MEDICAL PHYSIOLOGY

SECOND EDITION

MEDICAL PHYSIOLOGY

SECOND EDITION

EDITED BY

Rodney A. Rhoades, Ph.D.

Professor and Chairman, Department of Cellular and Integrative Physiology
Indiana University School of Medicine
Indianapolis, Indiana

George A. Tanner, Ph.D.

Professor of Physiology
Indiana University School of Medicine
Indianapolis, Indiana

ILLUSTRATIONS BY

Victoria Heim, CMI

Loganville, Georgia

LIPPINCOTT WILLIAMS & WILKINS
A **Wolters Kluwer** Company
Philadelphia • Baltimore • New York • London
Buenos Aires • Hong Kong • Sydney • Tokyo

Acquisitions Editor: Betty Sun
Developmental Editor: Kathleen Scogna
Marketing Manager: Joe Schott
Production Editor: Bill Cady
Designer: Carmen DiBartolomeo
Compositor: Maryland Composition
Printer: Quebecor World

351 West Camden Street
Baltimore, MD 21201

530 Walnut Street
Philadelphia, PA 19106

The publisher is not responsible (as a matter of product liability, negligence, or otherwise) for any injury resulting from any material contained herein. This publication contains information relating to general principles of medical care that should not be construed as specific instructions for individual patients. Manufacturers' product information and package inserts should be reviewed for current information, including contraindications, dosages, and precautions.

Library of Congress Cataloging-in-Publication Data

Medical physiology / edited by Rodney A. Rhoades, George A. Tanner.—2nd ed.
p.; cm.
Includes bibliographical references and index.
ISBN 0-7817-1936-4
1. Human physiology. 2. Physiology. I. Rhoades, Rodney A. II. Tanner, George A. 3. Title.
[DNLM: 1. Physiology. QT 104 M4892 2003]
QP34.5 .M473 2003
612—dc21

2002034060

Printed in the United States of America

First Edition, 1995

The publishers have made every effort to trace the copyright holders for borrowed material. If they have inadvertently overlooked any, they will be pleased to make the necessary arrangements at the first opportunity.

To purchase additional copies of this book, call our customer service department at (800) 638-3030 or fax orders to (301) 824-7390. International customers should call (301) 714-2324.

Visit Lippincott Williams & Wilkins on the Internet: http://www.LWW.com. Lippincott Williams & Wilkins customer service representatives are available from 8:30 am to 6:00 pm, EST.

05 06 07
3 4 5 6 7 8 9 10

PREFACE

The goal of this second edition of *Medical Physiology* is to provide a clear, accurate, and up-to-date introduction to medical physiology for medical students and students in the allied health sciences. Physiology, the study of normal function, is key to understanding pathophysiology and pharmacology and is essential to the everyday practice of clinical medicine.

Level. The level of the book is meant to be midway between an oversimplified review book and an encyclopedic textbook of physiology. Each chapter is written by medical school faculty members who have had many years of experience teaching physiology and who are experts in their field. They have selected material that is important for medical students to know and have presented this material in a concise, uncomplicated, and understandable fashion. We have purposely avoided discussion of research laboratory methods or historical material because most medical students are too busy to be burdened by such information. We have also avoided topics that are unsettled, recognizing that new research constantly provides fresh insights and sometimes challenges old ideas.

Key Changes. Many changes have been instituted in this second edition. All chapters were rewritten, in some cases by new contributors, and most illustrations have been redrawn. The new illustrations are clearer and make better use of color. An effort has also been made to institute more conceptual illustrations, rather than including more graphs and tables of data. These conceptual diagrams help students understand the general underpinnings of physiology. Another key change is the book's size: It is more compact because of deletions of extraneous material and shortening of some of the sections, most notably the gastrointestinal physiology section. We also overhauled many of the features in the book. Each chapter now contains a list of key concepts. The clinical focus boxes have been updated; they are more practical and less research-oriented. Each chapter includes a case study, with questions and answers. All of the review questions at the end of each chapter are now of the United States Medical Licensing Examination (USMLE) type. Lists of common abbreviations in physiology and of normal blood values have been added.

Content. This book begins with a discussion of basic physiological concepts, such as homeostasis and cell signaling, in Chapter 1. Chapter 2 covers the cell membrane, membrane transport, and the cell membrane potential. Most of the remaining chapters discuss the different organ systems: nervous, muscle, cardiovascular, respiratory, renal, gastrointestinal, endocrine, and reproductive physiology. Special chapters on the blood and the liver are included. Chapters on acid-base regulation, temperature regulation, and exercise discuss these complex, integrated functions. The order of presentation of topics follows that of most United States medical school courses in physiology. After the first two chapters, the other chapters can be read in any order, and some chapters may be skipped if the subjects are taught in other courses (e.g., neurobiology or biochemistry).

Material on pathophysiology is included throughout the book. This not only reinforces fundamental physiological principles but also demonstrates the relevance of physiology to an understanding of numerous medically important conditions.

Pedagogy. This second edition incorporates many features that should aid the student in his or her study of physiology:

- **Chapter outline.** The outline at the beginning of each chapter gives a preview of the chapter and is a useful study aid.
- **Key concepts.** Each chapter starts with a short list of key concepts that the student should understand after reading the chapter.
- **Text.** The text is easy to read, and topics are developed logically. Difficult concepts are explained clearly, often with the help of figures. Minutiae or esoteric topics are avoided.
- **Topic headings.** Second-level topic headings are active full-sentence statements. For example, instead of heading a section "Homeostasis," the heading is "Homeostasis is the maintenance of steady states in the body by coordinated physiological mechanisms." In this way, the key idea in a section is immediately obvious.
- **Boldfacing.** Key terms are boldfaced upon their first appearance in a chapter.
- **Illustrations and tables.** The figures have been selected to illustrate important concepts. The illustrations often show interrelationships between different variables or components of a system. Many of the figures are flow diagrams, so that students can appreciate the sequence of events that follow when a factor changes. Tables often provide useful summaries of material explained in more detail in the text.
- **Clinical focus boxes.** Each chapter contains one or two clinical focus boxes that illustrate the relevance of the physiology discussed in the chapter to an understanding of medicine.
- **Case studies.** Each section concludes with a set of case studies, one for each chapter, with questions and answers. These case studies help to reinforce how an un-

derstanding of physiology is important in dealing with clinical conditions.

- **Review questions and answers.** Students can use the review questions at the end of each chapter to test whether they have mastered the material. These USMLE-type questions should help students prepare for the Step 1 examination. Answers to the questions are provided at the end of the book and include explanations as to why the choices are correct or incorrect.
- **Suggested reading.** Each chapter provides a short list of recent review articles, monographs, book chapters, classic papers, or Web sites where students can obtain additional information.
- **Abbreviations and normal values.** This second edition includes a table of common abbreviations in physiology and a table of normal blood, plasma, or serum values. All abbreviations are defined when first used in the text, but the table of abbreviations in the appendix serves as a useful reminder of abbreviations commonly used in physiology and medicine. Normal values for blood are also embedded in the text, but the table on the inside front and back covers provides a more complete and easily accessible reference.
- **Index.** A complete index allows the student to easily look up material in the text.

Design. The design of this second edition has been completely overhauled. The new design makes navigating the text easier. Likewise, the design highlights the pedagogical features, making them easier to find and use.

Image Bank. A unique feature of this edition is an image bank containing virtually all of the figures in the book, in both pdf and jpeg formats, available for download from our Connections web site. The image bank is free to instructors and can be accessed at http://connection.lww.com, under "Medical Education/Physiology."

We thank the contributors for their patience and for following directions so that we could achieve a textbook of reasonably uniform style. Dr. James McGill was kind enough to write the clinical focus boxes and case studies for Chapters 26 and 27. We thank Marlene Brown for her secretarial assistance, Betsy Dilernia for her critical editing of each chapter, and Kathleen Scogna, our development editor, without whose encouragement and support this revised edition would not have been possible.

Rodney A. Rhoades, Ph.D.
George A. Tanner, Ph.D.

CONTRIBUTORS

H. Glenn Bohlen, Ph.D.
Professor of Physiology
Indiana University School of Medicine
Indianapolis, Indiana

Robert V. Considine, Ph.D.
Associate Professor of Medicine and Physiology
Indiana University School of Medicine
Indianapolis, Indiana

Denis English, Ph.D.
Director, Bone Marrow Transplant Laboratory
Methodist Hospital of Indiana
Indianapolis, Indiana

Cynthia J. Forehand, Ph.D.
Professor of Anatomy/Neurobiology
University of Vermont College of Medicine
Burlington, Vermont

Patricia J. Gallagher, Ph.D.
Associate Professor of Physiology
Indiana University School of Medicine
Indianapolis, Indiana

Stephen A. Kempson, Ph.D.
Professor of Physiology
Indiana University School of Medicine
Indianapolis, Indiana

John C. Kincaid, M.D.
Professor of Neurology and Physiology
Indiana University School of Medicine
Indianapolis, Indiana

Bruce J. Martin, Ph.D.
Associate Professor of Physiology
Indiana University School of Medicine
Bloomington, Indiana

Richard A. Meiss, Ph.D.
Professor of Obstetrics and Gynecology and Physiology
Indiana University School of Medicine
Indianapolis, Indiana

Daniel E. Peavy, Ph.D.
Associate Professor of Physiology
Indiana University School of Medicine
Indianapolis, Indiana

Rodney A. Rhoades, Ph.D.
Professor and Chairman
Department of Cellular and Integrative Physiology
Indiana University School of Medicine
Indianapolis, Indiana

Thom W. Rooke, M.D.
Director, Vascular Medicine Section
Vascular Center
Mayo Clinic
Rochester, Minnesota

Harvey V. Sparks, Jr., M.D.
University Distinguished Professor
Michigan State University
East Lansing, Michigan

George A. Tanner, Ph.D.
Professor of Physiology
Indiana University School of Medicine
Indianapolis, Indiana

Paul F. Terranova, Ph.D.
Director, Center for Reproductive Sciences
University of Kansas Medical Center
Kansas City, Kansas

Patrick Tso, Ph.D.
Professor of Pathology
University of Cincinnati School of Medicine
Cincinnati, Ohio

C. Bruce Wenger, M.D., Ph.D. (deceased)
Research Pharmacologist, Military Ergonomics Division
USARIEM
Natick, Massachusetts

Jackie D. Wood, Ph.D.
Professor of Physiology
Ohio State University College of Medicine
Columbus, Ohio

CONTENTS

PART I Cellular Physiology

Homeostasis and Cellular Signaling

Patricia J. Gallagher, Ph.D.
George A. Tanner, Ph.D.

CHAPTER OUTLINE

KEY CONCEPTS

1. Physiology is the study of the functions of living organisms and how they are regulated and integrated.
2. A stable internal environment is necessary for normal cell function and survival of the organism.
3. Homeostasis is the maintenance of steady states in the body by coordinated physiological mechanisms.
4. Negative and positive feedback are used to modulate the body's responses to changes in the environment.
5. Steady state and equilibrium are distinct conditions. Steady state is a condition that does not change over time, while equilibrium represents a balance between opposing forces.
6. Cellular communication is essential to integrate and coordinate the systems of the body so they can participate in different functions.
7. Different modes of cell communication differ in terms of distance and speed.
8. Chemical signaling molecules (first messengers) provide the major means of intercellular communication; they include ions, gases, small peptides, protein hormones, metabolites, and steroids.
9. Receptors are the receivers and transmitters of signaling molecules; they are located either on the plasma membrane or within the cell.
10. Second messengers are important for amplification of the signal received by plasma membrane receptors.
11. Steroid and thyroid hormone receptors are intracellular receptors that participate in the regulation of gene expression.

Physiology is the study of processes and functions in living organisms. It is a broad field that encompasses many disciplines and has strong roots in physics, chemistry, and mathematics. Physiologists assume that the same chemical and physical laws that apply to the inanimate world govern processes in the body. They attempt to describe functions in chemical, physical, or engineering terms. For example, the distribution of ions across cell membranes is described in thermodynamic terms, muscle contraction is analyzed in terms of forces and velocities, and regulation in the body is described in terms of control systems theory. Because the functions of living systems are carried out by their constituent structures, knowledge of structure from gross anatomy to the molecular level is germane to an understanding of physiology.

The scope of physiology ranges from the activities or functions of individual molecules and cells to the interaction of our bodies with the external world. In recent years, we have seen many advances in our understanding of physiological processes at the molecular and cellular levels. In higher organisms, changes in cell function always occur in the context of a whole organism, and different tissues and organs obviously affect one another. The independent activity of an organism requires the coordination of function at all levels, from molecular and cellular to the organism as a whole. An important part of physiology is understanding how different parts of the body are controlled, how they interact, and how they adapt to changing conditions.

For a person to remain healthy, physiological conditions in the body must be kept at optimal levels and closely regulated. Regulation requires effective communication between cells and tissues. This chapter discusses several topics related to regulation and communication: the internal environment, homeostasis of extracellular fluid, intracellular homeostasis, negative and positive feedback, feedforward control, compartments, steady state and equilibrium, intercellular and intracellular communication, nervous and endocrine systems control, cell membrane transduction, and important signal transduction cascades.

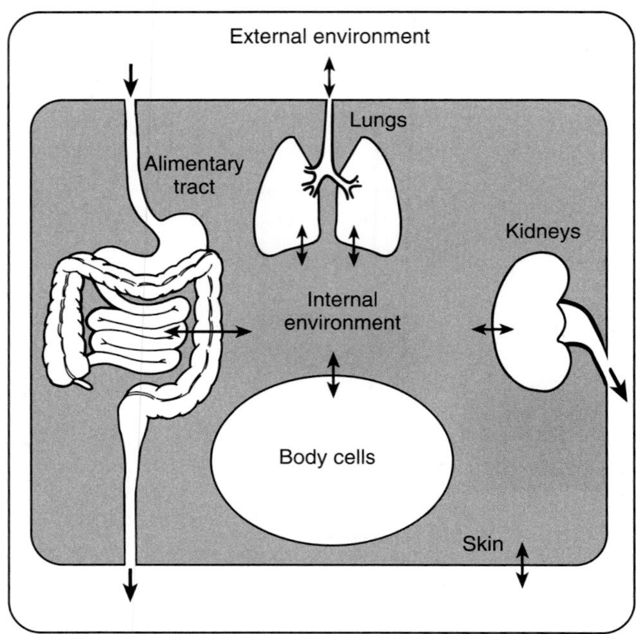

FIGURE 1.1 The living cells of our body, surrounded by an internal environment (extracellular fluid), communicate with the external world through this medium. Exchanges of matter and energy between the body and the external environment (indicated by arrows) occur via the gastrointestinal tract, kidneys, lungs, and skin (including the specialized sensory organs).

THE BASIS OF PHYSIOLOGICAL REGULATION

Our bodies are made up of incredibly complex and delicate materials, and we are constantly subjected to all kinds of disturbances, yet we keep going for a lifetime. It is clear that conditions and processes in the body must be closely controlled and regulated, i.e., kept at appropriate values. Below we consider, in broad terms, physiological regulation in the body.

A Stable Internal Environment Is Essential for Normal Cell Function

The nineteenth-century French physiologist Claude Bernard was the first to formulate the concept of the internal environment (*milieu intérieur*). He pointed out that an external environment surrounds multicellular organisms (air or water), but the cells live in a liquid internal environment (extracellular fluid). Most body cells are not directly exposed to the external world but, rather, interact with it through the internal environment, which is continuously renewed by the circulating blood (Fig. 1.1).

For optimal cell, tissue, and organ function in animals, several conditions in the internal environment must be maintained within narrow limits. These include but are not limited to (1) oxygen and carbon dioxide tensions, (2) concentrations of glucose and other metabolites, (3) osmotic pressure, (4) concentrations of hydrogen, potassium, calcium, and magnesium ions, and (5) temperature. Departures from optimal conditions may result in disordered functions, disease, or death.

Bernard stated that "stability of the internal environment is the primary condition for a free and independent existence." He recognized that an animal's independence from changing external conditions is related to its capacity to maintain a relatively constant internal environment. A good example is the ability of warm-blooded animals to live in different climates. Over a wide range of external temperatures, core temperature in mammals is maintained constant by both physiological and behavioral mechanisms. This stability has a clear survival value.

Homeostasis Is the Maintenance of Steady States in the Body by Coordinated Physiological Mechanisms

The key to maintaining stability of the internal environment is the presence of regulatory mechanisms in the body. In the first half of the twentieth century, the American physiologist Walter B. Cannon introduced a concept describing this capacity for self-regulation: **homeostasis**, the maintenance of steady states in the body by coordinated physiological mechanisms.

The concept of homeostasis is helpful in understanding and analyzing conditions in the body. The existence of steady conditions is evidence of regulatory mechanisms in the body that maintain stability. To function optimally under a variety of conditions, the body must sense departures from normal and must engage mechanisms for restoring conditions to normal. Departures from normal may be in the direction of too little or too much, so mechanisms exist for opposing changes in either direction. For example, if blood glucose concentration is too low, the hormone glucagon, from alpha cells of the pancreas, and epinephrine, from the adrenal medulla, will increase it. If blood glucose concentration is too high, insulin from the beta cells of the pancreas will lower it by enhancing the cellular uptake, storage, and metabolism of glucose. Behavioral responses also contribute to the maintenance of homeostasis. For example, a low blood glucose concentration stimulates feeding centers in the brain, driving the animal to seek food.

Homeostatic regulation of a physiological variable often involves several cooperating mechanisms activated at the same time or in succession. The more important a variable, the more numerous and complicated are the mechanisms that keep it at the desired value. Disease or death is often the result of dysfunction of homeostatic mechanisms.

The effectiveness of homeostatic mechanisms varies over a person's lifetime. Some homeostatic mechanisms are not fully developed at the time of birth. For example, a newborn infant cannot concentrate urine as well as an adult and is, therefore, less able to tolerate water deprivation. Homeostatic mechanisms gradually become less efficient as people age. For example, older adults are less able to tolerate stresses, such as exercise or changing weather, than are younger adults.

Intracellular Homeostasis Is Essential for Normal Cell Function

The term *homeostasis* has traditionally been applied to the internal environment—the extracellular fluid that bathes our tissues—but it can also be applied to conditions within cells. In fact, the ultimate goal of maintaining a constant internal environment is to promote intracellular homeostasis, and toward this end, conditions in the cytosol are closely regulated.

The many biochemical reactions within a cell must be tightly regulated to provide metabolic energy and proper rates of synthesis and breakdown of cellular constituents. Metabolic reactions within cells are catalyzed by enzymes and are therefore subject to several factors that regulate or influence enzyme activity.

- First, the final product of the reactions may inhibit the catalytic activity of enzymes, **end-product inhibition**. End-product inhibition is an example of negative-feedback control (see below).
- Second, intracellular regulatory proteins, such as the calcium-binding protein calmodulin, may control enzyme activity.
- Third, enzymes may be controlled by covalent modification, such as phosphorylation or dephosphorylation.
- Fourth, the ionic environment within cells, including hydrogen ion concentration ($[H^+]$), ionic strength, and calcium ion concentration, influences the structure and activity of enzymes.

Hydrogen ion concentration or pH affects the electrical charge of protein molecules and, hence, their configuration and binding properties. pH affects chemical reactions in cells and the organization of structural proteins. Cells regulate their pH via mechanisms for buffering intracellular hydrogen ions and by extruding H^+ into the extracellular fluid (see Chapter 25).

The structure and activity of cellular proteins are also affected by ionic strength. Cytosolic ionic strength depends on the total number and charge of ions per unit volume of water within cells. Cells can regulate their ionic strength by maintaining the proper mixture of ions and un-ionized molecules (e.g., organic osmolytes, such as sorbitol).

Many cells use calcium as an intracellular signal or "messenger" for enzyme activation, and, therefore, must possess mechanisms for regulating cytosolic $[Ca^{2+}]$. Such fundamental activities as muscle contraction, the secretion of neurotransmitters, hormones, and digestive enzymes, and the opening or closing of ion channels are mediated via transient changes in cytosolic $[Ca^{2+}]$. Cytosolic $[Ca^{2+}]$ in resting cells is low, about 10^{-7} M, and far below extracellular fluid $[Ca^{2+}]$ (about 2.5 mM). Cytosolic $[Ca^{2+}]$ is regulated by the binding of calcium to intracellular proteins, transport is regulated by adenosine triphosphate (ATP)-dependent calcium pumps in mitochondria and other organelles (e.g., sarcoplasmic reticulum in muscle), and the extrusion of calcium is regulated via cell membrane Na^+/Ca^{2+} exchangers and calcium pumps. Toxins or diminished ATP production can lead to an abnormally elevated cytosolic $[Ca^{2+}]$. A high cytosolic $[Ca^{2+}]$ activates many enzyme pathways, some of which have detrimental effects and may cause cell death.

Negative Feedback Promotes Stability; Feedforward Control Anticipates Change

Engineers have long recognized that stable conditions can be achieved by negative-feedback control systems (Fig. 1.2). Feedback is a flow of information along a closed loop. The components of a simple negative-feedback control system include a regulated variable, sensor (or detector), controller (or comparator), and effector. Each component controls the next component. Various disturbances may arise within or

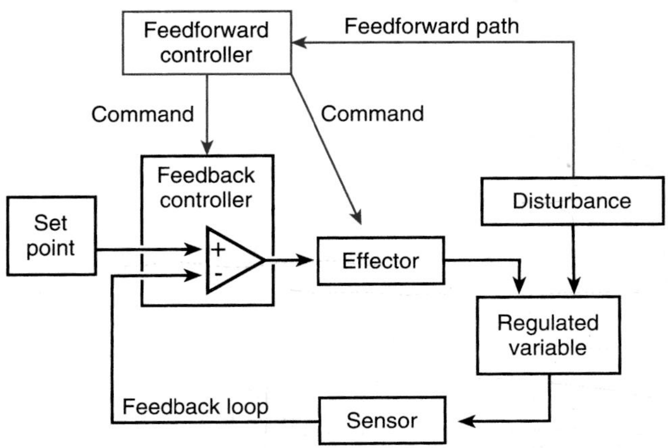

FIGURE 1.2 Elements of negative feedback and feedforward control systems (red). In a negative-feedback control system, information flows along a closed loop. The regulated variable is sensed, and information about its level is fed back to a feedback controller, which compares it to a desired value (set point). If there is a difference, an error signal is generated, which drives the effector to bring the regulated variable closer to the desired value. A feedforward controller generates commands without directly sensing the regulated variable, although it may sense a disturbance. Feedforward controllers often operate through feedback controllers.

outside the system and cause undesired changes in the regulated variable. With **negative feedback**, a regulated variable is sensed, information is fed back to the controller, and the effector acts to oppose change (hence, the term *negative*).

A familiar example of a negative-feedback control system is the thermostatic control of room temperature. Room temperature (regulated variable) is subjected to disturbances. For example, on a cold day, room temperature falls. A thermometer (sensor) in the thermostat (controller) detects the room temperature. The thermostat is set for a certain temperature (set point). The controller compares the actual temperature (feedback signal) to the set point temperature, and an error signal is generated if the room temperature falls below the set temperature. The error signal activates the furnace (effector). The resulting change in room temperature is monitored, and when the temperature rises sufficiently, the furnace is turned off. Such a negative-feedback system allows some fluctuation in room temperature, but the components act together to maintain the set temperature. Effective communication between the sensor and effector is important in keeping these oscillations to a minimum.

Similar negative-feedback systems maintain homeostasis in the body. One example is the system that regulates arterial blood pressure (see Chapter 18). This system's sensors (arterial baroreceptors) are located in the carotid sinuses and aortic arch. Changes in stretch of the walls of the carotid sinus and aorta, which follow from changes in blood pressure, stimulate these sensors. Afferent nerve fibers transmit impulses to control centers in the medulla oblongata. Efferent nerve fibers send impulses from the medullary centers to the system's effectors, the heart and blood vessels. The output of blood by the heart and the resistance to blood flow are altered in an appropriate direction to maintain blood pressure, as measured at the sensors, within a given range of values. This negative-feedback control system compensates for any disturbance that affects blood pressure, such as changing body position, exercise, anxiety, or hemorrhage. Nerves accomplish continuous rapid communication between the feedback elements. Various hormones are also involved in regulating blood pressure, but their effects are generally slower and last longer.

Feedforward control is another strategy for regulating systems in the body, particularly when a change with time is desired. In this case, a command signal is generated, which specifies the target or goal. The moment-to-moment operation of the controller is "open loop"; that is, the regulated variable itself is not sensed. Feedforward control mechanisms often sense a disturbance and can, therefore, take corrective action that anticipates change. For example, heart rate and breathing increase even before a person has begun to exercise.

Feedforward control usually acts in combination with negative-feedback systems. One example is picking up a pencil. The movements of the arm, hand, and fingers are directed by the cerebral cortex (feedforward controller); the movements are smooth, and forces are appropriate only in part because of the feedback of visual information and sensory information from receptors in the joints and muscles. Another example of this combination occurs during exercise. Respiratory and cardiovascular adjustments closely match muscular activity, so that arterial blood oxygen and carbon dioxide tensions hardly change during all but exhausting exercise. One explanation for this remarkable behavior is that exercise simultaneously produces a centrally generated feedforward signal to the active muscles and the respiratory and cardiovascular systems; feedforward control, together with feedback information generated as a consequence of increased movement and muscle activity, adjusts the heart, blood vessels, and respiratory muscles. In addition, control system function can adapt over a period of time. Past experience and learning can change the control system's output so that it behaves more efficiently or appropriately.

Although homeostatic control mechanisms usually act for the good of the body, they are sometimes deficient, inappropriate, or excessive. Many diseases, such as cancer, diabetes, and hypertension, develop because of a defective control mechanism. Homeostatic mechanisms may also result in inappropriate actions, such as autoimmune diseases, in which the immune system attacks the body's own tissue. Scar formation is one of the most effective homeostatic mechanisms of healing, but it is excessive in many chronic diseases, such as pulmonary fibrosis, hepatic cirrhosis, and renal interstitial disease.

Positive Feedback Promotes a Change in One Direction

With **positive feedback**, a variable is sensed and action is taken to reinforce a change of the variable. Positive feedback does not lead to stability or regulation, but to the opposite—a progressive change in one direction. One example of positive feedback in a physiological process is the upstroke of the action potential in nerve and muscle (Fig. 1.3). Depolarization of the cell membrane to a value greater than threshold leads to an increase in sodium (Na^+) permeability. Positively charged Na^+ ions rush into the cell through membrane Na^+ channels and cause further membrane depolarization; this leads to a further increase in Na^+ permeability and more Na^+ entry. This snowballing event, which occurs in a fraction of a mil-

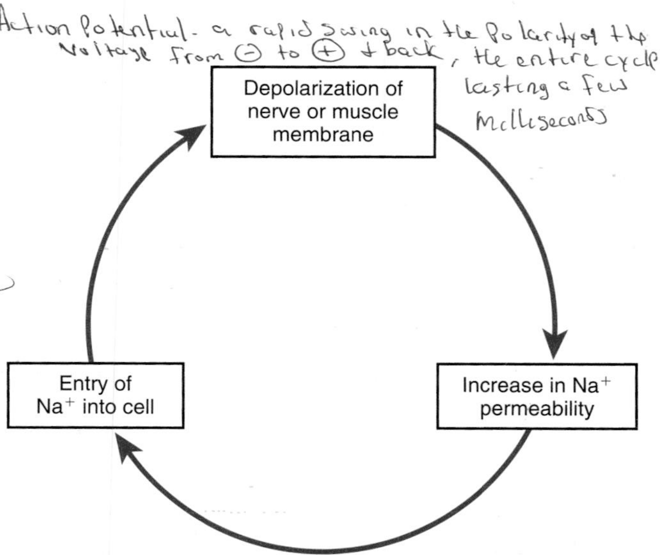

FIGURE 1.3 A positive-feedback cycle involved in the upstroke of an action potential.

lisecond, leads to an actual reversal of membrane potential and an electrical signal (action potential) conducted along the nerve or muscle fiber membrane. The process is stopped by inactivation (closure) of the Na^+ channels.

Another example of positive feedback occurs during the follicular phase of the menstrual cycle. The female sex hormone estrogen stimulates the release of luteinizing hormone, which in turn causes further estrogen synthesis by the ovaries. This positive feedback culminates in ovulation.

A third example is calcium-induced calcium release, which occurs with each heartbeat. Depolarization of the cardiac muscle plasma membrane leads to a small influx of calcium through membrane calcium channels. This leads to an explosive release of calcium from the muscle's sarcoplasmic reticulum, which rapidly increases the cytosolic calcium level and activates the contractile machinery. Many other examples are described in this textbook.

Positive feedback, if unchecked, can lead to a vicious cycle and dangerous situations. For example, a heart may be so weakened by disease that it cannot provide adequate blood flow to the muscle tissue of the heart. This leads to a further reduction in cardiac pumping ability, even less coronary blood flow, and further deterioration of cardiac function. The physician's task is sometimes to interrupt or "open" such a positive-feedback loop.

Steady State and Equilibrium Are Separate Ideas

Physiology often involves the study of exchanges of matter or energy between different defined spaces or **compartments**, separated by some type of limiting structure or **membrane**. The whole body can be divided into two major compartments: extracellular fluid and intracellular fluid. These two compartments are separated by cell plasma membranes. The extracellular fluid consists of all the body fluids outside of cells and includes the interstitial fluid, lymph, blood plasma, and specialized fluids, such as cerebrospinal fluid. It constitutes the internal environment of the body. Ordinary extracellular fluid is subdivided into interstitial fluid—lymph and plasma; these fluid compartments are separated by the endothelium, which lines the blood vessels. Materials are exchanged between these two compartments at the blood capillary level. Even withi[illegible] cells there is compartmentalization. The interiors of organ[illegible] separated from the cytosol by membranes, which restr[illegible]es and substrates to structures such as mitochondria a[illegible]mes and allow for the fine regulation of enzymatic re[illegible] and a greater variety of metabolic processes.

When two compartments are in **equilibrium**, o[illegible] forces are balanced, and there is no net transfer of a p[illegible]ular substance or energy from one compartment to [illegible] other. Equilibrium occurs if sufficient time for exchange h[illegible] been allowed and if no physical or chemical driving force would favor net movement in one direction or the other. For example, in the lung, oxygen in alveolar spaces diffuses into pulmonary capillary blood until the same oxygen tension is attained in both compartments. Osmotic equilibrium between cells and extracellular fluid is normally present in the body because of the high water permeability of most cell membranes. An equilibrium condition, if undisturbed, remains stable. No energy expenditure is required to maintain an equilibrium state.

Equilibrium and steady state are sometimes confused with each other. A **steady state** is simply a condition that does not change with time. It indicates that the amount or concentration of a substance in a compartment is constant. In a steady state, there is no net gain or net loss of a substance in a compartment. Steady state and equilibrium both suggest stable conditions, but a steady state does not necessarily indicate an equilibrium condition, and energy expenditure may be required to maintain a steady state. For example, in most body cells, there is a steady state for Na^+ ions; the amounts of Na^+ entering and leaving cells per unit time are equal. But intracellular and extracellular Na^+ ion concentrations are far from equilibrium. Extracellular $[Na^+]$ is much higher than intracellular $[Na^+]$, and Na^+ tends to move into cells down concentration and electrical gradients. The cell continuously uses metabolic energy to pump Na^+ out of the cell to maintain the cell in a steady state with respect to Na^+ ions. In living systems, conditions are often displaced from equilibrium by the constant expenditure of metabolic energy.

Figure 1.4 illustrates the distinctions between steady state and equilibrium. In Figure 1.4A, the fluid level in the

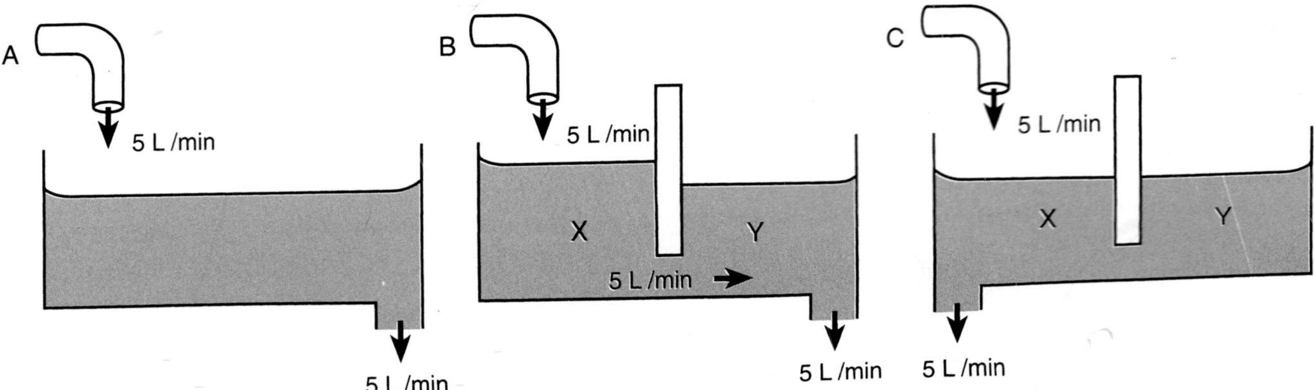

FIGURE 1.4 **Models of the concepts of steady state and equilibrium.** A, B, and C, Depiction of a steady state. In C, compartments X and Y are in equilibrium. (Modified from Riggs DS. The Mathematical Approach to Physiological Problems. Cambridge, MA: MIT Press, 1970;169.)

sink is constant (steady state) because the rates of inflow [illegible] are equal. If we were to increase the rate of in- [illegible] the tap), the fluid level would rise, and with [illegible] and new steady state might be established at a higher [illegible] In Figure 1.4B, the fluids in compartments X and Y [illegible] not in equilibrium (the fluid levels are different), but [illegible] system as a whole and each compartment are in a steady state, since inputs and outputs are equal. In Figure 1.4C, the system is in a steady state and compartments X and Y are in equilibrium. Note that the term *steady state* can apply to a single or several compartments; the term *equilibrium* describes the relation between at least two adjacent compartments that can exchange matter or energy with each other.

Coordinated Body Activity Requires Integration of Many Systems

Body functions can be analyzed in terms of several systems, such as the nervous, muscular, cardiovascular, respiratory, renal, gastrointestinal, and endocrine systems. These divisions are rather arbitrary, however, and all systems interact and depend on each other. For example, walking involves the activity of many systems. The nervous system coordinates the movements of the limbs and body, stimulates the muscles to contract, and senses muscle tension and limb position. The cardiovascular system supplies blood to the muscles, providing for nourishment and the removal of metabolic wastes and heat. The respiratory system supplies oxygen and removes carbon dioxide. The renal system maintains an optimal blood composition. The gastrointestinal system supplies energy-yielding metabolites. The endocrine system helps adjust blood flow and the supply of various metabolic substrates to the working muscles. Coordinated body activity demands the integration of many systems.

Recent research demonstrates that many diseases can be explained on the basis of abnormal function at the molecular level. This reductionist approach has led to incredible advances in our knowledge of both normal and abnormal function. Diseases occur within the context of a whole organism, however, and it is important to understand how all cells, tissues, organs, and organ systems respond to a disturbance (disease process) and interact. The saying, "The whole is more than the sum of its parts," certainly applies to what happens in living organisms. The science of physiology has the unique challenge of trying to make sense of the complex interactions that occur in the body. Understanding the body's processes and functions is clearly fundamental to the intelligent practice of medicine.

MODES OF COMMUNICATION AND SIGNALING

The human body has several means of transmitting information between cells. These mechanisms include direct communication between adjacent cells through gap junctions, autocrine and paracrine signaling, and the release of neurotransmitters and hormones produced by endocrine and nerve cells (Fig. 1.5).

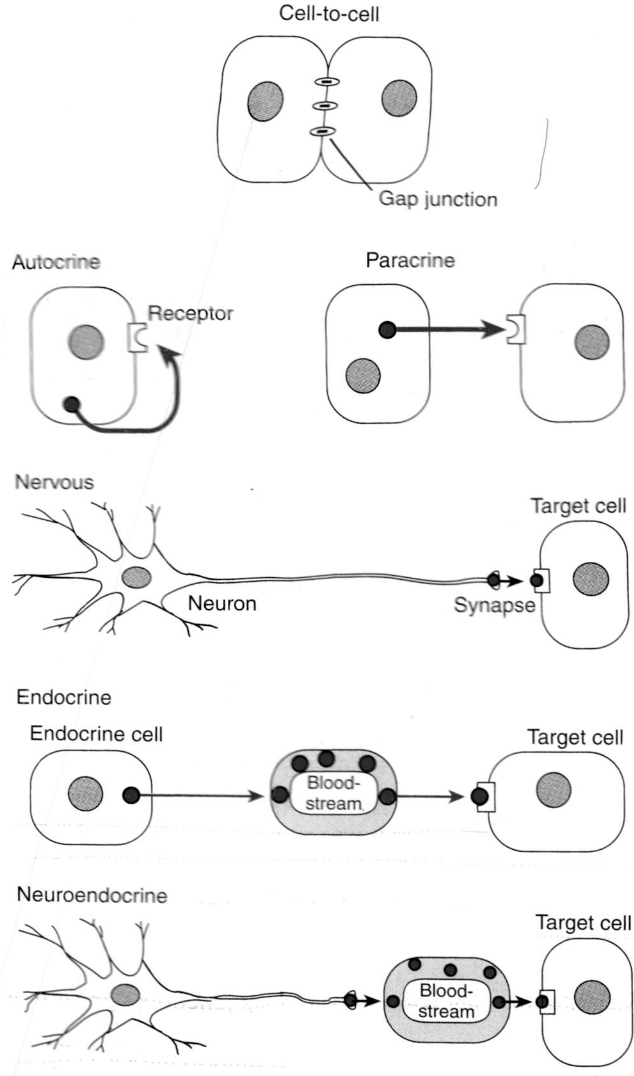

FIGURE 1.5 **Modes of intercellular signaling.** Cells may communicate with each other directly via gap junctions or chemical messengers. With autocrine and paracrine signaling, a chemical messenger diffuses a short distance through the extracellular fluid and binds to a receptor on the same cell or a nearby cell. Nervous signaling involves the rapid transmission of action potentials, often over long distances, and the release of a neurotransmitter at a synapse. Endocrine signaling involves the release of a hormone into the bloodstream and the binding of the hormone to specific target cell receptors. Neuroendocrine signaling involves the release of a hormone from a nerve cell and the transport of the hormone by the blood to a distant target cell.

Gap Junctions Provide a Pathway for Direct Communication Between Adjacent Cells

Adjacent cells sometimes communicate directly with each other via **gap junctions**, specialized protein channels made of the protein **connexin** (Fig. 1.6). Six connexins form a half-channel called a **connexon**. Two connexons join end to end to form an intercellular channel between adjacent cells. Gap junctions allow the flow of ions (hence, electrical current) and small molecules between the cytosol of

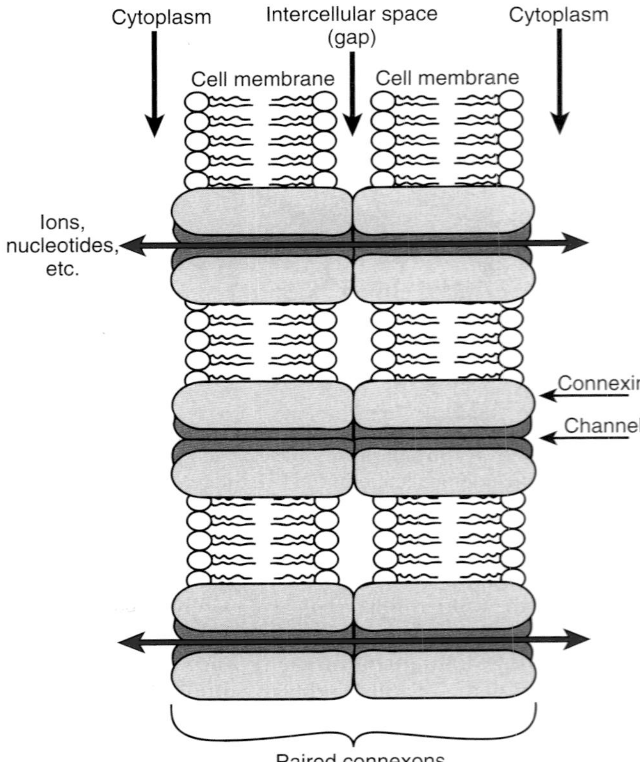

FIGURE 1.6 **The structure of gap junctions.** The channel connects the cytosol of adjacent cells. Six molecules of the protein connexin form a half-channel called a connexon. Ions and small molecules, such as nucleotides, can flow through the pore formed by the joining of connexons from adjacent cells.

neighboring cells (see Fig. 1.5). Gap junctions appear to be important in the transmission of electrical signals between neighboring cardiac muscle cells, smooth muscle cells, and some nerve cells. They may also functionally couple adjacent epithelial cells. Gap junctions are thought to play a role in the control of cell growth and differentiation by allowing adjacent cells to share a common intracellular environment. Often when a cell is injured, gap junctions close, isolating a damaged cell from its neighbors. This isolation process may result from a rise in calcium and a fall in pH in the cytosol of the damaged cell.

Cells May Communicate Locally by Paracrine and Autocrine Signaling

Cells may signal to each other via the local release of chemical substances. This means of communication, present in primitive living forms, does not depend on a vascular system. With **paracrine signaling**, a chemical is liberated from a cell, diffuses a short distance through interstitial fluid, and acts on nearby cells. Paracrine signaling factors affect only the immediate environment and bind with high specificity to cell receptors. They are also rapidly destroyed by extracellular enzymes or bound to extracellular matrix, thus preventing their widespread diffusion. **Nitric oxide** (**NO**), originally called "endothelium-derived relaxing factor (EDRF)," is an example of a paracrine signaling molecule. Although most cells can produce NO, it has major roles in mediating vascular smooth muscle tone, facilitating central nervous system neurotransmission activities, and modulating immune responses (see Chapters 16 and 26).

The production of NO results from the activation of **nitric oxide synthase** (**NOS**), which deaminates arginine to citrulline (Fig. 1.7). NO, produced by endothelial cells, regulates vascular tone by diffusing from the endothelial cell to the underlying vascular smooth muscle cell, where it activates its effector target, a cytoplasmic enzyme **guanylyl cyclase**. The activation of cytoplasmic guanylyl cyclase results in increased intracellular **cyclic guanosine monophosphate** (**cGMP**) levels and the activation of **cGMP-dependent protein kinase**. This enzyme phosphorylates potential target substrates, such as calcium pumps in the sarcoplasmic reticulum or sarcolemma, leading to reduced cytoplasmic levels of calcium. In turn, this deactivates the contractile machinery in the vascular smooth muscle cell and produces relaxation or a decrease of tone (see Chapter 16).

In contrast, during **autocrine signaling**, the cell releases a chemical into the interstitial fluid that affects its own activity by binding to a receptor on its own surface (see Fig. 1.5). Eicosanoids (e.g., prostaglandins), are examples of signaling molecules that often act in an autocrine manner. These molecules act as local hormones to influence a variety of physiological processes, such as uterine smooth muscle contraction during pregnancy.

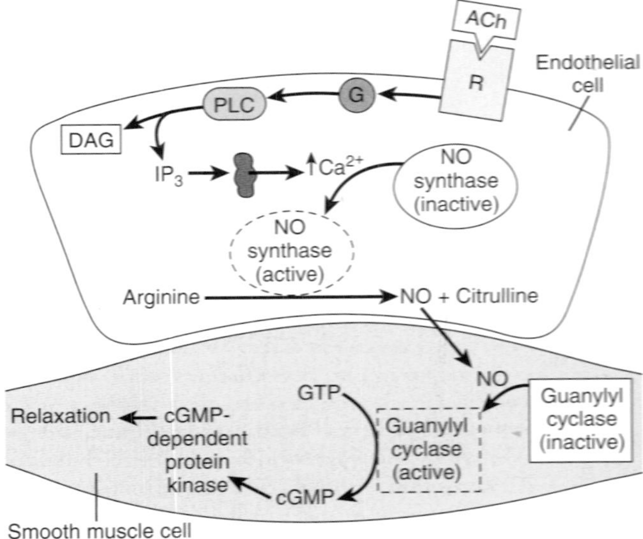

FIGURE 1.7 **Paracrine signaling by nitric oxide (NO) after stimulation of endothelial cells with acetylcholine (ACh).** The NO produced diffuses to the underlying vascular smooth muscle cell and activates its effector, cytoplasmic guanylyl cyclase, leading to the production of cGMP. Increased cGMP leads to the activation of cGMP-dependent protein kinase, which phosphorylates target substrates, leading to a decrease in cytoplasmic calcium and relaxation. Relaxation can also be mediated by nitroglycerin, a pharmacological agent that is converted to NO in smooth muscle cells, which can then activate guanylyl cyclase. G, G protein; PLC, phospholipase C; DAG, diacylglycerol; IP_3, inositol trisphosphate.

The Nervous System Provides for Rapid and Targeted Communication

The nervous system provides for rapid communication between body parts, with conduction times measured in milliseconds. This system is also organized for discrete activities; it has an enormous number of "private lines" for sending messages from one distinct locus to another. The conduction of information along nerves occurs via action potentials, and signal transmission between nerves or between nerves and effector structures takes place at a **synapse.** Synaptic transmission is almost always mediated by the release of specific chemicals or **neurotransmitters** from the nerve terminals (see Fig. 1.5). Innervated cells have specialized protein molecules (receptors) in their cell membranes that selectively bind neurotransmitters. Chapter 3 discusses the actions of various neurotransmitters and how they are synthesized and degraded. Chapters 4 to 6 discuss the role of the nervous system in coordinating and controlling body functions.

The Endocrine System Provides for Slower and More Diffuse Communication

The endocrine system produces **hormones** in response to a variety of stimuli. In contrast to the effects of nervous system stimulation, responses to hormones are much slower (seconds to hours) in onset, and the effects often last longer. Hormones are carried to all parts of the body by the bloodstream (see Fig. 1.5). A particular cell can respond to a hormone only if it possesses the specific receptor ("receiver") for the hormone. Hormone effects may be discrete. For example, arginine vasopressin increases the water permeability of kidney collecting duct cells but does not change the water permeability of other cells. They may also be diffuse, influencing practically every cell in the body. For example, thyroxine has a general stimulatory effect on metabolism. Hormones play a critical role in controlling such body functions as growth, metabolism, and reproduction.

Cells that are not traditional endocrine cells produce a special category of chemical messengers called **tissue growth factors**. These growth factors are protein molecules that influence cell division, differentiation, and cell survival. They may exert effects in an autocrine, paracrine, or endocrine fashion. Many growth factors have been identified, and probably many more will be recognized in years to come. **Nerve growth factor** enhances nerve cell development and stimulates the growth of axons. **Epidermal growth factor** stimulates the growth of epithelial cells in the skin and other organs. **Platelet-derived growth factor** stimulates the proliferation of vascular smooth muscle and endothelial cells. **Insulin-like growth factors** stimulate the proliferation of a wide variety of cells and mediate many of the effects of growth hormone. Growth factors appear to be important in the development of multicellular organisms and in the regeneration and repair of damaged tissues.

The Nervous and Endocrine Control Systems Overlap

The distinction between nervous and endocrine control systems is not always clear. First, the nervous system exerts important controls over endocrine gland function. For example, the hypothalamus controls the secretion of hormones from the pituitary gland. Second, specialized nerve cells, called **neuroendocrine cells**, secrete hormones. Examples are the hypothalamic neurons, which liberate releasing factors that control secretion by the anterior pituitary gland, and the hypothalamic neurons, which secrete arginine vasopressin and oxytocin into the circulation. Third, many proven or potential neurotransmitters found in nerve terminals are also well-known hormones, including arginine vasopressin, cholecystokinin, enkephalins, norepinephrine, secretin, and vasoactive intestinal peptide. Therefore, it is sometimes difficult to classify a particular molecule as either a hormone or a neurotransmitter.

THE MOLECULAR BASIS OF CELLULAR SIGNALING

Cells communicate with one another by many complex mechanisms. Even unicellular organisms, such as yeast cells, utilize small peptides called **pheromones** to coordinate mating events that eventually result in haploid cells with new assortments of genes. The study of intercellular communication has led to the identification of many complex signaling systems that are used by the body to network and coordinate functions. These studies have also shown that these signaling pathways must be tightly regulated to maintain cellular homeostasis. Dysregulation of these signaling pathways can transform normal cellular growth into uncontrolled cellular proliferation or cancer (see Clinical Focus Box 1.1).

Signaling systems consist of **receptors** that reside either in the plasma membrane or within cells and are activated by a variety of extracellular signals or **first messengers**, including peptides, protein hormones and growth factors, steroids, ions, metabolic products, gases, and various chemical or physical agents (e.g., light). Signaling systems also include **transducers** and **effectors** that are involved in conversion of the signal into a physiological response. The pathway may include additional intracellular messengers, called **second messengers** (Fig. 1.8). Examples of second messengers are cyclic nucleotides such as cyclic adenosine monophosphate (cAMP) and cyclic guanosine monophosphate (cGMP), inositol 1,4,5-trisphosphate (IP_3) and diacylglycerol (DAG), and calcium.

A general outline for a signaling system is as follows: The signaling cascade is initiated by binding of a hormone to its appropriate ligand-binding site on the outer surface domain of its cognate membrane receptor. This results in activation of the receptor; the receptor may adopt a new conformation, form aggregates (multimerize), or become phosphorylated. This change usually results in an association of adapter signaling molecules that transduce and amplify the signal through the cell by activating specific effector molecules and generating a second messenger. The outcome of the signal transduction cascade is a physiological response, such as secretion, movement, growth, division, or death.

CLINICAL FOCUS BOX 1.1

From Signaling Molecules to Oncoproteins and Cancer

Cancer may result from defects in critical signaling molecules that regulate many cell properties, including cell proliferation, differentiation, and survival. Normal cellular regulatory proteins or **proto-oncogenes** may become altered by mutation or abnormally expressed during cancer development. **Oncoproteins**, the altered proteins that arise from proto-oncogenes, in many cases are signal transduction proteins that normally function in the regulation of cellular proliferation. Examples of signaling molecules that can become oncogenic span the entire signal transduction pathway and include ligands (e.g., growth factors), receptors, adapter and effector molecules, and transcription factors.

There are many examples of how normal cellular proteins can be converted into oncoproteins. One occurs in chronic myeloid leukemia (CML). This disease usually results from an acquired chromosomal abnormality that involves translocation between chromosomes 9 and 22. This translocation results in the fusion of the *bcr* gene, whose function is unknown, with part of the *cellular abl* (c-*abl*) gene. The c-*abl* gene encodes a protein tyrosine kinase whose normal substrates are unknown. The chimeric (composed of fused parts of *bcr* and c-*abl*) Bcr-Abl fusion protein has unregulated tyrosine kinase activity and, through the Abl SH2 and SH3 domains, binds to and phosphorylates many signal transduction proteins that the wild-type Abl tyrosine kinase would not normally activate. Increased expression of the unregulated Bcr-Abl protein activates many growth regulatory genes in the absence of normal growth factor signaling.

The chromosomal translocation that results in the formation of the Bcr-Abl oncoprotein occurs during the development of hematopoietic stem cells and is observed as the diagnostic, shorter, Philadelphia chromosome 22. It results in chronic myeloid leukemia that is characterized by a progressive leukocytosis (increase in number of circulating white blood cells) and the presence of circulating immature blast cells. Other secondary mutations may spontaneously occur within the mutant stem cell and can lead to acute leukemia, a rapidly progressing disease that is often fatal. Understanding of the molecules and signaling pathways that regulate normal cell physiology can help us understand what happens in some types of cancer.

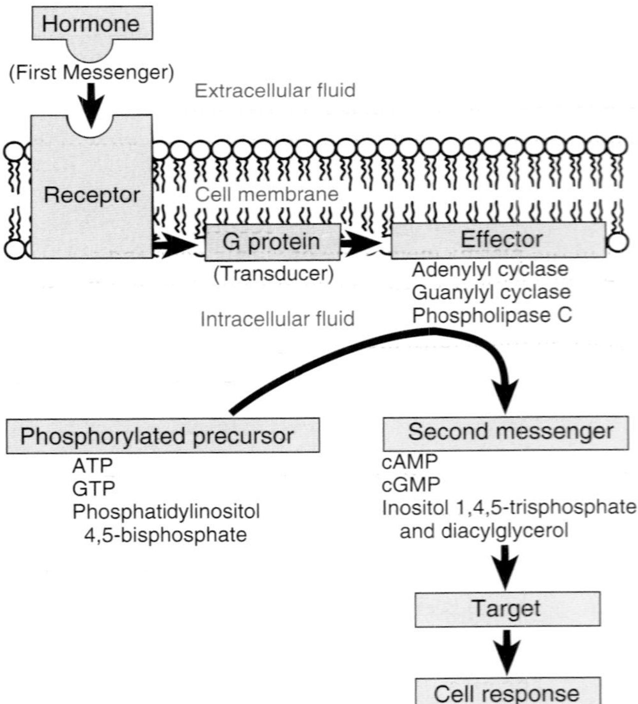

FIGURE 1.8 **Signal transduction patterns common to second messenger systems.** A protein or peptide hormone binds to a plasma membrane receptor, which stimulates or inhibits a membrane-bound effector enzyme via a G protein. The effector catalyzes the production of many second messenger molecules from a phosphorylated precursor (e.g., cAMP from ATP, cGMP from GTP, or inositol 1,4,5-trisphosphate and diacylglycerol from phosphatidylinositol 4,5-bisphosphate). The second messengers, in turn, activate protein kinases (targets) or cause other intracellular changes that ultimately lead to the cell response.

SIGNAL TRANSDUCTION BY PLASMA MEMBRANE RECEPTORS

As mentioned above, the molecules that are produced by one cell to act on itself (autocrine signaling) or other cells (paracrine, neural, or endocrine signaling) are ligands or first messengers. Many of these ligands bind directly to receptor proteins that reside in the plasma membrane, and others cross the plasma membrane and interact with cellular receptors that reside in either the cytoplasm or the nucleus. Thus, cellular receptors are divided into two general types, **cell-surface receptors** and **intracellular receptors**. Three general classes of cell-surface receptors have been identified: G protein-coupled receptors, ion channel-linked receptors, and enzyme-linked receptors. Intracellular receptors include steroid and thyroid hormone receptors and are discussed in a later section in this chapter.

G Protein-Coupled Receptors Transmit Signals Through the Trimeric G Proteins

G protein-coupled receptors (GPCRs) are the largest family of cell-surface receptors, with more than 1,000 members. These receptors indirectly regulate their effector targets, which can be ion channels or plasma membrane-bound effector enzymes, through the intermediary activity of a separate membrane-bound adapter protein complex called the **trimeric GTP-binding regulatory protein** or **G protein** (see Clinical Focus Box 1.2). GPCRs mediate cellular responses to numerous types of first messenger signaling molecules, including proteins, small peptides, amino acids, and fatty acid derivatives. Many first messenger ligands can activate several different GPCRs. For example, serotonin can activate at least 15 different GPCRs.

G protein-coupled receptors are structurally and functionally similar molecules. They have a ligand-binding ex-

CLINICAL FOCUS BOX 1.2

G Proteins and Disease

G proteins function as key transducers of information across cell membranes by coupling receptors to effectors such as adenylyl cyclase (AC) or phospholipase C (see Fig. 1.9). They are part of a large family of proteins that bind and hydrolyze guanosine triphosphate (GTP) as part of an "on" and "off" switching mechanism. G proteins are heterotrimers, consisting of Gα, Gβ, and Gγ subunits, each of which is encoded by a different gene.

Some strains of bacteria have developed toxins that can modify the activity of the α subunit of G proteins, resulting in disease. For example, **cholera toxin**, produced by the microorganism that causes cholera, *Vibrio cholerae*, causes ADP ribosylation of the stimulatory (Gα_s) subunit of G proteins. This modification abolishes the GTPase activity of Gα_s and results in an α_s subunit that is always in the "on" or active state. Thus, cholera toxin results in continuous stimulation of AC. The main cells affected by this bacterial toxin are the epithelial cells of the intestinal tract, and the excessive production of cAMP causes them to secrete chloride ions and water. This causes severe diarrhea and dehydration and may result in death.

Another toxin, **pertussis toxin**, is produced by *Bordatella pertussis* bacteria and causes whooping cough. The pertussis toxin alters the activity of Gα_i by ADP ribosylation. This modification inhibits the function of the α_i subunit by preventing association with an activated receptor. Thus, the α_i subunit remains GDP-bound and in an "off" state, unable to inhibit the activity of AC. The molecular mechanism by which pertussis toxin causes whooping cough is not understood.

The understanding of the actions of cholera and pertussis toxins highlights the importance of normal G-protein function and illustrates that dysfunction of this signaling pathway can cause acute disease. In the years since the discovery of these proteins, there has been an explosion of information on G proteins and several chronic human diseases have been linked to genetic mutations that cause abnormal function or expression of G proteins. These mutations can occur either in the G proteins themselves or in the receptors to which they are coupled.

Mutations in G protein-coupled receptors (GPCRs) can result in the receptor being in an active conformation in the absence of ligand binding. This would result in sustained stimulation of G proteins. Mutations of G-protein subunits can result in either constitutive activation (e.g., continuous stimulation of effectors such as AC) or loss of activity (e.g., loss of cAMP production).

Many factors influence the observed manifestations resulting from defective G-protein signaling. These include the specific GPCRs and the G proteins that associate with them, their complex patterns of expression in different tissues, and whether the mutation is germ-line or somatic. Mutation of a ubiquitously expressed GPCR or G protein results in widespread manifestations, while mutation of a GPCR or G protein with restricted expression will result in more focused manifestations.

Somatic mutation of Gα_s during embryogenesis can result in the dysregulated activation of this G protein and is the source of several diseases that have multiple pleiotropic or local manifestations, depending on when the mutation occurs. For example, early somatic mutation of Gα_s and its overactivity can lead to McCune-Albright syndrome (MAS). The consequences of the mutant Gα_s in MAS are manifested in many ways, with the most common being a triad of features that includes polyostotic (affecting many bones) fibrous dysplasia, *café-au-lait* skin hyperpigmentation, and precocious puberty. A later mutation of Gα_s can result in a more restricted focal syndrome, such as monostotic (affecting a single bone) fibrous dysplasia.

The complexity of the involvement of GPCR or G proteins in the pathogenesis of many human diseases is beginning to be appreciated, but already this information underscores the critical importance of understanding the molecular events involved in hormone signaling so that rational therapeutic interventions can be designed.

tracellular domain on one end of the molecule, separated by a seven-pass transmembrane-spanning region from the cytosolic regulatory domain at the other end, where the receptor interacts with the membrane-bound G protein. Binding of ligand or hormone to the extracellular domain results in a conformational change in the receptor that is transmitted to the cytosolic regulatory domain. This conformational change allows an association of the ligand-bound, activated receptor with a trimeric G protein associated with the inner leaflet of the plasma membrane. The interaction between the ligand-bound, activated receptor and the G protein, in turn, activates the G protein, which dissociates from the receptor and transmits the signal to its effector enzyme or ion channel (Fig. 1.9).

The trimeric G proteins are named for their requirement for GTP binding and hydrolysis and have been shown to have a broad role in linking various seven-pass transmembrane receptors to membrane-bound effector systems that generate intracellular messengers. G proteins are tethered to the membrane through lipid linkage and are heterotrimeric, that is, composed of three distinct subunits. The subunits of a G protein are an α subunit, which binds and hydrolyzes GTP, and β and γ subunits, which form a stable, tight noncovalent-linked dimer. When the α subunit binds GDP, it associates with the $\beta\gamma$ subunits to form a trimeric complex that can interact with the cytoplasmic domain of the GPCR (Fig. 1.10). The conformational change that occurs upon ligand binding causes the GDP-bound trimeric (α, β, γ complex) G protein to associate with the ligand-bound receptor. The association of the GDP-bound trimeric complex with the GPCR activates the exchange of GDP for GTP. Displacement of GDP by GTP is favored in cells because GTP is in higher concentration.

The displacement of GDP by GTP causes the α subunit to dissociate from the receptor and from the $\beta\gamma$ subunits of the G protein. This exposes an effector binding site on the α subunit, which then associates with an effector molecule (e.g., adenylyl cyclase or phospholipase C) to result in the generation of second messengers (e.g., cAMP or IP_3 and DAG). The hydrolysis of GTP to GDP by the α subunit re-

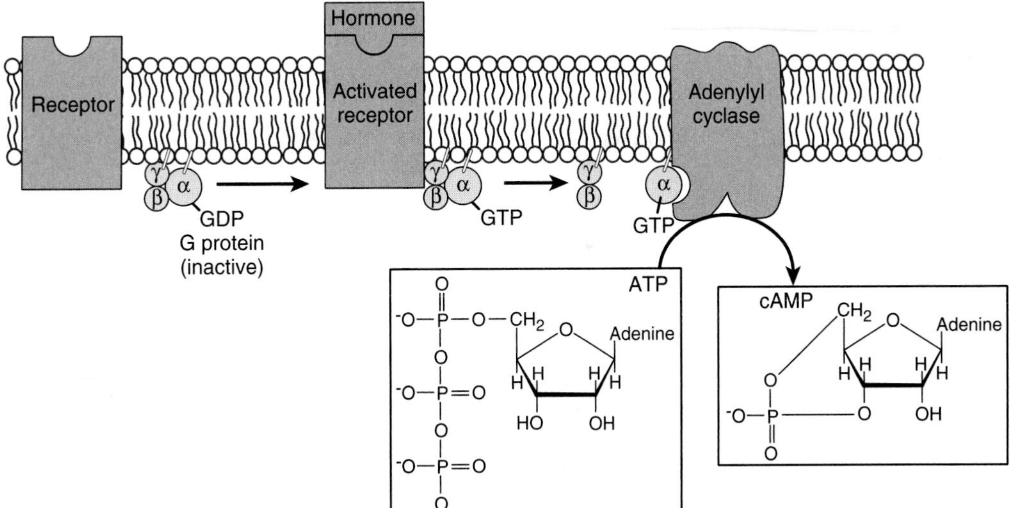

FIGURE 1.9 **Activation of a G protein-coupled receptor and the production of cAMP.** Binding of a hormone causes the interaction of the activated receptor with the inactive, GDP-bound G protein. This interaction results in activation of the G protein through GDP to GTP exchange and dissociation of the α and βγ subunits. The activated α subunit of the G protein can then interact with and activate the membrane protein adenylyl cyclase to catalyze the conversion of ATP to cAMP.

sults in the reassociation of the α and βγ subunits, which are then ready to repeat the cycle.

The cycling between inactive (GDP-bound) and active forms (GTP-bound) places the G proteins in the family of **molecular switches,** which regulate many biochemical events. When the switch is "off," the bound nucleotide is GDP. When the switch is "on," the hydrolytic enzyme (G protein) is bound to GTP, and the cleavage of GTP to GDP will reverse the switch to an "off" state. While most of the signal transduction produced by G proteins is due to the activities of the α subunit, a role for βγ subunits in activating effectors during signal transduction is beginning to be appreciated. For example, βγ subunits can activate K^+ channels. Therefore, both α and βγ subunits are involved in regulating physiological responses.

The catalytic activity of a G protein, which is the hydrolysis of GTP to GDP, resides in its Gα subunit. Each Gα subunit within this large protein family has an intrinsic rate of GTP hydrolysis. The intrinsic catalytic activity rate of G proteins is an important factor contributing to the amplifi-

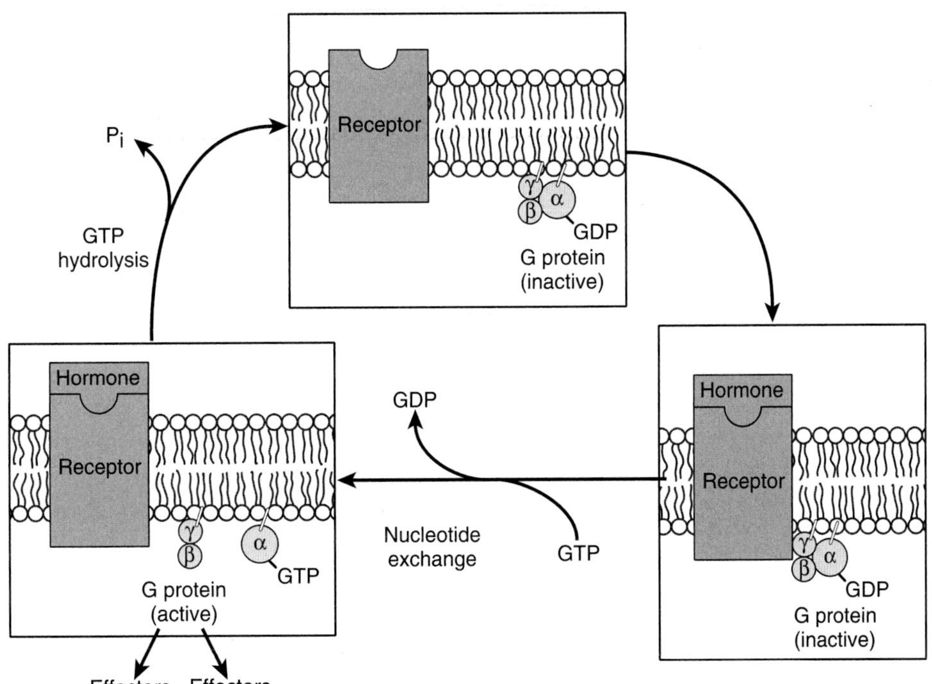

FIGURE 1.10 **Activation and inactivation of G proteins.** When bound to GDP, G proteins are in an inactive state and are not associated with a receptor. Binding of a hormone to a G protein-coupled receptor results in an association of the inactive, GDP-bound G protein with the receptor. The interaction of the GDP-bound G protein with the activated receptor results in activation of the G protein via the exchange of GDP for GTP by the α subunit. The α and βγ subunits of the activated GTP-bound G protein dissociate and can then interact with their effector proteins. The intrinsic GTPase activity in the α subunit of the G protein hydrolyzes the bound GTP to GDP. The GDP-bound α subunit reassociates with the βγ subunit to form an inactive, membrane-bound G-protein complex.

cation of the signal produced by a single molecule of ligand binding to a G protein-coupled receptor. For example, a Gα subunit that remains active longer (slower rate of GTP hydrolysis) will continue to activate its effector for a longer period and result in greater production of second messenger.

The G proteins functionally couple receptors to several different effector molecules. Two major effector molecules that are regulated by G-protein subunits are **adenylyl cyclase** (AC) and phospholipase C (PLC). The association of an activated Gα subunit with AC can result in either the stimulation or the inhibition of the production of cAMP. This disparity is due to the two types of α subunit that can couple AC to cell-surface receptors. Association of an α_s subunit (*s* for stimulatory) promotes the activation of AC and production of cAMP. The association of an α_i (*i* for inhibitory) subunit promotes the inhibition of AC and a decrease in cAMP. Thus, bidirectional regulation of adenylyl cyclase is achieved by coupling different classes of cell-surface receptors to the enzyme by either G_s or G_i (Fig. 1.11).

In addition to α_s and α_i subunits, other isoforms of G-protein subunits have been described. For example, α_q activates PLC, resulting in the production of the second messengers diacylglycerol and inositol trisphosphate. Another Gα subunit, α_T or **transducin**, is expressed in photoreceptor tissues, and has an important role in signaling in rod cells by activation of the effector **cGMP phosphodiesterase**, which degrades cGMP to 5′GMP (see Chapter 4). All three subunits of G proteins belong to large families that are expressed in different combinations in different tissues. This tissue distribution contributes to both the specificity of the transduced signal and the second messenger produced.

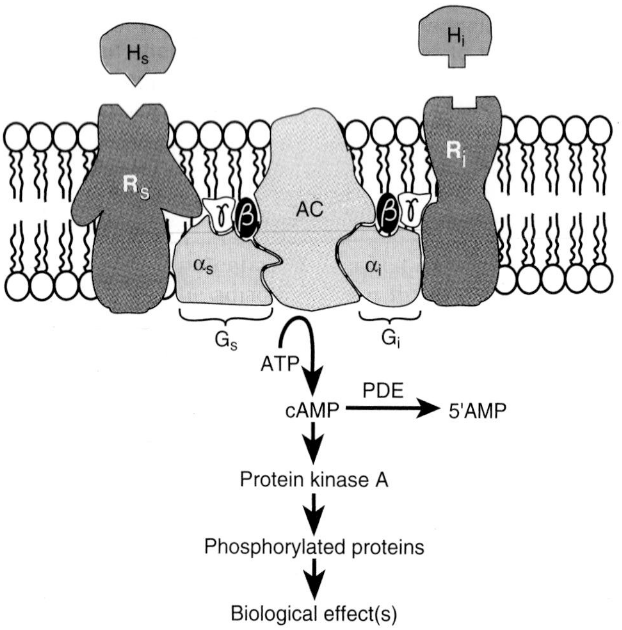

FIGURE 1.11 **Stimulatory and inhibitory coupling of G proteins to adenylyl cyclase (AC).** Stimulatory (G_s) and inhibitory (G_i) G proteins couple hormone binding to the receptor with either activation or inhibition of AC. Each G protein is a trimer consisting of Gα, Gβ, and Gγ subunits. The Gα subunits in G_s and G_i are distinct in each and provide the specificity for either AC activation or AC inhibition. Hormones (H_s) that stimulate AC interact with "stimulatory" receptors (R_s) and are coupled to AC through stimulatory G proteins (G_s). Conversely, hormones (H_i) that inhibit AC interact with "inhibitory" receptors (R_i) that are coupled to AC through inhibitory G proteins (G_i). Intracellular levels of cAMP are modulated by the activity of phosphodiesterase (PDE), which converts cAMP to 5′AMP and turns off the signaling pathway by reducing the level of cAMP.

The Ion Channel-Linked Receptors Help Regulate the Intracellular Concentration of Specific Ions

Ion channels, found in all cells, are transmembrane proteins that cross the plasma membrane and are involved in regulating the passage of specific ions into and out of cells.

Ion channels may be opened or closed by changing the membrane potential or by the binding of ligands, such as neurotransmitters or hormones, to membrane receptors. In some cases, the receptor and ion channel are one and the same molecule. For example, at the neuromuscular junction, the neurotransmitter acetylcholine binds to a muscle membrane nicotinic cholinergic receptor that is also an ion channel. In other cases, the receptor and an ion channel are linked via a G protein, second messengers, and other downstream effector molecules, as in the muscarinic cholinergic receptor on cells innervated by parasympathetic postganglionic nerve fibers. Another possibility is that the ion channel is directly activated by a cyclic nucleotide, such as cGMP or cAMP, produced as a consequence of receptor activation. This mode of ion channel control is predominantly found in the sensory tissues for sight, smell, and hearing. The opening or closing of ion channels plays a key role in signaling between electrically excitable cells.

The Tyrosine Kinase Receptors Signal Through Adapter Proteins to the Mitogen-Activated Protein Kinase Pathway

Many hormones, growth factors, and cytokines signal their target cells by binding to a class of receptors that have tyrosine kinase activity and result in the phosphorylation of tyrosine residues in the receptor and other target proteins. Many of the receptors in this class of plasma membrane receptors have an intrinsic tyrosine kinase domain that is part of the cytoplasmic region of the receptor (Fig. 1.12). Another group of related receptors lacks an intrinsic tyrosine kinase but, when activated, becomes associated with a cytoplasmic tyrosine kinase (see Fig. 1.12). This family of tyrosine kinase receptors utilizes similar signal transduction pathways, and we discuss them together.

The **tyrosine kinase receptors** consist of a hormone-binding region that is exposed to the extracellular fluid. Typical agonists for these receptors include hormones (e.g., insulin), growth factors (e.g., epidermal, fibroblast, and platelet-derived growth factors), or cytokines. The cytokine receptors include receptors for interferons, interleukins (e.g., IL-1 to IL-17), tumor necrosis factor, and colony-stimulating factors (e.g., granulocyte and monocyte colony-stimulating factors).

The signaling cascades generated by the activation of tyrosine kinase receptors can result in the amplification of

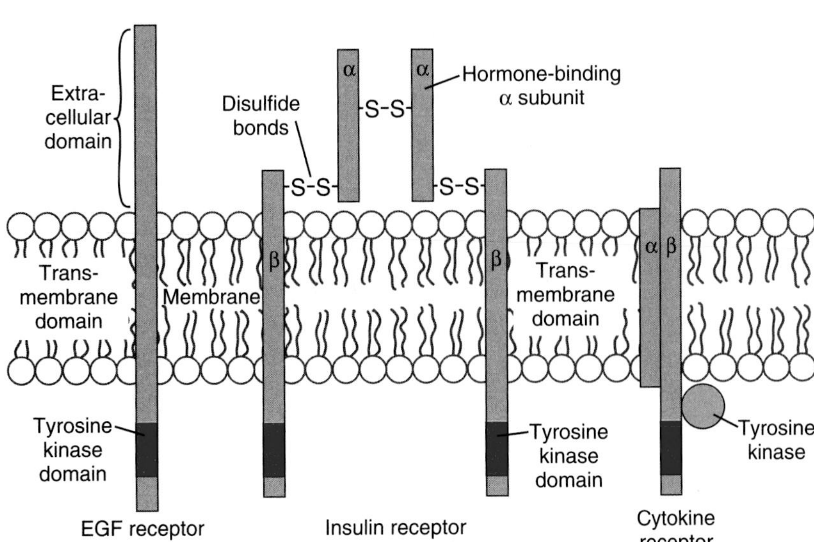

FIGURE 1.12 General structures of the tyrosine kinase receptor family. Tyrosine kinase receptors have an intrinsic protein tyrosine kinase activity that resides in the cytoplasmic domain of the molecule. Examples are the epidermal growth factor (EGF) and insulin receptors. The EGF receptor is a single-chain transmembrane protein consisting of an extracellular region containing the hormone-binding domain, a transmembrane domain, and an intracellular region that contains the tyrosine kinase domain. The insulin receptor is a heterotetramer consisting of two α and two β subunits held together by disulfide bonds. The α subunits are entirely extracellular and involved in insulin binding. The β subunits are transmembrane proteins and contain the tyrosine kinase activity within the cytoplasmic domain of the subunit. Some receptors become associated with cytoplasmic tyrosine kinases following their activation. Examples can be found in the family of cytokine receptors, which generally consist of an agonist-binding subunit and a signal-transducing subunit that become associated with a cytoplasmic tyrosine kinase.

gene transcription and *de novo* transcription of genes involved in growth, cellular differentiation, and movements such as crawling or shape changes. The general scheme for this signaling pathway begins with the agonist binding to the extracellular portion of the receptor (Fig. 1.13). The binding of agonists causes two of the agonist-bound receptors to associate or **dimerize**, and the associated or intrinsic tyrosine kinases become activated. The tyrosine kinases then phosphorylate tyrosine residues in the other subunit of the dimer. The phosphorylated tyrosine residues in the cytoplasmic domains of the dimerized receptor serve as "docking sites" for additional signaling molecules or adapter proteins that have a specific sequence called an **SH2 domain**. The SH2-containing adapter proteins may be serine/threonine protein kinases, phosphatases, or other proteins that help in the assembly of signaling complexes that transmit the signal from an activated receptor to many signaling pathways, resulting in a cellular response.

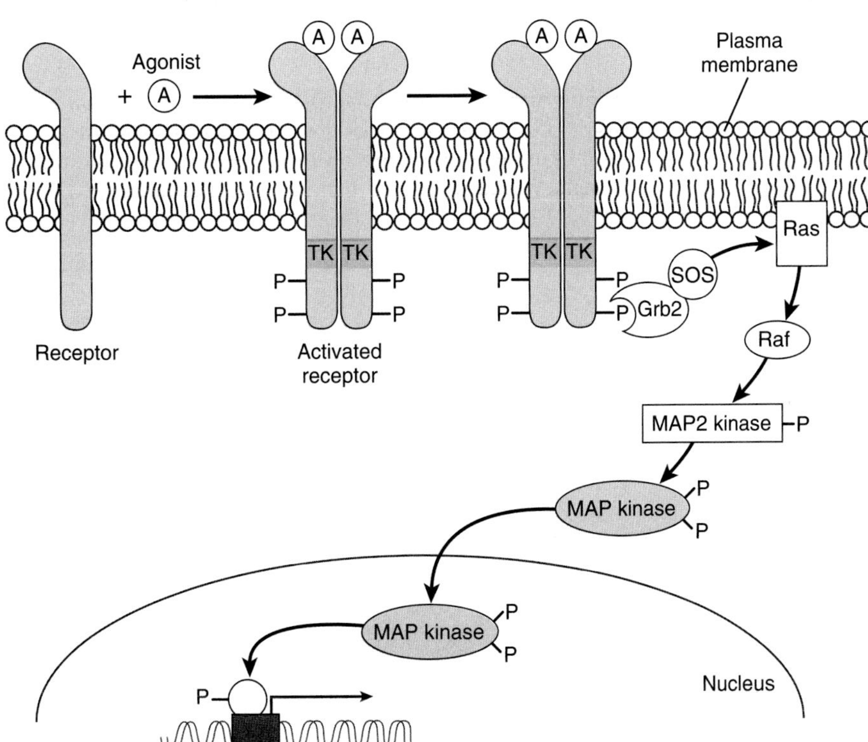

FIGURE 1.13 A signaling pathway for tyrosine kinase receptors. Binding of agonist to the tyrosine kinase receptor (TK) causes dimerization, activation of the intrinsic tyrosine kinase activity, and phosphorylation of the receptor subunits. The phosphotyrosine residues serve as docking sites for intracellular proteins, such as Grb2 and SOS, which have SH2 domains. Ras is activated by the exchange of GDP for GTP. Ras-GTP (active form) activates the serine/threonine kinase Raf, initiating a phosphorylation cascade that results in the activation of MAP kinase. MAP kinase translocates to the nucleus and phosphorylates transcription factors to modulate gene transcription.

One of these signaling pathways includes the activation of another GTPase that is related to the trimeric G proteins. Members of the **ras** family of **monomeric G proteins** are activated by many tyrosine kinase receptor growth factor agonists and, in turn, activate an intracellular signaling cascade that involves the phosphorylation and activation of several protein kinases called **mitogen-activated protein kinases** (MAP kinases). Activated MAP kinase translocates to the nucleus, where it activates the transcription of genes involved in the transcription of other genes, the **immediate early genes**.

SECOND MESSENGER SYSTEMS AND INTRACELLULAR SIGNALING PATHWAYS

Second messengers transmit and amplify the first messenger signal to signaling pathways inside the cell. Only a few second messengers are responsible for relaying these signals within target cells, and because each target cell has a different complement of intracellular signaling pathways, the physiological responses can vary. Thus, it is useful to keep in mind that every cell in our body is programmed to respond to specific combinations of messengers and that the same messenger can elicit a distinct physiological response in different cell types. For example, the neurotransmitter acetylcholine can cause heart muscle to relax, skeletal muscle to contract, and secretory cells to secrete.

cAMP Is an Important Second Messenger in All Cells

As a result of binding to specific G protein-coupled receptors, many peptide hormones and catecholamines produce an almost immediate increase in the intracellular concentration of cAMP. For these ligands, the receptor is coupled to a stimulatory G protein ($G\alpha_s$), which upon activation and exchange of GDP for GTP can diffuse in the membrane to interact with and activate adenylyl cyclase (AC), a large transmembrane protein that converts intracellular ATP to the second messenger, cAMP.

In addition to those hormones that stimulate the production of cAMP through a receptor coupled to $G\alpha_s$, some hormones act to decrease cAMP formation and, therefore, have opposing intracellular effects. These hormones bind to receptors that are coupled to an inhibitory ($G\alpha_i$) rather than a stimulatory ($G\alpha_s$) G protein. cAMP is perhaps the most widely distributed second messenger and has been shown to mediate various cellular responses to both hormonal and nonhormonal stimuli, not only in higher organisms but also in various primitive life forms, including slime molds and yeasts. The intracellular signal provided by cAMP is rapidly terminated by its hydrolysis to 5′AMP by a group of enzymes known as **phosphodiesterases**, which are also regulated by hormones in some instances.

Protein Kinase A Is the Major Mediator of the Signaling Effects of cAMP

cAMP activates an enzyme, **protein kinase A** (or **cAMP-dependent protein kinase**), which in turn catalyzes the phosphorylation of various cellular proteins, ion channels, and transcription factors. This phosphorylation alters the activity or function of the target proteins and ultimately leads to a desired cellular response. However, in addition to activating protein kinase A and phosphorylating target proteins, in some cell types, cAMP directly binds to and affects the activity of ion channels.

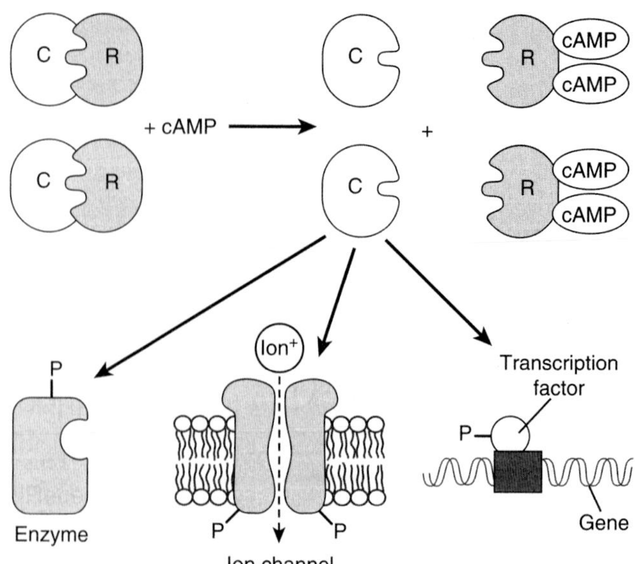

FIGURE 1.14 **Activation and targets of protein kinase A.** Inactive protein kinase A consists of two regulatory subunits complexed with two catalytic subunits. Activation of adenylyl cyclase results in increased cytosolic levels of cAMP. Two molecules of cAMP bind to each of the regulatory subunits, leading to the release of the active catalytic subunits. These subunits can then phosphorylate target enzymes, ion channels, or transcription factors, resulting in a cellular response. R, regulatory subunit; C, catalytic subunit; P, phosphate group.

Protein kinase A consists of catalytic and regulatory subunits, with the kinase activity residing in the catalytic subunit. When cAMP concentrations in the cell are low, two regulatory subunits bind to and inactivate two catalytic subunits, forming an inactive tetramer (Fig. 1.14). When cAMP is formed in response to hormonal stimulation, two molecules of cAMP bind to each of the regulatory subunits, causing them to dissociate from the catalytic subunits. This relieves the inhibition of catalytic subunits and allows them to catalyze the phosphorylation of target substrates and produce the resultant biological response to the hormone (see Fig. 1.14).

cGMP Is an Important Second Messenger in Smooth Muscle and Sensory Cells

cGMP, a second messenger similar and parallel to cAMP, is formed, much like cAMP, by the enzyme **guanylyl cyclase**. Although the full role of cGMP as a second messenger is not as well understood, its importance is finally being appreciated with respect to signal transduction in sensory tissues (see Chapter 4) and smooth muscle tissues (see Chapters 9 and 16).

One reason for its less apparent role is that few substrates for **cGMP-dependent protein kinase**, the main target of cGMP production, are known. The production of cGMP is mainly regulated by the activation of a cytoplasmic form of guanylyl cyclase, a target of the paracrine mediator nitric oxide (NO) that is produced by endothelial as well as other cell types and can mediate smooth muscle relaxation (see Chapter 16). Atrial natriuretic peptide and guanylin (an intestinal hormone) also use cGMP as a second messenger, and in these cases, the plasma membrane receptors for these hormones express guanylyl cyclase activity.

Second Messengers 1,2-Diacylglycerol (DAG) and Inositol Trisphosphate (IP_3) Are Generated by the Hydrolysis of Phosphatidylinositol 4,5-Bisphosphate (PIP_2)

Some G protein-coupled receptors are coupled to a different effector enzyme, **phospholipase C (PLC)**, which is localized to the inner leaflet of the plasma membrane. Similar to other GPCRs, binding of ligand or agonist to the receptor results in activation of the associated G protein, usually $G\alpha_q$ (or G_q). Depending on the isoform of the G protein associated with the receptor, either the α or the $\beta\gamma$ subunit may stimulate PLC. Stimulation of PLC results in the hydrolysis of the membrane phospholipid, phosphatidylinositol 4,5-bisphosphate (PIP_2), into 1,2-diacylglycerol (DAG) and inositol trisphosphate (IP_3). Both DAG and IP_3 serve as second messengers in the cell (Fig. 1.15).

In its second messenger role, DAG accumulates in the plasma membrane and activates the membrane-bound calcium- and lipid-sensitive enzyme **protein kinase C** (see Fig. 1.15). When activated, this enzyme catalyzes the phosphorylation of specific proteins, including other enzymes and transcription factors, in the cell to produce appropriate physiological effects, such as cell proliferation. Several tumor-promoting phorbol esters that mimic the structure of DAG have been shown to activate protein kinase C. They can, therefore, bypass the receptor by passing through the plasma membrane and directly activating protein kinase C, causing the phosphorylation of downstream targets to result in cellular proliferation.

IP_3 promotes the release of calcium ions into the cytoplasm by activation of endoplasmic or sarcoplasmic reticulum IP_3-gated calcium release channels (see Chapter 9). The concentration of free calcium ions in the cytoplasm of most cells is in the range of 10^{-7} M. With appropriate stimulation, the concentration may abruptly increase 1,000 times or more. The resulting increase in free cytoplasmic calcium synergizes with the action of DAG in the activation of some forms of protein kinase C and may also activate many other calcium-dependent processes.

Mechanisms exist to reverse the effects of DAG and IP_3 by rapidly removing them from the cytoplasm. The IP_3 is dephosphorylated to inositol, which can be reused for phosphoinositide synthesis. The DAG is converted to phosphatidic acid by the addition of a phosphate group to carbon number 3. Like inositol, phosphatidic acid can be used for the resynthesis of membrane inositol phospholipids (see Fig.1.15). On removal of the IP_3 signal, calcium is quickly pumped back into its storage sites, restoring cytoplasmic calcium concentrations to their low prestimulus levels.

A

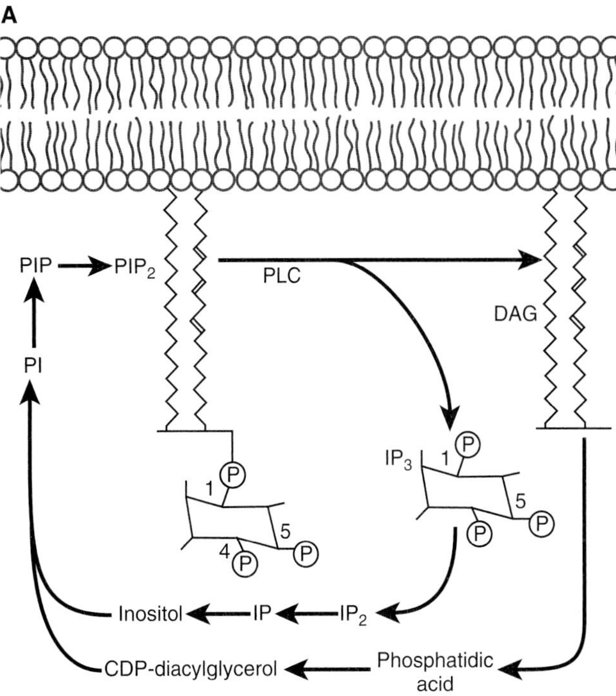

B

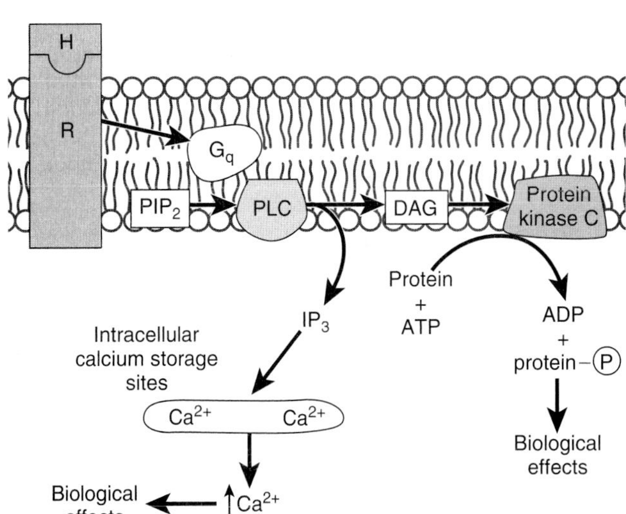

FIGURE 1.15 **The phosphatidylinositol second messenger system.** **A**, The pathway leading to the generation of inositol trisphosphate and diacylglycerol. The successive phosphorylation of phosphatidylinositol (PI) leads to the generation of phosphatidylinositol 4,5-bisphosphate (PIP_2). Phospholipase C (PLC) catalyzes the breakdown of PIP_2 to inositol trisphosphate (IP_3) and diacylglycerol (DAG), which are used for signaling and can be recycled to generate phosphatidylinositol. **B**, The generation of IP_3 and DAG and their intracellular signaling roles. The binding of hormone (H) to a G protein-coupled receptor (R) can lead to the activation of PLC. In this case, the $G\alpha$ subunit is G_q, a G protein that couples receptors to PLC. The activation of PLC results in the cleavage of PIP_2 to IP_3 and DAG. IP_3 interacts with calcium release channels in the endoplasmic reticulum, causing the release of calcium to the cytoplasm. Increased intracellular calcium can lead to the activation of calcium-dependent enzymes. An accumulation of DAG in the plasma membrane leads to the activation of the calcium- and phospholipid-dependent enzyme protein kinase C and phosphorylation of its downstream targets.

In addition to IP_3, other, perhaps more potent phosphoinositols, such as IP_4 or IP_5, may also be produced in response to stimulation. These are formed by the hydrolysis of appropriate phosphatidylinositol phosphate precursors found in the cell membrane. The precise role of these phosphoinositols is unknown. Evidence suggests that the hydrolysis of other phospholipids, such as phosphatidylcholine, may play an analogous role in hormone-signaling processes.

Cells Use Calcium as a Second Messenger by Keeping Resting Intracellular Calcium Levels Low

The levels of cytosolic calcium in an unstimulated cell are about 10,000 times less (10^{-7} M versus 10^{-3} M) than in the extracellular fluid. This large gradient of calcium is maintained by the limited permeability of the plasma membrane to calcium, by calcium transporters in the plasma membrane that extrude calcium, by calcium pumps in intracellular organelles, and by cytoplasmic and organellar proteins that bind calcium to buffer its free cytoplasmic concentration. Several plasma membrane ion channels serve to increase cytosolic calcium levels. Either these ion channels are voltage-gated and open when the plasma membrane depolarizes, or they may be controlled by phosphorylation by protein kinase A or protein kinase C.

In addition to the plasma membrane ion channels, the endoplasmic reticulum has two other main types of ion channels that, when activated, release calcium into the cytoplasm, causing an increase in cytoplasmic calcium. The small water-soluble molecule IP_3 activates the **IP_3-gated calcium release channel** in the endoplasmic reticulum. The activated channel opens to allow calcium to flow down a concentration gradient into the cytoplasm. The IP_3-gated channels are structurally similar to the second type of calcium release channel, the **ryanodine receptor**, found in the sarcoplasmic reticulum of muscle cells. Ryanodine receptors release calcium to trigger muscle contraction when an action potential invades the transverse tubule system of skeletal or cardiac muscle fibers (see Chapter 8). Both types of channels are regulated by positive feedback, in which the released cytosolic calcium can bind to the receptor to enhance further calcium release. This causes the calcium to be released suddenly in a spike, followed by a wave-like flow of the ion throughout the cytoplasm.

Increasing cytosolic free calcium activates many different signaling pathways and leads to numerous physiological events, such as muscle contraction, neurotransmitter secretion, and cytoskeletal polymerization. Calcium acts as a second messenger in two ways:

- It binds directly to an effector molecule, such as protein kinase C, to participate in its activation.
- It binds to an intermediary cytosolic calcium-binding protein, such as calmodulin.

Calmodulin is a small protein (16 kDa) with four binding sites for calcium. The binding of calcium to calmodulin causes calmodulin to undergo a dramatic conformational change and increases the affinity of this intracellular calcium "receptor" for its effectors (Fig. 1.16). Calcium-calmodulin complexes bind to and activate a variety of cellular proteins, including protein kinases that are important in many physiological processes, such as smooth muscle contraction (myosin light-chain kinase; see Chapter 9) and hormone synthesis (aldosterone synthesis; see Chapter 34), and ultimately result in altered cellular function.

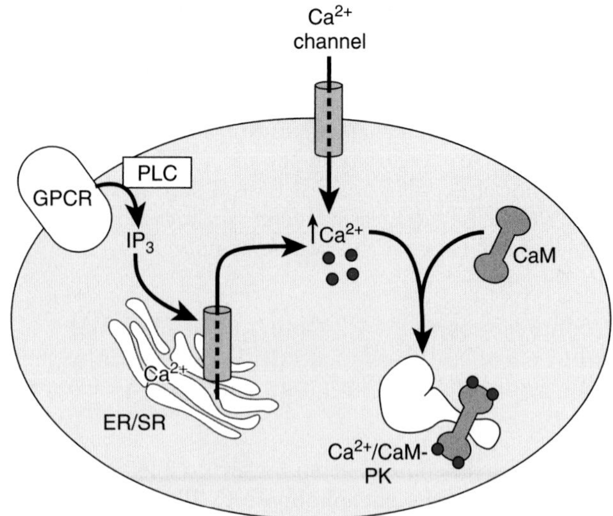

FIGURE 1.16 The role of calcium in intracellular signaling and activation of calcium-calmodulin-dependent protein kinases. Levels of intracellular calcium are regulated by membrane-bound ion channels that allow the entry of calcium from the extracellular space or release calcium from internal stores (e.g., endoplasmic reticulum, sarcoplasmic reticulum in muscle cells, and mitochondria). Calcium can also be released from intracellular stores via the G-protein-mediated activation of PLC and the generation of IP_3. IP_3 causes the release of calcium from the endoplasmic or sarcoplasmic reticulum in muscle cells by interaction with calcium ion channels. When intracellular calcium rises, four calcium ions complex with the dumbbell-shaped calmodulin protein (CaM) to induce a conformational change. Ca^{2+}/CaM can then bind to a spectrum of target proteins including Ca^{2+}/CaM-PKs, which then phosphorylate other substrates, leading to a response. IP_3, inositol trisphosphate; PLC, phospholipase C; CaM, calmodulin; Ca^{2+}/CaM-PK, calcium-calmodulin-dependent protein kinases; ER/SR, endoplasmic/sarcoplasmic reticulum.

Two mechanisms operate to terminate calcium action. The IP_3 generated by the activation of PLC can be dephosphorylated and, thus, inactivated by cellular phosphatases. In addition, the calcium that enters the cytosol can be rapidly removed. The plasma membrane, endoplasmic reticulum, sarcoplasmic reticulum, and mitochondrial membranes all have ATP-driven calcium pumps that drive the free calcium out of the cytosol to the extracellular space or into an intracellular organelle. Lowering cytosolic calcium concentrations shifts the equilibrium in favor of the release of calcium from calmodulin. Calmodulin then dissociates from the various proteins that were activated, and the cell returns to its basal state.

INTRACELLULAR RECEPTORS AND HORMONE SIGNALING

The intracellular receptors, in contrast to the plasma membrane-bound receptors, can be located in either the cytosol or the nucleus and are distinguished by their mode of activation and function. The ligands for these receptors must be lipid soluble because of the plasma membranes that must be traversed for the ligand to reach its receptor. The main result of activation of the intracellular receptors is altered gene expression.

Steroid and Thyroid Hormone Receptors Are Intracellular Receptors Located in the Cytoplasm or Nucleus

For the activation of intracellular receptors to occur, ligands must cross the plasma membrane. The hormone ligands that belong to this group include the steroids (e.g., estradiol, testosterone, progesterone, cortisone, and aldosterone), 1,25-dihydroxyvitamin D_3, thyroid hormone, and retinoids. These hormones are typically delivered to their target cells bound to specific carrier proteins. Because of their lipid solubility, these hormones freely diffuse through both plasma and nuclear membranes. These hormones bind to specific receptors that reside either in the cytoplasm or the nucleus. Steroid hormone receptors are located in the cytoplasm and are usually found complexed with other proteins that maintain the receptor in an inactive conformation. In contrast, the thyroid hormones and retinoic acid bind to receptors that are already bound to response elements in the DNA of specific genes. The unoccupied receptors are inactive until the hormone binds, and they serve as repressors in the absence of hormone. These receptors are discussed in Chapters 31 and 33. The model of steroid hormone action shown in Figure 1.17 is generally applicable to all steroid and thyroid hormones.

All **steroid hormone receptors** have similar structures, with three main domains. The N-terminal regulatory domain regulates the transcriptional activity of the receptor and may have sites for phosphorylation by protein kinases that may also be involved in modifying the transcriptional activity of the receptor. There is a centrally located DNA-binding domain and a carboxyl-terminal hormone-binding and dimerization domain. When hormones bind, the hormone-receptor complex moves to the nucleus, where it binds to specific DNA sequences in the gene regulatory (promoter) region of specific hormone-responsive genes. The targeted DNA sequence in the promoter is called a **hormone response element** (HRE). Binding of the hormone-receptor complex to the HRE can either activate or repress transcription. The end result of stimulation by steroid hormones is a change in the readout or transcription of the genome. While most effects involve increased production of specific proteins, repressed production of certain proteins by steroid hormones can also occur. These newly synthesized proteins and/or enzymes will affect cellular metabolism with responses attributable to that particular steroid hormone.

Hormones Bound to Their Receptors Regulate Gene Expression

The interaction of hormone and receptor leads to the activation (or transformation) of receptors into forms with an increased affinity for binding to specific HRE or acceptor sites on the chromosomes. The molecular basis of activation *in vivo* is unknown but appears to involve a decrease in apparent molecular weight or in the aggregation state of receptors, as determined by density gradient centrifugation. The binding of hormone-receptor complexes to chromatin results in alterations in RNA polymerase activity that lead to either increased or decreased transcription of specific portions of the genome. As a result, mRNA is produced, leading to the production of new cellular proteins or changes in the rates of synthesis of preexisting proteins (see Fig. 1.17).

The molecular mechanism of steroid hormone-receptor activation and/or transformation, how the hormone-receptor complex activates transcription, and how the hormone-receptor complex recognizes specific response elements of the genome are not well understood but are under active investigation. Steroid hormone receptors are also known to undergo phosphorylation/dephosphorylation reactions. The effect of this covalent modification is also an area of active research.

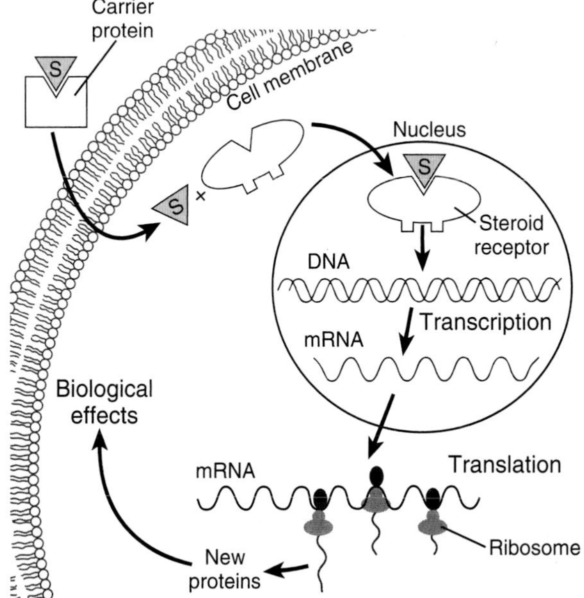

FIGURE 1.17 The general mechanism of action of steroid hormones. Steroid hormones (S) are lipid soluble and pass through the plasma membrane, where they bind to a cognate receptor in the cytoplasm. The steroid hormone-receptor complex then moves to the nucleus and binds to a hormone response element in the promoter-regulatory region of specific hormone-responsive genes. Binding of the steroid hormone-receptor complex to the response element initiates transcription of the gene, to form messenger RNA (mRNA). The mRNA moves to the cytoplasm, where it is translated into a protein that participates in a cellular response. Thyroid hormones are thought to act by a similar mechanism, although their receptors are already bound to a hormone response element, repressing gene expression. The thyroid hormone-receptor complex forms directly in the nucleus and results in the activation of transcription from the thyroid hormone-responsive gene.

REVIEW QUESTIONS

DIRECTIONS: Each of the numbered items or incomplete statements in this section is followed by answers or by completions of the statement. Select the ONE lettered answer or completion that is BEST in each case.

1. If a region or compartment is in a steady state with respect to a particular substance, then
   (A) The amount of the substance in the compartment is increasing
   (B) The amount of the substance in the compartment is decreasing
   (C) The amount of the substance in the compartment does not change with respect to time
   (D) There is no movement into or out of the compartment
   (E) The compartment must be in equilibrium with its surroundings
2. A 62-year-old woman eats a high carbohydrate meal. Her plasma glucose concentration rises, and this results in increased insulin secretion from the pancreatic islet cells. The insulin response is an example of
   (A) Chemical equilibrium
   (B) End-product inhibition
   (C) Feedforward control
   (D) Negative feedback
   (E) Positive feedback
3. In animal models of autosomal recessive polycystic kidney disease, epidermal growth factor (EGF) receptors may be abnormally expressed on the urine side of kidney epithelial cells and may be stimulated by EGF in the urine, causing excessive cell proliferation and formation of numerous kidney cysts. What type of drug might be useful in treating this condition?
   (A) Adenylyl cyclase stimulator
   (B) EGF agonist
   (C) Phosphatase inhibitor
   (D) Phosphodiesterase inhibitor
   (E) Tyrosine kinase inhibitor
4. Second messengers
   (A) Are extracellular ligands
   (B) Are always available for signal transduction
   (C) Always produce the same cellular response
   (D) Include nucleotides, ions, and gases
   (E) Are produced only by tyrosine kinase receptors
5. The second messengers cyclic AMP and cyclic GMP
   (A) Activate the same signal transduction pathways
   (B) Are generated by the activation of cyclases
   (C) Activate the same protein kinase
   (D) Are important only in sensory transduction
   (E) Can activate phospholipase C
6. Binding of estrogen to its steroid hormone receptor
   (A) Stimulates the GTPase activity of the trimeric G protein coupled to the estrogen receptor
   (B) Stimulates the activation of the IP_3 receptor in the sarcoplasmic reticulum to increase intracellular calcium
   (C) Stimulates phosphorylation of tyrosine residues in the cytoplasmic domain of the receptor
   (D) Stimulates the movement of the hormone-receptor complex to the nucleus to cause gene activation
   (E) Stimulates the activation of the MAP kinase pathway and results in the regulation of several transcription factors
7. A single cell within a culture of freshly isolated cardiac muscle cells is injected with a fluorescent dye that cannot cross cell membranes. Within minutes, several adjacent cells become fluorescent. The most likely explanation for this observation is the presence of
   (A) Ryanodine receptors
   (B) IP_3 receptors
   (C) Transverse tubules
   (D) Desmosomes
   (E) Gap junctions
8. Many signaling pathways involve the generation of inositol trisphosphate (IP_3) and diacylglycerol (DAG). These molecules
   (A) Are first messengers
   (B) Activate phospholipase C
   (C) Can activate tyrosine kinase receptors
   (D) Can activate calcium calmodulin-dependent protein kinases
   (E) Are derived from PIP_2
9. Tyrosine kinase receptors
   (A) Have constitutively active tyrosine kinase domains
   (B) Phosphorylate and activate ras directly
   (C) Mediate cellular processes involved in growth and differentiation
   (D) Are not phosphorylated upon activation
   (E) Are monomeric receptors upon activation
10. A pituitary tumor is removed from a 40-year-old man with acromegaly resulting from excessive secretion of growth hormone. It is known that G proteins and adenylyl cyclase normally mediate the stimulation of growth hormone secretion produced by growth hormone-releasing hormone (GHRH). Which of the following problems is most likely to be present in the patient's tumor cells?
    (A) Adenylyl cyclase activity is abnormally low
    (B) The $G\alpha_s$ subunit is unable to hydrolyze GTP
    (C) The $G\alpha_s$ subunit is inactivated
    (D) The $G\alpha_i$ subunit is activated
    (E) The cells lack GHRH receptors

SUGGESTED READING

Conn PM, Means AR, eds. Principles of Molecular Regulation. Totowa, NJ: Humana Press, 2000.

Farfel Z, Bourne HR, Iiri T. The expanding spectrum of G protein diseases. N Engl J Med 1999;340:1012–1020.

Heldin C-H, Purton M, eds. Signal Transduction. London: Chapman & Hall, 1996.

Krauss G. Biochemistry of Signal Transduction and Regulation. New York: Wiley-VCH, 1999.

Lodish H, Berk A, Zipursky S, et al. Molecular Cell Biology. 4th Ed. New York: WH Freeman, 2000.

Schultz SG. Homeostasis, humpty dumpty, and integrative biology. News Physiol Sci 1996;1:238–246.

The Plasma Membrane, Membrane Transport, and the Resting Membrane Potential

Stephen A. Kempson, Ph.D.

CHAPTER OUTLINE

- **THE STRUCTURE OF THE PLASMA MEMBRANE**
- **MECHANISMS OF SOLUTE TRANSPORT**
- **THE MOVEMENT OF WATER ACROSS THE PLASMA MEMBRANE**
- **THE RESTING MEMBRANE POTENTIAL**

KEY CONCEPTS

1. The two major components of the plasma membrane of a cell are proteins and lipids, present in about equal proportions.
2. Membrane proteins are responsible for most of the functions of the plasma membrane, including the transport of water and solutes across the membrane and providing specific binding sites for extracellular signaling molecules such as hormones.
3. Carrier-mediated transport systems allow the rapid transport of polar molecules, reach a maximum rate at high substrate concentration, exhibit structural specificity, and are competitively inhibited by molecules of similar structure.
4. Voltage-gated channels are opened by a change in the membrane potential, and ligand-gated channels are opened by the binding of a specific agonist.
5. The Na^+/K^+-ATPase pump is an example of primary active transport, and Na^+-coupled glucose transport is an example of secondary active transport.
6. The polarized organization of epithelial cells produces a directional movement of solutes and water across the epithelium.
7. Many cells regulate their volume when exposed to osmotic stress by activating transport systems that allow the exit or entry of solute so that water will follow.
8. The Goldman equation gives the value of the membrane potential when all the permeable ions are accounted for.
9. In most cells, the resting membrane potential is close to the Nernst potential for K^+.

The intracellular fluid of living cells, the **cytosol**, has a composition very different from that of the extracellular fluid. For example, the concentrations of potassium and phosphate ions are higher inside cells than outside, whereas sodium, calcium, and chloride ion concentrations are much lower inside cells than outside. These differences are necessary for the proper functioning of many intracellular enzymes; for instance, the synthesis of proteins by the ribosomes requires a relatively high potassium concentration. The cell membrane or **plasma membrane** creates and maintains these differences by establishing a permeability barrier around the cytosol. The ions and cell proteins needed for normal cell function are prevented from leaking out; those not needed by the cell are unable to enter the cell freely. The cell membrane also keeps metabolic intermediates near where they will be needed for further synthesis or processing and retains metabolically expensive proteins inside the cell.

The plasma membrane is necessarily selectively permeable. Cells must receive nutrients in order to function, and they must dispose of metabolic waste products. To function in coordination with the rest of the organism, cells receive and send information in the form of hormones and neurotransmitters. The plasma membrane has mechanisms that allow specific molecules to cross the barrier around the cell. A selective barrier surrounds not only the cell but also every intracellular organelle that requires an internal milieu different from that of the cytosol. The cell nucleus, mitochondria, endoplasmic reticulum, Golgi apparatus, and lysosomes are delimited by membranes similar in composition to the plasma membrane. This chapter describes the specific types of membrane transport mechanisms for ions and other solutes, their relative contributions to the resting membrane potential, and how their activities are coordinated to achieve directional transport from one side of a cell layer to the other.

THE STRUCTURE OF THE PLASMA MEMBRANE

The first theory of membrane structure proposed that cells are surrounded by a double layer of lipid molecules, a **lipid bilayer**. This theory was based on the known tendency of lipid molecules to form lipid bilayers with low permeability to water-soluble molecules. However, the lipid bilayer theory did not explain the selective movement of certain water-soluble compounds, such as glucose and amino acids, across the plasma membrane. In 1972, Singer and Nicolson proposed the **fluid mosaic model** of the plasma membrane (Fig. 2.1). With minor modifications, this model is still accepted as the correct picture of the structure of the plasma membrane.

The Plasma Membrane Has Proteins Inserted in the Lipid Bilayer

Proteins and lipids are the two major components of the plasma membrane, present in about equal proportions by weight. The various lipids are arranged in a lipid bilayer, and two different types of proteins are associated with this bilayer. **Integral proteins** (or intrinsic proteins) are embedded in the lipid bilayer; many span it completely, being accessible from the inside and outside of the membrane. The polypeptide chain of these proteins may cross the lipid bilayer once or may make multiple passes across it. The membrane-spanning segments usually contain amino acids with nonpolar side chains and are arranged in an ordered α-helical conformation. **Peripheral proteins** (or extrinsic proteins) do not penetrate the lipid bilayer. They are in contact with the outer side of only one of the lipid layers—either the layer facing the cytoplasm or the layer facing the extracellular fluid (see Fig. 2.1). Many membrane proteins have carbohydrate molecules, in the form of specific sugars, attached to the parts of the proteins that are exposed to the extracellular fluid. These molecules are known as **glycoproteins**. Some of the integral membrane proteins can move in the plane of the membrane, like small boats floating in the "sea" formed by the bilayer arrangement of the lipids. Other membrane proteins are anchored to the cytoskeleton inside the cell or to proteins of the extracellular matrix.

The proteins in the plasma membrane play a variety of roles. Many peripheral membrane proteins are enzymes, and many membrane-spanning integral proteins are carriers or channels for the movement of water-soluble molecules and ions into and out of the cell. Another important role of membrane proteins is structural; for example, certain membrane proteins in the erythrocyte help maintain the biconcave shape of the cell. Finally, some membrane proteins serve as highly specific recognition sites or receptors on the outside of the cell membrane to which extracellular molecules, such as hormones, can bind. If the receptor is a membrane-spanning protein, it provides a mechanism for converting an extracellular signal into an intracellular response.

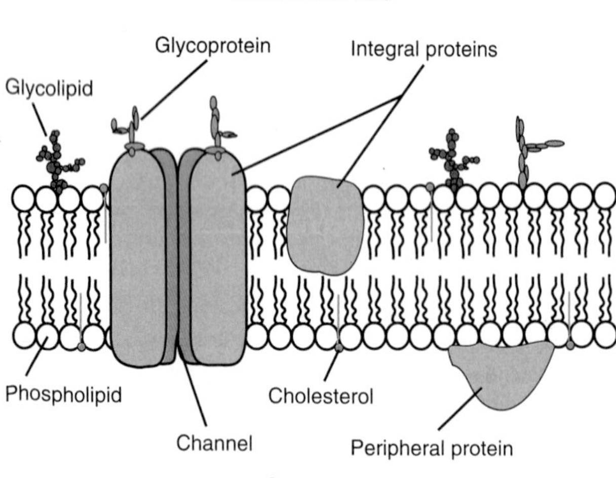

FIGURE 2.1 The fluid mosaic model of the plasma membrane. Lipids are arranged in a bilayer. Integral proteins are embedded in the bilayer and often span it. Some membrane-spanning proteins form channels. Peripheral proteins do not penetrate the bilayer.

There Are Different Types of Membrane Lipids

Lipids found in cell membranes can be classified into two broad groups: those that contain fatty acids as part of the lipid molecule and those that do not. Phospholipids are an example of the first group, and cholesterol is the most important example of the second group.

Phospholipids. The fatty acids present in **phospholipids** are molecules with a long hydrocarbon chain and a carboxyl terminal group. The hydrocarbon chain can be saturated (no double bonds between the carbon atoms) or unsaturated (one or more double bonds present). The composition of fatty acids gives them some peculiar characteristics. The long hydrocarbon chain tends to avoid contact with water and is described as **hydrophobic**. The carboxyl group at the other end is compatible with water and is termed **hydrophilic**. Fatty acids are said to be **amphipathic** because both hydrophobic and hydrophilic regions are present in the same molecule.

Phospholipids are the most abundant complex lipids found in cell membranes. They are amphipathic molecules formed by two fatty acids (normally, one saturated and one unsaturated) and one phosphoric acid group substituted on the backbone of a glycerol or sphingosine molecule. This arrangement produces a hydrophobic area formed by the two fatty acids and a polar hydrophilic head. When phospholipids are arranged in a bilayer, the polar heads are on the outside and the hydrophobic fatty acids on the inside. It is difficult for water-soluble molecules and ions to pass directly through the hydrophobic interior of the lipid bilayer.

The phospholipids, with a backbone of sphingosine (a long amino alcohol), are usually called sphingolipids and are present in all plasma membranes in small amounts. They are especially abundant in brain and nerve cells.

Glycolipids are lipid molecules that contain sugars and sugar derivatives (instead of phosphoric acid) in the polar head. They are located mainly in the outer half of the lipid bilayer, with the sugar molecules facing the extracellular fluid.

Cholesterol. **Cholesterol** is an important component of mammalian plasma membranes. The proportion of cholesterol in plasma membranes varies from 10% to 50% of total lipids. Cholesterol has a rigid structure that stabilizes the cell membrane and reduces the natural mobility of the complex lipids in the plane of the membrane. Increasing amounts of cholesterol make it more difficult for lipids and proteins to move in the membrane. Some cell functions, such as the response of immune system cells to the presence of an antigen, depend on the ability of membrane proteins to move in the plane of the membrane to bind the antigen. A decrease in membrane fluidity resulting from an increase in cholesterol will impair these functions.

MECHANISMS OF SOLUTE TRANSPORT

All cells need to import oxygen, sugars, amino acids, and some small ions and to export carbon dioxide, metabolic wastes, and secretions. At the same time, specialized cells require mechanisms to transport molecules such as enzymes, hormones, and neurotransmitters. The movement of large molecules is carried out by endocytosis and exocytosis, the transfer of substances into or out of the cell, respectively, by vesicle formation and vesicle fusion with the plasma membrane. Cells also have mechanisms for the rapid movement of ions and solute molecules across the plasma membrane. These mechanisms are of two general types: **passive movement**, which requires no direct expenditure of metabolic energy, and **active movement**, which uses metabolic energy to drive solute transport.

Macromolecules Cross the Plasma Membrane by Vesicle Fusion

Phagocytosis and Endocytosis. **Phagocytosis** is the ingestion of large particles or microorganisms, usually occurring only in specialized cells such as macrophages (Fig. 2.2). An important function of macrophages in humans is to remove invading bacteria. The phagocytic vesicle (1 to 2 μm in diameter) is almost as large as the phagocytic cell itself. Phagocytosis requires a specific stimulus. It occurs only after the extracellular particle has bound to the extracellular surface. The particle is then enveloped by expansion of the cell membrane around it.

Endocytosis is a general term for the process in which a region of the plasma membrane is pinched off to form an endocytic vesicle inside the cell. During vesicle formation, some fluid, dissolved solutes, and particulate material from the extracellular medium are trapped inside the vesicle and internalized by the cell. Endocytosis produces much smaller endocytic vesicles (0.1 to 0.2 μm in diameter) than phagocytosis. It occurs in almost all cells and is termed a constitutive process because it occurs continually and specific stimuli are not required. In further contrast to phagocytosis, endocytosis originates with the formation of depressions in the cell membrane. The depressions pinch off

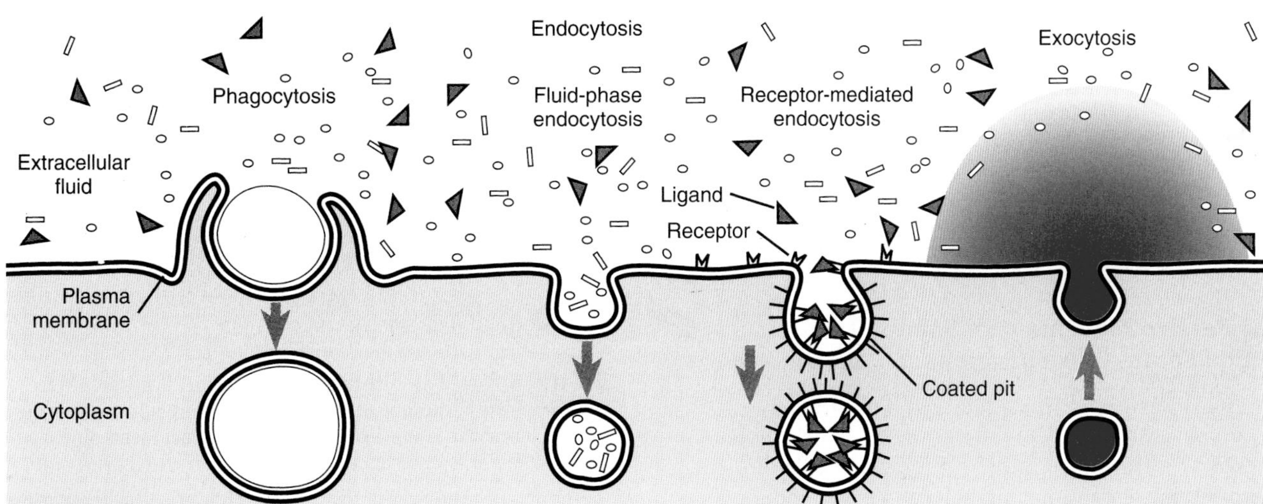

FIGURE 2.2 The transport of macromolecules across the plasma membrane by the formation of vesicles. Particulate matter in the extracellular fluid is engulfed and internalized by phagocytosis. During fluid-phase endocytosis, extracellular fluid and dissolved macromolecules enter the cell in endocytic vesicles that pinch off at depressions in the plasma membrane. Receptor-mediated endocytosis uses membrane receptors at coated pits to bind and internalize specific solutes (ligands). Exocytosis is the release of macromolecules destined for export from the cell. These are packed inside secretory vesicles that fuse with the plasma membrane and release their contents outside the cell. (Modified from Dautry-Varsat A, Lodish HF. How receptors bring proteins and particles into cells. Sci Am 1984;250(5):52–58.)

within a few minutes after they form and give rise to endocytic vesicles inside the cell.

Two main types of endocytosis can be distinguished (see Fig. 2.2). **Fluid-phase endocytosis** is the nonspecific uptake of the extracellular fluid and all its dissolved solutes. The material is trapped inside the endocytic vesicle as it is pinched off inside the cell. The amount of extracellular material internalized by this process is directly proportional to its concentration in the extracellular solution. **Receptor-mediated endocytosis** is a more efficient process that uses receptors on the cell surface to bind specific molecules. These receptors accumulate at specific depressions known as **coated pits**, so named because the cytosolic surface of the membrane at this site is covered with a coat of several proteins. The coated pits pinch off continually to form endocytic vesicles, providing the cell with a mechanism for rapid internalization of a large amount of a specific molecule without the need to endocytose large volumes of extracellular fluid. The receptors also aid the cellular uptake of molecules present at low concentrations outside the cell. Receptor-mediated endocytosis is the mechanism by which cells take up a variety of important molecules, including hormones; growth factors; and serum transport proteins, such as **transferrin** (an iron carrier). Foreign substances, such as diphtheria toxin and certain viruses, also enter cells by this pathway.

Exocytosis. Many cells synthesize important macromolecules that are destined for **exocytosis** or export from the cell. These molecules are synthesized in the endoplasmic reticulum, modified in the Golgi apparatus, and packed inside transport vesicles. The vesicles move to the cell surface, fuse with the cell membrane, and release their contents outside the cell (see Fig. 2.2).

There are two exocytic pathways—constitutive and regulated. Some proteins are secreted continuously by the cells that make them. Secretion of mucus by **goblet cells** in the small intestine is a specific example. In this case, exocytosis follows the **constitutive pathway**, which is present in all cells. In other cells, macromolecules are stored inside the cell in secretory vesicles. These vesicles fuse with the cell membrane and release their contents only when a specific extracellular stimulus arrives at the cell membrane. This pathway, known as the **regulated pathway**, is responsible for the rapid "on-demand" secretion of many specific hormones, neurotransmitters, and digestive enzymes.

The Passive Movement of Solutes Tends to Equilibrate Concentrations

Simple Diffusion. Any solute will tend to uniformly occupy the entire space available to it. This movement, known as **diffusion**, is due to the spontaneous Brownian (random) movement that all molecules experience and that explains many everyday observations. Sugar diffuses in coffee, lemon diffuses in tea, and a drop of ink placed in a glass of water will diffuse and slowly color all the water. The net result of diffusion is the movement of substances according to their difference in concentrations, from regions of high concentration to regions of low concentration. Diffusion is an effective way for substances to move short distances.

The speed with which the diffusion of a solute in water occurs depends on the difference of concentration, the size of the molecules, and the possible interactions of the diffusible substance with water. These different factors appear in Fick's law, which describes the diffusion of any solute in water. In its simplest formulation, Fick's law can be written as:

$$J = DA\,(C_1 - C_2)/\Delta X \qquad (1)$$

where J is the flow of solute from region 1 to region 2 in the solution, D is the diffusion coefficient of the solute and takes into consideration such factors as solute molecular size and interactions of the solute with water, A is the cross-sectional area through which the flow of solute is measured, C is the concentration of the solute at regions 1 and 2, and ΔX is the distance between regions 1 and 2. J is expressed in units of amount of substance per unit area per unit time, for example, mol/cm^2 per hour, and is also referred to as the solute **flux**.

Diffusive Membrane Transport. Solutes can enter or leave a cell by diffusing passively across the plasma membrane. The principal force driving the diffusion of an uncharged solute is the difference of concentration between the inside and the outside of the cell (Fig. 2.3). In the case of an electrically charged solute, such as an ion, diffusion is also driven by the membrane potential, which is the electrical gradient across the membrane. The membrane potential of most living cells is negative inside the cell relative to the outside.

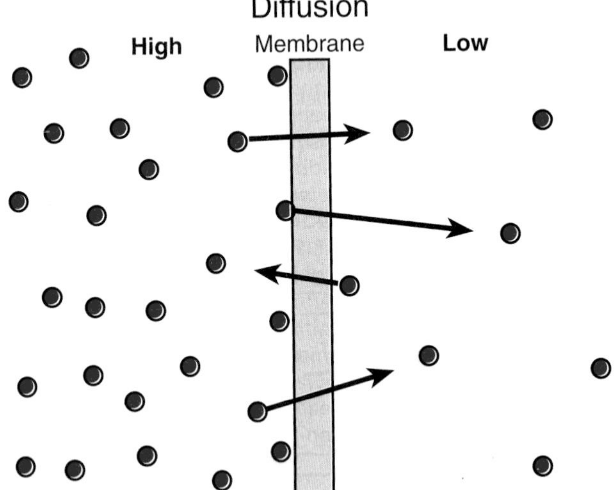

FIGURE 2.3 **The diffusion of gases and lipid-soluble molecules through the lipid bilayer.** In this example, the diffusion of a solute across a plasma membrane is driven by the difference in concentration on the two sides of the membrane. The solute molecules move randomly by Brownian movement. Initially, random movement from left to right across the membrane is more frequent than movement in the opposite direction because there are more molecules on the left side. This results in a net movement of solute from left to right across the membrane until the concentration of solute is the same on both sides. At this point, equilibrium (no net movement) is reached because solute movement from left to right is balanced by equal movement from right to left.

Diffusion across a membrane has no preferential direction; it can occur from the outside of the cell toward the inside or from the inside of the cell toward the outside. For any substance, it is possible to measure the **permeability coefficient (P)**, which gives the speed of the diffusion across a unit area of plasma membrane for a defined driving force. Fick's law for the diffusion of an uncharged solute across a membrane can be written as:

$$J = PA\,(C_1 - C_2) \qquad (2)$$

which is similar to equation 1. P includes the membrane thickness, diffusion coefficient of the solute within the membrane, and solubility of the solute in the membrane. Dissolved gases, such as oxygen and carbon dioxide, have high permeability coefficients and diffuse across the cell membrane rapidly. Since diffusion across the plasma membrane usually implies that the diffusing solute enters the lipid bilayer to cross it, the solute's solubility in a lipid solvent (e.g., olive oil or chloroform) compared with its solubility in water is important in determining its permeability coefficient.

A substance's solubility in oil compared with its solubility in water is its **partition coefficient**. Lipophilic substances that mix well with the lipids in the plasma membrane have high partition coefficients and, as a result, high permeability coefficients; they tend to cross the plasma membrane easily. Hydrophilic substances, such as ions and sugars, do not interact well with the lipid component of the membrane, have low partition coefficients and low permeability coefficients, and diffuse across the membrane more slowly.

For solutes that diffuse across the lipid part of the plasma membrane, the relationship between the rate of movement and the difference in concentration between the two sides of the membrane is linear (Fig. 2.4). The higher the difference in concentration ($C_1 - C_2$), the greater the amount of substance crossing the membrane per unit time.

Facilitated Diffusion via Carrier Proteins. For many solutes of physiological importance, such as sugars and amino acids, the relationship between transport rate and concentration difference follows a curve that reaches a plateau (Fig. 2.5). Furthermore, the rate of transport of these hydrophilic substances across the cell membrane is much faster than expected for simple diffusion through a lipid bilayer. Membrane transport with these characteristics is often called **carrier-mediated transport** because an integral membrane protein, the carrier, binds the transported solute on one side of the membrane and releases it at the other side. Although the details of this transport mechanism are unknown, it is hypothesized that the binding of the solute causes a conformational change in the carrier protein, which results in translocation of the solute (Fig. 2.6). Because there are limited numbers of these carriers in any cell membrane, increasing the concentration of the solute initially uses the existing "spare" carriers to transport the solute at a higher rate than by simple diffusion. As the concentration of the solute increases further and more solute molecules bind to carriers, the transport system eventually reaches **saturation**, when all the carriers are involved in translocating molecules of solute. At this point, additional increases in solute concentration do not increase the rate of solute transport (see Fig. 2.5).

The types of carrier-mediated transport mechanisms considered here can transport a solute along its concentration gradient only, as in simple diffusion. Net movement

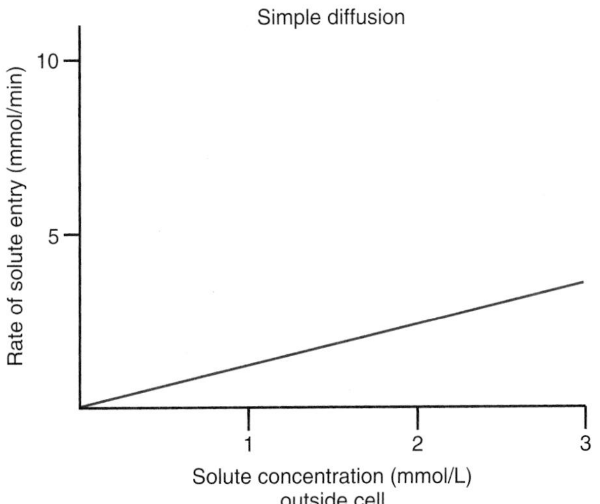

FIGURE 2.4 **A graph of solute transport across a plasma membrane by simple diffusion.** The rate of solute entry increases linearly with extracellular concentration of the solute. Assuming no change in intracellular concentration, increasing the extracellular concentration increases the gradient that drives solute entry.

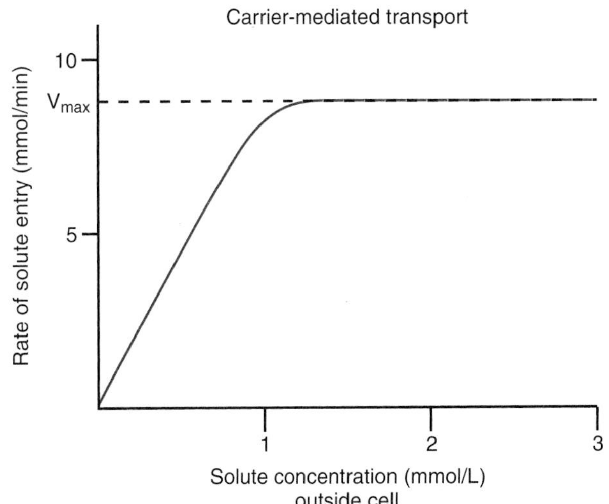

FIGURE 2.5 **A graph of solute transport across a plasma membrane by carrier-mediated transport.** The rate of transport is much faster than that of simple diffusion (see Fig. 2.4) and increases linearly as the extracellular solute concentration increases. The increase in transport is limited, however, by the availability of carriers. Once all are occupied by solute, further increases in extracellular concentration have no effect on the rate of transport. A maximum rate of transport (V_{max}) is achieved that cannot be exceeded.

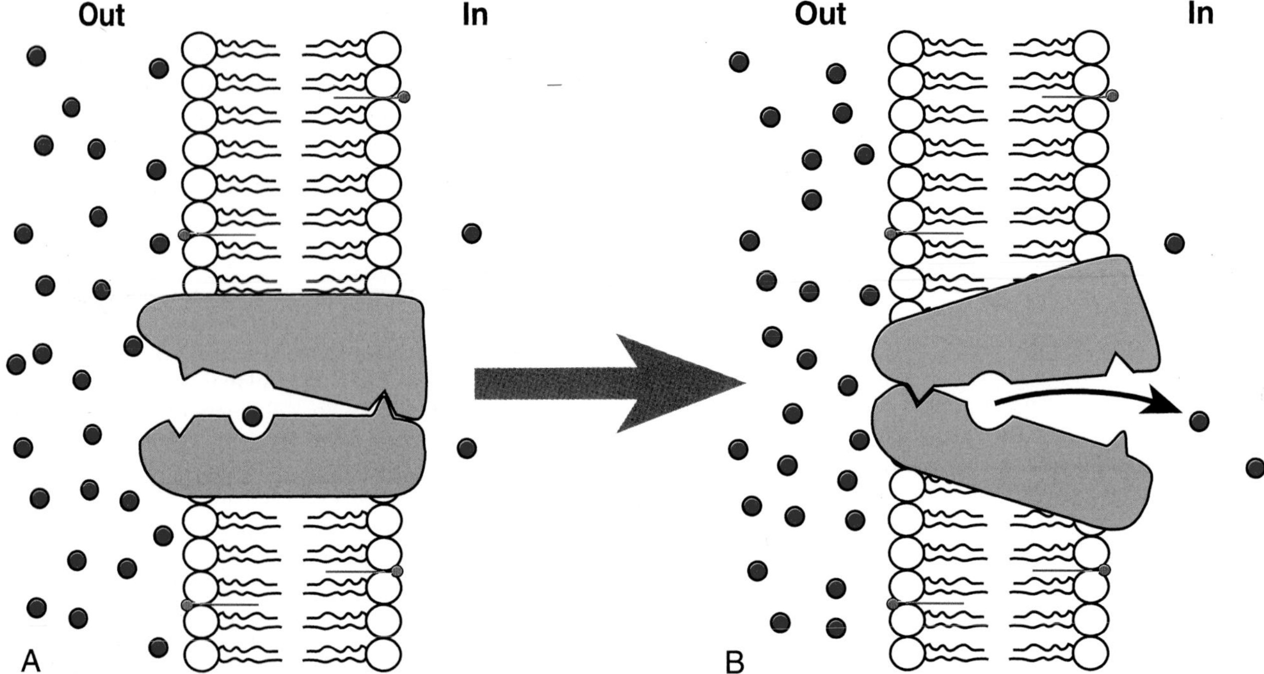

FIGURE 2.6 The role of a carrier protein in facilitated diffusion of solute molecules across a plasma membrane. In this example, solute transport into the cell is driven by the high solute concentration outside compared to inside. **A,** Binding of extracellular solute to the carrier, a membrane-spanning integral protein, may trigger a change in protein conformation that exposes the bound solute to the interior of the cell. **B,** Bound solute readily dissociates from the carrier because of the low intracellular concentration of solute. The release of solute may allow the carrier to revert to its original conformation (A) to begin the cycle again.

stops when the concentration of the solute has the same value on both sides of the membrane. At this point, with reference to equation 2, $C_1 = C_2$ and the value of J is 0. The transport systems function until the solute concentrations have **equilibrated.** However, equilibrium is attained much faster than with simple diffusion.

Equilibrating carrier-mediated transport systems have several characteristics:

- They allow the transport of polar (hydrophilic) molecules at rates much higher than expected from the partition coefficient of these molecules.
- They eventually reach saturation at high substrate concentration.
- They have structural specificity, meaning each carrier system recognizes and binds specific chemical structures (a carrier for D-glucose will not bind or transport L-glucose).
- They show competitive inhibition by molecules with similar chemical structure. For example, carrier-mediated transport of D-glucose occurs at a slower rate when molecules of D-galactose also are present. This is because galactose, structurally similar to glucose, competes with glucose for the available glucose carrier proteins.

A specific example of this type of carrier-mediated transport is the movement of glucose from the blood to the interior of cells. Most mammalian cells use blood glucose as a major source of cellular energy, and glucose is transported into cells down its concentration gradient. The transport process in many cells, such as erythrocytes and the cells of fat, liver, and muscle tissues, involves a plasma membrane protein called GLUT 1 (glucose transporter 1). The erythrocyte GLUT 1 has an affinity for D-glucose that is about 2,000-fold greater than the affinity for L-glucose. It is an integral membrane protein that contains 12 membrane-spanning α-helical segments.

Equilibrating carrier-mediated transport, like simple diffusion, does not have directional preferences. It functions equally well in bringing its specific solutes into or out of the cell, depending on the concentration gradient. Net movement by equilibrating carrier-mediated transport ceases once the concentrations inside and outside the cell become equal.

The anion exchange protein (AE1), the predominant integral protein in the mammalian erythrocyte membrane, provides a good example of the reversibility of transporter action. AE1 is folded into at least 12 transmembrane α-helices and normally permits the one-for-one exchange of Cl^- and HCO_3^- ions across the plasma membrane. The direction of ion movement is dependent only on the concentration gradients of the transported ions. AE1 has an important role in transporting CO_2 from the tissues to the lungs. The erythrocytes in systemic capillaries pick up CO_2 from tissues and convert it to HCO_3^-, which exits the cells via AE1. When the erythrocytes enter pulmonary capillaries, the AE1 allows plasma HCO_3^- to enter erythrocytes, where it is converted back to CO_2 for expiration by the lungs (see Chapter 21).

Facilitated Diffusion Through Ion Channels. Small ions, such as Na^+, K^+, Cl^-, and Ca^{2+}, also cross the plasma membrane faster than would be expected based on their partition coefficients in the lipid bilayer. An ion's electrical charge makes it difficult for the ion to move across the lipid bilayer. The rapid movement of ions across the membrane, however, is an aspect of many cell functions. The nerve action potential, the contraction of muscle, the pacemaker function of the heart, and many other physiological events are possible because of the ability of small ions to enter or leave the cell rapidly. This movement occurs through selective ion channels.

Ion channels are integral proteins spanning the width of the plasma membrane and are normally composed of several polypeptide subunits. Certain specific stimuli cause the protein subunits to open a **gate**, creating an aqueous channel through which the ions can move (Fig. 2.7). In this way, ions do not need to enter the lipid bilayer to cross the membrane; they are always in an aqueous medium. When the channels are open, the ions move rapidly from one side of the membrane to the other by facilitated diffusion. Specific interactions between the ions and the sides of the channel produce an extremely rapid rate of ion movement; in fact, ion channels permit a much faster rate of solute transport (about 10^8 ions/sec) than carrier-mediated systems.

Ion channels are often selective. For example, some channels are selective for Na^+, for K^+, for Ca^{2+}, for Cl^-, and for other anions and cations. It is generally assumed that some kind of ionic selectivity filter must be built into the structure of the channel (see Fig. 2.7). No clear relation between the amino acid composition of the channel protein and ion selectivity of the channel has been established.

A great deal of information about the characteristic behavior of channels for different ions has been revealed by the **patch clamp technique.** The small electrical current caused by ion movement when a channel is open can be detected with this technique, which is so sensitive that the opening and closing of a single ion channel can be observed (Fig. 2.8). In general, ion channels exist either fully open or completely closed, and they open and close very rapidly. The frequency with which a channel opens is variable, and the time the channel remains open (usually a few milliseconds) is also variable. The overall rate of ion transport across a membrane can be controlled by changing the frequency of a channel opening or by changing the time a channel remains open.

Most ion channels usually open in response to a specific stimulus. Ion channels can be classified according to their gating mechanisms, the signals that make them open or close. There are voltage-gated channels and ligand-gated channels. Some ion channels are always open and these are referred to as nongated channels (see Chapter 3).

Voltage-gated ion channels open when the membrane potential changes beyond a certain threshold value. Channels of this type are involved in the conduction of action potentials along nerve axons and they include sodium and potassium channels (see Chapter 3). Voltage-gated ion channels are found in many cell types. It is thought that some charged amino acids located in a membrane-spanning α-helical segment of the channel protein are sensitive to the transmembrane potential. Changes in the membrane potential cause these amino acids to move and induce a conformational change of the protein that opens the way for the ions.

Ligand-gated (or, **chemically gated**) **ion channels** cannot open unless they first bind to a specific agonist. The opening of the gate is produced by a conformational change in the protein induced by the ligand binding. The ligand can be a neurotransmitter arriving from the extracellular medium. It also can be an intracellular second messenger, produced in response to some cell activity or hormone action, that reaches the ion channel from the inside of the cell. The nicotinic acetylcholine receptor channel found in the postsynaptic neuromuscular junction (see Chapters 3 and 9) is a ligand-gated ion channel that is opened by an extracellular ligand (acetylcholine). Examples of ion channels gated by intracel-

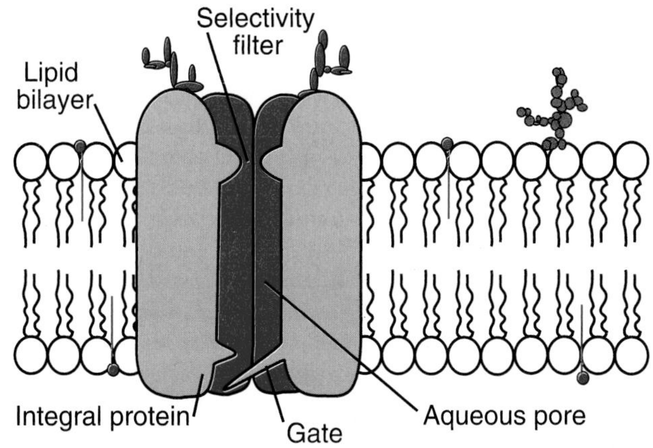

FIGURE 2.7 **An ion channel.** Ion channels are formed between the polypeptide subunits of integral proteins that span the plasma membrane, providing an aqueous pore through which ions can cross the membrane. Different types of gating mechanisms are used to open and close channels. Ion channels are often selective for a specific ion.

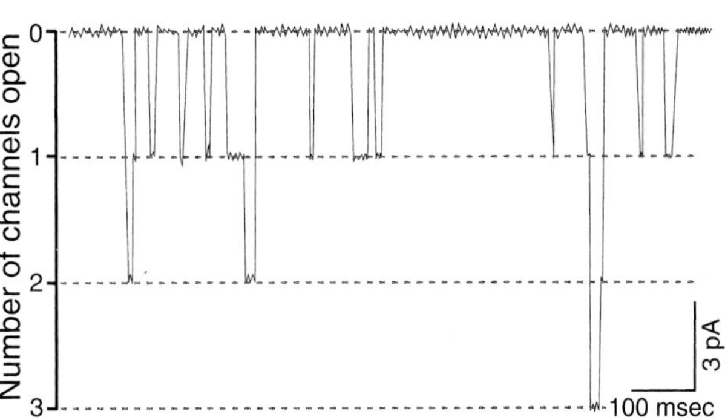

FIGURE 2.8 **A patch clamp recording from a frog muscle fiber.** Ions flow through the channel when it opens, generating a current. The current in this experiment is about 3 pA and is detected as a downward deflection in the recording. When more than one channel opens, the current and the downward deflection increase in direct proportion to the number of open channels. This record shows that up to three channels are open at any instant. (Modified from Kandel ER, Schwartz JH, Jessell TM. Principles of Neural Science. 3rd Ed. New York: Elsevier, 1991.)

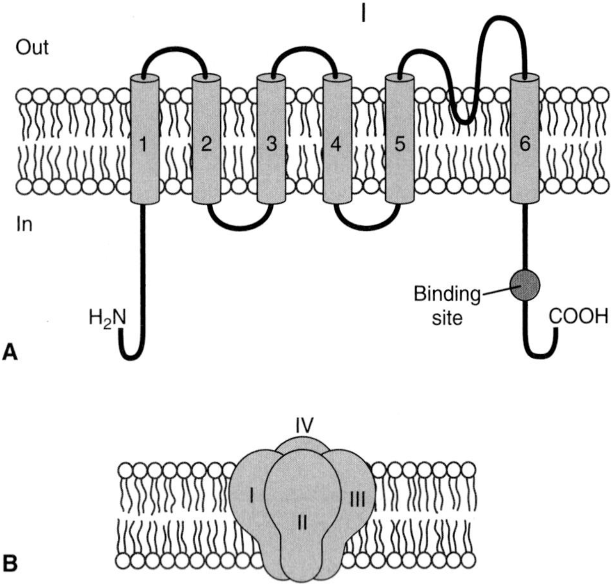

FIGURE 2.9 Structure of a cyclic nucleotide-gated ion channel. A, The secondary structure of a single subunit has six membrane-spanning regions and a binding site for cyclic nucleotides on the cytosolic side of the membrane. B, Four identical subunits (I–IV) assemble together to form a functional channel that provides a hydrophilic pathway across the plasma membrane.

lular messengers also abound in nature. This type of gating mechanism allows the channel to open or close in response to events that occur at other locations in the cell. For example, a sodium channel gated by intracellular cyclic GMP is involved in the process of vision (see Chapter 4). This channel is located in the rod cells of the retina and it opens in the presence of cyclic GMP. The generalized structure of one subunit of an ion channel gated by cyclic nucleotides is shown in Figure 2.9. There are six membrane-spanning regions and a cyclic nucleotide-binding site is exposed to the cytosol. The functional protein is a tetramer of four identical subunits. Other cell membranes have potassium channels that open when the intracellular concentration of calcium ions increases. Several known channels respond to inositol 1,4,5-trisphosphate, the activated part of G proteins, or ATP. The gating of the epithelial chloride channel by ATP is described in the Clinical Focus Box 2.1 in this chapter.

Solutes Are Moved Against Gradients by Active Transport Systems

The passive transport mechanisms discussed all tend to bring the cell into equilibrium with the extracellular fluid. Cells must oppose these equilibrating systems and preserve intracellular concentrations of solutes, particularly ions, that are compatible with life.

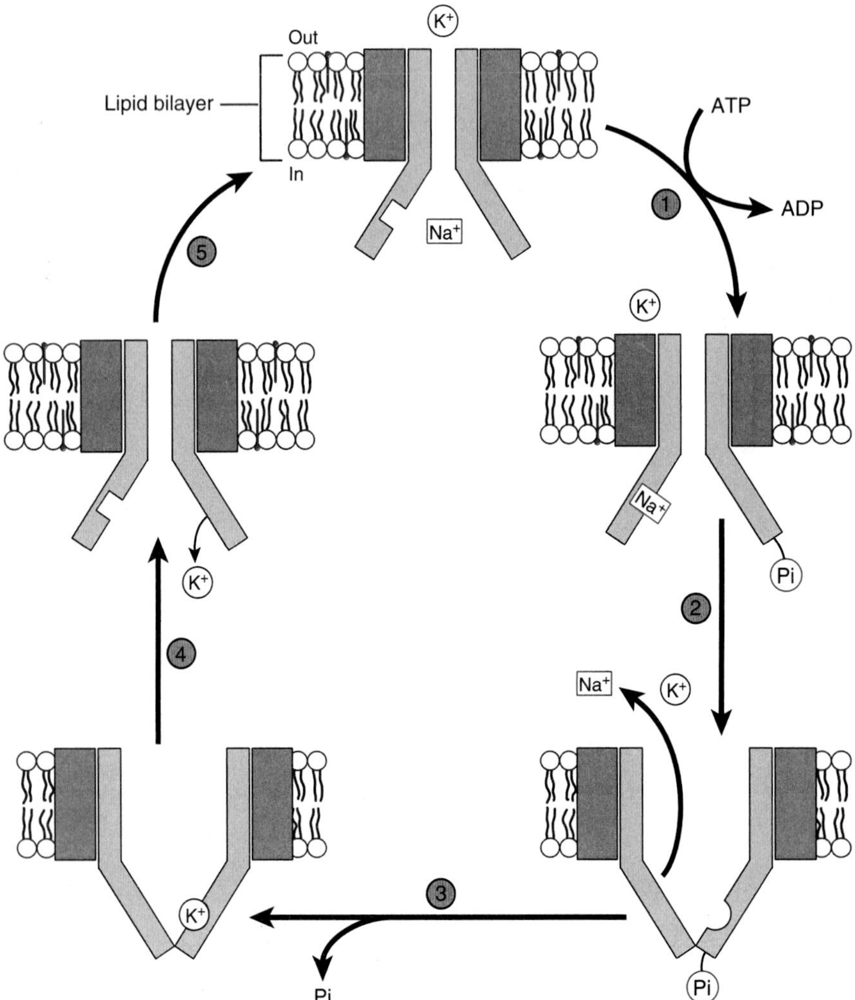

FIGURE 2.10 The possible sequence of events during one cycle of the sodium-potassium pump. The functional form may be a tetramer of two large catalytic subunits and two smaller subunits of unknown function. Binding of intracellular Na^+ and phosphorylation by ATP inside the cell may induce a conformational change that transfers Na^+ to the outside of the cell (steps 1 and 2). Subsequent binding of extracellular K^+ and dephosphorylation return the protein to its original form and transfer K^+ into the cell (steps 3, 4, and 5). There are thought to be three Na^+ binding sites and two K^+ binding sites. During one cycle, three Na^+ are exchanged for two K^+, and one ATP molecule is hydrolyzed.

CLINICAL FOCUS BOX 2.1

Cystic Fibrosis

Cystic fibrosis is one of the most common lethal genetic diseases of Caucasians. In northern Europe and the United States, for example, about 1 child in 2,500 is born with the disease. It was first recognized clinically in the 1930s, when it appeared to be a gastrointestinal problem because patients usually died from malnutrition during the first year of life. Survival has improved as management has improved; afflicted newborns now have a life expectancy of about 40 years. Cystic fibrosis affects several organ systems, with the severity varying enormously among individuals. Clinical features can include deficient secretion of digestive enzymes by the pancreas; infertility in males; increased concentration of chloride ions in sweat; intestinal and liver disease; and airway disease, leading to progressive lung dysfunction. Involvement of the lungs determines survival: 95% of cystic fibrosis patients die from respiratory failure.

The basic defect in cystic fibrosis is a failure of chloride transport across epithelial plasma membranes, particularly in the epithelial cells that line the airways. Much of the information about defective chloride transport was obtained by studying individual chloride channels using the patch clamp technique. One hypothesis is that all the pathophysiology of cystic fibrosis is a direct result of chloride transport failure. In the lungs, for example, reduced secretion of chloride ion is usually accompanied by a reduced secretion of sodium and bicarbonate ions. These changes retard the secretion of water, so the mucus secretions that line airways become thick and sticky and the smaller airways become blocked. The thick mucus also traps bacteria, which may lead to bacterial infection. Once established, bacterial infection is difficult to eradicate from the lungs of a patient with cystic fibrosis.

It was predicted that the flawed gene in patients with cystic fibrosis would normally encode either a chloride channel protein or a membrane protein that regulates chloride channels. The gene was identified in 1989 and encodes a protein of 1,480 amino acids, the **cystic fibrosis transmembrane conductance regulator** (CFTR). Evidence indicates that CFTR contains both a chloride channel and a channel regulator. Although it functions as an ion channel, it has structural similarities to adenosine triphosphate (ATP)-driven ion pumps that are integral membrane proteins. The CFTR protein is anchored in the plasma membrane by 12 membrane-spanning segments that also form a channel. A large regulatory domain is exposed to the cytosol and contains several sites that can be phosphorylated by various protein kinases, such as cyclic adenosine monophosphate (AMP)-dependent protein kinase. Two nucleotide-binding domains (NBD) control channel activity through interactions with nucleotides, such as ATP, present in the cell cytosol. A two-step process controls the gating of CFTR: (1) phosphorylation of specific sites within the regulatory domain, and (2) binding and hydrolysis of ATP at the NBD. After initial phosphorylation, gating between the closed and open states is controlled by ATP hydrolysis. It is believed that the channel is opened by ATP hydrolysis at one NBD and closed by subsequent ATP hydrolysis at the other NBD.

A common mutation in CFTR, found in 70% of cystic fibrosis patients, results in the loss of the amino acid phenylalanine from one of the NBD. This mutation produces severe symptoms because it results in defective targeting of newly synthesized CFTR proteins to the plasma membrane. The number of functional CFTR proteins at the correct location is decreased to an inadequate level.

Increased understanding of the pathophysiology of airway disease in cystic fibrosis has given rise to new therapies, and a definitive solution may be close at hand. Two approaches are undergoing clinical trials. One approach is to design pharmacological agents that will either regulate (open or close) defective CFTR chloride channels or bypass CFTR and stimulate other membrane chloride channels in the same cells. The other approach is the use of gene therapy to insert a normal gene for CFTR into affected airway epithelial cells. This has the advantage of restoring both the known and unknown functions of the gene. The field of gene therapy is in its infancy, and although there have been no "cures" for cystic fibrosis, much has been learned about the problems presented by the inefficient and short-lived transfer of genes *in vivo*. The next phase of gene therapy will focus on improving the technology for gene delivery. Gene therapy may become a reality for many lung diseases during this century.

Primary Active Transport. Integral membrane proteins that directly use metabolic energy to transport ions against a gradient of concentration or electrical potential are known as **ion pumps**. The direct use of metabolic energy to carry out transport defines a **primary active transport mechanism**. The source of metabolic energy is ATP synthesized by mitochondria, and the different ion pumps hydrolyze ATP to ADP and use the energy stored in the third phosphate bond to carry out transport. Because of this ability to hydrolyze ATP, ion pumps also are called **ATPases**.

The most abundant ion pump in higher organisms is the **sodium-potassium pump** or **Na^+/K^+-ATPase**. It is found in the plasma membrane of practically every eukaryotic cell and is responsible for maintaining the low sodium and high potassium concentrations in the cytoplasm. The sodium-potassium pump is an integral membrane protein consisting of two subunits, one large and one small. Sodium ions are transported out of the cell and potassium ions are brought in. It is known as a **P-type ATPase** because the protein is phosphorylated during the transport cycle (Fig. 2.10). The pump counterbalances the tendency of sodium ions to enter the cell passively and the tendency of potassium ions to leave passively. It maintains a high intracellular potassium concentration necessary for protein synthesis. It also plays a role in the resting membrane potential by maintaining ion gradients. The sodium-potassium pump can be inhibited either by metabolic poisons that stop the synthesis and supply of ATP or by specific pump blockers, such as the cardiac glycoside **digitalis**.

Calcium pumps, Ca^{2+}-ATPases, are found in the plasma membrane, in the membrane of the endoplasmic reticulum, and, in muscle cells, in the sarcoplasmic reticulum membrane. They are also P-type ATPases. They pump calcium ions from the cytosol of the cell either into the extracellular space or into the lumen of these organelles. The organelles store calcium and, as a result, help maintain a low cytosolic concentration of this ion (see Chapter 1).

The **H^+/K^+-ATPase** is another example of a P-type ATPase. It is present in the luminal membrane of the parietal cells in oxyntic (acid-secreting) glands of the stomach. By pumping protons into the lumen of the stomach in exchange for potassium ions, this pump maintains the low pH in the stomach that is necessary for proper digestion (see Chapter 28). It is also found in the colon and in the collecting ducts of the kidney. Its role in the kidney is to secrete H^+ ions into the urine and to reabsorb K^+ ions (see Chapter 25).

Proton pumps, H^+-ATPases, are found in the membranes of the lysosomes and the Golgi apparatus. They pump protons from the cytosol into these organelles, keeping the inside of the organelles more acidic (at a lower pH) than the rest of the cell. These pumps, classified as **V-type ATPases** because they were first discovered in intracellular vacuolar structures, have now been detected in plasma membranes. For example, the proton pump in the luminal plasma membrane of kidney cells is characterized as a V-type ATPase. By secreting protons, it plays an important role in acidifying the tubular urine.

Mitochondria have **F-type ATPases** located in the inner mitochondrial membrane. This type of proton pump normally functions in reverse. Instead of using the energy stored in ATP molecules to pump protons, its principal function is to synthesize ATP by using the energy stored in a gradient of protons. The proton gradient is generated by the respiratory chain.

Secondary Active Transport. The net effect of ion pumps is maintenance of the various environments needed for the proper functioning of organelles, cells, and organs. Metabolic energy is expended by the pumps to create and maintain the differences in ion concentrations. Besides the importance of local ion concentrations for cell function, differences in concentrations represent stored energy. An ion releases potential energy when it moves down an electrochemical gradient, just as a body releases energy when falling to a lower level. This energy can be used to perform work. Cells have developed several carrier mechanisms to transport one solute against its concentration gradient by using the energy stored in the favorable gradient of another solute. In mammals, most of these mechanisms use sodium as the driver solute and use the energy of the sodium gradient to carry out the "uphill" transport of another important solute (Fig. 2.11). Because the sodium gra-

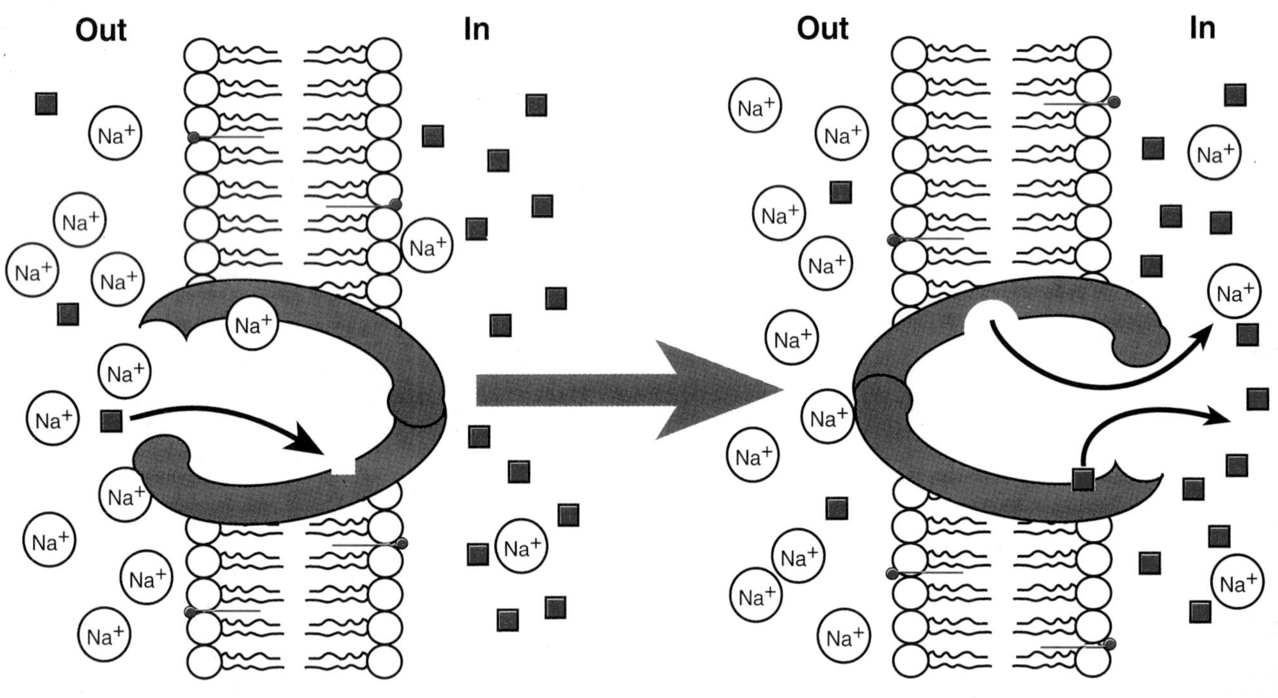

FIGURE 2.11 **A possible mechanism of secondary active transport.** A solute is moved against its concentration gradient by coupling it to Na^+ moving down a favorable gradient. Binding of extracellular Na^+ to the carrier protein may increase the affinity of binding sites for solute, so that solute also can bind to the carrier, even though its extracellular concentration is low. A conformational change in the carrier protein may expose the binding sites to the cytosol, where Na^+ readily dissociates because of the low intracellular Na^+ concentration. The release of Na^+ decreases the affinity of the carrier for solute and forces the release of the solute inside the cell, where solute concentration is already high.

dient is maintained by the action of the sodium-potassium pump, the function of these transport systems also depends on the function of the pump. Although they do not directly use metabolic energy for transport, these systems ultimately depend on the proper supply of metabolic energy to the sodium-potassium pump. They are called **secondary active transport mechanisms**. Disabling the pump with metabolic inhibitors or pharmacological blockers causes these transport systems to stop when the sodium gradient has been dissipated.

Similar to passive carrier-mediated systems, secondary active transport systems are integral membrane proteins; they have specificity for the solute they transport and show saturation kinetics and competitive inhibition. They differ, however, in two respects. First, they cannot function in the absence of the driver ion, the ion that moves along its electrochemical gradient and supplies energy. Second, they transport the solute against its own concentration or electrochemical gradient. Functionally, the different secondary active transport systems can be classified into two groups: **symport** (cotransport) systems, in which the solute being transported moves in the same direction as the sodium ion; and **antiport** (exchange) systems, in which sodium moves in one direction and the solute moves in the opposite direction (Fig. 2.12).

Examples of symport mechanisms are the sodium-coupled sugar transport system and the several sodium-coupled amino acid transport systems found in the small intestine and the renal tubule. The symport systems allow efficient absorption of nutrients even when the nutrients are present at very low concentrations. The Na^+-glucose cotransporter in the human intestine has been cloned and sequenced. It is called sodium-dependent glucose transporter (SGLT). The protein contains 664 amino acids, and the polypeptide chain is thought to contain 14 membrane-spanning segments (Fig. 2.13). Another example of a symport system is the family of sodium-coupled phosphate transporters (termed NaPi, types I and II) in the intestine and renal proximal tubule. These transporters have 6 to 8 membrane-spanning segments and contain 460 to 690 amino acids. Sodium-coupled chloride transporters in the kidney are targets for inhibition by specific diuretics. The Na^+-Cl^- cotransporter in the distal tubule, known as NCC, is inhibited by thiazide diuretics, and the Na^+-K^+-$2Cl^-$ cotransporter in the ascending limb of the loop of Henle, referred to as NKCC, is inhibited by bumetanide.

The most important examples of antiporters are the Na^+/H^+ exchange and Na^+/Ca^{2+} exchange systems, found mainly in the plasma membrane of many cells. The first uses the sodium gradient to remove protons from the cell, controlling the intracellular pH and counterbalancing the production of protons in metabolic reactions. It is an **electroneutral system** because there is no net movement of charge. One Na^+ enters the cell for each H^+ that leaves. The second antiporter removes calcium from the cell and, together with the different calcium pumps, helps maintain a low cytosolic calcium concentration. It is an **electrogenic system** because there is a net movement of charge. Three Na^+ enter the cell and one Ca^{2+} leaves during each cycle.

The structures of the symport and antiport protein transporters that have been characterized (see Fig. 2.13) share a common property with ion channels (see Fig. 2.9) and equilibrating carriers, namely the presence of multiple membrane-spanning segments within the polypeptide chain. This supports the concept that, regardless of the mechanism, the membrane-spanning regions of a transport protein form a hydrophilic pathway for rapid transport of ions and solutes across the hydrophobic interior of the membrane lipid bilayer.

Out
In
Symport
Na^+ → Na^+
S → **S**
Na^+ → Na^+
S ← S
Antiport

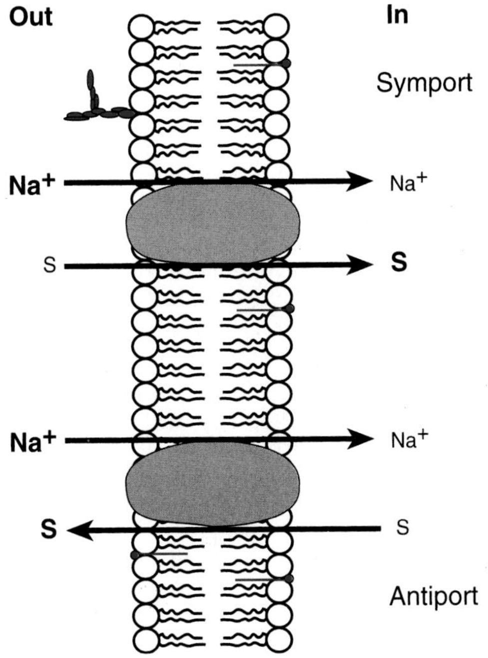

FIGURE 2.12 Secondary active transport systems. In a symport system (top), the transported solute (S) is moved in the same direction as the Na^+ ion. In an antiport system (bottom), the solute is moved in the opposite direction to Na^+. Large and small type indicate high and low concentrations, respectively, of Na^+ ions and solute.

The Movement of Solutes Across Epithelial Cell Layers. Epithelial cells occur in layers or sheets that allow the directional movement of solutes not only across the plasma membrane but also from one side of the cell layer to the other. Such regulated movement is achieved because the plasma membranes of epithelial cells have two distinct regions with different morphology and different transport systems. These regions are the **apical membrane**, facing the lumen, and the **basolateral membrane**, facing the blood supply (Fig. 2.14). The specialized or polarized organization of the cells is maintained by the presence of **tight junctions** at the areas of contact between adjacent cells. Tight junctions prevent proteins on the apical membrane from migrating to the basolateral membrane and those on the basolateral membrane from migrating to the apical membrane. Thus, the entry and exit steps for solutes can be localized to opposite sides of the cell. This is the key to **transcellular transport** across epithelial cells.

An example is the absorption of glucose in the small intestine. Glucose enters the intestinal epithelial cells by active transport using the electrogenic Na^+-glucose cotransporter system (SGLT) in the apical membrane. This

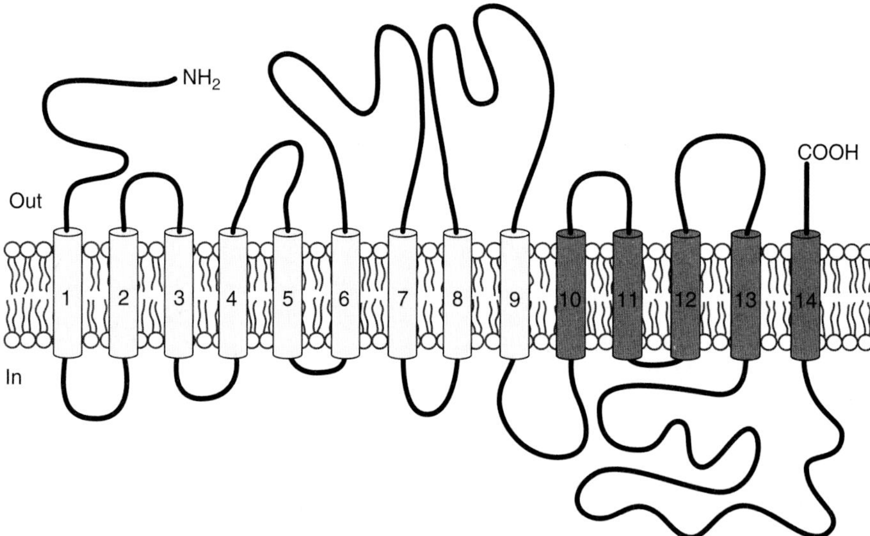

FIGURE 2.13 A model of the secondary structure of the Na^+-glucose cotransporter protein (SGLT) in the human intestine. The polypeptide chain of 664 amino acids passes back and forth across the membrane 14 times. Each membrane-spanning segment consists of 21 amino acids arranged in an α-helical conformation. Both the NH_2 and the COOH ends are located on the extracellular side of the plasma membrane. In the functional protein, it is likely that the membrane-spanning segments are clustered together to provide a hydrophilic pathway across the plasma membrane. The N-terminal portion of the protein, including helices 1 to 9, is required to couple Na^+ binding to glucose transport. The five helices (10 to 14) at the C-terminus may form the transport pathway for glucose. (Modified from Panayotova-Heiermann M, Eskandari S, Turk E, et al. Five transmembrane helices form the sugar pathway through the Na^+-glucose cotransporter. J Biol Chem 1997;272:20324–20327.)

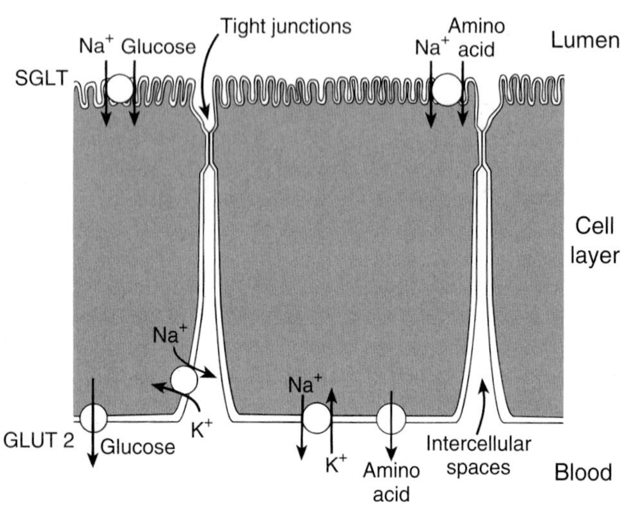

FIGURE 2.14 The localization of transport systems to different regions of the plasma membrane in epithelial cells of the small intestine. A polarized cell is produced, in which entry and exit of solutes, such as glucose, amino acids, and Na^+, occur at opposite sides of the cell. Active entry of glucose and amino acids is restricted to the apical membrane and exit requires equilibrating carriers located only in the basolateral membrane. For example, glucose enters on SGLT and exits on GLUT 2. Na^+ that enters via the apical symporters is pumped out by the Na^+/K^+-ATPase on the basolateral membrane. The result is a net movement of solutes from the luminal side of the cell to the basolateral side, ensuring efficient absorption of glucose, amino acids, and Na^+ from the intestinal lumen.

increases the intracellular glucose concentration above the blood glucose concentration, and the glucose molecules move passively out of the cell and into the blood via an equilibrating carrier mechanism (GLUT 2) in the basolateral membrane (see Fig. 2.14). The intestinal GLUT 2, like the erythrocyte GLUT 1, is a sodium-independent transporter that moves glucose down its concentration gradient. Unlike GLUT 1, the GLUT 2 transporter can accept other sugars, such as galactose and fructose, that are also absorbed in the intestine. The sodium ions that enter the cell with the glucose molecules on SGLT are pumped out by the Na^+/K^+-ATPase that is located in the basolateral membrane only. The polarized organization of the epithelial cells and the integrated functions of the plasma membrane transporters form the basis by which cells accomplish transcellular movement of both glucose and sodium ions.

THE MOVEMENT OF WATER ACROSS THE PLASMA MEMBRANE

Since the lipid part of the plasma membrane is very hydrophobic, the movement of water across it is too slow to explain the speed at which water can move in and out of the cells. The partition coefficient of water into lipids is very low; therefore, the permeability of the lipid bilayer for water is also very low. Specific membrane proteins that function as **water channels** explain the rapid movement of water across the plasma membrane. These water channels are small (molecular weight about 30 kDa) integral membrane proteins known as **aquaporins.** Ten different forms have been discovered so far in mammals. At least six forms are expressed in cells in the kidney and seven forms in the gastrointestinal tract, tissues where water movement across plasma membranes is particularly rapid.

In the kidney, aquaporin-2 (AQP2) is abundant in the collecting duct and is the target of the hormone **arginine vasopressin,** also known as antidiuretic hormone. This hormone increases water transport in the collecting duct by stimulating the insertion of AQP2 proteins into the apical plasma membrane. Several studies have shown that AQP2 has a critical role in inherited and acquired disorders of water reabsorption by the kidney. For example, **diabetes insipidus** is a condition in which the kidney loses its ability to reabsorb water properly, resulting in excessive loss of water and excretion of a large volume of very dilute urine (polyuria). Although inherited forms of diabetes insipidus are relatively rare, it can develop in patients receiving chronic lithium therapy for psychiatric disorders, giving rise to the term **lithium-induced polyuria**. Both of these conditions are associated with a decrease in the number of AQP2 proteins in the collecting ducts of the kidney.

The Movement of Water Across the Plasma Membrane Is Driven by Differences in Osmotic Pressure

The spontaneous movement of water across a membrane driven by a gradient of water concentration is the process known as **osmosis**. The water moves from an area of high concentration of water to an area of low concentration. Since concentration is defined by the number of particles per unit of volume, a solution with a high concentration of solutes has a low concentration of water, and vice versa. Osmosis can, therefore, be viewed as the movement of water from a solution of high water concentration (low concentration of solute) toward a solution with a lower concentration of water (high solute concentration). Osmosis is a passive transport mechanism that tends to equalize the total solute concentrations of the solutions on both sides of every membrane.

If a cell that is normally in osmotic equilibrium is transferred to a more dilute solution, water will enter the cell, the cell volume will increase, and the solute concentration of the cytoplasm will be reduced. If the cell is transferred to a more concentrated solution, water will leave the cell, the cell volume will decrease, and the solute concentration of the cytoplasm will increase. As we will see below, many cells have regulatory mechanisms that keep cell volume within a certain range. Other cells, such as mammalian erythrocytes, do not have volume regulatory mechanisms and large volume changes occur when the solute concentration of the extracellular fluid is changed.

The driving force for the movement of water across the plasma membrane is the difference in water concentration between the two sides of the membrane. For historical reasons, this driving force is not called the chemical gradient of water but the difference in osmotic pressure. The **osmotic pressure** of a solution is defined as the pressure necessary to stop the net movement of water across a selectively permeable membrane that separates the solution from pure water. When a membrane separates two solutions of different osmotic pressure, water will move from the solution with low osmotic pressure (high water concentration) to the solution of high osmotic pressure (low water concentration). In this context, the term *selectively permeable* means that the membrane is permeable to water but not solutes. In reality, most biological membranes contain membrane transport proteins that permit solute movement.

The osmotic pressure of a solution depends on the number of particles dissolved in it, the total concentration of all solutes. Many solutes, such as salts, acids, and bases, dissociate in water, so the number of particles is greater than the molar concentration. For example, NaCl dissociates in water to give Na^+ and Cl^-, so one molecule of NaCl will produce two osmotically active particles. In the case of $CaCl_2$, there are three particles per molecule. The equation giving the osmotic pressure of a solution is:

$$\pi = n\,R\,T\,C \tag{3}$$

where π is the osmotic pressure of the solution, n is the number of particles produced by the dissociation of one molecule of solute (2 for NaCl, 3 for $CaCl_2$), R is the universal gas constant (0.0821 L·atm/mol·K), T is the absolute temperature, and C is the concentration of the solute in mol/L. Osmotic pressure can be expressed in **atmospheres (atm)**. Solutions with the same osmotic pressure are called **isosmotic**. A solution is **hyperosmotic** with respect to another solution if it has a higher osmotic pressure and **hypoosmotic** if it has a lower osmotic pressure.

Equation 3, called the **van't Hoff equation**, is valid only when applied to very dilute solutions, in which the particles of solutes are so far away from each other that no interactions occur between them. Generally, this is not the case at physiological concentrations. Interactions between dissolved particles, mainly between ions, cause the solution to behave as if the concentration of particles is less than the theoretical value (nC). A correction coefficient, called the **osmotic coefficient** (ϕ) of the solute, needs to be introduced in the equation. Therefore, the osmotic pressure of a solution can be written more accurately as:

$$\pi = n\,R\,T\,\phi\,C \tag{4}$$

The osmotic coefficient varies with the specific solute and its concentration. It has values between 0 and 1. For example, the osmotic coefficient of NaCl is 1.00 in an infinitely dilute solution but changes to 0.93 at the physiological concentration of 0.15 mol/L.

At any given T, since R is constant, equation 4 shows that the osmotic pressure of a solution is directly proportional to the term $n\phi C$. This term is known as the **osmolality** or **osmotic concentration** of a solution and is expressed in osm/kg H_2O. Most physiological solutions, such as blood plasma, contain many different solutes, and each contributes to the total osmolality of the solution. The osmolality of a solution containing a complex mixture of solutes is usually measured by freezing point depression. The freezing point of an aqueous solution of solutes is lower than that of pure water and depends on the total number of solute particles. Compared with pure water, which freezes at 0°C, a solution with an osmolality of 1 osm/kg H_2O will freeze at −1.86°C. The ease with which osmolality can be measured has led to the wide use of this parameter for comparing the osmotic pressure of different solutions. The osmotic pressures of physiological solutions are not trivial. Consider blood plasma, for example, which usually has an osmolality of 0.28 osm/kg H_2O, determined by freezing point depression. Equation 4 shows that the osmotic pressure of plasma at 37°C is 7.1 atm, about 7 times greater than atmospheric pressure.

Many Cells Can Regulate Their Volume

Cell volume changes can occur in response to changes in the osmolality of extracellular fluid in both normal and pathophysiological situations. Accumulation of solutes also can produce volume changes by increasing the intracellular osmolality. Many cells can correct these volume changes.

Volume regulation is particularly important in the brain, for example, where cell swelling can have serious consequences because expansion is strictly limited by the rigid skull.

Osmolality and Tonicity. A solution's osmolality is determined by the total concentration of all the solutes present. In contrast, the solution's **tonicity** is determined by the concentrations of only those solutes that do not enter ("penetrate") the cell. Tonicity determines cell volume, as illustrated in the following examples. Na^+ behaves as a nonpenetrating solute because it is pumped out of cells by the Na^+/K^+-ATPase at the same rate that it enters. A solution of NaCl at 0.2 osm/kg H_2O is hypoosmotic compared to cell cytosol at 0.3 osm/kg H_2O. The NaCl solution is also **hypotonic** because cells will accumulate water and swell when placed in this solution. A solution containing a mixture of NaCl (0.3 osm/kg H_2O) and urea (0.1 osm/kg H_2O) has a total osmolality of 0.4 osm/kg H_2O and will be hyperosmotic compared to cell cytosol. The solution is **isotonic**, however, because it produces no permanent change in cell volume. The reason is that cells shrink initially as a result of loss of water but urea is a penetrating solute that rapidly enters the cells. Urea entry increases the intracellular osmolality so water also enters and increases the volume. Entry of water ceases when the urea concentration is the same inside and outside the cells. At this point, the total osmolality both inside and outside the cells will be 0.4 osm/kg H_2O and the cell volume will be restored to normal.

Volume Regulation. When cell volume increases because of extracellular hypotonicity, the response of many cells is rapid activation of transport mechanisms that tend to decrease the cell volume (Fig. 2.15A). Different cells use different **regulatory volume decrease (RVD) mechanisms** to move solutes out of the cell and decrease the number of particles in the cytosol, causing water to leave the cell. Since cells have high intracellular concentrations of potassium, many RVD mechanisms involve an increased efflux of K^+, either by stimulating the opening of potassium channels or by activating symport mechanisms for KCl. Other cells activate the efflux of some amino acids, such as taurine or proline. The net result is a decrease in intracellular solute content and a reduction of cell volume close to its original value (see Fig. 2.15A).

When placed in a **hypertonic** solution, cells rapidly lose water and their volume decreases. In many cells, a decreased volume triggers **regulatory volume increase (RVI) mechanisms,** which increase the number of intracellular particles, bringing water back into the cells. Because Na^+ is the main extracellular ion, many RVI mechanisms involve an influx of sodium into the cell. Na^+-Cl^- symport, Na^+-K^+-$2Cl^-$ symport, and Na^+/H^+ antiport are some of the mechanisms activated to increase the intracellular concentration of Na^+ and increase the cell volume toward its original value (Fig. 2.15B).

Mechanisms based on an increased Na^+ influx are effective for only a short time because, eventually, the sodium pump will increase its activity and reduce intracellular Na^+

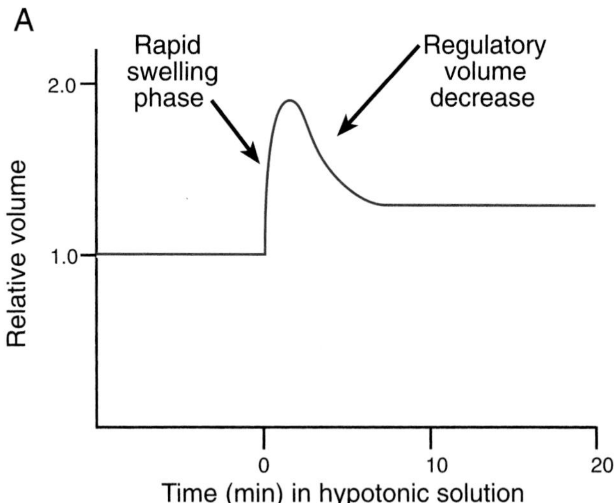

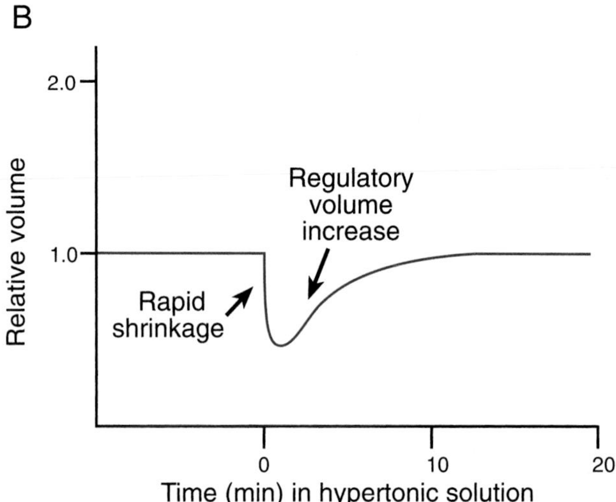

FIGURE 2.15 **The effect of tonicity changes on cell volume.** Cell volume changes when a cell is placed in either a hypotonic or a hypertonic solution. **A,** In a hypotonic solution, the reversal of the initial increase in cell volume is known as a regulatory volume decrease. Transport systems for solute exit are activated, and water follows movement of solute out of the cell. **B,** In a hypertonic solution, the reversal of the initial decrease in cell volume is a regulatory volume increase. Transport systems for solute entry are activated, and water follows solute into the cell.

to its normal value. Cells that regularly encounter hypertonic extracellular fluids have developed additional mechanisms for maintaining normal volume. These cells can synthesize specific organic solutes, enabling them to increase intracellular osmolality for a long time and avoiding altering the concentrations of ions they must maintain within a narrow range of values. The organic solutes are usually small molecules that do not interfere with normal cell function when they accumulate inside the cell. For example, cells of the medulla of the mammalian kidney can increase the level of the enzyme aldose reductase when subjected to elevated extracellular osmolality. This enzyme converts glucose to an osmotically active solute, sorbitol. Brain cells can synthesize and store inositol. Synthesis of sorbitol and inositol represents different answers to the problem of increasing the total intracellular osmolality, allowing normal cell volume to be maintained in the presence of hypertonic extracellular fluid.

Oral Rehydration Therapy

Oral administration of rehydration solutions has dramatically reduced the mortality resulting from cholera and other diseases that involve excessive losses of water and solutes from the gastrointestinal tract. The main ingredients of rehydration solutions are glucose, NaCl, and water. The glucose and Na^+ ions are reabsorbed by SGLT and other transporters in the epithelial cells lining the lumen of the small intestine (see Fig. 2.14). Deposition of these solutes on the basolateral side of the epithelial cells increases the osmolality in that region compared with the intestinal lumen and drives the osmotic absorption of water. Absorption of glucose increases the absorption of NaCl and water and helps to compensate for excessive diarrheal losses of salt and water.

THE RESTING MEMBRANE POTENTIAL

The different passive and active transport systems are coordinated in a living cell to maintain intracellular ions and other solutes at concentrations compatible with life. Consequently, the cell does not equilibrate with the extracellular fluid, but rather exists in a **steady state** with the extracellular solution. For example, intracellular Na^+ concentration (10 mmol/L in a muscle cell) is much lower than extracellular Na^+ concentration (140 mmol/L), so Na^+ enters the cell by passive transport through nongated Na^+ channels. The rate of Na^+ entry is matched, however, by the rate of active transport of Na^+ out of the cell via the sodium-potassium pump (Fig. 2.16). The net result is that intracellular Na^+ is maintained constant and at a low level, even though Na^+ continually enters and leaves the cell. The reverse is true for K^+, which is maintained at a high concentration inside the cell relative to the outside. The passive exit of K^+ through nongated K^+ channels is matched by active entry via the pump (see Fig. 2.16). Maintenance of this steady state with ion concentrations inside the cell different from those outside the cell is the basis for the difference in electrical potential across the plasma membrane or the **resting membrane potential**.

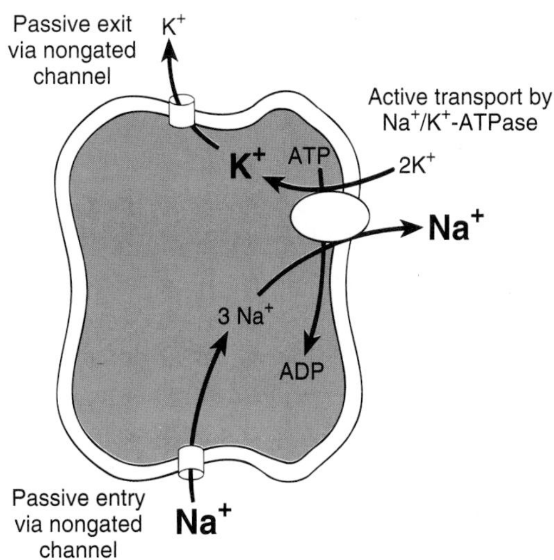

FIGURE 2.16 **The concept of a steady state.** Na^+ enters a cell through nongated Na^+ channels, moving passively down the electrochemical gradient. The rate of Na^+ entry is matched by the rate of active transport of Na^+ out of the cell via the Na^+/K^+-ATPase. The intracellular concentration of Na^+ remains low and constant. Similarly, the rate of passive K^+ exit through nongated K^+ channels is matched by the rate of active transport of K^+ into the cell via the pump. The intracellular K^+ concentration remains high and constant. During each cycle of the ATPase, two K^+ are exchanged for three Na^+ and one molecule of ATP is hydrolyzed to ADP. Large type and small type indicate high and low ion concentrations, respectively.

Ion Movement Is Driven by the Electrochemical Potential

If there are no differences in temperature or hydrostatic pressure between the two sides of a plasma membrane, two forces drive the movement of ions and other solutes across the membrane. One force results from the difference in the concentration of a substance between the inside and the outside of the cell and the tendency of every substance to move from areas of high concentration to areas of low concentration. The other force results from the difference in electrical potential between the two sides of the membrane, and it applies only to ions and other electrically charged solutes. When a difference in electrical potential exists, positive ions tend to move toward the negative side, while negative ions tend to move toward the positive side.

The sum of these two driving forces is called the gradient (or difference) of **electrochemical potential** across the membrane for a specific solute. It measures the tendency of that solute to cross the membrane. The expression of this force is given by:

$$\Delta\mu = RT \ln \frac{C_i}{C_o} + zF\,(E_i - E_o) \qquad (5)$$

where μ represents the electrochemical potential ($\Delta\mu$ is the difference in electrochemical potential between two sides of the membrane); C_i and C_o are the concentrations of the solute inside and outside the cell, respectively; E_i is the electrical potential inside the cell measured with respect to the electrical potential outside the cell (E_o); R is the universal gas constant (2 cal/mol·K); T is the absolute temperature (K); z is the valence of the ion; and F is the Faraday constant (23 cal/mV·mol). By inserting these units in equation 5 and simplifying, the electrochemical potential will be expressed in cal/mol, which are units of energy. If the solute is not an ion and has no electrical charge, then z = 0 and the last term of the equation becomes zero. In this case, the electrochemical potential is defined only by the different concentrations of the uncharged solute, called the **chemical potential**. The driving force for solute transport becomes solely the difference in chemical potential.

Net Ion Movement Is Zero at the Equilibrium Potential

Net movement of an ion into or out of a cell continues as long as the driving force exists. Net movement stops and equilibrium is reached only when the driving force of electrochemical potential across the membrane becomes zero. The condition of equilibrium for any permeable ion will be $\Delta\mu = 0$. Substituting this condition into equation 5, we obtain:

$$0 = RT \ln \frac{C_i}{C_o} + zF\,(E_i - E_o)$$

$$E_i - E_o = -\frac{RT}{zF} \ln \frac{C_i}{C_o} \qquad (6)$$

$$E_i - E_o = \frac{RT}{zF} \ln \frac{C_o}{C_i}$$

Equation 6, known as the **Nernst equation**, gives the value of the electrical potential difference ($E_i - E_o$) necessary for a specific ion to be at equilibrium. This value is known as the **Nernst equilibrium potential** for that particular ion and it is expressed in millivolts (mV), units of voltage. At the equilibrium potential, the tendency of an ion to move in one direction because of the difference in concentrations is exactly balanced by the tendency to move in the opposite direction because of the difference in electrical potential. At this point, the ion will be in equilibrium and there will be no net movement. By converting to $\log_{10}$ and assuming a physiological temperature of 37°C and a value of +1 for z (for Na^+ or K^+), the Nernst equation can be expressed as:

$$E_i - E_o = 61 \log_{10} \frac{C_o}{C_i} \qquad (7)$$

Since Na^+ and K^+ (and other ions) are present at different concentrations inside and outside a cell, it follows from equation 7 that the equilibrium potential will be different for each ion.

The Resting Membrane Potential Is Determined by the Passive Movement of Several Ions

The resting membrane potential is the electrical potential difference across the plasma membrane of a normal living cell in its unstimulated state. It can be measured directly by the insertion of a microelectrode into the cell with a reference electrode in the extracellular fluid. The resting membrane potential is determined by those ions that can cross the membrane and are prevented from attaining equilibrium by active transport systems. Potassium, sodium, and chloride ions can cross the membranes of every living cell, and each of these ions contributes to the resting membrane potential. By contrast, the permeability of the membrane of most cells to divalent ions is so low that it can be ignored in this context.

The **Goldman equation** gives the value of the membrane potential (in mV) when all the permeable ions are accounted for:

$$E_i - E_o = \frac{RT}{F} \ln \frac{P_K[K^+]_o + P_{Na}[Na^+]_o + P_{Cl}[Cl^-]_i}{P_K[K^+]_i + P_{Na}[Na^+]_i + P_{Cl}[Cl^-]_o} \qquad (8)$$

where P_K, P_{Na}, and P_{Cl} represent the permeability of the membrane to potassium, sodium, and chloride ions, respectively; and brackets indicate the concentration of the ion inside (i) and outside (o) the cell. If a certain cell is not permeable to one of these ions, the contribution of the impermeable ion to the membrane potential will be zero. If a specific cell is permeable to an ion other than the three considered in equation 8, that ion's contribution to the membrane potential must be included in the equation.

It can be seen from equation 8 that the contribution of any ion to the membrane potential is determined by the membrane's permeability to that particular ion. The higher the permeability of the membrane to one ion relative to the others, the more that ion will contribute to the membrane potential. The plasma membranes of most living cells are much more permeable to potassium ions than to any other ion. Making the assumption that P_{Na} and P_{Cl} are zero relative to P_K, equation 8 can be simplified to:

$$E_i - E_o = \frac{RT}{F} \ln \frac{P_K[K^+]_o}{P_K[K^+]_i}$$

$$E_i - E_o = \frac{RT}{F} \ln \frac{[K^+]_o}{[K^+]_i} \qquad (9)$$

which is the Nernst equation for the equilibrium potential for K^+ (see equation 6). This illustrates two important points:

- In most cells, the resting membrane potential is close to the equilibrium potential for K^+.
- The resting membrane potential of most cells is dominated by K^+ because the plasma membrane is more permeable to this ion compared to the others.

As a typical example, the K^+ concentrations outside and inside a muscle cell are 3.5 mmol/L and 155 mmol/L, respectively. Substituting these values in equation 7 gives an equilibrium potential for K^+ of −100 mV, negative inside the cell relative to the outside. The resting membrane potential in a muscle cell is −90 mV (negative inside). This value is close to, although not the same as, the equilibrium potential for K^+.

The reason the resting membrane potential in the muscle cell is less negative than the equilibrium potential for K^+ is as follows. Under physiological conditions, there is passive entry of Na^+ ions. This entry of positively charged ions has a small but significant effect on the negative potential inside the cell. Assuming intracellular Na^+ to be 10 mmol/L and extracellular Na^+ to be 140 mmol/L, the Nernst equation gives a value of +70 mV for the Na^+ equilibrium potential (positive inside the cell). This is far from the resting membrane potential of −90 mV. Na^+ makes only a small contribution to the resting membrane potential because membrane permeability to Na^+ is very low compared to that of K^+.

The contribution of Cl^- ions need not be considered because the resting membrane potential in the muscle cell is the same as the equilibrium potential for Cl^-. Therefore, there is no net movement of chloride ions.

In most cells, as shown above using a muscle cell as an example, the equilibrium potentials of K^+ and Na^+ are different from the resting membrane potential, which indicates that neither K^+ ions nor Na^+ ions are at equilibrium. Consequently, these ions continue to cross the plasma membrane via specific nongated channels, and these passive ion movements are *directly* responsible for the resting membrane potential.

The Na^+/K^+-ATPase is important *indirectly* for maintaining the resting membrane potential because it sets up the gradients of K^+ and Na^+ that drive passive K^+ exit and Na^+ entry. During each cycle of the pump, two K^+ ions are moved into the cell in exchange for three Na^+, which are moved out (see Fig. 2.16). Because of the unequal exchange mechanism, the pump's activity contributes slightly to the negative potential inside the cell.

REVIEW QUESTIONS

DIRECTIONS: Each of the numbered items or incomplete statements in this section is followed by answers or by completions of the statement. Select the ONE lettered answer or completion that is BEST in each case.

1. Which one of the following is a common property of all phospholipid molecules?
(A) Hydrophilic
(B) Steroid structure
(C) Water-soluble
(D) Amphipathic
(E) Hydrophobic
2. Select the true statement about membrane phospholipids.
(A) A phospholipid contains cholesterol
(B) Phospholipids move rapidly in the plane of the bilayer
(C) Specific phospholipids are always present in equal proportions in the two halves of the bilayer
(D) Phospholipids form ion channels through the membrane
(E) Na^+-glucose symport is mediated by phospholipids
3. Several segments of the polypeptide chain of integral membrane proteins usually span the lipid bilayer. These segments frequently
(A) Adopt an α-helical configuration
(B) Contain many hydrophilic amino acids
(C) Form covalent bonds with cholesterol
(D) Contain unusually strong peptide bonds
(E) Form covalent bonds with phospholipids
4. The electrical potential difference necessary for a single ion to be at equilibrium across a membrane is best described by the
(A) Goldman equation
(B) van't Hoff equation
(C) Fick's law
(D) Nernst equation
(E) Permeability coefficient
5. The ion present in highest concentration inside most cells is
(A) Sodium
(B) Potassium
(C) Calcium
(D) Chloride
(E) Phosphate
6. Solute movement by active transport can be distinguished from solute transport by equilibrating carrier-mediated transport because active transport
(A) Is saturable at high solute concentration
(B) Is inhibited by other molecules with structures similar to that of the solute
(C) Moves the solute against its electrochemical gradient
(D) Allows movement of polar molecules
(E) Is mediated by specific membrane proteins
7. A sodium channel that opens in response to an increase in intracellular cyclic GMP is an example of
(A) A ligand-gated ion channel
(B) An ion pump
(C) Sodium-coupled solute transport
(D) A peripheral membrane protein
(E) Receptor-mediated endocytosis
8. During regulatory volume decrease, many cells will increase
(A) Their volume
(B) Influx of Na^+
(C) Efflux of K^+
(D) Synthesis of sorbitol
(E) Influx of water
9. At equilibrium the concentrations of Cl^- inside and outside a cell are 8 mmol/L and 120 mmol/L, respectively. The equilibrium potential for Cl^- at 37°C is calculated to be
(A) +4.07 mV
(B) −4.07 mV
(C) +71.7 mV
(D) −71.7 mV
(E) +91.5 mV
(F) −91.5 mV
10. What is the osmotic pressure (in atm) of an aqueous solution of 100 mmol/L $CaCl_2$ at 27 °C? (Assume the osmotic coefficient is 0.86 and the gas constant is 0.0821 L·atm/mol·K).
(A) 738 atm
(B) 635 atm
(C) 211 atm
(D) 7.38 atm
(E) 6.35 atm
(F) 2.11 atm

SUGGESTED READING

Barrett KE, Keely SJ. Chloride secretion by the intestinal epithelium: Molecular basis and regulatory aspects. Annu Rev Physiol 2000;62:535–572.

Barrett MP, Walmsley AR, Gould GW. Structure and function of facilitative sugar transporters. Curr Opin Cell Biol 1999;11:496–502.

DeWeer P. A century of thinking about cell membranes. Annu Rev Physiol 2000;62:919–926.

Giebisch G. Physiological roles of renal potassium channels. Semin Nephrol 1999;19:458–471.

Hebert SC. Molecular mechanisms. Semin Nephrol 1999;19:504–523.

Hwang TC, Sheppard DN. Molecular pharmacology of the CFTR Cl^- channel. Trends Pharmacol Sci 1999;20:448–453.

Kanai Y. Family of neutral and acidic amino acid transporters: Molecular biology, physiology and medical implications. Curr Opin Cell Biol 1997;9:565–572.

(continued)

Ma T, Verkman AS. Aquaporin water channels in gastrointestinal physiology. J Physiol (London) 1999;517:317–326.

Nielsen S, Kwon TH, Christensen BM, et al. Physiology and pathophysiology of renal aquaporins. J Am Soc Nephrol 1999;10:647–663.

O'Neill WC. Physiological significance of volume-regulatory transporters. Am J Physiol 1999;276:C995–C1011.

Pilewski JM, Frizzell RA. Role of CFTR in airway disease. Physiol Rev 1999;79(Suppl):S215–S255.

Reuss L. One hundred years of inquiry: The mechanism of glucose absorption in the intestine. Annu Rev Physiol 2000;62:939–946.

Rojas CV. Ion channels and human genetic diseases. News Physiol Sci 1996;11:36–42.

Saier MH. Families of proteins forming transmembrane channels. J Membr Biol 2000;175:165–180.

Wright EM. Glucose galactose malabsorption. Am J Physiol 1998;275:G879–G882.

Yeaman C, Grindstaff KK, Nelson WJ. New perspectives on mechanisms involved in generating epithelial cell polarity. Physiol Rev 1999;79:73–98.

The Action Potential, Synaptic Transmission, and Maintenance of Nerve Function

Cynthia J. Forehand, Ph.D.

CHAPTER OUTLINE

- **PASSIVE MEMBRANE PROPERTIES, THE ACTION POTENTIAL, AND ELECTRICAL SIGNALING BY NEURONS**
- **SYNAPTIC TRANSMISSION**
- **NEUROCHEMICAL TRANSMISSION**
- **THE MAINTENANCE OF NERVE CELL FUNCTION**

KEY CONCEPTS

1. Nongated ion channels establish the resting membrane potential of neurons; voltage-gated ion channels are responsible for the action potential and the release of neurotransmitter.
2. Ligand-gated ion channels cause membrane depolarization or hyperpolarization in response to neurotransmitter.
3. Nongated ion channels are distributed throughout the neuronal membrane; voltage-gated channels are largely restricted to the axon and its terminals, while ligand-gated channels predominate on the cell body (soma) and dendritic membrane.
4. Membrane conductance and capacitance affect ion flow in neurons.
5. An action potential is a transient change in membrane potential characterized by a rapid depolarization followed by a repolarization; the depolarization phase is due to a rapid activation of voltage-gated sodium channels and the repolarization phase to an inactivation of the sodium channels and the delayed activation of voltage-gated potassium channels.
6. Initiation of an action potential occurs when an axon hillock is depolarized to a threshold for rapid activation of a large number of voltage-gated sodium channels.
7. Propagation of an action potential depends on local current flow derived from the inward sodium current depolarizing adjacent regions of an axon to threshold.
8. Conduction velocity depends on the size of an axon and the thickness of its myelin sheath, if present.
9. Following an action potential in one region of an axon, that region is temporarily refractory to the generation of another action potential because of the inactivation of the voltage-gated sodium channels.
10. When an action potential invades the nerve terminal, voltage-gated calcium channels open, allowing calcium to enter the terminal and start a cascade of events leading to the release of neurotransmitter.
11. Synaptic transmission involves a relatively small number of neurotransmitters that activate specific receptors on their postsynaptic target cells.
12. Most neurotransmitters are stored in synaptic vesicles and released upon nerve stimulation by a process of calcium-mediated exocytosis; once released, the neurotransmitter binds to and stimulates its receptors briefly before being rapidly removed from the synapse.
13. Metabolic maintenance of neurons requires specialized functions to match their specialized morphology and complex interconnections.

The nervous system coordinates the activities of many other organ systems. It activates muscles for movement, controls the secretion of hormones from glands, regulates the rate and depth of breathing, and is involved in modulating and regulating a multitude of other physiological processes. To perform these functions, the nervous system relies on neurons, which are designed for the rapid transmission of information from one cell to another by conducting electrical impulses and secreting chemical neurotransmitters. The electrical impulses propagate along the length of nerve fiber processes to their terminals, where they initiate a series of events that cause the release of chemical neurotransmitters. The release of neurotransmitters occurs at sites of synaptic contact between two nerve cells. Released neurotransmitters bind with their receptors on the postsynaptic cell membrane. The activation of these receptors either excites or inhibits the postsynaptic neuron.

The propagation of action potentials, the release of neurotransmitters, and the activation of receptors constitute the means whereby nerve cells communicate and transmit information to one another and to nonneuronal tissues. In this chapter, we examine the specialized membrane properties of nerve cells that endow them with the ability to produce action potentials, explore the basic mechanisms of synaptic transmission, and discuss aspects of neuronal structure necessary for the maintenance of nerve cell function.

PASSIVE MEMBRANE PROPERTIES, THE ACTION POTENTIAL, AND ELECTRICAL SIGNALING BY NEURONS

Neurons communicate by a combination of electrical and chemical signaling. Generally, information is integrated and transmitted along the processes of a single neuron electrically and then transmitted to a target cell chemically. The chemical signal then initiates an electrical change in the target cell. Electrical signals that depend on the passive properties of the neuronal cell membrane spread electrotonically over short distances. These potentials are initiated by local current flow and decay with distance from their site of initiation. Alternatively, an **action potential** is an electrical signal that propagates over a long distance without a change in amplitude. Action potentials depend on a regenerative wave of channel openings and closings in the membrane.

Special Anatomic Features of Neurons Adapt Them for Communicating Information

The shape of a nerve cell is highly specialized for the reception and transmission of information. One region of the neuron is designed to receive and process incoming information; another is designed to conduct and transmit information to other cells. The type of information that is processed and transmitted by a neuron depends on its location in the nervous system. For example, nerve cells associated with visual pathways convey information about the external environment, such as light and dark, to the brain; neurons associated with motor pathways convey information to control the contraction and relaxation of muscles for walking. Regardless of the type of information transmitted by neurons, they transduce and transmit this information via similar mechanisms. The mechanisms depend mostly on the specialized structures of the neuron and the electrical properties of their membranes.

Emerging from the **soma** (cell body) of a neuron are processes called **dendrites** and **axons** (Fig. 3.1). Many neurons in the central nervous system (CNS) also have knoblike structures called **dendritic spines** that extend from the dendrites. The dendritic spines, dendrites, and soma receive information from other nerve cells. The axon conducts and transmits information and may also receive information. Some axons are coated with **myelin**, a lipid structure formed by **glial cells** (oligodendrocytes in the CNS or Schwann cells in the peripheral nervous system, the PNS). Regular intermittent gaps in the myelin sheath are called **nodes of Ranvier**. The speed with which an axon conducts information is directly proportional to the size of the axon and the thickness of the myelin sheath. The end of the axon, the **axon terminal,** contains small **vesicles** packed with **neurotransmitter** molecules. The site of contact between a neuron and its target cell is called a **synapse.** Synapses are classified according to their site of contact as axospinous, axodendritic, axosomatic, or axoaxonic (Fig. 3.2). When a neuron is activated, an action potential is generated in the **axon hillock** (or initial segment) and conducted along the axon. The action potential causes the release of a neurotransmitter from the terminal. These neurotransmitter molecules bind to **receptors** located on target cells.

The binding of a neurotransmitter to its receptor typically causes a flow of ions across the membrane of the postsynaptic cell. This temporary redistribution of ionic charge can lead to the generation of an action potential, which itself is mediated by the flow of specific ions across the membrane. These electrical charges, critical for the transmission of information, are the result of ions moving through ion channels in the plasma membrane (see Chapter 2).

Channels Allow Ions to Flow Through the Nerve Cell Membrane

Ions can flow across the nerve cell membrane through three types of ion channels: nongated (leakage), ligand-gated, and voltage-gated (Fig. 3.3). **Nongated ion channels** are always open. They are responsible for the influx of Na^+ and efflux of K^+ when the neuron is in its resting state. **Ligand-gated ion channels** are directly or indirectly activated by chemical neurotransmitters binding to membrane receptors. In this type of channel, the receptor itself forms part of the ion channel or may be coupled to the channel via a G protein and a second messenger. When chemical transmitters bind to their receptors, the associated ion channels can either open or close to permit or block the movement of specific ions across the cell membrane. **Voltage-gated ion channels** are sensitive to the voltage difference across the membrane. In their initial resting state, these channels are typically closed; they open when a critical voltage level is reached.

Each type of ion channel has a unique distribution on the nerve cell membrane. Nongated ion channels, important for the establishment of the resting membrane potential, are found throughout the neuron. Ligand-gated channels, located at sites of synaptic contact, are found predominantly on dendritic spines, dendrites, and somata. Voltage-gated channels, required for the initiation and propagation of action potentials or for neurotransmitter release, are found predominantly on axons and axon terminals.

In the unstimulated state, nerve cells exhibit a **resting membrane potential** that is approximately -60 mV relative to the extracellular fluid. The resting membrane potential reflects a steady state that can be described by the Goldman equation (see Chapter 2). One should remember that the extracellular concentration of Na^+ is much greater than the

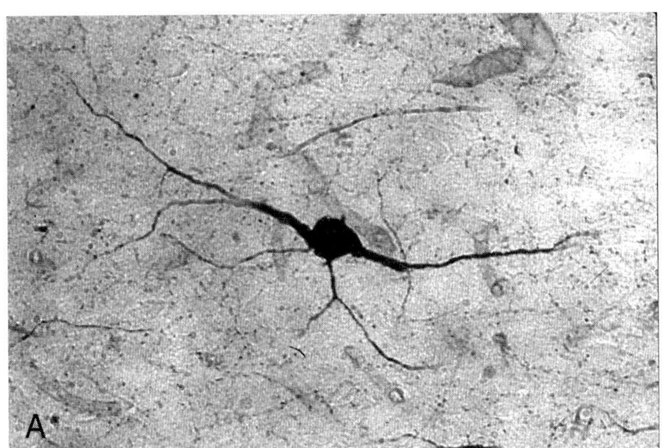

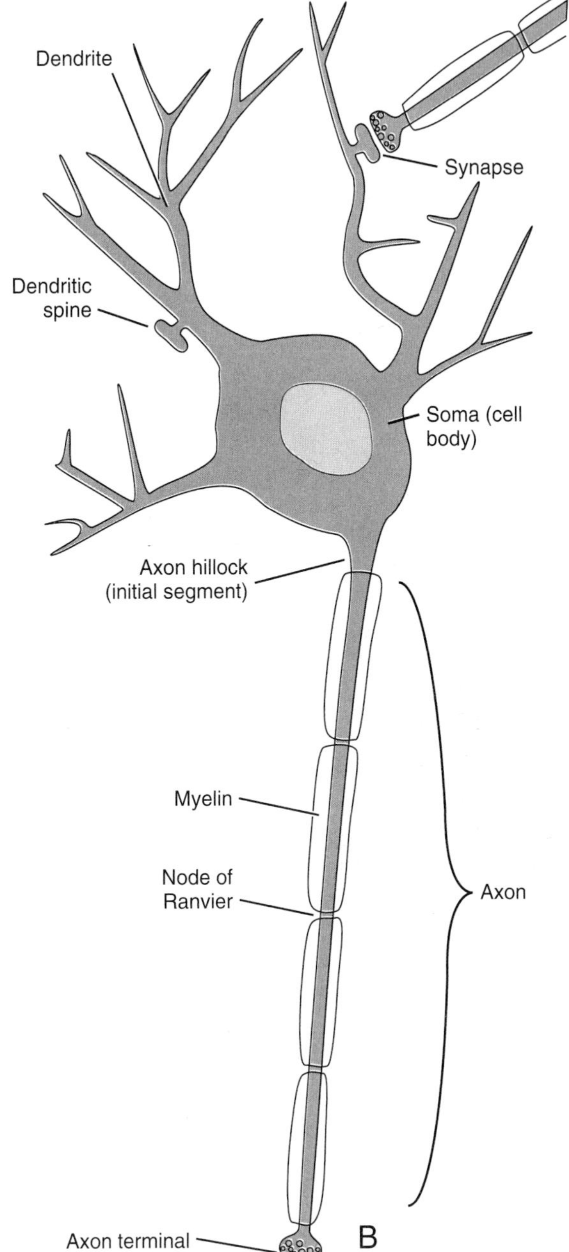

FIGURE 3.1 **The structure of a neuron.** A, A light micrograph. B, The structural components and a synapse.

intracellular concentration of Na^+, while the opposite is true for K^+. Moreover, the permeability of the membrane to potassium (P_K) is much greater than the permeability to sodium (P_{Na}) because there are many more leakage (nongated) channels in the membrane for K^+ than in the membrane for Na^+; therefore, the resting membrane potential is much closer to the equilibrium potential for potassium (E_K) than it is for sodium (see Chapter 2). Typical values for equilibrium potentials in neurons are +70 mV for sodium and −100 mV for potassium. Because sodium is far from its equilibrium potential, there is a large driving force on sodium, so sodium ions move readily whenever a voltage-gated or ligand-gated sodium channel opens in the membrane.

Electrical Properties of the Neuronal Membrane Affect Ion Flow

The electrical properties of the neuronal membrane play important roles in the flow of ions through the membrane, the initiation and conduction of action potentials along the axon, and the integration of incoming information at the dendrites and the soma. These properties include membrane conductance and capacitance.

The movement of ions across the nerve membrane is driven by ionic concentration and electrical gradients (see Chapter 2). The ease with which ions flow across the membrane through their channels is a measure of the membrane's **conductance**; the greater the conductance, the greater the flow of ions. Conductance is the inverse of **resistance**, which is measured in ohms. The conductance (g) of a membrane or single channel is measured in siemens. For an individual ion channel and a given ionic solution, the conductance is a constant value, determined in part by such factors as the relative size of the ion with respect to that of the channel and the charge distribution within the channel. Ohm's law describes the relationship between a single channel conductance, ionic current, and the membrane potential:

$$I_{ion} = g_{ion}(E_m - E_{ion})$$

or

$$g_{ion} = I_{ion}/(E_m - E_{ion}) \quad (1)$$

where I_{ion} is the ion current flow, E_m is the membrane potential, E_{ion} is the equilibrium (Nernst) potential for a specified ion, and g_{ion} is the channel conductance for an ion. Notice that if $E_m = E_{ion}$, there is no net movement of the ion and $I_{ion} = 0$. The conductance for a nerve membrane is the summation of all of its single channel conductances.

Another electrical property of the nerve membrane that influences the movement of ions is **capacitance**, the membrane's ability to store an electrical charge. A capacitor consists of two conductors separated by an insulator. Positive charge accumulates on one of the conductive plates while negative charge accumulates on the other plate. The biological capacitor is the lipid bilayer of the plasma membrane, which separates two conductive regions, the extracellular and intracellular fluids. Positive charge accumulates on the extracellular side while negative charge accumulates

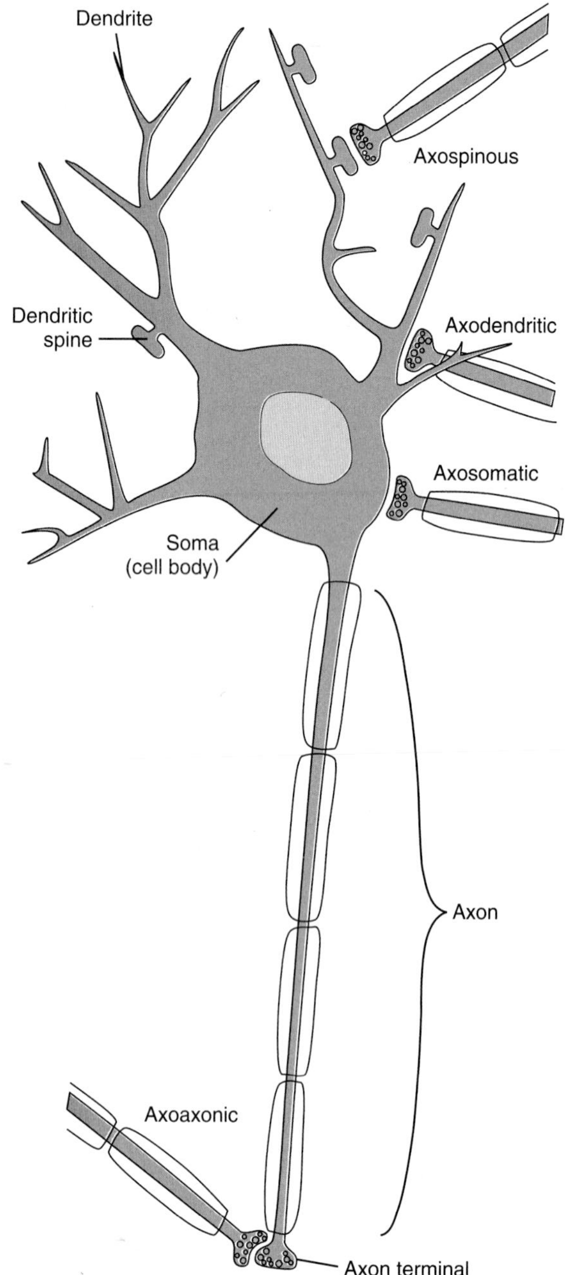

FIGURE 3.2 **Types of synapses.** The dendritic and somatic areas of the neuron, where most synapses occur, integrate incoming information. Synapses can also occur on the axon, which conducts information in the form of electrical impulses.

on the intracellular side. Membrane capacitance is measured in units of farads (F).

One factor that contributes to the amount of charge a membrane can store is its surface area; the greater the surface area, the greater the storage capacity. Large-diameter dendrites can store more charge than small-diameter dendrites of the same length. The speed with which the charge accumulates when a current is applied depends on the resistance of the circuit. Charge is delivered more rapidly when resistance is low. The time required for the mem-

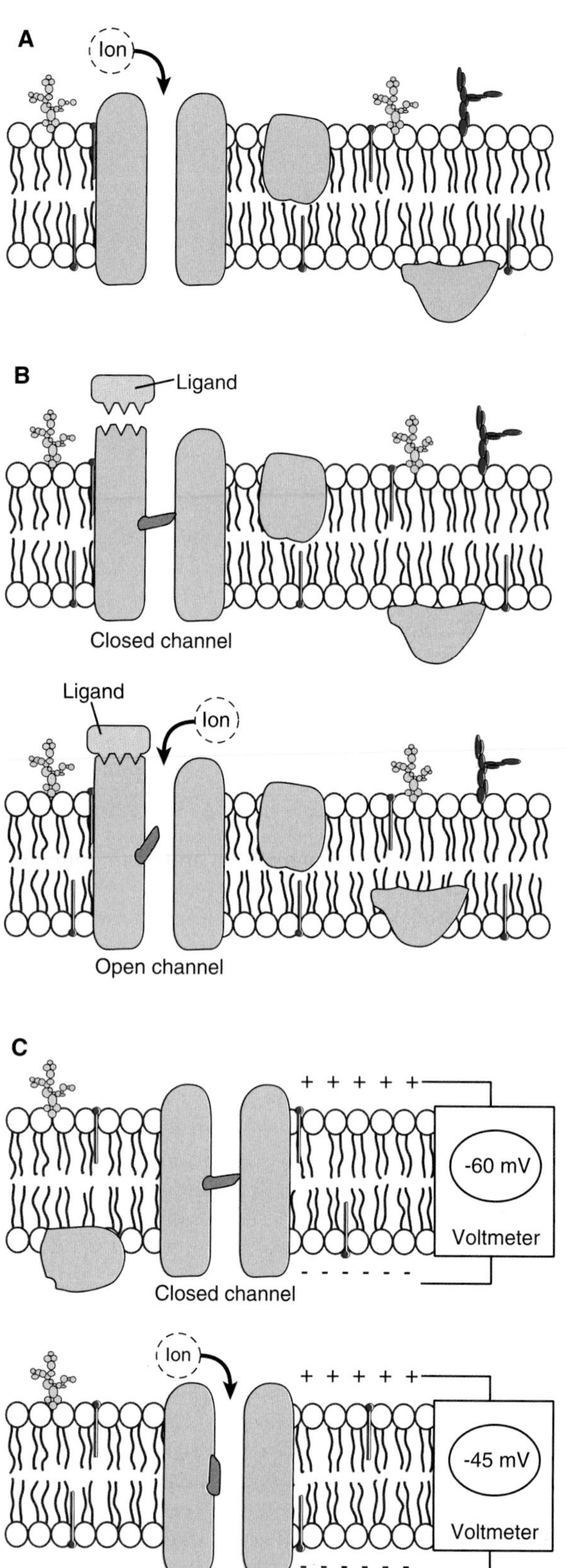

FIGURE 3.3 **The three types of ion channels.** **A**, The nongated channel remains open, permitting the free movement of ions across the membrane. **B**, The ligand-gated channel remains closed (or open) until the binding of a neurotransmitter. **C**, The voltage-gated channel remains closed until there is a change in membrane potential.

brane potential to change after a stimulus is applied is called the **time constant** or τ, and its relationship to capacitance (C) and resistance (R) is defined by the following equation:

$$\tau = RC \quad (2)$$

In the absence of an action potential, a stimulus applied to the neuronal membrane results in a local potential change that decreases with distance away from the point of stimulation. The voltage change at any point is a function of current and resistance as defined by Ohm's law. If a ligand-gated channel opens briefly and allows positive ions to enter the neuron, the electrical potential derived from that current will be greatest near the channels that opened, and the voltage change will steadily decline with increasing distance away from that point. The reason for the decline in voltage change with distance is that some of the ions back-leak out of the membrane because it is not a perfect insulator, and less charge reaches more distant sites. Since membrane resistance is a stable property of the membrane, the diminished current with distance away from the source results in a diminished voltage change. The distance at which the initial transmembrane voltage change has fallen to 37% of its peak value is defined as the **space constant** or λ. The value of the space constant depends on the internal axoplasmic resistance (R_a) and on the transmembrane resistance (R_m) as defined by the following equation:

$$\lambda = \sqrt{R_m / R_a} \quad (3)$$

R_m is usually measured in ohm-cm and R_a in ohm/cm. R_a decreases with increasing diameter of the axon or dendrite; thus, more current will flow farther along inside the cell, and the space constant is larger. Similarly, if R_m increases, less current leaks out and the space constant is larger. The larger the space constant, the farther along the membrane a voltage change is observed after a local stimulus is applied.

Membrane capacitance and resistance, and the resultant time and space constants, play an important role in both the propagation of the action potential and the integration of incoming information.

An Action Potential Is Generated at the Axon Hillock and Conducted Along the Axon

An action potential depends on the presence of voltage-gated sodium and potassium channels that open when the neuronal membrane is depolarized. These voltage-gated channels are restricted to the axon of most neurons. Thus, neuronal dendrites and cell bodies do not conduct action potentials. In most neurons, the axon hillock of the axon has a very high density of these voltage-gated channels. This region is also known as the **trigger zone** for the action potential. In sensory neurons that convey information to the CNS from distant peripheral targets, the trigger zone is in the region of the axon close to the peripheral target.

When the axon is depolarized slightly, some voltage-gated sodium channels open; as Na^+ ions enter and cause more depolarization, more of these channels open. At a critical membrane potential called the **threshold**, incoming Na^+ exceeds outgoing K^+ (through leakage channels), and the resulting explosive opening of the remaining voltage-gated sodium channels initiates an action potential. The action potential then propagates to the axon terminal, where the associated depolarization causes the release of neurotransmitter. The initial depolarization to start this process derives from synaptic inputs causing ligand-gated channels to open on the dendrites and somata of most neurons. For peripheral sensory neurons, the initial depolarization results from a generator potential initiated by a variety of sensory receptor mechanisms (see Chapter 4).

Characteristics of the Action Potential. Depolarization of the axon hillock to threshold results in the generation and propagation of an action potential. The action potential is a transient change in the membrane potential characterized by a gradual depolarization to threshold, a rapid rising phase, an overshoot, and a repolarization phase. The repolarization phase is followed by a brief **afterhyperpolarization (undershoot)** before the membrane potential again reaches resting level (Fig. 3.4A).

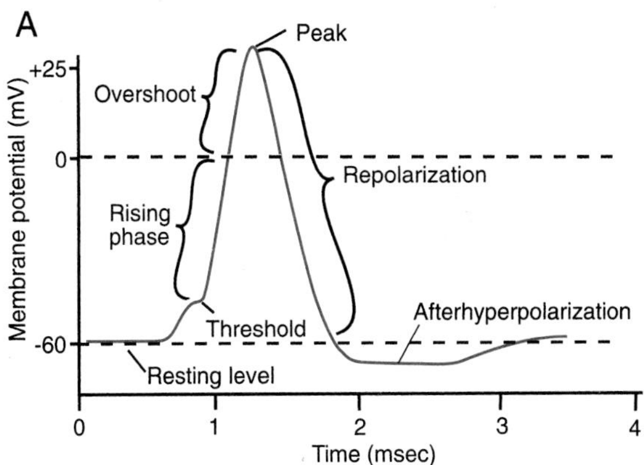

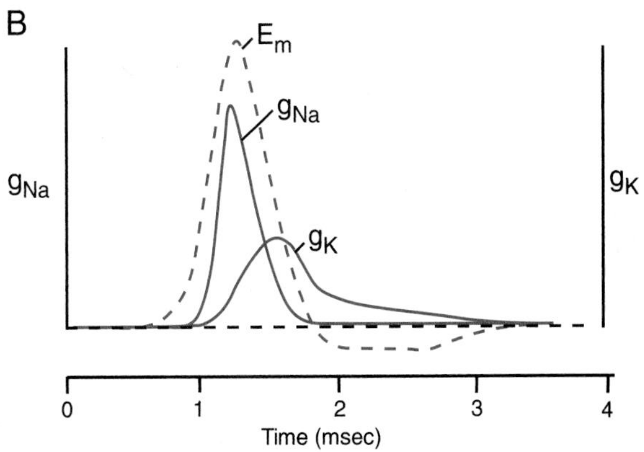

FIGURE 3.4 **The phases of an action potential.** A, Depolarization to threshold, the rising phase, overshoot, peak, repolarization, afterhyperpolarization, and return to the resting membrane potential. B, Changes in sodium (g_{Na}) and potassium (g_K) conductances associated with an action potential. The rising phase of the action potential is the result of an increase in sodium conductance, while the repolarization phase is a result of a decrease in sodium conductance and a delayed increase in potassium conductance.

The action potential may be recorded by placing a microelectrode inside a nerve cell or its axon. The voltage measured is compared to that detected by a reference electrode placed outside the cell. The difference between the two measurements is a measure of the membrane potential. This technique is used to monitor the membrane potential at rest, as well as during an action potential.

Action Potential Gating Mechanisms. The depolarizing and repolarizing phases of the action potential can be explained by relative changes in membrane conductance (permeability) to sodium and potassium. During the rising phase, the nerve cell membrane becomes more permeable to sodium; as a consequence, the membrane potential begins to shift more toward the equilibrium potential for sodium. However, before the membrane potential reaches E_{Na}, sodium permeability begins to decrease and potassium permeability increases. This change in membrane conductance again drives the membrane potential toward E_K, accounting for repolarization of the membrane (Fig. 3.4B).

The action potential can also be viewed in terms of the flow of charged ions through selective ion channels. These voltage-gated channels are closed when the neuron is at rest (Fig. 3.5A). When the membrane is depolarized, these channels begin to open. The Na^+ channel quickly opens its **activation gate** and allows Na^+ ions to flow into the cell (Fig. 3.5B). The influx of positively charged Na^+ ions causes the membrane to depolarize. In fact, the membrane potential actually reverses, with the inside becoming positive; this is called the **overshoot.** In the initial stage of the action potential, more Na^+ than K^+ channels are opened because the K^+ channels open more slowly in response to depolarization. This increase in Na^+ permeability compared to that of K^+ causes the membrane potential to move toward the equilibrium potential for Na^+.

At the peak of the action potential, the sodium conductance begins to fall as an **inactivation gate** closes. Also, more K^+ channels open, allowing more positively charged K^+ ions to leave the neuron. The net effect of inactivating Na^+ channels and opening additional K^+ channels is the repolarization of the membrane (Fig. 3.5C).

As the membrane continues to repolarize, the membrane potential becomes more negative than its resting level. This **afterhyperpolarization** is a result of K^+ channels remaining open, allowing the continued efflux of K^+ ions. Another way to think about afterhyperpolarization is that the membrane's permeability to K^+ is higher than when the neuron is at rest. Consequently, the membrane potential is driven even more toward the K^+ equilibrium potential (Fig. 3.5D).

The changes in membrane potential during an action potential result from selective alterations in membrane conductance (see Fig. 3.4B). These membrane conductance changes reflect the summated activity of individual voltage-gated sodium and potassium ion channels. From the temporal relationship of the action potential and the membrane conductance changes, the depolarization and rising phase of the action potential can be attributed to the increase in sodium ion conductance, the repolarization phases to both the decrease in sodium conductance and the increase in potassium conductance, and afterhyperpolarization to the sustained increase of potassium conductance.

Alterations in voltage-gated sodium and potassium channels, as well as in voltage-gated calcium and chloride channels, are now known to be the basis of several diseases of nerve and muscle. These diseases are collectively known as **channelopathies** (see Clinical Focus Box 3.1).

Initiation of the Action Potential. In most neurons, the axon hillock (initial segment) is the trigger zone that generates the action potential. The membrane of the initial segment contains a high density of voltage-gated sodium and potassium ion channels. When the membrane of the initial segment is depolarized, voltage-gated sodium channels are opened, permitting an influx of sodium ions. The influx of these positively charged ions further depolarizes the membrane, leading to the opening of other voltage-gated sodium channels. This cycle of membrane depolarization, sodium channel activation, sodium ion influx, and membrane depolarization is an example of positive feedback, a regenerative process (Fig. 1.3) that results in the explosive activation of many sodium ion channels when the threshold membrane potential is reached. If the depolarization of the initial segment does not reach threshold, then not enough sodium channels are activated to initiate the regenerative process. The initiation of an action potential is, therefore, an "all-or-none" event; it is generated completely or not at all.

Propagation and Speed of the Action Potential. After an action potential is generated, it propagates along the axon toward the axon terminal; it is conducted along the axon with no decrement in amplitude. The mode in which action potentials propagate and the speed with which they are conducted along an axon depend on whether the axon is myelinated. The diameter of the axon also influences the speed of action potential conduction: larger-diameter axons have faster action potential conduction velocities than smaller-diameter axons.

In unmyelinated axons, voltage-gated Na^+ and K^+ channels are distributed uniformly along the length of the axonal membrane. An action potential is generated when the axon hillock is depolarized by the passive spread of synaptic potentials along the somatic and dendritic membrane (see below). The hillock acts as a "sink" where Na^+ ions enter the cell. The "source" of these Na^+ ions is the extracellular space along the length of the axon. The entry of Na^+ ions into the axon hillock causes the adjacent region of the axon to depolarize as the ions that entered the cell, during the peak of the action potential, flow away from the sink. This local spread of the current depolarizes the adjacent region to threshold and causes an action potential in that region. By sequentially depolarizing adjacent segments of the axon, the action potential propagates or moves along the length of the axon from point to point, like a traveling wave (Fig. 3.6A).

Just as large-diameter tubes allow a greater flow of water than small-diameter tubes because of their decreased resistance, large-diameter axons have less cytoplasmic resistance, thereby permitting a greater flow of ions. This increase in ion flow in the cytoplasm causes greater lengths of the axon to be depolarized, decreasing the time needed for the action potential to travel along the axon. Recall

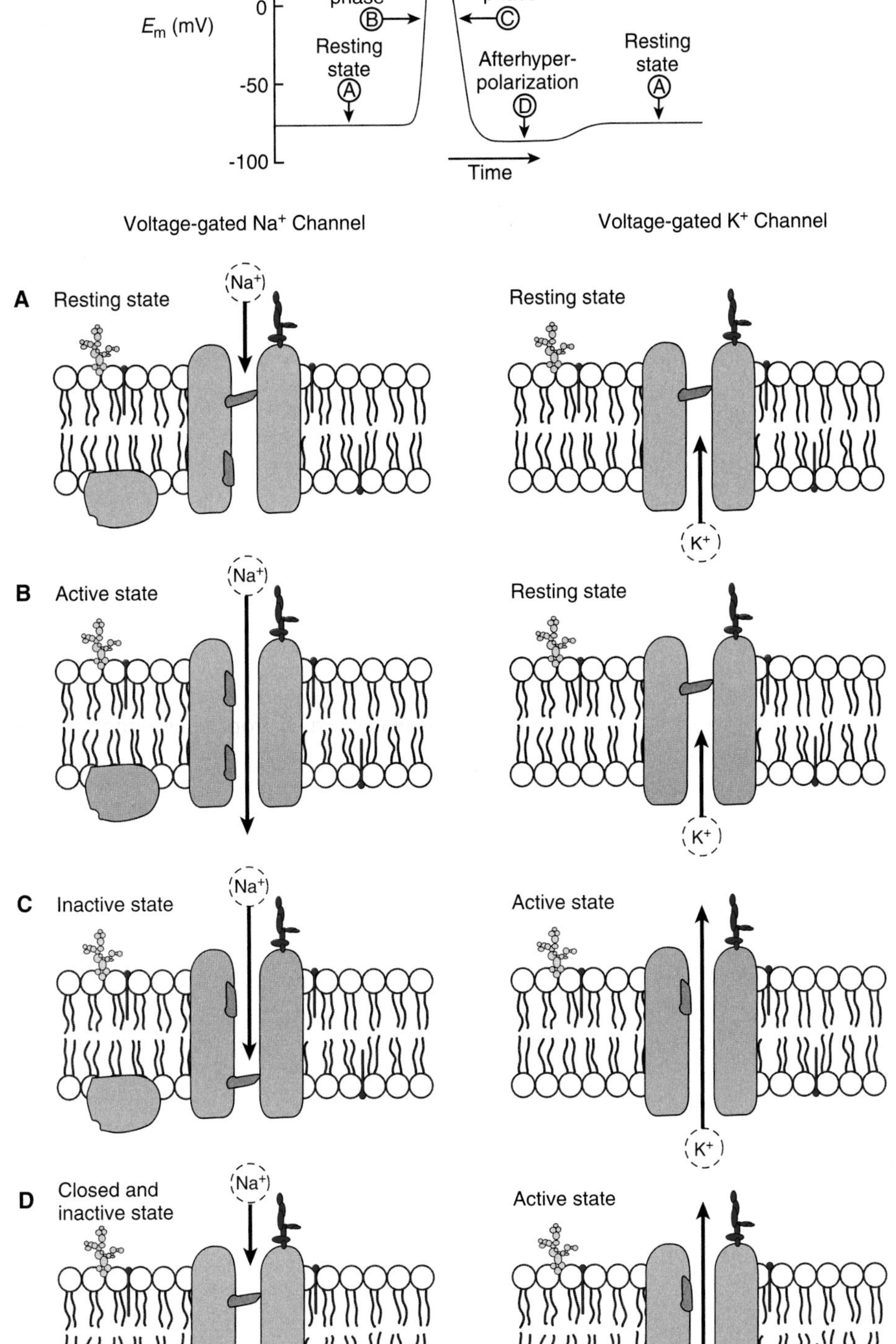

FIGURE 3.5 **The states of voltage-gated sodium and potassium channels correlated with the course of the action potential.** A, At the resting membrane potential, both channels are in a closed, resting state. B, During the depolarizing phase of the action potential the voltage-gated sodium channels are activated (open), but the potassium channels open more slowly and, therefore, have not yet responded to the depolarization. C, During the repolarizing phase, sodium channels become inactivated, while the potassium channels become activated (open). D, During the afterhyperpolarization, the sodium channels are both closed and inactivated, and the potassium channels remain in their active state. Eventually, the potassium channels close and the sodium channel inactivation is removed, so that both channels are in their resting state and the membrane potential returns to resting membrane potential. Note that the voltage-gated potassium channel does not have an inactivated state. (Modified from Matthews GG. Neurobiology: Molecules, Cells and Systems. Malden, MA: Blackwell Science, 1998.)

that the space constant, λ, determines the length along the axon that a voltage change is observed after a local stimulus is applied. In this case, the local stimulus is the inward sodium current that accompanies the action potential. The larger the space constant, the farther along the membrane a voltage change is observed after a local stimulus is applied. The space constant increases with axon diameter because the internal axoplasmic resistance, R_a, decreases, allowing the current to spread farther down the inside of the axon before leaking back across the membrane. Therefore,

CLINICAL FOCUS BOX 3.1

Channelopathies

Voltage-gated channels for sodium, potassium, calcium, and chloride are intimately associated with excitability in neurons and muscle cells and in synaptic transmission. Until the early 1990s, most of our knowledge about channel properties derived from biophysical studies of isolated cells or their membranes. The advent of molecular approaches resulted in the cloning of the genes for a variety of channels and the subsequent expression of these genes in a large cell, such as the *Xenopus* oocyte, for further characterization.

This approach also allowed experimental manipulation of the channels by expressing genes that were altered in known ways. In this way, researchers could determine which parts of channel molecules were responsible for particular properties, including voltage sensitivity, ion specificity, activation, inactivation, kinetics, and interaction with other cellular components. This genetic understanding of the control of channel properties led to the realization that many unexplained diseases may be caused by alterations in the genes for ion channels. Diseases based on altered ion channel function are now collectively called **channelopathies.** These diseases affect neurons, skeletal muscle, cardiac muscle, and even nonexcitable cells, such as kidney tubular cells.

One of the best-known sets of channelopathies is a group of channel mutations that lead to the Long Q-T (LQT) syndrome in the heart. The QT interval on the electrocardiogram is the time between the beginning of ventricular depolarization and the end of ventricular repolarization. In patients with LQT, the QT interval is abnormally long because of defective membrane repolarization, which can lead to ventricular arrhythmia and sudden death. Affected individuals generally have no cardiovascular disease other than that associated with electrical abnormality. The defect in membrane repolarization could be a result of a prolonged inward sodium current or a reduced outward potassium current. In fact, mutations in potassium channels account for two different LQT syndromes, and a third derives from a sodium channel mutation.

Myotonia is a condition characterized by a delayed relaxation of muscle following contraction. There are several types of myotonias, all related to abnormalities in muscle membrane. Some myotonias are associated with a skeletal muscle sodium channel, and others are associated with a skeletal muscle chloride channel.

Channelopathies affecting neurons include episodic and spinocerebellar ataxias, some forms of epilepsy, and familial hemiplegic migraine. Ataxias are a disruption in gait mediated by abnormalities in the cerebellum and spinal motor neurons. One specific ataxia associated with an abnormal potassium channel is episodic ataxia with myokymia. In this disease, which is autosomal-dominant, cerebellar neurons have abnormal excitability and motor neurons are chronically hyperexcitable. This hyperexcitability causes indiscriminant firing of motor neurons, observed as the twitching of small groups of muscle fibers, akin to worms crawling under the skin (myokymia). It is likely that many other neuronal (and muscle) disorders of currently unknown pathology will be identified as channelopathies.

when an action potential is generated in one region of the axon, more of the adjacent region that is depolarized by the inward current accompanying the action potential reaches the threshold for action potential generation. The result is that the speed at which action potentials are conducted, or **conduction velocity**, increases as a function of increasing axon diameter and concomitant increase in the space constant.

Several factors act to increase significantly the conduction velocity of action potentials in myelinated axons. Schwann cells in the PNS and oligodendrocytes in the CNS wrap themselves around axons to form myelin, layers of lipid membrane that insulate the axon and prevent the passage of ions through the axonal membrane (Fig. 3.6B). Between the myelinated segments of the axon are the nodes of Ranvier, where action potentials are generated.

The signal that causes these glial cells to myelinate the axons apparently derives from the axon, and its potency is a function of axon size. In general, axons larger than approximately 1 μm in diameter are myelinated, and the thickness of the myelin increases as a function of axon diameter. Since the smallest myelinated axon is bigger than the largest unmyelinated axon, conduction velocity is faster for myelinated axons based on size alone. In addition, the myelin acts to increase the effective resistance of the axonal membrane, R_m, since ions that flow across the axonal membrane must also flow through the tightly wrapped layers of myelin before they reach the extracellular fluid. This increase in R_m increases the space constant. The layers of myelin also decrease the effective capacitance of the axonal membrane because the distance between the extracellular and intracellular conducting fluid compartments is increased. Because the capacitance is decreased, the time constant is decreased, increasing the conduction velocity.

While the effect of myelin on R_m and capacitance are important for increasing conduction velocity, there is an even greater factor at play—an alteration in the mode of conduction. In myelinated axons, voltage-gated Na^+ channels are highly concentrated in the nodes of Ranvier, where the myelin sheath is absent, and are in low density beneath the segments of myelin. When an action potential is initiated at the axon hillock, the influx of Na^+ ions causes the adjacent node of Ranvier to depolarize, resulting in an action potential at the node. This, in turn, causes depolarization of the next node of Ranvier and the eventual initiation of an action potential. Action potentials are successively generated at neighboring nodes of Ranvier; therefore, the action potential in a myelinated axon appears to jump from one node to the next, a process called **saltatory conduction** (Fig. 3.6C). This process results in a faster conduction velocity for myelinated than unmyelinated axons. The conduction velocity in mammals ranges from 3 to 120 m/sec for myelinated axons and 0.5 to 2.0 m/sec for unmyelinated axons.

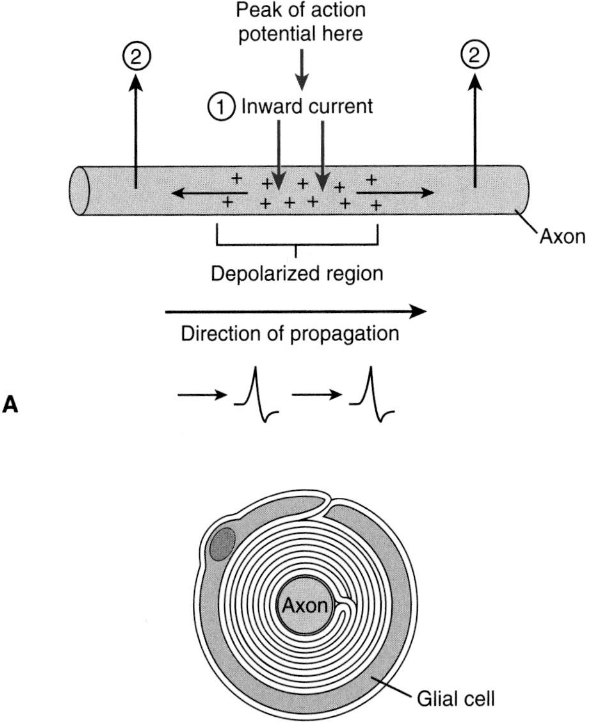

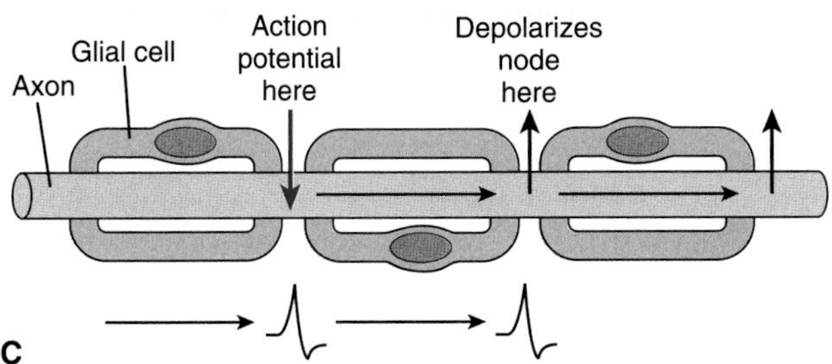

FIGURE 3.6 **Myelinated axons and saltatory conduction.** **A**, Propagation of an action potential in an unmyelinated axon. The initiation of an action potential in one segment of the axon depolarizes the immediately adjacent section, bringing it to threshold and generating an action potential. **B**, A sheath of myelin surrounding an axon. **C**, The propagation of an action potential in a myelinated axon. The initiation of an action potential in one node of Ranvier depolarizes the next node. Jumping from one node to the next is called saltatory conduction. (Modified from Matthews GG. Neurobiology: Molecules, Cells and Systems. Malden, MA: Blackwell Science, 1998.)

Refractory Periods. After the start of an action potential, there are periods when the initiation of additional action potentials requires a greater degree of depolarization and when action potentials cannot be initiated at all. These are called the **relative** and **absolute refractory periods**, respectively (Fig. 3.7).

The inability of a neuronal membrane to generate an action potential during the absolute refractory period is primarily due to the state of the voltage-gated Na^+ channel. After the inactivation gate closes during the repolarization phase of an action potential, it remains closed for some time; therefore, another action potential cannot be generated no matter how much the membrane is depolarized. The importance of the absolute refractory period is that it limits the rate of firing of action potentials. The absolute refractory period also prevents action potentials from traveling in the wrong direction along the axon.

In the relative refractory period, the inactivation gate of a portion of the voltage-gated Na^+ channels is open. Since these channels have returned to their initial resting state, they can now respond to depolarizations of the membrane. Consequently, when the membrane is depolarized, many of the channels open their activation gates and permit the influx of Na^+ ions. However, because only a portion of the Na^+ channels have returned to the resting state, depolarization of the membrane to the original threshold level activates an insufficient number of channels to initiate an action potential. With greater levels of depolarization, more channels are activated, until eventually an action potential is generated. The K^+ channels are maintained in the open state during the relative refractory period, leading to membrane hyperpolarization. By these two mechanisms, the action potential threshold is increased during the relative refractory period.

SYNAPTIC TRANSMISSION

Neurons communicate at synapses. Two types of synapses have been identified: electrical and chemical. At **electrical synapses**, passageways known as **gap junctions** connect the cytoplasm of adjacent neurons (see Fig. 1.6) and permit the bidirectional passage of ions from one cell to another. Electrical synapses are uncommon in the adult mammalian nervous system. Typically, they are found at dendrodendritic sites of contact; they are thought to synchronize the activity of neuronal populations. Gap junctions are more common in the embryonic nervous system, where they may act to aid the development of appropriate synaptic connections based on synchronous firing of neuronal populations.

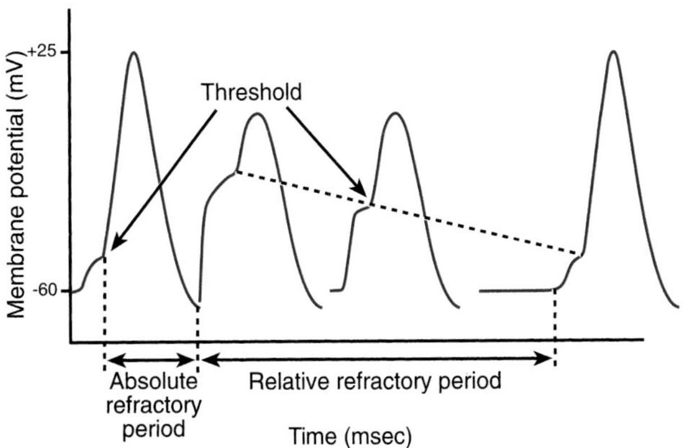

FIGURE 3.7 **Absolute and relative refractory periods.** Immediately after the start of an action potential, a nerve cell is incapable of generating another impulse. This is the absolute refractory period. With time, the neuron can generate another action potential, but only at higher levels of depolarization. The period of increased threshold for impulse initiation is the relative refractory period. Note that action potentials initiated during the relative refractory period have lower-than-normal amplitude.

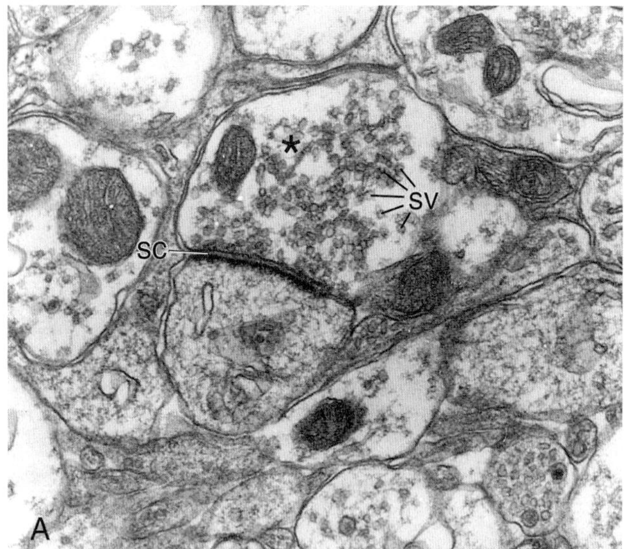

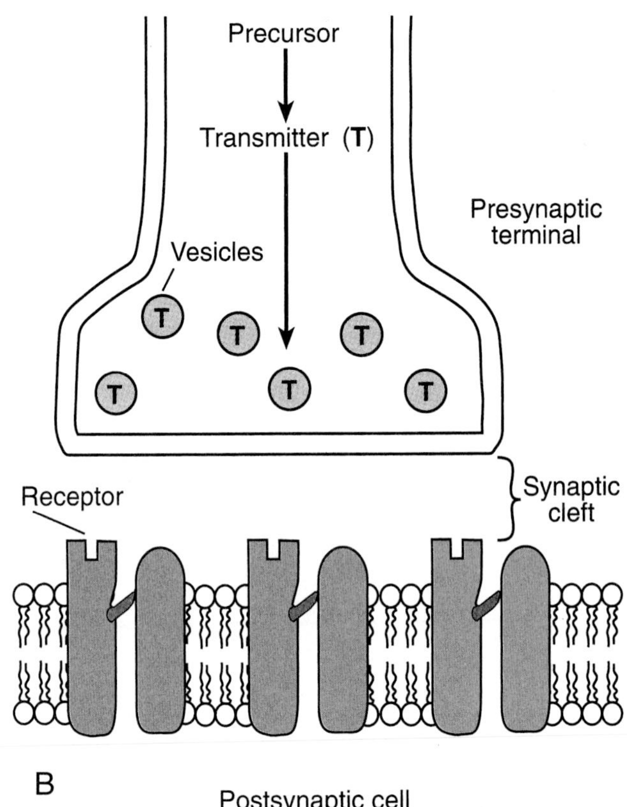

FIGURE 3.8 **A chemical synapse.** A, This electron micrograph shows a presynaptic terminal (asterisk) with synaptic vesicles (SV) and synaptic cleft (SC) separating presynaptic and postsynaptic membranes (magnification 60,000×) (Courtesy of Dr. Lazaros Triarhou, Indiana University School of Medicine.) **B**, The main components of a chemical synapse.

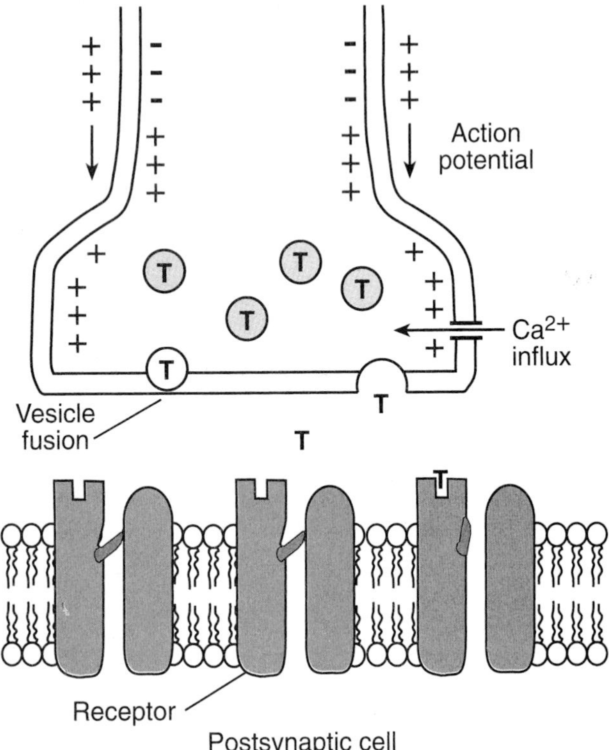

FIGURE 3.9 **The release of neurotransmitter.** Depolarization of the nerve terminal by the action potential opens voltage-gated calcium channels. Increased intracellular Ca^{2+} initiates fusion of synaptic vesicles with the presynaptic membrane, resulting in the release of neurotransmitter molecules into the synaptic cleft and binding with postsynaptic receptors.

Synaptic Transmission Usually Occurs via Chemical Neurotransmitters

At **chemical synapses**, a space called the **synaptic cleft** separates the presynaptic axon terminal from the postsynaptic cell (Fig. 3.8). The presynaptic terminal is packed with vesicles containing chemical neurotransmitters that are released into the synaptic cleft when an action potential enters the terminal. Once released, the chemical neurotransmitter diffuses across the synaptic cleft and binds to receptors on the postsynaptic cell. The binding of the transmitter to its receptor leads to the opening (or closing) of specific ion channels, which, in turn, alter the membrane potential of the postsynaptic cell.

The release of neurotransmitters from the presynaptic terminal begins with the invasion of the action potential into the axon terminal (Fig. 3.9). The depolarization of the terminal by the action potential causes the activation of voltage-gated Ca^{2+} channels. The electrochemical gradients for Ca^{2+} result in forces that drive Ca^{2+} into the terminal. This increase in intracellular ionized calcium causes a fusion of vesicles, containing neurotransmitters, with the presynaptic membrane at **active zones**. The neurotransmitters are then released into the cleft by **exocytosis.** Increasing the amount of Ca^{2+} that enters the terminal increases the amount of transmitter released into the synaptic cleft. The number of transmitter molecules released by any one exocytosed vesicle is called a **quantum**, and the total number of quanta released when the synapse is activated is called the **quantum content**. Under normal conditions, quanta are fixed in size but quantum content varies, particularly with the amount of Ca^{2+} that enters the terminal.

The way in which the entry of Ca^{2+} leads to the fusion of the vesicles with the presynaptic membrane is still being elucidated. It is clear that there are several proteins involved in this process. One hypothesis is that the vesicles are anchored to cytoskeletal components in the terminal by **synapsin**, a protein surrounding the vesicle. The entry of Ca^{2+} ions into the terminal is thought to result in phosphorylation of this protein and a decrease in its binding to the cytoskeleton, releasing the vesicles so they may move to the synaptic release sites.

Other proteins (rab GTP-binding proteins) are involved in targeting synaptic vesicles to specific docking sites in the presynaptic terminal. Still other proteins cause the vesicles to dock and bind to the presynaptic terminal membrane; these proteins are called **SNARES** and are found on both the vesicle and the nerve terminal membrane (called v-SNARES or t-SNARES, respectively). Tetanus toxin and botulinum toxin exert their devastating effects on the nervous system by disrupting the function of SNARES, preventing synaptic transmission. Exposure to these toxins can be fatal because the failure of neurotransmission between neurons and the muscles involved in breathing results in respiratory failure. To complete the process begun by Ca^{2+} entry into the nerve terminal, the docked and bound vesicles must fuse with the membrane and create a pore through which the transmitter may be released into the synaptic cleft. The vesicle membrane is then removed from the terminal membrane and recycled within the nerve terminal.

Once released into the synaptic cleft, neurotransmitter molecules exert their actions by binding to receptors in the postsynaptic membrane. These receptors are of two types. In some, the receptor forms part of an ion channel; in others, the receptor is coupled to an ion channel via a G protein and a second messenger system. In receptors associated with a specific G protein, a series of enzyme steps is initiated by binding of a transmitter to its receptor, producing a second messenger that alters intracellular functions over a longer time than for direct ion channel opening. These membrane-bound enzymes and the second messengers they produce inside the target cells include **adenylyl cyclase,** which produces cAMP; **guanylyl cyclase,** which produces cGMP; and **phospholipase C,** which leads to the formation of two second messengers, diacylglycerol and inositol trisphosphate (see Chapter 1).

When a transmitter binds to its receptor, membrane conductance changes occur, leading to depolarization or hyperpolarization. An increase in membrane conductance to Na^+ depolarizes the membrane. An increase in membrane conductance that permits the efflux of K^+ or the influx of Cl^- hyperpolarizes the membrane. In some cases, membrane hyperpolarization can occur when a decrease in membrane conductance reduces the influx of Na^+. Each of these effects results from specific alterations in ion channel function, and there are many different ligand-gated and voltage-gated channels.

Integration of Postsynaptic Potentials Occurs in the Dendrites and Soma

The transduction of information between neurons in the nervous system is mediated by changes in the membrane potential of the postsynaptic cell. These membrane depolarizations and hyperpolarizations are integrated or summated and can result in activation or inhibition of the postsynaptic neuron. Alterations in the membrane potential that occur in the postsynaptic neuron initially take place in the dendrites and the soma as a result of the activation of afferent inputs.

Since depolarizations can lead to the excitation and activation of a neuron, they are commonly called **excitatory postsynaptic potentials** (EPSPs). In contrast, hyperpolarizations of the membrane prevent the cell from becoming activated and are called **inhibitory postsynaptic potentials** (IPSPs). These membrane potential changes are caused by the influx or efflux of specific ions (Fig. 3.10).

The rate at which the membrane potential of a postsynaptic neuron is altered can greatly influence the efficiency of transducing information from one neuron to the next. If the activation of a synapse leads to the influx of positively charged ions, the postsynaptic membrane will depolarize. When the influx of these ions is stopped, the membrane will repolarize back to the resting level. The rate at which it repolarizes depends on the membrane time constant, τ, which is a function of membrane resistance and capacitance and represents the time required for the membrane potential to decay to 37% of its initial peak value (Fig. 3.11).

The decay rate for repolarization is slower for longer time constants because the increase in membrane resistance and/or capacitance results in a slower discharge of the membrane. The slow decay of the repolarization allows additional time for the synapse to be reactivated and depolarize the membrane. A second depolarization of the mem-

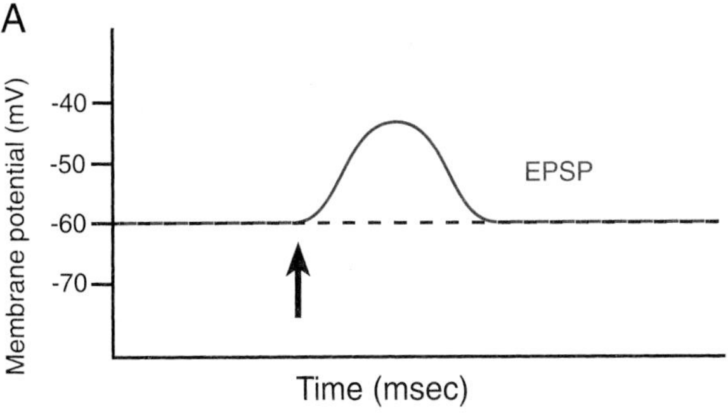

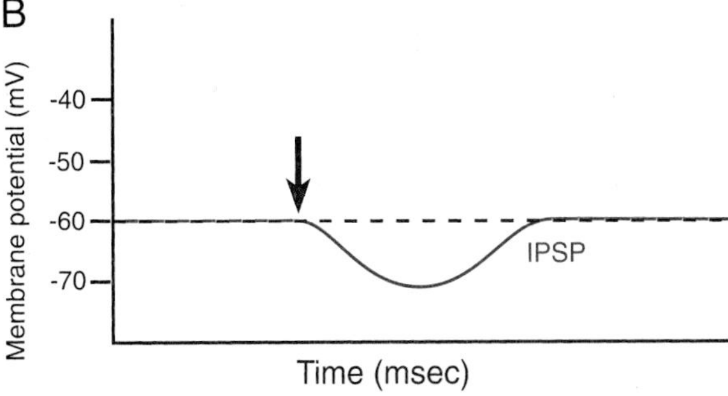

FIGURE 3.10 **Excitatory and inhibitory postsynaptic potentials.** A, The depolarization of the membrane (arrow) brings a nerve cell closer to the threshold for the initiation of an action potential and produces an excitatory postsynaptic potential (EPSP). **B,** The hyperpolarization of the membrane produces an inhibitory postsynaptic potential (IPSP).

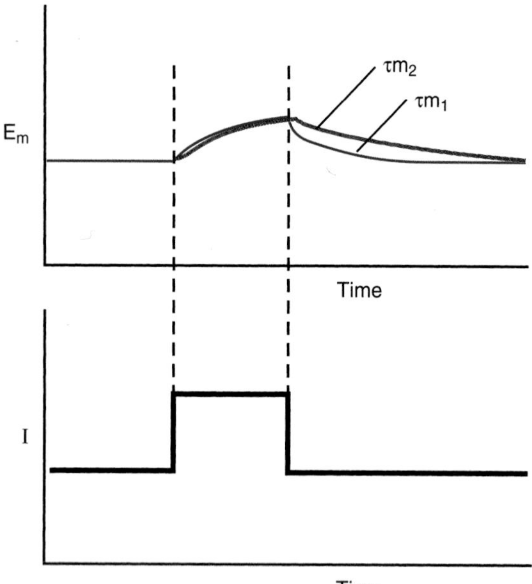

FIGURE 3.11 Membrane potential decay rate and time constant. The rate of decay of membrane potential (E_m) varies with a given neuron's membrane time constant. The responses of two neurons to a brief application of depolarizing current (I) are shown here. Each neuron depolarizes to the same degree, but the time for return to the baseline membrane potential differs for each. Neuron 2 takes longer to return to baseline than neuron 1 because its time constant is longer ($\tau_{m2} > \tau_{m1}$).

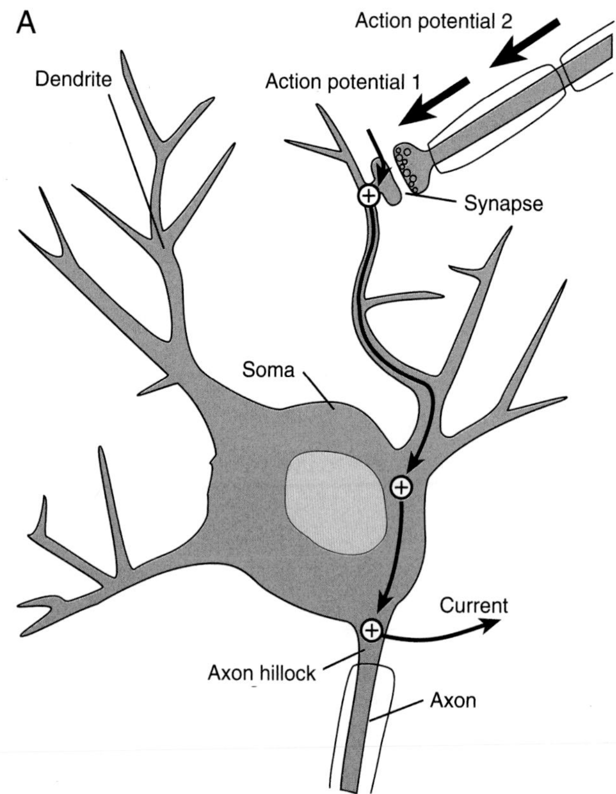

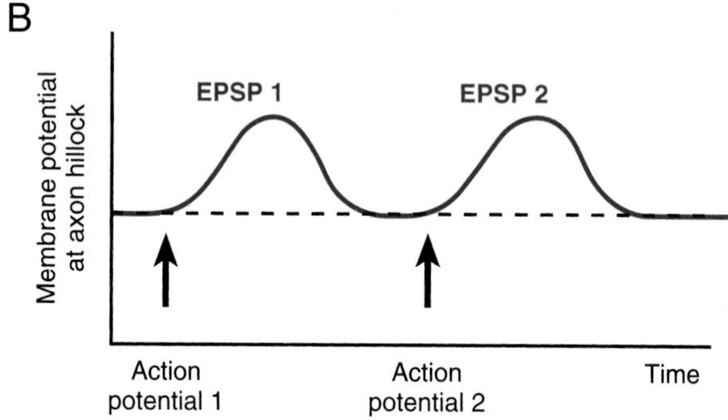

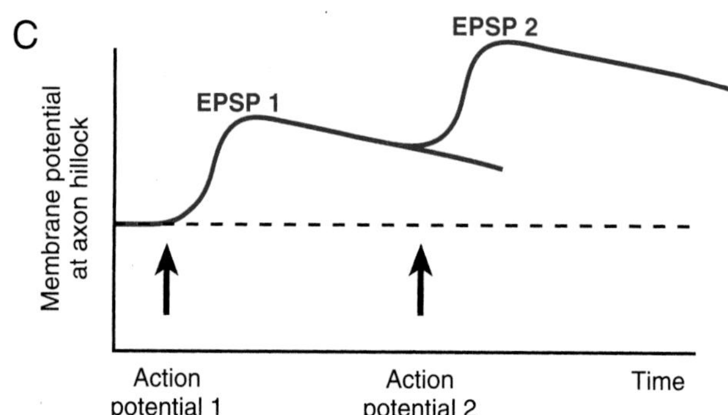

FIGURE 3.12 A model of temporal summation. **A,** Depolarization of a dendrite by two sequential action potentials. **B,** A dendritic membrane with a short time constant is unable to summate postsynaptic potentials. **C,** A dendritic membrane with a long time constant is able to summate membrane potential changes.

brane can be added to that of the first depolarization. Consequently, longer periods of depolarization increase the likelihood of summating two postsynaptic potentials. The process in which postsynaptic membrane potentials are added with time is called **temporal summation** (Fig. 3.12). If the magnitude of the summated depolarizations is above a threshold value, as detected at the axon hillock, it will generate an action potential.

The summation of postsynaptic potentials also occurs with the activation of several synapses located at different sites of contact. This process is called **spatial summation**. When a synapse is activated, causing an influx of positively charged ions, a depolarizing **electrotonic potential** develops, with maximal depolarization occurring at the site of synaptic activation. The electrotonic potential is due to the passive spread of ions in the dendritic cytoplasm and across the membrane. The amplitude of the electrotonic potential decays with distance from the synapse activation site (Fig. 3.13). The decay of the electrotonic potential per unit length along the dendrite is determined by the length or space constant, λ, which represents the length required for the membrane potential depolarization to decay to 37% of its maximal value. The larger the space constant value, the smaller the decay per unit length; thus, more charge is delivered to more distant membrane patches.

By depolarizing distal patches of membrane, other electrotonic potentials that occur by activating synaptic inputs at other sites can summate to produce even greater depolarization, and the resulting postsynaptic potentials

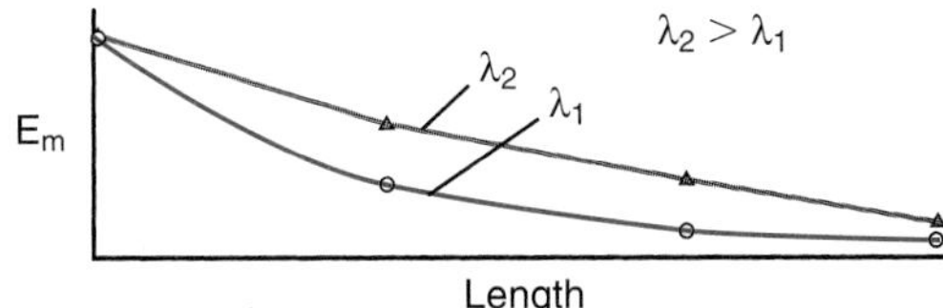

FIGURE 3.13 A profile of the electrotonic membrane potential produced along the length of a dendrite. The decay of the membrane potential, E_m, as it proceeds along the length of the dendrite is affected by the space constant, λ_m. Long space constants cause the electrotonic potential to decay more gradually. Profiles are shown for two dendrites with different space constants, λ_1 and λ_2. The electrotonic potential of dendrite 2 decays less steeply than that of dendrite 1 because its space constant is longer.

are added along the length of the dendrite. As with temporal summation, if the depolarizations resulting from spatial summation are sufficient to cause the membrane potential in the region of axon hillock to reach threshold, the postsynaptic neuron will generate an action potential (Fig. 3.14).

Because of the spatial decay of the electrotonic potential, the location of the synaptic contact strongly influences whether a synapse can activate a postsynaptic neuron. For example, axodendritic synapses, located in distal segments of the dendritic tree, are far removed from the axon hillock, and their activation has little impact on the membrane potential near this trigger zone. In contrast, axosomatic synapses have a greater effect in altering the membrane potential at the axon hillock because of their proximal location.

NEUROCHEMICAL TRANSMISSION

Neurons communicate with other cells by the release of chemical neurotransmitters, which act transiently on postsynaptic receptors and then must be removed from the synaptic cleft (Fig. 3.15). Transmitter is stored in synaptic vesicles and released on nerve stimulation by the process of exocytosis, following the opening of voltage-gated calcium ion channels in the nerve terminal. Once released, the neurotransmitter binds to and stimulates its receptors briefly before being rapidly removed from the synapse, thereby allowing the transmission of a new neuronal message. The most common mode of removal of the neurotransmitter following release is called **high-affinity reuptake** by the presynaptic terminal. This is a carrier-mediated, sodium-dependent, secondary active transport that uses energy from the Na^+/K^+- ATPase pump. Other removal mechanisms include enzymatic degradation into a nonactive metabolite in the synapse or diffusion away from the synapse into the extracellular space.

The details of synaptic events in chemical transmission were originally described for PNS synapses. CNS synapses appear to use similar mechanisms, with the important difference that muscle and gland cells are the targets of transmission in peripheral nerves, whereas neurons make up the postsynaptic elements at central synapses. In the central nervous system, glial cells also play a crucial role in remov-

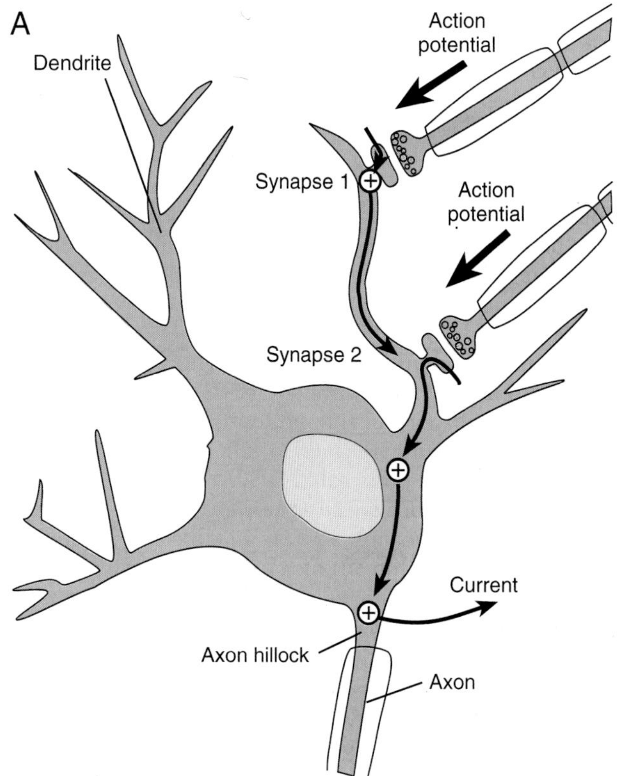

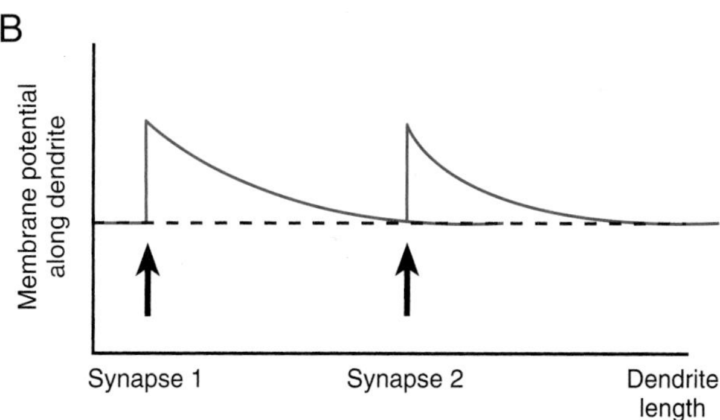

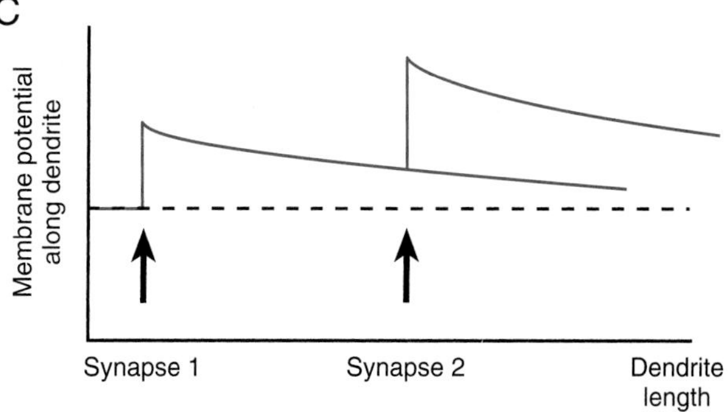

FIGURE 3.14 A model of spatial summation. A, The depolarization of a dendrite at two spatially separated synapses. **B**, A dendritic membrane with a short space constant is unable to summate postsynaptic potentials. **C**, A dendritic membrane with a long space constant is able to summate membrane potential changes.

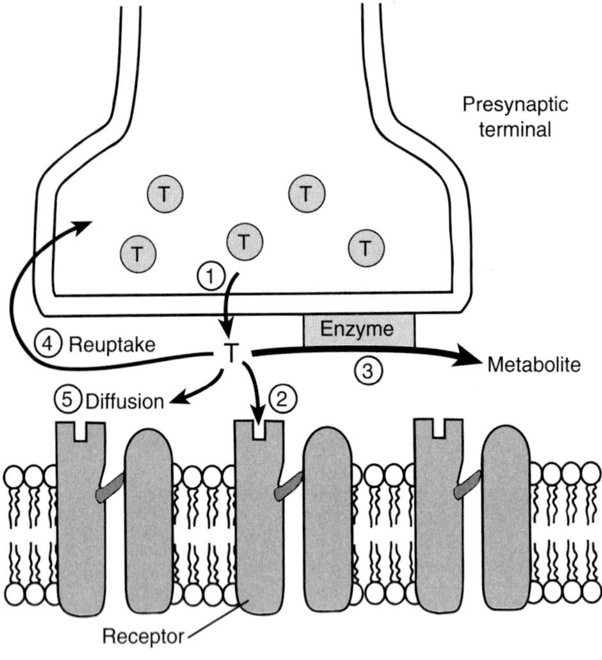

FIGURE 3.15 **The basic steps in neurochemical transmission.** Neurotransmitter molecules (T) are released into the synaptic cleft (1), reversibly bind to receptors on the postsynaptic cell (2), and are removed from the cleft by enzymatic degradation (3), reuptake into the presynaptic nerve terminal (4), or diffusion (5).

ing some neurotransmitters from the synaptic cleft via high-affinity reuptake.

There Are Several Classes of Neurotransmitters

The first neurotransmitters described were acetylcholine and norepinephrine, identified at synapses in the peripheral nervous system. Many others have since been identified, and they fall into three main classes: amino acids, monoamines, and polypeptides. Amino acids and monoamines are collectively termed **small-molecule transmitters.** The monoamines (or biogenic amines) are so named because they are synthesized from a single, readily available amino acid precursor. The polypeptide transmitters (or neuropeptides) consist of an amino acid chain, varying in length from three to several dozen. Recently, a novel set of neurotransmitters has been identified; these are membrane-soluble molecules that may act as both anterograde and retrograde signaling molecules between neurons.

Examples of amino acid transmitters include the excitatory amino acids glutamate and aspartate and the inhibitory amino acids glycine and γ-aminobutyric. (Note that γ-aminobutyric is biosynthetically a monoamine, but it has the features of an amino acid transmitter, not a monoaminergic one.) Examples of monoaminergic neurotransmitters are acetylcholine, derived from choline; the catecholamine transmitters dopamine, norepinephrine, and epinephrine, derived from the amino acid tyrosine; and an indoleamine, serotonin or 5-hydroxytryptamine, derived from tryptophan. Examples of polypeptide transmitters are the opioids and substance P. The best known membrane-soluble neurotransmitters are nitric oxide and arachidonic acid.

The human nervous system has some 100 billion neurons, each of which communicates with postsynaptic targets via chemical neurotransmission. As noted above, there are essentially only a handful of neurotransmitters. Even counting all the peptides known to act as transmitters, the number is well less than 50. Peptide transmitters can be colocalized, in a variety of combinations, with nonpeptide and other peptide transmitters, increasing the number of different types of chemical synapses. However, the specific neuronal signaling that allows the enormous complexity of function in the nervous system is due largely to the specificity of neuronal connections made during development.

There is a pattern to neurotransmitter distribution. Particular sets of pathways use the same neurotransmitter; some functions are performed by the same neurotransmitter in many places (Table 3.1). This redundant use of neurotransmitters is problematic in pathological conditions affecting one anatomic pathway or one neurotransmitter type. A classic example is Parkinson's disease, in which a particular set of dopaminergic neurons in the brain degenerates, resulting in a specific movement disorder. Therapies for Parkinson's disease, such as L-DOPA, that increase dopamine signaling do so globally, so other dopaminergic pathways become overly active. In some cases, patients receiving L-DOPA develop psychotic reactions because of excess dopamine signaling in limbic system pathways. Conversely, antipsychotic medications designed to decrease dopamine signaling in the limbic system may cause parkinsonian side effects. One strategy for decreasing the adverse effects of medications that affect neurotransmission is to target the therapies to specific types of receptors that may be preferentially distributed in one of the pathways that use the same neurotransmitter.

Acetylcholine. Neurons that use **acetylcholine** (ACh) as their neurotransmitter are known as **cholinergic neurons.** Acetylcholine is synthesized in the cholinergic neuron from choline and acetate, under the influence of the enzyme **choline acetyltransferase** or choline acetylase. This enzyme is localized in the cytoplasm of cholinergic neurons, especially in the vicinity of storage vesicles, and it is an identifying marker of the cholinergic neuron.

TABLE 3.1 **General Functions of Neurotransmitters**

Neurotransmitter	Function
Dopamine	Affect, reward, control of movement
Norepinephrine	Affect, alertness
Serotonin	Mood, arousal, modulation of pain
Acetylcholine	Control of movement, cognition
GABA	General inhibition
Glycine	General inhibition
Glutamate	General excitation, sensation
Substance P	Transmission of pain
Opioid peptides	Control of pain
Nitric oxide	Vasodilation, metabolic signaling

All the components for the synthesis, storage, and release of ACh are localized in the terminal region of the cholinergic neuron (Fig. 3.16). The storage vesicles and choline acetyltransferase are produced in the soma and are transported to the axon terminals. The rate-limiting step in ACh synthesis in the nerve terminals is the availability of choline, of which specialized mechanisms ensure a continuous supply. Acetylcholine is stored in vesicles in the axon terminals, where it is protected from enzymatic degradation and packaged appropriately for release upon nerve stimulation.

The enzyme **acetylcholinesterase** (AChE) hydrolyzes ACh back to choline and acetate after the release of ACh. This enzyme is found in both presynaptic and postsynaptic cell membranes, allowing rapid and efficient hydrolysis of extracellular ACh. This enzymatic mechanism is so efficient that normally no ACh spills over from the synapse into the general circulation. The choline generated from ACh hydrolysis is taken back up by the cholinergic neuron by a high-affinity, sodium-dependent uptake mechanism, which ensures a steady supply of the precursor for ACh synthesis. An additional source of choline is the low-affinity transport used by all cells to take up choline from the extracellular fluid for use in the synthesis of phospholipids.

The receptors for ACh, known as **cholinergic receptors,** fall into two categories, based on the drugs that mimic or antagonize the actions of ACh on its many target cell types. In classical studies dating to the early twentieth century, the drugs **muscarine**, isolated from poisonous mushrooms, and **nicotine**, isolated from tobacco, were used to distinguish two separate receptors for ACh. Muscarine stimulates some of the receptors and nicotine stimulates all the others, so receptors were designated as either **muscarinic** or **nicotinic.** It should be noted that ACh has the actions of both muscarine and nicotine at cholinergic receptors (Fig. 3.16); however, these two drugs cause fundamental differences that ACh cannot distinguish.

The **nicotinic acetylcholine receptor** is composed of five components: two α subunits and a β, γ, and δ subunit (Fig. 3.17). The two α subunits are binding sites for ACh. When ACh molecules bind to both α subunits, a conformational change occurs in the receptor, which results in an increase in channel conductance for Na^+ and K^+, leading to depolarization of the postsynaptic membrane. This depolarization is due to the strong inward electrical and chemical gradient for Na^+, which predominates over the outward gradient for K^+ ions and results in a net inward flux of positively charged ions.

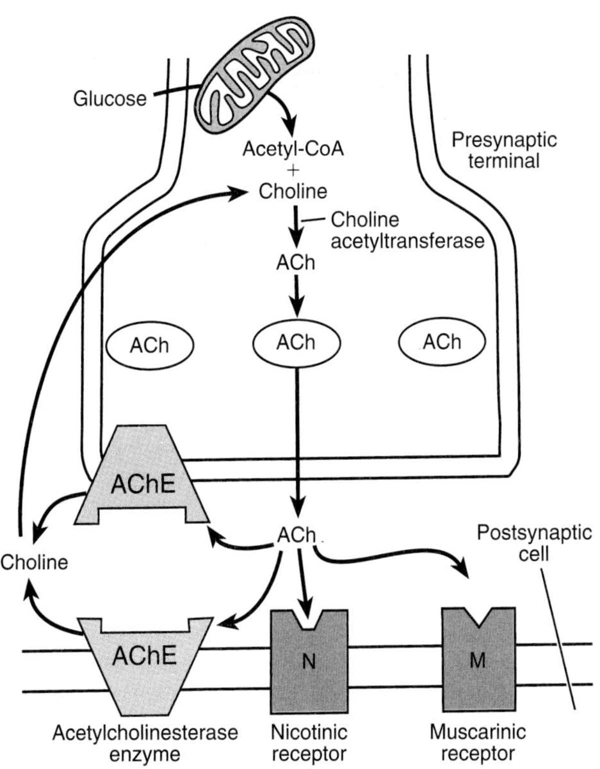

FIGURE 3.16 **Cholinergic neurotransmission.** When an action potential invades the presynaptic terminal, ACh is released into the synaptic cleft and binds to receptors on the postsynaptic cell to activate either nicotinic or muscarinic receptors. ACh is also hydrolyzed in the cleft by the enzyme acetylcholinesterase (AChE) to produce the metabolites choline and acetate. Choline is transported back into the presynaptic terminal by a high-affinity transport process to be reused in ACh resynthesis.

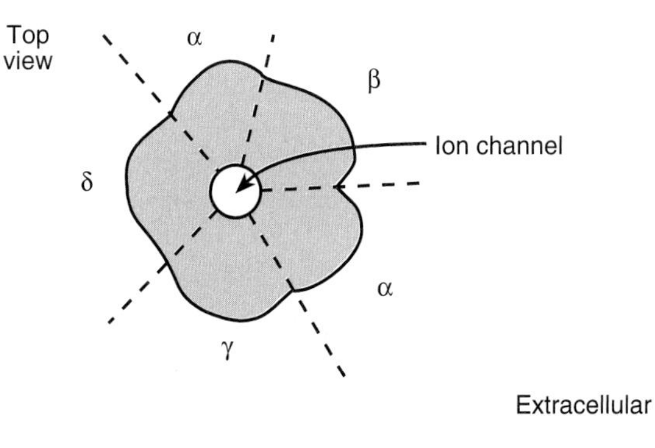

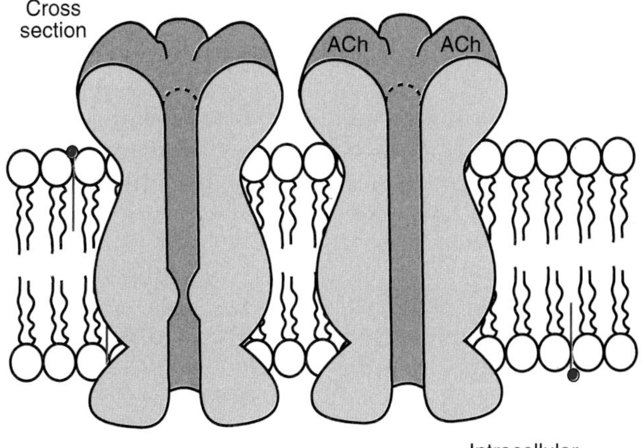

FIGURE 3.17 **The structure of a nicotinic acetylcholine receptor.** The nicotinic receptor is composed of five subunits: two α subunits and β, γ, and δ subunits. The two α subunits serve as binding sites for ACh. Both binding sites must be occupied to open the channel, permitting sodium ion influx and potassium ion efflux.

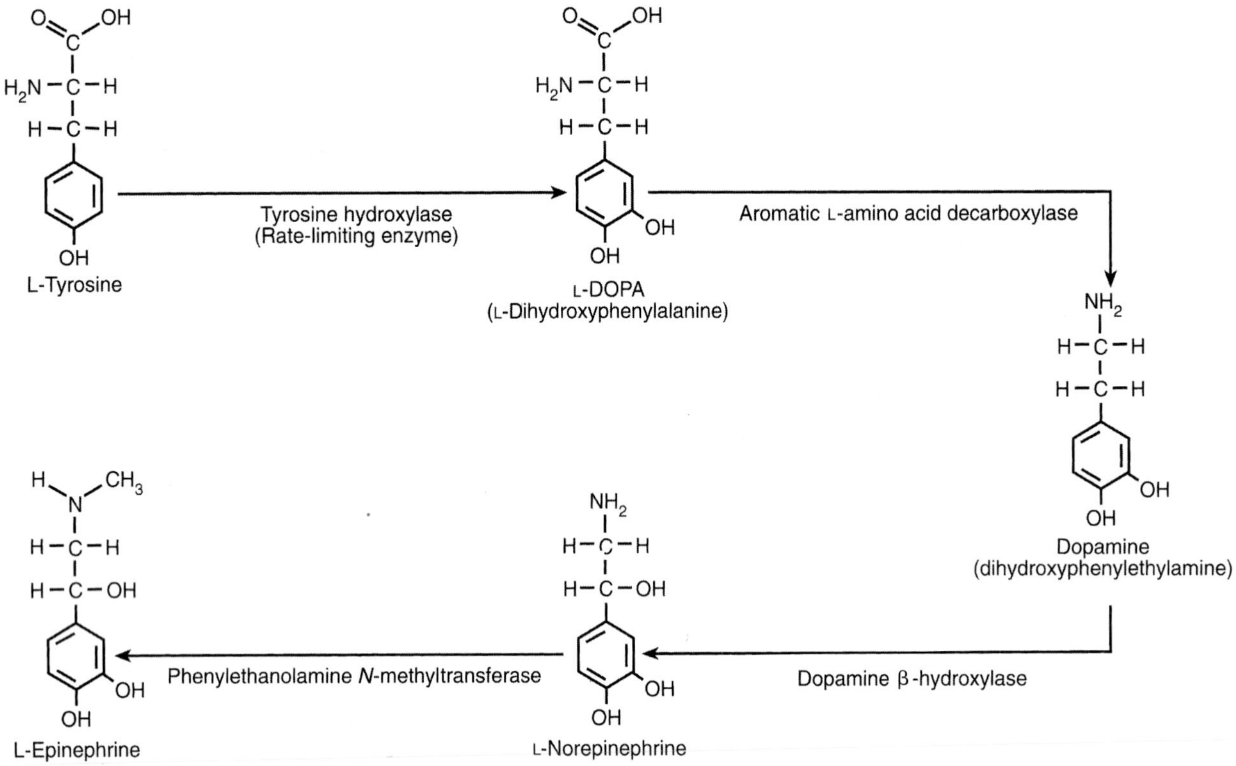

FIGURE 3.18 **The synthesis of catecholamines.** The catecholamine neurotransmitters are synthesized by way of a chain of enzymatic reactions to produce L-DOPA, dopamine, L-norepinephrine, and L-epinephrine.

The structure and the function of the **muscarinic acetylcholine receptor** are different. Five subtypes of muscarinic receptors have been identified. The M_1 and M_2 receptors are composed of seven membrane-spanning domains, with each exerting action through a G protein. The activation of M_1 receptors results in a decrease in K^+ conductance via phospholipase C, and activation of M_2 receptors causes an increase in K^+ conductance by inhibiting adenylyl cyclase. As a consequence, when ACh binds to an M_1 receptor, it results in membrane depolarization; when ACh binds to an M_2 receptor, it causes hyperpolarization.

Catecholamines. The catecholamines are so named because they consist of a catechol moiety (a phenyl ring with two attached hydroxyl groups) and an ethylamine side chain. The catecholamines **dopamine** (DA), **norepinephrine** (NE), and **epinephrine** (EPI) share a common pathway for enzymatic biosynthesis (Fig. 3.18). Three of the enzymes involved—tyrosine hydroxylase (TH), dopamine β-hydroxylase (DBH), and phenylethanolamine *N*-methyltransferase (PNMT)—are unique to catecholamine-secreting cells, and all are derived from a common ancestral gene. **Dopaminergic neurons** express only TH, **noradrenergic neurons** express both TH and DBH, and epinephrine-secreting cells express all three. **Epinephrine-secreting cells** include a small population of CNS neurons, as well as the hormonal cells of the adrenal medulla, **chromaffin cells,** which secrete EPI during the fight-or-flight response (see Chapter 6).

The rate-limiting enzyme in catecholamine biosynthesis is **tyrosine hydroxylase**, which converts L-tyrosine to L-3,4-dihydroxyphenylalanine (L-DOPA). Tyrosine hydroxylase is regulated by short-term activation and long-term induction. Short-term excitation of dopaminergic neurons results in an increase in the conversion of tyrosine to DA. This phenomenon is mediated by the phosphorylation of TH via a cAMP-dependent protein kinase, which results in an increase in functional TH activity. Long-term induction is mediated by the synthesis of new TH.

A nonspecific cytoplasmic enzyme, aromatic L-amino acid decarboxylase, catalyzes the formation of dopamine from L-DOPA. Dopamine is then taken up in storage vesicles and protected from enzymatic attack. In NE- and EPI-synthesizing neurons, DBH, which converts DA to NE, is found within vesicles, unlike the other synthetic enzymes, which are in the cytoplasm. In EPI-secreting cells, PNMT is localized in the cytoplasm. The PNMT adds a methyl group to the amine in NE to form EPI.

Two enzymes are involved in degrading the catecholamines following vesicle exocytosis. **Monoamine oxidase** (MAO) removes the amine group, and **catechol-O-methyltransferase** (COMT) methylates the 3-OH group on the catechol ring. As shown in Figure 3.19, MAO is localized in mitochondria, present in both presynaptic and postsynaptic cells, whereas COMT is localized in the cytoplasm and only postsynaptically. At synapses of noradrenergic neurons in the PNS (i.e., postganglionic sympathetic neurons of the autonomic nervous system) (see Chapter 6), the postsynaptic COMT-containing cells are the muscle and gland cells and other nonneuronal tissues that receive sympathetic stimulation. In the CNS, on the other hand, most of the COMT is localized in glial cells (especially astrocytes) rather than in postsynaptic target neurons.

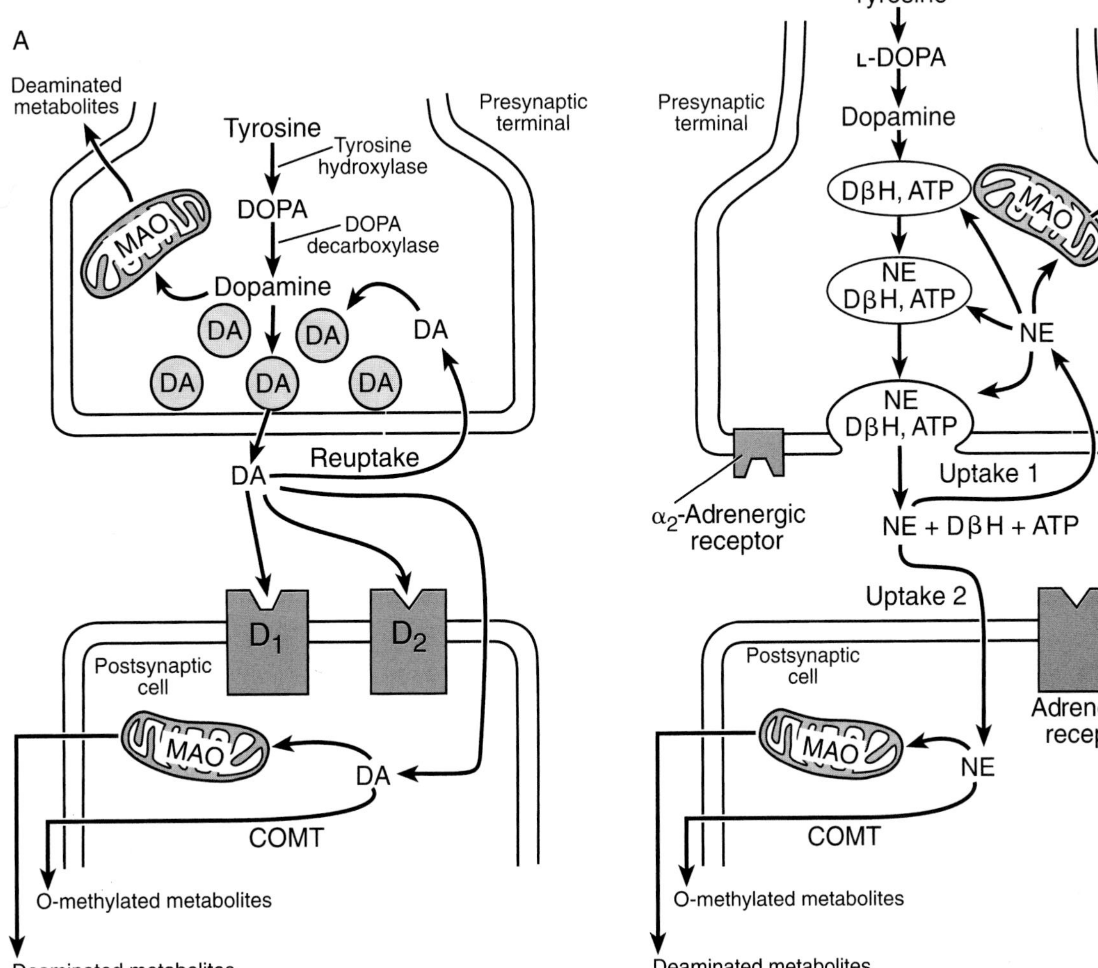

FIGURE 3.19 **Catecholaminergic neurotransmission.** A, In dopamine-producing nerve terminals, dopamine is enzymatically synthesized from tyrosine and taken up and stored in vesicles. The fusion of DA-containing vesicles with the terminal membrane results in the release of DA into the synaptic cleft and permits DA to bind to dopamine receptors (D_1 and D_2) in the postsynaptic cell. The termination of DA neurotransmission occurs when DA is transported back into the presynaptic terminal via a high-affinity mechanism. **B**, In norepinephrine (NE)-producing nerve terminals, DA is transported into synaptic vesicles and converted into NE by the enzyme dopamine β-hydroxylase (DBH). On release into the synaptic cleft, NE can bind to postsynaptic α- or β-adrenergic receptors and presynaptic α_2-adrenergic receptors. Uptake of NE into the presynaptic terminal (uptake 1) is responsible for the termination of synaptic transmission. In the presynaptic terminal, NE is repackaged into vesicles or deaminated by mitochondrial MAO. NE can also be transported into the postsynaptic cell by a low-affinity process (uptake 2), in which it is deaminated by MAO and O-methylated by catechol-O-methyltransferase (COMT).

Most of the catecholamine released into the synapse (up to 80%) is rapidly removed by uptake into the presynaptic neuron. Once inside the presynaptic neuron, the transmitter enters the synaptic vesicles and is made available for recycling. In peripheral noradrenergic synapses (the sympathetic nervous system), the neuronal uptake process described above is referred to as **uptake 1**, to distinguish it from a second uptake mechanism, **uptake 2**, localized in the target cells (smooth muscle, cardiac muscle, and gland cells) (Fig. 3.19B). In contrast with uptake 1, an active transport, uptake 2 is a facilitated diffusion mechanism, which takes up the sympathetic transmitter NE, as well as the circulating hormone EPI, and degrades them enzymatically by MAO and COMT localized in the target cells. In the CNS, there is little evidence of an uptake 2 of NE, but glia serve a comparable role by taking up catecholamines and degrading them enzymatically by glial MAO and COMT. Unlike uptake 2 in the PNS, glial uptake of catecholamines has many characteristics of uptake 1.

The catecholamines differ substantially in their interactions with receptors; DA interacts with DA receptors and NE and EPI interact with adrenergic receptors. Up to five subtypes of DA receptors have been described in the CNS. Of these five, two have been well characterized. **D_1 receptors** are coupled to stimulatory G proteins (G_s), which activate adenylyl cyclase, and **D_2 receptors** are coupled to inhibitory G proteins (G_i), which inhibit adenylyl cyclase. Activation of D_2 receptors hyperpolarizes the postsynaptic membrane by increasing potassium conductance. A third subtype of DA receptor postulated to modulate the release of DA is localized on the cell membrane of the nerve terminal that releases DA; accordingly, it is called an **autoreceptor.**

Adrenergic receptors, stimulated by EPI and NE, are located on cells throughout the body, including the CNS and the peripheral target organs of the sympathetic nervous system (see Chapter 6). Adrenergic receptors are classified as either α or β, based on the rank order of potency of catecholamines and related analogs in stimulating each type. The analogs used originally in distinguishing α- from β-adrenergic receptors are NE, EPI, and the two synthetic compounds isoproterenol (ISO) and phenylephrine (PE). Ahlquist, in 1948, designated α as those receptors in which EPI was highest in potency and ISO was least potent (EPI > NE > > ISO). β-Receptors exhibited a different rank order: ISO was most potent and EPI either more potent or equal in potency to NE. Studies with PE further distinguished these two classes of receptors: α-receptors were stimulated by PE, whereas β-receptors were not.

Serotonin. **Serotonin** or **5-hydroxytryptamine** (5-HT) is the transmitter in **serotonergic neurons.** Chemical transmission in these neurons is similar in several ways to that described for catecholaminergic neurons. Tryptophan hydroxylase, a marker of serotonergic neurons, converts tryptophan to 5-hydroxytryptophan (5-HTP), which is then converted to 5-HT by decarboxylation (Fig. 3.20).

5-Hydroxytryptamine is stored in vesicles and is released by exocytosis upon nerve depolarization. The major mode of removal of released 5-HT is by a high-affinity, sodium-dependent, active uptake mechanism. There are several receptor subtypes for serotonin. The 5-HT-3 receptor contains an ion channel. Activation results in an increase in sodium and potassium ion conductances, leading to EPSPs. The remaining well-characterized receptor subtypes appear to operate through second messenger systems. The 5-HT-1A receptor, for example, uses cAMP. Activation of this receptor results in an increase in K^+ ion conductance, producing IPSPs.

Glutamate and Aspartate. Both **glutamate** (GLU) and **aspartate** (ASP) serve as excitatory transmitters of the CNS. These dicarboxylic amino acids are important substrates for transaminations in all cells; but, in certain neurons, they also serve as neurotransmitters—that is, they are sequestered in high concentration in synaptic vesicles, released by exocytosis, stimulate specific receptors in the synapse, and are removed by high-affinity uptake. Since GLU and ASP are readily interconvertible in transamination reactions in cells, including neurons, it has been difficult to distinguish neurons that use glutamate as a transmit-

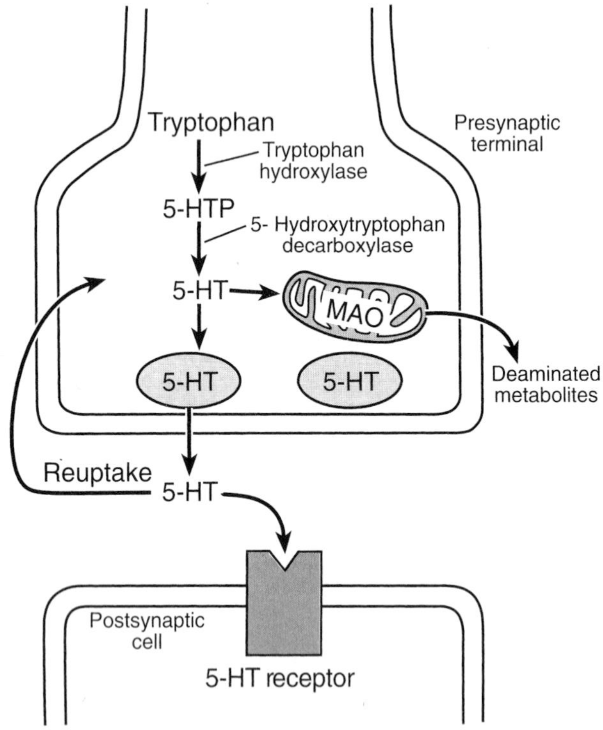

FIGURE 3.20 **Serotonergic neurotransmission.** Serotonin (5-HT) is synthesized by the hydroxylation of tryptophan to form 5-hydroxytryptophan (5-HTP) and the decarboxylation of 5-HTP to form 5-HT. On release into the synaptic cleft, 5-HT can bind to a variety of serotonergic receptors on the postsynaptic cell. Synaptic transmission is terminated when 5-HT is transported back into the presynaptic terminal for repackaging into vesicles.

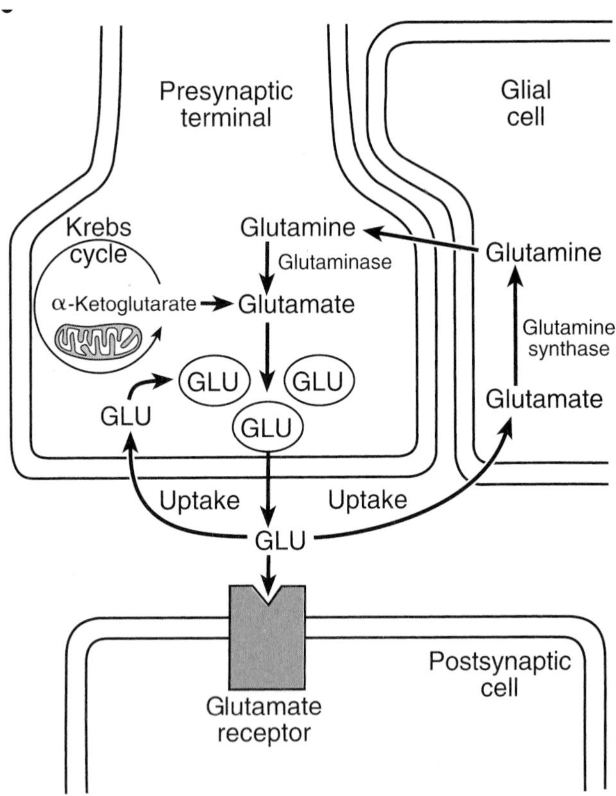

FIGURE 3.21 **Glutamatergic neurotransmission.** Glutamate (GLU) is synthesized from α-ketoglutarate by enzymatic amination. Upon release into the synaptic cleft, GLU can bind to a variety of receptors. The removal of GLU is primarily by transport into glial cells, where it is converted into glutamine. Glutamine, in turn, is transported from glial cells to the nerve terminal, where it is converted to glutamate by the enzyme glutaminase.

ter from those that use aspartate. This difficulty is further compounded by the fact that GLU and ASP stimulate common receptors. Accordingly, it is customary to refer to both as **glutamatergic neurons.**

Sources of GLU for neurotransmission are the diet and mitochondrial conversion of α-ketoglutarate derived from the Krebs cycle (Fig. 3.21). Glutamate is stored in vesicles and released by exocytosis, where it activates specific receptors to depolarize the postsynaptic neuron. Two efficient active transport mechanisms remove GLU rapidly from the synapse. Neuronal uptake recycles the transmitter by re-storage in vesicles and re-release. Glial cells (particularly astrocytes) contain a similar, high-affinity, active transport mechanism that ensures the efficient removal of excitatory neurotransmitter molecules from the synapse (see Fig. 3.21). Glia serves to recycle the transmitter by converting it to **glutamine,** an inactive storage form of GLU containing a second amine group. Glutamine from glia readily enters the neuron, where glutaminase removes the second amine, regenerating GLU for use again as a transmitter.

At least five subtypes of GLU receptors have been described, based on the relative potency of synthetic analogs in stimulating them. Three of these, named for the synthetic analogs that best activate them—kainate, quisqualate, and *N*-methyl-D-aspartate (NMDA) receptors—are associated with cationic channels in the neuronal membrane. Activation of the **kainate** and **quisqualate receptors** produces EPSPs by opening ion channels that increase Na^+ and K^+ conductance. Activation of the **NMDA receptor** increases Ca^{2+} conductance. This receptor, however, is blocked by Mg^{2+} when the membrane is in the resting state and becomes unblocked when the membrane is depolarized. Thus, the NMDA receptor can be thought of as both a ligand-gated and a voltage-gated channel. Calcium gating through the NMDA receptor is crucial for the development of specific neuronal connections and for neural processing related to learning and memory. In addition, excess entry of Ca^{2+} through NMDA receptors during ischemic disorders of the brain is thought to be responsible for the rapid death of neurons in stroke and hemorrhagic brain disorders (see Clinical Focus Box 3.2).

γ-Aminobutyric Acid and Glycine. The inhibitory amino acid transmitters **γ-aminobutyric acid** (GABA) and **glycine** (GLY) bind to their respective receptors, causing hyperpolar-

CLINICAL FOCUS BOX 3.2

The Role of Glutamate Receptors in Nerve Cell Death in Hypoxic/Ischemic Disorders

Excitatory amino acids (EAA), GLU and ASP, are the neurotransmitters for more than half the total neuronal population of the CNS. Not surprisingly, most neurons in the CNS contain receptors for EAA. When transmission in glutamatergic neurons functions normally, very low concentrations of EAA appear in the synapse at any time, primarily because of the efficient uptake mechanisms of the presynaptic neuron and neighboring glial cells.

In certain pathological states, however, extraneuronal concentrations of EAA exceed the ability of the uptake mechanisms to remove them, resulting in cell death in a matter of minutes. This can be seen in severe **hypoxia,** such as during respiratory or cardiovascular failure, and in **ischemia,** where the blood supply to a region of the brain is interrupted, as in stroke. In either condition, the affected area is deprived of oxygen and glucose, which are essential for normal neuronal functions, including energy-dependent mechanisms for the removal of extracellular EAA and their conversion to glutamine.

The consequences of prolonged exposure of neurons to EAA has been described as **excitotoxicity**. Much of the cytotoxicity can be attributed to the destructive actions of high intracellular calcium brought about by stimulation of the various subtypes of glutamatergic receptors. One subtype, a presynaptic kainate receptor, opens voltage-gated calcium channels and promotes the further release of GLU. Several postsynaptic receptor subtypes depolarize the nerve cell and promote the rise of intracellular calcium via ligand-gated and voltage-gated channels and second messenger-mediated mobilization of intracellular calcium stores. The spiraling consequences of increased extracellular GLU, leading to the further release of GLU, and of increased calcium entry, leading to the further mobilization of intracellular calcium, bring about cell death, resulting from the inability of ischemic/hypoxic conditions to meet the high metabolic demands of excited neurons and the triggering of destructive changes in the cell by increased free calcium.

Intracellular free calcium is an activator of calcium-dependent proteases, which destroy microtubules and other structural proteins that maintain neuronal integrity. Calcium activates phospholipases, which break down membrane phospholipids and lead to lipid peroxidation and the formation of oxygen-free radicals, which are toxic to cells. Another consequence of activated phospholipase is the formation of arachidonic acid and metabolites, including prostaglandins, some of which constrict blood vessels and further exacerbate hypoxia/ischemia. Calcium activates cellular endonucleases, leading to DNA fragmentation and the destruction of chromatin. In mitochondria, high calcium induces swelling and impaired formation of ATP via the Krebs cycle. Calcium is the primary toxic agent in EAA-induced cytotoxicity.

In addition to calcium, nitric oxide (NO) is known to mediate EAA-induced cytotoxicity. Nitric oxide synthase (NOS) activity is enhanced by NMDA receptor activation. Neurons that exhibit NOS and, therefore, synthesize NO are protected from NO, but NO released from NOS-expressing neurons in response to NMDA receptor activation kills adjacent neurons.

Proposed new treatment strategies promise to enhance survival of neurons in brain ischemic/hypoxic disorders. These therapies include drugs that block specific subtypes of glutamatergic receptors, such as the NMDA receptor, which is most responsible for promoting high calcium levels in the neuron. Other strategies include drugs that destroy oxygen-free radicals, calcium ion channel blocking agents, and NOS antagonists.

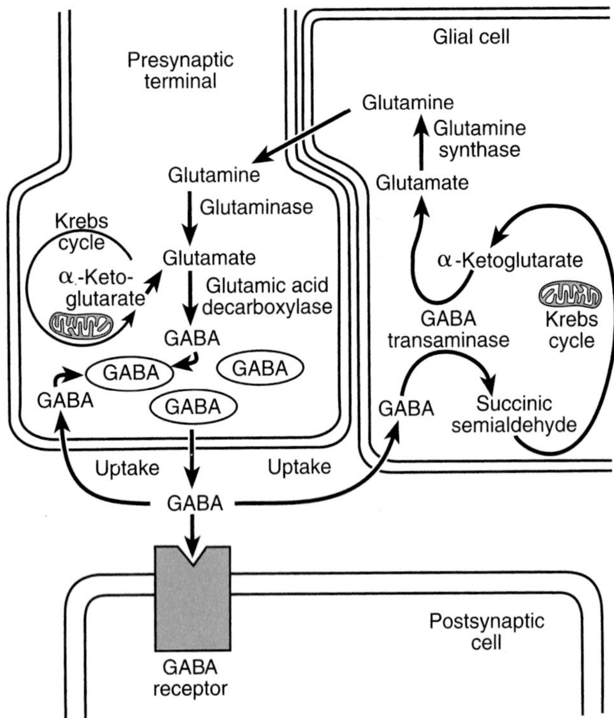

FIGURE 3.22 **GABAergic neurotransmission.** γ-Aminobutyric acid (GABA) is synthesized from glutamate by the enzyme glutamic acid decarboxylase. Upon release into the synaptic cleft, GABA can bind to GABA receptors ($GABA_A$, $GABA_B$). The removal of GABA from the synaptic cleft is primarily by uptake into the presynaptic neuron and surrounding glial cells. The conversion of GABA to succinic semialdehyde is coupled to the conversion of α-ketoglutarate to glutamate by the enzyme GABA-transaminase. In glia, glutamate is converted into glutamine, which is transported back into the presynaptic terminal for synthesis into GABA.

ization of the postsynaptic membrane. **GABAergic neurons** represent the major inhibitory neurons of the CNS, whereas **glycinergic neurons** are found in limited numbers, restricted only to the spinal cord and brainstem. Glycinergic transmission has not been as well characterized as transmission using GABA; therefore, only GABA will be discussed here.

The synthesis of GABA in neurons is by decarboxylation of GLU by the enzyme glutamic acid decarboxylase, a marker of GABAergic neurons. GABA is stored in vesicles and released by exocytosis, leading to the stimulation of postsynaptic receptors (Fig. 3.22).

There are two types of GABA receptors: $GABA_A$ and $GABA_B$. The **$GABA_A$ receptor** is a ligand-gated Cl^- channel, and its activation produces IPSPs by increasing the influx of Cl^- ions. The increase in Cl^- conductance is facilitated by **benzodiazepines,** drugs that are widely used to treat anxiety. Activation of the **$GABA_B$ receptor** also produces IPSPs, but the IPSP results from an increase in K^+ conductance via the activation of a G protein. Drugs that inhibit GABA transmission cause seizures, indicating a major role for inhibitory mechanisms in normal brain function.

GABA is removed from the synaptic cleft by transport into the presynaptic terminal and glial cells (astrocytes) (Fig. 3.22). The GABA enters the Krebs cycle in both neuronal and glial mitochondria and is converted to succinic semialdehyde by the enzyme GABA-transaminase. This enzyme is also coupled to the conversion of α-ketoglutarate to glutamate. The glutamate produced in the glial cell is converted to glutamine. As in the recycling of glutamate, glutamine is transported into the presynaptic terminal, where it is converted into glutamate.

Neuropeptides. Neurally active peptides are stored in synaptic vesicles and undergo exocytotic release in common with other neurotransmitters. Many times, vesicles containing neuropeptides are colocalized with vesicles containing another transmitter in the same neuron, and both can be shown to be released during nerve stimulation. In these colocalization instances, release of the peptide-containing vesicles generally occurs at higher stimulation frequencies than release of the vesicles containing nonpeptide neurotransmitters.

The list of candidate peptide transmitters continues to grow; it includes well-known gastrointestinal hormones, pituitary hormones, and hypothalamic-releasing factors. As a class, the neuropeptides fall into several families of peptides, based on their origins, homologies in amino acid composition, and similarities in the response they elicit at common or related receptors. Table 3.2 lists some members of each of these families.

TABLE 3.2 Some Recognized Neuropeptide Neurotransmitters

Neuropeptide	Amino Acid Composition
Opioids	
Met-enkephalin	Tyr-Gly-Gly-Phe-Met-OH
Leu-enkephalin	Tyr-Gly-Gly-Phe-Leu-OH
Dynorphin	Tyr-Gly-Gly-Phe-Leu-Arg-Arg-Ile
β-Endorphin	Tyr-Gly-Gly-Phe-Met-Thr-Glu-Lys-Ser-Gln-Thr-Pro-Leu-Val-Thr-Leu-Phe-Lys-Asn-Ala-Ile-Val-Lys-Asn-His-Lys-Gly-Gln-OH
Gastrointestinal peptides	
Cholecystokinin octapeptide (CCK-8)	Asp-Tyr-Met-Gly-Trp-Met-Asp-Phe-NH_2
Substance P	Arg-Pro-Lys-Pro-Gln-Gln-Phe-Phe-Gly-Leu-Met
Vasoactive intestinal peptide	His-Ser-Asp-Ala-Val-Phe-Thr-Asp-Asn-Tyr-Thr-Arg-Leu-Arg-Lys-Gln-Met-Ala-Val-Lys-Lys-Tyr-Leu-Asn-Ser-Ile-Leu-Asn-NH_2
Hypothalamic and pituitary peptides	
Thyrotropin-releasing hormone (TRH)	Pyro-Glu-His-Pro-NH_2
Somatostatin	Ala-Gly-Cys-Asn-Phe-Phe-Trp-Lys-Thr-Phe-Thr-Ser-Cys
Luteinizing hormone-releasing hormone (LHRH)	Pyro-Glu-His-Trp-Ser-Tyr-Gly-Leu-Arg-Pro-Gly
Vasopressin	Cys-Tyr-Phe-Gln-Asn-Cys-Pro-Arg-Gly-NH_2
Oxytocin	Cys-Tyr-Ile-Gln-Asn-Cys-Pro-Leu-Gly-NH_2

Peptides are synthesized as large prepropeptides in the endoplasmic reticulum and are packaged into vesicles that reach the axon terminal by **axoplasmic transport.** While in transit, the prepropeptide in the vesicle is posttranslationally modified by proteases that split it into small peptides and by other enzymes that alter the peptides by hydroxylation, amidation, sulfation, and so on. The products released by exocytosis include a neurally active peptide fragment, as well as many unidentified peptides and enzymes from within the vesicles.

The most common removal mechanism for synaptically released peptides appears to be diffusion, a slow process that ensures a longer-lasting action of the peptide in the synapse and in the extracellular fluid surrounding it. Peptides are degraded by proteases in the extracellular space; some of this degradation may occur within the synaptic cleft. There are no mechanisms for the recycling of peptide transmitters at the axon terminal, unlike more classical transmitters, for which the mechanisms for recycling, including synthesis, storage, reuptake, and release, are contained within the terminals. Accordingly, classical transmitters do not exhaust their supply, whereas peptide transmitters can be depleted in the axon terminal unless replenished by a steady supply of new vesicles transported from the soma.

Peptides can interact with specific peptide receptors located on postsynaptic target cells and, in this sense, are considered to be true neurotransmitters. However, peptides can also modify the response of a coreleased transmitter interacting with its own receptor in the synapse. In this case, the peptide is said to be a **modulator** of the actions of other neurotransmitters.

Opioids are peptides that bind to opiate receptors. They appear to be involved in the control of pain information. Opioid peptides include met-enkephalin, leu-enkephalin, dynorphins, and β-endorphin. Structurally, they share homologous regions consisting of the amino acid sequence Tyr-Gly-Gly-Phe. There are several opioid receptor subtypes: β-endorphin binds preferentially to **μ receptors**, enkephalins bind preferentially to μ and **δ receptors**; and dynorphin binds preferentially to **κ receptors**.

Originally isolated in the 1930s, **substance P** was found to have the properties of a neurotransmitter four decades later. Substance P is a polypeptide consisting of 11 amino acids, and is found in high concentrations in the spinal cord and hypothalamus. In the spinal cord, substance P is localized in nerve fibers involved in the transmission of pain information. It slowly depolarizes neurons in the spinal cord and appears to use inositol 1,4,5-trisphosphate as a second messenger. Antagonists that block the action of substance P produce an analgesic effect. The opioid enkephalin also diminishes pain sensation, probably by presynaptically inhibiting the release of substance P.

Many of the other peptides found throughout the CNS were originally discovered in the hypothalamus as part of the neuroendocrine system. Among the hypothalamic peptides, somatostatin has been fairly well characterized in its role as a transmitter. As part of the neuroendocrine system, this peptide inhibits the release of growth hormone by the anterior pituitary (see Chapter 32). About 90% of brain somatostatin, however, is found outside the hypothalamus. Application of somatostatin to target neurons inhibits their electrical activity, but the ionic mechanisms mediating this inhibition are unknown.

Nitric Oxide and Arachidonic Acid. Recently a novel type of neurotransmission has been identified. In this case, membrane-soluble molecules diffuse through neuronal membranes and activate "postsynaptic" cells via second messenger pathways. **Nitric oxide (NO)** is a labile free-radical gas that is synthesized on demand from its precursor, L-arginine, by nitric oxide synthase (NOS). Because NOS activity is exquisitely regulated by Ca^{2+}, the release of NO is calcium-dependent even though it is not packaged into synaptic vesicles.

Nitric oxide was first identified as the substance formed by macrophages that allow them to kill tumor cells. NO was also identified as the endothelial-derived relaxing factor in blood vessels before it was known to be a neurotransmitter. It is a relatively common neurotransmitter in peripheral autonomic pathways and **nitrergic neurons** are also found throughout the brain, where the NO they produce may be involved in damage associated with hypoxia (see Clinical Focus Box 3.2). The effects of NO are mediated through its activation of second messengers, particularly guanylyl cyclase.

Arachidonic acid is a fatty acid released from phospholipids in the membrane when phospholipase A2 is activated by ligand-gated receptors. The arachidonic acid then diffuses retrogradely to affect the presynaptic cell by activating second messenger systems. Nitric oxide can also act in this retrograde fashion as a signaling molecule.

THE MAINTENANCE OF NERVE CELL FUNCTION

Neurons are highly specialized cells and, thus, have unique metabolic needs compared to other cells, particularly with respect to their axonal and dendritic extensions. The axons of some neurons can exceed 1 meter long. Consider the control of toe movement in a tall individual. Neurons in the motor cortex of the brain have axons that must connect with the appropriate motor neurons in the lumbar region of the spinal cord; these motor neurons, in turn, have axons that connect the spinal cord to muscles in the toe. An enormous amount of axonal membrane and intraaxonal material must be supported by the cell bodies of neurons; additionally, a typical motor neuron soma may be only 40 μm in diameter and support a total dendritic arborization of 2 to 5 mm.

Another specialized feature of neurons is their intricate connectivity. Mechanisms must exist to allow the appropriate connections to be made during development.

Proteins Are Synthesized in the Soma of Neurons

The nucleus of a neuron is large, and a substantial portion of the genetic information it contains is continuously transcribed. Based on hybridization studies, it is estimated that one third of the genome in brain cells is actively transcribed, producing more mRNA than any other kind of cell in the body. Because of the high level of transcriptional activity, the nuclear chromatin is dispersed. In contrast, the chromatin in nonneuronal cells in the brain, such as glial cells, is found in clusters on the internal face of the nuclear membrane.

Most of the proteins formed by free ribosomes and polyribosomes remain within the soma, whereas proteins formed by rough endoplasmic reticulum (rough ER) are exported to the dendrites and the axon. Polyribosomes and rough ER are found predominantly in the soma of neurons. Axons contain no rough ER and are unable to synthesize proteins. The smooth ER is involved in the intracellular storage of calcium. Smooth ER in neurons binds calcium and maintains the intracellular cytoplasmic concentration at a low level, about 10^{-7} M. Prolonged elevation of intracellular calcium leads to neuronal death and degeneration (see Clinical Focus Box 3.2).

The Golgi apparatus in neurons is found only in the soma. As in other types of cells, this structure is engaged in the terminal glycosylation of proteins synthesized in the rough ER. The Golgi apparatus forms export vesicles for proteins produced in the rough ER. These vesicles are released into the cytoplasm, and some are carried by axoplasmic transport to the axon terminals.

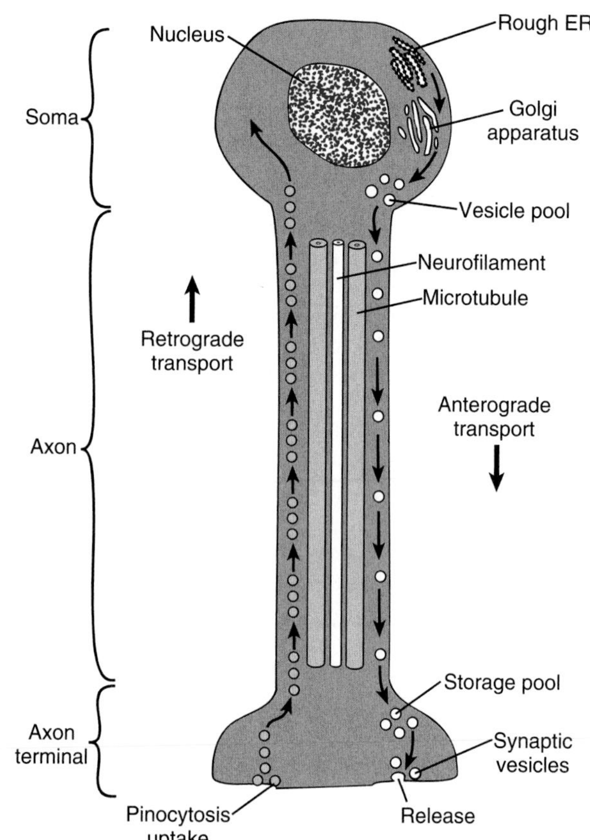

FIGURE 3.23 Anterograde and retrograde axoplasmic transport. Transport of molecules in vesicles along microtubules is mediated by kinesin for anterograde transport and by dynein for retrograde transport.

The Cytoskeleton Is the Infrastructure for Neuron Form

The transport of proteins from the Golgi apparatus and the highly specialized form of the neuron depend on the internal framework of the cytoskeleton. The neuronal cytoskeleton is made of microfilaments, neurofilaments, and microtubules. **Microfilaments** are composed of actin, a contractile protein also found in muscle. They are 4 to 5 nm in diameter and are found in dendritic spines. **Neurofilaments** are found in both axons and dendrites and are thought to provide structural rigidity. They are not found in the growing tips of axons and dendritic spines, which are more dynamic structures. Neurofilaments are about the size of intermediate filaments found in other types of cells (10 nm in diameter). In other cell types, however, intermediate filaments consist of one protein, whereas neurofilaments are composed of three proteins. The core of neurofilaments consists of a 70 kDa protein, similar to intermediate filaments in other cells. The two other neurofilament proteins are thought to be side arms that interact with microtubules.

Microtubules are responsible for the rapid movement of material in axons and dendrites. They are 23 nm in diameter and are composed of tubulin. In neurons, microtubules have accessory proteins, called **microtubule-associated proteins** (MAPs), thought to be responsible for the specific distribution of material to dendrites or axons.

Mitochondria Are Important for Synaptic Transmission

Mitochondria in neurons are highly concentrated in the region of the axon terminals. They produce ATP, which is required as a source of energy for many cellular processes. In the axon terminal, mitochondria provide both a source of energy for processes associated with synaptic transmission and substrates for the synthesis of certain neurotransmitter chemicals, such as the amino acid glutamate. In addition, mitochondria contain enzymes for degrading neurotransmitter molecules, such as MAO, which degrades catecholamines and 5-HT, and GABA-transaminase, which degrades GABA.

Transport Mechanisms Distribute Material Needed by the Neuron and Its Fiber Processes

The shape of most cells in the body is relatively simple, compared to the complexity of neurons, with their elaborate axons and dendrites. Neurons have mechanisms for transporting the proteins, organelles, and other cellular materials needed for the maintenance of the cell along the length of axons and dendrites. These transport mechanisms are capable of moving cellular components in an **anterograde** direction, away from the soma, or in a **retrograde** direction, toward the soma (Fig. 3.23). **Kinesin,** an MAP, is involved in anterograde transport of organelles and vesicles via the hydrolysis of ATP. Retrograde transport of organelles and vesicles is mediated by **dynein,** another MAP.

In the axon, anterograde transport occurs at both slow and fast rates. The rate of **slow axoplasmic transport** is 1 to 2 mm/day. Structural proteins, such as actin, neurofilaments, and microtubules, are transported at this speed. Slow axonal transport is rate limiting for the regeneration of axons following neuronal injury. The rate of **fast axoplasmic transport** is about 400 mm/day. Fast transport mechanisms are used for organelles, vesicles, and membrane glycoproteins needed at the axon terminal. In dendrites, anterograde transport occurs at a rate of approximately 0.4 mm/day. Dendritic transport also moves ribosomes and RNA, suggesting that protein synthesis occurs within dendrites.

In retrograde axoplasmic transport, material is moved from terminal endings to the cell body. This provides a mechanism for the cell body to sample the environment around its synaptic terminals. In some neurons, maintenance of synaptic connections depends on the **transneuronal transport** of trophic substances, such as nerve growth factor, across the synapse. After retrograde transport to the soma, nerve growth factor activates mechanisms for protein synthesis.

Nerve Fibers Migrate and Extend During Development and Regeneration

One of the major features that distinguishes differentiation and growth in nerve cells from these processes in other types of cells is the outgrowth of the axon that extends along a specific pathway to form synaptic connections with appropriate targets. Axonal growth is determined largely by interactions between the growing axon and the tissue environment. At the leading edge of a growing axon is the **growth cone**, a flat structure that gives rise to protrusions called **filopodia**. Growth cones contain actin and are motile, with filopodia extending and retracting at a velocity of 6 to 10 μm/min. Newly synthesized membranes in the form of vesicles are also found in the growth cone and fuse with the growth cone as it extends. As the growth cone elongates, microtubules and neurofilaments are added to the distal end of the fiber and partially extend into the growth cone. They are transported to the growth cone by slow axoplasmic transport.

The direction of axonal growth is dictated, in part, by **cell adhesion molecules** (CAMs), plasma membrane glycoproteins that promote cell adhesion. Neuron-glia-CAM (N-CAM) is expressed in postmitotic neurons and is particularly prominent in growing axons and dendrites, which migrate along certain types of glial cells that provide a guiding path to target sites. The secretion of tropic factors by target cells also influences the direction of axon growth. When the proper target site is reached and synaptic connections are formed, the processes of growth cone elongation and migration are terminated.

During the formation and maturation of specific neuronal connections, the initial connections made are more widespread than the final outcome. Some connections are lost, concomitant with a strengthening of other connections. This pruning of connections is a result of a selection process in which the most electrically active inputs predominate and survive and the less active contacts are lost. While the number of connections between different neurons decreases during this process, the total number of synapses increases dramatically as the remaining connections grow stronger.

Growth cones are also present in axons that regenerate following injury. When axons are severed, the distal portion—that is, the portion cut off from the cell body—degenerates. The proximal portion of the axon then develops a growth cone and begins to elongate. The signal to the cell body that injury has occurred is the loss of retrogradely transported signaling molecules normally derived from the axon terminal. The success of neuronal regeneration depends on the severity of the damage, the proximity of the damage to the cell body, and the location of the neurons. Axons in the CNS regenerate less successfully than axons in the PNS. Neurons damaged close to the cell body often die rather than regenerate because so much of their membrane and cytoplasm is lost.

REVIEW QUESTIONS

DIRECTIONS: Each of the numbered items or incomplete statements in this section is followed by answers or by completions of the statement. Select the ONE lettered answer or completion that is BEST in each case.

1. A pharmacological or physiological perturbation that increases the resting P_K/P_{Na} ratio for the plasma membrane of a neuron would
(A) Lead to depolarization of the cell
(B) Lead to hyperpolarization of the cell
(C) Produce no change in the value of the resting membrane potential
2. The afterhyperpolarization phase of the action potential is caused by
(A) An outward calcium current
(B) An inward chloride current
(C) An outward potassium current
(D) An outward sodium current
3. Saltatory conduction in myelinated axons results from the fact that
(A) Salt concentration is increased beneath the myelin segments
(B) Nongated ion channels are present beneath the segments of myelin
(C) Membrane resistance is decreased beneath the segments of myelin
(D) Voltage-gated sodium channels are concentrated at the nodes of Ranvier
(E) Capacitance is decreased at the nodes of Ranvier
4. In individuals with multiple sclerosis, regions of CNS axons lose their myelin sheath. When this happens, the space constant of these unmyelinated regions would
(A) Not change
(B) Increase
(C) Decrease
5. Tetanus toxin and botulinum toxin exert their effects by disrupting the function of SNARES, inhibiting
(A) Propagation of the action potential
(B) The function of voltage-gated ion channels
(C) The docking and binding of synaptic vesicles to the presynaptic membrane
(D) The binding of transmitter to the postsynaptic receptor
(E) The reuptake of neurotransmitter by the presynaptic cell
6. What property of the postsynaptic neuron would optimize the effectiveness of two closely spaced axodendritic synapses?
(A) A high membrane resistance
(B) A high dendritic cytoplasmic resistance
(C) A small cross-sectional area

(continued)

(D) A small space constant
(E) A small time constant

7. A gardener was accidentally poisoned by a weed killer that inhibits acetylcholinesterase. Which of the following alterations in neurochemical transmission at brain cholinergic synapses is the most likely result of this poisoning?
(A) Blockade of cholinergic receptors
(B) A pileup of choline outside the cholinergic neuron (in the synaptic cleft)
(C) A pileup of acetylcholine outside the cholinergic neuron (in the synaptic cleft)
(D) Up-regulation of postsynaptic cholinergic receptors
(E) Increased synthesis of choline acetyltransferase

8. The major mode of removal of catecholamines from the synaptic cleft is
(A) Diffusion
(B) Breakdown by MAO
(C) Reuptake by the presynaptic nerve terminal
(D) Breakdown by COMT
(E) Endocytosis by the postsynaptic neuron

9. A patient in the emergency department exhibits psychosis. Pharmacological intervention to decrease the psychosis would most likely involve
(A) Blockade of dopaminergic neurotransmission
(B) Stimulation of dopaminergic neurotransmission
(C) Blockade of nitrergic neurotransmission
(D) Stimulation of nitrergic neurotransmission
(E) Blockade of cholinergic neurotransmission
(F) Stimulation of cholinergic transmission

10. Which class of neurotransmitter would be most affected by a toxin that disrupted microtubules within neurons?
(A) Amino acid transmitters
(B) Catecholamine transmitters
(C) Membrane-soluble transmitters
(D) Peptide transmitters
(E) Second messenger transmitters

11. A teenager in the emergency department exhibits convulsions. The friend who accompanied her indicated that she does not have a seizure disorder. The friend also indicated that the patient had ingested an unknown substance at a party. From her symptoms, you suspect the substance interfered with
(A) Epinephrine receptors
(B) GABA receptors
(C) Nicotinic receptors
(D) Opioid receptors
(E) Serotonin receptors

12. A 45-year-old lawyer complains of nausea, vomiting, and a tingling feeling in his extremities. He had dined on red snapper with a client at a fancy seafood restaurant the night before. His client also became ill with similar symptoms. Which of the following is the most likely cause of his problem?
(A) Chronic demyelinating disorder
(B) Ingestion of a toxin that activates sodium channels
(C) Ingestion of a toxin that blocks sodium channels
(D) Ingestion of a toxin that blocks nerve-muscle transmission
(E) Cerebral infarct (stroke)

13. A summated (compound) action potential is recorded from the affected peripheral nerve of a patient with a demyelinating disorder. Compared to a recording from a normal nerve, the recording from the patient will have a
(A) Greater amplitude
(B) Increased rate of rise
(C) Lower conduction velocity
(D) Shorter duration afterhyperpolarization

14. A syndrome of muscle weakness associated with certain types of lung cancer is caused by antibodies against components of the cancer plasma membrane that cross-react with voltage-gated calcium channels. The interaction of the antibodies impairs ion channel opening and would likely cause
(A) Decreased nerve conduction velocity
(B) Delayed repolarization of axon membranes
(C) Impaired release of acetylcholine from motor nerve terminals
(D) More rapid upstroke of the nerve action potential
(E) Repetitive nerve firing

SUGGESTED READING

Cooper EC, Jan LW. Ion channel genes and human neurological disease: Recent progress, prospects, and challenges. Proc Natl Acad Sci U S A 1999;4759–4766.

Geppert M, Sudhof TC. RAB3 and synaptotagmin: The yin and yang of synaptic membrane fusion. Annu Rev Neurosci 1998;21:75–95.

Kandel ER, Schwartz JH, Jessell TM. Principles of Neural Science. 4th Ed. New York: McGraw-Hill, 2000.

Lehmann-Horn F, Rüdel R. Channelopathies: Their contribution to our knowledge about voltage-gated ion channels. News Physiol Sci 1997;12:105–112.

Matthews GG. Neurobiology: Molecules, Cells and Systems. Malden, MA: Blackwell Science, 1998.

Sattler R, Tymianski M. Molecular mechanisms of calcium-dependent excitotoxicity. J Mol Med 2000;78:3–13.

Schulz JB, Matthews RT, Klockgether T, Dichgans J, Beal MF. The role of mitochondrial dysfunction and neuronal nitric oxide in animal models of neurodegenerative diseases. Mol Cell Biochem 1997;174:193–197.

Snyder SH, Jaffrey SR, Zakhary R. Nitric oxide and carbon monoxide: Parallel roles as neural messengers. Brain Res Rev 1998;26:167–175.

CASE STUDIES FOR PART I

CASE STUDY FOR CHAPTER 1

Severe, Acute Diarrhea

A 29-year-old woman had spent the past 2 weeks visiting her family in southern Louisiana. On the last night of her visit, she consumed a dozen fresh oysters. Twenty-four hours later, following her return home, she awoke with nausea, vomiting, abdominal pain, and profuse watery diarrhea. She went into shock and was transported to the emergency department, where she was found to be dehydrated and lethargic. She does not have an elevated temperature, but her abdomen is distended. There is no tenderness to the abdomen, and her bowel sounds are hyperactive. Laboratory results show she is hypokalemic, with a plasma potassium level of 1.4 mEq/L (normal values, 3.5 to 5.0 mEq/L). Plasma sodium and chloride levels are slightly lower than normal, and plasma bicarbonate is 11 mEq/L (normal values, 22 to 28 mEq/L). After oral rehydration and antibiotic therapy, she rapidly improves and is discharged on the fourth hospital day.

Questions

1. What disease is consistent with this patient's symptoms?
2. Describe the pathophysiology associated with this disease.

Answers to Case Study Questions for Chapter 1

1. The disease consistent with the symptoms of this patient is cholera. Cholera is a self-limiting disease characterized by acute diarrhea and dehydration without febrile symptoms (no fever). The microorganism responsible for this disease is *Vibrio cholerae*. The ingestion of water or food that has been contaminated with feces or vomitus of an individual transmits the bacterium, causing the disease.
2. The pathophysiology associated with this disease is related to the production of a toxin by the *V. cholerae* bacterium. The toxin has two subunits (α and β). The α subunit causes the activation of adenylyl cyclase (AC) and the β subunit recognizes and binds to an apical (facing the lumen of the intestine) membrane component of intestinal epithelial cells, causing the toxin to become engulfed into the cell. Inside the cell, the toxin is transported to the basolateral membrane, where the α subunit ADP-ribosylates the G_s protein. ADP-ribosylation of G_s results in inhibition of the GTPase activity of the G_s subunit and the stabilization of the G protein in an active or "on" conformation. The continuous stimulation of AC and concomitant sustained production of cAMP result in opening of a chloride channel in the apical plasma membrane. This produces net chloride secretion, with sodium and water following. Bicarbonate and potassium ions are also lost in the stool. The loss of water and electrolytes in diarrheal fluid can be so severe (20 L/day) that it may be fatal.

CASE STUDY FOR CHAPTER 2

Cystic Fibrosis

A 12-month-old baby is brought to a pediatrician's office because the parents are concerned about a recurrent cough and frequent foul-smelling stools. The doctor has followed the child from birth and notices that the baby's weight has remained below the normal range. A chest X-ray reveals hyperinflation consistent with the obstruction of small airways.

Questions

1. What is the explanation for the frequent stools and poor growth?
2. What is causing obstruction of the small airways?
3. What is the fundamental defect at the molecular level that underlies these symptoms?

Answers to Case Study Questions for Chapter 2

1. Impaired secretion of chloride ions by epithelial cells of pancreatic ducts limits the function of a Cl^-/HCO_3^- exchanger to secrete bicarbonate. Secretion of Na^+ is also impaired, and the resultant failure to secrete $NaHCO_3$ retards water movement into the ducts. Mucus in the ducts becomes dehydrated and thick and blocks the delivery of pancreatic enzymes. The deficiency of pancreatic enzymes in the intestinal lumen leads to malabsorption of protein and fats, hence, the malnutrition and frequent malodorous stools.
2. An analogous mechanism in the epithelial cells of small airways results in reduced secretion of NaCl and retardation of water movement. The dehydrated mucus cannot be cleared from the small airways and not only obstructs them but also traps bacteria that initiate localized infections.
3. The defect in chloride transport is a result of mutations in the gene for the chloride channel known as the **cystic fibrosis transmembrane regulator** (CFTR). Some mutated forms of the CFTR protein are destroyed in the epithelial cell before they reach the apical plasma membrane; other mutations result in a CFTR protein that is inserted in the plasma membrane but functions abnormally.

Reference

Quinton PM. Physiological basis of cystic fibrosis: A historical perspective. Physiol Rev 1999;79(Suppl):S3–S22.

CASE STUDY FOR CHAPTER 3

Episodic Ataxia

A 3-year-old child was brought to the pediatrician because of visible muscle twitching. The parents described the twitches as looking like worms crawling under the skin. The child also periodically complained that her legs hurt, and the mother reported she could feel that the child's leg muscles were somewhat rigid at these times. Occasionally, the child would exhibit a loss of motor coordination (ataxia) that lasted 20 to 30 minutes; these episodes sometimes followed exertion or startle. Neurological function seemed normal between these episodes; the parents reported that the child's motor development seemed similar to that of their older child. The neurological examination confirms the parents' perception. Electromyographic analysis of the child's leg muscles indicates no abnormality in muscle membrane responses, and a muscle biopsy is histologically normal. Spinal anesthesia eliminated the muscle twitching. The child's mother indicates that one of the child's sisters also had frequent muscle twitches as a child, but did not have episodes of ataxia.

Questions

1. What is the likely source of the abnormal muscle activity?
2. What information in the presentation supports your answer to question 1?
3. Spontaneous muscle twitches indicate hyperexcitability of nerve or muscle. If this hyperexcitability is a result of an abnormality in action potential repolarization, what channels associated with the nerve action potential might lead to this condition?

Answers to Case Study Questions for Chapter 3

1. The abnormal muscle activity derives from the motor neurons.
2. Spontaneous muscle twitching could be a result of a defect in the muscle, the motor neurons that control the muscle, the neuromuscular junction (synapse), or the central nervous system elements that control spinal motor neurons. The description of muscle twitches that look like worms crawling under the skin indicates that individual motor units are firing randomly and spontaneously. (A motor unit is one motor neuron and all of the muscle fibers it innervates.) The muscle biopsy and electromyographic studies indicate it is not the muscle. Spinal anesthesia eliminates the muscle twitching indicating that the defect is at the level of the motor neurons.
3. The nerve action potential may fail to repolarize properly if there is a defect in the inactivation of voltage-gated sodium channels or in the activation of voltage-gated potassium channels. Genetic analysis in this individual, whose diagnosis is episodic ataxia with myokymia, would indicate a mutation in the potassium channel.

References

Adelman JP, Bond CT, Pessia M, Maylie J. Episodic ataxia results from voltage-dependent potassium channels with altered functions. Neuron 1995;15:1449–1454.

Browne DL, Gancher ST, Nutt JG, et al. Episodic ataxia/myokymia syndrome is associated with point mutations in the human potassium channel gene, KCNA1. Nat Genet 1994;8:136–140.

PART II Neurophysiology

Sensory Physiology

Richard A. Meiss, Ph.D.

CHAPTER OUTLINE

- THE GENERAL PROBLEM OF SENSATION
- SPECIFIC SENSORY RECEPTORS

KEY CONCEPTS

1. Sensory transduction takes place in a series of steps, starting with stimuli from the external or internal environment and ending with neural processing in the central nervous system.
2. The structure of sensory organs optimizes their response to the preferred types of stimuli.
3. A stimulus gives rise to a generator potential, which, in turn, causes action potentials to be produced in the associated sensory nerve.
4. The speeds of adaptation of particular sensory receptors are related to their biological roles.
5. Specific sensory receptors for a variety of types of tactile stimulation are located in the skin.
6. Somatic pain is associated with the body surface and the musculature; visceral pain is associated with the internal organs.
7. The sensory function of the eyeball is determined by structures that form and adjust images and by structures that transform images into neural signals.
8. The retina contains several cell types, each with a specific role in the process of visual transduction.
9. The rod cells in the retina have a high sensitivity to light but produce indistinct images without color, while the cones provide sharp color vision with less sensitivity to light.
10. The visual transduction process requires many steps, beginning with the absorption of light and ending with an electrical response.
11. The outer ear receives sound waves and passes them to the middle ear; they are modified and passed to the inner ear, where the process of sound transduction takes place.
12. The transmission of sound through the middle ear greatly increases the efficiency of its detection, while its protective mechanisms guard the inner ear from damage caused by extremely loud sounds. Disturbances in this transmission process can lead to hearing impairments.
13. Sound vibrations enter the cochlea through the oval window and travel along the basilar membrane, where their energy is transformed into neural signals in the organ of Corti.
14. Displacements of the basilar membrane cause deformation of the hair cells, the ultimate transducers of sound. Different sites along the basilar membrane are sensitive to different frequencies.
15. The vestibular apparatus senses the position of the head and its movements by detecting small deflections of its sensory structures.
16. Taste is mediated by sensory epithelial cells in the taste buds. There are five fundamental taste sensations: sweet, sour, salty, bitter, and umami.
17. Smell is detected by nerve cells in the olfactory mucosa. Thousands of different odors can be detected and distinguished.

The survival of any organism, human included, depends on having adequate information about the external environment, where food is to be found and where hazards abound. Equally important for maintaining the function of a complex organism is information about the state of numerous internal bodily processes and functions. Events in our external and internal worlds must first be translated into signals that our nervous systems can process. Despite the wide range of types of information to be sensed and acted on, a small set of common principles underlie all sensory processes.

This chapter discusses the functions of the organs that permit us to gather this information, the sensory receptors. The discussion emphasizes somatic sensations, those dealing with the external aspect of the body, and does not specifically treat visceral sensations, those that come from internal organs.

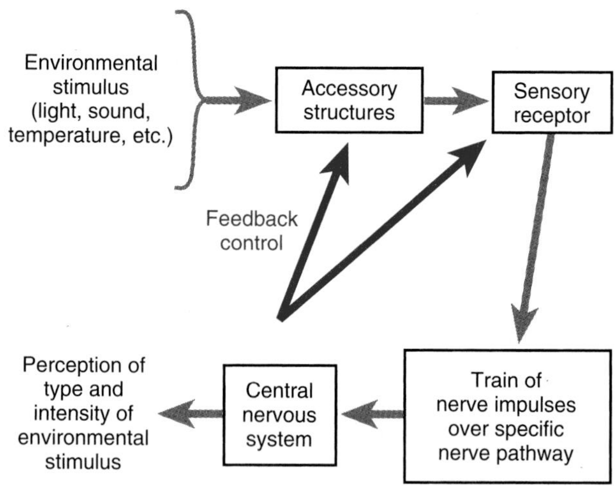

FIGURE 4.1 **A basic model for the translation of an environmental stimulus into a perception.** While the details vary with each type of sensory modality, the overall process is similar.

THE GENERAL PROBLEM OF SENSATION

While the human body contains a very large number of different sensory receptors, they have many functional features in common. Some basic themes are shared by almost all receptors, and the wide variety of specialized functions is a result of structural and physiological adaptations that adapt a particular receptor for its role in the overall economy of an organism.

Sensory Receptors Translate Energy From the Environment Into Biologically Useful Information

The process of sensation essentially involves sampling selected small amounts of energy from the environment and using it to control the generation of action potentials or nerve impulses (see Fig. 4.1). This process is the function of **sensory receptors,** biological structures that can be as simple as a free nerve ending or as complicated as the human eye or ear. The pattern of sensory action potentials, along with the specific nature of the sensory receptor and its nerve pathways in the brain, provide an internal representation of a specific component of the external world. The process of sensation is a portion of the more complex process of **perception,** in which sensory information is integrated with previously learned information and other sensory inputs, enabling us to make judgments about the quality, intensity, and relevance of what is being sensed.

The Nature of Environmental Stimuli. A factor in the environment that produces an effective response in a sensory receptor is called a **stimulus.** Stimuli involve exchanges of energy between the environment and the receptors. Typical stimuli include electromagnetic quantities, such as radiant heat or light; mechanical quantities, such as pressure, sound waves, and other vibrations; and chemical qualities, such as acidity and molecular shape and size. Common to all these types of stimuli is the property of **intensity,** a measure of the energy content (or concentration, in the case of chemical stimuli) available to interact with the sensory receptor. It is not surprising, therefore, that a fundamental property of receptors is their ability to respond to different intensities of stimulation with an appropriate output. Also related to receptor function is the concept of **sensory modality.** This term refers to the kind of sensation, which may range from the relatively general modalities of taste, smell, touch, sight, and hearing (the traditional five senses), to more complex sensations, such as slipperiness or wetness. Many sensory modalities are a combination of simpler sensations; the sensation of wetness is composed of sensations of pressure and temperature. (Try placing your hand in a plastic bag and immersing it in cold water. Although the skin will remain dry, the perception will be one of wetness.)

It is often difficult to communicate a precise definition of a sensory modality because of the subjective perception or **affect** that accompanies it. This property has to do with the psychological feeling attached to the stimulus. Some stimuli may give rise to an impression of discomfort or pleasure apart from the primary sensation of, for example, cold or touch. Previous experience and learning play a role in determining the affect of a sensory perception.

Some sensory receptors are classified by the nature of the signals they sense. For example, **photoreceptors** sense light and serve a visual function. **Chemoreceptors** detect chemical signals and serve the senses of taste and smell, as well as detecting the presence of specific substances in the body. **Mechanoreceptors** sense physical deformation, serve the senses of touch and hearing, and can detect the amount of stress in a tendon or muscle; and **thermal receptors** detect heat (or its relative lack). Other sensory receptors are classified by their "vantage point" in the body. Among these, **exteroceptors** detect stimuli from outside the body; **enteroceptors** detect internal stimuli; **proprioceptors** (receptors of "one's own") provide information about the positions of joints and about muscle activity and the orientation of the body in space. **Nociceptors** (pain receptors) detect noxious agents, both internally and externally.

The Specificity of Sensory Receptors. Most sensory receptors respond preferentially to a single kind of environmental stimulus. The usual stimulus for the eye is light; that for the ear is sound. This specificity is due to several features that match a receptor to its preferred stimulus. In many cases, **accessory structures,** such as the lens of the eye or the structures of the outer and middle ears, enhance the specific sensitivity of the receptor or exclude unwanted stimuli. Often these accessory structures are a control system that adjusts their sensitivity according to the information being received (Fig. 4.1). The usual and appropriate stimulus for a receptor is called its **adequate stimulus**. For the adequate stimulus, the receptor has the lowest **threshold**, the lowest stimulus intensity that can be reliably detected. A threshold is often difficult to measure because it can vary over time and with the presence of interfering stimuli or the action of accessory structures. Although most receptors will respond to stimuli other than the adequate stimulus, the threshold for inappropriate stimuli is much higher. For example, gently pressing the outer corner of the eye will produce a visual sensation caused by pressure, not light; extremes of temperature may be perceived as pain. Almost all receptors can be stimulated electrically to produce sensations that mimic the one usually associated with that receptor.

The central nervous pathway over which sensory information travels is also important in determining the nature of the perception; information arriving by way of the optic nerve, for example, is always perceived as light and never as sound. This is known as the concept of the **labeled line**.

The Process of Sensory Transduction Changes Stimuli Into Biological Information

This section focuses on the actual function of the sensory receptor in translating environmental energy into action potentials, the fundamental units of information in the nervous system. A device that performs such a translation is called a **transducer**; sensory receptors are biological transducers. The sequence of electrical events in the sensory transduction process is shown in Figure 4.2.

The Generator Potential. The sensory receptor in this example is a mechanoreceptor. Deformation or deflection of the tip of the receptor gives rise to a series of action potentials in the sensory nerve fiber leading to the central nervous system (CNS). The stimulus (1) is applied at the tip of the receptor, and the deflection (2) is held constant (dotted lines). This deformation of the receptor causes a

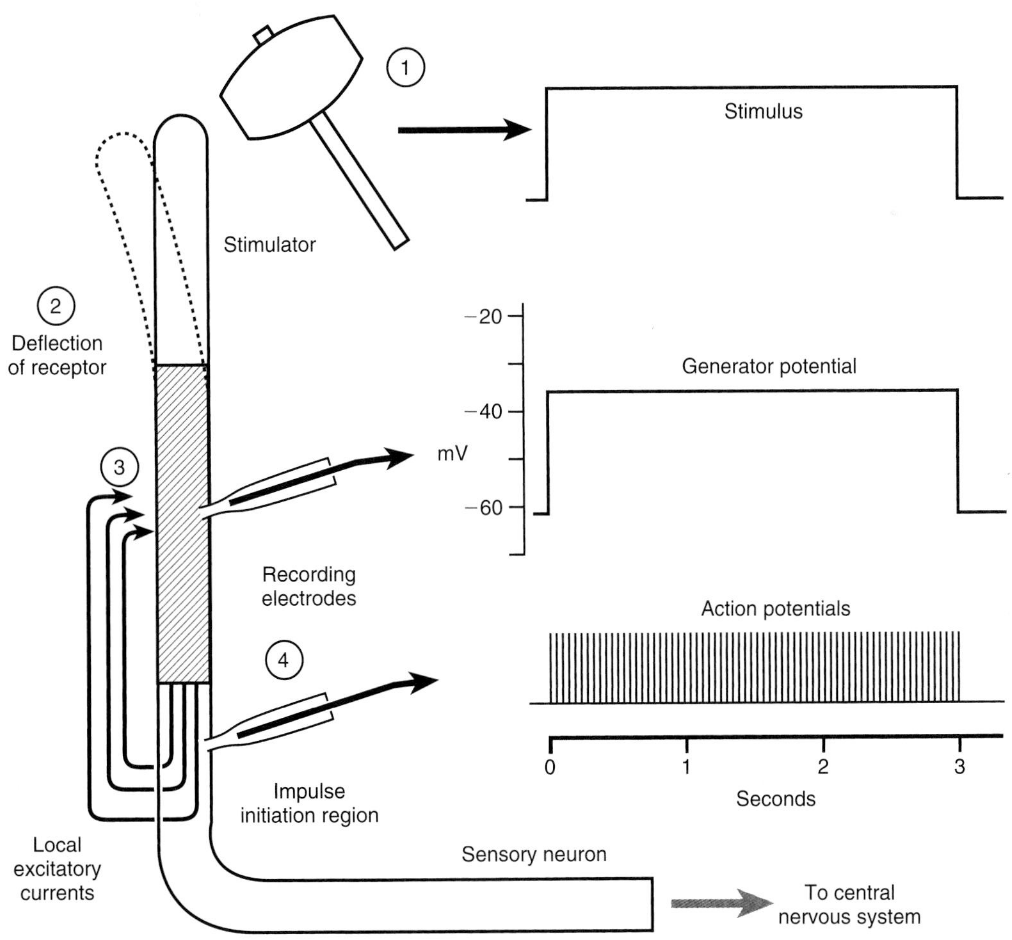

FIGURE 4.2 **The relation between an applied stimulus and the production of sensory nerve action potentials.** (See text for details.)

portion of its cell membrane (shaded region [3]) to become more permeable to positive ions (especially sodium). The increased permeability of the membrane leads to a localized depolarization, called the **generator potential.** At the depolarized region, sodium ions enter the cell down their electrochemical gradient, causing a current to flow in the extracellular fluid. Because current is flowing into the cell at one place, it must flow out of the cell in another place. It does this at a region of the receptor membrane (4) called the **impulse initiation region** (or coding region) because here the flowing current causes the cell membrane to produce action potentials at a frequency related to the strength of the current caused by the stimulus. These currents, called **local excitatory currents,** provide the link between the formation of the generator potential and the excitation of the nerve fiber membrane.

In complex sensory organs that contain a great many individual receptors, the generator potential may be called a **receptor potential,** and it may arise from several sources within the organ. Often the receptor potential is given a special name related to the function of the receptor; for example, in the ear it is called the cochlear microphonic, while an electroretinogram may be recorded from the eye. Note that in the eye the change in receptor membrane potential associated with the stimulus of light is a **hyperpolarization,** not a depolarization.

The production of the generator potential is of critical importance in the transduction process because it is the step in which information related to stimulus intensity and duration is transduced. The strength (intensity) of the stimulus applied (in Fig. 4.2, the amount of deflection) determines the size of the generator potential depolarization.

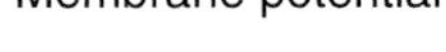

Varying the intensity of the stimulation will correspondingly vary the generator potential, although the changes will not usually be directly proportional to the intensity. This is called a **graded response,** in contrast to the all-or-none response of an action potential, and it causes a similar gradation of the strength of the local excitatory currents. These, in turn, determine the amount of depolarization produced in the impulse initiation region (4) of the receptor, and events in this region constitute the next important link in the process.

The Initiation of Nerve Impulses

Figure 4.3 shows a variety of possible events in the impulse initiation region. The threshold (colored line) is a critical level of depolarization; membrane potential changes below this level are caused by the local excitatory currents and vary in proportion to them, while the membrane activity above the threshold level consists of locally produced **action potentials.** The lower trace shows a series of different stimuli applied to the receptor, and the upper trace shows the resulting electrical events in the impulse initiation region.

No stimulus is given at **A,** and the membrane voltage is at the resting potential. At **B,** a small stimulus is applied, producing a generator potential too small to bring the impulse initiation region membrane to threshold, and no action potential activity results. (Such a stimulus would not be sensed at all.) A brief stimulus of greater intensity is given at **C**; the resulting generator potential displacement is of sufficient amplitude to trigger a single action potential. As in all excitable and all-or-none nerve membranes, the action potential is immediately followed by repolarization, often to a level that transiently hyperpolarizes the membrane potential because of temporarily high potassium conductance. Since the brief stimulus has been removed by this time, no further action potentials are produced. A longer stimulus of the same intensity (**D**) produces repetitive action potentials because as the membrane repolarizes from the action potential, local excitatory currents are still flowing. They bring

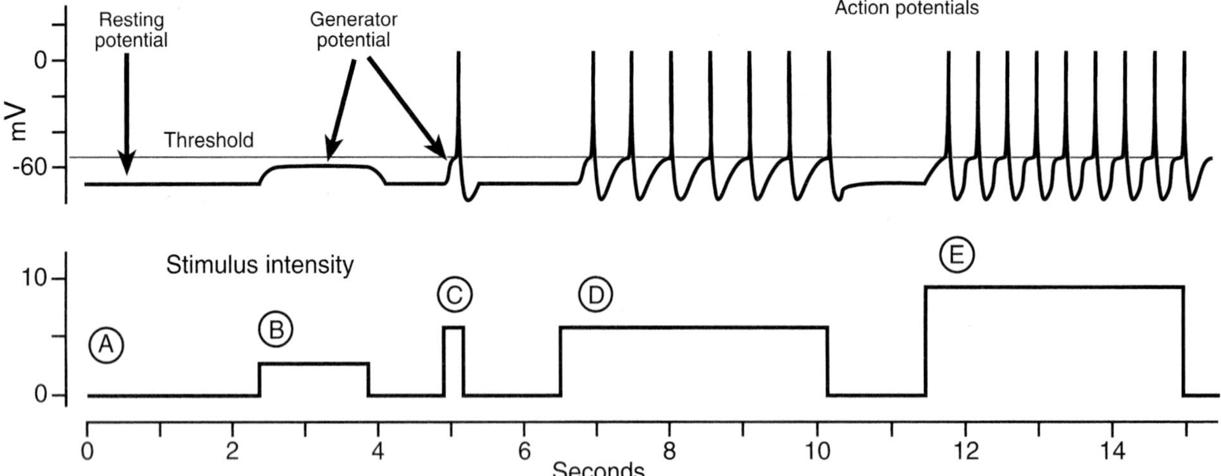

FIGURE 4.3 **Sensory nerve activity with different stimulus intensities and durations.** A, With no stimulus, the membrane is at rest. B, A subthreshold stimulus produces a generator potential too small to cause membrane excitation. C, A brief, but intense, stimulus can cause a single action potential. D, Maintaining this stimulus leads to a train of action potentials. E, Increasing the stimulus intensity leads to an increase in the action potential firing rate.

the repolarized membrane to threshold at a rate proportional to their strength. During this time interval, the fast sodium channels of the membrane are being reset, and another action potential is triggered as soon as the membrane potential reaches threshold. As long as the stimulus is maintained, this process will repeat itself at a rate determined by the stimulus intensity. If the intensity of the stimulus is increased (E), the local excitatory currents will be stronger and threshold will be reached more rapidly. This will result in a reduction of the time between each action potential and, as a consequence, a higher action potential frequency. This change in action potential frequency is critical in communicating the intensity of the stimulus to the CNS.

Adaptation. The discussion thus far has depicted the generator potential as though it does not change when a constant stimulus is applied. Although this is approximately correct for a few receptors, most will show some degree of **adaptation.** In an adapting receptor, the generator potential and, consequently, the action potential frequency will decline even though the stimulus is maintained. Part A of Figure 4.4 shows the output from a receptor in which there is no adaptation. As long as the stimulus is maintained, there is a steady rate of action potential firing. Part B shows **slow adaptation**; as the generator potential declines, the interval between the action potentials increases correspondingly. Part C demonstrates **rapid adaptation**; the action potential frequency falls rapidly and then maintains a constant slow rate that does not show further adaptation. Responses in which there is little or no adaptation are called **tonic,** whereas those in which significant adaptation occurs are called **phasic.** In some cases, tonic receptors may be called **intensity receptors,** and phasic receptors called **velocity receptors.** Many receptors—**muscle spindles**, for example—show a combination of responses; on application of a stimulus, a rapidly adapting phasic response is followed by a steady tonic response. Both of these responses may be graded by the intensity of the stimulus. As a receptor adapts, the sensory input to the CNS is reduced, and the sensation is perceived as less intense.

The phenomenon of adaptation is important in preventing "sensory overload," and it allows less important or unchanging environmental stimuli to be partially ignored. When a change occurs, however, the phasic response will occur again, and the sensory input will become temporarily more noticeable. Rapidly adapting receptors are also important in sensory systems that must sense the rate of change of a stimulus, especially when its intensity can vary over a range that would overload a tonic receptor.

Receptor adaptation can occur at several places in the transduction process. In some cases, the receptor's sensitivity is changed by the action of accessory structures, as in the constriction of the pupil of the eye in the presence of bright light. This is an example of feedback-controlled adaptation; in the sensory cells of the eye, light-controlled changes in the amounts of the visual pigments also can change the basic sensitivity of the receptors and produce adaptation. As mentioned above, adaptation of the generator potential can produce adaptation of the overall sensory response. Finally, the phenomenon of **accommodation** in the impulse initiation region of the sensory nerve fiber can slow the rate of action potential production even though the generator potential may show no change. Accommodation refers to a gradual increase in threshold caused by prolonged nerve depolarization, resulting from the inactivation of sodium channels.

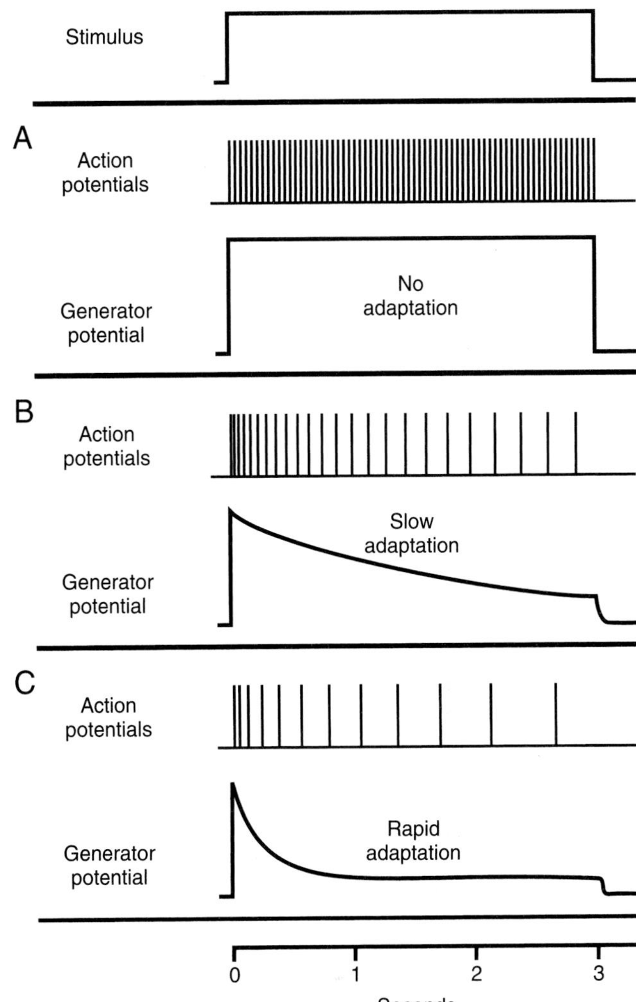

FIGURE 4.4 **Adaptation.** Adaptation in a sensory receptor is often related to a decline in the generator potential with time. **A,** The generator potential is maintained without decline, and the action potential frequency remains constant. **B,** A slow decline in the generator potential is associated with slow adaptation. **C,** In a rapidly adapting receptor, the generator potential declines rapidly.

The Perception of Sensory Information Involves Encoding and Decoding

After the acquisition of sensory stimuli, the process of perception involves the subsequent encoding and transmission of the sensory signal to the central nervous system. Further processing or decoding yields biologically useful information.

Encoding and Transmission of Sensory Information. Environmental stimuli that have been partially processed by a sensory receptor must be conveyed to the CNS in such a way that the complete range of the intensity of the stimulus is preserved.

Compression. The first step in the encoding process is **compression.** Even when the receptor sensitivity is modified by accessory structures and adaptation, the range of input intensities is quite large, as shown in Figure 4.5. At the left is a 100-fold range in the intensity of a stimulus. At the right is an intensity scale that results from events in the sensory receptor. In most receptors, the magnitude of the generator potential is not exactly proportional to the stimulus intensity; it increases less and less as the stimulus intensity increases. The frequency of the action potentials produced in the impulse initiation region is also not proportional to the strength of the local excitatory currents; there is an upper limit to the number of action potentials per second because of the refractory period of the nerve membrane. These factors are responsible for the process of compression; changes in the intensity of a small stimulus cause a greater change in action potential frequency than the same change would cause if the stimulus intensity were high. As a result, the 100-fold variation in the stimulus is compressed into a threefold range after the receptor has processed the stimulus. Some information is necessarily lost in this process, but integrative processes in the CNS can restore the information or compensate for its absence. Physiological evidence for compression is based on the observed nonlinear (logarithmic or power function) relation between the actual intensity of a stimulus and its perceived intensity.

Information Transfer. The next step is to transfer the sensory information from the receptor to the CNS. The encoding processes in the receptors have already provided the basis for this transfer by producing a series of action potentials related to the stimulus intensity. A special process is necessary for the transfer because of the nature of the conduction of action potentials. As an action potential travels along a nerve fiber, it is sequentially recreated at a sequence of locations along the nerve. Its duration and amplitude do not change. The only information that can be conveyed by a single action potential is its presence or absence. However, relationships between and among action potentials can convey large amounts of information, and this is the system found in the sensory transmission process. This biological process can be explained by analogy to a physical system such as that used for transmission of signals in communications systems.

Figure 4.6 outlines a hypothetical **frequency-modulated (FM)** encoding, transmission, and reception system. An input signal provided by some physical quantity (1) is continuously measured and converted into an electrical signal (2), analogous to the generator potential, whose amplitude is proportional to the input signal. This signal then controls the frequency of a pulse generator (3), as in the impulse initiation region of a sensory nerve fiber. Like action potentials, these pulses are of a constant height and duration, and the amplitude information of the original input signal is now contained in the intervals between the pulses. The resulting signals may be sent along a transmission line (analogous to a nerve pathway) to some distant point, where they produce an electrical voltage (4) proportional to the frequency of the arriving pulses. This voltage is a replica of the input voltage (2) and is not affected by changes in the amplitude of the pulses as they travel along the transmission line. Further processing can produce a graphic record (5) of the input data. In a biological system, these latter functions are accompanied by processing and interpretation in the CNS.

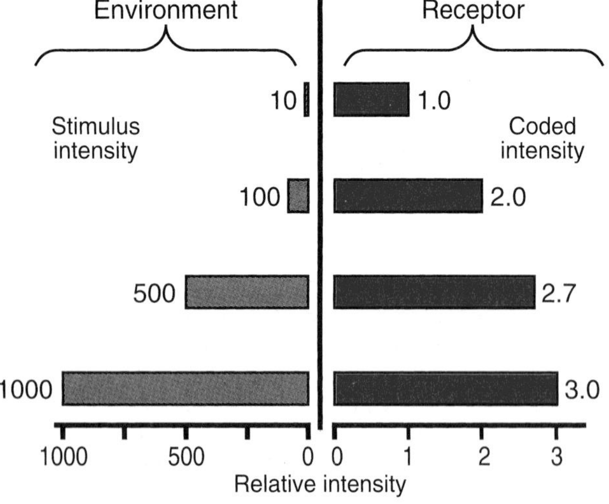

FIGURE 4.5 **Compression in sensory process.** By a variety of means, a wide range of input intensities is coded into a much narrower range of responses that can be represented by variations in action potential frequency.

The Interpretation of Sensory Information. The interpretation of encoded and transmitted information into a perception requires several other factors. For instance, the interpretation of sensory input by the CNS depends on the neural pathway it takes to the brain. All information arriving on the **optic nerves** is interpreted as light, even though the signal may have arisen as a result of pressure applied to the eyeball. The **localization** of a cutaneous sensation to a particular part of the body also depends on the particular pathway it takes to the CNS. Often a sensation (usually pain) arising in a visceral structure (e.g., heart, gallbladder) is perceived as coming from a portion of the body surface, because developmentally related nerve fibers come from these anatomically different regions and converge on the same spinal neurons. Such a sensation is called **referred pain.**

SPECIFIC SENSORY RECEPTORS

The remainder of this chapter surveys specific sensory receptors, concentrating on the **special senses.** These traditionally include cutaneous sensation (touch, temperature, etc.), sight, hearing, taste, and smell.

Cutaneous Sensation Provides Information From the Body Surface

The skin is richly supplied with sensory receptors serving the modalities of touch (light and deep pressure), temperature (warm and cold), and pain, as well as the more compli-

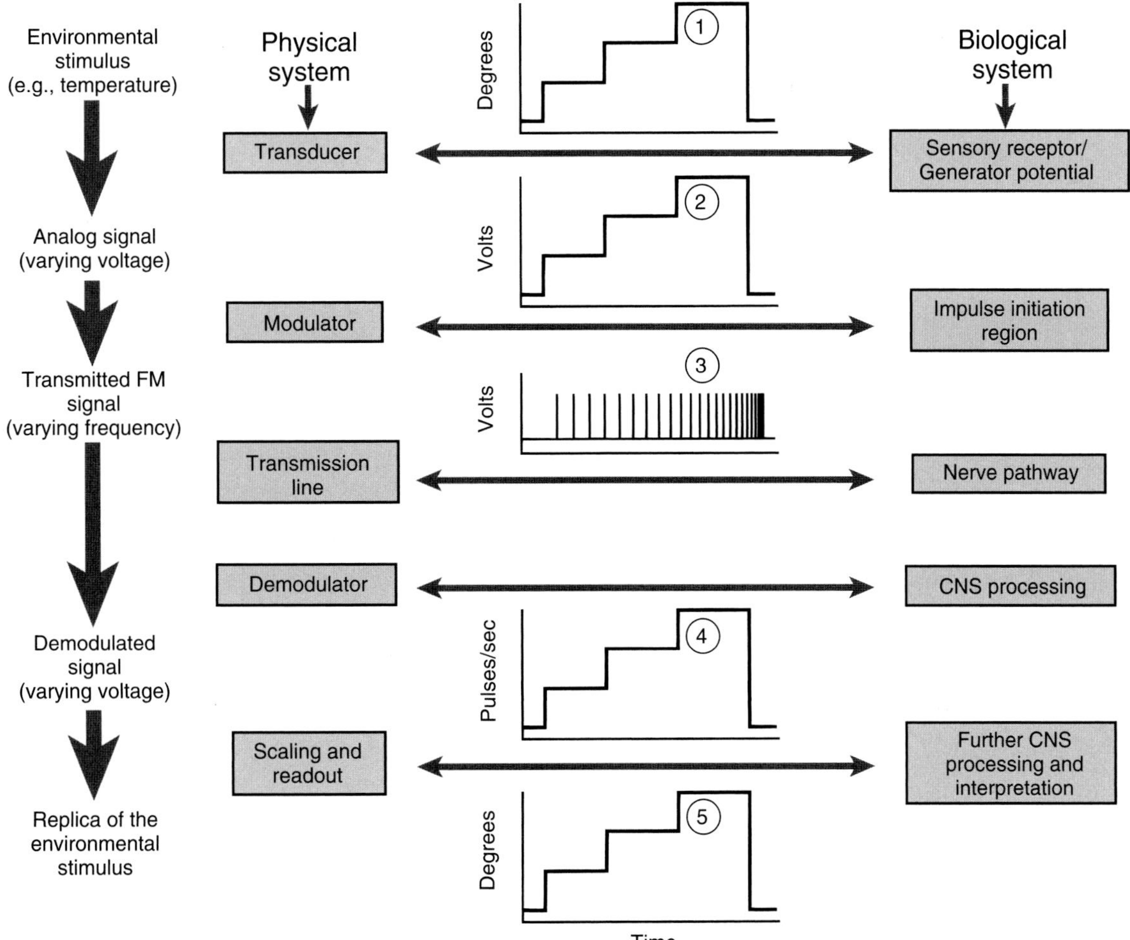

FIGURE 4.6 **The transmission of sensory information.** Because signals of varying amplitude cannot be transmitted along a nerve fiber, specific intensity information is transformed into a corresponding action potential frequency, and CNS processes decode the nerve activity into biologically useful information. The steps in the process are shown at the left, with the parts of a physical system that perform them (FM, frequency modulation). At the right are the analogous biological steps involved in the same process.

cated composite modalities of itch, tickle, wet, and so on. By using special probes that deliver highly localized stimuli of pressure, vibration, heat, or cold, the distribution of cutaneous receptors over the skin can be mapped. In general, areas of skin used in tasks requiring a high degree of spatial localization (e.g., fingertips, lips) have a high density of specific receptors, and these areas are correspondingly well represented in the somatosensory areas of the cerebral cortex (see Chapter 7).

Tactile Receptor. Several receptor types serve the sensations of touch in the skin (Fig. 4.7). In regions of hairless skin (e.g., the palm of the hand) are found Merkel's disks, Meissner's corpuscles, and pacinian corpuscles. **Merkel's disks** are intensity receptors (located in the lowest layers of the epidermis) that show slow adaptation and respond to steady pressure. **Meissner's corpuscles** adapt more rapidly to the same stimuli and serve as velocity receptors. **Pacinian corpuscles** are very rapidly adapting (acceleration) receptors. They are most sensitive to fast-changing stimuli, such as vibration. In regions of hairy skin, small hairs serve as accessory structures for **hair-follicle receptors,** mechanoreceptors that adapt more slowly. **Ruffini endings** (located in the dermis) are also slowly adapting receptors. Merkel's disks in areas of hairy skin are grouped into **tactile disks.** Pacinian corpuscles also sense vibrations in hairy skin. Nonmyelinated nerve endings, also usually found in hairy skin, appear to have a limited tactile function and may sense pain.

Temperature Sensation. From a physical standpoint, warm and cold represent values along a temperature continuum and do not differ fundamentally except in the amount of molecular motion present. However, the familiar subjective differentiation of the temperature sense into "warm" and "cold" reflects the underlying physiology of the two populations of receptors responsible for thermal sensation.

Temperature receptors (**thermoreceptors**) appear to be naked nerve endings supplied by either thin myelinated fibers (cold receptors) or nonmyelinated fibers (warm receptors) with low conduction velocity. Cold receptors form a population with a broad response peak at about

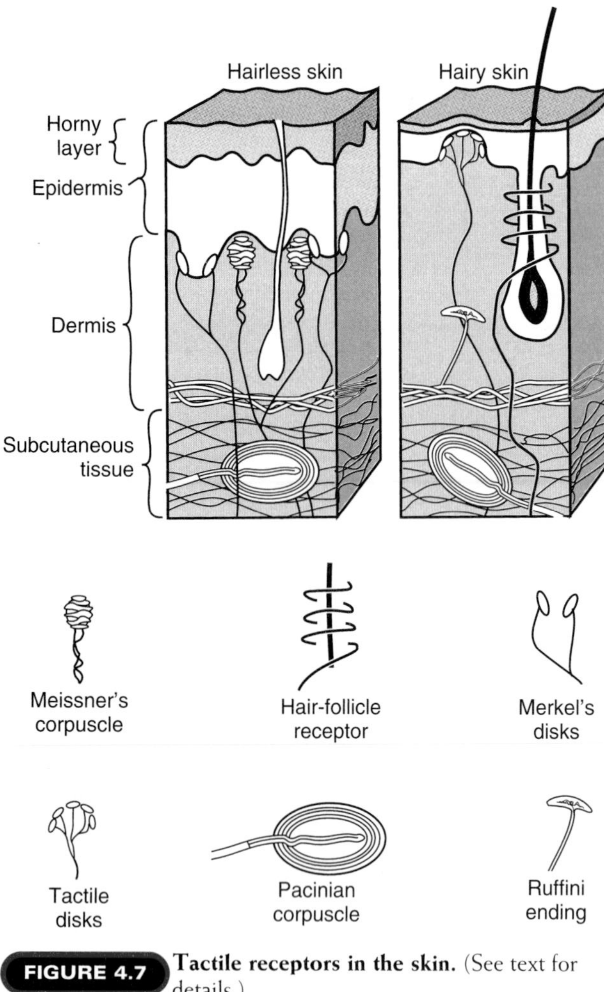

FIGURE 4.7 **Tactile receptors in the skin.** (See text for details.)

30°C; the warm receptor population has its peak at about 43°C (Fig. 4.8). Both sets of receptors share some common features:

- They are sensitive only to thermal stimulation.
- They have both a phasic response that is rapidly adapting and responds only to temperature changes (in a fashion roughly proportional to the rate of change) and a tonic (intensity) response that depends on the local temperature.

The density of temperature receptors differs at different places on the body surface. They are present in much lower numbers than cutaneous mechanoreceptors, and there are many more cold receptors than warm receptors.

The perception of temperature stimuli is closely related to the properties of the receptors. The phasic component of the response is apparent in our adaptation to sudden immersion in, for example, a warm bath. The sensation of warmth, apparent at first, soon fades away, and a less intense impression of the steady temperature may remain. Moving to somewhat cooler water produces an immediate sensation of cold that soon fades away. Over an intermediate temperature range (the "comfort zone"), there is no appreciable temperature sensation. This range is approximately 30 to 36°C for a small area of skin; the range is narrower when the whole body is exposed. Outside this range, steady temperature sensation depends on the ambient (skin) temperature. At skin temperatures lower than 17°C, **cold pain** is sensed, but this sensation arises from pain receptors, not cold receptors. At very high skin temperatures (above 45°C), there is a sensation of **paradoxical cold,** caused by activation of a part of the cold receptor population.

Temperature perception is subject to considerable processing by higher centers. While the perceived sensations reflect the activity of specific receptors, the phasic component of temperature perception may take many minutes to be completed, whereas the adaptation of the receptors is complete within seconds.

Pain. The familiar sensation of pain is not limited to cutaneous sensation; pain coming from stimulation of the body surface is called **superficial pain,** while that arising from within muscles, joints, bones, and connective tissue is called **deep pain.** These two categories comprise **somatic pain. Visceral pain** arises from internal organs and is often due to strong contractions of visceral muscle or its forcible deformation.

Pain is sensed by a population of specific receptors called nociceptors. In the skin, these are the free endings of thin myelinated and nonmyelinated fibers with characteristically low conduction velocities. They typically have a high threshold for mechanical, chemical, or thermal stimuli (or a combination) of intensity sufficient to cause tissue destruction. The skin has many more points at which pain can be elicited than it has mechanically or thermally sensitive sites. Because of the high threshold of pain receptors (compared with that of other cutaneous receptors), we are usually unaware of their existence.

Superficial pain may often have two components: an immediate, sharp, and highly localizable **initial pain**; and, after a latency of about 1 second, a longer-lasting and more diffuse **delayed pain.** These two submodalities appear to be mediated by different nerve fiber endings. In addition to

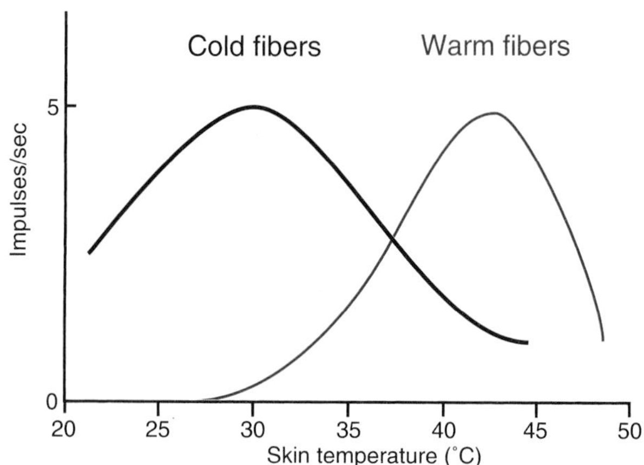

FIGURE 4.8 **Responses of cold and warm receptors in the skin.** The skin temperature was held at different values while nerve impulses were recorded from representative fibers leading from each receptor type. (Modified from Kenshalo. In: Zotterman Y. Sensory Functions of Skin in Primates. Oxford: Pergamon, 1976.)

their normally high thresholds, both cutaneous and deep pain receptors show little adaptation, a fact that is unpleasant but biologically necessary. Deep and visceral pain appear to be sensed by similar nerve endings, which may also be stimulated by local metabolic conditions, such as ischemia (lack of adequate blood flow, as may occur during the heart pain of angina pectoris).

The free nerve endings mediating pain sensation are anatomically distinct from other free nerve endings involved in the normal sensation of mechanical and thermal stimuli. The functional differences are not microscopically evident and are likely to relate to specific elements in the molecular structure of the receptor cell membrane.

The Eye Is a Sensor for Vision

The eye is an exceedingly complex sensory organ, involving both sensory elements and elaborate accessory structures that process information both before and after it is detected by the light-sensitive cells. A satisfactory understanding of vision involves a knowledge of some of the basic physics of light and its manipulation, in addition to the biological aspects of its detection.

The Properties of Light and Lenses. The adequate stimulus for human visual receptors is **light**, which may be defined as electromagnetic radiation between the wavelengths of 770 nm (red) and 380 nm (violet). The familiar colors of the spectrum all lie between these limits. A wide range of intensities, from a single photon to the direct light of the sun, exists in nature.

As with all such radiation, light rays travel in a straight line in a given medium. Light rays are **refracted** or bent as they pass between media (e.g., glass, air) that have different **refractive indices.** The amount of bending is determined by the angle at which the ray strikes the surface; if the angle is 90°, there is no bending, while successively more oblique rays are bent more sharply. A simple prism (Fig. 4.9A) can, therefore, cause a light ray to deviate from its path and travel in a new direction. An appropriately chosen pair of prisms can turn parallel rays to a common point (Fig. 4.9B). A **convex lens** may be thought of as a series of such prisms with increasingly more bending power (Fig. 4.9C, D), and such a lens, called a **converging lens** or **positive lens,** will bring an infinite number of parallel rays to a common point, called the focal point. A converging lens can form a **real image.** The distance from the lens to this point is its **focal length (FL)**, which may be expressed in meters. A convex lens with less curvature has a longer focal length (Fig. 4.9E). Often the **diopter (D)**, which is the inverse of the focal length (1/FL), is used to describe the power of a lens. For example, a lens with a focal length of 0.5 meter has a power of 2 D. An advantage of this system is that dioptric powers are additive; two convex lenses of 25 D each will function as a single lens with a power of 50 D when placed next to each other (Fig. 4.9F).

A **concave lens** causes parallel rays to diverge (Fig. 4.9G). Its focal length (and its power in diopters) is **negative,** and it cannot form a real image. A concave lens placed before a positive lens lengthens the focal length (Fig. 4.9H) of the lens system; the diopters of the two lenses are added algebraically. External lenses (eyeglasses or contact lenses) are used to compensate for optical defects in the eye.

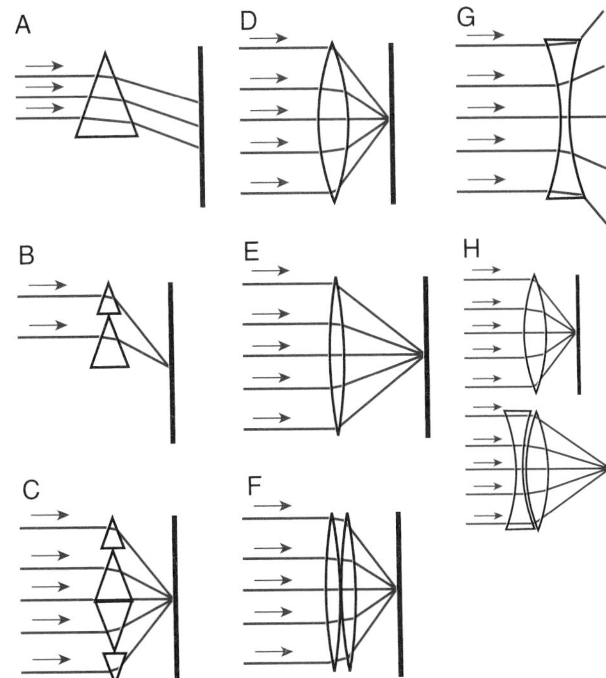

FIGURE 4.9 **How lenses control the refraction of light.** A, A prism bends the path of parallel rays of light. B, The amount of bending varies with the prism shape. C, A series of prisms can bring parallel rays to a point. D, The limiting case of this arrangement is a convex (converging) lens. E, Such a lens with less curvature has a longer focal length. F, Placing two such lenses together produces a shorter focal length. G, A concave (negative) lens causes rays to diverge. H, A negative lens can effectively increase the focal length of a positive lens.

The Structure of the Eye. The human eyeball is a roughly spherical organ consisting of several layers and structures (Fig. 4.10). The outermost of these consists of a tough, white, connective tissue layer, the **sclera,** and a transparent layer, the **cornea.** Six **extraocular muscles** that control the direction of the eyeball insert on the sclera. The next layer is the **vascular coat;** its rear portion, the **choroid,** is pigmented and highly vascular, supplying blood to the outer portions of the retina. The front portion contains the **iris,** a circular smooth muscle structure that forms the **pupil,** the neurally controlled aperture through which light is admitted to the interior of the eye. The iris also gives the eye its characteristic color.

The transparent **lens** is located just behind the iris and is held in place by a radial arrangement of **zonule fibers,** suspensory ligaments that attach it to the **ciliary body,** which contains smooth muscle fibers that regulate the curvature of the lens and, hence, its focal length. The lens is composed of many thin, interlocking layers of fibrous protein and is highly elastic.

Between the cornea and the iris/lens is the **anterior chamber,** a space filled with a thin clear liquid called the **aqueous humor,** similar in composition to cerebrospinal

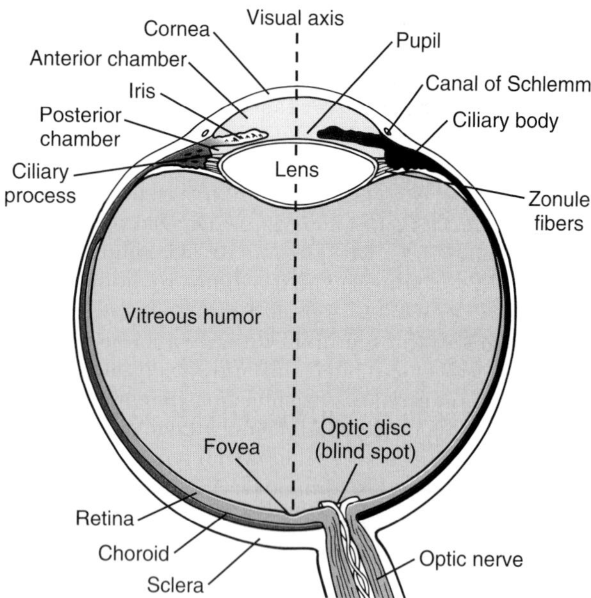

FIGURE 4.10 **The major parts of the human eye.** This is a view from above, showing the relative positions of its optical and structural parts.

fluid. This liquid is continuously secreted by the epithelium of the **ciliary processes,** located behind the iris. As the fluid accumulates, it is drained through the **canal of Schlemm** into the venous circulation. (Drainage of aqueous humor is critical. If too much pressure builds up in the anterior chamber, the internal structures are compressed and **glaucoma,** a condition that can cause blindness, results.) The **posterior chamber** lies behind the iris; along with the anterior chamber, it makes up the **anterior cavity**. The **vitreous humor** (or **vitreous body**), a clear gelatinous substance, fills the large cavity between the rear of the lens and the front surface of the retina. This substance is exchanged much more slowly than the aqueous humor.

The innermost layer of the eyeball is the **retina,** where the optical image is formed. This tissue contains the photoreceptor cells, called **rods** and **cones,** and a complex multilayered network of nerve fibers and cells that function in the early stages of image processing. The rear of the retina is supplied with blood from the choroid, while the front is supplied by the central artery and vein that enter the eyeball with the **optic nerve,** the fiber bundle that connects the retina with structures in the brain. The vascular supply to the front of the retina, which ramifies and spreads over the retinal surface, is visible through the lens and affords a direct view of the microcirculation; this window is useful for diagnostic purposes, even for conditions not directly related to ocular function.

At the optical center of the retina, where the image falls when one is looking straight ahead (i.e., along the **visual axis**), is the **macula lutea,** an area of about 1 mm^2 specialized for very sharp color vision. At the center of the macula is the **fovea centralis,** a depressed region about 0.4 mm in diameter, the **fixation point** of direct vision. Slightly off to the nasal side of the retina is the **optic disc,** where the optic nerve leaves the retina. There are no photoreceptor cells here, resulting in a **blind spot** in the field of vision. However, because the two eyes are mirror images of each other, information from the overlapping visual field of one eye "fills in" the missing part of the image from the other eye.

The Optics of the Eye. The image that falls on the retina is real and inverted, as in a camera. Neural processing restores the upright appearance of the field of view. The image itself can be modified by optical adjustments made by the lens and the iris. Most of the refractive power (about 43 D) is provided by the curvature of the cornea, with the lens providing an additional 13 to 26 D, depending on the focal distance. The muscle of the ciliary body has primarily a parasympathetic innervation, although some sympathetic innervation is present. When it is fully relaxed, the lens is at its flattest and the eye is focused at infinity (actually, at anything more than 6 meters away) (Fig. 4.11A). When the ciliary muscle is fully contracted, the lens is at its most curved and the eye is focused at its nearest point of distinct vision (Fig. 4.11B). This adjustment of the eye for close vision is called **accommodation.** The **near point** of vision for the eye of a young adult is about 10 cm. With age, the lens loses its elasticity and the near point of vision moves farther away, becoming approximately 80 cm at age 60. This condition is called **presbyopia;** supplemental refractive power,

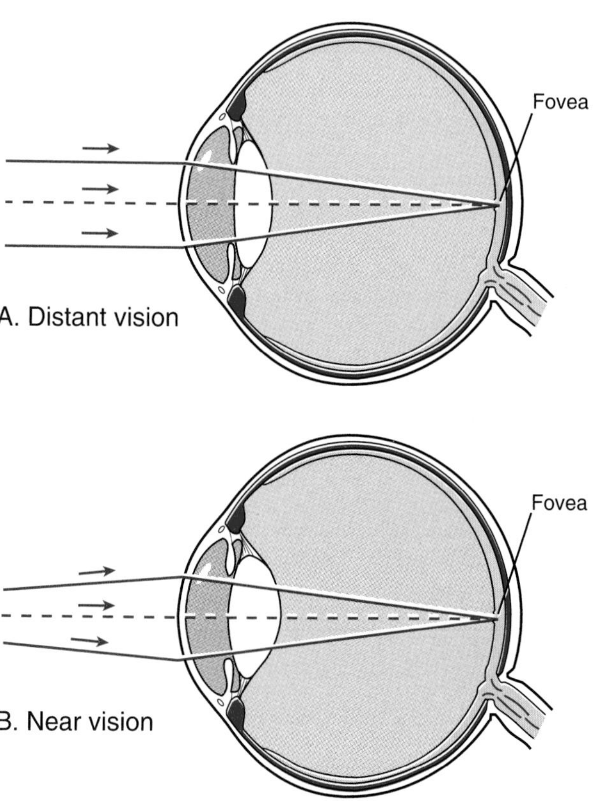

FIGURE 4.11 **The eye as an optical device.** During fixation the center of the image falls on the fovea. **A,** With the lens flattened, parallel rays from a distant object are brought to a sharp focus. **B,** Lens curvature increases with accommodation, and rays from a nearby object are focused.

in the form of external lenses (reading glasses), is required for distinct near vision.

Errors of refraction are common (Fig. 4.12). They can be corrected with external lenses (eyeglasses or contact lenses). Farsightedness or **hyperopia** is caused by an eyeball that is physically too short to focus on distant objects. The natural accommodation mechanism may compensate for distance vision, but the near point will still be too far away; the use of a positive (converging) lens corrects this error. If the eyeball is too long, nearsightedness or **myopia** results. In effect, the converging power of the eye is too great; close vision is clear, but the eye cannot focus on distant objects. A negative (diverging) lens corrects this defect. If the curvature of the cornea is not symmetric, **astigmatism** results. Objects with different orientations in the field of view will have different focal positions. Vertical lines may appear sharp, while horizontal structures are blurred. This condition is corrected with the use of a **cylindrical lens,** which has different radii of curvature at the proper orientations along its surfaces. Normal vision (i.e., the absence of any refractive errors) is termed **emmetropia** (literally, "eye in proper measure").

Normally the lens is completely transparent to visible light. Especially in older adults, there may be a progressive increase in its opacity, to the extent that vision is obscured. This condition, called a **cataract,** is treated by surgical removal of the defective lens. An artificial lens may be implanted in its place, or eyeglasses may be used to replace the refractive power of the lens.

The iris, which has both sympathetic and parasympathetic innervation, controls the diameter of the pupil. It is capable of a 30-fold change in area and in the amount of light admitted to the eye. This change is under complex reflex control, and bright light entering just one eye will cause the appropriate constriction response in both eyes. As with a camera, when the pupil is constricted, less light enters, but the image is focused more sharply because the more poorly focused peripheral rays are cut off.

Eye Movements. The extraocular muscles move the eyes. These six muscles, which originate on the bone of the **orbit** (the eye socket) and insert on the sclera, are arranged in three sets of antagonistic pairs. They are under visually compensated feedback control and produce several types of movement:

- Continuous activation of a small number of motor units produces a small tremor at a rate of 30 to 80 cycles per second. This movement and a slow drift cause the image to be in constant motion on the retina, a necessary condition for proper visual function.
- Larger movements include rapid flicks, called **saccades,** which suddenly change the orientation of the eyeball, and large, slow movements, used in following moving objects.

Organized movements of the eyes include:

- **Fixation,** the training of the eyes on a stationary object
- **Tracking movements,** used to follow the course of a moving target

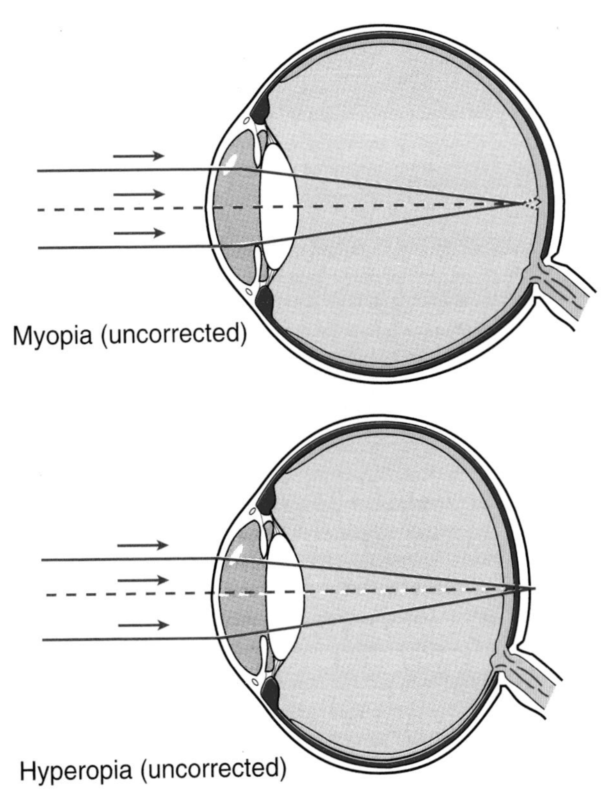

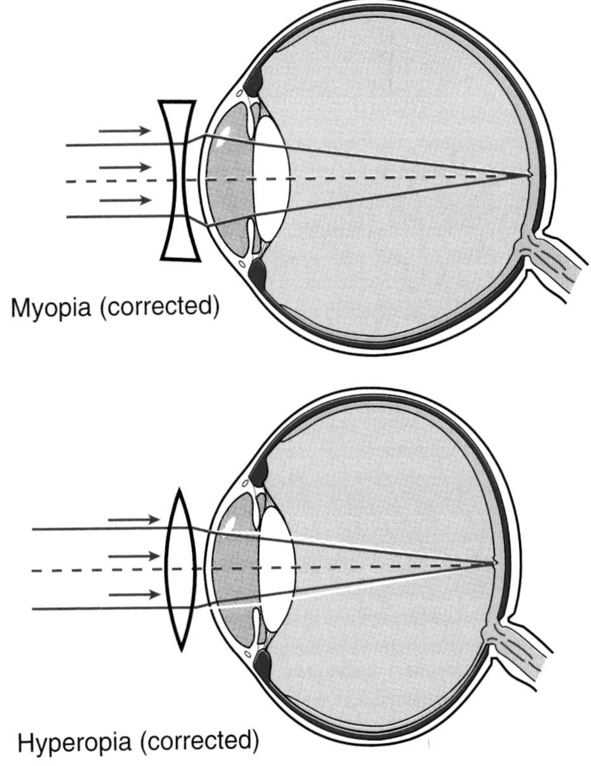

FIGURE 4.12 **The use of external lenses to correct refractive errors.** The external optical corrections change the effective focal length of the natural optical components.

- **Convergence** adjustments, in which both eyes turn inward to fix on near objects
- **Nystagmus,** a series of slow and saccadic movements (part of a vestibular reflex) that serves to keep the retinal image steady during rotation of the head

Because the eyes are separated by some distance, each receives a slightly different image of the same object. This property, **binocular vision,** along with information about the different positions of the two eyes, allows **stereoscopic** vision and its associated **depth perception,** abilities that are largely lost in the case of blindness in one eye. Many abnormalities of eye movement are types of **strabismus** ("squinting"), in which the two eyes do not work together properly. Other defects include **diplopia** (double vision), when the convergence mechanisms are impaired, and **amblyopia,** when one eye assumes improper dominance over the other. Failure to correct this latter condition can lead to loss of visual function in the subordinate eye.

The Retina and Its Photoreceptors.

The retina is a multilayered structure containing the photoreceptor cells and a complex web of several types of nerve cells (Fig. 4.13). There are 10 layers in the retina, but this discussion employs a simpler four-layer scheme: pigment epithelium, photoreceptor layer, neural network layer, and ganglion cell layer. These four layers are discussed in order, beginning with the deepest layer (pigment epithelium) and moving toward the layer nearest to the inner surface of the eye (ganglion cell layer). Note that this is the direction in which visual signal processing takes place, but it is opposite to the path taken by the light entering the retina.

Pigment Epithelium.

The **pigment epithelium** (Fig. 4.13B) consists of cells with high **melanin** content. This opaque material, which also extends between portions of individual rods and cones, prevents the scattering of stray light, thereby greatly sharpening the resolving power of the retina. Its presence ensures that a tiny spot of light (or a tiny portion of an image) will excite only those receptors on which it falls directly. People with albinism lack this pigment and have blurred vision that cannot be corrected effectively with external lenses. The pigment epithelial cells also phagocytose bits of cell membrane that are constantly shed from the outer segments of the photoreceptors.

Photoreceptor Layer.

In the photoreceptor layer (Fig. 4.13C), the rods and cones are packed tightly side by side, with a density of many thousands per square millimeter, depending on the region of the retina. Each eye contains about 125 million rods and 5.5 million cones. Because of the eye's mode of embryologic development, the photoreceptor cells occupy a deep layer of the retina, and light must pass through several overlying layers to reach them. The photoreceptors are divided into two classes. The cones are responsible for **photopic** (daytime) **vision,** which is in color (chromatic), and the rods are responsible for **scotopic** (nighttime) **vision,** which is not in color. Their functions are basically similar, although they have important structural and biochemical differences.

Cones have an **outer segment** that tapers to a point (Fig. 4.14). Three different photopigments are associated with cone cells. The pigments differ in the wavelength of light that optimally excites them. The peak spectral sensitivity for the **red-sensitive pigment** is 560 nm; for the **green-sensitive pigment,** it is about 530 nm; and for the **blue-sensitive pigment,** it is about 420 nm. The corresponding photoreceptors are called red, green, and blue cones, respectively. At wavelengths away from the optimum, the pigments still absorb light but with reduced sensitivity. Because of the interplay between light intensity and wavelength, a retina with only one class of cones would not be able to detect colors unambiguously. The presence of two of the three pigments in each cone removes this uncertainty. Colorblind individuals, who have a genetic lack of one or more of the pigments or lack an associated transduction mechanism, cannot distinguish between the af-

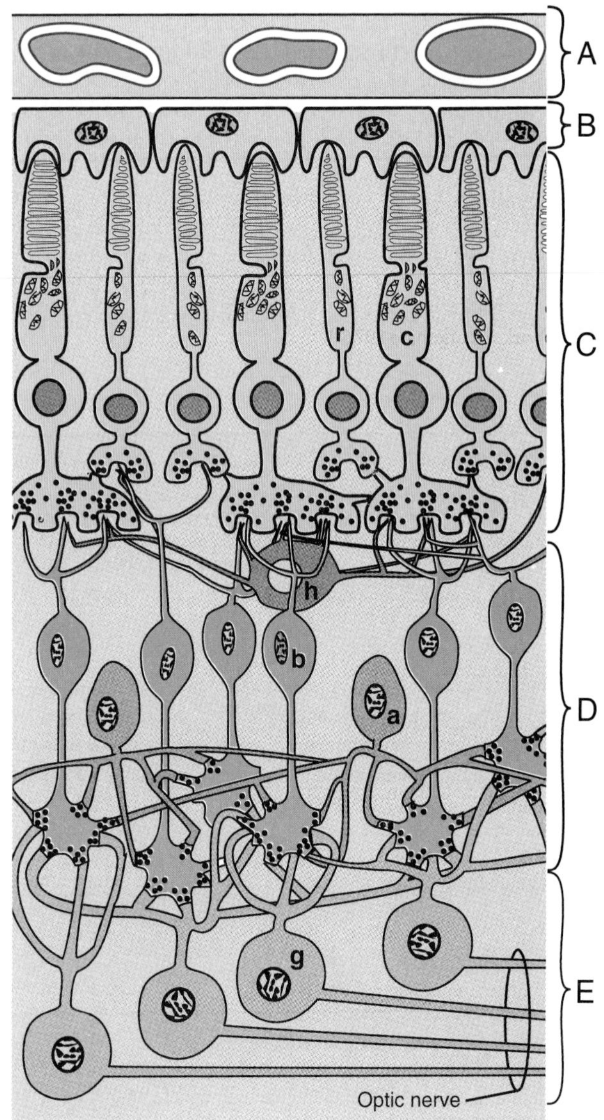

FIGURE 4.13 **Organization of the human retina.** A, Choroid. B, Pigment epithelium. C, Photoreceptor layer. D, Neural network layer. E, Ganglion cell layer. r, rod; c, cone; h, horizontal cell; b, bipolar cell; a, amacrine cell; g, ganglion cell. (See text for details.) (Modified from Dowling JE, Boycott BB. Organization of the primate retina: Electron microscopy. Proc R Soc Lond 1966:166:80–111.)

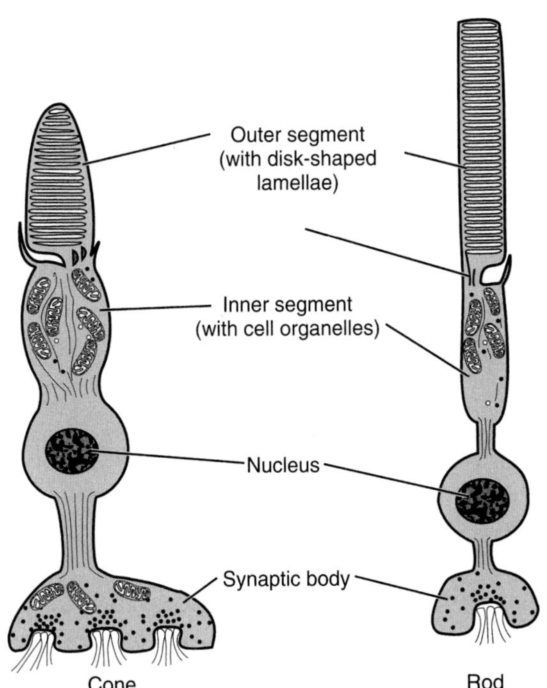

FIGURE 4.14 **Photoreceptors of the human retina.** Cone and rod receptors are compared. (Modified from Davson H, ed. The Eye: Visual Function in Man. 2nd Ed. New York: Academic, 1976.)

fected colors. Loss of a single color system produces **dichromatic vision** and lack of two of the systems causes **monochromatic vision**. If all three are lacking, vision is monochromatic and depends only on the rods.

A **rod cell** is long, slender, and cylindrical and is larger than a cone cell (Fig. 4.14). Its outer segment contains numerous photoreceptor disks composed of cellular membrane in which the molecules of the photopigment **rhodopsin** are embedded. The lamellae near the tip are regularly shed and replaced with new membrane synthesized at the opposite end of the outer segment. The **inner segment,** connected to the outer segment by a modified **cilium,** contains the cell nucleus, many mitochondria that provide energy for the phototransduction process, and other cell organelles. At the base of the cell is a **synaptic body** that makes contact with one or more bipolar nerve cells and liberates a transmitter substance in response to changing light levels.

The visual pigments of the photoreceptor cells convert light to a nerve signal. This process is best understood as it occurs in rod cells. In the dark, the pigment rhodopsin (or visual purple) consists of a light-trapping **chromophore** called **scotopsin** that is chemically conjugated with **11-*cis*-retinal**, the aldehyde form of vitamin A_1. When struck by light, rhodopsin undergoes a series of rapid chemical transitions, with the final intermediate form **metarhodopsin II** providing the critical link between this reaction series and the electrical response. The end-products of the light-induced transformation are the original scotopsin and an all-*trans* form of retinal, now dissociated from each other. Under conditions of both light and dark, the all-*trans* form of retinal is isomerized back to the 11-*cis* form, and the rhodopsin is reconstituted. All of these reactions take place in the highly folded membranes comprising the outer segment of the rod cell.

The biochemical process of visual signal transduction is shown in Figure 4.15. The coupling of the light-induced reactions and the electrical response involves the activation of **transducin,** a G protein; the associated exchange of GTP for GDP activates a **phosphodiesterase.** This, in turn, catalyzes the breakdown of cyclic GMP (cGMP) to 5'-GMP. When cellular cGMP levels are high (as in the dark), membrane sodium channels are kept open, and the cell is relatively depolarized. Under these conditions, there is a tonic release of neurotransmitter from the synaptic body of the rod cell. A decrease in the level of cGMP as a result of light-induced reactions causes the cell to close its sodium channels and hyperpolarize, thus, reducing the release of neurotransmitter; this change is the signal that is further processed by the nerve cells of the retina to form the final response in the optic nerve. An active sodium pump main-

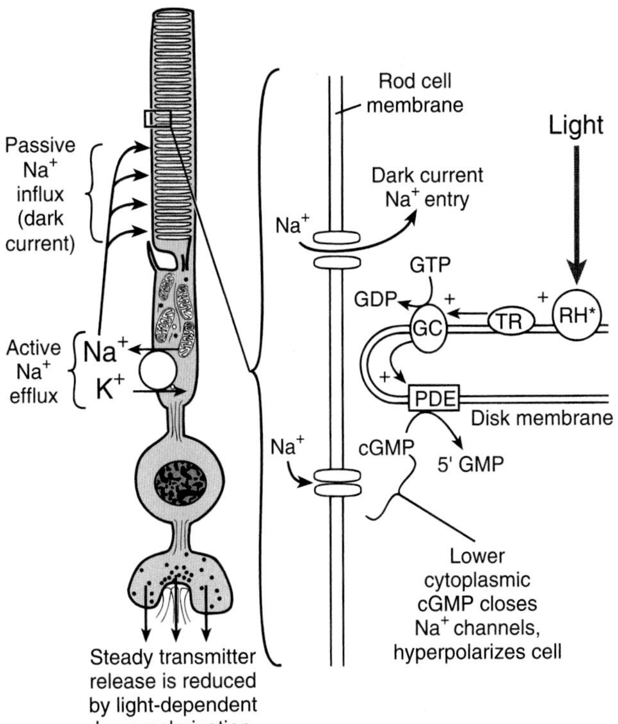

FIGURE 4.15 **The biochemical process of visual signal transduction.** Left: An active Na^+/K^+ pump maintains the ionic balance of a rod cell, while Na^+ enters passively through channels in the plasma membrane, causing a maintained depolarization and a dark current under conditions of no light. Right: The amplifying cascade of reactions (which take place in the disk membrane of a photoreceptor) allows a single activated rhodopsin molecule to control the hydrolysis of 500,000 cGMP molecules. (See text for details of the reaction sequence.) In the presence of light, the reactions lead to a depletion of cGMP, resulting in the closing of cell membrane Na^+ channels and the production of a hyperpolarizing generator potential. The release of neurotransmitter decreases during stimulation by light. RH*, activated rhodopsin; TR, transducin; GC, guanylyl cyclase; PDE, phosphodiesterase.

tains the cellular concentration at proper levels. A large amplification of the light response takes place during the coupling steps; one activated rhodopsin molecule will activate approximately 500 transducins, each of which activates the hydrolysis of several thousand cGMP molecules. Under proper conditions, a rod cell can respond to a single photon striking the outer segment. The processes in cone cells are similar, although there are three different opsins (with different spectral sensitivities) and the specific transduction mechanism is also different. The overall sensitivity of the transduction process is also lower.

In the light, much rhodopsin is in its unconjugated form, and the sensitivity of the rod cell is relatively low. During the process of **dark adaptation,** which takes about 40 minutes to complete, the stores of rhodopsin are gradually built up, with a consequent increase in sensitivity (by as much as 25,000 times). Cone cells adapt more quickly than rods, but their final sensitivity is much lower. The reverse process, **light adaptation,** takes about 5 minutes.

Neural Network Layer. **Bipolar cells, horizontal cells,** and **amacrine cells** comprise the **neural network layer.** These cells together are responsible for considerable initial processing of visual information. Because the distances between neurons here are so small, most cellular communication involves the **electrotonic spread** of cell potentials, rather than propagated action potentials. Light stimulation of the photoreceptors produces hyperpolarization that is transmitted to the bipolar cells. Some of these cells respond with a depolarization that is excitatory to the ganglion cells, whereas other cells respond with a hyperpolarization that is inhibitory. The horizontal cells also receive input from rod and cone cells but spread information laterally, causing inhibition of the bipolar cells on which they synapse. Another important aspect of retinal processing is **lateral inhibition.** A strongly stimulated receptor cell can, via lateral inhibitory pathways, inhibit the response of neighboring cells that are less well-illuminated. This has the effect of increasing the apparent contrast at the edge of an image. Amacrine cells also send information laterally but synapse on ganglion cells.

Ganglion Cell Layer. In the **ganglion cell layer** (Fig. 4.13E) the results of retinal processing are finally integrated by the **ganglion cells,** whose axons form the **optic nerve.** These cells are tonically active, sending action potentials into the optic nerve at an average rate of five per second, even when unstimulated. Input from other cells converging on the ganglion cells modifies this rate up or down.

Many kinds of information regarding color, brightness, contrast, and so on are passed along the optic nerve. The output of individual photoreceptor cells is **convergent** on the ganglion cells. In keeping with their role in visual acuity, relatively few cone cells converge on a ganglion cell, especially in the fovea, where the ratio is nearly 1:1. Rod cells, however, are highly convergent, with as many as 300 rods converging on a single ganglion cell. While this mechanism reduces the sharpness of an image, it allows for a great increase in light sensitivity.

Central Projections of the Retina. The optic nerves, each carrying about 1 million fibers from each retina, enter the rear of the orbit and pass to the underside of the brain to the **optic chiasma,** where about half the fibers from each eye "cross over" to the other side. Fibers from the temporal side of the retina do not cross the midline, but travel in the **optic tract** on the same side of the brain. Fibers originating from the nasal side of the retina cross the optic chiasma and travel in the optic tract to the opposite side of the brain. Hence, information from right and left visual fields is transmitted to opposite sides of the brain. The divided output goes through the optic tract to the paired **lateral geniculate bodies** (part of the thalamus) and then via the **geniculocalcarine tract** (or **optic radiation**) to the **visual cortex** in the **occipital lobe** of the brain (Fig. 4.16). Specific portions of each retina are mapped to specific areas of the cortex; the foveal and macular regions have the greatest representation, while the peripheral areas have the least. Mechanisms in the visual cortex detect and integrate visual information, such as shape, contrast, line, and intensity, into a coherent visual perception.

Information from the optic nerves is also sent to the **suprachiasmatic nucleus** of the hypothalamus, where it participates in the regulation of circadian rhythms; the **pretectal nuclei,** concerned with the control of visual fixation and pupillary reflexes; and the **superior colliculus,** which

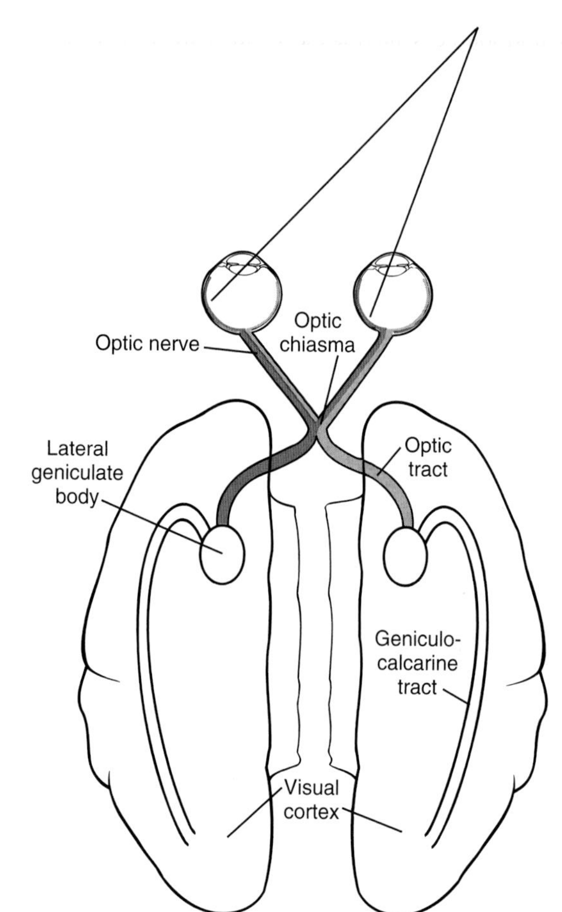

FIGURE 4.16 **The CNS pathway for visual information.** Fibers from the right visual field will stimulate the left half of each retina, and nerve impulses will be transmitted to the left hemisphere.

coordinates simultaneous bilateral eye movements, such as tracking and convergence.

The Ear Is Sensor for Hearing and Equilibrium

The human ear has a degree of complexity probably as great as that of the eye. Understanding our sense of hearing requires familiarity with the physics of sound and its interactions with the biological structures involved in hearing.

The Nature of Sound. **Sound waves** are mechanical disturbances that travel through an elastic medium (usually air or water). A sound wave is produced by a mechanically vibrating structure that alternately compresses and rarefies the air (or water) in contact with it. For example, as a loudspeaker cone moves forward, air molecules in its path are forced closer together; this is called **compression** or **condensation.** As the cone moves back, the space between the disturbed molecules is increased; this is known as **rarefaction**. The compression (or rarefaction) of air molecules in one region causes a similar compression in adjacent regions. Continuation of this process causes the disturbance (the sound wave) to spread away from the source.

The speed at which the sound wave travels is determined by the elasticity of the air (the tendency of the molecules to spring back to their original positions). Assuming the sound source is moving back and forth at a constant rate of alternation (i.e., at a constant **frequency**), a propagated compression wave will pass a given point once for every cycle of the source. Because the propagation speed is constant in a given medium, the compression waves are closer together at higher frequencies; that is, more of them pass the given point every second.

The distance between the compression peaks is called the **wavelength** of the sound, and it is inversely related to the frequency. A tone of 1,000 cycles per second, traveling through the air, has a wavelength of approximately 34 cm, while a tone of 2,000 cycles per second has a wavelength of 17 cm. Both waves, however, travel at the same speed through the air. Because the elastic forces in water are greater than those in air, the speed of sound in water is about 4 times as great, and the wavelength is correspondingly increased. Since the wavelength depends on the elasticity of the medium (which varies according to temperature and pressure), it is more convenient to identify sound waves by their frequency. Sound frequency is usually expressed in units of Hertz (Hz or cycles per second).

Another fundamental characteristic of a sound wave is its intensity or amplitude. This may be thought of as the relative amount of compression or rarefaction present as the wave is produced and propagated; it is related to the amount of energy contained in the wave. Usually the intensity is expressed in terms of **sound pressure**, the pressure the compressions and rarefactions exert on a surface of known area (expressed in dynes per square centimeter). Because the human ear is sensitive to sounds over a millionfold range of sound pressure levels, it is convenient to express the intensity of sound as the logarithm of a ratio referenced to the **absolute threshold of hearing** for a tone of 1,000 Hz. This reference level has a value of 0.0002 dyne/cm^2, and the scale for the measurements is the **decibel (dB)** scale. In the expression

$$dB = 20 \log (P/P_{ref}), \quad (1)$$

the sound pressure (P) is referred to the absolute reference pressure (P_{ref}). For a sound that is 10 times greater than the reference, the expression becomes

$$dB = 20 \log (0.002 / 0.0002) = 20. \quad (2)$$

Thus, any two sounds having a tenfold difference in intensity have a decibel difference of 20; a 100-fold difference would mean a 40 dB difference and a 1,000-fold difference would be 60 dB. Usually the reference value is assumed to be constant and standard, and it is not expressed when measurements are reported.

Table 4.1 lists the sound pressure levels and the decibel levels for some common sounds. The total range of 140 dB shown in the table expresses a relative range of 10 millionfold. Adaptation and compression processes in the human auditory system allow encoding of most of this wide range into biologically useful information.

Sinusoidal sound waves contain all of their energy at one frequency and are perceived as **pure tones**. Complex sound waves, such as those in speech or music, consist of the addition of several simpler waveforms of different frequencies and amplitudes. The human ear is capable of hearing sounds over the range of 20 to 16,000 Hz, although the upper limit decreases with age. Auditory sensitivity varies with the frequency of the sound; we hear sounds most readily in the range of 1,000 to 4,000 Hz and at a sound pressure level of around 60 dB. Not surprisingly, this is the frequency and intensity range of human vocalization. The ear's sensitivity is also affected by **masking**: In the presence of background sounds or noise, the auditory threshold for a given tone rises. This may be due to refractoriness induced by the masking sound, which would reduce the number of available receptor cells.

The Outer Ear. An overall view of the human ear is shown in Figure 4.17. The **pinna**, the visible portion of the outer ear, is not critical to hearing in humans, although it does

TABLE 4.1 The Relative Pressures of Some Common Sounds

Pressure (dynes/cm^2)	Sound Pressure Level (dB)	Sound Source	Relative Pressure
0.0002	0	Absolute threshold	1
0.002	+ 20	Faint whisper	10
0.02	+ 40	Quiet office	100
0.2	+ 60	Conversation	1,00
2	+ 80	City bus	10,000
20	+100	Subway train	100,000
200	+120	Loud thunder	1,000,000
2,000	+140	Pain and damage	10,000,000

Modified from Gulick WL, Gescheider GA, Frisina RD. Hearing: Physiological Acoustics, Neural Coding, and Psychoacoustics. New York: Oxford University Press, 1989;51, Table 2.2.

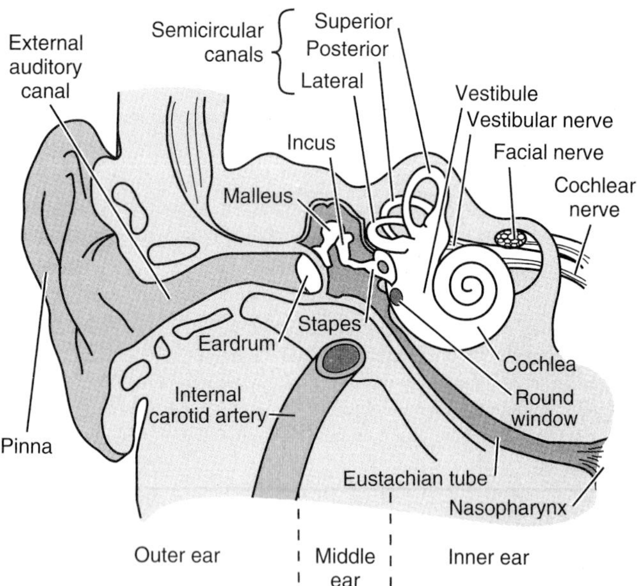

FIGURE 4.17 **The overall structure of the human ear.** The structures of the middle and inner ear are encased in the temporal bone of the skull.

slightly emphasize frequencies in the range of 1,500 to 7,000 Hz and aids in the localization of sources of sound. The **external auditory canal** extends inward through the **temporal bone.** Wax-secreting glands line the canal, and its inner end is sealed by the **tympanic membrane** or **eardrum,** a thin, oval, slightly conical, flexible membrane that is anchored around its edges to a ring of bone. An incoming pressure wave traveling down the external auditory canal causes the eardrum to vibrate back and forth in step with the compressions and rarefactions of the sound wave. This is the first mechanical step in the transduction of sound. The overall acoustic effect of the outer ear structures is to produce an amplification of 10 to 15 dB in the frequency range broadly centered around 3,000 Hz.

The Middle Ear. The next portion of the auditory system is an air-filled cavity (volume about 2 mL) in the mastoid region of the temporal bone. The middle ear is connected to the pharynx by the **eustachian tube.** The tube opens briefly during swallowing, allowing equalization of the pressures on either side of the eardrum. During rapid external pressure changes (such as in an elevator ride or during takeoff or descent in an airplane), the unequal forces displace the eardrum; such physical deformation may cause discomfort or pain and, by restricting the motion of the tympanic membrane, may impair hearing. Blockages of the eustachian tube or fluid accumulation in the middle ear (as a result of an infection) can also lead to difficulties with hearing.

Bridging the gap between the tympanic membrane and the inner ear is a chain of three very small bones, the **ossicles** (Fig. 4.18). The **malleus (hammer)** is attached to the eardrum in such a way that the back-and-forth movement of the eardrum causes a rocking movement of the malleus. The **incus (anvil)** connects the head of the malleus to the third bone, the **stapes (stirrup).** This last bone, through its oval **footplate,** connects to the **oval window** of the inner ear and is anchored there by an annular ligament.

Four separate **suspensory ligaments** hold the ossicles in position in the middle ear cavity. The superior and lateral ligaments lie roughly in the plane of the ossicular chain and anchor the head and shaft of the malleus. The anterior ligament attaches the head of the malleus to the anterior wall of the middle ear cavity, and the posterior ligament runs from the head of the incus to the posterior wall of the cavity. The suspensory ligaments allow the ossicles sufficient freedom to function as a lever system to transmit the vibrations of the tympanic membrane to the oval window. This mechanism is especially important because, although the eardrum is suspended in air, the oval window seals off a fluid-filled chamber. Transmission of sound from air to liquid is inefficient; if sound waves were to strike the oval window directly, 99.9% of the energy would be reflected away and lost.

Two mechanisms work to compensate for this loss. Although it varies with frequency, the ossicular chain has a lever ratio of about 1.3:1, producing a slight gain in force. In addition, the relatively large area of the tympanic membrane is coupled to the smaller area of the oval window (approximately a 17:1 ratio). These conditions result in a pressure gain of around 25 dB, largely compensating for the potential loss. Although the efficiency depends on the frequency, approximately 60% of the sound energy that strikes the eardrum is transmitted to the oval window.

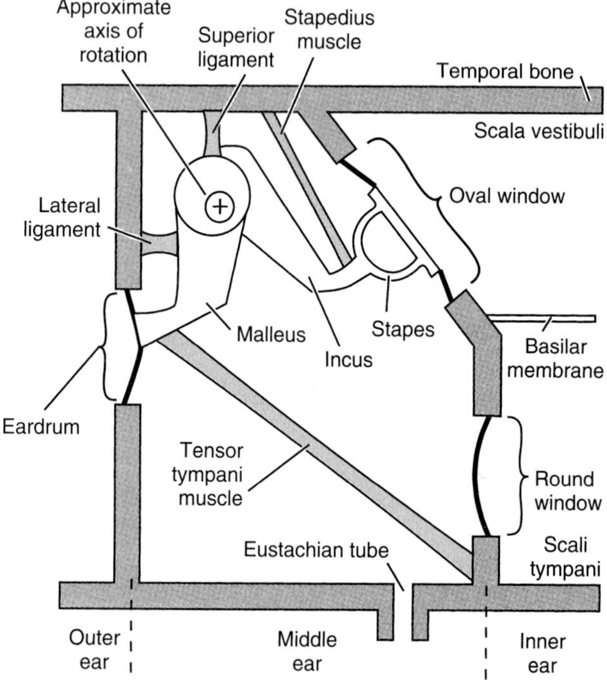

FIGURE 4.18 **A model of the middle ear.** Vibrations from the eardrum are transmitted by the lever system formed by the ossicular chain to the oval window of the scala vestibuli. The anterior and posterior ligaments, part of the suspensory system for the ossicles, are not shown. The combination of the four suspensory ligaments produces a virtual pivot point (marked by a cross); its position varies with the frequency and intensity of the sound. The stapedius and tensor tympani muscles modify the lever function of the ossicular chain.

Sound transmission through the middle ear is also affected by the action of two small muscles that attach to the ossicular chain and help hold the bones in position and modify their function (see Fig. 4.18). The **tensor tympani muscle** inserts on the malleus (near the center of the eardrum), passes diagonally through the middle ear cavity, and enters the **tensor canal,** in which it is anchored. Contraction of this muscle limits the vibration amplitude of the eardrum and makes sound transmission less efficient. The **stapedius muscle** attaches to the stapes near its connection to the incus and runs posteriorly to the mastoid bone. Its contraction changes the axis of oscillation of the ossicular chain and causes dissipation of excess movement before it reaches the oval window. These muscles are activated by a reflex (simultaneously in both ears) in response to moderate and loud sounds; they act to reduce the transmission of sound to the inner ear and, thus, to protect its delicate structures. Because this **acoustic reflex** requires up to 150 msec to operate (depending on the loudness of the stimulus), it cannot provide protection from sharp or sudden bursts of sound.

The process of sound transmission can bypass the ossicular chain entirely. If a vibrating object, such as a tuning fork, is placed against a bone of the skull (typically the mastoid), the vibrations are transmitted mechanically to the fluid of the inner ear, where the normal processes act to complete the hearing process. Bone conduction is used as a means of diagnosing hearing disorders that may arise because of lesions in the ossicular chain. Some hearing aids employ bone conduction to overcome such deficits.

The Inner Ear. The actual process of sound transduction takes place in the inner ear, where the sensory receptors and their neural connections are located. The relationship between its structure and function is a close and complex one. The following discussion includes the most significant aspects of this relationship.

Overall Structure. The auditory structures are located in the **cochlea** (Fig. 4.19), part of a cavity in the temporal bone called the bony labyrinth. The cochlea (meaning "snail shell") is a fluid-filled spiral tube that arises from a

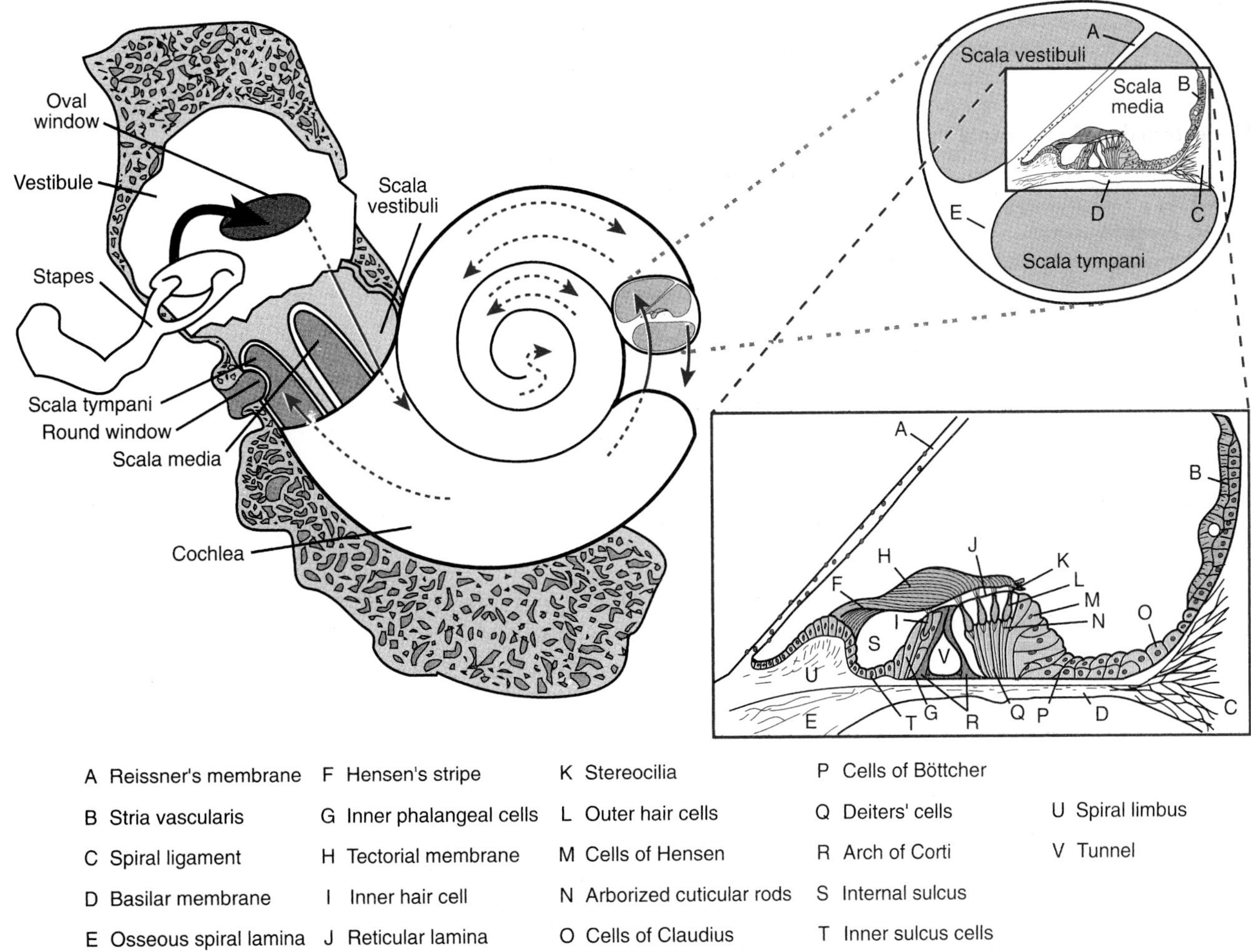

FIGURE 4.19 **The cochlea and the organ of Corti.** Left: An overview of the membranous labyrinth of the cochlea. Upper right: A cross section through one turn of the cochlea, showing the canals and membranes that make up the structures involved in the final processes of auditory sensation. Lower right: An enlargement of a cross section of the organ of Corti, showing the relationships among the hair cells and the membranes. (Modified from Gulick WL, Gescheider GA, Frisina RD. Hearing: Physiological Acoustics, Neural Coding, and Psychoacoustics. New York: Oxford University Press, 1989.)

cavity called the **vestibule,** with which the organs of balance also communicate. In the human ear, the cochlea is about 35 mm long and makes about $2^3/_4$ turns. Together with the vestibule it contains a total fluid volume of 0.1 mL. It is partitioned longitudinally into three divisions (canals) called the **scala vestibuli** (into which the oval window opens), the **scala tympani** (sealed off from the middle ear by the **round window**), and the **scala media** (in which the sensory cells are located). Arising from the bony center axis of the spiral (the **modiolus**) is a winding shelf called the **osseous spiral lamina**; opposite it on the outer wall of the spiral is the **spiral ligament**, and connecting these two structures is a highly flexible connective tissue sheet, the **basilar membrane,** that runs for almost the entire length of the cochlea. The basilar membrane separates the scala tympani (below) from the scala media (above). The **hair cells,** which are the actual sensory receptors, are located on the upper surface of the basilar membrane. They are called hair cells because each has a bundle of hair-like **cilia** at the end that projects away from the basilar membrane.

Reissner's membrane, a delicate sheet only two cell layers thick, divides the scala media (below) from the scala vestibuli (above) (see Fig. 4.19). The scala vestibuli communicates with the scala tympani at the apical (distal) end of the cochlea via the **helicotrema,** a small opening where a portion of the basilar membrane is missing. The scala vestibuli and scala tympani are filled with **perilymph,** a fluid high in sodium and low in potassium. The scala media contains **endolymph,** a fluid high in potassium and low in sodium. The endolymph is secreted by the **stria vascularis,** a layer of fibrous vascular tissue along the outer wall of the scala media. Because the cochlea is filled with incompressible fluid and is encased in hard bone, pressure changes caused by the in-and-out motion at the oval window (driven by the stapes) are relieved by an out-and-in motion of the flexible round window membrane.

Sensory Structures. The **organ of Corti** is formed by structures located on the upper surface of the basilar membrane and runs the length of the scala media (see Fig. 4.19). It contains one row of some 3,000 **inner hair cells**; the **arch of Corti** and other specialized supporting cells separate the inner hair cells from the three or four rows of **outer hair cells** (about 12,000) located on the stria vascularis side. The rows of inner and outer hair cells are inclined slightly toward each other and covered by the **tectorial membrane,** which arises from the **spiral limbus,** a projection on the upper surface of the osseous spiral lamina.

Nerve fibers from cell bodies located in the **spiral ganglia** form **radial bundles** on their way to synapse with the inner hair cells. Each nerve fiber makes synaptic connection with only one hair cell, but each hair cell is served by 8 to 30 fibers. While the inner hair cells comprise only 20% of the hair cell population, they receive 95% of the afferent fibers. In contrast, many outer hair cells are each served by a single external spiral nerve fiber. The collected afferent fibers are bundled in the **cochlear nerve,** which exits the inner ear via the **internal auditory meatus.** Some efferent fibers also innervate the cochlea. They may serve to enhance pitch discrimination and the ability to distinguish sounds in the presence of noise. Recent evidence suggests that efferent fibers to the outer hair cells may cause them to shorten (contract), altering the mechanical properties of the cochlea.

The Hair Cells. The hair cells of the inner and the outer rows are similar anatomically. Both sets are supported and anchored to the basilar membrane by **Deiters' cells** and extend upward into the scala media toward the tectorial membrane. Extensions of the outer hair cells actually touch the tectorial membrane, while those of the inner hair cells appear to stop just short of contact. The hair cells make synaptic contact with afferent neurons that run through channels between Deiters' cells. A chemical transmitter of unknown identity is contained in synaptic vesicles near the base of the hair cells; as in other synaptic systems, the entry of calcium ions (associated with cell membrane depolarization) is necessary for the migration and fusion of the synaptic vesicles with the cell membrane prior to transmitter release.

At the apical end of each inner hair cell is a projecting bundle of about 50 **stereocilia,** rod-like structures packed in three, parallel, slightly curved rows. Minute strands link the free ends of the stereocilia together, so the bundle tends to move as a unit. The height of the individual stereocilia increases toward the outer edge of the cell (toward the stria vascularis), giving a sloping appearance to the bundle. Along the cochlea, the inner hair cells remain constant in size, while the stereocilia increase in height from about 4 μm at the basal end to 7 μm at the apical end. The outer hair cells are more elongated than the inner cells, and their size increases along the cochlea from base to apex. Their stereocilia (about 100 per hair cell) are also arranged in three rows that form an exaggerated *W* figure. The height of the stereocilia also increases along the length of the cochlea, and they are embedded in the tectorial membrane. The stereocilia of both types of hair cells extend from the **cuticular plate** at the apex of the cell. The diameter of an individual stereocilium is uniform (about 0.2 μm) except at the base, where it decreases significantly. Each stereocilium contains cross-linked and closely packed **actin filaments,** and, near the tip, is a cation-selective **transduction channel.**

Mechanical transduction in hair cells is shown in Figure 4.20. When a hair bundle is deflected slightly (the threshold is less than 0.5 nm) toward the stria vascularis, minute mechanical forces open the transduction channels, and cations (mostly potassium) enter the cells. The resulting **depolarization,** roughly proportional to the deflection, causes the release of transmitter molecules, generating afferent nerve action potentials. Approximately 15% of the transduction channels are open in the absence of any deflection, and bending in the direction of the modiolus of the cochlea results in hyperpolarization, increasing the range of motion that can be sensed. Hair cells are quite insensitive to movements of the stereocilia bundles at right angles to their preferred direction.

The response time of hair cells is remarkable; they can detect repetitive motions of up to 100,000 times per second. They can, therefore, provide information throughout the course of a single cycle of a sound wave. Such rapid response is also necessary for the accurate localization of sound sources. When a sound comes from directly in front of a listener, the waves arrive simultaneously at both ears. If the sound originates off to one side, it reaches one ear

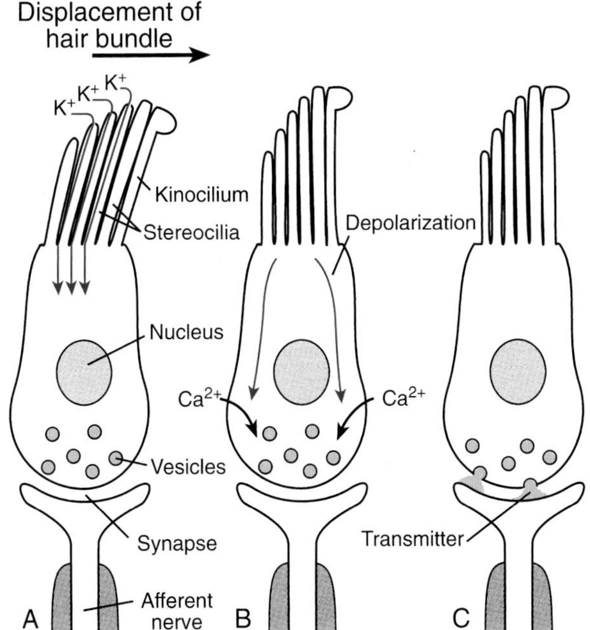

FIGURE 4.20 Mechanical transduction in the hair cells of the ear. A, Deflection of the stereocilia opens apical K^+ channels. B, The resulting depolarization allows the entry of Ca^{2+} at the basal end of the cell. This causes the release of the neurotransmitter, thereby exciting the afferent nerve. (Adapted from Hudspeth AJ. The hair cells of the inner ear. Sci Am 1983;248(1):54–64.)

sooner than the other and is slightly more intense at the nearer ear. The difference in arrival time is on the order of tenths of a millisecond, and the rapid response of the hair cells allows them to provide temporal input to the auditory cortex. The timing and intensity information are processed in the auditory cortex into an accurate perception of the location of the sound source.

Integrated Function of the Organ of Corti. The actual transduction of sound requires an interaction among the tectorial membrane, the arches of Corti, the hair cells, and the basilar membrane. When a sound wave is transmitted to the oval window by the ossicular chain, a pressure wave travels up the scala vestibuli and down the scala tympani (Fig. 4.21). The canals of the cochlea, being encased in bone, are not deformed, and movements of the round window allow the small volume change needed for the transmission of the pressure wave. Resulting eddy currents in the cochlear fluids produce an undulating distortion in the basilar membrane. Because the stiffness and width of the membrane vary with its length (it is wider and less stiff at the apex than at the base), the membrane deformation takes the form of a "**traveling wave,**" which has its maximal amplitude at a position along the membrane corresponding to the particular frequency of the sound wave (Fig. 4.22). Low-frequency sounds cause a maximal displacement of the membrane near its apical end (near the helicotrema), whereas high-frequency sounds produce their maximal effect at the basal end (near the oval window). As the basilar membrane moves, the arches of Corti transmit the movement to the tectorial membrane, the stereocilia of the outer hair cells (embedded in the tectorial membrane) are subjected to lateral shearing forces that stimulate the cells, and action potentials arise in the afferent neurons.

Because of the tuning effect of the basilar membrane, only hair cells located at a particular place along the membrane are maximally stimulated by a given frequency (pitch). This localization is the essence of the **place theory** of pitch discrimination, and the mapping of specific tones (pitches) to specific areas is called **tonotopic organization.** As the signals from the cochlea ascend through the complex pathways of the auditory system in the brain, the tonotopic organization of the neural elements is at least partially preserved, and pitch can be spatially localized throughout the system. The sense of pitch is further sharpened by the resonant characteristics of the different-length stereocilia along the length of the cochlea and by the frequency-response selectivity of neurons in the auditory pathway. The cochlea acts as both a transducer for sound waves and a frequency analyzer that sorts out the different pitches so they

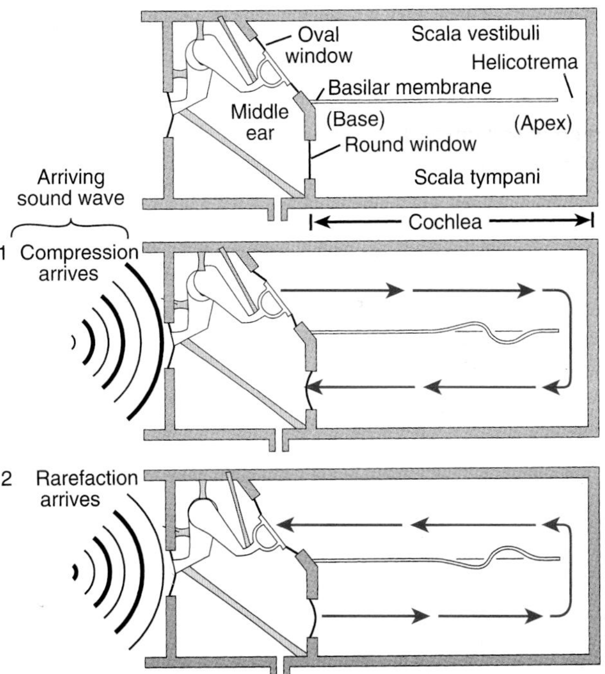

FIGURE 4.21 The mechanics of the cochlea, showing the action of the structures responsible for pitch discrimination (with only the basilar membrane of the organ of Corti shown). When the compression phase of a sound wave arrives at the eardrum, the ossicles transmit it to the oval window, which is pushed inward. A pressure wave travels up the scala vestibuli and (via the helicotrema) down the scala tympani. To relieve the pressure, the round window membrane bulges outward. Associated with the pressure waves are small eddy currents that cause a traveling wave of displacement to move along the basilar membrane from base to apex. The arrival of the next rarefaction phase reverses these processes. The frequency of the sound wave, interacting with the differences in the mass, width, and stiffness of the basilar membrane along its length, determines the characteristic position at which the membrane displacement is maximal. This localization is further detailed in Figure 4.22.

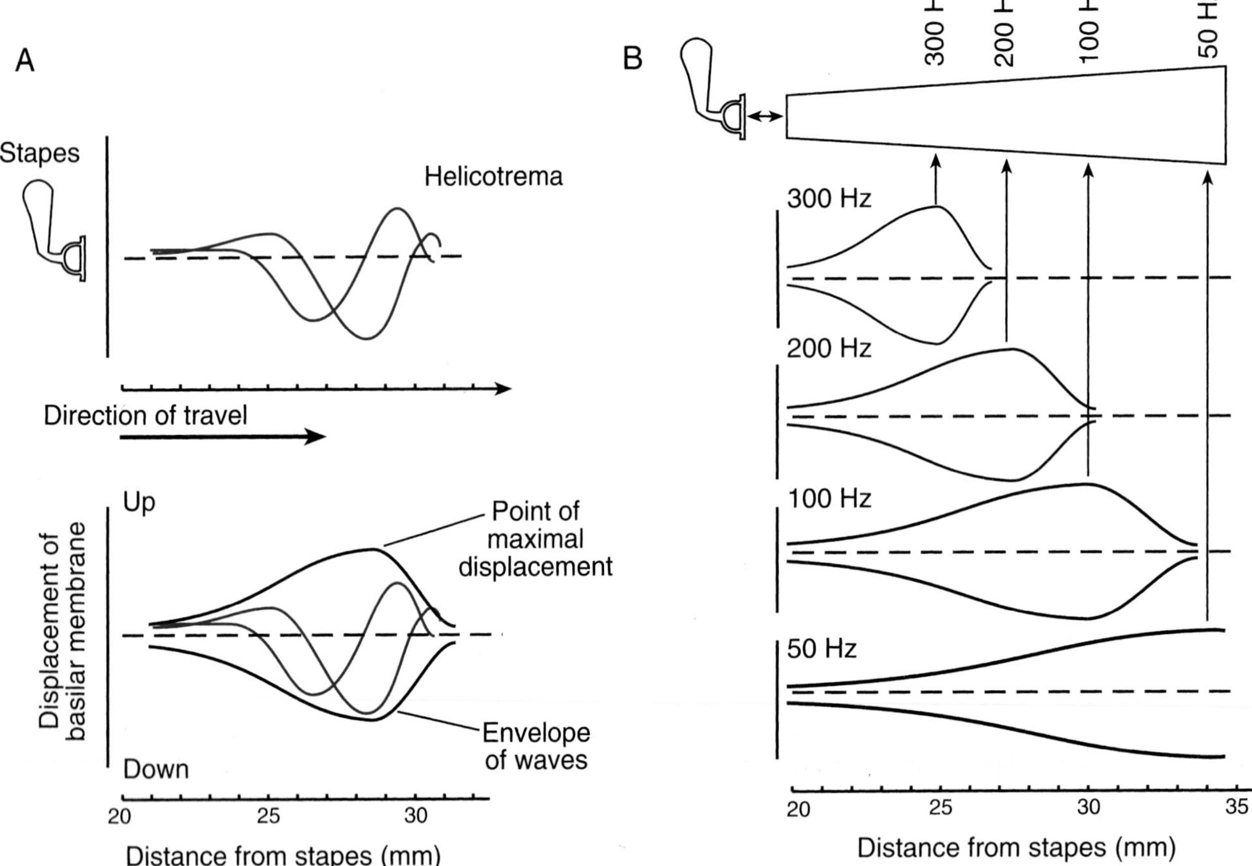

FIGURE 4.22 **Membrane localization of different frequencies.** **A**, The upper portion shows a traveling wave of displacement along the basilar membrane at two instants. Over time, the peak excursions of many such waves form an envelope of displacement with a maximal value at about 28 mm from the stapes (lower portion); at this position, its stimulating effect on the hair cells will be most intense. **B**, The effect of frequency. Lower frequencies produce a maximal effect at the apex of the basilar membrane, where it is the widest and least stiff. Pure tones affect a single location; complex tones affect multiple loci. (Modified from von Békésy G. Experiments in Hearing. New York: McGraw-Hill, 1960.)

can be separately distinguished. In the midrange of hearing (around 1,000 Hz), the human auditory system can sense a difference in frequency of as little as 3 Hz. The tonotopic organization of the basilar membrane has facilitated the invention of prosthetic devices whose aim is to provide some replacement of auditory function to people suffering from deafness that arises from severe malfunction of the middle or inner ear (see Clinical Focus Box 4.1).

Central Auditory Pathways. Nerve fibers from the cochlea enter the spiral ganglion of the organ of Corti; from there, fibers are sent to the **dorsal** and **ventral cochlear nuclei.** The complex pathway that finally ends at the **auditory cortex** in the superior portion of the **temporal lobe** of the brain involves several sets of synapses and considerable crossing over and intermediate processing. As with the eye, there is a spatial correlation between cells in the sensory organ and specific locations in the primary auditory cortex. In this case, the representation is called a **tonotopic map,** with different pitches being represented by different locations, even though the firing rates of the cells no longer correspond to the frequency of sound originally presented to the inner ear.

The Function of the Vestibular Apparatus. The ear also has important nonauditory sensory functions. The sensory receptors that allow us to maintain our equilibrium and balance are located in the **vestibular apparatus,** which consists (on each side of the head) of three **semicircular canals** and two **otolithic organs,** the **utricle** and the **saccule** (Fig. 4.23). These structures are located in the bony labyrinth of the temporal bone and are sometimes called the **membranous labyrinth.** As with hearing, the basic sensing elements are hair cells.

The semicircular canals, hoop-like tubular membranous structures, sense rotary acceleration and motion. Their interior is continuous with the scala media and is filled with endolymph; on the outside, they are bathed by perilymph. The three canals on each side lie in three mutually perpendicular planes. With the head tipped forward by about 30 degrees, the **horizontal** (**lateral**) **canal** lies in the horizontal plane. At right angles to this are the planes of the **anterior**

CLINICAL FOCUS BOX 4.1

Cochlear Implants

Disorders of hearing are broadly divided into the categories of **conductive hearing loss,** related to structures of the outer and middle ear; **sensorineural hearing loss** ("nerve deafness"), dealing with the mechanisms of the cochlea and peripheral nerves; and **central hearing loss,** concerning processes that lie in higher portions of the central nervous system.

Damage to the cochlea, especially to the hair cells of the organ of Corti, produces sensorineural hearing loss by several means. Prolonged exposure to loud occupational or recreational noises can lead to hair cell damage, including mechanical disruption of the stereocilia. Such damage is localized in the outer hair cells along the basilar membrane at a position related to the pitch of the sound that produced it. Antibiotics such as streptomycin and certain diuretics can cause rapid and irreversible damage to hair cells similar to that caused by noise, but it occurs over a broad range of frequencies. Diseases such as meningitis, especially in children, can also lead to sensorineural hearing loss.

In carefully selected patients, the use of a **cochlear implant** can restore some function to the profoundly deaf. The device consists of an external microphone, amplifier, and speech processor coupled by a plug-and-socket connection, magnetic induction, or a radio frequency link to a receiver implanted under the skin over the mastoid bone. Stimulating wires then lead to the cochlea. A single **extracochlear electrode,** applied to the round window, can restore perception of some environmental sounds and aid in lip-reading, but it will not restore pitch or speech discrimination. A **multielectrode intracochlear implant** (with up to 22 active elements spaced along it) can be inserted into the basal turn of the scala tympani. The linear spatial arrangement of the electrodes takes advantage of the tonotopic organization of the cochlea, and some pitch (frequency) discrimination is possible. The external processor separates the speech signal into several frequency bands that contain the most critical speech information, and the multielectrode assembly presents the separated signals to the appropriate locations along the cochlea. In some devices the signals are presented in rapid sequence, rather than simultaneously, to minimize interference between adjacent areas.

When implanted successfully, such a device can restore much of the ability to understand speech. Considerable training of the patient and fine-tuning of the speech processor are necessary. The degree of restoration of function ranges from recognition of critical environmental sounds to the ability to converse over a telephone. Cochlear implants are most successful in adults who became deaf after having learned to speak and hear naturally. Success in children depends critically on their age and linguistic ability; currently, implants are being used in children as young as age 2.

Infrequent problems with infection, device failure, and natural growth of the auditory structures may limit the usefulness of cochlear implants for some patients. In certain cases, psychological and social considerations may discourage the advisability of using of auditory prosthetic devices in general. From a technical standpoint, however, continual refinements in the design of implantable devices and the processing circuitry are extending the range of subjects who may benefit from cochlear implants. Research directed at external stimulation of higher auditory structures may eventually lead to even more effective treatments for profound hearing loss.

vertical (superior) canal and the **posterior vertical canal,** which are perpendicular to each other. The planes of the anterior vertical canals are each at approximately 45° to the midsagittal section of the head (and at 90° to each other). Thus, the anterior canal on one side lies in a plane parallel with the posterior canal on the other side, and the two function as a pair. The horizontal canals also lie in a common plane.

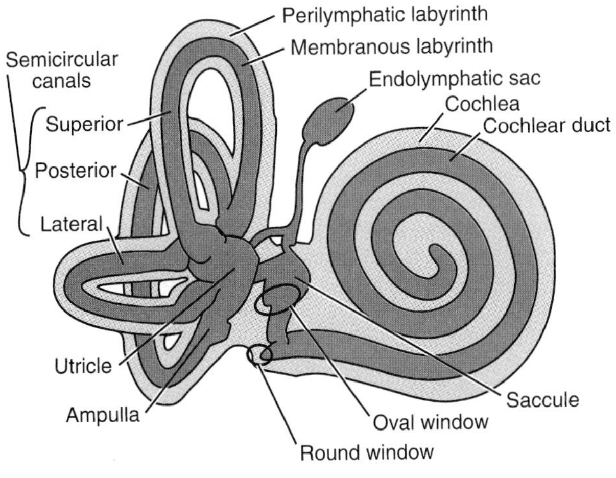

Near its junction with the utricle, each canal has a swollen portion called the **ampulla.** Each ampulla contains a **crista ampullaris,** the sensory structure for that semicircular canal; it is composed of hair cells and supporting cells encapsulated by a **cupula,** a gelatinous mass (Fig. 4.24). The cupula extends to the top of the ampulla and is moved back and forth by movements of the endolymph in the canal. This movement is sensed by displacement of the stereocilia of the hair cells. These cells are much like those of the organ of Corti, except that at the "tall" end of the stereocilia array there is one larger cilium, the **kinocilium.** All the hair cells have the same orientation. When the stereocilia are bent toward the kinocilium, the frequency of action potentials in the afferent neurons leaving the ampulla increases; bending in the other direction decreases the action potential frequency.

The role of the semicircular canals in sensing rotary acceleration is shown on the left side of Figure 4.25. The mechanisms linking stereocilia deflection to receptor potentials and action potential generation are quite similar to those in the auditory hair cells. Because of the inertia of the endolymph in the canals, when the position of the head is changed, fluid currents in the canals cause the deflection of

FIGURE 4.23 **The vestibular apparatus in the bony labyrinth of the inner ear.** The semicircular canals sense rotary acceleration and motion, while the utricle and saccule sense linear acceleration and static position.

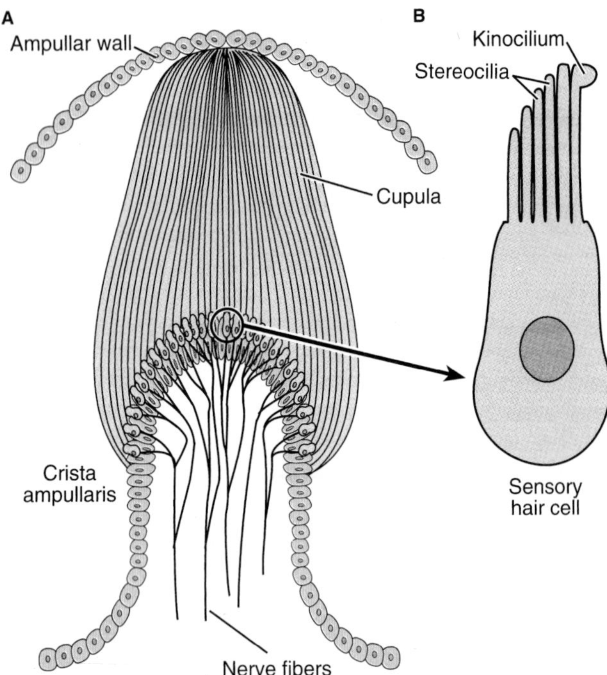

FIGURE 4.24 The sensory structure of the semicircular canals. A, The crista ampularis contains the hair (receptor) cells, and the whole structure is deflected by motion of the endolymph. B, An individual hair cell.

the cupula and the hair cells are stimulated. The fluid currents are roughly proportional to the rate of change of velocity (i.e., to the rotary acceleration), and they result in a proportional increase or decrease (depending on the direction of head rotation) in action potential frequency. As a result of the bilateral symmetry in the vestibular system, canals with opposite pairing produce opposite neural effects. The vestibular division of cranial nerve VIII passes the impulses first to the **vestibular ganglion,** where the cell bodies of the primary sensory neurons lie. The information is then passed to the **vestibular nuclei** of the brainstem and from there to various locations involved in sensing, correcting, and compensating for changes in the motions of the body.

The remaining vestibular organs, the saccule and the utricle, are also part of the membranous labyrinth. They communicate with the semicircular canals, the cochlear duct, and the endolymphatic duct. The sensory structures in these organs, called **maculae,** also employ hair cells, similar to those of the ampullar cristae (Fig. 4.26). The macular hair cells are covered with the **otolithic membrane,** a gelatinous substance in which are embedded numerous small crystals of calcium carbonate called **otoliths** (otoconia). Because the otoliths are heavier than the endolymph, tilting of the head results in gravitational movement of the otolithic membrane and a corresponding change in sensory neuron action potential frequency. As in the ampulla, the action potential frequency increases or decreases depending on the direction of displacement. The maculae are adapted to provide a steady signal in response to displacement; in addition, they are located away from the semicircular canals and are not subject to motion-induced currents in the endolymph. This allows them to monitor the position of the head with respect to a steady gravitational field. The maculae also respond proportionally to **linear acceleration.**

The vestibular apparatus is an important component in several reflexes that serve to orient the body in space and maintain that orientation. Integrated responses to

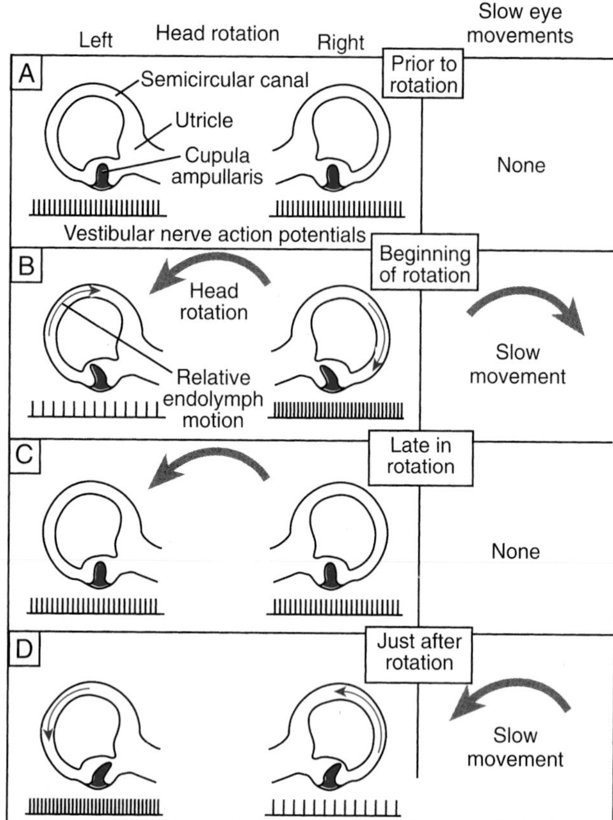

FIGURE 4.25 The role of the semicircular canals in sensing rotary acceleration. This sensation is linked to compensatory eye movements by the vestibuloocular reflex. Only the horizontal canals are considered here. This pair of canals is shown as if one were looking down through the top of a head looking toward the top of the page. Within the ampulla of each canal is the cupula, an extension of the crista ampullaris, the structure that senses motion in the endolymph fluid in the canal. Below each canal is the action potential train recorded from the vestibular nerve. **A,** The head is still, and equal nerve activity is seen on both sides. There are no associated eye movements (right column). **B,** The head has begun to rotate to the left. The inertia of the endolymph causes it to lag behind the movement, producing a fluid current that stimulates the cupulae (arrows show the direction of the relative movements). Because the two canals are mirror images, the neural effects are opposite on each side (the cupulae are bent in relatively opposite directions). The reflex action causes the eyes to move slowly to the right, opposite to the direction of rotation (right column); they then snap back and begin the slow movement again as rotation continues. The fast movement is called rotatory nystagmus. **C,** As rotation continues, the endolymph "catches up" with the canal because of fluid friction and viscosity, and there is no relative movement to deflect the cupulae. Equal neural output comes from both sides, and the eye movements cease. **D,** When the rotation stops, the inertia of the endolymph causes a current in the same direction as the preceding rotation, and the cupulae are again deflected, this time in a manner opposite to that shown in part B. The slow eye movements now occur in the same direction as the former rotation.

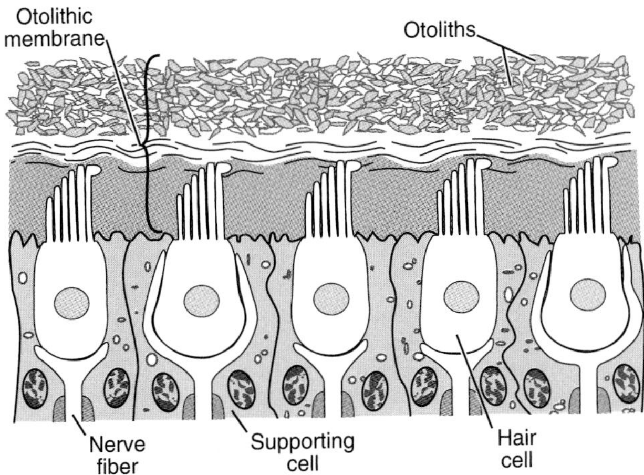

FIGURE 4.26 The relation of the otoliths to the sensory cells in the macula of the utricle and saccule. The gravity-driven movement of the otoliths stimulates the hair cells.

vestibular sensory input include balancing and steadying movements controlled by skeletal muscles, along with specific reflexes that automatically compensate for bodily motions. One such mechanism is the **vestibuloocular reflex.** If the body begins to rotate and, thereby, stimulate the horizontal semicircular canals, the eyes will move slowly in a direction opposite to that of the rotation and then suddenly snap back the other way (see Fig. 4.25, right). This movement pattern, called **rotatory nystagmus,** aids in visual fixation and orientation and takes place even with the eyes closed. It functions to keep the eyes fixed on a stationary point (real or imaginary) as the head rotates. By convention, the direction of the rapid eye movement is used to label the direction of the nystagmus, and this movement is in the same direction as the rotation. As rotation continues, the relative motion of the endolymph in the semicircular canals ceases, and the nystagmus disappears. When rotation stops, the inertia of the endolymph causes it to continue in motion and again the cupulae are displaced, this time from the opposite direction. The slow eye movements are now in the same direction as the prior rotation; the **postrotatory nystagmus** (fast phase) that develops is in a direction opposite to the previous rotation. As long as the endolymph continues its relative movement, the nystagmus (and the sensation of rotary motion) persists. Irrigation of the ear with water above or below body temperature causes convection currents in the endolymph. The resulting unilateral **caloric stimulation** of the semicircular canal produces symptoms of vertigo, nystagmus, and nausea. Disturbances of the labyrinthine function produce the symptoms of **vertigo,** a disorder that can significantly affect daily activities (see Clinical Focus Box 4.2).

Related mechanisms involving the otolithic organs produce automatic compensations (via the postural and extraocular musculature) when the otolithic organs are stimulated by transient or maintained changes in the position of the head. If the otolithic organs are stimulated rhythmically, as by the motion of a ship or automobile, the distressing symptoms of motion sickness (vertigo, nausea, sweating, etc.) may appear. Over time, these symptoms lessen and disappear.

The Special Chemical Senses Detect Molecules in the Environment

Chemical sensation includes not only the special chemical senses described below, but also internal sensory receptor functions that monitor the concentrations of gases and other chemical substances dissolved in the blood or other body fluids. Since we are seldom aware of these internal chemical sensations, they are treated throughout this book as needed; the discussion here covers only taste and smell.

Gustatory Sensation. The sense of taste is mediated by multicellular receptors called **taste buds,** several thousand of which are located on folds and projections on the dorsal tongue, called **papillae.** Taste buds are located mainly on the tops of the numerous **fungiform papillae** but are also located on the sides of the less numerous **foliate** and **vallate papillae.** The **filiform papillae,** which cover most of the tongue, usually do not bear taste buds. An individual taste bud is a spheroid collection of about 50 individual cells that is about 70 μm high and 40 μm in diameter (Fig. 4.27). The cells of a taste bud lie mostly buried in the surface of the tongue, and materials access the sensory cells by way of the **taste pore.**

Most of the cells of a taste bud are **sensory cells.** At their apical ends, they are connected laterally by tight junctions, and they bear **microvilli** that greatly increase the surface area they present to the environment. At their basal ends, they form synapses with the facial (VII) and glossopharyngeal (IX) cranial nerves. This arrangement indicates that the sensory cells are actually **secondary receptors** (like the hair cells of the ear), since they are anatomically separate from the afferent sensory nerves. About 50 afferent fibers enter each taste bud, where they branch so that each axon synapses with more than one sensory cell. Among the sensory cells are elongated **supporting cells** that do not have synaptic connections. The sensory cells typically have a lifespan of 10 days. They are continually replenished by new sensory cells formed from the **basal cells** of the lower part of the taste buds. When a sensory cell is replaced by a maturing basal cell, the old synaptic connections are broken, and new ones must be formed.

From the point of view of their receptors, the traditional four modalities of taste—**sweet, sour, salty,** and **bitter**—are well defined, and the areas of the tongue where they are located are also rather specific, although the degree of localization depends on the concentration of the stimulating substance. In general, the receptors for sweetness are located just behind the tip of the tongue, sour receptors are located along the sides, the salt sensation is localized at the tip, and the bitter sensation is found across the rear of the tongue. (The two "accessory qualities" of taste sensation are **alkaline** [soapy] and **metallic.**) The broad surface of the tongue is not as well supplied with taste buds. Most taste experiences involve several different sensory modalities, including taste, smell, mechanoreception (for texture), and temperature; artificially confining the taste sensation to only the four modalities found on the tongue (e.g., by

CLINICAL FOCUS BOX 4.2

Vertigo

A common medical complaint is dizziness. This symptom may be a result of several factors, such as cerebral ischemia ("feeling faint"), reactions to medication, disturbances in gait, or disturbances in the function of the vestibular apparatus and its central nervous system connections. Such disturbances can produce the phenomenon of **vertigo**, which may be defined as the illusion of motion (usually rotation) when no motion is actually occurring. Vertigo is often accompanied by autonomic nervous system symptoms of nausea, vomiting, sweating, and pallor.

The body uses three integrated systems to establish its place in space: the vestibular system, which senses position and rotation of the head; the visual system, which provides spatial information about the external environment; and the somatosensory system, which provides information from joint, skin, and muscle receptors about limb position. Several forms of vertigo can arise from disturbances in these systems. **Physiological vertigo** can result when there is discordant input from the three systems. Seasickness results from the unaccustomed repetitive motion of a ship (sensed via the vestibular system). Rapidly changing visual fields can cause visually-induced motion sickness, and space sickness is associated with multiple-input disturbances. **Central positional vertigo** can arise from lesions in cranial nerve VIII (as may be associated with multiple sclerosis or some tumors), vertebrovascular insufficiency (especially in older adults), or from impingement of vascular loops on neural structures. It is commonly present with other CNS symptoms. **Peripheral vertigo** arises from disturbances in the vestibular apparatus itself. The problem may be either unilateral or bilateral. Causes include trauma, physical defects in the labyrinthine system, and pathological syndromes such as Ménière's disease. As in the cochlea, aging produces considerable hair cell loss in the cristae and maculae of the vestibular system. Caloric stimulation can be used as an indicator of the degree of vestibular function.

The most common form of peripheral vertigo is **benign paroxysmal positional vertigo (BPPV)**. This is a severe vertigo, with incidence increasing with age. Episodes appear rapidly and are limited in duration (from minutes to days). They are usually brought on by assuming a particular position of the head, such as one might do when painting a ceiling. BPPV is thought to be due to the presence of **canaliths**, debris in the lumen of one of the semicircular canals. The offending particles are usually clumps of otoconia (otoliths) that have been shed from the maculae of the saccule and utricle, whose passages are connected to the semicircular canals. These clumps act as gravity-driven pistons in the canals, and their movement causes the endolymph to flow, producing the sensation of rotary motion. Because they are in the lowest position, the posterior canals are the most frequently affected. In addition to the rotating sensation, this input gives rise, via the vestibuloocular system, to a pattern of nystagmus (eye movements) appropriate to the spurious input.

The specific site of the problem can be determined by using the **Dix-Hallpike maneuver**, which is a series of physical maneuvers (changes in head and body position). By observing the resulting pattern of nystagmus and reported symptoms, the location of the defect can be deduced. Another set of maneuvers known as the **canalith repositioning procedure of Epley** can cause gravity to collect the loose canaliths and deposit them away from the lumen of the semicircular canal. This procedure is highly effective in cases of true BPPV, with a cure rate of up to 85% on the first attempt and nearly 100% on a subsequent attempt. Patients can be taught to perform the procedure on themselves if the problem returns.

Ménière's disease is a syndrome of uncertain (but peripheral) origin associated with vertigo. Its cause(s) and precipitating factors are not well understood. Typical associated findings include fluctuating hearing loss and tinnitus (ringing in the ears). Episodes involve increased fluid pressure in the labyrinthine system, and symptoms may decrease in response to salt restriction and diuretics. Other cases of peripheral vertigo may be caused by trauma (usually unilateral) or by toxins or drugs (such as some antibiotics); this type is often bilateral.

Central and peripheral vertigo may often be differentiated on the basis of their specific symptoms. Peripheral vertigo is more severe, and its nystagmus shows a delay (latency) in appearing after a position change. Its nystagmus fatigues and can be reduced by visual fixation. Position sensitive and of finite duration, the condition usually involves a horizontal orientation. Central vertigo, usually less severe, shows a vertically oriented nystagmus without latency and fatigability; it is not suppressed by visual fixation and may be of long duration.

Treatment for vertigo, beyond that mentioned above, can involve bed rest and vestibular inhibiting drugs (such as some antihistamines). However, these treatments are not always effective and may delay the natural compensation that can be aided by physical motion, such as walking (unpleasant as that may be). In severe cases that require surgical intervention (labyrinthectomy, etc.), patients can often achieve a workable position sense via the other sensory inputs involved in maintaining equilibrium. Some activities, such as underwater swimming, must be avoided by those with an impaired sense of orientation, since false cues may lead to moving in inappropriate directions and increase the risk of drowning.

References

Baloh RW. Vertigo. Lancet 1998;352:1841–1846.

Furman JM, Cass SP. Primary care: Benign paroxysmal positional vertigo. N Engl J Med 1999;341:1590–1596.

blocking the sense of smell) greatly diminishes the range of taste perceptions.

Recent studies have provided evidence for a fifth taste modality, one that is called **umami**, or savoriness. Its receptors are stimulated quite specifically by glutamate ions, which are contained in naturally occurring dietary protein and are responsible for a "meaty" taste. Glutamate ions can also be provided as a flavor-enhancer in the well-known food additive MSG, monosodium glutamate.

While the functional receptor categories are well defined, it is much more difficult to determine what kind of stimulating chemical will produce a given taste sensation. Chemicals that produce a sour sensation are usually acids, and the intensity of the perception depends on the degree

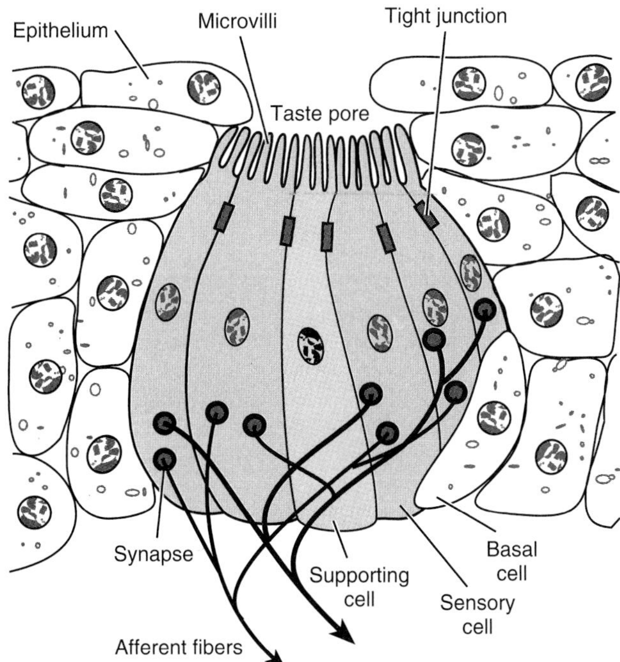

FIGURE 4.27 **The sensory and supporting cells in a taste bud.** The afferent nerve synapse with the basal areas of the sensory cells. (Modified from Schmidt RF, ed. Fundamentals of Sensory Physiology. 2nd Ed. New York: Springer-Verlag, 1981.)

of dissociation of the acid (i.e., the number of free hydrogen ions). Most sweet substances are organic; sugars, especially, tend to produce a sweet sensation, although thresholds vary widely. For example, sucrose is about 8 times as sweet as glucose. By comparison, the apparent sweetness of saccharin, an artificial sweetener, is 600 times as great as that of sucrose, although it is not a sugar. Unfortunately, the salts of lead are also sweet, which can lead to ingestion of toxic levels of this poisonous metal. Substances producing a bitter taste form a heterogeneous group. The classic bitter substance is quinine; nicotine and caffeine are also bitter, as are many of the salts of calcium, magnesium, and ammonium, the bitter taste being due to the cation portion of the salt. Sodium ions produce a salty sensation; some organic compounds, such as lysyltaurine, are even more potent in this regard than sodium chloride.

The intensity of a taste sensation depends on the concentration of the stimulating substance, but application of the same concentration to larger areas of the tongue produces a more intense sensation; this is probably due to facilitation involving a greater number of afferent fibers. Some taste sensations also increase with time, although taste receptors show a slow but definite adaptation. Elevated temperature, over some ranges, tends to increase the perceived taste intensity, while dilution by saliva and serous secretions from the tongue decreases the intensity. The specificity of the taste sensation arising from a particular stimulating substance results from the effects of specific receptor molecules on the microvilli of the sensory cells. Salty substances probably depolarize sensory cells directly, while sour substances may produce depolarization by blocking potassium channels with hydrogen ions. Bitter substances bind to specific G protein-coupled receptors and activate phospholipase C to increase the cell concentration of inositol trisphosphate, which promotes calcium release from the endoplasmic reticulum. Sweet substances also act through G protein-coupled receptors and cause increases in adenylyl cyclase activity, increasing cAMP, which, in turn, promotes the phosphorylation of membrane potassium channels. The resulting decrease in potassium conductance leads to depolarization. In the case of the umami taste, there is evidence of specific G protein-coupled receptors in the cell membranes of sensory taste cells.

Olfactory Sensation. Compared with that of many other animals, the human sense of smell is not particularly acute. Nevertheless, we can distinguish 2,000 to 4,000 different odors that cover a wide range of chemical species. The receptor organ for olfaction is the **olfactory mucosa**, an area of approximately 5 cm^2 located in the roof of the nasal cavity. Normally there is little air flow in this region of the nasal tract, but sniffing serves to direct air upward, increasing the likelihood of an odor being detected.

The olfactory mucosa contains about 10 to 20 million receptor cells. In contrast to the taste sensory cells, the olfactory cells are neurons and, as such, are **primary receptors.** These cells are interspersed among **supporting** (**sustentacular**) **cells**, and **tight junctions** bind the cells together at their sensory ends (Fig. 4.28). The receptor

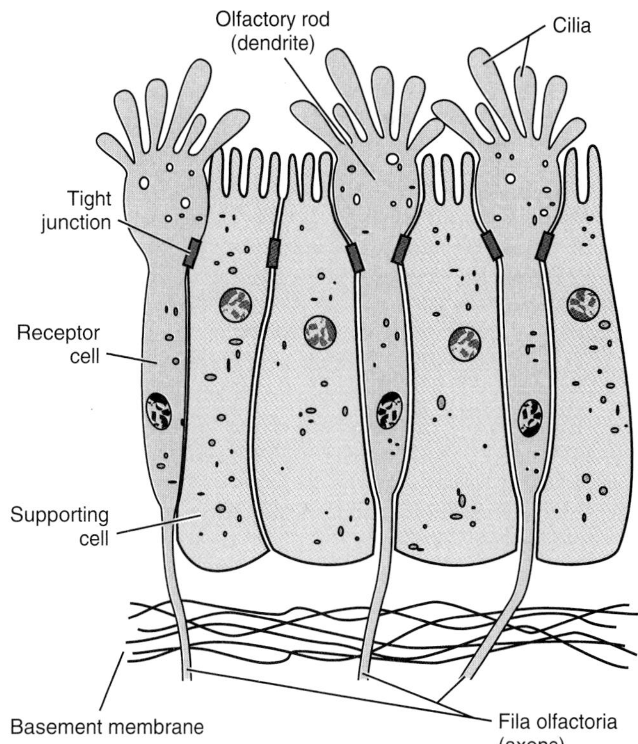

FIGURE 4.28 **The sensory cells in the olfactory mucosa.** The fila olfactoria, the axons leading from the receptor cells, are part of the sensory cells, in contrast to the situation in taste receptors. (Modified from Ganong WF. Review of Medical Physiology. 20th Ed. Stamford, CT: McGraw-Hill, 2001.)

cells terminate at their apical ends with short, thick dendrites called **olfactory rods**, and each cell bears 10 to 20 cilia that extend into a thin covering of mucus secreted by **Bowman's glands** located throughout the olfactory mucosa. Molecules to be sensed must be dissolved in this mucous layer. The basal ends of the receptor cells form axonal processes called **fila olfactoria** that pass through the **cribriform plate** of the **ethmoid bone.** These short axons synapse with the **mitral cells** in complex spherical structures called **olfactory glomeruli** located in the **olfactory bulb,** part of the brain located just above the olfactory mucosa. Here the complex afferent and efferent neural connections for the olfactory tract are made. Approximately 1,000 fila olfactoria synapse on each mitral cell, resulting in a highly convergent relationship. Lateral connections are also plentiful in the olfactory bulb, which also contains efferent fibers thought to have an inhibitory function.

The olfactory mucosa also contains sensory fibers from the trigeminal (V) cranial nerve. They are sensitive to certain odorous substances, such as peppermint and chlorine, and play a role in the initiation of reflex responses (e.g., sneezing) that result from irritation of the nasal tract.

The modalities of smell are numerous and do not fall into convenient classes, though some general categories, such as flowery, sweaty, or rotten, may be distinguished. Olfactory thresholds vary widely from substance to substance; the threshold concentration for the detection of ethyl ether is around 5.8 mg/L air, while that for methyl mercaptan (the odor of garlic) is approximately 0.5 ng/L. This represents a 10 million-fold difference in sensitivity. The basis for odor discrimination is not well understood. It is not likely that there is a receptor molecule for every possible odor substance located in the membranes of the olfactory cilia, and it appears that complex odor sensations arise from unique spatial patterns of activation throughout the olfactory mucosa.

Signal transduction appears to involve the binding of a molecule of an odorous substance to a G protein-coupled receptor on a cilium of a sensory cell. This binding causes the production of cAMP that binds to, and opens, sodium channels in the ciliary membrane. The resulting inward sodium current depolarizes the cell to produce a generator potential, which causes action potentials to arise in the initial segments of the fila olfactoria. The sense of smell shows a high degree of adaptation, some of which takes place at the level of the generator potential and some of which may be due to the action of efferent neurons in the olfactory bulb. Discrimination between odor intensities is not well defined; detectable differences may be about 30%.

REVIEW QUESTIONS

DIRECTIONS: Each of the numbered items or incomplete statements in this section is followed by answers or completions of the statement. Select the ONE lettered answer or completion that is BEST in each case.

1. An increase in the action potential frequency in a sensory nerve usually signifies
   (A) Increased intensity of the stimulus
   (B) Cessation of the stimulus
   (C) Adaptation of the receptor
   (D) A constant and maintained stimulus
   (E) An increase in the action potential conduction velocity
2. Why is the blind spot on the retina not usually perceived?
   (A) It is very small, below the ability of the sensory cells to detect
   (B) It is present only in very young children
   (C) Its location in the visual field is different in each eye
   (D) Constant eye motion prevents the spot from remaining still
   (E) Lateral input from adjacent cells fills in the missing information
3. The condition known as presbyopia is due to
   (A) Change in the shape of the eyeball as a result of age
   (B) An age-related loss of cells in the retina
   (C) Change in the elasticity of the lens as a result of age
   (D) A loss of transparency in the lens
   (E) Increased opacity of the vitreous humor
4. What external aids can be used to help a myopic eye compensate for distance vision?
   (A) A positive (converging) lens placed in front of the eye
   (B) A negative (diverging) lens placed in front of the eye
   (C) A cylindrical lens placed in front of the eye
   (D) Eyeglasses that are partially opaque, to reduce the light intensity
   (E) No help is needed because the eye itself can accommodate
5. At which location along the basilar membrane are the highest-frequency sounds detected?
   (A) Nearest the oval window
   (B) Farthest from the oval window, near the helicotrema
   (C) Uniformly along the basilar membrane
   (D) At the midpoint of the membrane
   (E) At a series of widely-spaced locations along the membrane
6. Motion of the endolymph in the semicircular canals when the head is held still will result in the perception of
   (A) Being upside-down
   (B) Moving in a straight line
   (C) Continued rotation
   (D) Being upright and stationary
   (E) Lying on one's back
7. A decrease in sensory response while a stimulus is maintained constant is due to the phenomenon of
   (A) Adaptation
   (B) Fatigue
   (C) The graded response
   (D) Compression
8. Sensory receptors that adapt rapidly are well suited to sensing
   (A) The weight of an object held in the hand
   (B) The rate at which an extremity is being moved
   (C) Resting body orientation in space
   (D) Potentially hazardous chemicals in the environment
   (E) The position of an extended limb
9. Adaptation in a sensory receptor is associated with a
   (A) Decline in the amplitude of action potentials in the sensory nerve
   (B) Reduction in the intensity of the applied stimulus
   (C) Decline in the conduction velocity of sensory nerve action potentials

(continued)

(D) Decline in the amplitude of the generator potential
(E) Reduction in the duration of the sensory action potentials

10. Which of the following is the principal function of the bones (ossicles) of the middle ear?
(A) They provide mechanical support for the flexible membranes to which they are attached (i.e., the eardrum and the oval window)
(B) They reduce the amplitude of the vibrations reaching the oval window, protecting it from mechanical damage
(C) They increase the efficiency of vibration transfer through the middle ear
(D) They control the opening of the eustachian tubes and allow pressures to be equalized
(E) They have little effect on the process of hearing in humans, since they are essentially passive structures

11. On a moonlit night, human vision is monochromatic and less acute than vision during the daytime. This is because
(A) Objects are being illuminated by monochromatic light, and there is no opportunity for color to be produced
(B) The cone cells of the retina, while more closely packed than the rod cells, have a lower sensitivity to light of all colors
(C) Light rays of low intensity do not carry information as to color
(D) Retinal photoreceptor cells that have become dark-adapted can no longer respond to varying wavelengths of light
(E) At low light levels, the lens cannot accommodate to sharpen vision

SUGGESTED READING

Ackerman D. A Natural History of the Senses. New York: Random House, 1990.

Gulick WL, Gescheider GA, Frisina RD. Hearing: Physiological Acoustics, Neural Coding, and Psychoacoustics. New York: Oxford University Press, 1989.

Hudspeth AJ. How hearing happens. Neuron 1997;19:947–950.

Spielman AI. Chemosensory function and dysfunction. Crit Rev Oral Biol Med 1998;9:267–291.

The Motor System

John C. Kincaid, M.D.

CHAPTER OUTLINE

KEY CONCEPTS

1. The contraction of skeletal muscle produces movement by acting on the skeleton.
2. Motor neurons activate the skeletal muscles.
3. Sensory feedback from muscles is important for precise control of contraction.
4. The output of sensory receptors like the muscle spindle can be adjusted.
5. The spinal cord is the source of reflexes that are important in the initiation and control of movement.
6. Spinal cord function is influenced by higher centers in the brainstem.
7. The highest level of motor control comes from the cerebral cortex.
8. The basal ganglia and the cerebellum provide feedback to the motor control areas of the cerebral cortex and brainstem.

The finger movements of a neurosurgeon manipulating microsurgical instruments while repairing a cerebral aneurysm, and the eye-hand-body control of a professional basketball player making a rimless three-point shot, are two examples of the motor control functions of the nervous system operating at high skill levels. The coordinated contraction of the hip flexors and ankle extensors to clear a slight pavement irregularity encountered during walking is a familiar example of the motor control system working at a seemingly automatic level. The stiff-legged stride of a patient who experienced a stroke and the swaying walk plus slurred speech of an intoxicated person are examples of perturbed motor control.

Although our understanding of the anatomy and physiology of the motor system is still far from complete, a significant fund of knowledge exists. This chapter will proceed through the constituent parts of the motor system, beginning with the skeleton and ending with the brain.

THE SKELETON AS THE FRAMEWORK FOR MOVEMENT

Bones are the body's framework and system of levers. They are the elements that move. The way adjacent bones articulate determines the motion and range of movement at a joint. Ligaments hold the bones together across the joint. Movements are described based on the anatomic planes through which the skeleton moves and the physical structure of the joint. Most joints move in only one plane, but some permit movement in multiple anatomic reference planes (Fig. 5.1).

Hinge joints, such as the elbow, are uniaxial, permitting movements in the sagittal plane. The wrist is an example of a biaxial joint. The shoulder is a multiaxial joint; movement can occur in oblique planes as well as the three major planes of that joint. Flexion and extension describe movements in the sagittal plane. **Flexion** movements decrease the angle between the moving body segments. **Extension** describes movement in the opposite direction. **Abduction** moves the

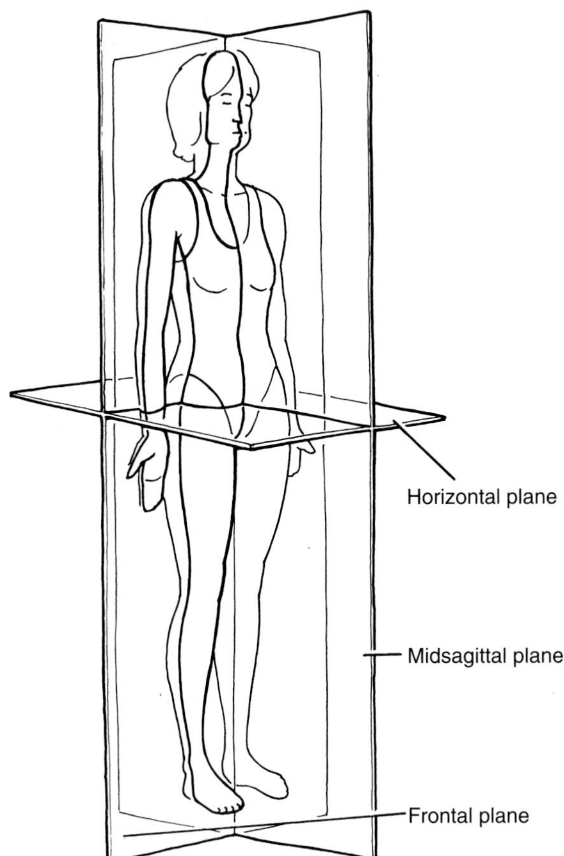

FIGURE 5.1 **Anatomic reference planes.** The figure is shown in the standard anatomic position with the associated primary reference planes.

body part away from the midline, while **adduction** moves the body part toward midline.

MUSCLE FUNCTION AND BODY MOVEMENT

Muscles span joints and are attached at two or more points to the bony levers of the skeleton. The muscles provide the power that moves the body's levers. Muscles are described in terms of their origin and insertion attachment sites. The **origin** tends to be the more fixed, less mobile location, while the **insertion** refers to the skeletal site that is more mobile. Movement occurs when a muscle generates force on its attachment sites and undergoes shortening. This type of action is termed an **isotonic** or **concentric contraction**. Another form of muscular action is a controlled lengthening while still generating force. This is an **eccentric contraction**. A muscle may also generate force but hold its attachment sites static, as in **isometric contraction**.

Because muscle contraction can produce movement in only one direction, at least two muscles opposing each other at a joint are needed to achieve motion in more than one direction. When a muscle produces movement by shortening, it is an **agonist**. The **prime mover** is the muscle that contributes most to the movement. Muscles that oppose the action of the prime mover are **antagonists**. The quadriceps and hamstring muscles are examples of agonist-antagonist pairs in knee extension and flexion. During both simple and light-load skilled movements, the antagonist is relaxed. Contraction of the agonist with concomitant relaxation of the antagonist occurs by the nervous system function of **reciprocal inhibition**. Co-contraction of agonist and antagonist occurs during movements that require precise control.

A muscle functions as a **synergist** if it contracts at the same time as the agonist while cooperating in producing the movement. Synergistic action can aid in producing a movement (e.g., the activity of both flexor carpi ulnaris and extensor carpi ulnaris are used in producing ulnar deviation of the wrist); eliminating unwanted movements (e.g., the activity of wrist extensors prevents flexion of the wrist when finger flexors contract in closing the hand); or stabilizing proximal joints (e.g., isometric contractions of muscles of the forearm, upper arm, shoulder, and trunk accompany a forceful grip of the hand).

PERIPHERAL NERVOUS SYSTEM COMPONENTS FOR THE CONTROL OF MOVEMENT

We can identify the components of the nervous system that are predominantly involved in the control of motor function and discuss the probable roles for each of them. It is important to appreciate that even the simplest reflex or voluntary movement requires the interaction of multiple levels of the nervous system (Fig. 5.2).

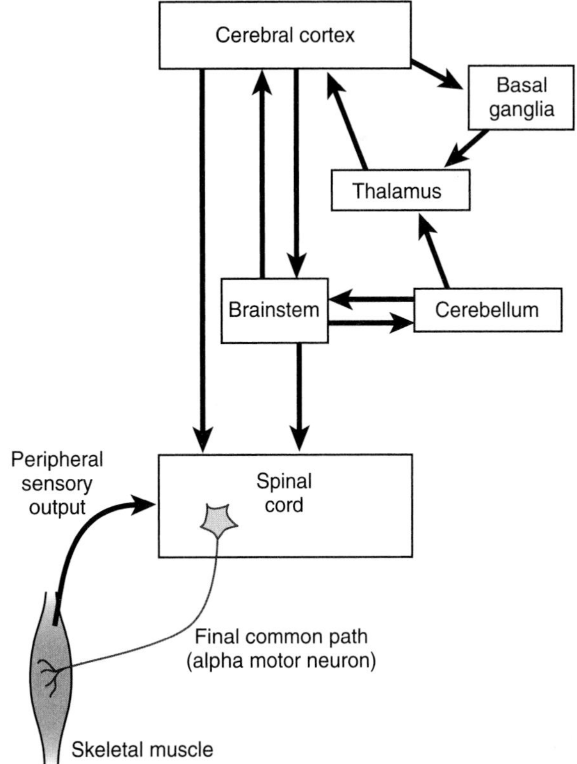

FIGURE 5.2 **Motor control system.** Alpha motor neurons are the final common path for motor control. Peripheral sensory input and spinal cord tract signals that descend from the brainstem and cerebral cortex influence the motor neurons. The cerebellum and basal ganglia contribute to motor control by modifying brainstem and cortical activity.

The motor neurons in the spinal cord and cranial nerve nuclei, plus their axons and muscle fibers, constitute the **final common path**, the route by which all central nervous activity influences the skeletal muscles. The motor neurons located in the ventral horns of the spinal gray matter and brainstem nuclei are influenced by both local reflex circuitry and by pathways that descend from the brainstem and cerebral cortex. The brainstem-derived pathways include the rubrospinal, vestibulospinal, and reticulospinal tracts; the cortical pathways are the corticospinal and corticobulbar tracts. Although some of the cortically derived axons terminate directly on motor neurons, most of the axons of the cortical and the brainstem-derived tracts terminate on interneurons, which then influence motor neuron function. The outputs of the basal ganglia of the brain and cerebellum provide fine-tuning of cortical and brainstem influences on motor neuron functions.

Alpha Motor Neurons Are the Final Common Path for Motor Control

Motor neurons segregate into two major categories, alpha and gamma. **Alpha motor neurons** innervate the **extrafusal muscle fibers,** which are responsible for force generation. **Gamma motor neurons** innervate the **intrafusal muscle fibers,** which are components of the muscle spindle. An alpha motor neuron controls several muscle fibers, 10 to 1,000, depending on the muscle. The term **motor unit** describes a motor neuron, its axon, the branches of the axon, the neuromuscular junction synapses at the distal end of each axon branch, and all of the extrafusal muscle fibers innervated by that motor neuron (Fig. 5.3). When a motor neuron generates an action potential, all of its muscle fibers are activated.

Alpha motor neurons can be separated into two populations according to their cell body size and axon diameter. The larger cells have a high threshold to synaptic stimulation, have fast action potential conduction velocities, and are active in high-effort force generation. They innervate fast-twitch, high-force but fatigable muscle fibers. The smaller alpha motor neurons have lower thresholds to synaptic stimulation, conduct action potentials at a somewhat slower velocity, and innervate slow-twitch, low-force, fatigue-resistant muscle fibers (see Chapter 9). The muscle fibers of each motor unit are homogeneous, either fast-twitch or slow-twitch. This property is ultimately determined by the motor neuron. Muscle fibers that are denervated secondary to disease of the axon or nerve cell body may change twitch type if reinnervated by an axon sprouted from a different twitch-type motor neuron.

The organization into different motor unit types has important functional consequences for the production of smooth, coordinated contractions. The smallest neurons have the lowest threshold and are, therefore, activated first when synaptic activity is low. These produce sustainable, relatively low-force tonic contractions in slow-twitch, fatigue-resistant muscle fibers. If additional force is required, synaptic drive from higher centers increases the action potential firing rate of the initially activated motor neurons and then activates additional motor units of the same type. If yet higher force levels are needed, the larger motor neurons are recruited, but their contribution is less sustained as a result of fatigability. This orderly process of motor unit **recruitment** obeys what is called the **size principle**—the smaller motor neurons are activated first. A logical corollary of this arrangement is that muscles concerned with endurance, such as antigravity muscles, contain predominantly slow-twitch muscle fibers in accordance with their function of continuous postural support. Muscles that contain predominantly fast-twitch fibers, including many physiological flexors, are capable of producing high-force contractions.

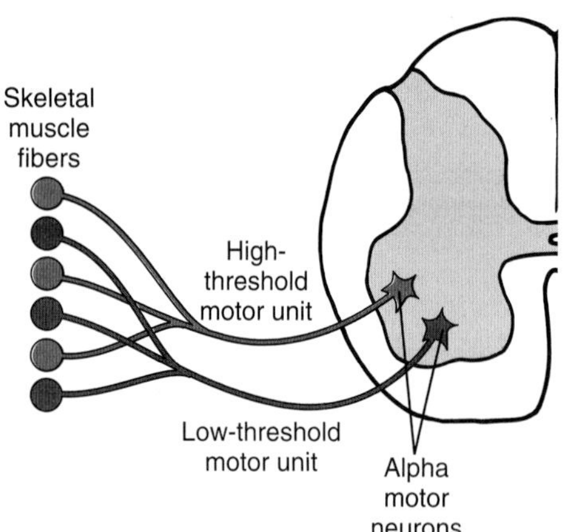

FIGURE 5.3 **Motor unit structure.** A motor unit consists of an alpha motor neuron and the group of extrafusal muscle fibers it innervates. Functional characteristics, such as activation threshold, twitch speed, twitch force, and resistance to fatigue, are determined by the motor neuron. Low- and high-threshold motor units are shown.

Afferent Muscle Innervation Provides Feedback for Motor Control

The muscles, joints, and ligaments are innervated with sensory receptors that inform the central nervous system about body position and muscle activity. Skeletal muscles contain muscle spindles, Golgi tendon organs, free nerve endings, and some Pacinian corpuscles. Joints contain Ruffini endings and Pacinian corpuscles; joint capsules contain nerve endings; ligaments contain Golgi tendon-like organs. Together, these are the **proprioceptors,** providing sensation from the deep somatic structures. These sensations, which may not reach a conscious level, include the position of the limbs and the force and speed of muscle contraction. They provide the feedback that is necessary for the control of movements.

Muscle spindles provide information about the muscle length and the velocity at which the muscle is being stretched. Golgi tendon organs provide information about the force being generated. Spindles are located in the mass of the muscle, in parallel with the extrafusal muscle fibers. Golgi tendon organs are located at the junction of the muscle and its tendons, in series with the muscle fibers (Fig. 5.4).

Muscle Spindles. Muscle spindles are sensory organs found in almost all of the skeletal muscles. They occur in

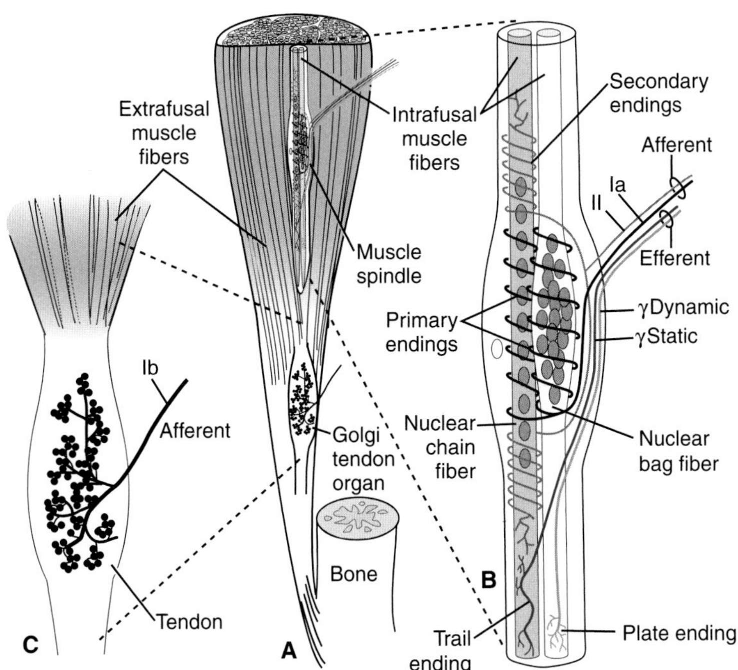

FIGURE 5.4 **Muscle spindle and Golgi tendon organ structure.** A, Muscle spindles are located parallel to extrafusal muscle fibers; Golgi tendon organs are in series. B, This enlarged spindle shows nuclear bag and nuclear chain types of intrafusal fibers; afferent innervation by Ia axons, which provide primary endings to both types of fibers; type II axons, which have secondary endings mainly on chain fibers; and motor innervation by the two types of gamma motor axons, static and dynamic. C, An enlarged Golgi tendon organ. The sensory receptor endings interdigitate with the collagen fibers of the tendon. The axon is type Ib.

greatest density in small muscles serving fine movements, such as those of the hand, and in the deep muscles of the neck. The muscle spindle, named for its long fusiform shape, is attached at both ends to extrafusal muscle fibers. Within the spindle's expanded middle portion is a fluid-filled capsule containing 2 to 12 specialized striated muscle fibers entwined by sensory nerve terminals. These intrafusal muscle fibers, about 300 μm long, have contractile filaments at both ends. The noncontractile midportion contains the cell nuclei (Fig. 5.4B). Gamma motor neurons innervate the contractile elements. There are two types of intrafusal fibers: **nuclear bag fibers**, named for the large number of nuclei packed into the midportion, and **nuclear chain fibers,** in which the nuclei are arranged in a longitudinal row. There are about twice as many nuclear chain fibers as nuclear bag fibers per spindle. The nuclear bag type fibers are further classified as bag_1 and bag_2, based on whether they respond best in the dynamic or static phase of muscle stretch, respectively.

Sensory axons surround both the noncontractile midportion and paracentral region of the contractile ends of the intrafusal fiber. The sensory axons are categorized as **primary (type Ia)** and **secondary (type II)**. The axons of both types are myelinated. Type Ia axons are larger in diameter (12 to 20 μm) than type II axons (6 to 12 μm) and have faster conduction velocities. Type Ia axons have spiral shaped endings that wrap around the middle of the intrafusal muscle fiber (see Fig. 5.4B). Both nuclear bag and nuclear chain fibers are innervated by type Ia axons. Type II axons innervate mainly nuclear chain fibers and have nerve endings that are located along the contractile components on either side of the type Ia spiral ending. The nerve endings of both primary and secondary sensory axons of the muscle spindles respond to stretch by generating action potentials that convey information to the central nervous system about changes in muscle length and the velocity of length change (Fig 5.5). The primary endings temporarily cease generating action potentials during the release of a muscle stretch (Fig. 5.6).

Golgi Tendon Organs. **Golgi tendon organs** (GTOs) are 1-mm-long, slender receptors encapsulated within the tendons of the skeletal muscles (see Fig. 5.4A and C). The distal pole of a GTO is anchored in collagen fibers of the tendon. The proximal pole is attached to the ends of the extrafusal muscle fibers. This arrangement places the GTO in series with the extrafusal muscle fibers such that contractions of the muscle stretch the GTO.

A large-diameter, myelinated type Ib afferent axon arises from each GTO. These axons are slightly smaller in diameter than the type Ia variety, which innervate the muscle spindle. Muscle contraction stretches the GTO and generates action potentials in type Ib axons. The GTO output provides information to the central nervous system about the force of the muscle contraction.

Information entering the spinal cord via type Ia and Ib axons is directed to many targets, including the spinal interneurons that give rise to the **spinocerebellar tracts.** These tracts convey information to the cerebellum about the status of muscle length and tension.

Gamma Motor Neurons. Alpha motor neurons innervate the extrafusal muscle fibers, and gamma motor neurons innervate the intrafusal fibers. Cells bodies of both alpha and gamma motor neurons reside in the ventral horns of the spinal cord and in nuclei of the cranial motor nerves. Nearly one third of all motor nerve axons are destined for intrafusal muscle fibers. This high number reflects the complex role of the spindles in motor system control. Intrafusal muscle fibers likewise constitute a significant portion of the total number of muscle cells, yet they contribute little or nothing to the total force generated when the muscle con-

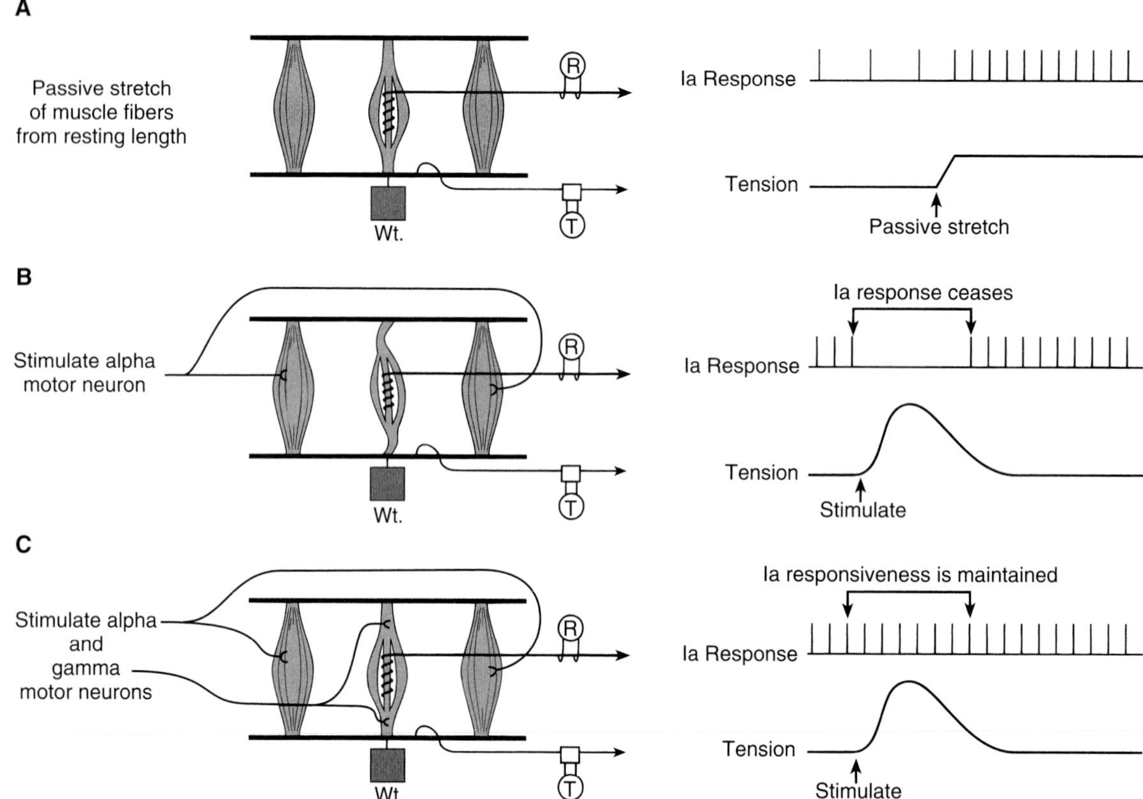

FIGURE 5.5 **Action potential recording (R) from type Ia endings and muscle tension (T).** A, The Ia sensory endings from the muscle spindles discharge at a slow rate when the muscle is at its resting length and show an increased firing rate when the muscle is stretched. B, Alpha motor neuron activation shortens the muscle and releases tension on the muscle spindle. Ia activity ceases temporarily during the tension release. C, Concurrent alpha and gamma motor neuron activation, as occurs in normal, voluntary muscle contraction, shortens the muscle spindle along with the extrafusal fibers, maintaining the spindle's responsiveness to the stretch.

tracts. Rather, the contractions of intrafusal fibers play a modulating role in sensation, as they alter the length and, thereby, the sensitivity of the muscle spindles.

Even when the muscle is at rest, the muscle spindles are slightly stretched, and type Ia afferent nerves exhibit a slow discharge of action potentials. Contraction of the muscle increases the firing rate in type Ib axons from Golgi tendon organs, whereas type Ia axons temporarily cease or reduce firing because the shortening of the surrounding extrafusal fibers unloads the intrafusal muscle fibers. If a load on the spindle were reinstituted, the Ia nerve endings would resume their sensitivity to stretch. The role of the gamma motor neurons is to "reload" the spindle during muscle contraction by activating the contractile elements of the intrafusal fibers. This is accomplished by coordinated activation of the alpha and gamma motor neurons during muscle contraction (see Fig. 5.5).

The gamma motor neurons and the intrafusal fibers they innervate are traditionally referred to as the **fusimotor system**. Axons of the gamma neurons terminate in one of two

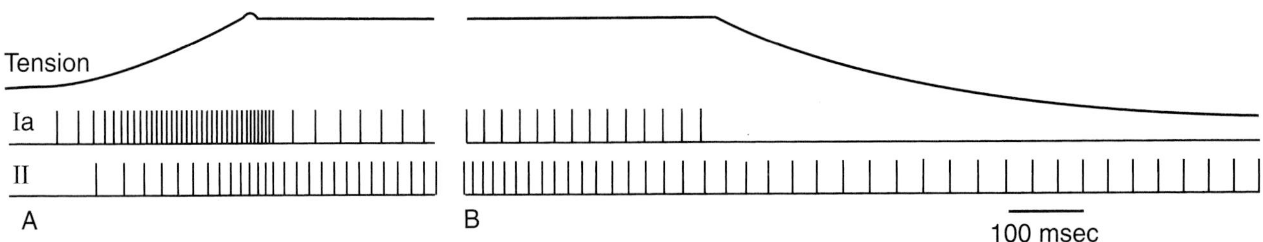

FIGURE 5.6 **Response of types Ia and II sensory endings to a muscle stretch.** A, During rapid stretch, type Ia endings show a greater firing rate increase, while type II endings show only a modest increase. B, With the release of the stretch, Ia endings cease firing, while firing of type II endings slows. Ia endings report both the velocity and the length of muscle stretch; type II endings report length.

types of endings, each located distal to the sensory endings on the striated poles of the spindle's muscle fibers (see Fig. 5.4B). The nerve terminals are either plate endings or trail endings; each intrafusal fiber has only one of these two types of endings. **Plate endings** occur predominantly on bag_1 fibers (dynamic), whereas **trail endings**, primarily on chain fibers, are also seen on bag_2 (static) fibers. This arrangement allows for largely independent control of the nuclear bag and nuclear chain fibers in the spindle.

Gamma motor neurons with plate endings are designated **dynamic** and those with trail endings are designated **static**. This functional distinction is based on experimental findings showing that stimulation of gamma neurons with plate endings enhanced the response of type Ia sensory axons to stretch, but only during the dynamic (muscle length changing) phase of a muscle stretch. During the static phase of the stretch (muscle length increase maintained) stimulation of the gamma neurons with trail endings enhanced the response of type II sensory axons. Static gamma neurons can affect the responses of both types Ia and II sensory axons; dynamic gamma neurons affect the response of only type Ia axons. These differences suggest that the motor system has the ability to monitor muscle length more precisely in some muscles and the speed of contraction in others.

THE SPINAL CORD IN THE CONTROL OF MOVEMENT

Muscles interact extensively in the maintenance of posture and the production of coordinated movement. The circuitry of the spinal cord automatically controls much of this interaction. Sensory feedback from muscles reaches motor neurons of related muscles and, to a lesser degree, of more distant muscles. In addition to activating local circuits, muscles and joints transmit sensory information up the spinal cord to higher centers. This information is processed and can be relayed back to influence spinal cord circuits.

The Structural Arrangement of Spinal Motor Systems Correlates With Function

The cell bodies of the spinal cord motor neurons are grouped into pools in the ventral horns. A **pool** consists of the motor neurons that serve a particular muscle. The number of motor neurons that control a muscle varies in direct proportion to the delicacy of control required. The motor neurons are organized so that those innervating the axial muscles are grouped medially and those innervating the limbs are located laterally (Fig. 5.7). The lateral limb motor neuron areas are further organized so that proximal actions, such as girdle movements, are controlled from relatively medial locations, while distal actions, such as finger movements, are located the most laterally. Neurons innervating flexors and extensors are also segregated. A motor neuron pool may extend over several spinal segments in the form of a column of motor neurons. This is mirrored by the innervation serving a single muscle emerging from the spinal cord in two or even three adjacent spinal nerve root levels. A physiological advantage to such an arrangement is that injury to a single nerve root, as could be produced by herniation of an intervertebral disk, will not completely paralyze a muscle.

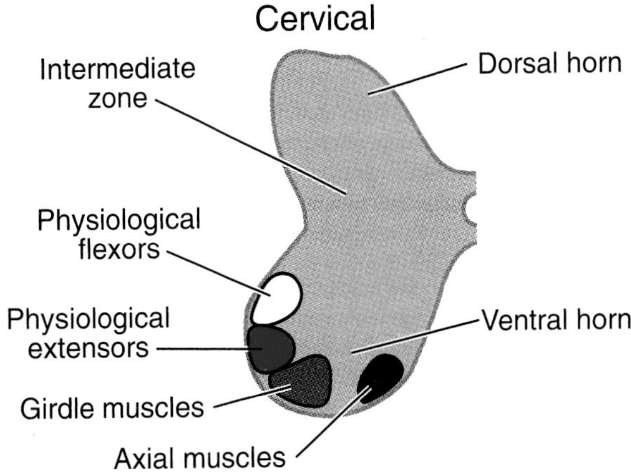

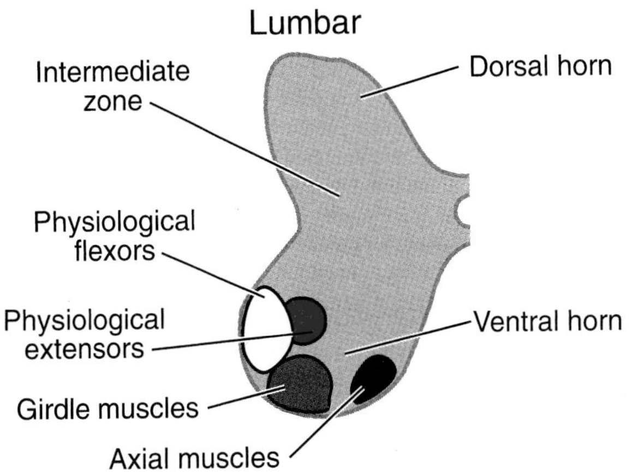

FIGURE 5.7 **Spinal cord motor neuron pools.** Motor neurons controlling axial, girdle, and limb muscles are grouped in pools oriented in a medial-to-lateral fashion. Limb flexor and extensor motor neurons also segregate into pools.

A zone between the medial and lateral pools contains interneurons that project to limb motor neuron pools ipsilaterally and axial pools bilaterally. Between the spinal cord's dorsal and ventral horns lies the intermediate zone, which contains an extensive network of interneurons that interconnect motor neuron pools (see Fig. 5.7). Some interneurons make connections in their own cord segment; others have longer axon projections that travel in the white matter to terminate in other segments of the spinal cord. These longer axon interneurons, termed **propriospinal cells**, carry information that aids coordinated movement. The importance of spinal cord interneurons is reflected in the fact that they comprise the majority of neurons in the spinal cord and provide the majority of the motor neuron synapses.

The Spinal Cord Mediates Reflex Activity

The spinal cord contains neural circuitry to generate **reflexes**, stereotypical actions produced in response to a peripherally applied stimulus. One function of a reflex is to generate a rapid response. A familiar example is the rapid, involuntary withdrawal of a hand after touching a dangerously hot object well before the heat or pain is perceived. This type of reflex protects the organism before higher CNS levels identify the problem. Some reflexes are simple, others much more complex. Even the simplest requires coordinated action in which the agonist contracts while the antagonist relaxes. The functional unit of a reflex consists of a sensor, an afferent pathway, an integrating center, an efferent pathway, and an effector. The sensory receptors for spinal reflexes are the proprioceptors and cutaneous receptors. Impulses initiated in these receptors travel along afferent nerves to the spinal cord, where interneurons and motor neurons constitute the integrating center. The final common path, or motor neurons, make up the efferent pathway to the effector organs, the skeletal muscles. The responsiveness of such a functional unit can be modulated by higher motor centers acting through descending pathways to facilitate or inhibit its activation.

Study of the three types of spinal reflexes—the myotatic, the inverse myotatic, and the flexor withdrawal—provides a basis for understanding the general mechanism of reflexes.

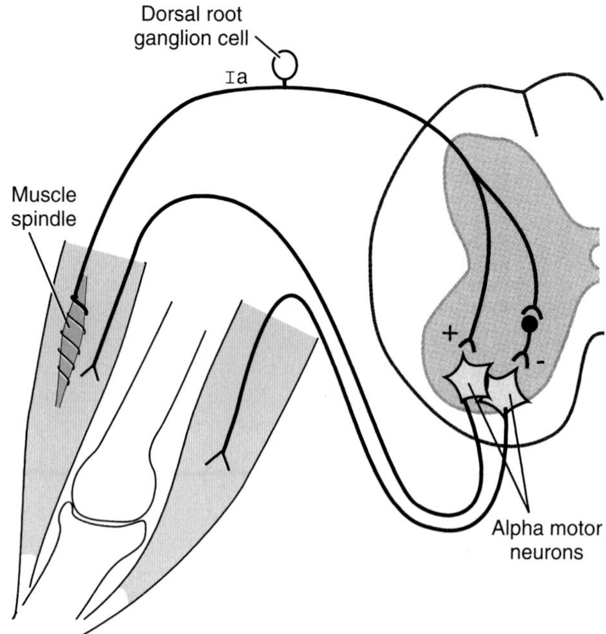

FIGURE 5.8 **Myotatic reflex circuitry.** Ia afferent axons from the muscle spindle make excitatory monosynaptic contact with homonymous motor neurons and with inhibitory interneurons that synapse on motor neurons of antagonist muscles. The plus sign indicates excitation; the minus sign indicates inhibition.

The Myotatic (Muscle Stretch) Reflex. Stretching or elongating a muscle—such as when the patellar tendon is tapped with a reflex hammer or when a quick change in posture is made—causes it to contract within a short time period. The period between the onset of a stimulus and the response, the **latency period**, is on the order of 30 msec for a knee-jerk reflex in a human. This response, called the **myotatic** or **muscle stretch reflex**, is due to monosynaptic circuitry, where an afferent sensory neuron synapses directly on the efferent motor neuron (Fig 5.8). The stretch activates muscle spindles. Type Ia sensory axons from the spindle carry action potentials to the spinal cord, where they synapse directly on motor neurons of the same (homonymous) muscle that was stretched and on motor neurons of synergistic (heteronymous) muscles. These synapses are excitatory and utilize glutamate as the neurotransmitter. Monosynaptic type Ia synapses occur predominantly on alpha motor neurons; gamma motor neurons seemingly lack such connections.

Collateral branches of type Ia axons also synapse on interneurons, whose action then inhibits motor neurons of antagonist muscles (see Fig 5.8). This synaptic pattern, called **reciprocal inhibition**, serves to coordinate muscles of opposing function around a joint. Secondary (type II) spindle afferent fibers also synapse with homonymous motor neurons, providing excitatory input through both monosynaptic and polysynaptic pathways. Golgi tendon organ input via type Ib axons has an inhibitory influence on homonymous motor neurons.

The myotatic reflex has two components: a phasic part, exemplified by tendon jerks, and a tonic part, thought to be important for maintaining posture. The phasic component is more familiar. These components blend together, but either one may predominate, depending on whether other synaptic activity, such as from cutaneous afferent neurons or pathways descending from higher centers, influences the motor response. Primary spindle afferent fibers probably mediate the tendon jerk, with secondary afferent fibers contributing mainly to the tonic phase of the reflex.

The myotatic reflex performs many functions. At the most general level, it produces rapid corrections of motor output in the moment-to-moment control of movement. It also forms the basis for postural reflexes, which maintain body position despite a varying range of loads and/or external forces on the body.

The Inverse Myotatic Reflex. The active contraction of a muscle also causes reflex inhibition of the contraction. This response is called the **inverse myotatic reflex** because it produces an effect that is opposite to that of the myotatic reflex. Active muscle contraction stimulates Golgi tendon organs, producing action potentials in the type Ib afferent axons. Those axons synapse on inhibitory interneurons that influence homonymous and heteronymous motor neurons and on excitatory interneurons that influence motor neurons of antagonists (Fig 5.9).

The function of the inverse myotatic reflex appears to be a tension feedback system that can adjust the strength of contraction during sustained activity. The inverse myotatic reflex does not have the same function as reciprocal inhibition. Reciprocal inhibition acts primarily on the antagonist, while the inverse myotatic reflex acts on the agonist.

The inverse myotatic reflex, like the myotatic reflex, has a more potent influence on the physiological extensor muscles than on the flexor muscles, suggesting that the two reflexes act together to maintain optimal responses in the antigravity muscles during postural adjustments. Another hypothesis about the conjoint function is that both of these

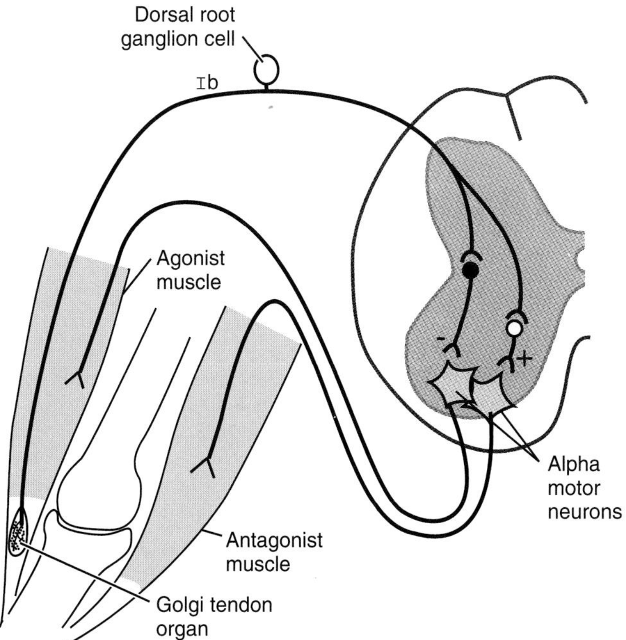

FIGURE 5.9 **Inverse myotatic reflex circuitry.** Contraction of the agonist muscle activates the Golgi tendon organ and Ib afferents, which synapse on interneurons that inhibit agonist motor neurons and excite the motor neurons of the antagonist muscle.

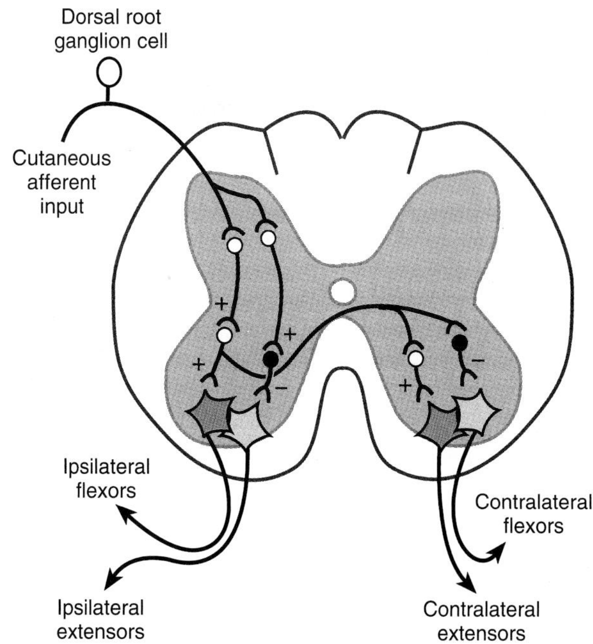

FIGURE 5.10 **Flexor withdrawal reflex circuitry.** Stimulation of cutaneous afferents activates ipsilateral flexor muscles via excitatory interneurons. Ipsilateral extensor motor neurons are inhibited. Contralateral extensor motor neuron activation provides postural support for withdrawal of the stimulated limb.

reflexes contribute to the smooth generation of tension in muscle by regulating muscle stiffness.

The Flexor Withdrawal Reflex. Cutaneous stimulation—such as touch, pressure, heat, cold, or tissue damage—can elicit a **flexor withdrawal reflex**. This reflex consists of a contraction of flexors and a relaxation of extensors in the stimulated limb. The action may be accompanied by a contraction of the extensors on the contralateral side. The axons of cutaneous sensory receptors synapse on interneurons in the dorsal horn. Those interneurons act ipsilaterally to excite the motor neurons of flexor muscles and inhibit those of extensor muscles. Collaterals of interneurons cross the midline to excite contralateral extensor motor neurons and inhibit flexors (Fig. 5.10).

There are two types of flexor withdrawal reflexes: those that result from innocuous stimuli and those that result from potentially injurious stimulation. The first type produces a localized flexor response accompanied by slight or no limb withdrawal; the second type produces widespread flexor contraction throughout the limb and abrupt withdrawal. The function of the first type of reflex is less obvious, but may be a general mechanism for adjusting the movement of a body part when an obstacle is detected by cutaneous sensory input. The function of the second type is protection of the individual. The endangered body part is rapidly removed, and postural support of the opposite side is strengthened if needed (e.g., if the foot is being withdrawn).

Collectively, these reflexes provide for stability and postural support (the myotatic and inverse myotatic) and mobility (flexor withdrawal). The reflexes provide a foundation of automatic responses on which more complicated voluntary movements are built.

The Spinal Cord Can Produce Basic Locomotor Actions

For locomotion, muscle action must occur in the limbs, but the posture of the trunk must also be controlled to provide a foundation from which the limb muscles can act. For example, when a human takes a step forward, not only must the advancing leg flex at the hip and knee, the opposite leg and bilateral truncal muscles must also be properly activated to prevent collapse of the body as weight is shifted from one leg to the other. Responsibility for the different functions that come together in successful locomotion is divided between several levels of the central nervous system.

Studies in experimental animals, mostly cats, have demonstrated that the spinal cord contains the capability for generating basic locomotor movements. This neural circuitry, called a **central pattern generator**, can produce the alternating contraction of limb flexors and extensors that is needed for walking. It has been shown experimentally that application of an excitatory amino acid like glutamate to the spinal cord produces rhythmic action potentials in motor neurons. Each limb has its own pattern generator, and the actions of different limbs are then coordinated. The normal strategy for generating basic locomotion engages central pattern generators and uses both sensory feedback and efferent impulses from higher motor control centers for the refinement of control.

Spinal Cord Injury Alters Voluntary and Reflex Motor Activity

When the spinal cord of a human or other mammal is severely injured, voluntary and reflex movements are immediately lost caudal to the level of injury. This acute impairment of function is called **spinal shock**. The loss of voluntary motor control is termed **plegia,** and the loss of reflexes is termed **areflexia**. Spinal shock may last from days to months, depending on the severity of cord injury. Reflexes tend to return, as may some degree of voluntary control. As recovery proceeds, myotatic reflexes become hyperactive, as demonstrated by an excessively vigorous response to tapping the muscle tendon with a reflex hammer. Tendon tapping, or even limb repositioning that produces a change in the muscle length, may also provoke **clonus**, a condition characterized by repetitive contraction and relaxation of a muscle in an oscillating fashion every second or so, in response to a single stimulus. Flexor withdrawal reflexes may also reappear and be provoked by lesser stimuli than would be normally required. The acute loss and eventual overactivity of all of these reflexes results from the lack of influence of the neural tracts that descend from higher motor control centers to the motor neurons and associated interneuron pools.

SUPRASPINAL INFLUENCES ON MOTOR CONTROL

Descending signals from the cervical spinal cord, brainstem, and cortex can influence the rate of motor neuron firing and the recruitment of additional motor neurons to increase the speed and force of muscle contraction. The influence of higher motor control centers is illustrated by a walking dog whose right and left limbs show alternating contractions and then change to a running pattern in which both sides contract in synchrony.

The brainstem contains the neural circuitry for initiating locomotion and for controlling posture. The maintenance of posture requires coordinated activity of both axial and limb muscles in response to input from proprioceptors and spatial position sensors, such as the inner ear. Cerebral cortex input through the corticospinal system is necessary for the control of fine individual movements of the distal limbs and digits. Each higher level of the nervous system acts on lower levels to produce appropriate, more refined movements.

The Brainstem Is the Origin of Three Descending Tracts That Influence Movement

Three brainstem nuclear groups give rise to descending motor tracts that influence motor neurons and their associated interneurons. These consist of the **red nucleus**, the **vestibular nuclear complex**, and the **reticular formation** (Fig. 5.11). The other major descending influence on the motor neurons is the corticospinal tract, the only volitional control pathway in the motor system. In most cases, the descending pathways act through synaptic connections on interneurons. The connection is less commonly made directly with motor neurons.

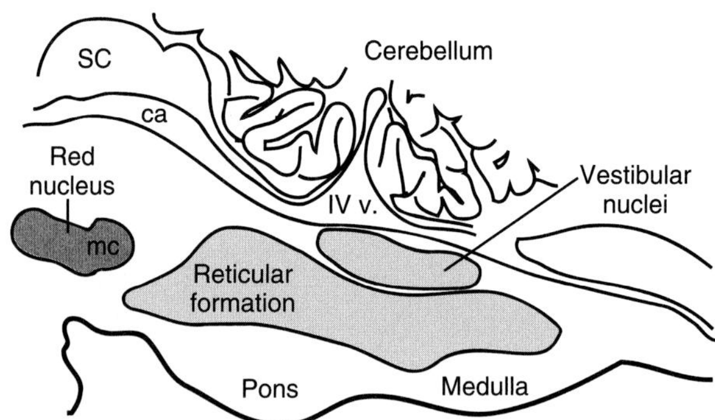

FIGURE 5.11 **Brainstem nuclei of descending motor pathways.** The magnocellular portion of the red nucleus is the origin of the rubrospinal tract. The lateral vestibular nucleus is the source of the vestibulospinal tract. The reticular formation is the source of two tracts, one from the pontine portion and one from the medulla. Structures illustrated are from the monkey. SC, superior colliculus; ca, cerebral aqueduct; IV v., fourth ventricle; Red nucleus mc, red nucleus magnocellular area.

The Rubrospinal Tract. The red nucleus of the mesencephalon receives major input from both the cerebellum and the cerebral cortical motor areas. Output via the **rubrospinal tract** is directed predominantly to contralateral spinal motor neurons that are involved with movements of the distal limbs. The axons of the rubrospinal tract are located in the lateral spinal white matter, just anterior to the corticospinal tract. Rubrospinal action enhances the function of motor neurons innervating limb flexor muscles while inhibiting extensors. This tract may also influence gamma motor neuron function.

Electrophysiological studies reveal that many rubrospinal neurons are active during locomotion, with more than half showing increased activity during the swing phase of stepping, when the flexors are most active. This system appears to be important for the production of movement, especially in the distal limbs. Experimental lesions that interrupt rubrospinal axons produce deficits in distal limb flexion, with little change in more proximal muscles. In higher animals, the corticospinal tract supersedes some of the function of the rubrospinal tract.

The Vestibulospinal Tract. The vestibular system regulates muscular function for the maintenance of posture in response to changes in the position of the head in space and accelerations of the body. There are four major nuclei in the vestibular complex: the **superior**, **lateral**, **medial**, and **inferior vestibular nuclei**. These nuclei, located in the pons and medulla, receive afferent action potentials from the vestibular portion of the ear, which includes the semicircular canals, the utricle, and the saccule (see Chapter 4). Information about rotatory and linear motions of the head and body are conveyed by this system. The vestibular nuclei are reciprocally connected with the superior colliculus on the dorsal surface of the mesencephalon. Input from the retina is received there and is utilized in adjusting eye position during movement of the head. Reciprocal connections to the vestibular nuclei are also made with the cerebellum and reticular formation.

The chief output to the spinal cord is the **vestibulospinal tract**, which originates predominantly from the lateral vestibular nucleus. The tract's axons are located in the anterior-lateral white matter and carry excitatory action potentials to ipsilateral extensor motor neuron pools, both alpha and gamma. The extensor motor neurons and their musculature are important in the maintenance of posture. Lesions in the brainstem secondary to stroke or trauma may abnormally enhance the influence of the vestibulospinal tract and produce dramatic clinical manifestations (see Clinical Focus Box 5.1).

The Reticulospinal Tract. The reticular formation in the central gray matter core of the brainstem contains many axon bundles interwoven with cells of various shapes and sizes. A prominent characteristic of reticular formation neurons is that their axons project widely in ascending and descending pathways, making multiple synaptic connections throughout the neuraxis. The medial region of the reticular formation contains large neurons that project upward to the thalamus, as well as downward to the spinal cord. Afferent input to the reticular formation comes from the spinal cord, vestibular nuclei, cerebellum, lateral hypothalamus, globus pallidus, tectum, and sensorimotor cortex.

Two areas of the reticular formation are important in the control of motor neurons. The descending tracts arise from the **nucleus reticularis pontis oralis and nucleus reticularis pontis caudalis** in the pons, and from the **nucleus reticularis gigantocellularis** in the medulla. The pontine reticular area gives rise to the ipsilateral **pontine reticulospinal tract**, whose axons descend in the medial spinal cord white matter. These axons carry excitatory action potentials to interneurons that influence alpha and gamma motor neuron pools of axial muscles. The medullary area gives rise to the **medullary reticulospinal tract**, whose axons descend mostly ipsilateral in the anterior spinal white matter. These axons have inhibitory influences on interneurons that modulate extensor motor neurons.

The Terminations of the Brainstem Motor Tracts Correlate With Their Functions

The vestibulospinal and reticulospinal tracts descend medially in the spinal cord and terminate in the ventromedial part of the intermediate zone, an area in the gray matter containing propriospinal interneurons (Fig. 5.12). There are also some direct connections with motor neurons of the neck and back muscles and the proximal limb muscles. These tracts are the main CNS pathways for maintaining posture and head position during movement.

The rubrospinal tract descends laterally in the spinal cord and terminates mostly on interneurons in the lateral spinal intermediate zone, but it also has some monosynaptic connections directly on motor neurons to muscles of the distal extremities. This tract supplements the medial descending pathways in postural control and the corticospinal tract for independent movements of the extremities.

In accordance with their medial or lateral distributions to spinal motor neurons, the reticulospinal and vestibulospinal tracts are thought to be most important for the control of axial and proximal limb muscles, whereas the rubrospinal (and corticospinal) tracts are most important for the control of distal limb muscles, particularly the flexors.

Sensory and Motor Systems Work Together to Control Posture

The maintenance of an upright posture in humans requires active muscular resistance against gravity. For movement to occur, the initial posture must be altered by flexing some body parts against gravity. Balance must be maintained during movement, which is achieved by postural reflexes initiated by several key sensory systems. Vision, the vestibular system, and the somatosensory system are important for postural reflexes.

CLINICAL FOCUS BOX 5.1

Decerebrate Rigidity

A patient with a history of poorly controlled hypertension, a result of noncompliance with his medication, is brought to the emergency department because of sudden collapse and subsequent unresponsiveness. A neurological examination performed about 30 minutes after onset of the collapse shows no response to verbal stimuli. No spontaneous movements of the limbs are observable. A mildly painful stimulus, compression of the soft tissue of the supraorbital ridge, causes immediate extension of the neck and both arms and legs. This posture relaxes within a few seconds after the stimulation is stopped. After the patient is stabilized medically, he undergoes a magnetic resonance imaging (MRI) study of the brain. The study demonstrates a large area of hemorrhage bilaterally in the upper pons and lower mesencephalon.

The posture this patient demonstrated in response to a noxious stimulus is termed **decerebrate rigidity**. Its presence is associated with lesions of the mesencephalon that isolate the portions of the brainstem below that level from the influence of higher centers. The abnormal posture is a result of extreme activation of the antigravity extensor muscles by the unopposed action of the lateral vestibular nucleus and the vestibulospinal tract. A model of this condition can be produced in experimental animals by a surgical lesion located between the mesencephalon and pons. It can also be shown in experimental animals that a destructive lesion of the lateral vestibular nucleus relieves the rigidity on that side.

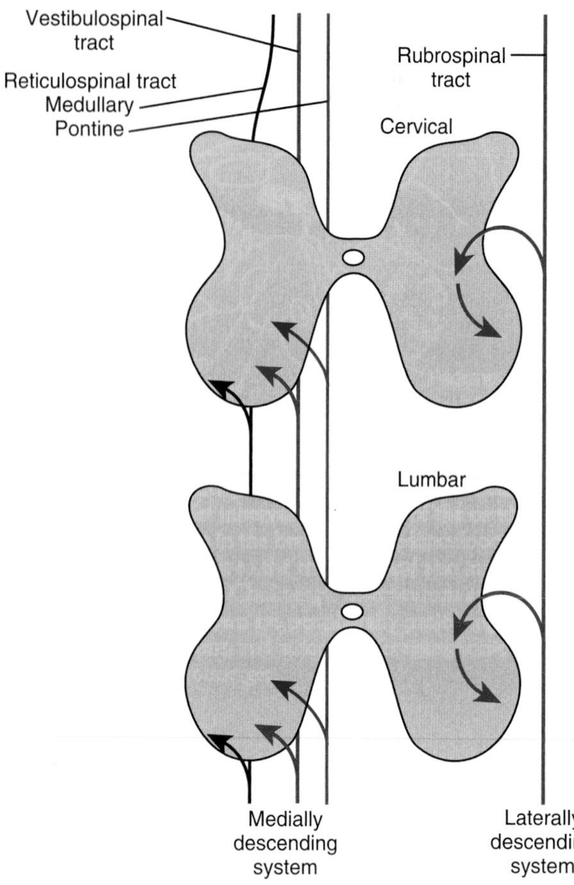

FIGURE 5.12 **Brainstem motor control tracts.** The vestibulospinal and reticulospinal tracts influence motor neurons that control axial and proximal limb muscles. The rubrospinal tract influences motor neurons controlling distal limb muscles. Excitatory pathways are shown in red.

Somatosensory input provides information about the position and movement of one part of the body with respect to others. The vestibular system provides information about the position and movement of the head and neck with respect to the external world. Vision provides both types of information, as well as information about objects in the external world. Visual and vestibular reflexes interact to produce coordinated head and eye movements associated with a shift in gaze. Vestibular reflexes and somatosensory neck reflexes interact to produce reflex changes in limb muscle activity. The quickest of these compensations occurs at about twice the latency of the monosynaptic myotatic reflex. These response types are termed **long loop reflexes**. The extra time reflects the action of other neurons at different anatomic levels of the nervous system.

THE ROLE OF THE CEREBRAL CORTEX IN MOTOR CONTROL

The cerebral cortical areas concerned with motor function exert the highest level of motor control. It is difficult to formulate an unequivocal definition of a **cortical motor area**, but three criteria may be used. An area is said to have a motor function if

- Stimulation using very low current strengths elicits movements.
- Destruction of the area results in a loss of motor function.
- The area has output connections going directly or relatively directly (i.e., with a minimal number of intermediate connections) to the motor neurons.

Some cortical areas fulfill all of these criteria and have exclusively motor functions. Other areas fulfill only some of the criteria yet are involved in movement, particularly volitional movement.

Distinct Cortical Areas Participate in Voluntary Movement

The **primary motor cortex** (MI), Brodmann's area 4, fulfills all three criteria for a motor area (Fig. 5.13). The **supplementary motor cortex** (MII), which also fulfills all three criteria, is rostral and medial to MI in Brodmann's area 6. Other areas that fulfill some of the criteria include the rest of Brodmann's area 6; areas 1, 2, and 3 of the postcentral

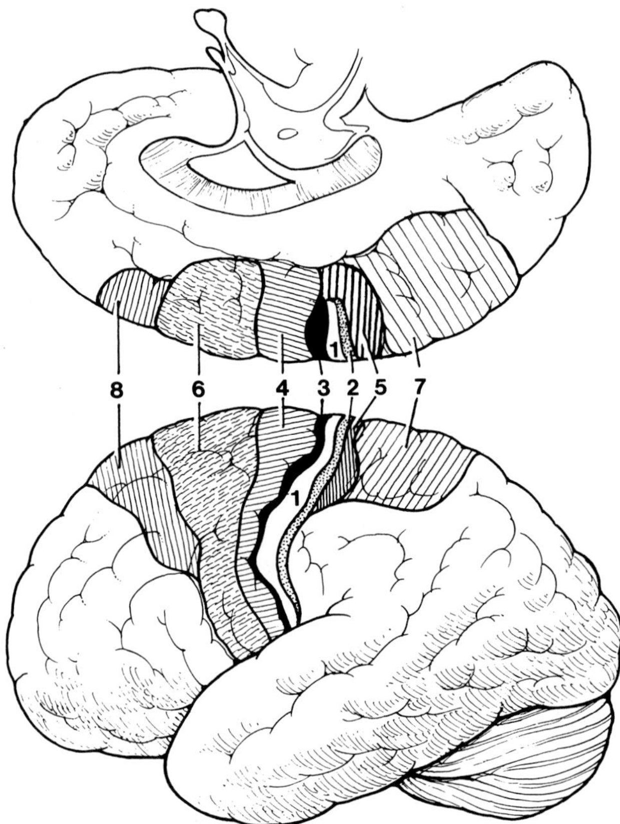

FIGURE 5.13 **Brodmann's cytoarchitectural map of the human cerebral cortex.** Area 4 is the primary motor cortex (MI); area 6 is the premotor cortex and includes the supplementary motor area (MII) on the medial aspect of the hemisphere; area 8 influences voluntary eye movements; areas 1, 2, 3, 5, and 7 have sensory functions but also contribute axons to the corticospinal tract.

gyrus; and areas 5 and 7 of the parietal lobe. All of these areas contribute fibers to the **corticospinal tract,** the efferent motor pathway from the cortex.

The Primary Motor Cortex (MI). This cortical area corresponds to Brodmann's area 4 in the precentral gyrus. Area 4 is structured in six well-defined layers (I to VI), with layer I being closest to the pial surface. Afferent fibers terminate in layers I to V. Thalamic afferent fibers terminate in two layers; those that carry somatosensory information end in layer IV, and those from nonspecific nuclei end in layer I. Cerebellar afferents terminate in layer IV. Efferent axons arise in layers V and VI to descend as the corticospinal tract. Body areas are represented in an orderly manner, as **somatotopic maps**, in the motor and sensory cortical areas (Fig 5.14). Those parts of the body that perform fine movements, such as the digits and the facial muscles, are controlled by a greater number of neurons that occupy more cortical territory than the neurons for the body parts only capable of gross movements.

Low-level electrical stimulation of MI produces twitch-like contraction of a few muscles or, less commonly, a single muscle. Slightly stronger stimuli also produce responses in adjacent muscles. Movements elicited from area 4 have the lowest stimulation thresholds and are the most discrete of any movements elicited by stimulation. Stimulation of MI limb areas produces contralateral movement, while cranial cortical areas may produce bilateral motor responses. Destruction of any part of the primary motor cortex leads to immediate paralysis of the muscles controlled by that area. In humans, some function may return weeks to months later, but the movements lack the fine degree muscle control of the normal state. For example, after a lesion in the arm area of MI, the use of the hand recovers, but the capacity for discrete finger movements does not.

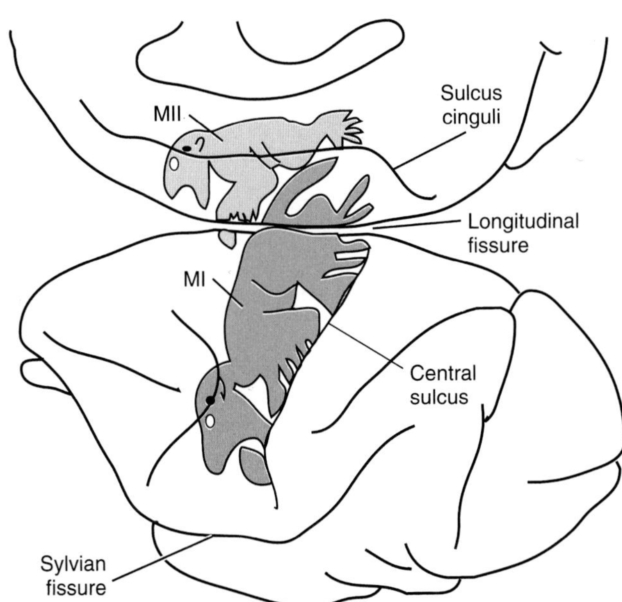

FIGURE 5.14 **A cortical map of motor functions.** Primary motor cortex (MI) and supplementary motor cortex (MII) areas in the monkey brain. MII is on the medial aspect of the hemisphere.

Neurons in MI encode the capability to control muscle force, muscle length, joint movement, and position. The area receives somatosensory input, both cutaneous and proprioceptive, via the ventrobasal thalamus. The cerebellum projects to MI via the red nucleus and ventrolateral thalamus. Other afferent projections come from the nonspecific nuclei of the thalamus, the contralateral motor cortex, and many other ipsilateral cortical areas. There are many axons between the precentral (motor) and postcentral (somatosensory) gyri and many connections to the visual cortical areas. Because of their connections with the somatosensory cortex, the cortical motor neurons can also respond to sensory stimulation. For example, cells innervating a particular muscle may respond to cutaneous stimuli originating in the area of skin that moves when that muscle is active, and they may respond to proprioceptive stimulation from the muscle to which they are related. Many efferent fibers from the primary motor cortex terminate in brain areas that contribute to ascending somatic sensory pathways. Through these connections, the motor cortex can control the flow of somatosensory information to motor control centers.

The close coupling of sensory and motor functions may play a role in two cortically controlled reflexes that were originally described in experimental animals as being important for maintaining normal body support during locomotion—the placing and hopping reactions. The **placing reaction** can be demonstrated in a cat by holding it so that its limbs hang freely. Contact of any part of the animal's foot with the edge of a table provokes immediate placement of the foot on the table surface. The **hopping reaction** is demonstrated by holding an animal so that it stands on one leg. If the body is moved forward, backward, or to the side, the leg hops in the direction of the movement so that the foot is kept directly under the shoulder or hip, stabilizing the body position. Lesions of the contralateral precentral or postcentral gyrus abolish placing. Hopping is abolished by a contralateral lesion of the precentral gyrus.

The Supplementary Motor Cortex (MII). The MII cortical area is located on the medial surface of the hemispheres, above the cingulate sulcus, and rostral to the leg area of the primary motor cortex (see Fig. 5.14). This cortical region within Brodmann's area 6 has no clear cytoarchitectural boundaries; that is, the shapes and sizes of cells and their processes are not obviously compartmentalized, as in the layers of MI.

Electrical stimulation of MII produces movements, but a greater strength of stimulating current is required than for MI. The movements produced by stimulation are also qualitatively different from MI; they last longer, the postures elicited may remain after the stimulation is over, and the movements are less discrete. Bilateral responses are common. MII is reciprocally connected with MI, and receives input from other motor cortical areas. Experimental lesions in MI eliminate the ability of MII stimulation to produce movements.

Current knowledge is insufficient to adequately describe the unique role of MII in higher motor functions. MII is thought to be active in bimanual tasks, in learning and preparing for the execution of skilled movements, and in the control of muscle tone. The mechanisms that underlie the more complex aspects of movement, such as thinking about and performing skilled movements and using complex sensory information to guide movement, remain incompletely understood.

The Primary Somatosensory Cortex and Superior Parietal Lobe. The **primary somatosensory cortex** (Brodmann's areas 1, 2, and 3) lies on the postcentral gyrus (see Fig. 5.13) and has a role in movement. Electrical stimulation here can produce movement, but thresholds are 2 to 3 times higher than in MI. The somatosensory cortex is reciprocally interconnected with MI in a somatotopic pattern—for example, arm areas of sensory cortex project to arm areas of motor cortex. Efferent fibers from areas 1, 2, and 3 travel in the corticospinal tract and terminate in the dorsal horn areas of the spinal cord.

The **superior parietal lobe** (Brodmann's areas 5 and 7) also has important motor functions. In addition to contributing fibers to the corticospinal tract, it is well connected to the motor areas in the frontal lobe. Lesion studies in animals and humans suggest this area is important for the utilization of complex sensory information in the production of movement.

The Corticospinal Tract Is the Primary Efferent Path From the Cortex

Traditionally, the descending motor tract originating in the cerebral cortex has been called the **pyramidal tract** because it traverses the medullary pyramids on its way to the spinal cord (Fig. 5.15). This path is the **corticospinal tract**. All other descending motor tracts emanating from the brainstem were generally grouped together as the extrapyramidal system. Cells in Brodmann's area 4 (MI) contribute 30% of the corticospinal fibers; area 6 (MII) is the origin of 30% of the fibers; and the parietal lobe, especially Brodmann's areas 1, 2, and 3, supplies 40%. In primates, 10 to 20% of corticospinal fibers ends directly on motor neurons; the others end on interneurons associated with motor neurons.

From the cerebral cortex, the corticospinal tract axons descend through the brain along a path located between the basal ganglia and the thalamus, known as the **internal capsule**. They then continue along the ventral brainstem as the **cerebral peduncles** and on through the pyramids of the medulla. Most of the corticospinal axons cross the midline in the medullary pyramids; thus, the motor cortex in each hemisphere controls the muscles on the contralateral side of the body. After crossing in the medulla, the corticospinal axons descend in the dorsal lateral columns of the spinal cord and terminate in lateral motor pools that control distal muscles of the limbs. A smaller group of axons do not cross in the medulla and descend in the ventral spinal columns. These axons terminate in the motor pools and adjacent intermediate zones that control the axial and proximal musculature.

The corticospinal tract is estimated to contain about 1 million axons at the level of the medullary pyramid. The largest-diameter, heavily myelinated axons are between 9 and 20 μm in diameter, but that population accounts for only a small fraction of the total. Most corticospinal axons are small, 1 to 4 μm in diameter, and half are unmyelinated.

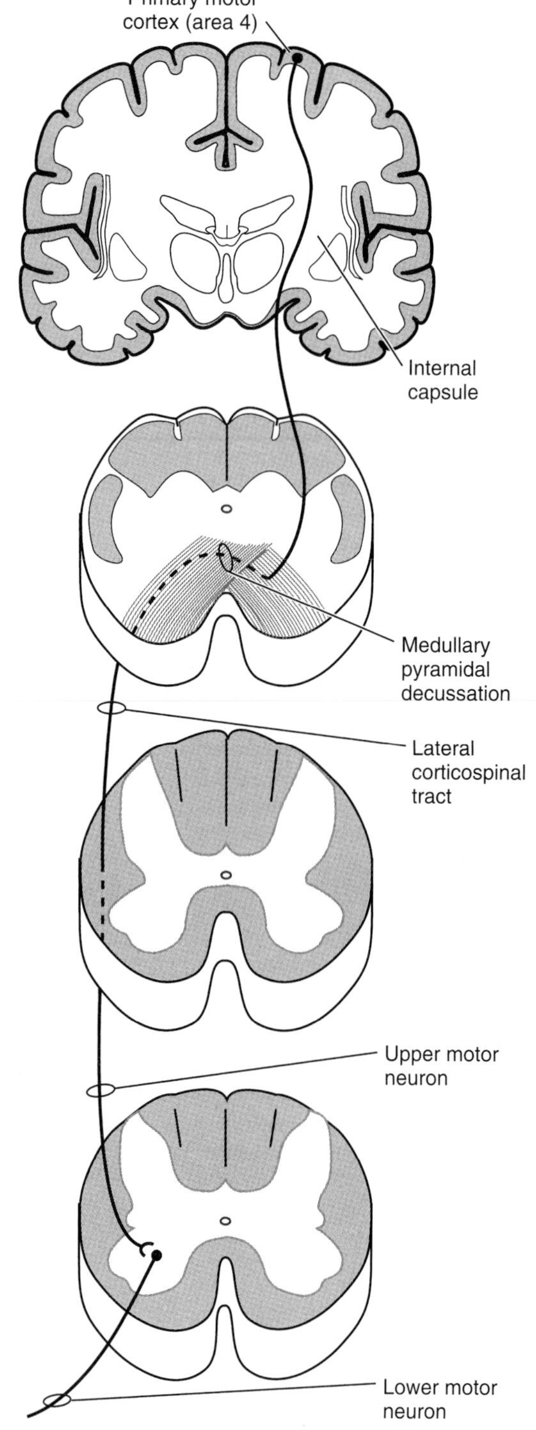

FIGURE 5.15 **The corticospinal tract.** Axons arising from cortical neurons, including the primary motor area, descend through the internal capsule, decussate in the medulla, travel in the lateral area of the spinal cord as the lateral corticospinal tract, and terminate on motor neurons and interneurons in the ventral horn areas of the spinal cord. Note the upper and lower motor neuron designations.

In addition to the direct corticospinal tract, there are other indirect pathways by which cortical fibers influence motor function. Some cortical efferent fibers project to the reticular formation, then to the spinal cord via the reticulospinal tract; others project to the red nucleus, then to the spinal cord via the rubrospinal tract. Despite the fact that these pathways involve intermediate neurons on the way to the cord, volleys relayed through the reticular formation can reach the spinal cord motor circuitry at the same time as, or earlier than, some volleys along the corticospinal tract.

THE BASAL GANGLIA AND MOTOR CONTROL

The **basal ganglia** are a group of subcortical nuclei located primarily in the base of the forebrain, with some in the diencephalon and upper brainstem. The striatum, globus pallidus, subthalamic nucleus, and substantia nigra comprise the basal ganglia. Input is derived from the cerebral cortex and output is directed to the cortical and brainstem areas concerned with movement. Basal ganglia action influences the entire motor system and plays a role in the preparation and execution of coordinated movements.

The forebrain (telencephalic) components of the basal ganglia consist of the **striatum,** which is made up of the **caudate nucleus** and the **putamen**, and the **globus pallidus**. The caudate nucleus and putamen are histologically identical but are separated anatomically by fibers of the anterior limb of the internal capsule. The globus pallidus has two subdivisions: the **external segment** (GPe), adjacent to the medial aspect of the putamen, and the **internal segment** (GPi), medial to the GPe. The other main nuclei of the basal ganglia are the **subthalamic nucleus** in the diencephalon and the **substantia nigra** in the mesencephalon.

The Basal Ganglia Are Extensively Interconnected

Although the circuitry of the basal ganglia appears complex at first glance, it can be simplified into input, output, and internal pathways (Fig. 5.16). Input is derived from the cerebral cortex and is directed to the striatum and the subthalamic nucleus. The predominant nerve cell type in the striatum is termed the **medium spiny neuron,** based on its cell body size and dendritic structure. This type of neuron receives input from all of the cerebral cortex except for the primary visual and auditory areas. The input is roughly somatotopic and is via neurons that use glutamate as the neurotransmitter. The putamen receives the majority of the cortical input from sensorimotor areas. Input to the subthalamus is from the cortical areas concerned with motor function, including eye movement, and is also via glutamate-releasing neurons.

Basal ganglia output is from the internal segment of the globus pallidus (GPi) and one segment of the substantia nigra. The GPi output is directed to ventrolateral and ventral anterior nuclei of the thalamus, which feed back to the cortical motor areas. The output of the GPi is also directed to a region in the upper brainstem termed the **midbrain extrapyramidal area**. This latter area then projects to the neurons of the reticulospinal tract. The substantia nigra output arises from the **pars reticulata** (SNr), which is histologically

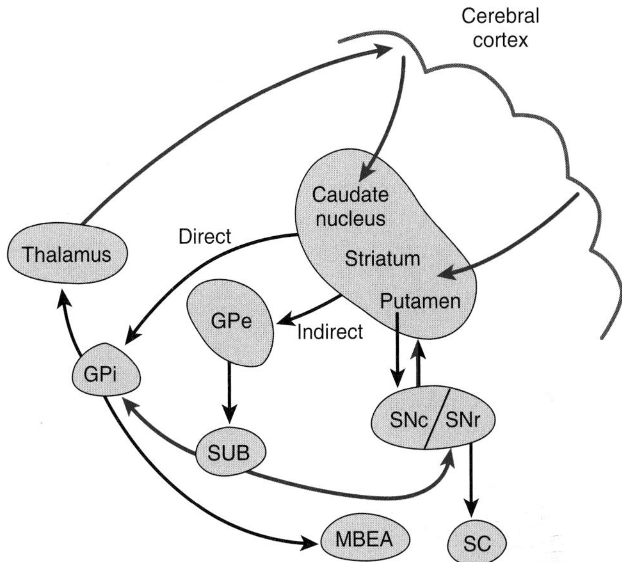

FIGURE 5.16 **Basal ganglia nuclei and circuitry.** The circuit of cerebral cortex to striatum to GPi to thalamus and back to the cortex is the main pathway for basal ganglia influence on motor control. Note the direct and indirect pathways involving the striatum, GPi, GPe, and subthalamic nucleus. GPi output is also directed to the midbrain extrapyramidal area (MBEA). The SNr to SC pathway is important in eye movements. Excitatory pathways are shown in red, inhibitory pathways are in black. GPe and GPi, globus pallidus externa and interna; SUB, subthalamic nucleus; SNc and SNr, substantia nigra pars compacta and pars reticulata; SC, superior colliculus.

similar to the GPi. The output is directed to the superior colliculus of the mesencephalon, which is involved in eye movement control. The GPi and SNr output is inhibitory via neurons that use GABA as the neurotransmitter.

The internal pathway circuits link the various nuclei of the basal ganglia. The globus pallidus externa (GPe), the subthalamic nucleus, and the pars compacta region of the substantia nigra (SNc) are the nuclei in these pathways. The GPe receives inhibitory input from the striatum via GABA-releasing neurons. The output of the GPe is also inhibitory via GABA release and is directed to the GPi and the subthalamic nucleus. The subthalamic nucleus output is excitatory and is directed to the GPi and the SNr. This striatum-GPe-subthalamic nucleus-GPi circuit has been termed the **indirect pathway** in contrast to the **direct pathway** of striatum to Gpi (see Fig. 5.16). The SNc receives inhibitory input from the striatum and produces output back to the striatum via dopamine-releasing neurons. The output can be either excitatory or inhibitory depending on the receptor type of the target neurons in the striatum. The action of the SNc may modulate cortical input to the striatum.

The Functions of the Basal Ganglia Are Partially Revealed by Disease

Basal ganglia diseases produce profound motor dysfunction in humans and experimental animals. The disorders can result in reduced motor activity, **hypokinesis**, or abnormally enhanced activity, **hyperkinesis**. Two well-known neurological conditions that show histological abnormality in basal ganglia structures, Parkinson's disease and Huntington's disease, illustrate the effects of basal ganglia dysfunction. Patients with **Parkinson's disease** show a general slowness of initiation of movement and paucity of movement when in motion. The latter takes the form of reduced arm swing and lack of truncal swagger when walking. These patients also have a resting tremor of the hands, described as "pill rolling." The tremor stops when the hand goes into active motion. At autopsy, patients with Parkinson's disease show a severe loss of dopamine-containing neurons in the SNc region. Patients with **Huntington's disease** have uncontrollable, quick, brief movements of individual limbs. These movements are similar to what a normal individual might show when flicking a fly off a hand or when quickly reaching up to scratch an itchy nose. At autopsy, a severe loss of striatal neurons is found.

The function of the basal ganglia in normal individuals remains unclear. One theory is that the primary action is to inhibit undesirable movements, thereby, allowing desired motions to proceed. Neuronal activity is increased in the appropriate areas of the basal ganglia prior to the actual execution of movement. The basal ganglia act as a brake on undesirable motion by the inhibitory output of the GPi back to the cortex through the thalamus. Enhanced output from the GPi increases this braking effect. The loss of dopamine-releasing neurons in Parkinson's disease is thought to produce this type of result by reducing inhibitory influence on the striatum and, thereby, increasing the excitatory action of the subthalamic nucleus on the GPi through the indirect basal ganglia pathway (see Clinical Focus Box 5.2). Hyperkinetic disorders like Huntington's disease are thought to result from decreased GPi output secondary to the loss of inhibitory influence of the striatum through the direct pathway.

THE CEREBELLUM IN THE CONTROL OF MOVEMENT

The **cerebellum**, or "little brain," lies caudal to the occipital lobe and is attached to the posterior aspect of the brainstem through three paired fiber tracts: the **inferior**, **middle**, and **superior cerebellar peduncles**. Input to the cerebellum comes from peripheral sensory receptors, the brainstem, and the cerebral cortex. The inferior, middle and, to a lesser degree, superior cerebellar peduncles carry the input. The output projections are mainly, if not totally, to other motor control areas of the central nervous system and are mostly carried in the superior cerebellar peduncle. The cerebellum contains three pairs of intrinsic nuclei: the **fastigial**, **interpositus** (interposed), and **dentate**. In some classification schemes, the interposed nucleus is further divided into the **emboliform** and **globose** nuclei.

The Structural Divisions of the Cerebellum Correlate With Function

The cerebellar surface is arranged in multiple, parallel, longitudinal folds termed **folia**. Several deep fissures divide the cerebellum into three main morphological components—the **anterior**, **posterior**, and **flocculonodular lobes**, which also correspond with the functional subdivisions of the cerebellum (Fig. 5.17). The functional divisions are the vestibulocerebellum, the spinocerebellum, and the cerebrocerebellum. These divisions appear in sequence during evo-

CLINICAL FOCUS BOX 5.2

Stereotactic Neurosurgery for Parkinson's disease

Parkinson's disease is a CNS disorder producing a generalized slowness of movement and resting tremor of the hands. Loss of dopamine-producing neurons in the substantia nigra pars compacta is the cause of the condition. Treatment with medications that stimulate an increased production of dopamine by the surviving substantia nigra neurons has revolutionized the management of Parkinson's disease. Unfortunately, the benefit of the medications tends to lessen after 5 to 10 years of treatment. Increasing difficulty in initiating movement and worsening slowness of movement are features of a declining responsiveness to medication. Improved knowledge of basal ganglia circuitry has enabled neurosurgeons to develop surgical procedures to ameliorate some of the effects of the advancing disease.

Degeneration of the dopamine-releasing cells of the substantia nigra reduces excitatory input to the putamen. Inhibitory output of the putamen to the GPe greatly increases via the indirect pathway. This results in decreased inhibitory GPe output to the subthalamic nucleus, which, in turn, acts unrestrained to stimulate the GPi. Stimulation of the GPi enhances its inhibitory influence on the thalamus and results in decreased excitatory drive back to the cerebral cortex.

Stereotactic neurosurgery is a technique in which a small probe can be precisely placed into a target within the brain. Magnetic resonance imaging (MRI) of the brain defines the three-dimensional location of the GPi. The surgical probe is introduced into the brain through a small hole made in the skull and is guided to the target by the surgeon using the MRI coordinates. The correct positioning of the probe into the GPi can be further confirmed by recording the electrical activity of the GPi neurons with an electrode located at the tip of the probe. GPi neurons have a continuous, high-frequency firing pattern that, when amplified and presented on a loudspeaker, sounds like heavy rain striking a metal roof. When the target location is reached, the probe is heated to a temperature that destroys a precisely controllable amount of the GPi. The inhibitory outflow of the GPi is reduced and movement improves.

The use of implantable stimulators to modify activity of the basal ganglia nuclei is also being investigated to improve function in patients with Parkinson's disease and other types of movement disorders.

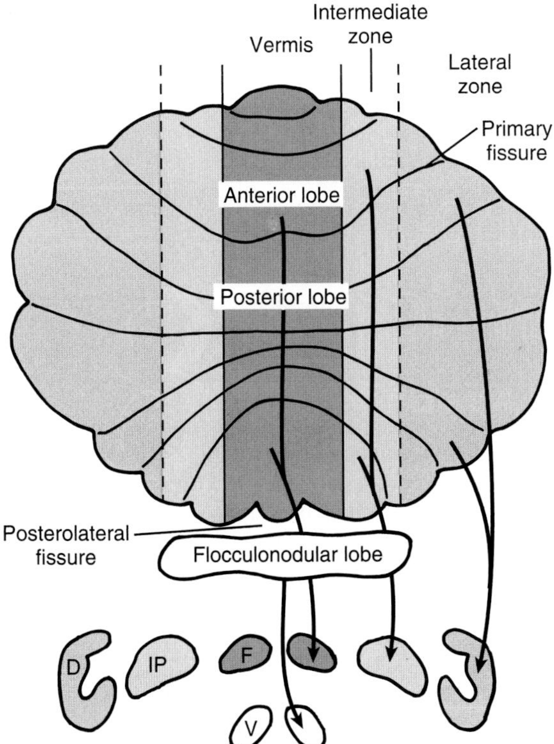

FIGURE 5.17 **The structure of the cerebellum.** The three lobes are shown: anterior, posterior, and flocculonodular. The functional divisions are demarcated by color. The vestibulocerebellum (white) is the flocculonodular lobe and projects to the vestibular (V) nuclei. The spinocerebellum includes the vermis (dark pink) and intermediate zone (pink), which project to the fastigial (F) and interposed (IP) nuclei, respectively. The cerebrocerebellum (gray) projects to the dentate nuclei (D).

lution. The lateral cerebellar hemispheres increase in size along with expansion of the cerebral cortex. The three divisions have similar intrinsic circuitry; thus, the function of each depends on the nature of the output nucleus to which it projects.

The **vestibulocerebellum** is composed of the flocculonodular lobe. It receives input from the vestibular system and visual areas. Output goes to the vestibular nuclei, which can, in a sense, be considered as an additional pair of intrinsic cerebellar nuclei. The vestibulocerebellum functions to control equilibrium and eye movements.

The medially placed **spinocerebellum** consists of the midline **vermis** plus the medial portion of the lateral hemispheres, called the **intermediate zones**. Spinocerebellar pathways carrying somatosensory information terminate in the vermis and intermediate zones in somatotopic arrangements. The auditory, visual, and vestibular systems and sensorimotor cortex also project to this portion of the cerebellum. Output from the vermis is directed to the fastigial nuclei, which project through the inferior cerebellar peduncle to the vestibular nuclei and reticular formation of the pons and medulla. Output from the intermediate zones goes to the interposed nuclei and from there to the red nucleus and, ultimately, to the motor cortex via the ventrolateral nucleus of the thalamus. It is believed that both the fastigial and interposed nuclei contain a complete representation of the muscles of the body. The fastigial output system controls antigravity muscles in posture and locomotion, while the interposed nuclei, perhaps, act on stretch reflexes and other somatosensory reflexes.

The **cerebrocerebellum** occupies the lateral aspects of the cerebellar hemispheres. Input comes exclusively from the cerebral cortex, relayed through the middle cerebellar peduncles of the pons. The cortical areas that are prominent in motor control are the sources for most of this input. Output is directed to the dentate nuclei and from there via the ventrolateral thalamus back to the motor and premotor cortices.

The Intrinsic Circuitry of the Cerebellum Is Very Regular

The cerebellar cortex is composed of five types of neurons arranged into three layers (Fig. 5.18). The molecular layer is the outermost and consists mostly of axons and dendrites plus two types of interneurons, **stellate cells** and **basket cells**. The next layer contains the dramatic **Purkinje cells,** whose dendrites reach upward into the molecular layer in a fan-like array. The Purkinje cells are the only efferent neurons of the cerebellar cortex. Their action is inhibitory via GABA as the neurotransmitter. Deep to the Purkinje cells is the **granular layer**, containing **Golgi cells,** and small local circuit neurons, the **granule cells**. The granule cells are numerous; there are more granule cells in the cerebellum than neurons in the entire cerebral cortex!

Afferent axons to the cerebellar cortex are of two types: mossy fibers and climbing fibers. **Mossy fibers** arise from the spinal cord and brainstem neurons, including those of the pons that receive input from the cerebral

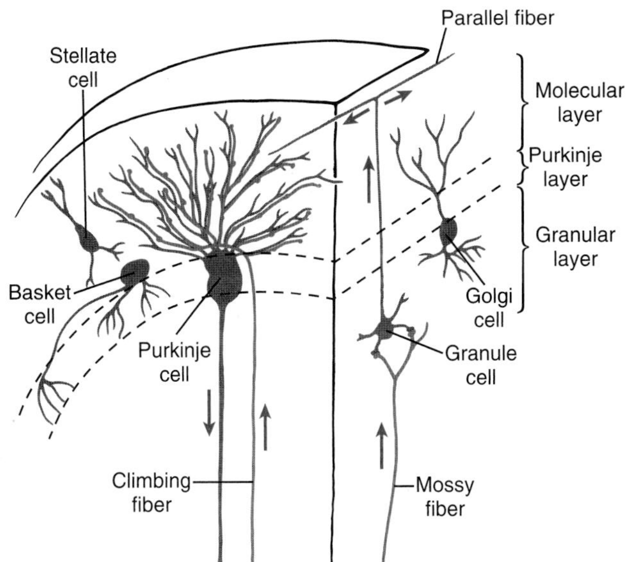

FIGURE 5.18 **Cerebellar circuitry.** The cell types and action potential pathways are shown. Mossy fibers bring afferent input from the spinal cord and the cerebral cortex. Climbing fibers bring afferent input from the inferior olive nucleus in the medulla and synapse directly on the Purkinje cells. The Purkinje cells are the efferent pathways of the cerebellum.

cortex. Mossy fibers make complex multicontact synapses on granule cells. The granule cell axons then ascend to the molecular layer and bifurcate, forming the **parallel fibers**. These travel perpendicular to and synapse with the dendrites of Purkinje cells, providing excitatory input via glutamate. Mossy fibers discharge at high tonic rates, 50 to 100 Hz, which increases further during voluntary movement. When mossy fiber input is of sufficient strength to bring a Purkinje cell to threshold, a single action potential results.

Climbing fibers arise from the **inferior olive,** a nucleus in the medulla. Each climbing fiber synapses directly on the dendrites of a Purkinje cell and exerts a strong excitatory influence. One action potential in a climbing fiber produces a burst of action potentials in the Purkinje cell called a complex spike. Climbing fibers also synapse with basket, Golgi, and stellate interneruons, which then make inhibitory contact with adjacent Purkinje cells. This circuitry allows a climbing fiber to produce excitation in a single Purkinje cell and inhibition in the surrounding ones.

Mossy and climbing fibers also give off excitatory collateral axons to the deep cerebellar nuclei before reaching the cerebellar cortex. The cerebellar cortical output (Purkinje cell efferents) is inhibitory to the cerebellar and vestibular nuclei, but the ultimate output of the cerebellar nuclei is mostly excitatory. A smaller population of neurons of the deep cerebellar nuclei produces inhibitory outflow directed mainly back to the inferior olive.

Lesions Reveal the Function of the Cerebellum. Lesions of the cerebellum produce impairment in the coordinated action of agonists, antagonists, and synergists. This impairment is clinically known as **ataxia**. The control of limb, axial, and cranial muscles may be impaired depending on the site of the cerebellar lesion. Limb ataxia might manifest as the coarse jerking motions of an arm and hand during reaching for an object instead of the expected, smooth actions. This jerking type of motion is also referred to as **action tremor**. The swaying walk of an intoxicated individual is a vivid example of truncal ataxia.

Cerebellar lesions can also produce a reduction in muscle tone, **hypotonia**. This condition is manifest as a notable decrease in the low level of resistance to passive joint movement detectable in normally relaxed individuals. Myotatic reflexes produced by tapping a tendon with a reflex hammer reverberate for several cycles (pendular reflexes) because of impaired damping from the reduced muscle tone. The hypotonia is likely a result of impaired processing of cerebellar afferent action potentials from the muscle spindles and Golgi tendon organs.

While these lesions establish a picture of the absence of cerebellar function, we are left without a firm idea of what the cerebellum does in the normal state. Cerebellar function is sometimes described as comparing the intended with the actual movement and adjusting motor system output in ongoing movements. Other putative functions include a role in learning new motor and even cognitive skills.

REVIEW QUESTIONS

DIRECTIONS: Each of the numbered items or incomplete statements in this section is followed by answers or by completions of the statement. Select the ONE lettered answer or completion that is BEST in each case.

1. Which type of motor unit is of prime importance in generating the muscle power necessary for the maintenance of posture?
(A) Low threshold, fatigue-resistant
(B) High threshold, fatigable
(C) Intrafusal, gamma controlled
(D) High threshold, high force
(E) Extrafusal, gamma controlled
2. Which type of sensory receptor provides information about the force of muscle contraction?
(A) Nuclear bag fiber
(B) Nuclear chain fiber
(C) Golgi tendon organ
(D) Bare nerve ending
(E) Type Ia ending
3. If a patient experiences enlargement of the normally rudimentary central canal of the spinal cord in the midcervical region, which, if any, muscular functions would become abnormal first?
(A) Finger flexion
(B) Elbow flexion
(C) Shoulder abduction
(D) Truncal extension
(E) No muscles would become abnormal
4. Tapping the patellar tendon with a reflex hammer produces a brief contraction of the knee extensors. What is the cause of the muscle contraction?
(A) Elastic rebound of muscle connective tissue
(B) Golgi tendon organ response
(C) Muscle spindle activation
(D) Muscle spindle unloading
(E) Gamma motor neuron discharge
5. The cyclical flexion and extension motions of a leg during walking result from activity at which level of the nervous system?
(A) Cerebral cortex
(B) Cerebellum
(C) Globus pallidus
(D) Red nucleus
(E) Spinal cord
6. Which brainstem-derived descending tract produces action similar to the corticospinal tract?
(A) Vestibulospinal
(B) Reticulospinal
(C) Spinocerebellar
(D) Rubrospinal
(E) None
7. What is the location of the primary motor area of the cerebral cortex?
(A) Upper parietal lobe
(B) Superior temporal lobe
(C) Precentral gyrus
(D) Postcentral gyrus
(E) Medial aspect of the hemisphere
8. Concurrent flexion of both wrists in response to electrical stimulation is characteristic of which area of the nervous system?
(A) Postcentral gyrus
(B) Vestibulospinal tract
(C) Dentate nucleus
(D) Primary motor cortex
(E) Supplementary motor cortex
9. If you could histologically examine the spinal cord of a patient who had experienced a viral illness 10 years before in which only the neurons of the primary motor area of the cerebral cortex were destroyed, what findings would you expect?
(A) The corticospinal tract would be completely degenerated
(B) The rubrospinal tract would show an increased number of axons

(continued)

(C) The corticospinal tract would be about one-third depleted of axons
(D) The alpha motor neurons would be atrophic
(E) The corticospinal tract would be normal

10. A disease that produces decreased inhibitory input to the internal segment of the globus pallidus should have what effect on the motor area of the cerebral cortex?
(A) Increased excitatory feedback directly to the cortex
(B) No effect
(C) Decreased excitatory output from the thalamus to the cortex
(D) Increased excitatory output from the putamen to the cortex
(E) Increased excitatory output from the thalamus to the cortex

11. Which cerebellar component would be abnormal in a degenerative disease that affected spinal sensory neurons?
(A) Purkinje cells
(B) Mossy fibers
(C) Parallel fibers
(D) Climbing fibers
(E) Granule cells

SUGGESTED READING

Alexander G, Crutcher M, DeLong M. Basal ganglia-thalamocortical circuits: Parallel substrates for motor, oculomotor, prefrontal, and limbic functions. Progr Brain Res 1990;85:119–146.

Kandel E, Schwartz J, Jessel T, eds. Principles of Neural Science. 4th Ed. New York: McGraw-Hill, 2000.

Parent A. Carpenter's Human Neuroanatomy. 9th Ed. Media, PA: Williams & Wilkins, 1996.

Wichmann T, DeLong MR. Functional and pathophysiological models of the basal ganglia. Curr Opin Neurobiol 1996;6:751–758.

Zigmond M, Bloom F, Landis S, et al. Fundamentals of Neuroscience. San Diego: Academic Press, 1999.

The Autonomic Nervous System

John C. Kincaid, M.D.

CHAPTER OUTLINE

KEY CONCEPTS

1. The autonomic nervous system regulates the involuntary functions of the body.
2. The autonomic nervous system has three divisions: sympathetic, parasympathetic, and enteric.
3. A two-neuron efferent path is utilized by the autonomic nervous system.
4. The three divisions have neurochemical differences.
5. The sympathetic and parasympathetic divisions differ in anatomic origin and function.
6. The central nervous system controls autonomic function through a hierarchy of reflexes and integrative centers.

The sweating sunbather lying quietly in the summer sun or the racing heart and "hair-standing-on-end" sensations experienced by a person suddenly frightened by a horror movie are familiar examples of the body responding automatically to changes in the physical or emotional environment. These responses occur as a result of the actions of the autonomic portion of the nervous system and take place without conscious action on the part of the individual. The term *autonomic* is derived from the root *auto* (meaning "self") and *nomos* (meaning "law"). Our concept of the autonomic part of the nervous system has evolved during several centuries. The recognition of anatomic differences between the spinal cord and peripheral nerve pathways that control visceral functions from those that control skeletal muscles was a major step. Observations on the effects of the substance released by the vagus nerve on heart rate helped define unique biochemical features.

The functions of the **autonomic nervous system** (ANS) fall into three major categories:

- Maintaining homeostatic conditions within the body
- Coordinating the body's responses to exercise and stress
- Assisting the endocrine system to regulate reproduction

The ANS regulates the functions of the involuntary organs, which include the heart, the blood vessels, the exocrine glands, and the visceral organs. In some organs, the actions of the ANS are joined by circulating endocrine hormones and by locally produced chemical mediators to complete the control process.

AN OVERVIEW OF THE AUTONOMIC NERVOUS SYSTEM

On the basis of anatomic, functional, and neurochemical differences, the ANS is usually subdivided into three divisions: sympathetic, parasympathetic, and enteric. The enteric nervous system is concerned with the regulation of gastrointestinal function and covered in more detail in Chapter 26. The sympathetic and parasympathetic divisions are the primary focus of this chapter.

Coordination of the body's activities by the nervous system was the process of **sympathy** in classical anatomic and physiological thinking. Regulation of the involuntary organs came to be associated with the portions of the nervous system that were located, at least in part, outside the standard spinal cord and peripheral nerve pathways. The ganglia, located along either side of the spine in the thorax and abdominal regions and somewhat detached from the nerve trunks destined for the limbs, were found to be associated with involuntary bodily functions and, therefore, designated the **sympathetic division**. This collection of structures was also termed the **thoracolumbar division** of the ANS because of the location of the ganglia and the neuron cell bodies that supply axons to the ganglia. Nuclei and their axons that controlled internal functions were also found in the brainstem and associated cranial nerves, as well as in the most caudal part of the spinal cord. Those pathways were somewhat distinct from the sympathetic system and were designated the **parasympathetic division**. The term **craniosacral** was applied to this portion of the ANS because of the origin of cell bodies and axons.

Neurochemical differences were recognized between these two divisions, leading to the designation of the sympathetic system as **adrenergic**, for the adrenaline-like actions resulting from sympathetic nerve activation; and the parasympathetic system as **cholinergic**, for the acetylcholine-like actions of nerve stimulation.

The functions of the sympathetic and parasympathetic divisions are often simplified into a two-part scheme. The sympathetic division is said to preside over the utilization of metabolic resources and emergency responses of the body. The parasympathetic division presides over the restoration and buildup of the body's reserves and the elimination of waste products. In reality, most of the organs supplied by the ANS receive both sympathetic and parasympathetic innervation. In many instances, the two divisions are activated in a reciprocal fashion, so that if the firing rate in one division is increased, the rate is decreased in the other. An example is controlling the heart rate: Increased firing in the sympathetic nerves and simultaneous decreased firing in the parasympathetic nerves result in increased heart rate.

In some organs, the two divisions work synergistically. For example, during secretion by exocrine glands of the gastrointestinal tract, the parasympathetic nerves increase volume and enzyme content at the same time that sympathetic activation contributes mucus to the total secretory product. Some organs, such as the skin and blood vessels, receive only sympathetic innervation and are regulated by a decrease or increase in a baseline firing rate of the sympathetic nerves.

A Two-Neuron Efferent Path Is Utilized by the Autonomic Nervous System

The nervous system supplies efferent innervation to all organs via the motor system (see Chapter 5) or the ANS. In the motor system, there is an uninterrupted path from the cell body of the motor neuron, located in either the ventral horn of the spinal cord or a brainstem motor nucleus, to the skeletal muscle cells. In the ANS, the efferent path consists of a two-neuron chain with a synapse interposed between the CNS and the effector cells (Fig. 6.1). The cell bodies of the autonomic motor neurons are located in the spinal cord or specific brainstem nuclei. An efferent fiber emerges as the **preganglionic axon** and then synapses with neurons located in a peripheral ganglion. The neuron in the ganglion then projects a **postganglionic axon** to the autonomic effector cells.

The Primary Neurotransmitters of the ANS Are Acetylcholine and Norepinephrine

In the somatic nervous system, neurotransmitter is released from specialized nerve endings that make intimate contact with the target structure. The mammalian motor endplate, with one nerve terminal to one skeletal muscle fiber, illustrates this principle. This arrangement contrasts with the ANS, where postganglionic axons terminate in varicosities, swellings enriched in synaptic vesicles, which release the transmitter into the extracellular space surrounding the effector cells (see Fig. 6.1). The response to the ANS output originates in some of the effector cells and then propagates to the remainder via gap junctions.

Acetylcholine. **Acetylcholine** (ACh) is the transmitter released by the preganglionic nerve terminals of both the sympathetic and the parasympathetic divisions (Fig 6.2). The synapse at those sites utilizes a **nicotinic receptor** similar in structure to the receptor at the neuromuscular junction. Parasympathetic postganglionic neurons release ACh at the synapse with the effectors. The postganglionic sympathetic neurons to the sweat glands and to some blood

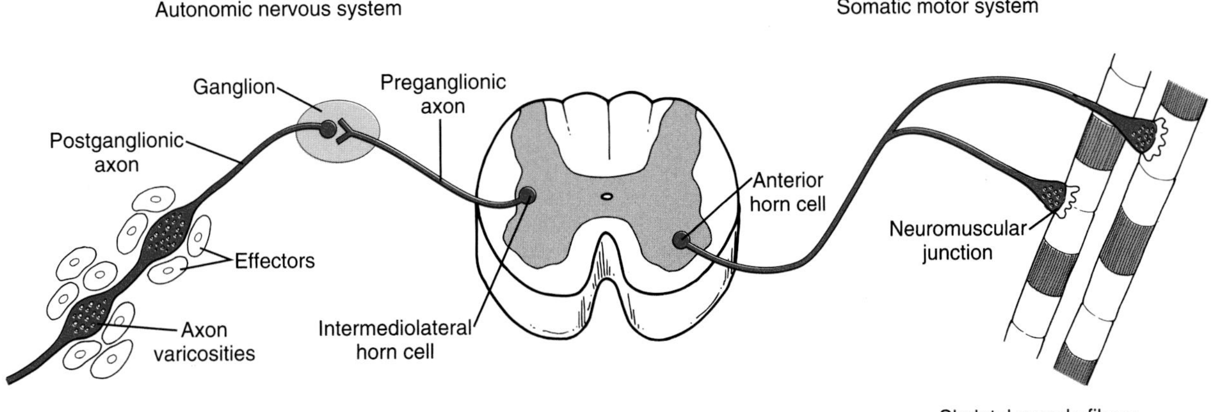

FIGURE 6.1 **The efferent path of the ANS as contrasted with the somatic motor system.** The ANS uses a two-neuron pathway. Note the structural differences between the synapses at autonomic effectors and skeletal muscle cells.

vessels in skeletal muscle also use ACh as the neurotransmitter. The synapse between the postganglionic neuron and the target tissues utilizes a **muscarinic receptor**. This receptor classification scheme is based on the response of the synapses to the alkaloids nicotine and muscarine, which act as agonists at their respective type of synapse. The nicotinic receptor of the ANS is blocked by the antagonist hexamethonium, in contrast to the neuromuscular junction receptor, which is blocked by curare. The muscarinic receptor is blocked by atropine.

The nicotinic receptor is of the direct ligand-gated type, meaning that the receptor and the ion channel are contained in the same structure. The muscarinic receptor is of the indirect ligand-gated type and uses a G protein to link receptor and effector functions (see Chapter 3). The action of ACh is terminated by the enzyme acetylcholinesterase. Choline released by the enzyme action is taken back into the nerve terminal and resynthesized into ACh.

Norepinephrine. The catecholamine **norepinephrine** (NE) is the neurotransmitter for postganglionic synapses of the sympathetic division (see Fig. 6.2). The synapses that utilize NE receptors can also be activated by the closely related compound epinephrine (adrenaline), which is released into the general circulation by the adrenal medulla—hence, the original designation of these type receptors as adrenergic. Adrenergic receptors are classified as either α or β, based on their responses to pharmacological agents that mimic or block the actions of NE and related compounds. Alpha receptors respond best to epinephrine, less well to NE, and least well to the synthetic compound isoproterenol. Beta receptors respond best to isoproterenol, less well to epinephrine, and least well to NE. Propranolol is a drug that acts as an antagonist at β receptors but has no action on α receptors. Each class of receptors is further classified as α_1 or α_2, and β_1, β_2, or β_3 on the basis of responses to additional pharmacological agents.

The adrenergic receptors are of the indirect, ligand-gated, G protein-linked type. They share a general structural similarity with the muscarinic type of ACh receptor. The α_1 receptors activate phospholipase C and increase the intracellular concentrations of diacylglycerol and inositol trisphosphate. The α_2 receptors inhibit adenylyl cyclase, while the β types stimulate it. The action of NE and epinephrine at a synapse is terminated by diffusion of the molecule away from the synapse and reuptake into the nerve terminal.

Other Neurotransmitters. Neurally active peptides are often colocalized with small molecule transmitters and are released simultaneously during nerve stimulation in the CNS. This is the same in the ANS, especially in the intrinsic plexuses of the gut, where amines, amino acid transmitters, and neurally active peptides are widely distributed. In the ANS, examples of a colocalized amine and peptide are seen in the sympathetic division, where NE and neuropeptide Y are coreleased by vasoconstrictor nerves. Vasoactive intestinal polypeptide (VIP) and calcitonin-gene-related peptide (CGRP) are released along with ACh from nerve terminals innervating the sweat glands.

Nitric oxide is another type of neurotransmitter produced by some autonomic nerve endings. The term **nonadrenergic noncholinergic** (NANC) has been applied to such nerves. Nitric oxide is a highly diffusible substance important in the regulation of smooth muscle contraction, (see Chapter 1).

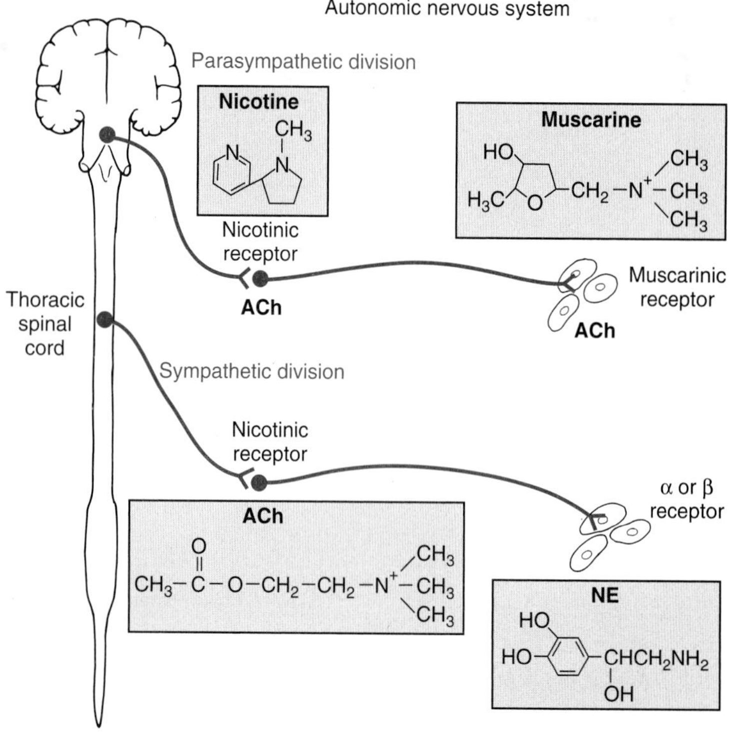

FIGURE 6.2 **The neurochemistry of the autonomic paths.** The structures of the neurotransmitters and the agonists for which the synapses were originally named are shown.

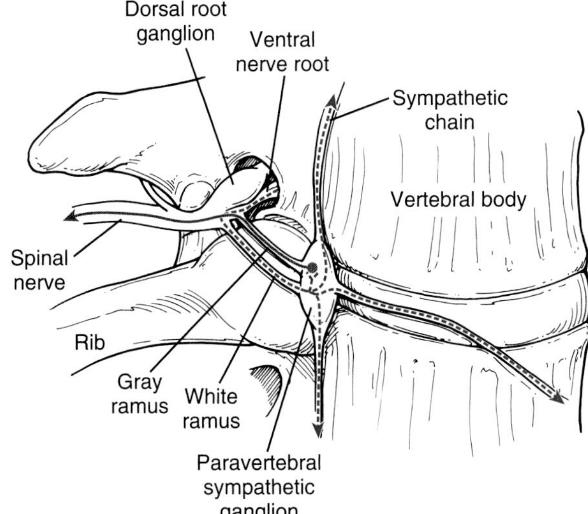

FIGURE 6.3 **Peripheral sympathetic anatomy.** The preganglionic axons course through the spinal nerve and white ramus to the paravertebral ganglion. Synapse with the postganglionic neuron may occur at the same spinal level, or at levels above or below. Postganglionic axons rejoin the spinal nerve through the gray ramus to innervate structures in the limbs or proceed to organs, such as the lungs or heart, in discrete nerves. Preganglionic axons may also pass to a prevertebral ganglion without synapsing in a paravertebral ganglion.

THE SYMPATHETIC NERVOUS SYSTEM

Preganglionic neurons of the sympathetic division originate in the intermediolateral horn of the thoracic (T1 to T12) and upper lumbar (L1 to L3) spinal cord. The preganglionic axons exit the spinal cord in the ventral nerve roots. Immediately after the ventral and dorsal roots merge to form the spinal nerve, the sympathetic axons leave the spinal nerve via the **white ramus** and enter the **paravertebral sympathetic ganglia** (Fig. 6.3). The paravertebral ganglia form an interconnected chain located on either side of the vertebral column. These ganglia extend above and below the thoracic and lumbar spinal levels, where preganglionic fibers emerge, to provide postganglionic sympathetic axons to the cervical and lumbosacral spinal nerves (Fig. 6.4). The preganglionic axons that ascend to the cervical levels arise from T1 to T5 and form three major ganglia: the **superior**, the **middle**, and the **inferior cervical ganglia**. Preganglionic axons descend below L3, forming two additional lumbar and at least four sacral ganglia. The preganglionic axons may synapse with postganglionic neurons in the paravertebral ganglion at the same level, ascend or descend up to several spinal levels and then synapse, or pass through the paravertebral ganglia en route to a **prevertebral ganglion**.

Postganglionic axons that are destined for somatic structures—such as sweat glands, pilomotor muscles, or blood vessels of the skin and skeletal muscles—leave the paravertebral ganglion in the **gray ramus** and rejoin the spinal nerve for distribution to the target tissues. Postganglionic axons to the head, heart, and lungs originate in the cervical or upper thoracic paravertebral ganglia and make their way to the specific organs as identifiable, separate nerves (e.g., the cardiac nerves), as small-caliber individual nerves that may group together, or as perivascular plexuses of axons that accompany arteries.

The superior cervical ganglion supplies sympathetic axons that innervate the structures of the head. These axons travel superiorly in the perivascular plexus along the carotid arteries. Structures innervated include the radial muscle of the iris, responsible for dilation of the pupil; Müller's muscle, which assists in elevating the eyelid; the lacrimal gland; and the salivary glands. Lesions that interrupt this pathway produce easily detectable clinical signs (see Clinical Focus Box 6.1). The middle and inferior cervical ganglia innervate organs of the chest, including the trachea, esophagus, heart, and lungs.

Postsynaptic axons destined for the abdominal and pelvic visceral organs arise from the prevertebral ganglia (see Fig. 6.4). The three major prevertebral ganglia, also called **collateral ganglia**, overlie the celiac, superior mesenteric, and inferior mesenteric arteries at their origin from the aorta and are named accordingly. The **celiac ganglion** provides sympathetic innervation to the stomach, liver, pancreas, gallbladder, small intestine, spleen, and kidneys. Preganglionic axons originate in the T5 to T12 spinal levels. The

CLINICAL FOCUS BOX 6.1

Horner's Syndrome

Lesions of the sympathetic pathway to the head produce abnormalities that are easily detectable on physical examination. The deficits of function occur ipsilateral to the lesion and include:

- Partial constriction of the pupil as a result of loss of sympathetic pupillodilator action
- Drooping of the eyelid, termed *ptosis*, as a result of loss of sympathetic activation of Müller's muscle of the eyelid
- Dryness of the face as a result of the lack of sympathetic activation of the facial sweat glands

A pattern of historical or physical examination findings that is consistent from patient to patient is often termed a ***syndrome***. Johann Horner, a 19th century Swiss ophthalmologist, described this pattern of eye and facial abnormalities in patients, and these are referred to as Horner's syndrome. Etiologies for Horner's syndrome include:

- Brainstem lesions, such as produced by strokes, which interrupt the tracts that descend to the sympathetic neurons in the spinal cord
- Upper thoracic nerve root lesions, such as those produced by excessive traction on the arm or from infiltration of the nerve roots by cancer spreading from the lung
- Cervical paravertebral ganglia lesions from accidental or surgical trauma, or metastatic cancer
- Arterial injury in the neck, from neck hyperextension, or direct trauma, which interrupt the postganglionic axons traveling in the carotid periarterial plexus

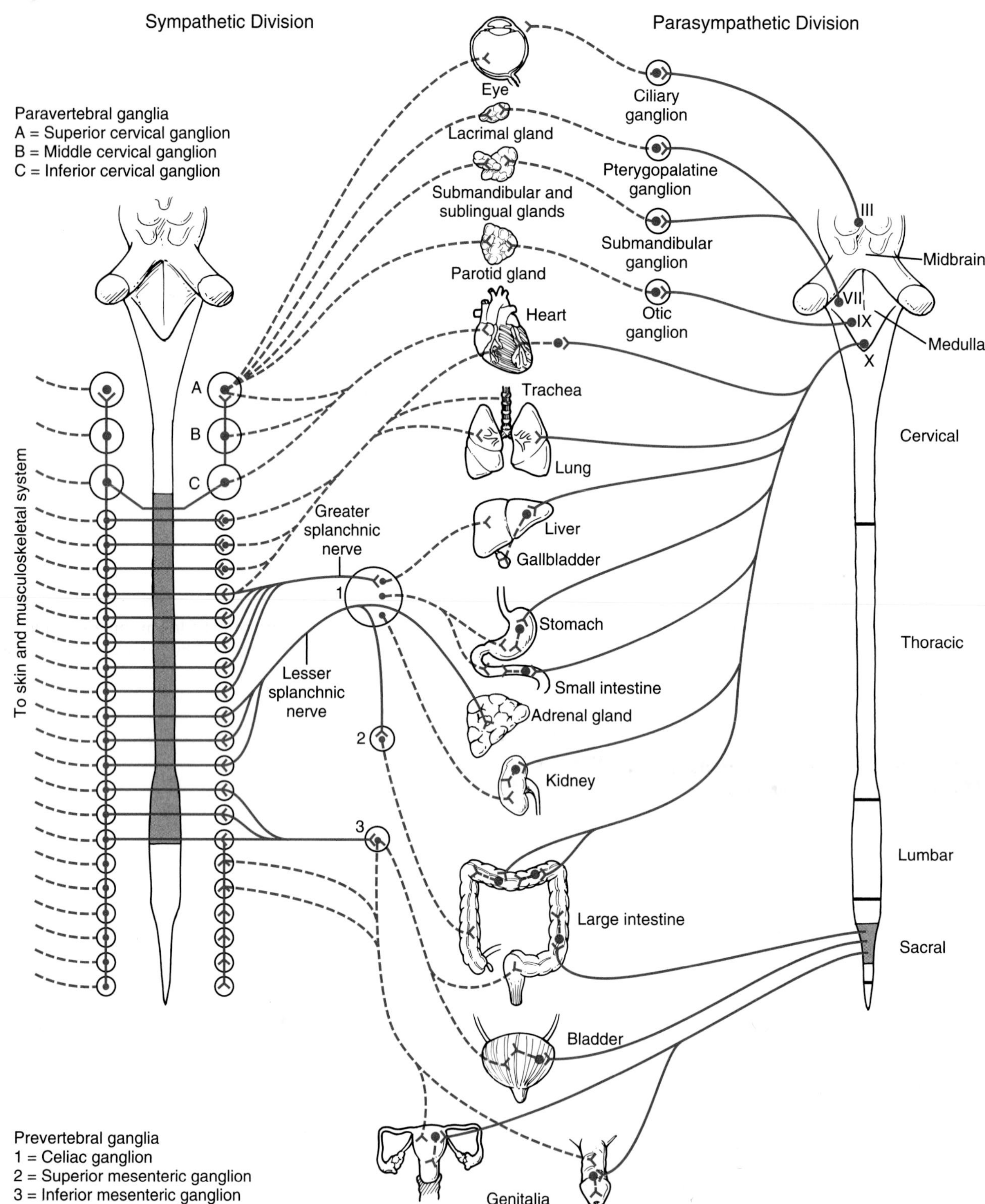

FIGURE 6.4 **The organ-specific arrangement of the ANS.** Preganglionic axons are indicated by solid lines, postganglionic axons by dashed lines. Sympathetic axons destined for the skin and musculoskeletal system are shown on the left side of the spinal cord. Note the named paravertebral and prevertebral ganglia.

superior mesenteric ganglion innervates the small and large intestines. Preganglionic axons originate primarily in T10 to T12. The **inferior mesenteric ganglion** innervates the lower colon and rectum, urinary bladder, and reproductive organs. Preganglionic axons originate in L1 to L3.

The Sympathetic Division Can Produce Local or Widespread Responses

The sympathetic division exerts a continuous influence on the organs it innervates. This continuous level of control is called **sympathetic tone**, and it is accomplished by a persistent, low rate of discharge of the sympathetic nerves. When the situation dictates, the rate of firing to a particular organ can be increased or decreased, such as an increased firing rate of the sympathetic neurons supplying the iris to produce pupillary dilation in dim light or a decreased firing rate and pupillary constriction during drowsiness.

The number of postganglionic axons emerging from the paravertebral ganglia is greater than the number of preganglionic neurons that originate in the spinal cord. It is estimated that postganglionic sympathetic neurons outnumber preganglionic neurons by 100:1 or more. This spread of influence, termed **divergence**, is accomplished by collateral branching of the presynaptic sympathetic axons, which then make synaptic connections with postganglionic neurons both above and below their original level of emergence from the spinal cord. Divergence enables the sympathetic division to produce widespread responses of many effectors when physiologically necessary.

The Adrenal Medulla Is a Mediator of Sympathetic Function

In addition to divergence, the sympathetic division has a hormonal mechanism to activate target tissues endowed with adrenergic receptors, including those innervated by the sympathetic nerves. The hormone is the catecholamine **epinephrine**, which is secreted with much lesser amounts of norepinephrine by the adrenal medulla during generalized response to stress.

The adrenal medulla, a neuroendocrine gland, forms the inner core of the adrenal gland situated on top of each kidney. Cells of the adrenal medulla are innervated by the lesser splanchnic nerve, which contains preganglionic sympathetic axons originating in the lower thoracic spinal cord (see Fig. 6.4). These axons pass through the paravertebral ganglia and the celiac ganglion without synapsing and terminate on the **chromaffin** cells of the adrenal medulla (Fig. 6.5). The chromaffin cells are modified ganglion cells that synthesize both epinephrine and norepinephrine in a ratio of about 8:1 and store them in secretory vesicles. Unlike neurons, these cells possess neither axons nor dendrites but function as neuroendocrine cells that release hormone directly into the bloodstream in response to preganglionic axon activation.

Circulating epinephrine mimics the actions of sympathetic nerve stimulation but with greater efficacy because epinephrine is usually more potent than norepinephrine in stimulating both α-adrenergic and β-adrenergic receptors. Epinephrine can also stimulate adrenergic receptors on cells that receive little or no direct sympathetic innervation, such as liver and adipose cells for mobilizing glucose and fatty acids, and blood cells which participate in the clotting and immune responses.

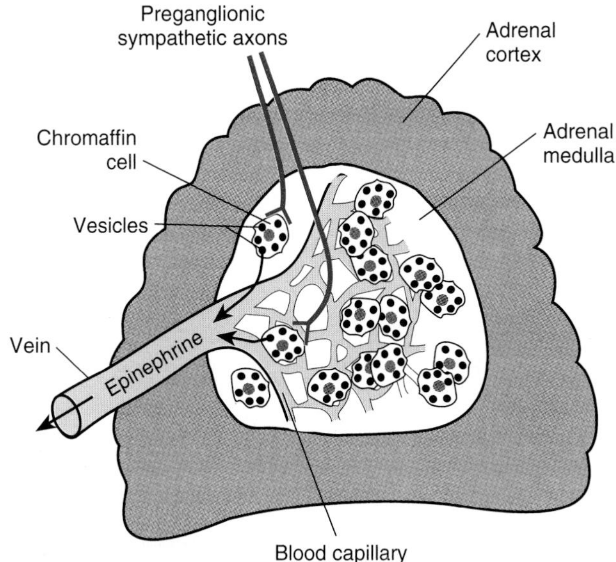

FIGURE 6.5 **Sympathetic innervation of the adrenal medulla.** Preganglionic sympathetic axons terminate on the chromaffin cells. When stimulated, the chromaffin cells release epinephrine into the circulation.

The Fight-or-Flight Response Is a Result of Widespread Sympathetic Activation

This response is the classic example of the sympathetic nervous system's ability to produce widespread activation of its effectors; it is activated when an organism's survival is in jeopardy and the animal may have to fight or flee. Some components of the response result from the direct effects of sympathetic activation, while the secretion of epinephrine by the adrenal medulla also contributes.

Sympathetic stimulation of the heart and blood vessels results in a rise in blood pressure because of increased cardiac output and increased total peripheral resistance. There is also a redistribution of the blood flow so that the muscles and heart receive more blood, while the splanchnic territory and the skin receive less. The need for an increased exchange of blood gases is met by acceleration of the respiratory rate and dilation of the bronchiolar tree. The volume of salivary secretion is reduced but the relative proportion of mucus increases, permitting lubrication of the mouth despite increased ventilation. The potential demand for an enhanced supply of metabolic substrates, like glucose and fatty acids, is met by the actions of the sympathetic nerves and circulating epinephrine on hepatocytes and adipose cells. Glycogenolysis mobilizes stored liver glycogen, increasing plasma levels of glucose. Lipolysis in fat cells converts stored triglycerides to free fatty acids that enter the bloodstream.

The skin plays an important role in maintaining body temperature in the face of increased heat production from contracting muscles. The sympathetic innervation of the skin vasculature can adjust blood flow and heat exchange by vasodilation to dissipate heat or by vasoconstriction to protect blood volume. The eccrine sweat glands are important structures that also can be activated to enhance heat loss. Sympathetic nerve stimulation of the sweat glands results in the secretion of a watery fluid, and evaporation then dissipates body heat. Constriction of the skin vasculature, concurrent with sweat gland activation, produces the cold, clammy skin of a frightened individual. Hair-standing-on-end sensations result from activation of the piloerector muscles associated with hair follicles. In humans, this action is likely a phylogenetic remnant from animals that use hair erection for body temperature preservation or to enhance the appearance of body size or ferocity.

THE PARASYMPATHETIC NERVOUS SYSTEM

The parasympathetic division is comprised of a cranial portion, emanating from the brainstem, and a sacral portion, originating in the intermediate gray zone of the sacral spinal cord (see Fig. 6.4). In contrast to the widespread activation pattern of the sympathetic division, the neurons of the parasympathetic division are activated in a more localized fashion. There is also much less tendency for divergence of the presynaptic influence to multiple postsynaptic neurons—on average, one presynaptic parasympathetic neuron synapses with 15 to 20 postsynaptic neurons. An example of localized activation is seen in the vagus nerve, where one portion of its outflow can be activated to slow the heart rate without altering the vagal control to the stomach.

Ganglia in the parasympathetic division are located either close to the organ innervated or embedded within its walls. The organs of the gastrointestinal system demonstrate the latter pattern. Because of this arrangement, preganglionic axons are much longer than postganglionic axons.

Brainstem Parasympathetic Neurons Innervate Structures in the Head, Chest, and Abdomen

Four of the twelve cranial nerves—numbers III, VII, IX, and X—contain parasympathetic axons. The nuclei of these nerves, which occupy areas of the tectum in the midbrain, pons, and medulla, are the centers for the initiation and integration of autonomic reflexes for the organ systems they innervate. Parasympathetic and sympathetic activities are coordinated by these nuclei.

Cranial Nerve III. The **oculomotor nerve** originates from nuclei in the tectum of the midbrain, where synaptic connections with the axons of the optic nerves provide input for ocular reflexes. The parasympathetic neurons are located in the **Edinger-Westphal nucleus**. The presynaptic axons travel in the superficial aspect of cranial nerve III to the **ciliary ganglion**, located inside the orbit where the synapse occurs. The postganglionic axons enter the eyeball near the optic nerve and travel between the sclera and the choroid. These axons supply the sphincter muscle of the iris; the ciliary muscle, which focuses the lens; and the choroidal blood vessels. About 90% of the axons are destined for the ciliary muscle, while only about 3 to 4% innervate the iris sphincter.

Cranial Nerve VII. The parasympathetic presynaptic axons of the **facial nerve** arise from the **superior salivatory nuclei** in the rostral medulla. Presynaptic axons pass from the facial nerve into the greater superficial petrosal nerve and synapse in the **pterygopalatine ganglion**. The postsynaptic axons from that ganglion innervate the lacrimal gland and the glands of the nasal and palatal mucosa. Other facial nerve presynaptic axons travel via the chorda tympani and synapse in the **submandibular ganglion**. These postsynaptic axons stimulate the production of saliva by the submandibular and sublingual glands. Parasympathetic activation can also produce dilation of the vasculature within the areas supplied by the facial nerve.

Cranial Nerve IX. The parasympathetic presynaptic axons of the **glossopharyngeal nerve** arise from the **inferior salivatory nuclei** of the medulla. The axons follow a circuitous course through the lesser petrosal nerve to reach the **otic ganglion**, where they synapse. From the otic ganglion, the postsynaptic axons join the auriculotemporal branch of cranial nerve V and arrive at the parotid gland, where they stimulate secretion of saliva.

Sensory axons that are important for autonomic function are also conveyed in cranial nerve IX. The carotid bodies sense the concentrations of oxygen and carbon dioxide in blood flowing in the carotid arteries and transmit that chemosensory information to the medulla via glossopharyngeal afferents. The carotid sinus, which is located in the proximal internal carotid artery, monitors blood pressure and transmits this baroreceptor information to the **tractus solitarius** in the medulla.

Cranial Nerve X. The **vagus nerve** has an extensive autonomic component, which arises from the **nucleus ambiguus** and the **dorsal motor nuclei** in the medulla. It has been estimated that vagal output comprises up to 75% of total parasympathetic activity. Long preganglionic axons travel in the vagus trunks to ganglia in the heart and lungs and to the intrinsic plexuses of the gastrointestinal tract. Sympathetic postsynaptic axons also intermingle with the parasympathetic presynaptic axons in these plexuses and travel together to the target tissues.

The right vagus nerve supplies axons to the sinoatrial node of the heart, and the left vagus nerve supplies the atrioventricular node. Vagal activation slows the heart rate and reduces the force of contraction. The vagal efferents to the lung control smooth muscle that constricts bronchioles, and also regulate the action of secretory cells. Vagal input to the esophagus and stomach regulates motility and influences secretory function in the stomach. Acetylcholine plus vasoactive intestinal peptide (VIP) are the transmitters of the postsynaptic neurons.

There is also vagal innervation to the kidneys, liver, spleen, and pancreas, but the role of these inputs is not yet fully established.

Sacral Spinal Cord Parasympathetic Neurons Innervate Structures in the Pelvis

Preganglionic fibers of the sacral division originate in the intermediate gray matter of the sacral spinal cord, emerging from segments S2, S3, and S4 (see Fig. 6.4). These preganglionic fibers synapse in ganglia in or near the pelvic organs, including the lower portion of the gastrointestinal tract (the sigmoid colon, rectum, and internal anal sphincter), the urinary bladder, and the reproductive organs.

SPECIFIC ORGAN RESPONSES TO AUTONOMIC ACTIVITY

As noted earlier, most involuntary organs are dually innervated by the sympathetic and parasympathetic divisions, often with opposing actions. A list of these organs and a summary of their responses to sympathetic and parasympathetic stimulation is given in Table 6.1. The type of receptor at the synapse with the effectors is also indicated. More detailed discussions of the effects of autonomic nerve activation are found in the chapters on the specific organ systems.

TABLE 6.1 Responses of Effectors to Parasympathetic and Sympathetic Stimulation

Effector	Parasympathetic	Sympathetic
Eye		
Pupil	Constriction	Dilation (α_1)
Ciliary muscle	Contraction	Relaxation (β_2)
Müller's muscle	None	Contraction (α_1)
Lacrimal gland	Secretion	None
Nasal glands	Secretion	Inhibition (α_1)
Salivary glands	Secretion	Amylase secretion (β)
Skin		
Sweat glands	None	Secretion (cholinergic muscarinic)
Piloerector muscles	None	Contraction (α_1)
Blood vessels		
Skin	None	Constriction (α)
Skeletal muscle	None	Dilation (β_2), constriction (α)
Viscera	None	Constriction (α_1)
Heart		
Rate	Decrease	Increase (β_1, β_2)
Force	Decrease	Increase (β_1, β_2)
Lungs		
Bronchioles	Constriction	Dilation (β_2)
Glands	Secretion	Decreased (α_1), incr. (β_2) secretion
Gastrointestinal tract		
Wall muscles	Contraction	Relaxation (α, β_2)
Sphincters	Relaxation	Contraction (α_1)
Glands	Secretion	Inhibition
Liver	None	Glycogenolysis and gluconeogenesis (α_1, β_2)
Pancreas (insulin)	None	Decreased secretion (α_2)
Adrenal medulla	None	Secretion of epinephrine (cholinergic nicotinic)
Urinary system		
Ureter	Relaxation	Contraction (α_1)
Detrusor	Contraction	Relaxation (β_2)
Sphincter	Relaxation	Contraction (α_1)
Reproductive system		
Uterus	Variable	Contraction (α_1)
Genitalia	Erection	Ejaculation/vaginal contraction (α)
Adipose cells	None	Lipolysis (β)

CONTROL OF THE AUTONOMIC NERVOUS SYSTEM

The autonomic nervous system utilizes a hierarchy of reflexes to control the function of autonomic target organs. These reflexes range from local, involving only a part of one neuron, to regional, requiring mediation by the spinal cord and associated autonomic ganglia, to the most complex, requiring action by the brainstem and cerebral centers. In general, the higher the level of complexity, the more likely the reflex will require coordination of both sympathetic and parasympathetic responses. Somatic motor neurons and the endocrine system may also be involved.

Sensory Input Contributes to Autonomic Function

The ANS is traditionally regarded as an efferent system, and the sensory neurons innervating the involuntary organs are not considered part of the ANS. Sensory input, however, is important for autonomic functioning. The sensory innervation to the visceral organs, blood vessels, and skin forms the afferent limb of autonomic reflexes (Fig. 6.6). Most of the sensory axons from ANS-innervated structures are unmyelinated C fibers.

Sensory information from these pathways may not reach the level of consciousness. Sensations that are perceived may be vaguely localized or may be felt in a somatic structure rather than the organ from which the afferent action potentials originated. The perception of pain in the left arm during a myocardial infarction is an example of pain being referred from a visceral organ.

Local Axon Reflexes Are Paths for Autonomic Activation

A sensory neuron may have several terminal branches peripherally that enlarge the receptive area and innervate multiple receptors. As a sensory action potential which originated in one of the terminal branches propagates afferently, or **orthodromically,** it may also enter some other branches of that same axon and then conduct efferently, or **antidromically**, for short distances. The distal ends of the sensory axons may release neurotransmitters in response to the antidromic action potentials. The process of action potential spread can result in a more wide-ranging reaction than that produced by the initial stimulus. If the sensory neuron innervates blood vessels or sweat glands, the response can produce reddening of the skin as a result of vasodilation, local sweating as a result of sweat gland activation, or pain as a result of the action of the released neurotransmitter. This process is called a **local axon reflex** (see Fig. 6.6). It differs from the usual reflex pathway in that a synapse with an efferent neuron in the spinal cord or peripheral ganglion is not required to produce a response. The neurotransmitter producing this local reflex is likely the same as that released

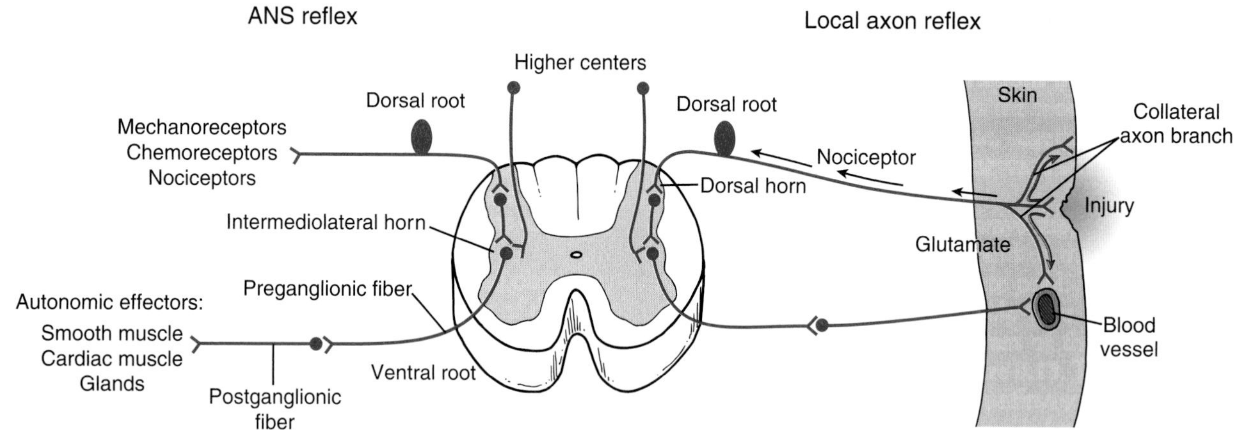

FIGURE 6.6 Sensory components of autonomic function. Left, Sensory action potentials from mechanical, chemical, and nociceptive receptors that propagate to the spinal cord can trigger ANS reflexes. Right, Local axon reflexes occur when an orthodromic action potential from a sensory nerve ending propagates antidromically into collateral branches of the same neuron. The antidromic action potentials may provoke release of the same neurotransmitters, like substance P or glutamate, from the nerve endings as would be released at the synapse in the spinal cord. Local axon reflexes may perpetuate pain, activate sweat glands, or cause vasomotor actions.

at the synapse in the spinal cord—substance P or glutamate for sensory neurons or ACh and NE at the target tissues for autonomic neurons. Local axon reflexes in nociceptive nerve endings that become persistently activated after local trauma can produce dramatic clinical manifestations (see Clinical Focus Box 6.2).

The Autonomic Ganglia Can Modify Reflexes

Although the paravertebral ganglia may serve merely as relay stations for synapse of preganglionic and postganglionic sympathetic neurons, evidence suggests that synaptic activity in these ganglia may modify efferent activity. Input from other preganglionic neurons provides the modifying influence. Prevertebral ganglia also serve as integrative centers for reflexes in the gastrointestinal tract. Chemoreceptors and mechanoreceptors located in the gut produce afferent action potentials that pass to the spinal cord and then to the celiac or mesenteric ganglia where changes in motility and secretion may be instituted during digestion. The integrative actions of these ganglia are also responsible for halting motility and secretion in the gastrointestinal tract during a generalized stress reaction (the fight-or-flight response).

The intrinsic plexuses of the gastrointestinal visceral wall are reflex integrative centers where input from presynaptic parasympathetic axons, postganglionic sympathetic axons, and the action of intrinsic neurons may all participate in reflexes that influence motility and secretion. The intrinsic plexuses also participate in centrally mediated gastrointestinal reflexes (see Chapter 26).

The Spinal Cord Coordinates Many Autonomic Reflexes

Reflexes coordinated by centers in the lumbar and sacral spinal cord include micturition (emptying the urinary bladder), defecation (emptying the rectum), and sexual response (engorgement of erectile tissue, vaginal lubrication, and ejaculation of semen). Sensory action potentials from receptors in the wall of the bladder or bowel report about degrees of distension. Sympathetic, parasympathetic, and somatic efferent actions require coordination to produce many of these responses.

CLINICAL FOCUS BOX 6.2

Reflex Sympathetic Dystrophy (RSD)

RSD is a clinical syndrome that includes spontaneous pain, painful hypersensitivity to nonnoxious stimuli such as light touch or moving air, and evidence of ANS dysfunction in the form of excessive coldness and sweating of the involved body part. The foot, knee, hand, and forearm are the more common sites of involvement. Local trauma, which may be minor in degree, and surgical procedures on joints or bones are common precipitating events. The term *dystrophy* applies to atrophic changes that may occur in the skin, soft tissue, and bone in the painful areas if the condition goes untreated for many months. A full explanation of the pathogenetic mechanisms is still lacking. Local axon reflexes in traumatized nociceptive neurons and reflex activation of the sympathetic nervous system are thought to be contributors. Repeated blockade of sympathetic neuron action by local anesthetic injection into the paravertebral ganglia serving the involved body part, followed by mobilization of the body part in a physical therapy program, are the mainstays of treatment. RSD is now termed *Complex Regional Pain Syndrome Type I.*

Higher centers provide facilitating or inhibiting influences to the spinal cord reflex centers. The ability to voluntarily suppress the urge to urinate when the sensation of bladder fullness is perceived is an example of higher CNS centers inhibiting a spinal cord reflex. Following injury to the cervical or upper thoracic spinal cord, micturition may occur involuntarily or be provoked at much lower than normal bladder volumes. Episodes of hypertension and piloerection in spinal cord injury patients are another example of uninhibited autonomic reflexes arising from the spinal cord.

The Brainstem Is a Major Control Center for Autonomic Reflexes

Areas within all three levels of the brainstem are important in autonomic function (Fig. 6.7). The **periaqueductal gray matter** of the midbrain coordinates autonomic responses to painful stimuli and can modulate the activity of the sensory tracts that transmit pain. The **parabrachial nucleus** of the pons participates in respiratory and cardiovascular control. The **locus ceruleus** may have a role in micturition reflexes. The medulla contains several key autonomic areas. The **nucleus of the tractus solitarius** receives afferent input from cardiac, respiratory, and gastrointestinal receptors. The **ventrolateral medullary area** is the major center for control of the preganglionic sympathetic neurons in the spinal cord. Vagal efferents arise from this area also. Neurons that control specific functions like blood pressure and heart rate are clustered within this general region. The descending paths for regulation of the preganglionic sympathetic and spinal parasympathetic neurons are not yet fully delineated. The reticulospinal tracts may carry some of these axons. Autonomic reflexes coordinated in the brainstem include pupillary reaction to light, lens accommodation, salivation, tearing, swallowing, vomiting, blood pressure regulation, and cardiac rhythm modulation.

The Hypothalamus and Cerebral Hemispheres Provide the Highest Levels of Autonomic Control

The **periventricular**, **medial**, and **lateral areas** of the **hypothalamus** in the diencephalon control circadian rhythms, and homeostatic functions such as thermoregulation, appetite, and thirst. Because of the major role of the hypothalamus in autonomic function, it has at times been labeled the "head ganglion of the ANS." The **insular** and **medial prefrontal areas** of the cerebral cortex are the respective sensory and motor areas involved with the regulation of autonomic function. The **amygdala** in the temporal lobe coordinates the autonomic components of emotional responses.

The areas of the cerebral hemispheres, diencephalon, brainstem, and central path to the spinal cord that are involved with the control of autonomic functions are collectively termed the **central autonomic network** (see Fig. 6.7).

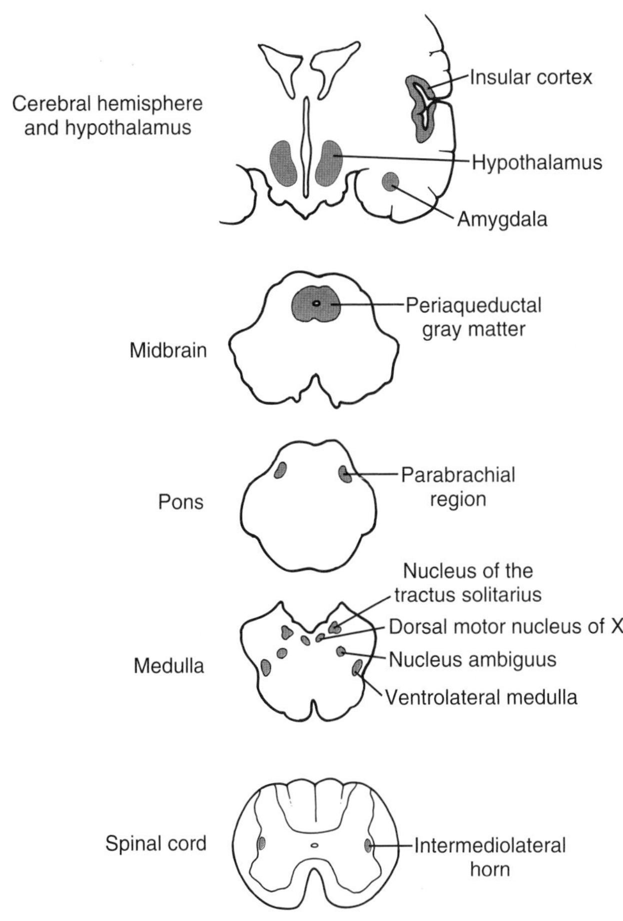

FIGURE 6.7 **The central autonomic network.** Note the cerebral, hypothalamic, brainstem, and spinal cord components. A hierarchy of reflexes initiated from these different levels regulates autonomic function.

REVIEW QUESTIONS

DIRECTIONS: Each of the numbered items or incomplete statements in this section is followed by answers or by completions of the statement. Select the ONE lettered answer or completion that is BEST in each case.

1. Impaired dilation of the pupil when entering a dark room is due to deficient functioning of
(A) Presynaptic axons that travel in the oculomotor nerve
(B) Postsynaptic axons that travel in the facial nerve
(C) Acetylcholine delivered by the circulatory system
(D) Postsynaptic axons arising from paravertebral ganglia
(E) Postsynaptic axons arising from prevertebral ganglia

2. Which effects would destruction of the lumbar paravertebral ganglia by a gunshot cause in the ipsilateral leg?
(A) It would be cold and clammy
(B) It would be weak
(C) There would be decreased sensation for painful stimuli
(D) It would be warm and dry
(E) There would be no detectable change

(continued)

3. Which of these is not a neurotransmitter in the autonomic nervous system?
(A) Acetylcholine
(B) Norepinephrine
(C) Epinephrine
(D) Muscarine
(E) Neuropeptide Y
4. With which other entity do the receptors of the parasympathetic postganglionic target tissue synapse share general structural similarity?
(A) The receptor of the sympathetic postganglionic target tissue synapse
(B) The receptor of the sympathetic preganglionic synapse
(C) The receptor of the parasympathetic preganglionic synapse
(D) The voltage-gated calcium channel
(E) The receptor at the neuromuscular junction
5. By which route are the sweat glands supplied with parasympathetic innervation?
(A) Vagal preganglionics to paravertebral ganglion to cutaneous nerve
(B) Vagal preganglionics to prevertebral ganglion to cutaneous nerve
(C) Spinal preganglionics to paravertebral ganglion to cutaneous nerve
(D) Spinal gray ramus to cutaneous nerve
(E) There is no parasympathetic innervation to the sweat glands
6. Which statement correctly describes the relationship between preganglionic and postganglionic sympathetic axons?
(A) The number of presynaptic axons is much greater than the number of postsynaptic axons
(B) The number of postsynaptic axons is much greater than the number of presynaptic axons
(C) The number of presynaptic and postsynaptic axons is equal
(D) Convergence of presynaptic influence onto the postsynaptic neurons is the rule
(E) Presynaptic and postsynaptic neurons are joined by gap junctions
7. A patient who is being treated with a medication complains of the adverse effect of difficulty adjusting his eyes to bright lights. How is the medication modifying autonomic function?
(A) Enhancing cholinergic activity
(B) Enhancing adrenergic activity
(C) Mimicking the action of epinephrine
(D) Inhibiting cholinergic activity
(E) Inhibiting adrenergic activity
8. The activation of which type of synapse could alter cyclic AMP levels in the postsynaptic cell?
(A) Preganglionic to postganglionic sympathetic
(B) Preganglionic to postganglionic parasympathetic
(C) Postganglionic axon-target tissue nicotinic
(D) Postganglionic axon-target tissue muscarinic
(E) Postganglionic-target tissue curare-sensitive
9. A concurrent increase in parasympathetic and decrease in sympathetic outflow to the heart would be coordinated at which level of the nervous system?
(A) Insular cortex
(B) Axon reflexes in cardiac sensory nerves
(C) Periaqueductal gray matter of the mesencephalon
(D) Gray matter of the upper thoracic spinal cord
(E) Reticular formation of the medulla

SUGGESTED READING

Low PA. Clinical Autonomic Disorders. 2nd Ed. Philadelphia: Lippincott-Raven, 1997.

Parent A. Carpenter's Human Neuroanatomy. 9th Ed. Media, PA: Williams & Wilkins, 1996.

Siegel GJ, Agranoff BW, Albers RW, Fisher SK, Uhler MD. Basic Neurochemistry. 6th Ed. Philadelphia: Lippincott Williams & Wilkins, 1999.

Integrative Functions of the Nervous System

Cynthia J. Forehand, Ph.D.

CHAPTER OUTLINE

- THE HYPOTHALAMUS
- THE RETICULAR FORMATION
- THE FOREBRAIN

KEY CONCEPTS

1. Homeostatic functions are regulated by the hypothalamus.
2. Homeostatically regulated functions fluctuate in a daily pattern.
3. The reticular formation serves as the activating system of the forebrain.
4. Sleep occurs in stages that exhibit different EEG patterns.
5. The limbic system receives distributed monoaminergic and cholinergic innervation.
6. Limbic structures play a role in the brain's reward system.
7. The limbic system regulates aggression and sexual activity.
8. Affective disorders and schizophrenia are disruptions in limbic function.
9. The cerebral cortex and hippocampus are involved in learning and memory.
10. Language is a lateralized function of association cortex.

The **central nervous system** (CNS) receives sensory stimuli from the body and the outside world and processes that information in neural networks or centers of integration to mediate an appropriate response or learned experience. Centers of integration are hierarchical in nature. In a caudal-to-rostral sequence, the more rostral it is placed, the greater the complexity of the neural network. This chapter considers functions integrated within the diencephalon and telencephalon, where emotionally motivated behavior, appetitive drive, consciousness, sleep, language, memory, and cognition are coordinated.

THE HYPOTHALAMUS

The **hypothalamus** coordinates autonomic reflexes of the brainstem and spinal cord. It also activates the endocrine and somatic motor systems when responding to signals generated either within the hypothalamus or brainstem or in higher centers, such as the **limbic system**, where the emotions and motivations are generated. The hypothalamus can accomplish this by virtue of its unique location at the interface between the limbic system and the endocrine and autonomic nervous systems.

As a major regulator of homeostasis, the hypothalamus receives input about the internal environment of the body via signals in the blood. In most of the brain, capillary endothelial cells are connected by tight junctions that prevent substances in the blood from entering the brain. These tight junctions are part of the **blood-brain barrier**. The blood-brain barrier is missing in several small regions of the brain called **circumventricular organs**, which are adjacent to the fluid-filled ventricular spaces. Several circumventricular organs are in the hypothalamus. Capillaries in these regions, like those in other organs, are fenestrated ("leaky"), allowing the cells of hypothalamic nuclei to sample freely, from moment to moment, the composition of the blood. Neurons in the hypothalamus then initiate the mechanisms necessary to maintain levels of constituents at a given set point, fixed within narrow limits by a specific hypothalamic nucleus. Homeostatic functions regulated by the hypothalamus include body temperature, water and electrolyte balance, and blood glucose levels.

The hypothalamus is the major regulator of endocrine function because of its connections with the pituitary gland, the master gland of the endocrine system. These connections include direct neuronal innervation of the posterior pituitary lobe by specific hypothalamic nuclei and a direct hormonal connection between specific hypothalamic nuclei and the anterior pituitary. **Hypothalamic hormones**, designated as **releasing factors**, reach the anterior pituitary lobe by a portal system of capillaries. Releasing factors then regulate the secretion of most hormones of the endocrine system.

The Hypothalamus Is Composed of Anatomically Distinct Nuclei

The **diencephalon** includes the hypothalamus, thalamus, and subthalamus (Fig. 7.1). The rostral border of the hypothalamus is at the optic chiasm, and its caudal border is at the mammillary body.

On the basal surface of the hypothalamus, exiting the median eminence, the pituitary stalk contains the **hypothalamo-hypophyseal portal blood vessels** (see Fig. 32.3). Neurons within specific nuclei of the hypothalamus secrete releasing factors into these portal vessels. The releasing factors are then transported to the anterior pituitary, where they stimulate secretion of hormones that are trophic to other glands of the endocrine system (see Chapter 32).

The pituitary stalk also contains the axons of magnocellular neurons whose cell bodies are located in the supraoptic and paraventricular hypothalamic nuclei. These axons form the **hypothalamo-hypophyseal tract** within the pituitary stalk and represent the efferent limbs of neuroendocrine reflexes that lead to the secretion of the hormones vasopressin and oxytocin into the blood. These hormones are made in the magnocellular neurons and released by their axon terminals next to the blood vessels within the posterior pituitary.

The nuclei of the hypothalamus have ill-defined boundaries, despite their customary depiction (Fig. 7.2). Many are named according to their anatomic location (e.g., anterior hypothalamic nuclei, ventromedial nucleus) or for the structures they lie next to (e.g., the periventricular nucleus surrounds the third ventricle, the suprachiasmatic nucleus lies above the optic chiasm).

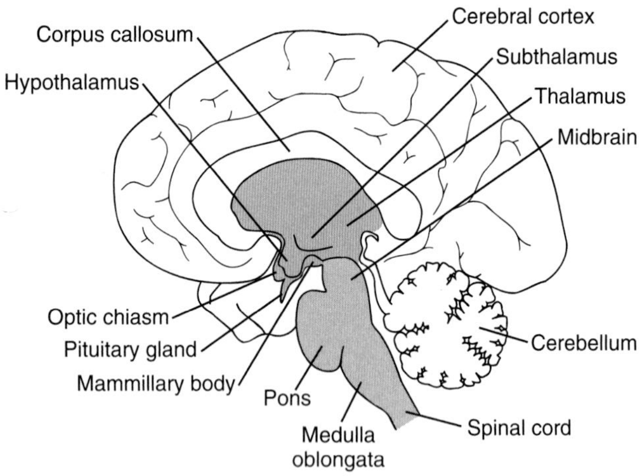

FIGURE 7.1 **A midsagittal section through the human brain, showing the most prominent structures of the brainstem (gray), diencephalon (red), and forebrain (white).** The cerebellum is also shown. (Modified from Kandel ER, Schwartz JH, Jessel TM. Principles of Neural Science. 3rd Ed. New York: Elsevier, 1991.)

The hypothalamus receives afferent inputs from all levels of the CNS. It makes reciprocal connections with the limbic system via fiber tracts in the **fornix**. The hypothalamus also makes extensive reciprocal connections with the brainstem, including the reticular formation and the medullary centers of cardiovascular, respiratory, and gastrointestinal regulation. Many of these connections travel within the **medial forebrain bundle**, which also connects the brainstem with the cerebral cortex.

Several major connections of the hypothalamus are one-way rather than reciprocal. One of these, the **mammillothalamic tract**, carries information from the mammillary bodies of the hypothalamus to the anterior nucleus of the **thalamus**, from where information is relayed to limbic regions of the cerebral cortex. A second one-way pathway carries visual information from the retina to the **suprachiasmatic nucleus** of the hypothalamus via the optic nerve. Through this retinal input, the light cues of the day/night cycle entrain or synchronize the "biological clock" of the brain to the external clock. A third one-way connection is the hypothalamo-hypophyseal tract from the supraoptic and paraventricular nuclei to the posterior pituitary gland. The hypothalamus also projects directly to the spinal cord to activate sympathetic and parasympathetic preganglionic neurons (see Chapter 6).

Hypothalamic Nuclei Are Centers of Physiological Regulation

The nuclei of the hypothalamus contain groups of neurons that regulate several important physiological functions:

1) Water and electrolyte balance in magnocellular cells of the supraoptic and paraventricular nuclei (see Chapter 32)

2) Secretion of hypothalamic releasing factors in the arcuate and periventricular nuclei and in parvocellular cells of the paraventricular nucleus (see Chapters 32 and 33)

3) Temperature regulation in the anterior and posterior hypothalamic nuclei (see Chapter 29)

4) Activation of the sympathetic nervous system and adrenal medullary hormone secretion in the dorsal and posterior hypothalamus (see Chapter 34)

5) Thirst and drinking regulation in the lateral hypothalamus (see Chapter 24)

6) Hunger, satiety, and the regulation of eating behavior in the arcuate nucleus, ventromedial nucleus, and lateral hypothalamic area

7) Regulation of sexual behavior in the anterior and preoptic areas

8) Regulation of circadian rhythms in the suprachiasmatic nucleus

The Hypothalamus Regulates Eating Behavior

Classically, the hypothalamus has been considered a grouping of regulatory centers governing homeostasis. With respect to eating, the ventromedial nucleus of the hypothalamus serves as a **satiety center** and the lateral hypothalamic area serves as a **feeding center**. Together, these areas coordinate the processes that govern eating behavior and the subjective perception of satiety. These hypothalamic areas also influence the secretion of hormones, partic-

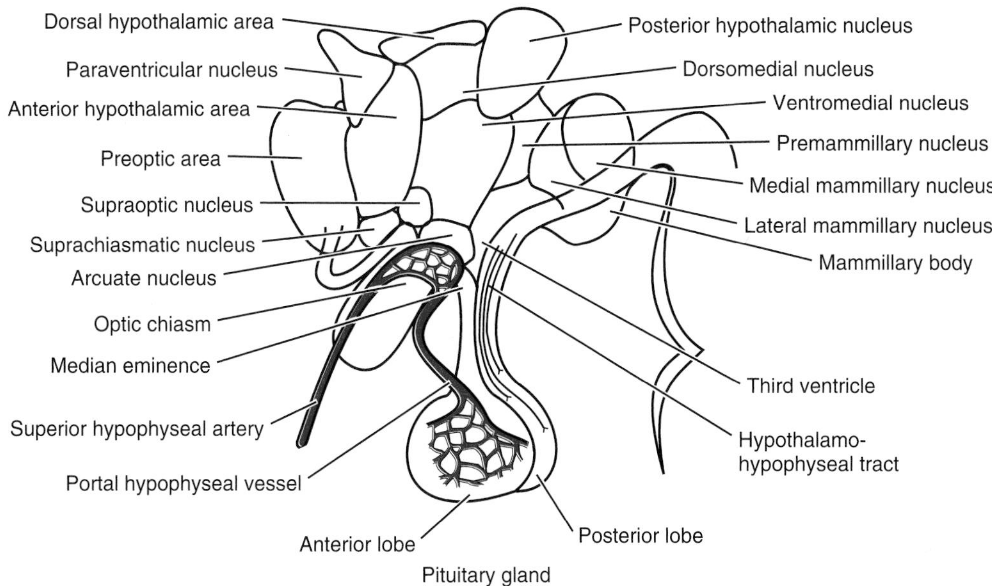

FIGURE 7.2 **The hypothalamus and its nuclei.** The connections between the hypothalamus and the pituitary gland are also shown. (Modified from Ganong WF. Review of Medical Physiology. 16th Ed. Norwalk, CT: Appleton & Lange, 1993.)

ularly from the thyroid gland, adrenal gland, and pancreatic islet cells, in response to changing metabolic demands.

Lesions in the ventromedial nucleus in experimental animals lead to morbid obesity as a result of unrestricted eating (hyperphagia). Conversely, electrical stimulation of this area results in the cessation of eating (hypophagia). Destructive lesions in the lateral hypothalamic area lead to hypophagia, even in the face of starvation; electrical stimulation of this area initiates feeding activity, even when the animal has already eaten.

The regulation of eating behavior is part of a complex pathway that regulates food intake, energy expenditure, and reproductive function in the face of changes in nutritional state. In general, the hypothalamus regulates caloric intake, utilization, and storage in a manner that tends to maintain the body weight in adulthood. The presumptive set point around which it attempts to stabilize body weight, however, is poorly defined or maintained, as it changes readily with changes in physical activity, composition of the diet, emotional states, stress, pregnancy, and so on.

A key player in the regulation of body weight is the hormone **leptin**, which is released by white fat cells (adipocytes). As fat stores increase, plasma leptin levels increase; conversely, as fat stores are depleted, leptin levels decrease. Cells in the arcuate nucleus of the hypothalamus appear to be the sensors for leptin levels. Physiological responses to low leptin levels (starvation) are initiated by the hypothalamus to increase food intake, decrease energy expenditure, decrease reproductive function, decrease body temperature, and increase parasympathetic activity. Physiological responses to high leptin levels (obesity) are initiated by the hypothalamus to decrease food intake, increase energy expenditure, and increase sympathetic activity. Hypothalamic pathways involving neuropeptide Y are important for the starvation response, while pathways involving the melanocyte-stimulating hormone are important for the obesity response.

In addition to long-term regulation of body weight, the hypothalamus also regulates eating behavior more acutely. Factors that limit the amount of food ingested during a single feeding episode originate in the gastrointestinal tract and influence the hypothalamic regulatory centers. These include sensory signals carried by the vagus nerve that signify stomach filling and chemical signals giving rise to the sensation of satiety, including absorbed nutrients (glucose, certain amino acids, and fatty acids) and gastrointestinal hormones, especially cholecystokinin.

The Hypothalamus Controls the Gonads and Sexual Activity

The anterior and preoptic hypothalamic areas are sites for regulating gonadotropic hormone secretion and sexual behavior. Neurons in the preoptic area secrete gonadotropin-releasing hormone (GnRH), beginning at puberty, in response to signals that are not understood. These neurons contain receptors for gonadal steroid hormones, testosterone and/or estradiol, which regulate GnRH secretion in either a cyclic (female) or a continual (male) pattern following the onset of puberty.

At a critical period in fetal development, circulating testosterone secreted by the testes of a male fetus changes the characteristics of cells in the preoptic area that are destined later in life to secrete GnRH. These cells, which would secrete GnRH *cyclically* at puberty, had they not been exposed to androgens prenatally, are transformed into cells that secrete GnRH *continually* at a homeostatically regulated level. As a result, males exhibit a steady-state secretion rate for gonadotropic hormones and, consequently, for testosterone (see Chapter 37).

In the absence of androgens in fetal blood during development, the preoptic area remains unchanged, so that at puberty the GnRH-secreting cells begin to secrete in a cyclic pattern. This pattern is reinforced and synchronized throughout female reproductive life by the cyclic feedback of ovarian steroids, estradiol and progesterone, on secretion of GnRH by the hypothalamus during the menstrual cycle (see Chapter 38).

Steroid levels during prenatal and postnatal development are known to mediate differentiation of sexually dimorphic regions of the brain of most vertebrate species. Sexually dimorphic brain anatomy, behavior, and susceptibility to neurological and psychiatric illness are evident in humans; however, with the exception of the GnRH-secreting cells, it has been difficult to definitively show a steroid dependency for sexually dimorphic differentiation in the human brain.

The Hypothalamus Contains the "Biological Clock"

Many physiological functions, including body temperature and sleep/wake cycles, vary throughout the day in a pattern that repeats itself daily. Others, such as the female menstrual cycle, repeat themselves approximately every 28 days. Still others, such as reproductive function in seasonal breeders, repeat annually. The hypothalamus is thought to play a major role in regulating all of these biological rhythms. Furthermore, these rhythms appear to be endogenous (within the body) because they persist even in the absence of time cues, such as day/night cycles for light and dark periods, lunar cycles for monthly rhythms, or changes in temperature and day length for seasonal change. Accordingly, most organisms, including humans, are said to possess an endogenous timekeeper, a so-called **biological clock** that times the body's regulated functions.

Most homeostatically regulated functions exhibit peaks and valleys of activity that recur approximately daily. These are called **circadian rhythms** or diurnal rhythms. The circadian rhythms of the body are driven by the **suprachiasmatic nucleus (SCN),** a center in the hypothalamus that serves as the brain's biological clock. The SCN, which influences many hypothalamic nuclei via its efferent connections, has the properties of an oscillator whose spontaneous firing patterns change dramatically during a day/night cycle. This diurnal cycle of activity is maintained *in vitro* and is an internal property of SCN cells. The molecular basis of the cellular rhythm is a series of transcriptional/translational feedback loops. The genes involved in these loops are apparently conserved from prokaryotes to humans. An important pathway influencing the SCN is the afferent **retinohypothalamic tract** of the optic nerve, which originates in the retina and enters the brain through the optic chiasm and terminates in the SCN. This pathway is the principal means by which light signals from the outside world transmit the day/night rhythm to the brain's internal clock, thereby entraining the endogenous oscillator to the external clock.

Figure 7.3 illustrates some of the circadian rhythms of the body. One of the most vivid is alertness, which peaks in the afternoon and is lowest in the hours preceding and following sleep. Another, body temperature, ranges approximately 1°C (about 2°F) throughout the day, with the low point occurring during sleep. Plasma levels of growth hormone increase greatly during sleep, in keeping with this hormone's metabolic role as a glucose-sparing agent during the nocturnal fast. Cortisol, on the other hand, has its highest daily plasma level prior to arising in the morning. The mechanism by which the SCN can regulate diverse functions is related to its control of the production of **melatonin** by the **pineal gland**. Melatonin levels increase with decreasing light as night ensues.

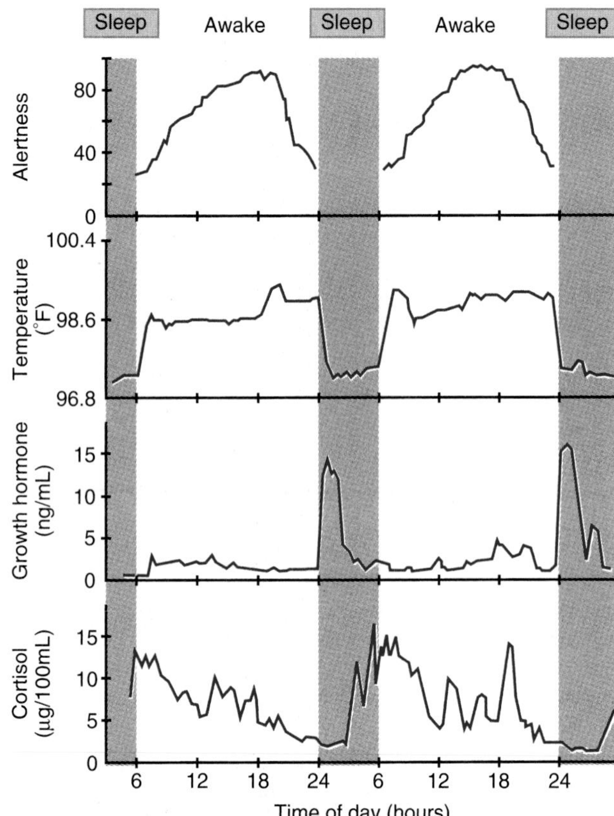

FIGURE 7.3 **Circadian rhythms in some homeostatically regulated functions during two 24-hour periods.** Alertness is measured on an arbitrary scale between sleep and most alert. (Modified from Coleman RM. Wide Awake at 3:00 AM. New York: WH Freeman, 1986.)

Other homeostatically regulated functions exhibit diurnal patterns as well; when they are all in synchrony, they function harmoniously and impart a feeling of well-being. When there is a disruption in rhythmic pattern, such as by sleep deprivation or when passing too rapidly through several time zones, the period required for reentrainment of the SCN to the new day/night pattern is characterized by a feeling of malaise and physiological distress. This is commonly experienced as jet lag in travelers crossing several time zones or by workers changing from day shift to night shift or from night shift to day shift. In such cases, the hypothalamus requires time to "reset its clock" before the regular rhythms are restored and a feeling of well-being ensues. The SCN uses the new pattern of light/darkness, as perceived in the retina, to entrain its firing rate to a pattern consistent with the external world. Resetting the clock may be facilitated by the judicious use of exogenous melatonin and by altering exposure to light.

THE RETICULAR FORMATION

The brainstem contains anatomic groupings of cell bodies clearly identified as the nuclei of cranial sensory and motor nerves or as relay cells of ascending sensory or descending motor systems. The remaining cell groups of the brainstem, located in the central core, constitute a diffuse-appearing system of neurons with widely branching axons, known as the **reticular formation**.

Neurons of the Reticular Formation Exert Widespread Modulatory Influence in the CNS

As neurochemistry and cytochemical localization techniques improve, it is becoming increasingly clear that the reticular formation is not a diffuse, undefined system; it contains highly organized clusters of transmitter-specific cell groups that influence functions in specific areas of the CNS. For example, the nuclei of monoaminergic neuronal systems are located in well-defined cell groups throughout the reticular formation.

A unique characteristic of neurons of the reticular formation is their widespread system of axon collaterals, which make extensive synaptic contacts and, in some cases, travel over long distances in the CNS. A striking example is the demonstration, using intracellular labeling of individual cells and their processes, that one axon branch descends all the way into the spinal cord, while the collateral branch projects rostrally all the way to the forebrain, making myriad synaptic contacts along both axonal pathways.

The Ascending Reticular Activating System Mediates Consciousness and Arousal

Sensory neurons bring peripheral sensory information to the CNS via specific pathways that ascend and synapse with specific nuclei of the thalamus, which, in turn, innervate primary sensory areas of the cerebral cortex. These pathways involve three to four synapses, starting from a receptor that responds to a specific sensory modality—such as touch, hearing, or vision. Each modality has, in addition, a nonspecific form of sensory transmission, in that axons of the ascending fibers send collateral branches to cells of the reticular formation (Fig. 7.4). The latter, in turn, send their axons to the **intralaminar nuclei of thalamus,** which innervate wide areas of the cerebral cortex and limbic system. In the cerebral cortex and limbic system, the influence of the nonspecific projections from the reticular formation is arousal of the organism. This series of connections from the reticular formation through the intralaminar nuclei of the thalamus and on to the forebrain is termed the **ascending reticular activating system.**

The reticular formation also houses the neuronal systems that regulate sleep/wake cycles and consciousness. So important is the ascending reticular activating system to the state of arousal that a malfunction in the reticular formation, particularly the rostral portion, can lead to a loss of consciousness and coma.

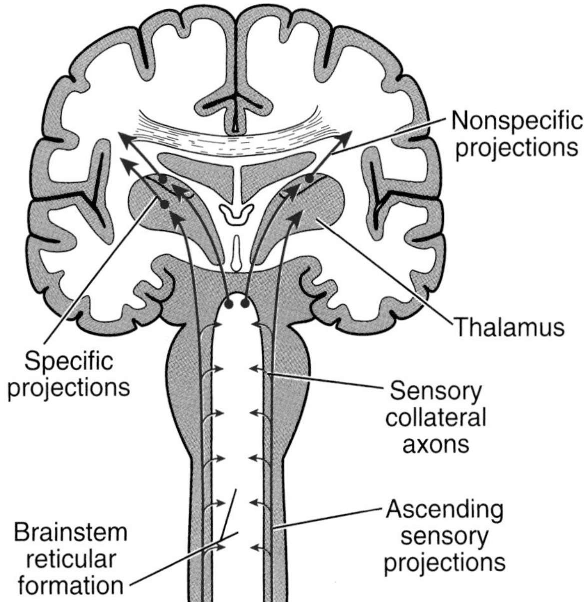

FIGURE 7.4 **The brainstem reticular formation and reticular activating system.** Ascending sensory tracts send axon collateral fibers to the reticular formation. These give rise to fibers synapsing in the intralaminar nuclei of the thalamus. From there, these nonspecific thalamic projections influence widespread areas of the cerebral cortex and limbic system.

An Electroencephalogram Records Electrical Activity of the Brain's Surface

The influence of the ascending reticular activating system on the brain's activity can be monitored via **electroencephalography.** The electroencephalograph is a sensitive recording device for picking up the electrical activity of the brain's surface through electrodes placed on designated sites on the scalp. This noninvasive tool measures simultaneously, via multiple leads, the electrical activity of the major areas of the cerebral cortex. It is also the best diagnostic tool available for detecting abnormalities in electrical activity, such as in epilepsy, and for diagnosing sleep disorders.

The detected electrical activity reflects the extracellular recording of the myriad postsynaptic potentials in cortical neurons underlying the electrode. The summated electrical potentials recorded from moment to moment in each lead are influenced greatly by the input of sensory information from the thalamus via specific and nonspecific projections to the cortical cells, as well as inputs that course laterally from other regions of the cortex.

EEG Waves. The waves recorded on an **electroencephalogram** (EEG) are described in terms of frequency, which usually ranges from less than 1 to about 30 Hz, and amplitude or height of the wave, which usually ranges from 20 to 100 μV. Since the waves are a summation of activity in a complex network of neuronal processes, they are highly variable. However, during various states of consciousness, EEG waves have certain characteristic patterns. At the highest state of alertness, when sensory input is greatest, the waves are of high frequency and low amplitude, as many

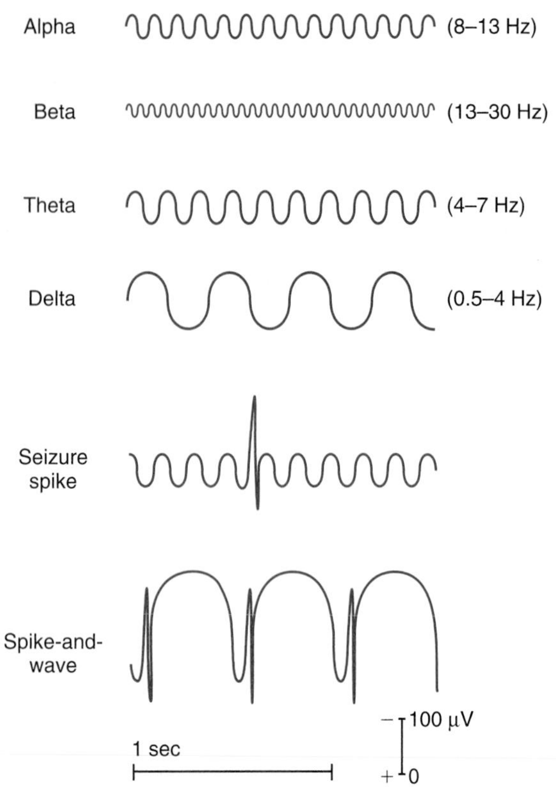

FIGURE 7.5 Patterns of brain waves recorded on an EEG. Wave patterns are designated alpha, beta, theta, or delta waves, based on frequency and relative amplitude. In epilepsy, abnormal spikes and large summated waves appear as many neurons are activated simultaneously.

units discharge asynchronously. At the opposite end of the alertness scale, when sensory input is at its lowest, in deep sleep, a synchronized EEG has the characteristics of low frequency and high amplitude. An absence of EEG activity is the legal criterion for death in the United States.

EEG wave patterns are classified according to their frequency (Fig. 7.5). **Alpha waves**, a rhythm ranging from 8 to 13 Hz, are observed when the person is awake but relaxed with the eyes closed. When the eyes are open, the added visual input to the cortex imparts a faster rhythm to the EEG, ranging from 13 to 30 Hz and designated **beta waves**. The slowest waves recorded occur during sleep: **theta waves** at 4 to 7 Hz and **delta waves** at 0.5 to 4 Hz, in deepest sleep.

Abnormal wave patterns are seen in **epilepsy**, a neurological disorder of the brain characterized by spontaneous discharges of electrical activity, resulting in abnormalities ranging from momentary lapses of attention, to seizures of varying severity, to loss of consciousness if both brain hemispheres participate in the electrical abnormality. The characteristic waveform signifying seizure activity is the appearance of spikes or sharp peaks, as abnormally large numbers of units fire simultaneously. Examples of spike activity occurring singly and in a spike-and-wave pattern are shown in Figure 7.5.

Sleep and the EEG. Sleep is regulated by the reticular formation. The ascending reticular activating system is periodically shut down by influences from other regions of the reticular formation. The EEG recorded during sleep reveals a persistently changing pattern of wave amplitudes and frequencies, indicating that the brain remains continually active even in the deepest stages of sleep. The EEG pattern recorded during sleep varies in a cyclic fashion that repeats approximately every 90 minutes, starting from the time of falling asleep to awakening 7 to 8 hours later (Fig. 7.6). These cycles are associated with two different forms of sleep, which follow each other sequentially:

1. **Slow-wave sleep:** four stages of progressively deepening sleep (i.e., it becomes harder to wake the subject)
2. **Rapid eye movement (REM) sleep:** back-and-forth movements of the eyes under closed lids, accompanied by autonomic excitation

EEG recordings of sleeping subjects in laboratory settings reveal that the brain's electrical activity varies as the

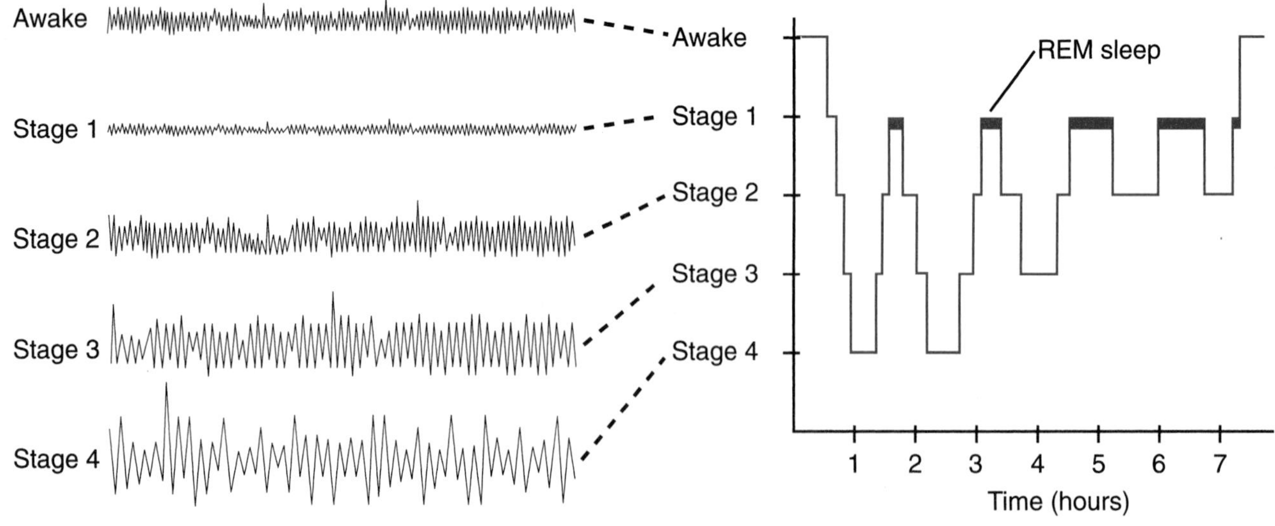

FIGURE 7.6 The brain wave patterns during a normal sleep cycle. (See text for details.) (Modified from Kandel ER, Schwartz JH, Jessel TM. Principles of Neural Science. 3rd Ed. New York: Elsevier, 1991.)

subject passes through cycles of slow-wave sleep, then REM sleep, on through the night.

A normal sleep cycle begins with slow-wave sleep, four stages of increasingly deep sleep during which the EEG becomes progressively slower in frequency and higher in amplitude. Stage 4 is reached at the end of about an hour, when delta waves are observed (see Fig. 7.6). The subject then passes through the same stages in reverse order, approaching stage 1 by about 90 minutes, when a REM period begins, followed by a new cycle of slow-wave sleep. Slow-wave sleep is characterized by decreased heart rate and blood pressure, slow and regular breathing, and relaxed muscle tone. Stages 3 and 4 occur only in the first few sleep cycles of the night. In contrast, REM periods increase in duration with each successive cycle, so that the last few cycles consist of approximately equal periods of REM sleep and stage 2 slow-wave sleep.

REM sleep is also known as **paradoxical sleep,** because of the seeming contradictions in its characteristics. First, the EEG exhibits unsynchronized, high-frequency, low-amplitude waves (i.e., a beta rhythm), which is more typical of the awake state than sleep, yet the subject is as difficult to arouse as when in stage 4 slow-wave sleep. Second, the autonomic nervous system is in a state of excitation; blood pressure and heart rate are increased and breathing is irregular. In males, autonomic excitation in REM sleep includes penile erection. This reflex is used in diagnosing impotence, to determine whether erectile failure is based on a neurological or a vascular defect (in which case, erection does not accompany REM sleep).

When subjects are awakened during a REM period, they usually report dreaming. Accordingly, it is customary to consider REM sleep as dream sleep. Another curious characteristic of REM sleep is that most voluntary muscles are temporarily paralyzed. Two exceptions, in addition to the muscles of respiration, include the extraocular muscles, which contract rhythmically to produce the rapid eye movements, and the muscles of the middle ear, which protect the inner ear (see Chapter 4). Muscle paralysis is caused by an active inhibition of motor neurons mediated by a group of neurons located close to the locus ceruleus in the brainstem. Many of us have experienced this muscle paralysis on waking from a bad dream, feeling momentarily incapable of running from danger. In certain sleep disorders in which skeletal muscle contraction is not temporarily paralyzed in REM sleep, subjects act out dream sequences with disturbing results, with no conscious awareness of this happening.

Sleep in humans varies with developmental stage. Newborns sleep approximately 16 hours per day, of which about 50% is spent in REM sleep. Normal adults sleep 7 to 8 hours per day, of which about 25% is spent in REM sleep. The percentage of REM sleep declines further with age, together with a loss of the ability to achieve stages 3 and 4 of slow-wave sleep.

THE FOREBRAIN

The forebrain contains the **cerebral cortex** and the **subcortical structures** rostral to the diencephalon. The cortex, a few-millimeters-thick outer shell of the cerebrum, has a rich, multilayered array of neurons and their processes forming columns perpendicular to the surface. The axons of cortical neurons give rise to descending fiber tracts and intrahemispheric and interhemispheric fiber tracts, which, together with ascending axons coursing toward the cortex, make up the prominent white matter underlying the outer cortical gray matter. A deep sagittal fissure divides the cortex into a right and left hemisphere, each of which receives sensory input from and sends its motor output to the opposite side of the body. A set of **commissures** containing axonal fibers interconnects the two hemispheres, so that processed neural information from one side of the forebrain is transmitted to the opposite hemisphere. The largest of these commissures is the **corpus callosum,** which interconnects the major portion of the hemispheric regions (Fig. 7.7).

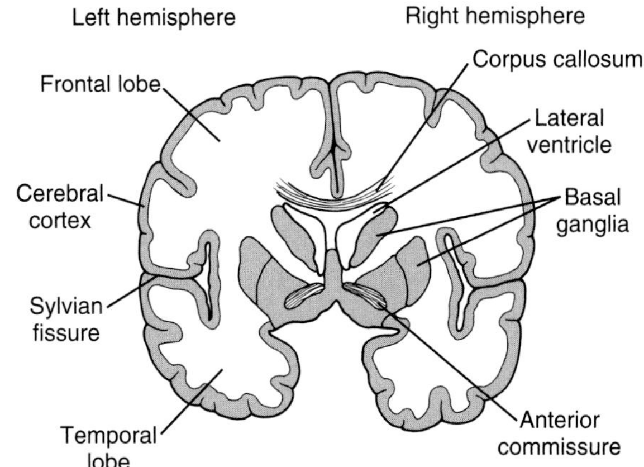

FIGURE 7.7 The cerebral hemispheres and some deep structures in a coronal section through the rostral forebrain. The corpus callosum is the major commissure that interconnects the right and left hemispheres. The anterior commissure connects rostral components of the right and left temporal lobes. The cortex is an outer rim of gray matter (neuronal cell bodies and dendrites); deep to the cortex is white matter (axonal projections) and then subcortical gray matter.

Among the subcortical structures located in the forebrain are the components of the limbic system, which regulates emotional response, and the **basal ganglia** (caudate, putamen, and globus pallidus), which are essential for coordinating motor activity (see Chapter 5).

The Cerebral Cortex Is Functionally Compartmentalized

In the human brain, the surface of the cerebral cortex is highly convoluted, with **gyri** (singular, *gyrus*) and **sulci** (singular, *sulcus*), which are akin to hills and valleys, respectively. Deep sulci are also called **fissures.** Two deep fissures form prominent landmarks on the surface of the cortex; the **central sulcus** divides the **frontal lobe** from the **parietal lobe,** and the **sylvian fissure** divides the parietal lobe from the **temporal lobe** (Fig. 7.8). The **occipital lobe** has less prominent sulci separating it from the parietal and temporal lobes.

Topographically, the cerebral cortex is divided into areas of specialized functions, including the primary sensory areas for vision (occipital cortex), hearing (temporal cortex), somatic sensation (postcentral gyrus), and primary

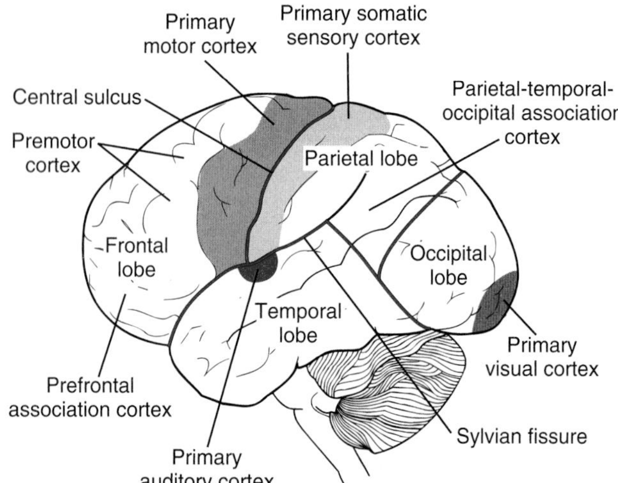

FIGURE 7.8 The four lobes of the cerebral cortex, containing primary sensory and motor areas and major association areas. The central sulcus and sylvian fissure are prominent landmarks used in defining the lobes of the cortex. Imaginary lines are drawn in to indicate the boundaries between the occipital, temporal, and parietal lobes. (Modified from Kandel ER, Schwartz JH, Jessel TM. Principles of Neural Science. 3rd Ed. New York: Elsevier, 1991.)

motor area (precentral gyrus) (see Chapters 4 and 5). As shown in Figure 7.8, these well-defined areas comprise only a small fraction of the surface of the cerebral cortex. The majority of the remaining cortical area is known as **association cortex,** where the processing of neural information is performed at the highest levels of which the organism is capable; among vertebrates, the human cortex contains the most extensive association areas. The association areas are also sites of long-term memory, and they control such human functions as language acquisition, speech, musical ability, mathematical ability, complex motor skills, abstract thought, symbolic thought, and other cognitive functions.

Association areas interconnect and integrate information from the primary sensory and motor areas via intrahemispheric connections. The **parietal-temporal-occipital association cortex** integrates neural information contributed by visual, auditory, and somatic sensory experiences. The **prefrontal association cortex** is extremely important as the coordinator of emotionally motivated behaviors, by virtue of its connections with the limbic system. In addition, the prefrontal cortex receives neural input from the other association areas and regulates motivated behaviors by direct input to the **premotor area,** which serves as the association area of the motor cortex.

Sensory and motor functions are controlled by cortical structures in the contralateral hemisphere (see Chapters 4 and 5). Particular cognitive functions or components of these functions may be lateralized to one side of the brain (see Clinical Focus Box 7.1).

The Limbic System Is the Seat of the Emotions

The limbic system comprises large areas of the forebrain where the emotions are generated and the responses to emotional stimuli are coordinated. Understanding its functions is particularly challenging because it is a complex system of numerous and disparate elements, most of which have not been fully characterized. A compelling reason for studying the limbic system is that the major psychiatric disorders—including bipolar disorder, major depression, schizophrenia, and dementia—involve malfunctions in the limbic system.

Anatomy of the Limbic System. The limbic system comprises specific areas of the cortex and subcortical structures interconnected via circuitous pathways that link the cerebrum with the diencephalon and brainstem (Fig. 7.9). Originally the limbic system was considered to be restricted to a ring of structures surrounding the corpus callosum, including the olfactory system, the **cingulate gyrus, parahippocampal gyrus,** and **hippocampus,** together with the fiber tracts that interconnect them with the diencephalic components of the limbic system, the hypothalamus and anterior thalamus. Current descriptions of the limbic system also include the **amygdala** (deep in the temporal lobe), **nucleus accumbens** (the limbic portion of the basal ganglia), **septal nuclei** (at the base of the forebrain), the **prefrontal cortex** (anterior and inferior components of the frontal lobe) and the **habenula** (in the diencephalon).

Circuitous loops of fiber tracts interconnect the limbic structures. The main circuit links the hippocampus to the mammillary body of the hypothalamus by way of the fornix, the hypothalamus to the anterior thalamic nuclei via the mammillothalamic tract, and the anterior thalamus to the cingulate gyrus by widespread, anterior thalamic projections (Fig. 7.10). To complete the circuit, the cingulate gyrus connects with the hippocampus, to enter the circuit again. Other structures of the limbic system form smaller loops within this major circuit, forming the basis for a wide range of emotional behaviors.

The fornix also connects the hippocampus to the base of the forebrain where the septal nuclei and nucleus accumbens reside. Prefrontal cortex and other areas of association cortex provide the limbic system with information based on previous learning and currently perceived needs. Inputs from the brainstem provide visceral and somatic sensory signals, including tactile, pressure, pain, and temperature information from the skin and sexual organs and pain information from the visceral organs.

At the caudal end of the limbic system, the brainstem has reciprocal connections with the hypothalamus (see Fig. 7.10). As noted above, all ascending sensory systems in the brainstem send axon collaterals to the reticular formation, which, in turn, innervates the limbic system, particularly via monoaminergic pathways. The reticular formation also forms the ascending reticular activating system, which serves not only to arouse the cortex but also to impart an emotional tone to the sensory information transmitted nonspecifically to the cerebral cortex.

Monoaminergic Innervation. **Monoaminergic neurons** innervate all parts of the CNS via widespread, divergent pathways starting from cell groups in the reticular formation. The limbic system and basal ganglia are richly innervated by catecholaminergic (noradrenergic and dopamin-

CLINICAL FOCUS BOX 7.1

The Split Brain

Patients with life-threatening, intractable epileptic seizures were treated in the past by surgical commissurotomy or cutting of the corpus callosum (see Fig. 7.7). This procedure effectively cut off most of the neuronal communication between the left and right hemispheres and vastly improved patient status because seizure activity no longer spread back and forth between the hemispheres.

There was a remarkable absence of overt signs of disability following commissurotomy; patients retained their original motor and sensory functions, learning and memory, personality, talents, emotional responding, and so on. This outcome was not unexpected because each hemisphere has bilateral representation of most known functions; moreover, those ascending (sensory) and descending (motor) neuronal systems that crossed to the opposite side were known to do so at levels lower than the corpus callosum.

Notwithstanding this appearance of normalcy, following commissurotomy, patients were shown to be impaired to the extent that one hemisphere literally did not know what the other was doing. It was further shown that each hemisphere processes neuronal information differently from the other, and that some cerebral functions are confined exclusively to one hemisphere.

In an interesting series of studies by Nobel laureate Roger Sperry and colleagues, these patients with a so-called split brain were subjected to psychophysiological testing in which each disconnected hemisphere was examined independently. Their findings confirmed what was already known: Sensory and motor functions are controlled by cortical structures in the contralateral hemisphere. For example, visual signals from the left visual field were perceived in the right occipital lobe, and there were contralateral controls for auditory, somatic sensory, and motor functions. (Note that the olfactory system is an exception, as odorant chemicals applied to one nostril are perceived in the olfactory lobe on the same side.) However, the scientists were surprised to find that language ability was controlled almost exclusively by the left hemisphere. Thus, if an object was presented to the left brain via any of the sensory systems, the subject could readily identify it by the spoken word. However, if the object was presented to the right hemisphere, the subject could not find words to identify it. This was not due to an inability of the right hemisphere to perceive the object, as the subject could easily identify it among other choices by nonverbal means, such as feeling it while blindfolded. From these and other tests it became clear that the right hemisphere was mute; it could not produce language.

In accordance with these findings, anatomic studies show that areas in the temporal lobe concerned with language ability, including Wernicke's area, are anatomically larger in the left hemisphere than in the right in a majority of humans, and this is seen even prenatally. Corroborative evidence of language ability in the left hemisphere is shown in persons who have had a stroke, where aphasias are most severe if the damage is on the left side of the brain. Analysis of people who are deaf who communicated by sign language prior to a stroke has shown that sign language is also a left-hemisphere function. These patients show the same kinds of grammatical and syntactical errors in their signing following a left-hemisphere stroke as do speakers.

In addition to language ability, the left hemisphere excels in mathematical ability, symbolic thinking, and sequential logic. The right hemisphere, on the other hand, excels in visuospatial ability, such as three-dimensional constructions with blocks and drawing maps, and in musical sense, artistic sense, and other higher functions that computers seem less capable of emulating. The right brain exhibits some ability in language and calculation, but at the level of children ages 5 to 7. It has been postulated that both sides of the brain are capable of all of these functions in early childhood, but the larger size of the language area in the left temporal lobe favors development of that side during language acquisition, resulting in nearly total specialization for language on the left side for the rest of one's life.

ergic) and serotonergic nerve terminals emanating from brainstem nuclei that contain relatively few cell bodies compared to their extensive terminal projections. From neurochemical manipulation of monoaminergic neurons in the limbic system, it is apparent that they play a major role in determining emotional state.

Dopaminergic neurons are located in three major pathways originating from cell groups in either the midbrain (the substantia nigra and ventral tegmental area) or the hypothalamus (Fig. 7.11). The **nigrostriatal system** consists of neurons with cell bodies in the substantia nigra (pars compacta) and terminals in the neostriatum (caudate and putamen) located in the basal ganglia. This dopaminergic pathway is essential for maintaining normal muscle tone and initiating voluntary movements (see Chapter 5). The **tuberoinfundibular system** of dopaminergic neurons is located entirely within the hypothalamus, with cell bodies in the arcuate nucleus and periventricular nuclei and terminals in the median eminence on the ventral surface of the hypothalamus. The tuberoinfundibular system is responsible for the secretion of hypothalamic releasing factors into a portal system that carries them through the pituitary stalk into the anterior pituitary lobe (see Chapter 32).

The **mesolimbic/mesocortical system** of dopaminergic neurons originates in the ventral tegmental area of the midbrain region of the brainstem and innervates most structures of the limbic system (olfactory tubercles, septal nuclei, amygdala, nucleus accumbens) and limbic cortex (frontal and cingulate cortices). This dopaminergic system plays an important role in motivation and drive. For example, dopaminergic sites in the limbic system, particularly the more ventral structures such as the septal nuclei and nucleus accumbens, are associated with the brain's **reward system**. Drugs that increase dopaminergic transmission, such as cocaine, which inhibits dopamine reuptake, and amphetamine, which promotes dopamine release and inhibits its reuptake, lead to repeated administration and abuse presumably because they stimulate the brain's reward system. The mesolimbic/mesocortical dopaminergic system is also the site of action of **neuroleptic drugs**, which

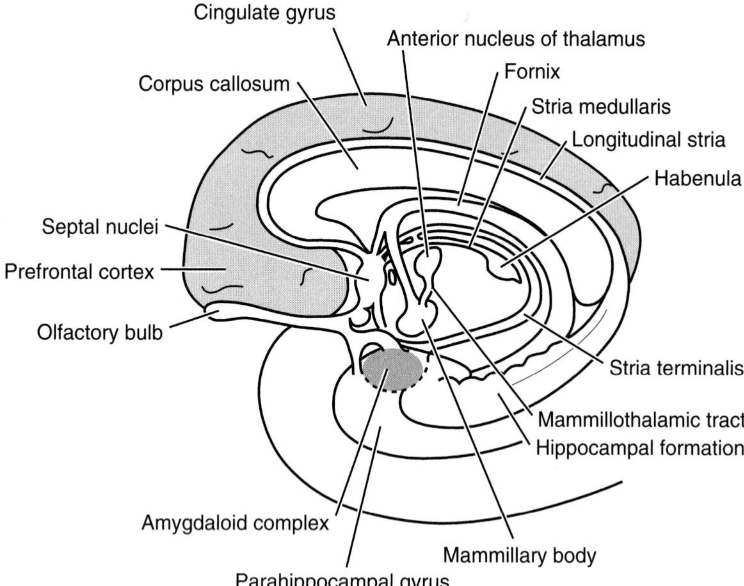

FIGURE 7.9 The cortical and subcortical structures of the limbic system extending from the cerebral cortex to the diencephalon. The fiber tracts that interconnect the structures of the limbic system are also shown. (Modified from Truex RC, Carpenter MB. Strong and Elwyn's Human Neuroanatomy. 5th Ed. Baltimore: Williams & Wilkins, 1964.)

are used to treat schizophrenia (discussed later) and other psychotic conditions.

Noradrenergic neurons (containing norepinephrine) are located in cell groups in the medulla and pons (Fig. 7.12). The medullary cell groups project to the spinal cord, where they influence cardiovascular regulation and other autonomic functions. Cell groups in the pons include the **lateral system,** which innervates the basal forebrain and hypothalamus, and the **locus ceruleus,** which sends efferent fibers to nearly all parts of the CNS.

Noradrenergic neurons innervate all parts of the limbic system and the cerebral cortex, where they play a major role in setting **mood** (sustained emotional state) and **affect** (the emotion itself; e.g., euphoria, depression, anxiety). Drugs that alter noradrenergic transmission have profound effects on mood and affect. For example, reserpine, which depletes brain norepinephrine (NE), induces a state of depression. Drugs that enhance NE availability, such as monoamine oxidase inhibitors (MAOIs) and inhibitors of reuptake, reverse this depression. Amphetamines and cocaine have effects on boosting noradrenergic transmission similar to those described for dopaminergic transmission; they inhibit reuptake and/or promote the release of norepinephrine. Increased noradrenergic transmission results in an elevation of mood, which further contributes to the po-

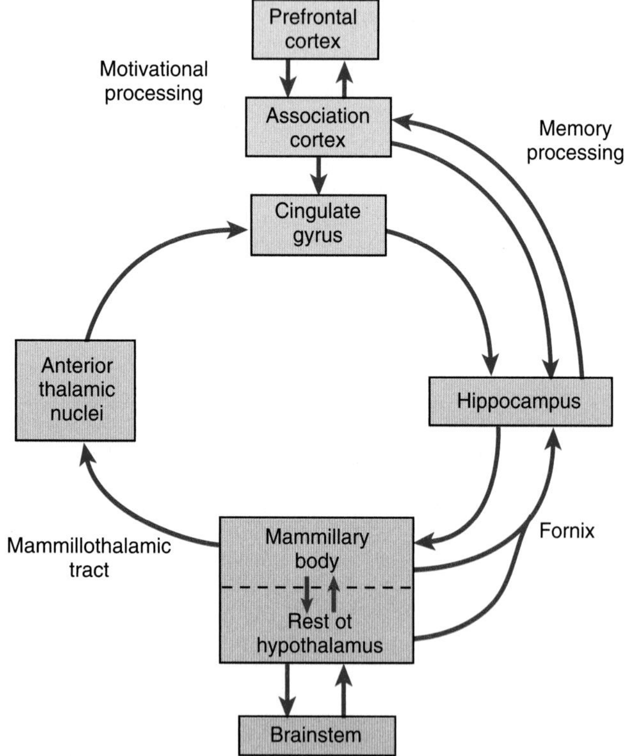

FIGURE 7.10 The main circuit of the limbic system.

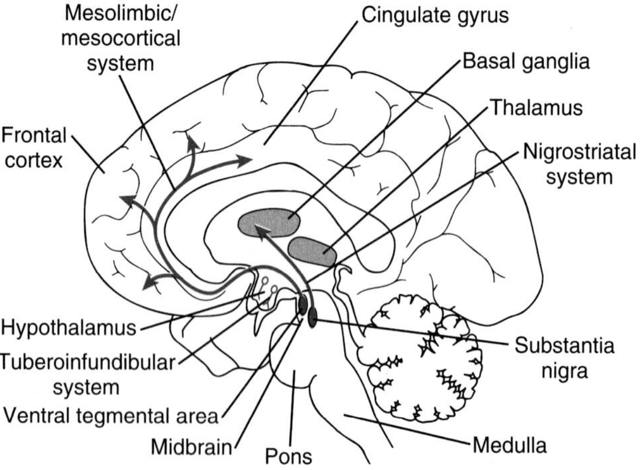

FIGURE 7.11 The origins and projections of the three major dopaminergic systems. (Modified from Heimer L. The Human Brain and Spinal Cord. New York: Springer-Verlag, 1983.)

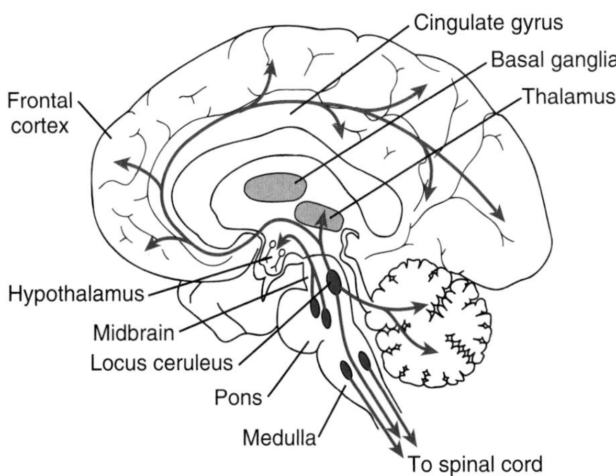

FIGURE 7.12 The origins and projections of five of seven cell groups of noradrenergic neurons of the brain. The depicted groups originate in the medulla and pons. Among the latter, the locus ceruleus in the dorsal pons innervates most parts of the CNS. (Modified from Heimer L. The Human Brain and Spinal Cord. New York: Springer-Verlag, 1983.)

tential for abusing such drugs, despite the depression that follows when drug levels fall. Some of the unwanted consequences of cocaine or amphetamine-like drugs reflect the increased noradrenergic transmission, in both the periphery and the CNS. This can result in a hypertensive crisis, myocardial infarction, or stroke, in addition to marked swings in affect, starting with euphoria and ending with profound depression.

Serotonergic neurons also innervate most parts of the CNS. Cell bodies of these neurons are located at the midline of the brainstem (the raphe system) and in more laterally placed nuclei, extending from the caudal medulla to the midbrain (Fig. 7.13). Serotonin plays a major role in the defect underlying affective disorders (discussed later). Drugs

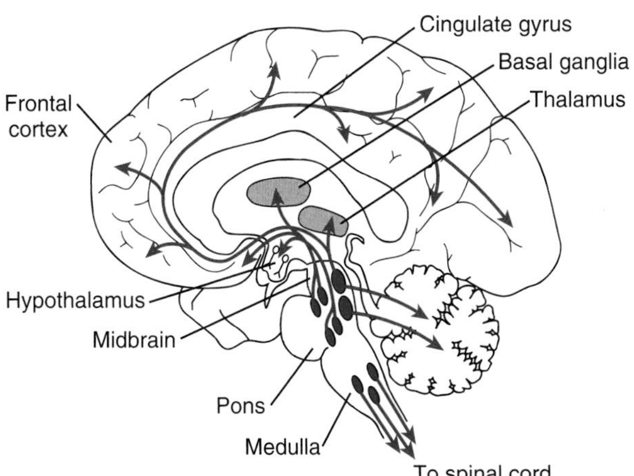

FIGURE 7.13 The origins and projections of the nine cell groups of the serotonergic system of the brain. The depicted groups originate in the caudal medulla, pons, and midbrain and send projections to most regions of the brain. (Modified from Heimer L. The Human Brain and Spinal Cord. New York: Springer-Verlag, 1983.)

that increase serotonin transmission are effective antidepressant agents.

The Brain's Reward System. Experimental studies beginning early in the last century demonstrated that stimulating the limbic system or creating lesions in various parts of the limbic system can alter emotional states. Most of our knowledge comes from animal studies, but emotional feelings are reported by humans when limbic structures are stimulated during brain surgery. The brain has no pain sensation when touched, and subjects awakened from anesthesia during brain surgery have communicated changes in emotional experience linked to electrical stimulation of specific areas.

Electrical stimulation of various sites in the limbic system produces either pleasurable (rewarding) or unpleasant (aversive) feelings. To study these findings, researchers use electrodes implanted in the brains of animals. When electrodes are implanted in structures presumed to generate rewarding feelings and the animals are allowed to deliver current to the electrodes by pressing a bar, repeated and prolonged self-stimulation is seen. Other needs—such as food, water, and sleep—are neglected. The sites that provoke the highest rates of electrical self-stimulation are in the ventral limbic areas, including the septal nuclei and nucleus accumbens. Extensive studies of electrical self-stimulatory behavior indicate that dopaminergic neurons play a major role in mediating reward. The nucleus accumbens is thought to be the site of action of addictive drugs, including opiates, alcohol, nicotine, cocaine, and amphetamine.

Aggression and the Limbic System. A fight-or-flight response, including the autonomic components (see Chapter 6) and postures of rage and aggression characteristic of fighting behavior, can be elicited by electrical stimulation of sites in the hypothalamus and amygdala. If the frontal cortical connections to the limbic system are severed, rage postures and aggressiveness become permanent, illustrating the importance of the higher centers in restraining aggression and, presumably, in invoking it at appropriate times. By contrast, bilateral removal of the amygdala results in a placid animal that cannot be provoked.

Sexual Activity. The biological basis of human sexual activity is poorly understood because of its complexity and because findings derived from nonhuman animal studies cannot be extrapolated. The major reason for this limitation is that the cerebral cortex, uniquely developed in the human brain, plays a more important role in governing human sexual activity than the instinctive or olfactory-driven behaviors in nonhuman primates and lower mammalian species. Nevertheless, several parallels in human and nonhuman sexual activities exist, indicating that the limbic system, in general, coordinates sex drive and mating behavior, with higher centers exerting more or less overriding influences.

Copulation in mammals is coordinated by reflexes of the sacral spinal cord, including male penile erection and ejaculation reflexes and engorgement of female erectile tissues, as well as the muscular spasms of the orgasmic response. Copulatory behaviors and postures can be elicited in animals by stimulating parts of the hypothalamus, olfactory system, and other limbic areas, resulting in mounting behavior in males and lordosis (arching the back and raising the tail) in females. Ablation studies have shown that sexual behavior also requires an intact connection of the limbic system with the frontal cortex.

Olfactory cues are important in initiating mating activity in seasonal breeders. Driven by the hypothalamus' endogenous seasonal clock, the anterior and preoptic areas of the hypothalamus initiate hormonal control of the gonads. Hormonal release leads to the secretion of odorants (pheromones) by the female reproductive tract, signaling the onset of estrus and sexual receptivity to the male. The odorant cues are powerful stimulants, acting at extremely low concentrations to initiate mating behavior in males. The olfactory system, by virtue of its direct connections with the limbic system, facilitates the coordination of behavioral, endocrine, and autonomic responses involved in mating.

Although human and nonhuman primates are not seasonal breeders (mating can occur on a continual basis), vestiges of this pattern remain. These include the importance of the olfactory and limbic systems and the role of the hypothalamus in cyclic changes in female ovarian function and the continuous regulation of male testicular function. More important determinants of human sexual activity are the higher cortical functions of learning and memory, which serve to either reinforce or suppress the signals that initiate sexual responding, including the sexual reflexes coordinated by the sacral spinal cord.

Psychiatric Disorders Involve the Limbic System

The major psychiatric disorders, including affective disorders and schizophrenia, are disabling diseases with a genetic predisposition and no known cure. The biological basis for these disorders remains obscure, particularly the role of environmental influences on individuals with a genetic predisposition to developing a disorder. Altered states of the brain's monoaminergic systems have been a major focus as possible underlying factors, based on extensive human studies in which neurochemical imbalances in catecholamines, acetylcholine, and serotonin have been observed. Another reason for focusing on the monoaminergic systems is that the most effective drugs used in treating psychiatric disorders are agents that alter monoaminergic transmission.

Affective Disorders. The **affective disorders** include **major depression,** which can be so profound as to provoke suicide, and **bipolar disorder** (or **manic-depressive disorder**), in which periods of profound depression are followed by periods of mania, in a cyclic pattern. Biochemical studies indicate that depressed patients show decreased use of brain NE. In manic periods, NE transmission increases. Whether in depression or in mania, all patients seem to have decreased brain serotonergic transmission, suggesting that serotonin may exert an underlying permissive role in abnormal mood swings, in contrast with norepinephrine, whose transmission, in a sense, titrates the mood from highest to lowest extremes.

The most effective treatments for depression, including antidepressant drugs such as MAOIs and selective serotonin reuptake inhibitors (SSRIs) and electroconvulsive therapy, have in common the ability to stimulate both noradrenergic and serotonergic neurons serving the limbic system. A therapeutic response to these treatments ensues only after treatment is repeated over time. Similarly, when treatment stops, symptoms may not reappear for several weeks. This time lag in treatment response is presumably due to alterations in the long-term regulation of receptor and second messenger systems in relevant regions of the brain.

The most effective long-term treatment for mania is lithium, although antipsychotic (neuroleptic) drugs, which block dopamine receptors, are effective in the acute treatment of mania. The therapeutic actions of lithium remain unknown, but the drug has an important action on a receptor-mediated second messenger system. Lithium interferes with regeneration of phosphatidylinositol in neuronal membranes by blocking the hydrolysis of inositol-1-phosphate. Depletion of phosphatidylinositol in the membrane renders it incapable of responding to receptors that use this second messenger system.

Schizophrenia. **Schizophrenia** is the collective name for a group of psychotic disorders that vary greatly in symptoms among individuals. The features most commonly observed are thought disorder, inappropriate emotional response, and auditory hallucinations. While the biochemical imbalance resulting in schizophrenia is poorly understood, the most troubling symptoms of schizophrenia are ameliorated by neuroleptic drugs, which block dopamine receptors in the limbic system.

Current research is focused on finding the subtype of dopamine receptor that mediates mesocortical/mesolimbic dopaminergic transmission but does not affect the nigrostriatal system, which controls motor function (see Fig. 7.12). So far, neuroleptic drugs that block one pathway almost always block the other as well, leading to unwanted neurological side effects, including abnormal involuntary movements (tardive dyskinesia) after long-term treatment or parkinsonism in the short term. Similarly, some patients with Parkinson's disease who receive L-DOPA to augment dopaminergic transmission in the nigrostriatal pathway must be taken off the medication because they develop psychosis.

Memory and Learning Require the Cerebral Cortex and Limbic System

Memory and learning are inextricably linked because part of the learning process involves the assimilation of new information and its commitment to memory. The most likely sites of learning in the human brain are the large association areas of the cerebral cortex, in coordination with subcortical structures deep in the temporal lobe, including the hippocampus and amygdala. The association areas draw on sensory information received from the primary visual, auditory, somatic sensory, and olfactory cortices and on emotional feelings transmitted via the limbic system. This information is integrated with previously learned skills and stored memory, which presumably also reside in the association areas.

The learning process itself is poorly understood, but it can be studied experimentally at the synaptic level in isolated slices of mammalian brain or in more simple invertebrate nervous systems. Synapses subjected to repeated presynaptic neuronal stimulation show changes in the excitability of postsynaptic neurons. These changes include the facilitation of neuronal firing, altered patterns of neurotransmitter release, second messenger formation, and, in intact organisms, evidence that learning occurred. The phenomenon of increased excitability and altered chemical state on repeated synaptic stimulation is known as **long-term potentiation,** a persistence beyond the cessation of electrical stimulation, as is expected of learning and memory. An early event in long-term potentiation is a series of protein phosphorylations induced by receptor-activated second messengers and leading to activation of a host of intracellular proteins and altered excitability. In addition to biochemical changes in synaptic efficacy associated with learning at the cellular level, structural alterations occur. The number of connections between sets of neurons increases as a result of experience.

Much of our knowledge about human memory formation and retrieval is based on studies of patients in whom stroke, brain injury, or surgery resulted in memory disorders. Such knowledge is then examined in more rigorous experiments in nonhuman primates capable of cognitive functions. From these combined approaches, we know that the prefrontal cortex is essential for coordinating the formation of memory, starting from a learning experience in the cerebral cortex, then processing the information and communicating it to the subcortical limbic structures. The prefrontal cortex receives sensory input from the parietal, occipital, and temporal lobes and emotional input from the limbic system. Drawing on skills such as language and mathematical ability, the prefrontal cortex integrates these inputs in light of previously acquired learning. The prefrontal cortex can thus be considered the site of **working memory**, where new experiences are processed, as opposed to sites that consolidate the memory and store it. The processed information is then transmitted to the hippocampus, where it is consolidated over several hours into a more permanent form that is stored in, and can be retrieved from, the association cortices.

Declarative and Procedural Memory. A remarkable finding from studies of surgical patients who had bilateral resections of the medial temporal lobe is that there are two fundamentally different memory systems in the brain. **Declarative memory** refers to memory of events and facts and the ability to consciously access them. Patients with bilateral medial temporal lobectomies lose their ability to form any new declarative memories. However, they retain their ability to learn and remember new skills and procedures. This type of memory is called **procedural memory** and involves several different regions of the brain, depending on the type of procedure. In contrast to declarative memory, structures in the medial temporal lobe are not involved in procedural memory. Learning and remembering new motor skills and habits requires the striatum, motor areas of the cortex, and the cerebellum. Emotional associations require the amygdala. Conditioned reflexes require the cerebellum.

An early demonstration of the dichotomy between declarative and procedural memory came from studies by Dr. Brenda Milner on a patient of Dr. Wilder Penfield in the mid-1950s. This patient (H.M.) had received a bilateral medial temporal lobectomy to treat severe epilepsy and, since that time, has been unable to form any new declarative memories. This deficit is called **anterograde amnesia**. Dr. Milner was quite surprised to learn that H.M. could learn a relatively difficult mirror-drawing task, in which (like anyone else) he got better with repeated trials and retained the skill over time. However, he could not remember ever having done the task before.

Short-Term Memory. Declarative memory can be divided into that which can be recalled for only a brief period (seconds to minutes) and that which can be recalled for weeks to years. Newly acquired learning experiences can be readily recalled for only a few minutes or more using **short-term memory**. An example of short-term memory is looking up a telephone number, repeating it mentally until you finish dialing the number, then promptly forgetting it as you focus your attention on starting the conversation. Short-term memory is a product of working memory; the decision to process information further for permanent storage is based on judgment as to its importance or on whether it is associated with a significant event or emotional state. An active process involving the hippocampus must be employed to make a memory more permanent.

Long-Term Memory. The conversion of short-term to **long-term memory** is facilitated by repetition, by adding more than one sensory modality to learn the new experience (e.g., writing down a newly acquired fact at the same time one hears it spoken) and, even more effective, by tying the experience (through the limbic system) to a strong, meaningful emotional context. The role of the hippocampus in consolidating the memory is reinforced by its participation in generating the emotional state with which the new experience is associated. As determined by studying patients such as H.M., the most important regions of the medial temporal lobe for long-term declarative memory formation are the hippocampus and parahippocampal cortex.

Once a long-term memory is formed, the hippocampus is not required for subsequent retrieval of the memory. Thus, H.M. showed no evidence of a loss of memories laid down prior to surgery; this type of memory loss is known as **retrograde amnesia**. Nor was there loss of intellectual capacity, mathematical skills, or other cognitive functions. An extreme example of H.M.'s memory loss is that Dr. Milner, who worked with him for years, had to introduce herself to her patient every time they met, even though he could readily remember people and events that had occurred before his surgery.

Cholinergic Innervation. The primacy of the hippocampus and its connections with the base of the forebrain for memory formation implicates acetylcholine as a major transmitter in cognitive function and learning and memory. The basal forebrain region contains prominent populations of cholinergic neurons that project to the hippocampus and to all regions of the cerebral cortex

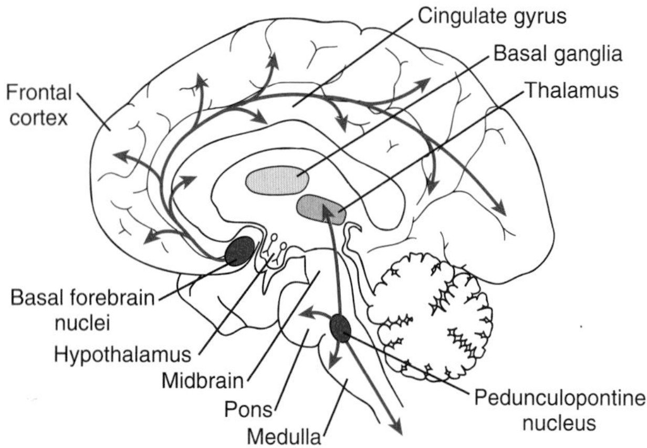

FIGURE 7.14 **The origins and projections of major cholinergic neurons.** Cholinergic neurons in the basal forebrain nuclei innervate all regions of the cerebral cortex. Cholinergic neurons in the brainstem's pedunculopontine nucleus provide a major input to the thalamus and also innervate the brainstem and spinal cord. Cholinergic interneurons are found in the basal ganglia. Not shown are peripherally projecting neurons, the somatic motor neurons, and autonomic preganglionic neurons, which also are cholinergic.

(Fig. 7.14). These cholinergic neurons are known generically as **basal forebrain nuclei** and include the septal nuclei, the nucleus basalis, and the nucleus accumbens. Another major cholinergic projection derives from a region of the brainstem reticular formation known as the **pedunculopontine nucleus**, which projects to the thalamus, spinal cord, and other regions of the brainstem. Roughly 90% of brainstem inputs to all nuclei of the thalamus are cholinergic.

Cortical cholinergic connections are thought to control selective attention, a function congruent with the cholinergic brainstem projections through the ascending reticular activating system. Loss of cholinergic function is associated with **dementia**, an impairment of memory, abstract thinking, and judgment (see Clinical Focus Box 7.2). Other cholinergic neurons include motor neurons and autonomic preganglionic neurons, as well as a major interneuronal pool in the striatum.

Language and Speech Are Coordinated in Specific Areas of Association Cortex

The ability to communicate by language, verbally and in writing, is one of the most difficult cognitive functions to study because only humans are capable of these skills. Thus, our knowledge of language processing in the brain has been inferred from clinical data by studying patients with **aphasias**—disturbances in producing or understanding the meaning of words—following brain injury, surgery, or other damage to the cerebral cortex.

Two areas appear to play an important role in language and speech: Wernicke's area, in the upper temporal lobe, and Broca's area, in the frontal lobe (Fig. 7.15). Both of these areas are located in association cortex, adjacent to cortical areas that are essential in language communication. **Wernicke's area** is in the parietal-temporal-occipital association cortex, a major association area for processing sensory information from the somatic sensory, visual, and auditory cortices. **Broca's area** is in the prefrontal association cortex, adjacent to the portion of the motor cortex that regulates movement of the muscles of the mouth, tongue, and throat (i.e., the structures used in the mechanical production of speech). A fiber tract, the **arcuate fasciculus**, connects Wernicke's area with Broca's area to

CLINICAL FOCUS BOX 7.2

Alzheimer's Disease

Alzheimer's disease (AD) is the most common cause of dementia in older adults. The cause of the disease still is unknown and there is no cure. In 1999, an estimated 4 million people in the United States suffered from AD. While the disease usually begins after age 65, and risk of AD goes up with age, it is important to note that AD is not a normal part of aging. The aging of the baby boom population has made AD one of the fastest growing diseases; estimates indicate that by the year 2040, some 14 million people in the United States will suffer from AD.

Cognitive deficits are the primary symptoms of AD. Early on, there is mild memory impairment; as the disease progresses, memory problems increase and difficulties with language are generally observed, including word-finding problems and decreased verbal fluency. Many patients also exhibit difficulty with visuospatial tasks. Personality changes are common, and patients become disoriented as the memory problems worsen. A progressive deterioration of function follows and, at late stages, the patient is bedridden, nearly mute, unresponsive, and incontinent. A definitive diagnosis of AD is not possible until autopsy, but the constellation of symptoms and disease progression allows a reasonably certain diagnosis.

Gross pathology consistent with AD is mild to severe cortical atrophy (depending on age of onset and death). Microscopic pathology indicates two classic signs of the disease even at the earliest stages: the presence of senile plaques (SPs) and neurofibrillary tangles (NFTs). As the disease progresses, synaptic and neuronal loss or atrophy and an increase in SPs and NFTs occur.

While many neurotransmitter systems are implicated in AD, the most consistent pathology is the loss or atrophy of cholinergic neurons in the basal forebrain. Medications that ameliorate the cognitive symptoms of AD are cholinergic function enhancers. These observations emphasize the importance of cholinergic systems in cognitive function.

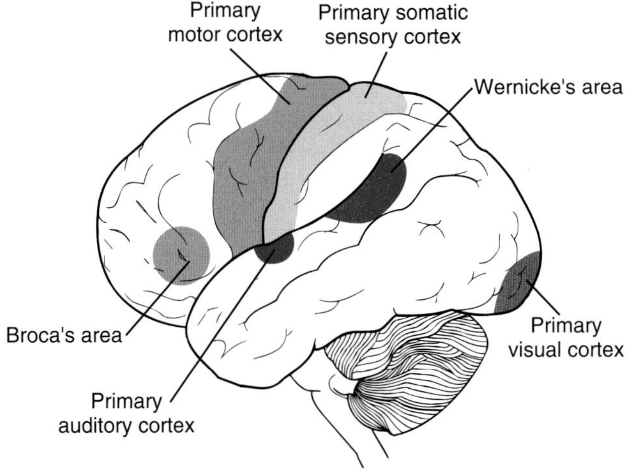

FIGURE 7.15 **Wernicke's and Broca's areas and the primary motor, visual, auditory, and somatic sensory cortices.**

coordinate aspects of understanding and executing speech and language skills.

Clinical evidence indicates that Wernicke's area is essential for the comprehension, recognition, and construction of words and language, whereas Broca's area is essential for the mechanical production of speech. Patients with a defect in Broca's area show evidence of comprehending a spoken or written word but they are not able to say the word. In contrast, patients with damage in Wernicke's area can produce speech, but the words they put together have little meaning.

Language is a highly lateralized function of the brain residing in the left hemisphere (see Clinical Focus Box 7.1). This dominance is observed in left-handed as well as right-handed individuals. Moreover, it is language that is lateralized, not the reception or production of speech. Thus native signers (individuals who use sign language) that have been deaf since birth still show left-hemisphere language function.

REVIEW QUESTIONS

DIRECTIONS: Each of the numbered items or incomplete statements in this section is followed by answers or by completions of the statement. Select the ONE lettered answer or completion that is BEST in each case.

1. An EEG technician can look at an electroencephalogram and tell that the subject was awake, but relaxed with eyes closed, during generation of the recording. She can tell this because the EEG recording exhibits
(A) Alpha rhythm
(B) Beta rhythm
(C) Theta rhythm
(D) Delta rhythm
(E) Variable rhythm
2. A patient's wife complains that, several times during the last few weeks, her husband struck her as he flailed around violently during sleep. The husband indicates that when he wakes up during one of these sessions, he has been dreaming. What is the likely cause of his problem?
(A) Increased muscle tone during stage 4 sleep
(B) Increased drive to the motor cortex during REM sleep
(C) Lack of behavioral inhibition by the prefrontal cortex during sleep
(D) Lack of abolished muscle tone during REM sleep
(E) Abnormal functioning of the amygdala during paradoxical sleep
3. The hormone secreted by the pineal gland under control of the suprachiasmatic nucleus is
(A) Adrenaline
(B) Leptin
(C) Melanocyte-stimulating hormone
(D) Melatonin
(E) Vasopressin
4. The basal forebrain nuclei and the pedunculopontine nuclei are similar in that neurons within them
(A) Are major inputs to the striatum
(B) Receive innervation from the cingulate gyrus
(C) Process information related to language construction
(D) Utilize acetylcholine as their neurotransmitter
(E) Are atrophied in patients with schizophrenia
5. A scientist develops a reagent that allows identification of leptin-sensing neurons in the CNS. The reagent is a fluorescent compound that binds to the plasma membrane of cells that sense leptin. Application of this reagent to sections of the brain would result in fluorescent staining located in the
(A) Arcuate nucleus of the hypothalamus
(B) Mammillary nuclei of the hypothalamus
(C) Paraventricular nucleus of the hypothalamus
(D) Preoptic nucleus of the hypothalamus
(E) Ventromedial nucleus of the thalamus
6. The hypothalamus receives cues concerning the cycle of sunlight and darkness in a 24-hour day. Following removal of a tumor and concomitant destruction of surrounding tissue, a patient's hypothalamus no longer received this information. The most likely location of this tumor was in
(A) The body's internal clock
(B) A direct neural pathway from the optic nerve to the suprachiasmatic nucleus
(C) The reticular formation
(D) A projection from the occipital lobe of the cerebral cortex to the hypothalamus
(E) The pineal gland
7. Posterior pituitary hormone secretion is mediated by
(A) A portal capillary system from the hypothalamus to the posterior pituitary
(B) The fight-or-flight response
(C) The hypothalamo-hypophyseal tract originating from magnocellular neurons in the supraoptic and paraventricular nuclei
(D) The reticular activating system's input to the hypothalamus
(E) The emotional state (i.e., mood and affect)
8. Language and speech require the participation of both Wernicke's area and Broca's area. These two regions of the brain communicate with each other via a fiber bundle called
(A) The thalamocortical tract
(B) The reticular activating system
(C) The prefrontal lobe
(D) The fornix
(E) The arcuate fasciculus
9. A chemist is trying to produce a new neuroleptic drug. To be an effective

(continued)

neuroleptic, the new compound must target
(A) Acetylcholine receptors
(B) Dopamine receptors
(C) Neuropeptide Y receptors
(D) Norepinephrine receptors
(E) Serotonin receptors

10. A patient suffered a stroke that destroyed the intralaminar nuclei of the thalamus. The location of the stroke was confirmed by magnetic resonance imaging of the brain; however, an indication that the stroke affected these nuclei was provided prior to imaging by an alteration in arousal in the patient. Which of the following alterations in arousal is most likely following destruction of these nuclei?
(A) Loss of consciousness
(B) Increased time spent in beta rhythm
(C) Increased attention to specific sensory inputs
(D) Alterations in paradoxical, but not slow-wave sleep
(E) Alteration in the period of the biological clock

11. A blindfolded subject is asked to verbally identify a common object presented to her left hand. She is not allowed to touch the object with her right hand. Which of the following structures must be intact for her to complete this task?
(A) The primary somatic sensory cortex on the left side of her brain
(B) The primary visual cortex on the right side of her brain
(C) The fornix
(D) The corpus callosum
(E) The hippocampus

12. A viral infection causes damage to both hippocampi in a patient. This damage would cause the patient to exhibit functional deficits in
(A) Recalling an old declarative memory
(B) Recalling an old procedural memory
(C) Forming a new short-term memory
(D) Forming a new long-term memory
(E) Forming a new procedural memory

13. An older gentleman is brought to the emergency department (ED) by his daughter. She had gone to his house for lunch, which she did on a daily basis. During her visit that day, she was alarmed because his speech did not make sense to her even though he talked a lot and the words themselves were clear. The physician in the ED informed the daughter that her father had most likely suffered a stroke that damaged
(A) Broca's area
(B) The corpus callosum
(C) The hippocampus
(D) The arcuate fasciculus
(E) Wernicke's area

14. A woman agreed to visit her physician because her husband was very worried about her behavior. She told the doctor she felt great and that she was going to run for governor of the state because she was smarter than the current governor and people would immediately agree to her plans. Her husband said she had been sleeping very little the last several days and had spent several thousand dollars in a shopping spree the day before. This was not typical behavior and had significantly affected their ability to meet their obligations for household expenses. The physician indicated a diagnosis of mania and started her on a course of a drug that would decrease neurotransmission in
(A) Cholinergic pathways
(B) Dopaminergic pathways
(C) Glutaminergic pathways
(D) Noradrenergic pathways
(E) Serotonergic pathways

15. Persons with mild cognitive impairments who smoke may experience a worsening of symptoms if they stop smoking. This worsening of symptoms is because nicotine acts as an agonist for receptors of a particular neurotransmitter. That neurotransmitter is
(A) Acetylcholine
(B) Dopamine
(C) Neuropeptide Y
(D) Nitric oxide
(E) Serotonin

SUGGESTED READING

Bavelier D, Corina DP, Neville HJ. Brain and language: A perspective from sign language. Neuron 1998;21:275–278.

Cooke B, Hegstrom CD, Villenueve LS, Breedlove SM. Sexual differentiation of the vertebrate brain: Principles and mechanisms. Front Neuroendocr 1998;19:323–362.

Dijk D-J, Duffy JF. Circadian regulation of human sleep and age-related changes in its timing, consolidation and EEG characteristics. Ann Med 1999;31:130–140.

Elmquist JK, Elias CF, Saper CB. From lesions to leptin: Hypothalamic control of food intake and body weight. Neuron 1999;22:221–232.

Gazzaniga MS. The split brain revisited. Sci Am 1998;279(1):50–55.

Kandel ER, Schwartz JH, Jessell TM. Principles of Neural Science. 4th Ed. New York: McGraw-Hill, 2000.

Milner B, Squire LR, Kandel ER. Cognitive neuroscience and the study of memory. Neuron 1998;20:445–468.

Perry E, Walker M, Grace J, Perry R. Acetylcholine in mind: A neurotransmitter correlate of consciousness? Trends Neurosci 1999;22:273–280.

CASE STUDIES FOR PART II

CASE STUDY FOR CHAPTER 4

Dizziness

A 35-year-old man consulted his family physician because of some recent episodes of what he described as dizziness. He was concerned that this complaint might be related to a fall from a stepladder that had occurred the previous month, although his symptoms did not begin immediately after the incident. At the time of his visit to the doctor, his symptoms are minimal, and he appears to be in good general health. He states that the feeling of dizziness, which also included sensations of nausea (without vomiting) and "ringing in the ears," makes him feel as though his surroundings were spinning around him. The episodes, which could last for several days at a time, are quite annoying and sufficiently severe to cause him concern for his safety on the job. When questioned, he indicates that he also may not be hearing as well as he should, but at other times he does not notice any hearing problems. He further indicates that he may have had occasional dizzy spells before the ladder incident, but that they now appear to be much more frequent. The only medication he takes is aspirin for an occasional headache. He has no difficulty in following a moving finger with his head held stationary, and on the day of the visit he walks with a normal gait. He reports no light-headedness with moderate and continued exertion.

Gentle irrigation of his external ear canals with warm water (at approximately 39°C) produces a feeling of dizziness and nausea accompanied by nystagmus. The subjective sensations appeared to be the same for each ear. He is further evaluated with the Dix-Hallpike maneuver, and no sensations of vertigo are elicited during the positional maneuvers. However, when he is rapidly rotated in a swivel chair, he reports dizziness that was more severe than his usual symptoms. Rotation in the opposite direction produced similar symptoms. His physician ad-

(continued)

vises him that there may be some appropriate specific medications for his condition, but he would first like him to try a salt-restricted diet for the next 4 weeks. He also prescribes a mild diuretic.

Upon his return visit 4 weeks later, the patient reports a gradual lessening of the frequency and duration of his spells of dizziness and accompanying symptoms.

Questions

1. What features of this case would indicate that trauma from the stepladder incident was not the precipitating cause of the symptoms?
2. What factors would tend to rule out a diagnosis of benign paroxysmal positional vertigo?
3. Would the use of water at body temperature yield the same diagnostic information as warmer or cooler water?
4. Is the patient's lack of light-headedness with moderate exercise relevant to the diagnosis of this problem?
5. Is it likely that the sensations produced by rapid rotation are mimicking those produced by his underlying disorder?
6. What is the purpose of the salt-restricted diet and diuretic therapy? Why was this tried before prescribing medication for his problem?

Answers to Case Study Questions for Chapter 4

1. Several features of this case suggest that trauma from the stepladder incident was not the precipitating cause. The caloric stimulation test and the rotation in the swivel chair indicate that his vestibular function is bilaterally symmetrical and of normal sensitivity. A defect arising from trauma would likely be localized to the injured side. His uncertainty of the timing of the onset of the symptoms indicates that the problem may have preceded the accident (and, perhaps, led to it), and the lack of immediate appearance of symptoms also tends to rule out trauma.
2. The relatively young age of the patient and the negative findings from the Dix-Hallpike test argue against positional vertigo and lend support to a tentative diagnosis of Ménière's disease, as would the presence of tinnitus and the fluctuating hearing loss. The patient's positive response to salt restriction and diuretic therapy is also indicative of this syndrome. (See answer to Question 6.)
3. The purpose of the application of water is to provide a thermal stimulus that will heat or cool the endolymph in the semicircular canals and cause convection currents that would stimulate the ampullae. Use of water at body temperature would not produce this effect, and no symptoms would be elicited. Warmer or cooler water would each produce symptoms of vertigo.
4. This observation tends to rule out cerebral ischemia as a result of circulatory (vascular) or heart problems, factors that would also be more likely in an older patient.
5. The symptoms produced by the rotation are severe because of the simultaneous involvement of both sets of vestibular apparatus and the resulting heavy neural input, which is likely to be greater than that produced by his underlying condition.
6. The use of salt restriction and diuretics would reduce the overall hydration state of his body and tend to reduce abnormal pressure within the labyrinthine system. The use of antimotion sickness drugs would interfere with the natural neural compensation that would, it is hoped, reduce the severity of the symptoms with time.

Reference

Drachman DA. A 69-year-old man with chronic dizziness. JAMA 1998;280:2111–2118.

CASE STUDY FOR CHAPTER 5

Upper Motor Neuron Lesion

A 50-year-old man comes for evaluation of persistent difficulty using his right arm and leg. The patient was well until 1 month previously when he had abrupt onset of weakness on the right side of his body while watching a television show. He was taken to the hospital by ambulance within 1 hour of onset of symptoms. The initial evaluation in the hospital emergency department showed elevated blood pressure with values of 200 mm Hg systolic and 150 mm Hg diastolic. The right arm and leg were severely weak. Activity of the myotatic reflexes on the right side was very reduced in comparison with the left side, where they were normal. Right side limb movements were slightly improved by 12 hours after onset but were still moderately impaired on the fourth hospital day.

A magnetic resonance imaging study (MRI) of the brain performed on the second day of hospitalization showed a stroke involving the left cerebral hemisphere in the region of the internal capsule. The blood pressure remained elevated, and medication to lower it was begun during the hospital stay. The patient was transferred to a rehabilitation hospital on the fourth day for extensive physical therapy to assist further recovery of neurological function.

At follow-up examination 1 month after onset of the stroke, the blood pressure remains normal on the medication that was started in the hospital. Neurological examination demonstrates mild weakness of the right arm and leg. There is still a slight but obvious delay between asking the patient to move those limbs and the movement actually beginning. Passive movement of the right arm and leg by the physician provokes involuntary contraction of the muscles in those limbs that seem to counteract the attempted movement. Right side myotatic reflexes are very hyperactive, compared with those obtained on the left. When the skin over the lateral plantar area of the right foot is stroked, the first toe extends involuntarily. When this maneuver is performed on the left, the toes flex.

Questions

1. Explain the neurophysiology of the muscular weakness, slowness of movement initiation, increased muscle resistance to passive movement, and overactive myotatic reflexes on the right side one month after stroke onset.
2. Explain why the toes extend on the right side and flex on the left in response to plantar stimulation.

Answers to Case Study Questions for Chapter 5

1. The motor pathways that descend to the spinal cord from higher CNS levels initiate voluntary muscle action and also regulate the sensitivity of the muscle stretch (myotatic) reflex. Impairment of corticospinal tract input to the alpha motor neuron pools results in weakness and slowness of initiation of voluntary movement. The corticospinal tract deficit also produces an increased sensitivity of the spinal reflex pathways, resulting in overly vigorous muscle stretch reflexes. Muscle tone, the normal slight resistance to passive movement that is detectable in a relaxed muscle, becomes greatly increased and demonstrates a pattern that is called **spasticity**. Spastic tone is most evident in the flexor muscles of the arm and the extensor muscles of the leg.
2. The extensor movement of the first toe in response to stroking the plantar aspect of the foot, termed ***Babinski sign***, is thought to occur because of modification of flexor withdrawal reflexes secondary to the impaired input of the corticospinal tract. The normal response is for the toes to flex when the plantar surface is stimulated.

The neurophysiological details of how the deficit in corticospinal input actually produces these commonly encountered abnormalities in muscle tone and reflex patterns are still not well understood. A current theory is that the disturbance of central control reduces the threshold of the stretch reflex but does not alter its gain.

References

Lance JW. The control of muscle tone, reflexes, and movement. Neurology 1980;30:1303–1313.

Powers RK, Marder-Meyer J, Rymer WZ. Quantitative relations between hypertonia and stretch reflex threshold in spastic hemiparesis. Ann Neurol 1988;23:11–124.

CASE STUDY FOR CHAPTER 6

Autonomic Dysfunction as a Result of CNS Disease

A 30-year-old patient came to the hospital emergency department because of a terrible headache that began several hours ago and did not improve. Previously he had experienced only mild, infrequent tension headaches associated with stressful days. Because of the intensity of this new headache, he is treated with injectable analgesics and is admitted to the hospital for further observation.

During the next several hours, the patient's level of consciousness declines to the point of responding only to painful stimuli. An emergency computed tomography (CT) scan of the brain demonstrates the presence of blood diffusely in the subarachnoid space. The source of the blood is thought to be a ruptured cerebral artery aneurysm.

During the next 24 hours, the patient's ECG begins to show abnormalities consisting of both tachycardia and changes in the configuration of the waves suggestive of a heart attack. The patient has no risk factors for premature cardiac disease. A cardiology consultation is requested.

Questions

1. What is the explanation for the cardiac abnormalities in this situation?
2. Describe two other scenarios in which there are prominent manifestations of autonomic activation produced by abnormalities in the central nervous system.

Answers to Case Study Questions for Chapter 6

1. The consulting cardiologist reviewed the situation and stated that the ECG abnormalities were all a result of subarachnoid blood and that an adrenergic antagonist medication should be administered.

   Blood released into the subarachnoid space by rupture of blood vessels or direct trauma to the brain can stimulate excessive activity of the sympathetic nervous system. Although a full explanation is still lacking, it is postulated that the subarachnoid blood irritates the hypothalamus and autonomic regulatory areas in the medulla, resulting in excessive activation of the sympathetic pathways. This activation causes the secretion of norepinephrine from sympathetic nerve endings and epinephrine by the adrenal medulla. Direct stimulation of the sympathetic pathways that supply the heart can produce the same ECG abnormalities in experimental animals as were found in this patient. The heightened release of norepinephrine and epinephrine stimulates the cardiac conducting system and may also produce direct damage of the myocardium. Treatment with medications that attenuate the effects of sympathetic neurotransmitters can be lifesaving.
2. The Cushing response (described by famous neurosurgeon Harvey Cushing) consists of the development of hypertension, bradycardia, and apnea in patients with increased intracranial pressure most often a result of tumors or other lesions, such as hemorrhage, that compress the brain. The pressure is transmitted downward to the brainstem and distorts the medulla, where the centers for blood pressure, heart rate, and respiratory drive originate. Correct interpretation of these abnormalities in vital signs permits beginning treatments that reduce intracranial pressure. These include elevating the head of the bed, placing the patient on an artificial respirator, and then instituting hyperventilation to lower the blood P_{CO_2} to produce cerebral vasoconstriction and giving mannitol to reduce the fluid content of the brain temporarily.

   Another autonomic reaction from the CNS that is utilized daily in hospitals is the response of fetal heart rate to compression of the head during labor. During uterine contractions, the fetal head is temporarily compressed. As the fetal skull is still malleable because the bones of the cranium are not yet fused, the pressure of the contraction is transmitted to the brain. The same mechanism of cardiac slowing as cited for the Cushing response is presumed to cause the temporary bradycardia. Slowing of greater than established normal limits indicates the fetus is suffering significant physiological distress. Additional factors, such as umbilical cord compression, may also produce patterns of slowing outside of the normal range.

Reference

Talman WT. The central nervous system and cardiovascular control in health and disease. In: Low PA. Clinical Autonomic Disorders. 2nd Ed. Philadelphia: Lippincott-Raven, 1997

CASE STUDY FOR CHAPTER 7

Stroke

A 67-year-old man was taken to see his physician by his wife. For the preceding 2 days, the patient's wife had noticed that he did not seem to make sense when he spoke. She also indicated that he seemed a little disoriented and did not respond appropriately to her questions. He has no obvious motor or somatic sensory deficits.

On examination, the physician concludes that the man had a stroke in a region of one of his cerebral hemispheres. As part of the diagnosis, the physician tests the man's visual fields and notices a decreased awareness of stimuli presented to one visual field.

Questions

1. Which side of the brain most likely suffered the stroke?
2. Which regions of the hemisphere suffered the stroke?
3. What information from the case history gives the answers to questions 1 and 2?
4. Which visual field is affected by the stroke?

Answers to Case Study Questions for Chapter 7

1. The stroke occurred on the left side of the brain.
2. The stroke involved the superior posterior temporal lobe encompassing Wernicke's area and the occipital lobe encompassing the primary visual cortex.
3. Language deficits indicate involvement of the left hemisphere. The fluent but nonsensical speech indicates involvement of Wernicke's area. The visual field deficit indicates a loss in the visual cortex. The lack of motor or somatic sensory deficits excludes the posterior frontal and anterior parietal lobes.
4. The right visual field would be affected, because visual fields are represented in the contralateral hemispheres.

PART III Muscle Physiology

CHAPTER 8 Contractile Properties of Muscle Cells

Richard A. Meiss, Ph.D.

CHAPTER OUTLINE

KEY CONCEPTS

1. Muscle is classified into three categories, based on anatomic location, histological structure, and mode of control. The categories may overlap.
2. Skeletal (striated) muscle is used for voluntary movement of the skeleton.
3. Smooth muscle controls and aids the function of visceral organs.
4. Cardiac muscle provides the motive power for circulation of the blood.
5. The contractile proteins of muscle are arranged into two overlapping sets of myofilaments, one predominantly myosin-containing (thick), and one predominantly actin-containing (thin).
6. In skeletal and cardiac muscle, the myofilaments are arranged into sarcomeres, the fundamental organizational unit of the contractile machinery.
7. Crossbridges are projections of myosin filaments that make mechanical contact with actin filaments.
8. The myofilament arrangement and crossbridge contacts in smooth muscle occur without an organized sarcomere structure.
9. Changes in the length of a skeletal muscle result in changes in the degree of overlap of the myofilaments.
10. The crossbridge cycle is a series of chemical reactions that transform the energy stored in ATP into mechanical energy that produces muscle contraction.
11. ATP has two functions in the crossbridge cycle: to provide the energy for contraction, and to allow the myosin crossbridges to release from the actin filaments.
12. Overall muscle force and shortening occur as a result of the cumulative effects of millions of crossbridges acting to move myofilaments past one another.
13. Crossbridge interaction and the events of the crossbridge cycle are regulated by the action of calcium ions, which are stored in the sarcoplasmic reticulum when the muscle is at rest.
14. The release and uptake of calcium ions by the sarcoplasmic reticulum of skeletal muscle are controlled by the membrane potential of the muscle fibers.
15. The energy for muscle contraction is derived from both aerobic and anaerobic metabolism; muscle can adapt its function depending on the availability of oxygen.

Muscle tissue is responsible for most of our interactions with the external world. These familiar functions include moving, speaking, and a host of other everyday actions. Less familiar, but no less important, are the internal functions of muscle. It pumps our blood and regulates its flow, it moves our food as it is being digested and causes the expulsion of wastes, and it serves as a critical regulator of numerous internal processes.

Muscle contraction is a cellular phenomenon. The shortening of a whole muscle results from the shortening of its individual cells, and the force a muscle produces is the sum of forces produced by its cells. Activation of a whole muscle involves activating its individual cells, and muscle relaxation involves a return of the cells to their resting state. The study of muscle function must, therefore, include an investigation of the cellular processes that cause and regulate muscle contraction.

As the great variety of its functions might imply, muscle is a highly diverse tissue. But in spite of its wide range of anatomic and physiological specializations, there is an underlying similarity in the way muscles are constructed and in their mechanism of contraction. This chapter discusses some fundamental aspects of muscle contraction expressed in all types of muscle. Chapters 9 and 10 consider the important specializations of structure and function that belong to particular kinds of muscle.

THE ROLES OF MUSCLE

Different types of muscle fall naturally into categories that are related to their anatomic and physiological properties. Within each major category are subclassifications that further specify differences among the muscle types. As with any classification scheme, some exceptions are inevitable and some categories overlap. For this reason, three sets of criteria are commonly used.

Muscles Are Grouped in Three Major Categories

Muscles may be grouped according to

- Their location in relation to other body structures
- Their histological (tissue) structure
- The way their action is controlled

These classifications are not mutually exclusive.

Throughout the three chapters on muscle, the highlighted categories in Figure 8.1 will be the preferred usage. The alternative categories are still useful, however, because in some instances they express more precisely the special attributes of a certain muscle type. The inconsistencies in classification are likewise useful in describing the characteristics of specific muscles.

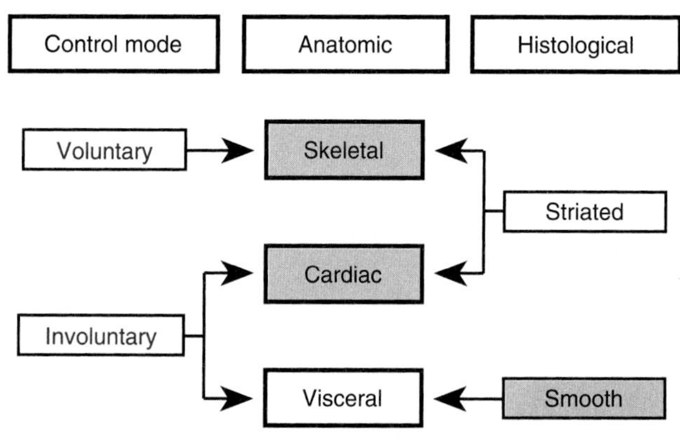

FIGURE 8.1 **Classification of types of muscles.** The categories overlap in different ways, depending on the criteria being used.

Skeletal Muscle: Interactions With the External Environment. As its name implies, **skeletal muscle** is usually associated with bones of the skeleton. It is responsible for large and forceful movements, such as those involved in walking, running, and lifting heavy objects, as well as for small and delicate movements that position the eyeballs or allow the manipulation of tiny objects. Some skeletal muscle is specialized for the long-term maintenance of tension; for example, muscles of the torso involved in maintaining an upright posture can be active for many hours without undue fatigue. Other skeletal muscles, such as those in the upper arm, are better adapted for making rapid and forceful movements, but these fatigue rather rapidly when required to lift and hold heavy loads.

Whatever its specialization, skeletal muscle serves as the link between the body and the external world. Much of this interaction, such as walking or speaking, is under *voluntary* control. Other actions, such as breathing or blinking the eyelids, are largely automatic, although they can be consciously suppressed for brief periods of time. All skeletal muscle is externally controlled; it cannot contract without a signal from the somatic nervous system.

Not all skeletal muscle is attached to the skeleton. The human tongue, for example, is made of skeletal muscle that does not move bones closer together. Among mammals, perhaps the most striking example of this exception is the trunk of the elephant, in which skeletal muscles are arranged in a structure capable of great dexterity even though no articulated bones are involved in its movement.

An important secondary function of skeletal muscle is the production of body heat. This may be desirable, as when one shivers to get warm. During heavy exercise, however, muscle contraction may be a source of excess heat that must be eliminated from the body.

All skeletal muscle has a striated appearance when viewed with a light microscope or an electron microscope (Fig. 8.2). The regular and periodic pattern of the cross-striations of skeletal muscle relates closely to the way it functions at a cellular level.

Smooth Muscle: Regulation of the Internal Environment. Of the many processes regulating the internal state of the human body, one of the most important is controlling the movement of fluids through the visceral organs and the circulatory system. Such regulation is the task of **smooth muscle**. Smooth muscle also has many individual specializations that suit it well to particular tasks. Some smooth muscle, such as that in **sphincters**, circular bands of muscle that can stop flow in tubular organs, can remain contracted for long periods while using its metabolic energy econom-

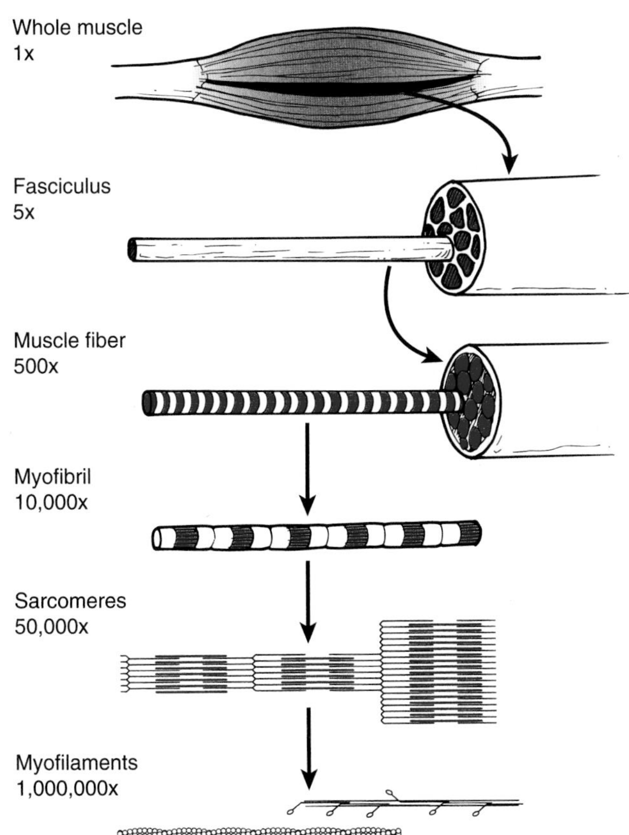

FIGURE 8.2 **Levels of complexity in the organization of skeletal muscle.** The approximate amount of magnification required to visualize each level is shown above each view.

ically. The muscle of the uterus, on the other hand, contracts and relaxes rapidly and powerfully during birth but is normally not very active during most of the rest of a woman's life. The economical use of energy is one of the most important general features of the physiology of smooth muscle.

The contraction of smooth muscle is *involuntary*. Although contraction may occur in response to a nerve stimulus, many smooth muscles are also controlled by circulating hormones or contracted under the influence of local hormonal or metabolic influences quite independent of the nervous system. Some indirect voluntary control of smooth muscle may be possible through mental processes such as biofeedback, but this ability is rare and is not an important aspect of smooth muscle function.

While one of the terms describing smooth muscle—*visceral*—implies its location in internal organs, much smooth muscle is located elsewhere. The muscles that control the diameter of the pupil of the eye and accommodate the eye for near vision, cause body hair to become erect (pilomotor muscles), and control the diameter of blood vessels are all examples of smooth muscles that are not visceral.

Cardiac Muscle: Motive Power for Blood Circulation. **Cardiac muscle** provides the force that moves blood throughout the body and is found only in the heart. It shares, with skeletal muscle, a striated cell structure, but its contractions are involuntary; the heartbeat arises from within the cardiac muscle and is not initiated by the nervous system. The nervous system, however, does participate in regulating the rate and strength of heart muscle contractions. Chapter 10 considers the special properties of cardiac muscle.

Muscles Have Specialized Adaptations of Structure and Function

All of the above should emphasize the varied and specialized nature of muscle function. Skeletal muscle, with its large and powerful contractions; smooth muscle, with its slow and economical contractions; and cardiac muscle, with its unceasing rhythm of contraction—all represent specialized adaptations of a basic cellular and biochemical system. An understanding of both the common features and the diversity of different muscles is important, and it is useful to emphasize particular types of muscle when investigating a general aspect of muscle function. Skeletal muscle is often used as the "typical" muscle for purposes of discussion, and this convention is followed in this chapter where appropriate, with an effort to point out those features relative to muscle in general. Important adaptations of the general features found in specific muscle types are considered in Chapters 9 and 10.

THE FUNCTIONAL ANATOMY AND ULTRASTRUCTURE OF MUSCLE

In biology, as in architecture, it can be said that form follows function. Nowhere is this truism more relevant than in the study of muscle. Investigations using light and electron microscopy, x-ray and light diffraction, and other modern visualization techniques have shown the complex and highly ordered internal structure of skeletal muscle. Elegant mechanical experiments have revealed how this structure determines the ways muscle functions.

Muscle Structure Provides a Key to Understanding the Mechanism of Contraction

Skeletal muscle is a highly organized tissue (Fig. 8.3). A whole skeletal muscle is composed of numerous muscle cells, also called **muscle fibers.** A cell can be up to 100 μm in diameter and many centimeters long, especially in larger muscles. The fibers are multinucleate, and the nuclei occupy positions near the periphery of the fiber. Skeletal muscle has an abundant supply of **mitochondria,** which are vital for supplying chemical energy in the form of ATP to the contractile system. The mitochondria lie close to the contractile elements in the cells. Mitochondria are especially plentiful in skeletal muscle fibers specialized for rapid and powerful contractions.

Each muscle fiber is further divided lengthwise into several hundred to several thousand parallel **myofibrils.** Electron micrographs show that each myofibril has alternating light and dark bands, giving the fiber a striated (striped) appearance. As shown in Figure 8.3, the bands repeat at regular intervals. Most prominent of these is a dark band

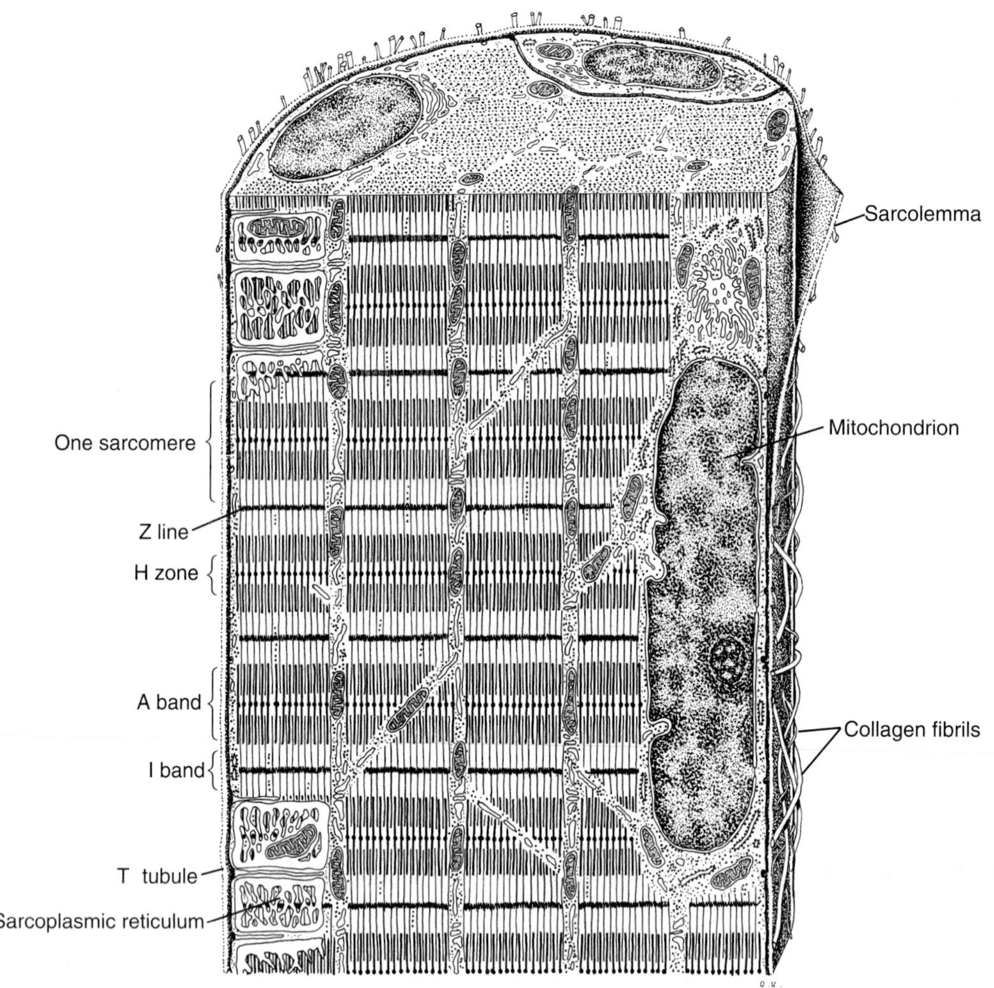

FIGURE 8.3 **The ultrastructure of skeletal muscle, a reconstruction based on electron micrographs.** (From Krstic RV. General Histology of the Mammal. New York: Springer-Verlag, 1984.)

called an **A band.** It is divided at its center by a narrow, lighter-colored region called an **H zone.** In many skeletal muscles, a prominent **M line** is found at the center of the H zone. Between the A bands lie the less dense **I bands.** (The letters *A* and *I* stand for anisotropic and isotropic; the bands are named for their appearance when viewed with polarized light.) Crossing the center of the I band is a dark structure called a **Z line** (sometimes termed a *Z disk* to emphasize its three-dimensional nature). The filaments of the I band attach to the Z line and extend in both directions into the adjacent A bands. This pattern of alternating bands is repeated over the entire length of the muscle fiber. The fundamental repeating unit of these bands is called a **sarcomere** and is defined as the space between (and including) two successive Z lines (Fig. 8.4).

Closer examination of a sarcomere shows the A and I bands to be composed of two kinds of parallel structures called **myofilaments.** The I band contains **thin filaments,** made primarily of the protein **actin,** and A bands contain **thick filaments** composed of the protein **myosin.**

Thin Myofilaments. Each thin (actin-containing) filament consists of two strands of macromolecular subunits entwined about each other (Fig. 8.5). The strands are composed of repeating subunits (monomers) of the globular protein **G-actin** (molecular weight, 41,700). These slightly ellipsoid molecules are joined front to back into long chains that wind about each other, forming a helical structure—**F-actin** (or filamentous actin)—that undergoes a half-turn every seven G-actin monomers. In the groove formed down the length of the helix, there is an end-to-end series of fibrous protein molecules (molecular weight, 50,000) called **tropomyosin.** Each tropomyosin molecule extends a distance of seven G-actin monomers along the F-actin groove. Near one end of each tropomyosin molecule is a protein complex called **troponin,** composed of three attached subunits: troponin-C (Tn-C), troponin-T (Tn-T), and troponin-I (Tn-I). The **Tn-C subunit** is capable of binding calcium ions, the **Tn-T subunit** attaches the complex to tropomyosin, and the **Tn-I subunit** has an inhibitory function. The troponin-tropomyosin complex regulates the contraction of skeletal muscle.

Thick Myofilaments. Thick (myosin-containing) filaments are also composed of macromolecular subunits (Fig. 8.6). The fundamental unit of a thick filament is

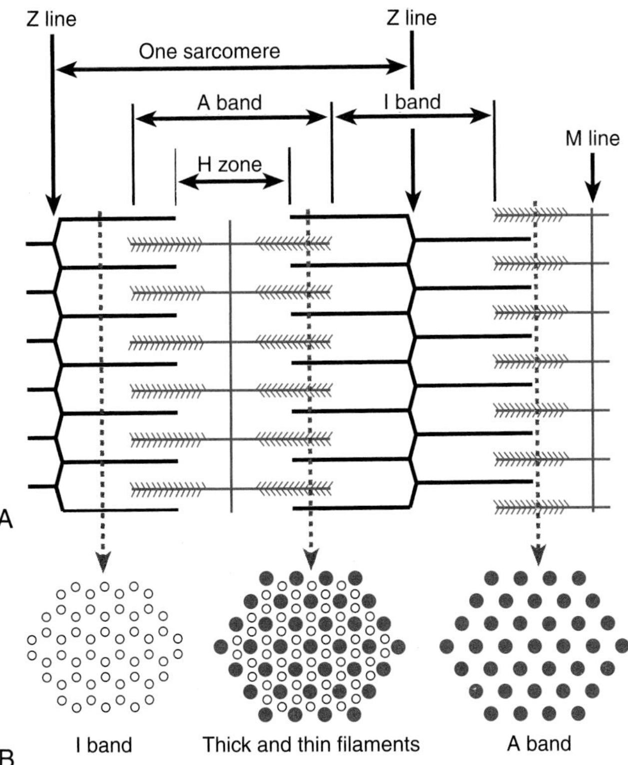

FIGURE 8.4 **Nomenclature of the skeletal muscle sarcomere.** **A**, The arrangement of the elements in a sarcomere. **B**, Cross sections through selected regions of the sarcomere, showing the overlap of myofilaments at different parts of the sarcomere.

myosin (molecular weight, approximately 500,000), a complex molecule with several distinct regions. Most of the length of the molecule consists of a long, straight portion, often called the "tail" region, composed of **light meromyosin** (LMM). The remainder of the molecule, **heavy meromyosin** (HMM), consists of a protein chain that terminates in a globular head portion. The head portion, called the **S1 region** (or subfragment 1), is responsible for the enzymatic and chemical activity that results in muscle contraction. It contains an **actin-binding site,** by which it can interact with the thin filament, and an **ATP-binding site** that is involved in the supply of energy for the actual process of contraction. The chain portion of HMM, the **S2 region** (or subfragment 2), serves as a flexible link between the head and tail regions. Associated with the S1 region are two loosely attached peptide chains of a much lower molecular weight. The **essential light chain** is necessary for myosin to function, and the **regulatory light chain** can be phosphorylated during muscle activity and modulates muscle function. Functional myosin molecules are paired; their tail and S2 regions are wound about each other along their lengths, and the two heads (each bearing its two light chains and its own ATP- and actin-binding sites) lie adjacent to each other. The molecule, with its attached light chains, exists as a functional dimer, but the degree of functional independence of the two heads is not yet known with certainty.

The assembly of individual myosin dimers into thick filaments involves close packing of the myosin molecules such that their tail regions form the "backbone" of the thick filament, with the head regions extending outward in a helical fashion. A myosin head projects every 60 degrees around the circumference of the filament, with each one displaced 14.4 nm further along the filament. The effect is like that of a bundle of golf clubs bound tightly by the handles, with the heads projecting from the bundle. The myosin molecules are packed so that they are tail-to-tail in the center of the thick filament and extend outward from the center in both directions, creating a bare zone (i.e., no heads protruding) in the middle of the filament (see Figs. 8.4 and 8.6).

Other Muscle Proteins. In addition to the proteins directly involved in the process of contraction, there are several other important structural proteins. **Titin**, a large filamentous protein, extends from the Z lines to the bare

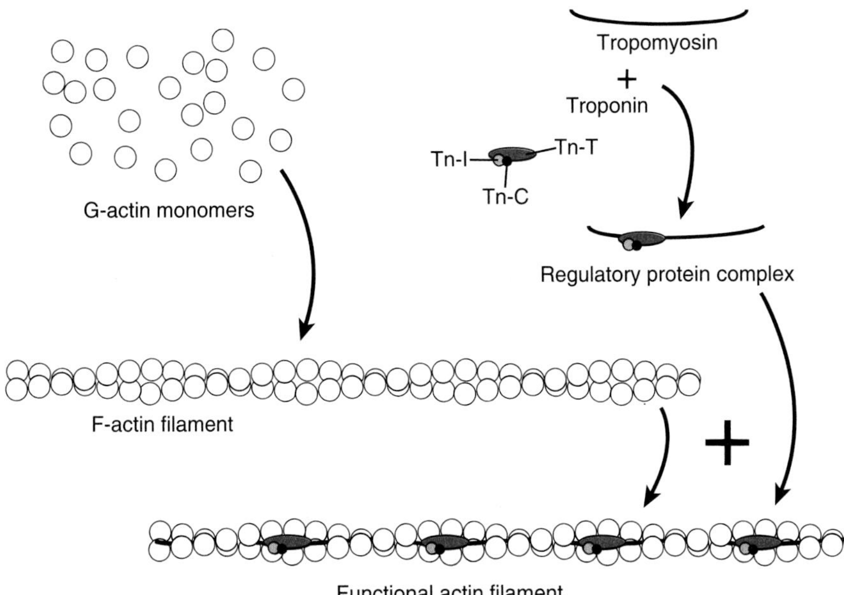

FIGURE 8.5 **The assembly of the thin (actin) filaments of skeletal muscle.** (See text for details.)

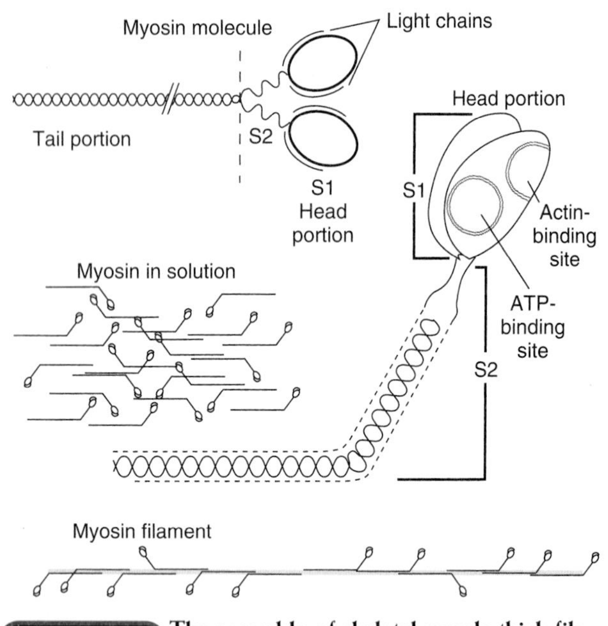

FIGURE 8.6 **The assembly of skeletal muscle thick filaments from myosin molecules.** (See text for details.)

portion of the myosin filaments and may help to prevent overextension of the sarcomeres and maintain the central location of the A bands. **Nebulin**, a filamentous protein that extends along the thin filaments, may play a role in stabilizing thin filament length during muscle development. The protein **α-actinin**, associated with the Z lines, serves to anchor the thin filaments to the structure of the Z line.

Dystrophin, which lies just inside the sarcolemma, participates in the transfer of force from the contractile system to the outside of the cells via membrane-spanning proteins called **integrins**. External to the cells, the protein **laminin** forms a link between integrins and the extracellular matrix. These proteins are disrupted in the group of genetic diseases collectively called **muscular dystrophy**, and their lack or malfunction leads to muscle degeneration and weakness and death (see Clinical Focus Box 8.1).

Polymyositis is an inflammatory disorder that produces damage to several or many muscles (Clinical Focus Box 8.2). The progressive muscle weakness in polymyositis usually develops more rapidly than in muscular dystrophy.

Skeletal Muscle Membrane Systems. Muscle cells, like other types of living cells, have a system of surface and in-

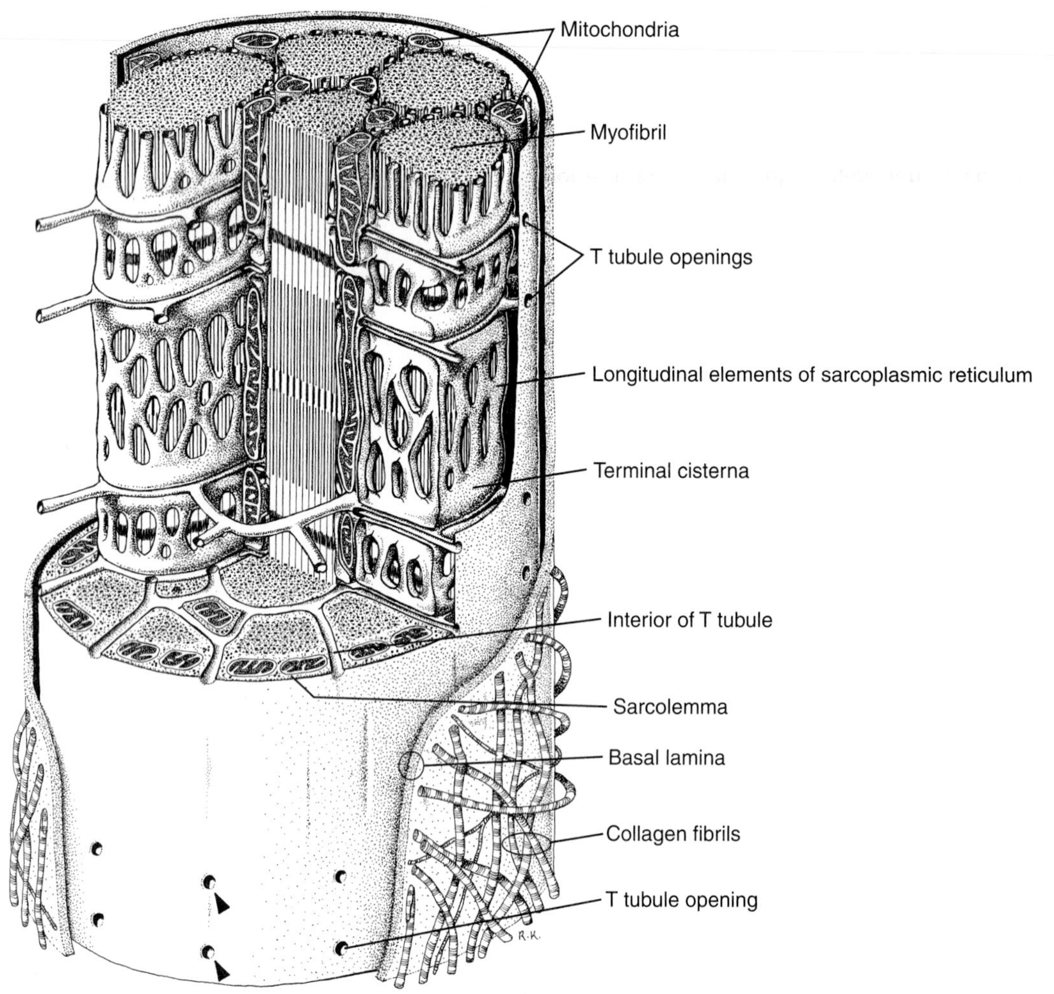

FIGURE 8.7 **The internal membrane system of skeletal muscle, responsible for communication between the surface membrane and contractile filaments.** This reconstruction is based on electron micrographs. (From Krstic RV. General Histology of the Mammal. New York: Springer-Verlag, 1984.)

CLINICAL FOCUS BOX 8.1

Muscular Dystrophy Research

The term ***muscular dystrophy*** (MD) encompasses a variety of degenerative muscle diseases. The most common of these diseases is **Duchenne's muscular dystrophy** (DMD) (also called pseudohypertrophic MD), which is an X-linked hereditary disease affecting mostly male children (1 of 3,500 live male births). DMD is manifested by progressive muscular weakness during the growing years, becoming apparent by age 4. A characteristic enlargement of the affected muscles, especially the calf muscles, is due to a gradual degeneration and necrosis of muscle fibers and their replacement by fibrous and fatty tissue. By age 12, most sufferers are no longer ambulatory, and death usually occurs by the late teens or early twenties. The most serious defects are in skeletal muscle, but smooth and cardiac muscle are affected as well, and many patients suffer from cardiomyopathy (see Chapter 10). A related (and rarer) disease, **Becker's muscular dystrophy** (BMD), has similar symptoms but is less severe; BMD patients often survive into adulthood. Some six other rarer forms of muscular dystrophy have their primary effect on particular muscle groups.

Using the genetic technique of chromosome mapping (using linkage analysis and positional cloning), researchers have localized the gene responsible for both DMD and BMD to the p21 region of the X chromosome, and the gene itself has been cloned. It is a large gene of some 2.5 million base pairs; apparently because of its great size, it has an unusually high mutation rate. About one third of DMD cases are due to new mutations and the other two thirds to sex-linked transmission of the defective gene. The BMD gene is a less severely damaged allele of the DMD gene.

The product of the DMD gene is dystrophin, a large protein that is absent in the muscles of DMD patients. Aberrant forms are present in BMD patients. The function of dystrophin in normal muscle appears to be that of a cytoskeletal component associated with the inside surface of the sarcolemma. Muscle also contains dystrophin-related proteins that may have similar functional roles. The most important of these is laminin 2, a protein associated with the basal lamina of muscle cells and concerned with mechanical connections between the exterior of muscle cells and the extracellular matrix. In several forms of muscular dystrophy, both laminin and dystrophin are lacking or defective.

A disease as common and devastating as DMD has long been the focus of intensive research. The recent identification of three animals—dog, cat, and mouse—in which genetically similar conditions occur promises to offer significant new opportunities for study. The manifestation of the defect is different in each of the three animals (and also differs in some details from the human condition). The **mdx mouse**, although it lacks dystrophin, does not suffer the severe debilitation of the human form of the disease. Research is underway to identify dystrophin-related proteins that may help compensate for the major defect. Mice, because of their rapid growth, are ideal for studying the normal expression and function of dystrophin. Progress has been made in transplanting normal muscle cells into mdx mice, where they have expressed the dystrophin protein. Such an approach has been less successful in humans and in dogs, and the differences may hold important clues. A gene expressing a truncated form of dystrophin, called **utrophin**, has been inserted into mice using transgenic methods and has corrected the myopathy.

The **mdx dog**, which suffers a more severe and human-like form of the disease, offers an opportunity to test new therapeutic approaches, while the cat dystrophy model shows prominent muscle fiber hypertrophy, a poorly understood phenomenon in the human disease. Taking advantage of the differences among these models promises to shed light on many missing aspects of our understanding of a serious human disease.

References

Burkin DJ, Kaufman SJ. The alpha7beta1 integrin in muscle development and disease. Cell Tissue Res 1999; 296: 183–190.

Tsao CY, Mendell JR. The childhood muscular dystrophies: Making order out of chaos. Semin Neurol 1999;19:9–23.

ternal membranes with several critical functions (see Fig. 8.7). A skeletal muscle fiber is surrounded on its outer surface by an electrically excitable cell membrane supported by an external meshwork of fine fibrous material. Together these layers form the cell's surface coat, the **sarcolemma.** In addition to the typical functions of any cell membrane, the sarcolemma generates and conducts action potentials much like those of nerve cells.

Contained wholly within a skeletal muscle cell is another set of membranes called the **sarcoplasmic reticulum** (SR), a specialization of the endoplasmic reticulum. The SR is specially adapted for the uptake, storage, and release of calcium ions, which are critical in controlling the processes of contraction and relaxation. Within each sarcomere, the SR consists of two distinct portions. The **longitudinal element** forms a system of hollow sheets and tubes that are closely associated with the myofibrils. The ends of the longitudinal elements terminate in a system of **terminal cisternae** (or lateral sacs). These contain a protein, calsequestrin, that weakly binds calcium, and most of the stored calcium is located in this region.

Closely associated with both the terminal cisternae and the sarcolemma are the **transverse tubules (T tubules)**, inward extensions of the cell membrane whose interior is continuous with the extracellular space. Although they traverse the muscle fiber, T tubules do not open into its interior. In many types of muscles, T tubules extend into the muscle fiber at the level of the Z line, while in others they penetrate in the region of the junction between the A and I bands. The association of a T tubule and the two terminal cisternae at its sides is called a **triad,** a structure important in linking membrane action potentials to muscle contraction.

CLINICAL FOCUS BOX 8.2

Polymyositis

Polymyositis is a skeletal muscle disease known as an inflammatory myopathy. Children (about 20% of cases) and adults may both be affected. Patients with the condition complain of muscle weakness initially associated with the proximal muscles of the limbs, making it hard to get up from a chair or use the stairs. They may have difficulty combing their hair or placing objects on a high shelf. Many patients have difficulty eating (dysphagia) because of the involvement of the muscles of the pharynx and the upper esophagus. A small percentage (about one third) of patients with polymyositis experience muscle tenderness or aching pain; a similar proportion of patients have some involvement of the heart muscle. The disease is progressive during a course of weeks or months.

Primary idiopathic polymyositis cases comprise approximately one third of the inflammatory myopathies. Twice as many women as men are affected. Another one third of polymyositis cases are associated with a closely related condition called dermatomyositis, symptoms of which include a mild heliotrope (light purple) rash around the eyes and nose and other parts of the body, such as knees and elbows. Nail bed abnormalities may also be present. Still other cases (approximately 8%) are associated with cancer present in the lung, breast, ovary, or gastrointestinal tract. This association occurs mostly in older patients. Finally, about one fifth of polymyositis cases are associated with other connective tissue disorders, such as rheumatoid arthritis and lupus erythematosus. Polymyositis can also occur in AIDS, as a result of either the disease itself or to a reaction to azidothymidine (AZT) therapy.

Polymyositis is thought to be primarily an autoimmune disease. Muscle histology shows infiltration by inflammatory cells such as lymphocytes, macrophages, and neutrophils. Muscle tissue destruction, which is almost always present, occurs by phagocytosis. The route of infiltration often follows the vascular supply. There may be elevated serum levels of enzymes normally present in muscle, such as creatine kinase (CK). These enzymes are released as muscle breaks down, and in severe cases, myoglobin may be found in the urine. The electrical activity of the affected muscle, as measured by electromyography, may show a characteristic pattern of abnormalities. In some cases, the weakness felt by the patient is greater than that suggested by the microscopic appearance of the tissue, and evidence indicates that diffusible factors produced by immune cells may have a direct effect on muscle contractile function. While the condition is not directly inherited, there is a strong familial component in its incidence. The cases of polymyositis associated with cancer (a paraneoplastic syndrome) are thought to be due to the altered immune status or tumor antigens that cross-react with muscle.

Several other disorders may present symptoms similar to polymyositis; these include neurological or neuromuscular junction conditions that result in muscle weakness without actual muscle pathology (see Chapter 9). Early stages of muscular dystrophy may mimic polymyositis, although the overall courses of the diseases differ considerably; the decline in function is much more rapid in untreated polymyositis. The parasitic infection trichinosis can produce symptoms of the disease, depending on the severity of the infection. A large number of commonly used drugs may produce the typical symptoms of muscle pain and weakness, and a careful drug history may suggest a specific cause. In cases in which dermatomyositis is combined with the typical symptoms of polymyositis, the diagnosis is quite certain.

Treatment of the disease usually involves high doses of glucocorticoids such as prednisone. Careful follow-up (by direct muscle strength testing and measurement of serum CK levels) is necessary to determine the ongoing effectiveness of treatment. After a course of treatment, the disease may become inactive, but relapses can occur, and other treatment approaches, such as the use of cytotoxic drugs, may be necessary. Long-term physical therapy and assistive devices are required when drug therapy is not sufficiently effective.

The Sliding Filament Theory Explains Muscle Contraction

The structure of skeletal muscle provides important clues to the mechanism of contraction. The width of the A bands (thick-filament areas) in striated muscle remains constant, regardless of the length of the entire muscle fiber, while the width of the I bands (thin-filament areas) varies directly with the length of the fiber. At the edges of the A band are fainter bands whose width also varies. These represent material extending into the A band from the I bands. The spacing between Z lines also depends directly on the length of the fiber. The lengths of the thin and thick myofilaments remain constant despite changes in fiber length.

The **sliding filament theory** proposes that changes in overall fiber length are directly associated with changes in the overlap between the two sets of filaments; that is, the thin filaments telescope into the array of thick filaments. This interdigitation accounts for the change in the length of the muscle fiber. It is accomplished by the interaction of the globular heads of the myosin molecules (**crossbridges**, which project from the thick filaments) with binding sites on the actin filaments. The crossbridges are the sites where force and shortening are produced and where the chemical energy stored in the muscle is transformed into mechanical energy. The total shortening of each sarcomere is only about 1 μm, but a muscle contains many thousands of sarcomeres placed end to end (in series). This arrangement has the effect of multiplying all the small sarcomere length changes into a large overall shortening of the muscle (Fig. 8.8). Similarly, the amount of force exerted by a single sarcomere is small (a few hundred micronewtons), but, again, there are thousands of sarcomeres side by side (in parallel), resulting in the production of considerable force.

The effects of sarcomere length on force generation are summarized in Figure 8.9. When the muscle is stretched beyond its normal resting length, decreased filament overlap occurs (3.65 μm and 3.00 μm, Fig. 8.9). This limits the

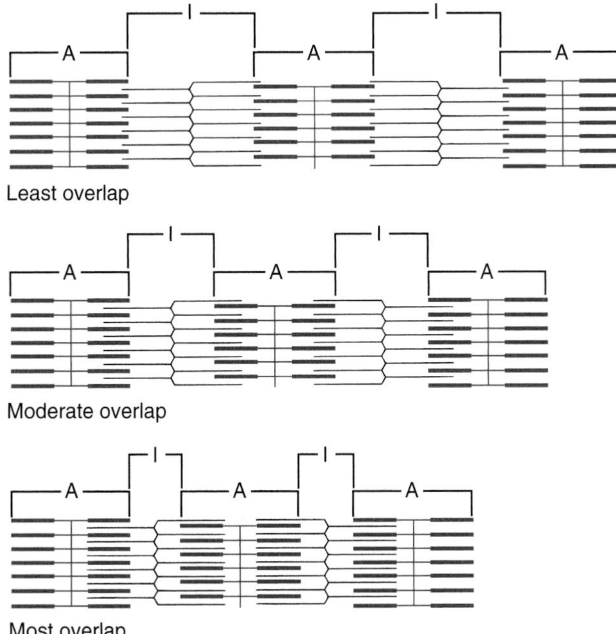

FIGURE 8.8 **Multiplying effect of sarcomeres placed in series.** The overall shortening is the sum of the shortening of the individual sarcomeres.

amount of force that can be produced, since a shorter length of thin filaments interdigitates with A band thick filaments and fewer crossbridges can be attached. Thus, over this region of lengths, force is directly proportional to the degree of overlap. At lengths near the normal **resting length** of the muscle (i.e., the length usually found in the body), the amount of force does not vary with the degree of overlap (2.25 μm and 1.95 μm, Fig. 8.10) because of the bare zone (the H zone) along the thick filaments at the center of the A band (where no myosin heads are present).

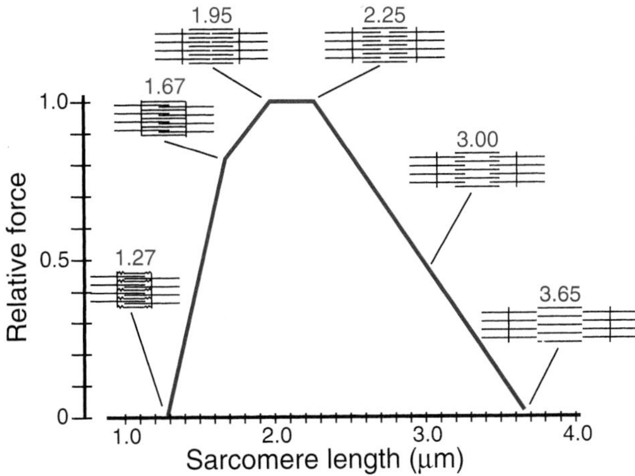

FIGURE 8.9 **Effect of filament overlap on force generation.** The force a muscle can produce depends on the amount of overlap between the thick and thin filaments because this determines how many crossbridges can interact effectively. (See text for details.)

Over this small region, further interdigitation does not lead to an increase in the number of attached crossbridges and the force remains constant.

At shorter lengths, additional geometric and physical factors play a role in myofilament interactions. Since muscle is a "telescoping" system, there is a physical limit to the amount of shortening. As thin myofilaments penetrate the A band from opposite sides, they begin to meet in the middle and interfere with each other (1.67 μm, Fig. 8.9). At the extreme, further shortening is limited by the thick filaments of the A band being forced against the structure of the Z lines (1.27 μm, Fig. 8.9).

The relationship between overlap and force at short lengths is more complex than that at longer lengths, since more factors are involved. It has also been shown that at very short lengths, the effectiveness of some of the steps in the excitation-contraction coupling process is reduced. These include reduced calcium binding to troponin and some loss of action potential conduction in the T tubule system. Some of the consequences for the muscle as a whole are apparent when the mechanical behavior of muscle is examined in more detail (see Chapter 9).

Events of the Crossbridge Cycle Drive Muscle Contraction

The process of contraction involves a cyclic interaction between the thick and thin filaments. The steps that comprise the **crossbridge cycle** are attachment of thick-filament crossbridges to sites along the thin filaments, production of a mechanical movement, crossbridge detachment from the thin filaments, and subsequent reattachment of the crossbridges at different sites along the thin filaments (Fig. 8.10). These mechanical changes are closely related to the biochemistry of the contractile proteins. In fact, the crossbridge association between actin and myosin actually functions as an enzyme, **actomyosin ATPase**, that catalyzes the breakdown of ATP and releases its stored chemical energy. Most of our knowledge of this process comes from studies on skeletal muscle, but the same basic steps are followed in all muscle types.

In resting skeletal muscle (Fig. 8.10, step 1), the interaction between actin and myosin (via the crossbridges) is weak, and the muscle can be extended with little effort. When the muscle is activated, the actin-myosin interaction becomes quite strong, and crossbridges become firmly attached (step 2). Initially, the crossbridges extend at right angles from each thick filament, but they rapidly undergo a change in angle of nearly 45 degrees. An ATP molecule bound to each crossbridge supplies the energy for this step. This ATP has been bound to the crossbridge in a partially broken-down form (ADP*P_i in step 1). The myosin head to which the ATP is bound is called "charged myosin" (M*ADP*P_i in step 1). When charged myosin interacts with actin, the association is represented as A*M*ADP*P_i (step 2).

The partial rotation of the angle of the crossbridge is associated with the final hydrolysis of the bound ATP and release of the hydrolysis products (step 3), an inorganic phosphate ion (P_i) and ADP. Since the myosin heads are temporarily attached to the actin filament, the partial rota-

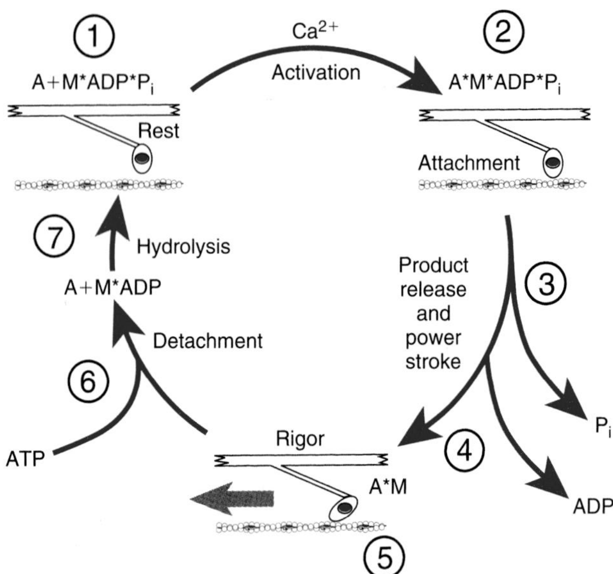

FIGURE 8.10 **Events of the crossbridge cycle in skeletal muscle.** ① At rest, ATP has been bound to the myosin head and hydrolyzed, but the energy of the reaction cannot be released until ② the myosin head can interact with actin. ③ The release of the hydrolysis products is associated with ④ the power stroke. ⑤ The rotated and still-attached crossbridge is now in the rigor state. ⑥ Detachment is possible when a new ATP molecule binds to the myosin head and is ⑦ subsequently hydrolyzed. These cyclic reactions can continue as long as the ATP supply remains and activation (via Ca^{2+}) is maintained. (See text for further details.) A, actin; M, myosin; *, chemical bond; +, a potential interaction.

tion pulls the actin filaments past the myosin filaments, a movement called the **power stroke** (step 4). Following this movement (which results in a relative filament displacement of around 10 nm), the actin-myosin binding is still strong and the crossbridge cannot detach; at this point in the cycle, it is termed a **rigor crossbridge** (A*M, step 5). For detachment to occur, a new molecule of ATP must bind to the myosin head (M*ATP, step 6) and undergo partial hydrolysis to M*ADP*P_i (step 7).

Once this new ATP binds, the newly recharged myosin head, momentarily not attached to the actin filament (step 1), can begin the cycle of attachment, rotation, and detachment again. This can go on as long as the muscle is activated, a sufficient supply of ATP is available, and the physiological limit to shortening has not been reached. If cellular energy stores are depleted, as happens after death, the crossbridges cannot detach because of the lack of ATP, and the cycle stops in an attached state (at step 5). This produces an overall stiffness of the muscle, which is observed as the **rigor mortis** that sets in shortly after death.

The crossbridge cycle obviously must be subject to control by the body to produce useful and coordinated muscular movements. This control involves several cellular processes that differ among the various types of muscle. Here, again, the case of skeletal muscle provides the basic description of the control process.

ACTIVATION AND INTERNAL CONTROL OF MUSCLE FUNCTION

Control of the contraction of skeletal muscle involves many steps between the arrival of the action potential in a motor nerve and the final mechanical activity. An important series of these steps, called **excitation-contraction coupling**, takes place deep within a muscle fiber. This is the subject of the remainder of this chapter; the very early events (communication between nerve and muscle) and the very late events (actual mechanical activity) are discussed in Chapter 9.

The Interaction Between Calcium and Specialized Proteins Is Central to Muscle Contraction

The most important chemical link in the control of muscle protein interactions is provided by calcium ions. The SR controls the internal concentration of these ions, and changes in the internal calcium ion concentration have profound effects on the actions of the contractile proteins of muscle.

Calcium and the Troponin-Tropomyosin Complex. The chemical processes of the crossbridge cycle in skeletal muscle are in a state of constant readiness, even while the muscle is relaxed. Undesired contraction is prevented by a specific inhibition of the interaction between actin and myosin. This inhibition is a function of the troponin-tropomyosin complex of the thin myofilaments. When a muscle is relaxed, calcium ions are at very low concentration in the region of the myofilaments. The long tropomyosin molecules, lying in the grooves of the entwined actin filaments, interfere with the myosin binding sites on the actin molecules. When calcium ion concentrations increase, the ions bind to the Tn-C subunit associated with each tropomyosin molecule. Through the action of Tn-I and Tn-T, calcium binding causes the tropomyosin molecule to change its position slightly, uncovering the myosin binding sites on the actin filaments. The myosin (already "charged" with ATP) is allowed to interact with actin, and the events of the crossbridge cycle take place until calcium ions are no longer bound to the Tn-C subunit.

The Switching Action of Calcium. An effective switching function requires the transition between the "off" and "on" states to be rapid and to respond to relatively small changes in the controlling element. The calcium switch in skeletal muscle satisfies these requirements well (Fig. 8.11). The curve describing the relationship between the relative force developed and the calcium concentration in the region of the myofilaments is very steep. At a calcium concentration of 1×10^{-8} M, the interaction between actin and myosin is negligible, while an increase in the calcium concentration to 1×10^{-5} M produces essentially full force development. This process is saturable, so that further increases in calcium concentration lead to little increase in force. In skeletal muscle, an excess of calcium ions is usually present during activation, and the contractile system is normally fully saturated. In cardiac and smooth muscle, however, only partial saturation occurs under normal conditions, and the

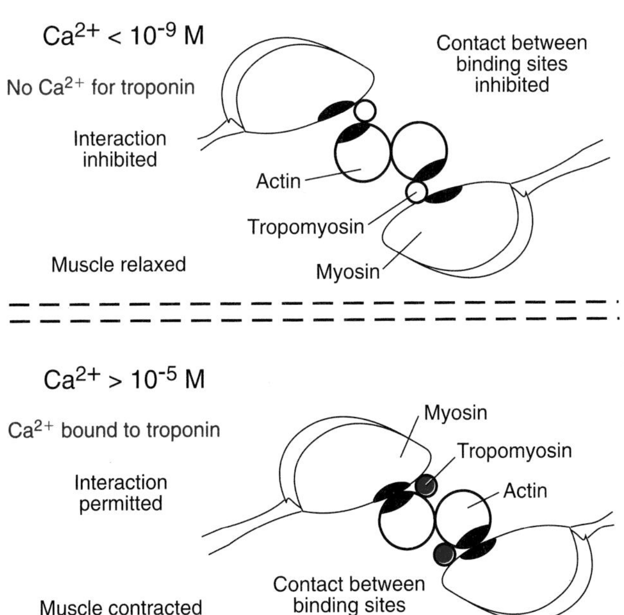

FIGURE 8.11 The calcium switch for controlling skeletal muscle contraction. Calcium ions, via the troponin-tropomyosin complex, control the unblocking of the interaction between the myosin heads (the crossbridges) and the active site on the thin filaments. The geometry of each tropomyosin molecule allows it to exert control over seven actin monomers.

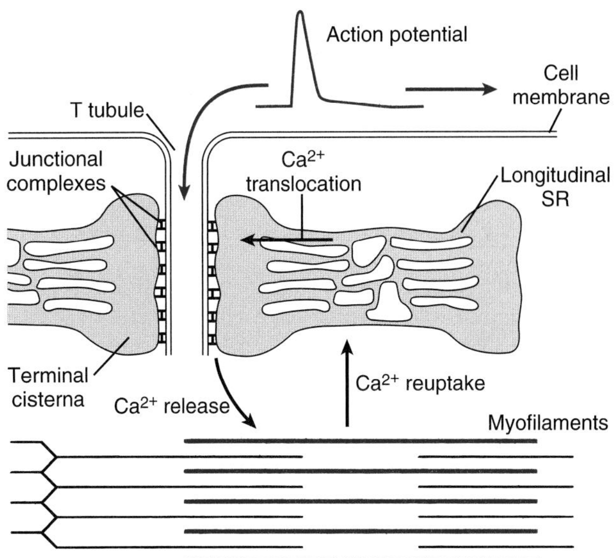

FIGURE 8.12 Excitation-contraction coupling and the cyclic movement of calcium. (See text for details of the process.)

degree of muscle activation can be adjusted by controlling the calcium concentration.

The switching action of the calcium-troponin-tropomyosin complex in skeletal and cardiac muscle is extended by the structure of the thin filaments, which allows one troponin molecule, via its tropomyosin connection, to control seven actin monomers. Since the calcium control in striated muscle is exercised through the thin filaments, it is termed **actin-linked regulation.** While the cellular control of smooth muscle contraction is also exercised by changes in calcium concentration, its effect is exerted on the thick (myosin) filaments. This is termed **myosin-linked regulation** and is described in Chapter 9.

Excitation-Contraction Coupling Links Electrical and Mechanical Events

When a nerve impulse arrives at the neuromuscular junction and its signal is transmitted to the muscle cell membrane, a rapid train of events carries the signal to the interior of the cell, where the contractile machinery is located. The large diameter of skeletal muscle cells places interior myofilaments out of range of the immediate influence of events at the cell surface, but the T tubules, SR, and their associated structures act as a specialized internal communication system that allows the signal to penetrate to interior parts of the cell. The end result of electrical stimulation of the cell is the liberation of calcium ions into regions of the sarcoplasm near the myofilaments, initiating the crossbridge cycle.

The process of excitation-contraction coupling, as outlined in Figure 8.12, begins in skeletal muscle with the electrical excitation of the surface membrane. An action potential sweeps rapidly down the length of the fiber. Its propagation is similar to that in nonmyelinated nerve fibers, in which successive areas of membrane are stimulated by local ionic currents flowing from adjacent areas of excited membrane. The lack of specialized conduction adaptations (e.g., myelination) makes this propagation slow compared with that in the motor nerve, but its speed is still sufficient to ensure the practically simultaneous activation of the entire fiber. When the action potential encounters the openings of T tubules, it propagates down the T tubule membrane. This propagation is also regenerative, resulting in numerous action potentials, one in each T tubule, traveling toward the center of the fiber. In the T tubules, the velocity of the action potentials is rather low, but the total distance to be traveled is quite short.

At some point along the T tubule, the action potential reaches the region of a triad. Here the presence of the action potential is communicated to the terminal cisternae of the SR. While the precise nature of this communication is not yet fully understood, it appears that the T tubule action potential affects specific protein molecules called **dihydropyridine receptors** (DHPRs). These molecules, which are embedded in the T tubule membrane in clusters of four, serve as **voltage sensors** that respond to the T tubule action potential. They are located in the region of the triad where the T tubule and SR membranes are the closest together, and each group of four is located in close proximity to a specific channel protein called a **ryanodine receptor** (**RyR**), which is embedded in the SR membrane. The RyR serves as a controllable channel (termed a **calcium release channel**) through which calcium ions can move readily when it is in the open state. DHPR and RyR form a functional unit called a **junctional complex** (Fig. 8.12).

When the muscle is at rest, the RyR is closed; when T tubule depolarization reaches the DHPR, some sort of linkage—most likely a mechanical connection—causes the

RyR to open and release calcium from the SR. In skeletal muscle, every other RyR is associated with a DHPR cluster; the RyRs without this connection open in response to calcium ions in a few milliseconds. This leads to rapid release of calcium ions from the terminal cisternae into the intracellular space surrounding the myofilaments. The calcium ions can now bind to the Tn-C molecules on the thin filaments. This allows the crossbridge cycle reactions to begin, and contraction occurs.

Even during calcium release from the terminal cisternae, the active transport processes in the membranes of the longitudinal elements of the SR pump free calcium ions from the myofilament space into the interior of the SR. The rapid release process stops very soon; there is only one burst of calcium ion release for each action potential, and the continuous **calcium pump** in the SR membrane reduces calcium in the region of the myofilaments to a low level (1×10^{-8} M). Because calcium ions are no longer available to bind to troponin, the contractile activity ceases and relaxation begins. The resequestered calcium ions are moved along the longitudinal elements to storage sites in the terminal cisternae, and the system is ready to be activated again. This entire process takes place in a few tens of milliseconds and may be repeated many times each second.

ENERGY SOURCES FOR MUSCLE CONTRACTION

Because contracting muscles perform work, cellular processes must supply biochemical energy to the contractile mechanism. Additional energy is required to pump the calcium ions involved in the control of contraction and for other cellular functions. In muscle cells, as in other cells, this energy ultimately comes from the universal high-energy compound, ATP.

Muscle Cells Obtain ATP From Several Sources

Although **ATP** is the immediate fuel for the contraction process, its concentration in the muscle cell is never high enough to sustain a long series of contractions. Most of the immediate energy supply is held in an "energy pool" of the compound **creatine phosphate** or **phosphocreatine (PCr)**, which is in chemical equilibrium with ATP. After a molecule of ATP has been split and yielded its energy, the resulting ADP molecule is readily rephosphorylated to ATP by the high-energy phosphate group from a creatine phosphate molecule. The creatine phosphate pool is restored by ATP from the various cellular metabolic pathways. These reactions (of which the last two are the reverse of each other) can be summarized as follows:

$$\text{ATP} \rightarrow \text{ADP} + \text{P}_i \text{ (Energy for contraction)} \quad (1)$$

$$\text{ADP} + \text{PCr} \rightarrow \text{ATP} + \text{Cr (Rephosphorylation of ATP)} \quad (2)$$

$$\text{ATP} + \text{Cr} \rightarrow \text{ADP} + \text{PCr (Restoration of PCr)} \quad (3)$$

Because of the chemical equilibria involved, the concentration of PCr can fall to very low levels before the ATP concentration shows a significant decline. It has been shown experimentally that when 90% of PCr has been used, the ATP concentration has fallen by only 10%. This situation results in a steady source of ATP for contraction that is maintained despite variations in energy supply and demand. Creatine phosphate is the most important storage form of high-energy phosphate; together with some other smaller sources, this energy reserve is sometimes called the **creatine phosphate pool**.

Two major metabolic pathways supply ATP to energy-requiring reactions in the cell and to the mechanisms that replenish the creatine phosphate pool. Their relative contributions depend on the muscle type and conditions of contraction. A simplified diagram of the energy relationships of muscle is shown in Figure 8.13. The first of the supply pathways is the **glycolytic pathway** or **glycolysis.** This is an **anaerobic** pathway; glucose is broken down without the use of oxygen to regenerate two molecules of ATP for every molecule of glucose consumed. Glucose for the glycolytic pathway may be derived from circulating blood glucose or from its storage form in muscle cells, the polymer **glycogen.** This reaction extracts only a small fraction of the energy contained in the glucose molecule.

The end product of anaerobic glycolysis is **lactic acid** or **lactate**. Under conditions of sufficient oxygen, this is converted to **pyruvic acid** or **pyruvate,** which enters another cellular (mitochondrial) pathway called the **Krebs cycle.** As a result of Krebs cycle reactions, substrates are made available for **oxidative phosphorylation**. The Krebs cycle and oxidative phosphorylation are **aerobic** processes that require a continuous supply of oxygen. In this pathway, an additional 36 molecules of ATP are regenerated from the energy in the original glucose molecule; the final products are carbon dioxide and water. While the oxidative phosphorylation pathway provides the greatest amount of energy, it cannot be used if the oxygen supply is insufficient; in this case, glycolytic metabolism predominates.

Glucose as an Energy Source. **Glucose** is the preferred fuel for skeletal muscle contraction at higher levels of exercise. At maximal work levels, almost all the energy used is derived from glucose produced by glycogen breakdown in muscle tissue and from bloodborne glucose from dietary sources. Glycogen breakdown increases rapidly during the first tens of seconds of vigorous exercise. This breakdown, and the subsequent entry of glucose into the glycolytic pathway, is catalyzed by the enzyme **phosphorylase a.** This enzyme is transformed from its inactive **phosphorylase b** form by a "cascade" of protein kinase reactions whose action is, in turn, stimulated by the increased Ca^{2+} concentration and metabolite (especially AMP) levels associated with muscle contraction. Increased levels of circulating epinephrine (associated with exercise), acting through cAMP, also increase glycogen breakdown. Sustained exercise can lead to substantial depletion of glycogen stores, which can restrict further muscle activity.

Other Important Energy Sources. At lower exercise levels (i.e., below 50% of maximal capacity) fats may provide 50 to 60% of the energy for muscle contraction. Fat, the major energy store in the body, is mobilized from adipose tissue to provide metabolic fuel in the form of **free fatty acids.** This process is slower than the liberation of glucose

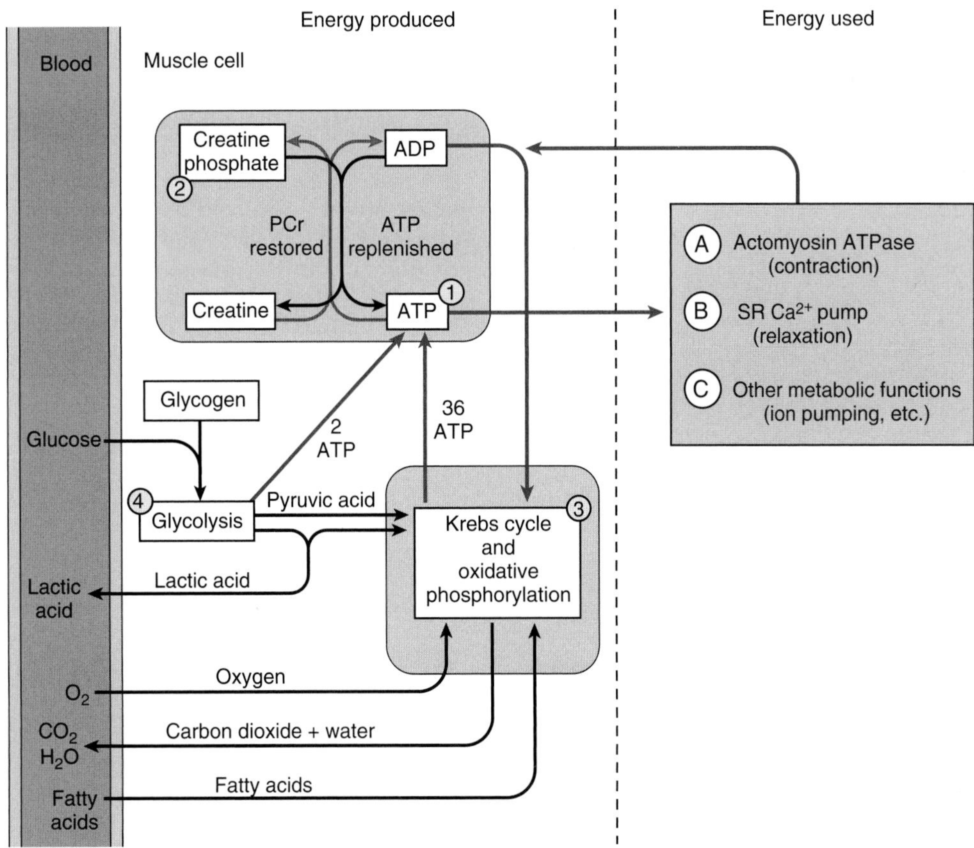

FIGURE 8.13 **The major metabolic processes of skeletal muscle.** These processes center on the supply of ATP for the actomyosin ATPase of the crossbridges. Energy sources are numbered in order of their proximity to the actual reactions of the crossbridge cycle. Energy is used by the cell in an *A*, *B*, and *C* order. The scheme shown here is typical for all types of muscle, although there are specific quantitative and qualitative variations.

from glycogen and cannot keep pace with the high demands of heavy exercise. Moderate activity, with brief rest periods, favors the consumption of fat as muscle fuel. Fatty acids enter the Krebs cycle at the acetyl-CoA-citrate step. Complete combustion of fat yields less ATP per mole of oxygen consumed than for glucose, but its high energy storage capacity (the equivalent of 138 moles of ATP per mole of a typical fatty acid) makes it an ideal energy store. The depletion of body fat reserves is almost never a limiting factor in muscle activity.

In the absence of other fuels, **protein** can serve as an energy source for contraction. However, protein is used by muscles for fuel mainly during dieting and starvation or during heavy exercise. Under such conditions, proteins are broken down into amino acids that provide energy for contraction and that can be resynthesized into glucose to meet other needs.

Many of the metabolic reactions and processes supplying energy for contraction and the recycling of metabolites (e.g., lactate, glucose) take place outside the muscle, particularly in the liver, and the products are transported to the muscle by the bloodstream. In addition to its oxygen- and carbon dioxide-carrying functions, the enhanced blood supply to exercising muscle provides for a rapid exchange of essential metabolic materials and the removal of heat.

Metabolic Adaptations Allow Contraction to Continue With an Inadequate Oxygen Supply

Glycolytic (anaerobic) metabolism can provide energy for sudden, rapid, and forceful contractions of some muscles. In such cases, the ready availability of glycolytic ATP compensates for the relatively low yield of this pathway, although a later adjustment must be made. In most muscles, especially under conditions of rest or moderate exercise, the supply of oxygen is adequate for aerobic metabolism (fed by fatty acids and by the end products of glycolysis) to supply the energy needs of the contractile system. As the level of exercise increases, several physiological mechanisms come into play to increase the blood supply (and, thus, the oxygen) to the working muscle. At some point, however, even these mechanisms fail to supply sufficient oxygen, and the end products of glycolysis begin to accumulate. The glycolytic pathway can continue to operate because the excess pyruvic acid that is produced is converted to **lactic acid,** which serves as a temporary storage medium. The formation of lactic acid, by preventing a buildup of pyruvic acid, also allows for the restoration of the enzyme cofactor NAD^+, needed for a critical step in the glycolytic pathway, so that the breakdown of glycogen can continue. Thus, ATP can continue to be produced under anaerobic conditions.

The accumulation of lactic acid is the largest contributor (more than 60%) to **oxygen deficit,** which allows short-term anaerobic metabolism to take place despite a relative lack of oxygen. Other depleted muscle oxygen stores have a smaller capacity but can still participate in oxygen deficit. The largest of these is the creatine phosphate pool (approximately 25%). Tissue fluids (including venous blood) account for another 7%, and the protein myoglobin can hold about 2.5%.

Eventually the lactic acid must be oxidized in the Krebs cycle and oxidative phosphorylation reactions, and the other energy stores (as listed above) must be replenished. This "repayment" of the oxygen deficit occurs over several minutes during recovery from heavy exercise, when the oxygen consumption and respiration rate remain high and depleted ATP is restored from the glucose breakdown products temporarily stored as lactic acid. As the cellular ATP levels return to normal, the energy stored in the creatine phosphate energy pool is also replenished.

Those muscles adapted for mostly aerobic metabolism contain significant amounts of the protein **myoglobin.** This iron-containing molecule, essentially a monomeric form of the blood protein hemoglobin (see Chapter 11), gives aerobic muscles their characteristic red color. The total oxygen storage capacity of myoglobin is quite low, and it does not make a significant direct contribution to the cellular stores; all the myoglobin-bound oxygen could support aerobic exercise for less than 1 second. However, because of its high affinity for oxygen even at low concentrations, myoglobin plays a major role in facilitating the diffusion of oxygen through exercising muscle tissue by binding and releasing oxygen molecules as they move down their concentration gradient.

Muscles of different types have varying capacities for sustaining an oxygen deficit; some skeletal muscles can sustain a considerable deficit, while cardiac muscle has an almost exclusively aerobic metabolism. Chapters 9 and 10 discuss metabolic adaptations that are specific to skeletal, smooth, and cardiac muscles.

REVIEW QUESTIONS

DIRECTIONS: Each of the numbered items or incomplete statements in this section is followed by answers or completions of the statement. Select the ONE lettered answer or completion that is BEST in each case.

1. Skeletal, smooth, and cardiac muscle all have which of the following in common?
(A) Their cellular structure is based on repeating sarcomeres
(B) The contractile cells are large relative to the size of the organ they comprise.
(C) The contractile system is based on an enzymatic interaction of actin and myosin.
(D) Initiation of contraction requires the binding of calcium ions to actin filaments
2. During the shortening of skeletal muscle,
(A) The distance between Z lines stays the same
(B) The width of the I band changes
(C) The width of the A band changes
(D) All internal spacings between repeating structures change proportionately
3. The compound ATP provides the energy for muscle contraction during the crossbridge cycle. A second important function for ATP in the cycle is to
(A) Provide the energy for relaxation
(B) Allow the thick and thin filaments to detach from each other during the crossbridge cycle
(C) Maintain the separation of thick and thin filaments when the muscle is at rest
(D) Promote the binding of calcium ions to the regulatory proteins
4. Calcium ions are required for the normal activation of all muscle types. Which statement below most closely describes the role of calcium ions in the control of skeletal muscle contraction?
(A) The binding of calcium ions to regulatory proteins on the thin filaments removes the inhibition of actin-myosin interaction
(B) The binding of calcium ions to the thick filament regulatory proteins activates the enzymatic activity of the myosin molecules
(C) Calcium ions serve as an inhibitor of the interaction of thick and thin filaments
(D) A high concentration of calcium ions in the myofilament space is required to maintain muscle in a relaxed state.
5. The normal process of relaxation in skeletal muscle depends on
(A) A sudden reduction in the amount of ATP available for the crossbridge interactions
(B) Metabolically supported pumping of calcium out of the cells when the membrane potential repolarizes
(C) A rapid reuptake of calcium into the sarcoplasmic reticulum
(D) An external force to separate the interacting myofilaments
6. When an isolated skeletal muscle is stretched beyond its optimal length (but not to the point where damage occurs), the reduction in contractile force is due to
(A) Lengthening of the myofilaments so that crossbridges become spaced farther apart and can interact less readily
(B) Decreased overlap between thick and thin filaments, which reduces the number of crossbridges that interact
(C) The thinning of the muscle, which reduces its cross-sectional area and, hence, the force that it can produce
(D) A proportional reduction in the amount of calcium released from the sarcoplasmic reticulum
7. The major immediate source of calcium for the initiation of skeletal muscle contraction is
(A) Calcium entry through the sarcolemma during the passage of an action potential
(B) A rapid release of calcium from its storage sites in the T tubules
(C) A rapid release of calcium from the terminal cisternae of the sarcoplasmic reticulum
(D) A release of calcium that is bound to cytoplasmic proteins in the region of the myofilaments
8. The relaxation of skeletal muscle is associated with a reduction in free intracellular calcium ion concentration. The effect of this reduction is
(A) A reestablishment of the inhibition of the actin-myosin interaction
(B) Deactivation of the enzymatic activity of the individual actin molecules

(continued)

(C) A change in the chemical nature of the myosin molecules, reducing their enzymatic activity
(D) Reduced contractile interaction by the binding of calcium to the active sites of the myosin molecules

9. The chemical energy source that most directly supports muscle contraction is
(A) Creatine phosphate
(B) Glucose
(C) ATP
(D) Free fatty acids

10. In the absence of an adequate supply of ATP for skeletal muscle contraction,
(A) Myofilament interaction ceases, and the muscle relaxes
(B) Actin and myosin filaments cannot separate, and the muscle stiffens
(C) Creatine phosphate can directly support myofilament interaction, although less efficiently
(D) The lower energy form, ADP, can support contraction at a reduced rate

11. In the face of insufficient oxygen to meet its current metabolic requirements, skeletal muscle
(A) Quickly loses its ability to contract and relaxes until oxygen is again available
(B) Maintains contraction by using metabolic pathways that do not require oxygen consumption
(C) Maintains contraction by using a large internal store of ATP that is kept in reserve
(D) Contracts more slowly at a given force, resulting in a saving of energy

12. If the calcium pumping ability of the sarcoplasmic reticulum were impaired (but not abolished),
(A) Muscles would relax more quickly because less calcium would be pumped
(B) Contraction would be slowed, but the muscle would relax normally
(C) The muscle would continue to develop force, but its relaxation would be slowed
(D) Activation of the muscle would no longer be possible

SUGGESTED READING

Bagshaw CR. Muscle Contraction. 2nd Ed. New York: Chapman & Hall, 1993.

Ford LE. Muscle Physiology and Cardiac Function. Carmel, IN: Biological Sciences Press-Cooper Group, 2000.

Matthews GG. Cellular Physiology of Nerve and Muscle. 2nd Ed. Boston: Blackwell, 1991.

Rüegg JC. Calcium in Muscle Contraction: Cellular and Molecular Physiology. 2nd Ed. New York: Springer-Verlag, 1992.

Squire JM, ed. Molecular Mechanisms in Muscle Contraction. Boca Raton: CRC Press, 1990.

Skeletal Muscle and Smooth Muscle

Richard A. Meiss, Ph.D.

CHAPTER OUTLINE

KEY CONCEPTS

1. The myoneural junction is a specialized synapse between the motor axon and a skeletal muscle fiber. A motor nerve and all of the muscle fibers it innervates is called a motor unit.
2. Neuromuscular transmission involves presynaptic transmitter release, diffusion of transmitter across the synaptic cleft, and binding to postsynaptic receptors.
3. The immediate postsynaptic electrical response to transmitter molecule binding is a local depolarization called the endplate potential, which is graded according to the relative number of channels that have been opened by the transmitter binding.
4. The endplate potential is localized to the endplate region and is not propagated. It causes current to flow into the muscle fiber at the endplate; the resulting outward current across adjacent areas of membrane leads to their depolarization and the generation of propagated nerve-like action potentials in the muscle cell membrane.
5. A twitch is a single muscle contraction, produced in response to a single action potential in the muscle cell membrane. A tetanus is a larger muscle contraction that results from repetitive stimulation (multiple action potentials) of the cell membrane. Its force represents the temporal summation of many twitch contractions.
6. Isometric contraction results when an activated muscle is prevented from shortening and force is produced without movement.
7. Isotonic contraction results when an activated muscle shortens against an external force (or load). The external load determines the force that the muscle will develop, and the developed force determines the velocity of shortening.
8. The length-tension curve describes the effect of the resting length of a muscle on the isometric force it can develop. This relationship, which passes through a maximum at the normal length of the muscle in the body, is determined largely by the molecular and cellular ultrastructure of the muscle.
9. The force-velocity curve describes the inverse relationship between the isotonic force and the shortening velocity in a fully activated muscle.
10. The power output of an isotonically contracting skeletal muscle is determined by the velocity of shortening, which is determined by the size of the load; it is maximal at approximately one-third of the maximal isometric force.
11. All muscles are arranged so that they may be extended by the action of antagonistic muscles or by an external force such as gravity. Muscles do not forcibly reextend themselves after shortening.
12. The control of skeletal muscle contraction is exercised through the thin filaments and is termed *actin-linked*. Smooth muscle contraction is controlled primarily via the thick filaments and is termed *myosin-linked*.
13. The links between cellular excitation and mechanical contraction in smooth muscle are varied and complex. In most of the pathways, the cellular concentration of free calcium ions is an important link in the process of activation and contraction.
14. The primary step in the regulation of smooth muscle contraction is the phosphorylation of the regulatory light chains of the myosin molecule, which is then free to interact with actin. Relaxation involves phosphatase-mediated dephosphorylation of the light chains.
15. The contractions of smooth muscle are considerably slower than those of skeletal muscle, but are much more economical in their use of cellular energy. A crossbridge mechanism called the "latch state" enables some smooth muscles to maintain contraction for extremely long periods of time.
16. Smooth muscle tissues, especially those in the walls of distensible organs, can operate over a wide range of lengths.

Chapter 8 dealt with the mechanics and activation of the internal cellular processes that produce muscle contraction. This chapter treats muscles as organized tissues, beginning with the events leading to membrane activation by nerve stimulation and continuing with the outward mechanical expression of internal processes.

ACTIVATION AND CONTRACTION OF SKELETAL MUSCLE

Skeletal muscle is controlled by the central nervous system (CNS), which provides a pattern of activation that is suited to the task at hand. The resulting contraction is further shaped by mechanical conditions external to the muscle. The connection between nerve and muscle has been studied for over a century, and a fairly clear picture of the process has emerged. While the process functions amazingly well, its complexity means that critical failures can lead to serious medical problems.

Impulse Transmission From Nerve to Muscle Occurs at the Neuromuscular Junction

The contraction of skeletal muscle occurs in response to action potentials that travel down somatic motor axons originating in the CNS. The transfer of the signal from nerve to muscle takes place at the **neuromuscular junction**, also called the **myoneural junction** or **motor endplate**. This special type of synapse has a close association between the membranes of nerve and muscle and a physiology much like that of excitatory neural synapses (see Chapter 3).

The Structure of the Neuromuscular Junction. On reaching a muscle cell, the axon of a motor neuron typically branches into several terminals, which constitute the **presynaptic portion** of the neuromuscular junction. The terminals lie in grooves or "gullies" in the surface of the muscle cell, outside the muscle cell membrane, and a Schwann cell covers them all (Fig. 9.1). Within the axoplasm of the nerve terminals are located numerous membrane-enclosed vesicles containing **acetylcholine** (**ACh**). Mitochondria, associated with the extra metabolic requirements of the terminal, are also plentiful.

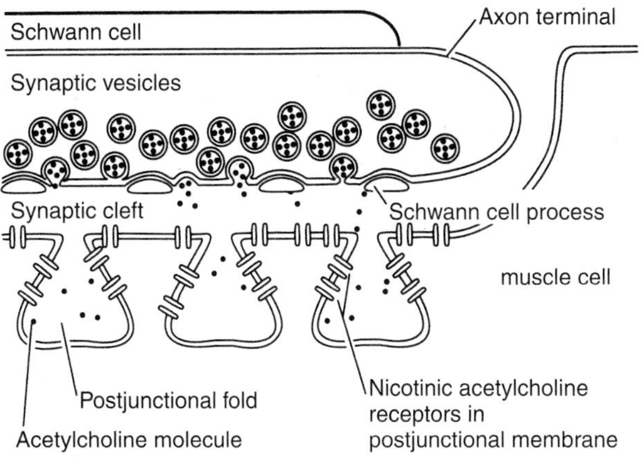

FIGURE 9.1 **Structural features of the neuromuscular junction.** Processes of the Schwann cell that overlie the axon terminal wrap around under it and divide the junctional area into active zones.

The postsynaptic portion of the junction or **endplate membrane** is that part of the muscle cell membrane lying immediately beneath the axon terminals. Here the membrane is formed into **postjunctional folds,** at the mouths of which are located many **nicotinic ACh receptor** molecules. These are *chemically gated* ion channels that increase the cation permeability of the postsynaptic membrane in response to the binding of ACh. Between the nerve and muscle is a narrow space called the **synaptic cleft.** Acetylcholine must diffuse across this gap to reach the receptors in the postsynaptic membrane. Also located in the synaptic cleft (and associated with the postsynaptic membrane) is the enzyme **acetylcholinesterase** (AChE).

Chemical Events at the Neuromuscular Junction. When the wave of depolarization associated with a nerve action potential spreads into the terminal of a motor axon, several processes are set in motion. The lowered membrane potential causes membrane channels to open and external calcium ions enter the axon. The rapid rise in intracellular calcium causes the cytoplasmic vesicles of ACh to migrate to the inner surface of the axon membrane, where they fuse with the membrane and release their contents. Because all the vesicles are of roughly the same size, they all release about the same amount—a **quantum**—of neurotransmitter. The transmitter release is called **quantal**; although so many vesicles are normally activated at once, their individual contributions are not separately identifiable.

When the ACh molecules arrive at the postsynaptic membrane after diffusing across the synaptic cleft, they bind to the ACh receptors. When two ACh molecules are bound to a receptor, it undergoes a configurational change that allows the relatively free passage of sodium and potassium ions down their respective electrochemical gradients. The binding of ACh to the receptor is reversible and rather loose. Soon ACh diffuses away and is hydrolyzed by AChE into choline and acetate, terminating its function as a transmitter molecule, and the membrane permeability returns to the resting state. The choline portion is taken up by the presynaptic terminal for resynthesis of ACh, and the acetate diffuses away into the extracellular fluid. These events take place over a few milliseconds and may be repeated many times per second without danger of fatigue.

Electrical Events at the Neuromuscular Junction. The binding of the ACh molecules to postsynaptic receptors initiates the electrical response of the muscle cell membrane, and what was a chemical signal becomes an electrical one. The stages of the development of the electrical signal are shown in Figure 9.2. With the opening of the postsynaptic ionic channels, sodium enters the muscle cell and potassium simultaneously leaves. Both ions share the same membrane channels; in this and several other respects, the endplate membrane is different from the general cell membrane of muscles and nerves. The opening of the channels depends only on the presence of neurotransmitter and not on mem-

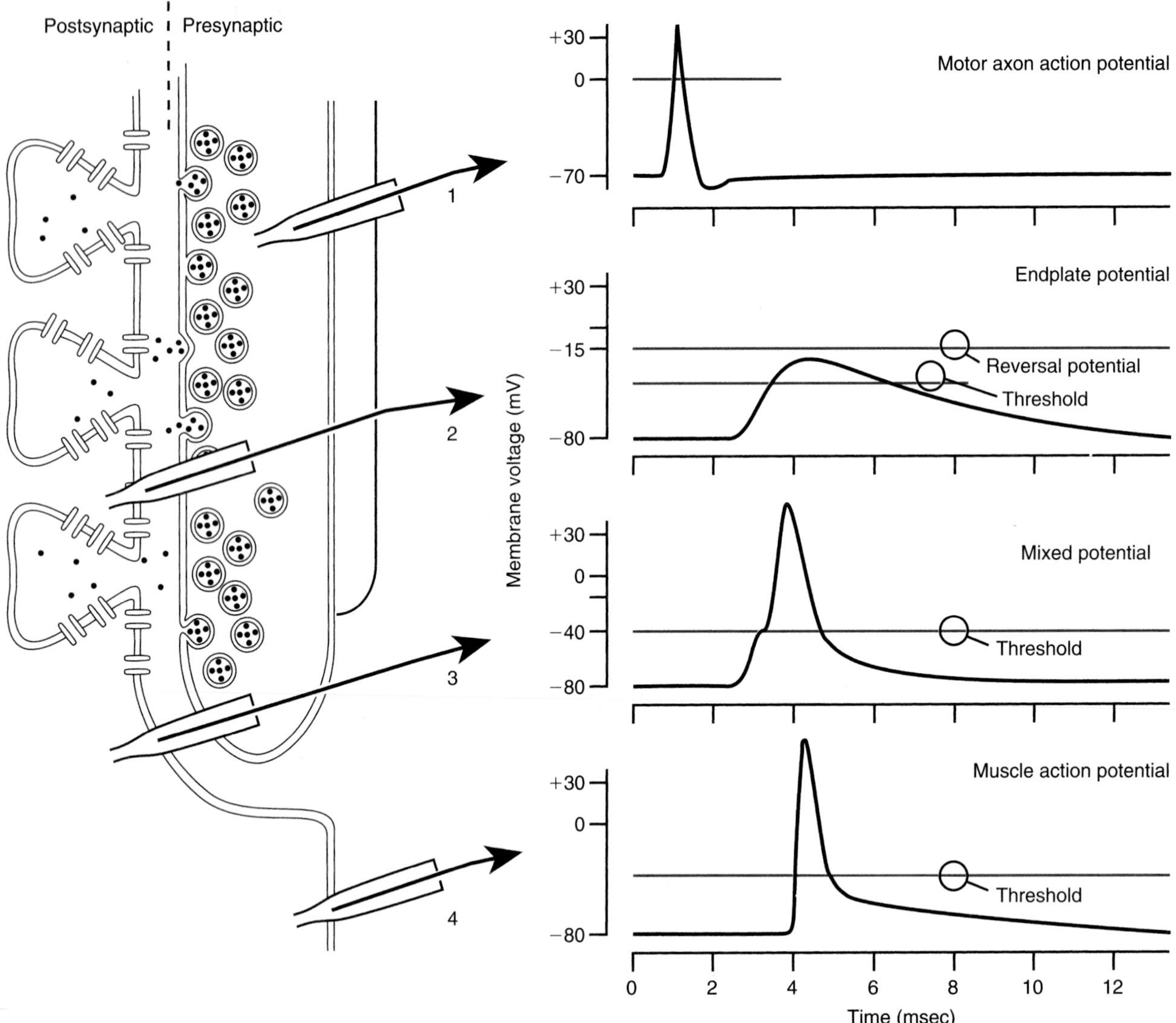

FIGURE 9.2 Electrical activity at the neuromuscular junction. The four microelectrodes sample membrane potentials at critical regions. (These are idealized records drawn to illustrate isolated portions of the response; in an actual recording, there would be considerable overlap of the responses because of the close spacing of the electrodes.) Note the time delays as a result of transmitter diffusion and endplate potential generation. The reversal potential is the membrane potential at which net current flow is zero (i.e., inward Na^+ and outward K^+ currents are equal).

brane voltage, and the sodium and potassium permeability changes occur simultaneously (rather than sequentially, as they do in nerve or in the general muscle membrane). As a result of the altered permeabilities, a net inward current, known as the **endplate current**, depolarizes the postsynaptic membrane. This voltage change is called the **endplate potential.** The voltage at which the net membrane current would become zero is called the **reversal potential** of the endplate (see Fig. 9.2), although time does not permit this condition to become established because the AChE is continuously inactivating transmitter molecules.

To complete the circuit, the current flowing inward at the postsynaptic membrane must be matched by a return current. This current flows through the local muscle cytoplasm (myoplasm), out across the adjacent muscle membrane and back through the extracellular fluid (Fig. 9.3). As this endplate current flows out across the muscle membrane in regions adjacent to the endplate, it depolarizes the membrane and causes voltage-gated sodium channels to open, bringing the membrane to threshold. This leads to an action potential in the muscle membrane. The muscle action potential is propagated along the muscle cell membrane by regenerative local currents similar to those in a nonmyelinated nerve fiber.

The endplate depolarization is **graded,** and its amplitude varies with the number of receptors with bound ACh. If some circumstance causes reduced ACh release, the amount of depolarization at the endplate could be correspondingly reduced. Under normal circumstances, however, the endplate potential is much more than sufficient to produce a muscle action potential; this reserve, referred to as a **safety factor,** can help preserve function under abnor-

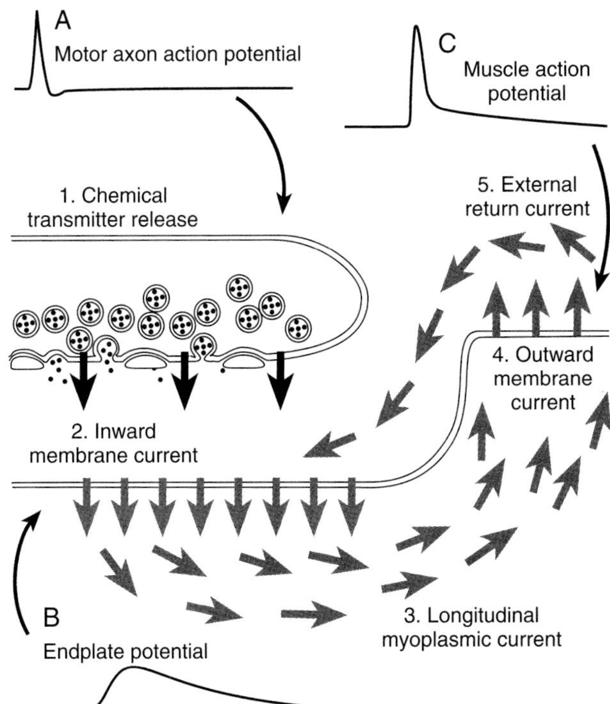

FIGURE 9.3 **Ionic currents at the neuromuscular junction.** **A,** The inward membrane current is carried by sodium ions through the channels associated with ACh receptors. The other currents are nonspecific and are carried by appropriately charged ions in the myoplasm and extracellular fluid. **B,** The endplate potential is localized to the endplate region. **C,** The muscle action potential is propagated along the surface of the muscle.

mal conditions. The rate of rise of the endplate potential is determined largely by the rate at which ACh binds to the receptors, and indirect clinical measurements of the size and rise time of the endplate potential are of considerable diagnostic importance. The rate of decay is determined by a combination of factors, including the rate at which the ACh diffuses away from the receptors, the rate of hydrolysis, and the electrical resistance and capacitance of the endplate membrane.

Neuromuscular Transmission Can Be Altered by Toxins, Drugs, and Trauma

The complex series of events making up neuromuscular transmission is subject to interference at several steps. **Presynaptic blockade** of the neuromuscular junction can occur if calcium does not enter the presynaptic terminal to participate in migration and emptying of the synaptic vesicles. The drug **hemicholinium** interferes with choline uptake by the presynaptic terminal and, thus, results in the depletion of ACh. **Botulinum toxin** interferes with ACh release. This bacterial toxin is used to treat focal dystonias (see Clinical Focus Box 9.1).

Postsynaptic blockade can result from a variety of circumstances. Drugs that partially mimic the action of ACh can be effective blockers. Derivatives of **curare,** originally used as arrow poison in South America, bind tightly to ACh receptors. This binding does not result in opening of the ion channels, however, and the endplate potential is reduced in proportion to the number of receptors occupied by curare. Muscle paralysis results. Although the muscle can be directly stimulated electrically, nerve stimulation is ineffective. The drug **succinylcholine** blocks the neuromuscular junction in a slightly different way; this molecule binds to the receptors and causes the channels to open. Because it is hydrolyzed very slowly by AChE, its action is long lasting and the channels remain open. This prevents resetting of the inactivation gates of muscle membrane sodium channels near the endplate region and blocks subsequent action potentials. Drugs that produce extremely long-lasting endplate potentials are referred to as **depolarizing blockers.**

Compounds such as **physostigmine (eserine)** are potent inhibitors of AChE and produce a depolarizing blockade. In carefully controlled doses, they can temporarily alleviate symptoms of **myasthenia gravis,** an autoimmune condition that results in a loss of postsynaptic ACh receptors. The principal symptom is muscular weakness caused by endplate potentials of insufficient amplitude. Partial inhibition of the enzymatic degradation of ACh allows ACh to remain effective longer and, thus, to compensate for the loss of receptor molecules.

Under normal conditions, ACh receptors are confined to the endplate region of a muscle. If accidental denervation occurs (e.g., by the severing of a motor nerve), the entire muscle becomes sensitive to direct application of ACh within several weeks. This extrasynaptic sensitivity is due to the synthesis of new ACh receptors, a process normally inhibited by the electrical activity of the motor axon. Artificial electrical stimulation has been shown experimentally to prevent the synthesis of new receptors, by regulating transcription of the genes involved. If reinnervation occurs, the extrasynaptic receptors gradually disappear. Muscle atrophy also occurs in the absence of functional innervation, which also can be at least partially reversed with artificial stimulation.

MECHANICAL PROPERTIES OF SKELETAL MUSCLE

The variety of controlled muscular movements that humans can make is remarkable, ranging from the powerful contractions of a weightlifter's biceps to the delicate movements of the muscles that position our eyes as we follow a moving object. In spite of this diversity, the fundamental mechanical events of the contraction process can be described by a relatively small set of specially defined functions that emphasize particular capabilities of muscle.

The Timing of Muscle Stimulation Is a Critical Determinant of Contractile Function

A skeletal muscle must be activated by the nervous system before it can begin contracting. Through the many processes previously described, a single nerve action potential arrives at each motor nerve axon terminal. A single muscle action potential then propagates along the length

CLINICAL FOCUS BOX 9.1

Focal Dystonias and Botulinum Toxin

Focal dystonias are neuromuscular disorders characterized by involuntary and repetitive or sustained skeletal muscle contractions that cause twisting, turning, or squeezing movements in a body part. Abnormal postures and considerable pain, as well as physical impairment, often result. Usually the abnormal contraction is limited to a small and specific region of muscles, hence, the term *focal* ("by itself"). *Dystonia* means "faulty contraction." **Spasmodic torticollis** and **cervical dystonia** (involving neck and shoulder muscles), **blepharospasm** (eyelid muscles), **strabismus** and **nystagmus** (extraocular muscles), **spasmodic dysphonia** (vocal muscles), **hemifacial spasm** (facial muscles), and **writer's cramp** (finger muscles in the forearm) are common dystonias. Such problems are neurological, not psychiatric, in origin, and sufferers can have severe impairment of daily social and occupational activities.

The specific cause is located somewhere in the central nervous system (CNS), but usually its exact nature is unknown. A genetic predisposition to the disorder may exist in some cases. Centrally acting drugs are of limited effectiveness, and surgical denervation, which carries a significant risk of permanent and irreversible paralysis, may provide only temporary relief. However, recent clinical trials using **botulinum toxin** to produce chemical denervation show significant promise in the treatment of these disorders.

Botulinum toxin is produced when the bacterium *Clostridium botulinum* grows anaerobically. It is one of the most potent natural toxins; a lethal dose for a human adult is about 2 to 3 μg. The active portion of the toxin is a protein with a molecular weight of about 150,000 that is conjugated with a variable number of accessory proteins. Type A toxin, the complex form most often used therapeutically, has a total molecular weight of 900,000 and is sold under the trade names Botox and Oculinum.

The toxin first binds to the cell membrane of presynaptic nerve terminals in skeletal muscles. The initial binding does not appear to produce paralysis until the toxin is actively transported into the cell, a process requiring more than an hour. Once inside the cell, the toxin disrupts calcium-mediated ACh release, producing an irreversible transmission block at the neuromuscular junction. The nerve terminals begin to degenerate, and the denervated muscle fibers atrophy. Eventually, new nerve terminals sprout from the axons of affected nerves and make new synaptic contact with the chemically denervated muscle fibers. During the period of denervation, which may be several months, the patient usually experiences considerable relief of symptoms. The relief is temporary, however, and the treatment must be repeated when reinnervation has occurred.

Clinically, highly diluted toxin is injected into the individual muscles involved in the dystonia. Often this is done in conjunction with electrical measurements of muscle activity (electromyography) to pinpoint the muscles involved. Patients typically begin to experience relief in a few days to a week. Depending on the specific disorder, relief may be dramatic and may last for several months or more. The abnormal contractions and associated pain are greatly reduced, speech can become clear again, eyes reopen and cease uncontrolled movements and, often, normal activities can be resumed.

The principal adverse effect is a temporary weakness of the injected muscles. A few patients develop antibodies to the toxin, which renders its further use ineffective. Studies have shown that the toxin's activity is confined to the injected muscles, with no toxic effects noted elsewhere. Long-term effects of the treatment, if any, are unknown.

of each muscle fiber innervated by that axon terminal. This leads to a single brief contraction of the muscle, a **twitch**. Though the contractile machinery may be fully activated (or nearly so) during a twitch, the amount of force produced is relatively low because the activation is so brief that the relaxation processes begin before contraction is fully established.

Effects of Repeated Stimulation. The duration of the action potential in a skeletal muscle fiber is short (about 5 msec) compared to the duration of a twitch (tens or hundreds of milliseconds, depending on muscle type, temperature, etc.). This means the absolute refractory period is also brief, and the muscle fiber membrane can be activated again long before the muscle has relaxed. Figure 9.4 shows the result of stimulating a muscle that is already active as a result of a prior stimulus. If the second stimulus is given during relaxation (Fig. 9.4B), well outside the refractory period caused by the first stimulus, significant additional force is developed. This additional force increment is associated with a second release of calcium ions from the SR, which adds to the calcium already there and reactivates actin and myosin interactions (see Chapter 8). When the second stimulus closely follows the first (even before force has begun to decline), the myoplasmic calcium concentration is still high (Fig. 9.4C), and the effect of the additional calcium ions is to increase the force and, to some extent, the duration of the twitch because a larger amount of calcium is present in the region of the myofilaments.

If stimuli are given repeatedly and rapidly, the result is a sustained contraction called a **tetanus**. When the contractions occur so close together that no fluctuations in force are observed, a fused tetanus results. The repetition rate at which this occurs is the **tetanic fusion frequency**, typically 20 to 60 stimuli per second, with the higher rates found in muscles that contract and relax rapidly. Figure 9.5 shows these effects in a special situation, in which the interval between successive stimuli is steadily reduced and the muscle responds at first with a series of twitches that become fused into a smooth tetanus at the highest stimulus frequency. Because it involves events that occur close together in time, a tetanus is a form of **temporal summation**.

Higher Forces Are Produced During a Tetanus. The amount of force produced in a tetanus is typically several times that of a twitch; the disparity is expressed as the

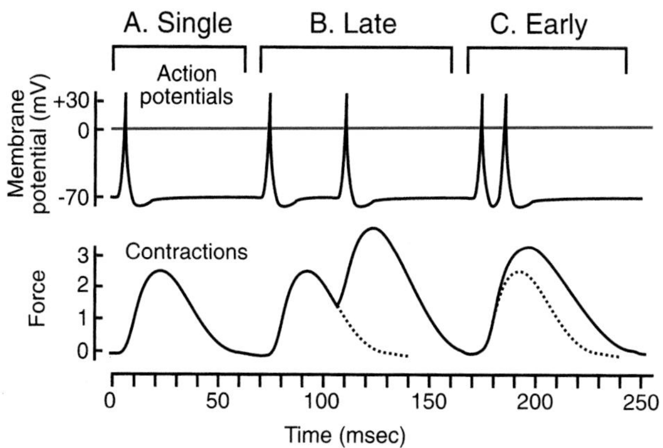

FIGURE 9.4 **Temporal summation of muscle twitches.** A, The first contraction is in response to a single action potential. B, The next contraction shows the summed response to a second stimulus given during relaxation; the two individual responses are evident. C, The last contraction is the result of two stimuli in quick succession. Though measured force was still rising when the second stimulus was given, the fact that there could be an added response shows that internal activation had begun to decline. In all cases, the solid line in the lower graph represents the actual summed tension.

tetanus-twitch ratio. The relaxation processes during a twitch, particularly the reuptake of calcium, begin to operate as soon as the muscle is activated, and full activation is brief (lasting less time than that required for the muscle to reach its peak force). Multiple stimuli, as in a tetanus, are needed for the full force to be expressed.

Another factor explaining the higher muscle force produced with repetitive stimulation is mechanical. Even if the ends of a muscle are held rigidly, internal dimensional changes take place on activation. Some of this internal motion is associated with the crossbridges, and the tendons at either end of the muscle make a considerable contribution.

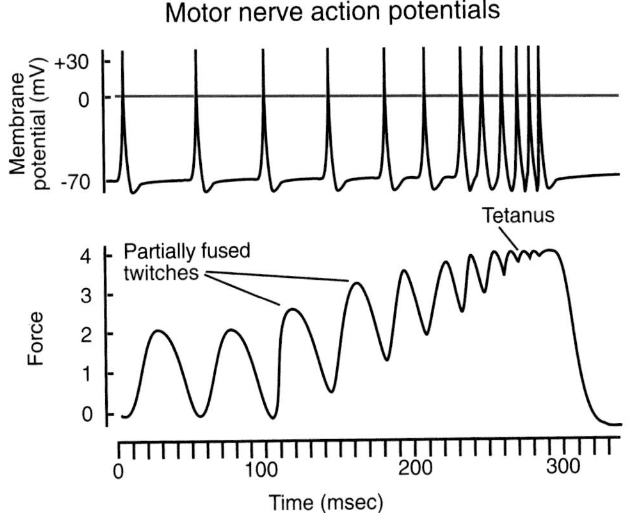

FIGURE 9.5 **Fusion of twitches into a smooth tetanus.** The interval between successive stimuli steadily decreases until no relaxation occurs between stimuli.

These deformable structures comprise the **series elastic component** of the muscle, and their extension takes a significant amount of time. The brief activation time of a twitch is not sufficient to extend the series elastic component fully, and not all of the potential force of the contraction is realized. Repeated activation in tetanus allows time for the internal "slack" to be more fully taken up, and more force is produced. Muscles with a large amount of series elasticity have a large tetanus-twitch ratio. The presence of series elasticity in human muscles provides some protection against sudden overloads of a muscle and allows for a small amount of mechanical energy storage. In jumping animals, such as kangaroos, a large fraction of muscular energy is stored in the elastic tendons and contributes significantly to the economy of locomotion.

Partial Activation of a Whole Muscle. Since a skeletal muscle consists of many fibers, each supplied by its own branch of a motor axon, it is possible (and usual) that only a portion of the muscle will be activated at any one time. The pattern of activation is determined by the CNS and by the distribution of the motor axons among the muscle fibers. A typical motor axon branches as it courses through the muscle, and each of its terminal branches innervates a single muscle fiber. All the fibers supplied by a single motor axon will contract together when a nerve action potential travels from the central nervous system and divides among the branches.

A single motor axon and all of the fibers it innervates are called a **motor unit.** Contractions in only some of the fibers in a motor unit are impossible, so the motor unit is normally the smallest functional unit of a muscle. In muscles adapted for fine and precise control, only a few muscle fibers are associated with a given motor axon; in muscles in which high force is more important, a single motor axon controls many more muscle fibers. The total force produced by a muscle is determined by the number of motor units active at any one time; as more motor units are brought into play, the force increases. This phenomenon, called **motor unit summation,** is illustrated in Figure 9.6. The force of contraction of the whole muscle is further modified by the degree of activation of each motor unit in the muscle; some may be fully tetanized, while others may be at rest or produce only a series of twitches. During a sustained contraction, the pattern of activity is continually changed by the CNS, and the burden of contraction is shared among the motor units. This results in a smooth contraction, with the force precisely controlled to produce the desired movement (or lack of it).

Externally Imposed Conditions Also Affect Contraction

Mechanical factors external to the muscle also influence the force and speed of contraction. For example, if a muscle is not allowed to shorten when it is stimulated, it will develop more force than it would if its length were allowed to change. If a muscle is in the process of lifting a load, its force of contraction is determined by the size of the load, not by the capabilities of the muscle. The speed with which a muscle shortens is likewise determined, at least in part, by external conditions.

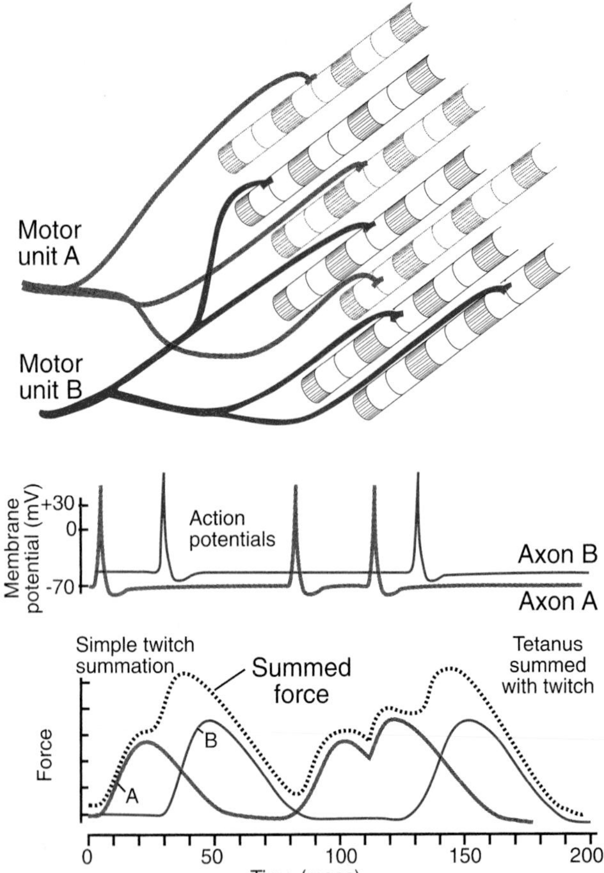

FIGURE 9.6 **Motor unit summation.** Two units are shown above; their motor nerve action potentials and muscle twitches are shown below. In the first contraction, there is a simple summation of two twitches; in the second, a brief tetanus in one motor unit sums with a twitch in the other.

Isometric Contraction. If a muscle is prevented from shortening when activated, the muscle will express its contractile activity by pulling against its attachments and developing force. This type of contraction is termed **isometric** (meaning "same length"). The forces developed during an isometric contraction can be studied by attaching a dissected muscle to an apparatus similar to that shown in Figure 9.7. This arrangement provides for setting the length of the muscle and tracing a record of force versus time. In a twitch, isometric force develops relatively rapidly, and subsequent isometric relaxation is somewhat slower. The durations of both contraction time and relaxation time are related to the rate at which calcium ions can be delivered to and removed from the region of the crossbridges, the actual sites of force development. During an isometric contraction, no actual physical work is done on the external environment because no movement takes place while the force is developed. The muscle, however, still consumes energy to fuel the processes that generate and maintain force.

Isotonic Contraction. When conditions are arranged so the muscle can shorten and exert a constant force while doing so, the contraction is called **isotonic** (meaning "same force"). In the simplest conditions, this constant force is

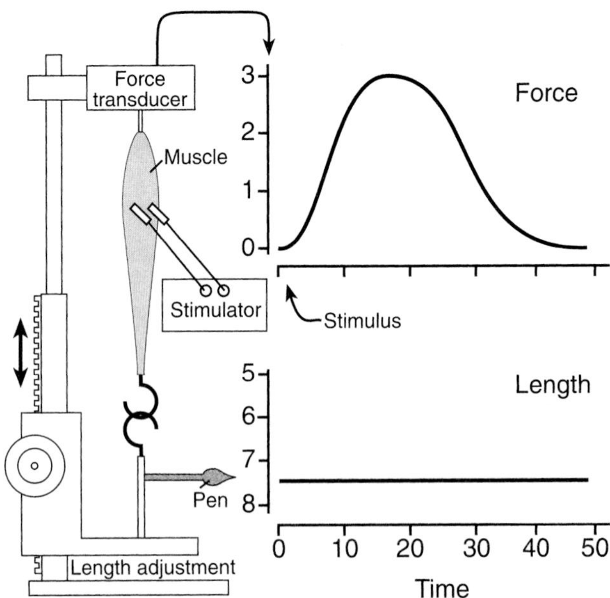

FIGURE 9.7 **A simple apparatus for recording isometric contractions.** The length of the muscle (marked on the graph by the pen attached near its lower end) is adjustable at rest but is held constant during contraction. The force transducer provides a record of the isometric force response to a single stimulus at a fixed length (isometric by definition). (Force, length, and time units are arbitrary.)

provided by the load a muscle lifts. This load is called an **afterload**, since its magnitude and presence are not apparent to the muscle until after it has begun to shorten.

Recording an isotonic contraction requires modification of the apparatus used to study isometric contraction (Fig. 9.8). Here the muscle is allowed to shorten while lifting an afterload, which is provided by the attached weight. This weight is chosen to present somewhat less than the peak force capability of the muscle. When the muscle is stimulated, it will begin to develop force without shortening, since it takes some time to build up enough force to begin to lift the weight. This means that early on, the contraction is isometric (phase 1; Fig. 9.8). After sufficient force has been generated, the muscle will begin to shorten and lift the load (phase 2). The contraction then becomes isotonic because the force exerted by the muscle exactly matches that of the weight, and the mass of the weight does not vary. Therefore, the upper tracing in Figure 9.8 shows a flat line representing constant force, while the muscle length (lower tracing) is free to change. As relaxation begins (phase 3), the muscle lengthens at constant force because it is still supporting the load; this phase of relaxation is isotonic, and the muscle is reextended by the weight. When the muscle has been extended sufficiently to return to its original length, conditions again become isometric (phase 4), and the remaining force in the muscle declines as it would in a purely isometric twitch. In almost all situations encountered in daily life, isotonic contraction is preceded by isometric force development; such contractions are called **mixed contractions** (isometric-isotonic-isometric).

The duration of the early isometric portion of the contraction varies, depending on the afterload. At low after-

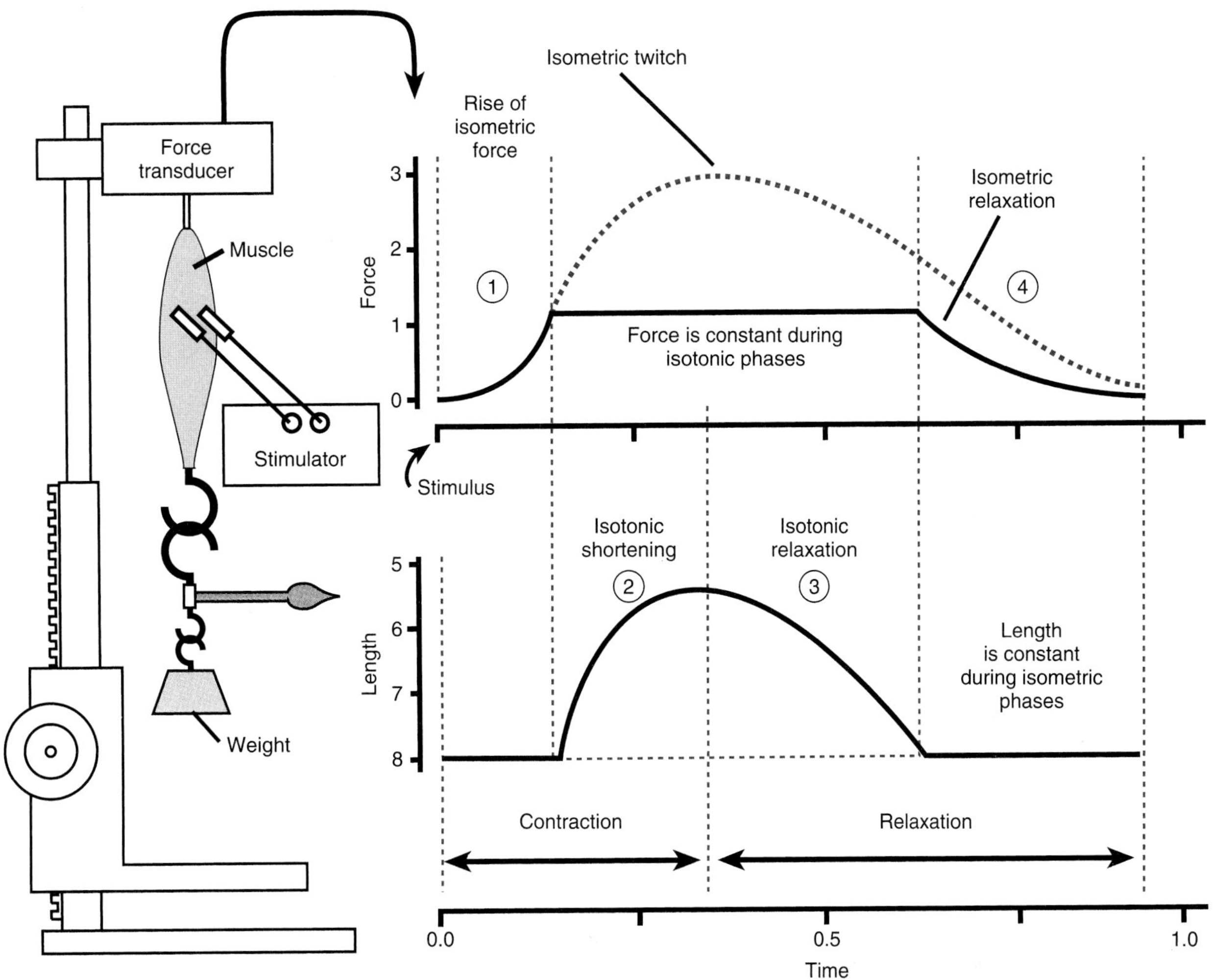

FIGURE 9.8 **A modified apparatus showing the recording of a single isotonic switch.** The pen at the lower end of the muscle marks its length, and the weight attached to the muscle provides the afterload, while the platform beneath the weight prevents the muscle from being overstretched at rest. The first part of the contraction, until sufficient force has developed to lift the weight, is isometric. During shortening and isotonic relaxation the force is constant (isotonic conditions), and during the final relaxation, conditions are again isometric because the muscle no longer lifts the weight. The dotted lines in the force and length traces show the isometric twitch that would have resulted if the force had been too large (greater than 3 units) for the muscle to lift. (Force, length, and time units are arbitrary.) (See text for details.)

loads, the muscle requires little time to develop sufficient force to begin to shorten, and conditions will be isotonic for a longer time. Figure 9.9 presents a series of three twitches. At the lowest afterload (weight A only), the isometric phase is the briefest and the isotonic phase is the longest with the lowest force. With the addition of weight B, the afterload is doubled and the isometric phase is longer, while the isotonic phase is shorter with twice the force. If weight C is added, the combined afterload represents more force that the muscle can exert, and the contraction is isometric for its entire duration. The speed and extent of shortening depend on the afterload in unique ways described shortly.

Other Types of Contraction. Other physical situations are sometimes encountered that modify the type of muscle contraction. When the force exerted by a shortening muscle continuously increases as it shortens, the contraction is said to be **auxotonic.** Drawing back a bowstring is an example of this type of contraction. If the force of contraction decreases as the muscle shortens, the contraction is called **meiotonic.**

In the body, a **concentric** contraction is one in which shortening (not necessarily isotonic) takes place. In an **eccentric** contraction, a muscle is extended (while active) by an external force. Activities such as descending stairs or landing from a jump utilize this type of contraction. Such contractions are potentially dangerous because the muscle can experience forces that are larger than it could develop on its own, and tearing (strain) injuries can result. A **static** contraction results in no movement, but this may be due to partial activation (fewer motor units active) opposing a load that is not maximal. (This is different from a true isometric contraction, in which shortening is physically impossible regardless of the degree of activation.)

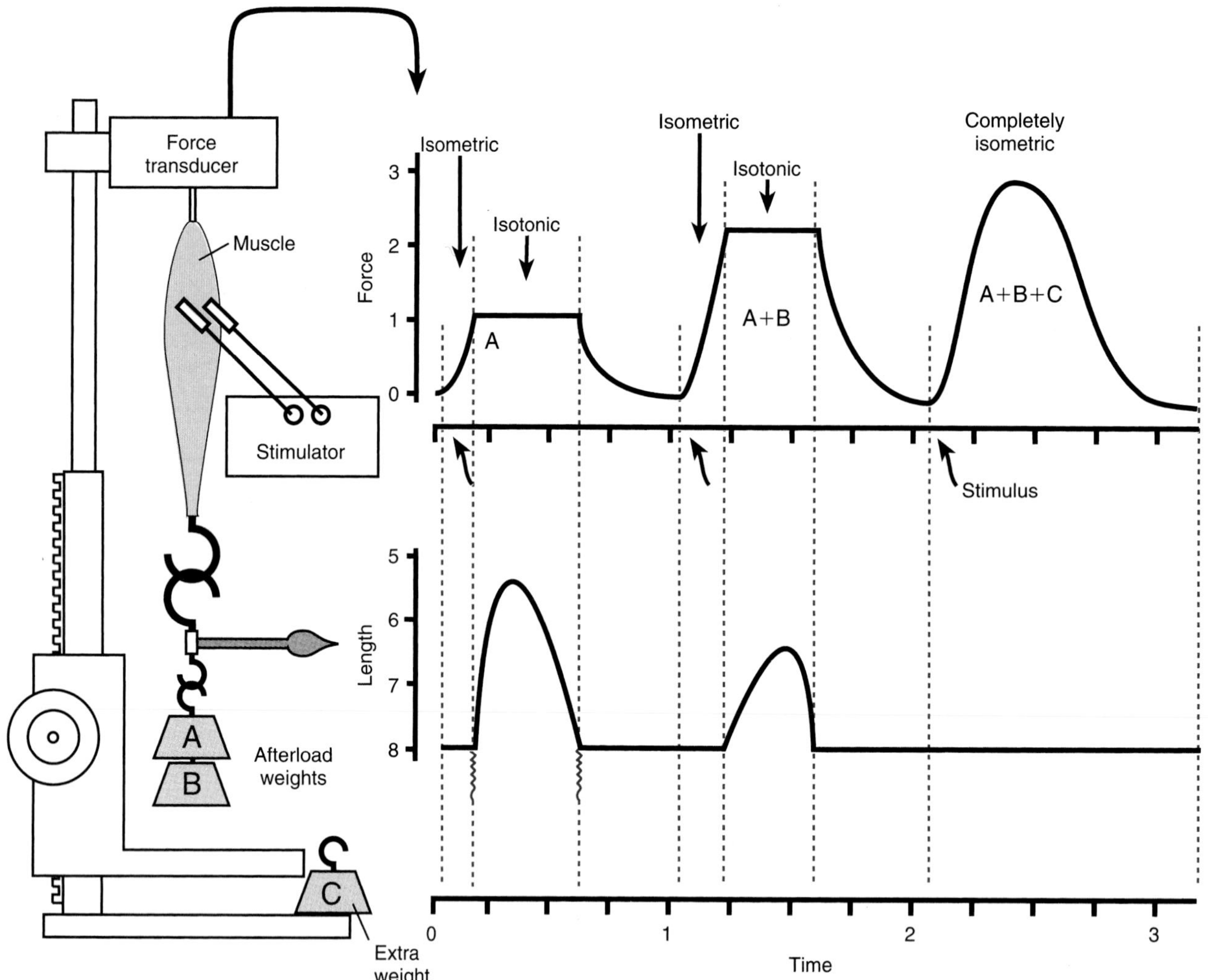

FIGURE 9.9 **A series of afterloaded isotonic contractions.** The curves labeled A and A + B correspond to the force and shortening records during the lifting of those weights. In each case, the adjustable platform prevents the muscle from being stretched by the attached weight, and all contractions start from the same muscle length. Note the lower force and greater shortening with the lower weight (A). If weight C (total weight = A + B + C) is added to the afterload, the muscle cannot lift it, and the entire contraction remains isometric. (Force, length, and time units are arbitrary.)

Special Mechanical Arrangements Allow a More Precise Analysis of Muscle Function

The types of contraction described above provide a basis for a better understanding of muscle function. The isometric and isotonic mechanical behavior of muscle can be described in terms of two important relationships:

- The length-tension curve, treating isometric contraction at different muscle lengths
- The force-velocity curve, concerned with muscle performance during isotonic contraction

Isometric Contraction and the Length-Tension Curve. Because it is made of contractile proteins and connective tissue, an isolated muscle can resist being stretched at rest. When it is very short, it is slack and will not resist passive extension. As it is made longer and longer, however, its resisting force increases more and more. Normally a muscle is protected against overextension by attachments to the skeleton or by other anatomic structures. If the muscle has not been stimulated, this resisting force is called **passive force** or **resting force**.

The relationship between force and length is much different in a stimulated muscle. The amount of active force or **active tension** a muscle can produce during an isometric contraction depends on the length at which the muscle is held. At a length roughly corresponding to the natural length in the body, the **resting length,** the maximum force is produced. If the muscle is set to a shorter length and then stimulated, it produces less force. At an extremely short length, it produces no force at all. If the muscle is made longer than its optimal length, it produces less force when stimulated. This behavior is summarized in the **length-tension curve** (Fig. 9.10).

In Figure 9.10, the left side of the top graph shows the force produced by a series of twitches made over the range

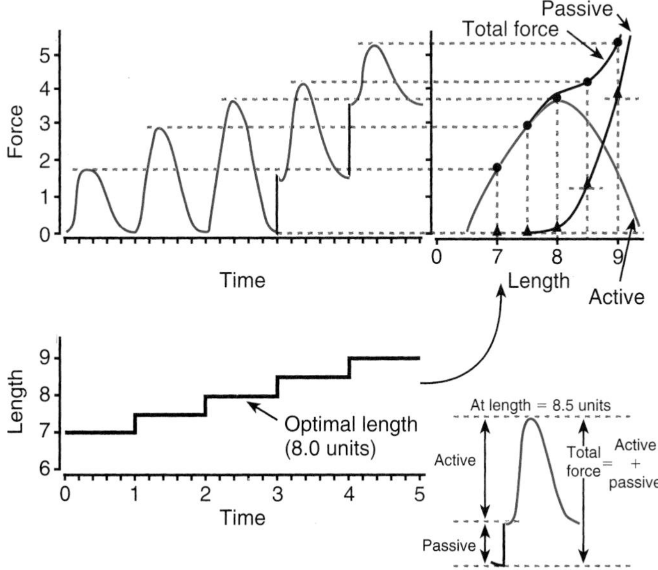

FIGURE 9.10 **A length-tension curve for skeletal muscle.** Contractions are made at several resting lengths, and the resting (passive) and peak (total) forces for each twitch are transferred to the graph at the right. Subtraction of the passive curve from the total curve yields the active force curve. These curves are further illustrated in the lower right corner of the figure. (Force, length, and time units are arbitrary.) (See text for details.)

of muscle lengths indicated at the left side of the bottom graph. Information from these traces is plotted at the right. The total peak force from each twitch is related to each length (dotted lines). The muscle length is changed only when the muscle is not stimulated, and it is held constant (isometric) during contraction. The difference between the total force and the passive force is called the active force (see inset; Fig. 9.10). The active force results directly from the active contraction of the muscle.

The length-tension curve shows that when the muscle is either longer or shorter than optimal length, it produces less force. Myofilament overlap is a primary factor in determining the active length-tension curve (see Chapter 8). However, studies have demonstrated that at very short lengths, the effectiveness of some steps in the excitation-contraction coupling process is reduced—binding of calcium to troponin is less and there is some loss of action potential conduction in the T tubule system.

The functional significance of the length-tension curve varies among the different muscle types. Many skeletal muscles are confined by their skeletal attachments to a relatively short region of the curve that is near the optimal length. In these cases, the lever action of the skeletal system, not the length-tension relationship, is of primary importance in determining the maximal force the muscle can exert. Cardiac muscle, however, normally works at lengths significantly less than optimal for force production, but its passive length-tension curve is shifted to shorter lengths (see Chapter 10). The length-tension relationship is, therefore, very important when considering the ability of cardiac muscle to adjust to changes in length (related to the volume of blood contained in the heart) to meet the body's changing needs. The role of the length-tension curve in smooth muscle is less clearly understood because of the great diversity among smooth muscles and their physiological roles. For all muscle types, however, the length-tension curve has provided important information about the cellular and molecular mechanisms of contraction.

Isotonic Contraction and the Force-Velocity Curve.

Everyday experience shows that the speed at which a muscle can shorten depends on the load that must be moved. Simply stated, light loads are lifted faster than heavy ones. Detailed analysis of this observation can provide insight into how the force and shortening of muscles are matched to the external tasks they perform, as well as how muscles function internally to liberate mechanical energy from their metabolic stores. The analysis is performed by arranging a muscle so that it can be presented with a series of afterloads (see Fig. 9.9; Fig. 9.11). When the muscle is maximally stimulated, lighter loads are lifted quickly and heavier loads more slowly. If the applied load is greater than the maximal force capability of the muscle, known as F_{max}, no shortening will result and the contraction will be isometric. If no load is applied, the muscle will shorten at its greatest possible speed, a velocity known as V_{max}.

The **initial velocity**—the speed with which the muscle begins to shorten—is measured at various loads. Initial velocity is measured because the muscle soon begins to slow down; as it gets shorter, it moves down its length-tension curve and is capable of less force and speed of shortening. When all the initial velocity measurements are related to each corresponding afterload lifted, an inverse relationship known as the **force-velocity curve** is obtained. The curve is steeper at low forces. When the measurements are made on a fully activated muscle, the force-velocity curve defines the upper limits of the muscle's isotonic capability. In practice, a completely unloaded contraction is very difficult to arrange, but mathematical extrapolation provides an accurate V_{max} value.

Figure 9.11 shows a force-velocity curve made from such a series of isotonic contractions. The initial velocity points (A–D) correspond to the contractions shown at the top. Factors that modify muscle performance, such as fatigue or incomplete stimulation (e.g., fewer motor units activated), result in operation *below* the limits defined by the force-velocity curve.

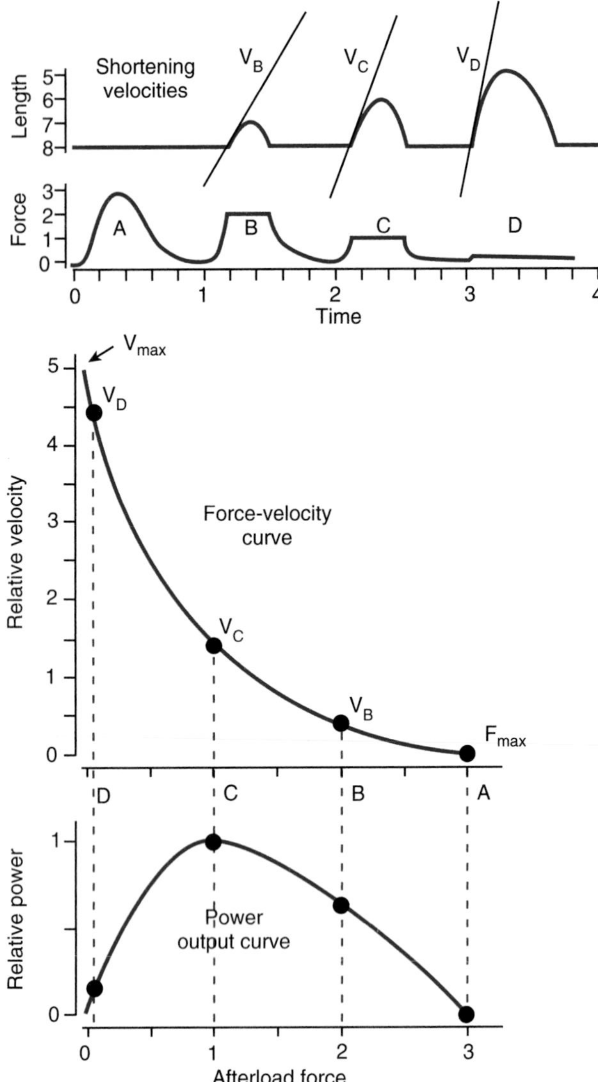

FIGURE 9.11 Force-velocity and power output curves for skeletal muscle. Contractions at four different afterloads (decreasing left to right) are shown in the top graphs. Note the differences in the amounts of shortening. The initial shortening velocity (slope) is measured (V_B, V_C, V_D) and the corresponding force and velocity points plotted on the axes in the bottom graph. Also shown is power output, the product of force and velocity. Note that it reaches a maximum at an afterload of about one-third of the maximal force. (Force, length, and time units are arbitrary.)

Consideration of the force-velocity relationship of muscle can provide insight into how it functions as a biological motor, its primary physiological role. For instance, V_{max} represents the maximal rate of crossbridge cycling; it is directly related to the biochemistry of the actin-myosin ATPase activity in a particular muscle type and can be used to compare the properties of different muscles.

Because isotonic contraction involves moving a force (the afterload) through a distance, the muscle does physical work. The rate at which it does this work is its **power output** (see Figure 9.11). The factors represented in the force-velocity curve are thus relevant to questions of muscle work and power. At the two extremes of the force-velocity curve (zero force, maximal velocity and maximal force, zero velocity), no work is done because, by definition, work requires moving a force through a distance. Between these two extremes, work and power output pass through a maximum at a point where the force is approximately one-third of its maximal value. The peak of the curve represents the combination of force and velocity at which the greatest power output is produced; at any afterload force greater or smaller than this, less power can be produced. It also appears in skeletal muscle that the optimal power output occurs under nearly the same conditions at which muscle **efficiency**, the amount of power produced for a given metabolic energy input, is greatest.

In terms of mechanical work, the chemical reactions of muscle are about 20% efficient; the energy from the remaining 80% of the fuel consumed (ATP) appears as heat. In some forms of locomotion, such as running, the measured efficiency is higher, approaching 40% in some cases. This apparent increase is probably due to the storage of mechanical energy (between strides) in elastic elements of the muscle and in the potential and kinetic energy of the moving body. This energy is then partly returned as work during the subsequent contraction. It has also been shown that stretching an active muscle (e.g., during running or descending stairs) can greatly reduce the breakdown of ATP, since the crossbridge cycle is disrupted when myofilaments are forced to slide in the lengthening direction.

These force-velocity and efficiency relationships are important when endurance is a significant concern. Athletes who are successful in long-term physical activity have learned to optimize their power output by "pacing" themselves and adjusting the velocity of contraction of their muscles to extend the duration of exercise. Such adjustments obviously involve compromises, as not all of the many muscles involved in a particular task can be used at optimal loading and rate and subjective factors, such as experience and training, enter into performance.

In rapid, short-term exercise, it is possible to work at an inefficient force-velocity combination to produce the most rapid or forceful movements possible. Such activity must necessarily be of more limited duration than that carried out under conditions of maximal efficiency. Examples of attempts at optimal matching of human muscles to varying loads can be found in the design of human-powered machinery, pedestrian ramps, and similar devices.

Interactions Between Isometric and Isotonic Contractions. The length-tension curve represents the effect of length on the isometric contraction of skeletal muscle. During isotonic shortening, however, muscle length does change while the force is constant. The limit of this shortening is also described by the length-tension curve. For example, a lightly loaded muscle will shorten farther than one starting from the same length and bearing a heavier load. If the muscle begins its shortening from a reduced length, its subsequent shortening will be reduced. These relationships are diagrammed in Figure 9.12. In the case of day-to-day skeletal muscle activity, these limits are not usually encountered because voluntary adjustments of the contracting muscle are usually made to accomplish a specific task. In the case of cardiac muscle, however, such interrelationships between force

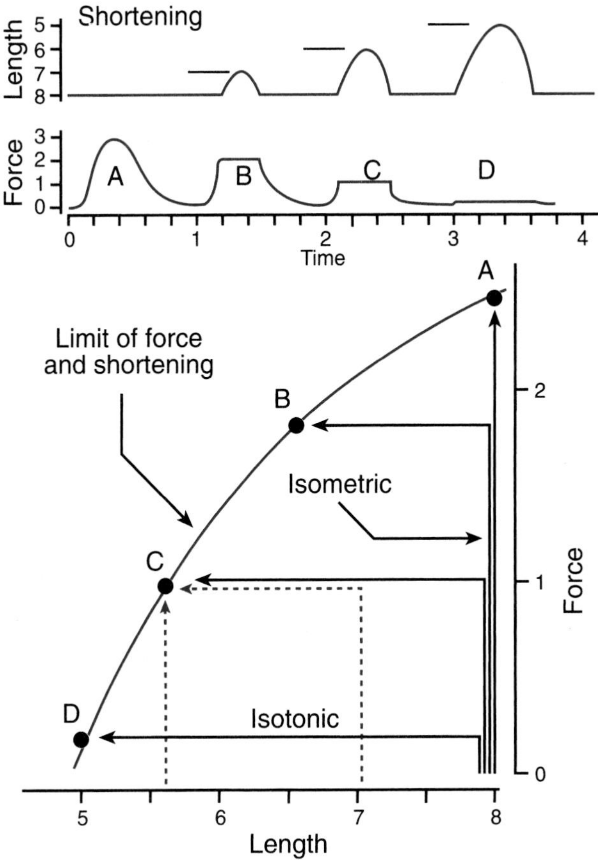

FIGURE 9.12 **The relationship between isotonic and isometric contractions.** The top graphs show the contractions from Figure 9.11, with different amounts of shortening. The bottom graph shows, for contractions B, C, and D, the initial portion is isometric (the line moves upward at constant length) until the afterload force is reached. The muscle then shortens at the afterload force (the line moves to the left) until its length reaches a limit determined (at least approximately) by the isometric length-tension curve. The dotted lines show that the same final force/length point can be reached by several different approaches. Relaxation data, not shown on the graph, would trace out the same pathways in reverse. (Force, length, and time units are arbitrary.)

and length are of critical importance in functional adjustment of the beating heart (see Chapter 10).

The Anatomic Arrangement of Muscle Is a Prime Determinant of Function

Anatomic location places restrictions on muscle function by limiting the amount of shortening or determining the kinds of loads encountered. Skeletal muscle is generally attached to bone, and bones are attached to each other. Because of the way the muscles are attached and the skeleton is articulated, the bones and muscles together constitute a **lever system**. This arrangement influences the physiology of the muscles and the functioning of the body as a whole. In most cases, the system works at a **mechanical disadvantage** with respect to the force exerted. The shortening capability of skeletal muscle by itself is rather limited, and the skeletal lever system multiplies the distance over which an extremity can be moved (Fig. 9.13). However, this means the muscle must exert a much greater force than the actual weight of the load being lifted (the muscle force is increased by the same ratio that the length change at the end of the extremity is increased). In the case of the human forearm, the biceps brachii, when moving a force applied to the hand, must exert a force at its insertion on the radius that is approximately 7 times as great. However, the resulting movement of the hand is approximately 7 times as far and 7 times as rapid as the shortening of the muscle itself. Muscles may be subject to large forces and this can lead to muscle injury (see Clinical Focus Box 9.2).

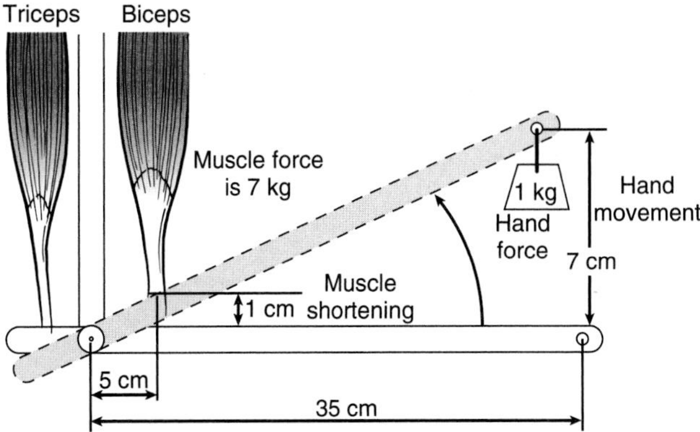

FIGURE 9.13 **Antagonistic pairs and the lever system of skeletal muscle.** Contraction of the biceps muscle lifts the lower arm (flexion) and elongates the triceps, while contraction of the triceps lowers the arm and hand (extension) and elongates the biceps. The bones of the lower arm are pivoted at the elbow joint (the fulcrum of the lever); the force of the biceps is applied through its tendon close to the fulcrum; the hand is 7 times as far away from the elbow joint. Thus, the hand will move 7 times as far (and fast) as the biceps shortens (lever ratio, 7:1), but the biceps will have to exert 7 times as much force as the hand is supporting.

Acting independently, a muscle can only shorten, and the force to relengthen it must be provided externally. These actions are achieved by the arrangement of muscles into **antagonistic pairs** of **flexors** and **extensors**. For example, the shortening of the biceps is countered by the action of the triceps; the triceps, in turn, is relengthened by contraction of the biceps. In some cases, gravity provides the restoring force.

Metabolic and Structural Adaptations Fit Skeletal Muscle for a Variety of Roles

Specific skeletal muscles are adapted for specialized functions. These adaptations involve primarily the structures and chemical reactions that supply the contractile system with energy. The enzymatic properties (i.e., the rate of ATP hydrolysis) of actomyosin ATPase also vary. The basic structural features of the sarcomeres and the thick/thin filament interactions are, however, essentially the same among the types of skeletal muscle.

Chapter 8 detailed the biochemical reactions responsible for providing ATP to the contractile system. Recall that

CLINICAL FOCUS BOX 9.2

Strain Injuries to Muscle

Skeletal muscle is subject to being damaged in several ways. In accidents that result in crushing or laceration, considerable muscle damage can occur. However, damage directly related to the contractile function of muscle is also possible. Such injuries are incidental to the muscle's primary function of exerting force and causing motion. In the areas of sports or physical labor, muscle strain is the most common type of injury.

The muscles most susceptible to injury are those of the limbs, especially those that go from joint to joint (e.g., the gastrocnemius or the rectus femoris) or that have a complex architecture (e.g., the adductor longus and, again, the rectus femoris). Often the injury will be confined to one muscle of a group used to perform a specific action. Injury can occur to a muscle that is overstretched while unstimulated, but most injuries occur during eccentric contraction, that is, during the forced extension of an activated muscle. Under such circumstances, the force in the muscle may rise to a level considerably higher than could be attained in an isometric contraction; relatively few injuries occur under isometric or isotonic (concentric) contraction conditions. The site of injury is most often at the myotendinous junction, a location that can be determined by physical examination and confirmed by magnetic resonance imaging (MRI) or by a computed tomography (CT) scan. There may also be extensive damage throughout the muscle itself. In some cases, there is complete disruption of the muscle (avulsion), although usually separation is not complete. Symptoms of a muscle strain injury include obvious soreness, weakness, delayed swelling, and "bunching up" in extreme cases.

Several predisposing factors may cause a muscle strain injury, including relative weakness of a given muscle, resulting from a lack of training early in a sports season, and fatigue, which leads to increased injury late in an athletic event. In general, factors that make a muscle less able to contract also predispose it to strain injury; laboratory experiments have shown that muscles in better physical condition are better able to safely absorb the energy that leads to injury. Retraining too rapidly or too soon after an injury or returning to activity too soon also make reinjury more likely.

Delayed-onset muscle soreness, as often experienced after unaccustomed exercise, also results from strain injury, but on a smaller scale. Muscle subjected to overload during eccentric contraction shows reduced contractile ability and ultrastructural damage to the contractile elements, especially at the Z lines. The pain peaks 1 to 2 days after exercise; as the healing progresses, the muscle becomes more able to withstand microinjury. Repeated bouts of exercise are tolerated increasingly well and are associated with the hypertrophy of the muscle; hence, the familiar phrase, "No pain, no gain."

Treatments for muscle strain injury are rather limited. They include the application of ice packs and enforced rest of the injured muscle. Nonsteroidal anti-inflammatory drugs (NSAIDs) can lessen the pain, but they also appear to delay healing somewhat. For injuries in which an actual separation of the muscle and tendon occurs, surgical repair is necessary. Massaging of an injured muscle does not appear to be as beneficial as light exercise, which may help to increase blood flow and promote healing. Recovery from strain injury is associated with the gradual regaining of strength, which will eventually reach near-normal levels if reinjury is avoided. Some muscle tissue is permanently replaced with scar tissue, which may change the geometry of the muscle. Most recovered muscles will have a somewhat increased susceptibility to injury for an extended period of time.

Precautions for avoiding strain injury include adequate physical conditioning and practiced expertise at the task at hand. Preexercise stretching and warm-up may be of some value in preventing strain injury, although the experimental evidence is equivocal.

muscle fibers contain both **glycolytic** (anaerobic) and **oxidative** (aerobic) metabolic pathways, which differ in their ability to produce ATP from metabolic fuels, particularly glucose and fatty acids. Among muscle fibers, the relative importance of each pathway and the presence or absence of associated supporting organelles and structures vary. These variations form the basis for the classification of skeletal muscle fiber types (Table 9.1). A typical skeletal muscle usually contains a mixture of fiber types, but in most muscles a particular type predominates. The major classification criteria are derived from mechanical measurements of muscle function and histochemical staining techniques in which dyes for specific enzymatic reactions are used to identify individual fibers in a muscle cross section.

Red Muscle Fibers and Aerobic Metabolism. The color differences of skeletal muscles arise from differences in the amount of **myoglobin** they contain. Similar to the related red blood cell protein hemoglobin, myoglobin can bind, store, and release oxygen. It is abundant in muscle fibers that depend heavily on aerobic metabolism for their ATP supply, where it facilitates oxygen diffusion (and serves as a minor auxiliary oxygen source) in times of heavy demand. Red muscle fibers are divided into **slow-twitch fibers** and **fast-twitch fibers** on the basis of their contraction speed (see Table 9.1). The differences in rates of contraction (shortening velocity or force development) arise from differences in actomyosin ATPase activity (i.e., in the basic crossbridge cycling rate). Mitochondria are abundant in these fibers because they contain the enzymes involved in aerobic metabolism.

White Muscle Fibers and Anaerobic Metabolism. White muscle fibers, which contain little myoglobin, are fast-twitch fibers that rely primarily on glycolytic metabolism. They contain significant amounts of stored **glycogen**, which can be broken down rapidly to provide a quick source of energy. Although they contract rapidly and powerfully, their endurance is limited by their ability to sustain an oxygen deficit (i.e., to tolerate the buildup of lactic acid). They require a period of recovery (and a supply of oxygen) after heavy use. White muscle fibers have fewer

TABLE 9.1 Classification of Skeletal Muscle Fiber Types

	Fast Twitch		Slow Twitch
Metabolic Type	Fast Glycolytic (White)	Fast Oxidative-Glycolytic (Red)	Slow Oxidative (Red)
Metabolic properties			
ATPase activity	High	High	Low
ATP source(s)	Anaerobic glycolysis	Anaerobic glycolysis/ Oxidative phosphorylation	Oxidative phosphorylation
Glycolytic enzyme content	High	Moderate	Low
Number of mitochondria	Low	High	High
Myoglobin content	Low	High	High
Glycogen content	High	Moderate	Low
Fatigue resistance	Low	Moderate	High
Mechanical properties			
Contraction speed	Fast	Fast	Slow
Force capability	High	Medium	Low
SR Ca^{2+}-ATPase activity	High	High	Moderate
Motor axon velocity	100 m/sec	100 m/sec	85 m/sec
Structural properties			
Fiber diameter	Large	Moderate	Small
Number of capillaries	Few	Many	Many
Functional role in body	Rapid and powerful movements	Medium endurance	Postural/endurance
Typical example	Latissimus dorsi	Mixed-fiber muscle, such as vastus lateralis	Soleus

mitochondria than red muscle fibers because the reactions of glycolysis take place in the myoplasm. There are indications that enzymes of the glycolytic pathway may be closely associated with the thin filament array.

Red and White Fibers and Muscle Function. The relative proportions of red and white muscle fibers fit muscles for different uses in the body. Muscles containing primarily slow-twitch oxidative red fibers are specialized for functions requiring slow movements and endurance, such as the maintenance of posture. Muscles containing a preponderance of fast-twitch red fibers support faster and more powerful contractions. They also typically contain varying numbers of fast-twitch white fibers; their resulting ability to use both aerobic and anaerobic metabolism increases their power and speed. Muscles containing primarily fast-twitch white fibers are suited for rapid, short, powerful contractions.

Fast muscles, both white and red, not only contract rapidly but also relax rapidly. Rapid relaxation requires a high rate of calcium pumping by the SR, which is abundant in these muscles. In such muscles, the energy used for calcium pumping can be as much as 30% of the total consumed. Fast muscles are supplied by large motor axons with high conduction velocities; this correlates with their ability to make quick and rapidly repeated contractions.

Muscle Fatigue. During a period of heavy exercise, especially when working above 70% of maximal aerobic capacity, skeletal muscle is subject to **fatigue.** The speed and force of contraction are diminished, relaxation time is prolonged, and a period of rest is required to restore normal function. While there is a close correlation between the oxidative capacity of a particular muscle fiber type and its fatigue resistance, chemical measurements of fatigued skeletal muscle specimens have shown that the ATP content, while reduced, is not completely exhausted. In well-motivated subjects, CNS factors do not appear to play an important role in fatigue, and transmission at the neuromuscular junction has such a large safety factor that impaired transmission also does not contribute to fatigue.

Studies on isolated muscle have distinguished two different mechanisms producing fatigue. Stimulation of the muscle at a rate far above that necessary for a fused tetanus quickly produces **high-frequency stimulation fatigue**; recovery from this condition is rapid (a few tens of seconds). In this type of fatigue, the principal defect seems to be a failure in T tubule action potential conduction, which leads to less Ca^{2+} release from the SR. Under most *in vivo* circumstances, feedback mechanisms in neural motor pathways work to reduce the stimulation to the minimum necessary for a smooth tetanus, and this type of fatigue is probably not often encountered.

Prolonged or repeated tetanic stimulation produces a longer-lasting fatigue with a longer recovery time. This type of fatigue—**low-frequency stimulation fatigue**—is related to the muscle's metabolic activities. The buildup of metabolites produced by crossbridge cycling, especially inorganic phosphate (P_i) and H^+ ions, reduces calcium sensitivity of the myofilaments and the contractile force generated per crossbridge. The reduced amount of metabolic energy available to the calcium transport system in the SR leads to reduced Ca^{2+} pumping. As a result, relaxation time increases and there is less Ca^{2+} available to activate the contraction with each stimulus, resulting in lowered peak force.

PROPERTIES OF SMOOTH MUSCLE

The properties of skeletal muscle described thus far apply in a general way to smooth muscle. Many of the basic muscle properties are highly modified in smooth muscle, however, because of the very different functional roles it plays in the body. The adaptations of smooth muscle structure and function are best understood in the context of the special requirements of the organs and systems of which smooth muscle is an integral component. Of particular importance are the high metabolic economy of smooth muscle, which allows it to remain contracted for long periods with little energy consumption, and the small size of its cells, which allows precise control of very small structures, such as blood vessels. Most smooth muscles are not discrete organs (like individual skeletal muscles) but are intimate components of larger organs. It is in the context of these specializations that the physiology of smooth muscle is best understood.

Structural Arrangements Equip Smooth Muscle for Its Special Roles

While there are major differences among the organs and systems in which smooth muscle plays a major part, the structure of smooth muscle is quite consistent at the tissue level and even more similar at the cellular level. Several typical arrangements of smooth muscle occur in a variety of locations.

The variety of smooth muscle tasks—regulating and promoting movement of fluids, expelling the contents of organs, moving visceral structures—is accomplished by a few basic types of tissue structures. All of these structures are subject, like skeletal muscle, to the requirement for antagonistic actions: If smooth muscle contracts, an external force must lengthen it again. The structures described below provide these restoring forces in a variety of ways.

Circular Organization: Blood Vessels. The simplest smooth muscle arrangement is found in the arteries and veins of the circulatory system. Smooth muscle cells are oriented in the circumference of a vessel so that shortening of the fibers results in reducing the vessel's diameter. This reduction may range from a slight narrowing to a complete obstruction of the vessel lumen, depending on the physiological needs of the body or organ. The orientation of the cells in the vessel walls is helical, with a very shallow pitch. In the larger muscular vessels, particularly arteries, there may be many layers of cells and the force of contraction may be quite high; in small arterioles, the muscle layer may consist of single cells wrapped around the vessel. The blood pressure provides the force to relengthen the cells in the vessel walls. This type of muscle organization is extremely important because the narrowing of a blood vessel has a powerful influence on the rate of blood flow through it (see Chapters 12 and 15). This circular arrangement is also prominent in the airways of the lungs, where it regulates the flow of air.

A further specialization of the circular muscle arrangement is a **sphincter,** a thickening of the muscular portion of the wall of a hollow or tubular organ, whose contraction has the effect of restricting flow or stopping it completely. Many sphincters, such as those in the gastrointestinal and urogenital tracts, have a special nerve supply and participate in complex reflex behavior. The muscle in sphincters is characterized by the ability to remain contracted for long periods with little metabolic cost.

Circular and Longitudinal Layers: The Small Intestine. Next, in order of complexity, is the combination of circular and longitudinal layers, as in the muscle of the small intestine. The outermost muscle layer, which is relatively thin, runs along the length of the intestine. The inner muscle layer, thicker and more powerful, has a circular arrangement. Coordinated alternating contractions and relaxations of these two layers propel the contents of the intestine, although most of the motive power is provided by circular muscle (see Chapter 26).

Complex Fiber Arrangements. The most complex arrangement of smooth muscle is found in organs such as the urinary bladder and uterus. Numerous layers and orientations of muscle fibers are present and the effect of their contraction is an overall reduction of the volume of the organ. Even with such a complex arrangement of fibers, coordinated and organized contractions take place. The relengthening force, in the case of these hollow organs, is provided by the gradual accumulation of contents. In the urinary bladder, for example, the muscle is gradually stretched as the emptied organ fills again.

In a few instances, smooth muscles are structurally similar to skeletal muscles in their arrangement. Some of the structures supporting the uterus, for example, are called ligaments; however, they contain large amounts of smooth muscle and are capable of considerable shortening. **Pilomotor muscles**, the small cutaneous muscles that erect the hairs, are also discrete structures whose shortening is basically unidirectional. Certain areas of mesentery also contain regions of linearly oriented smooth muscle fibers.

Small Cell Size Facilitates Precise Control

The most notable feature of smooth muscle tissue organization, in contrast to that of skeletal muscle, is the small size of the cells compared to the tissue they make up. Individual smooth muscle cells (depending somewhat on the type of tissue they compose) are 100 to 300 μm long and 5 to 10 μm in diameter. When isolated from the tissue, the cells are roughly cylindrical along most of their length and taper at the ends. The single nucleus is elongated and centrally located. Electron microscopy reveals that the cell margins contain many areas of small membrane invaginations, called **caveolae**, which may play a role in increasing the surface area of the cell (Fig. 9.14). Mitochondria are located at the ends of the nucleus and near the surface membrane. In some smooth muscle cells, the SR is abundant, although not to the extent found in skeletal muscle. In some cases, it closely approaches the cell membrane, but there is no organized T tubular system as in other types of muscle.

The bulk of the cell interior is occupied by three types of myofilaments: thick, thin, and intermediate. The thin filaments are similar to those of skeletal muscle but lack the

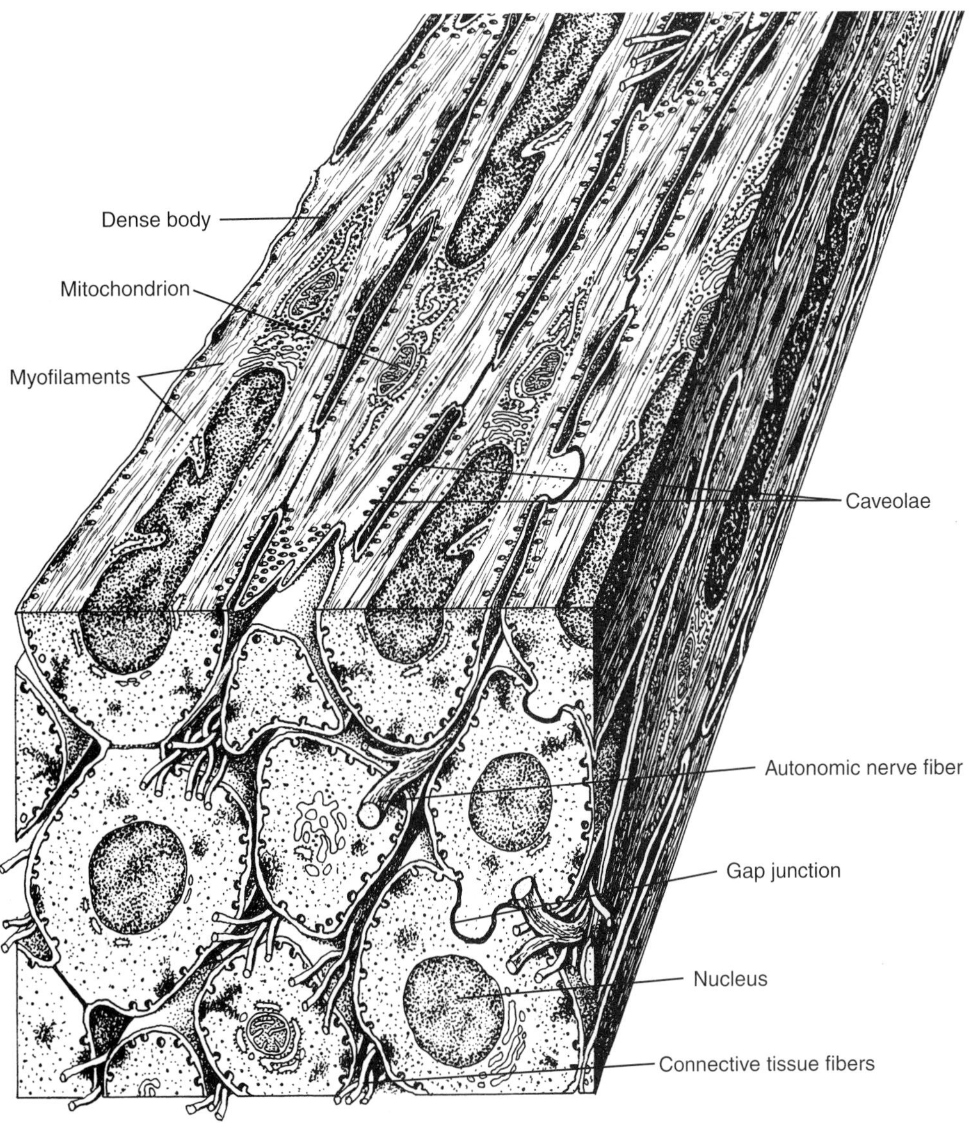

FIGURE 9.14 **A drawing from electron micrographs of smooth muscle, showing cells in cross section and longitudinal section.** (Adapted from Krstic RV. General Histology of the Mammal. New York: Springer-Verlag, 1984.)

troponin protein complex. The length of the individual filaments is not known with certainty because of their irregular organization. The thick filaments are composed of myosin molecules, as in skeletal muscle, but the details of the exact arrangement of the individual molecules into filaments are not completely understood. The thick filaments appear to be approximately 2.2 μm long, somewhat longer than in skeletal muscle (1.6 μm). The **intermediate filaments** are so named because their diameter of 10 nm is between that of the thick and thin filaments. Intermediate filaments appear to have a cytoskeletal, rather than a contractile, function. Prominent throughout the cytoplasm are small, dark-staining areas called **dense bodies.** They are associated with the thin and intermediate filaments and are considered analogous to the Z lines of skeletal muscle. Dense bodies associated with the cell margins are often called **membrane-associated dense bodies** (or **patches**) or **focal adhesions.** They appear to serve as anchors for thin filaments and to transmit the force of contraction to adjacent cells.

Smooth muscle lacks the regular sarcomere structure of skeletal muscle. Studies have shown some association among dense bodies down the length of a cell and a tendency of thick filaments to show a degree of lateral grouping. However, it appears that the lack of a strongly periodic arrangement of the contractile apparatus is an adaptation of smooth muscle associated with its ability to function over a wide range of lengths and to develop high forces despite a smaller cellular myosin content.

Mechanical Coupling. Because smooth muscle cells are so small compared to the whole tissue, some mechanical and electrical communication among them is necessary. Individual cells are coupled mechanically in several ways. A proposed arrangement of the smooth muscle contractile and force transmission system is shown in Figure 9.15. This

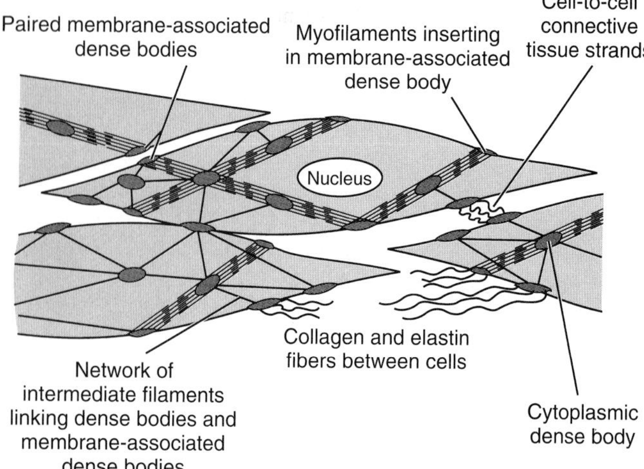

FIGURE 9.15 **The contractile system and cell-to-cell connections in smooth muscle.** Note regions of association between thick and thin filaments that are anchored by the cytoplasmic and membrane-associated dense bodies. A network of intermediate filaments provides some spatial organization (see, especially, the left side). Several types of cell-to-cell mechanical connections are shown, including direct connections and connections to the extracellular connective tissue matrix. Structures are not necessarily drawn to scale. (See text for details.)

picture represents a consensus from many researchers and areas of investigation. Note that assemblies of myofilaments are anchored within the cell by the dense bodies and at the cell margins by the membrane-associated dense bodies. The contractile apparatus lies oblique to the long axis of the cell. When single isolated smooth muscle cells contract, they undergo a "corkscrew" motion that is thought to reflect the off-axis orientation of the contractile filaments. In intact tissues, the connections to adjacent cells prevent this rotation.

Force appears to be transmitted from cell to cell and throughout the tissue in several ways. Many of the membrane-associated dense bodies are opposite one another in adjacent cells and may provide continuity of force transmission between the contractile apparatus in each cell. There are also areas of cell-to-cell contact, both lateral and end to end, where myofilament insertions are not apparent but where a direct transmission of force could occur. In some places, short strands of connective tissue link adjacent cells; in other places, cells are joined to the collagen and elastin fibers running throughout the tissue. These fibers, along with reticular connective tissue, comprise the **connective tissue matrix** or **stroma** found in all smooth muscle tissues. It serves to connect the cells and to give integrity to the whole tissue. In tissues that can resist considerable external force, this connective tissue matrix is well developed and may be organized into **septa**, which transmit the force of many cells.

Electrical Coupling. Smooth muscle cells are also coupled electrically. The structure most effective in this coupling is the **gap junction** (see Chapter 1). Gap junctions in smooth muscle appear to be somewhat transient structures that can form and disappear over time. In some tissues, this phenomenon is under hormonal control; in the uterus, for example, gap junctions are rare during most of pregnancy, and the contractions of the muscle are weak and lack coordination. However, just prior to the onset of labor, the number and size of gap junctions increase dramatically and the contractions become strong and well coordinated. Shortly after the cessation of labor, these gap junctions disappear and tissue function again becomes less coordinated.

Electrical coupling among smooth muscle cells is the basis for classifying smooth muscle into two major types:

- **Multiunit smooth muscle**, which has little cell-to-cell communication and depends directly on nerve stimulation for activation (like skeletal muscle). An example is the iris of the eye.
- **Unitary** or **single-unit smooth muscle**, which has a high degree of coupling among cells, so that large regions of tissue act as if they were a single cell. Its cells form a **functional syncytium** (an arrangement in which many cells behave as one). This type of smooth muscle makes up the bulk of the muscle in the visceral organs.

The Regulation and Control of Smooth Muscle Involve Many Factors

Smooth muscle is subject to a much more complex system of controls than skeletal muscle. In addition to contraction in response to nerve stimulation, smooth muscle responds to hormonal and pharmacological stimuli, the presence or lack of metabolites, cold, pressure, and stretch, or touch, and it may be spontaneously active as well. This multiplicity of controlling factors is vital for the integration of smooth muscle into overall body function. Skeletal muscle is primarily controlled by the CNS and by a relatively straightforward cellular control mechanism. The control of smooth muscle is much more closely related to the many factors that regulate the internal environment. It is not surprising, therefore, that many internal and external pathways have as their final effect the control of the interaction of smooth muscle contractile proteins.

Innervation of Smooth Muscle. Most smooth muscles have a nerve supply, usually from both divisions of the autonomic nervous system. There is much diversity in this area; the muscle response to a given neurotransmitter substance depends on the type of tissue and its physiological state. Smooth muscle does not contain the highly structured neuromuscular junctions found in skeletal muscle. Autonomic nerve axons run throughout the tissue; along the length of the axons are many swellings or **varicosities**, which are the sites of release of transmitter substances in response to nerve action potentials. Released molecules of excitatory or inhibitory transmitter diffuse from the nerve to the nearby smooth muscle cells, where they take effect. Since the cells are so small and numerous, relatively few are directly reached by the transmitters; those that are not reached are stimulated by cell-to-cell communication, as described above. Neuromuscular transmission in smooth muscle is a relatively slow process, and in many tissues, nerve stimulation serves mainly to modify (increase or decrease) spontaneous rhythmic mechanical activity.

Activation of Smooth Muscle Contraction. Chemical factors that control the function of smooth muscle cells most often have their first influence at the cell membrane. Some factors act by opening or closing cell membrane ion channels. Others result in production of a second messenger that diffuses to the interior of the cell, where it causes further changes (see Chapter 1). The final result of both mechanisms is usually a change in the intracellular concentration of Ca^{2+}, which, in turn, controls the contractile process itself.

The membrane potential of smooth muscle is subject to many external and internal influences, in contrast to the case in skeletal and cardiac muscle. In smooth muscle, the linkage between the electrical activity of the cell membrane and cellular functions, particularly contraction, is much more subtle and complex than in the other types of muscle.

The resting potential of most smooth muscles is approximately −50 mV. This is less negative than the resting potential of nerve and other muscle types, but here too it is determined primarily by the transmembrane potassium ion gradient. The smaller potential is due primarily to a greater resting permeability to sodium ions. In many smooth muscles, the resting potential varies periodically with time, producing a rhythmic potential change called a **slow wave** (see Chapter 26). Action potentials in smooth muscle also have a variety of forms. In many smooth muscles the action potential is a transient depolarization event lasting approximately 50 msec. At times, such action potentials will occur in rapid groups and produce repetitive membrane depolarizations that last for some time. Relatively rapid twitch-like contractions are usually the result of one or more action potentials. Sustained, low-level, partial contraction is often only loosely related to the electrical activity of the membrane.

The ionic basis of smooth muscle action potentials is complex because of the great variety of tissues, physiological conditions, and types of membrane channels. As a resting membrane potential of −50 mV results in the inactivation of typical fast sodium channels, sodium is usually not the major carrier of inward current during the action potential. In most cases, it has been shown that the rising (depolarizing) phase of a smooth muscle action potential is dominated by calcium, which enters through voltage-gated membrane channels. Repolarization current is carried by potassium ions, which leave through several types of channels, some voltage-controlled and others sensitive to the internal calcium concentration. These general ionic properties are typical of most smooth muscle types, although specific tissues may have variations within this general framework. The most important common feature is the entry of calcium ions during the action potential, since this inward flux is an important source of the calcium that controls the contractile process.

In addition to voltage-gated calcium channels, smooth muscle also contains receptor-activated calcium channels that are opened by the binding of hormones or neurotransmitters. One such ligand-gated channel in arterial smooth muscle is controlled by ATP, which acts as a transmitter substance in some types of smooth muscle tissues.

Smooth muscle can also be activated via the generation of second messengers, such as inositol 1,4,5-trisphosphate (IP_3) (see Chapter 1). This form of control involves chemical and hormonal activators and does not depend on membrane depolarization. The IP_3 causes the release of calcium from the SR, which initiates contraction.

The Role of Calcium in Smooth Muscle Contraction. All of the processes described above are ultimately concerned with the control of muscle contraction via the pool of intracellular calcium. Figure 9.16 summarizes these mechanisms in an overall picture of calcium regulation in smooth

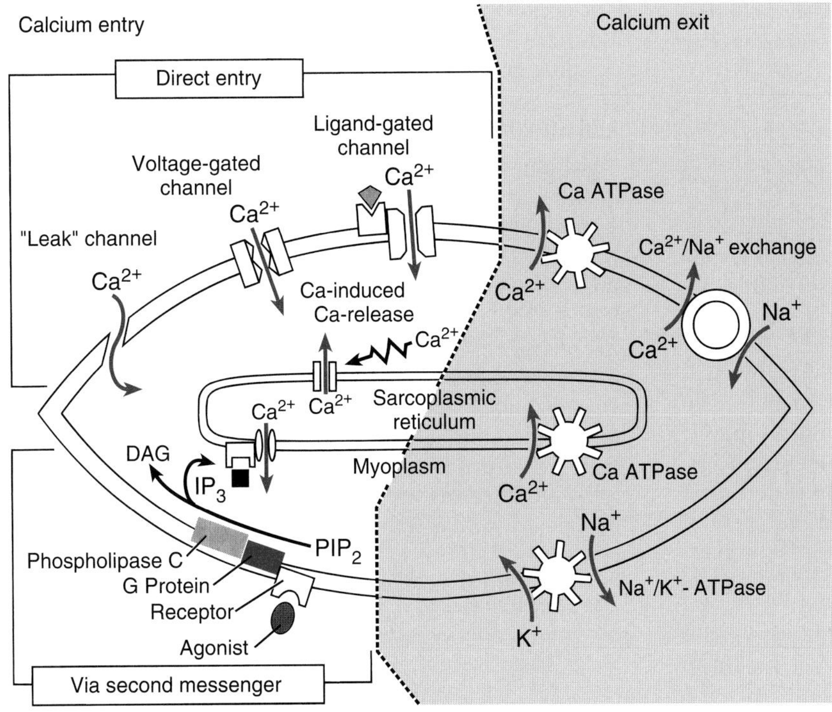

FIGURE 9.16 **Major routes of calcium entry and exit from the cytoplasm of smooth muscle.** The ATPase reactions are energy-consuming ion pumps. The processes on the left side increase cytoplasmic calcium and promote contraction; those on the right decrease internal calcium and cause relaxation. PIP_2, phosphatidylinositol 4,5-bisphosphate; IP_3, inositol 1,4,5-trisphosphate; DAG, diacylglycerol.

muscle. These processes may be grouped into those concerned with **calcium entry**, intracellular **calcium liberation**, and **calcium exit** from the cell. Calcium enters the cell through several pathways, including voltage-gated and ligand-gated channels and a relatively small number of unregulated "**leak**" **channels** that permit the continual passive entry of small amounts of extracellular calcium. Within the cell, the major storage site of calcium is the SR; in some types of smooth muscle, its capacity is quite small and these tissues are strongly dependent on extracellular calcium for their function. Calcium is released from the SR by at least two mechanisms, including IP_3-induced release and via **calcium-induced calcium release.** In this latter mechanism, calcium that has entered the cell via a membrane channel causes additional calcium release from the SR, amplifying its activating effect.

Studies in which internal calcium is continuously measured while the muscle is stimulated to contract typically reveal a high level of internal calcium early in the contraction; this activating burst most likely originates from internal SR storage. The level then decreases somewhat, although during the entire contraction it is maintained at a significantly elevated level. This sustained calcium level is the result of a balance between mechanisms allowing calcium entry and those favoring its removal from the cytoplasm. Calcium leaves the myoplasm in two directions: A portion of it is returned to storage in the SR by an active transport system (a Ca^{2+}-ATPase); and the rest is ejected from the cell by two principal means. The most important of these is another ATP-dependent active transport system located in the cell membrane. The second mechanism, also located in the plasma membrane, is sodium-calcium exchange, a process in which the entry of three sodium ions is coupled to the extrusion of one calcium ion. This mechanism derives its energy from the large sodium gradient across the plasma membrane; thus, it depends critically on the operation of the cell membrane Na^+/K^+-ATPase. (The sodium-calcium exchange mechanism, relatively unimportant in smooth muscle, is of much greater consequence in cardiac muscle; see Chapter 10.)

Biochemical Control of Contraction and Relaxation. The contractile proteins of smooth muscle, like those of skeletal and cardiac muscle, are controlled by changes in the intracellular concentration of calcium ions. Likewise, the general features of the actin-myosin contraction system are similar in all muscle types. It is in the control of the contractile proteins themselves that important differences exist. Because the control of contraction in skeletal and cardiac muscle is associated with thin filament proteins, it is called **actin-linked regulation.** The thin filaments of smooth muscle lack troponin; control of smooth muscle contraction relies instead on the thick filaments and is, therefore, called **myosin-linked regulation**. In actin-linked regulation, the contractile system is in a constant state of **inhibited readiness** and calcium ions remove the inhibition. In the myosin-linked regulation of smooth muscle, the role of calcium is to cause **activation** of a resting state of the contractile system. The general outlines of this process are well understood and appear to apply to all types of smooth muscle, although a variety of secondary regulatory mechanisms are being found in different tissue types. This general scheme is shown in Figure 9.17.

When smooth muscle is at rest, there is little cyclic interaction between the myosin and actin filaments because of a special feature of its myosin molecules. As in skeletal muscle, the S2 portion of each myosin molecule (the paired "head" portion) contains four protein **light chains.** Two of these have a molecular weight of 16,000 and are called **essential light chains**; their presence is necessary for actin-myosin interaction, but they do not appear to participate in the regulatory process. The other two light chains have a molecular weight of 20,000 and are called **regulatory light chains**; their role in smooth muscle is critical. These chains contain specific locations (amino acid residues) to which the terminal phosphate group of an ATP molecule can be attached via the process of **phosphorylation**; the enzyme responsible for promoting this reaction is **myosin light-chain kinase** (MLCK). When the regulatory light chains are phosphorylated, the myosin heads can interact in a cyclic fashion with actin, and the reactions of the **crossbridge cycle** (and its mechanical events) take place much as in skeletal muscle. It is important to note that the ATP molecule that phosphorylates a myosin light chain is separate and distinct from the one consumed as an energy source by the mechanochemical reactions of the crossbridge cycle.

For myosin phosphorylation to occur, the MLCK must be activated, and this step is also subject to control. Closely associated with the MLCK is **calmodulin** (CaM), a smaller protein that binds calcium ions. When four calcium ions are bound, the CaM protein activates its associated MLCK and light-chain phosphorylation can proceed. It is this MLCK-activating step that is sensitive to the cytoplasmic calcium concentration; at levels below 10^{-7} M Ca^{2+}, no calcium is bound to calmodulin and no contraction can take place. When cytoplasmic calcium concentration is greater than 10^{-4} M, the binding sites on calmodulin are fully occupied, light-chain phosphorylation proceeds at maximal rate, and contraction occurs. Between these extreme limits, variations in the internal calcium concentration can cause corresponding gradations in the contractile force. Such modulation of smooth muscle contraction is essential for its regulatory functions, especially in the vascular system.

Smooth Muscle Relaxation. The biochemical processes controlling relaxation in smooth muscle also differ from those in skeletal and cardiac muscle, in which a state of inhibition returns as calcium ions are withdrawn from being bound to troponin. In smooth muscle, the phosphorylation of myosin is reversed by the enzyme **myosin light-chain phosphatase** (MLCP). The activity of this phosphatase appears to be only partially regulated; that is, there is always some enzymatic activity, even while the muscle is contracting. During contraction, however, MLCK-catalyzed phosphorylation proceeds at a significantly higher rate, and phosphorylated myosin predominates. When the cytoplasmic calcium concentration falls, MLCK activity is reduced because the calcium dissociates from the calmodulin, and myosin dephosphorylation (catalyzed by the phosphatase) predominates. Because dephosphorylated myosin has a low affinity for actin, the reactions of the crossbridge cycle can no longer take place. Relaxation is, thus, brought about by mechanisms that lower cytoplasmic calcium concentrations

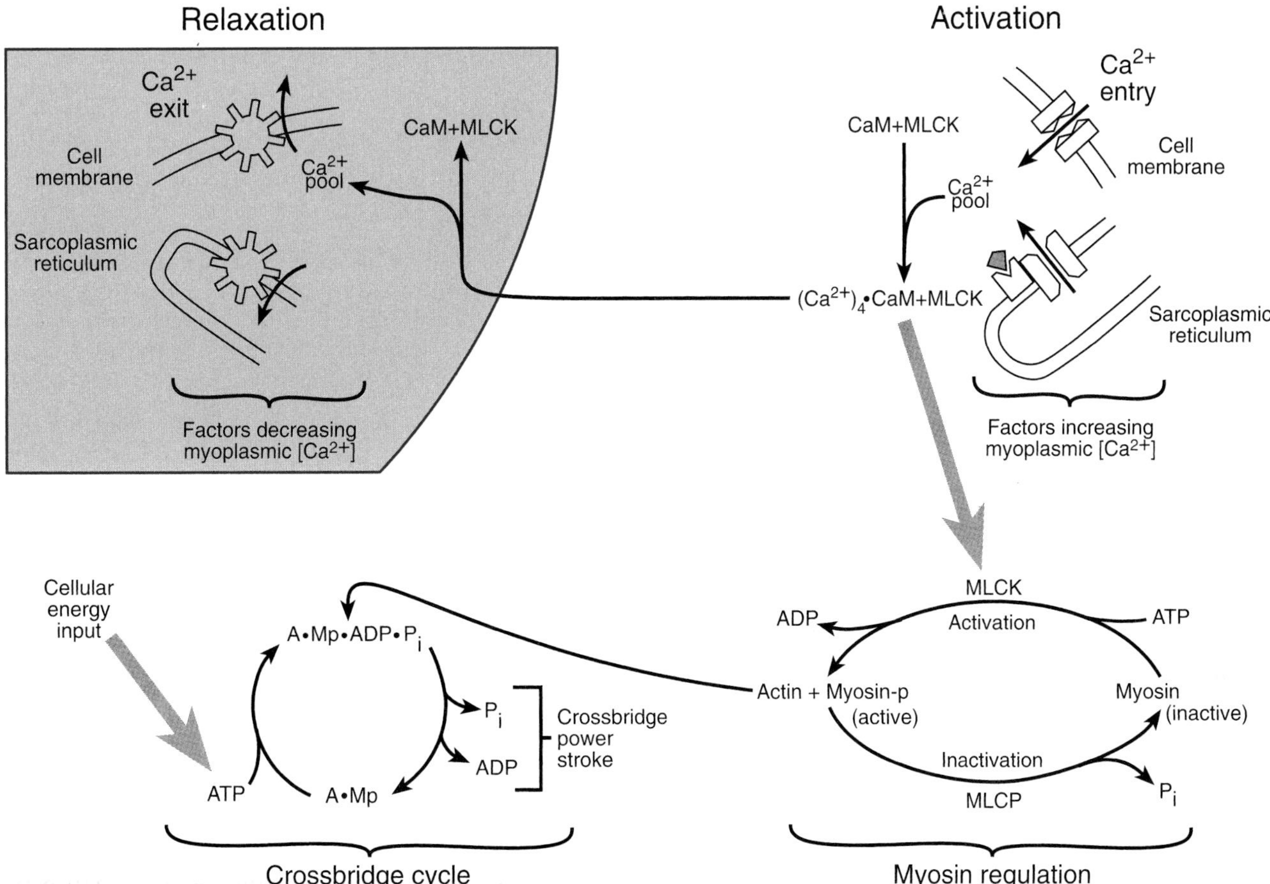

FIGURE 9.17 Reaction pathways involved in the basic regulation of smooth muscle contraction and relaxation. Activation begins (upper right) when cytoplasmic calcium levels are increased and calcium binds to calmodulin (CaM), activating the myosin light-chain kinase (MLCK). The kinase (lower right) catalyzes the phosphorylation of myosin, changing it to an active form (myosin-P or Mp). The phosphorylated myosin can then participate in a mechanical crossbridge cycle (lower left) much like that in skeletal muscle, although much slower. When calcium levels are reduced (upper left), calcium leaves calmodulin, the kinase is inactivated, and the myosin light-chain phosphatase (MLCP) dephosphorylates the myosin, making it inactive. The crossbridge cycle stops, and the muscle relaxes.

or decrease MLCK activity. Because of the importance of smooth muscle relaxation in physiological processes, this subject will be treated fully later in the chapter.

Secondary Mechanisms. In addition to myosin phosphorylation to control smooth muscle activation, secondary regulatory mechanisms are present in some types of smooth muscle. One of these provides long-term regulation of contraction in some tissues after the initial calcium-dependent myosin phosphorylation has activated the contractile system. For example, in vascular smooth muscle, the force of contraction may be maintained for long periods. This extended maintenance of force capability, called the **latch state,** appears to be related to a reduction in the cycling rate of crossbridges (possibly related to reduced phosphorylation) so that each remains attached for a longer portion of its total cycle. Even during the latch state, increased cytoplasmic calcium appears to be necessary for force to be maintained. Not all smooth muscle tissue can enter a latch state, however, and the details of the process are not completely understood.

Another possible secondary mechanism in some smooth muscle tissues involves the protein **caldesmon.** This molecule, also sensitive to the concentration of cytoplasmic calcium, is capable of binding to myosin at one of its ends and to actin and calmodulin at the other. While the process is not well understood, it is possible that caldesmon, under the control of calcium, could form crosslinks between actin and myosin filaments and, thus, aid in bearing force during a long-maintained contraction.

Other secondary regulatory mechanisms have been proposed. It is likely that several such mechanisms exist in various tissues, but the calcium-dependent phosphorylation of myosin light chains is the primary event in the activation of smooth muscle contraction.

Mechanical Activity in Smooth Muscle Is Adapted for Its Specialized Physiological Roles

The contraction of smooth muscle is much slower than that of skeletal or cardiac muscle; it can maintain contraction far longer and relaxes much more slowly. The source of these differences lies largely in the chemistry of the interaction between actin and myosin of smooth muscle. Recall that the crossbridges of muscle form an actin-myosin enzyme system (actomyosin ATPase) that releases energy from ATP so that it may be converted into a mechanical contraction (i.e., tension or shortening). The inherent rate of this ATPase correlates strongly with the velocity of shortening of the intact muscle. Most smooth muscles require several seconds (or even minutes) to develop maximal isometric force. A smooth muscle that contracts 100 times more slowly than a skeletal muscle will have an actomyosin ATPase that is 100 times as slow. The major source of this difference in rates is the myosin molecules; the actin found in smooth and skeletal muscles is rather similar. There is a close association in smooth muscle between maximal shortening velocity and degree of myosin light-chain phosphorylation.

A high economy of tension maintenance, typically 300 to 500 times greater than that in skeletal muscle, is vital to the physiological function of smooth muscle. Economy, as used here, means the amount of metabolic energy input compared to the tension produced. In smooth muscle, there is a direct relationship between isometric tension and the consumption of ATP. The economy is related to the basic cycling rate of the crossbridges: Early in a contraction (while tension is being developed and the crossbridges are cycling more rapidly), energy consumption is about 4 times as high as in the later steady-state phase of the contraction. Compared with skeletal muscle, the crossbridge cycle in smooth muscle is hundreds of times slower, and much more time is spent with the crossbridges in the attached phase of the cycle.

The cycling crossbridges are not the only energy-utilizing system in smooth muscle. Because the cells are so small and numerous, smooth muscle tissue contains a large cell membrane area. Maintenance of the proper ionic concentrations inside the cells requires the activity of the membrane-based ion pumps for sodium/potassium and calcium, and this ion pumping requires a significant portion of the cell's energy supply. Internal pumping of calcium ions into the SR during relaxation also requires energy, and the processes that result in phosphorylation of the myosin light chains consume a further portion of the cellular energy, as do the other processes of cellular maintenance and repair. Smooth muscle contains both glycolytic and oxidative metabolic pathways, with the oxidative pathway usually the most important; under some conditions, a transition may temporarily be made from oxidative to glycolytic metabolism. In terms of the entire body economy, the energy requirements of smooth muscle are small compared with those of skeletal muscle, but the critical regulatory functions of smooth muscle require that its energy supply not be interrupted.

Modes of Contraction. Smooth muscle contractile activity cannot be divided clearly into twitch and tetanus, as in skeletal muscle. In some cases, smooth muscle makes rapid **phasic contractions**, followed by complete relaxation. In other cases, smooth muscle can maintain a low level of active tension for long periods without cyclic contraction and relaxation; a long-maintained contraction is called **tonus** (rather than tetanus) or a **tonic contraction**. This is typical of smooth muscle activated by hormonal, pharmacological, or metabolic factors, whereas phasic activity is more closely associated with stimulation by neural activity.

Comparison With Skeletal Muscle. The force-velocity curve for smooth muscle reflects the differences in crossbridge functions described previously. Although smooth muscle contains one-third to one-fifth as much myosin as skeletal muscle, the longer smooth muscle myofilaments and the slower crossbridge cycling rate allow it to produce as much force per unit of cross-sectional area as does skeletal muscle. Thus, the maximum values for smooth muscle on the force axis would be similar, while the maximum (and intermediate) velocity values are very different (Fig. 9.18). Furthermore, smooth muscle can have a set of

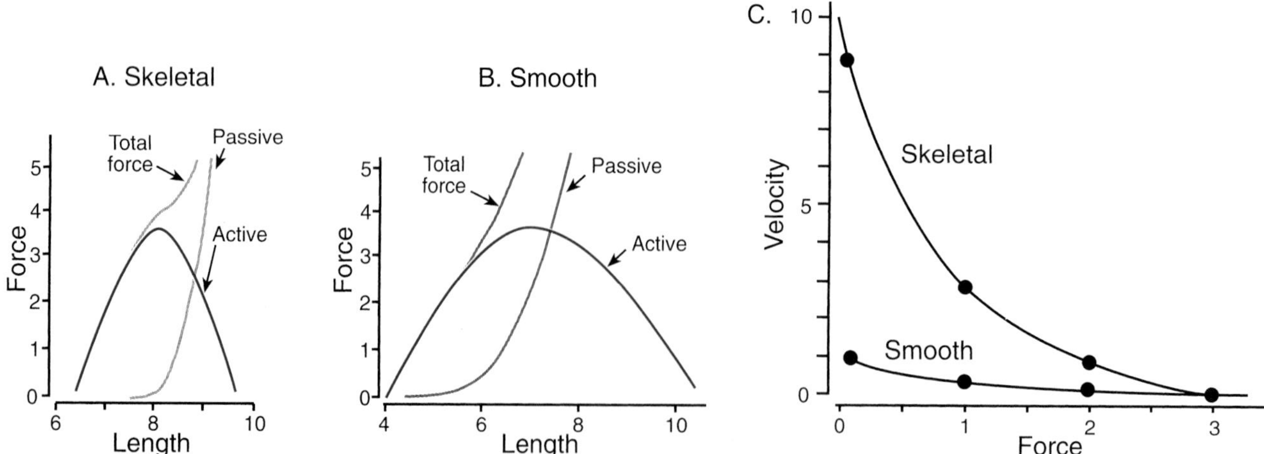

FIGURE 9.18 Smooth and skeletal muscle mechanical characteristics compared. A and B, Typical length-tension curves from skeletal and smooth muscle. Note the greater range of operating lengths for smooth muscle and the leftward shift of the passive (resting) tension curve. C, Skeletal and smooth muscle force-velocity curves. While the peak forces may be similar, the maximum shortening velocity of smooth muscle is typically 100 times lower than that of skeletal muscle. (Force and length units are arbitrary.)

force-velocity curves, each corresponding to a different level of myosin light-chain phosphorylation.

Other mechanical properties of smooth muscle are also related to its physiological roles. While its underlying cellular basis is uncertain, smooth muscle has a length-tension curve somewhat similar to that of skeletal muscle, although there are some significant differences (Fig. 9.18). At lengths at which the maximal isometric force is developed, many smooth muscles bear a substantial passive force. This is mostly a result of the network of connective tissue that supports the smooth muscle cells and resists overextension; in some cases, it may be partly a result of residual interaction between actin and attached but noncycling myosin cross-bridges. Compared to skeletal and cardiac muscle, smooth muscle can function over a significantly greater range of lengths. It is not constrained by skeletal attachments, and it makes up several organs that vary greatly in volume during the course of their normal functioning. The shape of the length-tension curve can also vary with time and the degree of distension. For example, when the urinary bladder is highly distended by its contents, the peak of the active length-tension curve can be displaced to longer muscle lengths. This means that as the muscle shortens to expel the organ's contents, it can reach lengths at which it can no longer exert active force. After a period of recovery at this shorter length, the muscle can again exert sufficient force to expel the contents.

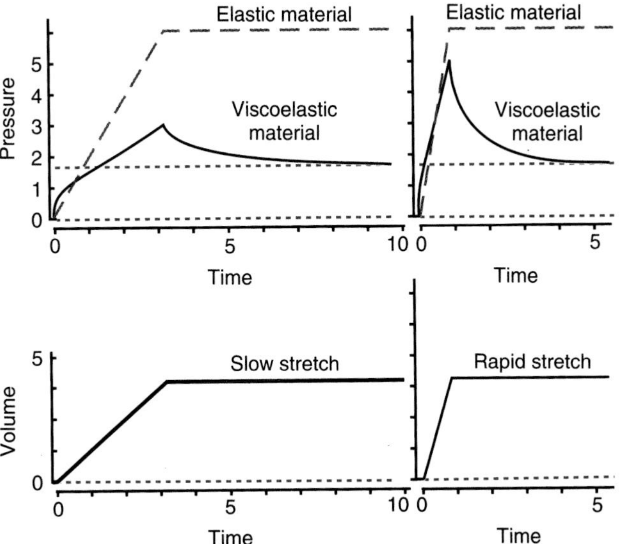

FIGURE 9.19 **Viscoelasticity.** The behavior of a viscoelastic material (e.g., the walls of a hollow, smooth muscle-containing organ) are subjected to slow (left) and rapid (right) elongation. The increase in force (or pressure) is proportional to the rate of extension, and at the end of the stretch, the force decays exponentially to a steady level. A purely elastic material (dashed line) maintains its force without stress relaxation.

Stress Relaxation and Viscoelasticity. These reversible changes in the length-tension relationship are, at least in part, the result of **stress relaxation,** which characterizes **viscoelastic materials** such as smooth muscle. When a viscoelastic material is stretched to a new length, it responds initially with a significant increase in force; this is an **elastic response**, and it is followed by a decline in force that is initially rapid and then continuously slows until a new steady force is reached. If a viscoelastic material is subjected to a constant force, it will elongate slowly until it reaches a new length. This phenomenon, the complement of stress relaxation, is called **creep.** In smooth muscle organs, the abundant connective tissue prevents overextension.

The viscoelastic properties of smooth muscle allow it to function well as a reservoir for fluids or other materials; if an organ is filled slowly, stress relaxation allows the internal pressure to adjust gradually, so that it rises much less than if the final volume had been introduced rapidly. This process is illustrated in Figure 9.19 for the case of a hollow smooth muscle organ subjected to both rapid and slow infusions of liquid (since this is a hollow structure, internal pressure and volume are directly related to the force and length of the muscle fibers in the walls). The dashed lines in the top graphs denote the pressure that would result if the material were simply elastic rather than having the additional property of viscosity.

Some of the viscoelasticity of smooth muscle is a property of the extracellular connective tissue and other materials, such as the hyaluronic acid gel, present between the cells; some of it is inherent in the smooth muscle cells, probably because of the presence of noncycling cross-bridges in resting tissue. One important feature of smooth muscle viscoelasticity is the tissue's ability to return to its original state following extreme extension. This capability is a result of the tonic contractile activity present in most smooth muscles under normal physiological conditions.

Other processes that are not yet well understood may also account for some of the length-dependent behavior of smooth muscle. In some smooth muscles, mechanical behavior in the later stages of a contraction depends strongly on the length at which the contraction began. This effect, called **plasticity** (not to be confused with nonrecoverable deformation), appears to arise from molecular rearrangements within the contractile protein array and may form the basis for both long- and short-term mechanical adaptation.

Modes of Relaxation. Relaxation is a complex process in smooth muscle. The central cause of relaxation is a reduction in the internal (cytoplasmic) calcium concentration, a process that is itself the result of several mechanisms. Electrical repolarization of the plasma membrane leads to a decrease in the influx of calcium ions, while the plasma membrane calcium pump and the sodium-calcium exchange mechanism (to a lesser extent) actively promote calcium efflux. Most important quantitatively is the uptake of calcium back into the SR. The net result of lowering the calcium concentration is a reduction in MLCK activity so that dephosphorylation of myosin can predominate over phosphorylation.

Biochemical Mechanisms. Both calcium uptake by the SR and the MLCK activity may be subject to another control mechanism called **β-adrenergic relaxation.** In some vascular smooth muscles, relaxation occurs in response to the presence of the hormone **norepinephrine.** Binding of this substance to cell membrane receptors causes the activation of **adenylyl cyclase** and the formation of cAMP (see

Chapter 1). Increased intracellular cAMP concentration is an effective promoter of relaxation in at least two major ways. The activity of the enzyme **cAMP-dependent protein kinase** increases as the concentration of cAMP rises. This enzyme (and perhaps also cAMP acting directly) enhances calcium uptake by the SR, resulting in a further lowering of the cytoplasmic calcium. At the same time, phosphorylation of MLCK (by the action of cAMP-dependent protein kinase) reduces its catalytic effectiveness, and myosin light-chain phosphorylation is decreased as if intracellular calcium had been lowered. Since many vascular muscles are continuously in a state of partial contraction, β-adrenergic relaxation is a physiologically important process in the adjustment of blood flow and pressure.

Another important relaxation pathway is present in the smooth muscle of small arteries (as well as other smooth muscle tissues). The lumen of arteries is lined with **endothelial cells**. In addition to their structural role, they serve as a controllable source of **nitric oxide** (NO), which was formerly known as endothelium-derived relaxing factor (EDRF) (see Chapter 1). The mechanical shearing effect of the flowing blood causes the endothelial cells to release NO; it is a small and highly diffusible molecule, and it quickly binds to membrane receptors on the vascular smooth muscle cells. This action results in a cascade of effects, the first of which is the stimulation of the enzyme **guanylyl cyclase**, which catalyzes the formation of **cyclic guanosine 3′,5′-monophosphate** (cGMP). By mechanisms similar to those in the case of cAMP, this leads to the activation of **cGMP-dependent protein kinase** (PKG), which affects several processes leading to relaxation. PKG promotes the reuptake of calcium ions, and it causes the opening of calcium-activated potassium ion channels in the cell membrane, leading to hyperpolarization and subsequent relaxation. PKG also blocks the activity of agonist-evoked **phospholipase C** (PLC), and this action reduces the liberation of stored calcium ions by IP_3. By mechanisms not well understood, cGMP reduces the calcium sensitivity of the myosin light-chain phosphorylation process, further promoting relaxation. Some drugs that relax vascular smooth muscle, such as **sodium nitroprusside**, work by mimicking the action of NO and causing similar intracellular events.

Mechanical Factors. Relaxation is obviously also a mechanical process. Contractile force decreases as cross-bridges detach and myofilaments become free again to slide past one another. Because most smooth muscle activity involves at least some shortening, relaxation must require elongation. As with other types of muscle, an external force must be applied for lengthening to occur. In the intestine, for example, material being propelled into a recently contracted region provides the extending force. Smooth muscle relaxation (or its absence) may have important indirect consequences. Hypertension, for example, can be caused by a failure of smooth muscle relaxation. In the uterus during labor, adequate relaxation between contractions is essential for the well-being of the fetus. During the contractions of labor, the muscular walls of the uterus become quite rigid and tend to compress the blood vessels that run through them. As a result, blood flow to the fetus is restricted, and failure of the muscle to relax adequately between contractions can result in fetal distress.

Adaptation to Changing Conditions. Several external influences, some not well understood, affect the growth and functional adaptation of smooth muscle. Some of these changes are vital for normal body function, while others can be part of a disease process.

Hormone-Induced Hypertrophy. The uterus and associated tissues are under the influence of the female sex hormones (see Chapters 38 and 39). During pregnancy, high levels of progesterone, later followed by high estrogen levels, promote significant changes in uterine growth and control. The mass of muscle layers, known as the **myometrium**, increases as much as 70-fold, primarily through an increase in muscle cell size—**hypertrophy**—associated with a large increase in content of contractile proteins and associated regulatory proteins. The distension caused by the growing fetus also promotes hypertrophy. Extracellular connective tissue also increases. The number of cells increases as well, a condition called **hyperplasia.**

Throughout most of pregnancy the cells are poorly coupled electrically and contractile activity is not well coordinated. As pregnancy nears term, the large increase in the number of gap junctions permits coordinated contractions that culminate in the birth process. Following delivery and the consequent hormonal and mechanical changes, the processes leading to hypertrophy are reversed and the muscle reverts to its nonpregnant state.

Other Forms of Hypertrophy. Chronic obstruction of hollow smooth muscle organs (e.g., the urinary bladder, small intestine, portal vein) produces a chronically elevated internal pressure. This acts as a stimulus for smooth muscle hypertrophy, although the cellular mechanisms involved are not well understood. In addition to structural changes, there may be alterations of the metabolic activities, contractile properties, and response to agonists. Hyperplasia is also present to some degree in these muscle adaptations, but its relative contribution is difficult to ascertain experimentally. Nonmuscular components of the organ wall (e.g., connective tissue) are also increased. These changes, especially those involving the muscle cells, usually revert to near normal when the mechanical cause of the hypertrophy is removed.

Vascular smooth muscle, especially of the arteries, is also subject to hypertrophy (and hyperplasia) when it encounters a sustained pressure overload. This is an important factor in **hypertension** or high blood pressure. An increase in blood pressure, perhaps a result of chronically elevated sympathetic nervous system activity, may be present before smooth muscle hypertrophy occurs. Enlargement of the smooth muscle layer is a response to this stimulus, and there may be a trophic effect of the sympathetic nervous system activity as well. The resulting thickening of the vascular wall further reduces the lumen diameter, aggravating the hypertension. Lowering the blood pressure by therapeutic means can result in a return of the vessel walls to a near-normal state. Hypertension in the pulmonary vasculature is also associated with increased smooth muscle growth and with the development of smooth muscle cells in areas of the arterial system that do not normally have smooth muscle in their walls.

Under some circumstances, smooth muscle cells can

lose most of their contractile function and become synthesizers of collagen and accumulators of low-density lipoproteins. The loss of contractile activity is accompanied by a significant loss in the number of myofilaments. Such a **phenotypic transformation** takes place, for example, in the formation of atherosclerotic lesions in artery linings. While the factors involved in initiating and sustaining this reversible transition are not well understood, they appear to involve growth-promoting substances released from platelets following endothelial injury, while circulating heparin-like substances block the transformation.

REVIEW QUESTIONS

DIRECTIONS: Each of the numbered items or incomplete statements in this section is followed by answers or completions of the statement. Select the ONE lettered answer or completion that is BEST in each case.

1. The endplate potential at the neuromuscular junction is the result of increased postsynaptic membrane permeability to
(A) Sodium first, then potassium
(B) Sodium and potassium simultaneously
(C) Sodium only
(D) Potassium only
2. The endplate potential differs from a muscle action potential in several ways. In which one of the following ways are they similar?
(A) They are both actively propagated down the length of the muscle fiber
(B) They both arise from changes in the permeability to sodium and potassium ions
(C) They are both initiated by the flow of electrical (ionic) current
(D) In both cases, the membrane potential becomes inside-positive
3. If transmission at the neuromuscular junction were blocked by the application of curare, which one of the events listed below would fail to occur when a motor nerve impulse arrived?
(A) Depolarization of the postsynaptic membrane
(B) Depolarization of the presynaptic membrane
(C) Entry of calcium ions into the presynaptic terminal
(D) Presynaptic release of transmitter substance
4. In a certain muscle, it takes 25 msec for a single twitch to develop its peak force in response to a single stimulus. If this muscle were stimulated with two stimuli spaced 15 msec apart, the result would be
(A) A single twitch identical to the one-stimulus twitch
(B) A contraction similar to a single twitch, but of higher amplitude
(C) Two distinct contractions of very short duration
(D) A failure of the muscle to contract at all
5. The factor most important in producing an isometric contraction is
(A) Keeping the muscle from changing its length
(B) Providing a stimulus adequate to activate all motor units
(C) Determining the resting length of the muscle
(D) Stimulating in a tetanic fashion to produce the maximal force
6. If a muscle is arranged so as to lift an afterload equal to one-half its maximal isometric capabilities, the ultimate force it develops is determined by the
(A) Length of the muscle
(B) Size of the afterload
(C) Strength of the stimulation
(D) Number of motor units activated
7. In a series of afterloaded isotonic twitches, as the load is increased, the
(A) Force developed by the muscle increases and the shortening velocity decreases
(B) Force developed by the muscle increases, while the velocity remains the same
(C) Velocity increases to compensate for the increased afterload
(D) Force developed is determined by the velocity of shortening
8. At which point along the isotonic force-velocity curve is the power output maximal?
(A) At the lowest force and highest velocity (V_{max})
(B) At the highest force and lowest velocity (F_{max})
(C) At a force that is about one third of F_{max}
(D) At a velocity that is about two thirds of V_{max}
9. Consider a load being lifted by a human hand. Because of the mechanical effects of the skeletal lever system, the biceps muscle exerts a force
(A) Less than the load, but shortens at a higher velocity
(B) Equal to the load, and shortens at a velocity equal to the load
(C) Greater than the load, but shortens at a lower velocity
(D) Independent of the load, and shortens at a velocity independent of the load
10. Muscles that are best suited for brief high-intensity exercise would contain which of the following types of fibers?
(A) Mostly glycolytic (white)
(B) Mostly slow-twitch oxidative (red)
(C) A mix of slow twitch (red) and fast twitch (red)
(D) A mix of glycolytic (white) and fast twitch (red)
11. Smooth muscles that are in the walls of hollow organs
(A) Can shorten without developing force
(B) Can develop force isometrically
(C) Have no contractile function, but resist lengthening
(D) Shorten as the volume of the organ increases
12. The relaxation of smooth muscle is associated with a reduction in free intracellular calcium ion concentration. The effect of the reduction is
(A) Reestablishment of the inhibition of the actin-myosin interaction
(B) Deactivation of the enzymatic activity of the individual actin molecules
(C) Decreased phosphorylation of myosin molecules
(D) Reduced contractile interaction by blocking the active sites of the myosin molecules
13. Which statement below most closely describes the role of calcium ions in the control of smooth muscle contraction?
(A) Binding of calcium ions to regulatory proteins on thin filaments removes the inhibition of actin-myosin interaction
(B) Binding of calcium ions to regulatory proteins associated with thick filaments, specifically calmodulin, activates the enzymatic activity of myosin molecules
(C) Calcium ions serve as a direct inhibitor of the interaction of thick and thin filaments
(D) A high concentration of calcium ions in the myofilament space is

(continued)

required to maintain muscle in a relaxed state

14. Compared with skeletal muscle, smooth muscle
(A) Contracts more slowly, but exerts considerably more force
(B) Contracts more rapidly, but exerts considerably less force
(C) Maintains long-duration contractions economically
(D) Exerts considerable force but can do little shortening

15. Compared with that of skeletal muscle, the crossbridge cycle of smooth muscle
(A) Is similar, but runs in the reverse direction
(B) Contains the same steps, but some of them are slower
(C) Does not have a step in which actin and myosin are bound together
(D) Can proceed without the consumption of ATP

16. Receptors in the smooth muscle cell membrane
(A) Function only in combination with electrical activation
(B) Cannot function if the cell is relaxed
(C) Play a variety of regulatory roles
(D) Control chemical activation, but do not affect electrical activation

SUGGESTED READING

Bagshaw CR. Muscle Contraction. 2nd Ed. New York: Chapman & Hall, 1993.

Barany M, ed. Biochemistry of Smooth Muscle Contraction. New York: Academic Press, 1996.

Barr L, Christ GJ, eds. A Functional View of Smooth Muscle. Stamford, CT: JAI Press, 2000.

Ford LE. Muscle Physiology and Cardiac Function. Carmel, IN: Biological Sciences Press-Cooper Group, 2000.

Kao CY, Carsten ME, eds. Cellular Aspects of Smooth Muscle Function. Cambridge, UK: Cambridge University Press, 1997.

Cardiac Muscle

Richard A. Meiss, Ph.D.

CHAPTER OUTLINE

- ANATOMIC SPECIALIZATIONS OF CARDIAC MUSCLE
- PHYSIOLOGICAL SPECIALIZATIONS OF CARDIAC MUSCLE

KEY CONCEPTS

1. Cardiac muscle is a striated muscle, with a sarcomere structure much like that of skeletal muscle. It has small cells (as in smooth muscle), firmly connected end-to-end at the intercalated disks.
2. The action potential in cardiac muscle is long lasting compared to the duration of the contraction, preventing a tetanic contraction.
3. Under normal circumstances, cardiac muscle operates at lengths somewhat less than the optimal length for peak force production, facilitating length-dependent regulation of the muscle activity.
4. A typical cardiac muscle contraction produces less than maximal force, allowing physiologically regulated changes in contractility to adjust the force of the muscle contraction to the body's current needs.
5. As in skeletal muscle, the speed of shortening of cardiac muscle is inversely related to the force being exerted, as expressed in the force-velocity curve.
6. The contractility of cardiac muscle is changed by inotropic interventions that include changes in the heart rate, the presence of circulating epinephrine, or sympathetic nerve stimulation.
7. Most changes in cardiac muscle contractility are associated with changes in the amount of calcium available to activate the contractile system.
8. Calcium enters a cardiac muscle cell during the plateau of the action potential. This entry promotes the release of internal calcium stores, which are located mainly in the sarcoplasmic reticulum (SR). Primary and secondary active transport systems remove calcium from the cytoplasm.
9. Cardiac muscle derives its energy primarily from the oxidative metabolism of lactic acid and free fatty acids. It has very little capacity for anaerobic metabolism.

The muscle mass of the heart, the **myocardium**, shares characteristics of both smooth muscle and skeletal muscle. The tissue is striated in appearance, as in skeletal muscle, and the structural characteristics of the sarcomeres and myofilaments are much like those of skeletal muscle. The regulation of contraction, involving calcium control of an actin-linked troponin-tropomyosin complex, is also quite like that of skeletal muscle. However, cardiac muscle is composed of many small cells, as is smooth muscle, and electrical and mechanical cell-to-cell communication is an essential feature of cardiac muscle structure and function. The mechanical properties of cardiac muscle relate more closely to those of skeletal muscle, although the mechanical performance is considerably more complex and subtle.

ANATOMIC SPECIALIZATIONS OF CARDIAC MUSCLE

The heart is composed of several varieties of cardiac muscle tissue. The **atrial myocardium** and **ventricular myocardium**, so named for their location, are similar structurally, although the electrical properties of these two areas differ significantly. The **conducting tissues** (e.g., Purkinje fibers) of the heart have a communicating function similar to nerve tissue, but they actually consist of muscle tissue that is highly adapted for the rapid and efficient conduction of action potentials, and their contractile ability is greatly reduced. Finally, there are the highly specialized tissues of the **sinoatrial** and **atrioventricular nodes**, muscle tissue that is greatly modified into structures concerned with the initiation and conduction of the heartbeat. The discussions that follow refer primarily to the ventricular myocardium, the tissue that makes up the greatest bulk of the muscle of the heart.

Cardiac Muscle Cells Are Structurally Distinct From Skeletal Muscle Cells

The small size of cardiac muscle cells is one of the critical aspects in determining the function of heart muscle. The cells are approximately 10 to 15 μm in diameter and about 50 μm long. Cardiac muscle tissue is a branching network of cells, also called **cardiac myocytes**, joined together at intercalated disks (Fig. 10.1). This arrangement aids in the spread of electrical activity. Cardiac myocytes have a single, centrally located nucleus, although many cells may contain two nuclei. The cell membrane and associated fine connective tissue structures form the sarcolemma, as in skeletal muscle. The sarcolemma of cardiac muscle supports the resting and action potentials and is the location of ion pumps and ion exchange mechanisms vital to cell function. Just inside the sarcolemma are components of the SR where significant amounts of calcium ions may be bound and kept from general access to the cytoplasm. This bound calcium can exchange rapidly with the extracellular space and can be rapidly freed from its binding sites by the passage of an action potential.

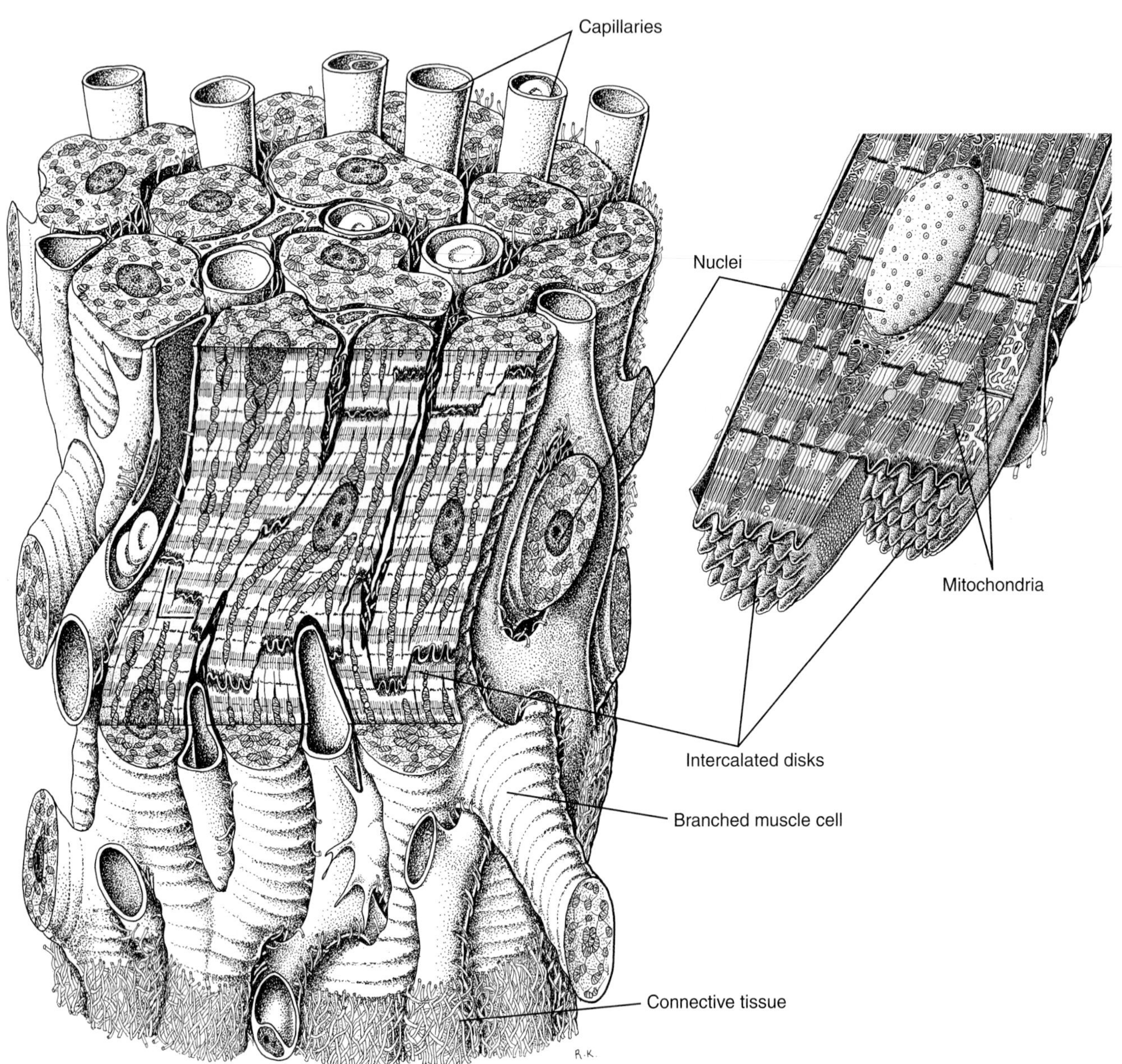

FIGURE 10.1 The structure of cardiac muscle tissue. Left: A small tissue sample in longitudinal and cross section. Note the branching nature of the cells and the large number of capillaries. Right: Two cardiac myocytes, showing the striated structure of the contractile filaments, the many mitochondria, and the three-dimensional structure of the intercalated disks. (Adapted from Krstic RV. General Histology of the Mammal. New York: Springer-Verlag, 1984.)

As in skeletal muscle, there is a system of transverse (T) tubules, but both it and the SR are not as extensive in cardiac muscle, together constituting less than 2% of the cell volume. This correlates with the small cell size and consequent reduction in diffusion distances between the cell surface membrane and contractile proteins. In cardiac muscle cells, the T tubules enter the cells at the level of the Z lines. In many cases, the link between a T tubule and the SR is not a triad, as in skeletal muscle, but rather a **dyad**, composed of the T tubule and the terminal cisterna of the SR of only one sarcomere. The small size of the SR also limits its calcium storage capacity, and the other source of calcium entry and exit, the sarcolemma, has an important role in the excitation-contraction coupling process in cardiac muscle.

The sarcomeres appear essentially like those of skeletal muscle, with similar A bands and I bands, Z lines, and M lines. Myofilaments make up almost half the cell volume and are bathed in the cytosol. Numerous mitochondria comprise another 30 to 40% of the cell volume, reflecting the highly aerobic nature of cardiac muscle function. The rest of the cell volume, about 15%, consists of cytosol, containing numerous enzymes and metabolic products and substrates.

Cardiac Muscle Cells Are Linked in a Functional Syncytium

Electron microscopy reveals that in the region of the intercalated disk, each cardiac myocyte sends processes deep into its neighboring cell to form an interdigitating junction with a large surface area. Gap junctions in the intercalated disks function like those of smooth muscle, allowing close electrical communication between cells. Also plentiful in the intercalated disk region are desmosomes, areas where there is a firm mechanical connection between cells. This mode of attachment, rather than an extensive extracellular connective tissue matrix as in smooth muscle, allows the transmission of force from cell to cell. The intercalated disk, therefore, allows cardiac muscle to form a **functional syncytium**, with cells acting in concert both mechanically and electrically.

The stimulus for cardiac muscle contraction arises entirely within the heart and is not dependent on its nerve supply (see Chapter 13). The conduction of action potentials is solely a function of the muscle tissue. Impulse propagation is aided by the branched nature of the cells, the intercalated disks, and specialized conducting tissue, such as **Purkinje fibers**—strands of myocytes, nerve-like in outward appearance, that are specialized for electrical conduction. Their contractile protein is only about 20% of the cell volume, and their large size optimizes their electrical characteristics for rapid action potential conduction. Innervation of cardiac muscle comes from both branches of the autonomic nervous system, allowing for external regulation of the heart rate and strength of contraction, as well as providing some degree of sensory feedback.

Cell and Tissue Structure Allow and Require Unique Adaptations

As a result of the small size of cardiac muscle cells, the communication system described above (and in Chapter 13) is necessary for organized function. The small cell size also makes each cell more critically dependent on the external environment, and cardiac function may be greatly altered by electrolyte and metabolic imbalances arising elsewhere in the body. Hormonal messengers, such as norepinephrine, also have quick access to cardiac muscle cells.

From a mechanical standpoint, the lack of skeletal attachments means cardiac muscle can function over a wide range of lengths. While the length-tension property is not of major importance in the functioning of many skeletal muscles, in cardiac muscle, it is the basis of the remarkable capacity of the heart to adjust to a wide range of physiological conditions and requirements.

PHYSIOLOGICAL SPECIALIZATIONS OF CARDIAC MUSCLE

Cardiac muscle is a striated muscle, but it functions rather differently from skeletal muscle. The lack of skeletal attachments provides a wider range of lengths over which it can operate. Special features of the excitation-contraction coupling process allow a subtle degree of control at the level of the muscle that is largely independent of the central nervous system (CNS).

Specialized Electrical and Metabolic Properties Control Cardiac Muscle Contraction

A more detailed treatment of the electrical properties of cardiac muscle is given in Chapter 13. The discussion here focuses on the electrical properties most closely related to controlling the mechanical function of cardiac ventricular muscle.

The Cardiac Action Potential. As in other types of muscle and in nerve, the muscle cells of the heart have an excitable and selectively permeable cell membrane that is responsible for both resting potentials and action potentials. These electrical phenomena are the result of ionic concentration differences and several ion-selective membrane channels, some of which are voltage- and time-dependent. In cardiac muscle, however, the membrane events are more diverse and complex than in skeletal muscles and are much more closely linked to the actual form of the mechanical contraction. The closer association of electrical and mechanical events is one key to the inherent properties of cardiac muscle that suit it to its role in an organ that is largely self-regulating.

Figure 10.2 illustrates some features of the cardiac muscle action potential that pertain directly to myocardial function. Note that the duration of the action potential is quite long; in fact, it lasts nearly as long as the muscle contraction. One consequence is that the absolute and relative refractory periods are likewise extended, and the muscle cannot be restimulated during any but the latest part of the contraction. During the repolarization phase of the action potential, there is a brief period in which the muscle actually shows an increased sensitivity to stimulation. This period of **supranormal excitability** is due to a lowered potassium conductance that persists late in the action potential (see Chapter 13). If the muscle is accidentally stimulated

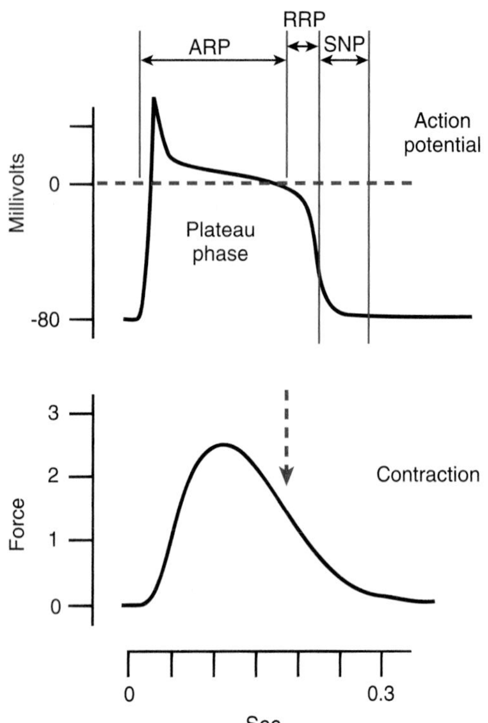

FIGURE 10.2 A cardiac muscle action potential and isometric twitch. Because of the duration of the action potential, an effective tetanic contraction cannot be produced, although a partial contraction can be elicited late in the twitch. ARP, absolute refractory period; RRP, relative refractory period; SNP, period of supranormal excitability.

during this period, the action potentials that are produced have reduced amplitude and duration and give rise to only small contractions. This period of supranormal excitability can lead to unwanted and untimely propagation of action potentials that can seriously interfere with the normal rhythm of the heart.

The long-lasting refractoriness of the cell membrane effectively prevents the development of a tetanic contraction (see Fig. 10.2); any failure of cardiac muscle to relax fully after every stimulus would make it quite unsuitable to function as a pump. When cardiac muscle is stimulated to contract more frequently (equivalent to an increase in the heart rate), the durations of the action potential and the contraction become less, and consecutive twitches remain separate contraction-and-relaxation events.

It must be emphasized that contraction in cardiac muscle is not the result of stimulation by motor nerves. Cells in some critical areas of the heart generate automatic and rhythmic action potentials that are conducted throughout the bulk of the tissue. These specialized cells are called **pacemaker cells** (see Chapter 13).

Excitation-Contraction Coupling in Cardiac Muscle. The rapid depolarization associated with the upstroke of the action potential is conducted down the T tubule system of the ventricular myocardium, where it causes the release of intracellular calcium ions from the SR. In cardiac muscle, a large part of the calcium released during rapid depolarization is from additional SR just inside the cell membrane. As in skeletal muscle, the principal role of the SR is in the rapid release, active uptake, storage, and buffering of cytosolic calcium. The action of calcium ions on the **troponin-tropomyosin complex** of the thin filaments is similar to that in skeletal muscle, but cardiac muscle differs in its cellular handling of the activator, calcium.

Along with the calcium ions released from the SR, a significant amount of calcium enters the cell from outside during the upstroke and **plateau phase** of the action potential (see Fig. 10.2). The principal cause of the sustained depolarization of the plateau phase is the presence of a population of voltage-gated membrane channels permeable to calcium ions. These channels open relatively slowly; while open, there is a net influx of calcium ions, called the **slow inward current**, moving down an electrochemical gradient. Although the calcium entering during an action potential does not directly affect that specific contraction, it can affect the *next* contraction, and it does increase the cellular calcium content over time because of the repeated nature of the cardiac muscle contraction.

In addition, even a small amount of Ca^{2+} entering through the sarcolemma causes the release of significant additional Ca^{2+} from the SR, a phenomenon known as **calcium-induced calcium release** (similar to that in smooth muscle). This constant influx of calcium requires that there be a cellular system that can rid the cell of excess calcium. Regulation of cellular calcium content has important consequences for cardiac muscle function, because of the close relationship between calcium and contractile activity.

Mechanical Properties of Cardiac Muscle Adapt It to Changing Physiological Requirements

The mechanical function of cardiac muscle differs somewhat from that of skeletal muscle contraction. Cardiac muscle, in its natural location, does not exist as separate strips of tissue with skeletal attachments at the ends. Instead, it is present as interwoven bundles of fibers in the heart walls, arranged so that shortening results in a reduction of the volume of the heart chamber, and its force or tension results in an increase in pressure in the chamber. Because of geometric complexities of the intact heart and the complex mechanical nature of the blood and aorta, shortening contractions of the intact heart muscle are more nearly auxotonic than truly isotonic (see Chapter 9).

The experimental basis for the present understanding of cardiac muscle mechanics comes largely from studies done on isolated **papillary muscles** from the ventricles of experimental animals. A papillary muscle is a relatively long, slender muscle that can serve as a representative of the whole myocardium. It can be arranged to function under the same sort of conditions as a skeletal muscle. Analysis of research results is aided by using simple afterloads to produce isotonic contractions. Despite the limitations these simplifications impose, many of the unique properties of the intact heart can be understood on the basis of studies of isolated muscle. As the various phenomena are explained here, substitute *volume changes* for *length changes* and *pressure* for *force*. You will then be able to relate the function of the heart as a pump to the properties of the muscle responsible for its operation (see Chapter 14).

The Length-Tension Curve. Some aspects of the cardiac muscle **length-tension curve** are associated with its specialized construction and physiological role (Fig. 10.3). Over the range of lengths that represent physiological ventricular volumes, there is an appreciable resting force that increases with length; at the length at which active force production is optimal, this can amount to 10 to 15% of the total force. Because this resting force exists before contraction occurs, it is known as **preload.** In the intact heart, the preload sets the resting fiber length according to the intracardiac blood pressure existing prior to contraction. The passive tension rises steeply beyond the optimal length and prevents overextension of the muscle (or overfilling of the heart). Note that the resting force curve is associated with the **diastolic (relaxed) phase** of the heart cycle, while the active force curve is associated with the **systolic (contraction) phase.**

The length-tension curve in Figure 10.3 describes isometric behavior; since the working heart never undergoes completely isometric contractions (see Chapter 14), other aspects of length-dependent behavior must be responsible for determining the effect length has on cardiac muscle function. One such aspect is the rate at which isometric force develops during a twitch. Notice the series of twitches shown in Figure 9.10; because of the constancy of the time required to reach peak force, the rate of rise of force also varies with muscle length. Other length-dependent aspects of contraction are encountered when we examine the complete contraction cycle of cardiac muscle.

The Contraction Cycle of Cardiac Muscle. A typical isotonic contraction of skeletal muscle (see Fig. 9.8) can be divided into four distinct phases:

1. **Isometric contraction**: the muscle force builds up to reach the afterload.
2. **Isotonic shortening**: the afterload is lifted.
3. **Isotonic lengthening**: the load stretches the muscle back to its starting length.
4. **Isometric relaxation**: the force dies away.

The isometric contraction and isotonic shortening phases of a typical cardiac muscle contraction are like those of skeletal muscle. However, in the intact heart at the peak of the shortening, the afterload is removed because of the closing of the aortic and pulmonary valves at the end of the cardiac ejection phase (see Chapter 14). Since the muscle is not allowed to lengthen (the inflow valves are still closed), it undergoes isometric relaxation at the shorter length. Some time later, the muscle is stretched back to its original length by an external force (the returning blood), producing an isotonic lengthening (isotonic relaxation) phase. Because the muscle has relaxed, only a small force is required for the reextension. In the intact heart, this force is supplied by the returning blood.

The principal difference between these two cycles is significant: in skeletal muscle, the work done on the afterload (by lifting it) is returned to the muscle. In cardiac muscle, the work done on the load is *not* returned to the muscle but is imparted to the afterload. The heart muscle is constrained by its anatomy and functional arrangements to follow different pathways during contraction and relaxation.

This pattern is seen clearly when the phases of the contraction-relaxation cycle are displayed on a length-tension curve. In Figure 10.4, at the beginning of the contraction (A), force increases without any change in length (isometric conditions); when the afterload is lifted (B), the muscle shortens at a constant force (isotonic conditions) to the shortest length possible for that afterload. The afterload is removed at the maximal extent of shortening, and the muscle relaxes (C) without any change in length (isometric conditions again, but at a reduced length). With sufficient force applied to the resting muscle by some external means (D), the mus-

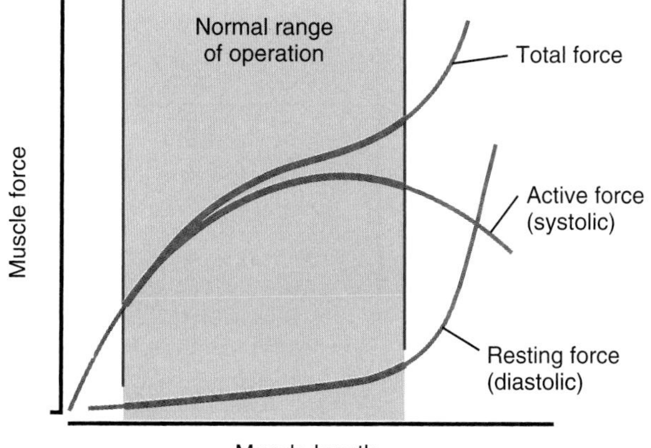

FIGURE 10.3 The isometric length-tension curve for isolated cardiac muscle. The total force at all physiologically significant lengths includes a resting force component.

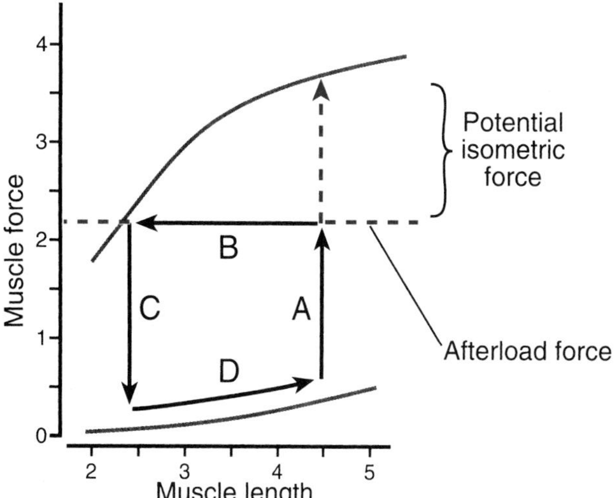

FIGURE 10.4 An afterloaded contraction of cardiac muscle, plotted in terms of the length-tension curve. The limit to force is provided by the afterload; the limit to shortening by the length-tension curve. A, isometric contraction phase; B, isotonic shortening phase; C, isometric relaxation phase; D, relengthening. (See text for details.)

cle is elongated back to its starting length. Because the muscle is unstimulated and its resting force rises somewhat during elongation, this phase is not strictly isotonic.

In physical terms, the area enclosed by this pathway represents work done by the muscle on the external load. If the afterload or the starting length (or both) is changed, then a different pathway will be traced (Fig. 10.5, left). The area enclosed will differ with changes in the conditions of contraction, reflecting differing amounts of external work delivered to the load. In a typical skeletal muscle contraction, as shown in Fig. 10.5 (right), steps A and B are reversed during relaxation. Such a contraction does no net external work, and no area is enclosed by the pathway.

Cardiac Muscle Self-Regulation. Each case in Figure 10.5 (center and left) demonstrates that the active portion of the length-tension curve provides the limit to shortening and, thus, interacts with the particular afterload chosen. With smaller afterloads, the muscle will shorten further than it would with a larger afterload. It is important to realize that during isotonic shortening, the muscle force is limited by the magnitude of the afterload and *not* by the length-tension capability of the muscle. It is the *extent of shortening* at a given afterload that is limited by the length-tension property of the muscle; this is a very important consideration when measuring cardiac performance under conditions of changing blood pressure and filling of the heart (see Chapter 14). This length- and force-dependent behavior is the key to **autoregulation** (self-regulation) by cardiac muscle and is the functional basis of **Starling's law of the heart** (see Chapter 14); when the muscle is set to a longer length at rest, the active contraction results in a greater shortening that is also more rapid and is preceded by a more rapid isometric phase. This allows the heart to adjust

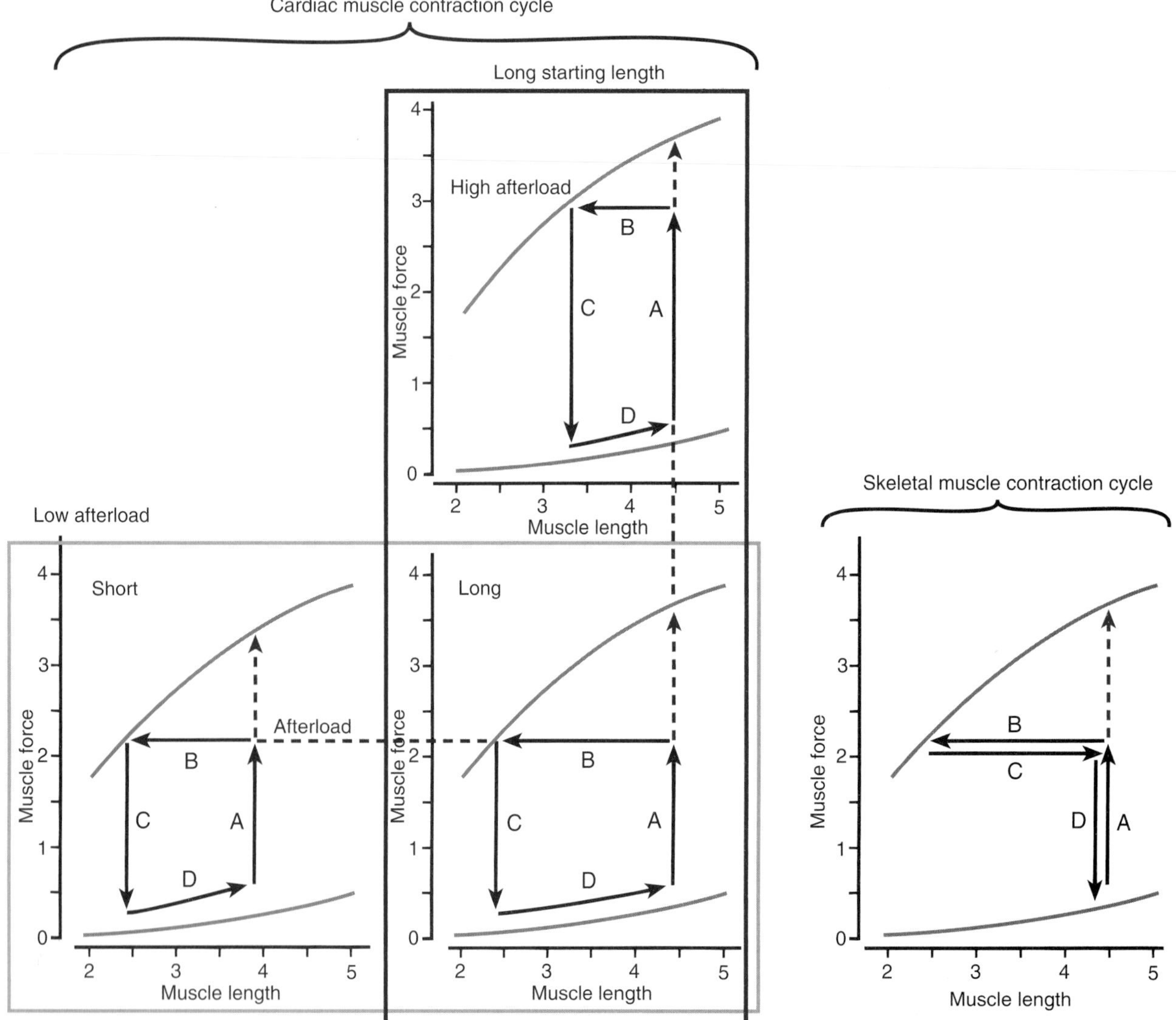

FIGURE 10.5 **Afterloaded contractions under a variety of conditions.** Left: Cardiac muscle contraction cycles. The horizontal box shows the effect of starting at two different initial lengths at the same afterload. The vertical box shows the effect of two different afterloads on shortening that begins at the same initial length. Increasing the afterload reduces the amount of shortening possible, as does decreasing the starting length; in both cases, the limit to shortening is determined by the length-tension curve. Right: The contraction cycle of skeletal muscle. Contraction and relaxation pathways are the same.

its pumping to exactly the amount required to keep the circulatory system in balance.

Variable Contractility Facilitates Essential Physiological Adjustments

Under a wide range of conditions, the contractile behavior of skeletal muscle is fixed and repeatable. The peak force and shortening velocity depend primarily on muscle length and afterload, and unless the muscle is worked to fatigue, these properties will not change from contraction to contraction. For this reason, skeletal muscle is said to possess **fixed contractility. Contractility** or the contractile state of muscle may be defined as a certain level of functional capability (as measured by a quantity such as isometric force, shortening velocity, etc.) when measured at a constant muscle length. (Length must be held constant to preclude the effects of the length-tension curve properties already discussed.) The regulation of skeletal muscle contraction to produce useful activity is primarily the task of the CNS, using the mechanisms of motor unit summation and partially fused tetani (see Chapter 9). Cardiac muscle has no motor innervation, but has a capacity for adjustment that is not solely accomplished by changes in afterload and starting length.

The **variable contractility** of cardiac muscle enables it to make adjustments to the varying demands of the circulatory system. Certain chemical and pharmacological agents, as well as physiological circumstances, affect cardiac contractility. The collective term for the influence of such agents is **inotropy**. Contractility is altered by **inotropic interventions,** agents or processes that change the functional state of cardiac muscle. **Positive inotropes** are inotropic interventions that increase contractility and include the action of adrenergic (sympathetic nervous system) stimulation, bloodborne catecholamine hormones, drugs such as the digitalis derivatives, and an increase in the rate of stimulation (i.e., increased heart rate). **Negative inotropes** include a decrease in heart rate, disease processes such as myocarditis or coronary artery disease, and certain drugs. Chronically reduced contractility can lead to the condition known as heart failure (see Clinical Focus Box 10.1).

Effects of Inotropic Interventions. Figure 10.6 shows an increase in contractility plotted on a length-tension graph. It has the effect of shifting the active length-tension curve upward and to the left; relaxation and the passive curve are minimally affected. Careful experiments have shown that one effect of short muscle length on muscle contraction is actually a reduction in contractility as a result of inefficiencies in the excitation-contraction coupling mechanism at these lengths. Such effects cannot be separated from other length-related effects on cardiac muscle functions, and they are usually included without mention in the more familiar length-dependent changes in muscle performance.

An example of the similarities and differences between changes in resting length and changes in contractility is shown in the force-velocity curves in Figure 10.7. The set of curves in Figure 10.7A represents the isotonic behavior of muscle at a constant level of contractility at three different muscle lengths. The maximum force point on each curve shows the isometric length-tension effect. When a particular afterload is chosen (in this case, 0.5 units), the initial shortening velocity varies with the starting length, although the curves do tend to converge at the lowest forces. The curves in Figure 10.7B represent contractions made at the same starting length but with the muscle operating at different levels of contractility. Again there is a difference in shortening velocity at a constant afterload, but there is no tendency for the curves to converge at the low forces. These examples show only one aspect of the effects of changing contractility; those not illustrated include changes in the rate of rise of isometric force and changes in the time required to reach peak force in a twitch.

Ultimately, any change in the muscle contraction will result in a change in the overall performance of the heart, but cardiac performance can change drastically even without changes in contractility because of length-tension effects. The need to distinguish such effects from changes in contractility (to guide treatment and therapy) has led to a search for aspects of muscle performance that are dependent on the state of contractility but independent of muscle length. The results of these studies (based on the properties of isolated muscle) are questionable because the complicated structure and function of the intact heart do not permit a reliable extrapolation of findings. Instead, several empirical measures have been developed from studies of the intact heart, some of which provide a reasonable and useful index of contractility (see Chapter 14).

The Cellular Basis of Contractility Changes. The basic determinant of the variable contractility of cardiac muscle is the calcium content in the myocardial cell. Under normal conditions, the contractile filaments of cardiac muscle are only partly activated. This is because, unlike with skeletal muscle, not enough calcium is released to occupy all of the troponin molecules, and not all potentially available crossbridges can attach and cycle. An increase in the availability of calcium would increase the number of crossbridges activated; thus, contractility would be increased. To understand the mechanisms of contractility change, it is necessary to consider the factors affecting cellular calcium handling.

The processes linking membrane excitation to contraction via calcium ions are illustrated in Figure 10.8. Since this involves many possible movements and locations of calcium, the processes are considered in the order in which they would be encountered during a single contraction.

The initial event is an action potential (1) traveling along the cell surface. As in skeletal muscle, the action potential enters a T tubule (2), where it can communicate with the SR (3) to cause calcium release. This mechanism for release of calcium in cardiac muscle is much less than in skeletal muscle and is insufficient to cause adequate activation of the contraction. To some extent, activation is aided by a calcium-induced calcium release mechanism (4) triggered by a rise in the cytoplasmic calcium concentration. The action potential (5) on the cell surface (sarcolemma) also causes the opening of calcium channels, through which strong inward calcium current flows. These calcium ions accumulate just inside the sarcolemma (6), although some probably diffuse rapidly into the cell interior. Calcium induces the rapid release of calcium from the subsarcolemmal SR, and the calcium then diffuses the short distance to the

CLINICAL FOCUS BOX 10.1

Heart Failure and Muscle Mechanics

Heart failure is evident when the heart is unable to maintain sufficient output to meet the body's normal metabolic needs. It is usually a progressively worsening condition. The condition is due to either deterioration of the heart muscle or worsening of the contributing factors external to the heart. The term *congestive heart failure* refers to fluid congestion of the lungs that often accompanies heart failure.

Patients suffering from heart failure may be unable to perform simple everyday tasks without fatigue or shortness of breath. In later stages, there may be significant distress even while resting. While many intrinsic and extrinsic factors contribute to the condition, this discussion will focus on those closely related to the mechanical properties of the heart muscle.

Much of poor cardiac function can be understood in terms of the mechanics of the heart muscle as it interacts with several external factors that determine the resting muscle length (or preload) or the load against which it must contract (the afterload). The most important aspects of the mechanical behavior are described by the length-tension and force-velocity curves, which, together with knowledge of the current state of contractility, can provide a complete picture of the muscle function.

Some heart failure is of the systolic type. If the heart has been damaged by a myocardial infarction (heart attack) or ischemia (impaired blood supply to the heart muscle) or by chronic overload (as with untreated high blood pressure), the muscle may become weakened and have reduced contractility. In this case, the load presented to the heart by the blood pressure is too high (relative to the weakened condition of the muscle), and (as the force-velocity curve describes) the rate of shortening (velocity) of the muscle will be reduced. The length-tension curve indicates that the larger the load, the less the shortening (see Fig. 10.5). Therefore, less blood will be pumped with each beat. Therapy for this type of failure involves improving the contractility of the muscle and/or reducing the load on the heart.

Heart failure can also be of the diastolic type (and may occur along with systolic failure). Here the relaxation is impaired, and the muscle is resistant to the stretch that must take place during its filling with blood. Some types of hypertrophy or connective tissue fibrosis also may contribute to diastolic failure. Because the muscle cannot be sufficiently lengthened during its rest period (diastole), it begins its contraction at too short a length. As the length-tension curve would predict, the muscle is unable to shorten sufficiently to pump an adequate volume of blood with each beat. Because the force-velocity curve is also length-dependent, the speed at which the muscle can shorten is also reduced.

Treatment of heart failure involves approaches that affect several areas of muscle mechanics. Drugs that increase the contractility of cardiac muscle, such as digitalis and its derivatives, may be used to cause more effective contraction and allow the muscle to operate along an improved force-velocity curve. Most contractility-increasing drugs work by increasing the amount of intracellular calcium available to the myofilaments, thereby increasing the number of crossbridges participating in the contractions. Care must be taken, however, that the increased contractility does not create a metabolic demand that would further weaken the muscle. Drugs that blunt the response of the heart to the excitatory action of the sympathetic nervous system (which affects both heart rate and muscle contractility) can protect against an increased workload. Drugs that lower blood pressure by their effects on the arterial muscle will reduce the load against which the heart muscle must contract, and the muscle can operate on a more efficient portion of the force-velocity curve. Drugs or dietary regimens that reduce blood volume (via increased renal excretion of salt and water) can also lower the load against which the muscle must contract; the same is true of drugs that cause relaxation of the muscle in the walls of the venous system. Lowering the blood volume also acts to decrease the over-distension of the heart during diastole. While it would seem that an increase in the resting muscle length would have a beneficial effect on the strength of contraction, geometric factors in the intact heart place the overstretched muscle at a mechanical disadvantage that the length-tension curve cannot adequately overcome.

Heart failure involves numerous interacting organ systems. The mechanical behavior of the heart muscle, as understood in the context of the length-tension and force-velocity curves, is only a part of the problem. Effective therapy must also consider factors external to the heart muscle itself.

---

myofilaments (7) and activates them. The amount of calcium in the cytoplasm, the **cytosolic calcium pool,** determines the magnitude of the myofilament activation and, hence, the level of contractility.

During relaxation, the cytoplasmic calcium concentration is rapidly lowered through several pathways. The SR membrane contains a vigorous Ca^{2+}-ATPase (8) that runs continuously and is further activated, through a protein phosphorylation mechanism, by high levels of cytoplasmic calcium. At the level of the sarcolemma, two additional mechanisms work to rid the cell of the calcium that entered via previous action potentials. A membrane Ca^{2+}-ATPase (9) actively extrudes calcium, ejecting one calcium ion for each ATP molecule consumed. Additional calcium is removed by a Na^+/Ca^{2+} exchange mechanism (10), also located in the cell membrane. This mechanism is part of a coupled transport system in which three sodium ions, entering the cell down their electrochemical gradient, are exchanged for the ejection of one calcium ion. Proper function of this exchange mechanism requires a steep sodium concentration gradient, maintained by the membrane Na^+/K^+-ATPase (11) located in the sarcolemma. Because the Na^+/Ca^{2+} exchange mechanism derives its energy from the sodium gradient, any reduction in the pumping action of the Na^+/K^+-ATPase leads to reduced calcium extrusion. Under normal conditions, these mechanisms can maintain a 10,000-fold Ca^{2+} concentration difference between the inside and outside of the cell. Since a cardiac cell contracts repeatedly many times per minute with each beat being accompanied by an influx of calcium, the extrusion mechanisms must also work continuously to balance the incoming calcium. The mito-

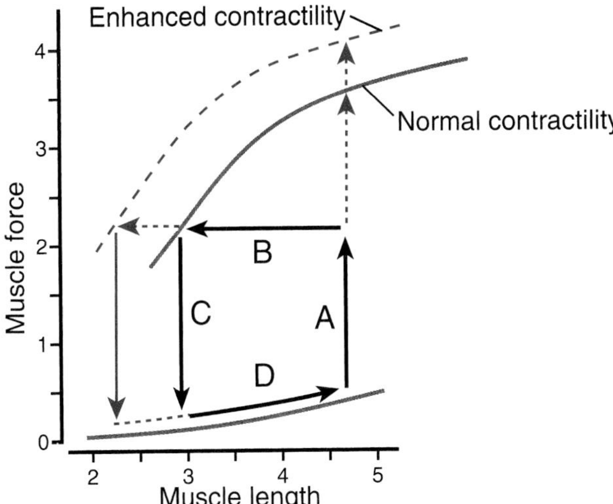

FIGURE 10.6 **Effect of enhanced contractility on the contraction cycle of cardiac muscle.** When contractility is increased, the rate of rise of force is increased, the time to afterload force is decreased, and potential force is increased. The muscle shortens faster and further (A) while isometric relaxation (B) and relengthening (C) are minimally affected (D).

chondria of cardiac muscle (12) are also capable of accumulating and releasing calcium, although this system does not appear to play a role in the normal functioning of the cell.

Calcium and the Function of Inotropic Agents. Inotropic agents usually work through changes in the internal calcium content of the cell. An increase in the heart rate, for instance, allows more separate influxes of calcium per minute, and the amount of releasable calcium in the subsarcolemmal space and SR increases. More crossbridges are activated, and the force of isometric contraction (and other indicators of contractility) increases. This is the basis of the **force-frequency relationship**, one of the principal means of changing myocardial contractility.

Cardiac glycosides are an important class of therapeutic agents used to increase the contractility of failing hearts. The drug **digitalis**, used for centuries for its effects on the circulation, is typical of these agents. While some details of its action are obscure, the drug has been shown to work by inhibiting the membrane Na^+/K^+-ATPase. This allows the cell to gain sodium and reduces the steepness of the sodium gradient. This makes the Na^+/Ca^{2+} exchange mechanism less effective, and the cell gains calcium. Since more calcium is available to activate the myofilaments, contractility increases. These effects, however, can lead to **digitalis toxicity** when the cell gains so much calcium that the capacity of the sarcoplasmic and sarcolemmal binding sites is exceeded. At this point, the mitochondria begin to take up the excess calcium; however, too much mitochondrial calcium interferes with ATP production. The cell, with its ATP needs already increased by enhanced contractility, is less able to pump out accumulated calcium, and the final result is a lowering of metabolic energy stores and a reduction in contractility. Some changes in the contractility of cardiac muscle may be permanent and life threatening. Many of these changes are due to disease or factors external to the heart and may be described by the general term **cardiomyopathy** (see Clinical Focus Box 10.2).

Sources of Energy for Cardiac Muscle Function

In contrast to skeletal muscle, cardiac muscle does not have the opportunity to rest from a period of intense activity to "pay back" an oxygen debt. As a result, the metabolism of cardiac muscle is almost entirely aerobic under basal conditions and uses free fatty acids and lactate as its primary substrates. This correlates with the high con-

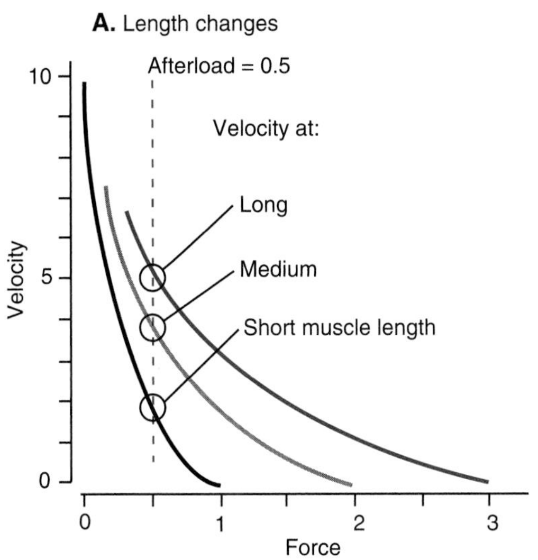

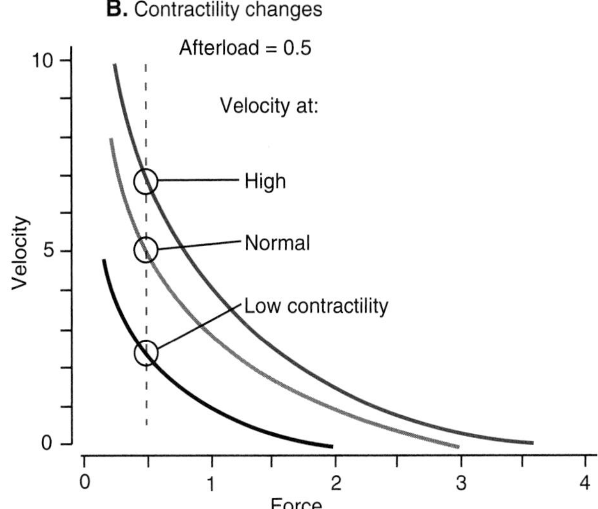

FIGURE 10.7 **Effect of length and contractility changes on the force-velocity curves of cardiac muscle.** A, Decreased starting length (with constant contractility) produces lower velocities of shortening at a given afterload. Because of the presence of resting force (characteristic of heart muscle), it is impossible to make a direct measure of a zero-force contraction at each length. There is a tendency for the curves to converge at the lower forces. B, Increased contractility produces increased velocity of shortening at a constant muscle length, but there is no tendency for the curves to converge at the low forces.

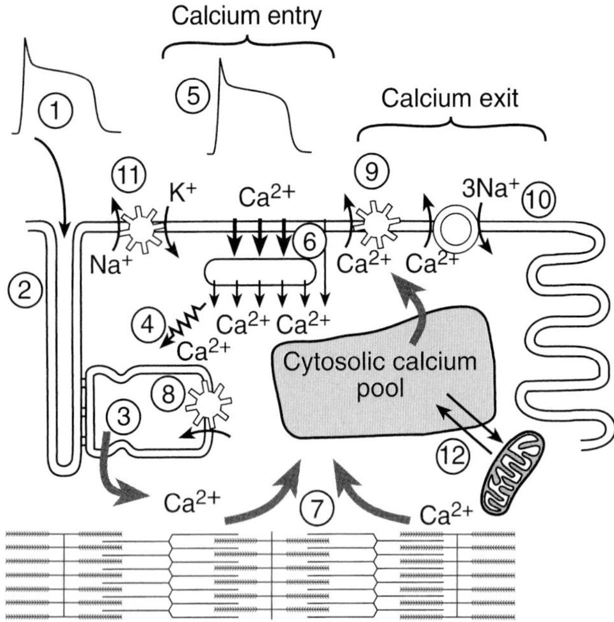

FIGURE 10.8 **The paths of calcium in and out of the cardiac muscle cell and its role in the regulation of contraction.** (See text for details.)

tent of mitochondria in the cells and with the high cellular content of myoglobin. Under conditions of **hypoxia** (lack of oxygen), the anaerobic component of the metabolism may approach 10% of the total, but beyond that limit, the supply of metabolic energy is insufficient to sustain adequate function.

The substrates that provide chemical energy input to the heart during periods of increased activity consist of carbohydrates (mostly in the form of lactic acid produced as a result of skeletal muscle exercise; see Chapters 8 and 9), fats (largely as free fatty acids), and, to a small degree, ketone body acids and amino acids. The relative amounts of the various metabolites vary according to the nutritional status of the body. Because of the highly aerobic nature of cardiac muscle metabolism, there is a strong correlation between the amount of work performed and the amount of oxygen consumed. Under most conditions, the contraction of cardiac muscle in the intact heart is approximately 20% efficient, with the remainder of the energy going to other cellular processes or wasted as heat. Regardless of the dietary or metabolic source of energy, ATP (as in all other muscle types) provides the immediate energy for contraction. As in skeletal muscle, cardiac muscle contains a "rechargeable" creatine phosphate buffering system that supplies the short-term ATP demands of the contractile system.

CLINICAL FOCUS BOX 10.2

Cardiomyopathies: Abnormalities of Heart Muscle

Heart disease takes many forms. While some of these are related to problems with the valves or the electrical conduction system (see Chapters 13 and 14), many are due to malfunctions of the cardiac muscle itself. These conditions, called **cardiomyopathy,** result in impaired heart function that may range from being essentially asymptomatic to malfunctions causing sudden death.

There are several types of cardiomyopathy, and they have several causes. In **hypertrophic cardiomyopathy,** an enlargement of the cardiac muscle fibers occurs because of a chronic overload, such as that caused by hypertension or a defective heart valve. Such muscle may fail because its high metabolic demands cannot be met, or fatal electrical arrhythmias may develop (see Chapter 13). **Congestive** or **dilated cardiomyopathy** refers to cardiac muscle so weakened that it cannot pump strongly enough to empty the heart properly with each beat. In **restrictive cardiomyopathy,** the muscle becomes so stiffened and inextensible that the heart cannot fill properly between beats. Chronic poisoning with heavy metals, such as cobalt or lead, can produce **toxic cardiomyopathy.** The skeletal muscle degeneration associated with **muscular dystrophy** is often accompanied by cardiomyopathy (see Chapter 8).

The cardiomyopathy arising from **viral myocarditis** is difficult to diagnose and may show no symptoms until death occurs. The action of some enteroviruses (e.g., coxsackievirus B) may cause an autoimmune response that does the actual damage to the muscle. This damage may occur at the subcellular level by interfering with energy metabolism while producing little apparent structural disruption. Such conditions, which can usually only be diagnosed by direct muscle biopsy, are difficult to treat effectively, although spontaneous recovery can occur. Excessive and chronic consumption of alcohol can also cause cardiomyopathy that is often reversible if total abstinence is maintained. In tropical regions, infection with a trypanosome **(Chagas' disease)** can produce chronic cardiomyopathy. The tick-borne spirochete infection called **Lyme disease** can cause heart muscle damage and lead to heart block, a conduction disturbance (see Chapter 13).

Another important kind of cardiomyopathy arises from ischemia, an inadequate oxygen (blood) supply to working cardiac muscle. An acute ischemic episode may be followed by a **stunned myocardium,** with reduced mechanical performance. Chronic ischemia can produce a **hibernating myocardium,** also with reduced mechanical performance. Ischemic tissue has impaired calcium handling, which can lead to destructively high levels of internal calcium. These conditions can be improved by reestablishing an adequate oxygen supply (e.g., following clot dissolution or coronary bypass surgery), but even this treatment is risky because rapidly restoring the blood flow to ischemic tissue can lead to the production of oxygen radicals that cause significant cellular damage. The use of calcium blockers and free radical scavengers, such as vitamin E, following ischemic episodes may limit this damage.

REVIEW QUESTIONS

DIRECTIONS: Each of the numbered items or incomplete statements in this section is followed by answers or completions of the statement. Select the ONE lettered answer or completion that is BEST in each case.

1. Which of the following sets of attributes best characterizes cardiac muscle?
(A) Large cells, electrically isolated, neurally stimulated
(B) Small cells, electrically coupled, chemically stimulated
(C) Small cells, electrically coupled, spontaneously active
(D) Small cells, electrically isolated, spontaneously active
2. Cardiac muscle functions as both an electrical and a mechanical syncytium. The structural basis of this ability is the
(A) T tubule system
(B) Intercalated disks
(C) Striated nature of the contractile system
(D) Extensive SR
3. The regulation of contraction in cardiac muscle is
(A) Most like that of smooth muscle (i.e., myosin-linked)
(B) Most like that of skeletal muscle (i.e., actin-linked)
(C) Independent of filament-related proteins
(D) Dependent on autonomic neural stimulation
4. What prevents cardiac muscle from undergoing a tetanic contraction?
(A) The rate of neural stimulation is limited by the CNS
(B) The muscle fatigues so quickly that it must relax fully between contractions
(C) The refractory period of the action potential lasts into the relaxation phase of the contraction
(D) The electrical activity is conducted too slowly for tetanus to occur
5. The contraction cycle for cardiac muscle differs in significant ways from that of skeletal muscle. Which situation below is more typical of cardiac muscle?
(A) The cycle involves only isometric contraction and relaxation
(B) Isometric relaxation occurs at a shorter length than isometric contraction
(C) The muscle relaxes along the same combination of lengths and forces that it took during contraction
(D) The complete cycle in cardiac muscle is isotonic
6. What is the physiological role of the passive length-tension curve in cardiac muscle?
(A) It ensures that force stays constant as the muscle is stretched
(B) It allows the muscle to be extended without limit when it is at rest
(C) It lets the resting muscle length to be set in proportion to the preload
(D) It prevents a contraction from having an isometric phase at shorter lengths
7. Why does cardiac muscle shorten less at higher afterloads?
(A) Higher loads cause a reduction in contractility and this limits the shortening
(B) Higher loads cause rapid fatigue, which limits the shortening
(C) Moving a heavy load causes premature relaxation
(D) It encounters the limit set by the length-tension curve with less shortening
8. What factor provides the most important limit to force production in cardiac muscle?
(A) The resting muscle length from which contraction begins
(B) The size of the preload, which sets the initial length
(C) The size of the afterload during isotonic shortening
(D) The rate (velocity) at which the muscle shortens
9. The factor common to most changes in cardiac muscle contractility is the
(A) Amplitude of the action potential
(B) Availability of cellular ATP
(C) Cytoplasmic calcium concentration
(D) Rate of neural stimulation
10. At a given muscle length, the velocity of contraction depends on
(A) Only the afterload
(B) Only the contractility of the muscle
(C) Both the contractility and the afterload
(D) Only the preload because the contractility is constant

SUGGESTED READING

American Heart Association. Website: http://www.americanheart.org.
Braunwald EG, Ross JR, Sonnenblick EH. Mechanisms of Contraction of the Normal and Failing Heart. Boston: Little, Brown, 1976.
Ford LE. Muscle Physiology and Cardiac Function. Carmel, IN: Biological Sciences Press-Cooper Group, 2000.
Heller LJ, Mohrman DE. Cardiovascular Physiology. New York: McGraw-Hill, 1981.
Katz AM. Physiology of the Heart. 2nd Ed. New York: Raven, 1992.
Noble D. The Initiation of the Heartbeat. Oxford: Oxford University Press, 1979.
WebMD. Website: http://www.webmd.org.

CASE STUDIES FOR PART III • • •

CASE STUDY FOR CHAPTER 8

Polymyositis in an Older Patient

A 67-year-old woman consulted her physician because of recent and progressive muscle weakness. She reported difficulty in rising out of a chair and had intermittent difficulty in swallowing. Physical examination reveals the presence of a light purple rash around her eyes and on her knuckles and elbows. Muscle weakness is noted in all four limbs, but the woman does not complain of muscular soreness. She is somewhat underweight, slightly short of breath, and speaks in a low voice. Laboratory tests show a moderately elevated creatine kinase level. There is no family history of muscle problems, and she is not currently taking any medication.

Because of the symptoms present, no muscle biopsy or electromyographic study is carried out. A tentative diagnosis of polymyositis/dermatomyositis is made. The woman

(continued)

is placed on high-dose prednisone, and arrangements are made for periodic tests for circulating muscle enzymes. Because of her age, she is referred to a cancer specialist to screen for a possible underlying malignancy, and physical therapy is strongly recommended.

In follow-up visits, the woman shows gradual improvement in muscle strength, and her rash is much less apparent. No malignancy is detected. She maintains a regimen of physical therapy and is able to have the prednisone dosage progressively reduced over the course of the next year.

Questions

1. What was a likely cause for the patient's underweight condition?
2. Could the shortness of breath also have been a result of polymyositis?
3. Does the pattern of recovery suggest that the diagnosis was correct?
4. What was the underlying cause of her disease?

Answers to Case Study Questions for Chapter 8

1. Patients with polymyositis involving the pharyngeal and esophageal muscles have difficulty swallowing. This leads to reduced nutritional intake, to the point where it may be life threatening.
2. Although several things could contribute to shortness of breath, weakness of the respiratory muscles can lead to hypoventilation; this, too, can be life threatening.
3. The response to therapy was what one would expect for a person suffering from polymyositis. Conditions such as muscular dystrophy would not have responded as well to the prednisone therapy.
4. Because a malignancy was ruled out, this case must be considered, like most cases of polymyositis, to be of idiopathic origin.

References

Dalakas MC, ed. Polymyositis and Dermatomyositis. London: Butterworth, 1988.

Maddison PJ, et al., eds. Oxford Textbook of Rheumatology. Vol 2. New York: Oxford University Press, 1993.

Case Study for Chapter 9

A Muscle-Pull Injury

A 35-year-old man visited his family physician early on a Monday morning. He walked into the waiting room with a pronounced limp, favoring his right leg, and was in obvious discomfort. When he arose from the waiting-room chair, it was with some difficulty and with considerable assistance from his arms and his left leg. He related that, during the weekend, he had been putting up a swing in a backyard tree for his children. At one point during the work, he jumped to the ground from a ladder leaning against the tree, a distance of about 4 feet. As he landed, he felt a sharp pain in the front of his right thigh, and he fell to his knees upon landing. He was immediately in considerable discomfort, and the pain did not lessen over the course of the weekend.

Physical examination reveals a somewhat swollen aspect to the lower part of the anterior surface of his right thigh. The area is tender to the touch, but the pain does not involve the knee joint. Using the left leg for comparison, he is considerably impaired in his ability to extend the lower portion of his right leg and doing so causes great discomfort.

After the physical examination, he is told that he has most likely experienced a strain (or "pull") of the rectus femoris muscle. He is given a few days' supply of a nonsteroidal drug to manage the pain and inflammation and is told to lessen the pain by applying ice packs to the affected region. He is advised to avoid stair climbing as much as possible during this time, but to begin walking as soon as he could do it without undue pain. On a follow-up visit 2 weeks later, he is experiencing little impairment in walking, although the strength of the leg is still less than normal and stair climbing is still somewhat of a problem. He is advised to return to regular activity, but to avoid any undue overloading of the affected leg for the foreseeable future.

Questions

1. What kind of contraction was the injured muscle undergoing at the time of the injury? Why does this kind of activity pose a special risk for injury?
2. What factors contributed to the occurrence and severity of this injury?
3. Why was the pain localized to the lower portion of the thigh?
4. What sort of activity would be most likely to reinjure the muscle?
5. What precautions should be taken to avoid reinjury?
6. Why was the patient given a limited supply of the pain medication?

Answers to Case Study Questions for Chapter 9

1. The muscle was undergoing an eccentric contraction; that is, the muscle was activated in order to break the fall upon landing, and the body weight extended it while it was active. Such a stretch can produce a force considerably in excess of the maximal isometric capability of a muscle.
2. The first factor was the sudden eccentric contraction (see above). Second, because the patient was not accustomed to the activity in question, the muscle was not conditioned to absorb the suddenly applied stretch. Third, the height from which the patient jumped could potentially generate a force considerably greater than the capability of the muscle.
3. The pain was localized in the general area of the myotendinous junction, the area where damage is most likely to occur.
4. Given the same conditions, a similar jump to the one causing the injury would be quite likely to result in reinjury. In general, any activity that would lead to an eccentric contraction of the muscle would put it at risk. This would explain the caution against stair climbing during the early stages of recovery.
5. There should be a gradual return to full activity, with adequate time for healing and repair, without any sudden increase in the use of the muscle. The initial precipitating behavior should be avoided.
6. The use of the anti-inflammatory medication should be limited because its continued use has been shown to delay the healing process, and it could also mask warning signs of reinjury.

References

Best TM. Soft-tissue injury and muscle tears. Clin Sports Med 1997;16:419–434.

Garrett WE. Muscle strain injuries. Am J Sports Med 1996;24:S2–S8.

Case Study for Chapter 10

Heart Failure

A 50-year-old man consulted his family physician with the principal complaint of shortness of breath and fatigue upon rather mild exertion and a recent weight gain. He appears to be rather pale, moderately overweight, and somewhat short of breath from walking from his car to the office. A careful history yields several pieces of information: He has been a light smoker for most of his adult life, although he has tried to quit; he attributes his morning cough, which resolves after being up for a while, to the smoking habit. He reports

(continued)

that sometimes he awakens suddenly during the night with a feeling of suffocation; sitting upright for a while makes this feeling go away. He has been treated for chronic hypertension, but is no longer taking his prescribed medication. Minor chest pain that he associates with heavy exertion quickly ceases on resting.

Physical examination notes some swelling of his ankles and feet, and palpation reveals a somewhat enlarged and tender liver. Distinct basilar rales (abnormal sounds that indicate pulmonary congestion) are heard during auscultation of the chest. A chest X-ray shows moderate enlargement of the heart, and the same finding (cardiomegaly) is apparent in an ultrasound examination.

The patient is placed on a mild diuretic, along with a drug designed to relax the smooth muscle in the walls of both arteries and veins. He is advised to limit salt intake to less than 4 grams per day; other dietary restrictions include a reduction in the amount of saturated fat and red meat. He is advised that moderate exercise, such as walking, would be beneficial if it is tolerated well. He is referred to a support program to help him quit smoking.

During the next few weeks, significant improvement in exercise tolerance is noted, and both systolic and diastolic blood pressures are reduced. His weight has decreased somewhat. The abnormal lung sounds are absent, and he has been able to quit smoking.

Questions

1. The X-ray and ultrasound data show an increase in the amount of heart muscle. If this was the case, why did the patient suffer from the problems reported above?
2. What effect would lowering the systolic blood pressure have on the ability of the heart muscle to shorten. Why?
3. This patient was not treated with contractility-enhancing drugs. Would such medication have been helpful in this case?
4. Did the result of the diuretic therapy relate most directly to the properties of the muscle at rest or during contraction?
5. Was the patient's morning cough most likely a result of smoking?
6. Did the complaint of fatigue during exercise relate more strongly to problems of the muscle at rest or during contraction?
7. Did the beneficial effects of his therapy relate more to changes in contractility or to changes in the mechanical situation of the heart muscle?
8. What is the benefit of a drug that tends to relax both arterial and venous smooth muscle?

Answers to Case Study Questions for Chapter 10

1. Because of the continuous overload of the heart muscle, it had hypertrophied. At this stage of the patient's disease, however, even the added muscle strength was not sufficient to handle the demands of the body during exercise.
2. With a lowered systolic pressure, the afterload during shortening would be reduced. An examination of the length-tension curve shows that more shortening would be possible, and the force-velocity curve would predict that the contraction would also be more rapid.
3. The use of drugs such as digitalis could have relieved the patient's symptoms sooner, but the risks of such drugs (heart rhythm disturbances, systemic and cardiac toxicity, etc.) make it advisable, if at all possible, to let the inherent properties of the muscle, when properly aided, to correct the problem.
4. The diuretic therapy reduced the blood volume, which meant that the heart muscle was less distended at rest. A lowered arterial volume would have also lowered the afterload on the muscle. Thus, both aspects of the problem were addressed.
5. Because the cough went away soon after arising, it was more likely a result of fluid accumulation in the lungs. The increased heart rate and contractility of the muscle associated with waking activity would have at least partly overcome this problem as the day progressed.
6. The feeling of fatigue is related to the lack of blood circulation in the skeletal muscle. This was most directly related to the weakened state of the heart muscle during contraction, which would reduce the amount of blood that could be pumped with each beat.
7. Because the patient was not given drugs that directly addressed the contractility of the muscle, the beneficial changes must have come about principally through the reduction of the preload and afterload on the muscle.
8. Such drugs can address problems of both excessive afterload and preload at the same time.

Reference

Poole-Wilson PA, Colucci WS, Massie BM, Chatterjee K, Coats AJS. Heart Failure: Scientific Principles and Clinical Practice. New York: Churchill Livingstone, 1997.

PART IV Blood and Cardiovascular Physiology

CHAPTER 11 Blood Components, Immunity, and Hemostasis

Denis English, Ph.D.

CHAPTER OUTLINE

- **THE COMPONENTS OF BLOOD**
- **THE IMMUNE SYSTEM**
- **HEMOSTASIS**

KEY CONCEPTS

1. Blood functions as a dynamic tissue.
2. Blood consists of erythrocytes, leukocytes, and platelets suspended within a solute-rich plasma.
3. Erythrocytes carry oxygen to the tissues.
4. Leukocytes protect the body against pathogens.
5. Platelets and plasma proteins control hemostasis, a process that stops blood loss after injury and promotes wound healing.
6. Blood cells are derived from bone marrow precursors.
7. In protecting the body against irritants and pathogens, the process of inflammation often results in the destruction of healthy tissue.
8. Adaptive immunity is specific and acquired.
9. During clotting, platelets release biologically active cofactors, which promote wound healing, inflammation, angiogenesis (blood vessel formation), and host defense.

Blood is a highly differentiated, complex living tissue that pulsates through the arteries to every part of the body, interacts with individual cells via an extensive capillary network, and returns to the heart through the venous system. Many of the functions of blood are undertaken in the capillaries, where the blood flow slows dramatically, allowing the efficient diffusion and transport of oxygen, glucose, and other molecules across the monolayer of endothelial cells that form the thin capillary walls. In addition to transport, blood and the cells within it mediate other essential aspects of immunity and hemostasis.

The human body is continually invaded by pathogenic microorganisms that enter through skin cuts, mucous membranes, and other sites of infection and tissue disruption. To oppose pathogenic microbes, the body has developed a highly sophisticated immune system. Cells of the immune system, the white blood cells, are derived from bone marrow precursors and are delivered to their sites of action by the blood. These cells, also known as leukocytes, exert their effects in conjunction with antibodies and protein cofactors in blood. In this chapter, we will see how certain leukocytes act without prior sensitization to neutralize offending pathogens, while others require a prior infectious insult to deal with invaders.

In addition to infectious assault, the body is continually threatened by the devastating consequences of vascular leak or hemorrhage as a result of even the most innocuous tissue injury. A highly organized clotting system, consisting of blood platelets that work in conjunction with blood plasma clotting factors, prevents excessive fluid loss by rapidly forming a hemostatic plug. In addition to physically constraining fluids within ruptured vessels, platelets release potent biological cofactors during the development of this hemostatic plug, which promote wound healing, prevent further infection, and promote the development and vascularization of new tissue.

THE COMPONENTS OF BLOOD

Blood is an opaque, red liquid consisting of several types of cells suspended in a complex, amber fluid known as **plasma.** When blood is allowed to clot or coagulate, the suspending medium is referred to as **serum**.

Blood Has a Higher Density and Viscosity than Water

Blood is normally confined to the circulation, including the heart and the pulmonary and systemic blood vessels. Blood accounts for 6 to 8% of the body weight of a healthy adult. The blood volume is normally 5.0 to 6.0 L in men and 4.5 to 5.5 L in women.

The **density** (or **specific gravity**) of blood is approximately 1.050 g/mL. Density depends on the number of blood cells present and the composition of the plasma. The density of individual blood cells varies according to cell type and ranges from 1.115 g/mL for erythrocytes to 1.070 g/mL for certain leukocytes.

While blood is only slightly heavier than water, it is certainly much thicker. The **viscosity** of blood, a measure of resistance to flow, is 3.5 to 5.5 times that of water. Blood's viscosity increases as the total number of cells present increases and when the concentration of large molecules (macromolecules) in plasma increases. At pathologically high viscosity, blood flows poorly to the extremities and internal organs.

The Erythrocyte Sedimentation Rate and Hematocrit Are Important Diagnostic Measurements

Erythrocytes are the red cells of blood. Since erythrocytes have only a slightly higher density than the suspending plasma, they normally settle out of whole blood very slowly. To determine the **erythrocyte sedimentation rate** (ESR), anticoagulated blood is placed in a long, thin, graduated cylinder (Fig. 11.1). As the red cells sink, they leave behind the less dense leukocytes and platelets in the suspending plasma. Erythrocytes in the blood of healthy men sediment at a rate of 2 to 8 mm/hr; those in the blood of healthy women sediment slightly faster (2 to 10 mm/hr).

The ESR can be an important diagnostic index, as values are often significantly elevated during infection, in patients with arthritis, and in patients with inflammatory diseases. In some diseases, such as **sickle cell anemia, polycythemia** (abnormal increase in red cell numbers), and **hyperglycemia** (elevated blood sugar levels), the ESR is slower than normal. The reasons for alterations in the ESR in disease states are not always clear, but the cells tend to sediment faster when the concentration of plasma proteins increases.

Blood cells can be quickly separated from the suspending fluid by simple centrifugation. When anticoagulated blood is placed in a tube that is rotated about a central point, centrifugal forces pull the blood cells from the suspending plasma. The **hematocrit** is the portion of the total blood volume that is made up of red cells. This value is determined by the centrifugation of small capillary tubes of anticoagulated blood to pack the cells.

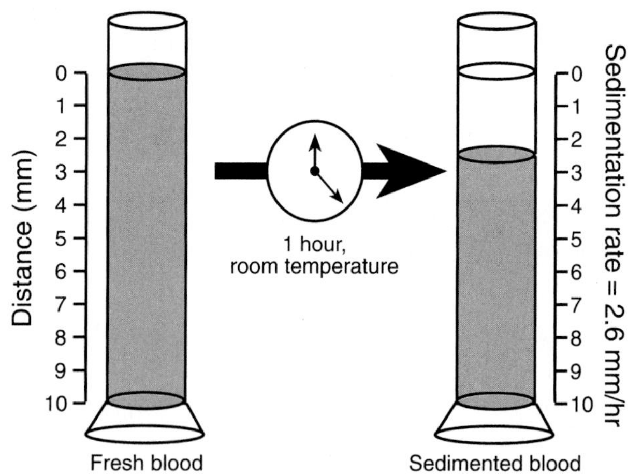

FIGURE 11.1 **Determination of the erythrocyte sedimentation rate (ESR).** Fresh, anticoagulated blood is allowed to settle at room temperature in a graduated cylinder. After a fixed time interval (1 hour), the distance (in millimeters) that the erythrocytes' sediment is measured.

Determination of hematocrit values is a simple and important screening diagnostic procedure in the evaluation of hematological disease. Hematocrit values of the blood of healthy adults are 47 ± 5% for men and 42 ± 5% for women. Decreased hematocrit values often reflect blood loss as a result of bleeding or deficiencies in blood cell production. Low hematocrit values indicate the presence of **anemia,** a reduction in the number of circulating erythrocytes. Increased hematocrit levels may likewise indicate a serious imbalance in the production and destruction of red cells. Increased production (or decreased rate of destruction) of erythrocytes results in polycythemia, as reflected by increased hematocrit values. Dehydration, which decreases the water content and, thus, the volume of plasma, also results in an increase in hematocrit.

Blood Functions as a Dynamic Tissue

While the cellular and plasma components of blood may act alone, they often work in concert to perform their functions. Working together, blood cells and plasma proteins play several important roles, including

- **Transport** of substances from one area of the body to another
- **Immunity,** the body's defense against disease
- **Hemostasis,** the arrest of bleeding
- **Homeostasis,** the maintenance of a stable internal environment

Transport. Blood carries several important substances from one area of the body to another, including oxygen, carbon dioxide, antibodies, acids and bases, ions, vitamins, cofactors, hormones, nutrients, lipids, gases, pigments, minerals, and water. Transport is one of the primary and most important functions of blood, and blood is the primary means of long-distance transport in the body. Substances can be transported free in plasma, bound to plasma proteins, or within blood cells.

Oxygen and carbon dioxide are two of the more important molecules transported by blood. Oxygen is taken up by the red cells as they pass through capillaries in the lung. In tissue capillaries, red cells release oxygen, which is then used by respiring tissue cells. These cells produce carbon dioxide and other wastes.

The blood also transports heat. By doing so, it maintains the proper temperature in different organs and tissues, and in the body as a whole.

Immunity. Blood leukocytes are involved in the body's battle against infection by microorganisms. While the skin and mucous membranes physically restrict the entry of infectious agents, microbes constantly penetrate these barriers and continuously threaten internal infection. Blood leukocytes, working in conjunction with plasma proteins, continuously patrol for microbial pathogens in the tissues and in the blood. In most cases, penetrating microbes are efficiently eliminated by the sophisticated and elaborate antimicrobial systems of the blood.

Hemostasis. Bleeding is controlled by the process of hemostasis. Complex and efficient hemostatic mechanisms have evolved to stop hemorrhage after injury, and their failure can quickly lead to fatal blood loss (exsanguination). Both physical and cellular mechanisms participate in hemostasis. These mechanisms, like those of the immune system, are complex, interrelated, and essential for survival.

Homeostasis. Homeostasis is a steady state that provides an optimal internal environment for cell function (see Chapter 1). By maintaining pH, ion concentrations, osmolality, temperature, nutrient supply, and vascular integrity, the blood system plays a crucial role in preserving homeostasis. Homeostasis is the result of normal functioning of the blood's transport, immune, and hemostatic systems.

Plasma Contains Many Important Solutes

Plasma is composed mostly of water (93%) with various dissolved solutes, including proteins, lipids (fats), carbohydrates, amino acids, vitamins, minerals, hormones, wastes, cofactors, gases, and electrolytes (Table 11.1). The solutes in plasma play crucial roles in homeostasis, such as maintaining normal plasma pH and osmolality.

TABLE 11.1 Some Components of Plasma

Class	Substance	Normal Concentration Range
Cations	Sodium (Na^+)	136–145 mEq/L
	Potassium (K^+)	3.5–5.0 mEq/L
	Calcium (Ca^{2+})	4.2–5.2 mEq/L
	Magnesium (Mg^{2+})	1.5–2.0 mEq/L
	Iron (Fe^{3+})	50–170 μg/dL
	Copper (Cu^{2+})	70–155 μg/dL
	Hydrogen (H^+)	35–45 nmol/L
Anions	Chloride (Cl^-)	95–105 mEq/L
	Bicarbonate (HCO_3^-)	22–26 mEq/L
	Lactate	0.67–1.8 mEq/L
	Sulfate ($SO_4^{2}-$)	0.9–1.1 mEq/L
	Phosphate ($HPO_4^{2}-/H_2PO_4-$)	3.0–4.5 mg/dL
Proteins	Total	6–8 g/dL
	Albumin	3.5–5.5 g/dL
	Globulin	2.3–3.5 g/dL
Fats	Cholesterol	150–200 mg/dL
	Phospholipids	150–220 mg/dL
	Triglycerides	35–160 mg/dL
Carbohydrates	Glucose	70–110 mg/dL
Vitamins, cofactors, and enzymes	Vitamin B_{12}	200–800 pg/mL
	Vitamin A	0.15–0.6 μg/mL
	Vitamin C	0.4–1.5 mg/dL
	2,3-Diphosphoglycerate (DPG)	3–4 mmol/L
	Transaminase (SGOT)	9–40 U/mL
	Alkaline phosphatase	20–70 U/L
	Acid phosphatase	0.5–2 U/L
Other substances	Creatinine	0.6–1.2 mg/dL
	Uric acid	0.18–0.49 mmol/L
	Blood urea nitrogen	7–18 mg/dL
	Iodine	3.5–8.0 μg/dL
	CO_2	23–30 mmol/L
	Bilirubin (total)	0.1–1.0 mg/dL
	Aldosterone	3–10 ng/dL
	Cortisol	5–18 μg/dL
	Ketones	0.2–2.0 mg/dL

There Are Three Types of Blood Cells

Blood cells include **erythrocytes** (**red blood cells**), **leukocytes** (**white blood cells**), and **platelets** (**thrombocytes**). Each microliter (a millionth of a liter) of blood contains 4 to 6 million erythrocytes, 4,500 to 10,000 leukocytes, and 150,000 to 400,000 platelets. There are several subtypes of leukocytes, defined by morphological differences (Fig. 11.2), each with vastly different functional characteristics and capabilities. Table 11.2 lists the normal circulating levels of different blood cell types.

Of the total leukocytes, 40 to 75% are neutrophilic, polymorphonuclear (multinucleated) cells, otherwise known as **neutrophils**. These phagocytic cells actively ingest and destroy invading microorganisms. **Eosinophils** and **basophils** are polymorphonuclear cells that are present in low numbers in blood (1 to 6% of total leukocytes) and participate in allergic hypersensitivity reactions. Mononuclear cells, including **monocytes** and **lymphocytes**, comprise 20 to 50% of the total leukocytes. These cells generate antibodies and mount cellular immune reactions against invading agents.

The number and relative proportion of the leukocyte subtypes can vary widely in different disease states. For example, the absolute neutrophil count often increases during infection, presumably in response to the infection. Eosinophil counts increase when allergic individuals are exposed to allergens. Lymphocyte counts decrease in AIDS and during some other viral infections. For this reason, in addition to a **blood cell count**, a **differential analysis** of leukocyte subtypes, performed by microscopic examination of stained slides, can provide important clues to the diagnosis of disease.

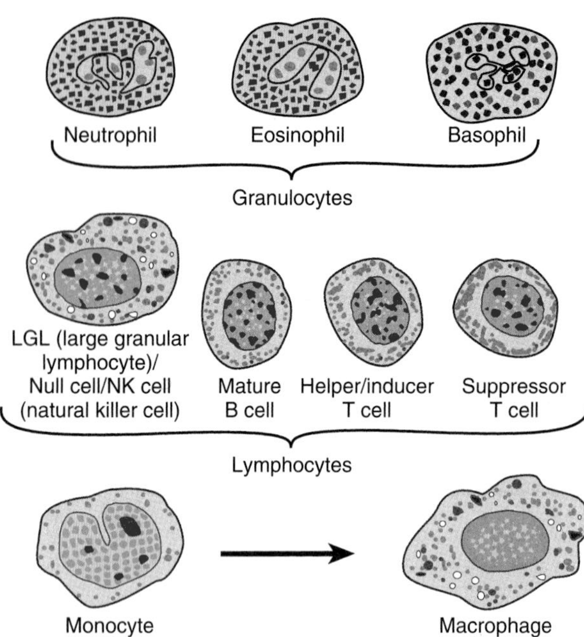

FIGURE 11.2 **Types of leukocytes in blood and tissues.** All of the cells shown here are found in the circulation except the macrophage, which differentiates from activated monocytes in tissue.

Erythrocytes Carry Oxygen to Tissues

Erythrocytes are the most numerous cells in blood. These biconcave disks lack a nucleus and have a diameter of about 7 μm and a maximum thickness of 2.5 μm. The shape of the erythrocyte optimizes its surface area, increasing the efficiency of gas exchange.

The erythrocyte maintains its shape by virtue of its complex membrane skeleton, which consists of an insoluble mesh of fibrous proteins attached to the inside of the plasma membrane. This structural arrangement allows the erythrocyte great flexibility as the cell twists and turns through small, curved vessels. In addition to structural proteins of the membrane, several functional proteins are found in the cytoplasm of erythrocytes. These include hemoglobin (the major oxygen-carrying protein), antioxidant enzymes, and glycolytic systems to provide cellular energy (ATP). The plasma membrane possesses ion pumps that maintain a high level of intracellular potassium and a low level of intracellular calcium and sodium.

Hemoglobin, the red, oxygen-transporting protein of erythrocytes, consists of a **globin** (or protein) portion and four **heme groups**, the iron-carrying portion. The molecular weight of hemoglobin is about 64,500. This complex protein possesses four polypeptide chains: two α-globin molecules of 141 amino acids each and two molecules of another type of globin chain (β, γ, δ, or ϵ), each containing 146 amino acid residues (Fig. 11.3).

Four types of hemoglobin molecules can be found in human erythrocytes: embryonic, fetal, and two different types found in adults (HbA, HbA_2). Each hemoglobin molecule is designated by its polypeptide composition. For example, the most prevalent adult hemoglobin, HbA, consists of two α chains and two β chains. Its formula is given as $\alpha_2\beta_2$. HbA_2, which makes up about 1.5 to 3% of total hemoglobin in an adult, has the subunit formula $\alpha_2\delta_2$. Fetal hemoglobin ($\alpha_2\gamma_2$) is the major hemoglobin component during intrauterine life. Its levels in circulating blood cells decrease rapidly during infancy and reach a concentration of 0.5% in adults. Embryonic hemoglobin is found earlier in development. It consists of two α chains and two ϵ chains ($\alpha_2\epsilon_2$). The production of ϵ chains ceases at about the third month of fetal development.

The production of each type of globin chain is controlled by an individual structural gene with five different loci. Mutations, which can occur anywhere in these five loci, have resulted in the production of over 550 types of abnormal hemoglobin molecules, most of which have no known clinical significance. Mutations can arise from a single substitution within the nucleic acid of the gene coding for the globin chain, a deletion of the codons, or gene rearrangement as a result of unequal crossing over between homologous chromosomes. Sickle cell anemia, for example, results from the presence of sickle cell hemoglobin (HbS), which differs from normal adult hemoglobin A be-

TABLE 11.2 **Circulating Blood Cell Levels**

Blood Cell Type	Approximate Normal Range
Erythrocytes (cells/μL)	
Men	$4.3–5.9 \times 10^6$
Women	$3.5–5.5 \times 10^6$
Leukocytes (cells/μL)	4,500–11,000
Neutrophils	4,000–7,000
Lymphocytes	2,500–5,000
Monocytes	100–1,000
Eosinophils	0–500
Basophils	0–100
Platelets (cells/μL)	150,000–400,000

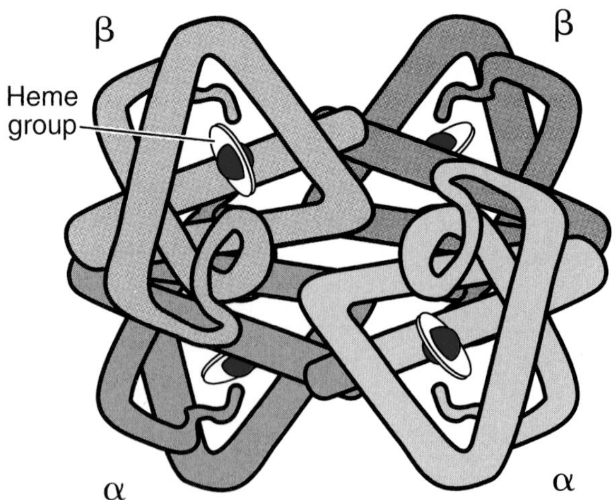

FIGURE 11.3 **Structure of hemoglobin A.** Each molecule of hemoglobin possesses four polypeptide chains, each containing iron bound to its heme group (Modified from Dickerson RE, Geis I. The Structure and Action of Proteins. New York: Harper & Row, 1969;3.)

cause of the substitution of a single amino acid in each of the two β chains.

Oxyhemoglobin (HbO_2), the oxygen-saturated form of hemoglobin, transports oxygen from the lungs to tissues, where the oxygen is released. When oxygen is released, HbO_2 becomes **reduced hemoglobin** (Hb). While oxygen-saturated hemoglobin is bright red, reduced hemoglobin is bluish-red, accounting for the difference in the color of blood in arteries and veins.

Certain chemicals readily block the oxygen-transporting function of hemoglobin. For example, carbon monoxide (CO) rapidly replaces oxygen in HbO_2, resulting in the formation of the stable compound **carboxyhemoglobin** (HbCO). The formation of HbCO accounts for the asphyxiating properties of CO. Nitrates and certain other chemicals oxidize the iron in Hb from the ferrous to the ferric state, resulting in the formation of **methemoglobin** (metHb). MetHb contains oxygen bound tightly to ferric iron; as such, it is useless in respiration. **Cyanosis,** the dark-blue coloration of skin associated with anoxia, becomes evident when the concentration of reduced hemoglobin exceeds 5 g/dL. Cyanosis may be rapidly reversed by oxygen if the condition is caused only by a diminished oxygen supply. However, cyanosis caused by the intestinal absorption of nitrates or other toxins, a condition known as **enterogenous cyanosis**, is due to the accumulation of stabilized methemoglobin and is not rapidly reversible by the administration of oxygen alone.

Normal Red Cell Values. In evaluating patients for hematological diseases, it is important to determine the hemoglobin concentration in the blood, the total number of circulating erythrocytes (the red cell count), and the hematocrit. From these values several other important blood values can be calculated, including **mean cell hemoglobin concentration** (MCHC), **mean cell hemoglobin** (MCH), **mean cell volume** (MCV), and blood **oxygen carrying capacity.**

The MCHC provides an index of the average hemoglobin content in the mass of circulating red cells. It is calculated as follows:

$$\text{MCHC} = \text{Hb (g/L)/hematocrit} \quad (1)$$

Example: 150 g/L ÷ 0.45 = 333 g/L

Low MCHC values indicate deficient hemoglobin synthesis. High MCHC values do not occur in erythrocyte disorders, because normally the hemoglobin concentration is close to the saturation point in red cells. Note that the MCHC value is easily obtained by a simple calculation from measurements that can be made without sophisticated instrumentation.

The MCH value is an estimate of the average hemoglobin content of each red cell. It is derived as follows:

$$\text{MCH} = \frac{\text{Blood hemoglobin (g/L)}}{\text{Red cell count (cells/L)}} \quad (2)$$

Example: 150 g/L ÷ (5×10^{12} cells/L) = 30×10^{-12} g/cell = 30 pg/cell

Since the red cell count is usually related to the hematocrit, the MCH is usually low when the MCHC is low. Exceptions to this rule yield important diagnostic clues.

The MCV value reflects the average volume of each red cell. It is calculated as follows:

$$\text{MCV} = \text{Hematocrit/Number of red cells} \quad (3)$$

Example: 0.450/(5×10^{12} cells/L) = 0.090×10^{-12} L/cell = 90 fL (1 fL = 10^{-15} L)

Each gram of hemoglobin can combine with and transport 1.34 mL of oxygen. Thus, the oxygen carrying capacity of 1 dL of normal blood containing 15 g of hemoglobin is 15 × 1.34 = 20.1 mL of oxygen.

Red Cell Morphology. In addition to revealing alterations in absolute values, stained blood films may provide valuable information based on the morphological appearance of blood cells. Erythrocytes are formed from precursor blast cells in the bone marrow (Fig. 11.4). This process, termed **erythropoiesis,** is regulated by **erythropoietin,** a hormone produced in the kidneys.

Changes in red cell appearance occur in a variety of pathological conditions (Fig. 11.5). Excessive variation in the size of cells is referred to as **anisocytosis.** Larger-than-

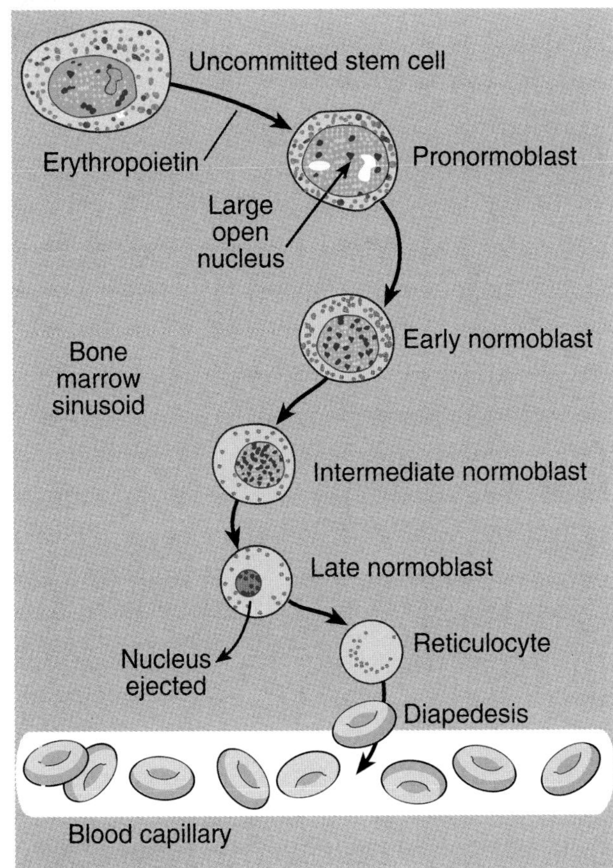

FIGURE 11.4 **Erythropoiesis.** Erythrocyte production in healthy adults occurs in marrow sinusoids. Driven by the hormone erythropoietin, the uncommitted stem cell differentiates along the erythrocyte lineage, forming normoblasts (also referred to as erythroblasts or burst-forming cells), reticulocytes and, finally, mature erythrocytes, which enter the bloodstream by the process of diapedesis.

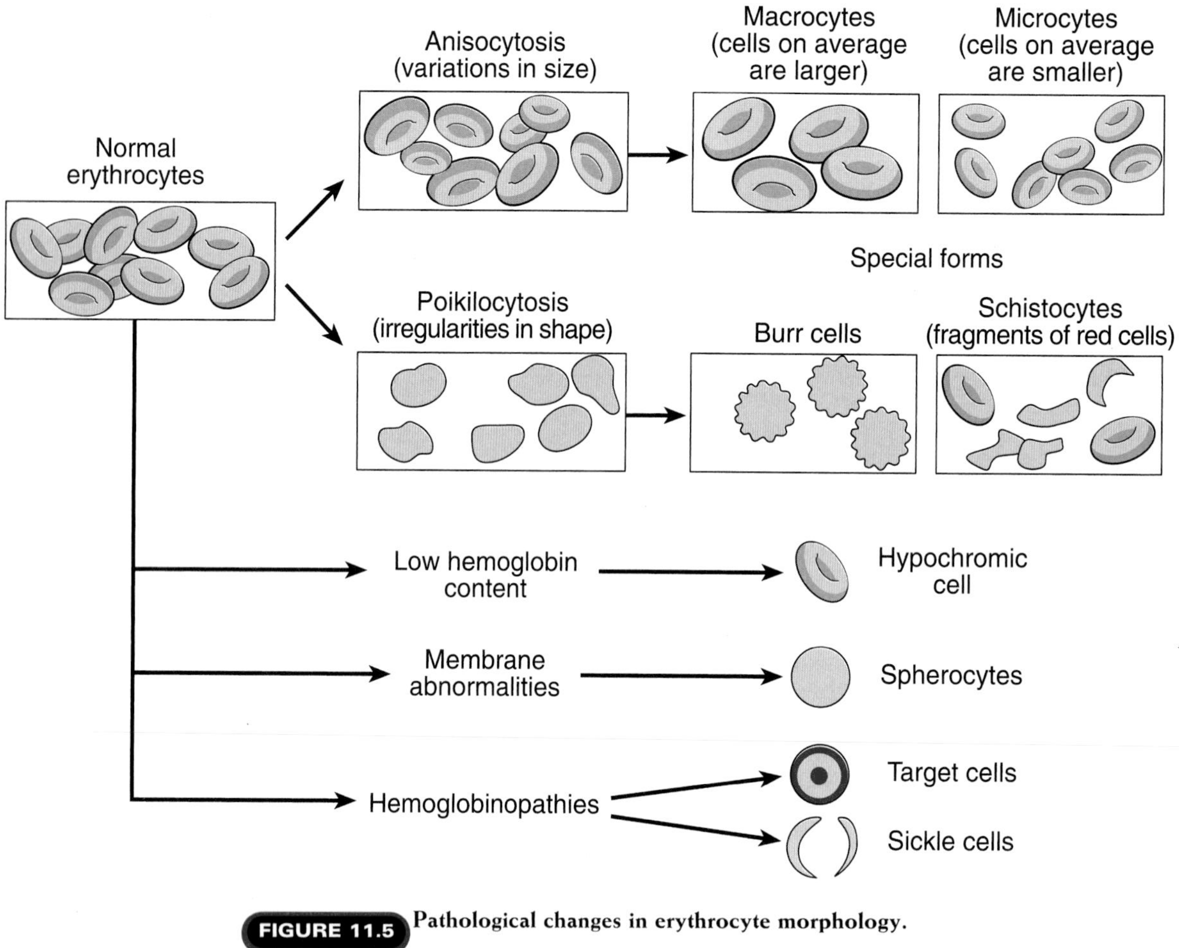

FIGURE 11.5 **Pathological changes in erythrocyte morphology.**

normal erythrocytes are termed **macrocytes**; smaller-than-normal erythrocytes are referred to as **microcytes. Poikilocytosis** is the presence of irregularly shaped erythrocytes. **Burr cells** are spiked erythrocytes generated by alterations in the plasma environment. **Schistocytes** are fragments of red cells damaged during blood flow through abnormal blood vessels or cardiac prostheses.

The hemoglobin content of erythrocytes is also reflected in the staining pattern of cells on dried films. Normal cells appear red-orange throughout, with a very slight central pallor as a result of the cell shape. **Hypochromic cells** appear pale with only a ring of deeply colored hemoglobin on the periphery. Other pathological variations in red cell appearance include **spherocytes**—small, densely staining red cells with loss of biconcavity as a result of congenital or acquired cell membrane abnormalities; and **target cells**—which have a densely staining central area with a pale surrounding area. Target cells are thin but bulge in the middle, unlike normal erythrocytes. This alteration is a consequence of **hemoglobinopathies**, mutations in the structure of hemoglobin. Target cells are observed in liver disease and after splenectomy.

Nucleated red cells are normally not seen in peripheral blood because their nuclei are lost before they move from the bone marrow into the blood. However, they appear in many blood and marrow disorders, and their presence can be of diagnostic significance. One type of nucleated red cell, the **normoblast** (see Fig. 11.4), is seen in several types of anemias, especially when the marrow is actively responding to demand for new erythrocytes. In seriously ill patients, the appearance of normoblasts in peripheral blood is a grave prognostic sign preceding death, often by several hours. Another nucleated erythrocyte, the **megaloblast,** is seen in peripheral blood in pernicious anemia and folic acid deficiency.

Erythrocyte Destruction. Red cells circulate for about 120 days after they are released from the marrow. Some of the senescent (old) red cells break up (hemolyze) in the bloodstream, but the majority are engulfed by macrophages in the monocyte-macrophage system. The hemoglobin released on destruction of red cells is metabolically catabolized and eventually reused in the synthesis of new hemoglobin. Hemoglobin released by red cells that lyse in the circulation either binds to **haptoglobin,** a protein in plasma, or is broken down to globin and heme. Heme binds a second plasma carrier protein, **hemopexin,** which, like haptoglobin, is cleared from the circulation by macrophages in the liver. In the macrophage, released hemoglobin is first broken into globin and heme. The globin portion is catabolized by proteases into constituent amino acids that are used in protein synthesis. Heme is broken down into free iron (Fe^{3+}) and **biliverdin**, a green substance that is further reduced to **bilirubin** (see Chapter 27).

Iron Recycling. Most of the iron needed for new hemoglobin synthesis is obtained from the heme of senescent red cells. Iron released by macrophages is transported in the ferric state in plasma bound to the iron transporting protein, **transferrin.** Cells that need iron (e.g., for heme synthesis) possess membrane receptors to which transferrin binds. The receptor-bound transferrin is then internalized. The iron is released, reduced intracellularly to the ferrous state, and either incorporated into heme or stored as **ferritin**, a complex of protein and ferrous hydroxide. Iron is also stored as ferritin by macrophages in the liver. A portion of the ferritin is catabolized to **hemosiderin**, an insoluble compound consisting of crystalline aggregates of ferritin. The accumulation of large amounts of hemosiderin formed during periods of massive hemolysis can result in damage to vital organs, including the heart, pancreas, and liver.

The recycling of iron is quite efficient, but small amounts are continuously lost. Iron loss increases substantially in women during menstruation. Iron stores must be replenished by dietary intake. The majority of iron in the diet is derived from heme in meat ("organic iron"), but iron can also be provided by the absorption of inorganic iron by intestinal epithelial cells. In these cells, iron attached to heme is released and reduced to the ferrous form (Fe^{2+}) by intracellular flavoprotein. The reduced iron (both released from heme and absorbed as the inorganic ion) is transported through the cytoplasm bound to a transferrin-like protein. When it is released to the plasma, it is oxidized to the ferric state and bound to transferrin for use in heme synthesis.

Platelets Participate in Clotting

Platelets are irregularly shaped, disk-like fragments of the membrane of their precursor cell, the **megakaryocyte.** Megakaryocytes shed platelets in the bone marrow **sinusoids.** From there the platelets are released to the blood, where they function in hemostasis. Several factors stimulate megakaryocytes to release platelets, including the hormone **thrombopoietin,** which is generated and released into the bloodstream when the number of circulating platelets drops. Platelets have no defined nucleus. They are one fourth to one third the size of erythrocytes. Platelets possess physiologically important proteins, stored in intracellular granules, which are secreted when the platelets are activated during coagulation. The role of platelets in blood clotting is discussed below.

Leukocytes Participate in Host Defense

Each of the three general types of leukocytes—**myeloid, lymphoid,** and **monocytic**—follows a separate line of development from primitive cells (see Fig. 11.2). Mature cells of the myeloid series are termed **granulocytes,** based on their appearance after staining with polychromatic dyes, such as Wright's stain. While monocytes and lymphocytes may also possess cytoplasmic granules, they are not clearly visualized with commonly used stains. Therefore, monocytes and lymphocytes are often referred to as **agranular leukocytes.**

The nuclei of most mature granulocytes are divided into two to five oval lobes connected by thin strands of chromatin. This nuclear separation imparts a multinuclear appearance to granulocytes, which are, therefore, also known as **polymorphonuclear leukocytes.** Three distinct types of granulocytes have been identified based on staining reactions of their cytoplasm with polychromatic dyes: neutrophils, eosinophils, and basophils.

Neutrophils. Neutrophils are usually the most prevalent leukocyte in peripheral blood. These dynamic cells respond instantly to microbial invasion by detecting foreign proteins or changes in host defense network proteins. Neutrophils provide an efficient defense against pathogens that have gotten past physical barriers such as the skin. Defects in neutrophil function quickly lead to massive infection—and, quite often, death.

Neutrophils are amoeba-like phagocytic cells. Invading bacteria induce neutrophil **chemotaxis**—migration to the site of infection. Chemotaxis is initiated by the release of **chemotactic factors** from the bacteria or by chemotactic factor generation in the blood plasma or tissues. Chemotactic factors are generated when bacteria or their products bind to circulating antibodies, by tissue cells when infected with bacteria, and by lymphocytes and platelets after interaction with bacteria.

After neutrophils migrate to the site of infection, they engulf the invading pathogen by the process of **phagocytosis**. Phagocytosis is facilitated when the bacteria are coated with the host defense proteins known as **opsonins.**

A burst of metabolic events occurs in the neutrophil after phagocytosis (Fig. 11.6). In the phagocytic vacuole or **phagosome**, the bacterium is exposed to enzymes that were originally positioned on the cell surface. Thus, phagocytosis involves invagination and then vacuolization of the segment of membrane to which a pathogen is bound. Membrane-bound enzymes, activated when the phagocytic vacuole closes, work in conjunction with enzymes secreted from intracellular **granules** into the phagocytic vacuole to destroy the invading pathogen efficiently. One important membrane-bound enzyme, **nicotinamide adenine dinucleotide phosphate (NADPH) oxidase,** produces **superoxide anion** (O_2^-). Superoxide is an unstable **free radical** that kills bacteria directly. Superoxide also participates in secondary free radical reactions to generate other potent antimicrobial agents, such as hydrogen peroxide. Superoxide generation in the phagocytic vacuole proceeds at the expense of reducing agents oxidized in the cytoplasm. The reducing agent, NADPH, is generated from glucose by the activity of the **hexose monophosphate shunt.** Aerobic cells generate reduced **nicotinamide adenine dinucleotide** (NADH) and ATP when glucose is oxidized to carbon dioxide. The hexose monophosphate shunt operates in neutrophils and other cells when large

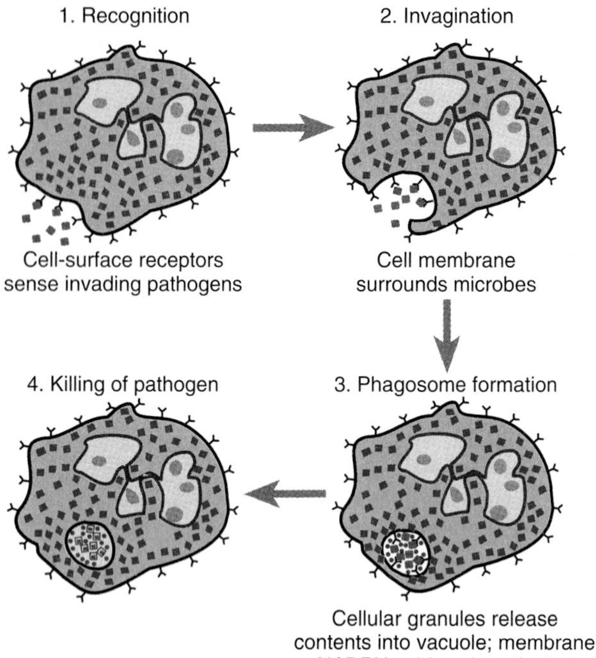
1. Recognition
2. Invagination
Cell-surface receptors sense invading pathogens
Cell membrane surrounds microbes
4. Killing of pathogen
3. Phagosome formation
Cellular granules release contents into vacuole; membrane NADPH oxidase is activated

FIGURE 11.6 Steps in phagocytosis and intracellular killing by neutrophils. *1,* Cell-surface receptors, including those for exposed opsonins, sense invading pathogens. *2,* The neutrophil plasma membrane invaginates to surround the organisms. *3,* A membrane-bounded vesicle formed from the invagination of the cell membrane, called a phagosome, traps the bacteria inside the neutrophil. *4,* Potent metabolic processes are activated to kill the ingested microbes, including activation of the respiratory burst, resulting in the generation of potent oxidants within the phagosome, and the secretion of bacteria-killing enzymes into the phagosome from neutrophil granules.

amounts of NADPH are needed to maintain intracellular reducing activity.

Oxygen reduction by the NADPH oxidase that generates superoxide in neutrophils is driven by an increased availability of NADPH after phagocytosis:

$$NADPH + 2O_2 \xrightarrow[\text{NADPH oxidase}]{} 2O_2^- + NADP^+ + H^+ \quad (4)$$

A complex set of biochemical events unfolds after phagocytosis to activate the neutrophil NADPH oxidase, which is dormant in resting cells. The oxidase is activated by its interaction with an activated **G protein** and cytosolic molecules that are generated during phagocytosis. The NADPH oxidase is activated in a manner that allows the enzyme to secrete the toxic free radical, superoxide, into the phagocytic vacuole while oxidizing NADPH in the cell's cytoplasm. This explosion of metabolic activity, collectively termed the **respiratory burst,** leads to the generation of potent, reactive agents not otherwise generated in biological systems. These agents are so reactive that they actually generate light (biological chemiluminescence) when they oxidize components in the bacterial cell wall.

Other bactericidal agents and processes operate in neutrophils to ensure efficient bacterial killing. Phagocytized bacteria encounter intracellular **defensins,** cationic proteins that bind to and inhibit the replication of bacteria. Defensins and other antibacterial agents pour into the phagocytic vacuole after phagocytosis. Agents stored in neutrophil granules include **lysozyme,** a bacteriolytic enzyme, and **myeloperoxidase,** which reacts with hydrogen peroxide to generate potent, bacteria-killing oxidants. One of the oxidants generated by the myeloperoxidase reaction is hypochlorous acid (HOCl), the killing agent found in household bleach. Granules also contain **collagenase** and other **proteases.**

Eosinophils. Eosinophils are rare in the circulation but are easily identified on stained blood films. As the name implies, the eosinophil takes on a deep eosin color during polychromatic staining; the large, refractile cytoplasmic granules of these cells stain orange-red to bright yellow. Like neutrophils, eosinophils migrate to sites where they are needed and exhibit a metabolic burst when activated. Eosinophils participate in defense against certain parasites, and they are involved in allergic reactions. The exposure of allergic individuals to an allergen often results in a transient increase in eosinophil count known as **eosinophilia.** Infection with parasites often results in a sustained overproduction of eosinophils.

Basophils. Basophils are polymorphonuclear leukocytes with multiple pleomorphic, coarse, deep-staining metachromatic granules throughout their cytoplasm. These granules contain **heparin** and **histamine,** which have anticoagulant and vasodilating properties, respectively. The release of these and other mediators by basophils increases regional blood flow, facilitating the transport of other leukocytes to areas of infection and allergic reactivity or other forms of hypersensitivity.

Monocytes and Lymphocytes. In contrast to granulocytes, monocytes and lymphocytes are mononuclear cells. Monocytes are phagocytic cells but lymphocytes are not; both participate in multiple aspects of immunity. Monocytes were originally differentiated from lymphocytes based on morphological characteristics. The cytoplasm of monocytes appears pale blue or blue-gray with Wright's stain. The cytoplasm contains multiple fine reddish-blue granules. The monocyte nucleus may be shaped like a kidney bean, indented, or shaped like a horseshoe. Frequently, however, it is rounded or ovoid. Upon activation, monocytes transform into **macrophages**—large, active mononuclear phagocytes.

Morphologically, circulating lymphocytes have been assigned to two broad categories: large and small lymphocytes. In blood, small lymphocytes are more numerous than larger ones; the latter closely resemble monocytes. Small lymphocytes possess a deeply stained, coarse nucleus that is large in relation to the remainder of the cell, so that often only a small rim of cytoplasm appears around parts of the nucleus. In contrast, a broad band of cytoplasm surrounds the nucleus of large lymphocytes; the nucleus of these cells is similar in size and appearance to that of small lymphocytes.

The morphological homogeneity of lymphocytes obscures their functional heterogeneity. As is discussed below, lymphocytes participate in multiple aspects of the immune response. Lymphocyte subtypes in blood (see Fig. 11.2) are often identified based on their reaction with fluorescent monoclonal antibodies. The majority of circulating lymphocytes are **T cells** or T lymphocytes (for "thymus-dependent lymphocytes"). These cells participate in certain types of immune responses that do not depend on antibody. T cells comprise 40 to 60% of the total circulating pool of lymphocytes.

Subtypes of T cells have been identified using fluorescent monoclonal antibodies to specific cell-surface antigens, known as CD antigens. All T cells possess the common CD3 antigen. So-called **helper T cells** possess the CD4 antigen cluster, while **suppressor T cells** lack CD4 but possess CD8. Patients with AIDS show decreased circulating levels of CD4-positive cells. **Natural killer (NK) cells** are T lymphocytes that possess the ability to kill tumor cells without prior exposure or priming.

Some 20 to 30% of circulating lymphocytes are **B cells,** which have immunoglobulin or antibody on their surface. B cells are bone marrow-derived lymphocytes; when immunologically activated, they transform into **plasma cells** that secrete immunoglobulin. Lymphocytes not characteristic of either T cells or B cells are called **null cells.** The entire scope of the function of null cells, which comprise only 1 to 5% of circulating lymphocytes, is unknown, but it has been established that null cells are capable of destroying tumor cells and virus-infected cells.

While B cells mediate immune responses by releasing antibody, T cells often exert their effects by synthesizing and releasing **cytokines,** hormone-like proteins that act by binding specific receptors on their target cells. Recent research has led to the discovery of many cytokines, with activities ranging from tumor destruction, a function of **tumor necrosis factor**, to the promotion of blood cell production. Cytokines that limit viral replication in cells, known as **interferons,** suppress or potentiate the function of T cells, stimulate macrophages, and activate neutrophils.

In some cases, cytokines, like other hormones, can exert potent effects when supplied exogenously. For example, colony-stimulating factors injected into cancer patients can prevent decreases in the production of leukocytes that result from the administration of chemotherapeutic drugs or radiation therapy. The technology of molecular biology is used to produce cytokines for therapy. In this process, sections of lymphocyte DNA containing the gene that codes for the specified cytokine are isolated and then **transfected** into a bacterial cell, fungus, or rapidly growing mammalian cell. These cells then produce the cytokine and release it into their culture supernatant, from which it can be purified, concentrated, and sterilized for injection. The biological diversity and potency of the cytokines has opened the door to the development of a variety of new pharmacological agents that have proved useful in the treatment of cancer, immune disorders, and other diseases.

Blood Cells Are Born in the Bone Marrow

Mature cells are transient residents of blood. Erythrocytes survive in the circulation for about 120 days, after which they are broken down and their components recycled, as discussed above. Platelets have an average lifespan of 15 to 45 days in the circulation; many, if not most, of these cells are consumed as they continuously participate in day-to-day hemostasis. The rate of platelet consumption accelerates rapidly during the repair of bleeding caused by trauma. Leukocytes have a variable lifespan. Some lymphocytes circulate for 1 year or longer after production. Neutrophils, constantly guarding body fluids and tissues against infection, have a circulating half-life of only a few hours. Neutrophils and other blood cells must, therefore, be continuously replenished.

As mentioned earlier, the process of blood cell generation, hematopoiesis, occurs in healthy adults only in the bone marrow. **Extramedullary hematopoiesis** (e.g., the generation of blood cells in the spleen) is observed only in some disease states, such as leukemia. Hematopoietic cells are found in high levels in the liver, spleen, and blood of the developing fetus. Shortly before birth, blood cell production gradually begins to shift to the marrow. In newborns, the hematopoietic cell content of the circulating blood is relatively high; hematopoietic cells are also found in the blood of adults, but in extremely low numbers. Large numbers of hematopoietic cells can be recovered from aspirates of the iliac crest, sternum, pelvic bones, long bones, and ribs of adults. Within the bones, hematopoietic cells germinate in extravascular sinuses, called **marrow stroma.** Circulating factors and factors released from capillary endothelial cells, stromal fibroblasts, and mature blood cells regulate the generation of immature blood cells from hematopoietic cells and the subsequent differentiation of newly formed immature cells.

Blood cell production begins with the proliferation of **pluripotent (uncommitted) stem cells**. Depending on the stimulating factors, the progeny of pluripotent stem cells may be other uncommitted stem cells or stem cells committed to development along a certain lineage. The committed stem cells include **myeloblasts,** which form cells of the myeloid series (neutrophils, basophils, and eosinophils); **erythroblasts**; **lymphoblasts**; and **monoblasts** (Fig. 11.7; see also Fig. 11.2). Promoted by hematopoietins and other cytokines, each of these blast cells differentiates further, a process that ultimately results in the formation of mature blood cells. This is a dynamic process; the hematopoietic cells of the bone marrow are among the most actively reproducing cells of the body. Interruption of hematopoiesis (e.g., by cancer treatment) results in the eventual disappearance of granulocytes from the blood, a condition known as **granulocytopenia**, or, when specific to neutrophils, **neutropenia**, in a matter of hours. Platelets disappear next—**thrombocytopenia**—followed by erythrocytes, a sequence that reflects

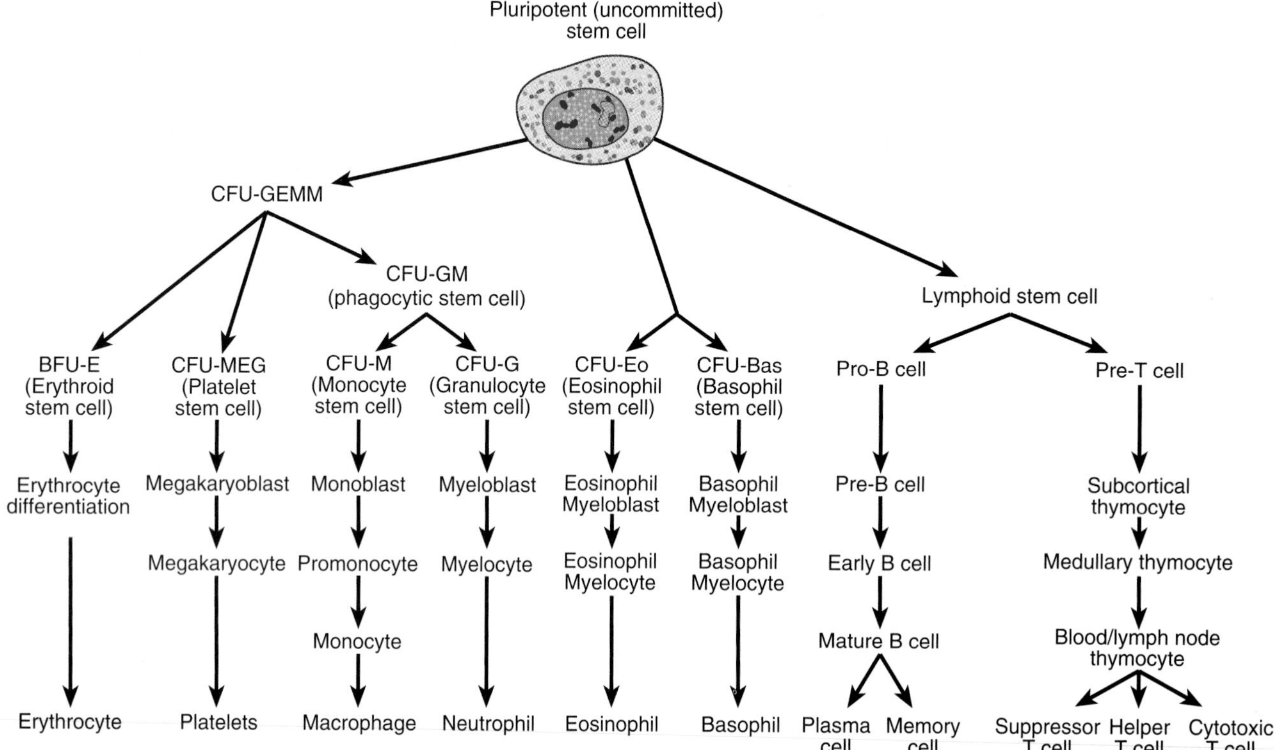

FIGURE 11.7 **Hematopoiesis.** All circulating blood cells are believed to be derived from a common, uncommitted bone marrow progenitor, the pluripotent stem cell. This cell differentiates along different lineages, depending on the conditions it encounters and the levels of individual hematopoietins available. CFU-GEMM, granulocyte-erythrocyte-macrophage-megakaryocyte colony-forming unit; CFU-GM, granulocyte-macrophage colony-forming unit; BFU-E, erythroid burst forming unit; CFU-MEG, megakaryocyte colony-forming unit; CFU-M, macrophage colony-forming unit; CFU-G, granulocyte colony-forming unit; CFU-Eo, eosinophil colony-forming unit; CFU-Bas, basophil colony-forming unit.

the circulating lifespan of each cell. Often, hematopoiesis can be restored after its interruption by an infusion of viable hematopoietic cells, e.g., a bone marrow transplant (see Clinical Focus Box 11.1). Administration of committed stem cells shows promise in treating specific blood cell defects (see Clinical Focus Box 11.2).

THE IMMUNE SYSTEM

Immunity or resistance to infection derives from the activity and intact functioning of two tightly interrelated systems, the **innate immune system** and the **adaptive immune system.** Elements of the innate or natural immune system include **exterior defenses,** such as skin and mucous membranes; **phagocytic leukocytes;** and serum proteins, which act nonspecifically and quickly against microbial invaders. Microbes that escape the onslaught of cells and molecules of the innate immune system face destruction by T cells and B cells of the adaptive immune system. Activation of the adaptive immune system results in the generation of antibodies and cells that specifically target the inducing organism or foreign molecule. Unlike the innate system, adaptive or acquired immune responses develop gradually but exhibit memory. Therefore, repeat exposure to the same infectious agent results in improved resistance mediated by the specific aspects of the adaptive immune system. Working together, elements of the innate and adaptive immune systems provide a considerable obstacle to the establishment and long-term survival of infectious agents.

The Innate Immune System Consists of Nonspecific Defenses

Infectious agents cannot easily penetrate intact skin, the first line of defense against infection. Infection is a major complication when the intact skin barrier is compromised, such as by burns or trauma. Even a small needle prick can result in a fatal infection.

Natural openings to body cavities and glands are an effective entry point of infectious agents. Usually, however, these openings are protected from invasion by pathogens in at least two ways. First, they are coated with mucus and other secretions that contain secretory immunoglobulins as well as antibacterial enzymes, such as lysozyme. Second, organisms that invade these openings cannot easily reach the blood but, instead, lodge in an organ that communicates with both the exterior and the interior of the body, such as a lung or the stomach. Many pathogens cannot survive the low pH of stomach acid. In the lungs, organisms face the efficient phagocytic activity of **alveolar macrophages.** These cells, derived from blood monocytes, are mobile but confined to the pulmonary capillary net-

CLINICAL FOCUS BOX 11.1

Bone Marrow Transplantation

When a patient has a terminal bone marrow disease, such as leukemia or aplastic anemia, often the only possibility for a cure is a **bone marrow transplant**. In this procedure, healthy bone marrow cells are used to replace the patient's diseased hematopoietic system. These cells are obtained from a donor who is usually a close relative of the patient. To identify a suitable donor, relatives' blood leukocytes are screened to determine whether their antigenic pattern matches that of the patient. The antigenic composition of leukocytes in bone marrow and peripheral blood are identical, so analysis of blood leukocytes usually provides enough information to determine whether the transplanted cells will engraft successfully. If significantly different from the recipient's tissue type, transplanted leukocytes may be recognized as foreign by the patient's immune system and, therefore, rejected.

More commonly, sufficient differences between the engrafted cells and the host's own tissue lead to debilitating consequences as a result of **graft-versus-host disease** (GVHD). In GVHD, functional T cells in the proliferating graft recognize host tissue as foreign and mount an immune response. The disease often begins with a skin rash, as transplanted lymphocytes invade the dermis, and ends in death as lymphocytes destroy every organ system in the marrow recipient.

Recent discoveries have led to useful ways to limit or prevent GVHD. These advances have decreased the morbidity of marrow transplants and have substantially increased the potential pool of bone marrow donors for a given patient. Immunosuppressive agents, including steroids, cyclosporine, and anti-T cell antiserum, effectively decrease the immune function of the transplanted lymphocytes. Another useful approach involves "purging"—the physical removal of T cells from bone marrow prior to transplantation. T cell-depleted bone marrow is much less capable of causing acute GVHD than untreated marrow. These techniques have enabled the successful transplantation of bone marrow obtained from unrelated donors. Unrelated transplants were never possible before these advances because GVHD would almost certainly develop, even when the antigenic type of the donor's leukocytes closely matched that of the recipient's. Thus, many patients died for lack of a related donor. Today, transplants of unrelated marrow are common.

Many problems remain, however. One of the most serious, and the most common, is donor identification. An unrelated transplant is successful only if the donor's leukocyte antigens closely match those of the recipient. Since there are several antigenic determinants and each can be occupied by any one of several genes, there are thousands of possible combinations of leukocyte antigens. The chance that any individual's cells will randomly match those of another is less than one in a million. Therefore, the identification of a suitable donor is a little more complicated than finding a needle in a haystack. On the other hand, these odds virtually guarantee that suitable donors are not only available but, in all probability, plentiful in the general population. Finding them is a formidable problem that often generates intense frustration when donors for terminally ill transplant candidates are not quickly identified.

To address this problem, bone marrow transplant registries have been established. In these registries, the results of extensive leukocyte antigen typing are stored in a computer. Typing is performed on leukocytes isolated from a small sample of blood, so the procedure does not significantly inconvenience prospective donors. For some registries, potential donors of a specific ethnic background are targeted; in others, blood samples are obtained from as many healthy individuals as possible, regardless of their heritage. The database is searched when an individual in need of a transplant cannot identify a suitable relative. In conjunction with continued development of methods to reduce or eliminate GVHD, the expanding bone marrow transplant registries may someday allow identification of a donor for anyone who needs a bone marrow transplant.

work. As efficient phagocytic cells, they continuously patrol the pulmonary vasculature to remove inhaled microbes.

Microbes that successfully break through these physical barriers face destruction by the **fixed macrophages** of the **monocyte-macrophage system**. These cells line the sinusoids and vasculature of many organs, including the liver, spleen, and bone marrow. The nonmobile, fixed phagocytic macrophages efficiently remove foreign particles, including bacteria, from the circulation.

Inflammation Is a Multifaceted Process

Microbial invaders that lodge in body tissues and begin to proliferate trigger an **inflammatory response** (Fig. 11.8). Inflammation provides a multifaceted defense against tissue invasion by pathogens. The inflammatory response is initiated by circulating proteins and blood cells when they contact invaders in a tissue. The response results in increased blood flow to the affected tissues, which accelerates the delivery of immune system elements to the site. The result is redness, heat, and swelling (edema) of the affected tissue. Blood cells participating in the inflammatory response release a variety of **inflammatory mediators** that perpetuate the response. If the pathogens persist, the inflammatory response may become chronic and may itself cause substantial tissue damage. Not only microbes, but also proteins, chemicals, and toxins the body recognized as foreign, can induce an inflammatory response.

Certain inflammatory mediators increase blood flow to the inflamed area. Other mediators increase capillary permeability, allowing diffusion of large molecules across the endothelium and into the infected site. These molecules may be plasma proteins, or they may be generated by plasma proteins or substances released by blood leukocytes. They often play important roles in eliminating the pathogenic agent or enhancing the inflammatory response. Finally, chemotactic factors produced by cells that arrived early in the inflammatory cascade cause polymorphonuclear leukocytes to migrate from the blood to the affected area. Neutrophils are an important participant in the in-

CLINICAL FOCUS BOX 11.2

Hematotherapy and Stem Cell Research: Clinical Tools of the Future

Many diseases result from a specific defect in the immune or hematopoietic system. These diseases may be effectively treated by infusion of specific precursors of the defective cells, a process termed **hematotherapy**. In a typical bone marrow transplant, the entire hematopoietic system (and, consequently, the immune system) of the recipient is ablated and restored with cells from the donor. In this situation, the most primitive stem cells of the immune or hematopoietic system are eliminated and replaced. In situations such as AIDS, thrombocytopenia, certain anemias, and genetic immunodeficiency, however, only specific committed progenitor cells of the hematopoietic or immune system are affected. We may soon be able to replace these and keep the healthy portion of the patient's hematopoietic system intact.

In recent years, much interest has focused on the isolation, identification, and propagation of the stem cells of various tissues. Hematopoietic stem cells have recently been grown in culture and may soon be used for therapeutic purposes. Hematopoietic stem cells are either committed or pluripotent. As such, they either are destined to generate a specific lineage of cells or are capable of generating further developed stem cells that can commit to development along any one of several lineages. Pluripotent stem cells are needed to reconstitute hematopoiesis after the complete disruption that occurs during whole-body irradiation or after the infusion of chemotherapeutic agents to treat leukemia and solid tumors.

Committed stem cells may be used for specific defects. For example, in AIDS, virus-laden T cells are rapidly eliminated, resulting in low circulating levels. Although pharmaceutical progress has resulted in extended survival for these patients, they are at high risk for life-threatening infection resulting from low T cell levels. It may be possible to support patients by periodic infusions of T cell precursors, generated in efficient bioreactors from the patient's own primitive stem cells. These bioreactors would be fueled by specific cytokines that direct the stem cells to specifically generate committed T cell progenitors. Stem cells used to initiate the culture would be obtained from the patient's marrow and grown under virus-free conditions. After sufficient T cell progenitors were generated, the cultures would be processed to isolate and concentrate the cells. Patients would receive an infusion whenever their T cell counts plummeted, protecting them against infection and allowing sustained survival.

In addition to AIDS, hematotherapy holds promise for several other diseases and conditions as well. Infusions of neutrophil progenitors may be useful for cancer patients during aggressive therapy. Red cell progenitors may be successfully cultivated and infused for those with certain anemias. Platelet progenitors may be used in patients with one of the many forms of inborn or acquired thrombocytopenia. In addition, this emerging therapeutic approach may soon be enhanced by genetic engineering. In this process, new or modified genes are inserted into the growing stem cells to replace defective or missing ones. For example, a patient may be unable to mount an appropriate immune response because of the lack of a specific enzyme secreted by healthy leukocytes. Stem cells of these patients may be modified in culture to eliminate this defect and infused back into the patient. If the infused cells take hold and generate sufficient progeny, the patient's immune defect may be reversed, resulting in a cure of a once fatal disease.

By far, the major clinical use of stem cells to date has been to restore the hematopoietic system of patients treated with radiation or chemotherapy for cancer. Less frequently, hematopoietic stem cells have been used to augment the defective immune system of patients born with genetic defects. New uses of stem cells appear to be on the horizon. In recent years, several groups have announced the successful isolation and culture of primitive, nonhematopoietic stem cells from human embryos and fetal tissue. In addition, current reports indicate that these primitive stem cells, cells that can be induced to differentiate into any type of cell in the body, can be successfully isolated from adult tissues, including tissues that would otherwise be discarded, such as fat obtained during liposuction. Stem cells could be potentially used for the regeneration and reconstruction of all types of damaged tissues.

References

Aldhous P. Stem cells. Panacea, or Pandora's box? Nature 2000;408:897–898.

Anonymous. Stem cells. Medicine's new frontier. Mayo Clin Health Lett 2000;18:1–3.

Asahara T, Kalka C, Isner JM. Stem cell therapy and gene transfer for regeneration. Gene Ther 2000;7:451–457.

Helmuth L. Neuroscience. Stem cells hear call of injured tissue. Science 2000;290:1479–1481.

Noble M. Can neural stem cells be used to track down and destroy migratory brain tumor cells while also providing a means of repairing tumor-associated damage? Proc Natl Acad Sci U S A 2000;97:12393–12395.

Spangrude GJ, Cooper DD. Paradigm shifts in stem-cell biology. Semin Hemat 2000;37(1 Suppl 2):3–10.

flammatory response. They can exert potent antimicrobial effects, as well as release a variety of agents that can further amplify and perpetuate the response.

The remarkable ability of the inflammatory response to sustain itself while it generates potent cytolytic agents can result in many undesirable effects, including extensive tissue damage and pain. A variety of **antiinflammatory agents** control some of these undesirable effects. These agents are designed to block some of the consequences of the inflammatory response without compromising its antimicrobial efficiency. They do this by neutralizing inflammatory mediators or by preventing inflammatory cells from releasing or responding to inflammatory mediators.

Defensive Mechanisms Are Integrated Systems

As discussed above, the innate and adaptive immune systems work together in ways that obscure their differences.

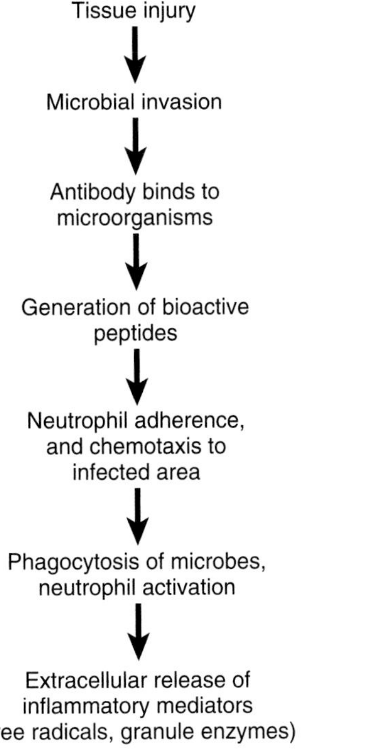

FIGURE 11.8 **Steps in the inflammatory response.** Inflammation can proceed along several divergent pathways, each involving inflammatory cells (e.g., neutrophils) and mediators. This shows a possible route of inflammation initiated by tissue injury.

Indeed, consideration of these two systems as distinct, individual entities is neither justified nor correct, owing to their extensive interdependence. They are described individually only as an aid to their presentation. In this respect, it is important to define the characteristics that differentiate each system (Table 11.3). In general, responses of the innate immune system are neither **specific** nor **inducible**; that is, the response is not programmed by or directed against a specific pathogen and is not amplified as a result of previous encounters with the pathogen. The adaptive response, in contrast, is both specific and inducible; the response is set in motion by a particular pathogen and develops against that specific pathogen.

TABLE 11.3 **Characteristics of the Innate and Adaptive Immune Systems**

	Innate	Adaptive
Resistance	Not improved by repeat infection	Improved by previous infection
Specificity	Not directed toward specific pathogen	Targeted response directed by specific elements of immune system
Soluble factors	Lysozyme, complement, acute phase proteins, interferon, cytokines	Antibodies
Cells	Phagocytic leukocytes, NK Cells	T cells, B cells

While characteristics of the innate and adaptive immune system differ, each system depends on elements of the other for optimal functioning. The initiation of responses by the innate system, as well as efficient phagocytosis by neutrophils in the tissues, often depends on the presence of a small amount of specific antibody in blood plasma. Antibody is generated by cells of the adaptive immune system in response to specific foreign molecules called **antigens.** In turn, the effective functioning of antibodies and other mediators of the adaptive immune system depends on neutrophils and other effector agents usually associated with the innate immune system. Thus, the innate and the adaptive systems depend on highly evolved, interactive, defensive mechanisms to kill and remove microbial intruders.

Adaptive Immunity Is Specific and Acquired

The adaptive immune system can be considered at three levels:

- The **afferent arm,** which gives the system its remarkable ability to recognize specific antigenic determinants of a wide range of infectious agents
- The **efferent arm,** which supplies a cellular and molecular assault on the invading pathogens
- **Immunological memory,** which specifically accelerates and potentiates subsequent responses to the same activating agent or antigen

The specificity of the recognition, effector, and memory aspects of the adaptive immune system derives from the specificity of antibody molecules as well as that of receptors on T cells and B cells. The lymphocytes of the immune system are capable of recognizing and specifically responding to hundreds of thousands of potential antigens, which may be presented, for example, as glycoproteins on the surface of bacteria, the coat protein of viruses, microbial toxins, or membranes of infected cells. Only a few circulating lymphocytes need to recognize an individual antigen initially. This initial recognition induces proliferation of the responsive cell, a process known as **clonal selection** (Fig. 11.9). Clonal selection amplifies the number of specific T cells or B cells (i.e., T or B lymphocytes programmed to respond to the inciting stimulus).

While all of the cells generated after a single clone has expanded are specific for the inducing antigen, they may not all possess the same functional characteristics. Some of the daughter lymphocytes may be effector cells. For example, when B cells are activated, their progeny **plasma cells** are capable of generating antibodies. Other progeny in the expanded clone may play an afferent recognition role and, thereby, function as **memory cells.** The increased number of these cells, which mimic the reactive specificity of the original lymphocytes that responded to the antigen, accelerate responsiveness when the antigen is encountered again. Memory cells thus account for one of the primary tenets of immunity: Resistance is increased after initial exposure to the infectious agent. Long-term immunity to many viruses—such as influenza, measles, smallpox, and polio—can be induced by **vaccination** with a killed or mutant form of the pathogen.

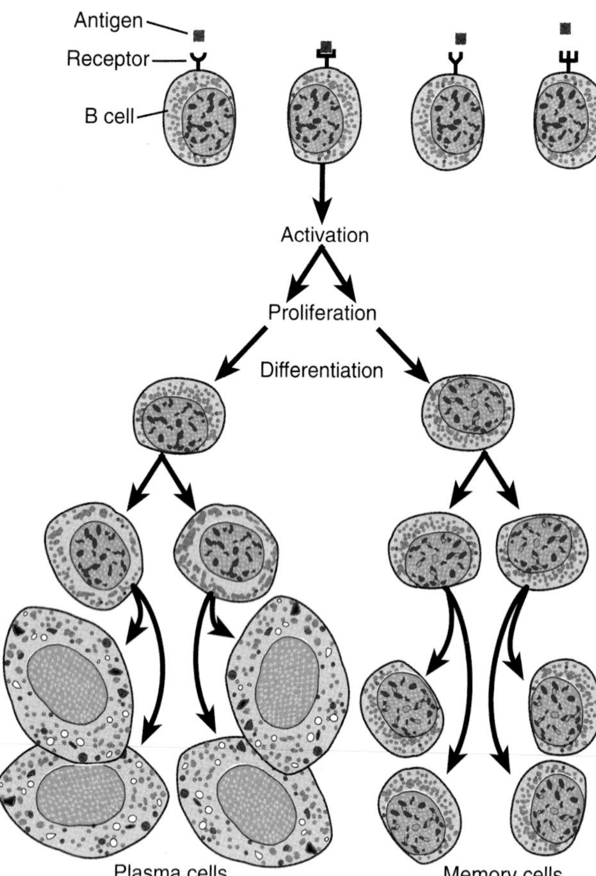

FIGURE 11.9 Clonal selection of committed lymphocytes. In this model, only the clone of lymphocytes that has the unique ability to recognize the antigen of interest proliferates, generating memory cells as well as effector cells specific to the inducing stimulus. This proliferation is initiated by the interaction of a specific recognition lymphocyte (afferent cell) with the antigen. Cells then proliferate and differentiate into either memory cells, which potentiate subsequent responses to the inciting antigen, or plasma cells, which secrete antibody.

The Adaptive Immune Response Involves Cellular and Humoral Components

Depending on the nature of the stimulus, its mode of presentation, and prior challenges to the immune system, an antigen may elicit either a cellular or humoral immune response. Both are ultimately mediated by lymphocytes, the cellular response by T cells and humoral response by B cells. As discussed above, stimulated B cells differentiate into plasma cells, which secrete antibody specific for the inciting stimulus. The antibody can be found in a variety of body fluids, including saliva, other secretions, and plasma.

Cell-Mediated Immunity. **Cell-mediated immunity (or cellular immunity)** is accomplished by activated T cells. The effector cells of this response do not secrete antibody but exert their influence by a variety of cellular mechanisms. These effector processes include direct cytotoxicity mediated by **cytotoxic T cells**; the suppression or activation of immune mechanisms in other cells—suppressor T cells or helper T cells, respectively; and the secretion of cytotoxic or immunomodulating cytokines, such as tumor necrosis factor and interleukin-2. T cells and their products may act directly or exert their effects in concert with other effector cells, such as neutrophils and macrophages.

The immune responses mediated by antibodies and T lymphocytes differ in several important respects. In general, antibodies are known to induce immediate responses to antigens and, thereby, provoke **immediate hypersensitivity reactions.** For example, **allergy** or **anaphylactic hypersensitivity** results when a certain type of antibody on the surface of fixed **mast cells** binds to its specific antigen. Antibody binding leads to the release of histamine and other mediators of the allergic response from intracellular granules.

Immediate hypersensitivity reactions also occur when circulating antibodies bind antigen in the tissues, thereby forming immune complexes that activate the **complement system,** a group of at least nine distinct proteins that circulate in plasma. A cascade of events occurs when the first protein recognizes preformed **immune complexes,** a large cross-linked mesh of antigen molecules bound to antibodies. In addition, complement can be activated when one of the proteins is exposed to the cell wall of certain bacteria. Initiation of this system results in edema, an influx of activated phagocytic cells (chemotaxis), and local inflammatory changes.

In contrast to the rapid onset of biological responses when antigen binds antibody, the consequences of T cell activation are not noticeable until 24 to 48 hours after antigen challenge. During this time, the T cells that initially recognize the antigen secrete factors that recruit and activate other cells (e.g., macrophages) and release factors that damage the antigen, cells possessing the antigen, or the surrounding tissue. A common example is the **delayed-type hypersensitivity reaction** to purified protein derivative (PPD), a response used to assess prior exposure to the bacteria that cause tuberculosis. Injected under the skin of sensitive individuals, PPD elicits the familiar inflammatory reaction characterized by local erythema and edema 1 to 2 days later.

Cell-mediated immune responses, while slow to develop, are potent and versatile. These delayed responses provide for defense against many pathogens, including viruses, fungi, and bacteria. T cells are responsible for the rejection of transplanted tissue grafts and containment of the growth of neoplastic cells. A deficiency in T cell immunity, such as that associated with AIDS, predisposes the affected patient to a wide array of serious, life-threatening infections.

Humoral Immunity. **Humoral immunity** consists of defense mechanisms carried out by soluble mediators in the blood plasma. **Antibodies** (also called **immunoglobulins**) are glycoproteins secreted by plasma cells. Antibodies are found in high levels in plasma and other body fluids. They have the ability to bind specifically to the antigenic determinant that induced their secretion.

Antibodies Bind Antigens

The primary structure of an antibody is illustrated in Figure 11.10. Each antibody molecule consists of four polypeptide

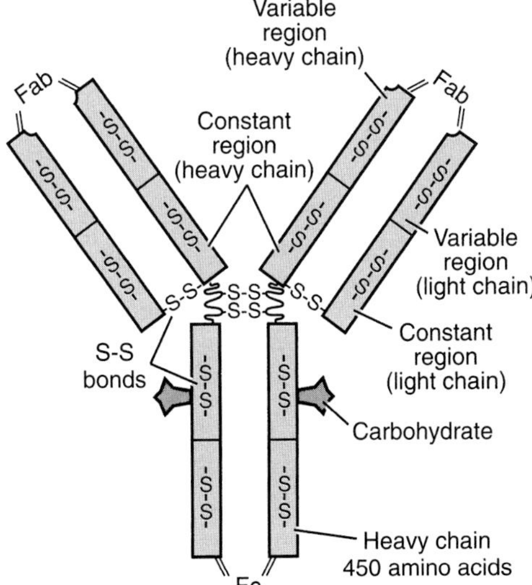

FIGURE 11.10 The structure of a typical antibody or immunoglobulin. Each molecule consists of two heavy chains and two light chains held together in a Y configuration by disulfide bonds. Each heavy chain and light chain possesses a constant region (where the amino acid sequence of individual molecules is similar) and a variable region, where alterations in the amino acid sequence convey to the antibody its individual antigen specificity.

chains (two **heavy chains** and two **light chains**) held together as a Y-shaped molecule by one or more disulfide bridges. Each polypeptide chain possesses both a conserved **constant region** and a **variable region**, where considerable amino acid sequence heterogeneity is found even within a single antibody class. This amino acid variability accounts for the widely diverse antigen-binding ability of antibody molecules, for it is the variable region that actually combines with the antigen, and there are millions of different antigens, ranging from viruses and proteins on bacterial cell walls to insect venom, pollen, and fluids secreted by plants.

The amino terminal portions of the variable regions, the **antigen-binding sites,** are known as the **Fab regions**. Each antibody unit possesses two identical antigen-binding sites, one at each end of the "Y." The carboxy terminal end of the heavy chain is termed the **Fc region**. Polypeptide fragments consisting of Fc and Fab regions of antibody molecules can be generated by protease digestion and separated by chromatography. Fc fragments can bind to cells such as neutrophils, monocytes, and mast cells through their **Fc receptors.** Fc receptor binding amplifies the biological activity of antigen-bound antibody. In addition to the ability to bind antigen, the antibody molecule may have a variety of other important biological functions, depending on its class.

Table 11.4 summarizes some characteristics and functions of the five major classes of antibodies; these classes are grouped based on differences in the amino acid composition of the constant region of the heavy chains. IgG is the most prevalent antibody in serum and is responsible for adaptive immunity to bacteria and other microorganisms. Bound to antigen, IgG can activate serum complement, which releases several inflammatory and bactericidal mediators. At the surface of bacteria, exposed Fc portions of IgG molecules facilitate the phagocytosis of bacteria by blood phagocytes, a process called **opsonization.** IgG exists in serum as a monomer. It can cross the placenta and is secreted into colostrum, protecting the fetus as well as the newborn from infection.

Unlike IgG, both IgM and IgA usually exist as polymers of the fundamental Y-shaped antibody unit. In most IgA molecules, two antibody units are held together by a **secretory piece** (**J chain**), a protein synthesized by epithelial cells. In this conformation, IgA is actively secreted into saliva, tears, colostrum, and mucus. IgA is thus known as **secretory immunoglobulin.** IgM is the first antibody secreted after an initial immune challenge and provides resistance early in the course of infection. IgM consists of five Y units. Its size and large number of antigen-binding sites provide the molecule with an excellent capacity for **agglutination**, the ability to clump particulate antigens, such as bacteria and blood cells. Clumped antigens are efficiently and quickly removed by fixed phagocytic cells of the monocyte-macrophage system.

IgE, a monomeric antibody slightly larger than IgG, avidly binds cells that store and release mediators of allergy and anaphylaxis, including mast cells and basophils. These cells are heavily granulated. The granules contain histamine, leukotrienes, and other biologically active agents that increase vascular permeability, dilate blood vessels (and, thereby, reduce blood pressure), and contract smooth muscle cells in lung airways. The granules are released when IgE, bound to mast cells at the Fc region, binds its specific anti-

TABLE 11.4 Characteristics of Different Antibody Classes

	IgG	IgA	IgM	IgD	IgE
Molecular weight ($\times 10^{-3}$)	150	150, 400	900	180	190
Y units/molecule	1	1–2	5	1	1
Serum concentration (mg/dL)	600–1500	85–300	50–400	<15	0.01–0.03
Crosses placenta	+	–	–	–	–
Enters secretions	+	+ +	–	–	–
Agglutinates particles	+	+	+ + +	–	–
Allergic reactions	+	–	–	–	+ + + +
Complement fixation	+	–	+ +	–	–
Fc receptor binding to monocytes and neutrophils	+ +	–	+	–	–

body. The ensuing allergic responses range from hay fever, hives, and bronchial asthma (induced by local or inhaled allergens) to systemic anaphylaxis, a potentially fatal response triggered when antigen is given systemically.

IgD, found in plasma and on the surface of some immature B cells, has no known function.

HEMOSTASIS

Circulating in a high-pressure, closed system that communicates with all tissues and cells in the body, blood exchanges oxygen, nutrients, and wastes and provides necessary components for host defense. This communication takes place largely in the complex and dynamic networks of capillary beds that provide oxygen to almost every cell in the body (only the cornea and intervertebral disks are avascular; these tissues receive oxygen by diffusion). Disruption of the integrity of the fragile capillaries may result from minor tissue injury associated with normal physical activity or from massive tissue trauma as a result of serious injury or infection, and may quickly lead to death. Any opening in the vascular network may lead to massive bruising or blood loss if left unrepaired.

To minimize bleeding and prevent blood loss after tissue injury, components of the hemostatic system are activated. The components of this dynamic, integrated system include blood platelets, endothelial cells, and plasma coagulation factors. They may be activated on exposure to foreign surfaces during bleeding, or by torn tissue at the site of injury, or by products released from the interior of damaged cells. Hemostasis can be viewed as four separate but interrelated events:

- Compression and vasoconstriction, which act immediately to stop the flow of blood
- Formation of a platelet plug
- Blood coagulation
- Clot retraction

Physical Factors Immediately Act to Constrain Bleeding

Immediately after tissue injury, blood flow through the disrupted vessel is slowed by the interplay of several important physical factors, including compression or back-pressure exerted by the tissue around the injured area, and vasoconstriction. The degree of compression varies in different tissues; for example, bleeding below the eye is not readily deterred because the skin in this area is easily distensible. Back-pressure increases as blood which leaks out of the disrupted capillaries accumulates. In some tissues, notably the uterus after childbirth, contraction of underlying muscles compresses blood vessels supplying the tissue and minimizes blood loss. Damaged cells at the site of tissue injury release potent substances that directly cause blood vessels to constrict, including **serotonin**, **thromboxane A_2**, **epinephrine**, and **fibrinopeptide B**.

Platelets Form a Hemostatic Plug

Platelets regulate bleeding in three stages. First, they form multicellular aggregates linked by protein strands at sites of openings in blood vessels. The aggregates form a physical barrier that begins to limit blood loss soon after the opening occurs. Second, **phospholipids** on the platelet plasma membrane activate the enzyme **thrombin**, which initiates a cascade of events ending in clot formation. Finally, platelets possess multiple storage granules, which they discharge (secrete) to enhance coagulation.

Platelet activation results in the sequential responses of **adherence**, **aggregation**, and **secretion**. Adherence is initiated when one or more substances, released from cells or activated in plasma at the site of a hemorrhage, bind to receptors in the platelet plasma membrane. Receptor binding results, via second messengers, in adherence (to other platelets and the inner, endothelial surface of blood vessels) and secretion.

Disruption of the endothelium at sites of tissue injury exposes a variety of proteins in the subendothelial matrix, such as **collagen** and **laminin**, which either induce or support platelet adherence. Endothelial cells also rapidly deploy cellular adherence proteins known as **integrins** on the outer surface of their plasma membranes during wound healing. These adherence proteins are deployed to the cell membrane by cellular processes set in motion by factors generated during coagulation or by factors released from platelets during clotting. In turn, activated endothelial cells release substances that participate in hemostasis. **von Willebrand factor**, a protein synthesized by endothelial cells and megakaryocytes, enhances platelet adherence by forming a bridge between cell surface receptors and collagen in the subendothelial matrix. The protein thrombin, which is generated by the plasma coagulation cascade, is a potent activator of platelet adherence and secretion. Ruptured cells at the site of tissue injury release adenosine diphosphate (ADP), which causes platelets to aggregate at the damaged site. These aggregates effectively stop the flow of blood from the ruptured vessels.

Blood Coagulation Results in the Production of Fibrin

Platelet aggregates are trapped in a highly organized, firm, and degradable network of **fibrin**, an insoluble protein generated in plasma as a consequence of activation of either the intrinsic or extrinsic clotting cascades, discussed below. The fibrin network traps red cells, leukocytes, platelets, and serum at sites of vascular damage, thereby forming a blood clot. The stable, fibrin-based blood clot eventually replaces the unstable platelet aggregate formed immediately after tissue injury. Fibrin is an insoluble polymer of proteolytic products of the plasma protein **fibrinogen**. Fibrin molecules are cleaved from fibrinogen by thrombin, which is generated in plasma during clotting. In the initial step of fibrin formation, thrombin cleaves four small peptides (fibrinopeptides) from each molecule of fibrinogen. The fibrinogen molecule devoid of these fibrinopeptides is called **fibrin monomer**. The fibrin monomers spontaneously assemble into ordered fibrous arrays of fibrin, resulting in an insoluble matrix of fibrous strands. At this stage, the clot is held together by noncovalent forces. A plasma enzyme, **fibrin stabilizing factor (Factor XIII)**, catalyzes the formation of covalent bonds between strands of polymerized fibrin, stabilizing and tightening the blood clot.

The Coagulation Cascade. Blood clotting is mediated by the sequential activation of a series of **coagulation factors,** proteins synthesized in the liver that circulate in the plasma in an inactive state. They are referred to by number (designated by a Roman numeral) in a sequence based on the order of the discovery of each factor. The plasma coagulation factors and their common names are listed in Table 11.5.

The sequential activation of a series of inactive molecules resulting in a biological response is called a **metabolic cascade.** The sequential activation of coagulation factors resulting in the conversion of fibrinogen to fibrin (and, hence, clotting) is called the **coagulation cascade.** The deficiency or deletion of any one factor of the cascade has severe consequences. Individuals deficient in factor VIII (antihemophilic factor), for example, display prolonged bleeding time on tissue injury, as a result of delayed clotting. Those who lack factor VIII have **hemophilia**, a condition resulting in severe coagulation defects.

Two separate coagulation cascades result in blood clotting in different circumstances. The two systems are the **intrinsic coagulation pathway** and the **extrinsic coagulation pathway** (Fig. 11.11). The final steps in fibrin formation are common to both pathways. In the intrinsic pathway, all the factors required for coagulation are present in the circulation. For initiation of the extrinsic pathway, a factor extrinsic to blood but released from injured tissue, called tissue thromboplastin or tissue factor (factor III), is required. Phospholipids are required for activation of both coagulation pathways. Phospholipids provide a surface for the efficient interaction of several factors. A component of tissue factor provides the necessary phospholipid for the extrinsic pathway. Phospholipids required for the activation of the intrinsic pathway are found on platelet membranes.

TABLE 11.5 Factors of the Coagulation Cascade

Scientific Name	Common Name	Other Names
Factor I	Fibrinogen	
Factor II	Prothrombin	
Factor III	Tissue thromboplastin	Tissue factor
Factor IV	Calcium	
Factor V	Proaccelerin	Labile factor
Factor VII	Proconvertin	Serum prothrombin conversion accelerator (SPCA)
Factor VIII	Antihemophilic factor	Platelet cofactor 1
Factor IX	Christmas factor	Platelet thromboplastin component
Factor X	Stuart factor	
Factor XI	Plasma thromboplastin antecedent	
Factor XII	Hageman factor	Contact factor
Factor XIII	Fibrin stabilizing factor	

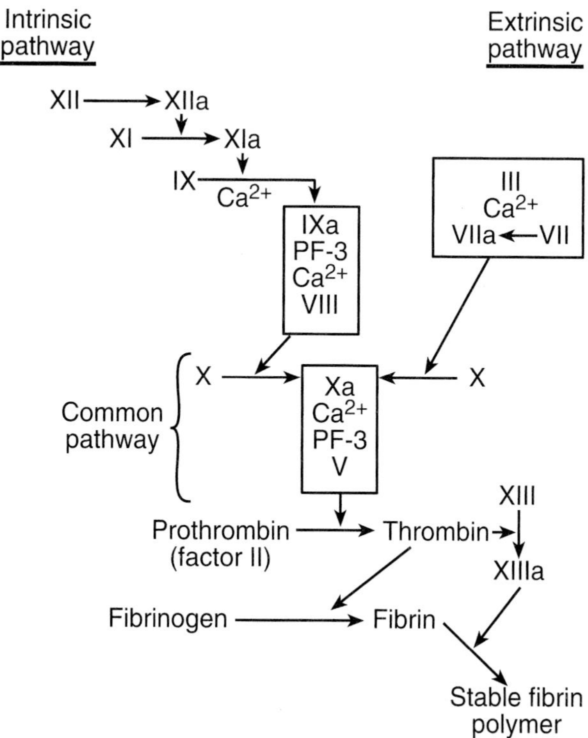

FIGURE 11.11 Steps in the coagulation cascade. The extrinsic pathway is initiated by tissue factor (factor III) released from damaged cells. In the presence of Ca^{2+}, factor III converts factor VII to factor VIIa, which then forms a complex with factor III and Ca^{2+}. This complex converts factor X to factor Xa. In the intrinsic system, factor XII is first converted to factor XIIa following its exposure to foreign surfaces, such as subendothelial matrix. Factor XIIa initiates a cascade of events, including activation of factor X, subsequent conversion of prothrombin to thrombin, and, finally, fibrin formation.

The final events leading to fibrin formation by both pathways result from the activation of the **common pathway.** The common pathway is initiated by the conversion of inactive clotting factor X to its active form, factor Xa (see Fig. 11.11) and results in the conversion of prothrombin to thrombin, thereby catalyzing the generation of fibrin. Thrombin also enhances the activity of clotting factors V and VIII, accelerating "upstream" events in the coagulation pathway. Finally, thrombin is a potent platelet and endothelial cell stimulus and enhances the participation of these cells in coagulation.

Factor X is activated during both the extrinsic and the intrinsic pathways. In the extrinsic pathway, factor X is activated by a complex consisting of activated factor VII, Ca^{2+}, and factor III (tissue factor). Activation of this complex by tissue factor bypasses the requirement for coagulation factors VIII, IX, XI, and XII used in the intrinsic pathway. In the intrinsic pathway, clotting is initiated by the activation of factor XII by contact to exposed surfaces, such as collagen in the subendothelial matrix. The activation of factor XII requires several cofactors, including **kallikrein** and **high-molecular-weight kininogen.** In this pathway, factor X is activated by a complex consisting of factor VIII, factor IXa, platelet factor 3, and Ca^{2+}.

Any attempt to describe a distinct division of coagulation into two separate pathways is an oversimplification, and the cascade theory has been extensively modified. There are many points of interaction between the two pathways, and no one pathway will account for hemostasis. For example, thrombin generated during activation of the extrinsic pathway is an essential cofactor for factor VIII of the intrinsic pathway. Factor VIIa of the extrinsic pathway directly activates factor IX of the intrinsic system. Factor VII can be activated by factors IXa, Xa, and XIIa and thrombin. The many additional points of interaction are beyond the scope of this discussion, but the concept of independently acting intrinsic versus extrinsic coagulation pathways has been abandoned. However, the activity of the intrinsic system and the extrinsic system are monitored individually in clinical coagulation tests for diagnostic purposes. The test used to monitor activity of the intrinsic system is the **partial thromboplastin time** (PTT). The extrinsic system is evaluated by determination of the **prothrombin time** (PT).

To a large extent, the interaction of coagulation factors occurs on the surfaces of platelets and endothelial cells. While plasma can eventually clot in the absence of surface contact, localization and assembly of coagulation factors on cell surfaces amplifies reaction rates by several orders of magnitude.

Clot retraction is a phenomenon that usually occurs within minutes or hours after clot formation. The clot draws together, extruding a very large fraction of the serum. The retraction requires platelets. Clot retraction decreases the breakdown of the clot and enhances wound healing.

Fibrinolysis and Wound Healing. Several important mechanisms exist to regulate and eventually reverse the final consequence of coagulation in order to allow healing to proceed. Platelet function is strongly inhibited, for example, by the endothelial cell metabolite **prostacyclin** (PGI_2), which is generated from arachidonic acid during cellular activation. Activated endothelial cells also release **tissue plasminogen activator** (TPA), which converts **plasminogen** to **plasmin,** a protein that hydrolyzes fibrin, resulting in dissolution of the fibrin clot in a process called **fibrinolysis**. Thrombin bound to **thrombomodulin** on the surface of endothelial cells converts **protein C** to an active protease. Activated protein C and its cofactor, **protein S**, restrain further coagulation by proteolysis of factors Va and VIIIa. Furthermore, activated protein C augments fibrinolysis by blocking an inhibitor of TPA. Finally, **antithrombin III** is a potent inhibitor of proteases involved in the coagulation cascade, such as thrombin. The activity of antithrombin III is accelerated by small amounts of heparin, a mucopolysaccharide present in the cells of many tissues. Deficiencies or abnormalities in proteins that regulate or constrain coagulation may result in **thrombotic** disorders, in which intravascular clot formation leads to severe problems, including embolism and stroke. Such disorders have been associated with abnormalities in protein C, protein S, antithrombin III, and plasminogen.

While the blood clot resolves, multiple factors participate in wound healing. Optimal wound healing requires the recruitment or generation of new tissue cells as well as new blood vessels to nourish the repairing tissue. Thus, secreted proteins and lipids that attract cells (chemoattractants), induce cells to proliferate (mitogens), and induce primitive cells to differentiate (growth factors) are called into play. These agents act in concert to induce the formation of new tissue and repair the injured area. The healing area is vascularized by a process known as **angiogenesis**, the formation of new blood vessels from preexisting ones. Platelets, activated during clotting, play an important role in the angiogenic response because they secrete factors that induce proliferation, migration, and differentiation of two of the major components of blood vessels, endothelial cells and smooth muscle cells.

Of the factors released from platelets involved in the angiogenic response, a novel lipid—**sphingosine 1-phosphate**—plays an important role in wound healing and angiogenesis. Released during clotting and acting in conjunction with protein growth factors, this lipid induces the proliferation of new tissue cells to replace damaged ones and drives the formation of new blood vessels until the healing process is complete. It does so by inducing the migration, proliferation, and differentiation of fibroblasts, smooth muscle cells, and endothelial cells at the site of tissue repair. Sphingosine 1-phosphate exerts its effects optimally when acting in conjunction with protein growth factors that possess angiogenic capabilities, including **vascular endothelial growth factor** (VEGF) and **fibroblast growth factor** (FGF). Recent research has been undertaken to define, in detail, the biochemical events that drive the angiogenic response because directed regulation of angiogenesis has profound clinical implications. For example, exogenously applied angiogenic factors may prove useful in accelerating repair of tissue damaged by thrombi in the pulmonary, cerebral, or cardiac circulation. In addition, angiogenic factors may assist in the repair of lesions that normally repair slowly—or not at all—such as skin ulcers in patients who are bedridden or diabetic.

Inhibition of angiogenesis may have profound clinical implications also, since unwanted tissues, such as growing tumors, require the development of blood vessels to survive. Therefore, agents which interfere with the angiogenic response, either by acting on the factors involved or the cells that respond to them, may prove particularly useful in the treatment of patients with cancer. Several novel pharmaceuticals are currently being evaluated for their use as regulators of angiogenesis, including **thrombospondin**, **angiostatin**, and **endostatin**, which block neovascularization in tumors and have shown great promise in laboratory testing. Further research will determine if these agents are effective in patients and will identify new, specific regulators of this fundamental process.

REVIEW QUESTIONS

DIRECTIONS: Each of the numbered items or incomplete statements in this section is followed by answers or by completion of the statement. Select the ONE lettered answer or completion that is BEST in each case.

1. Which type of hemoglobin is not normally found within human erythrocytes?
(A) HbA
(B) HbA_2
(C) HbCO
(D) HbO_2
(E) Reduced hemoglobin (Hb)
2. A reactant generated by neutrophils that plays an important role in bacterial killing is
(A) NADPH oxidase
(B) Hexose monophosphate shunt
(C) G proteins
(D) Superoxide anion
(E) Myeloperoxidase
3. Which cell type is defective in patients with AIDS?
(A) T cells
(B) B cells
(C) Neutrophils
(D) Monocytes
(E) Basophils
4. Which of the following would be expected to contain relatively high numbers of functional hematopoietic cells?
(A) Adult liver
(B) Umbilical cord blood
(C) Adult circulating blood
(D) Adult spleen
(E) Adult thymus
5. What is the process that amplifies the number of T cells or B cells programmed to respond to a specific infectious stimulus?
(A) Hematopoiesis
(B) Hematotherapy
(C) Inflammation
(D) Innate immunity
(E) Clonal selection
6. The response to the antigen used in the tuberculosis skin test, PPD, is not noticeable until 24 to 48 hours after injection because
(A) It takes that long for B cells to respond
(B) It takes that long for T cells to respond
(C) It takes that long for neutrophils to arrive at the site
(D) It takes that long for eosinophils to respond
(E) The skin test antigen is slowly converted to a more reactive antigen that quickly initiates the skin response
7. Antibody specificity is determined by the amino acid sequence within the
(A) Fc region
(B) Constant region
(C) Variable region
(D) Fc receptors
(E) J chain
8. The first step in the extrinsic coagulation pathway is
(A) Activation of factor X
(B) Activation of factor XII
(C) Conversion of prothrombin to thrombin
(D) Release of tissue thromboplastin
(E) Conversion of fibrinogen to fibrin

SUGGESTED READING

Browder T, Folkman J, Pirie-Shepherd S. The hemostatic system as a regulator of angiogenesis. J Biol Chem 2000;275:1521–1524.

Busslinger M, Nutt SL, Rolink AG. Lineage commitment in lymphopoiesis. Curr Opin Immunol 2000;12:151–158.

Claman HN. The biology of the immune response. JAMA 1992;268:2790–2796.

English D, Garcia JGN, Brindley DN. Platelet-released phospholipids link hemostasis and angiogenesis. Cardiovasc Res 2001;49:588–599.

Fischer A. Severe combined immunodeficiencies (SCID). Clin Exp Immunol 2000;122:143–149.

Fleisher TA, Bleesing JJ. Immune function. Pediatr Clin North Am 2000;4:1197–1209.

Grignani G, Maiolo A. Cytokines and hemostasis. Haematologica 2000;85:967–972.

Hoffman R, Benz EJ, Shattil SJ, et al. Hematology: Basic Principles and Practice. New York: Churchill Livingstone, 1991.

Lanier LL. The origin and functions of natural killer cells. Clin Immunol 2000;95:S14–S18.

Seaman WE. Natural killer cells and natural killer T cells. Arthritis Rheum 2000;43:1204–1217.

An Overview of the Circulation and Hemodynamics

Thom W. Rooke, M.D.
Harvey V. Sparks, Jr., M.D.

CHAPTER OUTLINE

- ONCE AROUND THE CIRCULATION
- HEMODYNAMIC PRINCIPLES OF THE CARDIOVASCULAR SYSTEM
- PRESSURES IN THE CARDIOVASCULAR SYSTEM
- SYSTOLIC AND DIASTOLIC PRESSURES
- TRANSPORT IN THE CARDIOVASCULAR SYSTEM
- THE LYMPHATIC CIRCULATION
- CONTROL OF THE CIRCULATION

KEY CONCEPTS

1. The circulatory system contributes to the maintenance of the internal environment by transporting nutrients to and waste products away from individual cells of the body. It also participates in the maintenance of the electrolyte and thermal environment of cells.
2. The circulatory system consists of two pumps in series. The right heart pumps blood into the lungs. The left heart pumps blood through the rest of the body.
3. The transport of nutrients and wastes over long distances (along the length of the blood vessels) occurs by bulk flow whereas transport over short distances (across the capillary walls) occurs via diffusion.
4. Pressure, flow, and resistance are related by Ohm's law.
5. Poiseuille's law shows how the radius and length of a vessel and blood viscosity contribute to vascular resistance.
6. The contractions of the heart generate the pressures that drive blood through the pulmonary and systemic circulations.

The physiological and medical importance of the cardiovascular system has been apparent since William Harvey first described the circulation of blood in 1628. A properly functioning, well-regulated cardiovascular system is essential to the maintenance of the internal environment of the body. Each cell must receive oxygen from the lungs and a variety of nutrients from the gastrointestinal tract. Each cell produces waste products that must be removed from its environment and taken to the lungs, kidneys, or other organs for metabolism and/or excretion. Cells in endocrine glands communicate with cells in other tissues by releasing hormones that are carried throughout the body by the circulation. Heat produced by the work of the body is brought to the surface of the body where it can be lost to the external environment by way of the circulation.

The circulatory system must perform all of these functions in the face of a variety of challenges, such as exercise, hot and cold environments, changes in posture, pregnancy and childbirth, and the hypoxia caused by high altitudes. Unfortunately, failure of the cardiovascular system to perform normally occurs all too often. In developed countries, the leading causes of death and morbidity include myocardial infarction, stroke, hypertension, congestive heart failure, and an assortment of other cardiovascular problems. Knowledge of the structure and function of the cardiovascular system is, therefore, crucial for understanding many aspects of health and disease.

ONCE AROUND THE CIRCULATION

An understanding of the circulation depends on knowledge of the physical principles governing blood flow. But first, we will briefly describe the cardiovascular system (Fig. 12.1). Contractions of the left ventricle propel blood into the aorta, the large arteries, and the vasculature beyond. Because of their elasticity, the aorta and large arteries are distended by each injection of blood from the heart. The aorta and large arteries recoil between ventricular contractions, continuing the flow of blood to the periphery.

Several regulatory mechanisms normally keep aortic pressure within a narrow range, providing a pulsatile but

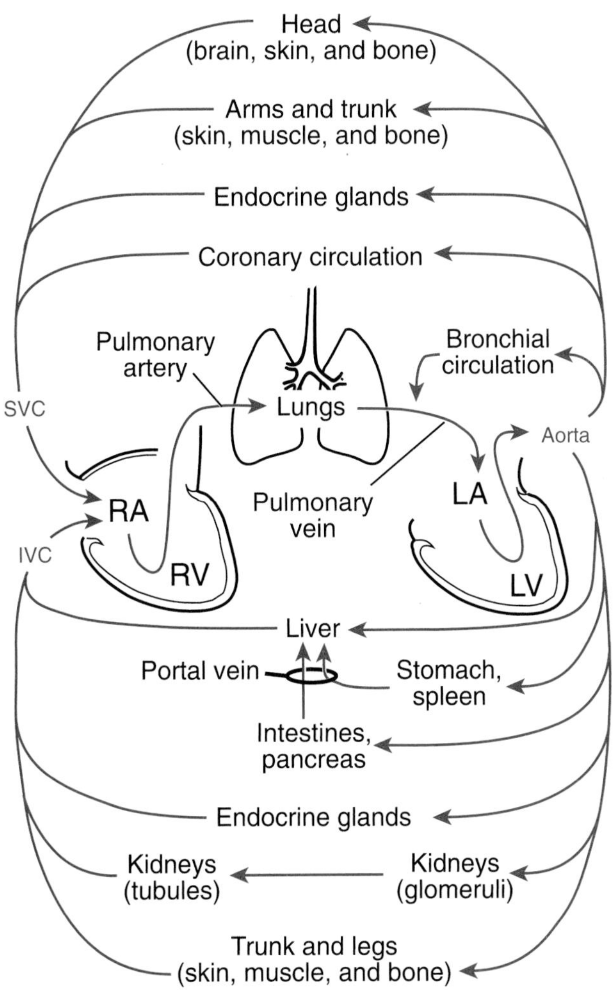

FIGURE 12.1 **A model of the cardiovascular system.** The right and left hearts are aligned in series, as are the systemic circulation and the pulmonary circulation. In contrast, the circulations of the organs other than the lungs are in parallel; that is, each organ receives blood from the aorta and returns it to the vena cava. Exceptions are the various "portal" circulations, which include the liver, kidney tubules, and hypothalamus. SVC, superior vena cava; IVC, inferior vena cava; RA, right atrium; RV, right ventricle; LA, left atrium; LV, left ventricle.

consistent pressure and driving blood to the small arteries and arterioles. Smooth muscle in the relatively thick walls of small arteries and arterioles can contract or relax, causing large changes in flow to a particular organ or tissue. Because of their ability to adjust their caliber, small arteries and arterioles are called **resistance vessels**. The prominent pressure pulsations in the aorta and large arteries are damped in the small arteries and arterioles. Pressure and flow are steady in the smallest arterioles.

Blood flows from arterioles into the capillaries. Capillaries are small enough that red blood cells flow through them in single file. They are numerous enough so that every cell in the body is close enough to a capillary to receive the nutrients it needs. The thin capillary walls allow rapid exchanges of oxygen, carbon dioxide, substrates, hormones, and other molecules and, for this reason, are called **exchange vessels**.

Blood flows from capillaries into venules and small veins. These vessels have larger diameters and thinner walls than the companion arterioles and small arteries. Because of their larger caliber they hold a larger volume of blood. When the smooth muscle in their walls contracts, the volume of blood they contain is reduced. These vessels, along with larger veins, are referred to as **capacitance vessels**. The pressure generated by the contractions of the left ventricle is largely dissipated by this point; blood flows through the veins to the right atrium at much lower pressures than are found on the arterial side of the circulation.

The right atrium receives blood from the largest veins, the superior and inferior vena cavae, which drain the entire body except the heart and lungs. The thin wall of the right atrium allows it to stretch easily to store the steady flow of blood from the periphery. Because the right ventricle can receive blood only when it is relaxing, this storage function of the right atrium is critical. The muscle in the wall of the right atrium contracts at just the right time to help fill the right ventricle. Contractions of the right ventricle propel blood through the lungs where oxygen and carbon dioxide are exchanged in the pulmonary capillaries. Pressures are much lower in the **pulmonary circulation** than in the **systemic circulation**. Blood then flows via the pulmonary vein to the left atrium, which functions much like the right atrium. The thick muscular wall of the left ventricle develops the high pressure necessary to drive blood around the systemic circulation.

The mechanisms that regulate all of the above anatomic elements of the circulation are the subject of the next few chapters. In this chapter, we consider the physical principles on which the study of the circulation is based.

HEMODYNAMIC PRINCIPLES OF THE CARDIOVASCULAR SYSTEM

Hemodynamics is the branch of physiology concerned with the physical principles governing pressure, flow, resistance, volume, and compliance as they relate to the cardiovascular system. These principles are used in the next few chapters to explain the performance of each part of the cardiovascular system.

Poiseuille's Law Describes the Relationship Between Pressure and Flow

Fluid flows when a pressure gradient exists. **Pressure** is force applied over a surface, such as the force applied to the cross-sectional surface of a fluid at each end of a rigid tube. The height of a column of fluid is often used as a measure of pressure. For example, the pressure at the bottom of a container containing a column of water 100 cm high is 100 cm of H_2O. The height of a column of mercury (Fig. 12.2) is frequently used for this purpose because it is dense (approximately 13 times more dense than water), and a relatively small column height can be used to measure physiological pressures. For example, mean arterial pressure is equal to the pressure at the bottom of a column of mercury approximately 93 mm high (abbreviated 93 mm Hg). If the same arterial pressure were measured

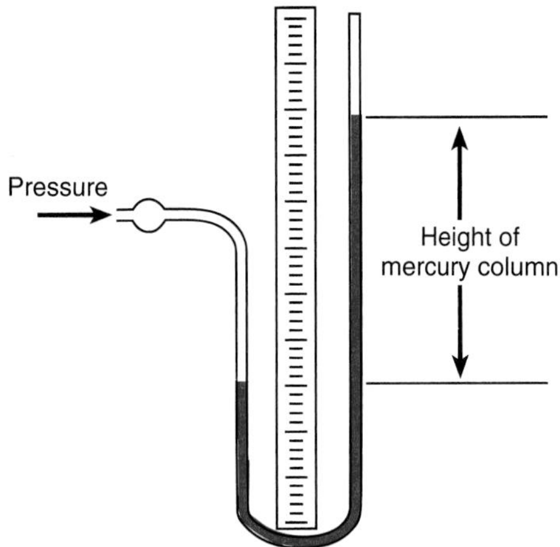

FIGURE 12.2 **Pressure expressed as the height of a column of fluid.** For the measurement of arterial pressures it is convenient to use mercury instead of water because its density allows the use of a relatively short column. A variety of electronic and mechanical transducers are used to measure blood pressure, but the convention of expressing pressure in mm Hg persists.

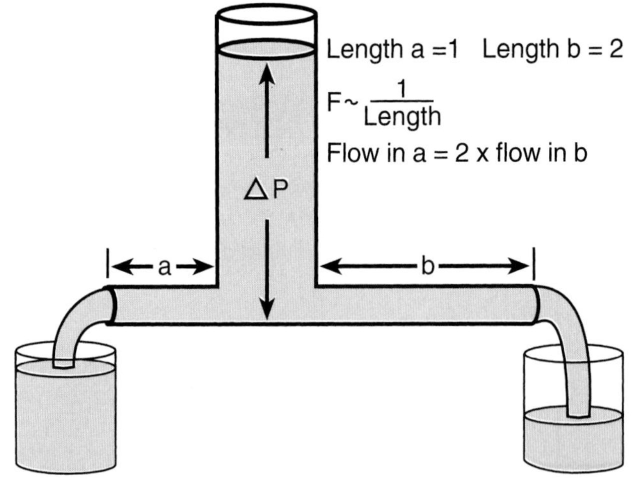

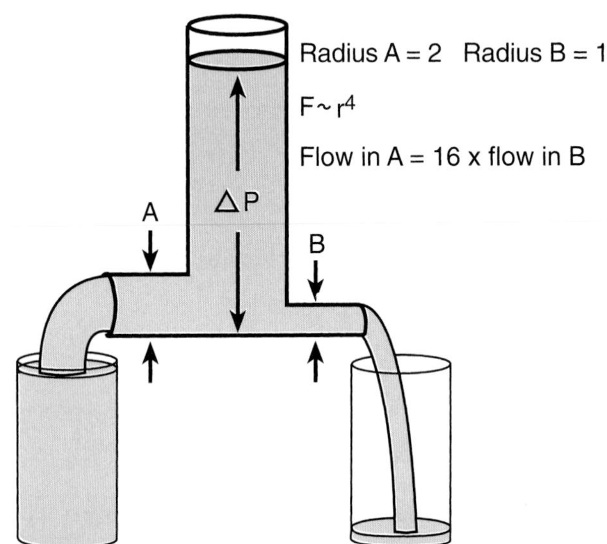

FIGURE 12.3 **The influence of tube length and radius on flow.** Because flow is determined by the fourth power of the radius, small changes in radius have a much greater effect than small changes in length. Furthermore, changes in blood vessel length do not occur over short periods of time and are not involved in the physiological control of blood flow. The pressure difference (ΔP) driving flow is the result of the height of the column of fluid above the openings of tubes A and B.

using a column of water, the column would be approximately 4 ft (or 1.3 m) high.

The **flow** of fluid through rigid tubes is governed by the pressure gradient and resistance to flow. Resistance depends on the radius and length of the tube as well as the viscosity of the fluid. All of this is summarized by **Poiseuille's law**. While not exactly descriptive of blood flow through elastic, tapering blood vessels, Poiseuille's law is useful in understanding blood flow. The volume of fluid flowing through a rigid tube per unit time ($\dot{Q}$) is proportional to the pressure difference (ΔP) between the ends of the tube and inversely proportional to the **resistance to flow** (R):

$$\dot{Q} = \Delta P/R \tag{1}$$

When fluid flows through a tube, the resistance to flow (R) is determined by the properties of both the fluid and the tube. Poiseuille found that the following factors determine resistance to steady, streamlined flow of fluid through a rigid, cylindrical tube:

$$R = 8\eta L/\pi r^4 \tag{2}$$

where r is the radius of the tube, L is its length, and η is the viscosity of the fluid; 8 and π are geometrical constants. Equation 2 shows that the resistance to blood flow increases proportionately with increases in fluid viscosity or tube length. In contrast, radius changes have a much greater influence because resistance is inversely proportional to the fourth power of the radius (Fig. 12.3). Equation 1 shows that if pressure and flow are expressed in units of mm Hg and mL/min, respectively, R is in mm Hg/(mL/min). The term **peripheral resistance unit** (PRU) is often used instead.

Poiseuille's law incorporates all of the factors influencing flow, so that

$$\dot{Q} = \Delta P \pi r^4/8\eta L \tag{3}$$

In the body, changes in radius are usually responsible for variations in blood flow. Length does not change. Although blood viscosity increases with hematocrit and with plasma protein concentration, blood viscosity only rarely changes enough to have a significant effect on resistance. Numerous control systems exist for the sole purpose of maintaining the arterial pressure relatively constant so there is a steady force to drive blood through the cardiovascular system. Small changes in arteriolar radius can cause large changes in flow to a tissue or organ because flow is related to the fourth power of the radius.

Conditions in the Cardiovascular System Deviate From the Assumptions of Poiseuille's Law

Despite the usefulness of Poiseuille's law, it is worthwhile to examine the ways the cardiovascular system does not strictly meet the criteria necessary to apply the law. First, the cardiovascular system is composed of tapering, branching, elastic tubes, rather than rigid tubes of constant diameter. These conditions, however, cause only small deviations from Poiseuille's law.

Application of Poiseuille's law requires that flow be steady rather than pulsatile, yet the contractions of the heart cause cyclical alterations in both pressure and flow. Despite this, Poiseuille's law gives a good estimate of the relationship between pressure and flow averaged over time.

Another criterion for applying Poiseuille's law is that flow be streamlined. **Streamline (laminar) flow** describes the movement of fluid through a tube in concentric layers that slip past each other. The layers at the center have the fastest velocity and those at the edge of the tube have the slowest. This is the most efficient pattern of flow velocities, in that the fluid exerts the least resistance to flow in this configuration. **Turbulent flow** has crosscurrents and eddies, and the fastest velocities are not necessarily in the middle of the stream. Several factors contribute to the tendency for turbulence: high flow velocity, large tube diameter, high fluid density, and low viscosity. All of these factors can be combined to calculate **Reynolds number** (N_R), which quantifies the tendency for turbulence:

$$N_R = \bar{v}d\rho/\eta \tag{4}$$

where $\bar{v}$ is the mean velocity, d is the tube diameter, ρ is the fluid density, and η is the fluid viscosity. Turbulent flow occurs when N_R exceeds a critical value. This value is hardly ever exceeded in a normal cardiovascular system, but high flow velocity is the most common cause of turbulence in pathological states.

Figure 12.4 shows that the relationship between pressure gradient along a tube and flow changes at the point that streamline flow breaks into eddies and crosscurrents (i.e., turbulent flow). Once turbulence occurs, a given increase in pressure gradient causes less increase in flow because the turbulence dissipates energy that would otherwise drive flow. Under normal circumstances, turbulent flow is found only in the aorta (just beyond the aortic valve) and in certain localized areas of the peripheral system, such as the carotid sinus. Pathological changes in the cardiac valves or a narrowing of arteries that raise flow velocity often induce turbulent flow. Turbulent flow generates vibrations that are transmitted to the surface of the body; these vibrations, known as **murmurs** and **bruits**, can be heard with a stethoscope.

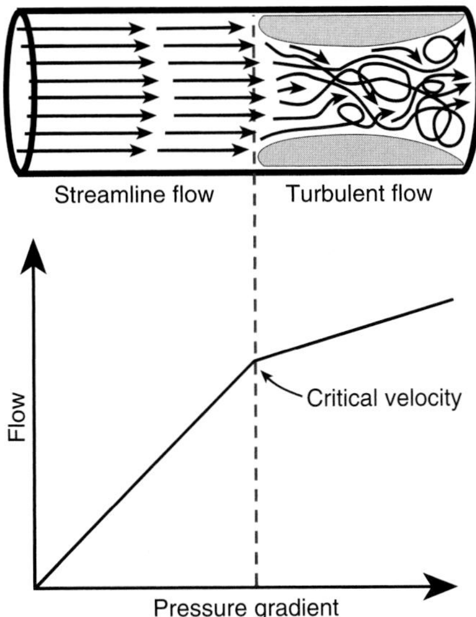

FIGURE 12.4 Streamline and turbulent blood flow. Blood flow is streamlined until a critical flow velocity is reached. When flow is streamlined, concentric layers of fluid slip past each other with the slowest layers at the interface between blood and vessel wall. The fastest layers are in the center of the blood vessel. When the critical velocity is reached, turbulent flow results. In the presence of turbulent flow, flow does not increase as much for a given rise in pressure because energy is lost in the turbulence. The Reynolds number defines critical velocity.

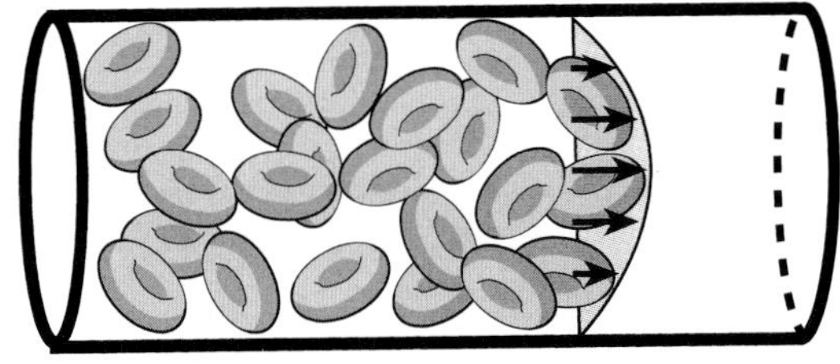

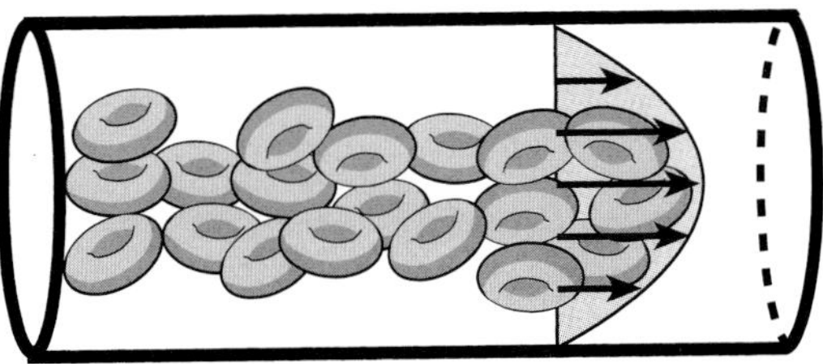

FIGURE 12.5 Axial streaming and flow velocity. The distribution of red blood cells in a blood vessel depends on flow velocity. As flow velocity increases, red blood cells move toward the center of the blood vessel (axial streaming), where velocity is highest. Axial streaming of red blood cells lowers the apparent viscosity of blood.

Finally, blood is not a strict **newtonian fluid**, a fluid that exhibits a constant viscosity regardless of flow velocity. When measured *in vitro*, the viscosity of blood decreases as the flow rate increases. This is because red cells tend to collect in the center of the lumen of a vessel as flow velocity increases, an arrangement known as **axial streaming** (Fig. 12.5). Axial streaming reduces the viscosity and, therefore, resistance to flow. Because this is a minor effect in the range of flow velocities in most blood vessels, we usually assume that the viscosity of blood (which is 3 to 4 times that of water) is independent of velocity.

PRESSURES IN THE CARDIOVASCULAR SYSTEM

Pressures in several regions of the cardiovascular system are readily measured and provide useful information. If arterial pressure is too high, it is a risk factor for cardiovascular diseases, including stroke and heart failure. When arterial pressure is too low, blood flow to vital organs is impaired.

Pressures in the various chambers of the heart are useful in evaluating cardiac function.

The Contractions of the Heart Produce Hemodynamic Pressure in the Aorta

The left ventricle imparts energy to the blood it ejects into the aorta, and this energy is responsible for the blood's circuit from the aorta back to the right side of the heart. Most of this energy is in the form of potential energy, which is the pressure referred to in Poiseuille's law. This is **hemodynamic pressure,** produced by contractions of the heart and stored in the elastic walls of the blood vessels. A much smaller component of the energy imparted by cardiac contractions is kinetic energy, which is the inertial energy associated with the movement of blood. The next section describes a third form of energy, hydrostatic pressure, derived from the force of gravity on the blood.

A Column of Fluid Exerts Hydrostatic Pressure

Fluid standing in a container exerts pressure proportional to the height of the fluid above it. The pressure at a given depth depends only on the height of the fluid and its density and not on the shape of the container. This **hydrostatic pressure** is caused by the force of gravity acting on the fluid. When a person stands, blood pressure is greater in the vessels of the legs than in analogous vessels in the arms because hydrostatic pressure is added to hemodynamic pressure. The hydrostatic pressure difference is proportional to the height of the column of blood between the arms and legs.

Two conventions are observed when measuring blood pressure. First, ambient atmospheric pressure is used as a zero reference, so the mean arterial pressure is actually about 93 mm Hg above atmospheric pressure. Second, all cardiovascular pressures are referred to the level of the heart. This takes into account the fact that pressures vary depending on position because of the addition of hydrostatic to hemodynamic pressure. (As we will see in Chapter 16, when capillary pressure is discussed, the term *hydrostatic pressure* is used to mean hemodynamic plus hydrostatic pressure. Although this is not strictly correct, it is the conventional usage.)

Transmural Pressure Stretches Blood Vessels in Proportion to Their Compliance

Thus far, we have discussed pressure and flow in the cardiovascular system as if blood vessels were rigid tubes. But blood vessels are elastic, and they expand when the blood in them is under pressure. The degree to which a distensible vessel or container expands when it is filled with fluid is determined by the transmural pressure and its compliance. **Transmural pressure** (P_{TM}) is the difference between the pressure inside and outside a blood vessel:

$$P_{TM} = P_{inside} - P_{outside} \quad (5)$$

Compliance (C) is defined by the equation:

$$C = \Delta V/\Delta P_{TM} \quad (6)$$

where ΔV is the change in volume and ΔP_{TM} is the change in transmural pressure.

A more compliant structure exhibits a greater change in volume for a given transmural pressure change. The lower the compliance of a vessel, the greater the pressure that will result when a given volume is introduced. For example, each time the left ventricle contracts and ejects blood into the aorta, the aorta expands; in doing so, it exerts an elastic force on the increased volume of blood it contains. This force is measured as the pressure in the aorta. With aging, the aorta becomes less compliant, and aortic pressure rises more for a given increase in aortic volume. Veins, which have thinner walls, are much more compliant than arteries. This means that, when we stand up and increased hydrostatic pressure is exerted on both the veins and the arteries of the legs, the volume of the veins expands much more than that of the arteries.

Mean Arterial Pressure Depends on Cardiac Output and Systemic Vascular Resistance

A simple model is useful in seeing how the pressures, flows, and volumes are established in the cardiovascular system. Imagine a circuit such as is shown in Figure 12.6. A pump propels fluid into stiff tubing that is of a large enough diameter to offer little resistance to flow. Midway around the circuit is a narrowing or stenosis of the tubing where almost all of the resistance to blood flow is located. The tubing downstream from the stenosis is 20 times more compliant than the tubing upstream from the stenosis. It has the same diameter as the upstream tubing and also offers almost no resistance to flow.

First imagine that the pump is turned off and the tubing is completely collapsed. At this point, enough fluid is infused into the circuit to fill all of the tubing and just begin to stretch the walls of the upstream and downstream tubing. Once the infused fluid comes to rest inside the tubing, the pressure inside the tubing is the same throughout because the pump is not adding energy to the circuit and there is no flow. The pressure inside the tubing is the pressure needed to "inflate" or fill the tubing in the resting state. The pressure outside the tubing is assumed to be atmospheric, and so the inside pressure equals the transmural pressure. Because the transmural pressure is the same throughout, and the left side of the circuit is made up of more compliant tubing, its volume is larger than the volume of the right side (see equation 6).

Imagine that the pump turns one cycle and shifts a small volume of fluid from the high-compliance tubing to the low-compliance tubing. The drop in volume on the left side has little effect on pressure because of its high compliance. However, an equivalent increase in volume on the low-compliance right side causes a 20-fold larger change in pressure. The pressure difference between the right and left side initiates flow from right to left. With only one stroke of the pump, the pressures on the two sides of the stenosis soon equalize as the volumes return to their resting values. At this point, flow ceases.

If the pump is turned on and left on, net volume is transferred from left to right until the pump has created

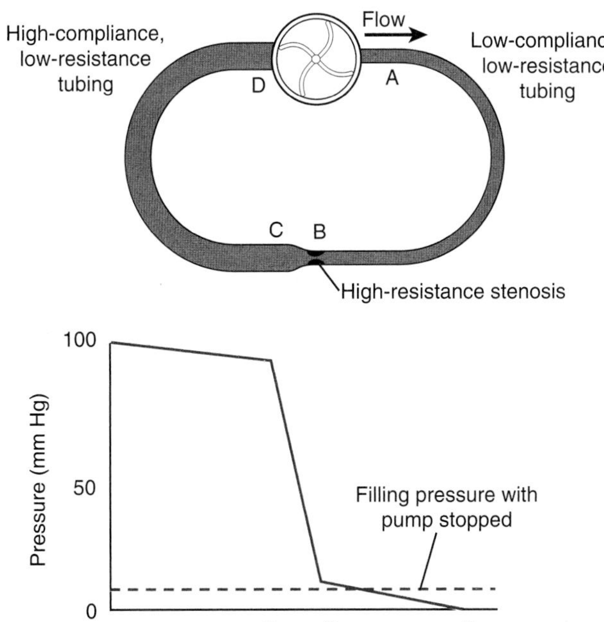

FIGURE 12.6 **A model of the systemic circulation.** When the pump is turned off, there is no flow and the pressures are the same everywhere in the circulation. This pressure is called the filling pressure, shown as a dotted line. When the pump is turned on, a small volume of fluid is transferred from the high compliance left-hand side (D) to the low compliance "arterial" side (A). This causes a small decrease in pressure in the left-hand tubing and a large increase in pressure in the right-hand tubing. The difference in the changes in pressures is because of the differences in compliance. Flow around the circulation occurs because of pressure difference established by transfer of fluid from the left- to the right-hand side of the model. Almost all of the resistance to flow is located at the high resistance stenosis between B and C. Because of this, almost all of the pressure drop occurs across the stenosis between B and C. This is shown by the pressures (solid line) observed when the pump is operating and the circulation is in a steady state.

a pressure difference sufficient to drive flow around the circuit equal to the output of the pump. In this new steady state, the pressure on the left side is slightly below the filling pressure and the pressure on the right side is much higher than the filling pressure. Although the volume removed from the left side exactly equals the volume added to the right side, the difference in the changes in pressures reflects the different compliances on the two sides of the pump.

The graph in Figure 12.6 shows that there is a small pressure drop from the outlet of the pump (A) to just before the stenosis (B), a large pressure drop occurs across the stenosis, and a very small pressure drop exists from just after the stenosis (C) to the inlet to the pump (D). This is because almost all of the resistance to flow is located at the stenosis between B and C.

In the steady state, flow ($\dot{Q}$) through the circuit equals the rate at which volume is transferred from D to A by the pump. In the steady state, $\dot{Q}$ is also equal to the pressure difference between point A (P_A) and point D (P_D) divided by the resistance (R) to flow (see equation 1):

$$\text{Rate of pump transfer of volume from D to A} = \dot{Q} = (P_A - P_D)/R \quad (7)$$

We can think about the coupling of the output of the left heart to the flow through the systemic circulation in an analogous fashion. The systemic circulation is filled by a volume of blood that inflates the blood vessels. The pressure required to fill the blood vessels is the **mean circulatory filling pressure.** This pressure can be observed experimentally by temporarily stopping the heart long enough to let blood flow out of the arteries into the veins, until pressure is the same everywhere in the systemic circulation and flow ceases. When this is done, the pressure measured throughout the systemic circulation is approximately 7 mm Hg.

Just as in the model, when the heart restarts after temporarily stopping, a net volume of blood is transferred to the arterial side from the venous side of the systemic circulation. Net transfer continues until the pressure difference builds up in the aorta and decreases in the right atrium enough to create a pressure difference to drive the blood to the venous side of the circulation at a flow rate equal to the output from the left ventricle. Because the venous side of the systemic circulation is approximately 20 times more compliant than the arterial side, the increase in pressure on the arterial side is 20 times the drop in pressure on the venous side.

The pumping action of the heart in combination with the elasticity of the aorta and large arteries make the aortic and arterial pressures pulsatile. In this discussion, we will concern ourselves with the **mean arterial pressure** ($\bar{P}_a$), the pulsatile pressure averaged over the cardiac cycle. Pressure in the aorta and large arteries is almost the same: there is only a 1 or 2 mm Hg pressure drop from the aorta to the large arteries. With vascular disease, the pressure drop in the large arteries can be much greater (see Clinical Focus Box 12.1). For most purposes, mean arterial pressure refers to the pressure measured in the aorta or any of the large arteries.

Flow through the aorta and large arteries ($\dot{Q}_{art}$), and on to the rest of the systemic circulation, is equal to the cardiac output in the steady state. It is proportional to the difference between mean arterial pressure and pressure in the right atrium (right atrial pressure, P_{ra}). It is inversely proportional to the resistance to flow offered by the systemic circulation, the **systemic vascular resistance** (SVR). As stated earlier, most of this resistance to flow is located in the small arteries, arterioles, and capillaries. Physiological changes in SVR are primarily caused by changes in radius of small arteries and arterioles, the resistance vessels of the systemic circulation. This is discussed in more detail in Chapter 15. The relationship between cardiac output, flow through the aorta and large arteries, mean arterial pressure, and systemic vascular resistance is analogous to the model (equation 7):

$$\text{Cardiac output} = \dot{Q}_{art} = (\bar{P}_a - P_{ra})/\text{SVR} \quad (8)$$

Systemic vascular resistance is calculated from cardiac output, mean arterial pressure, and right atrial pressure. Because right atrial pressure is normally close to zero and

CLINICAL FOCUS BOX 12.1

Effect of Vascular Disease on Arterial Resistance

The pressure gradient along large and medium-sized arteries, such as the aorta and renal arteries, is usually very small, due to the minimal resistance typically provided by these vessels. However, several disease processes can produce arterial narrowing and, thus, increase vascular resistance. Arterial narrowing exerts a profound effect on arterial blood flow because resistance varies inversely with the fourth power of the luminal radius.

The most common such disease is **atherosclerosis,** in which plaques composed of fatty substances (including cholesterol), fibrous tissue, and calcium form in the intimal layer of the artery. Atherosclerosis is the largest cause of morbidity and mortality in the United States: Myocardial infarction secondary to coronary atherosclerosis occurs more than 1 million times annually and accounts for over 700,000 deaths. Cerebrovascular infarction caused by carotid atherosclerosis is also a major cause of morbidity and mortality. Figure 12.A is an **arteriogram** from a patient with severe aortoiliac disease. The irregular luminal contour and focal narrowings of the iliac arteries (large arrowheads) and narrowing of the superior mesenteric artery (small arrowheads) are all caused by atherosclerosis.

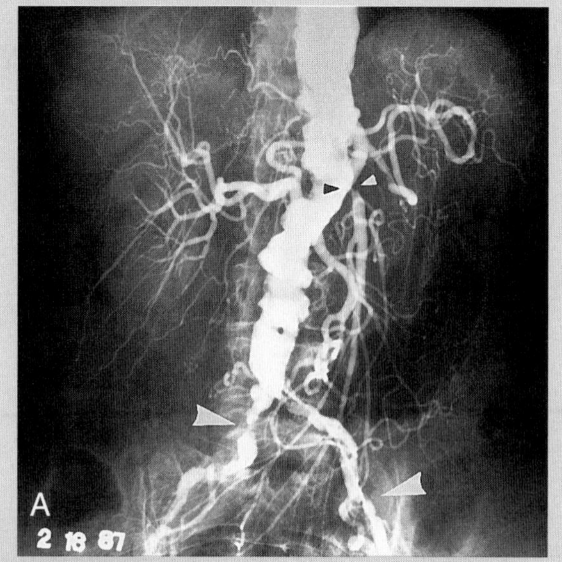

FIGURE 12.A An arteriogram of the abdominal aorta and iliac arteries, demonstrating atherosclerotic changes.

Other disease processes, such as inflammation, blunt trauma, and clotting abnormalities can also lead to significant arterial narrowing or occlusion. One such entity, **fibromuscular dysplasia,** is a condition in which the blood vessel wall develops structural irregularities. Fibromuscular dysplasia can affect people of any age or gender, but most commonly involves young women. The arteriogram in Figure 12.B shows a series of narrowings in the renal artery caused by this dysplastic disease.

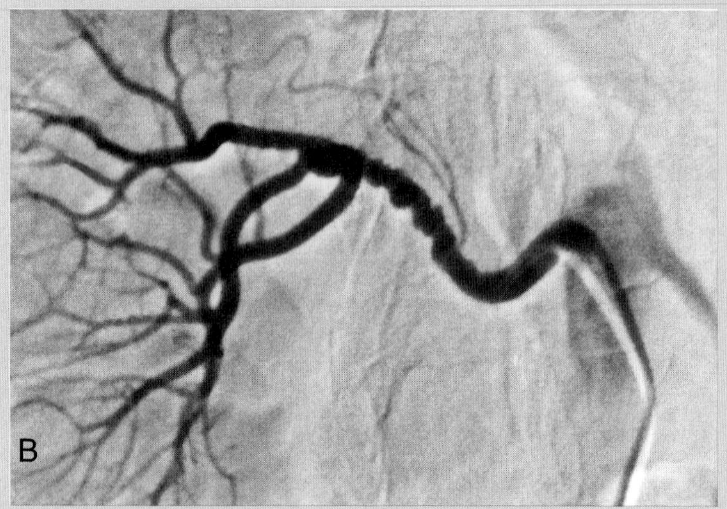

FIGURE 12.B An arteriogram of the left renal artery, demonstrating changes of fibromuscular dysplasia.

mean arterial pressure is much higher (e.g., 90 mm Hg), right atrial pressure is often ignored:

$$\text{Cardiac output} = \dot{Q}_{art} = \bar{P}_a/\text{SVR} \quad (9)$$

Cardiac output and systemic vascular resistance are controlled physiologically. Their control allows regulation of mean arterial pressure. Control of cardiac output and systemic vascular resistance is discussed in subsequent chapters.

An assumption in the above discussion is that the right heart and pulmonary circulation faithfully transfer blood flow from the systemic veins to the left heart. In fact, coupling of the output of the right heart and the pulmonary circulation can be analyzed in the same terms as our discussion of the systemic circulation (the pulmonary circulation is discussed in Chapter 20). Our assumption that in the steady state, the outputs of the right and left hearts are exactly equal is true. However, transient differences between the outputs of the left and right heart occur and are physiologically important (see Chapter 14).

SYSTOLIC AND DIASTOLIC PRESSURES

Thus far, we have discussed only mean arterial pressure, despite the fact that the pumping of blood by the heart is a cyclic event. In a resting individual, the heart ejects blood into the aorta about once every second (i.e., the heart rate is about 60 beats/min). The phase during which cardiac muscle contracts is called **systole**, from the Greek for "a drawing together." During atrial systole, the pressures in the atria increase and push blood into the ventricles. During ventricular systole, pressures in the ventricles rise and the blood is pushed into the pulmonary artery or aorta. During **diastole** ("a drawing apart"), the cardiac muscle relaxes and the chambers fill from the venous side. Because of the pulsatile nature of the cardiac pump, pressure in the arterial system rises and falls with each heartbeat. The large arteries are distended when the pressure within them is increased (during systole), and they recoil when the ejection of blood falls during the latter phase of systole and ceases entirely during diastole. This recoil of the arteries sustains the flow of blood into the distal vasculature when there is no ventricular input of blood into the arterial system. The peak in systemic arterial pressure occurs during ventricular systole and is called **systolic pressure**. The nadir of systemic arterial pressure is called **diastolic pressure**. The difference between systolic pressure and diastolic pressure is the **pulse pressure.** We will discuss these three pressures thoroughly in Chapter 15.

TRANSPORT IN THE CARDIOVASCULAR SYSTEM

The cardiovascular system depends on the energy provided by hemodynamic pressure gradients to move materials over long distances (bulk flow) and the energy provided by concentration gradients to move material over short distances (diffusion). Both types of movement are the result of differences in potential energy. As we have seen, bulk flow occurs because of differences in pressure. Diffusion occurs because of differences in chemical concentration.

Hemodynamic Pressure Gradients Drive Bulk Flow; Concentration Gradients Drive Diffusion

Blood circulation is an example of transport by **bulk flow**. This is an efficient means of transport over long distances, such as those between the legs and the lungs. **Diffusion** is accomplished by the random movement of individual molecules and is an effective transport mechanism over short distances. Diffusion occurs at the level of the capillaries, where the distances between blood and the surrounding tissue are short. The net transport of molecules by diffusion can occur within hundredths of a second or less when the distances involved are no more than a few microns. In contrast, minutes or hours would be needed for diffusion to occur over millimeters or centimeters.

Bulk Flow and Diffusion Are Influenced by Blood Vessel Size and Number

The aorta has the largest diameter of any artery, and the subsequent branches become progressively smaller down to the capillaries. Although the capillaries are the smallest blood vessels, there are several billion of them. For this reason, the total cross-sectional area of the lumens of all systemic capillaries (approximately 2,000 cm^2) greatly exceeds that of the lumen of the aorta (7 cm^2). In a steady state, the blood flow is equal at any two cross sections in series along the circulation. For example, the flow through the aorta is the same as the total flow through all of the systemic capillaries. Because the combined cross-sectional area of the capillaries is much greater and the total flow is the same, the velocity of flow in the capillaries is much lower. The slower movement of blood through the capillaries provides maximum opportunity for diffusional exchanges of substances between the blood and the tissue cells. In contrast, blood moves quickly in the aorta, where bulk flow, not diffusion, is important.

THE LYMPHATIC CIRCULATION

In vessels that are thin-walled and relatively permeable (e.g., capillaries and small venules), there is a net transfer of fluid out of the vessels and into the interstitial space. This fluid eventually returns from the interstitial space to the systemic circulation via another set of vessels, the **lymphatic vessels.** This movement of fluid from the systemic and pulmonary circulation into the interstitial space and then back to the systemic circulation via the lymphatic vessels is referred to as the **lymphatic circulation** (see Chapter 16). If the lymphatic circulation is interrupted, fluid accumulates in the interstitial space.

CONTROL OF THE CIRCULATION

The healthy cardiovascular system is capable of providing appropriate blood flow to each of the organs and tissues of the body under a wide range of conditions. This is done by

- Maintaining arterial blood pressure within normal limits
- Adjusting the output of the heart to the appropriate level
- Adjusting the resistance to blood flow in specific organs and tissues to meet special functional needs

The regulation of arterial pressure, cardiac output, and regional blood flow and capillary exchange is achieved by using a variety of neural, hormonal, and local mechanisms. In complex situations (e.g., standing or exercise), multiple mechanisms interact to regulate the cardiovascular response. In abnormal situations (e.g., heart failure), regulatory mechanisms that have evolved to handle normal events may be inadequate to restore proper function. The next few chapters describe these regulatory mechanisms in detail.

REVIEW QUESTIONS

DIRECTIONS: Each of the numbered items or incomplete statements in this section is followed by answers or by completions of the statement. Select the ONE lettered answer or completion that is BEST in each case.

1. Flow through a tube is proportional to the
   (A) Square of the radius
   (B) Square root of the length
   (C) Fourth power of the radius
   (D) Square of the length
   (E) Square root of the radius
2. Changes in transmural pressure
   (A) Can only be caused by changes in pressure inside a blood vessel
   (B) Cause changes in blood vessel volume, depending on the viscosity of the blood
   (C) Cause changes in blood vessel volume, depending on the compliance of the blood vessel
   (D) Cause proportional changes in blood flow
   (E) Are proportional to the length of a blood vessel
3. The pressure measured in either the arterial or the venous circulation when the heart has stopped long enough to allow the pressures to equalize is called the
   (A) Hemodynamic pressure
   (B) Mean arterial pressure
   (C) Transmural pressure
   (D) Mean circulatory filling pressure
   (E) Hydrostatic pressure
4. Blood flow becomes turbulent when
   (A) Flow velocity exceeds a certain value
   (B) Blood viscosity exceeds a certain value
   (C) Blood vessel diameter exceeds a certain value
   (D) The Reynolds number exceeds a certain value
5. The volume of an aorta is increased by 30 mL with an associated pressure increase from 80 to 120 mm Hg. The compliance of the aorta is
   (A) 1.33 mm Hg/mL
   (B) 4.0 mm Hg/mL
   (C) 0.75 mm Hg/mL
   (D) 1.33 mL/mm Hg
   (E) 0.75 mL/mm Hg
6. In the tube in the diagram to the right, the inlet pressure is 75 mm Hg and the outlet pressure at A and B is 25 mm Hg. The resistance to flow is
   (A) 2 PRU
   (B) 0.5 PRU
   (C) 2 (mL/min)/mm Hg
   (D) 0.75 mm Hg/(mL/min)
   (E) 0.5 (mL/min)/mm Hg

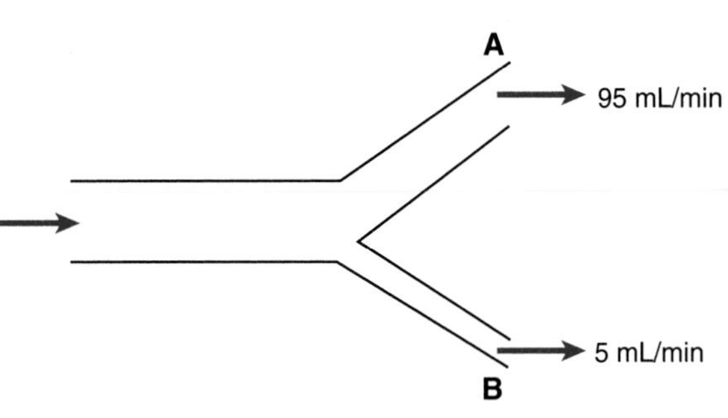

SUGGESTED READING

Fung YC. Biomechanics: Circulation. 2nd Ed. New York: Springer, 1997.

Janicki JS, Sheriff DD, Robotham JL, Wise RA. Cardiac output during exercise: Contributions of the cardiac, circulatory and respiratory systems. In: Rowell LB, Shepherd JT, eds. Handbook of Physiology. Exercise: Regulation and Integration of Multiple Systems. New York: Oxford University Press, 1996;649–704.

Li JK-J. The Arterial Circulation. Totowa, NJ: Humana Press, 2000.

Rowell LB. Human Cardiovascular Control. New York: Oxford University Press, 1993.

The Electrical Activity of the Heart

Thom W. Rooke, M.D.
Harvey V. Sparks, Jr., M.D.

CHAPTER OUTLINE

- **THE IONIC BASIS OF CARDIAC ELECTRICAL ACTIVITY: THE CARDIAC MEMBRANE POTENTIAL**
- **THE INITIATION AND PROPAGATION OF CARDIAC ELECTRICAL ACTIVITY**
- **THE ELECTROCARDIOGRAM**

KEY CONCEPTS

1. The electrical activity of cardiac cells is caused by the selective opening and closing of plasma membrane channels for sodium, potassium, and calcium ions.
2. Depolarization is achieved by the opening of sodium and calcium channels and the closing of potassium channels.
3. Repolarization is achieved by the opening of potassium channels and the closing of sodium and calcium channels.
4. Pacemaker potentials are achieved by the opening of channels for sodium and calcium ions and the closing of channels for potassium ions.
5. Electrical activity is normally initiated in the sinoatrial (SA) node where pacemaker cells reach threshold first.
6. Electrical activity spreads across the atria, through the atrioventricular (AV) node, through the Purkinje system, and to ventricular muscle.
7. Norepinephrine increases pacemaker activity and the speed of action potential conduction.
8. Acetylcholine decreases pacemaker activity and the speed of action potential conduction.
9. Voltage differences between repolarized and depolarized regions of the heart are recorded by an electrocardiogram (ECG).
10. The ECG provides clinically useful information about rate, rhythm, pattern of depolarization, and mass of electrically active cardiac muscle.

The heart beats in the absence of any nervous connections because the electrical (pacemaker) activity that generates the heartbeat resides within the cardiac muscle. After initiation, the electrical activity spreads throughout the heart, reaching every cardiac cell rapidly with the correct timing. This enables coordinated contraction of individual cells.

The electrical activity of cardiac cells depends on the ionic gradients across their plasma membranes and changes in permeability to selected ions brought about by the opening and closing of cation channels. This chapter describes how these ionic gradients and changes in membrane permeability result in the electrical activity of individual cells and how this electrical activity is propagated throughout the heart.

THE IONIC BASIS OF CARDIAC ELECTRICAL ACTIVITY: THE CARDIAC MEMBRANE POTENTIAL

The cardiac membrane potential is divided into 5 phases, phases 0 to 4 (Fig. 13.1). Phase 0 is the rapid upswing of the action potential; phase 1 is the small repolarization just after rapid depolarization; phase 2 is the plateau of the action potential; phase 3 is the repolarization to the resting membrane potential; and phase 4 is the resting membrane potential in atrial and ventricular cells and the pacemaker potential in nodal cells. In resting ventricular muscle cells, the potential inside the membrane is stable at approximately −90 mV relative to the outside of the cell (see phase 4, Fig. 13.1A). When the cell is brought to threshold, an action potential occurs (see Chapter 3). First, there is a rapid depolarization from −90 mV to +20 mV (phase 0). This is followed by a slight decline in membrane potential (phase 1) to a plateau (phase 2), at which time the membrane potential is close to 0 mV. Next, rapid repolarization (phase 3) returns the membrane potential to its resting value (phase 4).

In contrast to ventricular cells, cells of the **sinoatrial** (SA) **node** and **atrioventricular** (AV) **node** exhibit a progressive depolarization during phase 4 called the **pacemaker potential** (see Fig. 13.1B). When the membrane po-

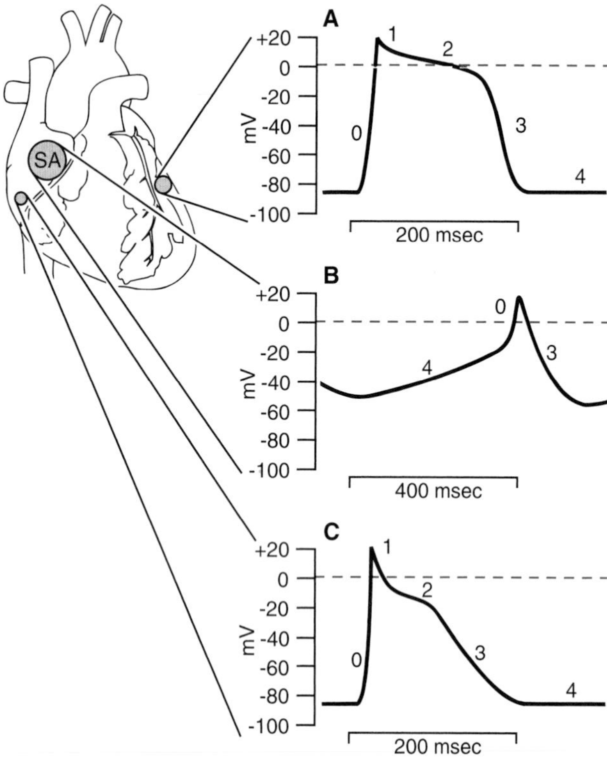

FIGURE 13.1 Cardiac action potentials (mV) recorded from A, ventricular, B, sinoatrial, and C, atrial cells. Note the difference in the time scale of the sinoatrial cell. Numbers 0 to 4 refer to the phases of the action potential (see text).

tential reaches threshold potential, there is a rapid depolarization (phase 0) to approximately +20 mV. The membrane subsequently repolarizes (phase 3) without going through a plateau phase, and the pacemaker potential resumes. Other myocardial cells combine various characteristics of the electrical activity of these two cell types. Atrial cells, for example (see Fig. 13.1C), have a steady diastolic resting membrane potential (phase 4) but lack a definite plateau (phase 2).

TABLE 13.1 Major Ion Channels Involved in Purkinje and Ventricular Myocyte Membrane Potentials

Name	Voltage (V)- or Ligand(L)- Gated	Functional Role
Voltage-gated Na^+ channel (fast, i_{Na})	V	Phase 0 of action potential (permits influx of Na^+)
Voltage-gated Ca^{2+} channel (long-lasting, i_{CaL})	V	Contributes to phase 2 of action potential (permits influx of Ca^{2+} when membrane is depolarized). β-Adrenergic agents increase the probability of channel opening and raise Ca^{2+} influx. ACh lowers the probability of channel opening.
Inward rectifying K^+ channel (i_{K1})	V	Maintains resting membrane potential (phase 4) by permitting outflux of K^+ at highly negative membrane potentials.
Outward (transient) rectifying K^+ channel (i_{to1})	V	Contributes briefly to phase 1 by transiently permitting outflux of K^+ at positive membrane potentials.
Outward (delayed) rectifying K^+ channels (i_{Kr}, i_{Ks})	V	Cause phase 3 of action potential by permitting outflux of K^+ after a delay when membrane depolarizes. I_{Kr} channel is also called HERG channel.
G protein-activated K^+ channel ($i_{K.G}$, $i_{K.ACh}$, $i_{K.ado}$)	L	G protein operated channel, opened by ACh and adenosine. This channel hyperpolarizes membrane during phase 4 and shortens during phase 2.

The Cardiac Membrane Potential Depends on Transmembrane Movements of Sodium, Potassium, and Calcium

The membrane potential of a cardiac cell depends on concentration differences in Na^+, K^+, and Ca^{2+} across the cell membrane and the opening and closing of channels that transport these cations. Some Na^+, K^+, and Ca^{2+} channels (voltage-gated channels) are opened and closed by changes in membrane voltage, and others (ligand-gated channels) are opened by a neurotransmitter, hormone, metabolite, and/or drug. Tables 13.1 and 13.2 list the major membrane channels responsible for conducting the ionic currents in cardiac cells.

The ion concentration gradients that determine transmembrane potentials are created and maintained by active transport. The transport of Na^+ and K^+ is accomplished by the plasma membrane Na^+/K^+-ATPase (see Chapter 2). Calcium is partially transported by means of a Ca^{2+}-ATPase and partially by an antiporter that uses energy derived from the Na^+ electrochemical gradient to remove Ca^{2+} from the cell. If the energy supply of myocardial cells is restricted by inadequate coronary blood flow, ATP synthesis (and, in turn, active transport) may be impaired. This situation leads to a reduction in ionic concentration gradients that eventually disrupts the electrical activity of the heart.

The magnitude of the intracellular potential depends on the relative permeability of the membrane to Na^+, Ca^{2+}, and K^+. The relative permeability to these cations at a particular time depends on which of the various cation channels are open. For example, in resting ventricular cells, mostly K^+ channels are open and the measured potential is close to that which would exist if the membrane were per-

TABLE 13.2 Major Ion Channels Involved in Nodal Membrane Potentials

Name	Voltage (V)- or Ligand(L)- Gated	Functional Role
Voltage-gated Ca^{2+} channel (long-lasting, i_{CaL})	V	Phase 0 of action potential of SA and AV nodal cells (carries influx of Ca^{2+} when membrane is depolarized); contributes to early pacemaker potential of nodal cells. β-Adrenergic agents increase the probability of channel opening and raise Ca^{2+} influx. ACh lowers the probability of channel opening.
Voltage-gated Ca^{2+} channel (transient, i_{CaT})	V	Contributes to the pacemaker potential.
Mixed cation channel (funny, i_f)	V	Carries Na^+ (mostly) and K^+ inward when activated by hyperpolarization. Contributes to pacemaker potential.
K^+ channel (delayed outward rectifier, i_K)	V	Contributes to phase 3 of action potential. Closing early in phase 4 contributes to pacemaker potential.
G protein-activated K^+ channel ($i_{K.G}$, $i_{K.ACh}$, $i_{K.ado}$)	L	G protein-operated channel, opened by ACh and adenosine. This channel hyperpolarizes membrane during phase 4, slowing pacemaker potential.

meable only to K^+ (potassium equilibrium potential). In contrast, when open Na^+ channels predominate (as occurs at the peak of phase 0 of the action potential), the measured potential is closer to the potential that would exist if the membrane were permeable only to Na^+ (sodium equilibrium potential) (see Fig. 13.2). The opening of Ca^{2+} channels causes the membrane potential to be closer to the calcium equilibrium potential, which is also positive; this occurs in phase 2. Specific changes in the number of open channels for these three cations are responsible for changes in membrane permeability and the different phases of the action potential.

The Opening and Closing of Cation Channels Causes the Ventricular Action Potential

In the normal heart, the sodium-potassium pump and calcium ion pump keep the ionic gradients constant. With constant ion gradients, the opening and closing of cation channels and the resulting changes in membrane permeability determine the membrane potential. Figures 13.3 and 13.4 depict the membrane changes that occur during an action potential in ventricular cells.

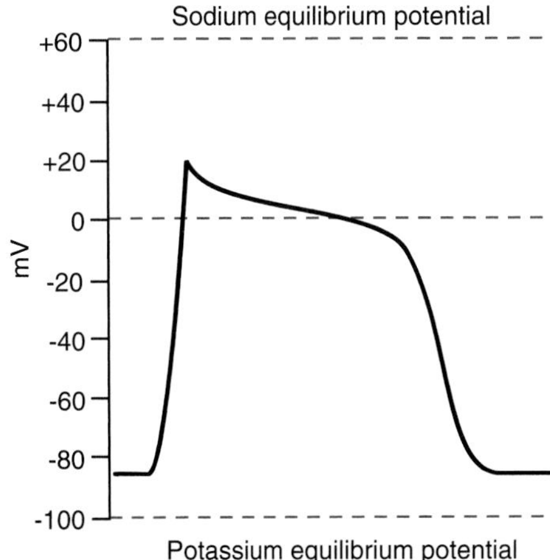

FIGURE 13.2 Effect of ionic permeability on membrane potential, primarily determined by the relative permeability of the membrane to Na^+, K^+, and Ca^{2+}. Relatively high permeability to K^+ places the membrane potential close to the K^+ equilibrium potential, and relatively high permeability to Na^+ places it close to the Na^+ equilibrium potential. The same is true for Ca^{2+}. An equilibrium potential is not shown for Ca^{2+} because, unlike Na^+ and K^+, it changes during the action potential. This is because cytosolic Ca^{2+} concentration changes approximately 5-fold during excitation. During the plateau of the action potential, the equilibrium potential for Ca^{2+} is approximately +90 mV. Membrane permeability to Na^+, K^+, and Ca^{2+} depends on ion channel proteins (see Table 13.1).

Depolarization Early in the Action Potential: Selective Opening of Sodium Channels. Depolarization occurs when the membrane potential moves away from the K^+ equilibrium potential and toward the Na^+ equilibrium potential. In ventricular cell membranes, this occurs passively at first, in response to the depolarization of adjacent membranes (discussed later). Once the ventricular cell membrane is brought to threshold, voltage-gated Na^+ channels open, causing the initial rapid upswing of the action potential (phase 0). The opening of Na^+ channels causes Na^+ permeability to increase. As permeability to Na^+ exceeds permeability to K^+, the membrane potential approaches the Na^+ equilibrium potential, and the inside of the cell becomes positively charged relative to the outside.

Phase 1 of the ventricular action potential is caused by a decrease in the number of open Na^+ channels and the opening of a particular type of K^+ channel (see Fig. 13.3 and Table 13.1). These changes tend to repolarize the membrane slightly.

Late Depolarization (Plateau): Selective Opening of Calcium Channels and Closing of Potassium Channels. The plateau of phase 2 results from a combination of the closing of K^+ channels (see Fig. 13.3 and Table 13.1) and the opening of voltage-gated Ca^{2+} channels. These chan-

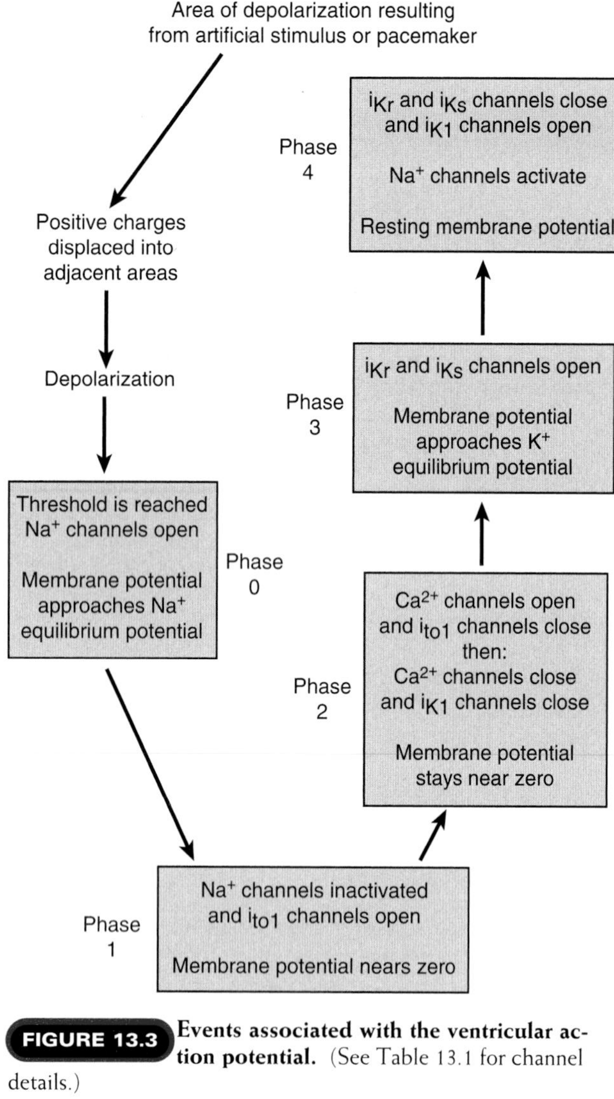

FIGURE 13.3 **Events associated with the ventricular action potential.** (See Table 13.1 for channel details.)

nels open more slowly than voltage-gated Na^+ channels and do not contribute to the rapid upswing of the ventricular action potential.

Repolarization: Selective Opening of Potassium Channels. The return of the membrane potential (phase 3, or repolarization) to the resting state is caused by the closing of Ca^{2+} channels and the opening of particular classes of K^+ channels (see Fig. 13.3 and Table 13.1). This relative increase in permeability to K^+ drives the membrane potential toward the K^+ equilibrium potential.

Resting Membrane Potential: Open Potassium Channels. The resting (diastolic) membrane potential (phase 4) of ventricular cells is maintained primarily by K^+ channels that are open at highly negative membrane potentials. They are called inward rectifying K^+ channels because, when the membrane is depolarized (e.g., by the opening of voltage-gated Na^+ channels), they do not permit outward movement of K^+. Other specialized K^+ channels help stabilize the resting membrane potential (see Table 13.1) and, in their absence, serious disorders of cardiac electrical activity can develop.

The Opening of Na^+ and Ca^{2+} and the Closing of K^+ Channels Causes the Pacemaker Potential of the SA and AV Nodes

When the electrical activity of a cell from the SA or AV node is compared with that of a ventricular muscle cell, three important differences are observed (see Fig. 13.1, Fig. 13.5): (1) the presence of a pacemaker potential, (2) the slow rise of the action potential, and (3) the lack of a well-defined plateau. The pacemaker potential results from changes in the permeability of the nodal cell membrane to all three of the major cations (see Table 13.2). First, K^+ channels, primarily responsible for repolarization, begin to close. Second, there is a steady increase in the membrane

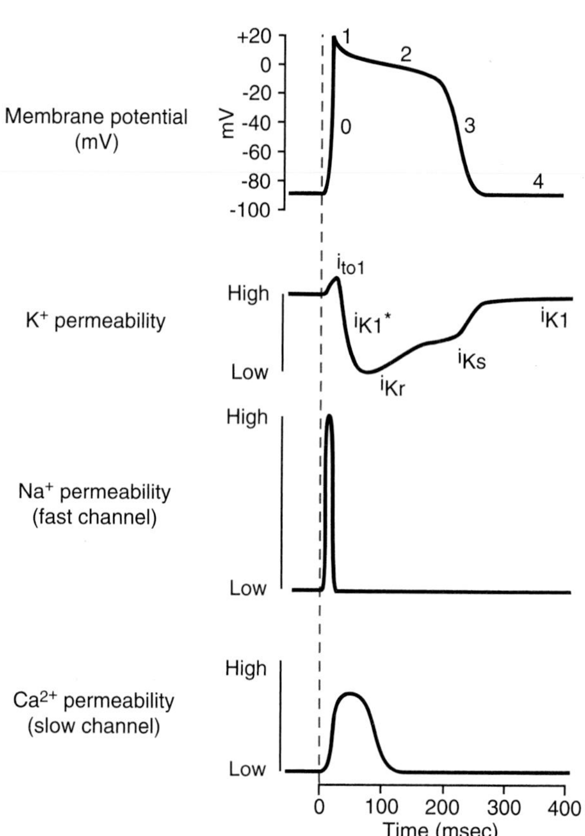

FIGURE 13.4 **Changes in cation permeabilities during a Purkinje fiber action potential (compare with Fig. 13.3).** The rise in action potential (phase 0) is caused by rapidly increasing Na^+ current carried by voltage-gated Na^+ channels. Na^+ current falls rapidly because voltage-gated Na^+ channels are inactivated. K^+ current rises briefly because of opening of i_{to1} channels and then falls precipitously because i_{K1} channels are closed by depolarization (*closing of i_{K1} channels). Ca^{2+} channels are opened by depolarization and are responsible, along with closed i_{K1} channels, for phase 2. K^+ current begins to increase because i_{Kr} and i_{Ks} channels are opened by depolarization, after a delay. Once repolarization occurs, Na^+ channels are activated. Reopened i_{K1} channels maintain phase 4.

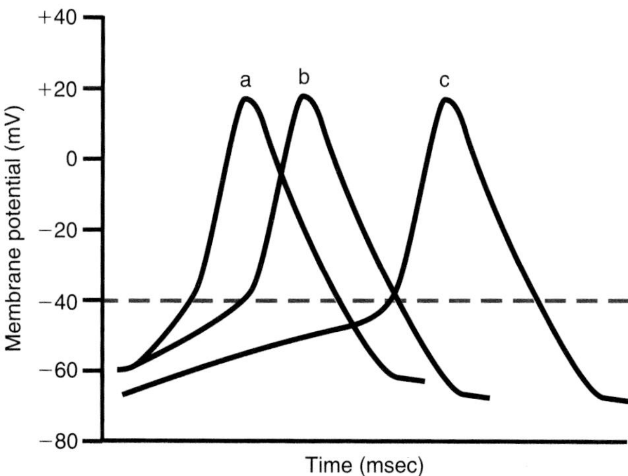

FIGURE 13.5 Sinoatrial plasma membrane potential as a function of time. Normal pacemaker potential (b) is affected by norepinephrine (a) and acetylcholine (c). The dashed line indicates threshold potential. The more rapidly rising pacemaker potential in the presence of norepinephrine (a) results from increased Na^+ permeability. The hyperpolarization and slower rising pacemaker potential in the presence of ACh results from decreased Na^+ permeability and increased K^+ permeability, due to the opening of ACh-activated K^+ channels.

permeability to Na^+ caused by the opening of a cation channel. Third, calcium moves in through the voltage-gated Ca^{2+} channel early in diastole. All three of these changes move the membrane potential in a positive direction toward the Na^+ and Ca^{2+} equilibrium potentials. An action potential is triggered when threshold is reached. This action potential rises more slowly than the ventricular action potential because the fast voltage-gated Na^+ channels play an insignificant role. Instead, the opening of slow voltage-gated Ca^{2+} channels is primarily responsible for the upstroke of the action potential in nodal cells. The absence of a well-defined plateau occurs because K^+ channels open and pull the membrane potential toward the K^+ equilibrium potential.

Purkinje fibers are also capable of pacemaker activity, but the rate of depolarization during phase 4 is much slower than that of the nodal cells. In the normal heart, phase 4 of Purkinje fibers is usually thought to be a stable resting membrane potential.

The Refractory Period Is Caused by a Delay in the Reactivation of Na^+ Channels

As discussed in Chapter 10, cardiac muscle cells display long refractory periods and, as a result, cannot be tetanized by fast, repeated stimulation. A prolonged refractory period eliminates the possibility that a sustained contraction might occur and prevent the cyclic contractions required to pump blood. The refractory period begins with depolarization and continues until nearly the end of phase 3 (see Fig. 10.2). This occurs because the Na^+ channels that open to cause phase 0 close and are inactive until the membrane repolarizes.

Neurotransmitters and Other Ligands Can Influence Membrane Ion Conductance

The normal pacemaker cells are under the influence of **parasympathetic nerves** (vagus) and **sympathetic nerves** (cardioaccelerator). The vagus nerves release acetylcholine (ACh) and the cardioaccelerator nerves release norepinephrine at their terminals in the heart. ACh slows the heart rate by reducing the rate of spontaneous depolarization of pacemaker cells (see Fig. 13.5), increasing the time required to reach threshold. Slowed heart rate is called **bradycardia** when the heart rate is below 60 beats/min. ACh exerts this effect by increasing the number of open K^+ channels and decreasing the number of open channels carrying Na^+ and Ca^{2+}; both actions hold the pacemaker potential closer to the K^+ equilibrium potential.

In contrast, norepinephrine causes an increase in the slope of the pacemaker potential so that the threshold is reached more rapidly and the heart rate increases. Increased heart rate is called **tachycardia** when the heart rate is above 100 beats/min. Norepinephrine increases the slope of the pacemaker potential by opening channels carrying Na^+ and Ca^{2+} and closing K^+ channels. Both effects result in faster movement of the pacemaker potential toward the Na^+ and Ca^{2+} equilibrium potentials. Norepinephrine and ACh exert these effects via G_s and G_i protein-mediated events.

Many other ligands, including metabolites (e.g., adenosine) and drugs (e.g., those which act on the autonomic nervous system), alter the heart rate by mechanisms similar to the ones outlined above.

THE INITIATION AND PROPAGATION OF CARDIAC ELECTRICAL ACTIVITY

Cardiac electrical activity is normally initiated and spread in an orderly fashion. The heart is said to be a **functional syncytium** because the excitation of one cardiac cell eventually leads to the excitation of all cells. The cellular basis for the functional syncytium is low-resistance areas of the intercalated disks (the end-to-end junctions of myocardial cells) called **gap junctions** (see Chapter 10). Gap junctions between adjacent cells allow small ions to move freely from one cell to the next, meaning that action potentials can be propagated from cell to cell, similar to the way an action potential is propagated along an axon (see Chapter 3).

Excitation Starts in the SA Node Because SA Cells Reach Threshold First

Excitation of the heart normally begins in the SA node because the pacemaker potential of this tissue (see Fig. 13.1) reaches threshold before the pacemaker potential of the AV node. The pacemaker rate of the SA node is normally 60 to 100 beats/min versus 40 to 55 beats/min for the AV node. Pacemaker activity in the bundle of His and the Purkinje system is even slower, at 25 to 40 beats/min. Normal atrial and ventricular cells do not exhibit pacemaker activity.

Many cells of the SA node reach threshold and depolarize almost simultaneously, creating a migration of ions between these depolarized SA nodal cells and nearby resting atrial cells. This leads to depolarization of the neighboring right atrial cells and a wave of depolarization begins to spread over the right and left atria.

The Action Potential Is Propagated by Local Currents Created During Depolarization

As Na^+ ions enter a cell during phase 0, their positive charge repels intracellular K^+ ions into nearby areas where depolarization has not yet occurred. Potassium is even driven into adjacent resting cells through gap junctions. The local buildup of K^+ depolarizes adjacent areas until threshold is reached. The cycle of depolarization to threshold, Na^+ entry, and subsequent displacement of positive charges into nearby areas explains the spread of electrical activity. Excitation proceeds as succeeding cycles of local ion current and action potential move out of the SA node and across the atria. This process is called the **propagation of the action potential.**

Excitation Usually Spreads From the SA Node to Atrial Muscle to the AV Node to the Purkinje System to Ventricular Muscle

A fibrous, nonconducting connective tissue ring separates the atria from the ventricles everywhere except at the AV node. For this reason, the transmission of electrical activity from the atria to the ventricles occurs only through the AV node. Action potentials in atrial muscle adjacent to the AV node produce local ion currents that invade the node and trigger intranodal action potentials.

Slow Conduction Through the AV Node. Excitation proceeds throughout the atria at a speed of approximately 1 m/sec. It requires 60 to 90 msec to excite all regions of the atria (Fig. 13.6). Propagation of the action potential continues within the AV node, but at a much slower velocity (0.05 to 0.1 m/sec). The slower conduction velocity is partially explained by the small size of the nodal cells. Less current is produced by the depolarization of a small nodal cell (compared with a large atrial or ventricular cell), and the relatively smaller current brings neighboring cells to threshold more slowly, decreasing the rate at which electrical activation spreads. Other significant factors are the slow upstroke of the action potential because it depends on slow voltage-gated Ca^{2+} channels and, possibly, weak electrical coupling as a result of relatively few gap junctions. Propagation of the action potential through the AV node takes approximately 120 msec. Excitation then proceeds through the AV bundle (bundle of His), the left and right bundle branches, and the Purkinje system.

The AV node is the weak link in the excitation of the heart. Inflammation, hypoxia, vagus nerve activity, and certain drugs (e.g., digitalis, beta blockers, and calcium entry blockers) can cause failure of the AV node to conduct some or all atrial depolarizations to the ventricles. On the other hand, its tendency to conduct slowly is sometimes of benefit in pathological situations in which atrial depolarizations are too frequent and/or uncoordinated, as in atrial flutter or fibrillation. In these conditions, not all of the electrical impulses that reach the AV node are conducted to the ventricles, and the ventricular rate tends to stay below the level at which diastolic filling is impaired (see Chapter 14). The benefit of slow AV nodal conduction in a normal heart is that it allows the ventricular filling associated with atrial systole to occur before the ventricles are excited.

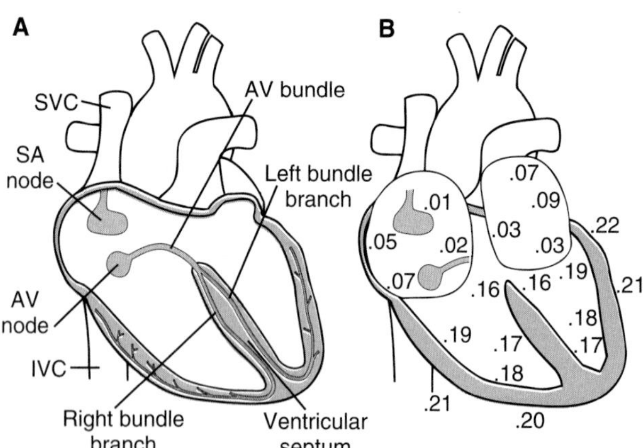

FIGURE 13.6 **The timing of excitation of various areas of the heart (in fractions of a second).** SVC, superior vena cava; IVC, inferior vena cava.

Rapid Conduction Through the Ventricles. The **Purkinje system** is composed of specialized cardiac muscle cells with large diameters. These cells rapidly conduct (conduction velocity up to 2 m/sec) action potentials throughout the subendocardium of both ventricles. Depolarization then proceeds from endocardium to epicardium (see Fig. 13.6). The conduction velocity through ventricular muscle is 0.3 m/sec; complete excitation of both ventricles takes approximately 75 msec. The rapid completion of excitation of the ventricles assures synchronized contraction of all ventricular muscle cells and maximal effectiveness in ejecting blood.

THE ELECTROCARDIOGRAM

The **electrocardiogram** (ECG) is a continuous record of cardiac electrical activity obtained by placing sensing electrodes on the surface of the body and recording the voltage differences generated by the heart. The equipment amplifies these voltages and causes a pen to deflect proportionally on a paper moving under it. This gives a plot of voltage as a function of time.

The ECG Records the Dipoles Produced by the Electrical Activity of the Heart

To understand the ECG, it is necessary to understand the behavior of electrical potentials in a three-dimensional conductor of electricity. Consider what happens when wires are run from the positive and negative terminals of a battery into a dish containing salt solution. Positively charged ions flow toward the negative wire (negative pole) and negatively charged ions simultaneously flow in the opposite direction toward the positive wire (positive pole).

The combination of two poles that are equal in magnitude and opposite in charge and located close to one another, is called a **dipole.** The flow of ions (current) is greatest in the region between the two poles, but some current flows at every point surrounding the dipole, reflecting the fact that voltage differences exist everywhere in the solution.

Measurement of the Voltage Associated With a Dipole. What points encircling the dipole in Figure 13.7 have the greatest voltage difference between them? Points A and B do because A is closest to the positive pole and B is closest to the negative pole. Positive charges are drawn from the area around point B by the negative end of the dipole, which is relatively near. The positive end of the dipole is relatively distant and, therefore, has little ability to attract negative charges from point B (although it can draw negative charges from point A). As positive charges are drawn away, point B is left with a negative charge (or negative voltage). The opposite happens between the positive end of the dipole and point A, leaving A with a net positive charge (or voltage). Points C and D have no voltage difference between them because they are equally distant from both poles and are, therefore, equally influenced by positive and negative charges. Any other two points on the circle, E and F, for example, have a voltage difference between them that is less than that between A and B and greater than that between C and D. This is also true of other combinations of points, such as A and C, B and D, and D and F. Voltage differences exist in all cases and are determined by the relative influences of the positive and negative ends of the dipole.

Changes in Dipole Magnitude and Direction. What would happen if the dipole were to change its orientation relative to points C and D? Figure 13.8 diagrams an apparatus in which electrodes from a voltmeter are placed at the edges of a dish of salt solution in which the dipole can be rotated. This solution is analogous to that depicted in Figure 13.7, except the dipole position is changed relative to the electrodes instead of the electrode being changed relative to the dipole. Figure 13.8 shows the changes in measured voltage that occur if the dipole is rotated 90 degrees. The measured voltage increases slowly as the dipole is turned and is maximal when the positive end of the dipole points to C and the negative end points to D. In each position, the dipole sets up current fields similar to those shown in Figure 13.7. The voltage measured depends on how the electrodes are positioned relative to those currents. Figure 13.8 also shows that the voltage between C and D will decrease to a new steady-state level as the voltage applied to the wires by the battery is decreased. These imaginary experiments illustrate two characteristics of a dipole that determine the voltage measured at distant points in a volume conductor: **direction** of the dipole relative to the measuring points and **magnitude** (voltage) of the dipole; this is another way of saying that a dipole is a **vector.**

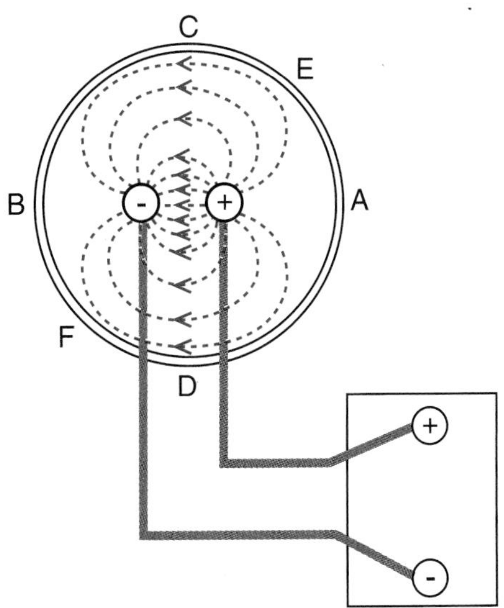

FIGURE 13.7 **Creating a dipole in a tub of salt solution.** The dashed lines indicate current flow; the current flows from the positive to the negative poles (see text for details).

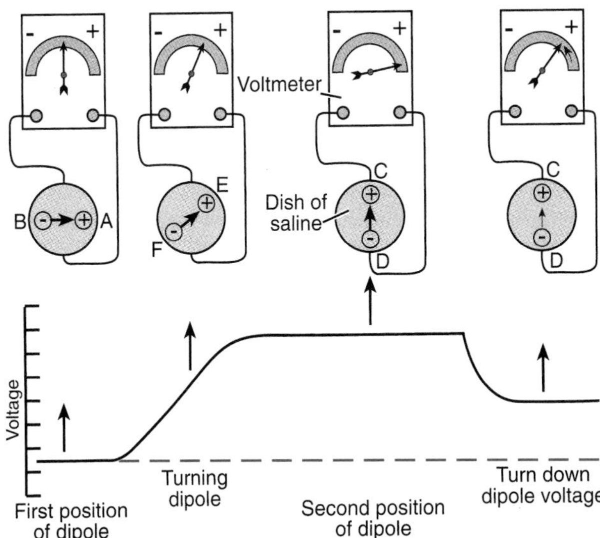

FIGURE 13.8 **Effect of dipole position and magnitude on recorded voltage.** In a salt solution, the dipole can be represented as a vector having a length and direction determined by the dipole magnitude and position, respectively. In this example, electrodes for the voltmeter are at points C and D. When a vector is directed parallel to a line between C and D, the voltage is maximum. If the magnitude of the vector is decreased, the voltage decreases.

Portions of the ECG Are Associated With Electrical Activity in Specific Cardiac Regions

We can use this analysis of a dipole in a volume conductor to rationalize the waveforms of the ECG. Of course, the actual case of the heart located in the chest is not as simple as the dipole in the tub of salt solution for two main reasons. First, excitation of the heart does not create one dipole; instead, there are many simultaneous dipoles. We will focus on the net dipole emerging as an average of all the individual dipoles. Second, the body is not a homogeneous volume conductor. The most significant problem is that the lungs are full of air, not salt solution. Despite these problems, the model is useful in an initial understanding of the generation of the ECG.

At rest, myocardial cells have a negative charge inside and a positive charge outside the cell membrane. As cells depolarize, the depolarized cells become negative on the outside, whereas the cells in the region ahead of the depolarized cells remain positive on the outside (Fig. 13.9). When the entire myocardium is depolarized, no voltage differences exist between any regions of myocardium because all cells are negative on the outside. When the cells in a given region depolarize during normal excitation, that portion of the heart generates a dipole. The depolarized portion constitutes the negative side, and the yet-to-be-depolarized portion constitutes the positive side of the dipole. The tub of salt solution is analogous to the rest of the body in that the heart is a dipole in a volume conductor. With electrodes located at various points around the volume conductor (i.e., the body), the voltage resulting from the dipole generated by the electrical activity of the heart can be measured.

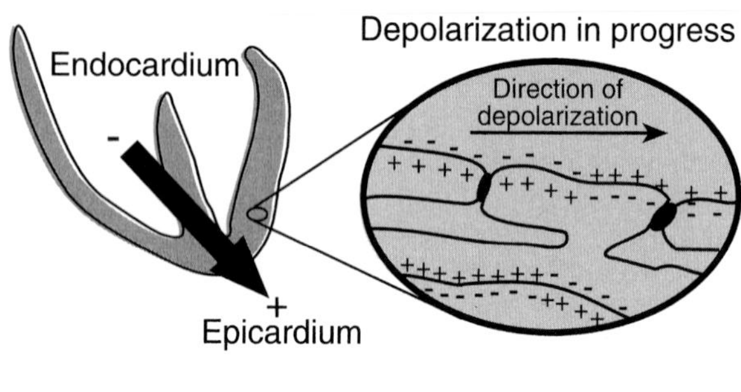

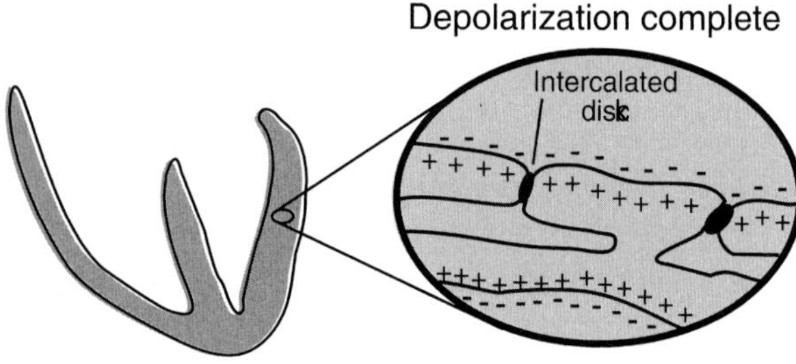

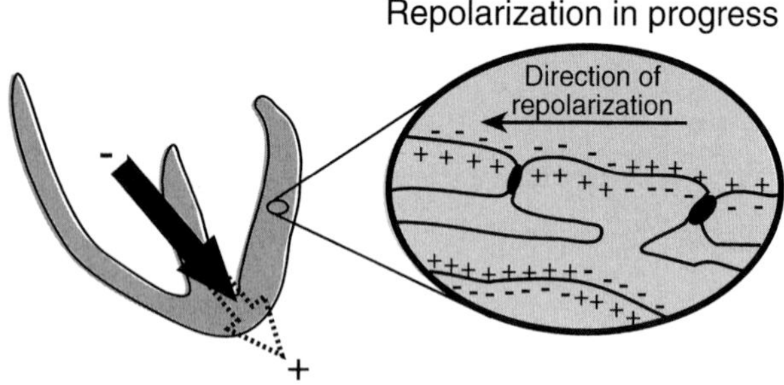

FIGURE 13.9 **Cardiac dipoles.** Partially depolarized or repolarized myocardium creates a dipole. Arrows show the direction of depolarization (or repolarization). Dipoles are present only when myocardium is undergoing depolarization or repolarization.

Consider the voltage changes produced by a two-dimensional model in which the body serves as a volume conductor and the heart generates a collection of changing dipoles (Fig. 13.10). An electrocardiographic recorder (a voltmeter) is connected between points A and B (lead I, see below). By convention, when point A is positive relative to point B, the ECG is deflected upward, and when B is positive relative to A, downward deflection results. The black arrows show (in two dimensions) the direction of the net dipole resulting from the many individual dipoles present at any one time. The lengths of the arrows are proportional to the magnitude (voltage) of the net dipole, which is related to the mass of myocardium generating the net dipole. The colored arrows show the magnitude of the dipole component that is parallel to the line between points A and B (the recorder electrodes); this component determines the voltage that will be recorded.

The P Wave and Atrial Depolarization. Atrial excitation results from a wave of depolarization that originates in the SA node and spreads over the atria, as indicated in panel 1 of Figure 13.10. The net dipole generated by this excitation has a magnitude proportional to the mass of the atrial muscle involved and a direction indicated by the black arrow. The head of the arrow points toward the positive end of the dipole, where the atrial muscle is not yet depolarized. The negative end of the dipole is located at the tail of the arrow, where depolarization has already occurred. Point A is, therefore, positive relative to point B, and there will be an upward deflection of the ECG as determined by the magnitude and direction of the dipole. Once the atria are completely depolarized, no voltage difference exists between A and B, and the voltage recording returns to 0. The voltage change associated with atrial excitation appears on the ECG as the **P wave.**

The PR Interval and Atrioventricular Conduction. After the P wave, the ECG returns to the baseline present before the P wave. The ECG is said to be **isoelectric** when there is no deflection from the baseline established before the P wave. During this time, the wave of depolarization moves slowly through the AV node, the AV bundle, the bundle branches, and the Purkinje system. The dipoles created by depolarization of these structures are too small to produce a deflection on the ECG. The isoelectric period between the end of the P wave and the beginning of the QRS complex, which signals ventricular depolarization, is called the **PR segment.** The P wave plus the PR segment is the **PR interval**. The duration of the PR interval is usually taken as an index of AV conduction time.

The QRS Complex and Ventricular Depolarization. The depolarization wave emerges from the AV node and travels along the AV bundle (bundle of His), bundle branches, and Purkinje system; these tracts extend down the interventricular septum. The net dipole that results from the initial depolarization of the septum is shown in panel 2 of Figure 13.10. Point B is positive relative to point A because the left side of the septum depolarizes before the right side. The small downward deflection produced on the ECG is the **Q wave.** The normal Q wave is often so small that it is not apparent.

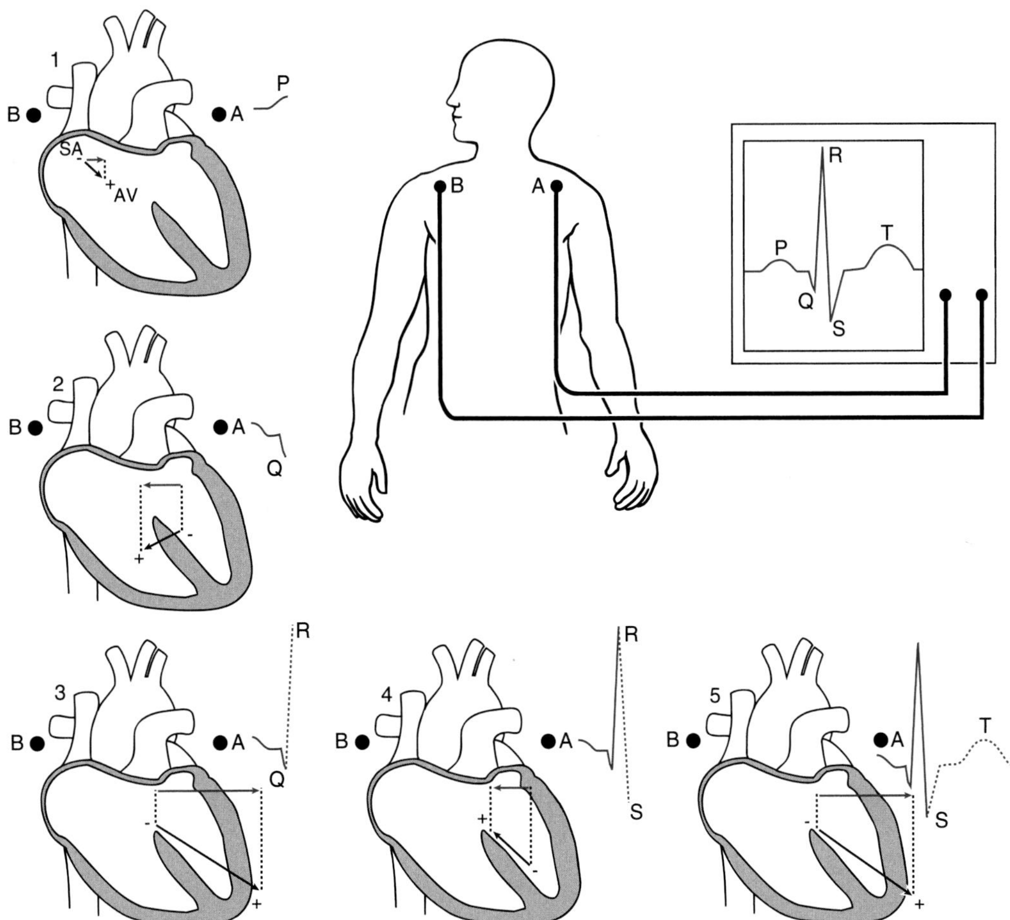

FIGURE 13.10 The sequence of major dipoles giving rise to ECG waveforms. The black arrows are vectors that represent the magnitude and direction of a major dipole. The magnitude is proportional to the mass of myocardium involved. The direction is determined by the orientation of depolarized and polarized regions of the myocardium. The vertical dashed lines project the vector onto the A-B coordinate (lead I); it is this component of the vector that is sensed and recorded (colored arrow). In panel 5, the tail of the vector (black arrow) shows the yet-to-be-repolarized region of the myocardium (negative) and the head points to the repolarized region (positive). The last areas of the ventricles to depolarize are the first to repolarize, i.e., repolarization appears to proceed in a direction opposite to that of depolarization. The projection of the vector (colored arrow) for repolarization points to the more positive electrode (A) as opposed to the less positive electrode (B), and so an upward deflection is recorded on this lead.

The wave of depolarization spreads via the Purkinje system across the inside surface of the free walls of the ventricles. Depolarization of free wall ventricular muscle proceeds from the innermost layers of muscle (subendocardium) to the outermost layers (subepicardium). Because the muscle mass of the left ventricle is much greater than that of the right ventricle, the net dipole during this phase has the direction indicated in panel 3. The deflection of the ECG is upward because point A is positive relative to point B, and it is large because of the great mass of tissue involved. This upward deflection is the **R wave.**

The last portions of the ventricle to depolarize generate a net dipole with the direction shown in panel 4. Point B is positive compared with point A, and the deflection on the ECG is downward. This final deflection is the **S wave.** The ECG tracing returns to baseline as all of the ventricular muscle becomes depolarized and all dipoles associated with ventricular depolarization disappear. The Q, R, and S waves together are known as the **QRS complex** and show the progression of ventricular muscle depolarization. The duration of the QRS complex is roughly equivalent to the duration of the P wave, despite the much greater mass of muscle of the ventricles. The relatively brief duration of the QRS complex is the result of the rapid, synchronous excitation of the ventricles.

The ST Segment and Phase 2 of the Ventricular Action Potential. The ST segment is the period between the end of the S wave and the beginning of the T wave. The ST segment is normally isoelectric, or nearly so. This indicates that no dipoles large enough to influence the ECG exist because all ventricular muscle is depolarized; that is, the action potentials of all ventricular cells are in phase 2 (Fig. 13.11).

The T Wave and Ventricular Repolarization. Repolarization, like depolarization, generates a dipole because the voltage of the depolarized area is different from that of the repolarized areas. The dipole associated with atrial repolarization does not appear as a separate deflection on the ECG because it generates a very low voltage and because it is

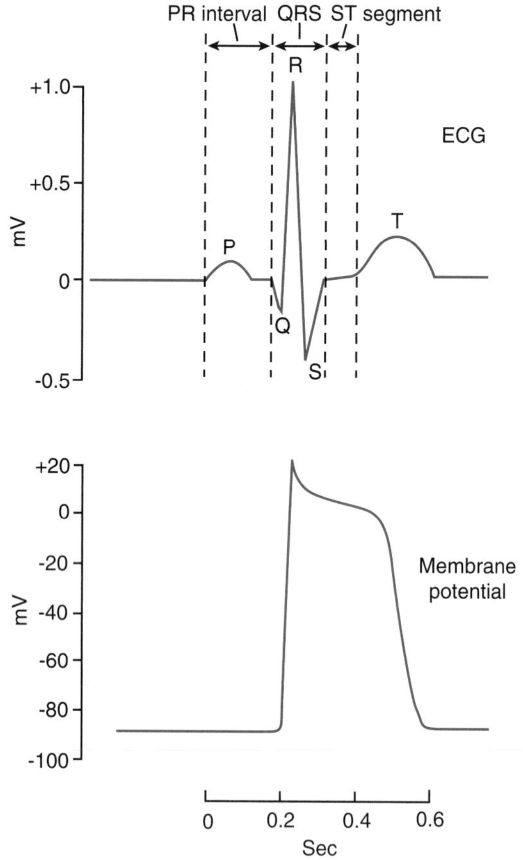

FIGURE 13.11 **The timing of the ventricular membrane potential and the ECG.** Note that the ST segment occurs during the plateau of the action potential.

masked by the much larger QRS complex that is present at the same time.

Ventricular repolarization is not as orderly as ventricular depolarization. The duration of ventricular action potentials is longer in subendocardial myocardium than in subepicardial myocardium. The longer duration of subendocardial action potentials means that even though subendocardial cells were the first to depolarize, they are the last to repolarize. Because subepicardial cells repolarize first, the subepicardium is positive relative to the subendocardium (see Fig. 13.9). That is, the polarity of the net dipole of repolarization is the same as the polarity of the dipole of depolarization. This results in an upward deflection because, as in depolarization, point A is positive with respect to point B. This deflection is the **T wave** (see panel 5, Fig. 13.10). The T wave has a longer duration than the QRS complex because repolarization does not proceed as a synchronized, propagated wave. Instead, the timing of repolarization is a function of properties of individual cells, such as numbers of particular K^+ channels.

The QT Interval. The **QT interval** is the time from the beginning of the QRS complex to the end of the T wave. If ventricular action potential and QT interval are compared, the QRS complex corresponds to depolarization, the ST segment to the plateau, and the T wave to repolarization (see Fig. 13.11). The relationship between a single ventricular action potential and the events of the QT interval are approximate because the events of the QT interval represent the combined influence of all of the ventricular action potentials.

The QT interval measures the total duration of ventricular activation. If ventricular repolarization is delayed, the QT interval is prolonged. Because delayed repolarization is associated with genesis of ventricular arrhythmias, this is clinically significant (see Clinical Focus Box 13.1).

ECG Leads Give the Voltages Measured Between Different Sites on the Body

An **electrocardiographic lead** is the pair of *electrical conductors* used to detect cardiac potential differences. An ECG lead is also used to refer to the *record* of potential differences made by the ECG machine. **Bipolar leads** give the potential difference between two electrodes placed at different sites. Elec-

CLINICAL FOCUS BOX 13.1

Long QT Syndrome

Some families have a rare inherited abnormality called **congenital long QT syndrome** (LQTS). Individuals with LQTS are often discovered because the individual or a family member presents to a physician with episodes of syncope (fainting) or because an otherwise healthy person dies suddenly and an alert physician suggests that their close relatives get an ECG. The ECG of affected individuals reveals either a long, irregular T wave, a prolonged ST segment, or both. Their hearts have delayed repolarization, which prolongs the ventricular action potential. In addition, when repolarization does occur, the freshly repolarized myocardium is subject to sudden, early depolarizations, called **afterdepolarizations**. These occur because the membrane potential in a small region of myocardium begins to depolarize before it has stabilized at the resting value. Afterdepolarizations may disrupt the normal, synchronized pattern of depolarization, and the ventricles may begin to depolarize in a chaotic pattern called **ventricular fibrillation**. With ventricular fibrillation, there is no synchronized contraction of ventricular muscle and the heart cannot pump the blood. Arterial pressure drops, blood flow to the brain and other parts of the body ceases, and sudden death occurs.

A single mutation of one of at least four genes, each of which codes for a particular cardiac muscle ion channel, causes LQTS. Mutations of three potassium channels have been discovered. The mutations decrease their function, decreasing potassium current and, thereby, increasing the tendency of the membrane to depolarize. A mutation of the sodium channel has also been found in some patients with LQTS. This mutation increases the sodium channel function, increasing sodium current and the tendency of the membrane to depolarize.

Individuals with congenital LQTS may be children or adults when the abnormality is identified. It is now apparent that at least one cause of sudden infant death syndrome (SIDS) involves a form of LQTS.

trodes of the traditional **bipolar limb leads** are placed on the left arm, right arm, and left leg (Fig. 13.12). The potential differences between each combination of two of these electrodes give leads I, II, and III. By convention, the left arm in lead I is the positive pole, and the left leg is the positive pole in leads II and III. A **unipolar lead** is the pair of electrical conductors giving the potential difference between an **exploring electrode** and a reference input, sometimes called the indifferent electrode. The reference input comes from a combination of electrodes at different sites, which is supposed to give roughly zero potential throughout excitation of the heart. Assuming this to be the case, the recorded electrical activity is the result of the influence of cardiac electrical activity on the exploring electrode. By convention, when the exploring electrode is positive relative to the reference input, an upward deflection is recorded.

The exploring electrode for the **precordial** or **chest leads** is the single electrode placed on the anterior and left lateral chest wall. For the chest leads, the reference input is obtained by connecting the three limb electrodes (Fig. 13.13). The observed ECGs recorded from the chest leads are each the result of voltage changes at a specified point on the surface of the chest. Unipolar chest leads are designated V_1 to V_6 and are placed over the areas of the chest

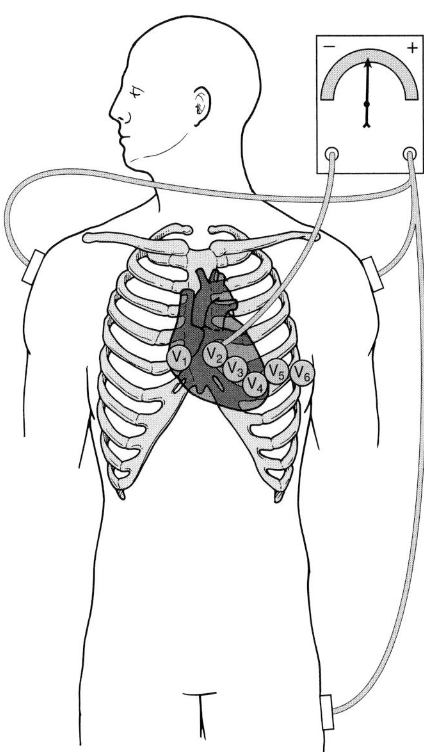

FIGURE 13.13 **Unipolar chest leads.** V_1 is just to the right of the sternum in the fourth intercostal space. V_2 is just to the left of the sternum in the fourth interspace. V_4 is in the fifth interspace in the midclavicular line. V_3 is midway between V_2 and V_4. V_5 is in the fifth interspace in the anterior axillary line. V_6 is in the fifth interspace in the midaxillary line. The three limb leads are combined to give the reference voltage (zero) for the unipolar chest lead (V).

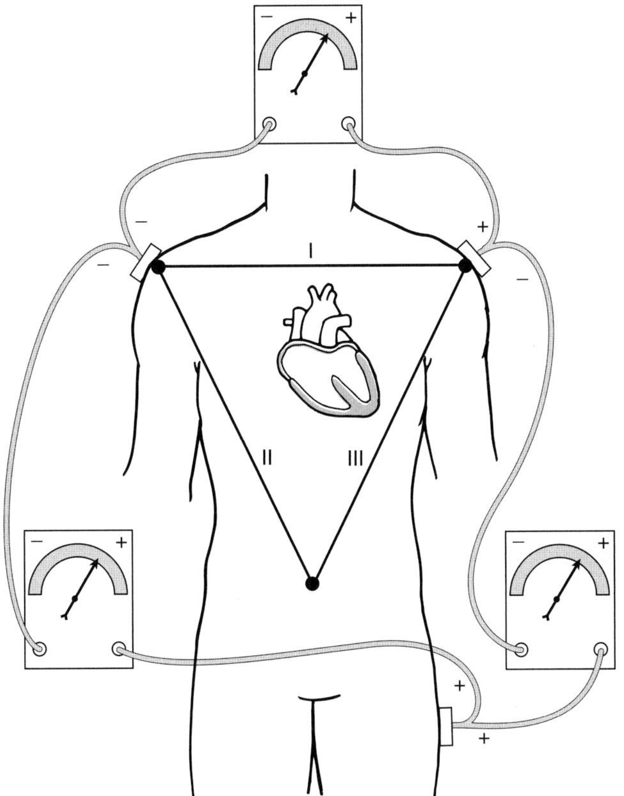

FIGURE 13.12 **Einthoven triangle.** Einthoven codified the analysis of electrical activity of the heart by proposing that certain conventions be followed. The heart is considered to be at the center of a triangle, each corner of which serves as the location for an electrode for two leads to the ECG recorder. The three resulting leads are I, II, and III. By convention, one electrode causes an upward deflection on the recorder when it is under the influence of a positive dipole relative to the other electrode.

shown in Figure 13.13. The generation of the QRS complex in the chest leads can be explained in a way similar to that for lead I.

The exploratory electrode for an **augmented limb lead** is an electrode on a single limb. The reference input is the two other limb electrodes connected together. Lead aVR gives the potential difference between the right arm (exploring electrode) and the combination of the left arm and the left leg (reference). Lead aVL gives the potential difference between the left arm and the combination of the right arm and left leg. Lead aVF gives the potential difference between the left leg and the combination of the left arm and right arm.

A standard 12-lead ECG, including six limb leads and six chest leads, is shown in Figure 13.14. The ECG is calibrated so that two dark horizontal lines (1 cm) represent 1 mV, and five dark vertical lines represent 1 second. This means that one light vertical line represents 0.04 sec.

The ECG Provides Information About Cardiac Dipoles as Vectors

Cardiac dipoles are vectors with both magnitude and direction. The net vector produced by all cardiac dipoles at a given time can be determined from the ECG. The direction of the vectors can be determined in the frontal and horizontal planes of the body.

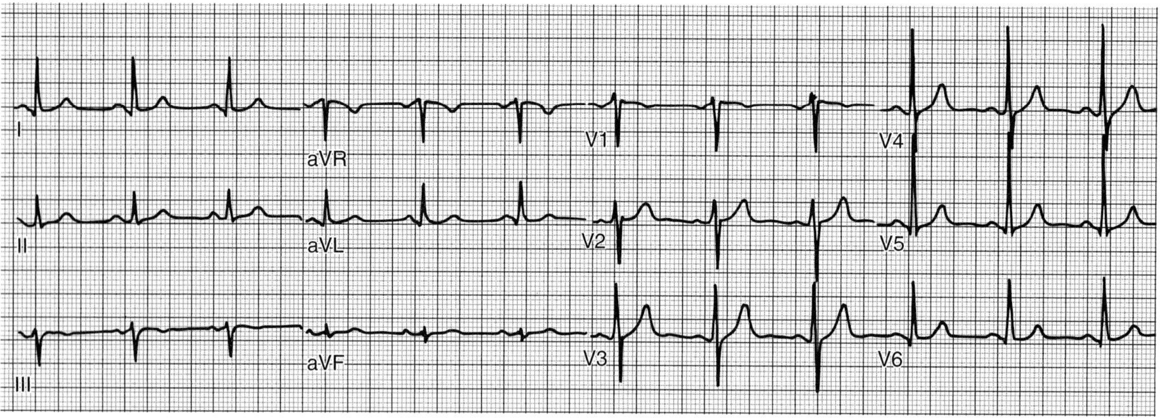

FIGURE 13.14 **Standard 12-lead ECG.** Six limb leads and six chest leads are shown. Two dark horizontal lines (10 mm) are calibrated to be 1 mV. Dark vertical lines represent 0.2 sec.

The bipolar limb leads (leads I, II, and III) and the augmented limb leads (aVR, aVL, and aVF) provide information about the electrical activity of the heart as observed in the frontal plane. As we have seen, lead I is the record of potential differences between the left and right arms. It records only the component of the electrical vector that is parallel to its **axis**. Lead I can be symbolized by a horizontal line (axis) going through the center of the chest (Fig. 13.15A) in the direction of right arm to left arm. Likewise, lead II can be symbolized by a 60° line drawn through the middle of the chest in the direction of right arm to left leg. The same type of representation can be done for lead III and for the augmented limb leads. The positive ends of the leads are shown by the arrowheads (see Fig. 13.15A). The diagram that results (see Fig. 13.15A) is called the **hexaxial reference system.**

A net cardiac dipole with its positive charge directed toward the positive end of the axis of a lead results in the recording of an upward deflection. A net cardiac dipole with its positive charge directed toward the negative end of the axis of a lead results in a downward deflection. A net cardiac dipole with its positive charge directed at a right angle to the axis of a lead results in no deflection. The hexaxial reference system can be used to predict the influence of a cardiac dipole on any of the six leads in the frontal plane. As we will see, this system is useful in understanding changes in the leads of the ECG associated with different diseases.

The unipolar chest leads provide information about cardiac dipoles generated in the horizontal plane (Figure 13.15B). Each chest lead can be represented as having an axis coming from the center of the chest to the site of the exploring electrode in the horizontal plane. The deflections recorded in each chest lead can be understood in terms of this axial system.

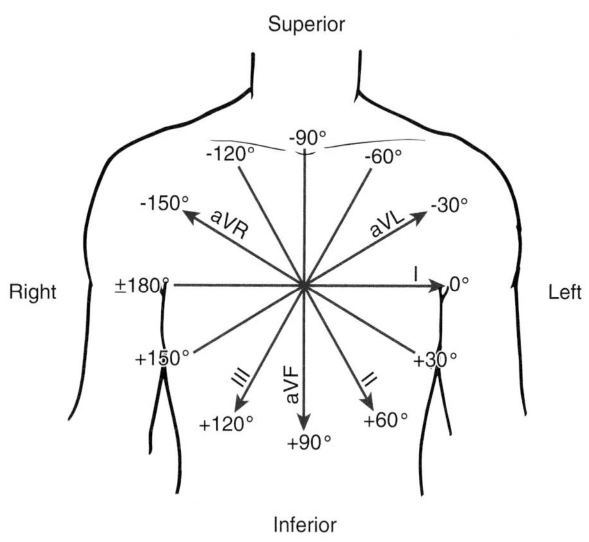

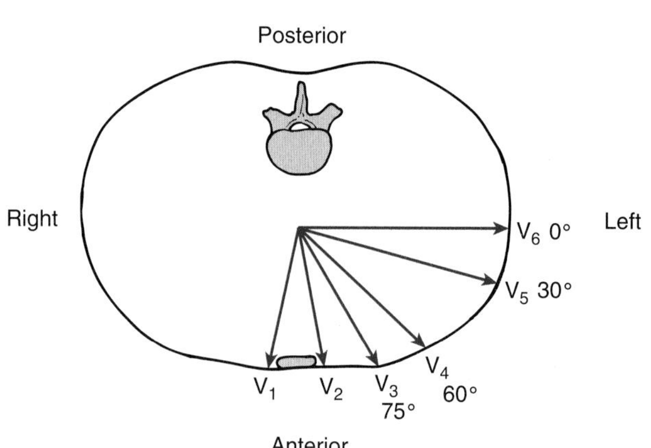

FIGURE 13.15 **Hexaxial reference system and chest leads.** A, The limb leads give information on cardiac dipole vectors in the frontal plane. **B,** Chest leads are influenced by dipole vectors in the horizontal plane.

The Mean QRS Electrical Axis Is Determined From the Limb Leads

As explained above, changes in the magnitude and direction of the cardiac dipole will cause changes in a given ECG lead, as predicted by the axial reference system. By examining the limb leads, the observer can determine the **mean electrical axis** during ventricular depolarization. One approach involves the use of **Einthoven's triangle.** Einthoven's triangle is an equilateral triangle with each side representing the axis of one of the bipolar limb leads (Fig. 13.16). The net magnitude of the QRS complex of any two of the three leads is measured and plotted on the appropriate axis. A perpendicular is dropped from each of the plotted points. A vector drawn between the center of the triangle and the intersection of the two perpendiculars gives the mean electrical axis. In this example, the data taken from the ECG in Figure 13.14 give a mean electrical axis of 3 degrees.

A second approach employs the hexaxial reference system (see Fig. 13.15A). First, the six limb leads are inspected to find the one in which the net QRS complex deflection is closest to zero. As discussed earlier, when the cardiac dipole is perpendicular to a particular lead, the net deflection is zero. Once the net QRS deflection closest to zero is identified, it follows that the mean electrical axis is perpendicular to that lead. The hexaxial reference system can be consulted to determine the angle of that axis. In Figure 13.14, the lead in which the net QRS deflection is closest to zero is lead aVF (the bipolar limb leads and lead aVF are enlarged in Figure 13.16). Lead I is perpendicular to the axis of lead aVF (see Fig. 13.15A). Because the QRS complex is upward in lead I, the mean electrical axis points to the left arm and is estimated to be about 0 degrees.

The mean QRS electrical axis is influenced by (a) the position of the heart in the chest, (b) the properties of the cardiac conduction system, and (c) the excitation and repolarization properties of the ventricular myocardium. Because the last two of these influences are most significant, the mean QRS electrical axis can provide valuable information about a variety of cardiac diseases.

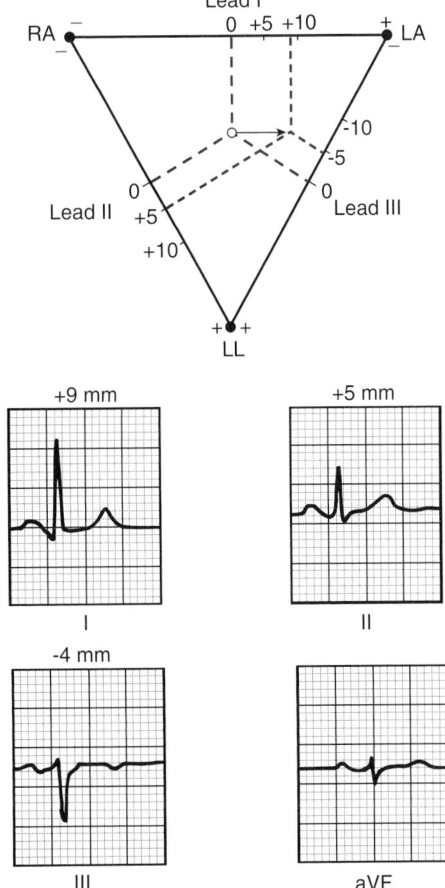

FIGURE 13.16 **Mean QRS electrical axis.** This axis can be estimated by using Einthoven's triangle and the net voltage of the QRS complex in any two of the bipolar limb leads. It can also be estimated by inspection of the six limb leads (see text for details). ECG tracings are from Figure 13.14.

The ECG Permits the Detection and Diagnosis of Irregularities in Heart Rate and Rhythm

The ECG provides information about the rate and rhythm of excitation, as well as the pattern of conduction of excitation throughout the heart. The following illustrations of cardiac rate and rhythm irregularities are not comprehensive; they were chosen to describe basic physiological principles. Disorders of cardiac rate and rhythm are referred to as **arrhythmias.**

Figure 13.14 shows the standard 12-lead ECG from an individual with normal sinus rhythm. We see that the P wave is always followed by a QRS complex of uniform shape and size. The PR interval (beginning of the P wave to the beginning of the QRS complex) is 0.16 sec (normal, 0.10 to 0.20 sec). This measurement indicates that the conduction velocity of the action potential from the SA node to the ventricular muscle is normal. The average time between R waves (successive heart beats) is about 0.84 sec, making the heart rate approximately 71 beats/min.

Figure 13.17A shows **respiratory sinus arrhythmia,** an increase in the heart rate with inspiration and a decrease with expiration. The presence of a P wave before each QRS complex indicates that these beats originate in the SA node. Intervals between successive R waves of 1.08, 0.88, 0.88, 0.80, 0.66, and 0.66 seconds correspond to heart rates of 56, 68, 68, 75, 91, and 91 beats/min. The interval between the beginning of the P wave and the end of the T wave is uniform, and the change in the interval between beats is primarily accounted for by the variation in time between the end of the T wave and the beginning of the P wave. Although the heart rate changes, the interval during which electrical activation of the atria and ventricles occurs does not change nearly as much as the interval between beats. Respiratory sinus arrhythmia is caused by cyclic changes in sympathetic and parasympathetic neural activity to the SA node that accompany respiration. It is observed in individuals with healthy hearts.

Figure 13.17B shows an ECG during excessive stimulation of the parasympathetic nerves. The stimulation releases ACh from nerve endings in the SA and AV nodes; ACh suppresses the pacemaker activity, slows the heart

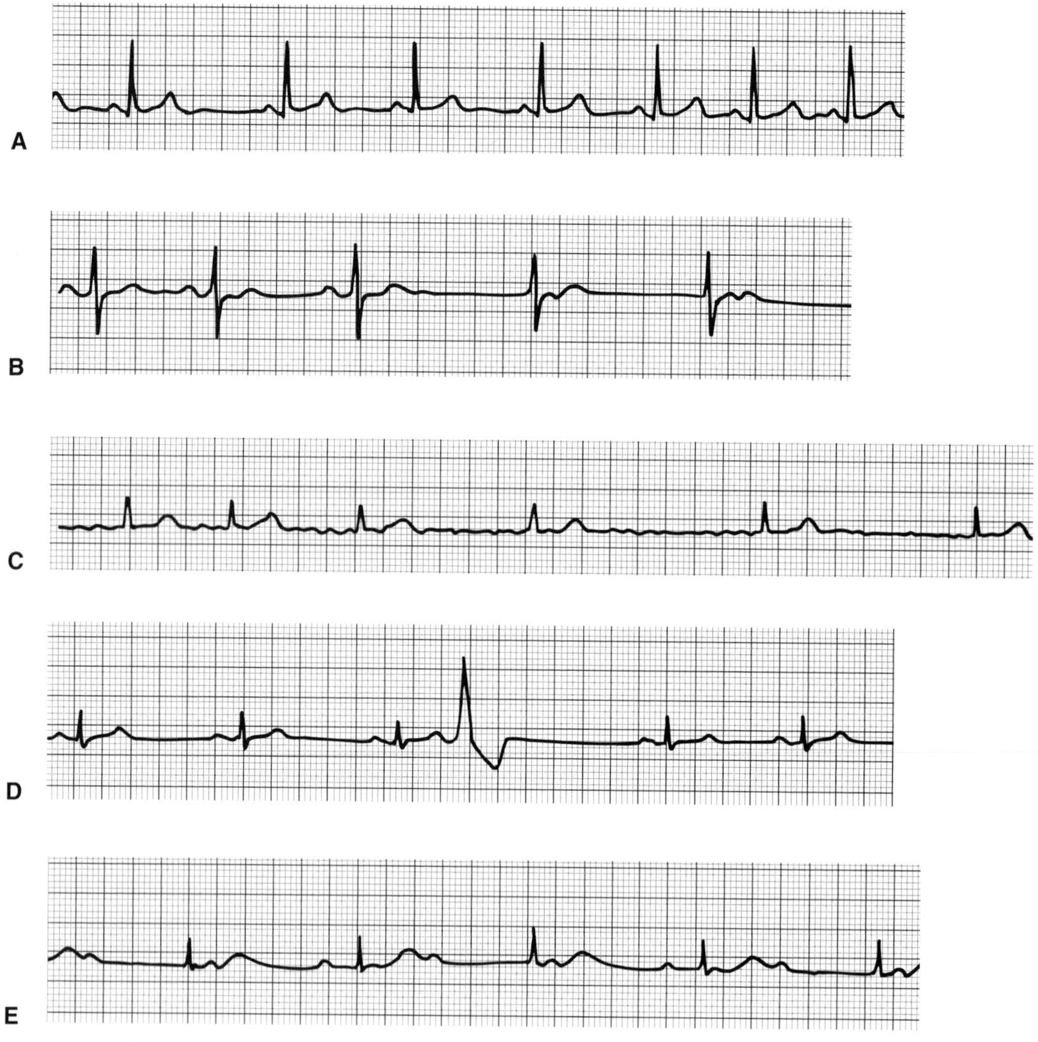

FIGURE 13.17 **ECGs (lead II) showing abnormal rhythms.** A, Respiratory sinus arrhythmia. B, Sinus arrest with vagal escape. C, Atrial fibrillation. D, Premature ventricular complex. E, Complete atrioventricular block.

rate, and increases the distance between P waves. The fourth and fifth QRS complexes are not preceded by P waves. When a QRS complex is recorded without a preceding P wave, it reflects the fact that ventricular excitation has occurred without a preceding atrial contraction, which means that the ventricles were excited by an impulse that originated below the atria. The normal configuration of the QRS complex suggests that the new pacemaker was in the AV node or bundle of His and that ventricular excitation proceeded normally from that point. This is called **junctional escape.**

The ECG in Figure 13.17C is from a patient with **atrial fibrillation.** In this condition, atrial systole does not occur because the atria are excited by many chaotic waves of depolarization. The AV node conducts excitation whenever it is not refractory and a wave of atrial excitation reaches it. Unless there are other abnormalities, conduction through the AV node and ventricles is normal and the resulting QRS complex is normal. The ECG shows QRS complexes that are not preceded by P waves. The ventricular rate is usually rapid and irregular. Atrial fibrillation is associated with numerous disease states, such as cardiomyopathy, pericarditis, hypertension, and hyperthyroidism, but it sometimes occurs in otherwise normal individuals.

The ECG in Figure 13.17D shows a **premature ventricular complex** (PVC). The first three QRS complexes are preceded by P waves; then after the T wave of the third QRS complex, a QRS complex of increased voltage and longer duration occurs. This premature complex is not preceded by a P wave and is followed by a pause before the next normal P wave and QRS complex. The premature ventricular excitation is initiated by an **ectopic focus,** an area of pacemaker activity in other than the SA node. In panel D, the focus is probably in the Purkinje system or ventricular muscle, where an aberrant pacemaker reaches threshold before being depolarized by the normal wave of excitation. Once the ectopic focus triggers an action potential, the excitation is propagated over the ventricles. The abnormal pattern of excitation accounts for the greater voltage, change of mean electrical axis, and longer duration (inefficient conduction) of the QRS complex. Although the abnormal wave of excitation reached the AV node, retrograde conduction usually dies out in the AV node. The next normal atrial excitation (P wave) occurs but is hidden by the inverted T wave associated with the abnormal QRS complex. This normal wave of atrial excitation does not result in ventricular excitation. Ventricular excitation does not occur because, when the impulse arrives, a portion of the AV node is still refractory following excitation by the premature complex. As a consequence, the next "scheduled" ventricular beat is missed. A prolonged interval following a premature ventricular beat is the **compensatory pause.**

Premature beats can also arise in the atria. In this case, the P wave is abnormal but the QRS complex is normal. Premature beats are often called extrasystoles, frequently a misnomer because there is no "extra" beat. However, in some cases, the premature beat is interpolated between two normal beats, and the premature beat is indeed "extra."

In Figure 13.17E, both P waves and QRS complexes are present, but their timing is independent of each other. This is **complete atrioventricular block** in which the AV node fails to conduct impulses from the atria to the ventricles. Because the AV node is the only electrical connection between these areas, the pacemaker activities of the two become entirely independent. In this example, the distance between P waves is about 0.8 sec, giving an atrial rate of 75 beats/min. The distance between R waves averages 1.2 sec, giving a ventricular rate of 50 beats/min. The atrial pacemaker is probably in the SA node, and the ventricular pacemaker is probably in a lower portion of the AV node or bundle of His.

AV block is not always complete. Sometimes the PR interval is lengthened, but all atrial excitations are eventually conducted to the ventricles. This is **first-degree atrioventricular block.** When some, but not all, of the atrial excitations are conducted by the AV node, it is **second-degree atrioventricular block.** If atrial excitation never reaches the ventricles, as in the example in Figure 13.17E, it is **third-degree (complete) atrioventricular block.**

The ECG Provides Three Types of Information About the Ventricular Myocardium

The ECG provides information about the pattern of excitation of the ventricles, changes in the mass of electrically active ventricular myocardium, and abnormal dipoles resulting from injury to the ventricular myocardium. It provides no direct information about the mechanical effectiveness of the heart; other tests are used to study the efficiency of the heart as a pump (see Chapter 14).

The Pattern of Ventricular Excitation. Disease or injury can affect the pattern of ventricular depolarization and produce an abnormality in the QRS complex. Figure 13.18 shows a normal QRS complex (Fig. 13.18A) and two examples of complexes that have been altered by impaired conduction. In Figure 13.18B, the AV bundle branch to the right side of the heart is not conducting (i.e., there is right bundle-branch block), and depolarization of right-sided myocardium, therefore, depends on delayed electrical activity coming from the normally depolarized left side of the heart. The resulting QRS complex has an abnormal shape because of aberrant electrical conduction and is prolonged because of the increased time necessary to fully depolarize the heart. In Figure 13.18C, the AV bundle branch to the left side of the heart is not conducting (i.e., there is left bundle-branch block), also resulting in a wide, deformed QRS complex.

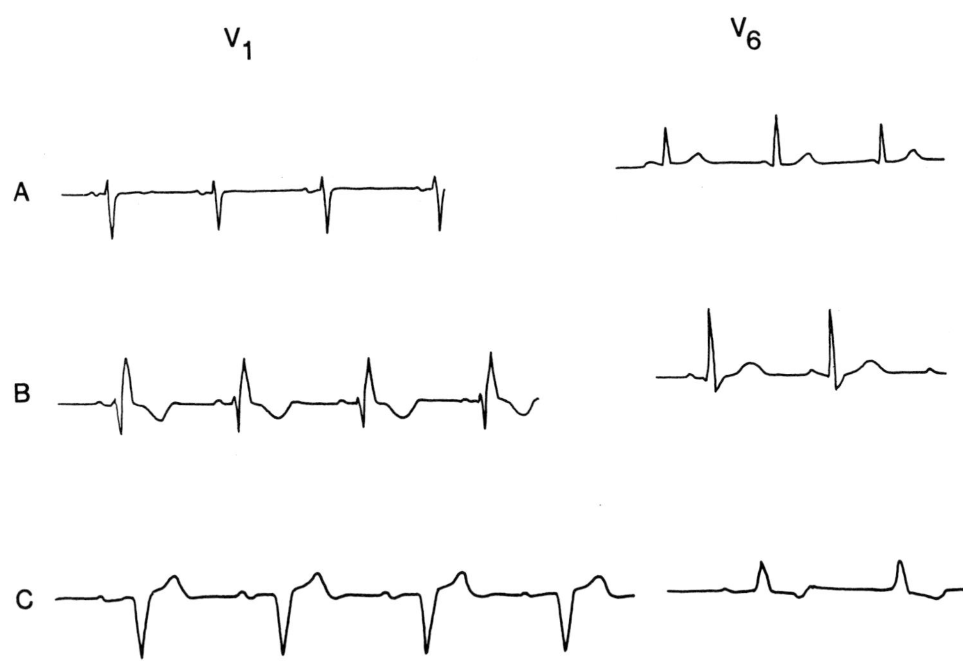

FIGURE 13.18 **ECGs (leads V_2 and V_6) of patients with various conditions.** A, patient with normal QRS complex. B, patient with right bundle-branch block. C, patient with left bundle-branch block.

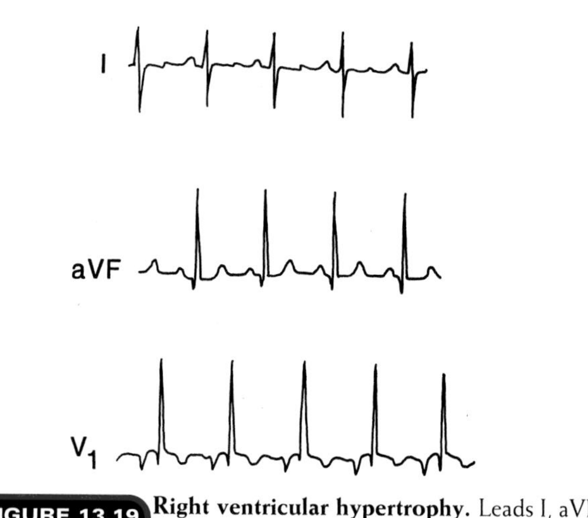

FIGURE 13.19 **Right ventricular hypertrophy.** Leads I, aVF, and V_1 of a patient are shown.

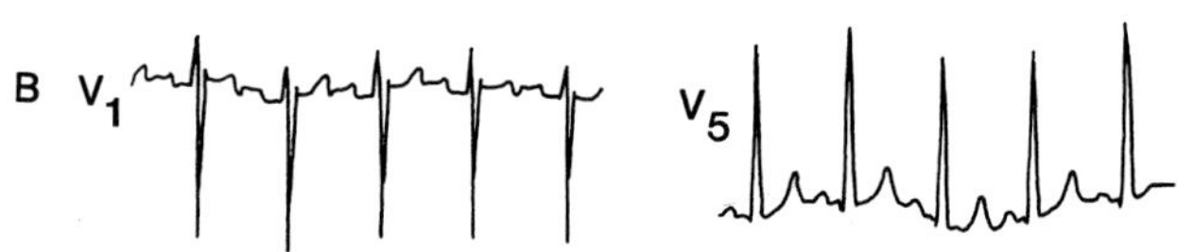

FIGURE 13.20 **Effects of cardiac hypertrophy.** A, Large P waves (lead III) caused by atrial hypertrophy. B, Altered QRS complex (leads V_1 and V_5) produced by left ventricular hypertrophy.

Changes in the Mass of Electrically Active Ventricular Myocardium. The recording in Figure 13.19 shows the effect of right ventricular enlargement on the ECG. The increased mass of right ventricular muscle changes the direction of the major dipole during ventricular depolarization, resulting in large R waves in lead V_1. The large S waves in lead I and the large R waves in lead aVF are also characteristic of a shift in the dipole of ventricular depolarization to the right. This illustrates how a change in the mass of excited tissue can affect the amplitude and direction of the QRS complex.

Figure 13.20 shows the effects of atrial hypertrophy on the P waves of lead III (see Fig. 13.20A) and the altered QRS complexes in leads V_1 and V_5 associated with left ventricular hypertrophy (see Fig. 13.20B). Left ventricular hypertrophy rotates the direction of the major dipole associated with ventricular depolarization to the left, causing large S waves in V_1 and large R waves in V_5.

Abnormal Dipoles Resulting From Ventricular Myocardial Injury. **Myocardial ischemia** is present when a portion of the ventricular myocardium fails to receive sufficient blood flow to meet its metabolic needs. In this case, the supply of ATP may decrease below the level required to maintain the active transport of ions across the cell membrane. The resulting alterations in the membrane potential in the ischemic region can affect the ECG. Normally, the ECG is at baseline (zero voltage) during

- The interval between the completion of the T wave and the onset of the P wave (the TP interval), during which all cardiac cells are at their resting membrane potential
- The ST segment, during which depolarization is complete and all ventricular cells are at the plateau (phase 2) of the action potential

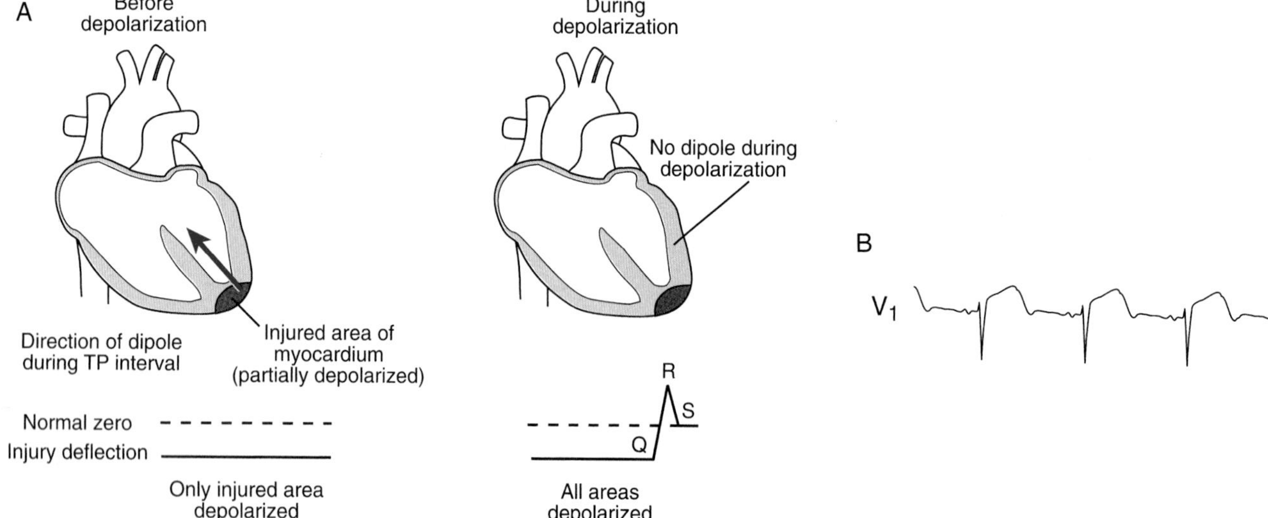

FIGURE 13.21 **Electrocardiogram changes in myocardial injury.** A, Dark shading depicts depolarized ventricular tissue. ST segment elevation can occur with myocardial injury. The apparent zero baseline of the ECG before depolarization is below zero because of partial depolarization of the injured area (shading). After depolarization (during the action potential plateau), all areas are depolarized and true zero is recorded. Because zero baseline is set arbitrarily (on the ECG recorder), a depressed diastolic baseline (TP segment) and an elevated ST segment cannot be distinguished. Regardless of the mechanism, this is referred to as an elevated ST segment. **B,** The ECG (lead V_1) of a patient with acute myocardial infarction.

With myocardial ischemia, the cells in the ischemic region partially depolarize to a lower resting membrane potential because of a lowering of the potassium ion concentration gradient, although they are still capable of action potentials. As a consequence, a dipole is present during the TP interval in injured hearts because of the voltage difference between normal (polarized) and abnormal (partially polarized) tissue. However, no dipole is present during the ST interval because depolarization is uniform and complete in both injured and normal tissue (this is the plateau period of ventricular action potentials). Because the ECG is designed so that the TP interval is recorded as zero voltage, the true zero during the ST interval is recorded as a positive or negative deflection (Fig. 13.21). These deflections during the ST interval are of major clinical utility in the diagnosis of cardiac injury.

REVIEW QUESTIONS

DIRECTIONS: Each of the numbered items or incomplete statements in this section is followed by answers or by completions of the statement. Select the ONE lettered answer or completion that is BEST in each case.

1. Rapid depolarization (phase 0) of the action potential of ventricular muscle results from opening of
   (A) Voltage-gated Ca^{2+} channels
   (B) Voltage-gated Na^+ channels
   (C) Acetylcholine-activated K^+ channels
   (D) Inward rectifying K^+ channels
   (E) ATP-sensitive K^+ channels
2. A 72-year-old man with an atrial rate of 80 beats/min develops third-degree (complete) AV block. A pacemaker site located in the AV node below the region of the block triggers ventricular activity, but at a rate of only 40 beats/min. What would be observed?
   (A) One P wave for each QRS complex
   (B) An inverted T wave
   (C) A shortened PR interval
   (D) A normal QRS complex
3. To ensure an adequate heart rate, a temporary electronic pacemaker lead is attached to the apex of the right ventricle, and the heart is paced by electrically stimulating this site at a rate of 70 beats/min. When the ECG during pacing is compared with the ECG before pacing, there would be a
   (A) Shortened PR interval
   (B) QRS complex similar to that seen with left bundle-branch block
   (C) QRS complex of shortened duration
   (D) P wave following each QRS complex
   (E) QRS complex similar to that seen with right bundle-branch block
4. What is most responsible for phase 0 of a cardiac nodal cell?
   (A) Voltage-gated Na^+ channels
   (B) Acetylcholine-activated K^+ channels
   (C) Inward rectifying K^+ channels
   (D) Voltage-gated Ca^{2+} channels
   (E) Pacemaker channels
5. Atrial repolarization normally occurs during the
   (A) P wave
   (B) QRS complex
   (C) ST segment
   (D) T wave
   (E) Isoelectric period
6. The P wave is normally positive in lead I of the ECG because
   (A) Depolarization of the ventricles proceeds from subendocardium to subepicardium
   (B) When the ECG electrode attached to the right arm is positive relative to the electrode attached to the left arm, an upward deflection is recorded
   (C) AV nodal conduction is slower than atrial conduction
   (D) Depolarization of the atria proceeds from right to left
   (E) When cardiac cells are depolarized, the inside of the cells is negative relative to the outside of the cells
7. Stimulation of the sympathetic nerves to the normal heart
   (A) Increases duration of the TP interval
   (B) Increases the duration of the PR interval
   (C) Decreases the duration of the QT interval
   (D) Leads to fewer P waves than QRS complexes
   (E) Decreases the frequency of QRS complexes
8. A drug that raises the heart rate from 70 to 100 beats per minute could
   (A) Be an adrenergic receptor antagonist
   (B) Cause the opening of acetylcholine-activated K^+ channels
   (C) Be a cholinergic receptor agonist
   (D) Be an adrenergic receptor agonist
   (E) Cause the closing of voltage-gated Ca^{2+} channels
9. Excitation of the ventricles
   (A) Always leads to excitation of the atria
   (B) Results from the action of norepinephrine on ventricular myocytes
   (C) Proceeds from the subendocardium to subepicardium
   (D) Is initiated during the plateau (phase 2) of the ventricular action potential
   (E) Results from pacemaker potentials of ventricular cells
10. AV nodal cells
   (A) Exhibit action potentials characterized by rapid depolarization (phase 0)
   (B) Exhibit increased conduction velocity when exposed to acetylcholine
   (C) Conduct impulses more slowly than either atrial or ventricular cells
   (D) Are capable of pacemaker activity at an intrinsic rate of 100 beats/min
   (E) Exhibit slowed conduction velocity when exposed to norepinephrine
11. Stimulation of the parasympathetic nerves to the normal heart can lead to complete inhibition of the SA node for several seconds. During that period
   (A) P waves would become larger
   (B) There would be fewer T waves than QRS complexes
   (C) There would be fewer P waves than T waves
   (D) There would be fewer QRS complexes than P waves
   (E) The shape of QRS complexes would change
12. The R wave in lead I of the ECG
   (A) Is larger than normal with right ventricular hypertrophy
   (B) Reflects a net dipole associated with ventricular depolarization
   (C) Reflects a net dipole associated with ventricular repolarization
   (D) Is largest when the mean electrical axis is directed perpendicular to a line drawn between the two shoulders
   (E) Is associated with atrial depolarization
13. The ST segment of the normal ECG
   (A) Occurs during a period when both ventricles are completely repolarized
   (B) Occurs when the major dipole is directed from subendocardium to subepicardium

(continued)

(C) Occurs during a period when both ventricles are completely depolarized
(D) Is absent in lead I of the ECG
(E) Occurs during depolarization of the Purkinje system

SUGGESTED READING

Fisch C. Electrocardiogram and mechanisms of arrhythmias. In: Podrid PJ, Kowey PR, eds. Cardiac Arrhythmia: Mechanisms, Diagnosis and Management. 2nd Ed. Baltimore: Lippincott Williams & Wilkins, 2001.

Katz AM. Physiology of the Heart. 3rd Ed. Philadelphia: Lippincott Williams & Wilkins, 2001.

Lauer MR, Sung RJ. Physiology of the conduction system. In: Podrid PJ, Kowey PR, eds. Cardiac Arrhythmia: Mechanisms, Diagnosis and Management. 2nd Ed. Baltimore: Lippincott Williams & Wilkins, 2001.

Lilly LS. Pathophysiology of Heart Disease. 2nd Ed. Baltimore: Williams & Wilkins, 1998.

Mirvis DM, Goldberger AL. Electrocardiography. In: Braunwald E, Zipes DP, Libby P, eds. Heart Disease. 6th Ed. Philadelphia: WB Saunders, 2001.

Rubart M, Zipes DP. Genesis of cardiac arrhythmias: Electrophysiological considerations. In: Braunwald E, Zipes DP, Libby P, eds. Heart Disease. 6th Ed. Philadelphia: WB Saunders, 2001.

CHAPTER

14

The Cardiac Pump

Thom W. Rooke, M.D.
Harvey V. Sparks, Jr., M.D.

CHAPTER OUTLINE

- THE CARDIAC CYCLE
- CARDIAC OUTPUT
- THE MEASUREMENT OF CARDIAC OUTPUT
- THE ENERGETICS OF CARDIAC FUNCTION

KEY CONCEPTS

1. Learning to correlate the ECG, pressures, volumes, flows, and heart sounds in time is fundamental to a working knowledge of the heart.
2. Cardiac output is the product of stroke volume times heart rate.
3. Stroke volume is determined by end-diastolic fiber length, contractility, afterload, and hypertrophy.
4. Heart rate influences ventricular filling time and stroke volume.
5. The influence of heart rate on cardiac output depends on simultaneous effects on ventricular contractility.
6. Cardiac output can be measured by methods that rely on mass balance or cardiac imaging.
7. Cardiac energy production depends primarily on the supply of oxygen to the heart.
8. Cardiac energy consumption depends on the work of the heart.
9. The external work of the heart depends on the volume of blood pumped and the pressure against which it is pumped.

The heart consists of a series of four separate chambers (two atria and two ventricles) that use one-way valves to direct blood flow. Its ability to pump blood depends on the integrity of the valves and the proper cyclic contraction and relaxation of the muscular walls of the four chambers. An understanding of the cardiac cycle is a prerequisite for understanding the performance of the heart as a pump.

THE CARDIAC CYCLE

The **cardiac cycle** refers to the sequence of electrical and mechanical events occurring in the heart during a single beat and the resulting changes in pressure, flow, and volume in the various cardiac chambers. The functional interrelationships of the cardiac cycle described below are represented in Figure 14.1.

Sequential Contractions of the Atria and Ventricles Pump Blood Through the Heart

The cycle of events described here occurs almost simultaneously in the right and left heart; the main difference is that the pressures are higher on the left side. The focus is on the left side of the heart, beginning with electrical activation of the atria.

Atrial Systole and Diastole. The P wave of the electrocardiogram (ECG) reflects atrial depolarization, which initiates **atrial systole.** Contraction of the atria "tops off" ventricular filling with a final, small volume of blood from the atria, producing the **a wave**. Under resting conditions, atrial systole is not essential for ventricular filling and, in its absence, ventricular filling is only slightly reduced. However, when increased cardiac output is required, as during exercise, the absence of atrial systole can limit ventricular filling and stroke volume. This happens in patients with atrial fibrillation, whose atria do not contract synchronously.

The P wave is followed by an electrically quiet period, during which atrioventricular (AV) node transmission occurs (the PR segment). During this electrical pause, the mechanical events of atrial systole and ventricular filling are concluded before excitation and contraction of the ventricles begin.

Atrial diastole follows atrial systole and occurs during ventricular systole. As the left atrium relaxes, blood enters the atrium from the pulmonary veins. Simultane-

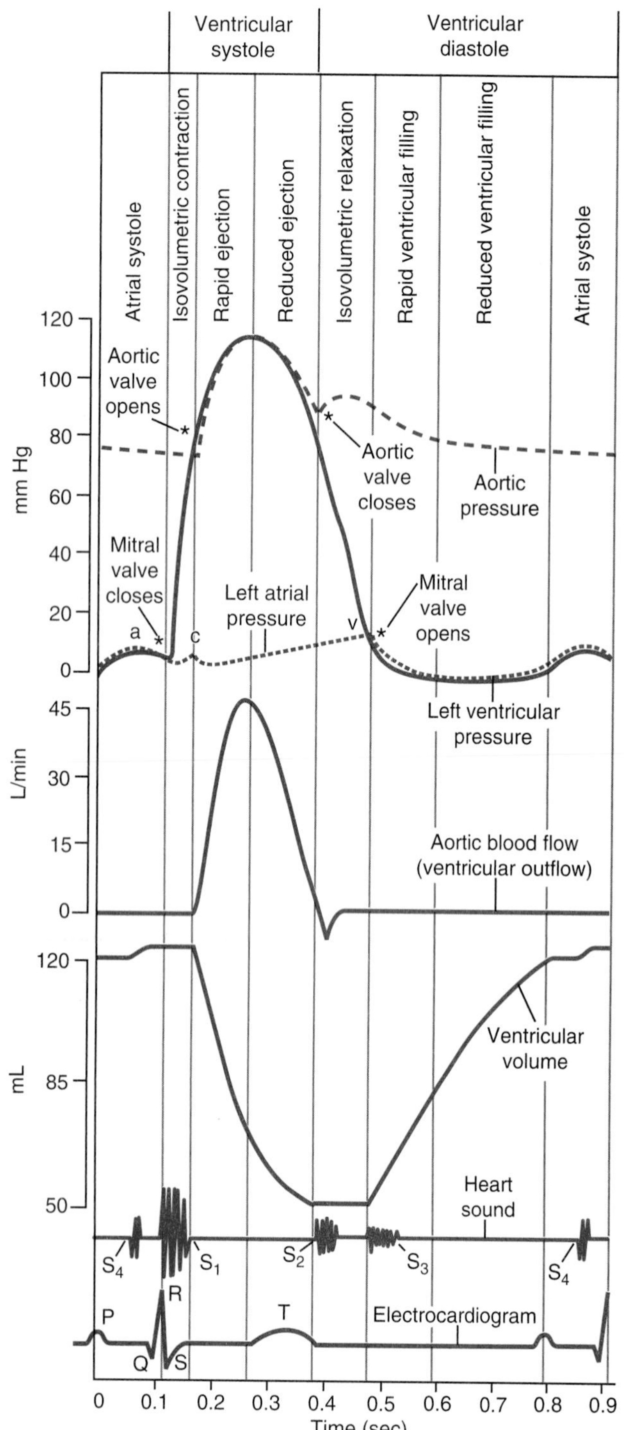

FIGURE 14.1 The timing of various events in the cardiac cycle.

ously, blood enters the right atrium from the superior and inferior vena cavae. The gradual rise in left atrial pressure during atrial diastole produces the **v wave** and reflects its filling. The small pressure oscillation early in atrial diastole, called the **c wave**, is caused by bulging of the mitral valve and movements of the heart associated with ventricular contraction.

Ventricular Systole. The QRS complex reflects excitation of ventricular muscle and the beginning of **ventricular systole** (see Fig. 14.1). As ventricular pressure rises above atrial pressure, the left **atrioventricular (mitral) valve** closes. Contraction of the papillary muscles prevents the mitral valve from everting into the left atrium and enables the valve to prevent the regurgitation of blood into the atrium as ventricular pressure rises. The aortic valve does not open until left ventricular pressure exceeds aortic pressure. During the interval when both mitral and aortic valves are closed, the ventricle contracts **isovolumetrically** (i.e., the ventricular volume does not change). The contraction causes ventricular pressure to rise, and when ventricular pressure exceeds aortic pressure (at approximately 80 mm Hg), the aortic valve opens and allows blood to flow from the ventricle into the aorta. At this point, ventricular muscle begins to shorten, reducing the volume of the ventricle.

When the rate of ejection begins to fall (see the aortic blood flow record in Fig. 14.1), the aortic and ventricular pressures decline. Ventricular pressure actually decreases slightly below aortic pressure prior to closure of the aortic valve, but flow continues through the aortic valve because of the inertia imparted to the blood by ventricular contraction. (Think of a rubber ball connected to a paddle by a rubber band. The ball continues to travel away from the paddle after you pull back because the inertial force on the ball exceeds the force generated by the rubber band.)

Ventricular Diastole. Ventricular repolarization (producing the T wave) initiates ventricular relaxation or **ventricular diastole.** When the ventricular pressure drops below the atrial pressure, the mitral valve opens, allowing the blood accumulated in the atrium during systole to flow rapidly into the ventricle; this is the rapid phase of ventricular filling. Both pressures continue to decrease—the atrial pressure because of emptying into the ventricle and the ventricular pressure because of continued ventricular relaxation (which, in turn, draws more blood from the atrium). About midway through ventricular diastole, filling slows as ventricular and atrial pressures converge. Finally, atrial systole tops off ventricular volume.

Pressures, Flows, and Volumes in the Cardiac Chambers, Aorta, and Great Veins Can Be Matched With the ECG and Heart Sounds

The pressures, flows, and volumes in the cardiac chambers, aorta, and great veins can be studied in conjunction with the ECG and heart sounds to yield an understanding of the coordinated activity of the heart. Ventricular diastole and systole can be defined in terms of both electrical and mechanical events. In electrical terms, ventricular systole is defined as the period between the QRS complex and the end of the T wave. In mechanical terms, it is the period between the closure of the mitral valve and the subsequent closure of the aortic valve. In either case, ventricular diastole comprises the remainder of the cycle.

The **first (S_1)** and **second (S_2) heart sounds** signal the beginning and end of mechanical systole. The first heart sound (usually described as a "lub") occurs as the ventricle contracts and ventricular pressure rises above atrial pressure, causing the atrioventricular valves to close. The relatively low-pitched sound associated with their closure is caused by vibrations of the valves and walls of the heart that occur as a result of their elastic properties when the flow of blood through the valves is suddenly stopped. In contrast, the aortic and pulmonic valves close at the end of ventricular systole, when the ventricles relax and pressures in the ventricles fall below those in the arteries. The elastic properties of the aortic and pulmonic valves produce the second heart sound, which is relatively high-pitched (typically described as a "dup"). Mechanical events other than vibrations of the valves and nearby structures contribute to these two sounds, especially S_1; these factors include movement of the great vessels and turbulence of the rapidly moving blood. The second heart sound often has two components—the first corresponds to aortic valve closure and the second to pulmonic valve closure. In normal individuals, **splitting** widens with inspiration and narrows or disappears with expiration.

A **third heart sound** (S_3) results from vibrations during the rapid phase of ventricular filling and is associated with ventricular filling that is too rapid. Although it may be heard in normal children and adolescents, its appearance in a patient older than age 35 usually signals the presence of a cardiac abnormality. A **fourth heart sound** (S_4) may be heard during atrial systole. It is caused by blood movement resulting from atrial contraction and, like S_3, is more common in patients with abnormal hearts.

CARDIAC OUTPUT

Cardiac output (CO) is defined as the volume of blood ejected from the heart per unit time. The usual resting values for adults are 5 to 6 L/min, or approximately 8% of body weight per minute. Cardiac output divided by body surface area is called the **cardiac index.** When it is necessary to normalize the value to compare the cardiac output among individuals of different sizes, either cardiac index or cardiac output divided by body weight can be used. Cardiac output is the product of heart rate (HR) and **stroke volume** (SV), the volume of blood ejected with each beat:

$$CO = SV \times HR \quad (1)$$

Stroke volume is the difference in the volume of blood in the ventricle at the end of diastole—**end-diastolic volume**—and the volume of blood in the ventricle at the end of systole—**end-systolic volume**. This is shown in Figure 14.1.

If heart rate remains constant, cardiac output increases in proportion to stroke volume, and stroke volume increases in proportion to cardiac output. Table 14.1 outlines the factors that influence cardiac output.

Ejection fraction (EF) is a commonly used measure of cardiac performance. It is the ratio of stroke volume to end-diastolic volume (EDV), expressed as a percentage:

$$EF = (SV/EDV) \times 100 \quad (2)$$

Ejection fraction is normally more than 55%. It is dependent on heart rate, preload, afterload, and contractility (all to be discussed below) and provides a nonspecific index of ventricular function. Still, it has proved to be valuable in predicting the severity of heart disease in individual patients.

TABLE 14.1 Factors Influencing Cardiac Output

I. Stroke volume
  A. Force of contraction
    1. End-diastolic fiber length (Starling's law, preload)
      a. End-diastolic pressure
      b. Ventricular diastolic compliance
    2. Contractility
      a. Sympathetic stimulation via norepinephrine acting on β_1 receptors
      b. Circulating epinephrine acting on β_1 receptors (minor)
      c. Intrinsic changes in contractility in response to changes in heart rate and afterload
      d. Drugs (positive inotropic drugs, e.g., digitalis; negative inotropic drugs, e.g., general anesthetics; toxins)
      e. Disease (coronary artery disease, myocarditis, cardiomyopathy, etc.)
    3. Hypertrophy
  B. Afterload
    1. Ventricular radius
    2. Ventricular systolic pressure
II. Heart rate (and pattern of electrical excitation)

Stroke Volume Is a Determinant of Cardiac Output

Stroke volume increases with increases in the force of contraction of ventricular muscle and decreases with increases in the afterload. The force of contraction is affected by end-diastolic fiber length, contractility, and hypertrophy. **Afterload**, the force against which the ventricle must contract to eject blood, is affected by the ventricular radius and ventricular systolic pressure. Because the pressure drop across the aortic valve is normally small, aortic pressure is often used as a substitute for ventricular pressure in such considerations.

Effect of End-Diastolic Fiber Length. The relationship between ventricular end-diastolic fiber length and stroke volume is known as **Starling's law of the heart.** Within limits, increases in the left ventricular end-diastolic fiber length augment the ventricular force of contraction, which increases the stroke volume. This reflects the relationship between the length of a muscle and the force of contraction (see Chapter 10). After reaching an optimal diastolic fiber length, stroke volume no longer increases with further stretching of the ventricle.

End-diastolic fiber length is determined by end-diastolic volume, which is dependent on end-diastolic pressure. End-diastolic pressure is the force that expands the ventricle to a particular volume. In Chapter 10, **preload** was defined as the passive force that establishes the muscle fiber length before contraction. For the intact heart, preload can be defined as end-diastolic pressure. For a given ventricular compliance (change in volume caused by a given change in pressure), a higher end-diastolic pressure (preload) increases both diastolic volume and fiber length. The end-diastolic pressure depends on the degree of ventricular filling during ventricular diastole, which is influenced largely by atrial pressure.

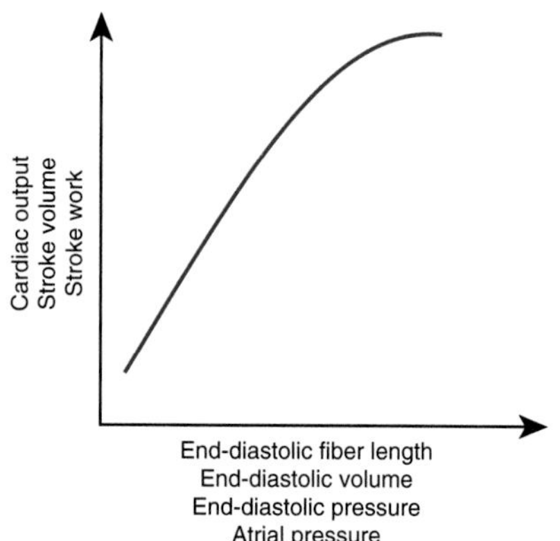

FIGURE 14.2 **A Starling (ventricular function) curve.** Stroke work increases with increased end-diastolic fiber length. Several other combinations of variables can be used to plot a Starling curve, depending on the assumptions made. For example, cardiac output can be substituted for stroke volume if heart rate is constant, and stroke volume can be substituted for stroke work if arterial pressure is constant. End-diastolic fiber length and volume are related by laws of geometry, and end-diastolic volume is related to end-diastolic pressure by ventricular compliance.

In heart disease, ventricular compliance can decrease because of impaired ventricular muscle relaxation or a buildup of connective tissue within the walls of the heart. In either case, the relationship between ventricular filling, end-diastolic pressure, and end-diastolic volume is altered. The effect is a decrease in end-diastolic fiber length and a resulting decrease in stroke volume.

The curve expressing the relationship between ventricular filling and ventricular contractile performance is called a **Starling curve** or a **ventricular function curve** (Fig. 14.2). This curve can be plotted with end-diastolic volume, end-diastolic pressure, or atrial pressure as the abscissa, as proxies for end-diastolic fiber length.

The ordinate on the plot of Starling's law (Fig. 14.2) can also be a variable other than stroke volume. For example, if heart rate remains constant, cardiac output can be substituted for stroke volume. The effect of arterial pressure on stroke volume can also be taken into account by plotting stroke work on the ordinate. **Stroke work** is stroke volume times mean arterial pressure. An increase in arterial pressure (afterload) decreases stroke volume by increasing the force that opposes the ejection of blood during systole. If stroke work is on the ordinate, any increase in the force of contraction that results in either increased arterial pressure or stroke volume shifts the stroke work curve upward and to the left. If stroke volume alone were the dependent variable, a change in the performance of the heart causing increased pressure would not be expressed by a change in the curve.

Starling's law explains the remarkable balancing of the output between the two ventricles. If the right heart were to pump 1% more blood than the left heart each minute without a compensatory mechanism, the entire blood volume of the body would be displaced into the pulmonary circulation in less than 2 hours. A similar error in the opposite direction would likewise displace all the blood volume into the systemic circuit. Fortunately, Starling's law prevents such an occurrence. If the right ventricle pumps slightly more blood than the left ventricle, left atrial filling (and pressure) will increase. As left atrial pressure increases, left ventricular pressure and left ventricular end-diastolic fiber length increase both the force of contraction and the stroke volume of the left ventricle. If the stroke volume rises too much, the left heart begins to pump more blood than the right heart and left atrial pressure drops; this decreases left ventricular filling and reduces stroke volume. The process continues until left heart output is exactly equal to right heart output.

The descending limb of the ventricular function curve, analogous to the descending limb of the length-tension curve (see Chapter 10), is probably never reached in a living heart because the resistance to stretch increases as the end-diastolic volume reaches the limit for optimum stroke volume. Further enlargement of the ventricle would require end-diastolic pressures that do not occur. As a result of increased resistance to stretch or decreased compliance, the atrial pressures necessary to produce further filling of the ventricles are probably never reached. The limited compliance, therefore, prevents optimal sarcomere length from being exceeded. In heart failure, the ventricles can dilate beyond the normal limit because they exhibit increased compliance. Even under these conditions, optimal sarcomere length is not exceeded. Instead, the sarcomeres appear to realign so that there are more of them in series, allowing the ventricle to dilate without stretching sarcomeres beyond their optimal length.

Effect of Changes in Contractility. Factors other than end-diastolic fiber length can influence the force of ventricular contraction. Different conditions produce different relationships between stroke volume (or work) and end-diastolic fiber length. For example, increased sympathetic nerve activity causes release of norepinephrine (see Chapter 3). Norepinephrine increases the force of contraction for a given end-diastolic fiber length (Fig. 14.3). The increase in force of contraction causes more blood to be ejected against a given aortic pressure and, thus, raises stroke volume. A change in

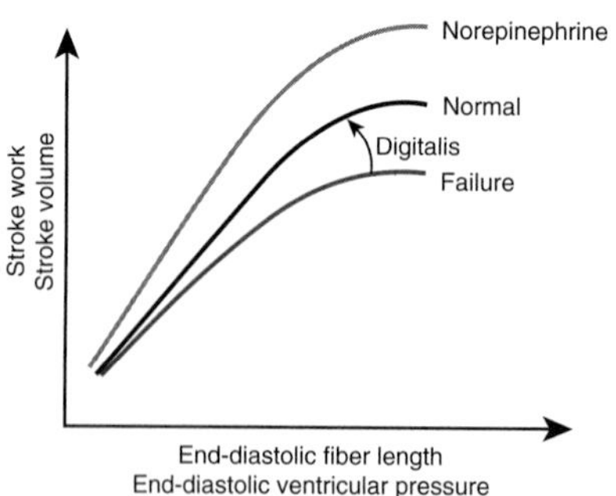

FIGURE 14.3 **Effect of norepinephrine and heart failure on the ventricular function curve.** Norepinephrine raises ventricular contractility (i.e., stroke volume and/or stroke work are elevated at a given end-diastolic fiber length). In heart failure, contractility is decreased, so that stroke volume and/or stroke work are decreased at a given end-diastolic fiber length. Digitalis raises the intracellular calcium ion concentration and restores the contractility of the failing ventricle.

the force of contraction at a constant end-diastolic fiber length reflects a change in the **contractility** of the heart. (The cellular mechanisms governing contractility are discussed in Chapter 10.) A shift in the ventricular function curve to the left indicates increased contractility (i.e., more force and/or shortening occurring at the same initial fiber length), and shifts to the right indicate decreased contractility. When an increase in contractility is accompanied by an increase in arterial pressure, the stroke volume may remain constant, and the increased contractility will not be evident by plotting the stroke volume against the end-diastolic fiber length. However, if stroke work is plotted, a leftward shift of the ventricular function curve is observed (see Fig. 14.3). A ventricular function curve with stroke volume on the ordinate can be used to indicate changes in contractility only when arterial pressure does not change.

During heart failure, the ventricular function curve is shifted to the right, causing a particular end-diastolic fiber length to be associated with less force of contraction and/or shortening and a smaller stroke volume. As described in Chapter 10, cardiac glycosides, such as digitalis, tend to normalize contractility; that is, they shift the ventricular curve of the failing heart back to the left (see Fig. 14.3).

The collection of ventricular function curves reflecting changes in contractility in a particular heart is known as a family of ventricular function curves.

Effect of Hypertrophy. In the normal heart, the force of contraction is also increased by **myocardial hypertrophy.** Regular, intense exercise results in increased synthesis of contractile proteins and enlargement of cardiac myocytes. The latter is the result of increased numbers of parallel myofilaments, increasing the number of actomyosin cross-bridges that can be formed. As each cell enlarges, the ventricular wall thickens and is capable of greater force development. The ventricular lumen may also increase in size, and this is accompanied by an increase in stroke volume. The hearts of appropriately trained athletes are capable of producing much greater stroke volumes and cardiac outputs than those of sedentary individuals. These changes are reversed if the athlete stops training. Myocardial hypertrophy also occurs in heart disease. In heart disease, although myocardial hypertrophy initially has positive effects, it ultimately has negative effects on myocardial force development. A thorough discussion of pathological hypertrophy is beyond the scope of this book.

Effect of Afterload. The second determinant of stroke volume is afterload (see Table 14.1), the force against which the ventricular muscle fibers must shorten. In normal circumstances, afterload can be equated to the aortic pressure during systole. If arterial pressure is suddenly increased, a ventricular contraction (at a given level of contractility and end-diastolic fiber length) produces a lower stroke volume. This decrease can be predicted from the force-velocity relationship of cardiac muscle (see Chapter 10). The shortening velocity of ventricular muscle decreases with increasing load, which means that for a given duration of contraction (reflecting the duration of the action potential), the lower velocity results in less shortening and a decrease in stroke volume (Fig. 14.4).

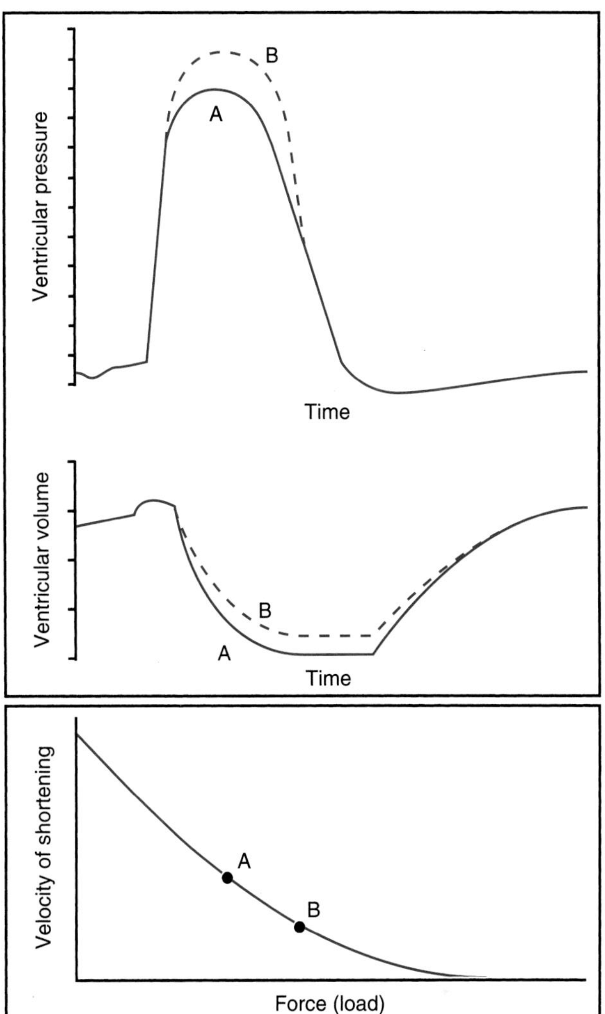

FIGURE 14.4 **Effect of aortic pressure on ventricular function.** Ventricular pressure, ventricular volume, and the force-velocity relationship are shown for (A) normal and (B) elevated aortic pressure. Increased afterload slows the velocity of shortening, decreasing ventricular emptying and stroke volume.

Fortunately, the heart can compensate for the decrease in left ventricular stroke volume produced by increased afterload. Although a sudden rise in systemic arterial pressure causes the left ventricle to eject less blood per beat, the output from the right heart remains constant. Left ventricular filling subsequently exceeds its output. As the end-diastolic volume and fiber length of the left ventricle increase, the ventricular force of contraction is enhanced. A new steady state is quickly reached in which the end-diastolic fiber length is increased and the previous stroke volume is maintained. Within limits, an additional compensation also occurs. During the next 30 seconds, the end-diastolic fiber length returns toward the control level, and the stroke volume is maintained despite the increase in aortic pressure. If arterial pressure times stroke volume (stroke work) is plotted against end-diastolic fiber length, it is apparent that stroke work has increased for a given end-diastolic fiber length. This leftward shift of the ventricular function curve indicates an increase in contractility.

Effect of the Ventricular Radius. The ventricular radius influences stroke volume because of the relationship between ventricular pressures (P_v) and ventricular wall tension (T). For a hollow structure, such as a ventricle, Laplace's law states that

$$P_v = T \times (1/r_1 + 1/r_2) \quad (3)$$

where r_1 and r_2 are the radii of curvature for the ventricular wall. Figure 14.5 shows this relationship for a simpler structure, in which curvature occurs in only one dimension (i.e., a cylinder). In this case, r_2 approaches infinity. Therefore:

$$P_v = T \times (1/r_1) \text{ or } T = P_v \times r_1 \quad (4)$$

The internal pressure expands the cylinder until it is exactly balanced by the wall tension. The larger the radius, the larger the tension needed to balance a particular pressure. For example, in a long balloon that has an inflated part with a large radius and an uninflated part with a much smaller radius, the pressure inside the balloon is the same everywhere, yet the tension in the wall is much higher in the inflated part because the radius is much greater (Fig. 14.6). This general principle also applies to noncylindrical objects, such as the heart and tapering blood vessels.

When the ventricular chamber enlarges, the wall tension required to balance a given intraventricular pressure increases. As a result, the force resisting ventricular wall shortening (afterload) likewise increases with ventricular size. Despite the effect of increased radius on afterload, an increase in ventricular size (within physiological limits) raises both wall tension and stroke volume. This occurs because the positive effects of adjustment in sarcomere length overcompensate for the negative effects of increasing ventricular radius. However, if a ventricle becomes pathologically dilated, the myocardial fibers may be unable to generate enough tension to raise pressure to the normal systolic level, and the stroke volume may fall.

Effect of Diastolic Compliance. Several diseases—including hypertension, myocardial ischemia, and cardiomyopathy—cause the left ventricle to be less compliant during diastole. In the presence of decreased diastolic compliance, a normal end-diastolic pressure stretches the ventricle less. Reduced stretch of the ventricle results in lowered stroke volume. In this situation, compensatory events increase central blood volume and end-diastolic pressure (see Chapter 18). A higher end-diastolic pressure stretches the stiffer ventricle and helps restore the stroke volume to normal. The physiological price for this compensation is higher left atrial and pulmonary pressures. Several pathological consequences, including pulmonary congestion and edema, can result.

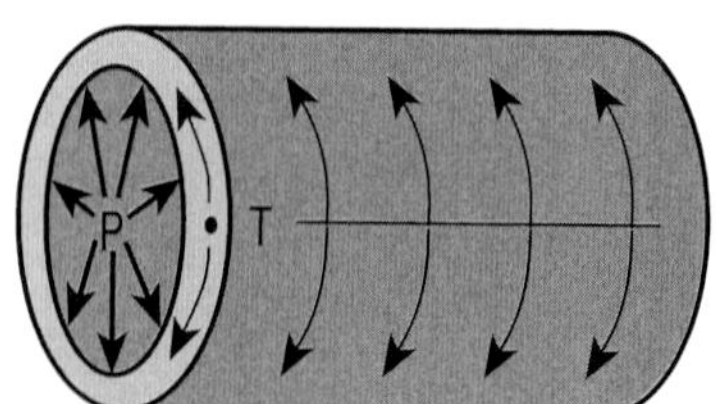

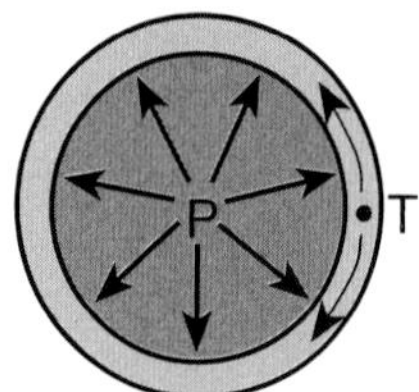

FIGURE 14.5 Pressure and tension in a cylindrical blood vessel. The tension tends to open an imaginary slit along the length of the blood vessel. The Laplace law relates pressure (P), radius, and tension (T), as described in the text.

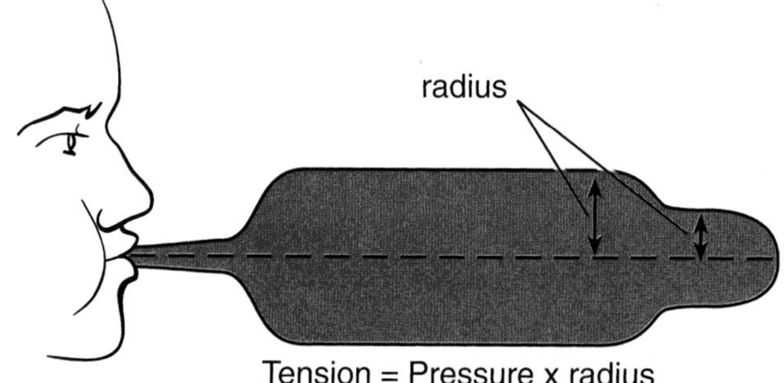

FIGURE 14.6 Effect of the radius of a cylinder on tension. The pressure inside an inflated balloon is the same everywhere. With the same inside pressure, the tension in the wall is proportional to the radius. The tension is lower in the portion of the balloon with the smaller radius.

Pressure-Volume Loops Provide Information Regarding Ventricular Performance

Figure 14.7A shows a plot of left ventricular pressure as a function of left ventricular volume. One cardiac cycle is represented by one counterclockwise circuit of the loop. At point 1, the mitral valve opens and the volume of the ventricle begins to increase. As it does, diastolic ventricular pressure rises a little, depending on the ventricular diastolic compliance. (Remember that compliance is $\Delta V/\Delta P$.) The less the pressure rises with the filling of the ventricle, the greater the compliance. The volume increase between point 1 and point 2 occurs during rapid and reduced ventricular filling and atrial systole (see Fig. 14.1). At point 2, the ventricle begins to contract and pressure rises rapidly. Because the mitral valve closes at this point and the aortic valve has not yet opened, the volume of the ventricle cannot change (isovolumetric contraction). At point 3, the aortic valve opens. As blood is ejected from the ventricle, ventricular volume falls. At first, ventricular pressure continues to rise because the ventricle continues to contract and build up pressure—this is the period of rapid ejection in Figure 14.1. Later, pressure begins to fall—this is the period of reduced ejection in Figure 14.1. The reduction in ventricular volume between points 3 and 4 is the difference between end-diastolic volume (3) and end-systolic volume (4) and equals stroke volume.

At point 4, ventricular pressure drops enough below aortic pressure to cause the aortic valve to close. The ventricle continues to relax after closure of the aortic valve, and this is reflected by the drop in ventricular pressure. Because the mitral valve has not yet opened, ventricular volume cannot change (isovolumetric relaxation). The loop returns to point 1 when the mitral valve opens and, once more, the ventricle begins to fill.

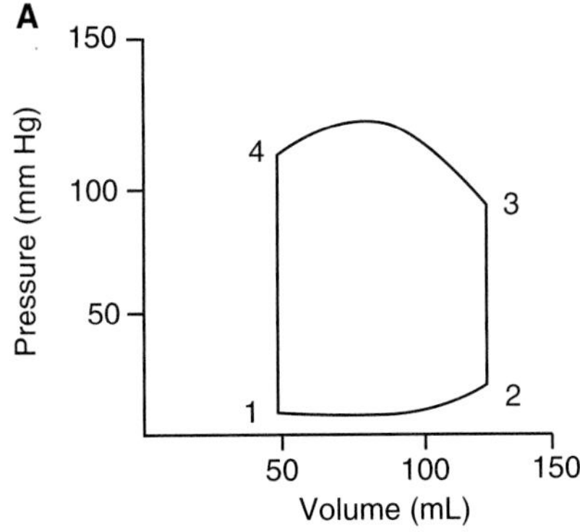

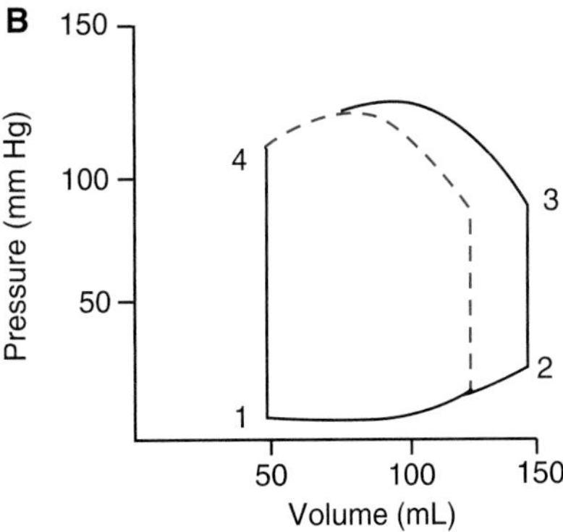

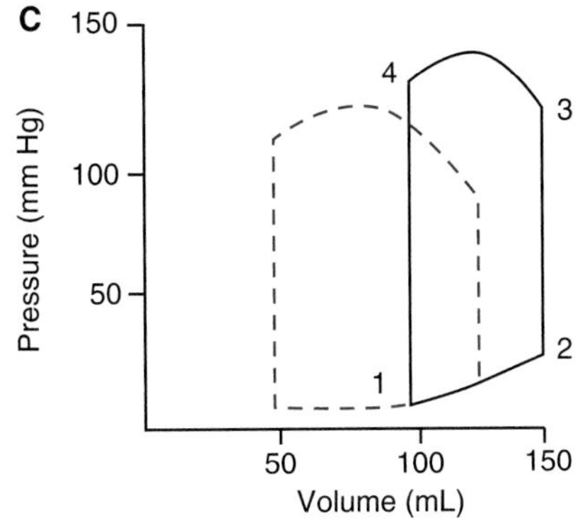

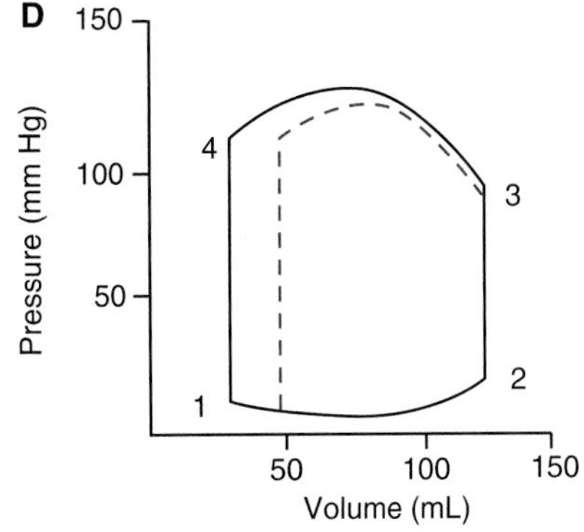

FIGURE 14.7 **Pressure-volume loops for the left ventricle.** 1. Mitral valve opens. 2. Mitral valve closes. 3. Aortic valve opens. 4. Aortic valve closes. **A,** The loop with normal values for ventricular volumes and pressures. **B,** The addition of a loop with increased preload. **C,** The addition of a loop with increased afterload. **D,** The addition of a loop with increased contractility.

Increased Preload. Figure 14.7B shows a pressure-volume loop from the same heart in the presence of increased preload. After opening of the mitral valve at point 1 in Figure 14.7B, diastolic pressure and volume increase to a higher value than in Figure14.7A. When isovolumetric contraction begins at point 2, end-diastolic volume is higher. Because afterload is unchanged, the aortic valve opens at the same pressure (point 3). In the idealized graph in Figure 14.7B, the greater force of contraction associated with higher preload causes the ventricle to eject all of the extra volume that entered during diastole. This means that, when the aortic valve closes at point 4, the volume and pressure of the ventricle are identical to the values in Figure 14.7A. The difference in volume between points 3 and 4 is larger, representing the larger stroke volume associated with increased preload.

Increased Afterload. Figure 14.7C shows the effect of an uncompensated increase in afterload on the pressure-volume loop. In this situation, the aortic valve opens (point 3) at a higher pressure because aortic pressure is increased, as compared with Figure 14.7A. The higher aortic pressure decreases stroke volume, and the aortic valve closes (point 4) at a higher pressure and volume. Mitral valve opening and ventricular filling (point 1) begin at a higher pressure and volume because more blood is left in the ventricle at the end of systole. Filling of the ventricle proceeds along the same diastolic pressure-volume curve from point 1 to point 2. Because the ventricle did not empty as much during systole and the atrium delivers as much blood during diastole, end-diastolic volume and pressure (preload) are increased.

Increased Contractility. Figure 14.7D shows the effect of increased contractility on the pressure-volume loop. In this idealized situation, there is no change in end-diastolic volume, and mitral valve closure occurs at the same pressure and volume (point 2). Afterload is also the same; therefore, the aortic valve opens at the same arterial pressure (point 3). The increased force of contraction causes the ventricle to eject more blood and the aortic valve closes at a lower end-systolic volume (point 4). This means that the mitral valve opens at a lower end-diastolic volume (point 1). Because diastolic compliance is unchanged, filling proceeds along the same pressure-volume curve from point 1 to point 2.

When there are changes in diastolic compliance, the pressure-volume curve between (1) and (2) is changed. This and other changes, such as heart failure, are beyond the scope of this text.

Heart Rate Interacts With Stroke Volume to Influence Cardiac Output

Heart rate can vary from less than 50 beats/min in a resting, physically fit individual to greater than 200 beats/min during maximal exercise. If stroke volume is held constant, increases in heart rate cause proportional increases in cardiac output. However, heart rate affects stroke volume; changes in heart rate do not necessarily cause proportional changes in cardiac output. In considering the influence of heart rate on cardiac output, it is important to recognize that as the heart rate increases and the duration of the cardiac cycle decreases, the duration of diastole decreases. As the duration of diastole decreases, the time for filling of the ventricles is diminished. Less filling of the ventricles leads to a reduced end-diastolic volume and decreased stroke volume.

Effect of Decreased Heart Rate on Cardiac Output. A consequence of the reciprocal relationship between heart rate and the duration of diastole is that, within limits, decreasing the rate of a normal resting heart does not decrease cardiac output. The lack of a decrease in cardiac output is because stroke volume increases as heart rate decreases. Stroke volume increases because as the heart rate falls, the duration of ventricular diastole increases, and the longer duration of diastole results in greater ventricular filling. The resulting elevated end-diastolic fiber length increases stroke volume, which compensates for the decreased heart rate. This balance works until the heart rate is below 20 beats/min. At this point, additional increases in end-diastolic fiber length cannot augment stroke volume further because the maximum of the ventricular function curve has been reached. At heart rates below 20 beats/min, cardiac output falls in proportion to decreases in heart rate.

Effect of Increased Heart Rate as a Result of Electronic Pacing. If an electronic pacemaker is attached to the right atrium and the heart rate is increased by electrical stimulation, surprisingly little increase in cardiac output results. This is because as the heart rate increases, the interval between beats shortens and the duration of diastole decreases. The decrease in diastole leaves less time for ventricular filling, producing a shortened end-diastolic fiber length, which subsequently reduces both the force of contraction and the stroke volume. The increased heart rate is, therefore, offset by the decrease in stroke volume. When the rate increases above 180 beats/min secondary to an abnormal pacemaker, stroke volume begins to fall as a result of poor diastolic filling. A person with abnormal tachycardia (e.g., caused by an ectopic ventricular pacemaker) may have a reduction in cardiac output despite an increased heart rate.

Events in the myocardium compensate to some degree for the decreased time available for filling. First, increases in heart rate reduce the duration of the action potential and, thus, the duration of systole, so the time available for diastolic filling decreases less than it would otherwise. Second, faster heart rates are accompanied by an increase in the force of contraction, which tends to maintain stroke volume. The increased contractility is sometimes called **treppe** or the **staircase phenomenon.** These internal adjustments are not very effective and, by themselves, would be insufficient to permit increases in heart rate to raise cardiac output.

Effects of Increased Heart Rate as a Result of Changes in Autonomic Nerve Activity. Increased heart rate usually occurs because of decreased parasympathetic and increased sympathetic neural activity. The release of norepinephrine by sympathetic nerves not only increases the heart rate (see Chapter 13) but also dramatically increases the force of contraction (see Fig. 14.3). Furthermore, norepinephrine increases conduction velocity in the heart, resulting in a more efficient and rapid ejection of blood from the ventricles. These effects, summarized in Figure 14.8, maintain the stroke volume as the heart rate increases. When the heart rate increases physiologically as a result of an increase in sympathetic nervous system activity (as during exercise), cardiac output increases proportionately over a broad range.

Influences on Stroke Volume and Heart Rate Regulate Cardiac Output

In summary, cardiac output is regulated by changing stroke volume and heart rate. Stroke volume is influenced by the contractile force of the ventricular myocardium and by the force opposing ejection (the aortic pressure or afterload). Myocardial contractile force depends on ventricular end-diastolic fiber length (Starling's law) and myocardial contractility. Contractility is influenced by four major factors:

1) Norepinephrine released from cardiac sympathetic nerves and, to a much lesser extent, circulating norepinephrine and epinephrine released from the adrenal medulla

2) Certain hormones and drugs, including glucagon, isoproterenol, and digitalis (which increase contractility) and anesthetics (which decrease contractility)

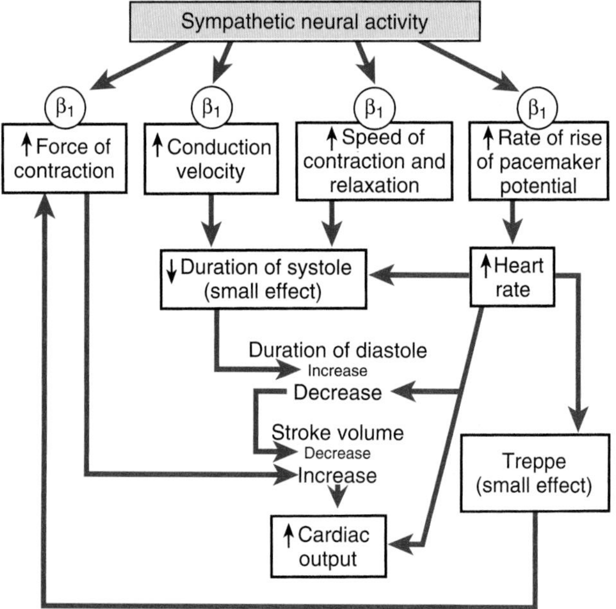

FIGURE 14.8 **Effects of increased sympathetic neural activity on heart rate, stroke volume, and cardiac output.** Various effects of norepinephrine on the heart compensate for the decreased duration of diastole and hold stroke volume relatively constant, so that cardiac output increases with increasing heart rate. The words "Increase" and "Decrease" in small type denote quantitatively less important effects than the same words in large type.

3) Disease states, such as coronary artery disease, myocarditis (see Chapter 10), bacterial toxemia, and alterations in plasma electrolytes and acid-base balance

4) Intrinsic changes in contractility with changes in heart rate and/or afterload

Heart rate is influenced primarily by sympathetic and parasympathetic nerves to the heart and, to a lesser extent, by circulating norepinephrine and epinephrine. The effect of heart rate on cardiac output depends on the extent of concomitant changes in ventricular filling and contractility.

Heart failure is a major problem in clinical medicine (see Clinical Focus Box 14.1).

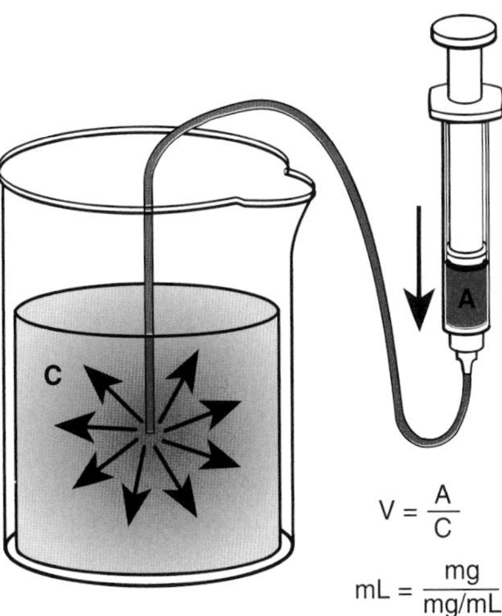

FIGURE 14.9 **The measurement of volume using the indicator dilution method.** The indicator is a dye. The volume (V) of liquid in the beaker equals the amount (A) of dye divided by the concentration (C) of the dye after it has dispersed uniformly in the liquid.

THE MEASUREMENT OF CARDIAC OUTPUT

The ability to measure output accurately is essential for performing physiological studies involving the heart and managing clinical problems in patients with heart disease or heart failure. Cardiac output is measured either by one of several applications of the Fick principle or by observing changes in the volume of the heart during the cardiac cycle.

Cardiac Output Can Be Measured Using Variations of the Principle of Mass Balance

The use of mass balance to measure cardiac output is best understood by considering the measurement of an unknown volume of liquid in a beaker (Fig. 14.9). The volume can be determined by dispersing a known quantity of dye throughout the liquid and then measuring the concentration of dye in a sample of liquid. Because mass is conserved, the quantity of dye (A) in the liquid is equal to the concentration of dye in the liquid (C) times the volume of liquid (V):

$$A = C \times V \quad (5)$$

Because A is known and C can be measured, V can be calculated:

$$V = A/C \quad (6)$$

When the principle of mass balance is applied to cardiac output, the goal is to measure the volume of blood flowing through the heart per unit of time. A known amount of dye or other indicator is injected and concentration of the dye or indicator is measured over time.

CLINICAL FOCUS BOX 14.1

Congestive Heart Failure

Heart failure occurs when the heart is unable to pump blood at a rate sufficient to meet the body's metabolic needs. One possible consequence of heart failure is that blood may "back up" on the atrial/venous side of the failing ventricle, leading to the engorgement and distension of veins (and the organs they drain) as the venous pressure rises. The signs and symptoms typically associated with this occurrence constitute **congestive heart failure** (CHF). This syndrome can be limited to the left ventricle (producing pulmonary venous distension, pulmonary edema, and symptoms such as dyspnea or cough) or the right ventricle (producing symptoms such as pedal edema, abdominal edema or ascites, and hepatic venous congestion), or it may affect both ventricles. Left heart failure (which increases pulmonary venous pressure) can eventually cause pulmonary artery pressure to rise and right heart failure to occur. Indeed, left heart failure is the most common reason for right heart failure.

The causes of CHF are numerous and include acquired and congenital conditions, such as valvular disease, myocardial infarction, assorted infiltrative processes (e.g., amyloid or hemochromatosis), inflammatory conditions (e.g., myocarditis), and various types of **cardiomyopathies** (a diverse assortment of conditions in which the heart becomes pathologically dilated, hypertrophied, or stiff).

The treatment of heart failure hinges on treating the underlying problem, when possible, and the judicious use of medical therapy. Medical treatment may include **diuretics** to reduce the venous fluid overload, **cardiac glycosides** (e.g., digitalis) to improve myocardial contractility, and **afterload reducing agents** (e.g., arterial vasodilators) to reduce the load against which the ventricle must contract. **Angiotensin converting enzyme inhibitors, aldosterone antagonists**, and **beta blockers** have all been shown to be effective in the treatment of CHF.

Heart transplantation is becoming an increasingly viable option for severe, intractable, unresponsive CHF. Although tens of thousands of patients worldwide have received new hearts for end-stage heart failure, the supply of donor hearts falls far below demand. For this reason, cardiac-assist devices, artificial hearts, and genetically modified animal hearts are undergoing intensive development and evaluation.

The Indicator Dilution Method. In the **indicator dilution method**, a known amount of indicator (A) is injected into the circulation, and the blood downstream is serially sampled after the indicator has had a chance to mix (Fig. 14.10). The indicator is usually injected on the venous side of the circulation (often into the right ventricle or pulmonary artery but, occasionally, directly into the left ventricle), and sampling is performed from a distal artery. The resulting concentration of indicator in the distal arterial blood (C) changes with time. First, the concentration rises as the portion of the indicator carried by the fastest-moving blood reaches the arterial sampling point. Concentration rises to a peak as the majority of indicator arrives and falls off as the indicator carried by the slower moving blood arrives. Before the last of the indicator arrives, the indicator carried by the blood flowing through the shortest pathways comes around again (recirculation). To correct for this recirculation, the downslope of the curve is assumed to be semilogarithmic, and the arterial value is extrapolated to zero indicator concentration. The average concentration of indicator can be determined by measuring the indicator concentration continuously from its first appearance (t_1) until its disappearance (t_2). The average concentration during that period ($\bar{C}$) is determined, and cardiac output is calculated:

$$CO = \frac{A}{\bar{C}\,(t_2 - t_1)} \tag{7}$$

Note the similarity between this equation and the one for calculating volume in a beaker. On the left is volume per minute (rather than volume, as in equation 6). In the numerator on the right is amount of indicator and in the denominator is the mean concentration over time (rather than concentration, as in equation 6). Concentration, volume, and amount appear in both equations 6 and 7, but time is present in the denominator on both sides in equation 7.

The Thermodilution Method. In most clinical situations, cardiac output is measured using a variation of the indicator dilution method called **thermodilution.** A Swan-Ganz catheter (a soft, flow-directed catheter with a balloon at the tip) is placed into a large vein and threaded through the right atrium and ventricle so that its tip lies in the pulmonary artery. The catheter is designed to allow a known amount of ice-cold saline solution to be injected into the right side of the heart via a side port in the catheter. This solution decreases the temperature of the surrounding blood. The magnitude of the decrease in temperature depends on the volume of blood that mixes with the solution, which depends on cardiac output. A thermistor on the catheter tip (located downstream in the pulmonary artery) measures the fall in blood temperature. The cardiac output can be determined using calculations similar to those described for the indicator dilution method.

The Fick Procedure. Another way the principle of mass balance is used to calculate cardiac output takes advantage of the continuous entry of oxygen into the blood via the

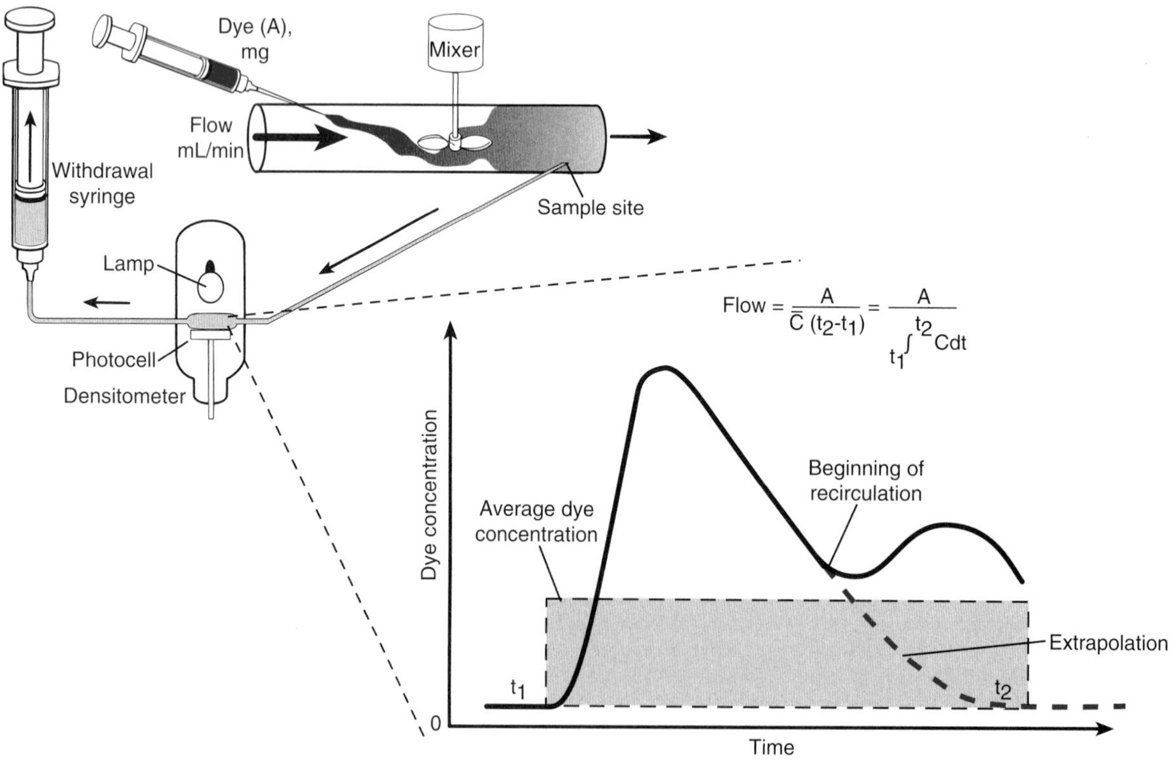

FIGURE 14.10 **The indicator dilution method for determining flow through a tube.** The volume per minute flowing in the tube equals the quantity of indicator (in this example, a dye) injected divided by the average dye concentration ($\bar{C}$) at the sample site, multiplied by the time between the appearance (t_1) and disappearance (t_2) of the dye. Note the analogy between this time-dependent measurement (volume/time) and the simple volume measurement in Figure 14.9. The downslope of the dye concentration curve shows the effects of recirculation of the dye (solid line) and the semilogarithmic extrapolation of the downslope (dashed line) used to correct for recirculation.

lungs (Fig. 14.11). In a steady state, the oxygen leaving the lungs (per unit time) via the pulmonary veins must equal the oxygen entering the lungs via the (mixed) venous blood and respiration (in a steady state, the amount of oxygen entering the blood through respiration is equal to the amount consumed by body metabolism):

$$O_2 \text{ in blood leaving the lungs} = O_2 \text{ in blood and air entering the lungs} \quad (8)$$

or

$$O_2 \text{ output via pulmonary veins} = O_2 \text{ input via pulmonary artery} + O_2 \text{ added by respiration} \quad (9)$$

The O_2 output via the pulmonary veins is equal to the pulmonary vein O_2 content multiplied by the cardiac output (CO). Because O_2 is neither added nor subtracted from the blood as it passes from the pulmonary veins through the left heart to the systemic arteries, the O_2 output via pulmonary veins is also equal to the arterial O_2 content (aO_2) multiplied by the cardiac output (CO). Similarly, O_2 input via the pulmonary artery is equal to mixed venous blood oxygen input to the right heart and is mixed venous blood O_2 content ($\bar{v}O_2$) multiplied by the cardiac output (CO). As indicated above, in the steady state, O_2 added by respiration is equal to oxygen consumption ($\dot{V}O_2$). By substitution in equation 9,

$$(CO)\,(aO_2) = [(CO)\,(\bar{v}O_2)] + \dot{V}O_2 \quad (10)$$

which rearranges to

$$CO = \dot{V}O_2/(aO_2 - \bar{v}O_2) \quad (11)$$

Systemic arterial blood oxygen content, pulmonary arterial (mixed venous) blood oxygen content, and oxygen consumption can all be measured and, therefore, cardiac output can be calculated. The theory behind this method is sounder than the theory behind the indicator dilution method because it avoids the need for extrapolation. However, because the cardiac catheterization required to measure pulmonary artery oxygen content is avoided, the indicator dilution method is more popular. The two methods agree well in a wide variety of circumstances.

Imaging Techniques Are Also Used for Measuring Cardiac Output

A variety of other techniques, many of which employ imaging modalities, can be used to measure or estimate cardiac output. All of them use time dependent images of the heart to estimate the difference between end-diastolic and end-systolic volumes. This difference gives stroke volume and, with heart rate, allows calculation of cardiac output.

Radionuclide Techniques. In **radionuclide techniques**, a radioactive substance (usually technetium-99) can be made to circulate throughout the vascular system by attaching (tagging) it to red blood cells or albumin. The radiation (gamma rays) emitted by the large pool(s) of blood in the cardiac chambers is measured using a specially designed **gamma camera.** The emitted radiation is proportional to the amount of technetium bound to the blood (easily determined by sampling the tagged blood) and the volume of blood in the heart. Using computerized analysis, the amount of radiation emitted by the left (or right) ventricle during various portions of the cardiac cycle can be determined (Fig. 14.12A and B). The amount

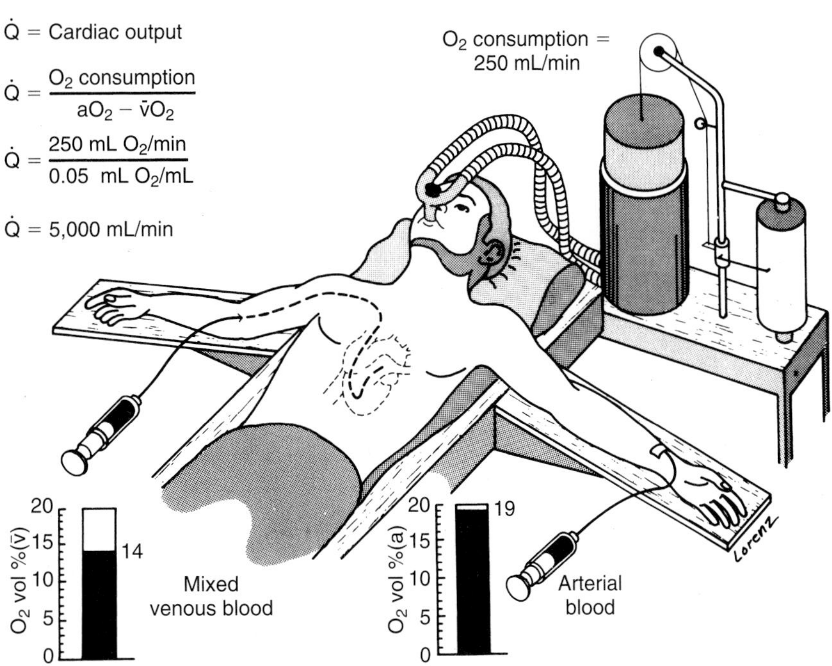

FIGURE 14.11 Calculating cardiac output using the oxygen uptake/consumption method. Oxygen is the "indicator" that is "added" to the mixed venous blood. For oxygen, 1 vol % = 1 mL oxygen/100 mL blood.

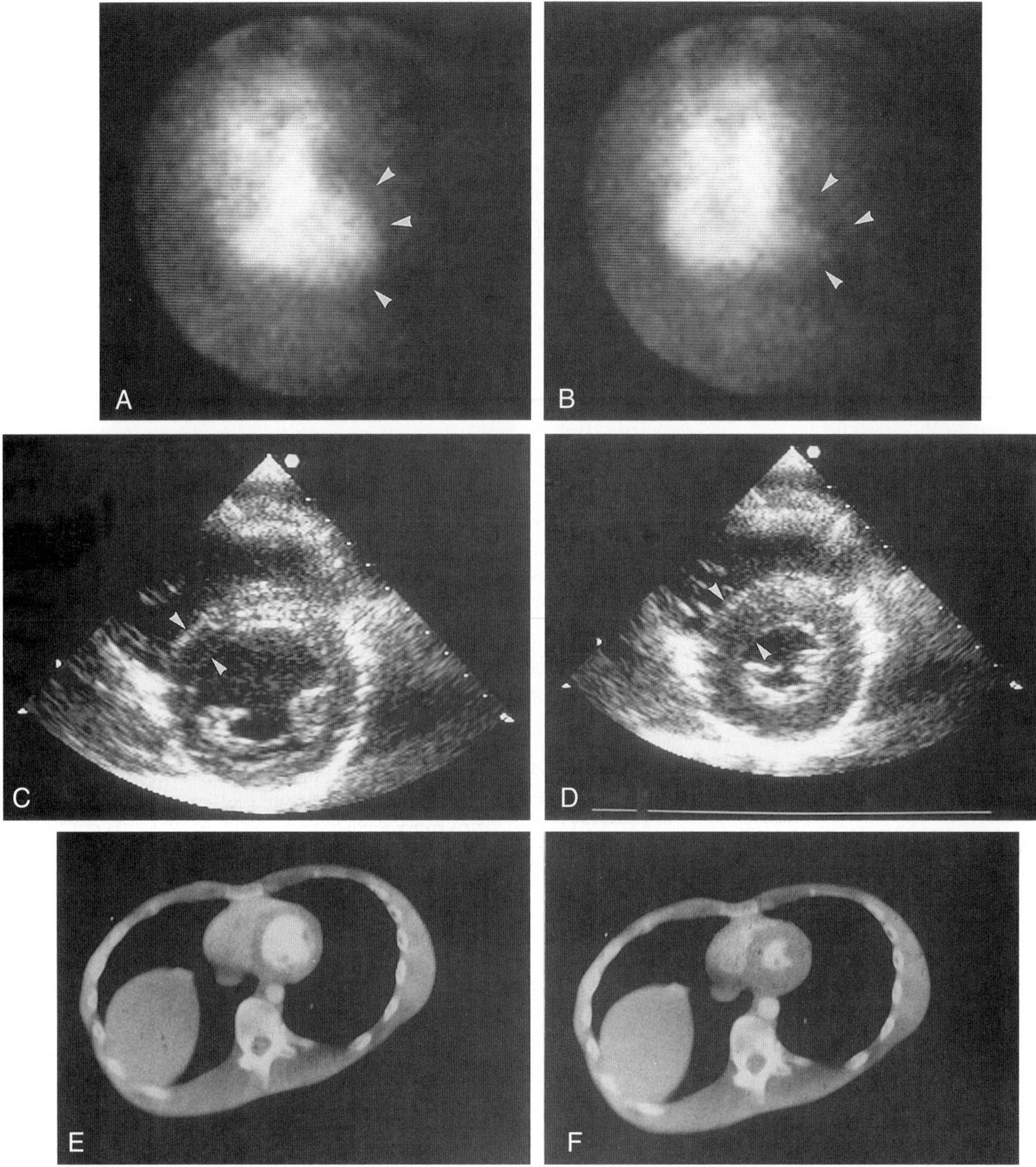

FIGURE 14.12 Imaging techniques for measuring cardiac output. A and B, Radionuclide angiograms. The white arrows in A show the boot-shaped left ventricle during cardiac diastole when it is maximally filled with radionuclide-labeled blood. In B, much of the apex appears to be missing (white arrows) because cardiac systole has caused the blood to be ejected as the intraventricular volume decreases. C and D, Two-dimensional echocardiograms. In this cross-sectional view, the left ventricle appears as a ring. White arrows indicate wall thickness. In diastole (C), the ventricle is large and the wall is thinned; during systole (D), the wall thickens and the ventricular size decreases. E and F, Ultrafast (cine) computed tomography. The ventricular size and wall thickness can be assessed during diastole and systole, and the change in ventricular size can be used to calculate cardiac output.

of blood ejected with each heartbeat (stroke volume) is determined by comparing the amount of radiation measured at the end of systole with that at the end of diastole; multiplying this number by the heart rate yields cardiac output.

Echocardiography. **Echocardiography** (ultrasound cardiography) provides two-dimensional, real-time images of the heart. In addition, the velocity of blood flow can be determined by measuring the Doppler shift (change in sound frequency) that occurs when the ultrasound wave is reflected off moving blood. Echocardiography can, therefore, be used to measure changes in ventricular chamber size (Fig. 14.12C and D), aortic diameter, and aortic blood flow velocity occurring throughout the cardiac cycle. With this information, cardiac output may be estimated in one of two ways. First, the change in ventricular volume occurring with each beat (stroke volume) can be determined and multiplied by the heart rate. Second, the average aortic blood flow velocity can be measured (just above or below the aortic valve) and multiplied by the measured aortic cross-sectional area to give aortic blood flow (which is nearly identical to cardiac output).

Computed Tomography. **Ultrafast (cine) computed tomography** and **magnetic resonance imaging** (MRI) provide cross-sectional views of the heart during different phases of the cardiac cycle (Fig. 14.12E and F). Stroke volume (and cardiac output) can be calculated using the same principles described for radionuclide techniques or echocardiography. When ventricular volume changes are estimated from cross-sectional data, assumptions are made about ventricular geometry. Although these assumptions can lead to errors in calculating cardiac output, these methods have proven to be highly useful.

THE ENERGETICS OF CARDIAC FUNCTION

The heart converts chemical energy in the form of ATP into mechanical work and heat. The relationship between the supply of oxygen and nutrients needed to synthesize ATP and the output of mechanical work by the heart is at the center of many clinical problems.

Cardiac Energy Production Depends Primarily on Oxidative Phosphorylation

The sources of energy for cardiac muscle function were described in Chapter 10. Although the major source of energy for the formation of ATP is oxidative phosphorylation, glycolysis can briefly compensate for a transient lack of aerobic production of ATP when a portion of the heart receives too little oxygen, as during brief coronary artery occlusion.

Oxidative phosphorylation in the heart can use either carbohydrates or fatty acids as metabolic substrates. The formation of ATP depends on a steady supply of oxygen via coronary blood flow. Oxygen delivery by coronary blood flow is, therefore, the most important determinant of an adequate supply of ATP for the mechanical, electrical, and metabolic energy needs of cardiac cells. Furthermore, cardiac oxygen consumption is an accurate measure of the use of energy by the heart. (Coronary blood flow is discussed in Chapter 17.)

As in skeletal muscle, ATP in cardiac muscle is in near equilibrium with phosphocreatine. The presence of phosphocreatine adds to the storage capacity of high-energy phosphate and speeds its transport from mitochondria to actomyosin crossbridges.

Cardiac Energy Consumption Is Required to Support External and Internal Cardiac Work

Cardiac energy consumption (which is equivalent to cardiac oxygen consumption) provides the energy for both **external work** and **internal work**.

Most of the external work of the heart involves the ejection of blood from the ventricles into the aorta and pulmonary artery. The work of ejecting blood from the ventricles is the stroke work. Stroke work, strictly speaking, is equal to the product of the volume of blood ejected (stroke volume, SV) and the pressure against which the blood is ejected (aortic and pulmonary artery pressure during systole). Because the systolic pressure in the pulmonary artery is about one sixth of the pressure in the aorta, more than 80% of external work is done by the left ventricle. Left ventricular stroke work (SW) is usually calculated as:

$$SW = SV \times \bar{P}_a \quad (12)$$

Mean arterial pressure ($\bar{P}_a$) is used instead of mean arterial pressure during systole because it is more readily available and is a reasonable index of mean systolic pressure.

A small additional component of external work (usually <10%) is kinetic work. Kinetic energy is the energy imparted to blood in the form of flow velocity as it is ejected with each heartbeat. We do not elaborate on this component of external work because it is of little importance in most situations.

Cardiac contractions involve many events that do not result in external work. These include internal mechanical events such as developing force by stretching series elasticity (see Chapter 10), overcoming internal viscosity, and rearranging the muscular architecture of the heart as it contracts. These activities, known as internal work, use far more energy (perhaps 5 times as much) than external work.

Cardiac Efficiency. The efficiency of the heart in performing external work can be estimated by dividing the external work of the heart by the energy equivalent of the oxygen consumed by the heart. Only 5 to 20% of the energy liberated by cardiac oxygen consumption is used for external work under most conditions. Therefore, changes in external work do not reveal much about changes in energy consumption in the heart. This is because internal work, the major determinant of oxygen consumption and, thereby, cardiac efficiency, varies independently of external work. As we shall see, large increases in internal work can occur in the absence of changes in external work. When this happens, oxygen consumption increases and efficiency decreases. The difference between pressure work and volume work illustrates this point.

"Pressure Work" Versus "Volume Work". Most of the cardiac energy devoted to internal work is used to maintain the force of contraction (and, thus, ventricular pressure) rather than to eject the blood. The importance of this is seen by comparing two tasks: lifting a 20-pound weight from the floor to a table and lifting the weight to the table height and continuing to hold it. The second task is clearly more difficult, even though the external work done (i.e., the force multiplied by the distance the object was moved) in each case is the same. The ventricles not only develop the pressure required to move the blood, but must maintain the pressure during systole. This takes far more energy than the external work alone as calculated from arterial pressure and stroke volume. In fact, if the external work of the heart is raised by increasing stroke volume but not mean arterial pressure, the oxygen consumption of the heart increases very little. Alternatively, if arterial pressure is increased, the oxygen consumption per beat goes up much more. In other words, pressure work by the heart is far more expensive in terms of oxygen consumption than volume work. This makes sense because internal work consumes far more energy than external work.

Afterload. The discussion of pressure work versus volume work emphasizes the importance of afterload as a determinant of energy use and oxygen consumption by the heart. Because of Laplace's law, an increase in ventricular radius is equivalent to an increase in arterial pressure. Thus, an increase in ventricular radius, as can occur with heart failure, also causes a proportional increase in internal work and energy use, independent of any change in external work.

Heart Rate. Thus far, we have considered only the energetic events associated with a single cardiac contraction. The energy consumed per unit time is equal to the energy consumed in a single heartbeat multiplied by the heart rate. It follows that the production of energy from oxidative phosphorylation per unit time must be sufficient to match the energy consumed in a single heartbeat multiplied by the heart rate.

There is another important consideration related to heart rate. Much of the internal work of the heart occurs during isovolumetric contraction, when force is being developed but no external work is being done. If cardiac output is increased by increasing heart rate, the energy expended in the internal work of isovolumetric contraction increases proportionately. By contrast, if cardiac output is increased by increasing stroke volume, there is a much smaller increase in internal work. This means that increasing cardiac output by increasing heart rate is more energetically costly than the same increase by means of stroke volume.

Contractility. Altered myocardial contractility has significant energetic consequences because of differential effects on external and internal work. Inotropic agents (e.g., norepinephrine) may increase pressure work by raising arterial pressure and, thereby, increase internal work. However, inotropic agents can also cause the heart to do the same stroke work at a smaller end-diastolic volume, reducing both afterload and internal work. During exercise, increased contractility causes end-diastolic volume to decrease despite the increase in venous return. This lowers the contribution of ventricular radius to afterload and avoids the inefficiency of an increase in end-diastolic volume.

The Double Product Is Used Clinically to Estimate the Energy Requirements of Cardiac Work

A useful index of the cardiac oxygen consumption is the product of aortic pressure and heart rate—the **double product**. This index includes one of the determinants of external work (pressure) and the determinant of energy use as a function of time, heart rate. The double product does not include the effect of changes in stroke volume on energy consumption, but these are less significant than changes in pressure. In addition, the double product does not take into account effect of changes in radius of the ventricle on energy consumption. The extra energy required by pathologically dilated hearts is not reflected in the double product.

REVIEW QUESTIONS

DIRECTIONS: Each of the numbered items or incomplete statements in this section is followed by answers or by completions of the statement. Select the ONE lettered answer or completion that is BEST in each case.

1. The figure below shows pressure-volume loops for two situations. When compared with loop A, loop B demonstrates

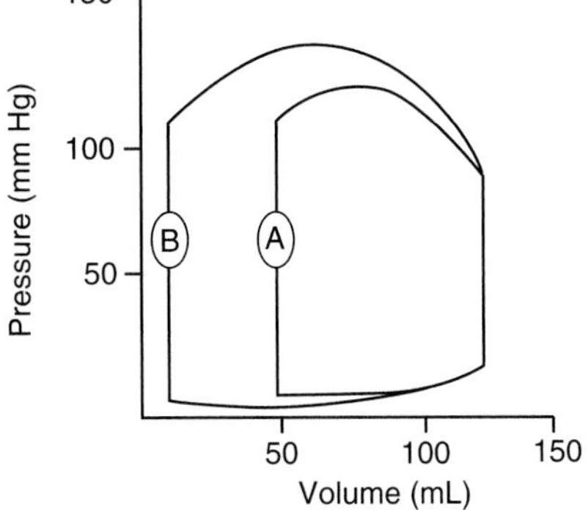

(A) Increased preload
(B) Decreased preload
(C) Increased contractility
(D) Increased afterload
(E) Decreased afterload

2. During the cardiac cycle,
(A) The aortic and mitral valves are never open at the same time
(B) The first heart sound is caused by the rapid ejection of blood from the ventricles
(C) The mitral valve is open throughout diastole
(D) Left ventricular pressure is always less than aortic pressure
(E) Ventricular filling occurs primarily during systole

3. During the cardiac cycle,
(A) The second heart sound is associated with opening of the aortic valve
(B) Left atrial pressure is always less than left ventricular pressure
(C) Aortic pressure reaches its lowest value during ventricular systole
(D) The ventricles eject blood during all of systole
(E) Ventricular end-systolic volume is greater than end-diastolic volume

4. Point Y in the figure below is the control point. Which point corresponds to a combination of increased contractility and increased ventricular filling?

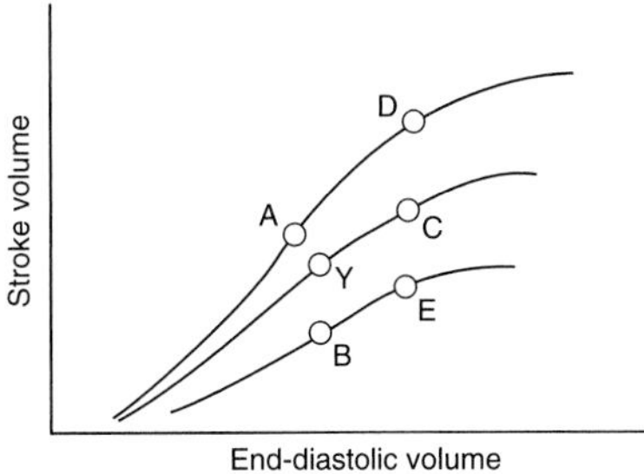

(A) Point A
(B) Point B
(C) Point C
(D) Point D
(E) Point E

5. Drug A causes a 33% increase in stroke volume and no change in systolic aortic blood pressure. Starting with the same baseline, drug B causes a 33% increase in systolic and mean aortic blood pressure and no change in stroke volume. Neither drug changes heart rate.
(A) Drug A increases the external work of the left ventricle more than drug B
(B) Drug B increases the internal work of the left ventricle more than drug A
(C) Drug A increases the oxygen consumption of the heart more than drug B
(D) The "double product" is greater for drug A than for drug B
(E) Cardiac efficiency is higher with drug B than with drug A

6. Using the data below, which is correct?
Volume in ventricle at end of diastole: 130 mL
Volume in ventricle at end of systole: 60 mL
Heart rate: 70 beats/min
Mean arterial blood pressure: 90 mm Hg
(A) Cardiac output is 9,100 mL/min
(B) Cardiac output is 4,200 mL/min
(C) Stroke work is 11,700 mL × mm Hg
(D) Stroke work is 6,300 mL × mm Hg
(E) Stroke work is 4,900 mL/min

7. The data below are from an athletic 70-kg man during heavy exercise. Which statement is correct?

Oxygen consumption:	4 L/min
Arterial oxygen content:	19 mL/100 mL blood
Mixed venous oxygen content:	3 mL/100 mL blood
Heart rate:	180 beats/min

(A) Cardiac output is 12 L/min
(B) Cardiac output is 25 L/min
(C) Stroke volume is 67 mL
(D) Stroke volume is 100 mL
(E) Stroke volume cannot be calculated without data on end-diastolic and end-systolic volume

8. Which of the following would cause a decrease in stroke volume, compared with the normal resting value?
(A) Reduction in afterload
(B) An increase in end-diastolic pressure
(C) Stimulation of the vagus nerves
(D) Electrical pacing to a heart rate of 200 beats/min
(E) Stimulation of sympathetic nerves to the heart

SUGGESTED READING

Davidson CJ, Bonow RO. Cardiac catheterization. In: Braunwald E, Zipes DP, Libby P, eds. Heart Disease. 6th Ed. Philadelphia: WB Saunders, 2001.

Fung YC. Biomechanics: Circulation. 2nd Ed. New York: Springer, 1997.

Katz AM. Physiology of the Heart. 3rd Ed. Philadelphia: Lippincott Williams & Wilkins, 2001.

LeWinter MM, Osol G. Normal physiology of the cardiovascular system. In: Fuster V, Alexander RW, O'Rourke RA, eds. Hurst's the Heart. 10th Ed. New York: McGraw-Hill, 2001.

Lilly LS. Pathophysiology of Heart Disease. 2nd Ed. Baltimore: Williams & Wilkins, 1998.

Opie LH. Mechanisms of cardiac contraction and relaxation. In: Braunwald E, Zipes DP, Libby P, eds. Heart Disease. 6th Ed. Philadelphia: WB Saunders, 2001.

The Systemic Circulation

Thom W. Rooke, M.D.
Harvey V. Sparks, Jr., M.D.

CHAPTER OUTLINE

- DETERMINANTS OF ARTERIAL PRESSURE
- THE MEASUREMENT OF ARTERIAL PRESSURE
- THE NORMAL RANGE OF ARTERIAL PRESSURE
- SYSTEMIC VASCULAR RESISTANCE (SVR)
- BLOOD VOLUME
- THE COUPLING OF VENOUS RETURN AND CARDIAC OUTPUT

KEY CONCEPTS

1. Cardiac output and systemic vascular resistance determine mean arterial pressure.
2. Stroke volume and arterial compliance are the main determinants of pulse pressure.
3. Arterial compliance decreases as arterial pressure increases.
4. Systolic and diastolic arterial pressure can be measured noninvasively.
5. Systemic vascular resistance is most influenced by the radius of arterioles.
6. The venous side of the systemic circulation contains a large fraction of the systemic blood volume.
7. Venous return and cardiac output are equal at a unique right atrial pressure.
8. Shifts in blood volume between the periphery (extrathoracic blood volume) and chest (central blood volume) influence preload and cardiac output.

An understanding of the major systemic hemodynamic variables—arterial pressure, systemic vascular resistance, and blood volume—is a prerequisite to understanding the regulation of arterial pressure and blood flow to individual tissues. The purpose of this chapter is to consider these variables in detail, in preparation for discussions of blood flow to specific regions of the body as well as the regulation of the circulation.

DETERMINANTS OF ARTERIAL PRESSURE

The key measures of systemic arterial pressure are mean arterial pressure, systolic and diastolic arterial pressures, and pulse pressure. These terms were introduced in Chapter 12 and, now that cardiac output, stroke volume, and heart rate have been discussed in Chapter 14, we can discuss them in more depth. For simplicity, mean arterial pressure, systolic pressure, and diastolic pressure are often presented as constant from moment-to-moment. Nothing could be further from the truth. Arterial pressures vary around average values from heartbeat to heartbeat and from minute to minute.

Mean Arterial Pressure Is Determined by Cardiac Output and Systemic Vascular Resistance

Mean arterial pressure ($\overline{P}_a$) is determined mathematically as indicated in Figure 15.1, but is often approximated from the equation,

$$\overline{P}_a = P_d + (P_s - P_d)/3 \text{ or } \overline{P}_a = (2P_d + P_s)/3 \qquad (1)$$

where P_d is the diastolic pressure, P_s is the systolic pressure, and $P_s - P_d$ is the pulse pressure. $\overline{P}_a$ is closer to P_d, instead of halfway between P_s and P_d, because the duration of diastole is about twice as long as systole.

The difference between mean arterial pressure and right atrial pressure (P_{ra}) is equal to the product of cardiac output (CO) and systemic vascular resistance (SVR):

$$\overline{P}_a - P_{ra} = CO \times SVR \qquad (2)$$

Because right atrial pressure is small compared to mean arterial pressure, cardiac output and SVR are usually considered to be the physiologically important determinants of mean arterial pressure.

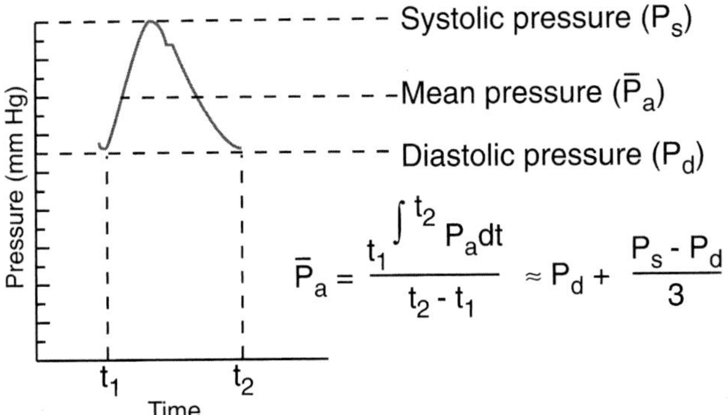

FIGURE 15.1 **Definition of mean arterial pressure.** Mean pressure is the area under the pressure curve divided by the time interval. This can be approximated as the diastolic pressure plus one-third the pulse pressure.

Pulse Pressure Is Determined Largely by Stroke Volume and Arterial Compliance

Arterial compliance is a nonlinear variable that depends on the volume of the aorta and major arteries. The volume of the aorta and major arteries is dependent on mean arterial pressure, meaning that pulse pressure is indirectly dependent on mean arterial pressure. Figure 15.2A shows the effect of a change in aortic volume on aortic pressure if aortic compliance were not a function of aortic volume. No matter what initial volume is present, the same change in volume causes the same change in pressure. In real life, however, aortic compliance decreases as aortic volume is increased, as shown in Figure 15.2B. Because of this, a given change in aortic volume at a low initial volume causes a relatively small change in pressure, but the same change in volume at a high initial volume causes a much larger change in pressure. The large arteries behave in an analogous manner.

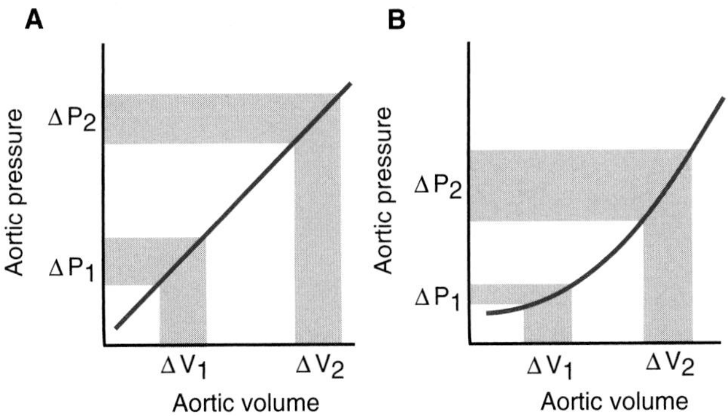

FIGURE 15.2 **Relationship between aortic volume and pressure.** **A**, aortic compliance is independent of aortic volume. The change in volume (ΔV_1) causes the change in pressure (ΔP_1). The same change in volume (ΔV_2) at a higher initial volume causes a change in pressure (ΔP_2) equal to ΔP_1. **B**, aortic compliance decreases as aortic volume increases. The change in volume (ΔV_1) causes the change in pressure (ΔP_1). The same change in volume (ΔV_2) at a higher initial volume causes a much larger change in pressure (ΔP_2).

The above discussion shows that the influence of stroke volume on pulse pressure depends on the mean arterial pressure. As mean arterial pressure increases, arterial compliance decreases. As arterial compliance decreases, a given stroke volume causes a larger pulse pressure.

Stroke Volume, Heart Rate, and Systemic Vascular Resistance Interact in Affecting Mean Arterial and Pulse Pressures

When cardiac output changes in the face of a constant SVR, mean arterial pressure is influenced according to the formula $\bar{P}_a = CO \times SVR$. The influence of a change in cardiac output on mean arterial pressure is independent of the cause of the change—heart rate or stroke volume (remember that $CO = SV \times HR$). In contrast, the effect of a change in cardiac output on pulse pressure greatly depends on whether stroke volume or heart rate changes. Below we consider the effects of changes in heart rate, stroke volume, cardiac output, SVR, and arterial compliance on pulse pressure and mean arterial pressure.

Effect of Changes in Heart Rate and Stroke Volume With No Change in Cardiac Output. If an increase in heart rate is balanced by a proportional and opposite change in stroke volume, *mean* arterial pressure does not change because cardiac output remains constant. However, the decrease in stroke volume that occurs in this situation results in a diminished pulse pressure; the diastolic pressure increases, while the systolic pressure decreases around an unchanged mean arterial pressure. An increase in stroke volume with no change in cardiac output likewise causes no change in mean arterial pressure. The increased stroke volume, however, produces a rise in pulse pressure; systolic pressure increases and diastolic pressure decreases.

Another way to think about these events is depicted in Figure 15.3A. The first two pressure waves have a diastolic pressure of 80 mm Hg, systolic pressure of 120 mm Hg, and mean arterial pressure of 93 mm Hg. Heart rate is 72 beats/min. After the second beat, the heart rate is slowed to 60 beats/min, but stroke volume is increased sufficiently to maintain the same cardiac output. The longer time interval between beats allows the diastolic pressure to fall to a new (lower) value of 70 mm Hg. The next systole, however, produces an increase in pulse pressure because of the ejection of a greater stroke volume, so systolic pressure rises to 130 mm Hg. The pressure then falls to the new (lower) diastolic pressure, and the cycle is repeated. Mean arterial pressure does not change because cardiac output and SVR are constant. The increased pulse pressure is distributed around the same mean arterial pressure.

If an increase in heart rate is balanced by a decrease in stroke volume so that there is no change in cardiac output, the result is no change in mean arterial pressure but a decrease in pulse pressure. Systolic pressure decreases and diastolic pressure increases.

Effect of Changes in Cardiac Output Balanced by Changes in Systemic Vascular Resistance. Mean arterial pressure may remain constant despite a change in cardiac output because of an alteration in SVR. A good exam-

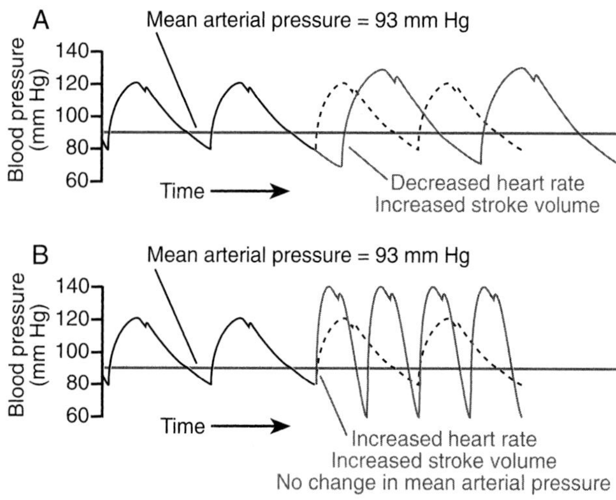

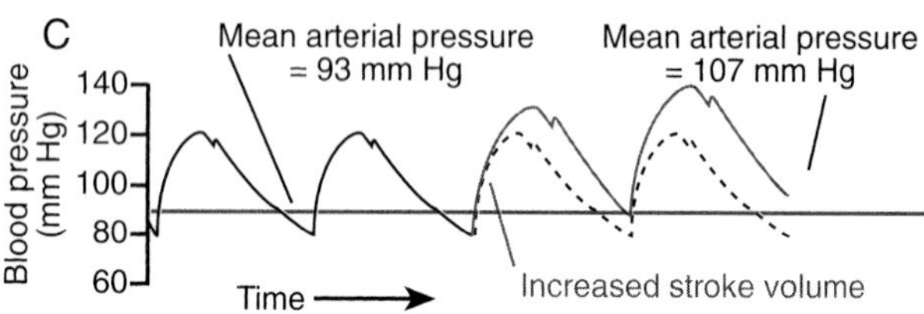

FIGURE 15.3 **Effects of changes in heart rate, stroke volume, and SVR on arterial pressure.** A, Effect of increased stroke volume on arterial pressure with constant cardiac output and SVR. When cardiac output is held constant by lowering heart rate, there is no change in mean arterial pressure (93 mm Hg) and systolic pressure increases while diastolic pressure decreases. B, Effect of increased heart rate *and* stroke volume with no change in mean arterial pressure because of decreased SVR. After the first two beats, stroke volume and heart rate are increased. Pulse pressure increases around an unchanged mean arterial pressure, and systolic pressure is higher and diastolic pressure is lower than the control. C, Effect of increased stroke volume, with constant heart rate and SVR. Cardiac output, mean arterial pressure, systolic pressure, diastolic pressure, and pulse pressure are all increased.

ple of this is **dynamic exercise** (e.g., running or swimming). Dynamic exercise often produces little change in mean arterial pressure because the increase in cardiac output is balanced by a decrease in SVR. The increase in cardiac output is caused by increases in both heart rate and stroke volume. The elevated stroke volume results in a higher pulse pressure. Systolic pressure is higher because of the elevated stroke volume. Diastolic pressure is lower because the fall in SVR increases flow from the aorta during diastole (Figs. 15.3B and 15.4). These examples demonstrate that when mean arterial pressure remains constant, moment-to-moment changes in pulse pressure can be predicted from changes in stroke volume.

Effect of Changes in Cardiac Output With Constant SVR. Figure 15.3C shows what happens if stroke volume is increased with no change in heart rate (cardiac output is increased). The increased stroke volume occurs at the time of the next expected beat, and the diastolic pressure is, as for previous beats, 80 mm Hg. After a transition beat, the increased stroke volume results in an elevation in systolic

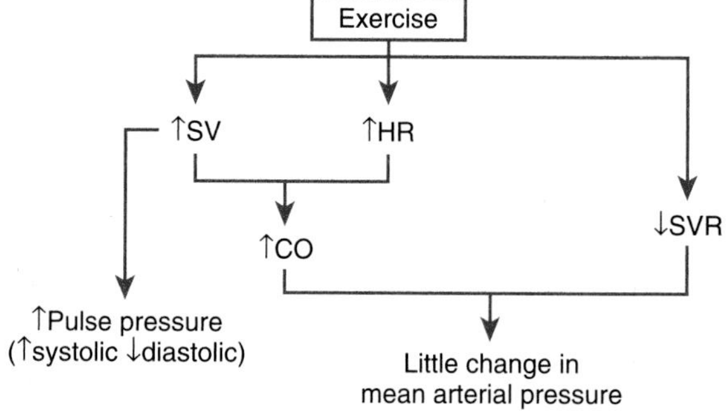

FIGURE 15.4 **Effect of dynamic exercise on mean arterial pressure and pulse pressure.** Heart rate (HR) and stroke volume (SV) increase, resulting in an increase in cardiac output (CO). However, dilation of resistance vessels in skeletal muscle lowers systemic vascular resistance (SVR), balancing the increase in cardiac output and causing little change in mean arterial pressure.

pressure to 140 mm Hg, after which the pressure falls to a new diastolic pressure of 90 mm Hg. In this new steady state, systolic, diastolic, and mean arterial pressures are all higher. The increase in mean arterial pressure (to 107 mm Hg) results in a decrease in arterial compliance (see Fig 15.2). The increase in pulse pressure results from both higher stroke volume and decreased arterial compliance.

Effect of Increased SVR. When SVR increases, flow out of the larger arteries transiently decreases. If cardiac output is unchanged, the volume in the aorta and large arteries increases (Fig. 15.5). Mean arterial pressure also increases, until it is sufficient to drive the blood out of the larger vessels and into the smaller vessels at the same rate as it enters from the heart (i.e., cardiac output). At a higher volume (and mean arterial pressure) arterial compliance is lower, and therefore pulse pressure is greater for a given stroke volume (see Fig. 15.2). The net result is an increase in mean arterial, systolic, and diastolic pressures. The extent of the increase in pulse pressure depends on how much arterial compliance decreases with the rise in mean arterial pressure and arterial volume.

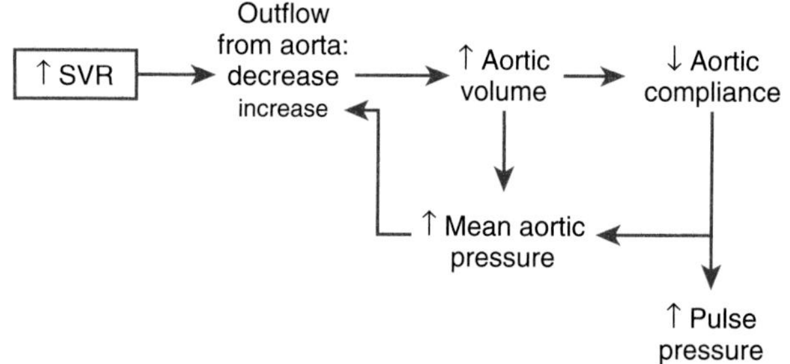

FIGURE 15.5 **Effect of increased SVR on mean arterial and pulse pressures.** Increased SVR impedes outflow from the aorta and large arteries, increasing their volume and pressure. The increase in aortic pressure brings the outflow from the aorta back to its original value, but at a higher aortic volume. The larger volume lowers aortic compliance and, thereby, raises pulse pressure at a constant stroke volume. The word "increase" in smaller type indicates a secondary change.

The compliance of the aorta decreases with age. The fall in compliance for a given increase in mean arterial pressure is greater in older than in younger individuals (Fig. 15.6). This explains the higher pulse and systolic pressures often observed in older individuals with modest elevations in SVR.

THE MEASUREMENT OF ARTERIAL PRESSURE

Arterial blood pressure can be measured by direct or indirect (noninvasive) methods. In the laboratory or hospital setting, a cannula can be placed in an artery and the pressure measured directly using electronic transducers. In clinical practice, however, blood pressure is usually measured indirectly.

The Routine Method for Measuring Human Blood Pressure Is by an Indirect Procedure Using a Sphygmomanometer

The **sphygmomanometer** uses an inflatable cuff that is wrapped around the patient's arm and inflated so that the pressure in it exceeds systolic blood pressure (Fig. 15.7). The external pressure compresses the artery and cuts off blood flow into the limb. The external pressure is measured by the height of a column of mercury in the manometer connected to the cuff or by means of a mechanical manometer calibrated by a column of mercury. The air in the cuff is slowly released until blood can leak past the occlusion at the peak of systole. Blood spurts past the point of partial occlusion at high velocity, resulting in turbulence. The vibrations associated with the turbulence are in the audible range, enabling a stethoscope (placed over the brachial artery) to detect noises caused by the turbulent flow of the blood pushing under the cuff; the noises are known as **Korotkoff sounds.** The pressure corresponding to the first appearance of blood pushing under the cuff is the systolic pressure. As pressure in the cuff continues to fall, the brachial artery returns toward its normal shape and both the turbulence and Korotkoff sounds cease. The pressure at which the Korotkoff sounds cease is the diastolic pressure.

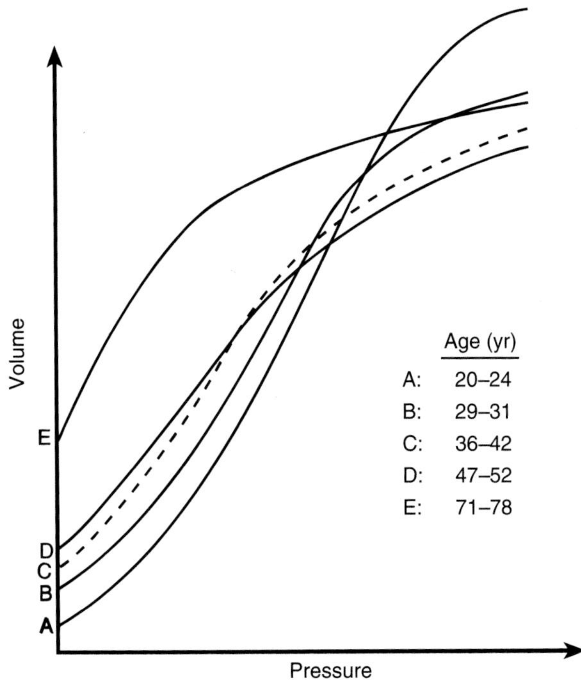

FIGURE 15.6 **Effect of aging on vascular compliance.** The curves illustrate the relationship between pressure and volume for aortas of humans in different age groups. In older aortas, because of decreased compliance, a given increase in volume causes a larger increase in pressure. (Modified from Hallock P, Benson IC. Studies on the elastic properties of human isolated aorta. J Clin Invest 1937;16:595–602.)

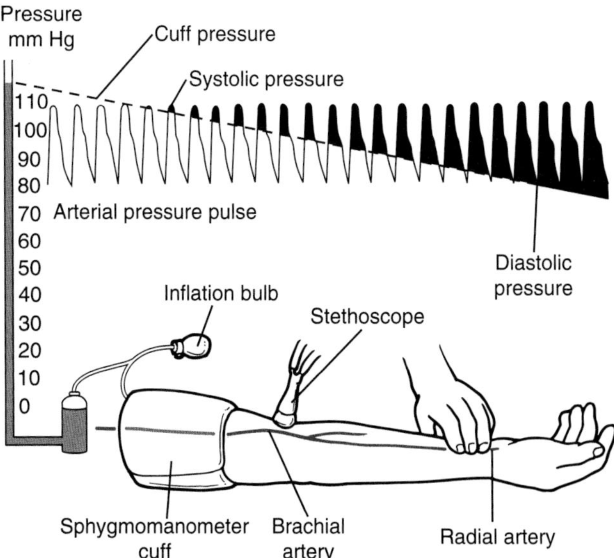

FIGURE 15.7 **The relationship between true arterial pressure and blood pressure as measured with a sphygmomanometer.** When cuff pressure falls just below systolic pressure, turbulent blood squirting through the partially occluded artery under the cuff produces the first Korotkoff sound, which can be heard via a stethoscope bell placed over the brachial artery (auscultatory method). Systolic pressure can also be estimated by palpating the radial artery and noting the cuff pressure at which the pulse is first felt at the wrist (palpatory method). When the cuff pressure falls just below diastolic pressure, the artery stays open, flow is no longer turbulent, and the sounds cease. The arterial pressure tracing is simplified in that systolic, diastolic, and mean arterial pressures vary around average values from moment-to-moment. For this reason, the production of sounds may vary from heartbeat to heartbeat. (From Rushmer RF. Cardiovascular Dynamics. 4th Ed. Philadelphia: WB Saunders, 1976;183.)

Indirect Methods of Measuring Arterial Pressure May Be Subject to Artifacts

The width of the inflatable cuff is an important factor that can affect pressure measurements. A cuff that is too narrow will give a falsely high pressure because the pressure in the cuff is not fully transmitted to the underlying artery. Ideally, cuff width should be approximately 1.5 times the diameter of the limb at the measurement site. In older adults (or those who have "stiff" or hard-to-compress blood vessels from other causes, such as arteriosclerosis), additional external pressure may be required to compress the blood vessels and stop the flow. This extra pressure gives a falsely high estimate of blood pressure. Obesity may contribute to an inaccurate assessment if the cuff used is too small.

THE NORMAL RANGE OF ARTERIAL PRESSURE

As with all physiological variables, values for individuals are distributed around a mean value. Although the range of blood pressures in the population as a whole is rather broad, changes in a given patient are of diagnostic importance. Normal arterial blood pressure in adults is approximately 120 mm Hg systolic and 80 mm Hg diastolic (usually written 120/80).

Age, Race, Gender, Diet and Body Weight, and Other Factors Affect Blood Pressure

In Western societies, arterial pressure is dependent on age. Systolic blood pressure rises throughout life, while diastolic blood pressure rises until the sixth decade of life after which it stays relatively constant. Blood pressure is higher among African Americans than Caucasian Americans. Blood pressure is higher among men than among women with functional ovaries. Dietary fat and salt, as well as obesity, are associated with higher blood pressures. Other factors that affect blood pressure are excessive alcohol intake, physical activity, psychosocial stress, potassium and calcium intake, and socioeconomic status.

Hypertension Is a Sustained Elevation in Blood Pressure

Epidemiological data show that chronically elevated blood pressure is associated with excess cardiovascular morbidity and mortality. In adults, hypertension is defined as sustained systolic blood pressure of 140 mm Hg or higher, sustained diastolic blood pressure of 90 mm Hg or higher, or taking antihypertensive medication. Hypertension causes damage to the arterial system, the myocardium, the kidneys, and the nervous system, including the retinas. Medical treatment that lowers blood pressure to normal values significantly reduces the risk of damage of these target tissues.

SYSTEMIC VASCULAR RESISTANCE (SVR)

SVR is the frictional resistance to blood flow provided by all of the vessels between the large arteries and right atrium, including the small arteries, arterioles, capillaries, venules, small veins, and veins.

Small Arteries, Arterioles, and Capillaries Account for 90% of Vascular Resistance

The relative importance of the various segments contributing to the systemic vascular resistance is appreciated by observing a profile of the pressure drop along the vascular tree (Fig. 15.8). Little change in pressure occurs in the aorta and large arteries. Approximately 70% of the pressure drop occurs in the small arteries and arterioles, and another 20% occurs in the capillaries. Contraction and relaxation of the smooth muscle in the walls of small arteries and arterioles cause changes in vessel diameter, which, in turn, influence blood flow.

When medium and large arteries are affected by disease, they may become major sources of increased resistance and significantly reduce blood flow to regions of the body (see Clinical Focus Box 15.1).

Blood Viscosity, Vessel Length, and Vessel Radius Affect Resistance

To understand the importance of smooth muscle in the control of SVR, we will consider the role of each factor expressed in Poiseuille's law (see Chapter 12):

$$R = 8\eta L/\pi r^4 \qquad (3)$$

Viscosity (η) increases with hematocrit, especially when the hematocrit is above the normal range of 38 to 54% (Fig. 15.9). An increase in viscosity raises vascular resistance and, thereby, limits flow. Because oxygen delivery depends on blood flow as well as blood oxygen content, a limited flow can negate the increase in oxygen content resulting from the increased number of red blood cells. In individuals with polycythemia (an increased number of red blood cells), less oxygen may actually be delivered to tissues because of increased viscosity; this occurs despite the enhanced oxygen-carrying capacity provided by the extra red blood cells. A normal hematocrit reflects a good balance between sufficient red blood cells for oxygen transport and the viscosity caused by red blood cells.

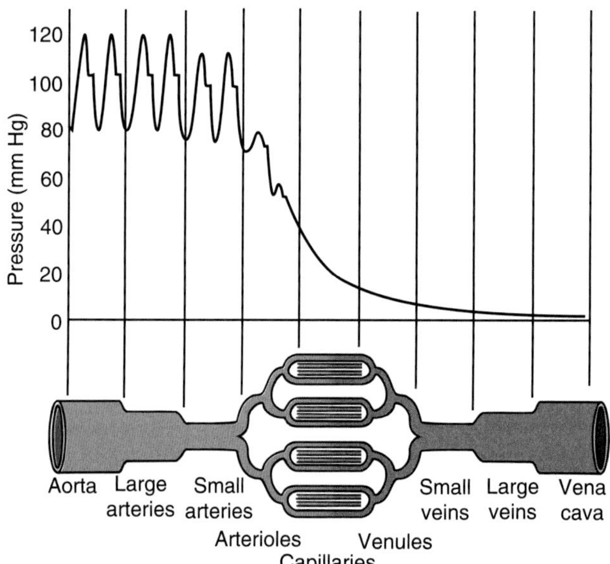

FIGURE 15.8 Pressures in different vessels of the systemic circulation. Pulse pressure is greatest in the aorta and large arteries. The greatest drop in pressure occurs in the arterioles.

CLINICAL FOCUS BOX 15.1

Arterial Disease

Disease processes such as **atherosclerosis** can reduce the diameter of most medium and large arteries, causing an increase in arterial resistance and a subsequent decrease in blood flow. The signs and symptoms resulting from atherosclerotic disease depend on which arteries are **stenotic** (narrowed) and the severity of the reduction in blood flow. Regions commonly affected by atherosclerosis include the heart, brain, and legs.

Coronary artery disease is the most common serious manifestation of atherosclerosis. When the stenotic lesions are relatively mild, blood flow may be inadequate only when the myocardial demand is high, such as during exercise. If blood flow is inadequate to meet the metabolic needs of a particular tissue, the tissue is said to be **ischemic.** In the heart, short periods of **ischemia** may produce chest pain known as **angina**. As the disease progresses and the coronary stenosis becomes more severe, ischemia tends to occur at progressively lower cardiac workloads, eventually resulting in angina at rest. In cases of severe stenosis and/or complete occlusion of the coronary arteries, blood flow may become inadequate to maintain myocardial viability, resulting in **infarction** (cell or tissue death). Millions of people in the United States are affected by coronary disease, with more than 1 million experiencing myocardial infarction each year and 700,000 ultimately dying from infarction, making this the leading cause of death in the nation.

Stenoses in the carotid or vertebral arteries can lead to ischemia and infarction—**stroke** or **cerebrovascular accident**—involving the brain. Strokes are the third leading cause of death in the United States and a leading cause of significant disability.

As with the heart, mild arterial disease involving the legs usually becomes symptomatic only when the demand for blood flow is high, such as during exercise involving the lower extremities. Muscle ischemia produces pain called **claudication,** which typically resolves rapidly when the patient rests. As the disease becomes more severe, symptoms may progress to include **rest pain** and, ultimately, limb infarction with gangrene.

In all of these cases, blood flow to the affected organ may be preserved by the development of collateral arteries, which can carry blood around the stenotic or occluded segments of arteries. When collateral flow is inadequate to meet needs, blood flow may be improved with **angioplasty** (using a balloon catheter, laser, etc.) or **bypass surgery** (using autologous vein or synthetic material to route blood around a blockage). More than 1 million revascularization procedures using these techniques are performed in the United States annually.

Returning to equation 3, despite the potential effect of blood viscosity on resistance, hematocrit normally does not change much and is usually not an important cause of changes in vascular resistance. Likewise, the length (L) of blood vessels does not change significantly (except with growth) and is, therefore, not important as a physiological determinant of vascular resistance. *The remaining influence, vessel radius (r), is the major determinant of changes in SVR.* Since resistance is inversely proportional to r^4, small changes in the radius cause relatively large changes in vascular resistance. For example, the vascular resistance to skeletal muscle during exercise may decrease 25-fold. This fall in resistance results from a 2.2-fold increase in resistance vessel radius (i.e., $2.2^4 = \sim 25$). Vessel radius is determined primarily by the contractile activity of smooth muscle in the vessel wall (see Chapter 16).

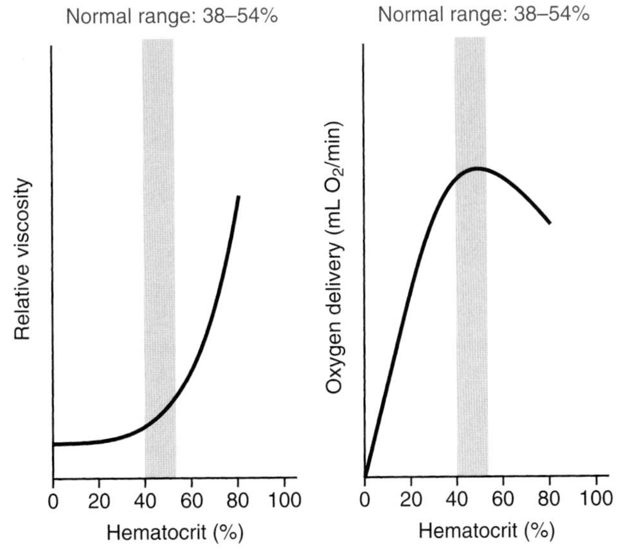

FIGURE 15.9 **Effect of hematocrit on blood viscosity.** Above-normal hematocrits produce a sharp increase in viscosity. Because increased viscosity raises vascular resistance, hemoglobin and oxygen delivery may fall when the hematocrit rises above the normal range.

Sources of Resistance in the Systemic Circulation Are Arranged in Series and in Parallel

Systemic vascular resistance is the net result of the resistance offered by many vessels arranged both in series and in parallel, and it is worth considering the effects of vessel arrangement on total resistance. Resistances in series are simply summed; for example:

$$SVR = R_{small\ arteries} + R_{arterioles} + R_{capillaries} + R_{venules} + R_{small\ veins} \quad (4)$$

For resistances in parallel, the reciprocals of the parallel resistances are summed, for example, for the various parallel blood flows in the body:

$$1/SVR = 1/R_{cerebral} + 1/R_{coronary} + 1/R_{splanchnic} + 1/R_{renal} + 1/R_{muscle} + 1/R_{skin} + 1/R_{other} \quad (5)$$

Resistances in hemodynamic circuits are treated the same way as in the analysis of electrical circuits.

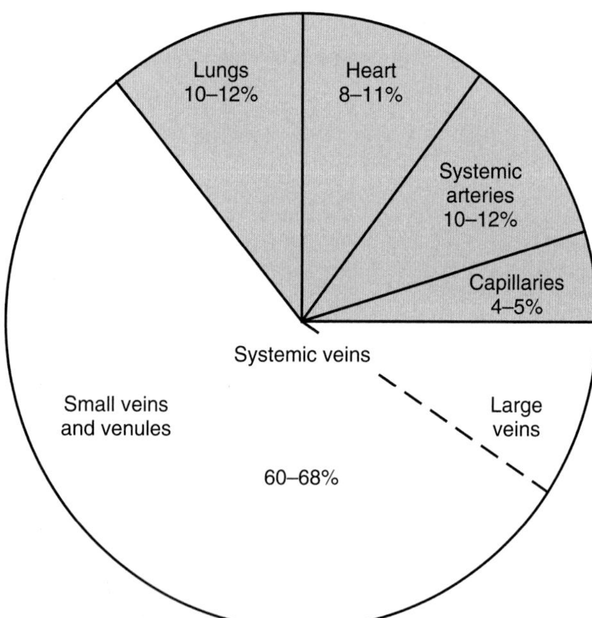

FIGURE 15.10 **Blood volumes of various elements of the circulation in a person at rest.** The majority of the blood volume is in systemic veins.

BLOOD VOLUME

The blood volume is distributed among the various portions of the circulatory system according to the pattern shown in Figure 15.10. Total blood volume in a 70-kg adult is 5.0 to 5.6 L.

Three Fourths of the Blood in the Systemic Circulation Is in the Veins

Approximately 80% of the total blood volume is located in the **systemic circulation** (i.e., the total volume minus the volume in the heart and lungs). About 60% of the total blood volume (or 75% of the systemic blood volume) is located on the venous side of the circulation. The blood present in the arteries and capillaries is only about 20% of the total blood volume. Because most of the systemic blood volume is in veins, it is not surprising that changes in systemic blood volume primarily reflect changes in venous volume.

Small Changes in Systemic Venous Pressure Can Cause Large Changes in Venous Volume

Systemic veins are approximately 20 times more compliant than systemic arteries; small changes in venous pressure are, therefore, associated with large changes in venous volume. If 500 mL of blood is infused into the circulation, about 80% (400 mL) locates in the systemic circulation. This increase in systemic blood volume raises mean circulatory filling pressure by a few mm Hg. This small rise in filling pressure, distributed throughout the systemic circulation, has a much larger effect on the volume of systemic veins than systemic arteries. Because of the much higher compliance of veins than arteries, 95% of the 400 mL (or 380 mL) is found in veins, and only 5% (20 mL) is found in arteries.

Central Blood Volume Is About One Fourth of Total Blood Volume

In considering the role of distribution of blood volume in filling the heart, it is useful to divide the blood volume into central (or intrathoracic) and extrathoracic portions. The **central blood volume** includes the blood in the superior vena cava and intrathoracic portions of the inferior vena cava, right atrium and ventricle, pulmonary circulation, and left atrium; this constitutes approximately 25% of the total blood volume. The central blood volume can be decreased or increased by shifts in blood to and from the extrathoracic blood volume. From a functional standpoint, the most important components of the **extrathoracic blood volume** are the veins of the extremities and abdominal cavity. Depending on several factors to be discussed below, blood shifts readily between these veins and the vessels containing the central blood volume. Although a part of the extrathoracic blood volume, the blood in the neck and head is less important because there is far less blood in these regions, and the blood volume inside the cranium cannot change much because the skull is rigid. Blood in the central and extrathoracic arteries can be ignored because the low compliance of these vessels means that little change in their volume occurs. The volume of blood in the veins of the abdomen and extremities is about equal to the central blood volume; therefore, about half of the total blood volume is involved in shifts in distribution that affect the filling of the heart.

The Measurement of Central Venous Pressure Provides Information on Central Blood Volume

Central venous pressure can be measured by placing the tip of a catheter in the right atrium. Changes in central venous pressures are a good indicator of central blood volume because the compliance of the intrathoracic vessels tends to be constant. In certain situations, however, the physiological meaning of central venous pressure is changed. For example, if the tricuspid valve is incompetent, right ventricular pressure is transmitted to the right atrium during ventricular systole. In general, the use of central venous pressure to assess changes in central blood volume depends on the assumption that the right heart is capable of pumping normally. Also, central venous pressure does not necessarily reflect left atrial or left ventricular filling pressure. Abnormalities in right or left heart function or in pulmonary vascular resistance can make it difficult to predict left atrial pressure from central venous pressure.

Unfortunately, measurements of the peripheral venous pressure, such as the pressure in an arm or leg vein, are subject to too many influences (e.g., partial occlusion caused by positioning or venous valves) to be helpful in most clinical situations.

Cardiac Output Is Sensitive to Changes in Central Blood Volume

Consider what happens if blood is steadily infused into the inferior vena cava of a normal individual. As this occurs, the volume of blood returning to the chest—venous return—is transiently greater than the volume leaving it—the cardiac output. This difference between the input and output of blood produces an increase in central blood volume. It will occur first in the right atrium where the accompanying increase in pressure enhances right ventricular filling, end-diastolic fiber length, and stroke volume. Increased flow into the lungs increases pulmonary blood volume and filling of the left atrium. Left cardiac output will increase according to Starling's law, so that the output of the two ventricles exactly matches. Cardiac output will increase until it equals the sum of the previous venous return to the heart plus the infusion of new blood.

Central Blood Volume Is Influenced by Total Blood Volume and Its Distribution

Changes in central blood volume initiate changes in filling of the ventricles, and therefore, central blood volume is an important influence on cardiac output. Central blood volume is altered by two events: changes in total blood volume and changes in the distribution of total blood volume between central and extrathoracic regions.

Changes in Total Blood Volume. An increase in total blood volume can occur as a result of an infusion of fluid, the retention of salt and water by the kidneys, or a shift in fluid from the interstitial space to plasma. A decrease in blood volume can occur as a result of hemorrhage, losses through sweat or other body fluids, or the transfer of fluid from plasma into the interstitial space. In the absence of compensatory events, changes in blood volume result in proportional changes in both central and extrathoracic blood volume. For example, a moderate hemorrhage (10% of blood volume) with no distribution shift would cause a 10% decrease in central blood volume. The reduced central blood volume would, in the absence of compensatory events, lead to decreased filling of the ventricles and diminished stroke volume and cardiac output.

Redistribution of Blood Volume. Central blood volume can be altered by a shift in blood volume to or away from the periphery. Shifts in the distribution of blood volume occur for two reasons: a change in transmural pressure or a change in venous compliance.

Changes in the transmural pressure of vessels in the chest or periphery enlarge or diminish their size. Because there is a finite volume of blood, it shifts in response to changes in transmural pressure in one or the other of these regions. Imagine a long balloon filled with water: If it is slowly turned end over end, the lower end of the balloon has the greatest transmural pressure because of the weight of the water pressing from above. As it is turned, the lower end of the balloon will bulge and the upper end will shrink.

The best physiological example of a change in transmural pressure occurs when a person stands up. Standing increases the transmural pressure in the blood vessels of the legs because it creates a vertical column of blood between the heart and the blood vessels of the legs. The arterial and venous pressures at the ankles during standing can easily be increased by 130 cm (4.3 ft) of water (blood), which is almost 100 mm Hg higher than in the recumbent position. The increased transmural pressure (outside pressure is still atmospheric) results in little distension of arteries because of their low compliance, but results in considerable distension of veins because of their high compliance. In fact, approximately 550 mL of blood is needed to fill the stretched veins of the legs and feet when an average person stands up. Filling of the veins of the buttocks and pelvis also increases, but to a lesser extent, because the increase in transmural pressure is less.

Blood is redistributed to the legs from the central blood volume by the following sequence of events. When a person stands, blood continues to be pumped by the heart at the same rate and stroke volume for one or two beats. However, much of the blood reaching the legs remains in the veins as they become passively stretched to their new size by the increased venous (transmural) pressure, decreasing the return of blood to the chest. As cardiac output exceeds venous return for a few beats, the central blood volume falls (as does the end-diastolic fiber length, stroke volume, and cardiac output). Once the veins of the legs reach their new steady-state volume, the venous return again equals cardiac output. The equality between venous return and cardiac output is reestablished even though the central blood volume is reduced by 550 mL. However, the new cardiac output and venous return are decreased (relative to what they were before standing) because of the reduction in central blood volume. Without compensation, the resulting decrease in systemic arterial pressure would cause a drop in brain blood flow and loss of consciousness. Compensatory events, including increased activity of the sympathetic nervous system, to be discussed in Chapter 18, are required to maintain arterial pressure in the face of decreased cardiac output.

When the smooth muscle of the systemic veins contracts, the compliance of the systemic veins decreases. This results in a redistribution of blood volume toward the central blood volume. Venoconstriction is an important compensatory mechanism following hemorrhage. The redistribution of blood toward the central blood volume helps to maintain ventricular filling and cardiac output.

THE COUPLING OF VENOUS RETURN AND CARDIAC OUTPUT

Because the blood moves in a closed circuit, **venous return**—the flow of blood from the periphery back to the right atrium—must equal cardiac output. The interplay between venous return and cardiac output can be analyzed from the viewpoint of the heart or the systemic circulation. From the viewpoint of the heart, venous return is kept equal to cardiac output by Starling's law. An increase in venous return raises diastolic filling of the ventricles and cardiac output rises to match the new venous return. The relation between cardiac output and right atrial pressure, shown in Figure 15.11, was presented earlier (see Fig. 14.2).

From the viewpoint of the systemic circulation, venous return to the heart is driven by the pressure gradient created by contractions of the left ventricle. The relationship between venous return and right atrial pressure is shown in Figure 15.11. If, in the absence of any reflex compensations, the heart fails and cardiac output falls below venous return, right atrial pressure rises. In Figure 15.11, the por-

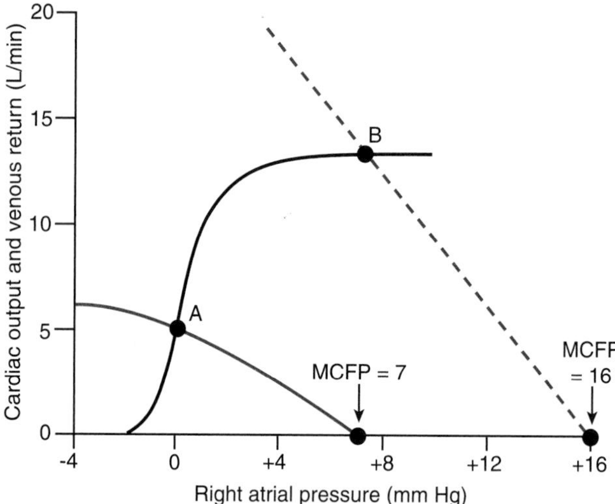

FIGURE 15.11 Interplay between venous return and cardiac output. The Starling curve relating cardiac output to right atrial pressure is shown in black. The normal curve showing venous return as a function of right atrial pressure is shown in solid red. Note that venous return is zero when right atrial pressure equals the mean circulatory filling pressure (7 mm Hg). The two curves intersect at point A where cardiac output and venous return are equal; the right atrial pressure in this case is 0 mm Hg. The dashed red line shows the venous return curve after transfusion of 1 L of blood. Filling of the cardiovascular system by the extra volume of blood raises mean circulatory filling pressure to 16 mm Hg. The slope of the venous return curve is also changed by the transfusion. The Starling curve is unchanged by the transfusion. The unique right atrial pressure that gives equal venous return and cardiac output (point B) is now 8 mm Hg. The transfusion raises cardiac output and venous return from 5 to 13 L/min. (From Guyton AC, Hall JE. Medical Physiology. 10th Ed. Philadelphia: WB Saunders, 2000;219.)

tion of the venous return curve for right atrial pressures above 0 mm Hg shows this. In this example, when right atrial pressure reaches 7 mm Hg, venous return stops. Generally, when right atrial pressure reaches the mean circulatory filling pressure, venous return stops. This is understandable because right atrial pressure and all other circulatory pressures will equal mean circulatory filling pressure when the heart stops (see Chapter 12 to review this point).

The venous return curve for right atrial pressures *below* 0 mm Hg is not sensitive to right atrial pressure (see Fig. 15.11). Instead of continuing to increase as right atrial pressure falls, venous return levels off. Venous return does not increase because, when right atrial pressure drops below zero (atmospheric pressure), the large veins collapse as they enter the chest. This is because the pressure in the lungs surrounding the veins is close to atmospheric pressure and the transmural pressure gradient favors collapse of the veins. The suction imposed by further drops in right atrial pressure further collapses the large veins, instead of sucking more blood into the chest. No matter how low right atrial pressure gets, venous return hardly increases.

Figure 15.11 shows a unique right atrial pressure at which a specific cardiac output curve and a specific venous return curve intersect. This is the right atrial pressure that provides a level of ventricular filling adequate to produce cardiac output that exactly matches the venous return.

The relationship between right atrial pressure, venous return, and cardiac output is not fixed. For example, Figure 15.11 shows the effects of transfusion of a liter of blood on these variables. Central blood volume participates in the increased blood volume, and filling of the heart is increased. This increases cardiac output from point A to point B, along an unchanged cardiac output curve. The increase in blood volume further fills the cardiovascular system and increases mean circulatory filling pressure. This changes the relationship between right atrial pressure and venous return, as shown by the dashed line. The curve is shifted to the right so that there is zero venous return at the new, elevated mean circulatory filling pressure (16 mm Hg). It also changes the slope of the venous return curve for reasons not discussed here. The unique right atrial pressure at which venous return is equal to cardiac output is now 8 mm Hg. Other factors that influence the relationship between cardiac output, venous return, and right atrial pressure include venous resistance to venous return, changes in sympathetic nervous system activity, and changes in SVR.

REVIEW QUESTIONS

DIRECTIONS: Each of the numbered items or incomplete statements in this section is followed by answers or by completions of the statement. Select the ONE lettered answer or completion that is BEST in each case.

1. Mean arterial pressure equals
   (A) Arterial compliance times stroke volume
   (B) Heart rate times stroke volume
   (C) Cardiac output times systemic vascular resistance
   (D) Cardiac output times arterial compliance
2. Mean arterial pressure changes if
   (A) Heart rate increases, with no changes in cardiac output or systemic vascular resistance
   (B) Stroke volume changes, with no changes in heart rate or systemic vascular resistance
   (C) Arterial compliance changes, with no changes in cardiac output or systemic vascular resistance
   (D) Heart rate doubles and systemic vascular resistance is halved, with no change in stroke volume
   (E) Cardiac output doubles and systemic vascular resistance is halved, with no change in heart rate
3. Blood pressure measured using a sphygmomanometer
   (A) May be falsely low with too narrow a cuff
   (B) May be falsely low in patients with badly stiffened arteries
   (C) May be falsely high in obese patients
   (D) Gives a direct reading of mean arterial pressure
   (E) Depends on the disappearance of sound to signal systolic pressure
4. In the systemic circulation, vascular resistance

(continued)

(A) Changes occur mainly in the aorta and large arteries
(B) Is altered more by changes in blood viscosity than radius
(C) Is altered more by changes in vessel radius than length
(D) Is altered more by changes in vessel length than radius

5. Standing up causes
(A) Decreased diameter of leg veins
(B) Decreased blood volume within the cranium
(C) Increased stroke volume
(D) Increased right atrial volume
(E) Decreased central blood volume
6. If a person has an arterial blood pressure of 125/75 mm Hg,
(A) The pulse pressure is 40 mm Hg
(B) The mean arterial pressure is 92 mm Hg
(C) Diastolic pressure is 80 mm Hg
(D) Systolic pressure is 120 mm Hg
(E) The mean arterial pressure is 100 mm Hg
7. A person with an arterial blood pressure of 150/90 mm Hg and a right atrial pressure of 3 mm Hg develops an incompetent tricuspid valve, and right atrial pressure rises to 13 mm Hg with no change in arterial pressure. The pressure gradient forcing blood through the systemic circulation
(A) Is unchanged
(B) Decreased from 107 to 97 mm Hg
(C) Increased from 103 to 113 mm Hg
(D) Decreased from 147 to 137 mm Hg
(E) Increased from 93 to 103 mm Hg
8. If mean arterial pressure increases (due to an increase in systemic vascular resistance) and stroke volume and heart rate remain constant, the pulse pressure
(A) Increases
(B) Decreases
(C) Does not change
9. If the compliance of veins were equal to that of arteries, the change in central blood volume with standing would be
(A) Less than normal
(B) Greater than normal
(C) The same as normal

SUGGESTED READING

Coleman TG, Hall JE. Systemic hemodynamics and regional blood flow regulation. In: Izzo JL, Black HR, eds. Hypertension Primer. Baltimore: Lippincott Williams & Wilkins, 1999.

Guyton AC, Hall JE. Medical Physiology. 10th Ed. Philadelphia: WB Saunders, 2000, Chapter 20.

Kaplan NM. Systemic hypertension: Mechanisms and diagnosis. In: Braunwald E, Zipes DP, Libby P, eds. Heart Disease. 6th Ed. Philadelphia: WB Saunders, 2001.

O'Rourke MF. Arterial stiffness and hypertension. In: Izzo JL, Black HR, eds. Hypertension Primer. Baltimore: Lippincott Williams & Wilkins, 1999.

Rowell LB. Human Cardiovascular Control. New York: Oxford University Press, 1993, Chapter 1.

CHAPTER

16 The Microcirculation and the Lymphatic System

H. Glenn Bohlen, Ph.D.

CHAPTER OUTLINE

- ARTERIAL MICROVASCULATURE
- THE CAPILLARIES
- VENOUS MICROVASCULATURE
- LYMPHATIC VASCULATURE
- VASCULAR AND TISSUE EXCHANGE OF SOLUTES
- TRANSCAPILLARY FLUID EXCHANGE
- REGULATION OF MICROVASCULAR PRESSURES
- REGULATION OF MICROVASCULAR RESISTANCE

KEY CONCEPTS

1. Arterioles regulate vascular resistance and microvascular pressures.
2. Capillaries are the primary sites for water and solute exchange.
3. Venules collect blood from the capillaries and act as reservoirs for blood volume.
4. Lymphatic vessels collect excess water and protein molecules from the interstitial space between cells.
5. Water-soluble materials pass through tiny pores between adjacent endothelial cells.
6. Lipid-soluble molecules pass through the endothelial cells.
7. The concentration difference of solutes across the capillary wall is the energy source for capillary exchange.
8. Plasma hydrostatic and colloid osmotic pressures are the primary forces for fluid filtration and absorption across capillary walls.
9. Tissue hydrostatic and colloid osmotic pressures are minor forces for absorption and filtration of fluid across capillary walls.
10. The ratio of postcapillary to precapillary resistance is a major determinant of capillary hydrostatic pressure.
11. Myogenic arteriolar regulation is a response to increased tension or stretch of the vessel wall muscle cells.
12. By-products of metabolism cause the dilation of arterioles.
13. The axons of the sympathetic nervous system release norepinephrine, which constricts the arterioles and venules.
14. Autoregulation of blood flow allows some organs to maintain nearly constant blood flow when arterial blood pressure is changed.

The **microcirculation** is the part of the circulation where nutrients, water, gases, hormones, and waste products are exchanged between the blood and cells. The microcirculation minimizes diffusion distances, facilitating exchange, its most important function. Virtually every cell in the body is in close contact with a microvessel. In fact, most cells are in direct contact with at least one microvessel. As a consequence, there are tens of thousands of microvessels per gram of tissue. The lens and cornea are exceptions because their nutrients are supplied by the fluids in the eye.

A second major function of the microcirculation is to regulate vascular resistance and thereby interact with cardiac output to maintain the arterial blood pressure (see Chapter 12). Normally, all microvessels, other than capillaries, are partially constricted by contraction of their vascular smooth muscle cells. If all microvessels were to dilate fully because of relaxation of their smooth muscle cells, the arterial blood pressure would plummet. Cerebral blood flow in a standing individual would be inadequate, resulting in fainting, or **syncope**. Regulation of vascular resistance in the microcirculation is an important aspect of total health.

There is a constant conflict between the regulation of vascular resistance to preserve the arterial pressure and simultaneously to allow each tissue to receive sufficient blood flow to sustain its metabolism. The compromise is to preserve the mean arterial pressure by increasing arterial resistance at the expense of reduced blood flow to most organs other than the heart and brain. The organs survive this conflict by increasing their extraction of oxygen and nutrients from blood in the microvessels as the blood flow is decreased.

The microvasculature is considered to begin where the smallest arteries enter the organs and to end where the smallest veins, the **venules,** exit the organs. In between are microscopic arteries, the **arterioles,** and the capillaries. Depending on an animal's size, the largest arterioles have an inner diameter of 100 to 400 μm, and the largest venules have a diameter of 200 to 800 μm. The arterioles divide into progressively smaller vessels so that each section of the tissue has its own specific microvessels. The branching pattern typical of the microvasculature of different major organs and how it relates to organ function are discussed in Chapter 17.

ARTERIAL MICROVASCULATURE

Large arteries have a low resistance to blood flow and function primarily as conduits (see Chapter 15). As arteries approach the organ they supply, they divide into many small arteries both just outside and within the organ. In most organs, these small arteries, which are 500 to 1,000 μm in diameter, control about 30 to 40% of the total vascular resistance. These smallest of arteries, combined with the arterioles of the microcirculation, constitute the **resistance blood vessels**; together they regulate about 70 to 80% of the total vascular resistance, with the remainder of the resistance about equally divided between the capillary beds and venules. Constriction of these vessels maintains the relatively high vascular resistance in organs. Constriction results from the release of norepinephrine by the sympathetic nervous system, from the myogenic mechanism (to be discussed later), and from other chemical and physical factors.

Arterioles Regulate Resistance by the Contraction of Vascular Smooth Muscle

The vast majority of arterioles, whether large or small, are tubes of **endothelial cells** surrounded by a connective tissue **basement membrane,** a single or double layer of vascular smooth muscle cells, and a thin outer layer of connective tissue cells, nerve axons, and mast cells (Fig. 16.1). The vascular smooth muscle cells around the arterioles are 70 to 90 μm long when fully relaxed. The muscle cells are anchored to the basement membrane and to each other in a way that any change in their length changes the diameter of the vessel. Vascular smooth muscle cells wrap around the arterioles at approximately a 90° angle to the long axis of the vessel. This arrangement is efficient because the tension developed by the vascular smooth muscle cell can be almost totally directed to maintaining or changing vessel diameter against the blood pressure within the vessel.

In the majority of organs, arteriolar muscle cells operate at about half their maximal length. If the muscle cells fully relax, the diameter of the vessel can nearly double to increase blood flow dramatically (flow increases as the fourth power of the vessel radius; see Chapter 12). When the muscle cells contract, the arterioles constrict, and with intense stimulation, the arterioles can literally shut for brief periods of time. A single muscle cell will not completely encircle a

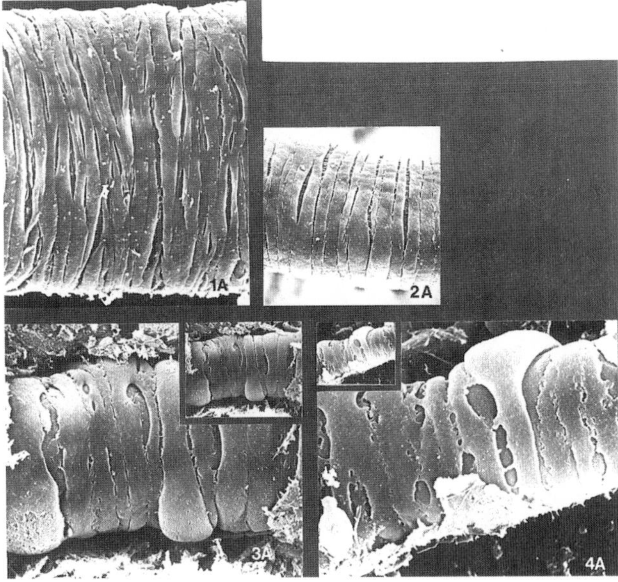

FIGURE 16.1 Scanning electron micrographs of smooth muscle cells wrapping around arterioles of various sizes. Each cell only partially passes around large-diameter (1A) and intermediate-diameter (2A) arterioles, but completely encircles the smaller arterioles (3A, 4A). 1A, 2A, and the small insets of 3A and 4A are at the same magnification. The enlarged views of 3A and 4A are at 4-times-greater magnification. (Modified from Miller BR, Overhage JM, Bohlen HG, Evan AP. Hypertrophy of arteriolar smooth muscle cells in the rat small intestine during maturation. Microvasc Res 1985;29:56–69.)

larger vessel, but may encircle a smaller vessel almost 2 times (see Fig. 16.1).

Vessel Wall Tension and Intravascular Pressure Interact to Determine Vessel Diameter

The smallest arteries and all arterioles are primarily responsible for regulating vascular resistance and blood flow. Vessel radius is determined by the transmural pressure gradient and wall tension, as expressed by Laplace's law (see Chapter 14). Changes in wall tension developed by arteriolar smooth muscle cells directly alter vessel radius. Most arterioles can dilate 60 to 100% from their resting diameter and can maintain a 40 to 50% constriction for long periods. Therefore, large decreases and increases in vascular resistance and blood flow are well within the capability of the microscopic blood vessels. For example, a 20-fold increase in blood flow can occur in contracting skeletal muscle during exercise, and blood flow in the same vasculature can be reduced to 20 to 30% of normal during reflex increases in sympathetic nerve activity.

THE CAPILLARIES

Exchanges Between Blood and Tissue Occur in Capillaries

Capillaries provide for most of the exchange between blood and tissue cells. The capillaries are supplied by the

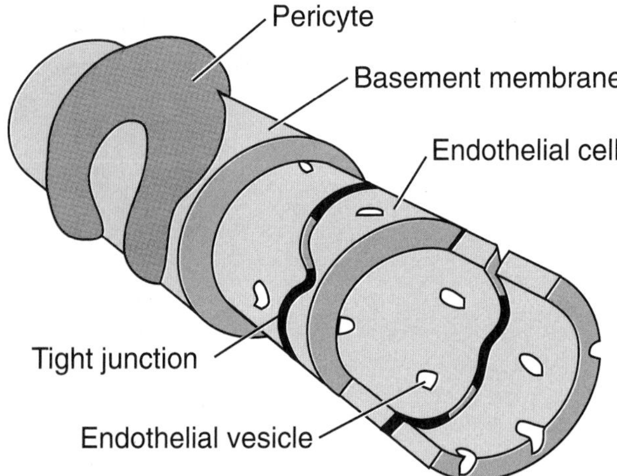

FIGURE 16.2 **The various layers of a mammalian capillary.** Adjacent endothelial cells are held together by tight junctions, which have occasional gaps. Water-soluble molecules pass through pores formed where tight junctions are imperfect. Vesicle formation and the diffusion of lipid-soluble molecules through endothelial cells provide other pathways for exchange.

smallest of arterioles, the **terminal arterioles,** and their outflow is collected by the smallest venules, **postcapillary venules**. A capillary is an endothelial tube surrounded by a basement membrane composed of dense connective tissue (Fig. 16.2). Capillaries in mammals do not have vascular smooth muscle cells and are unable to appreciably change their inner diameter. **Pericytes** (Rouget cells), wrapped around the outside of the basement membrane, may be a primitive form of vascular smooth muscle cell and may add structural integrity to the capillary.

Capillaries, with inner diameters of about 4 to 8 μm, are the smallest vessels of the vascular system. Although they are small in diameter and individually have a high vascular resistance, the parallel arrangement of many thousands of capillaries per mm^3 of tissue minimizes their collective resistance. For example, in skeletal muscle, the small intestine, and the brain, capillaries account for only about 15% of the total vascular resistance of each organ, even though a single capillary has a resistance higher than that of the entire organ's vasculature. The large number of capillaries arranged in hemodynamic parallel circuits allows their combined resistance to be quite low (see Chapter 15).

The capillary lumen is so small that red blood cells must fold into a shape resembling a parachute as they pass through and virtually fill the entire lumen. The small diameter of the capillary and the thin endothelial wall minimize the diffusion path for molecules from the capillary core to the tissue just outside the vessel. In fact, the diffusion path is so short that most gases and inorganic ions can pass through the capillary wall in less than 2 msec.

The Passage of Molecules Through the Capillary Wall Occurs Both Between Capillary Endothelial Cells and Through Them

The exchange function of the capillary is intimately linked to the structure of its endothelial cells and basement membrane. Lipid-soluble molecules, such as oxygen and carbon dioxide, readily pass through the lipid components of endothelial cell membranes. Water-soluble molecules, however, must diffuse through water-filled pathways formed in the capillary wall between adjacent endothelial cells. These pathways, known as **pores**, are not cylindrical holes but complex passageways formed by irregular tight junctions (see Fig. 16.2).

The capillaries of the brain and spinal cord have virtually continuous tight junctions between adjacent endothelial cells; consequently, only the smallest water-soluble molecules pass through their capillary walls. In all capillaries, there are sufficient open areas in adjacent tight junctions to provide pores filled with water for diffusion of small molecules. The pores are partially filled with a matrix of small fibers of submicron dimensions. The potential importance of this **fiber matrix** is that it acts partially to sieve the molecules approaching a water-filled pore. The combination of the fiber matrix and the small spaces in the basement membrane and between endothelial cells explains why the vessel wall behaves as if only about 1% of the total surface area were available for exchange of water-soluble molecules. The majority of pores permit only molecules with a radius less than 3 to 6 nm to pass through the vessel wall. These **small pores** only allow water and inorganic ions, glucose, amino acids, and similar small, water-soluble solutes to pass; they exclude large molecules, such as serum albumin and globular proteins.

A limited number of **large pores**, or possibly defects, allow virtually any large molecule in blood plasma to pass through the capillary wall. Even though few large pores exist, there are enough that nearly all the serum albumin molecules leak out of the cardiovascular system each day.

An alternative pathway for water-soluble molecules through the capillary wall is via **endothelial vesicles** (see Fig. 16.2). Membrane-bound vesicles form on either side of the capillary wall by pinocytosis, and exocytosis occurs when the vesicle reaches the opposite side of the endothelial cell. The vesicles appear to migrate randomly between the luminal and abluminal sides of the endothelial cell. Even the largest molecules may cross the capillary wall in this way. The importance of transport by vesicles to the overall process of transcapillary exchange remains unclear. Occasionally, continuous interconnecting vesicles have been found that bridge the endothelial cell. This open channel could be a random error or a purposeful structure, but in either case, it would function as a large pore to allow the diffusion of large molecules.

VENOUS MICROVASCULATURE

Venules Collect Blood From Capillaries

After the blood passes through the capillaries, it enters the venules, endothelial tubes usually surrounded by a monolayer of vascular smooth muscle cells. In general, the vascular muscle cells of venules are much smaller in diameter but longer than those of arterioles. The muscle size may reflect the fact that venules operate at intravascular pressures of 10 to 16 mm Hg, compared with 30 to 70 mm Hg in arterioles, and do not need a powerful muscular wall. The smallest venules are unique because they are more permeable than capillaries to large and small molecules. This increased permeability seems to exist because tight junctions between adjacent venular endothelial cells have more frequent and larger discontinuities or pores. It is probable that much of the exchange of large water-soluble molecules occurs as the blood passes through small venules.

The Venular Microvasculature Acts as a Blood Reservoir

In addition to their blood collection and exchange functions, the venules are an important component of the blood reservoir system in the venous circulation. At rest, approximately two thirds of the total blood volume is within the venous system, and perhaps more than half of this volume is within venules. Although the blood moves within the venous reservoir, it moves slowly, much like water in a reservoir behind a river dam. If venule radius is increased or decreased, the volume of blood in tissue can change up to 20 mL/kg of tissue; therefore, the volume of blood readily available for circulation would increase by more than 1 L in a 70-kg (154-pound) person. Such a large change in available blood volume can substantially improve the venous return of blood to the heart following depletion of blood volume caused by hemorrhage or dehydration. For example, the volume of blood typically removed from blood donors is about 500 mL, or about 10% of the total blood volume; usually no ill effects are experienced, in part because the venules and veins decrease their reservoir volume to restore the circulating blood volume.

LYMPHATIC VASCULATURE

Lymphatic Vessels Collect Excess Tissue Water and Plasma Proteins

Lymphatic vessels are microvessels that form an interconnected system of simple endothelial tubes within tissues. They do not carry blood, but transport fluid, serum proteins, lipids, and even foreign substances from the interstitial spaces back to the circulation. The gastrointestinal tract, the liver, and the skin have the most extensive lymphatic systems, and the central nervous system may not contain any lymph vessels. The lymphatic system typically begins as blind-ended tubes, or **lymphatic bulbs,** which drain into the meshwork of interconnected lymphatic vessels (Fig. 16.3). Although lymph collection begins in the lymphatic bulbs, lymph collection from tissue also occurs in the interconnected lymphatic vessels by the same mechanical processes.

A schematic drawing of the lymphatic system in the small intestine (Fig. 16.4) illustrates the complexity of lymphatic branching. The villus lacteals are lymphatic bulbs in individual villi of the small intestine. Note that lymph collection from the submucosal and muscle layers of this tissue must occur primarily in tubular lymphatic vessels because few, if any, lymphatic bulbs are present in these layers.

The lymphatic vessels coalesce into increasingly more developed and larger collection vessels. These larger vessels in the tissue and the macroscopic lymphatic vessels outside the organs have contractile cells similar to vascular smooth muscle cells. In connective tissues of the mesentery and skin, even the simplest of lymphatic vessels and bulbs spontaneously contract, perhaps as a result of contractile endothelial cells. Even if the lymphatic bulb or vessel cannot contract, compression of these lymphatic structures by movements of the organ (e.g., intestinal movements or skeletal muscle contractions) changes lymphatic vessel size. Forcing lymph from the organs is important because a volume of fluid equal to the plasma volume is filtered from the blood to tissues every day. It is absolutely essential that this fluid be returned by lymph flow to the venous system.

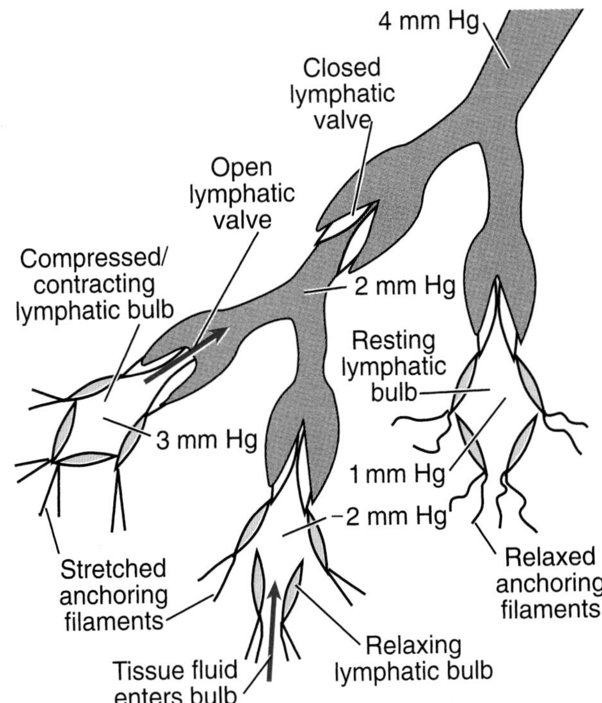

FIGURE 16.3 **Lymphatic vessels: basic structure and functions.** The contraction-relaxation cycle of lymphatic bulbs (bottom) is the fundamental process that removes excess water and plasma proteins from the interstitial spaces. Pressures along the lymphatics are generated by lymphatic vessel contractions and by organ movements.

Lymph Fluid Is Mechanically Collected Into Lymphatic Vessels From Tissue Fluid Between Cells

In all organ systems, more fluid is filtered than absorbed by the capillaries, and plasma proteins diffuse into the interstitial spaces through the large pore system. By removing the fluid, the lymphatic vessels also remove proteins. This function is essential because the protein concentration is higher in plasma than in tissue fluid and only some form of convective transport can return the protein to the plasma. The ability of lymphatic vessels to change diameter—whether initiated by the lymphatic vessel or by forces generated within a contractile organ—is important for lymph

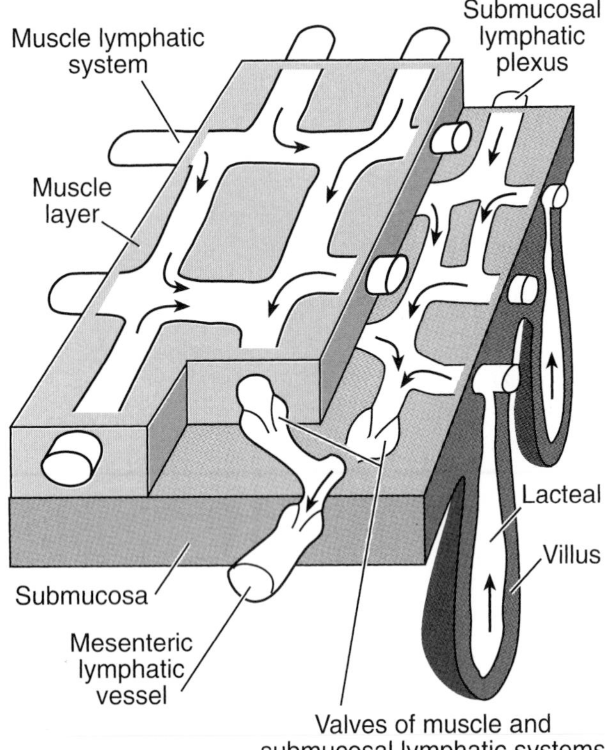

FIGURE 16.4 **The arrangement of lymphatic vessels in the small intestine.** The intestinal lymphatic vessels are unusual in that lymphatic valves are normally restricted to vessels about to exit the organ, whereas valves exist throughout the lymphatic system of the skin and skeletal muscles. (Modified from Unthank JL, Bohlen HG. Lymphatic pathways and role of valves in lymph propulsion from small intestine. Am J Physiol 1988;254:G389–G398.)

formation and protein removal. In the smallest lymphatic vessels and to some extent in the larger lymphatic vessels in a tissue, the endothelial cells are overlapped rather than fused together as in blood capillaries. The overlapped portions of the cells are attached to **anchoring filaments,** which extend into the tissue (Fig. 16.3). When stretched, anchoring filaments pull apart the free edges of the endothelial cells when the lymphatic vessels relax after a compression or contraction. The openings created in this process allow tissue fluid and molecules carried in the fluid to easily enter the lymphatic vessels.

The movement of fluid from tissue to the lymphatic vessel lumen is passive. When compressed or actively contracted lymphatic vessels are allowed to passively relax, the pressure in the lumen becomes slightly lower than in the interstitial space, and tissue fluid enters the lymphatic vessel. Once the interstitial fluid is in a lymphatic vessel, it is called **lymph.** When the lymphatic bulb or vessel again actively contracts or is compressed, the overlapped cells are mechanically sealed to hold the lymph. The pressure developed inside the lymphatic vessel forces the lymph into the next downstream segment of the lymphatic system. Because the anchoring filaments are stretched during this process, the overlapped cells can again be parted during relaxation of the lymphatic vessel.

The compression/relaxation cycle—whether controlled by lymphatic smooth muscle cells or the contractile lymphatic endothelial cells—increases in frequency and vigor when excess water is in the lymph vessels. Conversely, less fluid in the lymphatic vessels allows the vessels to become quiet and pump less fluid. This simple regulatory system ensures that the fluid status of the organ's interstitial environment is appropriate.

The active and passive compression of lymphatic bulbs and vessels also provides the force needed to propel the lymph back to the venous side of the blood circulation. To maintain directional lymph flow, microscopic lymphatic bulbs and vessels, as well as large lymphatic vessels, have one-way valves (see Fig. 16.3). These valves allow lymph to flow only from the tissue toward the progressively larger lymphatic vessels and, finally, into large veins in the chest cavity.

Lymphatic pressures are only a few mm Hg in the bulbs and smallest lymphatic vessels and as high as 10 to 20 mm Hg during contractions of larger lymphatic vessels. This progression from lower to higher lymphatic pressures is possible because, as each lymphatic segment contracts, it develops a slightly higher pressure than in the next lymphatic vessel and the lymphatic valve momentarily opens to allow lymph flow. When the activated lymphatic vessel relaxes, its pressure is again lower than that in the next vessel, and the lymphatic valve closes.

VASCULAR AND TISSUE EXCHANGE OF SOLUTES

The Large Number of Microvessels Provides a Large Vascular Surface Area for Exchange

The overall branching structure of the microvasculature is a tree-like system, with major trunks dividing into progressively smaller branches. This arrangement applies to both the arteriolar and the venular microvasculature; actually two "trees" exist—one to supply the tissue through arterioles and one to drain the tissue through venules. In general, there are four to five discrete branching steps from a small artery entering an organ to the capillary level and from the capillaries to the largest venules. These branching patterns are so consistent among like organ systems of various mammals, including humans, that they must be genetically determined.

The increasing numbers of vessels through successive branches dramatically increases the surface area of the microvasculature. The surface area is determined by the length, diameter, and number of vessels. In the small intestine, for example, the total surface area of the capillaries and smallest venules is more than 10 cm^2 for one cm^3 of tissue. The large surface area of the capillaries and smallest venules is important because the vast majority of exchange of nutrients, wastes, and fluid occurs across these tiny vessels.

The Large Number of Microvessels Minimizes the Diffusion Distance Between Cells and Blood

The spacing of microvessels in the tissues determines the distance molecules must diffuse from the blood to the interior of tissue cells. In the example shown in Figure 16.5A, a

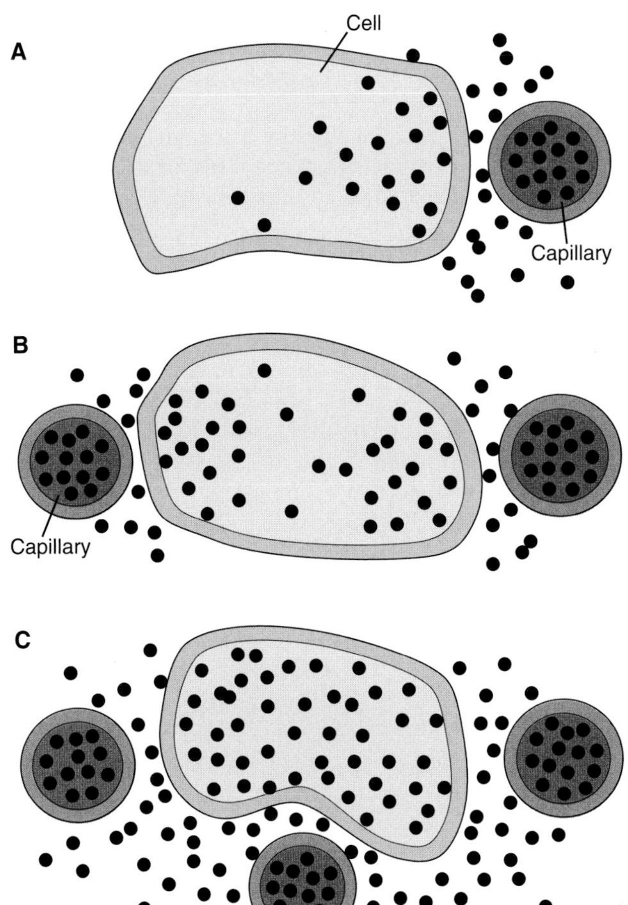

FIGURE 16.5 Effect of the number of perfused capillaries on cell concentration of bloodborne molecules (dots). A, With one capillary, the left side of the cell has a low concentration. B, The concentration can be substantially increased if a second capillary is perfused. C, The perfusion of three capillaries around the cell increases concentrations of bloodborne molecules throughout the cell.

single capillary provides all the nutrients to the cell. The concentration of bloodborne molecules across the cell interior is represented by the density of dots at various locations. Diffusion distances are important; as molecules travel farther from the capillary, their concentration decreases substantially because the volume into which diffusion proceeds increases as the square of the distance. In addition, some of the molecules may be consumed by different cellular components, which further reduces the concentration.

If there is a capillary on either side of a cell, as in Figure 16.5B, the cell has a higher internal concentration of molecules from the two capillaries. Therefore, increasing the number of microvessels reduces diffusion distances from a given point inside a cell to the nearest capillary. Doing so minimizes the dilution of molecules within the cells caused by large diffusion distances. At any given moment during resting conditions, only about 40 to 60% of the capillaries are perfused by red blood cells in most organs. The capillaries not in use do contain blood, but it is not moving. Exercise results in an increase in the number of perfused capillaries, decreasing diffusion distances. The arteriolar dilation during exercise allows arterioles to supply blood flow to nearly all of the available capillaries in muscle.

Regular exercise induces the growth of new capillaries in skeletal muscle. As shown in Figure 16.5C, three capillaries contribute to the nutrition of the cell and elevate cell concentrations of molecules derived from the blood. However, decreasing the number of capillaries perfused with blood by constricting arterioles or obliterating capillaries, as in diabetes mellitus, can lengthen diffusion distances and decrease exchange.

The Interstitial Space Between Cells Is a Complex Environment of Water- and Gel-Filled Areas

As molecules diffuse from the microvessels to the cells or from the cells to the microvessels, they must pass through the **interstitial space** that forms the extracellular environment between cells. This space contains strands of collagen and elastin together with hyaluronic acid (a high-molecular-weight unbranched polysaccharide) and proteoglycans (complex polysaccharides bound to polypeptides). These large molecules are arranged in complex, water-filled coils. To some extent, the large molecules and water may cause the interstitial space to behave as alternating regions of gel-like consistency and water-filled regions. The gel-like areas may restrict the diffusion of water-soluble solutes and may exclude solutes from their water.

An implication of the gel and water properties of the interstitial space is that the effective concentration of molecules in the free interstitial water is higher than expected because the molecules are restricted to readily accessible water-filled areas. The circuitous pathway a molecule must move in the maze of the interstitial gel- and water-filled spaces slows the diffusion of water-soluble molecules. It is also possible that the relative amounts of gel and water phases can be altered in a way that diffusion in the extracellular space is changed.

The Rate of Diffusion Depends on Permeability and Concentration Differences

Diffusion is by far the most important means for moving solutes across capillary walls. The rate of diffusion of a solute between blood and tissue is given by Fick's law (see Chapter 2):

$$J_s = P\,(C_b - C_t) \qquad (1)$$

J_s is the net movement of solute (often expressed in moles/min per 100 g tissue), P is the permeability coefficient, and C_b and C_t are, respectively, the blood and tissue concentrations of the solute.

The **permeability coefficient** is usually measured under conditions in which neither the surface area of the vasculature nor the diffusion distance is known, but the tissue mass can be determined. The permeability coefficient is directly related to the diffusion coefficient of the solute in the capillary wall and the vascular surface area available for exchange and is inversely related to the diffusion distance. The surface area and diffusion distance are determined, in part, by the number of microvessels with active blood flow. The diffusion coefficient is relatively constant, unless the capillaries are damaged, because it depends on the anatomic properties of the vessel wall (e.g., the size and abundance of pores) and the chemical nature of the material that is diffusing.

The number of perfused capillaries and blood and tissue concentrations of solutes are constantly changing, and chronic changes occur as well. Therefore, the diffusion distance and surface area for exchange can be influenced by physiological events. The same is true for concentrations in the tissue and blood. In this context, microvascular exchange is dynamically altered by many physiological events. For example, about half of the capillaries of the intestinal villus are perfused when the bowel lumen is empty. During absorption of foodstuff, all of the capillaries are perfused as arterioles dilate to provide a higher blood flow to support the increased metabolic rate of villus epithelial cells.

The magnitude of the difference in blood and tissue concentrations is influenced by many simultaneous and interacting processes. It is important to remember that the diffusion rate depends on the *difference* between the high and low concentrations, not the specific concentrations. For example, if the cell consumes a particular solute, the concentration in the cell will decrease, and for a constant concentration in blood plasma, the diffusion gradient will enlarge to increase the rate of diffusion. If the cell ceases to use as much of a given solute, the concentration in the cell will increase and the rate of diffusion will decrease. Both of these examples assume that more than sufficient blood flow exists to maintain a relatively constant concentration in the microvessel.

In many cases, the above scenario may not be true. For example, as blood passes through the tissues, the tissues extract approximately one fourth to one third of the oxygen contained in arterial blood before it reaches the capillaries. The oxygen diffuses directly through the walls of the arterioles and is readily available for any cells in the vicinity. There is usually ample oxygen in the capillary blood to maintain aerobic metabolism; however, if tissue metabolism is increased and blood flow is not appropriately elevated, the tissue will exhaust the available oxygen from the blood while it is in the microvessels. The result is that, although the cells have generated conditions to increase their aerobic metabolic rate, inadequate oxygen is exchanged for this increased need. To temporarily perform their functions, the active cells resort to anaerobic glycolysis to provide cell energy. This scenario routinely occurs when skeletal muscles begin to contract and blood flow has not yet been appropriately increased to meet the increased oxygen demand.

The Extraction of Molecules From Blood Is Influenced by Vascular Permeability, Surface Area, and Blood Flow

As a result of diffusional losses and gains of molecules as blood passes through the tissues, the concentrations of various molecules in venous blood can be very different from those in arterial blood. The **extraction** (E), or **extraction ratio**, of material from blood perfusing a tissue can be calculated from the arterial (C_a) and venous (C_v) blood concentration as:

$$E = (C_a - C_v)/C_a \qquad (2)$$

If the blood loses material to the tissue, the value of E is positive and has a maximum value of 1 if all material is removed from arterial blood ($C_v = 0$). An E value of 0 ($C_a = C_v$) indicates that no loss or gain occurred. A negative E value ($C_v > C_a$) indicates that the tissue added material to the blood. The total mass of material lost or gained by the blood can be calculated as:

$$\text{Amount lost or gained} = E \times \dot{Q} \times C_a \qquad (3)$$

E is extraction, $\dot{Q}$ is blood flow, and C_a is the arterial concentration. While this equation is useful for calculating the total amount of material exchanged between tissue and blood, it does not allow a direct determination of how changes in vascular permeability and exchange surface area influence the extraction process. The extraction can be related to the permeability (P) and surface area (A) available for exchange as well as the blood flow ($\dot{Q}$):

$$E = 1 - e^{-PA/\dot{Q}} \qquad (4)$$

The *e* is the base of the natural system of logarithms. This equation predicts that extraction increases when either permeability or exchange surface area increases or blood flow decreases. Extraction decreases when permeability and surface area decrease or blood flow increases. Consequently, physiologically induced changes in the number of perfused capillaries, which alters surface area, and changes in blood flow are important determinants of overall extraction and, therefore, exchange processes. The inverse effect of blood flow on extraction occurs because, if flow increases, less time is available for exchange. Conversely, a slowing of flow allows more time for exchange.

Ordinarily, the blood flow and total perfused surface area usually change in the same direction, although by different relative amounts. For example, surface area is usually able, at most, to double or be reduced by about half; however, blood flow can increase 3- to 5-fold or more in skeletal muscle, or decrease by about half in most organs, yet maintain viable tissue. The net effect is that extraction is rarely more than doubled or decreased by half relative to the resting value in most organs. This is still an important range because changes in extraction can compensate for reduced blood flow or enhance exchange when blood flow is increased.

TRANSCAPILLARY FLUID EXCHANGE

To force the blood through microvessels, the heart pumps blood into the elastic arterial system and provides the pressure needed to move the blood. This hemodynamic, or hydrostatic, pressure, while absolutely necessary, favors the pressurized filtration of water through pores because the hydrostatic pressure on the blood side of the pore is greater than on the tissue side. The capillary pressure is different in each organ, ranging from about 15 mm Hg in intestinal villus capillaries to 55 mm Hg in the kidney glomerulus. The interstitial hydrostatic pressure ranges from slightly negative to 8 to 10 mm Hg and, in most organs, is substantially less than capillary pressure.

The Osmotic Forces Developed by Plasma Proteins Oppose the Filtration of Fluid From Capillaries

The primary defense against excessive fluid filtration is the **colloid osmotic pressure,** also called plasma oncotic pressure, generated by plasma proteins. Plasma proteins are too large to pass readily through the vast majority of water-filled pores of the capillary wall. In fact, more than 90% of these large molecules are retained in the blood during its passage through the microvessels of most organs. Colloid osmotic pressure is conceptually similar to osmotic pressures for small molecules generated across selectively permeable cell membranes; both primarily depend on the number of molecules in solution. The major plasma protein that impedes filtration is serum albumin because it has the highest molar concentration of all plasma proteins. The colloid osmotic pressure of plasma proteins is typically 18 to 25 mm Hg in mammals when measured using a membrane that prevents the diffusion of all large molecules.

Colloid osmotic pressure offsets the capillary hydrostatic blood pressure to the extent that the net filtration force is only slightly positive or negative. If the capillary pressure is sufficiently low, the balance of colloid osmotic and hydrostatic pressures is negative, and tissue water is absorbed into the capillary blood. The majority of organs continuously form lymph, which indicates that capillary and venular filtration pressures generally are larger than absorption pressures. The balance of pressures is likely 1 to 2 mm Hg in most organs.

The Leakage of Plasma Proteins Into Tissues Increases the Filtration of Fluid From the Blood to the Tissues

A small amount of plasma protein enters the interstitial space; these proteins and, perhaps, native proteins of the space generate the tissue colloid osmotic pressure. This pressure of 2 to 5 mm Hg offsets part of the colloid osmotic pressure in the plasma. This is, in a sense, a filtration pressure that opposes the blood colloid osmotic pressure. As discussed earlier, the lymphatic vessels return plasma proteins in the interstitial fluid to the plasma.

Hydrostatic Pressure in Tissues Can Either Favor or Oppose Fluid Filtration From the Blood to the Tissues

The hydrostatic pressure on the tissue side of the endothelial pores is the tissue hydrostatic pressure. This pressure is determined by the water volume in the interstitial space and tissue distensibility. Tissue hydrostatic pressure can be increased by external compression, such as with support stockings, or by internal compression, such as in a muscle during contraction. The tissue hydrostatic pressure in various tissues during resting conditions is a matter of debate. Tissue pressure is probably slightly below atmospheric pressure (negative) to slightly positive ($<+3$ mm Hg) during normal hydration of the interstitial space and becomes positive when excess water is in the interstitial space. Tissue hydrostatic pressure is a filtration force when negative and an absorption force when positive.

Support stockings are routinely prescribed for people whose feet and lower legs swell during prolonged standing. Standing causes high capillary hydrostatic pressures from gravitational effects on blood in the arterial and venous vessels and results in excessive filtration. Support stockings compress the interstitial environment to raise hydrostatic tissue pressure and compress superficial veins, which helps lower venous pressure and, thereby, capillary pressure.

If water is removed from the interstitial space, the hydrostatic pressure becomes very negative and opposes further fluid loss (Fig. 16.6). If a substantial amount of water is added to the interstitial space, the tissue hydrostatic pressure is increased. However, a margin of safety exists over a wide range of tissue fluid volumes (see Fig. 16.6), and excessive tissue hydration or dehydration is avoided. If the tissue volume exceeds a certain range, swelling or **edema** occurs. In extreme situations, the tissue swells with fluid to the point that pressure dramatically increases and strongly opposes capillary filtration. The ability of tissues to allow substantial changes in interstitial volume with only small changes in pressure indicates that the interstitial space is distensible. As a general rule, about 500 to 1,000 mL of fluid can be withdrawn from the interstitial space of the entire body to help replace water losses due to sweating, diarrhea, vomiting, or blood loss.

The Balance of Filtration and Absorption Forces Regulates the Exchange of Fluid Between the Blood and the Tissues

The role of hydrostatic and colloid osmotic pressures in determining fluid movement across capillaries was first postulated by the English physiologist Ernest Starling at the end of the nineteenth century. In the 1920s, the American physiologist Eugene Landis obtained experimental proof

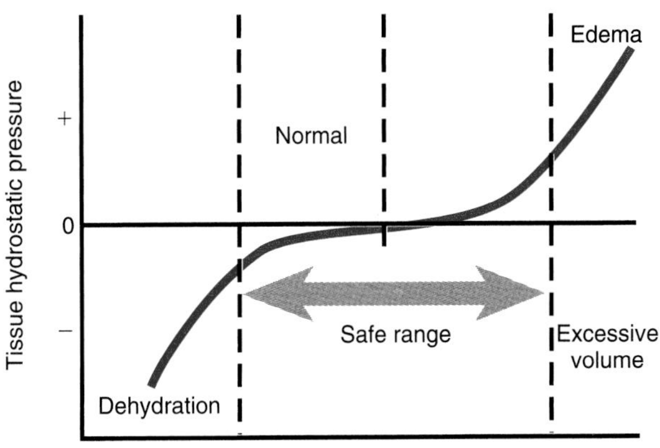

FIGURE 16.6 **Variations in tissue hydrostatic pressure as interstitial fluid volume is altered.** Under normal conditions, tissue pressure is slightly negative (subatmospheric), but an increase in volume can cause the pressure to be positive. If the interstitial fluid volume exceeds the "safe range," high tissue hydrostatic pressures and edema will be present. Tissue dehydration can cause negative tissue hydrostatic pressures.

for Starling's hypothesis. The relationship is defined for a single capillary by the **Starling-Landis equation:**

$$J_V = K_h A \{(P_c - P_t) - \sigma(COP_p - COP_t)\} \quad (5)$$

J_V is the net volume of fluid moving across the capillary wall per unit of time (μm^3/min). K_h is the **hydraulic conductivity for water,** which is the fluid permeability of the capillary wall. K_h is expressed as μm^3/min/(μm^2 of capillary surface area) per mm Hg pressure difference. The value of K_h increases up to 4-fold from the arterial to the venous end of a typical capillary. A is the vascular surface area, P_c is the capillary hydrostatic pressure, and P_t is the tissue hydrostatic pressure. COP_p and COP_t represent the plasma and tissue colloid osmotic pressures, respectively, and σ is the **reflection coefficient for plasma proteins.** This coefficient is included because the microvascular wall is slightly permeable to plasma proteins, preventing the full expression of the two colloid osmotic pressures. The value of σ is 1 when molecules cannot cross the membrane (i.e., they are 100% "reflected") and 0 when molecules freely cross the membrane (i.e., they are not reflected at all). Typical σ values for plasma proteins in the microvasculature exceed 0.9 in most organs other than the liver and spleen, which have capillaries that are very permeable to plasma proteins. The reflection coefficient is normally relatively constant but can be decreased dramatically by hypoxia, inflammatory processes, and tissue injury. This leads to increased fluid filtration because the effective colloid osmotic pressure is reduced when the vessel wall becomes more permeable to plasma proteins.

The capillary exchange of fluid is bidirectional because capillaries and venules may filter or absorb fluid, depending on the balance of hydrostatic and colloid osmotic pressures. It is possible that filtration occurs primarily at the arteriolar end of capillaries, where filtration forces exceed absorptive forces. It is equally likely that fluid absorption occurs in the venular end of the capillary and small venules because the friction of blood flow in the capillary has dissipated the hydrostatic blood pressure. Based on directly measured capillary hydrostatic and plasma colloid osmotic pressures, the entire length of the capillaries in skeletal muscle filters slightly all of the time, while the lower capillary pressures in the intestinal mucosa and brain primarily favor absorption along the entire capillary length. However, as each of these organs does filter fluid, some of the capillaries and, probably, the smaller arterioles are filtering fluid most of the time.

The extrapolation of fluid filtration or absorption for a single capillary to fluid exchange in a whole tissue is difficult. Within organs, there are regional variations in microvascular pressures, possible filtration and absorption of fluid in vessels other than capillaries, and physiologically and pathologically induced variations in the available surface area for capillary exchange. Therefore, for whole organs, a measurement of total fluid movement relative to the mass of the tissue is used. To take into account the various hydraulic conductivities and total surface areas of all vessels involved, the volume (mL) of fluid moved per minute for a change of 1 mm Hg in capillary pressure for each 100 g of tissue is determined. This value is called the **capillary filtration coefficient** (CFC), although it is likely that fluid exchange occurs in both venules and capillaries. CFC values in tissues such as skeletal muscle and the small intestine are typically in the range of 0.025 to 0.16 mL/min per mm Hg per 100 g.

The CFC replaces the hydraulic conductivity (K_h) and capillary surface area (A) in the Starling-Landis equation for filtration across a single capillary. The CFC can change if fluid permeability, the surface area (determined by the number of perfused microvessels), or both are altered. For example, during the intestinal absorption of foodstuff, particularly lipids, both capillary fluid permeability and perfused surface area increase, dramatically increasing CFC. In contrast, the skeletal muscle vasculature increases CFC primarily because of increased perfused capillary surface area during exercise and only small increases in fluid permeability occur.

The hydrostatic and colloid osmotic pressure differences across capillary walls—the **Starling forces**—cause the movement of water and dissolved solutes into the interstitial spaces. These movements are, however, normally quite small and contribute minimally to tissue nutrition. Most solutes transferred to the tissues move across capillary walls by simple diffusion, not by bulk flow of fluid.

REGULATION OF MICROVASCULAR PRESSURES

The microvascular pressures, both hydrostatic and colloid osmotic, involved in transcapillary fluid exchange depend on how the microvasculature dissipates the prevailing arterial and venous pressures and on the concentration of plasma proteins. Plasma protein concentration is determined largely by the rate of protein synthesis in the liver, where most of the plasma proteins are made. Disorders that impair protein synthesis—liver diseases and malnutrition—and kidney diseases—in which plasma proteins are filtered and lost into the urine—result in reduced plasma protein concentration. A lowered plasma colloid osmotic pressure favors the filtration of plasma water and gradually causes significant edema. Edema formation in the abdominal cavity, known as **ascites,** can allow large quantities of fluid to collect in and grossly distend the abdominal cavity.

Capillary Pressure Is Determined by the Resistance of and Blood Pressure in Arterioles and Venules

Capillary pressure (P_c) is not constant; it is influenced by four major variables: precapillary (R_{pre}) and postcapillary (R_{post}) resistances and arterial (P_a) and venous (P_v) pressures. Precapillary and postcapillary resistances can be calculated from the pressure dissipated across the respective vascular regions divided by the total tissue blood flow ($\dot{Q}$), which is essentially equal for both regions:

$$R_{pre} = (P_a - P_c)/\dot{Q} \quad (6)$$

$$R_{post} = (P_c - P_v)/\dot{Q} \quad (7)$$

In the majority of organ vasculatures, the precapillary resistance is 3 to 6 times higher than the postcapillary resistance. This has a substantial effect on capillary pressure.

To demonstrate the effect of precapillary and postcapillary resistances on capillary pressure, we use the equations for the precapillary and postcapillary resistances to solve for blood flow:

$$\dot{Q} = (P_a - P_c)/R_{pre} = (P_c - P_v)/R_{post} \quad (8)$$

The two equations to the right of the flow term can be solved for capillary pressure:

$$P_c = \frac{(R_{post}/R_{pre})P_a + P_v}{1 + (R_{post}/R_{pre})} \quad (9)$$

Equation 9 indicates that the ratio of postcapillary to precapillary resistance, rather than the absolute magnitude of either resistance, determines the effect of arterial pressure (P_a) on capillary pressure. In addition, venous pressure substantially influences capillary pressure. The denominator also influences both pressure effects. At a typical postcapillary to precapillary resistance ratio of 0.16:1, the denominator will be 1.16, which allows about 80% of a change in venous pressure to be reflected back to the capillaries. The postcapillary to precapillary resistance ratio increases during the arteriolar vasodilation that accompanies increased tissue metabolism; the decreased precapillary resistance and minimal change in postcapillary resistance increase capillary pressure. Because the balance of hydrostatic and colloid osmotic pressures is usually −2 to +2 mm Hg, a 10- to 15-mm Hg increase in capillary pressure during maximum vasodilation can cause a profound increase in filtration. The increased filtration associated with microvascular dilation is usually associated with a large increase in lymph production, which removes excess tissue fluid.

Capillary Pressure Is Reduced When the Sympathetic Nervous System Increases Arteriolar Resistance

When sympathetic nervous system stimulation causes a substantial increase in precapillary resistance and a proportionately smaller increase in postcapillary resistance, the capillary pressure can decrease up to 15 mm Hg and, thereby, greatly increase the absorption of tissue fluid. This process is important. As mentioned earlier, fluid taken from the interstitial space can compensate for vascular volume loss during sweating, vomiting, or diarrhea. As water is lost by any of these processes, the plasma proteins are concentrated because they are not lost.

REGULATION OF MICROVASCULAR RESISTANCE

The vascular smooth cells around arterioles and venules respond to a wide variety of physical and chemical stimuli, altering the diameter and resistance of the microvessels. Here we consider the various physical and chemical conditions in tissues that influence the muscle cells of the microvasculature.

Myogenic Vascular Regulation Allows Arterioles to Respond to Changes in Intravascular Pressure

Vascular smooth muscle can contract rapidly when stretched and, conversely, can reduce actively developed tension when passively shortened. In fact, vascular smooth muscle may be able to contract or relax when the load on the muscle is increased or decreased, respectively, even though the initial muscle length is not substantially changed. These responses are known to persist as long as the initial stimulus is present, unless vasoconstriction reduces blood flow to the extent that tissue becomes severely hypoxic. This process, called **myogenic regulation**, is activated when microvascular pressure is increased or decreased.

The cellular mechanisms responsible for myogenic regulation are not entirely understood, but several possibilities are likely involved. The first mechanism is a calcium ion-selective channel that is opened in response to increased membrane stretch or tension. Adding calcium to the cytoplasm would activate the smooth muscle cell and result in contraction. Limiting calcium entry would allow calcium pumps to remove calcium ions from the cytoplasm and favor relaxation. The second mechanism is a nonspecific cation channel that is opened in proportion to cell membrane stretch or tension. The entry of sodium ions through open channels would depolarize the cell and lead to the opening of voltage-activated calcium channels, followed by contraction as calcium ions flood into the cell. During reduced stretch or tension, the nonspecific channels would close and allow hyperpolarization to occur.

Other mechanisms are likely involved in myogenic regulation. What is clear is that vascular smooth muscle cells depolarize as the intravascular pressure is increased and hyperpolarize as the pressure is decreased. In addition, myogenic mechanisms are extremely fast and appear to be able to adjust to most rapid pressure changes.

Myogenic regulation has some benefits. First, and perhaps most important, blood flow can be regulated when the arterial pressure is too high or too low for appropriate tissue blood flow. Second, the myogenic response helps prevent tissue edema when venous pressure is elevated by more than about 5 to 10 mm Hg above the typical resting values. The elevation of venous pressure results in an increase in capillary and arteriolar pressures. Myogenic arteriolar constriction lowers the transmission of arterial pressure to the capillaries and small venules to minimize the risk of edema, but at the expense of a decreased blood flow. The myogenic response to elevated venous pressure may be due to venous pressures transmitted backward through the capillary bed to the arterioles and, perhaps, to some type of response initiated by venules and transmitted to arterioles, possibly through endothelial cells or local neurons.

Tissue Metabolism Influences Blood Flow

In almost all organs, an increase in metabolic rate is associated with increased blood flow and extraction of oxygen to meet the metabolic needs of the tissues. In addition, a reduction in oxygen within the blood is associated with dilation of the arterioles and increased blood flow, assuming neural reflexes to hypoxia are not activated. The local regulation of the microvasculature in response to the metabolic needs of tissues involves many different types of cellular mechanisms, one of which is linked to oxygen availability.

Oxygen is not stored in appreciable amounts in tissues, and the oxygen concentration will fall to nearly zero in about one minute if blood flow is stopped in any organ. An increase in metabolic rate would decrease the tissue oxygen concentration and possibly directly signal vascular muscle to relax by limiting the production of ATP for the contraction of smooth muscle cells. Figure 16.7 shows examples of the changes in oxygen partial pressure (tension) around arterioles (periarteriolar space), in the capillary bed, and around large venules during skeletal muscle contractions. At rest, venular blood oxygen tension is usually higher than in the capillary bed, possibly because venules acquire oxygen that diffuses out of nearby arterioles. Although both periarteriolar and capillary bed tissue oxygen tensions decrease at the onset of contractions, both are restored as arteriolar dilation occurs. The oxygen tension in venular blood rapidly and dramatically decreases at the onset of skeletal muscle contractions and demonstrates little recovery despite increased blood flow. The sustained decline in venular blood oxygen tension probably reflects increased extraction of oxygen from the blood. It is apparent from Figure 16.7 that the oxygen tension of venular blood in skeletal muscle is not a trustworthy indicator of the oxygen status of the capillary bed at rest or during contractions.

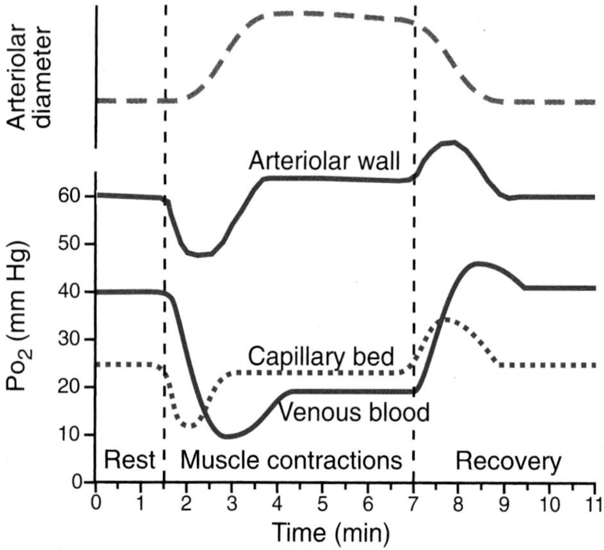

FIGURE 16.7 **Arteriolar dilation and tissue oxygen tensions during skeletal muscle contractions.** The decrease in arteriolar, capillary bed, and venous oxygen tensions at the start of contractions reflects increased oxygen use, which is not replenished by increased blood flow until the arterioles dilate. As arteriolar dilation occurs, arteriolar wall and capillary bed oxygen tensions are substantially restored, but venous blood has a low oxygen tension. During recovery, oxygen tensions transiently increase above resting values because blood flow remains temporarily elevated as oxygen use is rapidly lowered to normal. (Modified from Lash JM, Bohlen HG. Perivascular and tissue Po_2 in contracting rat spinotrapezius muscle. Am J Physiol 1987;252:H1192–H1202.)

In contrast to venules, many arterioles have a normal-to-slightly increased periarteriolar oxygen tension during skeletal muscle contractions because the increased delivery of oxygen through elevated blood flow offsets the increased use of oxygen by tissues immediately around the arteriole. Therefore, as long as blood flow is allowed to increase substantially, it is unlikely that oxygen availability at the arteriolar wall is a major factor in the sustained vasodilation that occurs during increased metabolism.

Recent studies indicate that vascular smooth muscle cells are not particularly responsive to a broad range of oxygen tensions. Only unusually low or high oxygen tensions seem to be associated with direct changes in vascular smooth muscle force. However, either oxygen depletion from an organ's cells or an increased metabolic rate does cause the release of adenine nucleotides, free adenosine, Krebs cycle intermediates, and, in hypoxic conditions, lactic acid. There is a large potential source of various molecules, most of which cause vasodilation at physiological concentrations, to influence the regulation of blood flow.

An increase in hydrogen ion concentration, resulting from accumulation of carbonic acid (formed from CO_2 and water) or acidic metabolites (such as lactic acid), causes vasodilation. However, usually only transient increases in venous blood and interstitial tissue acidity occur if blood flow through an organ with increased metabolism is allowed to increase appropriately.

Endothelial Cells Can Release Chemicals That Cause Relaxation or Constriction of Arterioles

An important contributor to local vascular regulation is released by endothelial cells. This substance, **endothelium-derived relaxing factor** (EDRF), is released from all arteries, microvessels, veins, and lymphatic endothelial cells. EDRF is **nitric oxide** (NO), which is formed by the action of **nitric oxide synthase** on the amino acid arginine. NO causes the relaxation of vascular smooth muscle by inducing an increase in cyclic guanosine monophosphate (cGMP). When cGMP is increased, the smooth muscle cell extrudes calcium ions and decreases calcium entry into the cell, inhibiting contraction and enzymatic processes that depend on calcium ions. Compounds such as acetylcholine, histamine, and adenine nucleotides (ATP, ADP) released into the interstitial space, as well as hypertonic conditions and hypoxia, cause the release of NO. Adenosine causes NO release from endothelial cells and directly relaxes vascular smooth muscle cells through adenosine receptors.

Another important mechanism to release NO is the **shear stress** generated by blood moving past the endothelial cells. Frictional forces between moving blood and the stationary endothelial cells distort the endothelial cells, opening special potassium channels and causing endothelial cell hyperpolarization. This increases calcium ion entry into the cell down the increased electrical gradient. The elevated cytosolic calcium ion concentration activates endothelial nitric oxide synthase to form more NO, and the blood vessels dilate.

This mechanism is used to coordinate various sized arterioles and small arteries. As small arterioles dilate in response to some signal from the tissue, the increased blood flow increases the shear stress in larger arterioles and small arteries, which prompts their endothelial cells to release NO and relax the smooth muscle. As larger arterioles and small arteries control much more of the total vascular resistance than do small arterioles, the cooperation of the larger resistance vessels is vital to adjusting blood flow to the needs of the tissue. Examples of this process, called **flow-mediated vasodilation**, have been observed in cerebral, skeletal muscle, and small intestinal vasculatures. Endothelial cells of arterioles also release vasodilatory prostaglandins when blood flow and shear stress are increased. However, NO appears to be the dominant vasodilator molecule for flow-dependent regulation. Clinical Focus Box 16.1 describes the defects in endothelial cell function and NO production that are a major contribution to the pathophysiology of diabetes mellitus.

Endothelial cells also release one of the most potent vasoconstrictor agents, the 21-amino acid-peptide **endothelin.** Extremely small amounts are released under natural conditions. Endothelin is the most potent biological constrictor of blood vessels yet to be found. The vasoconstriction occurs because of a cascade of events beginning with phospholipase C activation and leading to activation of protein kinase C (see Chapter 1). Two major types of endothelin receptors have been identified and others may exist. The constrictor function of endothelin is mediated by type B endothelin receptors. Type A endothelin receptors cause hyperplasia and hypertrophy of vascular muscle cells and the release of NO from endothelial cells. The precise function of endothelin in the normal vasculature is not clear; however, it is active during embryological development. In knockout mice, the absence of the endothelin A receptor results in serious cardiac defects so newborns are not viable. An absence of the type B receptor is associated with an enlarged colon, eventually leading to death. Endothelin clearly has functions other than vascular regulation.

In damaged heart tissue, such as after poor blood flow resulting in an infarct, cardiac endothelial cells increase endothelin production. The endothelin stimulates both vascular smooth muscle and cardiac muscle to contract more vigorously and induces the growth of surviving cardiac cells. However, excessive stimulation and hypertrophy of cells appears to contribute to heart failure, failure of contractility, and excessive enlargement of the heart. Part of the stimulation of endothelin production in the injured heart may be the damage per se. Also, increased formation of angiotensin II and norepinephrine during chronic heart disease stimulates endothelin production, probably at the gene expression level. Activation of protein kinase C (PKC) increases the expression of the c-*jun* proto-oncogene, which, in turn, activates the preproendothelin-1 gene. Endothelin has also been implicated as a contributor to renal vascular failure, both pulmonary hypertension and the systemic hypertension associated with insulin resistance, and the spasmodic contraction of cerebral blood vessels exposed to blood after a brain injury or stroke associated with blood loss to brain tissue.

The Sympathetic Nervous System Regulates Blood Pressure and Flow by Constricting the Microvessels

Although the microvasculature uses local control mechanisms to adjust vascular resistance based on the physical and chemical environment of the tissue and vasculature, the dominant regulatory system is the sympathetic nervous system. As Chapter 18 explains, the arterial pressure is monitored moment-to-moment by the baroreceptor system, and the brain adjusts the cardiac output and systemic vascular resistance as needed via the sympathetic and parasympathetic nervous systems. Sympathetic nerves communicate with the resistance vessels and venous system through the

CLINICAL FOCUS BOX 16.1

Diabetes Mellitus and Microvascular Function

More than 95% of persons with diabetes experience periods of elevated blood glucose concentration, or **hyperglycemia**, as a result of inadequate insulin action and the resulting decreased glucose transport into the muscle and fat tissues and increased glucose release from the liver. The most common cause of diabetes mellitus is obesity, which increases the requirement for insulin to the extent that even the high insulin concentrations provided by the pancreatic beta cells are insufficient. This overall condition is called **insulin resistance.**

Obesity independent of periods of hyperglycemia does not injure the microvasculature. However, periods of hyperglycemia over time cause reduced nitric oxide (NO) production by endothelial cells, increased reactivity of vascular smooth muscle to norepinephrine, accelerated atherosclerosis, and a reduced ability of microvessels to participate in tissue repair. The consequences are cerebrovascular accidents (stroke) and coronary artery disease as a result of endothelial cell abnormalities; loss of toes or whole legs as a result of microvascular and atherosclerotic pathology; and loss of retinal microvessels followed by a pathological overgrowth of capillaries, leading to blindness. The kidney glomerular capillaries are also damaged—this may lead to renal failure.

The mechanism of many of these abnormalities appears to stem from the fact that hyperglycemia activates protein kinase C (PKC) in endothelial cells. PKC inhibits nitric oxide synthase, so NO formation is gradually suppressed. This leads to loss of an important vasodilatory stimulus (NO) and vasoconstriction. PKC also activates phospholipase C, leading to increased diacylglycerol and arachidonic acid formation. The increased availability of arachidonic acid leads to increased prostaglandin synthesis and the generation of oxygen radicals that destroy part of the NO present. In addition, oxygen radicals damage cells of the microvasculature, and produce long-term problems caused by DNA breakage.

release of norepinephrine onto the surface of smooth muscle cells in vessel walls.

Because sympathetic nerves form an extensive meshwork of axons over the exterior of the microvessels, all vascular smooth muscle cells are likely to receive norepinephrine. Since the diffusion path is a few microns, norepinephrine rapidly reaches the vascular muscle and activates α-adrenergic receptors, and constriction begins within 2 to 5 seconds. Sympathetic nerve activation must occur quickly because rapid changes in body position or sudden exertion require immediate responses to maintain or increase arterial pressure. The sympathetic nervous system routinely overrides local regulatory mechanisms in most organs—except the heart and skeletal muscle—during exercise. But even in these, the sympathetic nervous system curtails somewhat the full increase in blood flow during submaximal contractions.

Certain Organs Control Their Blood Flow via Autoregulation and Reactive Hyperemia

If the arterial blood pressure to an organ is decreased to the extent that blood flow is compromised, the vascular resistance decreases and blood flow returns to approximately normal. If arterial pressure is elevated, flow is initially increased, but the vascular resistance increases and restores the blood flow toward normal; this is known as **autoregulation of blood flow.** Autoregulation appears to be primarily related to metabolic and myogenic control, as well as an increased release of NO if the tissue oxygen availability decreases. The cerebral and cardiac vasculatures, followed closely by the renal vasculature, are most able to autoregulate blood flow. Skeletal muscle and intestinal vasculatures exhibit less well-developed autoregulation.

A phenomenon related to autoregulation is **reactive hyperemia**. When blood flow to any organ is stopped or reduced by vascular compression for more than a few seconds, vascular resistance dramatically decreases. Absence of blood flow allows vasodilatory chemicals to accumulate as hypoxia occurs; the vessels also dilate due to decreased myogenic stimulation (low microvascular pressure). As soon as the vascular compression is removed, blood flow is dramatically increased for a few minutes. The excess blood in the part is called hyperemia; it is a reaction to the previous period of ischemia. A good example of reactive hyperemia is the redness of skin seen after a compression has been removed.

An example of autoregulation, based on data from the cerebral vasculature, is shown in Figure 16.8. Note that the arterioles continue to dilate at arterial pressures below 60 mm Hg, when blood flow begins to decrease significantly as arterial pressure is further lowered. The vessels clearly cannot dilate sufficiently to maintain blood flow at very low arterial pressures. At greater-than-normal arterial pressures, the arterioles constrict. If the mean arterial pressure is elevated appreciably above 150 to 160 mm Hg, the vessel walls cannot maintain sufficient tension to oppose passive distension by the high arterial pressure. The result is excessive blood flow and high microvascular pressures,

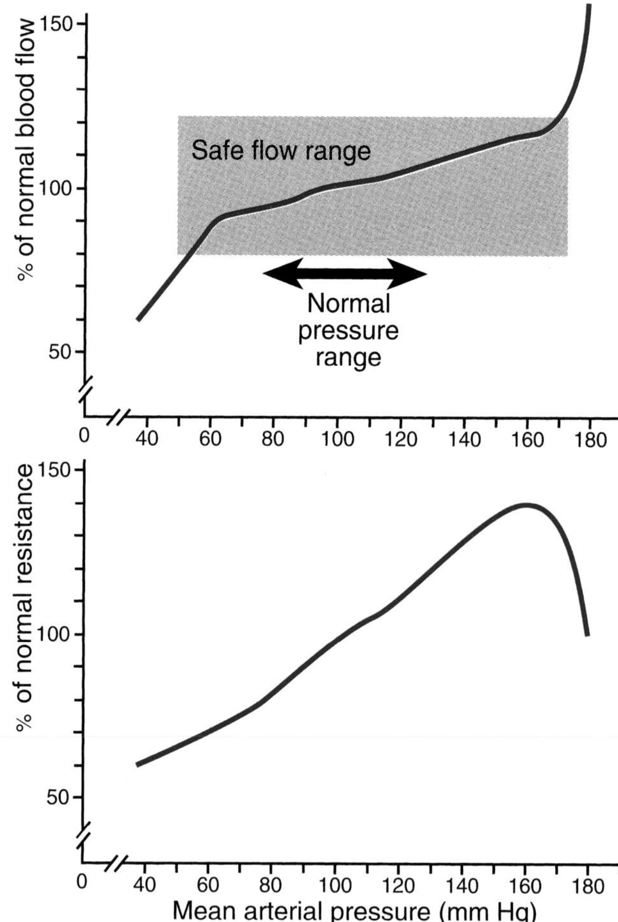

FIGURE 16.8 Autoregulation of blood flow and vascular resistance as mean arterial pressure is altered. The safe range for blood flow is about 80 to 125% of normal and usually occurs at arterial pressures of 60 to 160 mm Hg due to active adjustments of vascular resistance. At pressures above about 160 mm Hg, vascular resistance decreases because the pressure forces dilation to occur; at pressures below 60 mm Hg, the vessels are fully dilated, and resistance cannot be appreciably decreased further.

eventually leading to rupture of small vessels and excess fluid filtration into the tissue and edema.

Although the various mechanisms responsible for autoregulation are constantly interacting with the sympathetic nervous system, the actions of the sympathetic nervous system usually prevail in most organs. Only the cerebral and cardiac vasculatures exhibit impressive autoregulatory abilities because the sympathetic nervous system is incapable of causing large increases in resistance in the brain and heart. Sympathetic dominance of vascular control in the majority of organ systems is beneficial to the body as a whole. Maintenance of the arterial pressure by sustained constriction of most peripheral vascular beds and perfusion of the heart and brain at the expense of the other organs that can tolerate reduced blood flow for prolonged periods of time is lifesaving in an emergency.

REVIEW QUESTIONS

DIRECTIONS: Each of the numbered items or incomplete statements in this section is followed by answers or completions of the statement. Select the ONE lettered answer or completion that is BEST in each case.

1. The vessels most responsible for both controlling systemic vascular resistance and regulating blood flow to a particular organ are the
(A) Small arteries
(B) Arterioles
(C) Capillaries
(D) Venules
(E) Lymphatic vessels
2. The structures between adjacent capillary endothelial cells that primarily determine what size water-soluble molecules can enter the tissue are the
(A) Fiber matrices at the blood side of endothelial pores
(B) Molecular-sized openings within the tight junctions
(C) Basement membrane structures of the capillary
(D) Plasma proteins trapped in the spaces between cells
(E) Rare, large defects found between adjacent endothelial cells
3. The major pressures that determine filtration and absorption of fluid by capillaries are the
(A) Capillary hydrostatic pressure and plasma colloid osmotic pressure
(B) Plasma colloid osmotic pressure and interstitial hydrostatic pressure
(C) Interstitial hydrostatic pressure and tissue colloid osmotic pressure
(D) Capillary hydrostatic pressure and tissue colloid osmotic pressure
(E) Plasma colloid osmotic pressure and tissue colloid osmotic pressure
4. Myogenic vascular regulation is a cellular response initiated by
(A) A lack of oxygen in the tissue
(B) Nitric oxide release by vascular muscle cells
(C) Stretch or tension on vascular muscle cells
(D) Shear stress on the endothelial cells
(E) An accumulation of metabolites in the tissue
5. The most important function of the microcirculation is
(A) The exchange of nutrients and wastes between blood and tissue
(B) The filtration of water through capillaries
(C) The regulation of vascular resistance
(D) The autoregulation of blood flow
(E) Its role as a blood reservoir
6. When lipid-soluble molecules pass through a capillary wall, they primarily cross through
(A) The lipid component of cell membranes
(B) The water-filled spaces between cells
(C) The specialized transport proteins of the cell membranes
(D) The pinocytotic-exocytotic vesicles formed by endothelial cells
(E) Filtration through the capillary wall
7. Venules function to collect blood from the tissue and
(A) Act as a substantial source of resistance to regulate blood flow
(B) Serve as a reservoir for blood in the cardiovascular system
(C) Are virtually impermeable to both large and small molecules
(D) Are about the same diameter as arterioles
(E) Exchange a large amount of oxygen with the tissue
8. The interstitial space can best be described as a
(A) Water-filled space with a low plasma protein concentration
(B) Viscous space with a high plasma protein concentration
(C) Space with alternating gel and liquid areas with a low plasma protein concentration
(D) Space primarily filled with gel-like material and a small amount of liquid
(E) Major barrier to the diffusion of water and lipid-soluble molecules
9. An arteriole with a damaged endothelial cell layer will not
(A) Constrict when intravascular pressure is increased
(B) Dilate when adenosine is applied to the vessel wall
(C) Constrict in response to norepinephrine
(D) Dilate in response to adenosine diphosphate (ADP) or acetylcholine
(E) Dilate when blood flow is reduced
10. The first step for lymphatic vessels to remove excess fluid from interstitial tissue spaces is by
(A) Generating a lower intravascular than tissue hydrostatic pressure
(B) Contracting and forcing lymph into larger lymphatics
(C) Opening and closing one-way valves in the lymph vessels
(D) Lowering the colloid osmotic pressure inside the lymph vessel
(E) Closing the opening between adjacent lymphatic endothelial cells
11. When the sympathetic nervous system is activated,
(A) Norepinephrine is released by the vascular smooth muscle cells
(B) Acetylcholine is released onto vascular smooth muscle cells
(C) Norepinephrine is released from axons onto the arteriolar wall
(D) The arterioles constrict because nitric oxide production is suppressed
(E) The endothelial cells induce vascular smooth muscle cells to constrict
12. At a constant blood flow, an increase in the number of perfused capillaries improves the exchange between blood and tissue because of
(A) Greater surface area for the diffusion of molecules
(B) Faster flow velocity of plasma and red blood cells in capillaries
(C) Increased permeability of the microvasculature
(D) Decreased concentration of chemicals in the capillary blood
(E) Increased distances between the capillaries
13. For an arterial blood content of 20 mL oxygen per 100 mL blood and venous blood content of 15 mL oxygen per 100 mL of blood, how much oxygen is transferred from blood to tissue if the blood flow is 200 mL/min?
(A) 5 mL/min
(B) 10 mL/min
(C) 15 mL/min
(D) 20 mL/min
(E) 25 mL/min
14. Assume plasma proteins have a reflection coefficient of 0.9, plasma colloid osmotic pressure is 24 mm Hg, and tissue colloid osmotic pressure is 4 mm Hg. What is the net pressure available for filtration or absorption of fluid if capillary hydrostatic pressure is 23 mm Hg and tissue hydrostatic pressure is 1 mm Hg?
(A) 1 mm Hg
(B) 2 mm Hg
(C) 3 mm Hg
(D) 4 mm Hg
(E) 5 mm Hg

SUGGESTED READING

Davis MJ, Hill MA. Signaling mechanisms underlying the vascular myogenic response. Physiol Rev 1999;79:387–423.

Milnor WR. Hemodynamics. Baltimore: Williams & Wilkins, 1982;11–96.

Weinbaum S, Curry FE. Modelling the structural pathways for transcapillary exchange. Symp Soc Exp Biol 1995;49:323–345.

CHAPTER 17

Special Circulations

H. Glenn Bohlen, Ph.D.

CHAPTER OUTLINE

- CORONARY CIRCULATION
- CEREBRAL CIRCULATION
- SMALL INTESTINE CIRCULATION
- HEPATIC CIRCULATION
- SKELETAL MUSCLE CIRCULATION
- DERMAL CIRCULATION
- FETAL AND PLACENTAL CIRCULATIONS

KEY CONCEPTS

1. The ability of the heart to pump blood depends almost exclusively on oxygen supplied by the coronary microcirculation.
2. Brain blood flow increases when the neurons are active and require additional oxygen.
3. The regulation of intestinal blood flow during nutrient absorption depends on the elevated sodium chloride concentration in the tissue and the release of nitric oxide (NO).
4. The liver receives the portal venous blood from the gastrointestinal organs as its main blood supply, supplemented by hepatic arterial blood.
5. Skeletal muscle tissue receives minimal blood flow at rest because of its limited oxygen requirements, but flow and oxygen use can increase up to or beyond 20-fold during intense muscle activity.
6. The skin has a low oxygen requirement, but the high blood flow during warm temperatures or exercise supplies a large amount of heat for dissipation to the external environment.
7. The fetus obtains nutrients and oxygen from the mother's blood supply, using the combined maternal and fetal placental circulations.
8. The heart chambers have radically different roles in pumping blood in the fetus and adult.

This chapter discusses the anatomic and physiological properties of the vasculatures in the heart, brain, small intestine, liver, skeletal muscle, and skin. Table 17.1 presents data on blood flow and oxygen use by these different organs and tissues. The features of each vasculature, which are related to the specific functions and specialized needs of each organ or tissue, are described. The vascular anatomy and physiology of the fetus and placenta and the circulatory changes that occur at birth are also presented. The pulmonary and renal circulations are discussed in Chapters 20 and 23.

CORONARY CIRCULATION

The Work Done by the Heart Determines Its Oxygen Use and Blood Flow Requirements

The coronary circulation provides blood flow to the heart. During resting conditions, the heart muscle consumes about as much oxygen as does an equal mass of skeletal muscle during vigorous exercise (see Table 17.1). Coronary blood flow can normally increase about 4- to 5-fold, to provide more of the heart's oxygen needs, during heavy exercise. This increment in blood flow constitutes the **coronary blood flow reserve.** The ability to increase the blood flow to provide additional oxygen is imperative. Heart tissue extracts almost the maximum amount of oxygen from blood during resting conditions. Because the heart's ability to use anaerobic glycolysis to provide energy is limited, the only practical way to increase energy production is to increase blood flow and oxygen delivery. The production of lactic acid by the heart is an ominous sign of grossly inadequate oxygenation.

Cardiac Blood Flow Decreases During Systole and Increases During Diastole

Blood flow through the left ventricle decreases to a minimum when the muscle contracts because the small blood vessels

TABLE 17.1 Blood Flow and Oxygen Consumption of the Major Systemic Organs Estimated for a 70-kg Adult Man

Organ	Mass(kg)	Flow (mL/100g per min)	Total Flow (mL/min)	Oxygen Use (mL/100g per min)	Total Oxygen Use (mL/min)
Heart					
Rest	0.4–0.5	60–80	250	7.0–9.0	25–40
Exercise		200–300	1,000–1,200	25.0–40.0	65–85
Brain	1.4	50–60	750	4.0–5.0	50–60
Small intestine					
Rest	3	30–40	1,500	1.5–2.0	50–60
Absorption		45–70	2,200–2,600	2.5–3.5	80–110
Liver					
Total	1.8–2.0	100–300	1,400–1,500	13.0–14.0	180–200
Portal		70–90	1,100	5.0–7.0	
Hepatic Artery		30–40	350	5.0–7.0	
Muscle					
Rest	28	2–6	750–1,000	0.2–0.4	60
Exercise		40–100	15,000–20,000	8.0–15.0	2,400–4,200
Skin					
Rest	2.0–2.5	1–3	200–500	0.1–0.2	2–4
Exercise		5–15	1,000–2,500		

are compressed. Blood flow in the left coronary artery during cardiac systole is only 10 to 30% of that during diastole, when the heart musculature is relaxed and most of the blood flow occurs. The compression effect of systole on blood flow is minimal in the right ventricle, probably as a result of the lower pressures developed by a smaller muscle mass (Fig. 17.1). Changes in blood flow during the cardiac cycle in healthy people have no obvious deleterious effects even during maximal exercise; however, in people with compromised coronary arteries, an increased heart rate decreases the time spent in diastole, impairing coronary blood flow.

The heart musculature is perfused from the epicardial (outside) surface to the endocardial (inside) surface. Microvascular pressures are dissipated by blood flow friction as the vessels pass through the heart tissue. Therefore, the mechanical compression of systole has more effect on the blood flow through the endocardial layers where compressive forces are higher and microvascular pressures are lower. This problem occurs particularly in heart diseases of all types, and most kinds of tissue impairment affect the subendocardial layers.

Coronary Vascular Resistance Is Primarily Regulated by Responses to Heart Metabolism

Animal studies indicate that about 75% of total coronary vascular resistance occurs in vessels with inner diameters of less than about 200 μm. This observation is supported by clinical measurements in humans that show little arterial pressure dissipation in normal coronary arteries prior to their smaller branches entering the heart muscle tissue. The majority of the coronary resistance vessels—the small arteries and arterioles—are surrounded by cardiac muscle cells and are exposed to chemicals released by cardiac cells into the interstitial space. Many of these chemicals cause dilation of the coronary arterioles. For example, adenosine, derived from the breakdown of adenosine triphosphate (ATP) in cardiac cells, is a potent vasodilator, and its release increases whenever cardiac metabolism is increased or blood flow to the heart is experimentally or pathologically decreased. Blockade of the vasodilator actions of adenosine with theophylline, however, does not prevent coronary vasodilation when cardiac work is increased, blood flow is suppressed, or the arterial blood is depleted of oxygen. Therefore, while adenosine is likely an important contributor to cardiac vascular regulation, there are obviously other potent regulatory agents. Vasodilatory prostaglandins, H^+, CO_2, NO, and decreased availability of oxygen, as well as myogenic mechanisms, are capable of contributing to coronary vascular regulation. No single mechanism adequately explains the dilation of coronary arterioles and small arteries when the metabolic rate of the heart is increased, or when pathological or experimental means are used to restrict blood flow. However, the release of NO from endothelial cells—in response to blood flow-mediated dilation (see Chapter 16) and in response to ATP, adenosine diphosphate (ADP), tissue acidosis, and decreased oxygen availability—appears to be one of the most important mechanisms to produce vasodilation.

Coronary arteries and arterioles are innervated by the sympathetic nervous system and can be constricted by norepinephrine, whether released from nerves or carried in the arterial blood. The constrictor mechanism appears to be more important in equalizing blood flow through the layers of the heart than in reducing blood flow to the heart muscle in general. The coronary arteries and larger arterioles predominately have α_1-adrenergic receptors, which induce vascular constriction when activated by norepinephrine. Smaller arterioles predominately have β-adrenergic receptors, which cause vasodilation in response to epinephrine released by the adrenal medulla during sympathetic activity. In addition, epinephrine increases the

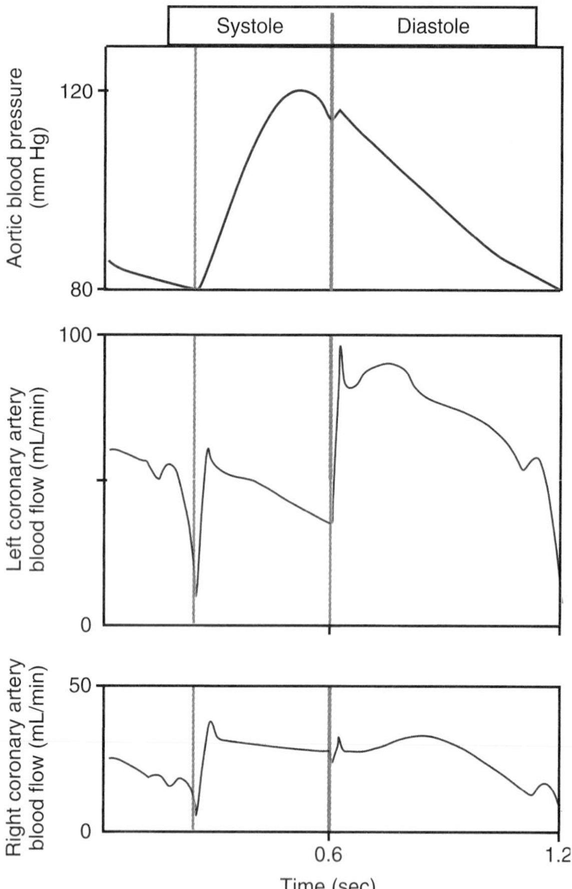

FIGURE 17.1 Aortic blood pressure and left and right coronary blood flows during the cardiac cycle. Note that left coronary artery blood flow decreases dramatically during the isovolumetric phase of systole, prior to opening of the aortic valve. Left coronary artery blood flow remains lower during systole than during diastole because of compression of the coronary blood vessels in the contracting myocardium. The left ventricle receives most of its arterial blood inflow during diastole. Right coronary artery blood flow tends to be sustained during both systole and diastole because lower intraventricular pressures are developed by the contracting right ventricle, resulting in less compression of coronary blood vessels. (Adapted from Gregg DE, Khouri EM, Rayford CR. Systemic and coronary energetics in the resting unanesthetized dog. Circ Res 1965;16:102–113; and Lowensohn HS, et al. Phasic right coronary artery blood flow in conscious dogs with normal and elevated right ventricular pressures. Circ Res 1976;39:760–766.)

metabolic rate of the heart via β_1-adrenergic receptors. This, in turn, leads to dilatory stimuli that potentially could overcome vasoconstriction.

The overall concept evolving from both human and animal studies is that the sympathetic nervous system suppresses the decrease in coronary vascular resistance during exercise despite the metabolic effects of epinephrine mentioned. The partial constriction of large coronary arterioles and most arteries by norepinephrine appears to limit the retrograde flow of blood during ventricular systole and, in doing so, prevents part of the decreased flow in the deep layers of the heart wall. In effect, the body trades a small decrease in flow, relative to what would exist without sympathetic effects on resistance vessels, for improved perfusion of the tissue at risk in the deeper layers of the heart.

Coronary Vascular Disease Limits Cardiac Blood Flow and Cardiac Work

Pathology of the coronary vasculature is the direct cause of death in about one third of the population in developed societies. Prior to death, most of these people have impaired cardiac function as a result of coronary artery disease, leading to heart failure with decreased quality of life. Progressive occlusion of coronary arteries by atherosclerotic plaques and acute occlusion as a result of the formation of blood clots in damaged coronary arteries are life-threatening because the metabolic needs of the cardiac muscle can no longer be met by the blood flow. Because the plaque or clot partially occludes the vessel lumen, vascular resistance is increased, and blood flow would decrease if smaller coronary vessels did not dilate to restore a relatively normal blood flow at rest. In doing so, the reserve for dilation of these vessels is compromised. While this usually has no effect at rest, when cardiac metabolism is increased, the decreased ability to increase blood flow can limit cardiac performance. In many cases, inadequate blood flow is first noticed as chest pain—known as **angina pectoris**—originating from the heart, and a feeling of shortness of breath during exercise or work. The vascular occlusion can cause conditions ranging from impaired contractile ability of the cardiac muscle, which limits cardiac output and tolerance to everyday work and exercise, to death of the muscle tissue, a **cardiac infarct.**

If the coronary occlusion is not severe, medication can be used to cause coronary vasodilation or decreased cardiac work, or both. If the arterial pressure is higher than normal, various approaches are used to lower the blood pressure, decreasing the heart's workload and oxygen needs. In addition to pharmacological treatment, mild to moderate exercise, depending on the status of the coronary disease, is often advised. Aerobic exercise stimulates the development of collateral vessels in the heart, improves the overall performance of the cardiovascular system, and increases the efficiency of the body during work and daily activities. This latter effect lowers the cardiac output needed for a given task, thereby decreasing the heart's metabolic energy requirement.

Significant changes in lifestyle—including strictly limiting dietary fat (especially saturated fat), strenuous and prolonged daily exercise, and reduced mental stress—have been shown to greatly slow and even slightly reverse coronary atherosclerosis. The goal is to lower blood levels of low-density lipoproteins (LDLs), which are known to accelerate the formation of cholesterol-containing arterial plaques. The LDL concentration should typically be lowered below 120 mg/dL, but some cardiologists favor lowering levels below 100 mg/dL. For most people, reductions in LDL below 120 mg/dL are not attainable with diet and exercise. In those persons, drugs, known as **statins**, which block the formation of cholesterol in the liver, appear to be highly effective in decreasing the risk and severity of coronary artery disease. Simultaneous treatment with an aerobic exercise program and large amounts of niacin, to increase high-density lipoproteins (HDLs), may help the body remove cholesterol for processing in the liver. (See Clinical Focus Box 17.1).

Collateral Vessels Interconnect Sections of the Cardiac Microvasculature

One of the likely contributing factors to compensate for slowly developing coronary vascular disease is the enlargement of **collateral blood vessels** between the left and right coronary arterial systems or among parts of each system. In the healthy heart of a sedentary person, collateral arterial vessels are rare, but arteriolar collaterals (internal diameter, <100 μm) do occur in small numbers. The expansion of existing collateral vessels and the limited formation of new collaterals provide a partial bypass for blood flow to areas of muscle whose primary supply vessels are impaired. Subendocardial arteriolar collaterals usually enlarge more than epicardial collaterals. In part, the greater collateral enlargement in the endocardium compared to the epicardium may be due to the lower pressure and blood flow in the endocardial vessels.

The exact mechanism responsible for the development of collateral vessels is unknown. However, periods of inadequate blood flow to the heart muscle caused by experimental flow reduction do stimulate collateral enlargement in healthy animals. It is assumed that in humans with coronary vascular disease who develop functional collateral vessels, the mechanism is related to occasional or even sustained periods of inadequate blood flow. Whether or not routine exercise aids in the development of collaterals in healthy humans is debatable; the benefits of exercise may be by other mechanisms, such as enlargement of the primary perfusion vessels and the reduction of atherosclerosis. However, there is no doubt that frequent and relatively intense aerobic exercise is beneficial to cardiac vascular function.

CEREBRAL CIRCULATION

The ultimate organ of life is the brain. Even the determination of death often depends upon whether or not the brain is viable. The most common cause of brain injury is some form of impaired brain blood flow. Such problems can develop as a result of accidents to arteries in the neck or brain, occlusion of vessels secondary to atherosclerotic processes, and, surprisingly frequently, **aneurysms** that occur as a result of vessel wall tearing. Fortunately, treatment of these problems is constantly improving.

Brain Blood Flow Is Virtually Constant Despite Changes in Arterial Blood Pressure

The cerebral circulation shares many of the physiological characteristics of the coronary circulation. The heart and brain have a high metabolic rate (see Table 17.1), extract a large amount of oxygen from blood, and have a limited ability to use anaerobic glycolysis for metabolism. Their vessels have a limited ability to constrict in response to sympathetic nerve stimulation. As described in Chapter 16, the brain and coronary vasculatures have an excellent ability to *autoregulate* blood flow at arterial pressures from about 50 to 60 mm Hg to about 150 to 160 mm Hg. The vasculature of the brainstem exhibits the most precise autoregulation, with good but less precise regulation of blood flow in the cerebral cortex. This regional variation in autoregulatory ability has clinical implications because the region of the brain most likely to suffer at low arterial pressure is the cortex, where consciousness will be lost long before the automatic cardiovascular and ventilatory regulatory functions of the brainstem are compromised.

A variety of mechanisms are responsible for cerebral vascular autoregulation. The identification of a specific chemical that causes cerebral autoregulation has not been possible. For example, when blood flow is normal, regardless of the arterial blood pressure, little extra adenosine, K^+, H^+, or other

CLINICAL FOCUS BOX 17.1

Coronary Vascular Disease

Approximately 45% of the adult population in the United States will, at some time during their lifetimes, require medical or surgical intervention because of atherosclerosis of the coronary arteries. The typical circumstance is rupture of the endothelial layer over an atherosclerotic plaque, followed by a clot that occludes or nearly occludes a coronary artery. About 10% of these incidents result in death before the patient reaches the hospital. For those who reach a coronary care facility, about 70% will be alive 1 year later, and about 50% will be alive in 5 years. If the patient does not have a risk of bleeding, the clot can be dissolved by administering tissue plasminogen activator or streptokinase. If the blood flow is quickly restored within a few hours, the damage to the heart muscle can be minimal. In some cases, advancing a catheter into the blocked artery to expand the vessel and remove the clot is the best approach. In a few cases, emergency replacement of the blocked artery is required; this is a much more invasive surgery and often requires several months of recovery.

Despite the multiple treatments available to deal with existing coronary artery blockage, the ideal treatment is to avoid the problem. Excessive intake of cholesterol-rich food, sedentary lifestyles that tend to raise low-density lipoproteins (LDLs) and lower high-density lipoproteins (HDLs), and obesity leading to insulin resistance are key problems leading to accelerated coronary heart disease. Two of the three can be addressed with a lowered cholesterol and calorie-restricted diet to promote loss of body fat. Aerobic exercise of any type for approximately 30 minutes, 3 days a week, has consistently been shown to lower LDL and raise HDL, as well as aid in body fat loss. Pharmacological blockade of cholesterol synthesis in the liver with the statin family of compounds is effective to both prevent second heart attacks and lower the risk of a first heart attack. These drugs are so effective that in the near future, most persons older than age 50 may be advised to follow a dietary and exercise plan complemented with statin therapy.

vasodilator metabolites are released, and brain tissue P_{O2} remains relatively constant. However, increasing concentrations of any of these chemicals causes vasodilation and increased blood flow. The brain vasculature does exhibit myogenic vascular responses and may use this mechanism as a major contributor to autoregulation. Animal studies indicate that both the cerebral arteries and cerebral arterioles are involved in cerebral vascular autoregulation and other types of vascular responses. In fact, the arteries can change their resistance almost proportionately to the arterioles during autoregulation. This may occur in part because cerebral arteries exhibit myogenic vascular responses and because they are partially to fully embedded in the brain tissues and would likely be influenced by the same vasoactive chemicals in the interstitial space as affect the arterioles.

Brain Microvessels Are Sensitive to CO_2 and H^+

The cerebral vasculature dilates in response to increased CO_2 and H^+ and constricts if either substance is decreased. Both of these substances are formed when cerebral metabolism is increased by nerve action potentials, such as during normal brain activation. In addition, interstitial K^+ is elevated when a large number of action potentials are fired. The cause of dilation in response to both K^+ and CO_2 involves the formation of nitric oxide (NO). However, the mechanism is not necessarily the typical endothelial formation of NO. The source of NO appears to be from nitric oxide synthase in neurons, as well as endothelial cells. The H^+ formed by the interaction of carbon dioxide and water or from acids formed by metabolism does not appear to cause dilation through a NO-dependent mechanism, but additional data are needed on this topic.

Reactions of cerebral blood flow to chemicals released by increased brain activity, such as CO_2, H^+, and K^+, are part of the overall process of matching the brain's metabolic needs to the blood supply of nutrients and oxygen. The 10 to 30% increase in blood flow in brain areas excited by peripheral nerve stimulation, mental activity, or visual activity may be related to these three substances released from active nerve cells. The cerebral vasculature also dilates when the oxygen content of arterial blood is reduced, but the vasodilatory effect of elevated CO_2 is much more powerful.

Cerebral Blood Flow Is Insensitive to Hormones and Sympathetic Nerve Activity

Circulating vasoconstrictor and vasodilator hormones and the release of norepinephrine by sympathetic nerve terminals on cerebral blood vessels do not play much of a role in moment-to-moment regulation of cerebral blood flow. The blood-brain barrier effectively prevents constrictor and dilator agents in blood plasma from reaching the vascular smooth muscle. Though the cerebral arteries and arterioles are fully innervated by sympathetic nerves, stimulation of these nerves produces only mild vasoconstriction in the majority of cerebral vessels. If, however, sympathetic activity to the cerebral vasculature is permanently interrupted, the cerebral vasculature has a decreased ability to autoregulate blood flow at high arterial pressures, and the integrity of the blood-brain barrier is more easily disrupted. Therefore, some aspect of sympathetic nerve activity other than the routine regulation of vascular resistance is important for the maintenance of normal cerebral vascular function. This may occur because of a trophic factor that promotes the health of endothelial and smooth muscle cells in the cerebral microvessels.

The Cerebral Vasculature Adapts to Chronic High Blood Pressure

In conditions of chronic hypertension, cerebral vascular resistance increases, thereby allowing cerebral blood flow and, presumably, capillary pressures to be normal. The adaptation of cerebral vessels to sustained hypertension lets them maintain vasoconstriction at arterial pressures that would overcome the contractile ability of a normal vasculature (Fig. 17.2).

The mechanisms that enable the cerebral vasculature to adjust the autoregulatory range upward appear to be hypertrophy of the vascular smooth muscle and a mechanical constraint to vasodilation, as a result of more muscle tissue,

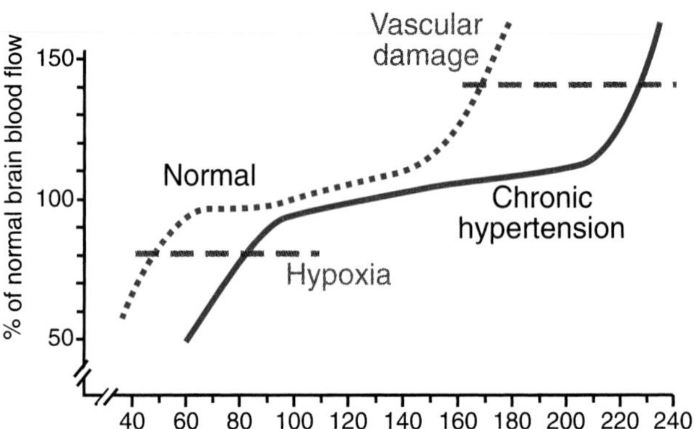

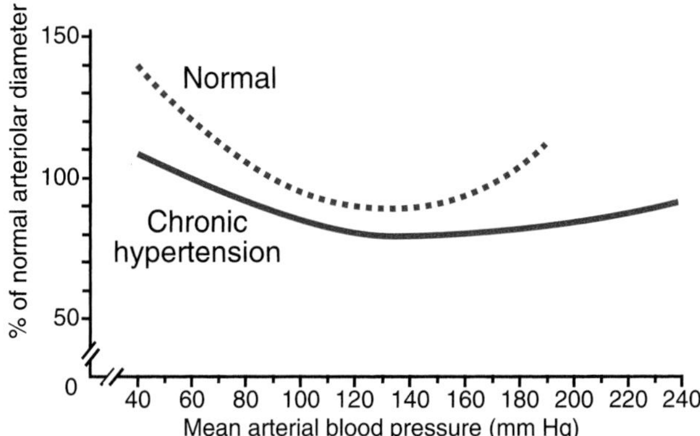

FIGURE 17.2 **Chronic hypertension.** This condition is associated with a rightward shift in the arterial pressure range over which autoregulation of cerebral blood flow occurs (upper panel) because, for any given arterial pressure, resistance vessels of the brain have smaller-than-normal diameters (lower panel). As a consequence, people with hypertension can tolerate high arterial pressures that would cause vascular damage in healthy people. However, they risk reduced blood flow and brain hypoxia at low arterial pressures that are easily tolerated by healthy people.

or more connective tissue, or both. The drawback to such adaptation is partial loss of the ability to dilate and regulate blood flow at low arterial pressures. This loss occurs because the passive structural properties of the resistance vessels restrict the vessel diameter at subnormal pressures and, in doing so, increase resistance. In fact, the lower pressure limit of constant blood flow (autoregulation) can be almost as high as the normal mean arterial pressure (see Fig. 17.2). This can be problematic if the arterial blood pressure is rapidly lowered to normal in a person whose vasculature has adapted to hypertension. The person may faint from inadequate brain blood flow, even though the arterial pressure is in the normal range. Fortunately, a gradual reduction in arterial pressure over weeks or months returns autoregulation to a more normal pressure range.

Cerebral Edema Impairs Blood Flow to the Brain

The brain is encased in a rigid bony case, the cranium. As such, should the brain begin to swell, the intracranial pressure will dramatically increase. There are many causes of **cerebral edema**—an excessive accumulation of fluid in the brain substance—including infection, tumors, trauma to the head that causes massive arteriolar dilation, and bleeding into the brain tissue after a stroke or trauma. In each case, the following approximate scenario occurs. As the intracranial pressure increases, the venules and veins are partially collapsed because their intravascular pressure is low. As these outflow vessels collapse, their resistance increases and capillary pressure rises (see Chapter 16). The increased capillary pressure favors increased filtration of fluid into the brain to further raise the intracranial pressure. The end result is a positive feedback system in which intracranial pressure will become so high as to begin to compress small arterioles and decrease blood flow.

Excessive intracranial pressure is a major clinical problem. Hypertonic mannitol can be given to promote water loss from swollen brain cells. Sometimes opening of the skull and drainage of cerebrospinal fluid or hemorrhaged blood, if any, may be necessary. Hemorrhaged blood is particularly a problem because clotted blood contains denatured hemoglobin that destroys nitric oxide. This in turn leads to inappropriate vasoconstriction of the arterioles in the area of the hemorrhage.

If blood flow to the pons and medulla of the brain is decreased, tissue hypoxia will activate the sympathetic nervous system control centers. This response—called **Cushing's reflex**—raises the arterial blood pressure, often dramatically. This can be viewed as an attempt to raise cerebral blood flow. While blood flow may improve, microvascular pressures are elevated, which worsens cerebral edema.

SMALL INTESTINE CIRCULATION

The small intestine completes the digestion of food and then absorbs the nutrients to sustain the remainder of the body. At rest, the intestine receives about 20% of the cardiac output and uses about 20% of the body's oxygen consumption. Both of these numbers nearly double after a large meal. Unless the blood flow can increase, food digestion and absorption simply do not occur. For example, if intense exercise is required in the midst of digesting a meal, blood flow through the small intestine can be reduced to half of normal by the sympathetic nervous system with no ill effects, other than delayed food absorption. Once the stress imposed on the body is over, intestinal blood flow again increases and the process of digesting and absorbing food resumes.

The Three Regions of the Intestinal Wall Are Supplied From a Common Set of Large Arterioles

Small arteries and veins penetrate the muscular wall of the bowel and form a microvascular distribution system in the submucosa (Fig. 17.3). The muscle layers receive small arterioles from the **submucosal vascular plexus**; other small arterioles continue into individual vessels of the deep submucosa around glands and to the villi of the mucosa. Small arteries and larger arterioles preceding the separate muscle and submucosal-mucosal vasculatures control about 70% of the intestinal vascular resistance. The small arterioles of the muscle, submucosal, and mucosal layers can partially adjust blood flow to meet the needs of small areas of tissue.

Compared with other major organ vasculatures, the circulation of the small intestine has a poorly developed autoregulatory response to locally decreased arterial pressure, and as a result, blood flow usually declines because resistance does not adequately decrease. However, elevation of venous pressure outside the intestine causes sustained myogenic constriction; in this regard, the intestinal circulation equals or exceeds similar regulation in other organ systems. Intestinal motility has little effect on the overall intestinal blood flow, probably because the increases in metabolic rate are so small. In contrast, the intestinal blood flow increases in approximate proportion to the elevated metabolic rate during food absorption.

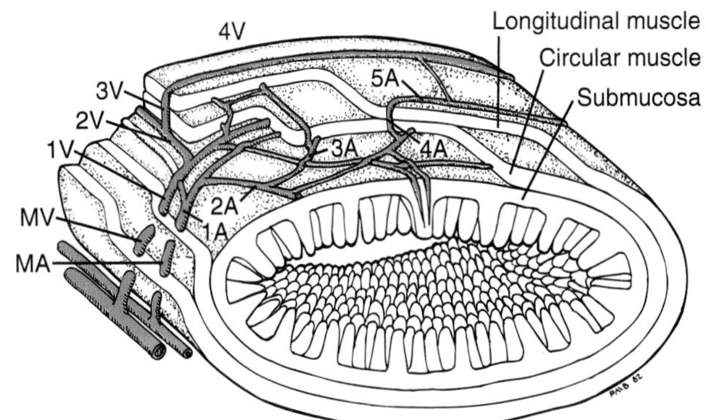

FIGURE 17.3 **The vasculature of the small intestine.** The intestinal vasculature is unusual because three very different tissues—the muscle layers, submucosa, and mucosal layer—are served by branches from a common vasculature located in the submucosa. Most of the intestinal vascular resistance is regulated by small arteries and arterioles preceding the separate muscle and submucosal and mucosal vasculatures. MA, muscular arteriole; 1A to 5A, successive branches of the arterioles; 1V to 4V, successive branches of the venules; MV, muscular venule. (Modified from Connors B. Quantification of the architectural changes observed in intestinal arterioles from diabetic rats. Ph.D. Dissertation, Indiana University, 1993.)

The Microvasculature of Intestinal Villi Has a High Blood Flow and Unusual Exchange Properties

The intestinal mucosa receives about 60 to 70% of the total intestinal blood flow. Blood flows of 70 to 100 mL/min per 100 g in this specialized tissue are probable and much higher than the average blood flow for the total intestinal wall (see Table 17.1). This blood flow can exceed the resting blood flow in the heart and brain.

The mucosa is composed of individual projections of tissue called **villi.** The interstitial space of the villi is mildly hyperosmotic (~400 mosm/kg H_2O) at rest as a result of NaCl. During food absorption, the interstitial osmolality increases to 600 to 800 mosm/kg H_2O near the villus tip, compared with 400 mosm/kg H_2O near the villus base. The primary cause of high osmolalities in the villi appears to be greater absorption than removal of NaCl and nutrient molecules. There is also a possible countercurrent exchange process in which materials absorbed into the capillary blood diffuse from the venules into the incoming blood in the arterioles.

Food Absorption Requires a High Blood Flow to Support the Metabolism of the Mucosal Epithelium

Lipid absorption causes a greater increase in intestinal blood flow, a condition known as **absorptive hyperemia,** and oxygen consumption than either carbohydrate or amino acid absorption. During absorption of all three classes of nutrients, the mucosa releases adenosine and CO_2 and oxygen is depleted. The hyperosmotic lymph and venous blood that leave the villus to enter the submucosal tissues around the major resistance vessels are also major contributors to absorptive hyperemia. By an unknown mechanism, hyperosmolality resulting from NaCl induces endothelial cells to release NO and dilate the major resistance arterioles in the submucosa. Hyperosmolality resulting from large organic molecules that do not enter endothelial cells does not cause appreciable increases in NO formation, producing much less of an increase in blood flow than equivalent hyperosmolality resulting from NaCl. These observations suggest that NaCl entering the endothelial cells is essential to induce NO formation.

The active absorption of amino acids and carbohydrates and the metabolic processing of lipids into chylomicrons by mucosal epithelial cells place a major burden on the microvasculature of the small intestine. There is an extensive network of capillaries just below the villus epithelial cells that contacts these cells. The villus capillaries are unusual in that portions of the cytoplasm are missing, so that the two opposing surfaces of the endothelial cell membranes appear to be fused. These areas of fusion, or **closed fenestrae,** are thought to facilitate the uptake of absorbed materials by capillaries. In addition, intestinal capillaries have a higher filtration coefficient than other major organ systems, which probably enhances the uptake of water absorbed by the villi (see Chapter 16). However, large molecules, such as plasma proteins, do not easily cross the fenestrated areas because the reflection coefficient for the intestinal vasculature is greater than 0.9, about the same as in skeletal muscle and the heart.

Low Capillary Pressures in Intestinal Villi Aid in Water Absorption

Although the mucosal layer of the small intestine has a high blood flow both at rest and during food absorption, the capillary blood pressure is usually 13 to 18 mm Hg and seldom higher than 20 mm Hg during food absorption. Therefore, plasma colloid osmotic pressure is higher than capillary blood pressure, favoring the absorption of water brought into the villi. During lipid absorption, the plasma protein reflection coefficient for the overall intestinal vasculature is decreased from a normal value of more than 0.9 to about 0.7. It is assumed that most of the decrease in reflection coefficient occurs in the mucosal capillaries. This lowers the ability of plasma proteins to counteract capillary filtration, with the net result that fluid is added to the interstitial space. Eventually, this fluid must be removed. Not surprisingly, the highest rates of intestinal lymph formation normally occur during fat absorption.

Sympathetic Nerve Activity Can Greatly Decrease Intestinal Blood Flow and Venous Volume

The intestinal vasculature is richly innervated by sympathetic nerve fibers. Major reductions in gastrointestinal blood flow and venous volume occur whenever sympathetic nerve activity is increased, such as during strenuous exercise or periods of pathologically low arterial blood pressure. Venoconstriction in the intestine during hemorrhage helps to mobilize blood and compensates for the blood loss. Gastrointestinal blood flow is about 25% of the cardiac output at rest; a reduction in this blood flow, by heightened sympathetic activity, allows more vital functions to be supported with the available cardiac output. However, gastrointestinal blood flow can be so drastically decreased by a combination of low arterial blood pressure (**hypotension**) and sympathetically mediated vasoconstriction that mucosal tissue damage can result.

HEPATIC CIRCULATION

The hepatic circulation perfuses one of the largest organs in the body, the liver. The liver is primarily an organ that maintains the organic chemical composition of the blood plasma. For example, all plasma proteins are produced by the liver, and the liver adds glucose from stored glycogen to the blood. The liver also removes damaged blood cells and bacteria and detoxifies many man-made or natural organic chemicals that have entered the body.

The Hepatic Circulation Is Perfused by Venous Blood From Gastrointestinal Organs and a Separate Arterial Supply

The human liver has a large blood flow, about 1.5 L/min or 25% of the resting cardiac output. It is perfused by both arterial blood through the **hepatic artery** and venous blood that has passed through the stomach, small intestine, pancreas, spleen, and portions of the large intestine.

The venous blood arrives via the **hepatic portal vein** and accounts for about 67 to 80% of the total liver blood flow (see Table 17.1). The remaining 20 to 33% of the total flow is through the hepatic artery. The majority of blood flow to the liver is determined by the flow through the stomach and small intestine.

About half of the oxygen used by the liver is derived from venous blood, even though the splanchnic organs have removed one third to one half of the available oxygen. The hepatic arterial circulation provides additional oxygen. The liver tissue efficiently extracts oxygen from the blood. The liver has a high metabolic rate and is a large organ; consequently, it has the largest oxygen consumption of all organs in a resting person. The metabolic functions of the liver are discussed in Chapter 28.

The Liver Acinus Is a Complex Microvascular Unit With Mixed Arteriolar and Venular Blood Flow

The liver vasculature is arranged into subunits that allow the arterial and portal blood to mix and provide nutrition for the liver cells. Each subunit, called an **acinus,** is about 300 to 350 μm long and wide. In humans, usually three acini occur together. The core of each acinus is supplied by a single **terminal portal venule; sinusoidal capillaries** originate from this venule (Fig. 17.4). The endothelial cells of the capillaries have fenestrated regions with discrete openings that facilitate exchange between the plasma and interstitial spaces. The capillaries do not have a basement membrane, which partially contributes to their high permeability.

The **terminal hepatic arteriole** to each acinus is paired with the terminal portal venule at the acinus core, and blood from the arteriole and blood from the venule jointly perfuse the capillaries. The intermixing of the arterial and portal blood tends to be intermittent because the vascular smooth muscle of the small arteriole alternately constricts and relaxes. This prevents arteriolar pressure from causing a sustained reversed flow in the sinusoidal capillaries, where pressures are 7 to 10 mm Hg. The best evidence is that hepatic artery and portal venous blood first mix at the level of the capillaries in each acinus. The sinusoidal capillaries are drained by the **terminal hepatic venules** at the outer margins of each acinus; usually at least two hepatic venules drain each acinus.

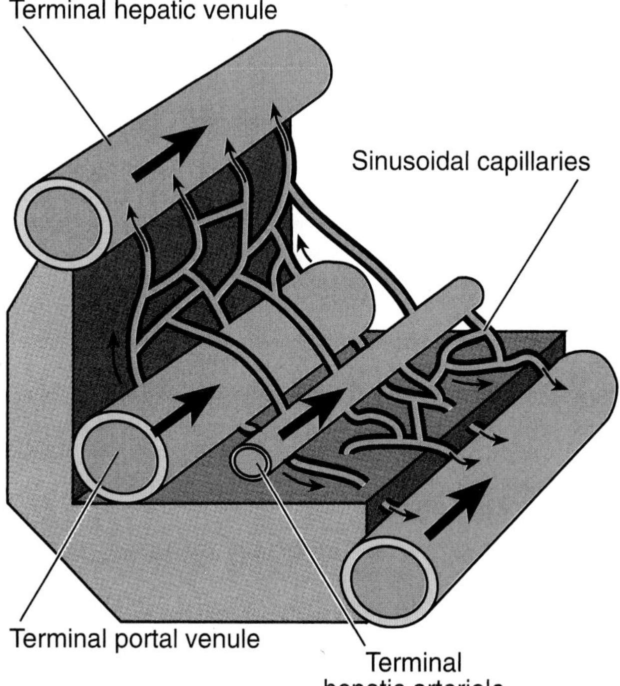

FIGURE 17.4 **Liver acinus microvascular anatomy.** A single liver acinus, the basic subunit of liver structure, is supplied by a terminal portal venule and a terminal hepatic arteriole. The mixture of portal venous and arterial blood occurs in the sinusoidal capillaries formed from the terminal portal venule. Usually two terminal hepatic venules drain the sinusoidal capillaries at the external margins of each acinus.

The Regulation of Hepatic Arterial and Portal Venous Blood Flows Requires an Interactive Control System

The regulation of portal venous and hepatic arterial blood flows is an interactive process: Hepatic arterial flow increases and decreases reciprocally with the portal venous blood flow. This mechanism, known as the **hepatic arterial buffer response,** can compensate or buffer about 25% of the decrease or increase in portal blood flow. Exactly how this is accomplished is still under investigation, but vasodilatory metabolite accumulation, possibly adenosine, during decreased portal flow, as well as increased metabolite removal during elevated portal flow, are thought to influence the resistance of the hepatic arterioles.

One might suspect that during digestion, when gastrointestinal blood flow and, therefore, portal venous blood flow are increased, the gastrointestinal hormones in portal venous blood would influence hepatic vascular resistance. However, at concentrations in portal venous blood equivalent to those during digestion, none of the major hormones appears to influence hepatic blood flow. Therefore, the increased hepatic blood flow during digestion would appear to be determined primarily by vascular responses of the gastrointestinal vasculatures.

The vascular resistances of the hepatic arterial and portal venous vasculatures are increased during sympathetic nerve activation, and the buffer mechanism is suppressed. When the sympathetic nervous system is activated, about half the blood volume of the liver can be expelled into the general circulation. Because up to 15% of the total blood volume is in the liver, constriction of the hepatic vasculature can significantly increase the circulating blood volume during times of cardiovascular stress.

SKELETAL MUSCLE CIRCULATION

The circulation of skeletal muscle involves the largest mass of tissue in the body: 30 to 40% of an adult's body weight. At rest, the skeletal muscle vasculature accounts for about 25% of systemic vascular resistance, even though individual muscles receive a low blood flow of about 2 to 6 mL/min per 100 g. The dominant mechanism controlling skeletal muscle resistance at rest is the sympathetic nervous system. Resting skeletal muscle has remarkably low oxygen consumption per 100 g of tissue, but its large mass makes its metabolic rate a major contributor to the total oxygen consumption in a resting person.

Skeletal Muscle Blood Flow and Metabolism Can Vary Over a Large Range

Skeletal muscle blood flow can increase 10- to 20-fold or more during the maximal vasodilation associated with high-performance aerobic exercise. Comparable increases in metabolic rate occur. Under such circumstances, total muscle blood flow may be equal to three or more times the resting cardiac output; obviously, cardiac output must increase during exercise to maintain the normal to increased arterial pressure (see Chapter 30).

With severe hemorrhage, which activates baroreceptor-induced reflexes, skeletal muscle vascular resistance can easily double as a result of increased sympathetic nerve activity, reducing blood flow. Skeletal muscle cells can survive long periods with minimal oxygen supply; consequently, low blood flow is not a problem. The increased vascular resistance helps preserve arterial blood pressure when cardiac output is compromised. In addition, contraction of the skeletal muscle venules and veins forces blood in these vessels to enter the general circulation and helps restore a depleted blood volume. In effect, the skeletal muscle vasculature can either place major demands on the cardiopulmonary system during exercise or perform as if expendable during a cardiovascular crisis, enabling absolutely essential tissues to be perfused with the available cardiac output.

The Regulation of Muscle Blood Flow Depends on Many Mechanisms to Provide Oxygen for Muscular Contractions

As discussed in Chapter 16, many potential local regulatory mechanisms adjust blood flow to the metabolic needs of the tissues. In fast-twitch muscles, which primarily depend on anaerobic metabolism, the accumulation of hydrogen ions from lactic acid is potentially a major contributor to the vasodilation that occurs. In slow-twitch skeletal muscles, which can easily increase oxidative metabolic requirements by more than 10 to 20 times during heavy exercise, it is not hard to imagine that whatever causes metabolically linked vasodilation is in ample supply at high metabolic rates.

During rhythmic muscle contractions, the blood flow during the relaxation phase can be high, and it is unlikely that the muscle becomes significantly hypoxic during submaximal aerobic exercise. Studies in humans and animals indicate that lactic acid formation, an indication of hypoxia and anaerobic metabolism, is present only during the first several minutes of submaximal exercise. Once the vasodilation and increased blood flow associated with exercise are established, after 1 to 2 minutes, the microvasculature is probably capable of maintaining ample oxygen for most workloads, perhaps up to 75 to 80% of maximum performance because remarkably little additional lactic acid accumulates in the blood. While the tissue oxygen content likely decreases as exercise intensity increases, the reduction does not compromise the high aerobic metabolic rate except with the most demanding forms of exercise. The changes in oxygen tensions before, during, and after a period of muscle contractions in an animal model were illustrated in Figure 16.7.

To ensure the best possible supply of nutrients, particularly oxygen, even mild exercise causes sufficient vasodilation to perfuse virtually all of the capillaries, rather than just 25 to 50% of them, as occurs at rest. However, near-maximum or maximum exercise exhausts the ability of the microvasculature to meet tissue oxygen needs and hypoxic conditions rapidly develop, limiting the performance of the muscles. The burning sensation and muscle fatigue during maximum exercise or at any time muscle blood flow is inadequate to provide adequate oxygen is partially a consequence of hypoxia. This type of burning sensation is particularly evident when a muscle must hold a weight in a steady position. In this situation, the contraction of the muscle compresses the microvessels, stopping the blood flow and, with it, the availability of oxygen.

The vasodilation associated with exercise is dependent upon NO. However, exactly which chemicals released or consumed by skeletal muscle induce the increased release of NO from endothelial cells is unknown. In addition, skeletal muscle cells can make NO and, although not yet tested, may produce a substantial fraction of the NO that causes the dilation of the arterioles. If endothelial production of NO is curtailed by the inhibition of endothelial nitric oxide synthase, the increased muscle blood flow during contractions is strongly suppressed. However, there is concern that the resting vasoconstriction caused by suppressed NO formation diminishes the ability of the vasculature to dilate in response a variety of mechanisms. Flow-mediated vasodilation, for example, appears to be used to dilate smaller arteries and larger arterioles to maximize the increase in blood flow initiated by the dilation of smaller arterioles in contact with active skeletal muscle cells. Studies in animals indicate these vessels make a major contribution to vascular regulation in skeletal muscle and must be participants in any significant increase in blood flow.

DERMAL CIRCULATION

The Skin Has a Microvascular Anatomy to Support Tissue Metabolism and Heat Dissipation

The structure of the skin vasculature differs according to location in the body. In all areas, an arcade of arterioles exists at the boundary of the dermis and the subcutaneous tissue over fatty tissues and skeletal muscles (Fig. 17.5). From this arteriolar arcade, arterioles ascend through the dermis into the superficial layers of the dermis, adjacent to the epidermal layers. These arterioles form a second network in the superficial dermal tissue and perfuse the extensive capillary loops that extend upward into the dermal papillae just beneath the epidermis.

The dermal vasculature also provides the vessels that surround hair follicles, sebaceous glands, and sweat glands.

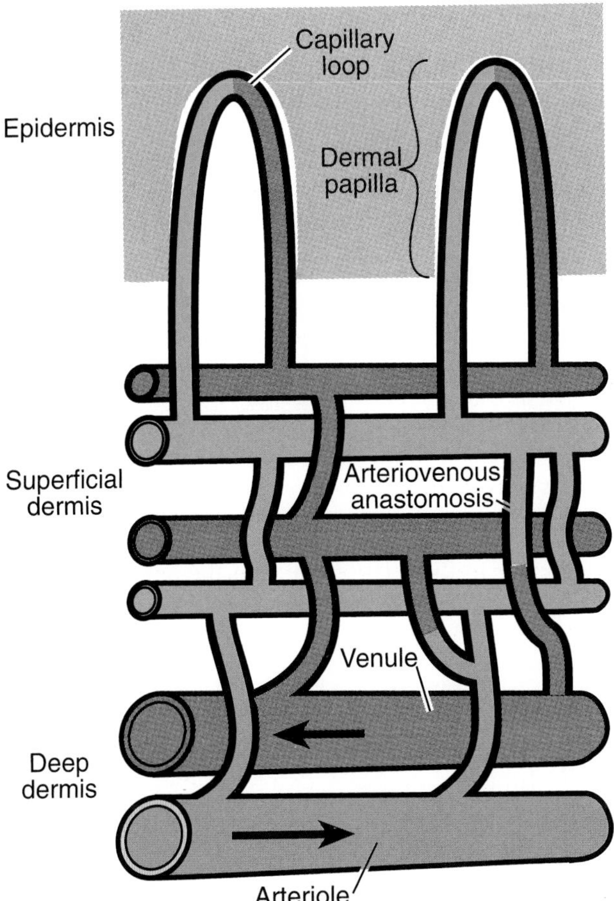

FIGURE 17.5 **The vasculature of the skin.** The skin vasculature is composed of a network of large arterioles and venules in the deep dermis, which send branches to the superficial network of smaller arterioles and venules. Arteriovenous anastomoses allow direct flow from arterioles to venules and greatly increase blood flow when dilated. The capillary loops into the dermal papillae beneath the epidermis are supplied and drained by microvessels of the superficial dermal vasculature.

Sweat glands derive virtually all sweat water from blood plasma and are surrounded by a dense capillary network in the deeper layers of the dermis. As explained in Chapter 29, neural regulation of the sweating mechanism not only causes the formation of sweat but also substantially increases skin blood flow. All the capillaries from the superficial skin layers are drained by venules, which form a venous plexus in the superficial dermis and eventually drain into many large venules and small veins beneath the dermis.

The vascular pattern just described is modified in the tissues of the hand, feet, ears, nose, and some areas of the face in that direct vascular connections between arterioles and venules, known as **arteriovenous anastomoses,** occur primarily in the superficial dermal tissues (see Fig. 17.5). By contrast, relatively few arteriovenous anastomoses exist in the major portion of human skin over the limbs and torso. If a great amount of heat must be dissipated, dilation of the arteriovenous anastomoses allows substantially increased skin blood flow to warm the skin, thereby increasing heat loss to the environment. This allows vasculatures of the hands and feet and, to a lesser extent, the face, neck, and ears to lose heat efficiently in a warm environment.

Skin Blood Flow Is Important in Body Temperature Regulation

The skin is a large organ, representing 10 to 15% of total body mass. The primary functions of the skin are protection of the body from the external environment and dissipation or conservation of heat during body temperature regulation.

The skin has one of the lowest metabolic rates in the body and requires relatively little blood flow for purely nutritive functions. Consequently, despite its large mass, its resting metabolism does not place a major flow demand on the cardiovascular system. However, in warm climates, body temperature regulation requires that warm blood from the body core be carried to the external surface, where heat transfer to the environment can occur. Therefore, at typical indoor temperatures and during warm weather, skin blood flow is usually far in excess of the need for tissue nutrition. The reddish color of the skin during exercise in a warm environment reflects the large blood flow and dilation of skin arterioles and venules (see Table 17.1).

The increase in the skin's blood flow probably occurs through two main mechanisms. First, an increase in body core temperature causes a reflex increase in the activity of sympathetic cholinergic nerves, which release acetylcholine. Acetylcholine release near sweat glands leads to the breakdown of a plasma protein (kininogen) to form bradykinin, a potent dilator of skin blood vessels, which increases the release of NO as a major component of the dilatory mechanism. Second, simply increasing skin temperature will cause the blood vessels to dilate. This can result from heat applied to the skin from the external environment, heat from underlying active skeletal muscle, or increased blood temperature as it enters the skin.

Total skin blood flows of 5 to 8 L/min have been estimated in humans during vigorous exercise in a hot environment. During mild to moderate exercise in a warm environment, skin blood flow can equal or exceed blood flow to the skeletal muscles. Exercise tolerance can, therefore, be lower in a warm environment because the vascular resistance of the skin and muscle is too low to maintain an appropriate arterial blood pressure, even at maximum cardiac output. One of the adaptations to exercise is an ability to increase blood flow in skin and dissipate more heat. In addition, aerobically trained humans are capable of higher sweat production rates; this increases heat loss and induces greater vasodilation of the skin arterioles.

The vast majority of humans live in cool to cold regions, where body heat conservation is imperative. The sensation of cool or cold skin, or a lowered body core temperature, elicits a reflex increase in sympathetic nerve activity, which causes vasoconstriction of blood vessels in the skin. Heat loss is minimized because the skin becomes a poorly perfused insulator, rather than a heat dissipator. As long as the skin temperature is higher than about 10 to 13°C (50 to 55°F), the neurally induced vasoconstriction is sustained. However, at lower tissue temperatures, the vascular smooth muscle cells progressively lose their contractile ability, and the vessels passively dilate to various extents. The reddish color of the hands, face, and ears on a cold day demonstrates increased blood flow and vasodilation as a result of low temperatures. To some extent, this cold-mediated vasodilation is useful because it lessens the chance of cold injury to exposed skin. However, if this process included most of the body surface, such as occurs when the body is submerged in cold water or inadequate clothing is worn, heat loss would be rapid and hypothermia would result. (Chapter 29 discusses skin blood flow and temperature regulation.)

FETAL AND PLACENTAL CIRCULATIONS

The Placenta Has Maternal and Fetal Circulations That Allow Exchange Between the Mother and Fetus

The development of a human fetus depends on nutrient, gas, water, and waste exchange in the maternal and fetal portions of the placenta. The human **fetal placenta** is supplied by two **umbilical arteries**, which branch from the internal iliac arteries, and is drained by a single **umbilical vein** (Fig. 17.6). The umbilical vein of the fetus returns oxygen and nutrients from the mother's body to the fetal cardiovascular system, and the umbilical arteries bring in blood laden with carbon dioxide and waste products to be transferred to the mother's blood. Although many liters of oxygen and carbon dioxide, together with hundreds of grams of nutrients and wastes, are exchanged between the mother and fetus each day, the exchange of red blood cells or white blood cells is a rare event. This large chemical exchange without cellular exchange is possible because the fetal and maternal blood are kept completely separate, or nearly so.

The fundamental anatomic and physiological structure for exchange is the **placental villus**. As the umbilical arteries enter the fetal placenta, they divide into many branches that penetrate the placenta toward the maternal system. These small arteries divide in a pattern similar to a fir tree, the placental villi being the small branches. The fetal capillaries bring in the fetal blood from the umbilical arteries

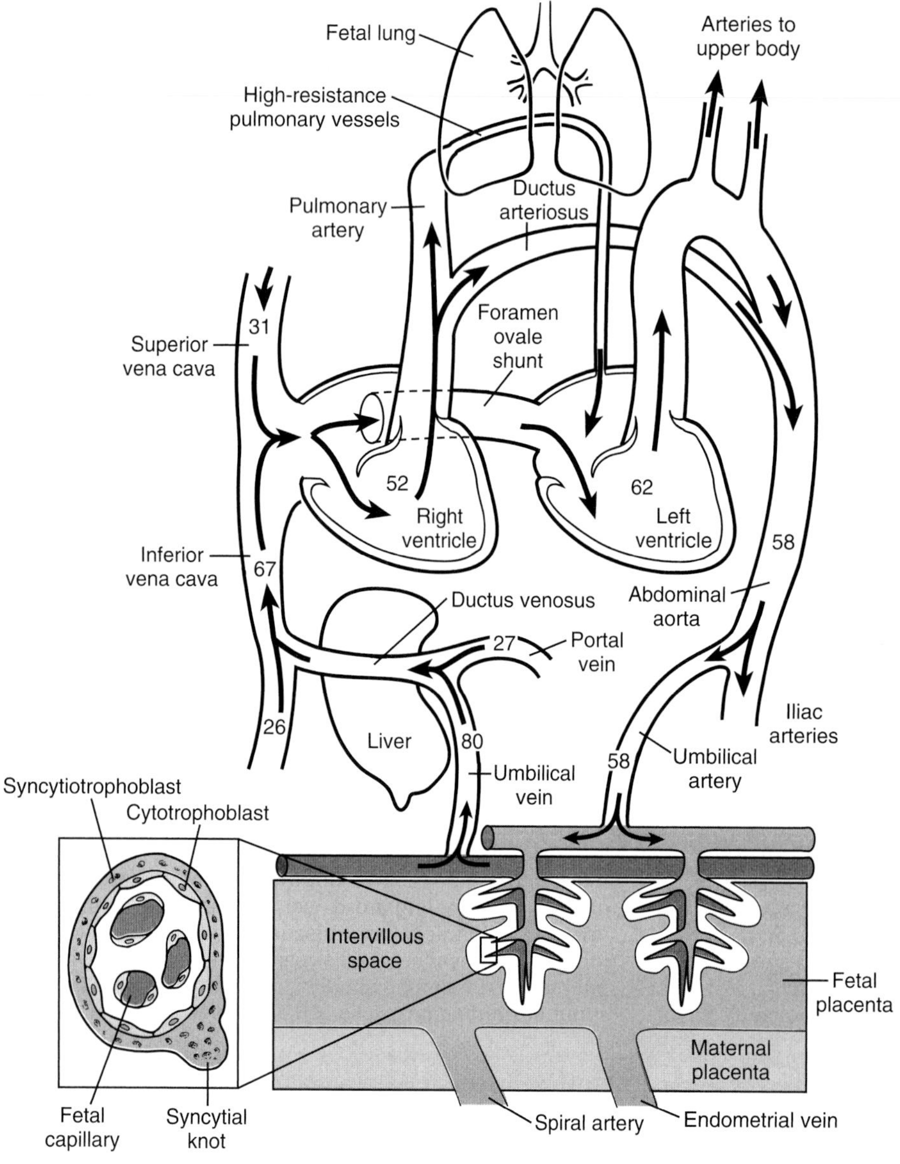

FIGURE 17.6 **The fetal and placental circulations.** Schematic representation of the left and right sides of the fetal heart are separated to emphasize the right-to-left shunt of blood through the open foramen ovale in the atrial septum and the right-to-left shunt through the ductus arteriosus. Arrows indicate the direction of blood flow. The numbers represent the percentage of saturation of blood hemoglobin with oxygen in the fetal circulation. Closure of the ductus venosus, foramen ovale, ductus arteriosus, and placental vessels at birth and the dilation of the pulmonary vasculature establish the adult circulation pattern. The insert is a cross-sectional view of a fetal placental villus, one of the branches of the tree-like fetal vascular system in the placenta. The fetal capillaries provide incoming blood, and the sinusoidal capillaries act as the venous drainage. The villus is completely surrounded by the maternal blood, and the exchange of nutrients and wastes occurs across the fetal syncytiotrophoblast.

and then blood leaves through sinusoidal capillaries to the umbilical venous system. Exchange occurs in the fetal capillaries and probably to some extent in the sinusoidal capillaries. The mother's vascular system forms a reservoir around the tree-like structure such that her blood envelops the placental villi.

As shown in Figure 17.6, the outermost layer of the placental villus is the **syncytiotrophoblast**, where exchange by passive diffusion, facilitated diffusion, and active transport between fetus and mother occurs through fully differentiated epithelial cells. The underlying **cytotrophoblast** is composed of less differentiated cells, which can form additional syncytiotrophoblast cells as required. As cells of the syncytiotrophoblast die, they form **syncytial knots**, and eventually these break off into the mother's blood system surrounding the fetal placental villi.

The placental vasculature of both the fetus and the mother adapt to the size of the fetus, as well as to the oxygen available within the maternal blood. For example, a minimal placental vascular anatomy will provide for a small fetus, but as the fetus develops and grows, a complex tree of placental vessels is essential to provide the surface area needed for the fetal-maternal exchange of gases, nutrients, and wastes. If the mother moves to a higher altitude where less oxygen is available, the complexity of the placental vascular tree increases, compensating with additional areas for exchange. If this type of adaptation does not take place, the fetus may be underdeveloped or die from a lack of oxygen.

During fetal development, the fetal tissues invade and cause partial degeneration of the maternal endometrial lining of the uterus. The result, after about 10 to 16 weeks gestation, is an **intervillous space** between fetal placental villi that is filled with maternal blood. Instead of microvessels, there is a cavernous blood-filled space. The intervillous space is supplied by 100 to 200 **spiral arteries** of the maternal endometrium and is drained by the **endometrial veins.** During gestation, the spiral arteries enlarge in diameter and simultaneously lose their vascular smooth muscle layer—it is the arteries preceding them that actually regulate blood flow through the placenta. At the end of gestation, the total maternal blood flow to the intervillous space is approximately 600 to 1,000 mL/min, which represents about 15 to 25% of the resting cardiac output. In comparison, the fetal placenta has a blood flow of about 600 mL/min, which represents about 50% of the fetal cardiac output.

The exchange of materials across the syncytiotrophoblast layer follows the typical pattern for all cells. Gases, primarily oxygen and carbon dioxide, and nutrient lipids move by simple diffusion from the site of highest concentration to the site of lowest concentration. Small ions are moved predominately by active transport processes. Glucose is passively transferred by the GLUT 1 transport protein, and amino acids require primarily facilitated diffusion through specific carrier proteins in the cell membranes, such as the system A transporter protein.

Large-molecular-weight peptides and proteins and many large, charged, water-soluble molecules used in pharmacological treatments do not readily cross the placenta. Part of the transfer of large molecules probably occurs between the cells of the syncytiotrophoblast layer and by pinocytosis and exocytosis. Lipid-soluble molecules diffuse through the lipid bilayer of cell membranes. For example, lipid-soluble anesthetic agents in the mother's blood do enter and depress the fetus. As a consequence, anesthesia during pregnancy is somewhat risky for the fetus.

The Placental Vasculature Permits Efficient Exchanges of O_2 and CO_2

Special fetal adaptations are required for gas exchange, particularly oxygen, because of the limitations of passive exchange across the placenta. The P_{O_2} of maternal arterial blood is about 80 to 100 mm Hg and about 20 to 25 mm Hg in the incoming blood in the umbilical artery. This difference in oxygen tension provides a large driving force for exchange; the result is an increase in the fetal blood P_{O_2} to 30 to 35 mm Hg in the umbilical vein. Fortunately, **fetal hemoglobin** carries more oxygen at a low P_{O_2} than adult hemoglobin carries at a P_{O_2} 2 to 3 times higher. In addition, the concentration of hemoglobin in fetal blood is about 20% higher than in adult blood. The net result is that the fetus has sufficient oxygen to support its metabolism and growth but does so at low oxygen tensions, using the unique properties of fetal hemoglobin. After birth, when much more efficient oxygen exchange occurs in the lung, the newborn gradually replaces the red cells containing fetal hemoglobin with red cells containing adult hemoglobin.

The Absence of Lung Ventilation Requires a Unique Circulation Through the Fetal Heart and Body

After the umbilical vein leaves the fetal placenta, it passes through the abdominal wall at the future site of the umbilicus (navel). The umbilical vein enters the liver's portal venous circulation, although the bulk of the oxygenated venous blood passes directly through the liver in the **ductus venosus** (see Fig. 17.6). The low-oxygen-content venous blood from the lower body and the high-oxygen-content placental venous blood mix in the inferior vena cava. The oxygen content of the blood returning from the lower body is about twice that of venous blood returning from the upper body in the superior vena cava. The two streams of blood from the superior and inferior vena cavae do not completely mix as they enter the right atrium. The net result is that oxygen-rich blood from the inferior vena cava passes through the open **foramen ovale** in the atrial septum to the left atrium, while the upper-body blood generally enters the right ventricle as in the adult. The preferential passage of oxygenated venous blood into the left atrium and the minimal amount of venous blood returning from the lungs to the left atrium allow blood in the left ventricle to have an oxygen content about 20% higher than that in the right ventricle. This relatively high-oxygen-content blood supplies the coronary vasculature, the head, and the brain.

The right ventricle actually pumps at least twice as much blood as the left ventricle during fetal life. In fact, the infant at birth has a relatively much more muscular right ventricular wall than the adult. Perfusion of the collapsed lungs of the fetus is minimal because the pulmonary vasculature has a high resistance. The elevated pulmonary resistance occurs because the lungs are not inflated and probably because the pulmonary vasculature has the unusual characteristic of vasoconstriction at low oxygen tensions. The right ventricle pumps blood into the systemic arterial circulation via a shunt—the **ductus arteriosus**—between the pulmonary artery and aorta (see Fig. 17.6). For ductus arteriosus blood to enter the initial part of the descending aorta, the right ventricle must develop a higher pressure than the left ventricle—the exact opposite of circumstances in the adult. The blood in the descending aorta has less oxygen content than that in the left ventricle and ascending aorta because of the mixture of less well-oxygenated blood from the right ventricle. This difference is crucial because about two thirds of this blood must be used to perfuse the placenta and pick up additional oxygen. In this situation, a lack of oxygen content is useful.

The Transition From Fetal to Neonatal Life Involves a Complex Sequence of Cardiovascular Events

After the newborn is delivered and the initial ventilatory movements cause the lungs to expand with air, pulmonary vascular resistance decreases substantially, as does pulmonary arterial pressure. At this point, the right ventricle can perfuse the lungs, and the circulation pattern in the newborn switches to that of an adult. In time, the reduced workload on the right ventricle causes its hypertrophy to subside.

The highly perfused, ventilated lungs allow a large amount of oxygen-rich blood to enter the left atrium. The increased oxygen tension in the aortic blood may provide the signal for closure of the ductus arteriosus, although suppression of vasodilator prostaglandins cannot be discounted. In any event, the ductus arteriosus constricts to virtual closure and over time becomes anatomically fused. Simultaneously, the increased oxygen to the peripheral tissues causes constriction in most body organs, and the sympathetic nervous system also stimulates the peripheral arterioles to constrict. The net result is that the left ventricle now pumps against a higher resistance. The combination of greater resistance and higher blood flow raises the arterial pressure and, in doing so, increases the mechanical load on the left ventricle. Over time, the left ventricle hypertrophies.

During all the processes just described, the open foramen ovale must be sealed to prevent blood flow from the left to right atrium. Left atrial pressure increases from the returning blood from the lungs and exceeds right atrial pressure. This pressure difference passively pushes the tissue flap on the left side of the foramen ovale against the open atrial septum. In time, the tissues of the atrial septum fuse; however, an anatomic passage that is probably only passively sealed can be documented in some adults. The ductus venosus in the liver is open for several days after birth but gradually closes and is obliterated within 2 to 3 months.

After the fetus begins breathing, the fetal placental vessels and umbilical vessels undergo progressive vasoconstriction to force placental blood into the fetal body, minimizing the possibility of fetal hemorrhage through the placental vessels. Vasoconstriction is related to increased oxygen availability and less of a signal for vasodilator chemicals and prostaglandins in the fetal tissue.

The final event of gestation is separation of the fetal and maternal placenta as a unit from the lining of the uterus. The separation process begins almost immediately after the fetus is expelled, but external delivery of the placenta can require up to 30 minutes. The separation occurs along the **decidua spongiosa**, a maternal structure, and requires that blood flow in the mother's spiral arteries be stopped. The cause of the placental separation may be mechanical, as the uterus surface area is greatly reduced by removal of the fetus and folds away from the uterine lining. Normally about 500 to 600 mL of maternal blood are lost in the process of placental separation. However, as maternal blood volume increases 1,000 to 1,500 mL during gestation, this blood loss is not of significant concern.

REVIEW QUESTIONS

DIRECTIONS: Each of the numbered items or incomplete statements in this section is followed by answers or completions of the statement. Select the ONE lettered answer or completion that is BEST in each case.

1. Which of the following would be an expected response by the coronary vasculature?
(A) Increased blood flow when the heart workload is increased
(B) Increased vascular resistance when the arterial blood pressure is increased
(C) Decreased blood flow when mean arterial pressure is reduced from 90 to 60 mm Hg by hemorrhage
(D) Decreased blood flow when blood oxygen content is reduced
(E) Increased vascular resistance during aerobic exercise
2. The intestinal blood flow during food digestion primarily increases because of
(A) Decreased sympathetic nervous system activity on intestinal arterioles
(B) Myogenic vasodilation associated with reduced arterial pressure after meals
(C) Tissue hypertonicity and the release of nitric oxide onto the arterioles
(D) Blood flow-mediated dilation by the major arteries of the abdominal cavity
(E) Increased parasympathetic nervous system activity associated with food absorption
3. Incoming arterial and portal venous blood mix in the liver
(A) As the hepatic artery and portal vein first enter the tissue
(B) In large arterioles and portal venules
(C) In the liver acinus capillaries
(D) In the terminal hepatic venules
(E) In the outflow venules of the liver
4. As arterial pressure is raised and lowered during the course of a day, blood flow through the brain would be expected to
(A) Change in the same direction as the arterial blood pressure because of the limited autoregulatory ability of the cerebral vessels
(B) Change in a direction opposite the change in mean arterial pressure

(continued)

(C) Remain about constant because cerebral vascular resistance changes in the same direction as arterial pressure
(D) Fluctuate widely, as both arterial pressure and brain neural activity status change
(E) Remain about constant because the cerebral vascular resistance changes in the opposite direction to the arterial pressure

5. Which of the following special circulations has the widest range of blood flows as part of its contributions to both the regulation of systemic vascular resistance and the modification of resistance to suit the organ's metabolic needs?
(A) Coronary
(B) Cerebral
(C) Small intestine
(D) Skeletal muscle
(E) Dermal

6. Which of the following sequences is a possible anatomic path for a red blood cell passing through a fetus and back to the placenta? (Some intervening structures are not included.)
(A) Umbilical vein, right ventricle, ductus arteriosus, pulmonary artery
(B) Ductus venosus, foramen ovale, right ventricle, ascending aorta
(C) Spiral artery, umbilical vein, left ventricle, umbilical artery
(D) Right ventricle, ductus arteriosus, descending aorta, umbilical artery
(E) Left ventricle, ductus arteriosus, pulmonary artery, left atrium

7. How does chronic hypertension affect the range of arterial pressure over which the cerebral circulation can maintain relatively constant blood flow?
(A) Very little change occurs
(B) The vasculature primarily adapts to higher arterial pressure
(C) The vasculature primarily loses regulation at low arterial pressure
(D) The entire range of regulation shifts to higher pressures
(E) The entire range of regulation shifts to lower pressures

8. Why is the oxygen content of blood sent to the upper body during fetal life higher than that sent to the lower body?
(A) Blood oxygenated in the fetal lungs enters the left ventricle
(B) Oxygenated blood passes through the foramen ovale to the left ventricle
(C) The upper body is perfused by the ductus arteriosus blood flow
(D) The heart takes less of the oxygen from the blood in the left ventricle
(E) The right ventricular stroke volume is greater than that of the left ventricle

SUGGESTED READING

Bohlen HG. Integration of intestinal structure, function and microvascular regulation. Microcirculation 1998;5:27–37.
Bohlen HG, Maass-Moreno R, Rothe CF. Hepatic venular pressures of rats, dogs, and rabbits. Am J Physiol 1991;261:G539–G547.
Delp MD, Laughlin MH. Regulation of skeletal muscle perfusion during exercise. Acta Physiol Scand 1998;162:411–419.
Fiegl EO. Neural control of coronary blood flow. J Vasc Res 1998;35:85–92.
Golding EM, Robertson CS, Bryan RM. The consequences of traumatic brain injury on cerebral blood flow and autoregulation: A review. Clin Exp Hypertens 1999;21:229–332.
Johnson JM. Physical training and the control of skin blood flow. Med Sci Sports Exerc 1998;30:382–386.

Control Mechanisms in Circulatory Function

Thom W. Rooke, M.D.
Harvey V. Sparks, Jr., M.D.

CHAPTER OUTLINE

KEY CONCEPTS

1. The sympathetic nervous system acts on the heart primarily via β-adrenergic receptors.
2. The parasympathetic nervous system acts on the heart via muscarinic cholinergic receptors.
3. The sympathetic nervous system acts on blood vessels primarily via α-adrenergic receptors.
4. Reflex control of the circulation is integrated primarily in pools of neurons in the medulla oblongata.
5. The integration of behavioral and cardiovascular responses occurs mainly in the hypothalamus.
6. Baroreceptors and cardiopulmonary receptors are key in the moment-to-moment regulation of arterial pressure.
7. The renin-angiotensin-aldosterone system, arginine vasopressin, and atrial natriuretic peptide are important in the long-term regulation of blood volume and arterial pressure.
8. Pressure diuresis is the mechanism that ultimately adjusts arterial pressure to a set level.
9. The defense of arterial pressure during standing involves the integration of multiple mechanisms.

The mechanisms controlling the circulation can be divided into neural control mechanisms, hormonal control mechanisms, and local control mechanisms. Cardiac performance and vascular tone at any time are the result of the integration of all three control mechanisms. To some extent, this categorization is artificial because each of the three categories affects the other two. This chapter deals with neural and hormonal mechanisms; local mechanisms are covered in Chapter 16.

Central blood volume and arterial pressure are normally maintained within narrow limits by neural and hormonal mechanisms. Adequate central blood volume is necessary to ensure proper cardiac output, and relatively constant arterial blood pressure maintains tissue perfusion in the face of changes in regional blood flow. Neural control involves sympathetic and parasympathetic branches of the autonomic nervous system (ANS). Blood volume and arterial pressure are monitored by stretch receptors in the heart and arteries. Afferent nerve traffic from these receptors is integrated with other afferent information in the medulla oblongata, which leads to activity in sympathetic and parasympathetic nerves that adjusts cardiac output and systemic vascular resistance (SVR) to maintain arterial pressure. Sympathetic nerve activity and, more importantly, hormones, such as arginine vasopressin (antidiuretic hormone), angiotensin II, aldosterone, and atrial natriuretic peptide, serve as effectors for the regulation of salt and water balance and blood volume. Neural control of cardiac output and SVR plays a larger role in the moment-by-moment regulation of arterial pressure, whereas hormones play a larger role in the long-term regulation of arterial pressure.

In some situations, factors other than blood volume and arterial pressure regulation strongly influence cardiovascular control mechanisms. These situations include the fight-or-flight response, diving, thermoregulation, standing, and exercise.

AUTONOMIC NEURAL CONTROL OF THE CIRCULATORY SYSTEM

Neural regulation of the cardiovascular system involves the firing of postganglionic parasympathetic and sympathetic neurons, triggered by preganglionic neurons in the brain (parasympathetic) and spinal cord (sympathetic and parasympathetic). Afferent input influencing these neurons comes from the cardiovascular system, as well as from other organs and the external environment.

Autonomic control of the heart and blood vessels was described in Chapter 6. Briefly, the heart is innervated by parasympathetic (vagus) and sympathetic (cardioaccelerator) nerve fibers (Fig. 18.1). Parasympathetic fibers release acetylcholine (ACh), which binds to muscarinic receptors of the sinoatrial node, the atrioventricular node, and specialized conducting tissues. Stimulation of parasympathetic fibers causes a slowing of the heart rate and conduction velocity. The ventricles are only sparsely innervated by parasympathetic nerve fibers, and stimulation of these fibers has little direct effect on cardiac contractility. Some cardiac parasympathetic fibers end on sympathetic nerves and inhibit the release of norepinephrine (NE) from sympathetic nerve fibers. Therefore, in the presence of sympathetic nervous system activity, parasympathetic activation reduces cardiac contractility.

Sympathetic fibers to the heart release NE, which binds to β_1-adrenergic receptors in the sinoatrial node, the atrioventricular node and specialized conducting tissues, and cardiac muscle. Stimulation of these fibers causes increased heart rate, conduction velocity, and contractility.

The two divisions of the autonomic nervous system tend to oppose each other in their effects on the heart, and activities along these two pathways usually change in a reciprocal manner.

Blood vessels (except those of the external genitalia) receive sympathetic innervation only (see Fig. 18.1). The neurotransmitter is NE, which binds to α_1-adrenergic receptors and causes vascular smooth muscle contraction and vasoconstriction. Circulating epinephrine, released from the adrenal medulla, binds to β_2-adrenergic receptors of vascular smooth muscle cells, especially coronary and skeletal muscle arterioles, producing vascular smooth muscle relaxation and vasodilation. Postganglionic parasympathetic fibers release ACh and nitric oxide (NO) to blood vessels in the external genitalia. ACh causes the further release of NO from endothelial cells; NO results in vascular smooth muscle relaxation and vasodilation.

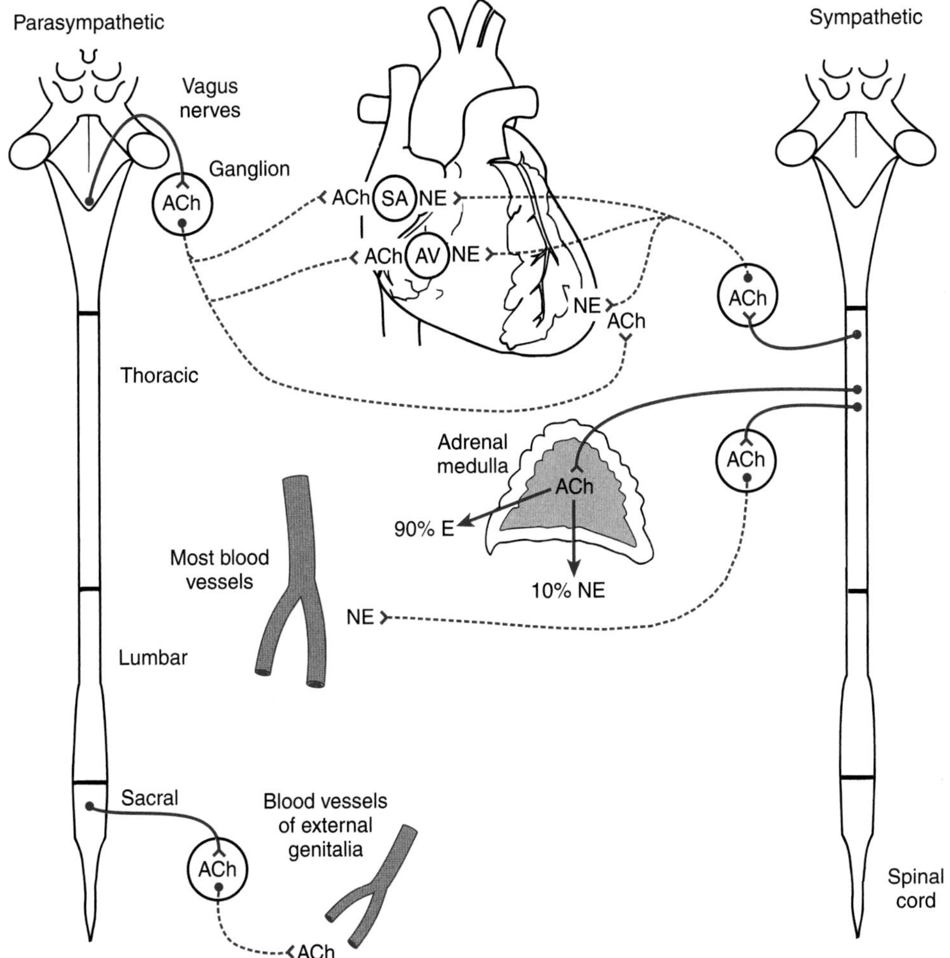

FIGURE 18.1 **Autonomic innervation of the cardiovascular system.** ACh, acetylcholine; NE, norepinephrine; E, epinephrine; SA, sinoatrial node; AV, atrioventricular node.

The Spinal Cord Exerts Control Over Cardiovascular Function

Preganglionic sympathetic neurons normally generate a steady level of background postganglionic activity (tone). This **sympathetic tone** produces a background level of sympathetic vasoconstriction, cardiac stimulation, and adrenal medullary catecholamine secretion, all of which contribute to the maintenance of normal blood pressure. This tonic activity is generated by excitatory signals from the medulla oblongata. When the spinal cord is acutely transected and these excitatory signals can no longer reach sympathetic preganglionic fibers, their tonic firing is reduced and blood pressure falls—an effect known as **spinal shock**.

Humans have spinal reflexes of cardiovascular significance. For example, the stimulation of pain fibers entering the spinal cord below the level of a chronic spinal cord transection can cause reflex vasoconstriction and increased blood pressure.

The Medulla Is a Major Area for Cardiovascular Reflex Integration

The medulla oblongata has three major cardiovascular functions:

- Generating tonic excitatory signals to spinal sympathetic preganglionic fibers
- Integrating cardiovascular reflexes
- Integrating signals from supramedullary neural networks and from circulating hormones and drugs

Specific pools of neurons are responsible for elements of these functions. Neurons in the **rostral ventrolateral nucleus** (RVL) are normally active and provide tonic excitatory activity to the spinal cord. Specific pools of neurons within the RVL have actions on heart and blood vessels. RVL neurons are critical in mediating reflex inhibition or activating sympathetic firing to the heart and blood vessels. The cell bodies of cardiac preganglionic parasympathetic neurons are located in the **nucleus ambiguus**; the activity of these neurons is influenced by reflex input, as well as input from respiratory neurons. Respiratory sinus arrhythmia, described in Chapter 13, is primarily the result of the influence of medullary respiratory neurons that inhibit firing of preganglionic parasympathetic neurons during inspiration and excite these neurons during expiration. Other inputs to the RVL and nucleus ambiguus will be described below.

The Baroreceptor Reflex Is Important in the Regulation of Arterial Pressure

The most important reflex behavior of the cardiovascular system originates in mechanoreceptors located in the aorta, carotid sinuses, atria, ventricles, and pulmonary vessels. These mechanoreceptors are sensitive to the stretch of the walls of these structures. When the wall is stretched by increased transmural pressure, receptor firing rate increases. Mechanoreceptors in the aorta and carotid sinuses are called **baroreceptors**. Mechanoreceptors in the atria, ventricles, and pulmonary vessels are referred to as low-pressure baroreceptors or **cardiopulmonary baroreceptors**.

Changes in the firing rate of the arterial baroreceptors and cardiopulmonary baroreceptors initiate reflex responses of the autonomic nervous system that alter cardiac output and SVR. The central terminals for these receptors are located in the **nucleus tractus solitarii** (NTS) in the medulla oblongata. Neurons from the NTS project to the RVL and nucleus ambiguus where they influence the firing of sympathetic and parasympathetic nerves.

Baroreceptor Reflex Effects on Cardiac Output and Systemic Vascular Resistance. Increased pressure in the carotid sinus and aorta stretches **carotid sinus baroreceptors** and **aortic baroreceptors** and raises their firing rate. Nerve fibers from carotid sinus baroreceptors join the glossopharyngeal (cranial nerve IX) nerves and travel to the NTS. Nerve fibers from the aortic baroreceptors, located in the wall of the arch of the aorta, travel with the vagus (cranial nerve X) nerves to the NTS.

The increased action potential traffic reaching the NTS leads to excitation of nucleus ambiguus neurons and inhibition of firing of RVL neurons. This results in increased parasympathetic neural activity to the heart and decreased sympathetic neural activity to the heart and resistance vessels (primarily arterioles) (Fig. 18.2), causing decreased cardiac output and SVR. Since mean arterial pressure is the product of SVR and cardiac output (see Chapter 12), mean arterial pressure is returned toward the normal level. This completes a negative-feedback loop by which increases in mean arterial pressure can be attenuated.

Conversely, decreases in arterial pressure (and decreased stretch of the baroreceptors) increase sympathetic neural activity and decrease parasympathetic neural activity, resulting in increased heart rate, stroke volume, and SVR; this

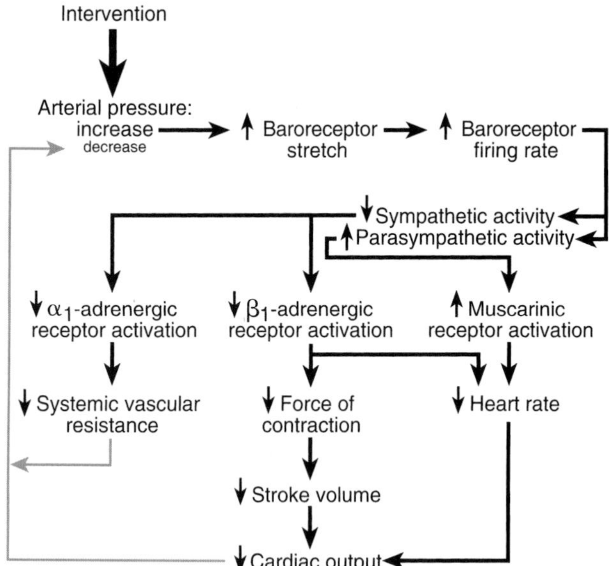

FIGURE 18.2 **Baroreceptor reflex response to increased arterial pressure.** An intervention elevates arterial pressure (either mean arterial pressure or pulse pressure), stretches the baroreceptors, and initiates the reflex. The resulting reduced systemic vascular resistance and cardiac output return arterial pressure toward the level existing before the intervention.

returns blood pressure toward the normal level. If the fall in mean arterial pressure is very large, increased sympathetic neural activity to veins is added to the above responses, causing contraction of the venous smooth muscle and reducing venous compliance. Decreased venous compliance shifts blood toward the central blood volume, increasing right atrial pressure and, in turn, stroke volume.

Baroreceptor Reflex Effects on Hormone Levels. The baroreceptor reflex influences hormone levels in addition to vascular and cardiac muscle. The most important influence is on the renin-angiotensin-aldosterone system (RAAS). A reduction in arterial pressure and baroreceptor firing results in increased sympathetic nerve activity to the kidneys, which causes the kidneys to release renin, activating the RAAS. The activation of this system causes the kidneys to save salt and water. Salt and water retention increases blood volume and, ultimately, causes blood pressure to rise. The details of the RAAS are discussed later in this chapter and in Chapter 24.

The information on the firing rate of the baroreceptors is also projected to the paraventricular nucleus of the hypothalamus where the release of arginine vasopressin (AVP) by the posterior pituitary is controlled (see Chapter 32). Decreased firing rate of the baroreceptors results in increased AVP release, causing the kidneys to save water. The result is an increase in blood volume. An increase in arterial pressure causes decreased AVP release and increased excretion of water by the kidneys.

Hormonal effects on salt and water balance and, ultimately, on cardiac output and blood pressure are powerful, but they occur more slowly (a timescale of many hours to days) than ANS effects (seconds to minutes).

Baroreceptor Reflex Effects on Specific Organs. The defense of arterial pressure by the baroreceptor reflex results in maintenance of blood flow to two vital organs: the heart and brain. If resistance vessels of the heart and brain participated in the sympathetically mediated vasoconstriction found in skeletal muscle, skin, and the splanchnic region, it would lower blood flow to these organs. This does not happen.

The combination of (1) a minimal vasoconstrictor effect of sympathetic nerves on cerebral blood vessels, and (2) a robust autoregulatory response keeps brain blood flow nearly normal despite modest decreases in arterial pressure (see Chapter 17). However, a large decrease in arterial pressure beyond the autoregulatory range causes brain blood flow to fall, accounting for loss of consciousness.

Activation of sympathetic nerves to the heart causes α_1-adrenergic receptor-mediated constriction of coronary arterioles and β_1-adrenergic receptor-mediated increases in cardiac muscle metabolism (see Chapter 17). The net effect is a marked increase in coronary blood flow, despite the increased sympathetic constrictor activity. In summary, when arterial pressure drops, the generalized vasoconstriction caused by the baroreflex spares the brain and heart, allowing flow to these two vital organs to be maintained.

Pressure Range for Baroreceptors. The effective range of the carotid sinus baroreceptor mechanism is approximately 40 mm Hg (when the receptor stops firing) to 180 mm Hg (when the firing rate reaches a maximum) (Fig. 18.3). Pulse pressure also influences the firing rate of the baroreceptors. For a given mean arterial pressure, the firing rate of the baroreceptors increases with pulse pressure.

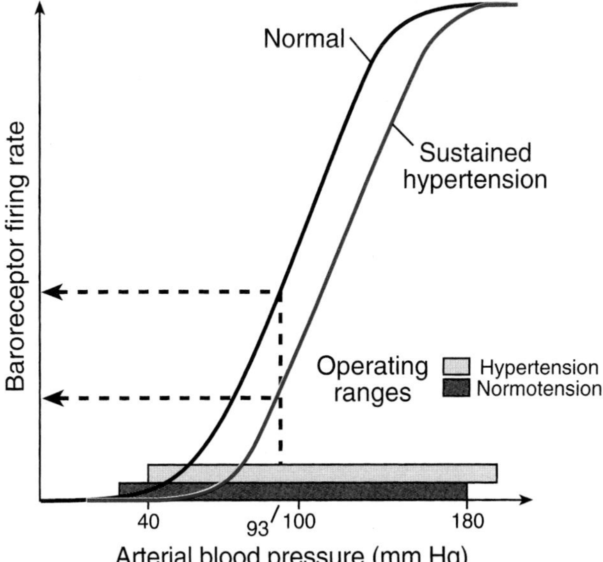

FIGURE 18.3 **Carotid sinus baroreceptor nerve firing rate and mean arterial pressure.** With normal conditions, a mean arterial pressure of 93 mm Hg is near the midrange of the firing rates for the nerves. Sustained hypertension causes the operating range to shift to the right, putting 93 mm Hg at the lower end of the firing range for the nerves.

Baroreceptor Adaptation. An important property of the baroreceptor reflex is that it adapts during a period of 1 to 2 days to the prevailing mean arterial pressure. When the mean arterial pressure is suddenly raised, baroreceptor firing increases. If arterial pressure is held at the higher level, baroreceptor firing declines during the next few seconds. Firing rate then continues to decline more slowly until it returns to the original firing rate, between 1 and 2 days. Consequently, if the mean arterial pressure is maintained at an elevated level, the tendency for the baroreceptors to initiate a decrease in cardiac output and SVR quickly disappears. This occurs, in part, because of the reduction in the rate of baroreceptor firing for a given mean arterial pressure mentioned above (see Fig. 18.3). This is an example of **receptor adaptation**. A "resetting" of the reflex in the central nervous system (CNS) occurs as well. Consequently, the baroreceptor mechanism is the "first line of defense" in the maintenance of normal blood pressure; it makes the rapid control of blood pressure needed with changes in posture or blood loss possible, but it does not provide for the long-term control of blood pressure.

Cardiopulmonary Baroreceptors Are Stretch Receptors That Sense Central Blood Volume

Cardiopulmonary baroreceptors are located in the cardiac atria, at the junction of the great veins and atria, in the ventricular myocardium, and in pulmonary vessels. Their nerve fibers run in the vagus nerve to the NTS, with projections to supramedullary areas as well. Unloading (i.e., decreasing the stretch) of the cardiopulmonary receptors by reducing central blood volume results in increased sympathetic nerve activity and decreased parasympathetic nerve activity to the heart and blood vessels. In addition, the cardiopulmonary reflex interacts with the baroreceptor reflex. Unloading of the cardiopulmonary receptors enhances the baroreceptor reflex, and loading the cardiopulmonary receptors, by increasing central blood volume, inhibits the baroreceptor reflex.

Like the arterial baroreceptors, decreased stretch of the cardiopulmonary baroreceptors activates the RAAS and increases the release of AVP.

Chemoreceptors Detect Changes in P_{CO_2}, pH, and P_{O_2}

The reflex response to changes in blood gases and pH begins with **chemoreceptors** located peripherally in the **carotid bodies** and **aortic bodies** and centrally in the medulla (see Chapter 22). The peripheral chemoreceptors of the carotid bodies and aortic bodies are specialized structures located in approximately the same areas as the carotid sinus and aortic baroreceptors. They send nerve impulses to the NTS and are sensitive to elevated P_{CO_2}, as well as decreased pH and P_{O_2}. Peripheral chemoreceptors exhibit an increased firing rate when (1) the P_{O_2} or pH of the arterial blood is low, (2) the P_{CO_2} of arterial blood is increased, (3) the flow through the bodies is very low or stopped, or (4) a chemical is given that blocks oxidative metabolism in the chemoreceptor cells. The central medullary chemoreceptors increase their firing rate primarily in response to elevated arterial P_{CO_2}, which causes a decrease in brain pH.

The increased firing of both peripheral and central chemoreceptors (via the NTS and RVL) leads to profound peripheral vasoconstriction. Arterial pressure is significantly elevated. If respiratory movements are voluntarily stopped, the vasoconstriction is more intense and a striking bradycardia and decreased cardiac output occur. This response pattern is typical of the diving response (discussed later). As in the case of the baroreceptor reflex, the coronary and cerebral circulations are not subject to the sympathetic vasoconstrictor effects and instead exhibit vasodilation, as a result of the combination of the direct effect of the abnormal blood gases and local metabolic effects.

In addition to its importance when arterial blood gases are abnormal, the chemoreceptor reflex is important in the cardiovascular response to severe hypotension. As blood pressure falls, blood flow through the carotid and aortic bodies decreases and chemoreceptor firing increases—probably because of changes in local P_{CO_2}, pH, and P_{O_2}.

Pain Receptors Produce Reflex Responses in the Cardiovascular System

Two reflex cardiovascular responses to pain occur. In the most common reflex, pain causes increased sympathetic activity to the heart and blood vessels, coupled with decreased parasympathetic activity to the heart. These events lead to increases in cardiac output, SVR, and mean arterial pressure. An example of this reaction is the **cold pressor response**—the elevated blood pressure that normally occurs when an extremity is placed in ice water. The increase in blood pressure produced by this challenge is exaggerated in several forms of hypertension.

A second type of response is produced by deep pain. The stimulation of deep pain fibers associated with crushing injuries, disruption of joints, testicular trauma, or distension of the abdominal organs results in diminished sympathetic activity and enhanced parasympathetic activity with decreased cardiac output, SVR, and blood pressure. This hypotensive response contributes to certain forms of cardiovascular shock.

Activation of Chemoreceptors in the Ventricular Myocardium Causes Reflex Bradycardia and Vasodilation

An injection of bradykinin, 5-hydroxytryptamine (serotonin), certain prostaglandins, or various other compounds into the coronary arteries supplying the posterior and inferior regions of the ventricles causes reflex bradycardia and hypotension. The chemoreceptor afferents are carried in the vagus nerves. The bradycardia results from increased parasympathetic tone. Dilation of systemic arterioles and veins is caused by withdrawal of sympathetic tone. This reflex is also elicited by myocardial ischemia and is responsible for the bradycardia and hypotension that can occur in response to acute infarction of the posterior or inferior myocardium.

INTEGRATED SUPRAMEDULLARY CARDIOVASCULAR CONTROL

The highest levels of organization in the ANS are the **supramedullary networks** of neurons with way stations in the limbic cortex, amygdala, and hypothalamus. These supramedullary networks orchestrate cardiovascular correlates of specific patterns of emotion and behavior by their projections to the ANS.

Unlike the medulla, supramedullary networks do not contribute to the tonic maintenance of blood pressure, nor are they necessary for most cardiovascular reflexes, although they modulate reflex reactivity.

The Fight-or-Flight Response Includes Specific Cardiovascular Changes

Upon stimulation of certain areas in the hypothalamus, cats demonstrate a stereotypical rage response, with spitting, clawing, tail lashing, back arching, and so on. This is accompanied by the autonomic **fight-or-flight response** described in Chapter 6. Cardiovascular responses include elevated heart rate and blood pressure.

The initial behavioral pattern during the fight-or-flight response includes increased skeletal muscle tone and general alertness. There is increased sympathetic neural activity to blood vessels and the heart. The result of this cardiovascular response is an increase in cardiac output (by increasing both heart rate and stroke volume), SVR, and arterial pressure. When the fight-or-flight response is consummated by fight or flight, arterioles in skeletal muscle dilate because of accumulation of local metabolites from the exercising muscles (see Chapter 17). This vasodilation may outweigh the sympathetic vasoconstriction in other organs and SVR may actually fall. With a fall in SVR, mean arterial pressure returns toward normal despite the increase in cardiac output.

Emotional situations often provoke the fight-or-flight response in humans, but it is usually not accompanied by muscle exercise (e.g., medical students taking an examination). It seems likely that repeated elevations in arterial pressure caused by dissociation of the cardiovascular component of the fight-or-flight response from muscular exercise component are harmful.

Fainting Can Be a Cardiovascular Correlate of Emotion

Vasovagal syncope (fainting) is a somatic and cardiovascular response to certain emotional experiences. Stimulation of specific areas of the cerebral cortex can lead to a sudden relaxation of skeletal muscles, depression of respiration, and loss of consciousness. The cardiovascular events accompanying these somatic changes include profound parasympathetic-induced bradycardia and withdrawal of resting sympathetic vasoconstrictor tone. There is a dramatic drop in heart rate, cardiac output, and SVR. The resultant decrease in mean arterial pressure results in unconsciousness because of lowered cerebral blood flow. Vasovagal syncope appears in lower animals as the "playing dead" response typical of the opossum.

The Cardiovascular Correlates of Exercise Require Integration of Central and Peripheral Mechanisms

Exercise causes activation of supramedullary neural networks that inhibit the activity of the baroreceptor reflex. The inhibition of medullary regions involved in the baroreceptor reflex is called **central command**. Central command results in withdrawal of parasympathetic tone to the heart with a resulting increase in heart rate and cardiac output. The increased cardiac output supplies the added requirement for blood flow to exercising muscle. As exercise intensity increases, central command adds sympathetic tone that further increases heart rate and contractility. It also recruits sympathetic vasoconstriction that redistributes blood flow away from splanchnic organs and resting skeletal muscle to exercising muscle. Finally, afferent impulses from exercising skeletal muscle terminate in the RVL where they further augment sympathetic tone.

During exercise, blood flow of the skin is largely influenced by temperature regulation, as described in Chapter 17.

The Diving Response Maintains Oxygen Delivery to the Heart and Brain

The **diving response** is best observed in seals and ducks, but it also occurs in humans. An experienced diver can exhibit intense slowing of the heart rate (parasympathetic) and peripheral vasoconstriction (sympathetic) of the extremities and splanchnic regions when his or her face is submerged in cold water. With breath holding during the dive, arterial P_{O_2} and pH fall and P_{CO_2} rises, and the chemoreceptor reflex reinforces the diving response. The arterioles of the brain and heart do not constrict and, therefore, cardiac output is distributed to these organs. This heart-brain circuit makes use of the oxygen stored in the blood that would normally be used by the other tissues, especially skeletal muscle. Once the diver surfaces, the heart rate and cardiac output increase substantially; peripheral vasoconstriction is replaced by vasodilation, restoring nutrient flow and washing out accumulated waste products.

Behavioral Conditioning Affects Cardiovascular Responses

Cardiovascular responses can be conditioned (as can other autonomic responses, such as those observed in Pavlov's famous experiments). Both classical and operant conditioning techniques have been used to raise and lower the blood pressure and heart rate of animals. Humans can also be taught to alter their heart rate and blood pressure, using a variety of behavioral techniques, such as biofeedback.

Behavioral conditioning of cardiovascular responses has significant clinical implications. Animal and human studies indicate that psychological stress can raise blood pressure, increase atherogenesis, and predispose to fatal cardiac arrhythmias. These effects are thought to result from an inappropriate fight-or-flight response. Other studies have shown beneficial effects of behavior patterns designed to introduce a sense of relaxation and well-being. Some clinical regimens for the treatment of cardiovascular disease take these factors into account.

Not All Cardiovascular Responses Are Equal

Supramedullary responses can override the baroreceptor reflex. For example, the fight-or-flight response causes the heart rate to rise above normal levels despite a simultaneous rise in arterial pressure. In such circumstances, the neurons connecting the hypothalamus to medullary areas inhibit the baroreceptor reflex and allow the corticohypothalamic response to predominate. Also, during exercise, input from supramedullary regions inhibits the baroreceptor reflex, promoting increased sympathetic tone and decreased parasympathetic tone despite an increase in arterial pressure.

Moreover, the various cardiovascular response patterns do not necessarily occur in isolation, as previously described. Many response patterns interact, reflecting the extensive neural interconnections between all levels of the CNS and interaction with various elements of the local control systems. For example, the baroreceptor reflex interacts with thermoregulatory responses. Cutaneous sympathetic nerves participate in body temperature regulation (see Chapter 29), but also serve the baroreceptor reflex. At moderate levels of heat stress, the baroreceptor reflex can cause cutaneous arteriolar constriction despite elevated core temperature. However, with severe heat stress, the baroreceptor reflex cannot overcome the cutaneous vasodilation; as a result, arterial pressure regulation may fail.

HORMONAL CONTROL OF THE CARDIOVASCULAR SYSTEM

Various hormones play a role in the control of the cardiovascular system. Important sites of hormone secretion include the adrenal medulla, posterior pituitary gland, kidney, and cardiac atrium.

Circulating Epinephrine Has Cardiovascular Effects

When the sympathetic nervous system is activated, the adrenal medulla releases epinephrine (> 90%) and norepinephrine (< 10%), which circulate in the blood (see Chapter 6). Changes in the circulating NE concentration are small relative to changes in NE resulting from the direct release from nerve endings close to vascular smooth muscle and cardiac cells. Increased circulating epinephrine, however, contributes to skeletal muscle vasodilation during the fight-or-flight response and exercise. In these cases, epinephrine binds to β_2-adrenergic receptors of skeletal muscle arteriolar smooth muscle cells and causes relaxation. In the heart, circulating epinephrine binds to cardiac cell β_1-adrenergic receptors and reinforces the effect of NE released from sympathetic nerve endings.

A comparison of the responses to infusions of epinephrine and norepinephrine illustrates not only the different effects of the two hormones but also the different reflex response each one elicits (Fig. 18.4). Epinephrine and norepinephrine have similar direct effects on the heart, but NE elicits a powerful baroreceptor reflex because it causes systemic vasoconstriction and increases mean arterial pressure. The reflex masks some of the direct cardiac effects of NE by significantly increasing cardiac parasympathetic tone. In contrast, epinephrine causes vasodilation in skeletal muscle and splanchnic beds. SVR may actually fall and mean arterial pressure does not rise. The baroreceptor reflex is not elicited, parasympathetic tone to the heart is not increased, and the direct cardiac effects of epinephrine are evident. At high concentrations, epinephrine binds to α_1-adrenergic receptors and causes peripheral vasoconstriction; this level of epinephrine is probably never reached except when it is administered as a drug.

Denervated organs, such as transplanted hearts, are very responsive to circulating levels of epinephrine and norepinephrine. This increased sensitivity to neurotransmitters is referred to as **denervation hypersensitivity**. Several factors contribute to denervation hypersensitivity, including the absence of sympathetic nerve endings to take up circulating norepinephrine and epinephrine actively, leaving more transmitter available for binding to receptors. In addition, denervation results in **up-regulation** of neurotransmitter receptors in target cells. During exercise, circulating levels of norepinephrine and epinephrine increase. Because of their enhanced response to circulating catecholamines, transplanted hearts can perform almost as well as normal hearts.

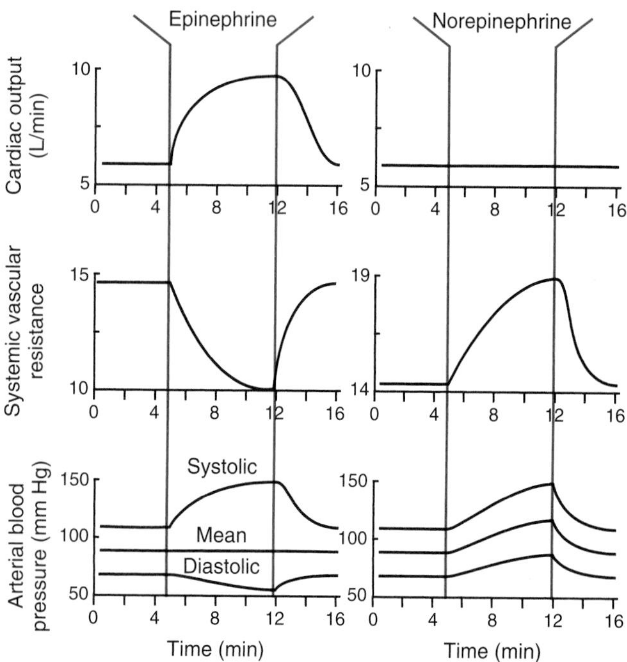

FIGURE 18.4 A comparison of the effects of intravenous infusions of epinephrine and norepinephrine. (See text for details.) (Modified from Rowell LB. Human Circulation: Regulation During Physical Stress. New York: Oxford University Press, 1986.)

The Renin-Angiotensin-Aldosterone System Helps Regulate Blood Pressure and Volume

The control of total blood volume is extremely important in regulating arterial pressure. Because changes in total blood volume lead to changes in central blood volume, the long-term influence of blood volume on ventricular end-diastolic volume and cardiac output is paramount. Cardiac output, in turn, strongly influences arterial pressure. Hormonal control of blood volume depends on hormones that regulate salt and water intake and output as well as red blood cell formation.

Reduced arterial pressure and blood volume cause the release of **renin** from the kidneys. Renin release is mediated by the sympathetic nervous system and by the direct effect of lowered arterial pressure on the kidneys. Renin is a proteolytic enzyme that catalyzes the conversion of angiotensinogen, a plasma protein, to angiotensin I (Fig. 18.5). Angiotensin I is then converted to angiotensin II by **angiotensin-converting enzyme** (ACE), primarily in the lungs. Angiotensin II has the following actions:

- It is a powerful arteriolar vasoconstrictor, and in some circumstances, it is present in plasma in concentrations sufficient to increase SVR.
- It reduces sodium excretion by increasing sodium reabsorption by proximal tubules of the kidney.
- It causes the release of **aldosterone** from the adrenal cortex.
- It causes the release of AVP from the posterior pituitary gland.

Angiotensin II is a significant vasoconstrictor in some circumstances. Angiotensin II directly stimulates contraction of vascular smooth muscle and also augments NE release from sympathetic nerves and sensitizes vascular smooth muscle to the constrictor effects of NE. It plays an

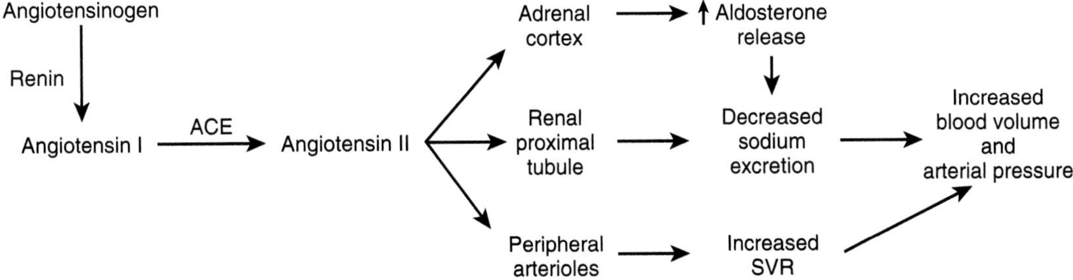

FIGURE 18.5 **Renin-angiotensin-aldosterone system.** This system plays an important role in the regulation of arterial blood pressure and blood volume. ACE, angiotensin-converting enzyme; SVR, systemic vascular resistance.

important role in increasing SVR, as well as blood volume, in individuals on a low-salt diet. If an ACE inhibitor is given to such individuals, blood pressure falls. Renin is released during blood loss, even before blood pressure falls, and the resulting rise in plasma angiotensin II increases the SVR.

One of the effects of aldosterone is to reduce renal excretion of sodium, the major cation of the extracellular fluid. Retention of sodium paves the way for increasing blood volume. Renin, angiotensin, aldosterone, and the factors that control their release and formation are discussed in Chapter 24. The RAAS is important in the normal maintenance of blood volume and blood pressure. It is critical when salt and water intake is reduced.

Rarely, renal artery stenosis causes hypertension that can be attributed solely to elevated renin and angiotensin II levels. In addition, the renin-angiotensin-aldosterone system plays an important (but not unique) role in maintaining elevated pressure in more than 60% of patients with essential hypertension. In patients with congestive heart failure, renin and angiotensin II are increased and contribute to elevated SVR as well as sodium retention.

Arginine Vasopressin Contributes to the Regulation of Blood Volume

Arginine vasopressin (AVP) is released by the posterior pituitary gland controlled by the hypothalamus. Three primary classes of stimuli lead to AVP release: increased plasma osmolality; decreased baroreceptor and cardiopulmonary receptor firing; and various types of stress, such as physical injury or surgery. In addition, circulating angiotensin II stimulates AVP release. Although AVP is a vasoconstrictor, it is not ordinarily present in plasma in high enough concentrations to exert an effect on blood vessels. However, in special circumstances (e.g., severe hemorrhage) it probably contributes to increased SVR. AVP exerts its major effect on the cardiovascular system by causing the retention of water by the kidneys (see Chapter 24)—an important part of the neural and humoral mechanisms that regulate blood volume.

Atrial Natriuretic Peptide Helps Regulate Blood Volume

Atrial natriuretic peptide (ANP) is a 28-amino acid polypeptide synthesized and stored in the atrial muscle cells and released into the bloodstream when the atria are stretched. By increasing sodium excretion, it decreases blood volume (see Chapter 24). It also inhibits renin release as well as aldosterone and AVP secretion. Increased ANP (along with decreased aldosterone and AVP) may be partially responsible for the reduction in blood volume that occurs with prolonged bed rest. When central blood volume and atrial stretch are increased, ANP secretion rises, leading to higher sodium excretion and a reduction in blood volume.

Erythropoietin Increases the Production of Erythrocytes

The final step in blood volume regulation is production of erythrocytes. **Erythropoietin** is a hormone released by the kidneys that causes bone marrow to increase production of red blood cells, raising the total mass of circulating red cells. The stimuli for erythropoietin release include hypoxia and reduced hematocrit. An increase in circulating AVP and aldosterone enhances salt and water retention and results in an elevated plasma volume. The increased plasma volume (with a constant volume of red blood cells) results in a lower hematocrit. The decrease in hematocrit stimulates erythropoietin release, which stimulates red blood cell synthesis and, therefore, balances the increase in plasma volume with a larger red blood cell mass.

COMPARISON OF SHORT-TERM AND LONG-TERM BLOOD PRESSURE CONTROL

Different mechanisms are responsible for the short-term and long-term control of blood pressure. Short-term control depends on activation of neural and hormonal responses by the baroreceptor reflexes (described earlier).

Long-term control depends on salt and water excretion by the kidneys. Excretion of salt and water by the kidneys is regulated by some neural and hormonal mechanisms, most of which have been mentioned earlier in this chapter. However, it is also regulated by arterial pressure. Increased arterial pressure results in increased excretion of salt and water—a phenomenon known as **pressure diuresis** (Fig. 18.6). Because of pressure diuresis, as long as mean arterial pressure is elevated, salt and water excretion will exceed the normal rate; this will tend to lower extracellular fluid vol-

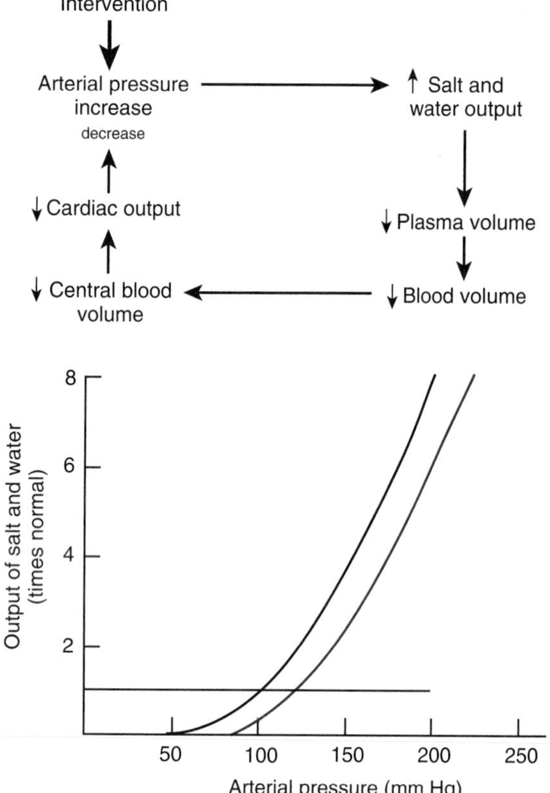

FIGURE 18.6 Regulation of arterial pressure by pressure diuresis. A higher output of salt and water in response to increased arterial pressure reduces blood volume. Blood volume is reduced until pressure returns to its normal level. The curve on the left shows the relationship in a person with normal blood pressure. The curve on the right shows the same relationship in an individual who is hypertensive. Note that the hypertensive individual has an elevated arterial pressure at a normal output of salt and water. (Modified from Guyton AC, Hall JE. Medical Physiology. 10th Ed. Philadelphia: WB Saunders, 2000;203.)

ume and, ultimately, blood volume. As discussed earlier in this chapter and in Chapter 15, a decrease in blood volume reduces stroke volume by lowering the end-diastolic filling of the ventricles. Decreased stroke volume lowers cardiac output and arterial pressure. Pressure diuresis persists until it lowers blood volume and cardiac output sufficiently to return mean arterial pressure to a set level. A decrease in mean arterial pressure has the opposite effect on salt and water excretion. Reduced pressure diuresis increases blood volume and cardiac output until mean arterial pressure is returned to a set level.

Pressure diuresis is a slow but persistent mechanism for regulating arterial pressure. Because it persists in altering salt and water excretion and blood volume as long as arterial pressure is above or below a set level, it will eventually return pressure to that level. In hypertensive patients, the curve shown in Figure 18.6 is shifted to the right, so that salt and water excretion are normal at a higher arterial pressure. If this were not the case, pressure diuresis would inexorably bring arterial pressure back to normal.

CARDIOVASCULAR CONTROL DURING STANDING

An integrated view of the cardiovascular system requires an understanding of the relationships among cardiac output, venous return, and central blood volume and how these relationships are influenced by interactions among various neural, hormonal, and other control mechanisms. Consideration of the responses to standing erect provides an opportunity to explore these elements in detail. Figure 18.7 compares venous pressures for the recumbent and standing positions. When a person is recumbent, pressure in the veins of the legs is only a few mm Hg above the pressure in the right atrium. The pressure distending the veins—transmural pressure—is equal to the pressure within the veins of the legs because the pressure outside the veins is atmospheric pressure (the zero-reference pressure).

When a person stands, the column of blood above the lower extremities raises venous pressure to about 50 mm Hg at the femoral level and 90 mm Hg at the foot. This is

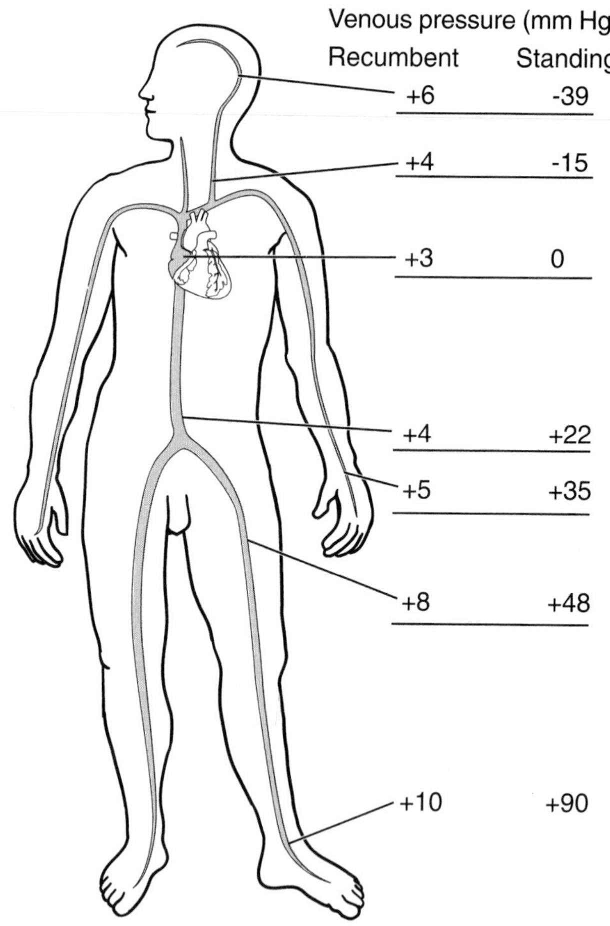

FIGURE 18.7 Venous pressures in the recumbent and standing positions. In this example, standing places a hydrostatic pressure of approximately 80 mm Hg on the feet. Right atrial pressure is lowered because of the reduction in central blood volume. The negative pressures above the heart with standing do not actually occur because once intravascular pressure drops below atmospheric pressure, the veins collapse. These are the pressures that would exist if the veins remained open.

the transmural (distending) pressure because the outside pressure is still zero (atmospheric). Because the veins are highly compliant, such a large increase in transmural pressure is accompanied by an increase in venous volume.

The arteries of the legs undergo exactly the same increases in transmural pressure. However, the increase in their volume is minimal because the compliance of the systemic arterial system is only 1/20th that of the systemic venous system. Standing increases pressure in the arteries and veins of the legs by exactly the same amount, so the added pressure has no influence on the difference in pressure driving blood flow from the arterial to the venous side of the circulation. It only influences the distension of the veins.

Standing Requires a Complex Cardiovascular Response

When a person stands and the veins of the legs are distended, blood that would normally be returned toward the right atrium remains in the legs, filling the expanding veins. For a few seconds after standing, venous return to the heart is lower than cardiac output and, during this time, there is a net shift of blood from the central blood volume to the veins of the legs.

When a 70-kg person stands, central blood volume is quickly reduced by approximately 550 mL. If no compensatory mechanisms existed, this would significantly reduce cardiac end-diastolic volume and cause a more than 60% decrease in stroke volume, cardiac output, and blood pressure; the resulting fall in cerebral blood flow would probably cause a loss of consciousness. If the individual continues to stand quietly for 30 minutes, 20% of plasma volume is lost by net filtration through the capillary walls of the legs. Therefore, quiet standing for half an hour without compensation is the hemodynamic equivalent of losing a liter of blood. It follows that an adequate cardiovascular response to the changes caused by upright posture—**orthostasis**—is absolutely essential to our lives as bipeds (see Clinical Focus Box 18.1).

The immediate cardiovascular adjustments to upright posture are the baroreceptor- and cardiopulmonary receptor-initiated reflexes, followed by the muscle and respiratory pumps and, later, adjustments in blood volume.

Standing Elicits Baroreceptor and Cardiopulmonary Reflexes

The decreased central blood volume caused by standing includes reduced atrial, ventricular, and pulmonary vessel volumes. These volume reductions unload the cardiopulmonary receptors and elicit a cardiopulmonary reflex. Reduced left ventricular end-diastolic volume decreases stroke volume and pulse pressure as well as cardiac output and mean arterial pressure, leading to decreased firing of aortic arch and carotid baroreceptors. The combined reduction in firing of cardiopulmonary receptors and baroreceptors results in a reflex decrease in parasympathetic nerve activity and an increase in sympathetic nerve activity to the heart.

When a person stands up, the heart rate generally increases by about 10 to 20 beats/min. The increased sympathetic nerve activity to the ventricular myocardium shifts the ventricle to a new function curve and, despite the lowered ventricular filling, stroke volume is decreased to only 50 to 60% of the recumbent value. In the absence of the compensatory increase in sympathetic nerve activity, stroke volume would fall much more. These cardiac adjustments maintain cardiac output at 60 to 80% of the recumbent value. An increase in sympathetic activity also causes arteriolar constriction and increased SVR. The effect of these compensatory changes in heart rate, ventricular con-

CLINICAL FOCUS BOX 18.1

Hypotension

Baroreceptors, volume receptors, chemoreceptors, and pain receptors all help maintain adequate blood pressure during various forms of hemodynamic stress, such as standing and exercise. However, in the presence of certain cardiovascular abnormalities, these mechanisms may fail to regulate blood pressure appropriately; when this occurs, a person may experience transient or sustained **hypotension**. As a practical definition, hypotension exists when symptoms are caused by low blood pressure and, in extreme cases, hypotension may cause weakness, lightheadedness, or even fainting.

Hypotension may be due to neurogenic or nonneurogenic factors. Neurogenic causes include autonomic dysfunction or failure, which can occur in association with other central nervous system abnormalities, such as Parkinson's disease, or may be secondary to systemic diseases that can damage the autonomic nerves, such as diabetes or amyloidosis; vasovagal hyperactivity; hypersensitivity of the carotid sinus; and drugs with sympathetic stimulating or blocking properties. Nonneurogenic causes of hypotension include vasodilation caused by alcohol, vasodilating drugs, or fever; cardiac disease (e.g., cardiomyopathy, valvular disease); or reduced blood volume secondary to hemorrhage, dehydration, or other causes of fluid loss. In many patients, multiple causative factors are involved.

The treatment of symptomatic hypotension is to eliminate the underlying cause whenever possible, which, in some cases, produces satisfactory results. When this approach is not possible, other adjunctive measures may be necessary, especially when the symptoms are disabling. Common treatment modalities include avoidance of factors that can precipitate hypotension (e.g., sudden changes in posture, hot environments, alcohol, certain drugs, large meals), volume expansion (using salt supplements and/or medications with salt-retaining/volume-expanding properties), and mechanical measures (including tight-fitting elastic compression stockings or pantyhose to prevent the blood from pooling in the veins of the legs upon standing). Unfortunately, even when these measures are employed, some patients continue to have severe, debilitating effects from hypotension.

tractility, and SVR is maintenance of mean arterial pressure. In fact, mean arterial pressure may be increased slightly above the recumbent value.

How is increased sympathetic nerve activity maintained if the mean arterial pressure reaches a value near or above that of the recumbent value? In other words, why doesn't the sympathetic nerve activity return to recumbent levels if the mean arterial pressure returns to the recumbent value? There are two reasons. First, although the mean arterial pressure returns to the same level (or even higher), pulse pressure remains reduced because the stroke volume is decreased to 50 to 60% of the recumbent value. As indicated earlier, the firing rate of the baroreceptors depends on both mean arterial and pulse pressures. Reduced pulse pressure means the baroreceptor firing rate is reduced even if the mean arterial pressure is slightly higher. Second, although mean arterial pressure is returned to the recumbent value, central blood volume remains low. Consequently, the cardiopulmonary receptors continue to discharge at a lower rate, leading to increased sympathetic activity. Some investigators believe it is the decreased stretch of the cardiopulmonary receptors that provides the primary steady state afferent information for the reflex cardiovascular response to standing.

The heart and brain do not participate in the arteriolar constriction caused by increased sympathetic nerve activity during standing; therefore, the blood flow and supply of oxygen and nutrients to these two vital organs are maintained.

Muscle and Respiratory Pumps Help Maintain Central Blood Volume

Although standing would appear to be a perfect situation for increased venoconstriction (which could return some of the blood from the legs to the central blood volume), reflex venoconstriction is a relatively minor part of the response to standing. A more powerful activation of the baroreceptor reflex, as occurs during severe hemorrhage, is required to cause significant venoconstriction. However, two other mechanisms return blood from the legs to the central blood volume. The more important mechanism is the **muscle pump** (Fig. 18.8). If the leg muscles periodically contract while an individual is standing, venous return is increased. Muscles swell as they shorten, and this compresses adjacent veins. Because of the venous valves in the limbs, the blood in the compressed veins can flow only toward the heart. The combination of contracting muscle and venous valves provides an effective pump that transiently increases venous return relative to cardiac output. This mechanism shifts blood volume from the legs to the central blood volume, and end-diastolic volume is increased. Even mild exercise, such as walking, returns the central blood volume and stroke volume to recumbent values (Fig. 18.9).

The **respiratory pump** is the other mechanism that acts to enhance venous return and restore central blood volume (Fig. 18.10). Quiet standing for 5 to 10 minutes invariably leads to sighing. This exaggerated respiratory movement lowers intrathoracic pressure more than usually occurs with inspiration. The fall in intrathoracic pressure raises the transmural pressure of the intrathoracic vessels, causing these vessels to expand. Contraction of the diaphragm simultaneously raises intraabdominal pressure, which compresses the abdominal veins. Because the venous valves prevent the backflow of blood into the legs, the raised intraabdominal pressure forces blood toward the intrathoracic vessels (which are expanding because of the lowered intrathoracic pressure). The seesaw action of the respiratory pump tends to displace extrathoracic blood volume toward the chest and raise right atrial pressure and stroke volume. Figure 18.11 provides an overview of the main cardiovascular events associated with a short period of standing.

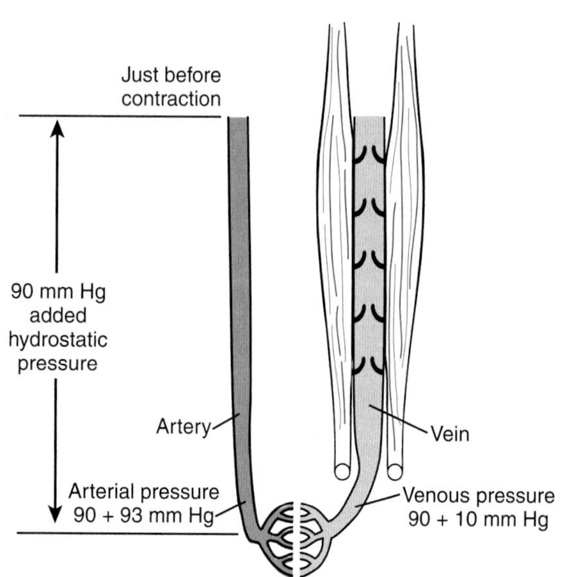

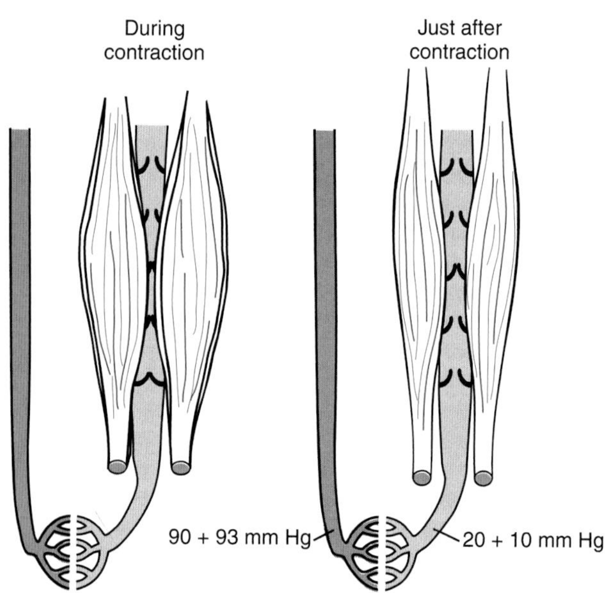

FIGURE 18.8 **Muscle pump.** This mechanism increases venous return and decreases venous volume. The valves (which are closed after contraction) break up the hydrostatic column of blood, lowering venous (and capillary) hydrostatic pressure.

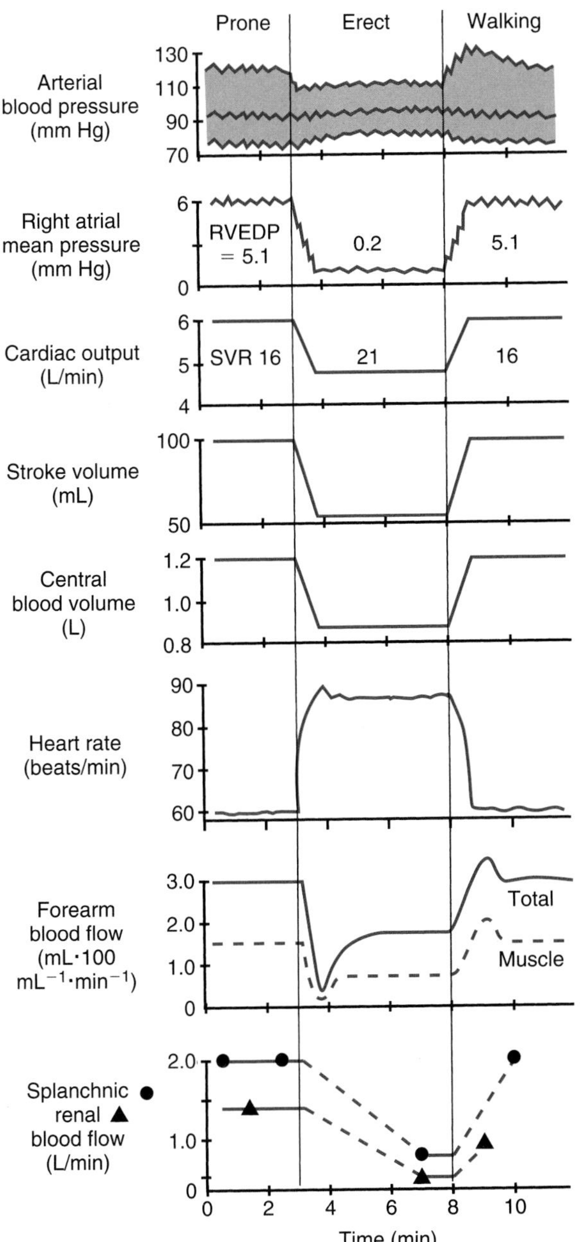

FIGURE 18.9 **Effect of the muscle pump on central blood volume and systemic hemodynamics.** The center section shows the effects of a shift from the prone to the upright position with quiet standing. The right panel shows the effect of activating the muscle pump by contracting leg muscles. Note that the muscle pump restores central blood volume and cardiac output to the levels in the prone position. The fall in heart rate and rise in peripheral blood flow (forearm, splanchnic, and renal) associated with activation of the muscle pump reflect the reduction in baroreceptor reflex activity associated with increased cardiac output. RVEDP, right ventricular end-diastolic pressure; SVR, systemic vascular resistance. (Modified from Rowell LB. Human Circulation: Regulation During Physical Stress. New York: Oxford University Press, 1986.)

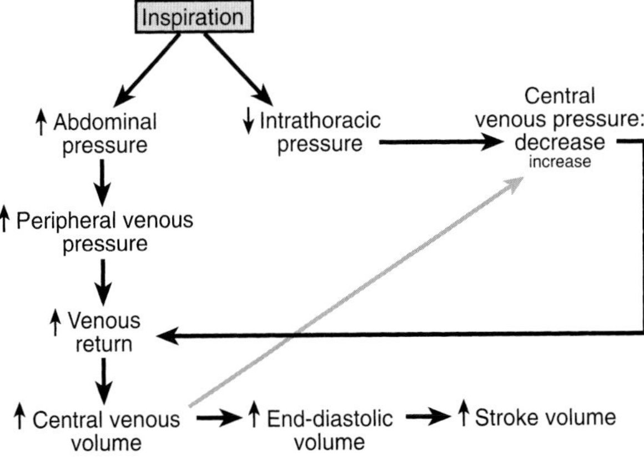

FIGURE 18.10 **Respiratory pump.** Inspiration leads to an increase in venous return and stroke volume. Small type represents a secondary change that returns variables toward the original values.

Capillary Filtration During Standing Further Reduces Central Blood Volume

During quiet (minimum muscular movement) standing for 10 to 15 minutes, the effects of the baroreceptor reflex on the heart and arterioles are insufficient to prevent a continued decline in arterial pressure. The decline in arterial pressure is caused by a steady loss of plasma volume, as fluid filters out of capillaries of the legs. The hydrostatic column of

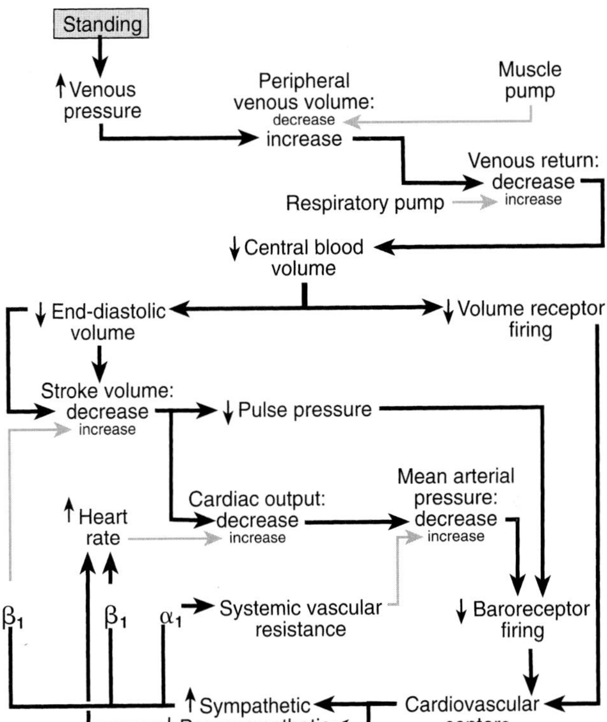

FIGURE 18.11 **Cardiovascular events associated with standing.** Small type represents compensatory changes that return variables toward the original values. α_1 and β_1 refer to adrenergic receptor types.

blood above the capillaries of the legs and feet raises capillary hydrostatic pressure and filtration. During a period of 30 minutes, a 10% loss of blood volume into the interstitial space can occur. This loss, coupled with the 550 mL displaced by redistribution from the central blood volume into the legs, causes central blood volume to fall to a level so low that reflex sympathetic nerve activity cannot maintain cardiac output and mean arterial pressure. Diminished cerebral blood flow and a loss of consciousness (fainting) result.

Arteriolar constriction due to the increased reflex sympathetic nerve activity tends to reduce capillary hydrostatic pressure. However, this alone does not bring capillary hydrostatic pressure back to normal because it does not affect the hydrostatic pressure exerted on the capillaries from the venous side. The muscle pump is the most important factor counteracting increased capillary hydrostatic pressure. The alternate compression and filling of the veins as the muscle pump works means the venous valves are closed most of the time. When the valves are closed, the hydrostatic column of blood in the leg veins at any point is only as high as the distance to the next valve.

The myogenic response of arterioles to increased transmural pressure also acts to oppose filtration. As discussed earlier, raising the transmural pressure stretches vascular smooth muscle and stimulates it to contract. This is especially true for the myocytes of precapillary arterioles. The elevated transmural pressure associated with standing causes a myogenic response and decreases the number of open capillaries. With fewer open capillaries, the filtration rate for a given capillary hydrostatic pressure imbalance is less.

In addition to the factors cited above, other safety factors against edema are important for preventing excessive translocation of plasma volume into the interstitial space (see Chapter 16). These factors, together with neural and myogenic responses and the muscle and respiratory pumps, play a significant role during the seconds and minutes following standing (Fig. 18.12). The combination of all of these factors minimizes net capillary filtration, making it possible to remain standing for long periods.

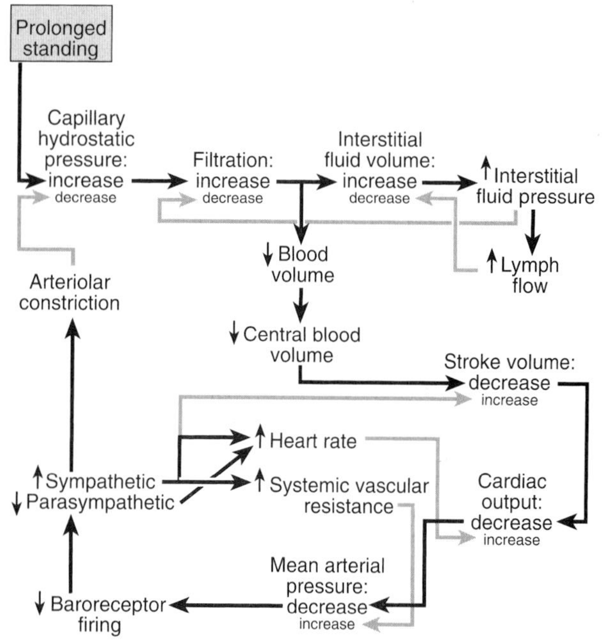

FIGURE 18.12 **Effects of prolonged standing.** With prolonged standing, capillary filtration reduces venous return. Without the compensatory events that result in the changes shown in small type, prolonged standing would inevitably lead to fainting.

Long-Term Responses Defend Venous Return During Prolonged Upright Posture

In addition to the relatively short-term cardiovascular responses, there are equally important long-term adjustments to orthostasis. These are observed in patients confined to bed (or astronauts not subject to the force of gravity). In people who are bedridden, intermittent upright posture does not shift the distribution of blood volume from the thorax to the legs. During the course of a day, average central blood volume (and pressure) is greater than in a person who is periodically standing up in the presence of gravity. The average increase in central blood volume caused by ex-

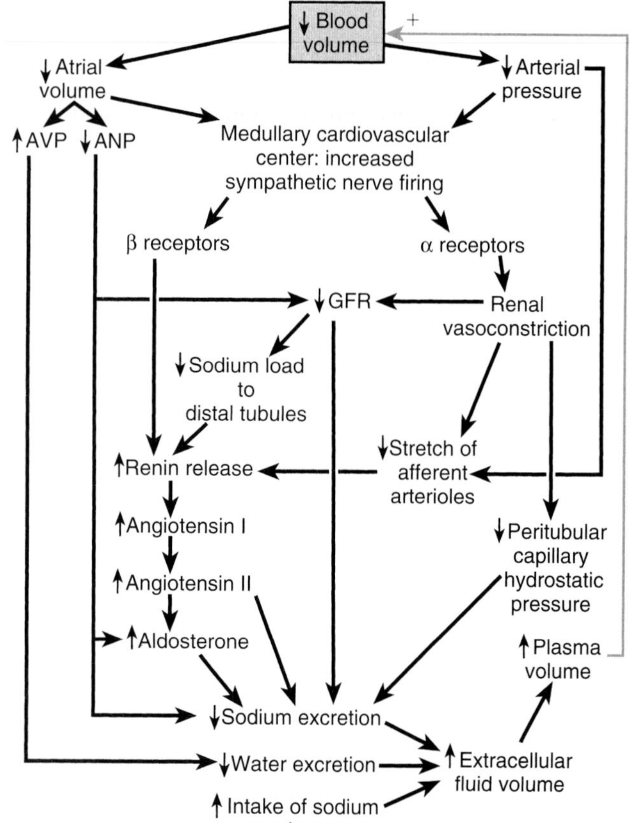

FIGURE 18.13 **Regulation of blood volume.** Blood loss influences sodium and water excretion by the kidney via several pathways. All these pathways, combined with an increased intake of salt and water, restore the extracellular fluid volume and, eventually, blood volume. These responses occur later than those shown in Figures 18.10, 18.11, and 18.12. The pathways responsible for stimulating an increased intake of salt and water are not shown. AVP, arginine vasopressin; ANP, atrial natriuretic peptide; GFR, glomerular filtration rate.

tended bed rest results in reduced activity of all of the pathways that increase blood volume in response to standing. The reduction in total blood volume begins during the first day and is quantitatively significant after a few days. At this point, standing becomes difficult because blood volume is not adequate to sustain a normal blood pressure. Looking at it another way, maintaining an erect posture in the Earth's gravitational field results in increased blood volume. This increase, proportioned between the extrathoracic and intrathoracic vessels, augments stroke volume during standing. If blood volume is not maintained by intermittent erect posture, standing becomes extremely difficult or impossible because of **orthostatic hypotension**—diminished blood pressure associated with standing.

The long-term regulation of blood volume is driven by changes in plasma volume accomplished by sympathetic nervous system effects on the kidneys; hormonal changes, including RAAS, AVP, and ANP; and alterations in pressure diuresis. Figure 18.13 depicts several components of plasma volume regulation by showing their response to a moderate (approximately 10%) blood loss, which is easily compensated for in healthy individuals.

Plasma is a part of the extracellular compartment and is subject to the factors that regulate the size of that space. The osmotically important electrolytes of the extracellular fluid are the sodium ion and its main partner, the chloride ion. The control of extracellular fluid volume is determined by the balance between the intake and excretion of sodium and water. This topic is discussed in depth in Chapter 24. Sodium excretion is much more closely regulated than sodium intake. Excretion of sodium is determined by the glomerular filtration rate, the plasma concentrations of aldosterone and ANP, and a variety of other factors, including angiotensin II.

Glomerular filtration rate is determined by glomerular capillary pressure, which is dependent on precapillary (afferent arteriolar) and postcapillary (efferent arteriolar) resistance and arterial pressure. Decreased mean arterial pressure and/or afferent arteriolar constriction tends to result in lowered glomerular capillary pressure, less filtration of fluid, and lower sodium excretion. Changes in glomerular capillary pressure are primarily the result of changes in sympathetic nerve activity and plasma angiotensin II and ANP concentrations.

Aldosterone acts on the distal nephron to cause increased reabsorption of sodium and, thereby, decrease its excretion. Aldosterone released from the adrenal cortex is increased by (among other things) angiotensin II. Water intake is determined by thirst and the availability of water.

The excretion of water is strongly influenced by AVP. Increased plasma osmolality, sensed by the hypothalamus, results in both thirst and increased AVP release. Thirst and AVP release are also increased by decreased stretch of baroreceptors and cardiopulmonary receptors.

Consider how these physiological variables are altered by an upright posture to produce an increase in the extracellular fluid volume. Renal arteriolar vasoconstriction associated with increased sympathetic nerve activity produced by standing reduces the glomerular filtration rate. This results in a decrease in filtered sodium and tends to decrease sodium excretion. The increased sympathetic nerve activity to the kidney also triggers the release of renin, which increases circulating angiotensin II and, in turn, aldosterone release. The decrease in central blood volume associated with standing reduces cardiopulmonary stretch receptor activity, causing an increased release of AVP from the posterior pituitary. Therefore, both sodium and water are retained and thirst is increased. Regulation of the precise quantities of water and sodium that are excreted maintains the correct osmolality of the plasma.

The distribution of extracellular fluid between plasma and interstitial compartments is determined by the balance of hydrostatic and colloid osmotic forces across the capillary wall. Retention of sodium and water tends to dilute plasma proteins, decreasing plasma colloid osmotic pressure and favoring the filtration of fluid from the plasma into the interstitial fluid. However, as increased synthesis of plasma proteins by the liver occurs, a portion of the retained sodium and water contributes to an increase in plasma volume.

Finally, the increase in plasma volume (in the absence of any change in total red cell volume) decreases hematocrit, which stimulates erythropoietin release and erythropoiesis. This helps total red blood cell volume keep pace with plasma volume.

REVIEW QUESTIONS

DIRECTIONS: Each of the numbered items or incomplete statements in this section is followed by answers or by completions of the statement. Select the ONE lettered answer or completion that is BEST in each case.

1. A person has cold, painful fingertips because of excessively constricted blood vessels in the skin. Which of the following alterations in autonomic function is most likely to be involved?
(A) Low concentration of circulating epinephrine
(B) High sensitivity of arterioles to norepinephrine
(C) High sensitivity of arterioles to nitric oxide
(D) Low parasympathetic nerve activity
(E) Arterioles insensitive to epinephrine
2. In the presence of a drug that blocks all effects of norepinephrine and epinephrine on the heart, the autonomic nervous system can
(A) Raise the heart rate above its intrinsic rate
(B) Lower the heart rate below its intrinsic rate
(C) Raise and lower the heart rate above and below its intrinsic rate
(D) Neither raise nor lower the heart rate from its intrinsic rate
3. The cold pressor response is initiated by stimulation of
(A) Baroreceptors
(B) Cardiopulmonary receptors
(C) Hypothalamic receptors
(D) Pain receptors
(E) Chemoreceptors

(continued)

4. Which of the following occurs when acetylcholine binds to muscarinic receptors?
(A) Heart rate slows
(B) Cardiac conduction velocity rises
(C) Norepinephrine release from sympathetic nerve terminals is enhanced
(D) Nitric oxide release from endothelial cells is inhibited
(E) Blood vessels of the external genitalia constrict
5. Carotid baroreceptors
(A) Are important in the rapid, short-term regulation of arterial blood pressure
(B) Do not fire until a pressure of approximately 100 mm Hg is reached
(C) Adapt over 1 to 2 weeks to the prevailing mean arterial pressure
(D) Stretch reflexly decreases cerebral blood flow
(E) Reflexly decrease coronary blood flow when blood pressure falls
6. Which of the following is true with respect to peripheral chemoreceptors?
(A) Activation is important in inhibiting the diving response
(B) Activity is increased by increased pH
(C) They are located in the medulla oblongata, but not the hypothalamus
(D) Activation is important in the cardiovascular response to hemorrhagic hypotension
(E) Activity is increased by lowering of the oxygen content, but not the Po_2, of arterial blood
7. Parasympathetic stimulation of the heart accompanied by a withdrawal of sympathetic tone to most of the blood vessels of the body is characteristic of
(A) The fight-or-flight response
(B) Vasovagal syncope
(C) Exercise
(D) The diving response
(E) The cold pressor response
8. A patient suffers a severe hemorrhage resulting in a lowered mean arterial pressure. Which of the following would be elevated *above normal* levels?
(A) Splanchnic blood flow
(B) Cardiopulmonary receptor activity
(C) Right ventricular end-diastolic volume
(D) Heart rate
(E) Carotid baroreceptor activity
9. A person stands up. Compared with the recumbent position, 1 minute after standing, the
(A) Skin blood flow increases
(B) Volume of blood in leg veins increases
(C) Cardiac preload increases
(D) Cardiac contractility decreases
(E) Brain blood flow decreases
10. Pressure diuresis lowers arterial pressure because it
(A) Lowers renal release of renin
(B) Lowers systemic vascular resistance
(C) Lowers blood volume
(D) Causes renal vasodilation
(E) Increases baroreceptor firing
11. Central blood volume is decreased by
(A) The muscle pump
(B) The respiratory pump
(C) Increased excretion of salt and water
(D) Lying down
(E) Living in a space station

SUGGESTED READING

Champleau MW. Arterial baroreflexes. In: Izzo JL, Black HR, eds. Hypertension Primer. Baltimore: Lippincott Williams & Wilkins, 1999.

Dampney RA. Functional organization of central pathways regulating the cardiovascular system. Physiol Rev 1994;74:323–364.

Hainsworth R, Mark AL, eds. Cardiovascular Reflex Control in Health and Disease. London: WB Saunders, 1993.

Katz AM. Physiology of the Heart. 3rd Ed. New York: Lippincott Williams & Wilkins, 2001.

Mohanty PK. Cardiopulmonary baroreflexes. In: Izzo JL, Black HR, eds. Hypertension Primer. Baltimore: Lippincott Williams & Wilkins, 1999.

Reis DJ. Functional neuroanatomy of central vasomotor control centers. In: Izzo JL, Black HR, eds. Hypertension Primer. Baltimore: Lippincott Williams & Wilkins, 1999.

Rowell LB. Human Cardiovascular Control. New York: Oxford University Press, 1993.

Waldrop TG, Eldridge FL, Iwamoto GA, Mitchell JH. Central neural control of respiration and circulation during exercise. In: Rowell LB, Shepherd JT, eds. Handbook of Physiology, Section 12. Exercise: Regulation and integration of multiple systems. New York: Oxford University Press, 1996.

CASE STUDIES FOR PART IV • • •

CASE STUDY FOR CHAPTER 11

Chronic Granulomatous Disease of Childhood

An 18-month-old boy, with a high fever and cough and with a history of frequent infections, is brought to the emergency department by his father. A blood examination shows elevated numbers of neutrophils but no other defects. A blood culture for bacteria is positive. The physician sends a sample of the boy's blood to a laboratory to test the ability of the patient's neutrophils to produce hydrogen peroxide. The ability of this patient's neutrophils to generate hydrogen peroxide is found to be completely absent.

Questions

1. What cellular defect may have led to the complete absence of hydrogen peroxide generation in this patient's neutrophils?
2. How might this disease be treated using hematotherapy?

Answers to Case Study Questions for Chapter 11

1. The disease, chronic granulomatous disease of childhood, results from a congenital lack of the superoxide-forming enzyme NADPH oxidase in this patient's neutrophils. The lack of this enzyme results in deficient hydrogen peroxide generation by these cells when they ingest or phagocytose bacteria, resulting in a compromised capacity to combat recurrent, life-threatening bacterial infections.
2. Normal neutrophil stem cells grown in culture may be infused to supplement the patient's own defective neutrophils. In addition, researchers are now trying to genetically reverse the defect in cultures of a patient's stem cells for subsequent therapeutic infusion.

Reference

Baehner RL. Chronic granulomatous disease of childhood: Clinical, pathological, biochemical, molecular, and genetic aspects of the disease. Pediatr Pathol 1990;10:143–153.

Case Study for Chapter 12

Congestive Heart Failure (Arteriovenous Fistula)

A 29-year-old man presented to his physician with fatigue, shortness of breath, and progressive ankle edema. These signs and symptoms had been worsening slowly for 3 months. His medical history included a motor vehicle accident 4 months ago, during which he sustained a deep puncture wound to the right thigh. The wound was closed with skin sutures on the day of the accident and had healed, although the area around the injury remained tender.

On physical examination, his resting blood pressure is 90/60 mm Hg, and his heart rate is 122 beats/min. He appears ill and has shortness of breath at rest. Bilateral lung crackles are present. Pitting edema is evident in both legs but is worse on the right. His pulses are intact, but the amplitude of the right femoral pulse is increased. A continuous bruit is present over the scar from his previous puncture injury. The superficial veins in the right thigh are prominent and appear distended.

Questions

1. What is the cause of the femoral bruit?
2. Why does the patient have fatigue, shortness of breath, leg edema, lung crackles, and an elevated heart rate?

Answers to Case Study Questions for Chapter 12

1. The patient has an arteriovenous (A-V) fistula caused by his previous puncture injury. During the injury, both the artery and the adjacent vein in the thigh were severed; the vessels healed but, during the healing process, a direct connection formed between the artery and the adjacent vein. The velocity of flow from the artery to the vein is very high; it produces turbulence and a bruit.
2. A large A-V fistula, such as this one, allows a substantial amount of the cardiac output to be shunted directly from the arterial system to the venous system, without passing through the resistance vessels. The lowered systemic vascular resistance leads to a lower arterial pressure. Compensatory mechanisms increase heart rate and cardiac output. However, continuous delivery of a high cardiac output for months causes the heart muscle to fail. As the heart muscle fails, the output of the heart cannot be maintained. This results in the accumulation of fluid in the lungs, causing crackles and shortness of breath, and in the legs, where it appears as pitting edema. Because so much blood is shunted directly to the venous circulation, there is reduced availability of arterial blood for many tissues, including skeletal muscle, thereby causing fatigue.

References

Schneider M, Creutzig A, Alexander K. Untreated arteriovenous fistula after World War II trauma. Vasa 1996;25:174–179.

Wang KT, Hou CJ, Hsieh JJ, et al. Late development of renal arteriovenous fistula following gunshot trauma—a case report. Angiology 1998;49:415–418.

Case Study for Chapter 13

Atrial Fibrillation

A 58-year-old woman arrived in the emergency department complaining of sudden onset of palpitations, light-headedness, and shortness of breath. These symptoms began approximately 2 hours previously. On examination, her blood pressure is 95/70 mm Hg, and her heart rate is 140 beats/min. An ECG demonstrates atrial fibrillation. The physical examination is otherwise unremarkable.

Questions

1. Explain why the patient has these symptoms.
2. Explain how medications could be useful in this setting.
3. While in the emergency department, the patient's symptoms worsened. What immediate action could be taken to stabilize or treat the patient?

Answers to Case Study Questions for Chapter 13

1. During atrial fibrillation, the AV node is incessantly stimulated. Depending upon the conduction velocity and refractory period of the node, the ventricular rate may be from 100 to more than 200 beats/min. When the ventricular rate is extremely rapid, there is little opportunity for ventricular filling to occur; despite the high heart rate, cardiac output falls in this setting (see Chapter 14). This leads to hypotension and associated symptoms such as light-headedness and shortness of breath.
2. Drugs that can slow down conduction through the AV node are useful in treating atrial fibrillation. These included digitalis, beta blockers, and calcium entry blockers. By slowing AV nodal conduction, these drugs reduce the rate of excitation of the ventricles. At a slower ventricular rate, there is more time for filling, and the output of the heart is increased.
3. Atrial fibrillation can be terminated by electrical cardioversion. In this procedure, a strong electrical current is passed through the heart to momentarily depolarize the entire heart. As repolarization occurs, a normal, coordinated rhythm is reestablished.

Reference

Shen W-K, Holmes DR Jr, Packer DL. Cardiac arrhythmias. In: Giuliani ER, Nishimura RA, Holmes DR Jr, eds. Mayo Clinic Practice of Cardiology. 3rd Ed. St. Louis: CV Mosby, 1996;727–747.

Case Study for Chapter 14

Left Ventricular Hypertrophy (Aortic Stenosis)

A 72-year-old woman presented to her physician with a complaint of poor exercise tolerance and dyspnea on exertion. Cardiac auscultation reveals a fourth heart sound and a loud systolic murmur heard best at the base of the heart. The murmur radiates into the region of the carotid artery. The carotid pulses are reduced in amplitude and feel "dampened." The ECG indicates left ventricular hypertrophy.

Questions

1. Why does the patient have a murmur?
2. Why has left ventricular hypertrophy developed?
3. How should this condition be managed?

Answers to Case Study Questions for Chapter 14

1. The aortic valve of this patient has become narrowed and calcified (aortic stenosis). Because blood must squeeze through the narrowed orifice, flow velocity increases and the blood flow becomes turbulent. This turbulence creates a murmur during cardiac systole (when blood is ejected through the valve).
2. To eject blood through the narrowed aortic valve, the ventricle must develop higher pressure during systole. In response to a sustained increase in afterload, hypertrophy of the muscle of the left ventricle occurs.
3. When symptoms develop and left ventricular enlargement is present, aortic stenosis is best treated with surgery. The valve can be replaced with a prosthetic valve.

Reference

Rahimtoola SH. Aortic stenosis. In: Fuster V, Alexander RW, O'Rourke FA , eds. Hurst's the Heart. 10th Ed. New York: McGraw-Hill, 2001.

CASE STUDY FOR CHAPTER 15

Pulmonary Embolism

A 68-year-old man receiving chemotherapy for colon cancer experienced the sudden onset of chest discomfort and shortness of breath. His blood pressure is 100/75 mm Hg, and his heart rate is 105 beats/min. The physical examination is unremarkable except for swelling and tenderness in the left leg, which began about 3 days earlier. The ECG shows no changes suggestive of cardiac ischemia.

Questions

1. How are the patient's chest discomfort, shortness of breath, arterial hypotension, tachycardia, and left leg symptoms explained?
2. Is right ventricular pressure likely to be increased or decreased? Why?
3. Would intravenous infusion of additional fluids (such as blood or plasma) help the patient's arterial blood pressure?

Answers to Case Study Questions for Chapter 15

1. The patient's symptoms are caused by pulmonary embolism. In this condition, a piece of blood clot located in a peripheral vein (in this case, a leg vein) breaks off and is carried through the right heart to a pulmonary artery where it lodges. Patients with certain medical problems, including cancer, have altered clotting mechanisms and are at risk of forming these clots. When this occurs, blood flow from the pulmonary artery to the left heart is obstructed (i.e., pulmonary vascular resistance increases), resulting in elevated pulmonary arterial pressure. The sudden rise in pressure causes distension of the artery, which may contribute to the sensation of chest discomfort. Increased pulmonary arterial pressure (pulmonary hypertension) leads to right heart failure. Because left atrial (and left ventricular) filling is reduced (as a result of lack of blood flow from the lungs), left-side cardiac output also falls. The fall in cardiac output causes a reflex increase in heart rate. The result is a combination of right- and left-side heart failure, producing the signs and symptoms seen in this patient.
2. The right ventricular pressure is likely to be increased because the blood clot in the pulmonary artery acts as a form of obstruction that raises the pulmonary artery resistance.
3. The problem here is increased afterload of the right ventricle caused by partial obstruction of the outflow tract. Because of this obstructed outflow, the diastolic volume of the right ventricle is already high. It is unlikely that infusing additional fluids into the veins will improve cardiac output because the extra filling of the right ventricle is unlikely to increase the force of contraction.

Reference

Brownell WH, Anderson FA Jr. Pulmonary embolism. In: Gloviczki P, Yao JST, eds. Handbook of Venous Disorders: Guidelines of the American Venous Forum. London: Chapman & Hall, 1996;274.

CASE STUDY FOR CHAPTER 16

Diabetic Microvascular Disease

A 48-year-old man went for a vision examination because his eyesight had been blurry for the past several months. His optometrist referred him to his family physician after seeing a few areas of dense clumps of capillaries over the retinas of both eyes.

The family physician finds fasting blood plasma glucose of 297 mg/dL. The man states he has had periods of tingling and numbness in his toes for a few weeks, which he attributes to gaining over 35 kg during the past 3 years.

Questions

1. Why were capillaries overgrowing the retina? Is this ever a normal finding?
2. Why does an elevated plasma glucose concentration during fasting indicate serious diabetes mellitus? Why does a large weight gain potentially lead to diabetes mellitus?
3. How might odd sensations in the feet be related to diabetes mellitus and microvascular disease?
4. What are the immediate and long-term treatments for minimizing further microvascular disease?

Answers to Case Study Questions for Chapter 16

1. The formation of clumps of capillaries over the retina is usually diagnostic for microvascular complications of diabetes mellitus and is rarely seen in other diseases. The capillaries probably overgrow the retina because they are attempting to replace capillaries that die off as a consequence of the disease.
2. A moderate elevation of blood glucose concentration after a carbohydrate meal can happen, but it should not exceed 140 to 150 mg/dL. Such a high blood glucose represents a major loss in the regulation of glucose metabolism. The patient is seriously overweight and is likely insulin-resistant. He has ample insulin but the cellular response to insulin is inadequate. The suppressed insulin response develops after repeated and sustained high insulin concentrations associated with excessive carbohydrate intake.
3. The peripheral sensory nerves of the body are nourished by microscopic blood vessels, and the loss of even a few vessels can alter the physiology of a nerve. An altered sensory nerve may fire too frequently, causing odd sensations, or not fire at all, causing numbness. Neuropathy or nerve impairment of the lower body is one of the most common problems in diabetes mellitus.
4. Even though this patient would likely have a high insulin concentration, additional insulin is required to stimulate the cells to take up glucose. However, pharmacological treatment could gradually be decreased or discontinued with a major change in diet, amount of body fat, and exercise level. Loss of body fat is associated with a progressive improvement in glucose metabolism. Exercise improves the ability of skeletal muscle cells to take up and burn glucose without the presence of insulin or at reduced insulin concentration.

Reference

Dahl-Jorgensen K. Diabetic microangiopathy. Acta Paediatr Suppl 1998;425:31–34.

CASE STUDY FOR CHAPTER 17

Coronary Artery Disease

A 57-year-old man experienced several months of vague pains in his left chest and shoulder when climbing stairs. During a touch football game at a family picnic, he had much more intense pain and had to rest. After about 45 minutes of intermittent pain, his family brought him to the emergency department.

His heart rate is 105 beats/min, his blood pressure is 105/85 mm Hg, and his hands and feet are cool to touch and somewhat bluish. He is sweating and is short of breath. An electrocardiogram indicates an elevated ST segment, which is most noticeable in leads V4 to V6. The attending cardiologist administers streptokinase intravenously.

One hour later, the ST segment abnormality is less noticeable. The heart rate is 87 beats/min, the arterial blood pressure is 120/85 mm Hg, and the patient's hands and feet are pink and warm. The patient is alert, not sweating, and does not complain of chest pain or shortness of breath.

During a 4-day stay in the hospital, percutaneous angioplasty is performed to open several partially blocked coronary arteries. The patient is told to take half of an adult aspirin pill every day and is given a prescription of a statin drug to lower blood lipids. In addition, he is assigned to a cardiac rehabilitation program designed to teach proper dietary habits and improve exercise performance and, together, to lower body fat gradually.

Questions

1. How did the left chest and shoulder pain during stair climbing predict some abnormality of coronary artery function?
2. Why was a 45-minute delay before going for medical intervention after intense pain started inappropriate for the man's health?
3. How does the lower than normal arterial pressure, smaller than normal arterial pulse pressure, and decreased blood flow to the hands and feet indicate impairment of the contractile function of the heart?
4. How did the streptokinase improve performance of the heart?
5. How is aspirin useful to protect the coronary vasculature from occlusions by blood clots?
6. How might lowering the low-density lipoproteins and raising the high-density lipoproteins with a combination of diet, exercise, and statin therapy lessen the chance of a second heart attack?

Answers to Case Study Questions for Chapter 17

1. The exercise of stair climbing imposed a substantial demand on the heart to pump blood, thereby, requiring more oxygen for the heart cells. Partially occluded arteries did not provide sufficient blood flow to provide the needed oxygen and hypoxia resulted. Coronary artery problems leading to mild hypoxia of the heart muscle typically cause a referred pain to the left chest and shoulder area. In some persons, the pain extends into the left arm and hand, as well as neck and jaw.
2. There is a major risk that cardiac hypoxia will initiate abnormal electrical activity in the heart. The results can range from mild disturbances of conduction to rapidly lethal ventricular fibrillation. In addition, the longer cardiac cells are without adequate blood flow, the more damage is done to the cells. The sooner oxygenation is restored, the less repair is needed in the heart tissue.
3. When the contractile ability of the heart is compromised, the typical result is a reduced stroke volume, which would explain the decreased pulse pressure. If cardiac output decreases, in spite of an increased heart rate, then arterial pressure tends to fall. The decreased blood flow to the hands and feet indicates that the sympathetic nervous system has been activated to constrict peripheral blood vessels, preserving the arterial pressure as much as possible in the presence of reduced cardiac function.
4. Streptokinase is a bacterial product that activates plasminogen, which leads to clot dissolution. Blood flow and oxygen supply to the downstream muscle will then be restored. If the muscle cells are not seriously injured, they will show prompt recovery of contractile function to restore the stroke volume and cardiac output.
5. Aspirin blocks the cyclooxygenase enzymes in all cells. With aspirin present, platelets are far less likely to be activated, limiting clot formation in areas of vessels with damaged endothelial cells. The production of prostaglandins by platelets is part of the clotting process. Also, thromboxane released by activated platelets will cause constriction of coronary arteries and arterioles, lowering blood flow in an already flow-deprived state.
6. Although regression of plaques is not dramatic when low-density lipoproteins are reduced, continued growth of the plaque is decreased and, in some cases, virtually stopped. This lowers the probability of a plaque rupturing and starting the formation of a new clot that will occlude the artery. In addition, lowering the LDL concentration will limit the formation of new plaques and, thereby, reduces the risk of vessel occlusion.

Reference

Lilly LS. Pathophysiology of Heart Disease. Baltimore: Williams & Wilkins, 1998.

CASE STUDY FOR CHAPTER 18

Hypertension

During a routine health assessment, a 52-year-old man is found to have a blood pressure of 180/95 mm Hg. He reports no significant health problems except "my blood pressure has always been a little high."

The physical examination, including an evaluation of the heart, eyes (including the blood vessels of the retina), and the peripheral pulses, is entirely normal. The resting heart rate is 87 beats/min.

Questions

1. How do changes in cardiac output or systemic vascular resistance affect arterial blood pressure?
2. Why did the physician examine the heart, eyes, and peripheral pulses?
3. Explain how drugs might lower the blood pressure by affecting β_1-adrenergic receptors, α_1-adrenergic receptors, intravascular fluid volume, the renin-angiotensin-aldosterone system, and intracellular calcium ion levels.

Answers to Case Study Questions for Chapter 18

1. Anything that increases cardiac output or SVR can cause an increase in arterial blood pressure. When this increase is sustained and significant, it is referred to as hypertension.
2. Chronic hypertension can damage many organs and tissues, some of which may be detected by physical examination. The heart can undergo left ventricular hypertrophy as a result of increased afterload. The blood vessels of the eye can become thickened and sclerotic. Because hypertension can contribute to atherosclerosis, the peripheral pulses may become diminished. Other organs, such as the kidneys, may also be damaged by hypertension, but these abnormalities require specific laboratory testing to evaluate and usually cannot be assessed by physical examination.
3. β_1-Adrenergic blockers reduce heart rate and contractility of the heart and lower cardiac output and blood pressure. They also block ability of the sympathetic nervous system to stimulate the release of renin. Drugs that block α_1-adrenergic receptors reduce peripheral vasoconstriction and thus lower SVR. Drugs that reduce intravascular fluid volume (diuretics such as furosemide or hydrochlorothiazide) reduce preload and, thereby, lower cardiac output and arterial pressure. Drugs that interfere with the RAAS (e.g., by blocking the effect of angiotensin-converting enzyme or by directly blocking the actions of angiotensin II) reduce blood pressure by preventing the vasoconstriction and sodium retention that would otherwise occur when the RAAS is acti-

vated. Calcium blockers diminish cardiac contractility (a determinant of cardiac output) and vascular smooth muscle contraction (a determinant of SVR). These drugs work by decreasing the cytosolic concentration of calcium ion by blocking either its entry or its release into the cytosol of cardiac or smooth muscle cells.

References

Izzo JL, Black HR, eds. Hypertension Primer. Baltimore: Lippincott Williams & Wilkins, 1999.

Vidt DG. Hypertension. In: Young JR, Olin JW, Bartholomew JR, eds. Peripheral Vascular Diseases. 2nd Ed. St. Louis: CV Mosby, 1996;189.

PART V Respiratory Physiology

CHAPTER 19 Ventilation and the Mechanics of Breathing

Rodney A. Rhoades, Ph.D.

CHAPTER OUTLINE

- STRUCTURAL AND FUNCTIONAL RELATIONSHIPS OF THE LUNGS
- PRESSURE CHANGES AND AIRFLOW DURING BREATHING
- SPIROMETRY AND LUNG VOLUMES
- ALVEOLAR VENTILATION
- MECHANICAL PROPERTIES OF THE LUNGS AND CHEST WALL
- AIRFLOW AND THE WORK OF BREATHING

KEY CONCEPTS

1. The primary function of the lungs is to bring fresh air into close contact with the pulmonary capillaries.
2. The pulmonary arterial tree accompanies the airway tree and branches with it.
3. Groups of alveolar ducts and their alveoli merge with pulmonary capillaries to form a terminal respiratory unit.
4. The alveolar-capillary membrane forms a large blood-gas interface for gas exchange.
5. Contraction of the diaphragm creates a small, negative pleural pressure, which causes lung inflation.
6. Spirometry is used to detect lung dysfunction.
7. Alveolar ventilation regulates carbon dioxide levels in the blood.
8. The portion of ventilation that is wasted is known as dead space ventilation.
9. Compliance is a measure of lung distensibility.
10. Surfactant and alveolar interdependence maintain alveolar stability.
11. Airway turbulence has a marked effect on airway resistance.
12. Lung compliance affects airway compression during forced expiration.
13. The work of breathing is required to expand the lungs and overcome airway resistance.

The lungs are the organs of gas exchange, beginning at birth when the newborn takes its first breath. With the first inrush of air, a series of events are set in motion that allow the newborn to switch from a dependent, placental life-support system to an independent air-breathing system. A breath in, a breath out, 12 to 15 times every minute—it may seem like a simple process on which to build the entire human gas exchange system that supplies oxygen to the trillions of cells in the body. This simplicity is deceptive, however, because breathing is amazingly responsive to small changes in blood chemistry, mood, level of alertness, and body activity.

The process of transferring oxygen to and removing carbon dioxide from cells in the body is known as **respiration**. Respiration takes place in two stages. The first stage, known as **gas exchange**, involves the transfer of oxygen and carbon dioxide between the atmosphere and blood in the pulmonary capillaries and then between the systemic blood and the metabolically active tissue. The human lungs are so efficiently designed to remove carbon dioxide and to supply oxygen to tissues that the gas exchange process rarely limits our activity. For example, a marathon runner who staggers across the 26-mile finish line in less than 3 hours or someone who swims the English Channel in record time is rarely limited by gas exchange in the lungs. The reason is that gas exchange can increase more than 20-fold to meet the body's energy demands. These examples of human activity not only underscore the functional capacity of the lungs but also illustrate the important role respiration plays in our extraordinary adaptability.

The second stage of respiration is known as **cellular respiration**, a series of complex metabolic reactions that break down molecules of food, releasing carbon dioxide and energy. Recall that oxygen is required in the final step of cellular respiration to serve as an electron acceptor in the process by which cells obtain energy.

The functions of the respiratory system can be divided into ventilation and the mechanics of breathing; blood flow to the lungs; gas transfer and transport; and the control of breathing. This chapter discusses ventilation and mechanical aspects of the respiratory system—the physical design of the lungs, lung volumes, airflow, alveolar ventilation, and the work of breathing. Chapter 20 discusses the pulmonary circulation and the matching of airflow with blood flow in the lungs. Chapter 21 discusses gas uptake and transport. Chapter 22 deals with basic breathing rhythms, breathing reflexes, and the integrated control of breathing.

STRUCTURAL AND FUNCTIONAL RELATIONSHIPS OF THE LUNGS

The movement of oxygen and carbon dioxide in and out of cells occurs by simple diffusion. In unicellular organisms, the process of gas exchange is simple and requires no special respiratory structures. In more complex organisms, however, cells deep in the body cannot exchange carbon dioxide and oxygen by simple diffusion between cells and the external environment because the diffusion distance is too long. As a result, special gas exchange organs have developed, including tracheal tubes in insects, gills in fish, and lungs in air-breathing animals.

The Airway Tree Divides Repeatedly to Increase Total Cross-Sectional Area

The human gas exchange organ consists of two lungs, each divided into several lobes. The lungs are comprised of two tree-like structures, the **vascular tree** and the **airway tree**, which are embedded in highly elastic connective tissue. The vascular tree consists of arteries and veins connected by capillaries (see Chapter 20). The airway tree consists of a series of hollow branching tubes that decrease in diameter at each branching (Fig. 19.1A). The main airway, the **trachea**, branches into two **bronchi**. Each bronchus enters a lung and branches many times into progressively smaller bronchi, which in turn form **bronchioles**.

A functional model of the airway tree is presented in Figure 19.1B. The trachea and the first 16 generations of airway branches make up the **conducting zone**. The trachea, bronchi, and bronchioles of the conducting zone have three important functions: (1) to warm and humidify inspired air; (2) to distribute air evenly to the deeper parts of the lungs; and (3) to serve as part of the body's defense system (removal of dust, bacteria, and noxious gases from the lungs). The first four generations of the conducting zone are subjected to changes in thoracic pressures and contain a considerable amount of cartilage to prevent airway collapse. In the trachea and main bronchi, the cartilage consists of U-shaped rings. Further down, in the lobar and segmental bronchi, the cartilaginous rings give way to small plates of cartilage. In the bronchioles, the cartilage disappears altogether. The smallest airways in the conducting zone are the **terminal bronchioles**. Bronchioles are suspended by elastic tissue in the lung parenchyma, and the elasticity of the lung tissue helps keep these airways open. The conducting zone has its own separate circulation, the **bronchial circulation**, which originates from the descending aorta and drains into the pulmonary veins. No gas exchange occurs in the conducting zone.

The last seven generations make up the **respiratory zone**, the site of gas exchange. The respiratory zone is composed almost entirely of alveolar ducts and alveoli. Like the conducting zone, the respiratory zone has its own separate and distinct circulation, the **pulmonary circulation.** The lungs have the most extensive capillary network of any organ in the body. Pulmonary capillaries occupy 70 to 80% of the alveolar surface area. The pulmonary circulation receives all of the cardiac output, and therefore, blood flow is very high. Red blood cells can pass through the pulmonary capillaries in less than 1 second. One pulmonary arterial branch accompanies each airway and branches with it.

An increase in internal surface area is accomplished by the formation of outpockets from the small airways to form **alveoli** (see Fig. 19.1). As mentioned above, a network of capillaries surrounds each alveolus and brings blood into close proximity with air inside the alveolus. Oxygen diffuses across the thin wall of the alveolus into the blood, and carbon dioxide diffuses from the blood into the alveolus. Adult lungs contain 300 to 500 million alveoli, with a combined internal surface area of approximately 75 m^2, approximately the size of a tennis court. This represents one of the largest biological membranes in the body. Alveolar surface area increases as the number and size of alveoli increase from birth to adolescence (Table 19.1). However, after adolescence, alveoli only increase in size and, if damaged, have limited ability to repair themselves.

The Vascular and Airway Trees Merge to Form a Blood-Gas Interface

In the respiratory zone, a group of alveolar ducts and their alveoli merge with pulmonary capillaries to form a **terminal respiratory unit**; there are approximately 60,000 of these units in both lungs. The **alveolar-capillary membrane** of these units forms a **blood-gas interface** that sep-

FIGURE 19.1 **Airway tree.** A, The airway tree consists of a series of highly branched hollow tubes that become narrower, shorter, and more numerous as they penetrate the deeper parts of the lung. B, The first 16 generations of branches conduct air to the deeper parts of the lungs and comprise the conducting zone. The last 7 generations participate in gas exchange and comprise the respiratory zone. (B is adapted from Weibel ER. Morphometry of the Human Lung. Berlin: Springer-Verlag, 1963.)

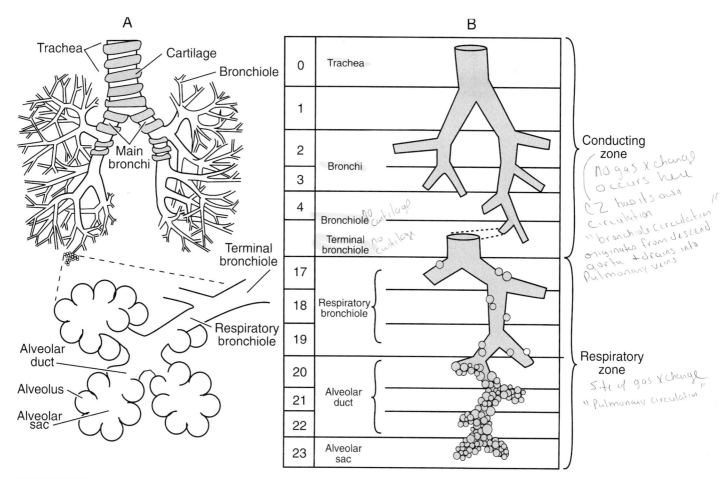

arates the blood in the pulmonary capillaries from the gas in the alveoli (Fig. 19.2). This interface is exceedingly thin (in some places less than 0.5 μm), and is composed of alveolar epithelium, interstitial fluid layer, and capillary endothelium. Air is brought to one side of the interface by **ventilation**—the movement of air to and from the alveoli. Blood is brought to the other side of the interface by the pulmonary circulation. As the blood perfuses the alveolar capillaries, oxygen and carbon dioxide cross the blood-gas interface by **diffusion**.

TABLE 19.1 **Age-Related Changes in Alveolar Number and Surface Area in the Human Lung**

Age	Number of Alveoli (10^6)	Alveolar Surface Area (m^2)	Skin Surface Area (m^2)
Birth	24	2.8	0.2
8 years old	300	32	0.9
Adult	300	75	1.8

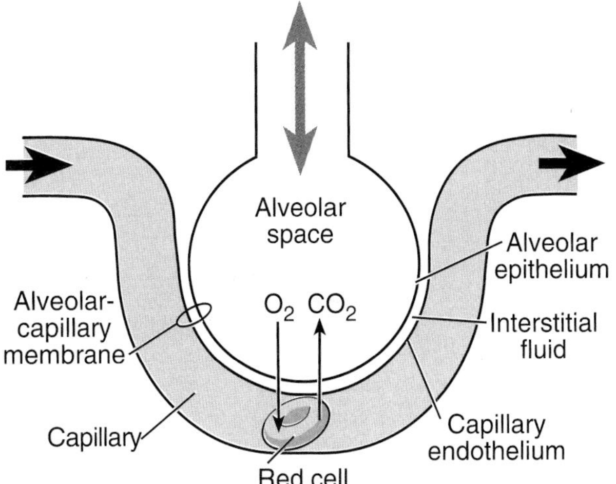

FIGURE 19.2 **Blood-gas interface.** The pulmonary capillaries and alveoli form a blood-gas interface. Thick arrows indicate direction of blood flow and ventilation, and thin arrows indicate the diffusion paths for O_2 and CO_2. The alveolar-capillary membrane is very thin (~0.5 μm) and, thus, the diffusion distance for gases is very small.

PRESSURE CHANGES AND AIRFLOW DURING BREATHING

A brief discussion of the pressure changes and muscles involved in breathing is useful before examining how air gets into the lungs. The lungs are housed in an airtight **thoracic cavity**, and they are separated from the abdomen by a large dome-shaped muscle, the **diaphragm.** The thoracic cavity is made up of 12 pairs of **ribs**, the **sternum**, and internal and external intercostal muscles that lie between the ribs. The **rib cage** is hinged to the vertebral column, allowing it to be raised and lowered during breathing. The space between the lungs and chest wall is the **pleural space**, which contains a thin layer of fluid (about 10 μm thick), which functions, in part, as a lubricant so the lungs can slide against the chest wall.

The Diaphragm Is the Main Muscle of Breathing

Inflation of the lungs is caused by contraction of the diaphragm, which is composed of skeletal muscle. Consider the breathing cycle (Fig. 19.3). When the diaphragm contracts, the thoracic cavity is expanded, and the lungs inflate automatically. This enlargement is accomplished in two ways. First, when the diaphragm (which is attached to the lower ribs and sternum) contracts, the abdominal contents are pushed down, enlarging the thoracic cavity in the vertical plane. Second, when the diaphragm descends and pushes down on the abdominal contents, it also pushes the rib cage outward, further enlarging the thoracic cavity.

The effectiveness of the diaphragm in changing thoracic volume is related to the strength of its contraction and its dome-shaped configuration when relaxed. With a normal breath, the diaphragm moves only about 1 to 2 cm, but with forced inspiration, a total excursion of 10 to 12 cm can occur. Obesity, pregnancy, and tight clothing around the abdominal wall can impede the effectiveness of the diaphragm in enlarging the thoracic cavity. Damage to the phrenic nerves (the diaphragm is innervated by two phrenic nerves, one to each lateral half) can lead to paralysis of the diaphragm. When a phrenic nerve is damaged, that portion of the diaphragm moves up rather than down during inspiration.

During forced inspiration, in which a large volume of air is taken in, the **accessory muscles** are also used. These include the **external intercostal muscles**, which raise the an-

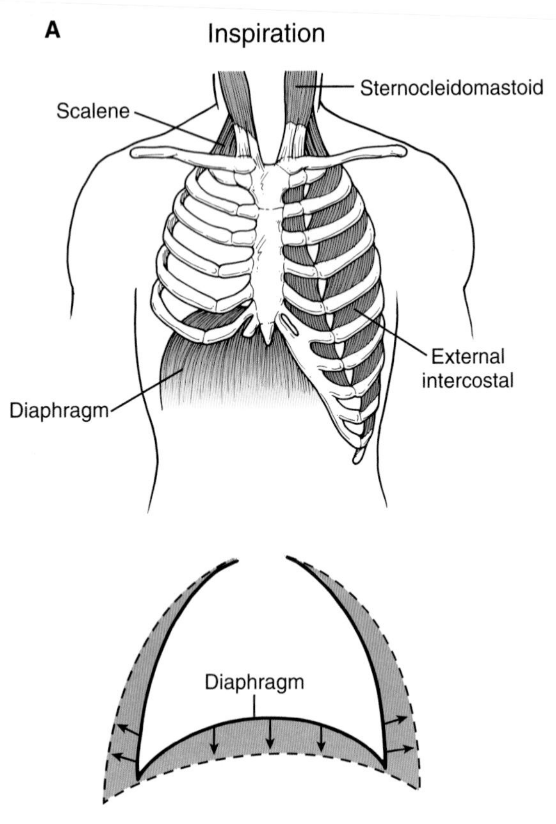

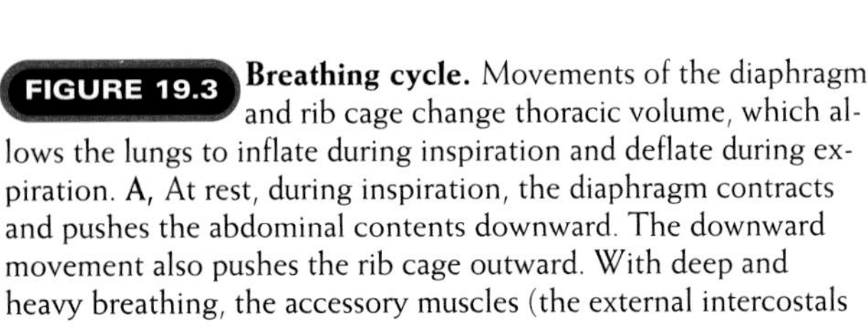

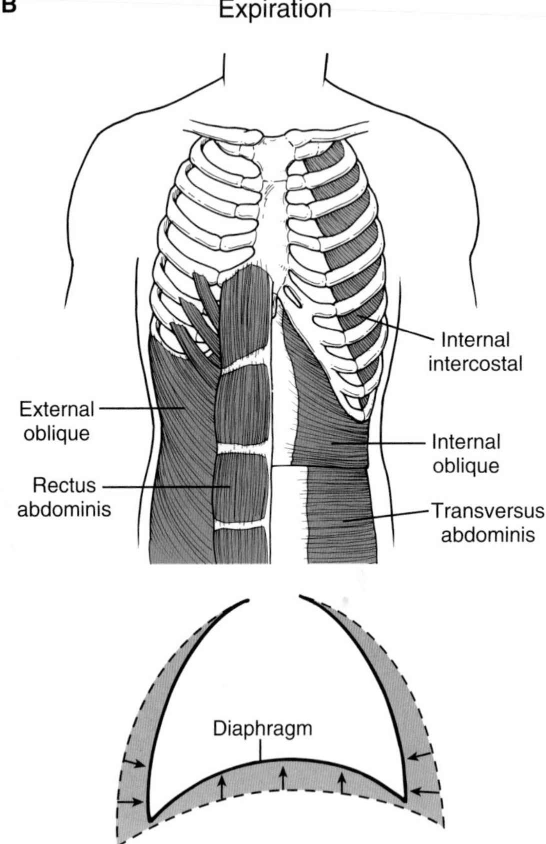

FIGURE 19.3 **Breathing cycle.** Movements of the diaphragm and rib cage change thoracic volume, which allows the lungs to inflate during inspiration and deflate during expiration. **A,** At rest, during inspiration, the diaphragm contracts and pushes the abdominal contents downward. The downward movement also pushes the rib cage outward. With deep and heavy breathing, the accessory muscles (the external intercostals and sternocleidomastoids) also contract and pull the rib cage upward and outward. **B,** Expiration is passive during resting conditions. The diaphragm relaxes and returns to its dome shape, and the rib cage is lowered. During forced expiration, however, the internal intercostal muscles contract and pull the rib cage downward and inward. The abdominal muscles also contract and help pull the rib cage downward, compressing thoracic volume.

terior end of the rib cage, causing the rib cage to be pulled upward and outward. To see how this works, stand sideways in front of a mirror and take a deep breath. The chest wall moves upward while the sternum moves outward, thereby enlarging the thoracic cavity. The **scalene muscles** of the neck and the **sternocleidomastoids**, which are inserted into the top of the sternum, are brought into play during deep and heavy breathing, such as exhaustive exercise, and are used to elevate the upper rib cage to further increase the thoracic volume.

Breathing out is a much simpler process than breathing in. At the end of a normal inspiration, the diaphragm relaxes; the rib cage drops, the thoracic volume decreases, and the lungs deflate. During normal breathing, expiration is purely passive. However, with exercise or forced expiration, the expiratory muscles become active. These muscles include those of the **abdominal wall** and the **internal intercostal muscles** (Fig. 19.3B). Contraction of the abdominal wall pushes the diaphragm upward into the chest and the internal intercostal muscles pull the rib cage down, reducing thoracic volume. These accessory respiratory muscles are necessary for such functions as coughing, straining, vomiting, and defecating. The expiratory muscles are extremely important in endurance running and are one of the reasons competitive long-distance runners, as part of their training program, often do exercises to strengthen their abdominal and chest muscles.

Different Pressures Are Found in the Lungs

The atmosphere, the air we breathe and live in, exerts a pressure (P) known as **barometric pressure** (P_B). At sea level, P_B is equal to 760 mm Hg. The total pressure or barometric pressure is the sum of individual **partial pressures** of the gases in the atmosphere. The relationship between the total pressure exerted by a mixture of gases and the pressure of individual gases is governed by **Dalton's law**, which states that the total barometric pressure (P_B) is equal to the sum of the partial pressures of the individual gases:

$$P_B = P_{N_2} + P_{O_2} + P_{H_2O} + P_{CO_2} \qquad (1)$$

where P_{N_2} equals partial pressure of nitrogen, P_{O_2} equals partial pressure of oxygen, P_{H_2O} equals partial pressure of water vapor, and P_{CO_2} equals partial pressure of carbon dioxide.

The partial pressures are the pressures that the individual gases would exert if each gas were present alone in the volume occupied by the whole mixture at the same temperature. Therefore, the partial pressure of oxygen (P_{O_2}), according to Dalton's law, is determined as $P_{O_2} = P_B \times F_{O_2}$, where F_{O_2} is the fractional concentration of oxygen. Because 21% of air is made up of oxygen, the partial pressure (P_{O_2}) exerted by oxygen is 160 mm Hg (760 × 0.21) at sea level. If all of the other gases in a container of air were removed, the remaining oxygen would still exert a pressure of 160 mm Hg. Partial pressure of a gas is often referred to as **gas tension**, and partial pressure and gas tension are used synonymously.

When air is inspired, it is warmed and humidified. The inspired air becomes saturated with water vapor at 37°C. The water vapor exerts a partial pressure that is a function of body temperature, not barometric pressure. At 37°C, water vapor exerts a partial pressure (P_{H_2O}) of 47 mm Hg. Water vapor pressure does not change the percentage of oxygen or nitrogen in a dry gas mixture; however, water vapor does lower the partial pressure of oxygen inside the lungs. The partial pressures of gases in the lungs are calculated based on a dry gas pressure; therefore, water vapor pressure is subtracted when the partial pressure of a gas is determined. The dry gas pressure in the trachea is 760 − 47 = 713 mm Hg, and the individual partial pressures of O_2 and N_2 are

$$P_{O_2} = 0.21 \times (760 - 47) = 150 \text{ mm Hg} \qquad (2)$$

and

$$P_{N_2} = 0.79 \times (760 - 47) = 563 \text{ mm Hg} \qquad (3)$$

Table 19.2 lists normal partial pressures of respiratory gases in different locations in the body.

In respiratory physiology, the pressures that affect airflow and lung volumes are small and are usually measured in centimeters of water (cm H_2O). A pressure of 1 cm H_2O is equal to 0.74 mm Hg (or 1 mm Hg = 1.36 cm H_2O). Changes in respiratory pressures during breathing are often expressed as **relative pressure**, a pressure relative to atmospheric pressure. For example, the pressure inside the alveoli can be −2 cm H_2O during inspiration. The negative sign indicates the pressure is 2 cm *below* P_B. Conversely, during expiration, the pressure inside the alveoli can be +3 cm

TABLE 19.2 Partial Pressures and Percentages of Respiratory Gases at Sea Level (P_B = 760 mm Hg)

	Ambient Dry Air		Moist Tracheal Air		Alveolar Air		Systemic Arterial Blood	Mixed Venous Blood
Gas	mm Hg	%	mm Hg	%	mm Hg	%	mm Hg	mm Hg
O_2	160	21	150	20	102	14	95	40
CO_2	0	0	0	0	40	5	40	46
Water vapor	0	0	47	6	47	6	47	47
N_2	600	79	563	74	571	75[a]	571	571
Total	760	100	760	100	760	100	760	704[b]

[a] Alveolar P_{N_2} increased by 1% because $R < 1$.
[b] Total pressure in venous blood is reduced because P_{O_2} decreases more than P_{CO_2} increases.

H_2O. This means the pressure is 3 cm H_2O *above* PB. A *positive* or *negative* pressure indicates that the pressure is relative to atmospheric pressure and is, respectively, above or below PB. When relative pressures are used, it is important to remember that PB is set at zero. If airway pressure is zero, pressure inside the airway equals atmospheric pressure. Unless otherwise specified, the pressures of breathing are relative pressures and the units are cm H_2O. A list of symbols and abbreviations used in respiratory physiology is shown in Table 19.3.

Since the thoracic cavity is airtight, an increase in thoracic volume causes the **pleural pressure** (Ppl), the pressure in the pleural space between the lung and chest wall, to fall. A decrease Ppl causes the lungs to expand and fill with air. This key pressure-volume relationship in breathing is based on two gas laws. **Boyle's law** states that, at a constant temperature, the pressure (P) of the gas varies inversely with the volume (V) of gas, or $P \propto 1/V$. If either pressure or volume changes and if temperature remains constant, the product of pressure and volume remains constant:

$$P_1V_1 = P_2V_2 \quad (4)$$

Charles' law states that if pressure is constant, the volume of a gas and its temperature vary proportionately, or $V \propto T$. If either temperature or volume changes and pressure remains constant, then

$$V_1/T_1 = V_2/T_2 \quad (5)$$

These two gas laws can be combined into the **general gas law**:

$$P_1V_1/T_1 = P_2V_2/T_2 \quad (6)$$

From the general gas law, at constant temperature, an increase in thoracic volume leads to a decrease in pleural pressure.

In addition to pleural pressure, several other pressures are associated with breathing and airflow (Fig. 19.4). The **alveolar pressure** (PA) is the pressure inside the alveoli. **Transmural pressure** (Ptm) is the pressure difference across an airway or across the lung wall or the pressure *inside* the wall minus the pressure *outside* the wall. Two major transmural pressures are involved in breathing. The first is **transpulmonary pressure** (P_L), which is the pressure difference across the lung wall. Transpulmonary pressure is measured by subtracting pleural pressure from alveolar pressure (PL = PA − Ppl). Transpulmonary pressure is the pressure that keeps the lungs inflated and prevents the lungs from collapsing. An increase in P_L is responsible for inflating the lungs above the resting volume. The second transmural pressure is **transairway pressure** (Pta), the pressure difference across the airways (Pta = Paw − Ppl), where Paw is the pressure inside the airway. Transairway pressure is important in keeping the airways open during forced expiration. One way to remember how to calculate transairway or transpulmonary pressure is "*in* minus *out*," where Ppl is the pressure outside the lung or airway.

Why is pleural pressure negative or subatmospheric? The reason is the **elasticity** of the lungs and chest wall—that is, their capability of being stretched and then recoiling to an unstretched configuration, like a spring. At the

TABLE 19.3 Symbols and Terminology Used in Respiratory Physiology

Symbol	Meaning
Primary Symbols	
C	Compliance
D	Diffusion
F	Fractional concentration of a gas
f	Frequency
P	Pressure or partial pressure
$\dot{Q}$	Volume of blood per unit time (blood flow or perfusion)
R	Resistance
S	Saturation
T	Time
V	Gas volume
$\dot{V}$	Volume of gas per unit time (airflow)
Secondary Symbols	
A	Alveolar
a	Arterial
aw	Airway
B	Barometric
D	Dead space
E	Expiratory
I	Inspiratory
L	Lung
pc'	Pulmonary end-capillary
pl	Pleural
pw	Pulmonary wedge
s	Shunt
T	Tidal
tp	Transpulmonary
v	Venous
Examples of Combinations	
CL	Lung compliance
DLCO	Lung diffusing capacity for carbon monoxide
FIO_2	Fractional concentration of inspired O_2
PB	Barometric pressure
PCO_2	Partial pressure of carbon dioxide
PA	Alveolar pressure
PO_2	Partial pressure of oxygen
$PaCO_2$	Partial pressure of carbon dioxide in arterial blood
$PACO_2$	Partial pressure of carbon dioxide in alveoli
PIO_2	Partial pressure of inspired O_2
$PECO_2$	Partial pressure of CO_2 in expired gas
$P(A\text{-}a)O_2$	Alveolar-arterial difference in partial pressure of O_2
Ppl	Pleural pressure
Raw	Airway resistance
SaO_2	Saturation of hemoglobin with oxygen in arterial blood
TI	Inspiratory time
TE	Expiratory time
$\dot{V}A$	Alveolar ventilation
$\dot{V}A/\dot{Q}$	Alveolar ventilation-perfusion ratio
VD	Dead space volume
$\dot{V}D$	Dead space ventilation
$\dot{V}E$	Expired minute ventilation
$\dot{V}O_2$	Oxygen consumption per minute

Note: A dot above a primary symbol denotes flow per unit time.

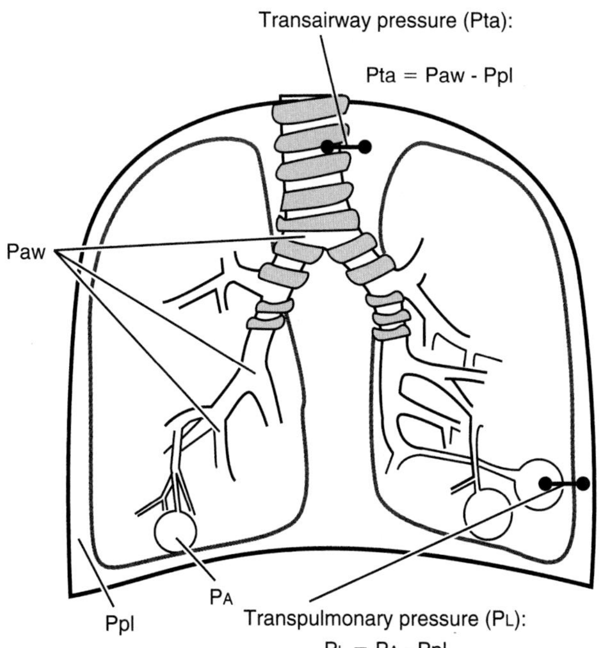

FIGURE 19.4 Pressures involved in breathing. Several important pressures are involved in breathing: airway pressure (Paw), alveolar pressure (PA), pleural pressure (Ppl), transpulmonary pressure (PL), and transairway pressure (Pta). Both transpulmonary pressure and transairway pressure can be defined as the pressure *inside* minus the pressure *outside*. In both cases, the pressure outside is pleural pressure (Ppl).

end of a normal expiration, the lungs and chest wall are stretched in equal but opposite directions (Fig. 19.5). The stretched lungs have the potential to recoil inwardly, and the stretched chest wall has the potential to recoil outwardly. These equal but opposing forces cause the pleural pressure to decrease below atmospheric pressure. Pleural pressure is negative during quiet breathing and becomes more negative with deep inspiration. Only during forced expiration does pleural pressure become positive.

The importance of pleural pressure is seen when the chest wall is punctured (Figure 19.5B) and air enters into the pleural space. The stretched lung collapses immediately (recoils inward), and the rib cage simultaneously expands outward (recoils outward). Since the normal pleural pressure is subatmospheric, air will rush into the pleural space any time the chest wall or lung is punctured, and the pleural pressure will become equal to atmospheric pressure because air moves from regions of high to low pressure. In this situation, transpulmonary pressure is zero (PTP = 0) because the pressure difference across the lung is eliminated. This condition, in which air or gas accumulates in the pleural space and the lung collapses, is known as **pneumothorax** (Fig. 19.5B, right side). A pneumothorax occurs with a knife or gunshot wound in which the chest wall is punctured or when the lung ruptures from an abscess or severe coughing. In the treatment of some lung disorders (e.g., tuberculosis), a pneumothorax is purposely created by inserting a sterile needle between the ribs and injecting nitrogen into the pleural fluid to rest the diseased lung.

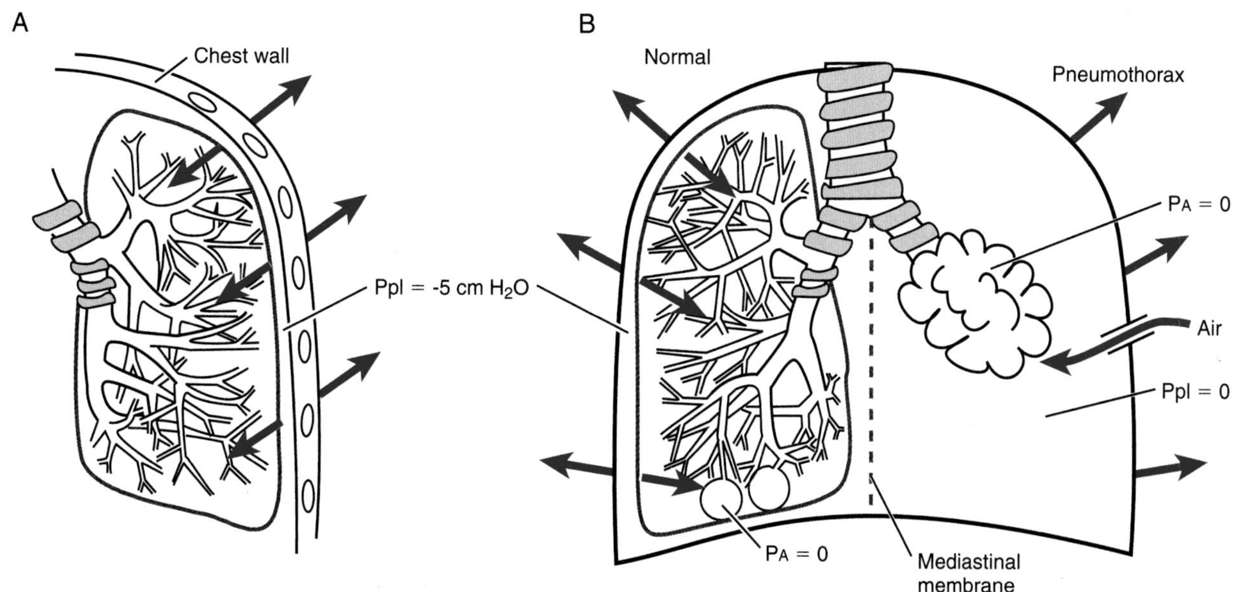

FIGURE 19.5 Negative pleural pressure resulting from elastic recoil of the chest wall and lungs. A, The stretched lung (at the end of a normal expiration) tends to recoil inward and the chest wall tends to recoil outward, but in *equal* and *opposite* directions. Consequently, pleural pressure (Ppl) becomes negative (i.e., less than atmospheric pressure). **B,** Rupture or puncture of the lung or chest wall results in a pneumothorax, during which the transpulmonary pressure becomes zero, and elastic recoil causes the lung to collapse. The mediastinal membrane prevents the other lung from collapsing.

Changes in Alveolar Pressure Move Air In and Out of the Lungs

Pressure changes during a normal breathing cycle are illustrated in Figure 19.6. At the end of expiration, the respiratory muscles are relaxed and there is no airflow. At this point, alveolar pressure is zero (equal to atmospheric pressure or P_B). Pleural pressure is -5 cm H_2O, and transpulmonary pressure is, therefore, 5 cm H_2O $[0 - (-5) = +5]$. Transpulmonary pressure is always positive in normal breathing; it is sometimes referred to as **distending pressure** because it keeps the lungs inflated. The more positive transpulmonary pressure becomes, the more the lungs are distended or inflated.

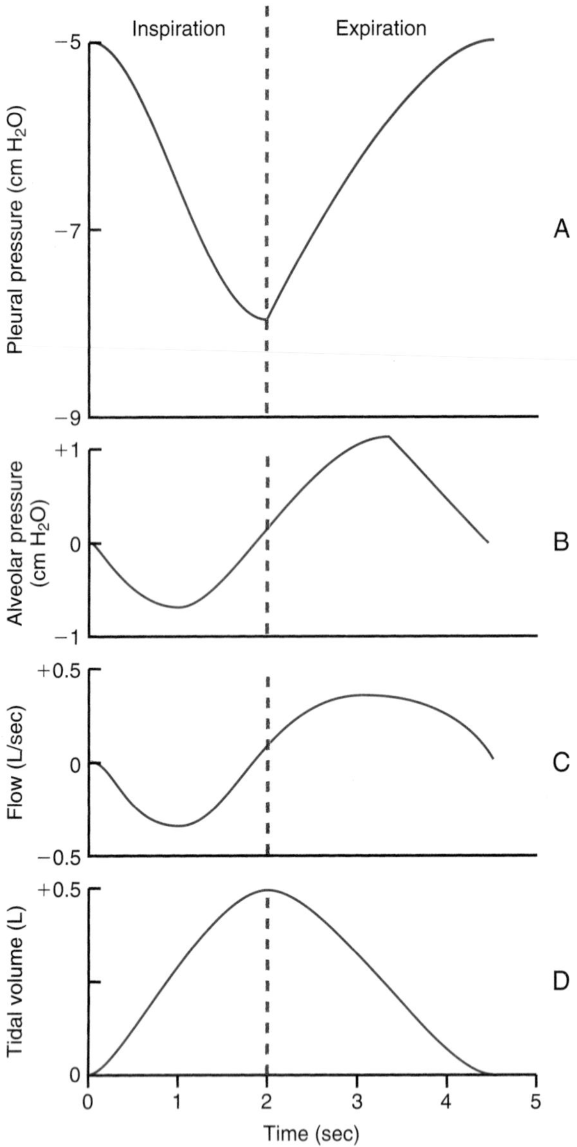

FIGURE 19.6 Changes in pressure and flow during a normal breathing cycle. The inspiratory time (T_I) is 2 seconds and is less than the expiratory time (T_E) of 3 seconds. This difference is a result of, in part, a higher airflow resistance during expiration, as is reflected by a higher alveolar pressure (P_A) change during expiration (1.2 cm H_2O) than during inspiration (0.8 cm H_2O). An increase in airway resistance will decrease the T_I/T_E ratio. Airflow occurs as a result of a pressure difference between the mouth and alveoli (See text for more details.). Only a small pressure change is required for a tidal volume.

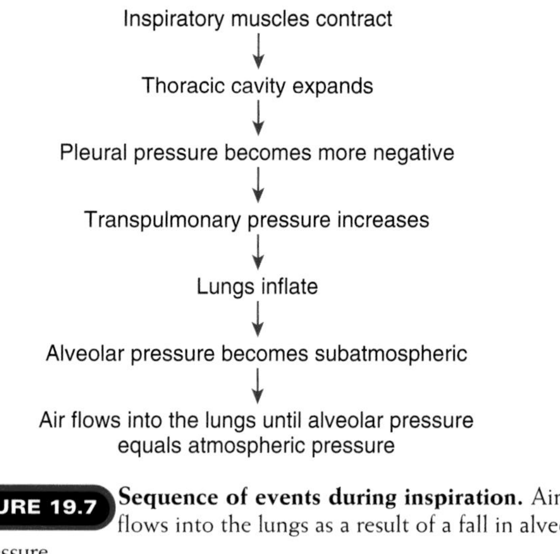

FIGURE 19.7 Sequence of events during inspiration. Air flows into the lungs as a result of a fall in alveolar pressure.

Inflation of the lungs is initiated by contraction of the diaphragm. If inspiration is started from the end of a maximal expiration, the chest wall can be felt to expand during inhalation. At no time, as our lungs fill, do we feel the need to close our epiglottis to keep the air in. This is because air is held in our lungs by only a slight pressure difference. In the example shown in Figure 19.6A, pleural pressure goes from -5 to -8 cm H_2O. One of the basic characteristics of gases, such as air, is that the pressures between two regions tend to equilibrate. Therefore, when pleural pressure decreases, transpulmonary pressure increases, and the lungs inflate. Inflation of the lungs causes the alveolar diameter to increase, and alveolar pressure to decrease below atmospheric pressure (Fig. 19.6B). This produces a pressure difference between the mouth and alveoli, which causes air to rush into the alveoli. Airflow stops at the end of inspiration because alveolar pressure again equals atmospheric pressure (Fig. 19.6C). The sequence of events is summarized in Figure 19.7.

During expiration, the inspiratory muscles relax, the rib cage drops, pleural pressure becomes less negative, transpulmonary pressure decreases, and the stretched lungs deflate. When alveolar diameter decreases during deflation, alveolar pressure becomes greater than atmospheric pressure and pushes air out of the lungs. Airflow out of the lungs occurs until alveolar pressure equals atmospheric pressure.

SPIROMETRY AND LUNG VOLUMES

The volume of air that is breathed in or out of the lungs is measured by the process of **spirometry** (Fig. 19.8). A **spirometer** is a volume recorder consisting of a double-walled cylinder in which an inverted bell is immersed in water to form a seal. A pulley attaches the bell to a marker that writes on a rotating drum. When air enters the spirometer from the lungs, the bell rises. Because of the pulley arrangement, the marker is lowered. Therefore, a downward deflection represents expiration and an upward deflection represents inspiration. The

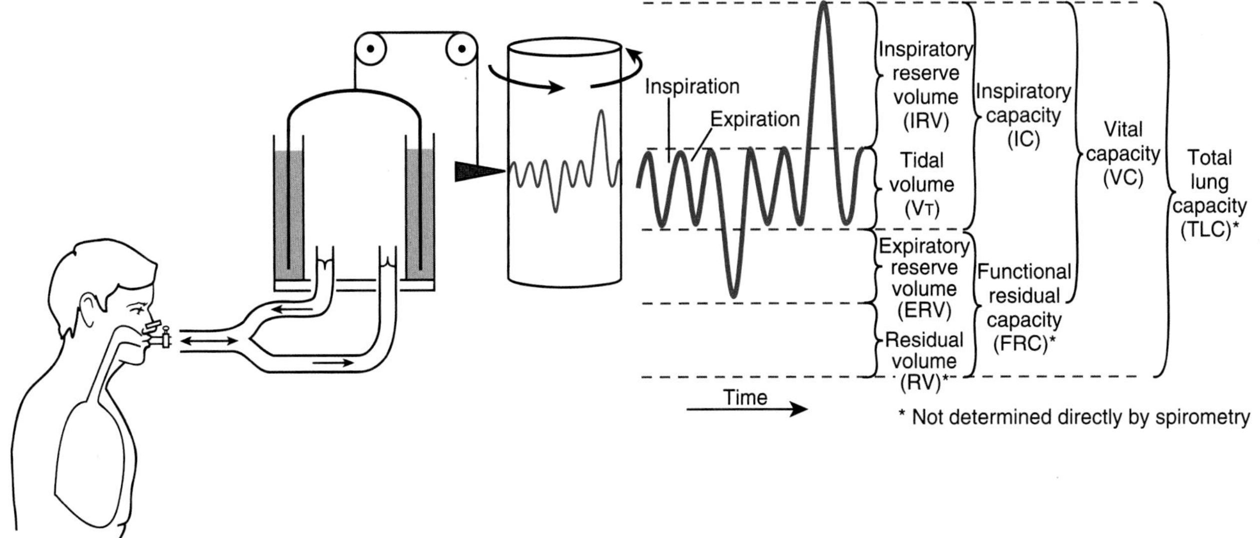

FIGURE 19.8 **Spirometry: the measurement of lung volumes and pulmonary function.** With expiration, the marker records a downward deflection. Note that residual volume (RV), functional residual capacity (FRC), and total lung capacity (TLC) cannot be measured directly by spirometry.

recording is known as a **spirogram**. The slope of the spirogram on the moving drum measures rate of airflow and the amplitude of the deflection measures the volume of air. Volume is plotted on the vertical axis (y-axis), and time is plotted on the horizontal axis (x-axis).

Spirometry Is Used to Measure Lung Volumes and Airflow

The volume of air leaving the lungs during a single breath is called **tidal volume** (VT). Under resting conditions, VT is approximately 500 mL and represents only a fraction of the air in the lungs. The maximum amount of air in the lungs at the end of a maximal inhalation is the **total lung capacity** (TLC) and is approximately 6 L in an adult man. Another important spirometry measurement is the **functional residual capacity** (FRC), the volume of air remaining in the lungs at the end of a normal tidal volume (end of expiration). Note the use of "volume" in the first term and "capacity" in the next two. Volume is used when only one volume is involved, and capacity is used when a volume can be broken down into two or more smaller volumes; for example, FRC equals **expiratory reserve volume** (ERV) plus **residual volume** (RV). The various lung volumes and capacities are summarized in Table 19.4.

Forced Vital Capacity Is One of the Most Useful Tests in Spirometry

The maximum volume of air that can be exhaled after a maximum inspiration is the **vital capacity** (VC). When expiration is performed as rapidly and as forcibly as possible into a spirometer, this volume is called **forced vital capacity** (FVC) and is about 5 L in an adult man (see Fig. 19.8). Vital capacity and forced vital capacity are the same volume. Vital capacity is the sum of expiratory reserve volume, tidal volume, and inspiratory reserve volume. Forced vital capacity is a direct volume measurement from spirometry and is one of the most useful measurements to assess ventilatory function of the lungs. To measure FVC, the individual inspires maximally and then exhales into the spirometer as forcefully, rapidly, and completely as possible.

Two additional determinations can be obtained from the FVC spirogram (Fig. 19.9). One is **forced expiratory volume** of air exhaled in 1 second (FEV_1). This volume has the least variability of the measurements obtained from a forced expiratory maneuver and is considered one of the most reliable spirometry measurements. Another useful way of expressing FEV_1 is as a percentage of FVC (i.e., $FEV_1/FVC \times 100$), which corrects for differences in lung size. Normally, the FEV_1/FVC ratio is 0.8, which means 80% of an individual's FVC can be exhaled in the first second of forced vital capacity. FVC and FEV_1 are important measurements in the diagnosis of certain types of lung diseases (see Clinical Focus Box 19.1).

A second measurement obtained from the FVC spirogram is **forced expiratory flow** (FEF_{25-75}); it has the greatest sensitivity in terms of detecting early airflow obstruction (see Fig. 19.9). This measurement represents the expiratory flow rate over the middle half of the forced vital capacity (between 25 and 75%). Forced expiratory flow is obtained by identifying the 25% and 75% volume points of the FVC, and then measuring the volume and time between these two points. The calculated flow rate is expressed in L/sec.

Not All Lung Volumes Can Be Measured Directly by Spirometry

Because the lungs cannot be emptied completely following forced expiration, neither RV nor FRC can be measured directly by simple spirometry. Instead, they are measured indirectly using a dilution technique involving helium, an in-

TABLE 19.4 Abbreviations and Definitions Used in Pulmonary Function

Abbreviation	Term	Definition	Normal Value[a]
EPP	*Equal pressure point*	The point at which the pressure inside the airway equals the pressure outside of the airway (i.e., Ppl).	
ERV	*Expiratory reserve volume*	The maximum volume of air exhaled at the end of the tidal volume.	1.2 L
FEF_{25-75}	*Forced expiratory flow*	The maximum midexpiratory flow rate, measured by drawing a line between points representing 25% and 75% of the forced vital capacity.	5 L/sec
FEV_1	*Forced expiratory volume*	The maximum volume of air forcibly exhaled in 1 second.	4.0 L
$FEV_1/FVC\%$	*Forced expired volume/forced vital capacity ratio*	The percentage of FVC forcibly exhaled in 1 second.	80%
FRC	*Functional residual capacity*	The volume of air remaining in the lungs at the end of a normal tidal volume.	2.4 L
FVC	*Forced vital capacity*	The maximum volume of air forcibly exhaled after a maximum inhalation.	4.8 L
IC	*Inspiratory capacity*	The maximum volume of air inhaled after a normal expiration.	3.6 L
IRV	*Inspiratory reserve volume*	The maximum volume of air inhaled at the end of a normal inspiration.	3.1 L
PEF	*Peak expiratory flow*	The maximal expiratory flow during a FVC maneuver.	7 L/sec
RV	*Residual volume*	The volume of air remaining in the lungs after maximum expiration.	1.2 L
RV/TLC%	*Residual volume/total lung capacity ratio*	The percentage of total lung capacity made up of residual volume.	20%
TLC	*Total lung capacity*	The volume of air in the lungs at the end of a maximum inspiration.	6.0 L
VC	*Vital capacity*	The maximum volume of air that can be exhaled. (Note that the values for FVC and VC are the same.) VC is calculated from static lung volumes ($VC = ERV + V_T + IRV$). FVC is determined from direct spirometry.	4.8 L
V_D/V_T	*Dead space–tidal volume ratio*	The fraction of tidal volume made up of dead space.	0.30
V_T	*Tidal volume*	The volume of air inhaled or exhaled with each breath.	0.5 L

[a] Values are for an average, healthy, young man.

CLINICAL FOCUS BOX 19.1

Chronic Obstructive Pulmonary Disease

Chronic bronchitis, emphysema, and asthma are major pathophysiological disorders of the lungs. **Bronchitis** is an inflammatory condition that affects one or more bronchi. **Emphysema** stems from an overdistension of alveoli and a loss of lung elastic recoil. **Asthma** is marked by spasmodic contractions of smooth muscle in the bronchi. Although the pathophysiology and etiology of chronic bronchitis, emphysema, and asthma are different, they all are classified as **chronic obstructive pulmonary disease** (COPD). The hallmark of COPD is the slowing down of air movement during forced expiration, which leads to a decrease in FVC and FEV_1.

COPD is characterized by chronic obstruction of the small airways. Airflow can become obstructed in three ways: excessive mucus production (as in bronchitis), airway narrowing caused by bronchial spasms (as in asthma), and airway collapse during expiration (as in emphysema). In the latter case, airway collapse stems from abnormally high compliance and concomitant loss of lung elastic recoil. In severe COPD, air is trapped in the lungs during forced expiration, leading to abnormally high residual volume.

Chronic bronchitis and emphysema often coexist. Bronchitis leads to excessive mucus production, which plugs the small airways. A cough is produced to clear the excess mucus from the airways; on repeated exposure to a bronchial irritant, such as tobacco smoke, a persistent cough develops. As a result, the alveoli become overdistended and can rupture from excessive pressure, especially in plugged airways.

Late in COPD, patients often experience hypoxemia (low blood oxygen) and hypercapnia (high blood carbon dioxide). These two conditions, especially hypoxemia, cause pulmonary arteries to constrict. Hypoxia-induced pulmonary vasoconstriction leads to pulmonary hypertension (high blood pressure in the pulmonary circulation), which causes right heart failure. Based on changes in blood oxygen levels, COPD patients can be categorized into two types. Those exhibiting predominantly emphysema are referred to as "pink puffers" because their oxygen levels are usually satisfactory and their skin remains pink. They develop a puffing style of breathing. Those manifesting predominantly chronic bronchitis are called "blue bloaters" because low oxygen levels give their skin a blue cast and fluid retention from heart failure gives them a bloated appearance.

COPD is by far the most common chronic lung disease in the United States. More than 10 million Americans suffer from COPD, and it is the fifth leading cause of death nationwide. The single most common cause of chronic bronchitis and emphysema is tobacco smoke.

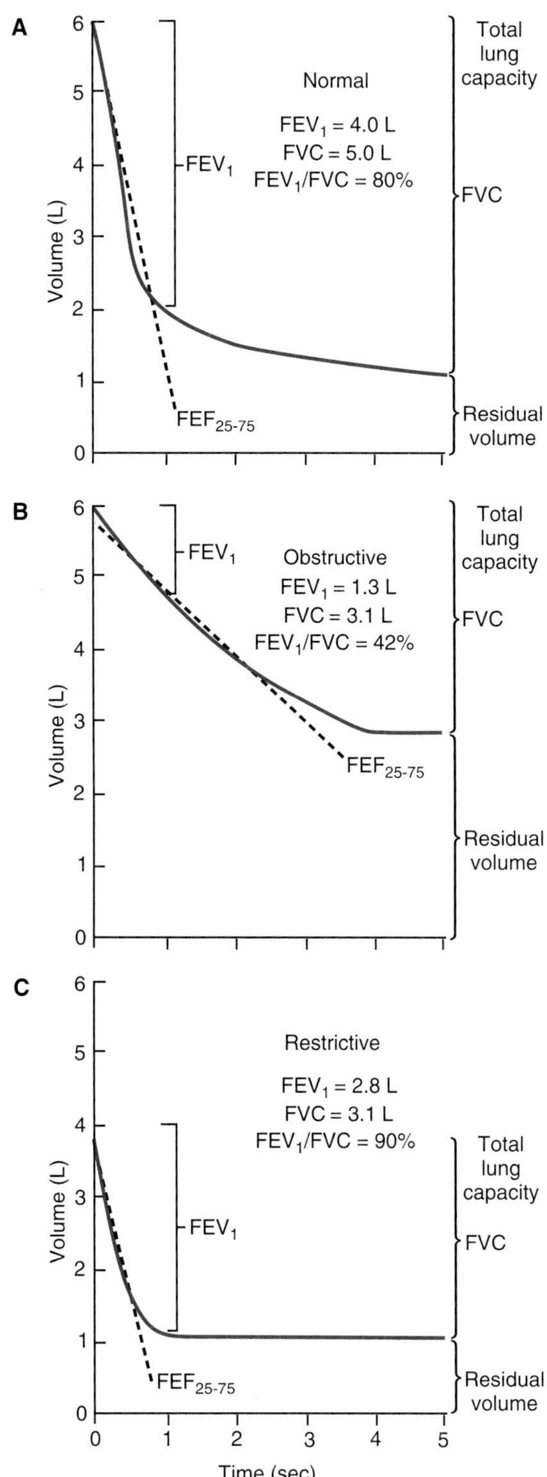

FIGURE 19.9 **Forced vital capacity measurement.** A, A healthy subject inspires maximally to total lung capacity and then exhales as forcefully and completely as possible into the spirometer. Two other measurements can be obtained from this maneuver: the forced expiratory volume in 1 second (FEV_1) and the flow rate over the middle half of the forced vital capacity (FEF_{25-75}). Measurements of FVC, FEV_1, FEV_1/FVC ratio, and FEF_{25-75} are used to detect obstructive and restrictive disorders. **B,** In an obstructive disorder, expiratory flow rate is significantly decreased, and the FEV_1/FVC ratio is low. **C,** In a restrictive disorder, lung inflation is decreased, resulting in reduced RV and TLC. Although FVC is decreased, it is important to note the FEV_1/FVC ratio is normal or increased in a restrictive disorder.

ert and relatively insoluble gas that is not readily taken up by blood in the lungs. The subject is connected to a spirometer filled with 10% helium in oxygen (Fig.19.10). The lungs initially contain no helium. After the subject rebreathes the helium-oxygen mixture and equilibrates with the spirometer, the helium concentration in the lungs will become the same as in the spirometer. From the conservation of mass principle, we can write

$$C_1 \times V_1 = C_2 \times (V_1 + V_2) \quad (7)$$

where C_1 equals the initial concentration of helium in the spirometer, V_1 equals the initial volume of helium-oxygen mixture in the spirometer, C_2 equals helium concentration after equilibration, and V_2 equals unknown volume in the lungs. Rearranging,

$$V_2 = \frac{V_1 (C_1 - C_2)}{C_2} \quad (8)$$

Starting the test at precisely the right time is important. If the test begins at the end of a normal tidal volume (end of expiration), the volume of air remaining in the lungs represents functional residual capacity. If the test begins at the end of a forced vital capacity, then the test will measure residual volume. Similarly, if the test starts after a maximal inspiration, then V_2 would equal total lung capacity. In practice, carbon dioxide is absorbed and oxygen is added to the spirometer to make up for the oxygen consumed by the individual during the test. Although the helium-dilution technique is an excellent test for the measurement of FRC and RV in normal individuals, it has a major limitation in patients whose lungs are poorly ventilated because of plugged airways or high airway resistance. In these diseased lungs, helium gives a falsely low FRC value.

An entirely different way to measure FRC that overcomes the problem of trapped gas in the lungs is the use of the **body plethysmograph** or body box (Fig. 19.11). The individual is seated comfortably in an airtight box in which changes in pressure and volume can be measured accurately. The subject takes a breath against a closed mouthpiece starting at FRC. During inspiration against a closed mouthpiece, the pressure in the lungs decreases, the chest volume increases, and the surrounding volume increases. An expiratory effort against a closed mouthpiece produces just the opposite effect: the pressure inside the lung increases as the chest volume decreases. In Figure 19.11, the lung volume is determined by applying Boyle's law.

ALVEOLAR VENTILATION

So far, ventilation has been described in terms of measurements of static lung volumes and forced lung volumes. However, breathing is a dynamic process involving how much air is brought in and out of the lungs in a minute. If 500 mL of air are inspired with each breath (V_T) and the breathing rate (f) is 14 times a minute, then the total **minute ventilation** ($\dot{V}$) is the amount of air that enters the lungs each minute (500 × 14 = 7,000 mL/min or 7 L/min). **Expired minute ventilation** ($\dot{V}_E$) is calculated from the amount of expired air per minute and can be represented by the equation:

$$\dot{V}_E = V_T \times f \quad (9)$$

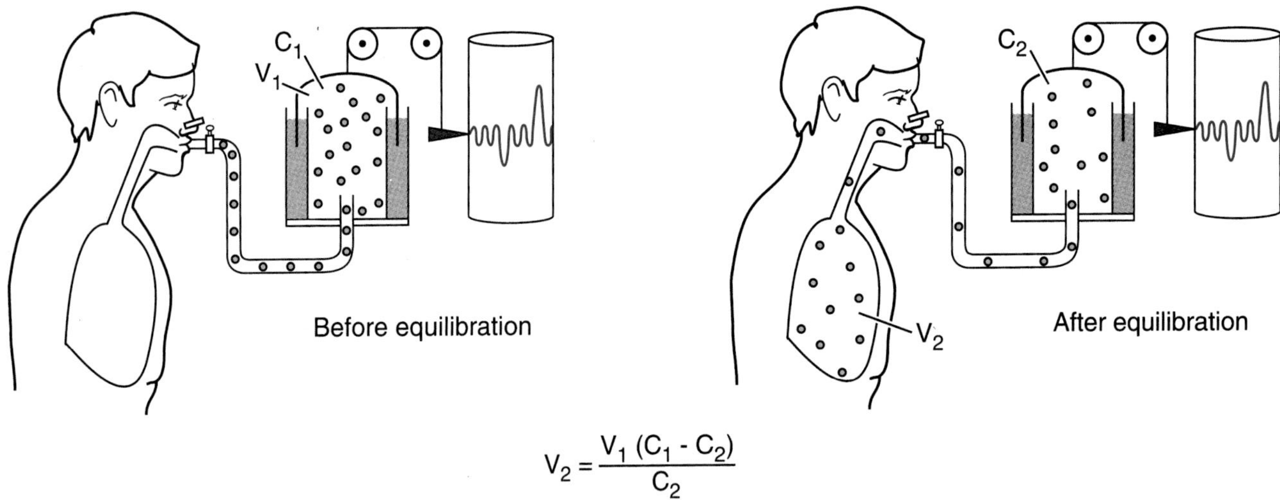

FIGURE 19.10 **Helium-dilution technique to measure FRC and RV.** Dots represent helium before and after equilibration. C = concentration; V = volume. (Modified from Rhoades R, Pflanzer R. Human Physiology. 3rd Ed. Fort Worth: Saunders College Publishing, 1996.)

Minute ventilation and expired minute ventilation are the same, based on the assumption that the volume of air inhaled equals the volume exhaled. This is not quite true because more oxygen is consumed than carbon dioxide is produced. This difference, for all practical purposes, is ignored.

Not All of the Inspired Air Participates in Gas Exchange

The tidal volume is distributed between the conducting airways and alveoli. Because gas exchange occurs only in the alveoli and not in the conducting airways, a fraction of the minute ventilation is wasted air. For each 500 mL of air inhaled, approximately 150 mL remain in the conducting airways and are not used in gas exchange (Fig. 19.12). This volume of wasted air is known as **dead space volume** (VD). Since VD is due to the anatomy of the airways, the volume is often referred to as **anatomic VD.**

Picture what occurs during a normal breathing cycle. A normal tidal volume of 500 mL is expired. During the next inspiration, another 500 mL are taken in, but the first 150 mL of air entering the alveoli is VD (old alveolar gas left be-

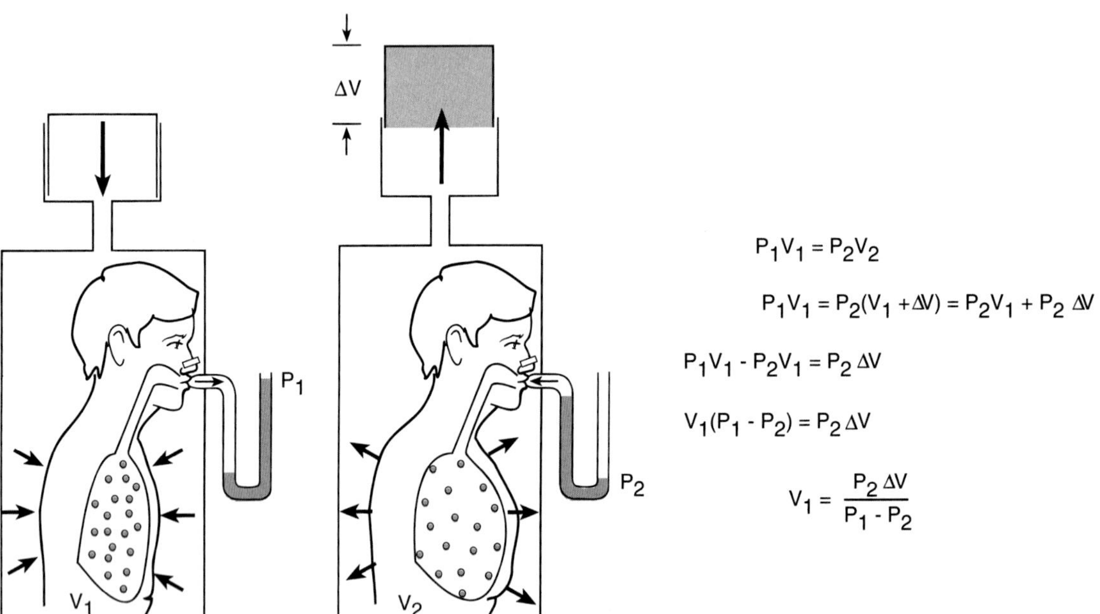

FIGURE 19.11 **Body plethysmograph for measuring functional residual capacity (FRC).** The subject sits in an airtight box and breathes air through a mouthpiece. The mouthpiece is closed at the end of a normal expiration, and the subject continues to try to breathe against a closed mouthpiece. P_1 is the lung pressure at the end of expiration and P_2 at the end of inspiration. ΔV is the volume change accompanying inspiration and expiration; it results from changes in thoracic volume against a closed mouthpiece. The body plethysmograph is used clinically for measuring FRC in patients with airway obstruction.

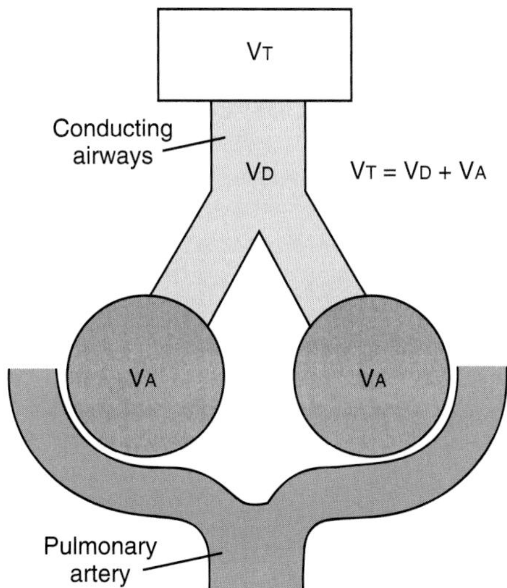

FIGURE 19.12 Distribution of tidal volume. Tidal volume (VT) is represented by the rectangle and is the volume of air that will be drawn in during inspiration. VT will be distributed between the conducting airways and alveoli. The volume of air in the conducting airways does not participate in gas exchange and constitutes dead space volume (VD). VA is the volume of fresh air added to the alveoli.

hind). Thus, only 350 mL of fresh air reach the alveoli and 150 mL are left in the conducting airways. The normal ratio of dead space volume to tidal volume (VD/VT) is in the range of 0.25 to 0.35. In this example, the ratio (150/500) is 0.30, which means 30% of the tidal volume or 30% of the minute ventilation does not participate in gas exchange and constitutes VD.

Dead space is not confined to the conducting airways alone. Any time gases in the alveoli do not participate in gas exchange, these gases also become part of the wasted air. For example, if inspired air is distributed to alveoli that have no blood flow, this constitutes dead space and is referred to as **alveolar dead space volume** (Fig. 19.13A). Alveolar dead

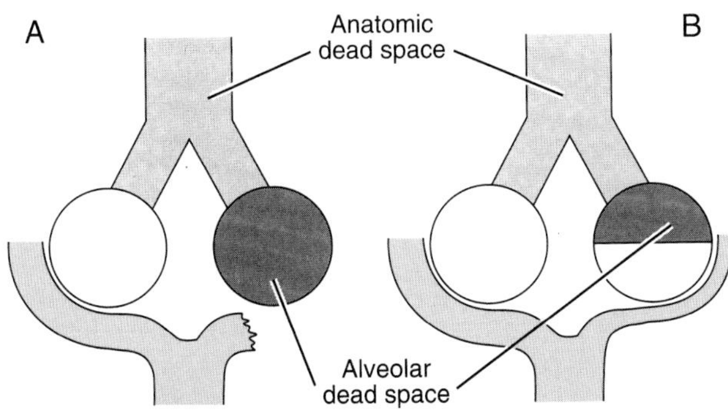

FIGURE 19.13 Anatomic and alveolar dead space. Dead space volume occurs in the conducting airways and in alveoli with poor capillary circulation. **A,** There is no blood flow to an alveolar region. **B,** There is reduced blood flow. In both cases, a portion of alveolar air does not participate in gas exchange and constitutes alveolar dead space volume. Note that physiological dead space is the sum of alveolar dead space plus anatomic dead space.

space volume is not confined to alveoli without blood flow. Alveoli that have reduced blood flow exchange less inspired air than normal (Fig. 19.13B); any portion of alveolar air in excess of that needed to maintain normal gas exchange constitutes alveolar dead space volume. Therefore, dead space volume (VD) may be either anatomic or alveolar in nature. The sum of the two types of dead space is **physiological dead space volume:**

$$\text{Physiological VD} = \text{Anatomic VD} + \text{alveolar VD} \quad (10)$$

Basically, anatomic VD, alveolar VD, and physiological VD are terms that denote the volume of inspired gas that does not participate in gas exchange. In one case, there is a fraction of VT that does not reach the alveoli (anatomic dead space volume). In another, a fraction of the VT reaches the alveoli, but there is reduced or no blood flow, leading to alveolar dead space volume. Physiological dead space volume represents the sum of anatomic and alveolar dead space volume. In normal individuals, physiological VD is approximately the same as anatomic dead space.

Alveolar Ventilation Is the Amount of Fresh Air That Reaches the Alveoli

The volume of fresh air per minute actually reaching the alveoli is known as **alveolar ventilation** ($\dot{V}A$). To determine how much fresh air reaches the alveoli per minute, dead space volume is subtracted from the tidal volume and the result is multiplied by breathing frequency, $\dot{V}_A$, and can be represented by

$$\dot{V}A = (VT - VD) \times f \quad (11)$$

where VT equals tidal volume, VD equals dead space volume, and f equals breathing frequency. For example, if an individual has a breathing rate of 14 breaths/min, a VT = 500 mL, and a V_D = 150 mL, then the volume of air entering the alveoli is 4.9 L/min [(500 − 150 mL) × 14 = 4,900 mL/min]. Only alveolar ventilation represents the amount of fresh air reaching the alveoli. For instance, a swimmer using a snorkel breathes through a tube that increases dead space volume. Similarly, a patient connected to a mechanical ventilator also has increased dead space volume. Indeed, if minute ventilation is held constant, then alveolar ventilation is decreased with snorkel breathing or with mechanical ventilation.

The significance of dead space volume, minute ventilation, and alveolar ventilation is shown in Table 19.5. Subject C's breathing is slow and deep, B's breathing is normal, and A's breathing is rapid and shallow. Note that each subject has the same minute ventilation (i.e., the total amount of expired air per minute), but each has marked differences in alveolar ventilation. Subject A has no alveolar ventilation and would die in a matter of minutes, while subject C has an alveolar ventilation greater than normal. The important lesson from the examples presented in Table 19.5 is that increasing the *depth* of breathing is far more effective in elevating alveolar ventilation than increasing the frequency or rate of breathing. A good example is exercise because in most exercise situations, increased alveolar ventilation is accomplished more by increases in the depth of breathing than in the rate. A well-trained athlete can often increase

TABLE 19.5 Effect of Breathing Patterns on Alveolar Ventilation

Subject	Tidal Volume (mL)	×	Frequency (breaths/min)	=	Minute Ventilation (mL/min)	−	Dead Space Ventilation[a] (mL/min)	=	Alveolar Ventilation (mL/min)
A	150	×	40	=	6000	−	150 × 40	=	0
B	500	×	12	=	6000	−	150 × 12	=	4200
C	1000	×	6	=	6000	−	150 × 6	=	5100

[a] Dead space volume is 150 mL in all three subjects.

alveolar ventilation during moderate exercise with little or no increase in breathing frequency.

Expired Carbon Dioxide Is Used to Measure Alveolar Ventilation

Alveolar ventilation is easy to calculate if dead space volume is known. However, dead space volume is not easily determined in a human subject. Often, dead space is approximated for a seated subject by assuming that dead space (in mL) is equal to the subject's weight in pounds (e.g., a subject who weighs 170 pounds would have a dead space volume of 170 mL). This assumption is fairly reliable for healthy individuals, but not in patients with respiratory problems. Alveolar ventilation is calculated in the pulmonary function laboratory from the volume of expired carbon dioxide per minute and fractional concentration of carbon dioxide in the alveolar gas (Fig. 19.14). Because no gas exchange occurs in the conducting airways and the inspired air contains essentially no carbon dioxide, all of the expired carbon dioxide originates from alveoli. Therefore,

$$\dot{V}_{ECO_2} = \dot{V}_A \times F_{ACO_2} \quad (12)$$

where $\dot{V}_{ECO_2}$ equals the volume of carbon dioxide expired per minute and F_{ACO_2} equals the fractional concentration of carbon dioxide in alveolar gas.

Rearranging yields the **alveolar ventilation equation**:

$$\dot{V}_A = \frac{\dot{V}_{ECO_2}}{F_{ACO_2}} \quad (13)$$

The carbon dioxide concentration in the alveoli can be obtained by sampling the last portion of the tidal volume during expiration, **end-tidal volume**, which contains alveolar gas.

Alveolar ventilation can also be determined from the partial pressure of carbon dioxide in the alveoli (P_{ACO_2}), based on the fact that P_{ACO_2} is equal to F_{ACO_2} times total alveolar gas pressure. Equation 13 can be written as

$$\dot{V}_A\ (\text{L/min}) = \frac{\dot{V}_{ECO_2}\ (\text{mL/min}) \times 0.863}{P_{ACO_2}\ (\text{mm Hg})} \quad (14)$$

where 0.863 is a constant with dimensions mm Hg × L/mL. Because P_{ACO_2} is in equilibrium with the partial pressure of carbon dioxide in the arterial blood (P_{aCO_2}), the latter value can be used to calculate $\dot{V}_A$, as follows:

$$\dot{V}_A = \frac{\dot{V}_{ECO_2} \times 0.863}{P_{aCO_2}} \quad (15)$$

It is important to recognize the inverse relationship between $\dot{V}_A$ and P_{aCO_2} (Fig. 19.15). If alveolar ventilation is halved, alveolar P_{CO_2} will double (assuming a steady state

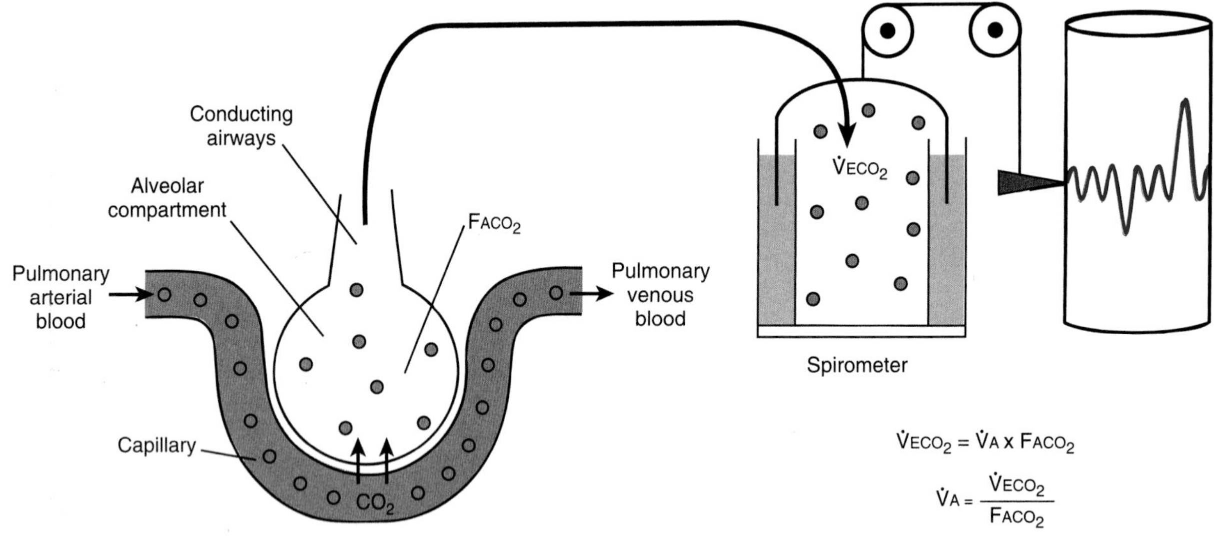

FIGURE 19.14 Calculation of alveolar ventilation using expired carbon dioxide. Because all of the CO_2 (represented by dots) in the expired air originates from the alveoli, the fractional concentration of CO_2 in alveolar gas (F_{ACO_2}) can be obtained at the end of a tidal volume (often referred to as end-tidal CO_2).

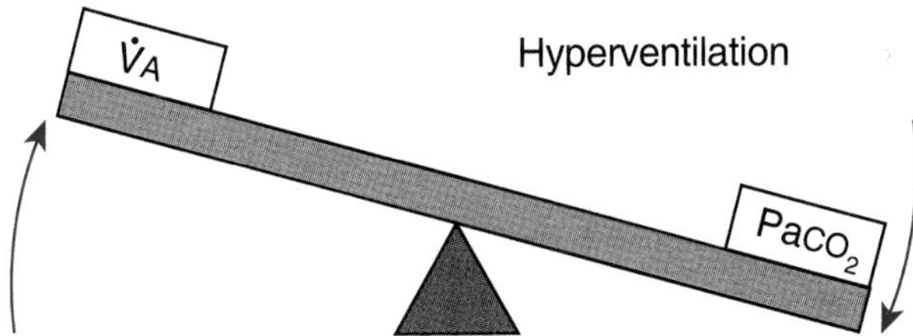

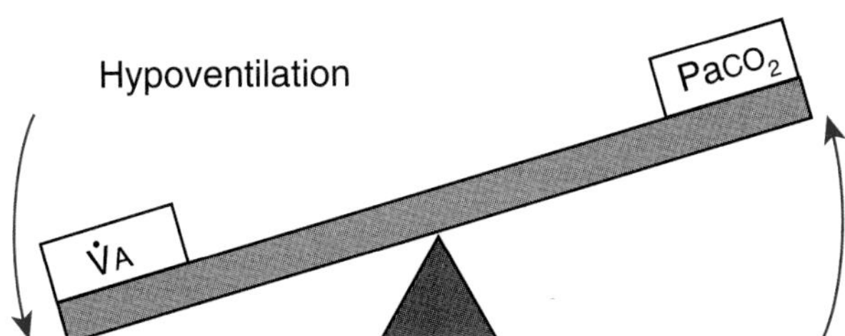

FIGURE 19.15 Relationship between arterial carbon dioxide and alveolar ventilation. Alveolar ventilation is a key determinant of arterial P_{CO_2}. Alveolar ventilation and arterial P_{CO_2} are inversely related. For example, if alveolar ventilation is halved, arterial P_{CO_2} will double, and if alveolar ventilation is doubled, arterial P_{CO_2} will be halved.

and constant carbon dioxide production). This decrease in alveolar ventilation is called **hypoventilation.** Conversely, increased ventilation, referred to as **hyperventilation**, leads to a fall in P_{ACO_2}. Clinically, the adequacy of alveolar ventilation is usually evaluated in terms of *arterial* P_{CO_2}. An increase in Pa_{CO_2} reflects hypoventilation and a decrease in Pa_{CO_2} reflects hyperventilation.

What happens to P_{AO_2} when alveolar ventilation increases? When alveolar ventilation increases, P_{AO_2} will also increase. However, doubling alveolar ventilation will not lead to a doubling of P_{AO_2}. The quantitative relationship between alveolar ventilation and P_{AO_2} is more complex than that for P_{ACO_2}, for two reasons. First, the inspired P_{O_2} is obviously not zero. Second, the **respiratory exchange ratio** (R), defined as the ratio of the volume of carbon dioxide exhaled to the volume of oxygen taken up ($\dot{V}_{CO_2}/\dot{V}_{O_2}$), is usually less than 1, which means more oxygen is removed from the alveolar gas per unit time than carbon dioxide is added.

The alveolar partial pressure of oxygen (P_{AO_2}) can be calculated by using the **alveolar gas equation:**

$$P_{AO_2} = P_{IO_2} - P_{ACO_2}\,[F_{IO_2} + (1 - F_{IO_2})/R] \quad (16)$$

where P_{IO_2} equals the partial pressure of inspired oxygen (moist tracheal air), P_{ACO_2} equals the partial pressure of carbon dioxide in the alveoli, F_{IO_2} equals the fractional concentration of oxygen in the inspired air, and R equals the respiratory exchange ratio. When R = 1, the complex term in the brackets equals 1. Also, if a subject breathes 100% oxygen (F_{IO_2} = 1.0) for a brief period, the correction factor in the brackets reduces to 1.

In a normal resting individual breathing air at sea level, with an R of 0.82, a P_{ACO_2} of 40 mm Hg, and a P_{IO_2} = 150 mm Hg, P_{AO_2} is calculated as follows:

$$P_{AO_2} = 150 \text{ mm Hg} - 40 \text{ mm Hg}\,[0.21 + (1 - 0.21)/0.82] \quad (17)$$

$$P_{AO_2} = 150 \text{ mm Hg} - 40 \text{ mm Hg} \times [1.2] = 102 \text{ mm Hg}$$

Since P_{IO_2} and F_{IO_2} stay fairly constant and P_{ACO_2} equals Pa_{CO_2}, the alveolar gas equation can be simplified to

$$P_{AO_2} = 150 - 1.2\,(Pa_{CO_2}) \quad (18)$$

MECHANICAL PROPERTIES OF THE LUNGS AND CHEST WALL

The lungs, airway tree, and vascular tree are embedded in elastic tissue. When the lungs are inflated, this elastic component is stretched. The degree of lung expansion at any given time in the breathing cycle is proportional to transpulmonary pressure. How well a lung inflates and deflates with a change in transpulmonary pressure depends on its elastic properties. An important feature of an elastic material is that once stretched, it will recoil to its unstretched position. Therefore, the lung is like a spring; it recoils when stretched.

Elastic Properties of the Lungs and Chest Wall Influence Breathing

The ease with which the lungs can be stretched or inflated is termed **distensibility.** Distensibility and elastic recoil are inversely related to each other. An analogy of this relationship is a coiled spring—the greater the distensibility, the less the elastic recoil; a lung that is easily inflated has less elastic recoil. Overstretching causes the lungs to lose elastic recoil. Lung distensibility and elastic recoil arise from the elastin and collagen fibers enmeshed around the alveolar walls, adjacent bronchioles, and small blood vessels. Elastin fibers are highly distensible and can be stretched to almost double their resting length. Collagen fibers, however, resist stretch and limit lung expansion at high lung volumes. As the lungs expand during inflation, the fiber network around alveoli, small blood vessels, and small airways unfolds and rearranges—similar to stretching a nylon stocking. When the stocking is stretched, there is not much change in individual fiber length, but the unfolding and rearrangement of the nylon mesh allows the stocking to be easily stretched out to fit the contour of the legs. However, if the nylon stocking is overstretched, it loses its elastic recoil and no longer fits the contour of the legs and sags or becomes "baggy." In the same way, lungs that lose their elastic recoil also become "baggy." In other words, "baggy lungs" are easy to inflate but are difficult to deflate because of their inability to recoil.

Lung Compliance Is Determined From a Pressure-Volume Curve

Lung distensibility and elastic recoil can be determined from a **pressure-volume curve.** A simple analogy is the inflation of a balloon with a syringe (Fig. 19.16). For each change in pressure, the balloon inflates to a new volume.

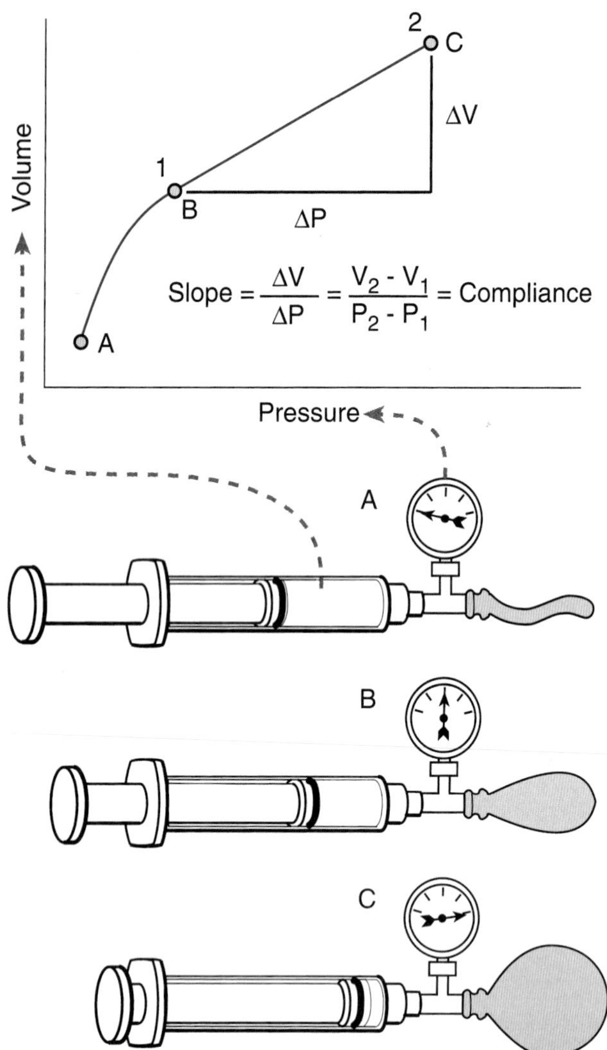

FIGURE 19.16 **Using a balloon to illustrate two elastic properties (distensibility and elastic recoil).** For each change in pressure (shown by movement of the arrow on the manometer dial), the balloon inflates to a new volume (plotted on the graph at points A, B, and C). Compliance is a measure of distensibility and is determined from the slope of the line between any two points on a pressure-volume curve. Recoil of an elastic material is inversely related to its compliance. (Modified from Rhoades R, Pflanzer R. Human Physiology. 3rd Ed. Fort Worth: Saunders College Publishing, 1996.)

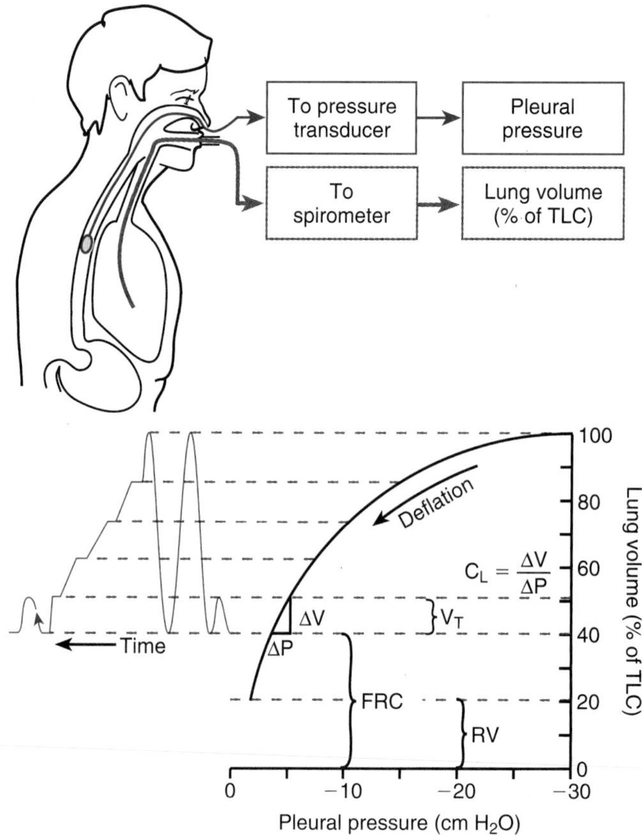

FIGURE 19.17 **Measuring lung compliance from a pressure-volume curve.** The subject first inspires maximally to total lung capacity (TLC) and then expires slowly, while airflow is periodically stopped to simultaneously measure pleural pressure and lung volume. Lung compliance (CL) is measured in L/cm H_2O. Note that pleural pressure is determined by measuring esophageal pressure.

The slope of the line of the pressure-volume curve is known as **compliance** (CL). Compliance is a measure of distensibility and is represented by

$$C_L = \Delta\text{volume}/\Delta\text{pressure} \qquad (19)$$

where Δvolume equals change in volume and Δpressure equals change in pressure (see Chapter 12) .

A similar pressure-volume curve can be generated for the human lungs by simultaneously measuring changes in lung volume with a spirometer and changes in pleural pressure with a pressure gauge (Fig. 19.17). Since the esophagus passes through the thorax, changes in pleural pressure can be obtained indirectly from the pressure changes in the esophagus by using a balloon catheter. In practice, a pressure-volume curve is obtained by having the individual first inspire maximally to total lung capacity (TLC) and then expire slowly. During the slow expiration, airflow is periodically interrupted (so that alveolar pressure is zero), and lung volume and pleural pressure are measured. Under these conditions, in which no airflow is occurring, the volume change per unit pressure change (ΔV/ΔP) is called **static compliance**.

In addition to distensibility and elastic recoil, lung compliance depends on lung volume, lung size, and surface tension inside the alveoli. Since the pressure-volume curve of the lung is nonlinear, compliance is not the same at all lung volumes; it is high at low lung volumes and low at high lung volumes. In the midrange of the pressure-volume curve, lung compliance is about 0.2 L/cm H_2O in adult humans. Lung size affects lung compliance. A mouse lung, for example, has a different compliance than an elephant lung. To allow comparisons among lungs of different sizes, lung compliance is normalized by dividing it by FRC to give a **specific compliance**.

What is the significance of an abnormally low or abnormally high compliance? Low lung compliance, indicating low distensibility, means more work is required to

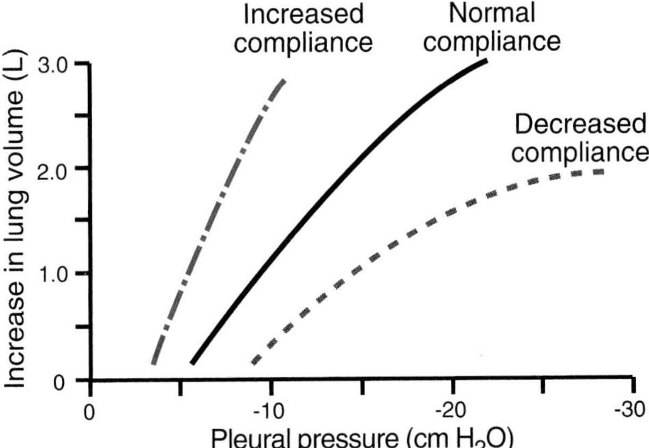

FIGURE 19.18 **Lung compliance in disease.** Lung compliance is abnormally high in patients with an obstructive disorder and abnormally low in patients with a restrictive disorder.

inflate the lungs to bring in a normal tidal volume. In a restrictive disorder, the compliance of the lungs is decreased (Fig. 19.18).

If the lungs are injured (by infection or by toxic environmental insults), the elastic properties are often altered, and the lungs lose their distensibility and recoil (see Clinical Focus Box 19.2). A high compliance is just as detrimental as a low compliance. Consider the disease **emphysema**, an obstructive disorder, with lungs being overstretched from chronic coughing and congested airways. Lungs of patients with emphysema have a high compliance (high distensibility) and are extremely easy to inflate. However, getting air out again is another matter. Lungs with abnormally high compliance have low elastic recoil, and additional effort is required to force air out of the lungs. Therefore, lungs with emphysema (which is strongly linked with smoking) retain an abnormally high residual volume of air.

Elastic Recoil of the Chest Wall Affects Lung Volumes

Just as the lungs have elastic properties, so does the chest wall. The outward elastic recoil of the chest wall aids lung expansion while the inward elastic recoil of the lungs pulls in the chest wall. The elastic recoil of the chest wall is such that if the chest were unopposed by the recoil of the lung, it would expand to about 70% of total lung capacity. This volume represents the resting position of the chest wall unopposed by the lung. If the chest wall is mechanically expanded beyond its resting position, it recoils inward. At volumes less than 70% of total lung capacity, the recoil of the chest wall is directed outward and is opposite the elastic recoil of the lung. Therefore, the outward elastic recoil of the chest wall is greatest at residual volume, whereas the inward elastic recoil of the lung is greatest at total lung capacity.

The lung volume at which lung and chest are at equilibrium (i.e., equal elastic recoil but in opposite directions) is represented by functional residual capacity (FRC). A change in the elastic properties of either the lungs or chest wall has a significant effect on FRC. For example, if the elastic recoil of the lungs is increased (i.e., lower C_L), a new equilibrium is established between the lungs and chest wall, resulting in a *decreased* FRC. Conversely, if the elastic recoil of the chest wall is increased, FRC is higher than normal. The elastic recoil of the chest wall at low lung volumes is a major determinant of residual volume in young people.

Differences in Regional Compliance Cause Uneven Ventilation

Lung compliance also affects ventilation and causes inspired air to be unevenly distributed in the lungs. In a normal upright individual, compliance at the top part of the lungs is less than at the base. This difference in compliance between the apex and base, known as **regional compliance**, is caused by gravity (Fig. 19.19). The gravitational effect

CLINICAL FOCUS BOX 19.2

Adult Respiratory Distress Syndrome

Diffuse lung injuries associated with a wide variety of conditions are collectively called **adult respiratory distress syndrome** (ARDS). Few patients have histories of previous lung disorders. The causes include trauma from head and chest injuries (e.g., car accidents), long bone injury, and pelvic injury. During the Vietnam War period, the trauma that caused ARDS was called Da Nang lung. Two other major causes are sepsis (the presence of a pathogen or toxin in the blood) and aspiration of gastric contents. The latter occurs with gastric reflux, and can occur at night during sleep or following surgery. Other causes are cardiopulmonary bypass, smoke inhalation, high altitude, and exposure to irritant gases. Exposure to irritant gases was responsible for the numerous ARDS cases that occurred at the disaster in Bhophal, India, in 1992.

Although the cause for ARDS is varied, the pathophysiology is nearly identical. ARDS is characterized by decreased lung compliance, pulmonary edema, focal atelectasis, hypoxemia, and an inflammatory reaction that leads to infiltration of neutrophils into the lung. Loss of lung surfactant leads to increased lung stiffness (i.e., decreased compliance) and edema.

Neutrophil aggregation is a key underlying mechanism of ARDS. Aggregation of neutrophils causes capillary endothelial damage by releasing several toxic substances, including oxygen free radicals, proteolytic enzymes, arachidonic acid metabolites (leukotrienes, thromboxane, prostaglandins), and platelet-activating factor.

Approximately 250,000 people develop ARDS each year. Despite improved diagnosis and treatment, the anticipated mortality for ARDS is still about 80%. Treatment requires long hospital stays and the medical costs often exceed $200,000. New approaches to therapy involve ways to reduce neutrophil chemoattraction and aggregation in the pulmonary capillaries and ways to reduce the amount of toxic substances released by neutrophils.

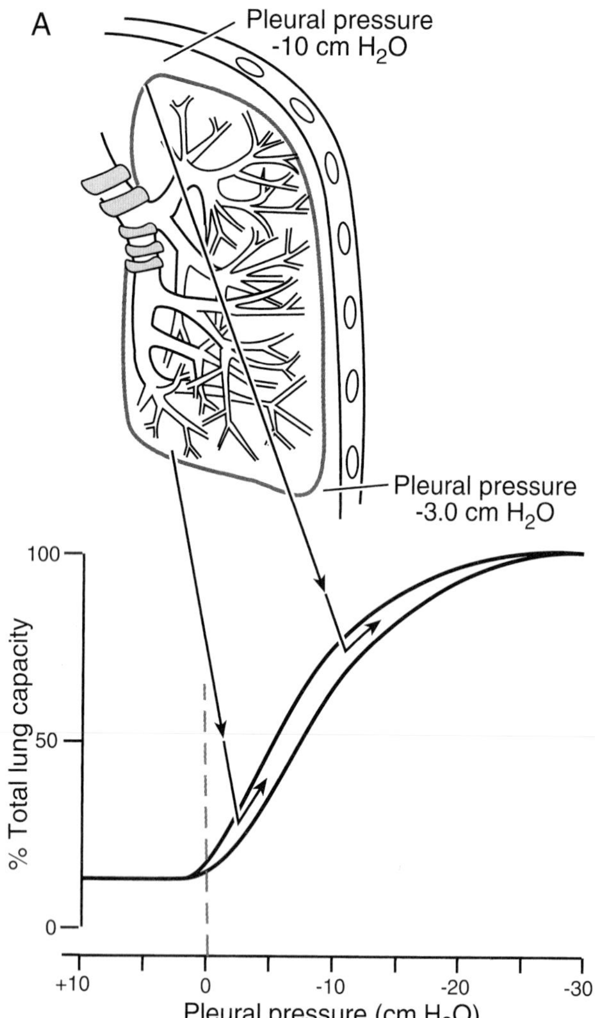

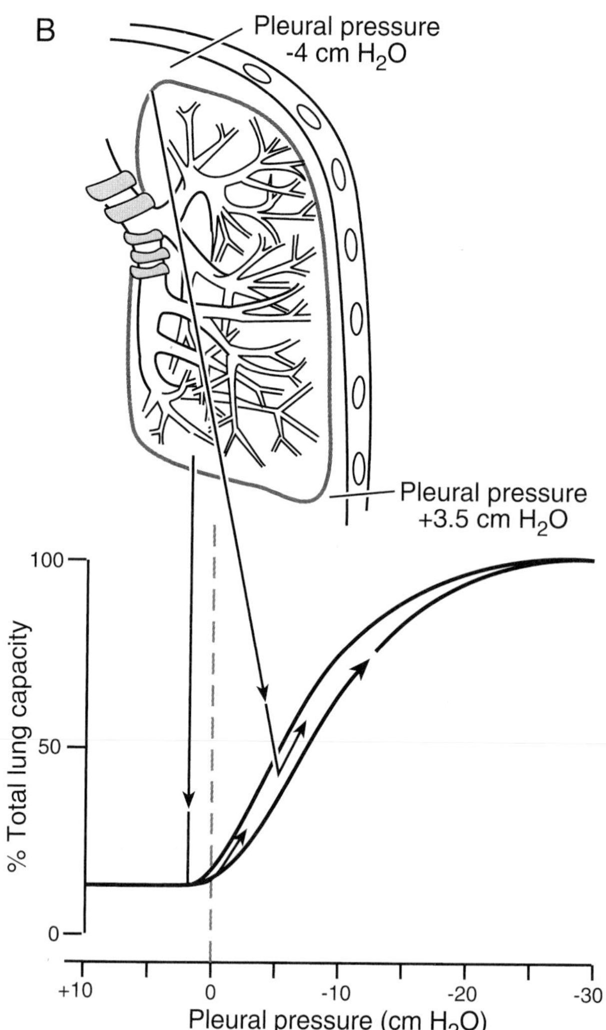

FIGURE 19.19 Effect of regional differences in lung compliance on the distribution of airflow. A, Because of gravity, pleural pressure in the upright person is more negative at the apex than at the base of the lungs. Consequently, the base of the lungs is more compliant at functional residual capacity. Therefore, more of the tidal volume will go to the base of the lungs. B, At very low lung volumes (residual volume), pleural pressure at the base becomes positive. As a result, the apical region is on the steeper part of the pressure-volume curve, is more compliant, and is initially filled with more of the inspired air than the base.

occurs because lung tissue is approximately 80% water and gravity exerts a "downward pull," resulting in a lower pleural pressure (i.e., more negative) at the apex than at the base of the lungs. As a result, there is a greater transpulmonary pressure at the apex (Fig. 19.19A). The higher transpulmonary pressure at the apex causes the alveoli to be more expanded and leads to regional differences in compliance. At any given volume from the FRC and above, the apex of the lung is less distensible (i.e., lower regional compliance) than at the base. This means the base of the lung has both a larger change in volume for the same pressure change and a smaller resting volume than at the apex. In other words, as one takes a breath in from FRC, a greater portion of the tidal volume will go to the base of the lungs, resulting in greater ventilation at the base.

At low lung volumes, alveoli at the apex of the lung are more compliant that at the base (Fig. 19.19B). At lung volumes approaching residual volume, the pleural pressure at the base of the lungs actually exceeds the pressure inside the airways, leading to airway closure. At residual volume, the base is compressed, and ventilation in the base is impossible until pleural pressure falls below atmospheric pressure. By contrast, the apex of the lung is in a more favorable portion of the compliance curve. Consequently, the first portion of the breath taken in from residual volume enters alveoli in the apex, and the distribution of ventilation is inverted at low lung volumes (i.e., the apex is better ventilated than the base).

Surface Tension Contributes to Lung Compliance

Another property that significantly affects lung compliance is surface tension at the air-liquid interface of the alveoli. The surface of the alveolar membrane is moist and is in contact with air, producing a large air-liquid interface. Surface tension (measured in dyne/cm) arises because water molecules are more strongly attracted to one another than to air molecules. In the alveoli, surface tension pro-

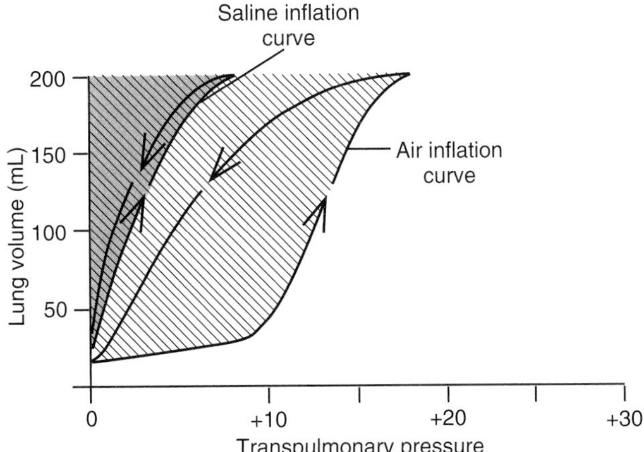

FIGURE 19.20 **Surface tension of the lungs.** Inflating lungs with saline instead of air eliminates surface tension. Two pressure-volume curves for lungs inflated to the same volume are shown, first with air and then with saline. The differences in the two curves occur because surface tension contributes significantly to lung compliance in the air-filled lungs. When lungs are inflated and deflated with saline, the lungs are more distensible and compliance increases. The shaded areas are equal to the work done in inflating the lungs.

duces an inwardly directed force that tends to reduce alveolar diameter.

Surface tension of the lungs can be studied by examining a pressure-volume curve. Figure 19.20 shows results from lungs that have been removed from the chest (excised lungs) and inflated and deflated in a stepwise fashion, first with air and then with saline. With air-filled lungs, the gas-liquid interface creates surface tension. However, with saline-filled lungs, the air-liquid interface is eliminated and the surface tension is eliminated.

Comparing these two pressure-volumes curves, two important observations can be made. First, the slope of the deflation limb of the saline curve is much steeper than that of the air curve. This means that when surface tension is eliminated, the lung is far more compliant (more distensible). Second, the different areas to the left of the saline and air inflation curves show that surface tension significantly contributes to the work required to inflate the lungs. Because the area to the left of each curve is equal to work, which can be defined as force (change in pressure) times distance (change in volume), the elastic forces and surface tension can be separated. The area to the left of the saline inflation curve is the work required to overcome the elastic recoil of the lung tissue. The area to the left of the air inflation curve is the work required to overcome both elastic tissue recoil and surface tension. Subtracting the area to the left of the saline curve from the area to the left of the air curve shows that approximately two thirds of the work required to inflate the lungs is needed to overcome surface tension. Lung distensibility and the work of breathing are significantly affected by surface tension.

Surface tension has important ramifications for maintaining alveolar stability. In a sphere such as an alveolus, surface tension produces a force that pulls inward and creates an internal pressure. The relationship between surface tension and pressure inside a sphere is shown in Figure 19.21. The pres-

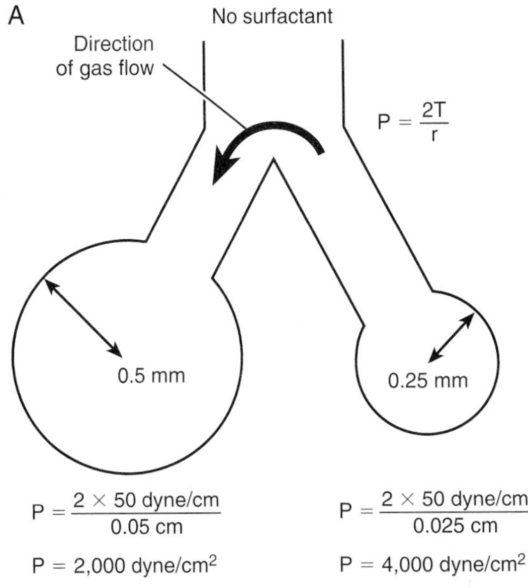

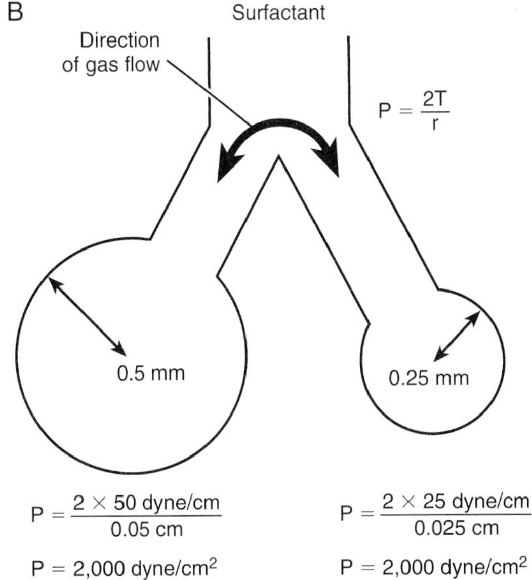

FIGURE 19.21 **Surface tension and alveolar stability.** **A,** If surface tension remains constant (50 dyne/cm), alveoli that are interconnected but differ in diameter become unstable and cannot coexist. Pressure in the smaller alveolus is greater than in the larger alveolus, which causes air from the smaller alveolus to empty into the larger alveolus. At low lung volumes, the smaller alveoli tend to collapse, a phenomenon known as atelectasis. **B,** Surfactant promotes alveolar stability by lowering surface tension proportionately more in the smaller alveolus. As a result, pressures in the two alveoli are equal, and alveoli of different diameters can coexist.

sure developed in an alveolus is given by the Laplace equation (see Chapter 12):

$$P = 2T/r \qquad (20)$$

where T equals surface tension and r equals radius. Because alveoli are interconnected and vary in diameter, the

Laplace equation assumes functional importance in the lung. In the example shown in Figure 19.21A, surface tension is constant at 50 dyne/cm, and an unstable condition results because pressure is greater in the smaller alveolus than in the larger one. Consequently, smaller alveoli tend to collapse, especially at low lung volumes, a phenomenon known as **atelectasis.** Larger alveoli become overdistended when atelectasis occurs. Therefore, two questions arise: How do alveoli of different sizes coexist when interconnected? How does the normal lung prevent atelectasis at low lung volumes? The answer lies, in part, in the fact that the alveolar surface tension is not constant at 50 dyne/cm, as in other biological fluids.

Surfactant Lowers Surface Tension and Stabilizes Alveoli

The alveolar lining is coated with a special surface-active agent, **pulmonary surfactant**, which not only lowers surface tension at the gas-liquid interface, but also changes surface tension with changes in alveolar diameter (Fig. 19.21B). Pulmonary surfactant is a lipoprotein rich in phospholipid. The principal agent responsible for its surface tension-reducing properties is **dipalmitoylphosphatidylcholine** (DPPC).

The functional importance of surfactant can be demonstrated by using a surface tension balance (Fig. 19.22). Surface tension is measured by placing a platinum flag connected to a force transducer into a trough of liquid. Surface tension creates a meniscus on each side of the platinum flag, and the greater the contact angle of the meniscus, the greater the surface tension. The surface is repeatedly expanded and compressed by a movable barrier that skims the surface of the liquid (simulating lung inflation and deflation). Surface tension of pure water is 72 dyne/cm, a value that is *independent* of the surface area of water in the balance. Therefore, when the surface is expanded and compressed, surface tension does not change. When a detergent is added, surface tension is reduced but again is independent of surface area. However, when a lung lavage is added that contains pulmonary surfactant, surface tension not only is reduced but also changes in a nonlinear fashion with surface area. Therefore, pulmonary surfactant makes it possible for alveoli of different diameters that are connected in parallel to coexist and be stable at low lung volumes, by lowering surface tension proportionately more in the smaller alveoli (see Fig 19.21B).

Surfactant works by reducing surface tension at the gas-liquid interface. When the molecules are compressed during lung deflation, surfactant causes a decrease in surface tension. At low lung volumes, when the molecules are tightly compressed, some surfactant is squeezed out of the surface and forms micelles (Fig. 19.23). On expansion (reinflation), new surfactant is required to form a new film that is spread on the alveolar surface lining.

When surface area remains fairly constant during quiet or shallow breathing, the spreading of surfactant is often impaired. A deep sigh or yawn causes the lungs to inflate to a larger volume and new surfactant molecules spread onto the gas-liquid interface. Patients recovering from anesthe-

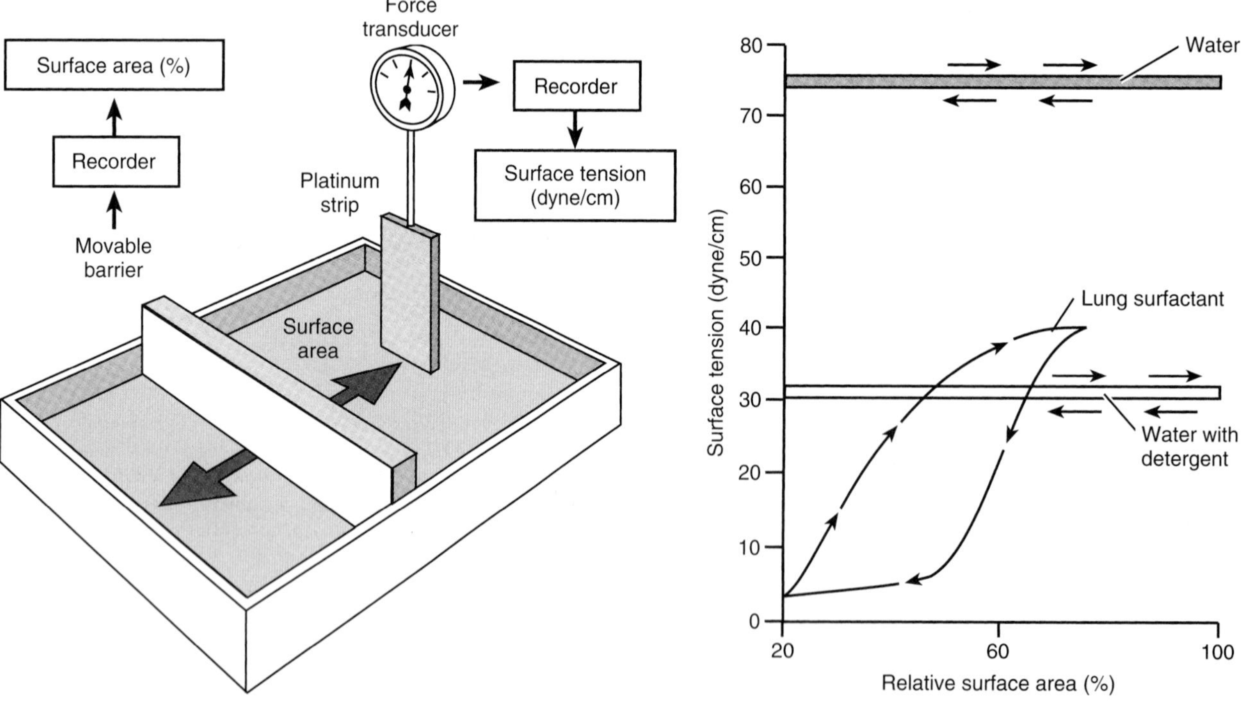

FIGURE 19.22 **Measurement of surface tension.** A surface tension balance is used to measure surface tension at the air-liquid interface. When distilled water is placed in the balance, surface tension is *independent* of surface area. The addition of a detergent reduces surface tension, but it is still independent of surface area. When lung surfactant (obtained from a lung lavage—a volume of saline flushed into the airways to rinse the alveoli) is placed in the balance, surface tension not only is decreased but also changes with surface area.

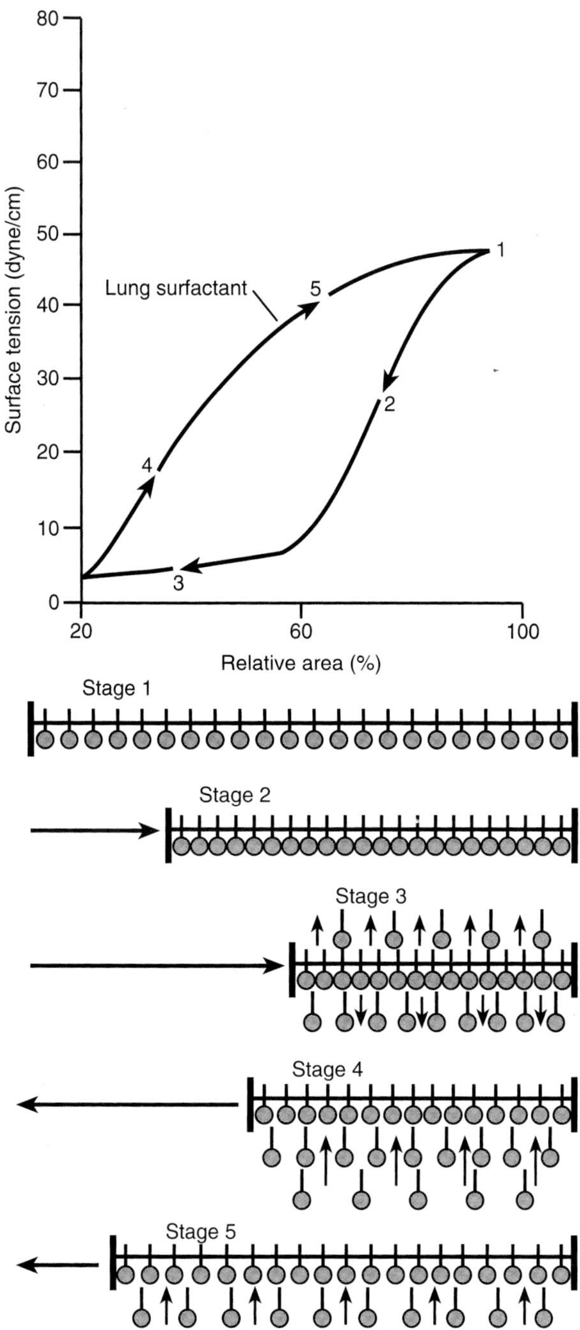

FIGURE 19.23 Lowering of the alveolar surface tension by lung surfactant. Surfactant molecules are compressed during lung deflation. At stage 3, surfactant molecules form micelles and are removed from the surface. On lung inflation, new surfactant is spread onto the surface film (stage 4). Turnover of lung surfactant is high because of continual replacement of surfactant during lung expansion.

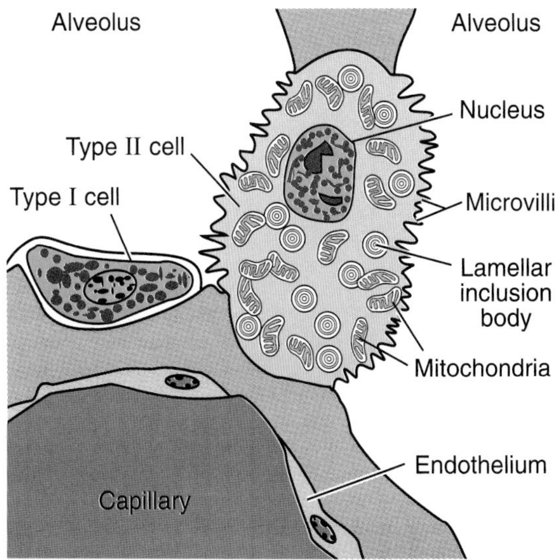

FIGURE 19.24 Cells of the alveolar epithelium. Alveolar type I cells occupy most of the alveolar surface, and alveolar type II cells are located in the corners between two adjacent alveoli. Alveolar type II cells produce lung surfactant. Also shown are endothelial cells that line the pulmonary capillaries.

sia are often encouraged to breathe deeply to enhance the spreading of surfactant. Patients who have undergone abdominal or thoracic surgery often find it too painful to breathe deeply; poor surfactant spreading results, causing part of their lungs to become atelectatic.

The alveolar epithelium consists basically of two cell types: alveolar type I and type II cells (Fig. 19.24). Alveolar type II cells are often referred to as type II pneumocytes. The ratio of these cells in the epithelial lining is about 1:1, but type I cells occupy approximately two thirds of the surface area. Type II cells seem to aggregate around the alveolar septa. Surfactant is synthesized in the alveolar type II cell. Compared to type I cells, they are rich in mitochondria and are metabolically active. Electron-dense **lamellar inclusion bodies** are a distinguishing feature of the type II cell. These lamellar inclusion bodies, rich in surfactant, are thought to be the storage sites for surfactant.

The process of surfactant synthesis is shown in Figure 19.25. Substrates (glucose, palmitate, and choline) are taken up from the circulating blood and synthesized into DPPC by alveolar type II cells. Stored surfactant from the lamellar inclusion bodies is discharged onto the alveolar surface. The turnover of surfactant is high because of the continual renewal of surfactant at the alveolar surface during each expansion of the lung. The high rate of replacement of surfactant probably accounts for the active lipid synthesis that occurs in the lung.

Because the lungs are among the last organs to develop, the synthesis of surfactant appears rather late in gestation. In humans, surfactant appears at about week 34 (a full-term pregnancy is 40 weeks). Regardless of the total duration of gestation in any mammalian species, the process of lung maturation seems to be "triggered" about the time gestation is 85 to 90% complete. Clearly, the fetal lung is endowed with a special regulatory mechanism to control the timing and appearance of surfactant. Failure of proper lung maturation during the perinatal period is still a major cause of death in newborns. The lung may be structurally intact but functionally immature if inadequate amounts of surfactant are available to reduce surface tension and stabilize surface forces during breathing.

Premature birth and certain hormonal disturbances (such as those seen in diabetic pregnancies) interfere with

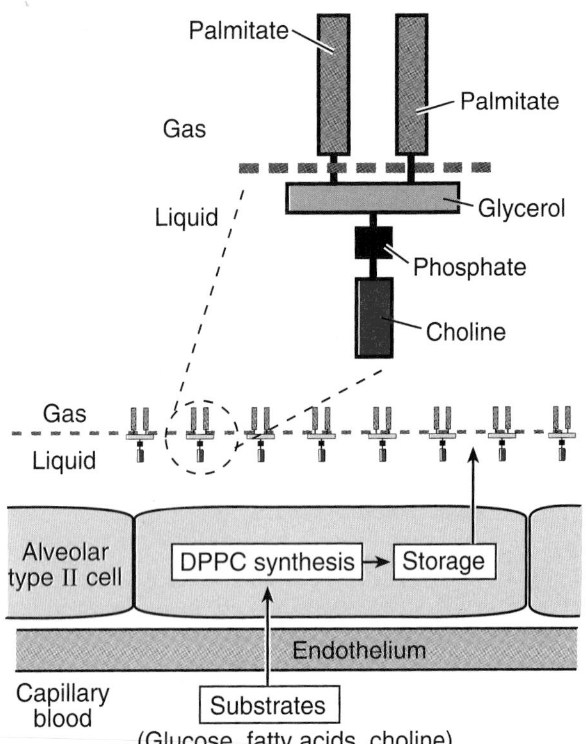

FIGURE 19.25 **Surfactant synthesis.** Substrates for lung surfactant synthesis are taken up by alveolar type II cells from the pulmonary capillary blood. Surfactant (dipalmitoylphosphatidylcholine or DPPC) is stored in lamellar inclusion bodies and subsequently discharged onto the alveolar surface. Surfactant is oriented perpendicular to the gas-liquid interface at the alveolar surface; the polar end is immersed in the liquid phase, while the nonpolar portion is in the gas phase.

the normal control and timing of lung maturation. These infants have immature lungs at birth, which often leads to **infant respiratory distress syndrome** (IRDS). Breathing is extremely labored because surface tension is high, making it difficult to inflate the lungs. Because of the high surface tension, these infants develop pulmonary edema and atelectasis. They are at high risk until the lungs become mature enough to secrete surfactant.

In addition to lowering alveolar surface tension and promoting alveolar stability, surfactant helps to prevent edema in the lung. The inward-contracting force that tends to collapse alveoli also tends to lower interstitial pressure, which "pulls" fluid from the capillaries. Pulmonary surfactant reduces this tendency by lowering surface forces. Some pulmonary physiologists think that keeping the lungs dry may be the major role of surfactant, especially in adults.

Alveolar Interdependence Promotes Alveolar Stability

Another mechanism that plays a role in promoting alveolar stability is interdependence or mutual support among adjacent alveoli. Because alveoli (except those next to the pleural surface) are interconnected with surrounding alveoli, they support each other. Studies have shown that this type of structural arrangement, with many connecting links, prevents the collapse of adjacent alveoli. For example, if alveoli tend to collapse, surrounding alveoli would develop large expanding forces. Therefore, interdependence can play a role in preventing atelectasis as well as in opening up lungs that have collapsed. Alveolar interdependence seems to be more important in adults than in newborns because newborns have fewer interconnecting links.

AIRFLOW AND THE WORK OF BREATHING

Two basic types of airflow occur in the lung. **Turbulent flow** occurs at high flow rates in the large airways (the trachea and large bronchi). Turbulent flow consists of completely disorganized patterns of airflow, resulting in a sound that can be heard with inspiration and expiration. The faster and deeper the breathing, the more noise produced from turbulence. In contrast, **laminar flow** is characterized by a streamlined flow that runs parallel to the sides of the airways. Laminar flow is silent because layers of air molecules slide over each other. This type of flow occurs in the small peripheral airways, where airflow is exceedingly slow.

Airway Resistance Decreases Airflow in the Lung

Both turbulent and laminar flow cause resistance to air moving in the airways. **Airway resistance** is expressed in cm H_2O/L per second and is defined as the ratio of driving pressure (ΔP) to airflow ($\dot{V}$). For total airway resistance (Raw), the driving pressure is the pressure difference between the mouth (P_{mouth}) and the alveoli (P_A).

The equation can be written as

$$\text{Raw} = \frac{P_{mouth} - P_A}{\dot{V}} \tag{21}$$

The major site of airway resistance is the medium bronchi (lobar and segmental) and bronchi down to about the seventh generation (Fig. 19.26). One would expect the major site of resistance, based on Poiseuille's law (see Chapter 12), to be located in the narrow airways (the bronchioles), which have the smallest radius. However, measurements show that only 10 to 20% of total airway resistance can be attributed to the small airways (those less than 2 mm in diameter). This apparent paradox results because so many small airways are arranged in parallel and their resistances are added as reciprocals. Resistance of each individual bronchiole is relatively high, but their great number results in a large total cross-sectional area, causing their total *combined resistance* to be low.

Many airway diseases often begin in the small airways. Diagnosing a disease in the small airways is difficult, however, because the small airways account for such a low percentage of the total Raw. Early detection is difficult because changes in airway resistance are not noticeable until the disease becomes severe.

Effect of Lung Volume. One of the major factors affecting airway diameter, especially that of bronchioles, is lung vol-

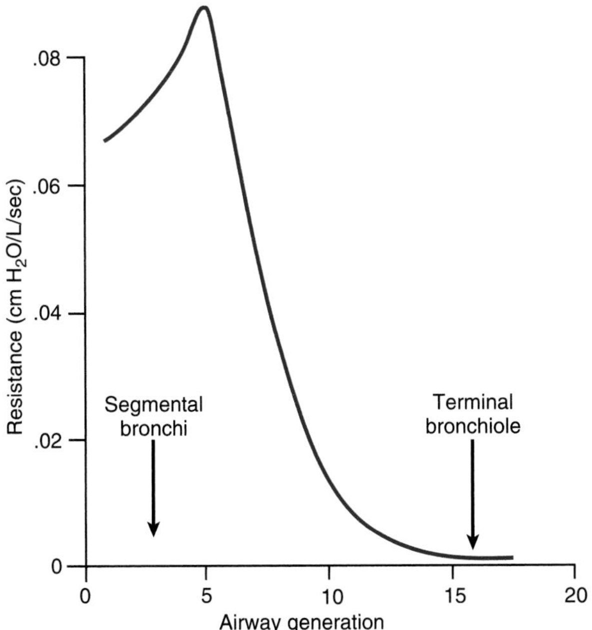

FIGURE 19.26 **Airway resistance.** Resistance to airflow is greatest in the large and medium-sized airways. The major site of resistance is the lobar and segmental bronchi, down to about the seventh generation of airway branches.

ume. Like lung tissue, the smaller airways are capable of being distended or compressed. Bronchi and smaller airways are embedded in lung parenchyma, and are connected by "guy wires" to surrounding tissue. As the lung enlarges, airway diameter increases, which results in a concomitant decrease in airway resistance during inspiration (Fig. 19.27). Conversely, at low lung volumes, airways are compressed, and airway resistance rises. Note that the inverse relationship between lung volume and airway resistance is nonlinear. At low lung volumes, airway resistance rises sharply.

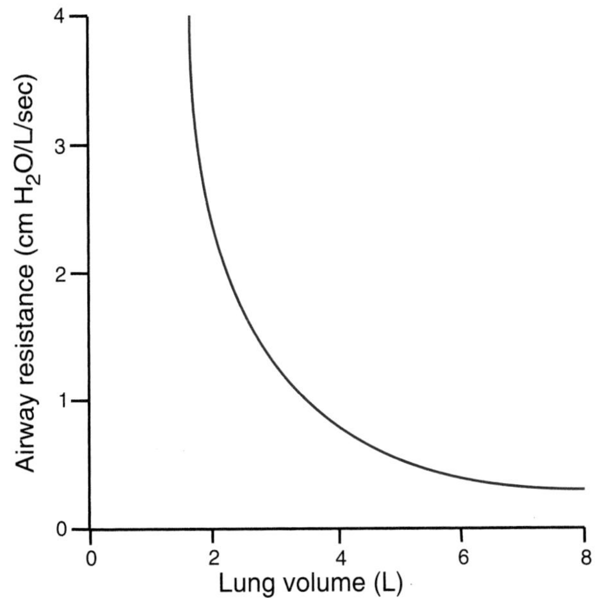

FIGURE 19.27 **Effect of lung volume on airway resistance.** Airways are compressed at low lung volumes, resulting in increased airway resistance.

Effect of Smooth Muscle Tone. Bronchial smooth muscle tone also affects airway diameter with a concomitant change in airway resistance. A change in smooth muscle tone will change airway diameter. The smooth muscles in the airway, from the trachea down to the terminal bronchioles, are under autonomic control. Stimulation of **parasympathetic** cholinergic postganglionic fibers causes bronchial constriction, as well as increased mucus secretion. Stimulation of **sympathetic** adrenergic fibers causes dilation of bronchial and bronchiolar airways and inhibition of glandular secretion. Drugs such as isoproterenol and epinephrine, which stimulate β_2-adrenergic receptors in the airways, cause dilation. These drugs alleviate bronchial constriction and are often used to treat asthmatic attacks. Environmental insults, such as breathing chemical irritants, dust, or smoke particles, can cause reflex constriction of the airways. Increased P_{CO_2} in the conducting airways can cause a local dilation. More important, a decrease in P_{CO_2} causes airway smooth muscle to contract.

Effect of Gas Density. The effect of gas density on airway resistance is seen most dramatically in deep-sea diving, in which the diver breathes air whose density may be greatly increased because of increased barometric pressure. The barometric pressure increases 1 atm for each 10 m or 33 feet underwater (e.g., at 10 m below the surface, P_B = 2 atm or 1,520 mm Hg). As a result of increased resistance, a large pressure gradient is required just to move a normal tidal volume. Instead of breathing air during diving, a helium-oxygen mixture is often used because helium is less dense than air and, consequently, makes breathing easier. The fact that density has a marked effect on airway resistance again indicates that the medium airways are the main site of resistance and that airflow here is primarily turbulent (see Chapter 12).

Airways Become Compressed During Forced Expiration

Airway resistance does not change much during normal quiet breathing. It is significantly increased, however, during forced expiration, such as in vigorous exercise. The marked change during forced expiration is a result of airway compression. This effect can be demonstrated with a **flow-volume curve**, which shows the relationship between airflow and lung volume during a forced inspiratory and expiratory effort (Fig. 19.28). The flow-volume curve is generated by taking the spirometer tracings of forced inspiratory and expiratory flow, and then closing them by connecting the end of maximum expiration back to the beginning of maximum inspiration. The small loop in Figure 19.28 is the flow-volume curve for a normal tidal volume. The large loop shows the maximal flow-volume curve. During forced inspiration, inspiratory flow is limited only by effort—that is, how hard the individual tries. During forced expiration (forced vital capacity), flow rises rapidly to a maximum value, **peak expiratory flow** (PEF), and then decreases linearly over most of expiration as lung volume decreases.

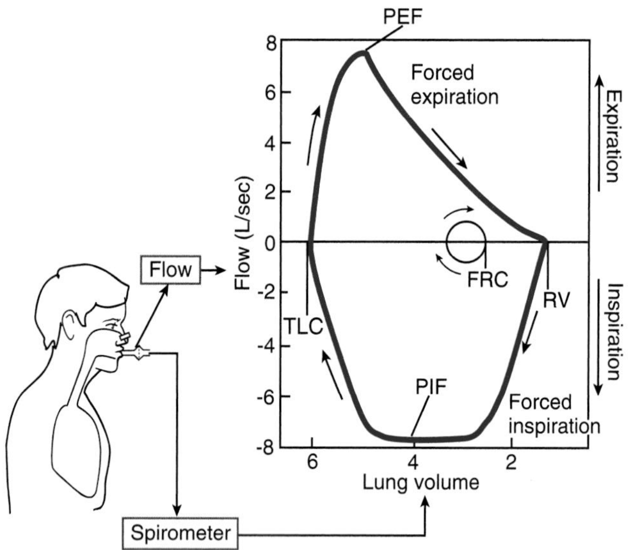

FIGURE 19.28 **Flow-volume curve.** The relationship between lung volume and airflow can be seen from a flow-volume curve. During tidal breathing, airflow velocity is very small (small loop labeled FRC [functional residual capacity]). However, during forced expiration, airflow rises rapidly and reaches a maximum at peak expiratory flow (PEF). PEF occurs before there is much change in lung volume. Once PEF is achieved, airflow rate decreases as lung volume decreases to residual volume (RV). Peak inspiratory flow (PIF) is maintained over a large change in lung volume because airways are distended and not compressed. The last part of the expiratory portion of the flow-volume curve (or a forced vital capacity) is *effort-independent* because of dynamic airway compression.

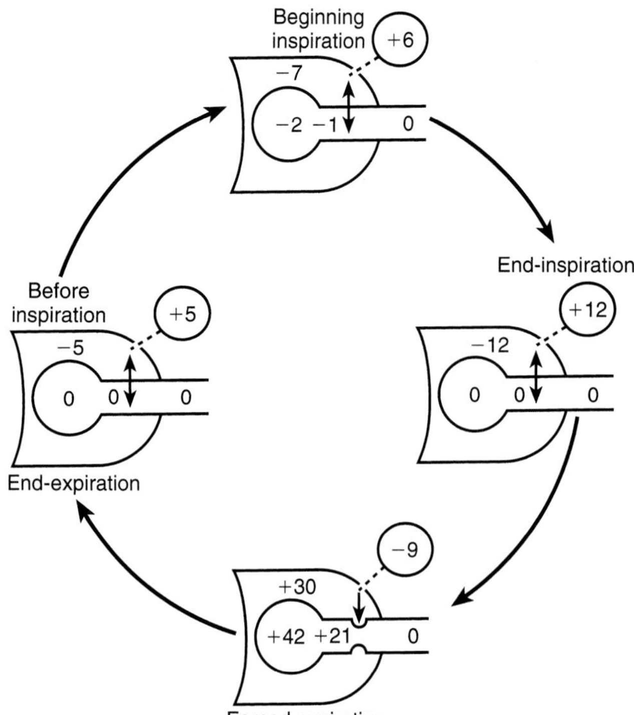

FIGURE 19.29 **Airway compression during forced expiration.** Transairway pressure is +5 cm H_2O before inspiration and reaches +12 cm H_2O at the end of inspiration. During forced expiration, transairway pressure becomes negative and the small airways are compressed.

The first part of the forced expiratory flow-volume curve is dependent on effort. Once PEF is achieved, flow is independent of effort in the last part of the flow-volume curve. Actually, over most of the expiratory flow-volume curve, flow is virtually independent of effort. The forced expiratory flow limitation illustrates the importance of dynamic airway compression. Dynamic airway compression increases airway resistance, which effectively limits the forced expiratory flow.

How does dynamic airway compression occur during forced expiration? The mechanism is related to changes in transairway pressure and the compressibility of the airways. For example, in Figure 19.29, pleural pressure (Ppl) is −5 cm H_2O and airway pressure (Paw) is zero before inspiration (no airflow). Transairway pressure (Paw − Ppl) is 5 cm H_2O [0 − (−5) = +5] and holds the airways open. At the start of maximum inspiration, pleural pressure decreases to −7 cm H_2O and alveolar pressure falls to −2 cm H_2O. The difference between alveolar pressure and pleural pressure is still 5 cm H_2O [–2 − (−7) = +5]. However, there is a pressure drop from mouth to alveoli because of resistance to airflow, and the transairway pressure will change along the airway. At the end of maximum inspiration, pleural pressure decreases further, to −12 cm H_2O, and airway pressure is again zero because of no airflow. During maximum inspiration, airway resistance actually decreases because transairway pressure increases, which enlarges the diameter of the airways, especially the small airways.

On forced expiration, pleural pressure is no longer negative but rises above atmospheric pressure and can increase up to +30 cm H_2O. The added pressure in the alveoli is due to the elastic recoil of the lungs and is termed **recoil pressure** (**precoil**) and can be written as

$$\text{Precoil} = P_A - \text{Ppl} \tag{22}$$

where P_A equals alveolar pressure and Ppl equals pleural pressure.

The recoil pressure in Figure 19.29 is 12 cm H_2O because at the beginning of expiration, lung volume has not appreciably decreased. Note that at the beginning of forced vital capacity, a pressure drop occurs along the airway because of airway resistance. Airway pressure falls progressively from the alveolar region to the airway opening (the mouth). The transairway pressure gradient along the airways reverses and tends to compress the airways. For example, at a point inside the airway where the pressure is 21 cm H_2O, the transairway pressure would be −9 cm H_2O, which would tend to close the airway.

At some point along the airway, the airway pressure equals pleural pressure, and transairway pressure is zero (Fig. 19.30). This is the **equal pressure point** (EPP). Theoretically, the EPP divides the airways into an upstream segment (from alveoli to EPP) and a downstream segment (EPP to the mouth). In the downstream segment, the airway pressure is below pleural pressure and the transairway pressure becomes negative. As a result, airways in the downstream segment are compressed or collapsed. The large air-

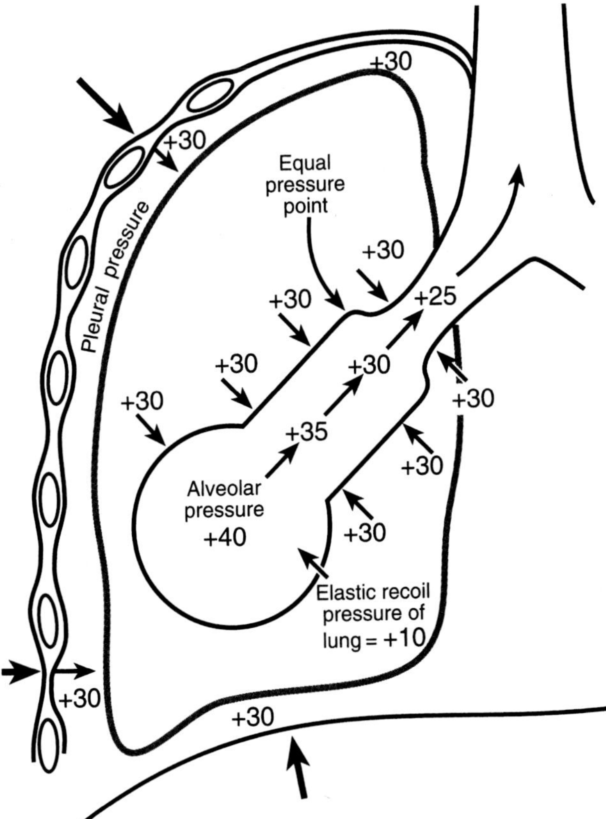

FIGURE 19.30 **Equal pressure point.** An equal pressure point (EPP) divides the airways into downstream and upstream segments. The EPP is established at peak expiratory flow (PEF) and occurs when the pressure inside the airway equals the pressure outside the airway (pleural pressure). The upstream segment is represented from the alveoli to the EPP, and the downstream segment is represented from the EPP to the mouth. The driving pressure for airflow is now alveolar pressure minus the pleural pressure. Airways are subjected to compression during forced expiration from the EPP to the trachea.

ways (the trachea and bronchi) are protected from collapse because they are supported by cartilage. However, small airways without this structural support are easily compressed and collapsed.

Lung Compliance Affects the Equal Pressure Point

The EPP is established after the peak expiratory flow is achieved (see Fig 19.28). As the forced expiratory effort continues, the EPP moves down the airways from larger to smaller airways because the recoil pressure decreases. A greater length of the downstream segment collapses. The driving pressure for airflow, once the EPP is established, is no longer the difference between alveolar pressure and mouth pressure but is alveolar pressure minus pleural pressure (see Fig. 19.30).

Two basic conclusions follow. First, regardless of the forcefulness of the expiratory effort, airflow cannot be increased because pleural pressure increases, causing more airway compression. This explains why the last part of the forced vital capacity is independent of effort. Second, elastic recoil of the lung determines maximum flow rates because it is the elastic recoil pressure that generates the alveolar-pleural pressure difference. As lung volume decreases, so does elastic recoil force. The decrease in elastic recoil is the main reason maximum flow falls so rapidly at low lung volumes.

The effect of elastic recoil on expiratory airflow can be demonstrated by comparing a normal lung to an emphysematous lung, in which the compliance is abnormally high. As seen in Figure 19.31, in both instances, pleural pressure rises to +30 cm H_2O with forced expiration. In the normal lung, the recoil pressure of 10 cm H_2O is added to produce an alveolar pressure of +40 cm H_2O.

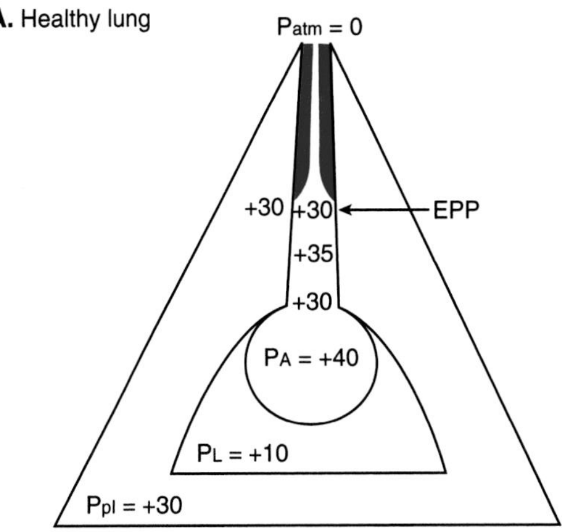

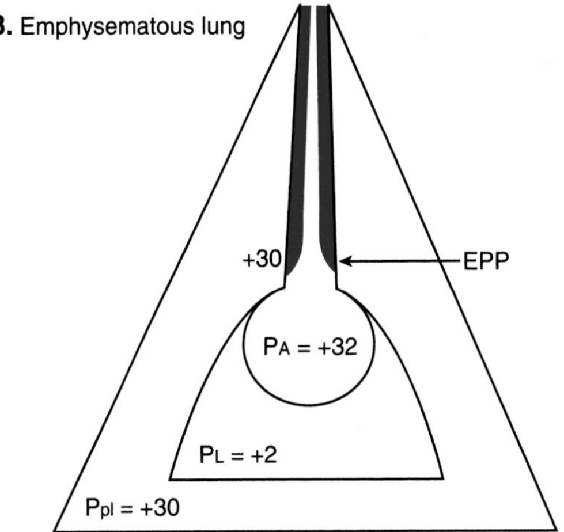

FIGURE 19.31 **Effect of elastic recoil on expiratory airflow. A,** In healthy lungs, elastic recoil adds 10 cm H_2O pressure, to produce an alveolar pressure of 40 cm H_2O at the beginning of a forced expiration. As a result, the equal pressure point (EPP) is established in the larger airways. Airway collapse is minimal in the large airways because cartilage supports the airways. **B,** A loss of elastic recoil causes the EPP to shift downward and to be established in the small airways. In emphysematous lungs, the elastic recoil pressure is low, and little recoil pressure is added to the alveolus. As a result, the EPP is shifted downward and is established in the smaller airways. The small airways are more easily compressed because they are thin and lack cartilaginous support.

With forced expiration, there is a progressive fall in airway pressure. Because of the elastic recoil, the normal lung has "added" pressure to the small airways that keeps their pressures *above* pleural pressure, and less collapse occurs. With emphysema, however, the lungs have diminished elastic recoil, resulting in less elastic recoil pressure. In the example shown in Figure 19.31, elastic recoil adds only 2 cm H_2O to alveolar pressure, resulting in an alveolar pressure of +32 cm H_2O. With expiratory effort, flow proceeds along the pressure gradient. But in the emphysematous lungs, the pressure inside the small airways falls below pleural pressure well before the large airways, resulting in more airway compression and collapsed airways at low lung volumes. As a result, flow stops temporarily in these collapsed airways. Airway pressure upstream to the collapsed segment then rises to equal alveolar pressure, causing the airways to open again. This process repeats itself continually, leading to the "wheeze" that is often heard in emphysematous patients.

In lungs with abnormally high compliance, the position of the EPP with forced expiration is established further down in the small airways, where there is no cartilage to keep them distended. Therefore, these patients are much more vulnerable to compression and collapse. The greatest problem for patients with emphysema is not getting air into the lungs but getting it out. Consequently, they tend to breathe at higher lung volumes, thereby increasing elastic recoil, which reduces airway resistance and facilitates expiration.

Work Is Required to Expand the Lungs and Overcome Airway Resistance

During inspiration, muscular work is involved in expanding the thoracic cavity, inflating the lungs, and overcoming airway resistance. Since work can be measured as force times distance, the amount of work involved in breathing can be expressed as a change in lung volume (distance) multiplied by the change in transpulmonary pressure (force). Thus, work (W) is equal to the product of pressure (P) and volume (V). With a volume change, the work involved in taking a breath is defined by this equation:

$$W = P \times \Delta V \quad (23)$$

where P equals transpulmonary pressure and ΔV equals change in lung volume.

During work, energy is expended with muscular contraction to create a force (transpulmonary pressure) to inflate the lungs. When a greater transpulmonary pressure is required to bring more air into the lungs, more muscular work and, hence, greater energy are required.

Figure 19.32 shows how a pressure-volume curve can be used to determine the work required for breathing. The shaded area in red represents the inspiratory work of breathing. In healthy individuals at rest, the energy needed for breathing represents approximately 5% of the body's total energy expenditure. During heavy exercise, about 20% of the total energy expenditure is involved in breathing. Breathing is efficient and is most economical when elastic and resistive forces yield the lowest work. Note the total inspiratory work of breathing in a restrictive lung disorder, compared with the normal lung, is increased and is a result of a greater inspiratory effort required. It is important to remember that lungs with a marked decrease in lung compliance (i.e., restrictive lung disorder) require more work to overcome increase in elastic recoil and surface forces. Patients with a restrictive disorder economize their ventilation by taking rapid and shallow breaths. In contrast, patients with severe airway obstruction tend to do the opposite; they take deeper breaths and breathe more slowly, to reduce their work due to the increased airway resistance. Despite this tendency, patients with obstructive disease still expend a considerable portion of their basal energy for breathing. The reason for this is the expiratory muscles must do additional work to overcome the increased airway resistance. These different breathing patterns help minimize the amount of work required for breathing.

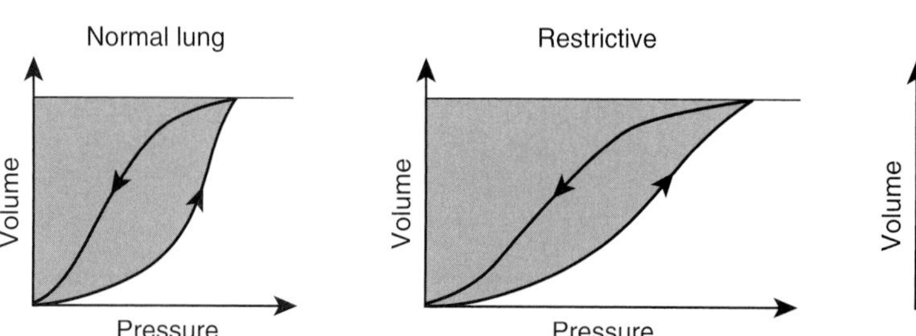

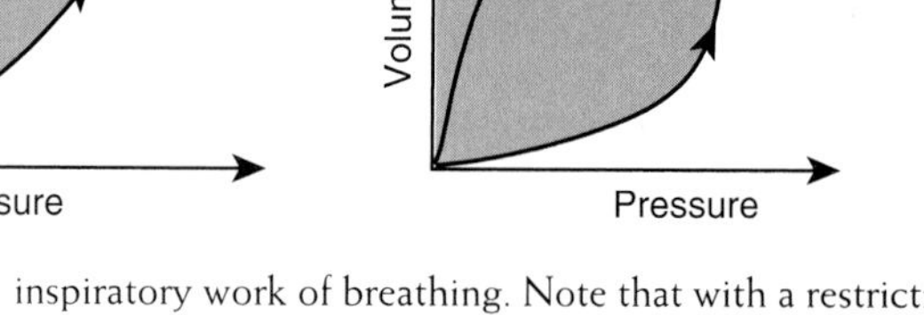

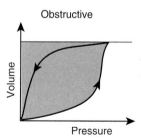

FIGURE 19.32 **Work of breathing.** A pressure-volume curve is used to determine the work of breathing. Pressure is plotted on the x-axis, and volume is plotted on the y-axis. The red area above the pressure-volume curve represents the inspiratory work of breathing. Note that with a restrictive disorder, inspiratory work of breathing is increased. Airway resistance is increased in an obstructive disorder, which requires more energy to force air out of the lungs during expiration.

REVIEW QUESTIONS

DIRECTIONS: Each of the numbered items or incomplete statements in this section is followed by answers or by completions of the statement. Select the ONE lettered answer or completion that is BEST in each case.

1. A 48-year-old patient undergoes a pulmonary function test. At which lung volume would her pleural pressure be most negative?
(A) Residual volume
(B) Functional residual capacity
(C) End of tidal volume
(D) Total lung capacity
(E) Middle of forced vital capacity
2. The same patient's transpulmonary pressure would be equal to
(A) Alveolar pressure minus pleural pressure
(B) Airway pressure minus pleural pressure
(C) Atmospheric pressure minus alveolar pressure
(D) Pleural pressure plus atmospheric pressure
(E) Pleural pressure plus alveolar pressure
3. The lungs and chest wall have opposite and equal recoil tendencies at what lung volume?
(A) Total lung capacity
(B) Functional residual capacity
(C) Residual volume
(D) Inspiratory reserve volume
(E) End of forced vital capacity
4. A patient's airways would most likely collapse under which of the following conditions of airway pressure (Paw) and pleural pressure (Ppl)?
(A) Paw = 35 cm H_2O, Ppl = 30 cm H_2O
(B) Paw = 15 cm H_2O, Ppl = 20 cm H_2O
(C) Paw = 0 cm H_2O, Ppl = −5 cm H_2O
(D) Paw = 20 cm H_2O, Ppl = 0 cm H_2O
5. A 21-year-old healthy athlete undergoes a pulmonary function test before spring training. Her expected ratio for FEV_1/FVC would be
(A) 0.20
(B) 0.40
(C) 0.60
(D) 0.80
(E) 1.00
6. In which of the following pulmonary disorders would a patient's lungs most likely have a pressure-volume curve that exhibits an abnormally high compliance?
(A) Pulmonary edema
(B) Fibrosis
(C) Emphysema
(D) Pulmonary congestion
(E) Respiratory distress syndrome
7. A 13-year-old girl suffers from asthma. She inhales a bronchodilator drug that causes a 3-fold increase in the radius of her small peripheral airways. Which of the following pulmonary function tests would show the greatest improvement?
(A) Total lung capacity
(B) Tidal volume
(C) Inspiratory capacity
(D) Forced vital capacity
(E) Functional residual capacity
8. A 62-year-old patient has been diagnosed with a restrictive pulmonary disease. Which of his following lung measurements is most likely to be increased?
(A) FEV_1
(B) FVC
(C) FEV_1/FVC
(D) FRC
(E) RV
9. Minute ventilation of a normal 65-year-old patient would be represented by
(A) Total expired air per minute
(B) Total air expired air per minute minus dead space ventilation
(C) Alveolar ventilation minus dead space volume
(D) Tidal volume divided by frequency of breathing
(E) Total lung capacity times frequency of breathing
10. A 52-year-old patient has a frequency of 10 breaths/min, vital capacity of 6 L, minute ventilation of 8 L/min, and functional residual capacity of 3 L. His tidal volume is
(A) 0.3 L
(B) 0.5 L
(C) 0.6 L
(D) 0.8 L
(E) 0.9 L
11. A 61-year-old patient has been diagnosed with pulmonary fibrosis. Compared with a healthy patient of the same age, his lungs would
(A) Be more compliant
(B) Be easier to deflate
(C) Require less work to inflate
(D) Have less elastic recoil
(E) Have greater airway resistance
12. A 25-year-old healthy subject holds his breath at residual volume with his glottis open. During this maneuver, his alveolar pressure is equal to
(A) Pleural pressure
(B) Transpulmonary pressure
(C) Airway pressure minus pleural pressure
(D) Atmospheric pressure
(E) Alveolar pressure minus pleural pressure
13. A patient inhales 500 mL of air from a spirometer. Pleural pressure before inspiration is −5 cm H_2O and −10 cm H_2O at the end of inspiration. What is the compliance of this patient's lungs?
(A) 5.0 L/cm H_2O
(B) 2.5 L/cm H_2O
(C) 0.5 L/cm H_2O
(D) 0.3 L/cm H_2O
(E) 0.1 L/cm H_2O
14. A first-year medical student undergoes a pulmonary function test. At one point during inspiration, her alveolar pressure is −1 cm H_2O, pleural pressure is −7 cm H_2O, and airway pressure is −4 cm H_2O. Atmospheric pressure at the time of her test was 750 mm Hg. What is her transpulmonary pressure?
(A) 8 cm H_2O
(B) 7 cm H_2O
(C) 6 cm H_2O
(D) 5 cm H_2O
(E) 3 cm H_2O
15. A 21-year-old athlete has a forced vital capacity of 5.0 L, functional residual capacity of 3.0 L, and residual volume of 1.2 L. What is her total lung capacity?
(A) 9.2 L
(B) 8.0 L
(C) 7.2 L
(D) 6.2 L
(E) 5.0 L
16. A 45-year-old patient has an alveolar ventilation of 5.0 L/min, a frequency of 10 breaths/min, and a tidal volume of 700 mL. Her vital capacity is 4 L. What is the patient's dead space ventilation?

(continued)

(A) 2.5 L/min
(B) 2.0 L/min
(C) 1.5 L/min
(D) 1.0 L/min
(E) 0.5 L/min

SUGGESTED READING

Avery ME. Surfactant deficiency in hyaline membrane disease: The story of discovery. Am J Respir Crit Care Med 2000;161:1074–1075.

Cotes JE. Lung Function: Assessment and Application in Medicine. 5th Ed. Boston: Blackwell Scientific, 1993.

Floros J, Padma K. Surfactant proteins: Molecular genetics of neonatal pulmonary diseases. Annu Rev Physiol 1998;60:365–384.

Hlastala MP, Berger AJ. Physiology of Respiration. 2nd Ed. New York: Oxford University Press, 2001.

Lumb AB. Nunn's Applied Respiratory Physiology. 5th Ed. Oxford, UK: Butterworth-Heinemann, 2000.

Massaro GD, Massaro D. Formation of pulmonary alveoli and gas exchange surface area: Quantitation and regulation. Annu Rev Physiol 1996;58:73–92.

Staub NC. Basic Respiratory Physiology. New York: Churchill Livingstone, 1991.

Pulmonary Circulation and the Ventilation-Perfusion Ratio

Rodney A. Rhoades, Ph.D.

CHAPTER OUTLINE

- FUNCTIONAL ORGANIZATION OF THE PULMONARY CIRCULATION
- PULMONARY VASCULAR RESISTANCE
- FLUID EXCHANGE IN PULMONARY CAPILLARIES
- BLOOD FLOW DISTRIBUTION IN THE LUNGS
- SHUNTS AND VENOUS ADMIXTURE
- THE BRONCHIAL CIRCULATION

KEY CONCEPTS

1. The pulmonary circulation is a high-flow, low-resistance, and low-pressure system.
2. Capillary recruitment and capillary distension cause the pulmonary vascular resistance to fall with increased cardiac output.
3. Alveolar oxygen tension (P_{AO_2}) regulates blood flow in the lungs.
4. High pulmonary capillary hydrostatic pressure leads to pulmonary edema.
5. Gravity causes lung perfusion to be better at the base than at the apex.
6. A mismatch of ventilation and blood flow occurs at both the base and the apex of the lungs.
7. Some of the blood that leaves the lungs is not fully oxygenated.
8. Poor regional ventilation is the major cause for a low ventilation-perfusion ratio in the lungs.
9. The bronchial circulation is part of the systemic circulation and does not participate in gas exchange.

FUNCTIONAL ORGANIZATION OF THE PULMONARY CIRCULATION

The heart drives two separate and distinct circulatory systems in the body: the pulmonary circulation and the systemic circulation. The pulmonary circulation is analogous to the entire systemic circulation. Similar to the systemic circulation, the pulmonary circulation receives all of the cardiac output. Therefore, the pulmonary circulation is not a regional circulation like the renal, hepatic, or coronary circulations. A change in pulmonary vascular resistance has the same implications for the right ventricle as a change in systemic vascular resistance has for the left ventricle.

The pulmonary arteries branch in the same tree-like manner as do the airways. Each time an airway branches, the arterial tree branches so that the two parallel each other (Fig. 20.1). More than 40% of lung weight is comprised of blood in the pulmonary blood vessels. The total blood volume of the pulmonary circulation (main pulmonary artery to left atrium) is approximately 500 mL or 10% of the total circulating blood volume (5,000 mL). The pulmonary veins contain more blood (270 mL) than the arteries (150 mL). The blood volume in the pulmonary capillaries is approximately equal to the stroke volume of the right ventricle (about 80 mL) under most physiological conditions.

The Pulmonary Circulation Functions in Gas Exchange and as a Filter, Metabolic Organ, and Blood Reservoir

The primary function of the pulmonary circulation is to bring venous blood from the superior and inferior vena cavae (i.e., mixed venous blood) into contact with alveoli for gas exchange. In addition to gas exchange, the pulmonary circulation has three secondary functions: it serves as a filter, a metabolic organ, and as a blood reservoir.

Pulmonary vessels protect the body against **thrombi** (blood clots) and **emboli** (fat globules or air bubbles) from entering important vessels in other organs. Thrombi and emboli often occur after surgery or injury and enter the systemic venous blood. Small pulmonary arterial vessels and capillaries trap the thrombi and emboli and prevent them from obstructing the vital coronary, cerebral, and renal vessels. Endothelial cells lining the pulmonary vessels release fibrinolytic substances that help dissolve thrombi. Emboli,

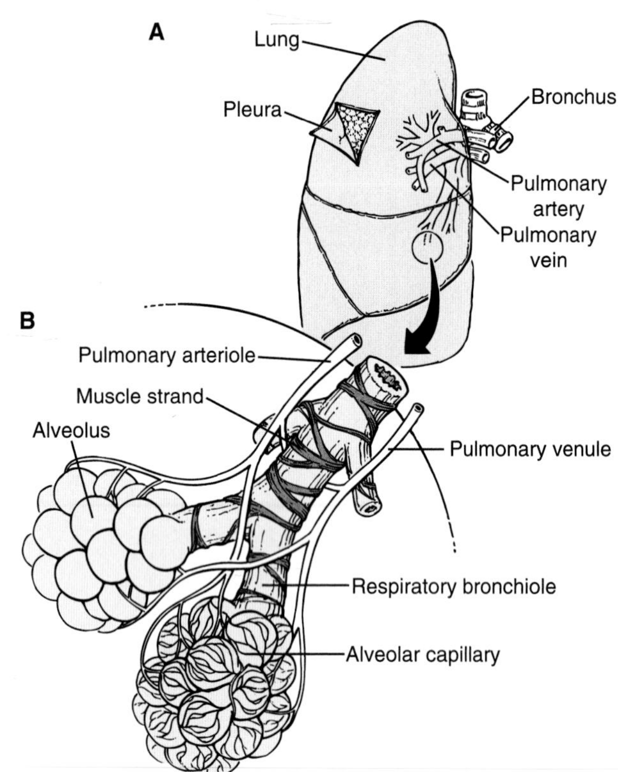

FIGURE 20.1 **Parallel structure of the vascular and airway trees.** **A,** Systemic venous blood flows through the pulmonary arteries into the alveolar capillaries and back to the heart via the pulmonary veins, to be pumped into the systemic circulation. **B,** A mesh of capillaries surrounds each alveolus. As the blood passes through the capillaries, it gives up carbon dioxide and takes up oxygen.

especially air emboli, are absorbed through the pulmonary capillary walls. If a large thrombus occludes a large pulmonary vessel, gas exchange can be severely impaired and can cause death. A similar situation occurs if emboli are extremely numerous and lodge all over the pulmonary arterial tree (see Clinical Focus Box 20.1).

Vasoactive hormones are metabolized in the pulmonary circulation. One such hormone is angiotensin I, which is activated and converted to angiotensin II in the lungs by **angiotensin-converting enzyme** (ACE) located on the surface of the pulmonary capillary endothelial cells. Activation is extremely rapid; 80% of angiotensin I (AI) can be converted to angiotensin II (AII) during a single passage through the pulmonary circulation. In addition to being a potent vasoconstrictor, AII has other important actions in the body (see Chapter 24). Metabolism of vasoactive hormones by the pulmonary circulation appears to be rather selective. Pulmonary endothelial cells inactivate bradykinin, serotonin, and the prostaglandins E_1, E_2, and $F_{2\alpha}$. Other prostaglandins, such as PGA_1 and PGA_2, pass through the lungs unaltered. Norepinephrine is inactivated, but epinephrine, histamine, and arginine vasopressin (AVP) pass through the pulmonary circulation unchanged. With acute lung injury (e.g., oxygen toxicity, fat emboli), the lungs can release histamine, prostaglandins, and leukotrienes, which can cause vasoconstriction of pulmonary arteries and pulmonary endothelial damage.

The lungs serve as a blood reservoir. Approximately 500 mL or 10% of the total circulating blood volume is in the pulmonary circulation. During hemorrhagic shock, some of this blood can be mobilized to improve the cardiac output.

The Pulmonary Circulation Has Unique Hemodynamic Features

In contrast to the systemic circulation, the pulmonary circulation is a high-flow, low-pressure, low-resistance system. The pulmonary artery and its branches have much thinner walls than the aorta and are more compliant. The pulmonary artery is much shorter and contains less elastin and smooth muscle in its walls. The pulmonary arterioles are thin-walled and contain little smooth muscle and, consequently, have less ability to constrict than the thick-walled, highly muscular systemic arterioles. The pulmonary veins are also thin-walled, highly compliant, and contain little smooth muscle compared with their counterparts in the systemic circulation.

The pulmonary capillary bed is also different. Unlike the systemic capillaries, which are often arranged as a network of tubular vessels with some interconnections, the pulmonary capillaries mesh together in the alveolar wall so that blood flows as a thin sheet. It is, therefore, misleading to refer to pulmonary capillaries as a capillary network; they comprise a dense **capillary bed.** The walls of the capillary bed are exceedingly thin, and a whole capillary bed can collapse if local alveolar pressure exceeds capillary pressure.

The systemic and pulmonary circulations differ strikingly in their pressure profiles (Fig. 20.2). Mean pulmonary arterial pressure is 15 mm Hg, compared with 93 mm Hg in the aorta. The driving pressure (10 mm Hg) for pulmonary flow is the difference between the mean pressure in the pulmonary artery (15 mm Hg) and the pressure in the left atrium (5 mm Hg). These pulmonary pressures are measured using a Swan-Ganz catheter, a thin, flexible tube with an inflatable rubber balloon surrounding the distal end. The balloon is inflated by injecting a small amount of air through the proximal end. Although the Swan-Ganz catheter is used for several pressure measurements, most useful is the **pulmonary wedge pressure** (Fig. 20.3). To measure wedge pressure, the catheter tip with balloon inflated is "wedged" into a small branch of the pulmonary artery. When the inflated balloon interrupts blood flow, the tip of the catheter measures downstream pressure. The downstream pressure in the occluded arterial branch represents pulmonary venous pressure, which, in turn, reflects left atrial pressure. Changes in pulmonary venous and left atrial pressures have a profound effect on gas exchange, and pulmonary wedge pressure provides an indirect measure of these important pressures.

PULMONARY VASCULAR RESISTANCE

The right ventricle pumps mixed venous blood through the pulmonary arterial tree, the alveolar capillaries (where oxygen is taken up and carbon dioxide is removed), the pulmonary veins, and then on to the left atrium. All of the cardiac output is pumped through the pulmonary circulation

CLINICAL FOCUS BOX 20.1

Pulmonary Embolism

Pulmonary embolism is clearly one of the more important disorders affecting the pulmonary circulation. The incidence of pulmonary embolism exceeds 500,000 per year with a mortality rate of approximately 10%. Pulmonary embolism is often misdiagnosed and, if improperly diagnosed, the mortality rate can exceed 30%.

The term **pulmonary embolism** refers to the movement of a blood clot or other plug from the systemic veins through the right heart and into the pulmonary circulation, where it lodges in one or more branches of the pulmonary artery. Although most pulmonary emboli originate from thrombosis in the leg veins, they can originate from the upper extremities as well. A thrombus is the major source of pulmonary emboli; however, air bubbles introduced during intravenous injections, hemodialysis, or the placement of central catheters can also cause emboli. Other sources of pulmonary emboli include fat emboli (a result of multiple long-bone fractures), tumor cells, amniotic fluid (secondary to strong uterine contractions), parasites, and various foreign materials in intravenous drug abusers.

The etiology of pulmonary emboli focuses on three factors that potentially contribute to the genesis of venous thrombosis: (1) hypercoagulability (e.g., a deficiency of antithrombin III, malignancies, the use of oral contraceptives, the presence of lupus anticoagulant); (2) endothelial damage (e.g., caused by atherosclerosis); and (3) stagnant blood flow (e.g., varicose veins). Several risk factors for thrombi include immobilization (e.g., prolonged bed rest, prolonged sitting during travel, or immobilization of an extremity after a fracture), congestive heart failure, obesity, underlying carcinoma, and chronic venous insufficiency.

When a thrombus migrates into the pulmonary circulation and lodges in pulmonary vessels, several pathophysiological consequences ensue. When a vessel is occluded, blood flow stops and perfusion to pulmonary capillaries ceases, and the ventilation-perfusion ratio in that lung unit becomes very high because ventilation is wasted. As a result, there is a significant increase in physiological dead space. Besides the direct mechanical effects of vessel occlusion, thrombi release vasoactive mediators that cause bronchoconstriction of small airways, which leads to hypoxemia. These vasoactive mediators also cause endothelial damage that leads to edema and atelectasis. If the pulmonary embolus is large and occludes a major pulmonary vessel, an additional complication occurs in the lung parenchyma distal to the site of the occlusion. The distal lung tissue becomes anoxic because it does not receive oxygen (either from airways or from the bronchial circulation). Oxygen deprivation leads to necrosis of lung parenchyma (pulmonary infarction). The parenchyma will subsequently contract and form a permanent scar.

Pulmonary emboli are difficult to diagnose because they do not manifest any specific symptoms. The most common clinical features include dyspnea and sometimes pleuritic chest pains. If the embolism is severe enough, a decreased arterial P_{O_2}, decreased P_{CO_2}, and increased pH result. The major screening test for pulmonary embolism is the perfusion scan, which involves the injection of aggregates of human serum albumin labeled with a radionuclide into a peripheral vein. These albumin aggregates (approximately 10 to 50 μm wide) travel through the right side of the heart, enter the pulmonary vasculature, and lodge in small pulmonary vessels. Only lung areas receiving blood flow will manifest an uptake of the tracer; the nonperfused region will not show any uptake of the tagged albumin. The aggregates fragment and are removed from the lungs in about a day.

at a much lower pressure than through the systemic circulation. As shown in Figure 20.2, the 10 mm Hg pressure gradient across the pulmonary circulation drives the same blood flow (5 L/min) as in the systemic circulation, where the pressure gradient is almost 100 mm Hg. Remember that vascular resistance (R) is equal to the pressure gradient (ΔP) divided by blood flow ($\dot{Q}$) (see Chapter 12):

$$R = \Delta P/\dot{Q} \quad (1)$$

Pulmonary vascular resistance is extremely low, about one-tenth that of systemic vascular resistance. The difference in resistances is a result, in part, of the enormous number of small pulmonary resistance vessels that are dilated. By contrast, systemic arterioles and precapillary sphincters are partially constricted.

Pulmonary Vascular Resistance Falls With Increased Cardiac Output

Another unique feature of the pulmonary circulation is the ability to decrease resistance when pulmonary arterial pressure rises, as seen with an increase in cardiac output. When pressure rises, there is a marked decrease in pulmonary vascular resistance (Fig. 20.4). Similarly, increasing pulmonary venous pressure causes pulmonary vascular resistance to fall. These responses are very different from those of the systemic circulation, where an increase in perfusion pressure increases vascular resistance. Two local mechanisms in the pulmonary circulation are responsible (Fig. 20.5). The first mechanism is known as **capillary recruitment**. Under normal conditions, some capillaries are partially or completely closed in the top part of the lungs because of the low perfusion pressure. As blood flow increases, the pressure rises and these collapsed vessels are opened, lowering overall resistance. This process of opening capillaries is the primary mechanism for the fall in pulmonary vascular resistance when cardiac output increases. The second mechanism is **capillary distension** or widening of capillary segments, which occurs because the pulmonary capillaries are exceedingly thin and highly compliant.

The fall in pulmonary vascular resistance with increased cardiac output has two beneficial effects. It opposes the tendency of blood velocity to speed up with increased flow rate, maintaining adequate time for pulmonary capillary blood to take up oxygen and dispose of carbon dioxide. It also results in an increase in capillary surface area, which

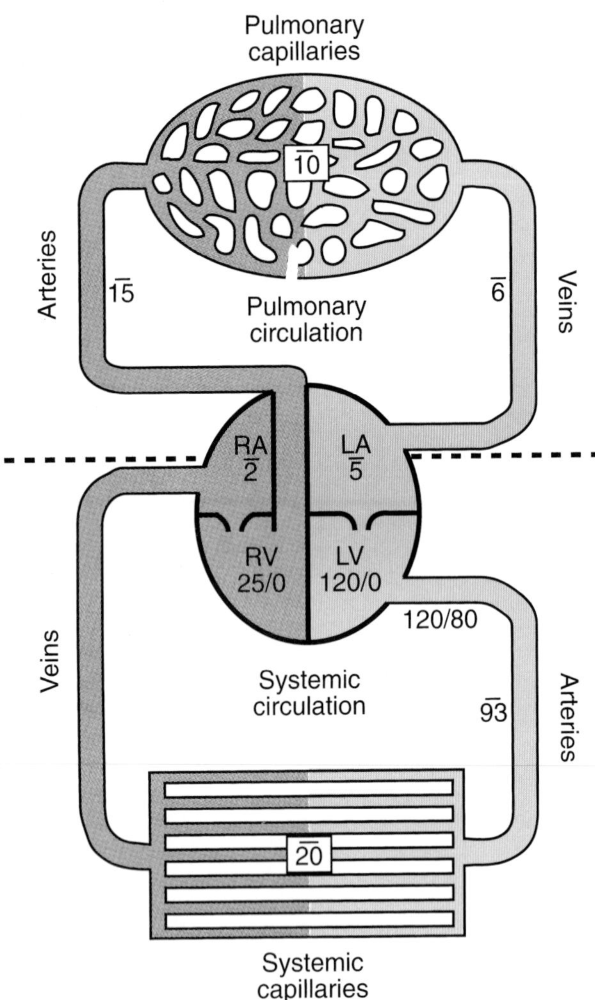

FIGURE 20.2 Pressure profiles of the pulmonary and systemic circulations. Unlike the systemic circulation, the pulmonary circulation is a low-pressure and low-resistance system. Pulmonary circulation is characterized as normally dilated, while the systemic circulation is characterized as normally constricted. Pressures are given in mm Hg; a bar over the number indicates mean pressure.

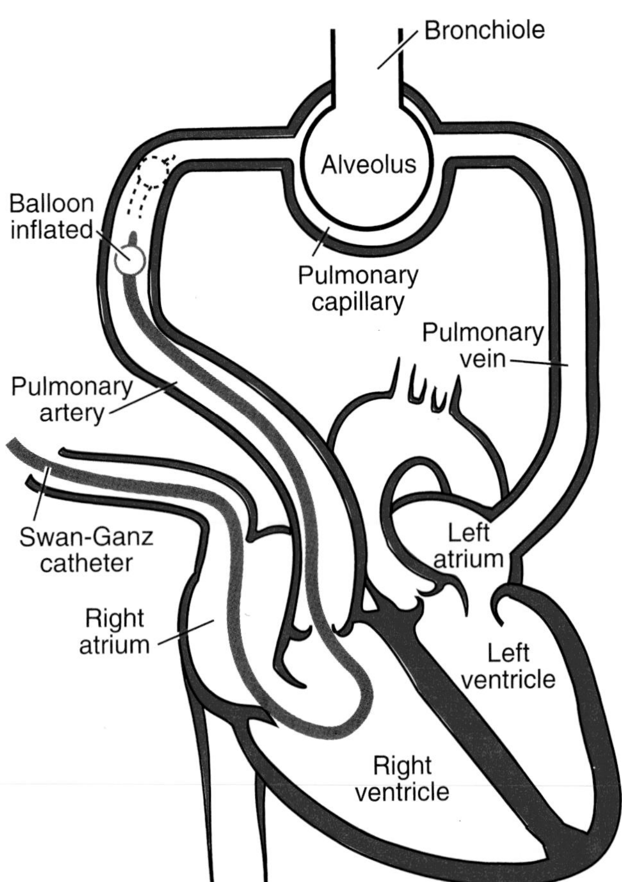

FIGURE 20.3 Measuring pulmonary wedge pressure. A catheter is threaded through a peripheral vein in the systemic circulation, through the right heart, and into the pulmonary artery. The wedged catheter temporarily occludes blood flow in a part of the vascular bed. The wedge pressure is a measure of downstream pressure, which is pulmonary venous pressure. Pulmonary venous pressure reflects left atrial pressure.

enhances the diffusion of oxygen into and carbon dioxide out of the pulmonary capillary blood.

Capillary recruitment and distension also have a protective function. High capillary pressure is a major threat to the lungs and can cause **pulmonary edema**, an abnormal accumulation of fluid, which can flood the alveoli and impair gas exchange. When cardiac output increases from a resting level of 5 L/min to 25 L/min with vigorous exercise, the decrease in pulmonary vascular resistance not only minimizes the load on the right heart but also keeps the capillary pressure low and prevents excess fluid from leaking out of the pulmonary capillaries.

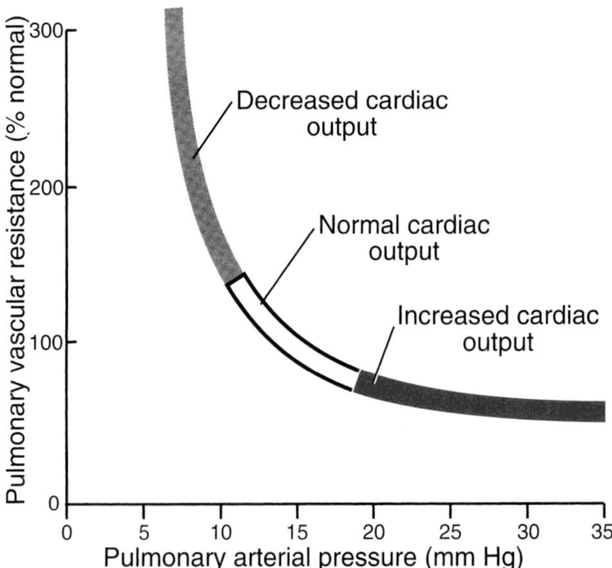

FIGURE 20.4 Effect of cardiac output on pulmonary vascular resistance. Pulmonary vascular resistance falls as cardiac output increases. Note that if pulmonary arterial pressure rises, pulmonary vascular resistance decreases.

Lung Volumes Affect Pulmonary Vascular Resistance

Pulmonary vascular resistance is also significantly affected by lung volume. Because pulmonary capillaries have little

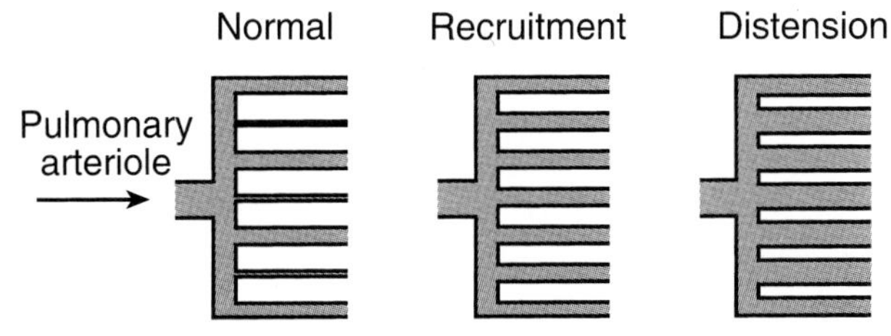

FIGURE 20.5 Capillary recruitment and capillary distension. These two mechanisms are responsible for decreasing pulmonary vascular resistance when arterial pressure increases. In the normal condition, not all capillaries are perfused. Capillary recruitment (the opening up of previously closed vessels) results in the perfusion of an increased number of vessels and a drop in resistance. Capillary distension (an increase in the caliber of vessels) also results in a lower resistance and higher blood flow.

structural support, they can be easily distended or collapsed, depending on the pressure surrounding them. It is the change in transmural pressure (pressure inside the capillary minus pressure outside the capillary) that influences vessel diameter. From a functional point of view, pulmonary vessels can be classified into two types: **extra-alveolar vessels** (pulmonary arteries and veins) and **alveolar vessels** (arterioles, capillaries, and venules). The extra-alveolar vessels are subjected to pleural pressure—any change in pleural pressure affects pulmonary vascular resistance in these vessels by changing transmural pressure. Alveolar vessels, however, are subjected primarily to alveolar pressure.

At high lung volumes, the pleural pressure is more negative. Transmural pressure in the extra-alveolar vessels increases, and they become distended (Fig. 20.6A). However, alveolar diameter increases at high lung volumes, causing transmural pressure in alveolar vessels to decrease. As the alveolar vessels become compressed, pulmonary vascular resistance increases. At low lung volumes, pulmonary vascular resistance also increases, as a result of more positive pleural pressure, which compresses the extra-alveolar vessels. Since alveolar and extra-alveolar vessels can be viewed as two groups of resistance vessels connected in series, their resistances are additive at any lung volume. Pulmonary vascular resistance is lowest at functional residual capacity (FRC) and increases at both higher and lower lung volumes (Fig. 20.6B).

Since smooth muscle plays a key role in determining the caliber of extra-alveolar vessels, drugs can also cause a change in resistance. Serotonin, norepinephrine, histamine, thromboxane A_2, and leukotrienes are potent vasoconstrictors, particularly at low lung volumes when the vessel walls are already compressed. Drugs that relax smooth muscle in the pulmonary circulation include adenosine, acetylcholine, prostacyclin (prostaglandin I_2), and isoproterenol. The pulmonary circulation is richly innervated with sympathetic nerves but, surprisingly, pulmonary vascular resistance is virtually unaffected by autonomic nerves under normal conditions.

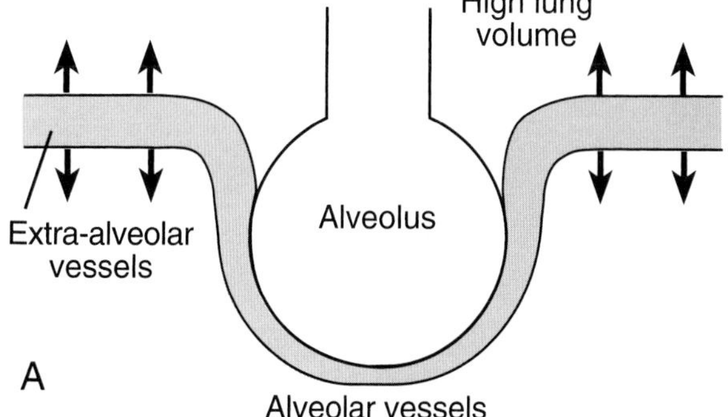

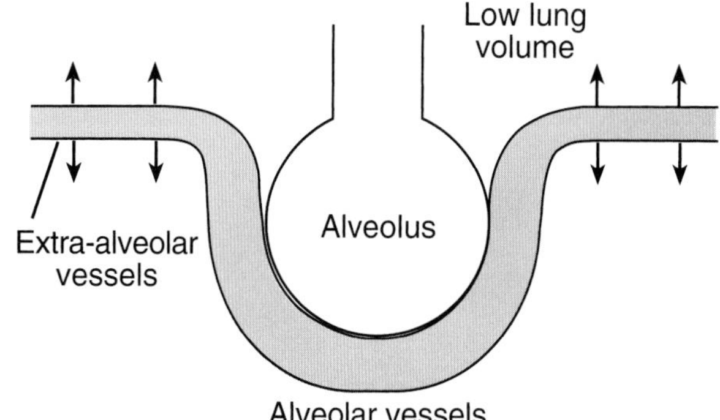

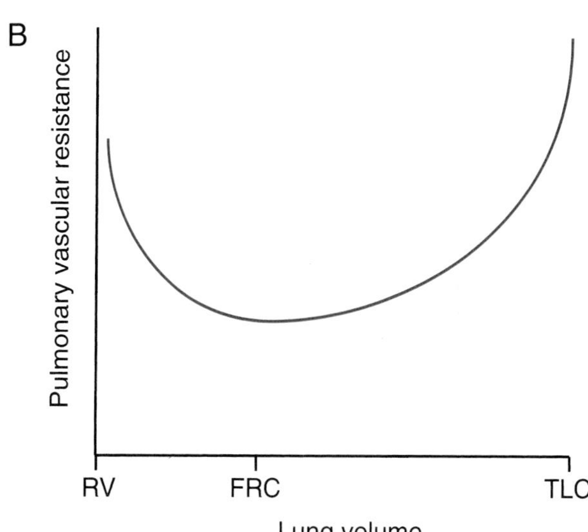

FIGURE 20.6 Effect of lung volume on pulmonary vascular resistance. A, At high lung volumes, alveolar vessels are compressed but extra-alveolar vessels are actually *distended* because of the lower pleural pressure. However, at low lung volumes, the extra-alveolar vessels are compressed from the pleural pressure and alveolar vessels are distended. **B,** Total pulmonary vascular resistance as a function of lung volumes follows a U-shaped curve, with resistance lowest at functional residual capacity (FRC).

Low Oxygen Tension Increases Pulmonary Vascular Resistance

Although changes in pulmonary vascular resistance are accomplished mainly by passive mechanisms, resistance can be increased by low oxygen in the alveoli, **alveolar hypoxia**, and low oxygen in the blood, **hypoxemia**. Hypoxemia causes vasodilation in systemic vessels but, in pulmonary vessels, hypoxemia or alveolar hypoxia causes vasoconstriction of small pulmonary arteries. This unique phenomenon of **hypoxia-induced pulmonary vasoconstriction** is accentuated by high carbon dioxide and low blood pH. The exact mechanism is not known, but hypoxia

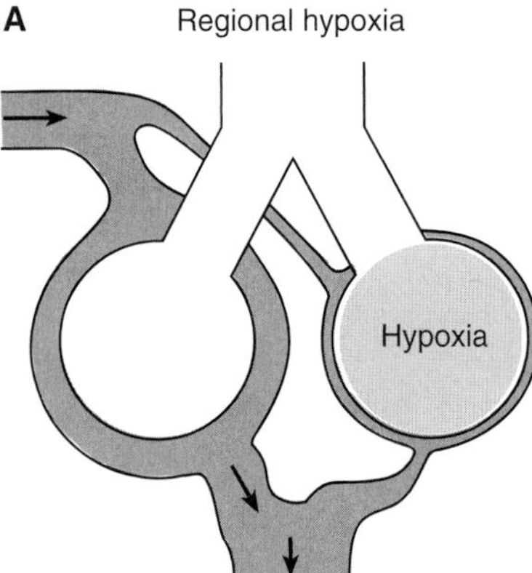

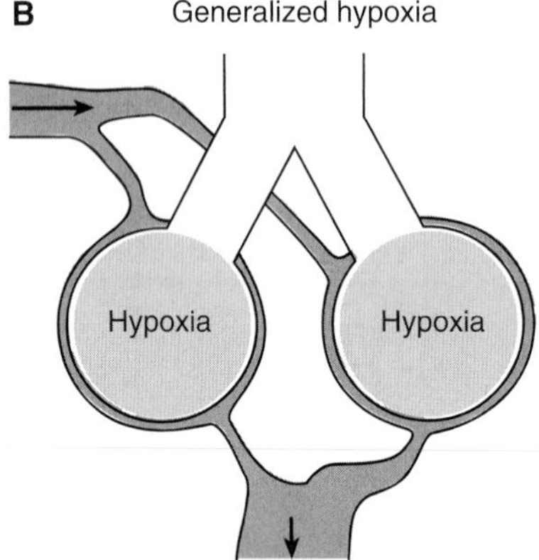

FIGURE 20.7 Effect of alveolar hypoxia on pulmonary arteries. Hypoxia-induced vasoconstriction is unique to vessels of the lungs and is the major mechanism regulating blood flow within normal lungs. **A,** With regional hypoxia, precapillary constriction diverts blood flow away from poorly ventilated regions; there is little change in pulmonary arterial pressure. **B,** In generalized hypoxia, which can occur with high altitude or with certain lung diseases, precapillary constriction occurs throughout the lungs and there is a marked increase in pulmonary arterial pressure.

can directly act on pulmonary vascular smooth muscle cells, independent of any agonist or neurotransmitter released by hypoxia.

Two types of alveolar hypoxia are encountered in the lungs, with different implications for pulmonary vascular resistance. In **regional hypoxia**, pulmonary vasoconstriction is localized to a specific region of the lungs and diverts blood away from a poorly ventilated region (e.g., caused by bronchial obstruction), minimizing effects on gas exchange (Fig. 20.7A). Regional hypoxia has little effect on pulmonary arterial pressure, and when alveolar hypoxia no longer exists, the vessels dilate and blood flow is restored. **Generalized hypoxia** causes vasoconstriction throughout both lungs, leading to a significant rise in resistance and pulmonary artery pressure (Fig 20.7B). Generalized hypoxia occurs when the partial pressure of alveolar oxygen (P_{AO_2}) is decreased with high altitude or with the chronic hypoxia seen in certain types of respiratory diseases (e.g., asthma, emphysema, and cystic fibrosis). Generalized hypoxia can lead to pulmonary hypertension (high pulmonary arterial pressure), which leads to pathophysiological changes (hypertrophy and proliferation of smooth muscle cells, narrowing of arterial lumens, and a change in contractile function). Pulmonary hypertension causes a substantial increase in workload on the right heart, often leading to right heart hypertrophy (see Clinical Focus Box 20.2). Generalized hypoxia plays an important nonpathophysiological role before birth. In the fetus, pulmonary vascular resistance is extremely high as a result of generalized hypoxia—less than 15% of the cardiac output goes to the lungs, and the remainder is diverted to the left side of the heart via the **foramen ovale** and to the aorta via the **ductus arteriosus**. When alveoli are oxygenated on the newborn's first breath, pulmonary vascular smooth muscle relaxes, the vessels dilate, and vascular resistance falls dramatically. The foramen ovale and ductus arteriosus close and pulmonary blood flow increases enormously.

FLUID EXCHANGE IN PULMONARY CAPILLARIES

Starling forces, which govern the exchange of fluid across capillary walls in the systemic circulation (see Chapter 16), also operate in the pulmonary capillaries. Net fluid transfer across the pulmonary capillaries depends on the *difference* between hydrostatic and colloid osmotic pressures inside and outside the capillaries. In the pulmonary circulation, two additional forces play a role in fluid transfer—surface tension and alveolar pressure. The force of alveolar surface tension (see Chapter 19) pulls inward, which tends to lower interstitial pressure and draw fluid into the interstitial space. By contrast, alveolar pressure tends to compress the interstitial space and interstitial pressure is increased (Fig. 20.8).

Low Capillary Pressure Enhances Fluid Removal

Mean pulmonary capillary hydrostatic pressure is normally 8 to 10 mm Hg, which is lower than the plasma colloid osmotic pressure (25 mm Hg). This is functionally important because the low hydrostatic pressure in the pulmonary capillaries favors the net absorption of fluid. Alveolar surface tension tends to offset this advantage and results in a net force that still favors a small continuous flux of fluid out of the capillaries and into the interstitial space. This excess fluid travels through the interstitium to the perivascular and peribronchial spaces in the lungs, where it then passes into the lymphatic channels (see Fig. 20.8). The lungs have a more extensive lymphatic system than most organs. The lymphatics are not found in the alveolar-capillary area but are strategically located near the terminal bronchioles to drain off excess fluid. Lymphatic channels, like small pulmonary blood vessels, are held open by tethers from surrounding connective tissue. Total lung lymph flow is about 0.5 mL/min, and the lymph is propelled by smooth muscle in the lymphatic walls and by ventilatory movements of the lungs.

Fluid Imbalance Leads to Pulmonary Edema

Pulmonary edema occurs when excess fluid accumulates in the lung interstitial spaces and alveoli, and usually results when capillary filtration exceeds fluid removal. Pulmonary edema can be caused by an increase in capillary hydrostatic

CLINICAL FOCUS BOX 20.2

Hypoxia-Induced Pulmonary Hypertension

Hypoxia has opposite effects on the pulmonary and systemic circulations. Hypoxia relaxes vascular smooth muscle in systemic vessels and elicits vasoconstriction in the pulmonary vasculature. Hypoxic pulmonary vasoconstriction is the major mechanism regulating the matching of regional blood flow to regional ventilation in the lungs. With regional hypoxia, the matching mechanism automatically adjusts regional pulmonary capillary blood flow in response to alveolar hypoxia and prevents blood from perfusing poorly ventilated regions in the lungs. Regional hypoxic vasoconstriction occurs without any change in pulmonary arterial pressure. However, when hypoxia affects all parts of the lung (generalized hypoxia), it causes pulmonary hypertension because all of the pulmonary vessels constrict. Hypoxia-induced pulmonary hypertension affects individuals who live at a high altitude (8,000 to 12,000 feet) and those with chronic obstructive pulmonary disease (COPD), especially patients with emphysema.

With chronic hypoxia-induced pulmonary hypertension, the pulmonary artery undergoes major remodeling during several days. An increase in wall thickness results from hypertrophy and hyperplasia of vascular smooth muscle and an increase in connective tissue. These structural changes occur in both large and small arteries. Also, there is abnormal extension of smooth muscle into peripheral pulmonary vessels where muscularization is not normally present; this is especially pronounced in precapillary segments. These changes lead to a marked increase in pulmonary vascular resistance. With severe, chronic hypoxia-induced pulmonary hypertension, the obliteration of small pulmonary arteries and arterioles, as well as pulmonary edema, eventually occur. The latter is caused, in part, by the hypoxia-induced vasoconstriction of pulmonary veins, which results in a significant increase in pulmonary capillary hydrostatic pressure.

A striking feature of the vascular remodeling is that both the pulmonary artery and the pulmonary vein constrict with hypoxia; however, only the arterial side undergoes major remodeling. The postcapillary segments and veins are spared the structural changes seen with hypoxia. Because of the hypoxia-induced vasoconstriction and vascular remodeling, pulmonary arterial pressure increases. Pulmonary hypertension eventually causes right heart hypertrophy and failure, the major cause of death in COPD patients.

pressure, capillary permeability, or alveolar surface tension or by a decrease in plasma colloid osmotic pressure. Increased capillary hydrostatic pressure is the most frequent cause of pulmonary edema and is often the result of an abnormally high pulmonary venous pressure (e.g., with mitral stenosis or left heart failure).

The second major cause of pulmonary edema is increased capillary permeability, which results in excess fluid and plasma proteins flooding the interstitial spaces and alveoli. Protein leakage makes pulmonary edema more severe because additional water is pulled from the capillaries to the alveoli when plasma proteins enter the interstitial spaces and alveoli. Increased capillary permeability occurs with pulmonary vascular injury, usually from oxidant damage (e.g., oxygen therapy, ozone toxicity), an inflammatory reaction (endotoxins), or neurogenic shock (e.g., head injury). High surface tension is the third major cause of pulmonary edema. Loss of surfactant causes high surface tension, lowering interstitial hydrostatic pressure and resulting in an increase of capillary fluid entering the interstitial space. A decrease in plasma colloid osmotic pressure occurs when plasma protein concentration is reduced (e.g., starvation).

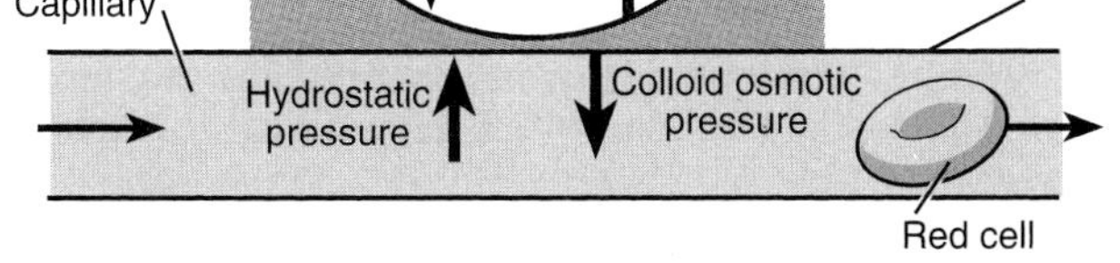

Pulmonary edema is a hallmark of adult respiratory distress syndrome (ARDS), and it is often associated with abnormally high surface tension. Pulmonary edema is a serious problem because it hinders gas exchange and, eventually, causes arterial P_{O_2} to fall below normal (i.e., $Pa_{O_2} < 85$ mm Hg) and arterial P_{CO_2} to rise above normal ($Pa_{CO_2} > 45$ mm Hg). As mentioned earlier, abnormally low arterial P_{O_2} produces hypoxemia and the abnormally high arterial P_{CO_2} produces hypercapnia. Pulmonary edema also obstructs small airways, thereby increasing airway resistance. Lung compliance is decreased with pulmonary edema because of interstitial swelling and the increase in alveolar surface tension. Decreased lung compliance, together with airway obstruction, greatly increases the work of breathing. The treatment of pulmonary edema is directed toward reducing pulmonary capillary hydrostatic pressure. This is accomplished by decreasing blood volume with a diuretic drug, increasing left ventricular function with digitalis, and administering a drug that causes vasodilation in systemic blood vessels.

FIGURE 20.8 Fluid exchange in pulmonary capillaries. Fluid movement in and out of capillaries depends on the net difference between hydrostatic and colloid osmotic pressures. Two additional factors involved in pulmonary fluid exchange are alveolar surface tension, which enhances filtration, and alveolar pressure, which opposes filtration. The relatively low pulmonary capillary hydrostatic pressure helps keep the alveoli "dry" and prevents pulmonary edema.

Although **fresh-water drowning** is often associated with aspiration of water into the lungs, the cause of death is not pulmonary edema but ventricular fibrillation. The low capillary pressure that normally keeps the alveolar-capillary membrane free of excess fluid becomes a severe disadvantage when fresh water accidentally enters the lungs. The aspirated water is rapidly pulled into the pulmonary capillary circulation via the alveoli because of the low capillary hydrostatic pressure and high colloid osmotic pressure. Consequently, the plasma is diluted and the hypotonic environment causes red cells to burst (hemolysis). The resulting elevation of plasma K^+ level and depression of Na^+ level alter the electrical activity of the heart. Ventricular fibrillation often occurs as a result of the combined effects of these electrolyte changes and hypoxemia. In **salt-water drowning**, the aspirated seawater is hypertonic, which leads to increased plasma Na^+ and pulmonary edema. The cause of death in this case is asphyxia.

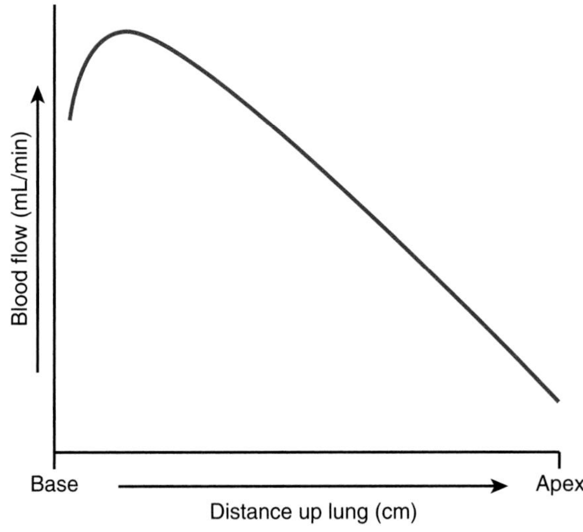

FIGURE 20.9 **Effect of gravity on pulmonary blood flow.** Gravity causes uneven pulmonary blood flow in the upright individual. The downward pull of gravity causes a lower blood pressure at the apex of the lungs. Consequently, pulmonary blood flow is very low at the apex and increases toward the base of the lungs.

BLOOD FLOW DISTRIBUTION IN THE LUNGS

As previously mentioned, blood accounts for approximately half the weight of the lungs. The effects of gravity on blood flow are dramatic and result in an uneven distribution of blood in the lungs. In an upright individual, the gravitational pull on the blood is downward. Since the vessels are highly compliant, gravity causes the blood volume and flow to be greater at the bottom of the lung (the base) than at the top (the apex). The pulmonary vessels can be compared with a continuous column of fluid. The difference in arterial pressure between the apex and base of the lungs is about 30 cm H_2O. Because the heart is situated midway between the top and bottom of the lungs, the arterial pressure is about 11 mm Hg less (15 cm $H_2O \div 1.36$ cm H_2O per mm Hg $= 11$ mm Hg) at the lungs' apex (15 cm above the heart) and about 11 mm Hg more than the mean pressure in the middle of the lungs at the lungs' base (15 cm below the heart). The low arterial pressure results in reduced blood flow in the capillaries at the lung's apex, while capillaries at the base are distended and blood flow is augmented.

Gravity Alters Capillary Perfusion

In an upright person, pulmonary blood flow increases almost linearly from apex to base (Fig. 20.9). Blood flow distribution is affected by gravity, and it can be altered by changes in body positions. For example, when an individual is lying down, blood flow is distributed relatively evenly from apex to base. The measurement of blood flow in a subject suspended upside-down would reveal an apical blood flow exceeding basal flow in the lungs. Exercise tends to offset the gravitational effects in an upright individual. As cardiac output increases with exercise, the increased pulmonary arterial pressure leads to capillary recruitment and distension in the lung's apex, resulting in increased blood flow and minimizing regional differences in blood flow in the lungs.

Since gravity causes capillary beds to be underperfused in the apex and overperfused in the base, the lungs are often divided into zones to describe the effect of gravity on pulmonary capillary blood flow (Fig. 20.10). **Zone 1** occurs when alveolar pressure is greater than pulmonary arterial pressure; pulmonary capillaries collapse and there is little or no blood flow. Pulmonary arterial pressure (Pa) is still greater than pulmonary venous pressure (Pv), hence, $P_A > Pa > Pv$. Because zone 1 is ventilated but not perfused (no blood flows through the pulmonary capillaries), alveolar dead space is increased (see Chapter 19). Zone 1 is usually very small or nonexistent in healthy individuals because the pulsatile pulmonary arterial pressure is sufficient to keep the capillaries partially open at the apex. Zone 1 may easily be created by conditions that elevate alveolar pressure or decrease pulmonary arterial pressure. For example, a zone 1 condition can be created when a patient is placed on a mechanical ventilator, which results in an increase in alveolar pressure with positive ventilation pressures. Hemorrhage or low blood pressure can create a zone 1 condition by lowering pulmonary arterial pressure. A zone 1 condition can also be created in the lungs of astronauts during a spacecraft launching. The rocket acceleration makes the gravitational pull even greater, causing arterial pressure in the top part of the lung to fall. To prevent or minimize a zone 1 from occurring, astronauts are placed in a supine position during blast-off.

A **zone 2** condition occurs in the middle of the lungs, where pulmonary arterial pressure, caused by the increased hydrostatic effect, is greater than alveolar pressure (see Fig. 20.10). Venous pressure is less than alveolar pressure. As a result, blood flow in a zone 2 condition is determined not by the arterial-venous pressure difference, but by the difference between arterial pressure and alveolar pressure. The pressure gradient in zone 2 is represented as $Pa > P_A > Pv$. The functional importance of this is that venous pressure in zone 2 has no effect on flow. In **zone 3**, venous pressure exceeds alveolar pressure and blood flow is determined by the usual arterial-venous pressure difference.

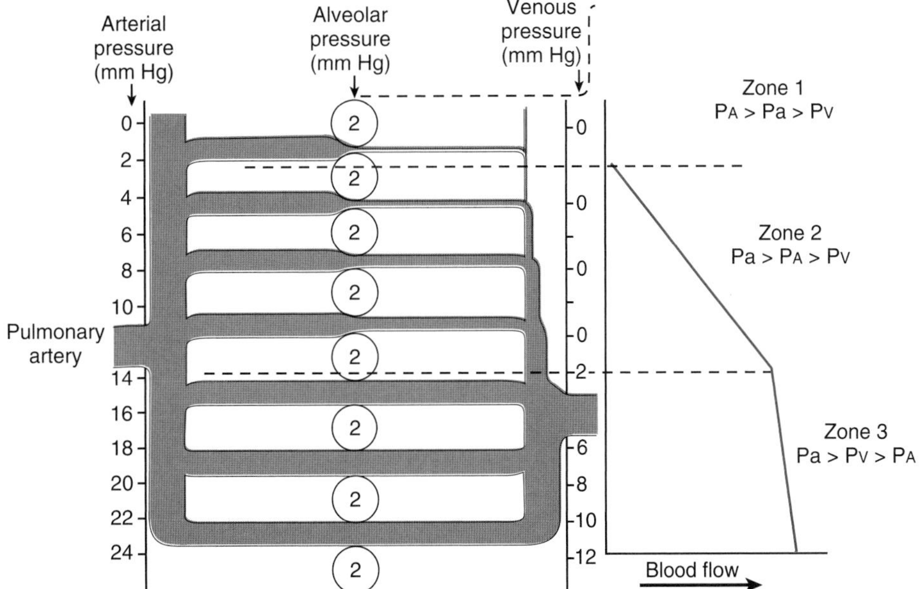

FIGURE 20.10 Zones of the lungs and the uneven distribution of pulmonary blood flow. The three zones depend on the relationship between pulmonary arterial pressure (Pa), pulmonary venous pressure (Pv), and alveolar pressure (PA). In zone 1, alveolar pressure exceeds arterial pressure and there is no blood flow. Zone 1 occurs only in abnormal conditions in which alveolar pressure is increased (e.g., positive pressure ventilation) or when arterial pressure is decreased below normal (e.g., the gravitational pull during the launching of a spacecraft). In zone 2, arterial pressure exceeds alveolar pressure, and blood flow depends on the difference between arterial and alveolar pressures. Blood flow is greater at the bottom than at the top of this zone. In zone 3, both arterial and venous pressures exceed alveolar pressure, and blood flow depends on the normal arterial-venous pressure difference. Note that arterial pressure increases down each zone, and transmural pressure also becomes greater, capillaries become more distended, and pulmonary vascular resistance falls.

The increase in blood flow down this region is primarily a result of capillary distension.

Regional Ventilation and Blood Flow Are Not Always Matched in the Lungs

Thus far, we have assumed that if ventilation and cardiac output are normal, gas exchange will also be normal. Unfortunately, this is not the case. Even though total ventilation and total blood flow (i.e., cardiac output) may be normal, there are regions in the lung where ventilation and blood flow are not matched, so that a certain fraction of the cardiac output is not fully oxygenated.

The matching of airflow and blood flow is best examined by considering the **ventilation-perfusion ratio**, which compares alveolar ventilation to blood flow in lung regions. Since resting healthy individuals have an alveolar ventilation ($\dot{V}A$) of 4 L/min and a cardiac output (pulmonary blood flow or perfusion) of 5 L/min, the ideal alveolar ventilation-perfusion ratio ($\dot{V}A/\dot{Q}$ ratio) should be 0.8 (there are no units, as this is a ratio). We have already seen that gravity can cause regional differences in blood flow and alveolar ventilation (see Chapter 19). In an upright person, the base of the lungs is better ventilated and better perfused than the apex.

Regional alveolar ventilation and blood flow are illustrated in Figure 20.11. Three points are apparent from this figure:

- Ventilation and blood flow are both gravity-dependent; airflow and blood flow increase down the lung.
- Blood flow shows about a 5-fold difference between the top and bottom of the lung, while ventilation shows about a 2-fold difference. This causes gravity-dependent regional variations in the $\dot{V}A/\dot{Q}$ ratio that range from 0.6 at the base to 3 or higher at the apex. Blood flow is proportionately greater than ventilation at the base, and ventilation is proportionately greater than blood flow at the apex.

The functional importance of lung ventilation-perfusion ratios is that the crucial factor in gas exchange is the matching of *regional ventilation and blood flow*, as opposed to total alveolar ventilation and total pulmonary blood flow. The distribution of $\dot{V}A/\dot{Q}$ in a healthy adult is shown in Figure 20.12. Even in healthy lungs, most of the ventilation and perfusion go to lung units with a $\dot{V}A/\dot{Q}$ ratio of about 1 instead of the ideal ratio of 0.8. At the apical region, where the $\dot{V}A/\dot{Q}$ ratio is high, there is overventilation relative to blood flow. At the base, where the ratio is low, the opposite occurs (i.e., overperfusion relative to ventilation). In the latter case, a fraction of the blood passes through the pulmonary capillaries at the base of the lungs without becoming fully oxygenated.

The effect of regional $\dot{V}A/\dot{Q}$ ratio on blood gases is shown in Figure 20.13. Because overventilation relative to blood flow (high $\dot{V}A/\dot{Q}$) occurs in the apex, the PAO_2 is high and the $PACO_2$ is low at the apex of the lungs. Oxygen tension (PO_2) in the blood leaving pulmonary capillaries at the base of the lungs is low because the blood is not fully oxygenated as a result of underventilation relative to blood flow. Regional differences in $\dot{V}_A/\dot{Q}$ ratios tend to localize some diseases to the top or bottom parts of the lungs. For example, tuberculosis tends to be localized in the apex because of a more favorable environment (i.e., higher oxygen levels for *Mycobacterium tuberculosis*).

SHUNTS AND VENOUS ADMIXTURE

Matching of the lung's airflow and blood flow is not perfect. On one side of the alveolar-capillary membrane there is "wasted air" (i.e., physiological dead space), and on the other side there is "wasted blood" (Fig. 20.14). Wasted blood refers to any fraction of the venous blood that does

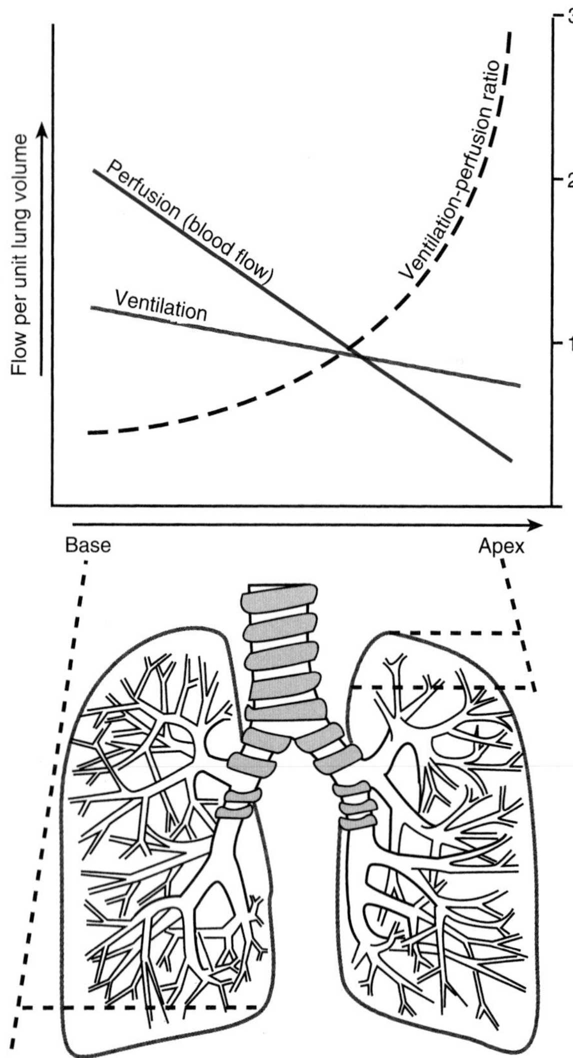

FIGURE 20.11 Regional alveolar ventilation and blood flow. Gravity causes a mismatch of blood flow and alveolar ventilation in the base and apex of the lungs. Both ventilation and perfusion are gravity-dependent. At the base of the lungs, blood flow exceeds alveolar ventilation, resulting in a low ventilation-perfusion ratio. At the apex, the opposite occurs; alveolar ventilation is greater than blood flow, resulting in a high ventilation-perfusion ratio.

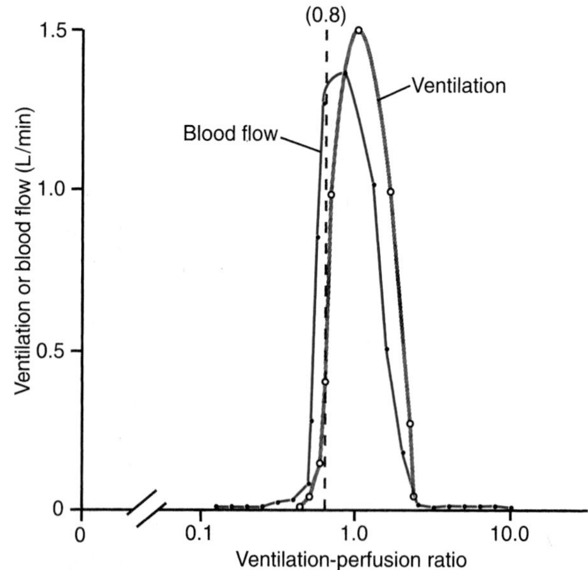

FIGURE 20.12 Profiles for alveolar ventilation and blood flow in healthy adults. The y-axis represents flow (either blood flow or airflow) in L/min. The ventilation-perfusion ratio is shown on the x-axis, plotted on a logarithmic scale. The optimal $\dot{V}_A/\dot{Q}$ ratio is 0.8 in healthy lungs. (Adapted from Lumb AB. Nunn's Applied Respiratory Physiology. 5th Ed. Oxford: Butterworth-Heinemann, 2000.)

	$\dot{V}_A$ (L/min)	$\dot{Q}$ (L/min)	$\dot{V}_A/\dot{Q}$	Pa_{O_2} (mm Hg)	Pa_{CO_2} (mm Hg)
Apex	0.25	0.07	3.6	130	28
Base	0.8	1.3	0.6	88	42

FIGURE 20.13 Effect of regional differences of ventilation-perfusion ratios on blood gases in the apex and base of the lungs.

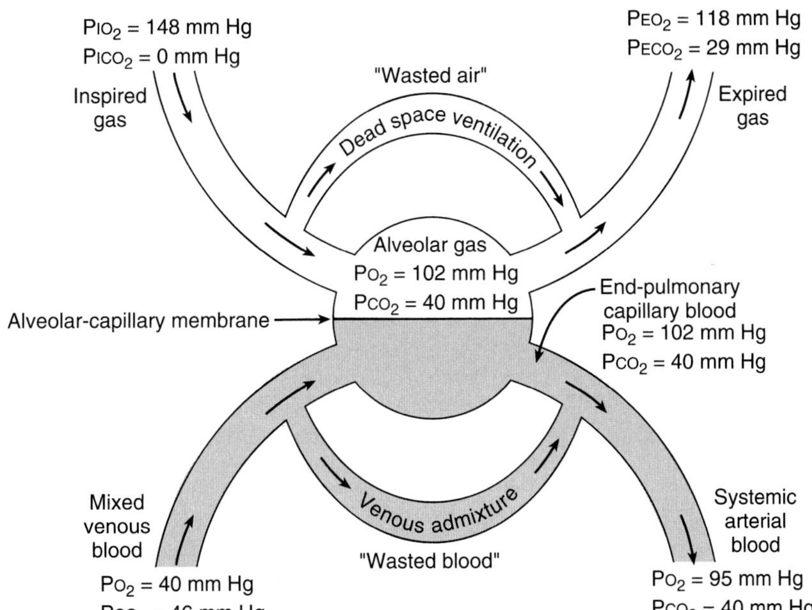

FIGURE 20.14 **"Wasted air" and "wasted blood."** The plumbing on both sides of the alveolar-capillary membrane is imperfect. On one side there is "wasted air" and on the other side there is "wasted blood." The total amount of wasted air constitutes physiological dead space and the total amount of wasted blood (venous admixture) constitutes physiological shunt.

not get fully oxygenated. The mixing of unoxygenated blood with oxygenated blood is known as **venous admixture**. There are two causes for venous admixture: a **shunt** and a low $\dot{V}_A/\dot{Q}$ ratio.

An **anatomic shunt** has a structural basis and occurs when blood bypasses alveoli through a channel, such as from the right to left heart through an atrial or ventricular septal defect or from a branch of the pulmonary artery connecting directly to the pulmonary vein. An anatomic shunt is often called a **right-to-left shunt**. The bronchial circulation also constitutes shunted blood because bronchial venous blood (deoxygenated blood) drains directly into the pulmonary veins that are carrying oxygenated blood.

The second cause for venous admixture is a low regional $\dot{V}_A/\dot{Q}$ ratio. This occurs when a portion of the cardiac output goes through the regular pulmonary capillaries but there is insufficient alveolar ventilation to fully oxygenate all of the blood. With a low regional $\dot{V}_A/\dot{Q}$ ratio, there is no abnormal anatomic connection and the blood does not bypass the alveoli. Rather, blood that passes through the alveolar capillaries is not completely oxygenated. In a healthy individual, a low $\dot{V}_A/\dot{Q}$ ratio occurs at the base of the lung (i.e., gravity dependent). A low regional $\dot{V}_A/\dot{Q}$ ratio can also occur with a partially obstructed airway (Fig. 20.15), in which underventilation with respect to blood flow results in **regional hypoventilation.** A fraction of the blood passing through a hypoventilated region is not fully oxygenated, resulting in an increase in venous admixture.

The total amount of venous admixture as a result of anatomic shunt and a low $\dot{V}_A/\dot{Q}$ ratio equals **physiological shunt** and represents the total amount of wasted blood that does not get fully oxygenated. Physiological shunt is analogous to physiological dead space; the two are compared in Table 20.1, in which one represents wasted blood flow and the other represents wasted air. It is important to remember that, in healthy individuals, there is some de-

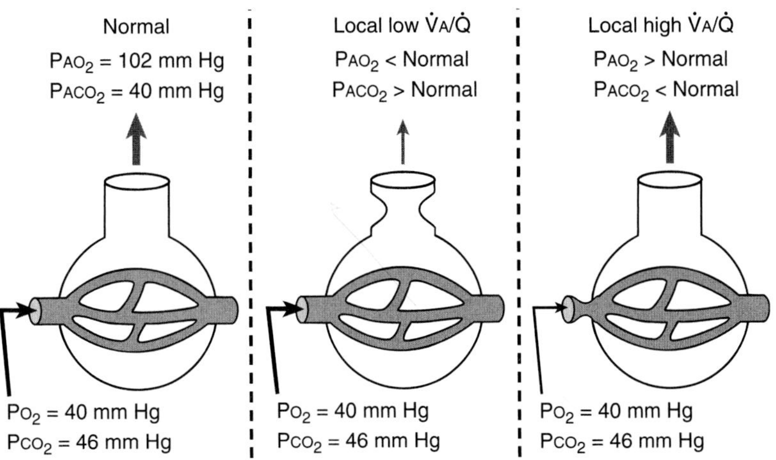

FIGURE 20.15 **Abnormal ventilation-perfusion ratios.** Airway obstruction (middle panel) causes a low regional ventilation-perfusion ($\dot{V}_A/\dot{Q}$) ratio. A partially blocked airway causes this region to be underventilated relative to blood flow. Note the alveolar gas composition. A low regional $\dot{V}_A/\dot{Q}$ ratio causes venous admixture and will increase the physiological shunt. A partially obstructed pulmonary arteriole (right panel) will cause an abnormally high $\dot{V}_A/\dot{Q}$ ratio in a lung region. Restricted blood flow causes this region to be overventilated relative to blood flow, which leads to an increase in physiological dead space.

TABLE 20.1 Shunts and Dead Spaces Compared

Shunt	Dead Space
Anatomic	Anatomic
+	+
Low $\dot{V}A/\dot{Q}$ ratio	Alveolar
=	=
Physiological shunt (calculated total "wasted blood")	Physiological dead space (calculated total "wasted air")

gree of physiological dead space as well as physiological shunt in the lungs.

In summary, venous admixture results from anatomic shunt and a low regional $\dot{V}A/\dot{Q}$ ratio. In healthy individuals, approximately 50% of the venous admixture comes from an anatomic shunt (e.g., bronchial circulation) and 50% from a low $\dot{V}A/\dot{Q}$ ratio at the base of the lungs as a result of gravity. Physiological shunt (i.e., total venous admixture) represents about 1 to 2% of cardiac output in healthy people. This amount can increase up to 15% of cardiac output with some bronchial diseases, and in certain congenital disorders, a right-to-left anatomic shunt can account for up to 50% of cardiac output. It is important to remember that any deviation of $\dot{V}A/\dot{Q}$ ratio from the ideal condition (0.8) impairs gas exchange and lowers oxygen tension in the arterial blood. A good way to remember the importance of a shunt is that it always leads to venous admixture and reduces the amount of oxygen carried in the systemic blood.

THE BRONCHIAL CIRCULATION

The conducting airways have a separate circulation known as the **bronchial circulation**, which is distinct from the pulmonary circulation. The primary function of the bronchial circulation is to nourish the walls of the conducting airways and surrounding tissues by distributing blood to the supporting structures of the lungs. Under normal conditions, the bronchial circulation does not supply blood to the terminal respiratory units (respiratory bronchioles, alveolar ducts, and alveoli); they receive their blood from the pulmonary circulation. Venous return from the bronchial circulation is by two routes: bronchial veins and pulmonary veins. About half of the bronchial blood flow returns to the right atrium by way of the bronchial veins, which empty into the azygos vein. The remainder returns through small bronchopulmonary anastomoses into the pulmonary veins.

Bronchial arterial pressure is approximately the same as aortic pressure, and bronchial vascular resistance is much higher than resistance in the pulmonary circulation. Bronchial blood flow is approximately 1 to 2% of cardiac output but, in certain inflammatory disorders of the airways (e.g., chronic bronchitis), it can be as high as 10% of cardiac output.

The bronchial circulation is the only portion of the circulation in the adult lung that is capable of undergoing **angiogenesis**, the formation of new vessels. This is extremely important in providing collateral circulation to the lung parenchyma, especially when the pulmonary circulation is compromised. When a clot or embolus obstructs pulmonary blood flow, the adjacent parenchyma is kept alive by the development of new blood vessels.

REVIEW QUESTIONS

DIRECTIONS: Each of the numbered items or incomplete statements in this section is followed by an answer or by completions of the statement. Select the ONE lettered answer or completion that is BEST in each case.

1. Which of the following best characterizes the pulmonary circulation?

	Flow	*Pressure*	*Resistance*	*Compliance*
(A)	Low	High	Low	High
(B)	High	Low	High	Low
(C)	Low	Low	Low	High
(D)	High	Low	High	High
(E)	High	Low	Low	High

2. Pulmonary vascular resistance is decreased
(A) At low lung volumes
(B) By breathing low oxygen
(C) At high lung volumes
(D) With increased pulmonary arterial pressure
3. In healthy individuals, the pulmonary and systemic circulations have the same
(A) Mean pressure
(B) Vascular resistance
(C) Compliance
(D) Flow per minute
(E) Capillary blood volume
4. The effect of gravity on the pulmonary circulation in an upright individual will cause
(A) Blood flow to be the greatest in the middle of the lung
(B) Capillary pressure to be greater at the base of the lung compared with the apex
(C) Alveolar pressure to be greater than capillary pressure at the base of the lung
(D) Lower vascular resistance at the apex of the lung compared with the base
(E) Venous pressure to be greater than alveolar pressure at the apex
5. A patient lying on his back and breathing normally has a mean left atrial pressure of 7 cm H_2O; a mean pulmonary arterial pressure of 15 cm H_2O; a cardiac output of 4 L/min; and an anteroposterior chest depth of 15 cm, measured at the xiphoid process. Most of his lung is perfused under which of the following conditions?
(A) Zone 1
(B) Zone 2
(C) Zone 3
(D) Zone 4
6. Lowering pulmonary venous pressure will have the greatest effect on regional blood flow in
(A) Zone 1
(B) Zone 2
(C) Zone 3
(D) Zones 1 and 2
(E) Zones 2 and 3
7. Which of the following best characterizes alveolar ventilation and blood flow at the base, compared with the apex, of the lungs of a healthy standing person?

	Ventilation	*Blood flow*	*Ventilation-perfusion ratio*
(A)	Higher	Higher	Lower
(B)	Lower	Higher	Higher
(C)	Lower	Lower	Lower
(D)	Higher	Lower	Higher
(E)	Lower	Lower	Higher

8. The regional changes seen in ventilation and perfusion in lungs of a

(continued)

healthy standing individual are largely brought about by
(A) Differences in alveolar surface tension
(B) The pyramidal shape of the lung
(C) The effects of gravity
(D) Differences in lung compliance
(E) Differences in lung elastic recoil

9. Regional differences in ventilation-perfusion ratios affect gas tensions in the pulmonary blood. Which of the following best describes the gas tensions in the blood leaving the alveolar capillaries of a healthy standing individual?

	O_2 *tension* (PO_2)	CO_2 *tension* (PCO_2)
(A)	Lowest at base	Highest at apex
(B)	Highest at base	Lowest at base
(C)	Highest at apex	Lowest at apex
(D)	Highest at apex	Lowest at base
(E)	Lowest at base	Lowest at base

10. A 26-year-old patient has a cardiac output of 5 L/min and mean pulmonary arterial and left atrial pressures of 20 and 5 mm Hg, respectively. What is her pulmonary vascular resistance?
(A) 4 mm Hg/L per minute
(B) 3 mm Hg/L per minute
(C) 0.33 mm Hg/L per minute
(D) 0.25 mm Hg/L per minute

11. The apex of the lungs of a 21-year-old subject is 20 cm above the heart. What pressure (in mm Hg) must his right ventricle produce to pump blood to the top of the lungs?
(A) 25 mm Hg
(B) 20 mm Hg
(C) 15 mm Hg
(D) 10 mm Hg
(E) 5 mm Hg

12. A 32-year-old patient has a pulmonary vascular resistance of 4 mm Hg/L per minute and a cardiac output of 5 L/min. What is her driving pressure for moving blood through the pulmonary circulation?
(A) 10 mm Hg
(B) 15 mm Hg
(C) 20 mm Hg
(D) 30 mm Hg
(E) 40 mm Hg

SUGGESTED READING

Cotes JE. Lung function: Assessment and Application in Medicine. 5th Ed. Boston: Blackwell Scientific, 1993.
Fishman AP. The Pulmonary Circulation. Philadelphia: University of Pennsylvania Press, 1989.
Lumb AB. Nunn's Applied Respiratory Physiology. 5th Ed. Oxford, UK: Butterworth-Heinemann, 2000.
Staub NC. Basic Respiratory Physiology. New York: Churchill Livingstone, 1991.
West JB. Pulmonary Pathophysiology: The Essentials. 5th Ed. Baltimore: Lippincott Williams & Wilkins, 1998.

Gas Transfer and Transport

Rodney A. Rhoades, Ph.D.

CHAPTER OUTLINE

- GAS DIFFUSION AND UPTAKE
- DIFFUSING CAPACITY
- GAS TRANSPORT BY THE BLOOD
- RESPIRATORY CAUSES OF HYPOXEMIA

KEY CONCEPTS

1. The diffusion of gases follows Fick's law.
2. Pulmonary blood flow limits the transfer of O_2 and CO_2 in the lungs.
3. Diffusing capacity depends on the diffusion properties of the lungs, hematocrit, and pulmonary capillary blood volume.
4. Most of the oxygen in the blood is carried by hemoglobin.
5. Arterial oxygen saturation is a measure of the percentage of hemoglobin loaded with oxygen.
6. Carbon monoxide has a strong affinity for hemoglobin and decreases the ability of the blood to carry O_2.
7. Most of the CO_2 in the blood is carried in the form of HCO_3^- in the plasma.
8. An alveolar-arterial oxygen (A-aO_2) gradient occurs because of venous admixture.
9. Hypoxemia is an abnormally low P_{O_2} or oxygen content in arterial blood.
10. A low $\dot{V}_A/\dot{Q}$ ratio is the major cause of hypoxemia.

GAS DIFFUSION AND UPTAKE

There are two types of gas movements in the lungs, **bulk flow** and **diffusion.** Gas moves in the airways, from the trachea down to the alveoli, by bulk flow, analogous to water coming out of a faucet, in which all molecules move as a unit. The driving pressure (P) for bulk flow in the airways is barometric pressure (P_B) at the mouth minus alveolar pressure (P_A).

Respiratory Gases Cross the Alveolar-Capillary Membrane by Diffusion

The movement of gases in the alveoli and across the alveolar-capillary membrane is by diffusion in response to partial pressure gradients (see Chapter 2). Recall that partial pressure or gas tension can be determined by measuring barometric pressure and the fractional concentration (F) of the gas (Dalton's law; see Chapter 19). At sea level, P_{O_2} is 160 mm Hg (760 mm Hg $\times$ 0.21). F_{O_2} does not change with altitude, which means that the percentage of O_2 in the atmosphere is essentially the same at 30,000 feet (about 9,000 m) as it is at sea level. Therefore, the decreased P_{O_2} at an altitude that makes it difficult to breathe is due to a decrease in the P_B, not to a decrease in F_{O_2} (Fig. 21.1).

Oxygen is taken up by blood in the lungs and is transported to the tissues. **Oxygen uptake** is the transfer of oxygen from the alveolar spaces to the blood in the pulmonary capillaries. Gas uptake is determined by three factors: the diffusion properties of the alveolar-capillary membrane, the partial pressure gradient, and pulmonary capillary blood flow.

The diffusion of gases is a function of the partial pressure difference of the individual gases. For example, oxygen diffuses across the alveolar-capillary membrane because of the difference in P_{O_2} between the alveoli and pulmonary capillaries (Fig. 21.2). The partial pressure difference for oxygen is referred to as the **oxygen diffusion gradient**; in the normal lung, the *initial* oxygen diffusion gradient, P_{AO_2} (102 mm Hg) minus $P_{\bar{V}O_2}$ (40 mm Hg), is 62 mm Hg. The *initial* diffusion gradient across the alveolar-capillary membrane for carbon dioxide ($P_{\bar{V}CO_2} - P_{ACO_2}$) is about 6 mm Hg, which is much smaller than that of oxygen.

When gases are exposed to a liquid such as blood plasma, gas molecules move into the liquid and exist in a dissolved state. The dissolved gases also exert a partial pressure. A gas will continue to dissolve in the liquid until the partial pressure of the dissolved gas equals the partial pressure above the liquid. **Henry's law** states that at equilibrium, the amount of gas dissolved in a liquid at a given tempera-

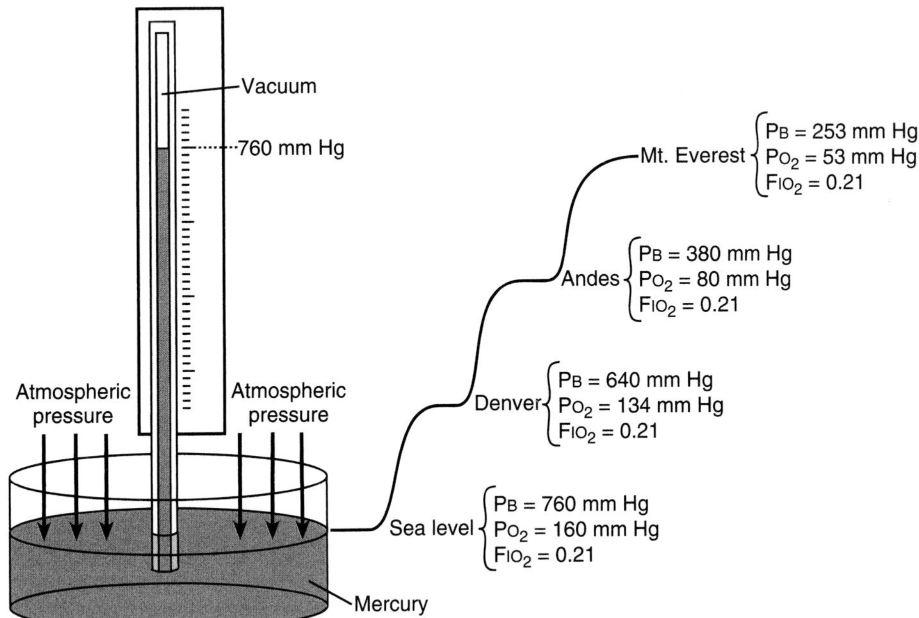

FIGURE 21.1 Changes in oxygen tension with altitude. The height of the column of mercury that is supported by air pressure decreases with increasing altitude and is a result of a fall in barometric pressure (PB). Because the fractional concentration of inspired O_2 (FIO_2) does not change with altitude, the decrease in PO_2 with altitude is caused entirely by a decrease in PB.

ture is directly proportional to the partial pressure and the solubility of the gas. Henry's law only accounts for the gas that is physically dissolved and not for chemically combined gases (e.g., oxygen bound to hemoglobin).

Gas diffusion in the lungs can be described by **Fick's law**, which states that the volume of gas diffusing per minute ($\dot{V}gas$) across a membrane is directly proportional to the membrane surface area (As), the diffusion coefficient of the gas (D), and the partial pressure difference (ΔP) of the gas and inversely proportional to membrane thickness (T) (Fig. 21.3):

$$\dot{V}gas = \frac{As \times D \times \Delta P}{T} \quad (1)$$

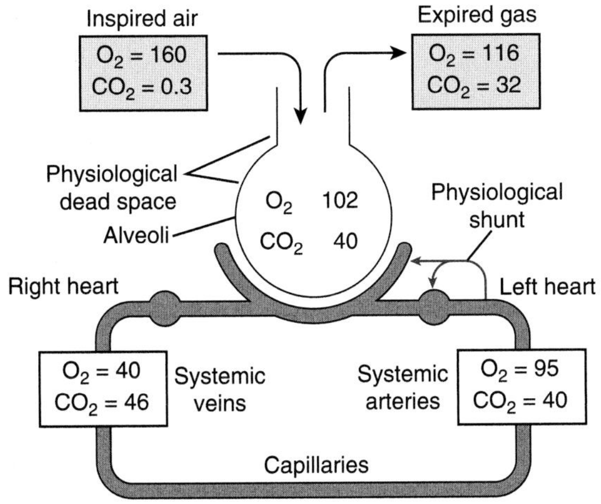

FIGURE 21.2 Partial pressures of oxygen (PO_2) and carbon dioxide (PCO_2) in the lungs and systemic circulation.

The diffusion coefficient of a gas is directly proportional to its solubility and inversely related to the square root of its molecular weight (MW):

$$D \propto \frac{\text{solubility}}{(MW)^{1/2}} \quad (2)$$

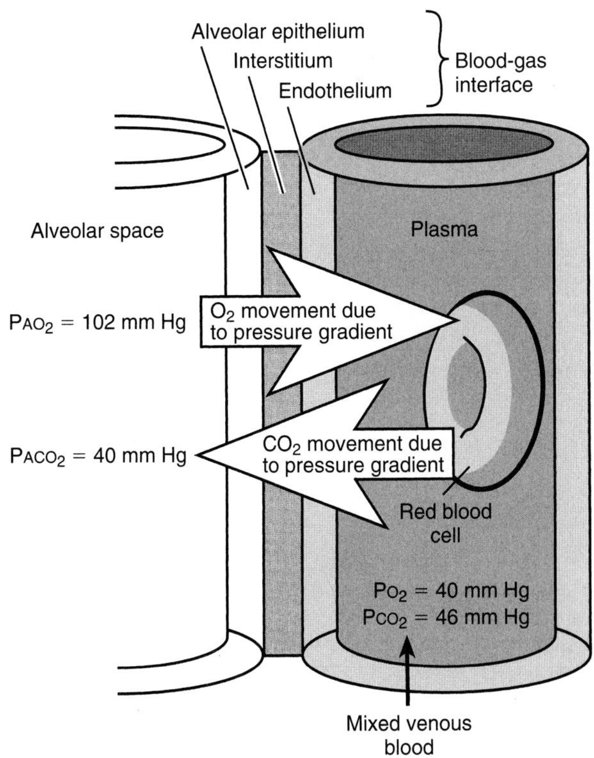

FIGURE 21.3 Diffusion path of O_2 and CO_2 in the lungs. Gases move across the blood-gas interface (alveolar-capillary membrane) by diffusion, following Fick's law.

Therefore, a small molecule or one that is very soluble will diffuse at a fast rate; for example, the diffusion coefficient of carbon dioxide in aqueous solutions is about 20 times greater than that of oxygen because of its higher solubility, even though it is a larger molecule than O_2.

Fick's law states that the rate of gas diffusion is inversely related to membrane thickness. This means that the diffusion of a gas will be halved if membrane thickness is doubled. Fick's law also states that the rate of diffusion is directly proportional to surface area (As). If two lungs have the same oxygen diffusion gradient and membrane thickness but one has twice the alveolar-capillary surface area, the rate of diffusion will differ by 2-fold.

Under steady state conditions, approximately 250 mL of oxygen per minute are transferred to the pulmonary circulation ($\dot{V}O_2$) while 200 mL of carbon dioxide per minute are removed ($\dot{V}CO_2$). The ratio $\dot{V}CO_2/\dot{V}O_2$ is the **respiratory exchange ratio** (R) and, in this case, is 0.8.

Capillary Blood Flow Limits Oxygen Uptake From Alveoli

Pulmonary capillary blood flow has a significant influence on oxygen uptake. The effect of blood flow on oxygen uptake is illustrated in Figure 21.4. The time required for the red cells to move through the capillary, referred to as **transit time**, is approximately 0.75 sec, during which time the gas tension in the blood equilibrates with the alveolar gas tension. Transit time can change dramatically with cardiac output. For example, when cardiac output increases, blood flow through the pulmonary capillaries increases, but transit time decreases (i.e., the time blood is in capillaries is less).

Figure 21.4 illustrates the effect of blood flow on the uptake of three test gases. In the first case, a trace amount of nitrous oxide (laughing gas), a common dental anesthetic, is breathed. Nitrous oxide (N_2O) is chosen because it diffuses across the alveolar-capillary membrane and dissolves in the blood, but does not combine with hemoglobin. The partial pressure in the blood rises rapidly and virtually reaches equilibrium with the partial pressure of N_2O in the alveoli by the time the blood is one tenth of the time in the capillary. At this point, the diffusion gradient for N_2O is zero. Once the pressure gradient becomes zero, no additional N_2O is transferred. The only way the transfer of N_2O can be increased is by increasing blood flow. The amount of N_2O that can be taken up is entirely limited by blood flow, not by diffusion of the gas. Therefore, the net transfer or uptake of N_2O is **perfusion-limited.**

When a trace amount of carbon monoxide (CO) is breathed, the transfer shows a different pattern (see Fig. 21.4). CO readily diffuses across the alveolar-capillary membrane but, unlike N_2O, CO has a strong affinity for hemoglobin. As the red cell moves through the pulmonary capillary, CO rapidly diffuses across the alveolar-capillary membrane into the blood and binds to hemoglobin. When a trace amount of CO is breathed, most is chemically bound in the blood, resulting in low partial pressure (P_{CO}). Consequently, equilibrium for CO across the alveolar-capillary membrane is never reached, and the transfer of CO to the blood is, therefore, **diffusion-limited** and not limited by the blood flow.

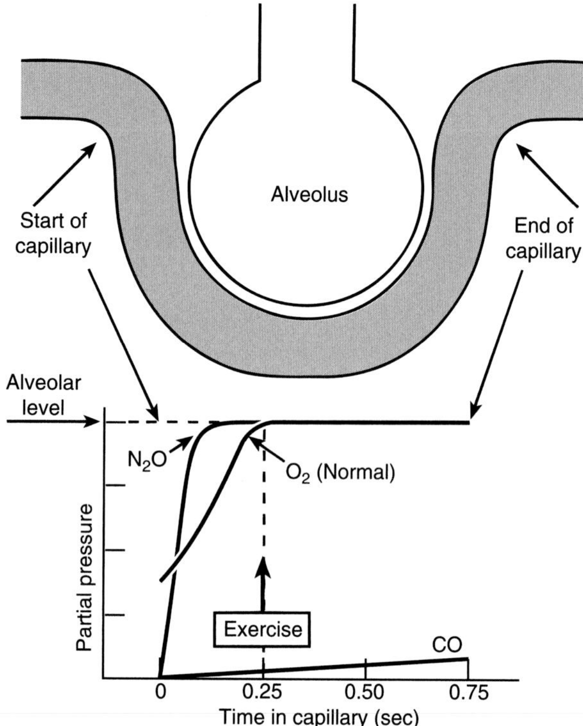

FIGURE 21.4 Uptake of N_2O, O_2, and CO by pulmonary capillary blood. Gas transfer is affected by pulmonary capillary blood flow. The horizontal axis shows time in the capillary. The average transit time it takes blood to pass through the pulmonary capillaries is 0.75 sec. The vertical axis indicates gas tension in the pulmonary capillary blood, and the top of the vertical axis indicates gas tension in the alveoli. Individual curves indicate the time it takes for the partial pressure of a specific gas in the pulmonary capillaries to equal the partial pressure in the alveoli. Nitrous oxide (N_2O) is used to illustrate how gas transfer is limited by blood flow; carbon monoxide (CO) illustrates how gas transfer is limited by diffusion. The profile for oxygen is more like that of N_2O, which means oxygen transfer is limited primarily by blood flow. Pulmonary capillary P_{O_2} equilibrates with the alveolar P_{O_2} in about 0.25 second (arrow).

Figure 21.4 shows that the equilibration curve for oxygen lies between the curves for N_2O and CO. Oxygen combines with hemoglobin, but not as readily as CO because it has a lower binding affinity. As blood moves along the pulmonary capillary, the rise in P_{O_2} is much greater than the rise in P_{CO} because of differences in binding affinity. Under resting conditions, the capillary P_{O_2} equilibrates with alveolar P_{O_2} when the blood is about one third of its time in the capillary. Beyond this point, there is no additional transfer of oxygen. Under normal conditions, oxygen transfer is more like that of N_2O and is limited primarily by blood flow in the capillary (perfusion-limited). Hence, an increase in cardiac output will increase oxygen uptake. Not only does cardiac output increase capillary blood flow, but it also increases capillary hydrostatic pressure. The latter increases the surface area for diffusion by opening up more capillary beds by recruitment.

The transit time at rest is normally about 0.75 sec, during which capillary oxygen tension equilibrates with alveolar oxygen tension. Ordinarily this process takes only about one third of the available time, leaving a wide safety margin to ensure that the end-capillary PO_2 is equilibrated with alveolar PO_2. With vigorous exercise, the transit time may be reduced to one third of a second (see Fig. 21.4). Thus, with vigorous exercise, there is still time to fully oxygenate the blood. Pulmonary end-capillary PO_2 still equals alveolar PO_2 and rarely falls with vigorous exercise. In abnormal situations, in which there is a thickening of the alveolar-capillary membrane so that oxygen diffusion is impaired, end-capillary PO_2 may not reach equilibrium with alveolar PO_2. In this case, there is measurable difference between alveolar and end-capillary PO_2.

DIFFUSING CAPACITY

In practice, direct measurements of As, T, and D in intact lungs are impossible to make. To circumvent this problem, Fick's law can be rewritten as shown in Figure 21.5, where the three terms are combined as **lung diffusing capacity** (DL).

Diffusing Capacity Is a Determinant of the Rate of Gas Transfer

The diffusing capacity provides a measure of the rate of gas transfer in the lungs per partial pressure gradient. For example, if 250 mL of O_2 per minute are taken up and the average alveolar-capillary PO_2 difference during a normal transit time is 14 mm Hg, then the DL for oxygen is 18 mL/min per mm Hg. Because the initial alveolar-capillary difference for oxygen cannot be measured and can only be estimated, CO is used to determine the lung diffusing capacity in patients. CO offers several advantages for measuring DL:

- Its uptake is limited by diffusion and not by blood flow.
- There is essentially no CO in the venous blood.
- The affinity of CO for hemoglobin is 210 times greater than that of oxygen, which causes the partial pressure of carbon monoxide to remain essentially zero in the pulmonary capillaries.

To measure the diffusing capacity in a patient with CO, the equation is

$$D_L = \frac{\dot{V}_{CO}}{P_{ACO}} \qquad (3)$$

where $\dot{V}_{CO}$ equals CO uptake in mL/min and P_{ACO} equals alveolar partial pressure of CO.

The most common technique for making this measurement is called the **single-breath test.** The patient inhales a single breath of a dilute mixture of CO and holds his or her breath for about 10 sec. By determining the percentage of CO in the alveolar gas at the beginning and the end of 10 sec and by measuring lung volume, one can calculate $\dot{V}_{CO}$. The single-breath test is very reliable. The normal resting value for DL_{CO} depends on age, sex, and body size. DL_{CO} ranges from 20 to 30 mL/min per mm Hg and decreases with pulmonary edema or a loss of alveolar membrane (e.g., emphysema).

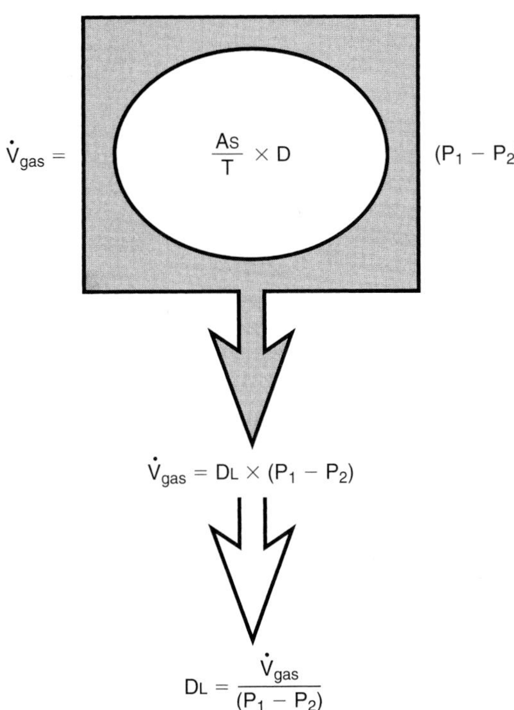

FIGURE 21.5 **Lung diffusing capacity.** Membrane surface area (As), gas diffusion coefficient (D), and membrane thickness (T) affect gas diffusion in the lungs. These properties are combined into one term, lung diffusing capacity (DL), which can be measured in a human subject. DL is equal to the volume of gas transferred/min ($\dot{V}gas$) divided by the mean partial pressure gradient for the gas.

Hemoglobin and Capillary Blood Volume Affect Lung Diffusing Capacity

Diffusing capacity does not depend solely on the diffusion properties of the lungs; it is also affected by blood hematocrit and pulmonary capillary blood volume. Both the hematocrit and capillary blood volume affect DL in the same direction (i.e., a decrease in either the hematocrit or capillary blood volume will lower the diffusing capacity in otherwise normal lungs). For example, if two individuals have the same pulmonary diffusion properties but one is anemic (reduced hematocrit), the anemic individual will have a decreased lung diffusing capacity. An abnormally low cardiac output lowers the pulmonary capillary blood volume, which decreases the alveolar capillary surface area and will, in turn, decrease the diffusing capacity in otherwise normal lungs.

GAS TRANSPORT BY THE BLOOD

The transport of O_2 and CO_2 by the blood, often referred to as **gas transport**, is an important step in the overall gas exchange process and is one of the important functions of the systemic circulation.

Oxygen Is Transported in Two Forms

Oxygen is transported to the tissues in two forms: combined with hemoglobin (Hb) in the red cell or physically dissolved in the blood. Approximately 98% of the oxygen is carried by hemoglobin and the remaining 2% is carried in the physically dissolved form. The amount of physically dissolved oxygen in the blood can be calculated from the following equation:

$$\text{Dissolved } O_2 \text{ (mL/dL)} = 0.003 \text{ (mL/dL per mm Hg)} \times PaO_2 \text{ (mm Hg)} \quad (4)$$

If PaO_2 equals 100 mm Hg, then dissolved O_2 = 0.3 mL/dL.

Binding Affinity of Hemoglobin for Oxygen. The hemoglobin molecule consists of four oxygen-binding heme sites and a globular protein chain. When hemoglobin binds with oxygen, it is called **oxyhemoglobin** (HbO_2). The hemoglobin that does not bind with O_2 is called **deoxyhemoglobin** (Hb). Each gram of hemoglobin can bind with 1.34 mL of oxygen. Oxygen binds rapidly and reversibly to hemoglobin: $O_2 + Hb \rightleftharpoons HbO_2$. The amount of oxyhemoglobin is a function of the partial pressure of oxygen in the blood. In the pulmonary capillaries, where PO_2 is high, the reaction is shifted to the right to form oxyhemoglobin. In tissue capillaries, where PO_2 is low, the reaction is shifted to the left; oxygen is unloaded from hemoglobin and becomes available to the cells. The maximum amount of oxygen that can be carried by hemoglobin is called the **oxygen carrying capacity**—about 20 mL O_2/dL blood in a healthy young adult. This value is calculated assuming a normal hemoglobin concentration of 15 g Hb/dL of blood (1.34 mL O_2/g Hb × 15 g Hb/dL blood = 20.1 mL O_2/dL blood).

Oxygen content is the amount of oxygen actually bound to hemoglobin (whereas capacity is the amount that can potentially be bound). The **percentage saturation of hemoglobin with oxygen** (SO_2) is calculated from the ratio of oxyhemoglobin content over capacity:

$$SO_2 = \frac{HbO_2 \text{ content}}{HbO_2 \text{ capacity}} \times 100 \quad (5)$$

Thus, the oxygen saturation is the ratio of the quantity of oxygen *actually bound* to the quantity that can be *potentially bound*. For example, if oxygen content is 16 mL O_2/dL blood and oxygen capacity is 20 mL O_2/dL blood, then the blood is 80% saturated. Arterial blood saturation of hemoglobin with oxygen (SaO_2) is normally about 98%.

Blood PO_2, O_2 saturation, and oxygen content are three closely related indices of oxygen transport. The relationship between PO_2, oxygen saturation, and oxygen content is illustrated by the **oxyhemoglobin equilibrium curve**, an S-shaped curve over a range of arterial oxygen tensions from 0 to 100 mm Hg (Fig. 21.6). The shape of the curve results because the hemoglobin affinity for oxygen increases progressively as blood PO_2 increases.

The shape of the oxyhemoglobin equilibrium curve reflects several important physiological advantages. The plateau region of the curve is the **loading phase**, in which oxygen is loaded onto hemoglobin to form oxyhemoglobin in the pulmonary capillaries. The plateau region illustrates how oxygen saturation and content remain fairly constant despite wide fluctuations in alveolar PO_2. For example, if PAO_2 were to rise from 100 to 120 mm Hg, hemoglobin would become only slightly more saturated (97 to 98%). For this reason, oxygen content cannot be raised appreciably by hyperventilation. The steep **unloading phase** of the curve allows large quantities of oxygen to be released or unloaded from hemoglobin in the tissue capillaries where a lower capillary PO_2 prevails. The S-shaped oxyhemoglobin equilibrium curve enables oxygen to saturate hemoglobin under high partial pressures in the lungs and to give up large amounts of oxygen with small changes in PO_2 at the tissue level.

A change in the binding affinity of hemoglobin for O_2 shifts the oxyhemoglobin-equilibrium curve to the right or left of normal (Fig. 21.7). The $\mathbf{P_{50}}$—the PO_2 at which 50% of the hemoglobin is saturated—provides a functional way to assess the binding affinity of hemoglobin for oxygen. The normal P_{50} for arterial blood is 26 to 28 mm Hg. A high P_{50} signifies a decrease in hemoglobin's affinity for oxygen and results in a rightward shift in the oxyhemoglobin equilibrium curve, whereas a low P_{50} signifies the opposite and shifts the curve to the left. A shift in the P_{50} in either direction has the greatest effect on the steep phase and only a small effect on the loading of oxygen in the normal lung, because loading occurs at the plateau.

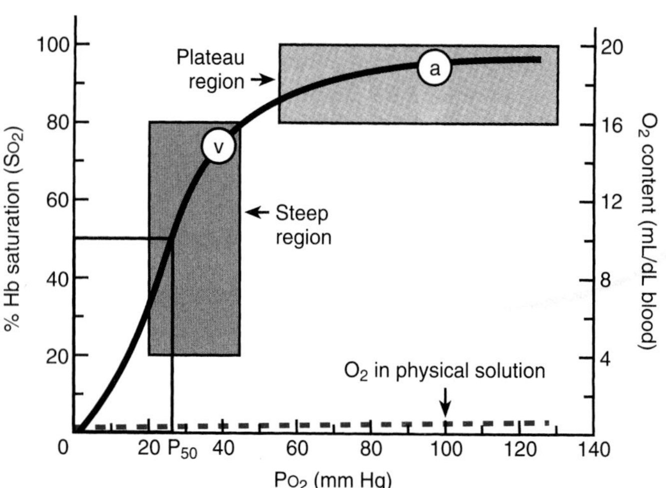

FIGURE 21.6 **An oxyhemoglobin equilibrium curve.** The oxygen saturation (left vertical axis) or oxygen content (right vertical axis) is plotted against partial pressure of oxygen (horizontal axis) to generate an oxyhemoglobin equilibrium curve. The curve is S-shaped and can be divided into a plateau region and a steep region. The dashed line indicates amount of oxygen dissolved in the plasma. a = arterial; v = venous; SO_2 = oxygen saturation; and P_{50} = partial pressure of O_2 required to saturate 50% of the hemoglobin with oxygen.

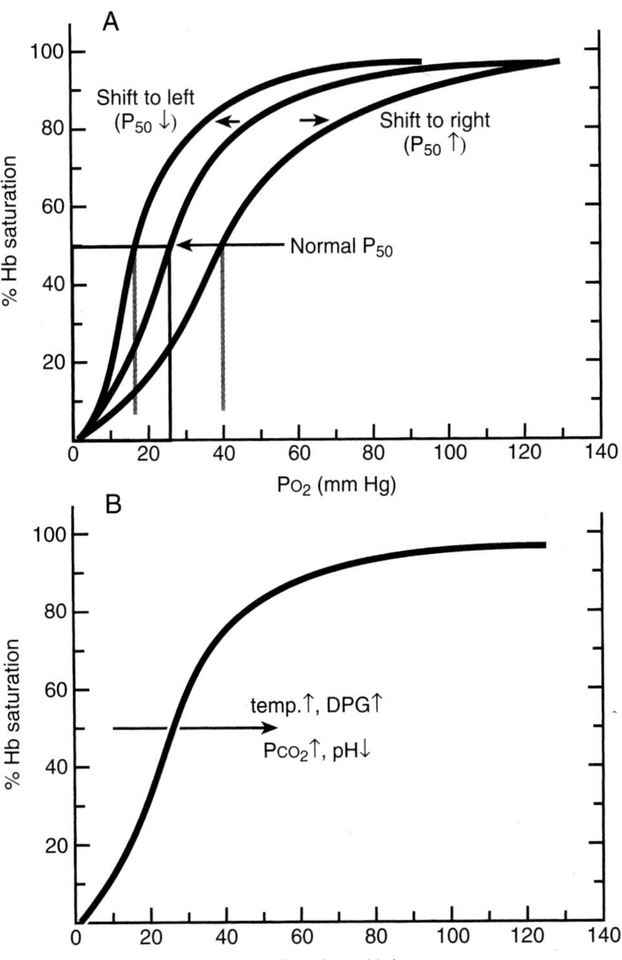

FIGURE 21.7 **Hemoglobin (Hb) binding affinity for O_2.** A, A shift in the oxyhemoglobin equilibrium curve affects the P_{50}. **B**, An increase in temperature, $[H^+]$, or arterial PCO_2 causes a rightward shift of the oxyhemoglobin equilibrium curve. A P_{50} increase indicates that binding affinity for oxygen decreases, which favors the unloading of O_2 from Hb at the tissue level. An increase in red cell levels of 2,3-diphosphoglycerate (DPG) will also shift the curve to the right. The increase in DPG occurs with hypoxemic conditions.

Effect of Blood Chemistry on Hemoglobin Binding Affinity. Several factors affect the binding affinity of hemoglobin for O_2, including temperature, arterial carbon dioxide tension, and arterial pH. A rise in PCO_2, a fall in pH, and a rise in temperature all shift the curve to the right (see Fig. 21.7). The effect of carbon dioxide and hydrogen ions on the affinity of hemoglobin for oxygen is known as the **Bohr effect.** A shift of the oxyhemoglobin equilibrium curve to the right is physiologically advantageous at the tissue level because the affinity is lowered (increased P_{50}). A rightward shift enhances the unloading of oxygen for a given PO_2 in the tissue, and a leftward shift increases the affinity of hemoglobin for oxygen, thereby lowering the ability to release oxygen to the tissues. A simple way to remember the functional importance of these shifts is that an exercising muscle is warm and acidic and produces large amounts of carbon dioxide (high PCO_2), all of which favor the unloading of more oxygen to metabolically active muscles.

Red blood cells contain 2,3-diphosphoglycerate (2,3-DPG), an organic phosphate compound that also can affect affinity of hemoglobin for oxygen. In red cells, 2,3-DPG levels are much higher than in other cells because erythrocytes lack mitochondria. An increase in 2,3-DPG facilitates unloading of oxygen from the red cell at the tissue level (shifts the curve to the right). An increase in red cell 2,3-DPG occurs with exercise and with hypoxia (e.g., high altitude, chronic lung disease).

Oxygen content, rather than PO_2 or SaO_2, is what keeps us alive and serves as a better gauge for oxygenation. For example, an individual can have a normal arterial PO_2 and SaO_2 but reduced oxygen content. This situation is seen in patients who have anemia (a decreased number of circulating red cells). A patient with anemia who has a hemoglobin concentration half of normal (7.5 g/dL instead of 15 g/dL) will have a normal arterial PO_2 and SaO_2, but oxygen content will be reduced to half of normal. A patient with anemia has a normal SaO_2 because that content and capacity are proportionally reduced. The usual oxyhemoglobin equilibrium curve does not show changes in blood oxygen content, since the vertical axis is saturation. If the vertical axis is changed to oxygen content (mL O_2/dL blood), then changes in content are seen (Fig. 21.8). The shape of the oxyhemoglobin equilibrium curve does not change, but the curve moves down to reflect the reduction in oxygen content. A good analogy for comparing an anemic patient with a normal patient is a bicycle tire and a truck tire: both can have the same air pressure, but the amount of air each tire holds is different.

Effect of Carbon Monoxide. Carbon monoxide interferes with oxygen transport by competing for the same binding sites on hemoglobin. Carbon monoxide binds to hemoglobin to form **carboxyhemoglobin** (HbCO). The reaction

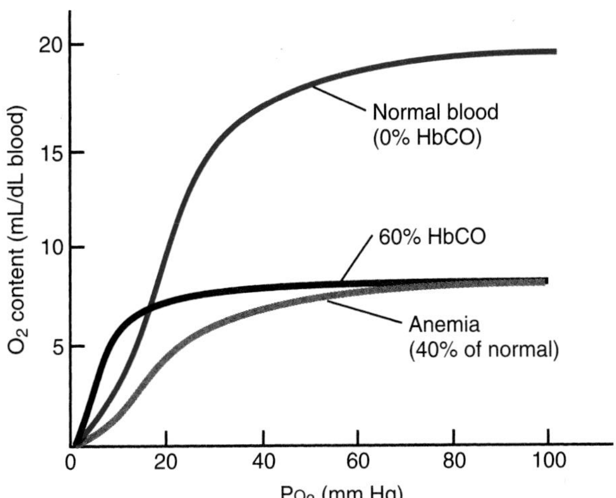

FIGURE 21.8 **Effect of blood hematocrit and CO on the oxyhemoglobin equilibrium curve.** Severe anemia can lower the O_2 content to 40% of normal. The blood O_2 content of an individual exposed to CO is shown for comparison. When the blood is 60% saturated with carbon monoxide (HbCO), O_2 content is reduced to about 8 mL/dL of blood. Note the leftward shift of the oxyhemoglobin equilibrium curve when CO binds with hemoglobin.

($Hb + CO \rightleftharpoons HbCO$) is reversible and is a function of P_{CO}. This means that breathing higher concentrations of CO will favor the reaction to the right. Breathing fresh air will favor the reaction to the left, which will cause CO to be released from the hemoglobin. A striking feature of CO is a binding affinity about 210 times that of oxygen. Consequently, CO will bind with the same amount of hemoglobin as oxygen at a partial pressure 210 times lower than that of oxygen. For example, breathing normal air (21% O_2) contaminated with 0.1% CO would cause half of the hemoglobin to be saturated with CO and half with O_2. With the high affinity of hemoglobin for CO, breathing a small amount CO can result in the formation of large amounts of HbCO. Arterial P_{O_2} in the plasma will still be normal because the oxygen diffusion gradient has not changed. However, oxygen content will be greatly reduced because oxygen cannot bind to hemoglobin. This is seen in Figure 21.8, which shows the effect of CO on the oxyhemoglobin equilibrium curve. When the blood is 60% saturated with CO (carboxyhemoglobin) the oxygen content is reduced to less than 10 mL/dL. The presence of CO also shifts the curve to the left, making it more difficult to unload or release oxygen to the tissues.

Carbon monoxide is dangerous for several reasons:

- It has a strong binding affinity for hemoglobin.
- As an odorless, colorless, and nonirritating gas, it is virtually undetectable.
- Pa_{O_2} is normal, and there is no feedback mechanism to indicate that oxygen content is low.
- There are no physical signs of hypoxemia (i.e., cyanosis or bluish color around the lips and fingers) because the blood is bright cherry red when CO binds with hemoglobin.

Therefore, a person can be exposed to CO and have oxygen content reduced to a level that becomes lethal, by causing tissue anoxia, without the individual being aware of the danger. The brain is one of the first organs affected by lack of oxygen. CO can alter reaction time, cause blurred vision and, if severe enough, cause unconsciousness.

The best treatment for CO poisoning is breathing 100% oxygen or a mixture of 95% O_2/5% CO_2. Since O_2 and CO compete for the same binding site on the hemoglobin molecule, breathing a high oxygen concentration will drive off the CO and favor the formation of oxyhemoglobin. The addition of 5% carbon dioxide to the inspired gas stimulates ventilation, which lowers the CO and enhances the

CLINICAL FOCUS BOX 21.1

Free Radical-Induced Lung Injury

Although an "oxygen paradox" has long been recognized in biology, only recently has it been well understood: Oxygen is essential for life, but too much oxygen or inappropriate oxygen metabolism can be harmful to both cells and the organism. The synthesis of ATP involves reactions in which molecular oxygen is reduced to form water. This reduction is accomplished by addition of four electrons by the mitochondrial electron transport system. About 98% of the oxygen consumed is reduced to water in the mitochondria. "Leaks" in the mitochondrial electron transport system, however, allow oxygen to accept less than four electrons, forming a free radical.

A **free radical** is any atom, molecule, or group of molecules with an unpaired electron in its outermost orbit. Free radicals include the **superoxide ion** ($O_2{\bullet}^-$) and the **hydroxyl radical** (•OH). The single unpaired electron in the free radical is denoted by a dot. The •OH radical is the most reactive and most damaging to cells. **Hydrogen peroxide** (H_2O_2), while not a free radical, is also reactive to tissues and has the potential to generate the hydroxyl radical (•OH). These three substances are collectively called **reactive oxygen species** (ROS). In addition to free radicals produced by leaks in the mitochondrial transport system, ROS also can be formed by cytochrome P450, in the production of NADPH and in arachidonic acid metabolism. Superoxide ion in the presence of NO will form **peroxynitrite**, another free radical that is also extremely toxic to cells. Under normal conditions, ROS are neutralized by the protective enzymes **superoxide dismutase**, **catalase**, and **peroxidases** and no damage occurs. However, when ROS are greatly increased, they overwhelm the protective enzyme systems and damage cells by oxidizing membrane lipids, cellular proteins, and DNA.

The lungs are a major organ for free radical injury, and the pulmonary vessels are the primary target site. Free radicals damage the pulmonary capillaries, causing them to become leaky, leading to pulmonary edema. In addition to intracellular production, ROS are produced during inflammation and episodes of oxidant exposure (i.e., oxygen therapy or breathing ozone and nitrogen dioxide from polluted air). During the inflammatory response, neutrophils become sequestered and activated; they undergo a respiratory burst (which produces free radicals) and release catalytic enzymes. This release of free radicals and catalytic enzymes functions to kill bacteria, but endothelial cells can become damaged in the process.

Paraquat, an agricultural herbicide, is another source of free radical-induced injury to the lungs. Crop dusters and migrant workers are particularly at risk because of exposure to paraquat through the lungs and skin. Tobacco or marijuana that has been sprayed with paraquat and subsequently smoked can also produce lung injury from ROS.

Ischemia-reperfusion, another cause of free radical-induced injury in the lungs, usually results from a blood clot that gets lodged in the pulmonary circulation. Tissues beyond the clot (or embolus) become ischemic, cellular ATP decreases, and hypoxanthine accumulates. When the clot dissolves, blood flow is reestablished. During the reperfusion phase, hypoxanthine, in the presence of oxygen, is converted to xanthine and then to urate. These reactions, catalyzed by the enzyme **xanthine oxidase** on the pulmonary endothelium, result in the production of superoxide ions. Neutrophils also become sequestered and activated in these vessels during the reperfusion phase. Thus, the pulmonary vasculature and surrounding lung parenchyma become damaged from a double hit of free radicals—those produced from the oxidation of hypoxanthine and those from activated neutrophils.

release of CO from hemoglobin. The loading and unloading of CO from hemoglobin is a function of P_{CO}.

Oxygen is not always beneficial. Oxygen metabolism can produce harmful products that injure tissues (see Clinical Focus Box 21.1).

Carbon Dioxide Is Transported in Three Forms

Figure 21.9 illustrates the processes involved in carbon dioxide transport. Carbon dioxide is carried in the blood in three forms:

- Physically dissolved in the plasma (10%)
- As bicarbonate ions in the plasma and in the red cells (60%)
- As **carbamino proteins** (30%)

The high P_{CO_2} in the tissues drives carbon dioxide into the blood, but only a small amount stays as dissolved CO_2 in the plasma. The bulk of the carbon dioxide diffuses into the red cell, where it forms either carbonic acid (H_2CO_3) or **carbaminohemoglobin.** In the red cell, carbonic acid is formed in the following reaction:

$$CO_2 + H_2O \underset{CA}{\rightleftharpoons} H_2CO_3 \rightleftharpoons H^+ + HCO_3^- \quad (6)$$

The hydration of CO_2 would take place very slowly if it were not accelerated about 1,000 times in red cells by the enzyme **carbonic anhydrase** (CA). This enzyme is also found in renal tubular cells, gastrointestinal mucosa, muscle, and other tissues, but its activity is highest in red blood cells.

Carbonic acid readily dissociates in red blood cells to form bicarbonate (HCO_3^-) and H^+. HCO_3^- leaves the red blood cells, and chloride diffuses in from the plasma to maintain electrical neutrality (see Fig. 21.9). The chloride movement is known as the **chloride shift** and is facilitated by a **chloride-bicarbonate exchanger** (**anion exchanger**) in the red blood cell membrane. The H^+ cannot readily move out because of the low permeability of the membrane to H^+. Most of the H^+ is buffered by hemoglobin: $H^+ + HbO_2^- \rightleftharpoons HHb + O_2$. As H^+ binds to hemoglobin, it decreases oxygen binding and shifts the oxyhemoglobin equilibrium curve to the right. This promotes the unloading of oxygen from hemoglobin in the tissues and favors the carrying of carbon dioxide. In the pulmonary capillaries, the oxygenation of hemoglobin favors the unloading of carbon dioxide.

Carbaminohemoglobin is formed in red cells from the reaction of carbon dioxide with free amine groups (NH_2) on the hemoglobin molecule:

$$CO_2 + HbNH_2 \rightleftharpoons HbNHCOOH \quad (7)$$

Deoxygenated hemoglobin can bind much more CO_2 in this way than oxygenated hemoglobin. Although major reactions related to CO_2 transport occur in the red cells, the bulk of the CO_2 is actually carried in the plasma in the form of bicarbonate.

A **carbon dioxide equilibrium curve** can be constructed in a fashion similar to that for oxygen (Fig. 21.10). The carbon dioxide equilibrium curve is nearly a straight-line function of P_{CO_2} in the normal arterial CO_2 range. Note that a higher P_{O_2} will shift the curve downward and to the right. This is known as the **Haldane effect**, and its advantage is that it allows the blood to load more CO_2 in the tissues and unload more CO_2 in the lungs.

Important differences are observed between the carbon dioxide and oxygen equilibrium curves (Fig. 21.11). First, one liter of blood can hold much more carbon dioxide than oxygen. Second, the CO_2 equilibrium curve is steeper and more linear, and because of the shape of the CO_2 equilibrium curve, large amounts of CO_2 can be loaded and un-

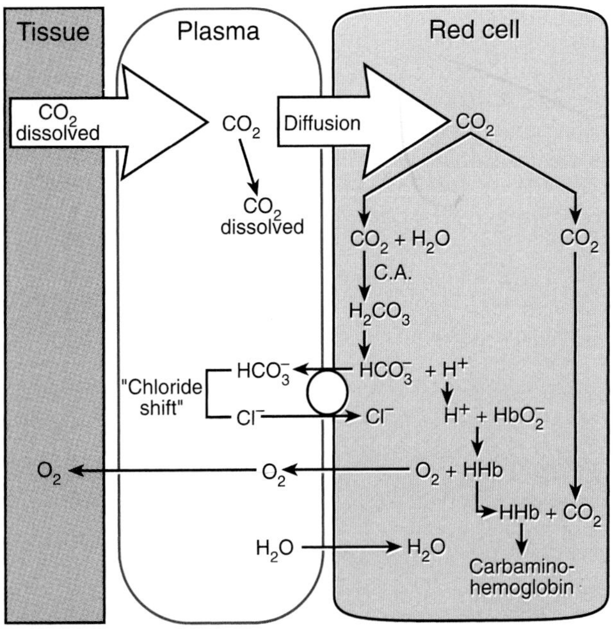

FIGURE 21.9 **Carbon dioxide transport.** CO_2 is transported in the blood in three forms: physically dissolved, as HCO_3^-, and as carbaminohemoglobin in the red cell (see text for details). The uptake of CO_2 favors the release of O_2.

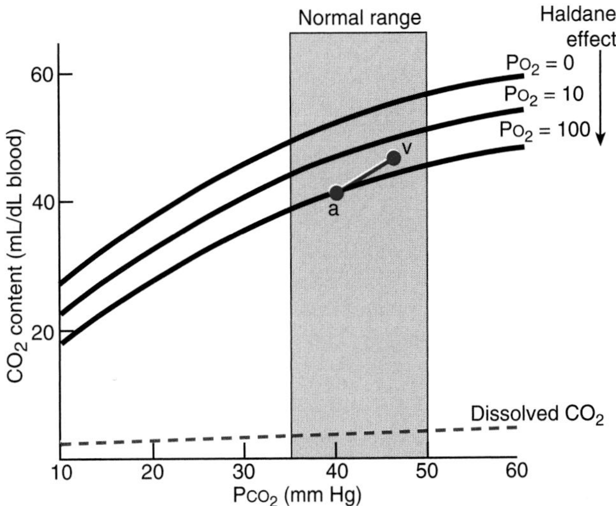

FIGURE 21.10 **Effect of O_2 on the carbon dioxide equilibrium curve.** The carbon dioxide equilibrium curve is relatively linear. An increase in P_{O_2} tension causes a rightward and downward shift of the curve. The P_{O_2} effect on the CO_2 equilibrium curve is known as the Haldane effect. The dashed line indicates the amount dissolved in plasma. a = CO_2 content in arterial blood; v = CO_2 content in mixed venous blood.

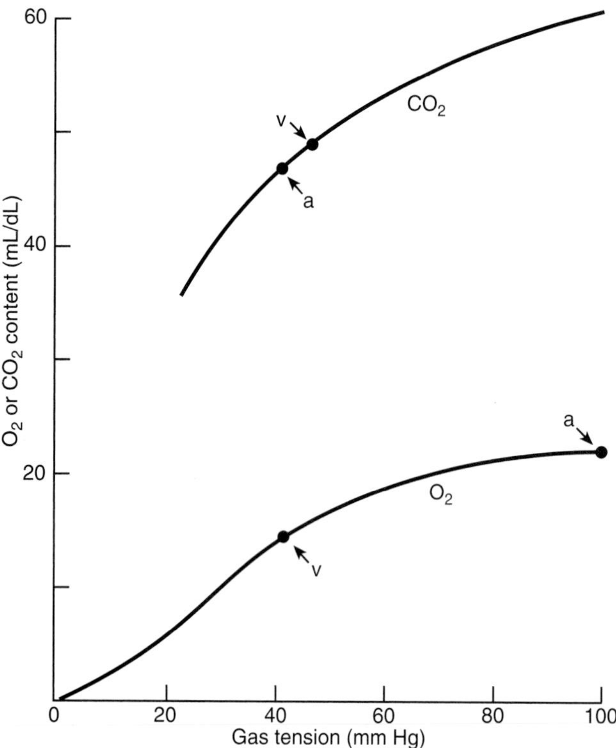

FIGURE 21.11 **Comparison of the oxyhemoglobin and CO_2 equilibrium curves.** The carrying capacity for CO_2 is much greater than for O_2. The increased steepness and linearity of the CO_2 equilibrium curve allow the lungs to remove large quantities of CO_2 from the blood with a small change in CO_2 tension. a = gas content in arterial blood; v = gas content in mixed venous blood.

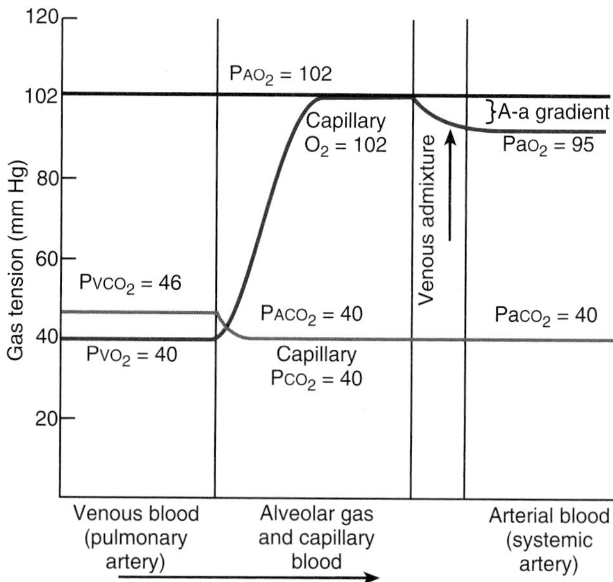

FIGURE 21.12 **The alveolar-arterial oxygen gradient.** The diagram shows O_2 and CO_2 tensions in blood in the pulmonary artery, pulmonary capillaries, and systemic arterial blood. The PO_2 leaving the pulmonary capillary has equilibrated with alveolar PO_2. However, systemic arterial PO_2 is below alveolar PO_2. Venous admixture results in the alveolar-arterial (A-a) oxygen gradient.

loaded from the blood with a small change in PCO_2. This is important not only in gas exchange and transport but also in the regulation of acid-base balance.

RESPIRATORY CAUSES OF HYPOXEMIA

Under normal conditions, hemoglobin is 100% saturated with oxygen when the blood leaves the pulmonary capillaries, and the end-capillary PO_2 equals alveolar PO_2. However, the blood that leaves the lungs (via the pulmonary veins) and returns to the left side of the heart has a lower PO_2 than pulmonary end-capillary blood. As a result, the systemic arterial blood has an average oxygen tension (PaO_2) of about 95 mm Hg and is only 98% saturated.

Venous Admixture Causes an Alveolar-Arterial Oxygen Gradient

The difference between alveolar oxygen tension (PAO_2) and arterial oxygen tension (PaO_2) is the **alveolar-arterial oxygen gradient** or **A-aO_2 gradient** (Fig. 21.12). Because alveolar PO_2 is normally 100 to 102 mm Hg and arterial PO_2 is 85 to 95 mm Hg, a normal A-aO_2 gradient is 5 to 15 mm Hg. The A-aO_2 gradient is obtained from blood gas measurements and the alveolar gas equation to determine PAO_2. Recall from Chapter 19 that the simplified equation is $PAO_2 = FIO_2 \times (PB - 47) - 1.2 \times PACO_2$.

The A-aO_2 gradient arises in the normal individual because of venous admixture as a result of a shunt (e.g., bronchial circulation) and regional variations of the $\dot{V}A/\dot{Q}$ ratio. Approximately half of the normal A-aO_2 gradient is caused by the bronchial circulation, and half is due to regional variations of the $\dot{V}A/\dot{Q}$ ratio. In some pathophysiological disorders, the A-aO_2 gradient can be greatly increased. A value greater than 15 mm Hg is considered abnormal and usually leads to low oxygen in the blood or hypoxemia. The normal ranges of blood gases are shown in Table 21.1. Values for PaO_2 below 85 mm Hg indicate hypoxemia. A $PaCO_2$ less than 35 mm Hg is called **hypocapnia,** and a $PaCO_2$ greater than 48 mm Hg is called **hypercapnia**. A pH value for arterial blood less than 7.35 or greater than 7.45 is called **acidemia** or **alkalemia**, respectively.

TABLE 21.1 **Arterial Blood Gases**

	Normal Range[a]
PaO_2	85–95 mm Hg
$PaCO_2$	35–48 mm Hg
SaO_2	94–98%
pH	7.35–7.45
HCO_3^-	23–28 mEq/L

[a] Normal range at sea level.

Respiratory Dysfunction Is the Major Cause of Hypoxemia

The causes of hypoxemia are classified as respiratory and nonrespiratory (Table 21.2). Respiratory dysfunction is by far the most common cause of hypoxemia in adults. Nonrespiratory causes include anemia, carbon monoxide poisoning, and a decreased inspired oxygen tension (as occurs at high altitude) (see Clinical Focus Box 21.2).

Regional Hypoventilation. The respiratory causes of hypoxemia are listed in order of frequency in Table 21.2. **Regional hypoventilation** is by far the most common cause of hypoxemia (about 90% of cases) and reflects a local $\dot{V}_A/\dot{Q}$ imbalance stemming from a partially obstructed airway. A fraction of the blood that passes through the lungs

TABLE 21.2 Pathophysiological Causes of Hypoxemia

Causes	Effect on A-aO_2 Gradient
Respiratory	
Regional low $\dot{V}_A/\dot{Q}$ ratio	Increased
Anatomic shunt	Increased
Generalized hypoventilation	Normal
Diffusion block	Increased
Nonrespiratory	
Intracardiac right-to-left shunt	Increased
Decreased P_{IO_2}, low P_B, low F_{IO_2}	Normal
Reduced oxygen content (anemia and CO poisoning)	Normal

CLINICAL FOCUS BOX 21.2

Anemia

Anemia, an abnormally low hematocrit or hemoglobin concentration, is by far the most common disorder affecting erythrocytes. The different causes of anemia can be grouped into three categories: decreased erythropoiesis by bone marrow, blood loss, and increased rate of red cell destruction (hemolytic anemia).

Several mechanisms lead to decreased production of red cells by the bone marrow, including aplastic anemia, malignant neoplasms, chronic renal disease, defective DNA synthesis, defective hemoglobin synthesis, and chronic liver disease. **Aplastic anemia** is the result of stem cell destruction in the bone marrow, which leads to decreased production of white cells, platelets, and erythrocytes. Malignant neoplasms (e.g., leukemia) cause an overproduction of immature red cells. Patients with chronic renal disease have a decreased production of erythropoietin, with a concomitant decrease in red cell production.

Patients with defective DNA synthesis have **megaloblastic anemia**, a condition in which red cell maturation in the bone marrow is abnormal; this may result from vitamin B_{12} or folic acid deficiency. These cofactors are essential for DNA synthesis. Vitamin B_{12} is present in high concentration in liver and, to some degree, in most meat, but it is absent in plants. Vitamin B_{12} deficiency is rare except in strict vegetarians. Folic acid is widely distributed in leafy vegetables; folic acid deficiency commonly occurs where malnutrition is prevalent. Pernicious anemia is a form of megaloblastic anemia resulting from vitamin B_{12} deficiency. Most commonly in adults over 60, it results not from deficient dietary intake but from a decreased vitamin B_{12} absorption by the small intestine. Pernicious anemia is linked to an autoimmune disease in which there is immunological destruction of the intestinal mucosa, particularly the gastric mucosa.

Iron-deficiency anemia is the most common cause of anemia worldwide. Although it occurs in both developed and undeveloped countries, the causes are different. In developed countries, the cause is usually due to pregnancy or chronic blood loss due to gastrointestinal ulcers or neoplasms. In undeveloped countries, hookworm infections account for most cases of iron-deficiency anemia.

Acute or chronic blood loss is another cause of anemia. With hemorrhage, red cells are lost and the hypovolemia causes the kidneys to retain water and electrolytes as a compensation. Retention of water and electrolytes restores the blood volume, but the concomitant dilution of the blood causes a further decrease in the red cell count, hemoglobin concentration, and hematocrit. Chronic bleeding is compensated by erythroid hyperplasia, which eventually depletes iron stores. Thus, chronic blood loss results in iron-deficiency anemia.

The last category, increased rate of red cell destruction, includes the Rh factor and sickle cell anemia. The Rhesus (Rh) blood group antigens are involved in maintaining erythrocyte structure. Patients who lack Rh antigens (Rh null) have severe deformation of the red cells.

Sickle cell anemia, associated with the abnormal hemoglobin HbS gene, is common in Africa, India, and among African Americans but rare in the Caucasian and Asian populations. In the sickle cell trait, which occurs in about 9% of African Americans, one abnormal gene is present. A single point mutation occurs in the hemoglobin molecule, causing the normal glutamic acid at position 6 of the beta chain to be replaced with valine, resulting in HbS. The amino acid substitution is on the surface, resulting in a tendency for the hemoglobin molecule to crystallize with anoxia. However, heterozygous individuals have no symptoms, and oxygen transport by fetal (HbF) and adult hemoglobin (HbA) is normal. Sickle cell trait (i.e., heterozygous individuals) offers protection against malaria, and this selective advantage is thought to have favored the persistence of the HbS gene, especially in regions where malaria is common. Sickle cell disease represents the homozygous condition (S/S) and occurs in about 0.2% of African Americans. The onset of sickle cell anemia occurs in infancy as HbS replaces HbF; death often occurs early in adult life. Patients with sickle cell anemia have >80% HbS in their blood with a decrease or an absence of normal HbA.

Whatever the cause of anemia, the pathophysiological effect is the same—hypoxemia. Symptoms include pallor of the lips and skin, weakness, fatigue, lethargy, dizziness, and fainting. If the anemia is severe, myocardial hypoxia can lead to angina pain.

does not get fully oxygenated, resulting in an increase in venous admixture. Only a small amount of venous admixture is required to lower systemic arterial P_{O_2}, due to the nature of the oxyhemoglobin equilibrium curve. This can be seen from Figure 21.13, which depicts oxygen content from three groups of alveoli with low, normal, and high $\dot{V}_A/\dot{Q}$ ratios. The oxygen content of the blood leaving these alveoli is 16.0, 19.5, and 20.0 mL/dL of blood, respectively. As Figure 21.13 shows, a low $\dot{V}_A/\dot{Q}$ ratio is far more serious because it has the greatest effect on lowering both the P_{O_2} and the O_2 content because of the nonlinear shape of the oxyhemoglobin equilibrium curve. Patients who have an abnormally low $\dot{V}_A/\dot{Q}$ ratio have a high A-aO_2 gradient, low P_{O_2}, and low O_2 content, but usually a normal or slightly elevated Pa_{CO_2}. Pa_{CO_2} does not change much because the CO_2 equilibrium curve is nearly linear, which allows excess CO_2 to be removed from the blood by the lungs.

Another cause for a regionally low $\dot{V}_A/\dot{Q}$ ratio is a large blood clot that occludes a major artery in the lungs. When a major pulmonary artery becomes occluded, a greater portion of the cardiac output is redirected to another part of the lungs, resulting in overperfusion with respect to alveolar ventilation. This causes a regionally low $\dot{V}_A/\dot{Q}$ ratio, and leads to an increase in venous admixture.

Shunts. The next most common cause of hypoxemia is a shunt, either a right-to-left anatomic shunt or an absolute intrapulmonary shunt. The latter occurs when an airway is totally obstructed by a foreign object (such as a peanut) or by tumors. Patients with hypoxemia stemming from a shunt also have a high A-aO_2 gradient, low P_{O_2}, and low O_2 content, and a normal or slightly elevated Pa_{CO_2}. A test that is often used to distinguish between an abnormally low $\dot{V}_A/\dot{Q}$ ratio and a shunt is to have the patient breathe 100% O_2 for 15 minutes. If the Pa_{O_2} is >150 mm Hg, the cause is a low $\dot{V}_A/\dot{Q}$ ratio. If the patient's Pa_{O_2} is <150 mm Hg, the cause of hypoxemia is a shunt. The principle for using 100% O_2 is illustrated in Figure 21.14. The patient with regional hypoventilation who breathes 100% O_2 compensates for the low $\dot{V}_A/\dot{Q}$ ratio, and because all of the blood leaving the pulmonary capillaries is now fully saturated, the venous admixture is eliminated. However, the low arterial P_{O_2} does not get corrected by breathing 100% O_2 in a patient with a shunt because the enriched oxygen mixture never comes into contact with the shunted blood.

Generalized Hypoventilation. **Generalized hypoventilation**, the third most common cause of hypoxemia, occurs when alveolar ventilation is depressed. This situation can

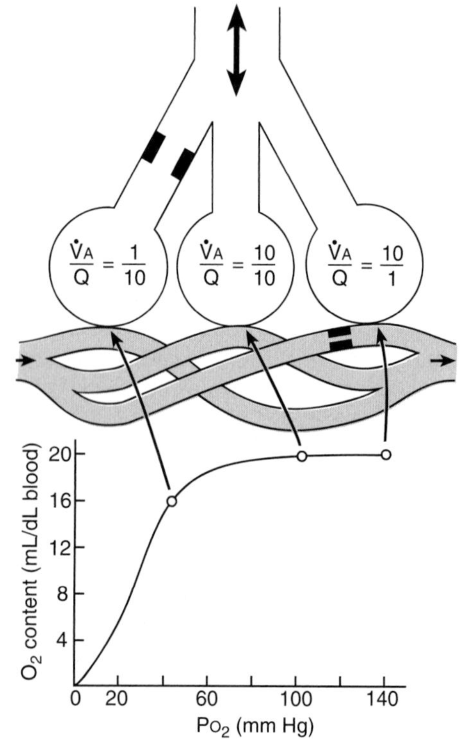

FIGURE 21.13 **Effect of venous admixture on O_2 content.** Because of the S-shaped oxyhemoglobin equilibrium curve, a high $\dot{V}_A/\dot{Q}$ ratio has little effect on arterial O_2 content. However, mixing with blood from a region with a low $\dot{V}_A/\dot{Q}$ ratio can dramatically lower P_{O_2} in blood leaving the lungs.

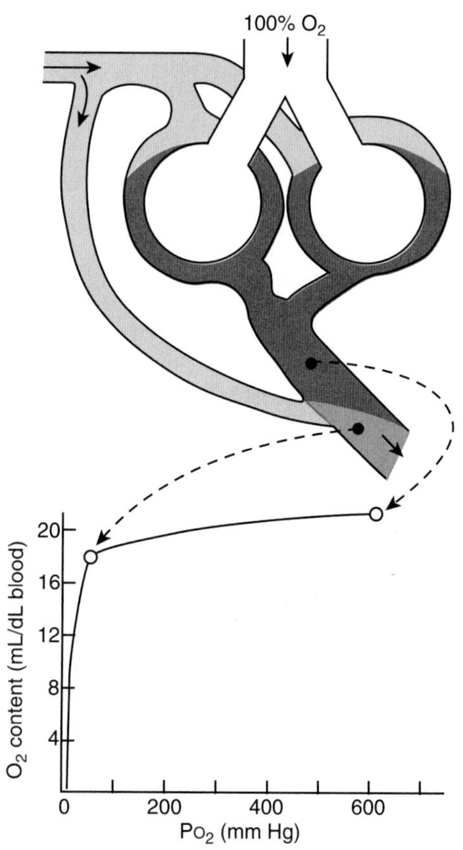

FIGURE 21.14 **Diagnosis of a shunt.** A shunt can be diagnosed by having the subject breathe 100% O_2 for 15 minutes. P_{O_2} in systemic arterial blood in a patient with a shunt does not increase above 150 mm Hg during the 15-minute period. The shunted blood is not exposed to 100% O_2, and the venous admixture reduces arterial P_{O_2}.

arise from a chronic obstructive pulmonary disorder (such as emphysema) or depressed respiration (as a result of a head injury or a drug overdose, for example). Since ventilation is depressed, there is also a significant increase in arterial PCO_2 with a concomitant decrease in arterial pH. In generalized hypoventilation, total ventilation is insufficient to maintain normal systemic arterial PO_2 and PCO_2. A feature that distinguishes generalized hypoventilation from the other causes of hypoxemia is a normal A-aO_2 gradient, as a result of the alveolar and arterial PO_2 being lowered equally. If a patient has a low PaO_2 and a normal A-aO_2 gradient, the cause of hypoxemia is entirely due to generalized hypoventilation. The best corrective measure for generalized hypoventilation is to place the patient on a mechanical ventilator, breathing room air. This treatment will return both arterial PO_2 and PCO_2 to normal. Administering supplemental oxygen to a patient with generalized hypoventilation will correct hypoxemia but not hypercapnia because ventilation is still depressed.

Diffusion Block. The least common cause of hypoxemia is a **diffusion block.** This condition occurs when the diffusion distance across the alveolar-capillary membrane is increased or the permeability of the alveolar-capillary membrane is decreased. It is characterized by a low PaO_2, high A-aO_2 gradient, and high $PaCO_2$. Pulmonary edema is one of the major causes of diffusion block.

In summary, there are four basic respiratory disturbances that cause hypoxemia. Examining the A-aO_2 gradient or $PaCO_2$ and/or breathing 100% oxygen distinguishes the four types. For example, if a patient has a low PaO_2, high $PaCO_2$, and normal A-aO_2 gradient, the cause of hypoxemia is generalized hypoventilation. If the PaO_2 is low and the A-aO_2 gradient is high, then the cause can be a shunt, regional low $\dot{V}A/\dot{Q}$ ratio, or a diffusion block. Breathing 100% O_2 will distinguish between a low $\dot{V}A/\dot{Q}$ ratio and a shunt. Diffusion impairment is the least likely cause and can be deduced if the other three causes have been eliminated.

REVIEW QUESTIONS

DIRECTIONS: Each of the numbered items or incomplete statements in this section is followed by answers or by completions of the statement. Select the one lettered answer or completion that is BEST in each case.

1. In healthy individuals, the cause of an A-aO_2 gradient is
(A) Low diffusing capacity for oxygen compared with that for carbon dioxide
(B) A high $\dot{V}A/\dot{Q}$ ratio in the apex of the lungs
(C) Overventilation in the base of the lung
(D) A bronchial circulation shunt
(E) A right-to-left shunt in the heart

2. Which of the following will *not* cause a low lung diffusing capacity (DL)?
(A) Decreased diffusion distance
(B) Decreased capillary blood volume
(C) Decreased surface area
(D) Decreased cardiac output
(E) Decreased hemoglobin concentration in the blood

3. With respect to oxygen and carbon dioxide transport,
(A) The slopes of the oxygen and carbon dioxide content curves are similar
(B) Equal amounts of oxygen and carbon dioxide can be carried in 100 mL of blood
(C) The presence of carbon dioxide decreases the P_{50} for O_2
(D) The presence of oxygen lowers carbon dioxide content in the blood
(E) Most of the O_2 and CO_2 are transported by the red blood cell

4. Which of the following best characterizes the blood oxygen of an otherwise healthy person who has lost enough blood to decrease his hemoglobin concentration from the normal 15g/dL of blood to 10 g/dL of blood?

	PaO_2	*SaO_2*	*O_2 content*
(A)	Normal	Normal	Decreased
(B)	Normal	Decreased	Decreased
(C)	Decreased	Decreased	Decreased
(D)	Decreased	Normal	Decreased
(E)	Decreased	Decreased	Normal

5. Which of the following would *not* favor the unloading of oxygen from hemoglobin in tissues?
(A) Increase in P_{50}
(B) Increase in tissue pH
(C) Increase in 2,3-DPG levels
(D) Increase in tissue PCO_2
(E) Increase in temperature

6. Which of the following sets of arterial blood gas data is consistent with the presence of an abnormally low $\dot{V}A/\dot{Q}$ ratio?
(A) PaO_2 = 130 mm Hg; $PaCO_2$ = 30 mm Hg
(B) PaO_2 = 98 mm Hg; $PaCO_2$ = 40 mm Hg
(C) PaO_2 = 95 mm Hg; $PaCO_2$ = 40 mm Hg
(D) PaO_2 = 60 mm Hg; $PaCO_2$ = 40 mm Hg
(E) PaO_2 = 50 mm Hg; $PaCO_2$ = 30 mm Hg

7. Which of the following ranges of hemoglobin O_2 saturation from systemic venous to systemic arterial blood represents a normal resting condition?
(A) 25 to 75%
(B) 40 to 75%
(C) 40 to 95%
(D) 60 to 98%
(E) 75 to 98%

8. A 54-year-old man sustains third-degree burns in a house fire. His respiratory rate is 30/min, Hb = 17 g/dL, arterial PO_2 is 95 mm Hg, and arterial O_2 saturation is 50%. The most likely cause of his low oxygen saturation is
(A) Airway obstruction from smoke inhalation
(B) Carbon monoxide poisoning
(C) Pulmonary edema
(D) Fever
(E) An abnormally high $\dot{V}A/\dot{Q}$ ratio

9. A patient's $PaCO_2$ is 68 mm Hg, PO_2 is 50 mm Hg, and A-aO_2 gradient is normal. These findings are most consistent with
(A) A shunt
(B) A low $\dot{V}A/\dot{Q}$ ratio
(C) A diffusion block
(D) Generalized hypoventilation
(E) A high $\dot{V}A/\dot{Q}$ ratio

10. A patient is breathing room air and has $PaCO_2$ of 45 mm Hg, PaO_2 of 70 mm Hg, pH of 7.30, and SaO_2 of 85%. What is her A-aO_2 gradient?
(A) 16 mm Hg
(B) 24 mm Hg
(C) 26 mm Hg
(D) 30 mm Hg
(E) 40 mm Hg

11. A patient inspired a gas mixture containing a trace amount of carbon monoxide and then held his breath for 10 sec. During breath holding, the

alveolar P_{CO} averaged 0.5 mm Hg and CO uptake was 10 mL/min. What is his pulmonary diffusing capacity ($D_{L_{CO}}$)?
(A) 2.0 mL/min per mm Hg
(B) 5.0 mL/min per mm Hg
(C) 10 mL/min per mm Hg
(D) 20 mL/min per mm Hg
(E) 200 mL/min per mm Hg

SUGGESTED READING

Cotes JE. Lung Function: Assessment and Application in Medicine. 5th Ed. Boston: Blackwell Scientific, 1993.

Fishman AP. The Fick principle and the steady state. Am J Respir Crit Care Med 2000;161:692–694.

Staub NC. Basic Respiratory Physiology. New York: Churchill Livingstone, 1991.

Wagner PD. Determinants of maximal oxygen transport and utilization. Annu Rev Physiol 1996;58:21–50.

West JB. Ventilation/Blood Flow and Gas Exchange. 4th Ed. Oxford, UK: Blackwell Scientific, 1985.

The Control of Ventilation

Rodney A. Rhoades, Ph.D.

CHAPTER OUTLINE

- **GENERATION OF THE BREATHING PATTERN**
- **REFLEXES FROM THE LUNGS AND CHEST WALL**
- **CONTROL OF BREATHING BY H^+, CO_2, AND O_2**
- **THE CONTROL OF BREATHING DURING SLEEP**
- **THE RESPONSE TO HIGH ALTITUDE**

KEY CONCEPTS

1. Ventilation is controlled by negative and positive feedback systems.
2. Normal arterial blood gases are maintained and the work of breathing is minimized despite changes in activity, the environment, and lung function.
3. The basic breathing rhythm is generated by neurons in the brainstem and can be modified by ventilatory reflexes.
4. The rate and depth of breathing are finely regulated by vagal nerve endings that are sensitive to lung stretch.
5. The autonomic nerves and vagal sensory nerves maintain local control of airway function.
6. Mechanical or chemical irritation of the airways and lungs induces coughing, bronchoconstriction, shallow breathing, and excess mucus production.
7. Arterial P_{CO_2} is the most important factor in determining the ventilatory drive in resting individuals.
8. Central chemoreceptors detect changes only in arterial P_{CO_2}; peripheral chemoreceptors detect changes in arterial P_{O_2}, P_{CO_2}, and pH.
9. The hypoxia-induced stimulation of ventilation is not great until the arterial P_{O_2} drops below 60 mm Hg.
10. Sleep is induced by the withdrawal of a wakefulness stimulus arising from the brainstem reticular formation and results in a general depression of breathing.
11. Chronic hypoxemia causes ventilatory acclimatization that increases breathing.

GENERATION OF THE BREATHING PATTERN

The control of breathing is critical for understanding of respiratory responses to activity, changes in the environment, and lung diseases. Breathing is an automatic process that occurs without any conscious effort while we are awake, asleep, or under anesthesia. Breathing is similar to the heartbeat in terms of an automatic rhythm. However, there is no single pacemaker that sets the basic rhythm of breathing and no single muscle devoted solely to the task of tidal air movement. Breathing depends on the cyclic excitation of many muscles that can influence the volume of the thorax. Control of that excitation is the result of multiple neuronal interactions involving all levels of the nervous system. Furthermore, the muscles used for breathing must often be used for other purposes as well. For example, talking while walking requires that some muscles simultaneously attend to the tasks of posturing, walking, phonation, and breathing. Because it is impossible to study extensively the subtleties of such a complex system in humans, much of what is known about the control of breathing has been obtained from the study of other species. Much, however, remains unexplained.

The control of upper and lower airway muscles that affect airway tone is integrated with control of the muscles that start tidal air movements. During quiet breathing, inspiration is brought about by a progressive increase in activation of inspiratory muscles, most importantly the diaphragm (Fig. 22.1). This nearly linear increase in activity with time causes the lungs to fill at a nearly constant rate until tidal volume has been reached. The end of inspiration is associated with a rapid decrease in excitation of inspiratory muscles, after which expiration occurs passively by elastic recoil of the lungs and chest wall. Some excitation of inspiratory muscles resumes during the first part of expiration, slowing the initial rate of expiration. As more ventilation is required—for example, during exercise—other inspiratory muscles (external intercostals, cervical muscles)

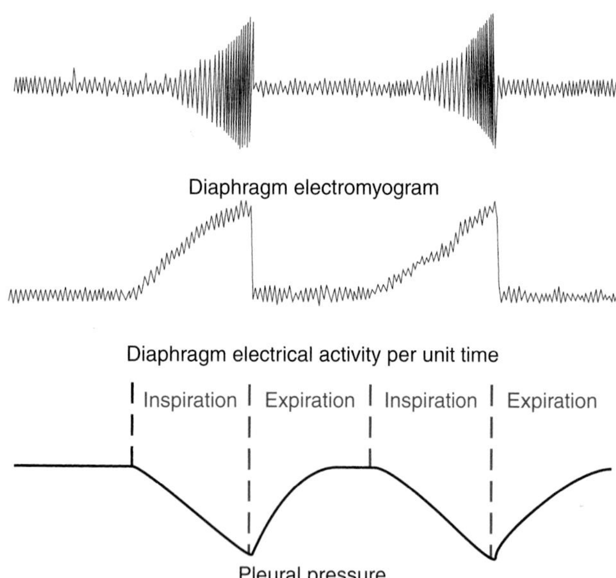

FIGURE 22.1 Relationship between electrical activity of the diaphragm and pleural pressure during quiet breathing. During inspiration, the number of active muscle fibers, and the frequency at which each fires, increases progressively, leading to a mirror-image fall in pleural pressure as the diaphragm descends.

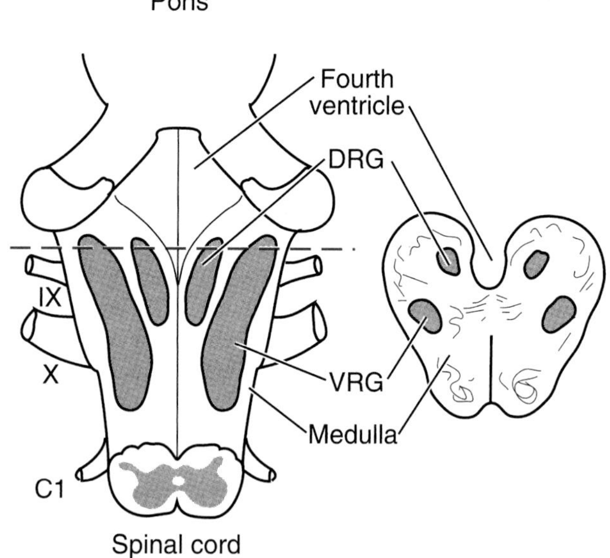

FIGURE 22.2 The general locations of the dorsal respiratory group (DRG) and ventral respiratory group (VRG). These drawings show the dorsal aspect of the medulla oblongata and a cross section in the region of the fourth ventricle. C1, first cervical nerve; X, vagus nerve; IX, glossopharyngeal nerve.

are recruited. In addition, expiration becomes an active process through the use, most notably, of the muscles of the abdominal wall. The neural basis of these breathing patterns depends on the generation and subsequent tailoring of cyclic changes in the activity of cells primarily located in the medulla oblongata in the brain.

Two Major Cell Groups in the Medulla Oblongata Control the Basic Breathing Rhythm

The central pattern for the basic breathing rhythm has been localized to fairly discrete areas in the medulla oblongata that discharge action potentials in a phasic pattern with respiration. Cells in the medulla oblongata associated with breathing have been identified by noting the correlation between their activity and mechanical events of the breathing cycle. Two different groups of cells have been found, and their anatomic locations are shown in Figure 22.2. The **dorsal respiratory group** (DRG), named for its dorsal location in the region of the nucleus tractus solitarii, predominantly contains cells that are active during inspiration. The **ventral respiratory group** (VRG) is a column of cells in the general region of the nucleus ambiguus that extends caudally nearly to the bulbospinal border and cranially nearly to the bulbopontine junction. The VRG contains both inspiration- and expiration-related neurons. Both groups contain cells projecting ultimately to the bulbospinal motor neuron pools. The DRG and VRG are bilaterally paired, but cross-communication enables them to respond in synchrony; as a consequence, respiratory movements are symmetric.

The neural networks forming the **central pattern generator for breathing** are contained within the DRG/VRG complex, but the exact anatomic and functional description remains uncertain. Central pattern generation probably does not arise from a single pacemaker or by reciprocal inhibition of two pools of cells, one having inspiratory- and the other expiratory-related activity. Instead, the progressive rise and abrupt fall of inspiratory motor activity associated with each breath can be modeled by the starting, stopping, and resetting of an integrator of background ventilatory drive. An integrator-based theoretical model, as described below, is suitable for a first understanding of respiratory pattern generation.

Integrator Neurons Synchronize the Onset of Inspiration

Many different signals (e.g., volition, anxiety, musculoskeletal movements, pain, chemosensor activity, and hypothalamic temperature) provide a background ventilatory drive to the medulla. Inspiration begins by the abrupt release from inhibition of a group of cells, **central inspiratory activity (CIA) integrator** neurons, located within the medullary reticular formation, that integrate this background drive (see Fig. 22.3). Integration results in a progressive rise in the output of the integrator neurons, which, in turn, excites a similar rise in activity of inspiratory premotor neurons of the DRG/VRG complex. The rate of rising activity of inspiratory neurons and, therefore, the rate of inspiration itself, can be influenced by changing the characteristics of the CIA integrator. Inspiration is ended by abruptly switching off the rising excitation of inspiratory neurons. The CIA integrator is reset before the beginning of each inspiration, so that activity of the inspiratory neurons begins each breath from a low level.

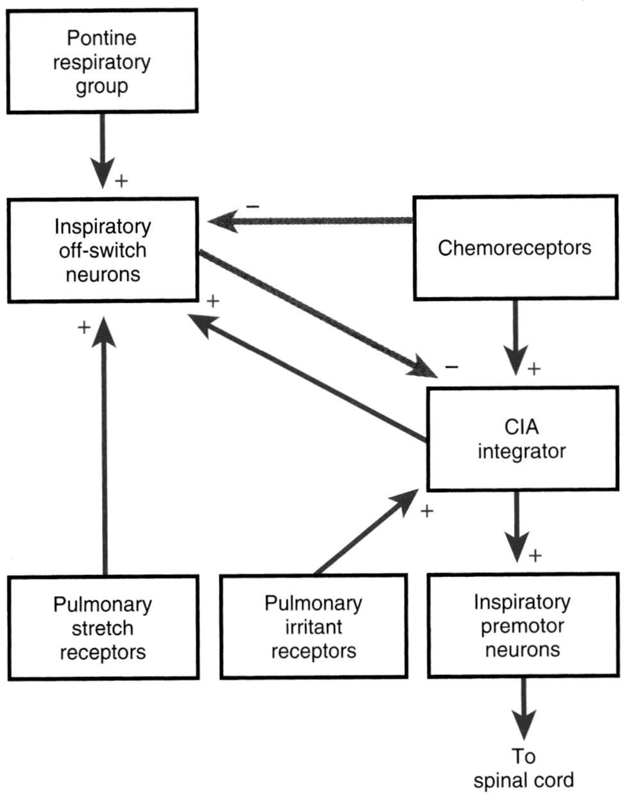

FIGURE 22.3 **The medullary inspiratory pattern generator.** CIA, central inspiratory activity.

Inspiratory Activity Is Switched Off to Initiate Expiration

Two groups of neurons, probably located within the VRG, seem to serve as an inspiratory off-switch (see Fig. 22.3). Switching occurs abruptly when the sum of excitatory inputs to the off-switch reaches a threshold. Adjustment of the threshold level is one of the ways in which depth of breathing can be varied. Two important excitatory inputs to the off-switch are a progressively increasing activity from the CIA integrator's rising output and an input from lung stretch receptors, whose afferent activity increases progressively with rising lung volume. (The first of these is what allows the medulla to generate a breathing pattern on its own; the second is one of many reflexes that influence breathing.) Once the critical threshold is reached, off-switch neurons apply a powerful inhibition to the CIA integrator. The CIA integrator is thus reset by its own rising activity. Other inputs, both excitatory and inhibitory, act on the off-switch and change its threshold. For example, chemical stimuli, such as hypoxemia and hypercapnia, are inhibitory, raising the threshold and causing larger tidal volumes.

An important excitatory input to the off-switch comes from a group of spatially dispersed neurons in the rostral pons called the **pontine respiratory group.** Electrical stimulation in this region causes variable effects on breathing, dependent not only on the site of stimulation but also on the phase of the respiratory cycle in which the stimulus is applied. It is believed that the pontine respiratory group may serve to integrate many different autonomic functions in addition to breathing.

Expiration Is Divided Into Two Phases

Shortly after the abrupt termination of inspiration, some activity of inspiratory muscles resumes. This activity serves to control expiratory airflow. This effect is greatest early in expiration and recedes as lung volume falls. Inspiratory muscle activity is essentially absent in the second phase of expiration, which includes continued passive recoil during quiet breathing or activation of expiratory muscles if more than quiet breathing is required.

The duration of expiration is determined by the intensity of inhibition of activity of inspiratory-related cells of the DRG/VRG complex. Inhibition is greatest at the start of expiration and falls progressively until it is insufficient to prevent the onset of inspiration. The progressive fall of inhibition amounts to a decline of threshold for initiating the switch from expiration to inspiration. The rate of decline of inhibition and the occurrence of events that trigger the onset of inspiration are subject to several influences. The duration of expiration can be controlled not only by neural information arriving during expiration but also in response to the pattern of the preceding inspiration. How the details of the preceding inspiration are stored and later recovered is unresolved.

Various Control Mechanisms Adjust Breathing to Meet Metabolic Demands

The basic pattern of breathing generated in the medulla is extensively modified by several control mechanisms. Multiple controls provide a greater capability for regulating breathing under a larger number of conditions. Their interactions modify each other and provide for backup in case of failure. The set of strategies for controlling a given variable, such as minute ventilation, typically includes individual schemes that differ in several respects, including choices of sensors and effectors, magnitudes of effects, speeds of action, and optimum operating points.

The use of multiple control mechanisms in breathing can be illustrated by considering some of the ways breathing changes in response to exercise. Perhaps the simplest strategies are feedforward mechanisms, in which breathing responds to some component of exercise but without recognition of how well the response meets the demand. One such mechanism would be for the central nervous system (CNS) to vary the activity of the medullary pattern generator in parallel with, and in proportion to, the excitation of the muscles used during exercise. Another prospective feedforward scheme involves sensing the magnitude of the carbon dioxide load delivered to the lungs by systemic venous return and then driving ventilation in response to the magnitude of that load. Experimental evidence supports this mechanism, but the identity of the required intrapulmonary sensor remains uncertain. Still another recognized feedforward mechanism is the enhancement of breathing in response to increased receptor activity in skeletal joints as joint motion increases with exercise.

Although feedforward methods bring about changes in the appropriate direction, they do not provide control in response to the difference between desired and prevailing conditions, as can be done with feedback control. For example, if $PaCO_2$ deviates from a reference point, say 40 mm Hg, ventilation could be adjusted by feedback control to reduce the discrepancy. This well-known control system, diagrammed according to the principles given in Chapter 1, is shown in Figure 22.4. Unlike feedforward control, feedback control requires a sensor, a reference (set point), and a comparator that together generate an error signal, which drives the effector. Negative-feedback systems provide good control in the presence of considerable variations of other properties of the system, such as lung stiffness or respiratory muscle strength. They can, if sufficiently sensitive, act quickly to reduce discrepancies from reference points to very low levels. Too much sensitivity, however, may lead to instability and undesirable excursions of the regulated variable.

Other mechanisms involve minimization or optimization. For example, evidence indicates that rate and depth of breathing are adjusted to minimize the work expenditure for ventilation of a given magnitude. In other words, the controller decides whether to use a large breath with its attendant large elastic load or more frequent smaller breaths with their associated higher resistive load. This strategy requires afferent neural information about lung volume, rate of volume change, and transpulmonary pressures, which can be provided by lung and chest wall mechanoreceptors. During exercise, such a controller would act in concert with, among other things, the feedback control of carbon dioxide described earlier. As a final example, an optimization model using two pieces of information is illustrated in Figure 22.5. Breathing is adjusted to minimize the sum of the muscle effort and the sensory "cost" of tolerating a raised $PaCO_2$.

Muscles of the Upper Airways Are Also Under Phasic Control

The same rhythm generator that controls the chest wall muscles also controls muscles of the nose, pharynx, and larynx. But unlike the inspiratory ramp-like rise of the stimulation of chest wall muscles, the excitation of upper airway muscles quickly reaches a plateau and is sustained until inspiration is ended. Flattening of the expected ramp excitation waveform probably results from progressive inhibition by the rising afferent activity of airway stretch reflexes as lung volume increases. Excitation during inspiration causes contractions of upper airway muscles, airway widening, and reduced resistance from the nostrils to the larynx.

During the first phase of quiet expiration, when expiration is slowed by renewed inspiratory muscle activity, there is also expiratory braking caused by active adduction of the vocal cords. However, during exercise-induced hyperpnea (increased depth and rate of breathing), the cords are separated during expiration and expiratory resistance is reduced.

REFLEXES FROM THE LUNGS AND CHEST WALL

Reflexes arising from the periphery provide feedback for fine-tuning, which adjusts frequency and tidal volume to minimize the work of breathing. Reflexes from the upper airways and lungs also act as defensive reflexes, protecting the lungs from injury and environmental insults. This section considers reflexes that arise from the lungs and chest wall. Among reflexes influencing breathing, the lung and chest wall mechanoreceptors and the chemoreflexes responding to blood pH and gas tension changes are the most widely recognized. Although many other less well-explored reflexes also influence breathing, most are not covered in this chapter. Examples are reflexes induced by changes in arterial blood pressure, cardiac stretch, epicardial irritation, sensations in the airway above the trachea, skin injury, and visceral pain.

Three Classes of Receptors Are Associated With Lung Reflexes

Pulmonary receptors can be divided into three groups: slowly adapting receptors, rapidly adapting receptors, and C fiber endings. Afferent fibers of all three types lie predominantly in the vagus nerves, although some pass with the sympathetic nerves to the spinal cord. The role of the sympathetic afferents is uncertain and is not considered further.

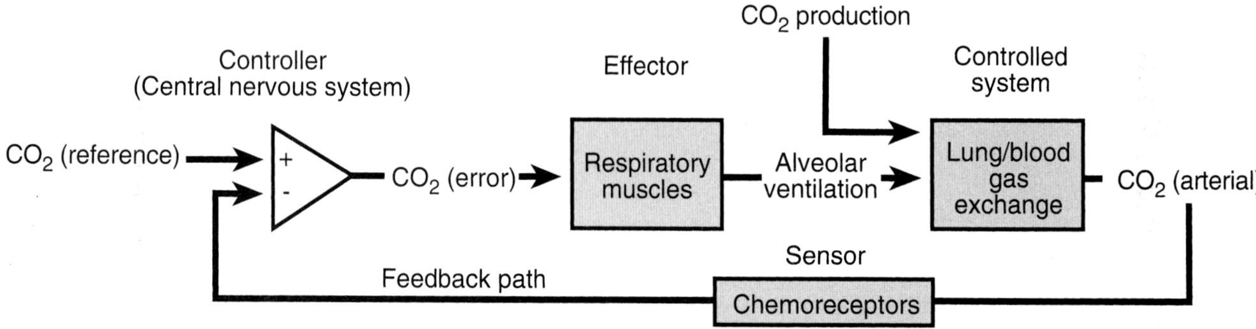

FIGURE 22.4 Negative-feedback control of arterial CO_2. Variations in CO_2 production lead to changes in arterial CO_2 that are sensed by chemoreceptors. The chemoreceptor signal is subtracted from a reference value. The absolute value of the difference is taken as an input by the CNS and passed on to respiratory muscles as new minute ventilation. The loop is completed as the new ventilation alters blood gas composition through the mechanism of lung-blood gas exchange.

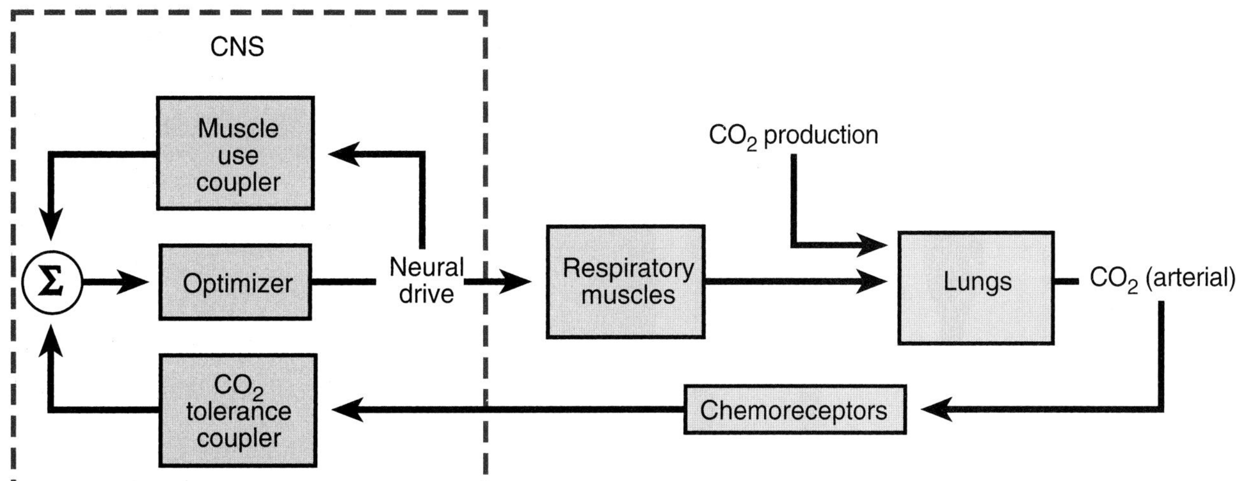

FIGURE 22.5 **An optimization controller.** The components inside the dashed box constitute the controller. In this strategy for breathing, the conflicting needs to maintain chemical homeostasis and to minimize respiratory effort are resolved by selecting an optimal ventilation. The muscle use and CO_2 tolerance couplers convert neural drive and the output of the chemoreceptors to a form interpreted by the neural optimizer as a cost to be minimized. (Modified from Poon CS. Ventilatory control in hypercapnia and exercise: Optimization hypothesis. J Appl Physiol 1987;62:2447–2459.)

Slowly Adapting Receptors. The **slowly adapting receptors** are sensory terminals of myelinated afferent fibers that lie within the smooth muscle layer of conducting airways. Because they respond to airway stretch, they are also called **pulmonary stretch receptors.** Slowly adapting receptors fire in proportion to applied airway transmural pressure, and their usual role is to sense lung volume. When stimulated, an increased firing rate is sustained as long as stretch is imposed; that is, they adapt slowly. Stimulation of these receptors causes an excitation of the inspiratory off-switch and a prolongation of expiration. Because of these two effects, inflating the lungs with a sustained pressure at the mouth terminates an inspiration in progress and prolongs the time before a subsequent inspiration occurs. This sequence is known as the **Hering-Breuer reflex** or **lung inflation reflex**.

The Hering-Breuer reflex probably plays a more important role in infants than in adults. In adults, particularly in the awake state, this reflex may be overwhelmed by more prominent central control. Because increasing lung volume stimulates slowly adapting receptors, which then excite the inspiratory off-switch, it is easy to see how they could be responsible for a feedback signal that results in cyclic breathing. However, as already mentioned, feedback from vagal afferents is not necessary for cyclic breathing to occur. Instead, feedback modifies a basic pattern established in the medulla. The effect may be to shorten inspiration when tidal volume is larger than normal. The most important role of slowly adapting receptors is probably their participation in regulating expiratory time, expiratory muscle activation, and functional residual capacity (FRC). Stimulation of slowly adapting receptors also relaxes airway smooth muscle, reduces systemic vasomotor tone, increases heart rate, and, as previously noted, influences laryngeal muscle activity.

Rapidly Adapting Receptors. The **rapidly adapting receptors** are sensory terminals of myelinated afferent fibers that are found in the larger conducting airways. They are frequently called **irritant receptors** because these nerve endings, which lie in the airway epithelium, respond to irritation of the airways by touch or by noxious substances, such as smoke and dust. Rapidly adapting receptors are stimulated by histamine, serotonin, and prostaglandins released locally in response to allergy and inflammation. They are also stimulated by lung inflation and deflation, but their firing rate rapidly declines when a volume change is sustained. Because of this rapid adaptation, bursts of activity occur that are in proportion to the change of volume and the rate at which that change occurs. Acute congestion of the pulmonary vascular bed also stimulates these receptors but, unlike the effect of inflation, their activity may be sustained when congestion is maintained.

Background activity of rapidly adapting receptors is inversely related to lung compliance, and they are thought to serve as sensors of compliance change. These receptors are probably nearly inactive in normal quiet breathing. Based on what stimulates them, their role would seem to be to sense the onset of pathological events. In spite of considerable information about what stimulates them, the effect of their stimulation remains controversial. As a general rule, stimulation causes excitatory responses such as coughing, gasping, and prolonged inspiration time.

C Fiber Endings. C fiber endings belong to unmyelinated nerves. These nerve endings are classified into two populations in the lungs. One group, **pulmonary C fibers**, is located adjacent to alveoli and is accessible from the pulmonary circulation. They are sometimes called **juxtapulmonary capillary receptors** or **J receptors.** A second group, **bronchial C fibers**, is accessible from the bronchial circulation and, consequently, is located in airways. Like rapidly adapting receptors, both groups play a protective role. They are both stimulated by lung injury, large inflation, acute pulmonary vascular congestion, and certain chemical agents.

Pulmonary C fibers are sensitive to mechanical events (e.g., edema, congestion, and pulmonary embolism), but are not as sensitive to products of inflammation, whereas the opposite is true of bronchial C fibers. Their activity excites breathing, and they probably provide a background excitation to the medulla. When stimulated, they cause rapid shallow breathing, bronchoconstriction, increased airway secretion, and cardiovascular depression (bradycardia, hypotension). **Apnea** (cessation of breathing) and a marked fall in systemic vascular resistance occur when they are stimulated acutely and severely. An abrupt reduction of skeletal muscle tone is an intriguing effect that follows intense stimulation of pulmonary C fibers, the homeostatic significance of which remains unexplained.

Chest Wall Proprioceptors Provide Information About Movement and Muscle Tension

Joint, tendon, and muscle spindle receptors—collectively called **proprioceptors**—may play a role in breathing, particularly when more than quiet breathing is called for or when breathing efforts are opposed by increased airway resistance or reduced lung compliance. Muscle spindles are present in considerable numbers in the intercostal muscles but are rare in the diaphragm. It has been proposed, but not fully verified, that muscle spindles may adjust breathing effort by sensing the discrepancy between tensions of the intrafusal and extrafusal fibers of the intercostal muscles. If a discrepancy exists, information from the spindle receptor alters the contraction of the extrafusal fiber, thereby minimizing the discrepancy. This mechanism provides increased motor excitation when movement is opposed. Evidence also shows that chest wall proprioceptors play a major role in the perception of breathing effort, but other sensory mechanisms may also be involved.

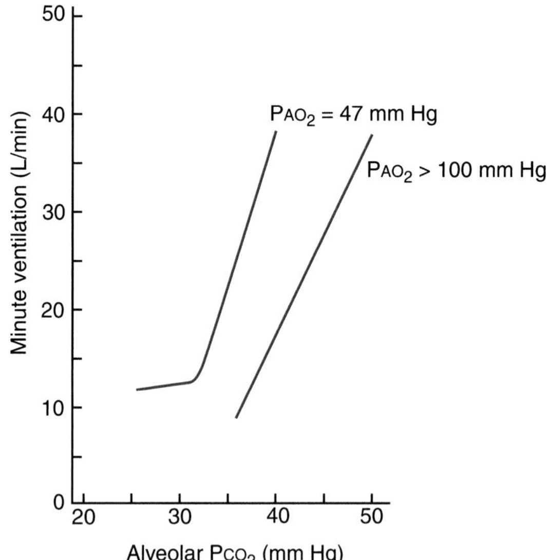

FIGURE 22.6 **Ventilatory responses to increasing alveolar CO_2 tension.** The line on the right represents the response when alveolar P_{O_2} was held at 100 mm Hg or greater to essentially eliminate O_2-dependent activity of the chemoreceptors. The line on the left represents the response when alveolar P_{O_2} was held at 47 mm Hg to provide an overlying hypoxic stimulus. Note that hypoxia increases the slope of the line in addition to changing its location. (Based on Nielsen M, Smith H. Studies on the regulation of respiration in acute hypoxia. Acta Physiol Scand 1952;24:293–313.)

CONTROL OF BREATHING BY H^+, P_{CO_2}, AND P_{O_2}

Breathing is profoundly influenced by the hydrogen ion concentration and respiratory gas composition of the arterial blood. The general rule is that breathing activity is inversely related to arterial blood P_{O_2} but directly related to P_{CO_2} and H^+. Figures 22.6 and 22.7 show the ventilatory responses of a typical person when alveolar P_{CO_2} and P_{O_2} are individually varied by controlling the composition of inspired gas. Responses to carbon dioxide and, to a lesser extent, blood pH depend on sensors in the brainstem and sensors in the carotid arteries and aorta. In contrast, responses to hypoxia are brought about only by the stimulation of arterial receptors.

Neuronal Cells of the Medulla Respond to Local H^+

Ventilatory drive is exquisitely sensitive to P_{CO_2} of blood perfusing the brain. The source of this chemosensitivity has been localized to bilaterally paired groups of cells just below the surface of the ventrolateral medulla immediately caudal to the pontomedullary junction. Each side contains a rostral and a caudal chemosensitive zone, separated by an intermediate zone in which the activities of the caudal and rostral groups converge and are integrated together with other autonomic functions. Exactly which cells exhibit chemosensitivity is unknown, but they are not the same as those of the DRG/VRG complex. Although specific cells have not been identified, the chemosensitive neurons that respond to the H^+ of the surrounding interstitial fluid are referred to as **central chemoreceptors**. The H^+ concentration in the interstitial fluid is a function of P_{CO_2} in the cerebral arterial blood and the bicarbonate concentration of cerebrospinal fluid.

Cerebrospinal Fluid pH Depends on Its Bicarbonate Concentration and P_{CO_2}

Cerebrospinal fluid (CSF) is formed mainly by the **choroid plexuses** of the ventricular cavities of the brain. The epithelium of the choroid plexus provides a barrier between blood and CSF that severely limits the passive movement of large molecules, charged molecules, and inorganic ions. However, choroidal epithelium actively transports several substances, including ions, and this active transport participates in determining the composition of CSF. Cerebrospinal fluid formed by the choroid plexuses is exposed to brain interstitial fluid across the surface of the brain and spinal cord, with the result that the composition of CSF away from the choroid plexuses is closer to that of interstitial fluid than it is to CSF as first formed. Brain interstitial fluid is also separated from blood

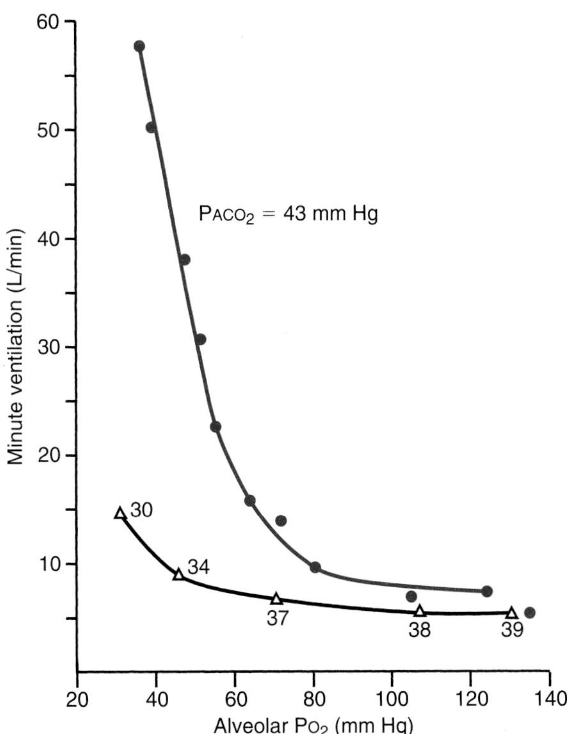

FIGURE 22.7 Ventilatory responses to hypoxia. Inspired oxygen was lowered while Pa_{O_2} was held at 43 mm Hg by adding CO_2 to the inspired air. If this had not been done (lower curve), hypocapnia secondary to the hypoxic hyperventilation would have reduced the ventilatory response. The numbers next to the lower curve are Pa_{CO_2} values measured at each point on the curve. (Based on Loeschke HH, Gertz KH. Einfluss des O_2-Druckes in der Einatmungsluft auf die Atemtätigkeit des Menschen, geprüft unter Konstanthaltung des alveolaren CO_2-Druckes. Pflügers Arch ges Physiol 1958;267:460–477.)

by the **blood-brain barrier** (capillary endothelium), which has its own transport capability.

Because of the properties of the limiting membranes, CSF is essentially protein-free, but it is not just a simple ultrafiltrate of plasma. CSF differs most notably from an ultrafiltrate by its lower bicarbonate and higher sodium and chloride ion concentrations. Potassium, magnesium, and calcium ion concentrations also differ somewhat from plasma; moreover they change little in response to marked changes in plasma concentrations of these cations. Bicarbonate serves as the only significant buffer in CSF, but the mechanism that controls bicarbonate concentration is controversial.

Most proposed regulatory mechanisms invoke the active transport of one or more ionic species by the epithelial and endothelial membranes. Because of the relative impermeabilities of the choroidal epithelium and capillary endothelium to H^+, changes in H^+ concentration of blood are poorly reflected in CSF. By contrast, molecular carbon dioxide diffuses readily; therefore, blood P_{CO_2} can influence the pH of CSF. The pH of CSF is primarily determined by its bicarbonate concentration and P_{CO_2}. The relative ease of movement of molecular carbon dioxide in contrast to hydrogen ions and bicarbonate is depicted in Figure 22.8.

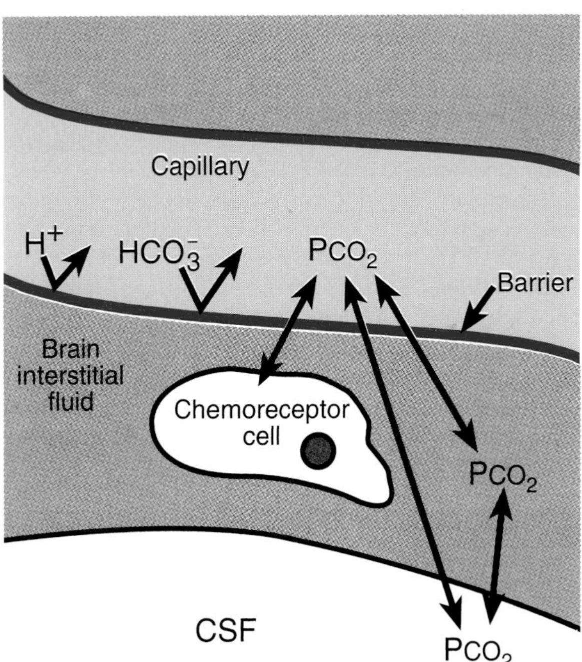

FIGURE 22.8 Movement of H^+, HCO_3^-, and molecular CO_2 between capillary blood, brain interstitial fluid, and CSF. The acid-base status of the chemoreceptors can be quickly changed only by changing Pa_{CO_2}.

In healthy people, the P_{CO_2} of CSF is about 6 mm Hg higher than that of arterial blood, approximating that of brain tissue. The pH of CSF, normally slightly below that of blood, is held within narrow limits. Cerebrospinal fluid pH changes little in states of metabolic acid-base disturbances (see Chapter 25)—about 10% of that in plasma. In respiratory acid-base disturbances, however, the change in pH of the CSF may exceed that of blood. During chronic acid-base disturbances, the bicarbonate concentration of CSF changes in the same direction as in blood, but the changes may be unequal. In metabolic disturbances, the CSF bicarbonate changes are about 40% of those in blood but, with respiratory disturbances, CSF and blood bicarbonate changes are essentially the same. When acute acid-base disturbances are imposed, CSF bicarbonate changes more slowly than does blood bicarbonate, and it may not reach a new steady state for hours or days. As already noted, the mechanism of bicarbonate regulation is unsettled. Irrespective of how it occurs, the bicarbonate regulation that occurs with acid-base disturbances is important because, by changing buffering, it influences the response to a given P_{CO_2}.

Peripheral Chemoreceptors Respond to P_{O_2}, P_{CO_2}, and pH

Peripheral chemoreceptors are located in the carotid and aortic bodies and detect changes in arterial blood P_{O_2}, P_{CO_2}, and pH. **Carotid bodies** are small (~ 2 mm wide) sensory organs located bilaterally near the bifurcations of the common carotid arteries near the base of the skull. Afferent nerves travel to the CNS from the carotid bodies in the glossopharyngeal nerves. **Aortic bodies** are located along the ascending aorta and are innervated by vagal afferents.

As with the medullary chemoreceptors, increasing $PaCO_2$ stimulates peripheral receptors. H^+ formed from H_2CO_3 within the peripheral chemoreceptors (glomus cells) is the stimulus and not molecular CO_2. About 40% of the effect of $PaCO_2$ on ventilation is brought about by peripheral chemoreceptors, while central chemoreceptors bring about the rest. Unlike the central sensor, peripheral chemoreceptors are sensitive to rising arterial blood H^+ and falling PO_2. They alone cause the stimulation of breathing by hypoxia; hypoxia in the brain has little effect on breathing unless severe, at which point breathing is depressed.

Carotid chemoreceptors play a more prominent role than aortic chemoreceptors; because of this and their greater accessibility, they have been studied in greater detail. The discharge rate of carotid chemoreceptors (and the resulting minute ventilation) is approximately linearly related to $PaCO_2$. The linear behavior of the receptor is reflected in the linear ventilatory response to carbon dioxide illustrated in Figure 22.6. When expressed using pH, the response curve is no longer linear but shows a progressively increasing effect as pH falls below normal. This occurs because pH is a logarithmic function of $[H^+]$, so the absolute change in $[H^+]$ per unit change in pH is greater when brought about at a lower pH.

The response of peripheral chemoreceptors to oxygen depends on arterial PaO_2, and *not* oxygen content. Therefore, anemia or carbon monoxide poisoning, two conditions that exhibit reduced oxygen content but have normal PaO_2, have little effect on the response curve. The shape of the response curve is not linear; instead, hypoxia is of increasing effectiveness as PO_2 falls below about 90 mm Hg. The behavior of the receptors is reflected in the ventilatory response to hypoxia illustrated in Figure 22.7. The shape of the curve relating ventilatory response to PO_2 resembles that of the oxyhemoglobin equilibrium curve when plotted upside down (see Chapter 21). As a result, the ventilatory response is inversely related in an approximately linear fashion to arterial blood oxygen saturation.

The nonlinearities of the ventilatory responses to PO_2 and pH, and the relatively low sensitivity across the normal ranges of these variables, cause ventilatory changes to be apparent only when PO_2 and pH deviate significantly from the normal range, especially toward hypoxemia or acidemia. By contrast, ventilation is sensitive to PCO_2 within the normal range, and carbon dioxide is normally the dominant chemical regulator of breathing through the use of both central and peripheral chemoreceptors (compare Figs. 22.6 and 22.7).

There is a strong interaction among stimuli, which causes the slope of the carbon dioxide response curve to increase if determined under hypoxic conditions (see Fig. 22.6), causing the response to hypoxia to be directly related to the prevailing PCO_2 and pH (see Fig. 22.7). As discussed in the next section, these interactions, and interaction with the effects of the central carbon dioxide sensor, profoundly influence the integrated chemoresponses to a primary change in arterial blood composition.

Carotid and aortic bodies also can be strongly stimulated by certain chemicals, particularly cyanide ion and other poisons of the metabolic respiratory chain. Changes in blood pressure have only a small effect on chemoreceptor activity, but responses can be stimulated if arterial pressure falls below about 60 mm Hg. This effect is more prominent in aortic bodies than in carotid bodies. Afferent activity of peripheral chemoreceptors is under some degree of efferent control capable of influencing responses by mechanisms that are not clear. Afferent activity from the chemoreceptors is also centrally modified in its effects by interactions with other reflexes, such as the lung stretch reflex and the systemic arterial baroreflex (see Chapter 18). Although the breathing interactions are not well understood in humans, they serve as examples of the complex interactions of cardiorespiratory regulation. Interactions among chemoreflexes, however, are easily demonstrated.

Significant Interactions Occur Among the Chemoresponses

The effect of PO_2 on the response to carbon dioxide and the effect of carbon dioxide on the response to PO_2 have already been noted. By virtue of this interdependence, a response to hypoxia is blunted by the subsequent increased ventilation, unless $PaCO_2$ is somehow maintained, because $PaCO_2$ ordinarily falls as ventilation is stimulated (see Fig. 22.7). The stimulating effect of hypoxia is blunted mainly by the central chemoreceptors, which respond more potently than the peripheral receptors to low $PaCO_2$.

The sequence of events in the response to hypoxia (e.g., ascent to high altitude) exemplifies interactions among chemoresponses. For example, if 100% oxygen is given to an individual newly arrived at high altitude, ventilation is quickly restored to its sea level value. During the next few days, ventilation in the absence of supplemental oxygen progressively rises further, but it is no longer restored to sea level value by breathing oxygen. Rising ventilation while acclimatizing to altitude could be explained by a reduction of blood and CSF bicarbonate concentrations. This would reduce the initial increase in pH created by the increased ventilation, and allow the hypoxic stimulation to be less strongly opposed. However, this mechanism is not the full explanation of altitude acclimatization. Cerebrospinal fluid pH is not fully restored to normal, and the increasing ventilation raises PaO_2 while further lowering $PaCO_2$, changes that should inhibit the stimulus to breathe. In spite of much inquiry, the reason for persistent hyperventilation in altitude-acclimatized subjects, the full explanation for altitude acclimatization, and the explanation for the failure of increased ventilation in acclimatized subjects to be relieved promptly by restoring a normal PaO_2 are still unknown.

Metabolic acidosis is caused by an accumulation of nonvolatile acids. The increase in blood $[H^+]$ initiates and sustains hyperventilation by stimulating the peripheral chemoreceptors. Because of the restricted movement of H^+ into CSF, the fall in blood pH cannot directly stimulate the central chemoreceptors. The central effect of the hyperventilation, brought about by decreased pH via the peripheral chemoreceptors, results in a paradoxical rise of CSF pH (i.e., an alkalosis as a result of reduced $PaCO_2$) that actually restrains the hyperventilation. With time, CSF bicarbonate concentration is adjusted downward, although it changes less than does that of blood, and the pH of CSF remains somewhat higher than blood pH. Ultimately, ventilation increases more than it did initially as the paradoxical CSF alkalosis is removed.

Respiratory acidosis (accumulation of carbon dioxide) is rarely a result of elevated environmental CO_2, although this occurs in submarine mishaps, while exploring wet limestone caves, and in physiology laboratories where responses to carbon dioxide are measured. Under these conditions, the response is a vigorous increase in minute ventilation proportional to the Pa_{CO_2}; Pa_{O_2} actually rises slightly and arterial pH falls slightly, but these have relatively little effect. If mild hypercapnia can be sustained for a few days, the intense hyperventilation subsides, probably as CSF bicarbonate is raised. More commonly, respiratory acidosis results from failure of the controller to respond to carbon dioxide (e.g., during anesthesia, following brain injury, and in some patients with chronic obstructive lung disease). Another cause of respiratory acidosis is a failure of the breathing apparatus to provide adequate ventilation at an acceptable effort, as may be the case in some patients with obstructive lung disease. When these subjects breathe room air, hypercapnia caused by reduced alveolar ventilation is accompanied by significant hypoxia and acidosis. If the hypoxic component alone is corrected—for example, by breathing oxygen-enriched air—a significant reduction in the ventilatory stimulus may result in greater underventilation, causing further hypercapnia and more severe acidosis. A more appropriate treatment is providing mechanical assistance for restoring adequate ventilation.

THE CONTROL OF BREATHING DURING SLEEP

We spend about one third of our lives asleep. Sleep disorders and disordered breathing during sleep are common and often have physiological consequences (see Clinical Focus Box 22.1). Chapter 7 described the two different neurophysiological sleep states: rapid eye movement (REM) sleep and slow-wave sleep. Sleep is a condition that results from withdrawal of the wakefulness stimulus that arises from the brainstem reticular formation. This wakefulness stimulus is one component of the tonic excitation of brainstem respiratory neurons, and one would predict correctly that sleep results in a general depression of breathing. There are, however, other changes, and the effects of REM and slow-wave sleep on breathing differ.

Sleep Changes the Breathing Pattern

During slow-wave sleep, breathing frequency and inspiratory flow rate are reduced, and minute ventilation falls. These responses partially reflect the reduced physical activity that accompanies sleep. However, because of the small rise in Pa_{CO_2} (about 3 mm Hg), there must also be a change in either the sensitivity or the set point of the carbon dioxide controller. In the deepest stage of slow-wave sleep (stage 4), breathing is slow, deep, and regular. But in stages 1 and 2, the depth of breathing sometimes varies periodically. The explanation is that during light sleep, withdrawal of the wakefulness stimulus varies over time in a periodic fashion. When the stimulus is removed, sleep is deepened and breathing is depressed; when returned, breathing is excited not only by the wakefulness stimulus but also by the carbon dioxide retained during the interval of sleep. This periodic pattern of breathing is known as **Cheyne-Stokes breathing** (Fig. 22.9).

In REM sleep, breathing frequency varies erratically while tidal volume varies little. The net effect on alveolar ventilation is probably a slight reduction, but this is achieved by averaging intervals of frank tachypnea (excessively rapid breathing) with intervals of apnea. Unlike slow-wave sleep, the variations during REM sleep do not reflect a changing wakefulness stimulus but instead represent responses to increased CNS activity of behavioral, rather than autonomic or metabolic, control systems.

CLINICAL FOCUS BOX 22.1

Sleep Apnea Syndrome

The analysis of multiple physiological variables recorded during sleep, known as **polysomnography**, is an important method for research into the control of breathing that has had increasing use in clinical evaluations of sleep disturbances. In normal sleep, reduced dilatory upper-airway muscle tone may be accompanied with brief intervals with no breathing movements. Some people, typically overweight and predominantly men, exhibit more severe disruption of breathing, referred to as **sleep apnea syndrome.** Sleep apnea is classified into two broad groups: obstructive and central.

In **central sleep apnea**, breathing movements cease for a longer than normal interval. In **obstructive sleep apnea**, the fault seems to lie in a failure of the pharyngeal muscles to open the airway during inspiration. This may be the result of decreased muscle activity, but the obstruction is worsened by an excessive amount of neck fat with which the muscles must contend. With obstructive sleep apnea, progressively larger inspiratory efforts eventually overcome the obstruction and airflow is temporarily resumed, usually accompanied by loud snoring.

Some patients exhibit both central and obstructive sleep apneas. In both types, hypoxemia and hypercapnia develop progressively during the apnea intervals. Frequent episodes of repeated hypoxia may lead to pulmonary and systemic hypertension and to myocardial distress; the accompanying hypercapnia is thought to be a cause of the morning headache these patients often experience. There may be partial arousal at the end of the periods of apnea, leading to disrupted sleep and resulting in drowsiness during the day. Daytime sleepiness, often leading to dangerous situations, is probably the most common and most debilitating symptom. The cause of this disorder is multivariate and often obscure, but mechanically assisted ventilation during sleep often results in significant symptomatic improvement.

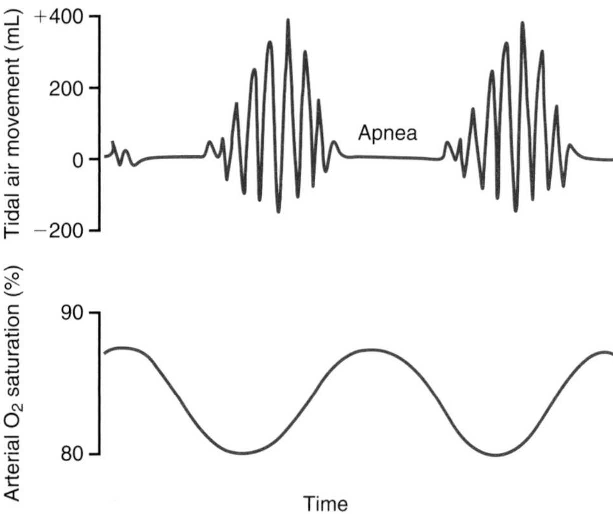

FIGURE 22.9 Cheyne-Stokes breathing and its effect on arterial O_2 saturation. Cheyne-Stokes breathing occurs frequently during sleep, especially in subjects at high altitude, as in this example. In the presence of preexisting hypoxemia secondary to high altitude or other causes, the periods of apnea may result in further falls of O_2 saturation to dangerous levels. Falling PO_2 and rising PCO_2 during the apnea intervals ultimately induce a response and breathing returns, reducing the stimuli and leading to a new period of apnea.

Sleep Changes the Responses to Respiratory Stimuli

Responsiveness to carbon dioxide is reduced during sleep. In slow-wave sleep, the reduction in sensitivity seems to be secondary to a reduction in the wakefulness stimulus and its tonic excitation of the brainstem rather than to a suppression of the chemosensory mechanisms. It is important to note that breathing remains responsive to carbon dioxide during slow-wave sleep, although at a less sensitive level, and that carbon dioxide stimulus may provide the major background brainstem excitation in the absence of the wakefulness stimulus or behavioral excitation. Hence, pathological alterations in the carbon dioxide chemosensory system may profoundly depress breathing during slow-wave sleep.

During intervals of REM sleep in which there is little sign of increased activity, the breathing response to carbon dioxide is slightly reduced, as in slow-wave sleep. However, during intervals of increased activity, responses to carbon dioxide during REM sleep are significantly reduced, and breathing seems to be regulated by the brain's behavioral control system. It is interesting that regulation of breathing during REM sleep by the behavioral control system, rather than by carbon dioxide, is similar to the way breathing is controlled during speech.

Ventilatory responses to hypoxia are probably reduced during both slow-wave and REM sleep, especially in individuals who have high sensitivity to hypoxia while awake. There does not seem to be a difference between the effects of slow-wave and REM sleep on hypoxic responsiveness, and the irregular breathing of REM sleep is unaffected by hypoxia.

Both slow-wave and REM sleep cause an important change in responses to airway irritation. Specifically, a stimulus that causes cough, tachypnea, and airway constriction during wakefulness will cause apnea and airway dilation during sleep unless the stimulus is sufficiently intense to cause arousal. The lung stretch reflex appears to be unchanged or somewhat enhanced during arousal from sleep, but the effect of stretch receptors on upper airways during sleep may be important.

Arousal Mechanisms Protect the Sleeper

Several stimuli cause arousal from sleep; less intense stimuli cause a shift to a lighter sleep stage without frank arousal. In general, arousal from REM sleep is more difficult than from slow-wave sleep. In humans, hypercapnia is a more potent arousal stimulus than hypoxia, the former requiring a $PaCO_2$ of about 55 mm Hg and the latter requiring a PaO_2 less than 40 mm Hg. Airway irritation and airway occlusion induce arousal readily in slow-wave sleep but much less readily during REM sleep.

All of these arousal mechanisms probably operate through the activation of a reticular arousal mechanism similar to the wakefulness stimulus. They play an important role in protecting the sleeper from airway obstruction, alveolar hypoventilation of any cause, and the entrance into the airways of irritating substances. Recall that coughing depends on the aroused state and without arousal airway irritation leads to apnea. Obviously, wakefulness altered by other than natural sleep—such as during drug-induced sleep, brain injury, or anesthesia—leaves the individual exposed to risk because arousal from those states is impaired or blocked. From a teleological point of view, the most important role of sensors of the respiratory system may be to cause arousal from sleep.

Upper Airway Tone May Be Compromised During Sleep

A prominent feature during REM sleep is a general reduction in skeletal muscle tone. Muscles of the larynx, pharynx, and tongue share in this relaxation, which can lead to obstruction of the upper airways. Airway muscle relaxation may be enhanced by the increased effectiveness of the lung inflation reflex.

A common consequence of airway narrowing during sleep is snoring. In many people, usually men, the degree of obstruction may at times be sufficient to cause essentially complete occlusion. In these people, an intact arousal mechanism prevents suffocation, and this sequence is not in itself unusual or abnormal. In some people, obstruction is more complete and more frequent, and the arousal threshold may be raised. Repeated obstruction leads to significant hypercapnia and hypoxemia, and repeated arousals cause sleep deprivation that leads to excessive daytime sleepiness, often interfering with normal daily activity.

THE RESPONSE TO HIGH ALTITUDE

Changes in activity and the environment initiate integrated ventilatory responses that involve changes in the cardiopulmonary system. Examples include the response to exercise (see Chapter 30) and the response to the low inspired oxygen tension at high altitudes. The importance of understanding integrated ventilatory responses is that similar interactions occur under pathophysiological conditions in patients with respiratory illnesses.

How the body responds to high altitude has fascinated physiologists for centuries. The French physiologist Paul Bert first recognized that the harmful effects of high altitude are caused by low oxygen tension. Recall from Chapter 21 that the percentage of oxygen does not change at high altitude but the barometric pressure decreases (see Fig. 21.1). So the hypoxic response at high altitude is caused by a decrease in inspired oxygen tension (P_{IO_2}). At high altitude, when the P_{IO_2} decreases and oxygen supply in the body is threatened, several compensations are made in an effort to deliver normal amounts of oxygen to the tissues. Chief among these responses to altitude is hyperventilation. Figure 22.7 shows that hypoxia-induced hyperventilation is not significantly increased until the alveolar P_{O_2} decreases below 60 mm Hg. In a healthy adult, a drop in alveolar P_{O_2} to 60 mm Hg occurs at an altitude of approximately 4,500 m (14,000 feet).

Figure 22.10 shows how ventilation and alveolar P_{CO_2} change with hypoxia. The hypoxia-induced hyperventilation appears in two stages. First, there is an immediate increase in ventilation, which is primarily a result of hypoxia-induced stimulation via the carotid bodies. However, the increase in ventilation seen in the first stage is small compared with the second stage, in which ventilation continues to rise slowly over the next 8 hours. After 8 hours of hypoxia, minute ventilation is sustained. The reason for the small rise in ventilation seen in the first stage is that the hypoxic stimulation is strongly opposed by the decrease in arterial P_{CO_2} as a result of excess carbon dioxide blown off with altitude-induced hyperventilation. The hypoxia-induced hyperventilation results in an increase in arterial pH. The decrease in arterial P_{CO_2} (hypocapnia) and the rise in blood pH work in concert to blunt the hypoxic drive.

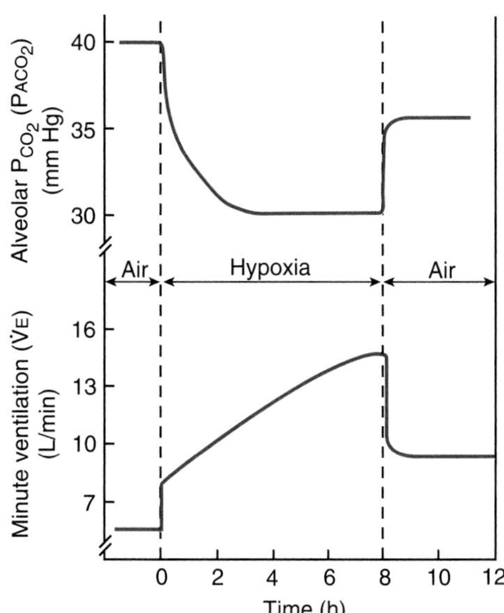

FIGURE 22.10 **Effect of hypoxia on minute ventilation and alveolar P_{CO_2}.** Hypoxia was induced by having a healthy subject breath 12% O_2 for 8 hours. With hypoxia-induced hyperventilation, excess CO_2 is blown off, resulting in a decrease in alveolar P_{CO_2}. Minute ventilation remains elevated for a while after the subject returns to room air.

Ventilatory Acclimatization Results in a Sustained Increase in Ventilation

The increased ventilation seen in the second stage is referred to as **ventilatory acclimatization**. Acclimatization occurs during prolonged exposure to hypoxia and is a physiological response, as opposed to a genetic or evolutionary change over generations leading to a permanent adaptation. Ventilatory acclimatization is defined as a time-dependent increase in ventilation that occurs over hours to days of continuous exposure to hypoxia. After 2 weeks, the hypoxia-induced hyperventilation reaches a stable plateau.

Although the physiological mechanisms responsible for ventilatory acclimatization are not completely understood, it is clear that two mechanisms are involved. One involves the chemoreceptors, and the second involves the kidneys. CSF pH, which becomes more alkaline when ventilation is stimulated by hypoxia, is brought closer to normal by the movement of bicarbonate out of the CSF. Also, during prolonged hypoxia, the carotid bodies increase their sensitivity to arterial P_{O_2}. These changes result in a further increase in ventilation.

The second mechanism responsible for ventilatory acclimatization involves the kidneys. The alkaline blood pH resulting from the hypoxia-induced hyperventilation is antagonistic to the hypoxic drive. Blood pH is regulated by both the lungs and the kidneys (see Chapter 25). The kidneys compensate by excreting more bicarbonate, which lowers the blood pH towards normal over 2 to 3 days; therefore, the antagonistic effect resulting from the hyperventilation-induced alkaline pH is minimized, allowing the hypoxic drive to increase minute ventilation further.

Cardiovascular Acclimatization Improves the Delivery of Oxygen to the Tissues

In addition to ventilatory acclimatization, the body undergoes other physiological changes to acclimatize to low oxygen levels. These include increased pulmonary blood flow, increased red cell production, and improved oxygen and carbon dioxide transport. There is an increase in cardiac output at high altitude resulting in increased blood flow to the lungs and other organs of the body. The increase in pulmonary blood flow reduces capillary transit time and results in an increase in oxygen uptake by the lungs. Low P_{O_2} causes vasodilation in the systemic circulation. The increase in blood flow resulting from the combined increased vasodilation and increased cardiac output sustains oxygen delivery to the tissues at high altitude.

Red cell production is also increased at high altitude, which improves oxygen delivery to the tissues. Hypoxia stimulates the kidneys to produce and release **erythropoietin**, a hormone that stimulates the bone marrow to produce erythrocytes, which are released into the circulation.

The increased hematocrit resulting from the hypoxia-induced **polycythemia** enables the blood to carry more oxygen to the tissues. However, the increased viscosity, as a result of the elevated hematocrit, increases the workload on the heart. In some cases, the polycythemia becomes so severe (hematocrit > 70%) at high altitude that blood has to be withdrawn periodically to permit the heart to pump effectively. Oxygen delivery to the cells is also favored by an increased concentration of 2,3-DPG in the red cells, which shifts the oxyhemoglobin equilibrium curve to the right, and favors the unloading of oxygen in the tissues (see Chapter 21).

Although the body undergoes many beneficial changes that allow acclimatization to high altitude, there are some undesirable effects. One of these is **pulmonary hypertension** (abnormally high pulmonary arterial blood pressure). Alveolar hypoxia causes pulmonary vasoconstriction. In addition, prolonged hypoxia causes vascular remodeling in which pulmonary arterial smooth muscle cells undergo hypertrophy and hyperplasia. The vascular remodeling results in narrowing of the small pulmonary arteries and leads to a significant increase in pulmonary vascular resistance and hypertension. With severe hypoxia, the pulmonary veins are also constricted. The increase in venous pressure elevates the filtration pressure in the alveolar capillary beds, leading to pulmonary edema. Pulmonary hypertension also increases the workload of the right heart, causing right heart hypertrophy, which, if severe enough, may lead to death.

REVIEW QUESTIONS

DIRECTIONS: Each of the numbered items or incomplete statements in this section is followed by answers or by completions of the statement. Select the ONE lettered answer or completion that is BEST in each case.

1. Generation of the basic cyclic pattern of breathing in the CNS requires participation of
(A) The pontine respiratory group
(B) Vagal afferent input to the pons
(C) Vagal afferent input to the medulla
(D) An inhibitory loop in the medulla
(E) An intact spinal cord
2. Quiet expiration is associated with
(A) A brief early burst by inspiratory neurons
(B) Active abduction of the vocal cords
(C) An early burst of activity by expiratory muscles
(D) Reciprocal inhibition of inspiratory and expiratory centers
(E) Increased activity of slowly adapting receptors
3. The ventilatory response to hypoxia
(A) Is independent of Pa_{CO_2}
(B) Is more dependent on aortic than carotid chemoreceptors
(C) Is exaggerated by hypoxia of the medullary chemoreceptors
(D) Bears an inverse linear relationship to arterial oxygen content
(E) Is a sensitive mechanism for controlling breathing in the normal range of blood gases
4. Which of the following is *not* a consequence of stimulation of lung C fiber endings?
(A) Bronchoconstriction
(B) Apnea
(C) Rapid shallow breathing
(D) Systemic vasoconstriction
(E) Skeletal muscle relaxation
5. Which of the following is true about cerebrospinal fluid?
(A) Its protein concentration is equal to that of plasma
(B) Its P_{CO_2} equals that of systemic arterial blood
(C) It is freely accessible to blood hydrogen ions
(D) Its composition is essentially that of a plasma ultrafiltrate
(E) Its pH is a function of Pa_{CO_2}
6. Slow-wave sleep is characterized by
(A) A fall in Pa_{CO_2}
(B) A tendency for breathing to vary in a periodic fashion
(C) Facilitation of the cough reflex
(D) Heightened ventilatory responsiveness to hypoxia
(E) Greater skeletal muscle relaxation than REM sleep
7. Which of the following is *not* true during sleep?
(A) Airway irritation evokes apnea
(B) Airway irritation evokes coughing
(C) Airway irritation evokes arousal
(D) Airway occlusion evokes arousal
(E) Hypercapnia evokes arousal
8. Negative-feedback control systems
(A) Would not apply to the regulation of Pa_{CO_2}
(B) Anticipate future events
(C) Give the best control when most sensitive
(D) Are ineffective if the properties of the controlled system change
(E) Are not necessarily stable
9. With regard to the control of minute ventilation by carbon dioxide
(A) About 80% of the effect of Pa_{CO_2} is mediated by the peripheral chemoreceptors
(B) Central effects are mediated by direct effects on cells of the DRG/VRG complex
(C) Sensitivity of the control system is inversely related to the prevailing Pa_{O_2}
(D) This mechanism is less sensitive than control in response to oxygen
(E) Transection of cranial nerves IX and X at the skull would have no effect
10. Which of the following relationships can be represented by a straight line sloping downward from left to right?
(A) Minute ventilation as a function of arterial pH
(B) Minute ventilation as a function of arterial oxygen percent saturation
(C) Carotid chemoreceptor firing frequency as a function of Pa_{CO_2}
(D) Minute ventilation as a function of Pa_{O_2} while Pa_{CO_2} is held constant
(E) Arterial pH as a function of arterial $[H^+]$

SUGGESTED READING

Cotes JE. Lung Function: Assessment and Application in Medicine. 5th Ed. Boston: Blackwell Scientific, 1993.

Haddad GG, Jian C. O_2-sensing mechanisms in excitable cells: Role of plasma membrane K^+ channels. Annu Rev Physiol 1997;59:23–41.

Lumb AB. Nunn's Applied Respiratory Physiology. Oxford, UK: Butterworth-Heinemann, 2000.

Patterson DJ. Potassium and breathing in exercise. Sports Med 1997; 23:149–163.

Schoene RB. Control of breathing at high altitude. Respiration 1997;64:407–415.

Thalhofer S, Dorow P. Central sleep apnea. Respiration 1997;64:2–9.

CASE STUDIES FOR PART V • • •

CASE STUDY FOR CHAPTER 19

Emphysema

A 65-year-old man went to the university hospital emergency department because of a 5-day history of shortness of breath and dyspnea on exertion. He also complained of a cough productive of green sputum. He appeared pale and said he felt feverish at home, but denied any shaking chills, sore throat, nausea, vomiting, or diarrhea. Having smoked two packs of cigarettes a day for the past 30 years, he had recently decreased his habit to one pack a day. He had not been previously hospitalized. He is a retired cab driver and lives with his wife; they have no pets. Although he has had dyspnea upon exertion for the past 2 years, he continues to maintain an active lifestyle. He still mows his lawn without much difficulty, and can walk 1 to 2 miles on a flat surface at a moderate pace. The patient said he rarely drinks alcohol. He denied having had any other significant past medical problems, including heart disease, hypertension, edema, childhood asthma, or any allergies. He did state that his father, also a heavy smoker, died of emphysema at age 55.

An initial exam shows that the patient is thin but has a large chest. He is in moderate respiratory distress. His blood pressure is 130/80 mm Hg; respiratory rate, 28 to 32 breaths/min; heart rate, 92/minute; and oral temperature, 37.9°C. His trachea is midline, and his chest expands symmetrically. He has decreased but audible breath sounds in both lung fields, with expiratory wheezing and a prolonged expiratory phase. Head, eyes, ears, nose, and throat findings are unremarkable. A pulse oximetry reading reveals his blood hemoglobin oxygen saturation is 91% when breathing room air.

Pulmonary function tests reveal severe limitation of airflow rates, particularly expiratory airflow. The patient is diagnosed with pulmonary emphysema.

Questions

1. What are the common spirometry findings associated with emphysema?
2. What are the mechanisms of airflow limitation in emphysema?
3. What is the most commonly held theory explaining the development of emphysema?

Answers to Case Study Questions for Chapter 19

1. The hallmark of emphysema is the limitation of airflow out of the lungs. In emphysema, expiratory flow rates (FVC, FEV_1, and FEV_1/FVC ratio) are significantly decreased. However, some lung volumes (TLC, FRC, and RV) are increased, and the increase is a result of the loss of lung elastic recoil (increased compliance).
2. The mechanisms that limit expiratory airflow in emphysema include hypersensitivity of airway smooth muscle, mucus hypersecretion, and bronchial wall inflammation and increased dynamic airway compression as a result of increased compliance.
3. Many of the pathophysiological changes in emphysema are a result of the loss of lung elastic recoil and destruction of the alveolar-capillary membrane. This is thought to be a result of an imbalance between the proteases and antiproteases (α_1-antitrypsin) in the lower respiratory tree. Normally, proteolytic enzyme activity is inactivated by antiproteases. In emphysema, excess proteolytic activity destroys elastin and collagen, the major extracellular matrix proteins responsible for maintaining the integrity of the alveolar-capillary membrane and the elasticity of the lung. Cigarette smoke increases proteolytic activity, which may arise through an increase in protease levels, a decrease in antiprotease activity, or a combination of the two.

Reference

Hogg JC. Chronic obstructive pulmonary disease: An overview of pathology and pathogenesis. Novartis Found Symp 2001;234:4–26.

CASE STUDY FOR CHAPTER 20

Chest Pain

A 27-year-old accountant recently drove cross-country to start a new job in Denver, Colorado. A week after her move, she started to experience chest pains. She drove to the emergency department after experiencing 24 hours of right-sided chest pain, which was worse with inspiration. She also experienced shortness of breath and stated that she felt warm. She denied any sputum production, hemoptysis, coughing, or wheezing. She is active and walks daily and never has experienced any swelling in her legs. She has never been treated for any respiratory problems and has never undergone any surgical procedures. Her medical history is negative, and she has no known drug allergies. Oral contraceptives are her only medication. She smokes a pack of cigarettes a day and consumes wine occasionally. She does not use intravenous drugs and has no other risk factors for HIV disease. Her family history is negative for asthma and any cardiovascular diseases.

Physical examination reveals a mildly obese woman in moderate respiratory distress. Her respiratory rate is 24 breaths/min and her pulse is 115 beats/min. Her blood pressure is 140/80 mm Hg, and no jugular vein distension is observed. Heart rate and rhythm are regular, with normal heart sounds and no murmurs. Her chest is clear, and her temperature is 38°C. Her extremities show signs of cyanosis, but no clubbing or edema is detected. Blood gases, obtained while she was breathing room air, reveal a P_{O_2} of 60 mm Hg and a P_{CO_2} of 32 mm Hg; her arterial blood pH is 7.49. Her alveolar-arterial $(A\text{-}a)_{O_2}$ gradient is 40 mm Hg. A Gram's stain sputum specimen exhibited a normal flora. A chest X-ray study reveals a normal heart shadow and clear lung fields, except for a small peripheral infiltrate in the left lower lobe. A lung scan reveals an embolus in the left lower lobe.

Case Study Questions

1. What is the cause of a widened alveolar-arterial gradient in patients with pulmonary embolism?
2. What causes the decreased arterial P_{CO_2} and elevated arterial pH?
3. Why do oral contraceptives induce hypercoagulability?

Answers to Case Study Questions for Chapter 20

1. A normal A-a_{O_2} gradient is 5 to 15 mm Hg. A pulmonary embolus will cause blood flow to be shunted to another region of the lung. Because cardiac output is unchanged, the shunting of blood causes overperfusion, which causes an abnormally low $\dot{V}_A/\dot{Q}$ ratio in another region of the lungs. Thus, blood leaving the lungs has a low P_{O_2}, resulting in hy-

poxemia (a low arterial Po_2). The decrease in arterial Po_2 accounts in part for the increase in the A-ao_2 gradient. However, ventilation is also stimulated as a compensatory mechanism to hypoxemia, which leads to hyperventilation with a concomitant increase in alveolar Po_2. The A-ao_2 gradient is, therefore, further increased because of the increased alveolar Po_2 caused by hyperventilation.

2. The decreased Pco_2 and increased pH are the result of hyperventilation as a result of the hypoxic drive (low Po_2) that stimulates ventilation.
3. The mechanisms by which oral contraceptives increase the risk of thrombus formation are not completely understood. The risk appears to be correlated best with the estrogen content of the pills. Hypotheses include increased endothelial cell proliferation, decreased rates of venous blood flow, and increased coagulability secondary to changes in platelets, coagulation factors, and the fibrinolytic system. Furthermore, there are changes in serum lipoprotein levels with an increase in LDL and VLDL and a variable effect on HDL. Driving cross-country, with long sedentary periods, may have exacerbated the patient's condition.

Reference

Cotes JE. Lung Function: Assessment and Application in Medicine. 5th Ed. Boston: Blackwell Scientific, 1993.

Case Study for Chapter 21

Anemia

A 68-year-old widow is seen by her physician because of complaints of fatigue and mild memory loss. The patient does not abuse alcohol and has not had a history of surgery in the past 5 years. Blood gases (Sao_2, Po_2, Pco_2, and pH) are normal. Blood analysis shows a white cell count of 5,200 cells/mm^3; Hb, 9.0 gm/dL; and a hematocrit of 27%. Her serum vitamin B_{12} is low, but her serum folate, thyroid-stimulating hormone (TSH), and liver enzymes are normal. Her peripheral blood smear is unremarkable.

Questions

1. Why are Sao_2 and arterial Po_2 normal in anemic patients who have hypoxemia?
2. How does anemia affect the oxygen diffusing capacity of the lungs?
3. Why might this patient be deficient in vitamin B_{12}?

Answers to Case Study Questions for Chapter 21

1. Hemoglobin increases the oxygen carrying capacity of the blood, but has no effect on arterial Po_2. By way of illustration, if 100 mL of blood are exposed to room air, the Po_2 in the blood will equal atmospheric Po_2 after equilibration. Removing the red cells, leaving only plasma, will not affect Po_2. An otherwise healthy patient with anemia will have a normal Sao_2 because both O_2 content and capacity are reduced proportionately. Hypoxemia in anemic patients is a result of low oxygen content, not a low Po_2.
2. DL_{CO} decreases with anemia because there is less hemoglobin available to bind CO.
3. There are several causes of vitamin B_{12} deficiency. In older individuals, especially those who live alone, insufficient dietary intake of animal protein may be the cause; other causes include loss of gastric mucosa or regional enteritis.

Reference

Wintrobe MM. Clinical Hematology. 9th Ed. Philadelphia: Lea & Febiger, 1993.

Case Study for Chapter 22

Pickwickian Syndrome

A 45-year-old man is referred to the pulmonary function laboratory because of polycythemia (hematocrit of 57%). At the time of referral, he weighs 142 kg (312 pounds), and his height is 175 cm (5 feet, 9 inches). A brief history reveals that he frequently falls asleep during the day. His blood gas values are Pao_2, 69 mm Hg; Sao_2, 94%; Pco_2, 35 mm Hg; and pH, 7.44. A few days later, he is admitted as an outpatient in the hospital's sleep center. He is connected to an ear oximeter and to a portable heart monitor. Within 30 minutes, the patient falls asleep, and within another 30 minutes, his Sao_2 decreases from 92% to 47%, and his heart rate increases from 92 to 108 beats/min, with two premature ventricular contractions. During this time, his chest wall continues to move, but airflow at the mouth and nose is not detected.

Questions

1. How would this patient's test results be interpreted?
2. What is the cause of the polycythemia?
3. How does hypoxia accelerate heart rate?

Answers to Case Study Questions for Chapter 22

1. This patient is suffering from what has been known as pickwickian syndrome, a disorder that occurs with severely obese individuals because of their excessive weight. The pickwickian syndrome was named after Joe, the fat boy who was always falling asleep in Charles Dickens' novel *The Pickwick Papers.* Pickwickian patients suffer from hypoventilation and often suffer from sleep apnea as well. Pickwickian syndrome is no longer an appropriate name because it does not indicate what type of sleep disorder is involved. About 80% of sleep apnea patients are obese, and 20% are of relatively normal weight.
2. Polycythemia is the result of chronic hypoxemia from hypoventilation, as well as from sleep apnea.
3. An increase in sympathetic discharge is often associated with sleep apnea and is responsible for the accelerated heart rate.

Reference

Martin RJ, ed. Cardiorespiratory Disorders During Sleep. 2nd Ed. Mt. Kisco, NY: Futura, 1990.

PART VI Renal Physiology and Body Fluids

Kidney Function

George A. Tanner, Ph.D.

CHAPTER OUTLINE

KEY CONCEPTS

1. The formation of urine involves glomerular filtration, tubular reabsorption, and tubular secretion.
2. The renal clearance of a substance is equal to its rate of excretion divided by its plasma concentration.
3. Inulin clearance provides the most accurate measure of glomerular filtration rate (GFR).
4. The clearance of *p*-aminohippurate (PAH) is equal to the effective renal plasma flow.
5. The rate of net tubular reabsorption of a substance is equal to its filtered load minus its excretion rate. The rate of net tubular secretion of a substance is equal to its excretion rate minus its filtered load.
6. The kidneys, especially the cortex, have a high blood flow.
7. Kidney blood flow is autoregulated; it is also profoundly influenced by nerves and hormones.
8. The glomerular filtrate is an ultrafiltrate of plasma.
9. GFR is determined by the glomerular ultrafiltration coefficient, glomerular capillary hydrostatic pressure, hydrostatic pressure in the space of Bowman's capsule, and glomerular capillary colloid osmotic pressure.
10. The proximal convoluted tubule reabsorbs about 70% of filtered Na^+, K^+, and water and nearly all of the filtered glucose and amino acids. It also secretes a large variety of organic anions and organic cations.
11. The transport of water and most solutes across tubular epithelia is dependent upon active reabsorption of Na^+.
12. The thick ascending limb is a water-impermeable segment that reabsorbs Na^+ via a Na-K-2Cl cotransporter in the apical cell membrane and a vigorous Na^+/K^+-ATPase in the basolateral cell membrane.
13. The distal convoluted tubule epithelium is water-impermeable and reabsorbs Na^+ via a thiazide-sensitive apical membrane Na-Cl cotransporter.
14. Cortical collecting duct principal cells reabsorb Na^+ and secrete K^+.
15. The kidneys save water for the body by producing urine with a total solute concentration (i.e., osmolality) greater than plasma.
16. The loops of Henle are countercurrent multipliers; they set up an osmotic gradient in the kidney medulla. Vasa recta are countercurrent exchangers; they passively help maintain the medullary gradient. Collecting ducts are osmotic equilibrating devices; they have a low water permeability, which is increased by arginine vasopressin (AVP).
17. Genetic defects in kidney epithelial cells account for several disorders.

The kidneys play a dominant role in regulating the composition and volume of the extracellular fluid (ECF). They normally maintain a stable internal environment by excreting appropriate amounts of many substances in the urine. These substances include not only waste products and foreign compounds, but also many useful substances that are present in excess because of eating, drinking, or metabolism. This chapter considers the basic renal processes that determine the excretion of various substances.

The kidneys perform a variety of important functions:

1) They regulate the osmotic pressure (osmolality) of the body fluids by excreting osmotically dilute or concentrated urine.

2) They regulate the concentrations of numerous ions in blood plasma, including Na^+, K^+, Ca^{2+}, Mg^{2+}, Cl^-, bicarbonate (HCO_3^-), phosphate, and sulfate.

3) They play an essential role in acid-base balance by excreting H^+, when there is excess acid, or HCO_3^-, when there is excess base.

4) They regulate the volume of the ECF by controlling Na^+ and water excretion.

5) They help regulate arterial blood pressure by adjusting Na^+ excretion and producing various substances (e.g., renin) that can affect blood pressure.

6) They eliminate the waste products of metabolism, including urea (the main nitrogen-containing end product of protein metabolism in humans), uric acid (an end product of purine metabolism), and creatinine (an end product of muscle metabolism).

7) They remove many drugs (e.g., penicillin) and foreign or toxic compounds.

8) They are the major production sites of certain hormones, including erythropoietin (see Chapter 11) and 1,25-dihydroxy vitamin D_3 (see Chapter 36).

9) They degrade several polypeptide hormones, including insulin, glucagon, and parathyroid hormone.

10) They synthesize ammonia, which plays a role in acid-base balance (see Chapter 25).

11) They synthesize substances that affect renal blood flow and Na^+ excretion, including arachidonic acid derivatives (prostaglandins, thromboxane A_2) and kallikrein (a proteolytic enzyme that results in the production of kinins).

When the kidneys fail, a host of problems ensue. Dialysis and kidney transplantation are commonly used treatments for advanced (end-stage) renal failure (see Clinical Focus Box 23.1).

CLINICAL FOCUS BOX 23.1

Dialysis and Transplantation

Chronic renal failure can result from a large variety of diseases but is most often due to inflammation of the glomeruli (glomerulonephritis) or urinary reflux and infections (pyelonephritis). Renal damage may occur over many years and may be undetected until a considerable loss of nephrons has occurred. When GFR has declined to 5% of normal or less, the internal environment becomes so disturbed that patients usually die within weeks or months if they are not dialyzed or provided with a functioning kidney transplant.

Most of the signs and symptoms of renal failure can be relieved by **dialysis,** the separation of smaller molecules from larger molecules in solution by diffusion of the small molecules through a selectively permeable membrane. Two methods of dialysis are commonly used to treat patients with severe, irreversible ("end-stage") renal failure.

In **continuous ambulatory peritoneal dialysis** (CAPD), the peritoneal membrane, which lines the abdominal cavity, acts as a dialyzing membrane. About 1 to 2 liters of a sterile glucose-salt solution are introduced into the abdominal cavity and small molecules (e.g., K^+ and urea) diffuse into the introduced solution, which is then drained and discarded. The procedure is usually done several times every day.

Hemodialysis is more efficient in terms of rapidly removing wastes. The patient's blood is pumped through an artificial kidney machine. The blood is separated from a balanced salt solution by a cellophane-like membrane, and small molecules can diffuse across this membrane. Excess fluid can be removed by applying pressure to the blood and filtering it. Hemodialysis is usually done 3 times a week (4 to 6 hours per session) in a medical facility or at home.

Dialysis can enable patients with otherwise fatal renal disease to live useful and productive lives. Several physiological and psychological problems persist, however, including bone disease, disorders of nerve function, hypertension, atherosclerotic vascular disease, and disturbances of sexual function. There is a constant risk of infection and, with hemodialysis, clotting and hemorrhage. Dialysis does not maintain normal growth and development in children. Anemia (primarily a result of deficient erythropoietin production by damaged kidneys) was once a problem but can now be treated with recombinant human erythropoietin.

Renal transplantation is the only real cure for patients with end-stage renal failure. It may restore complete health and function. In 1999, about 12,500 kidney transplant operations were performed in the United States. At present, 94% of kidneys grafted from living donors related to the recipient function for 1 year; about 90% of kidneys from unrelated donors (cadaver) function for 1 year.

Several problems complicate kidney transplantation. The immunological rejection of the kidney graft is a major challenge. The powerful drugs used to inhibit graft rejection compromise immune defensive mechanisms so that unusual and difficult-to-treat infections often develop. The limited supply of donor organs is also a major unsolved problem; there are many more patients who would benefit from a kidney transplant than there are donors. The median waiting time for a kidney transplant is currently more than 900 days. Finally, the cost of transplantation (or dialysis) is high. Fortunately for people in the United States, Medicare covers the cost of dialysis and transplantation, but these life-saving therapies are beyond the reach of most people in developing countries.

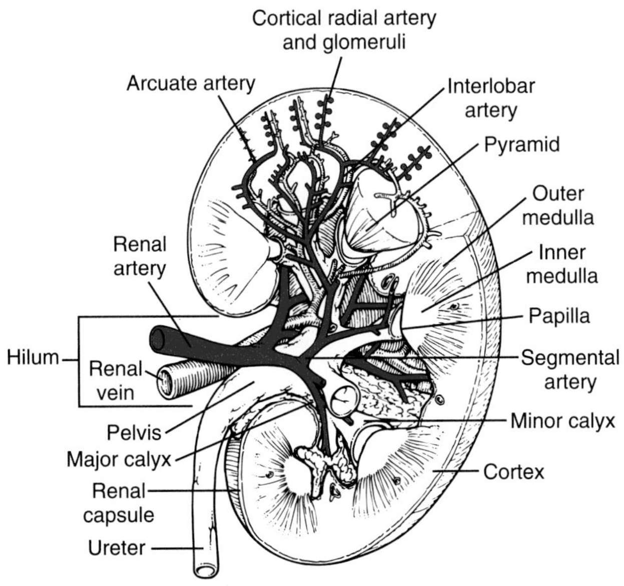

FIGURE 23.1 **The human kidney, sectioned vertically.** (From Smith HW. Principles of Renal Physiology. New York: Oxford University Press, 1956.)

FUNCTIONAL RENAL ANATOMY

Each kidney in an adult weighs about 150 g and is roughly the size of one's fist. If the kidney is sectioned (Fig. 23.1), two regions are seen: an outer part, called the **cortex,** and an inner part, called the **medulla.** The cortex typically is reddish brown and has a granulated appearance. All of the glomeruli, convoluted tubules, and cortical collecting ducts are located in the cortex. The medulla is lighter in color and has a striated appearance that results from the parallel arrangement of the loops of Henle, medullary collecting ducts, and blood vessels of the medulla. The medulla can be further subdivided into an **outer medulla,** which is closer to the cortex, and an **inner medulla,** which is farther from the cortex.

The human kidney is organized into a series of **lobes,** usually 8 to 10. Each lobe consists of a pyramid of medullary tissue and the cortical tissue overlying its base and covering its sides. The tip of the **medullary pyramid** forms a **renal papilla.** Each renal papilla drains its urine into a **minor calyx.** The minor calices unite to form a **major calyx,** and the urine then flows into the **renal pelvis.** The urine is propelled by peristaltic movements down the **ureters** to the **urinary bladder,** which stores the urine until the bladder is emptied. The medial aspect of each kidney is indented in a region called the **hilum,** where the ureter, blood vessels, nerves, and lymphatic vessels enter or leave the kidney.

The Nephron Is the Basic Unit of Renal Structure and Function

Each human kidney contains about one million **nephrons** (Fig. 23.2), which consist of a **renal corpuscle** and a **renal tubule.** The renal corpuscle consists of a tuft of capillaries, the **glomerulus,** surrounded by **Bowman's capsule.** The renal tubule is divided into several segments. The part of the tubule nearest the glomerulus is the **proximal tubule.** This is subdivided into a **proximal convoluted tubule** and **proximal straight tubule.** The straight portion heads toward the

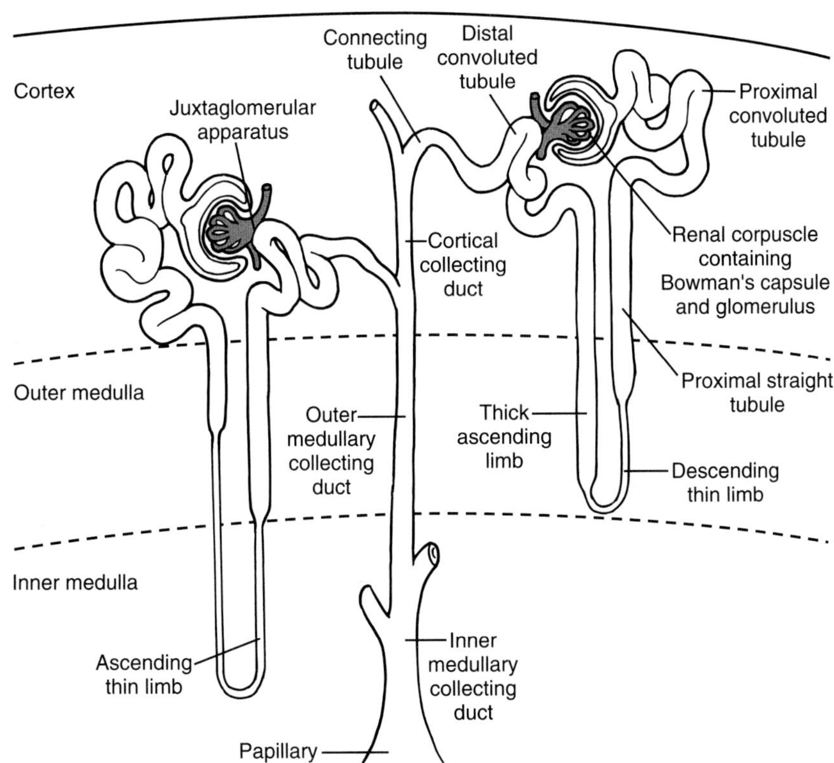

FIGURE 23.2 **Components of the nephron and the colle ing duct system.** On the left is a long-looped juxtamedullary nephron; on th is a superficial cortical nephron. (M from Kriz W, Bankir L. A standar clature for structures of the kidr Physiol 1988;254:F1–F8.)

medulla, away from the surface of the kidney. The **loop of Henle** includes the proximal straight tubule, **thin limb**, and **thick ascending limb**. The next segment, the short **distal convoluted tubule**, is connected to the collecting duct system by **connecting tubules**. Several nephrons drain into a **cortical collecting duct**, which passes into an **outer medullary collecting duct**. In the inner medulla, **inner medullary collecting ducts** unite to form large **papillary ducts**.

The collecting ducts perform the same types of functions as the renal tubules, so they are often considered to be part of the nephron. The collecting ducts and nephrons differ, however, in embryological origin, and because the collecting ducts form a branching system, there are many more nephrons than collecting ducts. The entire renal tubule and collecting duct system consists of a single layer of epithelial cells surrounding fluid (urine) in the tubule or duct lumen. Cells in each segment have a characteristic histological appearance. Each segment has unique transport properties (discussed later).

Not All Nephrons Are Alike

Three groups of nephrons are distinguished, based on the location of their glomeruli in the cortex: **superficial, midcortical,** and **juxtamedullary nephrons.** The juxtamedullary nephrons, whose glomeruli lie in the cortex next to the medulla, comprise about one-eighth of the nephron population. They differ in several ways from the other nephron types: they have a longer loop of Henle, longer thin limb (both descending and ascending portions), larger glomerulus, lower renin content, different tubular permeability and transport properties, and a different type of postglomerular blood supply. Figure 23.2 shows superficial and juxtamedullary nephrons; note the long loop of the juxtamedullary nephron.

The Kidneys Have a Rich Blood Supply and Innervation

Each kidney is typically supplied by a single **renal artery** that branches into anterior and posterior divisions, which give rise to a total of five **segmental arteries.** The segmental arteries branch into **interlobar arteries**, which pass toward the cortex between the kidney lobes (see Fig. 23.1). At the junction of cortex and medulla, the interlobar arteries branch to form **arcuate arteries.** These, in turn, give rise to smaller **cortical radial arteries,** which pass through the cortex toward the surface of the kidney. Several short, wide, muscular **afferent arterioles** arise from the cortical radial arteries. Each afferent arteriole gives rise to a **glomerulus.** The glomerular capillaries are followed by an **efferent arteriole.** The efferent arteriole hen divides into a second capillary network, the **per-**
ular capillaries, that surrounds the kidney tubules.
's vessels, in general, lie parallel to the arterial ves-
have similar names.
d supply to the medulla is derived from the ef-
of juxtamedullary glomeruli. These ves-
two patterns of capillaries: peritubular
re similar to those in the cortex, and
straight, long capillaries (Fig. 23.3).

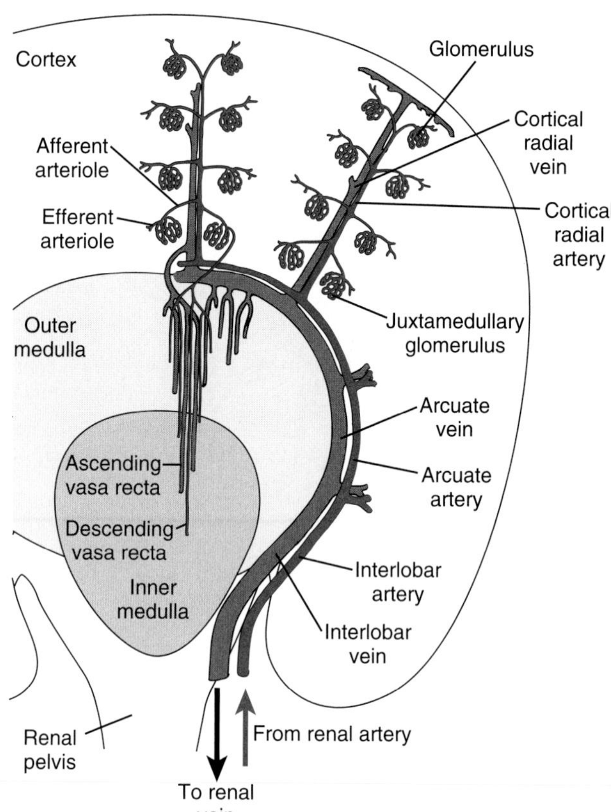

FIGURE 23.3 The blood vessels in the kidney; peritubular capillaries are not shown. (Modified from Kriz W, Bankir L. A standard nomenclature for structures of the kidney. Am J Physiol. 1988;254:F1–F8.)

Some vasa recta reach deep into the inner medulla. In the outer medulla, descending and ascending vasa recta are grouped in vascular bundles and are in close contact with each other. This arrangement greatly facilitates the exchange of substances between blood flowing in and out of the medulla.

The kidneys are richly innervated by **sympathetic nerve fibers**, which travel to the kidneys, mainly in thoracic spinal nerves T10, T11, and T12 and lumbar spinal nerve L1. Stimulation of sympathetic fibers causes constriction of renal blood vessels and a fall in renal blood flow. Sympathetic nerve fibers also innervate tubular cells and may cause an increase in Na^+ reabsorption by a direct action on these cells. In addition, stimulation of sympathetic nerves increases the release of renin by the kidneys. Afferent (sensory) renal nerves are stimulated by mechanical stretch or by various chemicals in the renal parenchyma.

Renal lymphatic vessels drain the kidneys, but little is known about their functions.

The Juxtaglomerular Apparatus Is the Site of Renin Production

Each nephron forms a loop, and the thick ascending limb touches the vascular pole of the glomerulus (see Fig. 23.2). At this site is the **juxtaglomerular apparatus**, a region com-

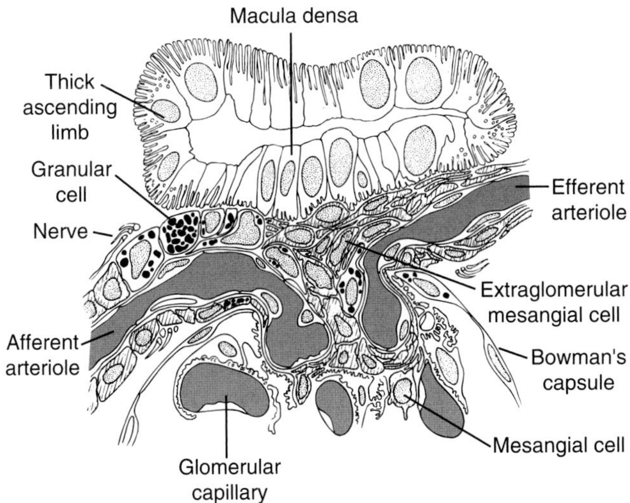

FIGURE 23.4 **Histological appearance of the juxtaglomerular apparatus.** A cross section through a thick ascending limb is on top and part of a glomerulus is below. The juxtaglomerular apparatus consists of the macula densa, extraglomerular mesangial cells, and granular cells. (From Taugner R, Hackenthal E. The Juxtaglomerular Apparatus: Structure and Function. Berlin: Springer, 1989.)

prised of the macula densa, extraglomerular mesangial cells, and granular cells (Fig. 23.4). The **macula densa** (dense spot) consists of densely crowded tubular epithelial cells on the side of the thick ascending limb that faces the glomerular tuft; these cells monitor the composition of the fluid in the tubule lumen at this point. The **extraglomerular mesangial cells** are continuous with mesangial cells of the glomerulus; they may transmit information from macula densa cells to the granular cells. The **granular cells** are modified vascular smooth muscle cells with an epithelioid appearance, located mainly in the afferent arterioles close to the glomerulus. These cells synthesize and release **renin**, a proteolytic enzyme that results in angiotensin formation (see Chapter 24).

AN OVERVIEW OF KIDNEY FUNCTION

Three processes are involved in forming urine: glomerular filtration, tubular reabsorption, and tubular secretion (Fig. 23.5). **Glomerular filtration** involves the ultrafiltration of plasma in the glomerulus. The filtrate collects in the urinary space of Bowman's capsule and then flows downstream through the tubule lumen, where its composition and volume are altered by tubular activity. **Tubular reabsorption** involves the transport of substances out of tubular urine; these substances are then returned to the capillary blood, which surrounds the kidney tubules. Reabsorbed substances include many important ions (e.g., Na^+, K^+, Ca^{2+}, Mg^{2+}, Cl^-, HCO_3^-, phosphate), water, important metabolites (e.g., glucose, amino acids), and even some waste products (e.g., urea, uric acid). **Tubular secretion** involves the transport of substances into the tubular urine. For example, many organic anions and cations are taken up by the tubular epithelium from the blood surrounding the tubules and added to the tubular urine. Some substances (e.g., H^+, ammonia) are produced in the tubular cells and secreted into the tubular urine. The terms *reabsorption* and *secretion* indicate movement out of and into tubular urine, respectively. Tubular transport (reabsorption, secretion) may be active or passive, depending on the particular substance and other conditions.

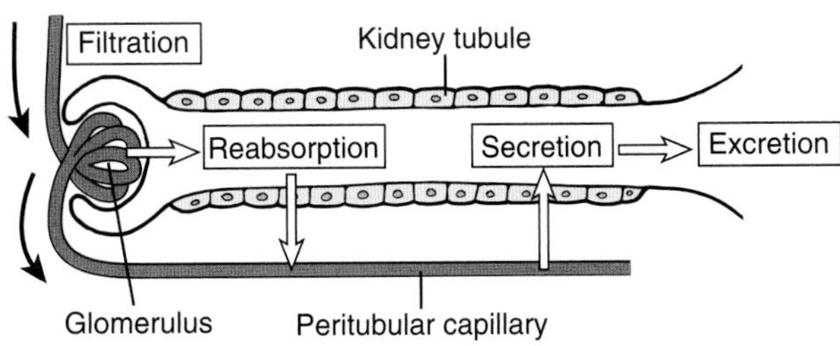

FIGURE 23.5 **Processes involved in urine formation.** This highly simplified drawing shows a nephron and its associated blood vessels.

Excretion refers to elimination via the urine. In general, the amount excreted is expressed by the following equation:

$$\text{Excreted} = \text{Filtered} - \text{Reabsorbed} + \text{Secreted} \quad (1)$$

The functional state of the kidneys can be evaluated using several tests based on the renal clearance concept. These tests measure the rates of glomerular filtration, renal blood flow, and tubular reabsorption or secretion of various substances. Some of these tests, such as the measurement of glomerular filtration rate, are routinely used to evaluate kidney function.

Renal Clearance Equals Urinary Excretion Rate Divided by Plasma Concentration

A useful way of looking at kidney function is to think of the kidneys as clearing substances from the blood plasma. When a substance is excreted in the urine, a certain volume of plasma is, in effect, freed (or cleared) of that substance. The **renal clearance** of a substance can be defined as the volume of plasma from which that substance is completely removed (cleared) per unit time. The clearance formula is:

$$C_x = \frac{U_x \times \dot{V}}{P_x} \quad (2)$$

where X is the substance of interest, C_X is the clearance of substance X, U_X is the urine concentration of substance, P_X is the plasma concentration of substance X, and $\dot{V}$ is the urine flow rate. The product $U_X \times \dot{V}$ equals the excretion rate and has dimensions of amount per unit time (e.g., mg/min or mEq/day). The clearance of a substance can easily be determined by measuring the concentrations of a substance in urine and plasma and the urine flow rate (urine volume/time of collection) and substituting these values into the clearance formula.

Inulin Clearance Equals the Glomerular Filtration Rate

An important measurement in the evaluation of kidney function is the **glomerular filtration rate** (GFR), the rate at

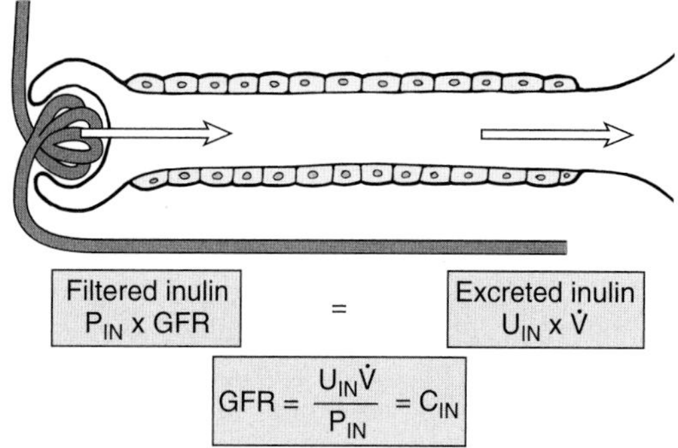

FIGURE 23.6 **The principle behind the measurement of glomerular filtration rate (GFR).** P_{IN} = plasma [inulin], U_{IN} = urine [inulin], $\dot{V}$ = urine flow rate, C_{IN} = inulin clearance.

which plasma is filtered by the kidney glomeruli. If we had a substance that was cleared from the plasma only by glomerular filtration, it could be used to measure GFR.

The ideal substance to measure GFR is **inulin,** a fructose polymer with a molecular weight of about 5,000. Inulin is suitable for measuring GFR for the following reasons:

- It is freely filterable by the glomeruli.
- It is not reabsorbed or secreted by the kidney tubules.
- It is not synthesized, destroyed, or stored in the kidneys.
- It is nontoxic.
- Its concentration in plasma and urine can be determined by simple analysis.

The principle behind the use of inulin is illustrated in Figure 23.6. The amount of inulin (IN) filtered per unit time, the **filtered load,** is equal to the product of the plasma [inulin] (P_{IN}) × GFR. The rate of inulin excretion is equal to $U_{IN} \times \dot{V}$. Since inulin is not reabsorbed, secreted, synthesized, destroyed, or stored by the kidney tubules, the filtered inulin load equals the rate of inulin excretion. The equation can be rearranged by dividing by the plasma [inulin]. The expression $U_{IN}\dot{V}/P_{IN}$ is defined as the **inulin clearance.** Therefore, inulin clearance equals GFR.

Normal values for inulin clearance or GFR (corrected to a body surface area of 1.73 m^2) are 110 ± 15 (SD) mL/min for young adult women and 125 ± 15 mL/min for young adult men. In newborns, even when corrected for body surface area, GFR is low, about 20 mL/min per 1.73 m^2 body surface area. Adult values (when corrected for body surface area) are attained by the end of the first year of life. After the age of 45 to 50 years, GFR declines, and is typically reduced by 30 to 40% by age 80.

If GFR is 125 mL plasma/min, then the volume of plasma filtered in a day is 180 L (125 mL/min × 1,440 min/day). Plasma volume in a 70-kg young adult man is only about 3 L, so the kidneys filter the plasma some 60 times in a day. The glomerular filtrate contains essential constituents (salts, water, metabolites), most of which are reabsorbed by the kidney tubules.

The Endogenous Creatinine Clearance Is Used Clinically to Estimate GFR

Inulin clearance is the gold standard for measuring GFR and is used whenever highly accurate measurements of GFR are desired. The clearance of iothalamate, an iodinated organic compound, also provides a reliable measure of GFR. It is not common, however, to use these substances in the clinic. They must be infused intravenously, and because short urine collection periods are used, the bladder is usually catheterized; these procedures are inconvenient. It would be simpler to use an endogenous substance (i.e., one native to the body) that is only filtered, is excreted in the urine, and normally has a stable plasma value that can be accurately measured. There is no such known substance, but creatinine comes close.

Creatinine is an end product of muscle metabolism, a derivative of muscle creatine phosphate. It is produced continuously in the body and is excreted in the urine. Long urine collection periods (e.g., a few hours) can be used because creatinine concentrations in the plasma are normally stable and creatinine does not have to be infused; consequently, there is no need to catheterize the bladder. Plasma and urine concentrations can be measured using a simple colorimetric method. The **endogenous creatinine clearance** is calculated from the formula:

$$C_{CREATININE} = \frac{U_{CREATININE} \times \dot{V}}{P_{CREATININE}} \quad (3)$$

There are two potential drawbacks to using creatinine to measure GFR. First, creatinine is not only filtered but also secreted by the human kidney. This elevates urinary excretion of creatinine, normally causing a 20% increase in the numerator of the clearance formula. The second drawback is due to errors in measuring plasma [creatinine]. The colorimetric method usually used also measures other plasma substances, such as glucose, leading to a 20% increase in the denominator of the clearance formula. Because both numerator and denominator are 20% too high, the two errors cancel, so the endogenous creatinine clearance fortuitously affords a good approximation of GFR when it is about normal. When GFR in an adult has been reduced to about 20 mL/min because of renal disease, the endogenous creatinine clearance may overestimate the GFR by as much as 50%. This results from higher plasma creatinine levels and increased tubular secretion of creatinine. Drugs that inhibit tubular secretion of creatinine or elevated plasma concentrations of chromogenic (color-producing) substances other than creatinine may cause the endogenous creatinine clearance to underestimate GFR.

Plasma Creatinine Concentration Can Be Used as an Index of GFR

Because the kidneys continuously clear creatinine from the plasma by excreting it in the urine, the GFR and plasma [creatinine] are inversely related. Figure 23.7 shows the steady state relationship between these variables—that is, when creatinine production and excretion are equal. Halving the GFR from a normal value of 180 L/day to 90 L/day results in a doubling of plasma [creatinine] from a normal value of 1 mg/dL to 2 mg/dL after a few days. Reducing GFR from 90 L/day to 45 L/day results in a greater increase in plasma creatinine, from 2 to 4 mg/dL. Figure 23.7 shows that with low GFR values, small absolute changes in GFR

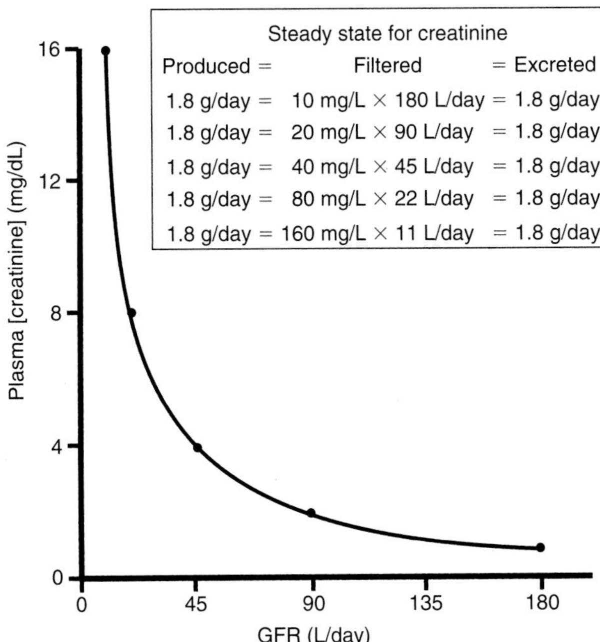

FIGURE 23.7 The inverse relationship between plasma [creatinine] and GFR. If GFR is decreased by half, plasma [creatinine] is doubled when the production and excretion of creatinine are in balance in a new steady state.

lead to much greater changes in plasma [creatinine] than occur at high GFR values.

The inverse relationship between GFR and plasma [creatinine] allows the use of plasma [creatinine] as an index of GFR, provided certain cautions are kept in mind:

1) It takes a certain amount of time for changes in GFR to produce detectable changes in plasma [creatinine].

2) Plasma [creatinine] is also influenced by muscle mass. A young, muscular man will have a higher plasma [creatinine] than an older woman with reduced muscle mass.

3) Some drugs inhibit tubular secretion of creatinine, leading to a raised plasma [creatinine] even though GFR may be unchanged.

The relationship between plasma [creatinine] and GFR is one example of how a substance's plasma concentration can depend on GFR. The same relationship is observed for several other substances whose excretion depends on GFR. For example, when GFR falls, the plasma [urea] (or blood urea nitrogen, BUN) rises in a similar fashion.

p-Aminohippurate Clearance Nearly Equals Renal Plasma Flow

Renal blood flow (RBF) can be determined from measurements of renal plasma flow (RPF) and blood hematocrit, using the following equation:

$$\text{RBF} = \text{RPF}/(1 - \text{Hematocrit}) \quad (4)$$

The hematocrit is easily determined by centrifuging a blood sample. Renal plasma flow is estimated by measuring the clearance of the organic anion *p*-aminohippurate (PAH), infused intravenously. PAH is filtered and vigorously secreted, so it is nearly completely cleared from all of the plasma flowing through the kidneys. The renal clearance of PAH, at low plasma PAH levels, approximates the renal plasma flow.

The equation for calculating the true value of the renal plasma flow is:

$$\text{RPF} = C_{PAH}/E_{PAH} \quad (5)$$

where C_{PAH} is the PAH clearance and E_{PAH} is the extraction ratio (see Chapter 16) for PAH—the arterial plasma [PAH] (P^a_{PAH}) minus renal venous plasma [PAH] (P^{rv}_{PAH}) divided by the arterial plasma [PAH]. The equation is derived as follows. In the steady state, the amounts of PAH per unit time entering and leaving the kidneys are equal. The PAH is supplied to the kidneys in the arterial plasma and leaves the kidneys in urine and renal venous plasma, or PAH entering kidneys is equal to PAH leaving kidneys:

$$\text{RPF} \times P^a_{PAH} = U_{PAH} \times \dot{V} + \text{RPF} \times P^{rv}_{PAH} \quad (6)$$

Rearranging, we get:

$$\text{RPF} = U_{PAH} \times \dot{V}/(P^a_{PAH} - P^{rv}_{PAH}) \quad (7)$$

If we divide the numerator and denominator of the right side of the equation by P^a_{PAH}, the numerator becomes C_{PAH} and the denominator becomes E_{PAH}.

If we assume extraction of PAH is 100% ($E_{PAH} = 1.00$), then the RPF equals the PAH clearance. When this assumption is made, the renal plasma flow is usually called the **effective renal plasma flow** and the blood flow calculated is called the **effective renal blood flow.** However, the extraction of PAH by healthy kidneys at suitably low plasma PAH concentrations is not 100% but averages about 91%. Assuming 100% extraction underestimates the true renal plasma flow by about 10%. To calculate the true renal plasma flow or blood flow, it is necessary to cannulate the renal vein to measure its plasma [PAH], a procedure not often done.

Net Tubular Reabsorption or Secretion of a Substance Can Be Calculated From Filtered and Excreted Amounts

The rate at which the kidney tubules reabsorb a substance can be calculated if we know how much is filtered and how much is excreted per unit time. If the filtered load of a substance exceeds the rate of excretion, the kidney tubules must have reabsorbed the substance. The equation is:

$$T_{reabsorbed} = P_x \times \text{GFR} - U_x \times \dot{V} \quad (8)$$

where T is the tubular transport rate.

The rate at which the kidney tubules secrete a substance is calculated from this equation:

$$T_{secreted} = U_x \times \dot{V} - P_x \times \text{GFR} \quad (9)$$

Note that the quantity excreted exceeds the filtered load because the tubules secrete X.

In equations 8 and 9, we assume that substance X is freely filterable. If, however, substance X is bound to the plasma proteins, which are not filtered, then it is necessary to correct the filtered load for this binding. For example, about 40% of plasma Ca^{2+} is bound to plasma proteins, so 60% of plasma Ca^{2+} is freely filterable.

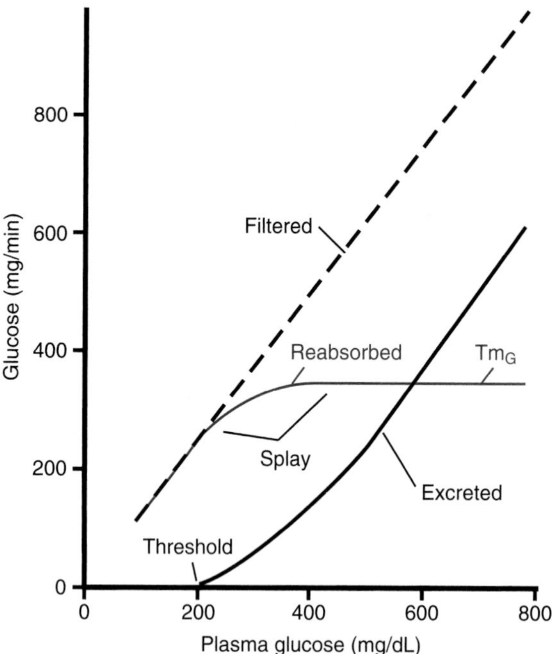

FIGURE 23.8 **Glucose titration study in a healthy man.** The plasma [glucose] was elevated by infusing glucose-containing solutions. The amount of glucose filtered per unit time (top line) is determined from the product of the plasma [glucose] and GFR (measured with inulin). Excreted glucose (bottom line) is determined by measuring the urine [glucose] and flow rate. Reabsorbed glucose is calculated from the difference between filtered and excreted glucose. Tm_G = tubular transport maximum for glucose.

Equations 8 and 9 for quantitating tubular transport rates yield the *net* rate of reabsorption or secretion of a substance. It is possible for a single substance to be both reabsorbed and secreted; the equations do not give unidirectional reabsorptive and secretory movements, but only the net transport.

The Glucose Titration Study Assesses Renal Glucose Reabsorption

Insights into the nature of glucose handling by the kidneys can be derived from a **glucose titration** study (Fig. 23.8). The plasma [glucose] is elevated to increasingly higher levels by the infusion of glucose-containing solutions. Inulin is infused to permit measurement of GFR and calculation of the filtered glucose load (plasma [glucose] × GFR). The rate of glucose reabsorption is determined from the difference between the filtered load and the rate of excretion. At normal plasma glucose levels (about 100 mg/dL), all of the filtered glucose is reabsorbed and none is excreted. When the plasma [glucose] exceeds a certain value (about 200 mg/dL, see Fig. 23.8), significant quantities of glucose appear in the urine; this plasma concentration is called the **glucose threshold.** Further elevations in plasma glucose lead to progressively more excreted glucose. Glucose appears in the urine because the filtered amount of glucose exceeds the capacity of the tubules to reabsorb it. At very high filtered glucose loads, the rate of glucose reabsorption reaches a constant maximal value, called the **tubular transport maximum** (Tm) for glucose (G). At Tm_G, the limited number of tubule glucose carriers are all saturated and transport glucose at the maximal rate.

The glucose threshold is not a fixed plasma concentration but depends on three factors: GFR, Tm_G, and amount of splay. A low GFR leads to an elevated threshold because the filtered glucose load is reduced and the kidney tubules can reabsorb all the filtered glucose despite an elevated plasma [glucose]. A reduced Tm_G lowers the threshold because the tubules have a diminished capacity to reabsorb glucose.

Splay is the rounding of the glucose reabsorption curve; Figure 23.8 shows that tubular glucose reabsorption does not abruptly attain Tm_G when plasma glucose is progressively elevated. One reason for splay is that not all nephrons have the same filtering and reabsorbing capacities. Thus, nephrons with relatively high filtration rates and low glucose reabsorptive rates excrete glucose at a lower plasma concentration than nephrons with relatively low filtration rates and high reabsorptive rates. A second reason for splay is that the glucose carrier does not have an infinitely high affinity for glucose, so glucose escapes in the urine even before the carrier is fully saturated. An increase in splay results in a decrease in glucose threshold.

In uncontrolled **diabetes mellitus,** plasma glucose levels are abnormally elevated, and more glucose is filtered than can be reabsorbed. Urinary excretion of glucose, **glucosuria,** produces an osmotic diuresis. A diuresis is an increase in urine output; in osmotic diuresis, the increased urine flow results from the excretion of osmotically active solute. Diabetes (from the Greek for "syphon") gets its name from this increased urine output.

The Tubular Transport Maximum for PAH Provides a Measure of Functional Proximal Secretory Tissue

p-Aminohippurate is secreted only by proximal tubules in the kidneys. At low plasma PAH concentrations, the rate of secretion increases linearly with the plasma [PAH]. At high plasma PAH concentrations, the secretory carriers are saturated and the rate of PAH secretion stabilizes at a constant maximal value, called the **tubular transport maximum for PAH** (Tm_{PAH}). The Tm_{PAH} is directly related to the number of functioning proximal tubules and, therefore, provides a measure of the mass of proximal secretory tissue. Figure 23.9 illustrates the pattern of filtration, secretion, and excretion of PAH observed when the plasma [PAH] is progressively elevated by intravenous infusion.

RENAL BLOOD FLOW

The kidneys have a very high blood flow. This allows them to filter the blood plasma at a high rate. Many factors, both intrinsic (autoregulation, local hormones) and extrinsic (nerves, bloodborne hormones), affect the rate of renal blood flow.

The Kidneys Have a High Blood Flow

In resting, healthy, young adult men, renal blood flow averages about 1.2 L/min. This is about 20% of the cardiac output (5 to 6 L/min). Both kidneys together weigh about

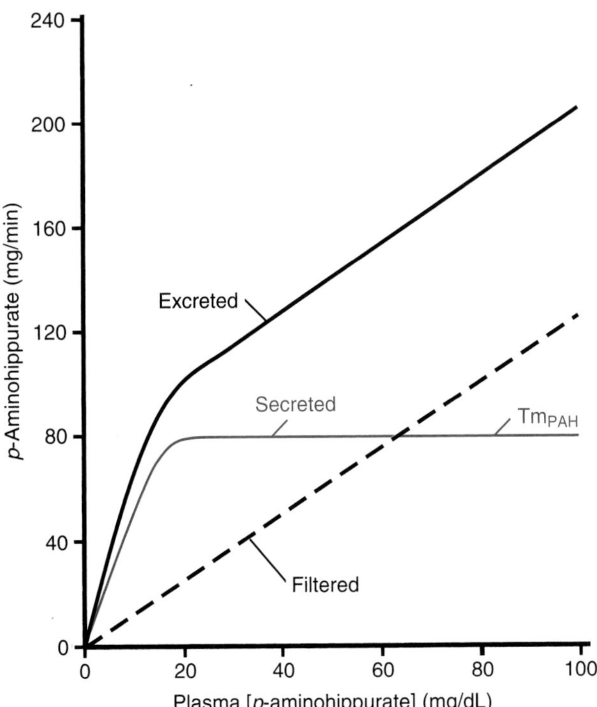

FIGURE 23.9 **Rates of excretion, filtration, and secretion of *p*-aminohippurate (PAH) as a function of plasma [PAH].** More PAH is excreted than is filtered; the difference represents secreted PAH.

300 g, so blood flow per gram of tissue averages about 4 mL/min. This rate of perfusion exceeds that of all other organs in the body, except the neurohypophysis and carotid bodies. The high blood flow to the kidneys is necessary for a high GFR and is not due to excessive metabolic demands.

The kidneys use about 8% of total resting oxygen consumption, but they receive much more oxygen than they need. Consequently, renal extraction of oxygen is low, and renal venous blood has a bright red color (because of a high oxyhemoglobin content). The anatomic arrangement of the vessels in the kidney permits a large fraction of the arterial oxygen to be shunted to the veins before the blood enters the capillaries. Therefore, the oxygen tension in the tissue is not as high as one might think, and the kidneys are certainly sensitive to ischemic damage.

Blood Flow Is Higher in the Renal Cortex and Lower in the Renal Medulla

Blood flow rates differ in different parts of the kidney (Fig. 23.10). Blood flow is highest in the cortex, averaging 4 to 5 mL/min per gram of tissue. The high cortical blood flow permits a high rate of filtration in the glomeruli. Blood flow (per gram of tissue) is about 0.7 to 1 mL/min in the outer medulla and 0.20 to 0.25 mL/min in the inner medulla. The relatively low blood flow in the medulla helps maintain a hyperosmolar environment in this region of the kidney.

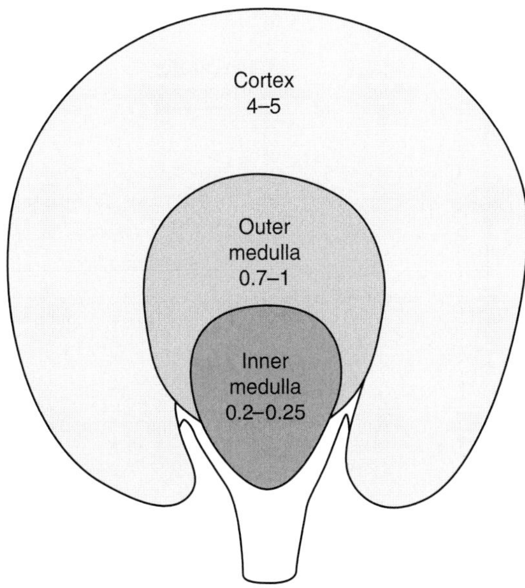

FIGURE 23.10 **Blood flow rates (in mL/min per gram of tissue) in different parts of the kidney.** (Modified from Brobeck JR, ed. Best & Taylor's Physiological Basis of Medical Practice. 10th Ed. Baltimore: Williams & Wilkins, 1979.)

The Kidneys Autoregulate Their Blood Flow

Despite changes in mean arterial blood pressure (from 80 to 180 mm Hg), renal blood flow is kept at a relatively constant level, a process known as **autoregulation** (see Chapter 16). Autoregulation is an intrinsic property of the kidneys and is observed even in an isolated, denervated, perfused kidney. GFR is also autoregulated (Fig. 23.11). When the blood pressure is raised or lowered, vessels upstream of the glomerulus (cortical radial arteries and afferent arterioles) constrict or dilate, respectively, maintaining relatively constant glomerular blood flow and capillary pressure. Below or above the autoregulatory range of pressures, blood flow and GFR change appreciably with arterial blood pressure.

Two mechanisms account for renal autoregulation: the myogenic mechanism and the tubuloglomerular feedback mechanism. In the **myogenic mechanism,** an increase in pressure stretches blood vessel walls and opens stretch-activated cation channels in smooth muscle cells. The ensuing membrane depolarization opens voltage-dependent Ca^{2+} channels and intracellular [Ca^{2+}] rises, causing smooth muscle contraction. Vessel lumen diameter decreases and vascular resistance increases. Decreased blood pressure causes the opposite changes.

In the **tubuloglomerular feedback mechanism,** the transient increase in GFR resulting from an increase in blood pressure leads to increased solute delivery to the macula densa (Fig. 23.12). This produces an increase in the tubular fluid [NaCl] at this site and increased NaCl reabsorption by macula densa cells. By mechanisms that are still uncertain, constriction of the nearby afferent arteriole results. The vasoconstrictor agent may be adenosine or ATP; it does not appear to be angiotensin II, although feedback sensitivity varies directly with the local concentration of angiotensin II. Blood flow and GFR are lowered to a more normal value. The tubuloglomerular feedback

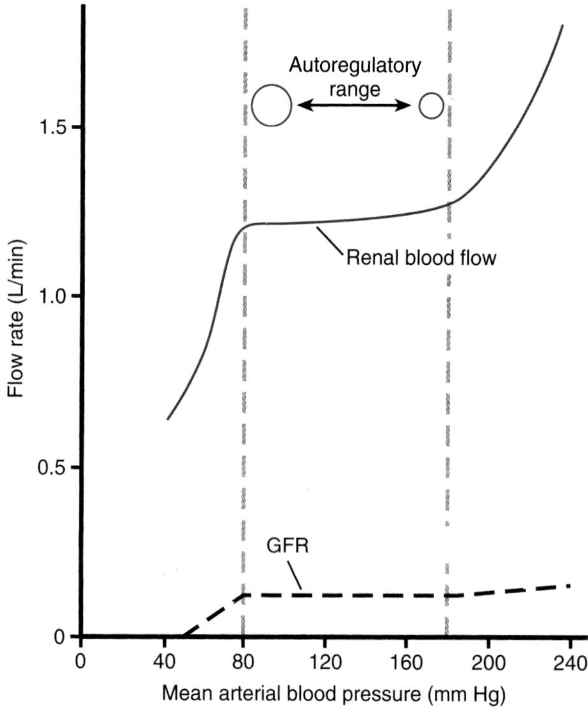

FIGURE 23.11 Renal autoregulation, based on measurements in isolated, denervated, and perfused kidneys. In the autoregulatory range, renal blood flow and GFR stay relatively constant despite changes in arterial blood pressure. This is accomplished by changes in the resistance (caliber) of preglomerular blood vessels. The circles indicate that vessel radius (r) is smaller when blood pressure is high and larger when blood pressure is low. Since resistance to blood flow varies as r^4, changes in vessel caliber are greatly exaggerated in this figure.

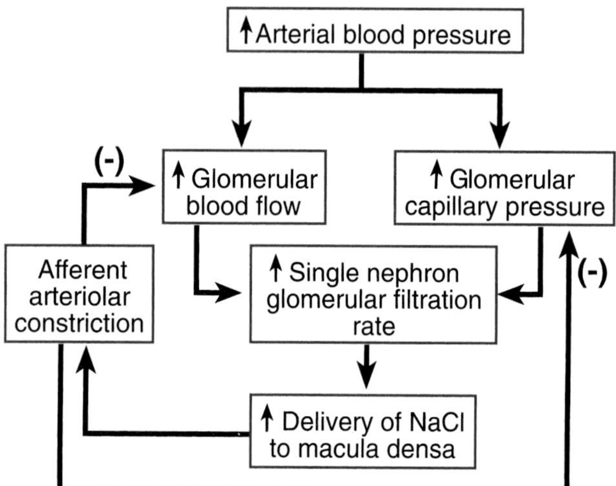

FIGURE 23.12 The tubuloglomerular feedback mechanism. When single nephron GFR is increased—for example, as a result of an increase in arterial blood pressure—more NaCl is delivered to and reabsorbed by the macula densa, leading to constriction of the nearby afferent arteriole. This negative-feedback system plays a role in renal blood flow and GFR autoregulation.

mechanism is a negative-feedback system that stabilizes renal blood flow and GFR.

If NaCl delivery to the macula densa is increased experimentally by perfusing the lumen of the loop of Henle, filtration rate in the perfused nephron decreases. This suggests that the purpose of tubuloglomerular feedback may be to control the amount of Na^+ presented to distal nephron segments. Regulation of Na^+ delivery to distal parts of the nephron is important because these segments have a limited capacity to reabsorb Na^+.

Renal autoregulation minimizes the impact of changes in arterial blood pressure on Na^+ excretion. Without renal autoregulation, increases in arterial blood pressure would lead to dramatic increases in GFR and potentially serious losses of NaCl and water from the ECF.

Renal Sympathetic Nerves and Various Hormones Change Renal Blood Flow

Renal blood flow may be changed by the stimulation of renal sympathetic nerves or by the release of various hormones. Sympathetic nerve stimulation causes renal vasoconstriction and a consequent decrease in renal blood flow. Renal sympathetic nerves are activated under stressful conditions, including cold temperatures, deep anesthesia, fearful situations, hemorrhage, pain, and strenuous exercise. In these conditions, the decrease in renal blood flow may be viewed as an emergency mechanism that makes more of the cardiac output available to perfuse other organs, such as the brain and heart, which are more important for short-term survival.

Several substances cause vasoconstriction in the kidneys, including adenosine, angiotensin II, endothelin, epinephrine, norepinephrine, thromboxane A_2, and vasopressin. Other substances cause vasodilation in the kidneys, including atrial natriuretic peptide, dopamine, histamine, kinins, nitric oxide, and prostaglandins E_2 and I_2. Some of these substances (e.g., prostaglandins E_2 and I_2) are produced locally in the kidneys. An increase in sympathetic nerve activity or plasma angiotensin II concentration stimulates the production of renal vasodilator prostaglandins. These prostaglandins then oppose the pure constrictor effect of sympathetic nerve stimulation or angiotensin II, reducing the fall in renal blood flow, preventing renal damage.

GLOMERULAR FILTRATION

Glomerular filtration involves the **ultrafiltration** of plasma. This term reflects the fact that the glomerular filtration barrier is an extremely fine molecular sieve that allows the filtration of small molecules but restricts the passage of macromolecules (e.g., the plasma proteins).

The Glomerular Filtration Barrier Has Three Layers

An ultrafiltrate of plasma passes from glomerular capillary blood into the space of Bowman's capsule through the **glomerular filtration barrier** (Fig. 23.13). This barrier consists of three layers. The first, the capillary **endothelium**, is called the lamina fenestra because it contains pores or win-

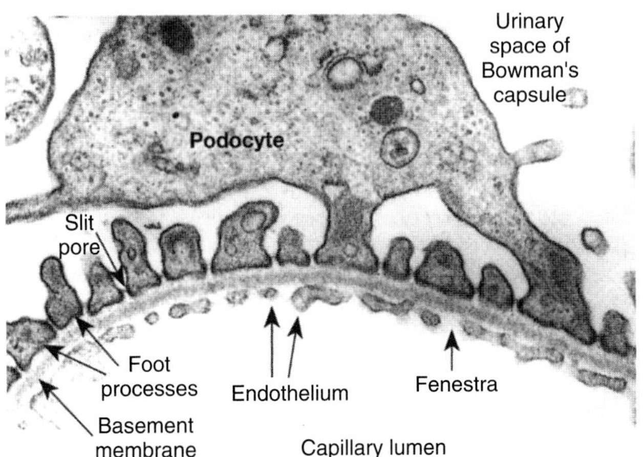

FIGURE 23.13 Electron micrograph showing the three layers of the glomerular filtration barrier: endothelium, basement membrane, and podocytes. (Courtesy of Dr. Andrew P. Evan, Indiana University.)

TABLE 23.1 Restrictions to the Glomerular Filtration of Molecules

Substance	Molecular Weight	Molecular Radius (nm)	[Filtrate]/ [Plasma Water]
Water	18	0.10	1.00
Glucose	180	0.36	1.00
Inulin	5,000	1.4	1.00
Myoglobin	17,000	2.0	0.75
Hemoglobin	68,000	3.3	0.03
Cationic dextran[a]		3.6	0.42
Neutral dextran		3.6	0.15
Anionic dextran		3.6	0.01
Serum albumin	69,000	3.6	0.001

[a]Dextrans are high-molecular-weight glucose polymers available in cationic (amine), neutral (uncharged), or anionic (sulfated) forms. (Adapted from Pitts RF. Physiology of the Kidney and Body Fluids. 3rd Ed. Chicago: Year Book, 1974; and Brenner BM, Bohrer MP, Baylis C, Deen WM. Determinants of glomerular permselectivity: Insights derived from observations *in vivo*. Kidney Int 1977;12:229–237.)

dows (fenestrae). At about 50 to 100 nm in diameter, these pores are too large to restrict the passage of plasma proteins. The second layer, the **basement membrane,** consists of a meshwork of fine fibrils embedded in a gel-like matrix. The third layer is composed of **podocytes,** which constitute the visceral layer of Bowman's capsule. Podocytes ("foot cells") are epithelial cells with extensions that terminate in foot processes, which rest on the outer layer of the basement membrane (see Fig. 23.13). The space between adjacent foot processes, called a **slit pore,** is about 20 nm wide and is bridged by a **filtration slit diaphragm.** A key component of the diaphragm is a molecule called **nephrin,** which forms a zipper-like structure; between the prongs of the zipper are rectangular pores. The nephrin is mutated in **congenital nephrotic syndrome,** a rare, inherited condition characterized by excessive filtration of plasma proteins. The glomerular filtrate normally takes an extracellular route, through holes in the endothelial cell layer, the basement membrane, and the pores between adjacent nephrin molecules.

Size, Shape, and Electrical Charge Affect the Filterability of Macromolecules

The permeability properties of the glomerular filtration barrier have been studied by determining how well molecules of different sizes pass through it. Table 23.1 lists several molecules that have been tested. Molecular radii were calculated from diffusion coefficients. The concentration of the molecule in the glomerular filtrate (fluid collected from Bowman's capsule) is compared to its concentration in plasma water. A ratio of 1 indicates complete filterability, and a ratio of zero indicates complete exclusion by the glomerular filtration barrier.

Molecular size is an important factor affecting filterability. All molecules with weights less than 10,000 are freely filterable, provided they are not bound to plasma proteins. Molecules with weights greater than 10,000 experience more restriction to passage through the glomerular filtration barrier. Very large molecules (e.g., molecular weight, 100,000) cannot get through at all. Most plasma proteins are large molecules, so they are not appreciably filtered. From studies with molecules of different sizes, it has been calculated that the glomerular filtration barrier behaves as though it were penetrated by cylindric pores of about 7.5 to 10 nm in diameter. However, no one has ever seen pores of this size in electron micrographs of the glomerular filtration barrier.

Molecular shape influences the filterability of macromolecules. For a given molecular weight, a slender and flexible molecule will pass through the glomerular filtration barrier more easily than a spherical, nondeformable molecule.

Electrical charge influences the passage of macromolecules through the glomerular filtration barrier because the barrier bears fixed negative charges. Glomerular endothelial cells and podocytes have a negatively charged surface coat (glycocalyx), and the glomerular basement membrane contains negatively charged sialic acid, sialoproteins, and heparan sulfate. These negative charges impede the passage of negatively charged macromolecules by electrostatic repulsion and favor the passage of positively charged macromolecules by electrostatic attraction. This is supported by the finding that the filterability of dextran is lowest for anionic dextran, intermediate for neutral dextran, and highest for cationic dextran (see Table 23.1).

In addition to its large molecular size, the net negative charge on serum albumin at physiological pH is an important factor that reduces its filterability. In some glomerular diseases, a loss of fixed negative charges from the glomerular filtration barrier causes increased filtration of serum albumin. **Proteinuria,** abnormal amounts of protein in the urine, results. Proteinuria is the hallmark of glomerular disease (see Clinical Focus Box 23.2 and the Case Study).

The layer of the glomerular filtration barrier primarily responsible for limiting the filtration of macromolecules is a matter of debate. The basement membrane is probably the principal size-selective barrier, and the filtration slit diaphragm forms a second barrier. The major electrostatic

CLINICAL FOCUS BOX 23.2

Glomerular Disease

The kidney glomeruli may be injured by several immunological, toxic, hemodynamic, and metabolic disorders. Glomerular injury impairs filtration barrier function and, consequently, increases the filtration and excretion of plasma proteins (proteinuria). Red cells may appear in the urine, and sometimes GFR is reduced. Three general syndromes are encountered: nephritic diseases, nephrotic diseases (nephrotic syndrome), and chronic glomerulonephritis.

In the **nephritic diseases**, the urine contains red blood cells, red cell casts, and mild to modest amounts of protein. A red cell cast is a mold of the tubule lumen formed when red cells and proteins clump together; the presence of such casts in the final urine indicates that bleeding had occurred in the kidneys (usually in the glomeruli), not in the lower urinary tract. Nephritic diseases are usually associated with a fall in GFR, accumulation of nitrogenous wastes (urea, creatinine) in the blood, and hypervolemia (hypertension, edema). Most nephritic diseases are due to immunological damage. The glomerular capillaries may be injured by antibodies directed against the glomerular basement membrane, by deposition of circulating immune complexes along the endothelium or in the mesangium, or by cell-mediated injury (infiltration with lymphocytes and macrophages). A renal biopsy and tissue examination by light and electron microscopy and immunostaining are often helpful in determining the nature and severity of the disease and in predicting its most likely course.

Poststreptococcal glomerulonephritis is an example of a nephritic condition that may follow a sore throat caused by certain strains of streptococci. Immune complexes of antibody and bacterial antigen are deposited in the glomeruli, complement is activated, and polymorphonuclear leukocytes and macrophages infiltrate the glomeruli. Endothelial cell damage, accumulation of leukocytes, and the release of vasoconstrictor substances reduce the glomerular surface area and fluid permeability and lower glomerular blood flow, causing a fall in GFR.

Nephrotic syndrome is a clinical state that can develop as a consequence of many different diseases causing glomerular injury. It is characterized by heavy proteinuria (>3.5 g/day per 1.73 m^2 body surface area), hypoalbuminemia (<3 g/dL), generalized edema, and hyperlipidemia. Abnormal glomerular leakiness to plasma proteins leads to increased catabolism of the reabsorbed proteins in the kidney proximal tubules and increased protein excretion in the urine. The loss of protein (mainly serum albumin) leads to a fall in plasma [protein] (and colloid osmotic pressure). The edema results from the hypoalbuminemia and renal Na^+ retention. Also, a generalized increase in capillary permeability to proteins (not just in the glomeruli) may lead to a decrease in the effective colloid osmotic pressure of the plasma proteins and may contribute to the edema. The hyperlipidemia (elevated serum cholesterol and, in severe cases, elevated triglycerides) is probably a result of increased hepatic synthesis of lipoproteins and decreased lipoprotein catabolism. Most often, nephrotic syndrome in young children cannot be ascribed to a specific cause; this is called idiopathic nephrotic syndrome. Nephrotic syndrome in children or adults can be caused by infectious diseases, neoplasia, certain drugs, various autoimmune disorders (such as lupus), allergic reactions, metabolic disease (such as diabetes mellitus), or congenital disorders.

The distinctions between nephritic and nephrotic diseases are sometimes blurred, and both may result in **chronic glomerulonephritis.** This disease is characterized by proteinuria and/or hematuria (blood in the urine), hypertension, and renal insufficiency that progresses over years. Renal biopsy shows glomerular scarring and increased numbers of cells in the glomeruli and scarring and inflammation in the interstitial space. The disease is accompanied by a progressive loss of functioning nephrons and proceeds relentlessly even though the initiating insult may no longer be present. The exact reasons for disease progression are not known, but an important factor may be that surviving nephrons hypertrophy when nephrons are lost. This leads to an increase in blood flow and pressure in the remaining nephrons, a situation that further injures the glomeruli. Also, increased filtration of proteins causes increased tubular reabsorption of proteins, and the latter results in production of vasoactive and inflammatory substances that cause ischemia, interstitial inflammation, and renal scarring. Dietary manipulations (such as a reduced protein intake) or antihypertensive drugs (such as angiotensin-converting enzyme inhibitors) may slow the progression of chronic glomerulonephritis. Glomerulonephritis in its various forms is the major cause of renal failure in people.

Reference

Falk RJ, Jennette JC, Nachman PH. Primary glomerular diseases. In: Brenner BM, ed. Brenner & Rector's The Kidney. 6th Ed. Philadelphia: WB Saunders, 2000;1263–1349.

barriers are probably the layers closest to the capillary lumen, the lamina fenestra and the innermost part of the basement membrane.

GFR Is Determined by Starling Forces

Glomerular filtration rate depends on the balance of hydrostatic and colloid osmotic pressures acting across the glomerular filtration barrier, the Starling forces (see Chapter 16); therefore, it is determined by the same factors that affect fluid movement across capillaries in general. In the glomerulus, the driving force for fluid filtration is the glomerular capillary hydrostatic pressure (P_{GC}). This pressure ultimately depends on the pumping of blood by the heart, an action that raises the blood pressure on the arterial side of the circulation. Filtration is opposed by the hydrostatic pressure in the space of Bowman's capsule (P_{BS}) and by the colloid osmotic pressure (COP) exerted by plasma proteins in glomerular capillary blood. Because the glomerular filtrate is virtually protein-free, we neglect the colloid osmotic pressure of fluid in Bowman's capsule. The **net ultrafiltration pressure gradi-**

ent (UP) is equal to the difference between the pressures favoring and opposing filtration:

$$GFR = K_f \times UP = K_f \times (P_{GC} - P_{BS} - COP) \quad (10)$$

where K_f is the glomerular ultrafiltration coefficient. Estimates of average, normal values for pressures in the human kidney are: P_{GC}, 55 mm Hg; P_{BS}, 15 mm Hg; and COP, 30 mm Hg. From these values, we calculate a net ultrafiltration pressure gradient of +10 mm Hg.

The Pressure Profile Along a Glomerular Capillary Is Unusual

Figure 23.14 shows how pressures change along the length of a glomerular capillary, in contrast to those seen in a capillary in other vascular beds (in this case, skeletal muscle). Note that average capillary hydrostatic pressure in the glomerulus is much higher (55 vs. 25 mm Hg) than in a skeletal muscle capillary. Also, capillary hydrostatic pressure declines little (perhaps 1 to 2 mm Hg) along the length of the glomerular capillary because the glomerulus contains many (30 to 50) capillary loops in parallel, making the resistance to blood flow in the glomerulus very low. In the skeletal muscle capillary, there is a much higher resistance to blood flow, resulting in an appreciable fall in capillary hydrostatic pressure with distance. Finally, note that in the glomerulus, the colloid osmotic pressure increases substantially along the length of the capillary because a large volume of filtrate (about 20% of the entering plasma flow) is pushed out of the capillary and the proteins remain in the circulation. The increase in colloid osmotic pressure opposes the outward movement of fluid.

In the skeletal muscle capillary, the colloid osmotic pressure hardly changes with distance, since little fluid moves across the capillary wall. In the "average" skeletal muscle capillary, outward filtration occurs at the arterial end and absorption occurs at the venous end. At some point along the skeletal muscle capillary, there is no net fluid movement; this is the point of so-called **filtration pressure equilibrium.** Filtration pressure equilibrium probably is not attained in the normal human glomerulus; in other words, the outward filtration of fluid probably occurs all along the glomerular capillaries.

Several Factors Can Affect GFR

The GFR depends on the magnitudes of the different terms in equation 10. Therefore, GFR varies with changes in K_f, hydrostatic pressures in the glomerular capillaries and Bow-

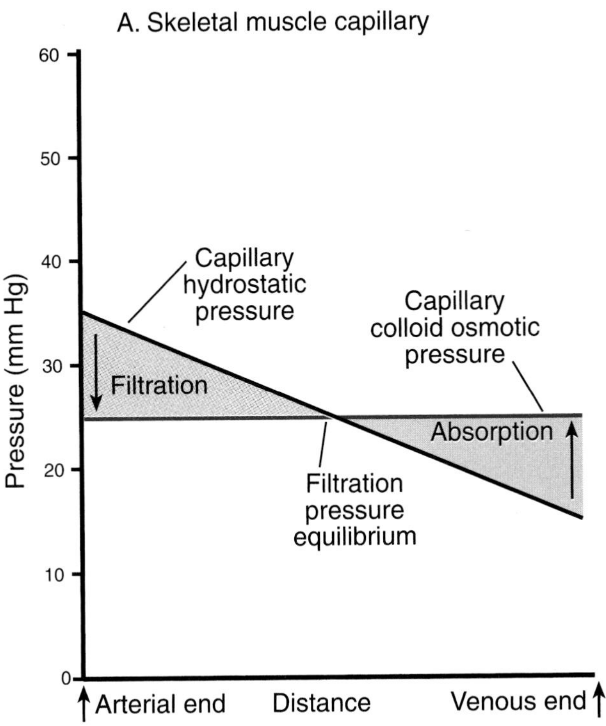

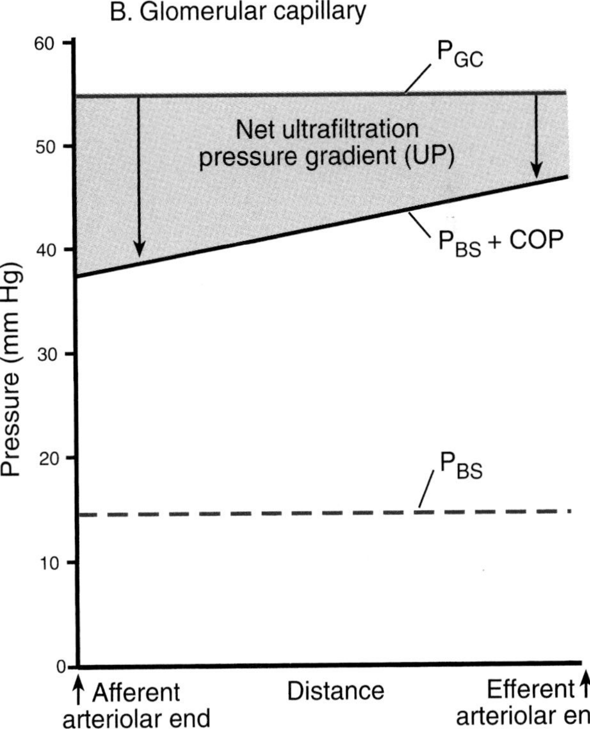

FIGURE 23.14 **Pressure profiles along a skeletal muscle capillary and a glomerular capillary.** A, In the typical skeletal muscle capillary, filtration occurs at the arterial end and absorption occurs at the venous end of the capillary. Interstitial fluid hydrostatic and colloid osmotic pressures are neglected here because they are about equal and counterbalance each other. **B**, In the glomerular capillary, glomerular hydrostatic pressure (P_{GC}) (top line) is high and declines only slightly with distance. The bottom line represents the hydrostatic pressure in Bowman's capsule (P_{BS}). The middle line is the sum of P_{BS} and the glomerular capillary colloid osmotic pressure (COP). The difference between P_{GC} and P_{BS} + COP is equal to the net ultrafiltration pressure gradient (UP). In the normal human glomerulus, filtration probably occurs along the entire capillary. Assuming that K_f is uniform along the length of the capillary, filtration rate would be highest at the afferent arteriolar end and lowest at the efferent arteriolar end of the glomerulus.

man's capsule, and the glomerular capillary colloid osmotic pressure. These factors are discussed next.

The Glomerular Ultrafiltration Coefficient. The **glomerular ultrafiltration coefficient** (K_f) is the glomerular equivalent of the capillary filtration coefficient encountered in Chapter 16. It depends on both the hydraulic conductivity (fluid permeability) and surface area of the glomerular filtration barrier. In chronic renal disease, functioning glomeruli are lost, leading to a reduction in surface area available for filtration and a fall in GFR. Acutely, a variety of drugs and hormones appear to change glomerular K_f and, thus, alter GFR, but the mechanisms are not completely understood.

Glomerular Capillary Hydrostatic Pressure. Glomerular capillary hydrostatic pressure (P_{GC}) is the driving force for filtration; it depends on the arterial blood pressure and the resistances of upstream and downstream blood vessels. Because of autoregulation, P_{GC} and GFR are maintained at relatively constant values when arterial blood pressure is varied from 80 to 180 mm Hg. Below a pressure of 80 mm Hg, however, P_{GC} and GFR decrease, and GFR ceases at a blood pressure of about 40 to 50 mm Hg. One of the classic signs of hemorrhagic or cardiogenic shock is an absence of urine output, which is due to an inadequate P_{GC} and GFR.

The caliber of afferent and efferent arterioles can be altered by a variety of hormones and by sympathetic nerve stimulation, leading to changes in P_{GC}, glomerular blood flow, and GFR. Some hormones act preferentially on afferent or efferent arterioles. Afferent arteriolar dilation increases glomerular blood flow and P_{GC} and, therefore, produces an increase in GFR. Afferent arteriolar constriction produces the exact opposite effects. Efferent arteriolar dilation increases glomerular blood flow but leads to a fall in GFR because P_{GC} is decreased. Constriction of efferent arterioles increases P_{GC} and decreases glomerular blood flow. With modest efferent arteriolar constriction, GFR increases because of the increased P_{GC}. With extreme efferent arteriolar constriction, however, GFR decreases because of the marked decrease in glomerular blood flow.

Hydrostatic Pressure in Bowman's Capsule. Hydrostatic pressure in Bowman's capsule (P_{BS}) depends on the input of glomerular filtrate and the rate of removal of this fluid by the tubule. This pressure opposes filtration. It also provides the driving force for fluid movement down the tubule lumen. If there is obstruction anywhere along the urinary tract—for example, stones, ureteral obstruction, or prostate enlargement—then pressure upstream to the block is increased, and GFR consequently falls. If tubular reabsorption of water is inhibited, pressure in the tubular system is increased because an increased pressure head is needed to force a large volume flow through the loops of Henle and collecting ducts. Consequently, a large increase in urine output caused by a diuretic drug may be associated with a tendency for GFR to fall.

Glomerular Capillary Colloid Osmotic Pressure. The COP opposes glomerular filtration. Dilution of the plasma proteins (e.g., by intravenous infusion of a large volume of isotonic saline) lowers the plasma COP and leads to an increase in GFR. Part of the reason glomerular blood flow has important effects on GFR is that the COP profile is changed along the length of a glomerular capillary. Consider, for example, what would happen if glomerular blood flow were low. Filtering a small volume out of the glomerular capillary would lead to a sharp rise in COP early along the length of the glomerulus. As a consequence, filtration would soon cease and GFR would be low. On the other hand, a high blood flow would allow a high rate of filtrate formation with a minimal rise in COP. In general, renal blood flow and GFR change hand in hand, but the exact relation between GFR and renal blood flow depends on the magnitude of the other factors that affect GFR.

Several Factors Contribute to the High GFR in the Human Kidney

The rate of plasma ultrafiltration in the kidney glomeruli (180 L/day) far exceeds that in all other capillary beds, for several reasons:

1) The filtration coefficient is unusually high in the glomeruli. Compared with most other capillaries, the glomerular capillaries behave as though they had more pores per unit surface area; consequently, they have an unusually high hydraulic conductivity. The total glomerular filtration barrier area is large, about 2 m^2.

2) Capillary hydrostatic pressure is higher in the glomeruli than in any other capillaries.

3) The high rate of renal blood flow helps sustain a high GFR by limiting the rise in colloid osmotic pressure, favoring filtration along the entire length of the glomerular capillaries.

In summary, glomerular filtration is high because the glomerular capillary blood is exposed to a large porous surface and there is a high transmural pressure gradient.

TRANSPORT IN THE PROXIMAL TUBULE

Glomerular filtration is a rather nonselective process, since both useful and waste substances are filtered. By contrast, tubular transport is selective; different substances are transported by different mechanisms. Some substances are reabsorbed, others are secreted, and still others are both reabsorbed and secreted. For most, the amount excreted in the urine depends in large measure on the magnitude of tubular transport. Transport of various solutes and water differs in the various nephron segments. Here we describe transport along the nephron and collecting duct system, starting with the proximal convoluted tubule.

The proximal convoluted tubule comprises the first 60% of the length of the proximal tubule. Because the proximal straight tubule is inaccessible to study *in vivo*, most quantitative information about function in the living animal is confined to the convoluted portion. Studies on isolated tubules *in vitro* indicate that both segments of the proximal tubule are functionally similar. The proximal tubule is responsible for reabsorbing all of the filtered glucose and amino acids; reabsorbing the largest fraction of the filtered Na^+, K^+, Ca^{2+}, Cl^-, HCO_3^-, and water and secreting various organic anions and organic cations.

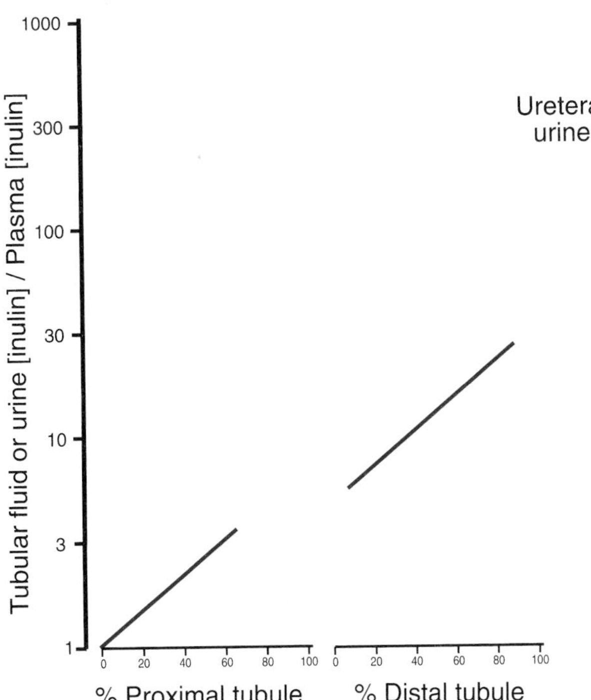

FIGURE 23.15 Tubular fluid (or urine) [inulin]/plasma [inulin] ratio as a function of collection site (data from micropuncture experiments in rats). The increase in this ratio depends on the extent of tubular water reabsorption. The distal tubule is defined in these studies as beginning at the macula densa and ending at the junction of the tubule and a collecting duct and it includes distal convoluted tubule, connecting tubule, and initial part of the collecting duct. (Modified from Giebisch G, Windhager E. Renal tubular transfer of sodium, chloride, and potassium. Am J Med 1964;36:643–669.)

The Proximal Convoluted Tubule Reabsorbs About 70% of the Filtered Water

The percentage of filtered water reabsorbed along the nephron has been determined by measuring the degree to which inulin is concentrated in tubular fluid, using the **kidney micropuncture** technique in laboratory animals. Samples of tubular fluid from surface nephrons are collected and analyzed, and the site of collection is identified by nephron microdissection. Because inulin is filtered but not reabsorbed by the kidney tubules, as water is reabsorbed, the inulin becomes increasingly concentrated. For example, if 50% of the filtered water is reabsorbed by a certain point along the tubule, the [inulin] in tubular fluid (TF_{IN}) will be twice the plasma [inulin] (P_{IN}). The percentage of filtered water reabsorbed by the tubules is equal to $100 \times (SNGFR - \dot{V}_{TF})/SNGFR$, where SN (single nephron) GFR gives the rate of filtration of water and $\dot{V}_{TF}$ is the rate of tubular fluid flow at a particular point. The SNGFR can be measured from the single nephron inulin clearance and is equal to $TF_{IN} \times \dot{V}_{TF}/P_{IN}$. From these relations:

$$\% \text{ of filtered water} = [1 - 1/(TF_{IN}/P_{IN})] \times 100 \quad (11)$$

Figure 23.15 shows how the TF_{IN}/P_{IN} ratio changes along the nephron in normal rats. In fluid collected from Bowman's capsule, the [inulin] is identical to that in plasma (inulin is freely filterable), so the concentration ratio starts at 1. By the end of the proximal convoluted tubule, the ratio is a little higher than 3, indicating that about 70% of the filtered water was reabsorbed in the proximal convoluted tubule. The ratio is about 5 at the beginning of the distal tubule, indicating that 80% of the filtered water was reabsorbed up to this point. From these measurements, we can conclude that the loop of Henle reabsorbed 10% of the filtered water. The urine/plasma inulin concentration ratio in the ureter is greater than 100, indicating that more than 99% of the filtered water was reabsorbed. These percentages are not fixed; they can vary widely, depending on conditions.

Proximal Tubular Fluid Is Essentially Isosmotic to Plasma

Samples of fluid collected from the proximal convoluted tubule are always essentially isosmotic to plasma, a consequence of the high water permeability of this segment (Fig. 23.16). Overall, 70% of filtered solutes and water are reabsorbed along the proximal convoluted tubule.

Na^+ salts are the major osmotically active solutes in the plasma and glomerular filtrate. Since osmolality does not change appreciably with proximal tubule length, it is

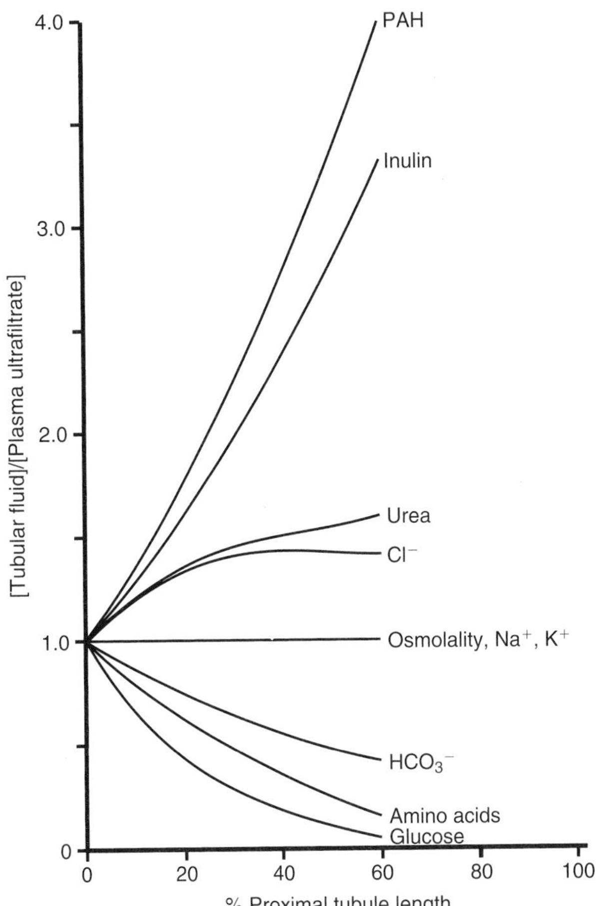

FIGURE 23.16 Tubular fluid-plasma ultrafiltrate concentration ratios for various solutes as a function of proximal tubule length. All values start at a ratio of 1, since the fluid in Bowman's capsule (0% proximal tubule length) is a plasma ultrafiltrate.

not surprising that [Na^+] also does not change under ordinary conditions.

If an appreciable quantity of nonreabsorbed solute is present (e.g., the sugar alcohol mannitol), proximal tubular fluid [Na^+] falls to values below the plasma concentration. This is evidence that Na^+ can be reabsorbed against a concentration gradient and is an active process. The fall in proximal tubular fluid [Na^+] increases diffusion of Na^+ into the tubule lumen and results in reduced net Na^+ and water reabsorption, leading to increased excretion of Na^+ and water, an osmotic diuresis.

Two major anions, Cl^- and HCO_3^-, accompany Na^+ in plasma and glomerular filtrate. HCO_3^- is preferentially reabsorbed along the proximal convoluted tubule, leading to a fall in tubular fluid [HCO_3^-], mainly because of H^+ secretion (see Chapter 25). The Cl^- lags behind; as water is reabsorbed, [Cl^-] rises (see Fig. 23.16). The result is a tubular fluid-to-plasma concentration gradient that favors Cl^- diffusion out of the tubule lumen. Outward movement of Cl^- in the late proximal convoluted tubule creates a small (1–2 mV), lumen-positive transepithelial potential difference that favors the passive reabsorption of Na^+.

Figure 23.16 shows that the [K^+] hardly changes along the proximal convoluted tubule. If K^+ were not reabsorbed, its concentration would increase as much as that of inulin. The fact that the concentration ratio for K^+ remains about 1 in this nephron segment indicates that 70% of filtered K^+ is reabsorbed along with 70% of the filtered water.

The concentrations of glucose and amino acids fall steeply in the proximal convoluted tubule. This nephron segment and the proximal straight tubule are responsible for complete reabsorption of these substances. Separate, specific mechanisms reabsorb glucose and various amino acids.

The concentration ratio for urea rises along the proximal tubule, but not as much as the inulin concentration ratio because about 50% of the filtered urea is reabsorbed. The concentration ratio for PAH in proximal tubular fluid increases more steeply than the inulin concentration ratio because of PAH secretion.

In summary, though the osmolality (total solute concentration) does not detectably change along the proximal convoluted tubule, it is clear that the concentrations of individual solutes vary widely. The concentrations of some substances fall (glucose, amino acids, HCO_3^-), others rise (inulin, urea, Cl^-, PAH), and still others do not change (Na^+, K^+). By the end of the proximal convoluted tubule, only about one-third of the filtered Na^+, water, and K^+ remain; almost all of the filtered glucose, amino acids, and HCO_3^- have been reabsorbed, and several solutes destined for excretion (PAH, inulin, urea) have been concentrated in the tubular fluid.

Na^+ Reabsorption Is the Major Driving Force for Reabsorption of Solutes and Water in the Proximal Tubule

Figure 23.17 is a model of a proximal tubule cell. Na^+ enters the cell from the lumen across the apical cell membrane and is pumped out across the basolateral cell membrane by the Na^+/K^+-ATPase. The Na^+ and accompanying anions and water are then taken up by the blood

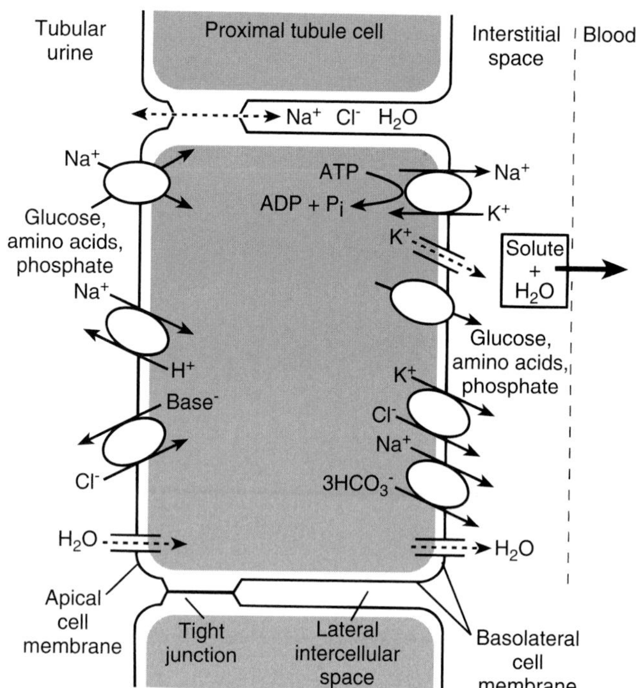

FIGURE 23.17 A cell model for transport in the proximal tubule. The luminal (apical) cell membrane in this nephron segment has a large surface area for transport because of the numerous microvilli that form the brush border (not shown). Glucose, amino acids, phosphate, and numerous other substances are transported by separate carriers.

surrounding the tubules, and filtered Na^+ salts and water are returned to the circulation.

At the luminal cell membrane (brush border) of the proximal tubule cell, Na^+ enters the cell down combined electrical and chemical potential gradients. The inside of the cell is about −70 mV compared to tubular fluid, and intracellular [Na^+] is about 30 to 40 mEq/L compared with a tubular fluid concentration of about 140 mEq/L. Na^+ entry into the cell occurs via several cotransporter and antiport mechanisms. Na^+ is reabsorbed together with glucose, amino acids, phosphate, and other solutes by way of separate, specific cotransporters. The downhill (energetically speaking) movement of Na^+ into the cell drives the uphill transport of these solutes. In other words, glucose, amino acids, phosphate, and so on are reabsorbed by secondary active transport. Na^+ is also reabsorbed across the luminal cell membrane in exchange for H^+. The Na^+/H^+ exchanger, an antiporter, is also a secondary active transport mechanism; the downhill movement of Na^+ into the cell energizes the uphill secretion of H^+ into the lumen. This mechanism is important in the acidification of urine (see Chapter 25). Cl^- may enter the cells by way of a luminal cell membrane Cl^--base (formate or oxalate) exchanger.

Once inside the cell, Na^+ is pumped out the basolateral side by a vigorous Na^+/K^+-ATPase that keeps intracellular [Na^+] low. This membrane ATPase pumps three Na^+ out of the cell and two K^+ into the cell and splits one ATP molecule for each cycle of the pump. K^+ pumped into the cell diffuses out the basolateral cell membrane mostly through a K^+ channel. Glucose, amino acids, and phosphate, accumulated in the cell because of active transport across the luminal cell membrane, exit across the basolateral cell membrane by way of separate, Na^+-independent facilitated diffusion mechanisms. HCO_3^- exits together with Na^+ by an electrogenic mechanism; the carrier transports three HCO_3^- for each Na^+. Cl^- may leave the cell by way of an electrically neutral K-Cl cotransporter.

The reabsorption of Na^+ and accompanying solutes establishes an osmotic gradient across the proximal tubule epithelium that is the driving force for water reabsorption. Because the water permeability of the proximal tubule epithelium is extremely high, only a small gradient (a few mosm/kg H_2O) is needed to account for the observed rate of water reabsorption. Some experimental evidence indicates that proximal tubular fluid is slightly hypoosmotic to plasma; since the osmolality difference is so small, it is still proper to consider the fluid as essentially isosmotic to plasma. Water crosses the proximal tubule epithelium through the cells via water channels (aquaporin-1) in the cell membranes and between the cells (tight junctions and lateral intercellular spaces).

The final step in the overall reabsorption of solutes and water is uptake by the peritubular capillaries. This mechanism involves the usual Starling forces that operate across capillary walls. Recall that blood in the peritubular capillaries was previously filtered in the glomeruli. Because a protein-free filtrate was filtered out of the glomeruli, the [protein] (hence, colloid osmotic pressure) of blood in the peritubular capillaries is high, providing an important driving force for the uptake of reabsorbed fluid. The hydrostatic pressure in the peritubular capillaries (a pressure that opposes the capillary uptake of fluid) is low because the blood has passed through upstream resistance vessels. The balance of pressures acting across peritubular capillaries favors the uptake of reabsorbed fluid from the interstitial spaces surrounding the tubules.

TABLE 23.2 Some Organic Compounds Secreted by Proximal Tubules[a]

Compound	Use
Organic Anions	
Phenol red (phenolsulfonphthalein)	pH indicator dye
p-Aminohippurate (PAH)	Measurement of renal plasma flow and proximal tubule secretory mass
Penicillin	Antibiotic
Probenecid (Benemid)	Inhibitor of penicillin secretion and uric acid reabsorption
Furosemide (Lasix)	Loop diuretic drug
Acetazolamide (Diamox)	Carbonic anhydrase inhibitor
Creatinine[b]	Normal end product of muscle metabolism
Organic Cations	
Histamine	Vasodilator, stimulator of gastric acid secretion
Cimetidine	Drug for treatment of gastric and duodenal ulcers
Cisplatin	Cancer chemotherapeutic agent
Norepinephrine	Neurotransmitter
Quinine	Antimalarial drug
Tetraethylammonium (TEA)	Ganglion blocking drug
Creatinine[b]	Normal end product of muscle metabolism

[a]This list includes only a few of the large variety of organic anions and cations secreted by kidney proximal tubules.

[b]Creatinine is an unusual compound because it is secreted by both organic anion and cation mechanisms. The creatinine molecule bears negatively charged and positively charged groups at physiological pH (it is a zwitterion), and this property may enable it to interact with both secretory mechanisms.

The Proximal Tubule Secretes Organic Ions

The proximal tubule, both convoluted and straight portions, secretes a large variety of organic anions and organic cations (Table 23.2). Many of these substances are endogenous compounds, drugs, or toxins. The organic anions are mainly carboxylates and sulfonates (carboxylic and sulfonic acids in their protonated forms). A negative charge on the molecule appears to be important for secretion of these compounds. Examples of organic anions actively secreted in the proximal tubule include penicillin and PAH. Organic anion transport becomes saturated at high plasma organic anion concentrations (see Fig. 23.9), and the organic anions compete with each other for secretion.

Figure 23.18 shows a cell model for active secretion. Proximal tubule cells actively take up PAH from the blood

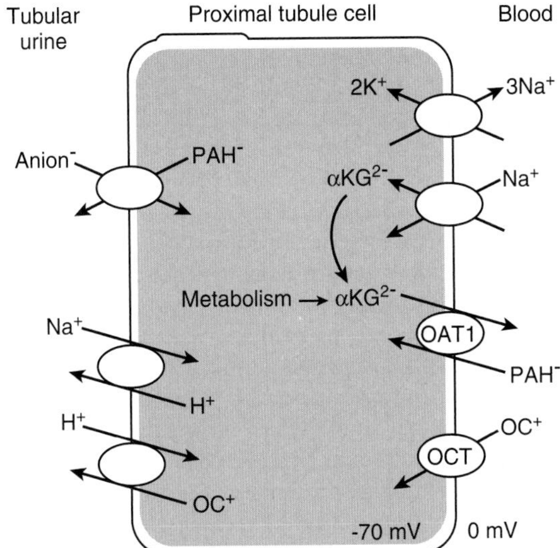

FIGURE 23.18 A cell model for the secretion of organic anions (PAH) and organic cations in the proximal tubule. Upward pointing arrows indicate transport against an electrochemical gradient (energetically uphill transport) and downward pointing arrows indicate downhill transport. There are two steps in the transcellular secretion of an organic anion or organic cation (OC^+): the active (uphill) transport step occurs in the basolateral membrane for PAH and in the luminal (brush border) membrane for the OC^+. There are actually more transporters for these molecules than are depicted in this figure. α-KG^{2-}, α-ketoglutarate; OAT1, organic anion transporter 1; OCT, organic cation transporter.

side by exchange for cell α-ketoglutarate. This exchange is mediated by an organic anion transporter (OAT) called OAT1. The cells accumulate α-ketoglutarate from metabolism and because of cell membrane Na^+-dependent dicarboxylate transporters. PAH accumulates in the cells at a high concentration and then moves downhill into the tubular urine in an electrically neutral fashion, by exchanging for an inorganic anion (e.g., Cl^-) or an organic anion.

The organic cations are mainly amine and ammonium compounds and are secreted by other transporters. Entry into the cell across the basolateral membrane is favored by the inside negative membrane potential and occurs via facilitated diffusion, mediated by an organic cation transporter (OCT). The exit of organic cations across the luminal membrane is accomplished by an organic cation/H^+ antiporter (exchanger) and is driven by the lumen-to-cell $[H^+]$ gradient established by Na^+/H^+ exchange. The transporters for organic anions and organic cations show broad substrate specificity and accomplish the secretion of a large variety of chemically diverse compounds.

In addition to being actively secreted, some organic compounds passively diffuse across the tubular epithelium. Organic anions can accept H^+ and organic cations can release H^+, so their charge is influenced by pH. The nonionized (uncharged) form, if it is lipid-soluble, can diffuse through the lipid bilayer of cell membranes down concentration gradients. The ionized (charged) form passively penetrates cell membranes with difficulty.

Consider, for example, the carboxylic acid probenecid ($pK_a = 3.4$). This compound is filtered by the glomeruli and secreted by the proximal tubule. When H^+ is secreted into the tubular urine (see Chapter 25), the anionic form (A^-) is converted to the nonionized acid (HA). The concentration of nonionized acid is also increased because of water reabsorption. A concentration gradient for passive reabsorption across the tubule wall is created, and appreciable quantities of probenecid are passively reabsorbed. This occurs in most parts of the nephron, but particularly in those where pH gradients are largest and where water reabsorption has resulted in the greatest concentration (i.e., the collecting ducts). The excretion of probenecid is enhanced by making the urine more alkaline (by administering $NaHCO_3$) and by increasing urine output (by drinking water).

Finally, a few organic anions and cations are also actively reabsorbed. For example, uric acid is both secreted and reabsorbed in the proximal tubule. Normally, the amount of uric acid excreted is equal to about 10% of the filtered uric acid, so reabsorption predominates. In **gout**, plasma levels of uric acid are increased. One treatment for gout is to promote urinary excretion of uric acid by administering drugs that inhibit its tubular reabsorption.

TUBULAR TRANSPORT IN THE LOOP OF HENLE

The loop of Henle includes several distinct segments with different structural and functional properties. As noted earlier, the proximal straight tubule has transport properties similar to those of the proximal convoluted tubule. The thin descending, thin ascending, and thick ascending limbs of the loop of Henle all display different permeability and transport properties.

Descending and Ascending Limbs Differ in Water Permeability

Tubular fluid entering the loop of Henle is isosmotic to plasma, but fluid leaving the loop is distinctly hypoosmotic. Fluid collected from the earliest part of the distal convoluted tubule has an osmolality of about 100 mosm/kg H_2O, compared with 285 mosm/kg H_2O in plasma because more solute than water is reabsorbed by the loop of Henle. The loop of Henle reabsorbs about 20% of filtered Na^+, 25% of filtered K^+, 30% of filtered Ca^{2+}, 65% of filtered Mg^{2+}, and 10% of filtered water. The descending limb of the loop of Henle (except for its terminal portion) is highly water-permeable. The ascending limb is water-impermeable. Because solutes are reabsorbed along the ascending limb and water cannot follow, fluid along the ascending limb becomes more and more dilute. Deposition of these solutes (mainly Na^+ salts) in the interstitial space of the kidney medulla is critical in the operation of the urinary concentrating mechanism.

The Luminal Cell Membrane of the Thick Ascending Limb Contains a Na-K-2Cl Cotransporter

Figure 23.19 is a model of a thick ascending limb cell. Na^+ enters the cell across the luminal cell membrane by an electrically neutral Na-K-2Cl cotransporter that is specifically inhibited by the "loop" diuretic drugs bumetanide and furosemide. The downhill movement of Na^+ into the cell results in secondary active transport of one K^+ and two Cl^-. Na^+ is pumped out the basolateral cell membrane by a vigorous Na^+/K^+-ATPase. K^+ recycles back into the lumen via a luminal cell membrane K^+ channel. Cl^- leaves through the basolateral side by a K-Cl cotransporter or Cl^- channel. The luminal cell membrane is predominantly permeable to K^+, and the basolateral cell membrane is pre-

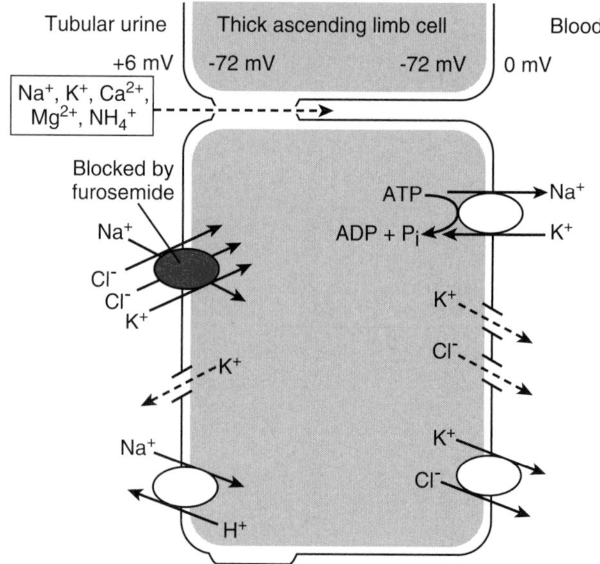

FIGURE 23.19 A cell model for ion transport in the thick ascending limb.

dominantly permeable to Cl^-. Diffusion of these ions out of the cell produces a transepithelial potential difference, with the lumen about +6 mV compared with interstitial space around the tubules. This potential difference drives small cations (Na^+, K^+, Ca^{2+}, Mg^{2+}, and NH_4^+) out of the lumen, between the cells. The tubular epithelium is extremely impermeable to water; there is no measurable water reabsorption along the ascending limb despite a large transepithelial gradient of osmotic pressure.

TUBULAR TRANSPORT IN THE DISTAL NEPHRON

The so-called **distal nephron** includes several distinct segments: distal convoluted tubule; connecting tubule; and cortical, outer medullary, and inner medullary collecting ducts (see Fig. 23.2). Note that the distal nephron includes the collecting duct system, which, strictly speaking, is not part of the nephron, but from a functional perspective, this is justified. Transport in the distal nephron differs from that in the proximal tubule in several ways:

1) The distal nephron reabsorbs much smaller amounts of salt and water. Typically, the distal nephron reabsorbs 9% of the filtered Na^+ and 19% of the filtered water, compared with 70% for both substances in the proximal convoluted tubule.

2) The distal nephron can establish steep gradients for salt and water. For example, the [Na^+] in the final urine may be as low as 1 mEq/L (versus 140 mEq/L in plasma) and the urine osmolality can be almost one-tenth that of plasma. By contrast, the proximal tubule reabsorbs Na^+ and water along small gradients, and the [Na^+] and osmolality of its tubule fluid are normally close to that of plasma.

3) The distal nephron has a "tight" epithelium, whereas the proximal tubule has a "leaky" epithelium (see Chapter 2). This explains why the distal nephron can establish steep gradients for small ions and water, whereas the proximal tubule cannot.

4) Na^+ and water reabsorption in the proximal tubule are normally closely coupled because epithelial water permeability is always high. By contrast, Na^+ and water reabsorption can be uncoupled in the distal nephron because water permeability may be low and variable.

Proximal reabsorption overall can be characterized as a coarse operation that reabsorbs large quantities of salt and water along small gradients. By contrast, distal reabsorption is a finer process.

The collecting ducts are at the end of the nephron system, and what happens there largely determines the excretion of Na^+, K^+, H^+, and water. Transport in the collecting ducts is finely tuned by hormones. Specifically, aldosterone increases Na^+ reabsorption and K^+ and H^+ secretion, and arginine vasopressin increases water reabsorption at this site.

The Luminal Cell Membrane of the Distal Convoluted Tubule Contains a Na-Cl Cotransporter

Figure 23.20 is a model of a distal convoluted tubule cell. In this nephron segment, Na^+ and Cl^- are transported from the lumen into the cell by a Na-Cl cotransporter that is inhibited by thiazide diuretics. Na^+ is pumped out the basolateral side by the Na^+/K^+-ATPase. Water permeability of the distal convoluted tubule is low and is not changed by arginine vasopressin.

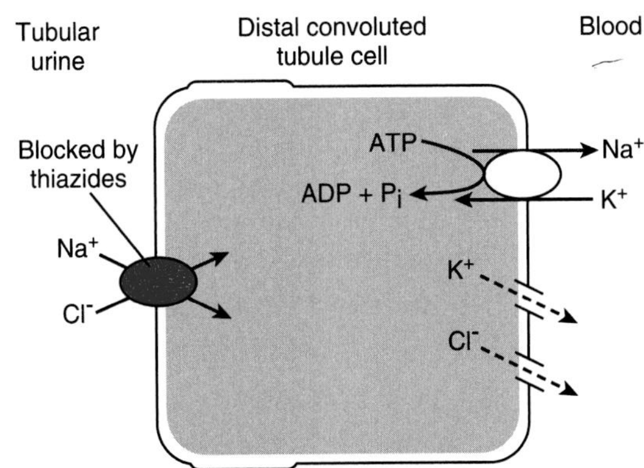

FIGURE 23.20 **A cell model for ion transport in the distal convoluted tubule.**

The Cortical Collecting Duct Is an Important Site Regulating K^+ Excretion

Under normal circumstances, most of the excreted K^+ comes from K^+ secreted by the cortical collecting ducts. With great K^+ excess (e.g., a high-K^+ diet), the cortical collecting ducts may secrete so much K^+ that more K^+ is excreted than was filtered. With severe K^+ depletion, the cortical collecting ducts reabsorb K^+.

K^+ secretion appears to be a function primarily of the collecting duct principal cell (Fig. 23.21). K^+ secretion involves active uptake by a Na^+/K^+-ATPase in the basolateral cell membrane, followed by diffusion of K^+ through luminal membrane K^+ channels. Outward diffusion of K^+ from the cell is favored by concentration gradients and opposed by electrical gradients. Note that the electrical gradient opposing exit from the cell is smaller across the luminal cell membrane than across the basolateral cell membrane, favoring movement of K^+ into the lumen rather than back into the blood. The luminal cell membrane potential difference is low (e.g., 20 mV, cell inside negative) because this membrane has a high Na^+ permeability and is depolarized by Na^+ diffusing into the cell. Recall that the entry of Na^+ into a cell causes membrane depolarization (see Chapter 3).

The magnitude of K^+ secretion is affected by several factors (see Fig. 23.21):

1) The activity of the basolateral membrane Na^+/K^+-ATPase is a key factor affecting secretion; the greater the pump activity, the higher the rate of secretion. A high plasma [K^+] promotes K^+ secretion. Increased amounts of Na^+ in the collecting duct lumen (e.g., a result of inhibition of Na^+ reabsorption by a loop diuretic drug) result in increased entry of Na^+ into principal cells, increased activity of the Na^+/K^+-ATPase, and increased K^+ secretion.

2) The lumen-negative transepithelial electrical potential promotes K^+ secretion.

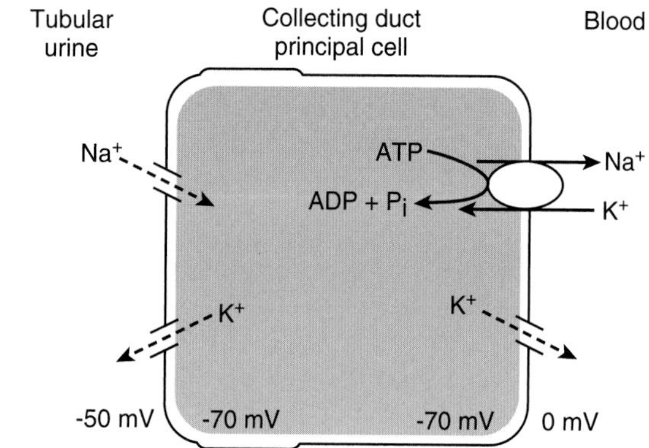

FIGURE 23.21 A model for ion transport by a collecting duct principal cell.

3) An increase in permeability of the luminal cell membrane to K^+ favors secretion.

4) A high fluid flow rate through the collecting duct lumen maintains the cell-to-lumen concentration gradient, which favors K^+ secretion.

The hormone aldosterone promotes K^+ secretion by several actions (see Chapter 24).

Na^+ entry into the collecting duct cell is by diffusion through a Na^+ channel (see Fig. 23.21). This channel has been cloned and sequenced and is known as **ENaC,** for **epithelial sodium (Na) channel.** The entry of Na^+ through this channel is rate-limiting for overall Na^+ reabsorption and is increased by aldosterone.

Intercalated cells are scattered among collecting duct principal cells; they are important in acid-base transport (see Chapter 25). A H^+/K^+-ATPase is present in the luminal cell membrane of α-intercalated cells and contributes to renal K^+ conservation when dietary intake of K^+ is deficient.

URINARY CONCENTRATION AND DILUTION

The human kidney can form urine with a total solute concentration greater or lower than that of plasma. Maximum and minimum urine osmolalities in humans are about 1,200 to 1,400 mosm/kg H_2O and 30 to 40 mosm/kg H_2O. We next consider the mechanisms involved in producing osmotically concentrated or dilute urine.

The Ability to Concentrate Urine Osmotically Is an Important Adaptation to Life on Land

When the kidneys form osmotically concentrated urine, they save water for the body. The kidneys have the task of getting rid of excess solutes (e.g., urea, various salts), which requires the excretion of solvent (water). Suppose, for example, we excrete 600 mosm of solutes per day. If we were only capable of excreting urine that is isosmotic to plasma (approximately 300 mosm/kg H_2O), we would need to excrete 2.0 L H_2O/day. If we can excrete the solutes in urine that is 4 times more concentrated than plasma (1,200 mosm/kg H_2O), only 0.5 L H_2O/day would be required. By excreting solutes in osmotically concentrated urine, the kidneys, in effect, saved $2.0 - 0.5 = 1.5$ L H_2O for the body. The ability to concentrate the urine decreases the amount of water we are obliged to find and drink each day.

Arginine Vasopressin Promotes the Excretion of an Osmotically Concentrated Urine

Changes in urine osmolality are normally brought about largely by changes in plasma levels of **arginine vasopressin** (AVP), also known as **antidiuretic hormone** (ADH) (see Chapter 32). In the absence of AVP, the kidney collecting ducts are relatively water-impermeable. Reabsorption of solute across a water-impermeable epithelium leads to osmotically dilute urine. In the presence of AVP, collecting duct water permeability is increased. Because the medullary interstitial fluid is hyperosmotic, water reabsorption in the medullary collecting ducts can lead to the production of an osmotically concentrated urine.

A model for the action of AVP on cells of the collecting duct is shown in Figure 23.22. When plasma osmolality is increased, plasma AVP levels increase. The hormone binds to a specific vasopressin (V_2) receptor in the basolateral cell membrane. By way of a guanine nucleotide stimulatory protein (G_s), the membrane-bound enzyme adenylyl cyclase is activated. This enzyme catalyzes the formation of cyclic AMP (cAMP) from ATP. Cyclic AMP then activates a cAMP-dependent protein kinase (protein kinase A [PKA]) that phosphorylates other proteins. This leads to the insertion, by exocytosis, of intracellular vesicles that contain water channels (aquaporin-2) into the luminal cell membrane. The resulting increase in number of luminal membrane water channels leads to an increase in water permeability. Water can then move out of the duct lumen through the cells, and the urinary solutes become concentrated. This response to AVP occurs in minutes. AVP also has delayed effects on collecting ducts; it increases the transcription of aquaporin-

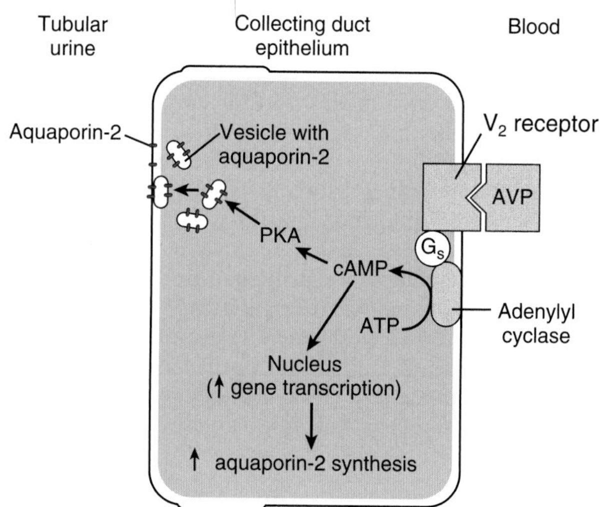

FIGURE 23.22 A model for the action of AVP on the epithelium of the collecting duct. The second messenger for AVP is cyclic AMP (cAMP). AVP has both prompt effects on luminal membrane water permeability (the movement of aquaporin-2-containing vesicles to the luminal cell membrane) and delayed effects (increased aquaporin-2 synthesis).

2 genes and produces an increase in the total number of aquaporin-2 molecules per cell.

The Loops of Henle Are Countercurrent Multipliers, and the Vasa Recta Are Countercurrent Exchangers

It has been known for longer than 50 years that there is a gradient of osmolality in the kidney medulla, with the highest osmolality present at the tips of the renal papillae. This gradient is explained by the **countercurrent hypothesis**. Two countercurrent processes occur in the kidney medulla—countercurrent multiplication and countercurrent exchange. The term *countercurrent* indicates a flow of fluid in opposite directions in adjacent structures (Fig. 23.23). The loops of Henle are **countercurrent multipliers.** Fluid flows toward the tip of the papilla along the descending limb of the loop and toward the cortex along the ascending limb of the loop. The loops of Henle set up the osmotic gradient in the medulla. Establishing a gradient requires work; the energy source is metabolism, which powers the active transport of Na^+ out of the thick ascending limb. The vasa recta are **countercurrent exchangers.** Blood flows in opposite directions along juxtaposed descending (arterial) and ascending (venous) vasa recta, and solutes and water are exchanged passively between these capillary blood vessels. The vasa recta help maintain the gradient in the medulla. The collecting ducts act as **osmotic equilibrating devices**; depending on the plasma level of AVP, the collecting duct urine is allowed to equilibrate more or less with the hyperosmotic medullary interstitial fluid.

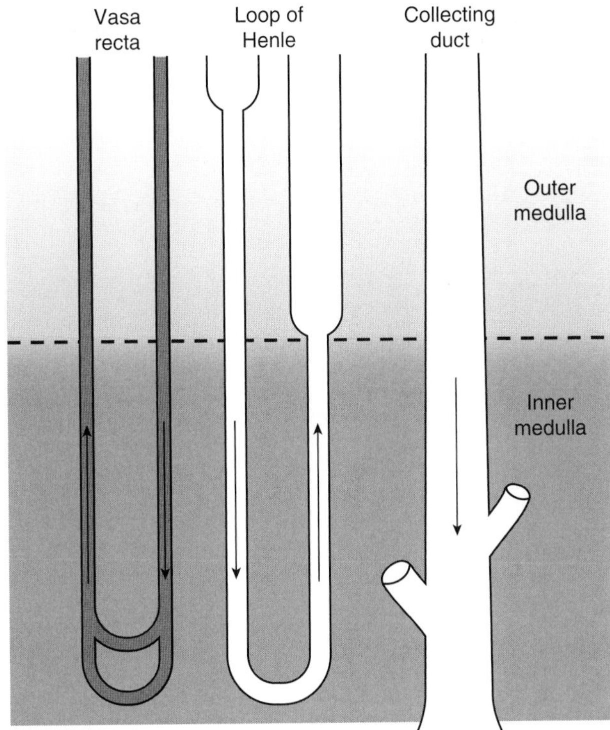

FIGURE 23.23 **Elements of the urinary concentrating mechanism.** The vasa recta are countercurrent exchangers, the loops of Henle are countercurrent multipliers, and the collecting ducts are osmotic equilibrating devices. Note that most loops of Henle and vasa recta do not reach the tip of the papilla, but turn at higher levels in the outer and inner medulla. Also, there are no thick ascending limbs in the inner medulla.

Countercurrent multiplication is the process in which a small gradient established at any level of the loop of Henle is increased (multiplied) into a much larger gradient along the axis of the loop. The osmotic gradient established at any level is called the **single effect.** The single effect involves movement of solute out of the water-impermeable ascending limb, solute deposition in the medullary interstitial fluid, and withdrawal of water from the descending limb. Because the fluid entering the next, deeper level of the loop is now more concentrated, repetition of the same process leads to an **axial gradient** of osmolality along the loop. The extent to which countercurrent multiplication can establish a large gradient along the axis of the loop depends on several factors, including the magnitude of the single effect, the rate of fluid flow, and the length of the loop of Henle. The larger the single effect, the larger the axial gradient. Impaired solute removal, as from the inhibition of active transport by thick ascending limb cells, leads to a reduced axial gradient. If flow rate through the loop is too high, not enough time is allowed for establishing a significant single effect, and consequently, the axial gradient is reduced. Finally, if the loops are long, there is more opportunity for multiplication and a larger axial gradient can be established.

Countercurrent exchange is a common process in the vascular system. In many vascular beds, arterial and venous vessels lie close to each other, and exchanges of heat or materials can occur between these vessels. For example, because of the countercurrent exchange of heat between blood flowing toward and away from its feet, a penguin can stand on ice and yet maintain a warm body (core) temperature. Countercurrent exchange between descending and ascending vasa recta in the kidney reduces dissipation of the solute gradient in the medulla. The descending vasa recta tend to give up water to the more concentrated interstitial fluid; this water is taken up by the ascending vasa recta, which come from more concentrated regions of the medulla. In effect, much of the water in the blood short-circuits across the tops of the vasa recta and does not flow deep into the medulla, where it would tend to dilute the accumulated solute. The ascending vasa recta tend to give up solute as the blood moves toward the cortex. Solute enters the descending vasa recta and, therefore, tends to be trapped in the medulla. Countercurrent exchange is a purely passive process; it helps maintain a gradient established by some other means.

Operation of the Urinary Concentrating Mechanism Requires an Integrated Functioning of the Loops of Henle, Vasa Recta, and Collecting Ducts

Figure 23.24 summarizes the mechanisms involved in producing osmotically concentrated urine. Maximally concentrated urine, with an osmolality of 1,200 mosm/kg H_2O and a low urine volume (1% of the original filtered water), is being excreted.

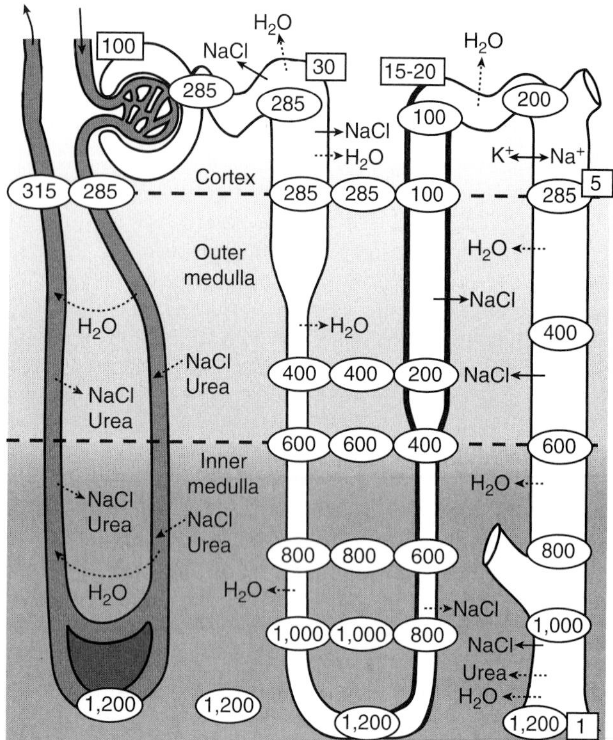

FIGURE 23.24 Osmotically concentrated urine. This diagram summarizes movements of ions, urea, and water in the kidney during production of maximally concentrated urine (1,200 mosm/kg H_2O). Numbers in ovals represent osmolality in mosm/kg H_2O. Numbers in boxes represent relative amounts of water present at each level of the nephron. Solid arrows indicate active transport; dashed arrows indicate passive transport. The heavy outlining along the ascending limb of the loop of Henle indicates relative water-impermeability.

About 70% of filtered water is reabsorbed along the proximal convoluted tubule, so 30% of the original filtered volume enters the loop of Henle. As discussed earlier, proximal reabsorption of water is essentially an isosmotic process, so fluid entering the loop is isosmotic. As the fluid moves along the descending limb of the loop Henle in the medulla, it becomes increasingly concentrated. This rise in osmolality, in principle, could be due to one of two processes:

1) The movement of water out of the descending limb because of the hyperosmolality of the medullary interstitial fluid

2) The entry of solute from the medullary interstitial fluid

The relative importance of these processes may depend on the species of animal. For most efficient operation of the concentrating mechanism, water removal should be predominant, so only this process is depicted in Figure 23.24. The removal of water along the descending limb leads to a rise in [NaCl] in the loop fluid to a value higher than in the interstitial fluid.

When the fluid enters the ascending limb, it enters water-impermeable segments. NaCl is transported out of the ascending limb and deposited in the medullary interstitial fluid. In the thick ascending limb, Na^+ transport is active and is powered by a vigorous Na^+/K^+-ATPase. In the thin ascending limb, NaCl reabsorption appears to be mainly passive. It occurs because the [NaCl] in the tubular fluid is higher than in the interstitial fluid and because the passive permeability of the thin ascending limb to Na^+ is high. There is also some evidence for a weak active Na^+ pump in the thin ascending limb. The net addition of solute to the medulla by the loops is essential for the osmotic concentration of urine in the collecting ducts.

Fluid entering the distal convoluted tubule is hypoosmotic compared to plasma (see Fig. 23.24) because of the removal of solute along the ascending limb. In the presence of AVP, the cortical collecting ducts become water-permeable and water is passively reabsorbed into the cortical interstitial fluid. The high blood flow to the cortex rapidly carries away this water, so there is no detectable dilution of cortical tissue osmolality. Before the tubular fluid reenters the medulla, it is isosmotic and reduced to about 5% of the original filtered volume. The reabsorption of water in the cortical collecting ducts is important for the overall operation of the urinary concentrating mechanism. If this water were not reabsorbed in the cortex, an excessive amount would enter the medulla. It would tend to wash out the gra-

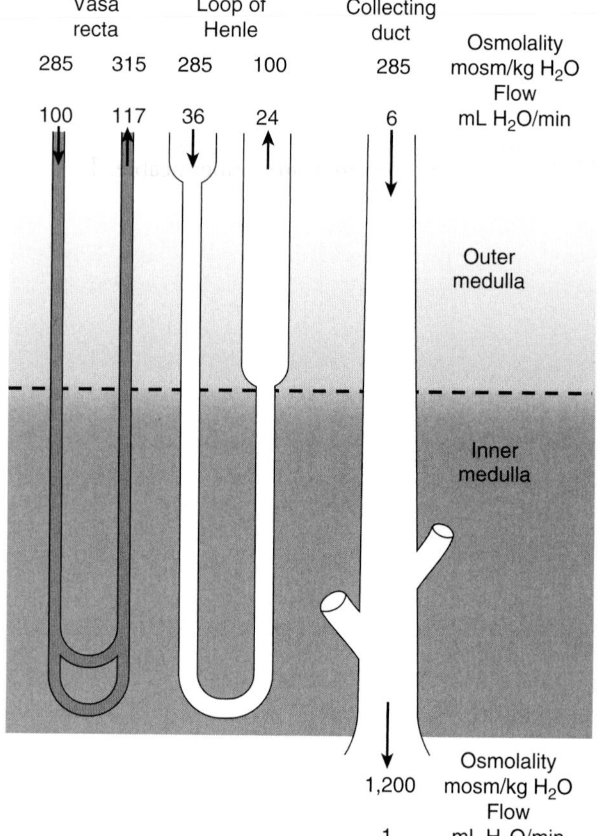

FIGURE 23.25 Mass balance considerations for the medulla as a whole. In the steady state, the inputs of water and solutes must equal their respective outputs. Water input into the medulla from the cortex (100 + 36 + 6 = 142 mL/min) equals water output from the medulla (117 + 24 + 1 = 142 mL/min). Solute input (28.5 + 10.3 + 1.7 = 40.5 mosm/min) is likewise equal to solute output (36.9 + 2.4 + 1.2 = 40.5 mosm/min).

dient in the medulla, leading to an impaired ability to concentrate the urine maximally.

All nephrons drain into collecting ducts that pass through the medulla. In the presence of AVP, the medullary collecting ducts are permeable to water. Water moves out of the collecting ducts into the more concentrated interstitial fluid. At high levels of AVP, the fluid equilibrates with the interstitial fluid, and the final urine becomes as concentrated as the tissue fluid at the tip of the papilla.

Many different models for the countercurrent mechanism have been proposed; each must take into account the principle of conservation of matter (mass balance). In the steady state, the inputs of water and every nonmetabolized solute must equal their respective outputs. This principle must be obeyed at every level of the medulla. Figure 23.25 presents a simplified scheme that applies the mass balance principle to the medulla as a whole. It provides some additional insight into the countercurrent mechanism. Notice that fluids entering the medulla (from the proximal tubule, descending vasa recta, and cortical collecting ducts) are isosmotic; they all have an osmolality of about 285 mosm/kg H_2O. Fluid leaving the medulla in the urine is hyperosmotic. It follows from mass balance considerations that somewhere a hypoosmotic fluid has to leave the medulla; this occurs in the ascending limb of the loop of Henle.

The input of water into the medulla must equal its output. Because water is added to the medulla along the descending limbs of the loops of Henle and the collecting ducts, this water must be removed at an equal rate. The ascending limbs of the loops of Henle cannot remove the added water, since they are water-impermeable. The water is removed by the vasa recta; this is why ascending exceeds descending vasa recta blood flow (see Fig. 23.25). The blood leaving the medulla is hyperosmotic because it drains a region of high osmolality and does not instantaneously equilibrate with the medullary interstitial fluid.

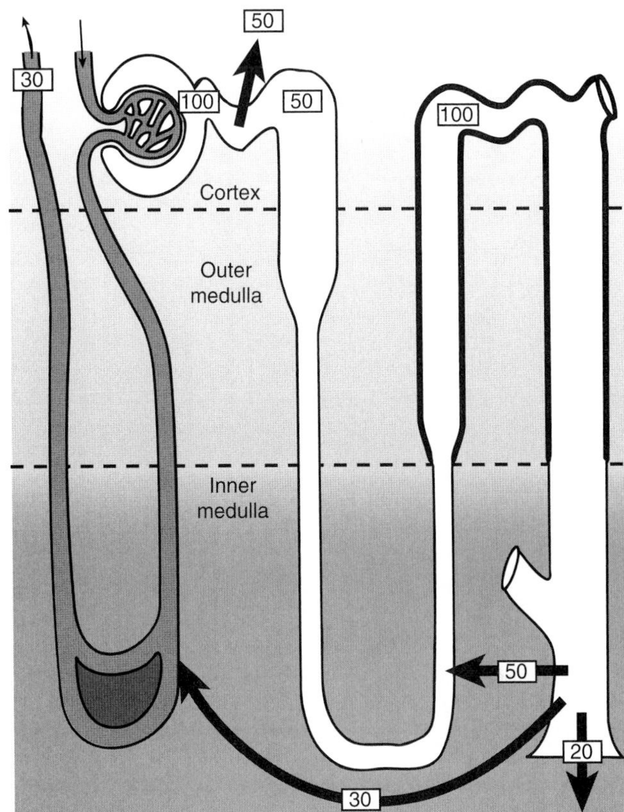

FIGURE 23.26 **Movements of urea along the nephron.** The numbers indicate relative amounts (100 = filtered urea), not concentrations. The heavy outline from the thick ascending limb to the outer medullary collecting duct indicates relatively urea-impermeable segments. Urea is added to the inner medulla by its collecting ducts; most of this urea reenters the loop of Henle, and some is removed by the vasa recta.

Urea Plays a Special Role in the Concentrating Mechanism

It has long been known that animals or humans on low-protein diets have an impaired ability to maximally concentrate the urine. A low-protein diet is associated with a decreased [urea] in the kidney medulla.

Figure 23.26 shows how urea is handled along the nephron. The proximal convoluted tubule is fairly permeable to urea and reabsorbs about 50% of the filtered urea. Fluid collected from the distal convoluted tubule, however, has as much urea as the amount filtered. Therefore, urea is secreted in the loop of Henle.

The thick ascending limb, distal convoluted tubule, connecting tubule, cortical collecting duct, and outer medullary collecting duct are relatively urea-impermeable. As water is reabsorbed along cortical and outer medullary collecting ducts, the [urea] rises. The result is the delivery to the inner medulla of a concentrated urea solution. A concentrated solution has chemical potential energy and can do work.

The inner medullary collecting duct has a facilitated urea transporter, which is activated by AVP and favors urea diffusion into the interstitial fluid of the inner medulla. Urea may reenter the loop of Henle and be recycled (see Fig. 23.26), building up its concentration in the inner medulla. Urea is also added to the inner medulla by diffusion from the urine surrounding the papillae (calyceal urine). Urea accounts for about half of the osmolality in the inner medulla. The urea in the interstitial fluid of the inner medulla counterbalances urea in the collecting duct urine, allowing the other solutes (e.g., NaCl) in the interstitial fluid to counterbalance osmotically the other solutes (e.g., creatinine, various salts) that need to be concentrated in the urine.

A Dilute Urine Is Excreted When Plasma AVP Levels Are Low

Figure 23.27 depicts kidney osmolalities during excretion of a dilute urine, as occurs when plasma AVP levels are low. Tubular fluid is diluted along the ascending limb and becomes more dilute as solute is reabsorbed across the relatively water-impermeable distal portions of the nephron and collecting ducts. Since as much as 15% of filtered water is not reabsorbed, a high urine flow rate results. In these circumstances, the osmotic gradient in the medulla is reduced but not abolished. The decreased gradient results from several factors, including an increased medullary blood flow,

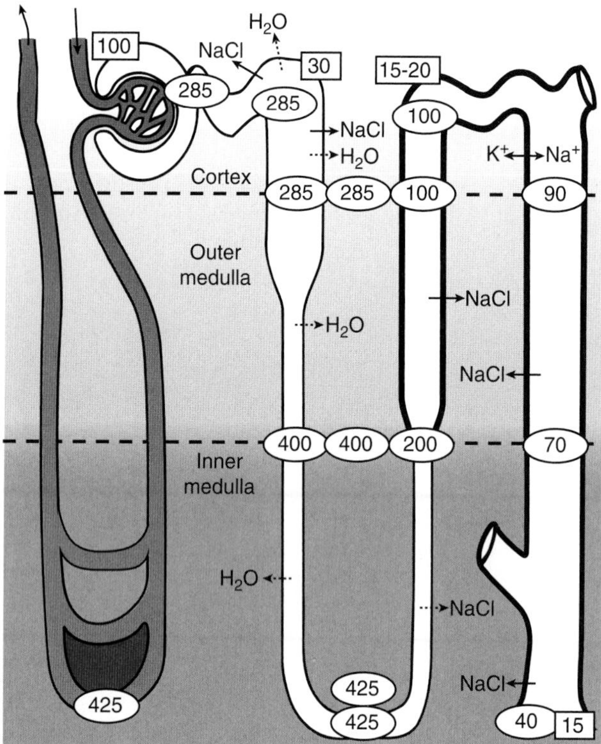

FIGURE 23.27 Osmotic gradients during excretion of osmotically very dilute urine. The collecting ducts are relatively water-impermeable (heavy outlining) because AVP is absent. Note that the medulla is still hyperosmotic, but less so than in a kidney producing osmotically concentrated urine.

TABLE 23.3 Inherited Defects in Kidney Tubule Epithelial Cells

Condition	Molecular Defect	Clinical Features
Renal glucosuria	Na^+-dependent glucose cotransporter	Glucosuria, polyuria, polydipsia, polyphagia
Cystinuria	Amino acid transporter	Kidney stone disease
Bartter's syndrome	Na-K-2Cl cotransporter, K channel or Cl channel in thick ascending limb	Salt wasting, hypokalemic metabolic alkalosis
Gitelman's syndrome	Thiazide-sensitive Na-Cl cotransporter in distal convoluted tubule	Salt wasting, hypokalemic metabolic alkalosis, hypocalciuria
Liddle's syndrome (pseudohyperaldosteronism)	Increased open time and number of principal cell epithelial sodium channels	Hypertension, hypokalemic metabolic alkalosis
Pseudohypoaldosteronism type 1	Decreased activity of epithelial sodium channels	Salt wasting, hyperkalemic metabolic acidosis
Distal renal tubular acidosis type 1	α-Intercalated cell Cl^-/HCO_3^- exchanger, H^+-ATPase	Metabolic acidosis, osteomalacia
Nephrogenic diabetes insipidus	Vasopressin-2 (V_2) receptor or aquaporin-2	Polyuria, polydipsia

reduced addition of urea, and the addition of too much water to the inner medulla by the collecting ducts.

INHERITED DEFECTS IN KIDNEY TUBULE EPITHELIAL CELLS

Recent studies have elucidated the molecular basis of several inherited kidney disorders. In many cases, the normal and mutated molecules have been cloned and sequenced. It appears that inherited defects in kidney tubule receptors (e.g., the vasopressin-2 receptor), ion channels, or carriers may explain the disturbed physiological processes of these conditions.

Table 23.3 lists some of these inherited disorders. Specific molecular defects have been identified in the proximal tubule (renal glucosuria, cystinuria), thick ascending limb (Bartter's syndrome), distal convoluted tubule (Gitelman's syndrome), and collecting duct (Liddle's syndrome, pseudohypoaldosteronism type 1, distal renal tubular acidosis, nephrogenic diabetes insipidus). Although these disorders are rare, they shed light on the pathophysiology of disease in general. For example, the finding that increased epithelial Na^+ channel activity in Liddle's syndrome leads to hypertension strengthens the view that excessive salt retention leads to high blood pressure.

REVIEW QUESTIONS

DIRECTIONS: Each of the numbered items of incomplete statements in this section is followed by answers or by completions of the statement. Select the ONE lettered answer or completion that is BEST in each case.

1. The dimensions of renal clearance are
   (A) mg/mL
   (B) mg/min
   (C) mL plasma/min
   (D) mL urine/min
   (E) mL urine/mL plasma
2. A luminal cell membrane Na^+ channel is the main pathway for Na^+ reabsorption in
   (A) Proximal tubule cells
   (B) Thick ascending limb cells
   (C) Distal convoluted tubule cells
   (D) Collecting duct principal cells
   (E) Collecting duct intercalated cells
3. A man needs to excrete 570 mosm of solute per day in his urine and his maximum urine osmolality is 1,140 mosm/kg H_2O. What is the minimum urine volume per day that he needs to excrete in order to stay in solute balance?
   (A) 0.25 L/day
   (B) 0.5 L/day

(continued)

(C) 2.0 L/day
(D) 4.0 L/day
(E) 180 L/day

4. Which of the following results in an increased osmotic gradient in the medulla of the kidney?
(A) Administration of a diuretic drug that inhibits Na^+ reabsorption by thick ascending limb cells
(B) A low GFR (e.g., 20 mL/min in an adult)
(C) Drinking a liter of water
(D) Long loops of Henle
(E) Low dietary protein intake

5. Dilation of efferent arterioles results in an increase in
(A) Glomerular blood flow
(B) Glomerular capillary pressure
(C) GFR
(D) Filtration fraction
(E) Hydrostatic pressure in the space of Bowman's capsule

6. The main driving force for water reabsorption by the proximal tubule epithelium is
(A) Active reabsorption of amino acids and glucose
(B) Active reabsorption of Na^+
(C) Active reabsorption of water
(D) Pinocytosis
(E) The high colloid osmotic pressure in the peritubular capillaries

7. The following clearance measurements were made in a man after he took a diuretic drug. What percentage of filtered Na^+ did he excrete?

Plasma [inulin]	1 mg/mL
Urine [inulin]	10 mg/mL
Plasma [Na^+]	140 mEq/L
Urine [Na^+]	70 mEq/L
Urine flow rate	10 mL/min

(A) 1%
(B) 5%
(C) 10%
(D) 50%
(E) 99%

8. Renal autoregulation
(A) Is associated with increased renal vascular resistance when arterial blood pressure is lowered from 100 to 80 mm Hg
(B) Mainly involves changes in the caliber of efferent arterioles
(C) Maintains a normal renal blood flow during severe hypotension (blood pressure, 50 mm Hg)
(D) Minimizes the impact of changes in arterial blood pressure on renal Na^+ excretion
(E) Requires intact renal nerves

9. In a kidney producing urine with an osmolality of 1,200 mosm/kg H_2O, the osmolality of fluid collected from the end of the cortical collecting duct is about
(A) 100 mosm/kg H_2O
(B) 300 mosm/kg H_2O
(C) 600 mosm/kg H_2O
(D) 900 mosm/kg H_2O
(E) 1,200 mosm/kg H_2O

10. An older woman with diabetes arrives at the hospital in a severely dehydrated condition, and she is breathing rapidly. Blood plasma [glucose] is 500 mg/dL (normal, ~100 mg/dL) and the urine [glucose] is zero (dipstick test). What is the most likely explanation for the absence of glucose in the urine?
(A) The amount of splay in the glucose reabsorption curve is abnormally increased
(B) GFR is abnormally low
(C) The glucose Tm is abnormally high
(D) The glucose Tm is abnormally low
(E) The renal plasma glucose threshold is abnormally low

11. In a suicide attempt, a nurse took an overdose of the sedative phenobarbital. This substance is a weak, lipid-soluble organic acid that is reabsorbed by nonionic diffusion in the kidneys. Which of the following would promote urinary excretion of this substance?
(A) Abstain from all fluids
(B) Acidify the urine by ingesting NH_4Cl tablets
(C) Administer a drug that inhibits tubular secretion of organic anions
(D) Alkalinize the urine by infusing a $NaHCO_3$ solution intravenously

12. Which of the following provides the most accurate measure of GFR?
(A) Blood urea nitrogen (BUN)
(B) Endogenous creatinine clearance
(C) Inulin clearance
(D) PAH clearance
(E) Plasma [creatinine]

13. Hypertension was observed in a young boy since birth. Which of the following disorders may be present?
(A) Bartter's syndrome
(B) Gitelman's syndrome
(C) Liddle's syndrome
(D) Nephrogenic diabetes insipidus
(E) Renal glucosuria

14. In a person with severe central diabetes insipidus (deficient production or release of AVP), urine osmolality and flow rate is typically about
(A) 50 mosm/kg H_2O, 18 L/day
(B) 50 mosm/kg H_2O, 1.5 L/day
(C) 300 mosm/kg H_2O, 1.5 L/day
(D) 300 mosm/kg H_2O, 18 L/day
(E) 1,200 mosm/kg H_2O, 0.5 L/day

15. Which of the following substances has the highest renal clearance?
(A) Creatinine
(B) Inulin
(C) PAH
(D) Na^+
(E) Urea

16. If the plasma concentration of a freely filterable substance is 2 mg/mL, GFR is 100 mL/min, urine concentration of the substance is 10 mg/mL, and urine flow rate is 5 mL/min, we can conclude that the kidney tubules
(A) reabsorbed 150 mg/min
(B) reabsorbed 200 mg/min
(C) secreted 50 mg/min
(D) secreted 150 mg/min
(E) secreted 200 mg/min

17. A clearance study was done on a young woman with suspected renal disease:

Arterial [PAH]	0.02 mg/mL
Renal vein [PAH]	0.01 mg/mL
Urine [PAH]	0.60 mg/mL
Urine flow rate	5.0 mL/min
Hematocrit, % cells	40

What is her true renal blood flow?
(A) 150 mL/min
(B) 300 mL/min
(C) 500 mL/min
(D) 750 mL/min
(E) 1,200 mL/min

18. A man has progressive, chronic kidney disease. Which of the following indicates the greatest absolute decrease in GFR?
(A) A fall in plasma creatinine from 4 mg/dL to 2 mg/dL
(B) A fall in plasma creatinine from 2 mg/dL to 1 mg/dL
(C) A rise in plasma creatinine from 1 mg/dL to 2 mg/dL
(D) A rise in plasma creatinine from 2 mg/dL to 4 mg/dL
(E) A rise in plasma creatinine from 4 mg/dL to 8 mg/dL

19. Renin in synthesized by
(A) Granular cells
(B) Intercalated cells
(C) Interstitial cells
(D) Macula densa cells
(E) Mesangial cells

20. The following determinations were made on a single glomerulus of a rat kidney: GFR, 42 nL/min; glomerular capillary hydrostatic pressure, 50 mm Hg; hydrostatic pressure in Bowman's space, 12 mm Hg; average glomerular capillary colloid osmotic pressure, 24 mm Hg. What is the glomerular ultrafiltration coefficient?
(A) 0.33 mm Hg per nL/min
(B) 0.49 nL/min per mm Hg
(C) 0.68 nL/min per mm Hg
(D) 1.48 mm Hg per nL/min
(E) 3.0 nL/min per mm Hg

SUGGESTED READING

Brooks VL, Vander AJ, eds. Refresher course for teaching renal physiology. Adv Physiol Educ 1998;20:S114–S245.

Burckhardt G, Bahn A, Wolff NA. Molecular physiology of renal *p*-aminohippurate secretion. News Physiol Sci 2001;16:113–118.

(continued)

Koeppen BM, Stanton BA. Renal Physiology. 3rd Ed. St. Louis: Mosby, 2001.
Kriz W, Bankir L. A standard nomenclature for structures of the kidney. Am J Physiol 1988;254:F1–F8.
Rose BD, Post TW. Clinical Physiology of Acid-Base and Electrolyte Disorders. 5th Ed. New York: McGraw-Hill, 2001.
Scheinman SJ, Guay-Woodford LM, Thakker RV, Warnock DG. Genetic disorders of renal electrolyte transport. N Engl J Med 1999;340:1177–1187.
Seldin DW, Giebisch G, eds. The Kidney: Physiology and Pathophysiology. 3rd Ed. Philadelphia: Lippincott Williams & Wilkins, 2000.
Valtin H, Schafer JA. Renal Function. 3rd Ed. Boston: Little, Brown, 1995.
Vander AJ. Renal Physiology. 5th Ed. New York: McGraw-Hill, 1995.

The Regulation of Fluid and Electrolyte Balance

George A. Tanner, Ph.D.

CHAPTER OUTLINE

- FLUID COMPARTMENTS OF THE BODY
- WATER BALANCE
- SODIUM BALANCE
- POTASSIUM BALANCE
- CALCIUM BALANCE
- MAGNESIUM BALANCE
- PHOSPHATE BALANCE
- URINARY TRACT

KEY CONCEPTS

1. Total body water is distributed in two major compartments: intracellular water and extracellular water. In an average young adult man, total body water, intracellular water, and extracellular water amount to 60%, 40%, and 20% of body weight, respectively. The corresponding figures for an average young adult woman are 50%, 30%, and 20% of body weight.
2. The volumes of body fluid compartments are determined by using the indicator dilution method and the equation: Volume = Amount of indicator ÷ Concentration of indicator at equilibrium.
3. Electrical neutrality is present in solutions of electrolytes; that is, the sum of the cations is equal to the sum of the anions (both expressed in milliequivalents).
4. Sodium (Na^+) is the major osmotically active solute in extracellular fluid (ECF), and potassium (K^+) has the same role in the intracellular fluid (ICF) compartment. Cells are typically in osmotic equilibrium with their external environment. The amount of water in (and, hence, the volume of) cells depends on the amount of K^+ they contain and, similarly, the amount of water in (and, hence, the volume of) the ECF is determined by its Na^+ content.
5. Plasma osmolality is closely regulated by arginine vasopressin (AVP), which governs renal excretion of water, and by habit and thirst, which govern water intake.
6. AVP is synthesized in the hypothalamus, released from the posterior pituitary gland, and acts on the collecting ducts of the kidney to increase their water permeability. The major stimuli for the release of AVP are an increase in effective plasma osmolality (detected by osmoreceptors in the anterior hypothalamus) and a decrease in blood volume (detected by stretch receptors in the left atrium, carotid sinuses, and aortic arch).
7. The kidneys are the primary site of control of Na^+ excretion. Only a small percentage (usually about 1%) of the filtered Na^+ is excreted in the urine, but this amount is of critical importance in overall Na^+ balance.
8. Multiple factors affect Na^+ excretion, including glomerular filtration rate, angiotensin II and aldosterone, intrarenal physical forces, natriuretic hormones and factors such as atrial natriuretic peptide, and renal sympathetic nerves. Changes in these factors may account for altered Na^+ excretion in response to excess Na^+ or Na^+ depletion. Estrogens, glucocorticoids, osmotic diuretics, poorly reabsorbed anions in the urine, and diuretic drugs also affect renal Na^+ excretion.
9. The effective arterial blood volume (EABV) depends on the degree of filling of the arterial system and determines the perfusion of the body's tissues. A decrease in EABV leads to Na^+ retention by the kidneys and contributes to the development of generalized edema in pathophysiological conditions, such as congestive heart failure.
10. The kidneys play a major role in the control of K^+ balance. K^+ is reabsorbed by the proximal convoluted tubule and the loop of Henle and is secreted by cortical collecting duct principal cells. Inadequate renal K^+ excretion produces hyperkalemia, and excessive K^+ excretion produces hypokalemia.
11. Calcium balance is regulated on both input and output sides. The absorption of Ca^{2+} from the small intestine is controlled by $1,25(OH)_2$ vitamin D_3, and the excretion of Ca^{2+} by the kidneys is controlled by parathyroid hormone (PTH).
12. Magnesium in the body is mostly in bone, but it is also an important intracellular ion. The kidneys regulate the plasma $[Mg^{2+}]$.
13. Filtered phosphate usually exceeds the maximal reabsorptive capacity of the kidney tubules for phosphate (Tm_{PO_4}), and about 5 to 20% of filtered phosphate is usually excreted. Phosphate reabsorption occurs mainly in the proximal tubules and is inhibited by PTH. Phosphate is an important pH buffer in the urine. Hyperphosphatemia is a significant problem in chronic renal failure.
14. The urinary bladder stores urine until it can be conveniently emptied. Micturition is a complex act involving both autonomic and somatic nerves.

A major function of the kidneys is to regulate the volume, composition, and osmolality of the body fluids. The fluid surrounding our body cells (the ECF) is constantly renewed and replenished by the circulating blood plasma. The kidneys constantly process the plasma; they filter, reabsorb, and secrete substances and, in health, maintain the internal environment within narrow limits. In this chapter, we begin with a discussion of the fluid compartments of the body—their location, magnitude, and composition. Then we consider water, sodium, potassium, calcium, magnesium, and phosphate balance, with special emphasis on the role of the kidneys in maintaining our fluid and electrolyte balance. Finally, we consider the role of the ureters, urinary bladder, and urethra in the transport, storage, and elimination of urine.

FLUID COMPARTMENTS OF THE BODY

Water is the major constituent of all body fluid compartments. **Total body water** averages about 60% of body weight in young adult men and about 50% of body weight in young adult women (Table 24.1). The percentage of body weight water occupies depends on the amount of adipose tissue (fat) a person has. A lean person has a high percentage and an obese individual a low percentage of body weight that is water because adipose tissue contains a low percentage of water (about 10%), whereas most other tissues have a much higher percentage of water. For example, muscle is about 75% water. Newborns have a low percentage of body weight as water because of a relatively large ECF volume and little fat (see Table 24.1). Adult women have relatively less water than men because, on average, they have more subcutaneous fat and less muscle mass. As people age, they tend to lose muscle and add adipose tissue; hence, water content declines with age.

Body Water Is Distributed in Several Fluid Compartments

Total body water can be divided into two compartments or spaces: **intracellular fluid** (ICF) and **extracellular fluid** (ECF). The ICF is comprised of the fluid within the trillions of cells in our body. The ECF is comprised of fluid outside of the cells. In a young adult man, two thirds of the body water is in the ICF, and one third is in the ECF (Fig. 24.1). These two fluids differ strikingly in terms of their electrolyte composition. However, their total solute concentrations (osmolalities) are normally equal, because of the high water permeability of most cell membranes, so that an osmotic difference between cells and ECF rapidly disappears.

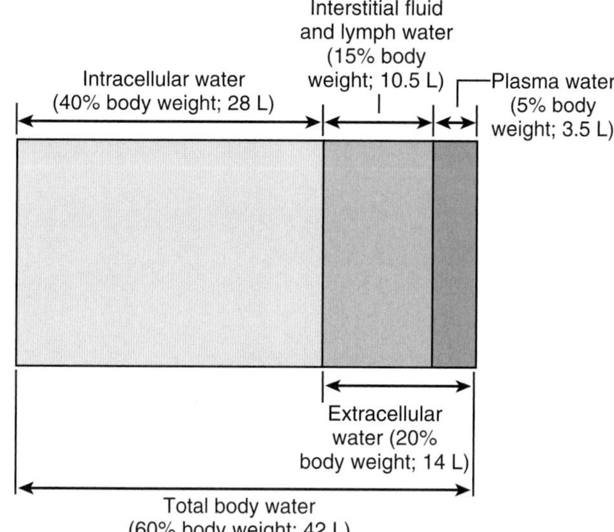

FIGURE 24.1 **Water distribution in the body.** This diagram is for an average young adult man weighing 70 kg. In an average young adult woman, total body water is 50% of body weight, intracellular water is 30% of body weight, and extracellular water is 20% of body weight.

The ECF can be further subdivided into two major subcompartments, which are separated from each other by the endothelium of blood vessels. The **blood plasma** is the ECF found within the vascular system; it is the fluid portion of the blood in which blood cells and platelets are suspended. The blood plasma water comprises about one fourth of the ECF or about 3.5 L for an average 70-kg man (see Fig. 24.1). The interstitial fluid and lymph are considered together because they cannot be easily separated. The water of the **interstitial fluid** and **lymph** comprises three fourths of the ECF. The interstitial fluid directly bathes most body cells, and the lymph is the fluid within lymphatic vessels. The blood plasma, interstitial fluid, and lymph are nearly identical in composition, except for the higher protein concentration in the plasma.

An additional ECF compartment (not shown in Fig. 24.1), the **transcellular fluid**, is small but physiologically important. Transcellular fluid amounts to about 1 to 3% of body weight. Transcellular fluids include cerebrospinal fluid, aqueous humor of the eye, secretions of the digestive tract and associated organs (saliva, bile, pancreatic juice), renal tubular fluid and bladder urine, synovial fluid, and sweat. In these cases, the fluid is separated from the blood plasma by an epithelial cell layer in addition to a capillary endothelium. The epithelial layer modifies the electrolyte composition of the fluid, so that transcellular fluids are not plasma ultrafiltrates (as is interstitial fluid and lymph); they have a distinct ionic composition. There is a constant turnover of transcellular fluids; they are continuously formed and absorbed or removed. Impaired for-

TABLE 24.1 Average Total Body Water as a Percentage of Body Weight

Age	Men	Both Sexes	Women
0–1 month		76	
1–12 months		65	
1–10 years		62	
10–16 years	59		57
17–39 years	61		50
40–59 years	55		52
60 years and older	52		46

From Edelman IS, Leibman J. Anatomy of body water and electrolytes. Am J Med 1959;27:256–277.

mation, abnormal loss from the body, or blockage of fluid removal can have serious consequences.

The Indicator Dilution Method Measures Fluid Compartment Size

The indicator dilution method can be used to determine the size of body fluid compartments (see Chapter 14). A known amount of a substance (the indicator), which should be confined to the compartment of interest, is administered. After allowing sufficient time for uniform distribution of the indicator throughout the compartment, a plasma sample is collected. The concentration of the indicator in the plasma at equilibrium is measured, and the distribution volume is calculated from this formula

$$\text{Volume} = \frac{\text{Amount of indicator}}{\text{Concentration of indicator}} \quad (1)$$

If there was loss of indicator from the fluid compartment, the amount lost is subtracted from the amount administered.

To measure total body water, heavy water (deuterium oxide), tritiated water (HTO), or antipyrene (a drug that distributes throughout all of the body water) is used as an indicator. For example, suppose we want to measure total body water in a 60-kg woman. We inject 30 mL of deuterium oxide (D_2O) as an isotonic saline solution into an arm vein. After a 2-hour equilibration period, a blood sample is withdrawn, and the plasma is separated and analyzed for D_2O. A concentration of 0.001 mL D_2O/mL plasma water is found. Suppose during the equilibration period, urinary, respiratory, and cutaneous losses of D_2O are 0.12 mL. Substituting these values into the indicator dilution equation, we get

$$\text{Total body water} = \frac{(30 - 0.12 \text{ mL } D_2O)}{0.001 \text{ mL } D_2O/\text{mL water}}$$

$$= 29{,}880 \text{ mL or } 30 \text{ L} \quad (2)$$

Therefore, total body water as a percentage of body weight equals 50% in this woman.

To measure extracellular water volume, the ideal indicator should distribute rapidly and uniformly outside the cells and should not enter the cell compartment. Unfortunately, there is no such ideal indicator, so the exact volume of the ECF cannot be measured. A reasonable estimate, however, can be obtained using two different classes of substances: impermeant ions and inert sugars. ECF volume has been determined from the volume of distribution of these ions: radioactive Na^+, radioactive Cl^-, radioactive sulfate, thiocyanate (SCN^-), and thiosulfate ($S_2O_3^{2-}$); radioactive sulfate ($^{35}SO_4^{2-}$) is probably the most accurate. However, ions are not completely impermeant; they slowly enter the cell compartment, so measurements tend to lead to an overestimate of ECF volume. Measurements with inert sugars (such as mannitol, sucrose, and inulin) tend to lead to an underestimate of ECF volume because they are excluded from some of the extracellular water—for example, the water in dense connective tissue and cartilage. Special techniques are required when using these sugars because they are rapidly filtered and excreted by the kidneys after their intravenous injection.

Cellular water cannot be determined directly with any indicator. It can, however, be calculated from the difference between measurements of total body water and extracellular water.

Plasma water is determined by using Evans blue dye, which avidly binds serum albumin, or radioiodinated serum albumin (RISA), and by collecting and analyzing a blood plasma sample. In effect, the plasma volume is measured from the distribution volume of serum albumin. The assumption is that serum albumin is completely confined to the vascular compartment, but this is not entirely true. Indeed, serum albumin is slowly (3 to 4% per hour) lost from the blood by diffusive and convective transport through capillary walls. To correct for this loss, repeated blood samples can be collected at timed intervals, and the concentration of albumin at time zero (the time at which no loss would have occurred) can be determined by extrapolation. Alternatively, the plasma concentration of indicator 10 minutes after injection can be used; this value is usually close to the extrapolated value. If plasma volume and hematocrit are known, total circulating blood volume can be calculated (see Chapter 11).

Interstitial fluid and lymph volume cannot be determined directly. It can be calculated as the difference between ECF and plasma volumes.

Body Fluids Differ in Electrolyte Composition

Body fluids contain many uncharged molecules (e.g., glucose and urea), but quantitatively speaking, **electrolytes** (ionized substances) contribute most to the total solute concentration (or osmolality) of body fluids. Osmolality is of prime importance in determining the distribution of water between intracellular and ECF compartments.

The importance of ions (particularly Na^+) in determining the plasma osmolality (P_{osm}) is exemplified by an equation that is of value in the clinic:

$$P_{osm} = 2 \times [Na^+] + \frac{[\text{glucose}] \text{ in mg/dL}}{18} + \frac{[\text{blood urea nitrogen}] \text{ in mg/dL}}{2.8} \quad (3)$$

If the plasma $[Na^+]$ is 140 mmol/L, blood glucose is 100 mg/dL (5.6 mmol/L), and blood urea nitrogen is 10 mg/dL (3.6 mmol/L), the calculated osmolality is 289 mosm/kg H_2O. The equation indicates that Na^+ and its accompanying anions (mainly Cl^- and HCO_3^-) normally account for more than 95% of the plasma osmolality. In some special circumstances (e.g., alcohol intoxication), plasma osmolality calculated from the above equation may be much lower than the true, measured osmolality as a result of the presence of unmeasured osmotically active solutes (e.g., ethanol).

The concentrations of various electrolytes in plasma, interstitial fluid, and ICF are summarized in Table 24.2. The ICF values are based on determinations made in skeletal muscle cells. These cells account for about two thirds of the cell mass in the human body. Concentrations are expressed in terms of milliequivalents per liter or per kg H_2O.

An **equivalent** contains one mole of univalent ions, and a **milliequivalent** (mEq) is 1/1,000th of an equivalent. Equiv-

TABLE 24.2 Electrolyte Composition of the Body Fluids

	Plasma Electrolyte (mEq/L)	Plasma Water (mEq/kg H_2O)	Interstitial Fluid (mEq/kg H_2O)	Intracellular Fluid[a] (mEq/kg H_2O)
Cations				
Na^+	142	153	145	10
K^+	4	4.3	4	159
Ca^{2+}	5	5.4	3	1
Mg^{2+}	2	2.2	2	40
Total	153	165	154	210
Anions				
Cl^-	103	111	117	3
HCO_3^-	25	27	28	7
Protein	17	18	—	45
Others	8	9	9	155
Total	153	165	154	210

[a] Skeletal muscle cells.

alents are calculated as the product of moles times valence and represent the concentration of charged species. For singly charged (univalent) ions, such as Na^+, K^+, Cl^-, or HCO_3^-, 1 mmol is equal to 1 mEq. For doubly charged (divalent) ions, such as Ca^{2+}, Mg^{2+}, or SO_4^{2-}, 1 mmol is equal to 2 mEq. Some electrolytes, such as proteins, are polyvalent, so there are several mEq/mmol. The usefulness of expressing concentrations in terms of mEq/L is based on the fact that in solutions, we have electrical neutrality; that is

$$\Sigma \text{ cations} = \Sigma \text{ anions} \tag{4}$$

If we know the total concentration (mEq/L) of all cations in a solution and know only some of the anions, we can easily calculate the concentration of the remaining anions. This was done in Table 24.2 for the anions labeled "Others." Plasma concentrations are listed in the first column of Table 24.2. Na^+ is the major cation in plasma, and Cl^- and HCO_3^- are the major anions. The plasma proteins (mainly serum albumin) bear net negative charges at physiological pH. The electrolytes are actually dissolved in the plasma water, so the second column in Table 24.2 expresses concentrations per kg H_2O. The water content of plasma is usually about 93%; about 7% of plasma volume is occupied by solutes, mainly the plasma proteins. To convert concentration in plasma to concentration in plasma water, we divided the plasma concentration by the plasma water content (0.93 L H_2O/L plasma). Therefore, 142 mEq Na^+/L plasma becomes 153 mEq/L H_2O or 153 mEq/kg H_2O (since 1 L of water weighs 1 kg).

Interstitial fluid (Column 3 of Table 24.2) is an ultrafiltrate of plasma. It contains all of the small electrolytes in essentially the same concentration as in plasma, but little protein. The proteins are largely confined to the plasma because of their large molecular size. Differences in small ion concentrations between plasma and interstitial fluid (compare Columns 2 and 3) occur because of the different protein concentrations in these two compartments. Two factors are involved. The first is an electrostatic effect: Because the plasma proteins are negatively charged, they cause a redistribution of small ions, so that the concentrations of diffusible cations (such as Na^+) are lower in interstitial fluid than in plasma and the concentrations of diffusible anions (such as Cl^-) are higher in interstitial fluid than in plasma. Second, Ca^{2+} and Mg^{2+} are bound to some extent (about 40% and 20%, respectively) by plasma proteins, and it is only the unbound ions that can diffuse through capillary walls. Hence, the total plasma Ca^{2+} and Mg^{2+} concentrations are higher than in interstitial fluid.

ICF composition (Table 24.2, Column 4) is different from ECF composition. The cells have a higher K^+, Mg^{2+}, and protein concentration than in the surrounding interstitial fluid. The intracellular Na^+, Ca^{2+}, Cl^-, and HCO_3^- levels are lower than outside the cell. The anions in skeletal muscle cells labeled "Others" are mainly organic phosphate compounds important in cell energy metabolism, such as creatine phosphate, ATP, and ADP. As described in Chapter 2, the high intracellular $[K^+]$ and low intracellular $[Na^+]$ are a consequence of plasma membrane Na^+/K^+-ATPase activity; this enzyme extrudes Na^+ from the cell and takes up K^+. The low intracellular $[Cl^-]$ and $[HCO_3^-]$ in skeletal muscle cells are primarily a consequence of the inside negative membrane potential (−90 mV), which favors the outward movement of these small, negatively charged ions. The intracellular $[Mg^{2+}]$ is high; most is not free, but is bound to cell proteins. Intracellular $[Ca^{2+}]$ is low; as discussed in Chapter 1, the cytosolic $[Ca^{2+}]$ in resting cells is about 10^{-7} M (0.0002 mEq/L). Most of the cell Ca^{2+} is sequestered in organelles, such as the sarcoplasmic reticulum in skeletal muscle.

Intracellular and Extracellular Fluids Are Normally in Osmotic Equilibrium

Despite the different compositions of ICF and ECF, the total solute concentration (osmolality) of these two fluid compartments is normally the same. ICF and ECF are in osmotic equilibrium because of the high water permeability of cell membranes, which does not permit an osmolality difference to be sustained. If the osmolality changes in one compartment, water moves to restore a new osmotic equilibrium (see Chapter 2).

The volumes of ICF and ECF depend primarily on the

volume of water present in these compartments. But the latter depends on the amount of solute present and the osmolality. This fact follows from the definition of the term *concentration*: concentration = amount/volume; hence, volume = amount/concentration. The main osmotically active solute in cells is K^+; therefore, a loss of cell K^+ will cause cells to lose water and shrink (see Chapter 2). The main osmotically active solute in the ECF is Na^+; therefore, a gain or loss of Na^+ from the body will cause the ECF volume to swell or shrink, respectively.

The distribution of water between intracellular and extracellular compartments changes in a variety of circumstances. Figure 24.2 provides some examples. Total body water is divided into the two major compartments, ICF and ECF. The y-axis represents total solute concentration and the x-axis the volume; the area of a box (concentration times volume) gives the amount of solute present in a compartment. Note that the height of the boxes is always equal, since osmotic equilibrium (equal osmolalities) is achieved.

In the normal situation (shown in Figure 24.2A), two thirds (28 L for a 70-kg man) of total body water is in the ICF, and one third (14 L) is in the ECF. The osmolality of both fluids is 285 mosm/kg H_2O. Hence, the cell compartment contains 7,980 mosm and the ECF contains 3,990 mosm.

In Figure 24.2B, 2.0 L of pure water were added to the ECF (e.g., by drinking water). Plasma osmolality is lowered, and water moves into the cell compartment along the osmotic gradient. The entry of water into the cells causes them to swell, and intracellular osmolality falls until a new equilibrium (solid lines) is achieved. Since 2 L of water were added to an original total body water volume of 42 L, the new total body water volume is 44 L. No solute was added, so the new osmolality at equilibrium is (7,980 + 3,990 mosm)/44 kg = 272 mosm/kg H_2O. The volume of the ICF at equilibrium, calculated by solving the equation, 272 mosm/kg H_2O × volume = 7,980 mosm, is 29.3 L. The volume of the ECF at equilibrium is 14.7 L. From these calculations, we conclude that two thirds of the added water ends up in the cell compartment and one third stays in the ECF. This description of events is artificial because, in reality, the kidneys would excrete the added water over the course of a few hours, minimizing the fall in plasma osmolality and cell swelling.

In Figure 24.2C, 2.0 L of isotonic saline (0.9% NaCl solution) were added to the ECF. Isotonic saline is isosmotic to plasma or ECF and, by definition, causes no change in cell volume. Therefore, all of the isotonic saline is retained in the ECF and there is no change in osmolality.

Figure 24.2D shows the effect of infusing intravenously 1.0 L of a 5% NaCl solution (osmolality about 1,580 mosm/kg H_2O). All the salt stays in the ECF. The cells are exposed to a hypertonic environment, and water leaves the cells. Solutes left behind in the cells become more concentrated as water leaves. A new equilibrium will be established, with the final osmolality higher than normal but equal inside and outside the cells. The final osmolality can be calculated from the amount of solute present (7,980 + 3,990 + 1,580 mosm) divided by the final volume (28 + 14 + 1 L); it is equal to 315 mosm/kg H_2O. The final volume of the ICF equals 7,980 mosm divided by 315 mosm/kg H_2O or 25.3 L, which is 2.7 L less than the initial volume. The final

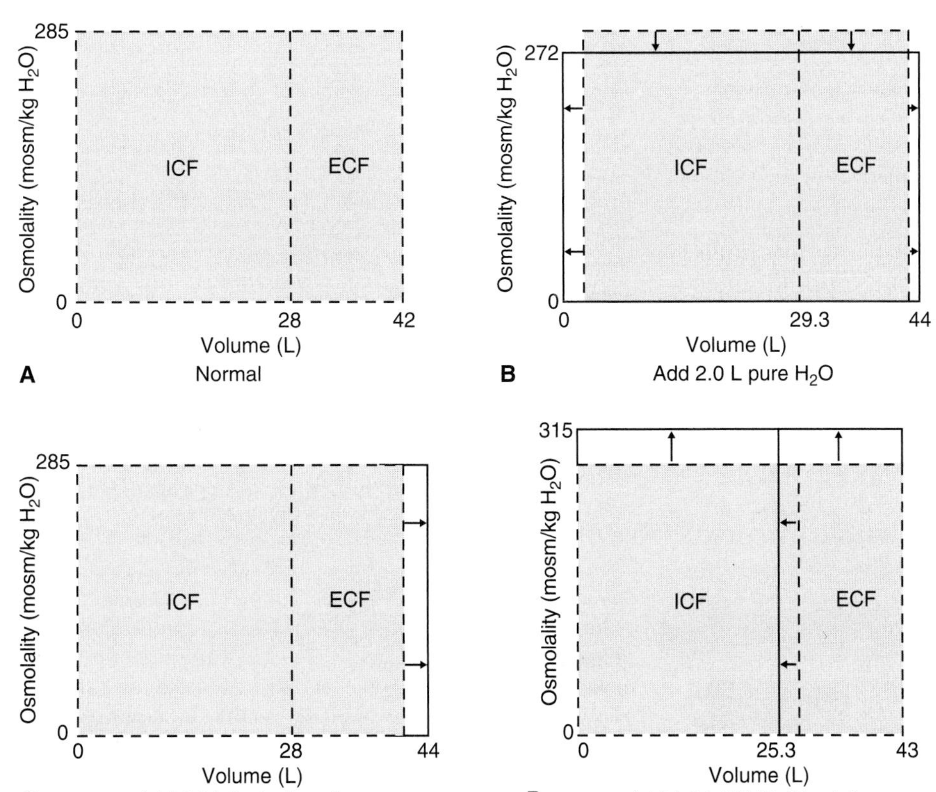

FIGURE 24.2 Effects of various disturbances on the osmolalities and volumes of intracellular fluid (ICF) and extracellular fluid (ECF). The dashed lines indicate the normal condition; the solid lines, the situation after a new osmotic equilibrium has been attained. (See text for details.)

volume of the ECF is 17.7 L, which is 3.7 L more than its initial value. The addition of hypertonic saline to the ECF, therefore, led to its considerable expansion mostly because of loss of water from the cell compartment.

WATER BALANCE

People normally stay in a stable water balance; that is, water input and output are equal. There are three major aspects to the control of water balance: arginine vasopressin, excretion of water by the kidneys, and habit and thirst.

Water Input and Output Are Equal

A balance chart for water for an average 70-kg man is presented in Table 24.3. The person is in a stable balance (or steady state) because the total input and total output of water from the body are equal (2,500 mL/day). On the input side, water is found in the beverages we drink and in the foods we eat. Solid foods, which consist of animal or vegetable matter, are, like our own bodies, mostly water. Water of oxidation is produced during metabolism; for example, when 1 mol of glucose is oxidized, 6 mol of water are produced. In a hospital setting, the input of water as a result of intravenous infusions would also need to be considered. On the output side, losses of water occur via the skin, lungs, gastrointestinal tract, and kidneys. We always lose water by simple evaporation from the skin and lungs; this is called **insensible water loss**.

Appreciable water loss from the skin, in the form of sweat, occurs at high temperatures or with heavy exercise. As much as 4 L of water per hour can be lost in sweat. Sweat, which is a hypoosmotic fluid, contains NaCl; excessive sweating can lead to significant losses of salt. Gastrointestinal losses of water are normally small (see Table 24.3), but with diarrhea, vomiting, or drainage of gastrointestinal secretions, massive quantities of water and electrolytes may be lost from the body.

The kidneys are the sites of adjustment of water output from the body. Renal water excretion changes to maintain balance. If there is a water deficiency, the kidneys diminish the excretion of water and urine output falls. If there is water excess, the kidneys increase water excretion and urine flow to remove the extra water. The renal excretion of water is controlled by arginine vasopressin.

The water needs of an infant or young child, per kg body weight, are several times higher than that of an adult. Children have, for their body weight, a larger body surface area and higher metabolic rate. They are much more susceptible to volume depletion.

TABLE 24.3 Daily Water Balance in an Average 70-kg Man

Input		Output	
Water in beverages	1,000 mL	Skin and lungs	900 mL
Water in food	1,200 mL	Gastrointestinal tract (feces)	100 mL
Water of oxidation	300 mL	Kidneys (urine)	1,500 mL
Total	2,500 mL	Total	2,500 mL

Arginine Vasopressin Is Critical in the Control of Renal Water Output and Plasma Osmolality

Arginine vasopressin (AVP), also known as **antidiuretic hormone** (ADH), is a nonapeptide synthesized in the body of nerve cells located in the supraoptic and paraventricular nuclei of the anterior hypothalamus (Fig. 24.3) (see Chapter 32). The hormone travels by axoplasmic flow down the hypothalamic-neurohypophyseal tract and is stored in vesicles in nerve terminals in the median eminence and, mostly, the posterior pituitary. When the cells are brought to threshold, they rapidly fire action potentials, Ca^{2+} enters the nerve terminals, the AVP-containing vesicles release their contents into the interstitial fluid surrounding the nerve terminals, and AVP diffuses into nearby capillaries. The hormone is carried by the bloodstream to its target tissue, the collecting ducts of the kidneys, where it increases water reabsorption (see Chapter 23).

Factors Affecting AVP Release. Many factors influence the release of AVP, including pain, trauma, emotional stress, nausea, fainting, most anesthetics, nicotine, morphine, and angiotensin II. These conditions or agents produce a decline in urine output and more concentrated urine. Ethanol and atrial natriuretic peptide inhibit AVP release, leading to the excretion of a large volume of dilute urine.

The main factor controlling AVP release under ordinary circumstances is a change in plasma osmolality. Figure 24.4 shows how plasma AVP concentrations vary as a function of plasma osmolality. When plasma osmolality rises, neurons called **osmoreceptor cells**, located in the anterior hypothalamus, shrink. This stimulates the nearby neurons in

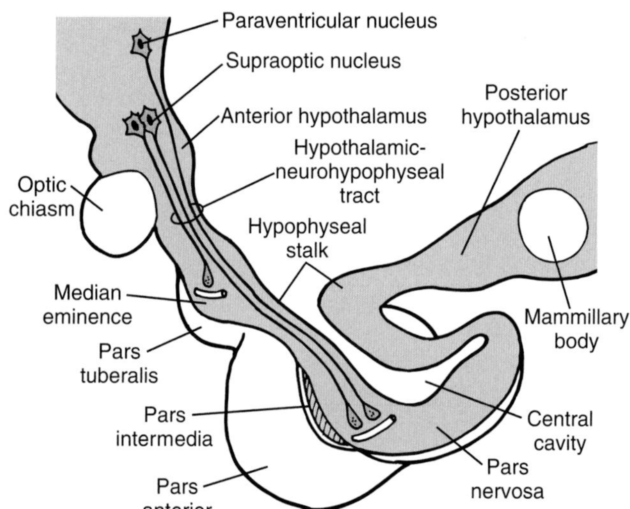

FIGURE 24.3 The pituitary and hypothalamus. AVP is synthesized primarily in the supraoptic nucleus and to a lesser extent in the paraventricular nuclei in the anterior hypothalamus. It is then transported down the hypothalamic-neurohypophyseal tract and stored in vesicles in the median eminence and posterior pituitary, where it can be released into the blood.

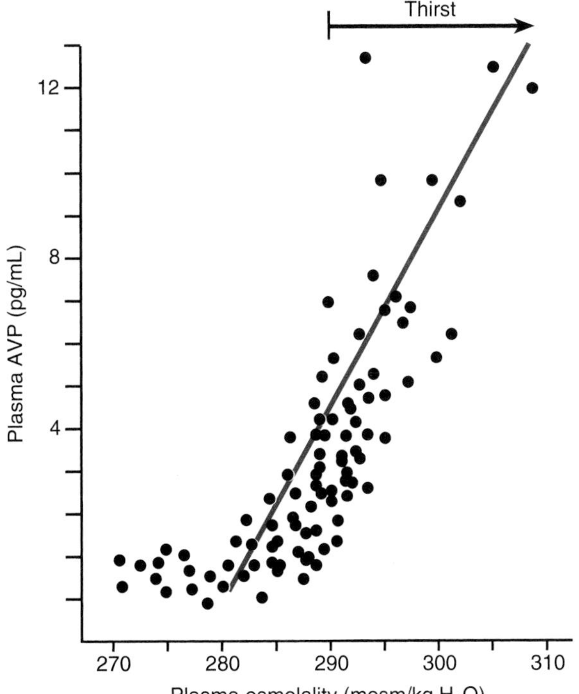

FIGURE 24.4 The relationship between plasma AVP level and plasma osmolality in healthy people. Decreases in plasma osmolality were produced by drinking water and increases by fluid restriction. Plasma AVP levels were measured by radioimmunoassay. At plasma osmolalities below 280 mosm/kg H_2O, plasma AVP is decreased to low or undetectable levels. Above this threshold, plasma AVP increases linearly with plasma osmolality. Normal plasma osmolality is about 285 to 287 mosm/kg H_2O, so we live above the threshold for AVP release. The thirst threshold is attained at a plasma osmolality of 290 mosm/kg H_2O, so the thirst mechanism "kicks in" only when there is an appreciable water deficit. Changes in plasma AVP and consequent changes in renal water excretion are normally capable of maintaining a normal plasma osmolality below the thirst threshold. (From Robertson GL, Aycinena P, Zerbe RL. Neurogenic disorders of osmoregulation. Am J Med 1982;72:339–353.)

the paraventricular and supraoptic nuclei to release AVP, and plasma AVP concentration rises. The result is the formation of osmotically concentrated urine. Not all solutes are equally effective in stimulating the osmoreceptor cells; for example, urea, which can enter these cells and, therefore, does not cause the osmotic withdrawal of water, is ineffective. Extracellular NaCl, however, is an effective stimulus for AVP release. When plasma osmolality falls in response to the addition of excess water, the osmoreceptor cells swell, AVP release is inhibited, and plasma AVP levels fall. In this situation, the collecting ducts express their intrinsically low water permeability, less water is reabsorbed, a dilute urine is excreted, and plasma osmolality can be restored to normal by elimination of the excess water. Figure 24.5 shows that the entire range of urine osmolalities, from dilute to concentrated urines, is a linear function of plasma AVP in healthy people.

A second important factor controlling AVP release is the blood volume—more precisely, the effective arterial blood volume. An increased blood volume inhibits AVP release, whereas a decreased blood volume (hypovolemia) stimulates AVP release. Intuitively, this makes sense, since with excess volume, a low plasma AVP level would promote the excretion of water by the kidneys. With hypovolemia, a high plasma AVP level would promote conservation of water by the kidneys.

The receptors for blood volume include stretch receptors in the left atrium of the heart and in the pulmonary veins within the pericardium. More stretch results in more impulses transmitted to the brain via vagal afferents and inhibition of AVP release. The common experiences of producing a large volume of dilute urine, a **water diuresis**—when lying down in bed at night, when exposed to cold weather, or when immersed in a pool during the summer—may be related to activation of this pathway. In all of these situations, the atria are stretched by an increased central blood volume. Arterial baroreceptors in the carotid sinuses and aortic arch also reflexly change AVP release; a fall in pressure at these sites stimulates AVP release. Finally, a decrease in renal blood flow stimulates renin release, which leads to increased angiotensin II production. Angiotensin II stimulates AVP release by acting on the brain.

Relatively large blood losses (more than 10% of blood volume) are required to increase AVP release (Fig. 24.6). With a loss of 15 to 20% of blood volume, however, large increases in plasma AVP are observed. Plasma levels of AVP may rise to levels much higher (e.g., 50 pg/mL) than are needed to concentrate the urine maximally (e.g., 5 pg/mL). (Compare Figures 24.5 and 24.6.) With severe hemorrhage, high circulating levels of AVP exert a significant vasoconstrictor effect, which helps compensate by raising the blood pressure.

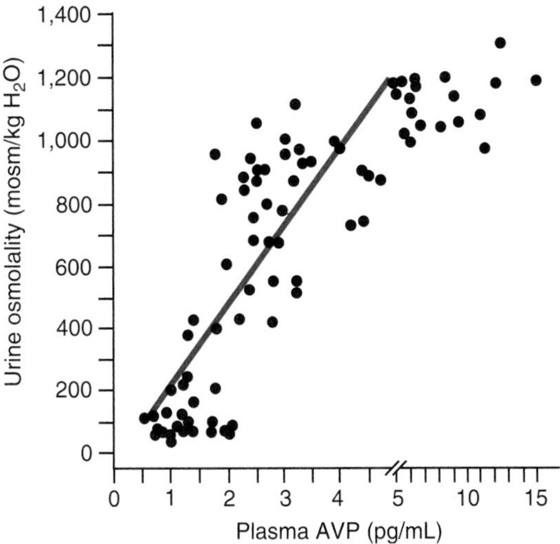

FIGURE 24.5 The relationship between urine osmolality and plasma AVP levels. With low plasma AVP levels, a hypoosmotic (compared to plasma) urine is excreted and, with high plasma AVP levels, a hyperosmotic urine is excreted. Note that maximally concentrated urine (1,200 to 1,400 mosm/kg H_2O) is produced when the plasma AVP level is about 5 pg/mL. (From Robertson GL, Aycinena P, Zerbe RL. Neurogenic disorders of osmoregulation. Am J Med 1982;72:339–353.)

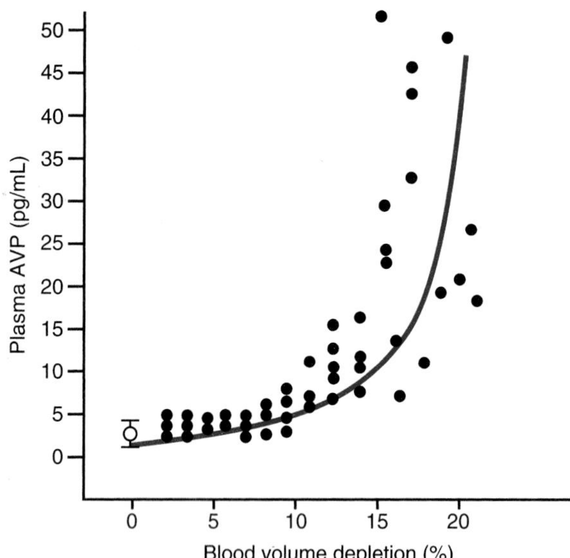

FIGURE 24.6 **The relationship between plasma AVP and blood volume depletion in the rat.** Note that severe hemorrhage (a loss of 20% of blood volume) causes a striking increase in plasma AVP. In this situation, the vasoconstrictor effect of AVP becomes significant and counteracts the low blood pressure. (From Dunn FL, Brennan TJ, Nelson AE, Robertson GL. The role of blood osmolality and volume in regulating vasopressin secretion in the rat. J Clin Invest 1973;52:3212–3219.)

Interaction Between Stimuli Affecting AVP Release. The two stimuli, plasma osmolality and blood volume, most often work synergistically to increase or decrease AVP release. For example, a great excess of water intake in a healthy person will inhibit AVP release because of both the fall in plasma osmolality and increase in blood volume. In certain important clinical circumstances, however, there is a conflict between these two inputs. For example, severe congestive heart failure is characterized by a decrease in the effective arterial blood volume, even though total blood volume is greater than normal. This condition results because the heart does not pump sufficient blood into the arterial system to maintain adequate tissue perfusion. The arterial baroreceptors signal less volume, and AVP release is stimulated. The patient will produce osmotically concentrated urine and will also be thirsty from the decreased effective arterial blood volume, with consequent increased water intake. The combination of decreased renal water excretion and increased water intake leads to hypoosmolality of the body fluids, which is reflected in a low plasma [Na^+] or **hyponatremia**. Despite the hypoosmolality, plasma AVP levels remain elevated and thirst persists. It appears that maintaining an effective arterial blood volume is of overriding importance, so osmolality may be sacrificed in this condition. The hypoosmolality creates new problems, such as the swelling of brain cells. Hyponatremia is discussed in Clinical Focus Box 24.1.

Clinical AVP Disorders. **Neurogenic diabetes insipidus** (central, hypothalamic, pituitary) is a condition characterized by a deficient production or release of AVP. Plasma AVP levels are low, and a large volume of dilute urine (up to 20 L/day) is excreted. In **nephrogenic diabetes insipidus**, the collecting ducts are partially or completely unresponsive to AVP. Urine output is increased, but the plasma AVP level is usually higher than normal (secondary to excessive loss of dilute fluid from the body). Nephrogenic diabetes insipidus may be acquired (e.g., via drugs such as lithium) or inherited. Mutations in the collecting duct AVP receptor gene or in the water channel (aquaporin-2) gene have now been identified in some families. In the **syndrome of inappropriate secretion of ADH** (SIADH), plasma AVP levels are inappropriately high for the existing osmolality. Plasma osmolality is low because the kidneys form concentrated urine and save water. This condition is sometimes caused by a bronchogenic tumor that produces AVP in an uncontrolled fashion.

Habit and Thirst Govern Water Intake

People drink water largely from habit, and this water intake normally covers an individual's water needs. Most of the time, we operate below the threshold for thirst. **Thirst**, a conscious desire to drink water, is mainly an emergency mechanism that comes into play when there is a perceived water deficit. Its function is obviously to encourage water intake to repair the water deficit. The **thirst center** is located in the anterior hypothalamus, close to the neurons that produce and control AVP release. This center relays impulses to the cerebral cortex, so that thirst becomes a conscious sensation.

Several factors affect the thirst sensation (Fig. 24.7). The major stimulus is an increase in osmolality of the blood, which is detected by osmoreceptor cells in the hypothalamus. These cells are distinct from those that affect AVP release. Ethanol and urea are not effective stimuli for the osmoreceptors because they readily penetrate these cells and do not cause them to shrink. NaCl is an effective stimulus. An increase in plasma osmolality of 1 to 2% (i.e., about 3 to 6 mosm/kg H_2O) is needed to reach the thirst threshold.

Hypovolemia or a decrease in the effective arterial blood volume stimulates thirst. Blood volume loss must be considerable for the thirst threshold to be reached; most blood donors do not become thirsty after donating 500 mL of blood (10% of blood volume). A larger blood loss (15 to 20% of blood volume), however, evokes intense thirst. A decrease in effective arterial blood volume as a result of severe diarrhea, vomiting, or congestive heart failure may also provoke thirst.

The receptors for blood volume that stimulate thirst include the arterial baroreceptors in the carotid sinuses and aortic arch and stretch receptors in the cardiac atria and great veins in the thorax. The kidneys may also act as volume receptors. When blood volume is decreased, the kidneys release renin into the circulation. This results in production of angiotensin II, which acts on neurons near the third ventricle of the brain to stimulate thirst.

The thirst sensation is reinforced by dryness of the mouth and throat, which is caused by a reflex decrease in secretion by salivary and buccal glands in a water-deprived person. The gastrointestinal tract also monitors water intake. Moistening of the mouth or distension of the stomach,

CLINICAL FOCUS BOX 24.1

Hyponatremia

Hyponatremia, defined as a plasma [Na^+] < 135 mEq/L, is the most common disorder of body fluid and electrolyte balance in hospitalized patients. Most often it reflects too much water, not too little Na^+, in the plasma. Since Na^+ is the major solute in the plasma, it is not surprising that hyponatremia is usually associated with hypoosmolality. Hyponatremia, however, may also occur with a normal or even elevated plasma osmolality.

Drinking large quantities of water (20 L/day) rarely causes frank hyponatremia because of the large capacity of the kidneys to excrete dilute urine. If, however, plasma AVP is not decreased when plasma osmolality is decreased or if the ability of the kidneys to dilute the urine is impaired, hyponatremia may develop even with a normal water intake.

Hyponatremia with hypoosmolality can occur in the presence of a decreased, normal, or even increased total body Na^+. Hyponatremia and decreased body Na^+ content may be seen with increased Na^+ loss, such as with vomiting, diarrhea, and diuretic therapy. In these instances, the decrease in ECF volume stimulates thirst and AVP release. More water is ingested, but the kidneys form osmotically concentrated urine and plasma hypoosmolality and hyponatremia result. Hyponatremia and a normal body Na^+ content are seen in hypothyroidism, cortisol deficiency, and the syndrome of inappropriate secretion of antidiuretic hormone (SIADH). SIADH occurs with neurological disease, severe pain, certain drugs (such as hypoglycemic agents), and with some tumors. For example, a bronchogenic tumor may secrete AVP without control by plasma osmolality. The result is renal conservation of water. Hyponatremia and increased total body Na^+ are seen in edematous states, such as congestive heart failure, hepatic cirrhosis, and nephrotic syndrome. The decrease in effective arterial blood volume stimulates thirst and AVP release. Excretion of a dilute urine may also be impaired because of decreased delivery of fluid to diluting sites along the nephron and collecting ducts. Although Na^+ and water are retained by the kidneys in the edematous states, relatively more water is conserved, leading to a dilutional hyponatremia.

Hyponatremia and hypoosmolality can cause a variety of symptoms, including muscle cramps, lethargy, fatigue, disorientation, headache, anorexia, nausea, agitation, hypothermia, seizures, and coma. These symptoms, mainly neurological, are a consequence of the swelling of brain cells as plasma osmolality falls. Excessive brain swelling may be fatal or may cause permanent damage. Treatment requires identifying and treating the underlying cause. If Na^+ loss is responsible for the hyponatremia, isotonic or hypertonic saline or NaCl by mouth is usually given. If the blood volume is normal or the patient is edematous, water restriction is recommended. Hyponatremia should be corrected slowly and with constant monitoring because too rapid correction can be harmful.

Hyponatremia in the presence of increased plasma osmolality is seen in hyperglycemic patients with uncontrolled diabetes mellitus. In this condition, the high plasma [glucose] causes the osmotic withdrawal of water from cells, and the extra water in the ECF space leads to hyponatremia. Plasma [Na^+] falls by 1.6 mEq/L for each 100 mg/dL rise in plasma glucose.

Hyponatremia and a normal plasma osmolality are seen with so-called **pseudohyponatremia**. This occurs when plasma lipids or proteins are greatly elevated. These molecules do not significantly elevate plasma osmolality. They do, however, occupy a significant volume of the plasma, and because the Na^+ is dissolved only in the plasma water, the [Na^+] measured in the entire plasma is low.

for example, inhibit thirst, preventing excessive water intake. For example, if a dog is deprived of water for some time and is then presented with water, it will commence drinking but will stop before all of the ingested water has been absorbed by the small intestine. Monitoring of water intake by the mouth and stomach in this situation limits water intake, preventing a dip in plasma osmolality below normal.

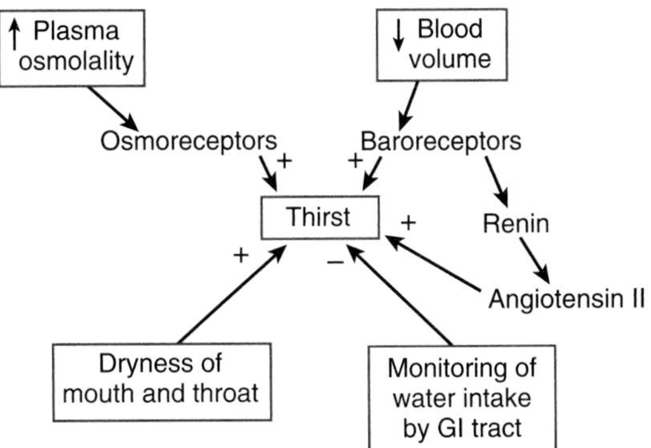

FIGURE 24.7 **Factors affecting the thirst sensation.** A plus sign indicates stimulation of thirst, the minus sign indicates an inhibitory influence.

SODIUM BALANCE

Na^+ is the most abundant cation in the ECF and, with its accompanying anions Cl^- and HCO_3^-, largely determines the osmolality of the ECF. Because the osmolality of the ECF is closely regulated by AVP, the kidneys, and thirst, the amount of water in (and, hence, the volume of) the ECF compartment is mainly determined by its Na^+ content. The kidneys are primarily involved in the regulation of Na^+ balance. We consider first the renal mechanisms involved in Na^+ excretion and then overall Na^+ balance.

The Kidneys Excrete Only a Small Percentage of the Filtered Na^+ Load

Table 24.4 shows the magnitude of filtration, reabsorption, and excretion of ions and water for a healthy adult man on an average American diet. The amount of Na^+ filtered was calculated from the product of the plasma [Na^+] and

TABLE 24.4 Magnitude of Daily Filtration, Reabsorption, and Excretion of Ions and Water in a Healthy Young Man on a Typical American Diet

	[Plasma] (mEq/L)	GFR (L/day)	Filtered (mEq/day)	Excreted (mEq/day)	Reabsorbed (mEq/day)	% Reabsorbed
Sodium	140	180	25,200	100	25,100	99.6
Chloride	105	180	18,900	100	18,800	99.5
Bicarbonate	24	180	4,320	2	4318	99.9
Potassium	4	180	720	100	620	86.1
Water	0.93[a]	180	167 L/day	1.5 L/day	165.5 L/day	99.1

[a] Plasma contains about 0.93 L H_2O per L.

glomerular filtration rate (GFR). The quantity of Na^+ reabsorbed was calculated from the difference between filtered and excreted amounts. Note that 99.6% (25,100 ÷ 25,200) of the filtered Na^+ was reabsorbed or, in other words, percentage excretion of Na^+ was only 0.4% of the filtered load. In terms of overall Na^+ balance for the body, the quantity of Na^+ excreted by the kidneys is of key importance because ordinarily about 95% of the Na^+ we consume is excreted by way of the kidneys. Tubular reabsorption of Na^+ must be finely regulated to keep us in Na^+ balance.

Figure 24.8 shows the percentage of filtered Na^+ reabsorbed in different parts of the nephron. Seventy percent of filtered Na^+, together with the same percentage of filtered water, is reabsorbed in the proximal convoluted tubule. The loop of Henle reabsorbs about 20% of filtered Na^+, but only 10% of filtered water. The distal convoluted tubule reabsorbs about 6% of filtered Na^+ (and no water), and the collecting ducts reabsorb about 3% of the filtered Na^+ (and 19% of the filtered water). Only about 1% of the filtered Na^+ (and water) is usually excreted. The distal nephron (distal convoluted tubule, connecting tubule, and collecting duct) has a lower capacity for Na^+ transport than more proximal segments and can be overwhelmed if too much Na^+ fails to be reabsorbed in proximal segments. The distal nephron is of critical importance in determining the final excretion of Na^+.

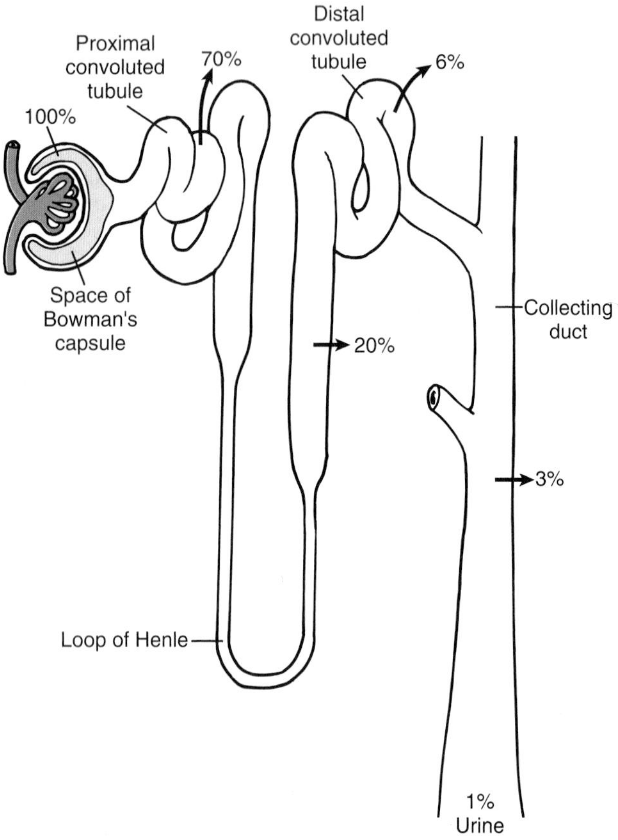

FIGURE 24.8 The percentage of the filtered load of Na^+ reabsorbed along the nephron. About 1% of the filtered Na^+ is usually excreted.

Many Factors Affect Renal Na^+ Excretion

Multiple factors affect renal Na^+ excretion; these are discussed below. A factor may promote Na^+ excretion either by increasing the amount of Na^+ filtered by the glomeruli or by decreasing the amount of Na^+ reabsorbed by the kidney tubules or, in some cases, by affecting both processes.

Glomerular Filtration Rate. Na^+ excretion tends to change in the same direction as GFR. If GFR rises—for example, from an expanded ECF volume—the tubules reabsorb the increased filtered load less completely, and Na^+ excretion increases. If GFR falls—for example, as a result of blood loss—the tubules can reabsorb the reduced filtered Na^+ load more completely, and Na^+ excretion falls. These changes are of obvious benefit in restoring a normal ECF volume.

Small changes in GFR could potentially lead to massive changes in Na^+ excretion, if it were not for a phenomenon called **glomerulotubular balance** (Table 24.5). There is a balance between the amount of Na^+ filtered and the amount of Na^+ reabsorbed by the tubules, so the tubules increase the rate of Na^+ reabsorption when GFR is increased and decrease the rate of Na^+ reabsorption when GFR is decreased. This adjustment is a function of the proximal convoluted tubule and the loop of Henle, and it reduces the impact of changes in GFR on Na^+ excretion.

The Renin-Angiotensin-Aldosterone System. **Renin** is a proteolytic enzyme produced by granular cells, which are located in afferent arterioles in the kidneys (see Fig. 23.4). There are three main stimuli for renin release:

TABLE 24.5 Glomerulotubular Balance[a]

Period	Filtered Na^+ (mEq/min)	− Reabsorbed Na^+ (mEq/min)	= Excreted Na^+ (mEq/min)
1	6.00	5.95	0.05
Increase GFR by one third			
2	8.00	7.90	0.10

[a] Results from an experiment performed on a 10-kg dog. Note that in response to an increase in GFR (produced by infusing a drug that dilated afferent arterioles), tubular reabsorption of Na^+ increased, so that only a modest increase in Na^+ excretion occurred. If there had been no glomerulotubular balance and if tubular Na^+ reabsorption had stayed at 5.95 mEq/min, the kidneys would have excreted 2.05 mEq/min in period 2. If we assume that the ECF volume in the dog is 2 L (20% of body weight) and if plasma $[Na^+]$ is 140 mEq/L, an excretion rate of 2.05 mEq/min would result in excretion of the entire ECF Na^+ (280 mEq) in a little more than 2 hours. The dog would have been dead long before this could happen, which underscores the importance of glomerulotubular balance.

1) A decrease in pressure in the afferent arteriole, with the granular cells being sensitive to stretch and functioning as an **intrarenal baroreceptor**

2) Stimulation of sympathetic nerve fibers to the kidneys via β_2-adrenergic receptors on the granular cells

3) A decrease in fluid delivery to the macula densa region of the nephron, resulting, for example, from a decrease in GFR

All three of these pathways are activated and reinforce each other when there is a decrease in the effective arterial blood volume—for example, following hemorrhage, transudation of fluid out of the vascular system, diarrhea, severe sweating, or a low salt intake. Conversely, an increase in the effective arterial blood volume inhibits renin release. Long-term stimulation causes vascular smooth muscle cells in the afferent arteriole to differentiate into granular cells and leads to further increases in renin supply. Renin in the blood plasma acts on a plasma α_2-globulin produced by the liver, called **angiotensinogen** (or renin substrate) and splits off the decapeptide **angiotensin I** (Fig. 24.9). Angiotensin I is converted to the octapeptide **angiotensin II** as the blood courses through the lungs. This reaction is catalyzed by the **angiotensin-converting enzyme** (ACE), which is present on the surface of endothelial cells. All the components of this system (renin, angiotensinogen, angiotensin-converting enzyme) are present in some organs (e.g., the kidneys and brain), so that angiotensin II may also be formed and act locally.

The **renin-angiotensin-aldosterone system** (RAAS) is a salt-conserving system. Angiotensin II has several actions related to Na^+ and water balance:

1) It stimulates the production and secretion of **aldosterone** from the zona glomerulosa of the adrenal cortex (see Chapter 36). This mineralocorticoid hormone then acts on the distal nephron to increase Na^+ reabsorption.

2) Angiotensin II directly stimulates tubular Na^+ reabsorption.

3) Angiotensin stimulates thirst and the release of AVP by the posterior pituitary.

Angiotensin II is also a potent vasoconstrictor of both resistance and capacitance vessels; increased plasma levels following hemorrhage, for example, help sustain blood pressure. Inhibiting angiotensin II production by giving an ACE inhibitor lowers blood pressure and is used in the treatment of hypertension.

The RAAS plays an important role in the day-to-day control of Na^+ excretion. It favors Na^+ conservation by the kidneys when there is a Na^+ or volume deficit in the body. When there is an excess of Na^+ or volume, diminished RAAS activity permits enhanced Na^+ excretion. In the absence of aldosterone (e.g., in an adrenalectomized individual) or in a person with adrenal cortical insufficiency—**Addison's disease**—excessive amounts of Na^+ are lost in the urine. Percentage reabsorption of Na^+ may decrease from a normal value of about 99.6% to a value of 98%. This change (1.6% of the filtered Na^+ load) may not seem like much, but if the kidneys filter 25,200 mEq/day (see Table 24.4) and excrete an extra $0.016 \times 25{,}200 = 403$ mEq/day, this is the amount of Na^+ in almost 3 L of ECF (assuming a $[Na^+]$ of 140 mEq/L). Such a loss of Na^+ would lead to a decrease in plasma and blood volume, circulatory collapse, and even death.

When there is an extra need for Na^+, people and many animals display a **sodium appetite**, an urge for salt intake, which can be viewed as a brain mechanism, much like thirst, that helps compensate for a deficit. Patients with Addison's disease often show a well-developed sodium appetite, which helps keep them alive.

Large doses of a potent mineralocorticoid will cause a person to retain about 200 to 300 mEq Na^+ (equivalent to about 1.4 to 2 L of ECF), and the person will "escape" from the salt-retaining action of the steroid. Retention of this amount of fluid is not sufficient to produce obvious edema. The fact that the person will not continue to accumulate Na^+ and water is due to the existence of numerous factors that are called into play when ECF volume is expanded; these factors promote renal Na^+ excretion and overpower the salt-retaining action of aldosterone. This phenomenon is called **mineralocorticoid escape**.

Intrarenal Physical Forces (Peritubular Capillary Starling Forces). An increase in the hydrostatic pressure or a decrease in the colloid osmotic pressure in peritubular capillaries (the so-called "physical" or Starling forces) results in reduced fluid uptake by the capillaries. In turn, an accumulation of the reabsorbed fluid in the kidney interstitial spaces results. The increased interstitial pressure causes a widening of the tight junctions between proximal tubule cells, and the epithelium becomes even more leaky than normal. The result is increased back-leak of salt and water into the tubule lumen and an overall reduction in net reabsorption. These changes occur, for example, if a large volume of isotonic saline is infused intravenously. They also occur if the **filtration fraction** (GFR/RPF) is lowered from the dilation of efferent arterioles, for example. In this case, the protein concentration (or colloid osmotic pressure) in efferent arteriolar blood and peritubular capillary blood is lower than normal because a smaller proportion of the plasma is filtered in the glomeruli. Also, with upstream vasodilation of efferent arterioles, hydrostatic pressure in the

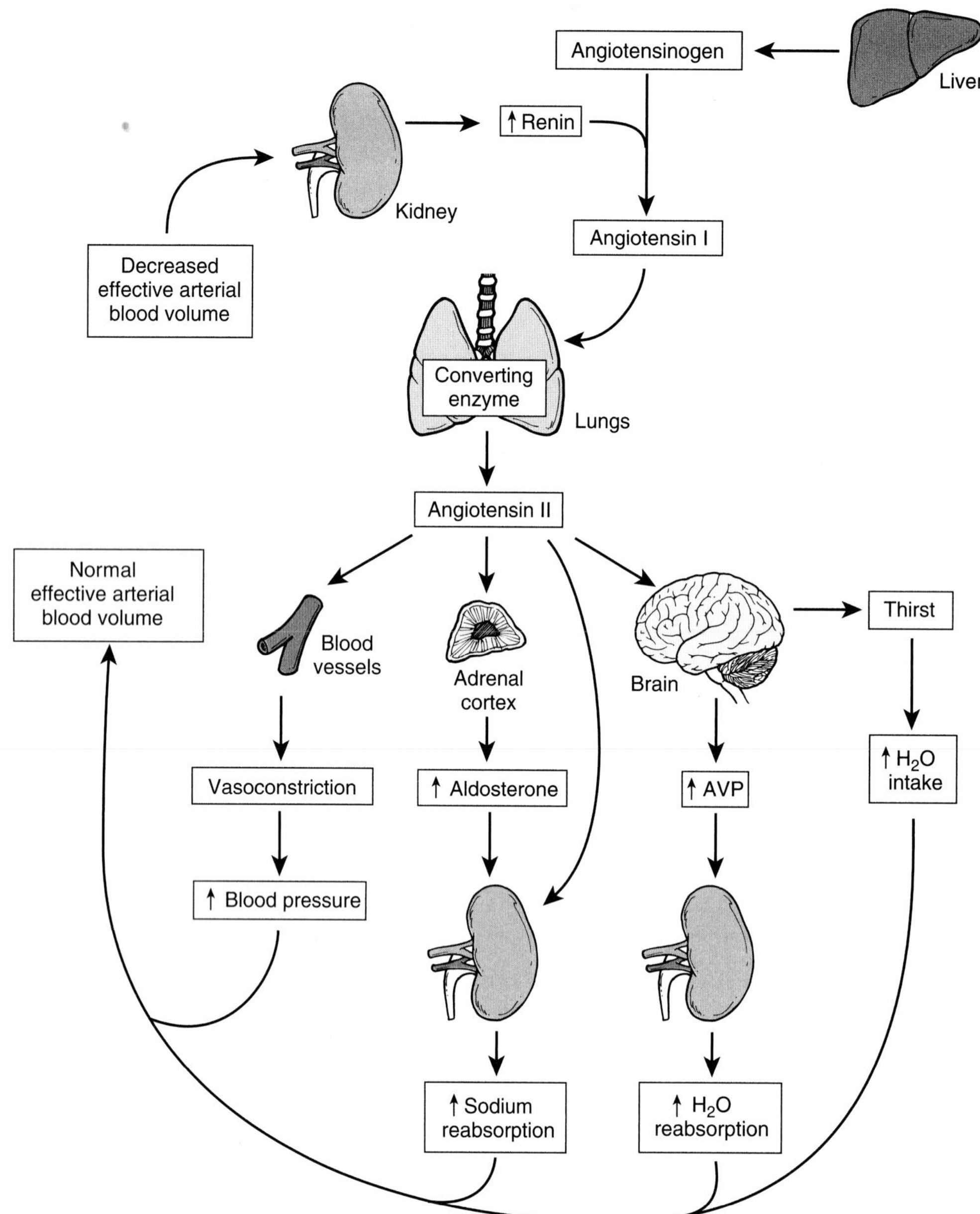

FIGURE 24.9 **Components of the renin-angiotensin-aldosterone system.** This system is activated by a decrease in the effective arterial blood volume (e.g., following hemorrhage) and results in compensatory changes that help restore arterial blood pressure and blood volume to normal.

peritubular capillaries is increased, leading to a **pressure natriuresis** and **pressure diuresis**. The term *natriuresis* means an increase in Na^+ excretion.

Natriuretic Hormones and Factors. **Atrial natriuretic peptide** (ANP) is a 28 amino acid polypeptide synthesized and stored in myocytes of the cardiac atria (Fig. 24.10). It is released upon stretch of the atria—for example, following volume expansion. This hormone has several actions that increase Na^+ excretion. ANP acts on the kidneys to increase glomerular blood flow and filtration rate and inhibits Na^+ reabsorption by the inner medullary collecting ducts. The second messenger for ANP in the collecting duct is cGMP. ANP directly inhibits aldosterone secretion by the adrenal cortex; it also indirectly inhibits aldosterone secretion by diminishing renal renin release. ANP is a vasodilator and, therefore, lowers blood pressure. Some evidence suggests that ANP inhibits AVP secretion. The actions of ANP are, in many respects, just the opposite of those of the RAAS; ANP promotes salt and water loss by the kidneys and lowers blood pressure.

Several other natriuretic hormones and factors have been described. **Urodilatin** (kidney natriuretic peptide) is a 32-amino acid polypeptide derived from the same prohormone as ANP. It is synthesized primarily by intercalated cells in the cortical collecting duct and secreted into the tubule lu-

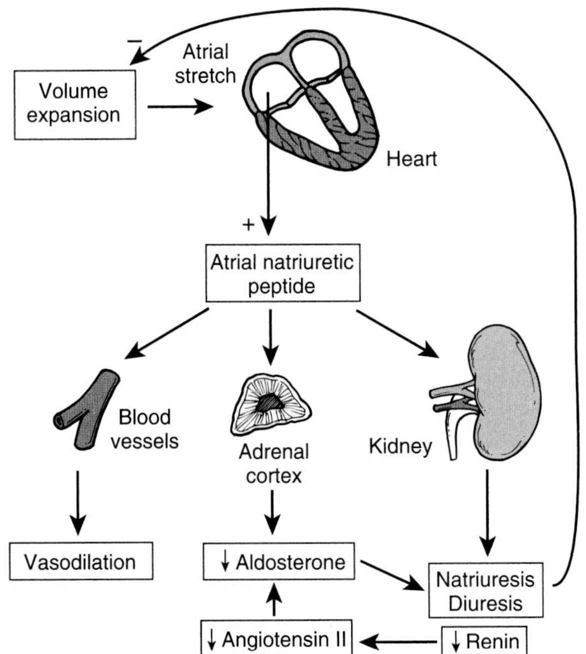

FIGURE 24.10 **Atrial natriuretic peptide and its actions.** ANP release from the cardiac atria is stimulated by blood volume expansion, which stretches the atria. ANP produces effects that bring blood volume back toward normal, such as increased Na^+ excretion.

men, inhibiting Na^+ reabsorption by inner medullary collecting ducts via cGMP. There is also a **brain natriuretic peptide**. **Guanylin** and **uroguanylin** are polypeptide hormones produced by the small intestine in response to salt ingestion. Like ANP and urodilatin, they activate guanylyl cyclase and produce cGMP as a second messenger, as their names suggest. **Adrenomedullin** is a polypeptide produced by the adrenal medulla; its physiological significance is still not certain. **Endoxin** is an endogenous digitalis-like substance produced by the adrenal gland. It inhibits Na^+/K^+-ATPase activity and, therefore, inhibits Na^+ transport by the kidney tubules. **Bradykinin** is produced locally in the kidneys and inhibits Na^+ reabsorption.

Prostaglandins E_2 and **I_2** (prostacyclin) increase Na^+ excretion by the kidneys. These locally produced hormones are formed from arachidonic acid, which is liberated from phospholipids in cell membranes by the enzyme phospholipase A_2. Further processing is mediated by a **cyclooxygenase** (**COX**) enzyme that has two isoforms, COX-1 and COX-2. In most tissues, COX-1 is constitutively expressed, while COX-2 is generally induced by inflammation. In the kidney, COX-1 and COX-2 are both constitutively expressed in cortex and medulla. In the cortex, COX-2 may be involved in macula densa-mediated renin release. COX-1 and COX-2 are present in high amounts in the renal medulla, where the main role of the prostaglandins is to inhibit Na^+ reabsorption. Because the prostaglandins (PGE_2, PGI_2) are vasodilators, the inhibition of Na^+ reabsorption occurs via direct effects on the tubules and collecting ducts and via hemodynamic effects (see Chapter 23). Inhibition of the formation of prostaglandins with common nonsteroidal anti-inflammatory drugs (NSAIDs), such as aspirin, may lead to a fall in renal blood flow and to Na^+ retention.

Renal Sympathetic Nerves. The stimulation of renal sympathetic nerves reduces renal Na^+ excretion in at least three ways:

1) It produces a decline in GFR and renal blood flow, leading to a decreased filtered Na^+ load and peritubular capillary hydrostatic pressure, both of which favor diminished Na^+ excretion.

2) It has a direct stimulatory effect on Na^+ reabsorption by the renal tubules.

3) It causes renin release, which results in increased plasma angiotensin II and aldosterone levels, both of which increase tubular Na^+ reabsorption.

Activation of the sympathetic nervous system occurs in several stressful circumstances (such as hemorrhage) in which the conservation of salt and water by the kidneys is of clear benefit.

Estrogens. Estrogens decrease Na^+ excretion, probably by the direct stimulation of tubular Na^+ reabsorption. Most women tend to retain salt and water during pregnancy, which may be partially related to the high plasma estrogen levels during this time.

Glucocorticoids. Glucocorticoids, such as cortisol (see Chapter 34), increase tubular Na^+ reabsorption and also cause an increase in GFR, which may mask the tubular effect. Usually a decrease in Na^+ excretion is seen.

Osmotic Diuretics. Osmotic diuretics are solutes that are excreted in the urine and increase urinary excretion of Na^+ and K^+ salts and water. Examples are urea, glucose (when the reabsorptive capacity of the tubules for glucose has been exceeded), and mannitol (a six-carbon sugar alcohol used in the clinic to promote Na^+ excretion or cell shrinkage). Osmotic diuretics decrease the reabsorption of Na^+ in the proximal tubule. This response results from the development of a Na^+ concentration gradient (lumen $[Na^+]$ $<$ plasma $[Na^+]$) across the proximal tubular epithelium in the presence of a high concentration of unreabsorbed solute in the tubule lumen. When this occurs, there is significant back-leak of Na^+ into the tubule lumen, down the concentration gradient. This back-leak results in decreased net Na^+ reabsorption. Because the proximal tubule is where most of the filtered Na^+ is normally reabsorbed, osmotic diuretics, by interfering with this process, can potentially cause the excretion of large amounts of Na^+. Osmotic diuretics may also increase Na^+ excretion by inhibiting distal Na^+ reabsorption (similar to the proximal inhibition) and by increasing medullary blood flow.

Poorly Reabsorbed Anions. Poorly reabsorbed anions result in increased Na^+ excretion. Solutions are electrically neutral; whenever there are more anions in the urine, there must also be more cations. If there is increased excretion of phosphate, ketone body acids (as occurs in uncontrolled diabetes mellitus), HCO_3^-, or SO_4^{2-}, more Na^+ is also excreted. To some extent, the Na^+ in the urine can be replaced by other cations, such as K^+, NH_4^+, and H^+.

Diuretic Drugs. Most of the diuretic drugs used today are specific Na^+ transport inhibitors. For example, the loop diuretic drugs (furosemide, bumetanide) inhibit the Na-K-2Cl cotransporter in the thick ascending limb, the thiazide diuretics inhibit the Na-Cl cotransporter in the distal convoluted tubule, and amiloride blocks the epithelial Na^+ channel in the collecting ducts (see Chapter 23). Spironolactone promotes Na^+ excretion by competitively inhibiting the binding of aldosterone to the mineralocorticoid receptor. The diuretic drugs are really natriuretic drugs; they produce an increased urine output (diuresis) because water reabsorption is diminished whenever Na^+ reabsorption is decreased. Diuretics are commonly prescribed for treating hypertension and edema.

The Kidneys Play a Dominant Role in Regulating Na^+ Balance

Figure 24.11 summarizes Na^+ balance throughout the body. Dietary intake of Na^+ varies and, in a typical American diet, amounts to about 100 to 300 mEq/day, mostly in the form of NaCl. Ingested Na^+ is mainly absorbed in the small intestine and is added to the ECF, where it is the major determinant of the osmolality and the amount of water in (or volume of) this fluid compartment. About 50% of the body's Na^+ is in the ECF, about 40% in bone, and about 10% within cells.

Losses of Na^+ occur via the skin, gastrointestinal tract, and kidneys. Skin losses are usually small, but can be considerable with sweating, burns, or hemorrhage. Likewise, gastrointestinal losses are usually small, but they can be large and serious with vomiting, diarrhea, or iatrogenic suction or drainage of gastrointestinal secretions. The kidneys are ordinarily the major routes of Na^+ loss from the body, excreting about 95% of the ingested Na^+ in a healthy person. Thus, the kidneys play a dominant role in the control of Na^+ balance. The kidneys can adjust Na^+ excretion over a wide range, reducing it to low levels when there is a Na^+ deficit and excreting more Na^+ when there is Na^+ excess in the body. Adjustments in Na^+ excretion occur by engaging many of the factors previously discussed.

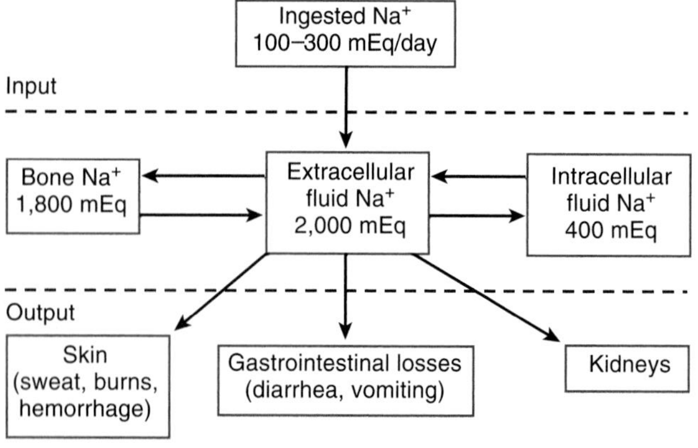

FIGURE 24.11 **Na^+ balance.** Most of the Na^+ consumed in our diets is excreted by the kidneys.

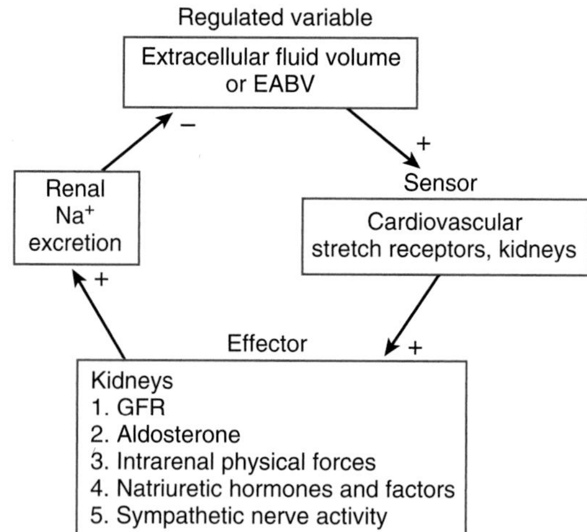

FIGURE 24.12 **The regulation of ECF volume or effective arterial blood volume (EABV) by a negative-feedback control system.** Arterial baroreceptors and the kidneys sense the degree of fullness of the arterial system. The kidneys are the effectors, and they change Na^+ excretion to restore EABV to normal.

In a healthy individual, one can think of the ECF volume as the regulated variable in a negative-feedback control system (Fig. 24.12). The kidneys are the effectors, and they change Na^+ excretion in an appropriate manner. An increase in ECF volume promotes renal Na^+ loss, which restores a normal volume. A decrease in ECF volume leads to decreased renal Na^+ excretion, and this Na^+ retention (with continued dietary Na^+ intake) leads to the restoration of a normal ECF volume. Closer examination of this concept, particularly when considering pathophysiological states, however, suggests that it is of limited usefulness. A more considered view suggests that the effective arterial blood volume (EABV) is actually the regulated variable. In a healthy individual, ECF volume and EABV usually change together in the same direction. In an abnormal condition such as congestive heart failure, however, EABV is low when the ECF volume is abnormally increased. In this condition, there is a potent stimulus for renal Na^+ retention that clearly cannot be the ECF volume.

When EABV is diminished, the degree of fullness of the arterial system is less than normal and tissue blood flow is inadequate. Arterial baroreceptors in the carotid sinuses and aortic arch sense the decreased arterial stretch. This will produce reflex activation of sympathetic nerve fibers to the kidneys, with consequently decreased GFR and renal blood flow and increased renin release. These changes favor renal Na^+ retention. Reduced EABV is also "sensed" in the kidneys in three ways:

1) A low pressure at the level of the afferent arteriole stimulates renin release via the intrarenal baroreceptor mechanism.

2) Decreases in renal perfusion pressure lead to a reduced GFR and, hence, diminished Na^+ excretion.

3) Decreases in renal perfusion pressure will also reduce peritubular capillary hydrostatic pressure, increasing the uptake of reabsorbed fluid and diminishing Na^+ excretion.

When kidney perfusion is threatened, the kidneys retain salt and water, a response that tends to improve their perfusion.

In several important diseases, including heart and liver and some kidney diseases, abnormal renal retention of Na^+ contributes to the development of **generalized edema**, a widespread accumulation of salt and water in the interstitial spaces of the body. The condition is often not clinically evident until a person has accumulated more than 2.5 to 3 L of ECF in the interstitial space. Expansion of the interstitial space has two components: (1) an altered balance of Starling forces exerted across capillaries, and (2) the retention of extra salt and water by the kidneys. Total plasma volume is only about 3.5 L; if edema fluid were derived solely from the plasma, hemoconcentration and circulatory shock would ensue. Conservation of salt and water by the kidneys is clearly an important part of the development of generalized edema.

Patients with congestive heart failure may accumulate many liters of edema fluid, which is easily detected as weight gain (since 1 L of fluid weighs 1 kg). Because of the effect of gravity, the ankles become swollen and pitting edema develops. As a result of heart failure, venous pressure is elevated, causing fluid to leak out of the capillaries because of their elevated hydrostatic pressure. Inadequate pumping of blood by the heart leads to a decrease in EABV, so the kidneys retain salt and water. Alterations in many of the factors discussed above—decreased GFR, increased RAAS activity, changes in intrarenal physical forces, and increased sympathetic nervous system activity—contribute to the renal salt and water retention. To minimize the accumulation of edema fluid, patients are often placed on a reduced Na^+ intake and given diuretic drugs.

Hypertension may often be a result of a disturbance in NaCl (salt) balance. Excessive dietary intake of NaCl or inadequate renal excretion of salt tends to increase intravascular volume; this change translates into an increase in blood pressure. A reduced salt intake, ACE inhibitors, diuretic drugs, or drugs that more directly affect the cardiovascular system (e.g., Ca^{2+} channel blockers or β-adrenergic blockers) are useful therapies in controlling hypertension in many people.

POTASSIUM BALANCE

Potassium (K^+) is the most abundant ion in the ICF compartment. It has many important effects in the body, and its plasma concentration is closely regulated. The kidneys play a dominant role in regulating K^+ balance.

K^+ Influences Cell Volume, Excitability, Acid-Base Balance, and Metabolism

As the major osmotically active solute in cells, the amount of cellular K^+ is the major determinant of the amount of water in (and, therefore, the volume of) the ICF compartment, in the same way that extracellular Na^+ is a major determinant of ECF volume. When cells lose K^+ (and accompanying anions), they also lose water and shrink; the converse is also true.

The distribution of K^+ across plasma membranes—that is, the ratio of intracellular to extracellular K^+ concentrations—is the major determinant of the resting membrane potential of cells and, hence, their excitability (see Chapter 3). Disturbances of K^+ balance often produce altered excitability of nerves and muscles. Low plasma [K^+] leads to membrane hyperpolarization and reduced excitability; muscle weakness is a common symptom. Excessive plasma K^+ levels lead to membrane depolarization and increased excitability. High plasma K^+ levels cause cardiac arrhythmias and, eventually, ventricular fibrillation, usually a lethal event.

K^+ balance is linked to acid-base balance in complex ways (see Chapter 25). K^+ depletion, for example, can lead to metabolic alkalosis, and K^+ excess to metabolic acidosis. A primary disturbance in acid-base balance can also lead to abnormal K^+ balance.

K^+ affects the activity of enzymes involved in carbohydrate metabolism and electron transport. K^+ is needed for tissue growth and repair. Tissue breakdown or increased protein catabolism result in a loss of K^+ from cells.

Most of the Body's K^+ Is in Cells

Total body content of K^+ in a healthy, young adult, 70-kg man is about 3,700 mEq. About 2% of this, about 60 mEq, is in the functional ECF (blood plasma, interstitial fluid, and lymph); this number was calculated by multiplying the plasma [K^+] of 4 mEq/L times the ECF volume (20% of body weight or 14 L). About 8% of the body's K^+ is in bone, dense connective tissue, and cartilage, and another 1% is in transcellular fluids. Ninety percent of the body's K^+ is in the cell compartment.

A normal plasma [K^+] is 3.5 to 5.0 mEq/L. By definition, plasma [K^+] below 3.5 mEq/L is **hypokalemia** and plasma [K^+] above 5.0 mEq/L is **hyperkalemia**. The [K^+] in skeletal muscle cells is about 150 mEq/L cell water. Skeletal muscle cells constitute the largest fraction of the cell mass in the human body and contain about two thirds of the body's K^+. One can easily appreciate that abnormal leakage of K^+ from muscle cells, for example, as a result of trauma, may lead to dangerous hyperkalemia.

A variety of factors influence the distribution of K^+ between cells and ECF (Fig. 24.13):

1) A key factor is the Na^+/K^+-ATPase, which pumps K^+ into cells. If this enzyme is inhibited—as a result of inadequate tissue oxygen supply or digitalis overdose, for example—hyperkalemia may result.

2) A decrease in ECF pH (an increase in ECF [H^+]) tends to produce a rise in ECF [K^+]. This results from an exchange of extracellular H^+ for intracellular K^+. When a mineral acid such as HCl is added to the ECF, a fall in blood pH of 0.1 unit leads to about a 0.6 mEq/L rise in plasma [K^+]. When an organic acid (which can penetrate plasma membranes) is added, the rise in plasma K^+ for a given fall in blood pH is considerably less. The fact that blood pH influences plasma [K^+] is sometimes used in the emergency treatment of hyperkalemia; intravenous infusion of a $NaHCO_3$ solution (which makes the blood more alkaline) will cause H^+ to move out of cells and K^+, in exchange, to move into cells.

3) Insulin promotes the uptake of K^+ by skeletal muscle and liver cells. This effect appears to be a result of stim-

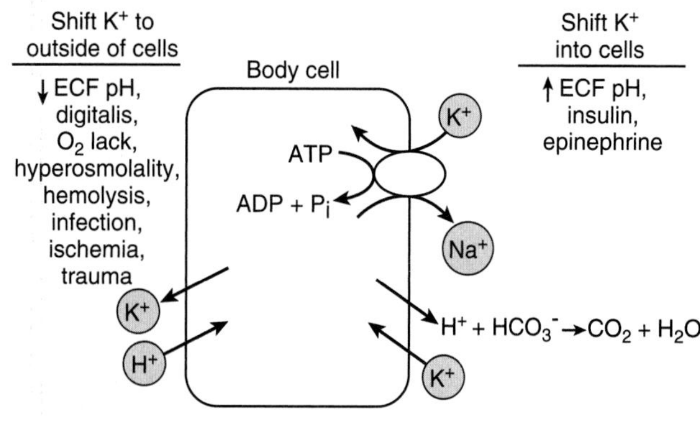

FIGURE 24.13 Factors influencing the distribution of K^+ between intracellular and extracellular fluids.

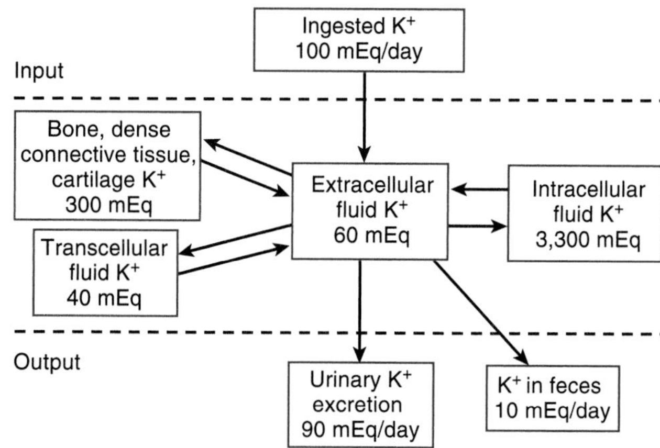

FIGURE 24.14 K^+ balance for a healthy adult. Most K^+ in the body is in the cell compartment. Renal K^+ excretion is normally adjusted to keep a person in balance.

ulation of plasma membrane Na^+/K^+-ATPase pumps. Insulin (administered with glucose) is also used in the emergency treatment of hyperkalemia.

4) Epinephrine increases K^+ uptake by cells, an effect mediated by β_2-receptors.

5) Hyperosmolality (e.g., a result of hyperglycemia) tends to raise plasma [K^+]; hyperosmolality causes cells to shrink and raises intracellular [K^+], which then favors outward diffusion of K^+ into the ECF.

6) Tissue trauma, infection, ischemia, hemolysis, or severe exercise release K^+ from cells and can cause significant hyperkalemia. An artifactual increase in plasma [K^+], **pseudohyperkalemia**, results if blood has been mishandled and red cells have been injured and allowed to leak K^+.

The plasma [K^+] is sometimes taken as an approximate guide to total body K^+ stores. For example, if a condition is known to produce an excessive loss of K^+ (such as taking a diuretic drug), a decrease in plasma [K^+] of 1 mEq/L may correspond to a loss of 200 to 300 mEq K^+. Clearly, however, many factors affect the distribution of K^+ between cells and ECF; in many circumstances, the plasma [K^+] is not a good index of the amount of K^+ in the body.

The Kidneys Normally Maintain K^+ Balance

Figure 24.14 depicts K^+ balance for a healthy adult man. Most of the food we eat contains K^+. K^+ intake (50 to 150 mEq/day) and absorption by the small intestine are unregulated. On the output side, gastrointestinal losses are normally small, but they can be large, especially with diarrhea. Diarrheal fluid may contain as much as 80 mEq K^+/L. K^+ loss in sweat is clinically unimportant. Normally, 90% of the ingested K^+ is excreted by the kidneys. The kidneys are the major sites of control of K^+ balance; they increase K^+ excretion when there is too much K^+ in the body and conserve K^+ when there is too little.

Abnormal Renal K^+ Excretion. The major cause of K^+ imbalances is abnormal renal K^+ excretion. The kidneys may excrete too little K^+; if the dietary intake of K^+ continues, hyperkalemia can result. For example, in Addison's disease, a low plasma aldosterone level leads to deficient K^+ excretion. Inadequate renal K^+ excretion also occurs with acute renal failure; the hyperkalemia caused by inadequate renal excretion is often compounded by tissue trauma, infection, and acidosis, all of which raise plasma [K^+]. In chronic renal failure, hyperkalemia usually does not develop until GFR falls below 15 to 20 mL/min because of the remarkable ability of the kidney collecting ducts to adapt and increase K^+ secretion.

Excessive loss of K^+ by the kidneys leads to hypokalemia. The major cause of renal K^+ wasting is iatrogenic, an unwanted side effect of diuretic drug therapy. Hyperaldosteronism causes excessive K^+ excretion. In uncontrolled diabetes mellitus, K^+ loss is increased because of the osmotic diuresis caused by glucosuria and an elevated rate of fluid flow in the cortical collecting ducts. Several rare inherited defects in tubular transport, including Bartter, Gitelman, and Liddle syndromes also lead to excessive renal K^+ excretion and hypokalemia (see Table 23.3).

Changes in Diet and K^+ Excretion. As was discussed in Chapter 23, K^+ is filtered, reabsorbed, and secreted in the kidneys. Most of the filtered K^+ is reabsorbed in the proximal convoluted tubule (70%) and the loop of Henle (25%), and the majority of K^+ excreted in the urine is usually the result of secretion by cortical collecting duct principal cells. The percentage of filtered K^+ excreted in the urine is typically about 15% (Fig. 24.15). With prolonged K^+ depletion, the kidneys may excrete only 1% of the filtered load. However, excessive K^+ intake may result in the excretion of an amount of K^+ that exceeds the amount filtered; in this case, there is greatly increased K^+ secretion by cortical collecting ducts.

When the dietary intake of K^+ is changed, renal excretion changes in the same direction. An important site for this adaptive change is the cortical collecting duct. Figure 24.16 shows the response to an increase in dietary K^+ intake. Two pathways are involved. First, an elevated plasma [K^+] leads to increased K^+ uptake by the basolateral plasma membrane Na^+/K^+-ATPase in collecting duct principal cells, resulting in increased intracellular [K^+], K^+ secretion, and K^+ excretion. Second, elevated plasma [K^+] has a direct effect (i.e., not mediated by renin and an-

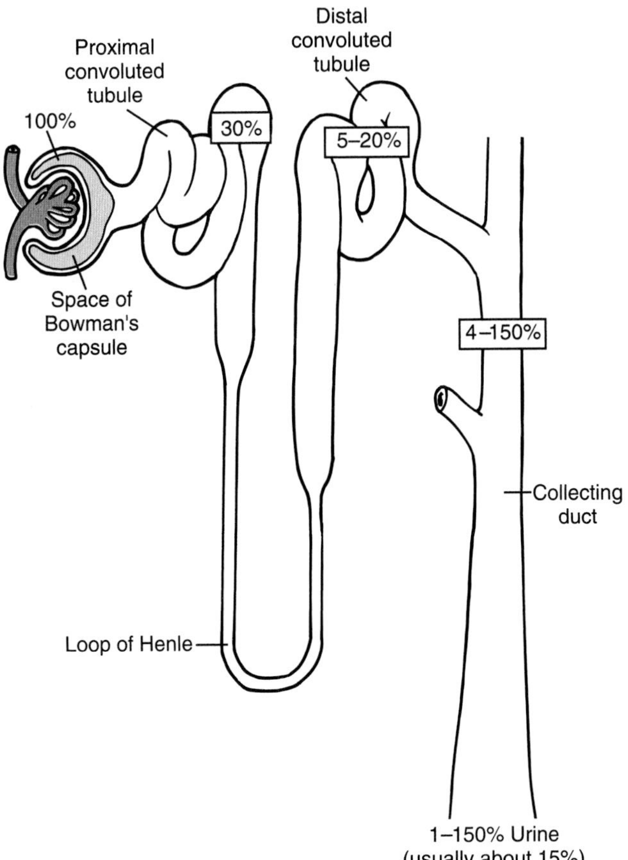

FIGURE 24.15 The percentage of the filtered load of K^+ remaining in tubular fluid as it flows down the nephron. K^+ is usually secreted in the cortical collecting duct. With K^+ loading, this secretion is so vigorous that the amount of K^+ excreted may actually exceed the filtered load. With K^+ depletion, K^+ is reabsorbed by the collecting ducts.

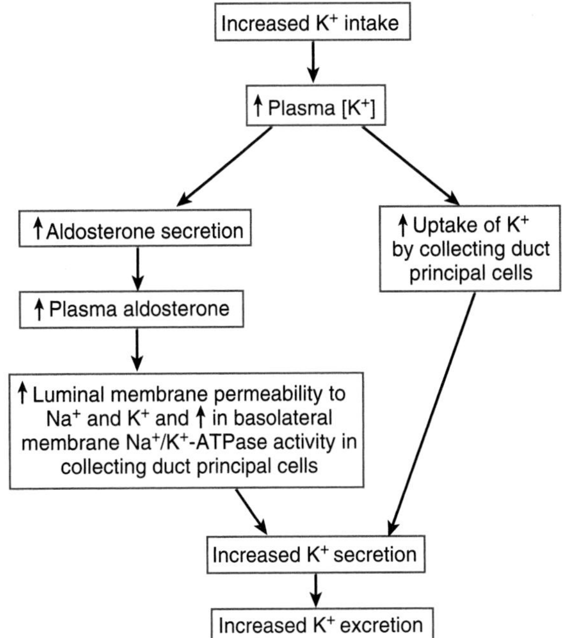

FIGURE 24.16 Effect of increased dietary K^+ intake on K^+ excretion. K^+ directly stimulates aldosterone secretion and leads to an increase in cell $[K^+]$ in collecting duct principal cells. Both of these lead to enhanced secretion and, hence, excretion, of K^+.

giotensin) on the adrenal cortex to stimulate the synthesis and release of aldosterone. Aldosterone acts on collecting duct principal cells to (1) increase the Na^+ permeability of the luminal plasma membrane, (2) increase the number and activity of basolateral plasma membrane Na^+/K^+-ATPase pumps, (3) increase the luminal plasma membrane K^+ permeability, and (4) increase cell metabolism. All of these changes result in increased K^+ secretion.

In cases of decreased dietary K^+ intake or K^+ depletion, the activity of the luminal plasma membrane H^+/K^+-ATPase found in α-intercalated cells is increased. This promotes K^+ reabsorption by the collecting ducts. The collecting ducts can greatly diminish K^+ excretion, but it takes a couple of weeks for K^+ loss to reach minimal levels.

Counterbalancing Influences on K^+ Excretion. Considering that aldosterone stimulates both Na^+ reabsorption and K^+ secretion, why is it that Na^+ deprivation, a stimulus that raises plasma aldosterone levels, does not lead to enhanced K^+ excretion? The explanation is related to the fact that Na^+ deprivation tends to lower GFR and increase proximal Na^+ reabsorption (Fig. 24.17). This response leads to a fall in Na^+ delivery and a decreased fluid flow rate in the cortical collecting ducts, which diminishes K^+ secretion and counterbalances the stimulatory effect of aldosterone. Consequently, K^+ excretion is unaltered.

Another puzzling question is: Why is it that K^+ excretion does not increase during water diuresis? In Chapter 23, we mentioned that an increase in fluid flow through the cortical collecting ducts increases K^+ secretion. AVP, in addition to its effects on water permeability, stimulates K^+ secretion by increasing the activity of luminal membrane K^+ channels in cortical collecting duct principal cells. Since plasma AVP levels are low during water diuresis, this will reduce K^+ secretion, oppos-

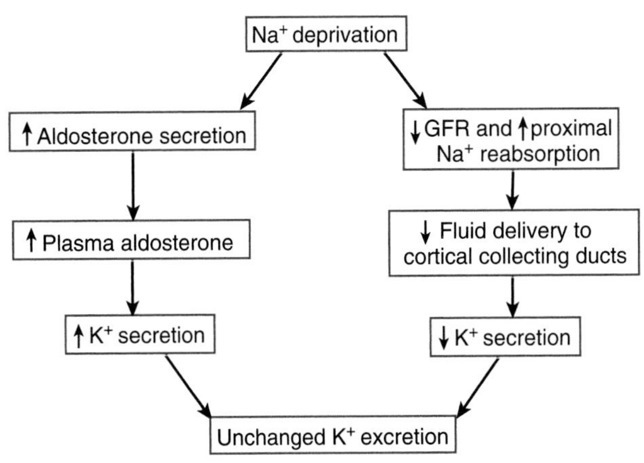

FIGURE 24.17 Why Na^+ depletion does not lead to enhanced K^+ excretion.

ing the effects of increased flow, with the result that K^+ excretion hardly changes.

CALCIUM BALANCE

The kidneys play an important role in the maintenance of Ca^{2+} balance. Ca^{2+} intake is about 1,000 mg/day and mainly comes from dairy products in the diet. About 300 mg/day are absorbed by the small intestine, a process controlled by 1,25$(OH)_2$ vitamin D_3. About 150 mg Ca^{2+}/day are secreted into the gastrointestinal tract (via saliva, gastric juice, pancreatic juice, bile, and intestinal secretions), so that net absorption is only about 150 mg/day. Fecal Ca^{2+} excretion is about 850 mg/day and urinary excretion about 150 mg/day.

A normal plasma [Ca^{2+}] is about 10 mg/dL, which is equal to 2.5 mmol/L (since the atomic weight of calcium is 40) or 5 mEq/L. About 40% of plasma Ca^{2+} is bound to plasma proteins (mainly serum albumin), 10% is bound to small diffusible anions (such as citrate, bicarbonate, phosphate, and sulfate) and 50% is free or ionized. It is the ionized Ca^{2+} in the blood that is physiologically important and closely regulated (see Chapter 36). Most of the Ca^{2+} in the body is in bone (99%), which constantly turns over. In a healthy adult, the rate of release of Ca^{2+} from old bone exactly matches the rate of deposition of Ca^{2+} in newly formed bone (500 mg/day).

Ca^{2+} that is not bound to plasma proteins (i.e., 60% of the plasma Ca^{2+}) is freely filterable in the glomeruli. About 60% of the filtered Ca^{2+} is reabsorbed in the proximal convoluted tubule (Fig. 24.18). Two thirds is reabsorbed via a paracellular route in response to solvent drag and the small lumen positive potential (+3 mV) found in the late proximal convoluted tubule. One third is reabsorbed via a transcellular route that includes Ca^{2+} channels in the apical plasma membrane and a primary Ca^{2+}-ATPase or 3 Na^+/1 Ca^{2+} exchanger in the basolateral plasma membrane. About 30% of filtered Ca^{2+} is reabsorbed along the loop of Henle. Most of the Ca^{2+} reabsorbed in the thick ascending limb is by passive transport through the tight junctions, propelled by the lumen positive potential.

Reabsorption continues along the distal convoluted tubule. Reabsorption here is increased by thiazide diuretics, which may be prescribed in cases of excessive Ca^{2+} in the urine, **hypercalciuria**, and kidney stone disease (see Clinical Focus Box 24.2). Thiazides inhibit the luminal membrane Na-Cl cotransporter in distal convoluted tubule cells, which leads to a fall in intracellular [Na^+]. This, in turn, promotes Na^+-Ca^{2+} exchange and increased basolateral extrusion of Ca^{2+} and increased Ca^{2+} reabsorption.

The late distal tubule (connecting tubule and initial part of the cortical collecting duct) is an important site of control of Ca^{2+} excretion because this is where parathyroid hormone (PTH) increases Ca^{2+} reabsorption. Ca^{2+} diffuses into the cells, primarily through an epithelial Ca^{2+} channel (ECaC) in the apical membrane, is transported through the cytoplasm by a 1,25$(OH)_2$ vitamin D_3-dependent calcium-binding protein, called calbindin, and is extruded by a Na^+/Ca^{2+} exchanger or Ca^{2+}-ATPase in the basolateral plasma membrane. Only about 0.5 to 2% of the filtered Ca^{2+} is usually excreted. (Chapter 34 discusses Ca^{2+} balance and its control by several hormones in more detail.)

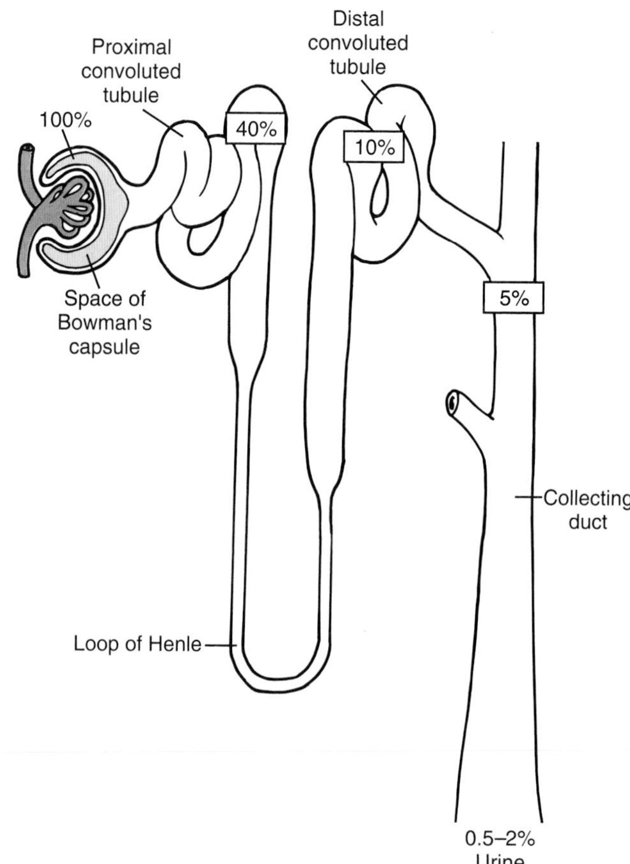

FIGURE 24.18 The percentage of the filtered load of Ca^{2+} remaining in tubular fluid as it flows down the nephron. The kidneys filter about 10,800 mg/day (0.6 × 100 mg/L × 180 L/day) and excrete only about 0.5 to 2% of the filtered load, that is, about 50 to 200 mg/day. Thiazides increase Ca^{2+} reabsorption by the distal convoluted tubule, and PTH increases Ca^{2+} reabsorption by the connecting tubule and cortical collecting duct.

MAGNESIUM BALANCE

An adult body contains about 2,000 mEq of Mg^{2+}, of which about 60% is present in bone, about 39% in cells, and about 1% in the ECF. Mg^{2+} is the second most abundant cation in cells, after K^+ (see Table 24.2). The bulk of intracellular Mg^{2+} is not free, but is bound to a variety of organic compounds, such as ATP. Mg^{2+} is present in the plasma at a concentration of about 1 mmol/L (2 mEq/L). About 20% of plasma Mg^{2+} is bound to plasma proteins, 20% is complexed with various anions, and 60% is free or ionized.

About 25% of the Mg^{2+} filtered by the glomeruli is reabsorbed in the proximal convoluted tubule (Fig. 24.19); this is a lower percentage than for Na^+, K^+, Ca^{2+}, or water. The proximal tubule epithelium is rather impermeable to Mg^{2+} under normal conditions, so there is little passive Mg^{2+} reabsorption. The major site of Mg^{2+} reabsorption is the loop of Henle (mainly the thick ascending limb), which

CLINICAL FOCUS BOX 24.2

Kidney Stone Disease (Nephrolithiasis)

A kidney stone is a hard mass that forms in the urinary tract. At least 1% of Americans develop kidney stones at some time during their lives. **Nephrolithiasis** or kidney stone disease occurs more commonly in men than in women and usually strikes men between the ages of 30 and 60. A stone lodged in the ureter will cause bleeding and intense pain. Kidney stone disease causes considerable suffering and loss of time from work, and it may lead to kidney damage. Once a stone forms in a person, stone formation often recurs.

Stones form when poorly soluble substances in the urine precipitate out of solution, causing crystals to form, aggregate, and grow. Most kidney stones (75 to 85%) are made up of insoluble Ca^{2+} salts of oxalate and phosphate. There may be excessive amounts of Ca^{2+} or oxalate in the urine as a result of diet, a genetic defect, or unknown causes. Stones may also form from precipitated ammonium magnesium phosphate (struvite), uric acid, and cystine. Struvite stones (10 to 15% of all stones) are the result of infection with bacteria, usually *Proteus* species. Uric acid stones (5 to 8% of all stones) may form in patients with excessive uric acid production and excretion, as occurs in some patients with gout. Defective tubular reabsorption of cystine (in patients with **cystinuria**) leads to cystine stones (1% of stones). The rather insoluble amino acid cystine was first isolated from a urinary bladder stone by Wollaston in 1810, hence, its name. Because low urine flow rate raises the concentration of all poorly soluble substances in the urine, favoring precipitation, a key to prevention of kidney stones is to drink plenty of water and maintain a high urine output day and night.

Fortunately, most stones are small enough to be passed down the urinary tract and spontaneously eliminated. Microscopic and chemical examination of the eliminated stones is used to determine the nature of the stone and help guide treatment. Sometimes a change in diet is recommended to reduce the amount of potential stone-forming material (e.g., Ca^{2+}, oxalate, or uric acid) in the urine. Thiazide diuretics are useful in reducing Ca^{2+} excretion if excessive urinary Ca^{2+} excretion (hypercalciuria) is the problem. Potassium citrate is useful in treating most stone disease because citrate complexes Ca^{2+} in the urine and inhibits the crystallization of Ca^{2+} salts. It also makes the urine more alkaline (since citrate is oxidized to HCO_3^- in the body). This is helpful in reducing the risk of uric acid stones because urates (favored in an alkaline urine) are more soluble than uric acid (the form favored in an acidic urine). Administering an inhibitor of uric acid synthesis, such as allopurinol, can help reduce the amount of uric acid in the urine.

If the stone is not passed, several options are available. Surgery to remove the stone can be done, but **extracorporeal shock wave lithotripsy** is more common, using a device called a **lithotriptor**. The patient is placed in a tub of water, and the stone is localized by X-ray imaging. Shock waves are generated in the water by high-voltage electric discharges and are focused on the stone through the body wall. The shock waves fragment the stone so that it can be passed down the urinary tract and eliminated. As some renal injury is produced by this procedure, it may not be entirely innocuous. Other procedures include passing a tube with an ultrasound transducer through the skin into the renal pelvis; stone fragments can be removed directly. A ureteroscope with a laser can also be used to break up stones.

reabsorbs about 65% of filtered Mg^{2+}. Reabsorption here is mainly passive and occurs through the tight junctions, driven by the lumen positive potential. Recent studies have identified a tight junction protein that is a channel that facilitates Mg^{2+} movement. Changes in Mg^{2+} excretion result mainly from changes in loop transport. More distal portions of the nephron reabsorb only a small fraction of filtered Mg^{2+} and, under normal circumstances, appear to play a minor role in controlling Mg^{2+} excretion.

An abnormally low plasma $[Mg^{2+}]$ is characterized by neuromuscular and CNS hyperirritability. Abnormally high plasma Mg^{2+} levels have a sedative effect and may cause cardiac arrest. Dietary intake of Mg^{2+} is usually 20 to 50 mEq/day; two thirds is excreted in the feces, and one third is excreted in the urine. The kidneys are mainly responsible for regulating the plasma $[Mg^{2+}]$. Excess amounts of Mg^{2+} are rapidly excreted by the kidneys. In Mg^{2+}-deficient states, Mg^{2+} virtually disappears from the urine.

PHOSPHATE BALANCE

A normal plasma concentration of inorganic phosphate is about 1 mmol/L. At a normal blood pH of 7.4, 80% of the phosphate is present as HPO_4^{2-} and 20% is present as $H_2PO_4^-$. Phosphate plays a variety of roles in the body: It is an important constituent of bone; it plays a critical role in cell metabolism, structure, and regulation (as organic phosphates); and it is a pH buffer.

Phosphate is mainly unbound in the plasma and freely filtered by the glomeruli. About 60 to 70% of filtered phosphate is actively reabsorbed in the proximal convoluted tubule and another 15% is reabsorbed by the proximal straight tubule via a Na^+-phosphate cotransporter in the luminal plasma membrane (Fig. 24.20). The remaining portions of the nephron and collecting ducts reabsorb little, if any, phosphate. The proximal tubule is the major site of phosphate reabsorption. Only about 5 to 20% of filtered phosphate is usually excreted. Phosphate in the urine is an important pH buffer and contributes to titratable acid excretion (see Chapter 25). Phosphate reabsorption is Tm-limited (see Chapter 23), and the amounts of phosphate filtered usually exceed the maximum reabsorptive capacity of the tubules for phosphate. This is different from the situation for glucose, in which normally less glucose is filtered than can be reabsorbed. If more phosphate is ingested and absorbed by the intestine, plasma [phosphate] rises, more phosphate is filtered, and the filtered load exceeds the Tm more than usual and the extra phosphate is excreted. Thus, the kidneys participate in regulating the plasma phosphate

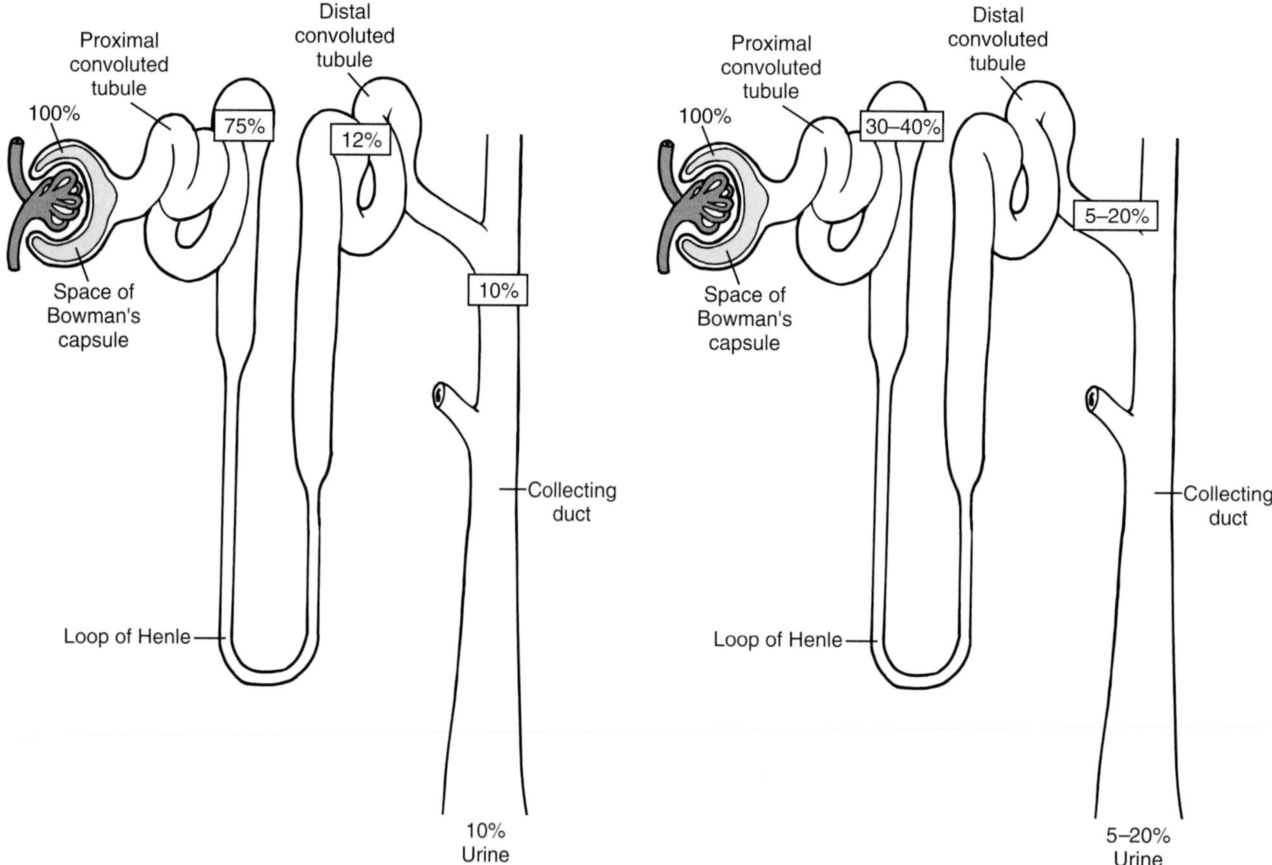

FIGURE 24.19 The percentage of the filtered load of Mg^{2+} remaining in tubular fluid as it flows down the nephron. The loop of Henle, specifically the thick ascending limb, is the major site of reabsorption of filtered Mg^{2+}.

FIGURE 24.20 The percentage of the filtered load of phosphate remaining in tubular fluid as it flows down the nephron. The proximal tubule is the major site of phosphate reabsorption, and downstream nephron segments reabsorb little, if any, phosphate.

by an "overflow" type mechanism. When there is an excess of phosphate in the body, they automatically increase phosphate excretion. In cases of phosphate depletion, the kidneys filter less phosphate and the tubules reabsorb a larger percentage of the filtered phosphate.

Phosphate reabsorption in the proximal tubule is controlled by a variety of factors. PTH is of particular importance; it decreases the phosphate Tm, increasing phosphate excretion.

Patients with chronic renal disease often develop an elevated plasma [phosphate] or **hyperphosphatemia**, depending on the severity of the disease. When GFR falls, the filtered phosphate load is diminished, and the tubules reabsorb phosphate more completely. Phosphate excretion is inadequate in the face of continued intake of phosphate in the diet. Hyperphosphatemia is dangerous because of the precipitation of calcium phosphate in soft tissue. For example, when calcium phosphate precipitates in the walls of blood vessels, blood flow will be impaired. Hyperphosphatemia can lead to myocardial failure and pulmonary insufficiency.

When plasma [phosphate] rises, the plasma ionized [Ca^{2+}] tends to fall, for two reasons. First, phosphate forms a complex with Ca^{2+}. Second, hyperphosphatemia decreases production of 1,25$(OH)_2$ vitamin D_3 in the kidneys by inhibiting the 1α-hydroxylase enzyme that forms this hormone. With decreased plasma levels of 1,25$(OH)_2$ vitamin D_3, there is less Ca^{2+} absorption by the small intestine and a tendency for hypocalcemia.

Low plasma ionized [Ca^{2+}] stimulates hyperplasia of the parathyroid glands and increased secretion of PTH. High plasma [phosphate] also stimulates PTH secretion directly. PTH then inhibits phosphate reabsorption by the proximal tubules, promotes phosphate excretion, and helps return plasma [phosphate] back to normal. Elevated PTH levels, however, also cause mobilization of Ca^{2+} and phosphate from bone. Increased bone resorption results, and the bone minerals are replaced with fibrous tissue that renders the bone more susceptible to fracture.

Patients with advanced chronic renal failure are often advised to restrict phosphate intake and consume substances (such as Ca^{2+} salts) that bind phosphate in the intestines, so as to avoid the many problems caused by hyperphosphatemia. Administration of synthetic 1,25$(OH)_2$ vitamin D_3 may compensate for deficient renal production of this hormone. This hormone opposes hypocalcemia and inhibits PTH synthesis and secretion. Parathyroidectomy is sometimes necessary in patients with advanced chronic renal failure.

URINARY TRACT

The kidneys form urine all of the time. The urine is transported by the ureters to the urinary bladder. The bladder is specialized to fill with urine at a low pressure and to empty its contents when appropriate. Contractions of the bladder and its sphincters are controlled by the nervous system.

The Ureters Convey Urine to the Bladder

The **ureters** are muscular tubes that propel the urine from the pelvis of each kidney to the urinary bladder. Peristaltic movements originate in the region of the calyces, which contain specialized smooth muscle cells that generate spontaneous pacemaker potentials. These pacemaker potentials trigger action potentials and contractions in the muscular regions of the renal pelvis that propagate distally to the ureter. Peristaltic waves sweep down the ureters at a frequency of one every 10 seconds to one every 2 to 3 minutes. The ureters enter the base of the bladder obliquely, forming a valvular flap that passively prevents the reflux of urine during contractions of the bladder. The ureters are innervated by sympathetic and parasympathetic nerve fibers. Sensory fibers mediate the intense pain that is felt when a stone distends or blocks a ureter.

The Bladder Stores Urine Until It Can Be Conveniently Emptied

The **urinary bladder** is a distensible hollow vessel containing smooth muscle in its wall (Fig. 24.21). The muscle is called the *detrusor* (from Latin for "that which pushes down"). The neck of the bladder, the involuntary **internal sphincter**, also contains smooth muscle. The bladder body and neck are innervated by parasympathetic **pelvic nerves** and sympathetic **hypogastric nerves**. The external sphincter, the **compressor urethrae**, is composed of skeletal muscle and innervated by somatic nerve fibers that travel in the **pudendal nerves**. Pelvic, hypogastric, and pudendal nerves contain both motor and sensory fibers.

The bladder has two functions: to serve as a distensible reservoir for urine and to empty its contents at appropriate intervals. When the bladder fills, it adjusts its tone to its capacity, so that minimal increases in bladder pressure occur. The external sphincter is kept closed by discharges along the pudendal nerves. The first sensation of bladder filling is experienced at a volume of 100 to 150 mL in an adult, and the first desire to void is elicited when the bladder contains about 150 to 250 mL of urine. A person becomes uncomfortably aware of a full bladder when the volume is 350 to 400 mL; at this volume, hydrostatic pressure in the bladder is about 10 cm H_2O. With further volume increases, bladder pressure rises steeply, partly as a result of reflex contractions of the detrusor. An increase in volume to 700 mL creates pain and often loss of control. The sensations of bladder filling, of conscious desire to void, and painful distension are mediated by afferents in the pelvic nerves.

Micturition Involves Autonomic and Somatic Nerves

Micturition (urination), the periodic emptying of the bladder, is a complex act involving both autonomic and somatic nerve pathways and several reflexes that can be either inhibited or facilitated by higher centers in the brain. The basic reflexes occur at the level of the sacral spinal cord and are modified by centers in the midbrain and cerebral cortex. Distension of the bladder is sensed by stretch receptors in the bladder wall; these induce reflex contraction of the detrusor and relaxation of the internal and external sphincters. This reflex is released by removing inhibitory influences from the cerebral cortex. Fluid flow through the urethra reflexively causes further contraction of the detrusor and relaxation of the external sphincter. Increased parasympathetic nerve activity stimulates contraction of the detrusor and relaxation of the internal sphincter. Sympathetic innervation is not essential for micturition. During micturition, the perineal and levator ani muscles relax, shortening the urethra and decreasing urethral resistance. Descent of the diaphragm and contraction of abdominal muscles raises intra-abdominal pressure, and aids in the expulsion of urine from the bladder.

Micturition is fortunately under voluntary control in healthy adults. In the young child, however, it is purely re-

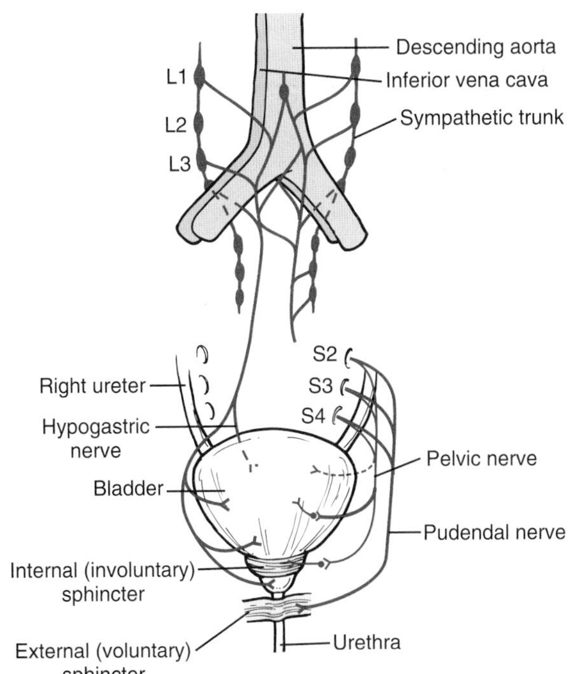

FIGURE 24.21 **The innervation of the urinary bladder.** The parasympathetic pelvic nerves arise from spinal cord segments S2 to S4 and supply motor fibers to the bladder musculature and internal (involuntary) sphincter. Sympathetic motor fibers supply the bladder via the hypogastric nerves, which arise from lumbar segments of the spinal cord. The pudendal nerves supply somatic motor innervation to the external (voluntary) sphincter. Sensory afferents (dashed lines) from the bladder travel mainly in the pelvic nerves but also to some extent in the hypogastric nerves.

flex and occurs whenever the bladder is sufficiently distended. At about 2½ years of age, it begins to come under cortical control, and in most children, complete control is achieved by age 3. Damage to the nerves that supply the bladder and its sphincters can produce abnormalities of micturition and incontinence. An increased resistance of the upper urethra commonly occurs in older men and is a result of enlargement of the surrounding prostate gland. This condition is called **benign prostatic hyperplasia**, and it results in decreased urine stream, overdistension of the bladder as a result of incomplete emptying, and increased urgency and frequency of urination.

REVIEW QUESTIONS

DIRECTIONS: Each of the numbered items or incomplete statements in this section is followed by answers or completions of the statement. Select the ONE lettered answer or completion that is BEST in each case.

1. Which of the following body fluid volumes cannot be directly determined with a single indicator?
(A) Extracellular fluid volume
(B) Intracellular fluid volume
(C) Plasma volume
(D) Total body water
2. Which of the following results in thirst?
(A) Cardiac failure
(B) Decreased plasma levels of angiotensin II
(C) Distension of the cardiac atria
(D) Distension of the stomach
(E) Hypotonic volume expansion
3. Arginine vasopressin (AVP) is synthesized in the
(A) Adrenal cortex
(B) Anterior hypothalamus
(C) Anterior pituitary
(D) Collecting ducts of the kidneys
(E) Posterior pituitary
4. A 60-kg woman is given 10 microcuries (μCi) (370 kilobecquerels) of radioiodinated serum albumin (RISA) intravenously. Ten minutes later, a venous blood sample is collected, and the plasma RISA activity is 4 μCi/L. Her hematocrit ratio is 0.40. What is her blood volume?
(A) 417 mL
(B) 625 mL
(C) 2.5 L
(D) 4.17 L
(E) 6.25 L
5. Which of the following leads to decreased Na^+ reabsorption by the kidneys?
(A) An increase in central blood volume
(B) An increase in colloid osmotic pressure in the peritubular capillaries
(C) An increase in GFR
(D) An increase in plasma aldosterone level
(E) An increase in renal sympathetic nerve activity
6. The nephron segment that reabsorbs the largest percentage of filtered Mg^{2+} is the
(A) Proximal convoluted tubule
(B) Thick ascending limb
(C) Distal convoluted tubule
(D) Cortical collecting duct
(E) Medullary collecting duct
7. Which of the following causes decreased renin release by the kidneys?
(A) Decreased fluid and solute delivery to the macula densa
(B) Hemorrhage
(C) Intravenous infusion of isotonic saline
(D) Narrowing (stenosis) of the renal artery
(E) Stimulation of renal sympathetic nerves
8. Which of the following may cause hyperkalemia?
(A) Epinephrine injection
(B) Hyperaldosteronism
(C) Insulin administration
(D) Intravenous infusion of a $NaHCO_3$ solution
(E) Skeletal muscle injury
9. Parathyroid hormone (PTH)
(A) Decreases tubular reabsorption of Ca^{2+}
(B) Decreases tubular reabsorption of phosphate
(C) Inhibits bone resorption
(D) Secretion is decreased in patients with chronic renal failure
(E) Secretion is stimulated by a rise in plasma ionized Ca^{2+}
10. Aldosterone acts on cortical collecting ducts to
(A) Decrease K^+ secretion
(B) Decrease Na^+ reabsorption
(C) Decrease water permeability
(D) Increase K^+ secretion
(E) Increase water permeability
11. In response to an increase in GFR, the proximal tubule and the loop of Henle demonstrate an increase in the rate of Na^+ reabsorption. This phenomenon is called
(A) Autoregulation
(B) Glomerulotubular balance
(C) Mineralocorticoid escape
(D) Saturation of tubular transport
(E) Tubuloglomerular feedback
12. A 60-year-old woman is always thirsty and wakes up several times during the night to empty her bladder. Plasma osmolality is 295 mosm/kg H_2O (normal range, 281 to 297 mosm/kg H_2O), urine osmolality is 100 mosm/kg H_2O, and plasma AVP levels are higher than normal. The urine is negative for glucose. The most likely diagnosis is
(A) Diabetes mellitus
(B) Diuretic drug abuse
(C) Nephrogenic diabetes insipidus
(D) Neurogenic diabetes insipidus
(E) Primary polydipsia
13. The volume of the extracellular fluid is most closely related to the amount of which solute in this compartment?
(A) HCO_3^-
(B) Glucose
(C) K^+
(D) Serum albumin
(E) Na^+
14. A homeless man was found comatose, lying in the doorway of a downtown department store at night. His plasma osmolality was 370 mosm/kg H_2O (normal, 281 to 297 mosm/kg H_2O), plasma [Na^+] was 140 mEq/L (normal, 136 to 145 mEq/L), plasma [glucose] 100 mg/dL (normal fasting level, 70 to 110 mg/dL), and BUN 15 mg/dL (normal, 7 to 18 mg/dL). His most likely problem is
(A) Alcohol intoxication
(B) Dehydration
(C) Diabetes insipidus
(D) Diabetes mellitus
(E) Renal failure
15. A hypertensive patient is given an angiotensin-converting enzyme (ACE) inhibitor. Which of the following changes would be expected?
(A) Plasma aldosterone level will rise
(B) Plasma angiotensin I level will rise
(C) Plasma angiotensin II level will rise
(D) Plasma bradykinin level will fall
(E) Plasma renin level will fall
16. If a person consumes a high-K^+ diet, the majority of K^+ excreted in the urine is derived from
(A) Glomerular filtrate
(B) K^+ that is not reabsorbed in the proximal tubule

(continued)

(C) K^+ secreted in the loop of Henle
(D) K^+ secreted by the cortical collecting duct
(E) K^+ secreted by the inner medullary collecting duct

17. Which of the following set of values would lead you to suspect that a person has syndrome of inappropriate secretion of ADH (SIADH)?

	Plasma Osmolality (mosm/ kg H_2O)	*Plasma $[Na^+]$ (mEq/L)*	*Urine Osmolality (mosm/ kg H_2O)*
(A)	300	145	100
(B)	270	130	50
(C)	285	140	600
(D)	270	130	450
(E)	285	140	1,200

18. A dehydrated hospitalized patient with uncontrolled diabetes mellitus has a plasma $[K^+]$ of 4.5 mEq/L (normal, 3.5 to 5.0 mEq/L), a plasma [glucose] of 500 mg/dL, and an arterial blood pH of 7.00 (normal, 7.35 to 7.45). These data suggest that the patient has
(A) A decreased total body store of K^+
(B) A normal total body store of K^+
(C) An increased total body store of K^+
(D) Hypokalemia
(E) Hyperkalemia

19. Intravenous infusion of 2.0 L of isotonic saline (0.9% NaCl) results in increased
(A) Intracellular fluid volume
(B) Plasma aldosterone level
(C) Plasma arginine vasopressin (AVP) concentration
(D) Plasma atrial natriuretic peptide (ANP) concentration
(E) Plasma volume, but no change in other body fluid compartments

20. The kidneys of a person with congestive heart failure avidly retain Na^+. The best explanation for this is that the
(A) Effective arterial blood volume is decreased
(B) Extracellular fluid volume is decreased
(C) Extracellular fluid volume is increased
(D) Total blood volume is decreased
(E) Total blood volume is increased

SUGGESTED READING

Adrogue HJ, Madias NE. Hypernatremia. N Engl J Med 2000;342:1493–1499.
Adrogue HJ, Madias NE. Hyponatremia. N Engl J Med 2000;342:1581–1589.
Braunwald E. Edema. In: Braunwald E, et al., eds. Harrison's Principles of Internal Medicine. 15th Ed. New York: McGraw-Hill, 2001;217–222.
Brooks VL, Vander AJ, eds. Refresher course for teaching renal physiology. Adv Physiol Education 1998;20:S114–S245.
Giebisch G. Renal potassium transport: Mechanisms and regulation. Am J Physiol 1998;274:F817–F833.
Hoenderop JGJ, Willems PHGM, Bindels RJM. Toward a comprehensive molecular model of active calcium reabsorption. Am J Physiol 2000;278:F352–F360.
Koeppen BM, Stanton BA. Renal Physiology. 3rd Ed. St. Louis: Mosby-Year Book, 2001.
Kumar R. New concepts concerning the regulation of renal phosphate excretion. News Physiol Sci 1997;12:211–214.
Quamme GA. Renal magnesium handling: New insights in understanding old problems. Kidney Int 1997;52:1180–1195.
Rose BD, Post TW. Clinical Physiology of Acid-Base and Electrolyte Disorders. 5th Ed. New York: McGraw-Hill, 2001.
Valtin H, Schafer JA. Renal Function. 3rd Ed. Boston: Little, Brown, 2001.
Vander AJ. Renal Physiology. 5th Ed. New York: McGraw-Hill, 1995.
Weiner ID, Wingo CS. Hyperkalemia: A potential silent killer. J Am Soc Nephrol 1998;9:1535–1543.

Acid-Base Balance

George A. Tanner, Ph.D.

CHAPTER OUTLINE

KEY CONCEPTS

1. The body is constantly threatened by acid resulting from diet and metabolism. The stability of blood pH is maintained by the concerted action of chemical buffers, the lungs, and the kidneys.
2. Numerous chemical buffers (e.g., HCO_3^-/CO_2, phosphates, proteins) work together to minimize pH changes in the body. The concentration ratio (base/acid) of any buffer pair, together with the pK of the acid, automatically defines the pH.
3. The bicarbonate/CO_2 buffer pair is effective in buffering in the body because its components are present in large amounts and the system is open.
4. The respiratory system influences plasma pH by regulating the P_{CO_2} by changing the level of alveolar ventilation. The kidneys influence plasma pH by getting rid of acid or base in the urine.
5. Renal acidification involves three processes: reabsorption of filtered HCO_3^-, excretion of titratable acid, and excretion of ammonia. New HCO_3^- is added to the plasma and replenishes depleted HCO_3^- when titratable acid (normally mainly $H_2PO_4^-$) and ammonia (as NH_4^+) are excreted.
6. The stability of intracellular pH is ensured by membrane transport of H^+ and HCO_3^-, by intracellular buffers (mainly proteins and organic phosphates), and by metabolic reactions.
7. Respiratory acidosis is an abnormal process characterized by an accumulation of CO_2 and a fall in arterial blood pH. The kidneys compensate by increasing the excretion of H^+ in the urine and adding new HCO_3^- to the blood, thereby diminishing the severity of the acidemia.
8. Respiratory alkalosis is an abnormal process characterized by an excessive loss of CO_2 and a rise in pH. The kidneys compensate by increasing the excretion of filtered HCO_3^-, thereby diminishing the alkalemia.
9. Metabolic acidosis is an abnormal process characterized by a gain of acid (other than H_2CO_3) or a loss of HCO_3^-. Respiratory compensation is hyperventilation, and renal compensation is an increased excretion of H^+ bound to urinary buffers (ammonia, phosphate).
10. Metabolic alkalosis is an abnormal process characterized by a gain of strong base or HCO_3^- or a loss of acid (other than H_2CO_3). Respiratory compensation is hypoventilation, and renal compensation is an increased excretion of HCO_3^-.
11. The plasma anion gap is equal to the plasma $[Na^+] - [Cl^-] - [HCO_3^-]$ and is most useful in narrowing down possible causes of metabolic acidosis.

Every day, metabolic reactions in the body produce and consume many moles of hydrogen ions (H^+s). Yet, the $[H^+]$ of most body fluids is very low (in the nanomolar range) and is kept within narrow limits. For example, the $[H^+]$ of arterial blood is normally 35 to 45 nmol/L (pH 7.45 to 7.35). Normally the body maintains acid-base balance; inputs and outputs of acids and bases are matched so that $[H^+]$ stays relatively constant both outside and inside cells.

Most of this chapter discusses the regulation of $[H^+]$ in extracellular fluid because ECF is easier to analyze than intracellular fluid and is the fluid used in the clinical evaluation of acid-base balance. In practice, systemic arterial blood is used as the reference for this purpose. Measurements on whole blood with a pH meter give values for the $[H^+]$ of plasma and, therefore, provide an ECF pH measurement.

A REVIEW OF ACID-BASE CHEMISTRY

In this section, we briefly review some principles of acid-base chemistry. We define acid, base, acid dissociation constant, weak and strong acids, pK_a, pH, and the Henderson-Hasselbalch equation and explain buffering. Students who already feel comfortable with these concepts can skip this section.

Acids Dissociate to Release Hydrogen Ions in Solution

An **acid** is a substance that can release or donate H^+; a **base** is a substance that can combine with or accept H^+. When an acid (generically written as HA) is added to water, it dissociates reversibly according to the reaction, $HA \rightleftharpoons H^+ + A^-$. The species A^- is a base because it can combine with a H^+ to form HA. In other words, when an acid dissociates, it yields a free H^+ and its conjugate (meaning "joined in a pair") base.

The Acid Dissociation Constant K_a Shows the Strength of an Acid

At equilibrium, the rate of dissociation of an acid to form $H^+ + A^-$, and the rate of association of H^+ and base A^- to form HA, are equal. The equilibrium constant (K_a), which is also called the ionization constant or acid **dissociation constant**, is given by the expression

$$K_a = \frac{[H^+] \times [A^-]}{[HA]} \qquad (1)$$

The higher the acid dissociation constant, the more an acid is ionized and the greater is its strength. Hydrochloric acid (HCl) is an example of a **strong acid**. It has a high K_a and is almost completely ionized in aqueous solutions. Other strong acids include sulfuric acid (H_2SO_4), phosphoric acid (H_3PO_4), and nitric acid (HNO_3).

An acid with a low K_a is a **weak acid**. For example, in a 0.1 M solution of acetic acid ($K_a = 1.8 \times 10^{-5}$) in water, most (99%) of the acid is nonionized and little (1%) is present as acetate$^-$ and H^+. The acidity (concentration of free H^+) of this solution is low. Other weak acids are lactic acid, carbonic acid (H_2CO_3), ammonium ion (NH_4^+), and dihydrogen phosphate ($H_2PO_4^-$).

pK_a Is a Logarithmic Expression of K_a

Acid dissociation constants vary widely and often are small numbers. It is convenient to convert K_a to a logarithmic form, defining pK_a as

$$pK_a = \log_{10}(1/K_a) = -\log_{10}K_a \qquad (2)$$

In aqueous solution, each acid has a characteristic pK_a, which varies slightly with temperature and the ionic strength of the solution. Note that pK_a is *inversely* proportional to acid strength. A strong acid has a high K_a and a low pK_a. A weak acid has a low K_a and a high pK_a.

pH Is Inversely Related to $[H^+]$

$[H^+]$ is often expressed in pH units. The following equation defines pH:

$$pH = \log_{10}(1/[H^+]) = -\log_{10}[H^+] \qquad (3)$$

where $[H^+]$ is in mol/L. Note that pH is *inversely* related to $[H^+]$. Each whole number on the pH scale represents a 10-fold (logarithmic) change in acidity. A solution with a pH of 5 has 10 times the $[H^+]$ of a solution with a pH of 6.

The Henderson-Hasselbalch Equation Relates pH to the Ratio of the Concentrations of Conjugate Base and Acid

For a solution containing an acid and its conjugate base, we can rearrange the equilibrium expression (equation 1) as

$$[H^+] = \frac{K_a \times [HA]}{[A^-]} \qquad (4)$$

If we take the negative logarithms of both sides,

$$-\log[H^+] = -\log K_a + \log\frac{[A^-]}{[HA]} \qquad (5)$$

Substituting pH for $-\log[H^+]$ and pK_a for $-\log K_a$, we get

$$pH = pK_a + \log\frac{[A^-]}{[HA]} \qquad (6)$$

This equation is known as the **Henderson-Hasselbalch equation**. It shows that the pH of a solution is determined by the pK_a of the acid and the ratio of the concentration of conjugate base to acid.

Buffers Promote the Stability of pH

The stability of pH is protected by the action of buffers. A **pH buffer** is defined as something that *minimizes* the change in pH produced when an acid or base is added. Note that a buffer *does not prevent* a pH change. A **chemical pH buffer** is a mixture of a weak acid and its conjugate base (or a weak base and its conjugate acid). Following are examples of buffers:

Weak Acid		Conjugate Base		
H_2CO_3 (carbonic acid)	$\rightleftharpoons$	HCO_3^- (bicarbonate)	$+ H^+$	(7)
$H_2PO_4^-$ (dihydrogen phosphate)	$\rightleftharpoons$	HPO_4^{2-} (monohydrogen phosphate)	$+ H^+$	(8)
NH_4^+ (ammonium ion)	$\rightleftharpoons$	NH_3 (ammonia)	$+ H^+$	(9)

Generally, the equilibrium expression for a buffer pair can be written in terms of the Henderson-Hasselbalch equation:

$$pH = pK_a + \log \frac{[\text{conjugate base}]}{[\text{acid}]} \qquad (10)$$

For example, for $H_2PO_4^-/HPO_4^{2-}$

$$pH = 6.8 + \log \frac{[HPO_4^{2-}]}{[H_2PO_4^-]} \qquad (11)$$

The effectiveness of a buffer—how well it reduces pH changes when an acid or base is added—depends on its concentration and its pK_a. A good buffer is present in high concentrations and has a pK_a close to the desired pH.

Figure 25.1 shows a titration curve for the phosphate buffer system. As a strong acid or strong base is progressively added to the solution (shown on the x-axis), the resulting pH is recorded (shown on the y-axis). Going from right to left as strong acid is added, H^+ combines with the basic form of phosphate: $H^+ + HPO_4^{2-} \rightleftharpoons H_2PO_4^-$. Going from left to right as strong base is added, OH^- combines with H^+ released from the acid form of the phosphate buffer: $OH^- + H_2PO_4^- \rightleftharpoons HPO_4^{2-} + H_2O$. These reactions lessen the fall or rise in pH.

At the pK_a of the phosphate buffer, the ratio $[HPO_4^{2-}]/[H_2PO_4^-]$ is 1 and the titration curve is flattest (the change in pH for a given amount of an added acid or base is at a minimum). In most cases, pH buffering is effective when the solution pH is within plus or minus one pH unit of the buffer pK_a. Beyond that range, the pH shift that a given amount of acid or base produces may be large, so the buffer becomes relatively ineffective.

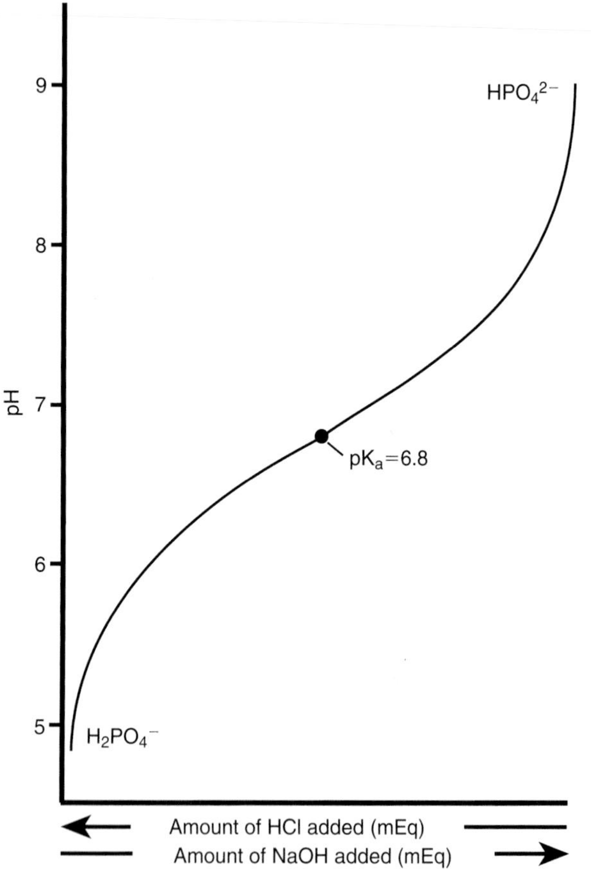

FIGURE 25.1 **A titration curve for a phosphate buffer.** The pK_a for $H_2PO_4^-$ is 6.8. A strong acid (HCl) (right to left) or strong base (NaOH) (left to right) was added and the resulting solution pH recorded (y-axis). Notice that buffering is best (i.e., the change in pH upon the addition of a given amount of acid or base is least) when the solution pH is equal to the pK_a of the buffer.

PRODUCTION AND REGULATION OF HYDROGEN IONS IN THE BODY

Acids are continuously produced in the body and threaten the normal pH of the extracellular and intracellular fluids. Physiologically speaking, acids fall into two groups: (1) H_2CO_3 (carbonic acid), and (2) all other acids (noncarbonic; also called "nonvolatile" or "fixed" acids). The distinction between these groups occurs because H_2CO_3 is in equilibrium with the volatile gas CO_2, which can leave the body via the lungs. The concentration of H_2CO_3 in arterial blood is, therefore, set by respiratory activity. By contrast, noncarbonic acids in the body are not directly affected by breathing. Noncarbonic acids are buffered in the body and excreted by the kidneys.

Metabolism Is a Constant Source of Carbon Dioxide

A normal adult produces about 300 L of CO_2 daily from metabolism. CO_2 from tissues enters the capillary blood, where it reacts with water to form H_2CO_3, which dissociates instantly to yield H^+ and HCO_3^-: $CO_2 + H_2O \rightleftharpoons H_2CO_3 \rightleftharpoons H^+ + HCO_3^-$. Blood pH would rapidly fall to lethal levels if the H_2CO_3 formed from CO_2 were allowed to accumulate in the body.

Fortunately, H_2CO_3 produced from metabolic CO_2 is only formed transiently in the transport of CO_2 by the blood and does not normally accumulate. Instead, it is converted to CO_2 and water in the pulmonary capillaries and the CO_2 is expired. In the lungs, the reactions reverse:

$$H^+ + HCO_3^- \rightleftharpoons H_2CO_3 \rightleftharpoons H_2O + CO_2 \qquad (12)$$

As long as CO_2 is expired as fast as it is produced, arterial blood CO_2 tension, H_2CO_3 concentration, and pH do not change.

Incomplete Carbohydrate and Fat Metabolism Produces Nonvolatile Acids

Normally, carbohydrates and fats are completely oxidized to CO_2 and water. If carbohydrates and fats are *incompletely* oxidized, nonvolatile acids are produced. Incomplete oxidation of carbohydrates occurs when the tissues do not receive enough oxygen, as during strenuous exercise or hemorrhagic or cardiogenic shock. In such states, glucose metabolism yields lactic acid ($pK_a = 3.9$), which dissociates into lactate and H^+, lowering the blood pH. Incomplete fatty acid oxidation occurs in uncontrolled diabetes mellitus, starvation, and alcoholism and produces ketone body acids (acetoacetic and β-hydroxybutyric acids). These acids have pK_a values around 4 to 5. At blood pH, they mostly dissociate into their anions and H^+, making the blood more acidic.

Protein Metabolism Generates Strong Acids

The metabolism of dietary proteins is a major source of H^+. The oxidation of proteins and amino acids produces strong acids such as H_2SO_4, HCl, and H_3PO_4. The oxidation of sulfur-containing amino acids (methionine, cysteine, cystine) produces H_2SO_4, and the oxidation of cationic amino acids (arginine, lysine, and some histidine residues) produces HCl. H_3PO_4 is produced by the oxidation of phosphorus-containing proteins and phosphoesters in nucleic acids.

On a Mixed Diet, Net Acid Gain Threatens pH

A diet containing both meat and vegetables results in a net production of acids, largely from protein oxidation. To some extent, acid-consuming metabolic reactions balance H^+ production. Food also contains basic anions, such as citrate, lactate, and acetate. When these are oxidized to CO_2 and water, H^+ ions are consumed (or, amounting to the same thing, HCO_3^- is produced). The balance of acid-forming and acid-consuming metabolic reactions results in a net production of about 1 mEq H^+/kg body weight/day in an adult person who eats a mixed diet. Persons who are vegetarians generally have less of a dietary acid burden and a more alkaline urine pH than nonvegetarians because most fruits and vegetables contain large amounts of organic anions that are metabolized to HCO_3^-. The body generally has to dispose of more or less nonvolatile acid, a function performed by the kidneys.

Whether a particular food has an acidifying or an alkalinizing effect depends on if and how its constituents are metabolized. Cranberry juice has an acidifying effect because of its content of benzoic acid, an acid that cannot be broken down in the body. Orange juice has an alkalinizing effect, despite its acidic pH of about 3.7, because it contains citrate, which is metabolized to HCO_3^-. The citric acid in orange juice is converted to CO_2 and water and has only a transient effect on blood pH and no effect on urine pH.

Many Buffering Mechanisms Protect and Stabilize Blood pH

Despite constant threats to acid-base homeostasis, a healthy person maintains a normal blood pH. Figure 25.2 shows some of the ways in which blood pH is kept at normal levels despite the daily net acid gain. The key buffering agents are chemical buffers, along with the lungs and kidneys.

1) Chemical buffering. Chemical buffers in extracellular and intracellular fluids and in bone are the first line of defense of blood pH. Chemical buffering minimizes a change in pH but does not remove acid or base from the body.

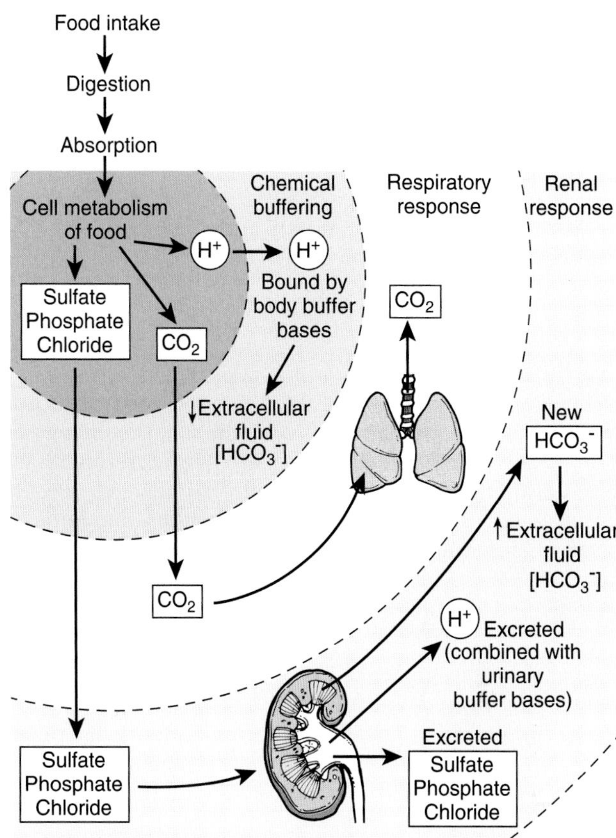

FIGURE 25.2 **The maintenance of normal blood pH by chemical buffers, the respiratory system, and the kidneys.** On a mixed diet, pH is threatened by the production of strong acids (sulfuric, hydrochloric, and phosphoric) mainly as a result of protein metabolism. These strong acids are buffered in the body by chemical buffer bases, such as ECF HCO_3^-. The kidneys eliminate hydrogen ions (combined with urinary buffers) and anions in the urine. At the same time, they add new HCO_3^- to the ECF, to replace the HCO_3^- consumed in buffering strong acids. The respiratory system disposes of CO_2.

2) Respiratory response. The respiratory system is the second line of defense of blood pH. Normally, breathing removes CO_2 as fast as it forms. Large loads of acid stimulate breathing (respiratory compensation), which removes CO_2 from the body and lowers the $[H_2CO_3]$ in arterial blood, reducing the acidic shift in blood pH.

3) Renal response. The kidneys are the third line of defense of blood pH. Although chemical buffers in the body can bind H^+ and the lungs can change $[H_2CO_3]$ of blood, the burden of removing excess H^+ falls directly on the kidneys. Hydrogen ions are excreted in combination with urinary buffers. At the same time, the kidneys add new HCO_3^- to the ECF to replace HCO_3^- used to buffer strong acids. The kidneys also excrete the anions (phosphate, chloride, sulfate) that are liberated from strong acids. The kidneys affect blood pH more slowly than other buffering mechanisms in the body; full renal compensation may take 1 to 3 days.

CHEMICAL REGULATION OF pH

The body contains many conjugate acid-base pairs that act as chemical buffers (Table 25.1). In the ECF, the main chemical buffer pair is HCO_3^-/CO_2. Plasma proteins and inorganic phosphate are also ECF buffers. Cells have large buffer stores, particularly proteins and organic phosphate compounds. HCO_3^- is present in cells, although at a lower concentration than in ECF. Bone contains large buffer stores, specifically phosphate and carbonate salts.

Chemical Buffers Are the First to Defend pH

When an acid or base is added to the body, the buffers just mentioned bind or release H^+, minimizing the change in pH. Buffering in ECF occurs rapidly, in minutes. Acids or bases also enter cells and bone, but this generally occurs more slowly, over hours, allowing cell buffers and bone to share in buffering.

A pK_a of 6.8 Makes Phosphate a Good Buffer

The pK_a for phosphate, $H_2PO_4^- \rightleftharpoons H^+ + HPO_4^{2-}$, is 6.8, close to the desired blood pH of 7.4, so phosphate is a good buffer. In the ECF, phosphate is present as inorganic phosphate. Its concentration, however, is low (about 1 mmol/L), so it plays a minor role in extracellular buffering.

Phosphate is an important intracellular buffer, however, for two reasons. First, cells contain large amounts of phosphate in such organic compounds as adenosine triphosphate (ATP), adenosine diphosphate (ADP), and creatine phosphate. Although these compounds primarily function in energy metabolism, they also act as pH buffers. Second, intracellular pH is generally lower than the pH of ECF and is closer to the pK_a of phosphate. (The cytosol of skeletal muscle, for example, has a pH of 6.9.) Phosphate is, thus, more effective in this environment than in one with a pH of 7.4. Bone has large phosphate salt stores, which also help in buffering.

TABLE 25.1 Major Chemical pH Buffers in the Body

Buffer	Reaction
Extracellular fluid	
Bicarbonate/CO_2	$CO_2 + H_2O \rightleftharpoons H_2CO_3 \rightleftharpoons H^+ + HCO_3^-$
Inorganic phosphate	$H_2PO_4^- \rightleftharpoons H^+ + HPO_4^{2-}$
Plasma proteins (Pr)	$HPr \rightleftharpoons H^+ + Pr^-$
Intracellular fluid	
Cell proteins (e.g., hemoglobin, Hb)	$HHb \rightleftharpoons H^+ + Hb^-$
Organic phosphates	Organic-$HPO_4^- \rightleftharpoons H^+ +$ organic-PO_4^{2-}
Bicarbonate/CO_2	$CO_2 + H_2O \rightleftharpoons H_2CO_3 \rightleftharpoons H^+ + HCO_3^-$
Bone	
Mineral phosphates	$H_2PO_4^- \rightleftharpoons H^+ + HPO_4^{2-}$
Mineral carbonates	$HCO_3^- \rightleftharpoons H^+ + CO_3^{2-}$

Proteins Are Excellent Buffers

Proteins are the largest buffer pool in the body and are excellent buffers. Proteins can function as both acids and bases, so they are **amphoteric**. They contain many ionizable groups, which can release or bind H^+. Serum albumin and plasma globulins are the major extracellular protein buffers, present mainly in the blood plasma. Cells also have large protein stores. Recall that the buffering properties of hemoglobin play an important role in the transport of CO_2 and O_2 by the blood (see Chapter 21).

The Bicarbonate/Carbon Dioxide Buffer Pair Is Crucial in pH Regulation

For several reasons, the HCO_3^-/CO_2 buffer pair is especially important in acid-base physiology:

1) Its components are abundant; the concentration of HCO_3^- in plasma or ECF normally averages 24 mmol/L. Although the concentration of dissolved CO_2 is lower (1.2 mmol/L), metabolism provides a nearly limitless supply.

2) Despite a pK of 6.10, a little far from the desired plasma pH of 7.40, it is effective because the system is "open."

3) It is controlled by the lungs and kidneys.

Forms of Carbon Dioxide. CO_2 exists in the body in several different forms: as gaseous CO_2 in the lung alveoli, and as dissolved CO_2, H_2CO_3, HCO_3^-, carbonate (CO_3^{2-}), and carbamino compounds in the body fluids. CO_3^{2-} is present at appreciable concentrations only in rather alkaline solutions, and so we will ignore it. We will also ignore any CO_2 that is bound to proteins in the carbamino form. The most important forms are gaseous CO_2, dissolved CO_2, H_2CO_3, and HCO_3^-.

The $CO_2/H_2CO_3/HCO_3^-$ Equilibria. Dissolved CO_2 in pulmonary capillary blood equilibrates with gaseous CO_2 in the lung alveoli. Consequently, the partial pressures of CO_2 (P_{CO_2}) in alveolar air and systemic arterial blood are normally identical. The concentration of dissolved CO_2 ($[CO_{2(d)}]$) is related to the P_{CO_2} by Henry's law (see Chapter 21). The solubility coefficient for CO_2 in plasma at 37°C is 0.03 mmol CO_2/L per mm Hg P_{CO_2}. Therefore, $[CO_{2(d)}] = 0.03 \times P_{CO_2}$. If P_{CO_2} is 40 mm Hg, then $[CO_{2(d)}]$ is 1.2 mmol/L.

In aqueous solutions, $CO_{2(d)}$ reacts with water to form H_2CO_3: $CO_{2(d)} + H_2O \rightleftharpoons H_2CO_3$. The reaction to the right is called the **hydration reaction**, and the reaction to the left is called the **dehydration reaction**. These reactions are slow if uncatalyzed. In many cells and tissues, such as the kidneys, pancreas, stomach, and red blood cells, the reactions are catalyzed by **carbonic anhydrase**, a zinc-containing enzyme. At equilibrium, $CO_{2(d)}$ is greatly favored; at body temperature, the ratio of $[CO_{2(d)}]$ to $[H_2CO_3]$ is about 400:1. If $[CO_{2(d)}]$ is 1.2 mmol/L, then $[H_2CO_3]$ equals 3 μmol/L. H_2CO_3 dissociates instantaneously into H^+ and HCO_3^-: $H_2CO_3 \rightleftharpoons H^+ + HCO_3^-$. The Henderson-Hasselbalch expression for this reaction is

$$pH = 3.5 + \log \frac{[HCO_3^-]}{[H_2CO_3]} \quad (13)$$

Note that H_2CO_3 is a fairly strong acid ($pK_a = 3.5$). Its low concentration in body fluids lessens its impact on acidity.

The Henderson-Hasselbalch Equation for HCO_3^-/CO_2. Because $[H_2CO_3]$ is so low and hard to measure and because $[H_2CO_3] = [CO_{2(d)}]/400$, we can use $[CO_{2(d)}]$ to represent the acid in the Henderson-Hasselbalch equation:

$$
\begin{aligned}
pH &= 3.5 + \log \frac{[HCO_3^-]}{[CO_{2(d)}]/400} \\
&= 3.5 + \log 400 + \log \frac{[HCO_3^-]}{[CO_{2(d)}]} \qquad (14)\\
&= 6.1 + \log \frac{[HCO_3^-]}{[CO_{2(d)}]}
\end{aligned}
$$

We can also use $0.03 \times P_{CO_2}$ in place of $[CO_{2(d)}]$:

$$pH = 6.1 + \log \frac{[HCO_3^-]}{0.03\ P_{CO_2}} \qquad (15)$$

This form of the Henderson-Hasselbalch equation is useful in understanding acid-base problems. Note that the "acid" in this equation appears to be $CO_{2(d)}$, but is really H_2CO_3 "represented" by CO_2. Therefore, this equation is valid only if $CO_{2(d)}$ and H_2CO_3 are in equilibrium with each other, which is usually (but not always) the case.

Many clinicians prefer to work with $[H^+]$ rather than pH. The following expression results if we take antilogarithms of the Henderson-Hasselbalch equation:

$$[H^+] = 24\ P_{CO_2}/[HCO_3^-] \qquad (16)$$

In this expression, $[H^+]$ is expressed in nmol/L, $[HCO_3^-]$ in mmol/L or mEq/L, and P_{CO_2} in mm Hg. If P_{CO_2} is 40 mm Hg and plasma $[HCO_3^-]$ is 24 mmol/L, $[H^+]$ is 40 nmol/L.

An "Open" Buffer System. As previously noted, the pK of the HCO_3^-/CO_2 system (6.10) is far from 7.40, the normal pH of arterial blood. From this, one might view this as a rather poor buffer pair. On the contrary, it is remarkably effective because it operates in an open system; that is, the two buffer components can be added to or removed from the body at controlled rates.

The HCO_3^-/CO_2 system is open in several ways:

1) Metabolism provides an endless source of CO_2, which can replace any H_2CO_3 consumed by a base added to the body.

2) The respiratory system can change the amount of CO_2 in body fluids by hyperventilation or hypoventilation.

3) The kidneys can change the amount of HCO_3^- in the ECF by forming new HCO_3^- when excess acid has been added to the body or excreting HCO_3^- when excess base has been added.

How the kidneys and respiratory system influence ECF pH by operating on the HCO_3^-/CO_2 system is described below. For now, the advantages of an open buffer system are best explained by an example (Fig. 25.3). Suppose we have 1 L of ECF containing 24 mmol of HCO_3^- and 1.2 mmol of dissolved $CO_{2(d)}$ (P_{CO_2} = 40 mm Hg). Using the special form of the Henderson-Hasselbalch equation described above, we find that the pH of the ECF is 7.40:

$$
\begin{aligned}
pH &= 6.10 + \log \frac{[HCO_3^-]}{0.03\ P_{CO_2}} \\
&= 6.10 + \log \frac{[24]}{[1.2]} = 7.40
\end{aligned} \qquad (17)
$$

Suppose we now add 10 mmol of HCl, a strong acid. HCO_3^- is the major buffer base in the ECF (we will neglect the contributions of other buffers). From the reaction $H^+ + HCO_3^- \rightleftharpoons H_2CO_3 \rightleftharpoons H_2O + CO_2$, we predict that the $[HCO_3^-]$ will fall by 10 mmol and that 10 mmol of $CO_{2(d)}$ will form. If the system were closed and no CO_2 could escape, the new pH would be

$$pH = 6.10 + \log \frac{[24 - 10]}{[1.2 + 10]} = 6.20 \qquad (18)$$

This is an intolerably low—indeed a fatal—pH.

Fortunately, however, the system is open and CO_2 can escape via the lungs. If all of the extra CO_2 is expired and the $[CO_{2(d)}]$ is kept at 1.2 mmol/L, the pH would be

$$pH = 6.10 + \log \frac{[24 - 10]}{[1.2]} = 7.17 \qquad (19)$$

Although this pH is low, it is compatible with life.

Still another mechanism promotes the escape of CO_2. In the body, an acidic blood pH stimulates breathing, which can make the P_{CO_2} lower than 40 mm Hg. If P_{CO_2} falls to 30 mm Hg ($[CO_{2(d)}]$ = 0.9 mmol/L) the pH would be

$$pH = 6.10 + \log \frac{[24 - 10]}{[0.9]} = 7.29 \qquad (20)$$

The system is also open at the kidneys and new HCO_3^- can be added to the plasma to correct the ECF $[HCO_3^-]$. Once the pH of the blood is normal, the stimulus for hyperventilation disappears.

Changes in Acid Production May Help Protect Blood pH

Another way in which blood pH may be protected is by changes in endogenous acid production (Fig. 25.4). An increase in blood pH caused by the addition of base to the body results in increased production of lactic acid and ketone body acids, which then reduces the alkaline shift in pH. A decrease in blood pH results in decreased production of lactic acid and ketone body acids, which opposes the acidic shift in pH.

This scenario is especially important when the endogenous production of these acids is high, as occurs during strenuous exercise or other conditions of circulatory inadequacy (lactic acidosis) or during ketosis as a result of uncontrolled diabetes, starvation, or alcoholism. These effects of pH on endogenous acid production result from changes in enzyme

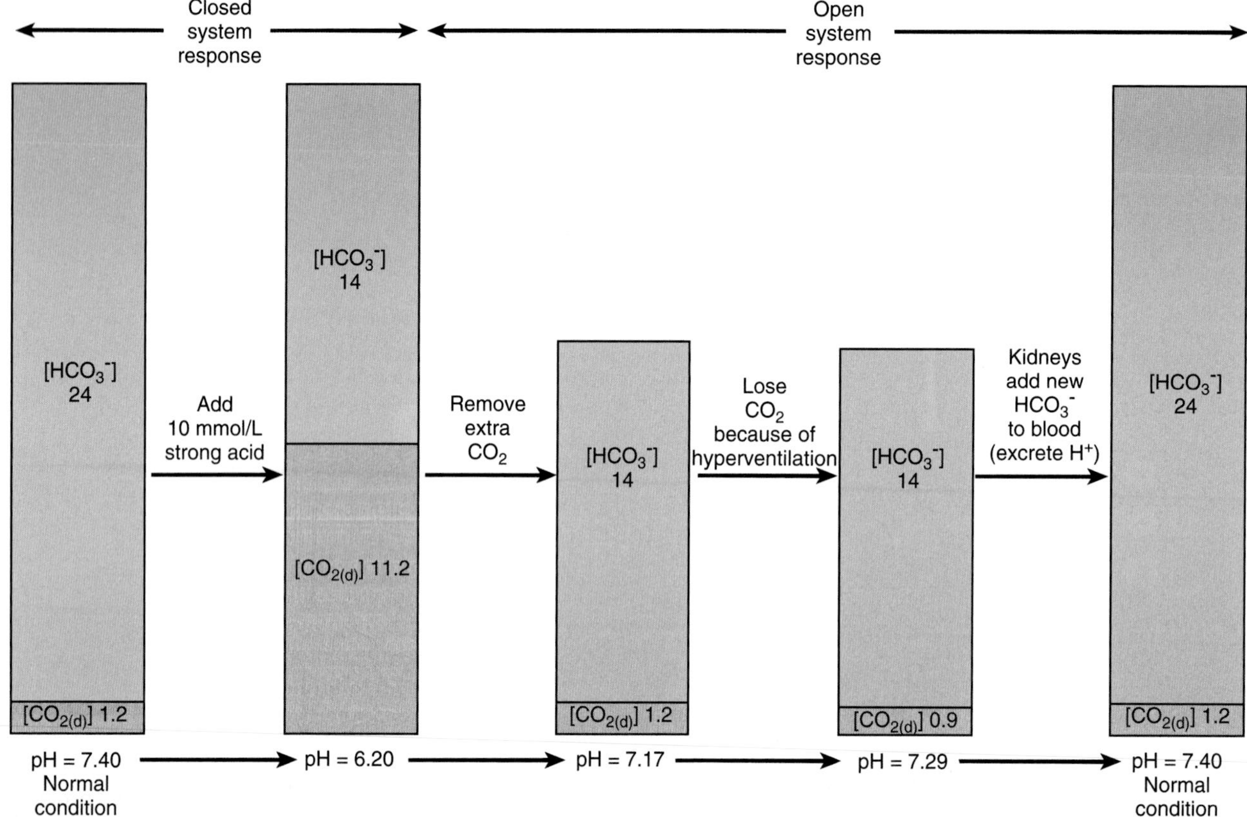

FIGURE 25.3 **The HCO_3^-/CO_2 system.** This system is remarkably effective in buffering added strong acid in the body because it is open. $[HCO_3^-]$ and $[CO_{2(d)}]$ are in mmol/L. See text for details. (Adapted from Pitts RF. Physiology of the Kidney and Body Fluids. 3rd Ed. Chicago: Year Book, 1974.)

activities brought about by the pH changes, and they are part of a negative-feedback mechanism regulating blood pH.

All Buffers Are in Equilibrium With the Same $[H^+]$

We have discussed the various buffers separately but, in the body, they all work together. In a solution containing multiple buffers, all are in equilibrium with the same $[H^+]$. This idea is known as the **isohydric principle** (*isohydric* meaning "same H^+"). For plasma, for example, we can write

$$\begin{aligned} pH &= 6.80 + \log \frac{[HPO_4^{2-}]}{[H_2PO_4^-]} \\ &= 6.10 + \log \frac{[HCO_3^-]}{0.03\ P_{CO_2}} \qquad (21) \\ &= pK_{protein} + \log \frac{[proteinate^-]}{[H\text{-}protein]} \end{aligned}$$

If an acid or a base is added to such a complex mixture of buffers, all buffers take part in buffering and shift from one form (base or acid) to the other. The relative importance of each buffer depends on its amount, pK, and availability.

The isohydric principle underscores the fact that it is the concentration ratio for any buffer pair, together with its pK, that sets the pH. We can focus on the concentration ratio for one buffer pair and all other buffers will automatically adjust their ratios according to the pH and their pK values.

The rest of this chapter emphasizes the role of the HCO_3^-/CO_2 buffer pair in setting the blood pH. Other buffers, however, are present and active. The HCO_3^-/CO_2 system is emphasized because physiological mechanisms (lungs and kidneys) regulate pH by acting on components of this buffer system.

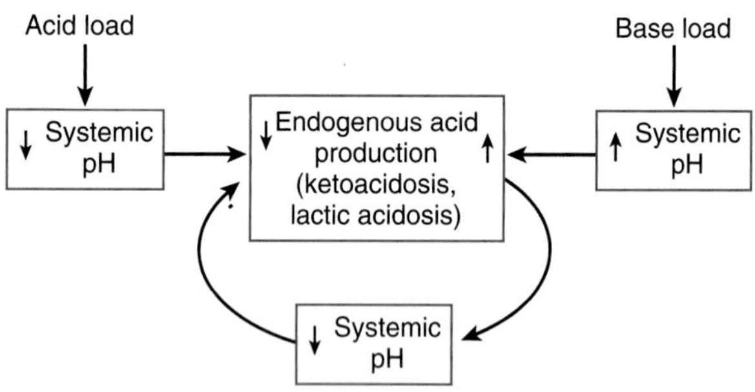

FIGURE 25.4 **Negative-feedback control of endogenous acid production.** The addition of an exogenous acid load or increased endogenous acid production result in a fall in pH, which, in turn, inhibits the production of ketone body acids and lactic acid. A base load, by raising pH, stimulates the endogenous production of acids. This negative-feedback mechanism attenuates changes in blood pH. (From Hood VL, Tannen RL. Protection of acid-base balance by pH regulation of acid production. N Engl J Med 1998;339:819–826.)

RESPIRATORY REGULATION OF pH

Reflex changes in ventilation help to defend blood pH. By changing the P_{CO_2} and, hence, $[H_2CO_3]$ of the blood, the respiratory system can rapidly and profoundly affect blood pH. As discussed in Chapter 22, a fall in blood pH stimulates ventilation, primarily by acting on peripheral chemoreceptors. An elevated arterial blood P_{CO_2} is a powerful stimulus to increase ventilation; it acts on both peripheral and central chemoreceptors, but primarily on the latter. CO_2 diffuses into brain interstitial and cerebrospinal fluids, where it causes a fall in pH that stimulates chemoreceptors in the medulla oblongata. When ventilation is stimulated, the lungs blow off more CO_2, making the blood less acidic. Conversely, a rise in blood pH inhibits ventilation; the consequent rise in blood $[H_2CO_3]$ reduces the alkaline shift in blood pH. Respiratory responses to disturbed blood pH begin within minutes and are maximal in about 12 to 24 hours.

RENAL REGULATION OF pH

The kidneys play a critical role in maintaining acid-base balance. If there is excess acid in the body, they remove H^+, or if there is excess base, they remove HCO_3^-. The usual challenge is to remove excess acid. As we have learned, strong acids produced by metabolism are first buffered by body buffer bases, particularly HCO_3^-. The kidneys then must eliminate H^+ in the urine and restore the depleted HCO_3^-.

Little of the H^+ excreted in the urine is present as free H^+. For example, if the urine has its lowest pH value (pH = 4.5), $[H^+]$ is only 0.03 mEq/L. With a typical daily urine output of 1 to 2 L, the amount of acid the body must dispose of daily (about 70 mEq) obviously is not excreted in the free form. Most of the H^+ combines with urinary buffers to be excreted as titratable acid and as NH_4^+.

Titratable acid is measured from the amount of strong base (NaOH) needed to bring the urine pH back to the pH of the blood (usually, 7.40). It represents the amount of H^+ ions that are excreted, combined with urinary buffers such as phosphate, creatinine, and other bases. The largest component of titratable acid is normally phosphate, that is, $H_2PO_4^-$.

Hydrogen ions secreted by the renal tubules also combine with the free base NH_3 and are excreted as NH_4^+. Ammonia (a term that collectively includes both NH_3 and NH_4^+) is produced by the kidney tubule cells and is secreted into the urine. Because the pK_a for NH_4^+ is high (9.0), most of the ammonia in the urine is present as NH_4^+. For this reason, too, NH_4^+ is not appreciably titrated when titratable acid is measured. Urinary ammonia is measured by a separate, often chemical, method.

Renal Net Acid Excretion Equals the Sum of Urinary Titratable Acid and Ammonia Minus Urinary Bicarbonate

In stable acid-base balance, net acid excretion by the kidneys equals the net rate of H^+ addition to the body by metabolism or other processes, assuming that other routes of loss of acid or base (e.g., gastrointestinal losses) are small and can be neglected, which normally is the case. The net loss of H^+ in the urine can be calculated from the following equation, which shows typical values in the parentheses:

$$\text{Renal net acid excretion (70 mEq/day)} = \text{urinary titratable acid (24 mEq/day)} + \text{urinary ammonia (48 mEq/day)} - \text{urinary } HCO_3^- \text{ (2 mEq/day)} \quad (22)$$

Urinary ammonia (as NH_4^+) ordinarily accounts for about two thirds of the excreted H^+, and titratable acid for about one third. Excretion of HCO_3^- in the urine represents a loss of base from the body. Therefore, it must be subtracted in the calculation of net acid excretion. If the urine contains significant amounts of organic anions, such as citrate, that potentially could have yielded HCO_3^- in the body, these should also be subtracted. Since the amount of free H^+ excreted is negligible, this is omitted from the equation.

Hydrogen Ions Are Added to Urine as It Flows Along the Nephron

As the urine flows along the tubule, from Bowman's capsule on through the collecting ducts, three processes occur: filtered HCO_3^- is reabsorbed, titratable acid is formed, and ammonia is added to the tubular urine. All three processes involve H^+ secretion (urinary acidification) by the tubular epithelium. The nature and magnitude of these processes vary in different nephron segments. Figure 25.5 summarizes measurements of tubular fluid pH along the nephron and shows ammonia movements in various nephron segments.

Acidification in the Proximal Convoluted Tubule. The pH of the glomerular ultrafiltrate, at the beginning of the proximal tubule, is identical to that of the plasma from which it is derived (7.4). H^+ ions are secreted by the proximal tubule epithelium into the tubule lumen; about two thirds of this is accomplished by a Na^+/H^+ exchanger and about one third by H^+-ATPase in the brush border membrane. Tubular fluid pH falls to a value of about 6.7 by the end of the proximal convoluted tubule (see Fig. 25.5).

The drop in pH is modest for two reasons: buffering of secreted H^+ and the high permeability of the proximal tubule epithelium to H^+. The glomerular filtrate and tubule fluid contain abundant buffer bases, especially HCO_3^-, which soak up secreted H^+, minimizing a fall in pH. The proximal tubule epithelium is rather leaky to H^+, so that any gradient from urine to blood, established by H^+ secretion, is soon limited by the diffusion of H^+ out of the tubule lumen into the blood surrounding the tubules.

Most of the H^+ ions secreted by the nephron are secreted in the proximal convoluted tubule and are used to bring about the reabsorption of filtered HCO_3^-. Secreted H^+ ions are also buffered by filtered phosphate to form titratable acid. Ammonia is produced by proximal tubule cells, mainly from glutamine. It is secreted into the tubular urine by the diffusion of NH_3, which then combines with a secreted H^+ to form NH_4^+, or via the brush border membrane Na^+/H^+ exchanger, which can operate in a Na^+/NH_4^+ exchange mode.

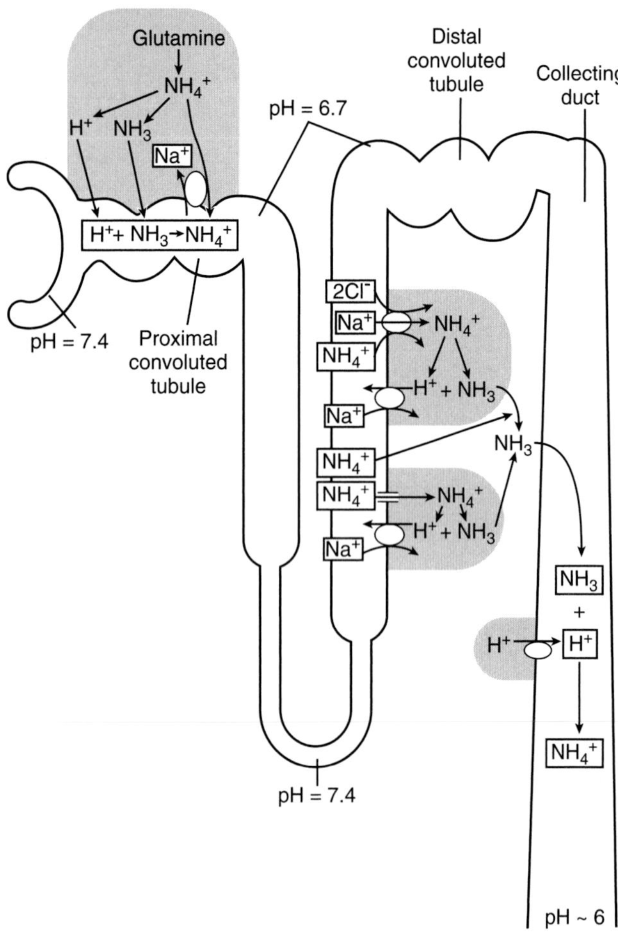

FIGURE 25.5 **Acidification along the nephron.** The pH of tubular urine decreases along the proximal convoluted tubule, rises along the descending limb of the loop of Henle, falls along the ascending limb, and reaches its lowest values in the collecting ducts. Ammonia ($=NH_3 + NH_4^+$) is chiefly produced in proximal tubule cells and is secreted into the tubular urine. NH_4^+ is reabsorbed in the thick ascending limb and accumulates in the kidney medulla. NH_3 diffuses into acidic collecting duct urine, where it is trapped as NH_4^+.

Acidification in the Loop of Henle. Along the descending limb of the loop of Henle, the pH of tubular fluid rises (from 6.7 to 7.4). This rise is explained by an increase in intraluminal [HCO_3^-] caused by water reabsorption. Ammonia is secreted along the descending limb.

The tubular fluid is acidified by secretion of H^+ along the ascending limb via a Na^+/H^+ exchanger. Along the thin ascending limb, ammonia is passively reabsorbed. Along the thick ascending limb, NH_4^+ is mostly actively reabsorbed by the Na-K-2Cl cotransporter in the luminal plasma membrane (NH_4^+ substitutes for K^+). Some NH_4^+ can be reabsorbed via a luminal plasma membrane K^+ channel. Also, some NH_4^+ can be passively reabsorbed between cells in this segment; the driving force is the lumen positive transepithelial electrical potential difference. Ammonia may undergo countercurrent multiplication in the loop of Henle, leading to an ammonia concentration gradient in the kidney medulla. The highest concentrations are at the tip of the papilla.

Acidification in the Distal Nephron. The distal nephron (distal convoluted tubule, connecting tubule, and collecting duct) differs from the proximal portion of the nephron in its H^+ transport properties. It secretes far fewer H^+ ions, and they are secreted primarily via an electrogenic H^+-ATPase or an electroneutral H^+/K^+-ATPase. The distal nephron is also lined by "tight" epithelia, so little secreted H^+ diffuses out of the tubule lumen, making steep urine-to-blood pH gradients possible (see Fig. 25.5). Final urine pH is typically about 6, but may be as low as 4.5.

The distal nephron usually almost completely reabsorbs the small quantities of HCO_3^- that were not reabsorbed by more proximal nephron segments. Considerable titratable acid forms as the urine is acidified. Ammonia, which was reabsorbed by the ascending limb of the loop of Henle and has accumulated in the medullary interstitial space, diffuses as lipid-soluble NH_3 into collecting duct urine and combines with secreted H^+ to form NH_4^+. The collecting duct epithelium is impermeable to the lipid-insoluble NH_4^+, so ammonia is trapped in an acidic urine and excreted as NH_4^+ (see Fig. 25.5).

The intercalated cells of the collecting duct are involved in acid-base transport and are of two major types: an acid-secreting **α-intercalated cell** and a bicarbonate-secreting **β-intercalated cell**. The α-intercalated cell has a vacuolar type of H^+-ATPase (the same kind as is found in lysosomes, endosomes, and secretory vesicles) and a H^+/K^+-ATPase (similar to that found in stomach and colon epithelial cells) in the luminal plasma membrane and a Cl^-/HCO_3^- exchanger in the basolateral plasma membrane (Fig. 25.6). The β-intercalated cell has the opposite polarity.

A more acidic blood pH results in the insertion of cytoplasmic H^+ pumps into the luminal plasma membrane of α-intercalated cells and enhanced H^+ secretion. If the blood is made alkaline, HCO_3^- secretion by β-intercalated cells is increased. Because the amounts of HCO_3^- secreted are ordinarily small compared to the amounts filtered and reabsorbed, HCO_3^- secretion will not be included in the remaining discussion.

The Reabsorption of Filtered HCO_3^- Restores Lost HCO_3^- to the Blood

HCO_3^- is freely filtered at the glomerulus, about 4,320 mEq/day (180 L/day × 24 mEq/L). Urinary loss of even a small portion of this HCO_3^- would lead to acidic blood and impair the body's ability to buffer its daily load of metabolically produced H^+. The kidney tubules have the important task of recovering the filtered HCO_3^- and returning it to the blood.

Figure 25.7 shows how HCO_3^- filtration, reabsorption, and excretion normally vary with plasma [HCO_3^-]. This type of graph should be familiar (see Fig. 23.8). The y-axis of the graph is unusual, however, because amounts of HCO_3^- per minute are factored by the GFR. The data are expressed in this way because the maximal rate of tubular reabsorption of HCO_3^- varies with GFR. The amount of HCO_3^- excreted in the urine per unit time is calculated as the difference between filtered and reabsorbed amounts. At low plasma concentrations of HCO_3^- (below about 26 mEq/L), all of the filtered HCO_3^- is reabsorbed. Because

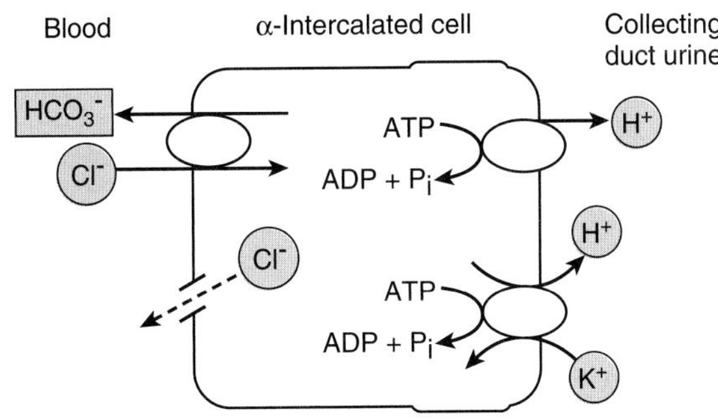

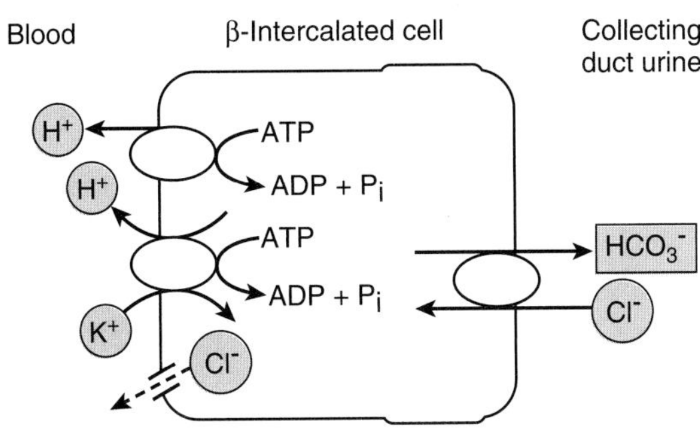

FIGURE 25.6 **Collecting duct intercalated cells.** The α-intercalated cell secretes H^+ via an electrogenic, vacuolar H^+-ATPase and electroneutral H^+/K^+-ATPase and adds HCO_3^- to the blood via a basolateral plasma membrane Cl^-/HCO_3^- exchanger. The β-intercalated cell, which is located in cortical collecting ducts, has the opposite polarity and secretes HCO_3^-.

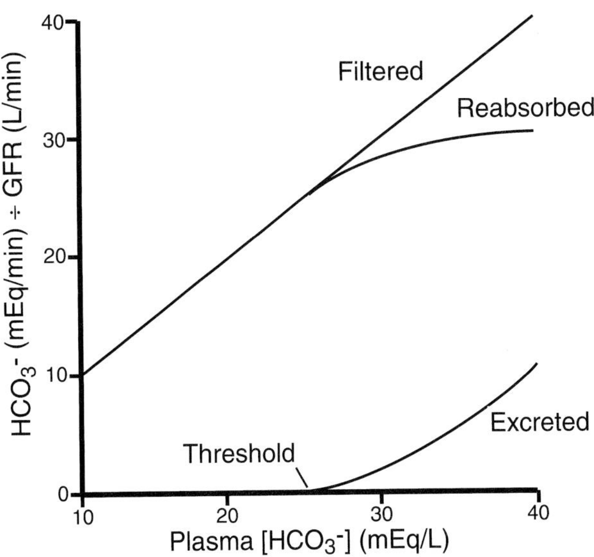

FIGURE 25.7 **The filtration, reabsorption, and excretion of HCO_3^-.** Decreases in plasma [HCO_3^-] were produced by ingestion of NH_4Cl and increases were produced by intravenous infusion of a solution of $NaHCO_3$. All the filtered HCO_3^- was reabsorbed below a plasma concentration of about 26 mEq/L. Above this value ("threshold"), appreciable quantities of filtered HCO_3^- were excreted in the urine. (Adapted from Pitts RF, Ayer JL, Schiess WA. The renal regulation of acid-base balance in man. III. The reabsorption and excretion of bicarbonate. J Clin Invest 1949;28:35–44.)

the plasma [HCO_3^-] and pH were decreased by ingestion of an acidifying salt (NH_4Cl), it makes good sense that the kidneys conserve filtered HCO_3^- in this situation.

If the plasma [HCO_3^-] is raised to high levels because of intravenous infusion of solutions containing $NaHCO_3$ for example, filtered HCO_3^- exceeds the reabsorptive capacity of the tubules and some HCO_3^- will be excreted in the urine (see Fig. 25.7). This also makes good sense. If the blood is too alkaline, the kidneys excrete HCO_3^-. This loss of base would return the pH of the blood to its normal value.

At the cellular level (see Fig. 25.8), filtered HCO_3^- is not reabsorbed directly across the tubule's luminal plasma membrane as, for example, is glucose. Instead, filtered HCO_3^- is reabsorbed indirectly via H^+ secretion in the following way. About 90% of the filtered HCO_3^- is reabsorbed in the proximal convoluted tubule, and we will emphasize events at this site. H^+ is secreted into the tubule lumen mainly via the Na^+/H^+ exchanger in the luminal membrane. It combines with filtered HCO_3^- to form H_2CO_3. Carbonic anhydrase (CA) in the luminal membrane (brush border) of the proximal tubule catalyzes the dehydration of H_2CO_3 to CO_2 and water in the lumen. The CO_2 diffuses back into the cell.

Inside the cell, the hydration of CO_2 (catalyzed by intracellular CA) yields H_2CO_3, which instantaneously forms H^+ and HCO_3^-. The H^+ is secreted into the lumen, and the HCO_3^- ion moves into the blood surrounding the tubules. In proximal tubule cells, this movement is favored by the inside negative membrane potential of the cell and by an electrogenic cotransporter in the basolateral membrane that simultaneously transports three HCO_3^- and one Na^+.

The reabsorption of filtered HCO_3^- does not result in H^+ excretion or the formation of any "new" HCO_3^-. The secreted H^+ is not excreted because it combines with filtered HCO_3^- that is, indirectly, reabsorbed. There is no net addition of HCO_3^- to the body in this operation. It is simply a recovery or reclamation process.

Excretion of Titratable Acid and Ammonia Generates New Bicarbonate

When H^+ is excreted as titratable acid and ammonia, new HCO_3^- is formed and added to the blood. New HCO_3^- replaces the HCO_3^- used to buffer the strong acids produced by metabolism.

The formation of new HCO_3^- and the excretion of H^+ are like two sides of the same coin. This fact is apparent if we assume that H_2CO_3 is the source of H^+:

$$CO_2 + H_2O \rightleftharpoons H_2CO_3 \begin{matrix} \nearrow H^+ \text{ (urine)} \\ \searrow HCO_3^- \text{ (blood)} \end{matrix} \qquad (23)$$

A loss of H^+ in the urine is equivalent to adding new HCO_3^- to the blood. The same is true if H^+ is lost from the body via another route, such as by vomiting of acidic gastric juice. This process leads to a rise in plasma [HCO_3^-]. *Conversely, a loss of HCO_3^- from the body is equivalent to adding H^+ to the blood.*

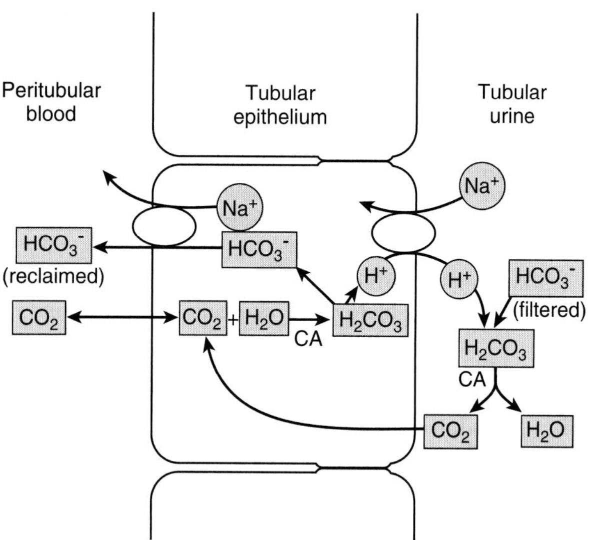

FIGURE 25.8 **A cell model for HCO_3^- reabsorption.** Filtered HCO_3^- combines with secreted H^+ and is reabsorbed indirectly. Carbonic anhydrase (CA) is present in the cells and in the proximal tubule on the brush border.

Titratable Acid Excretion. Figure 25.9 shows a cell model for the formation of titratable acid. In this figure, $H_2PO_4^-$ is the titratable acid formed. H^+ and HCO_3^- are produced in the cell from H_2CO_3. The secreted H^+ combines with the basic form of the phosphate (HPO_4^{2-}) to form the acid phosphate ($H_2PO_4^-$). The secreted H^+ replaces one of the Na^+ ions accompanying the basic phosphate. The new HCO_3^- generated in the cell moves into the blood, together with Na^+. For each mEq of H^+ excreted in the urine as titratable acid, a mEq of new HCO_3^- is added to the blood. This process eliminates H^+ in the urine, replaces ECF HCO_3^-, and restores a normal blood pH.

The amount of titratable acid excreted depends on two factors: the pH of the urine and the availability of buffer. If the urine pH is lowered, more titratable acid can form. The supply of phosphate and other buffers is usually limited. To excrete large amounts of acid, the kidneys must rely on increased ammonia excretion.

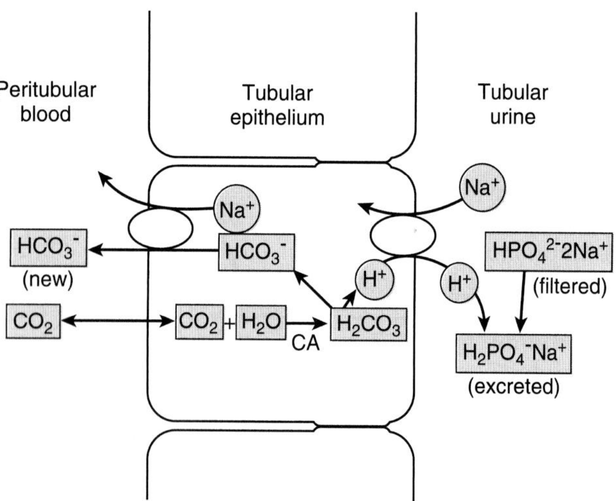

FIGURE 25.9 **A cell model for the formation of titratable acid.** Titratable acid (e.g., $H_2PO_4^-$) is formed when secreted H^+ is bound to a buffer base (e.g., HPO_4^{2-}) in the tubular urine. For each mEq of titratable acid excreted, a mEq of new HCO_3^- is added to the peritubular capillary blood.

Ammonia Excretion. Figure 25.10 shows a cell model for the excretion of ammonia. Most ammonia is synthesized in proximal tubule cells by deamidation and deamination of the amino acid glutamine:

$$\text{Glutamine} \xrightarrow[\text{Glutaminase}]{\nearrow NH_4^+} \text{Glutamate}^- \xrightarrow[\text{Glutamate dehydrogenase}]{\nearrow NH_4^+} \alpha\text{-Ketoglutarate}^{2-} \quad (24)$$

As discussed earlier, ammonia is secreted into the urine by two mechanisms. As NH_3, it diffuses into the tubular urine; as NH_4^+, it substitutes for H^+ on the Na^+/H^+ exchanger. In the lumen, NH_3 combines with secreted H^+ to form NH_4^+, which is excreted.

For each mEq of H^+ excreted as NH_4^+, one mEq of new HCO_3^- is added to the blood. The hydration of CO_2 in the tubule cell produces H^+ and HCO_3^-, as described earlier. Two H^+s are consumed when the anion α-ketoglutarate^{2-} is converted into CO_2 and water or into glucose in the cell. The new HCO_3^- returns to the blood along with Na^+.

If excess acid is added to the body, urinary ammonia excretion is increased for two reasons. First, a more acidic urine traps more ammonia (as NH_4^+) in the urine. Second, renal ammonia synthesis from glutamine increases over sev-

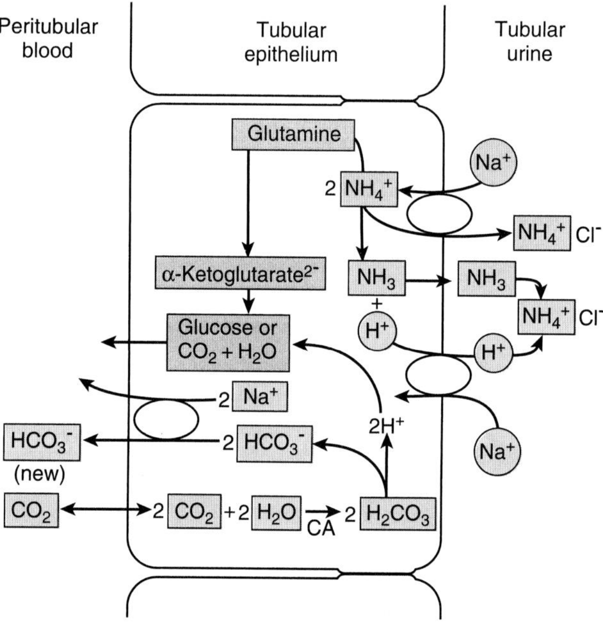

FIGURE 25.10 **A cell model for renal synthesis and excretion of ammonia.** Ammonium ions are formed from glutamine in the cell and are secreted into the tubular urine (top). H^+ from H_2CO_3 (bottom) is consumed when α-ketoglutarate is converted into glucose or CO_2 and H_2O. New HCO_3^- is added to the peritubular capillary blood—1 mEq for each mEq of NH_4^+ excreted in the urine.

eral days. Enhanced renal ammonia synthesis and excretion is a lifesaving adaptation because it allows the kidneys to remove large H^+ excesses and add more new HCO_3^- to the blood. Also, the excreted NH_4^+ can substitute in the urine for Na^+ and K^+, diminishing the loss of these cations. With severe metabolic acidosis, ammonia excretion may increase almost 10-fold.

Several Factors Influence Renal Excretion of Hydrogen Ions

Several factors influence the renal excretion of H^+, including intracellular pH, arterial blood P_{CO_2}, carbonic anhydrase activity, Na^+ reabsorption, plasma $[K^+]$, and aldosterone (Fig. 25.11).

Intracellular pH. The pH in kidney tubule cells is a key factor influencing the secretion and, therefore, the excretion of H^+. A fall in pH (increased $[H^+]$) enhances H^+ secretion. A rise in pH (decreased $[H^+]$) lowers H^+ secretion.

Arterial Blood P_{CO_2}. An increase in P_{CO_2} increases the formation of H^+ from H_2CO_3, leading to enhanced renal H^+ secretion and excretion—a useful compensation for any condition in which the blood contains too much H_2CO_3. (This will be discussed later, when we consider respiratory acidosis.) A decrease in P_{CO_2} results in lowered H^+ secretion and, consequently, less complete reabsorption of filtered HCO_3^- and a loss of base in the urine (a useful compensation for respiratory alkalosis, also discussed later).

Carbonic Anhydrase Activity. The enzyme carbonic anhydrase catalyzes two key reactions in urinary acidification:

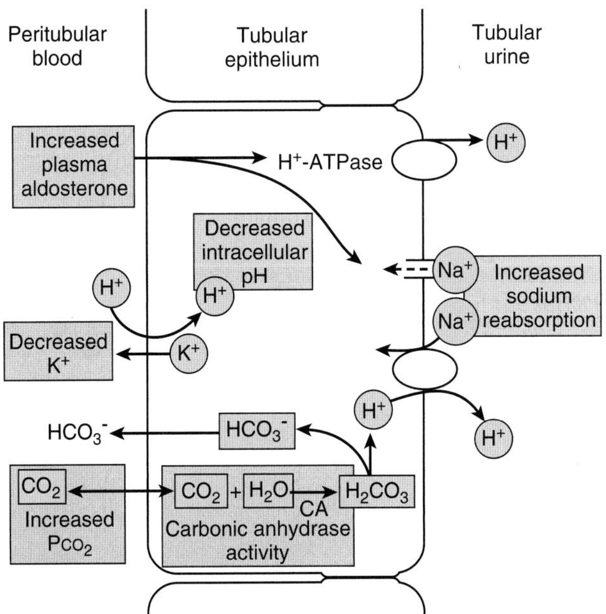

FIGURE 25.11 **Factors leading to increased H^+ secretion by the kidney tubule epithelium.** (See text for details.)

1) Hydration of CO_2 in the cells, forming H_2CO_3 and yielding H^+ for secretion

2) Dehydration of H_2CO_3 to H_2O and CO_2 in the proximal tubule lumen, an important step in the reabsorption of filtered HCO_3^-

If carbonic anhydrase is inhibited (usually by a drug), large amounts of filtered HCO_3^- may escape reabsorption. This situation leads to a fall in blood pH.

Sodium Reabsorption. Na^+ reabsorption is closely linked to H^+ secretion. In the proximal tubule, the two ions are directly linked, both being transported by the Na^+/H^+ exchanger in the luminal plasma membrane. The relation is less direct in the collecting ducts. Enhanced Na^+ reabsorption in the ducts leads to a more negative intraluminal electrical potential, which favors H^+ secretion by its electrogenic H^+-ATPase. The avid renal reabsorption of Na^+ observed in states of volume depletion is accompanied by a parallel rise in urinary H^+ excretion.

Plasma Potassium Concentration. Changes in plasma $[K^+]$ influence the renal excretion of H^+. A fall in plasma $[K^+]$ favors the movement of K^+ from body cells into interstitial fluid (or blood plasma) and a reciprocal movement of H^+ into cells. In the kidney tubule cells, these movements lower intracellular pH and increase H^+ secretion. K^+ depletion also stimulates ammonia synthesis by the kidneys. The result is the complete reabsorption of filtered HCO_3^- and the enhanced generation of new HCO_3^- as more titratable acid and ammonia are excreted. Consequently, hypokalemia (or a decrease in body K^+ stores) leads to increased plasma $[HCO_3^-]$ (metabolic alkalosis). Hyperkalemia (or excess K^+ in the body) results in the opposite changes: an increase in intracellular pH, decreased H^+ secretion, incomplete reabsorption of filtered HCO_3^-, and a fall in plasma $[HCO_3^-]$ (metabolic acidosis).

Aldosterone. Aldosterone stimulates the collecting ducts to secrete H^+ by three actions:

1) It directly stimulates the H^+-ATPase in collecting duct α-intercalated cells.

2) It enhances collecting duct Na^+ reabsorption, which leads to a more negative intraluminal potential and, consequently, promotes H^+ secretion by the electrogenic H^+-ATPase.

3) It promotes K^+ secretion. This response leads to hypokalemia, which increases renal H^+ secretion.

Hyperaldosteronism results in enhanced renal H^+ excretion and an alkaline blood pH; the opposite occurs with hypoaldosteronism.

pH Gradient. The secretion of H^+ by the kidney tubules and collecting ducts is gradient-limited. The collecting ducts cannot lower the urine pH below 4.5, corresponding to a urine/plasma $[H^+]$ gradient of $10^{-4.5}/10^{-7.4}$ or 800/1 when the plasma pH is 7.4. If more buffer base (NH_3, HPO_4^{2-}) is available in the urine, more H^+ can be secreted before the limiting gradient is reached. In some kidney tubule disorders, the secretion of H^+ is gradient-limited (see Clinical Focus Box 25.1).

CLINICAL FOCUS BOX 25.1

Renal Tubular Acidosis

Renal tubular acidosis (RTA) is a group of kidney disorders characterized by chronic metabolic acidosis, a normal plasma anion gap, and the absence of renal failure. The kidneys show inadequate H^+ secretion by the distal nephron, excessive excretion of HCO_3^-, or reduced excretion of NH_4^+.

In classic **type 1 (distal) RTA**, the ability of the collecting ducts to lower urine pH is impaired. This condition can be caused by inadequate secretion of H^+ (defective H^+-ATPase or H^+/K^+-ATPase) or abnormal leakiness of the collecting duct epithelium so that secreted H^+ ions diffuse back from lumen to blood. Because the urine pH is inappropriately high, titratable acid excretion is diminished and trapping of ammonia in the urine (as NH_4^+) is decreased. Type 1 RTA may be the result of an inherited defect, autoimmune disease, treatment with lithium or the antibiotic amphotericin B, or the result of diseases of the kidney medulla. A diagnosis of this form of RTA is established by challenging the subject with a standard oral dose of NH_4Cl and measuring the urine pH for the next several hours. This results in a urine pH below 5.0 in healthy people. In subjects with type 1 RTA, however, urine pH will not decrease below 5.5. Treatment of type 1 RTA involves daily administration of modest amounts of alkali (HCO_3^-, citrate) sufficient to cover daily metabolic acid production.

In **type 2 (proximal) RTA**, HCO_3^- reabsorption by the proximal tubule is impaired, leading to excessive losses of HCO_3^- in the urine. As a consequence, the plasma [HCO_3^-] falls and chronic metabolic acidosis ensues. In the new steady state, the tubules are able to reabsorb the filtered HCO_3^- load more completely because the filtered load is reduced. The distal nephron is no longer overwhelmed by HCO_3^- and the urine pH is acidic. In type 2 RTA, the administration of an NH_4Cl challenge results in a urine pH below 5.5. This disorder may be inherited, may be associated with several acquired conditions that result in a generalized disorder of proximal tubule transport, or may result from the inhibition of proximal tubule carbonic anhydrase by drugs such as acetazolamide. Treatment requires the daily administration of large amounts of alkali because when the plasma [HCO_3^-] is raised, excessive urinary excretion of filtered HCO_3^- occurs.

Type 4 RTA (there is no type 3 RTA) is also known as **hyperkalemic distal RTA**. Collecting duct secretion of both K^+ and H^+ is reduced, explaining the hyperkalemia and metabolic acidosis. Hyperkalemia reduces renal ammonia synthesis, resulting in reduced net acid excretion and a fall in plasma [HCO_3^-]. The urine pH can go below 5.5 after an NH_4Cl challenge because there is little ammonia in the urine to buffer secreted H^+. The underlying disorder is a result of inadequate production of aldosterone or impaired aldosterone action. Treatment of type 4 RTA requires lowering the plasma [K^+] to normal; if this therapy is successful, alkali may not be needed.

REGULATION OF INTRACELLULAR pH

The intracellular and extracellular fluids are linked by exchanges across plasma membranes of H^+, HCO_3^-, various acids and bases, and CO_2. By stabilizing ECF pH, the body helps to protect intracellular pH.

If H^+ ions were passively distributed across plasma membranes, intracellular pH would be lower than what is seen in most body cells. In skeletal muscle cells, for example, we can calculate from the Nernst equation (see Chapter 2) and a membrane potential of −90 mV that cytosolic pH should be 5.9 if ECF pH is 7.4; actual measurements, however, indicate a pH of 6.9. From this discrepancy, two conclusions are clear: H^+ ions are not at equilibrium across the plasma membrane, and the cell must use active mechanisms to extrude H^+.

Cells are typically threatened by acidic metabolic end products and by the tendency for H^+ to diffuse into the cell down the electrical gradient (Fig. 25.12). H^+ is extruded by Na^+/H^+ exchangers, which are present in nearly all body cells. Five different isoforms of these exchangers (designated NHE1, NHE2, etc.), with different tissue distributions, have been identified. These transporters exchange one H^+ for one Na^+ and, therefore, function in an electrically neutral fashion. Active extrusion of H^+ keeps the internal pH within narrow limits.

The activity of the Na^+/H^+ exchanger is regulated by intracellular pH and a variety of hormones and growth factors (Fig. 25.13). Not surprisingly, an increase in intracellu-

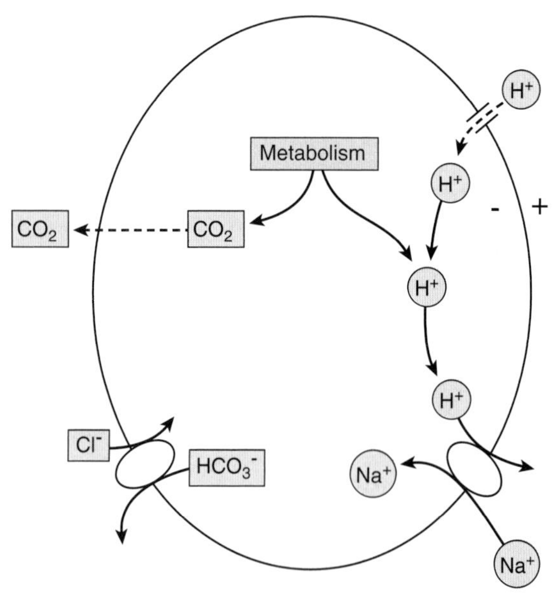

FIGURE 25.12 **Cell acid-base balance.** Body cells usually maintain a constant intracellular pH. The cell is acidified by the production of H^+ from metabolism and the influx of H^+ from the ECF (favored by the inside negative plasma membrane potential). To maintain a stable intracellular pH, the cell must extrude hydrogen ions at a rate matching their input. Many cells also possess various HCO_3^- transporters (not depicted), which defend against excess acid or base.

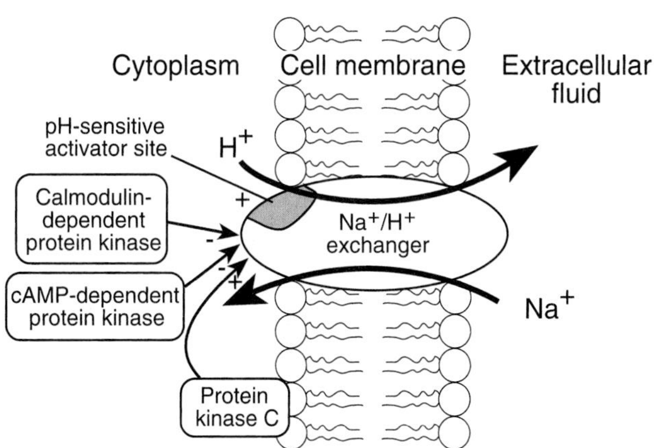

FIGURE 25.13 **The plasma membrane Na^+/H^+ exchanger.** This exchanger plays a key role in regulating intracellular pH in most body cells and is activated by a decrease in cytoplasmic pH. Many hormones and growth factors, acting via intracellular second messengers and protein kinases, can increase (+) or decrease (−) the activity of the exchange.

lar $[H^+]$ stimulates the exchanger but not only because of more substrate (H^+) for the exchanger. H^+ also stimulates the exchanger by protonating an activator site on the cytoplasmic side of the exchanger, making the exchanger more effective in dealing with the threat of intracellular acidosis. Many hormones and growth factors, via intracellular second messengers, activate various protein kinases that stimulate or inhibit the Na^+/H^+ exchanger. In this way, they produce changes in intracellular pH, which may lead to changes in cell activity.

Besides extruding H^+, the cell can deal with acids and bases in other ways. In some cells, various HCO_3^- transporting systems (e.g., Na^+-dependent and Na^+-independent Cl^-/HCO_3^- exchangers) may be present in plasma membranes. These exchangers may be activated by changes in intracellular pH. Cells have large stores of protein and organic phosphate buffers, which can bind or release H^+. Various chemical reactions in cells can also use up or release H^+. For example, the conversion of lactic acid to CO_2 and water to glucose effectively disposes of acid. In addition, various cell organelles may sequester H^+. For example, a H^+-ATPase in endosomes and lysosomes pumps H^+ out of the cytosol into these organelles. In summary, ion transport, buffering mechanisms, and metabolic reactions all ensure a relatively stable intracellular pH.

DISTURBANCES OF ACID-BASE BALANCE

Table 25.2 lists the normal values for the pH (or $[H^+]$), P_{CO_2}, and $[HCO_3^-]$ of arterial blood plasma. A blood pH of less than 7.35 ($[H^+] > 45$ nmol/L) indicates **acidemia**. A blood pH above 7.45 ($[H^+] < 35$ nmol/L) indicates **alkalemia**. The range of pH values compatible with life is approximately 6.8 to 7.8 ($[H^+] = 160$ to 16 nmol/L).

Four **simple acid-base disturbances** may lead to an abnormal blood pH: respiratory acidosis, respiratory alkalosis, metabolic acidosis, and metabolic alkalosis. The word "simple" indicates a single primary cause for the disturbance. **Acidosis** is an abnormal process that tends to produce acidemia. **Alkalosis** is an abnormal process that tends to produce alkalemia. If there is too much or too little CO_2, a **respiratory disturbance** is present. If the problem is too much or too little HCO_3^-, a **metabolic** (or **nonrespiratory**) **disturbance** of acid-base balance is present. Table 25.3 summarizes the changes in blood pH (or $[H^+]$), plasma $[HCO_3^-]$, and P_{CO_2} that occur in each of the four simple acid-base disturbances.

TABLE 25.2 **Normal Arterial Blood Plasma Acid-Base Values**

	Mean	Range[a]
pH	7.40	7.35–7.45
$[H^+]$, nmol/L	40	45–35
P_{CO_2}, mm Hg	40	35–45
$[HCO_3^-]$, mEq/L	24	22–26

[a] The range extends from 2 standard deviations below to 2 standard deviations above the mean and encompasses 95% of the healthy population.

In considering acid-base disturbances, it is helpful to recall the Henderson-Hasselbalch equation for HCO_3^-/CO_2:

$$pH = 6.10 + \log \frac{[HCO_3^-]}{0.03\ P_{CO_2}} \qquad (25)$$

If the primary problem is a change in $[HCO_3^-]$ or P_{CO_2}, the pH can be brought closer to normal by changing the other member of the buffer pair *in the same direction*. For example, if P_{CO_2} is primarily decreased, a decrease in plasma $[HCO_3^-]$ will minimize the change in pH. In various acid-base disturbances, the lungs adjust the blood P_{CO_2} and the kidneys adjust the plasma $[HCO_3^-]$ to reduce departures of pH from normal; these adjustments are called **compensations** (Table 25.4). Compensations generally do not bring about normal blood pH.

Respiratory Acidosis Results From an Accumulation of Carbon Dioxide

Respiratory acidosis is an abnormal process characterized by CO_2 accumulation. The CO_2 build-up pushes the following reactions to the right:

$$CO_2 + H_2O \rightleftharpoons H_2CO_3 \rightleftharpoons H^+ + HCO_3^- \qquad (26)$$

Blood $[H_2CO_3]$ increases, leading to an increase in $[H^+]$ or a fall in pH. Respiratory acidosis is usually caused by a failure to expire metabolically produced CO_2 at an adequate rate, leading to accumulation of CO_2 in the blood and a fall in blood pH. This disturbance may be a result of a decrease in overall alveolar ventilation (hypoventilation) or, as occurs commonly in lung disease, a mismatch between ventilation and perfusion. Respiratory acidosis also occurs if a person breathes CO_2-enriched air.

Chemical Buffering. In respiratory acidosis, more than 95% of the chemical buffering occurs within cells. The cells

TABLE 25.3 Directional Changes in Arterial Blood Plasma Values in the Four Simple Acid-Base Disturbances[a]

Disturbance	Arterial Plasma pH	$[H^+]$	$[HCO_3^-]$	P_{CO_2}	Compensatory Response
Respiratory acidosis	↓	↑	↑	⇑	Kidneys increase H^+ excretion
Respiratory alkalosis	↑	↓	↓	⇓	Kidneys increase HCO_3^- excretion
Metabolic acidosis	↓	↑	⇓	↓	Alveolar hyperventilation; kidneys increase H^+ excretion
Metabolic alkalosis	↑	↓	⇑	↑	Alveolar hypoventilation; kidneys increase HCO_3^- excretion

[a] Heavy arrows indicate the main effect.

contain many proteins and organic phosphates that can bind H^+. For example, hemoglobin (Hb) in red blood cells combines with H^+ from H_2CO_3, minimizing the increase in free H^+. Recall from Chapter 21 the buffering reaction:

$$H_2CO_3 + HbO_2^- \rightleftharpoons HHb + O_2 + HCO_3^- \quad (27)$$

This reaction raises the plasma $[HCO_3^-]$. In *acute* respiratory acidosis, such chemical buffering processes in the body lead to an increase in plasma $[HCO_3^-]$ of about 1 mEq/L for each 10 mm Hg increase in P_{CO_2} (see Table 25.4). Bicarbonate is not a buffer for H_2CO_3 because the reaction

$$H_2CO_3 + HCO_3^- \rightleftharpoons HCO_3^- + H_2CO_3 \quad (28)$$

is simply an exchange reaction and does not affect the pH.

An example illustrates how chemical buffering reduces a fall in pH during respiratory acidosis. Suppose P_{CO_2} increased from a normal value of 40 mm Hg to 70 mm Hg ($[CO_{2(d)}] = 2.1$ mmol/L). If there were no body buffer bases that could accept H^+ from H_2CO_3 (i.e., if there was no measurable increase in $[HCO_3^-]$), the resulting pH would be 7.16:

$$pH = 6.10 + \log \frac{[24]}{[2.1]} = 7.16 \quad (29)$$

In acute respiratory acidosis, a 3 mEq/L increase in plasma $[HCO_3^-]$ occurs with a 30 mm Hg rise in P_{CO_2} (see Table 25.4). Therefore, the pH is 7.21:

$$pH = 6.10 + \log \frac{[24 + 3]}{[2.1]} = 7.21 \quad (30)$$

The pH of 7.21 is closer to a normal pH because body buffer bases (mainly intracellular buffers) such as proteins and phosphates combined with H^+ ions liberated from H_2CO_3.

Respiratory Compensation. Respiratory acidosis produces a rise in P_{CO_2} and a fall in pH and is often associated with hypoxia. These changes stimulate breathing (see Chapter 22) and diminish the severity of the acidosis. In other words, a person would be worse off if the respiratory system did not reflexly respond to the abnormalities in blood P_{CO_2}, pH, and P_{O_2}.

Renal Compensation. The kidneys compensate for respiratory acidosis by adding more H^+ to the urine and adding new HCO_3^- to the blood. The increased P_{CO_2} stimulates renal H^+ secretion, which allows the reabsorption of all filtered HCO_3^-. Excess H^+ is excreted as titratable acid and NH_4^+; these processes add new HCO_3^- to the blood, causing plasma $[HCO_3^-]$ to rise. This compensation takes several days to fully develop.

With *chronic* respiratory acidosis, plasma $[HCO_3^-]$ increases, on average, by 4 mEq/L for each 10 mm Hg rise in P_{CO_2} (see Table 25.4). This rise exceeds that seen with acute respiratory acidosis because of the renal addition of HCO_3^- to the blood. One would expect a person with chronic respiratory acidosis and a P_{CO_2} of 70 mm Hg to have an increase in plasma HCO_3^- of 12 mEq/L. The blood pH would be 7.33:

$$pH = 6.10 + \log \frac{[24 + 12]}{[2.1]} = 7.33 \quad (31)$$

TABLE 25.4 Compensatory Responses in Acid-Base Disturbances[a]

Respiratory acidosis	
Acute	1 mEq/L increase in plasma $[HCO_3^-]$ for each 10 mm Hg increase in P_{CO_2}[b]
Chronic	4 mEq/L increase in plasma $[HCO_3^-]$ for each 10 mm Hg increase in P_{CO_2}[c]
Respiratory alkalosis	
Acute	2 mEq/L decrease in plasma $[HCO_3^-]$ for each 10 mm Hg decrease in P_{CO_2}[b]
Chronic	4 mEq/L decrease in plasma $[HCO_3^-]$ for each 10 mm Hg decrease in P_{CO_2}[c]
Metabolic acidosis	1.3 mm Hg decrease in P_{CO_2} for each 1 mEq/L decrease in plasma $[HCO_3^-]$[d]
Metabolic alkalosis	0.7 mm Hg increase in P_{CO_2} for each 1 mEq/L increase in plasma $[HCO_3^-]$[d]

From Valtin H, Gennari FJ. Acid-Base Disorders. Basic Concepts and Clinical Management. Boston: Little, Brown, 1987.

[a] Empirically determined average changes measured in people with simple acid-base disorders.

[b] This change is primarily a result of chemical buffering.

[c] This change is primarily a result of renal compensation.

[d] This change is a result of respiratory compensation.

With chronic respiratory acidosis, time for renal compensation is allowed, so blood pH (in this example, 7.33) is much closer to normal than is observed during acute respiratory acidosis (pH 7.21).

Respiratory Alkalosis Results From an Excessive Loss of Carbon Dioxide

Respiratory alkalosis is most easily understood as the opposite of respiratory acidosis; it is an abnormal process causing the loss of too much CO_2. This loss causes blood $[H_2CO_3]$ and, thus, $[H^+]$ to fall (pH rises). Alveolar hyperventilation causes respiratory alkalosis. Metabolically produced CO_2 is flushed out of the alveolar spaces more rapidly than it is added by the pulmonary capillary blood. This situation causes alveolar and arterial P_{CO_2} to fall. Hyperventilation and respiratory alkalosis can be caused by voluntary effort, anxiety, direct stimulation of the medullary respiratory center by some abnormality (e.g., meningitis, fever, aspirin intoxication), or hypoxia caused by severe anemia or high altitude.

Chemical Buffering. As with respiratory acidosis, during respiratory alkalosis more than 95% of chemical buffering occurs within cells. Cell proteins and organic phosphates liberate H^+ ions, which are added to the ECF and lower the plasma $[HCO_3^-]$, reducing the alkaline shift in pH.

With *acute* respiratory alkalosis, plasma $[HCO_3^-]$ falls by about 2 mEq/L for each 10 mm Hg drop in P_{CO_2} (see Table 25.4). For example, if P_{CO_2} drops from 40 to 20 mm Hg ($[CO_{2(d)}] = 0.6$ mmol/L) plasma $[HCO_3^-]$ falls by 4 mEq/L, and the pH will be 7.62:

$$pH = 6.10 + \log \frac{[24 - 4]}{[0.6]} = 7.62 \qquad (32)$$

If plasma $[HCO_3^-]$ had not changed, the pH would have been 7.70:

$$pH = 6.10 + \log \frac{[24]}{[0.6]} = 7.70 \qquad (33)$$

Respiratory Compensation. Although hyperventilation causes respiratory alkalosis, hyperventilation also causes changes (a fall in P_{CO_2} and a rise in blood pH) that inhibit ventilation and, therefore, limit the extent of hyperventilation.

Renal Compensation. The kidneys compensate for respiratory alkalosis by excreting HCO_3^- in the urine, thereby getting rid of base. A reduced P_{CO_2} reduces H^+ secretion by the kidney tubule epithelium. As a result, some of the filtered HCO_3^- is not reabsorbed. When the urine becomes more alkaline, titratable acid excretion vanishes and little ammonia is excreted. The enhanced output of HCO_3^- causes plasma $[HCO_3^-]$ to fall.

Chronic respiratory alkalosis is accompanied by a 4 mEq/L fall in plasma $[HCO_3^-]$ for each 10 mm Hg drop in P_{CO_2} (see Table 25.4). For example, in a person with chronic hyperventilation and a P_{CO_2} of 20 mm Hg, the blood pH is

$$pH = 6.10 + \log \frac{[24 - 8]}{[0.6]} = 7.53 \qquad (34)$$

This pH is closer to normal than the pH of 7.62 of acute respiratory alkalosis. The difference between the two situations is largely a result of renal compensation.

Metabolic Acidosis Results From a Gain of Noncarbonic Acid or a Loss of Bicarbonate

Metabolic acidosis is an abnormal process characterized by a gain of acid (other than H_2CO_3) or a loss of HCO_3^-. Either causes plasma $[HCO_3^-]$ and pH to fall. If a strong acid is added to the body, the reactions

$$H^+ + HCO_3^- \rightleftharpoons H_2CO_3 \rightleftharpoons H_2O + CO_2 \qquad (35)$$

are pushed to the right. The added H^+ consumes HCO_3^-. If a lot of acid is infused rapidly, P_{CO_2} rises, as the equation predicts. This increase occurs only transiently, however, because the body is an open system, and the lungs expire CO_2 as it is generated. P_{CO_2} actually falls below normal because an acidic blood pH stimulates ventilation (see Fig. 25.3).

Many conditions can produce metabolic acidosis, including renal failure, uncontrolled diabetes mellitus, lactic acidosis, the ingestion of acidifying agents such as NH_4Cl, abnormal renal excretion of HCO_3^-, and diarrhea. In renal failure, the kidneys cannot excrete H^+ fast enough to keep up with metabolic acid production and, in uncontrolled diabetes mellitus, the production of ketone body acids increases. Lactic acidosis results from tissue hypoxia. Ingested NH_4Cl is converted into urea and a strong acid, HCl, in the liver. Diarrhea causes a loss of alkaline intestinal fluids. Clinical Focus Box 25.2 discusses the metabolic acidosis seen in uncontrolled diabetes mellitus.

Chemical Buffering. Excess acid is chemically buffered in extracellular and intracellular fluids and bone. In metabolic acidosis, roughly half the buffering occurs in cells and bone. HCO_3^- is the principal buffer in the ECF.

Respiratory Compensation. The acidic blood pH stimulates the respiratory system to lower blood P_{CO_2}. This action lowers blood $[H_2CO_3]$ and tends to alkalinize the blood, opposing the acidic shift in pH. Metabolic acidosis is accompanied on average by a 1.3 mm Hg fall in P_{CO_2} for each 1 mEq/L drop in plasma $[HCO_3^-]$ (see Table 25.4). Suppose, for example, the infusion of a strong acid causes the plasma $[HCO_3^-]$ to drop from 24 to 12 mEq/L. If there was no respiratory compensation and the P_{CO_2} did not change from its normal value of 40 mm Hg, the pH would be 7.10:

$$pH = 6.10 + \log \frac{[12]}{[1.2]} = 7.10 \qquad (36)$$

CLINICAL FOCUS BOX 25.2

Metabolic Acidosis in Diabetes Mellitus

Diabetes mellitus is a common disorder characterized by an insufficient secretion of insulin or insulin-resistance by the major target tissues (skeletal muscle, liver, and adipocytes). A severe **metabolic acidosis** may develop in uncontrolled diabetes mellitus.

Acidosis occurs because insulin deficiency leads to decreased glucose utilization, a diversion of metabolism toward the utilization of fatty acids, and an overproduction of ketone body acids (acetoacetic acid and β-hydroxybutyric acids). Ketone body acids are fairly strong acids (pK_a 4 to 5); they are neutralized in the body by HCO_3^- and other buffers. Increased production of these acids leads to a fall in plasma $[HCO_3^-]$, an increase in plasma anion gap, and a fall in blood pH (acidemia).

Severe acidemia, whatever its cause, has many adverse effects on the body. It impairs myocardial contractility, resulting in a decrease in cardiac output. It causes arteriolar dilation, which leads to a fall in arterial blood pressure. Hepatic and renal blood flows are decreased. Reentrant arrhythmias and a decreased threshold for ventricular fibrillation can occur. The respiratory muscles show decreased strength and fatigue easily. Metabolic demands are increased due, in part, to activation of the sympathetic nervous system, but at the same time anaerobic glycolysis and ATP synthesis are reduced by acidemia. Hyperkalemia is favored and protein catabolism is enhanced. Severe acidemia causes impaired brain metabolism and cell volume regulation, leading to progressive obtundation and coma.

An increased acidity of the blood stimulates pulmonary ventilation, resulting in a compensatory lowering of alveolar and arterial blood P_{CO_2}. The consequent reduction in blood $[H_2CO_3]$ acts to move the blood pH back toward normal. The labored, deep breathing that accompanies severe uncontrolled diabetes is called **Kussmaul's respiration**.

The kidneys compensate for metabolic acidosis by reabsorbing all the filtered HCO_3^-. They also increase the excretion of titratable acid, part of which is comprised of ketone body acids. But these acids can only be partially titrated to their acid form in the urine because the urine pH cannot go below 4.5. Therefore, ketone body acids are excreted mostly in their anionic form; because of the requirement of electroneutrality in solutions, increased urinary excretion of Na^+ and K^+ results.

An important compensation for the acidosis is increased renal synthesis and excretion of ammonia. This adaptive response takes several days to fully develop, but it allows the kidneys to dispose of large amounts of H^+ in the form NH_4^+. The NH_4^+ in the urine can replace Na^+ and K^+ ions, resulting in conservation of these valuable cations.

The severe acidemia, electrolyte disturbances, and volume depletion that accompany uncontrolled diabetes mellitus may be fatal. Addressing the underlying cause, rather than just treating the symptoms best achieves correction of the acid-base disturbance. Therefore, the administration of a suitable dose of insulin is usually the key element of therapy. In some patients with marked acidemia (pH < 7.10), $NaHCO_3$ solutions may be infused intravenously to speed recovery, but this does not correct the underlying metabolic problem. Losses of Na^+, K^+, and water should be replaced.

With respiratory compensation, the P_{CO_2} falls by 16 mm Hg (12 × 1.3) to 24 mm Hg ($[CO_{2(d)}] = 0.72$ mmol/L) and pH is 7.32:

$$pH = 6.10 + \log \frac{[12]}{[0.72]} = 7.32 \quad (37)$$

This value is closer to normal than a pH of 7.10. The respiratory response develops promptly (within minutes) and is maximal after 12 to 24 hours.

Renal Compensation. The kidneys respond to metabolic acidosis by adding more H^+ to the urine. Since the plasma $[HCO_3^-]$ is primarily lowered, the filtered load of HCO_3^- drops, and the kidneys can accomplish the complete reabsorption of filtered HCO_3^- (see Fig. 25.7). More H^+ is excreted as titratable acid and NH_4^+. With chronic metabolic acidosis, the kidneys make more ammonia. The kidneys can, therefore, add more new HCO_3^- to the blood, to replace lost HCO_3^-. If the underlying cause of metabolic acidosis is corrected, then healthy kidneys can correct the blood pH in a few days.

The Plasma Anion Gap Is Calculated From Na^+, Cl^-, and HCO_3^- Concentrations

The anion gap is a useful concept, especially when trying to determine the possible cause of a metabolic acidosis. In any body fluid, the sums of the cations and anions are equal because solutions are electrically neutral. For blood plasma, we can write

$$\Sigma \text{ cations} = \Sigma \text{ anions} \quad (38)$$

or

$$[Na^+] + [\text{unmeasured cations}] = [Cl^-] + [HCO_3^-] + [\text{unmeasured anions}] \quad (39)$$

The unmeasured cations include K^+, Ca^{2+}, and Mg^{2+} ions and, because these are present at relatively low concentrations (compared to Na^+) and are usually fairly constant, we choose to neglect them. The unmeasured anions include plasma proteins, sulfate, phosphate, citrate, lactate, and other organic anions. If we rearrange the above equation, we get

$$[\text{unmeasured anions}] \text{ or anion gap} = [Na^+] - [Cl^-] - [HCO_3^-] \quad (40)$$

In a healthy person, the anion gap falls in the range of 8 to 14 mEq/L. For example, if plasma [Na^+] is 140 mEq/L, [Cl^-] is 105 mEq/L, and [HCO_3^-] is 24 mEq/L, the anion gap is 11 mEq/L. If an acid such as lactic acid is added to plasma, the reaction lactic acid + $HCO_3^- \rightleftharpoons$ lactate$^-$ + $H_2O + CO_2$ will be pushed to the right. Consequently, the plasma [HCO_3^-] will be decreased and because the [Cl^-] is not changed, the anion gap will be increased. The unmeasured anion in this case is lactate. In several types of metabolic acidosis, the low blood pH is accompanied by a high anion gap (Table 25.5). (These can be remembered from the mnemonic MULEPAKS formed from the first letters of this list.) In other types of metabolic acidosis, the low blood pH is accompanied by a normal anion gap (see Table 25.5). For example, with diarrhea and a loss of alkaline intestinal fluid, plasma [HCO_3^-] falls but plasma [Cl^-] rises, and the two changes counterbalance each other so the anion gap is unchanged. Again, the chief value of the anion gap concept is that it allows a clinician to narrow down possible explanations for metabolic acidosis in a patient.

Metabolic Alkalosis Results From a Gain of Strong Base or Bicarbonate or a Loss of Noncarbonic Acid

Metabolic alkalosis is an abnormal process characterized by a gain of a strong base or HCO_3^- or a loss of an acid (other than carbonic acid). Plasma [HCO_3^-] and pH rise; P_{CO_2} rises because of respiratory compensation. These changes are opposite to those seen in metabolic acidosis (see Table 25.3). A variety of situations can produce metabolic alkalosis, including the ingestion of antacids, vomiting of gastric acid juice, and enhanced renal H^+ loss (e.g., as a result of hyperaldosteronism or hypokalemia). Clinical Focus Box 25.3 discusses the metabolic alkalosis produced by vomiting of gastric juice.

Chemical Buffering. Chemical buffers in the body limit the alkaline shift in blood pH by releasing H^+ as they are titrated in the alkaline direction. About one third of the buffering occurs in cells.

Respiratory Compensation. The respiratory compensation for metabolic alkalosis is hypoventilation. An alkaline blood pH inhibits ventilation. Hypoventilation raises the blood P_{CO_2} and [H_2CO_3], reducing the alkaline shift in pH. A 1 mEq/L rise in plasma [HCO_3^-] caused by metabolic alkalosis is accompanied by a 0.7 mm Hg rise in P_{CO_2} (see Table 25.4). If, for example, the plasma [HCO_3^-] rose to 40 mEq/L, what would the plasma pH be with and without respiratory compensation? With respiratory compensation, the P_{CO_2} should rise by 11.2 mm Hg (0.7 × 16) to 51.2 mm Hg ([$CO_{2(d)}$] = 1.54 mmol/L). The pH is 7.51:

$$pH = 6.10 + \log \frac{[40]}{[1.54]} = 7.51 \tag{41}$$

Without respiratory compensation, the pH would be 7.62:

$$pH = 6.10 + \log \frac{[40]}{[1.2]} = 7.62 \tag{42}$$

Respiratory compensation for metabolic alkalosis is limited because hypoventilation leads to hypoxia and CO_2 retention, and both increase breathing.

Renal Compensation. The kidneys respond to metabolic alkalosis by lowering the plasma [HCO_3^-]. The plasma [HCO_3^-] is primarily raised and more HCO_3^- is filtered than can be reabsorbed (see Fig. 25.7); in addition, HCO_3^- is secreted in the collecting ducts. Both of these changes lead to increased urinary [HCO_3^-] excretion. If the cause of the metabolic alkalosis is corrected, the kidneys can often restore the plasma [HCO_3^-] and pH to normal in a day or two.

TABLE 25.5 High and Normal Anion Gap Metabolic Acidosis

Condition	Explanation
High anion gap metabolic acidosis	
Methanol intoxication	Methanol metabolized to formic acid
Uremia	Sulfuric, phosphoric, uric, and hippuric acids retained due to renal failure
Lactic acid	Lactic acid buffered by HCO_3^- and accumulates as lactate
Ethylene glycol intoxication	Ethylene glycol metabolized to glyoxylic, glycolic, and oxalic acids
p-Aldehyde intoxication	*p*-Aldehyde metabolized to acetic and chloroacetic acids
Ketoacidosis	Production of β-hydroxybutyric and acetoacetic acids
Salicylate intoxication	Impaired metabolism leads to production of lactic acid and ketone body acids; accumulation of salicylate
Normal anion gap metabolic acidosis	
Diarrhea	Loss of HCO_3^- in stool; kidneys conserve Cl^-
Renal tubular acidosis	Loss of HCO_3^- in urine or inadequate excretion of H^+; kidneys conserve Cl^-
Ammonium chloride ingestion	NH_4^+ is converted to urea in liver, a process that consumes HCO_3^-; excess Cl^- is ingested

CLINICAL FOCUS BOX 25.3

Vomiting and Metabolic Alkalosis

Vomiting of gastric acid juice results in **metabolic alkalosis** and fluid and electrolyte disturbances. Gastric acid juice contains about 0.1 M HCl. The acid is secreted by stomach parietal cells; these cells have a H^+/K^+-ATPase in their luminal plasma membrane and a Cl^-/HCO_3^- exchanger in their basolateral plasma membrane. When HCl is secreted into the stomach lumen and lost to the outside, there is a net gain of HCO_3^- in the blood plasma and no change in the anion gap. The HCO_3^-, in effect, replaces lost plasma Cl^-.

Ventilation is inhibited by the alkaline blood pH, resulting in a rise in P_{CO_2}. This respiratory compensation for the metabolic alkalosis, however, is limited because hypoventilation leads to a rise in P_{CO_2} and a fall in P_{O_2}, both of which stimulate breathing.

The logical renal compensation for metabolic alkalosis is enhanced excretion of HCO_3^-. In people with persistent vomiting, however, the urine is sometimes acidic and renal HCO_3^- reabsorption is enhanced, maintaining an elevated plasma $[HCO_3^-]$. This situation occurs because vomiting is accompanied by losses of ECF and K^+. Fluid loss leads to a decrease in effective arterial blood volume and engagement of mechanisms that reduce Na^+ excretion, such as decreased GFR and increased plasma renin, angiotensin, and aldosterone levels (see Chapter 24). Aldosterone stimulates H^+ secretion by collecting duct α-intercalated cells. Renal tubular Na^+/H^+ exchange is stimulated by volume depletion because the tubules reabsorb Na^+ more avidly than usual. With more H^+ secretion, more new HCO_3^- is added to the blood. The kidneys reabsorb filtered HCO_3^- completely, even though plasma HCO_3^- level is elevated, and maintain the metabolic alkalosis.

Vomiting results in K^+ depletion because of a loss of K^+ in the vomitus, decreased food intake and, most important quantitatively, enhanced renal K^+ excretion. Extracellular alkalosis results in a shift of K^+ into cells (including renal cells) and, thereby, promotes K^+ secretion and excretion. Elevated plasma aldosterone levels also favor K^+ loss in the urine.

Treatment of metabolic alkalosis primarily depends on eliminating the cause of vomiting. Correction of the alkalosis by administering an organic acid, such as lactic acid, does not make sense because this acid would simply be converted to CO_2 and H_2O; this approach also does not address the Cl^- deficit. The ECF volume depletion and the Cl^- and K^+ deficits can be corrected by administering isotonic saline and appropriate amounts of KCl. Because replacement of Cl^- is a key component of therapy, this type of metabolic alkalosis is said to be "chloride-responsive." After Na^+, Cl^-, water, and K^+ deficits have been replaced, excess HCO_3^- (accompanied by surplus Na^+) will be excreted in the urine, and the kidneys will return blood pH to normal.

Clinical Evaluation of Acid-Base Disturbances Requires a Comprehensive Study

Acid-base data should always be interpreted in the context of other information about a patient. A complete history and physical examination provide important clues to possible reasons for an acid-base disorder.

To identify an acid-base disturbance from laboratory values, it is best to look first at the pH. A low blood pH indicates acidosis; a high blood pH indicates alkalosis. If acidosis is present, for example, it could be either respiratory or metabolic. A low blood pH and elevated P_{CO_2} point to respiratory acidosis; a low pH and low plasma $[HCO_3^-]$ indicate metabolic acidosis. If alkalosis is present, it could be either respiratory or metabolic. A high blood pH and low plasma P_{CO_2} indicate respiratory alkalosis; a high blood pH and high plasma $[HCO_3^-]$ indicate metabolic alkalosis.

Whether the body is making an appropriate response for a simple acid-base disorder can be judged from the values in Table 25.4. Inappropriate values suggest that more than one acid-base disturbance may be present. Patients may have two or more of the four simple acid-base disturbances at the same time; in which case, they have a **mixed acid-base disturbance**.

REVIEW QUESTIONS

DIRECTIONS: Each of the numbered items or incomplete statements in this section is followed by answers or by completions of the statement. Select the ONE lettered answer or completion that is BEST in each case.

1. If the pK_a of NH_4^+ is 9.0, the ratio of NH_3 to NH_4^+ in a urine sample with a pH of 6.0 is
(A) 1:3
(B) 3:1
(C) 3:2
(D) 1:1,000
(E) 1,000:1
2. An arterial blood sample taken from a patient has a pH of 7.32 ($[H^+]$ = 48 nmol/L) and P_{CO_2} of 24 mm Hg. What is the plasma $[HCO_3^-]$?
(A) 6 mEq/L
(B) 12 mEq/L
(C) 20 mEq/L
(D) 24 mEq/L
(E) 48 mEq/L
3. Which segment can establish the steepest pH gradient (tubular fluid-to-blood)?
(A) Proximal convoluted tubule
(B) Thin ascending limb
(C) Thick ascending limb
(D) Distal convoluted tubule
(E) Collecting duct
4. Most of the hydrogen ions secreted by the kidney tubules are
(A) Consumed in the reabsorption of filtered bicarbonate
(B) Excreted in the urine as ammonium ions
(C) Excreted in the urine as free hydrogen ions
(D) Excreted in the urine as titratable acid

(continued)

5. The following measurements were made in a healthy adult:

Filtered bicarbonate	4,320 mEq/day
Excreted bicarbonate	2 mEq/day
Urinary titratable acid	30 mEq/day
Urinary ammonia (NH_4^+)	60 mEq/day
Urine pH	5

Net acid excretion by the kidneys is
(A) 28 mEq/day
(B) 30 mEq/day
(C) 88 mEq/day
(D) 90 mEq/day
(E) 92 mEq/day

6. If a patient with uncontrolled diabetes mellitus has a daily excretion rate of 200 mEq of titratable acid and 500 mEq of NH_4^+, how many mEq of new HCO_3^- have the kidney tubules added to the blood?
(A) 0 mEq
(B) 200 mEq
(C) 300 mEq
(D) 500 mEq
(E) 700 mEq

7. Which of the following causes increased tubular secretion of hydrogen ions?
(A) A decrease in arterial P_{CO_2}
(B) Adrenal cortical insufficiency
(C) Administration of a carbonic anhydrase inhibitor
(D) An increase in intracellular pH
(E) An increase in tubular sodium reabsorption

8. A homeless woman was found on a hot summer night lying on a park bench in a comatose condition. An arterial blood sample revealed a pH of 7.10, P_{CO_2} of 20 mm Hg, and plasma [HCO_3^-] of 6 mEq/L. Plasma glucose and blood urea nitrogen (BUN) values were normal. Plasma [Na^+] was 140 mEq/L and [Cl^-] was 105 mEq/L. Which of the following might explain her condition?
(A) Acute renal failure
(B) Diarrhea as a result of food poisoning
(C) Methanol intoxication
(D) Overdose with a drug that produces respiratory depression
(E) Uncontrolled diabetes mellitus

9. Which of the following arterial blood values might be expected in a mountain climber who has been residing at a high-altitude base camp below the summit of Mt. Everest for one week?

	pH	P_{O_2} *(mm Hg)*	P_{CO_2} *(mm Hg)*	*Plasma [HCO_3^-] (mEq/L)*
(A)	7.18	95	25	9
(B)	7.35	50	60	32
(C)	7.53	40	20	16
(D)	7.53	95	50	40
(E)	7.62	40	20	20

10. A 25-year-old nurse is brought to the emergency department shortly before midnight. Although somewhat drowsy, she was able to relate that she had attempted to kill herself by swallowing the contents of a bottle of aspirin tablets a few hours before. Which of the following set of arterial blood values is expected?

	pH	P_{O_2} *(mm Hg)*	P_{CO_2} *(mm Hg)*	*Plasma [HCO_3^-] (mEq/L)*
(A)	7.25	95	19	8
(B)	7.29	55	60	28
(C)	7.40	95	40	24
(D)	7.59	95	16	15
(E)	7.70	95	16	19

SUGGESTED READING

Abelow B. Understanding Acid-Base. Baltimore: Williams & Wilkins, 1998.
Adrogue HJ, Madias NE. Management of life-threatening acid base disorders. N Engl J Med 1998;338:26–34, 107–111.
Alpern RJ, Preisig PA. Renal acid-base transport. In: Schrier RW, Gottschalk CW, eds. Diseases of the Kidney. 6th Ed. Boston: Little, Brown, 1997;189–201.
Bevensee MO, Alper SL, Aronson PS, Boron WF. Control of intracellular pH. In: Seldin DW, Giebisch G, eds. The Kidney. Physiology and Pathophysiology. 3rd Ed. Philadelphia: Lippincott Williams & Wilkins, 2000;391–442.
Hood VL, Tannen RL. Protection of acid-base balance by pH regulation of acid production. N Engl J Med 1998; 339:819–826.
Knepper MA, Packer R, Good DW. Ammonium transport in the kidney. Physiol Rev 1989;69:179–249.
Lowenstein J. Acid and basics: A guide to understanding acid-base disorders. New York: Oxford University Press, 1993.
Rose BD, Post TW. Clinical Physiology of Acid-Base and Electrolyte Disorders. 5th Ed. New York: McGraw-Hill, 2001.
Valtin H, Gennari FJ. Acid-Base Disorders. Basic Concepts and Clinical Management. Boston: Little, Brown, 1987.

CASE STUDIES FOR PART VI • • •

CASE STUDY FOR CHAPTER 23

Nephrotic Syndrome

A 6-year-old boy is brought to the pediatrician by his mother because of a puffy face and lethargy. A few weeks before, he had an upper respiratory tract infection, probably caused by a virus. Body temperature is 36.8°C; blood pressure, 95/65; and heart rate, 90 beats/min. Puffiness around the eyes, abdominal swelling, and pitting edema in the legs are observed. A urine sample (dipstick) is negative for glucose but reveals 3+ protein. Microscopic examination of the urine reveals no cellular elements or casts. Plasma [Na^+] is 140 mEq/L; BUN, 10 mg/dL; [glucose], 100 mg/dL; creatinine, 0.8 mg/dL; serum albumin, 2.3 g/dL (normal, 3.0 to 4.5 g/dL); and cholesterol, 330 mg/dL. A 24-hour urine sample has a volume of 1.10 L and contains 10 mEq/L Na^+, 60 mg/dL creatinine, and 0.8 g/dL protein.

The child is treated with the corticosteroid prednisone, and the edema and proteinuria disappear in 2 weeks. Puffiness and proteinuria recur 4 months later, and a renal biopsy is performed. Glomeruli are normal by light microscopy, but effacement of podocyte foot processes and loss of filtration slits are seen with the electron microscope. No immune deposits or complement are seen after immunostaining. The biopsy indicates minimal change glomerulopathy. The podocyte cell surface and glomerular basement membrane show reduced staining with a cationic dye.

Questions

1. What features in this case would cause suspicion of nephrotic syndrome?
2. What is the explanation for the proteinuria?
3. Why does the abnormally high rate of urinary protein excretion underestimate the rate of renal protein loss?
4. What is the endogenous creatinine clearance, and is it normal? (The boy's body surface area is 0.86 m^2.)
5. What is the explanation for the edema?

Answers to Case Study Questions for Chapter 23

1. The child has the classical feature of nephrotic syndrome: heavy proteinuria (8.8 g/day), hypoalbuminemia (<3 g/dL), generalized edema, and hyperlipidemia (plasma cholesterol 330 mg/dL).

2. Proteinuria is a consequence of an abnormally high permeability of the glomerular filtration barrier to the normal plasma proteins. This condition might be a result of an increased size of "holes" or pores in the basement membrane and filtration slit diaphragms. The decreased staining with a cationic dye, however, suggests that there was a loss of fixed negative charges from the filtration barrier. Recall that serum albumin bears a net negative charge at physiological pH values, and that negative charges associated with the glomerular filtration barrier impede filtration of this plasma protein.
3. Proteins that have leaked across the glomerular filtration barrier are not only excreted in the urine but are reabsorbed by proximal tubules. The endocytosed proteins are digested in lysosomes to amino acids, which are returned to the circulation. Both increased renal catabolism by tubule cells and increased excretion of serum albumin in the urine contribute to the hypoalbuminemia. The liver, which synthesizes serum albumin, cannot keep up with the renal losses.
4. The endogenous creatinine (CR) clearance (an estimate of GFR) equals $(U_{CR} \times \dot{V})/P_{CR} = (60 \times 1.10)/0.8 = 82$ L/day. Normalized to a standard body surface area of 1.73 m^2, C_{CR} is 166 L/day -1.73 m^2, which falls within the normal range (150 to 210 L/day -1.73 m^2). Note that the permeability of the glomerular filtration barrier to macromolecules (plasma proteins) was abnormally high, but permeability to fluid was not increased. In some patients, a loss of filtration slits may be significant and may lead to a reduced fluid permeability and GFR.
5. The edema is a result of altered capillary Starling forces and renal retention of salt and water. The decline in plasma [protein] lowers the plasma colloid osmotic pressure, favoring fluid movement out of the capillaries into the interstitial compartment. The edema is particularly noticeable in the soft skin around the eyes (periorbital edema). The abdominal distension (in the absence of organ enlargement) suggests ascites (an abnormal accumulation of fluid in the abdominal cavity). The kidneys avidly conserve Na^+ (note the low urine [Na^+]) despite an expanded ECF volume. Although the exact reasons for renal Na^+ retention are controversial, a decrease in the effective arterial blood volume may be an important stimulus (see Chapter 24). This leads to activation of the renin-angiotensin-aldosterone system and stimulation of the sympathetic nervous system, both of which favor renal Na^+ conservation. In addition, distal segments of the nephron reabsorb more Na^+ than usual because of an intrinsic change in the kidneys.

Reference

Orth SR, Ritz E. The nephrotic syndrome. N Engl J Med 1998;338:1202–1211.

Case Study for Chapter 24

Water Intoxication

A 60-year-old woman with a long history of mental illness was institutionalized after a violent argument with her son. She experiences visual and auditory hallucinations and, on one occasion, ran naked through the ward screaming. She refuses to eat anything since admission, but maintains a good fluid intake. On the fifth hospital day, she complains of a slight headache and nausea and has three episodes of vomiting. Later in the day, she is found on the floor in a semiconscious state, confused and disoriented. She is pale and has cool extremities. Her pulse rate is 70/min, and her blood pressure is 150/100 mm Hg. She is transferred to a general hospital and, during transfer, has three grand mal seizures and arrives in a semiconscious, uncooperative state. A blood sample reveals a plasma [Na^+] of 103 mEq/L. Urine osmolality is 362 mosm/kg H_2O, and urine [Na^+] is 57 mEq/L. She is given an intravenous infusion of hypertonic saline (1.8% NaCl) and placed on water restriction. Several days after she had improved, bronchoscopy is performed.

Questions

1. What is the likely cause of the severe hyponatremia?
2. How much of an increase in plasma [Na^+] would an infusion of 1 L of 1.8% NaCl (308 mEq Na^+/L) produce? Assume that her total body water is 25 L (50% of her body weight). Why is the total body water used as the volume of distribution of Na^+, even though the administered Na^+ is limited to the ECF compartment?
3. Why is the brain so profoundly affected by hypoosmolality? Why should the hypertonic saline be administered slowly?
4. Why was the bronchoscopy performed?

Answers to Case Study Questions for Chapter 24

1. The problem started with ingestion of excessive amounts of water. Compulsive water drinking is a common problem in psychotic patients. The increased water intake, combined with an impaired ability to dilute the urine (note the inappropriately high urine osmolality), led to severe hyponatremia and water intoxication.
2. Addition of 1 L of 308 mEq Na^+/L to 25 L produces an increase in plasma [Na^+] of 12 mEq/L. The total body water is used in this calculation because when hypertonic NaCl is added to the ECF, it causes the movement of water out of the cell compartment, diluting the extracellular Na^+.
3. Because the brain is enclosed in a nondistensible cranium, when water moves into brain cells and causes them to swell, intracranial pressure can rise to very high values. This can damage nervous tissue directly or indirectly by impairing cerebral blood flow. The neurological symptoms seen in this patient (headache, semiconsciousness, grand mal seizures) are consequences of brain swelling. The increased blood pressure and cool and pale skin may be a consequence of sympathetic nervous system discharge resulting from increased intracranial pressure. Too rapid restoration of a normal plasma [Na^+] can produce serious damage to the brain (central pontine myelinolysis).
4. The physicians wanted to exclude the presence of a bronchogenic tumor, which is the most common cause of SIADH. No abnormality was detected. Today, a computed tomography (CT) scan would be performed first.

References

Goldman MB, Luchins DJ, Robertson GL. Mechanisms of altered water metabolism in psychotic patients with polydipsia and hyponatremia. N Engl J Med 1988;318:397–403.

Grainger DN. Rapid development of hyponatremic seizures in a psychotic patient. Psychol Med 1992;22:513–517.

Case Study for Chapter 25

Lactic Acidosis and Hemorrhagic Shock

During a violent argument over money, a 30-year-old man was stabbed in the stomach. The assailant escaped, but friends were able to rush the victim by car to the county hospital. The patient is unconscious, with a blood pressure (mm Hg) of 55/35 and heart rate of 165 beats/minute. Breathing is rapid and shallow. The subject is pale, with cool, clammy skin. On admission, about an hour after the stabbing, an arterial blood sample is taken, and the following data are reported:

	Patient	*Normal Range*
Glucose	125 mg/dL	70–110 mg/dL (3.9–6.1 mmol/L) (fasting values)
Na^+	140 mEq/L	136–145 mEq/L
K^+	4.8 mEq/L	3.5–5.0 mEq/L
Cl^-	103 mEq/L	95–105 mEq/L
HCO_3^-	4 mEq/L	22–26 mEq/L
BUN	23 mg/dL	7–18 mg/dL (1.2–3.0 mmol/L urea nitrogen)
Creatinine	1.1 mg/dL	0.6–1.2 mg/dL (53–106 μmol/L)
pH	7.08	7.35–7.45
$Pa{CO_2}$	14	35–45 mm Hg
$Pa{O_2}$	97 mm Hg	75–105 mm Hg
Hematocrit	35%	41–53%

Questions

1. What type of acid-base disturbance is present?
2. What is the reason for the low $Pa{CO_2}$?
3. Calculate the plasma anion gap and explain why it is high.
4. Why is the hematocrit low?
5. Discuss the status of kidney function.
6. What is the most appropriate treatment for the acid-base disturbance?

Answers to Case Study Questions for Chapter 25

1. The subject has a metabolic acidosis, with an abnormally low arterial blood pH and plasma [HCO_3^-].
2. The low $Pa{CO_2}$ is a result of respiratory compensation. Ventilation is stimulated by the low blood pH, sensed by the peripheral chemoreceptors.
3. The anion gap is = $[Na^+] - [Cl^-] - [HCO_3^-] = 140 - 103 - 4 = 33$ mEq/L, which is abnormally high. Considering the history and physical findings, the high anion gap is most likely caused by inadequate tissue perfusion, with resultant anaerobic metabolism and production of lactic acid. The lactic acid is buffered by HCO_3^- and lactate accumulates as the unmeasured anion. Note that tissue hypoxia can occur if blood flow is diminished, even when arterial $P{O_2}$ is normal.
4. The low hematocrit is a result of absorption of interstitial fluid by capillaries, consequent to the hemorrhage, low arterial blood pressure, and low capillary hydrostatic pressure.
5. In response to the blood loss and low blood pressure, kidney blood flow and GFR would be drastically reduced. The sympathetic nervous system, combined with increased plasma levels of AVP and angiotensin II, would produce intense renal vasoconstriction. The hydrostatic pressure in the glomeruli would be so low that practically no plasma would be filtered and little urine (oliguria) or no urine (anuria) would be excreted. Because of the short duration of renal shutdown, plasma [creatinine] is still in the normal range; the elevated BUN is probably mainly a result of bleeding into the gastrointestinal tract, digestion of blood proteins, and increased urea production.
6. Control of bleeding and administration of whole blood (or isotonic saline solutions and packed red blood cells) would help restore the circulation. With improved tissue perfusion, the lactate will be oxidized to HCO_3^-.

PART VII Gastrointestinal Physiology

Neurogastroenterology and Gastrointestinal Motility

Jackie D. Wood, Ph.D.

CHAPTER OUTLINE

KEY CONCEPTS

1. The musculature of the digestive tract is mainly smooth muscle.
2. Electrical slow waves and action potentials are the main forms of electrical activity in the gastrointestinal musculature.
3. Gastrointestinal smooth muscles have properties of a functional electrical syncytium.
4. A hierarchy of neural integrative centers in the central nervous system (CNS) and peripheral nervous system (PNS) determines moment-to-moment behavior of the digestive tract.
5. The digestive tract is innervated by the sympathetic, parasympathetic, and enteric divisions of the autonomic nervous system (ANS).
6. Vagus nerves transmit afferent sensory information to the brain and parasympathetic autonomic efferent signals to the digestive tract.
7. Splanchnic nerves transmit sensory information to the spinal cord and sympathetic autonomic efferent signals to the digestive tract.
8. The enteric nervous system (ENS) functions as a minibrain in the gut.
9. Fast and slow excitatory postsynaptic potentials, slow inhibitory postsynaptic potentials, presynaptic inhibition, and presynaptic facilitation are key synaptic events in the ENS.
10. Enteric motor neurons may be excitatory or inhibitory to the musculature.
11. Enteric inhibitory motor neurons to the intestinal circular muscle are continuously active and transiently inactivated to permit muscle contraction.
12. Enteric inhibitory motor neurons to the musculature of sphincters are inactive and transiently activated for timed opening and the passage of luminal contents.
13. A polysynaptic reflex circuit determines the behavior of the intestinal musculature during peristaltic propulsion.
14. Physiological ileus is the normal absence of contractile activity in the intestinal musculature.
15. Peristalsis and relaxation of the lower esophageal sphincter are the main motility events in the esophagus.
16. The gastric reservoir and antral pump have different motor behavior.
17. Vago-vagal reflexes are important in the control of gastric motor functions.
18. Feedback signals from the duodenum determine the rate of gastric emptying.
19. The migrating motor complex is the small intestinal motility pattern of the interdigestive state.

(continued)

20. Mixing movements are the small intestinal motility pattern of the digestive state.
21. Intestinal power propulsion is a protective response to harmful agents.
22. Cramping abdominal pain may be associated with intestinal power propulsion.
23. Motor functions of the large intestine are specialized for storage and dehydration of feces.
24. The physiology of the rectosigmoid region, anal canal, and pelvic floor musculature is important in maintaining fecal continence.

This chapter presents concepts and principles of neurogastroenterology in relation to motor functions of the specialized organs and muscle groups of the digestive tract. **Neurogastroenterology** is a subspecialty of clinical gastroenterology and digestive science. As such, it encompasses the investigative sciences dealing with functions, malfunctions, and malformations in the brain and spinal cord and the sympathetic, parasympathetic, and enteric divisions of the autonomic innervation of the digestive tract. Somatic motor systems are included insofar as pharyngeal phases of swallowing and pelvic floor involvement in defecation and continence are concerned. The basic physiology of smooth muscles, as it relates to enteric neural control of motor movements, is a part of neurogastroenterology. Psychological and psychiatric aspects of gastrointestinal disorders are significant components of the neurogastroenterological domain, especially in relation to projections of discomfort and pain to the digestive tract.

Gastrointestinal (GI) motility refers to wall movement or lack thereof in the digestive tract. The integrated function of multiple tissues and types of cells is necessary for generation of the various patterns of motility found in the organs of the digestive tract. Digestive motor movements involve the application of forces of muscle contraction to material that may be present in the mouth, pharynx, esophagus, stomach, gallbladder, or small and large intestines. The musculature is striated in the mouth, pharynx, upper esophagus, and pelvic floor and is visceral-type smooth muscle elsewhere. Specialized pacemaker cells, called interstitial cells of Cajal, are associated with the smooth musculature. The nervous system, with its different kinds of neurons and glial cells, organizes muscular activity into functional patterns of wall behavior. Functions of the nervous system are influenced by chemical signals released from enterochromaffin cells, enteroendocrine cells, and cells associated with the enteric immune system (e.g., mast cells and polymorphonuclear leukocytes).

Motility in the various organs of the digestive tract is organized to fulfill the specialized function of the individual organ. Esophageal motility, for example, differs from gastric motility, and gastric motility differs from small intestinal motility. The motility in the different organs reflects coordinated contractions and relaxations of the smooth muscle. Contractions are organized to produce the propulsive forces that move the contents along the tract, triturate large particles to smaller particles, mix ingested foodstuff with digestive enzymes, and bring nutrients into contact with the mucosa for efficient absorption. Relaxation of spontaneous tone in the smooth muscle allows sphincters to open and ingested material to be accommodated in reservoirs of the stomach and large intestine. The enteric nervous system (ENS), together with its input from the CNS, organizes motility into patterns of efficient behavior suited to differing digestive states (e.g., fasting and processing of a meal) as well as abnormal patterns such as occur during vomiting.

THE MUSCULATURE OF THE DIGESTIVE TRACT

The smooth muscles of the digestive tract are generally organized in distinct layers. Two important muscle layers for motility in the lower esophagus and small and large intestine are the longitudinal and circular layers (Fig. 26.1). The two layers form the intestinal **muscularis externa**. The stomach has an additional obliquely oriented muscle layer.

The Structure and Function of Circular and Longitudinal Muscles Differ

The circular muscle layer is thicker than the longitudinal layer and more powerful in exerting contractile forces on the contents of the lumen. The long axis of the muscle fibers of circular muscle is oriented in the circumferential direction. Consequently, contraction reduces the diameter of the lumen of an intestinal segment and increases its length. Because the long axis of the muscle fibers is oriented in the longitudinal direction, contraction of the longitudinal muscle coat shortens the segment of intestine where it occurs and expands the lumen.

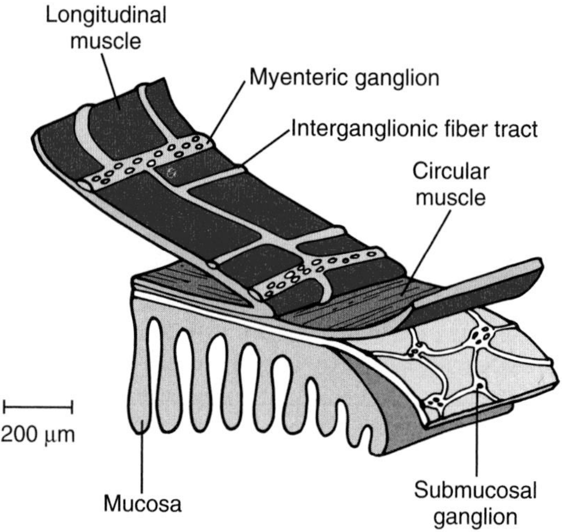

FIGURE 26.1 Structural relationship of the intestinal musculature and the enteric nervous system. Ganglia and interganglionic fiber tracts form the myenteric plexus between the longitudinal and the circular muscle layer and form the submucosal plexus between the mucosa and circular muscle layer.

Both longitudinal and circular muscle layers are innervated by motor neurons of the ENS. The longitudinal muscle layer is innervated mainly by excitatory motor neurons; the circular muscle layer by both excitatory and inhibitory motor neurons. Nonneural pacemaker cells and excitatory motor neurons activate contraction of the circular muscle, and excitatory motor neurons are the main triggers for contraction of the longitudinal muscle. More gap junctions between adjacent muscle fibers are found in the circular layer than in the longitudinal muscle layer. Calcium influx from outside the muscle cells is important for excitation-contraction coupling in longitudinal muscle fibers. Intracellular release from internal stores is more important for excitation-contraction coupling in the muscle fibers of the circular layer.

Smooth Muscles Are Classified as Unitary or Multiunit Types

Smooth muscles are classified based on their behavioral properties and associations with nerves (see Chapter 9). Muscles of the stomach and intestine behave like **unitary type** smooth muscle. These muscles contract spontaneously in the absence of neural or endocrine influence and contract in response to stretch. There are no structured neuromuscular junctions, and neurotransmitters travel over extended diffusion distances to influence relatively large numbers of muscle fibers. The smooth muscle of the esophagus and gallbladder is more like the **multiunit type**. These muscles do not contract spontaneously in the absence of nervous input and do not contract in response to stretch. Activation to contract is by nervous input to relatively small groups of muscle fibers.

Electromechanical and Pharmacomechanical Coupling Trigger Contractions in GI Muscles

GI smooth muscle differs from skeletal muscle in having two mechanisms that initiate the processes leading to contractile shortening and development of tension. In both skeletal muscle and GI smooth muscle, depolarization of the membrane electrical potential leads to the opening of voltage-gated calcium channels, followed by the elevation of cytosolic calcium, which, in turn, activates the contractile proteins. This mechanism is called **electromechanical coupling**. Smooth muscles have an additional mechanism in which the binding of a ligand to its receptor on the muscle membrane leads to the opening of calcium channels and the elevation of cytosolic calcium without any change in the membrane electrical potential. This mechanism is called **pharmacomechanical coupling**. The ligands may be chemical substances released as signals from nerves (neurocrine), from nonneural cells in close proximity to the muscle (paracrine), or from endocrine cells as hormones delivered to the muscle by the blood.

GI and Esophageal Smooth Muscles Have Properties of a Functional Electrical Syncytium

Smooth muscle fibers are connected to their neighbors by **gap junctions**, which are permeable to ions and, thereby, transmit electrical current from muscle fiber to muscle fiber. Ionic connectivity, without cytoplasmic continuity from fiber to fiber, accounts for the **electrical syncytial properties** of smooth muscle, which confers electrical behavior analogous to that of cardiac muscle (see Chapter 13). Electrical activity and associated contractions spread from a point of initiation (e.g., the pacemaker region) in three dimensions throughout the bulk of the muscle. The distance and the direction of electrical activity spread are controlled by the ENS. A failure of nervous control can lead to disordered motility that includes spasm and associated abdominal pain.

Slow Waves and Action Potentials Are Forms of Electrical Activity in GI Muscles

Electrical slow waves are omnipresent and responsible for triggering action potentials in some regions, whereas in other regions (e.g., the gastric antrum and large intestinal circular muscle) they represent the only form of electrical activity (Fig. 26.2). They are always present in the small intestine where they decrease in frequency along a gradient from the duodenum to the ileum. In the gastric antrum, the terms *slow wave* and *action potential* are used interchangeably for the same electrical event. When action potentials are associated with electrical slow waves, they occur during the plateau phase of the slow wave (see Fig. 26.2).

Action potentials in GI smooth muscle are mediated by changes in calcium and potassium conductances. The depolarization phase of the action potential is produced by an all-or-nothing increase in calcium conductance, with the inward calcium current carried by L-type calcium channels. The opening of potassium channels as the calcium channels are closing at or near the peak of the action potential accounts for the repolarization phase. The L-type calcium channels in GI smooth muscle are essentially the same as those found in cardiac and vascular smooth muscle. Therefore, disordered GI motility may be an adverse effect of treating cardiovascular disease with drugs that block L-type calcium channels.

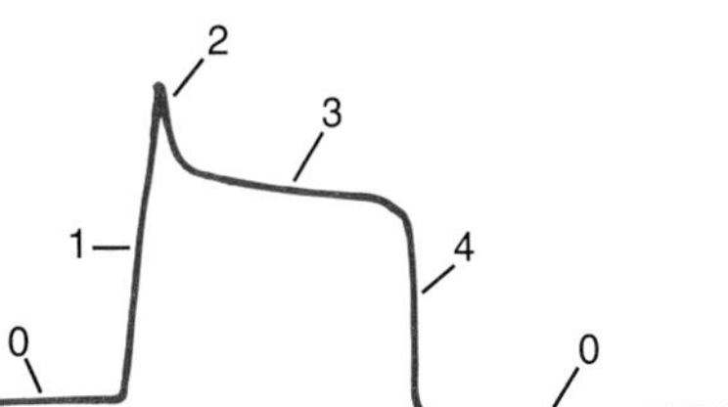

FIGURE 26.2 **Electrical slow waves.** In GI muscles, slow waves occur in four phases determined by specific ionic mechanisms. Phase 0: Resting membrane potential; outward potassium current. Phase 1, the rising phase (upstroke depolarization), activates voltage-gated calcium channels and voltage-gated potassium channels. Phase 2, partial repolarization, is due to the opening of voltage-gated potassium channels. Phase 3, the plateau phase, balances inward calcium current and outward potassium current. Phase 4, the falling phase (repolarization), inactivates voltage-gated calcium channels and activates calcium-gated potassium channels.

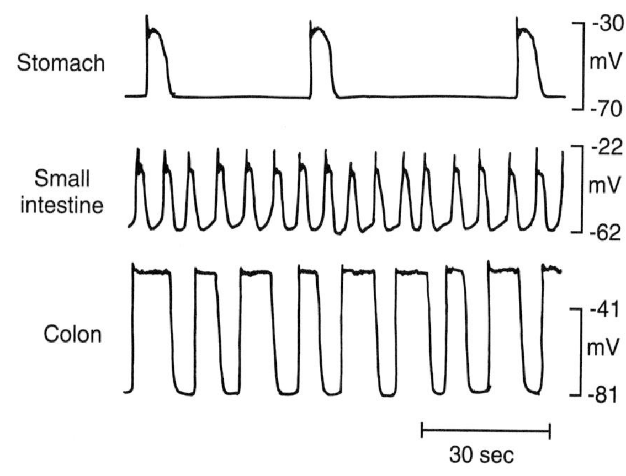

FIGURE 26.3 **Electrical slow wave frequencies.** Slow waves with similar waveforms occur at different frequencies in the stomach, small intestine, and colon.

Electrical Slow Wave Frequencies in the Stomach, Small Intestine, and Colon. Electrical slow waves with essentially the same waveform occur at different frequencies in the gastric antrum and small and large intestinal circular muscle when recorded with intracellular electrodes (Fig. 26.3). Slow waves occur at 3/min in the antrum, as high as 18/min in the duodenum, and 6 to 10/min in the colon. The maximum contractile frequency of the muscle does not exceed the frequency of the slow waves, but it may occur at a lower frequency because all slow waves may not trigger contractions. The nervous system determines the nature of the contractile response during each slow wave in the integrated functional state of the whole organ.

Electrical Slow Waves Without Action Potentials in the Small Intestine. As a general rule, slow waves in the small intestinal circular muscle trigger action potentials and action potentials trigger contractions. Slow waves are omnipresent in virtually all mammalian species and may or may not be accompanied by action potentials. Contractions do not occur in the absence of action potentials. The electrical slow waves in Figure 26.4 were recorded with an extracellular electrode attached to the serosal surface of the intestine. This method records from many circular muscle fibers. Shallow contractions appearing in the absence of action potentials on the slow waves reflect the responses of a few of the total population of muscle fibers under the electrode (Fig. 26.4A). In this case, the action potential currents from the small number of fibers are too small to be detected by the surface electrode. With this method of recording, the size of an action potential appears larger when larger numbers of the total population of muscle fibers are depolarized to action potential threshold by each slow wave. The amplitude of phasic contractions associated with each electrical slow wave increases in direct relation to the number of muscle fibers recruited to firing threshold by each slow wave cycle (Fig. 26.4B).

Electrical Slow Waves and Interstitial Cells of Cajal. **Interstitial cells of Cajal** (ICCs) are the generators of electrical slow waves in the stomach and small and large intestine (Fig. 26.5). The ICCs are interconnected into networks by gap junctions that impart the properties of a functional electrical syncytium to the network. Gap junctions also electrically connect the ICCs to the circular muscle. Electrical current flows from the ICC network across the gap junctions to depolarize the membrane potential of the circular muscle fibers to the threshold for action potential discharge.

Pacemaker networks of ICCs are located surrounding the small intestinal circular muscle at the border with the longitudinal muscle (myenteric border) and at its border with the submucosa. Slow waves generated by the ICC network at the submucosal border spread passively across gap junctions into the bulk of circular muscle, and those at the myenteric border spread passively into both longitudinal and circular muscle. Muscle fibers of the circular muscle are interconnected by gap junctions that transmit the slow wave electrical current from fiber to fiber throughout the bulk of the muscle.

CONTROL OF DIGESTIVE FUNCTIONS BY THE NERVOUS SYSTEM

The innervation of the digestive tract controls muscle contraction, secretion, and absorption across the mucosal lining and blood flow inside the walls of the esophagus, stomach, intestines, and gallbladder. Depending on the kind of neurotransmitter released, the neurons can activate or inhibit muscle contraction. The secretion of water, electrolytes, and mucus into the lumen and absorption from the lumen are determined by the innervation. The amount of blood flow within the wall and the distribution of flow between the muscle layers and mucosa are also controlled by nervous activity.

Sensory nerves transmit information on the state of the gut to the brain for processing. Sensory transmission and central processing account for sensations that are localized to the digestive tract. These include sensations of discomfort (such as upper abdominal fullness), abdominal pain, and chest pain (heartburn). Neural interactions include the sensory inflow of information from the gut to the brain and outflow from the brain to the gut. Outflow may originate in higher processing centers of the brain (the frontal cortex) and account for the projection of an individual's emotional state (psychogenic stress) to the gut. This kind of brain-gut interaction underlies the symptoms of diarrhea and lower abdominal discomfort often reported by students anticipating an examination.

A Hierarchy of Neural Integrative Centers Determines the Moment-to-Moment Motor Behavior of the Digestive Tract

The **sympathetic**, **parasympathetic**, and **enteric nervous systems** make up the divisions of the ANS that innervate the digestive tract. Figure 26.6 illustrates how neural control of the gut is hierarchical with five basic levels of integrative organization. Level 1 is the ENS, which behaves like a minibrain in the gut. Level 2 consists of the prevertebral ganglia of the sympathetic nervous system. Levels 3, 4, and 5 are within the CNS. Sympathetic and parasympa-

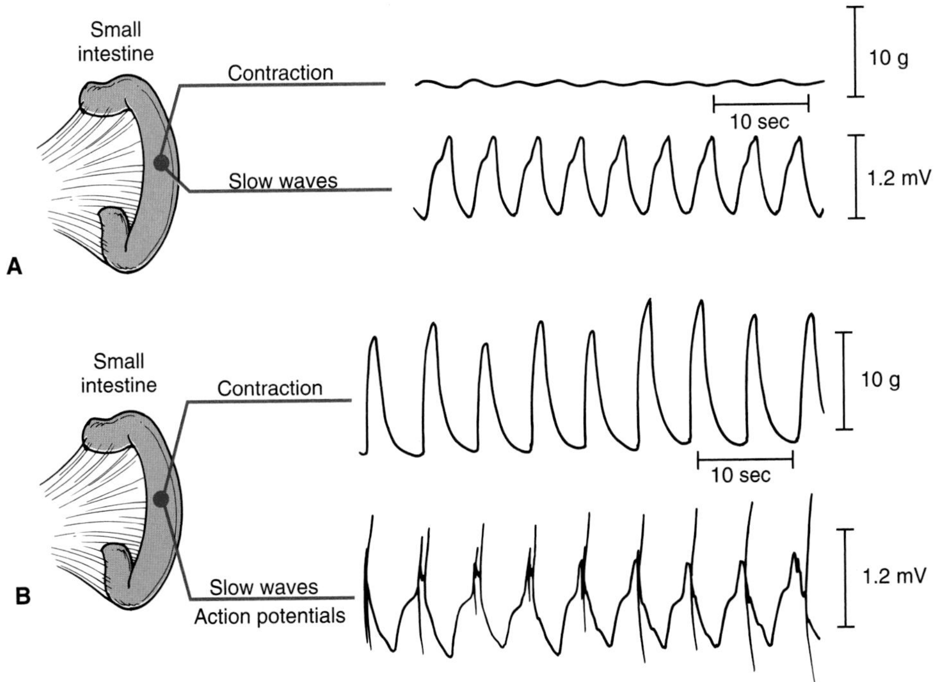

FIGURE 26.4 **Electrical slow waves in the small intestine.** A, No action potentials appear at the crests of the slow waves, and the muscle contractions associated with each slow wave are small. **B,** Muscle action potentials appear as sharp upward-downward deflections at the crests of the slow waves. Large-amplitude muscle contractions are associated with each slow wave when action potentials are present. Electrical slow waves trigger action potentials, and action potentials trigger contractions.

thetic signals to the digestive tract originate at levels 3 and 4 (central sympathetic and parasympathetic centers) in the medulla oblongata and represent the final common pathways for the outflow of information from the brain to the gut. Level 5 includes higher brain centers that provide input for integrative functions at levels 3 and 4.

Autonomic signals to the gut are carried from the brain and spinal cord by sympathetic and parasympathetic nervous pathways that represent the **extrinsic component** of innervation. Neurons of the enteric division form the local intramural control networks that make up the **intrinsic component** of the autonomic innervation. The parasympathetic and sympathetic subdivisions are identified by the positions of the ganglia containing the cell bodies of the postganglionic neurons and by the point of outflow from the CNS. Comprehensive autonomic innervation of the digestive tract consists of interconnections between the brain, the spinal cord, and the ENS.

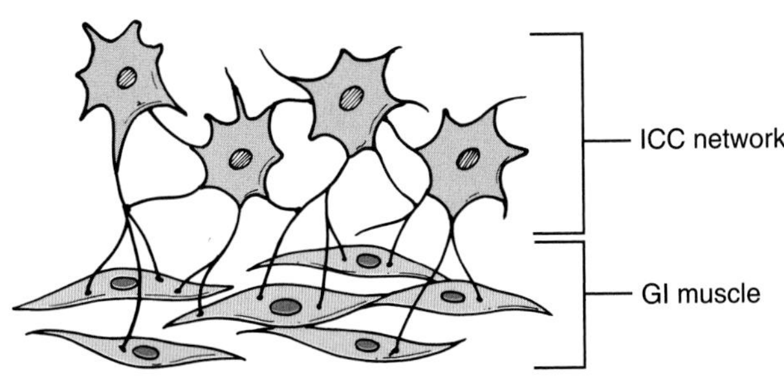

FIGURE 26.5 **Interstitial cells of Cajal.** ICCs form networks that contact the GI musculature. Electrical slow waves originate in the networks of ICCs. ICCs are the generators (pacemaker sites) of the slow waves. Gap junctions connect the ICCs to the circular muscle. Ionic current flows across the gap junctions to depolarize the membrane potential of the circular muscle fibers to the threshold for the discharge of action potentials.

Autonomic Parasympathetic Neurons Project to the Gut From the Medulla Oblongata and Sacral Spinal Cord

The origins of parasympathetic nerves to the gut are located in both the brainstem and sacral region of the spinal cord (Fig. 26.7). Projections to the digestive tract from these regions of the CNS are **preganglionic efferents**. Neu-

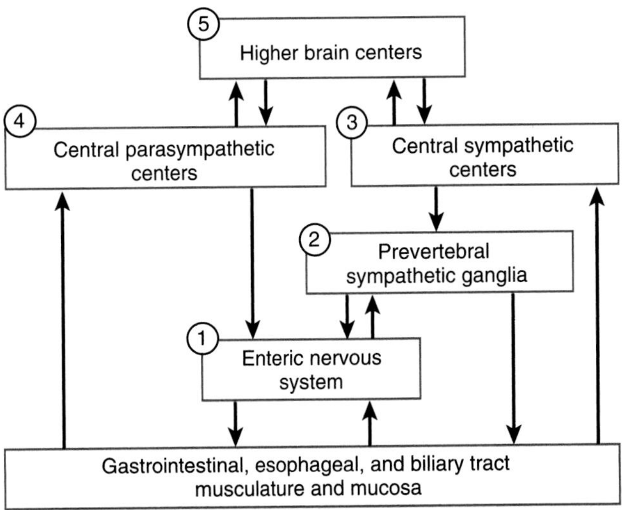

FIGURE 26.6 **A hierarchy of neural integrative centers.** Five levels of neural organization determine the moment-to-moment motor behavior of the digestive tract. (See text for details.)

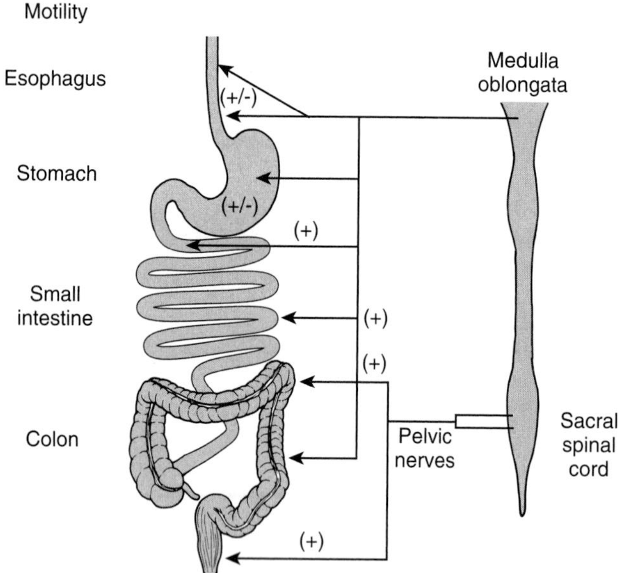

FIGURE 26.7 Parasympathetic innervation. Signals from parasympathetic centers in the CNS are transmitted to the enteric nervous system by the vagus and pelvic nerves. These signals may result in contraction (+) or relaxation (−) of the digestive tract musculature.

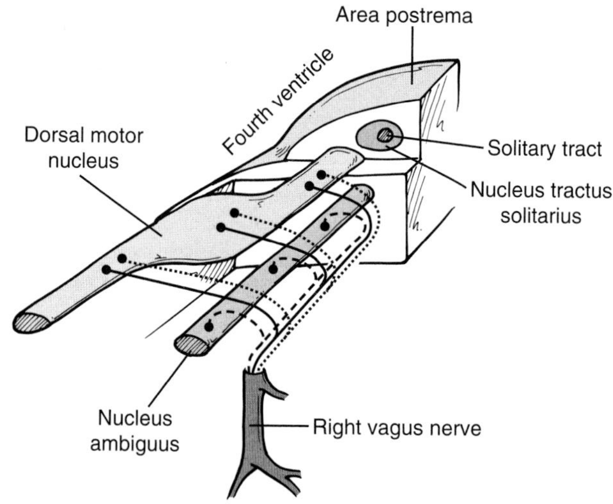

FIGURE 26.8 Dorsal vagal complex of medulla oblongata. Cell bodies of efferent vagal neurons are in the dorsal motor nucleus and nucleus ambiguus of the medulla oblongata. The dorsal motor nucleus and nucleus tractus solitarius form the dorsal vagal complex, which is the central vagal integrative center for the distal esophagus, stomach, and functional cluster of duodenum, gallbladder, and pancreas.

ronal cell bodies in the **dorsal motor nucleus** in the medulla oblongata project in the vagus nerves, and those in the sacral region of the spinal cord project in the pelvic nerves to the large intestine. Efferent fibers in the pelvic nerves make synaptic contact with neurons in ganglia located on the serosal surface of the colon and in ganglia of the ENS deeper within the large intestinal wall. Efferent vagal fibers synapse with neurons of the ENS in the esophagus, stomach, small intestine, and colon, as well as in the gallbladder and pancreas.

Efferent vagal nerves transmit signals to the enteric innervation of the GI musculature to control digestive processes both in anticipation of food intake and following a meal. This involves the stimulation and inhibition of contractile behavior in the stomach as a result of activation of the enteric circuits that control excitatory or inhibitory motor neurons, respectively. Parasympathetic efferents to the small and large intestinal musculature are predominantly stimulatory as a result of their input to the enteric microcircuits that control the activity of excitatory motor neurons.

The **dorsal vagal complex** consists of the dorsal motor nucleus of the vagus, **nucleus tractus solitarius**, **area postrema**, and **nucleus ambiguus**; it is the central vagal integrative center (Fig. 26.8). This center in the brain is more directly involved in the control of the specialized digestive functions of the esophagus, stomach, and the functional cluster of duodenum, gallbladder, and pancreas than the distal small intestine and large intestine. The circuits in the dorsal vagal complex and their interactions with higher centers are responsible for the rapid and more precise control required for adjustments to rapidly changing conditions in the upper digestive tract during anticipation, ingestion, and digestion of meals of varied composition.

Vago-Vagal Reflex Circuits Consist of Sensory Afferents, Second-Order Interneurons, and Efferent Neurons

A reflex circuit known as the **vago-vagal reflex** underlies moment-to-moment adjustments required for optimal digestive function in the upper digestive tract (see Clinical Focus Box 26.1). The afferent side of the reflex arc consists of vagal afferent neurons connected with a variety of sensory receptors specialized for the detection and signaling of mechanical parameters, such as muscle tension and mucosal brushing, or luminal chemical parameters, including glucose concentration, osmolality, and pH. Cell bodies of the vagal afferents are in the **nodose ganglia**. The afferent neurons are synaptically connected with neurons in the dorsal motor nucleus of the vagus and nucleus ambiguus. The nucleus of the tractus solitarius, which lies directly above the dorsal motor nucleus of the vagus (see Fig. 26.8), makes synaptic connections with the neuronal pool in the vagal motor nucleus. A synaptic meshwork formed by processes from neurons in both nuclei tightly links the two into an integrative center. The dorsal vagal neurons are second- or third-order neurons representing the efferent arm of the reflex circuit. They are the final common pathways out of the brain to the enteric circuits innervating the effector systems.

Efferent vagal fibers form synapses with neurons in the ENS to activate circuits that ultimately drive the outflow of signals in motor neurons to the effector systems. When the effector system is the musculature, its innervation consists of both inhibitory and excitatory motor neurons that participate in reciprocal control. If the effector systems are gastric glands or digestive glands, the secretomotor neurons are excitatory and stimulate secretory behavior.

The circuits for CNS control of the upper GI tract are organized much like those dedicated to the control of skeletal muscle movements (see Chapter 5), where fundamental re-

CLINICAL FOCUS BOX 26.1

Delayed Emptying and Rapid Emptying: Disorders of Gastric Motility

Disorders of gastric motility can be divided into the broad categories of delayed and rapid emptying. The generalized symptoms of both disorders overlap (Fig. 26.A).

Delayed gastric emptying is common in diabetes mellitus and may be related to disorders of the vagus nerves, as part of a spectrum of autonomic neuropathy. Surgical vagotomy results in a rapid emptying of liquids and a delayed emptying of solids. As mentioned later, vagotomy impairs adaptive relaxation and results in increased contractile tone in the reservoir (see Fig. 26.29). Increased pressure in the gastric reservoir more forcefully presses liquids into the antral pump. Paralysis with a loss of propulsive motility in the antrum occurs after a vagotomy. The result is **gastroparesis**, which can account for the delayed emptying of solids after a vagotomy. When selective vagotomy is performed as a treatment for peptic ulcer disease, the pylorus is enlarged surgically **(pyloroplasty)** to compensate for postvagotomy gastroparesis.

Delayed gastric emptying with no demonstrable underlying condition is common. Up to 80% of patients with anorexia nervosa have delayed gastric emptying of solids. Another such condition is **idiopathic gastric stasis**, in which no evidence of an underlying condition can be found. Motility-stimulating drugs are used successfully in treating these patients. In children, **hypertrophic pyloric stenosis** impedes gastric emptying. This is a thickening of the muscles of the pyloric canal associated with a loss of enteric neurons. The absence of inhibitory motor neurons and the failure of the circular muscles to relax account for the obstructive stenosis.

Rapid gastric emptying often occurs in patients who have had both vagotomy and gastric antrectomy for the treatment of peptic ulcer disease. These individuals have rapid emptying of solids and liquids. The pathological effects are referred to as the **dumping syndrome**, which results from the "dumping" of large osmotic loads into the proximal small intestine.

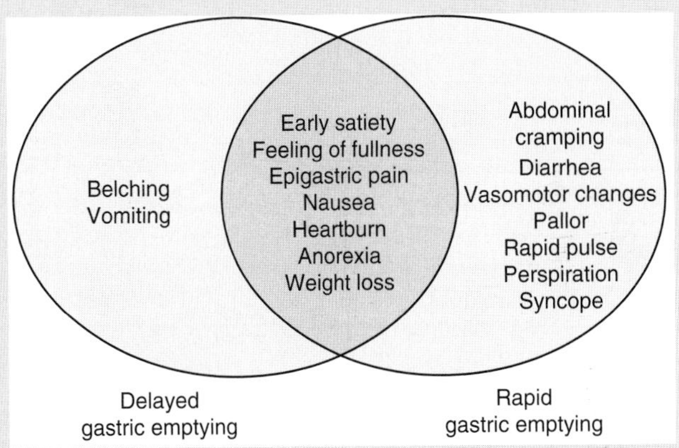

FIGURE 26.A Symptoms of disordered gastric emptying. Some of the symptoms of delayed and rapid gastric emptying overlap.

flex circuits are located in the spinal cord. Inputs to the spinal reflex circuits from higher order integrative centers in the brain (motor cortex and basal ganglia) complete the neural organization of skeletal muscle motor control. Memory, the processing of incoming information from outside the body, and the integration of proprioceptive information are ongoing functions of higher brain centers responsible for the logical organization of outflow to the skeletal muscles by way of the basic spinal reflex circuit. The basic connections of the vago-vagal reflex circuit are like somatic motor reflexes, in that they are "fine-tuned" from moment to moment by input from higher integrative centers in the brain.

Autonomic Sympathetic Neurons Project to the Gut From Thoracic and Upper Lumbar Segments of the Spinal Cord

Sympathetic innervation to the gut is located in thoracic and lumbar regions of the spinal cord (Fig. 26.9). The nerve cell bodies are in the intermediolateral columns. Efferent sympathetic fibers leave the spinal cord in the ventral roots to make their first synaptic connections with neurons in **prevertebral sympathetic ganglia** located in the abdomen. The prevertebral ganglia are the **celiac**, **superior mesenteric**, and **inferior mesenteric ganglia**. Cell bodies in the prevertebral ganglia project to the digestive tract where they synapse with neurons of the ENS in addition to innervating the blood vessels, mucosa, and specialized regions of the musculature.

Sympathetic input generally functions to shunt blood from the splanchnic to the systemic circulation during exercise and stressful environmental change, coinciding with the suppression of digestive functions, including motility and secretion. The release of **norepinephrine** (NE) from sympathetic postganglionic neurons is the principal mediator of these effects. NE acts directly on sphincteric muscles to increase tension and keep the sphincter closed. Presynaptic inhibitory action of NE at synapses in the control circuitry of the ENS is primarily responsible for inactivation of motility.

Suppression of synaptic transmission by the sympathetic nerves occurs at both fast and slow excitatory synapses in the neural networks of the ENS. This inactivates the neural circuits that generate intestinal motor behavior. Activation of the sympathetic inputs allows only continuous discharge of inhibitory motor neurons to the nonsphincteric muscles. The overall effect is a state of paralysis of intestinal motility in conjunction with reduced intestinal blood flow. When this state occurs transiently, it is called **physiological ileus** and, when it persists abnormally, is called **paralytic ileus**.

Splanchnic Nerves Transmit Sensory Information to the Spinal Cord and Efferent Sympathetic Signals to the Digestive Tract

The splanchnic nerves are mixed nerves that contain both sympathetic efferent and sensory afferent fibers. Sensory

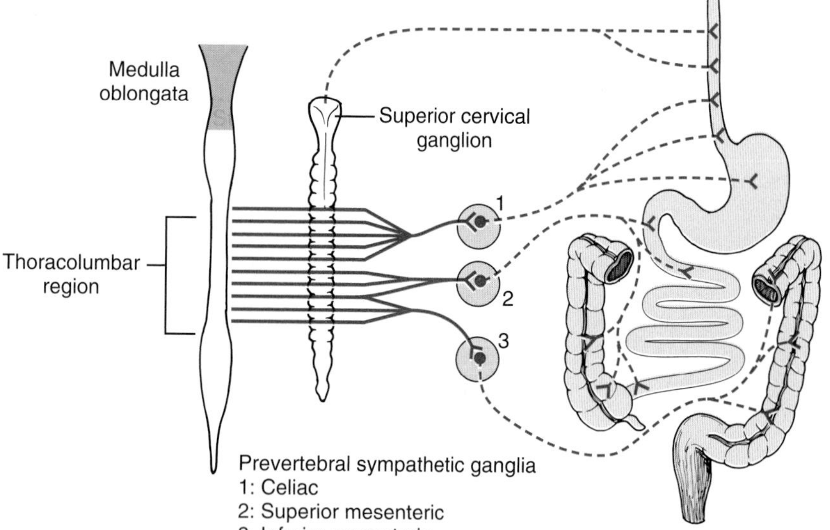

FIGURE 26.9 Sympathetic innervation. Neurons of the sympathetic division of the autonomic nervous system project to the gut from thoracic and first lumbar segments of the spinal cord. Efferent sympathetic fibers leave the spinal cord in the ventral roots to make their first synaptic connections with neurons in the prevertebral ganglia located in the abdomen. Cell bodies in the prevertebral ganglia project to the digestive tract, where they synapse with neurons of the enteric nervous system in addition to innervating the blood vessels, mucosa, and specialized regions of the musculature.

nerves course side by side with the sympathetic fibers; nevertheless, they are not part of the sympathetic nervous system. The term sympathetic afferent, which is sometimes used, is incorrect.

Sensory afferent fibers in the splanchnic nerves have their cell bodies in dorsal root spinal ganglia. They transmit information from the GI tract and gallbladder to the CNS for processing. These fibers transmit a steady stream of information to the local processing circuits in the ENS, to prevertebral sympathetic ganglia, and to the CNS. The gut has mechanoreceptors, chemoreceptors, and thermoreceptors. Mechanoreceptors sense mechanical events in the mucosa, musculature, serosal surface, and mesentery. They supply both the ENS and the CNS with information on stretch-related tension and muscle length in the wall and on the movement of luminal contents as they brush the mucosal surface. Mesenteric mechanoreceptors code for gross movements of the organ. Chemoreceptors generate information on the concentration of nutrients, osmolality, and pH in the luminal contents. Recordings of sensory information exiting the gut in afferent fibers reveal that most receptors are multimodal, in that they respond to both mechanical and chemical stimuli. The presence in the GI tract of pain receptors (nociceptors) equivalent to C fibers and A-delta fibers elsewhere in the body is likely, but not unequivocally confirmed, except for the gallbladder. The sensitivity of splanchnic afferents, including nociceptors, may be elevated when inflammation is present in intestine or gallbladder.

The Enteric Division of the ANS Functions as a Minibrain in the Gut

The ENS is a minibrain located close to the effector systems it controls. Effector systems of the digestive tract are the musculature, secretory glands, and blood vessels. Rather than crowding the vast numbers of neurons required for controlling digestive functions into the cranium as part of the cephalic brain and relying on signal transmission over long and unreliable pathways, the integrative microcircuits are located at the site of the effectors. The circuits at the effector sites have evolved as an organized array of different kinds of neurons interconnected by chemical synapses. Function in the circuits is determined by the generation of action potentials within single neurons and chemical transmission of information at the synapses.

The enteric microcircuits in the various specialized regions of the digestive tract are wired with large numbers of neurons and synaptic sites where information processing occurs. Multisite computation generates output behavior from the integrated circuits that could not be predicted from properties of their individual neurons and synapses. As in the brain and spinal cord, emergence of complex behaviors is a fundamental property of the neural networks of the ENS.

The processing of sensory signals is one of the major functions of the neural networks of the ENS. Sensory signals are generated by sensory nerve endings and coded in the form of action potentials. The code may represent the status of an effector system (such as tension in a muscle), or it may signal a change in an environmental parameter, such as luminal pH. Sensory signals are computed by the neural networks to generate output signals that initiate homeostatic adjustments in the behavior of the effector system.

The cell bodies of the neurons that make up the neural networks are clustered in ganglia that are interconnected by fiber tracts to form a plexus. The structure, function, and neurochemistry of the ganglia differ from other ANS ganglia. Unlike autonomic ganglia elsewhere in the body, where they function mainly as relay-distribution centers for signals transmitted from the brain and spinal cord, enteric ganglia are interconnected to form a nervous system with mechanisms for the integration and processing of information like those found in the CNS. This is why the ENS is sometimes referred to as the "minibrain-in-the-gut."

Myenteric and Submucous Plexuses Are Parts of the ENS

The ENS consists of ganglia, primary interganglionic fiber tracts, and secondary and tertiary fiber projections to the effector systems (i.e., musculature, glands, and blood vessels). These structural components of the ENS are interlaced to form a plexus. Two ganglionated plexuses are the most obvious constituents of the ENS (see Fig. 26.1). The **myenteric plexus**, also known as **Auerbach's plexus**, is located between the longitudinal and circular muscle layers of most of the digestive tract. The **submucous plexus**, also known as **Meissner's plexus**, is situated in the submucosal region between the circular muscle and mucosa. The submucous plexus is most prominent as a ganglionated network in the small and large intestines. It does not exist as a ganglionated plexus in the esophagus and is sparse in the submucosal space of the stomach.

Motor innervation of the intestinal crypts and villi originates in the submucous plexus. Neurons in submucosal ganglia send fibers to the myenteric plexus and also receive synaptic input from axons projecting from the myenteric plexus. The interconnections link the two networks into a functionally integrated nervous system.

Sensory Neurons, Interneurons, and Motor Neurons Form the Microcircuits of the ENS

The heuristic model for the ENS is the same as that for the brain and spinal cord (Fig. 26.10). In fact, the ENS has as many neurons as the spinal cord. Like the CNS, sensory neurons, interneurons, and motor neurons in the ENS are connected synaptically for the flow of information from sensory neurons to interneuronal integrative networks to motor neurons to effector systems. The ENS organizes and coordinates the activity of each effector system into meaningful behavior of the integrated organ. Bidirectional communication occurs between the central and enteric nervous systems.

SYNAPTIC TRANSMISSION

Multiple kinds of synaptic transmission occur in the microcircuits of the ENS. Both fast synaptic potentials with durations less than 50 msec and slow synaptic potentials lasting several seconds can be recorded in cell bodies of enteric ganglion cells. These synaptic events may be excitatory postsynaptic potentials (EPSPs) or inhibitory postsynaptic potentials (IPSPs). They can be evoked by experimental stimulation of presynaptic axons, or they may occur spontaneously. Presynaptic inhibitory and facilitatory events can involve axoaxonal, paracrine, or endocrine forms of transmission, and they occur at both fast and slow synaptic connections.

Figure 26.11 shows three kinds of synaptic events that occur in enteric neurons. The synaptic potentials in this illustration were evoked by placing fine stimulating electrodes on interganglionic fiber tracts of the myenteric or submucous plexus and applying electrical shocks to stimulate presynaptic axons and release the neurotransmitter at the synapse.

Enteric Slow EPSPs Have Specific Properties Mediated by Metabotropic Receptors

The slow EPSP in Figure 26.11 was evoked by repetitive shocks (5 Hz) applied to the fiber tract for 5 seconds. Slowly activating depolarization of the membrane potential with a time course lasting longer than 2 minutes after termination of the stimulus is apparent. Repetitive discharge of action potentials reflects enhanced neuronal excitability during the EPSP. The record shows hyperpolarizing after-potentials associated with the first four spikes of the train. As the slow EPSP develops, the hyperpolarizing after-potentials are suppressed and can be seen to recover at the end of the spike train as the EPSP subsides. Suppression of the after-potentials is part of the mechanism of slow synaptic excitation that permits the neuron to convert from low to high states of excitability.

Slow EPSPs are mediated by multiple chemical messengers acting at a variety of different **metabotropic receptors**. Different kinds of receptors, each of which mediates slow synaptic-like responses, are found in varied combinations

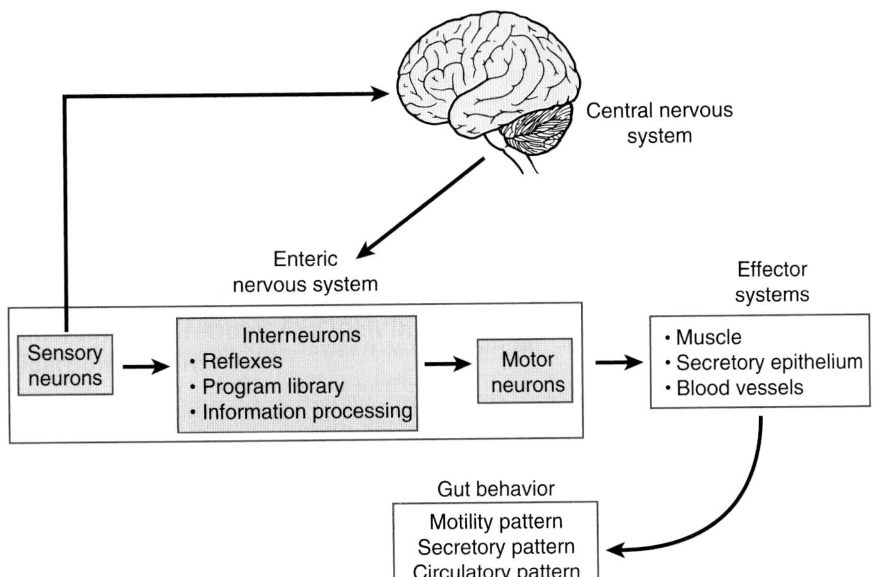

FIGURE 26.10 Enteric nervous system. Sensory neurons, interneurons, and motor neurons are synaptically interconnected to form the microcircuits of the ENS. As in the CNS, information flows from sensory neurons to interneuronal integrative networks to motor neurons to effector systems.

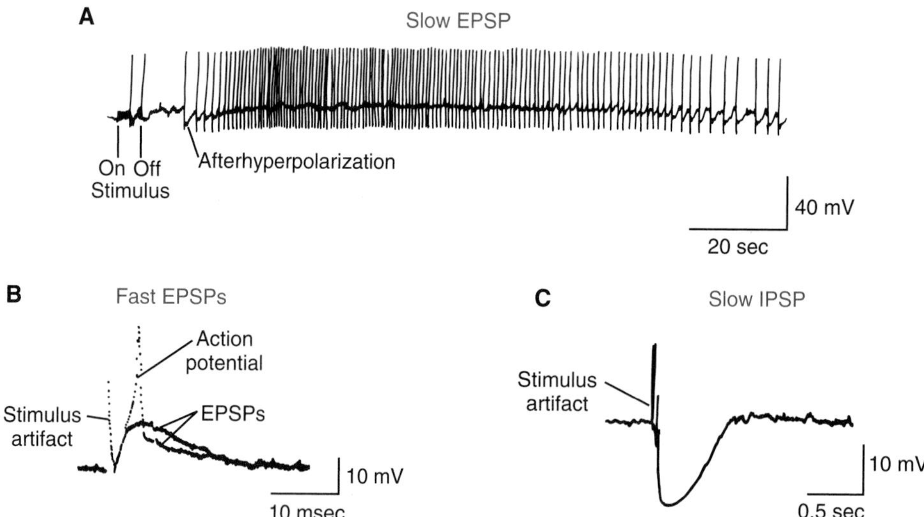

FIGURE 26.11 **Synaptic events in enteric neurons.** Slow EPSPs, fast EPSPs, and slow IPSPs all occur in enteric neurons. **A**, The slow EPSP was evoked by repetitive electrical stimulation of the synaptic input to the neuron. Slowly activating membrane depolarization of the membrane potential continues for almost 2 minutes after termination of the stimulus. During the slow EPSP, repetitive discharge of action potentials reflects enhanced neuronal excitability. **B**, The fast EPSPs were also evoked by single electrical shocks applied to the axon that synapsed with the recorded neuron. Two fast EPSPs were evoked by successive stimuli and are shown as superimposed records. Only one of the EPSPs reached the threshold for the discharge of an action potential. **C**, The slow IPSP was evoked by the stimulation of an inhibitory input to the neuron.

on each individual neuron. A common mode of signal transduction involves receptor activation of adenylyl cyclase and second messenger function of cAMP, which links several different chemical messages to the behavior of a common set of ionic channels responsible for generation of the slow EPSP responses. Serotonin, substance P, and acetylcholine (ACh) are examples of enteric neurotransmitters that evoke slow EPSPs. **Paracrine mediators** released from nonneural cells in the gut also evoke slow EPSP-like responses when released in the vicinity of the ENS. Histamine, for example, is released from mast cells during hypersensitivity reactions to antigens and acts at the histamine H_2 -receptor subtype to evoke slow EPSP-like responses in enteric neurons. Subpopulations of enteric neurons in specialized regions of the gut (e.g., the upper duodenum) have receptors for hormones, such as gastrin and cholecystokinin, that also evoke slow EPSP-like responses.

Slow EPSPs Are a Mechanism for Prolonged Neural Excitation or Inhibition of GI Effector Systems

The long-lasting discharge of spikes during the slow EPSP drives the release of neurotransmitter from the neuron's axon for the duration of the spike discharge. This may result in either prolonged excitation or inhibition at neuronal synapses and neuroeffector junctions in the gut wall.

Contractile responses within the musculature and secretory responses within the mucosal epithelium are slow events that span time courses of several seconds from start to completion. The train-like discharge of spikes during slow EPSPs is the neural correlate of long-lasting responses of the gut effectors during physiological stimuli. Figure 26.12 illustrates how the occurrence of slow EPSPs in excitatory motor neurons to the intestinal musculature or the mucosa results in prolonged contraction of the muscle or prolonged secretion from the crypts. The occurrence of slow EPSPs in inhibitory motor neurons to the musculature results in prolonged inhibition of contraction. This response is observed as a decrease in contractile tension.

Enteric Fast EPSPs Have Specific Properties Mediated by Ionotropic Receptors

Fast EPSPs (see Fig. 26.11B) are transient depolarizations of membrane potential that have durations of less than 50 msec. They occur in the enteric neural networks throughout the digestive tract. Most fast EPSPs are mediated by ACh acting at **ionotropic nicotinic receptors**. Ionotropic receptors are those coupled directly to ion channels. Fast EPSPs function in the rapid transfer and transformation of neurally coded information between the elements of the enteric microcircuits. They are "bytes" of information in the information-processing operations of the logic circuits.

Enteric Slow IPSPs Have Specific Properties Mediated by Multiple Chemical Receptors

The slow IPSP of Figure 26.11 was evoked by stimulation of an interganglionic fiber tract in the submucous plexus. This hyperpolarizing synaptic potential will suppress excitability (decrease the probability of spike discharge), compared with enhanced excitability during the slow EPSP.

Several different chemical messenger substances that may be peptidergic, purinergic, or cholinergic produce slow IPSP-like effects. Enkephalins, dynorphin, and morphine are all slow IPSP mimetics. This action is limited to subpopulations of neurons. Opiate receptors of the μ sub-

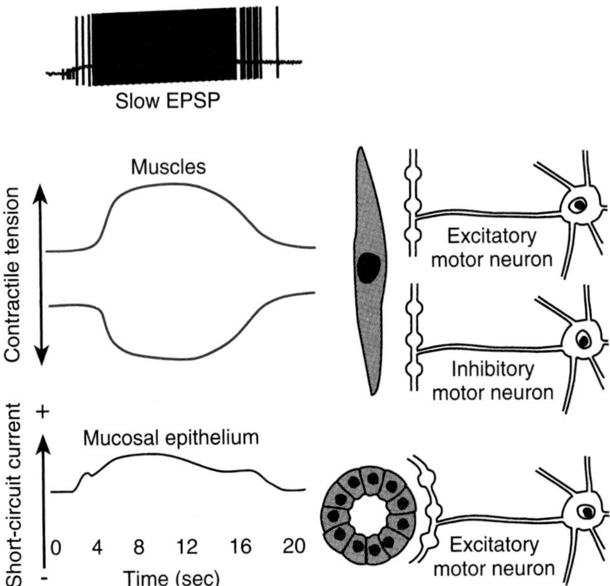

FIGURE 26.12 **The functional significance of slow EPSPs.** Slow EPSPs in excitatory motor neurons to the muscles or mucosal epithelium result in prolonged muscle contraction or mucosal crypt secretion. Stimulation of secretion in experiments is seen as an increase in ion movement (short-circuit current). Slow IPSPs in inhibitory motor neurons to the muscles result in prolonged inhibition of contractile activity, which is observed as decreased contractile tension.

type predominate on myenteric neurons in the small intestine; the receptors on neurons of the intestinal submucous plexus belong to the δ-opiate receptor subtype. The effects of opiates and opioid peptides are blocked by the antagonist naloxone. Addiction to morphine may be seen in enteric neurons, and withdrawal is observed as high-frequency spike discharge upon the addition of naloxone during chronic morphine exposure.

NE acts at β_2-adrenergic receptors to mimic slow IPSPs. This action occurs primarily in neurons of the submucous plexus that are involved in controlling mucosal secretion. The stimulation of sympathetic nerves evokes slow IPSPs that are blocked by α_2-adrenergic receptor antagonists in submucosal neurons. Slow IPSPs in submucosal neurons are a mechanism by which the sympathetic innervation suppresses intestinal secretion during physical exercise when blood is shunted from the splanchnic to systemic circulation.

Galanin is a 29-amino acid polypeptide that simulates slow synaptic inhibition when applied to any of the neurons of the myenteric plexus. The application of adenosine, ATP, or other purinergic analogs also mimics slow IPSPs. The inhibitory action of adenosine is at adenosine A_1 receptors. Inhibitory actions of adenosine A_1 agonists result from the suppression of the enzyme adenylyl cyclase and the reduction in intraneuronal cAMP.

Presynaptic Inhibitory Receptors Are Found at Enteric Synapses and Neuromuscular Junctions

Presynaptic inhibition (Fig. 26.13) is an important function at fast nicotinic synapses, at slow excitatory synapses, and at sympathetic inhibitory synapses in the neural networks of the submucous plexus and at excitatory neuromuscular junctions. It is a specialized form of neurocrine transmission whereby neurotransmitter released from an axon acts at receptors on a second axon to prevent the release of neurotransmitter from the second axon. Presynaptic inhibition, resulting from actions of paracrine or endocrine mediators on receptors at presynaptic release sites, is an alternative mechanism for modulating synaptic transmission.

Presynaptic inhibition in the ENS is mediated by multiple substances and their receptors, with variable combinations of the receptors involved at each release site. The chemical messenger substances may be peptidergic, aminergic, or cholinergic. NE acts at presynaptic α_2-adrenergic receptors to suppress fast EPSPs at nicotinic synapses, slow EPSPs, and cholinergic transmission at neuromuscular junctions. Serotonin suppresses both fast and slow EPSPs in the myenteric plexus. Opiates or opioid peptides suppress some fast EPSPs in the intestinal myenteric plexus.

ACh acts at muscarinic presynaptic receptors to suppress fast EPSPs in the myenteric plexus. This is a form of autoinhibition where ACh released at synapses with nicotinic postsynaptic receptors feeds back onto presynaptic

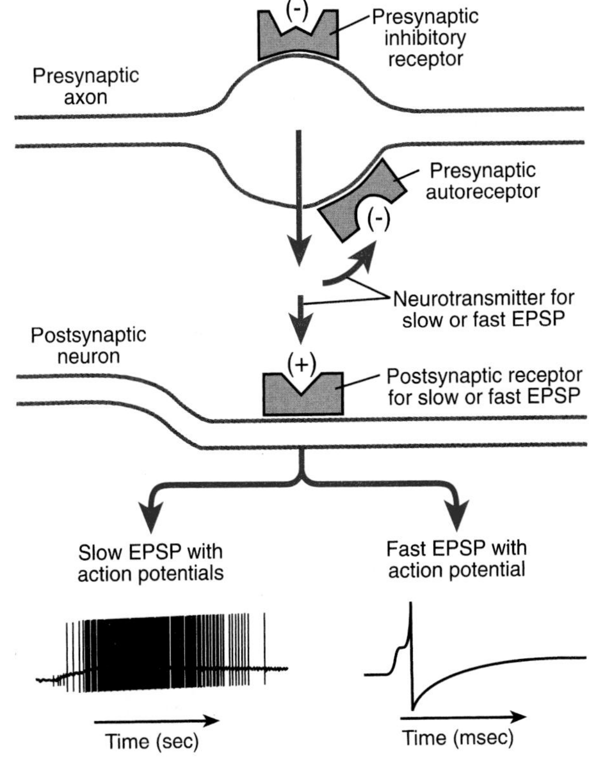

FIGURE 26.13 **Presynaptic inhibition.** Presynaptic inhibitory receptors are found on axons at neurotransmitter release sites for both slow and fast EPSPs. Different neurotransmitters act through the presynaptic inhibitory receptors to suppress axonal release of the transmitters for slow and fast EPSPs. Presynaptic autoreceptors are involved in a special form of presynaptic inhibition whereby the transmitter for slow or fast EPSPs accumulates at the synapse and acts on the autoreceptor to suppress further release of the neurotransmitter. (+), excitatory receptor; (−), inhibitory receptor.

CLINICAL FOCUS BOX 26.2

Chronic Intestinal Pseudoobstruction

Intestinal pseudoobstruction is characterized by symptoms of intestinal obstruction in the absence of a mechanical obstruction. The mechanisms for controlling orderly propulsive motility fail while the intestinal lumen is free from obstruction. This syndrome may result from abnormalities of the muscles or ENS. Its general symptoms of colicky abdominal pain, nausea and vomiting, and abdominal distension simulate mechanical obstruction.

Pseudoobstruction may be associated with degenerative changes in the ENS. Failure of propulsive motility reflects the loss of the neural networks that program and control the organized motility patterns of the intestine. This disorder can occur in varying lengths of intestine or in the entire length of the small intestine. Contractile behavior of the circular muscle is hyperactive but disorganized in the denervated segments. This behavior reflects the absence of inhibitory nervous control of the muscles, which are self-excitable when released from the braking action of enteric inhibitory motor neurons.

Paralytic ileus, another form of pseudoobstruction, is characterized by prolonged motor inhibition. The electrical slow waves are normal, but muscular action potentials and contractions are absent. Prolonged ileus commonly occurs after abdominal surgery. The ileus results from suppression of the synaptic circuits that organize propulsive motility in the intestine. A probable mechanism is presynaptic inhibition and the closure of synaptic gates (see Fig. 26.22).

Continuous discharge of the inhibitory motor neurons accompanies suppression of the motor circuits. This activity of the inhibitory motor neurons prevents the circular muscle from responding to electrical slow waves, which are undisturbed in ileus.

muscarinic receptors to suppress ACh release in negative-feedback fashion (see Fig. 26.13). Histamine acts at histamine H_3 presynaptic receptors to suppress fast EPSPs. Presynaptic inhibition mediated by paracrine or endocrine release of mediators is significant in pathophysiological states, such as inflammation. The release of histamine from intestinal mast cells in response to sensitizing allergens is an important example of paracrine-mediated presynaptic suppression in the enteric neural networks.

Presynaptic inhibition operates normally as a mechanism for selective shutdown or deenergizing of a microcircuit (see Clinical Focus Box 26.2). Preventing transmission among the neural elements of a circuit inactivates the circuit. For example, a major component of shutdown of gut function by the sympathetic nervous system involves the presynaptic inhibitory action of NE at fast nicotinic synapses.

Presynaptic Facilitation Enhances the Synaptic Release of Neurotransmitters and Increases the Amplitude of EPSPs

Presynaptic facilitation refers to an enhancement of synaptic transmission resulting from the actions of chemical mediators at neurotransmitter release sites on enteric axons (Fig. 26.14). The phenomenon is known to occur at fast excitatory synapses in the myenteric plexus of the small intestine and gastric antrum and at noradrenergic inhibitory synapses in the submucous plexus. It is also an action of cholecystokinin in the ENS of the gallbladder. Presynaptic facilitation is evident as an increase in amplitude of fast EPSPs at nicotinic synapses and reflects an enhanced ACh release from axonal release sites. At noradrenergic inhibitory synapses in the submucous plexus, it involves the elevation of cAMP in the postganglionic sympathetic fiber and appears as an enhancement of the slow IPSPs evoked by the stimulation of sympathetic postganglionic fibers.

Therapeutic agents that improve motility in the GI tract are known as **prokinetic drugs**. Presynaptic facilitation is the mechanism of action of some prokinetic drugs. Such drugs act to facilitate nicotinic transmission at the fast excitatory synapses in the enteric neural networks that control propulsive motor function. In both the stomach and the intestine, increases in EPSP amplitudes and rates of rise decrease the probability of transmission failure at the synapses, thereby increasing the speed of information transfer. This mechanism "energizes" the network circuits

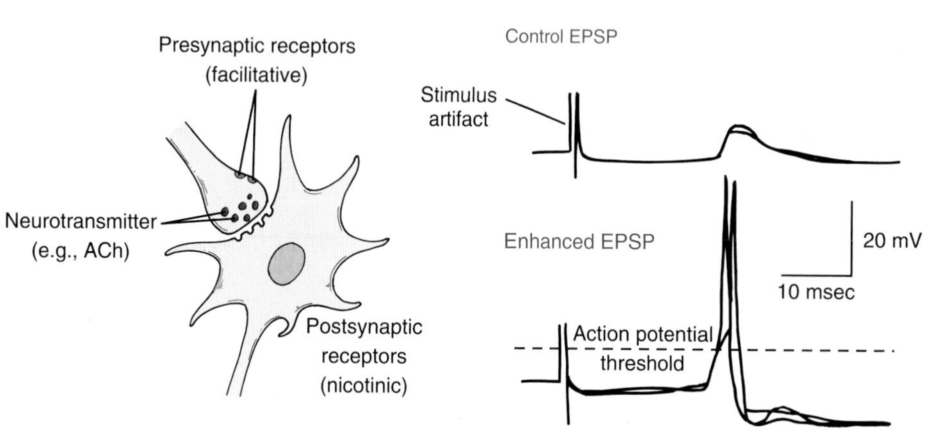

FIGURE 26.14 Presynaptic facilitation. Presynaptic facilitation enhances release of ACh and increases the amplitude of fast EPSPs at a nicotinic synapse.

and enhances propulsive motility (i.e., gastric emptying and intestinal transit).

ENTERIC MOTOR NEURONS

Motor neurons innervate the muscles of the digestive tract and, like spinal motor neurons, are the final pathways for signal transmission from the integrative microcircuits of the minibrain-in-the-gut (see Figs. 26.10 and 26. 15). The motor neuron pool of the ENS consists of excitatory and inhibitory neurons.

The **neuromuscular junction** is the site where neurotransmitters released from axons of motor neurons act on muscle fibers. Neuromuscular junctions in the digestive tract are simpler structures than the motor endplates of skeletal muscle (see Chapter 8). Most motor axons in the digestive tract do not release neurotransmitter from terminals as such; instead, release is from varicosities that occur along the axons. The neurotransmitter is released from the varicosities all along the axon during propagation of the action potential. Once released, the neurotransmitter diffuses over relatively long distances before reaching the muscle and/or interstitial cells of Cajal. This structural organization is an adaptation for the simultaneous application of a chemical neurotransmitter to a large number of muscle fibers from a small number of motor axons.

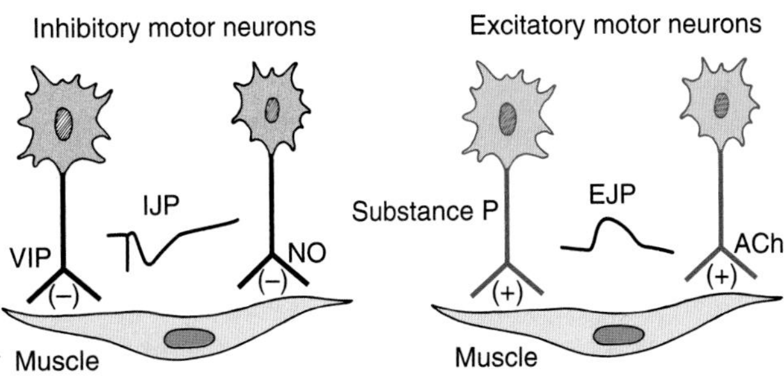

FIGURE 26.15 **Enteric motor neurons.** Motor neurons are final pathways from the ENS to the GI musculature. The motor neuron pool of the ENS consists of both excitatory and inhibitory neurons. Release of VIP or NO from inhibitory motor neurons evokes IJPs. Release of ACh or substance P from excitatory motor neurons evokes EJPs. VIP, vasoactive intestinal peptide; NO, nitric oxide; IJP, inhibitory junction potential; EJP, excitatory junction potential.

Excitatory Motor Neurons Evoke Muscle Contraction and Secretion in the Intestinal Crypts of Lieberkühn

Excitatory motor neurons release neurotransmitters that evoke contraction and increased tension in the GI muscles. ACh and substance P are the principal excitatory neurotransmitters released from enteric motor neurons to the musculature.

Two mechanisms of excitation-contraction coupling are involved in the neural initiation of muscle contraction in the GI tract. Transmitters from excitatory motor axons may trigger muscle contraction by depolarizing the muscle membrane to the threshold for the discharge of action potentials or by the direct release of calcium from intracellular stores. Neurally evoked depolarizations of the muscle membrane potential are called **excitatory junction potentials** (EJPs) (see Fig. 26.15). Direct release of calcium by the neurotransmitter fits the definition of pharmacomechanical coupling. In this case, occupation of receptors on the muscle plasma membrane by the neurotransmitter leads to the release of intracellular calcium, with calcium-triggered contraction independent of any changes in membrane electrical activity.

Cell bodies of the excitatory motor neurons are present in the myenteric plexus. In the small and large intestines, they project in the aboral direction to innervate the circular muscle.

Secretomotor neurons excite secretion of H_2O, electrolytes, and mucus from the crypts of Lieberkühn. ACh and VIP are the principal excitatory neurotransmitters. The cell bodies of secretomotor neurons are in the submucosal plexus. Excitation of these neurons, for example, by histamine release from mast cells during allergic responses, can lead to **neurogenic secretory diarrhea**. Suppression of excitability, for example, by morphine or other opiates, can lead to constipation.

Inhibitory Motor Neurons Suppress Muscle Contraction

Inhibitory neurotransmitters released from inhibitory motor neurons activate receptors on the muscle plasma membranes to produce **inhibitory junction potentials** (IJPs) (see Fig. 26.15). IJPs are hyperpolarizing potentials that move the membrane potential away from the threshold for the discharge of action potentials and, thereby, reduce the excitability of the muscle fiber. Hyperpolarization during IJPs prevents depolarization to the action potential threshold by the electrical slow waves and suppresses propagation of action potentials among neighboring muscle fibers within the electrical syncytium.

Early evidence suggested a purine nucleotide, possibly ATP, as the inhibitory transmitter released by enteric inhibitory motor neurons. Consequently, the term *purinergic neuron* temporarily became synonymous with enteric inhibitory motor neuron. The evidence for ATP as the inhibitory transmitter is now combined with evidence for vasoactive intestinal peptide (VIP), pituitary adenylyl cyclase–activating peptide, and nitric oxide (NO) as inhibitory transmitters. Enteric inhibitory motor neurons with VIP and/or NO synthase innervate the circular muscle of the stomach, intestines, gallbladder, and the various sphincters. Cell bodies of inhibitory motor neurons are present in the myenteric plexus. In the stomach and small and large intestines, they project in the aboral direction to innervate the circular muscle.

The longitudinal muscle layer of the small intestine does not appear to have inhibitory motor innervation. In contrast to the circular muscle, where inhibitory neural control is essential, enteric neural control of the longitudinal muscle during peristalsis may be exclusively excitatory.

Inhibitory Motor Neurons Control the Myogenic Intestinal Musculature

The need for inhibitory neural control is determined by the specialized physiology of the musculature. As mentioned earlier, the intestinal musculature behaves like a self-excitable electrical syncytium as a result of cell-to-cell communication across gap junctions and the presence of a pacemaker system. Action potentials triggered anywhere in the muscle will spread from muscle fiber to muscle fiber in three dimensions throughout the syncytium, which can be the entire length of the bowel. Action potentials trigger contractions as they spread. A nonneural pacemaker system of electrical slow waves (i.e., interstitial cells of Cajal) accounts for the self-excitable characteristic of the electrical syncytium. In the integrated system, the electrical slow waves are an extrinsic factor to which the circular muscle responds.

Why does the circular muscle fail to respond with action potentials and contractions to all slow wave cycles? Why don't action potentials and contractions spread in the syncytium throughout the entire length of intestine each time they occur? Answers to these questions lie in the functional significance of enteric inhibitory motor neurons.

Inhibitory Motor Neurons to the Circular Muscle. Figure 26.16A shows the spontaneous discharge of action potentials occurring in bursts, as recorded extracellularly from a neuron in the myenteric plexus of the small intestine. This kind of continuous discharge of action potentials by subsets of intestinal inhibitory motor neurons occurs in all mammals. The result is continuous inhibition of myogenic activity because, in intestinal segments where neuronal discharge in the myenteric plexus is prevalent, muscle action potentials and associated contractile activity are absent or occur only at reduced levels with each electrical slow wave. The continuous release of the inhibitory neurotransmitters VIP and NO can be detected in intestinal preparations in this case. When the inhibitory neuronal discharge is blocked experimentally with tetrodotoxin, every cycle of the electrical slow wave triggers an intense discharge of action potentials. Figure 26.16B shows how phasic contractions, occurring at slow wave frequency, progressively increase to maximal amplitude during a blockade of inhibitory neural activity after the application of tetrodotoxin in the small intestine. This response coincides with a progressive increase in baseline tension.

Tetrodotoxin is an effective pharmacological tool for demonstrating ongoing inhibition because it selectively blocks neural activity without affecting the muscle. This action is a result of a selective blockade of sodium channels in neurons. The rising phase of the muscle action potentials is caused by an inward calcium current that is unaffected by tetrodotoxin.

As a general rule, any treatment or condition that removes or inactivates inhibitory motor neurons results in tonic contracture and continuous, uncoordinated contractile activity of the circular musculature. Several circumstances that remove the inhibitory neurons are associated with conversion from a hypoirritable condition of the circular muscle to a hyperirritable state. These include the application of local anesthetics, hypoxia from restricted blood flow to an intestinal segment, an autoimmune attack on enteric neurons, congenital absence in Hirschsprung's disease, treatment with opiate drugs, and inhibition of NO synthase (see Clinical Focus Boxes 26.3 and 26.4).

Inhibitory Motor Neurons and the Strength of Contractions Evoked by Electrical Slow Waves. The strength of circular muscle contraction evoked by each slow wave cycle is a function of the number of inhibitory motor neurons in an active state. The circular muscle in an intestinal segment can respond to the electrical slow waves only when the inhibitory motor neurons are inactivated by inhibitory synaptic input from other neurons in the control circuits. This means that inhibitory neurons determine when the constantly running slow waves initiate a contraction, as well as the strength of the contraction that is initiated by each slow wave cycle. The strength of each contraction is determined by the proportion of muscle fibers in the population that can respond during a given slow wave cycle, which, in turn, is determined by the proportion exposed to inhibitory transmitters released by motor neurons. With maximum inhibition, no contractions can occur in response to a slow wave (see Fig.26.4A); contractions of maximum strength occur after all inhibition is removed and all of the muscle fibers in a segment are activated by each slow wave cycle (see Fig. 26.4B). Contractions between the two extremes are graded in strength according to the number of

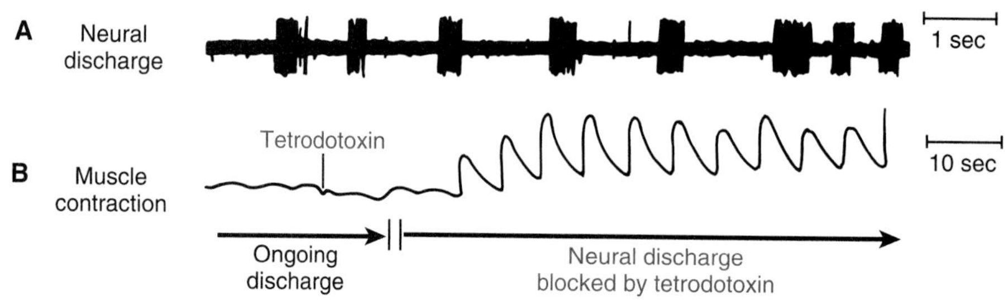

FIGURE 26.16 **Inhibitory motor neurons.** Ongoing firing of a subpopulation of inhibitory motor neurons to the intestinal circular muscle prevents electrical slow waves from triggering the action potentials that trigger contractions. When the inhibitory neural discharge is blocked with tetrodotoxin, every cycle of the electrical slow wave triggers discharge of action potentials and large-amplitude contractions. **A**, Electrical record of ongoing burst-like firing. **B**, Record of muscle contractile activity before and after application of tetrodotoxin.

CLINICAL FOCUS BOX 26.3

Hirschsprung's Disease and Incontinence: Motor Disorders of the Large Intestine and Anorectum

Hirschsprung's disease is a developmental disorder that is present at birth but may not be diagnosed until later childhood. It is characterized by defecation difficulty or failure. The disease is often called **congenital megacolon**, because the proximal colon may become grossly enlarged with impacted feces, or **congenital aganglionosis**, because the ganglia of the ENS fail to develop in the terminal region of the large intestine. Mutations in RET or endothelin genes account for the disease in some patients.

Enteric neurons may be absent in the rectosigmoid region only, in the descending colon, or in the entire colon. The aganglionic region appears constricted as a result of continuous contractile activity of the circular muscle, whereas the normally innervated intestine proximal to the aganglionic segment is distended with feces.

The constricted terminal segment of the large intestine in Hirschsprung's disease presents a functional obstruction to the forward passage of fecal material. Constriction and narrowing of the lumen of the segment reflects uncontrolled myogenic contractile activity in the absence of inhibitory motor neurons.

Incontinence is an inappropriate leakage of feces and flatus to a degree that it disables the patient by disrupting routine daily activities. As discussed later, the mechanisms for maintaining continence involve the coordinated interactions of several different components. Consequently, sensory malfunction, incompetence of the internal anal sphincter, or disorders of neuromuscular mechanisms of the external sphincter and pelvic floor muscles can be factors in the pathophysiology of incontinence.

Sensory malfunction renders the patient unaware of the filling of the rectum and stimulation of the anorectum, in which case he or she does not perceive the need for voluntary control over the muscular mechanisms of continence. This condition is tested clinically by distending an intrarectal balloon. The healthy subject will perceive the distension with an instilled volume of 15 mL or less, whereas the sensory-deprived patient either will not report any sensation at all or will require much larger volumes before becoming aware of the distension.

Incompetence of the internal anal sphincter is usually related to a surgical or mechanical factor or perianal disease, such as prolapsing hemorrhoids. Disorders of the neuromuscular mechanisms of the external sphincter and pelvic floor muscles may also result from surgical or mechanical trauma, such as during childbirth.

Physiological deficiencies of the skeletal motor mechanisms can be a significant factor in the common occurrence of incontinence in older adults. Whereas the resting tone of the internal anal sphincter does not seem to decrease with age, the strength of contraction of the external anal sphincter does weaken. Moreover, the striated muscles of the external anal sphincter and pelvic floor lose contractile strength with age. This condition occurs in parallel with a deterioration of nervous function, reflected by decreased conduction velocity in fibers of the pelvic nerves. Clinical examination with intra-anal manometry reveals a decreased ability of the patient with disordered voluntary muscle function to increase intra-anal pressure when asked to "squeeze" the intra-anal catheter.

inhibitory motor neurons that are inactivated by the ENS minibrain during each slow wave.

Control by Inhibitory Motor Neurons of the Length of Intestine Occupied by a Contraction and the Direction of Propagation of Contractions. The state of activity of inhibitory motor neurons determines the length of a contracting segment by controlling the distance of spread of action potentials within the three-dimensional electrical geometry of the muscular syncytium (Fig. 26.17). This occurs coincidently with control of contractile strength. Contractions can only occur in segments where ongoing inhibition has been inactivated, while it is prevented in adjacent segments where the inhibitory innervation is ac-

CLINICAL FOCUS BOX 26.4

Dysphagia, Diffuse Spasm, and Achalasia: Motor Disorders of the Esophagus

Failure of peristalsis in the esophageal body or failure of the lower esophageal sphincter to relax will result in **dysphagia** or difficulty in swallowing. Some people show abnormally high pressure waves as peristalsis propagates past the recording ports on manometric catheters. This condition, called **nutcracker esophagus**, is sometimes associated with chest pain that may be experienced as angina-like pain.

In **diffuse spasm**, organized propagation of the peristaltic behavioral complex fails to occur after a swallow. Instead, the act of swallowing results in simultaneous contractions all along the smooth muscle esophagus. On manometric tracings, this response is observed as a synchronous rise in intraluminal pressure at each of the recording sensors.

In **achalasia** of the lower esophageal sphincter, the sphincter fails to relax normally during a swallow. As a result, the ingested material does not enter the stomach and accumulates in the body of the esophagus. This leads to **megaesophagus**, in which distension and gross enlargement of the esophagus are evident. In advanced untreated cases of achalasia, peristalsis does not occur in response to a swallow.

Achalasia is a disorder of inhibitory motor neurons in the lower esophageal sphincter. The number of neurons in the lower esophageal sphincter is reduced, and the levels of the inhibitory neurotransmitter VIP and the enzyme NO synthase are diminished. This degenerative disease results in a loss of the inhibitory mechanisms for relaxing the sphincter with appropriate timing for a successful swallow.

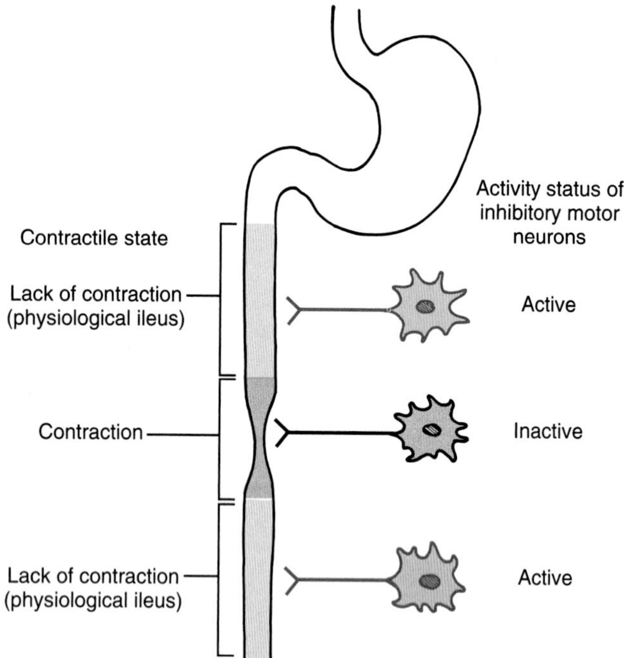

FIGURE 26.17 Inhibitory control of the intestinal musculature. Myogenic contraction occurs in segments of intestine where inhibitory motor neurons are inactive. Physiological ileus occurs in segments of intestine where the inhibitory neurons are actively firing.

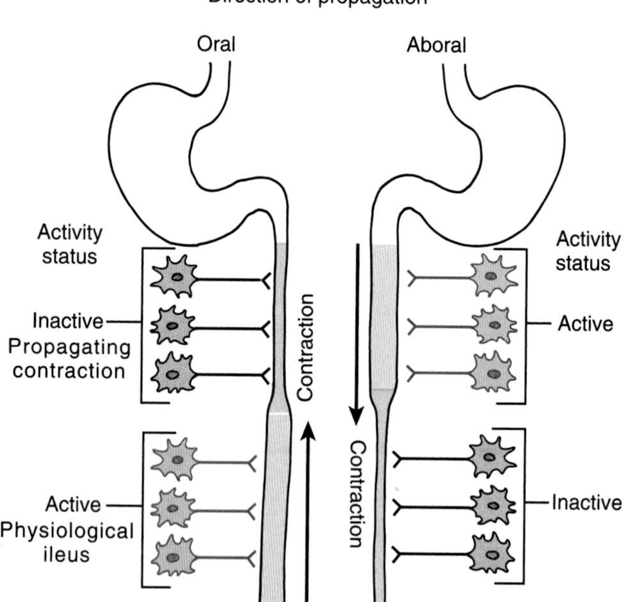

FIGURE 26.18 Inhibitory control of the direction of propagation of contractions. Contractions propagate into intestinal segments where inhibitory motor neurons are inactivated. Sequential inactivation in the oral direction permits oral propagation of contractions. Sequential inactivation in the aboral direction permits aboral propagation.

tive. The oral and aboral boundaries of a contracted segment reflect the transition zone from inactive to active inhibitory motor neurons. This is the mechanism by which the ENS generates short contractile segments during the digestive (mixing) pattern of small intestinal motility and longer contractile segments during propulsive motor patterns, such as "power propulsion" that travels over extended distances along the intestine.

As a result of the functional syncytial properties of the musculature, inhibitory motor neurons are necessary for control of the direction in which contractions travel along the intestine. The directional sequence in which inhibitory motor neurons are inactivated determines whether contractions propagate in the oral or aboral direction (Fig. 26.18). Normally, the neurons are inactivated sequentially in the aboral direction, resulting in contractile activity that propagates and moves the intraluminal contents distally. During vomiting, the integrative microcircuits of the ENS inactivate inhibitory motor neurons in a reverse sequence, allowing small intestinal propulsion to travel in the oral direction and propel the contents toward the stomach (see Clinical Focus Box 26.5).

The Inhibitory Innervation of GI Sphincters Is Transiently Activated for Timed Opening and the Passage of Luminal Contents

The circular muscle of sphincters remains tonically contracted to occlude the lumen and prevent the passage of contents between adjacent compartments, such as between stomach and esophagus. Inhibitory motor neurons are normally inactive in the sphincters and are switched on with timing appropriate to coordinate the opening of the sphincter with physiological events in adjacent regions

CLINICAL FOCUS BOX 26.5

Emesis

During **emesis** (vomiting), powerful propulsive peristalsis starts in the midjejunum and travels to the stomach. As a result, the small intestinal contents are propelled rapidly and continuously toward the stomach. As the propulsive complex advances, the gastroduodenal junction and the stomach wall relax, allowing passage of the intestinal contents into the stomach. At the same time, the longitudinal muscle of the esophagus and the gastroesophageal junction dilates. The overall result is the formation of a funnel-like cavity that allows the free flow of gastric contents into the esophagus as intra-abdominal pressure is increased by contraction of the diaphragm and abdominal muscles during retching.

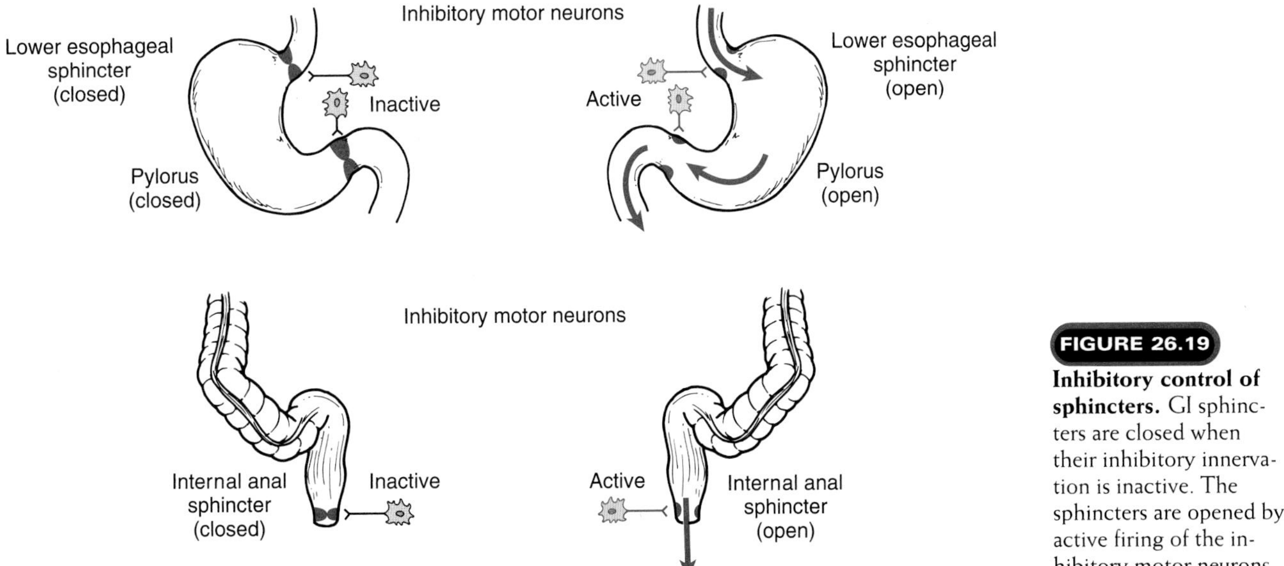

FIGURE 26.19 **Inhibitory control of sphincters.** GI sphincters are closed when their inhibitory innervation is inactive. The sphincters are opened by active firing of the inhibitory motor neurons.

(Fig. 26.19). When this occurs, the inhibitory neurotransmitter relaxes the ongoing muscle contraction in the sphincteric muscle and prevents excitation and contraction in the adjacent muscle from spreading into and closing the sphincter.

BASIC PATTERNS OF GI MOTILITY

Motility in the digestive tract accounts for the propulsion, mixing, and reservoir functions necessary for the orderly processing of ingested food and the elimination of waste products. **Propulsion** is the controlled movement of ingested foods, liquids, GI secretions, and sloughed cells from the mucosa through the digestive tract. It moves the food from the stomach into the small intestine and along the small intestine, with appropriate timing for efficient digestion and absorption. Propulsive forces move undigested material into the large intestine and eliminate waste through defecation. **Trituration**, the crushing and grinding of ingested food by the stomach, decreases particle size, increasing the surface area for action by digestive enzymes in the small intestine. **Mixing movements** blend pancreatic, biliary, and intestinal secretions with nutrients in the small intestine and bring products of digestion into contact with the absorptive surfaces of the mucosa. **Reservoir functions** are performed by the stomach and colon. The body of the stomach stores ingested food and exerts steady mechanical forces that are important determinants of gastric emptying. The colon holds material during the time required for the absorption of excess water and stores the residual material until defecation is convenient.

Each of the specialized organs along the digestive tract exhibits a variety of motility patterns. These patterns differ depending on factors such as time after a meal, awake or sleeping state, and the presence of disease. Motor patterns that accomplish propulsion in the esophagus and small and large intestines are derived from a basic peristaltic reflex circuit in the ENS.

Peristalsis Is a Stereotyped Propulsive Motor Reflex

Peristalsis is the organized propulsion of material over variable distances within the intestinal lumen. The muscle layers of the intestine behave in a stereotypical pattern during peristaltic propulsion (Fig. 26.20). This pattern is determined by the integrated circuits of the ENS. During peristalsis, the longitudinal muscle layer in the segment ahead of the advancing intraluminal contents contracts while the circular muscle layer simultaneously relaxes. The intestinal tube behaves like a cylinder with constant surface area. The shortening of the longitudinal axis of the cylinder is accompanied by a widening of the cross-sectional diameter. The simultaneous shortening of the longitudinal muscle and relaxation of the circular muscle results in expansion of the lumen, which prepares a **receiving segment** for the forward-moving intraluminal contents during peristalsis.

The second component of stereotyped peristaltic behavior is contraction of the circular muscle in the segment behind the advancing intraluminal contents. The longitudi-

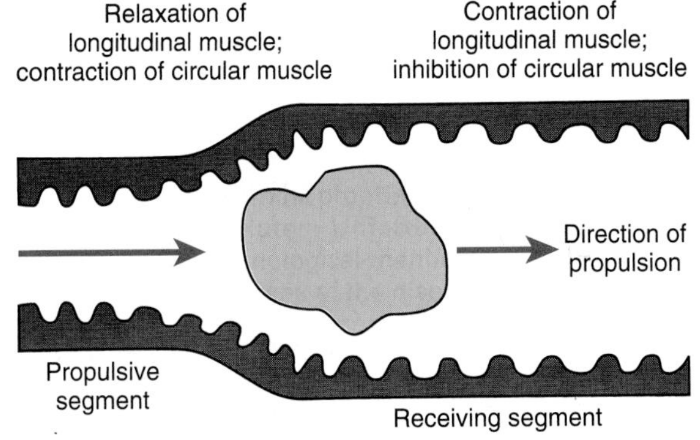

FIGURE 26.20 **Peristaltic propulsion.** Peristaltic propulsion involves formation of a propulsive and a receiving segment, mediated by reflex control of the intestinal musculature.

nal muscle layer in this segment relaxes simultaneously with contraction of the circular muscle, resulting in the conversion of this region to a **propulsive segment** that propels the luminal contents ahead, into the receiving segment. Intestinal segments ahead of the advancing front become receiving segments and then propulsive segments in succession as the peristaltic complex of propulsive and receiving segments travels along the intestine.

A Polysynaptic Reflex Circuit Determines Peristalsis

The peristaltic reflex (i.e., the formation of propulsive and receiving segments) can be triggered experimentally by distending the intestinal wall or by "brushing" the mucosa. Involvement of the reflex in the neural organization of peristaltic propulsion is similar to the reflex behavior mediated by the CNS for somatic movements of skeletal muscles. Reflex circuits with fixed connections in the spinal cord automatically reproduce a stereotypical pattern of behavior each time the circuit is activated (e.g., the myotatic reflex; see Chapter 5). Connections for the reflex remain, irrespective of the destruction of adjacent regions of the spinal cord. The peristaltic reflex circuit is similar, but the basic circuit is repeated along and around the intestine. Just as the monosynaptic reflex circuit of the spinal cord is the terminal circuit for the production of almost all skeletal muscle movements (see Chapter 5), the same basic peristaltic circuitry underlies all patterns of propulsive motility. Blocks of the same basic circuit are connected in series along the length of the intestine and repeated in parallel around the circumference. The basic peristaltic circuit consists of synaptic connections between sensory neurons, interneurons, and motor neurons. Distances over which peristaltic propulsion travels are determined by the number of blocks recruited in sequence along the bowel. Synaptic gates between blocks of the basic circuit determine whether or not recruitment occurs for the next circuit in the sequence.

The basic circuit for peristalsis is repeated serially along the intestine (Fig. 26.21). Synaptic gates connect the blocks of basic circuitry and provide a mechanism for controlling the distance over which the peristaltic behavioral complex travels. When the gates are opened, neural signals pass between successive blocks of the basic circuit, resulting in propagation of the peristaltic event over extended distances. Long-distance propulsion is prevented when all gates are closed (see Clinical Focus Box 26.1).

Presynaptic mechanisms are involved in gating the transfer of signals between sequentially positioned blocks of peristaltic reflex circuitry. Synapses between the neurons that carry excitatory signals to the next block of circuitry function as gating points for controlling the distance over which peristaltic propulsion travels (Fig. 26.22). Messenger substances that act presynaptically to inhibit the release of transmitter at the excitatory synapses close the gates to the transfer of information, determining the distance of propagation. Drugs that facilitate the release of neurotransmitters at the excitatory synapses have therapeutic application by increasing the probability of information transfer at the synaptic gates, enhancing propulsive motility.

Peristaltic Propulsion in the Upper Small Intestine During Vomiting. The enteric neural circuits can be programmed to produce peristaltic propulsion in either direction along the intestine. If forward passage of the intraluminal contents is impeded in the large intestine, reverse peristalsis propels the bolus over a variable distance away from the obstructed segment. **Retroperistalsis** then stops and forward peristalsis moves the bolus again in the direction of the obstruction. During the act of vomiting, retroperistalsis occurs in the small intestine. In this case, as well as in the obstructed intestine, the coordinated muscle behavior of peristalsis is the same except that it is organized by the nervous system to travel in the oral direction (see Clinical Focus Box 26.5).

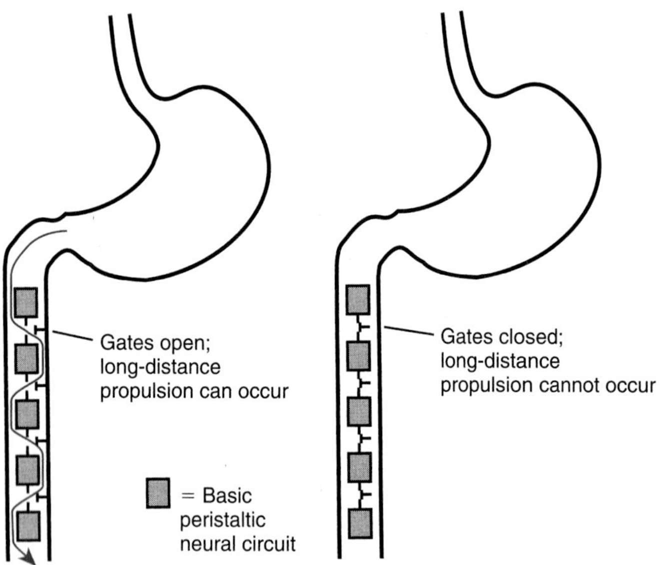

FIGURE 26.21 **Operation of synaptic gates between basic blocks of peristaltic circuitry.** Opening the gates between successive blocks of the basic circuit results in extended propagation of the propulsive event. Long-distance propulsion is prevented when all gates are closed.

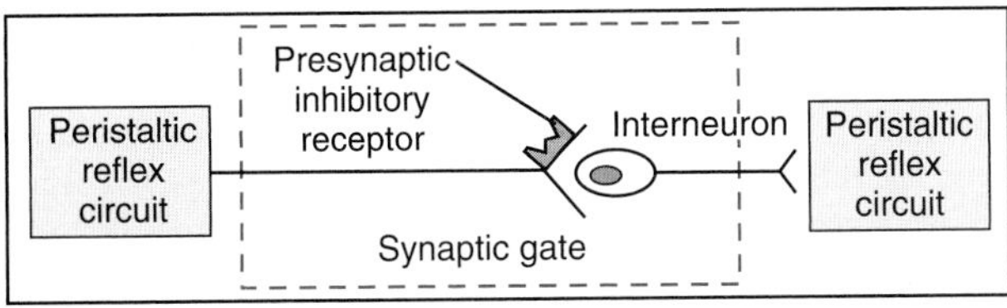

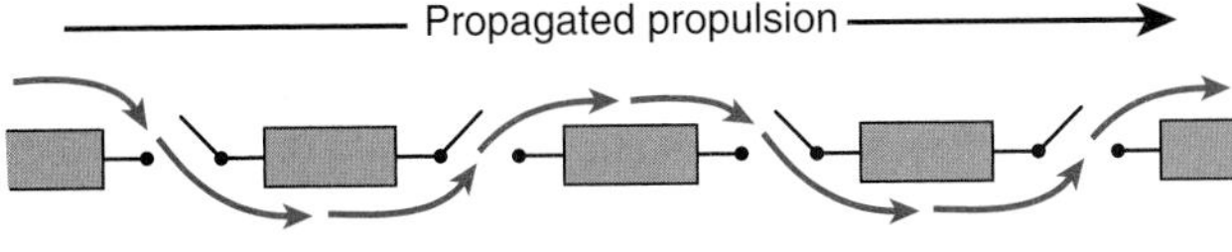

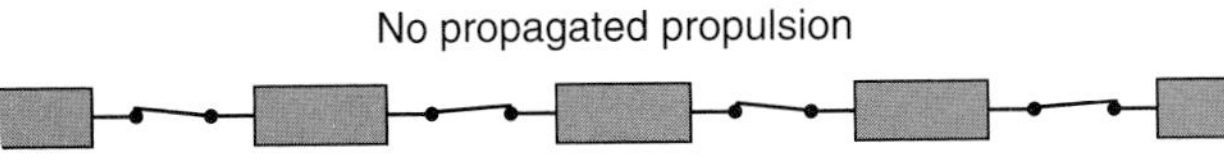

FIGURE 26.22 Control of the distance and direction of peristaltic propulsion. Synaptic gates determine distance and direction of propagation of propulsive motility. Presynaptic inhibitory receptors determine the open and closed states of the gates. When the gating synapses are uninhibited (i.e., no presynaptic inhibition), propagation proceeds in the direction in which the gates are open. The gates are closed by activation of presynaptic inhibitory receptors.

Ileus Reflects the Operation of a Program in the ENS

Physiological ileus is the absence of motility in the small and large intestine. It is a fundamental behavioral state of the intestine in which quiescence of motor function is neurally programmed. The state of physiological ileus disappears after ablation (removal) of the ENS. When enteric neural functions are destroyed by pathological processes, disorganized and nonpropulsive contractile behavior occurs continuously because of the myogenic electrical properties.

Quiescence of the intestinal circular muscle is believed to reflect the operation of a neural program in which all the gates within and between basic peristaltic circuits are held shut (see Fig. 26.22). In this state, the inhibitory motor neurons remain in a continuously active state and responsiveness of the circular muscle to the electrical slow waves is suppressed. This normal condition, physiological ileus, is in effect for varying periods of time in different intestinal regions, depending on such factors as the time after a meal.

The normal state of motor quiescence becomes pathological when the gates for the particular motor patterns are rendered inoperative for abnormally long periods. In this state of paralytic ileus, the basic circuits are locked in an inoperable state while unremitting activity of the inhibitory motor neurons suppresses myogenic activity (see Clinical Focus Box 26.2).

Sphincters Prevent the Reflux of Luminal Contents

Smooth muscle sphincters are found at the gastroesophageal junction, gastroduodenal junction, opening of the bile duct, ileocolonic junction, and termination of the large intestine in the anus. They consist of rings of smooth muscle that remain in a continuous state of contraction. The effect of the tonic contractile state is to occlude the lumen in a region that separates two specialized compartments. With the exception of the internal anal sphincter, sphincters function to prevent the backward movement of intraluminal contents.

The **lower esophageal sphincter** prevents the reflux of gastric acid into the esophagus. Incompetence results in chronic exposure of the esophageal mucosa to acid, which can lead to heartburn and dysplastic changes that may become cancerous. The **gastroduodenal sphincter** or **pyloric sphincter** prevents the excessive reflux of duodenal contents into the stomach. Incompetence of this sphincter can result in the reflux of bile acids from the duodenum. Bile acids are damaging to the protective barrier in the gastric mucosa; prolonged exposure can lead to gastric ulcers.

The **sphincter of Oddi** surrounds the opening of the bile duct as it enters the duodenum. It acts to prevent the reflux of intestinal contents into the ducts leading from the liver, gallbladder, and pancreas. Failure of this sphincter to open leads to distension, which is associated with the biliary tract pain that is felt in the right upper abdominal quadrant.

The **ileocolonic sphincter** prevents the reflux of colonic contents into the ileum. Incompetence can allow the entry of bacteria into the ileum from the colon, which may result in bacterial overgrowth. Bacterial counts are normally low in the small intestine. The **internal anal sphincter** prevents the uncontrolled movement of intraluminal contents through the anus.

The ongoing contractile tone in the smooth muscle sphincters is generated by **myogenic mechanisms**. The contractile state is an inherent property of the muscle and independent of the nervous system. Transient relaxation of the sphincter to permit the forward passage of material is accomplished by activation of inhibitory motor neurons (see Fig. 26.19). **Achalasia** is a pathological state in which smooth muscle sphincters fail to relax. Loss of the ENS and its complement of inhibitory motor neurons in the sphincters can underlie achalasia (see Clinical Focus Box 26.4).

MOTILITY IN THE ESOPHAGUS

The esophagus is a conduit for the transport of food from the pharynx to the stomach. Transport is accomplished by peristalsis, with propulsive and receiving segments produced by neurally organized contractile behavior of the longitudinal and circular muscle layers.

The esophagus is divided into three functionally distinct regions: the upper esophageal sphincter, the esophageal body, and the lower esophageal sphincter. Motor behavior of the esophagus involves striated muscle in the upper esophagus and smooth muscle in the lower esophagus.

Peristalsis and Relaxation of the Lower Esophageal Sphincter Are the Main Motility Events in the Esophagus

Esophageal peristalsis may occur as primary peristalsis or secondary peristalsis. **Primary peristalsis** is initiated by the voluntary act of swallowing, irrespective of the presence of food in the mouth. **Secondary peristalsis** occurs when the primary peristaltic event fails to clear the bolus from the body of the esophagus. It is initiated by activation of mechanoreceptors and can be evoked experimentally by distending a balloon in the esophagus.

When not involved in the act of swallowing, the muscles of the esophageal body are relaxed and the lower esophageal sphincter is tonically contracted. In contrast to the intestine, the relaxed state of the esophageal body is not produced by the ongoing activity of inhibitory motor neurons. Excitability of the muscle is low and there are no electrical slow waves to trigger contractions. The activation of excitatory motor neurons rather than myogenic mechanisms accounts for the coordinated contractions of the esophagus during a swallow.

Manometric Catheters Monitor Esophageal Motility and Diagnose Disordered Motility

Esophageal motor disorders are diagnosed clinically with **manometric catheters**, multiple small catheters fused into a single assembly with pressure sensors positioned at various levels (see Clinical Focus Box 26.4). They are placed into the esophagus via the nasal cavity. Manometric catheters record a distinctive pattern of motor behavior following a swallow (Fig. 26.23). At the onset of the swallow, the lower esophageal sphincter relaxes. This is recorded as a fall in pressure in the sphincter that lasts throughout the swallow and until the esophagus empties its contents into the stomach. Signals for relaxation of the lower esophageal sphincter are transmitted by the vagus nerves. The pressure-sensing ports along the catheter assembly show transient increases in pressure as the segment with the sensing port becomes the propulsive segment of the peristaltic pattern as it passes on its way to the stomach.

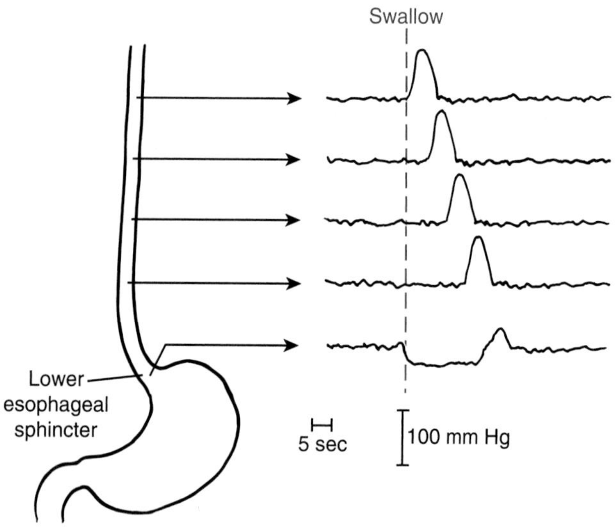

FIGURE 26.23 Manometric recordings of pressure events in the esophageal body and lower esophageal sphincter following a swallow. The propulsive segment of the peristaltic behavioral complex produces a positive pressure wave at each recording site in succession as it travels down the esophagus. Pressure falls in the lower esophageal sphincter shortly after the onset of the swallow, and the sphincter remains relaxed until the propulsive complex has transported the swallowed material into the stomach.

GASTRIC MOTILITY

The functional regions of the stomach do not correspond to the anatomic regions. The anatomic regions are the **fundus**, **corpus (body)**, **antrum**, and **pylorus** (Fig. 26.24). Functionally, the stomach is divided into a proximal **reservoir** and distal **antral pump** on the basis of distinct differences in motility between the two regions. The reservoir consists of the fundus and approximately one third of the corpus; the antral pump includes the caudal two thirds of the corpus, the antrum, and the pylorus.

Differences in motility between the reservoir and antral pump reflect adaptations for different functions. The muscles of the proximal stomach are adapted for maintaining continuous contractile tone (tonic contraction) and do not contract phasically. By contrast, the muscles of the antral pump contract phasically. The spread of phasic contractions in the region of the antral pump propels the gastric contents toward the gastroduodenal junction. Strong propulsive waves of this nature do not occur in the proximal stomach.

Motor Behavior of the Antral Pump Is Initiated by a Dominant Pacemaker

Gastric action potentials determine the duration and strength of the phasic contractions of the antral pump and

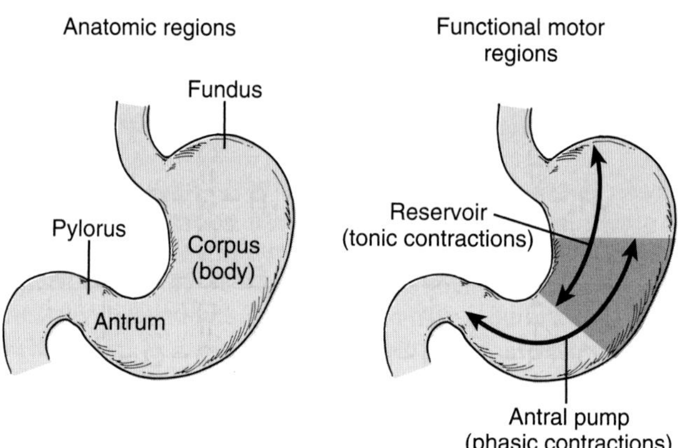

FIGURE 26.24 The stomach: three anatomic and two functional regions. The reservoir is specialized for receiving and storing a meal. The musculature in the region of the antral pump exhibits phasic contractions that function in the mixing and trituration of the gastric contents. No distinctly identifiable boundary exists between the reservoir and antral pump.

are initiated by a dominant pacemaker located in the corpus distal to the midregion. Once started at the pacemaker site, the action potentials propagate rapidly around the gastric circumference and trigger a ring-like contraction. The action potentials and associated ring-like contraction then travel more slowly toward the gastroduodenal junction.

Electrical syncytial properties of the gastric musculature account for the propagation of the action potentials from the pacemaker site to the gastroduodenal junction. The pacemaker region in humans generates action potentials and associated antral contractions at a frequency of 3/min. The gastric action potential lasts about 5 seconds and has a rising phase (depolarization), a plateau phase, and a falling phase (repolarization) (see Fig. 26.2).

The Gastric Action Potential Triggers Two Kinds of Contractions

The gastric action potential is responsible for two components of the propulsive contractile behavior in the antral pump. A **leading contraction**, with a relatively constant amplitude, is associated with the rising phase of the action potential, and a **trailing contraction**, of variable amplitude, is associated with the plateau phase (Fig. 26.25). Gastric action potentials are generated continuously by the pacemaker, but they do not trigger a trailing contraction when the plateau phase is reduced below threshold voltage. Trailing contractions appear when the plateau phase is above threshold. They increase in strength in direct relation to increases in the amplitude of the plateau potential above threshold.

The leading contractions produced by the rising phase of the gastric action potential have negligible amplitude as they propagate to the pylorus. As the rising phase reaches the terminal antrum and spreads into the pylorus, contraction of the pyloric muscle closes the orifice between the stomach and duodenum. The trailing contraction follows the leading contraction by a few seconds. As the trailing contraction approaches the closed pylorus, the gastric contents are forced into an antral compartment of ever-decreasing volume and progressively increasing pressure. This results in jet-like **retropulsion** through the orifice formed by the trailing contraction (Fig. 26.26). Trituration and reduction in particle size occur as the material is forcibly retropelled through the advancing orifice and back into the gastric reservoir to await the next propulsive cycle. Repetition at 3 cycles/min reduces particle size to the 1- to 7-mm range that is necessary before a particle can be emptied into the duodenum during the digestive phase of gastric motility.

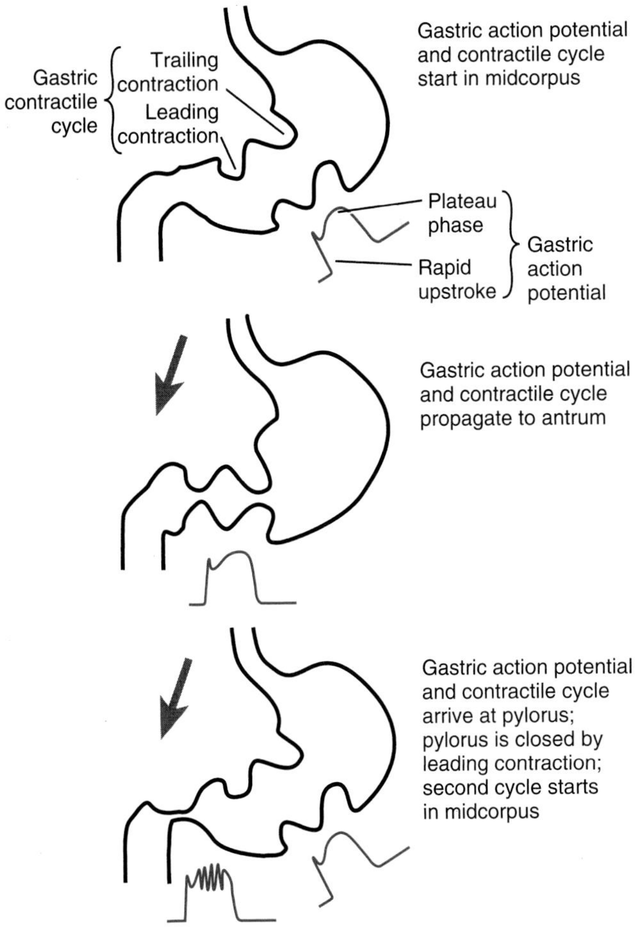

FIGURE 26.25 **Contractile cycle of the antral pump.** The rising phase of the gastric action potential accounts for the leading contraction that propagates toward the pylorus during one contractile cycle. The plateau phase accounts for the trailing contraction of the cycle.

Enteric Neurons Determine the Minute-to-Minute Strength of the Trailing Antral Contraction

The action potentials of the distal stomach are **myogenic** (i.e., an inherent property of the muscle) and occur in the absence of any neurotransmitters or other chemical messengers. The myogenic characteristics of the action potential are modulated by motor neurons in the gastric ENS. Neurotransmitters primarily affect the amplitude of the plateau phase of the action potential and, thereby, control the strength of the contractile event triggered by the plateau phase. Neurotransmitters, such as ACh from excitatory motor neurons, increase the amplitude of the plateau

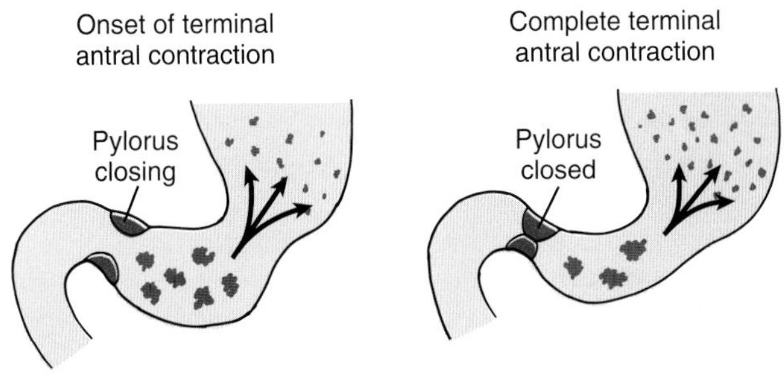

FIGURE 26.26 **Gastric retropulsion.** Jet-like retropulsion through the orifice of the antral compartment triturates solid particles in the stomach. The force for retropulsion is increased pressure in the terminal antrum as the trailing antral contraction approaches the closed pylorus.

phase and of the contraction initiated by the plateau. Inhibitory neurotransmitters, such as NE and VIP, decrease the amplitude of the plateau and the strength of the associated contraction.

The magnitude of the effects of neurotransmitters increases with increasing concentration of the transmitter substance at the gastric musculature. Higher frequencies of action potential discharged by motor neurons release greater amounts of neurotransmitter. In this way, motor neurons determine, through the actions of their neurotransmitters on the plateau phase, whether the trailing contraction of the propulsive complex of the distal stomach occurs. With sufficient release of transmitter, the plateau exceeds the threshold for contraction. Beyond threshold, the strength of contraction is determined by the amount of neurotransmitter released and present at receptors on the muscles.

The action potentials in the terminal antrum and pylorus differ somewhat in configuration from those in the more proximal regions. The principal difference is the occurrence of spike potentials on the plateau phase (see Fig. 26.25), which trigger short-duration phasic contractions superimposed on the phasic contraction associated with the plateau. These may contribute to the sphincteric function of the pylorus in preventing a reflux of duodenal contents back into the stomach.

Neural Control of Muscular Tone Determines Minute-to-Minute Volume and Pressure in the Gastric Reservoir

The gastric reservoir has two primary functions. One is to accommodate the arrival of a meal, without a significant increase in intragastric pressure and distension of the gastric wall. Failure of this mechanism can lead to the uncomfortable sensations of bloating, epigastric pain, and nausea. The second function is to maintain a constant compressive force on the contents of the reservoir. This pushes the contents into the 3 cycles/min motor activity of the antral pump. Drugs that relax the musculature of the gastric reservoir neutralize this function and suppress gastric emptying.

The musculature of the gastric reservoir is innervated by both excitatory and inhibitory motor neurons of the ENS. The motor neurons are controlled by the efferent vagus nerves and intramural microcircuits of the ENS. They function to adjust the volume and pressure of the reservoir to the amount of solid and/or liquid present while maintaining constant compressive forces on the contents. Continuous adjustments in the volume and pressure within the reservoir are required during both the ingestion and the emptying of a meal.

Increased activity of excitatory motor neurons, in coordination with decreased activity of inhibitory motor neurons, results in increased contractile tone in the reservoir, a decrease in its volume, and an increase in intraluminal pressure (Fig. 26.27). Increased activity of inhibitory motor neurons in coordination with decreased activity of excitatory motor neurons results in decreased contractile tone in the reservoir, expansion of its volume, and a decrease in intraluminal pressure.

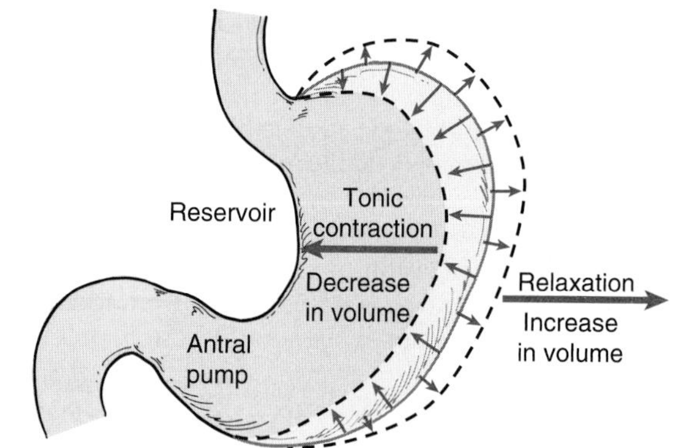

FIGURE 26.27 Muscular tone in the gastric reservoir. Tonic contraction of the musculature decreases the volume and exerts pressure on the contents. Tonic relaxation of the musculature expands the volume of the gastric reservoir. Neural mechanisms of feedback control determine intramural contractile tone in the reservoir.

Three Kinds of Relaxation Occur in the Gastric Reservoir

Neurally mediated decreases in tonic contraction of the musculature are responsible for relaxation in the gastric reservoir (i.e., increased volume). Three kinds of relaxation are recognized. **Receptive relaxation** is initiated by the act of swallowing. It is a reflex triggered by stimulation of mechanoreceptors in the pharynx followed by transmission over afferents to the dorsal vagal complex and activation of efferent vagal fibers to inhibitory motor neurons in the gastric ENS. **Adaptive relaxation** is triggered by distension of the gastric reservoir. It is a vago-vagal reflex triggered by stretch receptors in the gastric wall, transmission over vagal afferents to the dorsal vagal complex, and efferent vagal

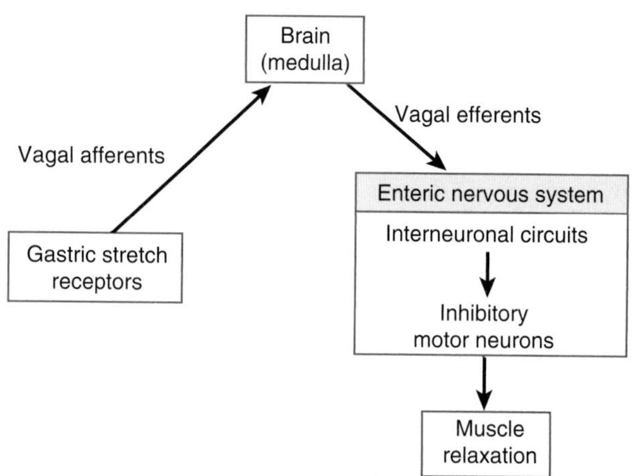

FIGURE 26.28 Adaptive relaxation in the gastric reservoir. Adaptive relaxation is a vago-vagal reflex in which information from gastric stretch receptors is the afferent component and outflow from the medullary region of the brain is the efferent component. Vagal efferents transmit to the ENS, which controls the activity of inhibitory motor neurons that relaxes contractile tone in the reservoir.

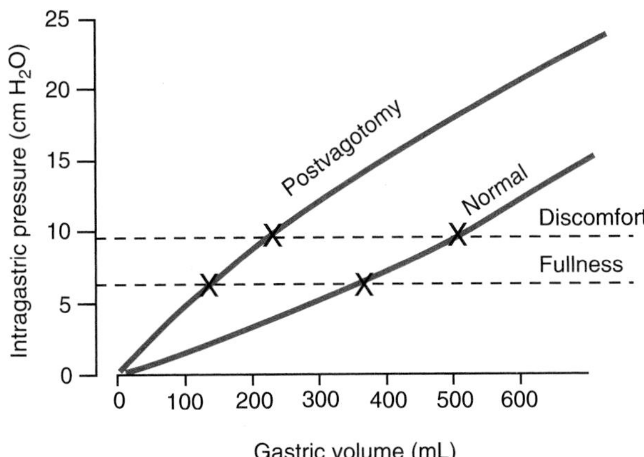

FIGURE 26.29 **Loss of adaptive relaxation following a vagotomy.** A loss of adaptive relaxation in the gastric reservoir is associated with a lowered threshold for sensations of fullness and epigastric pain.

fibers to inhibitory motor neurons in the gastric ENS (Fig. 26.28). **Feedback relaxation** is triggered by the presence of nutrients in the small intestine. It can involve both local reflex connections between receptors in the small intestine and the gastric ENS or hormones that are released from endocrine cells in the small intestine and transported by the blood to signal the gastric ENS.

Adaptive relaxation is lost in patients who have undergone a vagotomy as a treatment for gastric acid disease (e.g., peptic ulcer). Following a vagotomy, increased tone in the musculature of the reservoir decreases the wall compliance, which, in turn, affects the responses of gastric stretch receptors to distension of the reservoir. Pressure-volume curves before and after a vagotomy reflect the decrease in compliance of the gastric wall (Fig. 26.29). The loss of adaptive relaxation after a vagotomy is associated with a lowered threshold for sensations of fullness and pain. This response is explained by increased stimulation of the gastric mechanoreceptors that sense distension of the gastric wall. These effects of vagotomy may explain disordered gastric sensations in diseases with a component of vagus nerve pathology (e.g., autonomic neuropathy of diabetes mellitus) (see Clinical Focus Box 26.1).

The Rate of Gastric Emptying Is Determined by the Kind of Meal and Conditions in the Duodenum

In addition to storage in the reservoir and mixing and grinding by the antral pump, an important function of gastric motility is the orderly delivery of the gastric chyme to the duodenum at a rate that does not overload the digestive and absorptive functions of the small intestine (see Clinical Focus Box 26.1). The rate of gastric emptying is adjusted by neural control mechanisms to compensate for variations in the volume, composition, and physical state of the gastric contents.

The volume of liquid in the stomach is one of the important determinants of gastric emptying. The rate of emptying of isotonic noncaloric liquids (e.g., 0.9% NaCl) is proportional to the initial volume in the reservoir. The larger the initial volume, the more rapid the emptying.

With a mixed meal in the stomach, liquids empty faster than solids. If an experimental meal consisting of solid particles of various sizes suspended in water is instilled in the stomach, emptying of the particles lags behind emptying of the liquid (Fig. 26.30). With digestible particles (e.g., studies with isotopically labeled chunks of liver), the **lag phase** is the time required for the grinding action of the antral pump to reduce the particle size. If the particles are plastic spheres of various sizes, the smallest spheres are emptied first; however, spheres up to 7 mm in diameter empty at a slow but steady rate when digestible food is in the stomach. The selective emptying of smaller particles first is referred to as the **sieving action** of the distal stomach. Inert spheres larger than 7 mm in diameter are not emptied while food is in the stomach; they empty at the start of the first migrating motor complex as the digestive tract enters the interdigestive state.

Osmolality, acidity, and caloric content of the gastric chyme are major determinants of the rate of gastric emptying. Hypotonic and hypertonic liquids empty more slowly than isotonic liquids. The rate of gastric emptying decreases as the acidity of the gastric contents increases. Meals with a high caloric content empty from the stomach at a slower rate than meals with a low caloric content. The mechanisms of control of gastric emptying keep the rate of delivery of calories to the small intestine within a narrow range, regardless of whether the calories are presented as carbohydrate, protein, fat, or a mixture. Of all of these, fat is emptied the most slowly, or stated conversely, fat is the most potent inhibitor of gastric emptying. Part of the inhibition of gastric emptying by fats may involve the release of the hormone cholecystokinin, which itself is a potent inhibitor of gastric emptying.

The intraluminal milieu of the small intestine is extremely different from that of the stomach (see Chapter

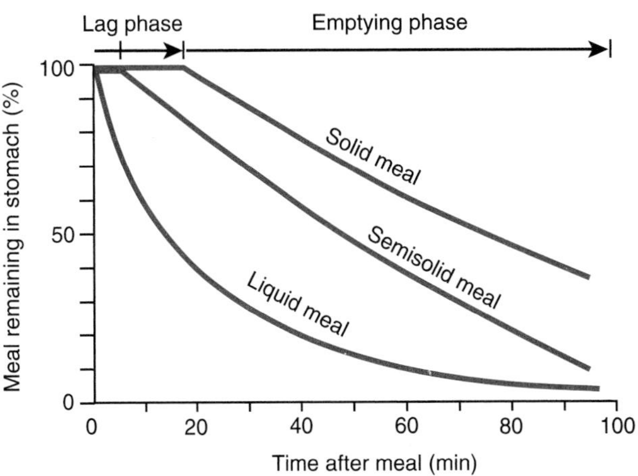

FIGURE 26.30 **Gastric emptying.** The rate of gastric emptying varies with the composition of the meal. Solid meals empty more slowly than semisolid or liquid meals. The emptying of a solid meal is preceded by a lag phase, the time required for particles to be reduced to sufficient size for emptying.

27). Undiluted stomach contents have a composition that is poorly tolerated by the duodenum. Mechanisms of control of gastric emptying automatically adjust the delivery of gastric chyme to an optimal rate for the small intestine. This guards against overloading the small intestinal mechanisms for the neutralization of acid, dilution to iso-osmolality, and enzymatic digestion of the foodstuff (see Clinical Focus Box 26.1).

MOTILITY IN THE SMALL INTESTINE

The time required for transit of experimentally labeled meals from the stomach to the small intestine to the large intestine is measured in hours (Fig. 26.31). Transit time in the stomach is most rapid of the three compartments; transit in the large intestine is the slowest. Three fundamental patterns of motility that influence the transit of material through the small intestine are the interdigestive pattern, the digestive pattern, and power propulsion. Each pattern is programmed by the small intestinal ENS.

The Migrating Motor Complex Is the Small Intestinal Motility Pattern of the Interdigestive State

The small intestine is in the **digestive state** when nutrients are present and the digestive processes are ongoing. It converts to the **interdigestive state** when the digestion and absorption of nutrients are complete, 2 to 3 hours after a meal. The pattern of motility in the interdigestive state is called the **migrating motor complex** (MMC). The MMC can be detected by placing pressure sensors in the lumen of the intestine or attaching electrodes to the intestinal surface (Fig. 26.32). Sensors in the stomach show the MMC starting as large-amplitude contractions at 3/min in the distal stomach. Elevated contraction of the lower esophageal sphincter coincides with the onset of the MMC in the stomach. Activity in the stomach appears to migrate into the duodenum and on through the small intestine to the ileum.

At a single recording site in the small intestine, the MMC consists of three consecutive phases:

- Phase I: a silent period having no contractile activity; corresponds to physiological ileus
- Phase II: irregularly occurring contractions
- Phase III: regularly occurring contractions

Phase I returns after phase III, and the cycle is repeated (Fig. 26.33). With multiple sensors positioned along the intestine, slow propagation of the phase II and phase III activity down the intestine becomes evident.

At a given time, the MMC occupies a limited length of intestine called the **activity front**, which has an upper and a lower boundary. The activity front slowly advances (migrates) along the intestine at a rate that progressively slows as it approaches the ileum. Peristaltic propulsion of luminal contents in the aboral direction occurs between the oral and aboral boundaries of the activity front. The frequency of the peristaltic waves within the activity front is the same as the frequency of electrical slow waves in that intestinal segment. Each peristaltic wave consists of propulsive and receiving segments, as described earlier (see Fig. 26.20). Successive peristaltic waves start, on average, slightly farther in the aboral direction and propagate, on average, slightly beyond the boundary where the previous one stopped. Thus, the entire activity front slowly migrates down the intestine, sweeping the lumen clean as it goes.

Phases II and III are commonly used descriptive terms of minimal value for understanding the MMC. Contractile activity described as phase II or phase III occurs because of the irregularity of the arrival of peristaltic waves at the aboral boundary of the activity front. On average, each consecutive peristaltic wave within the activity front propagates farther in the aboral direction than the previous wave. Nevertheless, at the lower boundary of the activity front, some waves terminate early and others travel farther (see Fig. 26.32). Therefore, as the lower boundary of the front passes the recording point, only the waves that reach the sensor are recorded, giving the appearance of irregular contractions. As propagation continues and the midpoint of the activity front reaches the recording point, the propulsive segment of every peristaltic wave is detected. Because the peristaltic waves occur with the same rhythmicity as the electrical slow waves, the contractions can be described as being "regular." The regular contractions that are seen

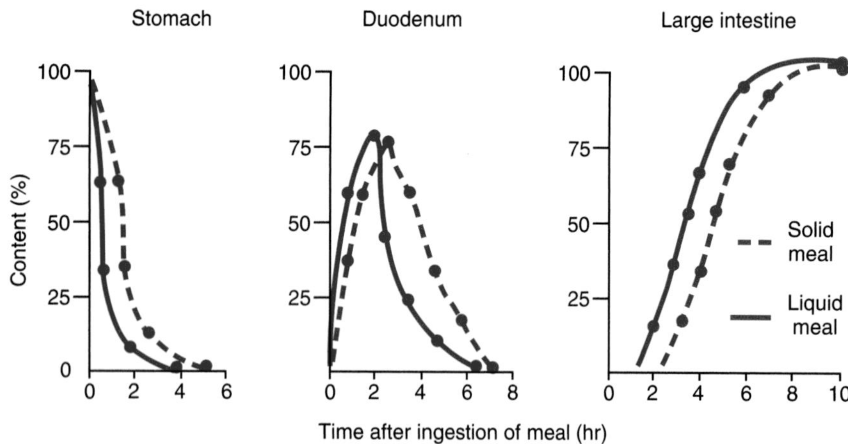

FIGURE 26.31 **GI transit times.** The time during which components of solid and liquid meals enter and leave the stomach, duodenum, and large intestine is measured in hours.

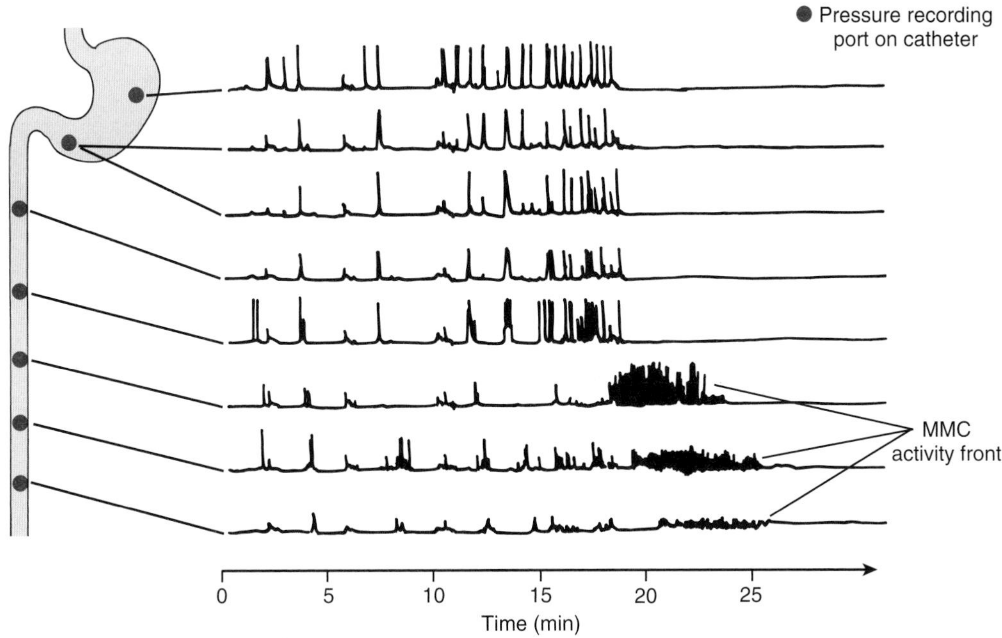

FIGURE 26.32 Migrating motor complex in the small intestine. The MMC consists of an activity front that starts in the gastric antrum and slowly migrates through the small intestine to the ileum. Repetitive peristaltic propulsion occurs within the activity front.

when the central region of the front passes a single recording site last for 8 to 15 minutes. This time is shortest in the duodenum and progressively increases as the MMC migrates toward the ileum.

The MMC is seen in most mammals, including humans, in conscious states and during sleep. It starts in the antrum of the stomach as an increase in the strength of the regularly occurring antral contractile complexes and accomplishes the emptying of indigestible particles (e.g., pills and capsules) greater than 7 mm. In humans, 80 to 120 minutes are required for the activity front of the MMC to travel from the antrum to the ileum. As one activity front terminates in the ileum, another begins in the antrum. In humans, the time between cycles is longer during the day than at night. The activity front travels at about 3 to 6 cm/min in the duodenum and progressively slows to about 1 to 2 cm/min in the ileum. It is important not to confuse the speed of travel of the activity front of the MMC with that of the electrical slow waves, action potentials, and peristaltic waves within the activity front. Slow waves with associated action potentials and associated contractions of circular muscle travel about 10 times faster.

Cycling of the MMC continues until it is ended by the ingestion of food. A sufficient nutrient load terminates the MMC simultaneously at all levels of the intestine. Termination requires the physical presence of a meal in the upper digestive tract; intravenous feeding does not end the fasting pattern. The speed with which the MMC is terminated at all levels of the intestine suggests a neural or hormonal mechanism. Gastrin and cholecystokinin, both of which are released during a meal, terminate the MMC in the stomach and upper small intestine but not in the ileum, when injected intravenously.

The MMC is organized by the microcircuits in the ENS. It continues in the small intestine after a vagotomy or sympathectomy but stops when it reaches a region of the intestine where the ENS has been interrupted. Presumably, command signals to the enteric neural circuits are necessary for initiating the MMC, but whether the commands are neural, hormonal, or both is unknown. Although levels of the hormone **motilin** increase in the blood at the onset of the MMC, it is unclear whether motilin is the trigger or is released as a consequence of its occurrence.

Adaptive Significance of the MMC.

Gallbladder contraction and delivery of bile to the duodenum is coordinated with the onset of the MMC in the intraduodenal region. After entering the duodenum, the activity front of the MMC propels the bile to the terminal ileum, where it is reabsorbed into the hepatic portal circulation. This mechanism minimizes the accumulation of concentrated bile in the gallbladder and increases the movement of bile acids in the enterohepatic circulation during the interdigestive state (see Chapter 27).

The adaptive significance of the MMC appears also to be a mechanism for clearing indigestible debris from the intestinal lumen during the fasting state. Large indigestible particles are emptied from the stomach only during the interdigestive state.

Bacterial overgrowth in the small intestine is associated with an absence of the MMC. This condition suggests that the MMC may play a housekeeper role in preventing the overgrowth of microorganisms that might occur in the small intestine if the contents were allowed to stagnate in the lumen.

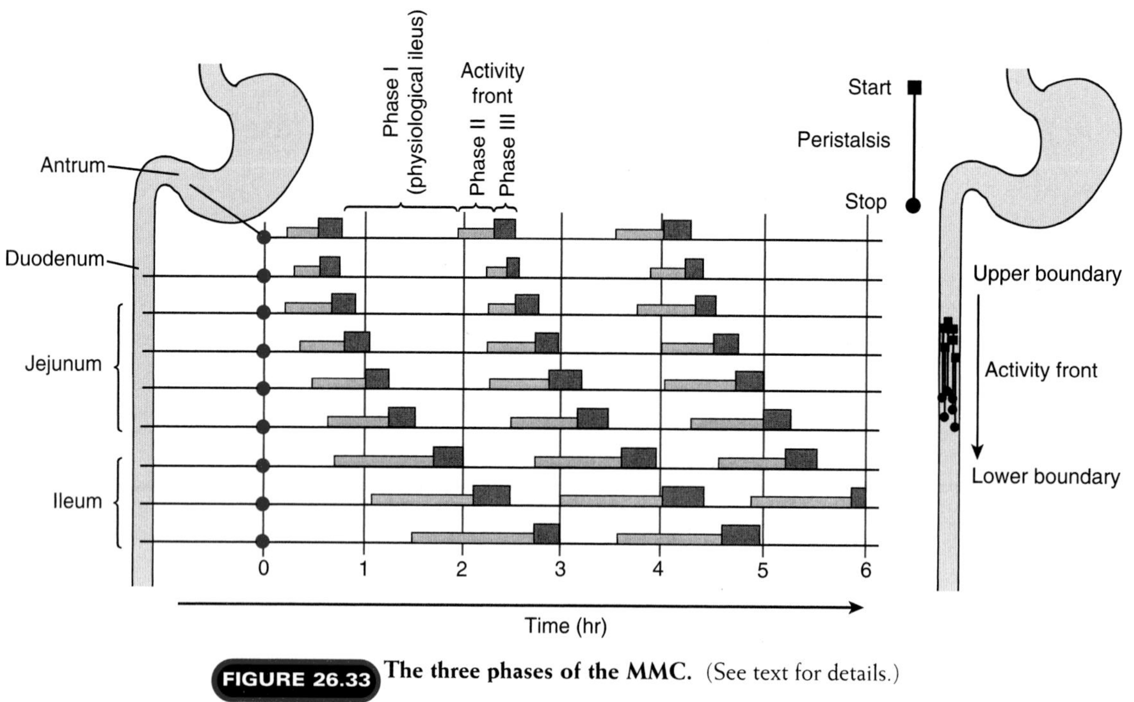

FIGURE 26.33 **The three phases of the MMC.** (See text for details.)

Mixing Movements Characterize the Digestive State

A mixing pattern of motility replaces the MMC when the small intestine is in the digestive state following ingestion of a meal. The mixing movements are sometimes called **segmenting movements** or **segmentation**, as a result of their appearance on X-ray films of the small intestine. Peristaltic contractions, which propagate for only short distances, account for the segmentation appearance. Circular muscle contractions in short propulsive segments are separated on either end by relaxed receiving segments (Fig. 26.34). Each propulsive segment jets the chyme in both directions into the relaxed receiving segments where stirring and mixing occur. This happens continuously at closely spaced sites along the entire length of the small intestine. The intervals of time between mixing contractions are the same as for electrical slow waves or are multiples of the shortest slow wave interval in the particular region of intestine. A higher frequency of electrical slow waves and associated contractions in more proximal regions and the peristaltic nature of the mixing movements result in a net aboral propulsion of the luminal contents over time.

The Role of the Vagus Nerves and ENS. The mixing pattern of small intestinal motility is programmed by the ENS. Signals transmitted by vagal efferent nerves to the ENS interrupt the MMC and initiate mixing motility during ingestion of a meal. After the vagus nerves are cut, a larger quantity of ingested food is necessary to interrupt the interdigestive motor pattern, and interruption of the MMCs is often incomplete. Evidence of vagal commands for the mixing pattern has been obtained in animals with cooling cuffs placed surgically around each vagus nerve. During the digestive state, cooling and blockade of impulse transmission in the nerves result in an interruption of the pattern of mixing movements. When the vagus nerves are blocked during the digestive state, MMCs reappear in the intestine but not in the stomach. With warming of the nerves and release of the neural blockade, the mixing motility pattern returns.

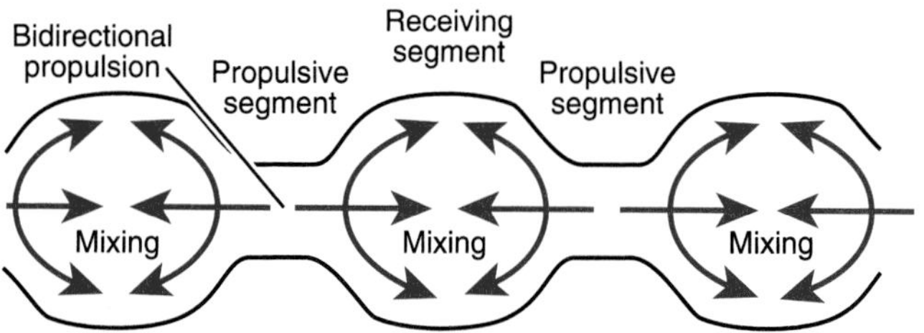

Same length of intestine later in time

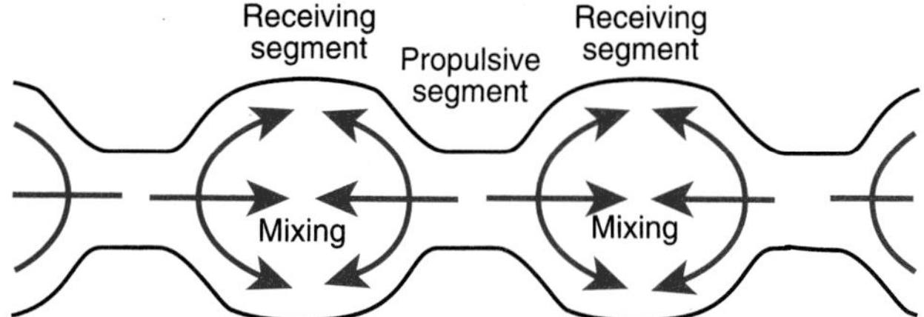

FIGURE 26.34 **Mixing movements.** The segmentation pattern of motility is characteristic of the digestive state. Propulsive segments separated by receiving segments occur randomly at many sites along the small intestine. Mixing of the luminal contents occurs in the receiving segments. Receiving segments convert to propulsive segments, while propulsive segments become receiving segments.

Power Propulsion Is a Defensive Response Against Harmful Agents

Power propulsion involves strong, long-lasting contractions of the circular muscle that propagate for extended distances along the small and large intestines. The giant migrating contractions are considerably stronger than the phasic contractions during the MMC or mixing pattern. Giant migrating contractions last 18 to 20 seconds and span several cycles of the electrical slow waves. They are a component of a highly efficient propulsive mechanism that rapidly strips the lumen clean as it travels at about 1 cm/sec over long lengths of intestine.

Intestinal power propulsion differs from peristaltic propulsion during the MMC and mixing movements, in that circular contractions in the propulsive segment are stronger and more open gates permit propagation over longer reaches of intestine. The circular muscle contractions are not time-locked to the electrical slow waves and probably reflect strong activation of the muscle by excitatory motor neurons.

Power propulsion occurs in the retrograde direction during emesis in the small intestine and in the orthograde direction in response to noxious stimulation in both the small and the large intestines. Abdominal cramping sensations and, sometimes, diarrhea are associated with this motor behavior. Application of irritants to the mucosa, the introduction of luminal parasites, enterotoxins from pathogenic bacteria, allergic reactions, and exposure to ionizing radiation all trigger the propulsive response. This suggests that power propulsion is a defensive adaptation for the rapid clearance of undesirable contents from the intestinal lumen. It may also accomplish mass movement of intraluminal material in normal states, especially in the large intestine.

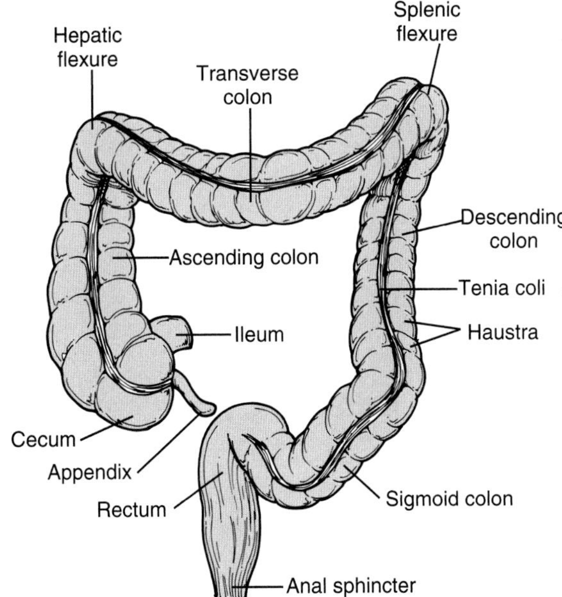

FIGURE 26.35 **Anatomy of the large intestine.** The main anatomic regions of the large intestine are the ascending colon, transverse colon, descending colon, sigmoid colon, and rectum. The hepatic flexure is the boundary between the ascending and the transverse colon; the splenic flexure is the boundary between the transverse and the descending colon. The sigmoid colon is so defined by its shape. The rectum is the most distal region. The cecum is the blind ending of the colon at the ileocecal junction. The appendix is an evolutionary vestige. Internal and external anal sphincters close the terminus of the large intestine. The longitudinal muscle layer is restricted to bundles of fibers called tenia coli.

MOTILITY IN THE LARGE INTESTINE

In the large intestine, contractile activity occurs almost continuously. Whereas the contents of the small intestine move through sequentially with no mixing of individual meals, the large bowel contains a mixture of the remnants of several meals ingested over 3 to 4 days. The arrival of undigested residue from the ileum does not predict the time of its elimination in the stool.

The large intestine is subdivided into functionally distinct regions corresponding approximately to the ascending colon, transverse colon, descending colon, rectosigmoid region, and internal anal sphincter (Fig. 26.35). The transit of small radiopaque markers through the large intestine occurs, on average, in 36 to 48 hours.

The Ascending Colon Is Specialized for Processing Chyme Delivered From the Terminal Ileum

Power propulsion in the terminal length of ileum may deliver relatively large volumes of chyme into the **ascending colon**, especially in the digestive state. Neuromuscular mechanisms analogous to adaptive relaxation in the stomach permit filling without large increases in intraluminal pressure. Chemoreceptors and mechanoreceptors in the cecum and ascending colon provide feedback information for controlling delivery from the ileum, analogous to the feedback control of gastric emptying from the small intestine.

Dwell-time of material in the ascending colon is found to be short when studied with gamma scintigraphic imaging of radiolabeled markers. When radiolabeled chyme is instilled into the human cecum, half of the instilled volume empties, on average, in 87 minutes. This period is long in comparison with an equivalent length of small intestine, but it is short in comparison with the transverse colon. It suggests that the ascending colon is not the primary site for the large intestinal functions of storage, mixing, and removal of water from the feces.

The motor pattern of the ascending colon consists of orthograde or retrograde peristaltic propulsion. The significance of backward propulsion in this region is uncertain; it may be a mechanism for temporary retention of the chyme in the ascending colon. Forward propulsion in this region is probably controlled by feedback signals on the fullness of the transverse colon.

The Transverse Colon Is Specialized for the Storage and Dehydration of Feces

Radioscintigraphy shows that the labeled material is moved relatively quickly into the **transverse colon** (Fig. 26.36),

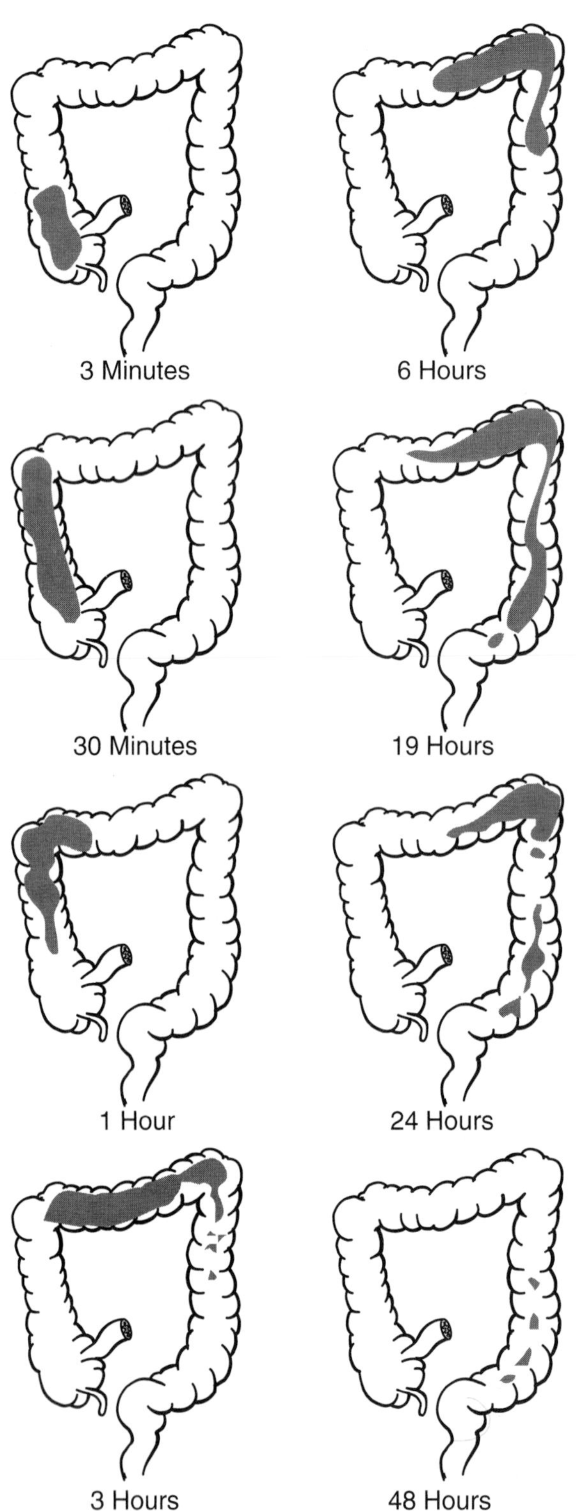

FIGURE 26.36 Colonic transit revealed by radioscintigraphy. Successive scintigrams reveal that the longest dwell-time for intraluminal markers injected initially into the cecum is in the transverse colon. The image is faint after 48 hours, indicating that most of the marker has been excreted with the feces.

where it is retained for about 24 hours. This suggests that the transverse colon is the primary location for the removal of water and electrolytes and the storage of solid feces in the large intestine.

A segmental pattern of motility programmed by the ENS accounts for the ultraslow forward movement of feces in the transverse colon. Ring-like contractions of the circular muscle divide the colon into pockets called **haustra** (Fig. 26.37). The motility pattern, called **haustration**, differs from segmental motility in the small intestine, in that the contracting segment and the receiving segments on either side remain in their respective states for longer periods. In addition, there is uniform repetition of the haustra along the colon. The contracting segments in some places appear to be fixed and are marked by a thickening of the circular muscle.

Haustrations are dynamic, in that they form and reform at different sites. The most common pattern in the fasting individual is for the contracting segment to propel the contents in both directions into receiving segments. This mechanism mixes and compresses the semiliquid feces in the haustral pockets and probably facilitates the absorption of water without any net forward propulsion.

Net forward propulsion occurs when sequential migration of the haustra occurs along the length of the bowel. The con-

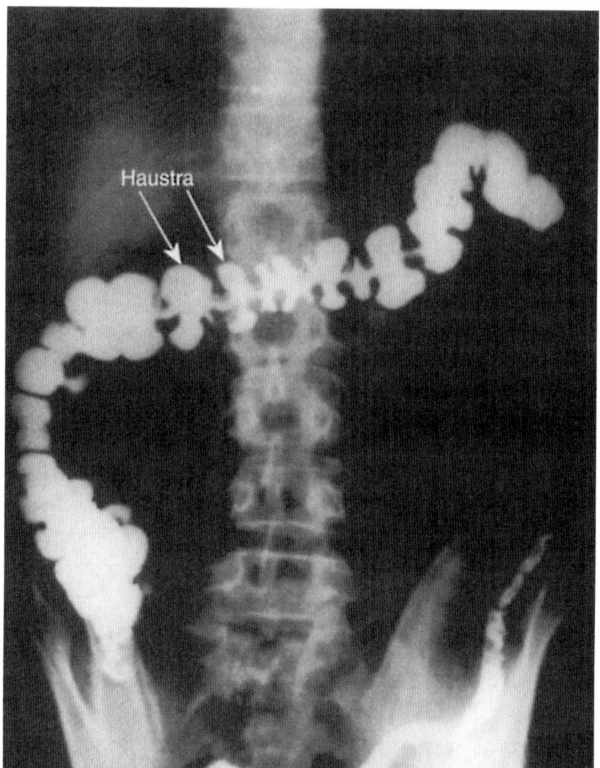

FIGURE 26.37 Haustra in the large intestine. This X-ray film shows haustral contractions in the ascending and the transverse colon. Between the haustral pockets are segments of contracted circular muscle. Ongoing activity of inhibitory motor neurons maintains the relaxed state of the circular muscle in the pockets. Inactivity of inhibitory motor neurons permits the contractions between the pockets.

tents of one haustral pocket are propelled into the next region, where a second pocket is formed, and from there to the next segment, where the same events occur. This pattern results in slow forward progression and is believed to be a mechanism for compacting the feces in storage.

Power propulsion is another programmed motor event in the transverse and the descending colon. This motor behavior fits the general pattern of neurally coordinated peristaltic propulsion and results in the mass movement of feces over long distances. Mass movements may be triggered by increased delivery of ileal chyme into the ascending colon following a meal. The increased incidence of mass movements and generalized increase in segmental movements following a meal is called the **gastrocolic reflex**. Irritant laxatives, such as castor oil, act to initiate the motor program for power propulsion in the colon. The presence of threatening agents in the colonic lumen, such as parasites, enterotoxins, and food antigens, can also initiate power propulsion.

Mass movement of feces (power propulsion) in the healthy bowel usually starts in the middle of the transverse colon and is preceded by relaxation of the circular muscle and the downstream disappearance of haustral contractions. A large portion of the colon may be emptied as the contents are propelled at rates up to 5 cm/min as far as the rectosigmoid region. Haustration returns after the passage of the power contractions.

The Descending Colon Is a Conduit Between the Transverse and Sigmoid Colon

Radioscintigraphic studies in humans show that feces do not have long dwell-times in the **descending colon**. Labeled feces begin to accumulate in the sigmoid colon and rectum about 24 hours after the label is instilled in the cecum. The descending colon functions as a conduit without long-term retention of the feces. This region has the neural program for power propulsion. Activation of the program is responsible for mass movements of feces into the sigmoid colon and rectum.

The Physiology of the Rectosigmoid Region, Anal Canal, and Pelvic Floor Musculature Maintains Fecal Continence

The **sigmoid colon** and **rectum** are reservoirs with a capacity of up to 500 mL in humans. Distensibility in this region is an adaptation for temporarily accommodating the mass movements of feces. The rectum begins at the level of the third sacral vertebra and follows the curvature of the sacrum and coccyx for its entire length. It connects to the anal canal surrounded by the internal and external anal sphincters. The pelvic floor is formed by overlapping sheets of striated fibers called **levator ani** muscles. This muscle group, which includes the **puborectalis muscle** and the striated external anal sphincter, comprise a functional unit that maintains continence. These skeletal muscles behave in many respects like the somatic muscles that maintain posture elsewhere in the body (see Chapter 5).

The pelvic floor musculature can be imagined as an inverted funnel consisting of the levator ani and external sphincter muscles in a continuous sheet from the bottom margins of the pelvis to the anal verge (the transition zone between mucosal epithelium and stratified squamous epithelium of the skin). After defecation, the levator ani contract to restore the perineum to its normal position. Fibers of the puborectalis join behind the anorectum and pass around it on both sides to insert on the pubis. This forms a U-shaped sling that pulls the anorectal tube anteriorly, such that the long axis of the anal canal lies at nearly a right angle to that of the rectum (Fig. 26.38). Tonic pull of the puborectalis narrows the anorectal tube from side to side at the bend of the angle, resulting in a physiological valve that is important in the mechanisms that control continence.

The puborectalis sling and the upper margins of the internal and external sphincters form the anorectal ring, which marks the boundary of the anal canal and rectum. Surrounding the anal canal for a length of about 2 cm are the internal and external anal sphincters. The **external anal sphincter** is skeletal muscle attached to the coccyx posteriorly and the perineum anteriorly. When contracted, it compresses the anus into a slit, closing the orifice. The **internal anal sphincter** is a modified extension of the circular muscle layer of the rectum. It is comprised of smooth muscle that, like other sphincteric muscles in the digestive tract, contracts tonically to sustain closure of the anal canal.

Sensory Innervation and Continence. Mechanoreceptors in the rectum detect distension and supply the enteric neural circuits with sensory information, similar to the innervation of the upper portions of the GI tract. Unlike the rectum, the anal canal in the region of skin at the anal verge is innervated by somatosensory nerves that transmit signals to the CNS. This region has sensory receptors that detect touch, pain, and temperature with high sensitivity. Processing of information from these receptors allows the in-

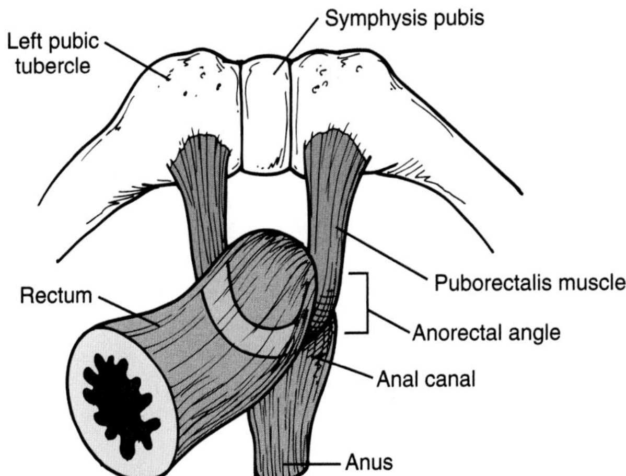

FIGURE 26.38 **Structural relationship of the anorectum and puborectalis muscle.** One end of the puborectalis muscle inserts on the left pubic tubercle, and the other inserts on the right pubic tubercle, forming a loop around the junction of the rectum and anal canal. Contraction of the puborectalis muscle helps form the anorectal angle, believed to be important in the maintenance of fecal continence.

dividual to discriminate consciously between the presence of gas, liquid, and solids in the anal canal. In addition, stretch receptors in the muscles of the pelvic floor detect changes in the orientation of the anorectum as feces are propelled into the region.

Contraction of the internal anal sphincter and the puborectalis muscles blocks the passage of feces and maintains continence with small volumes in the rectum (see Clinical Focus Box 26.3). When the rectum is distended, the **rectoanal reflex** or **rectosphincteric reflex** is activated to relax the internal sphincter. Like other enteric reflexes, this one involves a stretch receptor, enteric interneurons, and excitation of inhibitory motor neurons to the smooth muscle sphincter. Distension also results in the sensation of rectal fullness, mediated by the central processing of information from mechanoreceptors in the pelvic floor musculature.

Relaxation of the internal sphincter allows contact of the rectal contents with the sensory receptors in the lining of the anal canal, providing an early warning of the possibility of a breakdown of the continence mechanisms. When this occurs, continence is maintained by voluntary contraction of the external anal sphincter and the puborectalis muscle. The external sphincter closes the anal canal, and the puborectalis sharpens the anorectal angle. An increase in the anorectal angle works in concert with increases in intra-abdominal pressure to create a "flap" valve. The flap valve is formed by the collapse of the anterior rectal wall onto the upper end of the anal canal, occluding the lumen.

Whereas the rectoanal reflex is mediated by the ENS, synaptic circuits for the neural reflexes of the external anal sphincter and other pelvic floor muscles reside in the sacral portion of the spinal cord. The mechanosensory receptors are muscle spindles and Golgi tendon organs similar to those found in skeletal muscles elsewhere in the body. Sensory input from the anorectum and pelvic floor is transmitted over dorsal roots to the sacral cord, and motor outflow to these areas is in sacral root motor nerve fibers. The spinal circuits account for the reflex increases in contraction of the external sphincter and pelvic floor muscles by behaviors that raise intra-abdominal pressure, such as coughing, sneezing, and lifting weights.

Defecation Involves the Neural Coordination of Muscles in the Large Intestine and Pelvic Floor

Distension of the rectum by the mass movement of feces or gas results in an urge to defecate or release flatus. CNS processing of mechanosensory information from the rectum is the underlying mechanism for this sensation. Local processing of the mechanosensory information in the enteric neural circuits activates the motor program for relaxation of the internal anal sphincter. At this stage of rectal distension, voluntary contraction of the external anal sphincter and the puborectalis muscle prevents leakage. The decision to defecate at this stage is voluntary. When the decision is made, commands from the brain to the sacral cord shut off the excitatory input to the external sphincter and levator ani muscles. Additional skeletal motor commands contract the abdominal muscles and diaphragm to increase intra-abdominal pressure. Coordination of the skeletal muscle components of defecation results in a straightening of the anorectal angle, descent of the pelvic floor, and opening of the anus.

Programmed behavior of the smooth muscle during defecation includes shortening of the longitudinal muscle layer in the sigmoid colon and rectum, followed by strong contraction of the circular muscle layer. This behavior corresponds to the basic stereotyped pattern of peristaltic propulsion. It represents **terminal intestinal peristalsis**, in that the circular muscle of the distal colon and rectum becomes the final propulsive segment while the outside environment receives the forwardly propelled luminal contents.

A voluntary decision to resist the urge to defecate is eventually accompanied by relaxation of the circular muscle of the rectum. This form of adaptive relaxation accommodates the increased volume in the rectum. As wall tension relaxes, the stimulus for the rectal mechanoreceptors is removed, and the urge to defecate subsides. Receptive relaxation of the rectum is accompanied by a return of contractile tension in the internal anal sphincter, relaxation of tone in the external sphincter, and increased pull by the puborectalis muscle sling. When this occurs, the feces remain in the rectum until the next mass movement further increases the rectal volume and stimulation of mechanoreceptors again signals the neural mechanisms for defecation.

REVIEW QUESTIONS

DIRECTIONS: Each of the numbered items or incomplete statements in this section is followed by answers or by completions of the statement. Select the ONE lettered answer or completion that is BEST in each case.

1. A surgeon makes an incision in the jejunum starting at the serosal surface and ending in the lumen. What is the sequential order of bisected structures as the scalpel passes through the intestinal wall?
   (A) Circular muscle → longitudinal muscle → submucous plexus
   (B) Longitudinal muscle → myenteric plexus → circular muscle
   (C) Myenteric plexus → circular muscle → longitudinal muscle
   (D) Network of interstitial cells of Cajal → longitudinal muscle → circular muscle
   (E) Longitudinal muscle → network of interstitial cells of Cajal → submucous plexus
2. A mouse with a new genetic mutation is discovered not to have electrical slow waves in the small intestine. What cell type is most likely affected by the mutation?
   (A) Enteric neurons
   (B) Inhibitory motor neurons
   (C) Enterochromaffin cells
   (D) Interstitial cells of Cajal
   (E) Enteroendocrine cells
3. A patient with chronic intestinal pseudoobstruction has action potentials and large-amplitude contractions of the circular muscle associated with every electrical slow wave at all levels of the intestine in the interdigestive state. Dysplasia of which cell type most likely explains this patient's condition?
   (A) Unitary-type smooth muscle

(continued)

(B) Interstitial cells of Cajal
(C) Inhibitory motor neurons
(D) Sympathetic postganglionic neurons
(E) Vagal efferent neurons

4. A neural tracer technique labels the axon and cell body when it is applied to any part of a neuron. Where are labeled cell bodies most likely to be found after the tracer substance is injected into the wall of the stomach?
(A) Prefrontal cortex
(B) Intermediolateral horn of spinal cord
(C) Dorsal vagal complex
(D) Hypothalamus
(E) Gray matter of sacral spinal cord

5. An electrophysiological study of a neuron in the ENS detects a fast EPSP. Which is the most likely property associated with the EPSP?
(A) Acetylcholine (ACh) receptors
(B) Suppression of hyperpolarizing after-potentials
(C) Receptor activation of adenylyl cyclase
(D) Hyperpolarization of the membrane potential
(E) Mediation by a metabotropic receptor

6. The application of norepinephrine (NE) to the ENS suppresses cholinergically mediated EPSPs but has no effect on depolarizing responses to applied acetylcholine (ACh). This finding is best interpreted as
(A) Postsynaptic excitation
(B) Slow synaptic inhibition
(C) Presynaptic inhibition
(D) Postsynaptic facilitation
(E) Inhibitory junction potential

7. A 10-cm segment of small intestine is removed surgically and placed in a 37°C physiological solution containing tetrodotoxin. A stimulus at one end of the segment evokes an action potential and an accompanying contraction that travels to the opposite end of the segment. This finding is best explained by
(A) Electrical slow waves
(B) Varicose motor nerve fibers
(C) Interstitial cells of Cajal
(D) Functional electrical syncytial properties
(E) Release of neurotransmitters

8. A disease that results in the loss of enteric inhibitory motor neurons to the musculature of the digestive tract will most likely be expressed as
(A) Rapid intestinal transit
(B) Accelerated gastric emptying
(C) Gastroesophageal reflux
(D) Diarrhea
(E) Achalasia of the lower esophageal sphincter

9. The viewing of intestinal peristaltic propulsion in real time with magnetic resonance imaging shows the stereotyped formation of propulsive and receiving segments. What is the normal sequence of events in enteric neural programming of the propulsive and receiving segments?
(A) Relaxation of the longitudinal and circular muscles in the propulsive segment
(B) Relaxation of the circular and longitudinal muscles in the receiving segment
(C) Contraction of the longitudinal and circular muscles in the receiving segment
(D) Relaxation of the circular muscle and contraction of the longitudinal muscle in the receiving segment
(E) Contraction of the longitudinal muscle and relaxation of the circular muscle in the propulsive segment

10. Examination of the properties of a normal sphincter in the digestive tract will show that
(A) Primary flow across the sphincter is unidirectional
(B) The lower esophageal sphincter is relaxed at the onset of a migrating motor complex in the stomach
(C) Blockade of the sphincteric innervation by a local anesthetic causes the sphincter to relax
(D) The manometric pressure in the lumen of the sphincter is less than the pressure detected in the lumen on either side of the sphincter
(E) The inhibitory motor neurons to the sphincter muscle stop firing during a swallow

11. The absence of intestinal motility in the normal small intestine is best described as
(A) A migrating motor complex
(B) An interdigestive state
(C) Segmentation
(D) Physiological ileus
(E) Power propulsion

12. The best description of the lag phase of gastric emptying is the time required for
(A) Conversion from the interdigestive to the digestive enteric motor program
(B) Maximal stimulation of gastric secretion
(C) Return of the emptying curve to baseline
(D) Reduction of particle size to occur
(E) Emptying of half of a liquid meal

13. Increased strength of the trailing component of the contractile complex in the gastric antral pump is most likely to occur when
(A) Excitatory motor neurons are activated to release ACh at the antral musculature
(B) Sympathetic postganglionic neurons decrease the amplitude of the plateau phase of the gastric action potential
(C) Frequency of the gastric action potential increases beyond 3/min
(D) The pyloric sphincter opens
(E) Excitatory motor neurons to the musculature of the gastric reservoir are activated

14. When elevated in an ingested meal, the factor with the greatest effect in slowing gastric emptying is
(A) pH
(B) Carbohydrate
(C) Protein
(D) Lipid
(E) H_2O

15. On a return visit after receiving a diagnosis of functional dyspepsia, a 35-year-old woman reports sensations of early satiety and discomfort in the epigastric region after a meal. These symptoms are most likely a result of
(A) Malfunction of adaptive relaxation in the gastric reservoir
(B) Elevated frequency of contractions in the antral pump
(C) An incompetent lower esophageal sphincter
(D) Premature onset of the interdigestive phase of gastric motility
(E) Bile reflux from the duodenum

16. A 46-year-old university professor with an allergy to shellfish must be cautious when eating in restaurants because a trace of shrimp in any form of food triggers an allergic reaction, including abdominal cramping and diarrhea. Which kind of contractile behavior is the most likely intestinal motility pattern during the professor's allergic reaction to shellfish?
(A) Physiological ileus
(B) Migrating motor complex
(C) Retrograde peristaltic propulsion
(D) Segmentation
(E) Power propulsion

17. The instillation of markers in the large intestine is used to evaluate transit time in the large intestine and diagnose motility disorders. In healthy subjects, dwell-times for instilled markers in the large intestine are greatest in the
(A) Ascending colon
(B) Sigmoid colon
(C) Descending colon
(D) Transverse colon
(E) Anorectum

18. An 86-year-old woman has complaints of a compromised lifestyle because of fecal incontinence. Examination of this patient will most likely reveal the underlying cause of the incontinence to be
(A) Absence of the rectoanal reflex
(B) Elevated sensitivity to the presence of feces in the rectum

(continued)

(C) Loss of the ENS in the distal large intestine (adult Hirschsprung's disease)
(D) Weakness in the puborectalis and external anal sphincter muscles
(E) A myopathic form of chronic pseudoobstruction in the large intestine

SUGGESTED READING

Costa M, Glise H, Sjödal R. The enteric nervous system in health and disease. Gut 2000;47:1–88.

Costa M, Hennig GW, Brookes SJ. Intestinal peristalsis: A mammalian motor pattern controlled by enteric neural circuits. Ann N Y Acad Sci 1998;16:464–466.

Gershon MD. The Second Brain. New York: Harper Collins, 1998.

Krammer HJ, Enck P, Tack L. Neurogastroenterology—From the basics to the clinics. Z Gastroenterol (Suppl 2) 1997;3–68.

Kunze WA, Furness JB. The enteric nervous system and regulation of intestinal motility. Annu Rev Physiol 1999;61:117–142.

Makhlouf GM. Smooth muscle of the gut. In: Yamada T, Alpers DH, Owyang C, Powell DW, Silverstein FE, eds. Textbook of Gastroenterology. 2nd Ed. Philadelphia: Lippincott, 1995;86–111.

Sanders KM. A novel pacemaker mechanism drives gastrointestinal rhythmicity. News Physiol Sci 2000;15:291–298.

Szurszewski JH. A 100-year perspective on gastrointestinal motility. Am J Physiol 1998;274:G447–G453.

Wood JD. Enteric neuropathobiology. In: Phillips SF, Wingate DL, eds. Functional Disorders of the Gut: A Handbook for Clinicians. London: Harcourt Brace, 1998;19–42.

Wood JD. Physiology of the enteric nervous system. In: Johnson LR, Alpers DH, Christensen J, Jacobson ED, Walsh JH, eds. Physiology of the Gastrointestinal Tract. 3rd Ed. New York: Raven, 1994;423–482.

Wood JD, Alpers DH, Andrews PLR. Fundamentals of neurogastroenterology. Gut 1999;45:1–44.

Wood JD, Alpers DH, Andrews PLR. Fundamentals of neurogastroenterology: Basic science. In: Drossman DA, Talley NJ, Thompson WG, Corazziari E, eds. The Functional Gastrointestinal Disorders: Diagnosis, Pathophysiology and Treatment: A Multinational Consensus. McLean, VA: Degnon Associates, 2000;31–90.

Gastrointestinal Secretion, Digestion, and Absorption

Patrick Tso, Ph.D.

CHAPTER OUTLINE

KEY CONCEPTS

1. The major function of the GI tract is the digestion and absorption of nutrients.
2. Saliva assists in the swallowing of food, carbohydrate digestion, and the transport of immunoglobulins that combat pathogens.
3. Salivary secretion is mainly under the control of the autonomic nervous system. Parasympathetic and sympathetic nerves innervate the blood supply to the salivary glands. The parasympathetic nervous system increases the flow of saliva significantly, but the sympathetic nervous system only increases saliva flow marginally.
4. The stomach prepares chyme to aid in the digestion of food in the small intestine.
5. The gastric mucosa contains surface mucous cells that secrete mucus and bicarbonate ions, which protect the stomach from the acid in the stomach cavity.
6. Parietal cells secrete hydrochloric acid and intrinsic factor, and chief cells secrete pepsinogen.
7. Gastrin plays an important role in stimulating gastric acid secretion.
8. The acidity of gastric juice provides a barrier to microbial invasion of the GI tract.
9. Gastric secretion is under neural and hormonal control and consists of three phases: cephalic, gastric, and intestinal.
10. Gastric inhibitory peptide (GIP), secreted by intestinal endocrine cells, is a potent inhibitor of gastric acid secretion and enhances insulin release.
11. Pancreatic secretion neutralizes the acids in chyme and contains enzymes involved in the digestion of carbohydrates, fat, and protein.
12. Secretin stimulates the pancreas to secrete a bicarbonate-rich fluid, neutralizing acidic chyme.
13. CCK stimulates the pancreas to secrete an enzyme-rich fluid.
14. Pancreatic secretion is under neural and hormonal control and consists of three phases: cephalic, gastric, and intestinal.
15. Bile salts play an important role in the intestinal absorption of lipids.
16. Carbohydrates, when digested, form maltose, maltotriose, and α-limit dextrins, which are cleaved by brush border enzymes to monosaccharides and taken up by enterocytes.
17. Lipids absorbed by enterocytes are packaged and secreted as chylomicrons into lymph.
18. Protein is digested to form amino acids, dipeptides, and tripeptides that are taken up by enterocytes and transported in the blood.
19. The GI tract absorbs water-soluble vitamins and ions by different mechanisms.
20. Calcium-binding protein is involved in calcium absorption.
21. Heme and nonheme iron is absorbed in the small intestine by different mechanisms.
22. Most of the salt and water entering the intestinal tract, whether in the diet or in GI secretions, is absorbed in the small intestine.

The major function of the GI tract is the digestion and absorption of nutrients. Some absorption occurs in the stomach, including that of medium-chain fatty acids and some drugs, but most digestion and absorption of nutrients takes place in the small intestine. Secretions from the salivary glands, stomach, pancreas, and liver aid in the digestion and absorption process and protect the GI mucosa from the harmful effects of noxious agents. This chapter discusses the relevant anatomy, mechanism, composition, and regulation of GI secretion and the role the GI tract plays in the absorption of carbohydrate, fat, protein, fat-soluble and water-soluble vitamins, electrolytes, bile salts, and water.

GASTROINTESTINAL SECRETION

Secretions of the GI tract share several common features. A given secretion originates from individual groups of cells (e.g., acinar cells in the salivary gland) before pooling with other secretions. Secretions often empty into small ducts, which in turn empty into larger ducts, which empty into the lumen of the GI tract. Such a ductal system serves as a conduit for secretions from the salivary glands, pancreas, and liver, and modifies the primary secretion. Carbonic anhydrase, an enzyme present in gastric, pancreatic, and intestinal cells, is involved in the formation of GI secretions.

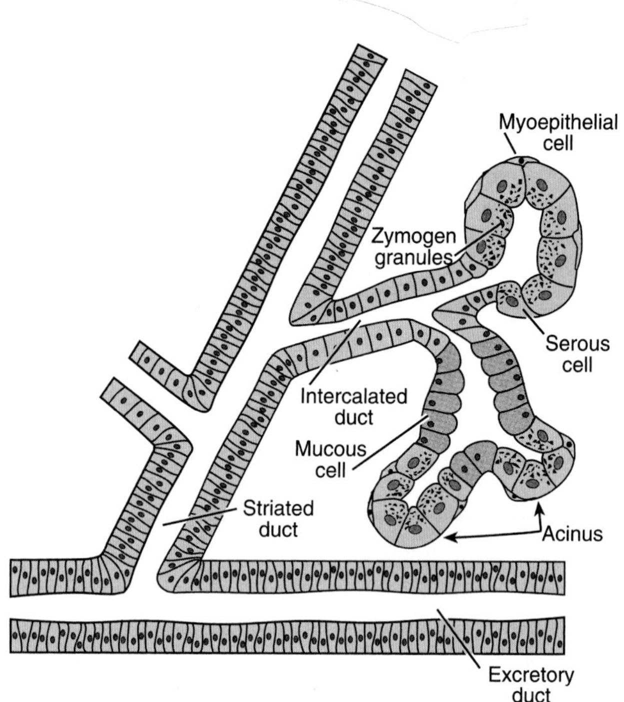

FIGURE 27.1 An acinus and associated ductal system from the human submandibular gland. (Modified from Johnson LR, Christensen J, Jackson MJ, et al. eds. Physiology of the Gastrointestinal Tract. New York: Raven, 1987.)

SALIVARY SECRETION

Salivary secretion is unique in that it is regulated almost exclusively by the nervous system. Saliva is produced by a heterogeneous group of exocrine glands called the **salivary glands.** Saliva performs several functions. It facilitates chewing and swallowing by lubricating food, carries immunoglobulins that combat pathogens, and assists in carbohydrate digestion.

The parotid, submandibular (submaxillary), and sublingual glands are the major salivary glands. They are drained by individual ducts into the mouth. The sublingual gland also has numerous small ducts that open into the floor of the mouth. The secretions of the major glands differ significantly. The parotid glands secrete saliva that is rich in water and electrolytes, whereas the submandibular and sublingual glands secrete saliva that is rich in mucin. There are also minor salivary glands located in the labial, palatine, buccal, lingual, and sublingual mucosae.

The salivary glands are endowed with a rich blood supply and are innervated by both the parasympathetic and sympathetic divisions of the autonomic nervous system. Although hormones may modify the composition of saliva, their physiological role is questionable, and it is generally believed that salivary secretion is mainly under autonomic control.

The Salivary Glands Consist of a Network of Acini and Ducts

A diagram of the human submandibular gland is shown in Figure 27.1. The basic unit, the **salivon**, consists of the acinus, the intercalated duct, the striated duct, and the excretory (collecting) duct. The **acinus** is a blind sac containing mainly pyramidal cells. Occasionally, there are stellate-shaped **myoepithelial cells** surrounding the large pyramidal cells. The cells of the acinus are not homogeneous. **Serous cells** secrete digestive enzymes, and **mucous cells** secrete mucin. Serous cells contain an abundance of rough endoplasmic reticulum (ER), reflecting active protein synthesis, and numerous **zymogen granules**. Salivary amylase is an important digestive enzyme synthesized and stored in the zymogen granules and secreted by the serous acinar cells.

Numerous mucin droplets are stored in the mucous acinar cells. Mucin is composed of glycoproteins of various molecular weights.

The **intercalated ducts** are lined with small cuboidal cells. The function of these cells is unclear, but they may be involved in the secretion of proteins, since secretory granules are occasionally observed in their cytoplasm. The intercalated ducts are connected to the striated duct, which eventually empties into the excretory duct. The **striated duct** is lined with columnar cells. Its major function is to modify the ionic composition of the saliva. The large **excretory ducts,** lined with columnar cells, also play a role in modifying the ionic composition of saliva. Although most proteins are synthesized and secreted by the acinar cells, the duct cells also synthesize several proteins, such as epidermal growth factor, ribonuclease, α-amylase, and proteases.

Saliva Contains Various Electrolytes and Proteins

The electrolyte composition of the **primary secretion** produced by the acinar cells resembles that of plasma. Micro-

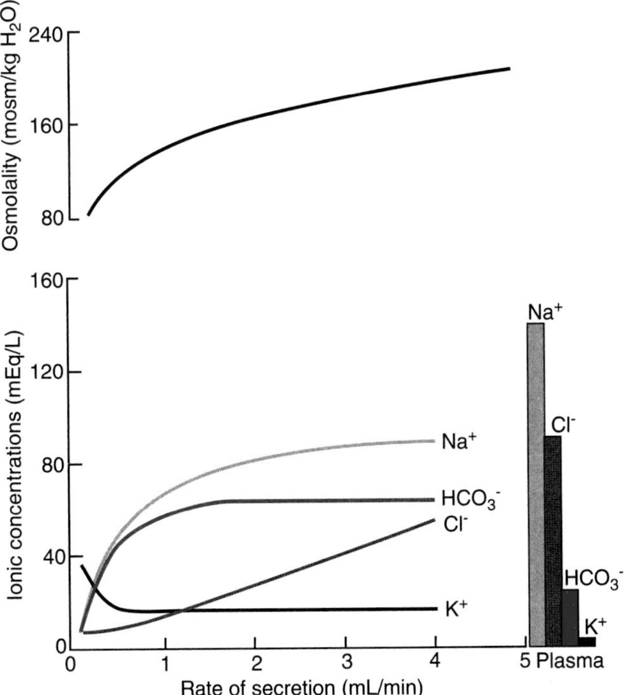

FIGURE 27.2 The osmolality and electrolyte composition of saliva at different secretion rates. (Modified from Granger DN, Barrowman JA, Kvietys PR. Clinical Gastrointestinal Physiology. Philadelphia: WB Saunders, 1985.)

puncture samples have revealed that there is little modification of the electrolyte composition of the primary secretion in the intercalated duct. However, samples from the excretory (collecting) ducts are hypotonic relative to plasma, indicating modification of the primary secretion in the striated and excretory ducts. As shown in Figure 27.2, there is less sodium (Na^+), less chloride (Cl^-), more potassium (K^+), and more bicarbonate (HCO_3^-) in saliva than in plasma. This is because Na^+ is actively absorbed from the lumen by the ductal cells, whereas K^+ and HCO_3^- ions are actively secreted into the lumen. Chloride ions leave the lumen either in exchange for HCO_3^- ions or by passive diffusion along the electrochemical gradient created by Na^+ absorption.

The electrolyte composition of saliva depends on the rate of secretion (see Fig. 27.2). As the secretion rate increases, the electrolyte composition of saliva approaches the ionic composition of plasma, but at low flow rates it differs significantly. At low secretion rates, the ductal epithelium has more time to modify and, thus, reduce the osmolality of the primary secretion, so the saliva has a much lower osmolality than plasma. The opposite is true at high secretion rates.

Although the absorption and secretion of ions may explain changes in the electrolyte composition of saliva, these processes do not explain why the osmolality of saliva is lower than that of the primary secretion of the acinar cells. Saliva is hypotonic to plasma because of a net absorption of ions by the ductal epithelium, a result of the action of a Na^+/K^+-ATPase in the basolateral cell membrane. The Na^+/K^+-ATPase transports three Na^+ ions out of the cell in place of two K^+ ions taken up by the cell. The epithelial lining of the duct is not permeable to water, so water does not follow the absorbed salt.

The two major proteins present in saliva are **amylase** and **mucin**. Salivary α-amylase (ptyalin) is produced predominantly by the parotid glands and mucin is produced mainly by the sublingual and submandibular salivary glands. Amylase catalyzes the hydrolysis of polysaccharides with α-1,4-glycosidic linkages. It is a hydrolytic enzyme involved in the digestion of starch. It is synthesized by the rough ER and transferred to the Golgi apparatus, where it is packaged into zymogen granules. The zymogen granules are stored at the apical region of the acinar cells and released with appropriate stimuli. Because some time usually passes before acids in the stomach can inactivate the amylase, a substantial amount of the ingested carbohydrate can be digested before reaching the duodenum. (The action of amylase is described later in the chapter.)

Mucin is the most abundant protein in saliva. The term describes a family of glycoproteins, each associated with different amounts of different sugars. Mucin is responsible for most of saliva's viscosity. Also present in saliva are small amounts of **muramidase**, a lysozyme that can lyse the muramic acid of certain bacteria (e.g., *Staphylococcus*); **lactoferrin**, a protein that binds iron; **epidermal growth factor**, which stimulates gastric mucosal growth; immunoglobulins (mainly IgA); and ABO blood group substances.

Saliva Has Protective Functions

Saliva's pH is almost neutral (pH = 7), and it contains HCO_3^- that can neutralize any acidic substance entering the oral cavity, including regurgitated gastric acid. Saliva plays an important role in the general hygiene of the oral cavity. The muramidase present in saliva combats bacteria by lysing the bacterial cell wall. The lactoferrin binds iron strongly, depriving microorganisms of sources of iron vital to their growth.

Saliva lubricates the mucosal surface, reducing the frictional damage caused by the rough surfaces of food. It helps small food particles stick together to form a bolus, which makes them easier to swallow. Moistening of the oral cavity with saliva facilitates speech. Saliva can dissolve flavorful substances, stimulating the different taste buds located on the tongue. Finally, saliva plays an important role in water intake; the sensation of dryness of the mouth due to low salivary secretion urges a person to drink.

Autonomic Nerves Are the Chief Modulators of Saliva Output and Content

As mentioned, salivary secretion is predominantly under the control of the autonomic nervous system. In the resting state, salivary secretion is low, amounting to about 30 mL/hr. The submandibular glands contribute about two thirds to resting salivary secretion, the parotid glands about one fourth, and the sublingual glands the remainder. Stimulation increases the rate of salivary secretion, most notably in the parotid glands, up to 400 mL/hr. The most potent stimuli for salivary secretion are acidic-tasting substances, such as citric acid. Other types of stimuli that induce salivary secretion include the smell of food and chewing. Secretion is inhibited by anxiety, fear, and dehydration.

Parasympathetic stimulation of the salivary glands results in increased activity of the acinar and ductal cells and increased salivary secretion. The parasympathetic nervous system plays an important role in controlling the secretion of saliva. The centers involved are located in the medulla oblongata. Preganglionic fibers from the inferior salivatory nucleus are contained in cranial nerve IX and synapse in the otic ganglion. They send postganglionic fibers to the parotid glands. Preganglionic fibers from the superior salivatory nucleus course with cranial nerve VII and synapse in the submandibular ganglion. They send postganglionic fibers to the submandibular and sublingual glands.

Blood flow to resting salivary glands is about 50 mL/min per 100 g tissue and can increase as much as 10-fold when salivary secretion is stimulated. This increase in blood flow is under parasympathetic control. Parasympathetic stimulation induces the acinar cells to release the protease **kallikrein**, which acts on a plasma globulin, **kininogen**, to release **lysyl-bradykinin**, which causes dilation of the blood vessels supplying the salivary glands (Fig. 27.3). Atropine, an anticholinergic agent, is a potent inhibitor of salivary secretion. Agents that inhibit acetylcholinesterase (e.g., pilocarpine) enhance salivary secretion. Some parasympathetic stimulation also increases blood flow to the salivary glands directly, apparently via the release of the neurotransmitter **vasoactive intestinal peptide** (VIP).

The salivary glands are also innervated by the sympathetic nervous system. Sympathetic fibers arise in the upper thoracic segments of the spinal cord and synapse in the **superior cervical ganglion.** Postganglionic fibers leave the superior cervical ganglion and innervate the acini, ducts, and blood vessels. Sympathetic stimulation tends to result in a short-lived and much smaller increase in salivary secretion than parasympathetic stimulation. The increase in salivary secretion observed during sympathetic stimulation is mainly via β-adrenergic receptors, which are more involved in stimulating the contraction of myoepithelial cells. Although both sympathetic and parasympathetic stimulation increases salivary secretion, the responses produced are different (Table 27.1).

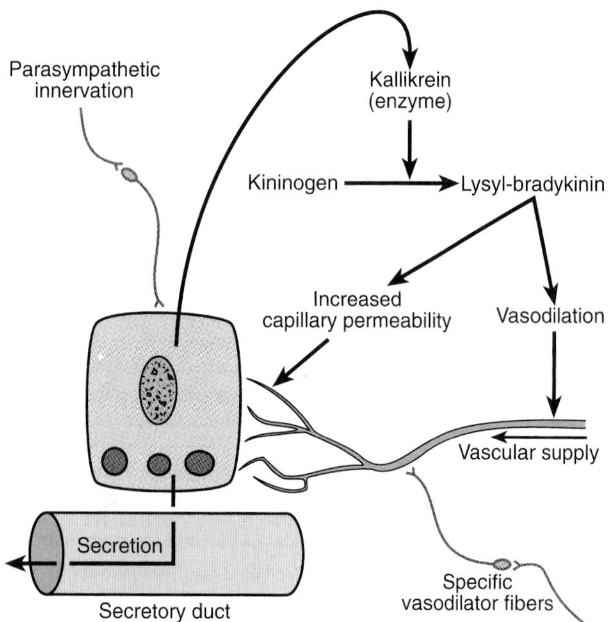

FIGURE 27.3 **The effect of parasympathetic innervation on blood flow to the salivary glands.** (Modified from Sanford PA. Digestive System Physiology. Baltimore: University Park Press, 1982.)

TABLE 27.1 Effects of Parasympathetic and Sympathetic Stimulation on Salivary Secretion Responses

Responses	Parasympathetic	Sympathetic
Saliva output	Copious	Scant
Temporal response	Sustained	Transient
Composition	Protein-poor, high K^+ and HCO_3^-	Protein-rich, low K^+ and HCO_3^-
Response to denervation	Decreased secretion, atrophy	Decreased secretion

Mineralocorticoid administration reduces the Na^+ concentration of saliva with a corresponding rise in K^+ concentration. Mineralocorticoids act mainly on the striated and excretory ducts. Arginine vasopressin (AVP) reduces the Na^+ concentration in saliva by increasing Na^+ reabsorption by the ducts. Some GI hormones (e.g., VIP and substance P) have been experimentally demonstrated to evoke salivary secretory responses.

GASTRIC SECRETION

The major function of the stomach is storage, but it also absorbs water-soluble and lipid-soluble substances (e.g., alcohol and some drugs). An important function of the stomach is to prepare the chyme for digestion in the small intestine. **Chyme** is the semi-fluid material produced by the gastric digestion of food. Chyme results partly from the conversion of large solid particles into smaller particles via the combined peristaltic movements of the stomach and contraction of the pyloric sphincter. The propulsive, grinding, and retropulsive movements associated with antral peristalsis are discussed in Chapter 26. A combination of the squirting of antral content into the duodenum, the grinding action of the antrum, and retropulsion provides much of the mechanical action necessary for the emulsification of dietary fat, which plays an important role in fat digestion.

Numerous Cell Types in the Stomach Contribute to Gastric Secretions

The fundus of the stomach is relatively thin-walled and can be expanded with ingested food (see Fig. 26.24). The main body (corpus) of the empty stomach is composed of many longitudinal folds called **rugae gastricae**. The stomach's mucosal lining, the **glandular gastric mucosa**, contains three main types of glands: cardiac, pyloric, and oxyntic. These glands contain mucous cells that secrete mucus and HCO_3^- ions, which protect the stomach from the acid in the stom-

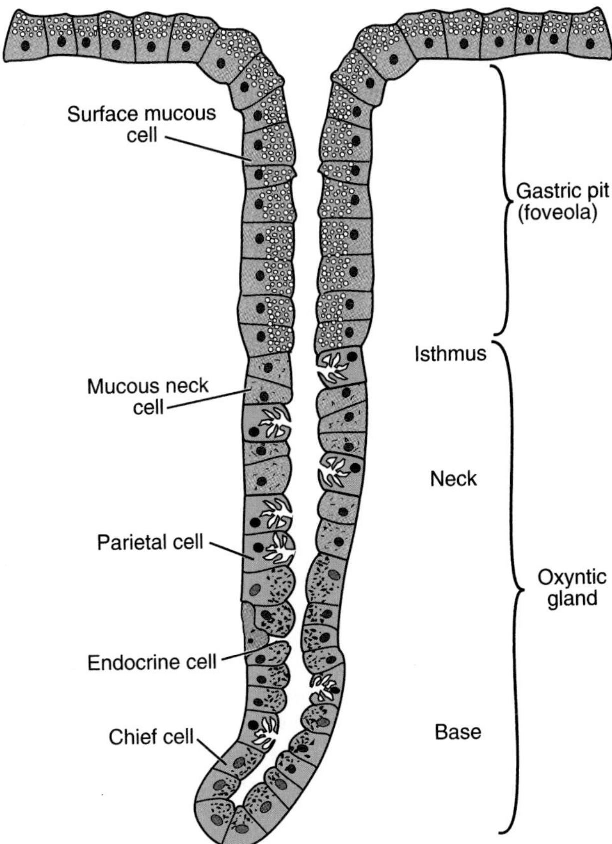

FIGURE 27.4 A simplified diagram of the oxyntic gland in the corpus of a mammalian stomach. One to several glands may open into a common gastric pit. (Modified from Ito S. Functional gastric morphology. In: Johnson LR, Christensen J, Jackson MJ, et al. eds. Physiology of the Gastrointestinal Tract. New York: Raven, 1987.)

ach lumen. The **cardiac glands** are located in a small area adjacent to the esophagus and are lined by mucus-producing columnar cells. The **pyloric glands** are located in a larger area adjacent to the duodenum. They contain cells similar to mucous neck cells but differ from cardiac and oxyntic glands in having many gastrin-producing cells called **G cells.** The **oxyntic glands,** the most abundant glands in the stomach, are found in the fundus and the corpus.

The oxyntic glands contain **parietal (oxyntic) cells, chief cells, mucous neck cells,** and some endocrine cells (Fig. 27.4). Surface mucous cells occupy the gastric pit (foveola); in the gland, most mucous cells are located in the neck region. The base of the oxyntic gland contains mostly chief cells, along with some parietal and endocrine cells. Mucous neck cells secrete mucus, parietal cells principally secrete **hydrochloric acid** (HCl) and **intrinsic factor,** and chief cells secrete **pepsinogen.** (Intrinsic factor and pepsinogen are discussed later in the chapter.)

Parietal cells are the most distinctive cells in the stomach. The structure of resting parietal cells is unique in that they have intracellular canaliculi as well as an abundance of mitochondria (Fig. 27.5A). This network consists of clefts and canals that are continuous with the lumen of the oxyntic gland. There is also an extensive smooth ER referred to as the **tubulovesicular membranes.** In active parietal cells (Fig. 27.5B), the tubulovesicular system is greatly diminished with a concomitant increase in the intracellular canaliculi. The mechanism for these morphological changes is not well understood. Hydrochloric acid is secreted across the parietal cell microvillar membrane and flows out of the intracellular canaliculi into the oxyntic gland lumen. As mentioned, surface mucous cells line the entire surface of the gastric mucosa and the openings of the cardiac, pyloric, and oxyntic glands. These cells secrete mucus and HCO_3^- to protect the gastric surface from the

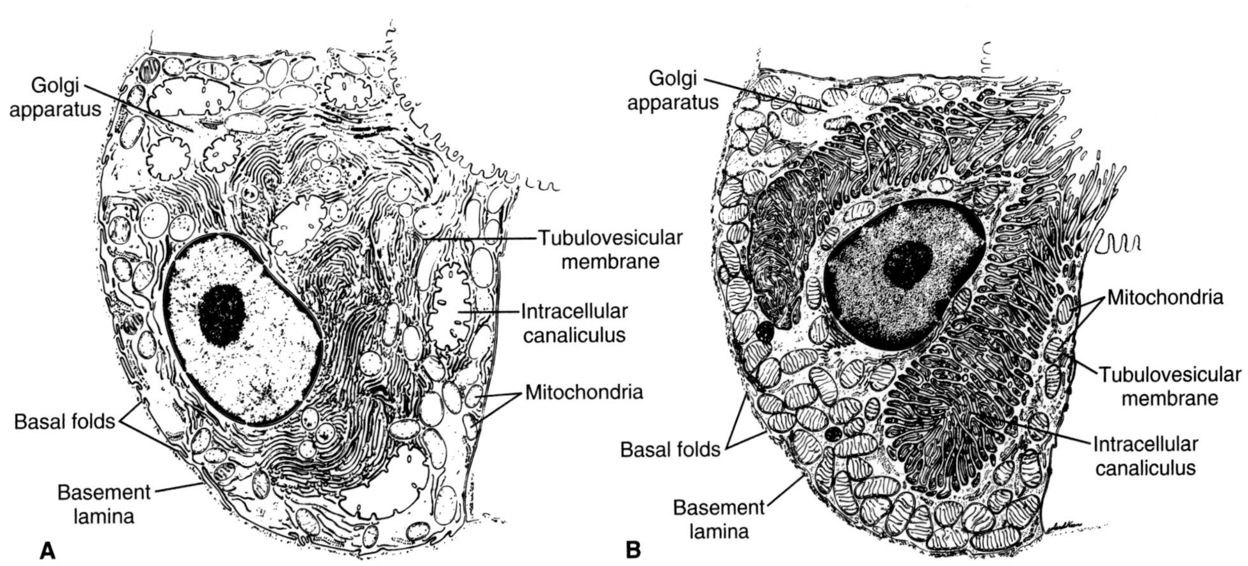

FIGURE 27.5 Parietal cells of the stomach. A, A nonsecreting parietal cell. The cytoplasm is filled with tubulovesicular membranes, and the intracellular canaliculi have become internalized, distended, and devoid of microvilli. **B,** An actively secreting parietal cell. Compared to the resting parietal cell, the most striking difference is the abundance of long microvilli and the paucity of the tubulovesicular system, making the mitochondria appear more numerous. (From Ito S. Functional gastric morphology. In: Johnson LR, Christensen J, Jackson MJ, et al. eds. Physiology of the Gastrointestinal Tract. New York: Raven, 1987.)

acidic environment of the stomach. The distinguishing characteristic of a surface mucous cell is the presence of numerous mucus granules at its apex. The number of mucus granules in storage varies depending on synthesis and secretion. The mucous neck cells of the oxyntic glands are similar in appearance to surface mucous cells.

Chief cells are morphologically distinguished primarily by the presence of zymogen granules in the apical region and an extensive ER. The zymogen granules contain pepsinogen, a precursor of the enzyme pepsin.

Also present in the stomach are various neuroendocrine cells, such as G cells, located predominantly in the antrum. These cells produce the hormone **gastrin,** which stimulates acid secretion by the stomach. An overabundance of gastrin secretion, a condition known as Zollinger-Ellison syndrome, results in gastric hypersecretion and peptic ulceration. **D cells,** also present in the antrum, produce **somatostatin,** another important GI hormone.

Gastric Juice Contains Hydrochloric Acid, Electrolytes, and Proteins

The important constituents of human gastric juice are HCl, electrolytes, pepsinogen, and intrinsic factor. The pH is low, about 0.7 to 3.8. This raises a question: How does the gastric mucosa protect itself from acidity? As mentioned earlier, the surface mucous cells secrete a fluid containing mucus and HCO_3^- ions. The mucus forms a **mucus gel layer** covering the surface of the gastric mucosa. Bicarbonate trapped in the mucus gel layer neutralizes acid, preventing damage to the mucosal cell surface.

Hydrochloric Acid Is Secreted by the Parietal Cells

The HCl present in the gastric lumen is secreted by the parietal cells of the corpus and fundus. The mechanism of HCl production is depicted in Figure 27.6. A **H^+/K^+-ATPase** in the apical (luminal) cell membrane of the parietal cell actively pumps H^+ out of the cell in exchange for K^+ entering the cell. The H^+/K^+-ATPase is inhibited by omeprazole. Omeprazole, an acid-activated prodrug that is converted in the stomach to the active drug, binds to two cysteines on the ATPase, resulting in an irreversible inactivation. Although the secreted H^+ is often depicted as being derived from carbonic acid (see Fig. 27.6), the source of H^+ is probably mostly from the dissociation of H_2O. Carbonic acid (H_2CO_3) is formed from carbon dioxide (CO_2) and H_2O in a reaction catalyzed by carbonic anhydrase. Carbonic anhydrase is inhibited by acetazolamide. The CO_2 is provided by metabolic sources inside the cell and from the blood.

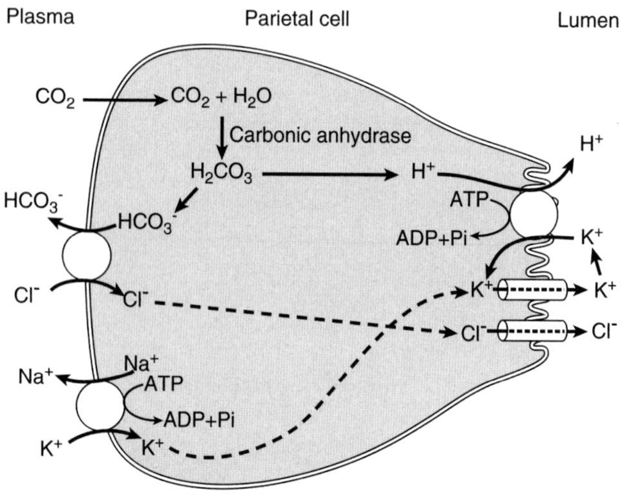

FIGURE 27.6 **The mechanism of HCl secretion by the gastric parietal cell.**

For the H^+/K^+-ATPase to work, an adequate supply of K^+ ions must exist outside the cell. Although the mechanism is still unclear, there is an increase in K^+ conductance (through K^+ channels) in the apical membrane of the parietal cells simultaneous with active acid secretion. This surge of K^+ conductance ensures plenty of K^+ in the lumen. The H^+/K^+-ATPase recycles K^+ ions back into the cell in exchange for H^+ ions. As shown in Figure 27.6, the basolateral cell membrane has an electroneutral Cl^-/HCO_3^- exchanger that balances the entry of Cl^- into the cell with an equal amount of HCO_3^- entering the bloodstream. The Cl^- inside the cell then leaks into the lumen through Cl^- channels, down an electrochemical gradient. Consequently, HCl is secreted into the lumen.

A large amount of HCl can be secreted by the parietal cells. This is balanced by an equal amount of HCO_3^- added to the bloodstream. The blood coming from the stomach during active acid secretion contains much HCO_3^-, a phenomenon called the **alkaline tide.** The osmotic gradient created by the HCl concentration in the gland lumen drives water passively into the lumen, thereby maintaining the iso-osmolality of the gastric secretion.

Gastric Juice Contains Various Electrolytes

Figure 27.7 depicts the changes in the electrolyte composition of gastric juice at different secretion rates. At a low secretion rate, gastric juice contains high concentrations of Na^+ and Cl^- and low concentrations of K^+ and H^+. When the rate of secretion increases, the concentration of Na^+ decreases while that of H^+ increases significantly. Also coupled with this increase in gastric secretion is an increase in Cl^- concentration. To understand the changes in electrolyte composition of gastric juice at different secretion rates, it is important to remember that gastric juice is derived from the secretions of two major sources: parietal cells and nonparietal cells. Secretion from nonparietal cells is probably constant; therefore, it is parietal secretion (HCl secretion) that contributes mainly to the changes in electrolyte composition with higher secretion rates.

Gastric Secretion Performs Digestive, Protective, and Other Functions

Gastric juice contains several proteins: pepsinogens, pepsins, salivary amylase, gastric lipase, and intrinsic factor. The chief cells of the oxyntic glands release inactive pepsinogen. Pepsinogen is activated by acid in the gastric lumen to form the active enzyme **pepsin.** Pepsin also cat-

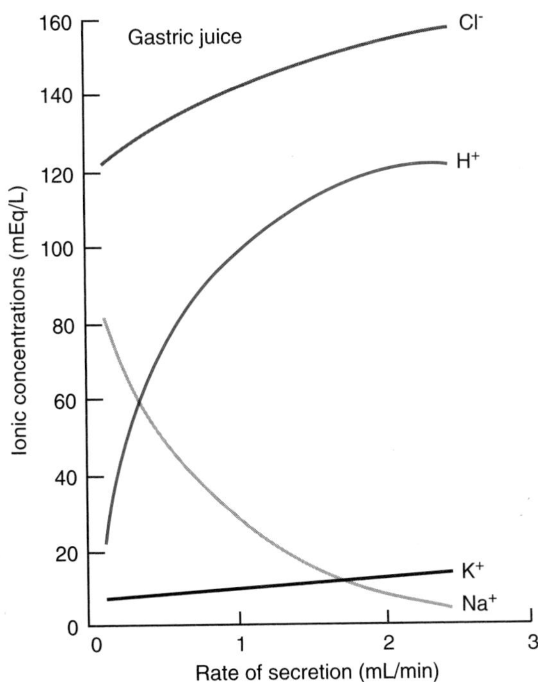

FIGURE 27.7 The concentration of electrolytes in the gastric juice of a healthy, young adult man as a function of the rate of secretion. (Modified from Davenport HW. Physiology of the Digestive Tract. Chicago: Year Book, 1977.)

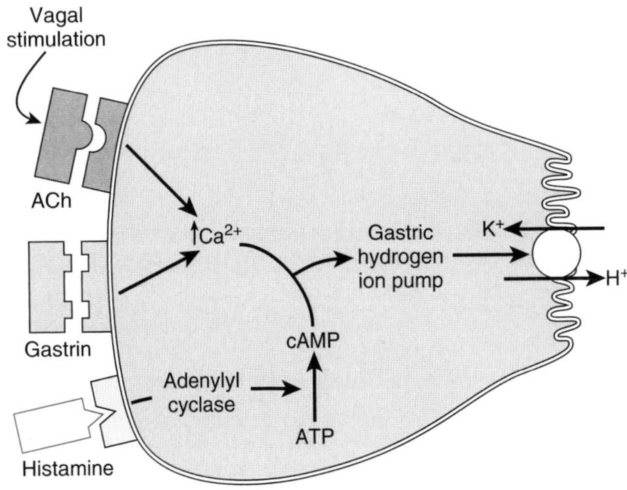

FIGURE 27.8 The stimulation of parietal cell acid secretion by histamine, gastrin, and acetylcholine (ACh), and potentiation of the process.

alyzes its own formation from pepsinogen. Pepsin, an **endopeptidase**, cleaves protein molecules from the inside, resulting in the formation of smaller peptides. The optimal pH for pepsin activity is 1.8 to 3.5; therefore, it is extremely active in the highly acidic medium of gastric juice.

The acidity of gastric juice poses a barrier to invasion of the GI tract by microbes and parasites. The intrinsic factor, produced by stomach parietal cells, is necessary for absorption of vitamin B_{12} in the terminal ileum.

Gastric Secretion Is Under Neural and Hormonal Control

Gastric acid secretion is mediated through neural and hormonal pathways. Vagus nerve stimulation is the neural effector; histamine and gastrin are the hormonal effectors (Fig. 27.8). Parietal cells possess special histamine receptors, **H_2 receptors**, whose stimulation results in increased acid secretion. Special endocrine cells of the stomach, known as **enterochromaffin-like (ECL) cells,** are believed to be the source of this histamine, but the mechanisms that stimulate them to release histamine are poorly understood. The importance of histamine as an effector of gastric acid secretion has been indirectly demonstrated by the effectiveness of cimetidine, a H_2 blocker, in reducing acid secretion. H_2 blockers are commonly used for the treatment of peptic ulcer disease or gastroesophageal reflux disease.

The effects of each of these three stimulants (ACh, gastrin, and histamine) augment those of the others, a phenomenon known as **potentiation.** Potentiation is said to occur when the effect of two stimulants is greater than the effect of either stimulant alone. For example, the interaction of gastrin and ACh molecules with their respective receptors results in an increase in intracellular Ca^{2+} concentration, and the interaction of histamine with its receptor results in an increase in cellular cAMP production. The increased intracellular Ca^{2+} and cAMP interact in numerous ways to stimulate the gastric H^+/K^+-ATPase, which brings about an increase in acid secretion (see Fig. 27.8). Exactly how the increase in intracellular Ca^{2+} and cAMP greatly enhances the effect of the other in stimulating gastric acid secretion is not well understood.

Acid Secretion Is Increased During a Meal

The stimulation of acid secretion resulting from the ingestion of food can be divided into three phases: the cephalic phase, the gastric phase, and the intestinal phase (Table 27.2). The **cephalic phase** involves the central nervous system. Smelling, chewing, and swallowing food (or merely the thought of food) send impulses via the vagus nerves to the parietal and G cells in the stomach. The nerve endings release ACh, which directly stimulates acid secretion from parietal cells. The nerves also release **gastrin-releasing peptide** (GRP), which stimulates G cells to release gastrin, indirectly stimulating parietal cell acid secretion. The fact that the effect of GRP is atropine-resistant indicates that it works through a noncholinergic pathway. The cephalic phase probably accounts for about 40% of total acid secretion.

The **gastric phase** is mainly a result of gastric distension and chemical agents such as digested proteins. Distension of the stomach stimulates mechanoreceptors, which stimulate the parietal cells directly through short local (enteric) reflexes and by long **vago-vagal reflexes.** Vago-vagal reflexes are mediated by afferent and efferent impulses traveling in the vagus nerves. Digested proteins in the stomach are also potent stimulators of gastric acid secretion, an ef-

TABLE 27.2 The Three Phases of Stimulation of Acid Secretion After Ingesting a Meal

Phase	Stimulus	Pathway	Stimulus to Parietal Cell
Cephalic	Thought of food, smell, taste, chewing, and swallowing	Vagus nerve to	
		Parietal cells	ACh
		G cells	Gastrin
Gastric	Stomach distension	Local (enteric) reflexes and vago-vagal reflexes to	
		Parietal cells	ACh
		G cells	Gastrin
	Digested peptides	G cells	Gastrin
Intestinal	Protein digestion products in duodenum	Amino acids in blood	Amino acids
	Distension	Intestinal endocrine cell	Enterooxyntin

fect mediated through gastrin release. Several other chemicals, such as alcohol and caffeine, stimulate gastric acid secretion through mechanisms that are not well understood. The gastric phase accounts for about 50% of total gastric acid secretion.

During the **intestinal phase,** protein digestion products in the duodenum stimulate gastric acid secretion through the action of the circulating amino acids on the parietal cells. Distension of the small intestine, probably via the release of the hormone **enterooxyntin** from intestinal endocrine cells, stimulates acid secretion. The intestinal phase accounts for only about 10% of total gastric acid secretion.

Gastric Acid Secretion Is Inhibited by Several Mechanisms

The inhibition of gastric acid secretion is physiologically important for two reasons. First, the secretion of acid is important only during the digestion of food. Second, excess acid can damage the gastric and the duodenal mucosal surfaces, causing ulcerative conditions (see Clinical Focus Box 27.1). The body has an elaborate system for regulating the amount of acid secreted by the stomach. Gastric luminal pH is a sensitive regulator of acid secretion. Proteins in food provide buffering in the lumen; consequently, the gastric luminal pH is usually above 3 after a meal. However, if the buffering capacity of protein is exceeded or if the stomach is empty, the pH of the gastric lumen will fall below 3. When this happens, the endocrine cells (D cells) in the antrum secrete somatostatin, which inhibits the release of gastrin and, thus, gastric acid secretion.

Another mechanism for inhibiting gastric acid secretion is acidification of the duodenal lumen. Acidification stimulates the release of **secretin,** which inhibits the release of gastrin, and several peptides, collectively known as **enterogastrones**, which are released by intestinal endocrine cells.

CLINICAL FOCUS BOX 27.1

Acid Secretion and Duodenal Ulcer

Ulcerative lesions of the gastroduodenal area are classified as **peptic ulcer disease.** Peptic ulcer disease is associated with a high rate of recurrence. The saying, "no acid, no ulcer," has withstood the test of time and is still accepted by most physicians and researchers as generally true. One possible cause of gastric and duodenal ulcers is reduced mucosal defense mechanisms. Human and animal data, however, have demonstrated that duodenal ulcers do not occur with reduced mucosal defense mechanisms alone but also require the presence of sufficient amounts of acid. In one study, patients suffering from duodenal ulcer had a significantly increased mean number of gastric parietal cells and appeared to have increased sensitivity to gastrin when compared with healthy subjects. Although the reason is unknown, the stomach emptying rate may be greatly increased in duodenal ulcer patients. Another abnormality in duodenal ulcer patients is decreased inhibition of gastrin release by acid and a reduced rate of duodenal bicarbonate secretion. It should be emphasized, however, that a significant number of patients with duodenal ulcer do not have excessive secretion of acid.

An exciting development in the field of peptic ulcer disease is the finding of a possible correlation between *Helicobacter pylori (H. pylori)* infection and the incidence of gastric and duodenal ulcers. The role of *H. pylori* infection in the genesis of peptic ulcers is unclear, but in a significant number of patients, eradication of the bacteria reduces the rate of ulcer recurrence. *H. pylori* produces large quantities of the enzyme urease, which hydrolyzes urea to produce ammonia. The ammonia neutralizes acid in the stomach, protecting the bacteria from the injurious effects of hydrochloric acid.

Although the mechanism has not been elucidated, the presence of *H. pylori* in the stomach enhances the secretion of gastrin by the gastric mucosa. Whether increased gastrin release by the presence of *H. pylori* is responsible for the increased recurrence of gastric and duodenal ulcers in patients has yet to be proven. It has been demonstrated that H_2 receptor antagonists (cimetidine and ranitidine) have no effect on *H. pylori* infection. In contrast, omeprazole (an inhibitor of the H^+/K^+-ATPase) appears to be bacteriostatic. A combined therapy using omeprazole and the antibiotic amoxicillin appears to be effective in the eradication of *H. pylori* in 50 to 80% of patients with peptic ulcer disease, resulting in a significant reduction of duodenal ulcer recurrence.

Acid, fatty acids, or hyperosmolar solutions in the duodenum stimulate the release of enterogastrones, which inhibit gastric acid secretion. **Gastric inhibitory peptide** (**GIP**), an enterogastrone produced by the small intestinal endocrine cells, inhibits parietal cell acid secretion. There are also several currently unidentified enterogastrones.

PANCREATIC SECRETION

One of the major functions of pancreatic secretion is to neutralize the acids in the chyme when it enters the duodenum from the stomach. This mechanism is important because pancreatic enzymes operate optimally near neutral pH. Another important function is the production of enzymes involved in the digestion of dietary carbohydrate, fat, and protein.

The Pancreas Consists of a Network of Acini and Ducts

The human pancreas is located in close apposition to the duodenum. It performs both endocrine and exocrine functions, but here we discuss only its exocrine function. (The endocrine functions are discussed in Chapter 35.)

The exocrine pancreas is composed of numerous small, sac-like dilatations called acini composed of a single layer of pyramidal **acinar cells** (Fig. 27.9). These cells are actively involved in the production of enzymes. Their cytoplasm is filled with an elaborate system of ER and Golgi apparatus. Zymogen granules are observed in the apical region of acinar cells. A few **centroacinar cells** line the lumen of the acinus. In contrast to acinar cells, these cells lack an elaborate ER and Golgi apparatus. Their major function seems to be modification of the electrolyte composition of the pancreatic secretion. Because the processes involved in the secretion or uptake of ions are active, centroacinar cells have numerous mitochondria in their cytoplasm.

The acini empty their secretions into intercalated ducts, which join to form intralobular and then interlobular ducts. The interlobular ducts empty into two pancreatic ducts: a major duct, the duct of Wirsung, and a minor duct, the duct of Santorini. The duct of Santorini enters the duodenum more proximally than the duct of Wirsung, which enters the duodenum usually together with the common bile duct. A ring of smooth muscle, the **sphincter of Oddi**, surrounds the opening of these ducts in the duodenum. The sphincter of Oddi not only regulates the flow of bile and pancreatic juice into the duodenum but also prevents the reflux of intestinal contents into the pancreatic ducts.

Pancreatic Secretions Are Rich in Bicarbonate Ions

The pancreas secretes about 1 L/day of HCO_3^--rich fluid. The osmolality of pancreatic fluid, unlike that of saliva, is equal to that of plasma at all secretion rates. The Na^+ and K^+ concentrations of pancreatic juice are the same as those in plasma, but unlike plasma, pancreatic juice is enriched with HCO_3^- and has a relatively low Cl^- concentration (Fig. 27.10). The HCO_3^- concentration increases with increases in secretion rate and reaches a maximal concentration of about 140 mEq/L, yielding a solution with a pH of 8.2. A reciprocal relationship exists between the Cl^- and HCO_3^- concentration in pancreatic juice. As the concentration of HCO_3^- increases with secretion rate, the Cl^- concentration falls accordingly, resulting in a combined total anion concentration that remains relatively constant (150 mEq/L) regardless of the pancreatic secretion rate.

Two separate mechanisms have been proposed to explain the secretion of a HCO_3^--rich juice by the pancreas and the HCO_3^- concentration changes. The first mechanism proposes that some cells, probably the acinar cells, secrete a plasma-like fluid containing predominantly Na^+ and Cl^-, while other cells, probably the centroacinar and duct cells, secrete a HCO_3^--rich solution when stimulated. Depending on the different rates of secretion from these three different cell types, the pancreatic juice can be rich in either HCO_3^- or Cl^-. The second mechanism depicts the primary secretion as rich in HCO_3^-. As the HCO_3^- solution moves down the ductal system, HCO_3^- ions are exchanged for Cl^- ions. When the flow is fast, there is little time for this exchange, so the concentration of HCO_3^- is high. The opposite is true when the flow is slow.

The secretion of electrolytes by pancreatic duct cells is depicted in Figure 27.11. A Na^+/H^+ exchanger is located in the basolateral cell membrane. The energy required to drive the exchanger is provided by the Na^+/K^+-ATPase-generated Na^+ gradient. Carbon dioxide diffuses into the cell and combines with H_2O to form H_2CO_3, a reaction catalyzed by carbonic anhydrase, which dissociates to H^+ and HCO_3^-. The H^+ is extruded by the Na^+/H^+ exchanger, and HCO_3^- is exchanged for luminal Cl^- via a Cl^-/HCO_3^- exchanger. Also located in the luminal cell membrane is a protein called **cystic fibrosis transmembrane conductance regulator** (CFTR). CFTR is an ion channel belonging to the ABC (ATP-binding cassette) family of proteins. Regulated by ATP, its major function is to secrete Cl^- ions out of the cells, providing Cl^- in the lumen for the Cl^-/HCO_3^- exchanger to work. The Na^+/K^+-ATPase removes cell Na^+ that enters through the Na^+/H^+ antiporter. Sodium from the interstitial space follows secreted HCO_3^- by diffusing through a paracellular path (between the cells). Movement of H_2O into the duct lumen is passive, driven by the osmotic gradient. The net result of pancreatic HCO_3^- secretion is the release of H^+ into the plasma; thus, pancreatic secretion is associated with an **acid tide** in the plasma.

Pancreatic Secretions Neutralize Luminal Acids and Digest Nutrients

As mentioned, one of the primary functions of pancreatic secretion is to neutralize the acidic chyme presented to the duodenum. The enzymes present in intestinal lumen work best at a pH close to neutral; therefore, it is crucial to increase the pH of the chyme. As described above, pancreatic juice is highly basic because of its HCO_3^- content. Thus, the acidic chyme presented to the duodenum is rapidly neutralized by pancreatic juice.

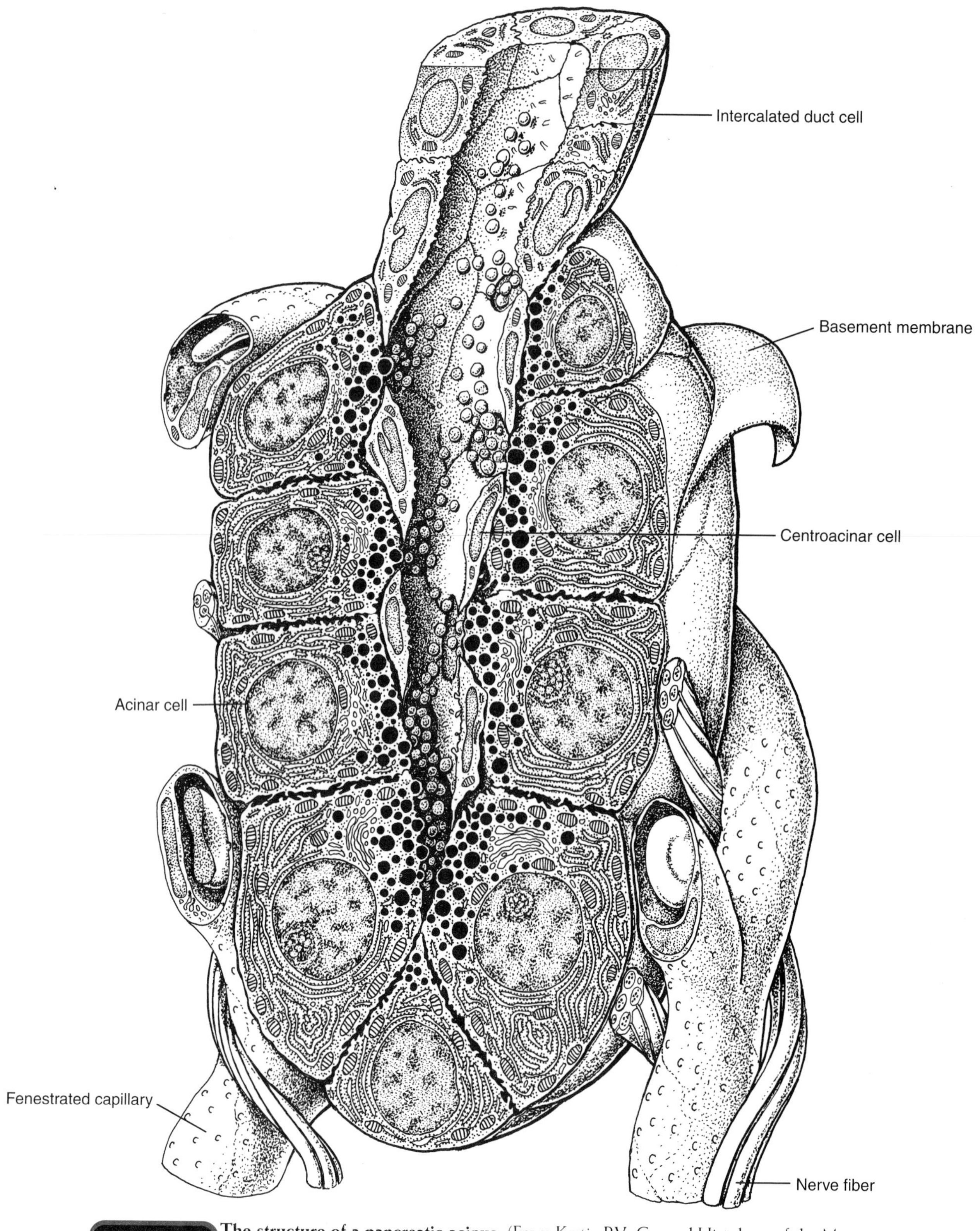

FIGURE 27.9 **The structure of a pancreatic acinus.** (From Krstic RV. General Histology of the Mammal. New York: Springer-Verlag, 1984.)

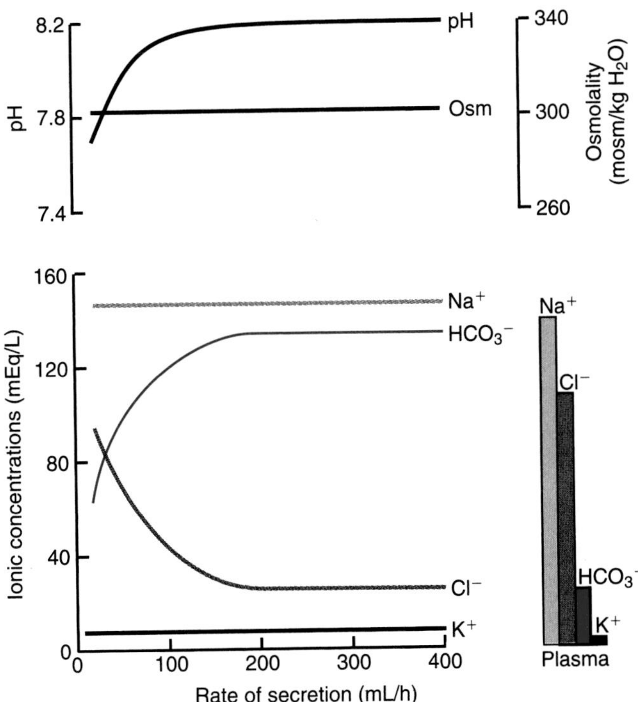

FIGURE 27.10 The pH, osmolality, and electrolyte composition of pancreatic juice at different secretion rates. Plasma electrolyte composition is provided for comparison. (Adapted from Granger DN, Barrowman JA, Kvietys PR. Clinical Gastrointestinal Physiology. Philadelphia: WB Saunders, 1985.)

The other major function of pancreatic secretion is the production of large amounts of pancreatic enzymes. Table 27.3 summarizes the various enzymes present in pancreatic juice. Some are secreted as proenzymes, which are activated in the duodenal lumen to form the active enzymes. (The digestion of nutrients by these enzymes is discussed later in the chapter.)

Pancreatic Secretion Is Under Neural and Hormonal Control

Pancreatic secretion is stimulated by parasympathetic fibers in the vagus nerve that release ACh. Stimulation of the vagus nerve results predominantly in an increase in enzyme secretion—fluid and HCO_3^- secretion are marginally stimulated or unchanged. Sympathetic nerve fibers mainly innervate the blood vessels supplying the pancreas, causing vasoconstriction. Stimulation of the sympathetic nerves neither stimulates nor inhibits pancreatic secretion, probably because of the reduction in blood flow.

The secretion of electrolytes and enzymes by the pancreas is greatly influenced by circulating GI hormones, particularly secretin and **cholecystokinin** (CCK). Secretin tends to stimulate a HCO_3^--rich secretion. CCK stimulates a marked increase in enzyme secretion. Both hormones are produced by the small intestine, and the pancreas has receptors for them.

Structurally similar hormones have effects similar to those of secretin and CCK. For example, VIP, structurally similar to secretin, stimulates the secretion of HCO_3^- and H_2O. However, because VIP is much weaker than secretin, it produces a weaker pancreatic response when given together with secretin than when secretin is given alone. Similarly, gastrin can stimulate pancreatic enzyme secretion because of its structural similarity to CCK, but unlike CCK, it is a weak agonist for pancreatic enzyme secretion.

Pancreatic Secretion Is Phasic

The regulation of pancreatic secretion by various hormonal and neural factors is summarized in Table 27.4. Seeing, smelling, tasting, chewing, swallowing, or thinking about food results in the secretion of a pancreatic juice rich in enzymes. In this cephalic phase, stimulation of pancreatic secretion is mainly mediated by direct efferent impulses sent by vagal centers in the brain to the pancreas and, to a minor extent, by the indirect effect of parasympathetic stimulation of gastrin release. The gastric phase is initiated when food enters the stomach and distends it. Pancreatic secretion is then stimulated by vago-vagal reflex. Gastrin may also be involved in this phase.

During the most important phase, the intestinal phase, the entry of acidic chyme from the stomach into the small intestine stimulates the release of secretin by the **S cells** (a type of endocrine cell) in the intestinal mucosa. When the pH of the lumen in the duodenum decreases, the secretin concentration in plasma increases. This response is followed by an increase in HCO_3^- output by the pancreas. The secretion of pancreatic enzymes is increased by circulating CCK and by parasympathetic stimulation through a vago-vagal reflex. The release of CCK by the **I cells** (a type of endocrine cell) in the intestinal mucosa is stimulated by exposure of the intestinal mucosa to long-chain fatty acids (lipid digestion products) and free amino acids.

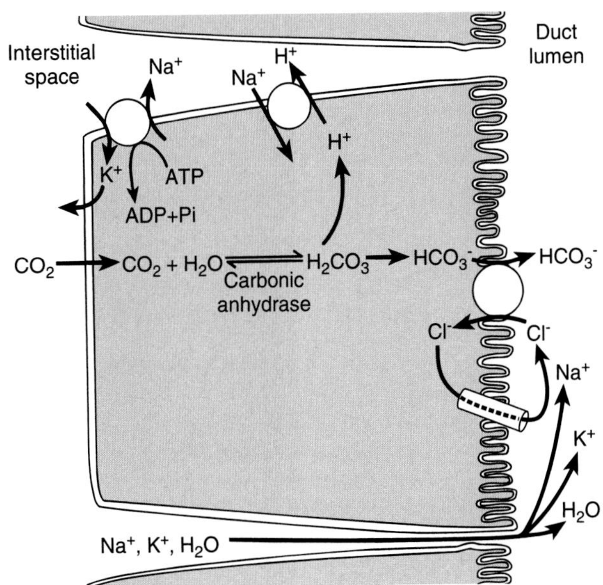

FIGURE 27.11 A model for electrolyte secretion by pancreatic duct cells. The luminal membrane Cl^- channel is CFTR (cystic fibrosis transmembrane conductance regulator).

TABLE 27.3 Characteristics of Pancreatic Enzymes

Enzyme	Specific Hydrolytic Activity
Proteolytic	
Endopeptidases	
Trypsin(ogen)	Cleaves peptide linkages in which the carboxyl group is either arginine or lysine
Chymotrypsin(ogen)	Cleaves peptides at the carboxyl end of hydrophobic amino acids, e.g., tyrosine or phenylalanine
(Pro)elastase	Cleaves peptide bonds at the carboxyl terminal of aliphatic amino acids
Exopeptidase	
(Pro)carboxypeptidase	Cleaves amino acids from the carboxyl end of the peptide
Amylolytic	
α-Amylase	Cleaves α-1,4-glycosidic linkages of glucose polymers
Lipases	
Lipase	Cleaves the ester bond at the 1 and 3 positions of triglycerides, producing free fatty acids and 2-monoglyceride
(Pro)phospholipase A_2	Cleaves the ester bond at the 2 position of phospholipids
Carboxylester hydrolase (cholesterol esterase)	Cleaves cholesteryl ester to free cholesterol
Nucleolytic	
Ribonuclease	Cleaves ribonucleic acids into mononucleotides
Deoxyribonuclease	Cleaves deoxyribonucleic acids into mononucleotides

The suffix *-ogen* or prefix *pro-* indicates the enzyme is secreted in an inactive form.

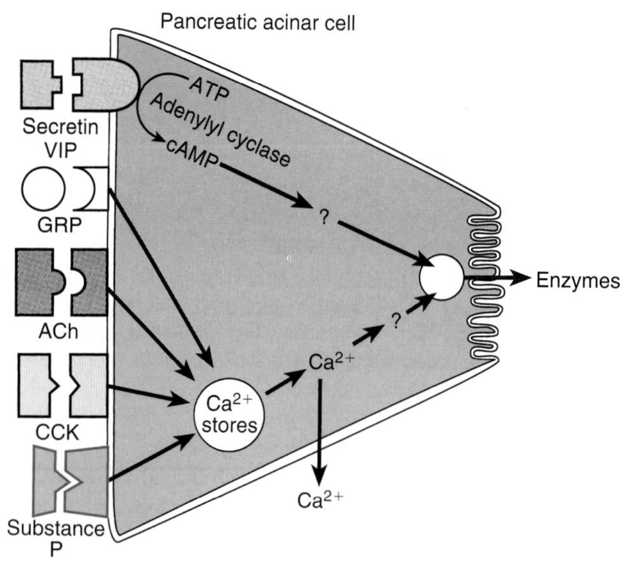

FIGURE 27.12 The stimulation of pancreatic secretion by hormones and neurotransmitters.

Potentiation, as previously described for gastric secretion, also exists in the pancreas. Its effect in pancreatic secretion is a result of the different receptors used for ACh, CCK, and secretin. Secretin binding triggers an increase in adenylyl cyclase activity, which, in turn, stimulates the formation of cAMP (Fig. 27.12). Acetylcholine (ACh), CCK, and the neuropeptides GRP and substance P bind to their respective receptors and trigger the release of Ca^{2+} from intracellular stores. The increase in intracellular Ca^{2+} release and cAMP formation results in an increase in pancreatic enzyme secretion. The mechanism by which this takes place is not well understood.

BILIARY SECRETION

The human liver secretes 600 to 1,200 mL/day of bile into the duodenum. Bile contains bile salts, bile pigments (e.g., bilirubin), cholesterol, phospholipids, and proteins and performs several important functions. For example, bile salts play an important role in the intestinal absorption of lipid. Bile salts are derived from cholesterol and, therefore, constitute a path for its excretion. Biliary secretion is an important route for the excretion of bilirubin from the body.

Bile canaliculi are fine tubular canals running between the hepatocytes. Bile flows through the canaliculi to the bile ducts, which drain into the gallbladder. During the interdigestive state, the sphincter of Oddi, which controls

TABLE 27.4 Factors Regulating Pancreatic Secretion After a Meal

Phase	Stimulus	Mediators	Response
Cephalic	Thought of food, smell, taste, chewing, and swallowing	Release of ACh and gastrin by vagal stimulation	Increased secretion, with a greater effect on enzyme output
Gastric	Protein in food	Gastrin	Increased secretion, with a greater effect on enzyme output
	Gastric distension	Vago-vagal reflex	Increased secretion, with a greater effect on enzyme output
Intestinal	Acid in chyme	Secretin	Increased H_2O and HCO_3^- secretion
	Long-chain fatty acids	CCK and vago-vagal reflex	Increased secretion, with a greater effect on enzyme output
	Amino acids and peptides	CCK and vago-vagal reflex	Increased secretion, with a greater effect on enzyme output

TABLE 27.5 Electrolyte Composition of Human Hepatic Bile

Constituent	Bile Concentration (mEq/L)	Plasma Concentration (mEq/L)
Na^+	140–170	145
K^+	4.0–6.0	4.5
Ca^{2+}	1.2–5.0	4.6
Mg^{2+}	1.5–3.0	1.6
Cl^-	95–125	105
HCO_3^-	15–60	24

the opening of the duct that carries biliary and pancreatic secretions, is contracted and the gallbladder is relaxed. Thus, most of the hepatic bile is stored in the gallbladder during this period. After the ingestion of a meal, CCK is released into the blood, causing contraction of the gallbladder and resulting in the delivery of bile into the duodenum.

The Major Components of Bile Are Electrolytes, Bile Salts, and Lipids

The electrolyte composition of human bile collected from the hepatic ducts is similar to that of blood plasma, except the HCO_3^- concentration may be higher, resulting in an alkaline pH (Table 27.5). **Bile acids** are formed in the liver from cholesterol. During the conversion, hydroxyl groups and a carboxyl group are added to the steroid nucleus. Bile acids are classified as primary or secondary (Fig. 27.13). The primary bile acids are synthesized by the hepatocytes and include **cholic acid** and **chenodeoxycholic acid.** Bile acids are secreted as conjugates of taurine or glycine. When bile enters the GI tract, bacteria present in the lumen act on the primary bile acids and convert them to secondary bile acids by dehydroxylation. Cholic acid is converted to **deoxycholic acid** and chenodeoxycholic acid to **lithocholic acid.**

At a neutral pH, the bile acids are mostly ionized and are referred to as **bile salts**. Conjugated bile acids ionize more readily than the unconjugated bile acids and, thus, usually exist as salts of various cations (e.g., sodium glycocholate). Bile salts are much more polar than bile acids, and have greater difficulty penetrating cell membranes. Consequently, bile salts are absorbed much more poorly by the small intestine than bile acids. This property of bile salts is important because they play an integral role in the intestinal absorption of lipid. Therefore, it is important that bile salts are absorbed by the small intestine only after all of the lipid has been absorbed.

The major lipids in bile are phospholipids and cholesterol. Of the phospholipids, the predominant species is phosphatidylcholine (lecithin). The phospholipid and cholesterol concentrations of hepatic bile are 0.3 to 11 mmol/L and 1.6 to 8.3 mmol/L, respectively. The concentrations of these lipids in the gallbladder bile are even higher because of the absorption of water by the gallbladder. Cholesterol in bile is responsible for the formation of cholesterol gallstones.

Total Bile Secretion Consists of Three Components, One of Which Depends on Bile Acids

The total bile flow is composed of the ductular secretion and the canalicular bile flow (Fig. 27.14). The ductular secretion is from the cells lining the bile ducts. These cells actively secrete HCO_3^- into the lumen, resulting in the movement of water into the lumen of the duct. Another mechanism that may contribute to ductular secretion of fluid is the presence of a cAMP-dependent Cl^- channel that secretes Cl^- into the

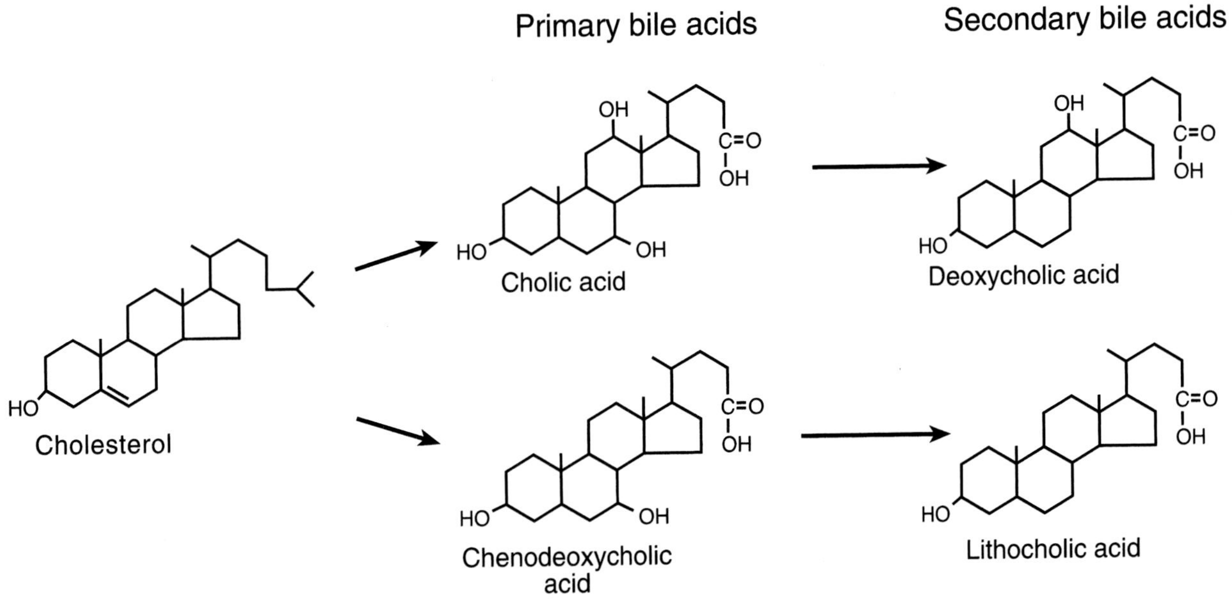

FIGURE 27.13 The formation of bile acids. Bile acids are conjugated with the amino acids glycine and taurine in the liver.

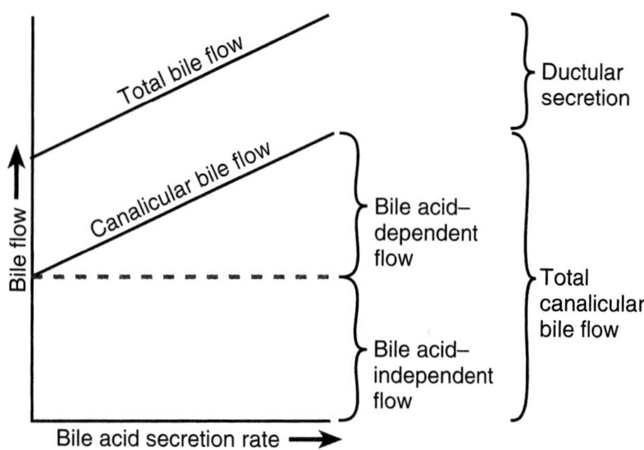

FIGURE 27.14 Components of total bile flow: canalicular bile flow and ductular secretion. Total canalicular bile flow is composed of bile acid–dependent flow and bile acid–independent flow. (Modified from Scharschmidt BF. In: Zakim D, Boyer T, eds. Hepatology. Philadelphia: WB Saunders, 1982.)

ductule lumen. Canalicular bile flow can be conceptually divided into two components: bile acid–dependent secretion and bile acid–independent secretion.

Canalicular Bile Acid–Dependent Flow. Hepatocyte uptake of free and conjugated bile salts is Na^+-dependent and mediated by **bile salt–sodium symport** (Fig. 27.15). The energy required is provided by the transmembrane Na^+ gradient generated by the Na^+/K^+-ATPase. This mechanism is a type of secondary active transport because the energy required for the active uptake of bile acid, or its conjugate, is not directly provided by ATP but by an ionic gradient. The free bile acids are reconjugated with taurine or glycine before secretion. Hepatocytes also make new bile acids from cholesterol. Bile salts are secreted by hepatocytes by a carrier located at the canalicular membrane. This secretion is not Na^+-dependent; instead, it is driven by the electrical potential difference between the hepatocyte and the canaliculus lumen.

Other major components of bile, such as phospholipid and cholesterol, are secreted in concert with bile salts. Bilirubin is secreted by hepatocytes via an active process. Although the secretion of cholesterol and phospholipid is not well understood, it is closely coupled to bile salt secretion. The osmotic pressure generated as a result of the secretion of bile salts draws water into the canaliculus lumen through the paracellular pathway.

Canalicular Bile Acid–Independent Flow. As the name implies, this component of canalicular flow is not dependent on the secretion of bile acids (see Figs. 27.14 and 27.15). The Na^+/K^+-ATPase plays an important role in bile acid-independent bile flow, a role that is clearly demonstrated by the marked reduction in bile flow when an inhibitor of this enzyme is applied. Another mechanism responsible for bile acid-independent flow is canalicular HCO_3^- secretion.

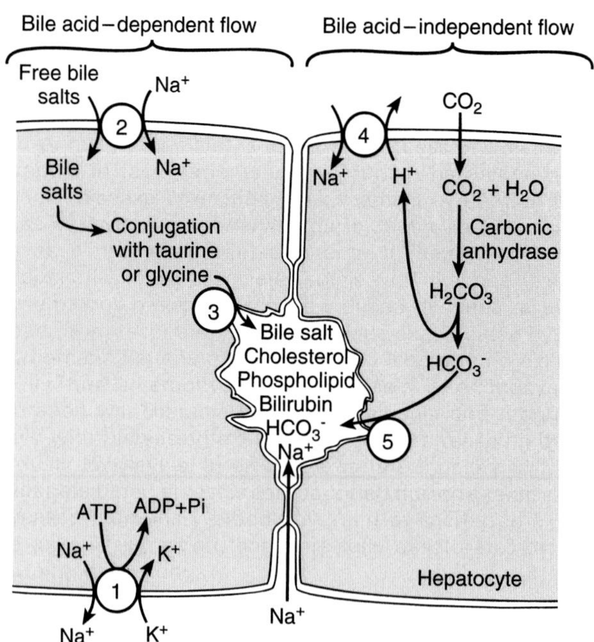

FIGURE 27.15 The mechanism of bile salt secretion and bile flow. (1) Na^+/K^+-ATPase. (2) Bile salt–sodium symport. (3) Canalicular bile salt carrier. (4) Na^+/H^+ exchanger. (5) HCO_3^- transport system.

Bile Secretion Is Primarily Regulated by a Feedback Mechanism, With Secondary Hormonal and Neural Controls

The major determinant of bile acid synthesis and secretion by hepatocytes is the bile acid concentration in hepatic portal blood, which exerts a negative-feedback effect on the synthesis of bile acids from cholesterol. The concentration of bile acids in portal blood also determines bile acid–dependent secretion. Between meals, the portal blood concentration of bile salts is usually extremely low, resulting in increased bile acid synthesis but reduced bile acid–dependent flow. After a meal, there is increased delivery of bile salts in the portal blood, which not only inhibits bile acid synthesis but also stimulates bile acid–dependent secretion.

CCK is secreted by the intestinal mucosa when fatty acids or amino acids are present in the lumen. CCK causes contraction of the gallbladder, which, in turn, causes increased pressure in the bile ducts. As the bile duct pressure rises, the sphincter of Oddi relaxes (another effect of CCK), and bile is delivered into the lumen.

When the mucosa of the small intestine is exposed to acid in the chyme, it releases secretin into the blood. Secretin stimulates HCO_3^- secretion by the cells lining the bile ducts. As a result, bile contributes to the neutralization of acid in the duodenum.

Gastrin stimulates bile secretion directly by affecting the liver and indirectly by stimulating increased acid production that results in increased secretin release. Steroid hormones (e.g., estrogen and some androgens) are inhibitors of bile secretion, and reduced bile secretion is a side effect associated with the therapeutic use of these hormones.

During pregnancy, the high circulating level of estrogen can reduce bile acid secretion.

The biliary system is supplied by parasympathetic and sympathetic nerves. Parasympathetic (vagal) stimulation results in contraction of the gallbladder and relaxation of the sphincter of Oddi, as well as increased bile formation. Bilateral vagotomy results in reduced bile secretion after a meal, suggesting that the parasympathetic nervous system plays a role in mediating bile secretion. By contrast, stimulation of the sympathetic nervous system results in reduced bile secretion and relaxation of the gallbladder.

Gallbladder Bile Differs From Hepatic Bile

Gallbladder bile has a very different composition from hepatic bile. The principal difference is that gallbladder bile is more highly concentrated. Water absorption is the major mechanism involved in concentrating hepatic bile by the gallbladder. Water absorption by the gallbladder epithelium is passive and is secondary to active Na^+ transport via a Na^+/K^+-ATPase in the basolateral membrane of the epithelial cells lining the gallbladder. As a result of isotonic fluid absorption from the gallbladder bile, the concentration of the various unabsorbed components of hepatic bile increases dramatically—as much as 20-fold.

The Enterohepatic Circulation Recycles Bile Salts Between the Small Intestine and the Liver

The **enterohepatic circulation** of bile salts is the recycling of bile salts between the small intestine and the liver. The total amount of bile acids in the body, primary or secondary, conjugated or free, at any time is defined as the **total bile acid pool.** In healthy people, the bile acid pool ranges from 2 to 4 g. The enterohepatic circulation of bile acids in this pool is physiologically extremely important. By cycling several times during a meal, a relatively small bile acid pool can provide the body with sufficient amounts of bile salts to promote lipid absorption. In a light eater, the bile acid pool may circulate 3 to 5 times a day; in a heavy eater, it may circulate 14 to 16 times a day. The intestine is normally extremely efficient in absorbing the bile salts by carriers located in the distal ileum. Inflammation of the ileum can lead to their malabsorption and result in the loss of large quantities of bile salts in the feces. Depending on the severity of illness, malabsorption of fat may result.

Bile salts in the intestinal lumen are absorbed via four pathways (Fig. 27.16). First, they are absorbed throughout the entire small intestine by passive diffusion, but only a small fraction of the total amount of bile salts is absorbed in this manner. Second, and most important, bile salts are absorbed in the terminal ileum by an active carrier-mediated process, an extremely efficient process in which usually less than 5% of the bile salts escape into the colon. Third, bacteria in the terminal ileum and colon deconjugate the bile salts to form bile acids, which are much more lipophilic than bile salts and, thus, can be absorbed passively. Fourth, these same bacteria are responsible for transforming the primary bile acids to secondary bile acids (deoxycholic and lithocholic acids) by dehydroxylation. Deoxycholic acid may be absorbed, but lithocholic acid is poorly absorbed.

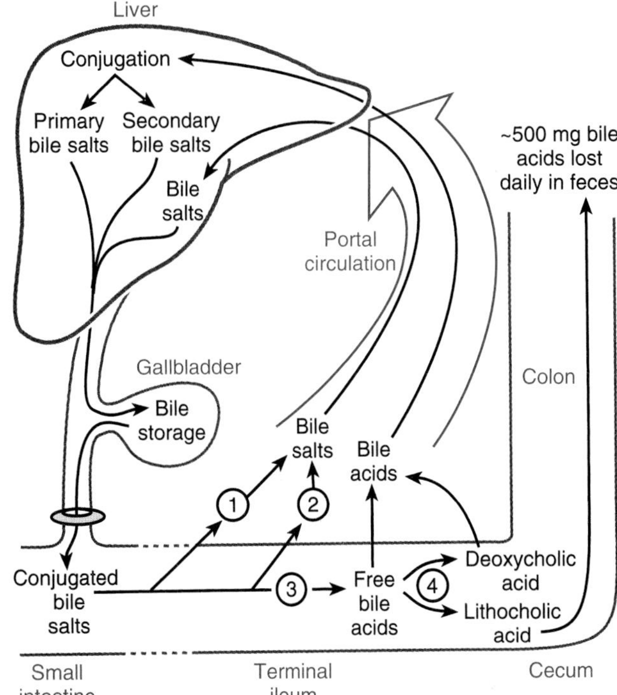

FIGURE 27.16 **The enterohepatic circulation of bile salts.** Bile salts are recycled out of the small intestine in four ways: (1) passive diffusion along the small intestine (plays a relatively minor role); (2) carrier-mediated active absorption in the terminal ileum (the most important absorption route); (3) deconjugation to primary bile acids before being absorbed either passively or actively; (4) conversion of primary bile acids to secondary bile acids with subsequent absorption of deoxycholic acid.

Although bile salt and bile acid absorption is extremely efficient, some salts and acids are nonetheless lost with every cycle of the enterohepatic circulation. About 500 mg of bile acids are lost daily. They are replenished by the synthesis of new bile acids from cholesterol. The loss of bile acid in feces is, therefore, an efficient way to excrete cholesterol.

Absorbed bile salts are transported in the portal blood bound to albumin or high-density lipoproteins (HDLs). The uptake of bile salts by hepatocytes is extremely efficient. In just one pass through the liver, more than 80% of the bile salts in the portal blood is removed. Once taken up by hepatocytes, bile salts are secreted into bile. The uptake of bile salts is a primary determinant of bile salt secretion by the liver.

The Liver Secretes Bile Pigments

The major pigment present in bile is the orange compound **bilirubin,** an end product of hemoglobin degradation in the monocyte-macrophage system in the spleen, bone marrow, and liver (Fig. 27.17). Hemoglobin is first converted to biliverdin with the release of iron and globin. Biliverdin is then converted into bilirubin, which is transported in blood bound to albumin. The liver removes bilirubin from the circulation rapidly and conjugates it with glucuronic acid. The glucuronide is secreted into the bile canaliculi through an active carrier-mediated process.

In the small intestine, bilirubin glucuronide is poorly ab-

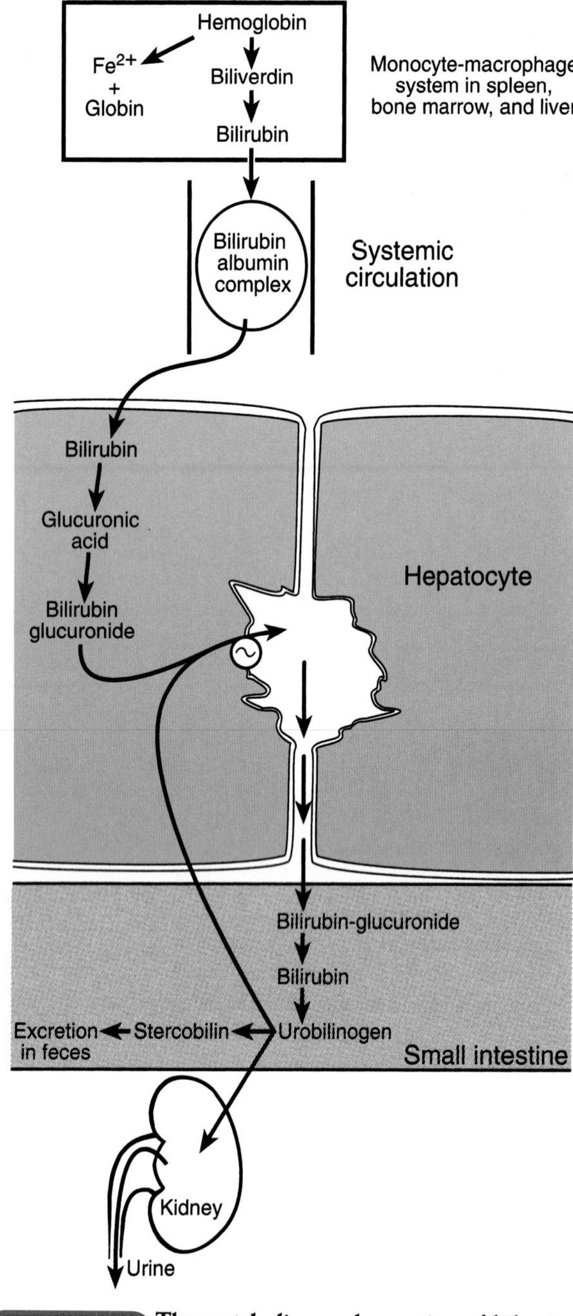

FIGURE 27.17 **The metabolism and excretion of bile pigment (bilirubin).**

sorbed. In the colon, however, bacteria deconjugate it, and part of the bilirubin released is converted to the highly soluble, colorless compound called urobilinogen. Urobilinogen can be oxidized in the intestine to stercobilin or absorbed by the small intestine. It is excreted in either urine or bile. Stercobilin is responsible for the brown color of the stool.

Cholesterol Gallstones Form When Cholesterol Supersaturates the Bile

Bile salts and lecithin in the bile help solubilize cholesterol. When the cholesterol concentration in bile increases to the point that it cannot be solubilized, it starts to crystallize, forming gallstones. Eventually, calcium deposits form in the stones, increasing their opacity and making them easily detectable on X-ray images of the gallbladder.

INTESTINAL SECRETION

The small intestine secretes 2 to 3 L/day of isotonic alkaline fluid. This secretion is derived mainly from cells in the **crypts of Lieberkühn**, tubular glands located at the base of intestinal villi. Of the three major cell types in the crypts of Lieberkühn—argentaffin cells, Paneth cells, and undifferentiated cells—the undifferentiated cells are responsible for intestinal secretions.

Intestinal secretion probably helps maintain the fluidity of the chyme and may also play a role in diluting noxious agents and washing away infectious microorganisms. The HCO_3^- in intestinal secretions protects the intestinal mucosa by neutralizing any H^+ present in the lumen. This is important in the duodenum and also in the ileum where bacteria degrade certain foods to produce acids (e.g., dietary fibers to short-chain fatty acids).

The fluid and electrolytes from intestinal secretions are usually absorbed by the small intestine and colon, but if secretion surpasses absorption (e.g., in cholera), watery diarrhea may result. If uncontrolled, this can lead to the loss of large quantities of fluid and electrolytes, which can result in dehydration and electrolyte imbalances and, ultimately, death. Several noxious agents, such as bacterial toxins (e.g., cholera toxin), can induce intestinal hypersecretion. Cholera toxin binds to the brush border membrane of crypt cells and increases intracellular adenylyl cyclase activity. The result is a dramatic increase in intracellular cAMP, which stimulates active Cl^- and HCO_3^- secretion into the lumen.

Also present in intestinal secretions are various mucins (mucoproteins) secreted by **goblet cells.** Mucins are glycoproteins high in carbohydrate, and they form gels in solution. They are extremely diverse in structure and are usually very large molecules. The mucus lubricates the mucosal surface and protects it from mechanical damage by solid food particles. It may also provide a physical barrier in the small intestine against the entry of microorganisms into the mucosa.

It is well documented that tactile stimulation, or an increase in intraluminal pressure, stimulates intestinal secretion. Other potent stimuli are certain noxious agents and the toxins produced by microorganisms. With the exception of toxin-induced secretion, our understanding of the normal control of intestinal secretion is meager. Vasoactive intestinal peptide is known to be a potent stimulator of intestinal secretion. This is demonstrated by a form of endocrine tumor of the pancreas that results in the secretion of large amounts of VIP into the circulation. In this condition, intestinal secretion rates are high.

DIGESTION AND ABSORPTION

To ensure the optimal absorption of nutrients, the GI tract has several unique features. For instance, after a meal, the small intestine undergoes rhythmic contractions called seg-

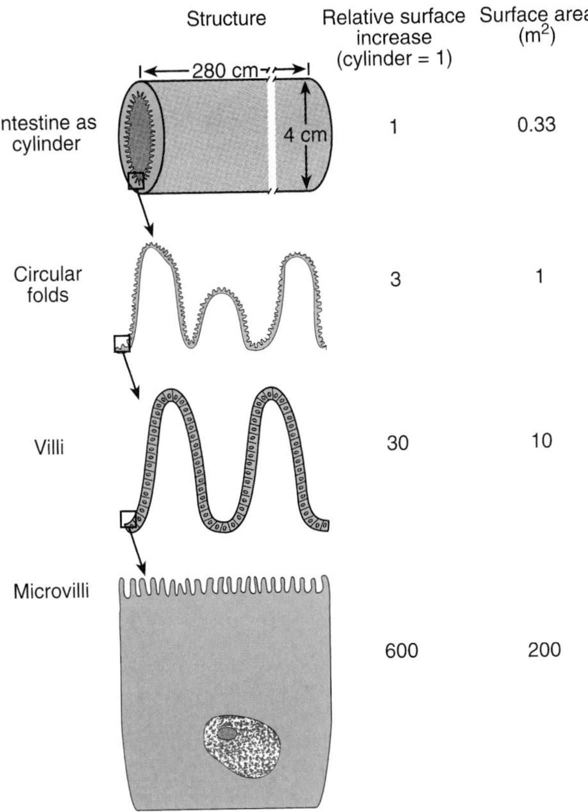

FIGURE 27.18 **Surface area amplification by the specialized features of the intestinal mucosa.** (Modified from Schmidt RF, Thews G. Human Physiology. Berlin: Springer-Verlag, 1993, p. 602.)

mentations (see Chapter 26), which ensure proper mixing of the small intestinal contents, exposure of the contents to digestive enzymes, and maximum exposure of digestion products to the small intestinal mucosa. The rhythmic segmentation has a gradient along the small intestine, with the highest frequency in the duodenum and the lowest in the ileum. This gradient ensures slow but forward movement of intestinal contents toward the colon.

Another unique feature of the small intestine is its architecture. Spiral or circular concentric folds increase the surface area of the intestine about 3 times (Fig. 27.18). Fingerlike projections of the mucosal surface called **villi** further increase the surface area of the small intestine about 30 times. To amplify the absorptive surface further, each epithelial cell, or **enterocyte**, is covered by numerous closely packed **microvilli**. The total surface area is increased to 600 times. The various nutrients, vitamins, bile salts, and water are absorbed by the GI tract by passive, facilitated, or active transport. (The site and mechanism of absorption will be discussed below.) The GI tract has a large reserve for the digestion and absorption of various nutrients and vitamins. Malabsorption of nutrients is usually not detected unless a large portion of the small intestine has been lost or damaged because of disease (see Clinical Focus Box 27.2).

Most nutrients and vitamins are absorbed by the duodenum and jejunum, but because bile salts are involved in the intestinal absorption of lipids, it is important that they not be absorbed prematurely. For effective fat absorption, the small intestine has adapted to absorb the bile salts in the terminal ileum through a bile salt transporter. The enterocytes along the villus that are involved in the absorption of nutrients are replaced every 2 to 3 days.

CLINICAL FOCUS BOX 27.2

Celiac Sprue (Gluten-Sensitive Enteropathy)

Celiac sprue, also called **gluten-sensitive enteropathy,** is a common disease involving a primary lesion of the intestinal mucosa. It is caused by the sensitivity of the small intestine to gluten. This disorder can result in the malabsorption of all nutrients as a result of the shortening or total loss of intestinal villi, which reduces the mucosal enzymes for nutrient digestion and the mucosal surface for absorption. Celiac sprue occurs in about 1 to 6 of 10,000 individuals in the Western world. The highest incidence is in western Ireland, where the prevalence is as high as 3 of 1,000 individuals. Although the disease may occur at any age, it is more common during the first few years and the third to fifth decades of life.

In patients with celiac sprue, the water-insoluble protein gluten (present in cereal grains such as wheat, barley, rye, and oats) or its breakdown product interacts with the intestinal mucosa and causes the characteristic lesion. Precisely how the binding of gluten to the intestinal mucosa causes mucosal injury is unclear. One hypothesis is that patients prone to celiac sprue may have a brush border peptidase deficiency and that the consequent incomplete digestion of gluten results in the production of a toxic substance, which injures the intestinal mucosa. This idea is probably incorrect, however, because the intestinal brush border peptidases revert to normal after the healing of damaged intestinal mucosa. Another hypothesis is that immune mechanisms are involved. This is supported by the fact that the number and activity of plasma cells and lymphocytes increase during the active phase of celiac sprue and that antigluten antibodies are usually present. It has been demonstrated that the small intestine makes a lymphokine-like substance, which inhibits the infiltration of leukocytes into the lamina propria of the intestinal mucosa when exposed to gluten. Unfortunately, it is not clear whether these immunological manifestations are primary or secondary phenomena of the disease.

The elimination of dietary gluten is a standard treatment for patients with celiac sprue. Occasionally, intestinal absorptive function and intestinal mucosal morphology of patients with celiac sprue are improved with glucocorticoid therapy. Presumably, such treatment is beneficial because of the immunosuppressive and anti-inflammatory actions of these hormones.

DIGESTION AND ABSORPTION OF CARBOHYDRATES

The digestion and absorption of dietary carbohydrates takes place in the small intestine. These are extremely efficient processes, in that essentially all of the carbohydrates consumed are absorbed. Carbohydrates are an extremely important component of food intake, since they constitute about 45 to 50% of the typical Western diet and provide the greatest and least expensive source of energy. Carbohydrates must be digested to monosaccharides before absorption.

The Diet Contains Both Digestible and Nondigestible Carbohydrates

Humans can digest most carbohydrates; those we cannot digest constitute the dietary fiber that forms roughage. Carbohydrate is present in food as monosaccharides, disaccharides, oligosaccharides, and polysaccharides. The monosaccharides are mainly hexoses (six-carbon sugars), and glucose is by far the most abundant of these. Glucose is obtained directly from the diet or from the digestion of disaccharides, oligosaccharides, or polysaccharides. The next most common monosaccharides are galactose, fructose, and sorbitol. Galactose is present in the diet only as milk lactose, a disaccharide composed of galactose and glucose. Fructose is present in abundance in fruit and honey and is usually present as disaccharides or polysaccharides. Sorbitol is derived from glucose and is almost as sweet as glucose, but sorbitol is absorbed much more slowly and, thus, maintains a high blood sugar level for a longer period when similar amounts are ingested. It has been used as a weight-reduction aid to delay the onset of hunger sensations.

The major disaccharides in the diet are sucrose, lactose, and maltose. Sucrose, present in sugar cane and honey, is composed of glucose and fructose. Lactose, the main sugar in milk, is composed of galactose and glucose. Maltose is composed of two glucose units.

The digestible polysaccharides are starch, dextrins, and glycogen. Starch, by far the most abundant carbohydrate in the human diet, is made of amylose and amylopectin. Amylose is composed of a straight chain of glucose units; amylopectin is composed of branched glucose units. Dextrins, formed from heating (e.g., toasting bread) or the action of the enzyme amylase, are intermediate products of starch digestion. Glycogen is a highly branched polysaccharide that stores carbohydrates in the body. The structure of glycogen is illustrated in Figure 27.19. Normally, about 300 to 400 g of glycogen is stored in the liver and muscle, with more stored in muscle than in the liver. Muscle glycogen is used exclusively by muscle, and liver glycogen is used to provide blood glucose during fasting.

Dietary fiber is made of polysaccharides that are usually poorly digested by the enzymes in the small intestine. They have an extremely important physiological function in that they provide the "bulk" that facilitates intestinal motility and function. Many vegetables and fruits are rich in fibers, and their frequent ingestion greatly decreases intestinal transit time.

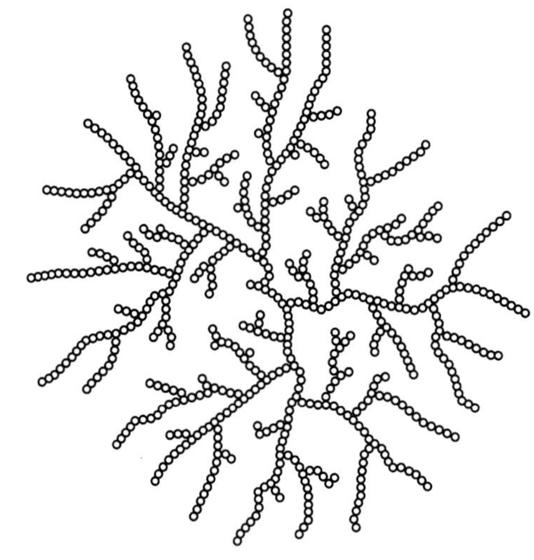

FIGURE 27.19 **The structure of glycogen.**

Carbohydrates Are Digested in Different Parts of the GI Tract

The digestion of carbohydrates starts when food is mixed with saliva during chewing. The enzyme salivary amylase acts on the α-1,4-glycosidic linkage of amylose and amylopectin of polysaccharides to release the disaccharide maltose and oligosaccharides maltotriose and α-limit dextrins (Fig. 27.20). Because salivary amylase works best at neutral pH, its digestive action terminates rapidly after the bolus mixes with acid in the stomach. However, if the food is thoroughly mixed with amylase during chewing, a substantial amount of complex carbohydrates is digested be-

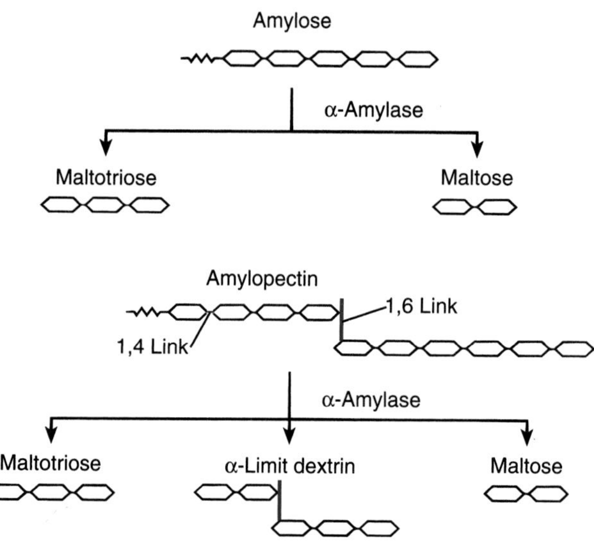

FIGURE 27.20 **The digestion products of starch after exposure to salivary or pancreatic α-amylase.** Sugar units are indicated by hexagons.

fore this point. **Pancreatic amylase** continues the digestion of the remaining carbohydrates. However, the chyme must first be neutralized by pancreatic secretions, since pancreatic amylase works best at neutral pH. The products of pancreatic amylase digestion of polysaccharides are also maltose, maltotriose, and α-limit dextrins.

The digestion products of starch and glycogen, together with disaccharides (sucrose and lactose), are further digested by enzymes located at the brush border membrane. Table 27.6 lists the enzymes involved in the digestion of disaccharides and oligosaccharides and the products of their action. The final products are glucose, fructose, and galactose.

Enterocytes Play an Important Role in Carbohydrate Absorption and Metabolism

Monosaccharides are absorbed by enterocytes either actively or by facilitated transport. Glucose and galactose are absorbed via secondary active transport by a symporter (see Chapter 2) that transports two Na^+ for every molecule of monosaccharide (Fig. 27.21). The movement of Na^+ into the cell, down concentration and electrical gradients, effects the uphill movement of glucose into the cell. The low intracellular Na^+ concentration is maintained by the basolateral membrane Na^+/K^+-ATPase. The osmotic effects of sugars increase the Na^+/K^+-ATPase activity and the K^+ conductance of the basolateral membrane. Sugars accumulate in the cell at a higher concentration than in plasma and leave the cell by Na^+-independent facilitated transport or passive diffusion through the basolateral cell membrane. Glucose and galactose share a common transporter at the brush border membrane of enterocytes and, thus, compete with each other during absorption.

Fructose is taken up by facilitated transport. Although facilitated transport is carrier-mediated, it is not an active process (see Chapter 2). Fructose absorption is much slower than glucose and galactose absorption and is not Na^+-dependent. Although in some animal species both galactose and fructose can be converted to glucose in enterocytes, this mechanism is probably not important in humans.

TABLE 27.6 The Digestion of Disaccharides and Oligosaccharides by Brush Border Enzymes

Enzyme	Substrate	Site of Action	Products
Sucrase	Sucrose	α-1,2-glycosidic linkage	Glucose and fructose
Lactase	Lactose	β-1,4-glycosidic linkage	Glucose and galactose
Isomaltase	α-Limit dextrins	α-1,6-glycosidic linkage	Glucose, maltose, and oligosaccharides
Maltase	Maltose, maltotriose	α-1,4-glycosidic linkage	Glucose

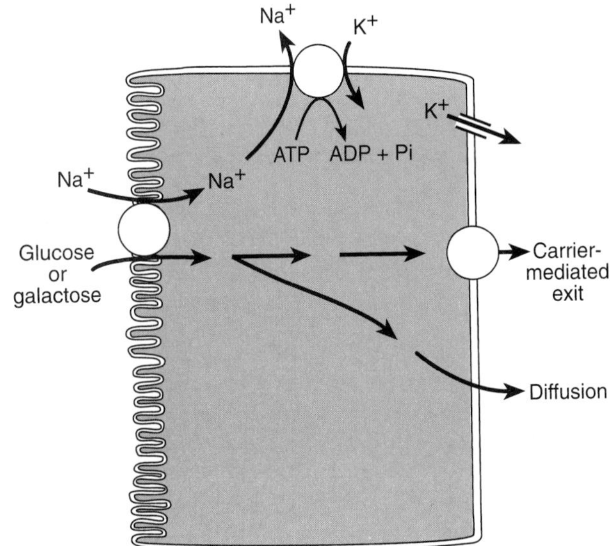

FIGURE 27.21 **The enterocyte Na^+-dependent carrier system for glucose and galactose.**

The sugars absorbed by enterocytes are transported by the portal blood to the liver where they are converted to glycogen or remain in the blood. After a meal, the level of blood glucose rises rapidly, usually peaking at 30 to 60 minutes. The concentration of glucose can be as high as 150 mg/dL. Although enterocytes can use glucose for fuel, glutamine is preferred. Both galactose and glucose can be used in the glycosylation of proteins in the Golgi apparatus of the enterocytes.

The Lack of Some Digestive Enzymes Impairs Carbohydrate Absorption

Impaired carbohydrate absorption caused by the absence of salivary or pancreatic amylase almost never occurs because these enzymes are usually present in great excess. However, impaired absorption due to a deficiency in membrane disaccharidases is rather common. Such deficiencies can be either genetic or acquired. Among congenital deficiencies, lactase deficiency is, by far, the most common. Affected individuals suffer from **lactose intolerance**, a condition in which the ingestion of milk products results in severe osmotic (watery) diarrhea. The mechanism responsible is depicted in Figure 27.22. Undigested lactose in the intestinal lumen increases the osmolality of the luminal contents. Osmolality is further increased by lactic acid produced from the action of intestinal bacteria on the lactose. Increased luminal osmolality results in net water secretion into the lumen. The accumulation of fluid distends the small intestine and accelerates peristalsis, eventually resulting in watery diarrhea.

Congenital sucrase deficiency results in symptoms similar to those of lactase deficiency. Sucrase deficiency can be inherited or acquired through disorders of the small intestine, such as tropical sprue or Crohn's disease.

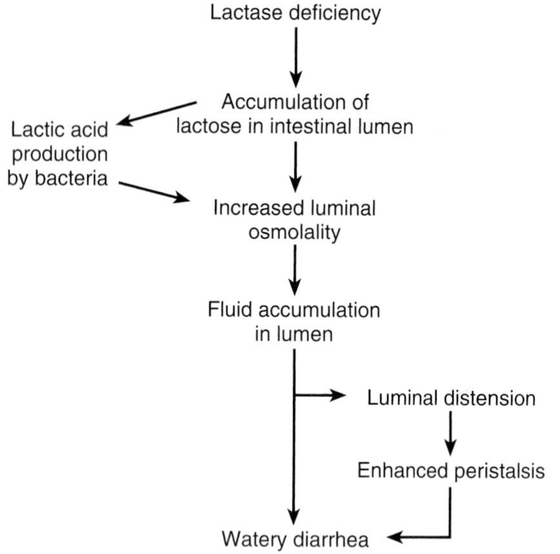

FIGURE 27.22 **The mechanism for osmotic diarrhea resulting from lactase deficiency.**

Dietary Fiber Plays an Important Role in GI Motility

Dietary fiber includes indigestible carbohydrates and carbohydrate-like components mainly found in fruits and vegetables. The most common are cellulose, hemicellulose, pectins, and gums. Cellulose and hemicellulose are insoluble in water and are poorly digested by humans, thus providing the bulkiness of stool.

Dietary fiber imparts bulk to the bolus and, therefore, greatly shortens transit time. It has been proposed that dietary fiber reduces the incidence of colon cancer by shortening GI transit time, which, in turn, reduces the formation of carcinogenic bile acids (e.g., lithocholic acid). Because dietary fiber also binds bile acids, which are formed from cholesterol, fiber consumption can result in a lowering of blood cholesterol by promoting excretion.

DIGESTION AND ABSORPTION OF LIPIDS

Lipids are a concentrated form of energy. They provide 30 to 40% of the daily caloric intake in the Western diet. Lipids are also essential for normal body functions, as they form part of cellular membranes and are precursors of bile acids, steroid hormones, prostaglandins, and leukotrienes. The human body is capable of synthesizing most of the lipids it requires with the exception of the **essential fatty acids** linoleic acid (C 18:2, an 18-carbon long fatty acid with two double bonds) and arachidonic acid (C 20:4). Both of these acids belong to the family of omega-6 fatty acids. Recently, researchers have provided convincing evidence that **eicosapentaenoic acid** (C 20:5) and **docosahexaenoic acid** (C 22:6) are also essential for the normal development of vision in newborns. Both of these acids are omega-3 fatty acids and are abundant in seafood and algae.

The Luminal Lipid Consists of Both Exogenous and Endogenous Lipids

Lipids are comprised of several classes of compounds that are insoluble in water but soluble in organic solvents. By far the most abundant dietary lipids are **triacylglycerols**, or **triglycerides**. They consist of a glycerol backbone esterified in the three positions with fatty acids (Fig. 27.23A). More than 90% of the daily dietary lipid intake is in the form of triglycerides.

The other lipids in the human diet are cholesterol and phospholipids. Cholesterol is a sterol derived exclusively from animal fat. Humans also ingest a small amount of plant sterols, notably β-sitosterol and campesterol. The phospholipid molecule is similar to a triglyceride with fatty acids occupying the first and second positions of the glycerol backbone (Fig. 27.23B). However, the third position of the glycerol backbone is occupied by a phosphate group coupled to a nitrogenous base (e.g., choline or ethanolamine), for which each type of phospholipid molecule is named.

Bile serves as an endogenous source of cholesterol and phospholipids. Bile contributes about 12 g/day of phospholipid to the intestinal lumen, most in the form of phosphatidylcholine, whereas dietary sources contribute 2 to 3 g/day. Another important endogenous source of lipid is desquamated intestinal villus epithelial cells.

Different Lipases Carry Out Lipid Hydrolysis

Lipid digestion mainly occurs in the lumen of the small intestine. Humans secrete an overabundance of **pancreatic lipase**. Depending on the substrate being digested, pancreatic lipase has an optimal pH of 7 to 8, allowing it to work well in the intestinal lumen after the acidic contents from the stomach have been neutralized by pancreatic HCO_3^- secretion. Pancreatic lipase hydrolyzes the triglyceride molecule to a 2-monoglyceride and two fatty acids (Fig. 27.24). It works on the triglyceride molecule at the oil-water interface; thus, the rate of lipolysis depends on the surface area of the interface. The products from the partial di-

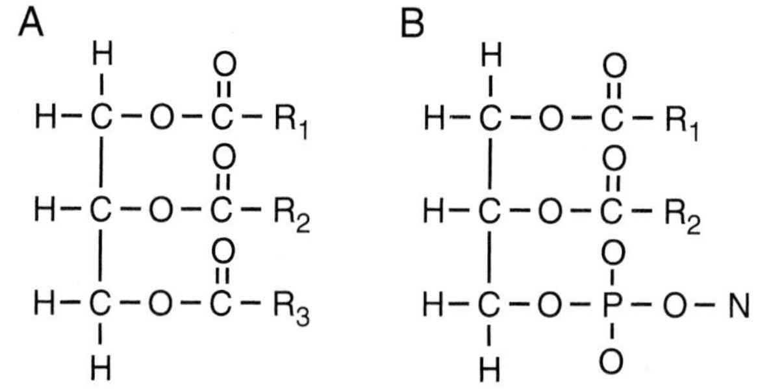

A

H–C(H)–O–C(=O)–R_1
H–C–O–C(=O)–R_2
H–C(H)–O–C(=O)–R_3

B

H–C(H)–O–C(=O)–R_1
H–C–O–C(=O)–R_2
H–C(H)–O–P(=O)(O)–O–N

FIGURE 27.23 **Dietary lipids. A**, A triglyceride molecule. R_1, R_2, and R_3 belong to different fatty acids. **B**, A phospholipid molecule. The fatty acid occupying the first position (R_1) is usually a saturated fatty acid and that in the second position (R_2) is usually an unsaturated or polyunsaturated fatty acid. The third position after the phosphate group is occupied by a nitrogenous base (N), such as choline or ethanolamine.

gestion of dietary triglyceride by gastric lipase and the churning action of the stomach produce a suspension of oil droplets (an emulsion) that help increase the area of the oil-water interface. Pancreatic juice also contains the peptide **colipase,** which is necessary for the normal digestion of fat by pancreatic lipase. Colipase binds lipase at a molar ratio of 1:1, thereby allowing the lipase to bind to the oil-water interface where lipolysis takes place. Colipase also counteracts the inhibition of lipolysis by bile salt, which, despite its importance in intestinal fat absorption, prevents the attachment of pancreatic lipase to the oil-water interface.

Phospholipase A_2 is the major pancreatic enzyme for digesting phospholipids, forming lysophospholipids and fatty acids. For instance, phosphatidylcholine (lecithin) is hydrolyzed to form lysophosphatidylcholine (lysolecithin) and fatty acid (see Fig. 27.24).

Dietary cholesterol is presented as a free sterol or as a sterol ester (cholesterol ester). The hydrolysis of cholesterol ester is catalyzed by the pancreatic enzyme carboxylester hydrolase, also called **cholesterol esterase** (see Fig. 27.24). The digestion of cholesterol ester is important because cholesterol can be absorbed only as the free sterol.

Bile Salt Plays an Important Role in Lipid Absorption

A layer of poorly stirred fluid called the **unstirred water layer** coats the surface of the intestinal villi (Fig. 27.25A). The unstirred water layer reduces the absorption of lipid digestion products because they are poorly soluble in water. They are rendered water-soluble by **micellar solubilization** by bile salts in the small intestinal lumen. This mechanism greatly enhances the concentration of these products in the unstirred water layer (Fig. 27.25B). The lipid digestion products are then absorbed by enterocytes, mainly by passive diffusion. Fatty acid and monoglyceride molecules are taken up individually. Similar mechanisms seem to operate for cholesterol and lysolecithin.

Bile salts are derived from cholesterol, but they are different from cholesterol in that they are water-soluble. They are essentially detergents—molecules that possess both hydrophilic and hydrophobic properties. Because bile salts are polar molecules, they penetrate cell membranes poorly. This is significant because it ensures their minimal absorption by the jejunum where most fat absorption takes place. At or above a certain concentration of bile salts, the **critical micellar concentration,** they aggregate to form micelles; the concentration of luminal bile salts is usually well above the critical micellar concentration. When bile salts alone are present in the micelle, it is called a **simple micelle.** Simple micelles incorporate the lipid digestion products—monoglyceride and fatty acids—to form **mixed micelles.** This renders the lipid digestion products water-soluble by incorporation into mixed micelles. Mixed micelles diffuse across the unstirred water layer and deliver lipid digestion products to the enterocytes for absorption.

Enterocytes Process Absorbed Lipid to Form Lipoproteins

After entering the enterocytes, the fatty acids and monoglycerides migrate to the smooth ER. A fatty acid–binding protein may be involved in the intracellular transport of fatty acids, but whether or not a protein carrier is involved in the intracellular transport of monoglycerides is unknown. In the smooth ER, monoglycerides and fatty acids

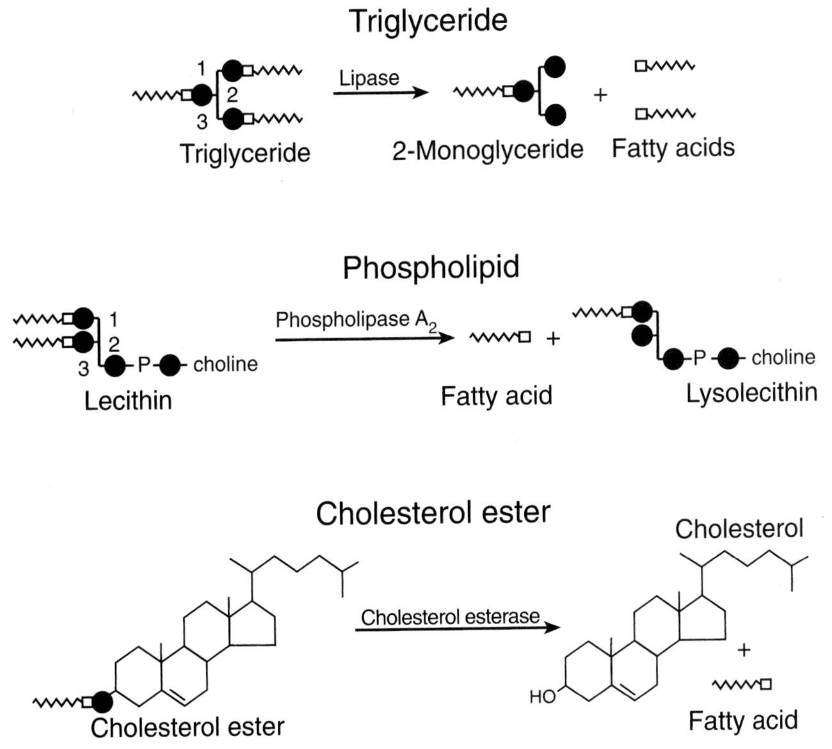

FIGURE 27.24 The digestion of dietary lipids by pancreatic enzymes in the small intestine. Solid circles represent oxygen atoms.

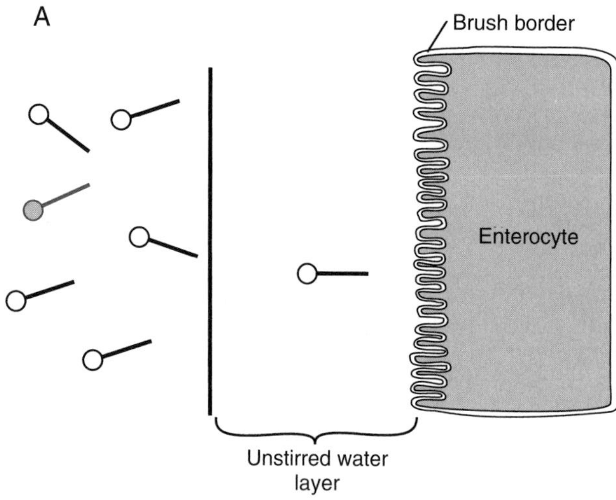

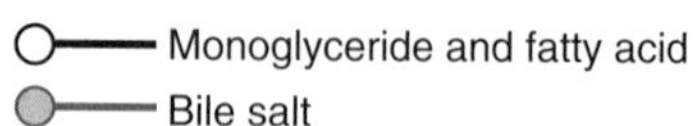

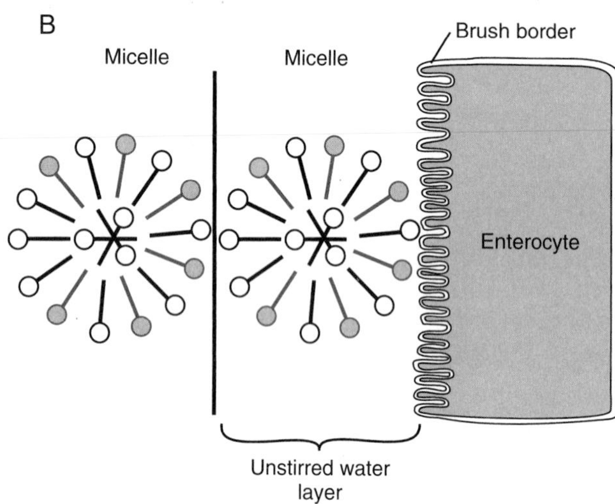

FIGURE 27.25 **The micellar solubilization of lipids.** Micellar solubilization enhances the delivery of lipid to the brush border membrane. **A,** In the absence of bile salts. **B,** In the presence of bile salts.

are rapidly reconstituted to form triglycerides (Fig. 27.26). Fatty acids are first activated to form acyl-CoA, which is then used to esterify monoglyceride to form diglyceride, which is transformed into triglyceride. The lysolecithin absorbed by the enterocytes can be reesterified in the smooth ER to form lecithin.

Cholesterol can be transported out of the enterocytes as free cholesterol or as esterified cholesterol. The enzyme responsible for the esterification of cholesterol to form cholesterol ester is **acyl-CoA cholesterol acyltransferase** (ACAT).

Enterocytes Secrete Chylomicrons and Very Low Density Lipoproteins

The reassembled triglycerides, lecithin, cholesterol, and cholesterol esters are then packaged into **lipoproteins** and

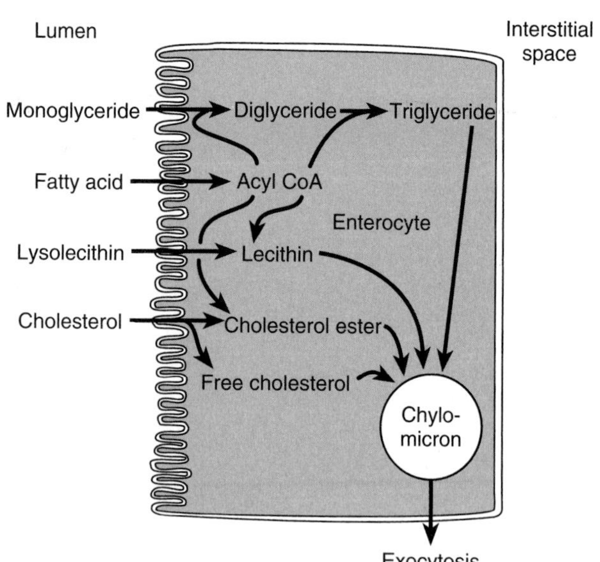

FIGURE 27.26 **The intracellular metabolism of absorbed lipid digestion products to form chylomicrons.**

exported from the enterocytes. The intestine produces two major classes of lipoproteins: chylomicrons and very low density lipoproteins (VLDLs). Both are triglyceride-rich lipoproteins with densities less than 1.006 g/mL. **Chylomicrons** are made exclusively by the small intestine, and their primary function is to transport the large amount of dietary fat absorbed by the small intestine from the enterocytes to the lymph. Chylomicrons are large, spherical lipoproteins with diameters of 80 to 500 nm. They contain less protein and phospholipid than VLDLs and are, therefore, less dense than VLDLs. **VLDLs** are made continuously by the small intestine during both fasting and feeding, although the liver contributes significantly more VLDLs to the circulation.

Apoproteins—apo A-I, apo A-IV, and apo B—are among the major proteins associated with the production of chylomicrons and VLDLs. Apo B is the only protein that seems to be necessary for the normal formation of intestinal chylomicrons and VLDLs. This protein is made in the small intestine. It has a molecular weight of 250,000 and it is extremely hydrophobic. Apo A-I is involved in a reaction catalyzed by the plasma enzyme **lecithin cholesterol acyltransferase** (LCAT). Plasma LCAT is responsible for the esterification of cholesterol in the plasma to form cholesterol ester with the fatty acid derived from the 2-position of lecithin. After the chylomicrons and VLDLs enter the plasma, apo A-I is rapidly transferred from chylomicrons and VLDLs to high-density lipoproteins (HDLs). Apo A-I is the major protein present in plasma HDLs. Apo A-IV is made by the small intestine and the liver. Recently, it was shown that apo A-IV, secreted by the small intestine, may be an important factor contributing to anorexia after fat feeding.

Newly synthesized lipoproteins in the smooth ER are transferred to the Golgi apparatus, where they are packaged in vesicles. Chylomicrons and VLDLs are released into the intercellular space by exocytosis (Fig. 27.27). From

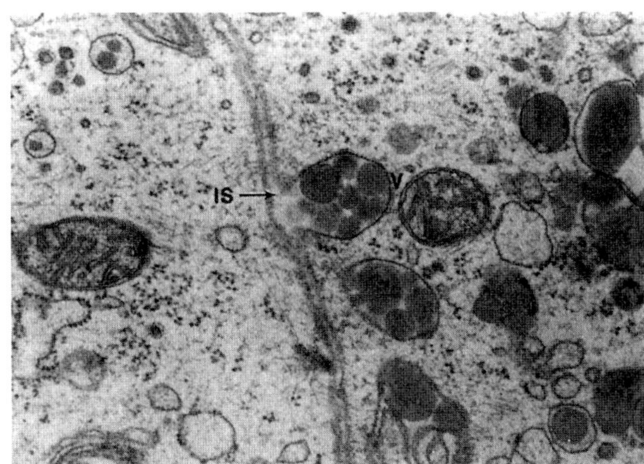

FIGURE 27.27 **Exocytosis of chylomicrons.** The exocytosis of chylomicrons is evident in this electron micrograph. The nascent chylomicrons in the secretory vesicle (V) are similar in size and morphology to those already present in the intercellular space (IS). (From Sabesin SM, Frase S. Electron microscopic studies of the assembly, intracellular transport, and secretion of chylomicrons by rat intestine. J Lipid Res 1977;18:496–511.)

there, they are transferred to the central lacteals (the beginnings of lymphatic vessels) by a process that is not well understood. Experimental evidence seems to indicate that the transfer probably occurs mostly by diffusion. Intestinal lipid absorption is associated with a marked increase in lymph flow called the **lymphagogic effect** of fat feeding. This increase in lymph flow plays an important role in the transfer of lipoproteins from the intercellular spaces to the central lacteal.

Fatty acids can also travel in the blood bound to albumin. While most of the long-chain fatty acids are transported from the small intestine as triglycerides packaged in chylomicrons and VLDLs, some are transported in the portal blood bound to serum albumin. Most of the medium-chain (8 to 12 carbons) and all of the short-chain fatty acids are transported by the hepatic portal route.

The Lack of Pancreatic Lipases or Bile Salts Can Impair Lipid Absorption

In several clinical conditions, lipid digestion and absorption are impaired, resulting in the malabsorption of lipids and other nutrients and fatty stools. Abnormal lipid absorption can result in numerous problems because the body requires certain fatty acids (e.g., linoleic and arachidonic acid, precursors of prostaglandins) to function normally. These are called essential fatty acids because the human body cannot synthesize them and is, therefore, totally dependent on the diet to supply them. Recent studies suggest that the human body may also require omega-3 fatty acids in the diet during development. These include linolenic, docosahexaenoic, and eicosapentaenoic acids. Linolenic acid is abundant in plants, and docosahexaenoic and eicosapentaenoic acids are abundant in fish. Docosahexaenoic acid is an important fatty acid present in the retina and other parts of the brain.

Pancreatic deficiency significantly reduces the ability of the exocrine pancreas to produce digestive enzymes. Because the pancreas normally produces an excess of digestive enzymes, enzyme production has to be reduced to about 10% of normal before symptoms of malabsorption develop. One characteristic of pancreatic deficiency is **steatorrhea** (fatty stool), resulting from the poor digestion of fat by the pancreatic lipase. Normally about 5 g/day of fat are excreted in human stool. With steatorrhea, as much as 50 g/day can be excreted.

Fat absorption subsequent to the action of pancreatic lipase requires solubilization by bile salt micelles. Acute or chronic liver disease can cause defective biliary secretion, resulting in bile salt concentrations lower than necessary for micelle formation. The normal absorption of fat is thereby inhibited.

Abetalipoproteinemia, an autosomal recessive disorder, is characterized by a complete lack of apo B, which is required for the formation and secretion of chylomicrons and VLDLs. Apo B–containing lipoproteins in the circulation—including chylomicrons, VLDLs, and low-density lipoproteins (LDLs)—are absent. Plasma LDLs are absent because they are derived mainly from the metabolism of VLDLs. Since individuals with abetalipoproteinemia do not produce any chylomicrons or VLDLs in the small intestine, they are unable to transport absorbed fat, resulting in an accumulation of lipid droplets in the cytoplasm of enterocytes. They also suffer from a deficiency of fat-soluble vitamins.

DIGESTION AND ABSORPTION OF PROTEINS

Proteins form the fundamental structure of cells and are the most abundant of all organic compounds in the body. Most proteins are found in muscle, with the remainder in other cells, blood, body fluids, and body secretions. Enzymes and many hormones are proteins. Proteins are composed of amino acids and have molecular weights of a few thousand to a few hundred thousand. More than 20 common amino acids form the building blocks for proteins (Table 27.7). Of these, nine are considered essential and must be provided by the diet. Although the **nonessential amino acids** are also required for normal protein synthesis, the body can synthesize them from other amino acids.

Complete proteins are those that can supply all of the essential amino acids in amounts sufficient to support normal growth and body maintenance. Examples are eggs, poultry, and fish. The proteins in most vegetables and grains are called **incomplete proteins** because they do not provide all of the essential amino acids in amounts sufficient to sustain normal growth and body maintenance. Vegetarians need to eat a variety of vegetables and soy protein to avoid amino acid deficiencies.

Luminal Protein Is Derived From the Diet, GI Secretions, and Enterocytes

The average American adult takes in 70 to 110 g/day of protein. The minimum daily protein requirement for adults is about 0.8 g/kg body weight (e.g., 56 g for a 70-kg per-

TABLE 27.7 The Amino Acids Found in Proteins

Essential	Nonessential
Histidine	Alanine
Isoleucine	Arginine
Leucine	Asparagine
Lysine	Aspartic acid
Methionine	Cysteine
Phenylalanine	Glutamic acid
Threonine	Glutamine
Tryptophan	Glycine
Valine	Hydroxyproline
	Proline
	Serine
	Tyrosine

son). Pregnant or lactating women require 20 to 30 g above the recommended daily allowance to meet the extra demand for protein. A lactating woman can lose as much as 12 to 15 g of protein per day as milk protein. Children need more protein for body growth; the recommended daily allowance for infants is about 2 g/kg body weight.

While most of the protein entering the GI tract is dietary protein, there are also proteins derived from endogenous sources such as pancreatic, biliary, and intestinal secretions, and the cells shed from the intestinal villi. About 20 to 30 g/day of protein enters the intestinal lumen in pancreatic juice and about 10 g/day in bile. Enterocytes of the intestinal villi are continuously shed into the intestinal lumen, and as much as 50 g/day of enterocyte proteins enter the intestinal lumen. An average of 150 to 180 g/day of total protein is presented to the small intestine, of which more than 90% is absorbed.

Proteins Are Digested in the GI Tract, Yielding Amino Acids and Peptides

Most of the protein in the intestinal lumen is completely digested into either amino acids or dipeptides or tripeptides before it is taken up by the enterocytes. Protein digestion starts in the stomach with the action of **pepsin,** which is secreted as a proenzyme and activated by acid in the stomach. Pepsin hydrolyzes protein to form smaller polypeptides. It is classified as an endopeptidase because it attacks specific peptide bonds inside the protein molecule. This phase of protein digestion is normally not important other than in individuals suffering from pancreatic exocrine deficiency.

Most of the digestion of proteins and polypeptides takes place in the small intestine. Most proteases are secreted in the pancreatic juice as inactive proenzymes. When the pancreatic juice enters the duodenum, trypsinogen is converted to trypsin by **enteropeptidase** (also known as enterokinase), an enzyme found on the luminal surface of enterocytes. The active **trypsin** then converts the other proenzymes to active enzymes.

The pancreatic proteases are classified as endopeptidases or exopeptidases (Table 27.3). **Endopeptidases** hydrolyze certain internal peptide bonds of proteins or polypeptides to release the smaller peptides. The three endopeptidases present in pancreatic juice are trypsin, chymotrypsin, and elastase. Trypsin splits off basic amino acids from the carboxyl terminal of a protein, **chymotrypsin** attacks peptide bonds with an aromatic carboxyl terminal, and **elastase** attacks peptide bonds with a neutral aliphatic carboxyl terminal. The **exopeptidases** in pancreatic juice are carboxypeptidase A and carboxypeptidase B. Like the endopeptidases, the exopeptidases are specific in their action. **Carboxypeptidase A** attacks polypeptides with a neutral aliphatic or aromatic carboxyl terminal. **Carboxypeptidase B** attacks polypeptides with a basic carboxyl terminal. The final products of protein digestion are amino acids and small peptides.

Specific Transporters in the Small Intestine Take Up Amino Acids and Peptides

Amino acids are taken up by enterocytes via secondary active transport. Six major amino acid carriers in the small intestine have been identified; they transport related groups of amino acids. The amino acid transporters favor the L form over the D form. As in the uptake of glucose, the uptake of amino acids is dependent on a Na^+ concentration gradient across the enterocyte brush border membrane.

The absorption of peptides by enterocytes was once thought to be less efficient than amino acid absorption. However, subsequent studies in humans clearly demonstrated that dipeptides and tripeptide uptake is significantly more efficient than the uptake of amino acids. Dipeptides and tripeptides use different transporters than those used by amino acids. The peptide transporter prefers dipeptides and tripeptides with either glycine or lysine residues. Furthermore, tetrapeptides and more complex peptides are only poorly transported by the peptide transporter. These peptides can be further broken down to dipeptides and tripeptides by the **peptidases** (exopeptidases) located on the brush border of the enterocytes. Dipeptides and tripeptides are given to individuals suffering from malabsorption because they are absorbed more efficiently and are more palatable than free amino acids. Another advantage of peptides over amino acids is the smaller osmotic stress created as a result of delivering them.

In adults, a negligible amount of protein is absorbed as undigested protein. In some individuals, however, intact or partially digested proteins are absorbed, resulting in anaphylactic or hypersensitivity reactions. The pulmonary and cardiovascular systems are the major organs involved in anaphylactic reactions. For the first few weeks after birth, the newborn's small intestine absorbs considerable amounts of intact proteins. This is possible because of low proteolytic activity in the stomach, low pancreatic secretion of peptidases, and poor development of intracellular protein degradation by lysosomal proteases.

The absorption of immunoglobulins (predominantly IgG) plays an important role in the transmission of **passive immunity** from the mother's milk to the newborn in several animal species (e.g., ruminants and rodents). In humans, the absorption of intact immunoglobulins does not appear to be an important mode of transmission of antibodies for two reasons. First, passive immunity in humans is derived almost entirely from the intrauterine transport of maternal antibodies. Second, human **colostrum**, the thin, yellowish, milky fluid secreted by the mammary glands a few days before or after parturition, contains mainly IgA, which is poorly absorbed by the small intestine. The ability to absorb intact proteins is rapidly lost as the gut matures—a process called **closure.** Colostrum contains a factor that promotes the closure of the small intestine.

After dipeptides and tripeptides are taken up by the enterocytes, they are further broken down to amino acids by peptidases in the cytoplasm. The amino acids are transported in the portal blood. The small amount of protein that is taken up by the adult intestine is largely degraded by lysosomal proteases, although some proteins escape degradation.

Defects in Digestion and Transport Can Impair Protein Absorption

Although pancreatic deficiency has the potential to affect protein digestion, it only does so in severe cases. Pancreatic deficiency seems to affect lipid digestion more than protein digestion. There are several extremely rare genetic disorders of amino acid carriers. In **Hartnup's disease,** the membrane carrier for neutral amino acids (e.g., tryptophan) is defective. **Cystinuria** involves the carrier for basic amino acids (e.g., lysine and arginine) and the sulfur-containing amino acids (e.g., cystine). Cystinuria was once thought to involve only the kidneys because of the excretion of amino acids such as cystine in urine, but the small intestine is involved as well.

Because the peptide transport system remains unaffected, disorders of some amino acid transporters can be treated with supplemental dipeptides containing these amino acids. However, this treatment alone is not effective if the kidney transporter is also involved, as in cystinuria.

ABSORPTION OF VITAMINS

Vitamins are organic substances from both animal and plant sources needed in small quantities for normal metabolic function and the growth and maintenance of the body. Because most of these organic compounds are not manufactured in the body, adequate dietary intake and efficient intestinal absorption are important. Vitamins are classified in many ways, but in terms of absorption, they are classified according to whether they are lipid-soluble or water-soluble.

The Fat-Soluble Vitamins Include A, D, E, and K

The only feature shared by the fat-soluble vitamins is their lipid solubility. Otherwise, they are structurally very different. Most are absorbed passively. The fat-soluble vitamins are summarized in Table 27.8.

Vitamin A. The principal form of **vitamin A** is **retinol**; the aldehyde (retinal) and the acid (retinoic acid) are also active forms of vitamin A. Retinol can be derived directly from animal sources or through conversion from β-carotene (found abundantly in carrots) in the small intestine. Vitamin A is rendered water-soluble by micellar solubilization and is absorbed by the small intestine passively. It is converted in the small intestinal mucosa to an ester, **retinyl ester**, which is incorporated in chylomicrons and taken up by the liver. Vitamin A is stored in the liver and released to the circulation bound to **retinol-binding protein** only when needed.

Vitamin A is important in the production and regeneration of rhodopsin of the retina and in the normal growth of the skin. Vitamin A–deficient individuals develop night blindness and skin lesions.

TABLE 27.8 Fat-Soluble Vitamins

Vitamin	RDA	Sources	Site and Mode of Absorption	Role
A	1,000 RE	Liver, kidney, butter, whole milk, cheese, and β-carotene (yields two molecules of retinol)	Small intestine; passive	Vision, bone development, epithelial development, and reproduction
D	200 IU	Liver, butter, cream, vitamin D fortified milk, conversion from 7-dehydrocholesterol by UV light	Small intestine; passive	Growth and development, formation of bones and teeth, stimulation of intestinal Ca^{2+} and phosphate absorption, mobilization of Ca^{2+} from bones
E	10 mg	Wheat germ, green plants, egg yolk, milk, butter, meat	Small intestine; passive	Antioxidant
K	70–100 μg	Green vegetables, intestinal flora	Phylloquinones from green vegetables are absorbed actively from the proximal small intestine; menaquinones from gut flora are absorbed passively	Blood clotting

RDA, recommended daily allowances; RE, retinol equivalent; IU, international unit: 1 IU = 0.025 μg

Vitamin D. **Vitamin D** is a group of fat-soluble compounds collectively known as the calciferols. Vitamin D_3 (also called cholecalciferol or activated dehydrocholesterol) in the human body is derived from two main sources: the skin, which contains a rich source of 7-dehydrocholesterol that is rapidly converted to cholecalciferol when exposed to UV light, and dietary vitamin D_3. Like vitamin A, vitamin D_3 is absorbed by the small intestine passively and is incorporated into chylomicrons. During the metabolism of chylomicrons, vitamin D_3 is transferred to a binding protein in plasma called the **vitamin D–binding protein.**

Unlike vitamin A, vitamin D is not stored in the liver but is distributed among the various organs depending on their lipid content. In the liver, vitamin D_3 is converted to 25-hydroxycholecalciferol, which is subsequently converted to the active hormone **1,25-dihydroxycholecalciferol** in the kidneys. The latter enhances Ca^{2+} and phosphate absorption by the small intestine and mobilizes Ca^{2+} and phosphate from bones.

Vitamin D is essential for normal development and growth and the formation of bones and teeth. Vitamin D deficiency can result in **rickets**, a disorder of normal bone ossification manifested by distorted bone movements during muscular action.

Vitamin E. The major dietary **vitamin E** is α-tocopherol. Vegetable oils are rich in vitamin E. It is absorbed by the small intestine by passive diffusion and incorporated into chylomicrons. Unlike vitamins A and D, vitamin E is transported in the circulation associated with lipoproteins and erythrocytes.

Vitamin E is a potent antioxidant and therefore prevents lipid peroxidation. Tocopherol deficiency is associated with increased red cell susceptibility to lipid peroxidation, which may explain why the red cells are more fragile in vitamin E–deficient individuals than in healthy individuals.

Vitamin K. **Vitamin K** can be derived from green vegetables in the diet or the gut flora. The vitamin K derived from green vegetables is in **phylloquinones**. Vitamin K derived from bacteria in the small intestine is in **menaquinones.** Phylloquinones are taken up by the small intestine via an energy-dependent process from the proximal small intestine. In contrast, menaquinones are absorbed from the small intestine passively, dependent only on the micellar solubilization of these compounds by bile salts. Vitamin K is incorporated into chylomicrons. It is rapidly taken up by the liver and secreted together with VLDLs. No carrier protein for vitamin K has been identified.

Vitamin K is essential for the synthesis of various clotting factors by the liver. Vitamin K deficiency is associated with bleeding disorders.

The Water-Soluble Vitamins Are C, B_1, B_2, B_6, B_{12}, Niacin, Biotin, and Folic Acid

Most of the water-soluble vitamins are absorbed by the small intestine by both passive and active processes. The water-soluble vitamins are summarized in Table 27.9.

Vitamin C. The major source of **vitamin C (ascorbic acid)** is green vegetables and fruits. It plays an important role in many oxidative processes by acting as a coenzyme or cofactor. It is absorbed mainly by active transport in the ileum. Vitamin C deficiency is associated with **scurvy**, a disorder characterized by weakness, fatigue, anemia, and bleeding gums.

Vitamin B_1. **Vitamin B_1 (thiamine)** plays an important role in carbohydrate metabolism. Thiamine is absorbed by the jejunum passively as well as by an active, carrier-mediated process. Thiamine deficiency results in **beriberi**, characterized by anorexia and disorders of the nervous system and heart.

Vitamin B_2. **Vitamin B_2 (riboflavin)** is a component of the two groups of flavoproteins—flavin adenine dinucleotide (FAD) and flavin mononucleotide (FMN). Riboflavin plays an important role in metabolism. Riboflavin is absorbed by a specific, saturable, active transport system located in the proximal small intestine. Riboflavin deficiency is associated with anorexia, impaired growth, impaired use of food, and nervous disorders.

Niacin. **Niacin** plays an important role as a component of the coenzymes NAD(H) and NADP(H), which participate in a wide variety of oxidation-reduction reactions involving H^+ transfer.

At low concentrations, niacin is absorbed by the small intestine by Na^+-dependent, carrier-mediated facilitated transport. At high concentrations, it is absorbed by passive diffusion. Niacin has been used to treat hypercholesterolemia, for the prevention of coronary artery disease. It decreases plasma total cholesterol and LDL cholesterol, yet increases plasma HDL cholesterol.

Niacin deficiency is characterized by many clinical symptoms, including anorexia, indigestion, muscle weakness, and skin eruptions. Severe deficiency leads to **pellagra**, a disease characterized by dermatitis, dementia, and diarrhea.

Vitamin B_6. **Vitamin B_6 (pyridoxine)** is involved in amino acid and carbohydrate metabolism. Vitamin B_6 is absorbed throughout the small intestine by simple diffusion. A deficiency of this vitamin is often associated with anemia and CNS disorders.

Biotin. **Biotin** acts as a coenzyme for carboxylase, transcarboxylase, and decarboxylase enzymes, which play an important role in the metabolism of lipids, glucose, and amino acids. At low luminal concentrations, biotin is absorbed by the small intestine by Na^+-dependent active transport. At high concentrations, biotin is absorbed by simple diffusion. Biotin is so common in food that deficiency is rarely observed.

Folic Acid. **Folic acid** is usually found in the diet as polyglutamyl conjugates (pteroylpolyglutamates). It is required for the formation of nucleic acids, the maturation of red blood cells, and growth. An enzyme on the brush border degrades pteroylpolyglutamates to yield a monoglutamylfolate, which is taken up by enterocytes by facilitated transport. Inside enterocytes, the monoglutamylfolate is released directly into the bloodstream or converted to

TABLE 27.9 Water-Soluble Vitamins

Vitamin	RDA	Sources	Site and Mode of Absorption	Role
C	60 mg/day	Fruits, vegetables, organ (liver and kidney) meat	Active transport by the ileum	Coenzyme or cofactor in many oxidative processes
B_1 (thiamine)	1 mg/day	Yeast, liver, cereal grains	At low luminal concentrations, by active, carrier-mediated process; at high luminal concentrations by passive diffusion	Carbohydrate metabolism
B_2 (riboflavin)	1.7 mg/day	Dairy products	Active transport in proximal small intestine	Metabolism
Niacin	19 mg/day	Brewer's yeast, meat	At low luminal concentrations, by Na^+-dependent, carrier-mediated, facilitated transport	Component of coenzymes NAD(H) and NADP(H); metabolism of carbohydrates, fats, and proteins; synthesis of fatty acid and steroid
B_6 (pyridoxine)	2.2 mg/day	Brewer's yeast, wheat germ, meat, whole grain cereals, dairy products	By passive diffusion in small intestine	Amino acid and carbohydrate metabolism
Biotin	200 μg /day	Brewer's yeast, milk, liver, egg yolk	At low luminal concentrations, by Na^+-dependent active transport; at high luminal concentrations, by simple diffusion	Coenzyme for carboxylase, transcarboxylase, and decarboxylase enzymes, metabolism of lipids, glucose, and amino acids
Folic acid	0.5 mg/day	Liver, beans, dark green leafy vegetables	By Na^+-dependent facilitated transport	Nucleic acid biosynthesis, maturation of red blood cells, promotion of growth
B_{12}	3 μg/day	Liver, kidney, dairy products, eggs, fish	Absorbed in terminal ileum by active transport involving binding to intrinsic factor	Normal cell division; bone marrow, intestinal mucosa, and nervous system most affected in deficiency state, characterized by pernicious anemia

5-methyltetrahydrofolate before exiting the cell. A folate-binding protein binds the free and methylated forms of folic acid in plasma. Folic acid deficiency causes a fall in plasma and red cell folic acid content and, in its most severe form, the development of megaloblastic anemia, dermatological lesions, and poor growth.

Vitamin B_{12}. The discovery of **vitamin B_{12} (cobalamin)** followed from the observation that patients with **pernicious anemia** who ate large quantities of raw liver recovered from the disease. Subsequent analysis of liver components isolated the cobalt-containing vitamin, which plays an important role in the production of red blood cells. A glycoprotein secreted by the parietal cells in the stomach called the **intrinsic factor** binds strongly with vitamin B_{12} to form a complex that is then absorbed in the terminal ileum through a receptor-mediated process (Fig. 27.28). Vitamin B_{12} is transported in the portal blood bound to the protein **transcobalamin.** Individuals who lack the intrinsic factor fail to absorb vitamin B_{12} and develop pernicious anemia.

ELECTROLYTE AND MINERAL ABSORPTION

Nearly all of the dietary nutrients and approximately 95 to 98% of the water and electrolytes that enter the upper small intestine are absorbed. The absorption of electrolytes and minerals involves both passive and active processes, resulting in the movement of electrolytes, water, and metabolic substrates into the blood for distribution and use throughout the body.

Sodium. The GI system is well equipped to handle the large amount of Na^+ entering the GI lumen daily—on average, about 25 to 35 g of Na^+ every day. Around 5 to 8 g are derived from the diet, and the rest from salivary, gastric, biliary, pancreatic, and small intestinal secretions. The GI tract is extremely efficient in conserving Na^+: only 0.5% of intestinal Na^+ is lost in the feces. The jejunum absorbs more than half of the total Na^+, and the ileum and colon absorb the remainder. The small intestine absorbs the bulk of the Na^+ presented to it, but the colon is most efficient in conserving Na^+.

Sodium is absorbed by several different mechanisms operating at varying degrees in different parts of the GI tract. When a meal that is hypotonic to plasma is ingested, considerable absorption of water from the lumen to the blood takes place, predominantly through tight junctions and intercellular spaces between the enterocytes, resulting in the absorption of small solutes such as Na^+ and Cl^- ions. This mode of absorption, called **solvent drag,** is responsible for a significant amount of the Na^+ absorption by the duode-

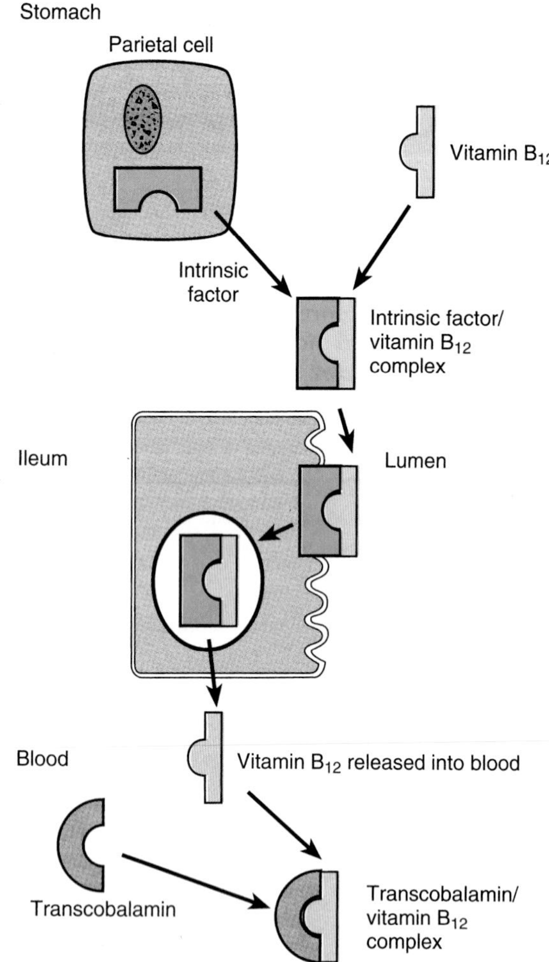

FIGURE 27.28 **The intestinal absorption of vitamin B_{12}.**

num and jejunum, but it probably plays a minor role in Na^+ absorption by the ileum and colon because more distal regions of the intestine are lined by a "tight" epithelium (see Chapter 2).

In the jejunum, Na^+ is actively pumped out of the basolateral surface of enterocytes by a Na^+/K^+-ATPase (Fig. 27.29A). The result is low intracellular Na^+ concentration, and the luminal Na^+ enters enterocytes down the electrochemical gradient, providing energy for the extrusion of H^+ into the lumen (via a Na^+/H^+ exchanger). The H^+ then reacts with HCO_3^- in bile and pancreatic secretions in the intestinal lumen to form H_2CO_3. Carbonic acid dissociates to form CO_2 and H_2O. The CO_2 readily diffuses across the small intestine into the blood. Another mode of Na^+ uptake is via a carrier located in the enterocyte brush border membrane, which transports Na^+ together with a monosaccharide (e.g., glucose) or an amino acid molecule (a symport type of transport).

In the ileum, the presence of a Na^+/K^+-ATPase at the basolateral membrane also creates a low intracellular Na^+ concentration, and luminal Na^+ enters enterocytes down the electrochemical gradient. Sodium absorption by Na^+-coupled symporters is not as great as in the jejunum because most of the monosaccharides and amino acids have already been absorbed by the small intestine (Fig. 27.29B). Sodium chloride is transported via two exchangers located at the brush border membrane. One is a Cl^-/HCO_3^- exchanger, and the other is a Na^+/H^+ exchanger. The downhill movement of Na^+ into the cell provides the energy required for the uphill movement of the H^+ from the cell to the lumen. Similarly, the downhill movement of HCO_3^- out of the cell provides the energy for the uphill entry of Cl^- into the enterocytes. The Cl^- then leaves the cell through facilitated transport. This mode of Na^+ uptake is called Na^+/H^+-Cl^-/HCO_3^- countertransport.

In the colon, the mechanisms for Na^+ absorption are mostly similar to those described for the ileum. There is no sugar- or amino acid–coupled Na^+ transport because most sugars and amino acids have already been absorbed. Sodium is also absorbed here via Na^+-selective ion channels in the apical cell membrane (electrogenic Na^+ absorption).

Potassium. The average daily intake of K^+ is about 4 g. Absorption takes place throughout the intestine by passive

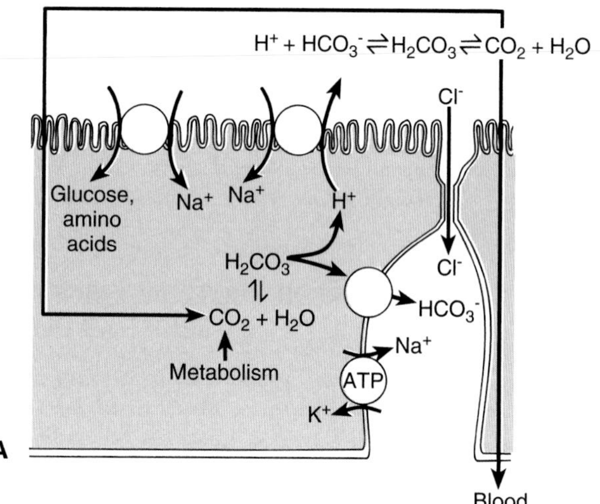

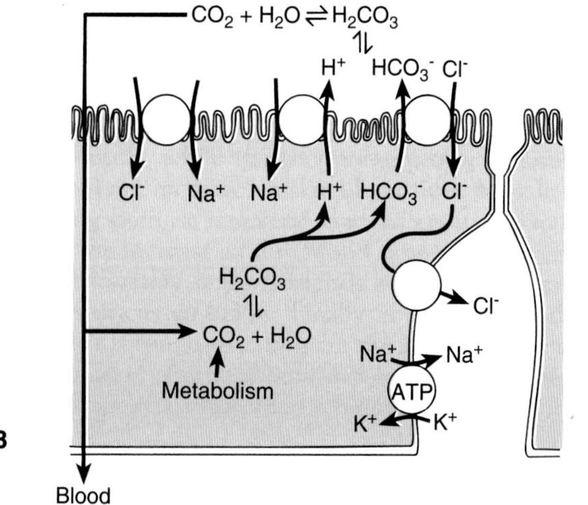

FIGURE 27.29 **The intestinal absorption of sodium. A,** Na^+ absorption by the jejunum. **B,** Na^+ absorption by the ileum.

diffusion through the tight junctions and lateral intercellular spaces of the enterocytes. The driving force for K^+ absorption is the difference between luminal and blood K^+ concentration. The absorption of water results in an increase in luminal K^+ concentration, resulting in K^+ absorption by the intestine. In the colon, K^+ can be absorbed or secreted depending on the luminal K^+ concentration. With diarrhea, considerable K^+ can be lost. Prolonged diarrhea can be life-threatening, because the dramatic fall in extracellular K^+ concentration can cause complications such as cardiac arrhythmias.

Chloride. Most of the Cl^- ions added to the GI tract from the diet and from the various secretions of the GI system are absorbed. Intestinal chloride absorption involves both passive and active processes. In the jejunum, active Na^+ absorption generates a potential difference across the small intestinal mucosa, with the serosal side more positive than the lumen. Chloride ions follow this potential difference and enter the bloodstream via the tight junctions and lateral intercellular spaces. In the ileum and colon, Cl^- is taken up actively by enterocytes via Cl^-/HCO_3^- exchange, as discussed above. This absorption of Cl^- is inhibited by the presence of other halides.

Bicarbonate. Bicarbonate ions are absorbed in the jejunum together with Na^+. In humans, the absorption of HCO_3^- by the jejunum stimulates the absorption of Na^+ and H_2O (see Fig. 27.29A). Through a Na^+/H^+ exchanger, H^+ is secreted into the intestinal lumen where H^+ and HCO_3^- react to form H_2CO_3, which then dissociates to form CO_2 and H_2O. The CO_2 diffuses into the enterocytes, where it reacts with H_2O to form H_2CO_3 (catalyzed by carbonic anhydrase). H_2CO_3 dissociates into HCO_3^- and H^+ and the HCO_3^- then diffuses into the blood.

In the ileum and colon, HCO_3^- is actively secreted into the lumen in exchange for Cl^-. This secretion of HCO_3^- is important in buffering the decrease in pH resulting from the short-chain fatty acids produced by bacteria in the distal ileum and colon.

Calcium. The amount of Ca^{2+} entering the GI tract is about 1 g/day, approximately half of which is derived from the diet. Most dietary Ca^{2+} is derived from meat and dairy products. Of the Ca^{2+} presented to the GI tract, about 40% is absorbed. Several factors affect Ca^{2+} absorption. For instance, the presence of fatty acid can retard Ca^{2+} absorption by the formation of Ca^{2+} soap. In contrast, bile salt molecules form complexes with Ca^{2+} ions, which facilitates Ca^{2+} absorption.

Calcium absorption takes place predominantly in the duodenum and jejunum, is mainly active, and involves three steps: (1) Calcium is taken up by enterocytes by passive diffusion through a Ca^{2+} channel, since there is a large Ca^{2+} concentration gradient; the luminal Ca^{2+} is about 5 to 10 mM, whereas free intracellular Ca^{2+} is about 100 nM. (2) Once inside the cell, Ca^{2+} is complexed with **Ca^{2+}-binding protein, calbindin D (CaBP)**. (3) At the basolateral membrane, Ca^{2+} is extruded from the enterocyte via the Ca^{2+}-ATPase pump. Calcium uptake by enterocytes, the level of CaBP in the cells, and transport by Ca^{2+}-ATPase pumps are increased by 1,25-dihydroxyvitamin D_3. Once inside the cell, the Ca^{2+} ions are sequestered in the ER and Golgi membranes by binding to the CaBP in these organelles.

Calcium absorption by the small intestine is regulated by the circulating plasma Ca^{2+} concentration. Lowering of the Ca^{2+} concentration stimulates the release of parathyroid hormone, which stimulates the conversion of vitamin D to its active metabolite—1,25-dihydroxyvitamin D_3—in the kidney. This in turn stimulates the synthesis of CaBP and the Ca^{2+}-ATPase by the enterocytes (Fig. 27.30). Because protein synthesis is involved in the stimulation of Ca^{2+} uptake by parathyroid hormone, a lapse of a few hours usually occurs between the release of parathyroid hormone and the increase in Ca^{2+} absorption by the enterocytes.

Magnesium. Humans ingest about 0.4 to 0.5 g/day of Mg^{2+}. The absorption of Mg^{2+} seems to take place along the entire small intestine, and the mechanism involved seems to be passive.

Zinc. The average daily zinc intake is 10 to 15 mg, about half of which is absorbed, primarily in the ileum. A carrier located in the brush border membrane actively transports zinc from the lumen into the cell, where it can be stored or transferred into the bloodstream. Zinc plays an important role in several metabolic activities. For example, a group of

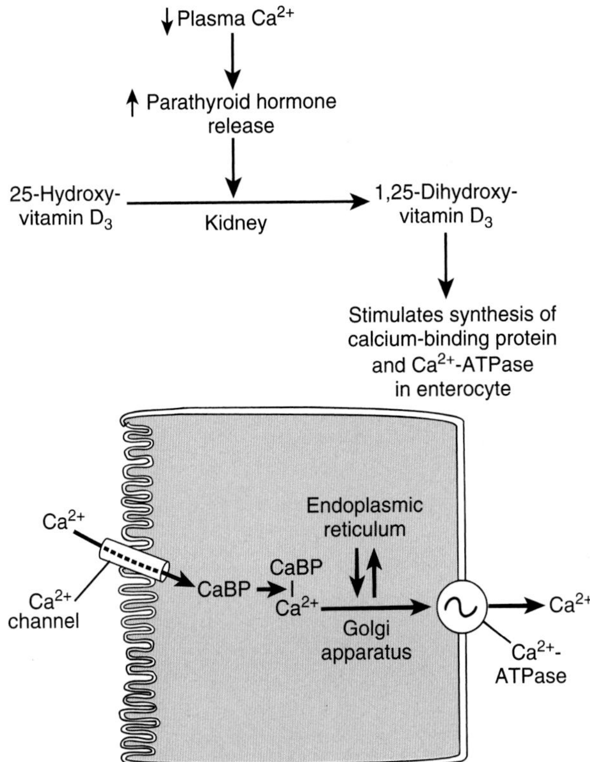

FIGURE 27.30 **Calcium absorption by enterocytes.** Parathyroid hormone stimulates the conversion of vitamin D_3 in the kidney to its active metabolite 1,25-dihydroxyvitamin D_3 (1,25-dihydroxycholecalciferol), which stimulates Ca^{2+} uptake via the Ca^{2+} channels. It also stimulates the synthesis of both Ca^{2+}-binding protein (CaBP) and the Ca^{2+}-ATPase.

metalloenzymes (e.g., alkaline phosphatase, carbonic anhydrase, and lactic dehydrogenase) requires zinc to function.

Iron. Iron plays an important role not only as a component of heme but also as a participant in many enzymatic reactions. About 12 to 15 mg/day of iron enter the GI tract, where it is absorbed mainly by the duodenum and upper jejunum (Fig. 27.31). There are two forms of dietary iron: heme and nonheme. The heme iron is absorbed intact by enterocytes. Nonheme iron absorption depends on both pH and concentration. Ferric (Fe^{3+}) salts are not soluble at pH 7, whereas ferrous (Fe^{2+}) salts are. Consequently, in the duodenum and upper jejunum, unless Fe^{3+} ion is chelated, it forms a precipitate. Several compounds, such as tannic acid in tea and phytates in vegetables, form insoluble complexes with iron, preventing absorption. Iron is absorbed by an active process via a carrier(s) located in the brush border membrane. One such transporter, the divalent metal transporter (DMT-1), is expressed abundantly in the duodenum.

Once inside the cell, heme iron is released by the action of heme oxygenase and mixed with the intracellular free iron pool. Iron is either stored in the enterocyte cytoplasm bound to the storage protein **apoferritin** to form **ferritin**, or transported across the cell bound to transport proteins, which carry the iron across the cytoplasm and release it into the intercellular space. Iron is bound and transported in the blood by **transferrin**, a β-globulin synthesized by the liver.

Iron absorption is closely regulated by iron storage in enterocytes and iron concentration in the plasma. Enterocytes are continuously shed into the lumen, and the ferritin contained within is also lost. Normally, iron in enterocytes is derived from the lumen and the blood (Fig. 27.32). The amount of iron absorbed is regulated by the amount stored

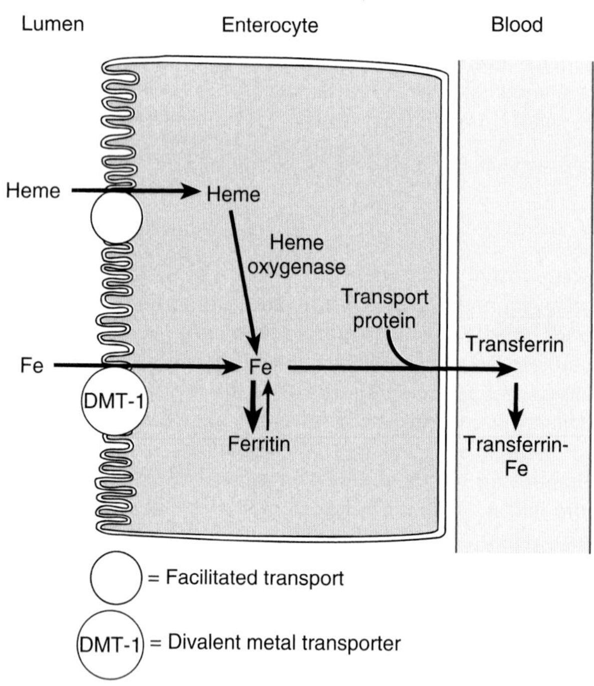

FIGURE 27.31 **Iron absorption.**

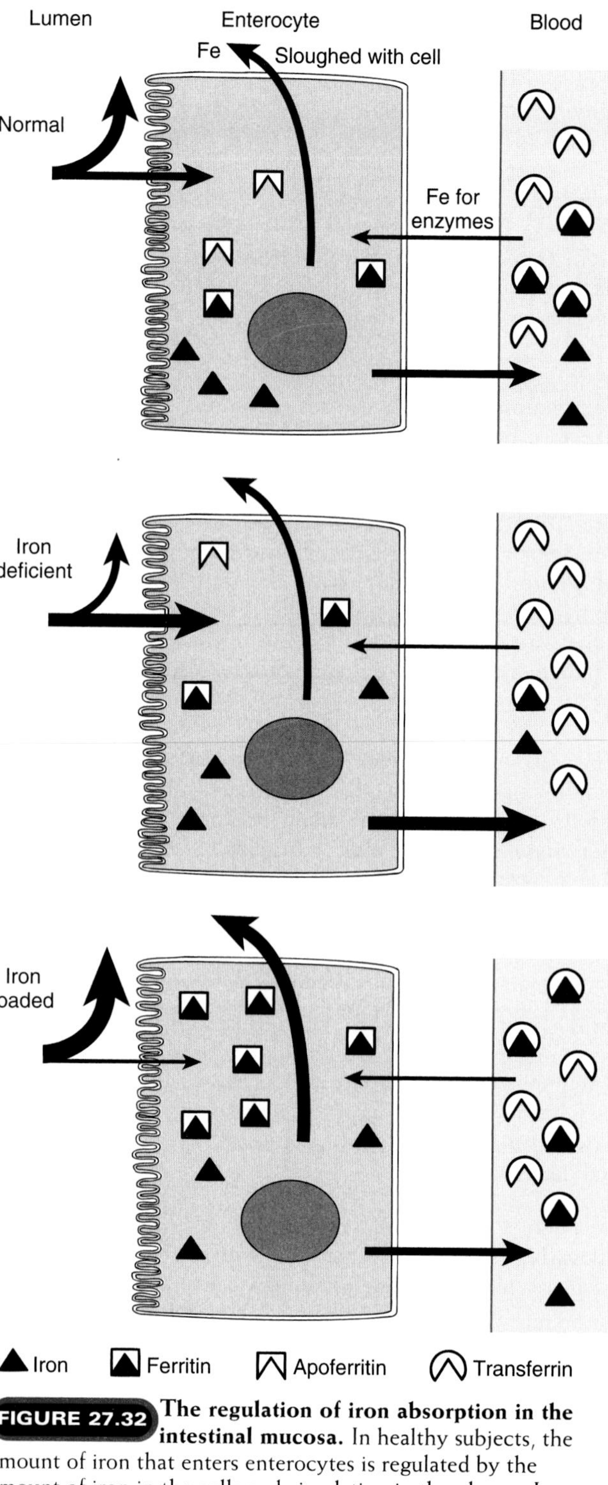

FIGURE 27.32 **The regulation of iron absorption in the intestinal mucosa.** In healthy subjects, the amount of iron that enters enterocytes is regulated by the amount of iron in the cells and circulating in the plasma. In iron-deficient subjects, little iron is incorporated into enterocytes and less is circulating in the plasma; therefore, absorption is increased and excretion is decreased. In iron-loaded subjects, the mucosal cells and transferrin are more highly saturated, limiting absorption and increasing excretion. (Modified from Krause MV, Mahan LK, eds. Food, Nutrition, and Diet Therapy. Philadelphia: WB Saunders, 1984.)

in enterocytes. In iron deficiency, the circulating plasma iron concentration is low, which stimulates the absorption of iron from the lumen and the transport of iron into the blood. Moreover, in a deficient state, less iron is stored as ferritin in the enterocytes, so the loss of iron through this means is significantly reduced. In iron-loaded patients, there is less absorption of iron because of the large amount of mucosal iron storage, which increases iron loss as a result of enterocyte shedding. Furthermore, because of the high level of circulating plasma iron, the transfer of iron from enterocytes to the blood is reduced. Through a combination of various mechanisms, body iron homeostasis is maintained.

TABLE 27.10 Water Intake, Absorption, and Excretion by the GI Tract

Water added to GI tract	
Food and beverages	2,000 mL
Salivary secretion	1,000 mL
Biliary secretion	1,000 mL
Gastric secretion	2,000 mL
Pancreatic secretion	1,000 mL
Intestinal mucosal secretion	2,000 mL
Water absorbed or lost in feces	
Water absorbed	
Duodenum and jejunum	4,000 mL
Ileum	3,500 mL
Colon	1,400 mL
Water loss in feces	100 mL

ABSORPTION OF WATER

In human adults, the average daily intake of water is about 2 L. As shown in Table 27.10, secretions from the salivary glands, pancreas, liver, and GI tract make up most of the fluid entering the GI tract (about 7 L). Despite this large volume of fluid, only 100 mL are lost in the feces. Therefore, the GI tract is extremely efficient in absorbing water. Water absorption by the GI tract is passive. The rate of absorption depends on both the region of the intestinal tract and the luminal osmolality. The duodenum, jejunum, and ileum absorb the bulk of the water that enters the GI tract daily. The colon normally absorbs about 1.4 L of water and excretes about 100 mL. It is capable of absorbing considerably more water (about 4.5 L), however, and watery diarrhea occurs only if this capacity is exceeded.

Because water absorption is determined by the osmolality difference of the lumen and the blood, water can move both ways in the intestinal tract (i.e., secretion and absorption). The osmolality of blood is about 300 mosm/kg H_2O. The ingestion of a hypertonic meal (e.g., 600 mosm/kg H_2O) initially leads to net water movement from blood to lumen; however, as the various nutrients and electrolytes are absorbed by the small intestine, the luminal osmolality falls, resulting in the net water movement from lumen to blood. The water of a hypertonic meal is therefore absorbed mainly in the ileum and colon. In contrast, if a hypotonic meal is ingested (e.g., 200 mosm/kg H_2O), net water movement is immediately from the lumen to the blood, resulting in the absorption of most of the water in the duodenum and jejunum.

REVIEW QUESTIONS

DIRECTIONS: Each of the numbered items or incomplete statements in this section is followed by answers or by completions of the statement. Select the ONE lettered answer or completion that is BEST in each case.

1. Most of the following GI secretions have a basal output during the interdigestive period (between meals). However, the sight and smell of a tasty meal stimulates GI secretions. Of the various GI secretions, which is the most stimulated?
(A) Gastric secretion
(B) Intestinal secretion
(C) Pancreatic secretion
(D) Salivary secretion
(E) Biliary secretion
2. Bile acid uptake by hepatocytes is dependent on
(A) Calcium
(B) Iron
(C) Sodium
(D) Potassium
(E) Chloride
3. Parietal cells in the stomach secrete a protein crucial for the absorption of vitamin B_{12} by the ileum. What is this protein?
(A) Intrinsic factor
(B) Gastrin
(C) Somatostatin
(D) Cholecystokinin (CCK)
(E) Chylomicrons
4. Gastric acid secretion is stimulated during several phases associated with the ingestion and digestion of a meal. Which phase is associated with the bulk of acid secretion?
(A) Cephalic
(B) Esophageal
(C) Gastric
(D) Intestinal
(E) Colonic
5 Carbonic anhydrase is an enzyme that occurs in plants, bacteria, and animals and is involved in the formation of which chemical?
(A) Carbon dioxide from carbon and oxygen
(B) Carbonic acid from carbon dioxide and water
(C) Bicarbonate ion from carbonic acid
(D) Hydrochloric acid
(E) Hypochlorous acid
6. Parasympathetic stimulation induces salivary acinar cells to release the protease
(A) Bradykinin
(B) Kallikrein
(C) Kininogen
(D) Kinin
(E) Aminopeptidase
7. Which protein is absent in saliva?
(A) Lactoferrin
(B) Amylase
(C) Mucin
(D) Intrinsic factor
(E) Muramidase
8. After the ingestion of a meal, the pH in the stomach lumen increases in response to the dilution and buffering of gastric acid by the arrival of food. The pH in the stomach lumen in the fasting state is usually between
(A) 0.1 to 0.5
(B) 1 to 2
(C) 4 to 5
(D) 6 to 7
(E) 9 to 10

(continued)

9. Unlike other GI secretions, salivary secretion is controlled almost exclusively by the nervous system and is significantly inhibited by
(A) Atropine
(B) Pilocarpine
(C) Cimetidine
(D) Aspirin
(E) Omeprazole
10. The chief cells of the stomach secrete
(A) Intrinsic factor
(B) Hydrochloric acid
(C) Pepsinogen
(D) Gastrin
(E) CCK
11. The interaction of histamine with its H_2 receptor in the parietal cell results in
(A) An increase in intracellular sodium concentration
(B) An increase in intracellular cAMP production
(C) An increase in intracellular cGMP production
(D) A decrease in intracellular calcium concentration
(E) A decrease in intracellular cAMP production
12. When the pH of the stomach lumen falls below 3, the antrum of the stomach releases a peptide that acts locally to inhibit gastrin release. This peptide is
(A) Enterogastrone
(B) Intrinsic factor
(C) Secretin
(D) Somatostatin
(E) CCK
13. Which hormone stimulates pancreatic secretion that is rich in bicarbonate?
(A) Somatostatin
(B) Secretin
(C) CCK
(D) Gastrin
(E) Insulin
14. A patient suffering from Zollinger-Ellison syndrome would be expected to have
(A) Excessive acid reflux into the esophagus, resulting in esophagitis
(B) Excessive secretion of CCK, causing continuous contraction of the gallbladder
(C) A gastrin-secreting tumor causing excessive stomach acid secretion and peptic ulcers
(D) Low plasma lipid levels, due to failure of the liver to secrete VLDLs
(E) Inadequate secretion of bicarbonate by the pancreas
15. Lactase is a brush border enzyme involved in the digestion of lactose. The digestion product or products of lactose is/are
(A) Glucose
(B) Glucose and galactose
(C) Glucose and fructose
(D) Galactose and fructose
(E) Fructose
16. Maltase hydrolyzes maltose to form
(A) Glucose
(B) Glucose and galactose
(C) Glucose and fructose
(D) Galactose and fructose
(E) Galactose
17. Which sugar is taken up by enterocytes by facilitated diffusion?
(A) Glucose
(B) Galactose
(C) Fructose
(D) Xylose
(E) Sucrose
18. Dietary triglyceride is a major source of nutrient for the human body. It is digested mostly in the intestinal lumen by pancreatic lipase to release
(A) Lysophosphatidylcholine and fatty acids
(B) Glycerol and fatty acids
(C) Diglyceride and fatty acids
(D) 2-Monoglyceride and fatty acids
(E) Lysophosphatidylcholine and diglyceride
19. After a meal of pizza, dietary lipid is absorbed by the small intestine and transported in the lymph mainly as
(A) VLDLs
(B) Free fatty acids bound to albumin
(C) Chylomicrons
(D) LDLs
(E) HDLs
20. Hartnup's disease is an inherited autosomal recessive disorder involving the malabsorption of amino acids, particularly tryptophan, by the small intestine. Feeding dipeptides and tripeptides containing tryptophan to patients with this disease improves their clinical condition because
(A) Dipeptides and tripeptides, unlike free amino acids, can be taken up passively by enterocytes in the small intestine
(B) Peptides, unlike free amino acids, can be taken up by defective amino acid transporters
(C) Dipeptides and tripeptides use transporters that are different from the defective amino acid transporters
(D) The presence of dipeptides and tripeptides in the intestinal lumen enhances the uptake of amino acids by the transporters
(E) Dipeptides and tripeptides, unlike amino acids, can be taken up passively by the colon
21. What would you expect to find in a sample of hepatic portal blood after protein has been digested and absorbed by the GI tract?
(A) Free amino acids
(B) Dipeptides and tripeptides
(C) Free amino acids and dipeptides
(D) Free amino acids and tripeptides
(E) Free amino acids, dipeptides, and tripeptides
22. Which vitamin is water-soluble?
(A) Vitamin A
(B) Vitamin D
(C) Vitamin K
(D) Vitamin B_1
(E) Vitamin E
23. Which one of the following vitamins stimulates calcium absorption by the GI tract?
(A) Vitamin E
(B) Vitamin D
(C) Vitamin A
(D) Vitamin K
(E) Vitamin C
24. Which vitamin is transported in chylomicrons as an ester?
(A) Vitamin E
(B) Vitamin D
(C) Vitamin A
(D) Vitamin K
(E) Vitamin B_{12}
25. Potassium is absorbed in the jejunum by
(A) Active transport
(B) Facilitated transport
(C) Passive transport
(D) Active and passive transport
(E) Coupling to sodium absorption
26. Ascorbic acid is a potent enhancer of iron absorption because it
(A) Enhances the absorption of heme iron
(B) Enhances the activity of heme oxygenase
(C) Is a reducing agent, thereby helping to keep iron in the ferrous state
(D) Decreases the production of ferritin by enterocytes
(E) Stimulates production of transferrin

SUGGESTED READING

Alpers DH. Digestion and absorption of carbohydrates and proteins. In: Johnson LR, ed. Physiology of the Gastrointestinal Tract. 3rd Ed. New York: Raven, 1994;1723–1749.

Boyer JL, Graf J, Meier PJ. Hepatic transport systems regulating pH, cell volume and bile secretion. Annu Rev Physiol 1992;54:415–438.

Choudari CP, Lehman GA, Sherman S. Pancreatitis and cystic fibrosis gene mutations. Gastroenterol Clin North Am 1999;28:543–549.

(continued)

Davenport HW. Physiology of the Digestive Tract. 5th Ed. Chicago: Year Book, 1982.

Hagenbuch B, Stieger B, Foguet M, Lubbert H, Meier PJ. Functional expression cloning and characterization of the hepatocyte Na^+/bile acid cotransport system. Proc Natl Acad Sci U S A 1991;88:10,629–10,633.

Ito S. Functional gastric morphology. In: Johnson LR, ed. Physiology of the Gastrointestinal Tract. 2nd Ed. New York: Raven, 1987;817–851.

Johnson LR. Gastrointestinal Physiology. 6th Ed. St. Louis: CV Mosby, 2001.

Phan CT, Tso P. Intestinal lipid absorption and transport. Front Biosci 2001;6:D299–D319.

Rose RC. Intestinal absorption of water-soluble vitamins. In: Johnson LR, ed. Physiology of the Gastrointestinal Tract. 2nd Ed. New York: Raven 1987; 1581–1596.

Scott D, Weeks D, Melchers K, Sachs G. The life and death of *Helicobacter pylori*. Gut 1998;43:S56–S60.

The Physiology of the Liver

Patrick Tso, Ph.D.

CHAPTER OUTLINE

KEY CONCEPTS

1. The liver sinusoid is lined with sinusoidal cells (endothelial cells), Kupffer cells, and fat storage cells (also called stellate or Ito cells), which perform important metabolic functions and defend the liver.
2. The liver plays an important role in maintaining blood glucose levels and in metabolizing drugs and toxic substances.
3. The liver has a remarkable capacity to regenerate.
4. The liver is extremely important in maintaining an adequate supply of nutrients for metabolism.
5. The liver synthesizes glucose from noncarbohydrate sources, a process called gluconeogenesis.
6. The liver is the first organ to experience and respond to changes in plasma insulin levels.
7. The liver is one of the main organs involved in fatty acid synthesis.
8. The liver aids in the elimination of cholesterol from the body.
9. The liver is a storage area for fat-soluble vitamins and iron.
10. The liver modifies the action of hormones released by other organs.

The liver is the largest internal organ in the body, constituting about 2.5% of an adult's body weight. During rest, it receives 25% of the cardiac output via the hepatic portal vein and hepatic artery. The hepatic portal vein carries the absorbed nutrients from the GI tract to the liver, which takes up, stores, and distributes nutrients and vitamins. The liver plays an important role in maintaining blood glucose levels. It also regulates the circulating blood lipids by the amount of very low density lipoproteins (VLDLs) it secretes. Many of the circulating plasma proteins are synthesized by the liver. In addition, the liver takes up numerous toxic compounds and drugs from the portal circulation. It is well equipped to deal with the metabolism of drugs and toxic substances. The liver also serves as an excretory organ for bile pigments, cholesterol, and drugs. Finally, it performs important endocrine functions.

THE ANATOMY OF THE LIVER

The liver is essential to the normal physiology of many organs and systems of the body. It interacts with the cardiovascular and immune systems, it secretes important substances into the GI tract, and it stores, degrades, and detoxifies many substrates.

The Arrangement of Hepatocytes Along Liver Sinusoids Aids the Rapid Exchange of Molecules

Hepatocytes are highly specialized cells. The bile canaliculus is usually lined by two hepatocytes and is separated from the pericellular space by tight junctions, which are impermeable and, thus, prevent the mixing of contents between the bile canaliculus and the pericellular space (Fig. 28.1). The bile from the bile canaliculus drains into a series of ducts, and it may eventually join the pancreatic duct near where it enters the duodenum. Drainage of bile into the duodenum is partly regulated by a sphincter located at the junction between the bile duct and the duodenum, the sphincter of Oddi (see Chapter 27).

The pericellular space, the space between two hepatocytes, is continuous with the perisinusoidal space (see Fig. 28.1). The **perisinusoidal space**, also known as the **space of Disse**, is separated from the sinusoid by a layer of **sinusoidal endothelial cells**. Hepatocytes possess numerous,

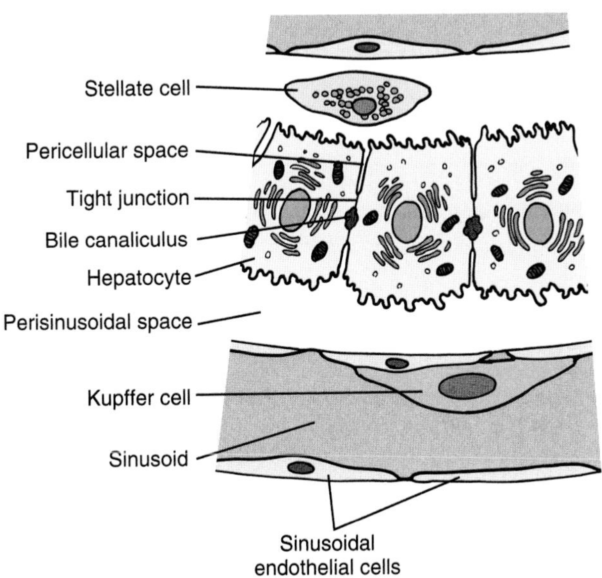

FIGURE 28.1 **The relationship between hepatocytes, the perisinusoidal space, and the sinusoid.**

finger-like projections that extend into the perisinusoidal space, greatly increasing the surface area over which hepatocytes contact the perisinusoidal fluid.

Endothelial cells of the liver, unlike those in other parts of the cardiovascular system, lack a basement membrane. Furthermore, they have sieve-like plates that permit the ready exchange of materials between the perisinusoidal space and the sinusoid. Electron microscopy has demonstrated that even particles as big as chylomicrons (80 to 500 nm in diameter) can penetrate these porous plates. Although the barrier between the perisinusoidal space and the sinusoid is permeable, it does have some sieving properties. For example, the protein concentration of hepatic lymph, assumed to derive from the perisinusoidal space, is lower than that of plasma by about 10%.

Kupffer cells also line the hepatic sinusoids. These are resident macrophages of the **fixed monocyte-macrophage system** that play an extremely important role in removing unwanted material (e.g., bacteria, virus particles, fibrin-fibrinogen complexes, damaged erythrocytes, and immune complexes) from the circulation. Endocytosis is the mechanism by which these materials are removed.

Some perisinusoidal cells contain distinct lipid droplets in the cytoplasm. These fat-storage cells are called **stellate cells** or **Ito cells**. The lipid droplets contain vitamin A. Through complex and typically inflammatory processes, stellate cells become transformed to myofibroblasts, which then become capable of both secreting collagen into the space of Disse and regulating sinusoidal portal pressure by their contraction or relaxation. Stellate cells may be involved in the pathological fibrosis of the liver.

The Liver Receives Venous Blood Through the Portal Vein and Arterial Blood Through the Hepatic Artery

Circulation to the liver is discussed in detail in Chapter 17; here, we briefly describe some of its unique features. The hepatic portal vein provides about 70 to 80% of the liver's blood supply, and the hepatic artery provides the rest. Hepatic portal blood is poorly oxygenated unlike that from the hepatic artery. The portal vein branches repeatedly, forming smaller venules that eventually empty into the sinusoids. The hepatic artery branches to form arterioles and then capillaries, which also drain into the sinusoids. Liver sinusoids can be considered specialized capillaries. As mentioned earlier, the hepatic sinusoid is extremely porous and allows the rapid exchange of materials between the perisinusoidal space and the sinusoid. The sinusoids empty into the central veins, which subsequently join to form the hepatic vein, which then joins the inferior vena cava.

Hepatic blood flow varies with activity, increasing after eating and decreasing during sleep. Blood flow to the intestines and spleen and, in turn, in the portal vein is predominantly regulated by the splanchnic arterioles. In this way, eating results in increased blood flow to the intestines followed by increased liver blood flow. Portal vein pressure is normally low. Increased resistance to portal blood flow results in **portal hypertension**. Portal hypertension is the most common complication of chronic liver disease and accounts for a large percentage of the morbidity and mortality associated with chronic liver diseases (see Clinical Focus Box 28.1).

The Liver Has an Important Lymphatic System

The hepatic lymphatic system is present in three main areas: adjacent to the central veins, adjacent to the portal veins, and coursing along the hepatic artery. As in other organs, it is through these channels that fluid and proteins are drained. The protein concentration is highest in lymph from the liver.

In the liver, the largest space drained by the lymphatic system is the perisinusoidal space. Disturbances in the balance of filtration and drainage are the primary causes of **ascites**, the accumulation of serous fluid in the peritoneal cavity. Ascites is another common cause of morbidity in patients with chronic liver disease.

The Liver Can Regenerate

Of the solid organs, the liver is the only one that can regenerate. There appears to be a critical ratio between functioning liver mass and body mass. Deviations in this ratio trigger a modulation of either hepatocyte proliferation or apoptosis, in order to maintain the liver's optimal size. Peptide growth factors—such as transforming growth factor-α (TGF-α), hepatocyte growth factor (HGF), and epidermal growth factor (EGF)—have been the best-studied stimuli of hepatocyte DNA synthesis. After these peptides bind to their receptors on the remaining hepatocytes and work their way through myriad transcription factors, gene transcription is accelerated, resulting in increased cell number and increased liver mass.

Alternatively, a decrease in liver volume is achieved by enhanced hepatocyte apoptosis rates. Apoptosis is a carefully programmed process by which cells kill themselves while maintaining the integrity of their cellular membranes. In contrast, cell death that results from necroinflammatory

CLINICAL FOCUS BOX 28.1

Esophageal Varices, a Common Manifestation of Portal Hypertension

Chronic liver injury can lead to a sequence of changes that terminates with fatal bleeding from esophageal blood vessels. In most forms of chronic liver injury, stellate cells are transformed into collagen-secreting myofibroblasts. These cells deposit collagen into the sinusoids, interfering with the exchange of compounds between the blood and hepatocytes and increasing resistance to portal venous flow. The resistance appears to be further increased when stellate cells contract. The increased resistance results in increased hepatic portal pressure and decreased liver blood flow. This disorder is seen in approximately 80% of patients with **cirrhosis**. In a compensatory effort, new channels are formed or dormant venous tributaries are expanded, resulting in the formation of varicose (unnaturally swollen) veins in the abdomen. Although varicose veins develop in many areas, portal pressure increases are least opposed in the esophagus because of the limited connective tissue support at the base of the esophagus. This structural condition, along with the negative intrathoracic pressure, favors the formation and rupture of **esophageal varices**. Approximately 30% of patients who develop an esophageal variceal hemorrhage die during the episode of bleeding, making it one of the most lethal medical illnesses.

Currently there are no well-recognized treatments to reverse cirrhosis, but numerous strategies are employed to reduce portal hypertension and bleeding. Chief among these is the use of nonselective beta blockers, which enhance splanchnic arteriolar vasoconstriction and thereby reduce portal venous pressure. Bleeding esophageal varices are frequently treated by endoscopic ligation of the varices. Shunts can be placed radiologically or surgically between the portal and systemic venous systems to reduce the portal pressure.

processes is characterized by a loss of cell membrane integrity and the activation of inflammatory reactions. Liver cell suicide is mediated by proapoptotic signals, such as tumor necrosis factor (TNF).

THE METABOLISM OF DRUGS AND XENOBIOTICS

Hepatocytes play an extremely important role in the metabolism of drugs and **xenobiotics**—compounds that are foreign to the body, some of which are toxic. Most drugs and xenobiotics are introduced into the body with food. The kidneys ultimately dispose of these substances, but for effective elimination, the drug or its metabolites must be made hydrophilic (polar, water-soluble). This is because reabsorption of a substance by the renal tubules is dependent on its hydrophobicity. The more hydrophobic (nonpolar, lipid-soluble) a substance is, the more likely it will be reabsorbed. Many drugs and metabolites are hydrophobic, and the liver converts them into hydrophilic compounds.

The Liver Converts Hydrophobic Drugs and Xenobiotics to Hydrophilic Compounds

Two reactions (phase I and II), catalyzed by different enzyme systems, are involved in the conversion of xenobiotics and drugs into hydrophilic compounds. In **phase I reactions**, the parent compound is biotransformed into more polar compounds by the introduction of one or more polar groups. The common polar groups are hydroxyl (OH) and carboxyl (COOH). Most phase I reactions involve oxidation of the parent compound. The enzymes involved are mostly located in the smooth ER; some are located in the cytoplasm. For example, alcohol dehydrogenase is located in the cytoplasm of hepatocytes and catalyzes the rapid conversion of alcohol to acetaldehyde. It may also play a role in the dehydrogenation of steroids.

The enzymes involved in phase I reactions of drug biotransformation are present as an enzyme complex composed of the **NADPH-cytochrome P450 reductase** and a series of hemoproteins called **cytochrome P450** (Fig. 28.2). The drug combines with the oxidized cytochrome $P450^{+3}$ to form the cytochrome $P450^{+3}$-drug complex. This complex is then reduced to the cytochrome $P450^{+2}$-drug complex, catalyzed by the enzyme NADPH-cytochrome P450 reductase. The reduced complex combines with molecular oxygen to form an oxygenated intermediate. One atom of the molecular oxygen

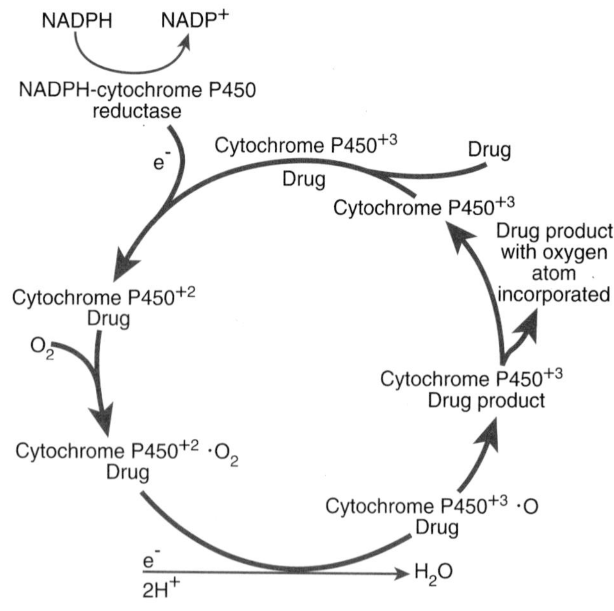

FIGURE 28.2 Phase I reactions in the metabolism of drugs.

then combines with two H^+ and an electron to form water. The other oxygen atom remains bound to the cytochrome $P450^{+3}$-drug complex and is transferred from the cytochrome $P450^{+3}$ to the drug molecule. The drug product with an oxygen atom incorporated is released from the complex. The cytochrome $P450^{+3}$ released can then be recycled for the oxidation of other drug molecules.

In **phase II reactions**, the phase I reaction products undergo conjugation with several compounds to render them more hydrophilic. Glucuronic acid is the substance most commonly used for conjugation, and the enzymes involved are the glucuronosyltransferases. Other molecules used in conjugation are glycine, taurine, and sulfates.

Aging, Nutrition, and Genetics Influence Drug Metabolism

The enzyme systems in phase I and II reactions are age-dependent. These systems are poorly developed in human newborns because their ability to metabolize any given drug is lower than that of adults. Older adults also have a lower capacity than young adults to metabolize drugs.

Nutritional factors can also affect the enzymes involved in phase I and II reactions. Insufficient protein in the diet to sustain normal growth results in the production of fewer of the enzymes involved in drug metabolism.

It is well known that drug-metabolizing enzymes can be induced by certain factors, such as polycyclic aromatic hydrocarbons. Persons who smoke inhale polycyclic aromatic hydrocarbons, increasing the metabolism of certain drugs, such as caffeine.

The role of genetics in the regulation of drug metabolism by the liver is less well understood. Briefly, drug metabolism by the liver can be controlled by a single gene or several genes (polygenic control). Careful study of the metabolism of a certain drug by the population can provide important clues as to whether its metabolism is under single gene or polygenic control. Genetic variability combined with the induction or inhibition of P450 enzymes by other drugs or compounds can have a profound effect on what is a safe and effective dose of a medicine.

ENERGY METABOLISM IN THE LIVER

The liver is pivotal in regulating the metabolism of carbohydrates, lipids, and proteins. It also helps to maintain a constant blood glucose concentration by converting other substances, such as amino acids, into glucose.

The Intestine Supplies Nutrients to the Liver

Most water-soluble nutrients and water-soluble vitamins and minerals absorbed from the small intestine are transported via the portal blood to the liver. The nutrients transported in portal blood include amino acids, monosaccharides, and fatty acids (predominantly short- and medium-chain forms). Short-chain fatty acids are largely derived from the fermentation of dietary fibers by bacteria in the colon. Some dietary fibers, such as pectin, are almost completely digested to form short-chain fatty acids (or volatile fatty acids), whereas cellulose is not well digested by the bacteria. Only a small amount of long-chain fatty acids, bound to albumin, is transported by the portal blood; most is transported in intestinal lymph as triglyceride-rich lipoproteins (chylomicrons).

The Liver Is Important in Carbohydrate Metabolism

The liver is extremely important in maintaining an adequate supply of nutrients for cell metabolism and regulating blood glucose concentration (Fig. 28.3). After the ingestion of a meal, the blood glucose increases to a concentration of 120 to 150 mg/dL, usually in 1 to 2 hours. Glucose is taken up by hepatocytes by a facilitated carrier-mediated process and is converted to glucose 6-phosphate and then UDP-glucose. UDP-glucose can be used for glycogen synthesis, or **glycogenesis**. It is generally believed that blood glucose is the major precursor of glycogen. However, recent evidence seems to indicate that the lactate in blood (from the peripheral metabolism of glucose) is also a major precursor of glycogen. Amino acids (e.g., alanine) can supply pyruvate to synthesize glycogen.

Glycogen is the main carbohydrate store in the liver, and may amount to as much as 7 to 10% of the weight of a normal, healthy liver. The glycogen molecule resembles a tree with many branches (see Fig. 27.19). Glucose units are linked via α-1,4 (to form a straight chain) or α-1,6 (to form a branched chain) glycosidic bonds. The advantage of such a configuration is that the glycogen chain can be broken down at multiple sites, making the release of glucose much more efficient than would be the case with a straight-chain polymer.

During fasting, glycogen is broken down by **glycogenolysis**. The enzyme **glycogen phosphorylase** catalyzes the cleavage of glycogen into glucose 1-phosphate. Glycogen phosphorylase acts only on the α-1,4-glycosidic bond,

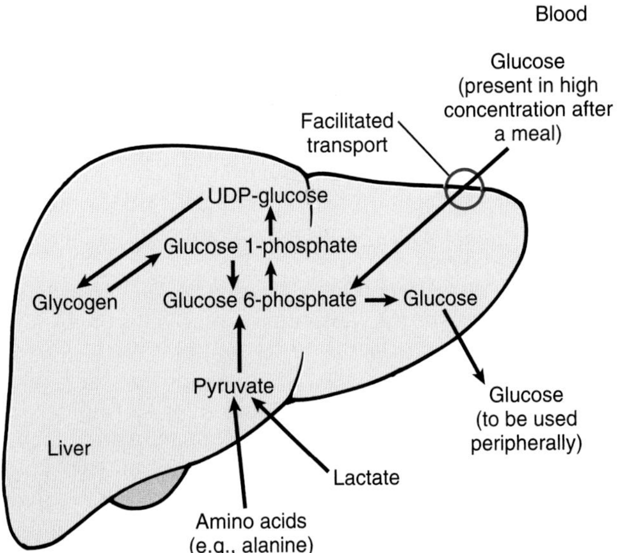

FIGURE 28.3 The regulation of carbohydrate metabolism in the liver.

and the enzyme α-1,6-glucosidase is used to break the α-1,6-glycosidic bonds.

Glucose 1-phosphate is converted to glucose 6-phosphate by the enzyme phosphoglucomutase. The enzyme **glucose-6-phosphatase**, which is present in the liver but not in muscle or brain, converts glucose 6-phosphate to glucose. This last reaction enables the liver to release glucose into the circulation. Glucose 6-phosphate is an important intermediate in carbohydrate metabolism because it can be channeled either to provide blood glucose or for glycogen formation.

Both glycogenolysis and glycogenesis are hormonally regulated. The pancreas secretes insulin into the portal blood. Therefore, the liver is the first organ to respond to changes in plasma insulin levels, to which it is extremely sensitive. For instance, a doubling of portal insulin concentration completely shuts down hepatic glucose production. About half the insulin in portal blood is removed in its first pass through the liver. Insulin tends to lower blood glucose by stimulating glycogenesis and suppressing glycogenolysis and gluconeogenesis. Glucagon, in contrast, stimulates glycogenolysis and gluconeogenesis, raising blood sugar levels. Epinephrine stimulates glycogenolysis.

The liver regulates the blood glucose concentrations within a narrow limit, 70 to 100 mg/dL. Although one might expect patients with liver disease to have difficulty regulating blood glucose, this is usually not the case because of the relatively large reserve of hepatic function. However, those with chronic liver disease occasionally have reduced glycogen synthesis and reduced gluconeogenesis. Some patients with advanced liver disease develop portal hypertension, which induces the formation of portosystemic shunting, resulting in elevated arterial blood levels of insulin and glucagon.

The Metabolism of Monosaccharides. Monosaccharides are first phosphorylated by a reaction catalyzed by the enzyme hexokinase. In the liver (but not in the muscle), there is a specific enzyme (glucokinase) for the phosphorylation of glucose to form glucose 6-phosphate. Depending on the energy requirement, the glucose 6-phosphate is channeled to glycogen synthesis or used for energy production by the glycolytic pathway.

Fructose is taken up by the liver and phosphorylated by fructokinase to form fructose 1-phosphate. This molecule is either isomerized to form glucose 6-phosphate or metabolized by the glycolytic pathway. Fructose 1-phosphate is used by the glycolytic pathway more efficiently than glucose 6-phosphate.

Galactose is an important sugar used not only to provide energy but also in the biosynthesis of glycoproteins and glycolipids. When galactose is taken up by the liver, it is phosphorylated to form galactose 1-phosphate, which then reacts with uridine diphosphate-glucose, or **UDP-glucose**, to form UDP-galactose and glucose 1-phosphate. The UDP-galactose can be used for glycoprotein and glycolipid biosynthesis or converted to UDP-glucose, which can then be recycled.

Gluconeogenesis. **Gluconeogenesis** is the production of glucose from noncarbohydrate sources such as fat, amino acids, and lactate. The process is energy-dependent, and the starting substrate is pyruvate. The energy required seems to be derived predominantly from the β-oxidation of fatty acids. Pyruvate can be derived from lactate and the metabolism of glucogenic amino acids—those that can contribute to the formation of glucose. The two major organs involved in the production of glucose from noncarbohydrate sources are the liver and the kidneys. However, because of its size, the liver plays a far more important role than the kidney in the production of sugar from noncarbohydrate sources.

Gluconeogenesis is important in maintaining blood glucose concentrations especially during fasting. The red blood cells and renal medulla are totally dependent on blood glucose for energy, and glucose is the preferred substrate for the brain. Most amino acids can contribute to the carbon atoms of the glucose molecule, and alanine from muscle is the most important. The rate-limiting factor in gluconeogenesis is not the liver enzymes but the availability of substrates. Gluconeogenesis is stimulated by epinephrine and glucagon but greatly suppressed by insulin. Thus, in type 1 diabetics, gluconeogenesis is greatly stimulated, contributing to the hyperglycemia observed in these patients (see Chapter 35).

The Liver Plays an Important Role in the Metabolism of Lipids

The liver plays a pivotal role in lipid metabolism (Fig. 28.4). It takes up free fatty acids and lipoproteins (complexes of lipid and protein) from the plasma. Lipid is circulated in the plasma as lipoproteins because lipid and water are not mis-

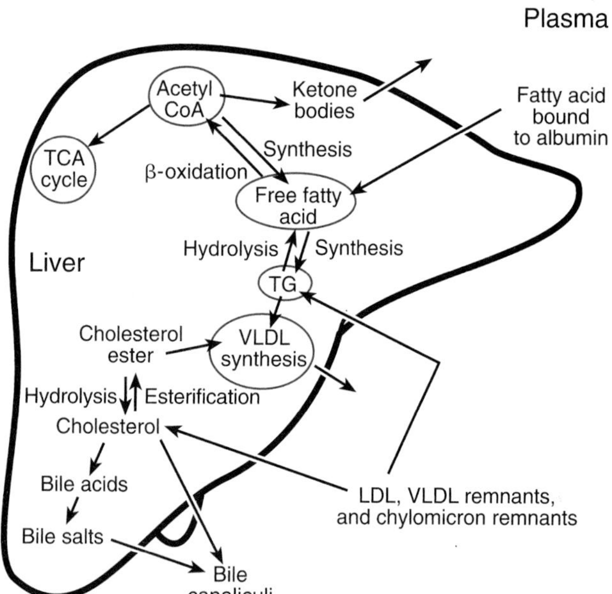

FIGURE 28.4 **The regulation of lipid metabolism in the liver.** LDL, low-density lipoprotein; VLDL, very low density lipoprotein; TG, triglycerides; TCA, tricarboxylic acid.

cible; the lipid droplets coalesce in an aqueous medium. The protein and phospholipid on the surface of the lipoprotein particles stabilize the hydrophobic triglyceride center of the particle.

During fasting, fatty acids are mobilized from adipose tissue and are taken up by the liver. They are used by the hepatocytes to provide energy via β-oxidation, for the generation of ketone bodies, and to synthesize the triglyceride necessary for VLDL formation. After feeding, chylomicrons from the small intestine are metabolized peripherally, and the chylomicron remnants formed are rapidly taken up by the liver. The fatty acids derived from the triglycerides of the chylomicron remnants are used for the formation of VLDLs or for energy production via β-oxidation.

Fatty Acid Oxidation and Synthesis. Fatty acids derived from the plasma can be metabolized in the mitochondria of hepatocytes by β-oxidation to provide energy. Fatty acids are broken down to form acetyl-CoA, which can be used in the tricarboxylic acid cycle for ATP production, in the synthesis of fatty acids, and in the formation of ketone bodies. Because fatty acids are synthesized from acetyl-CoA, any substances that contribute to acetyl-CoA, such as carbohydrate and protein sources, enhance fatty acid synthesis.

The liver is one of the main organs involved in fatty acid synthesis. Palmitic acid is synthesized in the hepatocellular cytosol; the other fatty acids synthesized in the body are derived by shortening, elongating, or desaturating the palmitic acid molecule.

Lipoprotein Synthesis. One of the major functions of the liver in lipid metabolism is lipoprotein synthesis. The four major classes of circulating plasma lipoproteins are chylomicrons, very low density lipoproteins (VLDLs), low-density lipoproteins (LDLs), and high-density lipoproteins (HDLs) (Table 28.1). These lipoproteins, which differ in chemical composition, are usually isolated from plasma according to their flotation properties.

Chylomicrons are the lightest of the four lipoprotein classes, with a density of less than 0.95 g/mL. They are made only by the small intestine and are produced in large quantities during fat ingestion. Their major function is to transport the large amount of absorbed fat to the bloodstream.

Very low density lipoproteins (**VLDLs**) are denser and smaller than chylomicrons. The liver synthesizes about 10 times more circulating VLDLs than the small intestine. Like chylomicrons, VLDLs are triglyceride-rich and carry most of the triglyceride from the liver to the other organs. The triglyceride of VLDLs is broken down by **lipoprotein lipase** to yield fatty acids, which can be metabolized to provide energy. The human liver normally has a considerable capacity to produce VLDLs, but in acute or chronic liver disorders, this ability is significantly compromised. Liver VLDLs are associated with an important class of proteins, the **apo B proteins**. The two forms of circulating apo B are B_{48} and B_{100}. The human liver makes only apo B_{100}, which has a molecular weight of about 500,000. Apo B_{100} is important for the hepatic secretion of VLDL. In **abetalipoproteinemia**, apo B synthesis and, therefore, the secretion of VLDLs is blocked. Large lipid droplets can be seen in the cytoplasm of the hepatocytes of abetalipoproteinemic patients.

Although considerable amounts of circulating plasma LDLs and HDLs are produced in the plasma, the liver also produces a small amount of these two cholesterol-rich lipoproteins. LDLs are denser than VLDLs, and HDLs are denser than LDLs. The function of LDLs is to transport cholesterol ester from the liver to the other organs. HDLs are believed to remove cholesterol from the peripheral tissue and transport it to the liver.

The formation and secretion of lipoproteins by the liver is regulated by precursors and hormones, such as estrogen and thyroid hormones. For instance, during fasting, the fatty acids in VLDLs are derived mainly from fatty acids mobilized from adipose tissue. In contrast, during fat feeding, fatty acids in VLDLs produced by the liver are largely derived from chylomicrons.

As noted earlier, the fatty acids taken up by the liver can be used for β-oxidation and ketone body formation. The relative amounts of fatty acid channeled for these various purposes are largely dependent on the individual's nutritional and hormonal status. More fatty acid is channeled to ketogenesis or β-oxidation when the supply of carbohydrate is short (during fasting) or under conditions of high circulating glucagon or low circulating insulin (diabetes mellitus). In contrast, more of the fatty acid is used for synthesis of triglyceride for lipoprotein export when the supply of carbohydrate is abundant (during feeding) or under conditions of low circulating glucagon or high circulating insulin.

Lipoprotein Catabolism. The importance of the liver in lipoprotein metabolism is exemplified by **familial hypercholesterolemia**, a disorder in which the liver fails to produce the LDL receptor. When LDL binds its receptor, it is internalized and catabolized in the hepatocyte. Consequently, the LDL receptor is crucial for the removal of LDL from the plasma. Individuals suffering from familial hypercholesterolemia usually have very high plasma LDLs,

TABLE 28.1 Characteristics of Human Plasma Lipoproteins

Lipoprotein	Source	Density (g/mL)	Size (nm)	Protein	Lipid
Chylomicron	Intestine	< 0.95	80–500	1%	99%
VLDL	Intestine and liver	0.95–1.006	30–80	7–10%	90–93%
LDL	Chylomicron and VLDL	1.019–1.063	18–28	20–22%	78–80%
HDL	Chylomicron and VLDL	1.063–1.21	5–14	35–60%	40–65%

VLDL, very low density lipoprotein; LDL, low-density lipoprotein; HDL, high-density lipoprotein.

which predisposes them to early coronary heart disease. Often the only effective treatment is a liver transplant.

The liver also plays an important role in the uptake of chylomicrons after their metabolism. After the chylomicrons produced by the small intestine enter the circulation, lipoprotein lipase on the endothelial cells of blood vessels acts on them to liberate fatty acids and glycerol from the triglycerides. As metabolism progresses, the chylomicrons shrink, resulting in the detachment of free cholesterol, phospholipid, and proteins, and the formation of HDL. Chylomicrons are converted to **chylomicron remnants** during metabolism, and chylomicron remnants are rapidly taken up by the liver via chylomicron remnant receptors.

The Production of Ketone Bodies. Most organs, except the liver, can use ketone bodies as fuel. For example, during prolonged fasting, the brain shifts to use ketone bodies for energy, although glucose is the preferred fuel for the brain. The two ketone bodies are acetoacetate and β-hydroxybutyrate. Their formation by the liver is normal and physiologically important. For instance, during fasting a rapid depletion of the glycogen stores in the liver occurs resulting in a shortage of substrates (e.g., oxaloacetate) for the citric acid cycle. There is also a rapid mobilization of fatty acids from adipose tissues to the liver. Under these circumstances, the acetyl-CoA formed from β-oxidation is channeled to ketone bodies.

The liver is efficient in producing ketone bodies. In humans, it can produce half of its equivalent weight of ketone bodies per day. However, it lacks the ability to metabolize the ketone bodies formed because it lacks the necessary enzyme ketoacid-CoA transferase.

The level of ketone bodies circulating in the blood is usually low, but during prolonged starvation and in diabetes mellitus it is highly elevated, a condition known as **ketosis**. In patients with diabetes, large amounts of β-hydroxybutyric acid can make the blood pH acidic, a state called **ketoacidosis**.

Cholesterol Metabolism. The liver plays an important role in cholesterol homeostasis. Liver cholesterol is derived from both *de novo* synthesis and the lipoproteins taken up by the liver. Hepatic cholesterol can be used in the formation of bile acids, biliary cholesterol secretion, the synthesis of VLDLs, and the synthesis of liver membranes. Because the absorption of biliary cholesterol and bile acids by the GI tract is incomplete, this method of eliminating cholesterol from the body is essential and efficient. However, patients with high plasma cholesterol levels might be given additional drugs, such as statins, to lower their plasma cholesterol levels. Statins act by inhibiting enzymes that play an essential role in cholesterol synthesis. VLDLs secreted by the liver provide cholesterol to organs that need it for the synthesis of steroid hormones (e.g., the adrenal glands, ovaries, and testes).

PROTEIN AND AMINO ACID METABOLISM IN THE LIVER

The liver is one of the major organs involved in synthesizing nonessential amino acids from the essential amino acids. The body can synthesize all but nine of the amino acids necessary for protein synthesis.

The Liver Produces Most of the Circulating Plasma Proteins

The liver synthesizes many of the circulating plasma proteins, albumin being the most important (Fig. 28.5). It synthesizes about 3 g of albumin a day. Albumin plays an important role in preserving plasma volume and tissue fluid balance by maintaining the colloid osmotic pressure of plasma. This important function of plasma proteins is illustrated by the fact that both liver disease and long-term starvation result in generalized edema and ascites. Plasma albumin plays a pivotal role in the transport of many substances in blood, such as free fatty acids and certain drugs, including penicillin and salicylate.

The other major plasma proteins synthesized by the liver are components of the complement system, components of the blood clotting cascade (fibrinogen and prothrombin), and proteins involved in iron transport (transferrin, haptoglobin, and hemopexin) (see Chapter 11).

The Liver Produces Urea

Ammonia, derived from protein and nucleic acid catabolism, plays a pivotal role in nitrogen metabolism and is needed in the biosynthesis of nonessential amino acids and nucleic acids. Ammonia metabolism is a major function of the liver. The liver has an ammonia level 10 times higher than the plasma ammonia level. High circulating ammonia levels are highly neurotoxic, and a deficiency in hepatic function can lead to several distinct neurological disorders, including coma in severe cases.

The liver synthesizes most of the urea in the body. The enzymes involved in the urea cycle are regulated by protein intake. In humans, starvation stimulates these enzymes.

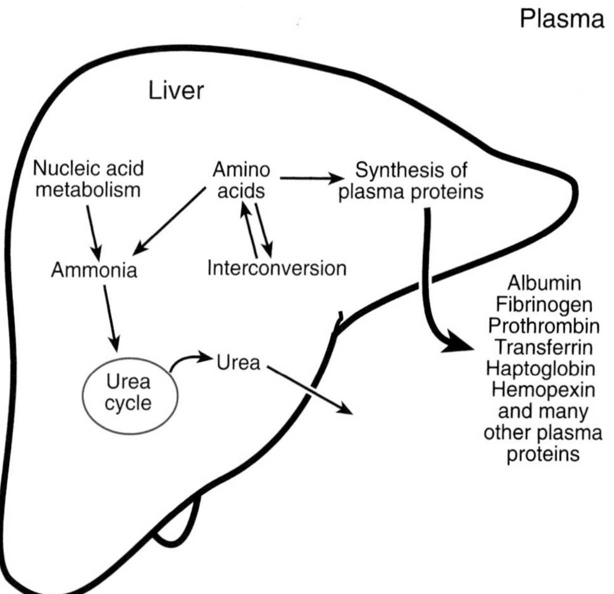

FIGURE 28.5 The regulation of protein and amino acid metabolism in the liver.

The Liver Plays an Important Role in the Synthesis and Interconversion of Amino Acids

The **essential amino acids** (see Table 27.7) must be supplied in the diet. The liver can form nonessential amino acids from the essential amino acids. For instance, tyrosine can be synthesized from phenylalanine and cysteine can be synthesized from methionine.

Glutamic acid and glutamine play an important role in the biosynthesis of certain amino acids in the liver. Glutamic acid is derived from the amination of α-ketoglutarate by ammonia. This reaction is important because ammonia is used directly in the formation of the α-amino group and constitutes a mechanism for shunting nitrogen from wasteful urea-forming products. Glutamic acid can be used in the amination of other α-keto acids to form the corresponding amino acids. It can also be converted to glutamine by coupling with ammonia, a reaction catalyzed by glutamine synthetase. After urea, glutamine is the second most important metabolite of ammonia in the liver. It plays an important role in the storage and transport of ammonia in the blood. Through the action of various transaminases, glutamine can be used to aminate various keto acids to their corresponding amino acids. It also acts as an important oxidative substrate, and in the small intestine it is the primary substrate for providing energy.

THE LIVER AS A STORAGE ORGAN

Another important role of the liver is the storage and metabolism of fat-soluble vitamins and iron. Some water-soluble vitamins, particularly vitamin B_{12}, are also stored in the liver. These stored vitamins are released into the circulation when a need for them arises.

The Liver Has a Central Role in Regulating Coagulation

Liver cells are important both in the production and the clearance of coagulation proteins. Most of the known clotting factors and inhibitors are secreted by hepatocytes, some of them exclusively. In addition, several coagulation and anticoagulation proteins require a vitamin K–dependent modification following synthesis, specifically factors II, VII, IX, and X and proteins C and S, to make them effective.

The monocyte-macrophage system of the liver, predominantly Kupffer cells, is an important system for clearing clotting factors and factor-inhibitor complexes. Disturbances in liver perfusion and function result in the ineffective clearance of activated coagulation proteins, so patients with advanced liver failure may be predisposed to developing disseminated intravascular coagulation.

Fat-Soluble Vitamins Are Stored in the Liver

Vitamin A comprises a family of compounds related to retinol. Vitamin A is important in vision, growth, the maintenance of epithelia, and reproduction. The liver plays a pivotal role in the uptake, storage, and maintenance of circulating plasma vitamin A levels by mobilizing its vitamin A store (Fig. 28.6). Retinol (an alcohol) is transported in chylomicrons mainly as an ester of long-chain fatty acids (see Chapter 27). When chylomicrons enter the circulation, the triglyceride is rapidly acted on by lipoprotein lipase; the triglyceride content of the particles is significantly reduced, while the retinyl ester content remains unchanged. Receptors in the liver mediate the rapid uptake of chylomicron remnants, which are degraded, and the retinyl ester is stored.

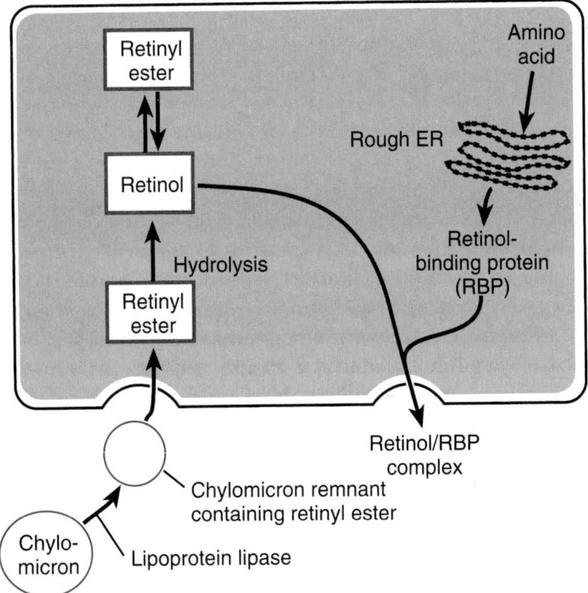

FIGURE 28.6 The metabolism of vitamin A (retinol) by the hepatocyte.

When the vitamin A level in blood falls, the liver mobilizes the vitamin A store by hydrolyzing the retinyl ester (see Fig. 28.6). The retinol formed is bound with **retinol-binding protein** (RBP), which is synthesized by the liver before it is secreted into the blood. The amount of RBP secreted into the blood is dependent on vitamin A status. Vitamin A deficiency significantly inhibits the release of RBP, whereas vitamin A loading stimulates its release.

Hypervitaminosis A develops when massive quantities of vitamin A are consumed. Since the liver is the storage organ for vitamin A, hepatotoxicity is often associated with hypervitaminosis A. The continued ingestion of excessive amounts of vitamin A eventually leads to portal hypertension and cirrhosis.

Vitamin D is thought to be stored mainly in skeletal muscle and adipose tissue. However, the liver is responsible for the initial activation of vitamin D by converting vitamin D_3 to 25-hydroxyvitamin D_3, and it synthesizes vitamin D-binding protein.

Vitamin K is a fat-soluble vitamin important in the hepatic synthesis of prothrombin. Prothrombin is synthesized as a precursor that is converted to the mature prothrombin, a reaction that requires the presence of vitamin K (Fig. 28.7). Vitamin K deficiency, therefore, leads to impaired blood clotting.

The largest vitamin K store is in skeletal muscle, but the physiological significance of this and other body stores is

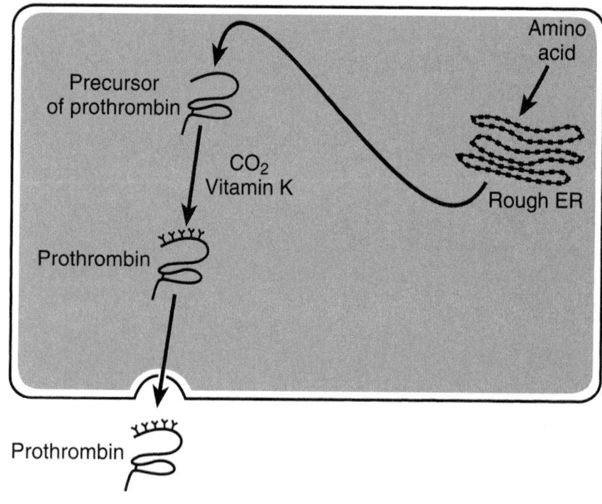

FIGURE 28.7 **The formation and secretion of prothrombin by the hepatocyte.**

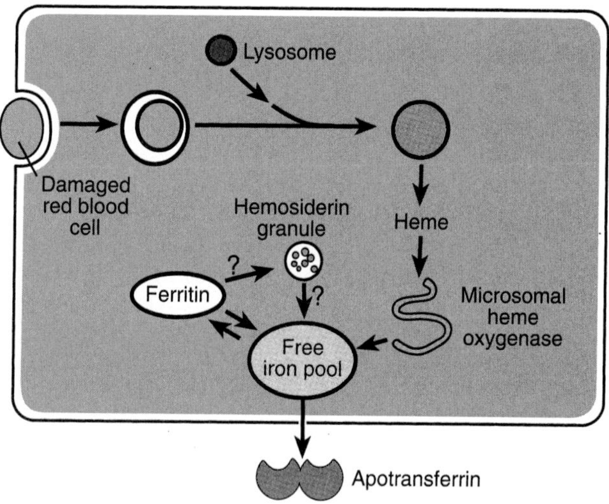

FIGURE 28.8 **The possible pathways following phagocytosis of damaged red blood cells by Kupffer cells.** (Modified from Young SP, Aisen P. The liver and iron. In: Arias I, Jakoby WB, Popper H, et al., eds. The Liver: Biology and Pathobiology. New York: Raven, 1988.)

unknown. The dietary vitamin K requirement is extremely small and is adequately supplied by the average North American diet. Bacteria in the GI tract also provide vitamin K. This appears to be an important source of vitamin K because prolonged administration of wide-spectrum antibiotics sometimes results in hypoprothrombinemia. Because vitamin K absorption is dependent on normal fat absorption, any prolonged malabsorption of lipid can result in its deficiency. The vitamin K store in the liver is relatively limited, and therefore, hypoprothrombinemia can develop within a few weeks. Vitamin K deficiency is not uncommon in the Western world. Parenteral administration of vitamin K usually provides a cure.

The Liver Is Important in the Storage and Homeostasis of Iron

The liver is the major site for the synthesis of several proteins involved in iron transport and metabolism. The protein **transferrin** plays a critical role in the transport and homeostasis of iron in the blood. The circulating plasma transferrin level is inversely proportional to the iron load of the body—the higher the concentration of ferritin in the hepatocyte, the lower the rate of transferrin synthesis. During iron deficiency, liver synthesis of transferrin is significantly stimulated, enhancing the intestinal absorption of iron. **Haptoglobin**, a large glycoprotein with a molecular weight of 100,000, binds free hemoglobin in the blood. The hemoglobin-haptoglobin complex is rapidly removed by the liver, conserving iron in the body. **Hemopexin** is another protein synthesized by the liver that is involved in the transport of free heme in the blood. It forms a complex with free heme, and the complex is removed rapidly by the liver.

The spleen is the organ that removes red blood cells that are slightly altered. Kupffer cells of the liver also have the capacity to remove damaged red blood cells, especially those that are moderately damaged (Fig. 28.8). The red cells taken up by Kupffer cells are rapidly digested by secondary lysosomes to release heme. Microsomal **heme oxygenase** releases iron from the heme, which then enters the free iron pool and is stored as ferritin or released into the bloodstream (bound to apotransferrin). Some of the ferritin iron may be converted to **hemosiderin granules**. It is unclear whether the iron from the hemosiderin granules is exchangeable with the free iron pool.

It was long believed that Kupffer cells were the only cells involved in iron storage, but recent studies suggest that hepatocytes are the major sites of long-term iron storage. Transferrin binds to receptors on the surface of hepatocytes, and the entire transferrin-receptor complex is internalized and processed (Fig. 28.9). The apotransferrin (not containing iron) is recycled back to the plasma, and the released iron enters a labile iron pool. The iron from transferrin is probably the major source of iron for the hepatocytes, but they also derive iron from haptoglobin-hemoglobin and hemopexin-heme complexes. When hemoglobin is released inside the hepatocytes, it is degraded in the secondary lysosomes, and heme is released. Heme is processed in the smooth ER and free iron released enters the labile iron pool. A significant portion of the free iron in the cytosol probably combines rapidly with apoferritin to form ferritin. Like Kupffer cells, hepatocytes may transfer some of the iron in ferritin to hemosiderin.

Iron is absolutely essential for survival, but iron overload can be extremely toxic, especially to the liver where it can cause **hemochromatosis**, a condition characterized by excessive amounts of hemosiderin in the hepatocytes. The hepatocytes in patients with hemochromatosis are defective and fail to perform many normal functions.

ENDOCRINE FUNCTIONS OF THE LIVER

The liver is important in regulating the endocrine functions of hormones. It can amplify the action of some hormones. It is also the major organ for the removal of peptide hormones.

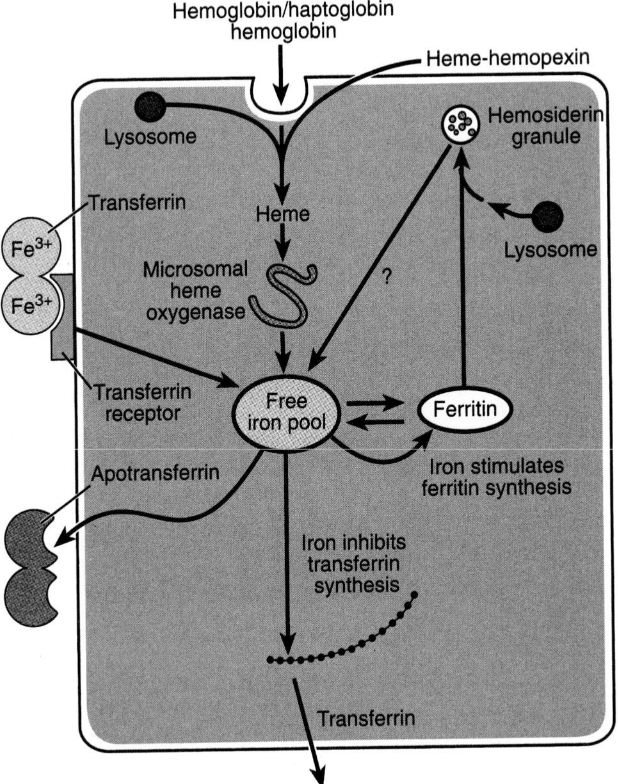

FIGURE 28.9 **The possible pathways followed by iron in the hepatocyte.** (Modified from Young SP, Aisen P. The liver and iron. In: Arias I, Jakoby WB, Popper H, et al., eds. The Liver: Biology and Pathobiology. New York: Raven, 1988.)

The Liver Can Modify or Amplify Hormone Action

As discussed before, the liver converts vitamin D_3 to 25-hydroxyvitamin D_3, an essential step before conversion to the active hormone 1,25-dihydroxyvitamin D_3 in the kidneys. The liver is also a major site of conversion of the thyroid hormone thyroxine (T_4) to the biologically more potent hormone triiodothyronine (T_3). The regulation of the hepatic T_4 to T_3 conversion occurs at both the uptake step and the conversion step. Because of the liver's relatively large reserve in converting T_4 to T_3, hypothyroidism is uncommon in patients with liver disease. In advanced chronic liver disease, however, signs of hypothyroidism may be evident.

The liver modifies the function of growth hormone (GH) secreted by the pituitary gland. Some growth hormone actions are mediated by **insulin-like growth factors** made by the liver (see Chapter 32).

The Liver Removes Circulating Hormones

The liver helps to remove and degrade many circulating hormones. Insulin is degraded in many organs, but the liver and kidneys are by far most important. The presence of insulin receptors on the surface of hepatocytes suggests that the binding of insulin to these receptors results in degradation of some insulin molecules. There is also degradation of insulin by proteases of hepatocytes that do not involve the insulin receptor.

Glucagon and growth hormone are degraded mainly by the liver and the kidneys. The liver may also degrade various GI hormones (e.g., gastrin), but the kidneys and other organs probably contribute more significantly to inactivating these hormones.

REVIEW QUESTIONS

DIRECTIONS: Each of the numbered items or incomplete statements in this section is followed by answers or by completions of statements. Select the ONE lettered answer or completion that is BEST in each case.

1. The first step in alcohol metabolism by the liver is the formation of acetaldehyde from alcohol, a chemical reaction catalyzed by
(A) Cytochrome P450
(B) NADPH-cytochrome P450 reductase
(C) Alcohol oxygenase
(D) Alcohol dehydrogenase
(E) Glycogen phosphorylase
2. The arterial blood glucose concentration in normal humans after a meal is in the range of
(A) 30 to 50 mg/dL
(B) 50 to 70 mg/dL
(C) 120 to 150 mg/dL
(D) 220 to 250 mg/dL
(E) 300 to 350 mg/dL
3. Both the liver and muscle contain glycogen, yet, unlike the liver, muscle is not capable of contributing glucose to the circulation because muscle
(A) Does not have the enzyme glucose-6-phosphatase
(B) Glycolytic activity consumes all of the glucose it generates
(C) Does not have the enzyme glucose-1-phosphatase
(D) Does not have the enzyme glycogen phosphorylase
(E) Is not as capable of gluconeogenesis as is the liver
4. The hepatocyte is compartmentalized to carry out specific functions. In which subcellular compartment does fatty acid synthesis occur?
(A) Cytoplasm
(B) Mitochondria
(C) Nucleus
(D) Endosomes
(E) Golgi apparatus
5. The small intestine secretes various triglyceride-rich lipoproteins, but the liver secretes only
(A) Chylomicrons
(B) VLDLs
(C) LDLs
(D) HDLs
(E) Chylomicron remnants
6. Because free ammonia in the blood is toxic to the body, it is transported in which of the following non-toxic forms?
(A) Histidine and urea
(B) Phenylalanine and methionine
(C) Glutamine and urea
(D) Lysine and glutamine
(E) Methionine and urea
7. In patients with a portacaval shunt (connection between the portal vein and vena cava), the circulating glucagon level is extremely high because the
(A) Pancreas produces more glucagon in these patients
(B) Kidney is less efficient in removing the circulating glucagon in these patients

(continued)

(C) Liver normally is the major site for the removal of glucagon
(D) Small intestine produces more glucagon in these patients
(E) Blood flow to the small intestine is compromised

8. Which protein is made by the liver and carries iron in the blood?
(A) Hemosiderin
(B) Haptoglobin
(C) Transferrin
(D) Ceruloplasmin
(E) Lactoferrin

9. The level of drug metabolizing enzymes in the liver determines how fast a drug is removed from the circulation. Therefore, it would be expected to find drug metabolizing enzymes
(A) Higher in smokers than in nonsmokers
(B) Similar in smokers and nonsmokers
(C) Lower in smokers than in nonsmokers
(D) Stimulated by malnutrition
(E) Higher in newborns than in adults

10. Phase I reactions of drug metabolism refer to the
(A) Conjugation of drugs with glucuronic acid
(B) Conjugation of drugs with glycine or taurine
(C) Introduction of one or more polar groups to the drug molecule
(D) Introduction of one or more hydrophobic groups to the drug molecule
(E) Conjugation of drugs with sulfate

11. The level of circulating 1,25-dihydroxycholecalciferol is significantly reduced in patients with chronic liver disease because
(A) The liver can no longer efficiently convert 25-hydroxycholecalciferol to 1,25-dihydroxycholecalciferol
(B) The liver can no longer efficiently convert vitamin D to cholecalciferol
(C) The liver can no longer efficiently convert vitamin D to 25-hydroxycholecalciferol
(D) The liver can no longer efficiently convert cholecalciferol to 1,25-dihydroxycholecalciferol
(E) The intestine has impaired absorption of 1,25-dihydroxy-cholecalciferol

12. The liver removes LDLs in the blood by the LDLs binding to
(A) LDL receptors and then internalizing them
(B) HDL receptors and then internalizing them
(C) The albumin present on LDLs and then internalizing them
(D) The transferrin present on LDL and then internalizing them
(E) The ceruloplasmin on LDLs and then internalizing them

SUGGESTED READING

Arias IM. The Liver: Biology and Pathobiology. 3rd Ed. New York: Lippincott-Raven, 1994.

Black ER. Diagnostic strategies and test algorithms in liver disease. Clin Chem 1997;43:1555–1560.

Chang EB, Sitrin MD, Black DD. Gastrointestinal, Hepatobiliary, and Nutritional Physiology. Philadelphia: Lippincott-Raven, 1996.

Liska DJ. The detoxification enzyme systems. Altern Med Rev 1998;3:187–198.

MacMathuna PM. Mechanisms and consequences of portal hypertension. Drugs 1992;44(Suppl 2):1–13, 70–72.

Oka K, Davis AR, Chan L. Recent advances in liver-directed gene therapy: Implications for the treatment of dyslipidemia. Curr Opin Lipidol 2000;11:179–186.

CASE STUDIES FOR PART VII

CASE STUDY FOR CHAPTER 26

Dysphagia

A 51-year-old woman is evaluated for difficulty in swallowing solid foods. She experiences chest pain while attempting to eat and often regurgitates swallowed food. Fluoroscopic examination of a barium swallow reveals a dilated lower esophagus with considerable residual barium remaining after the swallow. A manometric motility study of esophageal motility following a swallow reveals an absence of primary peristalsis in the distal third, without relaxation of contractile tone in the lower esophageal sphincter.

Questions

1. What is the explanation for the woman's dysphagia?
2. What is the most likely explanation for the failure of the lower esophageal sphincter relaxation during the swallow?
3. What are the possible treatments for the woman's condition?

Answers to Case Study Questions for Chapter 26

1. The best explanation for the patient's dysphagia is failure of the lower esophageal sphincter to relax (achalasia).
2. Loss of the ENS in the region of the lower esophageal sphincter and gastric cardia is the histoanatomic hallmark of lower esophageal sphincter achalasia. Failure of the sphincter to relax reflects the loss of inhibitory motor innervation of the sphincteric muscle.
3. There are several possible treatments. The time-tested treatment is pneumatic dilation of the lower esophageal sphincter, by placing a balloon in the lumen of the sphincter. Pharmacological approaches include calcium channel blockers (e.g., nifedipine) to relax the smooth muscle of the sphincter, and local endoscopic injection of botulinum toxin, an inhibitor of ACh release from nerve terminals.

Reference

Richter JE. Motility disorders of the esophagus. In: Yamada T, Alpers DH, Owyang C, Powell DW, Silverstein FE, eds. Textbook of Gastroenterology. 2nd Ed. Philadelphia: Lippincott, 1995;1174–1213.

CASE STUDY FOR CHAPTER 27

Lactose Intolerance

A 9-year-old Chinese American boy regularly complains of abdominal cramps, abdominal distension, and diarrhea after drinking milk. A gastroenterologist administers 50 g of lactose by mouth to the child and measures an increase in the boy's expired hydrogen gas.

Questions

1. How is lactose digested and absorbed in the small intestine?
2. Explain the symptoms that accompany lactose intolerance.
3. Why was the lactose breath test done?
4. How common is lactose intolerance?
5. What can be done about lactose intolerance?

Answers to Case Study Questions for Chapter 27

1. Lactose is hydrolyzed by a brush border enzyme called lactase to glucose and galactose. The monosaccharides are then absorbed by sodium-dependent secondary active transport.

2. If the lactase enzyme is deficient, lactose will not be broken down and will remain in the intestinal lumen. The osmotic activity of the lactose draws water into the intestinal lumen and results in a watery diarrhea. In the colon, bacteria metabolize the lactose to lactic acid, carbon dioxide, and hydrogen gas. The extra fluid and gas in the intestine result in distension and increased motility (cramps).
3. The child might have had an allergy to proteins in milk. The lactose breath test results indicate lactose intolerance.
4. In most of the world's population, intestinal lactase activity is high during childhood, but falls after ages 5 to 7 to low adult levels. The prevalence of lactose intolerance in adults is about 100% in Asian Americans, 95% in Native Americans, 81% in African Americans, 56% in Mexican Americans, and 24% in white Americans. Lactose intolerance is common (about 50 to 70%) in adult Americans of Mediterranean descent, but is low (0 to only a few %) in those of northern European ancestry.
5. Avoiding foods that contain lactose (milk, dairy products) is recommended for persons who are lactose-intolerant; however, calcium and caloric intake should not be compromised. Milk can be pretreated with an enzyme obtained from bacteria or yeasts that digests lactose, or lactase pills can be taken with meals.

Case Study for Chapter 28

Budd-Chiari Syndrome

A 51-year-old woman complained of 4 days of epigastric abdominal pain. She reported having been healthy all her life. She admitted to having gained approximately 9 kg (20 lb) over the preceding 6 months, which was unusual. Upon examination by her physician, she is found to have a distended abdomen that is tender in the area between her ribs at the top of her abdomen.

An exploratory laparotomy reveals an enlarged liver and no other disease. A liver biopsy is taken and reportedly shows no significant abnormalities. For unstated reasons, the patient is later taken for a venogram and is found to have thrombosis of her hepatic veins, Budd-Chiari syndrome. She is subsequently referred to a tertiary hospital. Initially, the patient is treated with diuretic medication (spironolactone and furosemide to increase renal excretion of sodium and water) and intermittent paracentesis (insertion of a needle into the peritoneal space, evacuating fluid, which relieves the abdominal distension and discomfort). She subsequently undergoes placement of a transjugular intrahepatic portosystemic shunt (TIPS), which serves to lower portal pressure by shunting blood into systemic veins. She is also given warfarin, an anticoagulant.

Questions

1. What is the probable explanation for her abdominal pain, distension, and weight gain over 6 months?
2. What is the rationale for giving an anticoagulant, and how does warfarin work?

Answers to Case Study Questions for Chapter 28

1. A common explanation for abdominal discomfort, distension, and weight gain in women is pregnancy. Her age makes this unlikely but not impossible. Any disorder that results in fluid retention may present with these symptoms. Common causes of marked abdominal fluid retention are nephrotic syndrome (the kidneys fail to adequately remove excess water), congestive heart failure (the heart fails to adequately pump blood to the kidneys, reducing their ability to remove excess water), and liver dysfunction (usually from an excess pressure in the sinusoids resulting in increased fluid loss into the abdomen). The general term to describe excess fluid in the abdominal cavity is ascites. Alternatively, symptoms may be due to intraabdominal malignancies, such as malignant ascites or large tumors. In a woman of this age, ovarian cancer would be considered a likely cause.
2. The anticoagulant warfarin was given to treat the patient's hypercoagulable disorder and to maintain shunt patency. Clotting factors, mostly produced in the liver, have a series of glutamic acid residues that must be carboxylated by a vitamin K–dependent carboxylase in order for them to bind to endothelial cells and activate platelets necessary for clot formation. The reduced form of vitamin K is a necessary cofactor for the carboxylation. During carboxylation of the clotting factor, vitamin K becomes an epoxide. Warfarin is thought to disrupt the vitamin K cycle, thereby preventing the necessary carboxylation of clotting factors. The liver continues to synthesize these factors, but they lack effect, and therefore, clotting is limited.

PART VIII Temperature Regulation and Exercise Physiology

CHAPTER 29 The Regulation of Body Temperature*

C. Bruce Wenger, M.D., Ph.D.†

CHAPTER OUTLINE

- BODY TEMPERATURES AND HEAT TRANSFER IN THE BODY
- THE BALANCE BETWEEN HEAT PRODUCTION AND HEAT LOSS
- HEAT DISSIPATION
- THERMOREGULATORY CONTROL
- THERMOREGULATORY RESPONSES DURING EXERCISE
- HEAT ACCLIMATIZATION
- RESPONSES TO COLD
- CLINICAL ASPECTS OF THERMOREGULATION

KEY CONCEPTS

1. The body is divided into an inner core and an outer shell; temperature is relatively uniform in the core and is regulated within narrow limits, while shell temperature is permitted to vary.
2. The body produces heat through metabolic processes and exchanges energy with the environment as mechanical work and heat; it is in thermal balance when the sum of metabolic energy production plus energy gain from the environment equals energy loss to the environment.
3. In humans, the chief physiological thermoregulatory responses are the secretion of sweat, which removes heat from the skin as it evaporates; the control of skin blood flow, which governs the flow of heat to the skin from the rest of the body; and increasing metabolic heat production in the cold.
4. The thermoregulatory set point (the setting of the body's "thermostat") varies cyclically with the circadian rhythm and the menstrual cycle, and is elevated during fever.
5. Core and whole-body skin temperatures govern the reflex control of physiological thermoregulatory responses, which are graded according to disturbances in the body's thermal state.
6. The control of thermoregulatory responses is accomplished through reflex signals generated in the central nervous system (CNS) according to the level of the thermoregulatory set point, as well as signals from temperature-sensitive CNS neurons and nerve endings elsewhere, chiefly in the skin. The response of sweat glands and superficial blood vessels to these signals is modified by local skin temperature.
7. Acclimatization to heat can dramatically increase the body's ability to dissipate heat, maintain cardiovascular homeostasis in hot temperatures, and conserve salt while sweating profusely. Acclimatization to cold has only modest effects, depending on how the acclimatization was produced, and may include increased tissue insulation and variable metabolic responses.
8. Adverse systemic effects of excessive heat stress include circulatory instability, fluid-electrolyte imbalance, exertional heat injury, and heatstroke. Exertional heat injury and heatstroke involve organ and tissue injury produced in several ways, some of which are not well understood. The primary adverse systemic effect of excessive cold stress is hypothermia.

*The views, opinions, and findings contained in this chapter are those of the author and should not be construed as official Department of the Army position, policy, or decision unless so designated by other official documentation. Approved for public release; distribution unlimited.

†Dr. Wenger died November 22, 2002.

Humans, like other mammals, are **homeotherms**, or warm-blooded animals, and regulate their internal body temperatures within a narrow range near 37°C, in spite of wide variations in environmental temperature (Fig. 29.1). Internal body temperatures of **poikilotherms**, or cold-blooded animals, by contrast, are governed by environmental temperature. The range of temperatures that living cells and tissues can tolerate without harm extends from just above freezing to nearly 45°C—far wider than the limits within which homeotherms regulate body temperature. What biological advantage do homeotherms gain by maintaining a stable body temperature? As we shall see, tissue temperature is important for two reasons.

First, temperature extremes injure tissue directly. High temperatures alter the configuration and overall structure of protein molecules, even though the sequence of amino acids is unchanged. Such alteration of protein structure is called **denaturation**. A familiar example of denaturation by heat is the coagulation of albumin in the white of a cooked egg. Since the biological activity of a protein molecule depends on its configuration and charge distribution, denaturation inactivates a cell's proteins and injures or kills the cell. Injury occurs at tissue temperatures higher than about 45°C, which is also the point at which heating the skin becomes painful. The severity of injury depends on the temperature to which the tissue is heated and how long the heating lasts.

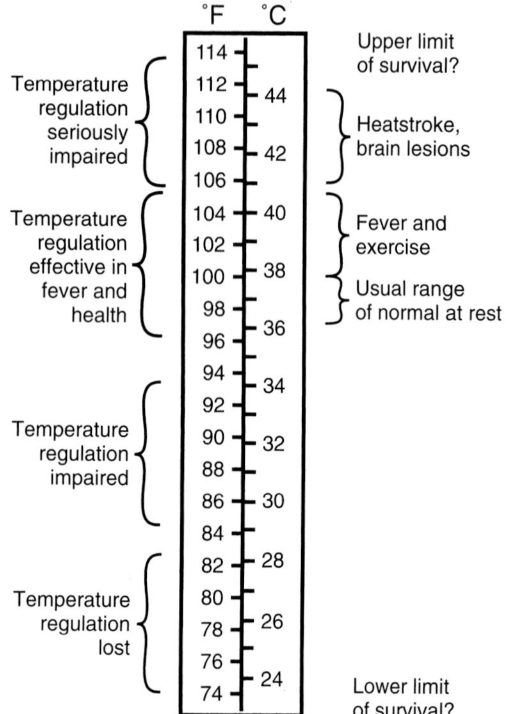

FIGURE 29.1 Rectal temperature ranges in healthy people, patients with fever, and people with impaired or failed thermoregulation. (Modified from Wenger CB, Hardy JD. Temperature regulation and exposure to heat and cold. In: Lehmann JF, ed. Therapeutic Heat and Cold. 4th Ed. Baltimore: Williams & Wilkins, 1990;150–178. Based on DuBois EF. Fever and the Regulation of Body Temperature. Springfield, IL: CC Thomas, 1948.)

Cold also can injure tissues. As a water-based solution freezes, ice crystals consisting of pure water form, so that all dissolved substances in the solution are left in the unfrozen liquid. Therefore, as more ice forms, the remaining liquid becomes more and more concentrated. Freezing damages cells through two mechanisms. Ice crystals probably injure the cell mechanically. In addition, the increase in solute concentration of the cytoplasm as ice forms denatures the proteins by removing their water of hydration, increasing the ionic strength of the cytoplasm, and causing other changes in the physicochemical environment in the cytoplasm.

Second, temperature changes profoundly alter biological function through specific effects on such specialized functions as electrical properties and fluidity of cell membranes, and through a general effect on most chemical reaction rates. In the physiological temperature range, most reaction rates vary approximately as an exponential function of temperature (T); increasing T by 10°C increases the reaction rate by a factor of 2 to 3. For any particular reaction, the ratio of the rates at two temperatures 10°C apart is called the $\mathbf{Q_{10}}$ for that reaction, and the effect of temperature on reaction rate is called the **$\mathbf{Q_{10}}$ effect**. The notion of Q_{10} may be generalized to apply to a group of reactions that have some measurable overall effect (such as O_2 consumption) in common and are, thus, thought of as comprising a physiological process. The Q_{10} effect is clinically important in managing patients who have high fevers and are receiving fluid and nutrition intravenously. A commonly used rule is that a patient's fluid and calorie needs are increased 13% above normal for each 1°C of fever.

The profound effect of temperature on biochemical reaction rates is illustrated by the sluggishness of a reptile that comes out of its burrow in the morning chill and becomes active only after being warmed by the sun. Homeotherms avoid such a dependence of metabolic rate on environmental temperature by regulating their internal body temperatures within a narrow range. A drawback of homeothermy is that, in most homeotherms, certain vital processes cannot function at low levels of body temperature that poikilotherms tolerate easily. For example, shipwreck victims immersed in cold water die of respiratory or circulatory failure (through disruption of the electrical activity of the brainstem or heart) at body temperatures of about 25°C, even though such a temperature produces no direct tissue injury and fish thrive in the same water.

BODY TEMPERATURES AND HEAT TRANSFER IN THE BODY

The body is divided into a warm internal core and a cooler outer shell (Fig. 29.2). Because the temperature of the shell is strongly influenced by the environment, its temperature is not regulated within narrow limits as the internal body temperature is, even though thermoregulatory responses strongly affect the temperature of the shell, especially its outermost layer, the skin. The thickness of the shell depends on the environment and the body's need to conserve heat. In a warm environment, the shell may be less than 1 cm thick, but in a subject conserving heat in a cold environment, it may extend several centimeters below the skin.

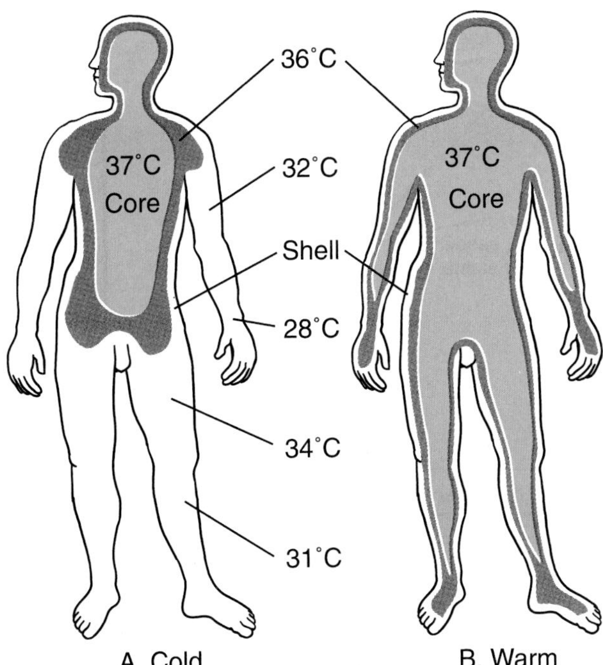

FIGURE 29.2 **Distribution of temperatures in the body's core and shell.** A, During exposure to cold. B, In a warm environment. Since the temperatures of the surface and the thickness of the shell depend on environmental temperature, the shell is thicker in the cold and thinner in the heat.

The internal body temperature that is regulated is the temperature of the vital organs inside the head and trunk, which, together with a variable amount of other tissue, comprise the warm internal core.

Heat is produced in all tissues of the body but is lost to the environment only from tissues in contact with the environment—predominantly from the skin and, to a lesser degree, from the respiratory tract. We, therefore, need to consider heat transfer within the body, especially heat transfer (1) from major sites of heat production to the rest of the body, and (2) from the core to the skin. Heat is transported within the body by two means: conduction through the tissues and convection by the blood, a process in which flowing blood carries heat from warmer tissues to cooler tissues.

Heat flow by conduction varies directly with the thermal conductivity of the tissues, the change in temperature over the distance the heat travels, and the area (perpendicular to the direction of heat flow) through which the heat flows. It varies inversely with the distance the heat must travel. As Table 29.1 shows, the tissues are rather poor heat conductors.

Heat flow by convection depends on the rate of blood flow and the temperature difference between the tissue and the blood supplying the tissue. Because the vessels of the microvasculature have thin walls and, collectively, a large total surface area, the blood comes to the temperature of the surrounding tissue before it reaches the capillaries. Changes in skin blood flow in a cool environment change the thickness of the shell. When skin blood flow is reduced in the cold, the affected skin becomes cooler, and the underlying tissues—

TABLE 29.1 **Thermal Conductivities and Rates of Heat Flow**

Material	Conductivity kcal/(s·m·°C)	Rate of Heat Flow[a] kcal/hr	Watts
Copper	0.092	33,120	38,474
Epidermis	0.00005	18	21
Dermis	0.00009	32	38
Fat	0.00004	14	17
Muscle	0.00011	40	46
Oak (across grain)	0.00004	14	17
Glass fiber insulation	0.00001	3.6	4.2

[a] Values are calculated for slabs 1 m^2 in area and 1 cm thick, with a 1°C temperature difference between the two faces of the slab.

which in the cold may include most of the limbs and the more superficial muscles of the neck and trunk—become cooler as they lose heat by conduction to cool overlying skin and, ultimately, to the environment. In this way, these underlying tissues, which in the heat were part of the body core, now become part of the shell. In addition to the organs in the trunk and head, the core includes a greater or lesser amount of more superficial tissue—mostly skeletal muscle—depending on the body's thermal state.

Because the shell lies between the core and the environment, all heat leaving the body core, except heat lost through the respiratory tract, must pass through the shell before being given up to the environment. Thus, the shell insulates the core from the environment. In a cool subject, the skin blood flow is low, so core-to-skin heat transfer is dominated by conduction; the shell is also thicker, providing more insulation to the core, since heat flow by conduction varies inversely with the distance the heat must travel. Changes in skin blood flow, which directly affect core-to-skin heat transfer by convection, also indirectly affect core-to-skin heat transfer by conduction by changing the thickness of the shell. In a cool subject, the subcutaneous fat layer contributes to the insulation value of the shell because the fat layer increases the thickness of the shell and because fat has a conductivity about 0.4 times that of dermis or muscle (see Table 29.1). Thus, fat is a correspondingly better insulator. In a warm subject, however, the shell is relatively thin, and provides little insulation. Furthermore, a warm subject's skin blood flow is high, so heat flow from the core to the skin is dominated by convection. In these circumstances the subcutaneous fat layer, which affects conduction but not convection, has little effect on heat flow from the core to the skin.

Core Temperature Is Close to Central Blood Temperature

Core temperature varies slightly from one site to another depending on such local factors as metabolic rate, blood supply, and the temperatures of neighboring tissues. However, temperatures at different places in the core are all close to the temperature of the central blood and tend to change together. The notion of a single uniform core temperature, although not strictly correct, is a useful approximation. The value of 98.6°F often given as the normal level of body temperature may give the misleading impression that body temperature is regulated so precisely that it is not allowed to deviate even a few tenths of a degree. In fact, 98.6°F is simply the Fahrenheit equivalent of 37°C, and body temperature does vary somewhat (see Fig. 29.1). The effects of heavy exercise and fever are familiar; variation among individuals and such factors as time of day (Fig. 29.3), phase of the menstrual cycle, and acclimatization to heat can also cause differences of up to about 1°C in core temperature at rest.

To maintain core temperature within a narrow range, the thermoregulatory system needs continuous information about the level of core temperature. Temperature-sensitive neurons and nerve endings in the abdominal viscera, great veins, spinal cord, and, especially, the brain provide this information. We discuss how the thermoregulatory system processes and responds to this information later in the chapter.

Core temperature should be measured at a site whose temperature is not biased by environmental temperature. Sites used clinically include the rectum, the mouth and, occasionally, the axilla. The rectum is well insulated from the environment; its temperature is independent of environmental temperature and is a few tenths of 1°C warmer than arterial blood and other core sites. The tongue is richly supplied with blood; oral temperature under the tongue is usually close to blood temperature (and 0.4 to 0.5°C below rectal temperature), but cooling the face, neck, or mouth can make oral temperature misleadingly low. If a patient holds his or her upper arm firmly against the chest to close the axilla, axillary temperature will eventually come reasonably close to core temperature. However, as this may take 30 minutes or more, axillary temperature is infrequently used. Infrared ear thermometers are convenient and widely used in the clinic, but temperatures of the tympanum and external auditory meatus are loosely related to more accepted indices of core temperature, and ear temperature in collapsed hyperthermic runners may be 3 to 6°C below rectal temperature.

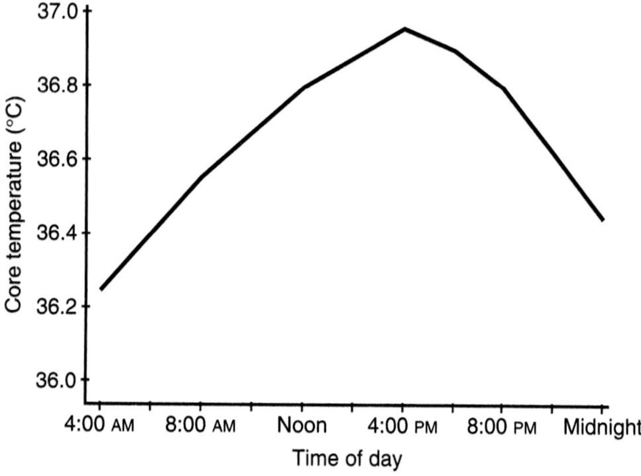

FIGURE 29.3 **Effect of time of day on internal body temperature of healthy resting subjects.** (Drawn from data of Mackowiak PA, Wasserman SS, Levine MM. A critical appraisal of 98.6°F, the upper limit of normal body temperature, and other legacies of Carl Reinhold August Wunderlich. JAMA 1992;268:1578–1580; and Stephenson LA, Wenger CB, O'Donovan BH, et al. Circadian rhythm in sweating and cutaneous blood flow. Am J Physiol 1984;246:R321–R324.)

Skin Temperature Is Important in Heat Exchange and Thermoregulatory Control

Most heat is exchanged between the body and the environment at the skin surface. Skin temperature is much more variable than core temperature; it is affected by thermoregulatory responses such as skin blood flow and sweat secretion, the temperatures of underlying tissues, and environmental factors such as air temperature, air movement, and thermal radiation. Skin temperature is one of the major factors determining heat exchange with the environment. For these reasons, it provides the thermoregulatory system with important information about the need to conserve or dissipate heat.

Many bare nerve endings just under the skin are sensitive to temperature. Depending on the relation of discharge rate to temperature, they are classified as either warm or cold receptors (see Chapter 4). Cold receptors are about 10 times more numerous than warm receptors. Furthermore, as the skin is heated, warm receptors respond with a transient burst of activity and cold receptors respond with a transient suppression; the reverse happens as the skin is cooled. These transient responses at the beginning of heating or cooling give the central thermoregulatory controller almost immediate information about changes in skin temperature and may explain, for example, the intense, brief sensation of being chilled that occurs during a plunge into cold water.

Since skin temperature usually is not uniform over the body surface, mean skin temperature ($\overline{T}_{sk}$) is frequently calculated from temperatures at several skin sites, usually weighting each temperature according to the fraction of body surface area it represents. $\overline{T}_{sk}$ is used to summarize the input to the CNS from temperature-sensitive nerve endings in the skin. $\overline{T}_{sk}$ also is commonly used, along with core temperature, to calculate a mean body temperature and to estimate the quantity of heat stored in the body, since the direct measurement of shell temperature would be difficult and invasive.

THE BALANCE BETWEEN HEAT PRODUCTION AND HEAT LOSS

All animals exchange energy with the environment. Some energy is exchanged as mechanical work, but most is exchanged as heat (Fig. 29.4). Heat is exchanged by conduction, convection, and radiation and as latent heat through evaporation or (rarely) condensation of water. If the sum of energy production and energy gain from the environment does not equal energy loss, the extra heat is "stored" in, or lost from, the body. This relationship is summarized in the heat balance equation:

$$M = E + R + C + K + W + S \qquad (1)$$

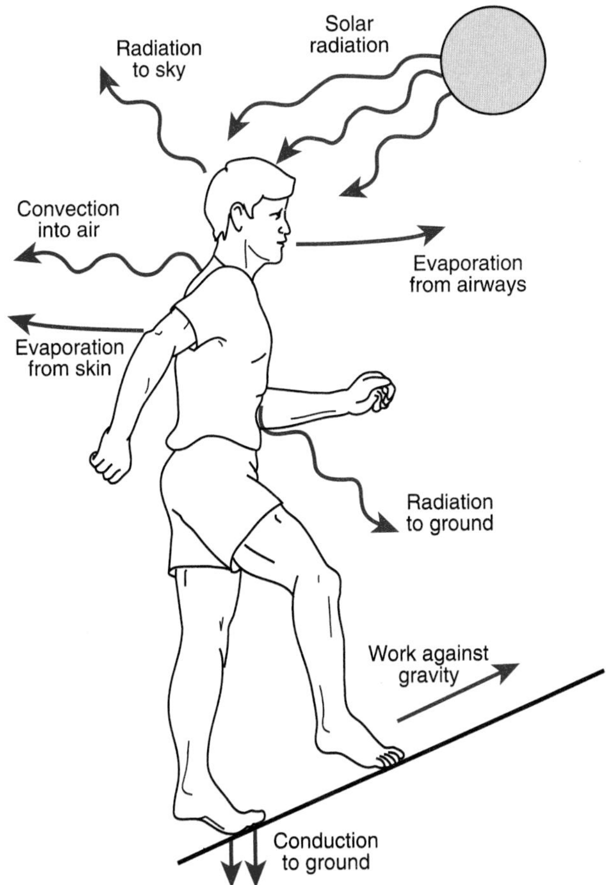

FIGURE 29.4 **Exchange of energy with the environment.** This hiker gains heat from the sun by radiation and loses heat by conduction to the ground through the soles of his feet, convection into the air, radiation to the ground and sky, and evaporation of water from his skin and respiratory passages. In addition, some of the energy released by his metabolic processes is converted into mechanical work, rather than heat, since he is walking uphill.

where M is metabolic rate; E is rate of heat loss by evaporation; R and C are rates of heat loss by radiation and convection, respectively; K is the rate of heat loss by conduction; W is rate of energy loss as mechanical work; and S is rate of heat storage in the body, manifested as changes in tissue temperatures.

M is always positive, but the terms on the right side of equation 1 represent energy exchange with the environment and storage and may be either positive or negative. E, R, C, K, and W are positive if they represent energy losses from the body and negative if they represent energy gains. When $S = 0$, the body is in heat balance and body temperature neither rises nor falls. When the body is not in heat balance, its mean tissue temperature increases if S is positive and decreases if S is negative. This situation commonly lasts only until the body's responses to the temperature changes are sufficient to restore balance. However, if the thermal stress is too great for the thermoregulatory system to restore balance, the body will continue to gain or lose heat until either the stress diminishes sufficiently or the animal dies.

The traditional units for measuring heat are a potential source of confusion, because the word calorie refers to two units differing by a 1,000-fold. The *calorie* used in chemistry and physics is the quantity of heat that will raise the temperature of 1 g of pure water by 1°C; it is also called the small calorie or gram calorie. The *Calorie* (capital C) used in physiology and nutrition is the quantity of heat that will raise the temperature of 1 kg of pure water by 1°C; it is also called the large calorie, kilogram calorie, or (the usual practice in thermal physiology) the **kilocalorie** (kcal). Because heat is a form of energy, it is now often measured in joules, the unit of work (1 kcal = 4,186 J), and rate of heat production or heat flow in watts, the unit of power (1 W = 1 J/sec). This practice avoids confusing calories and Calories. However, kilocalories are still used widely enough that it is necessary to be familiar with them, and there is a certain advantage to a unit based on water because the body itself is mostly water.

Heat Is a By-product of Energy-Requiring Metabolic Processes

Metabolic energy is used for active transport via membrane pumps, for energy-requiring chemical reactions, such as the formation of glycogen from glucose and proteins from amino acids, and for muscular work. Most of the metabolic energy used in these processes is converted into heat within the body. This conversion may occur almost immediately, as with energy used for active transport or heat produced as a by-product of muscular activity. Other energy is converted to heat only after a delay, as when the energy used in forming glycogen or protein is released as heat when the glycogen is converted back into glucose or the protein is converted back into amino acids.

Metabolic Rate and Sites of Heat Production at Rest.

Among subjects of different body size, metabolic rate at rest varies approximately in proportion to body surface area. In a resting and fasting young adult man it is about 45 W/m^2 (81 W or 70 kcal/hr for 1.8 m^2 body surface area), corresponding to an O_2 consumption of about 240 mL/min. About 70% of energy production at rest occurs in the body core—trunk viscera and the brain—even though they comprise only about 36% of the body mass (Table 29.2). As a by-product of their metabolic processes, these organs produce most of the heat needed to maintain heat balance at comfortable environmental temperatures; only in the cold must such by-product heat be supplemented by heat produced expressly for thermoregulation.

Factors other than body size that affect metabolism at rest include age and sex (Fig. 29.5), and hormones and digestion. The ratio of metabolic rate to surface area is highest in infancy and declines with age, most rapidly in childhood and adolescence and more slowly thereafter. Children have high metabolic rates in relation to surface area because of the energy used to synthesize the fats, proteins, and other tissue components needed to sustain growth. Similarly, a woman's metabolic rate increases during pregnancy to supply the energy needed for the growth of the fetus. However, a nonpregnant woman's metabolic rate is 5 to 10% lower than that of a man of the same age and surface area, proba-

TABLE 29.2 Relative Masses and Metabolic Heat Production Rates During Rest and Heavy Exercise

	% of Body Mass	% of Heat Production	
		Rest	Exercise
Brain	2	16	1
Trunk viscera	34	56	8
Muscle and skin	56	18	90
Other	8	10	1

bly because a higher proportion of the female body is composed of fat, a tissue with low metabolism.

The catecholamines and thyroxine are the hormones that have the greatest effect on metabolic rate. Catecholamines cause glycogen to break down into glucose and stimulate many enzyme systems, increasing cellular metabolism. Hypermetabolism is a clinical feature of some cases of pheochromocytoma, a catecholamine-secreting tumor of the adrenal medulla. Thyroxine magnifies the metabolic response to catecholamines, increases protein synthesis, and stimulates oxidation by the mitochondria. The metabolic rate is typically 45% above normal in hyperthyroidism (but up to 100% above normal in severe cases) and 25% below normal in hypothyroidism (but 45% below normal with complete lack of thyroid hormone). Other hormones have relatively minor effects on metabolic rate.

A resting person's metabolic rate increases 10 to 20% after a meal. This effect of food, called the **thermic effect of food** (formerly known as specific dynamic action), lasts several hours. The effect is greatest after eating protein and less after carbohydrate and fat; it appears to be associated with processing the products of digestion in the liver.

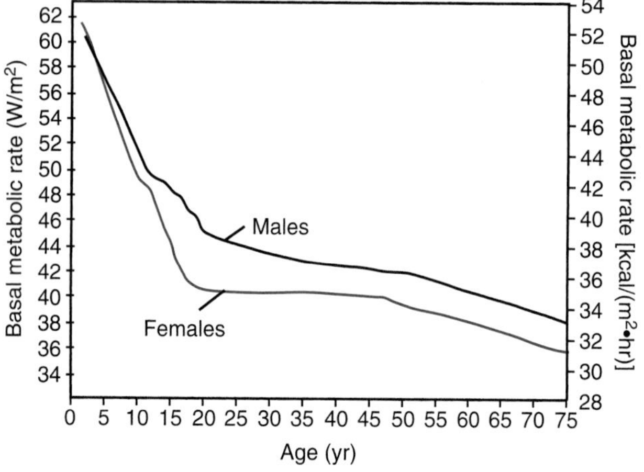

FIGURE 29.5 Effects of age and sex on the basal metabolic rate of healthy subjects. Metabolic rate here is expressed as the ratio of energy consumption to body surface area.

Measurement of Metabolic Rate. Because so many factors affect metabolism at rest, metabolic rate is often measured under a set of standard conditions to compare it with established norms. Metabolic rate measured under these conditions is called **basal metabolic rate** (BMR). The commonly accepted conditions for measuring BMR are that the person must have fasted for 12 hours; the measurement must be made in the morning after a good night's sleep, beginning after the person has rested quietly for at least 30 minutes; and the air temperature must be comfortable, about 25°C (77°F). Basal metabolic rate is "basal" only during wakefulness, since metabolic rate during sleep is somewhat less than BMR.

Heat exchange with the environment can be measured directly by using a human calorimeter. In this insulated chamber, heat can exit only in the air ventilating the chamber or in water flowing through a heat exchanger in the chamber. By measuring the flow of air and water and their temperatures as they enter and leave the chamber, one can determine the subject's heat loss by conduction, convection, and radiation. And by measuring the moisture content of air entering and leaving the chamber, one can determine heat loss by evaporation. This technique is called **direct calorimetry**, and though conceptually simple, it is cumbersome and costly.

Metabolic rate is often estimated by **indirect calorimetry**, which is based on measuring a person's rate of O_2 consumption, since virtually all energy available to the body depends ultimately on reactions that consume O_2. Consuming 1 L of O_2 is associated with releasing 21.1 kJ (5.05 kcal) if the fuel is carbohydrate, 19.8 kJ (4.74 kcal) if the fuel is fat, and 18.6 kJ (4.46 kcal) if the fuel is protein. An average value often used for the metabolism of a mixed diet is 20.2 kJ (4.83 kcal) per liter of O_2. The ratio of CO_2 produced to O_2 consumed in the tissues is called the **respiratory quotient** (RQ). The RQ is 1.0 for the oxidation of carbohydrate, 0.71 for the oxidation of fat, and 0.80 for the oxidation of protein. In a steady state where CO_2 is exhaled from the lungs at the same rate it is produced in the tissues, RQ is equal to the respiratory exchange ratio, R (see Chapter 19). One can improve the accuracy of indirect calorimetry by also determining R and either estimating the amount of protein oxidized—which usually is small compared to fat and carbohydrate—or calculating it from urinary nitrogen excretion.

Skeletal Muscle Metabolism and External Work. Even during mild exercise, the muscles are the principal source of metabolic heat, and during intense exercise, they may account for up to 90%. Moderately intense exercise by a healthy, but sedentary, young man may require a metabolic rate of 600 W (in contrast to about 80 W at rest), and intense activity by a trained athlete, 1,400 W or more. Because of their high metabolic rate, exercising muscles may be almost 1°C warmer than the core. Blood perfusing these muscles is warmed and, in turn, warms the rest of the body, raising the core temperature.

Muscles convert most of the energy in the fuels they consume into heat rather than mechanical work. During phosphorylation of ADP to form ATP, 58% of the energy released from the fuel is converted into heat, and only about 42% is captured in the ATP that is formed in the process. When a muscle contracts, some of the energy in the ATP that was hydrolyzed is converted into heat rather than mechanical work. The efficiency at this stage varies enormously; it is zero in isometric muscle contraction, in which a muscle's length does not change while it develops tension, so that no work is done even though metabolic energy is required. Finally, some of the mechanical work produced is converted by friction into heat within the body. (This is, for example, the fate of all of the mechanical work done by the heart in pumping blood.) At best, no more than 25% of the metabolic energy released during exercise is converted into mechanical work outside the body, and the other 75% or more is converted into heat within the body.

Convection, Radiation, and Evaporation Are the Main Avenues of Heat Exchange With the Environment

Convection is the transfer of heat resulting from the movement of a fluid, either liquid or gas. In thermal physiology, the fluid is usually air or water in the environment or blood, in the case of heat transfer inside the body. To illustrate, consider an object immersed in a fluid that is cooler than the object. Heat passes from the object to the immediately adjacent fluid by conduction. If the fluid is stationary, conduction is the only means by which heat can pass through the fluid, and over time, the rate of heat flow from the body to the fluid will diminish as the fluid nearest the object approaches the temperature of the object. In practice, however, fluids are rarely stationary. If the fluid is moving, heat will still be carried from the object into the fluid by conduction, but once the heat has entered the fluid, it will be carried by the movement of the fluid—by convection. The same fluid movement that carries heat away from the surface of the object constantly brings fresh cool fluid to the surface, so the object gives up heat to the fluid much more rapidly than if the fluid were stationary. Although conduction plays a role in this process, convection so dominates the overall heat transfer that we refer to the heat transfer as if it were entirely convection. Therefore, the conduction term (K) in the heat balance equation is restricted to heat flow between the body and other solid objects, and it usually represents only a small part of the total heat exchange with the environment.

Every surface emits energy as electromagnetic radiation, with a power output proportional to the area of the surface, the fourth power of its absolute temperature (i.e., measured from absolute zero), and the **emissivity** (e) of the surface, a number between 0 and 1 that depends on the nature of the surface and the wavelength of the radiation. (In this discussion, the term *surface* is broadly defined, so that a flame and the sky, for example, are surfaces.) Such radiation, called thermal radiation, has a characteristic distribution of power as a function of wavelength, which depends on the temperature of the surface. The emissivity of any surface is equal to the **absorptivity**—the fraction of incident radiant energy the surface absorbs. (For this reason, an ideal emitter, with an emissivity of 1, is called a **black body**.) If two bodies exchange heat by thermal radiation, radiation travels in both directions, but since each body emits radiation with an intensity that depends on its temperature, the net heat flow is from the warmer to the cooler body.

At ordinary tissue and environmental temperatures, virtually all thermal radiation is in a region of the infrared range where most surfaces, other than polished metals, have emissivities near 1 and emit with a power output near the theoretical maximum. However, bodies that are hot enough to glow, such as the sun, emit large amounts of radiation in the visible and near-infrared range, in which light-colored surfaces have lower emissivities and absorptivities than dark ones. Therefore, colors of skin and clothing affect heat exchange only in sunlight or bright artificial light.

When 1 g of water is converted into vapor at 30°C, it absorbs 2,425 J (0.58 kcal), the **latent heat of evaporation**, in the process. Evaporation of water is, thus, an efficient way of losing heat, and it is the body's only means of losing heat when the environment is hotter than the skin, as it usually is when the environment is warmer than 36°C. Evaporation must then dissipate both the heat produced by metabolic processes and any heat gained from the environment by convection and radiation. Most water evaporated in the heat comes from sweat, but even in cold temperatures, the skin loses some water by the evaporation of **insensible perspiration**, water that diffuses through the skin rather than being secreted. In equation 1, E is nearly always positive, representing heat loss from the body. However, E is negative in the rare circumstances in which water vapor gives up heat to the body by condensing on the skin (as in a steam room).

Heat Exchange Is Proportional to Surface Area and Obeys Biophysical Principles

Animals exchange heat with their environment through both the skin and the respiratory passages, but only the skin exchanges heat by radiation. In panting animals, respiratory heat loss may be large and may be an important means of achieving heat balance. In humans, however, respiratory heat exchange is usually relatively small and (though hyperthermic subjects may hyperventilate) is not predominantly under thermoregulatory control. Therefore, we do not consider it further here.

Convective heat exchange between the skin and the environment is proportional to the difference between skin and ambient air temperatures, as expressed by this equation:

$$C = h_c \times A \times (\overline{T}_{sk} - T_a) \quad (2)$$

where A is the body surface area, $\overline{T}_{sk}$ and T_a are mean skin and ambient temperatures, respectively, and h_c is the convective heat transfer coefficient.

The value of h_c includes the effects of the factors other than temperature and surface area that influence convective heat exchange. For the whole body, air movement is the most important of these factors, and convective heat exchange (and, thus, h_c) varies approximately as the square root of the air speed, except when air movement is slight (Fig. 29.6). Other factors that affect h_c include the direction of air movement and the curvature of the skin surface. As the radius of curvature decreases, h_c increases, so the hands and fingers are effective in convective heat exchange disproportionately to their surface area.

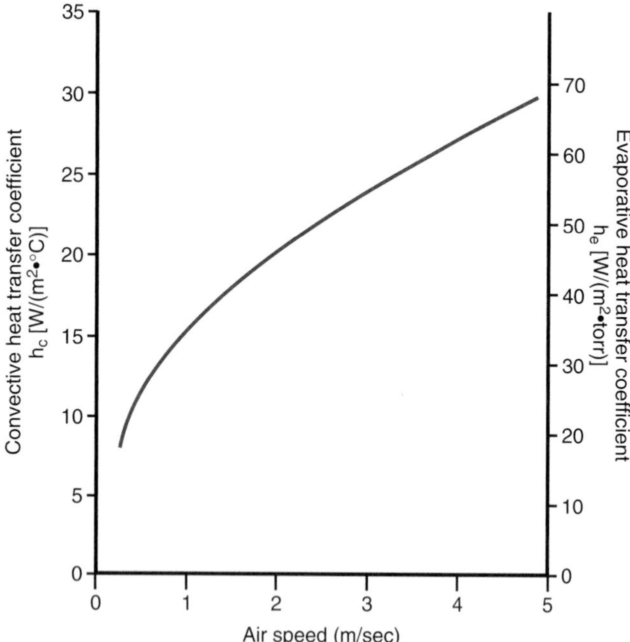

FIGURE 29.6 **Dependence of convection and evaporation on air movement.** This figure shows the convective heat transfer coefficient, h_c (left), and the evaporative heat transfer coefficient, h_e (right) for a standing human as a function of air speed. The convective and evaporative heat transfer coefficients are related by the equation $h_e = h_c \times 2.2°C/torr$. The horizontal axis can be converted into English units by using the relation 5 m/sec = 16.4 ft/sec = 11.2 miles/hr.

Radiative heat exchange is proportional to the difference between the fourth powers of the absolute temperatures of the skin and of the radiant environment (T_r) and to the emissivity of the skin (e_{sk}): $R \propto e_{sk} \times (\overline{T}_{sk}^4 - T_r^4)$. However, if T_r is close enough to $\overline{T}_{sk}$ that $\overline{T}_{sk} - T_r$ is much smaller than the absolute temperature of the skin, R is nearly proportional to $e_{sk} \times (\overline{T}_{sk} - T_r)$. Some parts of the body surface (e.g., the inner surfaces of the thighs and arms) exchange heat by radiation with other parts of the body surface, so the body exchanges heat with the environment as if it had an area smaller than its actual surface area. This smaller area, called the **effective radiating surface area** (A_r), depends on the body's posture, and it is closest to the actual surface area in a spread-eagle position and least in a curled-up position. Radiative heat exchange can be represented by the equation

$$R = h_r \times e_{sk} \times A_r \times (\overline{T}_{sk} - T_r) \quad (3)$$

where h_r is the radiant heat transfer coefficient, 6.43 W/(m²·°C) at 28°C.

Evaporative heat loss from the skin to the environment is proportional to the difference between the water vapor pressure at the skin surface and the water vapor pressure in the ambient air. These relations are summarized as:

$$E = h_e \times A \times (P_{sk} - P_a) \quad (4)$$

where P_{sk} is the water vapor pressure at the skin surface, P_a is the ambient water vapor pressure, and h_e is the evaporative heat transfer coefficient.

Water vapor, like heat, is carried away by moving air, so geometric factors and air movement affect E and h_e in the same way they affect C and h_c. If the skin is completely wet, the water vapor pressure at the skin surface is the saturation water vapor pressure at the temperature of the skin (Fig. 29.7), and evaporative heat loss is E_{max}, the maximum possible for the prevailing skin temperature and environmental conditions. This condition is described as:

$$E_{max} = h_e \times A \times (P_{sk,sat} - P_a) \quad (5)$$

where $P_{sk,sat}$ is the saturation water vapor pressure at skin temperature. When the skin is not completely wet, it is impractical to measure P_{sk}, the actual average water vapor pressure at the skin surface. Therefore, a coefficient called skin *wettedness* (w) is defined as the ratio E/E_{max}, with $0 < w < 1$. Skin wettedness depends on the hydration of the epidermis and the fraction of the skin surface that is wet. We can now rewrite equation 4 as:

$$E = h_e \times A \times w \times (P_{sk,sat} - P_a) \quad (6)$$

Wettedness depends on the balance between secretion and evaporation of sweat. If secretion exceeds evaporation, sweat accumulates on the skin and spreads out to wet more of the space between neighboring sweat glands, increasing wettedness and E; if evaporation exceeds secretion, the reverse occurs. If sweat rate exceeds E_{max}, once wettedness becomes 1, the excess sweat drips from the body, since it cannot evaporate.

Note that P_a, on which evaporation from the skin directly depends, is proportional to the actual moisture content in the air. By contrast, the more familiar quantity **relative humidity** (rh) is the ratio between the actual moisture content in the air and the maximum moisture content possible at the temperature of the air. It is important to recog-

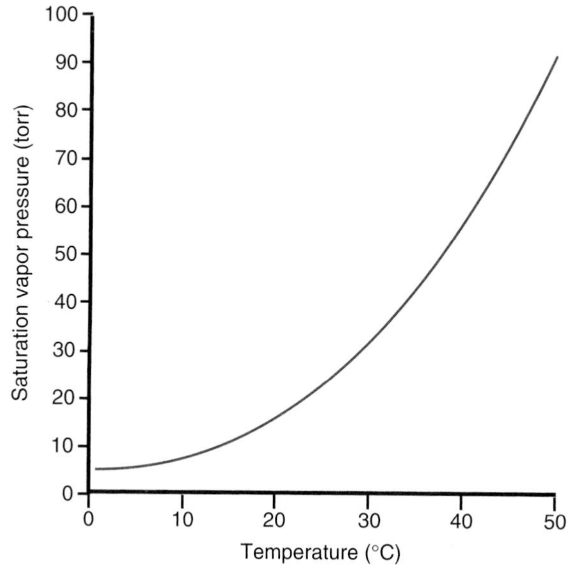

FIGURE 29.7 **Saturation vapor pressure of water as a function of temperature.** For any given temperature, the water vapor pressure is at its saturation value when the air is "saturated" with water vapor (i.e., holds the maximum amount possible at that temperature). At 37°C, PH_2O equals 47 torr.

nize that rh is only indirectly related to evaporation from the skin. For example, in a cold environment, P_a will be low enough that sweat can easily evaporate from the skin even if rh equals 100%, since the skin is warm and $P_{sk,sat}$, which depends on the temperature of the skin, will be much greater than P_a.

Heat Storage Is a Change in the Heat Content of the Body

The rate of **heat storage** is the difference between heat production and net heat loss (equation 1). (In the unusual circumstances in which there is a net heat gain from the environment, such as during immersion in a hot bath, storage is the sum of heat production and net heat gain.) It can be determined experimentally from simultaneous measurements of metabolism by indirect calorimetry and heat gain or loss by direct calorimetry. Storage of heat in the tissues changes their temperature, and the amount of heat stored is the product of body mass, the body's mean specific heat, and a suitable mean body temperature (T_b). The body's mean specific heat depends on its composition, especially the proportion of fat, and is about 3.55 kJ/(kg·°C) [0.85 kcal/(kg·°C)]. Empirical relations of T_b to core temperature (T_c) and $\overline{T}_{sk}$, determined in calorimetric studies, depend on ambient temperature, with T_b varying from $0.65 \times T_c + 0.35 \times \overline{T}_{sk}$ in the cold to $0.9 \times T_c + 0.1 \times \overline{T}_{sk}$ in the heat. The shift from cold to heat in the relative weighting of T_c and $\overline{T}_{sk}$ reflects the accompanying change in the thickness of the shell (see Fig. 29.2).

HEAT DISSIPATION

Figure 29.8 shows rectal and mean skin temperatures, heat losses, and calculated core-to-skin (shell) conductances for nude resting men and women at the end of 2-hour exposures in a calorimeter to ambient temperatures of 23 to 36°C. Shell conductance represents the sum of heat transfer by two parallel modes: conduction through the tissues of the shell, and convection by the blood. It is calculated by dividing heat flow through the skin (HF_{sk}) (i.e., total heat loss from the body less heat loss through the respiratory tract) by the difference between core and mean skin temperatures:

$$C = HF_{sk}/(T_c - \overline{T}_{sk}) \quad (7)$$

where C is shell conductance and T_c and $\overline{T}_{sk}$ are core and mean skin temperatures.

From 23 to 28°C, conductance is minimal because the skin is vasoconstricted and its blood flow is low. The minimal level of conductance attainable depends largely on the thickness of the subcutaneous fat layer, and the women's thicker layer allows them to attain a lower conductance than men. At about 28°C, conductance begins to increase, and above 30°C, conductance continues to increase and sweating begins.

For these subjects, 28 to 30°C is the zone of **thermoneutrality**, the range of comfortable environmental temperatures in which thermal balance is maintained without either shivering or sweating. In this zone, heat balance is maintained entirely by controlling conductance and $\overline{T}_{sk}$ and, thus, *R* and *C*. As equations 2 to 4 show, *C*, *R*, and *E* all depend on skin temperature, which, in turn, depends partly on skin blood flow. *E* depends also, through skin wettedness, on sweat secretion. Therefore, all these modes of heat exchange are partly under physiological control.

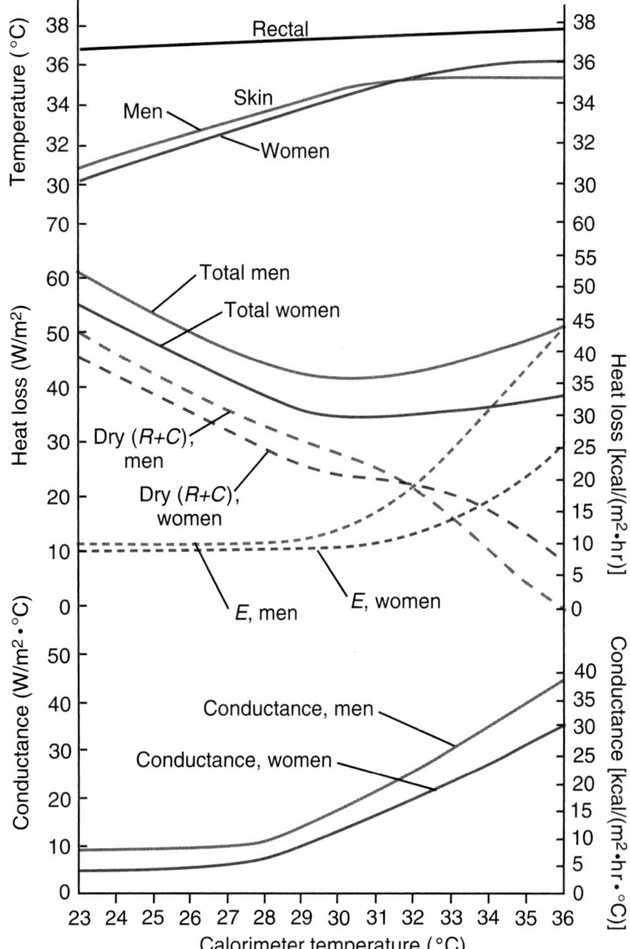

FIGURE 29.8 **Heat dissipation.** These graphs show the average values of rectal and mean skin temperatures, heat loss, and core-to-skin thermal conductance for nude resting men and women near steady state after 2 hours at different environmental temperatures in a calorimeter. (All energy exchange quantities in this figure have been divided by body surface area to remove the effect of individual body size.) Total heat loss is the sum of dry heat loss, by radiation (*R*) and convection (*C*), and evaporative heat loss (*E*). Dry heat loss is proportional to the difference between skin temperature and calorimeter temperature and decreases with increasing calorimeter temperature. (Based on data from Hardy JD, DuBois EF. Differences between men and women in their response to heat and cold. Proc Natl Acad Sci U S A 1940;26:389–398.)

The Evaporation of Sweat Can Dissipate Large Amounts of Heat

In Figure 29.8, evaporative heat loss is nearly independent of ambient temperature below 30°C and is 9 to 10 W/m², corresponding to evaporation of about 13 to 15 g/(m²·h), of which about half is moisture lost in breathing and half is insensible perspiration. This evaporation occurs independent of thermoregulatory control. As the ambient temperature increases, the body depends more and more on the evaporation of sweat to achieve heat balance.

The two histological types of sweat glands are **eccrine** and **apocrine**. In northern Europeans, apocrine glands are found mostly in the axilla and pigmented skin, such as the lips, but they are more widely distributed in some other populations. Eccrine sweat is essentially a dilute electrolyte solution, but apocrine sweat also contains fatty material. Eccrine sweat glands, the dominant type in all human populations, are more important in human thermoregulation and number about 2,500,000. They are controlled through postganglionic sympathetic nerves that release acetylcholine (ACh) rather than norepinephrine. A healthy man unacclimatized to heat can secrete up to 1.5 L/hr of sweat. Although the number of functional sweat glands is fixed before the age of 3, the secretory capacity of the individual glands can change, especially with endurance exercise training and heat acclimatization; men well acclimatized to heat can attain peak sweat rates greater than 2.5 L/hr. Such rates cannot be maintained, however; the maximum daily sweat output is probably about 15 L.

The sodium concentration of eccrine sweat ranges from less than 5 to 60 mmol/L (versus 135 to 145 mmol/L in plasma). In producing sweat that is hypotonic to plasma, the glands reabsorb sodium from the sweat duct by active transport. As sweat rate increases, the rate at which the glands reabsorb sodium increases more slowly, so that sodium concentration in the sweat increases. The sodium concentration of sweat is affected also by heat acclimatization and by the action of mineralocorticoids.

Skin Circulation Is Important in Heat Transfer

Heat produced in the body must be delivered to the skin surface to be eliminated. When skin blood flow is minimal, shell conductance is typically 5 to 9 W/°C per m^2 of body surface. For a lean resting subject with a surface area of 1.8 m^2, minimal whole body conductance of 16 W/°C [i.e., 8.9 $W/(°C \cdot m^2) \times 1.8\ m^2$] and a metabolic heat production of 80 W, the temperature difference between the core and the skin must be 5°C (i.e., 80 W ÷ 16 W/°C) for the heat produced to be conducted to the surface. In a cool environment, $\overline{T}_{sk}$ may easily be low enough for this to occur. However, in an ambient temperature of 33°C, $\overline{T}_{sk}$ is typically about 35°C, and without an increase in conductance, core temperature would have to rise to 40°C—a high, although not yet dangerous, level—for the heat to be conducted to the skin. If the rate of heat production were increased to 480 W by moderate exercise, the temperature difference between core and skin would have to rise to 30°C—and core temperature to well beyond lethal levels—to allow all the heat produced to be conducted to the skin. In the latter circumstances, the conductance of the shell must increase greatly for the body to reestablish thermal balance and continue to regulate its temperature. This is accomplished by increasing the skin blood flow.

Effectiveness of Skin Blood Flow in Heat Transfer. Assuming that blood on its way to the skin remains at core temperature until it reaches the skin, reaches skin temperature as it passes through the skin, and then stays at skin temperature until it returns to the core, we can compute the rate of heat flow (HF_b) as a result of convection by the blood as

$$HF_b = SkBF \times (T_c - T_{sk}) \times 3.85\ kJ/(L \cdot °C) \quad (8)$$

where SkBF is the rate of skin blood flow, expressed in L/sec rather than the usual L/min to simplify computing HF in W (i.e., J/sec); and 3.85 kJ/(L·°C) [0.92 kcal/(L·°C)] is the volume-specific heat of blood. Conductance as a result of convection by the blood (C_b) is calculated as:

$$C_b = HF_b/(T_c - T_{sk}) = SkBF \times 3.85\ kJ/(L \cdot °C) \quad (9)$$

Of course, heat continues to flow by conduction through the tissues of the shell, so total conductance is the sum of conductance as a result of convection by the blood, plus that resulting from conduction through the tissues. Total heat flow is given by

$$HF = (C_b + C_0) \times (T_c - T_{sk}) \quad (10)$$

in which C_0 is thermal conductance of the tissues when skin blood flow is minimal and, thus, is predominantly due to conduction through the tissues.

The assumptions made in deriving equation 8 are somewhat artificial and represent the conditions for maximum efficiency of heat transfer by the blood. In practice, blood exchanges heat also with the tissues through which it passes on its way to and from the skin. Heat exchange with these other tissues is greatest when skin blood flow is low; in such cases, heat flow to the skin may be much less than predicted by equation 8, as discussed further below. However, equation 8 is a reasonable approximation in a warm subject with moderate to high skin blood flow. Although measuring whole-body SkBF directly is not possible, it is believed to reach several liters per minute during heavy exercise in the heat. The maximum obtainable is estimated to be nearly 8 L/min. If SkBF = 1.89 L/min (0.0315 L/sec), according to equation 9, skin blood flow contributes about 121 W/°C to the conductance of the shell. If conduction through the tissues contributes 16 W/°C, total shell conductance is 137 W/°C, and if T_c = 38.5°C and T_{sk} = 35°C, this will produce a core-to-skin heat transfer of 480 W, the heat production in our earlier example of moderate exercise. Therefore, even a moderate rate of skin blood flow can have a dramatic effect on heat transfer.

When a person is not sweating, raising skin blood flow brings skin temperature nearer to blood temperature and lowering skin blood flow brings skin temperature nearer to ambient temperature. Under such conditions, the body can control dry (convective and radiative) heat loss by varying skin blood flow and, thus, skin temperature. Once sweating begins, skin blood flow continues to increase as the person becomes warmer. In these conditions, however, the tendency of an increase in skin blood flow to warm the skin is approximately balanced by the tendency of an increase in sweating to cool the skin. Therefore, after sweating has begun, further increases in skin blood flow usually cause little change in skin temperature or dry heat exchange and serve primarily to deliver to the skin the heat that is being removed by the evaporation of sweat. Skin blood flow and sweating work in tandem to dissipate heat under such conditions.

Sympathetic Control of Skin Circulation. Blood flow in human skin is under dual vasomotor control. In most of the skin, the vasodilation that occurs during heat exposure depends on sympathetic nerve signals that cause the blood vessels to dilate, and this vasodilation can be prevented or reversed by regional nerve block. Because it depends on the action of nerve signals, such vasodilation is sometimes referred to as active vasodilation. Active vasodilation occurs in almost all the skin, except in so-called acral regions—hands, feet, lips, ears, and nose. In skin areas where active vasodilation occurs, vasoconstrictor activity is minimal at thermoneutral temperatures, and active vasodilation during heat exposure does not begin until close to the onset of sweating. Therefore, skin blood flow in these areas is not much affected by small temperature changes within the thermoneutral range.

The neurotransmitter or other vasoactive substance responsible for active vasodilation in human skin has not been identified. Active vasodilation operates in tandem with sweating in the heat, and is impaired or absent in **anhidrotic ectodermal dysplasia**, a congenital disorder in which sweat glands are sparse or absent. For these reasons, the existence of a mechanism linking active vasodilation to the sweat glands has long been suspected, but never established. Earlier suggestions that active vasodilation is cholinergic or is caused by the release of bradykinin from activated sweat glands have not gained general acceptance. More recently, however, nerve endings containing both ACh and vasoactive peptides have been found near eccrine sweat glands in human skin, suggesting that active vasodilation may be mediated by a vasoactive cotransmitter that is released along with ACh from the endings of nerves that innervate sweat glands.

Reflex vasoconstriction, occurring in response to cold and as part of certain nonthermal reflexes such as baroreflexes, is mediated primarily through adrenergic sympathetic fibers distributed widely over most of the skin. Reducing the flow of impulses in these nerves allows the blood vessels to dilate. In the acral regions and superficial veins (whose role in heat transfer is discussed below), vasoconstrictor fibers are the predominant vasomotor innervation, and the vasodilation that occurs during heat exposure is largely a result of the withdrawal of vasoconstrictor activity. Blood flow in these skin regions is sensitive to small temperature changes even in the thermoneutral range, and may be responsible for "fine-tuning" heat loss to maintain heat balance in this range.

THERMOREGULATORY CONTROL

In discussions of control systems, the words "regulation" and "regulate" have meanings distinct from those of the word "control" (see Chapter 1). The variable that a control system acts to maintain within narrow limits (e.g., temperature) is called the *regulated* variable, and the quantities it controls to accomplish this (e.g., sweating rate, skin blood flow, metabolic rate, and thermoregulatory behavior) are called *controlled* variables.

Humans have two distinct subsystems for regulating body temperature: behavioral thermoregulation and physiological thermoregulation. Behavioral thermoregulation—through the use of shelter, space heating, air conditioning, and clothing—enables humans to live in the most extreme climates in the world, but it does not provide fine control of body heat balance. In contrast, physiological thermoregulation is capable of fairly precise adjustments of heat balance but is effective only within a relatively narrow range of environmental temperatures.

Behavioral Thermoregulation Is Governed by Thermal Sensation and Comfort

Sensory information about body temperatures is an essential part of both behavioral and physiological thermoregulation. The distinguishing feature of behavioral thermoregulation is the involvement of consciously directed efforts to regulate body temperature. Thermal discomfort provides the necessary motivation for thermoregulatory behavior, and behavioral thermoregulation acts to reduce both the discomfort and the physiological strain imposed by a stressful thermal environment. For this reason, the zone of thermoneutrality is characterized by both thermal comfort and the absence of shivering and sweating.

Warmth and cold on the skin are felt as either comfortable or uncomfortable, depending on whether they decrease or increase the physiological strain—a shower temperature that feels pleasant after strenuous exercise may be uncomfortably chilly on a cold winter morning. The processing of thermal information in behavioral thermoregulation is not as well understood as it is in physiological thermoregulation. However, perceptions of thermal sensation and comfort respond much more quickly than core temperature or physiological thermoregulatory responses to changes in environmental temperature and, thus, appear to anticipate changes in the body's thermal state. Such an anticipatory feature would be advantageous, since it would reduce the need for frequent small behavioral adjustments.

Physiological Thermoregulation Operates Through Graded Control of Heat-Production and Heat-Loss Responses

Familiar inanimate control systems, such as most refrigerators and heating and air-conditioning systems, operate at only two levels: on and off. In a steam heating system, for example, when the indoor temperature falls below the desired level, the thermostat turns on the burner under the boiler; when the temperature is restored to the desired level, the thermostat turns the burner off. Rather than operating at only two levels, most physiological control systems produce a graded response according to the size of the disturbance in the regulated variable. In many instances, changes in the controlled variables are proportional to displacements of the regulated variable from some threshold value; such control systems are called **proportional control systems**.

The control of heat-dissipating responses is an example of a proportional control system. Figure 29.9 shows how reflex control of two heat-dissipating responses, sweating and skin blood flow, depends on body core temperature and mean skin temperature. Each response has a core temperature threshold—a temperature at which the response starts to increase—and this threshold depends on mean skin tem-

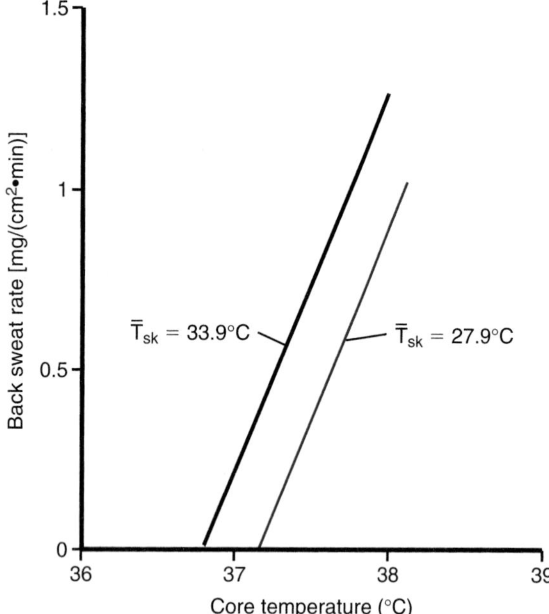

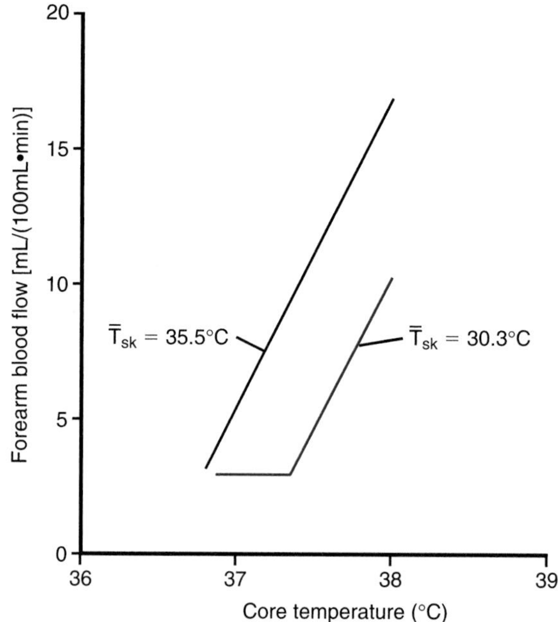

FIGURE 29.9 **Control of heat-dissipating responses.** These graphs show the relations of back (scapular) sweat rate (left) and forearm blood flow (right) to core temperature and mean skin temperatures ($\bar{T}_{sk}$). In these experiments, core temperature was increased by exercise. (Left: Based on data from Sawka MN, Gonzalez RR, Drolet LL, et al. Heat exchange during upper- and lower-body exercise. J Appl Physiol 1984;57:1050–1054. Right: Modified from Wenger CB, Roberts MF, Stolwijk JAJ, et al. Forearm blood flow during body temperature transients produced by leg exercise. J Appl Physiol 1975;38:58–63.)

perature. At any given skin temperature, the change in each response is proportional to the change in core temperature, and increasing the skin temperature lowers the threshold level of core temperature and increases the response at any given core temperature. In humans, a change of 1°C in core temperature elicits about 9 times as great a thermoregulatory response as a 1°C change in mean skin temperature. (Besides its effect on the reflex signals, skin temperature has a local effect that modifies the response of the blood vessels and sweat glands to the reflex signal, discussed later.)

Cold stress elicits increases in metabolic heat production through shivering and nonshivering thermogenesis. **Shivering** is a rhythmic oscillating tremor of skeletal muscles. The **primary motor center for shivering** lies in the dorsomedial part of the posterior hypothalamus and is normally inhibited by signals of warmth from the preoptic area of the hypothalamus. In the cold, these inhibitory signals are withdrawn, and the primary motor center for shivering sends impulses down the brainstem and lateral columns of the spinal cord to anterior motor neurons. Although these impulses are not rhythmic, they increase muscle tone, thereby increasing metabolic rate somewhat. Once the tone exceeds a critical level, the contraction of one group of muscle fibers stretches the muscle spindles in other fiber groups in series with it, eliciting contractions from those groups of fibers via the stretch reflex, and so on; thus, the rhythmic oscillations that characterize frank shivering begin.

Shivering occurs in bursts, and the "shivering pathway" is inhibited by signals from the cerebral cortex, so that voluntary muscular activity and attention can suppress shivering. Since the limbs are part of the shell in the cold, trunk and neck muscles are preferentially recruited for shivering—the **centralization of shivering**—to help retain the heat produced during shivering within the body core; and the familiar experience of teeth chattering is one of the earliest signs of shivering. As with heat-dissipating responses, the control of shivering depends on both core and skin temperatures, but the details of its control are not precisely understood.

The CNS Integrates Thermal Information From the Core and the Skin

Temperature receptors in the body core and skin transmit information about their temperatures through afferent nerves to the brainstem and, especially, the hypothalamus, where much of the integration of temperature information occurs. The sensitivity of the thermoregulatory system to core temperature enables it to adjust heat production and heat loss to resist disturbances in core temperature. Sensitivity to mean skin temperature lets the system respond appropriately to mild heat or cold exposure with little change in body core temperature, so that changes in body heat as a result of changes in environmental temperature take place almost entirely in the peripheral tissues (see Fig. 29.2). For example, the skin temperature of someone who enters a hot environment may rise and elicit sweating even if there is no change in core temperature. On the other hand, an increase in heat production within the body, as during exercise, elicits the appropriate heat-dissipating responses through a rise in core temperature.

Core temperature receptors involved in controlling thermoregulatory responses are unevenly distributed and are concentrated in the hypothalamus. In experimental mammals, temperature changes of only a few tenths of 1°C in the anterior preoptic area of the hypothalamus elicit changes in the thermoregulatory effector responses, and this area contains many neurons that increase their firing rate in response to either warming or cooling. Thermal receptors have been reported elsewhere in the core of laboratory animals, including the heart, pulmonary vessels, and spinal cord, but the thermoregulatory role of core thermal receptors outside the CNS is unknown.

Consider what happens when some disturbance—say, an increase in metabolic heat production resulting from exercise—upsets the thermal balance. Additional heat is stored in the body, and core temperature rises. The central thermoregulatory controller receives information about these changes from the thermal receptors and elicits appropriate heat-dissipating responses. Core temperature continues to rise, and these responses continue to increase until they are sufficient to dissipate heat as fast as it is being produced, restoring heat balance and preventing further increases in body temperatures. In the language of control theory, the rise in core temperature that elicits heat-dissipating responses sufficient to reestablish thermal balance during exercise is an example of a **load error**. A load error is characteristic of any proportional control system that is resisting the effect of some imposed disturbance or "load." Although the disturbance in this example is exercise, the same principle applies if the disturbance is a decrease in metabolic rate or a change in the environment. However, if the disturbance is in the environment, most of the temperature change will be in the skin and shell rather than in the core; if the disturbance produces a net loss of heat, the body will restore heat balance by decreasing heat loss and increasing heat production.

Relation of Controlling Signal to Thermal Integration and Set Point. Both sweating and skin blood flow depend on core and skin temperatures in the same way, and changes in the threshold for sweating are accompanied by similar changes in the threshold for vasodilation. We may, therefore, think of the central thermoregulatory controller as generating one thermal command signal for the control of both sweating and skin blood flow (Fig. 29.10). This signal is based on the information about core and skin temperatures that the controller receives and on the thermoregulatory **set point**—the target level of core temperature, or the setting of the body's "thermostat." In the operation of the thermoregulatory system, it is a reference point that determines the thresholds of all of the thermoregulatory responses. Shivering and thermal comfort are affected by changes in the set point in the same way as sweating and

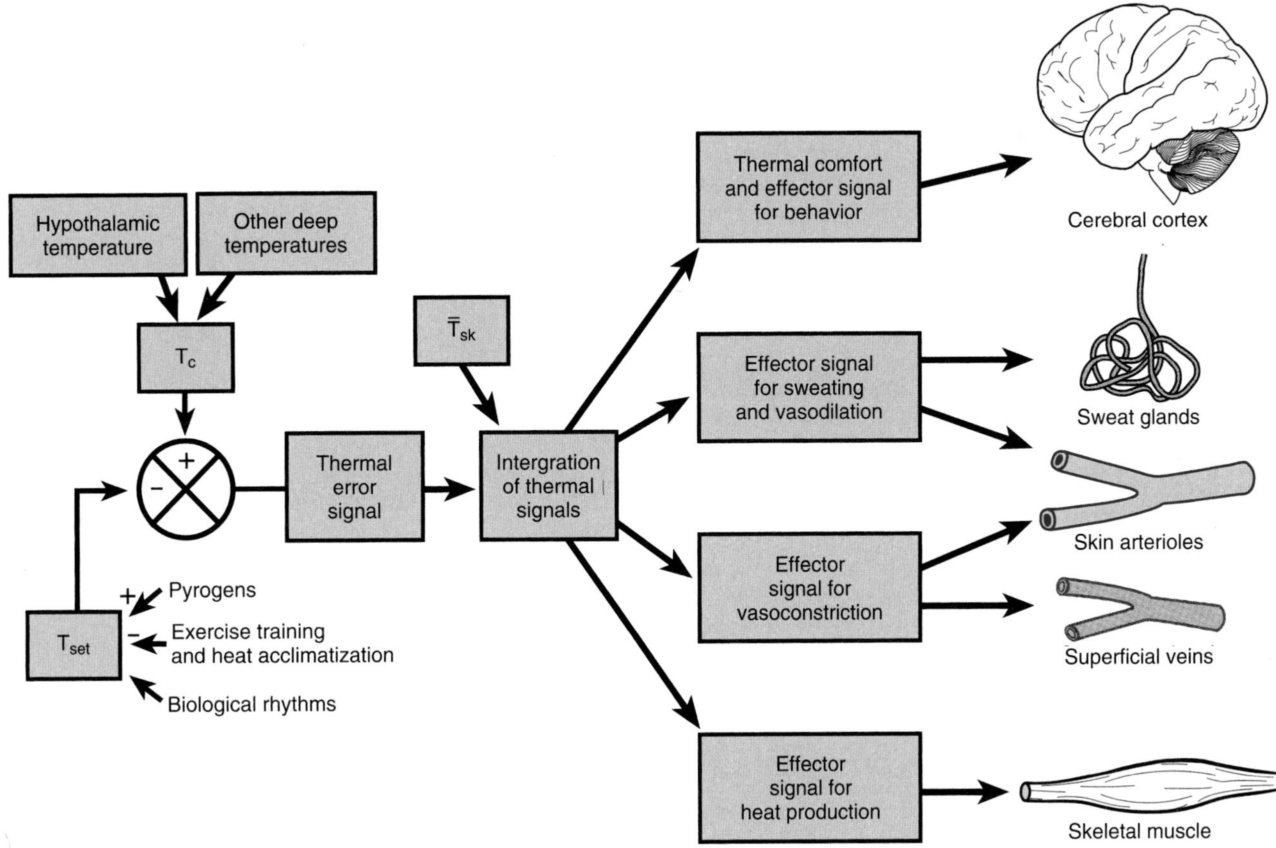

FIGURE 29.10 **Control of human thermoregulatory responses.** The plus and minus signs next to the inputs to T_{set} indicate that pyrogens raise the set point and heat acclimatization lowers it. Core temperature (T_c) is compared with the set point (T_{set}) to generate an error signal, which is integrated with thermal input from the skin to produce effector signals for the thermoregulatory responses.

skin blood flow. However, our understanding of the control of shivering is insufficient to say whether it is controlled by the same command signal as sweating and skin blood flow. (Thermal comfort, as we saw earlier, seems not to be controlled by the same command signal.)

Effect of Nonthermal Inputs on Thermoregulatory Responses. Each thermoregulatory response may be affected by inputs other than body temperatures and factors that influence the set point. We have already noted that voluntary activity affects shivering and certain hormones affect metabolic heat production. In addition, nonthermal factors may produce a burst of sweating at the beginning of exercise, and emotional effects on sweating and skin blood flow are matters of common experience. Skin blood flow is the thermoregulatory response most influenced by nonthermal factors because of its potential involvement in reflexes that function to maintain cardiac output, blood pressure, and tissue O_2 delivery under a variety of disturbances, including heat stress, postural changes, hemorrhage, and exercise.

Several Factors May Change the Thermoregulatory Set Point

Fever elevates core temperature at rest, heat acclimatization decreases it, and time of day and (in women) the phase of the menstrual cycle change it in a cyclic fashion. Core temperature at rest varies in an approximately sinusoidal fashion with time of day. The minimum temperature occurs at night, several hours before awaking, and the maximum, which is 0.5 to 1°C higher, occurs in the late afternoon or evening (see Fig. 29.3). This pattern coincides with patterns of activity and eating but does not depend on them, and it occurs even during bed rest in fasting subjects. This pattern is an example of a **circadian rhythm**, a rhythmic pattern in a physiological function with a period of about 1 day. During the menstrual cycle, core temperature is at its lowest point just before ovulation; during the next few days, it rises 0.5 to 1°C to a plateau that persists through most of the luteal phase. Each of these factors—fever, heat acclimatization, the circadian rhythm, and the menstrual cycle—change the core temperature at rest by changing the thermoregulatory set point, producing corresponding changes in the thresholds for all of the thermoregulatory responses.

Peripheral Factors Modify the Responses of Skin Blood Vessels and Sweat Glands

The skin is the organ most directly affected by environmental temperature. Skin temperature influences heat loss responses not only through reflex actions (see Fig. 29.9), but also through direct effects on the skin blood vessels and sweat glands.

Skin Temperature and Cutaneous Vascular and Sweat Gland Responses. Local temperature changes act on skin blood vessels in at least two ways. First, local cooling potentiates (and heating weakens) the constriction of blood vessels in response to nerve signals and vasoconstrictor substances. (At very low temperatures, however, cold-induced vasodilation increases skin blood flow, as discussed later.) Second, in skin regions where active vasodilation occurs, local heating causes vasodilation (and local cooling causes vasoconstriction) through a direct action on the vessels, independent of nerve signals. The local vasodilator effect of skin temperature is especially strong above 35°C; and, when the skin is warmer than the blood, increased blood flow helps cool the skin and protect it from heat injury, unless this response is impaired by vascular disease. Local thermal effects on sweat glands parallel those on blood vessels, so local heating potentiates (and local cooling diminishes) the local sweat gland response to reflex stimulation or ACh, and intense local heating elicits sweating directly, even in skin whose sympathetic innervation has been interrupted surgically.

Skin Wettedness and the Sweat Gland Response. During prolonged heat exposure (lasting several hours) with high sweat output, sweating rates gradually decline and the response of sweat glands to local cholinergic drugs is reduced. This reduction of sweat gland responsiveness is sometimes called sweat gland "fatigue." Wetting the skin makes the stratum corneum swell, mechanically obstructing the sweat gland ducts and causing a reduction in sweat secretion, an effect called **hidromeiosis**. The glands' responsiveness can be at least partly restored if air movement increases or humidity is reduced, allowing some of the sweat on the skin to evaporate. Sweat gland fatigue may involve processes besides hidromeiosis, since prolonged sweating also causes histological changes, including the depletion of glycogen, in the sweat glands.

THERMOREGULATORY RESPONSES DURING EXERCISE

Intense exercise may increase heat production within the body 10-fold or more, requiring large increases in skin blood flow and sweating to reestablish the body's heat balance. Although hot environments also elicit heat-dissipating responses, exercise ordinarily is responsible for the greatest demands on the thermoregulatory system for heat dissipation. Exercise provides an important example of how the thermoregulatory system responds to a disturbance in heat balance. In addition, exercise and thermoregulation impose competing demands on the circulatory system because exercise requires large increases in blood flow to exercising muscle, while the thermoregulatory responses to exercise require increases in skin blood flow. Muscle blood flow during exercise is several times as great as skin blood flow, but the increase in skin blood flow is responsible for disproportionately large demands on the cardiovascular system, as discussed below. Finally, if the water and electrolytes lost through sweating are not replaced, the resulting reduction in plasma volume will eventually create a further challenge to cardiovascular homeostasis.

Core Temperature Rises During Exercise, Triggering Heat-Loss Responses

As previously mentioned, the increased heat production during exercise causes an increase in core temperature, which in turn elicits heat-loss responses. Core temperature continues to rise until heat loss has increased enough to match heat production, and core temperature and the heat-loss responses reach new steady-state levels. Since the heat-loss responses are proportional to the increase in core temperature, the increase in core temperature at steady state is proportional to the rate of heat production and, thus, to the metabolic rate.

A change in ambient temperature causes changes in the levels of sweating and skin blood flow necessary to maintain any given level of heat dissipation. However, the change in ambient temperature also elicits, via direct and reflex effects of the accompanying skin temperature changes, altered responses in the right direction. For any given rate of heat production, there is a certain range of environmental conditions within which an ambient temperature change elicits the necessary changes in heat-dissipating responses almost entirely through the effects of skin temperature changes, with virtually no effect on core temperature. (The limits of this range of environmental conditions depend on the rate of heat production and such individual factors as skin surface area and state of heat acclimatization.) Within this range, the core temperature reached during exercise is nearly independent of ambient temperature; for this reason, it was once believed that the increase in core temperature during exercise is caused by an increase in the thermoregulatory set point, as during fever. As noted, however, the increase in core temperature with exercise is an example of a load error rather than an increase in set point.

This difference between fever and exercise is shown in Figure 29.11. Note that, although heat production may increase substantially (through shivering), when core temperature is rising early during fever, it need not stay high to maintain the fever; in fact, it returns nearly to prefebrile levels once the fever is established. During exercise, however, an increase in heat production not only causes the elevation in core temperature but is necessary to sustain it. Also, while core temperature is rising during fever, the rate of heat loss is, if anything, lower than it was before the fever began. During exercise, however, the heat-dissipating responses and the rate of heat loss start to increase early and continue increasing as core temperature rises.

Exercise in the Heat Can Threaten Cardiovascular Homeostasis

The rise in core temperature during exercise increases the temperature difference between the core and the skin somewhat, but not nearly enough to match the increase in metabolic heat production. Therefore, as we saw earlier, skin blood flow must increase to carry all of the heat that is produced to the skin. In a warm environment, where the temperature difference between core and skin is relatively small, the necessary increase in skin blood flow may be several liters per minute.

Impaired Cardiac Filling During Exercise in the Heat. The work of providing the skin blood flow required for thermoregulation in the heat may impose a heavy burden on a diseased heart, but in healthy people, the major car-

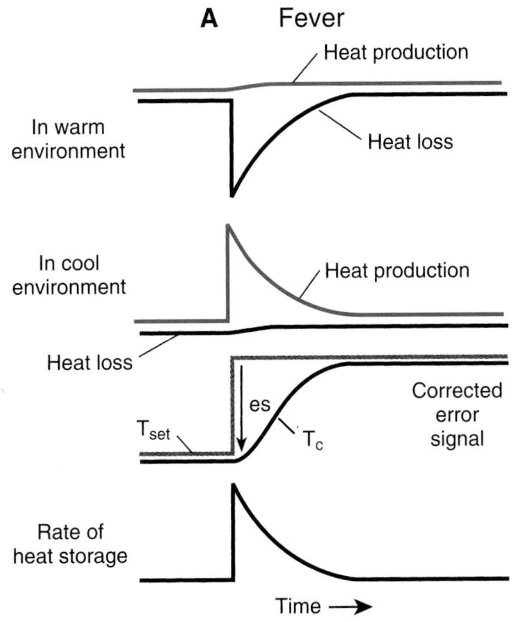

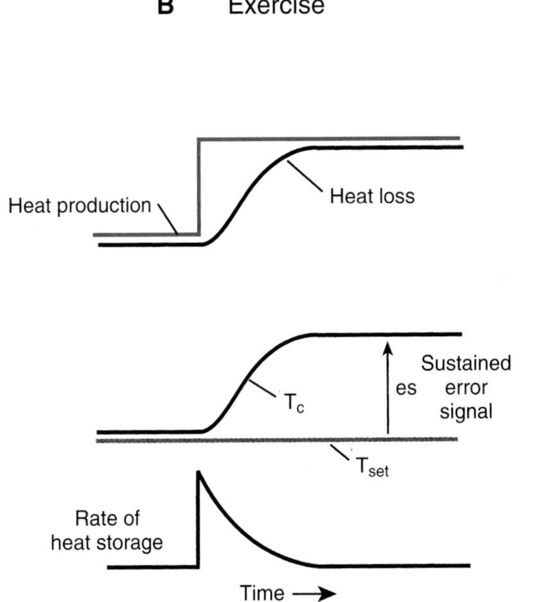

FIGURE 29.11 **Thermal events during fever and exercise.** A, The development of fever. **B**, The increase in core temperature (T_c) during exercise. The error signal is the difference between core temperature (T_c) and the set point (T_{set}). At the start of a fever, T_{set} has risen, so that T_{set} is higher than T_c and es is negative. At steady state, T_c has risen to equal the new level of T_{set} and es is corrected (i.e., it returns to zero.) At the start of exercise, $T_c = T_{set}$, so that es = 0. At steady state, T_{set} has not changed but T_c has increased and is greater than T_{set}, producing a sustained error signal, which is equal to the load error. (The error signal, or load error, is here represented with an arrow pointing downward for $T_c < T_{set}$ and with an arrow pointing upward for $T_c > T_{set}$.) (Modified from Stitt JT. Fever versus hyperthermia. Fed Proc 1979;38:39–43.)

diovascular burden of heat stress results from impaired venous return. As skin blood flow increases, the dilated vascular bed of the skin becomes engorged with large volumes of blood, reducing central blood volume and cardiac filling (Fig. 29.12). Stroke volume is decreased, and a higher heart rate is required to maintain cardiac output. These effects are aggravated by a decrease in plasma volume if the large amounts of salt and water lost in the sweat are not replaced. Since the main cation in sweat is sodium, disproportionately much of the body water lost in sweat is at the expense of extracellular fluid, including plasma, although this effect is mitigated if the sweat is dilute.

Compensatory Responses During Exercise in the Heat. Several reflex adjustments help maintain cardiac filling, cardiac output, and arterial pressure during exercise and heat stress. The most important of these is constriction of the renal and splanchnic vascular beds. A reduction in blood flow through these beds allows a corresponding diversion of cardiac output to the skin and the exercising muscles. In addition, since the splanchnic vascular beds are compliant, a decrease in their blood flow reduces the amount of blood pooled in them (see Fig. 29.12), helping compensate for decreases in central blood volume caused by reduced plasma volume and blood pooling in the skin.

The degree of vasoconstriction is graded according to the levels of heat stress and exercise intensity. During strenuous exercise in the heat, renal and splanchnic blood flows may fall to 20% of their values in a cool resting subject. Such intense splanchnic vasoconstriction may produce mild ischemic injury to the gut, helping explain the intestinal symptoms some athletes experience after endurance events. The cutaneous veins constrict during exercise; since most of the vascular volume is in the veins, constriction makes the cutaneous vascular bed less easily distensible and reduces peripheral pooling. Because of the essential role of skin blood flow in thermoregulation during exercise and heat stress, the body preferentially compromises splanchnic and renal flow for the sake of cardiovascular homeostasis. Above a certain level of cardiovascular strain, however, skin blood flow, too, is compromised.

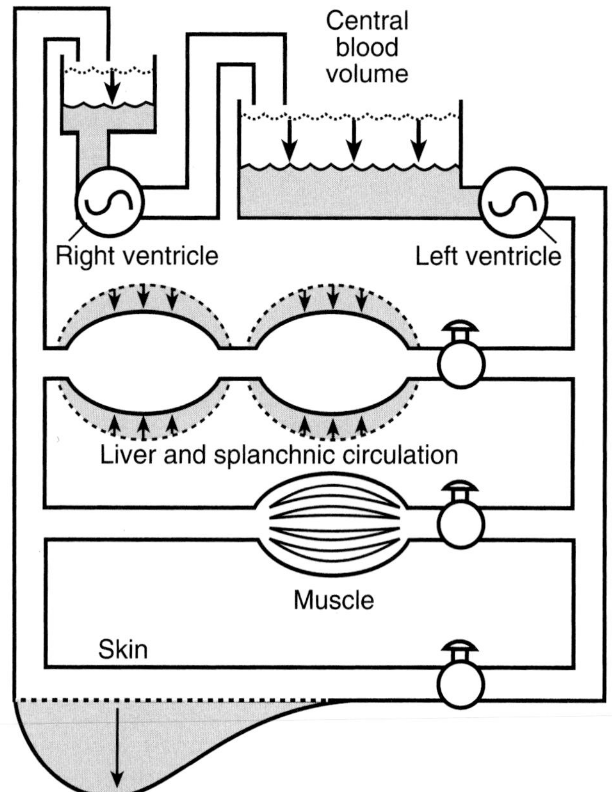

FIGURE 29.12 **Cardiovascular strain and compensatory responses during heat stress.** This figure first shows the effects of skin vasodilation on peripheral pooling of blood and the thoracic reservoirs from which the ventricles are filled; and second, the effects of compensatory vasomotor adjustments in the splanchnic circulation. The valves on the right represent the resistance vessels that control blood flow through the liver/splanchnic, muscle, and skin vascular beds. Arrows show the direction of the changes during heat stress. (Modified from Rowell LB. Cardiovascular aspects of human thermoregulation. Circ Res 1983;52:367–379.)

HEAT ACCLIMATIZATION

Prolonged or repeated exposure to stressful environmental conditions elicits significant physiological changes, called **acclimatization**, that reduce the resulting strain. (Such changes are often referred to as *acclimation* when produced in a controlled experimental setting.) Some degree of heat acclimatization occurs either by heat exposure alone or by regular strenuous exercise, which raises core temperature and provokes heat-loss responses. Indeed, the first summer heat wave produces enough heat acclimatization that most people notice an improvement in their level of energy and general feeling of well-being after a few days. However, the acclimatization response is greater if heat exposure and exercise are combined, causing a greater rise of internal temperature and more profuse sweating. Evidence of acclimatization appears in the first few days of combined exercise and heat exposure, and most of the improvement in heat tolerance occurs within 10 days. The effect of heat acclimatization on performance can be dramatic, and acclimatized subjects can easily complete exercise in the heat that earlier was difficult or impossible.

Heat Acclimatization Includes Adjustments in Heart Rate, Temperatures, and Sweat Rate

Cardiovascular adaptations that reduce the heart rate required to sustain a given level of activity in the heat appear quickly and reach nearly their full development within 1 week. Changes in sweating develop more slowly. After acclimatization, sweating begins earlier and at a lower core temperature (i.e., the core temperature threshold for sweating is reduced). The sweat glands become more sensitive to cholinergic stimulation, and a given elevation in core temperature elicits a higher sweat rate; in addition, the glands become resistant to hidromeiosis and fatigue, so higher sweat rates can be sustained. These changes reduce the levels of core and skin temperatures reached during a period of exer-

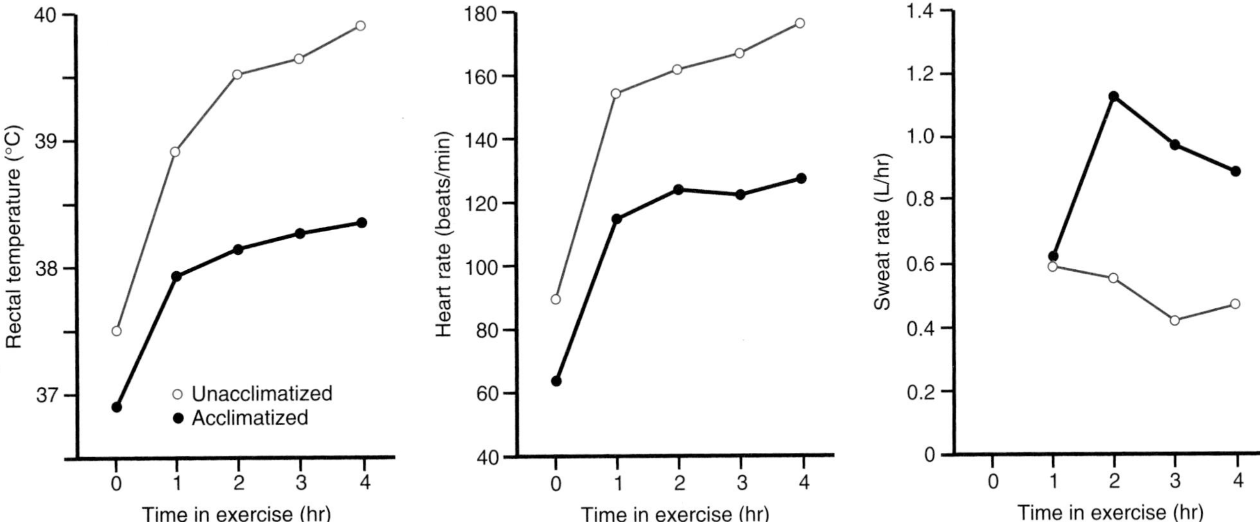

FIGURE 29.13 **Heat acclimatization.** These graphs show rectal temperatures, heart rates, and sweat rates during 4 hours' exercise (bench stepping, 35 W mechanical power) in humid heat (33.9°C dry bulb, 89% relative humidity, 35 torr ambient vapor pressure) on the first and last days of a 2-week program of acclimatizaton to humid heat. (Modified from Wenger CB. Human heat acclimatization. In: Pandolf KB, Sawka MN, Gonzalez RR, eds. Human Performance Physiology and Environmental Medicine at Terrestrial Extremes. Indianapolis: Benchmark, 1988;153–197. Based on data from Wyndham CH, Strydom NB, Morrison JF, et al. Heat reactions of Caucasians and Bantu in South Africa. J Appl Physiol 1964;19:598–606.)

cise in the heat, increase the sweat rate, and enable one to exercise longer. The threshold for cutaneous vasodilation is reduced along with the threshold for sweating, so heat transfer from the core to the skin is maintained. The lower heart rate and core temperature and the higher sweat rate are the three classical signs of heat acclimatization (Fig. 29.13).

Changes in Fluid and Electrolyte Balance Also Occur With Heat Acclimatization

During the first week, total body water and, especially, plasma volume increase. These changes likely contribute to the cardiovascular adaptations. Later, the fluid changes seem to diminish or disappear, although the cardiovascular adaptations persist. In an unacclimatized person, sweating occurs mostly on the chest and back, but during acclimatization, especially in humid heat, the fraction of sweat secreted on the limbs increases to make better use of the skin surface for evaporation. An unacclimatized person who is sweating profusely can lose large amounts of sodium. With acclimatization, the sweat glands become able to conserve sodium by secreting sweat with a sodium concentration as low as 5 mmol/L. This effect is mediated through aldosterone, which is secreted in response to sodium depletion and to exercise and heat exposure. The sweat glands respond to aldosterone more slowly than the kidneys, requiring several days; unlike the kidneys, the sweat glands do not escape the influence of aldosterone when sodium balance has been restored, but continue to conserve sodium for as long as acclimatization persists.

The cell membranes are freely permeable to water, so that any osmotic imbalance between the intracellular and extracellular compartments is rapidly corrected by the movement of water across the cell membranes (see Chapter 2). One important consequence of the salt-conserving response of the sweat glands is that the loss of a given volume of sweat causes a smaller decrease in the volume of the extracellular space than if the sodium concentration of the sweat is high (Table 29.3). Other consequences are discussed in Clinical Focus Box 29.1.

Heat acclimatization is transient, disappearing in a few weeks if not maintained by repeated heat exposure. The components of heat acclimatization are lost in the order in which they were acquired; the cardiovascular changes decay more quickly than the reduction in exercise core temperature and sweating changes.

RESPONSES TO COLD

The body maintains core temperature in the cold by minimizing heat loss and, when this response is insufficient, increasing heat production. Reducing shell conductance is the chief physiological means of heat conservation in humans. Furred or hairy animals also can increase the thickness of their coat and, thus, its insulating properties by making the hairs stand on end. This response, called **piloerection**, makes a negligible contribution to heat conservation in humans, but manifests itself as gooseflesh.

Blood Vessels in the Shell Constrict to Conserve Heat

The constriction of cutaneous arterioles reduces skin blood flow and shell conductance. Constriction of the superficial limb veins further improves heat conservation by diverting venous blood to the deep limb veins, which lie close to the major arteries of the limbs and do not constrict in the cold.

TABLE 29.3 **Effect of Sweat Secretion on Body Fluid Compartments and Plasma Sodium Concentration**[a]

		Extracellular Space		Intracellular Space		Total Body Water			
Subject	Condition	Volume (L)	Osmotic Content (mosm)	Volume (L)	Osmotic Content (mosm)	Volume (L)	Osmotic Content (mosm)	Osmolality (mosm/kg)	Plasma $[Na^+]$ (mmol/L)
	Initial	15	4,350	25	7,250	40	11,600	290	140
A	Loss of 5 L of sweat, 120 mosm/L, 60 mmol Na^+/L	11.9	3,750	23.1	7,250	35	11,000	314	151
	Above condition accompanied by intake of 5 L water	13.6	3,750	26.4	7,250	40	11,000	275	132
B	Loss of 5 L of sweat, 20 mosm/L, 10 mmol Na^+/L	12.9	4,250	22.1	7,250	35	11,500	329	159
	Above condition accompanied by intake of 5 L water	14.8	4,250	25.2	7,250	40	11,500	288	139

[a] Each subject has total body water of 40 L. The sweat of subject A has a relatively high $[Na^+]$ of 60 mmol/L, while that of subject B has a relatively low $[Na^+]$ of 10 mmol/L. Volumes of the extracellular and intracellular spaces are calculated assuming that water moves between the two spaces as needed to maintain osmotic balance.

CLINICAL FOCUS BOX 29.1

Water and Salt Depletion as a Result of Sweating

Changes in fluid and electrolyte balance are probably the most frequent physiological disturbances associated with sustained exercise and heat stress. Water loss via the sweat glands can exceed 1 L/hr for many hours. Salt loss in the sweat is variable; however, since sweat is more dilute than plasma, sweating always results in an increase in the osmolality of the fluid remaining in the body, and increased plasma $[Na^+]$ and $[Cl^-]$, as long as the lost water is not replaced.

Because people who secrete large volumes of sweat usually replace at least some of their losses by drinking water or electrolyte solutions, the final effect on body fluids may vary. In Table 29.3, the second and third conditions (subject A) represent the effects on body fluids of sweat losses alone and combined with replacement by an equal volume of plain water, respectively, for someone producing sweat with a $[Na^+]$ and $[Cl^-]$ in the upper part of the normal range. By contrast, the fourth and fifth conditions (subject B) represent the corresponding effects for a heat-acclimatized person secreting dilute sweat. Comparing the effects on these two individuals, we note: (1) The more dilute the sweat that is secreted, the greater the increase in osmolality and plasma $[Na^+]$ if no fluid is replaced; (2) Extracellular fluid volume, a major determinant of plasma volume (see Chapter 18), is greater in subject B (secreting dilute sweat) than in subject A (secreting saltier sweat), whether or not water is replaced; and (3) Drinking plain water allowed subject B to maintain plasma sodium and extracellular fluid volume almost unchanged while secreting 5 L of sweat. In subject A, however, drinking the same amount of water reduced plasma $[Na^+]$ by 8 mmol/L, and failed to prevent a decrease of almost 10% in extracellular fluid volume. In 5 L of sweat, subject A lost 17.5 g of salt, somewhat more than the daily salt intake in a normal Western diet, and he is becoming salt-depleted.

Thirst is stimulated by increased osmolality of the extracellular fluid, and by decreased plasma volume via a reduction in the activity of the cardiovascular stretch receptors (see Chapter 18). When sweating is profuse, however, thirst usually does not elicit enough drinking to replace fluid as rapidly as it is lost, so that people exercising in the heat tend to become progressively dehydrated—in some cases losing as much as 7 to 8% of body weight—and restore normal fluid balance only during long periods of rest or at meals. Depending on how much of his fluid losses he replaces, subject B may either be hypernatremic and dehydrated or be in essentially normal fluid and electrolyte balance. (If he drinks fluid well in excess of his losses, he may become overhydrated and **hyponatremic**, but this is an unlikely occurrence.) However, subject A, who is somewhat salt depleted, may be very dehydrated and hypernatremic, normally hydrated but hyponatremic, or somewhat dehydrated with plasma $[Na^+]$ anywhere in between these two extremes. Once subject A replaces all the water lost as sweat, his extracellular fluid volume will be about 10% below its initial value. If he responds to the accompanying reduction in plasma volume by continuing to drink water, he will become even more hyponatremic than shown in Table 29.3.

The disturbances shown in Table 29.3, while physiologically significant and useful for illustration, are not likely to require clinical attention. Greater disturbances, with correspondingly more severe clinical effects, may occur. The consequences of the various possible disturbances of salt and water balance can be grouped as effects of decreased plasma volume secondary to decreased extracellular fluid volume, effects of hypernatremia, and effects of hyponatremia.

(continued)

(Many penetrating veins connect the superficial veins to the deep veins, so that venous blood from anywhere in the limb potentially can return to the heart via either superficial or deep veins.) In the deep veins, cool venous blood returning to the core can take up heat from the warm blood in the adjacent deep limb arteries. Therefore, some of the heat contained in the arterial blood as it enters the limbs takes a "short circuit" back to the core. When the arterial blood reaches the skin, it is already cooler than the core, so it loses less heat to the skin than it otherwise would. (When the superficial veins dilate in the heat, most venous blood returns via superficial veins so as to maximize core-to-skin heat flow.) The transfer of heat from arteries to veins by this short circuit is called **countercurrent heat exchange**. This mechanism can cool the blood in the radial artery of a cool but comfortable subject to as low as 30°C by the time it reaches the wrist.

As we saw earlier, the shell's insulating properties increase in the cold as its blood vessels constrict and its thickness increases. Furthermore, the shell includes a fair amount of skeletal muscle in the cold, and although muscle blood flow is believed not to be affected by thermoregulatory reflexes, it is reduced by direct cooling. In a cool subject, the resulting reduction in muscle blood flow adds to the shell's insulating properties. As the blood vessels in the shell constrict, blood is shifted to the central blood reservoir in the thorax. This shift produces many of the same effects as an increase in blood volume, including so-called **cold diuresis** as the kidneys respond to the increased central blood volume.

Once skin blood flow is near minimal, metabolic heat production increases—almost entirely through shivering in human adults. Shivering may increase metabolism at rest by more than 4-fold—that is, to 350 to 400 W. Although it is often stated that shivering diminishes substantially after several hours and is impaired by exhaustive exercise, such effects are not well understood. In most laboratory mammals, chronic cold exposure also causes **nonshivering thermogenesis**, an increase in metabolic rate that is not due to muscle activity. Nonshivering thermogenesis appears to be elicited through sympathetic stimulation and circulating catecholamines. It occurs in many tissues, especially the liver and **brown fat**, a tissue specialized for nonshivering thermogenesis whose color is imparted by high concentrations of iron-containing respiratory enzymes. Brown fat is found in human infants, and nonshivering thermogenesis is important for their thermoregulation. The existence of brown fat and nonshivering thermogenesis in human adults is controversial, but there is some evi-

The circulatory effects of decreased volume are nearly identical to the effects of peripheral pooling of blood (see Fig. 29.12), and the combined effects of peripheral pooling and decreased volume will be greater than the effects of either alone. These effects include impairment of cardiac filling and cardiac output, and compensatory reflex reductions in renal, splanchnic, and skin blood flow. Impaired cardiac output leads to fatigue during exertion and decreased exercise tolerance; if skin blood flow is reduced, heat dissipation will be impaired. Exertional **rhabdomyolysis,** the injury of skeletal muscle fibers, is a frequent result of unaccustomed intense exercise. Myoglobin released from injured skeletal muscle cells appears in the plasma, rapidly enters the glomerular filtrate, and is excreted in the urine, producing **myoglobinuria** and staining the urine brown if enough myoglobin is present. This process may be harmless to the kidneys if urine flow is adequate; however, a reduction in renal blood flow reduces urine flow, increasing the likelihood that the myoglobin will cause renal tubular injury.

Hypernatremic dehydration is believed to predispose to heatstroke. Dehydration is often accompanied by both hypernatremia and reduced plasma volume. Hypernatremia impairs the heat-loss responses (sweating and increased skin blood flow) independently of any accompanying reduction in plasma volume and elevates the thermoregulatory set point. Hypernatremic dehydration promotes the development of high core temperature in multiple ways through the combination of hypernatremia and reduced plasma volume.

Even in the absence of sodium loss, overdrinking that exceeds the kidneys' ability to compensate dilutes all the body's fluid compartments, producing **dilutional hyponatremia,** which is also called **water intoxication** if it causes symptoms. The development of water intoxication requires either massive overdrinking, or a condition, such as the inappropriate secretion of arginine vasopressin, that impairs the excretion of free water by the kidneys. Overdrinking sufficient to cause hyponatremia may occur in patients with psychiatric disorders or disturbance of the thirst mechanism, or may be done with a mistaken intention of preventing or treating dehydration. However, individuals who secrete copious amounts of sweat with a high sodium concentration, like subject A or people with cystic fibrosis, may easily lose enough salt to become hyponatremic because of sodium loss. Some healthy young adults who come to medical attention for salt depletion after profuse sweating are found to have genetic variants of cystic fibrosis, which cause these individuals to have salty sweat without producing the characteristic digestive and pulmonary manifestations of cystic fibrosis.

As sodium concentration and osmolality of the extracellular space decrease, water moves from the extracellular space into the cells to maintain osmotic balance across the cell membranes. Most of the manifestations of hyponatremia are due to the resulting swelling of the brain cells. Mild hyponatremia is characterized by nonspecific symptoms such as fatigue, confusion, nausea, and headache, and may be mistaken for heat exhaustion. Severe hyponatremia can be a life-threatening medical emergency and may include seizures, coma, herniation of the brainstem (which occurs if the brain swells enough to exceed the capacity of the cranium) and death. In the setting of prolonged exertion in the heat, symptomatic hyponatremia is far less common than heat exhaustion, but potentially far more dangerous. Therefore, it is important not to treat a presumed case of heat exhaustion with large amounts of low-sodium fluids without first ruling out hyponatremia.

dence for functioning brown fat in the neck and mediastinum of outdoor workers.

Human Cold Acclimatization Confers a Modest Thermoregulatory Advantage

The pattern of human cold acclimatization depends on the nature of the cold exposure. It is partly for this reason that the occurrence of cold acclimatization in humans was controversial for a long time. Our knowledge of human cold acclimatization comes from both laboratory studies and studies of populations whose occupation or way of life exposes them repeatedly to cold temperatures.

Metabolic Changes in Cold Acclimatization. At one time it was believed that humans must acclimatize to cold as laboratory mammals do—by increasing their metabolic rate. There are a few reports of increased basal metabolic rate and, sometimes, thyroid activity in the winter. More often, however, increased metabolic rate has not been observed in studies of human cold acclimatization. In fact, several reports indicate the opposite response, consisting of a lower core temperature threshold for shivering, with a greater fall in core temperature and a smaller metabolic response during cold exposure. Such a response would spare metabolic energy and might be advantageous in an environment that is not so cold that a blunted metabolic response would allow core temperature to fall to dangerous levels.

Increased Tissue Insulation in Cold Acclimatization. A lower core-to-skin conductance (i.e., increased insulation by the shell) has often been reported in studies of cold acclimatization in which a reduction in the metabolic response to cold occurred. This increased insulation is not due to subcutaneous fat (in fact, it has been observed in very lean subjects), but apparently results from lower blood flow in the limbs or improved countercurrent heat exchange in the acclimatized subjects. In general, the cold stresses that elicit a lower shell conductance after acclimatization involve either cold water immersion or exposure to air that is chilly but not so cold as to risk freezing the vasoconstricted extremities.

Cold-Induced Vasodilation and the Lewis Hunting Response. As the skin is cooled below about 15°C, its blood flow begins to increase somewhat, a response called **cold-induced vasodilation** (CIVD). This response is elicited most easily in comfortably warm subjects and in skin rich in arteriovenous anastomoses (in the hands and feet). The mechanism has not been established but may involve a direct inhibitory effect of cold on the contraction of vascular smooth muscle or on neuromuscular transmission. The CIVD response varies greatly among individuals, and is usually rudimentary in hands and feet unaccustomed to cold exposure. After repeated cold exposure, CIVD begins earlier during cold exposure, produces higher levels of blood flow, and takes on a rhythmic pattern of alternating vasodilation and vasoconstriction. This is called the **Lewis hunting response** because the rhythmic pattern of blood flow suggests that it is "hunting" for its proper level. This response is often well developed in workers whose hands are exposed to cold, such as fishermen working with nets in cold water. Since the Lewis hunting response increases heat loss from the body somewhat, whether or not it is truly an example of acclimatization to cold is debatable. However, the response is advantageous because it keeps the extremities warmer, more comfortable, and functional and probably protects them from cold injury.

CLINICAL ASPECTS OF THERMOREGULATION

Temperature is important clinically because of the presence of fever in many diseases, the effects of many factors on tolerance to heat or cold stress, and the effects of heat or cold stress in causing or aggravating certain disorders.

Fever Enhances Defense Mechanisms

Fever may be caused by infection or noninfectious conditions (e.g., inflammatory processes such as collagen vascular diseases, trauma, neoplasms, acute hemolysis, immunologically-mediated disorders). **Pyrogens** are substances that cause fever and may be either exogenous or endogenous. Exogenous pyrogens are derived from outside the body; most are microbial products, microbial toxins, or whole microorganisms. The best studied of these is the lipopolysaccharide endotoxin of gram-negative bacteria. Exogenous pyrogens stimulate a variety of cells, especially monocytes and macrophages, to release endogenous pyrogens, polypeptides that cause the thermoreceptors in the hypothalamus (and perhaps elsewhere in the brain) to alter their firing rate and input to the central thermoregulatory controller, raising the thermoregulatory set point. This effect of endogenous pyrogens is mediated by the local synthesis and release of **prostaglandin E_2**. Aspirin and other drugs that inhibit the synthesis of prostaglandins also reduce fever.

Fever accompanies disease so frequently and is such a reliable indicator of the presence of disease that body temperature is probably the most commonly measured clinical index. Many of the body's defenses against infection and cancer are elicited by a group of polypeptides called **cytokines**; the endogenous pyrogen is usually a member of this group, **interleukin-1**. However, other cytokines, particularly **tumor necrosis factor**, **interleukin-6**, and the **interferons**, are also pyrogenic in certain circumstances. Elevated body temperature enhances the development of these defenses. If laboratory animals are prevented from developing a fever during experimentally induced infection, survival rates may be dramatically reduced. (Although, in this chapter, fever specifically means an elevation in core temperature resulting from pyrogens, some authors use the term more generally to mean any significant elevation of core temperature.)

Many Factors Affect Thermoregulatory Responses and Tolerance to Heat and Cold

Regular physical exercise and heat acclimatization increase heat tolerance and the sensitivity of the sweating response. Aging has the opposite effect; in healthy 65-year-old men, the sensitivity of the sweating response is half of that in 25-

year-old men. Many drugs inhibit sweating, most obviously those used for their anticholinergic effects, such as atropine and scopolamine. In addition, some drugs used for other purposes, such as glutethimide (a sleep-inducing drug), tricyclic antidepressants, phenothiazines (tranquilizers and antipsychotic drugs), and antihistamines, also have some anticholinergic action. All of these and several others have been associated with heatstroke. Congestive heart failure and certain skin diseases (e.g., ichthyosis and anhidrotic ectodermal dysplasia) impair sweating, and in patients with these diseases, heat exposure and especially exercise in the heat may raise body temperature to dangerous levels. Lesions that affect the thermoregulatory structures in the brainstem can also alter thermoregulation. Such lesions can produce **hypothermia** (abnormally low core temperature) if they impair heat-conserving responses. However, **hyperthermia** (abnormally high core temperature) is a more usual result of brainstem lesions and is typically characterized by a loss of both sweating and the circadian rhythm of core temperature.

Certain drugs, such as barbiturates, alcohol, and phenothiazines, and certain diseases, such as hypothyroidism, hypopituitarism, congestive heart failure, and septicemia, may impair the defense against cold. (Septicemia, especially in debilitated patients, may be accompanied by hypothermia, instead of the usual febrile response to infection.) Furthermore, newborns and many healthy older adults are less able than older children and younger adults to maintain adequate body temperature in the cold. This failing appears to be due to an impaired ability to conserve body heat by reducing heat loss and to increase metabolic heat production in the cold.

Heat Stress Causes or Aggravates Several Disorders

The harmful effects of heat stress are exerted through cardiovascular strain, fluid and electrolyte loss and, especially in heatstroke, tissue injury whose mechanism is uncertain. In a patient suspected of having hyperthermia secondary to heat stress, temperature should be measured in the rectum, since hyperventilation may render oral temperature spuriously low.

Heat Syncope. **Heat syncope** is circulatory failure resulting from a pooling of blood in the peripheral veins, with a consequent decrease in venous return and diastolic filling of the heart, resulting in decreased cardiac output and a fall of arterial pressure. Symptoms range from light-headedness and giddiness to loss of consciousness. Thermoregulatory responses are intact, so core temperature typically is not substantially elevated, and the skin is wet and cool. The large thermoregulatory increase in skin blood flow in the heat is probably the primary cause of the peripheral pooling. Heat syncope affects mostly those who are not acclimatized to heat, presumably because the plasma-volume expansion that accompanies acclimatization compensates for the peripheral pooling of blood. Treatment consists in laying the patient down out of the heat, to reduce the peripheral pooling of blood and improve the diastolic filling of the heart.

Heat Exhaustion. **Heat exhaustion**, also called heat collapse, is probably the most common heat disorder, and represents a failure of cardiovascular homeostasis in a hot environment. Collapse may occur either at rest or during exercise, and may be preceded by weakness or faintness, confusion, anxiety, ataxia, vertigo, headache, and nausea or vomiting. The patient has dilated pupils and usually sweats profusely. As in heat syncope, reduced diastolic filling of the heart appears to have a primary role in the pathogenesis of heat exhaustion. Although blood pressure may be low during the acute phase of heat exhaustion, the baroreflex responses are usually sufficient to maintain consciousness and may be manifested in nausea, vomiting, pallor, cool or even clammy skin, and rapid pulse. Patients with heat exhaustion usually respond well to rest in a cool environment and oral fluid replacement. In more severe cases, however, intravenous replacement of fluid and salt may be required. Core temperature may be normal or only mildly elevated in heat exhaustion. However, heat exhaustion accompanied by hyperthermia and dehydration may lead to heatstroke. Therefore, patients should be actively cooled if rectal temperature is 40.6°C (105°F) or higher.

The reasons underlying the reduced diastolic filling in heat exhaustion are not fully understood. Hypovolemia contributes if the patient is dehydrated, but heat exhaustion often occurs without significant dehydration. In rats heated to the point of collapse, compensatory splanchnic vasoconstriction develops during the early part of heating, but is reversed shortly before the maintenance of blood pressure fails. A similar process may occur in heat exhaustion.

Heatstroke. The most severe and dangerous heat disorder is characterized by high core temperature and the development of serious neurological disturbances with a loss of consciousness and, frequently, convulsions. **Heatstroke** occurs in two forms, classical and exertional. In the classical form, the primary factor is environmental heat stress that overwhelms an impaired thermoregulatory system, and most patients have preexisting chronic disease. In exertional heatstroke, the primary factor is high metabolic heat production. Patients with exertional heatstroke tend to be younger and more physically fit (typically, soldiers and athletes) than patients with the classical form. Rhabdomyolysis, hepatic and renal injury, and disturbances of blood clotting are frequent accompaniments of exertional heatstroke. The traditional diagnostic criteria of heatstroke—coma, hot dry skin, and rectal temperature above 41.3°C (106°F)—are characteristic of the classical form; however, patients with exertional heatstroke may have somewhat lower rectal temperatures and often sweat profusely. Heatstroke is a medical emergency, and prompt appropriate treatment is critically important to reducing morbidity and mortality. The rapid lowering of core temperature is the cornerstone of treatment, and it is most effectively accomplished by immersion in cold water. With prompt cooling, vigorous hydration, maintenance of a proper airway, avoidance of aspiration, and appropriate treatment of complications, most patients will survive, especially if they were previously healthy.

The pathogenesis of heatstroke is not well understood, but it seems clear that factors other than hyperthermia are involved, even if the action of these other factors partly depends on the hyperthermia. Exercise may contribute more to the pathogenesis than simply metabolic heat production. Elevated plasma levels of several inflammatory cytokines have been reported in patients presenting with heatstroke, suggesting a systemic inflammatory component. No trigger for such an inflammatory process has been established, although several possible candidates exist. One possible trigger is some product(s) of the bacterial flora in the gut, perhaps including lipopolysaccharide endotoxins. Several lines of evidence suggest that sustained splanchnic vasoconstriction may produce a degree of intestinal ischemia sufficient to allow these products to "leak" into the circulation and activate inflammatory responses.

The preceding diagnostic categories are traditional. However, they are not entirely satisfactory for heat illness associated with exercise because many patients have laboratory evidence of tissue and cellular injury, but are classified as having heat exhaustion because they do not have the serious neurological disturbances that characterize heatstroke. Some more recent literature uses the term **exertional heat injury** for such cases. The boundaries of exertional heat injury, with heat exhaustion on one hand and heatstroke on the other, are not clearly and consistently defined, and these categories probably represent parts of a continuum.

Malignant hyperthermia, a rare process triggered by depolarizing neuromuscular blocking agents or certain inhalational anesthetics, was once thought to be a form of

CLINICAL FOCUS BOX 29.2

Hypothermia

Hypothermia is classified according to the patient's core temperature as mild (32 to 35°C), moderate (28 to 32°C), or severe (below 28°C). Shivering is usually prominent in mild hypothermia, but diminishes in moderate hypothermia and is absent in severe hypothermia. The pathophysiology is characterized chiefly by the depressant effect of cold (via the Q_{10} effect) on multiple physiological processes and differences in the degree of depression of each process.

Other than shivering, the most prominent features of mild and moderate hypothermia are due to depression of the central nervous system. Beginning with mood changes (commonly, apathy, withdrawal, and irritability), they progress to confusion and lethargy, followed by ataxia and speech and gait disturbances, which may mimic a cerebrovascular accident (stroke). In severe hypothermia, voluntary movement, reflexes, and consciousness are lost and muscular rigidity appears. Cardiac output and respiration decrease as core temperature falls. Myocardial irritability increases in severe hypothermia, causing a substantial danger of ventricular fibrillation, with the risk increasing as cardiac temperature falls. The primary mechanism presumably is that cold depresses conduction velocity in Purkinje fibers more than in ventricular muscle, favoring the development of circus-movement propagation of action potentials. Myocardial hypoxia also contributes. In more profound hypothermia, cardiac sounds become inaudible and pulse and blood pressure are unobtainable because of circulatory depression; the electrical activity of the heart and brain becomes unmeasurable; and extensive muscular rigidity may mimic **rigor mortis.** The patient may appear clinically dead, but patients have been revived from core temperatures as low as 17°C, so that "no one is dead until warm and dead." The usual causes of death during hypothermia are respiratory cessation and the failure of cardiac pumping, because of either ventricular fibrillation or direct depression of cardiac contraction.

Depression of renal tubular metabolism by cold impairs the reabsorption of sodium, causing a diuresis and leading to dehydration and hypovolemia. Acid-base disturbances in hypothermia are complex. Respiration and cardiac output typically are depressed more than metabolic rate, and a mixed respiratory and metabolic acidosis results, because of CO_2 retention and lactic acid accumulation and the cold-induced shift of the hemoglobin-O_2 dissociation curve to the left. Acidosis aggravates the susceptibility to ventricular fibrillation.

Treatment consists of preventing further cooling and restoring fluid, acid-base, and electrolyte balance. Patients in mild to moderate hypothermia may be warmed solely by providing abundant insulation to promote the retention of metabolically produced heat; those who are more severely affected require active rewarming. The most serious complication associated with treating hypothermia is the development of ventricular fibrillation. Vigorous handling of the patient may trigger this process, but an increase in the patient's circulation (e.g., associated with warming or skeletal muscle activity) may itself increase the susceptibility to such an occurrence, as follows. Peripheral tissues of a hypothermic patient are, in general, even cooler than the core, including the heart, and acid products of anaerobic metabolism will have accumulated in underperfused tissues while the circulation was most depressed. As the circulation increases, a large increase in blood flow through cold, acidotic peripheral tissue may return enough cold, acidic blood to the heart to cause a transient drop in the temperature and pH of the heart, increasing its susceptibility to ventricular fibrillation.

The diagnosis of hypothermia is usually straightforward in a patient rescued from the cold but may be far less clear in a patient in whom hypothermia is the result of a serious impairment of physiological and behavioral defenses against cold. A typical example is the older person, living alone, who is discovered at home, cool and obtunded or unconscious. The setting may not particularly suggest hypothermia, and when the patient comes to medical attention, the diagnosis may easily be missed because standard clinical thermometers are not graduated low enough (usually only to 34.4°C) to detect hypothermia and, in any case, do not register temperatures below the level to which the mercury has been shaken. Because of the depressant effect of hypothermia on the brain, the patient's condition may be misdiagnosed as cerebrovascular accident or other primary neurological disease. Recognition of this condition depends on the physician's considering it when examining a cool patient whose mental status is impaired and obtaining a true core temperature with a low-reading glass thermometer or other device.

heatstroke but is now known to be a distinct disorder that occurs in people with a genetic predisposition. In 90% of susceptible individuals, biopsied skeletal muscle tissue contracts on exposure to caffeine or halothane in concentrations having little effect on normal muscle. Susceptibility may be associated with any of several myopathies, but most susceptible individuals have no other clinical manifestations. The control of free (unbound) calcium ion concentration in skeletal muscle cytoplasm is severely impaired in susceptible individuals; and when an attack is triggered, calcium concentration rises abnormally, activating myosin ATPase and leading to an uncontrolled hypermetabolic process that rapidly increases core temperature. Treatment with dantrolene sodium, which appears to act by reducing the release of calcium ions from the sarcoplasmic reticulum, has dramatically reduced the mortality rate of this disorder.

Aggravation of Disease States by Heat Exposure. Other than producing specific disorders, heat exposure aggravates several other diseases. Epidemiological studies show that during unusually hot weather, mortality may be 2 to 3 times that normally expected for the months in which heat waves occur. Deaths ascribed to specific heat disorders account for only a small fraction of the excess mortality (i.e., the increase above the expected mortality). Most of the excess mortality is accounted for by deaths from diabetes, various diseases of the cardiovascular system, and diseases of the blood-forming organs.

Hypothermia Occurs When the Body's Defenses Against Cold Are Disabled or Overwhelmed

Hypothermia reduces metabolic rate via the Q_{10} effect and prolongs the time tissues can safely tolerate a loss of blood flow. Since the brain is damaged by ischemia soon after circulatory arrest, controlled hypothermia is often used to protect the brain during surgical procedures in which its circulation is occluded or the heart is stopped. Much of our knowledge about the physiological effects of hypothermia comes from observations of surgical patients.

During the initial phases of cooling, stimulation of shivering through thermoregulatory reflexes overwhelms the Q_{10} effect. Metabolic rate, therefore, increases, reaching a peak at a core temperature of 30 to 33°C. At lower core temperatures, however, metabolic rate is dominated by the Q_{10} effect, and thermoregulation is lost. A vicious circle develops, wherein a fall in core temperature depresses metabolism and allows core temperature to fall further, so that at 17°C, the O_2 consumption is about 15%, and cardiac output 10%, of precooling values.

Hypothermia that is not induced for therapeutic purposes is called **accidental hypothermia** (Clinical Focus Box 29.2). It occurs in individuals whose defenses are impaired by drugs (especially ethanol, in the United States), disease, or other physical conditions and in healthy individuals who are immersed in cold water or become exhausted working or playing in the cold.

REVIEW QUESTIONS

DIRECTIONS: Each of the numbered items or incomplete statements in this section is followed by answers or by completions of the statement. Select the ONE lettered answer or completion that is BEST in each case.

1. Antipyretics such as aspirin effectively lower core temperature during fever, but they are not used to counteract the increase in core temperature that occurs during exercise. Which of the following best explains why it is inappropriate to use antipyretics for this purpose?
(A) The increase in core temperature during exercise stimulates metabolism via the Q_{10} effect, helping to support the body's increased metabolic energy demands
(B) A moderate increase in core temperature during exercise is harmless, so there is no benefit in preventing it
(C) Antipyretics are ineffective during exercise because they act on a mechanism that operates during fever, but not to a significant degree during exercise
(D) Antipyretics increase skin blood flow so as to dissipate more heat, increasing circulatory strain during exercise
(E) The increased heat production during exercise greatly exceeds the ability of antipyretics to stimulate the responses for heat loss
2. A surgical sympathectomy has completely interrupted the sympathetic nerve supply to a patient's arm. How would one expect the thermoregulatory skin blood flow and sweating responses on that arm to be affected?

	Vasoconstriction in the Cold	Vasodilation in the Heat	Sweating
(A)	Abolished	Intact	Intact
(B)	Abolished	Intact	Abolished
(C)	Abolished	Abolished	Intact
(D)	Abolished	Abolished	Abolished
(E)	Intact	Abolished	Abolished

3. A person resting in a constant ambient temperature is tested in the early morning at 4:00 AM, and again in the afternoon at 4:00 PM. Compared to measurements made in the morning, one would expect to find in the afternoon:

	Core Temperature	Sweating Threshold	Threshold for Cutaneous Vasodilation
(A)	Unchanged	Higher	Lower
(B)	Unchanged	Unchanged	Unchanged
(C)	Higher	Higher	Higher
(D)	Higher	Unchanged	Lower
(E)	Lower	Lower	Lower

4. Compared to an unacclimatized person, one who is acclimatized to cold has
(A) Higher metabolic rate in the cold, to produce more heat
(B) Lower metabolic rate in the cold, to conserve metabolic energy
(C) Lower peripheral blood flow in the cold, to retain heat
(D) Higher blood flow in the hands and feet in the cold, to preserve their function
(E) Various combinations of the above, depending on the environment that produced acclimatization
5. Which statement best describes how the elevated core temperature during fever affects the outcome of most bacterial infections?
(A) Fever benefits the patient because most pathogens thrive best at the host's normal body temperature

(continued)

(B) Fever is beneficial because it helps stimulate the immune defenses against infection
(C) Fever is harmful because the accompanying protein catabolism reduces the availability of amino acids for the immune defenses
(D) Fever is harmful because the patient's higher temperature favors growth of the bacteria responsible for infection
(E) Fever has little overall effect either way

6. A manual laborer moves in March from Canada to a hot, tropical country and becomes acclimatized by working outdoors for a month. Compared with his responses on the first few days in the tropical country, for the same activity level after acclimatization one would expect higher
(A) Core temperature
(B) Heart rate
(C) Sweating rate
(D) Sweat salt concentration
(E) Thermoregulatory set point

In questions 7 to 8, assume a 70-kg young man with the following baseline characteristics: total body water (TBW) = 40 L, extracellular fluid (ECF) volume = 15 L, plasma volume = 3 L, body surface area = 1.8 m^2, plasma $[Na^+]$ = 140 mmol/L. Heat of evaporation of water = 2,425 kJ/kg = 580 kcal/kg.

7. Our subject begins an 8-hour hike in the desert carrying 5 L of water in canteens. During the hike, he sweats at a rate of 1 L/hr, his sweat $[Na^+]$ is 50 mmol/L, and he drinks all his water. After the end of his hike he rests and consumes 3 L of water. (For simplicity in calculations, assume that the plasma osmolality equals 2 times the plasma $[Na^+]$.) What are his plasma sodium concentration and ECF volume after he has replaced all the water that he lost?

	Plasma $[Na^+]$ (mmol/L)	ECF Volume (L)
(A)	140.5	12.1
(B)	130	13.1
(C)	122.3	13.9
(D)	113.3	15.0
(E)	113.3	13.9

8. Our subject is bicycling on a long road with a slight upward grade. His metabolic rate (*M* in the heat-balance equation) is 800 W (48 kJ/min). He performs mechanical work (against gravity, friction, and wind resistance) at a rate of 140 W. Air temperature is 20°C and h_c, the convective heat transfer coefficient, is 15 W/(m^2·°C). Assume that his mean skin temperature is 34°C, all the sweat he secretes is evaporated, respiratory water loss can be ignored, and net heat exchange by radiation is negligible. How rapidly must he sweat to achieve heat balance? (Remember that 1 W = 1 J/sec = 60 J/min.)
(A) 3.9 g/min
(B) 7.0 g/min
(C) 11.1 g/min
(D) 13.9 g/min
(E) 15.0 g/min

SUGGESTED READING

Boulant JA. Hypothalamic neurons regulating body temperature. In: Fregly MJ, Blatteis CM, eds. Handbook of Physiology. Section 4. Environmental Physiology. New York: Oxford University Press, 1996;105–126.

Danzl DF. Hypothermia and frostbite. In: Braunwald E, Fauci AS, Kasper DL, et al., eds. Harrison's Principles of Internal Medicine. 15th Ed. New York: McGraw-Hill, 2001;107–111.

Dinarello CA. Cytokines as endogenous pyrogens. J Infect Dis 1999;179(Suppl 2):S294–S304.

Dinarello CA, Gelfand JA. Fever and hyperthermia. In: Braunwald E, Fauci AS, Kasper DL, et al., eds. Harrison's Principles of Internal Medicine. 15th Ed. New York: McGraw-Hill, 2001;91–94.

Gagge AP, Gonzalez RR. Mechanisms of heat biophysics and physiology. In: Fregly MJ, Blatteis CM, eds. Handbook of Physiology. Section 4. Environmental Physiology. New York: Oxford University Press, 1996;45–84.

Jessen C. Interaction of body temperatures in control of thermoregulatory effector mechanisms. In: Fregly MJ, Blatteis CM, eds. Handbook of Physiology. Section 4. Environmental Physiology. New York: Oxford University Press, 1996;127–138.

Johnson JM, Proppe DW. Cardiovascular adjustments to heat stress. In: Fregly MJ, Blatteis CM, eds. Handbook of Physiology. Section 4. Environmental Physiology. New York: Oxford University Press, 1996;215–243.

Knochel JP, Reed G: Disorders of heat regulation. In: Narins RG, ed. Maxwell & Kleeman's Clinical Disorders of Fluid and Electrolyte Metabolism. 5th Ed. New York: McGraw-Hill, 1994;1549–1590.

Pandolf KB, Sawka MN, Gonzalez RR, eds. Human Performance Physiology and Environmental Medicine at Terrestrial Extremes. Indianapolis: Benchmark, 1988.

Exercise Physiology

Bruce J. Martin, Ph.D.

CHAPTER OUTLINE

- THE QUANTIFICATION OF EXERCISE
- CARDIOVASCULAR RESPONSES
- RESPIRATORY RESPONSES
- MUSCLE AND BONE RESPONSES
- GASTROINTESTINAL, METABOLIC, AND ENDOCRINE RESPONSES
- AGING, IMMUNE, AND PSYCHIATRIC RESPONSES

KEY CONCEPTS

1. Exercise must be accurately defined before acute or chronic physiological responses can be predicted.
2. Maximal oxygen uptake predicts work performance and the physiological responses to exercise.
3. Substantial regional blood flow shifts occur during dynamic and isometric exercise.
4. Training affects both myocardial muscle and the coronary circulation.
5. The respiratory system responds predictably to increased O_2 consumption and CO_2 production with exercise.
6. In healthy individuals, muscle fatigue during exercise is linked to ADP accumulation.
7. Chronic physical activity enhances insulin sensitivity and glucose entry into cells.

Exercise, or physical activity, is a ubiquitous physiological state, so common in its many forms that true physiological "rest" is indeed rarely achieved. Defined ultimately in terms of skeletal muscle contraction, exercise involves every organ system in coordinated response to increased muscular energy demands.

THE QUANTIFICATION OF EXERCISE

Exercise is as varied as it is ubiquitous. A single episode of exercise, or "acute" exercise, may provoke responses different from the adaptations seen when activity is chronic—that is, during **training**. The forms of exercise vary as well. The amount of muscle mass at work (one finger? one arm? both legs?), the intensity of the effort, its duration, and the type of muscle contraction (isometric, rhythmic) all influence the body's responses and adaptations.

These many aspects of exercise imply that its interaction with disease is multifaceted. There is no simple answer as to whether exercise promotes health. In fact, physical activity can be healthful, harmful, or irrelevant, depending on the patient, the disease, and the specific exercise in question.

Measuring Maximal Oxygen Uptake Is the Most Common Method of Quantifying Dynamic Exercise

Dynamic exercise is defined as skeletal muscle contractions at changing lengths and with rhythmic episodes of relaxation. Fundamental to any discussion of dynamic exercise is a description of its intensity. Since dynamically exercising muscle primarily generates energy from oxidative metabolism, a traditional standard is to measure, by mouth, the oxygen uptake ($\dot{V}O_2$) of an exercising subject. This measurement is limited to dynamic exercise and usually to the steady state, when exercise intensity and oxygen consumption are stable and no net energy is provided from nonoxidative sources. Three implications of the original oxygen consumption measurements deserve mention. First, the centrality of oxygen usage to work output gave rise to the now-standard term "aerobic" exercise. Second, the apparent excess in oxygen consumption during the first minutes of recovery has been termed the **oxygen debt** (Fig. 30.1). The "excess" oxygen consumption of recovery results from a multitude of physiological processes and little usable information is obtained from its measurement. Third, and more

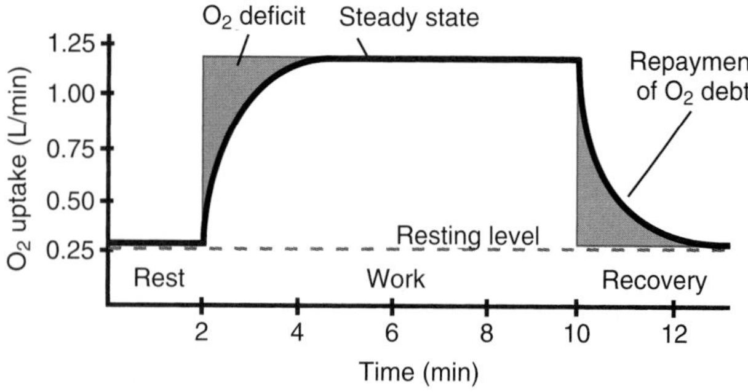

FIGURE 30.1 **Oxygen uptake before, during, and after light steady-state exercise.**

useful, during dynamic exercise that uses a large muscle mass, each person has a **maximal oxygen uptake**, a ceiling up to 20 times basal consumption, that cannot be exceeded, although it can be increased by appropriate training. This maximal oxygen uptake is a useful but imperfect predictor of the ability to perform prolonged dynamic external work or, more specifically, of endurance athletic performance. Maximal oxygen uptake is decreased, all else being equal, by age, bed rest, or increased body fat.

Maximal oxygen uptake is also used to express relative work capacity. A world champion cross-country skier obviously has a greater capacity to consume oxygen than a novice. However, when both are exercising at intensities requiring two thirds of their respective maximal oxygen uptakes (the world champion is moving much faster in doing this, as a result of higher capacity), both become exhausted at roughly the same time and for the same physiological reasons (Fig. 30.2). In the discussion that follows, relative as well as absolute (expressed as L/min of oxygen uptake) work levels are used to explain physiological responses. The energy costs and relative demands of some familiar activities are listed in Table 30.1.

What causes oxygen uptake to reach a ceiling? Historically, many arguments claim primacy for either cardiac output (oxygen delivery) or muscle metabolic capacity (oxygen use) limitations. However, it may be that every link in the chain taking oxygen from the atmosphere to the mitochondrion reaches its capacity at about the same time. In practical terms, this means that any lung, heart, vascular, or musculoskeletal illness that reduces oxygen flow capacity will diminish a patient's functional capacity.

In **isometric exercise**, force is generated at constant muscle length and without rhythmic episodes of relaxation. Isometric work intensity is usually described as a percentage of the **maximal voluntary contraction** (MVC), the peak isometric force that can be briefly generated for that specific exercise. Analogous to work levels relative to maximal oxygen uptake, the ability to endure isometric effort, and many physiological responses to that effort, are predictable when the percentage of MVC among individuals is held constant.

CARDIOVASCULAR RESPONSES

Increased energy expenditure with exercise demands more energy production. For prolonged work, this energy is supplied by the oxidation of foodstuff, with the oxygen carried to working muscles by the cardiovascular system.

Blood Flow Is Preferentially Directed to Working Skeletal Muscle During Exercise

Local control of blood flow ensures that only working muscles with increased metabolic demands receive increased blood and oxygen delivery. If the legs alone are active, leg muscle blood flow should increase while arm muscle blood flow remains unchanged or is reduced. At rest, skeletal muscle receives only a small fraction of the cardiac output. In dynamic exercise, both total cardiac output and relative and absolute output directed to working skeletal muscle increase dramatically (Table 30.2).

Cardiovascular control during exercise involves systemic regulation (cardiovascular centers in the brain, with their autonomic nervous output to the heart and systemic resistance vessels) in tandem with local control. For millennia our ancestors successfully used exercise both to escape being eaten and to catch food; therefore, it is no surprise that cardiovascular control in exercise is complex and unique. It's as if a brain software program entitled "Exercise"

TABLE 30.1 **Absolute and Relative Costs of Daily Activities**

Activity	Energy Cost (kcal/min)	% Maximal Oxygen Uptake: Sedentary 22-Year-Old	% Maximal Oxygen Uptake: Sedentary 70-Year-Old
Sleeping	1	6	8
Sitting	2	12	17
Standing	3	19	25
Dressing, undressing	3	19	25
Walking (3 miles/hr)	4	25	33
Making a bed	5	31	42
Dancing	7	44	58
Gardening/shoveling	8	50	67
Climbing stairs	11	69	92
Crawl swimming (50 m/min)	16	100	
Running (8 miles/hr)	16	100	

TABLE 30.2 Blood Flow Distribution During Rest and Dynamic Exercise in an Athlete

	Rest		Heavy Exercise	
Area	mL/min	%	mL/min	%
Splanchnic	1,400	24	300	1
Renal	1,100	19	900	4
Brain	750	13	750	3
Coronary	250	4	1,000	4
Skeletal muscle	1,200	21	22,000	86
Skin	500	9	600	2
Other	600	10	100	0.5
Total Cardiac Output	5,800	100	25,650	100

were inserted into the brain as work begins. Initially, the motor cortex is activated: The total neural activity is roughly proportional to the muscle mass and its work intensity. This neural activity communicates with the cardiovascular control centers, reducing vagal tone on the heart (which raises heart rate) and resetting the arterial baroreceptors to a higher level. As work rate is increased further, lactic acid is formed in actively contracting muscles, which stimulates muscle afferent nerves to send information to the cardiovascular center that increases sympathetic outflow to the heart and systemic resistance vessels. However, despite this **muscle chemoreflex** activity, within these same working muscles, low PO_2, increased nitric oxide, vasodilator prostanoids, and associated local vasoactive factors dilate arterioles despite rising sympathetic vasoconstrictor tone. Increased sympathetic drive does elevate heart rate and cardiac contractility, resulting in increased cardiac output; local factors in the coronary vessels mediate coronary vasodilation. Increased sympathetic vasoconstrictor tone in the renal and splanchnic vascular beds, and in inactive muscle, reduces blood flow to these tissues. Blood flow to these inactive regions can fall 75% if exercise is strenuous. Increased vascular resistance and decreased blood volume in these tissues helps maintain blood pressure during dynamic exercise. In contrast to blood flow reductions in the viscera and in inactive muscle, the brain autoregulates blood flow at constant levels independent of exercise. The skin remains vasoconstricted only if thermoregulatory demands are absent. Table 30.3 shows how a profound fall in systemic vascular resistance matches the enormous rise in cardiac output during dynamic exercise.

TABLE 30.3 Cardiac Output, Mean Arterial Pressure, and Systemic Vascular Resistance Changes With Exercise

	Rest	Strenuous Dynamic Exercise
Cardiac output (L/min)	6	21
Mean arterial pressure (mm Hg)	90	105
Systemic vascular resistance (mm Hg · min/L)	15	5

Dynamic exercise, at its most intense level, forces the body to choose between maximum muscle vascular dilation and defense of blood pressure. Blood pressure is, in fact, maintained. During strenuous exercise, sympathetic drive can begin to limit vasodilation in active muscle. When exercise is prolonged in the heat, increased skin blood flow and sweating-induced reduction in plasma volume both contribute to the risk of hyperthermia and hypotension (heat exhaustion). Although chronic exercise provides some heat acclimatization, even highly trained people are at risk for hyperthermia and hypotension if work is prolonged and water is withheld in demanding environmental conditions.

Isometric exercise causes a somewhat different cardiovascular response. Muscle blood flow increases relative to the resting condition, as does cardiac output, but the higher mean intramuscular pressure limits these flow increases much more than when exercise is rhythmic. Because the blood flow increase is blunted inside a statically contracting muscle, the fruits of hard work with too little oxygen appear quickly: a shift to anaerobic metabolism, the production of lactic acid, a rise in the ADP/ATP ratio, and fatigue. Maintaining just 50% of the MVC is agonizing after about 1 minute and usually cannot be continued after 2 minutes. A long-term sustainable level is only about 20% of maximum. These percentages are much less than the equivalent for dynamic work, as defined in terms of maximal oxygen uptake. Rhythmic exercise requiring 70% of the maximal oxygen uptake can be maintained in healthy individuals for about an hour, while work at 50% of the maximal oxygen uptake may be prolonged for several hours (see Fig. 30.2).

The reliance on anaerobic metabolism in isometric exercise triggers muscle ischemic chemoreflex responses that raise blood pressure more and cardiac output and heart rate

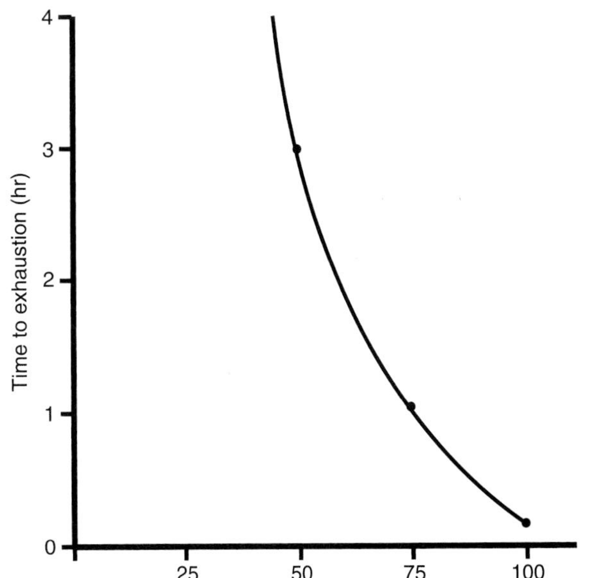

FIGURE 30.2 Time to exhaustion during dynamic exercise. Exhaustion is predictable on the basis of relative demand upon the maximal oxygen uptake.

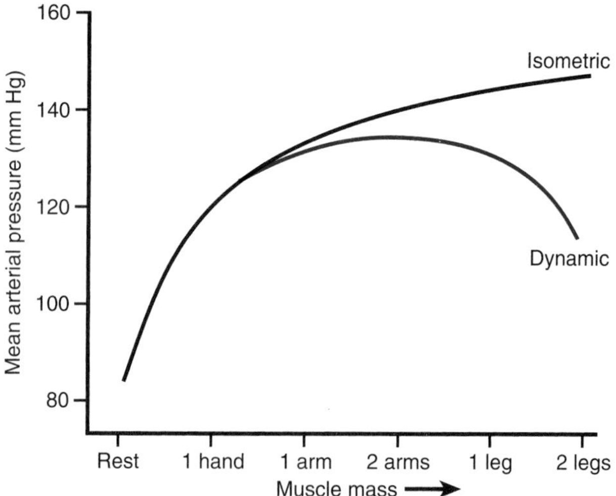

FIGURE 30.3 Effect of active muscle mass on mean arterial pressure during exercise. The highest pressures during dynamic exercise occur when an intermediate muscle mass is involved; pressure continues to rise in isometric exercise as more muscle is added.

less than in dynamic work (Fig. 30.3). Oddly, for dynamic exercise, the elevation of blood pressure is most pronounced when a medium muscle mass is working (see Fig. 30.3). This response results from the combination of a small, dilated active muscle mass with powerful central sympathetic vasoconstrictor drive. Typically, the arms exemplify a medium muscle mass; shoveling snow is a good example of primarily arm and heavily isometric exercise. Shoveling snow can be risky for people in danger of stroke or heart attack because it substantially raises systemic arterial pressure. The elevated pressure places compromised cerebral arteries at risk and presents an ischemic or failing heart with a greatly increased afterload.

Acute and Chronic Responses of the Heart and Blood Vessels to Exercise Differ

In acute dynamic exercise, vagal withdrawal and increases in sympathetic outflow elevate heart rate and contractility in proportion to exercise intensity (Table 30.4). Cardiac output is also aided in dynamic exercise by factors enhancing venous return. These include the "muscle pump," which compresses veins as muscles rhythmically contract, and the "respiratory pump," which increases breath-by-breath oscillations in intrathoracic pressure (see Chapter 18). The importance of these factors is clear in patients with heart transplants who lack extrinsic cardiac innervation. Stroke volume rises in cardiac transplant patients with increasing exercise intensity as a result of increased venous return that enhances cardiac preload. In addition, circulating epinephrine and norepinephrine from the adrenal medulla and norepinephrine from sympathetic nerve spillover augment heart rate and contractility.

Maximal dynamic exercise yields a maximal heart rate: further vagal blockade (e.g., via pharmacological means) cannot elevate heart rate further. Stroke volume, in contrast, reaches a plateau in moderate work and is unchanged as exercise reaches its maximum intensity (see Table 30.4). This plateau occurs in the face of ever-shortening filling time, testimony to the increasing effectiveness of the mechanisms that enhance venous return and those that promote cardiac contractility. Sympathetic stimulation decreases left ventricular volume and pressure at the onset of cardiac relaxation (as a result of increased ejection fraction), leading to more rapid ventricular filling early in diastole. This helps maintain stroke volume as diastole shortens. Even in untrained individuals, the ejection fraction (stroke volume as a percentage of end-diastolic volume) reaches 80% in strenuous exercise.

The increased blood pressure, heart rate, stroke volume, and cardiac contractility seen in exercise all increase myocardial oxygen demands. These demands are met by a linear increase in coronary blood flow during exercise that can reach a value 5 times the basal level. This increase in flow is driven by local, metabolically linked factors (nitric oxide, adenosine, and the activation of ATP-sensitive K^+ channels) acting on coronary resistance vessels in defiance of sympathetic vasoconstrictor tone. Coronary oxygen extraction, high at rest, increases further with exercise (up to 80% of delivered oxygen). In healthy people, there is no evidence of myocardial ischemia under any exercise condition, and there may be a coronary vasodilator reserve in even the most intense exercise (Clinical Focus Box 30.1).

Over longer periods of time, the heart adapts to exercise overload much as it does to high-demand pathological states: by increasing left ventricular volume when exercise requires high blood flow, and by left ventricular hypertrophy when exercise creates high systemic arterial pressure (high afterload). Consequently, the hearts of individuals adapted to prolonged, rhythmic exercise that involves relatively low arterial pressure exhibit large left ventricular

TABLE 30.4 Acute Cardiac Response to Graded Exercise in a 30-Year-Old Untrained Woman

Exercise Intensity	Oxygen Uptake (L/min)	Heart Rate (beats/min)	Stroke Volume (mL/beat)	Cardiac Output (L/min)
Rest	0.25	72	70	5
Walking	1.0	110	90	10
Jogging	1.8	150	100	15
Running fast	2.5	190	100	19

CLINICAL FOCUS BOX 30.1

Stress Testing

To detect coronary artery disease, physicians often record an electrocardiogram (ECG), but at rest, many disease sufferers have a normal ECG. To increase demands on the heart and coronary circulation, an ECG is performed while the patient walks on a treadmill or rides a stationary bicycle. It is sometimes called a **stress test.**

Exercise increases the heart rate and the systemic arterial blood pressure. These changes increase cardiac work and the demand for coronary blood flow. In many patients, coronary blood flow is adequate at rest, but because of coronary arterial blockage, cannot rise sufficiently to meet the increased demands of exercise. During a stress test, specific ECG changes can indicate that cardiac muscle is not receiving sufficient blood flow and oxygen delivery.

As heart rate increases during exercise, the distance between any portion of the ECG (for example, the R wave) on the ECG becomes shorter (Fig. 30.A, 1 and 2). In patients suffering from ischemic heart disease, however, other changes occur. Most common is an abnormal depression between the S and T waves, known as ST segment depression (see Fig. 30.A2). Depression of the ST segment arises from changes in cardiac muscle electrical activity secondary to lack of blood flow and oxygen delivery.

During the stress test, the ECG is continuously analyzed for changes while blood pressure and arterial blood oxygen saturation are monitored. At the start of the test, the exercise load is mild. The load is increased at regular intervals, and the test ends when the patient becomes exhausted, the heart rate safely reaches a maximum, significant pain occurs, or abnormal ECG changes are noted. With proper supervision, the stress test is a safe method for detecting coronary artery disease. Because the exercise load is gradually increased, the test can be stopped at the first sign of problems.

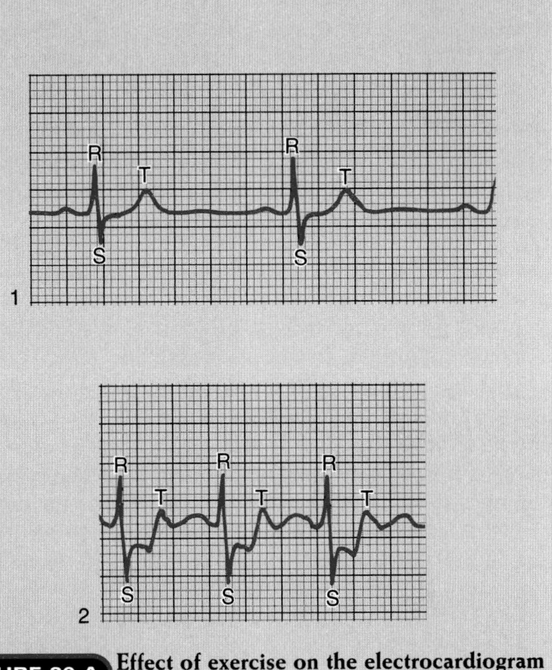

FIGURE 30.A **Effect of exercise on the electrocardiogram (ECG) in a patient with ischemic heart disease. 1,** The ECG is normal at rest. **2,** During exercise, the interval between R waves is reduced, and the ECG segment between the S and T waves is depressed.

volumes with normal wall thickness, while wall thickness is increased at normal volume in those adapted to activities involving isometric contraction and greatly elevated arterial pressure, such as lifting weights.

The larger left ventricular volume in people chronically active in dynamic exercise leads directly to larger resting and exercise stroke volume. A simultaneous increase in vagal tone and decrease in β-adrenergic sensitivity enhance the resting and exercise bradycardia seen after training, so that in effect the trained heart operates further up the ascending limb of its length-tension relationship (see Fig. 10.3). Nonetheless, resting bradycardia is a poor index of endurance fitness because genetic factors explain a much larger proportion of the individual variation in resting heart rate than does training.

The effects of endurance training on coronary blood flow are partly mediated through changes in myocardial oxygen uptake. Since myocardial oxygen consumption is roughly proportional to the rate-pressure product (heart rate × mean arterial pressure), and since heart rate falls after training at any absolute exercise intensity, coronary flow at a fixed submaximal workload is reduced in parallel. The peak coronary blood flow is, however, increased by training, as are cardiac muscle capillary density and peak capillary exchange capacity. Training also improves endothelium-mediated regulation, responsiveness to adenosine, and control of intracellular free calcium ions within coronary vessels. Preserving endothelial vasodilator function may be the primary benefit of chronic physical activity on the coronary circulation.

The Blood Lipid Profile Is Influenced by Exercise Training

Chronic, dynamic exercise is associated with increased circulating levels of high-density lipoproteins (HDLs) and reduced low-density lipoproteins (LDLs), such that the ratio of HDL to total cholesterol is increased. These changes in cholesterol fractions occur at any age if exercise is regular. Weight loss and increased insulin sensitivity, which typically accompany increased chronic physical activity in sedentary individuals, undoubtedly contribute to these changes in plasma lipoproteins. Nonetheless, in people with lipoprotein levels that place them at high risk for coronary heart disease, exercise appears to be an essential adjunct to dietary restriction and weight loss for lowering

LDL cholesterol levels. Because exercise acutely and chronically enhances fat metabolism and cellular metabolic capacities for β-oxidation of free fatty acids, it is not surprising that regular activity increases both muscle and adipose tissue lipoprotein lipase activity. Changes in lipoprotein lipase activity, in concert with increased lecithin-cholesterol acyltransferase activity and apo A-I synthesis, enhance the levels of circulating HDLs.

Exercise Has a Role in Preventing and Recovering From Several Cardiovascular Diseases

Changes in the ratio of HDL to total cholesterol that take place with regular physical activity reduce the risk of atherogenesis and coronary artery disease in active people, as compared with those who are sedentary. A lack of exercise is now established as a risk factor for coronary heart disease similar in magnitude to hypercholesterolemia, hypertension, and smoking. A reduced risk grows out of the changes in lipid profiles noted above, reduced insulin requirements and increased insulin sensitivity, and reduced cardiac β-adrenergic responsiveness and increased vagal tone. When coronary ischemia does occur, increased vagal tone may reduce the risk of fibrillation.

Regular exercise often, but not always, reduces resting blood pressure. Why some people respond to chronic activity with a resting blood pressure decline and others do not remains unknown. Responders typically show diminished resting sympathetic tone, so that systemic vascular resistance falls. In obesity-linked hypertension, declining insulin secretion and increasing insulin sensitivity with exercise may explain the salutary effects of combining training with weight loss. Nonetheless, because some obese people who exercise and lose weight show no blood pressure changes, exercise remains adjunctive therapy for hypertension.

Pregnancy Shares Many Cardiovascular Characteristics With the Trained State

The physiological demands and adaptations of pregnancy in some ways are similar to those of chronic exercise. Both of them increase blood volume, cardiac output, skin blood flow, and caloric expenditure. Exercise clearly has the potential to be deleterious to the fetus. Acutely, it increases body core temperature, causes splanchnic (hence, uterine and umbilical) vasoconstriction, and alters the endocrinological milieu; chronically, it increases caloric requirements. This last demand may be devastating if food shortages exist: the superimposed caloric demands of successful pregnancy and lactation are estimated at 80,000 kcal. Given adequate nutritional resources, however, there is little evidence of other damaging effects of maternal exercise on fetal development. The failure of exercise to harm well-nourished pregnant women may relate in part to the increased maternal and fetal mass and blood volume, which reduces specific heat loads, moderates vasoconstriction in the uterine and umbilical circulations, and diminishes the maternal exercise capacity.

At least in previously active women, even the most intense concurrent exercise regimen (unless associated with excessive weight loss) does not alter fertility, implantation, or embryogenesis, although the combined effects of exercise on insulin sensitivity and central obesity can restore ovulation in anovulatory obese women suffering from polycystic ovary disease.

Regular exercise may reduce the risk of spontaneous abortion of a chromosomally normal fetus. Continued exercise throughout pregnancy characteristically results in normal-term infants after relatively brief labor. These infants are usually normal in length and lean body mass but reduced in fat. The risk of large infant size for gestational age, increased in diabetic mothers, is reduced by maternal exercise through improved glucose tolerance. The incidence of umbilical cord entanglement, abnormal fetal heart rate during labor, stained amniotic fluid, and low fetal responsiveness scores may all be reduced in women who are active throughout pregnancy. Further, when examined 5 days after birth, newborns of exercising women perform better in their ability to orient to environmental stimuli and their ability to quiet themselves after sound and light stimuli than weight-matched children of nonexercising mothers.

RESPIRATORY RESPONSES

Increased breathing is perhaps the single most obvious physiological response to acute dynamic exercise. Figure 30.4 shows that minute ventilation (the product of breathing frequency and tidal volume) initially rises linearly with work intensity and then supralinearly beyond that point. Ventilation of the lungs in exercise is linked to the twin goals of oxygen intake and carbon dioxide removal.

Ventilation in Exercise Matches Metabolic Demands, but the Exact Control Mechanisms Are Unknown

Exercise increases oxygen consumption and carbon dioxide production by working muscles, and the pulmonary response is precisely calibrated to maintain homeostasis of these gases in arterial blood. In mild or moderate work, ar-

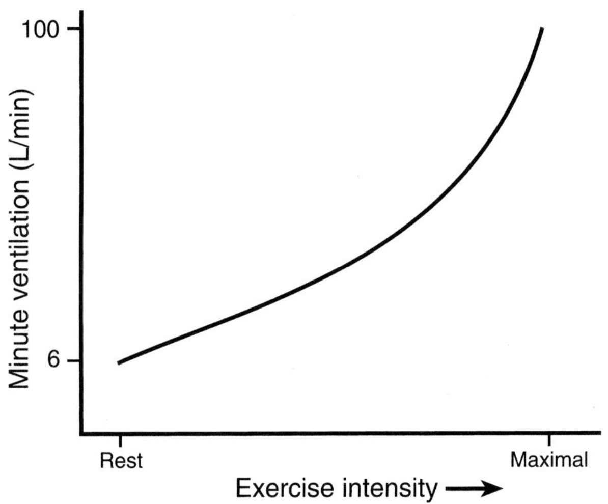

FIGURE 30.4 **The dependence of minute ventilation on the intensity of dynamic exercise.** Ventilation rises linearly with intensity until exercise nears maximal levels.

terial PO_2 (and, hence, oxygen content), PCO_2, and pH all remain unchanged from resting levels (Table 30.5). The respiratory muscles accomplish this severalfold increase in ventilation primarily by increasing tidal volume, without provoking sensations of dyspnea.

More intense exercise presents the lungs with tougher challenges. Near the halfway point from rest to maximal dynamic work, lactic acid formed in working muscles begins to appear in the circulation. This point, which depends on the type of work involved and the person's training status, is called the **lactate threshold**. Lactate concentration gradually rises with work intensity, as more and more muscle fibers must rely on anaerobic metabolism. Almost fully dissociated, lactic acid causes metabolic acidosis. During exercise, healthy lungs respond to lactic acidosis by further increasing ventilation, lowering the arterial PCO_2, and maintaining arterial blood pH at normal levels; it is the response to acidosis that spurs the supralinear ventilation rise seen in strenuous exercise (see Fig. 30.4). Through a range of exercise levels, the pH effects of lactic acid are fully compensated by the respiratory system; however, eventually in the hardest work—near-exhaustion—ventilatory compensation becomes only partial, and both pH and arterial PCO_2 may fall well below resting values (see Table 30.5). Tidal volume continues to increase until pulmonary stretch receptors limit it, typically at or near half of vital capacity. Frequency increases at high tidal volume produce the remainder of the ventilatory volume increases.

Hyperventilation relative to carbon dioxide production in heavy exercise helps maintain arterial oxygenation. The blood returned to the lungs during exercise is more thoroughly depleted of oxygen because active muscles with high oxygen extraction receive most of the cardiac output. Because the pulmonary arterial PO_2 is reduced in exercise, blood shunted past ventilated areas can profoundly depress systemic arterial oxygen content. Other than having a diminished oxygen content, pulmonary arterial blood flow (cardiac output) rises during exercise. In compensation, ventilation rises faster than cardiac output: The ventilation-perfusion ratio of the lung rises from near 1 at rest to greater than 4 with strenuous exercise (see Table 30.5). Healthy people maintain nearly constant arterial PO_2 with acute exercise, although the alveolar-to-arterial PO_2 gradient does rise. This increase shows that, despite the increase in the ventilation-perfusion ratio, areas of relative pulmonary underventilation and, possibly, some mild diffusion limitation exist even in highly trained, healthy individuals.

The ventilatory control mechanisms in exercise remain undefined. Where there are stimuli—such as in mixed venous blood, which is hypercapnic and hypoxic in proportion to exercise intensity—there are seemingly no receptors. Conversely, where there are receptors—the carotid bodies, the lung parenchyma or airways, the brainstem bathed by cerebrospinal fluid—no stimulus proportional to the exercise demand exists. Paradoxically, the central chemoreceptor is immersed in increasing alkalinity as exercise intensifies, a consequence of blood-brain barrier permeability to CO_2 but not hydrogen ions. Perhaps exercise respiratory control parallels cardiovascular control, with a central command proportional to muscle activity directly stimulating the respiratory center and feedback modulation from the lung, respiratory muscles, chest wall mechanoreceptors, and carotid body chemoreceptors.

The Respiratory System Is Largely Unchanged by Training

The effects of training on the pulmonary system are minimal. Lung diffusing capacity, lung mechanics, and even lung volumes change little, if at all, with training. The widespread assumption that training improves vital capacity is false; even exercise designed specifically to increase inspiratory muscle strength elevates vital capacity by only 3%. The demands placed on respiratory muscles increase their endurance, an adaptation that may reduce the sensation of dyspnea in exercise. Nonetheless, the primary respiratory changes with training are secondary to lower lactate production that reduces ventilatory demands at previously heavy absolute work levels.

In Lung Disease, Respiratory Limitations May Be Evidenced by Shortness of Breath or Decreased Oxygen Content of Arterial Blood

Any compromise of lung or chest wall function is much more apparent during exercise than at rest. One hallmark of lung disease is **dyspnea** (difficult or labored breathing) during exertion, when this exertion previously was unproblematic. Restrictive lung diseases limit tidal volume, reducing the ventilatory reserve volumes and exercise capacity. Obstructive lung diseases increase the work of breathing, exaggerating dyspnea and limiting work output. Lung diseases that compromise oxygen diffusion from alveolus to blood exaggerate exercise-induced widening of the alveolar-to-arterial PO_2 gradient. This effect contributes to po-

TABLE 30.5 Acute Respiratory Response to Graded Dynamic Exercise in a 30-Year-Old Untrained Woman

Exercise Intensity	Ventilation (L/min)	Ventilation-Perfusion Ratio	Alveolar PO_2 (mm Hg)	Arterial PO_2 (mm Hg)	Arterial PCO_2 (mm Hg)	Arterial pH
Rest	5	1	103	100	40	7.40
Walking	20	2	103	100	40	7.40
Jogging	45	3	106	100	36	7.40
Running fast	75	4	110	100	25	7.32

CLINICAL FOCUS BOX 30.2

Exercise in Patients with Emphysema

Normally, the respiratory system does not limit exercise tolerance. In healthy individuals, arterial blood saturation with oxygen, which averages 98% at rest, is maintained at or near 98% in even the most strenuous dynamic or isometric exercise. The healthy response includes the ability to augment ventilation more than cardiac output; the resulting rise in the ventilation-perfusion ratio counterbalances the falling oxygen content of mixed venous blood.

In patients with **emphysema**, ventilatory limitations to exercise occur long before ceilings are imposed by either skeletal muscle oxidative capacity or by the ability of the cardiovascular system to deliver oxygen to exercising muscle. These limitations are manifest during a stress test on the basis of three primary measurements. First, patients with ventilatory limitations typically cease exercise at relatively low heart rate, indicating that exhaustion is due to factors unrelated to cardiovascular limitations. Second, their primary complaint is usually shortness of breath, or dyspnea. In fact, patients with chronic obstructive pulmonary disease often first seek medical evaluation because of dyspnea experienced during such routine activities as climbing a flight of stairs. In healthy people, exhaustion is rarely associated solely with dyspnea. In emphysematous patients, exercise-induced dyspnea results, in part, from respiratory muscle fatigue exacerbated by diaphragmatic flattening brought on by loss of lung elastic recoil. Third, in emphysematous patients, arterial oxygen saturation will characteristically fall steeply and progressively with increasing exercise, sometimes reaching dangerously low levels. In emphysema, the inability to fully oxygenate blood at rest is compounded during exercise by increased pulmonary blood flow, and by increased exercise oxygen extraction that more fully desaturates blood returning to the lungs.

tentially dangerous systemic arterial hypoxia during exercise. The signs and symptoms of a respiratory limitation to exercise include exercise cessation with low maximal heart rate, oxygen desaturation of arterial blood, and severe shortness of breath (Clinical Focus Box 30.2). The prospects of training-based rehabilitation are modest, although locomotor muscle-based adaptations can reduce lactate production and ventilatory demands in exercise. Specific training of respiratory muscles to increase their strength and endurance is of minimal benefit to patients with compromised lung function.

Exercise causes bronchoconstriction in nearly every asthmatic patient and is the sole provocative agent for asthma in many people. In healthy individuals, catecholamine release from the adrenal medulla and sympathetic nerves dilates the airways during exercise. Sympathetic bronchodilation in people with asthma is outweighed by constrictor influences, among them heat loss from airways (cold, dry air is a potent bronchoconstrictor), release of inflammatory mediators, and increases in airway tissue osmolality. Leukotriene-receptor antagonists block exercise-induced symptoms in most people. The effects of exercise on airways are due to increased ventilation per se; the exercise is incidental. Individuals with exercise-induced bronchoconstriction are simply the most sensitive people along a continuum; for example, breathing high volumes of cold, dry air provokes at least mild bronchospasm in everyone.

MUSCLE AND BONE RESPONSES

Events within exercising skeletal muscle are a primary factor in fatigue. These same events, when repeated during training, lead to adaptations that increase exercise capacity and retard fatigue during similar work. Skeletal muscle contraction also increases stresses placed on bone, leading to specific bone adaptations.

Muscle Fatigue Is Independent of Lactic Acid

Although strenuous exercise can reduce intramuscular pH to values as low as 6.8 (arterial blood pH may fall to 7.2), there is little evidence that elevations in hydrogen ion concentration are the sole cause of fatigue. The best correlate of fatigue in healthy individuals is ADP accumulation in the face of normal or slightly reduced ATP, such that the ADP/ATP ratio is very high. Because the complete oxidation of glucose, glycogen, or free fatty acids to carbon dioxide and water is the major source of energy in prolonged work, people with defects in glycolysis or electron transport exhibit a reduced ability to sustain exercise.

These metabolic defects are distinct from another group of disorders exemplified by the various muscular dystrophies. In these illnesses, the loss of active muscle mass as a result of fat infiltration, cellular necrosis, or atrophy reduces exercise tolerance despite normal capacities (in healthy fibers) for ATP production. It is unclear whether fatigue in health ever occurs centrally (pain from fatigued muscle may feed back to the brain to lower motivation and, possibly, to reduce motor cortical output) or at the level of the motor neuron or the neuromuscular junction.

Endurance Activity Enhances Muscle Oxidative Capacity

Within skeletal muscle, adaptations to training are specific to the form of muscle contraction. Increased activity with low loads results in increased oxidative metabolic capacity without hypertrophy; increased activity with high loads produces muscle hypertrophy. Increased activity without overload increases capillary and mitochondrial density, myoglobin concentration, and virtually the entire enzymatic machinery for energy production from oxygen (Table 30.6). Coordination of energy-producing and energy-utilizing systems in muscle ensures that even after atrophy the remaining contractile proteins are adequately

TABLE 30.6 Effects of Training and Immobilization on the Human Biceps Brachii Muscle in a 22-Year-Old Woman

	Sedentary	After Endurance Training	After Strength Training	After 4 Months Immobilization
Total number of cells	300,000	300,000	300,000	300,000
Total cross-sectional area (cm^2)	10	10	13	6
Isometric strength (% control)	100	100	200	60
Fast-twitch fibers (% by number)	50	50	50	50
Fast-twitch fibers, average area ($\mu m^2 \times 10^2$)	67	67	87	40
Capillaries/fiber	0.8	1.3	0.8	0.6
Succinate dehydrogenase activity/unit area (% control)	100	150	77	100

Modified from Gollnick PD, Saltin B. Skeletal muscle physiology. In: Teitz CC, ed. Scientific Foundations of Sports Medicine. Toronto: BC Decker, 1989;185–242.

supported metabolically. In fact, the easy fatigability of atrophied muscle is due to the requirement that more motor units be recruited for identical external force; the fatigability per unit cross-sectional area is normal. The magnitude of the skeletal muscle endurance training response is limited by factors outside the muscle, since cross-innervation or chronic stimulation of muscles in animals can produce adaptations 5 times larger than those created by the most intense and prolonged exercise.

Local adaptations of skeletal muscle to endurance activity reduce reliance on carbohydrate as a fuel and allow more metabolism of fat, prolonging endurance and decreasing lactic acid accumulation. Decreased circulating lactate, in turn, reduces the ventilatory demands of heavier work. Because metabolites accumulate less rapidly inside trained muscle, there is reduced chemosensory feedback to the central nervous system at any absolute workload. This reduces sympathetic outflow to the heart and blood vessels, reducing cardiac oxygen demands at a fixed exercise level.

Muscle Hypertrophies in Response to Eccentric Contractions

Everyone knows it is easier to walk downhill than uphill, but the mechanisms underlying this commonplace phenomenon are complex. Muscle forces are identical in the two situations. However, moving the body uphill against gravity involves muscle shortening, or **concentric contractions**. In contrast, walking downhill primarily involves muscle tension development that resists muscle lengthening, or **eccentric contractions**. All routine forms of physical activity, in fact, involve combinations of concentric, eccentric, and isometric contractions. Because less ATP is required for force development during a contraction when external forces lengthen the muscle, the number of active motor units is reduced and energy demands are less for eccentric work. However, perhaps because the force per active motor unit is greater in eccentric exercise, eccentric contractions can readily cause muscle damage. These include weakness (apparent the first day), soreness and edema (delayed 1 to 3 days in peak magnitude), and elevated plasma levels of intramuscular enzymes (delayed 2 to 6 days). Histological evidence of damage may persist for 2 weeks. Damage is accompanied by an acute phase reaction that includes complement activation, increases in circulating cytokines, neutrophil mobilization, and increased monocyte cell adhesion capacity. Training adaptation to the eccentric components of exercise is efficient; soreness after a second episode is minimal if it occurs within two weeks of a first episode.

Eccentric contraction-induced muscle damage and its subsequent response may be the essential stimulus for muscle hypertrophy. While standard resistance exercise involves a mixture of contraction types, careful studies show that when one limb works purely concentrically and the other purely eccentrically at equivalent force, only the eccentric limb hypertrophies. The immediate changes in actin and myosin production that lead to hypertrophy are mediated at the posttranslational level; after a week of loading, mRNA for these proteins is altered. Although its precise role remains unclear, the activity of the 70-kDa S6 protein kinase is tightly linked with long-term changes in muscle mass. The cellular mechanisms for hypertrophy include the induction of insulin-like growth factor I, and upregulation of several members of the fibroblast growth factor family.

Exercise Plays a Role in Calcium Homeostasis

Skeletal muscle contraction applies force to bone. Because the architecture of bone remodeling involves osteoblast and osteoclast activation in response to loading and unloading, physical activity is a major site-specific influence on bone mineral density and geometry. Repetitive physical activity can create excessive strain, leading to inefficiency in bone remodeling and stress fracture; however, extreme inactivity allows osteoclast dominance and bone loss.

The forces applied to bone during exercise are related both to the weight borne by the bone during activity and to the strength of the involved muscles. Consequently, bone strength and density appear to be closely related to applied gravitational forces and to muscle strength. This suggests that exercise programs to prevent or treat **osteoporosis** should emphasize weight-bearing activities and strength as well as endurance training. Adequate dietary calcium is essential for any exercise effect: weight-bearing activity enhances spinal bone mineral density in post-

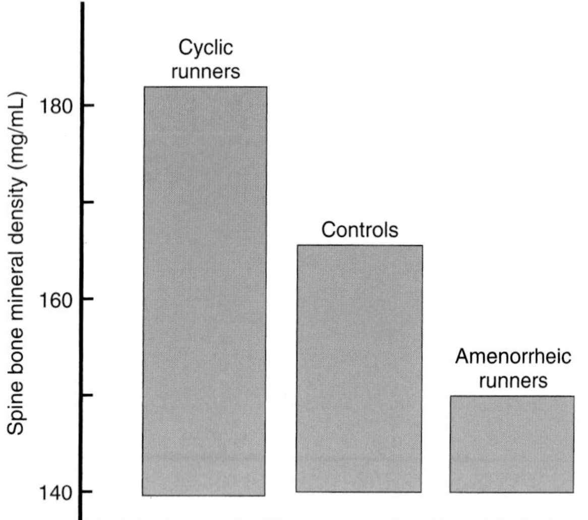

FIGURE 30.5 **Exercise and bone density.** This graph shows spine bone density in young adult women who are nonathletes (controls), distance runners with regular menstrual cycles (cyclic runners), and distance runners with amenorrhea (amenorrheic runners). Differences from controls indicate the roles that exercise and estrogen play in determination of bone mineral density.

menopausal women only when calcium intakes exceed 1 g/day. Because exercise may also improve gait, balance, coordination, proprioception, and reaction time, even in older and frail persons, the risk of falls and osteoporosis are reduced by chronic activity. In fact, the incidence of hip fracture is reduced nearly 50% when older adults are involved in regular physical activity. However, even when activity is optimal, it is apparent that genetic contributions to bone mass are greater than exercise. Perhaps 75% of the population variance is genetic, and 25% is due to different levels of activity. In addition, the predominant contribution of estrogen to homeostasis of bone in young women is apparent when amenorrhea occurs secondary to chronic heavy exercise. These exceptionally active women are typically very thin and exhibit low levels of circulating estrogens, low trabecular bone mass, and a high fracture risk (Fig. 30.5).

Exercise also plays a role in the treatment of **osteoarthritis**. Controlled clinical trials find that appropriate, regular exercise decreases joint pain and degree of disability, although it fails to influence the requirement for anti-inflammatory drug treatment. In **rheumatoid arthritis**, exercise also increases muscle strength and functional capacity without increasing pain or medication requirements. Whether or not exercise alters disease progression in either rheumatoid arthritis or osteoarthritis is not known.

GASTROINTESTINAL, METABOLIC, AND ENDOCRINE RESPONSES

The effects of exercise on gastrointestinal (GI) function remain poorly understood. However, chronic physical activity plays a major role in the control of obesity and type 2 diabetes mellitus.

Exercise Can Modify the Rate of Gastric Emptying and Intestinal Absorption

Dynamic exercise must be strenuous (demanding more than 70% of the maximal oxygen uptake) to slow gastric emptying of liquids. Little is known of the neural, hormonal, or intrinsic smooth muscle basis for this effect. Although gastric acid secretion is unchanged by acute exercise of any intensity, nothing is known about the effects of exercise on other factors relevant to the development or healing of peptic ulcers. There is some evidence that strenuous postprandial dynamic exercise provokes gastroesophageal reflux by altering esophageal motility.

Chronic physical activity accelerates gastric emptying rates and small intestinal transit. These adaptive responses to chronically increased energy expenditure lead to more rapid processing of food and increased appetite. Animal models of hyperphagia show specific adaptations in the small bowel (increased mucosal surface area, height of microvilli, content of brush border enzymes and transporters) that lead to more rapid digestion and absorption; these same effects likely take place in humans rendered hyperphagic by regular physical activity.

Blood flow to the gut decreases in proportion to exercise intensity, as sympathetic vasoconstrictor tone rises. Water, electrolyte, and glucose absorption may be slowed in parallel, and acute diarrhea is common in endurance athletes during competition. However, these effects are transient, and malabsorption as a consequence of acute or chronic exercise does not occur in healthy people. While exercise may not improve symptoms or disease progression in inflammatory bowel disease, there is some evidence that repetitive dynamic exercise may reduce the risk for this illness.

Although exercise is often recommended as treatment for postsurgical ileus, uncomplicated constipation, or irritable bowel syndrome, little is known in these areas. However, chronic dynamic exercise does substantially decrease the risk for colon cancer, possibly via increases in food and fiber intake, with consequent acceleration of colonic transit.

Chronic Exercise Increases Appetite Slightly Less Than Caloric Expenditure in Obese People

Obesity is common in sedentary societies. Obesity increases the risk for hypertension, heart disease, and diabetes and is characterized, at a descriptive level, as an excess of caloric intake over energy expenditure. Because exercise enhances energy expenditure, increasing physical activity is a mainstay of treatment for obesity.

The metabolic cost of exercise averages 100 kcal/mile walked. For exceptionally active people, exercise expenditure can exceed 3,000 kcal/day added to the basal energy expenditure, which for a 55-kg woman averages about 1,400 kcal/day. At high levels of activity, appetite and food intake match caloric expenditure (Fig. 30.6). The biological factors that allow this precise balance have never been defined. In obese people, modest increases in physical activity increase energy expenditure more than food intake, so progressive weight loss can be instituted if exercise can be regularized (see Fig. 30.6). This method of weight control is superior to dieting alone, since substantial caloric re-

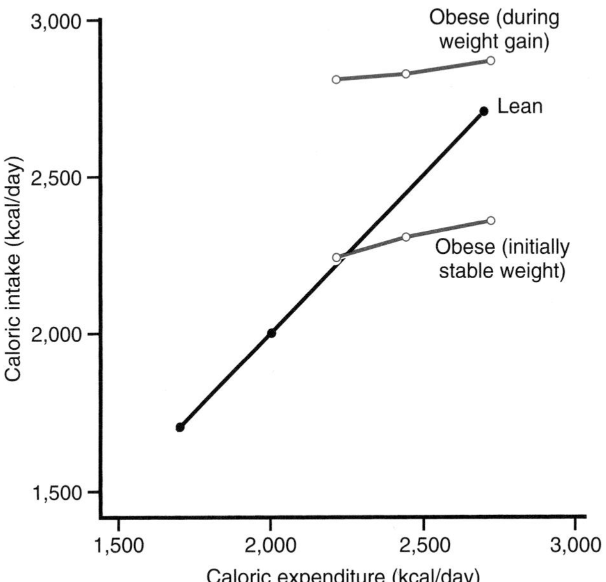

FIGURE 30.6 Caloric intake as a function of exercise-induced increases in daily caloric expenditure. For lean individuals, intake matches expenditure over a wide range. For obese individuals during periods of weight gain or periods of stable weight, increases in expenditure are not matched by increases in caloric intake. (Modified from Pi-Sunyer FX. Exercise effects on calorie intake. Annals NY Acad Sci 1987;499:94–103.)

striction (>500 kcal/day) results in both a lowered BMR and a substantial loss of fat-free body mass.

Exercise has other, subtler, positive effects on the energy balance equation as well. A single exercise episode may increase basal energy expenditure for several hours and may increase the thermal effect of feeding. The greatest practical problem remains compliance with even the most precise exercise "prescription"; patient dropout rates from even short-term programs typically exceed 50%.

Acute and Chronic Exercise Increases Insulin Sensitivity, Insulin Receptor Density, and Glucose Transport Into Muscle

Though skeletal muscle is omnivorous, its work intensity and duration, training status, inherent metabolic capacities, and substrate availability determine its energy sources. For very short-term exercise, stored phosphagens (ATP and creatine phosphate) are sufficient for crossbridge interaction between actin and myosin; even maximal efforts lasting 5 to 10 seconds require little or no glycolytic or oxidative energy production. When work to exhaustion is paced to be somewhat longer in duration, glycolysis is driven (particularly in fast glycolytic fibers) by high intramuscular ADP concentrations, and this form of anaerobic metabolism, with its by-product lactic acid, is the major energy source. The carbohydrate provided to glycolysis comes from stored, intramuscular glycogen or blood-borne glucose. Exhaustion from work in this intensity range (50 to 90% of the maximal oxygen uptake) is associated with carbohydrate depletion. Accordingly, factors that increase carbohydrate availability improve fatigue resistance. These include prior high dietary carbohydrate, cellular training adaptations that increase the enzymatic potential for fatty acid oxidation (thereby sparing carbohydrate stores), and oral carbohydrate intake during exercise. Frank hypoglycemia rarely occurs in healthy people during even the most prolonged or intense physical activity. When it does, it is usually in association with the depletion of muscle and hepatic stores and a failure to supplement carbohydrate orally.

Exercise suppresses insulin secretion by increasing sympathetic tone at the pancreatic islets. Despite acutely falling levels of circulating insulin, both non-insulin-dependent and insulin-dependent muscle glucose uptake increase during exercise. Exercise recruits glucose transporters from their intracellular storage sites to the plasma membrane of active skeletal muscle cells. Because exercise increases insulin sensitivity, patients with **type 1 diabetes** (insulin-dependent) require less insulin when activity increases. However, this positive result can be treacherous because exercise can accelerate hypoglycemia and increase the risk of insulin coma in these individuals. Chronic exercise, through its reduction of insulin requirements, up-regulates insulin receptors. This effect appears to be due less to training than simply to a repeated acute stimulus; the effect is full-blown after 2 to 3 days of regular physical activity and can be lost as quickly. Consequently, healthy active people show strikingly greater insulin sensitivity than do their sedentary counterparts (Fig. 30.7). In addition, up-

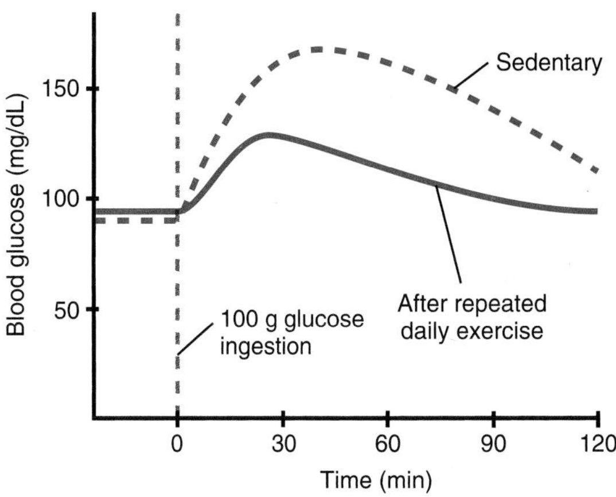

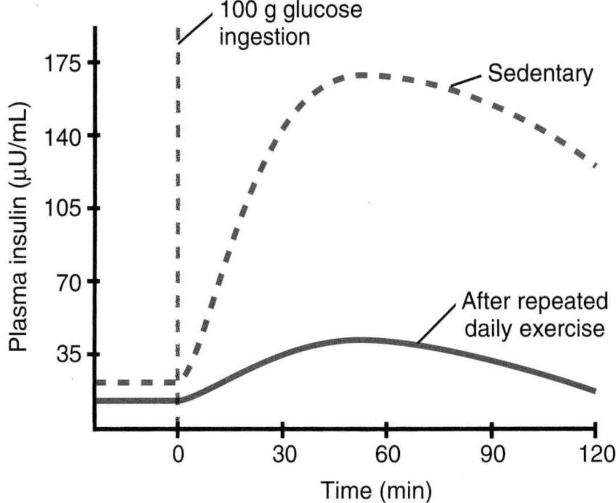

FIGURE 30.7 Repeated daily exercise and the blood glucose and insulin response to glucose ingestion. Both responses are blunted by repeated exercise, demonstrating increased insulin sensitivity.

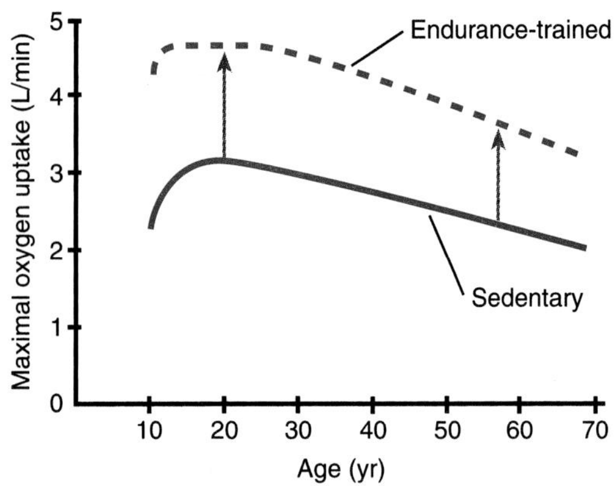

FIGURE 30.8 **Maximal oxygen uptake, endurance training, and age.** Endurance-trained subjects possess greater maximal oxygen uptake than sedentary subjects, regardless of age.

regulation of insulin receptors and reduced insulin release after chronic exercise is ideal therapy in **type 2 diabetes** (non–insulin-dependent), a disease characterized by high insulin secretion and low receptor sensitivity. In persons with type 2 diabetes, a single episode of exercise results in substantial glucose transporter translocation to the plasma membrane in skeletal muscle.

AGING, IMMUNE, AND PSYCHIATRIC RESPONSES

Maximal dynamic and isometric exercise capacities are lower at age 70 than at age 20. There is overwhelming evidence, however, that declines in strength and endurance with advancing age can be substantially mitigated by training. Changes in functional capacity, as well as protection against heart disease and diabetes, do increase longevity in active persons. However, it remains controversial if chronic exercise enhances lifespan, or if exercise boosts the immune system, prevents insomnia, or enhances mood.

As People Age, the Effects of Exercise on Functional Capacity Are More Profound Than Their Effect on Longevity

The influence of exercise on strength and endurance at any age is dramatic. Although the ceiling for oxygen uptake during work gradually falls with age, the ability to train toward an age-appropriate ceiling is as intact at age 70 as it is at age 20 (Fig. 30.8). In fact, a highly active 70-year-old, otherwise healthy, will typically display an absolute exercise capacity greater than a sedentary 20-year-old. Aging affects all the links in the chain of oxygen transport and use, so aging-induced declines in lung elasticity, lung diffusing capacity, cardiac output, and muscle metabolic potential take place in concert. Consequently, the physiological mechanisms underlying fatigue are similar at all ages.

Regular dynamic exercise, compared with inactivity, increases longevity in rats and humans. In descriptive terms, the effects of exercise are modest; all-cause mortality is reduced, but only in amounts sufficient to increase longevity by 1 to 2 years. These facts leave open the possibility that exercise might alter biological aging. While physical activity increases cellular oxidative stress, it simultaneously increases antioxidant capacity. Food-restricted rats experience increased life span, and exhibit elevated spontaneous activity levels, but the role exercise may play in the apparent delay of aging in these animals remains unclear.

Acute Exercise Transiently Alters Many Circulating Immune System Markers, but the Long-Term Effects of Training on Immune Function Are Unclear

In protein-calorie malnutrition, the catabolism of protein for energy lowers immunoglobulin levels and compromises the body's resistance to infection. Clearly, in this circumstance, exercise merely speeds the starvation process by increasing daily caloric expenditure and would be expected to diminish the immune response further. Nazi labor camps of the early 1940s became death camps, partly, by severe food restrictions and incessant demands for physical work—a combination guaranteed to cause starvation.

If nutrition is adequate, it is less clear whether adopting an active versus a sedentary lifestyle alters immune responsivity. In healthy people, an acute episode of exercise briefly increases blood leukocyte concentration and transiently enhances neutrophil production of microbicidal reactive oxygen species and natural killer cell activity. However, it remains unproven that regular exercise over time can lower the frequency or reduce the intensity of, for example, upper respiratory tract infections. In HIV-positive men and in men with AIDS and advanced muscle wasting, strength and endurance training yield normal gains. There is also incomplete evidence that training may slow progression to AIDS in HIV-positive men, with a corresponding increase in CD4 lymphocytes.

Exercise May Help Relieve Depression, but Its Efficacy and Neurochemical Effects Are Uncertain

In healthy people, prolonged exercise increases subsequent deep sleep, defined as stages 3 and 4 of slow-wave sleep (see Chapter 7). This effect is apparently mediated entirely through the thermal effects of exercise, since equivalent passive heating produces the same result. Whether or not exercise can improve sleep in patients with insomnia is not known.

Clinical depression is characterized by sleep and appetite dysfunction and profound changes in mood. Whether acute or chronic exercise can help relieve depression remains unproven. The two most prominent biological theories of depression—the dysregulation of central monoamine activity and dysfunction of the hypothalamic-pituitary-adrenal axis—have received almost no study with regard to the impact of exercise.

Panic disorder patients, often characterized by agoraphobia, have reduced exercise capacity. Although sodium lactate infusion does provoke panic in these patients, the anxiety mediator appears to be hypernatremia, not lactate; even strenuous exercise with substantial lactic acidosis will not trigger panic attacks in these individuals.

REVIEW QUESTIONS

DIRECTIONS: Each of the numbered items or incomplete statements in this section is followed by answers or by completions of the statement. Select the ONE lettered answer or completion that is BEST in each case.

1. In an effort to strengthen selected muscles after surgery and immobilization has led to muscle atrophy, isometric exercise is recommended. The intensity of isometric exercise is best quantified
(A) Relative to the maximal oxygen uptake
(B) As mild, moderate, or strenuous
(C) As percentage of the maximum voluntary contraction
(D) In terms of anaerobic metabolism
(E) On the basis of the total muscle mass involved
2. Two people, one highly trained and one not, each exercising at 75% of the maximal oxygen uptake, become fatigued
(A) For similar physiological reasons
(B) Very slowly
(C) At different times
(D) While performing equally well for at least a short period of time
(E) Despite much higher circulating lactic acid levels in the trained person
3. A patient completes a graded, dynamic exercise test on a treadmill while showing a modest rise (25%) in mean arterial blood pressure. In contrast, during the highest level of exercise at the end of the test, an indirect method shows that cardiac output has risen 300% from rest. These results indicate that during graded, dynamic exercise to exhaustion, systemic vascular resistance
(A) Is constant
(B) Rises slightly
(C) Falls only if work is prolonged
(D) Falls dramatically
(E) Cannot be measured
4. A patient with inflammatory bowel disease and compromised kidney function asks if exercise will alter blood flow to either the gastrointestinal tract or to the kidneys. The answer is that vasoconstriction in both the renal and splanchnic vascular beds during exercise
(A) Rarely occurs
(B) Occurs only after prolonged training
(C) Helps maintain arterial blood pressure
(D) Allows renal and splanchnic flows to parallel cerebral blood flow
(E) Will be balanced by local dilation in these vascular beds
5. A young, healthy, highly trained individual enters a marathon (40 km) run on a warm, humid day (32°C, 70% humidity). The best medical advice for this individual is to be concerned about the possibility for
(A) Heat exhaustion
(B) Coronary ischemia
(C) Renal ischemia and anoxia
(D) Hypertension
(E) Gastric mucosal ischemia and increased risk for gastric ulceration
6. An individual with hypertension has been advised to increase physical activity. At the same time, this person has been counseled to avoid activities that substantially increase the systemic arterial blood pressure. In terms of dynamic exercise, this individual should avoid exercise that
(A) Causes fatigue
(B) Is prolonged
(C) Uses untrained muscle groups
(D) Is substituted for isometric exercise
(E) Involves an intermediate muscle mass
7. In a patient with heart disease, a treadmill test involving graded dynamic exercise results in falling blood pressure at each exercise level. Eventually, faintness and dizziness cause termination of the test. These results arise from inadequate cardiac output during exercise because the baroreceptors, during exercise,
(A) Reset blood pressure to a lower level
(B) Are "turned off"
(C) Are increased in sensitivity by training
(D) Are decreased in sensitivity by training
(E) Reset blood pressure to a higher level
8. A man with a family history of heart disease has both diabetes and hypertension. His total serum cholesterol is 270 mg/dL. In addition, his LDL cholesterol is elevated and his HDL cholesterol is reduced, compared with individuals with low cardiovascular disease risk. When exercise and diet are recommended, this individual asks what effect a long-term exercise program will have on the blood lipid profile. The answer is that exercise, over time, will
(A) Have no independent effect on blood cholesterol levels
(B) Elevate both HDL and LDL
(C) Lower HDL and LDL, thereby lowering total cholesterol
(D) Reduce risk of myocardial infarction despite elevated total cholesterol levels
(E) Elevate HDL and lower LDL
9. A healthy individual, aged 60, completes a 500 m freestyle swim at an age-group competition. Breathing hard after the race, she explains that her increased ventilation is a normal response to heavy, dynamic exercise. Her increased ventilation results in
(A) Clinically significant systemic arterial hypoxemia
(B) Normal or reduced arterial P_{CO_2}
(C) Respiratory alkalosis
(D) Respiratory acidosis
(E) Dizziness and decreased cerebral blood flow
10. A 33-year-old woman embarks on an extensive program of daily exercise, with both strenuous dynamic and isometric exercise included. After two years, her maximal voluntary contraction of many major muscle groups and her maximal oxygen uptake, are both increased 30%. Predictably, pulmonary function tests show
(A) A 30% rise in vital capacity
(B) No effect on lung elasticity, inspiratory or expiratory flow rates, or vital capacity
(C) An increase in resting pulmonary diffusing capacity of 30 to 50%
(D) A 25% increase in maximal forced expiratory flow rate
(E) Decreases in residual volume and airways resistance at rest
11. In older adults at risk for falls, osteoporosis, and fractures, a program of weight-bearing exercise
(A) Increases the risk of hip fracture
(B) Decreases bone mineral density
(C) Leaves gait, coordination, proprioception, and reaction time unaltered
(D) Reduces the risk of osteoporosis, falls, and fractures
(E) Is less valuable than dynamic exercise during water immersion
12. A 57-year-old woman, told that she is at risk for osteoporosis, starts an exercise class that emphasizes weight-bearing activities and development of muscle strength. She develops extensive muscle soreness after the first two sessions, indicating that the exercise that she performed
(A) Involved isometric contractions
(B) Produced muscle ischemia
(C) Was actually most effective for increasing muscle endurance

(continued)

(D) Involved eccentric contractions
(E) Required at least 50% of the maximum voluntary contractile force

13. A high-school football player injures a knee early in the season. The knee requires immobilization for six weeks, after which time the athlete undergoes rehabilitation before joining the team. Immediately after rehabilitation begins, the individual notices that the flexors and extensors of the knee are much weaker than before the injury because during contraction at a fixed force
(A) Fewer motor units are involved
(B) There is a relative excess of contractile protein
(C) Muscle cells are small, so more cells are required to perform the same work
(D) Oxidative energy-producing systems are up-regulated
(E) Eccentric work is less, while concentric work is increased

14. A tenth-grade distance runner finishes in the top five of her statewide high school cross-country championships. Encouraged, she redoubles her training intensity, only to find that her menstrual periods cease for nearly a year. After finally visiting her doctor, her serum estrogen levels are found to be well below normal. In addition, it is predictable that this young woman will be found to have
(A) Dynamic exercise endurance less than an untrained person
(B) Weak leg muscles
(C) Normal body weight
(D) No risk for fractures as a result of her young age
(E) Low trabecular bone mass

15. A man with recently diagnosed type 2 diabetes asks for advice about exercise. His specific concern is the impact that an acute episode of exercise will have on his blood glucose levels and insulin requirements. He is correctly informed that during exercise, an important factor to consider is that
(A) Muscle glucose uptake decreases in patients with either type 1 or type 2 diabetes
(B) The pancreas will release increased amounts of both insulin and glucagon
(C) Muscle glucose uptake will increase only if endogenous or exogenous insulin levels rise
(D) Muscle glucose transporters will be translocated to the plasma membrane, increasing insulin-dependent and insulin-independent glucose uptake
(E) Insulin-independent glucose uptake is reduced in active muscles

16. A highly active woman is pregnant for the first time. She asks what benefits might ensue from continued physical activity during pregnancy. Which of the following is a predictable effect of chronic, dynamic exercise during pregnancy?
(A) Increased average gestational length
(B) Increased fetal weight at term
(C) Decreased risk of maternal gestational diabetes
(D) Increased risk of spontaneous abortion during the first trimester
(E) Decreased neonatal responsiveness scores

SUGGESTED READING

Beck LH. Update in preventive medicine. Ann Intern Med 2001;134:128–135.

Berchtold MW, Brinkmeier H, Muntener M. Calcium ion in skeletal muscle: Its crucial role for muscle function, plasticity, and disease. Physiol Rev 2000; 80:1215–1265.

Booth FW, Gordon SE, Carlson CJ, et al. Waging war on modern chronic diseases: Primary prevention through exercise biology. J Appl Physiol 2000; 88:774–787.

Bray MS. Genomics, genes, and environmental interaction: the role of exercise. J Appl Physiol 2000;88:788–792.

Clapp JF 3rd. Exercise during pregnancy. A clinical update. Clin Sports Med 2000;19:273–286.

Fairfield WP, Treat M, Rosenthal DI, et al. Effects of testosterone and exercise on muscle leanness in eugonadal men with AIDS wasting. J Appl Physiol 2001; 90:2166–2171.

Gielen S, Schuler G, Hambrecht R. Exercise training in coronary artery disease and coronary vasomotion. Circulation 2001;103:E1–E6.

Jones NL, Killian KJ. Exercise limitation in health and disease. N Engl J Med 2000;343:632–641.

Marcus R. Role of exercise in preventing and treating osteoporosis. Rheum Dis Clin North Am 2001;27:131–141.

Pedersen BK, Hoffman-Goetz L. Exercise and the immune system: regulation, integration, and adaptation. Physiol Rev 2000;80:1055–1081.

Peters HP, De Vries WR, Vanberge-Henegouwen GP, et al. Potential benefits and hazards of physical activity and exercise on the gastrointestinal tract. Gut 2001;48:435–439.

Ryder JW, Chibalin AV, Zierath JR. Intracellular mechanisms underlying increases in glucose uptake in response to insulin or exercise in skeletal muscle. Acta Physiol Scand 2001; 171:249–257.

CASE STUDIES FOR PART VIII • • •

Case Study for Chapter 29

Heat Exhaustion With Dehydration

A Michigan National Guard infantry unit was sent at the end of May to Louisiana for a field training exercise. Spring in Michigan was cool, but during the exercise in Louisiana, the temperature reached at least 30°C (86°F) every afternoon. At 3:30 PM on the second day of the exercise, a 70-kg infantryman became unsteady and, after a few more steps, sat on the ground. He told his comrades that he was dizzy and had a headache. When they urged him to drink from his canteen, he took a few swallows and said that he was sick in his stomach.

At the field aid station, he is observed to be sweating, his rectal temperature is 38.5°C, and his pulse is rapid. He appears dazed, and his answers to questions are coherent but slow. He cannot produce a urine sample. Blood samples are drawn, and an intravenous drip is started. The laboratory report shows serum [Na^+] of 156 mmol/L (normal range,135 to 145 mmol/L). Two liters of normal saline (0.9% NaCl) are infused over 45 minutes. Well before the end of the infusion, the patient is alert, his nausea disappears, and he asks for, and is given, water to drink. After the end of the infusion he is sent back to his unit with instructions to consume salt with dinner, drink at least three quarts of fluid before going to bed, and to return for follow-up in the morning.

Questions

1. What is the likely basis of the patient's nausea, which also contributes to his inability to produce a urine specimen?

2. If we assume that the patient's total body water was 36 L when he came for treatment, it can be shown that giving the patient 3 L of water without salt (by mouth and/or as an intravenous infusion of glucose in water) would reduce serum [Na^+] to 144 mmol/L. Such treatment would improve the patient's condition considerably. How might the medical officer argue the case for giving 2 L of normal saline?
3. What other (and relatively unusual) condition could produce the patient's symptoms? Did the medical officer rule this possibility out by appropriate means?

Answers to Case Study Questions for Chapter 29

1. The patient's nausea is probably a result of constriction of the splanchnic vascular beds, which is part of the homeostatic cardiovascular response that helps maintain cardiac output and blood pressure when central blood volume is reduced. Central blood volume, in turn, was reduced by the loss of body water and pooling of blood in the peripheral vascular beds. This homeostatic response also includes constriction of the renal vascular beds, which, in turn, contributes (along with the release of vasopressin and activation of the renin-angiotensin system) to scanty urine production.
2. Because the weather was cool back home, the patient probably was probably not acclimatized to heat and was not conserving salt in his sweat. He was probably secreting large amounts of sweat, and losing correspondingly large amounts of salt because of the weather and the activity involved in the exercise. If the patient returns to training the next morning without correcting the salt deficit, he is likely to have further difficulties in the heat. Even if the medical officer has guessed incorrectly about the patient's salt balance, a patient with normal renal function and adequate fluid intake should be able to excrete any excess salt resulting from the treatment.
3. Hyponatremia can produce symptoms similar to the patient's symptoms. However, the medical officer was able to exclude hyponatremia (although not necessarily some degree of salt deficit) on the basis of elevated serum [Na^+]. Giving a hyponatremic patient large volumes of fluid without an equivalent of salt (which would have been a reasonable alternative treatment for the patient in this example) would worsen the hyponatremia, perhaps to a dangerous degree.

Reference

Knochel JP. Clinical complications of body fluid and electrolyte balance. In: Buskirk ER, Puhl SM, eds. Body Fluid Balance: Exercise and Sport. Boca Raton, FL: CRC Press, 1996;297–317.

CASE STUDY FOR CHAPTER 30

A Patient With Dyspnea During Exercise

A 56-year-old man complained of shortness of breath and chest pain when climbing stairs or mowing the lawn. He is subjected to a stress test, with noninvasive monitoring of heart rate, blood pressure, arterial blood oxygen saturation, and cardiac electrical activity. His resting heart rate is 73 beats/min; blood pressure, 118/75 mm Hg; arterial blood oxygen saturation, 96%; and the ECG, normal. After 3.5 minutes of increasingly intense exercise, the test is terminated because of the subject's severe dyspnea. His heart rate is 119 beats/min (his age and sex-adjusted predicted maximal heart rate is 168 beats/min), blood pressure is 146/76 mm Hg, arterial blood oxygen saturation is 88%, and the ECG is normal.

Questions

1. What are three lines of evidence for ventilatory limitation to this subject's exercise?
2. Why did arterial blood oxygen saturation fall during exercise?
3. Why did exhaustion occur before maximal heart rate was reached?
4. Why did the pulse pressure rise in exercise?
5. Why would endurance exercise training likely increase this individual's exercise capacity?

Answers to Case Study Questions for Chapter 30

1. Ventilatory limitation is evidenced by severe dyspnea as a primary symptom in exercise, falling arterial blood oxygenation, and exercise termination at relatively low heart rate.
2. Arterial blood oxygen saturation fell during exercise because increased cardiac output (increased pulmonary blood flow) and decreased pulmonary arterial blood oxygen content (a result of increased skeletal muscle oxygen extraction) increase demands for oxygenation in lungs with inadequate diffusing capacity.
3. Exhaustion occurred before a maximal heart rate was reached because lung disease creates severe dyspnea even in mild exercise.
4. The pulse pressure rose during exercise because sympathetic stimulation and enhanced venous return increase the stroke volume at constant arterial compliance.
5. Endurance exercise training would have little effect on any aspect of lung function. However, training would cause adaptations within exercising muscle that would increase muscle oxidative capacity and reduce lactic acid production. By reducing the ventilatory demands of exercise, these changes would increase exercise capacity in this individual.

PART IX Endocrine Physiology

CHAPTER 31 Endocrine Control Mechanisms

Daniel E. Peavy, Ph.D.

CHAPTER OUTLINE

- GENERAL CONCEPTS OF ENDOCRINE CONTROL
- THE NATURE OF HORMONES
- MECHANISMS OF HORMONE ACTION

KEY CONCEPTS

1. Hormones are chemical substances involved in cell-to-cell communication that promote the maintenance of homeostasis.
2. There are six classes of steroid hormones, based on their primary actions.
3. Most polypeptide hormones are initially synthesized as preprohormones.
4. Steroid hormones and thyroid hormones are generally transported in the bloodstream bound to carrier proteins, whereas most peptide and protein hormones are soluble in the plasma and are carried free in solution.
5. RIA and ELISA have provided major advancements in the field of endocrinology, but each type of assay has limitations.
6. Altered hormone-receptor interactions may lead to endocrine abnormalities.

Endocrinology is the branch of physiology concerned with the description and characterization of processes involved in the regulation and integration of cells and organ systems by a group of specialized chemical substances called hormones. The diagnosis and treatment of a large number of endocrine disorders is an important aspect of any general medical practice. Certain endocrine disease states, such as diabetes mellitus, thyroid disorders, and reproductive disorders, are fairly common in the general population; therefore, it is likely that they will be encountered repeatedly in the practice of medicine.

In addition, because hormones either directly or indirectly affect virtually every cell or tissue in the body, a number of other prominent diseases not primarily classified as endocrine diseases may have an important endocrine component. Atherosclerosis, certain forms of cancer, and even certain psychiatric disorders are examples of conditions in which an endocrine disturbance may contribute to the progression or severity of disease.

GENERAL CONCEPTS OF ENDOCRINE CONTROL

Hormones are bloodborne substances involved in regulating a variety of processes. The word "hormone" is derived from the Greek *hormaein,* which means to "excite" or to "stir up." The endocrine system forms an important communication system that serves to regulate, integrate, and coordinate a variety of different physiological processes. The processes that hormones regulate fall into four areas: (1) the digestion, utilization, and storage of nutrients; (2) growth and development; (3) ion and water balance; and (4) reproductive function.

Hormones Regulate and Coordinate Many Functions

It is difficult to describe hormones in absolute terms. As a working definition, however, it can be said that hormones serve as regulators and coordinators of various biological functions in the animals in which they are produced. They are highly potent, specialized, organic molecules produced by endocrine cells in response to specific stimuli and exert their actions on specific target cells. These target cells are equipped with receptors that bind hormones with high affinity and specificity; when bound, they initiate characteristic biological responses by the target cells.

In the past, definitions or descriptions of hormones usually included a phrase indicating that these substances were secreted into the bloodstream and carried by the blood to a distant target tissue. Although many hormones travel by this mechanism, we now realize that there are many hormones or hormone-like substances that play important roles in cell-to-cell communication that are not secreted directly into the bloodstream. Instead, these substances reach their target cells by diffusion through the interstitial fluid. Recall the discussion of autocrine and paracrine mechanisms in Chapter 1.

Hormone Receptors Determine Whether a Cell Will Respond to a Hormone

In the endocrine system, a hormone molecule secreted into the blood is free to circulate and contact almost any cell in the body. However, only **target cells**, those cells that possess specific receptors for the hormone, will respond to that hormone. A **hormone receptor** is the molecular entity (usually a protein or glycoprotein) either outside or within a cell that recognizes and binds a particular hormone. When a hormone binds to its receptor, biological effects characteristic of that hormone are initiated. Therefore, in the endocrine system, the basis for specificity in cell-to-cell communication rests at the level of the receptor. Similar concepts apply to autocrine and paracrine mechanisms of communication.

A certain degree of specificity is ensured by the restricted distribution of some hormones. For example, several hormones produced by the hypothalamus regulate hormone secretion by the anterior pituitary. These hormones are carried via small blood vessels directly from the hypothalamus to the anterior pituitary, prior to entering the general systemic circulation. The anterior pituitary is, therefore, exposed to considerably higher concentrations of these hypothalamic hormones than the rest of the body; as a result, the actions of these hormones focus on cells of the anterior pituitary. Another mechanism that restricts the distribution of active hormone is the local transformation of a hormone within its target tissue from a less active to a more active form. An example is the formation of dihydrotestosterone from testosterone, occurring in such androgen target tissues as the prostate gland. Dihydrotestosterone is a much more potent androgen than testosterone. Because the enzyme that catalyzes this conversion is found only in certain locations, its cell or tissue distribution partly localizes the actions of the androgens to these sites. Therefore, while receptor distribution is the primary factor in determining the target tissues for a specific hormone, other factors may also focus the actions of a hormone on a particular tissue.

Feedback Regulation Is an Important Part of Endocrine Function

The endocrine system, like many other physiological systems, is regulated by feedback mechanisms. The mechanism is usually negative feedback, although a few positive feedback mechanisms are known. Both types of feedback control occur because the endocrine cell, in addition to synthesizing and secreting its own hormone product, has the ability to sense the biological consequences of secretion of that hormone. This enables the endocrine cell to adjust its rate of hormone secretion to produce the desired level of effect, ensuring the maintenance of homeostasis.

Hormone secretion may be regulated via simple first-order feedback loops or more complex multilevel second- or third-order feedback loops. Since negative feedback is most prevalent in the endocrine system, only examples of this type are illustrated here.

Simple Feedback Loops. First-order feedback regulation is the simplest type and forms the basis for more complex modes of regulation. Figure 31.1A illustrates a simple first-order feedback loop. In this example, an endocrine cell secretes a hormone that produces a specific biological effect in its target tissue. It also senses the magnitude of the effect produced by the hormone. As the biological response increases, the amount of hormone secreted by the endocrine cell is appropriately decreased.

Complex Feedback Loops. More commonly, feedback regulation in the endocrine system is complex, involving second- or third-order feedback loops. For example, multiple levels of feedback regulation may be involved in regulating hormone production by various endocrine glands under the control of the anterior pituitary (Fig. 31.1B). The regulation of target gland hormone secretion, such as adrenal steroids or thyroid hormones, begins with production of a releasing hormone by the hypothalamus. The releasing hormone stimulates production of a trophic hormone by the anterior pituitary, which, in turn, stimulates the production of the target gland hormone by the target gland. As indicated by the dashed lines in Figure 31.1B, the target gland hormone may have negative-feedback effects to inhibit secretion of both the trophic hormone from the anterior pituitary and the releasing hormone from the hypothalamus. In addition, the trophic hormone may inhibit releasing hormone secretion from the hypothalamus, and in some cases, the releasing hormone may inhibit its own secretion by the hypothalamus.

The more complex multilevel form of regulation appears to provide certain advantages compared with the simpler system. Theoretically, it permits a greater degree of fine-tuning of hormone secretion, and the multiplicity of regulatory steps minimizes changes in hormone secretion in the event that one component of the system is not functioning normally.

It is important to bear in mind the normal feedback relationships that control the secretion of each individual hormone are discussed in the chapters that follow. Clinical diagnoses are often made based on the evaluation of

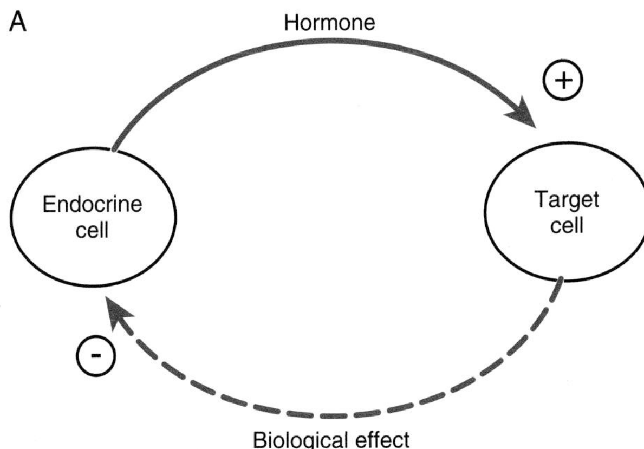

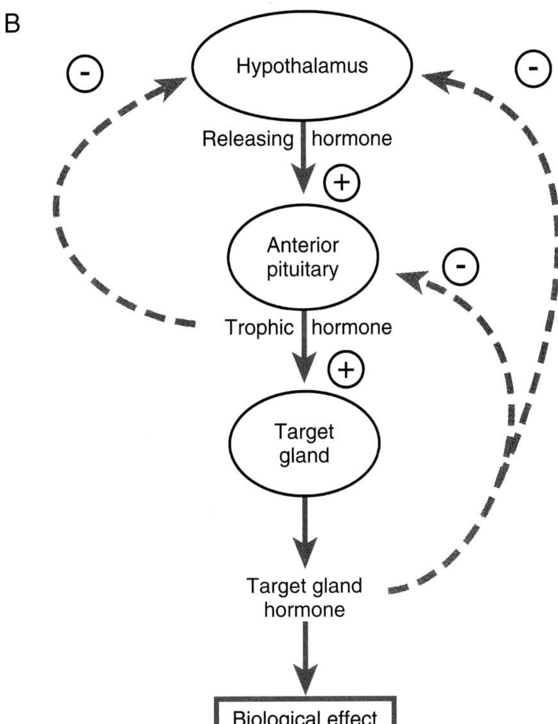

FIGURE 31.1 **Simple and complex feedback loops in the endocrine system. A,** A simple first-order feedback loop. **B,** A complex, multilevel feedback loop: the hypothalamic-pituitary-target gland axis. Solid lines indicate stimulatory effects; dashed lines indicate inhibitory, negative-feedback effects.

hormone-effector pairs relative to normal feedback relationships. For example, in the case of anterior pituitary hormones, measuring both the trophic hormone and the target gland hormone concentration provides important information to help determine whether a defect in hormone production exists at the level of the pituitary or at the level of the target gland. Furthermore, most dynamic tests of endocrine function performed clinically are based on our knowledge of these feedback relationships. Dynamic tests involve a prescribed perturbation of the feedback relationship(s); the range of response in a normal individual is well established, while a response outside the normal range is indicative of abnormal function at some level and greatly enhances information gained from static measurements of hormone concentrations (see Clinical Focus Box 31.1).

Signal Amplification Is an Important Characteristic of the Endocrine System

Another important feature of the endocrine system is **signal amplification**. Blood concentrations of hormones are exceedingly low, generally, 10^{-9} to 10^{-12} mol/L. Even at the higher concentration of 10^{-9} mol/L, only one hormone molecule would be present for roughly every 50 billion water molecules. Therefore, for hormones to be effective regulators of biological processes, amplification must be part of the overall mechanism of hormone action.

Amplification generally results from the activation of a series of enzymatic steps involved in hormone action. At each step, many times more signal molecules are generated than were present at the prior step, leading to a cascade of ever-increasing numbers of signal molecules. The self-multiplying nature of the hormone action pathways provides the molecular basis for amplification in the endocrine system.

Pleiotropic Hormone Effects and Multiplicity of Regulation Also Characterize the Endocrine System

Most hormones have multiple actions in their target tissues and are, therefore, said to have **pleiotropic effects**. For example, insulin exhibits pleiotropic effects in skeletal muscle, where it stimulates glucose uptake, stimulates glycolysis, stimulates glycogenesis, inhibits glycogenolysis, stimulates amino acid uptake, stimulates protein synthesis, and inhibits protein degradation.

In addition, some hormones are known to have different effects in several different target tissues. For example, testosterone, the male sex steroid, promotes normal sperm formation in the testes, stimulates growth of the accessory sex glands, such as the prostate and seminal vesicles, and promotes the development of several secondary sex characteristics, such as beard growth and deepening of the voice.

Multiplicity of regulation is also common in the endocrine system. The input of information from several sources allows a highly integrated response to a variety of stimuli, which is of ultimate benefit to the whole animal. For example, liver glycogen metabolism may be regulated or influenced by several different hormones, including insulin, glucagon, epinephrine, thyroid hormones, and adrenal glucocorticoids.

Hormones Are Often Secreted in Definable Patterns

The secretion of any particular hormone is either stimulated or inhibited by a defined set of chemical substances in

CLINICAL FOCUS BOX 31.1

Growth Hormone and Pulsatile Hormone Secretion

Growth hormone is a 191-amino acid protein hormone that is synthesized and secreted by somatotrophs of the anterior lobe of the pituitary gland. As described in Chapter 32, the hormone plays a role in regulating bone growth and energy metabolism in skeletal muscle and adipose tissue. A deficiency in growth hormone production during adolescence results in dwarfism and overproduction results in gigantism. Measurements of circulating growth hormone levels are, therefore, desirable in children whose growth rate is not appropriate for their age.

Like many other peptide hormones, growth hormone secretion occurs in a pulsatile fashion. The most consistent pulse occurs just after the onset of deep sleep and lasts for about 1 hour. There are usually 4 to 6 irregularly timed pulses throughout the remainder of the day. In order to obtain reliable information about growth hormone secretion, endocrinologists employ a dynamic test of growth hormone secretory capacity. There are several variations of this test that are used at different hospitals. In one test, a bolus of arginine, which is known to stimulate growth hormone secretion, is given and a blood sample is taken a short time later for the measurement of growth hormone concentrations. Another test makes use of the fact that hypoglycemia is a known stimulus for growth hormone secretion. Mild hypoglycemia is induced by an injection of insulin, and a blood sample is drawn a short time later. Regardless of which test is used, by perturbing the system in a well-prescribed fashion, the endocrinologist is able to gain important information about growth hormone secretion that would not be possible if a random blood sample were used.

the blood or environmental factors. In addition to these specific **secretagogues**, many hormones are secreted in a defined, rhythmic pattern. These rhythms can take several forms. For example, they may be pulsatile, episodic spikes in secretion lasting just a few minutes, or they may follow a daily, monthly, or seasonal change in overall pattern. Pulsatile secretion may occur in addition to other longer secretory patterns.

For these reasons, a single randomly drawn blood sample for determining a certain hormone concentration may be of little or no diagnostic value. A dynamic test of endocrine function in which hormone secretion is specifically stimulated by a known agent often provides much more meaningful information.

THE NATURE OF HORMONES

Hormones can be categorized by a number of criteria. Grouping them by chemical structure is convenient, since in many cases hormones with similar structures also use similar mechanisms to produce their biological effects. In addition, hormones with similar chemical structures are usually produced by tissues with similar embryonic origins. Hormones can generally be classed as one of three chemical types.

The Simplest Hormones, in Terms of Structure, Consist of One or Two Modified Amino Acids

Hormones derived from one or two amino acids are small in size and often hydrophilic. These hormones are formed by conversion from a commonly occurring amino acid; epinephrine and thyroxine, for example, are derived from tyrosine. Each of these hormones is synthesized by a particular sequence of enzymes that are primarily localized in the endocrine gland involved in its production. The synthesis of amino acid-derived hormones can, therefore, be influenced in a relatively specific fashion by a variety of environmental or pharmacological agents. The steps involved in the synthesis of these hormones are discussed in detail in later chapters.

Many Hormones Are Polypeptides

Hormones in the polypeptide group are quite diverse in size and complexity. They may be as small as the tripeptide thyrotropin-releasing hormone (TRH) or as large as human chorionic gonadotropin (hCG), which is composed of separate alpha and beta subunits, has a molecular weight of approximately 34 kDa, and is a glycoprotein comprised of 16% carbohydrate by weight.

Within the polypeptide class of hormones are a number of families of hormones, some of which are listed in Table 31.1. Hormones can be grouped into these families as a result of considerable homology with regard to amino acid sequence and structure. Presumably, the similarity of struc-

TABLE 31.1 Examples of Peptide Hormone Families

Insulin Family
Insulin
Insulin-like growth factor I
Insulin-like growth factor II
Relaxin
Glycoprotein Family
Luteinizing hormone (LH)
Follicle-stimulating hormone (FSH)
Thyroid-stimulating hormone (TSH)
Human chorionic gonadotropin (hCG)
Growth Hormone Family
Growth hormone (GH)
Prolactin (PRL)
Human placental lactogen (hPL)
Secretin Family
Secretin
Vasoactive intestinal peptide (VIP)
Glucagon
Gastric inhibitory peptide (GIP)

ture in these families resulted from the evolution of a single ancestral hormone into each of the separate and distinct hormones. In many cases, there is also considerable homology among receptors for the hormones within a family.

Steroid Hormones Are Derived From Cholesterol

Steroids are lipid-soluble, hydrophobic molecules synthesized from cholesterol. They can be classified into six categories, based on their primary biological activity. An example of each category is shown in Figure 31.2.

Glucocorticoids, such as cortisol, are primarily produced in cells of the adrenal cortex and regulate processes involved in glucose, protein, and lipid homeostasis. Glucocorticoids generally produce effects that are catabolic in nature. Aldosterone, a primary example of a **mineralocorticoid**, is produced in cells of the outermost portion of the adrenal cortex. Aldosterone is primarily involved in regulating sodium and potassium balance by the kidneys and is the principal mineralocorticoid in the body.

Androgens, such as testosterone, are primarily produced in the testes, but physiologically significant amounts can be synthesized by the adrenal cortex as well. The primary female sex hormone is estradiol, a member of the **estrogen** family, produced by the ovaries and placenta. **Progestins**, such as progesterone, are involved in maintenance of pregnancy and are produced by the ovaries and placenta.

The **calciferols**, such as 1,25-dihydroxycholecalciferol, are involved in the regulation of calcium homeostasis. 1,25-Dihydroxycholecalciferol is the hormonally active form of vitamin D and is formed by a sequence of reactions occurring in skin, liver, and kidneys.

Polypeptide and Protein Hormones Are Synthesized in Advance of Need and Stored in Secretory Vesicles

Steroid hormones are synthesized and secreted on demand, but polypeptide hormones are typically stored prior to secretion. Steroid hormone synthesis and secretion are dis-

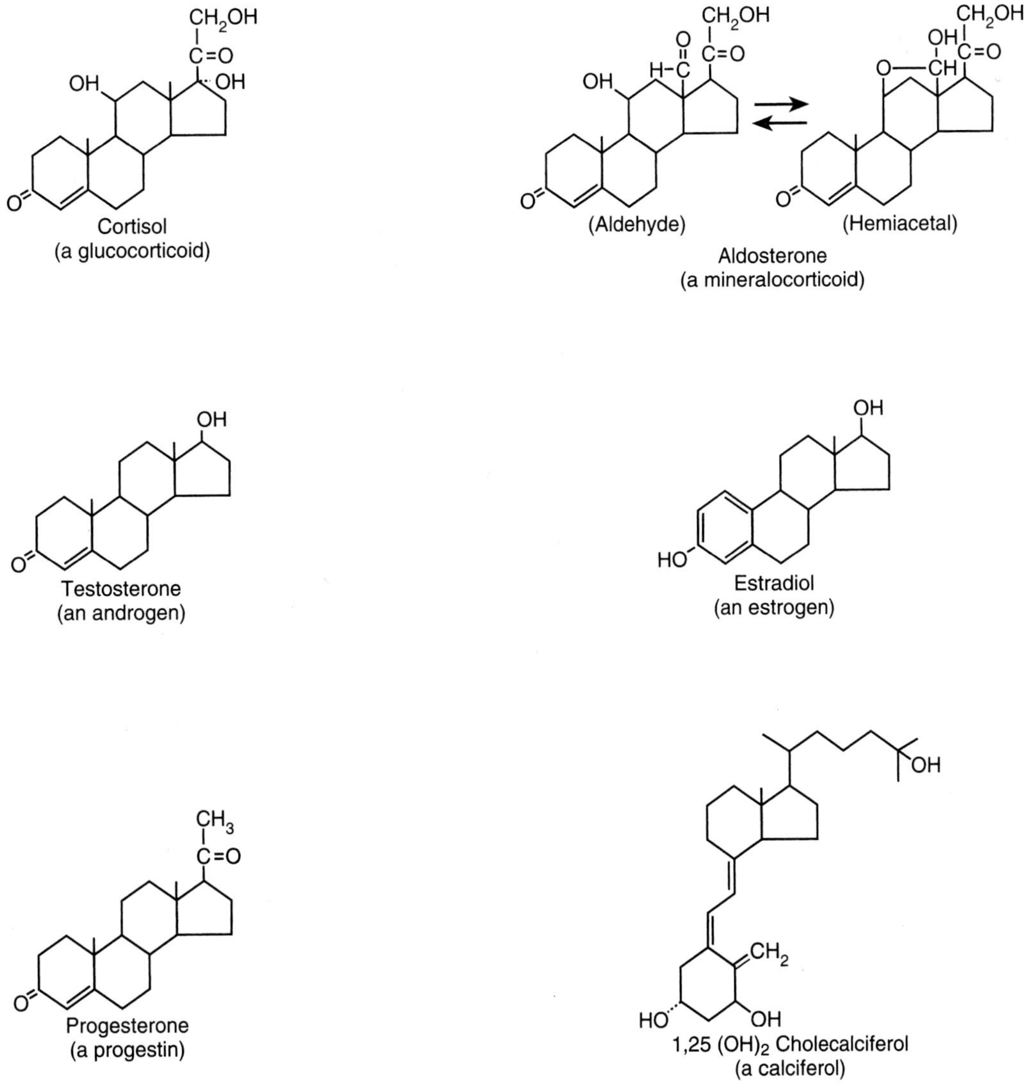

Cortisol
(a glucocorticoid)

(Aldehyde) (Hemiacetal)
Aldosterone
(a mineralocorticoid)

Testosterone
(an androgen)

Estradiol
(an estrogen)

Progesterone
(a progestin)

1,25 $(OH)_2$ Cholecalciferol
(a calciferol)

FIGURE 31.2 **Examples of the six types of naturally occurring steroids.**

cussed in Chapter 34; the discussion here is confined to the synthesis and secretion of polypeptide hormones.

Preprohormones and Prohormones. Like other proteins destined for secretion, polypeptide hormones are synthesized with a *pre-* or *signal* peptide at their amino terminal end that directs the growing peptide chain into the cisternae of the rough ER. Most, if not all, polypeptide hormones are synthesized as part of an even larger precursor or **preprohormone**. The prepeptide is cleaved off upon entry of the preprohormone into the rough ER, to form the **prohormone**. As the prohormone is processed through the Golgi apparatus and packaged into secretory vesicles, it is proteolytically cleaved at one or more sites to yield active hormone. In many cases, preprohormones may contain the sequences for several different biologically active molecules. These active elements may, in some cases, be separated by inactive spacer segments of peptide.

Examples of prohormones that are the precursors for polypeptide hormones, which illustrate the multipotent nature of these precursors, are shown schematically in Figure 31.3. Note, for example, that proopiomelanocortin (POMC) actually contains the sequences for several biologically active signal molecules. Propressophysin serves as the precursor for the nonapeptide hormone arginine vasopressin (AVP). The precursor for TRH contains five repeats of the TRH tripeptide in one single precursor molecule.

In general, two basic amino acid residues, either lys-arg or arg-arg, demarcate the point(s) at which the prohormone will be cleaved into its biologically active components. Presumably, these two basic amino acids serve as specific recognition sites for the trypsin-like endopeptidases thought to be responsible for cleavage of the prohormones. Although somewhat rare, there are documented cases of inherited diseases in which a point mutation involving an amino acid residue at the cleavage site results in an inability to convert the prohormone into active hormone, resulting in a state of hormone deficiency. Partially cleaved precursor molecules having limited biological activity may be found circulating in the blood in some of these cases.

In some disease states, large amounts of intact precursor molecules are found in the circulation. This situation may be the result of endocrine cell hyperactivity or even uncontrolled production of hormone precursor by nonendocrine tumor cells. Although precursors usually have relatively low biological activity, if they are secreted in sufficiently high amounts, they may still produce biological effects. In some cases, these effects may be the first recognized sign of neoplasia.

Tissue-specific differences in the processing of prohormones are well known. Although the same prohormone gene may be expressed in different tissues, tissue-specific differences in the way the molecule is cleaved give rise to different final secretory products. For example, within alpha cells of the pancreas, proglucagon is cleaved at two positions to yield three peptides, illustrated in Figure 31.4 (left). Glucagon, an important hormone in the regulation of carbohydrate metabolism, is the best characterized of the three peptides. In contrast, in other cells of the gastrointestinal (GI) tract in which proglucagon is also produced, the molecule is cleaved at three different positions such that glicentin, glucagon-like peptide-1 (GLP-1), and glucagon-like peptide-2 (GLP-2) are produced (Fig. 31.4, right).

Intracellular Movement of Secretory Vesicles and Exocytosis. Upon insertion of the preprohormone into the cisternae of the ER, the prepeptide or signal peptide is rapidly cleaved from the amino terminal end of the molecule. The resulting prohormone is translocated to the Golgi appara-

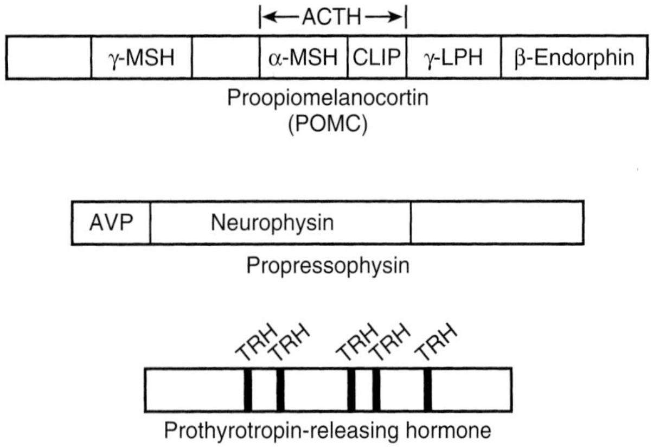

FIGURE 31.3 **The structure of three prohormones.** Relative sizes of individual peptides are only approximations. MSH = melanocyte-stimulating hormone; CLIP = corticotropin-like intermediate lobe peptide; LPH = lipotropin; AVP = arginine vasopressin; TRH = thyrotropin-releasing hormone.

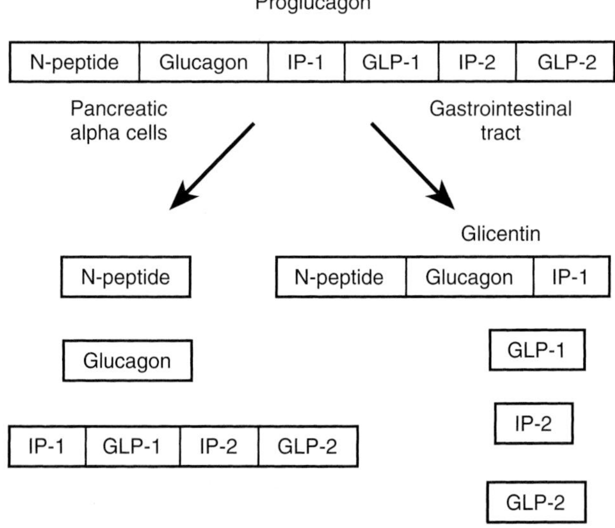

FIGURE 31.4 **The differential processing of prohormones.** In alpha cells of the pancreas (left), the major bioactive product formed from proglucagon is glucagon itself. It is not currently known whether the other peptides are processed to produce biologically active molecules. In intestinal cells (right), proglucagon is cleaved to produce the four peptides shown. Glicentin is the major glucagon-containing peptide in the intestine. IP-1, intervening peptide 1; IP-2, intervening peptide 2; GLP-1, glucagon-like peptide-1; GLP-2, glucagon-like peptide 2.

CLINICAL FOCUS BOX 31.2

Pancreatic Beta Cell Function and C-Peptide

Beta cells of the human pancreas produce and secrete **insulin**. The product of the insulin gene is a peptide known as preproinsulin. As with other secretory peptides, the prepeptide or signal peptide is cleaved off early in the biosynthetic process, yielding proinsulin. Proinsulin is an 86-amino acid protein that is subsequently cleaved at two sites to yield insulin and a 31-amino acid peptide known as **C-peptide**. Insulin and C-peptide are, therefore, localized within the same secretory vesicle and are co-secreted into the bloodstream.

For these reasons, measurements of circulating C-peptide levels can provide a valuable indirect assessment of beta cell insulin secretory capacity. In diabetic patients who are receiving exogenous insulin injections, the measurement of circulating insulin levels would not provide any useful information about their own pancreatic function because it would primarily be the injected insulin that would be measured. However, an evaluation of C-peptide levels in such patients would provide an indirect measure of how well the beta cells were functioning with regard to insulin production and secretion.

tus, where it is processed and packaged for export. After processing in the Golgi apparatus, peptide hormones are stored in membrane-enclosed secretory vesicles. Secretion of the peptide hormone occurs by exocytosis; the secretory vesicle is translocated to the cell surface, its membrane fuses with the plasma membrane, and its contents are released into the extracellular fluid. Movement of the secretory vesicle and membrane fusion are triggered by an increase in cytosolic calcium stemming from an influx of calcium into the cytoplasm from internal organelles or the extracellular fluid. In some cells, an increase in cAMP and the subsequent activation of protein kinases is also involved in the stimulus-secretion coupling process. Elements of the microtubule-microfilament system play a role in the movement of secretory vesicles from their intracellular storage sites toward the cell membrane.

The cleavage of prohormone into active hormone molecules typically takes place during transit through the Golgi apparatus or, perhaps, soon after entry into secretory vesicles. Secretory vesicles, therefore, contain not only active hormone but also the excised biologically inactive fragments. When active hormone is released into the blood, a quantitatively similar amount of inactive fragment is also released. In some instances, this forms the basis for an indirect assessment of hormone secretory activity (see Clinical Focus Box 31.2). Other types of processing of peptide hormones that may occur during transit through the Golgi apparatus include glycosylation and coupling of subunits.

Many Hormones Reach Their Target Cells by Transport in the Bloodstream

According to the classical definition, hormones are carried by the bloodstream from their site of synthesis to their target tissues. However, the manner in which different hormones are carried in the blood varies.

Transport of Amino Acid-Derived and Polypeptide Hormones. Most amino acid-derived and polypeptide hormones dissolve readily in the plasma, and thus no special mechanisms are required for their transport. Steroid and thyroid hormones are relatively insoluble in plasma. Mechanisms are present to promote their solubility in the aqueous phase of the blood and ultimate delivery to a target cell.

Transport of Steroid and Thyroid Hormones. In most cases, 90% or more of steroid and thyroid hormones in the blood are bound to plasma proteins. Some of the plasma proteins that bind hormones are specialized, in that they have a considerably higher affinity for one hormone over another, whereas others, such as serum albumin, bind a variety of hydrophobic hormones. The extent to which a hormone is protein-bound and the extent to which it binds to specific versus nonspecific transport proteins vary from one hormone to another. The principal binding proteins involved in specific and nonspecific transport of steroid and thyroid hormones are listed in Table 31.2. These proteins are synthesized and secreted by the liver, and their production is influenced by changes in various nutritional and endocrine factors.

Typically, for hormones that bind to carrier proteins, only 1 to 10% of the total hormone present in the plasma exists free in solution. However, only this free hormone is biologically active. Bound hormone cannot directly interact with its receptor and, thus, is part of a temporarily inactive pool. However, free hormone and carrier-bound hormone are in a dynamic equilibrium with each other (Fig. 31.5). The size of the free hormone pool and, therefore, the amount available to receptors are influenced not only by changes in the rate of secretion of the hormone but also by the amount of carrier protein available for hormone binding and the rate of degradation or removal of the hormone from the plasma.

TABLE 31.2 Circulating Transport Proteins

Transport Protein	Principal Hormone(s) Transported
Specific	
Corticosteroid-binding globulin (CBG, transcortin)	Cortisol, aldosterone
Thyroxine-binding globulin (TBG)	Thyroxine, triiodothyronine
Sex hormone-binding globulin (SHBG)	Testosterone, estrogen
Nonspecific	
Serum albumin	Most steroids, thyroxine, triiodothyronine
Transthyretin (prealbumin)	Thyroxine, some steroids

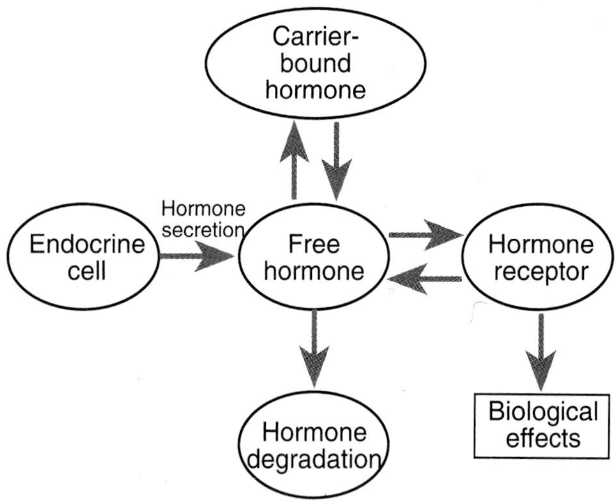

FIGURE 31.5 The relationship between hormone secretion, carrier protein binding, and hormone degradation. This relationship determines the amount of free hormone available for receptor binding and the production of biological effects.

In addition to increasing the total amount of hormone that can be carried in plasma, transport proteins also provide a relatively large reservoir of hormone that buffers rapid changes in free hormone concentrations. As unbound hormone leaves the circulation and enters cells, additional hormone dissociates from transport proteins and replaces free hormone that is lost from the free pool. Similarly, following a rapid increase in hormone secretion or the therapeutic administration of a large dose of hormone, the majority of newly appearing hormone is bound to transport proteins, since under most conditions these are present in considerable excess.

Protein binding greatly slows the rate of clearance of hormones from plasma. It not only slows the entry of hormones into cells, slowing the rate of hormone degradation, but also prevents loss by filtration in the kidneys.

From a diagnostic standpoint, it is important to recognize that most hormone assays are reported in terms of total concentration (i.e., the sum of free and bound hormone), not just free hormone concentration. The amount of transport protein and the total plasma hormone content are known to change under certain physiological or pathological conditions, while the free hormone concentration may remain relatively normal. For example, increased concentrations of binding proteins are seen during pregnancy and decreased concentrations are seen with certain forms of liver or kidney disease. Assays of total hormone concentration might be misleading, since free hormone concentrations may be in the normal range. In such cases, it is helpful to determine the extent of protein binding, so free hormone concentrations can be estimated.

The proportion of a hormone that is free, bound to a specific transport protein, and bound to albumin varies depending on its solubility, its relative affinity for the two classes of transport proteins, and the relative abundance of the transport proteins. For example, the affinity of cortisol for corticosteroid-binding globulin (CBG) is more than 1,000 times greater than its affinity for albumin, but albumin is present in much higher concentrations than CBG. Therefore, about 70% of plasma cortisol is bound to CBG, 20% is bound to albumin, and the remaining 10% is free in solution. Aldosterone also binds to CBG, but with a much lower affinity, such that only 17% is bound to CBG, 47% associates with albumin, and 36% is free in solution.

As this example indicates, more than one hormone may be capable of binding to a specific transport protein. When several such hormones are present simultaneously, they compete for a limited number of binding sites on these transport proteins. For example, cortisol and aldosterone compete for CBG binding sites. Increases in plasma cortisol result in displacement of aldosterone from CBG, raising the unbound (active) concentration of aldosterone in the plasma. Similarly, prednisone, a widely used synthetic corticosteroid, can displace about 35% of the cortisol normally bound to CBG. As a result, with prednisone treatment, the free cortisol concentration is higher than might be predicted from measured concentrations of total cortisol and CBG.

Peripheral Transformation, Degradation, and Excretion of Hormones, in Part, Determine Their Activity

As a general rule, hormones are produced by their gland or tissue of origin in an active form. However, for a few notable exceptions, the peripheral transformation of a hormone plays a very important role in its action.

Peripheral Transformation of Hormones. Specific hormone transformations may be impaired because of a congenital enzyme deficiency or drug-induced inhibition of enzyme activity, resulting in endocrine abnormalities. Well-known transformations are the conversion of testosterone to dihydrotestosterone (see Chapter 37) and the conversion of thyroxine to triiodothyronine (see Chapter 33). Other examples are the formation of the octapeptide angiotensin II from its precursor, angiotensinogen (see Chapter 34), and the formation of 1,25-dihydroxycholecalciferol from cholecalciferol (see Chapter 36).

Mechanisms of Hormone Degradation and Excretion. As in any regulatory control system, it is necessary for the hormonal signal to dissipate or disappear once appropriate information has been transferred and the need for further stimulus has ceased. As described earlier, steady-state plasma concentrations of hormone are determined not only by the rate of secretion but also by the rate of degradation. Thus, any factor that significantly alters the degradation of a hormone can potentially alter its circulating concentration. Commonly, however, secretory mechanisms can compensate for altered degradation such that plasma hormone concentrations remain within the normal range. Processes of hormone degradation show little, if any, regulation; alterations in the rates of hormone synthesis or secretion in most cases provide the primary mechanism for altering circulating hormone concentrations.

For most hormones, the liver is quantitatively the most important site of degradation; for a few others, the kidneys play a significant role as well. Diseases of the liver and kidneys may, therefore, indirectly influence endocrine status as a result of altering the rates at which hormones are removed from the circulation. Various drugs also alter normal rates of hormone degradation; thus, the possibility of indirect drug-induced endocrine abnormalities also exists. In addition to the liver and kidneys, target tissues may take up and degrade quantitatively smaller amounts of hormone. In the case of peptide and protein hormones, this occurs via receptor-mediated endocytosis.

The nature of specific structural modification(s) involved in hormone inactivation and degradation differs for each hormone class. As a general rule, however, specific enzyme-catalyzed reactions are involved. Inactivation and degradation may involve complete metabolism of the hormone to entirely different products, or it may be limited to a simpler process involving one or two steps, such as a covalent modification to inactivate the hormone. Urine is the primary route of excretion of hormone degradation products, but small amounts of intact hormone may also appear in the urine. In some cases, measuring the urinary content of a hormone or hormone metabolite provides a useful, indirect, noninvasive means of assessing endocrine function.

The degradation of peptide and protein hormones has been studied only in a limited number of cases. However, it appears that peptide and protein hormones are inactivated in a variety of tissues by proteolytic attack. The first step appears to involve attack by specific peptidases, resulting in the formation of several distinct hormone fragments. These fragments are then metabolized by a variety of nonspecific peptidases to yield the constituent amino acids, which can be reused.

The metabolism and degradation of steroid hormones has been studied in much more detail. The primary organ involved is the liver, although some metabolism also takes place in the kidneys. Complete steroid metabolism generally involves a combination of one or more of five general classes of reactions: reduction, hydroxylation, side chain cleavage, oxidation, and esterification. Reduction reactions are the principal reactions involved in the conversion of biologically active steroids to forms that possess little or no activity. Esterification (or conjugation) reactions are also particularly important. Groups added in esterification reactions are primarily glucuronate and sulfate. The addition of such charged moieties enhances the water solubility of the metabolites, facilitating their excretion. Steroid metabolites are eliminated from the body primarily via the urine, although smaller amounts also enter the bile and leave the body in the feces.

At times, quantitative information concerning the rate of hormone metabolism is clinically useful. One index of the rate at which a hormone is removed from the blood is the **metabolic clearance rate** (MCR). The metabolic clearance of a hormone is analogous to that of renal clearance (see Chapter 23). The MCR is the volume of plasma cleared of the hormone in question per unit time. It is calculated from the equation:

$$\text{MCR} = \frac{\text{Hormone removed per unit time (mg/min)}}{\text{Plasma concentration (mg/mL)}} \quad (1)$$

and is expressed in mL plasma/min.

One approach to measuring MCR involves injecting a small amount of radioactive hormone into the subject and then collecting a series of timed blood samples to determine the amount of radioactive hormone remaining. Based on the rate of disappearance of hormone from the blood, its half-life and MCR can be calculated. The MCR and half-life are inversely related—the shorter the half-life, the greater the MCR. The half-lives of different hormones vary considerably, from 5 minutes or less for some to several hours for others. The circulating concentration of hormones with short half-lives can vary dramatically over a short period of time. This is typical of hormones that regulate processes on an acute minute-to-minute basis, such as a number of those involved in regulating blood glucose. Hormones for which rapid changes in concentration are not required, such as those with seasonal variations and those that regulate the menstrual cycle, typically have longer half-lives.

The Measurement of Hormone Concentrations Is an Important Tool in Endocrinology

The concentration of hormone present in a biological fluid is often measured to make a clinical diagnosis of a suspected endocrine disease or to study basic endocrine physiology. Substantial advancements have been made in measuring hormone concentrations.

Bioassay. Even before hormones were chemically characterized, they were quantitated in terms of biological responses they produced. Thus, early assays for measuring hormones were bioassays that depended on a hormone's ability to produce a characteristic biological response. As a result, hormones came to be quantitated in terms of units, defined as an amount sufficient to produce a response of specified magnitude under a defined set of conditions. A unit of hormone is, thus, arbitrarily determined. Although bioassays are rarely used today for diagnostic purposes, many hormones are still standardized in terms of **biological activity units**. For example, commercial insulin is still sold and dispensed based on the number of units in a particular preparation, rather than by the weight or the number of moles of insulin.

Bioassays in general suffer from a number of shortcomings, including a relative lack of specificity and a lack of sensitivity. In many cases, they are slow and cumbersome to perform, and often they are expensive, since biological variability often requires the inclusion of many animals in the assay.

Radioimmunoassay. Development of the **radioimmunoassay** (RIA) in the late 1950s and early 1960s was a major step forward in clinical and research endocrinology. Much of our current knowledge of endocrinology is based on this method. A RIA or closely related assay is now available for virtually every known hormone. In addition, RIAs have been developed to measure circulating concentrations of a variety of other biologically relevant proteins, drugs, and vitamins.

The RIA is a prototype for a larger group of assays termed **competitive binding assays**. These are modifications and adaptations of the original RIA, relying to a large degree on the principle of competitive binding on which the RIA is based. It is beyond the scope of this text to describe in detail the competitive binding assays currently used to measure hormone concentrations, but the principles are the same as those for the RIA.

The two key components of a RIA are a specific antibody (Ab) that has been raised against the hormone in question and a radioactively labeled hormone (H*). If the hormone being measured is a peptide or protein, the molecule is commonly labeled with a radioactive iodine atom (^{125}I or ^{131}I) that can be readily attached to tyrosine residues of the peptide chain. For substances lacking tyrosine residues, such as steroids, labeling may be accomplished by incorporating radioactive carbon (^{14}C) or hydrogen (^{3}H). In either case, the use of the radioactive hormone permits detection and quantification of very small amounts of the substance.

The RIA is performed *in vitro* using a series of test tubes. Fixed amounts of Ab and of H* are added to all tubes (Fig. 31.6A). Samples (plasma, urine, cerebrospinal fluid, etc.) to be measured are added to individual tubes. Varying known concentrations of unlabeled hormone (the standards) are added to a series of identical tubes. The principle of the RIA, as indicated in Figure 31.6B, is that labeled and unlabeled hormone compete for a limited number of antibody binding sites. The amount of each hormone that is bound to antibody is a proportion of that present in solution. In a sample containing a high concentration of hormone, less radioactive hormone will be able to bind to the antibody, and in a sample containing a low concentration of hormone, more radioactive hormone will be able to bind to the antibody. In each case, the amount of radioactivity present as antibody-bound H* is determined. The response produced by the standards is used to generate a **standard curve** (Fig. 31.7). Responses produced by the unknown samples are then compared to the standard curve to determine the amount of hormone present in the unknowns (see dashed lines in Fig. 31.7).

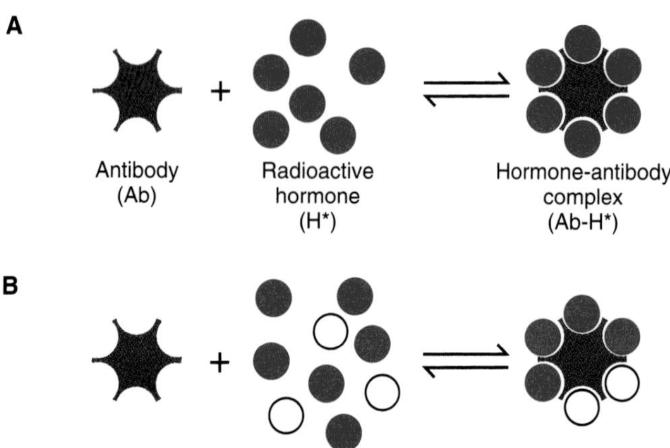

FIGURE 31.6 **The principles of radioimmunoassay (RIA).** A, Specific antibodies (Ab) bind with radioactive hormone (H*) to form hormone-antibody complexes (Ab-H*). **B**, When unlabeled hormone (open circles) is also introduced into the system, less radioactive hormone binds to the antibody. (Modified from Hedge GA, Colby HD, Goodman RL. Clinical Endocrine Physiology. Philadelphia: WB Saunders, 1987.)

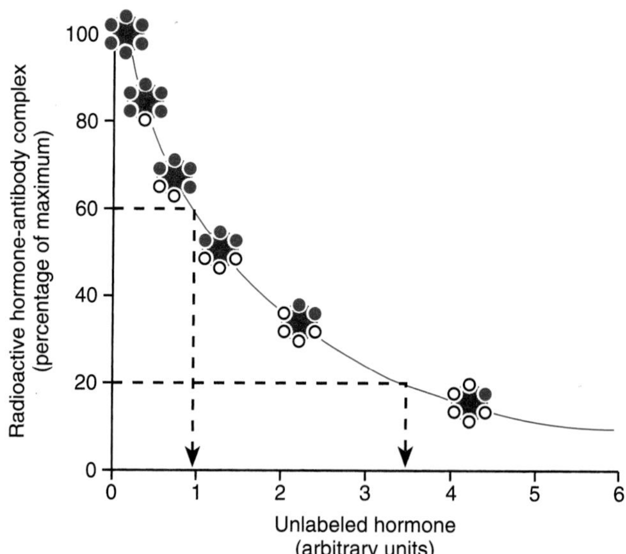

FIGURE 31.7 **A typical RIA standard curve.** As indicated by the dashed lines, the hormone concentration in unknown samples can be deduced from the standard curve. (Modified from Hedge GA, Colby HD, Goodman RL. Clinical Endocrine Physiology. Philadelphia: WB Saunders, 1987.)

One major limitation of RIAs is that they measure immunoreactivity, rather than biological activity. The presence of an immunologically related but different hormone or of heterogeneous forms of the same hormone can complicate the interpretation of the results. For example, POMC, the precursor of ACTH, is often present in high concentrations in the plasma of patients with bronchogenic carcinoma. Antibodies for ACTH may cross-react with POMC. The results of a RIA for ACTH in which such an antibody is used may suggest high concentrations of ACTH, when actually POMC is being detected. Because POMC has less than 5% of the biological potency of ACTH, there may be little clinical evidence of significantly elevated ACTH. If appropriate measures are taken, however, such possible pitfalls can be overcome in most cases, and reliable results from the RIA can be obtained.

One important modification of the RIA is the **radioreceptor assay**, which uses specific hormone receptors rather than antibodies as the hormone-binding reagent. In theory, this method measures biologically active hormone, since receptor binding rather than antibody recognition is assessed. However, the need to purify hormone receptors and the somewhat more complex nature of this assay limit its usefulness for routine clinical measurements. It is more likely to be used in a research setting.

ELISA. The **enzyme-linked immunosorbent assay** (ELISA) is a solid-phase, enzyme-based assay whose use and application have increased considerably over the past two decades. A typical ELISA is a colorimetric or fluorometric assay, and therefore, the ELISA, unlike the RIA, does not produce radioactive waste, which is an advantage, considering environmental concerns and the rapidly increasing cost of radioactive waste disposal. In addition, because it is

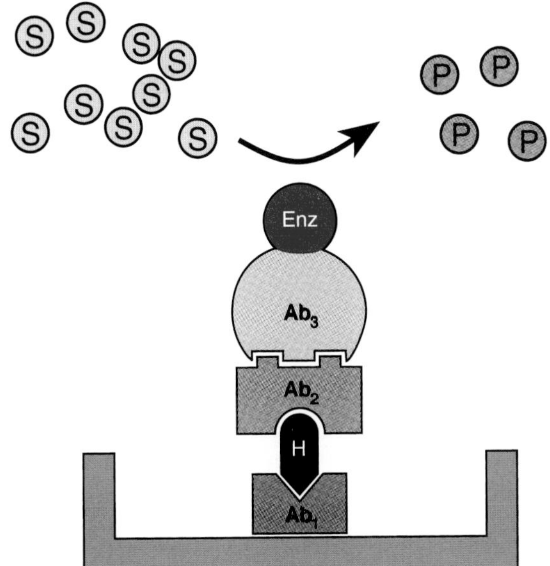

FIGURE 31.8 **The basic components of an ELISA.** A typical ELISA is performed in a 3 × 5-inch plastic plate containing 96 small wells. Each well is precoated with an antibody (Ab_1) that is specific for the hormone (H) being measured. Unknown samples or standards are introduced into the wells, followed by a second hormone-specific antibody (Ab_2). A third antibody (Ab_3), which recognizes Ab_2, is then added. Ab_3 is coupled to an enzyme that will convert an appropriate substrate (S) into a colored or fluorescent product (P). The amount of product formed can be determined using optical methods. After the addition of each antibody or sample to the wells, the plates are incubated for an appropriate period of time to allow antibodies and hormones to bind. Any unbound material is washed out of the well before the addition of the next reagent. The amount of colored product formed is directly proportional to the amount of hormone present in the standard or unknown sample. Concentrations are determined using a standard curve. For simplicity, only one Ab_1 molecule is shown in the bottom of the well, when, in fact, there is an excess of Ab_1 relative to the amount of hormone to be measured.

a solid-phase assay, the ELISA can be automated to a large degree, which reduces costs. Figure 31.8 shows a relatively simple version of an ELISA. More complex assays using similar principles have been developed to overcome a variety of technical problems, but the basic principle remains the same. In recent years the RIA has been the primary assay used clinically; its use has expanded considerably, and it will likely be the predominant assay in the future because of the advantages listed above.

MECHANISMS OF HORMONE ACTION

As indicated earlier, hormones are one mechanism by which cells communicate with one another. Fidelity of communication in the endocrine system depends on each hormone's ability to interact with a specific receptor in its target tissues. This interaction results in the activation (or inhibition) of a series of specific events in cells that results in precise biological responses characteristic of that hormone.

The binding of a hormone to its receptor with subsequent activation of the receptor is the first step in hormone action and also the point at which specificity is determined within the endocrine system. Abnormal interactions of hormones with their receptors are involved in the pathogenesis of a number of endocrine disease states, and therefore, considerable attention has been paid to this aspect of hormone action.

The Kinetics of Hormone-Receptor Binding Determines, in Part, the Biological Response

The probability that a hormone-receptor interaction will occur is related to both the abundance of cellular receptors and the receptor's affinity for the hormone relative to the ambient hormone concentration. The more receptors available to interact with a given amount of hormone, the greater the likelihood of a response. Similarly, the higher the affinity of a receptor for the hormone, the greater the likelihood that an interaction will occur. The circulating hormone concentration is, of course, a function of the rate of hormone secretion relative to hormone degradation.

The association of a hormone with its receptor generally behaves as if it were a simple, reversible chemical reaction that can be described by the following kinetic equation:

$$[H] + [R] \rightleftharpoons [HR] \qquad (2)$$

where [H] is the free hormone concentration, [R] is the unoccupied receptor concentration, and [HR] is the **hormone-receptor complex** (also referred to as bound hormone or occupied receptor).

Assuming a simple chemical equilibrium, it follows that

$$K_a = [HR]/[H] \times [R] \qquad (3)$$

where K_a is the association constant. If R_0 is defined as the total receptor number (i.e., [R] + [HR]), then after substituting and rearranging, we obtain the following relationship:

$$[HR]/[H] = -K_a[HR] + K_aR_0 \qquad (4)$$

Literally translated, this equation states:

$$\frac{\text{Bound hormone}}{\text{Free hormone}} = -K_a \times \text{Bound hormone} + K_a \times \text{Total receptor number} \qquad (5)$$

Notice that equations 4 and 5 have the general form of an equation for a straight line: $y = mx + b$.

To obtain information regarding a particular hormone-receptor system, a fixed number of cells (and, therefore, a fixed number of receptors) is incubated *in vitro* in a series of test tubes with increasing amounts of hormone. At each higher hormone concentration, the amount of receptor-bound hormone is increased until all receptors are occupied by hormone. Receptor number and affinity can be obtained by using the relationships given in equation 5 above and plotting the results as the ratio of receptor-bound hormone to free hormone ([HR]/[H]) as a function of the amount of bound hormone ([HR]). This type of analysis is known as a **Scatchard plot** (Fig. 31.9). In theory, a Scatchard plot of simple, reversible equilibrium binding is a straight line (Fig. 31.9A), with the slope of the line being equal to the negative of the association constant ($-K_a$) and the x-intercept being equal to the total receptor number (R_0). Other equally valid mathematical and graphic methods can be used to analyze hormone-receptor interactions, but the Scatchard plot is probably the most widely used.

In practice, Scatchard plots are not always straight lines but instead can be curvilinear (Fig. 31.9B). Insulin is a classic example of a hormone that gives curved Scatchard plots. One interpretation of this result is that cells contain two separate and distinct classes of receptors, each with a different binding affinity. Typically, one receptor population has a higher affinity but is fewer in number compared to the second population. Therefore, as indicated in Figure 31.9B, $Ka_1 > Ka_2$, but $R0_2 > R0_1$. Computer analysis is often required to fit curvilinear Scatchard plots accurately to a two-site model.

Another explanation for curvilinear Scatchard plots is that occupied receptors influence the affinity of adjacent, unoccupied receptors by **negative cooperativity**. According to this theory, when one hormone molecule binds to its receptor, it causes a decrease in the affinity of nearby unoccupied receptors, making it more difficult for additional hormone molecules to bind. The greater the amount of hormone bound, the lower the affinity of unoccupied receptors. Therefore, as shown in Figure 31.9B, as bound hormone increases, the affinity (slope) steadily decreases. Whether curvilinear Scatchard plots in fact result from two-site receptor systems or from negative cooperativity between receptors is unknown.

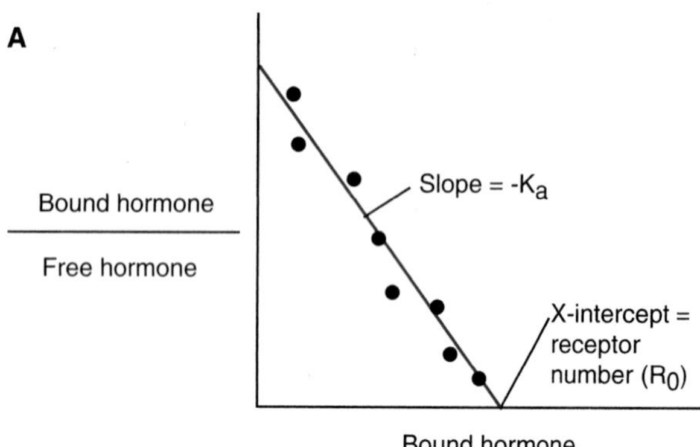

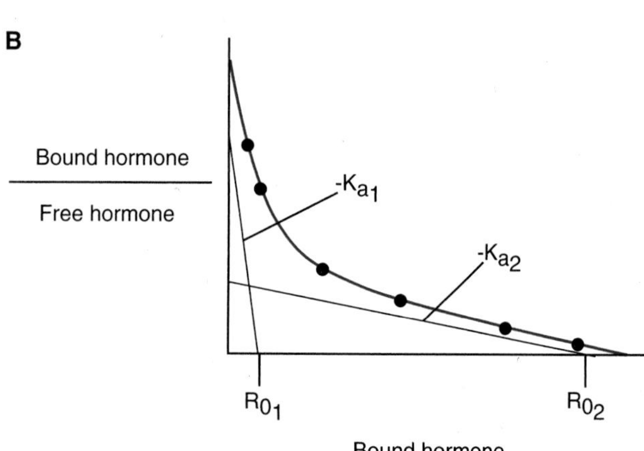

FIGURE 31.9 **Scatchard plots of hormone-receptor binding data. A,** A straight-line plot typical of hormone binding to a single class of receptors. **B,** A curvilinear Scatchard plot typical of some hormones. Several models have been proposed to account for nonlinearity of Scatchard plots. (See text for details.)

Dose-Response Curves Are Useful in Determining Whether There Has Been a Change in Responsiveness or Sensitivity

Hormone effects are generally not all-or-none phenomena—that is, they generally do not switch from totally off to totally on, and then back again. Instead, target cells exhibit graded responses proportional to the concentration of free hormone present.

The dose-response relationship for a hormone generally exhibits a sigmoid shape when plotted as the biological response on the y-axis versus the log of the hormone concentration on the x-axis (Fig. 31.10). Regardless of the biological pathway or process being considered, cells typically exhibit an intrinsic **basal level** of activity in the absence of added hormone, even well after any previous exposure to hormone. As the hormone concentration surrounding the cells increases, a minimal **threshold concentration** must be present before any measurable increase in the cellular response can be produced. At higher hormone concentrations, a **maximal response** by the target cell is produced, and no greater response can be elicited by increasing the hormone concentration. The concentration of hormone required to produce a response half-way between the maximal and basal responses, the **median effective dose** or **ED_{50}**, is a useful index of the **sensitivity** of the target cell for that particular hormone (see Fig. 31.10).

For some peptide hormones, the maximal response may occur when only a small percentage (5 to 10%) of the total receptor population is occupied by hormone. The remaining 90 to 95% of the receptors are called **spare receptors**

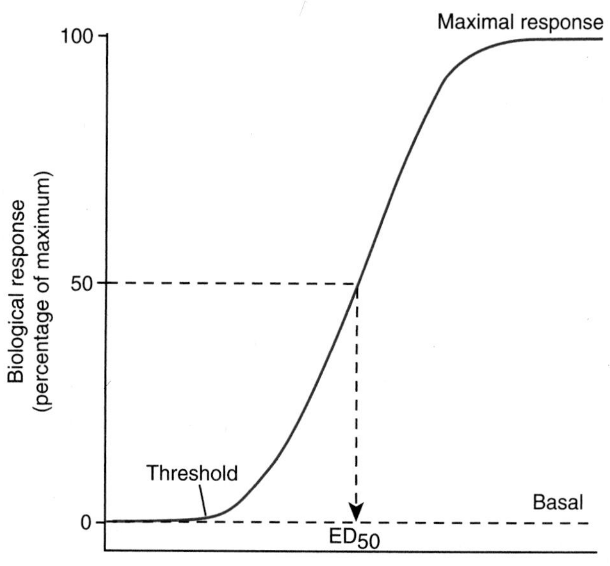

FIGURE 31.10 **A normal dose-response curve of hormone activity.**

because on initial inspection they do not appear necessary to produce a maximal response. This term is unfortunate, because the receptors are not "spare" in the sense of being unused. While at any one point in time only 5 to 10% of the receptors may be occupied, hormone-receptor interactions are an equilibrium process, and hormones continually dissociate and reassociate with their receptors. Therefore, from one point in time to the next, different subsets of the total population of receptors may be occupied, but presumably all receptors participate equally in producing the biological response.

Physiological or pathophysiological alterations in target tissue responses to hormones can take one of two general forms, as indicated by changes in their dose-response curves (Fig. 31.11). Although changes in dose-response curves are not routinely assessed in the clinical setting, they can serve to distinguish between a **receptor abnormality** and a **postreceptor abnormality** in hormone action, providing useful information regarding the underlying cause of a particular disease state. A change in responsiveness is indicated by an increase or decrease in the maximal response of the target tissue and may be the result of one or more factors (Fig. 31.11A). Altered responsiveness can be caused by a change in the number of functional target cells in a tissue, by a change in the number of receptors per cell for the hormone in question or, if receptor function itself is not rate-limiting for hormone action, by a change in the specific rate-limiting postreceptor step in the hormone action pathway.

A change in sensitivity is reflected as a right or left shift in the dose-response curve and, thus, a change in the ED_{50}; a right shift indicates decreased sensitivity and a left shift indicates increased sensitivity for that hormone (Fig. 31.11B). Changes in sensitivity reflect (1) an alteration in receptor affinity or, if submaximal concentrations of hormone are present, (2) a change in receptor number. Dose-response curves may also reflect combinations of changes in responsiveness and sensitivity in which there is both a right or left shift of the curve (a sensitivity change) and a change in maximal biological response to a lower or higher level (a change in responsiveness).

Cells can regulate their receptor number and/or function in several ways. Exposing cells to an excess of hormone for a sustained period of time typically results in a decreased number of receptors for that hormone per cell. This phenomenon is referred to as **down-regulation**. In the case of peptide hormones, which have receptors on cell surfaces, a redistribution of receptors from the cell surface to intracellular sites usually occurs as part of the process of down-regulation. Therefore, there may be fewer total receptors per cell, and a smaller percentage may be available for hormone binding on the cell surface. Although somewhat less prevalent than down-regulation, **up-regulation** may occur when certain conditions or treatments cause an increase in receptor number compared to normal. Changes in rates of receptor synthesis may also contribute to long-term down- or up-regulation.

In addition to changing receptor number, many target cells can regulate receptor function. Chronic exposure of cells to a hormone may cause the cells to become less responsive to subsequent exposure to the hormone by a process termed **desensitization**. If the exposure of cells to a hormone has a desensitizing effect on further action by that same hormone, the effect is termed **homologous desensitization**. If the exposure of cells to one hormone has a desensitizing effect with regard to the action of a different hormone, the effect is termed **heterologous desensitization**.

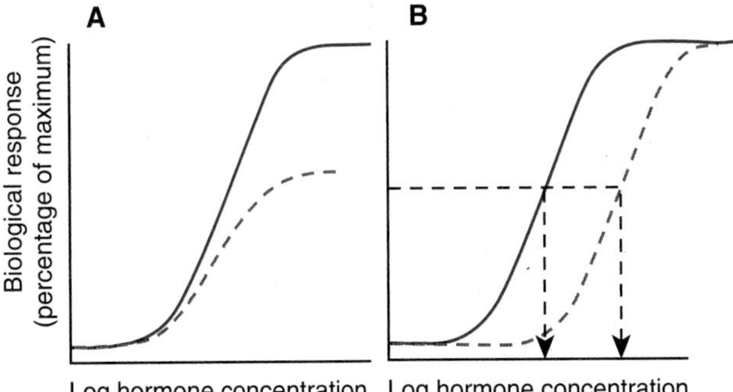

FIGURE 31.11 Altered target tissue responses reflected by dose-response curves. A, Decreased target tissue responsiveness. B, Decreased target tissue sensitivity.

REVIEW QUESTIONS

DIRECTIONS: Each of the numbered items or incomplete statements in this section is followed by answers or by completions of the statement. Select the ONE lettered answer or completion that is BEST in each case.

1. A shift to the right in the biological activity dose-response curve for a hormone with no accompanying change in the maximal response indicates
(A) Decreased responsiveness *and* decreased sensitivity
(B) Increased responsiveness
(C) Decreased sensitivity
(D) Increased sensitivity *and* decreased responsiveness
(E) Increased sensitivity
2. Within the endocrine system, specificity of communication is determined by
(A) The chemical nature of the hormone
(B) The distance between the endocrine cell and its target cell(s)
(C) The presence of specific receptors on target cells
(D) Anatomic connections between the endocrine and target cells
(E) The affinity of binding between the hormone and its receptor
3. The principal mineralocorticoid in the body is
(A) Aldosterone
(B) Testosterone
(C) Progesterone
(D) Prostaglandin E_2
(E) Cortisol
4. An index of the binding affinity of a hormone for its receptor can be obtained by examining the
(A) Y-intercept of a Scatchard plot
(B) Slope of a Scatchard plot

(continued)

(C) Maximum point on a biological dose-response curve
(D) X-intercept of a Scatchard plot
(E) The threshold point of a biological dose-response curve

5. Most peptide and protein hormones are synthesized as
(A) A secretagogue
(B) A pleiotropic hormone
(C) Proopiomelanocortin (POMC)
(D) A preprohormone
(E) Propressophysin

6. The primary form of cortisol in the plasma is that which is
(A) Bound to albumin
(B) Bound to transthyretin
(C) Free in solution
(D) Bound to cortisol receptors
(E) Bound to corticosteroid-binding globulin (CBG)

7. The ability of hormones to be effective regulators of biological function despite circulating at very low concentrations results from
(A) The multiplicity of their effects
(B) Transport proteins
(C) Pleiotropic effects
(D) Signal amplification
(E) Competitive binding

SUGGESTED READING

Goodman HM. Basic Medical Endocrinology. 2nd Ed. New York: Raven, 1994.

Griffin JE, Ojeda SR, eds. Textbook of Endocrine Physiology. 4th Ed. New York: Oxford University Press, 2000.

Hedge GA, Colby HD, Goodman RL. Clinical Endocrine Physiology. Philadelphia: WB Saunders, 1987.

Norman AW, Litwack G. Hormones. 2nd Ed. San Diego: Academic Press, 1997.

Scott JD, Pawson T. Cell communication: The inside story. Sci Am 2000; 282(6):72–79.

Wilson JD, Foster DW, Kronenberg HM, Larsen PR, eds. Williams Textbook of Endocrinology. 9th Ed. Philadelphia: WB Saunders, 1998.

The Hypothalamus and the Pituitary Gland

Robert V. Considine, Ph.D.

CHAPTER OUTLINE

- HYPOTHALAMIC-PITUITARY AXIS
- HORMONES OF THE POSTERIOR PITUITARY
- HORMONES OF THE ANTERIOR PITUITARY

KEY CONCEPTS

1. The hypothalamic-pituitary axis is composed of the hypothalamus, infundibular stalk, posterior pituitary, and anterior pituitary.
2. Arginine vasopressin (AVP) and oxytocin are synthesized in hypothalamic neurons whose axons terminate in the posterior pituitary.
3. AVP increases water reabsorption by the kidneys in response to a rise in blood osmolality or a fall in blood volume.
4. Oxytocin stimulates milk letdown in the breast in response to suckling and muscle contraction in the uterus in response to cervical dilation during labor.
5. The hormones ACTH, TSH, GH, FSH, LH, and PRL are synthesized in the anterior pituitary and secreted in response to hypothalamic releasing hormones carried in the hypophyseal portal circulation.
6. Hypothalamic CRH stimulates ACTH release from corticotrophs, which, in turn, stimulates glucocorticoid release from the adrenal cortex, to comprise the hypothalamic-pituitary-adrenal axis.
7. ACTH secretion is regulated by glucocorticoids, physical and emotional stress, AVP, and the sleep-wake cycle.
8. Hypothalamic TRH stimulates TSH release from thyrotrophs, which, in turn, stimulates T_3 and T_4 release from the thyroid follicles, to comprise the hypothalamic-pituitary-thyroid axis.
9. TSH secretion is regulated by the thyroid hormones, cold temperatures, and the sleep-wake cycle.
10. Hypothalamic GHRH increases and hypothalamic SRIF decreases GH secretion from somatotrophs.
11. GH secretion is regulated by GH itself, IGF-I, aging, deep sleep, stress, exercise, and hypoglycemia.
12. LHRH stimulates the secretion of FSH and LH from the anterior pituitary. These hormones, in turn, affect functions of the ovaries and testes.
13. Dopamine inhibits the secretion of prolactin.

The pituitary gland is a complex endocrine organ that secretes an array of peptide hormones that have important actions on almost every aspect of body function. Some pituitary hormones influence key cellular processes involved in preserving the volume and composition of body fluids. Others bring about changes in body function, which enable the individual to grow, reproduce, and respond appropriately to stress and trauma. The pituitary hormones produce these physiological effects by either acting directly on their target cells or stimulating other endocrine glands to secrete hormones, which, in turn, bring about changes in body function.

Stimuli that affect the secretion of pituitary hormones may originate within or outside the body. These stimuli are perceived and processed by the brain, which signals the pituitary gland to increase or decrease the rate of secretion of a particular hormone. Thus, the brain links the pituitary gland to events occurring within or outside the body, which call for changes in pituitary hormone secretion. This important functional connection between the brain and the pituitary, in which the hypothalamus plays a central role, is called the **hypothalamic-pituitary axis**.

HYPOTHALAMIC-PITUITARY AXIS

The human pituitary is composed of two morphologically and functionally distinct glands connected to the hypothalamus. The **pituitary gland** or **hypophysis** is located at the base of the brain and is connected to the hypothalamus by a stalk. It sits in a depression in the sphenoid bone of the skull called the **sella turcica**. The two morphologically and functionally distinct glands comprising the human pituitary are the adenohypophysis and the neurohypophysis (Fig. 32.1). The **adenohypophysis** consists of the **pars tuberalis**, which forms the outer covering of the pituitary

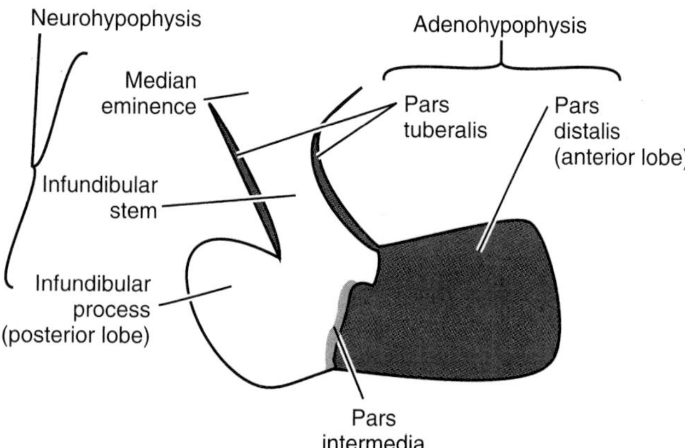

FIGURE 32.1 A midsagittal section of the human pituitary gland.

stalk, and the **pars distalis** or **anterior lobe**. The **neurohypophysis** is composed of the **median eminence** of the hypothalamus, the **infundibular stem**, which forms the inner part of the stalk, and the **infundibular process** or **posterior lobe**. In most vertebrates, the pituitary contains a third anatomically distinct lobe, the **pars intermedia** or **intermediate lobe**. In adult humans, only a vestige of the intermediate lobe is found as a thin diffuse region of cells between the anterior and posterior lobes.

The adenohypophysis and neurohypophysis have different embryological origins. The adenohypophysis is formed from an evagination of the oral ectoderm called **Rathke's pouch**. The neurohypophysis forms as an extension of the developing hypothalamus, which fuses with Rathke's pouch as development proceeds. The posterior lobe is, therefore, composed of neural tissue and is a functional part of the hypothalamus.

Posterior Pituitary Hormones Are Synthesized by Hypothalamic Neurons Whose Axons Terminate in the Posterior Lobe

The infundibular stem of the pituitary gland contains bundles of nonmyelinated nerve fibers, which terminate on the capillary bed in the posterior lobe. These fibers are the axons of neurons that originate in the **supraoptic nuclei** and **paraventricular nuclei** of the hypothalamus. The cell bodies of these neurons are large compared to those of other hypothalamic neurons; hence, they are called **magnocellular neurons**. The hormones arginine vasopressin (AVP) and oxytocin are synthesized as parts of larger precursor proteins (prohormones) in the cell bodies of these neurons. Prohormones are then packaged into granules and enzymatically processed to produce AVP and oxytocin. The granules are transported down the axons by axoplasmic flow; they accumulate at the axon terminals in the posterior lobe.

Stimuli for the secretion of posterior lobe hormones may be generated by events occurring within or outside the body. These stimuli are processed by the central nervous system (CNS), and the signal for the secretion of AVP or oxytocin is then transmitted to neurosecretory neurons in the hypothalamus. Secretory granules containing the hormone are then released into the nearby capillary circulation, from which the hormone is carried into the systemic circulation.

Anterior Pituitary Hormones Are Synthesized and Secreted in Response to Hypothalamic Releasing Hormones Carried in the Hypophyseal Portal Circulation

The anterior lobe contains clusters of histologically distinct types of cells closely associated with blood sinusoids that drain into the venous circulation. These cells produce anterior pituitary hormones and secrete them into the blood sinusoids. The six well-known anterior pituitary hormones are produced by separate kinds of cells. **Adrenocorticotropic hormone** (ACTH), also known as **corticotropin**, is secreted by **corticotrophs**, **thyroid-stimulating hormone** (TSH) by **thyrotrophs**, **growth hormone** (GH) by **somatotrophs**, **prolactin** (PRL) by **lactotrophs**, and **follicle-stimulating hormone** (FSH) and **luteinizing hormone** (LH) by **gonadotrophs**.

The cells that produce anterior pituitary hormones are not innervated and, therefore, are not under direct neural control. Rather, their secretory activity is regulated by **releasing hormones**, also called **hypophysiotropic hormones**, synthesized by neural cell bodies in the hypothalamus. Granules containing releasing hormones are stored in the axon terminals of these neurons, located in capillary networks in the median eminence of the hypothalamus and lower infundibular stem. These capillary networks give rise to the principal blood supply to the anterior lobe of the pituitary.

The blood supply to the anterior pituitary is shown in Figure 32.2. Arterial blood is brought to the hypothalamic-pituitary region by the superior and inferior hypophyseal arteries. The **superior hypophyseal arteries** give rise to a rich capillary network in the median eminence. The capillaries converge into long veins that run down the pituitary stalk and empty into the blood sinusoids in the anterior lobe. They are considered to be portal veins because they deliver blood to the anterior pituitary rather than joining the venous circulation that carries blood back to the heart; therefore, they are called **long hypophyseal portal vessels**. The **inferior hypophyseal arteries** provide arterial blood to the posterior lobe. They also penetrate into the lower infundibular stem, where they form another important capillary network. The capillaries of this network converge into **short hypophyseal portal vessels**, which also deliver blood into the sinusoids of the anterior pituitary. The special blood supply to the anterior lobe of the pituitary gland is known as the **hypophyseal portal circulation**.

When a neurosecretory neuron is stimulated to secrete, the releasing hormone is discharged into the hypophyseal portal circulation (see Fig. 32.2). Releasing hormones travel only a short distance before they come in contact with their target cells in the anterior lobe. Only the amount of releasing hormone needed to control anterior pituitary hormone secretion is delivered to the hypophyseal portal circulation by neurosecretory neurons. Consequently, releasing hormones are almost undetectable in systemic blood.

A releasing hormone either stimulates or inhibits the synthesis and secretion of a particular anterior pituitary

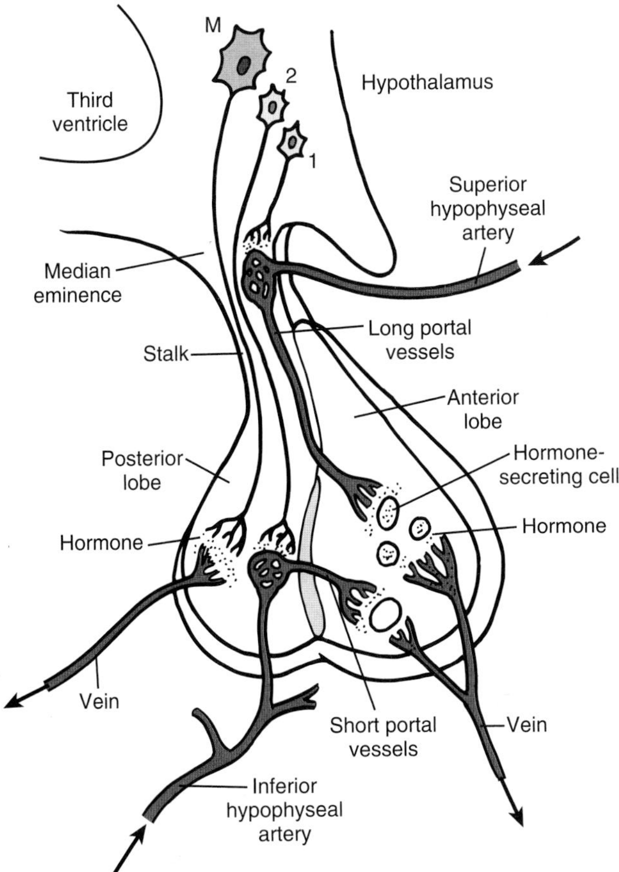

FIGURE 32.2 **The blood supply to the anterior pituitary.** This illustration shows the relationship of the pituitary blood supply to hypothalamic magnocellular neurons and to hypothalamic neurosecretory cells that produce releasing hormones. M represents a magnocellular neuron releasing AVP or oxytocin at its axon terminals into capillaries that give rise to the venous drainage of the posterior lobe. Neurons 1 and 2 are secreting releasing factors into capillary networks that give rise to the long and short hypophyseal portal vessels, respectively. Releasing hormones are shown reaching the hormone-secreting cells of the anterior lobe via the portal vessels.

hormone. **Corticotropin-releasing hormone** (CRH), **thyrotropin-releasing hormone** (TRH), and **growth hormone-releasing hormone** (GHRH) stimulate the synthesis and secretion of ACTH, TSH, and GH, respectively (Table 32.1). **Luteinizing hormone-releasing hormone** (LHRH), also known as **gonadotropin-releasing hormone** (GnRH), stimulates the synthesis and release of FSH and LH. In contrast, **somatostatin**, also called **somatotropin release inhibiting factor** (SRIF), inhibits GH secretion. All of the releasing hormones are peptides, with the exception of **dopamine**, which is a catecholamine that inhibits the synthesis and secretion of PRL. Releasing hormones can be produced synthetically, and several are currently under study for use in the diagnosis and treatment of diseases of the endocrine system. For example, synthetic GnRH is now used for treating infertility in women.

Releasing hormones are secreted in response to neural inputs from other areas of the CNS. These signals are generated by external events that affect the body or by changes occurring within the body itself. For example, sensory nerve excitation, emotional or physical stress, biological rhythms, changes in sleep patterns or in the sleep-wake cycle, and changes in circulating levels of certain hormones or metabolites all affect the secretion of particular anterior pituitary hormones. Signals generated in the CNS by such events are transmitted to the neurosecretory neurons in the hypothalamus. Depending on the nature of the event and the signal generated, the secretion of a particular releasing hormone may be either stimulated or inhibited. In turn, this response affects the rate of secretion of the appropriate anterior pituitary hormone. The neural pathways involved in transmitting these signals to the neurosecretory neurons in the hypothalamus are not well defined.

HORMONES OF THE POSTERIOR PITUITARY

Arginine vasopressin (AVP), also known as antidiuretic hormone (ADH), and **oxytocin** are produced by magnocellular neurons in the supraoptic and paraventricular nuclei of the hypothalamus. Individual neurons make either AVP or

TABLE 32.1 **Hypothalamic Releasing Hormones**

Hormone	Chemistry	Actions on Anterior Pituitary
Corticotropin-releasing hormone (CRH)	Single chain of 41 amino acids	Stimulates ACTH secretion by corticotrophs; stimulates expression of POMC gene in corticotrophs
Thyrotropin-releasing hormone (TRH)	Peptide of 3 amino acids	Stimulates TSH secretion by thyrotrophs; stimulates expression of genes for α and β subunits of TSH in thyrotrophs; stimulates PRL synthesis by lactotrophs
Growth hormone-releasing hormone (GHRH)	Two forms in human: single chain of 44 amino acids, single chain of 40 amino acids	Stimulates GH secretion by somatotrophs; stimulates expression of GH gene in somatotrophs
Luteinizing hormone-releasing hormone (LHRH), gonadotropin-releasing hormone (GnRH)	Single chain of 10 amino acids	Stimulates FSH and LH secretion by gonadotrophs
Somatostatin, somatotropin release inhibiting factor (SRIF)	Single chain of 14 amino acids	Inhibits GH secretion by somatotrophs; inhibits TSH secretion by thyrotrophs
Dopamine	Catecholamine	Inhibits PRL synthesis and secretion by lactotrophs

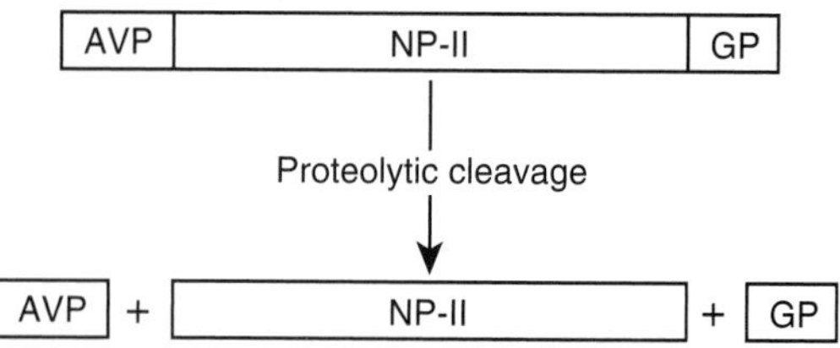

FIGURE 32.3 **The structural organization and proteolytic processing of AVP from its prohormone.** AVP, arginine vasopressin; NP-II, neurophysin II; GP, glycoprotein.

oxytocin, but not both. The axons of these neurons form the infundibular stem and terminate on the capillary network in the posterior lobe, where they discharge AVP and oxytocin into the systemic circulation.

AVP and oxytocin are closely related small peptides, each consisting of nine amino acid residues. Two forms of vasopressin, one containing arginine and the other containing lysine, are made by different mammals. Arginine vasopressin is made in humans. Although AVP and oxytocin differ by only two amino acid residues, the structural differences are sufficient to give these two molecules very different hormonal activities. They are similar enough, however, for AVP to have slight oxytocic activity and for oxytocin to have slight antidiuretic activity.

The genes for AVP and oxytocin are located near one another on chromosome 20. They code for much larger prohormones that contain the amino acid sequences for AVP or oxytocin and for a 93-amino acid peptide called **neurophysin** (Fig. 32.3). The neurophysin coded by the AVP gene has a slightly different structure than that coded by the oxytocin gene. Neurophysin is important in the processing and secretion of AVP, and mutations in the neurophysin portion of the AVP gene are associated with **central diabetes insipidus**, a condition in which AVP secretion is impaired. Prohormones for AVP and oxytocin are synthesized in the cell bodies of magnocellular neurons and transported in secretory granules to axon terminals in the posterior lobe, as described earlier. During the passage of the granules from the Golgi apparatus to axon terminals, prohormones are cleaved by proteolytic enzymes to produce AVP or oxytocin and their associated neurophysins.

When magnocellular neurons receive neural signals for AVP or oxytocin secretion, action potentials are generated in these cells, triggering the release of AVP or oxytocin and neurophysin from the axon terminals. These substances diffuse into nearby capillaries and then enter the systemic circulation.

AVP Increases the Reabsorption of Water by the Kidneys

Two physiological signals, a rise in the osmolality of the blood and a decrease in blood volume, generate the CNS stimulus for AVP secretion. Chemical mediators of AVP release include catecholamines, angiotensin II, and atrial natriuretic peptide (ANP). The main physiological action of AVP is to increase water reabsorption by the collecting ducts of the kidneys. The result is decreased water excretion and the formation of osmotically concentrated urine (see Chapter 23). This action of AVP works to counteract the conditions that stimulate its secretion. For example, reducing water loss in the urine limits a further rise in the osmolality of the blood and conserves blood volume. Low blood AVP levels lead to diabetes insipidus and the excessive production of dilute urine (see Chapter 24).

Oxytocin Stimulates the Contraction of Smooth Muscle in the Mammary Glands and Uterus

Two physiological signals stimulate the secretion of oxytocin by hypothalamic magnocellular neurons. Breast-feeding stimulates sensory nerves in the nipple. Afferent nerve impulses enter the CNS and eventually stimulate oxytocin-secreting magnocellular neurons. These neurons fire in synchrony and release a bolus of oxytocin into the bloodstream. Oxytocin stimulates the contraction of **myoepithelial cells**, which surround the milk-laden alveoli in the lactating mammary gland, aiding in milk ejection.

Oxytocin secretion is also stimulated by neural input from the female reproductive tract during childbirth. Cervical dilation before the beginning of labor stimulates stretch receptors in the cervix. Afferent nerve impulses pass through the CNS to oxytocin-secreting neurons. Oxytocin release stimulates the contraction of smooth muscle cells in the uterus during labor, aiding in the delivery of the newborn and placenta. The actions of oxytocin on the mammary glands and the female reproductive tract are discussed further in Chapter 39.

HORMONES OF THE ANTERIOR PITUITARY

The anterior pituitary secretes six protein hormones, all of which are small, ranging in molecular size from 4.5 to 29 kDa. Their chemical and physiological features are given in Table 32.2.

Four of the anterior pituitary hormones have effects on the morphology and secretory activity of other endocrine glands; they are called *tropic* (Greek meaning "to turn to") or *trophic* ("to nourish") hormones. For example, ACTH maintains the size of certain cells in the adrenal cortex and stimulates these cells to synthesize and secrete **glucocorticoids**, the hormones **cortisol** and **corticosterone**. Similarly, TSH maintains the size of the cells of the thyroid follicles and stimulates these cells to produce and secrete the thyroid hormones **thyroxine** (T_4) and **triiodothyronine** (T_3). The two other tropic hormones, FSH and LH, are called **gonadotropins** because both act on the ovaries and testes. FSH stimulates the development of follicles in the ovaries and regulates the process of **spermatogenesis** in the testes. LH causes **ovulation** and **luteinization** of the ovulated **graafian follicle** in the ovary of the human female and stimulates the production of the female sex hormones **estrogen** and **progesterone** by the ovary. In the male, LH stimulates the **Leydig cells** of the testis to produce and secrete the male sex hormone, **testosterone**.

The two remaining anterior pituitary hormones, GH and PRL, are not usually thought of as tropic hormones because their main target organs are not other endocrine

TABLE 32.2 Hormones of the Anterior Pituitary

Hormone	Chemistry	Physiological Actions
Adrenocorticotropic hormone (ACTH, corticotropin)	Single chain of 39 amino acids; 4.5 kDa	Stimulates production of glucocorticoids and androgens by adrenal cortex; maintains size of zona fasciculata and zona reticularis of cortex
Thyroid-stimulating hormone (TSH, thyrotropin)	Glycoprotein having two subunits, α and β; 28 kDa	Stimulates production of thyroid hormones, T_4 and T_3, by thyroid follicular cells; maintains size of follicular cells
Growth hormone (GH, somatotropin)	Single chain of 191 amino acids; 22 kDa	Stimulates postnatal body growth; stimulates triglyceride lipolysis; inhibits insulin action on carbohydrate and lipid metabolism
Follicle-stimulating hormone (FSH)	Glycoprotein having two subunits, α and β; 28–29 kDa	Stimulates development of ovarian follicles; regulates spermatogenesis in testes
Luteinizing hormone (LH)	Glycoprotein having two subunits, α and β; 28–29 kDa	Causes ovulation and formation of corpus luteum in ovaries; stimulates production of estrogen and progesterone by ovaries; stimulates testosterone production by testes
Prolactin (PRL)	Single chain of 199 amino acids	Essential for milk production by lactating mammary glands

glands. As discussed later, however, these two hormones have certain effects that can be regarded as "tropic." The main physiological action of GH is its stimulatory effect on the growth of the body during childhood. In humans, PRL is essential for the synthesis of milk by the mammary glands during **lactation**.

The following discussion focuses on ACTH, TSH, and GH. Regulation of the secretion of the gonadotropins and PRL, and descriptions of their actions, are given in greater detail in Chapters 37 to 39.

ACTH Regulates the Function of the Adrenal Cortex

The adrenal cortex produces the glucocorticoid hormones, cortisol and corticosterone, in the cells of its two inner zones, the **zona fasciculata** and the **zona reticularis**. These cells also synthesize **androgens** or male sex hormones. The main androgen synthesized is **dehydroepiandrosterone**.

Glucocorticoids act on many processes, mainly by altering gene transcription and, thereby, changing the protein composition of their target cells. Glucocorticoids permit metabolic adaptations during fasting, which prevent the development of **hypoglycemia** or low blood glucose level. They also play an essential role in the body's response to physical and emotional stress. Other actions of glucocorticoids include their inhibitory effect on inflammation, their ability to suppress the immune system, and their regulation of vascular responsiveness to norepinephrine.

Aldosterone, the other physiologically important hormone made by the adrenal cortex, is produced by the cells of the outer zone of the cortex, the **zona glomerulosa**. It acts to stimulate sodium reabsorption by the kidneys.

Adrenocorticotropic hormone (ACTH) is the physiological regulator of the synthesis and secretion of glucocorticoids by the zona fasciculata and zona reticularis. ACTH stimulates the synthesis of these steroid hormones and promotes the expression of the genes for various enzymes involved in steroidogenesis. It also maintains the size and functional integrity of the cells of the zona fasciculata and zona reticularis. ACTH is not an important regulator of aldosterone synthesis and secretion.

The actions of ACTH on glucocorticoid synthesis and secretion and details about the physiological effects of glucocorticoids are described in Chapter 34.

The Structure and Synthesis of ACTH. ACTH, the smallest of the six anterior pituitary hormones, consists of a single chain of 39 amino acids and has a molecular size of 4.5 kDa. ACTH is synthesized in corticotrophs as part of a larger 30-kDa prohormone called **proopiomelanocortin** (POMC). Enzymatic cleavage of POMC in the anterior pituitary results in ACTH, an amino terminal protein, and **β-lipotropin** (Fig. 32.4). β-Lipotropin has effects on lipid metabolism, but its physiological function in humans has not been estab-

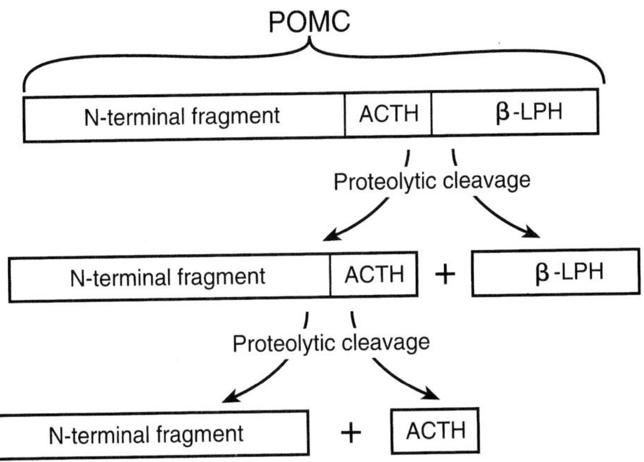

FIGURE 32.4 The proteolytic processing of proopiomelanocortin (POMC) by the human corticotroph. β-LPH, β-lipotropin.

lished. Although POMC can be cleaved into other peptides, such as β-endorphin, only ACTH and β-lipotropin are produced from POMC in the human corticotroph. Proteolytic processing of POMC occurs after it is packaged into secretory granules. Therefore, when the corticotroph receives a signal to secrete, ACTH and β-lipotropin are released into the bloodstream in a 1:1 molar ratio.

POMC is also synthesized by cells of the intermediate lobe of the pituitary gland and neurons in the hypothalamus. In the intermediate lobe, the ACTH sequence of POMC is cleaved to release a small peptide, **α-melanocyte-stimulating hormone** (α-MSH), and, therefore, very little ACTH is produced. α-MSH acts in lower vertebrates to produce temporary changes in skin color by causing the dispersion of melanin granules in pigment cells. As noted earlier, the adult human has only a vestigial intermediate lobe and does not produce and secrete significant amounts of α-MSH or other hormones derived from POMC. However, because ACTH contains the α-MSH amino acid sequence at its N-terminal end, it has melanocyte-stimulating activity when present in the blood at high concentrations. Humans who have high blood levels of ACTH, as a result of **Addison's disease** or an ACTH-secreting tumor, are often hyperpigmented. In the hypothalamus, α-MSH is important in the regulation of feeding behavior.

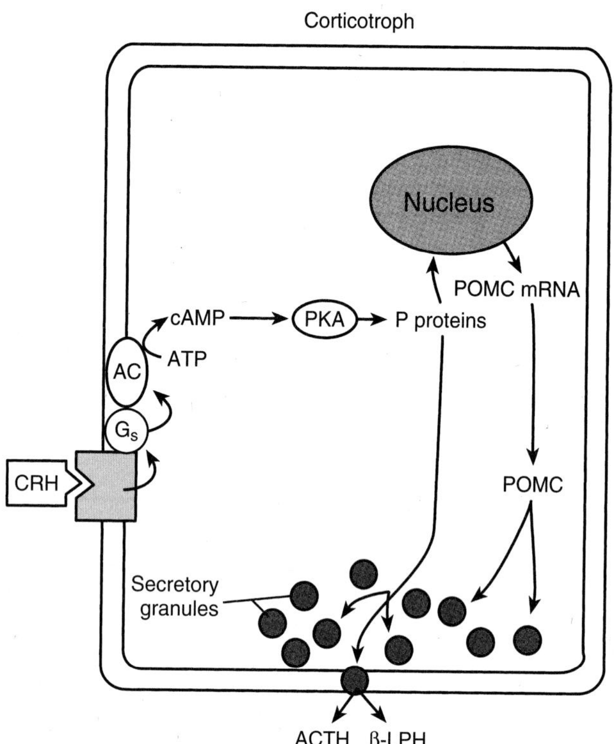

FIGURE 32.5 The main actions of corticotropin-releasing hormone (CRH) on a corticotroph. CRH binds to membrane receptors that are coupled to adenylyl cyclase (AC) by stimulatory G proteins (G_s). Adenylyl cyclase is stimulated, and cAMP rises in the cell. cAMP activates protein kinase A (PKA), which then phosphorylates proteins (P proteins) involved in stimulating ACTH secretion and the expression of the POMC gene.

CRH and ACTH Synthesis and Secretion. Corticotropin-releasing hormone is the main physiological regulator of ACTH secretion and synthesis. In humans, CRH consists of 41 amino acid residues in a single peptide chain.

CRH is synthesized in the paraventricular nuclei of the hypothalamus by a group of neurons with small cell bodies, called **parvicellular neurons**. The axons of parvicellular neurons terminate on capillary networks that give rise to hypophyseal portal vessels. Secretory granules containing CRH are stored in the axon terminals of these cells. Upon receiving the appropriate stimulus, these cells secrete CRH into the capillary network; CRH enters the hypophyseal portal circulation and is delivered to the anterior pituitary gland.

CRH binds to receptors on the plasma membranes of corticotrophs. These receptors are coupled to adenylyl cyclase by stimulatory G proteins. The binding of CRH to its receptor increases the activity of adenylyl cyclase, which catalyzes the formation of cAMP from ATP (Fig. 32.5). The rise in cAMP concentration in the corticotroph activates protein kinase A (PKA), which then phosphorylates cell proteins. PKA-mediated protein phosphorylation stimulates the corticotroph to secrete ACTH and β-lipotropin by unknown mechanisms.

Increased cAMP production in the corticotroph by CRH also stimulates expression of the gene for POMC, increasing the level of POMC mRNA in these cells (see Fig. 32.5). Thus, CRH not only stimulates ACTH secretion but also maintains the capacity of the corticotroph to synthesize the precursor for ACTH.

Glucocorticoids and ACTH Synthesis and Secretion. A rise in glucocorticoid concentration in the blood resulting from the action of ACTH on the adrenal cortex inhibits the secretion of ACTH. Thus, glucocorticoids have a negative-feedback effect on ACTH secretion, which, in turn, reduces the rate of secretion of glucocorticoids by the adrenal cortex. If the blood glucocorticoid level begins to fall for some reason, this negative-feedback effect is reduced, stimulating ACTH secretion and restoring the blood glucocorticoid level. This interactive relationship is called the **hypothalamic-pituitary-adrenal axis** (Fig. 32.6). This control loop ensures that the level of glucocorticoids in the blood remains relatively stable in the resting state, although there is a diurnal variation in glucocorticoid secretion. As discussed later, physical and emotional stress can alter the mechanism regulating glucocorticoid secretion.

The negative-feedback effect of glucocorticoids on ACTH secretion results from actions on both the hypothalamus and the corticotroph (see Fig. 32.6). When the concentration of glucocorticoids rises in the blood, CRH secretion from the hypothalamus is inhibited. As a result, the stimulatory effect of CRH on the corticotroph is reduced and the rate of ACTH secretion falls. Glucocorticoids act directly on parvicellular neurons to inhibit CRH release, and indirectly through neurons in the hippocampus that project to the hypothalamus, to affect the activity of parvicellular neurons. At the corticotroph, glucocorticoids inhibit the actions of CRH to stimulate ACTH secretion.

If the blood concentration of glucocorticoids remains high for a long period of time, expression of the gene for POMC is inhibited. As a result, the amount of POMC mRNA falls in the corticotroph, and gradually the produc-

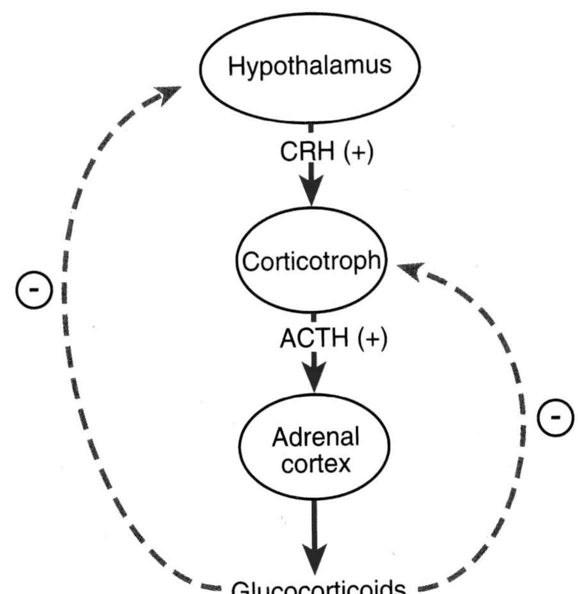

FIGURE 32.6 **The hypothalamic-pituitary-adrenal axis.** The negative-feedback actions of glucocorticoids on the corticotroph and the hypothalamus are indicated by dashed lines.

tion of ACTH and the other POMC peptides declines as well. Since CRH stimulates POMC gene expression and glucocorticoids inhibit CRH secretion, glucocorticoids inhibit POMC gene expression, in part, by suppressing CRH secretion. Glucocorticoids also act directly in the corticotroph itself to suppress POMC gene expression.

The negative-feedback actions of glucocorticoids are essential for the normal operation of the hypothalamic-pituitary-adrenal axis. This relationship is vividly illustrated by the disturbances that occur when blood glucocorticoid levels are changed drastically by disease or glucocorticoid administration. For example, if an individual's adrenal glands have been surgically removed or damaged by disease (e.g., Addison's disease), the resulting lack of glucocorticoids allows corticotrophs to secrete large amounts of ACTH. As noted earlier, this response may result in hyperpigmentation as a result of the melanocyte-stimulating activity of ACTH. Individuals with glucocorticoid deficiency caused by inherited genetic defects affecting enzymes involved in steroid hormone synthesis by the adrenal cortex have high blood ACTH levels resulting from the absence of the negative-feedback effects of glucocorticoids on ACTH secretion. Because a high blood concentration of ACTH causes hypertrophy of the adrenal glands, these genetic diseases are collectively called **congenital adrenal hyperplasia** (see Chapter 34). By contrast, in individuals treated chronically with large doses of glucocorticoids, the adrenal cortex atrophies because the high level of glucocorticoids in the blood inhibits ACTH secretion, resulting in the loss of its trophic influence on the adrenal cortex.

Stress and ACTH Secretion. The hypothalamic-pituitary-adrenal axis is greatly influenced by stress. When an individual experiences physical or emotional stress, ACTH secretion is increased. As a result, the blood level of glucocorticoids rises rapidly. Regardless of the blood glucocorticoid concentration, stress stimulates the hypothalamic-pituitary-adrenal axis because stress-induced neural activity generated at higher CNS levels stimulates parvicellular neurons in the paraventricular nuclei to secrete CRH at a greater rate. Thus, stress can override the normal operation of the hypothalamic-pituitary-adrenal axis. If the stress persists, the blood glucocorticoid level remains high because the glucocorticoid negative-feedback mechanism functions at a higher set point.

AVP and ACTH Secretion. Glucocorticoid deficiency and certain types of stress also increase the concentration of arginine vasopressin (AVP) in hypophyseal portal blood. The physiological significance is that AVP, like CRH, can stimulate corticotrophs to secrete ACTH. Acting along with CRH, AVP amplifies the stimulatory effect of CRH on ACTH secretion.

AVP interacts with a specific receptor on the plasma membrane of the corticotroph. These receptors are coupled to the enzyme **phospholipase C** (PLC) by G proteins. The interaction of AVP with its receptor activates PLC, which, in turn, hydrolyzes phosphatidylinositol 4,5-bisphosphate (PIP_2) present in the plasma membrane. This generates the intracellular second messengers **inositol trisphosphate** (IP_3) and **diacylglycerol** (DAG). IP_3 mobilizes intracellular calcium stores and DAG activates the phospholipid- and calcium-dependent protein kinase C (PKC) to mediate the stimulatory effect of AVP on ACTH secretion.

As noted earlier, AVP and oxytocin are produced by magnocellular neurons of the supraoptic and paraventricular nuclei of the hypothalamus. These neurons terminate in the posterior lobe, where they secrete AVP and oxytocin into capillaries that feed into the systemic circulation. However, parvicellular neurons in the paraventricular nuclei also produce AVP, which they secrete into hypophyseal portal blood. It appears that much of the AVP secreted by parvicellular neurons is made in the same cells that produce CRH. It is assumed that the AVP in hypophyseal portal blood comes from these cells and from a small number of AVP-producing magnocellular neurons whose axons pass through the median eminence of the hypothalamus on their way to the posterior lobe.

The Sleep-Wake Cycle and ACTH Secretion. Under normal circumstances, the hypothalamic-pituitary-adrenal axis in humans functions in a pulsatile manner, resulting in several bursts of secretory activity over a 24-hour period. This pattern appears to be due to rhythmic activity in the CNS, which causes bursts of CRH secretion and, in turn, bursts of ACTH and glucocorticoid secretion (Fig. 32.7). A diurnal oscillation in secretory activity of the axis is thought to be due to changes in the sensitivity of CRH-producing neurons to the negative-feedback action of glucocorticoids, altering their rate of CRH secretion. As a result, there is a diurnal oscillation in the rate of ACTH and glucocorticoid secretion. This **circadian rhythm** is reflected in the daily pattern of glucocorticoid secretion. In individuals who are awake during the day and sleep at night, the blood gluco-

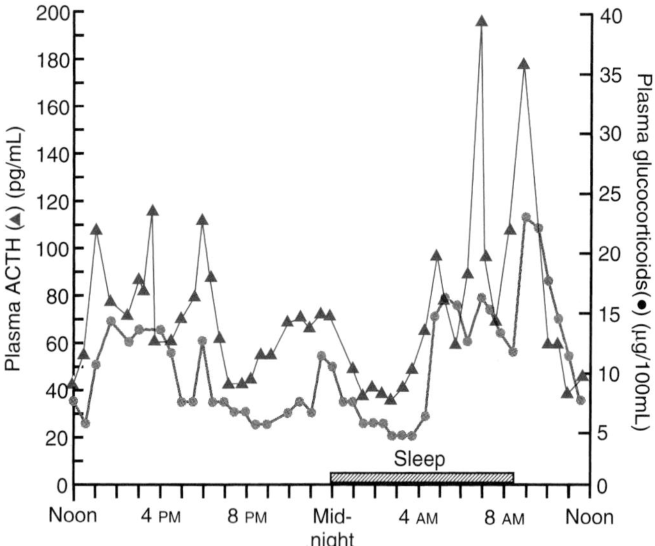

FIGURE 32.7 **ACTH secretion and the sleep-wake cycle.** Pulsatile changes in the concentrations of ACTH and glucocorticoids in the blood of a young woman over a 24-hour period. Note that the amplitude of the pulses in ACTH and glucocorticoids is lower during the evening hours and increases greatly during the early morning hours. This pattern is due to the diurnal oscillation of the hypothalamic-pituitary-adrenal axis. (Modified from Krieger DT. Rhythms in CRF, ACTH and corticosteroids. In: Krieger DT, ed. Endocrine Rhythms. New York: Raven, 1979.)

corticoid level begins to rise during the early morning hours, reaches a peak sometime before noon, and then falls gradually to a low level around midnight (see Fig. 32.7). This pattern is reversed in individuals who sleep during the day and are awake at night. This inherent biological rhythm is superimposed on the normal operation of the hypothalamic-pituitary-adrenal axis.

TSH Regulates the Function of the Thyroid Gland

The thyroid gland is composed of aggregates of **follicles**, which are formed from a single layer of cells. The follicular cells produce and secrete thyroxine (T_4) and triiodothyronine (T_3), thyroid hormones that are iodinated derivatives of the amino acid tyrosine. The thyroid hormones act on many cells by changing the expression of certain genes, changing the capacity of their target cells to produce particular proteins. These changes are thought to bring about the important actions of the thyroid hormones on the differentiation of the CNS, on body growth, and on the pathways of energy and intermediary metabolism.

Thyroid-stimulating hormone (TSH) is the physiological regulator of T_4 and T_3 synthesis and secretion by the thyroid gland. It also promotes nucleic acid and protein synthesis in the cells of the thyroid follicles, maintaining their size and functional integrity. The actions of TSH on thyroid hormone synthesis and secretion, and the physiological effects of the thyroid hormones, are described in detail in Chapter 33.

The Structure and Synthesis of TSH.

TSH is a glycoprotein consisting of two structurally different subunits. The α **subunit** of TSH is a single peptide chain of 92 amino acid residues with two carbohydrate chains linked to its structure. The β **subunit** is a single peptide chain of 112 amino acid residues, to which a single carbohydrate chain is linked. The α and β subunits are held together by noncovalent bonds. The two subunits combined give the TSH molecule a molecular weight of about 28,000. Neither subunit has significant TSH activity by itself. The two subunits must be combined in a 1:1 ratio to form an active hormone. The gonadotropins FSH and LH are also composed of two noncovalently combined subunits. The α subunits of TSH, FSH, and LH are derived from the same gene and are identical, but the β subunit gives each hormone its particular set of physiological activities.

Thyrotrophs synthesize the peptide chains of the α and β subunits of TSH from separate mRNA molecules, which are transcribed from two different genes. The peptide chains of the α and β subunits are combined and undergo glycosylation in the rough ER. These processes are completed as TSH molecules pass through the Golgi apparatus and are packaged into secretory granules. Normally, thyrotrophs make more α subunits than β subunits. As a result, secretory granules contain excess α subunits. When a thyrotroph is stimulated to secrete TSH, it releases both TSH and free α subunits into the bloodstream. In contrast, very little free TSH β subunit is in the blood.

TRH and TSH Synthesis and Secretion.

Thyrotropin-releasing hormone (TRH) is the main physiological stimulator of TSH secretion and synthesis by thyrotrophs. TRH is a small peptide consisting of three amino acid residues produced by neurons in the hypothalamus. These neurons terminate on the capillary networks that give rise to the hypophyseal portal vessels. Normally, these neurons secrete TRH into the hypophyseal portal circulation at a constant or tonic rate. It is assumed that the TRH concentration in the blood that perfuses the thyrotrophs does not change greatly; therefore, the thyrotrophs are continuously exposed to TRH.

TRH binds to receptors on the plasma membranes of thyrotrophs. These receptors are coupled to PLC by G proteins (Fig. 32.8). The interaction of TRH with its receptor activates PLC, causing the hydrolysis of PIP_2 in the membrane. This action releases the intracellular messengers IP_3 and DAG. IP_3 causes the concentration of Ca^{2+} in the cytosol to rise, which stimulates the secretion of TSH into the blood.

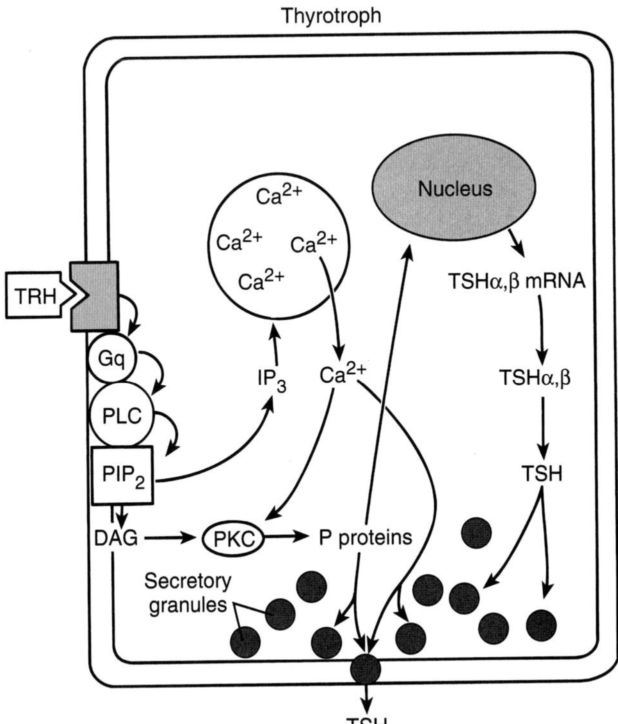

FIGURE 32.8 **The actions of TRH on a thyrotroph.** TRH binds to membrane receptors, which are coupled to phospholipase C (PLC) by G proteins (G_q). PLC hydrolyzes phosphatidylinositol 4,5-bisphosphate (PIP_2) in the plasma membrane, generating inositol trisphosphate (IP_3) and diacylglycerol (DAG). IP_3 mobilizes intracellular stores of Ca^{2+}. The rise in Ca^{2+} stimulates TSH secretion. Ca^{2+} and DAG activate protein kinase C (PKC), which phosphorylates proteins (P proteins) involved in stimulating TSH secretion and the expression of the genes for the α and β subunits of TSH.

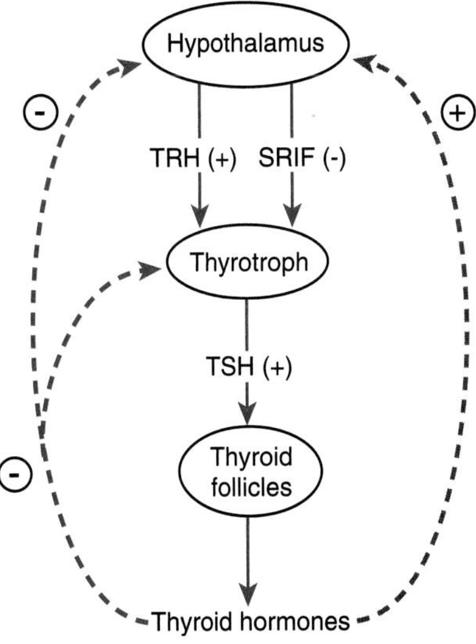

FIGURE 32.9 **The hypothalamic-pituitary-thyroid axis.** TRH stimulates and somatostatin (SRIF) inhibits TSH release by acting directly on the thyrotroph. The negative-feedback loops (-), shown in red, inhibit TRH secretion and action on the thyrotroph, causing a decrease in TSH secretion. The feedback loops (+), shown in gray, stimulate somatostatin secretion, causing a decrease in TRH secretion. SRIF, somatostatin, or somatotropin release inhibiting factor.

The rise in cytosolic Ca^{2+} and the increase in DAG activate PKC in thyrotrophs. PKC phosphorylates proteins that are in some way involved in stimulating TSH secretion.

TRH also stimulates the expression of the genes for the α and β subunits of TSH (see Fig. 32.8). As a result, the amount of mRNA for the α and β subunits is maintained in the thyrotroph and the production of TSH is fairly constant.

Thyroid Hormones and TSH Synthesis and Secretion.

The thyroid hormones exert a direct negative-feedback effect on TSH secretion. For example, when the blood concentration of thyroid hormones is high, the rate of TSH secretion falls. In turn, the stimulatory effect of TSH on the follicular cells of the thyroid is reduced, resulting in a decrease in T_4 and T_3 secretion. However, when the circulating levels of T_4 and T_3 are low, their negative-feedback effect on TSH release is reduced and more TSH is secreted from thyrotrophs, increasing the rate of thyroid hormone secretion. This control system is part of the **hypothalamic-pituitary-thyroid axis** (Fig. 32.9).

The thyroid hormones exert negative-feedback effects on both the hypothalamus and the pituitary. In the hypothalamic TRH-secreting neurons, thyroid hormones reduce TRH mRNA and TRH prohormone to decrease TRH release. The thyroid hormones also increase the release of somatostatin from the hypothalamus. Somatostatin (SRIF) inhibits the release of TSH from the thyrotroph (see Fig. 32.9). In the pituitary, thyroid hormones reduce the sensitivity of the thyrotroph to TRH and inhibit TSH synthesis.

The negative-feedback effects of the thyroid hormones on thyrotrophs are produced primarily through the actions of T_3. Both T_4 and T_3 circulate in the blood bound to plasma proteins, with only a small percentage (less than 1%) unbound or free (see Chapter 33). The free T_4 and T_3 molecules are taken up by thyrotrophs, and T_4 is converted to T_3 by the enzymatic removal of one iodine atom. The newly formed T_3 molecules and those taken up directly from the blood enter the nucleus, where they bind to thyroid hormone receptors in the chromatin. The interaction of T_3 with its receptors changes the expression of specific genes in the thyrotroph, which decreases the cell's ability to produce and secrete TSH. For example, T_3 inhibits the expression of the genes for the α and β subunits of TSH, directly decreasing the synthesis of TSH. Also, T_3 influences the expression of other unidentified genes that code for proteins that decrease thyrotroph sensitivity to TRH. The loss in sensitivity is thought to be partly due to a reduction in the number of TRH receptors in thyrotroph plasma membranes.

Other Factors Affecting TSH Secretion.

The exposure of certain animals to a cold environment stimulates TSH secretion. This makes sense from a physiological perspective because the thyroid hormones are important in regulating body heat production (see Chapter 33). Brief exposure of experimental animals to a cold environment stimulates the secretion of TSH, presumably a result of enhanced TRH secretion. Newborn humans behave much the same way, in that they respond to brief cold exposure with an increase in TSH secretion. This response to cold does not occur in adult humans.

The hypothalamic-pituitary-thyroid axis, like the hypothalamic-pituitary-adrenal axis, follows a diurnal circadian rhythm in humans. Peak TSH secretion occurs in the early morning and a low point is reached in the evening. Physical and emotional stress can alter TSH secretion but the effects of stress on the hypothalamic-pituitary-thyroid axis are not as pronounced as on the hypothalamic-pituitary-adrenal axis.

GH Regulates Growth During Childhood and Remains Important Throughout Life

As its name implies, growth hormone (GH) promotes the growth of the human body. It does not appear to stimulate fetal growth, nor is it an important growth factor during the first few months after birth. Thereafter, it is essential for the normal rate of body growth during childhood and adolescence.

Growth hormone (also called somatotropin) is secreted by the anterior pituitary throughout life and remains physiologically important even after growth has stopped. In addition to its growth-promoting action, GH has effects on many aspects of carbohydrate, lipid, and protein metabolism. For example, GH is thought to be one of the physiological factors that counteract and, thus, modulate some of the actions of insulin on the liver and peripheral tissues.

The Structure and Synthesis of Human GH.

Human GH is a globular 22 kDa protein consisting of a single chain of 191 amino acid residues with two intrachain disulfide bridges. Human GH has considerable structural similarity to human PRL and placental lactogen.

Growth hormone is produced in somatotrophs of the anterior pituitary. It is synthesized in the rough ER as a larger prohormone consisting of an N-terminal signal peptide and the 191-amino acid hormone. The signal peptide is then cleaved from the prohormone, and the hormone traverses the Golgi apparatus and is packaged in secretory granules.

Hypothalamic growth hormone-releasing hormone (GHRH) regulates the production of GH by stimulating the expression of the GH gene in somatotrophs. Expression of the GH gene is also stimulated by thyroid hormones. As a result, the normal rate of GH production depends on these hormones. For example, a thyroid hormone deficient individual is also GH-deficient. This important action of thyroid hormones is discussed further in Chapter 33.

Regulation of GH Secretion by GHRH and Somatostatin.

The secretion of GH is regulated by two opposing hypothalamic releasing hormones. GHRH stimulates GH secretion and somatostatin inhibits GH secretion by inhibiting the action of GHRH. The rate of GH secretion is determined by the net effect of these counteracting hormones on somatotrophs. When GHRH predominates, GH secretion is stimulated. When somatostatin predominates, GH secretion is inhibited.

Human GHRH is a peptide composed of a single chain of 44 amino acid residues. A slightly smaller version of GHRH consisting of 40 amino acid residues is also present in humans. GHRH is synthesized in the cell bodies of neurons in the **arcuate nuclei** and **ventromedial nuclei** of the hypothalamus. The axons of these cells project to the capillary networks giving rise to the portal vessels. When these neurons receive a stimulus for GHRH secretion, they discharge GHRH from their axon terminals into the hypophyseal portal circulation.

GHRH binds to receptors in the plasma membranes of somatotrophs (Fig. 32.10). These receptors are coupled to adenylyl cyclase by a stimulatory G protein, G_s. The interaction of GHRH with its receptors activates adenylyl cyclase, increasing the concentration of cyclic AMP (cAMP) in the somatotroph. The rise in cAMP activates protein kinase A (PKA), which, in turn, phosphorylates proteins that stimulate GH secretion and GH gene expression. GHRH binding to its receptor also increases intracellular Ca^{2+}, which stimulates GH secretion. In addition, some evidence suggests that GHRH may stimulate PLC, causing the hy-

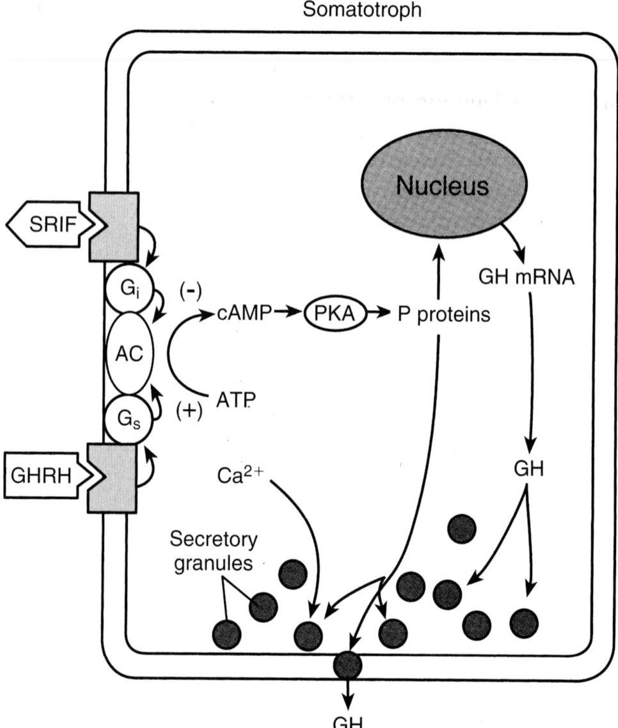

FIGURE 32.10 **The actions of GHRH and somatostatin on a somatotroph.** GHRH binds to membrane receptors that are coupled to adenylyl cyclase (AC) by stimulatory G proteins (G_s). Cyclic AMP (cAMP) rises in the cell and activates protein kinase A (PKA), which then phosphorylates proteins (P proteins) involved in stimulating GH secretion and the expression of the gene for GH. Ca^{2+} is also involved in the action of GHRH on GH secretion. The possible involvement of the phosphatidylinositol pathway in GHRH action is not shown. Somatostatin (SRIF) binds to membrane receptors that are coupled to adenylyl cyclase by inhibitory G proteins (G_i). This action inhibits the ability of GHRH to stimulate adenylyl cyclase, blocking its action on GH secretion.

drolysis of membrane PIP_2 in the somatotroph. The importance of this phospholipid pathway for the stimulation of GH secretion by GHRH is not established.

Somatostatin is a small peptide consisting of 14 amino acid residues. Although made by neurosecretory neurons in various parts of the hypothalamus, somatostatin neurons are especially abundant in the **anterior periventricular region** (close to the third ventricle). The axons of these cells terminate on the capillary networks giving rise to the hypophyseal portal circulation, where they release somatostatin into the blood.

Somatostatin binds to receptors in the plasma membranes of somatotrophs. These receptors, like those for GHRH, are also coupled to adenylyl cyclase, but they are coupled by an inhibitory G protein (see Fig. 32.10). The binding of somatostatin to its receptor decreases adenylyl cyclase activity, reducing intracellular cAMP. Somatostatin binding to its receptor also lowers intracellular Ca^{2+}, reducing GH secretion. When the somatroph is exposed to both somatostatin and GHRH, the effects of somatostatin are dominant and intracellular cAMP and Ca^{2+} are reduced. Thus, somatostatin has a negative modulating influence on the action of GHRH.

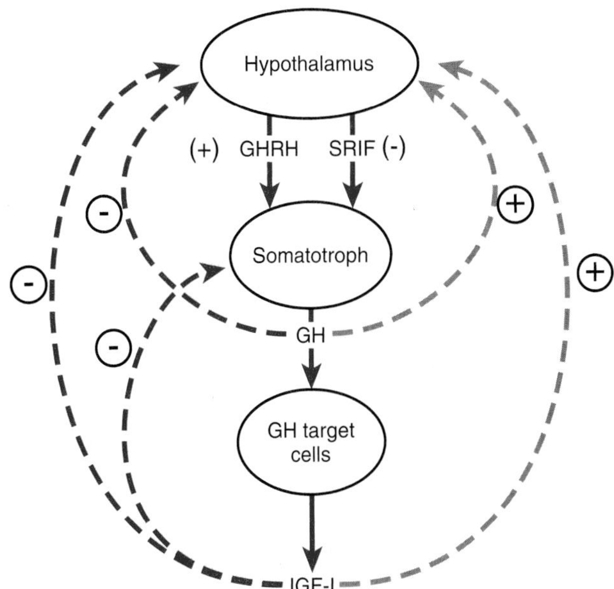

FIGURE 32.11 **The hypothalamic-pituitary-GH axis.** Growth hormone-releasing hormone (GHRH) stimulates, and somatostatin inhibits, GH secretion by acting directly on the somatotroph. The negative-feedback loops (−), shown in red, inhibit GHRH secretion and action on the somatotroph, causing a decrease in GH secretion. The feedback loops (+), shown in gray, stimulate somatostatin secretion, causing a decrease in GH secretion. IGF-I, insulin-like growth factor I.

GH and Insulin-Like Growth Factor I. GH is not considered a traditional trophic hormone; however, it does stimulate the production of a trophic hormone called **insulin-like growth factor I** (IGF-I). IGF-I is a potent **mitogenic agent** that mediates the growth-promoting action of GH. IGF-I was originally called **somatomedin C** or somatotropin-mediating hormone because of its role in promoting growth. Somatomedin C was renamed IGF-I because of its structural similarity to proinsulin.

Insulin-like growth factor II (IGF-II), an additional growth factor induced by GH, is structurally similar to IGF-I and has many of the same metabolic and mitogenic actions. However, IGF-I appears to be the more important mediator of GH action.

IGF-I is a 7.5 kDa protein consisting of a single chain of 70 amino acids. Because of its structural similarity to proinsulin, IGF-I can produce some of the effects of insulin. IGF-I is produced by many cells of the body; however, the liver is the main source of IGF-I in the blood. Most IGF-I in the blood is bound to specific IGF-I-binding proteins; only a small amount circulates in the free form. The bound form of circulating IGF-I has little insulin-like activity, so it does not play a physiological role in the regulation of blood glucose level.

GH increases the expression of the genes for IGF-I in various tissues and organs, such as the liver, and stimulates the production and release of IGF-I. Excessive secretion of GH results in a greater than normal amount of IGF-I in the blood. Individuals with GH deficiency have lower than normal levels of IGF-I, but there is still some present, since the production of IGF-I by cells is regulated by a variety of hormones and factors in addition to GH.

IGF-I has a negative-feedback effect on the secretion of GH (Fig. 32.11). It acts directly on somatotrophs to inhibit the stimulatory action of GHRH on GH secretion. It also inhibits GHRH secretion and stimulates the secretion of somatostatin by neurons in the hypothalamus. The net effect of these actions is the inhibition of GH secretion. By stimulating IGF-I production, GH inhibits its own secretion. This mechanism is analogous to the way ACTH and TSH regulate their own secretion through the respective negative-feedback effects of the glucocorticoid and thyroid hormones. This interactive relationship involving GHRH, somatostatin, GH, and IGF-I comprises the **hypothalamic-pituitary-GH axis**.

Feedback Effects of GH on Its Own Secretion. An increase in the blood concentration of GH has direct feedback effects on its own secretion, independent of the production of IGF-I. These effects of GH are due to the inhibition of GHRH secretion and the stimulation of somatostatin secretion by hypothalamic neurons (see Fig. 32.11). GH circulating in the blood can enter the interstitial spaces of the median eminence of the hypothalamus because there is no blood-brain barrier in this area.

Pulsatile Secretion of GH. In humans, GH is secreted in periodic bursts, which produce large but short-lived peaks in GH concentration in the blood. Between these episodes of high GH secretion, somatotrophs release little GH; as a result, the blood concentration of GH falls to very low levels. It is believed that these periodic bursts of GH secretion are caused by an increase in the rate of GHRH secretion and a fall in the rate of somatostatin secretion. The intervals between bursts, when GH secretion is suppressed, are thought to be caused by increased somatostatin secretion. These changes in GHRH and somatostatin secretion result from neural activity generated in higher levels of the CNS, which affects the secretory activity of GHRH and somatostatin producing neurons in the hypothalamus.

Bursts of GH secretion occur during both awake and sleep periods of the day; however, GH secretion is maximal at night. The bursts of GH secretion during sleep usually occur within the first hour after the onset of deep sleep (stages 3 and 4 of slow-wave sleep). Mean GH levels in the blood are highest during adolescence (peaking in late puberty) and decline in adults. The reduction in blood GH with aging is mainly due to decrease in the size of the GH secretory burst but not the number of pulses (Fig 32.12).

A variety of factors affect the rate of GH secretion in humans. These factors are thought to work by changing the secretion of GHRH and somatostatin by neurons in the hypothalamus. For example, emotional or physical stress causes a great increase in the rate of GH secretion. Vigorous exercise also stimulates GH secretion. Obesity results in reduced GH secretion.

Changes in the circulating levels of metabolites also affect GH secretion. A decrease in blood glucose concentration stimulates GH secretion, whereas hyperglycemia inhibits it. Growth hormone secretion is also stimulated by an

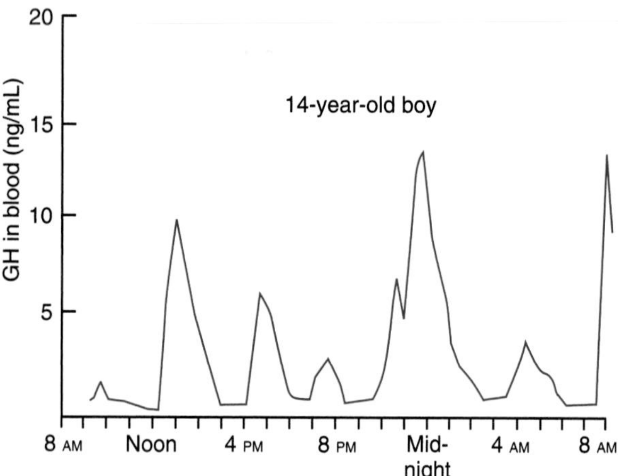

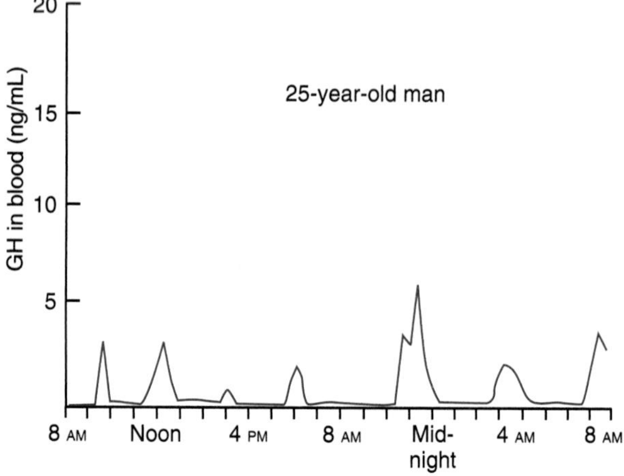

FIGURE 32.12 Pulsatile GH secretion in an adolescent boy and in an adult. In the adult, GH levels are reduced as a result of smaller pulse width and amplitude rather than a decrease in the number of pulses.

increase in the blood concentration of the amino acids arginine and leucine.

The Actions of GH. The cells of many tissues and organs of the body have receptors for GH in their plasma membranes. The interaction of GH with these receptors produces its growth-promoting and other metabolic effects, but the mechanisms that produce these effects are not fully understood. The binding of GH to its receptor activates a tyrosine kinase (JAK2), which initiates changes in the phosphorylation pattern of cytoplasmic and nuclear proteins. These phosphorylated proteins ultimately stimulate the transcription of specific genes, such as that for IGF-I.

Many of the mitogenic effects of GH are mediated by IGF-I; however, evidence indicates that GH has direct growth-promoting actions on **progenitor cells** or **stem cells**, such as **prechondrocytes** in the growth plates of bone and **satellite cells** of skeletal muscle. GH stimulates such progenitor cells to differentiate into cells with the capacity to undergo cell division. An important action of GH on the differentiation of progenitor cells is stimulation of the expression of the IGF-I gene; IGF-I is produced and released by these cells. IGF-I exerts an autocrine mitogenic action on the cells that produced it or a paracrine action on neighboring cells. In response to IGF-I, these cells undergo division, causing the tissue to grow mainly through cell replication.

As mentioned earlier, GH deficiency in childhood causes a decrease in the rate of body growth. If left untreated, the deficiency results in **pituitary dwarfism**. Individuals with this condition may be deficient in GH only, or they may have multiple anterior pituitary hormone deficiencies. GH deficiency can be caused by a defect in the mechanisms that control GH secretion or the production of GH by somatotrophs. In some individuals, the target cells for GH fail to respond normally to the hormone because of several different mutations in the GH receptor. See Clinical Focus Box 32.1 and the Case Study for further discussion of growth hormone deficiency, its detection, and treatment.

The excessive secretion of GH during childhood, caused by a defect in the mechanisms regulating GH secretion or a GH-secreting tumor, results in **gigantism**. Affected individuals may grow to a height of 7 to 8 feet (2.1 to 2.4 m). When excessive GH secretion occurs in an adult, further linear growth does not occur because the growth plates of the long bones have calcified. Instead, it causes the bones of the face, hands, and feet to become thicker and certain organs, such as the liver, to undergo hypertrophy. This condition, known as **acromegaly**, can also be caused by the chronic administration of excessive amounts of GH to adults.

Although the main physiological action of GH is on body growth, it also has important effects on certain aspects of fat and carbohydrate metabolism. Its main action on fat metabolism is to stimulate the mobilization of triglycerides from the fat depots of the body. This process, known as **lipolysis**, involves the hydrolysis of triglycerides to fatty acids and glycerol by the enzyme **hormone-sensitive lipase**. The fatty acids and glycerol are released from adipocytes and enter the bloodstream. How GH stimulates lipolysis is not understood, but most evidence suggests that it causes adipocytes to be more responsive to other lipolytic stimuli, such as fasting and catecholamines.

CLINICAL FOCUS BOX 32.1

Recombinant Human Growth Hormone and GH Deficiency

Growth hormone (GH) is species-specific, and humans do not respond to GH derived from animals. In the past, the only human GH available for treating children who were GH-deficient was a very limited amount made from human pituitaries obtained at autopsy, but there was never enough to meet the need. This problem was solved when the gene for human GH was cloned in 1979 and then expressed in bacteria. The production of large amounts of recombinant human GH, with all the activities of the natural substance, was now possible. During the 1980s, careful clinical trials established that recombinant human GH was safe to use in GH-deficient children to promote growth. The hormone was approved for clinical use and is now produced and sold worldwide.

Despite the availability of recombinant GH, the diagnosis of **GH deficiency** has remained controversial. GH is released in periodic bursts, the greatest of which occur in the early morning hours. Between pulses of secretion, the blood concentration of GH is nearly undetectable by most techniques. For these reasons, a random measure of GH in the blood is not useful for diagnosing GH deficiency. However, a random blood sample may be useful to detect GH resistance, a syndrome in which the patient exhibits symptoms of GH deficiency but presents with high GH levels in the blood.

An alternative means of diagnosing GH deficiency is to measure the levels of IGF-I, IGF-II, and the IGF-binding protein 3 (IGFBP3) in the blood. The IGFs mediate many of the mitogenic effects of GH on tissues in the body. IGF-I and IGF-II bind to IGFBP3 in the blood. IGFBP3 extends the half-life of the IGFs, transports them to target cells, and facilitates their interaction with IGF receptors. GH stimulates the production of all three molecules, which are present in the blood at fairly constant, readily detectable levels in normal individuals. In children with GH deficiency, the concentration of IGFs and IGFBP3 are low. Treatment with recombinant GH will increase IGF-I, IGF-II, and IGFBP3 in the blood, which will result in increased long bone growth. The epiphyseal growth plate in the bone becomes less responsive to GH and IGF-I several years after puberty, and long bone growth stops in adulthood (see Chapter 36).

GH is also thought to function as one of the counterregulatory hormones that limit the actions of insulin on muscle, adipose tissue, and the liver. For example, GH inhibits glucose use by muscle and adipose tissue and increases glucose production by the liver. These effects are opposite those of insulin. Also, GH makes muscle and fat cells resistant to the action of insulin itself. Thus, GH normally has a tonic inhibitory effect on the actions of insulin, much like the glucocorticoid hormones (see Chapter 34).

The insulin-opposing actions of GH can produce serious metabolic disturbances in individuals who secrete excessive amounts of GH (people with acromegaly) or are given large amounts of GH for an extended time. They may develop insulin resistance and an elevated insulin level in the blood. They may also have hyperglycemia caused by the underutilization and overproduction of glucose. These disturbances are much like those in individuals with non-insulin-dependent (type 2) diabetes mellitus. For this reason, this metabolic response to excess GH is called its **diabetogenic action**.

In GH-deficient individuals, GH has a transitory **insulin-like action**. For example, intravenous injection of GH in a person who is GH-deficient produces hypoglycemia. The hypoglycemia is caused by the ability of GH to stimulate the uptake and use of glucose by muscle and adipose tissue and to inhibit glucose production by the liver. After about 1 hour, the blood glucose level returns to normal. If this person is given a second injection of GH, hypoglycemia does not occur because the person has become insensitive or refractory to the insulin-like action of GH and remains so for some hours. Normal individuals do not respond to the insulin-like action of GH, presumably because they are always refractory from being exposed to their own endogenous GH. The actions of GH in humans are summarized in Table 32.3.

Gonadotropins Regulate Reproduction

The testes and ovaries have two essential functions in human reproduction. The first is to produce sperm cells and ova (egg cells), respectively. The second is to produce an array of steroid and peptide hormones, which influence virtually every aspect of the reproductive process. The gonadotropic hormones FSH and LH regulate both of these functions. The production and secretion of the gonadotropins by the anterior pituitary is, in turn, regulated by the hypothalamic releasing hormone LHRH and the hormones produced by the testes and ovaries in response to gonadotropic stimulation. The regulation of human reproduction by this **hypothalamic-pituitary-gonad axis** is dis-

TABLE 32.3 The Actions of Growth Hormone

Growth-promoting	Stimulates IGF-I gene expression by target cells; IGF-I produced by these cells has autocrine or paracrine stimulatory effect on cell division, resulting in growth
Lipolytic	Stimulates mobilization of triglycerides from fat deposits
Diabetogenic	Inhibits glucose use by muscle and adipose tissue and increases glucose production by the liver
	Inhibits the action of insulin on glucose and lipid metabolism by muscle and adipose tissue
Insulin-like	Transitory stimulatory effect on uptake and use of glucose by muscle and adipose tissue in GH-deficient individuals
	Transitory inhibitory effect on glucose production by liver of GH-deficient individuals

cussed in Chapters 37 and 38. Here, we describe the chemistry and formation of the gonadotropins.

Like TSH, human FSH and LH are composed of two structurally different glycoprotein subunits, called α and β, which are held together by noncovalent bonds. The β subunit of human FSH consists of a peptide chain of 111 amino acid residues, to which two chains of carbohydrate are attached. The β subunit of human LH is a peptide of 121 amino acid residues. It is also glycosylated with two carbohydrate chains. The combined α and β subunits of FSH and LH give these hormones a molecular size of about 28 to 29 kDa.

As with TSH, the individual subunits of the gonadotropins have no hormonal activity. They must be combined with each other in a 1:1 ratio in order to have activity. Again, it is the β subunit that gives the gonadotropin molecule either FSH or LH activity because the α subunits are identical.

FSH and LH are produced by the same gonadotrophs in the anterior pituitary. There are separate genes for the α and β subunits in the gonadotroph; hence, the peptide chains of these subunits are translated from separate mRNA molecules. Glycosylation of these chains begins as they are synthesized and before they are released from the ribosome. The folding of the subunit peptides into their final three-dimensional structure, the combination of an α subunit and a β subunit, and the completion of glycosylation all occur as these molecules pass through the Golgi apparatus and are packaged into secretory granules. As with the thyrotroph, the gonadotroph produces an excess of α subunits over FSH and LH β subunits. Therefore, the rate of β subunit production is considered to be the rate-limiting step in gonadotropin synthesis.

The synthesis of FSH and LH is regulated by the hormones of the hypothalamic-pituitary-gonad axis. For example, gonadotropin production is stimulated by LHRH. It is also affected by the steroid and peptide hormones produced by the gonads in response to stimulation by the gonadotropins. Such hormonally regulated changes in gonadotropin production are caused mainly by changes in the expression of the genes for the gonadotropin subunits. More information about the regulation of gonadotropin synthesis and secretion is found in Chapters 37 and 38.

Prolactin Regulates the Synthesis of Milk

Lactation is the final phase of the process of human reproduction. During pregnancy, **alveolar cells** of the mammary glands develop the capacity to synthesize milk in response to stimulation by a variety of steroid and peptide hormones. Milk synthesis by these cells begins shortly after childbirth. To continue to synthesize milk, these cells must be stimulated periodically by prolactin (PRL), and this is thought to be the main physiological function of PRL in the human female. What role, if any, PRL has in the human male is unclear. It is known to have some supportive effect on the action of androgenic hormones on the male reproductive tract, but whether this is an important physiological function of PRL is not established.

Human PRL is a globular protein consisting of a single peptide chain of 199 amino acid residues with three intrachain disulfide bridges. Its molecular size is about 23 kDa. Human PRL has considerable structural similarity to human GH and to a PRL-like hormone produced by the human placenta called **placental lactogen** (hPL). It is thought that these hormones are structurally related because their genes evolved from a common ancestral gene during the course of vertebrate evolution. Because of its structural similarity to human PRL, human GH has substantial PRL-like or **lactogenic activity**. However, PRL and hPL have little GH-like activity. Human placental lactogen is discussed further in Chapter 39.

Prolactin is synthesized and secreted by lactotrophs in the anterior pituitary. PRL is synthesized in the rough ER as a larger peptide. Its N-terminal signal peptide sequence is then removed and the 199-amino acid protein passes through the Golgi apparatus and is packaged into secretory granules.

The synthesis and secretion of PRL is stimulated by estrogens and other hormones, such as TRH, which increase the expression of the PRL gene. However, dopamine inhibits the synthesis of PRL. Dopamine produced by hypothalamic neurons plays a major role in the regulation of PRL synthesis and secretion by the hypothalamic-pituitary axis. The regulation of the synthesis and secretion of PRL and its physiological actions are discussed in Chapter 39.

REVIEW QUESTIONS

DIRECTIONS: Each of the numbered items or incomplete statements in this section is followed by answers or by completions of the statement. Select the ONE lettered answer or completion that is the BEST in each case.

1. Which of the following conditions is consistent with a decreased rate of ACTH secretion?
   (A) Hyperosmolality of the blood
   (B) Low serum glucocorticoid
   (C) Loss of hypothalamic neurons
   (D) Primary adrenal insufficiency
   (E) Stress as a result of emotional trauma
   (F) Increased PKA activity in corticotrophs
2. Which of the following statements most accurately describes the feedback effects of thyroid hormones?
   (A) They increase the sensitivity of thyrotrophs to TRH
   (B) They stimulate transcription of the α and β subunits of TSH in thyrotrophs
   (C) They increase the secretion of TSH by thyrotrophs
   (D) They stimulate the expression of the GH gene in somatotrophs
   (E) They increase IP_3 in thyrotrophs
   (F) They increase ACTH release
3. A 30-year-old woman completed a routine pregnancy with the uncomplicated delivery of a normal-sized baby girl 6 months ago. The woman is currently experiencing galactorrhea (persistent discharge of milk-like secretions from the breast) and has not yet resumed regular menstrual periods. The baby had been bottle-fed since birth. What is the

(continued)

most likely explanation of the galactorrhea?
(A) Normal postpartum response
(B) Excess PRL secretion
(C) Insufficient TSH secretion
(D) Reduced GH secretion
(E) Increased dopamine synthesis in the hypothalamus

4. A decrease in blood volume would result in an increase in the secretion of
(A) Neurophysin
(B) Oxytocin
(C) β-Lipotropin
(D) Somatostatin
(E) ACTH
(F) POMC

5. A 50-year-old man complains of decreased muscle strength, libido, and exercise intolerance. Examination reveals a 10% reduction in lean body mass and an increase in body fat, primarily localized to the abdominal region. Thyroid hormone levels are normal. Which diagnosis is most consistent with these symptoms?
(A) Glucocorticoid deficiency
(B) Addison's disease
(C) GH deficiency
(D) PRL deficiency
(E) Acromegaly

6. For evaluation of possible adrenocortical dysfunction in a middle-aged man, ACTH and cortisol were measured in blood samples taken at 8 AM, 8:30 AM, 8 PM, and 8:30 PM. The values obtained for ACTH were 110, 90, 120, and 200 pg/mL, respectively. The values obtained for cortisol were 10, 15, 25, and 20 μg/dL. These concentrations of ACTH demonstrate
(A) Normal circadian pulsatile release
(B) Primary adrenal insufficiency
(C) Inverted circadian pulsatile release
(D) Secondary adrenal insufficiency
(E) Normal circadian nonpulsatile release
(F) ACTH-secreting tumor

7. Which treatment would provide the greatest therapeutic benefit in patients with acromegaly?
(A) Glucocorticoid
(B) Somatostatin
(C) Growth hormone
(D) Insulin
(E) GHRH
(F) Thyroid hormone

8. Which of the following is mediated by a rise in cAMP?
(A) Inhibition of GH secretion by somatostatin
(B) Stimulation of GH gene expression by GHRH
(C) Stimulation of TSH secretion by TRH
(D) Inhibition of TSH α and β subunit gene expression by TRH
(E) Release of AVP
(F) Inhibition of ACTH synthesis in corticotrophs

SUGGESTED READING

Cuttler L. The regulation of growth hormone secretion. Endocrinol Metab Clin North Am 1996;3:541–571.

Fliers E, Wiersinga WM, Swaab DF. Physiological and pathophysiological aspects of thryotropin-releasing hormone gene expression in the human hypothalamus. Thyroid 1998;8:921–928.

Itoi K, Seasholtz AF, Watson SJ. Cellular and extracellular regulatory mechanisms of hypothalamic corticotropin-releasing hormone neurons. Endocr J 1998;45:13–33.

Reichlin S. Neuroendocrinology. In: Wilson JD, Foster DW, Kronenberg HM, Larsen PR, eds. Williams Textbook of Endocrinology. 9th Ed. Philadelphia: WB Saunders, 1998.

Zingg HH, Bourque CH, Bichet DG, eds. Vasopressin and oxytocin. Molecular, cellular and clinical advances. Adv Exp Med Biol 1998;449.

The Thyroid Gland

Robert V. Considine, Ph.D.

CHAPTER OUTLINE

KEY CONCEPTS

1. The thyroid gland consists of two lobes attached to either side of the trachea. Within the lobes of the thyroid gland are spherical follicles surrounded by a single layer of epithelial cells. Parafollicular cells that secrete calcitonin are also present within the walls of the follicles.
2. The major thyroid hormones are thyroxine (T_4) and triiodothyronine (T_3), both of which contain iodine.
3. Thyroid hormones are synthesized by iodination and the coupling of tyrosines in reactions catalyzed by the enzyme thyroid peroxidase.
4. Thyroid hormones are released from the thyroid gland by the degradation of thyroglobulin within the follicular cells.
5. The synthesis and release of thyroid hormones is regulated by thyroid-stimulating hormone (TSH), mainly via cAMP.
6. TSH release from the anterior pituitary is regulated by the concentration of thyroid hormones in the circulation.
7. In peripheral tissues, T_4 is deiodinated to the physiologically active hormone T_3 by 5′- deiodinase.
8. In target tissues, T_3 binds to the thyroid hormone receptor (TR), which then associates with a second TR or other nuclear receptor to regulate transcription.
9. TR regulates transcription by binding to specific thyroid hormone response elements (TRE) in target genes.
10. Thyroid hormones are important regulators of central nervous system development.
11. Thyroid hormones stimulate growth by regulating growth hormone release from the pituitary and by direct actions on target tissues, such as bone.
12. Thyroid hormones regulate the basal metabolic rate and intermediary metabolism through effects on mitochondrial ATP synthesis and the expression of genes encoding metabolic enzymes.
13. An excess of thyroid hormone (hyperthyroidism) is characterized by nervousness and increased metabolic rate, resulting in weight loss.
14. A deficiency of thyroid hormone (hypothyroidism) is characterized by decreased metabolic rate, resulting in weight gain.

The development of the human body, from embryo to adult, is an orderly, programmed process. The timing of developmental events is remarkably consistent from one individual to the next, with developmental milestones reached at about the same time in all of us. For example, the early development of motor skills, body growth, the start of puberty, and final sexual and physical maturation occur within rather narrow timeframes during the human life span.

At the level of the individual cell, the timing or rate of metabolic processes is also tightly regulated. For example, energy metabolism occurs at a rate needed to make the amount of ATP required for activities such as excitability, secretion, maintaining osmotic integrity, and countless biosynthetic processes. The cell not only meets its basic metabolic "housekeeping" needs but also remains poised to do its own special work in the body, such as conducting nerve impulses and contracting, absorbing, and secreting. During its life span, the cell continues to make the enzymatic and structural proteins that ensure the maintenance of an appropriate rate of metabolism.

The thyroid hormones, thyroxine and triiodothyronine, play key roles in the regulation of body development and govern the rate at which metabolism occurs in individual cells. Although these hormones are not essential for life, without them, life would lose its orderly nature. Without adequate levels of thyroid hormones, the body fails to develop on time. Cellular housekeeping moves at a slower pace, eventually influencing the ability of individual cells to carry out their physiological functions. The thyroid hormones exert their regulatory functions by influencing gene expression, affecting the developmental program and the amount of cellular constituents needed for the normal rate of metabolism.

FUNCTIONAL ANATOMY OF THE THYROID GLAND

The human thyroid gland consists of two lobes attached to either side of the trachea by connective tissue. The two lobes are connected by a band of thyroid tissue or isthmus, which lies just below the cricoid cartilage. A normal thyroid gland in a healthy adult weighs about 20 g.

Each lobe of the thyroid receives its arterial blood supply from a superior and an inferior thyroid artery, which arise from the external carotid and subclavian artery, respectively. Blood leaves the lobes of the thyroid by a series of thyroid veins that drain into the external jugular and innominate veins. This circulation provides a rich blood supply to the thyroid gland, giving it a higher rate of blood flow per gram than even that of the kidneys.

The thyroid gland receives adrenergic innervation from the cervical ganglia and cholinergic innervation from the vagus nerves. This innervation regulates vasomotor function to increase the delivery of TSH, iodide, and metabolic substrates to the thyroid gland. The adrenergic system can also affect thyroid function by direct effects on the cells.

Thyroxine and Triiodothyronine Are Synthesized and Secreted by the Thyroid Follicle

The lobes of the thyroid gland consist of aggregates of many spherical **follicles**, lined by a single layer of epithelial cells (Fig. 33.1). The apical membranes of the follicular cells, which face the lumen, are covered with microvilli. Pseudopods formed from the apical membrane extend into the lumen. The lateral membranes of the follicular cells are connected by tight junctions, which provide a seal for the contents of the lumen. The basal membranes of the follicular cells are close to the rich capillary network that penetrates the stroma between the follicles.

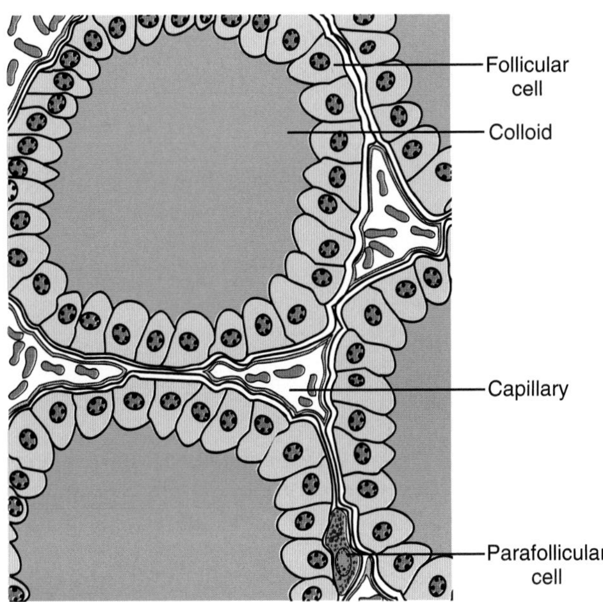

FIGURE 33.1 A cross-sectional view through a portion of the human thyroid gland.

The lumen of the follicle contains a thick, gel-like substance called **colloid** (see Fig. 33.1). The colloid is a solution composed primarily of **thyroglobulin**, a large protein that is a storage form of the thyroid hormones. The high viscosity of the colloid is due to the high concentration (10 to 25%) of thyroglobulin.

The thyroid follicle produces and secretes two thyroid hormones, **thyroxine** (T_4) and **triiodothyronine** (T_3). Their molecular structures are shown in Figure 33.2. Thyroxine and triiodothyronine are iodinated derivatives of the amino acid tyrosine. They are formed by the coupling of the phenyl rings of two iodinated tyrosine molecules in an ether linkage. The resulting structure is called an **iodothyronine**. The mechanism of this process is discussed in detail later.

Thyroxine contains four iodine atoms on the 3, 5, 3′, and 5′ positions of the thyronine ring structure, whereas triiodothyronine has only three iodine atoms, at ring positions 3, 5, and 3′ (see Fig. 33.2). Consequently, thyroxine is usually abbreviated as T_4 and triiodothyronine as T_3. Because T_4 and T_3 contain the element iodine, their synthesis by the thyroid follicle depends on an adequate supply of iodine in the diet.

Parafollicular Cells Are the Sites of Calcitonin Synthesis

In addition to the epithelial cells that secrete T_4 and T_3, the wall of the thyroid follicle contains small numbers of **parafollicular cells** (see Fig. 33.1). The parafollicular cell is usually embedded in the wall of the follicle, inside the basal lamina surrounding the follicle. However, its plasma membrane does not form part of the wall of the lumen. Parafollicular cells produce and secrete the hormone calcitonin. Calcitonin and its effects on calcium metabolism are discussed in Chapter 36.

SYNTHESIS, SECRETION, AND METABOLISM OF THE THYROID HORMONES

T_4 and T_3 are not directly synthesized by the thyroid follicle in their final form. Instead, they are formed by the chemical modification of tyrosine residues in the peptide structure of thyroglobulin as it is secreted by the follicular cells into the lumen of the follicle. Therefore, the T_4 and T_3 formed by this chemical modification are actually part of the amino acid sequence of thyroglobulin.

The high concentration of thyroglobulin in the colloid provides a large reservoir of stored thyroid hormones for later processing and secretion by the follicle. The synthesis of T_4 and T_3 is completed when thyroglobulin is retrieved through pinocytosis of the colloid by the follicular cells. Thyroglobulin is then hydrolyzed by lysosomal enzymes

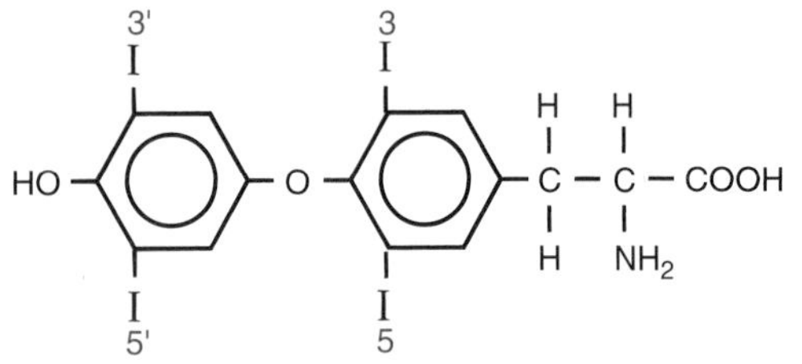

3' 3 I I H H HO O C C COOH H NH2 I I 5' 5

Thyroxine (T_4)

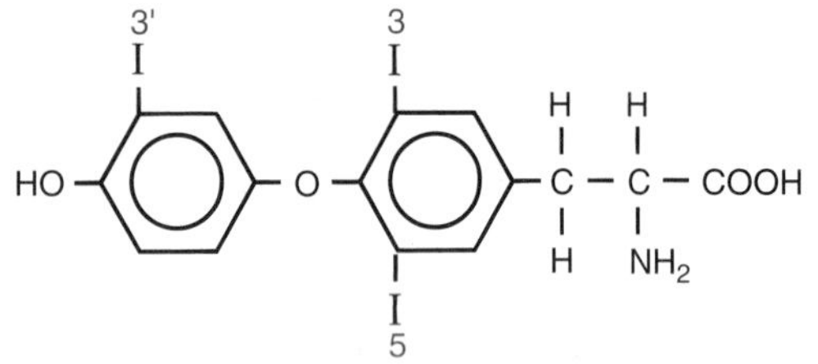

3' 3 I I H H HO O C C COOH H NH2 I 5

Triiodothyronine (T_3)

FIGURE 33.2 **The molecular structure of the thyroid hormones.** The numbering of the iodine atoms on the iodothyronine ring structure is shown in red.

to its constituent amino acids, releasing T_4 and T_3 molecules from their peptide linkage. T_4 and T_3 are then secreted into the blood.

Follicular Cells Synthesize Iodinated Thyroglobulin

The steps involved in the synthesis of iodinated thyroglobulin are shown in Figure 33.3. This process involves the synthesis of a thyroglobulin precursor, the uptake of iodide, and the formation of iodothyronine residues.

Synthesis and Secretion of the Thyroglobulin Precursor. The synthesis of the protein precursor for thyroglobulin is the first step in the formation of T_4 and T_3. This substance is a 660-kDa glycoprotein composed of two similar 330-kDa subunits held together by disulfide bridges. The subunits are synthesized by ribosomes on the rough ER and then undergo dimerization and glycosylation in the smooth ER. The completed glycoprotein is packaged into vesicles by the Golgi apparatus. These vesicles migrate to the apical membrane of the follicular cell and fuse with it. The thyroglobulin precursor protein is then extruded onto the apical surface of the cell, where iodination takes place.

Iodide Uptake. The iodide used for iodination of the thyroglobulin precursor protein comes from the blood perfusing the thyroid gland. The basal plasma membranes of follicular cells, which are near the capillaries that supply the follicle, contain **iodide transporters**. These transporters move iodide across the basal membrane and into the cytosol of the follicular cell. The iodide transporter is an active transport mechanism that requires ATP, is saturable, and can also transport certain other anions, such as bromide, thiocyanate, and perchlorate. It enables the follicular cell to concentrate iodide many times over the concentration of iodide present in the blood; therefore, follicular cells are efficient extractors of the small amount of iodide circulating in the blood. Once inside follicular cells, the iodide ions diffuse rapidly to the apical membrane, where they are used for iodination of the thyroglobulin precursor.

Formation of the Iodothyronine Residues. The next step in the formation of thyroglobulin is the addition of one or two iodine atoms to certain tyrosine residues in the precursor protein. The precursor of thyroglobulin contains 134 tyrosine residues, but only a small fraction of these become iodinated. A typical thyroglobulin molecule contains only 20 to 30 atoms of iodine.

The iodination of thyroglobulin is catalyzed by the enzyme **thyroid peroxidase**, which is bound to the apical membranes of follicular cells. Thyroid peroxidase binds an iodide ion and a tyrosine residue in the thyroglobulin precursor, bringing them in close proximity. The enzyme oxidizes the iodide ion and the tyrosine residue to short-lived free radicals, using hydrogen peroxide that has been generated within the mitochondria of follicular cells. The free radicals then undergo addition. The product formed is a **monoiodotyrosine** (MIT) residue, which remains in peptide linkage in the thyroglobulin structure. A second iodine atom may be added to a MIT residue by this same enzymatic process, forming a **diiodotyrosine** (DIT) residue (see Fig. 33.3).

Iodinated tyrosine residues that are close together in the thyroglobulin precursor molecule undergo a **coupling reaction**, which forms the iodothyronine structure. Thyroid peroxidase, the same enzyme that initially oxidizes iodine, is believed to catalyze the coupling reaction through the oxidation of neighboring iodinated tyrosine residues to short-lived free radicals. These free radicals undergo addition, as shown in Figure 33.4. The addition reaction produces an iodothyronine residue and a **dehydroalanine residue**, both of which remain in peptide linkage in the thyroglobulin structure. For example, when two neighboring DIT residues couple by this mechanism, T_4 is formed (see Fig. 33.4). After being iodinated, the thyroglobulin molecule is stored as part of the colloid in the lumen of the follicle.

Only about 20 to 25% of the DIT and MIT residues in the thyroglobulin molecule become coupled to form iodothyronines. For example, a typical thyroglobulin molecule contains five to six uncoupled residues of DIT and two to three residues of T_4. However, T_3 is formed in only about one of three thyroglobulin molecules. As a result, the thyroid secretes substantially more T_4 than T_3.

Thyroid Hormones Are Formed From the Hydrolysis of Thyroglobulin

When the thyroid gland is stimulated to secrete thyroid hormones, vigorous pinocytosis occurs at the apical membranes of follicular cells. Pseudopods from the apical membrane reach into the lumen of the follicle, engulfing bits of the colloid (see Fig. 33.3). Endocytotic vesicles or **colloid droplets** formed by this pinocytotic activity migrate toward the basal region of the follicular cell. Lysosomes, which are mainly located in the basal region of resting fol-

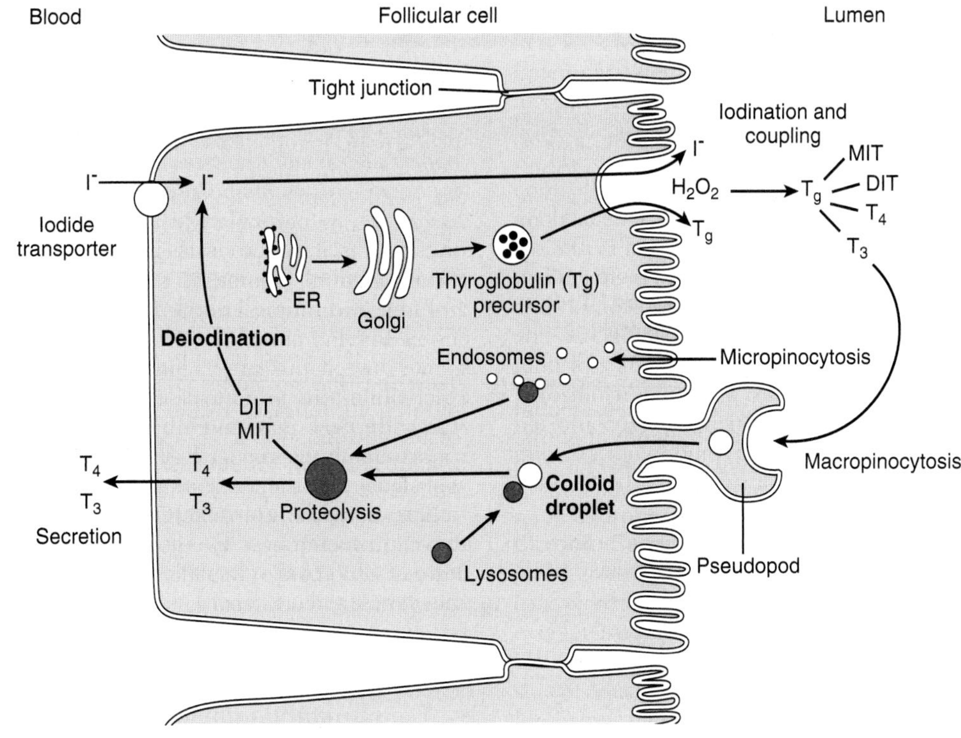

FIGURE 33.3 **Thyroid hormone synthesis and secretion.** (See text for details.) DIT, diiodotyrosine; MIT, monoiodotyrosine.

licular cells, migrate toward the apical region of the stimulated cells. The lysosomes fuse with the colloid droplets and hydrolyze the thyroglobulin to its constituent amino acids. As a result, T_4 and T_3 and the other iodinated amino acids are released into the cytosol.

Secretion of Free T_4 and T_3. T_4 and T_3 formed from the hydrolysis of thyroglobulin are released from the follicular cell and enter the nearby capillary circulation; however, the mechanism of transport of T_4 and T_3 across the basal plasma membrane has not been defined. The DIT and MIT generated by the hydrolysis of thyroglobulin are deiodinated in the follicular cell. The released iodide is then reutilized by the follicular cell for the iodination of thyroglobulin (see Fig. 33.3).

Binding of T_4 and T_3 to Plasma Proteins. Most of the T_4 and T_3 molecules that enter the bloodstream become bound to plasma proteins. About 70% of the T_4 and 80% of the T_3 are noncovalently bound to **thyroxine-binding globulin** (TBG), a 54-kDa glycoprotein that is synthesized and secreted by the liver. Each molecule of TBG has a single binding site for a thyroid hormone molecule. The remaining T_4 and T_3 in the blood are bound to **transthyretin** or to albumin. Less than 1% of the T_4 and T_3 in blood is in the free form, and it is in equilibrium with the large protein-bound fraction. It is this small amount of free thyroid hormone that interacts with target cells.

The protein-bound form of T_4 and T_3 represents a large reservoir of preformed hormone that can replenish the small amount of circulating free hormone as it is cleared from the blood. This reservoir provides the body with a buffer against drastic changes in circulating thyroid hormone levels as a result of sudden changes in the rate of T_4 and T_3 secretion. The protein-bound T_4 and T_3 molecules are also protected from metabolic inactivation and excretion in the urine. As a result of these factors, the thyroid hormones have long half-lives in the bloodstream. The half-life of T_4 is about 7 days; the half-life of T_3 is about 1 day.

Thyroid Hormones Are Metabolized by Peripheral Tissues

Thyroid hormones are both activated and inactivated by deiodination reactions in the peripheral tissues. The enzymes that catalyze the various deiodination reactions are regulated, resulting in different thyroid hormone concentrations in various tissues in different physiological and pathophysiological conditions.

Conversion of T_4 to T_3. As noted earlier, T_4 is the major secretory product of the thyroid gland and is the predominant thyroid hormone in the blood. However, about 40% of the T_4 secreted by the thyroid gland is converted to T_3 by enzymatic removal of the iodine atom at position 5′ of the thyronine ring structure (Fig. 33.5). This reaction is catalyzed by a **5′-deiodinase** (type 1) located in the liver, kidneys, and thyroid gland. The T_3 formed by this deiodination and that secreted by the thyroid react with thyroid hormone receptors in target cells; therefore, T_3 is the physiologically active form of the thyroid hormones. A second 5′-deiodinase (type 2) is

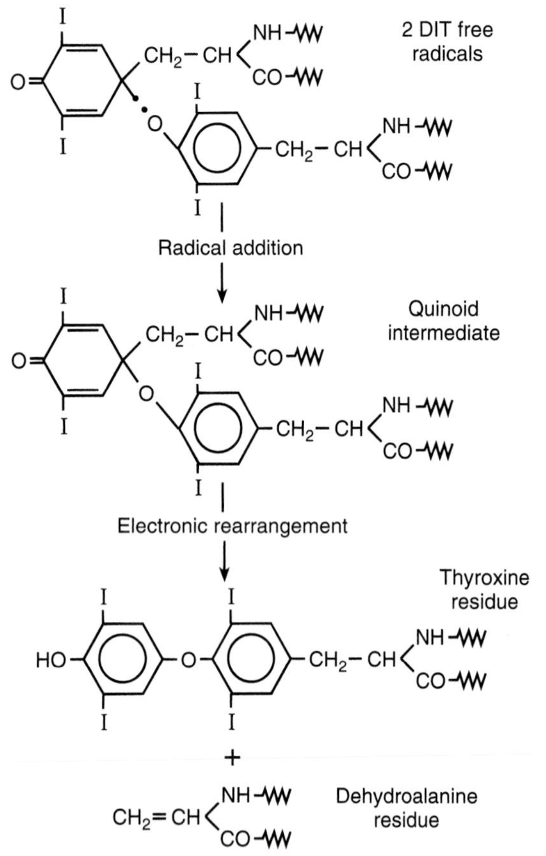

FIGURE 33.4 Theoretical model for the coupling reaction between two diiodotyrosine (DIT) residues in iodinated thyroglobulin. This model is based on free radical formation catalyzed by thyroid peroxidase. (Adapted from Taurog AM. Hormone synthesis: Thyroid iodine metabolism. In: Braverman LE, Utiger RD, eds. Werner & Ingbar's The Thyroid: A Fundamental and Clinical Text. 8th Ed. Philadelphia: Lippincott Williams & Wilkins, 2000;61–85.)

present in skeletal muscle, the CNS, the pituitary gland, and the placenta. Type 2 deiodinase is believed to function primarily to maintain intracellular T_3 in target tissues, but it may also contribute to the generation of circulating T_3. All of the deiodinases contain **selenocysteine** in the active center. This rare amino acid has properties that make it ideal for catalysis of oxidoreductive reactions.

Deiodinations That Inactivate T_4 and T_3. Whereas the 5′-deiodination of T_4 to produce T_3 can be viewed as a metabolic activation process, both T_4 and T_3 undergo enzymatic deiodinations, particularly in the liver and kidneys, which inactivate them. For example, about 40% of the T_4 secreted by the human thyroid gland is deiodinated at the 5 position on the thyronine ring structure by a 5-deiodinase. This produces reverse T_3 (see Fig. 33.5). Since reverse T_3 has little or no thyroid hormone activity, this deiodination reaction is a major pathway for the metabolic inactivation or disposal of T_4. Triiodothyronine and reverse T_3 also undergo deiodination to yield 3,3′-diiodothyronine. This inactivated metabolite may be further deiodinated before being excreted.

Regulation of 5′-Deiodination. The 5′-deiodination reaction is a regulated process influenced by certain physiological and pathological factors. The result is a change in the relative amounts of T_3 and reverse T_3 produced from T_4. For example, a human fetus produces less T_3 from T_4 than a child or adult because the 5′-deiodination reaction is less active in the fetus. Also, 5′-deiodination is inhibited during fasting, particularly in response to carbohydrate restriction, but it can be restored to normal when the individual is fed again. Trauma, as well as most acute and chronic illnesses, also suppresses the 5′-deiodination reaction. Under all of these circumstances, the amount of T_3 produced from T_4 is reduced and its blood concentration falls. However, the amount of reverse T_3 rises in the circulation, mainly because its conversion to 3,3′-diiodothyronine by 5′-deiodination is reduced. A rise in reverse T_3 in the blood may signal that the 5′-deiodination reaction is suppressed.

Note that during fasting or in the disease states mentioned above, the secretion of T_4 is usually not increased, despite the decrease of T_3 in the circulation. This response indicates that, under these circumstances, a T_3 decrease in the blood does not stimulate the hypothalamic-pituitary-thyroid axis.

Minor Degradative Pathways. T_4 and, to a lesser extent, T_3 are also metabolized by conjugation with glucuronic acid in the liver. The conjugated hormones are secreted into the bile and eliminated in the feces. Many tissues also metabolize thyroid hormones by modifying the three-carbon side chain of the iodothyronine structure. These modifications include decarboxylation and deamination. The derivatives formed from T_4, such as **tetraiodoacetic acid** (tetrac), may also undergo deiodinations before being excreted (see Fig. 33.5).

TSH Regulates Thyroid Hormone Synthesis and Secretion

When the concentrations of free T_4 and T_3 fall in the blood, the anterior pituitary gland is stimulated to secrete **thyroid-stimulating hormone** (TSH), raising the concentration of TSH in the blood. This action results in increased interactions between TSH and its receptors on thyroid follicular cells.

TSH Receptors and Second Messengers. The receptor for TSH is a transmembrane glycoprotein thought to be located on the basal plasma membrane of the follicular cell. These receptors are coupled by G_s proteins, mainly to the adenylyl cyclase-cAMP-protein kinase A pathway; however, there is also evidence for effects via phospholipase C (PLC), inositol trisphosphate, and diacylglycerol (see Chapter 1). The physiological importance of TSH-stimulated phospholipid metabolism in human follicular cells is unclear, since very high concentrations of TSH are needed to activate PLC.

TSH and Thyroid Hormone Formation and Secretion. TSH stimulates most of the processes involved in thyroid hormone synthesis and secretion by follicular cells. The rise in cAMP produced by TSH is believed to cause many of these effects. TSH stimulates the uptake of iodide by follicular cells, usually after a short interval during which io-

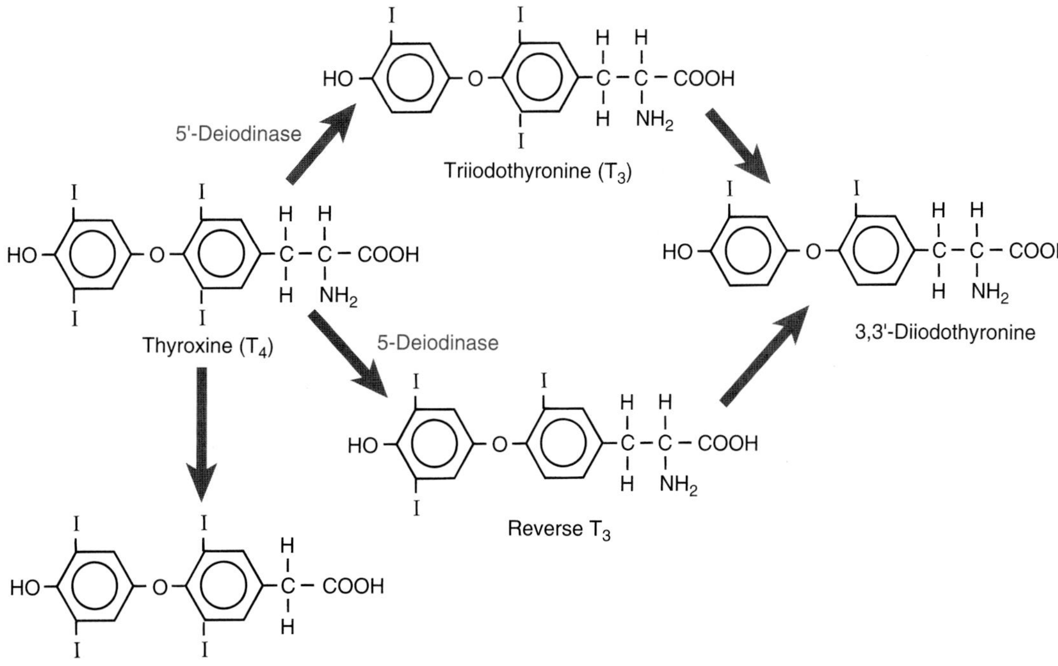

FIGURE 33.5 **The metabolism of thyroxine.** Thyroxine is deiodinated by 5′-deiodinase to form T_3, the physiologically active thyroid hormone. Some T_4 is also enzymatically deiodinated at the 5 position to form the inactive metabolite, reverse T_3. T_3 and reverse T_3 undergo additional deiodinations (e.g., to 3,3′-diiodothyronine) before being excreted. A small amount of T_4 is also decarboxylated and deaminated to form the metabolite, tetraiodoacetic acid (tetrac). Tetrac may then be deiodinated before being excreted.

dide transport is actually depressed. TSH also stimulates the iodination of tyrosine residues in the thyroglobulin precursor and the coupling of iodinated tyrosines to form iodothyronines. Moreover, it stimulates the pinocytosis of colloid by the apical membranes, resulting in a great increase in endocytosis of thyroglobulin and its hydrolysis. The overall result of these effects of TSH is an increased release of T_4 and T_3 into the blood. In addition to its effects on thyroid hormone synthesis and secretion, TSH rapidly increases energy metabolism in the thyroid follicular cell.

TSH and Thyroid Size. Over the long term, TSH promotes protein synthesis in thyroid follicular cells, maintaining their size and structural integrity. Evidence of this **trophic effect** of TSH is seen in a hypophysectomized patient, whose thyroid gland atrophies, largely as a result of a reduction in the height of follicular cells. However, the chronic exposure of an individual to excessive amounts of TSH causes the thyroid gland to increase in size. This enlargement is due to an increase in follicular cell height and number. Such an enlarged thyroid gland is called a **goiter**. These trophic and proliferative effects of TSH on the thyroid are primarily mediated by cAMP.

Dietary Iodide Is Essential for the Synthesis of Thyroid Hormones

Because iodine atoms are constituent parts of the T_4 and T_3 molecules, a continual supply of iodide is required for the synthesis of these hormones. If an individual's diet is severely deficient in iodide, as in some parts of the world, T_4 and T_3 synthesis is limited by the amount of iodide available to the thyroid gland. As a result, the concentrations of T_4 and T_3 in the blood fall, causing a chronic stimulation of TSH secretion, which, in turn, produces a goiter. Enlargement of the thyroid gland increases its capacity to accumulate iodide from the blood and to synthesize T_4 and T_3. However, the degree to which the enlarged gland can produce thyroid hormones to compensate for their deficiency in the blood depends on the severity of the deficiency of iodide in the diet. To prevent iodide deficiency and the consequent goiter formation in the human population, iodide is added to the table salt (iodized salt) sold in most developed countries.

THE MECHANISM OF THYROID HORMONE ACTION

Most cells of the body are targets for the action of thyroid hormones. The sensitivity or responsiveness of a particular cell to thyroid hormones correlates to some degree with the number of receptors for these hormones. The cells of the CNS appear to be an exception. As is discussed later, the thyroid hormones play an important role in CNS development during fetal and neonatal life, and developing nerve cells in the brain are important targets for thyroid hormones. In the adult, however, brain cells show little responsiveness to the metabolic regulatory action of thyroid hormones, although they have numerous receptors for these hormones. The reason for this discrepancy is unclear.

Thyroid hormone receptors (TR) are located in the nuclei of target cells bound to **thyroid hormone response elements** (TRE) in the DNA. TRs are protein molecules of about 50 kDa that are structurally similar to the nuclear receptors for steroid hormones and vitamin D. Thyroid receptors bound to the TRE in the absence of T_3 generally act to repress gene expression.

The free forms of T_3 and T_4 are taken up by target cells from the blood through a carrier-mediated process that requires ATP. Once inside the cell, T_4 is deiodinated to T_3, which enters the nucleus of the cell and binds to its receptor in the chromatin. The TR with bound T_3 forms a complex with other nuclear receptors (called a heterodimer) or with another TR (homodimer) to activate transcription. Other transcription factors may also complex with the TR heterodimer or homodimer. As a result, the production of mRNA for certain proteins is either increased or decreased, changing the cell's capacity to make these proteins (Fig. 33.6). T_3 can influence differentiation by regulating the kinds of proteins produced by its target cells and can influence growth and metabolism by changing the amounts of structural and enzymatic proteins present in the cells. The mechanisms by which T_3 alters gene expression continue to be investigated.

The gene expression response to T_3 is slow to appear. When T_3 is given to an animal or human, several hours elapse before its physiological effects can be detected. This delayed action undoubtedly reflects the time required for changes in gene expression and consequent changes in the synthesis of key proteins to occur. When T_4 is administered, its course of action is usually slower than that of T_3 because of the additional time required for the body to convert T_4 to T_3.

Thyroid hormones also have effects on cells that occur much faster and do not appear to be mediated by nuclear TR receptors, including effects on signal transduction pathways that alter cellular respiration, cell morphology, vascular tone, and ion homeostasis. The physiological relevance of these effects is currently being investigated.

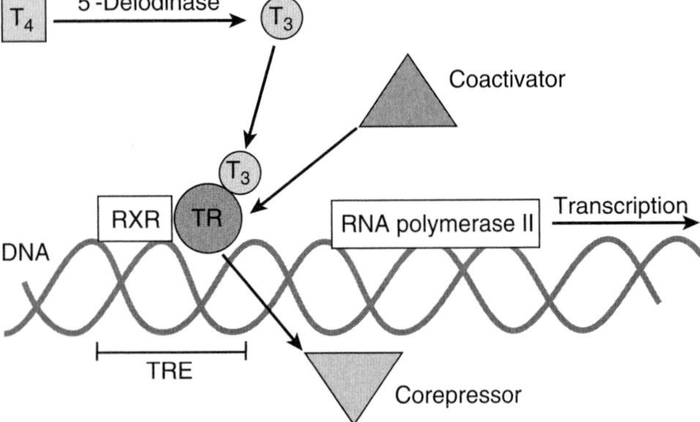

FIGURE 33.6 **The activation of transcription by thyroid hormone.** T_4 is taken up by the cell and deiodinated to T_3, which then binds to the thyroid hormone receptor (TR). The activated TR heterodimerizes with a second transcription factor, 9-*cis* retinoic acid receptor (RXR), and binds to the thyroid hormone response element (TRE). The binding of TR/RXR to the TRE displaces repressors of transcription and recruits additional coactivators. The final result is the activation of RNA polymerase II and the transcription of the target gene.

ROLE OF THE THYROID HORMONES IN DEVELOPMENT, GROWTH, AND METABOLISM

Thyroid hormones play a critical role in the development of the central nervous system (CNS). They are also essential for normal body growth during childhood, and in basal energy metabolism.

Thyroid Hormones Are Essential for Development of the CNS

The human brain undergoes its most active phase of growth during the last 6 months of fetal life and the first 6 months of postnatal life. During the second trimester of pregnancy, the multiplication of neuroblasts in the fetal brain reaches a peak and then declines. As pregnancy progresses and the rate of neuroblast division drops, neuroblasts differentiate into neurons and begin the process of synapse formation that extends into postnatal life.

Thyroid hormones first appear in the fetal blood during the second trimester of pregnancy, and levels continue to rise during the remaining months of fetal life. Thyroid hormone receptors increase about 10-fold in the fetal brain at about the time the concentrations of T_4 and T_3 begin to rise in the blood. These events are critical for normal brain development because thyroid hormones are essential for timing the decline in nerve cell division and the initiation of differentiation and maturation of these cells.

If thyroid hormones are deficient during these prenatal and postnatal periods of differentiation and maturation of the brain, mental retardation occurs. The cause is thought to be inadequate development of the neuronal circuitry of the CNS. Thyroid hormone therapy must be given to a thyroid hormone-deficient child during the first few months of postnatal life to prevent mental retardation. Starting thyroid hormone therapy after behavioral deficits have occurred cannot reverse the mental retardation (i.e., thyroid hormone must be present when differentiation normally occurs). Thyroid hormone deficiency during infancy causes both mental retardation and growth impairment, as discussed below. Fortunately, this occurs rarely today because thyroid hormone deficiency is usually detected in newborn infants and hormone therapy is given at the proper time.

The exact mechanism by which thyroid hormones influence differentiation of the CNS is unknown. Animal studies have demonstrated that thyroid hormones inhibit nerve cell replication in the brain and stimulate the growth of nerve cell bodies, the branching of dendrites, and the rate of myelinization of axons. These effects of thyroid hormones are presumably due to their ability to regulate the expression of genes involved in nerve cell replication and differentiation. However, the details, particularly in the human, are unclear.

Thyroid Hormones Are Essential for Normal Body Growth

The thyroid hormones are important factors regulating the growth of the entire body. For example, an individual who is deficient in thyroid hormones, who does not receive thyroid hormone therapy during childhood, will not grow to a normal adult height.

Thyroid Hormones and the Gene for GH. A major way thyroid hormones promote normal body growth is by stimulating the expression of the gene for **growth hormone** (GH) in the somatotrophs of the anterior pituitary gland. In a thyroid hormone-deficient individual, GH synthesis by the somatotrophs is greatly reduced and consequently GH secretion is impaired; therefore, a thyroid hormone-deficient individual will also be GH-deficient. If this condition occurs in a child, it will cause growth retardation, largely a result of the lack of the growth-promoting action of GH (see Chapter 32).

Other Effects of Thyroid Hormones on Growth. The thyroid hormones have additional effects on growth. In tissues such as skeletal muscle, the heart, and the liver, thyroid hormones have direct effects on the synthesis of a variety of structural and enzymatic proteins. For example, they stimulate the synthesis of structural proteins of mitochondria, as well as the formation of many enzymes involved in intermediary metabolism and oxidative phosphorylation.

Thyroid hormones also promote the calcification and, hence, the closure, of the cartilaginous growth plates of the bones of the skeleton. This action limits further linear body growth. How the thyroid hormones promote calcification of the growth plates of bones is not understood.

Thyroid Hormones Regulate the Basal Energy Economy of the Body

When the body is at rest, about half of the ATP produced by its cells is used to drive energy-requiring membrane transport processes. The remainder is used in involuntary muscular activity, such as respiratory movements, peristalsis, contraction of the heart, and in many metabolic reactions requiring ATP, such as protein synthesis. The energy required to do this work is eventually released as body heat.

Basal Oxygen Consumption and Body Heat Production. The major site of ATP production is the mitochondria, where the oxidative phosphorylation of ADP to ATP takes place. The rate of oxidative phosphorylation depends on the supply of ADP for electron transport. The ADP supply is, in turn, a function of the amount of ATP used to do work. For example, when more work is done per unit time, more ATP is used and more ADP is generated, increasing the rate of oxidative phosphorylation. The rate at which oxidative phosphorylation occurs is reflected in the amount of oxygen consumed by the body because oxygen is the final electron acceptor at the end of the electron transport chain.

Activities that occur when the body is not at rest, such as voluntary movements, use additional ATP for the work involved; the amounts of oxygen consumed and body heat produced depend on total body activity.

Thermogenic Action of the Thyroid Hormones. Thyroid hormones regulate the basal rate at which oxidative phosphorylation takes place in cells. As a result, they set the basal rate of body heat production and of oxygen consumed by the body. This is called the **thermogenic action** of thyroid hormones.

Thyroid hormone levels in the blood must be within normal limits for basal metabolism to proceed at the rate needed for a balanced energy economy of the body. For example, if thyroid hormones are present in excess, oxidative phosphorylation is accelerated, and body heat production and oxygen consumption are abnormally high. The converse occurs when the blood concentrations of T_4 and T_3 are lower than normal. The fact that thyroid hormones affect the amount of oxygen consumed by the body has been used clinically to assess the status of thyroid function. Oxygen consumption is measured under resting conditions and compared with the rate expected of a similar individual with normal thyroid function. This measurement is the **basal metabolic rate** (BMR) test.

Tissues Affected by the Thermogenic Action of Thyroid Hormones. Not all tissues are sensitive to the thermogenic action of thyroid hormones. Tissues and organs that give this response include skeletal muscle, the heart, the liver, and the kidneys. These are also tissues in which thyroid hormone receptors are abundant. The adult brain, skin, lymphoid organs, and gonads show little thermogenic response to thyroid hormones. With the exception of the adult brain, these tissues contain few thyroid hormone receptors, which may explain their poor response.

Molecular and Cellular Mechanisms. The thermogenic action of the thyroid hormones is poorly understood at the molecular level. The thermogenic effect takes many hours to appear after the administration of thyroid hormones to a human or animal, probably because of the time required for changes in the expression of genes involved. T_3 is known to stimulate the synthesis of cytochromes, cytochrome oxidase, and Na^+/K^+-ATPase in certain cells. This action suggests that T_3 may regulate the number of respiratory units in these cells, affecting their capacity to carry out oxidative phosphorylation. A greater rate of oxidative phosphorylation would result in greater heat production.

Thyroid hormone also stimulates the synthesis of **uncoupling protein-1** (UCP-1) in brown adipose tissue. ATP is synthesized by ATP synthase in the mitochondria when protons flow down their electrochemical gradient. UCP-1 acts as a channel in the mitochondrial membrane to dissipate the ion gradient without making ATP. As the protons move down their electrochemical gradient *uncoupled* from ATP synthesis, energy is released as heat. Adult humans have little brown adipose tissue, so it is not likely that UCP-1 makes a significant contribution to nutrient oxidation or body heat production. However, several uncoupling proteins (UCP-2 and UCP-3) have recently been discovered in many tissues, and their expression is regulated by thyroid hormones.

These novel uncoupling proteins may be involved in the thermogenic action of thyroid hormones.

Thyroid Hormones Stimulate Intermediary Metabolism

In addition to their ability to regulate the rate of basal energy metabolism, thyroid hormones influence the rate at which most of the pathways of intermediary metabolism operate in their target cells. When thyroid hormones are deficient, pathways of carbohydrate, lipid, and protein metabolism are slowed, and their responsiveness to other regulatory factors, such as other hormones, is decreased. However, these same metabolic pathways run at an abnormally high rate when thyroid hormones are present in excess. Thyroid hormones, therefore, can be viewed as amplifiers of cellular metabolic activity. The amplifying effect of thyroid hormones on intermediary metabolism is mediated through the activation of genes encoding enzymes involved in these metabolic pathways.

TABLE 33.1 The Physiological Actions of Thyroid Hormones

Development of CNS	Inhibit nerve cell replication Stimulate growth of nerve cell bodies Stimulate branching of dendrites Stimulate rate of axon myelinization
Body growth	Stimulate expression of gene for GH in somatotrophs Stimulate synthesis of many structural and enzymatic proteins Promote calcification of growth plates of bones
Basal energy economy of the body	Regulate basal rates of oxidative phosphorylation, body heat production, and oxygen consumption (thermogenic effect)
Intermediary metabolism	Stimulate synthetic and degradative pathways of carbohydrate, lipid, and protein metabolism
Thyroid-stimulating hormone (TSH) secretion	Inhibit TSH secretion by decreasing sensitivity of thyrotrophs to thyrotropin-releasing hormone (TRH)

Thyroid Hormones Regulate Their Own Secretion

An important action of the thyroid hormones is the ability to regulate their own secretion. As discussed in Chapter 32, T_3 exerts an inhibitory effect on TSH secretion by thyrotrophs in the anterior pituitary gland by decreasing thyrotroph sensitivity to thyrotropin-releasing hormone (TRH). Consequently, when the circulating concentration of free thyroid hormones is high, thyrotrophs are relatively insensitive to TRH, and the rate of TSH secretion decreases. The resulting fall of TSH levels in the blood reduces the rate of thyroid hormone release from the follicular cells in the thyroid. When the free thyroid hormone level falls in the blood, however, the negative-feedback effect of T_3 on thyrotrophs is reduced, and the rate of TSH secretion increases. The rise in TSH in the blood stimulates the thyroid gland to secrete thyroid hormones at a greater rate. This action of T_3 on thyrotrophs is thought to be due to changes in gene expression in these cells.

The physiological actions of the thyroid hormones described above are summarized in Table 33.1.

THYROID HORMONE DEFICIENCY AND EXCESS IN ADULTS

A deficiency or an excess of thyroid hormones produces characteristic changes in the body. These changes result from dysregulation of nervous system function and altered metabolism.

Thyroid Hormone Deficiency Causes Nervous and Metabolic Disorders

Thyroid hormone deficiency in humans has a variety of causes. For example, iodide deficiency may result in a reduction in thyroid hormone production. Autoimmune diseases, such as **Hashimoto's disease**, impair thyroid hormone synthesis (see Clinical Focus Box 33.1). Other causes of thyroid hormone deficiency include heritable diseases that affect certain steps in the biosynthesis of thyroid hormones and hypothalamic or pituitary diseases that interfere with TRH or TSH secretion. Obviously, radioiodine ablation or surgical removal of the thyroid gland also causes thyroid hormone deficiency. **Hypothyroidism** is the disease state that results from thyroid hormone deficiency.

Thyroid hormone deficiency impairs the functioning of most tissues in the body. As described earlier, a deficiency of thyroid hormones at birth that is not treated during the first few months of postnatal life causes irreversible mental retardation. Thyroid hormone deficiency later in life also influences the function of the nervous system. For example, all cognitive functions, including speech and memory, are slowed and body movements may be clumsy. These changes can usually be reversed with thyroid hormone therapy.

Metabolism is also reduced in thyroid hormone-deficient individuals. Basal metabolic rate is reduced, resulting in impaired body heat production. Vasoconstriction occurs in the skin as a compensatory mechanism to conserve body heat. Heart rate and cardiac output are reduced. Food intake is reduced, and the synthetic and degradative processes of intermediary metabolism are slowed. In severe hypothyroidism, a substance consisting of hyaluronic acid and chondroitin sulfate complexed with protein is deposited in the extracellular spaces of the skin, causing water to accumulate osmotically. This effect gives a puffy appearance to the face, hands, and feet called **myxedema**. All of the above disorders in adults can be normalized with thyroid hormone therapy.

An Excess of Thyroid Hormone Produces Nervous and Other Disorders

The most common cause of excessive thyroid hormone production in humans is **Graves' disease**, an autoimmune

CLINICAL FOCUS BOX 33.1

Autoimmune Thyroid Disease—Postpartum Thyroiditis

Certain diseases affecting the function of the thyroid gland occur when an individual's immune system fails to recognize particular thyroid proteins as "self" and reacts to the proteins as if they were foreign. This usually triggers both humoral and cellular immune responses. As a result, antibodies to these proteins are generated, which then alter thyroid function. Two common autoimmune diseases with opposite effects on thyroid function are Hashimoto's disease and Graves' disease. In Hashimoto's disease, the thyroid gland is infiltrated by lymphocytes, and elevated levels of antibodies against several components of thyroid tissue (e.g., antithyroid peroxidase and antithyroglobulin antibodies) are found in the serum. The thyroid gland is destroyed, resulting in hypothyroidism. In Graves' disease, stimulatory antibodies to the TSH receptor activate thyroid hormone synthesis, resulting in hyperthyroidism (see text for details).

A third, fairly common autoimmune disease is postpartum thyroiditis, which usually occurs within 3 to 12 months after delivery. The disease is characterized by a transient thyrotoxicosis (hyperthyroidism) often followed by a period of hypothyroidism lasting several months. Many patients eventually return to the euthyroid state. Often only the hypothyroid phase of the disease may be observed, occurring in more than 30% of women with antibodies to thyroid peroxidase detectable preconception. The disease is also observed in patients known to have Graves' disease. The postpartum occurrence of the disorder is likely due to increased immune system function following the suppression of its activity during pregnancy.

It has been estimated that 5 to 10% of women develop postpartum thyroiditis. Of these women, about 50% have transient thyrotoxicosis alone, 25% have transient hypothyroidism alone, and the remaining 25% have both phases of the disease. The prevalence of the disease has prompted a clinical recommendation suggesting that thyroid function (serum T_4, T_3, and TSH levels) be surveyed postpartum at 2, 4, 6, and 12 months in all women with thyroid peroxidase antibodies or symptoms suggestive of thyroid dysfunction. Patients who have experienced one episode of postpartum thyroiditis should also be considered at risk for recurrence after pregnancy.

Treatment for thyrotoxicosis commonly involves inhibiting thyroid hormone synthesis and secretion. Thionamides are a class of drugs that inhibit the oxidation and organic binding of thyroid iodide to reduce thyroid hormone production. Some drugs in this class also inhibit the conversion of T_4 to T_3 in the peripheral tissues. Thyroid hormone replacement is required to treat hypothyroidism.

---

disorder caused by antibodies directed against the TSH receptor in the plasma membranes of thyroid follicular cells. These antibodies bind to the TSH receptor, resulting in an increase in the activity of adenylyl cyclase. The consequent rise in cAMP in follicular cells produces effects similar to those caused by the action of TSH. The thyroid gland enlarges to form a **diffuse toxic goiter**, which synthesizes and secretes thyroid hormones at an accelerated rate, causing thyroid hormones to be chronically elevated in the blood. Feedback inhibition of thyroid hormone production by the thyroid hormones is also lost.

Less common conditions that cause chronic elevations in circulating thyroid hormones include adenomas of the thyroid gland that secrete thyroid hormones and excessive TSH secretion caused by malfunctions of the hypothalamic-pituitary-thyroid axis. The disease state that develops in response to excessive thyroid hormone secretion, called **hyperthyroidism** or **thyrotoxicosis**, is characterized by many changes in the functioning of the body that are the opposite of those caused by thyroid hormone deficiency.

Hyperthyroid individuals are nervous and emotionally irritable, with a compulsion to be constantly moving around. However, they also experience physical weakness and fatigue. Basal metabolic rate is increased and, as a result, body heat production is increased. Vasodilation in the skin and sweating occur as compensatory mechanisms to dissipate excessive body heat. Heart rate and cardiac output are increased. Energy metabolism increases, as does appetite. However, despite the increase in food intake, a net degradation of protein and lipid stores occurs, resulting in weight loss. All of these changes can be reversed by reducing the rate of thyroid hormone secretion with drugs or by removal of the thyroid gland by radioactive ablation or surgery.

REVIEW QUESTIONS

DIRECTIONS: Each of the numbered items or incomplete statements in this section is followed by answers or by completions of the statement. Select the ONE lettered answer or completion that is the BEST in each case.

1. The effects of TSH on thyroid follicular cells include
   (A) Stimulation of endocytosis of thyroglobulin stored in the colloid
   (B) Release of a large pool of T_4 and T_3 stored in secretory vesicles in the cell
   (C) Stimulation of the uptake of iodide from the thyroglobulin stored in the colloid
   (D) Increase in perfusion by the blood
   (E) Stimulation of the binding of T_4 and T_3 to thyroxine-binding globulin
   (F) Increased cAMP hydrolysis
2. A child is born with a rare disorder in which the thyroid gland does not respond to TSH. What would be the predicted effects on mental ability, body growth rate, and thyroid gland size when the child reaches 6 years of age?

(continued)

(A) Mental ability would be impaired, body growth rate would be slowed, and thyroid gland size would be larger than normal
(B) Mental ability would be unaffected, body growth rate would be slowed, and thyroid gland size would be smaller than normal
(C) Mental ability would be impaired, body growth rate would be slowed, and thyroid gland size would be smaller than normal
(D) Mental ability would be unaffected, body growth rate would be unaffected, and thyroid gland size would be smaller than normal
(E) Mental ability would be impaired, body growth rate would be slowed, and thyroid gland size would be normal
(F) Mental ability would be unaffected, body growth rate would be unaffected, and thyroid gland size would be unaffected

3. If the 6-year-old child described in the previous question is now treated with thyroid hormones, how would mental ability, body growth rate, and thyroid gland size be affected?
(A) Mental ability would remain impaired, body growth rate would be improved, and thyroid gland size would be smaller than normal
(B) Mental ability would be improved, body growth rate would be improved, and thyroid gland size would be normal
(C) Mental ability would remain impaired, body growth rate would be improved, and thyroid gland size would be normal
(D) Mental ability would remain impaired, body growth rate would be improved, and thyroid gland size would be larger than normal
(E) Mental ability would be improved, body growth rate would remain slowed, and thyroid gland size would be normal
(F) Mental ability would be improved, body growth rate would remain slowed, and thyroid gland size would be larger than normal

4. Uncoupling proteins
(A) Utilize the proton gradient across the mitochondrial membrane to facilitate ATP synthesis
(B) Are decreased by thyroid hormones
(C) Dissipate the proton gradient across the mitochondrial membrane to generate heat
(D) Are present exclusively in brown fat
(E) Uncouple fatty acid oxidation from glucose oxidation in mitochondria
(F) Are essential for maintaining body temperature in mammals

5. Triiodothyronine (T_3)
(A) Is produced in greater amounts by the thyroid gland than T_4
(B) Is bound by the thyroid receptor present in the cytosol of target cells
(C) Is formed from T_4 through the action of a 5-deiodinase
(D) Has a half-life of a few minutes in the bloodstream
(E) Is released from thyroglobulin through the action of thyroid peroxidase
(F) Can be produced by the deiodination of T_4 in pituitary thyrotrophs

6. A 40-year-old man complains of chronic fatigue, aching muscles, and occasional numbness in his fingers. Physical examination reveals a modest weight gain but no goiter is detected. Laboratory findings include TSH > 10 μU/L (normal range, 0.5 to 5 μU/L) and a free T_4 of low to low-normal. These findings are most consistent with a diagnosis of
(A) Hypothyroidism secondary to a hypothalamic-pituitary defect
(B) Hyperthyroidism secondary to a hypothalamic-pituitary defect
(C) Hyperthyroidism as a result of iodine excess
(D) Hypothyroidism as a result of autoimmune thyroid disease
(E) Hypothyroidism as a result of iodine deficiency
(F) Hyperthyroidism as a result of autoimmune thyroid disease

7. The reaction catalyzed by thyroid peroxidase
(A) Produces hydrogen peroxide as an end product
(B) Couples two iodotyrosine residues to form an iodothyronine residue
(C) Occurs on the basal membrane of the follicular cell
(D) Catalyzes the release of thyroid hormones into the circulation
(E) Couples MIT and DIT to thyroglobulin
(F) Couples dehydroalanine with a thyroxine residue

8. A 25-year-old woman complains of weight loss, heat intolerance, excessive sweating, and weakness. TSH and thyroid hormones are elevated, goiter is present, but no antithyroid antibodies are detected. Which of the following diagnoses is consistent with these symptoms?
(A) Graves' disease
(B) Resistance to thyroid hormone action
(C) Plummer's disease (thyroid gland adenoma)
(D) A 5′-deiodinase deficiency
(E) Acute Hashimoto's disease
(F) TSH-secreting pituitary tumor

SUGGESTED READING

Apriletti JW, Ribeiro RC, Wagner RL, et al. Molecular and structural biology of thyroid hormone receptors. Clin Exp Pharmacol Physiol Suppl 1998;25:S2–S11.

Braverman LE, Utiger RD. Werner and Ingbar's The Thyroid: A Fundamental and Clinical Text. 8th Ed. Philadelphia: Lippincott Williams & Wilkins, 2000.

Goglia F, Moreno M, Lanni A. Action of thyroid hormones at the cellular level: The mitochondrial target. FEBS Lett 1999;452:115–120.

Larsen PR, Davies TF, Hay ID. The thyroid gland. In: Wilson JD, Foster DW, Kronenberg HM, Larsen PR, eds: Williams Textbook of Endocrinology. 9th Ed. Philadelphia: WB Saunders, 1998.

Meier CA. Thyroid hormone and development: Brain and peripheral tissue. In: Hauser P, Rovet J, eds. Thyroid Diseases of Infancy and Childhood. Washington, DC: American Psychiatric Press, 1999.

Motomura K, Brent GA. Mechanisms of thyroid hormone action. Endocrinol Metab Clin North Am 1998;27:1–23.

Munoz A, Bernal J. Biological activities of thyroid hormone receptors. Eur J Endocrinol 1997;137:433–445.

Reitman ML, He Y, Gong D-W. Thyroid hormone and other regulators of uncoupling proteins. Int J Obes Relat Metab Disord 1999;23(Suppl 6): S56–S59.

The Adrenal Gland

Robert V. Considine, Ph.D.

CHAPTER OUTLINE

KEY CONCEPTS

1. The adrenal gland is comprised of an outer cortex surrounding an inner medulla. The cortex contains three histologically distinct zones (from outside to inside): the zona glomerulosa, zona fasciculata, and zona reticularis.
2. Hormones secreted by the adrenal cortex include glucocorticoids, aldosterone, and adrenal androgens.
3. The glucocorticoids cortisol and corticosterone are synthesized in the zona fasciculata and zona reticularis of the adrenal cortex.
4. The mineralocorticoid aldosterone is synthesized in the zona glomerulosa of the adrenal cortex.
5. Cholesterol, used in the synthesis of the adrenal cortical hormones, comes from cholesterol esters stored in the cells. Stored cholesterol is derived mainly from low-density lipoprotein particles circulating in the blood, but it can also be synthesized *de novo* from acetate within the adrenal gland.
6. The conversion of cholesterol to pregnenolone in mitochondria is the common first step in the synthesis of all adrenal steroids and occurs in all three zones of the cortex.
7. The liver is the main site for the metabolism of adrenal steroids, which are conjugated to glucuronic acid and excreted in the urine.
8. ACTH increases glucocorticoid and androgen synthesis in adrenal cortical cells in the zona fasciculata and zona reticularis by increasing intracellular cAMP. ACTH also has a trophic effect on these cells.
9. Angiotensin II and angiotensin III stimulate aldosterone synthesis in the cells of the zona glomerulosa by increasing cytosolic calcium and activating protein kinase C.
10. Glucocorticoids bind to glucocorticoid receptors in the cytosol of target cells. The glucocorticoid-bound receptor translocates to the nucleus and then binds to glucocorticoid response elements in the DNA to increase or decrease the transcription of specific genes.
11. Glucocorticoids are essential to the adaptation of the body to fasting, injury, and stress.
12. The catecholamines epinephrine and norepinephrine are synthesized and secreted by the chromaffin cells of the adrenal medulla.
13. Catecholamines interact with four adrenergic receptors (α_1, α_2, β_1, and β_2) that mediate the cellular effects of the hormones.
14. Stimuli such as injury, anger, pain, cold, strenuous exercise, and hypoglycemia generate impulses in the cholinergic preganglionic fibers innervating the chromaffin cells, resulting in the secretion of catecholamines.
15. To counteract hypoglycemia, catecholamines stimulate glucose production in the liver, lactate release from muscle, and lipolysis in adipose tissue.

To remain alive, the organs and tissues of the human body must have a finely regulated extracellular environment. This environment must contain the correct concentrations of ions to maintain body fluid volume and to enable excitable cells to function. The extracellular environment must also have an adequate supply of metabolic substrates for cells to generate ATP. Salts, water, and other organic substances are continually lost from the body as a result of perspiration, respiration, and excretion. Metabolic substrates are constantly used by cells. Under normal conditions, these critical constituents of the body's extracellular environment are replenished by the intake of food and liquids. However, a person can survive for weeks on little else but water because the body has a remarkable capacity for adjusting the functions of its organs and tissues to preserve body fluid volume and composition.

The adrenal glands play a key role in making these adjustments. This is readily apparent from the fact that an adrenalectomized animal, unlike its normal counterpart, cannot survive prolonged fasting. Its blood glucose supply diminishes, ATP generation by the cells becomes inadequate to support life, and the animal eventually dies. Even when fed a normal diet, an adrenalectomized animal typically loses body sodium and water over time, and eventually dies of circulatory collapse. Its death is caused by a lack of certain steroid hormones that are produced and secreted by the cortex of the adrenal gland.

The glucocorticoid hormones, **cortisol** and **corticosterone**, play essential roles in adjusting the metabolism of carbohydrates, lipids, and proteins in liver, muscle, and adipose tissues during fasting, which assures an adequate supply of glucose and fatty acids for energy metabolism despite the absence of food. The mineralocorticoid hormone **aldosterone**, another steroid hormone produced by the adrenal cortex, stimulates the kidneys to conserve sodium and, hence, body fluid volume.

The glucocorticoids also enable the body to cope with physical and emotional traumas or stresses. The physiological importance of this action of the glucocorticoids is emphasized by the fact that adrenalectomized animals lose their ability to cope with physical or emotional stresses. Even when given an appropriate diet to prevent blood glucose and body sodium depletion, an adrenalectomized animal may die when exposed to traumas that are not fatal to normal animals.

Hormones produced by the other endocrine component of the adrenal gland, the medulla, are also involved in compensatory reactions of the body to trauma or life-threatening situations. These hormones are the catecholamines, **epinephrine** and **norepinephrine**, which have widespread effects on the cardiovascular system and muscular system and on carbohydrate and lipid metabolism in liver, muscle, and adipose tissues.

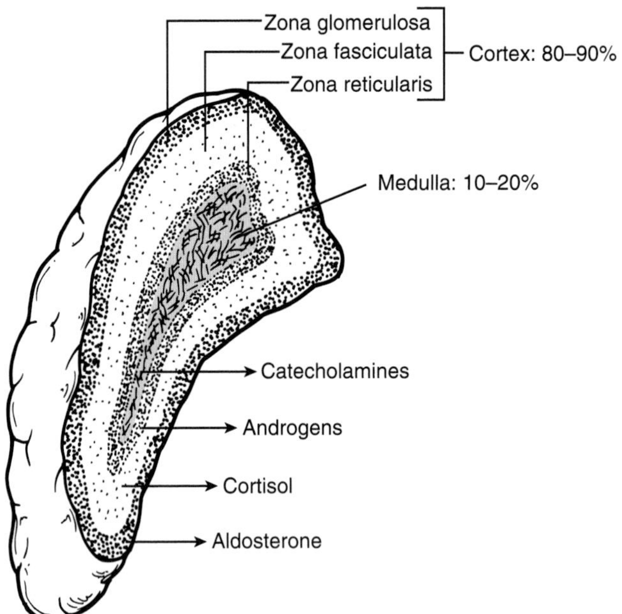

FIGURE 34.1 **The three zones of the adrenal cortex and corresponding hormone secretion.**

FUNCTIONAL ANATOMY OF THE ADRENAL GLAND

The human adrenal glands are paired, pyramid-shaped organs located on the upper poles of each kidney. The adrenal gland is actually a composite of two separate endocrine organs, one inside the other, each secreting separate hormones and each regulated by different mechanisms. The outer portion or **cortex** of the adrenal gland completely surrounds the inner portion or **medulla** and makes up most of the gland. During embryonic development, the cortex forms from mesoderm; the medulla arises from neural ectoderm.

The Adrenal Cortex Consists of Three Distinct Zones

In the adult human, the adrenal cortex consists of three histologically distinct zones or layers (Fig. 34.1). The outer zone, which lies immediately under the capsule of the gland, is called the **zona glomerulosa** and consists of small clumps of cells that produce the mineralocorticoid aldosterone. The **zona fasciculata** is the middle and thickest layer of the cortex and consists of cords of cells oriented radially to the center of the gland. The inner layer is comprised of interlaced strands of cells called the **zona reticularis**. The zona fasciculata and zona reticularis both produce the physiologically important glucocorticoids, cortisol and corticosterone. These layers of the cortex also produce the androgen **dehydroepiandrosterone**, which is related chemically to the male sex hormone **testosterone**. The molecular structures of these hormones are shown in Figure 34.2.

Like all endocrine organs, the adrenal cortex is highly vascularized. Many small arteries branch from the aorta and renal arteries and enter the cortex. These vessels give rise to capillaries that course radially through the cortex and terminate in venous sinuses in the zona reticularis and adrenal medulla; therefore, the hormones produced by the cells of the cortex have ready access to the circulation.

The cells of the adrenal cortex contain abundant lipid droplets. This stored lipid is functionally significant because cholesterol esters present in the droplets are an important source of the cholesterol used as a precursor for the synthesis of steroid hormones.

The Adrenal Medulla Is a Modified Sympathetic Ganglion

The adrenal medulla can be considered a modified sympathetic ganglion. The medulla consists of clumps and strands of **chromaffin cells** interspersed with venous sinuses. Chromaffin cells, like the modified postganglionic neurons that receive sympathetic preganglionic cholinergic innervation from the splanchnic nerves, produce catecholamine hormones, principally epinephrine and norepinephrine (NE). Epinephrine and NE are stored in granules in chromaffin cells and discharged into venous sinuses of the adrenal medulla when the adrenal branches of splanchnic nerves are stimulated (see Fig. 6.5).

HORMONES OF THE ADRENAL CORTEX

Only small amounts of the glucocorticoids, aldosterone, and adrenal androgens are found in adrenal cortical cells at

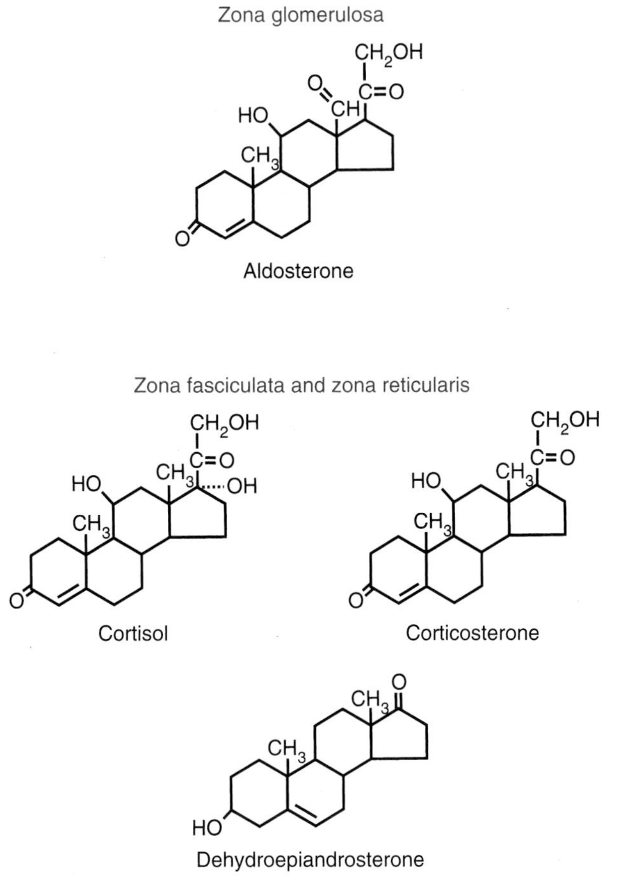

Zona glomerulosa

Aldosterone

Zona fasciculata and zona reticularis

Cortisol

Corticosterone

Dehydroepiandrosterone

FIGURE 34.2 **Molecular structures of the important hormones secreted by the adrenal cortex.**

TABLE 34.2 Comparison of Shared Activities of Adrenal Cortical Hormones

Hormone	Glucocorticoid Activity[a]	Mineralocorticoid Activity[b]
Cortisol	100	0.25
Corticosterone	20	0.5
Aldosterone	10	100

[a]Percentage activity, with cortisol being 100%
[b]Percentage activity, with aldosterone being 100%

a given time because those cells produce and secrete these hormones on demand, rather than storing them. Table 34.1 shows the daily production of adrenal cortex hormones in a healthy adult under resting (unstimulated) conditions. Because the molecular weights of these substances do not vary greatly, comparing the amounts secreted indicates the relative number of molecules of each hormone produced daily. Humans secrete about 10 times more cortisol than corticosterone during an average day, and corticosterone has only one fifth of the glucocorticoid activity of cortisol (Table 34.2). Cortisol is considered the physiologically important glucocorticoid in humans. Compared with the glucocorticoids, a much smaller amount of aldosterone is secreted each day.

Because of similarities in their structures, the glucocorticoids and aldosterone have overlapping actions. For example, cortisol and corticosterone have some mineralocorticoid activity; conversely, aldosterone has some glucocorticoid activity. However, given the amounts of these hormones secreted under normal circumstances and their relative activities, glucocorticoids are not physiologically important mineralocorticoids, nor does aldosterone function physiologically as a glucocorticoid.

TABLE 34.1 The Average Daily Production of Hormones by the Adrenal Cortex

Hormone	Amount Produced (mg/day)
Cortisol	20
Corticosterone	2
Aldosterone	0.1
Dehydroepiandrosterone	30

As discussed in detail later, the amounts of glucocorticoids and aldosterone secreted by an individual can vary greatly from those given in Table 34.1. The amount secreted depends on the person's physiological state. For example, in an individual subjected to severe physical or emotional trauma, the rate of cortisol secretion may be 10 times greater than the resting rate shown in Table 34.1. Certain diseases of the adrenal cortex that involve steroid hormone biosynthesis can significantly increase or decrease the amount of hormones produced.

The adrenal cortex also produces and secretes substantial amounts of androgenic steroids. Dehydroepiandrosterone (DHEA) in both the free form and the sulfated form (DHEAS) is the main androgen secreted by the adrenal cortex of both men and women (see Table 34.1). Lesser amounts of other androgens are also produced. The adrenal cortex is the main source of androgens in the blood in human females. In the human male, however, androgens produced by the testes and adrenal cortex contribute to the male sex hormones circulating in the blood. Adrenal androgens normally have little physiological effect other than a role in development before the start of puberty in both girls and boys. This is because the male sex hormone activity of the adrenal androgens is weak. Exceptions occur in individuals who produce inappropriately large amounts of certain adrenal androgens as a result of diseases affecting the pathways of steroid biosynthesis in the adrenal cortex.

Adrenal Steroid Hormones Are Synthesized From Cholesterol

Cholesterol is the starting material for the synthesis of steroid hormones. A cholesterol molecule consists of four interconnected rings of carbon atoms and a side chain of eight carbon atoms extending from one ring (Fig. 34.3). In all, there are 27 carbon atoms in cholesterol, numbered as shown in the figure.

Sources of Cholesterol. The immediate source of cholesterol used in the biosynthesis of steroid hormones is the abundant lipid droplets in adrenal cortical cells. The cho-

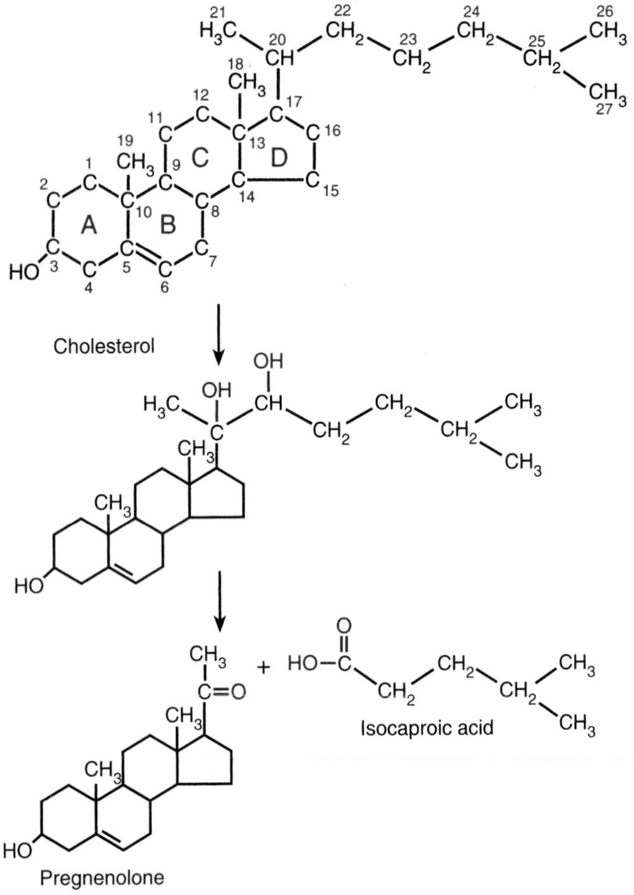

FIGURE 34.3 The formation of pregnenolone from cholesterol by the action of cholesterol side-chain cleavage enzyme (CYP11A1). Note the chemical structure of cholesterol, how the four rings are lettered (A to D), and how the carbons are numbered. The hydrogen atoms on the carbons composing the rings are omitted from the figure.

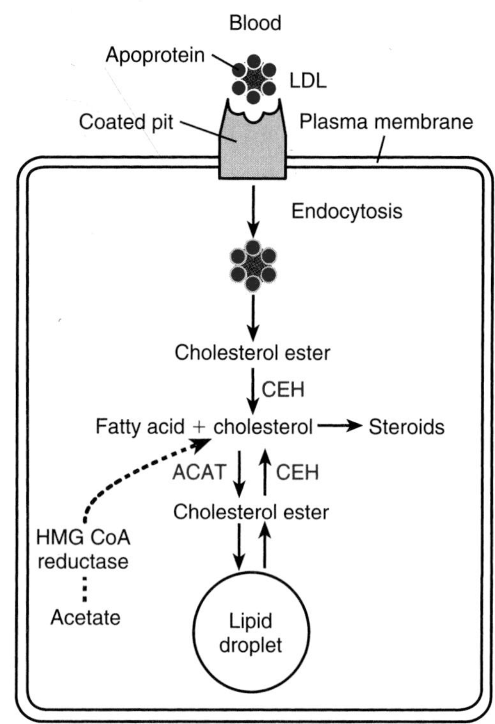

FIGURE 34.4 Sources of cholesterol for steroid biosynthesis by the adrenal cortex. Most cholesterol comes from low-density lipoprotein (LDL) particles in the blood, which bind to receptors in the plasma membrane and are taken up by endocytosis. The cholesterol in the LDL particle is used directly for steroidogenesis or stored in lipid droplets for later use. Some cholesterol is synthesized directly from acetate. CEH, cholesterol ester hydrolase; ACAT, acyl-CoA:cholesterol acyltransferase; HMG, 3-hydroxy-3-methylglutaryl.

lesterol present in these lipid droplets is mainly in the form of **cholesterol esters**, single molecules of cholesterol esterified to single fatty acid molecules. The free cholesterol used in steroid biosynthesis is generated from these cholesterol esters by the action of **cholesterol esterase (cholesterol ester hydrolase [CEH])**, which hydrolyzes the ester bond. The free cholesterol generated by that cleavage enters mitochondria located in close proximity to the lipid droplet. The process of remodeling the cholesterol molecule into steroid hormones is then initiated.

The cholesterol that has been removed from the lipid droplets for steroid hormone biosynthesis is replenished in two ways (Fig. 34.4). Most of the cholesterol converted to steroid hormones by the human adrenal gland comes from cholesterol esters contained in **low-density lipoprotein** (LDL) particles circulating in the blood. The LDL particles consist of a core of cholesterol esters surrounded by a coat of cholesterol and phospholipids. A 400-kDa protein molecule called **apoprotein B_{100}** is also present on the surface of the LDL particle; it is recognized by LDL receptors localized to coated pits on the plasma membrane of adrenal cortical cells (see Fig. 34.4). The apoprotein binds to the LDL receptor, and both the LDL particle and the receptor are taken up by the cell through endocytosis. The endocytic vesicle containing the LDL particles fuses with a lysosome and the particle is degraded. The cholesterol esters in the core of the particle are hydrolyzed to free cholesterol and fatty acid by the action of CEH.

Any cholesterol not immediately used by the cell is converted again to cholesterol esters by the action of the enzyme **acyl-CoA:cholesterol acyltransferase** (ACAT). The esters are then stored in the lipid droplets of the cell to be used later.

When steroid biosynthesis is proceeding at a high rate, cholesterol delivered to the adrenal cell may be diverted directly to mitochondria for steroid production rather than reesterified and stored. Accumulating evidence suggests that **high-density lipoprotein** (HDL) cholesterol may also be used as a substrate for adrenal steroidogenesis.

In humans, cholesterol that has been synthesized *de novo* from acetate by the adrenal glands is a significant but minor source of cholesterol for steroid hormone formation. The rate-limiting step in this process is catalyzed by the enzyme **3-hydroxy-3-methylglutaryl CoA reductase (HMG CoA reductase)**. The newly synthesized cholesterol is then incorporated into cellular structures, such as membranes, or converted to cholesterol esters through the action of ACAT and stored in lipid droplets (see Fig. 34.4).

Pathways for the Synthesis of Steroid Hormones. Adrenal steroid hormones are synthesized by four CYP enzymes. The CYPs are a large family of oxidative enzymes with a 450 nm absorbance maximum when complexed with carbon monoxide; hence, these molecules were once referred to as cytochrome P450 enzymes. The adrenal CYPs are more commonly known by their trivial names, which denote their function in steroid biosynthesis (see Table 34.3).

The conversion of cholesterol into steroid hormones begins with the formation of free cholesterol from the cholesterol esters stored in intracellular lipid droplets. Free cholesterol molecules enter the mitochondria, which are located close to the lipid droplets, by a mechanism that is not well understood. Evidence indicates that free cholesterol associates with a small protein called **sterol carrier protein 2**, which facilitates its entry into the mitochondrion in some manner. Several other proteins, as well as cAMP, appear to be involved in cholesterol transport into mitochondria, but the process is still unclear.

Once inside a mitochondrion, single cholesterol molecules bind to the **cholesterol side-chain cleavage enzyme** (CYP11A1), embedded in the inner mitochondrial membrane. This enzyme catalyzes the first and rate-limiting reaction in steroidogenesis, which remodels the cholesterol molecule into a 21-carbon steroid intermediate called **pregnenolone**. The reaction occurs in three steps, as shown in Figure 34.3. The first two steps consist of the hydroxylation of carbons 20 and 22 by cholesterol side-chain cleavage enzyme. Then the enzyme cleaves the side chain of cholesterol between carbons 20 and 22, yielding pregnenolone and **isocaproic acid**.

Once formed, pregnenolone molecules dissociate from cholesterol side-chain cleavage enzyme, leave the mitochondrion, and enter the smooth ER nearby. This mechanism is not understood. At this point, the further remodeling of pregnenolone into steroid hormones can vary, depending on whether the process occurs in the zona fasciculata and zona reticularis or the zona glomerulosa. We first consider what occurs in the zona fasciculata and zona reticularis. These biosynthetic events are summarized in Figure 34.5.

TABLE 34.3 Nomenclature for the Steroidogenic Enzymes

Common Name	Previous Form	Current Form	Gene
Cholesterol side-chain cleavage enzyme	$P450_{SCC}$	CYP11A1	*CYP11A1*
3β-Hydroxysteroid dehydrogenase	3β-HSD	3β-HSD II	*HSD3B2*
17α-Hydroxylase	$P450_{C17}$	CYP17	*CYP17*
21-Hydroxylase	$P450_{C21}$	CYP21A2	*CYP21A2*
11β-Hydroxylase	$P450_{C11}$	CYP11B1	*CYP11B1*
Aldosterone synthase	$P450_{C11AS}$	CYP11B2	*CYP11B2*

In cells of the zona fasciculata and zona reticularis, most of the pregnenolone is converted to cortisol and the main adrenal androgen dehydroepiandrosterone (DHEA). Pregnenolone molecules bind to the enzyme **17α-hydroxylase** (CYP17), embedded in the ER membrane, which hydroxylates pregnenolone at carbon 17. The product formed by this reaction is **17α-hydroxypregnenolone** (see Fig. 34.5).

The 17α-hydroxylase has an additional enzymatic action that becomes important at this step in the steroidogenic process. Once the enzyme has hydroxylated carbon 17 of pregnenolone to form 17α-hydroxypregnenolone, it has the ability to lyse or cleave the carbon 20–21 side chain from the steroid structure. Some molecules of 17α-hydroxypregnenolone undergo this reaction and are converted to the 19-carbon steroid DHEA. This action of 17α-hydroxylase is essential for the formation of androgens (19 carbon steroids) and estrogens (18 carbon steroids), which lack the carbon 20–21 side chain. Therefore, this **lyase** activity of 17α-hydroxylase is important in the gonads, where androgens and estrogens are primarily made. 17α-hydroxylase does not exert significant lyase activity in children before age 7 or 8. As a result, young boys and girls do not secrete significant amounts of adrenal androgens. The appearance of significant adrenal androgen secretion in children of both sexes is termed **adrenarche**. It is not related to the onset of puberty, since it normally occurs before the activation of the hypothalamic-pituitary-gonad axis, which initiates puberty. The adrenal androgens produced as a result of adrenarche are a stimulus for the growth of pubic and axillary hair.

Those molecules of 17α-hydroxypregnenolone that dissociate as such from 17α-hydroxylase bind next to another ER enzyme, **3β-hydroxysteroid dehydrogenase** (3β-HSD II). This enzyme acts on 17α-hydroxypregnenolone to isomerize the double bond in ring B to ring A and to dehydrogenate the 3β-hydroxy group, forming a 3-keto group. The product formed is **17α-hydroxyprogesterone** (see Fig. 34.5). This intermediate then binds to another enzyme, **21-hydroxylase** (CYP21A2), which hydroxylates it at carbon 21. The mechanism of this hydroxylation is similar to that performed by the 17α-hydroxylase. The product formed is **11-deoxycortisol**, which is the immediate precursor for cortisol.

To be converted to cortisol, 11-deoxycortisol molecules must be transferred back into the mitochondrion to be acted on by **11β-hydroxylase** (CYP11B1) embedded in the inner mitochondrial membrane. This enzyme hydroxylates 11-deoxycortisol on carbon 11, converting it into cortisol. The 11β-hydroxyl group is the molecular feature that confers glucocorticoid activity on the steroid. Cortisol is then secreted into the bloodstream.

Some of the pregnenolone molecules generated in cells of the zona fasciculata and zona reticularis first bind to 3β-hydroxysteroid dehydrogenase when they enter the endoplasmic reticulum. As a result, they are converted to **progesterone**. Some of these progesterone molecules are hydroxylated by 21-hydroxylase to form the mineralocorticoid **11-deoxycorticosterone** (DOC) (see Fig. 34.5). The 11-deoxycorticosterone formed may be either secreted or transferred back into the mitochondrion. There it is acted on by 11β-hydroxylase to form corticosterone, which is then secreted into the circulation.

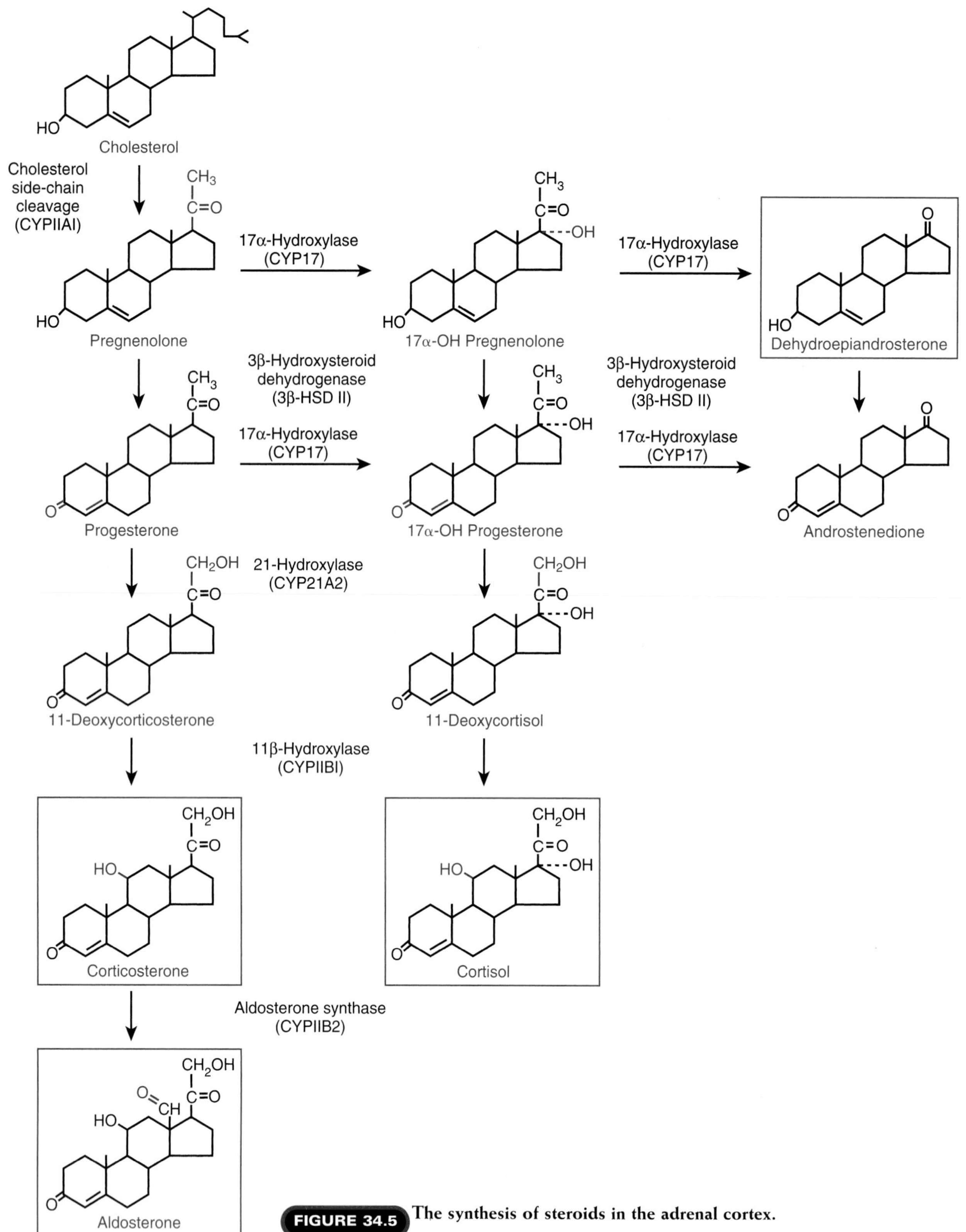

FIGURE 34.5 **The synthesis of steroids in the adrenal cortex.**

Progesterone may also undergo 17α-hydroxylation in the zona fasciculata and zona reticularis. It is then converted to either cortisol or the adrenal androgen **androstenedione**.

The 17α-hydroxylase is not present in cells of the zona glomerulosa; therefore, pregnenolone does not undergo 17α-hydroxylation in these cells, and cortisol and adrenal androgens are not formed by these cells. Instead, the enzymatic pathway leading to the formation of aldosterone is followed (see Fig. 34.5). Pregnenolone is converted by enzymes in the endoplasmic reticulum to progesterone and 11-deoxycorticosterone. The latter compound then moves into the mitochondrion, where it is converted to aldosterone. This conversion involves three steps: the hydroxylation of carbon 11 to form corticosterone, the hydroxylation of carbon 18 to form 18-hydroxycorticosterone, and the oxidation of the 18-hydroxymethyl group to form aldosterone. In humans, these three reactions are catalyzed by a single enzyme, **aldosterone synthase** (CYP11B2), an isozyme of 11β-hydroxylase (CYP11B1), expressed only in glomerulosa cells. The 11β-hydroxylase enzyme, which is expressed in the zona fasciculata and zona reticularis, although closely related to aldosterone synthase, cannot catalyze all three reactions involved in the conversion of 11-deoxycorticosterone to aldosterone; therefore, aldosterone is not synthesized in the zona fasciculata and zona reticularis of the adrenal cortex.

Genetic Defects in Adrenal Steroidogenesis. Inherited genetic defects can cause relative or absolute deficiencies in the enzymes involved in the steroid hormone biosynthetic pathways. The immediate consequences of these defects are changes in the types and amounts of steroid hormones secreted by the adrenal cortex. The end result is disease.

Most of the genetic defects affecting the steroidogenic enzymes impair the formation of cortisol. As discussed in Chapter 32, a drop in cortisol concentration in the blood stimulates the secretion of adrenocorticotropic hormone (ACTH) by the anterior pituitary. The consequent rise in ACTH in the blood exerts a trophic (growth-promoting) effect on the adrenal cortex, resulting in adrenal hypertrophy. Because of this mechanism, individuals with genetic defects affecting adrenal steroidogenesis usually have hypertrophied adrenal glands. These diseases are collectively called **congenital adrenal hyperplasia**.

In humans, inherited genetic defects occur that affect cholesterol side-chain cleavage enzyme, 17α-hydroxylase, 3β-hydroxysteroid dehydrogenase, 21-hydroxylase, 11β-hydroxylase, and aldosterone synthase. The most common defect involves mutations in the gene for 21-hydroxylase and occurs in 1 of 7,000 people. The gene for 21-hydroxylase may be deleted entirely, or mutant genes may code for forms of 21-hydroxylase with impaired enzyme activity. The consequent reduction in the amount of active 21-hydroxylase in the adrenal cortex interferes with the formation of cortisol, corticosterone, and aldosterone, all of which are hydroxylated at carbon 21. Because of the reduction of cortisol (and corticosterone) secretion in these individuals, ACTH secretion is stimulated. This, in turn, causes hypertrophy of the adrenal glands and stimulates the glands to produce steroids.

Because 21-hydroxylation is impaired, the ACTH stimulus causes pregnenolone to be converted to adrenal androgens in inappropriately high amounts. Thus, women afflicted with 21-hydroxylase deficiency exhibit **virilization** from the masculinizing effects of excessive adrenal androgen secretion. In severe cases, the deficiency in aldosterone production can lead to sodium depletion, dehydration, vascular collapse, and death, if appropriate hormone therapy is not given.

Addison's Disease. Glucocorticoid and aldosterone deficiency also occur as a result of pathological destruction of the adrenal glands by microorganisms or autoimmune disease. This disorder is called **Addison's disease**. If sufficient adrenal cortical tissue is lost, the resulting decrease in aldosterone production can lead to vascular collapse and death, unless hormone therapy is given (see Clinical Focus Box 34.1).

Transport of Adrenal Steroids in Blood. As noted earlier, steroid hormones are not stored to any extent by cells of the adrenal cortex but are continually synthesized and secreted. The rate of secretion may change dramatically, however, depending on stimuli received by the adrenal cortical cells. The process by which steroid hormones are secreted is not well studied. It has been assumed that the accumulation of the final products of the steroidogenic pathways creates a concentration gradient for steroid hormone between cells and blood. This gradient is thought to be the driving force for diffusion of the lipid-soluble steroids through cellular membranes and into the circulation.

A large fraction of the adrenal steroids that enter the bloodstream become bound noncovalently to certain plasma proteins. One of these is **corticosteroid-binding globulin** (CBG), a glycoprotein produced by the liver. CBG binds glucocorticoids and aldosterone, but has a greater affinity for the glucocorticoids. **Serum albumin** also binds steroid molecules. Albumin has a high capacity for binding steroids, but its interaction with steroids is weak. The binding of a steroid hormone to a circulating protein molecule prevents it from being taken up by cells or being excreted in the urine.

Circulating steroid hormone molecules not bound to plasma proteins are free to interact with receptors on cells and, therefore, are cleared from the blood. As this occurs, bound hormone dissociates from its binding protein and replenishes the circulating pool of free hormone. Because of this process, adrenal steroid hormones have long half-lives in the body, ranging from many minutes to hours.

Metabolism of Adrenal Steroids in the Liver. Adrenal steroid hormones are eliminated from the body primarily by excretion in the urine after they have been structurally modified to destroy their hormone activity and increase their water solubility. Although many cells are capable of carrying out these modifications, they primarily occur in the liver.

The most common structural modifications made in adrenal steroids involve reduction of the double bond in ring A and conjugation of the resultant hydroxyl group formed on carbon 3 with glucuronic acid. Figure 34.6 shows how cortisol is modified in this manner to produce a major excretable metabolite, **tetrahydrocortisol glucuronide**. Cortisol, and other 21-carbon steroids with a 17α-hydroxyl group and a 20-keto group, may undergo lysis of the carbon 20–21 side chain as well. The resultant metabolite, with a keto group on carbon 17, appears as one of the **17-ketosteroids** in the urine. Adrenal androgens are also 17-ketosteroids. They are usually conjugated with sulfuric acid or glucuronic acid before being excreted and normally comprise the bulk of the 17-ketosteroids in the urine. Before the development of specific methods to measure androgens and 17α-hydroxycorticosteroids in body fluids, the amount of 17-ketosteroids in urine was used clinically as a crude in-

CLINICAL FOCUS BOX 34.1

Primary Adrenal Insufficiency: Addison's Disease

Adrenal insufficiency may be caused by destruction of the adrenal cortex (primary adrenal insufficiency), low pituitary ACTH secretion (secondary adrenal insufficiency), or deficient hypothalamic release of CRH (tertiary adrenal insufficiency). **Addison's disease** (primary adrenal insufficiency) results from the destruction of the adrenal gland by microorganisms or autoimmune disease. When Addison first described primary adrenal insufficiency in the mid-1800s, bilateral adrenal destruction by tuberculosis was the most common cause of the disease. Today, autoimmune destruction accounts for 70 to 90% of all cases, with the remainder resulting from infection, cancer, or adrenal hemorrhage. The prevalence of primary adrenal insufficiency is about 40 to 110 cases per 1 million adults, with an incidence of about 6 cases per 1 million adults per year.

In primary adrenal insufficiency, all three zones of the adrenal cortex are usually involved. The result is inadequate secretion of glucocorticoids, mineralocorticoids, and androgens. Major symptoms are not usually detected until 90% of the gland has been destroyed. The initial symptoms generally have a gradual onset, with only a partial glucocorticoid deficiency resulting in inadequate cortisol increase in response to stress. Mineralocorticoid deficiency may only appear as a mild postural hypotension. Progression to complete glucocorticoid deficiency results in a decreased sense of well-being and abnormal glucose metabolism. Lack of mineralocorticoid leads to decreased renal potassium secretion and reduced sodium retention, the loss of which results in hypotension and dehydration. The combined lack of glucocorticoid and mineralocorticoid can lead to vascular collapse, shock, and death. Adrenal androgen deficiency is observed in women only (men derive most of their androgen from the testes) as decreased pubic and axillary hair and decreased libido.

Antibodies that react with all three zones of the adrenal cortex have been identified in autoimmune adrenalitis and are more common in women than in men. The presence of antibodies appears to precede the development of adrenal insufficiency by several years. Antiadrenal antibodies are mainly directed to the steroidogenic enzymes cholesterol side-chain cleavage enzyme (CYP11A1), 17α-hydroxylase (CYP17), and 21-hydroxylase (CYP21A2), although antibodies to other steroidogenic enzymes may also be present. In the initial stages of the disease, the adrenal glands may be enlarged with extensive lymphocyte infiltration. Genetic susceptibility to autoimmune adrenal insufficiency is strongly linked with the HLA-B8, HLA-DR3, and HLA-DR4 alleles of human leukocyte antigen (HLA). The earliest sign of adrenal insufficiency is an increase in plasma renin activity, with a low or normal aldosterone level, which suggests that the zona glomerulosa is affected first during disease progression.

Treatment for acute adrenal insufficiency should be directed at reversal of the hypotension and electrolyte abnormalities. Large volumes of 0.9% saline or 5% dextrose in saline should be infused as quickly as possible. Dexamethasone or a soluble form of injectable cortisol should also be given. Daily glucocorticoid and mineralocorticoid replacement allows the patient to lead a normal active life.

Reference

Orth DN, Kovacs WJ. The adrenal cortex. In: Wilson JD, Foster DW, Kronenberg HM, Larsen PR, eds. Williams Textbook of Endocrinology. 9th Ed. Philadelphia: WB Saunders, 1998;517–664.

dicator of the production of these substances by the adrenal gland.

ACTH Regulates the Synthesis of Adrenal Steroids

Adrenocorticotropic hormone (ACTH) is the physiological regulator of the synthesis and secretion of glucocorticoids and androgens by the zona fasciculata and zona reticularis. It has a very rapid stimulatory effect on steroidogenesis in these cells, which can result in a great rise in blood glucocorticoids within seconds. It also exerts several long-term trophic effects on these cells, all directed toward maintaining the cellular machinery necessary to carry out steroidogenesis at a high, sustained rate. These actions of ACTH are summarized in Figure 34.7.

Role of cAMP. When the level of ACTH in the blood rises, increased numbers of ACTH molecules interact with receptors on the plasma membranes of adrenal cortical cells. These ACTH receptors are coupled to the enzyme adenylyl cyclase by stimulatory guanine nucleotide-binding proteins (G_s proteins). The production of cAMP from ATP greatly increases, and the concentration of cAMP rises in the cell. cAMP activates protein kinase A (PKA), which phosphorylates proteins that regulate steroidogenesis.

The rapid rise in cAMP produced by ACTH stimulates the mechanism that transfers cholesterol into the inner mitochondrial membrane. This action provides abundant cholesterol for side-chain cleavage enzyme, which carries out the rate-limiting step in steroidogenesis. As a result, the rates of steroid hormone formation and secretion rise greatly.

Gene Expression for Steroidogenic Enzymes. Adrenocorticotropic hormone maintains the capacity of the cells of the zona fasciculata and zona reticularis to produce steroid hormones by stimulating the transcription of the genes for many of the enzymes involved in steroidogenesis. For example, transcription of the genes for side-chain cleavage enzyme, 17α-hydroxylase, 21-hydroxylase, and 11β-hydroxylase, is increased several hours after adrenal cortical cells have been stimulated by ACTH. Because normal individuals are continually exposed to episodes of ACTH secretion (see Fig. 32.7), the mRNA for these enzymes is well maintained in the cells. Again, this long-term or maintenance effect of ACTH is due to its ability to increase cAMP in the cells (see Fig. 34.7).

The importance of ACTH in gene transcription be-

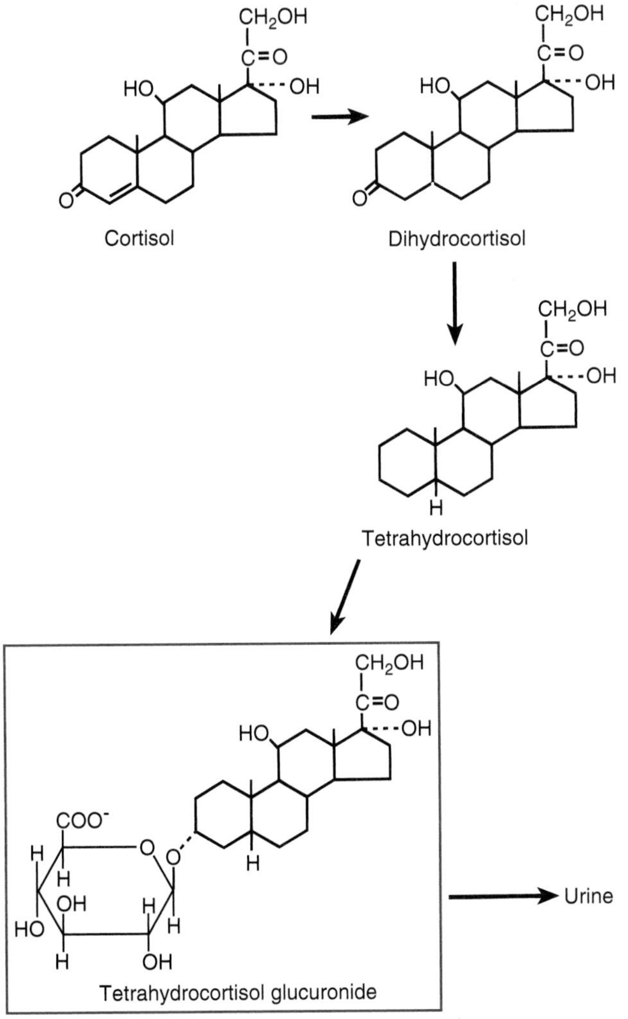

FIGURE 34.6 **The metabolism of cortisol to tetrahydrocortisol glucuronide in the liver.** The reduced and conjugated steroid is inactive. Because it is more water-soluble than cortisol, it is easily excreted in the urine.

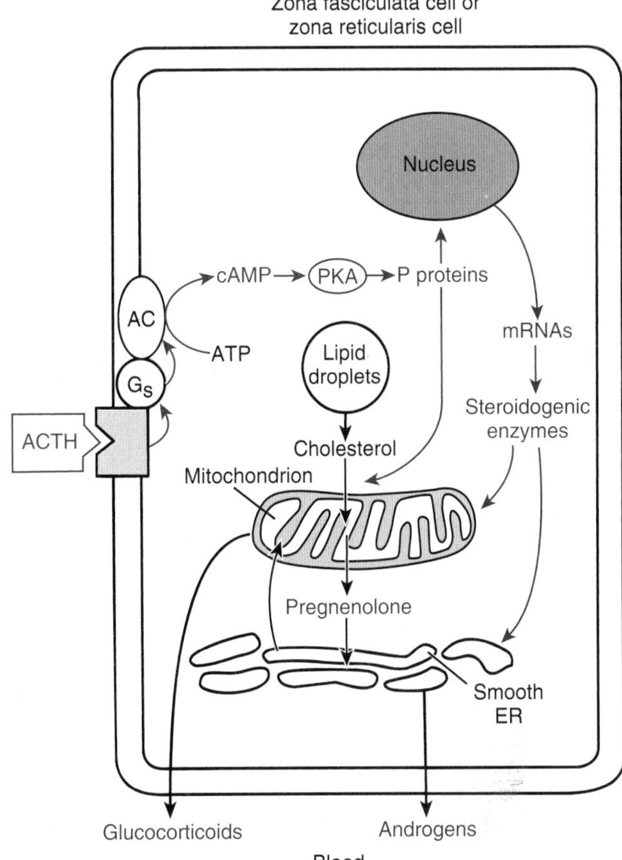

FIGURE 34.7 **The main actions of ACTH on steroidogenesis.** ACTH binds to plasma membrane receptors, which are coupled to adenylyl cyclase (AC) by stimulatory G proteins (G_s). cAMP rises in the cells and activates protein kinase A (PKA), which then phosphorylates certain proteins (P-Proteins). These proteins presumably initiate steroidogenesis and stimulate the expression of genes for steroidogenic enzymes.

comes evident in hypophysectomized animals or humans with ACTH deficiency. An example of the latter is a human treated chronically with large doses of cortisol or related steroids, which causes prolonged suppression of ACTH secretion by the anterior pituitary. The chronic lack of ACTH decreases the transcription of the genes for steroidogenic enzymes, causing a deficiency in these enzymes in the adrenals. As a result, the administration of ACTH to such an individual does not cause a marked increase in glucocorticoid secretion. Chronic exposure to ACTH is required to restore mRNA levels for the steroidogenic enzymes and, hence, the enzymes themselves, to obtain normal steroidogenic responses to ACTH. A patient receiving long-term treatment with glucocorticoid may suffer serious glucocorticoid deficiency if hormone therapy is halted abruptly; withdrawing glucocorticoid therapy gradually allows time for endogenous ACTH to restore steroidogenic enzyme levels to normal.

Effects on Cholesterol Metabolism. ACTH has several long-term effects on cholesterol metabolism that support steroidogenesis in the zona fasciculata and zona reticularis. It increases the abundance of LDL receptors and the activity of the enzyme HMG-CoA reductase in these cells. These actions increase the availability of cholesterol for steroidogenesis. It is not clear whether ACTH exerts these effects directly. The abundance of LDL receptors in the plasma membrane and the activity of HMG-CoA reductase in most cells are inversely related to the amount of cellular cholesterol. By stimulating steroidogenesis, ACTH reduces the amount of cholesterol in adrenal cells; therefore, the increased abundance of LDL receptors and high HMG-CoA reductase activity in ACTH-stimulated cells may merely result from the normal compensatory mechanisms that function to maintain cell cholesterol levels.

ACTH also stimulates the activity of cholesterol esterase in adrenal cells, which promotes the hydrolysis of the cholesterol esters stored in the lipid droplets of these cells, making free cholesterol available for steroidogenesis. The cholesterol esterase in the adrenal cortex appears to be identical to hormone-sensitive lipase, which is activated when it is phosphorylated by a cAMP-dependent protein kinase. The rise in cAMP concentration produced by ACTH might account for its effect on the enzyme.

Trophic Action on Adrenal Cortical Cell Size. ACTH maintains the size of the two inner zones of the adrenal cortex, presumably by stimulating the synthesis of structural elements of the cells; however, it does not affect the size of the cells of the zona glomerulosa. The trophic effect of ACTH is clearly evident in states of ACTH deficiency or excess. In hypophysectomized or ACTH-deficient individuals, the cells of the two inner zones atrophy. Chronic stimulation of these cells with ACTH causes them to hypertrophy. The mechanisms involved in this trophic action of ACTH are unclear.

ACTH and Aldosterone Production. The cells of the zona glomerulosa have ACTH receptors, which are coupled to adenylyl cyclase. In these cells, cAMP increases in response to ACTH, resulting in some increase in aldosterone secretion. However, angiotensin II is the important physiological regulator of aldosterone secretion, not ACTH. Other factors, such as an increase in serum potassium, can also stimulate aldosterone secretion, but normally, they play only a secondary role.

Formation of Angiotensin II. Angiotensin II is a short peptide consisting of eight amino acid residues. It is formed in the bloodstream by the proteolysis of the α_2-globulin **angiotensinogen**, which is secreted by the liver. The formation of angiotensin II occurs in two stages (Fig. 34.8). Angiotensinogen is first cleaved at its N-terminal end by the circulating protease **renin**, releasing the inactive decapeptide **angiotensin I**. Renin is produced and secreted by granular (juxtaglomerular) cells in the kidneys (see Chapter 23). A dipeptide is then removed from the C-terminal end of angiotensin I, producing angiotensin II. This cleavage is performed by the protease **angiotensin-converting enzyme** present on the endothelial cells lining the vasculature. This step usually occurs as angiotensin I molecules traverse the pulmonary circulation. The rate-limiting factor for the formation of angiotensin II is the renin concentration of the blood.

Cleavage of the N-terminal aspartate from angiotensin II results in the formation of angiotensin III, which circulates at a concentration of 20% that of angiotensin II. Angiotensin III is as potent a stimulator of aldosterone secretion as angiotensin II.

Action of Angiotensin II on Aldosterone Secretion. Angiotensin II stimulates aldosterone synthesis by promoting the rate-limiting step in steroidogenesis (i.e., the movement of cholesterol into the inner mitochondrial membrane and its conversion to pregnenolone). The primary mechanism is shown in Figure 34.9.

The stimulation of aldosterone synthesis is initiated when angiotensin II binds to its receptors on the plasma membranes of zona glomerulosa cells. The signal generated by the interaction of angiotensin II with its receptors is transmitted to phospholipase C (PLC) by a G protein, and the enzyme becomes activated. The PLC then hydrolyzes phosphatidylinositol 4,5 bisphosphate (PIP_2) in the plasma membrane, producing the intracellular second messengers inositol trisphosphate (IP_3) and diacylglycerol (DAG). The IP_3 mobilizes calcium, which is bound to intracellular structures, increasing the calcium concentration in the cytosol. This increase in intracellular calcium and DAG activates protein kinase C (PKC). The rise in intracellular calcium also activates **calmodulin-dependent protein kinase** (CMK). These enzymes phosphorylate proteins, which then become involved in initiating steroidogenesis.

Signals for Increased Angiotensin II Formation. Although angiotensin II is the final mediator in the physiological regulation of aldosterone secretion, its formation from angiotensinogen is dependent on the secretion of renin by the kidneys. The rate of renin secretion ultimately determines the rate of aldosterone secretion. Renin is secreted by the granular cells in the walls of the afferent arterioles of renal glomeruli. These cells are stimulated to secrete renin by three signals that indicate a possible loss of body fluid: a fall in blood pressure in the afferent arterioles of the glomeruli, a drop in sodium chloride concentration in renal tubular fluid at the macula densa, and an increase in renal sympathetic nerve activity (see Chapters 23 and 24).

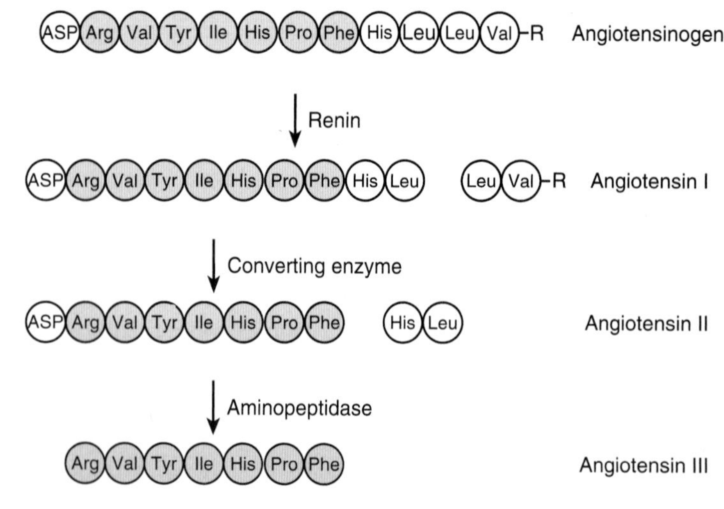

FIGURE 34.8 **The formation of angiotensins I, II, and III from angiotensinogen.**

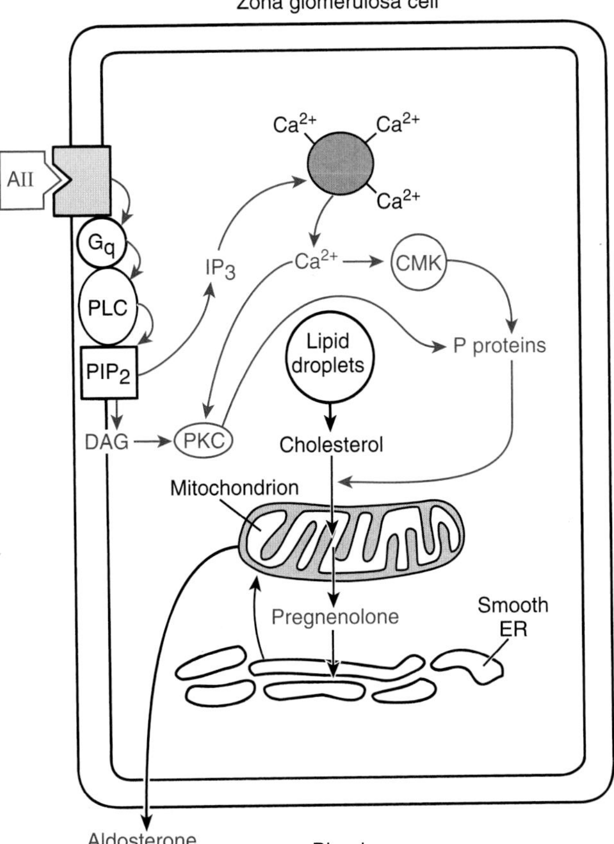

FIGURE 34.9 **The action of angiotensin II on aldosterone synthesis.** Angiotensin II (AII) binds to receptors on the plasma membrane of zona glomerulosa cells. This activates phospholipase C (PLC), which is coupled to the angiotensin II receptor by G proteins (G_q). PLC hydrolyzes phosphatidylinositol 4,5 bisphosphate (PIP_2) in the plasma membrane, producing inositol trisphosphate (IP_3) and diacylglycerol (DAG). IP_3 mobilizes intracellularly bound Ca^{2+}. The rise in Ca^{2+} and DAG activates protein kinase C (PKC) and calmodulin-dependent protein kinase (CMK). These enzymes phosphorylate proteins (P-Proteins) involved in initiating aldosterone synthesis.

Increased renin secretion results in an increase in angiotensin II formation in the blood, thereby stimulating aldosterone secretion by the zona glomerulosa. This series of events tends to conserve body fluid volume because aldosterone stimulates sodium reabsorption by the kidneys.

Extracellular Potassium Concentration and Aldosterone Secretion. Aldosterone secretion is also stimulated by an increase in the potassium concentration in extracellular fluid, caused by a direct effect of potassium on zona glomerulosa cells. Glomerulosa cells are sensitive to this effect of extracellular potassium and, therefore, increase their rate of aldosterone secretion in response to small increases in blood and interstitial fluid potassium concentration. This signal for aldosterone secretion is appropriate from a physiological point of view because aldosterone promotes the renal excretion of potassium (see Chapter 24).

A rise in extracellular potassium depolarizes glomerulosa cell membranes, activating voltage-dependent calcium channels in the membranes. The consequent rise in cytosolic calcium is thought to stimulate aldosterone synthesis by the mechanisms described above for the action of angiotensin II.

Aldosterone and Sodium Reabsorption by Kidney Tubules. The physiological action of aldosterone is to stimulate sodium reabsorption in the kidneys by the distal tubule and collecting duct of the nephron and to promote the excretion of potassium and hydrogen ions. The mechanism of action of aldosterone on the kidneys and its role in water and electrolyte balance are discussed in Chapter 24.

Glucocorticoids Play a Role in the Reactions to Fasting, Injury, and Stress

Glucocorticoids widely influence physiological processes. In fact, most cells have receptors for glucocorticoids and are potential targets for their actions. Consequently, glucocorticoids have been used extensively as therapeutic agents, and much is known about their pharmacological effects.

Actions on Transcription. Unlike many other hormones, glucocorticoids influence physiological processes slowly, sometimes taking hours to produce their effects. Glucocorticoids that are free in the blood diffuse through the plasma membranes of target cells; once inside, they bind tightly but noncovalently to receptor proteins present in the cytoplasm. The interaction between the glucocorticoid molecule and its receptor molecule produces an activated glucocorticoid-receptor complex, which translocates into the nucleus.

These complexes then bind to specific regions of DNA called **glucocorticoid response elements** (GREs), which are near glucocorticoid-sensitive target genes. The binding triggers events that either stimulate or inhibit the transcription of the target gene. As a result of the change in transcription, amounts of mRNA for certain proteins are either increased or decreased. This, in turn, affects the abundance of these proteins in the cell, which produces the physiological effects of the glucocorticoids. The apparent slowness of glucocorticoid action is due to the time required by the mechanism to change the protein composition of a target cell.

Glucocorticoids and the Metabolic Response to Fasting. During the fasting periods between food intake in humans, metabolic adaptations prevent hypoglycemia. The maintenance of sufficient blood glucose is necessary because the brain depends on glucose for its energy needs. Many of the adaptations that prevent hypoglycemia are not fully expressed in the course of daily life because the individual eats before they fully develop. Full expression of these changes is seen only after many days to weeks of fasting. Glucocorticoids are necessary for the metabolic adaptation to fasting.

At the onset of a prolonged fast, there is a gradual decline in the concentration of glucose in the blood. Within 1 to 2 days, the blood glucose level stabilizes at a concentration of 60 to 70 mg/dL, where it remains even if the fast is prolonged for many days (Fig. 34.10). The blood glucose

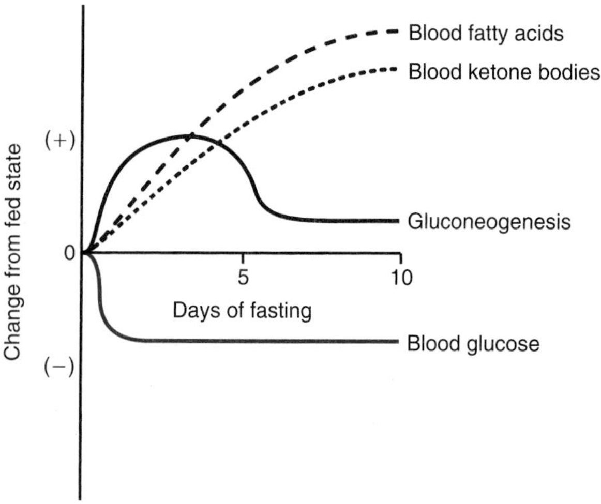

FIGURE 34.10 Metabolic adaptations during fasting. This graphs shows the changes in the concentrations of blood glucose, fatty acids, and ketone bodies and the rate of gluconeogenesis during the course of a prolonged fast. Only the *direction* of change over time is indicated: increase (+) or decrease (−).

level is stabilized by the production of glucose by the body and the restriction of its use by tissues other than the brain. Although a limited supply of glucose is available from glycogen stored in the liver, the more important source of blood glucose during the first days of a fast is gluconeogenesis in the liver and, to some extent, in the kidneys.

Gluconeogenesis begins several hours after the start of a fast. Amino acids derived from tissue protein are the main substrates. Fasting results in protein breakdown in the skeletal muscle and accelerated release of amino acids into the bloodstream. Protein breakdown and protein accretion in adult humans are regulated by two opposing hormones, insulin and glucocorticoids. During fasting, insulin secretion is suppressed and the inhibitory effect of insulin on protein breakdown is lost. As proteins are broken down, glucocorticoids inhibit the reuse of amino acids derived from tissue proteins for new protein synthesis, promoting the release of these amino acids from the muscle. Amino acids released into the blood by the skeletal muscle are extracted from the blood at an accelerated rate by the liver and kidneys. The amino acids then undergo metabolic transformations in these tissues, leading to the synthesis of glucose. The newly synthesized glucose is then delivered to the bloodstream.

The glucocorticoids are essential for the acceleration of gluconeogenesis during fasting. They play a permissive role in this process by maintaining gene expression and, therefore, the intracellular concentrations of many of the enzymes needed to carry out gluconeogenesis in the liver and kidneys. For example, glucocorticoids maintain the amounts of transaminases, pyruvate carboxylase, phosphoenolpyruvate carboxykinase, fructose-1,6-diphosphatase, fructose-6-phosphatase, and glucose-6-phosphatase needed to carry out gluconeogenesis at an accelerated rate. In an untreated, glucocorticoid-deficient individual, the amounts of these enzymes in the liver are greatly reduced. As a consequence, the individual cannot respond to fasting with accelerated gluconeogenesis and will die from hypoglycemia. In essence, the glucocorticoids maintain the liver and kidney in a state that enables them to carry out accelerated gluconeogenesis should the need arise.

The other important metabolic adaptation that occurs during fasting involves the mobilization and use of stored fat. Within the first few hours of the start of a fast, the concentration of free fatty acids rises in the blood (see Fig. 34.10). This action is due to the acceleration of lipolysis in the fat depots, as a result of the activation of **hormone-sensitive lipase** (HSL). HSL hydrolyzes the stored triglyceride to free fatty acids and glycerol, which are released into the blood.

HSL is activated when it is phosphorylated by cAMP-dependent protein kinase (PKA). As the level of insulin falls in the blood during fasting, the inhibitory effect of insulin on cAMP accumulation in the fat cell diminishes. There is a rise in the cellular level of cAMP, and HSL is activated. The glucocorticoids are essential for maintaining fat cells in an enzymatic state that permits lipolysis to occur during a fast. This is evident from the fact that accelerated lipolysis does not occur when a glucocorticoid-deficient individual fasts.

The abundant fatty acids produced by lipolysis are taken up by many tissues. The fatty acids enter mitochondria, undergo β-oxidation to acetyl CoA, and become the substrate for ATP synthesis. The enhanced use of fatty acids for energy metabolism spares the blood glucose supply. There is also significant gluconeogenesis in liver from the glycerol released from triglyceride by lipolysis. In prolonged fasting, when the rate of glucose production from body protein has declined, a significant fraction of blood glucose is derived from triglyceride glycerol.

Within a few hours of the start of a fast, the increased delivery to and oxidation of fatty acids in the liver results in the production of the ketone bodies. As a result of these events in the liver, a gradual rise in ketone bodies occurs in the blood as a fast continues over many days (Fig. 34.10). Ketone bodies become the principal energy source used by the CNS during the later stages of fasting.

The increased use of fatty acids for energy metabolism by skeletal muscle results in less use of glucose in this tissue, sparing blood glucose for use by the CNS. Two products resulting from the breakdown of fatty acids, acetyl CoA and citrate, inhibit glycolysis. As a result, the uptake and use of glucose from the blood is reduced.

In summary, the strategy behind the metabolic adaptation to fasting is to provide the body with glucose produced primarily from protein until the ketone bodies become abundant enough in the blood to be a principal source of energy for the brain. From that point on, the body uses mainly fat for energy metabolism, and it can survive until the fat depots are exhausted. Glucocorticoids do not trigger the metabolic adaptations to fasting but only provide the metabolic machinery necessary for the adaptations to occur.

Cushing's Syndrome. When present in excessive amounts, glucocorticoids can trigger many of the metabolic adaptations to the fasting state. **Cushing's syndrome** is the name of such pathological hypercortisolic states.

Cushing's syndrome may be ACTH-dependent or ACTH-independent. One type of ACTH-dependent syndrome (actually called **Cushing's disease**) is caused by a corticotroph adenoma, which secretes excessive ACTH and stimulates the adrenal cortex to produce large amounts of cortisol. ACTH-independent Cushing's syndrome is usually a result of an adrenocortical adenoma that secretes large amounts of cortisol. Whatever the cause, prolonged exposure of the body to large amounts of glucocorticoids causes the breakdown of skeletal muscle protein, increased glucose production by the liver, and mobilization of lipid from the fat depots. Despite the increased mobilization of lipid, there is also an abnormal deposition of fat in the abdominal region, between the shoulders, and in the face. The increased mobilization of lipid provides abundant fatty acids for metabolism, and the increased oxidation of fatty acids by tissues reduces their ability to use glucose. The underutilization of glucose by skeletal muscle, coupled with increased glucose production by the liver, results in hyperglycemia, which, in turn, stimulates the pancreas to secrete insulin. In this instance, however, the rise in insulin is not effective in reducing the blood glucose concentration because glucose uptake and use are decreased in the skeletal muscle and adipose tissue. Evidence also indicates that excessive glucocorticoids decrease the affinity of insulin receptors for insulin. The net result is that the individual becomes insensitive or resistant to the action of insulin and little glucose is removed from the blood, despite the high level of circulating insulin. The persisting hyperglycemia continually stimulates the pancreas to secrete insulin. The result is a form of "diabetes" similar to type 2 diabetes mellitus (see Chapter 35).

The opposite situation occurs in the glucocorticoid-deficient individual. Little lipid mobilization and use occur, so there is little restriction on the rate of glucose use by tissues. The glucocorticoid-deficient individual is sensitive to insulin in that a given concentration of blood insulin is more effective in clearing the blood of glucose than it is in a healthy person. The administration of even small doses of insulin to such individuals may produce hypoglycemia.

The Anti-inflammatory Action of Glucocorticoids. Tissue injury triggers a complex mechanism called **inflammation** that precedes the actual repair of damaged tissue. A host of chemical mediators are released into the damaged area by neighboring cells, adjacent vasculature, and phagocytic cells that migrate to the damaged site. Mediators released under these circumstances include **prostaglandins, leukotrienes, kinins, histamine, serotonin,** and **lymphokines**. These substances exert a multitude of actions at the site of injury and directly or indirectly promote the local vasodilation, increased capillary permeability, and edema formation that characterize the inflammatory response (see Chapter 11).

Because glucocorticoids inhibit the inflammatory response to injury, they are used extensively as therapeutic anti-inflammatory agents; however, the mechanisms are not clear. Their regulation of the production of prostaglandins and leukotrienes is the best understood. These substances play a major role in mediating the inflammatory reaction. They are synthesized from the unsaturated fatty acid arachidonic acid, which is released from plasma membrane phospholipids by the hydrolytic action of **phospholipase A_2**. Glucocorticoids stimulate the synthesis of a family of proteins called **lipocortins** in their target cells. Lipocortins inhibit the activity of phospholipase A_2, reducing the amount of arachidonic acid available for conversion to prostaglandins and leukotrienes.

Effects on the Immune System. Glucocorticoids have little influence on the human immune system under normal physiological conditions. When administered in large doses over a prolonged period, however, they can suppress antibody formation and interfere with cell-mediated immunity. Glucocorticoid therapy, therefore, is used to suppress the rejection of surgically transplanted organs and tissues.

Immature T cells in the thymus and immature B cells and T cells in lymph nodes can be killed by exposure to high concentrations of glucocorticoids, decreasing the number of circulating lymphocytes. The destruction of immature T and B cells by glucocorticoids also causes some reduction in the size of the thymus and lymph nodes.

Maintenance of the Vascular Response to Norepinephrine (NE). Glucocorticoids are required for the normal responses of vascular smooth muscle to the vasoconstrictor action of NE. NE is much less active on vascular smooth muscle in the absence of glucocorticoids and is another example of the permissive action of glucocorticoids.

Glucocorticoids and Stress. Perhaps the most interesting, but least understood, of all glucocorticoid action is the ability to protect the body against stress. All that is really known is that the body cannot cope successfully with even mild stresses in the absence of glucocorticoids. One must presume that the processes that enable the body to defend itself against physical or emotional trauma require glucocorticoids. This, again, emphasizes the permissive role they play in physiological processes.

Stress stimulates the secretion of ACTH, which increases the secretion of glucocorticoids by the adrenal cortex (see Chapter 32). In humans, this increase in glucocorticoid secretion during stress appears to be important for the appropriate defense mechanisms to be put into place. It is well known, for example, that glucocorticoid-deficient individuals receiving replacement therapy require larger doses of glucocorticoid to maintain their well-being during periods of stress.

Regulation of Glucocorticoid Secretion. An important physiological action of glucocorticoids is the ability to regulate their own secretion. This effect is achieved by a negative-feedback mechanism of glucocorticoids on the secretion of corticotropin-releasing hormone (CRH) and ACTH and on proopiomelanocortin (POMC) gene expression (see Chapter 32).

PRODUCTS OF THE ADRENAL MEDULLA

The catecholamines, epinephrine and norepinephrine (NE), are the two hormones synthesized by the chromaffin cells of the adrenal medulla. The human adrenal medulla produces and secretes about 4 times more epinephrine than NE. Postganglionic sympathetic neurons also produce and release NE from their nerve terminals but do not produce epinephrine.

Epinephrine and NE are formed in the chromaffin cells from the amino acid tyrosine. The pathway for the synthesis of catecholamines is illustrated in Figure 3.18.

Trauma, Exercise, and Hypoglycemia Stimulate the Medulla to Release Catecholamines

Epinephrine and some NE are released from chromaffin cells by the fusion of secretory granules with the plasma membrane. The contents of the granules are extruded into the interstitial fluid. The catecholamines diffuse into capillaries and are transported in the bloodstream.

Neural stimulation of the cholinergic preganglionic fibers that innervate chromaffin cells triggers the secretion of catecholamines. Stimuli such as injury, anger, anxiety, pain, cold, strenuous exercise, and hypoglycemia generate impulses in these fibers, causing a rapid discharge of the catecholamines into the bloodstream.

Catecholamines Have Rapid, Widespread Effects

Most cells of the body have receptors for catecholamines and, thus, are their target cells. There are four structurally related forms of catecholamine receptors, all of which are transmembrane proteins: α_1, α_2, β_1, and β_2. All can bind epinephrine or NE, to varying extents (see Chapter 3).

Fight-or-Flight Response. Epinephrine and NE produce widespread effects on the cardiovascular system, muscular system, and carbohydrate and lipid metabolism in liver, muscle, and adipose tissues. In response to a sudden rise in catecholamines in the blood, the heart rate accelerates, coronary blood vessels dilate, and blood flow to the skeletal muscles is increased as a result of vasodilation (but vasoconstriction occurs in the skin). Smooth muscles in the airways of the lungs, gastrointestinal tract, and urinary bladder relax. Muscles in the hair follicles contract, causing piloerection. Blood glucose level also rises. This overall reaction to the sudden release of catecholamines is known as the fight-or-flight response (see Chapter 6).

Catecholamines and the Metabolic Response to Hypoglycemia. Catecholamines secreted by the adrenal medulla and NE released from sympathetic postganglionic nerve terminals are key agents in the body's defense against hypoglycemia. Catecholamine release usually starts when the blood glucose concentration falls to the low end of the physiological range (60 to 70 mg/dL). A further decline in blood glucose concentration into the hypoglycemic range produces marked catecholamine release. Hypoglycemia can result from a variety of situations, such as insulin overdosing, catecholamine antagonists, or drugs that block fatty acid oxidation. Hypoglycemia is always a dangerous condition because the CNS will die of ATP deprivation in extended cases. The length of time profound hypoglycemia can be tolerated depends on its severity and the individual's sensitivity.

When the blood glucose concentration drops toward the hypoglycemic range, CNS receptors monitoring blood glucose are activated, stimulating the neural pathway leading to the fibers innervating the chromaffin cells. As a result, the adrenal medulla discharges catecholamines. Sympathetic postganglionic nerve terminals also release NE.

Catecholamines act on the liver to stimulate glucose production. They activate glycogen phosphorylase, resulting in the hydrolysis of stored glycogen, and stimulate gluconeogenesis from lactate and amino acids. Catecholamines also activate glycogen phosphorylase in skeletal muscle and adipose cells by interacting with β receptors, activating adenylyl cyclase and increasing cAMP in the cells. The elevated cAMP activates glycogen phosphorylase. The glucose 6-phosphate generated in these cells is metabolized, although glucose is not released into the blood, since the cells lack glucose-6-phosphatase. The glucose 6-phosphate in muscle is converted by glycolysis to lactate, much of which is released into the blood. The lactate taken up by the liver is converted to glucose via gluconeogenesis and returned to the blood.

In adipose cells, the rise in cAMP produced by catecholamines activates hormone-sensitive lipase, causing the hydrolysis of triglycerides and the release of fatty acids and glycerol into the bloodstream. These fatty acids provide an alternative substrate for energy metabolism in other tissues, primarily skeletal muscle, and block the phosphorylation and metabolism of glucose.

During profound hypoglycemia, the rapid rise in blood catecholamine levels triggers some of the same metabolic adjustments that occur more slowly during fasting. During fasting, these adjustments are triggered mainly in response to the gradual rise in the ratio of glucagon to insulin in the blood. The ratio also rises during profound hypoglycemia, reinforcing the actions of the catecholamines on glycogenolysis, gluconeogenesis, and lipolysis. The catecholamines released during hypoglycemia are thought to be partly responsible for the rise in the glucagon-to-insulin ratio by directly influencing the secretion of these hormones by the pancreas. Catecholamines stimulate the secretion of glucagon by the alpha cells and inhibit the secretion of insulin by beta cells (see Chapter 35). These catecholamine-mediated responses to hypoglycemia are summarized in Table 34.4.

TABLE 34.4 Catecholamine-Mediated Responses to Hypoglycemia

Liver	Stimulation of glycogenolysis
	Stimulation of gluconeogenesis
Skeletal muscle	Simulation of glycogenolysis
Adipose tissue	Simulation of glycogenolysis
	Stimulation of triglyceride lipolysis
Pancreatic islets	Inhibition of insulin secretion by beta cells
	Stimulation of glucagon secretion by alpha cells

REVIEW QUESTIONS

DIRECTIONS: Each of the numbered items or incomplete statements in this section is followed by answers or by completions of the statement. Select the ONE lettered answer or completion that is the BEST in each case.

1. Which of the following sources of cholesterol is most important for sustaining adrenal steroidogenesis when it occurs at a high rate for a long time?
(A) *De novo* synthesis of cholesterol from acetate
(B) Cholesterol in LDL particles
(C) Cholesterol in the plasma membrane
(D) Cholesterol in lipid droplets within adrenal cortical cells
(E) Cholesterol from the endoplasmic reticulum
(F) Cholesterol in lipid droplets within adrenal medullary cells
2. A 7-year-old boy comes to the pediatric endocrine unit for evaluation of excess body weight. Review of his growth charts indicates substantial weight gain over the previous 3 years but little increase in height. To differentiate between the development of obesity and Cushing's disease, blood and urine samples are taken. Which of the following would be most diagnostic of Cushing's disease?
(A) Increased serum ACTH, decreased serum cortisol, and increased urinary free cortisol
(B) Decreased serum ACTH, increased serum cortisol, and increased serum insulin
(C) Increased serum ACTH, increased serum cortisol, and increased serum insulin
(D) Increased serum ACTH, decreased serum cortisol, and decreased serum insulin
(E) Increased serum ACTH, decreased serum cortisol, and decreased urinary free cortisol
(F) Decreased serum ACTH, decreased serum cortisol, and increased serum insulin
3. Congenital adrenal hyperplasia is most likely a result of
(A) Defects in adrenal steroidogenic enzymes
(B) Addison's disease
(C) Defects in ACTH secretion
(D) Defects in corticosteroid-binding globulin
(E) Cushing's disease
(F) Defects in aldosterone synthase
4. What is the mechanism through which catecholamines stabilize blood glucose concentration in response to hypoglycemia?
(A) Catecholamines stimulate glycogen phosphorylase to release glucose from muscle
(B) Catecholamines inhibit glycogenolysis in the liver
(C) Catecholamines stimulate the release of insulin from the pancreas
(D) Catecholamines inhibit the release of fatty acids from adipose tissue
(E) Catecholamines stimulate gluconeogenesis in the liver
(F) Catecholamines inhibit the release of lactate from muscle
5. A patient receiving long-term glucocorticoid therapy plans to undergo hip replacement surgery. What would the physician recommend prior to surgery and why?
(A) Glucocorticoids should be decreased to prevent serious hypoglycemia during recovery
(B) Glucocorticoids should be increased to stimulate immune function and prevent possible infection
(C) Glucocorticoids should be decreased to minimize potential interactions with anesthetics
(D) Glucocorticoids should be increased to stimulate ACTH secretion during surgery to promote wound healing
(E) Glucocorticoids should be decreased to prevent inadequate vascular response to catecholamines during recovery
(F) Glucocorticoids should be increased to compensate for the increased stress associated with surgery
6. Which of the following is most likely to result in a decreased rate of aldosterone release?
(A) An increase in renin secretion by the kidney
(B) A rise in serum potassium
(C) A fall in blood pressure in the kidney
(D) A decrease in tubule fluid sodium concentration at the macula densa
(E) An increase in renal sympathetic nerve activity
(F) A decrease in IP_3 in cells of the zona glomerulosa
7. The rate-limiting step in the synthesis of cortisol is catalyzed by
(A) 21-Hydroxylase
(B) 3β-Hydroxysteroid dehydrogenase
(C) Cholesterol side-chain cleavage enzyme
(D) 11β-Hydroxylase
(E) 3-Hydroxy-3-methylglutaryl CoA reductase
(F) 17α-Hydroxylase
8. A patient complains of generalized weakness and fatigue, anorexia, and weight loss associated with gastrointestinal symptoms (nausea, vomiting). Physical examination notes hyperpigmentation and hypotension. Laboratory findings include hyponatremia (low plasma sodium) and hyperkalemia (high plasma potassium). The most likely diagnosis is
(A) Cushing's disease
(B) Addison's disease
(C) Primary hypoaldosteronism
(D) Congenital adrenal hyperplasia
(E) Hypopituitarism
(F) Glucocorticoid-suppressible hyperaldosteronism
9. Through what "permissive action" do glucocorticoids accelerate gluconeogenesis during fasting?
(A) Glucocorticoids stimulate the secretion of insulin, which activates gluconeogenic enzymes in the liver
(B) Glucocorticoids inhibit the use of glucose by skeletal muscle
(C) Glucocorticoids maintain the vascular response to norepinephrine
(D) Glucocorticoids inhibit glycogenolysis
(E) Glucocorticoids maintain the intracellular concentrations of many of the enzymes needed to carry out gluconeogenesis through effects on transcription
(F) Glucocorticoids inhibit the release of fatty acids from adipose tissue

SUGGESTED READING

Bornstein SR, Chrousos GP. Clinical review 104. Adrenocorticotropin (ACTH)- and non-ACTH-mediated regulation of the adrenal cortex: Neural and immune inputs. J Clin Endocrinol Metab 1999;84:1729–1736.

Lumbers ER. Angiotensin and aldosterone. Regul Pept 1999;80:91–100.

Miller WL. Early steps in androgen biosynthesis: From cholesterol to DHEA. Baillieres Clin Endocrinol Metab 1998;12:67–81.

Nordenstrom A, Thilen A, Hagenfeldt L, Larsson A, Wedell A. Genotyping is a valuable diagnostic complement to neonatal screening for congenital adrenal hyperplasia due to steroid 21-hydroxylase deficiency. J Clin Endocrinol Metab 1999;84:1505–1509.

(continued)

Orth DN, Kovacs WJ. The adrenal cortex. In: Wilson JD, Foster DW, Kronenberg HM, Larsen PR, eds. Williams Textbook of Endocrinology. 9th Ed. Philadelphia: WB Saunders, 1998.

Sapolsky RM, Romero LM, Munck AU. How do glucocorticoids influence stress responses? Integrating permissive, suppressive, stimulatory, and preparative actions. Endocr Rev 2000;21:55–89.

Young JB, Landsberg L. Catecholamines and the adrenal medulla. In: Wilson JD, Foster DW, Kronenberg HM, Larsen PR, eds. Williams Textbook of Endocrinology. 9th Ed. Philadelphia: WB Saunders, 1998.

CHAPTER 35

The Endocrine Pancreas

Daniel E. Peavy, Ph.D.

CHAPTER OUTLINE

- **SYNTHESIS AND SECRETION OF THE ISLET HORMONES**
- **METABOLIC EFFECTS OF INSULIN AND GLUCAGON**
- **DIABETES MELLITUS**

KEY CONCEPTS

1. The relative distribution of alpha, beta, and delta cells within each islet of Langerhans shows a distinctive pattern and suggests that there may be some paracrine regulation of secretion.
2. Plasma glucose is the primary physiological regulator of insulin and glucagon secretion, but amino acids, fatty acids, and some GI hormones also play a role.
3. Insulin has anabolic effects on carbohydrate, lipid, and protein metabolism in its target tissues, where it promotes the storage of nutrients.
4. Effects of glucagon on carbohydrate, lipid, and protein metabolism occur primarily in the liver and are catabolic in nature.
5. Type 1 diabetes mellitus results from the destruction of beta cells, whereas type 2 diabetes often results from a lack of responsiveness to circulating insulin.
6. Diabetes mellitus may produce both acute complications, such as ketoacidosis, and chronic secondary complications, such as peripheral vascular disease, neuropathy, and nephropathy.

The development of mechanisms for the storage of large amounts of metabolic fuel was an important adaptation in the evolution of complex organisms. The processes involved in the digestion, storage, and use of fuels require a high degree of regulation and coordination. The pancreas, which plays a vital role in these processes, consists of two functionally different groups of cells.

Cells of the **exocrine pancreas** produce and secrete digestive enzymes and fluids into the upper part of the small intestine. The **endocrine pancreas**, an anatomically small portion of the pancreas (1 to 2% of the total mass), produces hormones involved in regulating fuel storage and use.

For convenience, functions of the exocrine and endocrine portions of the pancreas are usually discussed separately. While this chapter focuses primarily on hormones of the endocrine pancreas, the overall function of the pancreas is to coordinate and direct a wide variety of processes related to the digestion, uptake, and use of metabolic fuels.

SYNTHESIS AND SECRETION OF THE ISLET HORMONES

The endocrine pancreas consists of numerous discrete clusters of cells, known as the islets of Langerhans, which are located throughout the pancreatic mass. The islets contain specific types of cells responsible for the secretion of the hormones insulin, glucagon, and somatostatin. Secretion of these hormones is regulated by a variety of circulating nutrients.

The Islets of Langerhans Are the Functional Units of the Endocrine Pancreas

The **islets of Langerhans** contain from a few hundred to several thousand hormone-secreting endocrine cells. The islets are found throughout the pancreas but are most abundant in the tail region of the gland. The human pancreas contains, on average, about 1 million islets, which vary in size from 50 to 300 μm in diameter. Each islet is separated from the surrounding acinar tissue by a connective tissue sheath.

Islets are composed of four hormone-producing cell types: insulin-secreting beta cells, glucagon-secreting alpha cells, somatostatin-secreting delta cells, and pancreatic polypeptide-secreting F cells. Immunofluorescent staining techniques have shown that the four cell types are arranged in each islet in a pattern suggesting a highly organized cellular community, in which paracrine influences may play an important role in determining hormone secretion rates (Fig. 35.1). Further evidence that cell-to-cell communica-

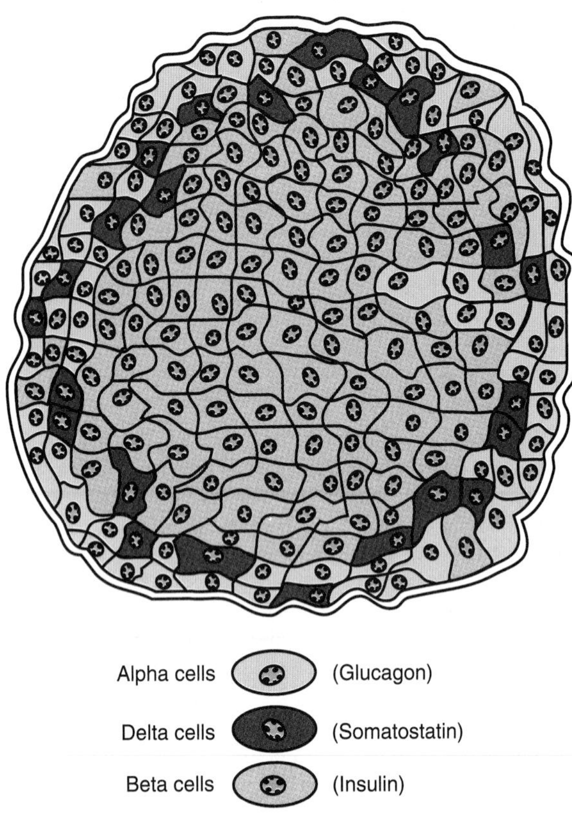

FIGURE 35.1 **Major cell types in a typical islet of Langerhans.** Note the distinct anatomic arrangement of the various cell types. (Modified from Orci L, Unger RH. Functional subdivision of islets of Langerhans and possible role of D cells. Lancet 1975;2:1243–1244.)

tion within the islet may play a role in regulating hormone secretion comes from the finding that islet cells have both gap junctions and tight junctions. Gap junctions link different cell types in the islets cells and potentially provide a means for the transfer of ions, nucleotides, or electrical current between cells. The presence of tight junctions between outer membrane leaflets of contiguous cells could result in the formation of microdomains in the interstitial space, which may also be important for paracrine communication. Although the existence of gap junctions and tight junctions in pancreatic islets is well documented, their exact function has not been fully defined.

The arrangement of the vascular supply to islets is also consistent with paracrine involvement in regulating islet secretion. Afferent blood vessels penetrate nearly to the center of the islet before branching out and returning to the surface of the islet. The innermost cells of the islet, therefore, receive arterial blood, while those cells nearer the surface receive blood containing secretions from inner cells. Since there is a definite anatomic arrangement of cells in the islet (see Fig. 35.1), one cell type could affect the secretion of others. In general, the effluent from smaller islets passes through neighboring pancreatic acinar tissue before entering into the hepatic portal venous system. By contrast, the effluent from larger islets passes directly into the venous system without first perfusing adjacent acinar tissue. Therefore, islet hormones arrive in high concentrations in some areas of the exocrine pancreas before reaching peripheral tissues. However, the exact physiological significance of these arrangements is unknown.

Neural inputs also influence islet cell hormone secretion. Islet cells receive sympathetic and parasympathetic innervation. Responses to neural input occur as a result of activation of various adrenergic and cholinergic receptors (described below). Neuropeptides released together with the neurotransmitters may also be involved in regulating hormone secretion.

Beta Cells. In the early 1900s, M. A. Lane established a histochemical method by which two kinds of islet cells could be distinguished. He found that alcohol-based fixatives dissolved the secretory granules in most of the islet cells but preserved them in a small minority of cells. Water-based fixatives had the opposite effect.He named cells containing alcohol-insoluble granules A cells or alpha cells and those containing alcohol-soluble granules B cells or beta cells. Many years later, other investigators used immunofluorescence techniques to demonstrate that beta cells produce insulin and alpha cells produce glucagon.

Insulin-secreting **beta cells** are the most numerous cell type of the islet, comprising 70 to 90% of the endocrine cells. Beta cells typically occupy the most central space of the islets (see Fig. 35.1). They are generally 10 to 15 μm in diameter and contain secretory granules that measure 0.25 μm.

Alpha Cells. **Alpha cells** comprise most of the remaining cells of the islets. They are generally located near the periphery, where they form a cortex of cells surrounding the more centrally located beta cells. Blood vessels pass through the outer zone of the islet before extensive branching occurs. Inward extensions of the cortex may be present along the axes of blood vessels toward the center of the islet, giving the appearance that the islet is subdivided into small lobules.

Delta Cells. **Delta cells** are the sites of production of somatostatin in the pancreas. These cells are typically located in the periphery of the islet, often between beta cells and the surrounding mantle of alpha cells. Somatostatin produced by pancreatic delta cells is identical to that previously described in a neurotransmitter role (see Chapter 3) and as a hypothalamic hormone that inhibits growth hormone secretion by the anterior pituitary (see Chapter 32).

F Cells. **F cells** are the least abundant of the hormone-secreting cells of islets, representing only about 1% of the total cell population. The distribution of F cells is generally similar to that of delta cells. F cells secrete **pancreatic polypeptide**.

Increased Blood Glucose Stimulates the Secretion of Insulin

A variety of factors, including other pancreatic hormones, are known to influence insulin secretion. The primary physiological regulator of insulin secretion, however, is the blood glucose concentration.

Proinsulin Synthesis. The gene for insulin is located on the short arm of chromosome 11 in humans. Like other hormones and secretory proteins, insulin is first synthesized by ribosomes of the rough ER as a larger precursor peptide that is then converted to the mature hormone prior to secretion (see Chapter 31).

The insulin gene product is a 110-amino acid peptide, preproinsulin. Proinsulin consists of 86 amino acids (Fig. 35.2); residues 1 to 30 constitute what will form the B chain of insulin, residues 31 to 65 form the connecting peptide, and residues 66 to 86 constitute the A chain. (Note that "connecting peptide" should not be confused with "C-peptide.") In the process of converting proinsulin to insulin, two pairs of basic amino acid residues are clipped out of the proinsulin molecule, resulting in the formation of insulin and **C-peptide**, which are ultimately secreted from the beta cell in equimolar amounts.

It is of clinical significance that insulin and C-peptide are co-secreted in equal amounts. Measurements of circulating C-peptide levels may sometimes provide important information regarding beta cell secretory capacity that could not be obtained by measuring circulating insulin levels alone.

Insulin Secretion. Table 35.1 lists the physiologically relevant regulators of insulin secretion. As indicated previously, an elevated blood glucose level is the most important regulator of insulin secretion. In humans, the threshold value for glucose-stimulated insulin secretion is a plasma glucose concentration of approximately 100 mg/dL (5.6 mmol/L).

Based on studies using isolated animal pancreas preparations maintained *in vitro*, it has been determined that insulin is secreted in a biphasic manner in response to a marked increase in blood glucose. An initial burst of insulin secretion may last 5 to 15 minutes, resulting from the secretion of preformed insulin secretory granules. This response is followed by more gradual and sustained insulin secretion that results largely from the synthesis of new insulin molecules.

In addition to glucose, several other factors serve as important regulators of insulin secretion (see Table 35.1). These include dietary constituents, such as amino acids and fatty acids, as well as hormones and drugs. Among the amino acids, arginine is the most potent secretagogue for insulin. Among the fatty acids, long-chain fatty acids (16 to 18 carbons) generally are considered the most potent stimulators of insulin secretion. Several hormones secreted by the gastrointestinal tract, including gastric inhibitory peptide (GIP), gastrin, and secretin, promote insulin secretion. An oral dose of glucose produces a greater increment in insulin secretion than an equivalent intravenous dose because oral glucose promotes the secretion of GI hormones that

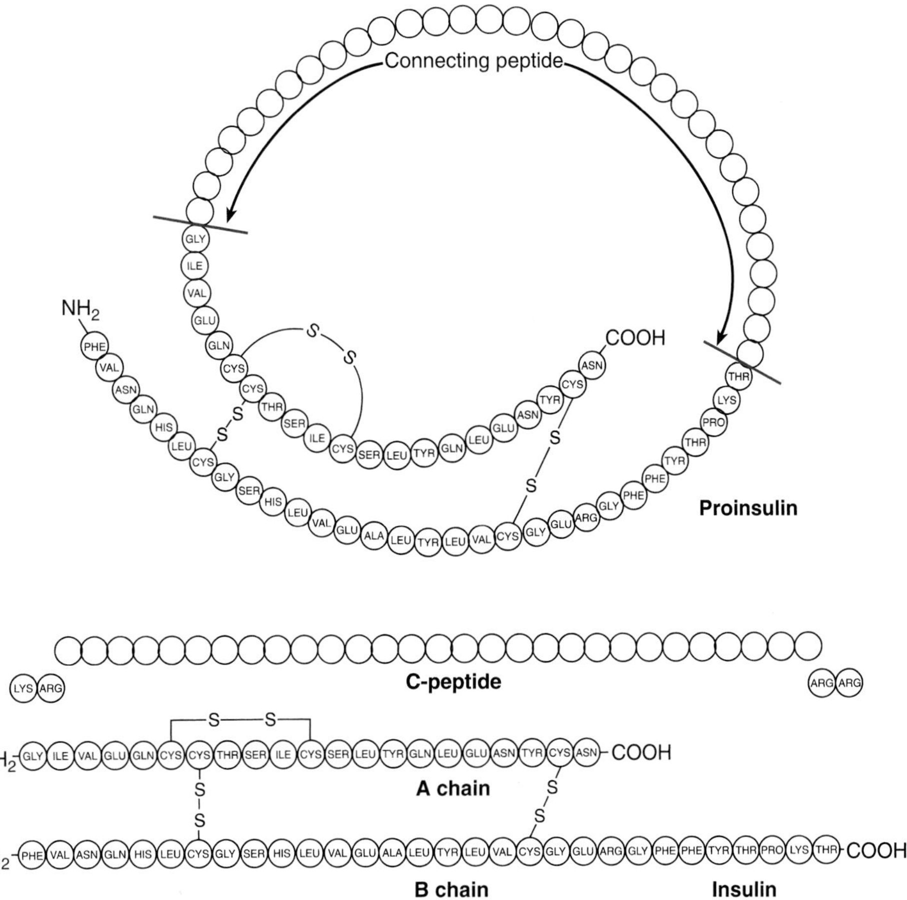

FIGURE 35.2 **The structure of proinsulin, C-peptide, and insulin.** Note that with the removal of two pairs of basic amino acids, proinsulin is converted into insulin and C-peptide.

TABLE 35.1 Factors Regulating Insulin Secretion From the Pancreas

Stimulatory agents or conditions	Hyperglycemia Amino acids Fatty acids, especially long-chain Gastrointestinal hormones, especially gastric inhibitory peptide (GIP), gastrin, and secretin Acetylcholine Sulfonylureas
Inhibitory agents or conditions	Somatostatin Norepinephrine Epinephrine

TABLE 35.2 Factors Regulating Glucagon Secretion From the Pancreas

Stimulatory agents or conditions	Hypoglycemia Amino acids Acetylcholine Norepinephrine Epinephrine
Inhibitory agents or conditions	Fatty acids Somatostatin Insulin

augment insulin secretion by the pancreas. Direct infusion of acetylcholine into the pancreatic circulation stimulates insulin secretion, reflecting the role of parasympathetic innervation in regulating insulin secretion. Sulfonylureas, a class of drugs used orally in the treatment of type 2 diabetes, promote insulin's action in peripheral tissues and directly stimulate insulin secretion.

In addition to factors that stimulate insulin secretion, there are several potent inhibitors. Exogenously administered somatostatin is a strong inhibitor. It is presumed that pancreatic somatostatin plays a role in regulating insulin secretion, but the importance of this effect has not been fully established. Epinephrine and norepinephrine, the primary catecholamines, are also potent inhibitors of insulin secretion. This response would appear appropriate because during periods of stress and high catecholamine secretion, the desired response is mobilization of glucose and other nutrient stores. Insulin generally promotes the opposite response, and by inhibiting insulin secretion, the catecholamines produce their full effect without the opposing actions of insulin.

Decreased Blood Glucose Stimulates the Secretion of Glucagon

Similar to insulin, glucagon is first synthesized as part of a larger precursor protein. Glucagon secretion is regulated by many of the factors that also regulate insulin secretion. In most cases, however, these factors have the opposite effect on glucagon secretion.

Synthesis of Proglucagon. Glucagon is a simple 29-amino acid peptide. The initial gene product for glucagon, preproglucagon, is a much larger peptide. Like other peptide hormones, the "pre" piece is removed in the ER, and the prohormone is converted into a mature hormone as it is packaged and processed in secretory granules (see Chapter 31).

Secretion of Glucagon. The principal factors that influence glucagon secretion are listed in Table 35.2. With a few exceptions, this table is nearly a mirror image of Table 35.1, the factors that regulate insulin secretion. The primary regulator of glucagon secretion is blood glucose; specifically, a decrease in blood glucose below about 100 mg/dL promotes glucagon secretion. As with insulin, amino acids, especially arginine, are potent stimulators of glucagon secretion. Somatostatin inhibits glucagon secretion, as it does insulin secretion.

Increased Blood Glucose and Glucagon Stimulate the Secretion of Somatostatin

Somatostatin is first synthesized as a larger peptide precursor, preprosomatostatin. The hypothalamus also produces this protein, but the regulation of somatostatin secretion from the hypothalamus is independent of that from the pancreatic delta cells. Upon insertion of preprosomatostatin into the rough ER, it is initially cleaved and converted to prosomatostatin. The prohormone is converted into active hormone during packaging and processing in the Golgi apparatus.

Factors that stimulate pancreatic somatostatin secretion include hyperglycemia, glucagon, and amino acids. Glucose and glucagon are generally considered the most important regulators of somatostatin secretion.

The exact role of somatostatin in regulating islet hormone secretion has not been fully established. Somatostatin clearly inhibits both glucagon and insulin secretion from the alpha and beta cells of the pancreas, respectively, when it is given exogenously. The anatomic and vascular relationships of delta cells to alpha and beta cells further suggest that somatostatin may play a role in regulating both glucagon and insulin secretion. Although many of the data are circumstantial, it is generally accepted that somatostatin plays a paracrine role in regulating insulin and glucagon secretion from the pancreas.

METABOLIC EFFECTS OF INSULIN AND GLUCAGON

The endocrine pancreas secretes hormones that direct the storage and use of fuels during times of nutrient abundance (fed state) and nutrient deficiency (fasting). Insulin is secreted in the fed state and is called the "hormone of nutrient abundance." By contrast, glucagon is secreted in response to an overall deficit in nutrient supply. These two hormones play an important role in directing the flow of metabolic fuels.

Insulin Affects the Metabolism of Carbohydrates, Lipids, and Proteins in Liver, Muscle, and Adipose Tissues

The primary targets for insulin are liver, skeletal muscle, and adipose tissues. Insulin has multiple individual actions in each of these tissues, the net result of which is fuel storage.

Mechanism of Insulin Action. Although insulin was one of the first peptide hormones to be identified, isolated, and characterized, its exact mechanism of action remains elusive. The **insulin receptor** is a heterotetramer, consisting of a pair of α/β subunit complexes held together by disulfide bonds (Fig. 35.3). The α subunit is an extracellular protein containing the insulin-binding component of the receptor. The β subunit is a transmembrane protein that couples the extracellular event of insulin binding to its intracellular actions.

Activation of the β subunit of the insulin receptor results in autophosphorylation, involving the phosphorylation of a few selected tyrosine residues in the intracellular portion of the receptor. This event further activates the tyrosine kinase portion of the β subunit, leading to tyrosine phosphorylation of specific intracellular substrates. A cascade of events follows, leading to the pleiotropic actions of insulin in its target cells. While tyrosine phosphorylation events appear to be the early steps in insulin action, serine/threonine phosphorylation or dephosphorylation is involved in many of the final steps of insulin action.

Insulin and Glucose Transport. Perhaps one of the most important functions of insulin is to promote the uptake of glucose from blood into cells. Glucose uptake into many cell types is by facilitated diffusion. A specific cell membrane carrier is involved but no energy is required, and the process cannot move glucose against a concentration gradient. The carriers shuttle glucose across the membrane faster than would occur by diffusion alone. Considerable recent work has revealed not just one transporter, but a family of about seven different glucose transporters (GLUT), commonly called GLUT 1 to GLUT 7. These transporters are expressed in different tissues and, in some cases, at different times during fetal development.

GLUT 4, the insulin-stimulated glucose transporter, is the primary form of the transporter present in skeletal muscle tissue and adipose tissue. It is present in plasma membranes and in intracellular vesicles of the smooth ER. In target cells, the effect of insulin is to promote the translocation of GLUT 4 transporters from the intracellular pool into plasma membranes. As a result, more transporters are available in the plasma membrane, and glucose uptake by target cells is, thereby, increased.

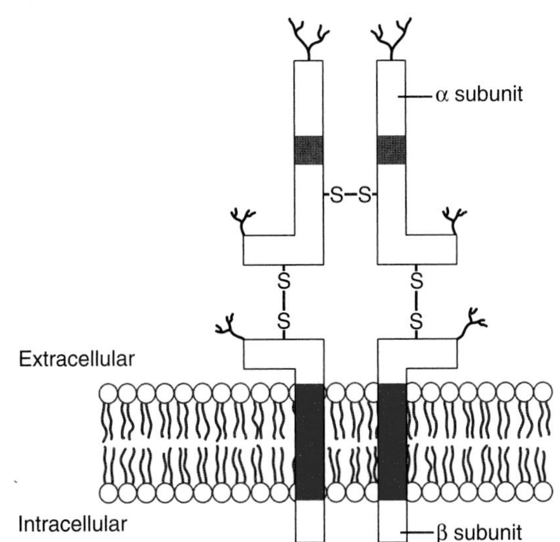

FIGURE 35.3 **The structure of the insulin receptor.** The insulin receptor is a heterotetramer consisting of two extracellular insulin-binding α subunits linked by disulfide bonds to two transmembrane β subunits. The β subunits contain an intrinsic tyrosine kinase that is activated upon insulin binding to the α subunit.

Insulin and the Synthesis of Glycogen. Besides promoting glucose uptake into target cells, insulin promotes its storage. Glucose carbon is stored in the body in two primary forms: as glycogen and (by metabolic conversion) as triglycerides. Glycogen is a short-term storage form that plays an important role in maintaining normal blood glucose levels. The primary glycogen storage sites are the liver and skeletal muscle; other tissues, such as adipose tissue, also store glycogen but in quantitatively small amounts. Insulin promotes glycogen storage primarily through two enzymes (Fig. 35.4). It activates **glycogen synthase** by promoting its dephosphorylation and concomitantly inactivates **glycogen phosphorylase**, also by promoting its dephosphorylation. The result is that glycogen synthesis is promoted and glycogen breakdown is inhibited.

Insulin and Glycolysis. Insulin also enhances glycolysis. In addition to increasing glucose uptake and providing a mass action stimulus for glycolysis, insulin activates the enzymes glucokinase and hexokinase and phosphofructokinase, pyruvate kinase, and pyruvate dehydrogenase of the glycolytic pathway (see Fig. 35.4).

Lipogenic and Antilipolytic Effects of Insulin. In adipose tissue and liver tissue, insulin promotes lipogenesis and inhibits lipolysis (Fig. 35.5). Insulin has similar actions in muscle, but since muscle is not a major site of lipid storage, the discussion here focuses on actions in adipose tissue and the liver. By promoting the flow of intermediates through glycolysis, insulin promotes the formation of α-glycerol phosphate and fatty acids necessary for triglyceride formation. In addition, it stimulates fatty acid synthase, leading directly to increased fatty acid synthesis. Insulin inhibits the breakdown of triglycerides by inhibiting hormone-sensitive lipase, which is activated by a variety of counterregulatory hormones, such as epinephrine and adrenal glucocorticoids. By inhibiting this enzyme, insulin promotes the accumulation of triglycerides in adipose tissue.

In addition to promoting *de novo* fatty acid synthesis in adipose tissue, insulin increases the activity of lipoprotein lipase, which plays a role in the uptake of fatty acids from the blood into adipose tissue. As a result, lipoproteins synthesized in the liver are taken up by adipose tissue, and fatty acids are ultimately stored as triglycerides.

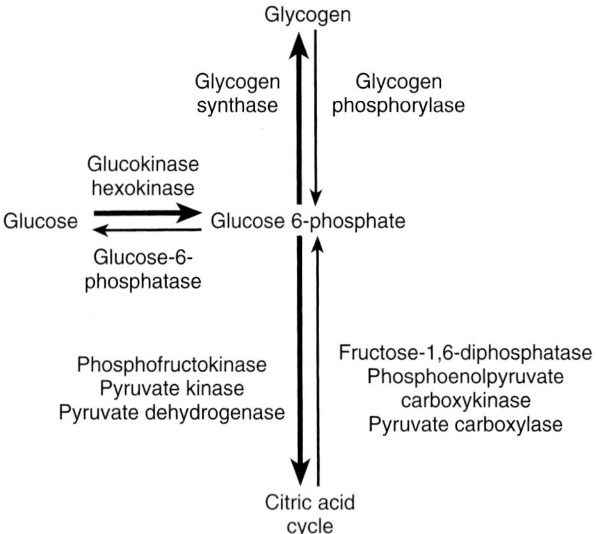

FIGURE 35.4 **Insulin stimulation of glycogen synthesis and glucose metabolism.** Insulin promotes glucose uptake into target tissues, stimulates glycogen synthesis, and inhibits glycogenolysis. In addition it promotes glycolysis in its target tissues. Heavy arrows indicate processes stimulated by insulin; light arrows indicate processes inhibited by insulin.

Effects of Insulin on Protein Synthesis and Protein Degradation. Insulin promotes protein accumulation in its primary target tissues—liver, adipose tissue, and muscle—in three specific ways (Fig. 35.6). First, it stimulates amino acid uptake. Second, it increases the activity of several factors involved in protein synthesis. For example, it increases the activity of protein synthesis initiation factors, promoting the start of translation and increasing the efficiency of protein synthesis. Insulin also increases the amount of protein synthesis machinery in cells by promoting ribosome synthesis. Third, insulin inhibits protein degradation by reducing lysosome activity and possibly other mechanisms as well.

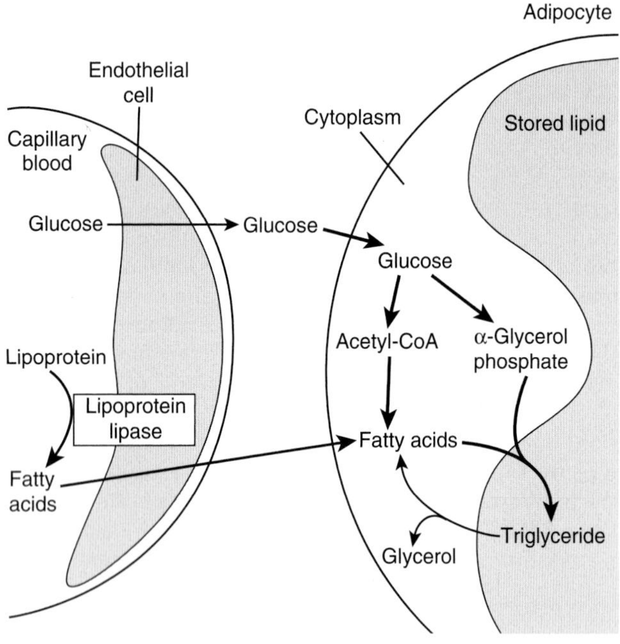

FIGURE 35.5 **Effects of insulin on lipid metabolism in adipocytes.** Insulin promotes the accumulation of lipid (triglycerides) in adipocytes by stimulating the processes shown by the heavy arrows and inhibiting the processes shown by the light arrows. Similar stimulatory and inhibitory effects occur in liver cells.

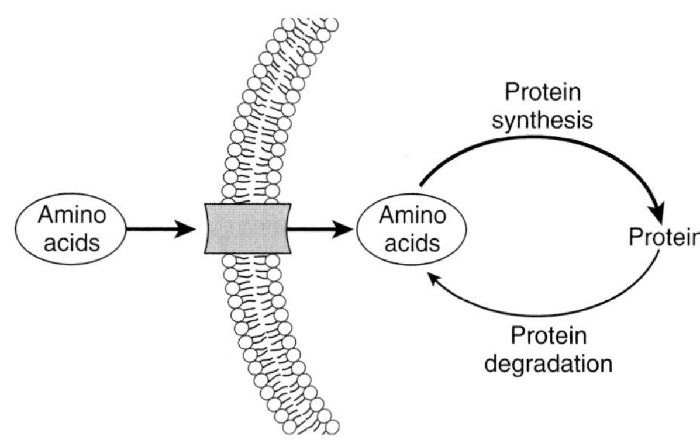

FIGURE 35.6 **Effects of insulin on protein synthesis and protein degradation.** Insulin promotes the accumulation of protein by stimulating (heavy arrows) amino acid uptake and protein synthesis and by inhibiting (light arrows) protein degradation in liver, skeletal muscle, and adipose tissue.

Glucagon Primarily Affects the Liver Metabolism of Carbohydrates, Lipids, and Proteins

The primary physiological actions of glucagon are exerted in the liver. Numerous effects of glucagon have been documented in other tissues, primarily adipose tissue, when the hormone has been added at high, nonphysiological concentrations in experimental situations. While these effects may play a role in certain abnormal situations, the normal daily effects of glucagon occur primarily in the liver.

Mechanism of Glucagon Action. Glucagon initiates its biological effects by interacting with one or more types of cell membrane receptors. **Glucagon receptors** are coupled to G proteins and promote increased intracellular cAMP, via the activation of adenylyl cyclase, or elevated cytosolic calcium as a result of phospholipid breakdown to form IP_3.

Glucagon and Glycogenolysis. Glucagon is an important regulator of hepatic glycogen metabolism. It produces a net effect of glycogen breakdown by increasing intracellular cAMP levels, initiating a cascade of phosphorylation events that ultimately results in the phosphorylation of phosphorylase b and its activation by conversion into phosphorylase a. Similarly, glucagon promotes the net breakdown of glycogen by promoting the inactivation of glycogen synthase (Fig. 35.7).

Glucagon and Gluconeogenesis. In addition to promoting hepatic glucose production by stimulating glycogenolysis, glucagon stimulates hepatic gluconeogenesis (Fig. 35.8). It does this principally by increasing the transcription of mRNA coding for the enzyme phosphoenolpyruvate carboxykinase (PEPCK), a key rate-limiting enzyme in gluconeogenesis. Glucagon also stimulates amino acid

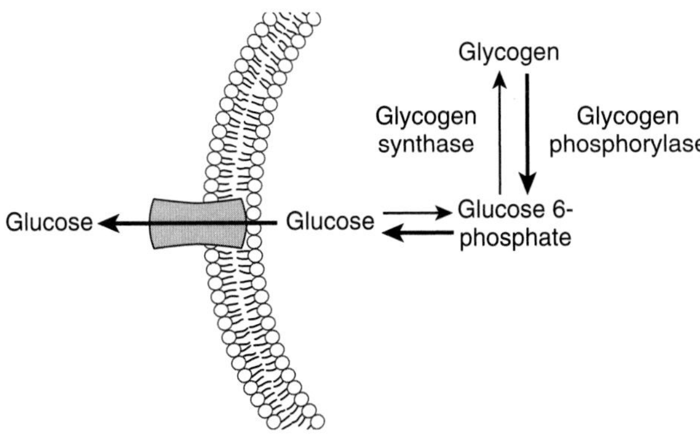

FIGURE 35.7 The role of glucagon in glycogenolysis and glucose production in liver cells. Heavy arrows indicate processes stimulated by glucagon; light arrows indicate processes inhibited by glucagon.

transport into liver cells and the degradation of hepatic proteins, helping provide substrates for gluconeogenesis.

Glucagon and Ureagenesis. The glucagon-enhanced conversion of amino acids into glucose leads to increased formation of ammonia. Glucagon assists in the disposal of ammonia by increasing the activity of the urea cycle enzymes in liver cells (see Fig. 35.8).

Glucagon and Lipolysis. Glucagon promotes lipolysis in liver cells (Fig. 35.9), although the quantity of lipids stored in liver is small compared to that in adipose tissue.

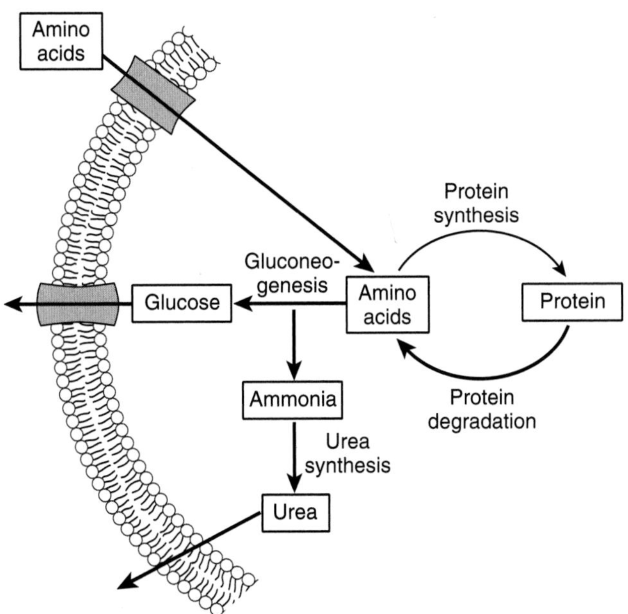

FIGURE 35.8 The role of glucagon in gluconeogenesis and ureagenesis in liver cells. Heavy arrows indicate processes stimulated by glucagon; the light arrow indicates a process inhibited by glucagon.

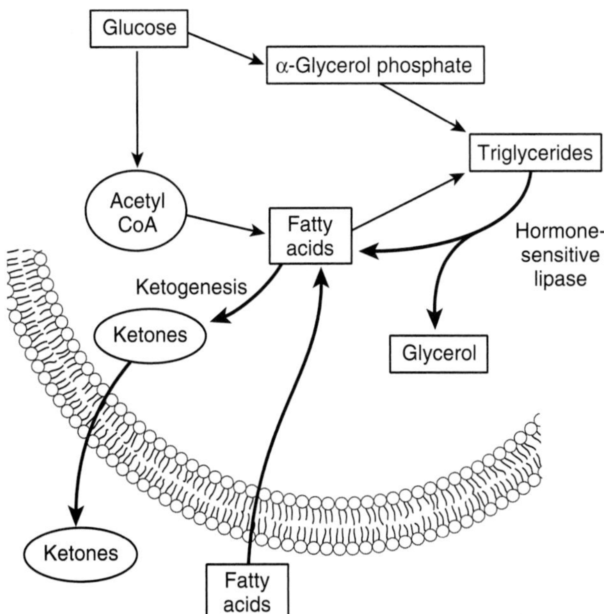

FIGURE 35.9 The role of glucagon in lipolysis and ketogenesis in liver cells. Heavy arrows indicate processes stimulated by glucagon; light arrows indicate processes inhibited by glucagon.

Glucagon and Ketogenesis. Glucagon promotes ketogenesis, the production of ketones, by lowering the levels of malonyl CoA, relieving an inhibition of palmitoyl transferase and allowing fatty acids to enter the mitochondria for oxidation to ketones (see Fig 35.9). Ketones are an important source of fuel for muscle cells and heart cells during times of starvation, sparing blood glucose for other tissues that are obligate glucose users, such as the central nervous system. During prolonged starvation, the brain adapts its metabolism to use ketones as a fuel source, lessening the overall need for hepatic glucose production (see Chapter 34).

The Insulin-Glucagon Ratio Determines Metabolic Status

In most instances, insulin and glucagon produce opposing effects. Therefore, the net physiological response is determined by the relative levels of both hormones in the blood plasma, the **insulin-glucagon ratio** (I/G ratio).

I/G Ratio in the Fed and Fasting States. The I/G ratio may vary 100-fold or more because the plasma concentration of each hormone can vary considerably in different nutritional states. In the fed state, the molar I/G ratio is approximately 30. After an overnight fast, it may fall to about 2, and with prolonged fasting, it may fall to as low as 0.5.

Inappropriate I/G Ratios in Diabetes. A good example of the profound influence of the I/G ratio on metabolic status is in insulin-deficient diabetes. Insulin levels are low, so pathways that insulin stimulates operate at a reduced level.

However, insulin is also necessary for alpha cells to sense blood glucose appropriately; in the absence of insulin, the secretion of glucagon is inappropriately elevated. The result is an imbalance in the I/G ratio and an accentuation of glucagon effects well above what would be seen in normal states of low insulin, such as in fasting.

DIABETES MELLITUS

Diabetes mellitus is a disease of metabolic dysregulation—most notably a dysregulation of glucose metabolism—accompanied by long-term vascular and neurological complications. Diabetes has several clinical forms, each of which has a distinct etiology, clinical presentation, and course. Insights into diabetes and its complications have been gleaned from extensive metabolic studies, the use of radioimmunoassays for insulin and glucagon, and the application of molecular biology strategies. Diabetes is the most common endocrine disorder. Some 16 million people may have the disease in the United States; the exact number is not known because many people have a borderline, subclinical form of the disorder. Many deaths attributed to cardiovascular disease are in fact the result of complications from diabetes.

Diagnosing diabetes mellitus is not difficult to do. Symptoms usually include frequent urination, increased thirst, increased food consumption, and weight loss. The standard criterion for a diagnosis of diabetes is an elevated plasma glucose level after an overnight fast on at least two separate occasions. A glucose value above 126 mg/dL (7.0 mmol/L) is often used as the diagnostic value.

Diabetes mellitus is a heterogeneous disorder. The causes, symptoms, and general medical outcomes are variable. Generally, the disease takes one of two forms, **type 1 diabetes** or **type 2 diabetes**. Other forms of diabetes, such as **gestational diabetes**, are also well known.

Most Forms of Type 1 Diabetes Mellitus Involve an Autoimmune Disorder

Type 1 diabetes is characterized by the inability of beta cells to produce physiologically appropriate amounts of insulin. In some instances, this may result from a mutation in the preproinsulin gene. However, the most common form of type 1 diabetes results from destruction of the pancreatic beta cells by the immune system. The initial pathological event is **insulitis,** involving a lymphocytic attack on beta cells. Antibodies to beta cell cell-surface antigens have also been found in the circulation of many persons with type 1 diabetes, but this is not a primary causative factor and probably results from the initial cellular damage.

Studies of identical twins have provided important information regarding the genetic basis of type 1 diabetes. If one twin develops type 1 diabetes, the odds that the second will develop the disease are much higher than for any random individual in the population, even when the twins are raised apart under different socioeconomic conditions. In addition, individuals with certain cell-surface HLA antigens bear a higher risk for the disease than others.

Environmental factors are involved as well because the development of type 1 diabetes in one twin predicts only a 50% or less chance that the second will develop the disease. The specific environmental factors have not been identified, although much evidence implicates viruses. Therefore, it appears that a combination of genetics and environment are strong contributing factors to the development of type 1 diabetes.

Because the primary defect in type 1 diabetes is the inability of beta cells to secrete adequate amounts of insulin, these patients must be treated with injections of insulin. In an attempt to match insulin concentrations in the blood with the metabolic requirements of the individual, various formulations of insulin with different durations of action have been developed. Patients inject an appropriate amount of these different insulin forms to match their dietary and lifestyle requirements.

The long-term control of type 1 diabetes depends on maintaining a balance between three factors: insulin, diet, and exercise. To strictly control their blood glucose, patients are advised to monitor their diet and level of physical activity, as well as their insulin dosage. Exercise per se, much like insulin, increases glucose uptake by muscle. Diabetic patients must take this into account and make appropriate adjustments in diet or insulin whenever general exercise levels change dramatically.

Type 2 Diabetes Mellitus Primarily Originates in the Target Tissue

Type 2 diabetes mellitus results primarily from impaired ability of target tissues to respond to insulin. There are multiple forms of the disease, each with a different etiology. In some cases, it is a permanent, lifelong disorder; in others, it results from the secretion of counterregulatory hormones in a normal (e.g., pregnant) or pathophysiological (e.g., Cushing's syndrome) state. **Gestational diabetes** occurs in 2 to 5% of all pregnancies but usually disappears after delivery. Women who have had gestational diabetes have an increased risk of developing type 2 diabetes later in life.

Insulin Resistance in Type 2 Diabetes. In most cases of type 2 diabetes, normal or higher-than-normal amounts of insulin are present in the circulation. Therefore, there is no impairment in the secretory capacity of pancreatic beta cells but only in the ability of target cells to respond to insulin. In some instances, it has been demonstrated that the fundamental defect is in the insulin receptor. In most cases, however, receptor function appears normal, and the impairment in insulin action is ascribed to a postreceptor defect. Since the exact mechanism of insulin action has not been determined, it is difficult to explore the causes of insulin resistance in much greater depth.

Genetics, Environment, and Type 2 Diabetes. As with type 1 diabetes, key information on the influence of genetics and environmental factors in type 2 diabetes comes from studies of identical twins. These studies indicate that there is a strong genetic component to the development of type 2 diabetes and that environmental factors, including diet, play a considerably lesser role. If one identical twin develops type 2 diabetes, chances are nearly 100% that the second will as well, even if they are raised apart under entirely different conditions.

Many persons with type 2 diabetes are overweight, and often the severity of their disease can be lessened simply by weight loss. However, no strict cause-and-effect relationship between these two conditions has been established. Clearly, not all persons with type 2 diabetes are obese, and not all obese individuals develop diabetes.

Treatment Options for Type 2 Diabetes. In milder forms of type 2 diabetes, dietary restriction leading to weight loss may be the only treatment necessary. Commonly, however, dietary restriction is supplemented by treatment with one of several orally active agents, most often of the sulfonylurea class. These drugs appear to act in two ways. First, they promote insulin action in target cells, lessening insulin resistance in tissues. Second, they correct or reverse a somewhat sluggish response of pancreatic beta cells often seen in type 2 diabetes, normalizing insulin secretory responses to glucose. The exact mechanisms of these effects are unknown. In some cases, persons with type 2 diabetes may also be treated with insulin, although in most cases a regimen of oral agents and dietary manipulation is sufficient.

Diabetes Mellitus Complications Present Major Health Problems

If left untreated or if glycemic control is poor, diabetes leads to acute complications that may prove fatal. However, even with reasonably good control of blood glucose, over a period of years, most diabetics develop secondary complications of the disease that result in tissue damage, primarily involving the cardiovascular and nervous systems.

Acute Complications of Diabetes. The nature of acute complications that develop in type 1 and type 2 diabetics differs. Persons with poorly controlled type 1 diabetes often exhibit hyperglycemia, glucosuria, dehydration, and **diabetic ketoacidosis**. As blood glucose becomes elevated above the renal plasma threshold, glucose appears in the urine. As a result of osmotic effects, water follows glucose, leading to polyuria, excessive loss of fluid from the body, and dehydration. With fluid loss, the circulating blood volume is reduced, compromising cardiovascular function, which may lead to circulatory failure.

Excessive ketone formation leads to acidosis and electrolyte imbalances in persons with type 1 diabetes. If uncontrolled, ketones may be elevated in the blood to such an extent that the odor of acetone (one of the ketones) is noticeable on the breath. Production of the primary ketones, β-hydroxybutyric acid and acetoacetic acid, results in the generation of excess hydrogen ions and a metabolic acidosis. Ketones may accumulate in the blood to such a degree that they exceed renal transport capacities and appear in the urine. As a result of osmotic effects, water is also lost in the urine. In addition, the pK of ketones is such that even with the most acidic urine a normal kidney can produce, about half of the excreted ketones are in the salt (or base) form. To ensure electrical neutrality, these must be accompanied by a cation, usually either sodium or potassium. The loss of ketones in the urine, therefore, also results in a loss of important electrolytes. Excessive ketone production in type 1 diabetes results in acidosis, a loss of cations, and a loss of fluids. Emergency department procedures are directed toward immediate correction of these acute problems and usually involve the administration of base, fluids, and insulin.

The complex sequence of events that can result from uncontrolled type 1 diabetes is shown in Figure 35.10. If left unchecked, many of these complications can have an additive effect to further the severity of the disease state.

Persons with type 2 diabetes are generally not ketotic and do not develop acidosis or the electrolyte imbalances characteristic of type 1 diabetes. Hyperglycemia leads to fluid loss and dehydration. Severe cases may result in hyperosmolar coma as a result of excessive fluid loss. The initial objective of treatment in these individuals is the administration of fluids to restore fluid volumes to normal and eliminate the hyperosmolar state.

Chronic Secondary Complications of Diabetes. With good control of their disease, most persons with diabetes can avoid the acute complications described above; however, it is rare that they will not suffer from some of the chronic secondary complications of the disease. In most instances, such complications will ultimately lead to reduced life expectancy.

Most lesions occur in the circulatory system, although the nervous system is also often affected. Large vessels often show changes similar to those in atherosclerosis, with the deposition of large fatty plaques in arteries. However, most of the circulatory complications in diabetes occur in microvessels. The common finding in affected vessels is a

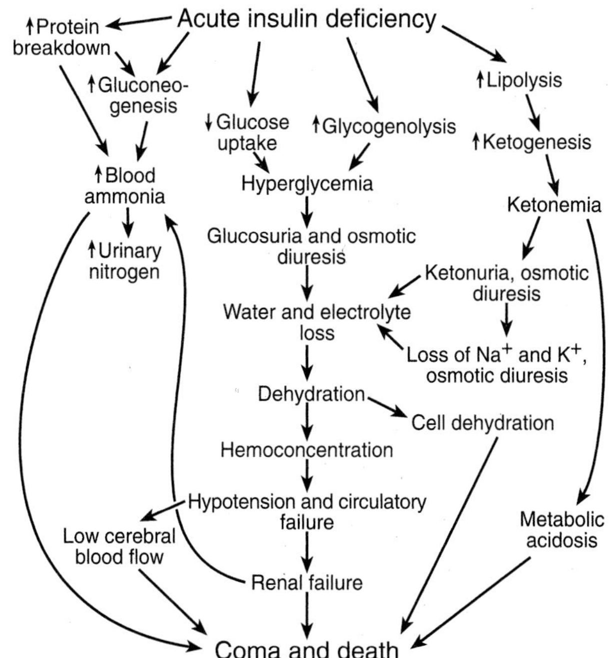

FIGURE 35.10 Events resulting from acute insulin deficiency in type 1 diabetes mellitus. If left untreated, insulin deficiency may lead to several complications, which may have additive or confounding effects that may ultimately result in death.

CLINICAL FOCUS BOX 35.1

The Diabetic Foot

Despite efforts to control their disease and maintain a normal glycemic state, most persons with diabetes eventually develop one or more secondary complications of the disease. These complications may be somewhat subtle in onset and slow in progression; however, they account for the high rates of morbidity and mortality. While the specific mechanisms involved remain areas of debate and research activity, most secondary complications are vascular or neural in nature.

Vascular complications may involve atherosclerotic-like lesions in the large blood vessels or impaired function in the microcirculation. Damage to the basement membrane of capillaries in the eye (**diabetic retinopathy**) or kidney (**diabetic nephropathy**) is commonly seen. Although there is no satisfactory direct treatment for diabetic vascular disease, its progression is often monitored closely as an indirect indicator of the overall diabetic state.

Diabetic neuropathy typically involves symmetric sensory loss in the distal lower extremities or autonomic neuropathy, leading to impotence, GI dysfunction, or anhidrosis (lack of sweating) in the lower extremities. The **diabetic foot** is an example of several complicating factors exacerbating one another. About 50 to 70% of nontraumatic amputations in the United States each year are due to diabetes. Breakdown of the foot in persons who are diabetic is commonly due to a combination of neuropathy, vascular impairment, and infection. In a typical scenario, small lesions on the foot result from dryness of the skin due to a combination of neural and vascular complications. Impairments in sensory nerve function may result in these small lesions going unnoticed by the patient until a severe infection or gangrene has become well established.

Loss of the affected foot or limb often can be avoided with patient and physician education. The focus in managing patients with diabetes is the maintenance of normal blood glucose levels; avoiding primary complications, such as diabetic ketoacidosis or hyperosmolar coma; and initial secondary complications, such as diabetic retinopathy. There is an increasing awareness of the importance of assessing the feet of a diabetic patient at each visit. The results of one study show that the likelihood of amputation is reduced by half if patients with diabetes simply remove their shoes for foot inspection during every outpatient clinic visit. Therefore, while the underlying physiological mechanisms of the problem may be complex, the problem can be relatively easily avoided.

thickening of the basement membrane. This condition leads to impaired delivery of nutrients and hormones to the tissues and inadequate removal of waste products, resulting in irreparable tissue damage.

Some of the more disabling consequences of diabetic circulatory impairment are deterioration of blood flow to the retina of the eye, causing retinopathy and blindness; deterioration of blood flow to the extremities, causing, in some cases, the need for foot or leg amputation; and deterioration of glomerular filtration in the kidneys, leading to renal failure.

Diabetic peripheral neuropathy is also a common complication of long-standing diabetes. This disorder usually involves sensory nerves and those of the autonomic nervous system. Many persons with diabetes experience diminished sensation in the extremities, especially in the feet and legs, which compounds the problem of diminished blood flow to these areas (see Clinical Focus Box 35.1). Often, impaired sensory nerve function results in lack of awareness of severe ulcerations of the feet caused by reduced blood flow. Men may develop impotence, and both men and women may have impaired bladder and bowel function.

REVIEW QUESTIONS

DIRECTIONS: Each of the numbered items or incomplete statements in this section is followed by answers or completions of the statement. Select the ONE lettered answer or completion that is BEST in each case.

1. Which of the following stimulate the secretion of *both* insulin and glucagon from the pancreas?
(A) Epinephrine
(B) Amino acids
(C) Acetylcholine
(D) Both amino acids and acetylcholine
2. The effects of insulin include
(A) Inhibition of amino acid uptake into skeletal muscle
(B) Stimulation of glucose uptake into all tissues in the body
(C) Inhibition of protein degradation in skeletal muscle
(D) Stimulation of hormone-sensitive lipase in adipose tissue
3. The effects of glucagon include
(A) Inhibition of insulin secretion by pancreatic beta cells
(B) Primary actions in adipose tissue
(C) Promotion of gluconeogenesis and urea synthesis in liver cells
(D) Indirect stimulation of ketogenesis in liver cells by the inhibition of pancreatic somatostatin secretion
4. A 55-year-old man was diagnosed with type 1 diabetes at the age of 8. Which would be most characteristic of his form of the disease?
(A) Insulin resistance
(B) Treatment with exogenous insulin
(C) Sulfonylurea treatment
(D) Virtual absence of secondary complications
5. Type 2 diabetes
(A) Has a strong genetic component to the development of the disease
(B) Is characterized by low or negligible circulating insulin
(C) Occurs only in obese individuals
(D) Is treated in the same manner as type 1 diabetes
6. Which of the following would you least likely see in a person with long-standing type 2 diabetes?

(continued)

(A) Neuropathy
(B) Nephropathy
(C) Retinopathy
(D) Ketoacidosis

7. Delta cells of the islets of Langerhans produce which hormone?
(A) Insulin
(B) Glucagon
(C) Acetylcholine
(D) Somatostatin

8. The insulin-glucagon ratio would be expected to be lowest
(A) Immediately after a high-carbohydrate meal
(B) Immediately after a high-protein meal
(C) After an overnight fast
(D) After a 3-day fast

SUGGESTED READING

American Diabetes Association Web site. Available at: http://www.diabetes.org.

Elmendorf JS, Pessin JE. Insulin signaling regulating the trafficking and plasma membrane fusion of GLUT4-containing intracellular vesicles. Exp Cell Res 1999:253:55–62.

Porte D Jr, Sherwin RS, eds. Ellenberg & Rifkin's Diabetes Mellitus. 5th Ed. Stamford, CT: Appleton & Lange, 1997.

Saltiel AR, Kahn CR. Insulin signaling and the regulation of glucose and lipid metabolism. Nature 2001;414:799–806.

Virkamaki A, Ueki K, Kahn CR. Protein-protein interaction in insulin signaling and the molecular mechanisms of insulin resistance. J Clin Invest 1999;103:931–943.

Wilson JD, Foster DW, Kronenberg HM, Larsen PR, eds. Williams Textbook of Endocrinology. 9th Ed. Philadelphia: WB Saunders, 1998.

Endocrine Regulation of Calcium, Phosphate, and Bone Homeostasis

Daniel E. Peavy, Ph.D.

CHAPTER OUTLINE

KEY CONCEPTS

1. When plasma calcium levels fall below normal, spontaneous action potentials can be generated in nerves, leading to tetany of muscles, which, if severe, can result in death.
2. About half of the circulating calcium is in the free or ionized form, about 10% is bound to small anions, and about 40% is bound to plasma proteins. Most of the phosphorus circulates free as orthophosphate.
3. The majority of ingested calcium is not absorbed by the GI tract and leaves the body via the feces; by contrast, phosphate is almost completely absorbed by the GI tract and leaves the body mostly via the urine.
4. Secretion of parathyroid hormone (PTH), a polypeptide hormone produced by the parathyroid glands, is stimulated by a decrease in plasma ionized calcium. PTH plays a vital role in calcium and phosphate homeostasis, and acts on bones, kidneys, and intestine to raise the plasma calcium concentration and lower the plasma phosphate concentration.
5. Vitamin D is converted to the active hormone 1,25-dihydroxycholecalciferol by sequential hydroxylation reactions in the liver and kidneys. This hormone stimulates intestinal calcium absorption and, thereby, raises the plasma calcium concentration.
6. Calcitonin, a polypeptide hormone produced by the thyroid glands, tends to lower the plasma calcium concentration, but its physiological importance in humans has been questioned.
7. Osteoporosis, osteomalacia and rickets, and Paget's disease are the most common forms of metabolic bone disease.

The plasma calcium concentration is among the most closely regulated of all physiological parameters in the body. Typically, it varies by only 1 to 2% daily or even weekly. Such stringent regulation in a biological system usually implies that the parameter plays an important role in one or more critical processes.

Phosphate also plays a variety of important roles in the body, although its concentration is not as tightly regulated as that of calcium. Many of the factors involved in regulating calcium also affect phosphate.

AN OVERVIEW OF CALCIUM AND PHOSPHORUS IN THE BODY

Calcium plays a key role in many physiologically important processes. A significant decrease in plasma calcium can rapidly lead to death. A chronic increase in plasma calcium can lead to soft tissue calcification and formation of stones. Phosphorus also plays important roles in the body.

Calcium Plays Key Roles in Nerve and Muscle Excitation, Muscle Contraction, Enzyme Function, and Bone Mineral Balance

Calcium affects nerve and muscle excitability, neurotransmitter release from axon terminals, and excitation-contraction coupling in muscle cells. It serves as a second or third messenger in several intracellular signal transduction pathways. Some enzymes use calcium as a cofactor, including some in the blood-clotting cascade. Finally, calcium is a major constituent of bone.

Of all these roles, the one that demands the most careful regulation of plasma calcium is the effect of calcium on nerve excitability. Calcium affects the sodium permeability of nerve membranes, which influences the ease with which action potentials are triggered. Low plasma calcium can lead to the generation of spontaneous action potentials in nerves. When motor neurons are affected, tetany of the muscles of the motor unit may occur; this condition is called **hypocalcemic tetany**. Latent tetany may be revealed by certain diagnostically important signs. **Trousseau's sign** is a characteristic spasm of the muscles of the forearm that causes flexion of the wrist and thumb and extension of the fingers. It may occur spontaneously or be elicited by inflation of a blood pressure cuff placed on the upper arm. **Chvostek's sign** is a unilateral spasm of the facial muscles that can be elicited by tapping the facial nerve at the point where it crosses the angle of the jaw.

Phosphate Participates in pH Buffering and Is a Major Constituent of Macromolecules and Bones

Phosphorus (usually as phosphate) also participates in many important metabolic processes. Phosphate serves as an important component of intracellular pH buffering and various metabolic intermediates. DNA, RNA, and phosphoproteins all contain phosphate as an integral part of their structure. Phosphate is also a major component of bones.

The Distributions of Calcium and Phosphorus Differ

Table 36.1 shows the relative distributions of calcium and phosphate in a healthy adult. The average adult body contains approximately 1 to 2 kg of calcium, roughly 99% of it in bones. Despite its critical role in excitation-contraction coupling, only about 0.3% of total body calcium is located in muscle. About 0.1% of total calcium is in extracellular fluid.

Of the roughly 600 g of phosphorus in the body, most is in bones (86%). Compared with calcium, a much larger percentage of phosphorus is located in cells (14%). The amount of phosphorus in extracellular fluid is rather low (0.08% of body content).

Bones also contain a relatively high percentage of the total body content of several other inorganic substances (Table 36.2). About 80% of the total carbonate in the body is located in bones. This carbonate can be mobilized into the blood to combat acidosis; thus, bone participates in pH buffering in the body. Long-standing uncorrected acidosis can result in considerable loss of bone mineral. Significant percentages of the body's magnesium and sodium and nearly 10% of its total water content are in bones.

TABLE 36.1 Body Content and Tissue Distribution of Calcium and Phosphorus in a Healthy Adult

	Calcium	Phosphorus
Total Body Content	1,300 g	600 g
Relative Tissue Distribution (% of total body content)		
Bones and teeth	99%	86%
Extracellular fluid	0.1%	0.08%
Intracellular fluid	1.0%	14%

TABLE 36.2 Inorganic Constituents of Bone

Constituent	Percentage of Total Body Content Present in Bone
Calcium	99
Phosphate	86
Carbonate	80
Magnesium	50
Sodium	35
Water	9

Calcium and Phosphorus Are Present in the Plasma in Several Forms

In humans, the normal plasma calcium concentration is 9.0 to 10.5 mg/dL. Plasma calcium exists in three forms: ionized or free calcium (50% of the total), protein-bound calcium (40%), and calcium bound to small diffusible anions, such as citrate, phosphate, and bicarbonate (10%). The association of calcium with plasma proteins is pH-dependent. At an alkaline pH, more calcium is bound; the opposite is true at an acidic pH.

Plasma phosphorus concentrations may fluctuate significantly during the course of a day, from 50 to 150% of the average value for any particular individual. In adults, the normal range of plasma concentrations is 3.0 to 4.5 mg/dL (expressed in terms of milligrams of phosphorus).

Phosphorus circulates in the plasma primarily as inorganic **orthophosphate** (PO_4). At a normal blood pH of 7.4, 80% of the phosphate is in the HPO_4^{2-} form and 20% is in the $H_2PO_4^-$ form. Nearly all plasma inorganic phosphate is ultrafilterable. In addition to free orthophosphate, phosphate is present in small amounts in the plasma in organic form, such as in hexose or lipid phosphates.

The Homeostatic Pathways for Calcium and Phosphorus Differ Quantitatively

Both calcium and phosphate are obtained from the diet. The ultimate fate of each substance is determined primarily by the gastrointestinal (GI) tract, the kidneys, and the bones.

Calcium Handling by the GI Tract, Kidneys, and Bones. The approximate tissue distribution and average daily flux of calcium among tissues in a healthy adult are shown in Figure 36.1. Dietary intakes may vary widely, but an "average" diet contains approximately 1,000 mg/day of calcium. Intakes up to twice that amount are usually well tolerated, but excessive calcium intake can result in soft tissue calcification or kidney stones. Only about one third of ingested calcium is actually absorbed from the GI tract; the remainder is excreted in the feces. The efficiency of calcium uptake from the GI tract varies with the individual's physiological status. The percentage uptake of calcium may be increased in young growing children and pregnant or nursing women; often it is reduced in older adults.

Figure 36.1 also indicates that approximately 150 mg/day of calcium actually enter the GI tract from the

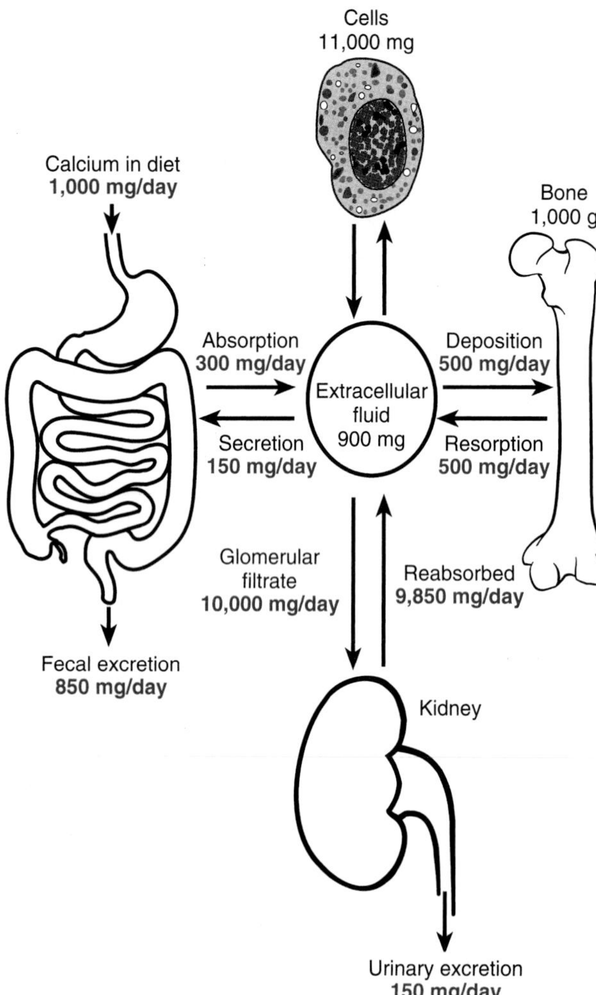

FIGURE 36.1 Typical daily exchanges of calcium between different tissue compartments in a healthy adult. Fluxes of calcium (mg/day) are shown in red. Total calcium content in each compartment is shown in black. Note that the majority of ingested calcium is eliminated from the body via the feces.

body. This component of the calcium flux partly results from sloughing of mucosal cells that line the GI tract and also from calcium that accompanies various secretions into the GI tract. This component of calcium metabolism is relatively constant, so the primary determinant of *net* calcium uptake from the GI tract is calcium absorption. Intestinal absorption is important in regulating calcium homeostasis.

Bone in an average individual contains approximately 1,000 g of calcium. Bone mineral is constantly resorbed and deposited in the remodeling process. As much as 500 mg/day of calcium may flow in and out of the bones (see Fig. 36.1). Since bone calcium serves as a reservoir, both bone resorption and bone formation are important in regulating plasma calcium concentration.

In overall calcium balance, the net uptake of calcium from the GI tract presents a daily load of calcium that will eventually require elimination. The primary route of elimination is via the urine, and therefore, the kidneys play an important role in regulating calcium homeostasis. The 150 mg/day of calcium excreted in the urine represent only about 1% of the calcium initially filtered by the kidneys; the remaining 99% is reabsorbed and returned to the blood. Therefore, small changes in the amount of calcium reabsorbed by the kidneys can have a dramatic impact on calcium homeostasis.

Phosphate Handling by the GI Tract, Kidneys, and Bones.

Figure 36.2 shows the overall daily flux of phosphate in the body. A typical adult ingests approximately 1,400 mg/day of phosphorus. In marked contrast to calcium, most (1,300 mg/day) of this phosphorus is absorbed from the GI tract, typically as inorganic phosphate. There is an obligatory contribution of phosphorus to the contents of the GI tract (about 200 mg/day), much like that for calcium, resulting in a net uptake of phosphorus of 1,100 mg/day and excretion of 300 mg/day via the feces. Thus, the majority of ingested phosphate is absorbed from the GI tract and little passes through to the feces.

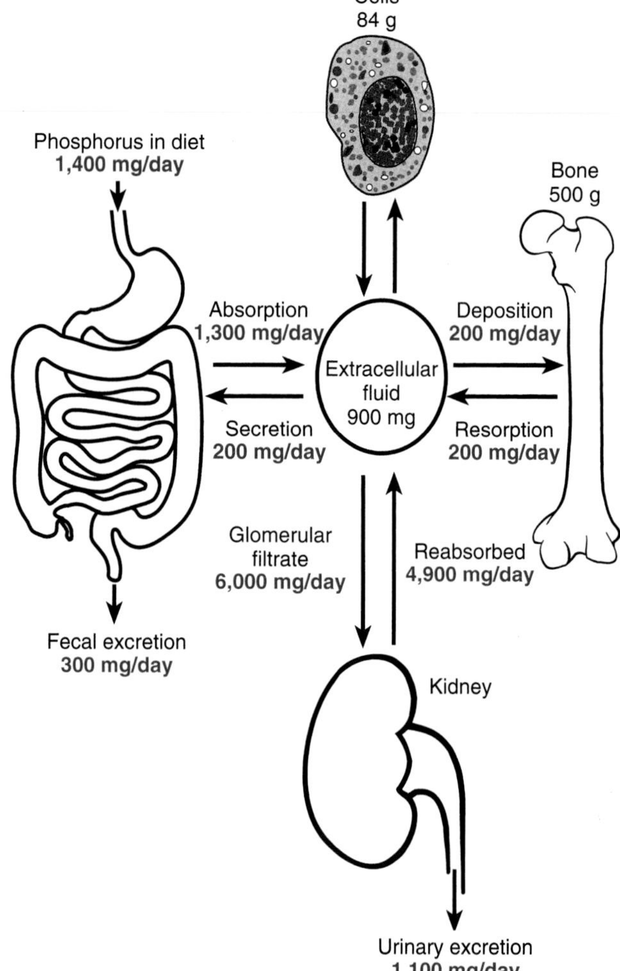

FIGURE 36.2 Typical daily exchanges of phosphorus between different tissue compartments in a healthy adult. Fluxes of phosphorus (mg/day) are shown in red. Total phosphorus content in each compartment is shown in black. Note that the majority of ingested phosphorus is absorbed and eventually eliminated from the body via the urine.

Because most ingested phosphate is absorbed, phosphate homeostasis is greatly influenced by renal excretory mechanisms. Since the majority of circulating phosphate is readily filtered in the kidneys, tubular phosphate reabsorption is a major process regulating phosphate homeostasis.

MECHANISMS OF CALCIUM AND PHOSPHATE HOMEOSTASIS

As indicated above, the GI tract, kidneys, and bone each play a role in the regulation of calcium and phosphate homeostasis.

Calcium and Phosphate Are Absorbed Primarily by the Small Intestine

Calcium absorption in the small intestine occurs by both active transport and diffusion. The relative contribution of each process varies with the region and with total calcium intake. Uptake of calcium by active transport predominates in the duodenum and jejunum; in the ileum, simple diffusion predominates. The relative importance of active transport in the duodenum and jejunum versus passive diffusion in the ileum depends on several factors. At very high levels of calcium intake, active transport processes are saturated and most of the uptake occurs in the ileum, partly because of its greater length, compared with other intestinal segments. With moderate or low calcium intake, however, active transport predominates because the gradient for diffusion is low.

Active transport is the regulated variable in controlling calcium uptake from the small intestine. Metabolites of vitamin D provide a regulatory signal to increase intestinal calcium absorption. Under the influence of 1,25-dihydroxycholecalciferol, calcium-binding proteins in intestinal mucosal cells increase in number, enhancing the capacity of these cells to transport calcium actively (see Chapter 27).

The small intestine is also a primary site for phosphate absorption. Uptake occurs by active transport and passive diffusion, but active transport is the primary mechanism. As indicated in Figure 36.2, phosphate is efficiently absorbed from the small intestine; typically, 80% or more of ingested phosphate is absorbed. However, phosphate absorption from the small intestine is regulated very little. To a minor extent, active transport of phosphate is coupled to calcium transport. Therefore, when active transport of calcium is low, as with vitamin D deficiency, phosphate absorption is also low.

The Kidneys Play an Important Role in Regulating Plasma Concentrations of Calcium and Phosphate

As a result of regulating the urinary excretion of calcium and phosphate, the kidneys are in a key position to regulate the total body balance of these two ions. Hormones are an important signal to the kidneys to direct the excretion or retention of calcium and phosphate.

Renal Handling of Calcium. As discussed in Chapter 24, filterable calcium comprises about 60% of the total calcium in the plasma. It consists of free calcium ions and calcium bound to small diffusible anions. The remaining 40% of the total calcium is bound to plasma proteins and is not filterable by the glomeruli. Ordinarily, 99% of the filtered calcium is reabsorbed by the kidney tubules and returned to the plasma. Reabsorption occurs both in the proximal and distal tubules and in the loop of Henle. Approximately 60% of filtered calcium is reabsorbed in the proximal tubule, 30% in the loop of Henle, and 9% in the distal tubule; the remaining 1% is excreted in the urine. Renal calcium excretion is controlled primarily in the late distal tubule; parathyroid hormone stimulates calcium reabsorption here, promoting calcium retention and lowering urinary calcium. Parathyroid hormone is an important regulator of plasma calcium concentration.

Renal Handling of Phosphate. Most ingested phosphate is absorbed from the GI tract, and the primary route of excretion of this phosphate is via the urine. Therefore, the kidneys play a key role in regulating phosphate homeostasis. Ordinarily about 85% of filtered phosphate is reabsorbed and 15% is excreted in the urine. Phosphate reabsorption occurs via active transport, mainly in the proximal tubule where 65 to 80% of filtered phosphate is reabsorbed. Parathyroid hormone inhibits phosphate reabsorption in the proximal tubule and has a major regulatory effect on phosphate homeostasis. It increases urinary phosphate excretion, leading to the condition of **phosphaturia**, with an accompanying decrease in the plasma phosphate concentration.

Substantial Amounts of Calcium and Phosphate Enter and Leave Bone Each Day

Although bone may be considered as being a relatively inert material, it is active metabolically. Considerable amounts of calcium and phosphate both enter and exit bone each day, and these processes are hormonally controlled.

Composition of Bone. Mature bone can be simply described as inorganic mineral deposited on an organic framework. The mineral portion of bone is composed largely of calcium phosphate in the form of **hydroxyapatite crystals**, which have the general chemical formula $Ca_{10}(PO_4)_6(OH)_2$. The mineral portion of bone typically comprises about 25% of its volume, but because of its high density, the mineral fraction is responsible for approximately half the weight of bone. Bone contains considerable amounts of the body's content of carbonate, magnesium, and sodium in addition to calcium and phosphate (see Table 36.2).

The organic matrix of bone on which the bone mineral is deposited is called **osteoid**. **Type I collagen** is the primary constituent of osteoid, comprising 95% or more. Collagen in bone is similar to that of skin and tendons, but bone collagen exhibits some biochemical differences that impart increased mechanical strength. The remaining noncollagen portion (5%) of organic matter is referred to as **ground substance**. Ground substance consists of a mixture of various **proteoglycans**, high-molecular-weight compounds consisting of different types of polysaccharides linked to a polypeptide backbone. Typically, they are 95% or more carbohydrate.

Electron microscopic study of bone reveals needle-like hydroxyapatite crystals lying alongside collagen fibers. This orderly association of hydroxyapatite crystals with the collagen fibers is responsible for the strength and hardness characteristic of bone. A loss of either bone mineral or organic matrix greatly affects the mechanical properties of bone. Complete demineralization of bone leaves a flexible collagen framework, and the complete removal of organic matrix leaves a bone with its original shape, but extremely brittle.

Cell Types Involved in Bone Formation and Bone Resorption. The three principal cell types involved in bone formation and bone resorption are osteoblasts, osteocytes, and osteoclasts (Fig. 36.3).

Osteoblasts are located on the bone surface and are responsible for osteoid synthesis. Like many cells that actively synthesize proteins for export, osteoblasts have an abundant rough ER and Golgi apparatus. Cells actively engaged in osteoid synthesis are cuboidal, while those less active are more flattened. Numerous cytoplasmic processes connect adjacent osteoblasts on the bone surface and connect osteoblasts with osteocytes deeper in the bone. Osteoid produced by osteoblasts is secreted into the space adjacent to the bone. Eventually, new osteoid becomes mineralized, and in the process, osteoblasts become surrounded by mineralized bone.

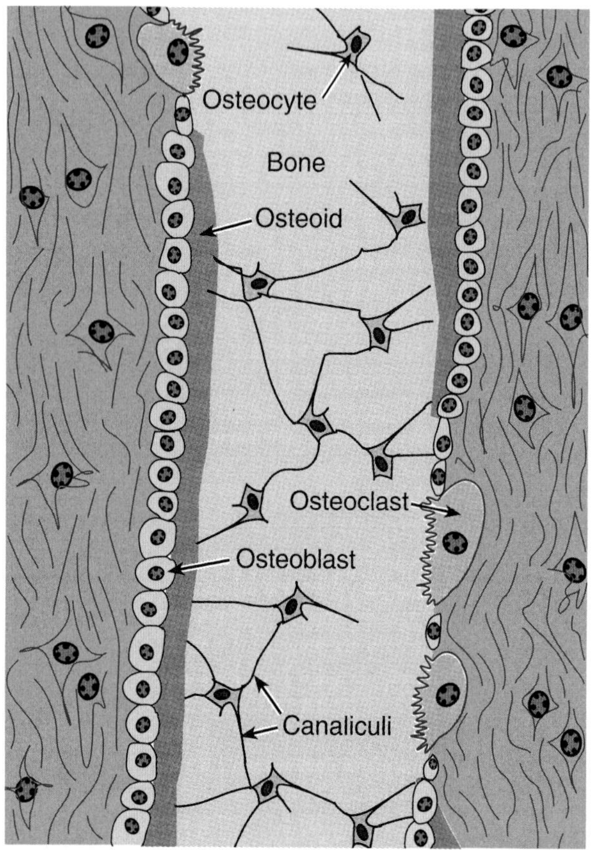

FIGURE 36.3 **The location and relationship of the three primary cell types involved in bone metabolism.**

As osteoblasts are progressively incorporated into mineralized bone, they lose much of their bone-forming ability and become quiescent. At this point they are called **osteocytes**. Many of the cytoplasmic connections in the osteoblast stage are maintained into the osteocyte stage. These connections become visible channels or **canaliculi** that provide direct contact for osteocytes deep in bone with other osteocytes and with the bone surface. It is generally believed that these canaliculi provide a mechanism for the transfer of nutrients, hormones, and waste products between the bone surface and its interior.

Osteoclasts are cells responsible for bone resorption. They are large, multinucleated cells located on bone surfaces. Osteoclasts promote bone resorption by secreting acid and proteolytic enzymes into the space adjacent to the bone surface. Surfaces of osteoclasts facing bone are ruffled to increase their surface area and promote bone resorption. Bone resorption is a two-step process. First, osteoclasts create a local acidic environment that increases the solubility of surface bone mineral. Second, proteolytic enzymes secreted by osteoclasts degrade the organic matrix of bone.

Bone Formation and Bone Remodeling. Early in fetal development, the skeleton consists of little more than a cartilaginous model of what will later form the bony skeleton. The process of replacing this cartilaginous model with mature, mineralized bone begins in the center of the cartilage and progresses toward the two ends of what will later form the bone. As mineralization progresses, the bone increases in thickness and in length.

The **epiphyseal plate** is a region of growing bone of particular interest because it is here that the elongation and growth of bones occurs after birth. Histologically, the epiphyseal plate shows considerable differences between its leading and trailing edges. The leading edge consists primarily of **chondrocytes**, which are actively engaged in the synthesis of cartilage of the epiphyseal plate. These cells gradually become engulfed in their own cartilage and are replaced by new cells on the cartilage surface, allowing the process to continue. The cartilage gradually becomes calcified, and the embedded chondrocytes die. The calcified cartilage begins to erode, and osteoblasts migrate into the area. Osteoblasts secrete osteoid, which eventually becomes mineralized, and new mature bone is formed. In the epiphyseal plate, therefore, the continuing processes of cartilage synthesis, calcification, erosion, and osteoblast invasion result in a zone of active bone formation that moves away from the middle or center of the bone toward its end.

Chondrocytes of epiphyseal plates are controlled by hormones. Insulin-like growth factor I (IGF-I), primarily produced by the liver in response to growth hormone, serves as a primary stimulator of chondrocyte activity and, ultimately, of bone growth. Insulin and thyroid hormones provide an additional stimulus for chondrocyte activity.

Beginning a few years after puberty, the epiphyseal plates in long bones (as in the legs and arms) gradually become less responsive to hormonal stimuli and, eventually, are totally unresponsive. This phenomenon is referred to as **closure of the epiphyses**. In most individuals, epiphyseal closure is complete by about age 20; adult height is reached at this point, since further linear growth is impossible. Not all bones undergo closure. For example, those in the fingers, feet, skull, and jaw remain responsive, which accounts for the skeletal changes seen in acromegaly, the condition of growth hormone overproduction (see Chapter 32).

The flux of calcium and phosphate into and out of bone each day reflects a turnover of bone mineral and changes in bone structure generally referred to as **remodeling**. Bone remodeling occurs along most of the outer surface of the bone, making it either thinner or thicker, as required. In long bones, remodeling can also occur along the inner surface of the bone shaft, next to the marrow cavity. Remodeling is an adaptive process that allows bone to be reshaped to meet changing mechanical demands placed on the skeleton. It also allows the body to store or mobilize calcium rapidly.

REGULATION OF PLASMA CALCIUM AND PHOSPHATE CONCENTRATIONS

Regulatory mechanisms for calcium include rapid nonhormonal mechanisms with limited capacity and somewhat slower hormonally regulated mechanisms with much greater capacity. There are also similar mechanisms involved in regulating plasma phosphate concentrations.

Nonhormonal Mechanisms Can Rapidly Buffer Small Changes in Plasma Concentrations of Free Calcium

The calcium bound to plasma proteins and a small fraction of that in bone mineral can help prevent a rapid decrease in the plasma calcium concentration.

Protein-Bound Calcium. The association of calcium with proteins is a simple, reversible, chemical equilibrium process. Protein-bound calcium, therefore, has the capacity to serve as a buffer of free plasma calcium concentrations. This effect is rapid and does not require complex signaling pathways; however, the capacity is limited, and the mechanism cannot serve a long-term role in calcium homeostasis.

A Readily Exchangeable Pool of Calcium in Bones. Recall that approximately 99% of total body calcium is present in bones, and a healthy adult body has about 1 to 2 kg of calcium. Most of the calcium in bones exists as mature, hardened bone mineral that is not readily exchangeable but can be moved into the plasma via hormonal mechanisms (described below). However, approximately 1% (or 10 g) of the calcium in bones is in a simple chemical equilibrium with plasma calcium. This readily exchangeable calcium source is primarily located on the surface of newly formed bones. Any change in free calcium in the plasma or extracellular fluid results in a shift of calcium either into or out of the bone mineral until a new equilibrium is reached. Although this mechanism, like that described above, provides for a rapid defense against changes in free calcium concentrations, it is limited in capacity and can provide for only short-term adjustments in calcium homeostasis.

Hormonal Mechanisms Provide High-Capacity, Long-Term Regulation of Plasma Calcium and Phosphate Concentrations

The hormonal mechanisms described here have a large capacity and the ability to make long-term adjustments in calcium and phosphate fluxes, but they do not respond instantaneously. It may take several minutes or hours for the response to occur and adjustments to be made. However, these are the principal mechanisms that regulate plasma calcium and phosphate concentrations.

The Chemistry of Parathyroid Hormone, Calcitonin, and 1,25-Dihydroxycholecalciferol and the Regulation of Their Production. One of the primary regulators of plasma calcium concentrations is **parathyroid hormone** (PTH). PTH is an 84-amino acid polypeptide produced by the parathyroid glands. Synthetic peptides containing the first 34 amino terminal residues appear to be as active as the native hormone.

There are two pairs of parathyroid glands, located on the dorsal surface of the left and right lobes of the thyroid gland. Because of this close proximity, damage to the parathyroid glands or to their blood supply may occur during surgical removal of the thyroid gland.

The primary physiological stimulus for PTH secretion is a *decrease* in plasma calcium. Figure 36.4 shows the relationship between serum parathyroid hormone concentration and total plasma calcium concentration. It is actually a decrease in the **ionized calcium** concentration that triggers an increase in PTH secretion. The net effect of PTH is to in-

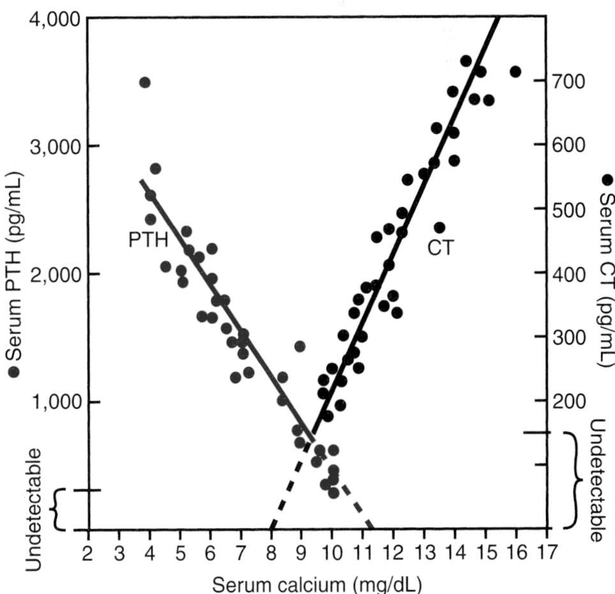

FIGURE 36.4 **Effect of changes in plasma calcium on parathyroid hormone (PTH) and calcitonin (CT) secretion.** (Modified from Arnaud, CD, Littledike T, Tsao HS. Simultaneous measurements of calcitonin and parathyroid hormone in the pig. In: Taylor S, Foster GV, eds. Proceedings of the Symposium on Calcitonin and C Cells. London: Heinemann, 1969;99.)

crease the flow of calcium into plasma, and return the plasma calcium concentration toward normal.

Calcitonin (CT) is a 32-amino acid polypeptide. Also known as thyrocalcitonin, CT is produced by **parafollicular cells** of the thyroid gland (see Fig. 33.1). Unlike PTH, for which only the initial amino terminal segment is required, the full polypeptide is required for CT activity. Salmon calcitonin differs from human calcitonin in 9 of 32 amino acid residues and is 10 times more potent than human CT in its hypocalcemic effect. The higher potency may be due to a greater affinity for receptors and slower degradation by peripheral tissues. CT is often used clinically as a synthetic peptide matching the sequence of salmon calcitonin.

In contrast to PTH, CT secretion is stimulated by an increase in plasma calcium (see Fig. 36.4). Hormones of the GI tract, especially gastrin, also promote CT secretion. Because the net effect of CT is to promote calcium deposition in bone, the stimulation of CT secretion by GI hormones provides an additional mechanism for facilitating the uptake of calcium into bone after the ingestion of a meal.

The third key hormone involved in regulating plasma calcium is **vitamin D_3** (cholecalciferol). More precisely, a metabolite of vitamin D_3 serves as a hormone in calcium homeostasis. The D vitamins, a group of lipid-soluble compounds derived from cholesterol, have long been known to be effective in the prevention of rickets. Research during the past 30 years indicates that vitamin D exerts it effects through a hormonal mechanism.

Figure 36.5 shows the structure of vitamin D_3 and the related compound **vitamin D_2** (ergocalciferol). Ergocalciferol is the form principally found in plants and yeasts and is commonly used to supplement human foods because of its relative availability and low cost. Although it is less potent on a mole-per-mole basis, vitamin D_2 undergoes the same metabolic conversion steps and, ultimately, produces the same biological effects as vitamin D_3. The physiological actions of vitamin D_3 also apply to vitamin D_2.

Vitamin D_3 can be provided by the diet or formed in the skin by the action of ultraviolet light on a precursor, **7-dehydrocholesterol**, derived from cholesterol (Fig. 36.6). In many countries where food is not systematically supplemented with vitamin D, this pathway provides the major source of vitamin D. Because of the number of variables involved, it is difficult to specify a minimum exposure time. However, exposure to moderately bright sunlight for 30 to 120 min/day usually provides enough vitamin D to satisfy the body's needs without any dietary supplementation.

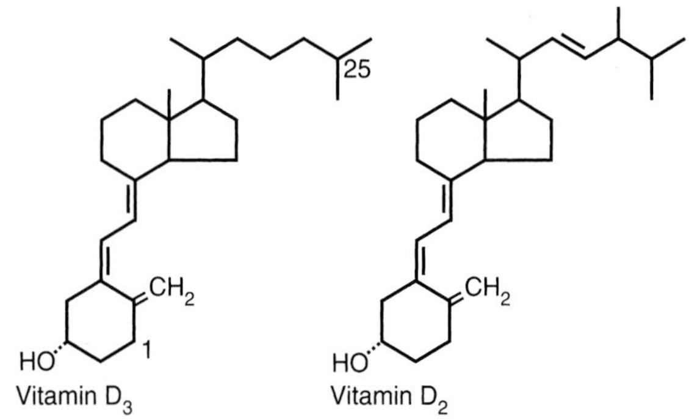

Vitamins D_3 and D_2 are by themselves relatively inactive. However, they undergo a series of transformations in the liver and kidneys that convert them into powerful calcium-regulatory hormones (see Fig. 36.6). The first step occurs in the liver and involves addition of a hydroxyl group to carbon 25, to form 25-hydroxycholecalciferol (25-OH D_3). This reaction is largely unregulated, although certain drugs and liver diseases may affect this step. Next, 25-hydroxycholecalciferol is released into the blood, and it undergoes a second hydroxylation reaction on carbon 1 in the kidney. The product is **1,25-dihydroxycholecalciferol**, also known as **1,25-dihydroxyvitamin D_3** or **calcitriol**, the principal hormonally active form of the vitamin. The biological activity of 1,25-dihydroxycholecalciferol is approximately 100 to 500 times greater than that of 25-hydroxycholecalciferol. The reaction in the kidney is catalyzed by the enzyme 1α-hydroxylase, which is located in tubule cells.

The final step in 1,25-dihydroxycholecalciferol formation is highly regulated. The activity of 1α-hydroxylase is regulated primarily by PTH, which stimulates its activity. Therefore, if plasma calcium levels fall, PTH secretion increases; in turn, PTH promotes the formation of 1,25-dihydroxycholecalciferol. In addition, enzyme activity increases in response to a decrease in plasma phosphate. This does not appear to involve any intermediate hormonal signals but apparently involves direct activation of either the enzyme or cells in which the enzyme is located. Both a decrease in plasma calcium, which triggers PTH secretion, and a decrease in circulating phosphate result in the activation of 1α-hydroxylase and an increase in 1,25-dihydroxycholecalciferol synthesis.

FIGURE 36.5 **The structures of vitamin D_3 and vitamin D_2.** Note that they differ only by a double bond between carbons 22 and 23 and a methyl group at position 24.

The Actions of Parathyroid Hormone, Calcitonin, and 1,25-Dihydroxycholecalciferol. Most hormones generally improve the quality of life and the chance for survival when an animal is placed in a physiologically challenging situation. However, PTH is essential for life. The complete absence of PTH causes death from hypocalcemic tetany within just a few days. The condition can be avoided with hormone replacement therapy.

The net effects of PTH on plasma calcium and phosphate and its sites of action are shown in Figure 36.7. PTH causes an increase in plasma calcium concentration while decreasing plasma phosphate. This decrease in phosphate concentration is important with regard to calcium homeostasis. At normal plasma concentrations, calcium and phosphate are at or near chemical saturation levels. If PTH were to increase *both* calcium and phosphate levels, they would simply crystallize in bone or soft tissues as calcium phosphate, and the necessary increase in plasma calcium concentration would not occur. Thus, the effect of PTH to lower plasma phosphate is an important aspect of its role in regulating plasma calcium.

Parathyroid hormone has several important actions in the kidneys (see Fig. 36.7). It stimulates calcium reabsorption in the thick ascending limb and late distal tubule, decreasing calcium loss in the urine and increasing plasma concentra-

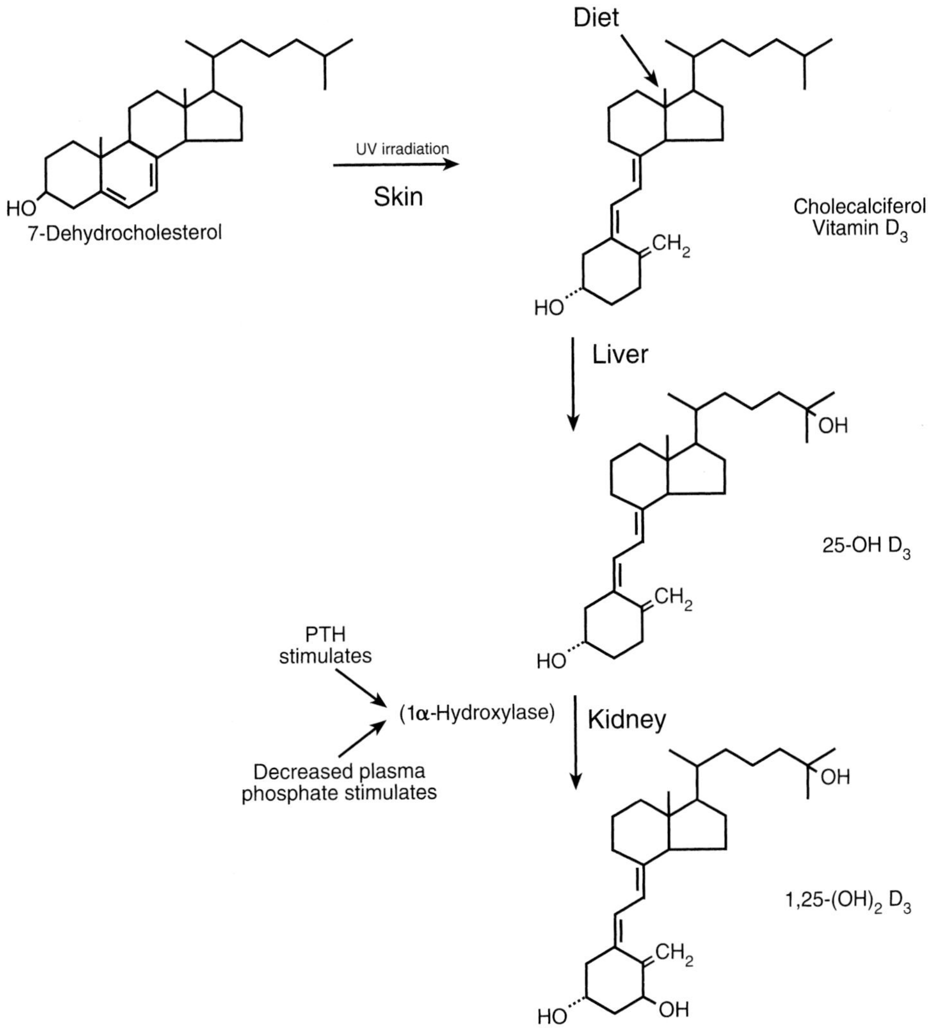

FIGURE 36.6 **The conversion pathway of vitamin D_3 into 1,25-dihydroxycholecalciferol [1,25-$(OH)_2$ D_3].**

tions. It also inhibits phosphate reabsorption in the proximal tubule, leading to increased urinary phosphate excretion and a decrease in plasma phosphate. Another important effect of PTH is to increase the activity of kidney 1α-hydroxylase, which is involved in forming active vitamin D.

In bone, PTH activates osteoclasts to increase bone resorption and the delivery of calcium from bone into plasma (see Fig. 36.7). In addition to stimulating active osteoclasts, PTH stimulates the maturation of immature osteoclasts into mature, active osteoclasts. PTH also inhibits collagen synthesis by osteoblasts, resulting in decreased bone matrix formation and decreased flow of calcium from plasma into bone mineral. The actions of PTH to promote bone resorption are augmented by 1,25-dihydroxycholecalciferol.

PTH does not appear to have any major direct effects on the GI tract. However, because it increases active vitamin D formation, it ultimately increases the absorption of both calcium and phosphate from the GI tract (see Fig. 36.7).

Calcitonin is important in several lower vertebrates, but despite its many demonstrated biological effects in humans, it appears to play only a minor role in calcium homeostasis. This conclusion mostly stems from two lines of evidence. First, CT loss following surgical removal of the thyroid gland (and, therefore, removal of CT-secreting parafollicular cells) does not lead to overt clinical abnormalities of calcium homeostasis. Second, CT hypersecretion, such as from thyroid tumors involving parafollicular cells, does not cause any overt problems. On a daily basis, calcitonin probably only fine-tunes the calcium regulatory system.

The overall action of calcitonin is to decrease *both* calcium and phosphate concentrations in plasma (Fig. 36.8). The primary target of CT is bone, although some lesser effects also occur in the kidneys. In the kidneys, CT decreases the tubular reabsorption of calcium and phosphate. This leads to an increase in urinary excretion of both calcium and phosphate and, ultimately, to decreased levels of both ions in the plasma. In bones, CT opposes the action of PTH on osteoclasts by inhibiting their activity. This leads to decreased bone resorption and an overall net transfer of calcium from plasma into bone. Calcitonin has little or no direct effect on the GI tract.

The net effect of 1,25-dihydroxycholecalciferol is to increase both calcium and phosphate concentrations in plasma (Fig. 36.9). The activated form of vitamin D primarily influences the GI tract, although it has actions in the kidneys and bones as well.

In the kidneys, 1,25-dihydroxycholecalciferol increases the tubular reabsorption of calcium and phosphate, pro-

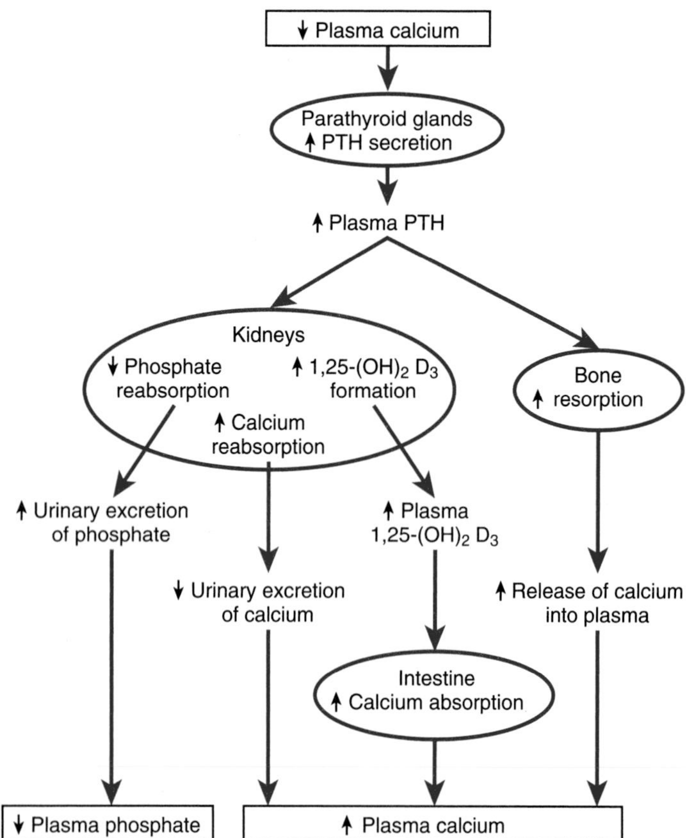

FIGURE 36.7 Effects of parathyroid hormone (PTH) on calcium and phosphate metabolism.

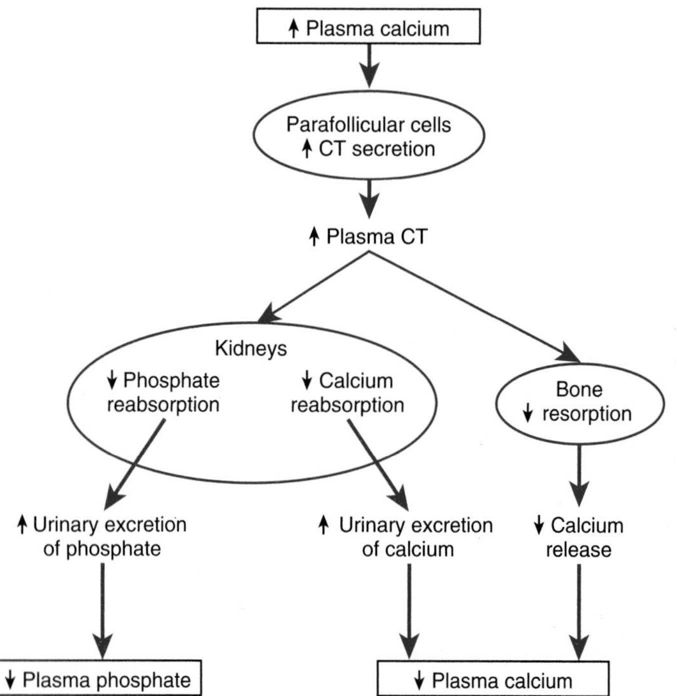

FIGURE 36.8 Effects of calcitonin (CT) on calcium and phosphate metabolism.

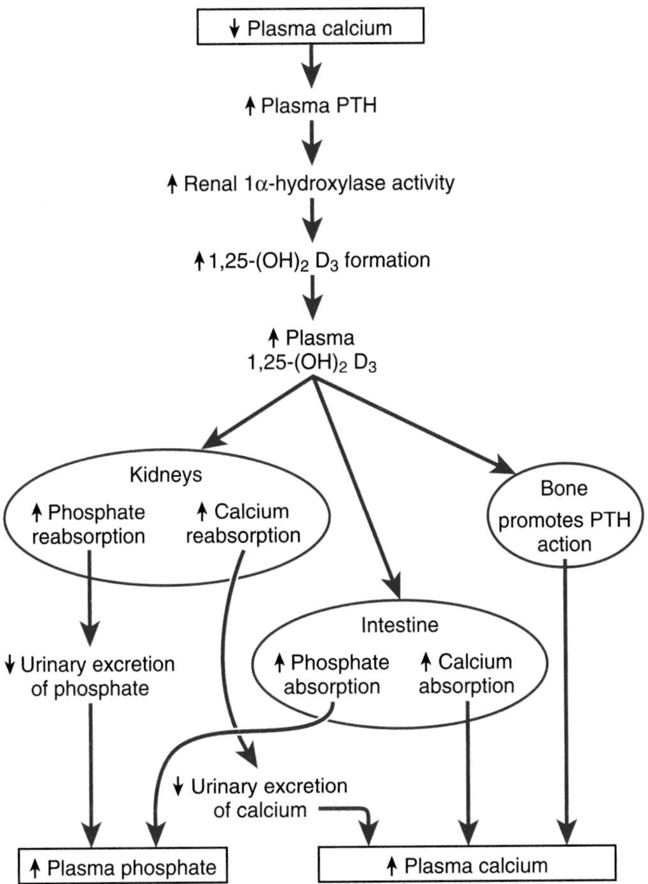

FIGURE 36.9 **Effects of 1,25-dihydroxycholecalciferol [1,25-$(OH)_2$ D_3] on calcium and phosphate metabolism.**

moting the retention of both ions in the body. However, this is a weak and probably only minor effect of the hormone. In bones, the hormone promotes actions of PTH on osteoclasts, increasing bone resorption (see Fig. 36.9).

In the gastrointestinal tract, 1,25-dihydroxycholecalciferol stimulates calcium and phosphate absorption by the small intestine, increasing plasma concentrations of both ions. This effect is mediated by increased production of calcium transport proteins resulting from gene transcription events and usually requires several hours to appear.

ABNORMALITIES OF BONE MINERAL METABOLISM

There are several **metabolic bone diseases**, all typified by ongoing disruption of the normal processes of either bone formation or bone resorption. The conditions most frequently encountered clinically are osteoporosis, osteomalacia, and Paget's disease.

Osteoporosis Is a Reduction in Bone Mass

Osteoporosis is a major health problem, particularly because older adults are more prone to this disorder and the average age of the population is increasing (see Clinical Focus Box 36.1). Osteoporosis involves a reduction in total bone mass with an equal loss of both bone mineral and organic matrix. Several factors are known to contribute directly to osteoporosis. Long-term dietary calcium deficiency can lead to osteoporosis because bone mineral is mobilized to maintain plasma calcium levels. Vitamin C deficiency also can result in a net loss of bone because vitamin C is required for normal collagen synthesis to occur. A defect in matrix production and the inability to produce new bone eventually result in a net loss of bones. For reasons that are not entirely understood, a reduction in the mechanical stress placed on bone can lead to bone loss. Immobilization or disuse of a limb, such as with a cast or paralysis, can result in localized osteoporosis of the affected limb. Space flight can produce a type of disuse osteoporosis resulting from the condition of weightlessness.

Most commonly, osteoporosis is associated with advancing age in both men and women, and it cannot be assigned to any specific definable cause. For several reasons, women are more prone to develop the disease than men. Figure 36.10 shows the average bone mineral content (as grams of calcium) for men and women versus age. Until about the time of puberty, males and females have similar bone mineral content. However, at puberty, males begin to acquire bone mineral at a greater rate; peak bone mass may be approximately 20% greater than that of

CLINICAL FOCUS BOX 36.1

The Toll of Osteoporosis

Osteoporosis is often called the "silent disease" because bone loss initially occurs without symptoms. People may not know that they have significant bone loss until their bones become so weak that a sudden strain, bump, or fall causes a fracture. Osteoporosis is a major public health threat in the United States because it affects some 28 million Americans. Some 10 million individuals have been diagnosed with the disease and another 18 million have low bone mass, placing them at increased risk for osteoporosis. Approximately 80% of those affected by osteoporosis are women. While osteoporosis is often thought of as an older person's disease, it can strike at any age. The seeds of osteoporosis are often sown in childhood, and it takes a lifetime of effort to prevent the disease. A frightening number of children do not get sufficient exercise, vitamin D, and calcium to ensure protection from developing the disease later in life.

Osteoporosis is responsible for more than 1.5 million fractures annually, including 300,000 hip fractures, 700,000 vertebral fractures, 250,000 wrist fractures, and 300,000 fractures at other sites. It is estimated that osteoporosis costs some 10 to 15 billion dollars a year for hospitalization and nursing home care. Additional losses in wages and productivity send these numbers far higher. Nearly one third of people who have hip fractures end up in nursing homes within a year; nearly 20% die within a year.

What causes osteoporosis, and what can be done to prevent or treat the disease? While it is known that a diet low in calcium or vitamin D, certain medications such as glucocorticoids and anticonvulsants, and excessive ingestion of aluminum-containing antacids can cause osteoporosis, in most cases, the exact cause is unknown. However, several identified risk factors associated with the disease are being a woman (especially a postmenopausal woman); being Caucasian or Asian; being of advanced age; having a family history of the disease; having low testosterone levels (in men); having an inactive lifestyle; cigarette smoking; and an excessive use of alcohol.

A comprehensive program to help prevent osteoporosis includes a balanced diet rich in calcium and vitamin D, regular weight-bearing exercise, a healthy lifestyle with no smoking or excessive alcohol use, and bone density testing and medication when appropriate. Although at present there is no cure for osteoporosis, there are five FDA-approved medications to either prevent or treat the disease in women: estrogens; alendronate and risedronate (both bisphosphonates); calcitonin; and raloxifene, a selective estrogen receptor modulator. Although 20% of all osteoporosis cases occur in men, only alendronate and risedronate are currently FDA approved for use in men and only for cases of corticosteroid-induced osteoporosis. Testosterone replacement therapy is often helpful in a man with a low testosterone level.

women. Maximum bone mass is attained between 30 and 40 years of age and then tends to decrease in both sexes. Initially this occurs at an approximately equivalent rate, but women begin to experience a more rapid bone mineral loss at the time of menopause (about age 45 to 50). This loss appears to result from the decline in estrogen secretion that occurs at menopause. Low-dose estrogen supplementation of postmenopausal women is usually effective in retarding bone loss without causing adverse effects. This condition of increased bone loss in women after menopause is called **postmenopausal osteoporosis** (see Clinical Focus Box 36.2).

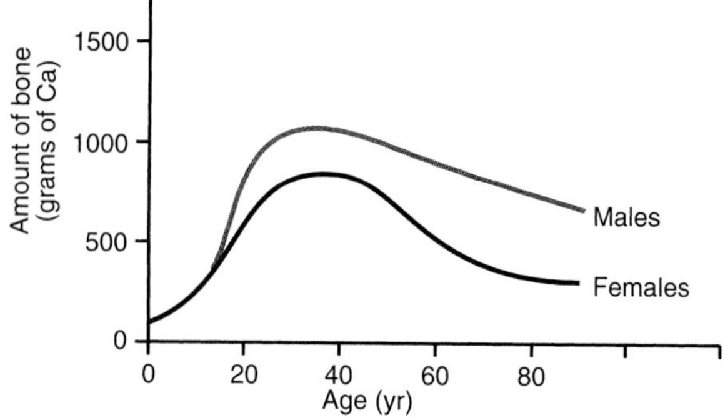

FIGURE 36.10 Changes in bone calcium content as a function of age in males and females. These changes can be roughly extrapolated into changes in bone mass and bone strength.

Osteomalacia and Rickets Result From Inadequate Bone Mineralization

Osteomalacia and **rickets** are characterized by the inadequate mineralization of new bone matrix, such that the ratio of bone mineral to matrix is reduced. As a result, bones may have reduced strength and are subject to distortion in response to mechanical loads. When the disease occurs in adults, it is called osteomalacia; when it occurs in children, it is called rickets. In children, the condition often produces a bowing of the long bones in the legs. In adults, it is often associated with severe bone pain.

TABLE 36.3 Causes of Osteomalacia and Rickets

Inadequate availability of vitamin D	Dietary deficiency or lack of exposure to sunlight Fat-soluble vitamin malabsorption
Defects in metabolic activation of vitamin D	25-Hydroxylation (liver) Liver disease Certain anticonvulsants, such as phenobarbital 1-Hydroxylation (kidney) Renal failure Hypoparathyroidism
Impaired action of 1,25-dihydroxycholecalciferol on target tissues	Certain anticonvulsants 1,25-Dihydroxycholecalciferol receptor defects Uremia

CLINICAL FOCUS BOX 36.2

Cytokines, Estrogens, and Osteoporosis

It is well established that a decline in circulating levels of 17β-estradiol is a major contributing factor in the development of osteoporosis in postmenopausal women. Until recently, specific mechanisms by which estradiol might influence bone metabolism were largely unknown. Recent studies suggest that estradiol influences the production and/or modulates the activity of several cytokines involved in regulating bone remodeling.

Normal bone remodeling involves a regulated balance between the processes of bone formation and bone resorption. Osteoclast-mediated bone resorption involves two processes: the activation of mature, functional osteoclasts and the recruitment and differentiation of osteoclast precursors. In addition to PTH, the cytokines interleukin-1 (IL-1) and tumor necrosis factor (TNF) are involved in the activation of mature osteoclasts to cause bone resorption. For maturation of osteoclast precursors, the cytokines macrophage-colony stimulating factor (M-CSF) and interleukin-6 (IL-6) appear to be involved. Estradiol plays a role in bone remodeling by suppressing the formation of these cytokines. As a result of its ability to interact with bone cells and their precursors to regulate local paracrine signaling mechanisms, estradiol produces anti-osteoporotic effects in bone.

When estradiol is present, as in a premenopausal state, it acts as a governor to reduce cytokine production and limit osteoclast activity. When estradiol levels are reduced, the governor is lost, secretion of these cytokines increases, and osteoclast formation and activity increase, resulting in increased bone resorption.

Current research efforts attempt to define more clearly the specific source(s) and roles of the cytokines involved. The elucidation of these factors might allow the development of diagnostic tools, such as the assessment of cytokine levels, to monitor osteoporosis. In addition, such knowledge should facilitate the development of drugs that might interfere with cytokine action and potentially be of value in the treatment of osteoporosis.

The primary cause of osteomalacia and rickets is a deficiency in vitamin D *activity*. Vitamin D may be deficient in the diet; it may not be adequately absorbed by the small intestine; it may not be converted into its hormonally active form; or target tissues may not adequately respond to the active hormone (Table 36.3). Dietary deficiency is generally not a problem in the United States, where vitamin D is added to many foods; however it is a major health problem in other parts of the world. Because the liver and kidneys are involved in converting vitamin D_3 into its hormonally active form, primary disease of either of these organs may result in vitamin D deficiency. Impaired vitamin D actions are somewhat rare but can be produced by certain drugs. In particular, some anticonvulsants used in the treatment of epilepsy may produce osteomalacia or rickets with prolonged treatment.

Paget's Disease Leads to Disordered Bone Formation

Paget's disease affects about 3% of people older than 40. It is typified by disordered bone formation and resorption (remodeling) and may occur at a single local site or at multiple sites in the body. Radiographs of affected bone often exhibit increased density, but the abnormal structure makes the bone weaker than normal. Often those with Paget's disease experience considerable pain, and in severe cases, crippling deformities may lead to serious neurological complications.

The cause of the disease is not well understood. Both genetic and environmental factors (probably viral) appear to be important. Several therapies are available for treating the disease, including treatment with CT, but these typically offer only temporary relief from pain and complications.

REVIEW QUESTIONS

DIRECTIONS: Each of the numbered items or incomplete statements in this section is followed by answers or by completions of the statement. Select the ONE lettered answer or completion that is BEST in each case.

1. As part of a routine physical exam, a patient's serum electrolyte levels were measured. Among the measurements, it was determined that total plasma calcium concentration was 10.2 mg/dL. What percentage of total plasma calcium is normally present as the free Ca^{2+} ion?
(A) 1%
(B) 50%
(C) 60%
(D) 100%
2. A healthy individual consumed 1,000 mg of calcium during a 24-hour period. What is the major route of calcium excretion from the body?
(A) Urine
(B) Sweat
(C) Feces
(D) Bile
3. The major route by which ingested phosphate leaves the body is via the
(A) Urine
(B) Sweat
(C) Feces
(D) Bile
4. 1,25-Dihydroxycholecalciferol can be formed in the body by metabolism of cholesterol. Which of the following is *not* either directly or indirectly involved in formation of 1,25-dihydroxycholecalciferol?
(A) Bone
(B) Skin
(C) Kidney
(D) Liver
5. A 42-year-old woman develops an autoimmune disease that damages her kidneys. Of the following conversions, which is most likely to be impaired in this person?

(continued)

(A) Cholecalciferol to 7-dehydrocholesterol
(B) Vitamin D_3 to vitamin D_2
(C) 25-Hydroxycholecalciferol to 1,25-dihydroxycholecalciferol
(D) Calcium to hydroxyapatite

6. A 62-year-old woman stumbles on a crack in the sidewalk, falls, and breaks her right hip. She suffers from a form of metabolic bone disease in which there is an equivalent loss of bone mineral and organic matrix. What is this disease?
(A) Paget's disease
(B) Rickets
(C) Osteoporosis
(D) Osteomalacia
7. Parathyroid hormone has effects on bone, kidney, and indirectly the GI tract. The net result of PTH actions is that it tends to
(A) Raise plasma calcium and phosphate
(B) Lower plasma calcium and phosphate
(C) Lower plasma calcium and raise plasma phosphate
(D) Raise plasma calcium and lower plasma phosphate

SUGGESTED READING

Aurbach GD, Marx SJ, Spiegel AM. Parathyroid hormone, calcitonin, and the calciferols. In: Wilson JD, Foster DW, Kronenberg HM, Larsen PR, eds. Williams Textbook of Endocrinology. 9th Ed. Philadelphia: WB Saunders, 1998;1397–1496.

Bilezikian JP, Kurland ES, Rosen CJ. Male skeletal health and osteoporosis. Trends Endocrinol Metab 1999;10:244–250.

Griffin JE, Ojeda SR, eds. Textbook of Endocrine Physiology. 4th Ed. Oxford: Oxford University Press, 2000.

Henry HL. The 25-hydroxyvitamin D 1α-hydroxylase. In: Feldman D, Glorieux FH, Pike JW, eds. Vitamin D. San Diego: Academic Press, 1997.

National Osteoporosis Foundation Web site. Available at: http://www.nof.org

Norman AW, Litwack G. Hormones. 2nd Ed. San Diego: Academic Press, 1997.

CASE STUDIES FOR PART IX • • •

CASE STUDY FOR CHAPTER 31

Diabetes Mellitus

Charlie was diagnosed with type 1 diabetes mellitus during the summer of his eighth year. Charlie's mother suspected he was drinking excessive amounts of fluids that summer; however, he was in and out of the house and visiting friends, and she couldn't be certain. During an afternoon at a friend's birthday party, Charlie drank nearly 3 quarts of fruit juice; his mother became alerted to a possible problem and took him to their family doctor.

Charlie's tests are normal, except that he tests positive for glucose in the urine (dipstick test) and his fasting blood sugar is elevated (620 mg/dL). Plasma insulin (5 μU/mL) and C-peptide (0.6 ng/mL) are reduced. Charlie is placed on a regimen of daily insulin injections, along with monitoring of blood and urine glucose concentrations. His mother is instructed about changes in Charlie's diet. During the next year, Charlie returns to the doctor for several follow-up visits and to adjust his insulin dosage. Data from his 1-year visit are as follows: fasting blood glucose, 120 mg/dL; C-peptide, 0.1 ng/mL.

Questions

1. What might be the reason for the decrease in Charlie's C-peptide after one year?
2. Why weren't plasma insulin values measured after one year?

Answers to Case Study Questions for Chapter 31

1. The decrease in C-peptide reflects further destruction of Charlie's insulin-producing pancreatic beta cells and indicates a further impairment in his own insulin production capacity.
2. Because Charlie is taking insulin injections, measurement of circulating insulin levels would not have provided any information about insulin secretion. This information is inferred from the C-peptide data.

CASE STUDY FOR CHAPTER 32

Growth Hormone

A 6-year-old boy was brought to the clinic to be evaluated for GH deficiency. The boy's height is between 2 and 3 standard deviations below the average height for his age. Initial physical examination rules out head trauma, chronic illness, and malnutrition. The patient's family history does not suggest similar short stature in immediate relatives. Thyroid hormones are normal.

Questions

1. Why would the doctor order a blood test for levels of IGFBP3 and IGF-I?
2. The levels of IGFBP3 and IGF-I are below the normal range in this patient. What does this finding suggest?
3. What is GH resistance, and what measurements would support the presence of this problem?
4. Why is it important to treat GH deficiency and short stature prior to the onset of puberty?
5. Why is resistance to insulin action a potential adverse effect of giving extremely high pharmacological doses of GH for a long time?

Answers to Case Study Questions for Chapter 32

1. GH is released in a pulsatile manner; between pulses of GH secretion the blood concentration may be undetectable in normal individuals. GH induces the synthesis and secretion of IGF-I and IGFBP3, both of which are easily detectable in the serum. Low IGF-I and IGFBP3 levels would indicate insufficient GH secretion.
2. In most cases, low levels of IGF-I and IGFBP3 in the blood would indicate insufficient GH release. However, low levels of IGF-I and IGFBP3 could also be due to a defect in the GH receptor, resulting in GH resistance.
3. GH resistance is characterized by impaired growth as a result of low levels of IGF-I and IGFBP3 in the blood. However, the blood concentration of GH is high. Defects in the GH receptor, which prevent GH from stimulating the production of IGF-I and IGFBP3, are a common cause of GH resistance.

Measurement of a GH in the blood should detect the extremely high levels of the hormone to confirm diagnosis.

4. GH and IGF-I stimulate the epiphyseal growth plate of the long bones to grow. The epiphyseal plate fuses several years after puberty, at which time GH and IGF-I can no longer stimulate the growth of the bone. Therefore, the earlier GH therapy is initiated, the greater will be the chance of achieving normal adult height before long bone growth stops.
5. GH has diabetogenic actions, which oppose the actions of insulin. Thus, chronic, high doses of GH can impair the actions of insulin. Insulin resistance is a condition in which tissues in the body do not respond very well to insulin (see Chapter 35).

Reference

Grumbach MM, Bin-Abbas BS, Kaplan SL. The growth hormone cascade: Progress and long-term results of growth hormone treatment in growth hormone deficiency. Horm Res 1998;49(Suppl 2):41–57.

Case Study for Chapter 33

Thyroiditis

A 35-year-old woman is seen in the Endocrine Clinic for evaluation of thyroid disease. The patient complains of weight loss, irritability, and restlessness. Physical examination reveals enlargement of the thyroid gland, weakness in maintaining the leg in an extended position, warm moist skin, and tachycardia. Family history indicates that the patient's mother had hypothyroidism after the birth of the patient's brother and an aunt had Hashimoto's disease.

Questions

1. Based on the history and physical examination, what would be a reasonable initial diagnosis?
2. From a blood sample, what hormone concentrations should the laboratory measure, and what would be the likely results?
3. What antibody titers should the laboratory determine? Which antibody titer is the most useful in the diagnosis of Hashimoto's disease?
4. Which antibody titer would be most useful in the diagnosis of Graves' disease?
5. The antibody titers indicate that the patient has Graves' disease. What treatment would be appropriate for this patient?

Answers to Case Study Questions for Chapter 33

1. The physical findings, including the presence of goiter, suggest that the patient may be hyperthyroid. However, goiter can also occur in hypothyroidism. Since autoimmune thyroid disease runs in families, the family history suggests that the thyroiditis might be due to an autoimmune response.
2. The laboratory should determine the blood levels of thyroid hormones (T_4 and T_3) and TSH. Thyroid hormones should be increased. TSH may be increased if it is early in the progression of Hashimoto's disease or decreased if the patient has Graves' disease.
3. The laboratory should measure antibodies to TSH receptor, thyroid peroxidase, and thyroglobulin. Antibodies to thyroid peroxidase are elevated to the greatest extent in Hashimoto's disease.
4. Antibodies to TSH receptor, thyroid peroxidase, and thyroglobulin can all be elevated in Graves' disease. However, the presence of TSH receptor antibodies is diagnostic.
5. A thionamide compound should first be used to inhibit thyroid hormone synthesis. This treatment will relieve the symptoms of hyperthyroidism and may result in a reduction in immune response. The drug may be withdrawn after several months of treatment to determine whether the disease is in remission. If thyroid hormone levels increase with cessation of the drug, ablation of the thyroid gland with ^{131}I (or less commonly with surgery) would be indicated.

Case Study for Chapter 34

Congenital Adrenal Hyperplasia

The pediatric endocrinologist is called in to consult on the case of a 1-week-old girl. The baby was born at home and is now in the emergency department because she appeared listless and has not nursed during the past 24 hours. On physical examination, the baby exhibits signs of virilization (growth of pubic hair) and volume depletion, and laboratory results indicate hyponatremia and hyperkalemia.

1. Based on the history, physical examination, and laboratory findings, what would be a reasonable initial hypothesis?
2. What are the two most likely congenital defects in adrenal steroidogenic enzymes that could explain the findings in this child?
3. From a blood sample, what hormones/metabolites should the laboratory measure, and what would be the likely results?
4. From the hormone/metabolite analysis, how would the two most likely causes for this case of congenital adrenal hyperplasia be distinguished?
5. A genetic screen utilizing DNA from the baby's white cells identifies an inactivating mutation in the gene (*CYP21A2)* for 21-hydroxylase. What would be appropriate hormone replacement for this patient?

Answers to Case Study Questions for Chapter 34

1. A reasonable initial hypothesis is that the baby has a form of congenital adrenal hyperplasia. The virilization (appearance of pubic hair) suggests the presence of excess androgen production by the adrenal gland. The hyponatremia, hyperkalemia, and volume depletion suggest a "salt wasting" syndrome.
2. Mutations in *CYP21A2*, which encodes 21-hydroxylase, account for more than 90% of all cases of adrenal hyperplasia associated with excess androgen production. Mutations in *CYP11B1*, which encodes 11β-hydroxylase, would also result in excess adrenal androgen production.
3. Adrenal androgens would be significantly elevated in patients with virilizing forms of congenital adrenal hyperplasia. Adrenal hyperplasia is usually due to defects in cortisol production. Therefore, the serum concentrations of precursors of cortisol biosythesis such as progesterone, 17α-hydroxyprogesterone, and 11-deoxycortisol could be elevated. In addition, serum ACTH would be elevated as a result of the lack of negative feedback from the absent cortisol.
4. Genetic defects in the gene for 11β-hydroxylase, resulting in a reduction in the activity of this enzyme, would result in increased 11-deoxycortisol. Defects in the gene for 21-hydroxylase, which impair the activity of the enzyme, would not lead to the production of 11-deoxycortisol. Since 11-deoxycortisol has significant mineralocorticoid activity, excess production of this steroid is usually associated with hypertension, rather than the volume depletion and hypotension observed in this patient.
5. Treatment would be directed toward replacement of gluco-

corticoids and mineralocorticoids. Glucocorticoids would replace the missing cortisol and also suppress ACTH secretion. With less ACTH stimulation of steroid production from the adrenal gland, the hyperandrogenemia should subside. Mineralocorticoids are given to treat the "salt wasting" that occurs in the absence of aldosterone.

Case Study for Chapter 35

Type 2 Diabetes

A 65-year-old semi-retired college professor was diagnosed with type 2 diabetes about 4 years ago during a routine physical examination at his family doctor's office. Treatment for the diabetes initially consisted of one tablet daily of an oral antidiabetic drug of the sulfonylurea class and two daily injections of insulin. The patient's doctor also recommended modest weight loss and a regular exercise program. With diligence to the treatment program, the patient was able to control his blood sugar levels adequately.

About 2 years ago, the patient developed gallstones, which required surgery to remove the gallbladder. For about one week after the surgery, the patient had to increase his insulin dosage to maintain normal blood glucose levels. He gradually returned to his presurgery insulin dose.

Because of the surgery, the patient vows to take better care of himself. He increases his physical activity and begins a diet that results in loss of 7 kg in 3 months. The weight loss and exercise result in the cessation of the patient's need for insulin injections, although he still takes his daily oral medication.

Questions

1. Why might the gallbladder disease and resulting surgery have increased the patient's need for insulin?
2. What might be the consequences if the patient were to regain the weight he lost after surgery?
3. Why is exercise an important part of the treatment regimen for type 2 diabetes?

Answers to Case Study Questions for Chapter 35

1. Stress, such as surgery, results in increased production of epinephrine and norepinephrine, both of which inhibit insulin secretion. The patient's pancreas will produce less insulin, and thus, more exogenous insulin will need to be provided.
2. If the patient were to regain weight, he would most likely have to go back to taking insulin injections.
3. Exercise not only helps to control weight, it stimulates glucose uptake in skeletal muscle, lessening the requirements for injected insulin.

Case Study for Chapter 36

Bone Fractures

A 38-year-old Caucasian man recently came to the attention of his physician when he suffered the second of two bone fractures in the past year and a half. He previously was in relatively good health, was not a smoker, and used alcohol only moderately. However, his only form of exercise was cutting the lawn on weekends during the summer months. He has not required any major surgeries during his lifetime, and had only minor bouts of the typical childhood illnesses. However, at age eight he was diagnosed with asthma after he suffered severe respiratory problems during a baseball game on a hot summer day. He has been treated ever since with a daily tablet of a synthetic glucocorticoid and the occasional use of an inhaler when needed to relieve acute symptoms of the disease.

The fractures that the patient experienced were to the left wrist and the right forearm. In both cases, the trauma that caused the fracture was relatively minor. Suspecting that there may be an underlying problem, his physician orders a series of bone density scans. Results of these studies show that the patient has a considerable reduction in bone mass compared with other men of the same age.

Questions

1. What is the most probable diagnosis?
2. What is the most probable underlying cause for the patient's problem?
3. What risk factors are present (or absent) in this case?

Answers to Case Study Questions for Chapter 36

1. Osteoporosis and, perhaps, glucocorticoid-induced osteoporosis.
2. Because the patient is young and has a relatively healthy lifestyle, the most probable cause of his osteoporosis is his 30-year history of treatment with glucocorticoids for asthma. Glucocorticoids increase bone loss by inhibiting osteoblasts, stimulating bone resorption, impairing intestinal calcium absorption, increasing urinary calcium loss, inhibiting secretion of sex hormones, and other effects.
3. The patient lacks the risk factors of smoking, excessive alcohol intake, and being female. He does appear, however, to have the risk factor of a somewhat sedentary lifestyle.

PART X Reproductive Physiology

CHAPTER 37 The Male Reproductive System

Paul F. Terranova, Ph.D.

CHAPTER OUTLINE

- AN OVERVIEW OF THE MALE REPRODUCTIVE SYSTEM
- REGULATION OF TESTICULAR FUNCTION
- THE MALE REPRODUCTIVE ORGANS
- SPERMATOGENESIS
- TESTICULAR STEROIDOGENESIS
- THE ACTIONS OF ANDROGENS
- REPRODUCTIVE DYSFUNCTIONS

KEY CONCEPTS

1. In the testes, luteinizing hormone (LH) controls the synthesis of testosterone by Leydig cells, and follicle-stimulating hormone (FSH) increases the production of androgen-binding protein, inhibin, and estrogen by Sertoli cells.
2. Spermatozoa are produced within the seminiferous tubules of both testes. Sperm develop from spermatogonia through a series of developmental stages that include spermatocytes and spermatids.
3. The sperm mature and are stored in the epididymis. At the time of ejaculation, sperm are moved by muscular contractions of the epididymis and vas deferens through the ejaculatory ducts into the prostatic urethra. The sperm are finally moved out of the body through the urethra in the penis.
4. LH and FSH secretion by the anterior pituitary are controlled by gonadotropin-releasing hormone (GnRH).
5. Testosterone mainly reduces LH secretion, whereas inhibin reduces the secretion of FSH. The testicular hormones complete a negative-feedback loop with the hypothalamic-pituitary axis.
6. Androgens have several target organs and have roles in regulating the development of secondary sex characteristics, the libido, and sexual behavior.
7. The most potent natural androgen is dihydrotestosterone, which is produced from the precursor, testosterone, by the action of the enzyme 5α-reductase.
8. Male reproductive dysfunction is often due to a lack of LH and FSH secretion or abnormal testicular morphology.

The testes have two primary functions, spermatogenesis, the process of producing mature sperm, and steroidogenesis, the synthesis of testosterone. Both processes are regulated by the pituitary gonadotropins LH and FSH. Testosterone is the primary sex hormone in the male and is responsible for primary and secondary sex characteristics. The primary sex characteristics include those structures responsible for promoting the development, preservation, and delivery of sperm. The secondary sex characteristics are those structures and behavioral features that make men externally different from women and include the typical male hair pattern, deep voice, and large muscle and bone masses.

AN OVERVIEW OF THE MALE REPRODUCTIVE SYSTEM

A diagram of reproduction regulation in the male is presented in Figure 37.1. The system is divided into factors affecting male function: brain centers, which control pituitary release of hormones and sexual behavior; gonadal

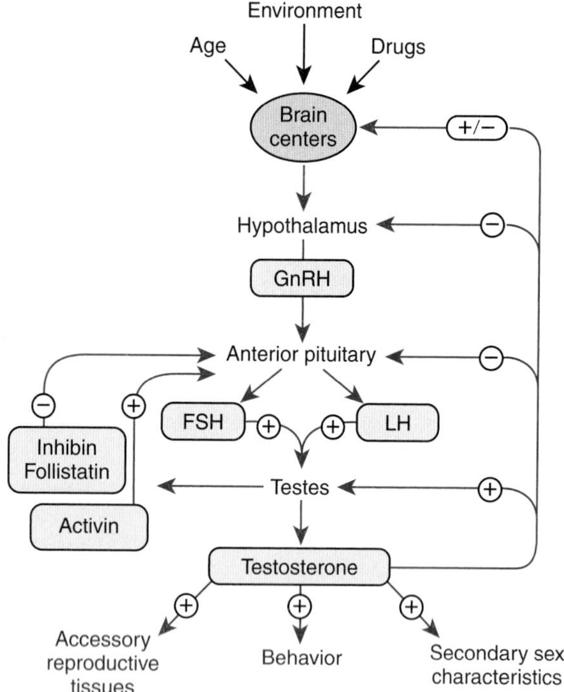

FIGURE 37.1 **Regulation of reproduction in the male.** The main reproductive hormones are shown in boxes. Positive and negative regulations are depicted by plus and minus signs, respectively.

structures, which produce sperm and hormones; a ductal system, which stores and transports sperm; and accessory glands, which support sperm viability.

The endocrine glands of the male reproductive system include the hypothalamus, anterior pituitary, and testes. The hypothalamus processes information obtained from the external and internal environment using neurotransmitters that regulate the secretion of gonadotropin-releasing hormone (GnRH). GnRH moves down the hypothalamic-pituitary portal system and stimulates the secretion of LH and FSH by the gonadotrophs of the anterior pituitary. LH binds to receptors on the Leydig cells and FSH binds to receptors on the Sertoli cells. Leydig cells reside in the interstitium of the testes, between seminiferous tubules, and produce testosterone. Sertoli cells are located within the seminiferous tubules, support spermatogenesis, contain FSH and testosterone receptors, and produce estradiol, albeit at low levels.

Testosterone belongs to a class of steroid hormones, the androgens, which promote "maleness." It carries out multiple functions, including feedback on the hypothalamus and anterior pituitary; the support of spermatogenesis; the regulation of behavior, including sexual behavior; and the development and maintenance of secondary sex characteristics. Sertoli cells also produce glycoprotein hormones—inhibin, activin, and follistatin—that regulate the secretion of FSH.

The duct system that transports sperm from the testis to the outside through the penis includes the epididymis, vas deferens, and urethra. The sperm acquire motility and the capability to fertilize in the epididymis; they are stored in the epididymis and in the vas deferens. They are transported via the urethra through the penis and are ultimately expelled by ejaculation. The accessory structures of the male reproductive tract include the prostate gland, seminal vesicles, and bulbourethral glands. These glands contribute several constituents to the seminal fluid that are necessary for maintaining functional sperm.

REGULATION OF TESTICULAR FUNCTION

Testicular function is regulated by LH and FSH. LH regulates the secretion of testosterone by the Leydig cells and FSH, in synergy with testosterone, regulates the production of spermatozoa.

Hypothalamic Neurons Produce Gonadotropin-Releasing Hormone

Hypothalamic neurons produce **gonadotropin-releasing hormone** (GnRH), a decapeptide, which regulates the secretion of **luteinizing hormone** (LH) and **follicle-stimulating hormone** (FSH). Although neurons that produce GnRH can be located in various areas of the brain, their highest concentration is in the medial basal hypothalamus, in the region of the infundibulum and arcuate nucleus. GnRH enters the hypothalamic-pituitary portal system and binds to receptors on the plasma membranes of pituitary cells, resulting in the synthesis and release of LH and FSH.

A variety of external cues and internal signals influence the secretion of GnRH, LH, and FSH. For example, the amount of GnRH, FSH, and LH secreted changes with age, stress levels, and hormonal state. In addition, various disease states lead to hyposecretion of GnRH. Little, if any, secretion of hypothalamic GnRH occurs in patients with prepubertal **hypopituitarism**, resulting in a failure of the development of the testes, primarily a result of a lack of LH, FSH, and testosterone.

Male patients with **Kallmann's syndrome** are hypogonadal from a deficiency in LH and FSH secretion because of a failure of GnRH neurons to migrate from the olfactory bulbs, their embryological site of origin. These patients do not have a sufficient hypothalamic source of GnRH to maintain secretion of LH and FSH, and the testes fail to undergo significant development.

GnRH originates from a large precursor molecule called preproGnRH (Fig. 37.2). PreproGnRH consists of a signal peptide, native GnRH, and a GnRH-associated peptide (GAP). The signal peptide (or leader sequence) allows the protein to cross the membrane of the rough ER. However, both the signal peptide and GAP are enzymatically cleaved at the rough ER prior to GnRH secretion.

Distinct Gonadotrophs Produce LH and FSH

Three distinct pituitary LH- and FSH-secreting cells have been identified. Gonadotrophs contain either LH or FSH, and some cells contain both LH and FSH. GnRH can induce the secretion of both hormones simultaneously because GnRH receptors are present on all of these cell types.

LH and FSH each contain two polypeptide subunits, referred to as alpha and beta chains, that are about 15 kDa in

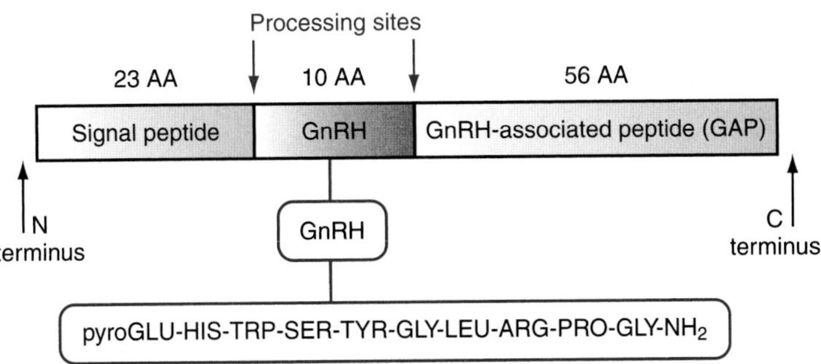

FIGURE 37.2 The precursor molecule, pre-proGnRH, that contains GnRH. The amino acid sequence of GnRH, a decapeptide, is indicated at the bottom.

size. Both hormones contain the same α subunit but different β subunits. Each hormone is glycosylated prior to release into the general circulation. Glycosylation regulates the half-life, protein folding for receptor recognition, and biological activity of the hormone.

LH and FSH bind membrane receptors on Leydig and Sertoli cells, respectively. The activation of LH and FSH receptors on these cells increases the intracellular second messenger cAMP. The two gonadotropin receptors are linked to G proteins and adenylyl cyclase for the production of cAMP from ATP. For the most part, cAMP can account for all of the actions of LH and FSH on testicular cells. cAMP binds to protein kinase A, which activates transcription factors such as **steroidogenic factor-1** (SF-1) and **cAMP response element binding protein** (CREB). These factors activate the promoter region of the genes of steroidogenic enzymes that control testosterone production by Leydig cells. Similar signal-transducing events occur in Sertoli cells that regulate the production of estradiol. The testis converts testosterone and some other androgens to estradiol by the process of aromatization, although estradiol production is low in males.

Another major function of the testis is the production of mature sperm, inhibin (a protein produced by Sertoli cells that suppresses FSH secretion), and androgen-binding protein. Activin and follistatin production by testicular cells in humans is currently being investigated.

GnRH Is Secreted in a Pulsatile Manner

GnRH in the hypothalamus is secreted in a pulsatile manner into the hypothalamic-hypophyseal portal blood. GnRH pulsatility is ultimately necessary for proper functioning of the testes because it regulates the secretion of FSH and LH, which are also released in a pulsatile fashion (Fig. 37.3). Continuous exposure of gonadotrophs to GnRH results in desensitization of GnRH receptors, leading to a decrease in LH and FSH release. Therefore, the pulsatile pattern of GnRH release serves an important physiological function. The administration of GnRH at an improper frequency results in a decrease in circulating concentrations of LH and FSH.

Most evidence for GnRH pulses has come from animal studies because GnRH must be measured in hypothalamic-hypophyseal portal blood, an extremely difficult area to obtain blood samples in humans. Since discrete pulses of GnRH are followed by distinct pulses of FSH and LH, measurements of the pulses of LH and FSH in serum indirectly indicate that GnRH pulses have occurred. Numerous human studies measuring pulsatile secretion of LH and FSH in peripheral blood at various times have provided much of the information regarding the role of LH and FSH in regulating testicular development and function. However, the exact relationship between endogenous GnRH pulses and LH and FSH secretion in humans is unknown.

Hypogonadal eunuchoid men exhibit low levels of LH in serum and do not exhibit pulsatile secretion of LH. Pulsatile injections of GnRH restore LH and FSH secretion and increase sperm counts. FSH pulses tend to be smaller in amplitude than LH pulses, mostly because FSH has a longer half-life than LH in the circulation.

Although the exact identity of the cells responsible for generating GnRH pulsatility is unknown, the presence of a **pulse generator** in the hypothalamus has been postulated. The putative pulse generator resides in the medial basal hypothalamus and is responsible for the synchronized and rhythmic firing of a population of neurons. The activity of the pulse generator is modified by several factors. For example, castration causes a large increase in basal LH levels in serum, as evidenced by an increase in frequency and amplitude of LH pulses. Therefore, the pulse generator may be tonically inhibited by testosterone. However, GnRH neurons lack receptors for gonadal steroids, suggesting that

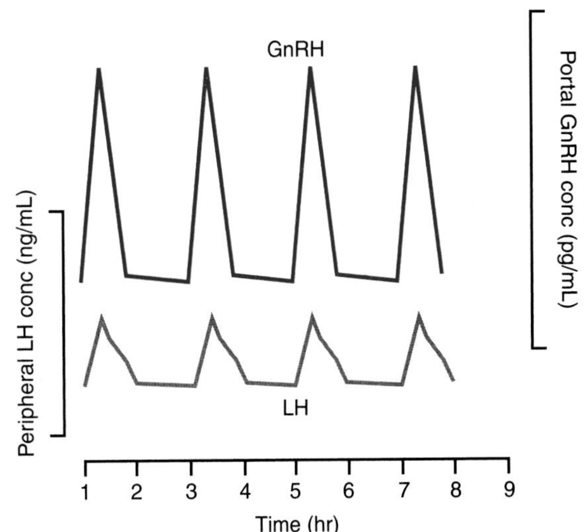

FIGURE 37.3 A diagram of the pulsatile release of GnRH in portal blood and LH in peripheral blood.

steroidal effects are mediated by other neurons whose neuropeptides, neurohormones, or vasoactive agents regulate the activity of the GnRH-producing neurons.

Steroids and Polypeptides From the Testis Regulate LH and FSH Secretion

Testosterone, estradiol, inhibin, activin, and follistatin are major testicular hormones that regulate the release of the gonadotropins LH and FSH. Generally, testosterone, estradiol, and inhibin reduce the secretion of LH and FSH in the male. Activin stimulates the secretion of FSH, whereas follistatin inhibits FSH secretion.

Testosterone inhibits LH release by decreasing the secretion of GnRH and, to a lesser extent, by reducing gonadotroph sensitivity to GnRH. Estradiol formed from testosterone by aromatase also has an inhibitory effect on GnRH secretion. Acute testosterone treatment does not alter pituitary responsiveness to GnRH, but prolonged exposure significantly reduces the secretory response to GnRH.

Removal of the testes results in increased circulating levels of LH and FSH. Replacement therapy with physiological doses of testosterone restores LH to precastration levels but does not completely correct FSH levels. This observation led to a search for a gonadal factor that specifically inhibits FSH release. The polypeptide hormone **inhibin** was eventually isolated from seminal fluid. Inhibin is produced by Sertoli cells, and has a molecular weight of 32 to 120 kDa, the 32-kDa form being the most prominent. Inhibin is composed of two dissimilar subunits, α and β, which are held together by disulfide bonds. There are two β subunit forms, called A and B. Inhibin B consists of the α subunit bound by a disulfide bridge to the βB subunit and is the physiologically important form of inhibin in the human male. Inhibin acts directly on the anterior pituitary and inhibits the secretion of FSH but not LH.

Activin is produced by Sertoli cells, stimulates the secretion of FSH, has an approximate molecular weight of 30 kDa and has multiple forms based on the βA and βB subunits of inhibin. The multiple forms of activin are called activin A (two βA subunits linked by a disulfide bridge), activin B (two βB subunits), and activin AB (one βA and one βB subunit). The major form of activin in the male is currently unknown although both Sertoli and Leydig cells have been implicated in its secretion.

Follistatin is a 31 to 45 kDa single-chain protein hormone, with several isoforms, that binds and deactivates activin. Thus, the deactivation of activin by binding to follistatin reduces FSH secretion. Follistatin is apparently produced by Sertoli cells and acts as a paracrine factor on the developing spermatogenic cells.

THE MALE REPRODUCTIVE ORGANS

The testes produce spermatozoa and transport them through a series of ducts in preparation for fertilization. The testes also produce testosterone that regulates development of the male gametes, male sex characteristics, and male behavior.

The Testis Is the Site of Sperm Formation

During embryonic stages of development, the testes lie attached to the posterior abdominal wall. As the embryo elongates, the testes move to the inguinal ring. Between the seventh month of pregnancy and birth, the testes descend through the inguinal canal into the scrotum. The location of the testes in the scrotum is important for sperm production, which is optimal at 2 to 3°C lower than core body temperature. Two systems help maintain the testes at a cooler temperature. One is the **pampiniform plexus** of blood vessels, which serves as a countercurrent heat exchanger between warm arterial blood reaching the testes and cooler venous blood leaving the testes. The second is the **cremasteric muscle**, which responds to changes in temperature by moving the testes closer or farther away from the body. Prolonged exposure of the testes to elevated temperature, fever, or thermoregulatory dysfunction can lead to temporary or permanent sterility as a result of a failure of spermatogenesis, whereas steroidogenesis is unaltered.

The testes are encapsulated by a thick fibrous connective tissue layer, the tunica albuginea. Each human testis contains hundreds of tightly packed **seminiferous tubules**, ranging from 150 to 250 μm in diameter and from 30 to 70 cm long. The tubules are arranged in lobules, separated by extensions of the tunica albuginea, and open on both ends into the rete testis. Examination of a cross section of a testis reveals distinct morphological compartmentalization. Sperm production is carried out in the avascular seminiferous tubules, whereas testosterone is produced by the **Leydig cells**, which are scattered in a vascular, loose connective tissue between the seminiferous tubules in the interstitial compartment.

Each seminiferous tubule is composed of two somatic cell types (myoid cells and Sertoli cells) and germ cells. The seminiferous tubule is surrounded by a basement membrane (basal lamina) with myoid cells on its perimeter, which define its outer limit. On the inside of the basement membrane are large, irregularly shaped **Sertoli cells**, which extend from the basement membrane to the lumen (Fig. 37.4). Sertoli cells are attached to one another near their base by tight junctions (Fig. 37.5). The tight junctions divide each tubule into a **basal compartment**, whose constituents are exposed to circulating agents, and an **adluminal compartment**, which is isolated from bloodborne elements. The tight junctions limit the transport of fluid and macromolecules from the interstitial space into the tubular lumen, forming the **blood-testis barrier**.

Located between the nonproliferating Sertoli cells are germ cells at various stages of division and differentiation. Mitosis of the **spermatogonia** (diploid progenitors of spermatozoa) occurs in the basal compartment of the seminiferous tubule (see Fig. 37.5). The early meiotic cells (primary spermatocytes) move across the junctional complexes into the adluminal compartment, where they mature into **spermatozoa** or **gametes** after meiosis. The adluminal compartment is an immunologically privileged site. Spermatozoa that develop in the adluminal compartment are not recognized as "self" by the immune system. Consequently, males can develop antibodies against their own sperm, resulting in infertility. Sperm antibodies neutralize the ability of sperm to function. Sperm antibodies are often present after vasec-

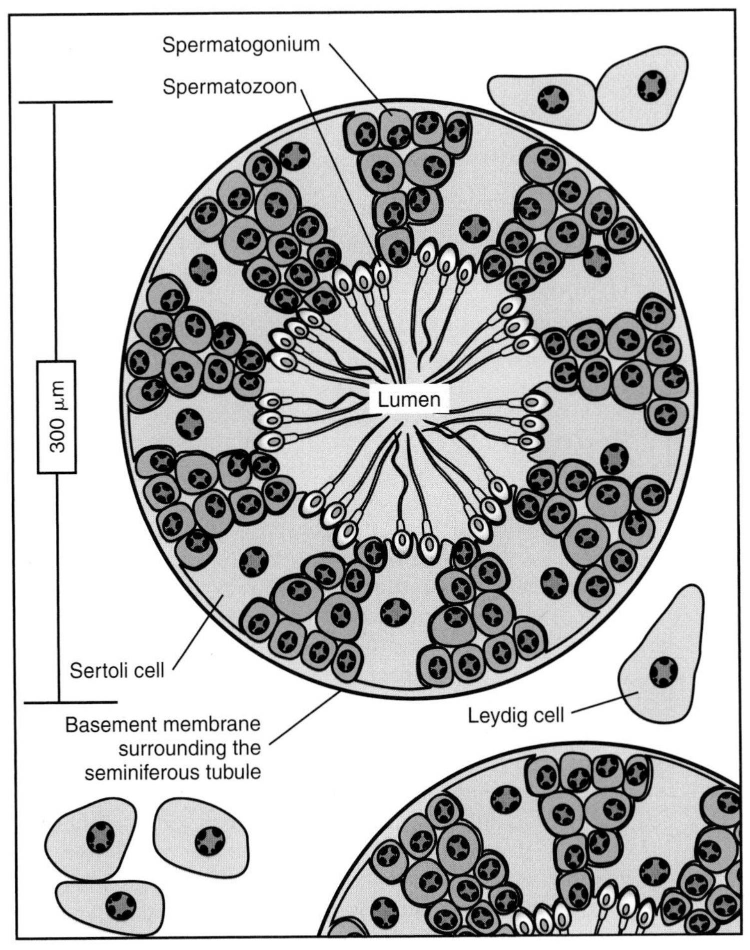

FIGURE 37.4 **The testis.** This cross-sectional view shows the anatomic relationship of the Leydig cells, basement membrane, seminiferous tubules, Sertoli cells, spermatogonia, and spermatozoa. (Modified from Alberts B, Bray D, Lewis M, et al. Molecular Biology of the Cell. 3rd Ed. New York: Garland, 1994.)

tomy or testicular injury and in some autoimmune diseases where the adluminal compartment is ruptured, allowing sperm to mingle with immune cells from the circulation.

Sertoli Cells Have Multiple Functions

Sertoli cells are critical to germ cell development, as indicated by their close contact. As many as 6 to 12 spermatids may be attached to a Sertoli cell. Sertoli cells phagocytose residual bodies (excess cytoplasm resulting from the transformation of spermatids to spermatozoa) and damaged germ cells, provide structural support and nutrition for germ cells, secrete fluids, and assist in **spermiation**, the final detachment of mature spermatozoa from the Sertoli cell into the lumen. Spermiation may involve **plasminogen activator**, which converts plasminogen to plasmin, a proteolytic enzyme that assists in the release of the mature sperm into the lumen. Sertoli cells also synthesize large amounts of transferrin, an iron-transport protein important for sperm development.

During the fetal period, Sertoli cells and gonocytes form the seminiferous tubules as Sertoli cells undergo numerous rounds of cell divisions. Shortly after birth, Sertoli cells cease proliferating, and throughout life, the number of sperm produced is directly related to the number of Sertoli cells. At puberty, the capacity of Sertoli cells to bind FSH and testosterone increases. Receptors for FSH, present only on the plasma membranes of Sertoli cells, are glycoproteins linked to adenylyl cyclase via G proteins. FSH exerts multiple effects on the Sertoli cell, most of which are mediated by cAMP and protein kinase A (Fig. 37.6). FSH stimulates the production of androgen-binding protein and plasminogen activator, increases secretion of inhibin, and induces aromatase activity for the conversion of androgens to estrogens. The testosterone receptor is within the nucleus of the Sertoli cell.

Androgen-binding protein (ABP) is a 90-kDa protein, made of a heavy and a light chain, that has a high binding affinity for dihydrotestosterone and testosterone. It is similar in function, with some homology in structure, to another binding protein, sex hormone-binding globulin (SHBG), synthesized in the liver. ABP is found at high concentrations in the human testis and epididymis. It serves as a carrier of testosterone in Sertoli cells, as a storage protein for androgens in the seminiferous tubules, and as a carrier of testosterone from the testis to the epididymis.

Other products of the Sertoli cell are inhibin, follistatin, and activin. Inhibin suppresses FSH release from the pituitary gonadotrophs. The pituitary gonadotrophs and testicular Sertoli cells form a classical negative-feedback loop in which FSH stimulates inhibin secretion and inhibin suppresses FSH release. Inhibin also functions as a paracrine

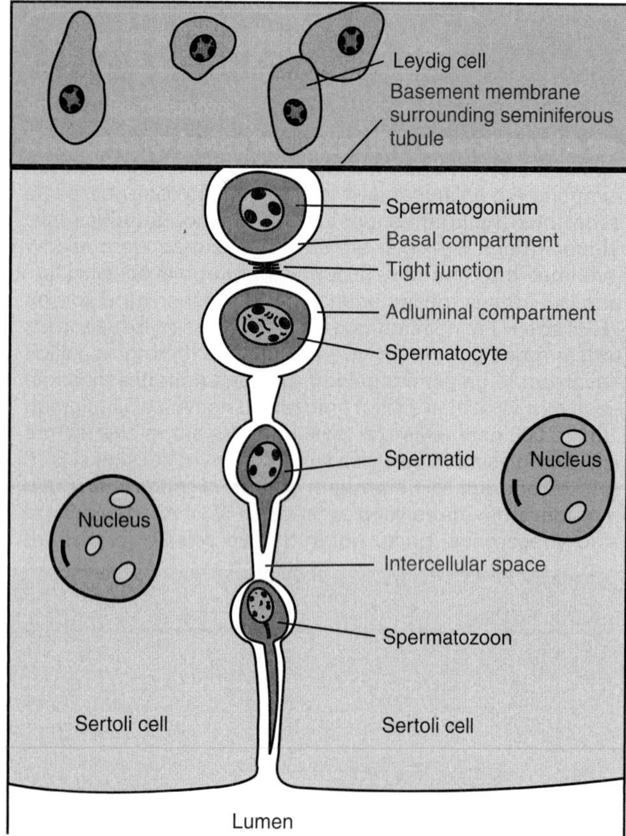

FIGURE 37.5 **Sertoli cells.** Sertoli cells are connected by tight junctions, which divide the intercellular space into a basal compartment and an adluminal compartment. Spermatogonia are located in the basal compartment and maturing sperm in the adluminal compartment. Spermatocytes are formed from the spermatogonia and cross the tight junctions into the adluminal compartment, where they mature into spermatozoa. (Modified from Alberts B, Bray D, Lewis M, et al. Molecular Biology of the Cell. 3rd Ed. New York: Garland, 1994.)

agent in the testes. Activin stimulates the release of FSH. Follistatin, an activin-binding protein, reduces FSH secretion induced by activin.

Leydig Cells Produce Testosterone

Leydig cells are large polyhedral cells that are often found in clusters near blood vessels in the interstitium between seminiferous tubules. They are equipped to produce steroids because they have numerous mitochondria, a prominent smooth ER, and conspicuous lipid droplets.

Leydig cells undergo significant changes in quantity and activity throughout life. This mechanism may depend on a nuclear transcription factor, steroidogenic factor-1 (SF-1), that recognizes a sequence in the promoter of all genes encoding CYP enzymes. In the human fetus, the period from weeks 8 to 18 is marked by active steroidogenesis, which is obligatory for differentiation of the male genital ducts. Leydig cells at this time are prominent and very active, reaching their maximal steroidogenic activity at about 14 weeks, when they constitute more than 50% of the testicular volume. Because the fetal hypothalamic-pituitary axis is still underdeveloped, steroidogenesis is controlled by **human chorionic gonadotropin (hCG)** from the placenta, rather than by LH from the fetal pituitary (see Chapter 39); LH and hCG bind the same receptor. After this period, Leydig cells slowly regress. At about 2 to 3 months of postnatal life, male infants have a significant rise in testosterone production (infantile testosterone surge), the regulation and function of which are unknown. Leydig cells remain quiescent throughout childhood but increase in number and activity at the onset of puberty.

Leydig cells do not have FSH receptors, but FSH can increase the number of developing Leydig cells by stimulating the production of growth stimulators from Sertoli cells that subsequently enhance the growth of the Leydig cells. In addition, androgens stimulate the proliferation of developing Leydig cells. Estrogen receptors are present on Leydig cells, and they reduce the proliferation and activity of these cells.

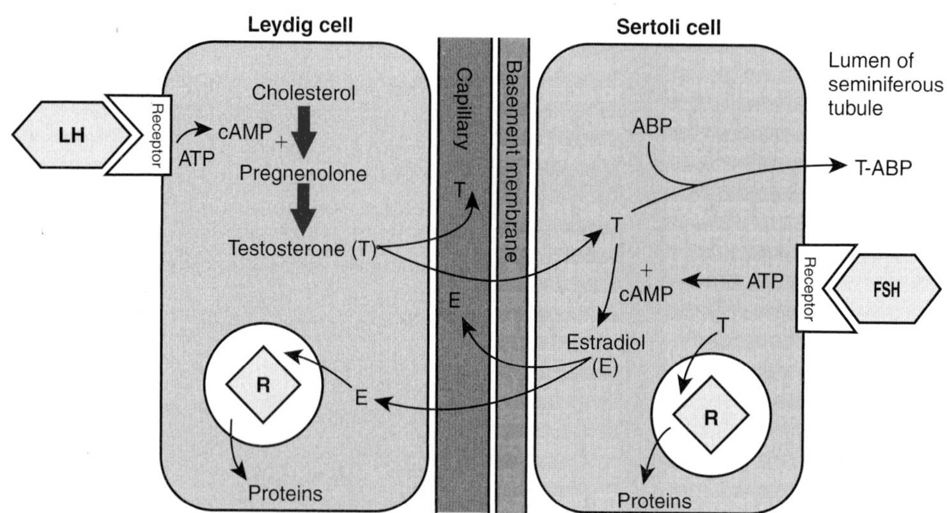

FIGURE 37.6 **Regulation, hormonal products, and interactions between Leydig and Sertoli cells.** ABP, androgen-binding protein; E, estradiol; T, testosterone; R, receptor.

Leydig cells have LH receptors, and the major effect of LH is to stimulate androgen secretion via a cAMP-dependent mechanism (see Fig. 37.6). The main product of Leydig cells is testosterone, but two other androgens of less biological activity, dehydroepiandrosterone (DHEA) and androstenedione, are also produced.

There are bidirectional interactions between Sertoli and Leydig cells (see Fig. 37.6). The Sertoli cell is incapable of producing testosterone but contains testosterone receptors as well as FSH-dependent aromatase. The Leydig cell does not produce estradiol but contains receptors for it, and estradiol can suppress the response of the Leydig cell to LH. Testosterone diffuses from the Leydig cells, crosses the basement membrane, enters the Sertoli cell, and binds to ABP. As a result, androgen levels can reach high local concentrations in the seminiferous tubules. Testosterone is obligatory for spermatogenesis and the proper functioning of Sertoli cells. In Sertoli cells, testosterone also serves as a precursor for estradiol production. The daily role of estradiol in the functioning of Leydig cells is unclear, but it may modulate responses to LH.

The Duct System Functions in Sperm Maturation, Storage, and Transport

After formation in the seminiferous tubules, spermatozoa are transported to the rete testes and from there through the efferent ductules to the **epididymis**. This movement of sperm is accomplished by ciliary movement in the efferent ductules, by muscle contraction, and by the flow of fluid.

The epididymis is a single, tightly coiled duct, 4 to 5 m long. It is composed of a head (caput), a body (corpus), and a tail (cauda) (Fig. 37.7). The functions of the epididymis are storage, protection, transport, and maturation of sperm cells. Maturation at this point includes a change in functional capacity as sperm make their way through the epididymis. The sperm become capable of forward mobility during migration through the body of the epididymis. A significant portion of sperm maturation is carried out in the caput, whereas sperm are stored in the cauda.

Frequent ejaculation results in reduced sperm numbers and increased numbers of immotile sperm in the ejaculate. The cauda connects to the vas deferens, which forms a dilated tube, the ampulla, prior to entering the prostate. The ampulla also serves as a storage site for sperm. Cutting and ligation of the vas deferens or vasectomy is an effective method of male contraception. Because sperm are stored in the ampulla, men remain fertile for 4 to 5 weeks after vasectomy.

Erection and Ejaculation Are Neurally Regulated

Erection is associated with sexual arousal emanating from sexually related psychic and/or physical stimuli. During sexual arousal, impulses from the genitalia, together with nerve signals originating in the limbic system, elicit motor impulses in the spinal cord. These neuronal impulses are carried by the parasympathetic nerves in the sacral region of the spinal cord via the cavernous nerve branches of the prostatic plexus that enter the penis. Those signals cause vasodilation of the arterioles and corpora cavernosa. The smooth muscles in those structures relax, and the blood vessels dilate and begin to engorge with blood. The thin-walled veins become compressed by the swelling of the blood-filled arterioles and cavernosa, restricting blood flow. The result is a reduction in the outflow of blood from the penis, and blood is trapped in the surrounding erectile tissue, leading to engorgement, rigidity, and elongation of the penis in an erect position.

Semen, consisting of sperm and the associated fluids, is expelled by a neuromuscular reflex that is divided into two sequential phases: emission and ejaculation. **Emission** moves sperm and associated fluids from the cauda epididymis and vas deferens into the urethra. The latter process involves efferent stimuli originating in the lumbar areas (L1 and L2) of the spinal cord and is mediated by adrenergic sympathetic (hypogastric) nerves that induce contraction of smooth muscles of the epididymis and vas deferens. This action propels sperm through the ejaculatory ducts and into the urethra. Sympathetic discharge also closes the internal urethral sphincter, which prevents retrograde ejaculation into the urinary bladder. **Ejaculation** is the expulsion of the semen from the penile urethra; it is initiated after emission. The filling of the urethra with sperm initiates sensory signals via the pudendal nerves that travel to the sacrospinal region of the cord. A spinal reflex mechanism that induces rhythmic contractions of the striated bulbospongiosus muscles surrounding the penile urethra results in propelling the semen out of the tip of the penis.

The secretions of the accessory glands promote sperm survival and fertility. The accessory glands that contribute to the secretions are the seminal vesicles, prostate gland, and bulbourethral glands. The semen contains only 10% sperm by volume, with the remainder consisting of the combined secretions of the accessory glands. The normal volume of semen is 3 mL with 20 to 50 million sperm per milliliter; normal is considered more than 20 million sperm per milliliter. The **seminal vesicles** contribute about 75% of the semen volume. Their secretion contains fructose (the principal substrate for glycolysis of ejaculated sperm), ascorbic acid, and prostaglandins. In fact, prostaglandin concentrations are high and were first discovered in semen but were mistakenly considered the product of the prostate. Seminal vesicle secretions are also responsible for coagulation of the semen seconds after ejaculation. Prostate gland secretions (~0.5 mL) include fibrinolysin, which is responsible for liquefaction of the coagulated semen 15 to 30 minutes after ejaculation, releasing sperm.

SPERMATOGENESIS

Spermatogenesis is a continual process involving mitosis of the male germ cells that undergo extensive morphological changes in cell shape and, ultimately, meiosis to produce the haploid spermatozoa. Sperm are produced throughout life beginning with puberty. Sperm production declines in the elderly.

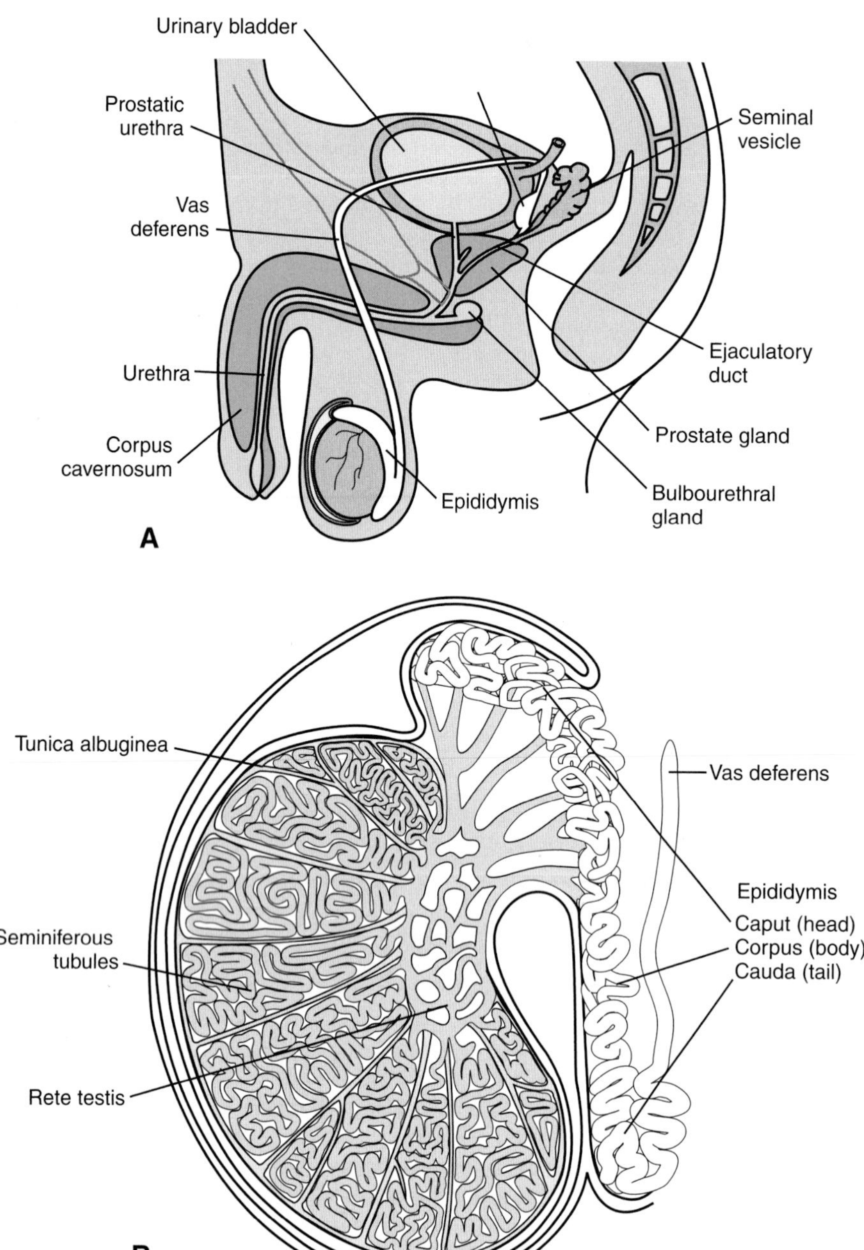

FIGURE 37.7 The male reproductive organs. The top drawing is a general side view. The bottom enlargement shows a sagittal section of the testis, epididymis, and vas deferens.

Spermatogenesis Is an Ongoing Process From Puberty to Senescence

Spermatogenesis is the process of transformation of male germ cells into spermatozoa. This process can be divided into three distinct phases. The phases include cellular proliferation by **mitosis**, two reduction divisions by **meiosis** to produce haploid spermatids, and cell differentiation by a process called **spermiogenesis**, in which the spermatids differentiate into spermatozoa (Fig. 37.8). Spermatogenesis begins at puberty, so the seminiferous tubules are quiescent throughout childhood. Spermatogenesis is initiated shortly before puberty, under the influence of the rising levels of gonadotropins and testosterone, and continues throughout life, with a slight decline during old age.

The time required to produce mature spermatozoa from the earliest stage of spermatogonia is 65 to 70 days. Because several developmental stages of spermatogenic cells occur during this time frame, the stages are collectively known as the **spermatogenic cycle**. There is synchronized development of spermatozoa within the seminiferous tubules, and each stage is morphologically distinct. A spermatogonium becomes a mature spermatozoon after going through several rounds of mitotic divisions, a couple of meiotic divisions, and a few weeks of differentiation. Hormones can alter the number of spermatozoa, but they generally do not affect the duration of the cycle. Spermatogenesis occurs along the length of each seminiferous tubule in successive cycles. New cycles are initiated at regular time intervals (every 2 to 3 weeks) before the previous ones are com-

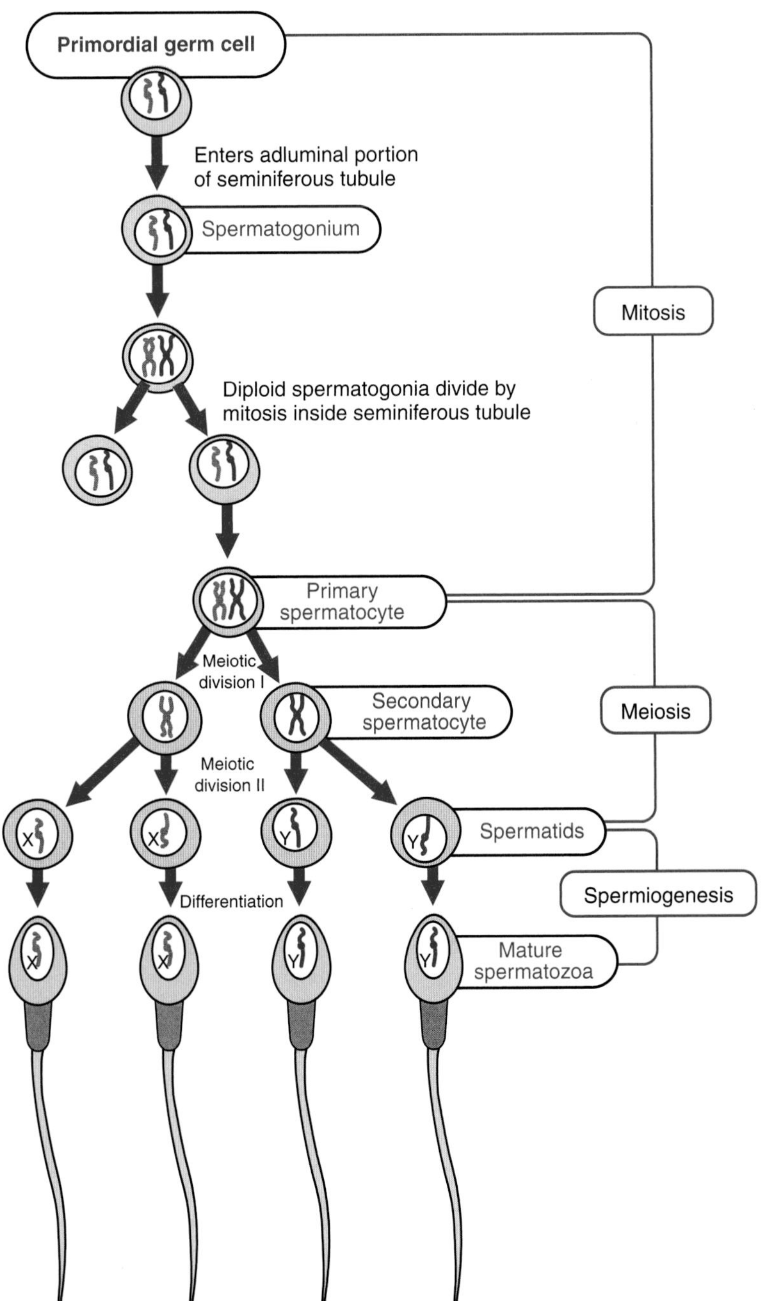

FIGURE 37.8 **The process of spermatogenesis, showing successive cell divisions and remodeling leading to the formation of haploid spermatozoa.** (Modified from Alberts B, Bray D, Lewis M, et al. Molecular Biology of the Cell. 3rd Ed. New York: Garland, 1994.)

pleted. Consequently, cells at different stages of development are spaced along each tubule in a "spermatogenic wave." Such a succession ensures the continuous production of fresh spermatozoa. Approximately 200 million spermatozoa are produced daily in the adult human testes, which is about the same number of sperm present in a normal ejaculate.

Since sperm cells are rapidly dividing and undergoing meiosis, they are sensitive to external agents that alter cell division. Chemical carcinogens, chemotherapeutic agents, certain drugs, environmental toxins, irradiation, and extreme temperatures are factors that can reduce the number of replicating germ cells or cause chromosomal abnormalities in individual cells. While defective somatic cells are normally detected and destroyed by the immune system, the blood-testis barrier isolates advanced germ cells from immune surveillance.

If the blood-testis barrier is ruptured by physical injury or infection and sperm cells within the barrier are exposed to circulating immune cells, it is possible that antibodies will develop to the sperm cells. In the past, it was thought that the development of antisperm antibodies could lead to male infertility. It appears that men with high levels of antisperm antibodies may exhibit some infertility problems. However, studies of men who have developed low or moderate levels of antisperm antibodies after vasectomy and who have had their vasa deferens reconnected have normal fertility if the vasectomy was for a relatively short time. Vasectomy does not appear to change hormone or sperm production by the testes. Nevertheless, in some cases, a high level of antisperm antibodies in men and women leads to infertility.

Spermatogonia Undergo Mitotic and Meiotic Divisions and Become Spermatids

Spermatogonia undergo several rounds of mitotic division prior to entering the meiotic phase (see Fig. 37.8). The spermatogonia remain in contact with the Sertoli cells, migrate away from the basal compartment near the walls of the seminiferous tubules and cross into the adluminal compartment of the tubule (see Fig. 37.5). After crossing into the adluminal compartment, the cells differentiate into spermatocytes prior to undergoing meiosis I. The first meiotic division of **primary spermatocytes** gives rise to diploid (2n chromosomes) **secondary spermatocytes**.

The second meiotic division produces haploid (1 set of chromosomes) cells called **spermatids**. Of every four spermatids emanating from a primary spermatocyte, two contain X chromosomes and two have Y chromosomes (see Fig. 37.8). Because of the numerous mitotic divisions and two rounds of meiosis, each spermatogonium committed to meiosis should have yielded 256 spermatids, if all cells survive.

There are numerous developmental disorders of spermatogenesis. The most frequent is **Klinefelter's syndrome**, which causes hypogonadism and infertility in men. Patients with this disorder have an accessory X chromosome caused by meiotic nondisjunction. The typical karyotype is 47 XXY, but there are other chromosomal mosaics. Testicular volume is reduced more than 75% and ejaculates contain few, if any, spermatozoa. Spermatogonic cell differentiation beyond the primary spermatocyte stage is rare.

The Formation of a Mature Spermatozoon Requires Extensive Cell Remodeling

Spermatids are small, round, and nondistinctive cells. During the second half of the spermatogenic cycle they undergo considerable restructuring to form mature spermatozoa. Notable changes include alterations in the nucleus, the formation of a tail, and a massive loss of cytoplasm. The nucleus becomes eccentric and decreases in size, and the chromatin becomes condensed. The **acrosome**, a lysosome-like structure unique to spermatozoa, buds from the Golgi apparatus, flattens, and covers most of the nucleus. The centrioles, located near the Golgi apparatus, migrate to the caudal pole and form a long axial filament made of nine peripheral doublet microtubules surrounding a central pair (9 + 2 arrangement). This becomes the **axoneme** or major portion of the tail. Throughout this reshaping process, the cytoplasmic content is redistributed and discarded. During spermiation, most of the remaining cytoplasm is shed in the form of residual bodies.

The reasons for this lengthy and metabolically costly process become apparent when the unique functions of this cell are considered. Unlike other cells, the spermatozoon serves no apparent purpose in the organism. Its only function is to reach, recognize, and fertilize an egg; hence, it must fulfill several prerequisites: It should possess an energy supply and means of locomotion, it should be able to withstand a foreign and even hostile environment, it should be able to recognize and penetrate an egg, and it must carry all the genetic information necessary to create a new individual.

The mature spermatozoon exhibits a remarkable degree of structural and functional specialization well adapted to carry out these functions. The cell is small, compact, and streamlined; it is about 1 to 2 μm in diameter and can exceed 50 μm in length in humans. It is packed with specialized organelles and long axial fibers but contains only a few of the normal cytoplasmic constituents, such as ribosomes, ER, and Golgi apparatus. It has a very prominent nucleus, a flexible tail, numerous mitochondria, and an assortment of proteolytic enzymes.

The spermatozoon consists of three main parts: a head, a middle piece, and a tail. The two major components in the head are the condensed chromatin and the acrosome. The haploid chromatin is transcriptionally inactive throughout the life of the sperm until fertilization, when the nucleus decondenses and becomes a pronucleus. The acrosome contains proteolytic enzymes, such as hyaluronidase, acrosin, neuraminidase, phospholipase A, and esterases. They are inactive until the **acrosome reaction** occurs upon contact of the sperm head with the egg (see Chapter 39). Their proteolytic action enables sperm to penetrate through the egg membranes. The middle piece contains spiral sheaths of mitochondria that supply energy for sperm metabolism and locomotion. The tail is composed of a 9 + 2 arrangement of microtubules, which is typical of cilia and flagella, and is surrounded by a fibrous sheath that provides some rigidity. The tail propels the sperm by a twisting motion, involving interactions between tubulin fibers and dynein side arms and requiring ATP and magnesium.

Testosterone Is Essential for Sperm Production and Maturation

Spermatogenesis requires high intratesticular levels of testosterone, secreted from the LH-stimulated Leydig cells. The testosterone diffuses across the basement membrane of the seminiferous tubule, crosses the blood-testis barrier, and complexes with ABP. Sertoli cells, but not spermatogenic cells, contain receptors for testosterone. Sertoli cells also contain FSH receptors. However, recent studies using mice, in which the β subunit of FSH has been mutated to an inactive form, reveal that the testes are small but do produce sperm. The absolute requirement for FSH in sperm production remains unknown. From these data, it appears that testosterone may be sufficient for spermatogenesis.

The actions of FSH and testosterone at each point of sperm cell production are unknown. Upon entering meiosis, spermatogenesis appears to depend on the availability of FSH and testosterone. In human males, FSH is thought to be required for the initiation of spermatogenesis before puberty. When adequate sperm production has been achieved, LH alone (through stimulation of testosterone production) or testosterone alone is sufficient to maintain spermatogenesis.

TESTICULAR STEROIDOGENESIS

Following spermatogenesis, the second primary function of the testes is steroidogenesis. **Steroidogenesis** is the production of the steroid hormones, mainly testosterone. Testosterone is then converted to dihydrotesterone (DHT), the most biologically active androgen, and to estradiol, the most biologically estrogen.

Testosterone Production Requires Two Intracellular Compartments and Several Enzymes

Steroid hormones are produced from cholesterol by the adrenal cortex, ovaries, testes, and placenta. Cholesterol, a 27-carbon (C27) steroid, can be obtained from the diet or synthesized within the body from acetate. Each organ uses a similar steroid biosynthetic pathway, but the relative amount of the final products depends on the particular subset of enzymes expressed in that tissue and the trophic hormones (LH, FSH, ACTH) stimulating specific cells within the organ. The major steroid produced by the testis is **testosterone**, but other androgens, such as androstenediol, androstenedione, and dehydroepiandrosterone (DHEA), as well as a small amount of estradiol, are also produced.

Cholesterol from low-density lipoprotein (LDL) and high-density lipoprotein (HDL) is released in the Leydig cell and transported from the outer mitochondrial membrane to the inner mitochondrial membrane, a process regulated by **steroidogenic acute regulatory protein** (StAR). Under the influence of LH, with cAMP as a second messenger, cholesterol is converted to pregnenolone (C21) by cholesterol side-chain cleavage enzyme (CYP11A1), which removes 6 carbons attached to the 21 position. Pregnenolone is a key intermediate for all steroid hormones in various steroidogenic organs (Fig. 37.9; see also Fig. 34.5). Pregnenolone is transported out of mitochondria by specific transport proteins. The pregnenolone then moves by diffusion to the smooth ER, where the remainder of sex hormone biosynthesis takes place.

Pregnenolone can be converted to testosterone via two pathways, the **delta 5 pathway** and the **delta 4 pathway**. In the delta 5 pathway, the double bond is in ring B; in the delta 4 pathway the double bond is in ring A (see Fig. 37.9). The delta 5 intermediates include 17α-hydroxypregnenolone, DHEA, and androstenediol, while the delta 4 intermediates are progesterone, 17α-hydroxyprogesterone, and androstenedione.

The conversion of C21 steroids (the progestins) to androgens (C19 steroids) proceeds in two steps: first, 17α-hydroxylation of pregnenolone (to form 17α-hydroxypregnenolone) and second, C17,20 cleavage; thus, two carbons are removed to form DHEA. This hydroxylation and cleavage is accomplished by a single enzyme, 17α-hydroxylase or 17,20-lyase (CYP17). DHEA is converted to androstenedione by another two-step enzymatic reaction: dehydrogenation in position 3 (catalyzed by 3β-hydroxysteroid dehydrogenase [3β-HSD]) and shifting of the double bond from ring B to ring A (catalyzed by delta 4,5-ketosteroid isomerase); these two may be the same enzyme. The final reaction yielding testosterone is carried out by 17-ketosteroid reductase (17β-hydroxysteroid dehydrogenase), which substitutes the keto group in position 17 with a hydroxyl group. Unlike all the preceding enzymatic reactions, this is a reversible step but tends to favor testosterone.

Although estrogens are only minor products of testicular steroidogenesis, they are normally found in low concentrations in men. Androgens (C19) are converted to estrogens (C18) by the action of the enzyme complex aromatase (CYP19). Aromatization involves the removal of the methyl group in position 19 and the rearrangement of ring A into an unsaturated aromatic ring. The products of aromatization of testosterone and androstenedione are estradiol and estrone, respectively (see Fig. 37.9). In the testis, the Sertoli cell is the main site of aromatization, which is stimulated by FSH; however, aromatization may also occur in peripheral tissues that lack FSH receptors (e.g., adipose tissue).

The Effects of LH on Leydig Cells Are Primarily Mediated by cAMP

The action of LH on Leydig cells is mediated through specific LH receptors on the plasma membrane. A Leydig cell has about 15,000 LH receptors, and occupancy of less than 5% of these is sufficient for maximal steroidogenesis. This is an example of "spare receptors" (see Chapter 31). Excess receptors increase target cell sensitivity to low circulating levels of hormones by increasing the probability that sufficient receptors will be occupied to induce a response. After exposure to a high LH concentration, the number of LH receptors and testosterone production decrease. However, in response to the initial high concentration of LH, testosterone production will increase and then decrease. Thereafter, subsequent challenges with LH lead to no response or decreased responses. This so-called desensitization involves a loss of surface LH receptors as a result of internalization and receptor modification by phosphorylation.

The LH receptor is a single 93-kDa glycoprotein composed of three functional domains: a glycosylated extracellular hormone-binding domain, a transmembrane spanning domain that contains seven noncontiguous segments, and an intracellular domain. The receptor is coupled to a stimulatory G protein (G_s) via a loop of one of the LH receptor transmembrane segments. The activation of G_s results in increased adenylyl cyclase activity, the production of cAMP, and the activation of protein kinase A (Fig. 37.10).

Low doses of LH can stimulate testosterone production without detectable changes in total cell cAMP concentration. However, the amount of cAMP bound to the regulatory subunit of protein kinase A (PKA) increases in response to such low doses of LH. This response emphasizes the importance of compartmentalization for both enzymes and substrates in mediating hormonal action. Other intracellular mediators, such as the phosphatidylinositol system or calcium, have roles in regulating Leydig cell steroidogenesis, but it appears that the PKA pathway may predominate.

The proteins phosphorylated by PKA are specific for each cell type. Some of these, such as cAMP response element binding protein (CREB), which functions as a DNA-binding protein, regulate the transcription of cholesterol

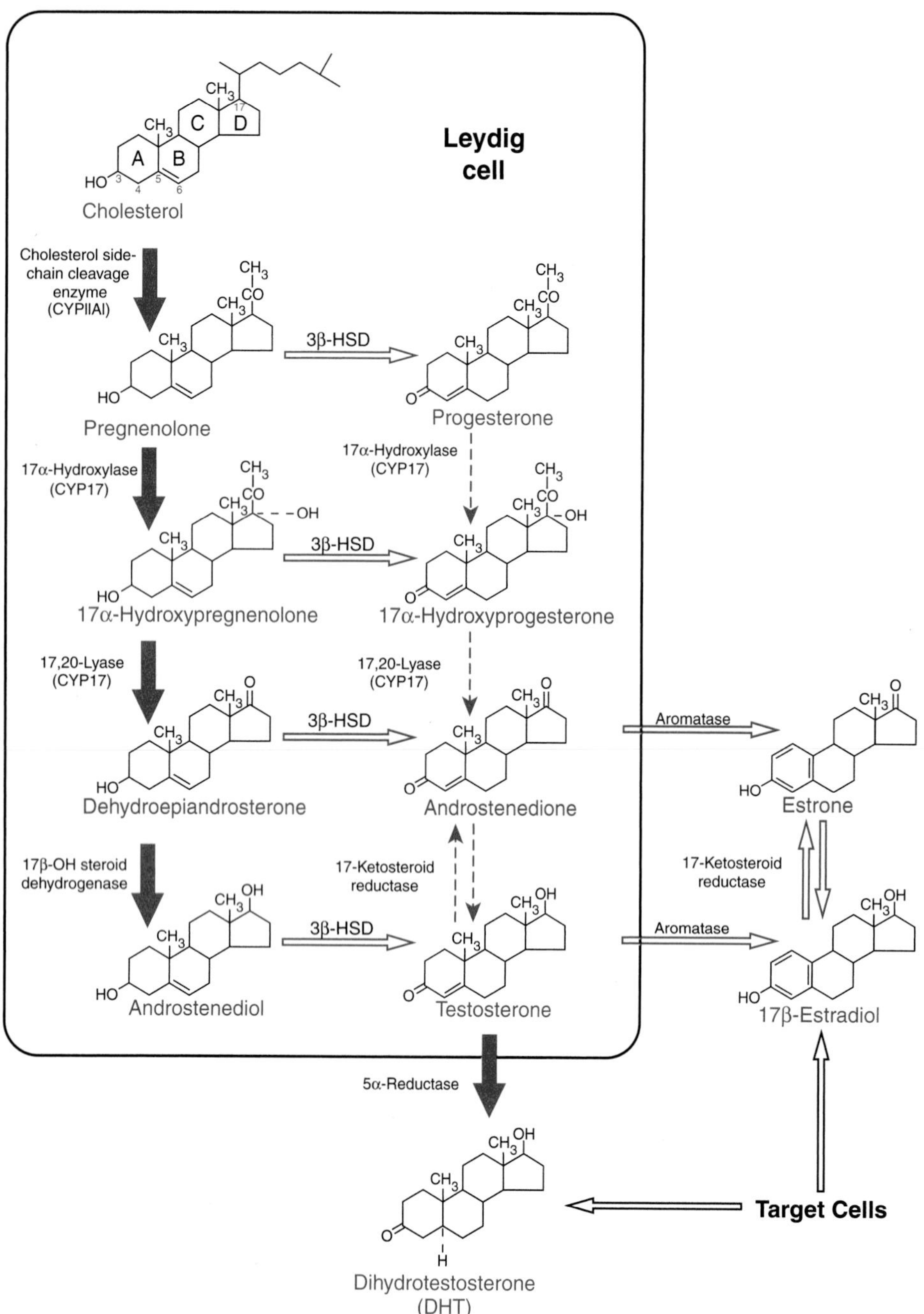

FIGURE 37.9 Steroidogenesis in Leydig cells and further modifications of androgens in target cells. Solid arrows represent the delta 5 pathway. Dashed arrows represent the delta 4 pathway.

side-chain cleavage enzyme (CYP11A1), the rate-limiting enzyme in the conversion of cholesterol to pregnenolone. cAMP is inactivated by **phosphodiesterase** to AMP. This enzyme plays a major role in regulating LH (and, possibly, FSH) responses because phosphodiesterase is activated by gonadotropin stimulation. The increase in phosphodiesterase reduces the response to LH (and FSH). Certain drugs can inhibit phosphodiesterase; gonadotropin hormone responses will increase dramatically in the presence of those drugs. Numerous isoforms of phosphodiesterase and adenylyl cyclase exist; specific types of each in the testis have not yet been revealed.

LH stimulates steroidogenesis by two principal activations. One is the phosphorylation of cholesterol esterase, which releases cholesterol from its intracellular stores. The other is the activation of CYP11A1.

Leydig cells also contain receptors for **prolactin** (PRL). Hyperprolactinemia in men with pituitary tumors, usually microadenomas, is associated with decreased testosterone levels. This condition is due to a direct effect of elevated circulating levels of PRL on Leydig cells, reducing the number of LH receptors or inhibiting downstream signaling events. In addition, hyperprolactinemia may decrease LH secretion by reducing the pulsatile nature of its release. Under nonpathological conditions, however, PRL may synergize with LH to stimulate testosterone production by increasing the number of LH receptors.

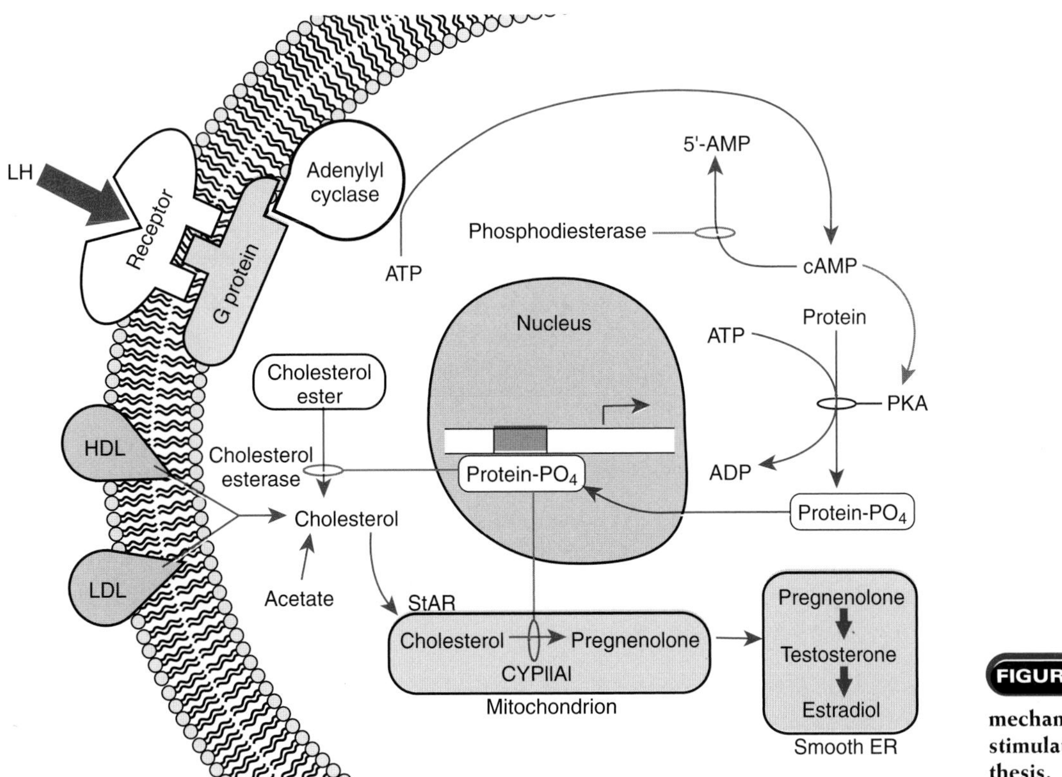

FIGURE 37.10 **A proposed intracellular mechanism by which LH stimulates testosterone synthesis.**

THE ACTIONS OF ANDROGENS

DHT enhances development of the male reproductive tract, accompanying accessory ducts and glands, and male sex characteristics, including behavior. A lack of androgen secretion or action causes feminization.

Peripheral Tissues Process and Metabolize Testosterone

Testosterone is not stored in Leydig cells but diffuses into the blood immediately after being synthesized. An adult man produces 6 to 7 mg testosterone per day. This amount slowly declines after age 50 and reaches about 4 mg/day in the seventh decade of life. Therefore, men do not undergo a sudden cessation of sex steroid production upon aging, as women do during their postmenopausal period, when the ova are completely depleted.

Testosterone circulates bound to plasma proteins, with only 2 to 3% present as the free hormone. About 30 to 40% is bound to albumin and the remainder to **sex hormone-binding globulin** (SHBG), a 94-kDa glycoprotein produced by the liver. SHBG binds both estradiol and testosterone, with a higher binding affinity for testosterone. Because its production is increased by estrogens and decreased by androgens, plasma SHBG concentration is higher in women than in men. SHBG serves as a reservoir for testosterone, and therefore, a sudden decline in newly formed testosterone may not be evident because of the large pool bound to proteins. SHBG, in effect, deactivates testosterone because only the unbound hormone can enter the cell. SHBG also prolongs the half-life of circulating testosterone because testosterone is cleared from the circulation much more slowly if bound to a protein. Any type of liver damage or disease will generally reduce SHBG production. The latter can upset the hormonal balance between LH and testosterone. For example, if SHBG declines acutely, then free testosterone may increase while the total amount of circulating testosterone would decrease. In response to the increase in free testosterone, LH levels would decline in a homeostatic attempt to reduce testosterone production.

Once testosterone is released into the circulation, its fate is variable. In most target tissues, testosterone functions as a prohormone and is converted to the biologically active derivatives DHT by **5α-reductase** or estradiol by aromatase (Fig. 37.11). Skin, hair follicles, and most of the male reproductive tract contain an active 5α-reductase. The enzyme irreversibly catalyzes the reduction of the double bond in ring A and generates DHT (see Fig. 37.9). DHT has a high binding affinity for the androgen receptor and is 2 to 3 times more potent than testosterone.

Congenital deficiency of 5α-reductase in males results in ambiguous genitalia containing female and male characteristics because DHT is critical for directing the normal development of male external genitalia during embryonic life (see Chapter 39). Without DHT, the female pathway may predominate, even though the genetic sex is male and small, undescended testes are present in the inguinal region. DHT is nonaromatizable and cannot be converted to estrogens.

Drugs that inhibit 5α-reductase are currently used to reduce prostatic hypertrophy because DHT induces hyperplasia of prostatic epithelial cells. In addition, analogs of GnRH, as either agonists or antagonists, can be given to patients to reduce the secretion of androgen in androgen-dependent

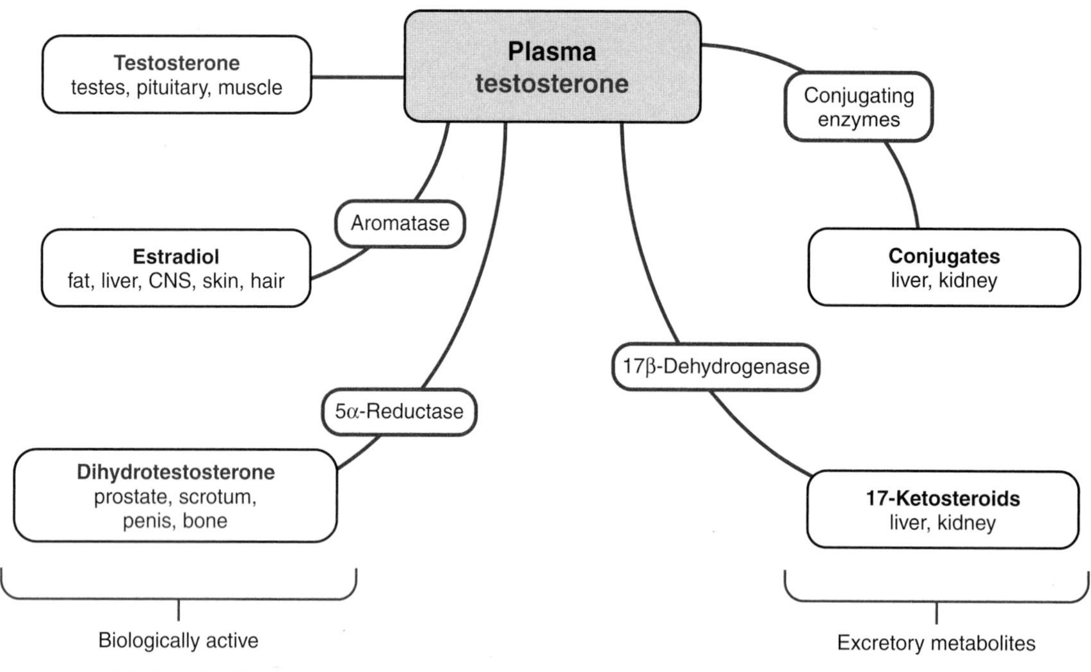

FIGURE 37.11 **Conversion of testosterone to different products in extratesticular sites.**

neoplasia or cancer (see Clinical Focus Box 37.1). In the case of the GnRH antagonist, this analog blocks the secretion of LH. In contrast, GnRH agonists given in large quantities initially induce the secretion of LH (and androgen). However, this response is followed by down-regulation of GnRH receptors on the pituitary gonadotrophs and, ultimately, a dramatic decline in circulating LH and androgen.

Aromatization of some androgens to estrogens occurs in fat, liver, skin, and brain cells. Circulating levels of total estrogens (estradiol plus estrone) in men can approach those of women in their early follicular phase. Men are protected from feminization as long as production of and tissue responsiveness to androgens are normal. The treatment of hypogonadal male patients with high doses of aromatizable testosterone analogs (or testosterone), the use of anabolic steroids by athletes, abnormal reductions in testosterone secretion, estrogen-producing testicular tumors, and tissue insensitivity to androgens can lead to **gynecomastia** or breast enlargement. All of these conditions are characterized by a decrease in the testosterone-to-estradiol ratio.

Androgens are metabolized in the liver to biologically inactive water-soluble derivatives suitable for excretion by the kidneys. The major products of testosterone metabolism are two 17-ketosteroids, androsterone and etiocholanolone. These, as well as native testosterone, are conjugated in position 3 to form sulfates and glucuronides, which are water-soluble and excreted into the urine (see Fig. 37.11).

Androgens Have Effects on Reproductive and Nonreproductive Tissues

An **androgen** is a substance that stimulates the growth of the male reproductive tract and the development of secondary sex characteristics. Androgens have effects on almost every tissue, including alteration of the primary sex structures (i.e., the testes and genital tract) and stimulation of the secondary sex structures (i.e., accessory glands) and development of secondary sex characteristics responsible for masculine phenotypic expression. Androgens also affect both sexual and nonsexual behavior. The relative potency ranking of androgens is DHT $>$ testosterone $>$ androstenedione $>$ DHEA. The action of sex steroid hormones on somatic tissue, such as muscle, is referred to as "anabolic" because the end result is increased muscle size. This action is mediated by the same molecular mechanisms that result in virilization.

Between 8 and 18 weeks of fetal life, androgens mediate differentiation of the male genitalia. The organogenesis of the wolffian (mesonephric) ducts into the epididymis, vas deferens, and seminal vesicles is directly influenced by testosterone, which reaches these target tissues by diffusion rather than by a systemic route. The differentiation of the urogenital sinus and the genital tubercle into the penis, scrotum, and prostate gland depends on testosterone being converted to DHT. Toward the end of fetal life, the descent of the testes into the scrotum is promoted by testosterone and insulin-like hormones from Leydig cells (see Chapter 39).

The onset of puberty is marked by enhanced androgenic activity. Androgens promote the growth of the penis and scrotum, stimulate the growth and secretory activity of the epididymis and accessory glands, and increase the pigmentation of the genitalia. Enlargement of the testes occurs under the influence of the gonadotropins (LH and FSH). Spermatogenesis, which is initiated during puberty, depends on adequate amounts of testosterone. Throughout adulthood, androgens are responsible for maintaining the structural and functional integrity of all reproductive tissues. Castration of adult men results in regression of the reproductive tract and involution of the accessory glands.

CLINICAL FOCUS BOX 37.1

Prostate Cancer

Some prostate cancers are highly dependent upon androgens for cellular proliferation; therefore, physicians attempt to totally ablate the secretion of androgens by the testes. Generally, two options for those patients are surgical castration and chemical castration. Surgical castration is irreversible and requires the removal of the testes, while chemical castration is reversible.

One option for chemical treatment of these patients is the use of analogs of GnRH, the hormone that regulates the secretion of LH and FSH. Long-acting GnRH agonists or antagonists reduce LH and FSH secretion by different mechanisms. GnRH agonists reduce gonadotropin secretion by desensitization of the pituitary gonadotrophs to GnRH, leading to a reduction of LH and FSH secretion. GnRH agonists initially stimulate GnRH receptors on pituitary cells but ultimately reduce their numbers. GnRH antagonists bind to GnRH receptors on the pituitary cells, prevent endogenous GnRH from binding to those receptors, and subsequently reduce LH and FSH secretion. Shortly after treatment, testicular concentrations of androgens decline because of the low levels of circulating LH and FSH. The expectation is that androgen-dependent cancer cells will cease or slow proliferation and, ultimately, die.

GnRH agonists (leuprolide acetate [trade name Lupron]) are usually used in combination with other drugs in order to block most effectively androgenic activity. For example, one of the androgen-blocking drugs includes 5α-reductase inhibitors that prevent the conversion of testosterone to the highly active androgen dihydrotestosterone (DHT). In addition, antiandrogens, such as flutamide, bind to the androgen receptor and prevent binding of endogenous androgen. Some prostate cancers are androgen-independent, and the treatment requires nonhormonal therapies, including chemotherapy and radiation.

Androgens Are Responsible for Secondary Sex Characteristics and the Masculine Phenotype

Androgens effect changes in hair distribution, skin texture, pitch of the voice, bone growth, and muscle development. Hair is classified by its sensitivity to androgens into nonsexual (eyebrows and extremities); ambisexual (axilla), which is responsive to low levels of androgens; and sexual (face, chest, upper pubic triangle), which is responsive only to high androgen levels. Hair follicles metabolize testosterone to DHT or androstenedione. Androgens stimulate the growth of facial, chest, and axillary hair; however, along with genetic factors, they also promote temporal hair recession and loss. Normal axillary and pubic hair growth in women is also under androgenic control, whereas excess androgen production in women causes the excessive growth of sexual hair (**hirsutism**).

The growth and secretory activity of the sebaceous glands on the face, upper back, and chest are stimulated by androgens, primarily DHT, and inhibited by estrogens. Increased sensitivity of target cells to androgenic action, especially during puberty, is the cause of **acne vulgaris** in both males and females. Skin derived from the urogenital ridge (e.g., the prepuce, scrotum, clitoris, and labia majora) remains sensitive to androgens throughout life and contains an active 5α-reductase. Growth of the larynx and thickening of the vocal cords are also androgen-dependent. Eunuchs maintain the high-pitched voice typical of prepubertal boys because they were castrated prior to puberty.

The growth spurt of adolescent males is influenced by a complex interplay between androgens, growth hormone (GH), nutrition, and genetic factors. The growth spurt includes growth of the vertebrae, long bones, and shoulders. The mechanism by which androgens (likely DHT) alter bone metabolism is unclear. Androgens accelerate closure of the epiphyses in the long bones, eventually limiting further growth. Because of the latter, precocious puberty is associated with a final short adult stature, whereas delayed puberty or eunuchoidism usually results in tall stature. Androgens have multiple effects on skeletal and cardiac muscle. Because 5α-reductase activity in muscle cells is low, the androgenic action is due to testosterone. Testosterone stimulates muscle hypertrophy, increasing muscle mass; however, it has minimal or no effect on muscle hyperplasia. Testosterone, in synergy with GH, causes a net increase in muscle protein.

Other nonreproductive organs and systems are affected, directly or indirectly, by androgens, including the liver, kidneys, adipose tissue, and hematopoietic and immune systems. The kidneys are larger in males, and some renal enzymes (e.g., β-glucuronidase and ornithine decarboxylase) are induced by androgens. HDL levels are lower and triglyceride concentrations higher in men, compared to premenopausal women, a fact that may explain the higher prevalence of atherosclerosis in men. Androgens increase red blood cell mass (and, hence, hemoglobin levels) by stimulating erythropoietin production and by increasing stem cell proliferation in the bone marrow.

The Brain Is a Target Site for Androgen Action

Many sites in the brain contain androgen receptors, with the highest density in the hypothalamus, preoptic area, septum, and amygdala. Most of those areas also contain aromatase and many of the androgenic actions in the brain result from the aromatization of androgens to estrogens. The pituitary also has abundant androgen receptors, but no aromatase. The enzyme 5α-reductase is widely distributed in the brain, but its activity is generally higher during the prenatal period than in adults. Sexual dimorphism in the size, number, and arborization of neurons in the preoptic area, amygdala, and superior cervical ganglia has been recently recognized in humans.

Unlike most species, which mate only to produce offspring, in humans, sexual activity and procreation are not tightly linked. Superimposed on the basic reproductive mechanisms dictated by hormones are numerous psychological and societal factors. In normal men, no correlation is found between circulating testosterone levels and sexual drive, frequency of intercourse, or sexual fantasies. Similarly, there is no correlation between testosterone levels and impotence or homosexuality. Castration of adult men results in a slow decline in, but not a complete elimination of, sexual interest and activity. See Clinical Focus Box 37.2 for a discussion of the effects of testosterone administration.

REPRODUCTIVE DYSFUNCTIONS

Male reproductive dysfunctions may by caused by endocrine disruption, morphological alterations in the reproductive tract, neuropathology, and genetic mutations. Several medical tests, including serum hormone levels, physical examination of the reproductive organs, and sperm count are important in ascertaining causes of reproductive dysfunctions.

Hypogonadism Can Result From Defects at Several Levels

Male **hypogonadism** may result from defects in spermatogenesis, steroidogenesis, or both. It may be a primary defect in the testes or secondary to hypothalamic-pituitary dysfunction, and determining whether the onset of gonadal failure occurred before or after puberty is important in establishing the cause. However, several factors must be considered. First, normal spermatogenesis almost never occurs with defective steroidogenesis, but normal steroidogenesis can be present with defective spermatogenesis. Second, primary testicular failure removes feedback inhibition from the hypothalamic-pituitary axis, resulting in elevated plasma gonadotropins. In contrast, hypothalamic and/or pituitary failure is almost always accompanied by decreased gonadotropin and steroid levels and reduced testicular size. Third, gonadal failure before puberty results in the absence of secondary sex characteristics, creating a distinctive clinical presentation called **eunuchoidism**. In contrast, men with a postpubertal testicular failure retain masculine features but exhibit low sperm counts or a reduced ability to produce functional sperm.

To establish the cause(s) of reproductive dysfunction, physical examination and medical history, semen analysis, hormone determinations, hormone stimulation tests, and genetic analysis are performed. Physical examination should establish whether eunuchoidal features (i.e., infantile appearance of external genitalia and poor or absent development of secondary sex characteristics) are present. In men with adult-onset reproductive dysfunction, physical examination can uncover problems such as cryptorchidism (nondescendent testes), testicular injury, varicocele (an abnormality of the spermatic vasculature), testicular tumors, prostatic inflammation, or gynecomastia. Medical and family history help determine delayed puberty, anosmia (an inability to smell, often associated with GnRH dysfunction), previous fertility, changes in sexual performance, ejaculatory disturbances, or impotence (an inability to achieve or maintain erection).

One step in the evaluation of fertility is semen analysis. Semen are analyzed on specimens collected after 3 to 5 days of sexual abstinence, as the number of sperm ejaculated remains low for a couple of days after ejaculation. Initial examination includes determination of viscosity, liquefaction, and semen volume. The sperm are then counted and the percentage of sperm showing forward motility is scored. The spermatozoa are evaluated morphologically, with attention to abnormal head configuration and defective tails. Chemical analysis can provide information on the secretory activity of the accessory glands, which is considered abnormal if semen volume is too low or sperm motility is impaired. Fructose and prostaglandin levels are determined to assess the function of the seminal vesicles and levels of zinc, magnesium, and acid phosphatase to evaluate the prostate. Terms used in evaluating fertility include aspermia (no semen), hypospermia and hyperspermia (too small or too large semen volume), azoospermia (no spermatozoa), and oligozoospermia (reduced number of spermatozoa).

Serum testosterone, estradiol, LH, and FSH analyses are performed using radioimmunoassays. Free and total testosterone levels should be measured; because of the pulsatile nature of LH release, several consecutive blood samples are needed. Dynamic hormone stimulation tests are most valuable for establishing the site of abnormality. A failure to increase LH release upon treatment with **clomiphene**, an

CLINICAL FOCUS BOX 37.2

Effects of Testosterone Administration

Although testosterone has a role in stimulating spermatogenesis, infertile men with a low sperm count do not benefit from testosterone treatment. Unless given at supraphysiological doses, exogenous testosterone cannot achieve the required local high concentration in the testis. One function of androgen-binding protein in the testis is to sequester testosterone, which significantly increases its local concentration.

Exogenous testosterone given to men would normally inhibit endogenous LH release through a negative-feedback effect on the hypothalamic-pituitary axis, and lead to a suppression of testosterone production by the Leydig cells and a further decrease in testicular testosterone concentrations. Ultimately, because LH levels decrease when exogenous testosterone is administered, testicular size decreases, as has been reported for men who abuse androgens.

High levels of androgens have an anabolic effect on muscle tissue, leading to increased muscle mass, strength, and performance, a desired result for body builders and athletes. Androgen abuse has been associated with abnormally aggressive behavior and the potential for increased incidence of liver and brain tumors.

antiestrogen, likely indicates a hypothalamic abnormality. Clomiphene blocks the inhibitory effects of estrogen and testosterone on endogenous GnRH release. An absence of or blunted testosterone rise after hCG injection suggests a primary testicular defect. Genetic analysis is used when congenital defects are suspected. The presence of the Y chromosome can be revealed by karyotyping of cultured peripheral lymphocytes or direct detection of specific Y antigens on cell surfaces.

Reproductive Disorders Are Associated With Hypogonadotropic or Hypergonadotropic States

Endocrine factors are responsible for approximately 50% of hypogonadal or infertility cases. The remainder is of unknown etiology or the result of injury, deformities, and environmental factors. Endocrine-related hypogonadism can be classified as hypothalamic-pituitary defects (hypogonadotropic because of the lack of LH and/or FSH), primary gonadal defects (hypergonadotropic because gonadotropins are high as a result of a lack of negative feedback from the testes), and defective androgen action (usually the result of absence of androgen receptor or 5α-reductase). Each of these is further subdivided into several categories, but only a few examples are discussed here.

Hypogonadotropic hypogonadism can be congenital, idiopathic, or acquired. The most common congenital form is Kallmann's syndrome, which results from decreased or absent GnRH secretion, as mentioned earlier. It is often associated with anosmia or hyposmia and is transmitted as an autosomal dominant trait. Patients do not undergo pubertal development and have eunuchoidal features. Plasma LH, FSH, and testosterone levels are low, and the testes are immature and have no sperm. There is no response to clomiphene, but intermittent treatment with GnRH can produce sexual maturation and full spermatogenesis.

Another category of hypogonadotropic hypogonadism, **panhypopituitarism** or **pituitary failure**, can occur before or after puberty and is usually accompanied by a deficiency of other pituitary hormones. **Hyperprolactinemia**, whether from hypothalamic disturbance or pituitary adenoma, often results in decreased GnRH production, hypogonadotropic state, impotence, and decreased libido. It can be treated with dopaminergic agonists (e.g., bromocriptine), which suppress PRL release (see Chapter 38). Excess androgens can also result in suppression of the hypothalamic-pituitary axis, resulting in lower LH levels and impaired testicular function. This condition often results from **congenital adrenal hyperplasia** and increased adrenal androgen production from 21-hydroxylase (CYP21A2) deficiency (see Chapter 34).

Hypergonadotropic hypogonadism usually results from impaired testosterone production, which can be congenital or acquired. The most common disorder is Klinefelter's syndrome discussed earlier.

Male Pseudohermaphroditism Often Results From Resistance to Androgens

A **pseudohermaphrodite** is an individual with the gonads of one sex and the genitalia of the other. One of the most interesting causes of male reproductive abnormalities is an end organ insensitivity to androgens. The best characterized syndrome is **testicular feminization**, an X-linked recessive disorder caused by a defect in the testosterone receptor. In the classical form, patients are male pseudohermaphrodites with a female phenotype and an XY male genotype. They have abdominal testes that secrete testosterone but no other internal genitalia of either sex (see Chapter 39). They commonly have female external genitalia, but with a short vagina ending in a blind pouch. Breast development is typical of a female (as a result of peripheral aromatization of testosterone), but axillary and pubic hair, which are androgen-dependent, are scarce or absent. Testosterone levels are normal or elevated, estradiol levels are above the normal male range, and circulating gonadotropin levels are high. The inguinally located testes usually have to be removed because of an increased risk of cancer. After orchiectomy, patients are treated with estradiol to maintain a female phenotype.

REVIEW QUESTIONS

DIRECTIONS: Each of the numbered items or incomplete statements in this section is followed by answers or by completions of the statement. Select the ONE lettered answer or completion that is BEST in each case.

1. A major causal factor in some cases of hypogonadism is
   (A) Reduced secretion of gonadotropin-releasing hormone (GnRH)
   (B) Hypersecretion of pituitary LH and FSH as the result of increased GnRH
   (C) Excess secretion of testicular activin by Sertoli cells
   (D) Failure of the hypothalamus to respond to testosterone
   (E) Increased number of FSH receptors in the testis
2. The major function of follistatin is
   (A) Bind FSH and increase FSH secretion
   (B) Inhibit the production of seminal fluid
   (C) Reduce testosterone secretion by Leydig cells
   (D) Stimulate the production of spermatogonia
   (E) Bind activin and thus decrease FSH secretion
3. A major function of the epididymis is
   (A) Storage and transport of mature sperm
   (B) Initiating the development of spermatozoa
   (C) Secretion of estrogens
   (D) Production of inhibin
   (E) Secretion of fluids that contribute to semen
4. The production of mature spermatozoa from spermatogonia
   (A) Takes 32 days
   (B) Takes 70 days
   (C) Takes 150 days
   (D) Is unaffected by Kallmann's syndrome
   (E) Is independent of testicular temperature

(continued)

5. The first enzymatic reaction, which is the rate-limiting step, in the production of testosterone
(A) Occurs in the mitochondria
(B) Occurs in the ribosomes
(C) Involves aromatization
(D) Generates progesterone as the immediate derivative
(E) Is stimulated by FSH
6. Testosterone is
(A) Bound to high-density lipoprotein (HDL)
(B) Bound to activin
(C) Converted to dihydrotestosterone in the prostate
(D) Converted to 17-hydroxyprogesterone in the liver
(E) Metabolized by cholesterol side-chain cleavage enzyme
7. Sex hormone-binding globulin (SHBG)
(A) Binds testosterone with a higher affinity than estradiol
(B) Reduces the total amount of circulating testosterone
(C) Decreases the half-life of testosterone
(D) Stimulates the secretion of inhibin
(E) Blocks the synthesis of androgen-binding protein
8. The production of estradiol by the testes requires
(A) Sertoli cell follistatin
(B) LH and Leydig cells
(C) Activin but not LH
(D) Leydig cells, Sertoli cells, LH, and FSH
(E) Leydig cells and FSH
9. Eunuchs are tall because
(A) Estrogens stimulate the growth of long bones
(B) Excess LH delays epiphyseal closure of the long bones
(C) Reduced androgen and estrogen delays epiphyseal closure in long bones
(D) The lack of testes stimulates closure of the epiphyses
(E) They secrete excess androgen

SUGGESTED READING

Burger H, DeKretser D. The Testis. New York: Raven Press, 1989.

Fawcett DW. A Textbook of Histology. 12th Ed. New York: Chapman & Hall, 1994;796–850.

Griswold MD, Russell LD. Sertoli cells, function. In: Knobil E, Neill JD, eds. The Encyclopedia of Reproduction. New York: Academic Press, 1999;371–380.

Johnson L, McGowen TA, Keillor GE. Testis, overview. In: Knobil E, Neill JD, eds. The Encyclopedia of Reproduction. New York: Academic Press, 1999;769–784.

Payne A, Hardy M, Russell L. The Leydig Cell. Vienna, IL: Cache River Press, 1996.

Redman JF. Male reproductive system, human. In: Knobil E, Neill JD, eds. The Encyclopedia of Reproduction. New York: Academic Press, 1999;30–41.

CHAPTER

38 The Female Reproductive System

Paul F. Terranova, Ph.D.

CHAPTER OUTLINE

KEY CONCEPTS

1. Pulses of hypothalamic GnRH regulate the secretion of LH and FSH, which enhance follicular development, steroidogenesis, ovulation, and formation of the corpus luteum.
2. LH and FSH, in coordination with ovarian theca and granulosa cells, regulate the secretion of follicular estradiol.
3. Ovulation occurs as the result of a positive feedback of follicular estradiol on the hypothalamic-pituitary axis that induces LH and FSH surges.
4. Follicular development occurs in distinct steps: primordial, primary, secondary, tertiary, and graafian follicle stages.
5. Follicular rupture (ovulation) requires the coordination of appropriately timed LH and FSH surges that induce inflammatory reactions in the graafian follicle, leading to dissolution at midcycle of the follicular wall by several ovarian enzymes.
6. Follicular atresia results from the withdrawal of gonadotropin support.
7. The formation of a functional corpus luteum requires the presence of an LH surge, adequate numbers of LH receptors, sufficient granulosa cells, and significant progesterone secretion.
8. The uterine cycle is regulated by estradiol and progesterone, such that estradiol induces proliferation of the uterine endometrium, whereas progesterone induces differentiation of the uterine endometrium and the secretion of distinct products.
9. During puberty, the hypothalamus begins to secrete increasing quantities of GnRH, which increases LH and FSH secretion, enhances ovarian function, and leads to the first ovulation.
10. Menopause ensues from the loss of numerous oocytes in the ovary and the subsequent failure of follicular development and estradiol secretion. LH and FSH levels rise from the lack of negative feedback by estradiol.

The fertility of the mature human female is cyclic. The release from the ovary of a mature female germ cell or ovum occurs at a distinct phase of the menstrual cycle. The secretion of ovarian steroid hormones, estradiol and progesterone, and the subsequent release of an ovum during the menstrual cycle are controlled by cyclic changes in LH and FSH from the pituitary gland, and estradiol and progesterone from the ovaries. The cyclic changes in steroid hormone secretion cause significant changes in the structure and function of the uterus in preparing it for the reception of a fertilized ovum. At different stages of the menstrual cycle, progesterone and estradiol exert negative- and positive-feedback effects on the hypothalamus and on pituitary gonadotrophs, generating the cyclic pattern of LH and FSH release characteristic of the female reproductive system. Since the hormonal events during the menstrual cycle are delicately synchronized, the menstrual cycle can be readily affected by stress and by environmental, psychological, and social factors.

The female cycle is characterized by monthly bleeding, resulting from the withdrawal of ovarian steroid hormone support of the uterus, which causes shedding of the superficial layers of the uterine lining at the end of each cycle. The first menstrual cycle occurs during puberty. Menstrual cycles are interrupted during pregnancy and lactation and cease at menopause. Menstruation signifies a failure to conceive and results from regression of the corpus luteum and subsequent withdrawal of luteal steroid support of the superficial endometrial layer of the uterus.

AN OVERVIEW OF THE FEMALE REPRODUCTIVE SYSTEM

An overview of the interactions of hormonal factors in female reproduction is shown in Figure 38.1. The female hormonal system consists of the brain, pituitary, ovaries, and reproductive tract (oviduct, uterus, cervix, and vagina). In the brain, the hypothalamus produces gonadotropin-releasing hormone (GnRH), which controls the secretion of luteinizing hormone (LH) and follicle-stimulating hormone (FSH).

The mature ovary has two major functions: the maturation of germ cells and steroidogenesis. Each germ cell is ultimately enclosed within a follicle, a major source of steroid hormones during the menstrual cycle. At ovulation, the ovum or egg is released and the ruptured follicle is transformed into a corpus luteum, which secretes progesterone as its main product. FSH is primarily involved in stimulating the growth of ovarian follicles, while LH induces ovulation. Both LH and FSH regulate follicular steroidogenesis and androgen and estradiol secretion, and LH regulates the secretion of progesterone from the corpus luteum. Ovarian steroids inhibit the secretion of LH and FSH with one exception: Just prior to ovulation (at midcycle), estradiol has a positive-feedback effect on the hypothalamic-pituitary axis and induces significant increases in the secretion of GnRH, LH, and FSH. The ovary also produces three polypeptide hormones. **Inhibin** suppresses the secretion of FSH. **Activin** (an inhibin-binding protein) increases the secretion of FSH, and **follistatin** (an activin-binding protein) reduces the secretion of FSH.

Shortly after fertilization, the embryo begins to develop placenta cells, which attach to the uterine lining and unite with the maternal placental cells. The **placenta** produces several pituitary-like and ovarian steroid-like hormones. These hormones support placental and fetal development throughout pregnancy and have a role in parturition. The mammary glands are also under the control of pituitary hormones and ovarian steroids, and provide the baby with immunological protection and nutritional support through lactation. Lactation is hormonally controlled by **prolactin** (PRL) from the anterior pituitary, which regulates milk production, and **oxytocin** from the posterior pituitary, which induces milk ejection from the breasts.

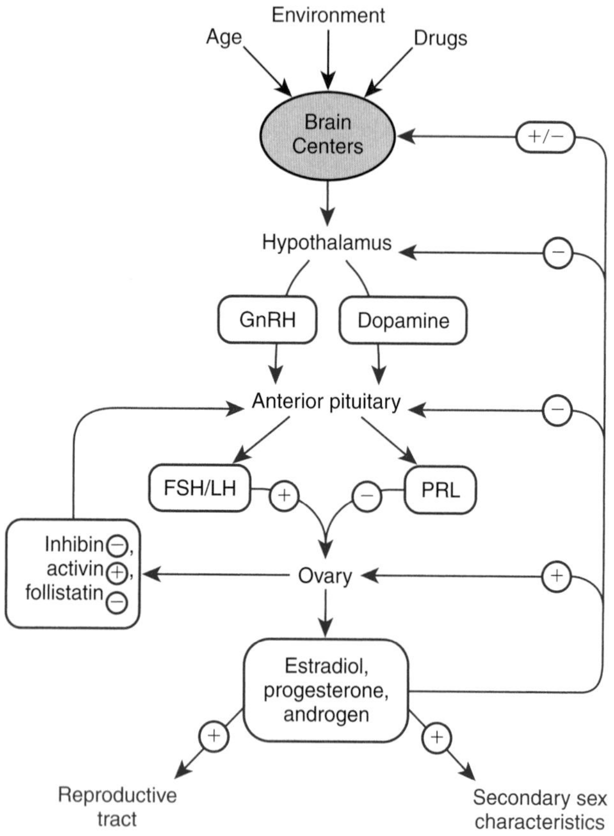

FIGURE 38.1 **Regulation of the reproductive tract in the female.** The main reproductive hormones are shown in boxes. Positive and negative regulations are depicted by plus and minus signs.

THE HYPOTHALAMIC-PITUITARY AXIS

The hypothalamic-pituitary axis has an important role in regulating the menstrual cycle. GnRH, a decapeptide produced in the hypothalamus and released in a pulsatile manner, controls the secretion of LH and FSH through a portal vascular system (see Chapter 32). Blockade of the portal system reduces the secretion of LH and FSH and leads to ovarian atrophy and a reduction in ovarian hormone secretion. The secretion of GnRH by the hypothalamus is regulated by neurons from other brain regions. Neurotransmitters, such as epinephrine and norepinephrine, stimulate the secretion of GnRH, whereas dopamine and serotonin inhibit secretion of GnRH. In addition, ovarian steroids and peptides and hypothalamic neuropeptides can regulate the secretion of GnRH. GnRH stimulates the pituitary gonadotrophs to secrete LH and FSH. GnRH binds to high-affinity receptors on the gonadotrophs and stimulates the secretion of LH and FSH through a phosphoinositide-protein kinase C-mediated pathway (see Chapter 1).

LH release throughout the female life span is depicted in Figure 38.2. During the neonatal period, LH is released at low and steady rates without pulsatility; this period coincides with lack of development of mature ovarian follicles and very low to no ovarian estradiol secretion. Pulsatile release begins with the onset of puberty and for several years is expressed only during sleep; this period coincides with increased but asynchronous follicular development and with increased secretion of ovarian estradiol. Upon the establishment of regular functional menstrual cycles associated with regular ovulation, LH pulsatility prevails throughout the 24-hour period, changing in a monthly cyclic manner. In postmenopausal women whose ovaries lack sustained follicular development and exhibit low ovar-

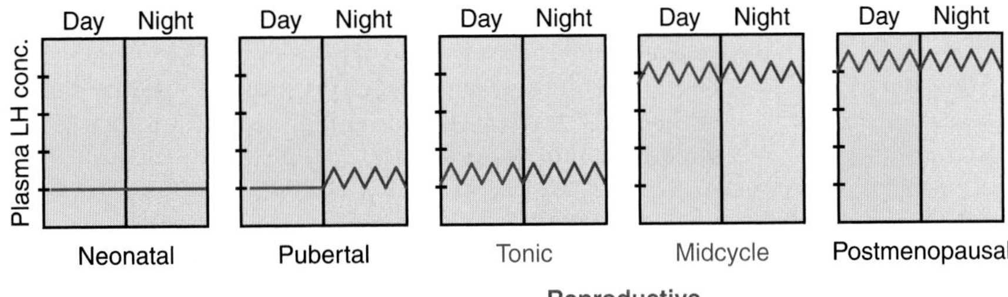

FIGURE 38.2 Relative levels of LH release in human females throughout life. (Modified from Yen SSC. In: Krieger DT, Hughes JC, eds. Neuroendocrinology. Sunderland, MA: Sinauer, 1980.)

ian estradiol secretion, mean circulating LH levels are high and pulses occur at a high frequency.

THE FEMALE REPRODUCTIVE ORGANS

The female reproductive tract has two major components: the ovaries, which produce the mature ovum and secrete progestins, androgens, and estrogens; and the ductal system, which transports the ovum, is the place of the union of the sperm and egg, and maintains the developing conceptus until delivery. The morphology and function of these structures change in a cyclic manner under the influence of the reproductive hormones.

The ovaries are in the pelvic portion of the abdominal cavity on both sides of the uterus and are anchored by ligaments (Fig. 38.3). An adult ovary weighs 8 to 12 g and consists of an outer cortex and an inner medulla, without a sharp demarcation. The cortex is surrounded by a fibrous tissue, the tunica albuginea, covered by a single layer of surface epithelium continuous with the mesothelium covering the other organs in the abdominal cavity. The cortex contains oocytes enclosed in **follicles** of various sizes, **corpora lutea**, **corpora albicantia**, and **stromal cells**. The medulla contains connective

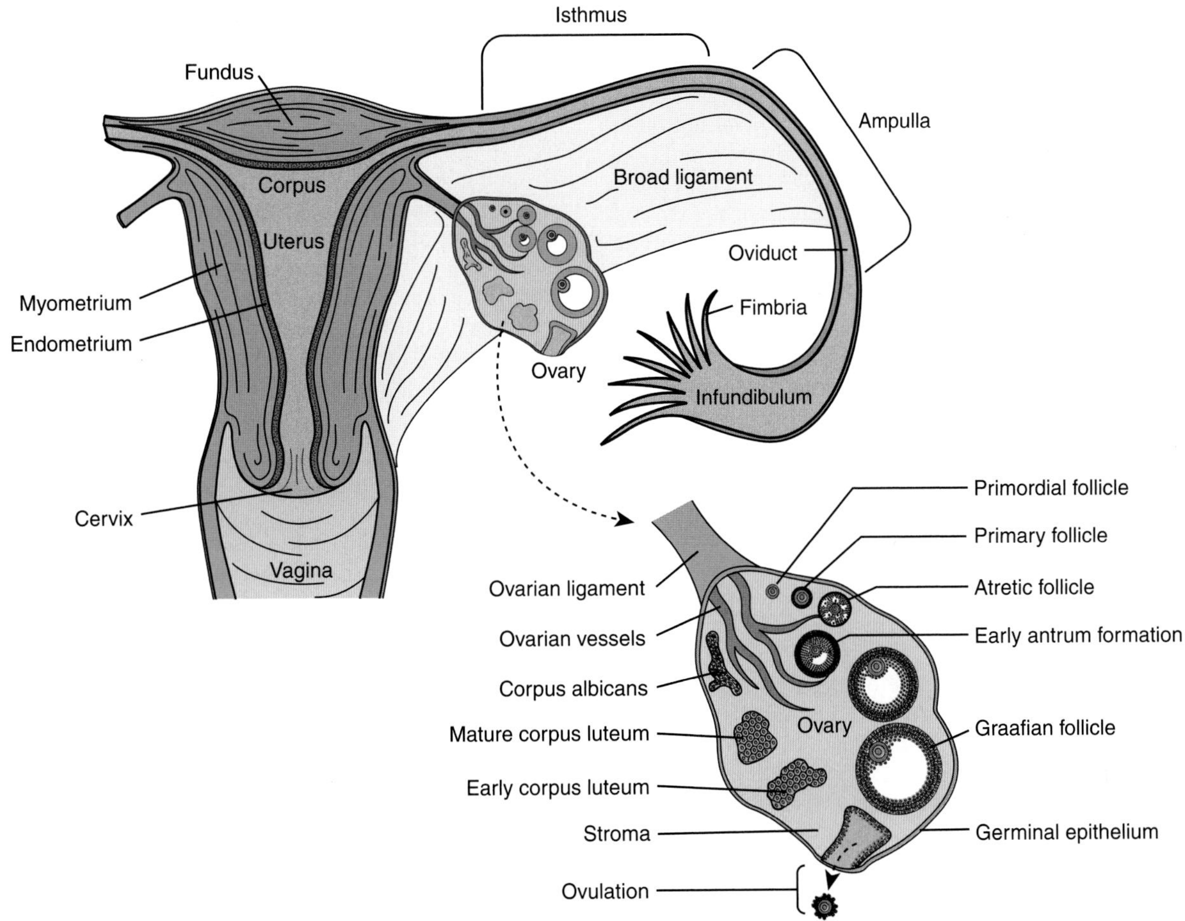

FIGURE 38.3 The female reproductive organs. (Modified from Patton BM. Human Embryology. New York: McGraw-Hill, 1976.)

and interstitial tissues. Blood vessels, lymphatics, and nerves enter the medulla of the ovary through the hilus.

On the side that ovulates, the **oviduct** (fallopian tube) receives the ovum immediately after ovulation. The oviducts are the site of fertilization and provide an environment for development of the early embryo. The oviducts are 10 to 15 cm long and composed of sequential regions called the **infundibulum**, **ampulla**, and **isthmus**. The infundibulum is adjacent to the ovary and opens to the peritoneal cavity. It is trumpet-shaped with finger-like projections called **fimbria** along its outer border that grasp the ovum at the time of follicular rupture. Its thin walls are covered with densely ciliated projections, which facilitate ovum uptake and movement through this region. The ampulla is the site of fertilization. It has a thin musculature and well-developed mucosal surface. The isthmus is located at the uterotubal junction and has a narrow lumen surrounded by smooth muscle. It has sphincter-like properties and can serve as a barrier to the passage of germ cells. The oviducts transport the germ cells in two directions: sperm ascend toward the ampulla and the zygote descends toward the uterus. This requires coordination between smooth muscle contraction, ciliary movement, and fluid secretion, all of which are under hormonal and neuronal control.

The **uterus** is situated between the urinary bladder and rectum. On each upper side, an oviduct opens into the uterine lumen, and on the lower side, the uterus connects to the vagina. The uterus is composed of two types of tissue. The outer part is the **myometrium**, composed of multiple layers of smooth muscle. The inner part, lining the lumen of the uterus, is the **endometrium**, which contains a deep **stromal layer** next to the myometrium and a superficial epithelial layer. The stroma is permeated by spiral arteries and contains much connective tissue. The epithelial layer is interrupted by uterine glands, which also penetrate the stromal layer and are lined by columnar secretory cells. The uterus provides an environment for the developing fetus, and eventually, the myometrium will generate rhythmic contractions that assist in expelling the fetus at delivery.

The **cervix** (neck) is a narrow muscular canal that connects the vagina and the body (corpus) of the uterus. It must dilate in response to hormones to allow the expulsion of the fetus. The cervix has numerous glands with a columnar epithelium that produces mucus under the control of estradiol. As more and more estradiol is produced during the follicular phase of the cycle, the cervical mucus changes from a scanty viscous material to a profuse watery and highly elastic substance called **spinnbarkeit**. The viscosity of the spinnbarkeit can be tested by touching it with a piece of paper and lifting vertically. The mucus can form a thread up to 6 cm under the influence of elevated estradiol. If a drop of the cervical mucus is placed on a slide and allowed to dry, it will form a typical **ferning** pattern when under the influence of estradiol.

The **vagina** is well innervated, and has a rich blood supply. It is lined by several layers of epithelium that change histologically during the menstrual cycle. When estradiol levels are low, as during the prepubertal or postmenopausal periods, the vaginal epithelium is thin and the secretions are scanty, resulting in a dry and infection-susceptible area. Estradiol induces proliferation and **cornification (keratinization)** of the vaginal epithelium, whereas progesterone opposes those actions and induces the influx of polymorphonuclear leukocytes into the vaginal fluids. Estradiol also activates vaginal glands that produce lubricating fluid during coitus.

FOLLICULOGENESIS, STEROIDOGENESIS, ATRESIA, AND MEIOSIS

Most follicles in the ovary will undergo atresia. However, some will develop into mature follicles, produce steroids, and ovulate. As follicles mature, oocytes will also mature by entering meiosis, which produces the proper number of chromosomes in preparation for fertilization.

The Primordial Follicle Contains an Oocyte Arrested in Meiosis

Female germ cells develop in the embryonic yolk sac and migrate to the genital ridge where they participate in the development of the ovary (Table 38.1). Without germ cells, the ovary does not develop. The germs cells, called **oogonia**, actively divide by mitosis. Oogonia undergo mitosis only during the prenatal period. By birth, the ovaries contain a finite number of oocytes, estimated to be about 1 million. Most of them will die by a process called atresia. By puberty, only 200,000 oocytes remain; by age 30, only 26,000 remain; and by the time of menopause, the ovaries are essentially devoid of oocytes.

When oogonia cease the process of mitosis, they are called **oocytes**. At that time they enter the meiotic cycle (or meiosis, to prepare for the production of a haploid ovum), become arrested in prophase of the first meiotic division, and remain arrested in that phase until they either die or grow into mature oocytes at the time of ovulation. The **primordial follicle** (Fig. 38.4) is 20 μm in diameter and contains an oocyte, which may or may not be surrounded by a single layer of flattened (squamous) **pregranulosa cells**. When pregranulosa cells surround the oocyte, a basement membrane develops, separating the granulosa from the ovarian stroma.

A Graafian Follicle Is the Final Stage of Follicle Development

Folliculogenesis (also called follicular development) is the process by which follicles develop and mature (see Fig. 38.3). Follicles are in one of the following physiological states: resting, growing, degenerating, or ready to ovulate. During each menstrual cycle, the ovaries produce a group of growing follicles of which most will fail to grow to maturity and will undergo follicular atresia (death) at some stage of development. However, one **dominant follicle** generally emerges from the cohort of developing follicles and it will ovulate, releasing a mature haploid ovum.

Primordial follicles are generally considered the non-growing resting pool of follicles, which gets progressively depleted throughout life; by the time of menopause, the ovaries are essentially devoid of all follicles. Primordial follicles are located in the ovarian cortex (peripheral regions of the ovary) beneath the tunica albuginea.

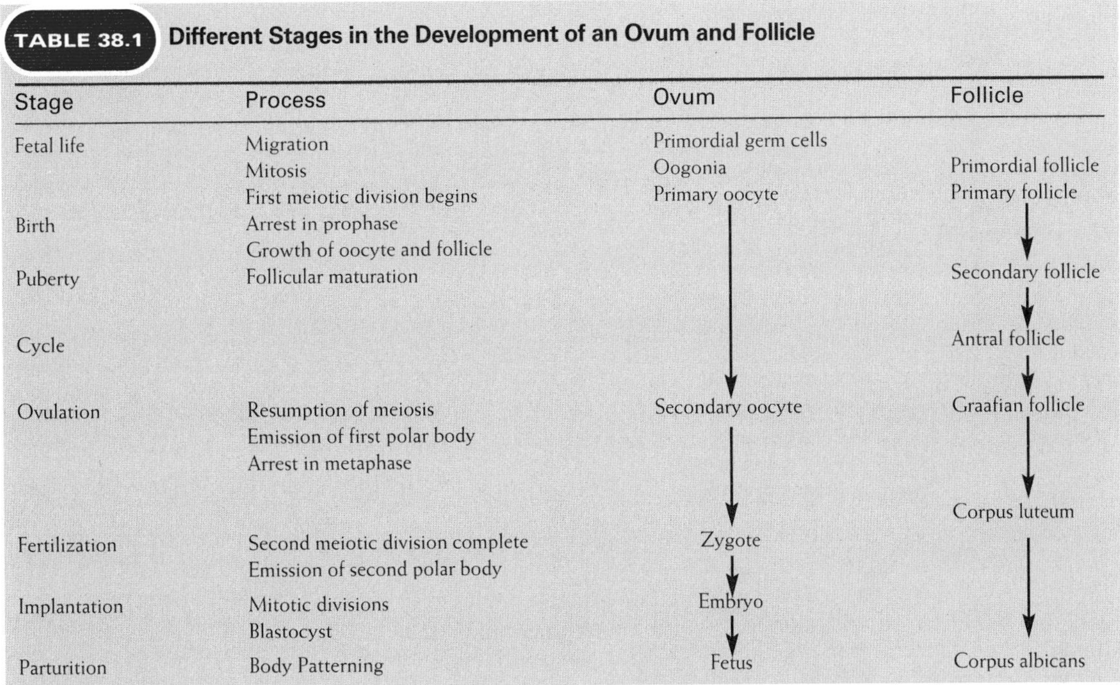

TABLE 38.1 Different Stages in the Development of an Ovum and Follicle

Stage	Process	Ovum	Follicle
Fetal life	Migration Mitosis First meiotic division begins	Primordial germ cells Oogonia Primary oocyte	Primordial follicle Primary follicle
Birth	Arrest in prophase Growth of oocyte and follicle		
Puberty	Follicular maturation		Secondary follicle
Cycle			Antral follicle
Ovulation	Resumption of meiosis Emission of first polar body Arrest in metaphase	Secondary oocyte	Graafian follicle
			Corpus luteum
Fertilization	Second meiotic division complete Emission of second polar body	Zygote	
Implantation	Mitotic divisions Blastocyst	Embryo	
Parturition	Body Patterning	Fetus	Corpus albicans

Progression from primordial to the next stage of follicular development, the primary stage, occurs at a relatively constant rate throughout fetal, juvenile, prepubertal, and adult life. Once primary follicles leave the resting pool, they are committed to further development or atresia. Most become atretic, and typically only one fully developed follicle will ovulate. The conversion from primordial to primary follicles is believed to be independent of pituitary gonadotropins. The exact signal that recruits a follicle from a resting to a growing pool is unknown; it could be programmed by the cell genome or influenced by local ovarian growth regulators.

The first sign that a primordial follicle is entering the growth phase is a morphological change of the flattened pregranulosa cells into cuboidal granulosa cells. The cuboidal granulosa cells proliferate to form a single continuous layer of cells surrounding the oocyte, which has enlarged from 20 μm in the primordial stage to 140 μm in diameter. At this stage, a glassy membrane, the **zona pellucida**, surrounds the oocyte and serves as means of attachment through which the granulosa cells communicate with the oocyte. This is the **primary follicular stage** of development, consisting of one layer of cuboidal granulosa cells and a basement membrane.

The follicle continues to grow, mainly through proliferation of its granulosa cells, so that several layers of granulosa cells exist in the **secondary follicular stage** of development (see Fig. 38.4). As the secondary follicle grows deeper into the cortex, stromal cells, near the basement membrane, begin to differentiate into cell layers called **theca interna** and **theca externa**, and a blood supply with lymphatics and nerves forms within the thecal component. The granulosa layer remains avascular.

The theca interna cells become flattened, epithelioid, and steroidogenic. The granulosa cells of secondary follicles acquire receptors for FSH and start producing small amounts of estrogen. The theca externa remains fibroblastic and provides structural support to the developing follicle.

Development beyond the primary follicle is gonadotropin-dependent, begins at puberty, and continues in a cyclic manner throughout the reproductive years. As the follicle continues to grow, theca layers expand, and fluid-filled spaces or **antra** begin to develop around the granulosa cells. This early antral stage of follicle development is referred to as the **tertiary follicular stage** (see Fig. 38.4). The critical hormone responsible for progression from the preantral to the antral stage is FSH. Mitosis of the granulosa cells is stimulated by FSH. As the number of granulosa cells increases, the production of estrogens, the binding capacity for FSH, the size of the follicle, and the volume of the follicular fluid all increase significantly.

As the antra increase in size, a single, large, coalesced antrum develops, pushing the oocyte to the periphery of the follicle and forming a large 2- to 2.5-cm-diameter **graafian follicle** (preovulatory follicle; see Fig. 38.4). Three distinct granulosa cell compartments are evident in the graafian follicle. Granulosa cells surrounding the oocyte are **cumulus granulosa cells** (collectively called cumulus oophorus). Those cells lining the antral cavity are called **antral granulosa cells** and those attached to the basement membrane are called **mural granulosa cells**. Mural and antral granulosa cells are more steroidogenically active than cumulus cells.

In addition to bloodborne hormones, antral follicles have a unique microenvironment in which the follicular fluid contains different concentrations of pituitary hormones, steroids, peptides, and growth factors. Some are present in the follicular fluid at a concentration 100 to 1,000 times higher than in the circulation. Table 38.2 lists some parameters of human follicles at successive stages of development in

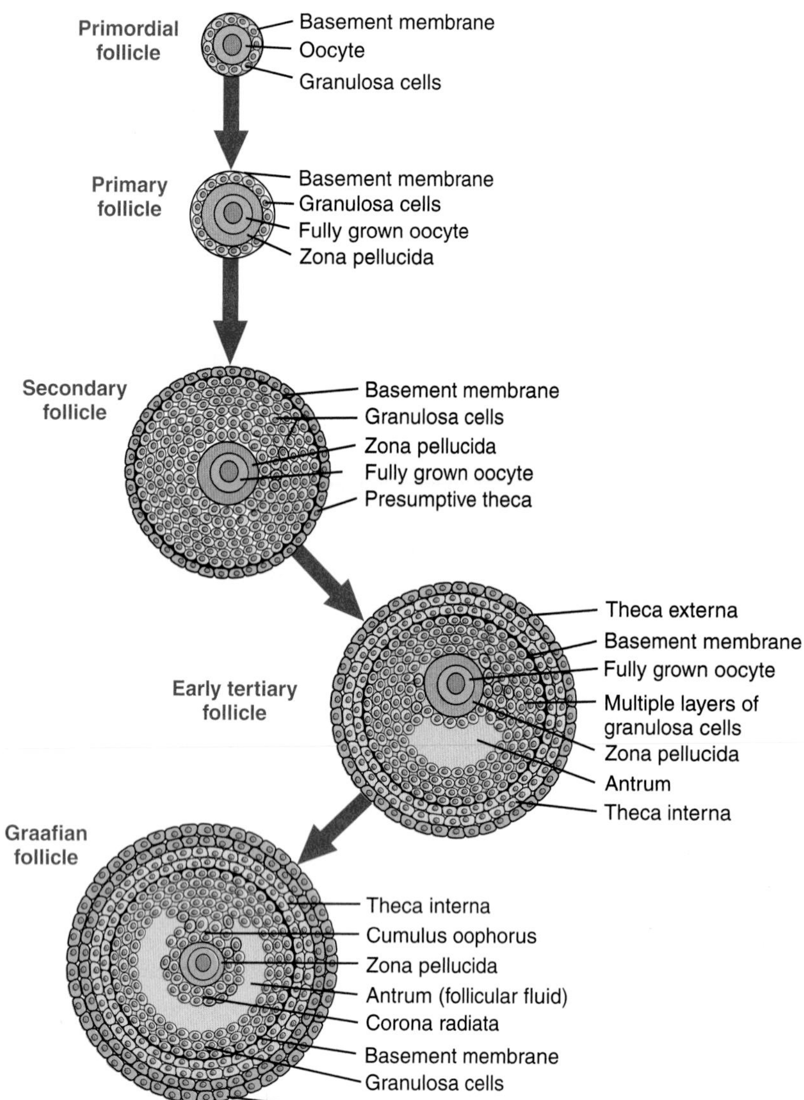

FIGURE 38.4 The developing follicle, from primordial through graafian. (Modified from Erickson GF. In: Sciarra JJ, Speroff L, eds. Reproductive Endocrinology, Infertility, and Genetics. New York: Harper & Row, 1981.)

the follicular phase. There is a 5-fold increase in follicular diameter and a 25-fold rise in the number of granulosa cells. As the follicle matures, the intrafollicular concentration of FSH does not change much, whereas that of LH increases and that of PRL declines. Of the steroids, the concentrations of estradiol and progesterone increase 20-fold, while androgen levels remain unchanged.

The follicular fluid contains other substances, including inhibin, activin, GnRH-like peptide, growth factors, opioid peptides, oxytocin, and plasminogen activator. Inhibin and activin inhibit and stimulate, respectively, the release of FSH from the anterior pituitary. Inhibin is secreted by granulosa cells. In addition to its effect on FSH secretion, inhibin also has a local effect on ovarian cells.

TABLE 38.2 Different Parameters of Follicles During the First Half of the Menstrual Cycle

Cycle (day)	Diameter (mm)	Volume (mL)	Granulosa Cells ($\times 10^6$)	Follicular Fluid FSH (ng/mL)	LH (ng/mL)	PRL (ng/mL)	A (ng/mL)	E2 (ng/mL)	P4 (ng/mL)
1	4	0.05	2	2.5	—	60	800	100	—
4	7	0.15	5	2.5	—	40	800	500	100
7	12	0.50	15	3.6	2.8	20	800	1,000	300
12	20	0.50	50	3.6	2.8	5	800	2,000	2,000

FSH, follicle-stimulating hormone; LH, luteinizing hormone; PRL, prolactin; A, androstenedione; E2, estradiol; P4, progesterone. (Modified from Erickson GF. An analysis of follicle development and ovum maturation. Semin Reprod Endocrinol 1986;4:233–254.)

Granulosa and Theca Cells Both Participate in Steroidogenesis

The main physiologically active steroid produced by the follicle is **estradiol**, a steroid with 18 carbons. **Steroidogenesis**, the process of steroid hormone production, depends on the availability of cholesterol, which originates from several sources and serves as the main precursor for all of steroidogenesis. Ovarian cholesterol can come from plasma lipoproteins, *de novo* synthesis in ovarian cells, and cholesterol esters within lipid droplets in ovarian cells. For ovarian steroidogenesis, the primary source of cholesterol is low-density lipoprotein (LDL).

The conversion of cholesterol to **pregnenolone** by **cholesterol side-chain cleavage enzyme** is a rate-limiting step regulated by LH using the second messenger cAMP (Fig. 38.5). LH binds to specific membrane receptors on theca cells, activates adenylyl cyclase through a G protein, and increases the production of cAMP. cAMP increases LDL receptor mRNA, the uptake of LDL cholesterol, and cholesterol ester synthesis. cAMP also increases the transport of cholesterol from the outer to the inner mitochondrial membrane, the site of pregnenolone synthesis, using a unique protein called **steroidogenic acute regulatory protein** (StAR). Pregnenolone, a 21-carbon steroid of the progestin family, diffuses out of the mitochondria and enters the ER, the site of subsequent steroidogenesis.

Two steroidogenic pathways may be used for subsequent steroidogenesis (see Fig. 37.9). In theca cells, the **delta 5 pathway** is predominant; in granulosa cells and the corpus luteum, the **delta 4 pathway** is predominant. Pregnenolone gets converted to either **progesterone** by 3β-hydroxysteroid dehydrogenase in the delta 4 pathway or to **17α-hydroxypregnenolone** by 17α-hydroxylase in the delta 5 pathway. In the delta 4 pathway, progesterone gets converted to **17α-hydroxyprogesterone** (by 17α-hydroxylase), which is subsequently converted to **androstenedione** and **testosterone** by 17,20-lyase and 17β-hydroxysteroid dehydrogenase (17-ketosteroid reductase), respectively. In the delta 5 pathway, 17α-hydroxypregnenolone gets converted to **dehydroepiandrosterone** (by 17,20-lyase), which is subsequently converted to androstenedione by 3β-hydroxysteroid dehydrogenase. The androgens contain 19 carbons. Testosterone and androstenedione diffuse from the thecal compartment, cross the basement membrane, and enter the granulosa cells.

In the granulosa cell, under the influence of FSH, with cAMP as a second messenger, testosterone and androstenedione are then converted to estradiol and **estrone**, respectively, by the enzyme aromatase, which aromatizes the A ring of the steroid and removes one carbon (see Fig. 38.5; see Fig. 37.9). Estrogens typically have 18 carbons. Estrone can then be converted to estradiol by 17β-hydroxysteroid dehydrogenase in granulosa cells.

In summary, estradiol secretion by the follicle requires cooperation between granulosa and theca cells and coordination between FSH and LH. An understanding of this **two-cell, two-gonadotropin hypothesis** requires recognition that the actions of FSH are restricted to granulosa cells because all other ovarian cell types lack FSH receptors. LH actions, on the other hand, are exerted on theca, granulosa, and stromal (interstitial) cells and the corpus luteum. The expression of LH receptors is time-dependent because theca cells acquire LH receptors at a relatively early stage, whereas LH receptors on granulosa cells are induced by FSH in the later stages of the maturing follicle.

The biosynthetic enzymes are differentially expressed in the two cells. Aromatase is expressed only in granulosa cells, and its activation and induction are regulated by FSH. Granulosa cells are deficient in 17α-hydroxylase and cannot proceed beyond the C-21 progestins to generate C-19 androgenic compounds (see Fig. 38.5). Consequently, estrogen production by granulosa cells depends on an adequate supply of exogenous aromatizable androgens, provided by theca cells. Under LH regulation, theca cells produce androgenic substrates, primarily androstenedione and testosterone, which reach the granulosa cells by diffusion. The androgens are then converted to estrogens by aromatization.

In follicles, theca and granulosa cells are exposed to different microenvironments. Vascularization is restricted to the theca layer because blood vessels do not penetrate the

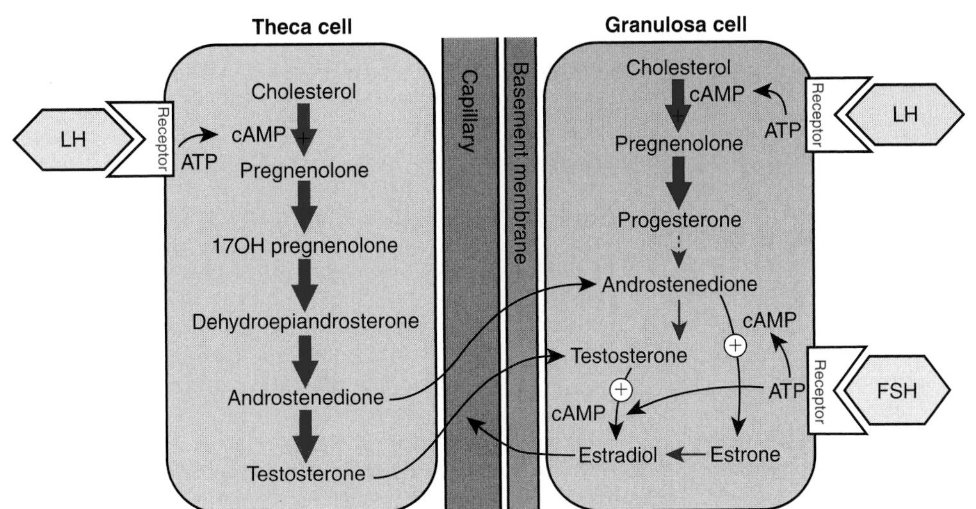

FIGURE 38.5 **The two-cell, two-gonadotropin hypothesis.** The follicular theca cells, under control of LH, produce androgens that diffuse to the follicular granulosa cells, where they are converted to estrogens via an FSH-supported aromatization reaction. The dashed arrow indicates that granulosa cells cannot convert progesterone to androstenedione because of the lack of the enzyme 17α-hydroxylase.

basement membrane. Theca cells, therefore, have better access to circulating cholesterol, which enters the cells via LDL receptors. Granulosa cells, on the other hand, primarily produce cholesterol from acetate, a less efficient process than uptake. In addition, granulosa cells are bathed in follicular fluid and exposed to autocrine, paracrine, and juxtacrine control by locally produced peptides and growth factors. "Juxtacrine" describes the interaction of a membrane-bound growth factor on one cell with its membrane-bound receptor on an adjacent cell.

FSH acts on granulosa cells by a cAMP-dependent mechanism and produces a broad range of activities, including increased mitosis and cell proliferation, the stimulation of progesterone synthesis, the induction of aromatase, and increased inhibin synthesis. As the follicle matures, the number of receptors for both gonadotropins increases. FSH stimulates the formation of its own receptors and induces the appearance of LH receptors. The combined activity of the two gonadotropins greatly amplifies estrogen production.

Androgens are produced by theca and stromal cells. They serve as precursors for estrogen synthesis and also have a distinct local action. At low concentrations, androgens enhance aromatase activity, promoting estrogen production. At high concentrations, androgens are converted by 5α-reductase to a more potent androgen, such as dihydrotestosterone (DHT). When follicles are overwhelmed by androgens, the intrafollicular androgenic environment antagonizes granulosa cell proliferation and leads to apoptosis of the granulosa cells and subsequent follicular atresia.

Follicular Atresia Probably Results From a Lack of Gonadotropin Support

Follicular **atresia**, the degeneration of follicles in the ovary, is characterized by the destruction of the oocyte and granulosa cells. Atresia is a continuous process and can occur at any stage of follicular development. During a woman's lifetime approximately 400 to 500 follicles will ovulate; those are the only follicles that escape atresia, and they represent a small percentage of the 1 to 2 million follicles present at birth. The cause of follicular atresia is likely due to lack of gonadotropin support of the growing follicle. For example, at the beginning of the menstrual cycle, several follicles are selected for growth but only one follicle, the dominant follicle, will go on to ovulate. Because the dominant follicle has a preferential blood supply, it gets the most FSH (and LH). Other reasons for the lack of gonadotropin support of nondominant follicles could be a lack of FSH and LH receptors or the inability of granulosa cells to transduce the gonadotropin signals.

During atresia, granulosa cell nuclei become **pyknotic** (referring to an apoptotic process characterized by DNA laddering), and/or the oocyte undergoes **pseudomaturation**, characteristic of meiosis. During the early stages of oocyte death, the nuclear membrane disintegrates, the chromatin condenses, and the chromosomes form a metaphase plate with a spindle; the term *pseudomaturation* is appropriate because these oocytes are not capable of successful fertilization. During atresia of follicles containing theca cells, the theca layer may undergo hyperplasia and hypertrophy and may remain in the ovary for extended periods of time.

Meiosis Resumes During the Periovulatory Period

All healthy oocytes in the ovary remain arrested in prophase of the first meiosis. When a graafian follicle is subjected to a surge of gonadotropins (LH and/or FSH), the oocyte within undergoes the final stages of meiosis, resulting in the production of a mature gamete. This maturation is accomplished by two successive cell divisions in which the number of chromosomes is reduced, producing haploid gametes. At fertilization, the diploid state is restored.

Primary oocytes arrested in meiotic prophase 1 (of the first meiosis) have duplicated their centrioles and DNA (4n DNA) so that each chromosome has two identical **chromatids**. Crossing over and chromatid exchange occur during this phase, producing genetic diversity. The resumption of meiosis, ending the first meiotic prophase and beginning of meiotic metaphase 1, is characterized by disappearance of the nuclear membrane, condensation of the chromosomes, nuclear dissolution (germinal vesicle breakdown), and alignment of the chromosomes on the equator of the spindle. At meiotic anaphase 1, the homologous chromosomes move in opposite directions under the influence of the retracting meiotic spindle at the cellular periphery. At meiotic telophase 1, an unequal division of the cell cytoplasm yields a large **secondary oocyte** (2n DNA) and a small, nonfunctional cell, the **first polar body** (2n DNA). Each cell contains half the original 4n number of chromosomes (only one member of each homologous pair is present, but each chromosome consists of two unique chromatids).

The secondary oocyte is formed several hours after the initiation of the LH surge but before ovulation. It rapidly begins the second meiotic division and proceeds through a short prophase to become arrested in metaphase. At this stage, the secondary oocyte is expelled from the graafian follicle. The second arrest period is relatively short. In response to penetration by a spermatozoon during fertilization, meiosis 2 resumes and is rapidly completed. A second unequal cell division soon follows, producing a small **second polar body** (1n DNA) and a large fertilized egg, the **zygote** (2n DNA, 1n from the mother and 1n from the father). The first and second polar bodies either degenerate or divide, yielding small nonfunctional cells. If fertilization does not occur, the secondary oocyte begins to degenerate within 24 to 48 hours.

FOLLICLE SELECTION AND OVULATION

The number of ovulating eggs is species-specific and is influenced by genetic, nutritional, and environmental factors. In humans, normally only one follicle will ovulate, but multiple ovulations in a single cycle (superovulation) can be induced by the timed administration of gonadotropins or antiestrogens. The mechanism by which one follicle is selected from a cohort of growing follicles is poorly understood. It occurs during the first few days of the cycle, immediately after the onset of menstruation. Once selected, the follicle begins to grow and differentiate at an exponential rate and becomes the dominant follicle.

In parallel with the growth of the dominant follicle, the rest of the preantral follicles undergo atresia. Two main factors contribute to atresia in the nonselected follicles. One is the suppression of plasma FSH in response to increased estradiol secretion by the dominant follicle. The decline in FSH support decreases aromatase activity and estradiol production and interrupts granulosa cell proliferation in those nondominant follicles. The dominant follicle is protected from a fall in circulating FSH levels because it has a healthy blood supply, FSH accumulated in the follicular fluid, and an increased density of FSH receptors on its granulosa cells. Another factor in selection is the accumulation of atretogenic androgens, such as DHT, in the nonselected follicles. The increase in DHT changes the intrafollicular ratio of estrogen to androgen and antagonizes the actions of FSH.

As the dominant follicle grows, vascularization of the theca layer increases. On day 9 or 10 of the cycle, the vascularity of the dominant follicle is twice that of the other antral follicles, permitting a more efficient delivery of cholesterol to theca cells and better exposure to circulating gonadotropins. At this time, the main source of circulating estradiol is the dominant follicle. Since estradiol is the primary regulator of LH and FSH secretion by positive and negative feedback, the dominant follicle ultimately determines its own fate.

The midcycle LH surge occurs as a result of rising levels of circulating estradiol, and it causes multiple changes in the dominant follicle, which occur within a relatively short time. These include the resumption of meiosis in the oocyte (as already discussed); granulosa cell differentiation and transformation into luteal cells; the activation of proteolytic enzymes that degrade the follicle wall and surrounding tissues; increased production of prostaglandins, histamine, and other local factors that cause localized hyperemia; and an increase in progesterone secretion. Within 30 to 36 hours after the onset of the LH surge, this coordinated series of biochemical and morphological events culminates in follicular rupture and ovulation. The midcycle FSH surge is not essential for ovulation because an injection of either LH or human chorionic gonadotropin (hCG) before the endogenous gonadotropin surge can induce normal ovulation. However, only follicles that have been adequately primed with FSH will ovulate because they contain sufficient numbers of LH receptors for ovulation and subsequent luteinization.

Four ovarian proteins are essential for ovulation: the progesterone receptor, the cyclooxygenase enzyme (which converts arachidonic acid to prostaglandins), cyclin D2 (a cell cycle regulator), and a transcription factor called C/EBPβ (CCAAT/enhancer binding protein). The mechanisms by which these proteins interact to regulate follicular rupture are largely unknown. However, mice with specific disruption of genes for any of these proteins fail to ovulate, and these proteins are likely to have a functional role in human ovulation.

The earliest responses of the ovary to the midcycle LH surge are the release of vasodilatory substances, such as histamine, bradykinin, and prostaglandins, which mediate increased ovarian and follicular blood flow. The highly vascularized dominant follicle becomes hyperemic and edematous and swells to a size of at least 20 to 25 mm in diameter. There is also an increased production of follicular fluid, disaggregation of granulosa cells, and detachment of the oocyte-cumulus complex from the follicular wall, moving it to the central portion of the follicle. The basement membrane separating theca cells from granulosa cells begins to disintegrate, granulosa cells begin to undergo luteinization, and blood vessels begin to penetrate the granulosa cell compartment.

Just prior to follicular rupture, the follicular wall thins by cellular deterioration and bulges at a specific site called the **stigma**, the point on the follicle that actually ruptures. As ovulation approaches, the follicle enlarges and protrudes from the surface of the ovary at the stigma. In response to the LH surge, **plasminogen activator** is produced by theca and granulosa cells of the dominant follicle and converts plasminogen to **plasmin**. Plasmin is a proteolytic enzyme that acts directly on the follicular wall and stimulates the production of **collagenase**, an enzyme that digests the connective tissue matrix. The thinning and increased distensibility of the wall facilitates the rupture of the follicle. The extrusion of the oocyte-cumulus complex is aided by smooth muscle contraction. At the time of rupture, the oocyte-cumulus complex and follicular fluid are ejected from the follicle.

The LH surge triggers the resumption of the first meiosis. Up to this point, the primary oocyte has been protected by unknown factors within the follicle from premature cell division. The LH surge also causes transient changes in plasma estradiol and a prolonged increase in plasma progesterone concentrations. Within a couple of hours after the initiation of the LH surge, the production of progesterone, androgens, and estrogens begins to increase. Progesterone, acting through the progesterone receptor on granulosa cells, promotes ovulation by releasing mediators that increase the distensibility of the follicular wall and enhance the activity of proteolytic enzymes. As LH levels reach their peak, plasma estradiol levels plunge because of down-regulation by LH of FSH receptors on granulosa cells and the inhibition of granulosa cell aromatase. Eventually, LH receptors on luteinizing granulosa cells escape the down-regulation, and progesterone production increases.

FORMATION OF THE CORPUS LUTEUM FROM THE POSTOVULATORY FOLLICLE

In response to the LH and FSH surges and after ovulation, the wall of the graafian follicle collapses and becomes convoluted, blood vessels course through the luteinizing granulosa and theca cell layers, and the antral cavity fills with blood. The granulosa cells begin to cease their proliferation and begin to undergo hypertrophy and produce progesterone as their main secretory product. The ruptured follicle develops a rich blood supply and forms a solid structure called the **corpus luteum** (yellow body). The mature corpus luteum develops as the result of numerous biochemical and morphological changes, collectively referred to as **luteinization**. The granulosa cells and theca cells in the corpus luteum are called **granulosa lutein cells** and **theca lutein cells**, respectively.

Continued stimulation by LH is needed to ensure morphological integrity (healthy luteal cells) and functionality (progesterone secretion). If pregnancy does not occur, the corpus luteum regresses, a process called **luteolysis** or **luteal regression**. Luteolysis occurs as a result of apoptosis and necrosis of the luteal cells. After degeneration, the luteinized cells are replaced by fibrous tissue, creating a nonfunctional structure, the **corpus albicans**. Therefore, the corpus luteum is a transient endocrine structure formed from the postovulatory follicle. It serves as the main source of circulating steroids during the luteal (postovulatory) phase of the cycle and is essential for maintaining pregnancy during the first trimester (see Case Study) as well as maintaining menstrual cycles of normal length.

The process of luteinization begins before ovulation. After acquiring a high concentration of LH receptors, granulosa cells respond to the LH surge by undergoing morphological and biochemical transformation. This change involves cell enlargement (hypertrophy) and the development of smooth ER and lipid inclusions, typical of steroid-secreting cells. Unlike the nonvascular granulosa cells in the follicle, luteal cells have a rich blood supply. Invasion by capillaries starts immediately after the LH surge and is facilitated by the dissolution of the basement membrane between theca and granulosa cells. Peak vascularization is reached 7 to 8 days after ovulation.

Differentiated theca and stroma cells, as well as granulosa cells, are incorporated into the corpus luteum, and all three classes of steroids—androgens, estrogens, and progestins—are synthesized. Although some progesterone is secreted before ovulation, peak progesterone production is reached 6 to 8 days after the LH surge. The life span of the corpus luteum is limited. Unless pregnancy occurs, it degenerates within about 13 days after ovulation. During the menstrual cycle, the function of the corpus luteum is maintained by LH; therefore, LH is referred to as a **luteotropic hormone**. Lack of LH can lead to luteal insufficiency (see Clinical Focus Box 38.1).

Regression of the corpus luteum at the end of the cycle is not understood. Luteal regression is thought to be induced by locally produced luteolytic agents that inhibit LH action. Several ovarian hormones, such as estrogen, oxytocin, prostaglandins, and GnRH, have been proposed, but their role as **luteolysins** is controversial. The corpus luteum is rescued from degeneration in the late luteal phase by the action of human chorionic gonadotropin (hCG), an LH-like hormone that is produced by the embryonic trophoblast during the implantation phase (see Chapter 39). This hormone binds the LH receptor and increases cAMP and progesterone secretion.

THE MENSTRUAL CYCLE

Under normal conditions, ovulation occurs at timed intervals. Sexual intercourse may occur at any time during the cycle, but fertilization occurs only during the postovulatory period. Once pregnancy occurs, ovulation ceases, and after parturition, lactation also inhibits ovulation. The first **menstrual cycle** occurs in adolescence, usually around age 12. The initial period of bleeding is called the **menarche**. The first few cycles are usually irregular and anovulatory, as the result of delayed maturation of the positive feedback by estradiol on a hypothalamus that fails to secrete significant GnRH. During puberty, LH secretion occurs more during periods of sleep than during periods of being awake, resulting in a diurnal cycle.

CLINICAL FOCUS BOX 38.1

Luteal Insufficiency

Occasionally, the corpus luteum will not produce sufficient progesterone to maintain pregnancy during its very early stages. Initial signs of early spontaneous pregnancy termination include pelvic cramping and the detection of blood, similar to indications of menstruation. If the corpus luteum is truly deficient, then fertilization may occur around the idealized day 14 (ovulation), pregnancy will terminate during the deficient luteal phase, and menses will start on schedule. Without measuring levels of hCG, the pregnancy detection hormone, the woman would not know that she is pregnant because of the continuation of regular menstrual cycles. **Luteal insufficiency** is a common cause of infertility. Women are advised to see their physician if pregnancy does not result after 6 months of unprotected intercourse.

Analysis of the regulation of progesterone secretion by the corpus luteum provides insights into this clinical problem. There are several reasons for luteal insufficiency. First, the number of luteinized granulosa cells in the corpus luteum may be insufficient because of the ovulation of a small follicle or the premature ovulation of a follicle that was not fully developed. Second, the number of LH receptors on the luteinized granulosa cells in the graafian follicle and developing corpus luteum may be insufficient. LH receptors mediate the action of LH, which stimulates progesterone secretion. An insufficient number of LH receptors could be due to insufficient priming of the developing follicle with FSH. It is well known that FSH increases the number of LH receptors in the follicle. Third, the LH surge could have been inadequate in inducing full luteinization of the corpus luteum, yet there was sufficient LH to induce ovulation. It has been estimated that only 10% of the LH surge is required for ovulation, but the amount required for full luteinization and adequate progesterone secretion to maintain pregnancy is not known.

If progesterone values are low in consecutive cycles at the midluteal phase and do not match endometrial biopsies, exogenous progesterone may be administered in order to prevent early pregnancy termination during a fertile cycle. Other options include the induction of follicular development and ovulation with clomiphene and hCG. This treatment would likely produce a large, healthy, estrogen-secreting graafian follicle with sufficient LH receptors for luteinization. The exogenous hCG is given to supplement the endogenous LH surge and to ensure full stimulation of the graafian follicle, ovulation, adequate progesterone, and luteinization of the developing corpus luteum.

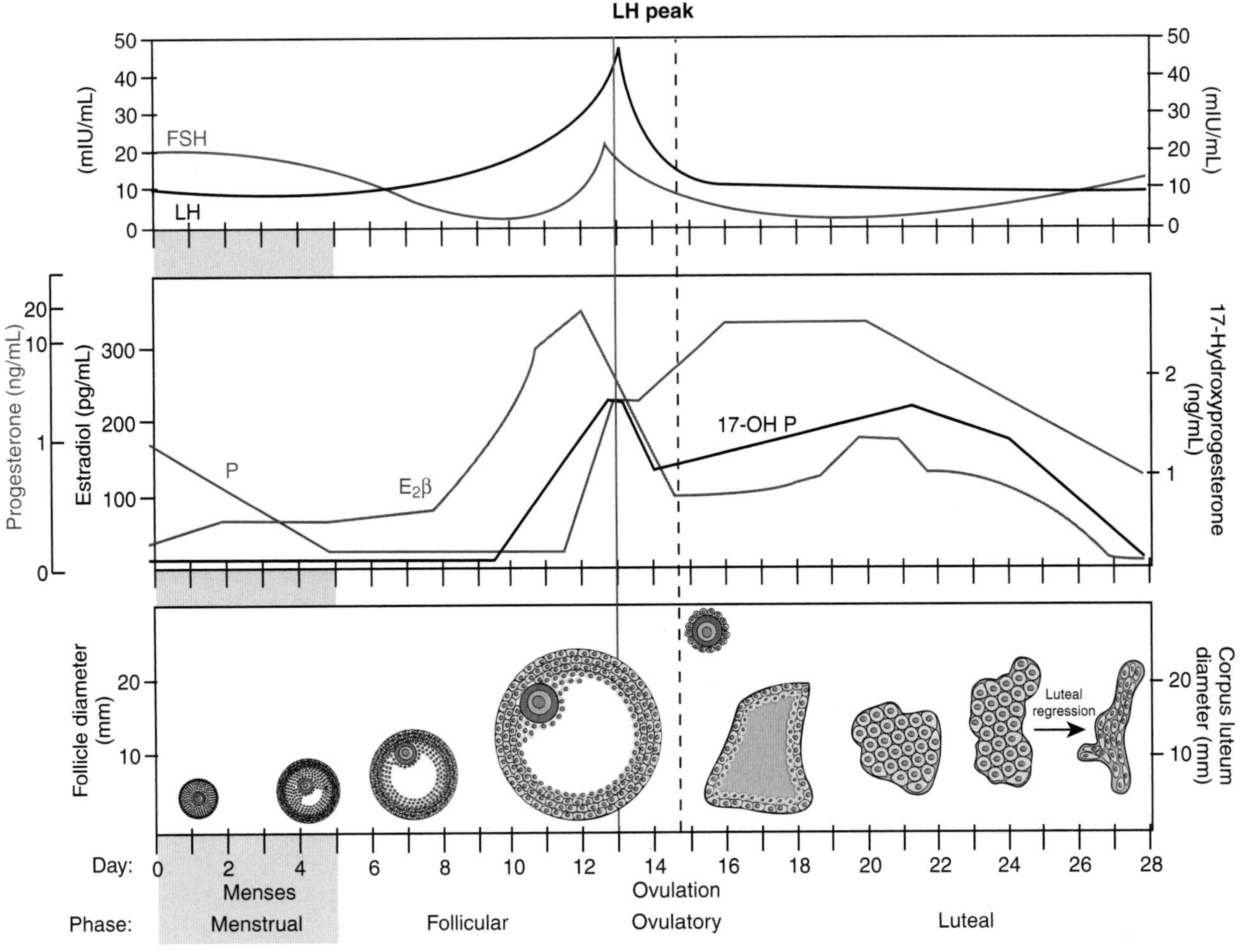

FIGURE 38.6 **Hormonal and ovarian events during the menstrual cycle.** P, progesterone; $E_2\beta$, estradiol; 17-OH P, 17-hydroxyprogesterone.

The average menstrual cycle length in adult women is 28 days, with a range of 25 to 35 days. The interval from ovulation to the onset of menstruation is relatively constant, averaging 14 days in most women and is dictated by the fixed life span of the corpus luteum. In contrast, the interval from the onset of menses to ovulation (the follicular phase) is more variable and accounts for differences in cycle lengths among ovulating women.

The menstrual cycle is divided into four phases (Fig. 38.6). The **menstrual phase**, also called **menses** or **menstruation**, is the bleeding phase and lasts about 5 days. The ovarian **follicular phase** lasts about 10 to 16 days; follicle development occurs, estradiol secretion increases, and the uterine endometrium undergoes proliferation in response to rising estrogen levels. The **ovulatory phase** lasts 24 to 48 hours, and the **luteal phase** lasts 14 days. In the luteal phase, progesterone is produced, and the endometrium secretes numerous proteins in preparation for implantation of an embryo.

The cycles become irregular as menopause approaches around age 50, and cycles cease thereafter. During the reproductive years, menstrual cycling is interrupted by conception and lactation and is subjected to modulation by physiological, psychological, and social factors.

The Menstrual Cycle Requires Synchrony Among the Ovary, Brain, and Pituitary

The menstrual cycle requires several coordinated elements: hypothalamic control of pituitary function, ovarian follicular and luteal changes, and positive and negative feedback of ovarian hormones at the hypothalamic-pituitary axis. We have discussed separately the mechanisms that regulate the synthesis and release of the reproductive hormones; now we put them together in terms of sequence and interaction. For this purpose, we use a hypothetical cycle of 28 days (see Fig. 38.6), divided into four phases as follows: menstrual (days 0 to 5), follicular (days 0 to 13), ovulatory (days 13 to 14), and luteal (days 14 to 28).

During menstruation, estrogen, progesterone, and inhibin levels are very low as a result of the luteal regression that has just occurred and the low estrogen synthesis by immature follicles. The plasma FSH levels are high while LH levels are low in response to the removal of negative feedback by estrogen, progesterone, and inhibin. A few days later, however, LH levels slowly begin to rise. FSH acts on a cohort of follicles recruited 20 to 25 days earlier from a resting pool of smaller follicles. The follicles on days 3 to 5 average 4 to 6 mm in diameter, and they are stimulated by FSH to grow into the preantral stages. In response to FSH, the granulosa cells proliferate, aromatase activity increases, and plasma estradiol levels rise slightly between days 3 and 7. The designated dominant follicle is selected between days 5 and 7, and increases in size and steroidogenic activity. Between days 8 and 10, plasma estradiol levels rise sharply, reaching peak levels above 200 pg/mL on day 12, the day before the LH surge.

During the early follicular phase, LH pulsatility is of low amplitude and high frequency (about every hour). Coinciding pulses of GnRH are released about every hour. As estradiol levels rise, the pulse frequency in GnRH further increases, without a change in amplitude. The mean plasma LH level increases and further supports follicular steroidogenesis, especially since FSH has increased the number of LH receptors on growing follicles. During the midfollicular to late follicular phase, rising estradiol and inhibin from the dominant follicle suppress FSH release. The decline in FSH, together with an accumulation of nonaromatizable androgens, induces atresia in the nonselected follicles. The dominant follicle is saved by virtue of its high density of FSH receptors, the accumulation of FSH in its follicular fluid (see Table 38.2), and the acquisition of LH receptors by the granulosa cells.

The midcycle surge of LH is rather short (24 to 36 hours) and is an example of positive feedback. For the LH surge to occur, estradiol must be maintained at a critical concentration (about 200 pg/mL) for a sufficient duration (36 to 48 hours) prior to the surge. Any reduction of the estradiol rise or a rise that is too small or too short eliminates or reduces the LH surge. In addition, in the presence of elevated progesterone, high concentrations of estradiol do not induce an LH surge. Paradoxically, although it exerts negative feedback on LH release most of the time, positive feedback by estradiol is required to generate the midcycle surge.

Estrogen exerts its effects directly on the anterior pituitary, with GnRH playing a permissive, albeit mandatory, role. This concept is derived from experiments in monkeys whose medial basal hypothalamus, including the GnRH-producing neurons, was destroyed by lesioning, resulting in a marked decrease in plasma LH levels. The administration of exogenous GnRH at a fixed frequency restored LH release. When estradiol was given at an optimal concentration for an appropriate time, an LH surge was generated, in spite of maintaining steady and unchanging pulses of GnRH.

The mechanism that transforms estradiol from a negative to a positive regulator of LH release is unknown. One factor involves an increase in the number of GnRH receptors on the gonadotrophs, increasing pituitary responsiveness to GnRH. Another factor is the conversion of a storage pool of LH (perhaps within a subpopulation of gonadotrophs) to a readily releasable pool. Estrogen may also increase GnRH release, serving as a fine-tuning or fail-safe mechanism. A small but distinct rise in progesterone occurs before the LH surge. This rise is important for augmenting the LH surge and, together with estradiol, promotes a concomitant surge in FSH. There are indications that the midcycle FSH surge is important for inducing enough LH receptors on granulosa cells for luteinization, stimulating plasminogen activator for follicular rupture, and activating a cohort of follicles destined to develop in the next cycle.

The LH surge reduces the concentration of 17α-hydroxylase and subsequently decreases androstenedione production by the dominant follicle. Estradiol levels decline, 17-hydroxyprogesterone increases, and progesterone levels plateau. The prolonged exposure to high LH levels during the surge down-regulates the ovarian LH receptors, accounting for the immediate postovulatory suppression of estradiol. As the corpus luteum matures, it increases progesterone production and reinitiates estradiol secretion. Both reach high plasma concentrations on days 20 to 23, about 1 week after ovulation.

During the luteal phase, circulating FSH levels are suppressed by the elevated steroids. The LH pulse frequency is reduced during the early luteal phase, but the amplitude is higher than that during the follicular phase. LH is important at this time for maintaining the function of the corpus luteum and sustaining steroid production. In the late luteal phase, both LH pulse frequency and amplitude are reduced by a progesterone-dependent, opioid-mediated suppression of the GnRH pulse generator.

After the demise of the corpus luteum on days 24 to 26, estradiol and progesterone levels plunge, causing the withdrawal of support of the uterine endometrium, culminating within 2 to 3 days in menstruation. The reduction in ovarian steroids acts centrally to remove feedback inhibition. The FSH level begins to rise and a new cycle is initiated.

Estradiol and Progesterone Influence Cyclic Changes in the Reproductive Tract

The female reproductive tract undergoes cyclic alterations in response to the changing levels of ovarian steroids. The most notable changes occur in the function and histology of the oviduct and uterine endometrium, the composition of cervical mucus, and the cytology of the vagina (Fig. 38.7). At the time of ovulation, there is also a small but detectable rise in basal body temperature, caused by progesterone. All of the above parameters are clinically useful for diagnosing menstrual dysfunction and infertility.

The oviduct is a muscular tube lined internally with a ciliated, secretory, columnar epithelium with a deeper stromal tissue. Fertilization occurs in the oviduct, after which the zygote enters the uterus; therefore, the oviduct is involved in transport of the gametes and provides a site for fertilization and early embryonic development. Estrogens maintain the ciliated nature of the epithelium, and ovariectomy causes a loss of the cilia. Estrogens also increase the motility of the oviducts. Exogenous estrogen given around the time of fertilization can cause premature expulsion of the fertilized egg, whereas extremely high doses of estrogen can cause "tube locking," the entrapment of the fertilized egg and an ectopic pregnancy. Progesterone opposes these actions of estrogen.

The endometrium (also called uterine mucosa) is composed of a superficial layer of epithelial cells and an underlying stromal layer. The epithelial layer contains glands that penetrate the stromal layer. The glands are lined by a secretory columnar epithelium.

The **endometrial cycle** consists of four phases. The **proliferative phase** coincides with the midfollicular to late follicular phase of the menstrual cycle. Under the influence of the rising plasma estradiol concentration, the stromal and epithelial layers of the uterine endometrium undergo hyperplasia and hypertrophy and increase in size and thickness. The endometrial glands elongate and are lined with columnar epithelium. The endometrium becomes vascularized, and more spiral arteries, a rich blood supply to this region, develop. Estradiol also induces the formation of progesterone receptors and increases myometrial excitability and contractility.

The **secretory phase** begins on the day of ovulation and coincides with the early to midluteal phase of the menstrual cycle. The endometrium contains numerous progesterone receptors. Under the combined action of progesterone and estrogen, the endometrial glands become coiled, store glycogen, and secrete large amounts of carbohydrate-rich mucus. The stroma increases in vascularity and becomes edematous, and the spiral arteries become tortuous (see Fig. 38.7). Peak secretory activity, edema formation, and overall thickness of the endometrium are reached on days 6 to 8 after ovulation in preparation for implantation of the blastocyst. Progesterone antagonizes the effect of estrogen on the myometrium and reduces spontaneous myometrial contractions.

The **ischemic phase**, generally not depicted graphically, occurs immediately before the menses and is initiated by the declining levels of progesterone and estradiol caused by regression of the corpus luteum. Necrotic changes and abundant apoptosis occur in the secretory epithelium as it collapses. The arteries constrict, reducing the blood supply to the superficial endometrium. Leukocytes and macrophages invade the stroma and begin to phagocytose the ischemic tissue. Leukocytes persist in large numbers throughout menstruation, providing resistance against infection to the denuded endometrial surface.

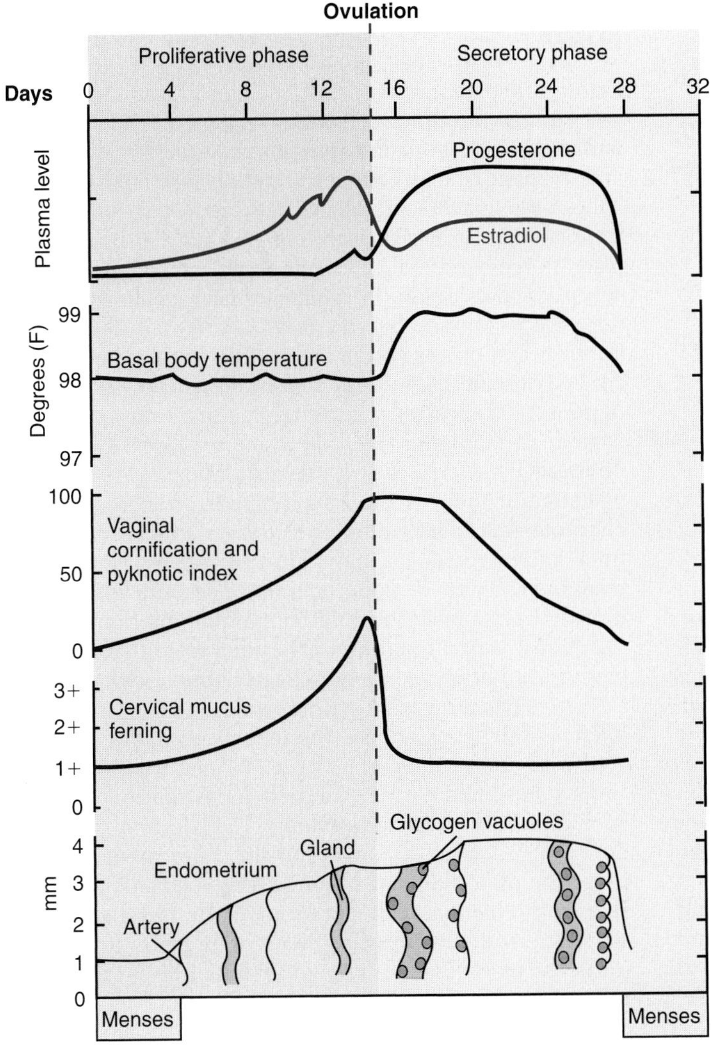

FIGURE 38.7 **Cyclic changes in the uterus, cervix, vagina, and body temperature in relationship to estradiol, progesterone, and ovulation during the menstrual cycle.** (Modified from Odell WD. The reproductive system in women. In: Degroot LJ, et al., eds. Endocrinology. Vol. 3. New York: Grune & Stratton, 1979.)

Desquamation and sloughing of the entire functional layer of the endometrium occurs during the **menstrual phase** (menses). The mechanism leading to necrosis is only partly understood. The reduction in steroids destabilizes lysosomal membranes in endometrial cells, resulting in the liberation of proteolytic enzymes and increased production of vasoconstrictor prostaglandins (e.g., $PGF_{2\alpha}$). The prostaglandins induce vasospasm of the spiral arteries, and the proteolytic enzymes digest the tissue. Eventually, the blood vessels rupture and blood is released, together with cellular debris. The endometrial tissue is expelled through the cervix and vagina, with blood from the ruptured arteries. The menstrual flow lasts 4 to 5 days and averages 30 to 50 mL in volume. It does not clot because of the presence of fibrinolysin, but the spiral arteries constrict, resulting in a reduction in bleeding.

Changes in the properties of the cervical mucus promote the survival and transport of sperm and, thus, can be important for normal fertility. The cervical mucus undergoes cyclic changes in composition and volume. During the follicular phase, estrogen increases the quantity, alkalinity, viscosity, and elasticity of the mucus. The cervical muscles relax, and the epithelium becomes secretory in response to estrogen. By the time of ovulation, elasticity of the mucus or spinnbarkeit is greatest. Sperm can readily pass through the estrogen-dominated mucus. With progesterone rising either after ovulation, during pregnancy, or with low-dose progestogen administration during the cycle, the quantity and elasticity of the mucus decline; it becomes thicker (low spinnbarkeit) and does not form a ferning pattern when dried on a microscope slide. With these conditions, the mucus provides better protection against infections and sperm do not easily pass through.

The vaginal epithelium proliferates under the influence of estrogen. Basophilic cells predominate early in the follicular phase. The columnar epithelium becomes cornified (keratinized) under the influence of estrogen and reaches its peak in the periovulatory period. During the postovulatory period, progesterone induces the formation of thick mucus, the epithelium becomes infiltrated with leukocytes, and cornification decreases (see Fig. 38.7).

ESTROGEN, PROGESTIN, AND ANDROGEN: TRANSPORT AND METABOLISM

The principal sex steroids in the female are estrogen, progestin, and androgen. Three **estrogens** are present in significant quantities—estradiol, estrone, and estriol. Estradiol is the most abundant and is 12 and 80 times more potent than estrone and estriol, respectively. Much of estrone is derived from peripheral conversion of either androstenedione or estradiol (see Fig 37.9). During pregnancy, large quantities of estriol are produced from dehydroepiandrosterone sulfate after 16α-hydroxylation by the fetoplacental unit (see Chapter 39). Most estrogens are bound to either **albumin** (~60%) with a low affinity or to **sex hormone-binding globulin** (SHBG) (~40%) with high affinity. Estrogens are metabolized in the liver through **oxidation** or conversion to **glucuronides** or **sulfates**. The metabolites are then excreted in the urine.

The most important **progestin** is progesterone. It is secreted in significant amounts during the luteal phase of the menstrual cycle. During pregnancy, the corpus luteum secretes progesterone throughout the first trimester, and the placenta continues progesterone production until parturition. Small amounts of 17-hydroxyprogesterone are secreted along with progesterone. Progesterone binds equally to albumin and to a plasma protein called **corticosteroid-binding protein (transcortin)**. Progesterone is metabolized in the liver to **pregnanediol** and, subsequently, excreted in the urine as a glucuronide conjugate.

Circulating **androgens** in the female originate from the ovaries and adrenals and from peripheral conversion. Androstenedione and dehydroepiandrosterone (DHEA) originate from the adrenal cortex (see Chapter 34), and ovarian theca and stroma cells. Peripheral conversion from androstenedione provides an additional source of testosterone. Testosterone can also be converted in peripheral tissues to **dihydrotestosterone** (DHT) by **5α-reductase**. However, the primary biologically active androgen in women is testosterone. Androgens bind primarily to SHBG and bind to albumin by about half as much. Androgens are also metabolized to water-soluble forms by oxidation, sulfation, or glucuronidation and excreted in the urine.

PUBERTY

During the prepubertal period, the hypothalamic-pituitary-ovarian axis becomes activated—an event known as **gonadarche**—and gonadotropins increase in the circulation and stimulate ovarian estrogen secretion. The increase in gonadotropins is a direct result of increased secretion of GnRH. Factors stimulating the secretion of GnRH include glutamate, norepinephrine, and neuropeptide Y emanating from synaptic inputs to GnRH-producing neurons. In addition, a decrease in γ-aminobutyric acid (GABA), an inhibitor of GnRH secretion, may occur at this time. It is also known that the response of the pituitary to GnRH increases at the time of **puberty**. Collectively, numerous factors control the rise in ovarian estradiol secretion that triggers the development of physical characteristics of sexual maturation.

Estradiol induces the development of **secondary sex characteristics**, including the breasts and reproductive tract, and increased fat in the hips. Estrogens also regulate the growth spurt at puberty, induce closure of the epiphyses, have a positive effect in maintaining bone formation, and can antagonize the degrading actions of parathyroid hormone on bone. Therefore, estrogens have a positive effect on bone maintenance, and later in life, exogenous estrogens oppose the osteoporosis often associated with menopause.

As mentioned earlier, the first menstruation is called menarche and occurs around age 12. The first ovulation does not occur until 6 to 9 months after menarche because the hypothalamic-pituitary axis is not fully responsive to the feedback effects of estrogen. During the pubertal period, the development of breasts, under the influence of estrogen, is known as **thelarche**. At this time, the appearance of axillary and pubic hair occurs, a development known as **pubarche**, controlled by adrenal androgens. The adrenals begin to produce significant amounts of androgens (dehydroepiandrosterone and androstenedione) 4 to 5 years prior to menarche, and this event is called **adrenarche**. The adrenal androgens are responsible in part for pubarche. Adrenarche is independent of gonadarche.

MENOPAUSE

Menopause is the time after which the final menses occurs. It is associated with the cessation of ovarian function and reproductive cycles. Generally, menstrual cycles and bleeding become irregular, and the cycles become shorter from the lack of follicular development (shortened follicular phases). The ovaries atrophy and are characterized by the presence of few, if any, healthy follicles.

The decline in ovarian function is associated with a decrease in estrogen secretion and a concomitant increase in LH and FSH, which is characteristic of menopausal women (Table 38.3). It is used as a diagnostic tool. The elevated LH stimulates ovarian stroma cells to continue producing androstenedione. Estrone, derived almost entirely from the peripheral conversion of adrenal and ovarian androstenedione, becomes the dominant estrogen (see Fig. 37.9). Because the ratio of estrogens to androgens decreases, some women exhibit hirsutism, which results from androgen excess. The lack of estrogen causes atrophic changes in the breasts and reproductive tract, accompanied by vaginal dryness, which often causes pain and irritation. Similar changes in the urinary tract may give rise to urinary disturbances. The epidermal layer of the skin becomes thinner and less elastic.

Hot flashes, as a result of the loss of vasomotor tone, osteoporosis, and an increased risk of cardiovascular disease are not uncommon. Hot flashes are associated with episodic increases in upper body and skin temperature, peripheral vasodilation, and sweating. They occur concurrently with LH pulses but are not caused by the gonadotropins because they are evident in hypophysectomized women. Hot flashes, consisting of episodes of sudden warmth and sweating, reflect temporary disturbances in the hypothalamic thermoregulatory centers, which are somehow linked to the GnRH pulse generator.

Osteoporosis increases the risk of hip fractures and estrogen replacement therapy reduces the risk. Estrogen antagonizes the effects of PTH on bone but enhances its effect on kidney, i.e., it stimulates retention of calcium. Estrogen also promotes the intestinal absorption of calcium through 1,25-dihydroxyvitamin D_3.

Menopausal symptoms are often treated with hormone replacement therapy (HRT), which includes estrogens and progestins. HRT is not an uncommon treatment to improve the quality of life. In some patients, treatment with estrogen can cause adverse effects, such as vaginal bleeding, nausea, and headache. Estrogen therapy is contraindicated in cases of existing reproductive tract carcinomas or hypertension and other cardiovascular disease. The prevailing opinion is that the benefit of treating postmenopausal women with estrogens for limited periods outweighs any risk of developing breast or endometrial carcinomas.

INFERTILITY

One of five women in the United States will be affected by infertility. A thorough understanding of female endocrinology, anatomy, and physiology are critical to gaining insights into solving this major health problem. Infertility can be caused by several factors. Environmental factors, disorders of the central nervous system, hypothalamic disease, pituitary disorders, and ovarian abnormalities can interfere with follicular development and/or ovulation. If a normal ovulation occurs, structural, pathological, and/or endocrine problems associated with the oviduct and/or uterus can prevent fertilization, impede the transport or implantation of the embryo, and, ultimately, interfere with the establishment or maintenance of pregnancy.

Amenorrhea Is Caused by Endocrine Disruption

Menstrual cycle disorders can be divided into two categories: **amenorrhea**, the absence of menstruation, and **oligomenorrhea**, infrequent or irregular menstruation. **Primary amenorrhea** is a condition in which menstruation has never occurred. An example is **Turner's syndrome**, also called gonadal dysgenesis, a congenital abnormality caused by a nondisjunction of one of the X chromosomes, resulting in a 45 XO chromosomal karyotype. Because the two X chromosomes are necessary for normal ovarian development, women with this condition have rudimentary gonads and do not have a normal puberty. Because of ovarian steroid deficiency (lack of estrogen), secondary sex characteristics remain prepubertal, and plasma LH and FSH are elevated. Other abnormalities include short stature, a webbed neck, coarctation of the aorta, and renal disorders.

Another congenital form of primary amenorrhea is **hypogonadotropism** with anosmia, similar to Kallmann's syn-

TABLE 38.3 Serum Gonadotropin and Steroid Levels in Premenopausal and Postmenopausal Women

		Menstrual Cycle			
Hormone	Units	Follicular	Preovulatory	Luteal	Postmenopausal
LH	mIU/mL	2.5–15	15–100	2.5–15	20–100
FSH	mIU/mL	2–10	10–30	2–6	20–140
Estradiol	pg/mL	70–200	200–500	75–300	10–60
Progesterone	ng/mL	≤0.5	≤1.5	4–20	≤0.5

drome in males (see Chapter 37). Patients do not progress through normal puberty and have low and nonpulsatile LH and FSH levels. However, they can have normal stature, female karyotype, and anosmia. The disorder is caused by a failure of olfactory lobe development and GnRH deficiency. Primary amenorrhea can also be caused by a congenital malformation of reproductive tract structures originating from the müllerian duct, including the absence or obstruction of the uterus, cervix, or upper vagina.

Secondary amenorrhea is the cessation of menstruation for longer than 6 months. Pregnancy, lactation, and menopause are common physiological causes of secondary amenorrhea. Other causes are premature ovarian failure, polycystic ovarian syndrome, hyperprolactinemia, and hypopituitarism.

Premature ovarian failure is characterized by amenorrhea, low estrogen levels, and high gonadotropin (LH and FSH) levels before age 40. The symptoms are similar to those of menopause, including hot flashes and an increased risk of osteoporosis. The etiology is variable, including chromosomal abnormalities; lesions resulting from irradiation, chemotherapy, or viral infections; and autoimmune conditions.

Polycystic ovarian syndrome, also called **Stein-Leventhal syndrome**, is a heterogeneous group of disorders characterized by amenorrhea or anovulatory bleeding, an elevated LH/FSH ratio, high androgen levels, hirsutism, and obesity. Although the etiology is unknown, the syndrome may be initiated by excessive adrenal androgen production, during puberty or following stress, that deranges the hypothalamic-pituitary axis secretion of LH. Androgens are converted peripherally to estrogens and stimulate LH release. Excess LH, in turn, increases ovarian stromal and thecal androgen production, resulting in impaired follicular maturation. The LH-stimulated ovaries are enlarged and contain many small follicles and hyperplastic and luteinized theca cells (the site of LH receptors). The elevated plasma androgen levels cause hirsutism, increased activity of sebaceous glands, and clitoral hypertrophy, which are signs of virilization in females.

Hyperprolactinemia is also a cause of secondary amenorrhea. **Galactorrhea**, a persistent milk-like discharge from the nipple in nonlactating individuals, is a frequent symptom and is due to the excess prolactin (PRL). The etiology of hyperprolactinemia is variable. Pituitary prolactinomas account for about 50% of cases. Other causes are hypothalamic disorders, trauma to the pituitary stalk, and psychotropic medications, all of which are associated with a reduction in dopamine release, resulting in an increased PRL secretion. Hypothyroidism, chronic renal failure, and hepatic cirrhosis are additional causes of hyperprolactinemia. In some forms of hypothyroidism, increased hypothalamic thyrotropin-releasing hormone (TRH) is thought to contribute to excess PRL secretion, as experimental studies reveal that exogenous TRH increases the secretion of PRL. The mechanism by which elevated PRL levels suppress ovulation is not entirely clear. It has been postulated that PRL may inhibit GnRH release, reduce LH secretion in response to GnRH stimulation, and act directly at the level of the ovary by inhibiting the action of LH and FSH on follicle development.

Oligomenorrhea can be caused by excessive exercise and by nutritional, psychological, and social factors. **Anorexia nervosa**, a severe behavioral disorder associated with the lack of food intake, is characterized by extreme malnutrition and endocrine changes secondary to psychological and nutritional disturbances. About 30% of patients develop amenorrhea that is not alleviated by weight gain. Strenuous exercise, especially by competitive athletes and dancers, frequently causes menstrual irregularities. Two main factors are thought to be responsible: a low level of body fat, and the effect of stress itself through endorphins that are known to inhibit the secretion of LH. Other types of stress, such as relocation, college examinations, general illness, and job-related pressures, have been known to induce some forms of oligomenorrhea.

Female Infertility Is Caused by Endocrine Malfunction and Abnormalities in the Reproductive Tract

The diagnosis and treatment of amenorrhea present a challenging problem. The amenorrhea must first be classified as primary or secondary, and menopause, pregnancy, and lactation must be excluded. The next step is to determine whether the disorder originates in one of the following areas: the hypothalamus and central nervous system, the anterior pituitary, the ovary, and/or the reproductive tract.

Several treatments can alleviate infertility problems; for example, some success has been achieved in hypothalamic disease with pulsatile administration of GnRH. When hypogonadotropism is the cause of infertility, sequential administration of FSH and hCG is a common treatment for inducing ovulation, although the risk of ovarian hyperstimulation and multiple ovulations is increased. Hyperprolactinemia can be treated surgically by removing the pituitary adenoma containing numerous lactotrophs (prolactin-secreting cells). It can also be treated pharmacologically with bromocriptine, a dopaminergic agonist that reduces the size and number of the lactotrophs and PRL secretion. Treatment with clomiphene, an antiestrogen that binds to and blocks estrogen receptors, can induce ovulation in women with endogenous estrogens in the normal range. Clomiphene reduces the negative feedback effects of estrogen and thus increases endogenous FSH and LH secretion. When reproductive tract lesions are the cause of infertility, corrective surgery or *in vitro* fertilization is the treatment of choice.

REVIEW QUESTIONS

DIRECTIONS: Each of the numbered items or incomplete statements in this section is followed by answers or by completions of the statement. Select the ONE lettered answer or completion that is BEST in each case.

1. Estradiol synthesis in the graafian follicle involves
(A) Activation of LH-stimulated granulosa production of androgen
(B) Stimulation of aromatase in the granulosa cell by FSH
(C) Decreased secretion of progesterone from the corpus luteum, resulting in increased LH
(D) Inhibition of the LH surge during the preovulatory period
(E) Synergy between FSH and progesterone
2. Granulosa cells do not produce estradiol from cholesterol because they do not have an active
(A) 17α-Hydroxylase
(B) Aromatase
(C) 5α-Reductase
(D) Sulfatase
(E) Steroidogenic acute regulatory protein
3. A clinical sign indicating the onset of the menopause is
(A) The onset of menses near age 50
(B) An increase in plasma FSH levels
(C) An excessive presence of corpora lutea
(D) An increased number of cornified cells in the vagina
(E) Regular menstrual cycles
4. Increased progesterone during the postovulatory period is associated with
(A) Proliferation of the uterine endometrium
(B) Enhanced development of graafian follicles
(C) Luteal regression
(D) An increase in basal body temperature by 0.5 to 1.0°C
(E) Increased secretion of FSH
5. The theca interna cells of the graafian follicle are distinguished by
(A) Their capacity to produce androgens from cholesterol
(B) The lack of cholesterol side-chain cleavage enzyme
(C) Aromatization of testosterone to estradiol
(D) The lack of a blood supply
(E) The production of inhibin
6. Disruption of the hypothalamic-pituitary portal system will lead to
(A) High circulating levels of PRL, low levels of LH and FSH, and ovarian atrophy
(B) Enhanced follicular development as a result of increased circulating levels of PRL
(C) Ovulation, followed by increased circulating levels of progesterone
(D) A reduction of ovarian inhibin levels, followed by increased circulating FSH
(E) Excessive androgen production by the ovaries
7. Inhibin is an ovarian hormone that
(A) Inhibits the secretion of LH and PRL
(B) Is produced by granulosa cells and inhibits the secretion of FSH
(C) Only has local ovarian effects and no effect on the secretion of FSH
(D) Has two forms, A and B, with the same β subunits but distinct α subunits
(E) Binds activin and increases FSH secretion
8. Spinnbarkeit formation is induced by
(A) Secretory endometrium
(B) Progesterone action on the uterus
(C) Androgen production from the ovaries
(D) Estrogen action on the vaginal secretions
(E) Prolactin secretion
9. Successful fertilization is most likely to occur when the oocyte is in
(A) The oviduct and has entered the second meiotic division
(B) The uterus and has completed the first meiotic division
(C) Metaphase of mitosis
(D) The graafian follicle, which then enters the oviduct
(E) The uterus, extruding the second polar body and implanting
10. The enzyme, 5α-reductase, is responsible for
(A) Conversion of cholesterol to pregnenolone and enhancing steroidogenesis
(B) Conversion of testosterone to dihydrotestosterone
(C) Aromatization of testosterone to estradiol
(D) Increasing the synthesis of LH
(E) Female secondary sex characteristics

SUGGESTED READING

Carr BR, Blackwell RE. Textbook of Reproductive Medicine. Norwalk, CT: Appleton & Lange, 1998.

Griffin JE, Ojeda SR. Textbook of Endocrine Physiology. 4th Ed. New York: Oxford University Press, 2000.

Johnson MH, Everitt BJ. Essential Reproduction. Oxford: Blackwell Science, 2000.

Kettyle WM, Arky RA. Endocrine Pathophysiology. Philadelphia: Lippincott-Raven, 1998.

Van Voorhis BJ. Follicular development. In: Knobil E, Neill JD, eds. The Encyclopedia of Reproduction. New York: Academic Press, 1999;376–389.

Van Voorhis BJ. Follicular steroidogenesis. In: Knobil E, Neill JD, eds. The Encyclopedia of Reproduction. New York: Academic Press, 1999;389–395.

Yen SSC, Jaffe RB, Barbieri RL. Reproductive Endocrinology. 4th Ed. Philadelphia: WB Saunders, 1999.

Fertilization, Pregnancy, and Fetal Development

Paul F. Terranova, Ph.D.

CHAPTER OUTLINE

- **OVUM AND SPERM TRANSPORT, FERTILIZATION, AND IMPLANTATION**
- **PREGNANCY**
- **FETAL DEVELOPMENT AND PARTURITION**
- **POSTPARTUM AND PREPUBERTAL PERIODS**

KEY CONCEPTS

1. Fertilization of the ovum occurs in the oviduct. Progesterone and estrogen released from the ovary prepare the oviduct and uterus for receiving the developing embryo.
2. The blastocyst enters the uterus, leaves the surrounding zona pellucida, and implants into the uterine wall on day 7 of gestation.
3. Human chorionic gonadotropin (hCG), produced by trophoblast cells of the developing embryo, activates the corpus luteum to continue producing progesterone and estradiol beyond its normal life span to maintain pregnancy.
4. Shortly after the embryo implants into the uterine wall, a placenta develops from embryonic and maternal cells and becomes the major steroid-secreting organ during pregnancy.
5. Major hormones produced by the fetoplacental unit are progesterone, estradiol, estriol, hCG, and human placental lactogen. Elevated estriol levels indicate fetal well-being, whereas low levels might indicate fetal stress. Human placental lactogen has a role in preparing the breasts for milk production.
6. The pregnant woman becomes insulin-resistant during the latter half of pregnancy in order to conserve maternal glucose consumption and make glucose available for the developing fetus.
7. The termination of pregnancy is initiated by strong uterine contractions induced by oxytocin. Estrogens, relaxin, and prostaglandins are involved in softening and dilating the uterine cervix so that the fetus may exit.
8. Lactogenesis is milk production, which requires prolactin (PRL), insulin, and glucocorticoids. Galactopoiesis is the maintenance of an established lactation and requires PRL and numerous other hormones. Milk ejection is the process by which stored milk is released; "milk letdown" is regulated by oxytocin, which contracts the myoepithelial cells surrounding the alveoli and ejects milk into the ducts.
9. Lactation is associated with the suppression of menstrual cycles and anovulation due to the inhibitory actions of PRL on GnRH release and the hypothalamic-pituitary-ovarian axis.
10. The hypothalamic-pituitary axis becomes activated during the late prepubertal period, resulting in increased frequency and amplitude of GnRH pulses, increased LH and FSH secretion, and increased steroid output by the gonads.
11. Most disorders of sexual development are caused by chromosomal or hormonal alterations, which may result in infertility, sexual dysfunction, or various degrees of intersexuality (hermaphroditism).

A mother is considered pregnant at the moment of fertilization—the successful union of a sperm and an egg. The life span of the sperm and an ovum is less than 2 days, so their rapid transport to the oviduct is required for fertilization to occur. Immediately after fertilization, the zygote or fertilized egg begins to divide and a new life begins. The cell division produces a morula, a solid ball of cells, which then forms a blastocyst. Because the early embryo contains a limited energy supply, the embryo enters the uterus within a short time and attaches to the uterine endometrium, a process that initiates the implantation phase. Implantation occurs only in a uterus that has been primed by gonadal steroids and is, therefore, receptive to accepting the blastocyst. At the time of implantation, the trophoblast cells of the early embryonic placenta begin to produce a hormone, human chorionic gonadotropin (hCG), which signals the ovary to continue to produce progesterone, the major hormone required for the maintenance of pregnancy. As a signal from the embryo to the mother to extend the life of the corpus luteum (and progesterone production), hCG prevents the onset of the next menstruation and ovulatory cycle. The placenta, an organ produced by the mother and fetus, exists only during pregnancy; it regulates the supply of oxygen and the removal of wastes and serves as an energy supply for the fetus. It also produces protein and steroid hormones, which duplicate, in part, the functions of the pituitary gland and gonads. Some of the fetal endocrine glands have important functions before birth, including sexual differentiation.

Parturition, the expulsion of the fully formed fetus from the uterus, is the final stage of gestation. The onset of parturition is triggered by signals from both the fetus and the mother and involves biochemical and mechanical changes in the uterine myometrium and cervix. After delivery, the mother's mammary glands must be fully developed and secrete milk in order to provide nutrition to the newborn baby. Milk is produced and secreted in response to suckling. The act of suckling, through neurohormonal signals, prevents new ovulatory cycles. Suckling acts as a natural contraceptive until the baby stops suckling. Thereafter, the mother regains metabolic balance, which has been reduced by the nutritional demands of pregnancy and lactation, and ovulatory cycles return. Sexual maturity of the offspring is attained during puberty, at approximately 12 years of age. The onset of puberty requires changes in the sensitivity, activity, and function of several endocrine organs, including those of the hypothalamic-pituitary-gonadal axis.

OVUM AND SPERM TRANSPORT, FERTILIZATION, AND IMPLANTATION

Sperm deposited in the female reproductive tract swim up the uterus and enter the oviduct where fertilization of the ovum occurs. The developing embryo transits the oviduct, enters the uterus, and implants into the endometrium.

The Egg and Sperm Enter the Oviduct

A meiotically active egg is released from the ovary in a cyclical manner in response to the LH surge. For a successful fertilization, fresh sperm must be present at the time the ovum enters the oviduct. To increase the probability that the sperm and egg will meet at an optimal time, the female reproductive tract facilitates sperm transport during the follicular phase of the menstrual cycle, prior to ovulation (see Chapter 38). However, during the luteal phase, after ovulation, sperm survival and access to the oviduct are decreased. If fertilization does not occur, the egg and sperm begin to exhibit signs of degeneration within 24 hours after release.

The volume of **semen** (ejaculatory fluids and sperm) in fertile men is 2 to 6 mL, and it contains some 20 to 30 million sperm per milliliter, which are deposited in the vagina. The liquid component of the semen, called **seminal plasma**, coagulates after ejaculation but liquefies within 20 to 30 minutes from the action of proteolytic enzymes secreted by the prostate gland. The coagulum forms a temporary reservoir of sperm, minimizing the expulsion of semen from the vagina. During intercourse, some sperm cells are immediately propelled into the cervical canal. Those remaining in the vagina do not survive long because of the acidic environment (pH 5.7), although some protection is provided by the alkalinity of the seminal plasma. The cervical canal constitutes a more favorable environment, enabling sperm survival for several hours. Under estrogen dominance, mucin molecules in the **cervical mucus** become oriented in parallel and facilitate sperm migration. Sperm stored in the cervical crypts constitute a pool for slow release into the uterus.

Sperm survival in the uterine lumen is short because of phagocytosis by leukocytes. The **uterotubal junction** also presents an anatomic barrier that limits the passage of sperm into the oviducts. Abnormal or dead spermatozoa may be prevented from entry to the oviduct. Of the millions of sperm deposited in the vagina, only 50 to 100, usually spaced in time, will reach the oviduct. Major losses of sperm occur in the vagina, uterus, and at the uterotubal junction. Spermatozoa that survive can reach the ampulla within 5 to 10 minutes after coitus. The motility of sperm largely accounts for this rapid transit. However, transport is assisted by muscular contractions of the vagina, cervix, and uterus; ciliary movement; peristaltic activity; and fluid flow in the oviducts. Semen samples with low sperm motility can be associated with male infertility.

There is no evidence for chemotactic interactions between the egg and sperm, although evidence exists for specific ligand-receptor binding between egg and sperm. Sperm arrive in the vicinity of the egg at random, and some exit into the abdominal cavity. Although sperm remain motile for up to 4 days, their fertilizing capacity is limited to 1 to 2 days in the female reproductive tract. Sperm can be cryopreserved for years, if agents such as glycerol are used to prevent ice crystal formation during freezing.

Freshly ejaculated sperm cannot immediately penetrate an egg. During maturation in the epididymis, the sperm acquire surface glycoproteins that act as stabilizing factors but also prevent sperm-egg interactions. To bind to and penetrate the zona pellucida, the sperm must undergo **capacitation**, an irreversible process that involves an increase in sperm motility, the removal of surface proteins, a loss of lipids, and merging of the acrosomal and plasma membranes of the sperm head. The uniting of these sperm membranes and change in acrosomal structure is called the **acrosome reaction**. The reaction occurs when the sperm cell binds to the zona pellucida of the egg. It involves a redistribution of membrane constituents, increased membrane fluidity, and a rise in calcium permeability. Capacitation takes place along the female genital tract and lasts 1 hour to several hours. Sperm can be capacitated in a chemically defined medium, a fact that has enabled *in vitro* fertilization (see Clinical Focus Box 39.1). *In vitro* fertilization may be used in female infertility as well.

Because the ovary is not entirely engulfed by the oviduct, an active "pickup" of the released ovum is required. The ovum is grasped by the **fimbria**, ciliated finger-like projects of the oviducts. The grasping of the egg is facilitated by ciliary movement and muscle contractions, under the influence of estrogen secreted during the periovulatory period. Because the oviduct opens into the peritoneal cavity, eggs that are not picked up by the oviducts can enter the abdominal cavity. An **ectopic pregnancy** may result if an abdominal ovum is fertilized. Egg transport from the fimbria to the **ampulla**, the swollen end of the oviduct, is accomplished by coordinated ciliary activity and depends on the presence of granulosa cells surrounding the egg.

CLINICAL FOCUS BOX 39.1

***In Vitro* Fertilization**

Candidates for *in vitro* fertilization (IVF) are women with disease of the oviducts, unexplained infertility, or endometriosis (occurrence of endometrial tissue outside the endometrial cavity, a condition that reduces fertility), and those whose male partners are infertile (e.g., low sperm count). Follicular development is induced with one or a combination of GnRH analogs, clomiphene, recombinant FSH, and menopausal gonadotropins (a combination of LH and FSH). Follicular growth is monitored by measuring serum estradiol concentration and by ultrasound imaging of the developing follicles. When the leading follicle is 16 to 17 mm in diameter and/or the estradiol level is greater than 300 pg/mL, hCG is injected to mimic an LH surge and induce final follicular maturation, including maturation of the oocyte. Approximately, 34 to 36 hours later, oocytes are retrieved from the larger follicles by aspiration using laparoscopy or a transvaginal approach. Oocyte maturity is judged from the morphology of the cumulus (granulosa) cells and the presence of the germinal vesicle and first polar body. The mature oocytes are then placed in culture media.

The donor's sperm are prepared by washing, centrifuging, and collecting those that are most motile. About 100,000 spermatozoa are added for each oocyte. After 24 hours, the eggs are examined for the presence of two pronuclei (male and female). Embryos are grown to the four- to eight-cell stage, about 60 to 70 hours after their retrieval from the follicles. Approximately three embryos are often deposited in the uterine lumen in order to increase the chance for a successful pregnancy. To ensure a receptive endometrium, daily progesterone administrations begin on the day of retrieval. A successful pregnancy rate of 15 to 25% has been reported by many groups, which compares favorably with that of natural human pregnancy.

The fertilizable life of the human ovum is about 24 hours, and fertilization occurs usually by 2 days after ovulation. The fertilized ovum remains in the oviduct for 2 to 3 days, develops into a solid ball of cells called a **morula**, and by day 3 or 4 enters the uterus. While in the uterus, the morula further develops into a **blastocyst**, the **zona pellucida** is shed, and the blastocyst implants into the wall of the uterus on day 7. The movement of the developing embryo from the oviduct to the uterus is largely regulated by progesterone and estrogen.

Fertilization Is Accompanied by a Multitude of Cellular Events

The initial stage of **fertilization** is the attachment of the sperm head to the zona pellucida of the egg. A successful fertilization restores the full complement of 46 chromosomes and subsequently initiates the development of an embryo. Fertilization involves several steps. Recognition of the egg by the sperm occurs first. The next step is the regulation of sperm entry into the egg. A series of key molecular events, collectively called **polyspermy block**, prevent multiple sperm from entering the egg. Coupled with fertilization is the completion of the **second meiotic division** of the egg, which extrudes the second polar body. At this point, the male and female pronuclei unite, followed by initiation of the **first mitotic cell division** (Fig. 39.1).

The zona pellucida contains specific glycoproteins that serve as sperm receptors. They selectively prevent the fusion of inappropriate sperm cells (e.g., from a different species) with the egg. Contact between the sperm and egg triggers the acrosome reaction, which is required for sperm penetration. Sperm proteolytic enzymes are released that dissolve the matrices of the cumulus (granulosa) cells surrounding the egg, enabling the sperm to move through this densely packed group of cells. The sperm penetrates the zona pellucida, aided by proteolytic enzymes and the propulsive force of the tail; this process may take up to 30 minutes. After entering the **perivitelline space**, the sperm head becomes anchored to the membrane surface of the egg, and microvilli protruding from the **oolemma** (plasma membrane of the egg) extend and clasp the sperm. The oolemma engulfs the sperm, and eventually, the whole head and then the tail are incorporated into the **ooplasm**.

Shortly after the sperm enters the egg, **cortical granules**, which are lysosome-like organelles located underneath the oolemma, are released. The cortical granules fuse with the oolemma. Fusion starts at the point of sperm attachment and propagates over the entire egg surface. The content of the granules is released into the perivitelline space and diffuses into the zona pellucida, inducing the **zona reaction**, which is characterized by sperm receptor inactivation and a hardening of the zona. Consequently, once the first spermatozoon triggers the zona reaction, other sperm cannot penetrate the zona, and therefore, polyspermia is prevented.

An increase in intracellular calcium initiated by sperm incorporation into the egg triggers the next event, which is the activation of the egg for completion of the second meiotic division. The chromosomes of the egg separate and half of the chromatin is extruded with the small **second polar body**. The remaining haploid nucleus with its 23 chromosomes is transformed into a **female pronucleus**. Soon after being incorporated into the ooplasm, the nuclear envelope of the sperm disintegrates; the **male pronucleus** is formed and increases 4 to 5 times in size. The two pronuclei, which are visible 2 to 3 hours after the entry of the sperm into the egg, are moved to the center of the cell by contractions of microtubules and microfilaments. Replication of the haploid chromosomes begins in both pronuclei. Pores are formed in their nuclear membranes, and the pronuclei fuse. The **zygote** (fertilized egg) then enters the first mitotic division (cleavage) producing two unequal sized cells called **blastomeres** within 24 to 36 hours after fertilization. Development proceeds with four-cell and eight-cell embryos and a morula, still in the oviduct, forming at approximately 48, 72, and 96 hours, respectively. The morula enters the uterine cavity at around 4 days after fertilization, and subsequently, a blastocyst develops at approximately 6 days after fertiliza-

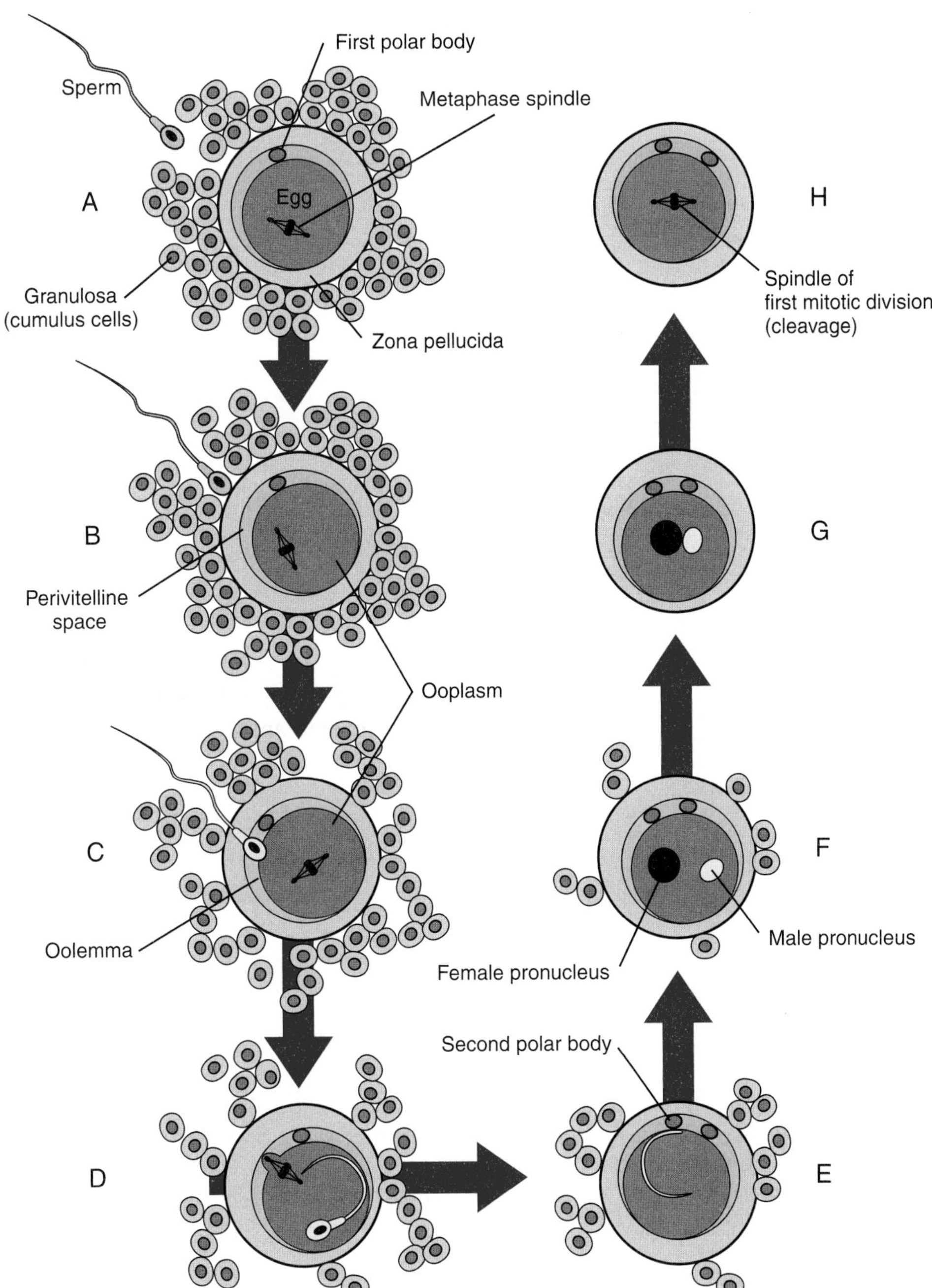

FIGURE 39.1 **The process of fertilization.** **A**, A sperm cell approaches an egg. **B**, Contact between the sperm and the zona pellucida. **C**, The entry of the sperm and contact with the oolemma. **D**, The resumption of the second meiotic division. **E**, The completion of meiosis. **F**, The formation of male and female pronuclei. **G**, The migration of the pronuclei to center of cell. **H**, The zygote is ready for the first mitotic division.

tion. The blastocyst implants into the uterine wall on approximately day 7 after fertilization.

Implantation Requires the Interaction of the Uterine Endometrium and the Embryo

Cell division of the fertilized egg occurs without growth. The cells of the early embryo become progressively smaller, reaching the dimension of somatic cells after several cell divisions. The embryonic cells continue to cleave as the embryo moves from the ampulla toward the uterus (Fig. 39.2). Until implantation, the embryo is enclosed in the zona pellucida. Retention of an intact zona is necessary for embryo transport, protection against mechanical damage or adhesion to the oviduct wall, and prevention of immunological rejection by the mother.

At the 20- to 30-cell stage, a fluid-filled cavity (blastocoele) appears and enlarges until the embryo becomes a hollow sphere, the blastocyst. The cells of the blastocyst have undergone significant differentiation. A single outer layer of the blastocyst consists of extraembryonic ectodermal cells called the **trophoblast**, which will participate in implantation, form the embryonic contribution to the placenta and embryonic membranes, produce hCG, and provide nutrition to the embryo. A cluster of smaller centrally located cells comprises the **embryoblast** or **inner cell mass** and will give rise to the fetus.

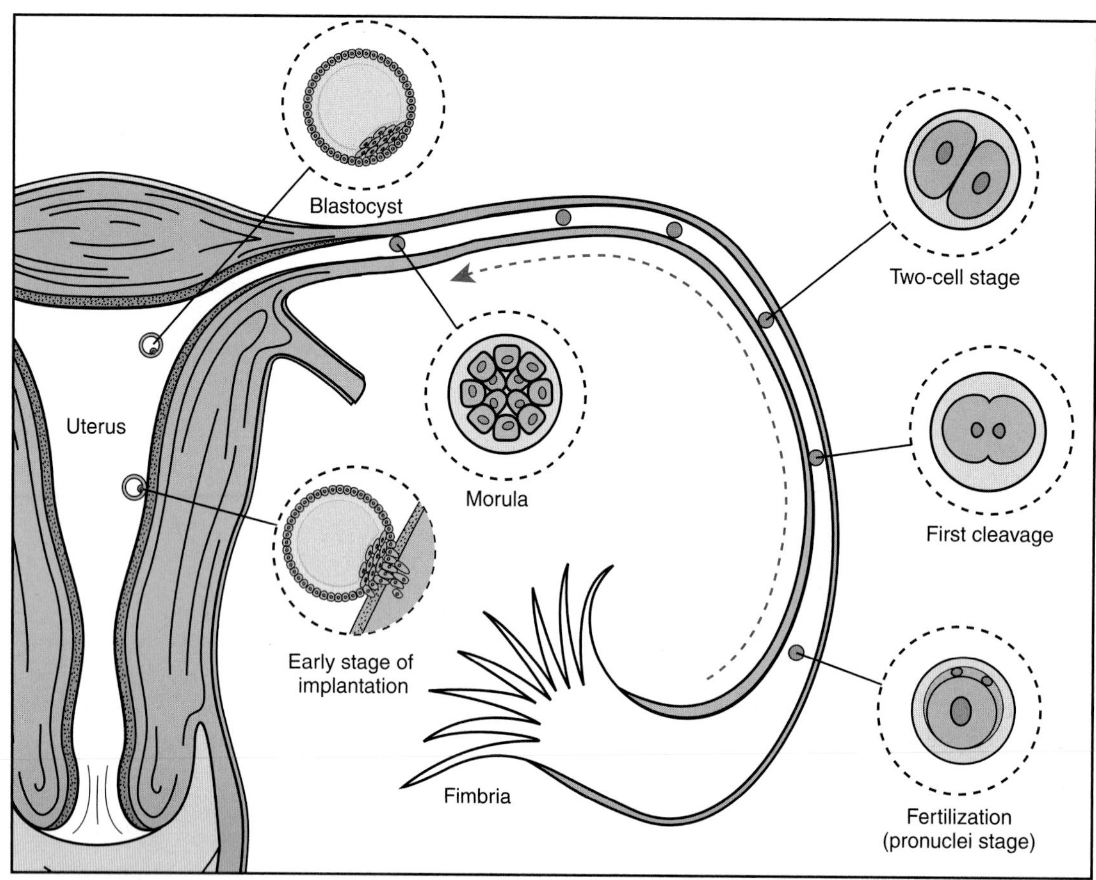

FIGURE 39.2 **Transport of the developing embryo from the oviduct, the site of fertilization, to the uterus, the site of implantation.**

The morula reaches the uterus about 4 days after fertilization. It remains suspended in the uterine cavity for 2 to 3 days while developing into a blastocyst and is nourished by constituents of the uterine fluid during that time. **Implantation** of the blastocyst, which is attachment to the surface endometrial cells of the uterine wall, begins on days 7 to 8 after fertilization and requires proper priming of the uterus by estrogen and progesterone. In preparing for implantation, the blastocyst escapes from the zona pellucida. The zona is ruptured by expansion of the blastocyst and lysed by enzymes. The denuded trophoblast cells become negatively charged and adhere to the endometrium via surface glycoproteins. Microvilli from the trophoblast cells interdigitate with and form junctional complexes with the uterine endometrial cells.

In the presence of progesterone emanating from the corpus luteum, the endometrium undergoes **decidualization**, which involves the hypertrophy of endometrial cells that contain large amounts of glycogen and lipid. In some cases, the cells are multinucleated. This group of decidualized cells is called the **decidua**, which is the site of implantation and the maternal contribution to the placenta. In the absence of progesterone, decidualization does not occur and implantation would fail. As the blastocyst implants into the decidualizing uterus, a **decidual reaction** occurs involving the dilation of blood vessels, increased capillary permeability, edema formation, and increased proliferation of endometrial glandular and epithelial cells (Fig. 39.3). The exact embryonic signals that trigger this reaction are unclear, but histamine, catechol estrogens, steroids, prostaglandins, leukemia inhibitory factor, epidermal growth factor, transforming growth factor β, platelet-derived growth factor, placental growth factor, and several other pregnancy-associated proteins have been proposed.

Invasion of the endometrium is mediated by the release of proteases produced by trophoblast cells adjacent to the uterine epithelium. By 8 to 12 days after ovulation, the human conceptus has penetrated the uterine epithelium and is embedded in the uterine stroma (see Fig. 39.3). The trophoblast cells have differentiated into large polyhedral **cytotrophoblasts**, surrounded by peripheral **syncytiotrophoblasts** lacking distinct cell boundaries. Maternal blood vessels in the endometrium dilate and spaces appear and fuse, forming blood-filled **lacunae**. Between weeks 2 and 3, villi, originating from the embryo, are formed that protrude into the lacunae, establishing a functional communication between the developing embryonic vascular system and the maternal blood (see Fig. 17.6). At this time, the embryoblast has differentiated into three layers:

- Ectoderm, destined to form the epidermis, its appendages (nails and hair), and the entire nervous system
- Endoderm, which will give rise to the epithelial lining of the digestive tract and associated structures
- Mesoderm, which will form the bulk of the body, in-

cluding connective tissue, muscle, bone, blood, and lymph

PREGNANCY

Pregnancy is maintained by protein and steroid hormones from the mother's ovary and the placenta. The maternal endocrine system adapts to allow optimum growth of the fetus.

The Mother and Fetus Contribute to the Placenta

In the human **placenta**, the maternal and fetal components are interdigitated. The functional units of the placenta, the **chorionic villi** (see Fig. 17.6), form on days 11 to 12 and extend tissue projections into the maternal lacunae that form from endometrial blood vessels immediately after implantation. By week 4, the villi are spread over the entire surface of the chorionic sac. As the placenta matures, it becomes discoid in shape. During the third month, the chorionic villi are confined to the area of the **decidua basalis**. The decidua basalis and **chorionic plate** together form the placenta proper (Fig. 39.4).

The **decidua capsularis** around the conceptus and the **decidua parietalis** on the uterine wall fuse and occlude the uterine cavity. The **yolk sac** becomes vestigial and the **amniotic sac** expands, pushing the chorion against the uterine wall. From the fourth month onward, the fetus is enclosed within the **amnion** and **chorion** and is connected to the placenta by the **umbilical cord**. Fetal blood flows through two umbilical arteries to capillaries in the villi, is brought into juxtaposition with maternal blood in the sinuses, and returns to the fetus through a single umbilical vein. The fetal and maternal circulations do not mix. The human placenta is a **hemochorial type**, in which the fetal endothelium and fetal connective tissues are surrounded by maternal blood. The chorionic villi aggregate into groups known as **cotyledons** and are surrounded by blood from the maternal **spiral arteries** that course through the decidua.

Major functions of the placenta are the delivery of nutrients to the fetus and the removal of its waste products. Oxygen diffuses from maternal blood to the fetal blood down an initial gradient of 60 to 70 mm Hg. The oxygen-transporting capacity of fetal blood is enhanced by **fetal hemoglobin**, which has a high affinity for oxygen. The P_{CO_2} of fetal arterial blood is 2 to 3 mm Hg higher than that of maternal blood, allowing the diffusion of carbon dioxide toward the maternal compartment. Other compounds, such as glucose, amino acids, free fatty acids, electrolytes, vitamins, and some hormones, are transported by diffusion, facilitated diffusion, or pinocytosis. Waste products, such as urea and creatinine, diffuse away from the fetus down their concentration gradients. Large proteins, including most polypeptide hormones, do not readily cross the placenta, whereas the lipid-soluble steroids pass through quite easily. The **blood-placental barrier** allows the transfer of some immunoglobulins, viruses, and drugs from the mother to the fetus (Fig. 39.5).

The Recognition and Maintenance of Pregnancy Depend on Maternal and Fetal Hormones

The placenta is an endocrine organ that produces **progesterone** and **estrogens**, hormones essential for the continuance of pregnancy. The placenta also produces protein hormones unique to pregnancy, such as **human placental**

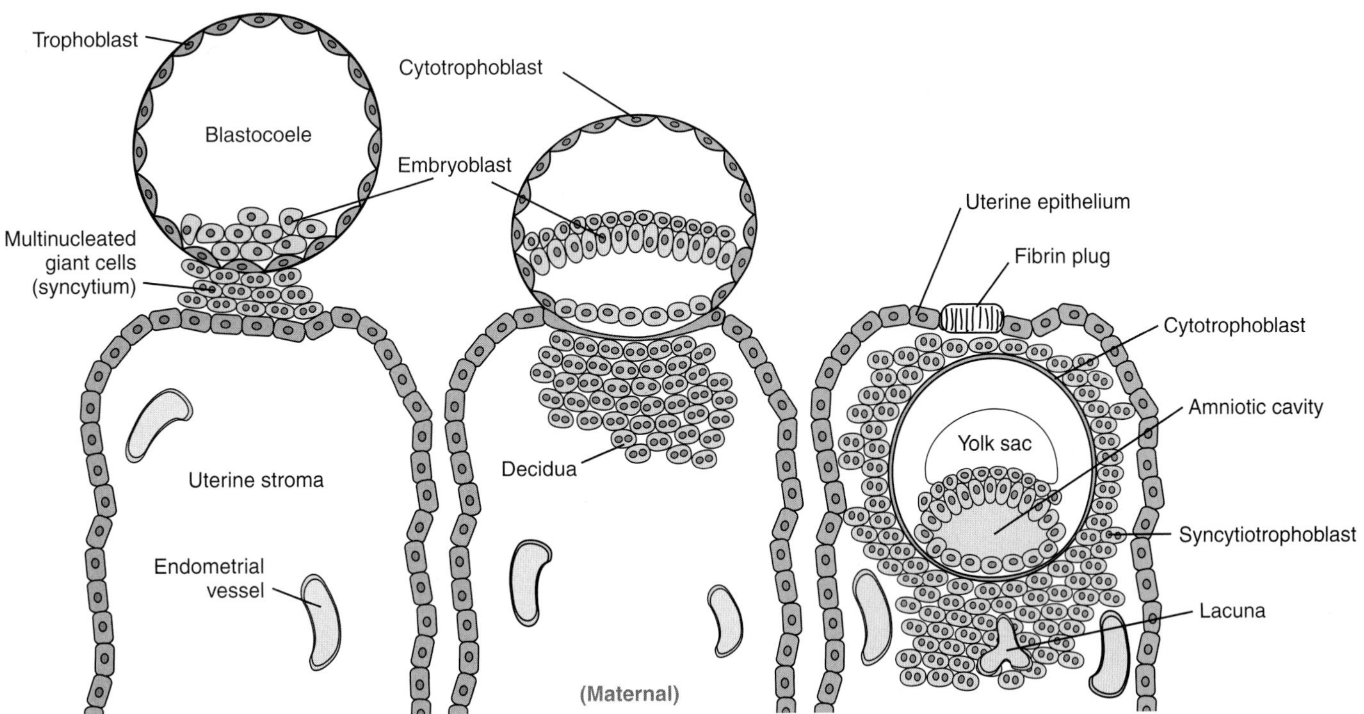

FIGURE 39.3 The process of embryo implantation and the decidual reaction.

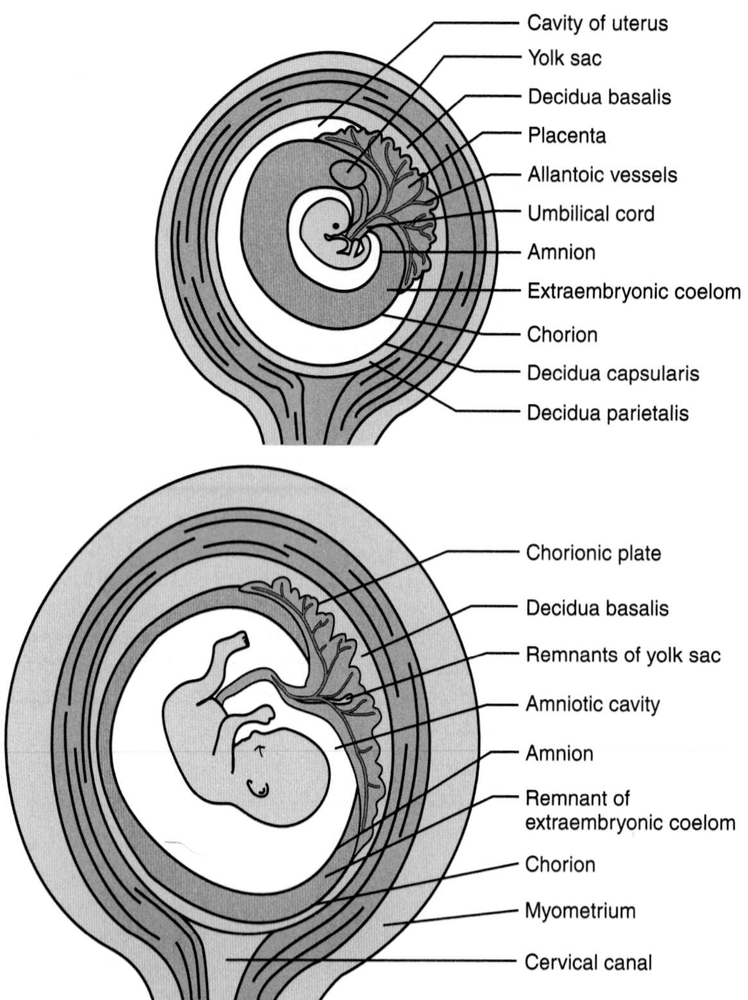

FIGURE 39.4 **Two stages in the development of the placenta, showing the origin of the membranes around the fetus.**

lactogen (hPL) and **human chorionic gonadotropin** (hCG). Several peptides and polypeptides, including corticotropin-releasing hormone (CRH), GnRH, and insulin-like growth factors, are also synthesized by the placenta and function as paracrine factors.

During the menstrual cycle, the corpus luteum forms shortly after ovulation and produces significant amounts of progesterone and estrogen to prepare the uterus for receiving a fertilized ovum. If the egg is not fertilized, the corpus luteum regresses at the end of the luteal phase, as indicated by declining levels of progesterone and estrogen in the circulation. After losing ovarian steroidal support, the superficial endometrial layer of the uterus is expelled, resulting in menstruation. If the egg is fertilized, the developing embryo signals its presence by producing hCG, which extends the life of the corpus luteum. This signaling process is called the maternal recognition of pregnancy. Syncytiotrophoblast cells produce hCG 6 to 8 days after ovulation (fertilization), and hCG enters the maternal and fetal circulations. Very similar to LH, hCG has a molecular weight of approximately 38 kDa, binds LH receptors on the corpus luteum, stimulates luteal progesterone production, and prevents menses at the end of the anticipated cycle. It can be detected in the pregnant woman's urine using commercial colorimetric kits.

Human chorionic gonadotropin is a glycoprotein made of two dissimilar subunits, α and β. It belongs to the same hormone family as luteinizing hormone (LH), follicle-stimulating hormone (FSH), and thyroid-stimulating hormone (TSH). The α subunit is made of the same 92 amino acids as the other glycoprotein hormones. The β subunit is made of 145 amino acids, with six N- and O-linked oligosaccharide units. It resembles the LH β subunit but has a 24-amino acid extension at the C-terminal end. Because of extensive glycosylation, the half-life of hCG in the circulation is longer than that of LH. Like LH, the major function of hCG in early pregnancy is the stimulation of luteal steroidogenesis. Both bind to the same or similar membrane receptors and increase the formation of pregnenolone from cholesterol by a cAMP-dependent mechanism.

The hCG level in plasma doubles about every 2 to 3 days in early pregnancy and reaches peak levels at about 10 to 15 weeks of gestation. It is reduced by about 75% by 25 weeks and remains at that level until term (Fig. 39.6). Fetal concentrations of hCG follow a similar pattern. The hCG levels are higher in pregnancies with multiple fetuses. During the first trimester, GnRH locally produced by cytotrophoblasts appears to regulate hCG production by a paracrine mechanism. The suppression of hCG release during the second half of pregnancy is attributed to negative

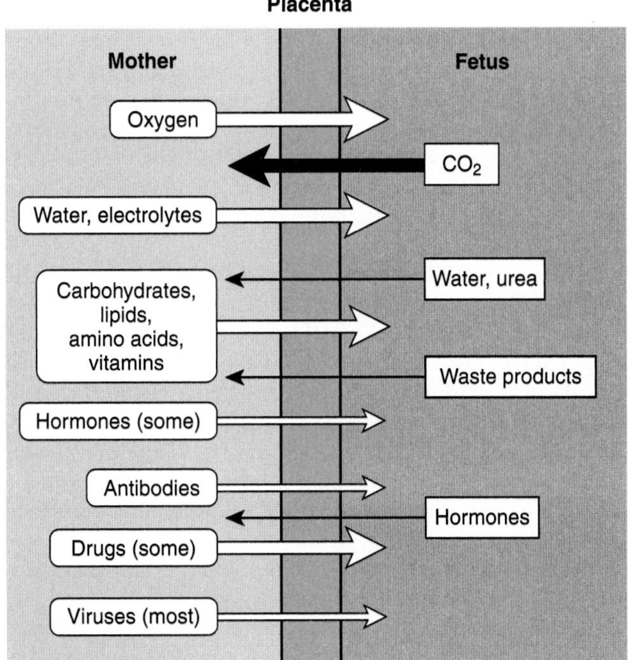

FIGURE 39.5 Role of the placenta in exchanges between the fetal and maternal compartments. The size of the arrows indicates the amount of exchange between the compartments.

feedback by placental progesterone or other steroids produced by the fetus. Progesterone secretion by the corpus luteum is maximal 4 to 5 weeks after conception and declines, although hCG levels are still rising. Corpus luteum refractoriness to hCG results from receptor desensitization and the rising levels of placental estrogens. From week 7 to 10 of gestation, steroid production by the corpus luteum is gradually replaced by steroid production by the placenta. Removal of the corpus luteum after week 10 does not terminate the pregnancy. Other placental-derived growth regulators affecting hCG production are activin, inhibin, and transforming growth factors α and β.

Human chorionic gonadotropin has been shown to increase progesterone production by the trophoblast. Therefore, hCG may have a critical role in maintaining placental steroidogenesis throughout pregnancy and replacing luteal progesterone secretion after week 10 when the ovaries are no longer needed to maintain pregnancy. Another important function of hCG is in sexual differentiation of the male fetus, which depends on testosterone production by the fetal testes. Peak production of testosterone occurs 11 to 17 weeks after conception. This timing coincides with peak hCG production and predates the functional maturity of the fetal hypothalamic-pituitary axis (fetal LH levels are low). Human chorionic gonadotropin appears to regulate fetal Leydig cell proliferation as well as testosterone biosynthesis, especially because LH/hCG receptors are present in the early fetal testes. The role of hCG in fetal ovarian development is less clear since LH/hCG receptors are not present on fetal ovaries. There are some indications that increased levels of hCG and thyroxine accompany maternal morning sickness, but a cause-and-effect relationship is not established.

Human placental lactogen (hPL) has lactogenic and growth hormone-like actions. As a result, it is also called human chorionic somatomammotropin and chorionic growth hormone. This hormone is synthesized by syncytiotrophoblasts and secreted into the maternal circulation, where its levels gradually rise from the third week of pregnancy until term. Although hPL is produced by the same cells as hCG, its pattern of secretion is different, indicating the possibility of control by different regulatory mechanisms. The hormone is composed of a single chain of 191 amino acids with two disulfide bridges and has a molecular weight of about 22 kDa. Its structure and function resemble those of prolactin (PRL) and growth hormone (GH).

Human placental lactogen promotes cell specialization in the mammary gland but is less potent than PRL in stimulating milk production and is much less potent than GH in stimulating growth. Its main function is to alter fuel availability by antagonizing maternal glucose consumption and enhancing fat mobilization. This ensures adequate fuel supplies for the fetus. Its effects on carbohydrate, protein, and fat metabolism are similar to those of GH. The amniotic fluid also contains large amounts of PRL produced mainly by the decidual compartments. Decidual PRL is indistinguishable from pituitary PRL, but its function and regulation are unclear.

Steroid Production During Pregnancy Involves the Ovary and Fetoplacental Unit

Progesterone is required to maintain normal human pregnancy. During the early stages of pregnancy (approximately the first 8 weeks), the ovaries produce most of the sex steroids; the corpus luteum produces primarily progesterone and estrogen. As the placenta develops, trophoblast cells gradually take over a major role in the production of progesterone and estrogen. Although the corpus luteum continues to secrete progesterone, the placenta secretes most of the progesterone. Progesterone levels gradually rise during early pregnancy and plateau during the transition period from corpus luteal to placental production (see Fig. 39.6). Thereafter, plasma progesterone levels continue

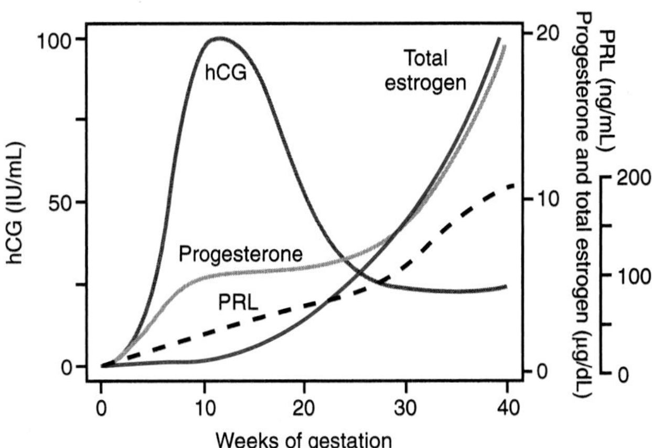

FIGURE 39.6 Profiles of hCG, progesterone, total estrogens, and PRL in the maternal blood throughout gestation.

to rise and reach about 150 ng/mL near the end of pregnancy. Two major estrogens, estradiol and estriol, gradually rise during the first half of pregnancy and steeply increase in the latter half of pregnancy to more than 25 ng/mL near term.

Progesterone and estrogen have numerous functions throughout gestation. Estrogens increase the size of the uterus and uterine blood flow, are critical in the timing of implantation of the embryo into the uterine wall, induce the formation of uterine receptors for progesterone and oxytocin, enhance fetal organ development, stimulate maternal hepatic protein production, and increase the mass of breast and adipose tissues. Progesterone is essential for maintaining the uterus and early embryo, inhibits myometrial contractions, and suppresses maternal immunological responses to fetal antigens. Progesterone also serves as a precursor for steroid production by the fetal adrenal glands and plays a role in the onset of parturition.

Beginning at approximately week 8 of gestation, progesterone production is carried out by the placenta, but its synthesis requires cholesterol, which is contributed from the mother. The placenta cannot make significant amounts of cholesterol from acetate and obtains it from the maternal blood via LDL cholesterol. Trophoblast cells have **LDL receptors**, which bind the LDL cholesterol and internalize it. Free cholesterol is released and used by cholesterol side-chain cleavage enzyme to synthesize pregnenolone. Pregnenolone is converted to progesterone by 3β-hydroxysteroid dehydrogenase.

The placenta lacks the 17α-hydroxylase for converting pregnenolone or progesterone to androgens (the precursors of the estrogens). Maternal 17α-hydroxyprogesterone can be measured during the first trimester and serves as a marker of corpus luteum function, since the placenta cannot make this steroid. The production of estrogens (estradiol, estrone, and estriol) during gestation requires cooperation between the maternal compartment and the placental and fetal compartments, referred to as the fetoplacental unit (Fig. 39.7). To produce estrogens, the placenta uses androgenic substrates derived from both the fetus and the mother. The primary androgenic precursor is **dehydroepiandrosterone sulfate** (DHEAS), which is produced by the fetal zone of the fetal adrenal gland. The fetal adrenal gland is extremely active in the production of steroid hormones, but because it lacks 3β-hydroxysteroid dehydrogenase, it cannot make progesterone. Therefore, the fetal adrenals use progesterone from the placenta to produce androgens, which are ultimately sulfated in the adrenal glands. The conjugation of androgenic precursors to sulfates ensures greater water solubility, aids in their transport, and reduces their biological activity while in the fetal circulation. DHEAS diffuses into the placenta and is cleaved by a **sulfatase** to yield a nonconjugated androgenic precursor. The placenta has an active aromatase that converts androgenic precursors to estradiol and estrone.

The major estrogen produced during human pregnancy is estriol, which has relatively weak estrogenic activity. **Estriol** is produced by a unique biosynthetic pathway (see Fig. 39.7). DHEAS from the fetal adrenal is converted to **16-hydroxydehydroepiandrosterone sulfate** by 16-hy-

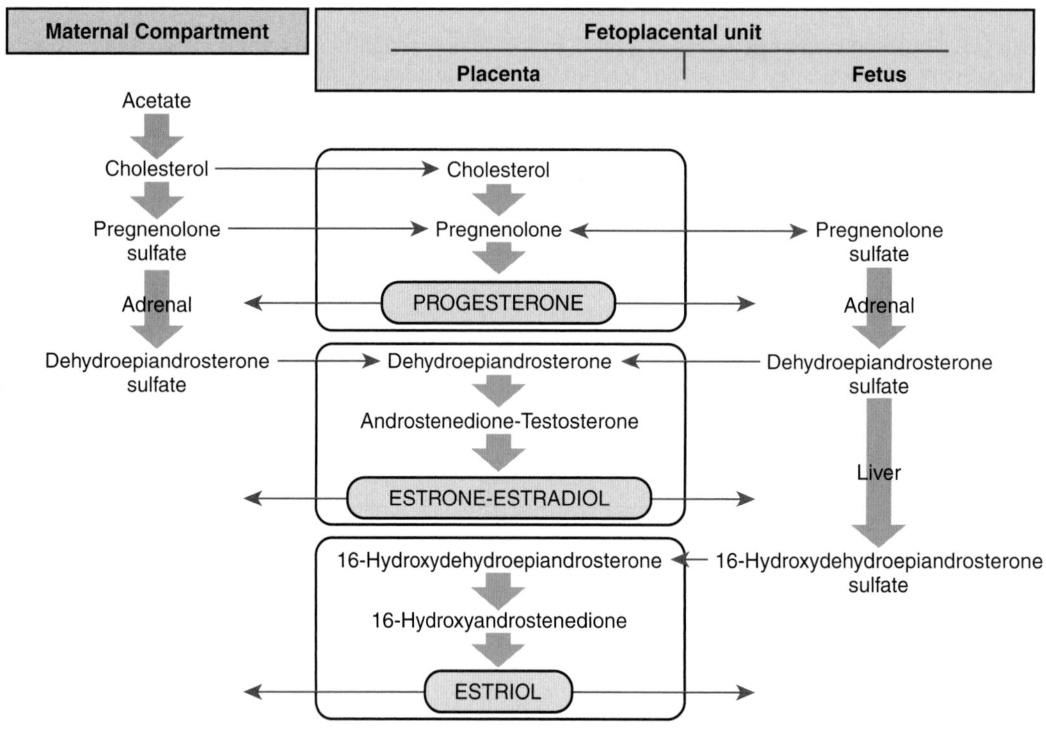

FIGURE 39.7 **The fetoplacental unit and steroidogenesis.** Note that estriol is the product of reactions occurring in the fetal adrenal, fetal liver, and placenta. (Modified from Goodman HM. Basic Medical Endocrinology. New York: Raven, 1988.)

droxylation in the fetal liver and, to a lesser extent, the fetal adrenal gland. This step is followed by desulfation using a placental sulfatase and conversion by 3β-hydroxysteroid dehydrogenase to 16-hydroxyandrostenedione, which is subsequently aromatized in the placenta to estriol. Although 16-OH-DHEAS can be made in the maternal adrenal from maternal DHEAS, the levels are low. It has been estimated that 90% of the estriol is derived from the fetal 16-OH-DHEAS. Therefore, the levels of estriol in plasma, amniotic fluid, or urine are used as an index of fetal well-being. Low levels of estriol would indicate potential fetal distress, although rare inherited sulfatase deficiencies can also lead to low estriol.

Maternal Physiology Changes Throughout Gestation

The pregnant woman provides nutrients for her growing fetus, is the sole source of fetal oxygen, and removes fetal waste products. These functions necessitate significant adjustments in her pulmonary, cardiovascular, renal, metabolic, and endocrine systems. Among the most notable changes during pregnancy are hyperventilation, reduced arterial blood P_{CO_2} and osmolality, increased blood volume and cardiac output, increased renal blood flow and glomerular filtration rate, and substantial weight gain. These are brought about by the rising levels of estrogens, progesterone, hPL, and other placental hormones and by mechanical factors, such as the expanding size of the uterus and the development of uterine and placental circulations.

The maternal endocrine system undergoes significant adaptations. The hypothalamic-pituitary-ovarian axis is suppressed by the high levels of sex steroids. Consequently, circulating gonadotropins are low, and ovulation does not occur during pregnancy. In contrast, the rising levels of estrogens stimulate PRL release. PRL levels begin to rise during the first trimester, increasing gradually to reach a level 10 times higher near term (see Fig. 39.6). Pituitary lactotrophs undergo hyperplasia and hypertrophy and mostly account for the enlargement of the pregnant woman's pituitary gland. However, somatotrophs that produce growth hormone are reduced, and GH levels are low throughout pregnancy.

The thyroid gland enlarges, but TSH levels are in the normal nonpregnant range. T_3 and T_4 increase, but thyroxine-binding globulin (TBG) also increases in response to the rising levels of estrogen, which are known to stimulate TBG synthesis. Therefore, the pregnant woman stays in an euthyroid state. The parathyroid glands and their hormone, PTH, increase mostly during the third trimester. PTH enhances calcium mobilization from maternal bone stores in response to the fetus's growing demands for calcium. The rate of adrenal secretion of mineralocorticoids and glucocorticoids increases, and plasma free cortisol is higher because of its displacement from transcortin, the cortisol-binding globulin, by progesterone, but hypercortisolism is not apparent during pregnancy.

Changes in maternal ACTH levels throughout pregnancy are variable, although there is a significant increase at the time of parturition. Current reports indicate that maternal pituitary secretion of ACTH may be suppressed by the high levels of steroids during pregnancy. However, the placenta can produce ACTH, so plasma levels tend to rise throughout pregnancy because placental secretion (unlike pituitary hormone secretion) is not regulated by the high level of steroids.

Maternal metabolism responds in several ways to the increasing nutritional demands of the fetus. The major net weight gain of the mother occurs during the first half of gestation, mostly resulting from fat deposition. This response is attributed to progesterone, which increases appetite and diverts glucose into fat synthesis. The extra fat stores are used as an energy source later in pregnancy, when the metabolic requirements of the fetus are at their peak, and also during periods of starvation. Several maternal and placental hormones act together to provide a constant supply of metabolic fuels to the fetus. Toward the second half of gestation, the mother develops a resistance to insulin. This is brought about by combined effects of hormones antagonistic to insulin action, such as GH, PRL, hPL, glucagon, and cortisol. As a result, maternal glucose use declines and gluconeogenesis increases, maximizing the availability of glucose to the fetus.

FETAL DEVELOPMENT AND PARTURITION

At fertilization, genetic sex is determined; subsequently, sexual differentiation is controlled by gonadal hormones. The fetal endocrine system participates in growth and development of the fetus, and parturition is regulated by interactions of fetal and maternal factors.

The Fetal Endocrine System Gradually Matures

The protective intrauterine environment postpones the initiation of some physiological functions that are essential for life after birth. For example, the fetal lungs and kidneys do not act as organs of gas exchange and excretion because their functions are carried out by the placenta. Constant isothermal surroundings alleviate the need to expend calories to maintain body temperature. The gastrointestinal tract does not carry out digestive activities, and fetal bones and muscles do not support weight or locomotion. Being exposed to low levels of external stimuli and environmental insults, the fetal nervous and immune systems develop slowly. Homeostasis in the fetus is regulated by hormones. The fetal endocrine system plays a vital role in fetal growth and development.

Given that most protein and polypeptide hormones are excluded from the fetus by the blood-placental barrier, the maternal endocrine system has little direct influence on the fetus. Instead, the fetus is almost self-sufficient in its hormonal requirements. Notable exceptions are some of the steroid hormones, which are produced by the fetoplacental unit; they cross easily between the different compartments and carry out integrated functions in both the fetus and the mother. By and large, fetal hormones perform the same functions as in the adult, but they also subserve unique processes, such as sexual differentiation and the initiation of labor.

The fetal hypothalamic nuclei, including their releasing hormones such as TRH, GnRH, and several of the neurotransmitters, are well developed by 12 weeks of gestation. At about week 4, the anterior pituitary begins its development from Rathke's pouch, an ectodermal evagination from the roof the fetal mouth (stomodeum), and by week 8, most anterior pituitary hormones can be identified. The posterior pituitary or neurohypophysis is an evagination from the floor of the primitive hypothalamus, and its nuclei, supraoptic and paraventricular with AVP and oxytocin, can be detected around week 14. The hypothalamic-pituitary axis is well developed by midgestation, and well-differentiated hormone-producing cells in the anterior pituitary are also apparent at this time. Whether the fetal pituitary is tightly regulated by hypothalamic hormones or possesses some autonomy is unclear. However, the release of pituitary hormones can occur prior to the establishment of the portal system, indicating that the hypothalamic-releasing hormones may diffuse down to the pituitary from the hypothalamic sites.

Experiments with long-term catheterization of monkey fetuses indicate that by the last trimester, both LH and testosterone increase in response to GnRH administration. In the adult, GH largely regulates the secretion of the insulin-like growth factors (IGF-I and IGF-II) from the liver. In the fetus, this may not be the case, since newborns with low GH have normal birth size; therefore, other mechanisms may control the secretion of IGFs in the fetus. GH levels increase in the fetus until midgestation and decline thereafter when fetal weight is increasing significantly, representing another dichotomy in GH and IGF in the fetus versus postnatal life. PRL levels increase in the fetus throughout gestation and can be inhibited by an exogenous dopamine agonist. Although the role of PRL in fetal growth is unclear, it has been implicated in adrenal and lung function, as well as in the regulation of amniotic fluid volume.

The fetal adrenal glands are unique in both structure and function. At month 4 of gestation, they are larger than the kidneys, as a result of the development of a fetal zone that constitutes 75 to 80% of the whole gland. The outer definitive zone will form the adult adrenal cortex, whereas the deeper fetal zone involutes after birth; the reason for the involution is unknown, but it is not caused by the withdrawal of ACTH support. The fetal zone produces large amounts of DHEAS and provides androgenic precursors for estrogen synthesis by the placenta (see Fig. 39.7). The definitive zone produces cortisol, which has multiple functions during fetal life, including the promotion of pancreas and lung maturation, the induction of liver enzymes, the promotion of intestinal tract cytodifferentiation and, possibly, the initiation of labor. ACTH is the main regulator of fetal adrenal steroidogenesis, partly evidenced by the observation that anencephalic fetuses have low ACTH and the fetal zone is small. The adrenal medulla develops by about week 10 and is capable of producing epinephrine and norepinephrine.

The rate of fetal growth increases significantly during the last trimester. Surprisingly, growth hormone of maternal, placental, or fetal origin has little effect on fetal growth, as judged by the normal weight of hypopituitary dwarfs or anencephalic fetuses. Fetal insulin is the most important hormone in regulating fetal growth. Glucose is the main metabolic fuel for the fetus. Fetal insulin, produced by the pancreas by week 12 of gestation, regulates tissue glucose use, controls liver glycogen storage, and facilitates fat deposition. It does not control the supply of glucose, however; this is determined by maternal gluconeogenesis and placental glucose transport. The release of insulin in the fetus is relatively constant, increasing only slightly in response to a rapid rise in blood glucose levels. When blood glucose levels are chronically elevated, as in diabetic women, the fetal pancreas becomes enlarged and circulating insulin levels increase. Consequently, fetal growth is accelerated, and infants of uncontrolled diabetic women are overweight (Fig. 39.8).

Calcium is in large demand because of the fetus's rapid growth and large amount of bone formation during pregnancy. Maternal calcium is highly important for meeting this fetal requirement. During pregnancy, maternal calcium intake increases, and 1,25-dihydroxyvitamin D_3 and PTH increase to meet the increased calcium demands of the fetus. In the mother, total plasma calcium and phosphate decline without affecting free calcium. The placenta has a specialized calcium pump that transfers calcium to the fetus, resulting in sustained increases in calcium and phosphate throughout pregnancy. Although PTH and calcitonin are evident in the fetus near week 12 of gestation, their role in regulating fetal calcium is unclear. In addition, the placenta has 1α-hydroxylase and can convert 25-hydroxyvitamin D_3 to 1,25-dihydroxyvitamin D_3. At the end of gestation, calcium and phosphate levels in the fetus are higher than in the mother. However, after delivery, neonatal calcium levels decrease and PTH levels rise to raise the levels of serum calcium.

The Sex Chromosomes Dictate the Development of the Fetal Gonads

Sexual differentiation begins at the time of fertilization by a random unification of an X-bearing egg with either an X- or Y-bearing spermatozoon and continues during early em-

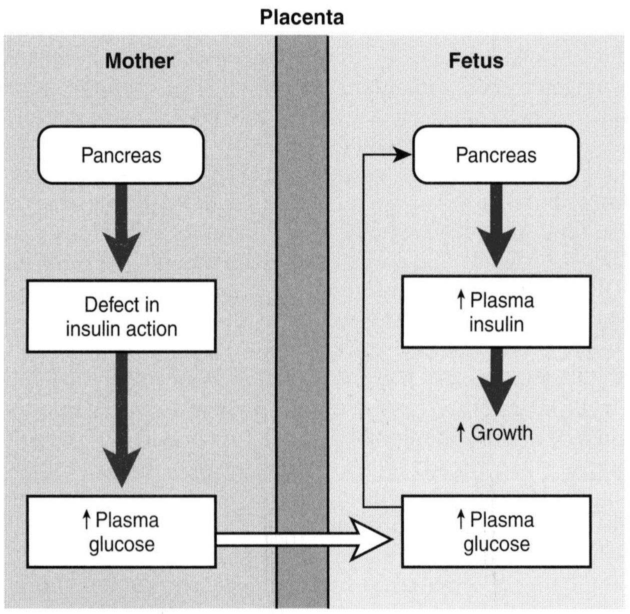

FIGURE 39.8 Effects of maternal diabetes on fetal growth.

bryonic life with the development of male or female gonads. Therefore, at the time of fertilization, **chromosomal sex** or **genetic sex** is determined. Sexual differentiation is controlled by gonadal hormones that act at critical times during organogenesis. Testicular hormones induce masculinization, whereas feminization does not require (female) hormonal intervention. The process of sexual development is incomplete at birth; the secondary sex characteristics and a functional reproductive system are not fully developed until puberty.

Human somatic cells have 44 autosomes and 2 sex chromosomes. The female is **homogametic** (having two X chromosomes) and produces similar X-bearing ova. The male is **heterogametic** (having one X and one Y chromosome) and generates two populations of spermatozoa, one with X chromosomes and the other with Y chromosomes. The X chromosome is large, containing 80 to 90 genes responsible for many vital functions. The Y chromosome is much smaller, carrying only few genes responsible for testicular development and normal spermatogenesis. Gene mutation of genes on an X chromosome results in the transmission of X-linked traits, such as hemophilia and color-blindness, to male offspring, which, unlike females, cannot compensate with an unaffected allele.

Theoretically, by having two X chromosomes, the female has an advantage over the male, who has only one. However, because one of the X chromosomes is inactivated at the morula stage, the advantage is lost. Each cell randomly inactivates either the paternally or the maternally derived X chromosome, and this continues throughout the cell's progeny. The inactivated X chromosome is recognized cytologically as the sex chromatin or **Barr body**. In males with more than one X chromosome or in females with more than two, extra X chromosomes are inactivated and only one remains functional. This does not apply to the germ cells. The single active X chromosome of the spermatogonium becomes inactivated during meiosis, and a functional X chromosome is not necessary for the formation of fertile sperm. The oogonium, however, reactivates its second X chromosome, and both are functional in oocytes and important for normal oocyte development.

Testicular differentiation requires a Y chromosome and occurs even in the presence of two or more X chromosomes. **Gonadal sex** determination is regulated by a testis-determining gene designated **SRY** (sex-determining region, Y chromosome). Located on the short arm of the Y chromosome, *SRY* encodes a DNA-binding protein, which binds to the target DNA in a sequence-specific manner. The presence or absence of *SRY* in the genome determines whether male or female gonadal differentiation takes place. Thus, in normal XX (female) fetuses, which lack a Y chromosome, ovaries, rather than testes, develop.

Whether possessing the XX or the XY karyotype, every embryo goes initially through an **ambisexual stage** and has the potential to acquire either masculine or feminine characteristics. A 4- to 6-week-old human embryo possesses indifferent gonads, and undifferentiated pituitary, hypothalamus, and higher brain centers.

The indifferent gonad consists of a **genital ridge**, derived from coelomic epithelium and underlying mesenchyme, and primordial germ cells, which migrate from the yolk sac to the genital ridges. Depending on genetic programming, the inner **medullary tissue** will become the testicular components, and the outer **cortical tissue** will develop into an ovary. The primordial germ cells will become oogonia or spermatogonia. In an XY fetus, the testes differentiate first. Between weeks 6 and 8 of gestation, the cortex regresses, the medulla enlarges, and the seminiferous tubules become distinguishable. Sertoli cells line the basement membrane of the tubules, and Leydig cells undergo rapid proliferation. Development of the ovary begins at weeks 9 to 10. Primordial follicles, composed of oocytes surrounded by a single layer of granulosa cells, are discernible in the cortex between weeks 11 and 12 and reach maximal development by weeks 20 to 25.

Differentiation of the Genital Ducts Is Determined by Hormones

During the indifferent stage, the primordial genital ducts are the paired **mesonephric (wolffian) ducts** and the paired **paramesonephric (müllerian) ducts**. In the normal male fetus, the wolffian ducts give rise to the epididymis, vas deferens, seminal vesicles, and ejaculatory ducts, while the müllerian ducts become vestigial. In the normal female fetus, the müllerian ducts fuse at the midline and develop into the oviducts, uterus, cervix, and upper portion of the vagina, while the wolffian ducts regress (Fig. 39.9). The mesonephros is the embryonic kidney.

The fetal testes differentiate between weeks 6 and 8 of gestation. Leydig cells, either autonomously or under regulation by hCG, start producing testosterone. Sertoli cells produce two nonsteroidal compounds. One is the **antimüllerian hormone** (AMH), also known as **müllerian inhibiting substance**, a large glycoprotein with a sequence homologous to inhibin and transforming growth factor β, which inhibits cell division of the müllerian ducts. The second is **androgen-binding protein** (ABP), which binds testosterone. Peak production of these compounds occurs between weeks 9 and 12, coinciding with the time of differentiation of the internal genitalia along the male line. The ovary, which differentiates later, does not produce hormones and has a passive role.

The primordial external genitalia include the genital tubercle, genital swellings, urethral folds, and urogenital sinus. Differentiation of the external genitalia also occurs between weeks 8 and 12 and is determined by the presence or absence of male sex hormones. Differentiation along the male line requires active **5α-reductase**, the enzyme that converts testosterone to DHT. Without DHT, regardless of the genetic, gonadal, or hormonal sex, the external genitalia develop along the female pattern. The structures that develop from the primordial structures are illustrated in Figure 39.10, and a summary of sexual differentiation during fetal life is shown in Figure 39.11. Androgen-dependent differentiation occurs only during fetal life and is thereafter irreversible. However, the exposure of females to high androgens either before or after birth can cause clitoral hypertrophy. Testicular descent into the scrotum, which occurs during the third trimester, is also controlled by androgens.

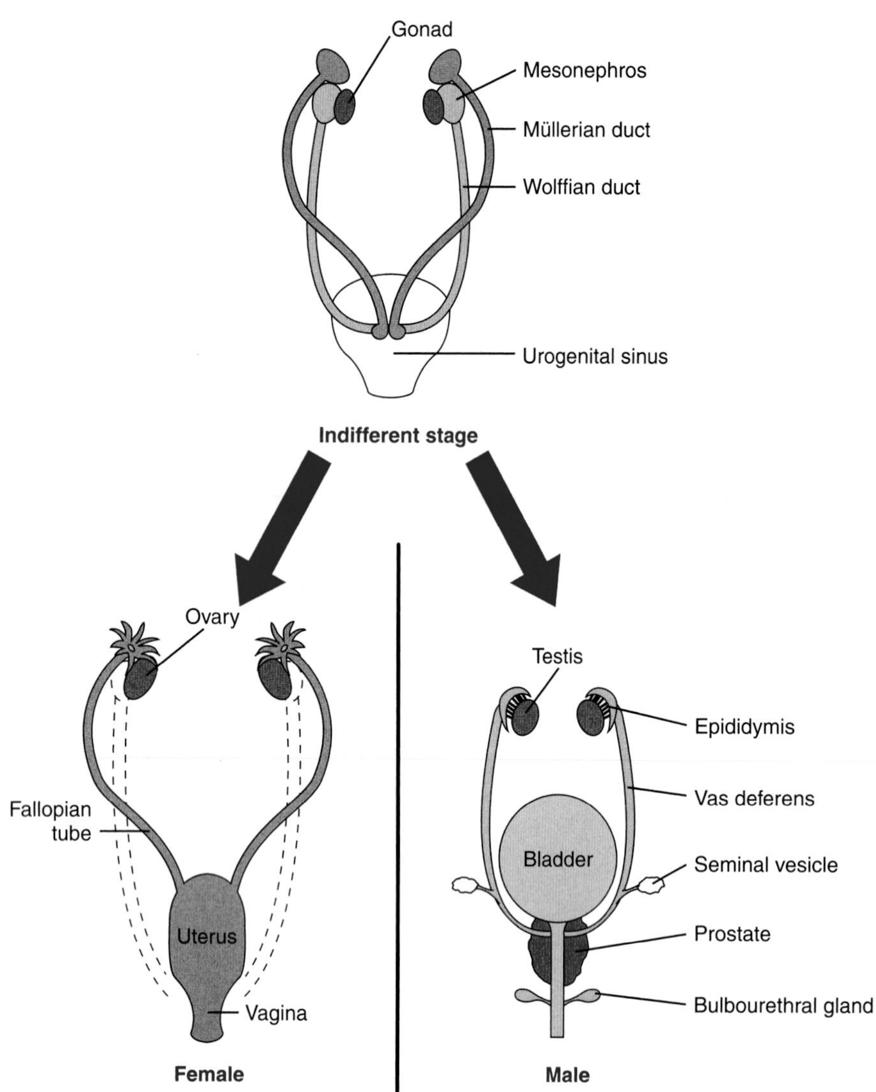

FIGURE 39.9 Differentiation of the internal genitalia and the primordial ducts. (Modified from George FW, Wilson JD. Embryology of the urinary tract. In: Walsh PC, Retik AB, Stamey TA, et al., eds. Campbell's Urology. 6th Ed. Philadelphia: WB Saunders, 1992;1496.)

A Complex Interplay Between Maternal and Fetal Factors Induces Parturition

The duration of pregnancy in women averages 270 ± 14 days from the time of fertilization. **Parturition** or the onset of birth is regulated by the interactions of fetal and maternal factors. Uncoordinated uterine contractions start about 1 month before the end of gestation. The termination of pregnancy is initiated by strong rhythmic contractions that may last several hours and eventually generate enough force to expel the conceptus. The contraction of the uterine muscle is regulated by hormones and by mechanical factors. The hormones include progesterone, estrogen, prostaglandins, oxytocin, and relaxin. The mechanical factors include distension of the uterine muscle and stretching or irritation of the cervix.

Progesterone hyperpolarizes myometrial cells, lowers their excitability, and suppresses uterine contractions. It also prevents the release of phospholipase A_2, the rate-limiting enzyme in prostaglandin synthesis. Estrogen, in general, has the opposite effects. The maintenance of uterine quiescence throughout gestation, preventing premature delivery, is called the **progesterone block**. In many species, a sharp decline in the circulating levels of progesterone and a concomitant rise in estrogen precede birth. In humans, progesterone does not fall significantly before delivery. However, its effective concentration may be altered by a rise in placental progesterone-binding protein or by a decline in the number of myometrial progesterone receptors.

Prostaglandins $F_{2\alpha}$ and E_2 are potent stimulators of uterine contractions and also cause significant ripening of the cervix and its dilation. They increase intracellular calcium concentrations of myometrial cells and activate the actin-myosin contractile apparatus. Shortly before the onset of parturition, the concentration of prostaglandins in amniotic fluid rises abruptly. Prostaglandins are produced by the myometrium, decidua, and chorion. Aspirin and indomethacin, inhibitors of prostaglandin synthesis, delay or prolong parturition.

Oxytocin is also a potent stimulator of uterine contractions, and its release from both maternal and fetal pituitaries increases during labor. Oxytocin is used clinically to

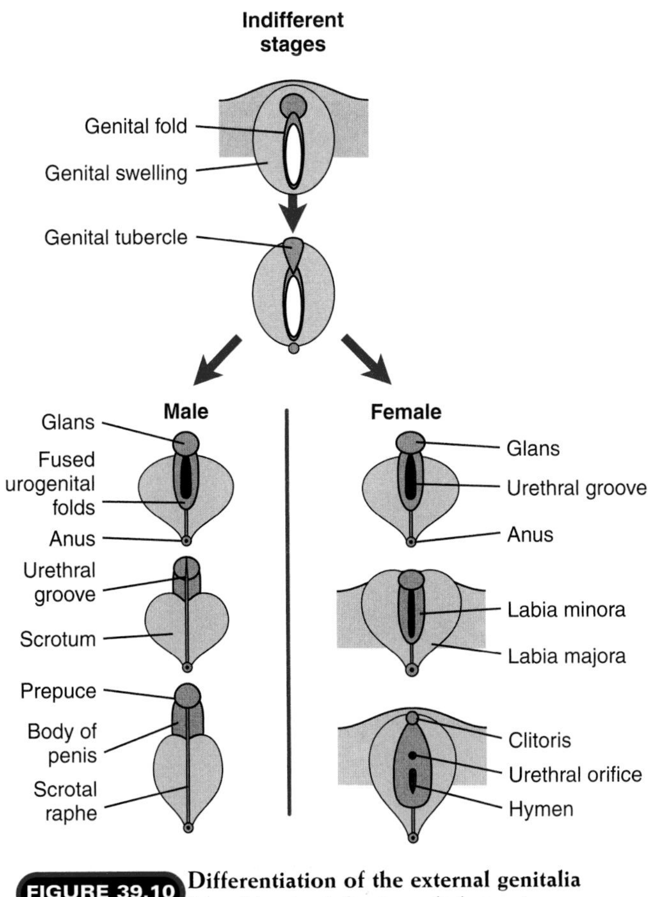

FIGURE 39.10 **Differentiation of the external genitalia from bipotential primordial structures.**

induce labor (see Clinical Focus Box 39.2). The functional significance of oxytocin is that it helps expel the fetus from the uterus, and by contracting uterine muscles, it reduces uterine bleeding when bleeding may be significant after delivery. Interestingly, oxytocin levels do not rise at the time of parturition.

Relaxin, a large polypeptide hormone produced by the corpus luteum and the decidua, assists parturition by softening the cervix, permitting the eventual passage of the fetus, and by increasing oxytocin receptors. However, the relative role of relaxin in parturition in humans is unclear, as its levels do not rise toward the end of gestation. Relaxin reaches its peak during the first trimester, declines to about half, and remains unchanged throughout the remainder of pregnancy.

The fetus may play a role in initiating labor. In sheep, the concentration of ACTH and cortisol in the fetal plasma rise during the last 2 to 3 days of gestation. Ablation of the fetal lamb pituitary or removal of the adrenals prolongs gestation, while administration of ACTH or cortisol leads to premature delivery. Cortisol enhances the conversion of progesterone to estradiol, changing the progesterone-to-estrogen ratio, and increases the production of prostaglandins. The role of cortisol and ACTH, however, has not been established in humans. Anencephalic or adrenal-deficient fetuses, which lack a pituitary and have atrophied adrenal glands, have an unpredictable length of gestation. Those pregnancies also exhibit low estrogen levels because of the lack of adrenal androgen precursors. Injections of ACTH and cortisol in late pregnancy do not induce labor. Interestingly, the administration of estrogens to the cervix causes ripening, probably by increasing the secretion of prostaglandins.

POSTPARTUM AND PREPUBERTAL PERIODS

Lactation is controlled by pituitary and ovarian hormones, requires suckling for continued milk production, and is the major source of nutrition for the newborn. As the child grows, puberty will occur around age 10 to 11 because the hypothalamus activates secretion of pituitary hormones that cause secretion of estrogens and androgens from the gonads and adrenals during that time. Alterations in hormone secretion lead to abnormal onset of puberty and gonadal development.

Mammogenesis and Lactogenesis Are Regulated by Multiple Hormones

Lactation (the secretion of milk) occurs at the final phase of the reproductive process. Several hormones participate in **mammogenesis**, the differentiation and growth of the mammary glands, and in the production and delivery of milk. **Lactogenesis** is milk production by alveolar cells. **Galactopoiesis**, the maintenance of lactation, is regulated by PRL. **Milk ejection** is the process by which stored milk is released from the mammary glands by the action of oxytocin.

Mammogenesis occurs at three distinct periods: embryonic, pubertal, and gestational. The **mammary glands** begin to differentiate in the pectoral region as an ectodermal thickening on the epidermal ridge during weeks 7 to 8 of fetal life. The prospective mammary glands lie along bilateral mammary ridges or milk lines extending from axilla to groin on the ventral side of the fetus. Most of the ridge disintegrates except in the axillary region. However, in mammals with serially repeated nipples, a distinct milk line with several nipples persists, accounting for the accessory nipples that can occur in both sexes, although rarely. Mammary buds are derived from surface epithelium, which invades the underlying mesenchyme. During the fifth month, the buds elongate, branch, and sprout, eventually forming the **lactiferous ducts**, the primary milk ducts. They continue to branch and grow throughout life. The ducts unite, grow, and extend to the site of the future nipple. The primary buds give rise to secondary buds, which are separated into lobules by connective tissue. These become surrounded by **myoepithelial cells** derived from epithelial progenitors. In response to oxytocin, myoepithelial cells will contract, and expel milk from the duct. The **nipple** and **areola**, which are first recognized as circular areas, are formed during the eighth month of gestation. The development of the mammary glands in utero appears to be independent of hormones but is influenced by paracrine interactions between the mesenchyme and epithelium.

The mammary glands of male and female infants are identical. Although underdeveloped, they have the capacity to respond to hormones, revealed by the secretion of small amounts of milk (witch's milk) in many newborns. Witch's milk results from the responsiveness of the fetal mammary

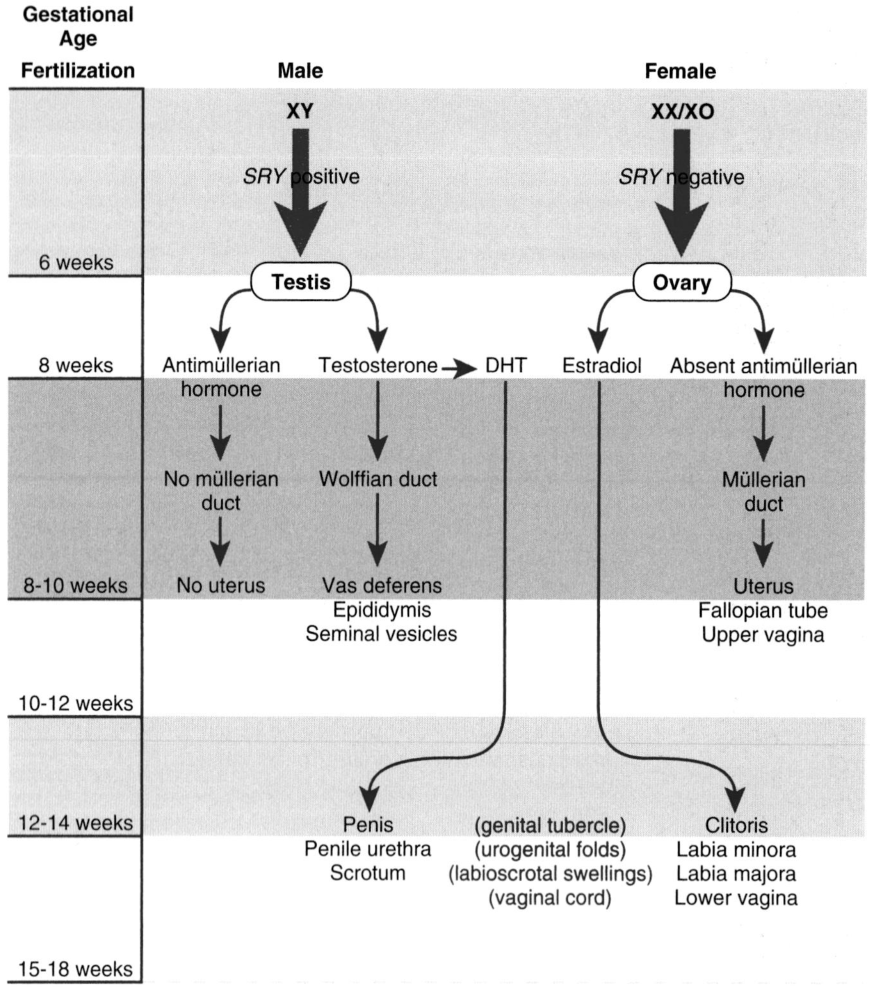

FIGURE 39.11 The process of sexual differentiation and its time course.

tissue to lactogenic hormones of pregnancy and the withdrawal of placental steroids at birth. Sexual dimorphism in breast development begins at the onset of puberty. The male breast is fully developed at about age 20 and is similar to the female breast at an early stage of puberty.

In females, estrogen exerts a major influence on breast growth at puberty. The first response to estrogen is an increase in size and pigmentation of the areola and accelerated deposition of adipose and connective tissues. In association with menstrual cycles, estrogen stimulates the growth and branching of the ducts, whereas progesterone acts primarily on the alveolar components. The action of both hormones, however, requires synergism with PRL, GH, insulin, cortisol, and thyroxine.

The mammary glands undergo significant changes during pregnancy. The ducts become elaborate during the first trimester, and new lobules and alveoli are formed in the second trimester. The terminal alveolar cells differentiate into

CLINICAL FOCUS BOX 39.2

Pharmacological Induction and Augmentation of Labor

Several drugs are currently used to assist in the therapeutic induction and augmentation of labor. Therapeutic induction implies that labor is initiated by the use of a drug. Augmentation indicates that labor has started and that the process is further stimulated by a therapeutic agent.

Oxytocin, the natural hormone produced from the posterior pituitary, is widely used to induce and augment labor. Several synthetic forms of oxytocin can be used by intravenous routes. Recently, the prostaglandins ($F_{2\alpha}$ and E_2) have also been used to induce and augment labor and cervical ripening. Prostaglandins promote dilatation and effacement of the cervix and can be used for various reasons intravaginally, intravenously, or intra-amniotically. Another therapeutic agent being tested for efficacy in labor induction and augmentation is mifepristone (RU-486), a progesterone receptor blocker. It is used to induce labor and to increase the sensitivity of the uterus to oxytocin and prostaglandins. An additional and interesting feature of these drugs is that they reduce postpartum hemorrhage by causing muscle contractions.

secretory cells, replacing most of the connective tissue. The development of the secretory capability requires estrogen, progesterone, PRL, and placental lactogen. Their action is supported by insulin, cortisol, and several growth factors. Lactogenesis begins during the fifth month of gestation, but only **colostrum** (initial milk) is produced. Full lactation during pregnancy is prevented by elevated progesterone levels, which antagonize the action of PRL. The ovarian steroids synergize with PRL in stimulating mammary growth but antagonize its actions in promoting milk secretion.

Lactogenesis is fully expressed only after parturition, on the withdrawal of placental steroids. Lactating women produce up to 600 mL of milk each day, increasing to 800 to 1,100 mL/day by the sixth postpartum month. Milk is isosmotic with plasma, and its main constituents include proteins, such as casein and lactalbumin, lipids, and lactose. The composition of milk changes with the stage of lactation. Colostrum, produced in small quantities during the first postpartum days, is higher in protein, sodium, and chloride content and lower in lactose and potassium than normal milk. Colostrum also contains immunoglobulin A, macrophages, and lymphocytes, which provide passive immunity to the infant by acting on its GI tract. During the first 2 to 3 weeks, the protein content of milk decreases, whereas that of lipids, lactose, and water-soluble vitamins increases.

The milk-secreting **alveolar cells** form a single layer of epithelial cells, joined by junctional complexes (Fig. 39.12). The bases of the cells abut on the contractile myoepithelial cells, and their luminal surface is enriched with microvilli. They have a well-developed endoplasmic reticulum and Golgi apparatus and numerous mitochondria and lipid droplets. Alveolar cells contain plasma membrane receptors for PRL, which can be internalized after binding to the hormone. In synergism with insulin and glucocorticoids, PRL is critical for lactogenesis, promotes mammary cell division and differentiation, and increases the synthesis of milk constituents. This hormone also stimulates the synthesis of casein by increasing its transcription rate and stabilizing its mRNA, and stimulates enzymes that regulate the production of lactose.

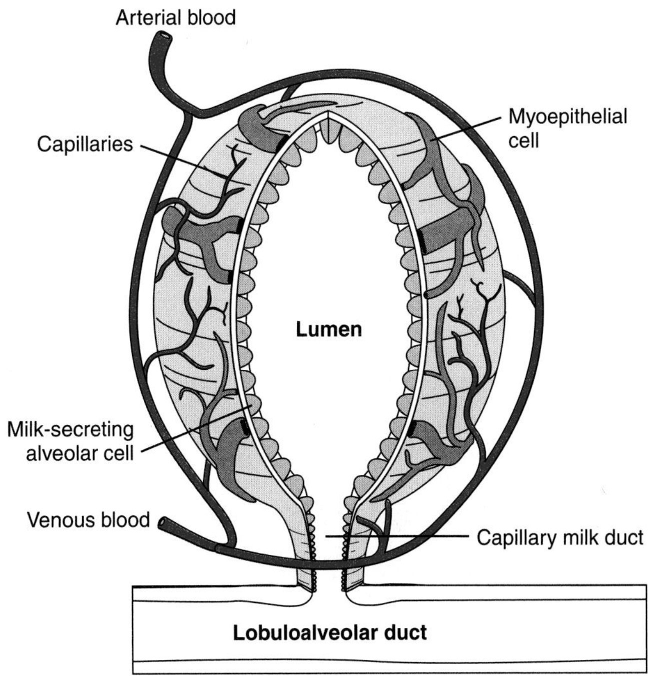

FIGURE 39.12 **The structure of a mammary alveolus.** Milk-producing cells are surrounded by a meshwork of contractile myoepithelial cells.

The Suckling Reflex Maintains Lactation and Inhibits Ovulation

The suckling reflex is central to the maintenance of lactation in that it coordinates the release of PRL and oxytocin and delays the onset of ovulation. Lactation involves two components, **milk secretion** (synthesis and release) and **milk removal**, which are regulated independently. Milk secretion is a continuous process, whereas milk removal is intermittent. Milk secretion involves the synthesis of milk constituents by the alveolar cells, their intracellular transport, and the subsequent release of formed milk into the alveolar lumen (see Fig. 39.12). PRL is the major regulator of milk secretion in women and most other mammals. Oxytocin is responsible for milk removal by activating **milk ejection** or **letdown**.

The stimulation of sensory nerves in the breast by the infant initiates the **suckling reflex**. Unlike ordinary reflexes with only neural components, the afferent arc of the suckling reflex is neural and the efferent arc is hormonal. The suckling reflex increases the release of PRL, oxytocin, and ACTH and inhibits the secretion of gonadotropins (Fig. 39.13). The neuronal component is composed of sensory receptors in the nipple that initiate nerve impulses in response to breast stimulation. These impulses reach the hypothalamus via ascending fibers in the spinal cord and then via the mesencephalon. Eventually, fibers terminating in the supraoptic and paraventricular nuclei trigger the release of oxytocin from the posterior pituitary into the general circulation (see Chapter 32). On reaching the mammary glands, oxytocin induces the contraction of myoepithelial cells, increasing intramammary pressure and forcing the milk into the main collecting ducts. The milk ejection reflex can be conditioned; milk ejection can occur because of anticipation or in response to a baby's cry.

PRL levels, which are elevated by the end of gestation, decline by 50% within the first postpartum week and decrease to near pregestation levels by 6 months. Suckling elicits a rapid and significant rise in plasma PRL. The amount released is determined by the intensity and duration of nipple stimulation. The exact mechanism by which suckling triggers PRL release is unclear, but the suppression of dopamine, the major inhibitor of PRL release, and the stimulation of prolactin-releasing factor(s) have been considered. Lactation can be terminated by dopaminergic agonists that reduce PRL or by the discontinuation of suckling. Swollen alveoli can depress milk production by exerting local pressure, resulting in vascular stasis and alveolar regression.

Lactation is associated with the suppression of cyclicity and **anovulation**. The contraceptive effect of lactation

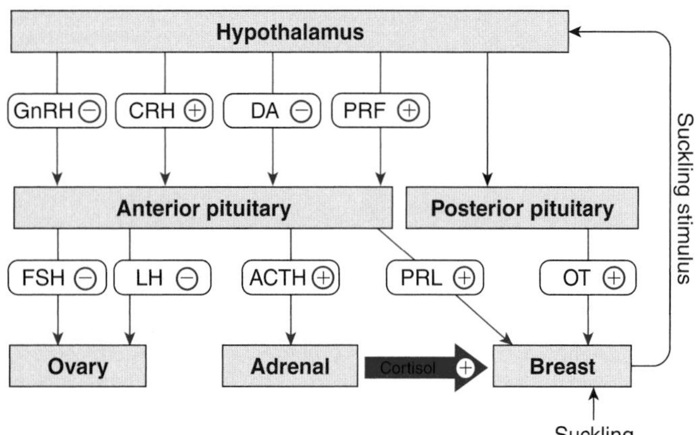

FIGURE 39.13 **Effect of suckling on hypothalamic, pituitary, and adrenal hormones.** GnRH, gonadotropin-releasing hormone; CRH, corticotropin-releasing hormone; DA, dopamine; PRF, prolactin-releasing factor; FSH, follicle-stimulating hormone; LH, luteinizing hormone; ACTH, adrenocorticotropic hormone; PRL, prolactin; OT, oxytocin. Plus and minus signs indicate positive and negative effects.

CLINICAL FOCUS BOX 39.3

Contraceptive Methods

Fertility can be controlled by interfering with the association between the sperm and ovum, by preventing ovulation or implantation, or by terminating an early pregnancy. Contraceptive methods may also be categorized as reversible and irreversible. Most current methods regulate fertility in women, with only a few contraceptives available for men (Table 39.A).

Methods based on preventing contact between the germ cells include coitus interruptus (withdrawal before ejaculation), the rhythm method (no intercourse at times of the menstrual cycle, especially when an ovum is present in the oviduct), and barriers. Barrier methods include condoms, diaphragms, and cervical caps. When combined with spermicidal agents, barrier methods approach the high success rate of oral contraceptives. Condoms are the most widely used reversible contraceptives for men. Because they also provide protection against the transmission of venereal diseases and AIDS, their use has increased in recent years. Diaphragms and cervical caps seal off the opening of the cervix. Spermicides are inserted into the vagina. Postcoital douching is not an effective contraceptive because some sperm enter the uterus and oviduct very rapidly.

Vasectomy is cutting of the two vasa deferentia, and it prevents sperm from passing into the ejaculate. An increased incidence of sperm antibodies occurs following vasectomy, but its consequences are unknown. Tubal ligation is the closure or ligation of the oviducts. Restorative surgery for the reversal of a tubal ligation and a vasectomy can be performed; its success is limited.

Oral contraceptive steroids prevent ovulation by reducing LH and FSH secretion through negative feedback. Reduced secretion of LH and FSH retards follicular development. The pill's effectiveness is also increased by adversely affecting the environment within the reproductive tract, making it unlikely for pregnancy to result even if fertilization were to occur. Exogenous estrogen and progesterone are likely to alter normal endometrial development and may contribute to their detrimental effects in the early establishment of pregnancy. Progesterone thickens cervical mucus and reduces oviductal peristalsis, impeding gamete transport.

Noncontraceptive benefits of the pill include a reduction in excessive menstrual bleeding, alleviation of premenstrual syndrome, and some protection against pelvic inflammatory disease. Adverse effects include nausea, headache, breast tenderness, water retention, and weight gain, some of which disappear after prolonged use. There is no evidence that fertility is reduced after discontinuation of the pill.

Several contraceptives act by interfering with zygote transport or implantation and cause early pregnancy termination. Among these are long-acting progesterone preparations, high doses of estrogen, and progesterone receptor antagonists, such as RU-486 (also called mifepristone). RU-486 blocks the action of the progesterone required for early pregnancy. Prostaglandins are given in combination with RU-486 to assist in the expulsion of the products of conception. The intrauterine device (IUD) also prevents implantation by provoking sterile inflammation of the endometrium and prostaglandin production. The contraceptive efficacy of IUDs, especially those impregnated with progestins, copper, or zinc, is high. The drawbacks include a high rate of expulsion, uterine cramps, excessive bleeding, perforation of the uterus, and increased incidence of ectopic pregnancy. Established pregnancy can be interrupted by surgical means (dilatation and curettage).

TABLE 39.A **Contraceptive Use and Efficacy Rates in the United States**

Method	Estimated Use (%)	Accidental Pregnancy in Year 1 (%)
Pill	32	3
Female sterilization	19	0.4
Condom	17	12
Male sterilization	14	0.15
Diaphragm	4–6	2–23
Spermicides	5	20
Rhythm	4	20
Intrauterine device	3	6

From Developing New Contraceptives: Obstacles and Opportunities. Washington, DC: National Academy Press, 1990.

is moderate in humans. In non-breast-feeding women, the menstrual cycle may return within 1 month after delivery, whereas *fully* lactating women have a period of several months of **lactational amenorrhea**, with the first few menstrual cycles being anovulatory. The cessation of cyclicity results from the combined effects of the act of suckling and elevated PRL levels. PRL suppresses ovulation by inhibiting pulsatile GnRH release, suppressing pituitary responsiveness to GnRH, reducing LH and FSH, and decreasing ovarian activity. It is also possible that PRL may inhibit the action of the low circulating levels of gonadotropins on ovarian cells. Thus, follicular development would be suppressed by a direct inhibitory action of PRL on the ovary. Although fertility is reduced by lactation, there are numerous other methods of contraception (see Clinical Focus Box 39.3).

The Onset of Puberty Depends on Maturation of the Hypothalamic GnRH Pulse Generator

The onset of puberty depends on a sequence of maturational processes that begin during fetal life. The hypothalamic-pituitary-gonadal axis undergoes a prolonged and multiphasic activation-inactivation process. By midgestation, LH and FSH levels in fetal blood are elevated, reaching near adult values. Experimental evidence suggests that the hypothalamic GnRH pulse generator is operative at this time, and gonadotropins are released in a pulsatile manner. The levels of FSH are lower in males than in females, probably because of suppression by fetal testosterone at midgestation. As the levels of placental steroids increase, they exert negative feedback on GnRH release, lowering LH and FSH to very low levels toward the end of gestation.

After birth, the newborn is deprived of maternal and placental steroids. The reduction in steroidal negative feedback stimulates gonadotropin secretion, which stimulates the gonads, resulting in transient increases in serum testosterone in male infants and estradiol in females. FSH levels in females are usually higher than those in males. At approximately 3 months of age, the levels of both gonadotropins and gonadal steroids are in the low-normal adult range. Circulating gonadotropins decline to low levels by 6 to 7 months in males and 1 to 2 years in females and remain suppressed until the onset of puberty.

Throughout childhood, the gonads are quiescent and plasma steroid levels are low. Gonadotropin release is also suppressed. The prepubertal restraint of gonadotropin secretion is explained by two mechanisms, both of which affect the hypothalamic GnRH pulse generator. One is a sex steroid-dependent mechanism that renders the pulse generator extremely sensitive to negative feedback by steroids. The other is an intrinsic central nervous system (CNS) inhibition of the GnRH pulse generator. Together, they suppress the amplitude, and probably the frequency, of GnRH pulses, resulting in diminished secretion of LH, FSH, and gonadal steroids. Throughout this period of quiescence, the pituitary and the gonads can respond to exogenous GnRH and gonadotropins, but at a relatively low sensitivity.

The hypothalamic-pituitary axis becomes reactivated during the late prepubertal period. This response involves a decrease in hypothalamic sensitivity to sex steroids and a reduction in the effectiveness of intrinsic CNS inhibition over the GnRH pulse generator. The mechanisms underlying these changes are unclear but might involve endogenous opioids. As a result of disinhibition, the frequency and amplitude of GnRH pulses increase. Initially, pulsatility is most prominent at night, entrained by deep sleep; later it becomes established throughout the 24-hour period. GnRH acts on the gonadotrophs of the anterior pituitary as a self-primer. It increases the number of GnRH receptors (up-regulation) and augments the synthesis, storage, and secretion of the gonadotropins. The increased responsiveness of FSH to GnRH in females occurs earlier than that of LH, accounting for a higher FSH/LH ratio at the onset of puberty than during late puberty and adulthood. A reversal of the ratio is seen again after menopause.

The increased pulsatile GnRH release initiates a cascade of events. The sensitivity of gonadotrophs to GnRH is increased, the secretion of LH and FSH is augmented, the gonads become more responsive to the gonadotropins, and the secretion of gonadal hormones is stimulated. The rising circulating levels of gonadal steroids induce progressive development of the secondary sex characteristics and establish an adult pattern of negative feedback on the hypothalamic-pituitary axis. Activation of the positive-feedback mechanism in females and the capacity to exhibit an estrogen-induced LH surge is a late event, expressed in midpuberty to late puberty.

The onset of puberty in humans begins at age 10 to 11. Lasting 3 to 5 years, the process involves the development of secondary sex characteristics, a growth spurt, and the ac-

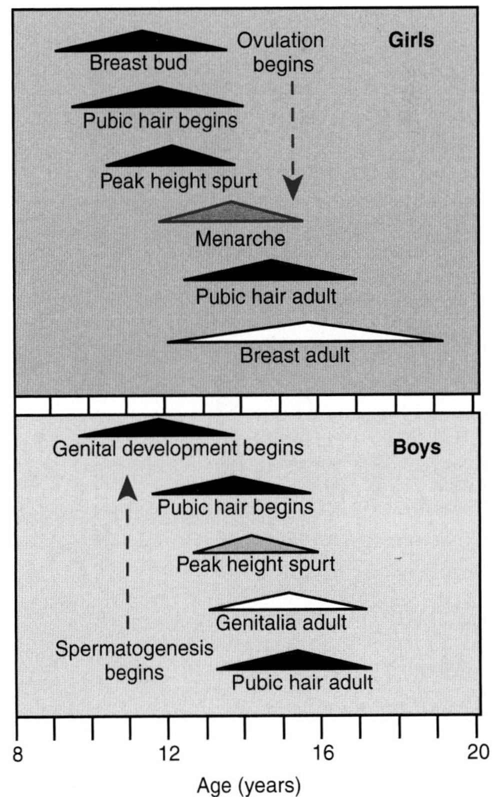

FIGURE 39.14 **Peripubertal maturation of secondary sex characteristics in girls and boys.**

quisition of fertility. The timing of puberty is determined by genetic, nutritional, climatic, and geographic factors. Over the last 150 years, the age of puberty has declined by 2 to 3 months per decade; this pattern appears to correlate with improvements in nutrition and general health in Americans.

The first physical signs of puberty in girls are breast budding, **thelarche**, and the appearance of pubic hair. Axillary hair growth and peak height spurt occur within 1 to 2 years. **Menarche**, the beginning of menstrual cycles, occurs at a median age of 12.8 years in American girls. The first few cycles are usually anovulatory. The first sign of puberty in boys is enlargement of the testes, followed by the appearance of pubic hair and enlargement of the penis. The peak growth spurt and appearance of axillary hair in boys usually occurs 2 years later than in girls. The growth of facial hair, deepening of the voice, and broadening of the shoulders are late events in male pubertal maturation (Fig. 39.14).

Puberty is also regulated by hormones other than gonadal steroids. The adrenal androgens DHEA and DHEAS are primarily responsible for the development of pubic and axillary hair. Adrenal maturation or **adrenarche** precedes gonadal maturation or **gonadarche** by 2 years. The pubertal growth spurt requires a concerted action of sex steroids and growth hormone. The principal mediator of GH is insulin-like growth factor-I (IGF-I). Plasma concentration of IGF-I increases significantly during puberty, with peak levels observed earlier in girls than in boys. IGF-I is essential for accelerated growth. The gonadal steroids appear to act primarily by augmenting pituitary growth hormone release, which stimulates the production of IGF-I in the liver and other tissues.

Disorders of Sexual Development Can Manifest Before or After Birth

Normal sexual development depends on a complex, orderly sequence of events that begins during early fetal life and is completed at puberty. Any deviation can result in infertility, sexual dysfunction, or various degrees of intersexuality or **hermaphroditism**. A true hermaphrodite possesses both ovarian and testicular tissues, either separate or combined as ovotestes. A **pseudohermaphrodite** has one type of gonads but a different degree of sexuality of the opposite sex. Sex is normally assigned according to the type of gonads. Disorders of sexual differentiation can be classified as gonadal dysgenesis, female pseudohermaphroditism, male pseudohermaphroditism, or true hermaphroditism. Se-

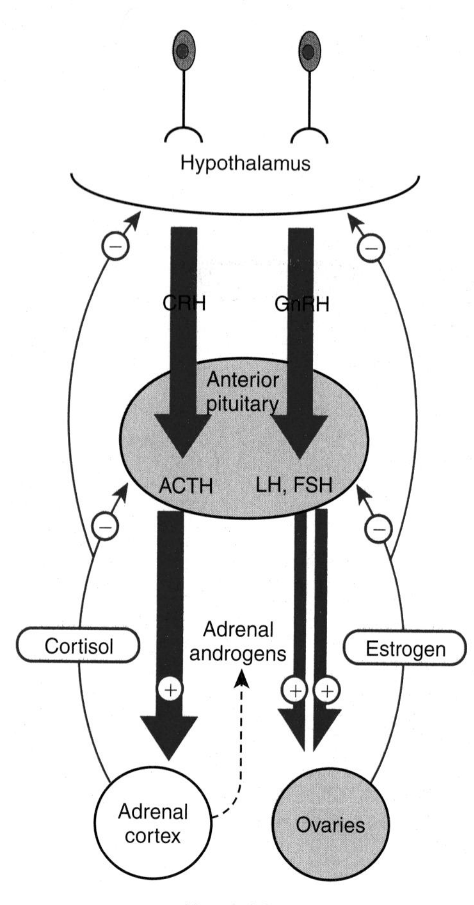

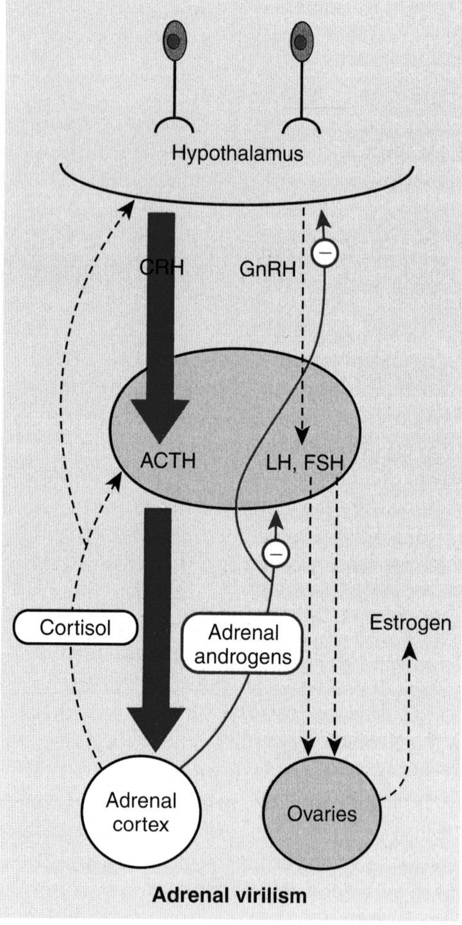

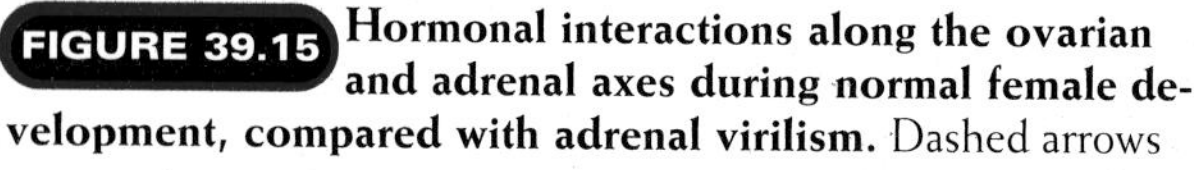

FIGURE 39.15 Hormonal interactions along the ovarian and adrenal axes during normal female development, compared with adrenal virilism. Dashed arrows indicate low production of the hormone. Heavy arrows indicate increased hormone production. Plus and minus signs indicate positive and negative effects.

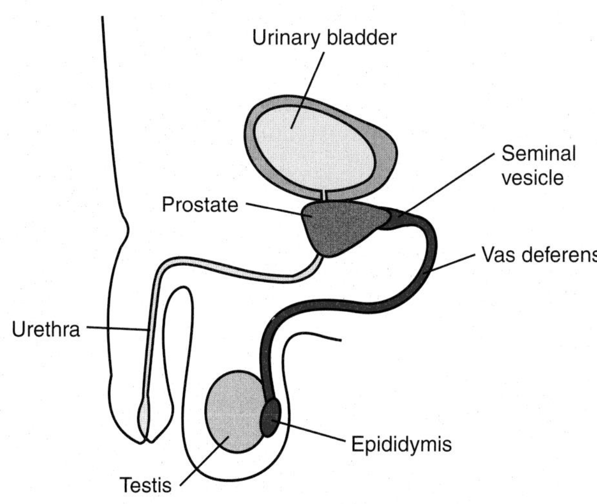

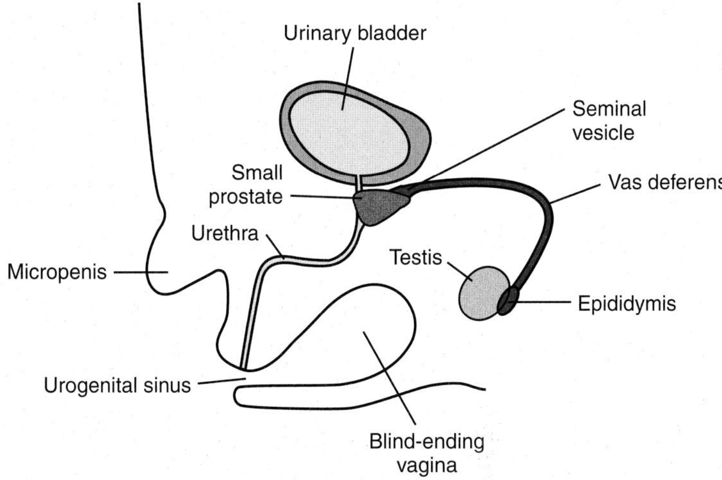

FIGURE 39.16 **Effects of 5α-reductase deficiency on differentiation of the internal and external genitalia.**

lected cases and their manifestations are briefly discussed here, as are disorders of pubertal development (see also Chapters 37 and 38).

Gonadal dysgenesis refers to incomplete differentiation of the gonads and is usually associated with sex chromosome abnormalities. These result from errors in the first or second meiotic division and occur by chromosomal nondisjunction, translocation, rearrangement, or deletion. The two most common disorders are Klinefelter's syndrome (47,XXY) and Turner's syndrome (45,XO). Because of a Y chromosome, an individual with a 47,XXY karyotype has normal testicular function in utero in terms of testosterone and AMH production and no ambiguity of the genitalia at birth. The extra X chromosome, however, interferes with the development of the seminiferous tubules, which show extensive hyalinization and fibrosis, whereas the Leydig cells are hyperplastic. Such males have small testes, are azoospermic, and often exhibit some eunuchoidal features. Because of having only one X chromosome, an individual with a 45,XO karyotype will have no gonadal development during fetal life and is presented at birth as a phenotypic female. Given the absence of ovarian follicles, such patients have very low levels of estrogens, primary amenorrhea, and do not undergo normal pubertal development.

Female pseudohermaphrodites are 46,XX females with normal ovaries and internal genitalia but a different degree of virilization of the external genitalia, resulting from exposure to excessive androgens in utero. The most common cause is **congenital adrenal hyperplasia**, an inherited abnormality in adrenal steroid biosynthesis, with most cases of virilization resulting from 21-hydroxylase or 11β-hydroxylase deficiency (see Chapter 34). In such cases, cortisol production is low, causing increased production of ACTH by activating the hypothalamic-pituitary axis (Fig. 39.15). The elevated ACTH levels induce adrenal hyperplasia and an abnormal production of androgens and corticosteroid precursors. These infants are born with ambiguous external genitalia (i.e., clitoromegaly, labioscrotal fusion, or phallic urethra). The degree of virilization depends on the time of onset of excess fetal androgen production. When aldosterone levels are also affected, a life-threatening salt-wasting disease results. Untreated patients with congenital adrenal virilism develop progressive masculinization, amenorrhea, and infertility.

Male pseudohermaphrodites are 46,XY individuals with differentiated testes but underdeveloped and/or absent wolffian-derived structures and inadequate virilization of the external genitalia. These effects result from defects in testosterone biosynthesis, metabolism, or action. The 5α-reductase deficiency is an autosomal recessive disorder caused by the inability to convert testosterone to DHT. Such infants have ambiguous or female external genitalia and normal male internal genitalia (Fig. 39.16). They are often raised as females but undergo a complete or partial testosterone-dependent puberty, including enlargement of the penis, testicular descent, and the development of male psychosexual behavior. Azoospermia is common.

The **testicular feminization** syndrome is an X-linked recessive disorder caused by end-organ insensitivity to androgens, usually because of absent or defective androgen receptors. These 46,XY males have abdominal testes that secrete normal testosterone levels. Because of androgen insensitivity, the wolffian ducts regress, and the external genitalia develop along the female line. The presence of AMH in utero causes regression of the müllerian ducts. These individuals have neither male nor female internal genitalia and phenotypic female external genitalia, with the vagina ending in a blind pouch. They are reared as females and undergo feminization during puberty because of the peripheral conversion of testosterone to estradiol.

Disorders of puberty are classified as **precocious puberty**, defined as sexual maturation before the age of 8 years, and **delayed puberty**, in which menses do not start by age 17 or testicular development is delayed beyond age 20. True precocious puberty results from premature activation of the hypothalamic-pituitary-gonadal axis, leading to the development of secondary sex characteristics as well as gametogenesis. The most frequent causes are CNS lesions or infections, hypothalamic disease, and hypothyroidism. **Pseudoprecocious puberty** is the early development of secondary sex characteristics without gametogenesis. It can result from the abnormal exposure of immature boys to androgens and of immature girls to estrogens. Augmented steroid production can be of gonadal or adrenal origin.

REVIEW QUESTIONS

DIRECTIONS: Each of the numbered items or incomplete statements in this section is followed by answers or by completions of the statement. Select the ONE lettered answer or completion that is BEST in each case.

1. The suckling reflex
(A) Has afferent hormonal and efferent neuronal components
(B) Increases placental lactogen secretion
(C) Increases the release of dopamine from the arcuate nucleus
(D) Triggers the release of oxytocin by stimulating the supraoptic nuclei
(E) Reduces PRL secretion from the pituitary
2. Implantation occurs
(A) On day 4 after fertilization
(B) After the endometrium undergoes a decidual reaction
(C) When the embryo is at the morula stage
(D) Only after priming of the uterine endometrium by progesterone and estrogen
(E) On the first day after entry of the embryo into the uterus
3. Upon contact between the sperm head and the zona pellucida, penetration of the sperm into the egg is allowed because of
(A) The acrosome reaction
(B) The zona reaction
(C) The perivitelline space
(D) Pronuclei formation
(E) Cumulus expansion
4. The next ovulatory cycle after implantation is postponed because of
(A) High levels of PRL
(B) The production of hCG by trophoblast cells
(C) The production of prostaglandins by the corpus luteum
(D) The depletion of oocytes in the ovary
(E) Low levels of progesterone
5. Polyspermy block occurs as a result of the
(A) Cortical reaction
(B) Enzyme reaction
(C) Acrosome reaction
(D) Decidual reaction
(E) Inflammatory reaction
6. Oral steroidal contraceptives are most effective in preventing pregnancy by
(A) Blocking ovulation
(B) Altering the uterine environment
(C) Thickening the cervical mucus
(D) Reducing sperm motility
(E) Inducing a premature LH surge
7. The maternal recognition of pregnancy occurs as a result of the
(A) Prolonged secretion of estrogen by the placenta
(B) Production of human placental lactogen
(C) Increased secretion of progesterone by the corpus luteum
(D) Secretion of hCG by the trophoblast
(E) Activation of an inflammatory reaction at implantation
8. Estriol production during pregnancy requires
(A) Androgens produced from cholesterol in the placenta
(B) Estradiol as a precursor from the mother's ovary
(C) Androgenic substrates from the fetus
(D) Androgens from the ovary of the mother
(E) Estradiol to be produced in the placenta
9. One benefit of insulin resistance in the mother during pregnancy is
(A) A reduction of her plasma glucose concentrations
(B) The blockage of the development of diabetes mellitus in later life
(C) The increased availability of glucose to the fetus
(D) A reduction of pituitary function
(E) Increased appetite
10. The primary reason that the female phenotype develops in an XY male is
(A) The secretion of progesterone
(B) Adrenal insufficiency
(C) The lack of testosterone action
(D) Increased inhibin secretion
(E) The secretion of antimüllerian hormone (AMH)

SUGGESTED READING

Carr BR. Fertilization, implantation, and endocrinology of pregnancy. In: Griffin JE, Ojeda SR, eds. Textbook of Endocrine Physiology. 4th Ed. New York: Oxford University Press, 2000;265–285.

Hay WW Jr. Metabolic changes in pregnancy. In: Knobil E, Neill JD, eds. The

Encyclopedia of Reproduction. New York: Academic Press, 1999;1016–1026.
Johnson MH, Everitt BJ. Essential Reproduction. Oxford, UK: Blackwell Science, 2000.
Parker CR Jr. The endocrinology of pregnancy. In: Carr BR, Blackwell RE, eds. Textbook of Reproductive Medicine. East Norwalk, CT: Appleton & Lange, 1993;17–40.
Regan CL. Overview, pregnancy in humans. In Knobil E, Neill JD, eds. The encyclopedia of reproduction. New York: Academic Press, 1999;986–991.
Spencer TE: Maternal recognition of pregnancy. In: Knobil E, Neill JD, eds. The Encyclopedia of Reproduction. New York: Academic Press, 1999;1006–1015.

CASE STUDIES FOR PART X • • •

CASE STUDY FOR CHAPTER 37

Steroid Abuse

A 30-year-old man and his 29-year-old wife have been trying to have a baby. She has been having regular menstrual cycles. They have intercourse 2 to 3 times a week, with no physical problems, and try to time intercourse around the time of her ovulation.

Physical examination and history for the wife are normal. The husband's physical examination reveals a muscular man with an excellent physique who works out regularly to build his body. His testes are small and soft. Laboratory results indicate that his plasma testosterone is 1850 ng/dL (normal, 300 to 100 ng/dL) and LH $< 2\ \mu$U/mL (normal, 3 to 18 μU/mL). His semen analysis reveals a sperm count of 1.2×10^6/mL (normal, $>20 \times 10^6$/mL).

Questions

1. What is the major reason for the failure of the wife to get pregnant?
2. What is a reasonable explanation for the abnormal hormone levels?

Answers to Case Study Questions for Chapter 37

1. He has an extremely low sperm count.
2. Since he is a body builder with small, soft testes, high testosterone levels, and low LH levels, the physician should suspect androgen abuse or possibly androgen-producing tumor (extremely rare). The high androgen levels would suppress LH secretion and reduce intratesticular testosterone levels. The low LH and intratesticular testosterone would correlate well with small testes and low sperm count, respectively.

CASE STUDY FOR CHAPTER 38

Early Spontaneous Pregnancy Termination

A 35-year-old woman visited her obstetrician/gynecologist and complained that she was unable to get pregnant. Upon taking a medical history, the physician notes that the patient had regular 28- to 30-day cycles during the past year, during which time she had regular unprotected intercourse. She does not smoke and does not use caffeine, drugs, or alcohol. She appears to be in good health. Her ovaries and uterus appear normal in size for her age. Laboratory tests indicate that her preovulatory (late follicular phase) estradiol is 40 pg/mL (normal, 200 to 500 pg/mL) and midluteal phase progesterone is 3 ng/mL (normal, 4 to 20 ng/mL). Her husband's sperm count is 30 million/mL.

Questions

1. What are the clinical indications of a fertility problem with this patient?
2. Based on the clinical signs, what basic physiological principles provide insight into the infertility?
3. What are some theoretical treatment options for this patient?

Answers to Case Study Questions for Chapter 38

1. There are two laboratory tests that indicate a problem, the low late follicular phase plasma estradiol concentration and the low midluteal phase plasma progesterone concentration.
2. The low estradiol could be due to the development of a small dominant graafian follicle with insufficient numbers of granulosa cells. The reduced number of granulosa cells would not contain sufficient aromatase to synthesize the high levels of estradiol required during the late follicular phase. In addition, low estradiol could be due to inadequate FSH receptors on the granulosa cells or inadequate FSH secretion. The low estradiol could also be explained by a lack of LH stimulation of thecal androgen production from the small dominant follicle, possibly the result of inadequate LH receptors on theca cells or low LH levels.

   The low progesterone during the luteal phase might be due to the ovulation of a small follicle or premature ovulation of a follicle that was not fully developed. The number of LH receptors on the luteinized granulosa cells in the graafian follicle and developing corpus luteum may be insufficient, or LH secretion may be deficient. LH receptors mediate the action of LH, which stimulates progesterone secretion. An insufficient number of LH receptors could be due to insufficient priming of the developing follicle with FSH. Finally, the LH surge may be insufficient for maximal progesterone secretion.
3. Theoretical treatment options for the patient include exogenous progesterone during the luteal phase, which would raise the overall circulating progesterone to levels compatible with maintaining pregnancy, allowing implantation of the embryo.

   Another option would be to use exogenous FSH to stimulate follicular development to produce larger follicle(s) with sufficient estradiol secretion and LH receptors. Follicles with adequate LH receptors would respond to an LH surge with increased progesterone in the normal range. Another option is the administration of hCG during the periovulatory period for inducing ovulation and full luteinization. The latter would overcome any deficiency in the endogenous LH surge. Finally, the use of clomiphene, an antiestrogen, would increase FSH (and LH) secretion in the follicular phase and, subsequently, induce follicular development with sufficient estradiol to induce a full LH surge. Exogenous hCG could be given during the ovulatory phase to ensure full luteinization of the corpus luteum with sufficient progesterone to maintain pregnancy.

CASE STUDY FOR CHAPTER 39

Female Infertility

A 25-year-old woman and her 29-year-old husband have been trying to have a baby for 1 year. She has regular menstrual cycles of 24 to 26 days in length. The couple has intercourse 3 times a week, with no physical problems, and they try to time intercourse around the time of her ovulation.

Physical examination and history for the wife are normal. The husband's physical examination is normal. The husband's semen analysis reveals a semen volume of 4 mL; pH 7.5; sperm count of 30 million/mL; and normal morphology and motility of the sperm. Because of the short cycles (24 to 26 days versus 28 days), the wife's plasma progesterone level during the midluteal phase is assessed and determined to be 10 ng/mL, which is considered normal (4 to 20 ng/mL, see Table 38.3).

Questions

1. What hormones can be measured in the blood to determine why the patient is not able to get pregnant?
2. Based on the hormone measurements, what treatment would likely result in a successful pregnancy?

Answers to Case Study Questions for Chapter 39

1. Estradiol should be measured at the end of the anticipated follicular phase. High serum levels of estradiol would indicate that a dominant follicle has been recruited and is active; low levels would indicate a subnormal dominant follicle or lack of a dominant follicle; this could be verified by ultrasound of the ovaries. Serum LH should be measured during the anticipated preovulatory period. High serum levels of LH would indicate that the dominant follicle is getting the signal to ovulate. Low levels of LH may lead to an unruptured dominant follicle that fails to ovulate but luteinizes, leading to progesterone levels in the normal range for the luteal phase. Plasma concentrations of hCG could be determined during the midluteal to late luteal phase to determine whether she was pregnant.
2. Low estradiol indicates a lack of development of a dominant follicle. Therapies such as gonadotropins and clomiphene (see Chapter 38) would be appropriate to stimulate follicular development, estradiol secretion, and ovulation. If a dominant follicle is present, then hCG can be given to induce follicular rupture. hCG binds to the LH receptor and is preferred over LH and GnRH for ovulation induction because of its longer half-life.

APPENDIX A

Answers to Review Questions

Chapter 1

1. **The answer is C.** In a steady state, the amount or concentration of a substance in a compartment does not change with respect to time. Although there may be considerable movements into and out of the compartment, there is no net gain or loss. Steady states in the body often do not represent an equilibrium condition, but they are displaced from equilibrium by the constant expenditure of metabolic energy.
2. **The answer is D.** The increase in plasma insulin lowers the plasma glucose concentration back to normal and is an example of negative feedback. Negative feedback opposes change and results in stability. Positive feedback would produce a further increase in plasma glucose concentration. Chemical equilibrium indicates a condition in which the rates of reactions in forward and backward directions are equal. End-product inhibition occurs when the products of a chemical reaction slow the reaction (for example, by inhibiting an enzyme) that produces them. Feedforward control involves a command signal and does not directly sense the regulated variable (plasma glucose concentration).
3. **The answer is E.** The EGF receptor is a tyrosine kinase receptor; therefore, an inhibitor of tyrosine kinases should have the desired effect. An adenylyl cyclase stimulator or phosphodiesterase inhibitor would increase intracellular levels of cAMP, but this is not the second messenger problem here. An EGF agonist would increase signaling along the EGF pathway and would increase the problem, causing an undesired effect. Likewise, a phosphatase inhibitor would slow the hydrolysis of phosphorylated intermediates and maintain the activated state of the EGF pathway.
4. **The answer is D.** Second messengers are a class of signaling molecules generated inside cells in response to the activation of a receptor. If second messengers were always available, signal transduction pathways could not be regulated. The response to a second messenger varies depending on the cell type because each cell type differs with respect to the number and complement of receptors, effectors, and downstream targets. Second messengers include nucleotides such as cAMP or cGMP, ions such as calcium, and gases such as nitric oxide. Many different plasma membrane receptors, not only tyrosine kinase receptors, are coupled to second messenger generating systems.
5. **The answer is B.** Cyclic nucleotides are generated by the action of either adenylyl or guanylyl cyclase on ATP or GTP, respectively. Cyclic AMP (cAMP) and cyclic GMP (cGMP) activate distinct signaling pathways. For example, cAMP can activate protein kinase A, which will phosphorylate its substrates; cGMP activates protein kinase G, which phosphorylates a different set of substrates. Although signal transduction in sensory tissues involves both cAMP and cGMP, cGMP has a more important role in signal transduction than cAMP. Phospholipase C activation is coupled to the activation of a G protein (G_q), not to cAMP or cGMP.
6. **The answer is D.** Steroid hormone receptors are transcriptional regulators found in the cytoplasm or in the nucleus. These receptors are activated by the binding of steroid ligands that diffuse through lipid bilayers and enter the cytosol. Activated steroid receptors mediate their effects by direct interaction with gene regulatory elements and do not activate G proteins or cause binding of IP_3 to the IP_3-gated calcium-release channel in the sarcoplasmic reticulum. Steroid hormone receptors do not have tyrosine kinase activity and do not cause the phosphorylation of tyrosine residues in these receptors. Steroid hormone receptors are not linked to activation of the MAP kinase pathway. Estrogen receptors are located in the cytoplasm of cells; upon binding of estrogen, they move to the nucleus to bind to estrogen response elements to activate gene transcription.
7. **The answer is E.** Cardiac muscle cells have many gap junctions that allow the rapid transmission of electrical activity and the coordination of heart muscle contraction. Gap junctions are pores composed of paired connexons that allow the passage of ions, nucleotides, and other small molecules between cells.
8. **The answer is E.** Inositol trisphosphate (IP_3) and diacylglycerol (DAG) are generated by the action of phospholipase C (PLC) on PIP_2, phosphatidylinositol 4,5-bisphosphate. IP_3 and DAG are second messengers, not first messengers. DAG is important for the activation of protein kinase C, not PLC. Tyrosine kinase receptors are activated by the binding of ligands, such as hormones or growth factors, not by IP_3 or DAG. IP_3 can indirectly activate calcium-calmodulin-dependent protein kinases by causing the release of calcium from intracellular stores; DAG has no direct effect on these kinases.
9. **The answer is C.** The activation of tyrosine kinase receptors often results in a cellular response that is involved in growth or differentiation. Tyrosine kinase receptors do not have constitutively active receptors; if

this were true, there could be no regulation of signaling. The activation of ras occurs indirectly by the activation of adapter molecules (Grb2 and SOS) that associate with phosphorylated tyrosine residues in the cytoplasmic tail of the receptor. The activation of tyrosine kinase receptors usually involves multimerization into dimers or trimers.

10. **The answer is B.** If it is unable to hydrolyze GTP, the $G\alpha_s$ subunit remains in its active form and results in increases in adenylyl cyclase activity, intracellular cAMP, and release of growth hormone. If $G\alpha_i$ is activated, adenylyl cyclase activity will be decreased. A lack of GHRH receptors should produce decreased, not increased, growth hormone secretion.

Chapter 2

1. **The answer is D.** Phospholipids have both a polar hydrophilic head group and a hydrophobic region because of the long hydrocarbon chains of the two fatty acids. Since the hydrophilic and hydrophobic regions are present within the same molecule, phospholipids are described as amphipathic. Phospholipids are not soluble in water and do not have a steroid structure.
2. **The answer is B.** Phospholipid molecules can rotate and move laterally within the plane of the lipid bilayer, but movement from one half of the bilayer to the other is slow because it is an energetically unfavorable process. Cholesterol is an example of a separate class of lipids that do not contain fatty acids. Many phospholipids are distributed unequally between the two halves of the lipid bilayer. In the red blood cell membrane, for example, most of the phosphatidylcholine is in the outer half. Both ion channels and symporters are membrane proteins, not phospholipids.
3. **The answer is A.** Membrane-spanning segments of integral proteins frequently adopt an α-helical conformation because this structure maximizes the opportunities for the polar peptide bonds to form hydrogen bonds with one another in the hydrophobic interior of the lipid bilayer. These segments are composed largely of amino acids with nonpolar hydrophobic side chains that interact with the surrounding lipids. There are no covalent bonds with cholesterol or phospholipids, and the peptide bonds are not unusually strong.
4. **The answer is D.** The Nernst equation calculates the membrane potential that develops when a single ion is distributed at equilibrium across a membrane. The Goldman equation gives the value of the membrane potential when all permeable ions are accounted for. The van't Hoff equation calculates the osmotic pressure of a solution, and Fick's law refers to the diffusional movement of solute. The permeability coefficient accounts for several factors that determine the ease with which a solute can cross a membrane.
5. **The answer is B.** Intracellular K^+ is high compared with all other intracellular ions.
6. **The answer is C.** Active transport always moves solute against its electrochemical gradient. All the other options are shared by both active transport and equilibrating carrier-mediated transport systems.
7. **The answer is A.** Cyclic GMP is the ligand that opens the ion channel in this example. Ion pumps and Na^+ solute-coupled transporters are examples of active transport systems, not Na^+ channels. The process is unrelated to receptor-mediated endocytosis.
8. **The answer is C.** K^+ is the major intracellular ion; efflux of K^+ will produce an osmotic flow of water out of the cell. Water exit will lead to a decrease in cell volume. An influx of Na^+ and synthesis of sorbitol do not occur during this process because both processes would increase intracellular osmolytes and drive water into the cell, increasing cell volume.
9. **The answer is D.** By substituting in the Nernst equation (equation 7):

$$\begin{aligned} E_i - E_o &= 61/{-1} \times \log_{10} 120/8 \\ &= -61 \times 1.176 \\ &= -71.7 \text{ mV inside the cell.} \end{aligned}$$

Note that for Cl^-, the value for z (valence) is -1.

10. **The answer is E.** By using the van't Hoff equation:

$$\begin{aligned} \pi &= n R T \phi C \\ &= 3 \times 0.0821 \times 300 \times 0.86 \times 0.1 \\ &= 6.35 \text{ atm} \end{aligned}$$

Chapter 3

1. **The answer is B.** Potassium ion concentration is high in the intracellular fluid relative to the extracellular fluid. The opposite is true for sodium ion concentration; therefore, there is a strong driving force for potassium to leave the cell and sodium to enter the cell. If the permeability to K^+ increases, more potassium would leave the cell, and the cell would become more negative (hyperpolarize). If the permeability to Na^+ decreases, less sodium would enter the cell, and the cell would become less positive (hyperpolarize).
2. **The answer is C.** Voltage-gated potassium channels open with a delay relative to voltage-gated sodium channels in response to depolarization. Concomitantly, there is a delay in their closing relative to the sodium channels. During the afterhyperpolarization phase of the action potential, the sodium channels are closed, but the potassium channels remain open. Because there is a strong driving force for K^+ to leave the cell, the cell hyperpolarizes. An outward calcium current, an outward sodium current, or an inward chloride current could conceivably hyperpolarize the cell, but these currents are not the basis for this phase of the action potential.
3. **The answer is D.** A specialization occurs in myelinated axons in which the voltage-gated sodium channels are preferentially distributed to the axonal membrane beneath the nodes of Ranvier. Since these channels are required for the generation of an action potential, the action potential jumps from node to node. This process is facilitated by an increased membrane resistance and a decreased capacitance associated with the myelinated regions of the axon, both of which promote the electrotonic spread of the positive charge that accumulates beneath one node of Ranvier at the

peak of the action potential. Nongated ion channels are not involved in the generation of action potential.

4. **The answer is C.** Myelin contributes substantially to the effective membrane resistance, R_m. The space constant increases as R_m increases because it is more difficult for ions to flow across the membrane relative to the ease with which they flow within the axon. When an axon demyelinates, its space constant decreases and conduction velocity is slowed. This slowing of the conduction velocity is the basis for the neurological deficits associated with multiple sclerosis.
5. **The answer is C.** SNARES are the group of proteins responsible for docking and binding synaptic vesicles to the presynaptic membrane to prepare them for release. If the vesicles cannot dock, they cannot fuse with the membrane to release their neurotransmitter. Disruption of SNARES has no direct effect on other components of neurochemical transmission, including action potential propagation, transmitter-receptor interaction, or uptake mechanisms.
6. **The answer is A.** Spatial summation of synaptic potentials can occur if they are close enough that the space constant spans the two synapses; therefore, properties of the cell that increase the space constant would optimize the effectiveness of the two synapses. The space constant increases with increasing membrane resistance or decreasing cytoplasmic resistance. Cytoplasmic resistance decreases as the cross-sectional area increases. Temporal summation could also increase the effectiveness of the two synapses; this would be facilitated by a large time constant.
7. **The answer is C.** Acetylcholinesterase is the enzyme that breaks acetylcholine down into acetate and choline. The acetate diffuses away, and choline is taken back up into the presynaptic nerve terminal for the synthesis of more ACh. Blocking the function of acetylcholinesterase would prevent the breakdown of ACh, which would accumulate in the cleft because there is no uptake mechanism for ACh and it diffuses away more slowly than acetate.
8. **The answer is C.** Catecholaminergic transmission is efficient, in part, because there is a significant reuptake of the catecholamines for repackaging into synaptic vesicles to use again. MAO and COMT are not found in the cleft and do not aid in the removal of the catecholamines from the cleft. The postsynaptic cell may have an uptake mechanism (not endocytosis) for the catecholamines, but the efficacy of this mechanism is substantially lower than the reuptake mechanism.
9. **The answer is A.** Dopamine plays a major role in two functional systems of the brain, the motor system and the limbic system. Within the limbic system, DA is associated with affect. Too much dopaminergic transmission can result in psychotic disorders, such as schizophrenia. A blockade of dopaminergic transmission ameliorates psychosis. Cholinergic transmission is involved in cognitive function and motor control. The role of nitrergic transmission in cognition and behavior is unknown.
10. **The answer is D.** Most neurotransmitters are synthesized locally within the axon terminals from precursors available at the terminals, using enzymes that reside in the terminals. However, peptides must be synthesized by ribosomes, which are not found in axons or terminals. The supply of peptide transmitters in the axon terminal must be continuously replenished via axoplasmic transport from the cell body. Microtubules are an essential component of axoplasmic transport; disrupting their integrity would diminish axonal transport and deplete the peptide transmitter from the terminal.
11. **The answer is B.** GABA is the major inhibitory transmitter in the brain. The activation of GABA receptors hyperpolarizes neurons. Activity of the GABA system is widespread in the brain, and a disruption of GABA signaling results in a hyperexcitability of neurons that can lead to seizures.
12. **The answer is B.** The acute onset of symptoms in both people suggests food poisoning and not a chronic disorder or a stroke. A toxin that blocked nerve-muscle transmission would produce muscle paralysis or weakness and no sensory disturbances. The tingling feeling suggests abnormally high excitability and firing of sensory nerves. Ciguatera toxin, the product of a dinoflagellate that sometimes contaminates red snapper and other reef fishes, is probably the cause of the sensory abnormality and gastrointestinal symptoms. Ciguatera toxin binds to voltage-gated sodium channels and results in their persistent activation.
13. **The answer is C.** Loss of myelin will result in a lower conduction velocity because the action potential will no longer "jump" from node to node. The compound action potential (the sum of many individual action potentials) will be more spread out and will have a slower rate of rise than normal. The afterhyperpolarization will last longer.
14. **The answer is C.** Release of transmitter depends on opening of voltage-gated calcium channels and entry of extracellular calcium into the nerve terminals. Deficient acetylcholine release by motor nerve terminals could explain muscle weakness. Nerve conduction velocity is not dependent on calcium channels. The repolarization phase of the nerve action potential depends on voltage-gated potassium channels. The upstroke of the nerve action potential depends on voltage-gated sodium channels. Nerve excitability (and, hence, nerve firing) is affected by extracellular calcium concentration (hypocalcemia results in increased excitability), but this is because of an effect on sodium channels, not calcium channels.

Chapter 4

1. **The answer is A.** The intensity of sensory information is encoded in the action potential frequency. Cessation of the stimulus would lead to a rapid decrease in the action potential frequency, and adaptation of the receptor would also lead to a decrease in frequency. With a constant and maintained stimulus, at least some adaptation would take place, and the frequency would fall somewhat (and certainly not increase). The action potential velocity is a property of the nerve—not the receptor—and it would not be affected.

2. **The answer is C.** Rods and cones are absent from the area of the retina where the optic nerve exits. The blind spot is of appreciable size, but because its location is off-center and the eyeballs are mirror images, each eye fills in the information missing from the other, even when the gaze is fixed at a point. There are no connections from lateral cells to the blind spot because nerves are exiting there and do not make synapses.
3. **The answer is C.** Presbyopia is the age-related inability of the eye to focus on close objects. The decreased compliance of the lens prevents it from assuming a sufficiently curved state, and the focal point is behind the retina. As a person ages, only minor changes occur in the shape of the eyeball. Age-related changes in the opacity of structures through which light must pass, while they can impair vision, have little effect on the focal point of the light rays.
4. **The answer is B.** A myopic eyeball is too long, and light rays coming from great distances focus in front of the retina. The range of motion available to the lens is not sufficient to provide accommodation regardless of the effort made. A negative lens placed in front of the eye corrects the eye's refractive power, and the rays will now focus on the retina. A positive lens would only worsen matters, and a cylindrical lens (which has two foci, depending on the orientation considered) would not compensate satisfactorily. Reducing the light intensity also would not help; the pupils would dilate and admit more peripheral rays that would be further out of focus.
5. **The answer is A.** The frequency response of the basilar membrane changes steadily from high to low along its length, so that high frequencies are detected close to the oval window and low frequencies are detected at the other end, near the helicotrema.
6. **The answer is C.** Relative motion between the endolymph and the cupulae of the semicircular canals is due to the inertia of the endolymph, whether the body motion is starting or stopping. As the fluid continues to move when the head has stopped moving, the cupulae will be stimulated, producing the sensation of rotary motion. Moving in a straight line, without acceleration, will produce no fluid movement and no sensation. The sensation of static body position is accomplished by the maculae, which are sensitive to gravity but not endolymph motion.
7. **The answer is A.** When a receptor adapts, the sensation decreases although the stimulus may be unchanged. Adaptation is largely a result of the fall in magnitude of the generator potential and is not due to fatigue. Sensory responses are graded in response to changes in stimulus intensity regardless of the level of adaptation, and the phenomenon of compression allows a wide range of environmental intensities to be translated into a much more narrow range of sensory responses.
8. **The answer is B.** Rapidly adapting sensory receptors are best suited for detecting motion and change. Actions such as holding a steady weight and sensing the resting position of the body or the position of an extended limb are best sensed by receptors that adapt slowly. Likewise, sensors that adapt quickly would not be well suited for detecting the continued presence of a chemical stimulus. (Our rapidly adapting olfactory sensors can sometimes fail to provide information about a continuing hazard.)
9. **The answer is D.** Reduction in the intensity of a sensation is largely the result of a decline in the generator potential. In this sense, it mimics the effects of a reduction in the stimulus intensity. Because the action potentials arising in a sensory nerve are all-or-none, their velocity of conduction, amplitude, and duration of depolarization are not affected by the stimulus intensity; rather, they are properties of the nerve cell.
10. **The answer is C.** The transfer of energy through the middle ear from the relatively large eardrum to the smaller oval window by the ossicular chain increases the efficiency of the mechanical transduction process. The bones do not support the membrane structures but allow them to move relatively freely. Interference with the ossicular transmission process by external influences (as by the stapedius and tensor tympani muscles) or by disease processes, acts to reduce the vibration transfer efficiency, a change that can be either protective or harmful. The function of the eustachian tube is independent of the ossicles. While the bones themselves are passive, they are essential to the process of sound conduction.
11. **The answer is B.** The cone cells, which are responsible for color vision, are located at the point of sharpest focus, but they do not function if the light intensity is too low. In such cases, the single-pigment rod cells (with greater sensitivity, but with less advantageous location and interconnection) provide monochromatic but diffuse vision. The color composition of light does not depend on its intensity, and dark adaptation does not change the spectral sensitivity of the individual pigments. While focusing mechanisms may be less effective with low light, they still function.

Chapter 5

1. **The answer is A.** The maintenance of posture requires continuous muscle action. The low threshold for activation, fatigue-resistant motor units are the type active in postural control. Intrafusal muscle fibers do not contribute to force generation.
2. **The answer is C.** The Golgi tendon organ senses the force of muscular contraction. The nuclear chain and bag fibers, along with type Ia endings, are all components of the muscle spindle which reports muscle length and velocity of muscle shortening.
3. **The answer is D.** Motor neurons controlling axial muscles are positioned most medially in the ventral horn area. An enlarged central canal would impinge on that pool of motor neurons first.
4. **The answer is C.** Muscle spindles monitor muscle length. If the muscle is suddenly stretched, the spindle produces action potentials that activate homonymous motor neurons to contract the stretched muscle and resist the length change.
5. **The answer is E.** The spinal cord has the intrinsic cir-

cuitry in the form of central pattern generators to produce the basic motions of walking. All the other listed areas may influence the local pattern generators.

6. **The answer is D.** The rubrospinal tract descends in the lateral spinal cord and influences distal muscle function. This is also the function of the corticospinal tract. The vestibulospinal and reticulospinal tracts descend medially and influence proximal muscle action. The spinocerebellar tract is an ascending pathway.
7. **The answer is C.** The primary motor area is located along the precentral gyrus. The supplementary motor area is located on the medial aspect of the hemisphere. The other areas have sensory and association functions that influence the motor areas.
8. **The answer is E.** The supplementary motor area tends to produce bilateral motor responses when stimulated. The other areas would tend to produce unilateral responses.
9. **The answer is C.** The neurons of the primary motor cortex contribute about one-third of the axons that make up the corticospinal tract. Other tracts, such as the rubrospinal, do not sprout additional axons. The alpha motor neurons do not atrophy if deprived of corticospinal input.
10. **The answer is C.** Decreased inhibitory input to the GPi from the putamen would enhance inhibitory output from the GPi to the thalamus. The result is inhibition of excitatory output from the thalamus back to the cortex.
11. **The answer is B.** Spinal input, such as from the spinocerebellar tracts, enters the cerebellum on the mossy fibers. The climbing fibers originate from the inferior olivary nucleus of the medulla. The other components are intrinsic to the cerebellum.

Chapter 6

1. **The answer is D.** Pupillary dilation is a function of the sympathetic innervation that originates from the upper thoracic spinal cord. The preganglionic axons pass up the paravertebral sympathetic chain to the superior cervical ganglion from which the postganglionic axons arise. These axons then ascend in the pericarotid plexus to the eye.
2. **The answer is D.** Destruction of the lumbar paravertebral ganglia would impair sympathetic function to the leg on that side. This would result in skin dryness from the absence of sweating and warmth from persistent vasodilation. There should be no alteration of sensation or skeletal muscle function.
3. **The answer is D.** Muscarine is an exogenous agonist that acts at postganglionic synapses. All the other agents are neurotransmitters used at autonomic synapses.
4. **The answer is A.** The muscarinic parasympathetic and adrenergic sympathetic receptors are both G protein-linked and share a seven-membrane-spanning segment configuration. The parasympathetic and sympathetic preganglionic synapses are both of the direct ligand-gated type, which is similar in configuration to the receptor at the neuromuscular junction.
5. **The answer is E.** Sweat glands are controlled by the sympathetic nervous system.
6. **The answer is B.** Postsynaptic neurons are about 100-fold more numerous than the presynaptic neurons. This divergence is why presynaptic neuron activation can produce a widespread sympathetic response.
7. **The answer is D.** Bright light would cause constriction of the pupil as a result of parasympathetic activation. Medications that inhibit the action of acetylcholine (anticholinergic drugs) could impair pupillary constriction. Inhibition of adrenergic action might assist in pupillary constriction, but the primary constrictor action is cholinergic.
8. **The answer is D.** Muscarinic receptors at the synapse between the postganglionic axon and the target tissue are of the indirect ligand-gated type, which utilizes a G protein. This type of synapse alters the function of adenylyl cyclase and produces changes in cyclic AMP levels. The preganglionic to postganglionic sympathetic and parasympathetic synapses are directly gated by acetylcholine. Curare blocks the receptor at the neuromuscular junction but not at the direct ligand-gated cholinergic autonomic synapses.
9. **The answer is E.** The medullary reticular formation is the anatomic site for coordination of cardiac sympathetic and parasympathetic activity.

Chapter 7

1. **The answer is A.** Alpha waves are noted in the EEG in a relaxed, awake person whose eyes are closed. An alert state is indicated by beta waves. Theta and delta waves are noted during sleep. Variability in the wave forms might indicate a seizure or damage locus.
2. **The answer is D.** Dreams are associated with REM sleep. Normally, a person does not "act out" his or her dream because all of the motor neurons to muscles other than those for respiration, the middle ear, and the extraocular eye muscles are inhibited, abolishing muscle tone. If this inhibition does not occur, a person exhibits marked and often dangerous movement during dreaming. Muscle tone is reduced but not abolished during slow-wave sleep; however, movements are not an issue, presumably because dreaming does not occur. (Note, however, that sleepwalking occurs in slow-wave sleep.) Increased activity in motor areas of the cortex (or other areas) during REM sleep normally would not cause movement because motor neurons are inhibited.
3. **The answer is D.** Melatonin is secreted by the pineal gland. Adrenaline is secreted by the adrenal medulla, leptin by adipocytes, and melanocyte-stimulating hormone and vasopressin by the pituitary.
4. **The answer is D.** The basal forebrain nuclei and the pedunculopontine nuclei are major sources of distributed cholinergic innervation in the CNS. These cell groups are functionally dissimilar. Neither is a major input to the striatum or involved in language construction. Only the basal forebrain nuclei receive input from the cingulate gyrus. Although not known for certain, it is unlikely that either of these cell groups is atrophied in

schizophrenia, which appears to be a disorder of dopaminergic function.

5. **The answer is A.** Circulating leptin levels are sensed by neurons in the arcuate nucleus, which does not possess a blood-brain barrier. While some other regions of the hypothalamus also lack a blood-brain barrier, these regions do not contain leptin-sensing cells.
6. **The answer is B.** Circadian rhythms are entrained by the SCN to the external day/night cycle. This external information reaches the SCN directly by an optic nerve projection from the retina. The internal clock resides in the SCN and regulates the production of melatonin by the pineal gland. Reticular formation and visual cortical inputs are not directly involved in the regulation of circadian rhythms.
7. **The answer is C.** Magnocellular neurons of the paraventricular and supraoptic nuclei of the hypothalamus, whose axons reach the posterior pituitary via the hypothalamo-hypophyseal tract, secrete the posterior pituitary hormones. The portal capillary system from the hypothalamus to the pituitary gland is associated with the anterior pituitary. While pituitary function may be altered by the fight-or-flight response, the reticular activating system, or emotional state, none of these directly mediates posterior pituitary hormone secretion.
8. **The answer is E.** The arcuate fasciculus is the fiber bundle connecting Broca's and Wernicke's areas. The fornix connects the hippocampus with the hypothalamus and basal forebrain. The thalamocortical tract connects the thalamus with the cortex and the reticular activating system connects the brainstem with the thalamus and cortex. The prefrontal cortex is not a fiber bundle.
9. **The answer is B.** Neuroleptic drugs ameliorate the symptoms of psychosis in disorders such as schizophrenia. While the etiology of schizophrenia is far from understood and many transmitter systems may be involved, all neuroleptics block dopamine receptors.
10. **The answer is A.** The intralaminar nuclei of the thalamus receive input from the brainstem reticular activating system and convey information to the cortex. These nuclei are critical for the maintenance of arousal and consciousness. Without the intralaminar nuclei, beta rhythms and attention would be severely compromised. Both slow-wave and REM sleep would be affected because the regulation of sleep is also driven by the reticular activating system and its input.
11. **The answer is D.** Somatic sensory information from the left hand would be perceived in the right cortex, which does not generate language. To verbally explain what the object is, the information must cross to the left hemisphere. This crossing occurs through the corpus callosum. The fornix and hippocampus would be involved in storing memories about particular items, not in retrieving the memory. Neither the primary somatic sensory cortex on the left side nor the visual cortex on either side plays a role in identifying an object placed in the left hand by tactile cues.
12. **The answer is D.** The hippocampus is crucial for the formation of long-term (declarative) memory. Without the hippocampus, short-term memory is intact but the conversion to long-term does not take place. The retrieval of stored declarative memory does not require the hippocampus. The hippocampus is not needed for the formation or retrieval of procedural memory.
13. **The answer is E.** Wernicke's area is responsible for the recognition and construction of words and language; when it is damaged, the individual speaks but the content is nonsensical. Damage to Broca's area results in an inability to speak clearly because it controls the motor patterns required to speak; the little speech that is produced is grammatically and syntactically correct. The hippocampus and corpus callosum are not involved in the generation of speech. Damage to the arcuate fasciculus would result in a loss of speech because language generated in Wernicke's area would not be conveyed to Broca's area.
14. **The answer is D.** Mania is an affective disorder characterized by increased transmission through noradrenergic pathways. Other transmitter systems may play a role, but effective treatments are targeted at the noradrenergic system.
15. **The answer is A.** Acetylcholine is critical for cognitive function because of the cholinergic neurons in the basal forebrain that relay hippocampal information to the rest of the cortex. Nicotine activates cholinergic receptors. The only effective drugs for the treatment of cognitive deficits in Alzheimer's disease are cholinergic, although cognition clearly involves neurons in many regions of the brain that utilize a variety of transmitters.

Chapter 8

1. **The answer is C.** In all muscle types, the interaction between actin and myosin provides the forces that result in shortening. Skeletal and cardiac muscle have repeating sarcomeres, but smooth muscle does not. Smooth and cardiac muscles have small cells, whereas skeletal muscle has large cells.
2. **The answer is B.** The width of the I band changes because the thin filaments enter farther into the A band. The Z lines move closer together. The decrease in I band width and the moving of the Z lines together are proportional, but there is no change in A band dimensions.
3. **The answer is B.** ATP must bind to the myosin heads to allow the crossbridges to detach and the cycle to continue.
4. **The answer is A.** Relaxed skeletal muscle is in a state of inhibited contraction. The enzymatic activity of myosin is greatly enhanced by its interaction with actin. The role of calcium is as an activator, not an inhibitor; at rest, the concentration of free calcium is low.
5. **The answer is C.** Removal of calcium from the myofilament space into the sarcoplasmic reticulum (not the extracellular space) is an absolute requirement for normal relaxation. A reduction of ATP would promote rigor, not relaxation.
6. **The answer is B.** When the myofilament overlap is decreased above the optimal length, fewer crossbridges

(borne on the myosin filaments) are able to interact with actin, and there is a proportionate decrease in the force produced. The filaments actually come closer together as the muscle becomes thinner.

7. **The answer is C.** The sarcoplasmic reticulum releases calcium rapidly and in close proximity to the myofilaments. Calcium is not stored in the T tubules and is not involved in action potential events at the sarcolemma in skeletal muscle.
8. **The answer is A.** Calcium diffuses away from the troponin complex because the intracellular concentration has become low and the gradient favors dissociation. Calcium does not bind to active sites on myosin molecules, and individual actin molecules do not have enzymatic activity.
9. **The answer is C.** ATP is the immediate source of energy. The other substances are in metabolic pathways that provide energy, via several routes, into the ATP pool. They are not used directly in the crossbridge cycle.
10. **The answer is B.** The condition called rigor mortis develops after death because the processes that generate ATP stop. ADP does not contain energy usable to support contraction.
11. **The answer is B.** By shifting its metabolism to anaerobic pathways (glycolysis), the muscle keeps functioning at the expense of generating end products that will eventually require oxygen consumption for their further processing. This condition is called oxygen deficit.
12. **The answer is C.** A reduction in the calcium-pumping ability of the sarcoplasmic reticulum would leave a higher concentration of calcium ions in the myofilament space for a longer time. The diffusion of calcium away from the regulatory proteins would be slower, and crossbridges would detach less rapidly; consequently, the muscle would relax more slowly. Activation processes would not be as affected because they do not directly depend on the effectiveness of the calcium pump.

Chapter 9

1. **The answer is B.** This is a chemically gated channel without a highly selective filter and voltage sensor mechanism. Both sodium and potassium pass through it simultaneously down their respective electrochemical gradients.
2. **The answer is B.** The endplate potential and the action potential are based on changed ionic permeabilities, but the postsynaptic channels in the endplate region are not voltage-sensitive. This means that the endplate potential cannot regenerate and be propagated. Because the channels do not select between sodium and potassium, the endplate potential is close to zero. As such, it can never assume a large inside-positive value.
3. **The answer is A.** The postsynaptic membrane channels are blocked by the bound curare molecules and will not allow ions to pass; therefore, this membrane will not depolarize. The other choices are all presynaptic events and will not be affected by the blocked postsynaptic membrane.
4. **The answer is B.** The contraction will be twitch-like, but it will have increased amplitude, reflecting the additional calcium released from the SR in response to the second stimulus. Its duration will also be increased for the same reason.
5. **The answer is A.** This is the definition of isometric. The three other responses address factors that might change the size of the contraction but have nothing to do with whether it is isometric.
6. **The answer is B.** As long as the muscle is actually lifting the afterload, this is the only factor that determines the force. The other factors may make the muscle shorten faster or slower, but they do not affect the force produced.
7. **The answer is A.** This is a statement of relationship that is graphically represented in the force-velocity curve. Regarding choice D, note that it is force that determines velocity and not the other way around.
8. **The answer is C.** This is a point at the maximum of the power output curve. F_{max} and V_{max} represent points of zero power. A velocity of two-thirds V_{max} corresponds to a force too small to produce maximal power.
9. **The answer is C.** The forearm/biceps combination, because of the proportions involved, operates at a mechanical disadvantage with regard to force, trading decreased hand force for increased hand velocity.
10. **The answer is D.** This mixture of fiber types is ideal for the stated exercise because it can mobilize energy quickly. Choice A is a possibility, but almost all muscles have some mixture of fiber types.
11. **The answer is B.** Isometric contraction is possible when the volume of the organ is prevented from changing, as by a closed sphincter. Any shortening of smooth muscle in a hollow organ would be against some sort of load.
12. **The answer is C.** The level of phosphorylation would decline because the myosin light chains dephosphorylated by the phosphatase could not be rephosphorylated by MLCK because it would no longer be calmodulin-activated, as a result of the lowered cellular calcium. Choice A represents the skeletal muscle condition.
13. **The answer is B.** This emphasizes the primary role of myosin-based regulation in smooth muscle. Choice A represents the skeletal muscle condition, while choices C and D do not reflect the actual physiological effects; in particular, choice D is the reverse of the truth.
14. **The answer is C.** While smooth and skeletal muscle can exert about the same amount of force per cross-sectional area, smooth muscle does it much more economically. It is capable of extreme shortening when conditions external to the muscle allow.
15. **The answer is B.** The crossbridge cycle of smooth muscle is similar to that of skeletal muscle, with the added complexity (in some smooth muscles) of the latch state of crossbridges.
16. **The answer is C.** Smooth muscle membrane receptors perform a wide variety of functions and are involved in

both chemical and electrical activities at the membrane.

Chapter 10

1. **The answer is C.** Cardiac muscle has small cells that must be coupled electrically for communication to occur. Because it receives no motor innervation, it must be spontaneously active.
2. **The answer is B.** The intercalated disk is the site of electrical coupling and mechanical linkage, both of which are necessary for the tissue to behave as a syncytium.
3. **The answer is B.** Cardiac muscle is similar to skeletal muscle in both the structure of its contractile apparatus and the means by which it is regulated. It is similar to smooth muscle in its small cell size and syncytial behavior.
4. **The answer is C.** Until the relative refractory period is over, the muscle cannot be restimulated. By this time, it has begun to relax, so a smooth (fused) tetanus can not occur.
5. **The answer is B.** Because the afterload is removed at the end of the isotonic portion of the contraction, it is not available to reextend the muscle and relaxation occurs isometrically at the shortened length.
6. **The answer is C.** Although the relationship is not linear, the muscle is extended in proportion to the preload. In the intact heart, when the heart is filled more at rest, the muscle will shorten a greater distance when it contracts.
7. **The answer is D.** Because the force is high, the muscle is nearer the limit set by the length-tension curve.
8. **The answer is C.** As is the case for skeletal muscle, when the muscle is shortening isotonically, the only factor that controls the force is the afterload. The other factors mentioned will affect the velocity or extent of the shortening, not the force.
9. **The answer is C.** Inotropic interventions of many types, including heart rate changes, epinephrine, and digitalis-like drugs, all affect the availability of calcium to the contractile proteins.
10. **The answer is C.** The force-velocity curve states the basic relationship between the speed of shortening and the afterload at a given level of contractility. This relationship can be modified (up or down) by changes in contractility.

Chapter 11

1. **The answer is C.** Adult erythrocytes normally do not contain any carboxyhemoglobin, which is formed when hemoglobin binds carbon monoxide. Adult erythrocytes possess two distinct types of hemoglobin, HbA and HbA_2. These hemoglobin molecules may be saturated with oxygen (HbO_2) or reduced to Hb when oxygen is released to cells within tissues.
2. **The answer is D.** Superoxide anion is generated when oxygen is reduced by cytoplasmic NADPH. The reduction is carried out by the enzyme NADPH oxidase, which is not a reactant but a catalyst activated in cells responding to bacteria. The hexose monophosphate shunt is an enzyme cascade (not a reactant) that functions to provide high levels of reduced NADPH to drive this reaction. G proteins are not reactants, but play an essential role in the activation of this cellular cascade. Similarly, the enzyme myeloperoxidase is not a reactant; it enhances the ability of reactants, such as hydrogen peroxide, to exert a lethal effect on invading bacteria.
3. **The answer is A.** T cells are infected by HIV in individuals who have AIDS. B cells, like T cells, are lymphocytes, but they are not targets for HIV. Neutrophils are not lymphocytes and are not infected by the AIDS virus. Monocytes and basophils similarly are not targets for the virus that causes AIDS.
4. **The answer is B.** Umbilical cord blood, derived from the circulating blood of newborn infants, possesses high levels of hematopoietic progenitors. Levels of circulating progenitors rapidly decrease after birth, depleting the progenitor content within the circulating blood of adults. The spleen of adult humans functions as a hematopoietic organ in certain disease states, such as leukemia. However, in other animals and in developing human fetuses, the spleen plays an important role in the hematopoietic response. While the liver and the thymus are important in hematopoiesis and immune reconstitution prior to birth, these organs are not involved in hematopoiesis in adult humans.
5. **The answer is E.** When specifically programmed T cells or B cells of the adaptive immune system first recognize specific antigens, they begin to divide rapidly, generating several copies of cells similarly programmed against the inciting stimulus. Hematopoiesis involves the nonspecific generation of all cells in blood, including leukocytes, erythrocytes, and platelets. Hematotherapy is a therapeutic process in which specifically amplified cells are infused in patients to increase resistance to infection or to restore hematopoiesis. Inflammation is not a specific response against individual antigenic determinants and does not require T cell or B cell amplification. Similarly, innate immunity does not require amplification of T cells or B cells as a result of interaction with an invading stimulus but is affected by cells present and programmed to respond to specific stimuli.
6. **The answer is B.** Delayed-type hypersensitivity reactions to PPD and other specific antigens develop slowly over 24 to 48 hours as T cells become activated and secrete factors that effect the skin response. B cells play no role in this type of reaction; instead, they produce antibodies involved in more immediate responses. Neutrophils do not arrive at sites of delayed-type hypersensitivity in large numbers. Eosinophils play a role in immediate hypersensitivity to many antigens that cause symptoms of allergy, such as sneezing and stuffy nose, but do not participate in the delayed response. Finally, the response to PPD is driven by cells programmed to respond specifically to this antigen derived from the bacteria that cause tuberculosis, and not by a metabolite of this protein.
7. **The answer is C.** Antibody specificity is dictated by the

sequence of amino acids within the variable regions of the light and heavy chains. The Fc region is a site for antibody docking to effector cells and does not play a role in antigen binding. The constant region has a similar structure in antibodies of widely divergent specificity and, therefore, does not dictate specificity. Fc receptors are sites on immune effector cells that interact with the Fc region of the antibody molecule and do not define an antibody's specificity. The J chain is a unique portion of secreted IgA molecules that allows the molecule to move from the circulation through mucous membranes.

8. **The answer is D.** The extrinsic coagulation pathway is activated when tissue thromboplastin (tissue factor) is released from injured tissues. Activation of factor X occurs later and is a step involved in the activation of both the intrinsic and the extrinsic pathways. Activation of factor XII is the first step in activation of the intrinsic coagulation pathway. Conversion of prothrombin to thrombin and conversion of fibrinogen to fibrin are the final steps that lead to clot formation by either the intrinsic or the extrinsic pathway.

Chapter 12

1. **The answer is C.** See equation 3 in the text.
2. **The answer is C.** See equation 6 in the text. Changes in transmural pressure can be caused by changes inside or outside of a vessel (see equation 5). The viscosity of blood does not directly affect transmural pressure. Resistance, not transmural pressure, is proportional to the length of a tube.
3. **The answer is D.** When the heart stops, blood continues to flow from the arteries to the veins until the pressures in the two sides of the circulation are equal. That pressure is mean circulatory filling pressure. Hemodynamic pressure is the potential energy that causes blood to flow. Mean arterial pressure is the average pressure in the aorta or a large artery over the cardiac cycle. Transmural pressure is the difference between the pressure inside and outside a blood vessel. Hydrostatic pressure is the pressure caused by the force of gravity acting on a fluid.
4. **The answer is D.** Although flow velocity, viscosity, and tube diameter all influence turbulence, it is the combination of these variables (plus the density of blood), expressed as the Reynolds number (equation 4 in the text), that determines whether flow is turbulent or laminar.
5. **The answer is E.**

$$\begin{aligned}\text{Compliance} &= \Delta V/\Delta P \\ &= 30 \text{ mL}/40 \text{ mm Hg} \\ &= 0.75 \text{ mL/mm Hg}\end{aligned}$$

6. **The answer is B.** See equation 1 in text. The tube is analogous to the systemic circulation in which there are many branches. The overall resistance can be calculated from the sum of the flows through the individual branches and ΔP, provided it is the same for all branches.

$$R = \Delta P/\dot{Q}$$

where $\dot{Q} = 95 + 5 = 100$ mL/min and $\Delta P = 75 - 25 = 50$ mm Hg.

$$R = 50/100 = 0.5 \text{ mm Hg/(mL/min)} = 0.5 \text{ PRU}$$

Chapter 13

1. **The answer is B.** Voltage-gated Na^+ channels are responsible for phase 0 in ventricular muscle. Voltage-gated Ca^{2+} channels are responsible for phase 0 in nodal cells. The potassium channels mentioned do not play a role in mediating depolarization.
2. **The answer is D.** The form of the QRS will be normal because electrical excitation of the ventricles occurs over essentially the normal pathway (i.e., AV node to bundle branches to Purkinje system to myocardium). The T wave will be normal as well. With complete heart block, P waves and QRS complexes are completely independent of each other. Some PR intervals could be shortened by chance, others will be very long; that is, there is no predictable PR interval. There will not be a consistent ratio of P waves to QRS complexes because the two are disassociated, but the average ratio would be 80/40 or 2:1.
3. **The answer is B.** The shape of the QRS complex will be significantly different from normal because depolarization now originates in the right ventricle and propagates in a retrograde fashion. Because the right side of the heart depolarizes before the left, the configuration of the QRS may resemble that seen with left bundle branch block, another situation in which the right side of the heart depolarizes before the left. The duration of the QRS complex will be increased because the specialized conducting system of the ventricles is not fully employed: Depolarization moves through more slowly conducting muscle instead of the rapidly conducting Purkinje system. Retrograde conduction through the AV node is extremely unlikely, so P waves will not follow each QRS complex. Because excitation of the atria and ventricles is still independent, there will be no predictable PR interval.
4. **The answer is D.** Voltage-gated Ca^{2+} channels are primarily responsible for the upswing of the action potential (phase 0) of nodal cells. Voltage-gated Na^+ channels are inactivated because the resting membrane potential in these cells never becomes sufficiently negative to allow reactivation. Acetylcholine-activated K^+ channels are important only in mediating the effect of ACh on the pacemaker potential of nodal cells. Inward rectifying K^+ channels are responsible for maintaining the resting membrane potential in nonnodal cells but have a less important role in cells with a pacemaker potential.
5. **The answer is B.** Atrial repolarization normally occurs during the QRS complex. A dipole is created by atrial repolarization but it is not observed on the ECG because the dipole created by ventricular depolarization is much larger.
6. **The answer is D.** Depolarization of the ventricles proceeds from subendocardium to subepicardium, but this

does not result in the P wave. In lead I, when the ECG electrode attached to the right arm is positive relative to the electrode attached to the left arm, a *downward* deflection is recorded. AV nodal conduction *is* slower than atrial conduction, but this does not cause the P wave. When cardiac cells are depolarized, the inside of the cells is positive or neutral relative to the outside of the cells.

7. **The answer is C.** Stimulation of the sympathetic nerves to the normal heart decreases the duration of the ventricular action potential and, therefore, decreases the QT interval. As heart rate increases, the duration of diastole and, therefore, the TP interval decreases. Increased conduction velocity in the AV node decreases the duration of the PR interval. Fewer P waves than QRS complexes are indicative of AV block. On the contrary, sympathetic stimulation may reverse AV block. The frequency of QRS complexes increases with the heart rate.
8. **The answer is D.** The drug could act on β_1-adrenergic receptors to increase the rate of depolarization of sinoatrial nodal cells. An adrenergic receptor antagonist would have the opposite effect, as would a cholinergic receptor agonist and the closing of voltage-gated Ca^{2+} channels. Opening of acetylcholine-activated K^+ channels would slow pacemaker depolarization by keeping the membrane potential closer to the K^+ equilibrium potential.
9. **The answer is C.** Excitation of the ventricles does not ordinarily lead to excitation of the atria because retrograde conduction in the AV node is unusual. Norepinephrine modulates the ventricular force of contraction and conduction velocity and lowers the threshold for excitation, but it does not, by itself, initiate excitation. Excitation of the ventricles is initiated by phase 0 of the action potential. Normal ventricular cells do not exhibit pacemaker potentials.
10. **The answer is C.** AV nodal cells exhibit action potentials characterized by slow depolarization (phase 0) because fast voltage-gated Na^+ channels do not participate. This is because the diastolic potential of these cells does not become sufficiently negative to allow reactivation of Na^+ channels. Acetylcholine slows and norepinephrine speeds conduction velocity. AV nodal cells are capable of pacemaker activity but at a rate of approximately 25 to 40 beats/min.
11. **The answer is C.** When stimulation of the parasympathetic nerves to the normal heart leads to complete inhibition of the SA node for several seconds, nodal escape usually occurs. In this situation, pacemaker activity usually is taken over by cells in the AV node or bundle of His. QRS complexes are normal because the pacemaker activity is high enough in the conducting system to lead to a normal pattern of ventricular excitation. T waves would be normal for the same reason. Because at least one beat begins without atrial excitation, there would be fewer P waves than either QRS complexes or T waves.
12. **The answer is B.** The R wave in lead I of the ECG reflects a net dipole associated with ventricular depolarization. Repolarization causes the T wave. The R wave is smallest when the mean axis is directed perpendicular to a line drawn between the two shoulders because both electrodes are equally influenced by the negative and positive sides of the dipole.
13. **The answer is C.** The ST segment of the normal ECG occurs during a period when both ventricles are completely depolarized. It is present in all leads.

Chapter 14

1. **The answer is C.** Loop B shows increased contractility because stroke volume is increased at a constant preload and afterload. When loop B is compared to loop A, preload is not increased or decreased because there is no change in the pressure or volume at which the mitral valve closes and isovolumetric contraction begins. Afterload is not changed because there is no change in the pressure or volume at which the aortic valve opens and ejection begins. The evidence that stroke volume is increased is the larger volume difference between the point at which the aortic valve opens and closes—that is, between isovolumetric contraction and relaxation.
2. **The answer is A.** The aortic and mitral valves are never open at the same time. This is the basic principle of the cardiac pump. The first heart sound is caused by closure of the mitral and tricuspid valves. The mitral valve is open throughout diastole *except* isovolumetric relaxation. Left ventricular pressure is less than aortic pressure during diastole and isovolumetric contraction but is greater than aortic pressure during a substantial period of ventricular ejection. Ventricular filling occurs during diastole.
3. **The answer is C.** Aortic pressure reaches its lowest value during the isovolumetric contraction phase of ventricular systole. The second heart sound is associated with closure of the aortic valve. Left atrial pressure is less than left ventricular pressure during ventricular systole and isovolumetric relaxation. The ventricles eject blood during all of systole except isovolumetric contraction. Ventricular end-diastolic volume is greater than end-systolic volume.
4. **The answer is D.** Increased ventricular filling means a larger end-diastolic volume. Of the three points representing increased end-diastolic volume, only choice D is on a higher ventricular function curve, signaling increased contractility. If you chose choice A, you recognized that the upper curve represented increased contractility, but missed the fact that end-diastolic volume would be increased as well. If you selected choices C or D, you recognized increased end-diastolic volume, but did not understand that increased contractility means that the ventricular function curve would be higher. Point B is the graphical definition of decreased contractility at an unchanged end-diastolic volume.
5. **The answer is B.** Drug B increases the internal work of the left ventricle more than drug A because it increases external work by increasing pressure. Drug A increases the external work of the left ventricle the same as drug B. External work is stroke volume multiplied by mean arterial pressure, so equivalent increases in stroke vol-

ume and pressure yield equivalent increases in stroke work. Because drug B increases internal work more than drug A, total work is more increased. For this reason, drug B increases the oxygen consumption of the heart more than drug A. The "double product" (aortic pressure times heart rate) is greater for drug B than for drug A. Cardiac efficiency is higher with drug A than with drug B because efficiency is a measure of the oxygen cost of external work. Because of the greater internal work, drug B increases oxygen consumption more than drug A. The ratio of external work to oxygen consumption would be higher for drug A than drug B.

6. **The answer is D.**

$$\begin{aligned} CO &= HR \times SV = HR \times (EDV - ESV) \\ &= 70 \text{ beats/min} \times (130 - 60) \text{ mL} \\ &= 4{,}900 \text{ mL/min} \end{aligned}$$

$$\begin{aligned} SW &= SV \times MAP \\ &= 70 \text{ mL} \times 90 \text{ mm Hg} \\ &= 6{,}300 \text{ mL} \times \text{mm Hg} \end{aligned}$$

7. **The answer is B.**

$$\begin{aligned} CO &= \dot{V}O_2/(aO_2 - \bar{v}O_2) \\ &= 4{,}000 \text{ mL } O_2\text{/min/}(190 - 30 \text{ mL } O_2\text{/L}) \\ &= 25 \text{ L/min} \end{aligned}$$

$$\begin{aligned} SV &= CO/HR \\ &= 25 \text{ (L/min)/(180 beats/min)} \\ &= 139 \text{ mL/beat} \end{aligned}$$

8. **The answer is D.** Electrical pacing to a heart rate of 200 beats/min would decrease time for filling and reduce end-diastolic volume. A reduction in afterload would make it easier for the ventricle to eject blood and would raise stroke volume. An increase in end-diastolic pressure will increase end-diastolic fiber length and increase the force of contraction and stroke volume. Stimulation of the vagus nerves slows the heart, increases the time for ventricular filling, and increases stroke volume. Stimulation of sympathetic nerves to the heart increases heart rate and contractility. Despite the decreased filling accompanying an increase in heart rate, stroke volume will stay the same or increase because of the increased contractility.

Chapter 15

1. **The answer is C.** Strictly speaking, mean arterial pressure minus right atrial pressure equals cardiac output times systemic vascular resistance. Right atrial pressure is often ignored because it is so much smaller than mean arterial pressure that it does not have much effect on the calculation.
2. **The answer is B.** A stroke volume change with no change in heart rate means that cardiac output is changed. If we assume that mean arterial pressure is determined by CO and SVR and SVR is constant, then mean arterial pressure must have changed. Heart rate changes with no changes in cardiac output or SVR will have no effect on mean arterial pressure. A doubling of heart rate with no change in stroke volume gives a doubling of cardiac output; if SVR is halved at the same time, then mean arterial pressure will not change. Arterial compliance influences pulse pressure but not mean arterial pressure.
3. **The answer is C.** If the cuff is too small, it takes a falsely high pressure in the cuff to transmit sufficient pressure to the vessel wall for total occlusion of the artery. Blood pressure may be falsely high in patients with badly stiffened arteries because of the extra pressure needed to compress the arteries. The measurement gives an indirect reading of systolic and diastolic pressure; mean arterial pressure must be calculated. The measurement depends on the *appearance* of sound to signal systolic pressure.
4. **The answer is C.** Vessel radius is the most important variable influencing vascular resistance. Resistance changes occur primarily in small arteries and arterioles. Blood viscosity and length are important determinants of underlying vascular resistance, but ordinarily do not change enough to be influential in altering vascular resistance.
5. **The answer is E.** Standing up causes a shift in blood from the chest to the periphery, lowering central blood volume. The diameter of the leg veins increases because of increased transmural pressure caused by the column of blood in the vessels above them. Right atrial pressure decreases and, therefore, decreases filling of the ventricles and stroke volume.
6. **The answer is B.** By convention, the first of the two numbers is the systolic pressure and the second is the diastolic pressure.

$$\begin{aligned} \text{Pulse pressure} &= 125 - 75 \text{ mm Hg} \\ &= 50 \text{ mm Hg} \end{aligned}$$

$$\begin{aligned} \text{Mean arterial pressure} &= 75 \text{ mm Hg} + 50 \text{ mm Hg/3} \\ &= 92 \text{ mm Hg} \end{aligned}$$

7. **The answer is B.** Mean arterial pressure both before and after the tricuspid valve becomes incompetent is 110 mm Hg. The pressure gradient before tricuspid insufficiency is $\bar{P}_a - P_{ra} = 107$ mm Hg. The pressure gradient after the valve becomes incompetent is $110 - 13 = 97$ mm Hg. If all other hemodynamic factors remain unchanged (which would be unlikely in this situation), systemic blood flow will fall in proportion to the decrease in pressure gradient.
8. **The answer is A.** Pulse pressure is determined by stroke volume and arterial compliance. Stroke volume is unchanged, and if arterial compliance were to remain constant, the pulse pressure would not change. However, an increase in mean arterial pressure will tend to stretch the aorta and decrease its compliance. Ejecting the same stroke volume into a less compliant aorta will result in an increased pulse pressure.
9. **The answer is A.** The increase in transmural pressure exerted by the column of blood above the veins would have little effect on their volume if they were as stiff as the arteries. In this situation, relatively little blood would accumulate in the veins and little would be displaced from the central blood volume.

Chapter 16

1. **The answer is B.** Although small arteries do have a significant resistance, arterioles dominate the total resistance.
2. **The answer is B.** Molecular-sized openings within the tight junctions are the most influential sites sieving the molecules that diffuse through the capillary wall. Large defects are highly permeable areas, but their occurrence is too infrequent to affect the total amount of material moved.
3. **The answer is A.** All the other possibilities include one minor force for filtration or absorption.
4. **The answer is C.** Myogenic mechanisms seem to involve only the physical loading of vascular muscle cells in the form of stretch and increased tension or in just increased tension.
5. **The answer is A.** Each of the choices is a function of the microcirculation, but its most important function by far is to provide tissue with nutrients and remove the wastes.
6. **The answer is A.** Lipids are not particularly water-soluble and must primarily diffuse through the lipid layers of cell membranes. A small amount of lipid does move through water-filled channels.
7. **The answer is B.** The cardiovascular system is designed to support a much higher metabolic rate than exists at rest. Only a fraction of the available blood flow is necessary for functioning at rest, and the remainder moves slowly through the venules and smallest veins.
8. **The answer is C.** The interstitial space consists of alternating gel and liquid areas with a low plasma protein concentration. It is permeable compared to the capillary wall.
9. **The answer is D.** Both adenosine diphosphate (ADP) and acetylcholine cause the release of NO from endothelial cells. The other choices involve mechanisms that function without endothelial cells.
10. **The answer is A.** Although all of the choices are events that happen in lymph vessels, the first key event is lowering the lymphatic hydrostatic pressure to enable tissue fluid to enter the lymphatic vessel.
11. **The answer is C.** Nerve fibers, not vascular smooth muscle, release norepinephrine. The norepinephrine from the sympathetic nerves simply diffuses from the axons and binds to specific receptors on smooth muscle cells.
12. **The answer is A.** More capillaries in use at a constant blood flow actually slows the flow velocity in individual capillaries. The distances between capillaries are decreased. The perfusion of additional capillaries does not influence the permeability of the individual capillaries.
13. **The answer is B.** The amount of oxygen exchanged is equal to the product of the blood flow and the arterial-venous oxygen content difference: 200 mL/min × (20 mL/100 mL − 15 mL/100 mL) = 10 mL/min.
14. **The answer is D.** Fluid will be filtered at a net pressure of +4 mm Hg. The balance of hydrostatic pressures is 22 mm Hg (capillary hydrostatic pressure − tissue hydrostatic pressure) and is greater than the balance of colloid osmotic pressures, which is 18 mm Hg [reflection coefficient × (plasma colloid osmotic pressure − tissue colloid osmotic pressure)].

Chapter 17

1. **The answer is A.** Increased arterial blood pressure or the increased cardiac output of exercise imposes an increased workload on the heart, and the coronary vessels dilate to improve oxygen delivery. When the blood pressure falls or the blood lacks oxygen, the autoregulatory mechanisms of the heart vasculature dilate the microvessels to maintain the blood flow.
2. **The answer is C.** Resting after a meal is associated with reduced sympathetic nervous system activity and reduced arterial pressure, but the expected increase in blood flow would not meet the substantially increased metabolic needs of the intestine during nutrient processing. Much more potent mechanisms are needed to increase blood flow, such as increased NO production. Although parasympathetic nervous system activity increases during food absorption, the effect on blood flow is minor.
3. **The answer is C.** The hepatic arterial and portal venous blood mix in the capillaries of the hepatic acinus.
4. **The answer is C.** Brain blood flow is constant despite large changes in the arterial blood pressure because vascular resistance usually changes in the same direction as the arterial pressure and by almost the same percentage.
5. **The answer is D.** The skeletal muscle vasculature has a 20-fold or greater range of blood flows, from minimal perfusion at rest to very high blood flows during intense exercise. No other organ system has appreciably more than a 4- to 5-fold change in blood flow from rest to maximum flow.
6. **The answer is D.** See Figure 17.6.
7. **The answer is D.** The autoregulatory range is shifted to higher pressures because the arteries and arterioles increase their resistance. The functional and structural changes increase the arterial pressure at which autoregulation of blood flow occurs, but increase the lowest pressure at which blood flow can be maintained.
8. **The answer is B.** Oxygenated blood from the placenta does not become fully mixed with blood returning from the superior vena cava and is diverted through the foramen ovale into the left atrium. Consequently, the oxygen content of blood in the ascending aorta is significantly greater than that in the ductus arteriosus. The upper body, brain, and coronary arteries are supplied by vessels that branch from the aorta before the ductus arteriosus. The ductus carries blood with lower oxygen content into the aorta, to perfuse the lower body and fetal placenta.

Chapter 18

1. **The answer is B.** Norepinephrine, the sympathetic postganglionic neurotransmitter, causes constriction of blood vessels in the skin. Increased sensitivity to NE would greatly reduce skin blood flow, which would

cause the skin to be cold and painful. Epinephrine constricts skin blood vessels; abnormally low epinephrine in the blood would allow skin vessels to dilate. An insensitivity of blood vessels to epinephrine would have the same effect. As there are no parasympathetic nerves to skin blood vessels, parasympathetic activity does not affect blood flow in the skin. Although acetylcholine causes nitric oxide release from skin blood vessels, this would cause vasodilation.

2. **The answer is B.** Activation of parasympathetic nerves to the heart would lower the heart rate below its intrinsic rate. However, with all effects of norepinephrine and epinephrine blocked, the sympathetic nervous system cannot raise the heart rate above its intrinsic rate. The withdrawal of parasympathetic nerve tone could only raise the heart rate to the intrinsic rate. (See Chapter 13 for a discussion of intrinsic heart rate.)
3. **The answer is D.** The cold pressor response is initiated by the stimulation of pain receptors by exposing the surface of the skin to ice water.
4. **The answer is A.** The release of acetylcholine from parasympathetic nerves to the sinoatrial node results in a slowing of diastolic depolarization of pacemaker cells and a slowing of the heart rate. ACh *slows* conduction velocity, *inhibits* NE release from sympathetic terminals, *enhances* NO release from endothelial cells, and *dilates* blood vessels of the external genitalia (via NO)—all by binding to muscarinic receptors.
5. **The answer is A.** The function of these baroreceptors is the rapid short-term regulation of arterial blood pressure. The receptors start firing at a pressure of approximately 40 mm Hg. They completely adapt over 1 to 2 *days*, not weeks. In general, changes in baroreceptor activity have little effect on cerebral blood flow. The sympathetic activity following a fall in blood pressure results in increased heart rate and contractility, which raises myocardial metabolism and coronary blood flow.
6. **The answer is D.** Peripheral chemoreceptor activation plays a significant role in *enhancing* the diving response by enhancing peripheral vasoconstriction and bradycardia. Activation is increased by a *decrease* in pH and by a lowering of arterial PO_2, not oxygen content. Peripheral chemoreceptors are located in the aortic and carotid bodies.
7. **The answer is B.** The fight-or-flight response and exercise are characterized by increased sympathetic tone and decreased parasympathetic tone. The diving response is associated with increased parasympathetic *and* sympathetic tone. The cold pressor response is characterized by increased sympathetic activity to the heart and blood vessels.
8. **The answer is D.** The hemorrhage has decreased arterial pressure below normal. The fall in blood volume would result in a fall in central blood volume, right ventricular end-diastolic volume, and cardiopulmonary receptor activity. Carotid baroreceptor activity would be lowered in the presence of a low mean arterial pressure. The resulting sympathetic activity would cause vasoconstriction in the splanchnic bed, and especially with a lowered arterial pressure, splanchnic blood flow would be decreased. The heart rate would be elevated by the increased sympathetic activity and decreased parasympathetic activity caused by the baroreceptor reflex.
9. **The answer is B.** Standing up increases the transmural pressure in the veins of the legs. Because the veins are highly compliant, their volume increases at the expense of central blood volume. A lower central blood volume means reduced cardiac filling pressure (preload). Within seconds, the decrease in preload decreases stroke volume, cardiac output, and arterial pressure. However, within the first minute, the arterial baroreflex and the cardiopulmonary reflex work together to increase sympathetic activity and decrease parasympathetic activity. As a result, cardiac contractility and heart rate increase, and cardiac output decreases less than it would have without compensation. "Noncritical" vascular regions, such as the splanchnic area and skin, constrict in response to increased sympathetic nervous system activity. Brain blood flow changes little because sympathetic nerve activation causes little vasoconstriction in the brain and autoregulation of blood flow prevents a fall in brain blood flow, even if mean arterial pressure decreases.
10. **The answer is C.** Pressure diuresis lowers arterial pressure by lowering blood volume and, thereby, lowering cardiac output. All of the other choices do lower arterial pressure, but are not caused by pressure diuresis.
11. **The answer is C.** By increasing the excretion of salt and water, blood volume is decreased. Central blood volume participates in this decrease, reducing ventricular filling, cardiac output, and venous return as well (remember that cardiac output equals venous return in the steady state). Both the muscle pump and the respiratory pump increase the pressure gradient toward the heart and increase venous return and central blood volume. Lying down increases venous return and central blood volume by reducing venous volume of the lower extremities. Going into space is similar to lying down, in that the force of gravity on blood is removed and blood is not held in the leg veins when a person is upright. Therefore, central blood volume is increased.

Chapter 19

1. **The answer is D.** Pleural pressure is the most negative at total lung capacity because of the elastic recoil of the lungs pulling inward. At residual volume, pleural pressure would be the least negative. Choice E is not an option because pleural pressure is positive during a forced vital capacity maneuver.
2. **The answer is A.** Transpulmonary pressure is equal to alveolar pressure minus pleural pressure.
3. **The answer is B.** The outward recoil of the chest wall and the inward recoil of the lungs reach equilibrium at FRC. At residual volume, the outward recoil of the chest is the greatest and the inward recoil of the lung is the smallest. At total lung capacity, the inward recoil of the lung is the greatest.
4. **The answer is B.** Transairway pressure is pressure across the airways and is measured by subtracting pleu-

ral pressure from airway pressure (Pta = Paw − Ppl). Transairway pressure is most negative in the conditions described in choice B. Transairway pressure is the most positive in the conditions described in choice D.

5. **The answer is D.** The ratio for FEV_1/FVC is 0.80 (80%) for healthy adults, including trained athletes. This value tends to decrease with age.
6. **The answer is C.** Emphysema is an obstructive disorder that leads to highly compliant lungs, while pulmonary edema, fibrosis, congestion, and respiratory distress syndrome are restrictive disorders that lead to stiff lungs with decreased compliance.
7. **The answer is D.** An increase in airway diameter lowers airway resistance, which has the greatest effect on forced expiration. Total lung capacity, inspiratory capacity, and tidal volume would not appreciably change. FRC is high with asthma and would decrease with a bronchodilator.
8. **The answer is C.** A restrictive lung disease causes a decrease in FEV_1, FVC, FRC, and RV. However, the ratio of FEV_1/FVC is likely to be increased.
9. **The answer is A.** Minute ventilation is equal to expired air per minute, tidal volume times frequency of breathing, or alveolar ventilation plus dead space ventilation.
10. **The answer is D.** Tidal volume = minute ventilation (8 L/min) ÷ frequency (10 breaths/min) = 0.8 L/breath.
11. **The answer is B.** Fibrosis leads to stiff lungs, resulting in reduced compliance and the need for more work to inflate the lungs. Stiffer lungs also have greater elastic recoil, so the lungs will deflate easier.
12. **The answer is D.** There is no airflow during breath holding with an open glottis. Under these conditions, alveolar pressure equals atmospheric pressure.
13. **The answer is E.** $C_L = \Delta V/\Delta P = 0.5$ L/5 cm H_2O = 0.1 L/cm H_2O.
14. **The answer is C.** $P_{TP} = P_A - Ppl = -1 - (-7)$ cm H_2O = + 6 cm H_2O.
15. **The answer is D.** TLC = VC + RV = 5.0 L + 1.2 L = 6.2 L.
16. **The answer is B.** $\dot{V}_D = \dot{V}_E - \dot{V}_A$ = 7.0 L/min − 5.0 L/min = 2.0 L/min.

Chapter 20

1. **The answer is E.** The pulmonary circulation is a high-flow, low-pressure, low-resistance, and high-compliance system.
2. **The answer is D.** Pulmonary vascular resistance decreases with an increase in pulmonary arterial pressure. The primary reason is capillary recruitment, but it is also due to capillary distension. Pulmonary vascular resistance is increased at low and high lung volumes (see Fig. 20.6) and by hypoxia.
3. **The answer is D.** The pulmonary and the systemic circulations both receive all of the cardiac output and have the same flow. Pressure, resistance, and compliance are different.
4. **The answer is B.** The gravitational effect on the pulmonary circulation causes blood flow to be greatest at the base. Vascular resistance is high at the apex because alveolar pressure exceeds capillary pressure.
5. **The answer is C.** In the supine position, the heart is in the middle of the chest. Pulmonary arterial pressure at the top of the chest is 15 cm H_2O minus 7.5 cm H_2O = 7.5 cm H_2O. Therefore, arterial pressure exceeds venous pressure (7 cm H_2O). Since alveolar pressure is less than venous pressure in a healthy individual, we have the situation that $Pa > Pv > P_A$, or a zone 3. There is no zone 4.
6. **The answer is C.** A drop in venous pressure has the greatest effect in zone 3 because the pressure gradient for flow is determined by the arterial-venous pressure difference. In zone 1 there is no flow, and the pressure gradient for flow in zone 2 is the arterial-alveolar pressure difference.
7. **The answer is A.** At the base, both airflow and blood flow are higher; however, blood flow exceeds airflow at the base, which results in a low $\dot{V}_A/\dot{Q}$ ratio. At the apex, blood flow and airflow are lower than at the base, but airflow is greater than blood flow, which leads to a high $\dot{V}_A/\dot{Q}$.
8. **The answer is C.** The regional differences in blood flow and airflow are the result of gravity.
9. **The answer is C.** The ventilation-perfusion ratio is highest at the apex and lowest at the base of the lung. As a result, the lungs are overventilated at the apex relative to blood flow; P_{O_2} is high and P_{CO_2} is low at the apex.
10. **The answer is B.** $R = \Delta P/\dot{Q}$ = (20 − 5 mm Hg)/5 L per min = 3 mm Hg/L per min.
11. **The answer is C.** 20 cm H_2O ÷ 1.36 cm H_2O per mm Hg = 15 mm Hg.
12. **The answer is C.** $\Delta P = R \times \dot{Q}$ = 4 mm Hg/L/min × 5 L/min = 20 mm Hg.

Chapter 21

1. **The answer is D.** The A-aO_2 gradient in a healthy person is due to both a low $\dot{V}_A/\dot{Q}$ ratio at the base of the lungs and a small shunt from the bronchial circulation.
2. **The answer is A.** A decrease in the diffusion distance will lead to an increase in D_L. A decrease in capillary blood volume, surface area, cardiac output, and blood hemoglobin concentration will decrease D_L.
3. **The answer is D.** The equilibrium curves are not similar; that for CO_2 is steeper and more linear. The blood carries more CO_2 than O_2. The presence of CO_2 will *increase* the P_{50}. Although red cells carry most of the O_2, the plasma carries the majority of the CO_2 (mainly as bicarbonate).
4. **The answer is A.** A decrease in hemoglobin concentration will decrease the O_2 content, but will not affect the oxygen saturation or P_{O_2}.
5. **The answer is B.** All will favor the unloading of oxygen from hemoglobin except a rise in pH.
6. **The answer is D.** A low $\dot{V}_A/\dot{Q}$ ratio will cause hypoxemia, but it will have little effect on arterial P_{CO_2} because of the linearity of the CO_2 equilibrium curve. Also, a low Pa_{O_2} stimulates ventilation, which promotes CO_2 loss.

7. **The answer is E.** In a normal resting condition, the blood leaving the lungs is 98% saturated with oxygen, and the blood returning to the lungs is 75% saturated with oxygen. With vigorous exercise, blood leaving the lungs is still 98% saturated, but blood returning is usually less than 75% saturated because more oxygen is unloaded from hemoglobin in exercising muscles.
8. **The answer is B.** Carbon monoxide will lower oxygen content and saturation, but the arterial PO_2 is unchanged. Airway obstruction or pulmonary edema will lower O_2 saturation and arterial PO_2.
9. **The answer is D.** A shunt, low $\dot{V}A/\dot{Q}$ ratio, and diffusion impairment all cause an increase in the A-aO_2 gradient. The reason the A-aO_2 gradient is normal with generalized hypoventilation is that both alveolar oxygen and arterial oxygen tension decrease together.
10. **The answer is C.** The alveolar gas equation is required to obtain the A-aO_2 gradient. The alveolar gas equation is $PAO_2 = 150$ mm Hg $- 1.2 \times PaCO_2$.
11. **The answer is D.** $DL = \dot{V}CO/PACO = 10$ mL/min $\div 0.5$ mm Hg $= 20$ mL/min per mm Hg.

Chapter 22

1. **The answer is D.** The basic rhythm exists in the absence of the pontine respiratory group, afferent vagal input to the pons and medulla, or an intact spinal cord. These can modify the rhythm of breathing but are not required.
2. **The answer is A.** A brief early burst by the inspiratory neurons occurs with expiration.
3. **The answer is D.** An inverse relationship exists between hypoxia-induced hyperventilation and oxygen content. Hypoxia-induced hyperventilation is dependent on $PaCO_2$ and more on carotid than aortic chemoreceptors.
4. **The answer is D.** Stimulation of lung C fibers will cause bronchoconstriction, apnea, rapid shallow breathing, and skeletal muscle relaxation.
5. **The answer is E.** CSF and plasma differ in protein concentration, PCO_2, and electrolyte composition (including the $[H^+]$).
6. **The answer is B.** Slow-wave sleep is characterized by periodic breathing, hypercapnia, and a decreased sensitivity to hypoxia. The cough reflex is suppressed, and skeletal muscle relaxation is less than in REM sleep.
7. **The answer is B.** During sleep, airway irritation *will not* evoke a cough, but will evoke apnea and arousal. Airway occlusion or hypercapnia will evoke arousal.
8. **The answer is E.** Negative-feedback systems are not necessarily the most stable.
9. **The answer is C.** The control of ventilation by $PaCO_2$ works primarily through the central chemoreceptors. However, the central effects are mediated indirectly through a change in CSF $[H^+]$, and the sensitivity is inversely related to PaO_2.
10. **The answer is B.** Minute ventilation is inversely related to SaO_2 and increases in linear fashion as SaO_2 decreases.

Chapter 23

1. **The answer is C.** Renal clearance is measured in volume of plasma per unit time.
2. **The answer is D.** Na^+ reabsorption by collecting duct principal cells occurs via a Na^+ channel called ENaC (epithelial sodium channel). Na^+ reabsorption in proximal tubule cells is coupled to transport of solutes via cotransport (e.g., Na-glucose) and antiport (e.g., Na^+/H^+ exchanger) mechanisms. Na^+ reabsorption in the thick ascending limb involves a Na-K-2Cl cotransporter and, in the distal convoluted tubule, a Na-Cl cotransporter. Collecting duct intercalated cells are primarily concerned with acid-base, not Na^+, transport.
3. **The answer is B.** Amount = concentration × volume or volume = amount/concentration. Volume = 570 mosm/day ÷ 1,140 mosm/kg H_2O = 0.5 kg H_2O/day (or, 0.5 L/day because urine is mostly water and a liter of urine weighs about 1 kg).
4. **The answer is D.** Long loops of Henle are associated with a steep gradient in the medulla because there is more opportunity for countercurrent multiplication. A drug that inhibits Na^+ reabsorption by the thick ascending limb will reduce the single effect, resulting in a loss of the medullary gradient. A very low GFR results in inadequate input of solute into the medulla and a diminished ability to concentrate the urine. Excess water intake causes the medullary gradient to fall because too much water is added to the medulla. A protein-deficient diet results in less urea production by the liver and less urea accumulation in the kidney medulla.
5. **The answer is A.** The decrease in vascular resistance leads to an increase in glomerular blood flow. Glomerular capillary pressure will fall, however, and consequently, GFR will fall. The filtration fraction (GFR/RPF) will fall because GFR falls and RPF rises. Less fluid is filtered into the space of Bowman's capsule, so the hydrostatic pressure there should fall.
6. **The answer is B.** Active reabsorption of Na^+, powered by the Na^+/K^+-ATPase, is the main driving force for water reabsorption. Reabsorption of amino acids and water is secondary to active Na^+ reabsorption. There is no active water reabsorption, and pinocytosis is too small to account for appreciable water reabsorption. The high colloid osmotic pressure in peritubular capillaries favors uptake of reabsorbed fluid from the renal interstitial fluid, but does not cause the removal of fluid from the proximal tubule lumen.
7. **The answer is B.** The percentage excretion is equal to $100 \times$ excreted Na^+/filtered $Na^+ = 100 \times (U_{Na} \times \dot{V}) \div (P_{Na} \times GFR) = 100 \times (U_{Na} \times \dot{V}) \div P_{Na} \times (U_{IN} \times \dot{V}/P_{IN}) = 100 \times U_{Na}/P_{Na} \div U_{IN}/P_{IN} = 7{,}000/140 \div 10/1 = 5$.
8. **The answer is D.** In the autoregulatory range, vascular resistance falls when arterial blood pressure falls. Changes in vessel caliber primarily occur in vessels upstream to the glomeruli (cortical radial arteries and afferent arterioles). Because autoregulatory range extends from an arterial blood pressure of about 80 to 180 mm Hg, renal blood flow is not maintained when blood pressure is low; in fact, the sympathetic nervous system will be activated and cause intense vasocon-

striction in the kidneys. Renal autoregulation does not depend on nerves.

9. **The answer is B.** When the kidney is producing maximally concentrated urine, fluid in the cortical collecting duct becomes isosmotic with the surrounding cortical interstitial fluid. Therefore, the osmolality will be about 300 mosm/kg H_2O; it cannot go above this value because hyperosmotic values (compared to systemic blood plasma) can be produced only in the kidney medulla.
10. **The answer is B.** The patient is older and severely dehydrated; the GFR can be expected to be low. Consequently, the proximal tubules may be able to reabsorb all of the filtered glucose (because the filtered load is reduced), even though the plasma [glucose] is elevated. If splay is increased, glucose Tm is low, or threshold is low, glucose should be present (not absent) from the urine. An abnormally high glucose Tm would reduce glucose excretion; however, in the scenario presented, this is not a likely cause of the absence of glucose in the urine.
11. **The answer is D.** Excretion of phenobarbital is promoted by increasing urine output and making the urine more alkaline. The latter would keep phenobarbital in its anionic form, which is not reabsorbed by the kidney tubules.
12. **The answer is C.** Inulin clearance is the standard for measuring GFR. PAH clearance is used to measure renal plasma flow, not GFR.
13. **The answer is C.** Liddle's syndrome is due to excessive activity of the Na^+ channel in collecting duct principal cells, leading to salt retention and hypertension. Bartter and Gitelman syndromes are salt-wasting disorders; blood pressure would tend to be low, not high. Diabetes insipidus and renal glucosuria produce excessive fluid loss and would not be likely causes of the patient's hypertension.
14. **The answer is A.** In the absence of arginine vasopressin, the kidneys produce a large volume of osmotically dilute urine.
15. **The answer is C.** The renal clearance of PAH is the highest (it is nearly equal to the renal plasma flow) because PAH is not only filtered by the glomeruli but is also secreted vigorously by proximal tubules. Creatinine is filtered and secreted, to a small extent only, in the human kidney. Inulin is only filtered. Urea is filtered and variably reabsorbed; its clearance is always below the inulin clearance in people. Na^+ has the lowest clearance of all because filtered Na^+ is extensively reabsorbed.
16. **The answer is A.** The filtered load of the substance is $P_x \times GFR = 2$ mg/mL $\times$ 100 mL/min $= 200$ mg/min. The rate of excretion is $U_x \times \dot{V} = 10$ mg/mL $\times$ 5 mL/min $= 50$ mg/min. Hence, more substance X was filtered than was excreted, and the difference, 200 mg/min $-$ 50 mg/min $=$ 150 mg/min, gives the rate of tubular reabsorption of substance X.
17. **The answer is C.** The true renal plasma flow (RPF) $= C_{PAH}/E_{PAH} = U_{PAH} \times \dot{V}/P_{PAH} \div (P^a_{PAH} - P^{rv}_{PAH})/P^a_{PAH} = 0.60 \times 5.0/0.02 \div (0.02 - 0.01)/0.02 = 300$ mL/min. The renal blood flow $=$ RPF/(1 $-$ hematocrit) $= 300/(1 - 0.40) = 500$ mL/min.
18. **The answer is C.** There is an inverse hyperbolic relationship between plasma [creatinine] and GFR and, therefore, a rise in plasma [creatinine] is associated with a fall in GFR (see Fig. 23.7). The greatest absolute change in GFR occurs when plasma [creatinine] doubles starting from a normal GFR and plasma [creatinine].
19. **A is the answer.** Granular cells (also known as juxtaglomerular cells) are located primarily in the wall of afferent arterioles and are the major site of renin synthesis and release.
20. **The answer is E.** $GFR = K_f (P_{GC} - P_{BS} - COP)$. Therefore, $K_f = 42$ nL/min $\div (50 - 12 - 24)$ mm Hg $= 3.0$ nL/min per mm Hg.

Chapter 24

1. **The answer is B.** ICF volume is calculated by subtracting ECF volume from the total body water. The other fluid volumes can be determined from the volume of distribution of a single indicator, such as radioactive sulfate for ECF volume, radioiodinated serum albumin for plasma volume, and deuterium oxide for total body water.
2. **The answer is A.** Cardiac failure results in a decrease in effective arterial blood volume, which stimulates thirst. Because angiotensin stimulates thirst, a low plasma level would have the opposite effect. Distension of the atria (increased blood volume) or stomach inhibits thirst. Volume expansion and a low plasma osmolality both inhibit thirst.
3. **The answer is B.** AVP is synthesized in the cell bodies of nerve cells located in the supraoptic and paraventricular nuclei of the anterior hypothalamus.
4. **The answer is D.** From the indicator dilution method, the plasma volume $= 10\ \mu Ci \div 4\ \mu Ci/L = 2.5$ L. If the hematocrit ratio is 0.4, then the blood volume $= 2.5$ L plasma $\div$ 0.6 L plasma per L blood $= 4.17$ L.
5. **The answer is A.** An increase in central blood volume will stretch the atria, cause the release of atrial natriuretic peptide, and result in diminished Na^+ reabsorption. All other choices produce increased tubular Na^+ reabsorption.
6. **The answer is B.** The loop of Henle (mostly the thick ascending limb) reabsorbs about 65% of the filtered Mg^{2+}.
7. **The answer is C.** Infusion of isotonic saline tends to raise blood pressure, decrease renal sympathetic nerve activity, and increase fluid delivery to the macula densa; all of these changes suppress renin release. All other choices result in increased renin release.
8. **The answer is E.** Skeletal muscle cells contain large amounts of K^+; injury of these cells can result in addition of large amounts of K^+ to the ECF. Insulin, epinephrine, and HCO_3^- promote the uptake of K^+ by cells. Hyperaldosteronism causes increased renal excretion of K^+ and a tendency to develop hypokalemia.
9. **The answer is B.** PTH inhibits tubular reabsorption of phosphate, stimulates tubular reabsorption of Ca^{2+}, and increases bone resorption. PTH secretion is in-

creased in patients with chronic renal failure. Its secretion is stimulated by a fall in plasma ionized Ca^{2+}.

10. **The answer is D.** Aldosterone increases K^+ secretion and Na^+ reabsorption by cortical collecting ducts. It does not affect water permeability.
11. **The answer is B.** Autoregulation refers to the relative constancy of renal blood flow and GFR despite changes in arterial blood pressure. Mineralocorticoid escape refers to the fact that the salt-retaining action of mineralocorticoids does not persist but is overpowered by factors that promote renal Na^+ excretion. Saturation of transport occurs when the maximal rate of tubular transport is reached. Tubuloglomerular feedback results in afferent arteriolar constriction when fluid delivery to the macula densa is increased; it contributes to renal autoregulation.
12. **The answer is C.** Nephrogenic diabetes insipidus is characterized by increased output of dilute urine. Plasma AVP is elevated because of the volume depletion. Plasma osmolality is on the high side of the normal range because of the loss of dilute fluid in the urine. The increased urine output is not due to diabetes mellitus because there is no glucose in the urine and the urine is very dilute. Diuretic drug abuse should not produce very dilute urine because Na^+ reabsorption is inhibited. Neurogenic diabetes insipidus is unlikely because the plasma AVP level is reduced in this case. Primary polydipsia produces output of a large volume of dilute urine, but plasma osmolality and AVP levels are decreased.
13. **The answer is E.** Na^+ is the major osmotically active solute in the ECF and is the major determinant of the amount of water in and, hence, volume of this compartment.
14. **The answer is A.** Although the plasma osmolality is extraordinarily high, the plasma Na^+, glucose, and BUN are normal. This indicates the presence of another solute (it could be ethanol) in the plasma. The calculated osmolality $= 2 \times [Na^+] + [glucose]/18 + [BUN]/2.8 = 280 + 5.6 + 5.4 = 291$ mosm/kg H_2O, a lot less than the measured osmolality (370 mosm/kg H_2O). Simple dehydration would cause a rise in plasma $[Na^+]$. Diabetes insipidus or diabetes mellitus cannot explain the high osmolality. The normal BUN does not support the existence of renal failure.
15. **The answer is B.** The inhibitor will block the conversion of angiotensin I to angiotensin II, and therefore, the plasma angiotensin I level will rise and the plasma angiotensin II and aldosterone levels will fall. The plasma bradykinin level will rise because the converting enzyme catalyzes the breakdown of this hormone. The plasma renin level will rise because (1) the fall in blood pressure stimulates renin release, and (2) angiotensin II directly inhibits renin release by acting on the granular cells of afferent arterioles, so that this inhibition is removed when less angiotensin II is present.
16. **The answer is D.** In response to an increase in dietary K^+ intake, the cortical collecting duct principal cells increase the rate of K^+ secretion, accounting for most of the K^+ excreted in the urine.
17. **The answer is D.** The subject in choice D has a low plasma osmolality but inappropriately concentrated urine. The subject in choice A may have diabetes insipidus. The subject in choice B has a low plasma osmolality, but the urine osmolality is appropriately low. The subjects in choices C and E are normal, although the subject in choice E is producing concentrated urine and may be water-deprived.
18. **The answer is A.** The low blood pH and hyperglycemia (or hyperosmolality) would tend to raise plasma $[K^+]$, yet the plasma $[K^+]$ is normal. These findings suggest that the total body store of K^+ is reduced. Remember that most of the body's K^+ is within cells. In uncontrolled diabetes mellitus, the osmotic diuresis (increased Na^+ and water delivery to the cortical collecting ducts), increased renal excretion of poorly reabsorbed anions (ketone body acids), and elevated plasma aldosterone level (secondary to volume depletion) would all favor enhanced excretion of K^+ by the kidneys. The subject has normokalemia, not hypokalemia or hyperkalemia.
19. **The answer is D.** Isotonic saline does not change cell volume. The plasma AVP level will fall because of volume expansion and cardiovascular stretch receptor inhibition of its release. The plasma aldosterone level will be low because of inhibited release of renin and less angiotensin II formation. The plasma ANP level will be increased from stretch of the cardiac atria. A large part of the infused isotonic saline will be filtered through capillary walls into the interstitial fluid.
20. **The answer is A.** ECF volume and blood volume are increased, but these should promote Na^+ excretion, not lead to Na^+ retention by the kidneys. A decrease in effective arterial blood volume is the best explanation for renal Na^+ retention.

Chapter 25

1. **The answer is D.** Using the Henderson-Hasselbalch equation, $6.0 = 9.0 + \log ([NH_3]/[NH_4^+])$, $[NH_3]/[NH_4^+] = 10^{-3.0} = 1{:}1{,}000$.
2. **The answer is B.** The plasma $[HCO_3^-]$ is easily calculated from the formula: $[H^+] = 24 \times P_{CO_2}/[HCO_3^-]$, so $[HCO_3^-] = 24 \times 24/48 = 12$ mEq/L. Alternatively, the Henderson-Hasselbalch equation could be used, but it requires the use of logarithms.
3. **The answer is E.** The collecting duct is lined by a tight epithelium and can lower the urine pH to 4.5 (a tubule fluid/plasma $[H^+]$ ratio of $10^{2.9}/1$ or about 800/1 if plasma pH is 7.4). The proximal convoluted tubule is lined by a leaky epithelium and can lower tubule fluid pH to about 6.7 (a tubule fluid/plasma $[H^+]$ ratio of $10^{0.7}/1$ or 5/1 when plasma pH is 7.4). Other nephron segments beyond the proximal convoluted tubule do not lower tubular fluid pH as much as the collecting ducts.
4. **The answer is A.** The kidneys filter about 4,320 mEq/day of HCO_3^- and usually reabsorb all but a few mEq/day; reabsorption of HCO_3^- occurs via H^+ secretion and consumes the bulk of secreted H^+. A typical excretion rate for NH_4^+ is about 50 mEq/day; for titratable acid about 25 mEq/day. The quantity of free

H^+ in a typical urine sample (e.g., 2 L/day, pH 6.0) is negligible (e.g., 0.002 mEq/day).

5. **The answer is C.** Net acid excretion is calculated from: urinary titratable acid + urinary NH_4^+ − urinary HCO_3^- excretion = 30 + 60 − 2 = 88 mEq/day, in this case.
6. **The answer is E.** In the process of excreting titratable acid and ammonia, the kidneys generate and add to the blood an equivalent amount of new HCO_3^-. Therefore, the answer is 200 + 500 = 700 mEq.
7. **The answer is E.** When Na^+ reabsorption is stimulated, Na^+/H^+ exchange is increased, resulting in greater H^+ secretion in the proximal tubule and loop of Henle. Additionally, increased Na^+ reabsorption in the collecting ducts renders the duct lumen more negative, which favors H^+ secretion. All of the other factors result in decreased H^+ secretion.
8. **The answer is C.** The subject has a severe metabolic acidosis. The anion gap (140 − 105 − 6 = 19 mEq/L) is high. Methanol intoxication (see Table 25.5) produces this type of acid-base disturbance, resulting mainly from formic acid production. Acute renal failure would also produce a high anion gap metabolic acidosis, but because the BUN is normal, this is unlikely. Uncontrolled diabetes mellitus also produces a high anion gap metabolic acidosis, but because the plasma glucose is normal, this is unlikely. Diarrhea produces a metabolic alkalosis. A drug that depresses breathing produces retention of CO_2 and respiratory acidosis.
9. **The answer is C.** Because of the low ambient barometric pressure and oxygen tension at high altitude, hypoxia develops. Therefore, we can immediately rule out choices A and D. Choice B is a subject with hypoxia that resulted from inadequate ventilation; this subject has CO_2 retention and a respiratory acidosis. Hypoxia stimulates ventilation and results in a low P_{CO_2} and respiratory alkalosis. Choice E shows values for an acute respiratory alkalosis; the plasma $[HCO_3^-]$ has been lowered by 4 mEq/L, corresponding to the 20 mm Hg decrease below normal in P_{CO_2} (see Table 25.4). Choice C shows typical values for a chronic (one week) respiratory alkalosis; the kidneys have further lowered the plasma $[HCO_3^-]$ and reduced the severity of the alkalemia.
10. **The answer is D.** Aspirin (salicylate) intoxication produces a mixed acid-base disturbance—respiratory alkalosis (as a result of stimulation of the respiratory center) and metabolic acidosis (as a result of inhibition of oxidative metabolism and accumulation of lactic and ketone body acids). The respiratory alkalosis predominates during the first several hours in adults; metabolic acidosis occurs at the same time and becomes overwhelming late in the course of the intoxication. Choice D shows the predominant respiratory alkalosis; the reduction in plasma $[HCO_3^-]$ is accounted for by the accumulation of organic acids in the blood and is too early to reflect significant renal compensation. Choice A represents metabolic acidosis with normal respiratory compensation. Choice B represents respiratory acidosis as a result of alveolar hypoventilation or a mismatch between alveolar ventilation and pulmonary capillary blood flow; note the abnormally low P_{O_2}. Choice C represents the normal condition for arterial blood. Choice E represents simple acute respiratory alkalosis.

Chapter 26

1. **The answer is B.** Successive small intestinal structures between the serosa and mucosa are longitudinal muscle, myenteric plexus, a network of interstitial cells of Cajal, circular muscle, submucous plexus, and muscularis mucosae.
2. **The answer is D.** Interstitial cells of Cajal are pacemaker cells that generate electrical slow waves. The other cell types do not generate electrical slow waves.
3. **The answer is C.** Inhibitory motor neurons determine when electrical slow waves trigger contractions. Damage to the enteric nervous system, including the inhibitory motor neurons, frees the musculature from inhibition. In the absence of inhibition, the muscle contracts continuously in a disorganized manner. Effective propulsion is impossible in the absence of the ENS.
4. **The answer is C.** Of the possible choices, only cell bodies in the dorsal vagal nucleus have axons ending in the wall of the stomach.
5. **The answer is A.** Fast EPSPs in the ENS are mediated mainly by nicotinic receptors for ACh. Hyperpolarizing after-potentials reduce excitability. Metabotropic receptors stimulate adenylyl cyclase. Fast EPSPs are not hyperpolarizing potentials.
6. **The answer is C.** Suppression of EPSPs by NE could be through an action at the presynaptic site of ACh release or an action at the postsynaptic membrane. The finding that NE does not affect the action of exogenously applied ACh while blocking the fast EPSP indicates that the mechanism of suppression of the EPSPs is suppression of ACh release at the synapse.
7. **The answer is D.** Once triggered by the stimulus, the action potential travels from muscle fiber to muscle fiber as the ionic current travels across gap junctions. Gap junctions account for the functional electrical syncytial properties of smooth muscle. Nerve fibers and the release of neurotransmitter cannot account for the spread of the action potential and associated contraction because tetrodotoxin blocked all neural function. Interstitial cells of Cajal is not correct because the action potential traveled from cell to cell in the bulk of the smooth muscle. Electrical slow waves are not correct because the action potential was triggered by a stimulus applied at one point, not slow waves originating along the segment of intestine.
8. **The answer is E.** Rapid transit is not likely because the loss of inhibitory motor neurons results in delayed transit (i.e., pseudoobstruction). Accelerated gastric emptying does not occur mainly because pseudoobstruction in the duodenum presents a high resistance to inflow from the stomach. Gastroesophageal reflux is not correct because in the absence of inhibition, the lower esophageal sphincter remains contracted and is a barrier to reflux. Diarrhea is unlikely because diarrhea

requires intestinal propulsion and this is compromised by the loss of inhibitory motor neurons. Inhibitory motor neurons are necessary for the relaxation of sphincters.

9. **The answer is D.** Longitudinal muscle is relaxed and circular muscle is contracted in the propulsive segment. Longitudinal muscle contracts and circular muscle is inhibited in the receiving segment.
10. **The answer is A.** Choice A is correct because sphincters function to prevent reflux; therefore, flow across a sphincter is generally unidirectional. Choice B is not correct because tone in the lower esophageal sphincter is increased during the MMC in the stomach. Choice C is incorrect because the sphincter cannot be relaxed after blockade of the inhibitory innervation by a local anesthetic. Choice D is incorrect because pressure in the sphincter is higher than in the two compartments it separates. Choice E is incorrect because inhibitory neurons fire to relax the sphincter during a swallow.
11. **The answer is D.** Physiological ileus is defined as the absence of contractile activity. It is a significant behavior pattern, requiring a functional ENS. Each of the other neurally programmed patterns involves contractile behavior and motility.
12. **The answer is D.** Gastric emptying of particles greater than about 7 mm does not occur during the digestive state. The lag phase is the time required for the stomach to grind large particles into smaller particles in this size range. Choice A is not correct because conversion from interdigestive to digestive states occurs immediately upon the first few swallows of a meal. Choice B is incorrect because cephalic and gastric phases of acid secretion reach maximum near the onset of the lag phase. Choice C is incorrect because the lag phase is at the beginning of the emptying curve, not at the end. Choice E is incorrect because the lag phase does not apply for a liquid meal.
13. **The answer is A.** The plateau phase of the gastric action potential and the associated trailing contraction increase in direct relation to the amount of ACh released by excitatory motor neurons to the antral musculature. The higher the firing frequency of the excitatory motor neurons, the more ACh is released. Choice B is not correct because the release of NE from sympathetic postganglionic neurons decreases the amplitude of the plateau phase of the gastric action potential. Choice C is incorrect because the firing frequency of the pacemaker does not affect the amplitude of the plateau phase. Opening of the pylorus cannot affect the trailing contraction; nevertheless, the pylorus is closed as the trailing contraction approaches. Choice E is incorrect because firing of the motor neurons to the gastric reservoir does not directly influence the innervation of the antral pump.
14. **The answer is D.** Lipids (fats) have the greatest effect in slowing gastric emptying because they have the highest caloric content. Decreased pH in the duodenum is also a powerful suppressant of gastric emptying. However, the question asks about an ingested meal, not conditions in the duodenum.
15. **The answer is A.** As the gastric reservoir fills during a meal, mechanoreceptors signal the CNS. When the limits of adaptive relaxation in the reservoir are reached, signals from the stretch receptors in the reservoir's walls account for the sensations of fullness and satiety. Overdistension is perceived as discomfort. Adaptive relaxation appears to malfunction in the forms of functional dyspepsia characterized by the symptoms described in this question. If adaptive relaxation is compromised (e.g., by an enteric neuropathy), mechanoreceptors are activated at lower distending volumes and the CNS wrongly interprets the signals as if the gastric reservoir were full. None of the other choices would be expected to activate mechanosensory signaling of the state of fullness of the gastric reservoir.
16. **The answer is E.** Power propulsion is the pattern of motility for defense of the intestinal tract. It occurs in the retrograde direction during emetic responses that empty the lumen of threatening material in the upper small intestine. It occurs in the orthograde direction in the lower small intestine and in the large intestine where it also functions to quickly eliminate threatening substances or organisms from the intestine. In the large intestine, secretion flushes the material from the mucosa and holds it in suspension in the lumen. This is followed by power propulsion, which rapidly clears the lumen of the material. This form of behavior is defensive but has the adverse effects of diarrhea and abdominal pain. None of the other choices evokes conscious sensations during daily occurrence.
17. **The answer is D.** Observations on the transit of markers after instillation in the human cecum show that the markers remain for the longest time in the transverse colon. Transit is significantly faster in the other parts of the large intestine.
18. **The answer is D.** Examination of older patients often reveals weakness in the pelvic floor musculature. Weakness in the puborectalis muscle allows the anorectal angle to straighten and lose its barrier function to the passage of feces into the anorectum. Choice A is incorrect because the rectoanal reflex (i.e., relaxation of the internal anal sphincter in response to distension of the rectum) does not weaken significantly in older persons. Choice B is incorrect because a deficit in sensory detection, not elevated sensitivity, can be a factor in fecal incontinence. Choice C is incorrect because adult Hirschsprung's disease results in constipation, not incontinence. The myopathic form of pseudoobstruction is not associated with fecal incontinence because propulsive motility is absent as a result of weakening of the intestinal smooth muscle.

Chapter 27

1. **The answer is D.** Salivary secretion is exclusively under neural control. The others need both neural and hormonal stimulation and are, therefore, only partially stimulated by the sight, smell, and chewing of food (cephalic phase). The sight, smell, and chewing of food stimulate the parasympathetic nervous system, which stimulates salivary secretion.

2. **The answer is C.** The uptake of bile acid by hepatocytes is sodium-dependent and is not dependent on calcium, iron, potassium, or chloride.
3. **The answer is A.** Intrinsic factor is critical for the absorption of vitamin B_{12} by the ileum. None of the other substances is secreted by parietal cells. Gastrin, somatostatin, and CCK are secreted by specialized GI endocrine cells, whereas chylomicrons are produced by enterocytes.
4. **The answer is C.** Although the cephalic and intestinal phases stimulate gastric secretion, the gastric phase is, by far, the most important.
5. **The answer is B.** Carbonic anhydrase catalyzes the formation of carbonic acid from carbon dioxide and water. It is not involved in the formation of carbon dioxide from carbon and oxygen, bicarbonate ion from carbonic acid, hydrochloric acid, or hypochlorous acid.
6. **The answer is B.** Parasympathetic stimulation induces the release of kallikrein by the salivary acinar cells, which converts kininogen to form lysyl-bradykinin (a potent vasodilator). Bradykinin is a vasoactive peptide. Kininogen is the precursor for kinins. Kinins include bradykinin and lysyl-bradykinin. Aminopeptidase releases amino acids from the amino end of peptides and is found in the brush border membrane and cytoplasm of enterocytes.
7. **The answer is D.** Intrinsic factor is secreted by the parietal cells of the stomach and is not secreted by the salivary glands. Lactoferrin, amylase, mucin, and muramidase are found in saliva.
8. **The answer is B.** In the fasting state, the pH of the stomach is low, between 1 and 2.
9. **The answer is A.** Salivary secretion is inhibited by atropine. Atropine is an anticholinergic drug that competitively inhibits ACh at postganglionic sites, inhibiting parasympathetic activity. Pilocarpine actually stimulates salivation because of its muscarinic action. Cimetidine is an antagonist for the histamine H_2 receptor. Aspirin is the most widely used analgesic (pain reducer), antipyretic (fever reducer), and anti-inflammatory drug. Omeprazole inhibits the H^+/K^+-ATPase and, thus, inhibits acid secretion.
10. **The answer is C.** The chief cells of the stomach secrete pepsinogen, and the parietal cells of the stomach secrete hydrochloric acid and intrinsic factor. Gastrin and CCK are secreted by specialized endocrine cells.
11. **The answer is B.** Histamine interacts with its receptor in parietal cells to increase the intracellular cAMP. Histamine does not cause an increase in intracellular sodium or cGMP or a decrease in intracellular calcium.
12. **The answer is D.** When the pH of the stomach falls below 3, the antrum secretes somatostatin, which acts locally to inhibit gastrin release; therefore, somatostatin inhibits gastric secretion. Enterogastrones are hormones produced by the duodenum that inhibit gastric secretion and motility. Intrinsic factor is involved in the absorption of vitamin B_{12} and is not involved in the release of gastrin. Secretin is a hormone secreted by the duodenal and jejunal mucosa when exposed to acidic chyme and is responsible for stimulating pancreatic secretion rich in bicarbonate. CCK stimulates the gallbladder to contract and the pancreas to secrete a juice rich in enzymes.
13. **The answer is B.** Secretin stimulates secretion of a bicarbonate-rich pancreatic juice. Somatostatin, gastrin, and insulin do not. CCK stimulates a pancreatic secretion rich in enzymes and potentiates the action of secretin.
14. **The answer is C.** Excessive production of gastrin results in acid hypersecretion and peptic ulcer disease. Patients with Zollinger-Ellison syndrome do not suffer from excessive acid reflux, excessive secretion of CCK, failure of the liver to secrete VLDLs, or failure to secrete a bicarbonate-rich pancreatic juice.
15. **The answer is B.** Lactase hydrolyzes lactose to form both glucose and galactose. None of the other combinations is correct.
16. **The answer is A.** Maltase hydrolyzes maltose to form glucose. Because maltose does not contain galactose or fructose, none of the other choices is correct.
17. **The answer is C.** Fructose is taken up by enterocytes by facilitated diffusion. Both glucose and galactose are taken up by enterocytes through a sodium-dependent transporter. Xylose and sucrose are not taken up by enterocytes.
18. **The answer is D.** Pancreatic lipase hydrolyzes triglyceride to form 2-monoglyceride and two fatty acids. The hydrolysis of phosphatidylcholine, not triglyceride, results in the formation of lysophosphatidylcholine. Although diglyceride is an intermediate in the hydrolysis of triglyceride by pancreatic lipase, the hydrolysis continues until 2-monoglyceride and fatty acids are formed. Pancreatic lipase does not hydrolyze triglyceride totally to form glycerol and fatty acids.
19. **The answer is C.** The small intestine transports dietary triglyceride as chylomicrons in lymph. VLDLs are secreted by the small intestine during fasting. Although some dietary fatty acids are transported in the portal blood bound to albumin, it is not the predominant pathway for the transport of dietary lipids to the circulation by the small intestine. The intestine does not secrete LDLs, and although it does secrete HDLs, they are not used as a vehicle for transporting dietary lipids to the blood by the small intestine.
20. **The answer is C.** Amino acids, as well as dipeptides and tripeptides, use different brush border transporters for their uptake. Dipeptides and tripeptides are not taken up passively by any part of the GI tract.
21. **The answer is A.** Dietary protein is transported in the portal blood as free amino acids. Although dipeptides and tripeptides are taken up by enterocytes, they are hydrolyzed by the brush border membrane, as well as by cytoplasmic peptidase to form free amino acids.
22. **The answer is D.** Vitamin B_1 is a water-soluble vitamin. Vitamins A, D, E, and K are all fat-soluble vitamins.
23. **The answer is B.** Vitamin D plays an important indirect role in the absorption of calcium by the GI tract. The other vitamins listed are not involved in the absorption of calcium.
24. **The answer is C.** Vitamin A is transported in chylomicrons as ester. Vitamins D, E, and K are transported in

the free form associated with chylomicrons. Vitamin B_{12}, a water-soluble vitamin, is transported in the blood bound to transcobalamin.

25. **The answer is C.** Potassium is passively absorbed by the jejunum. The other choices do not apply to the absorption of potassium by the small intestine.
26. **The answer is C.** Ascorbic acid enhances iron absorption mostly by its reducing capacity, keeping iron in the ferrous state. Ascorbic acid does not enhance heme iron absorption, nor does it affect heme oxygenase activity or the production of ferritin or transferrin.

Chapter 28

1. **The answer is D.** Alcohol dehydrogenase catalyzes the conversion of alcohol to acetaldehyde, which is then converted to acetate. Acetate is then metabolized by hepatocytes. Cytochrome P450 is a primary component of the oxidative enzyme system involved in the metabolism of drugs. NADPH-cytochrome P450 reductase is an enzyme involved in phase I reactions of drug metabolism. There is no such enzyme as alcohol oxygenase. Glycogen phosphorylase is an enzyme involved in glycogen breakdown, not alcohol metabolism.
2. **The answer is C.** Unlike patients who have diabetes, healthy humans are capable of keeping their blood glucose within a relatively narrow range after a meal, 120 to 150 mg/dL. Blood levels of 30 to 50 mg/dL and 50 to 70 mg/dL indicate hypoglycemia, and blood levels of 220 to 250 mg/dL and 300 to 350 mg/dL indicate hyperglycemia.
3. **The answer is A.** The liver has the enzyme glucose-6-phosphatase, but muscle does not. Consequently, muscle is incapable of releasing glucose from glucose 6-phosphate. Glucose undergoes reactions other than glycolysis. Both liver and muscle have glucose-1-phosphatase and glycogen phosphorylase enzymes. The synthesis of glucose, called gluconeogenesis, is carried out mostly in the liver and, to some extent, in the kidneys.
4. **The answer is A.** Fatty acid synthesis occurs only in the cytoplasm. Mitochondria are involved in fatty acid oxidation rather than synthesis. Fatty acid synthesis does not occur in the nucleus. Endosomes and the Golgi apparatus are not involved in fatty acid synthesis.
5. **The answer is B.** Although both chylomicrons and VLDLs are triglyceride-rich lipoproteins, the liver, unlike the small intestine, produces only VLDLs. LDLs and HDLs are not triglyceride-rich lipoproteins. Chylomicron remnants are generated in the circulation by the metabolism of chylomicrons.
6. **The answer is C.** Both urea and glutamine play an important role in the storage and transport of ammonia in the blood. Histidine, phenylalanine, methionine, and lysine are not involved in ammonia transport.
7. **The answer is C.** The liver is one of the major sites for the removal of hormones, including glucagon. Consequently, patients with a portacaval shunt have high levels of circulating glucagon and other hormones because portal blood bypasses the liver. Choice A is incorrect because the pancreas does not produce more glucagon in portacaval shunt patients. Choice B is incorrect because the kidneys are capable of removing glucagon in these patients. However, the kidneys are not nearly as important as the liver in removing glucagon in healthy individuals. Choice D is incorrect because the small intestine does not produce glucagon. Choice E is incorrect because blood flow to the small intestine is not compromised in portacaval shunt patients.
8. **The answer is C.** The liver makes transferrin to carry iron in the blood. Hemosiderin is an intracellular complex of ferric hydroxide, polysaccharides, and proteins. Haptoglobin binds free hemoglobin in the blood. Ceruloplasmin is a circulating plasma protein involved in the transport of copper. Lactoferrin is an iron-binding glycoprotein found in secretions (e.g., milk, saliva) and in neutrophil granules; it appears to contribute to antimicrobial host defenses.
9. **The answer is A.** Smokers inhale polycyclic aromatic hydrocarbons, which stimulate drug-metabolizing enzymes. Therefore, smokers have higher levels of hepatic drug-metabolizing enzymes than nonsmokers. The level of drug-metabolizing enzymes in the liver is lowered by malnutrition and is lower in the newborn.
10. **The answer is C.** Phase I reactions of drug metabolism refer to the addition of one or more polar groups to the drug molecule. Hydrophilic, not hydrophobic, groups are introduced into the drug molecule in a phase I reaction. The conjugation of drugs with glucuronic acid, glycine, taurine, or sulfate is a phase II reaction.
11. **The answer is C.** A healthy liver converts vitamin D (cholecalciferol) to form 25-hydroxycholecalciferol, but a diseased liver has a reduced capacity to do so. The kidney, not the liver, is responsible for the conversion of 25-hydroxycholecalciferol to 1,25-dihydroxycholecalciferol. Vitamin D, not 1,25-hydroxycholecalciferol, is absorbed by the small intestine.
12. **The answer is A.** LDLs are removed from the blood by the liver by binding to LDL receptors, followed by endocytosis of the LDL-receptor complex. LDLs do not bind to HDL receptors, albumin, transferrin, or ceruloplasmin.

Chapter 29

1. **The answer is C.** Antipyretics, such as aspirin, inhibit the synthesis of prostaglandin E_2, which mediates the elevation of the thermoregulatory set point during fever. Antipyretics cannot prevent the increase in core temperature during exercise because that increase is not produced by an elevated thermoregulatory set point (see Fig. 29.11). Therefore, considerations of the possible harm or benefits as a result of the increase, as in choices A and B, are irrelevant. As antipyretics do not directly stimulate heat loss responses, choices D and E are not applicable.
2. **The answer is D.** Blood vessels in the skin have a dual nervous control, but both vasoconstriction and active vasodilation are mediated by sympathetic fibers. The nerve endings that control sweating are also part of the

sympathetic nervous system, although they release acetylcholine. Sympathectomy abolishes sweating, vasoconstriction, and active vasodilation.

3. **The answer is C.** The core temperature of a resting person shows a circadian rhythm and is higher at 4:00 PM than at 4:00 AM. This rhythm in core temperature is the result of an underlying circadian rhythm in the thermoregulatory set point. Because a change in the thermoregulatory set point affects core temperature at rest and the thresholds for sweating and vasodilation all in the same way, these thresholds are also higher at 4:00 PM.
4. **The answer is E.** Acclimatization to cold produces several different (and contrasting) sets of changes, depending on the acclimatizing environment (and, perhaps, on characteristics of the population being acclimatized).
5. **The answer is B.** Fever enhances the body's defenses, partly by magnifying the responses of leukocytes and macrophages to the other stimuli that are operative during an immune response. Choice E reflects what was widely believed until the 1970s. Although a few pathogens do not flourish at temperatures above 37°C, they are so much the exception that A is not the best choice.
6. **The answer is C.** The classic changes observed in heat acclimatization are *lower* heart rate during exercise; an *increased* sweating response; and a *lower* core temperature during exercise, which is due to both the increased sweating response and a *lower* thermoregulatory set point. In addition, salt is conserved by a *reduced* salt concentration in sweat.
7. **The answer is B.** Before the hike, the total osmotic content of the body was 40 L × 280 mosm/L = 11,200 mosm (assuming that plasma osmolarity = 2 × plasma [Na^+]). The subject lost 400 mmol (50 mmol/L × 8 L) of NaCl or 800 mosm from the ECF in sweat and replaced all of his water losses. His plasma osmolarity after the hike and rest is 260 mosm/L [(11,200 − 800 mosm) ÷ 40 L], or plasma [Na^+] = 130 mmol/L. The reduced plasma osmolarity causes water to move from the ECF into the cells until a new osmotic equilibrium is established. The initial Na^+ content of the subject's ECF was 15 L × 140 mmol/L = 2,100 mmol. He lost 400 mmol Na^+ during the hike, and his ECF [Na^+] was lowered to 130 mmol/L. His new ECF volume = (2,100 − 400 mmol) ÷ 130 mmol/L = 13.1 L.
8. **The answer is B.** His metabolic rate is 800 W; however, he is performing external work at a rate of 140 W and needs to dissipate 660 W (= 800 W − 140 W) as heat. (It is true that he requires a higher metabolic rate to go uphill than if he were going on a level road, but we have already specified his metabolic rate.) His skin temperature is 14°C above air temperature, and the convective heat transfer coefficient is 15 W/(m^2·°C), so he loses heat by convection at a rate of 210 W/m^2 of surface area. Because his body surface area is 1.8 m^2, convection accounts for a loss of 378 W, leaving 282 W = 16,920 J/min to be dissipated by evaporation of sweat. Because it takes evaporation of 1 g of sweat to remove 2,425 J of heat, he must secrete and evaporate 7 g of sweat per min.

Chapter 30

1. **The answer is C.** A maximal voluntary contraction involving the identical muscles in an identical form of contraction provides the most readily quantified and accurate basis for normalization of isometric exercise intensity. Choice A is incorrect because the basis for comparison involves rhythmic, dynamic exercise. The other choices also contradict the principle that exercise can only be compared with other exercise involving the same muscles and the same types of muscle contraction.
2. **The answer is A.** The physiological responses to dynamic exercise are predictable when healthy individuals differing in endurance exercise capacity are compared at matched levels of relative oxygen transport demand. Exercise at 75% of the maximal oxygen uptake will lead to exhaustion in typically 1 to 2 hours, rendering choice B incorrect. The more highly trained person will show increased work output despite fatiguing at about the same time as the person with lower capacity, rendering choices C and D incorrect. Training lowers lactic acid production at any matched fractional use of the maximal oxygen uptake, making choice E incorrect.
3. **The answer is D.** Active muscle vasodilation during dynamic exercise is quantitatively much greater than the net vasoconstriction in the gut, skin, kidneys, and inactive muscle. Choices B, C, and D contradict this answer. Total systemic vascular resistance can be measured, albeit indirectly, from measurements of systemic arterial pressure and cardiac output.
4. **The answer is C.** This answer presumes that vasoconstriction occurs in these vascular beds and that its effect is to help balance vasodilation in active skeletal muscles and prevent exercise-induced systemic hypotension. This effect is ubiquitous across all individuals during all forms of dynamic exercise, making choices B, C, and E incorrect. Cerebral blood flow is held constant during all forms of exercise, unlike renal or splanchnic blood flow.
5. **The answer is A.** Even highly trained and heat-acclimatized individuals are at risk for heat-related illness if exercise is sufficiently prolonged and if environmental conditions are sufficiently severe. In healthy persons during exercise, coronary vasodilatory capacity is adequate, renal blood flow reductions in health are entirely safe, and gastric mucosal blood flow reductions are easily tolerated. In long-term exercise in a warm environment, hypotension, not hypertension, is the possible cardiovascular risk.
6. **The answer is E.** During dynamic exercise, the balance of active muscle vasodilation and sympathetically driven vasoconstriction in other organs provides the highest systemic arterial pressure when the involved muscle mass size is intermediate. Isometric exercise always causes blood pressure to increase more than matched dynamic exercise. Prolongation of work low-

ers blood pressure. The state of training, fatigue, and prolongation of activity are largely irrelevant or nonspecific as factors affecting blood pressure during exercise.

7. **The answer is E.** The baroreceptor blood pressure set point is increased during exercise, depending on exercise mode, intensity, and duration. Blood pressure only falls during exercise when there is preexisting cardiac disease or during prolonged work in the heat. Training has no apparent effects on the baroreflex.
8. **The answer is E.** In the broadest terms, the changes in cholesterol transport in response to chronically increased physical activity occur from prolonged enhancement of fat metabolism. The increase in HDL and decrease in LDL occur, at least in part, in response to enhanced lipoprotein lipase activity and increased apo A-I synthesis. These effects of long-term, regular, dynamic exercise are largely independent of diet and weight loss.
9. **The answer is B.** A normal or reduced arterial P_{CO_2} is a respiratory response that regulates arterial blood pH during exercise. Oxygen partial pressure significantly declines in arterial blood during exercise only when there is preexisting lung disease (choice A), while the respiratory control system allows ventilation to match, but not exceed, levels of CO_2 production (choice C). Exercise in healthy persons does not result in respiratory acidosis or dizziness resulting from decreased cerebral perfusion.
10. **The answer is B.** Exercise training has no effect on lung tissues other than the respiratory muscles. The incorrect choices represent aspects of lung function that are determined by lung tissues unaltered by any form of physical activity. Decreases in ventilation and dyspnea during exercise do occur with chronic increases in dynamic exercise, but these arise from adaptations localized in the active skeletal muscles (including the respiratory muscles).
11. **The answer is D.** Weight-bearing exercise and increased muscle strength reduce osteoporosis by increasing the forces applied to bone. These changes are augmented by exercise-linked improvements in motor coordination that reduce the risk of falls. These factors in combination sharply reduce the incidence of hip fracture in older persons. Activities that decrease gravitational forces on bone (e.g., water immersion), while valuable, decrease forces applied to bone and are less useful in the prevention of osteoporosis.
12. **The answer is D.** Eccentric contractions cause delayed muscle soreness. This muscle inflammatory response is a result of the greater force per active motor unit found during eccentric as compared with concentric exercise at the same force development. Soreness is not found after isometric exercise (choice A), solely on the basis of ischemia (which occurs in many forms of muscle contraction; choice B), in response to increased endurance (choice C), or in direct proportion to the percentage usage of the maximum voluntary contractile force (which is defined in terms of isometric contractions; choice E).
13. **The answer is C.** Motor unit rotation allows frequent rest and recovery for activated cells, delaying fatigue. Inactive muscle cells undergo atrophic changes that reduce cell cross-sectional area, reducing strength and increasing the number of mobilized cells and motor units required for a fixed external force development. These facts contradict choice A. Atrophy causes all systems required for force production to down-regulate in parallel, contradicting choice B, and lack of activity reduces, rather than increases, oxidative capacity, rendering choice D erroneous. Choice E is false because the form of exercise must be standardized for meaningful comparisons of strength or endurance.
14. **The answer is E.** Relative to weight-bearing activity, estrogen plays a more important role in the maintenance of bone mass in women. Reductions in bone mass, which increase the risk of fracture and are invariably associated with reduced body weight, occur despite increased dynamic exercise endurance and intramuscular adaptations that are appropriate for the high level of dynamic exercise training.
15. **The answer is D.** Increases in both insulin-dependent and insulin-independent glucose uptake in active muscles during exercise enhance the measured insulin sensitivity. These effects reduce the requirements for either insulin itself or for oral antiglycemic agents in persons with type 2 diabetes. In contrast, in type 1 diabetes, these same effects increase the risk for hypoglycemia, leading to requirements for careful monitoring of activity, as well as food intake, insulin administration, and stress in persons with this illness. All of the other choices directly contradict this principle, other than choice B, which is incorrect because increased sympathetic activity during exercise directly reduces pancreatic insulin release and blood insulin levels.
16. **The answer is C.** Maternal activity reduces the risk of maternal gestational diabetes as a result of the same mechanisms (increased insulin-dependent and insulin-independent muscle glucose uptake) that reduce the risk and severity of type 2 diabetes in all persons. There are no known negative effects of maternal exercise on either the course of pregnancy or its outcome, and maternal exercise does not alter the duration of gestation or fetal weight at term.

Chapter 31

1. **The answer is C.** Right or left shifts in dose-response curves indicate changes in sensitivity. Changes in maximal biological response indicate changes in responsiveness. Because there is no change in maximal response, the correct answer must relate to a change in sensitivity only. A right shift indicates decreased sensitivity.
2. **The answer is C.** Hormones produce their effects on target cells by interacting with specific receptors. Hormone binding to its receptor generally initiates a cascade of events that lead to biological effects in the target cells.
3. **The answer is A.** Aldosterone is a steroid and the primary mineralocorticoid in the body. Testosterone,

progesterone, and cortisol are steroid hormones having primarily androgen, progestin, and glucocorticoid activities, respectively. Prostaglandin E_2 is a local signaling molecule derived from arachidonic acid.

4. **The answer is B.** Scatchard plots of hormone-receptor binding data give information regarding the number of receptors and the affinity of the hormone for its receptor. The x-intercept provides data regarding total receptor number, and the slope is equal to the negative of the association constant ($-K_a$).
5. **The answer is D.** Preprohormones are the gene products for most peptide and protein hormones. These are rapidly cleaved to form prohormones. POMC and pro-pressophysin are two examples of specific prohormones.
6. **The answer is E.** Cortisol, like other steroid hormones, is carried in the blood largely bound to carrier proteins, although a small percentage exists free in solution. The majority of cortisol is bound to a specific carrier protein, corticosteroid-binding globulin (CBG), while smaller amounts are bound nonspecifically to albumin. Few, if any, cortisol receptors would be expected in the plasma and transthyretin binds primarily thyroxine.
7. **The answer is D.** Hormones generally circulate at concentrations from 10^{-9} to 10^{-12} M. They produce much larger changes in a variety of biological parameters as a result of signal amplification, in which the rather weak hormonal signal is amplified into a larger biological response.

Chapter 32

1. **The answer is C.** Destruction of the neurons in the paraventricular nuclei of the hypothalamus decreases CRH release, which causes decreased synthesis and secretion of ACTH. Hyperosmolality of the blood would lead to an increase in portal blood AVP, which increases ACTH secretion by corticotrophs. Physical or emotional stress increases ACTH release. Glucocorticoids feed back to the hypothalamus and anterior pituitary to decrease ACTH synthesis and secretion. Primary adrenal insufficiency is characterized by a lack of glucocorticoids in the blood, resulting in an increase in ACTH synthesis and secretion. Increased PKA activity in corticotrophs increases ACTH synthesis and secretion.
2. **The answer is D.** Thyroid hormones stimulate the expression of the GH gene in somatotrophs. Thyroid hormones exert a negative-feedback signal on the hypothalamic-pituitary-thyroid axis to inhibit their own synthesis and secretion. Therefore, thyroid hormones *decrease* the sensitivity of thyrotrophs to TRH, *decrease* the formation of IP_3 in thyrotrophs, *inhibit* the expression of the genes for the α and β subunits of TSH in thyrotrophs, and *decrease* the secretion of TSH by thyrotrophs. Thyroid hormones have no effect on ACTH release.
3. **The answer is B.** Galactorrhea is commonly associated with pituitary tumors secreting excess PRL. Prolactin is important in maintaining breast milk production in the woman after giving birth. Galactorrhea is diagnosed if present longer than 6 months postpartum in a nonnursing mother. The combination of both galactorrhea and amenorrhea is diagnostic of a PRL-secreting pituitary tumor. TSH generally has little effect on PRL secretion. GH has lactogenic activity when high, not low. Hypothalamic dopamine is an inhibitor of PRL release.
4. **The answer is A.** Neurophysin is the other product generated when the prohormone for AVP or oxytocin is cleaved. A decrease in blood volume would result in the release of AVP and neurophysin from magnocellular neurons. The hormones oxytocin, β-lipotropin, ACTH, and somatostatin are not involved in the regulation of blood volume.
5. **The answer is C.** Growth hormone deficiency in adults is characterized by decreased muscle strength and exercise intolerance and a reduced sense of well-being (including effects on libido). Lean body mass (muscle) is lost, and excess body fat deposition occurs in the abdominal region. GH replacement can reverse these effects. Thyroid dysfunction is ruled out by the normal thyroid hormone levels. Glucocorticoid deficiency usually results from primary adrenal insufficiency, as in Addison's disease. Clinical symptoms include a decreased sense of well-being, GI disturbances, and abnormal glucose metabolism. Primary adrenal insufficiency is also characterized by high blood levels of ACTH, which can result in hyperpigmentation as a result of the melanocyte-stimulating activity of the amino terminal portion of ACTH. Adrenal insufficiency is not usually associated with a redistribution of body fat to central stores. Prolactin does not appear to have a major physiological effect in human males. Acromegaly results from excessive GH secretion in an adult; the symptoms are not consistent with acromegaly.
6. **The answer is C.** The data demonstrate a higher average ACTH and higher average cortisol concentration in the evening hours. This is opposite the usual diurnal pattern in which ACTH and cortisol are high in the morning. It is possible that the subject works nights and has a reversed but normal diurnal rhythm of ACTH and cortisol release. There is no adrenal disease (primary or secondary) because both ACTH and cortisol are higher at the same time and then are lower at the same time. The diurnal pattern rules out an ACTH-secreting tumor because ACTH release would tend to be constant.
7. **The answer is B.** Somatostatin, given as a long-acting analog octreotide, is effective in reducing excess secretion of GH. It can also reduce tumor size, if one is present. Glucocorticoid would feed back to inhibit the hypothalamic-pituitary-adrenal axis but have little effect on GH release. Because acromegaly is characterized by excessive GH secretion, the administration of GH would be inappropriate. Insulin could be given to counter the diabetogenic effects of excess GH, but it would have little effect on tumor size (if present), bone thickening, or hypertrophy of the liver. GHRH and thyroid hormone would stimulate GH release in a situation of high GH.
8. **The answer is B.** GHRH increases cAMP and stimu-

lates GH synthesis and secretion; somatostatin decreases cAMP and inhibits GH synthesis and secretion from somatotrophs. TRH stimulates TSH secretion and the synthesis of the α and β subunits of TSH by increasing inositol trisphosphate and calcium in thyrotrophs. cAMP has no effect on AVP release, and it stimulates ACTH synthesis in corticotrophs.

Chapter 33

1. **The answer is A.** TSH stimulates the endocytosis of colloid by the apical membrane of the follicular cell. Thyroglobulin in the colloid is hydrolyzed in the lysosomal vesicles to release thyroid hormones. T_4 and T_3 are stored in thyroglobulin in the colloid, not in secretory vesicles in the follicular cell. TSH stimulates the uptake of iodide from the blood, not the colloid. It has no effect on blood flow to the thyroid gland and no direct effect on the binding of T_4 and T_3 to thyroxine-binding globulin. TSH stimulates an increase in cAMP, not an increase in the hydrolysis of this second messenger.
2. **The answer is C.** Thyroid hormones are important for normal development of the CNS and for body growth. TSH stimulates the synthesis and release of thyroid hormones, as well as the growth of the thyroid gland. In a disorder in which the thyroid gland does not respond to TSH, thyroid hormone production would be decreased, resulting in poor development of the CNS and poor body growth. TSH would also not be able to stimulate the growth of the thyroid, resulting in a small gland.
3. **The answer is A.** Giving thyroid hormones to the child would improve body growth but not mental ability because thyroid hormones are most important for CNS development in utero. Therefore, giving thyroid hormones after birth would be too late. Thyroid gland size would remain smaller than normal because thyroid hormones have no trophic effect on the gland; only TSH has a trophic effect.
4. **The answer is C.** Uncoupling proteins allow protons to flow down their electrochemical gradient across the mitochondrial membrane, uncoupled from the synthesis of ATP. The resulting energy generated is released as heat, and ATP is not synthesized. Uncoupling proteins are increased by thyroid hormones. The novel uncoupling proteins are found in many tissues, including muscle and adipose tissue. Oxidation of fatty acids and glucose is not coupled in mitochondria, and the uncoupling proteins are not the switch between oxidation of these two substrates. Uncoupling proteins have not been demonstrated to be essential to the maintenance of body temperature in mammals. However, UCP-1 is important in the ability of small mammals, such as rodents, to tolerate cold temperatures.
5. **The answer is F.** T_3 is produced from T_4 by 5′-deiodinase (type 2) in the anterior pituitary. The major thyroid hormone product of the thyroid gland is T_4. The thyroid hormone receptor (TR) is located in the nucleus. A 5-deiodinase acts on T_4 to make reverse T_3. The half-life of T_3 in the bloodstream is about 1 day because of the protective actions of the thyroid hormone-binding proteins. Thyroid peroxidase catalyzes the iodination of thyroglobulin to form MIT and DIT, precursor molecules for T_3.
6. **The answer is D.** The patient's symptoms of chronic fatigue, aching muscles, occasional numbness in the fingers, and weight gain are consistent with a hypothyroid state. The high TSH rules out a defect in the hypothalamic-pituitary axis and suggests an unresponsive thyroid gland, most likely a result of autoimmune thyroid disease. The presence of antibodies to thyroid peroxidase or thyroglobulin would confirm the diagnosis. The absence of a goiter rules out hypothyroidism as a result of iodine deficiency; low serum thyroid hormone levels would result in elevated TSH with subsequent trophic effects on thyroid growth. There is no growth of the thyroid in this patient because of the autoimmune attack on the gland.
7. **The answer is B.** Thyroid peroxidase catalyzes the coupling of two adjacent iodotyrosine residues in the thyroglobulin precursor to form iodothyronine and dehydroalanine. Thyroid peroxidase uses hydrogen peroxide produced by mitochondria to iodinate tyrosine residues and to couple adjacent iodotyrosine residues. Thyroid peroxidase is localized to the apical membrane of the follicular cell and catalyzes all reactions in this location. The release of thyroid hormone is mediated by lysosomal degradation of thyroglobulin. Thyroid peroxidase iodinates tyrosine residues in the thyroglobulin molecule to form MIT and DIT. Dehydroalanine is derived from the free-radical rearrangement of 2 DIT residues to form thyroxine. Thyroid peroxidase forms the free radicals necessary for this reaction.
8. **The answer is F.** A TSH secreting tumor of the pituitary would result in elevated thyroid hormone levels and symptoms of thyrotoxicosis. Graves' disease is characterized by elevated thyroid hormone levels and anti-TSH receptor antibodies. TSH would be low because of feedback inhibition of its release. Resistance to thyroid hormone action could result in elevated thyroid hormone levels but would not cause symptoms of thyrotoxicosis. Thyroid gland adenomas commonly result from a point mutation in the TSH receptor, resulting in chronic activation of signaling. This would increase thyroid hormones but should result in a reduction in TSH. A deficiency in 5′-deiodinase could result in increased thyroid hormone levels and symptoms of thyrotoxicosis, but would not be associated with elevated TSH. Early in the progression of Hashimoto's disease, symptoms of thyrotoxicosis may be present, but the absence of antithyroid antibodies rules out this condition.

Chapter 34

1. **The answer is B.** Cholesterol esters in LDL are the most important source of cholesterol for sustaining adrenal steroidogenesis when it occurs at a high rate over a long time. This cholesterol can be used directly after release from LDL and not stored. *De novo* synthesis of

cholesterol from acetate is a minor source of cholesterol in humans. Cholesterol from the plasma membrane or endoplasmic reticulum is not used for steroidogenesis. Cholesterol esters in lipid droplets within adrenal cortical cells would be used first and depleted during periods of high adrenal steroid hormone synthesis.

2. **The answer is C.** The increase in body weight with little linear growth suggests that the patient has Cushing's disease rather than general obesity because linear growth usually continues in obesity syndromes. Laboratory findings in Cushing's disease include elevated ACTH, serum cortisol, urinary cortisol, and serum insulin (as a result of the cortisol-induced resistance to insulin action in skeletal muscle and adipose tissue).
3. **The answer is A.** Congenital adrenal hyperplasia is the result of genetic defects that affect adrenal steroidogenic enzymes, producing an impaired formation of cortisol. Low serum cortisol is a stimulus for ACTH release from the hypothalamus. The increase in ACTH has a proliferative effect on the adrenal gland, resulting in hyperplasia. Addison's disease is the result of pathological destruction of the adrenal glands by microorganisms or autoimmune disease and would, therefore, not result in adrenal hyperplasia. ACTH stimulates the growth of the adrenal gland. A reduction in ACTH in the blood would result in atrophy of the adrenal gland. Corticosteroid-binding globulin noncovalently binds steroid hormones in plasma; defects in this protein are not associated with adrenal hyperplasia. Cushing's disease results from a pituitary ACTH-secreting tumor; adrenal hyperplasia is secondary, not congenital, in this disease. Aldosterone synthesis is regulated by the renin-angiotensin system. Defective aldosterone synthesis would, therefore, not lead to increased ACTH and adrenal hyperplasia.
4. **The answer is E.** Catecholamines stimulate glycogenolysis and gluconeogenesis in the liver, causing glucose to be synthesized and released into the blood. Catecholamines stimulate glycogen phosphorylase in muscle to free glucose for use by the muscle. Muscle cannot release glucose to the circulation because it lacks glucose-6-phosphatase. However, the muscle can release lactate, which can be used in gluconeogenesis by the liver. Catecholamines inhibit the release of insulin from the pancreas. Insulin would be counterproductive to attempts to increase blood glucose. Catecholamines increase the release of fatty acids from the adipose tissue, to be used in gluconeogenesis by the liver.
5. **The answer is F.** Patients on long-term glucocorticoid therapy should have the dose increased prior to undergoing surgery to minimize the effects of surgical stress. These patients cannot mount their own stress response because of the lack of adrenal cortisol release. Glucocorticoid-induced hypoglycemia or interactions with anesthetics are unlikely, and these concerns would be secondary to stimulating the response to surgical stress. Glucocorticoids inhibit ACTH release and the immune response. Glucocorticoids increase the response of the vasculature to catecholamines.
6. **The answer is F.** IP_3 is one of the second messengers in the cells of the zona glomerulosa that signals for aldosterone release. A decrease in IP_3 would result in less signal for aldosterone synthesis and release. The rate of aldosterone secretion would increase in response to an increase in renin release from the kidney. Renin catalyzes the rate-limiting step in the conversion of angiotensinogen to angiotensin II, which is a stimulus for aldosterone synthesis and release. A rise in serum potassium or renal sympathetic nerve activity, a fall in blood pressure in the kidney, or a decrease in tubule fluid sodium concentration at the macula densa would stimulate aldosterone synthesis and release.
7. **The answer is C.** The first and rate-limiting step in all steroid biosynthesis is catalyzed by cholesterol side-chain cleavage enzyme, resulting in pregnenolone and isocaproic acid. 17α-hydroxylase, 3β-hydroxysteroid dehydrogenase, 21-hydroxylase, and 11β-hydroxylase are all involved in the synthesis of cortisol, but are not rate-limiting. 3-Hydroxy-3-methylglutaryl CoA reductase catalyzes the rate-limiting step in *de novo* cholesterol synthesis.
8. **The answer is B.** Addison's disease results from the pathological destruction of the adrenal glands by microorganisms or by an autoimmune response; it is characterized by glucocorticoid and aldosterone deficiency. Hyperpigmentation is caused by a lack of cortisol production, which results in increased ACTH production. Hyponatremia and hyperkalemia occur in the absence of aldosterone, which normally stimulates sodium retention and potassium excretion by the kidneys. Cushing's disease produces excessive cortisol release from the adrenals, secondary to excessive anterior pituitary secretion of ACTH; patients with this disease do not have the symptoms of aldosterone deficiency. Primary hypoaldosteronism is characterized by a lack of aldosterone secretion. The hyperpigmentation indicates a more severe disease with lack of cortisol production as well. Congenital adrenal hyperplasia is the result of genetic defects that affect adrenal steroidogenic enzymes, resulting in impaired formation of cortisol. Low serum cortisol is a stimulus for ACTH release and hyperpigmentation. Congenital adrenal hyperplasia is usually associated with hypertension as a result of the excess production of steroidogenic intermediates such as deoxycorticosterone, which has substantial mineralocorticoid activity. Hypopituitarism is a condition in which pituitary function is suppressed, resulting in reduced ACTH secretion; this is not applicable because the patient presents with hyperpigmentation as a result of excess ACTH release. Patients with glucocorticoid-suppressible hyperaldosteronism are hypertensive.
9. **The answer is E.** Glucocorticoids maintain the transcription of genes and, therefore, the intracellular concentrations of many of the enzymes needed to carry out gluconeogenesis in the liver and kidneys. Glucocorticoids maintain the liver and kidneys in a state that makes them capable of accelerated gluconeogenesis when fasting occurs. Glucocorticoids inhibit insulin release. Insulin inhibits gluconeogenic enzymes in the liver. The glucocorticoid-induced inhibition of glu-

cose utilization by skeletal muscle does not stimulate gluconeogenesis but provides glucose for utilization by the CNS. Inhibition of glycogenolysis by glucocorticoids does not occur in fasting. Glucocorticoids do not inhibit, but actually permit, lipolysis and the release of fatty acids from adipose tissue.

Chapter 35

1. **The answer is D.** Epinephrine stimulates glucagon secretion but inhibits insulin secretion. Amino acids and acetylcholine both stimulate insulin and glucagon secretion.
2. **The answer is C.** Insulin inhibits protein degradation and stimulates amino acid uptake in skeletal muscle. It stimulates glucose uptake in many, but not all, tissues. It inhibits hormone-sensitive lipase in adipose tissue.
3. **The answer is C.** Glucagon stimulates gluconeogenesis and ureagenesis in the liver. Under certain conditions, glucagon can actually stimulate insulin secretion. Glucagon does not have its primary actions in adipose tissue. Somatostatin does not play a role in ketogenesis.
4. **The answer is B.** Persons with type 1 diabetes are insulin-deficient, not insulin resistant; they are treated with exogenous insulin. Persons with type 2 diabetes are treated with sulfonylureas. Secondary complications are difficult to avoid in any form of diabetes.
5. **The answer is A.** The development of type 2 diabetes has a strong genetic component. Persons with type 2 diabetes often have normal or elevated insulin levels. Although there is an association of type 2 diabetes and obesity, it is not true that it only occurs in obese individuals. Type 1 diabetes is a disease of insulin deficiency, and type 2 is a disease of insulin resistance.
6. **The answer is D.** Neuropathy, nephropathy, and retinopathy are chronic complications of type 2 diabetes. Ketoacidosis is an acute complication seen in type 1 diabetes.
7. **The answer is D.** Delta cells produce somatostatin.
8. **The answer is D.** The I/G ratio is highest after feeding and decreases progressively during fasting.

Chapter 36

1. **The answer is B.** Half (50%) of the total plasma calcium is free or ionized.
2. **The answer is C.** Most of the ingested calcium is not absorbed by the GI tract and leaves the body via the feces.
3. **The answer is A.** The majority of ingested phosphate is absorbed by the GI tract and leaves the body via the urine.
4. **The answer is A.** Skin, kidney, and liver can all be involved in forming the active metabolite of vitamin D, 1,25-dihydroxycholecalciferol. Bone does not form this hormone, but is a target for its actions.
5. **The answer is C.** The kidneys are the site of formation of 1,25-dihydroxycholecalciferol from 25-hydroxycholecalciferol, a reaction catalyzed by the 1α-hydroxylase enzyme. 7-Dehydrocholesterol is converted to cholecalciferol in the skin. Vitamin D_3 is not converted to vitamin D_2. Calcium is incorporated into hydroxyapatite in bone.
6. **The answer is C.** Osteoporosis is characterized by an equivalent loss of bone mineral and organic matrix. Paget's disease is characterized by disordered bone remodeling; rickets and osteomalacia are characterized by inadequate bone mineralization.
7. **The answer is D.** PTH stimulates bone resorption and renal calcium reabsorption and, via stimulated synthesis of 1,25-dihydroxycholecalciferol, intestinal calcium absorption, raising plasma calcium concentration. PTH inhibits renal phosphate reabsorption, leading to phosphaturia and hypophosphatemia.

Chapter 37

1. **The answer is A.** Reduced secretion of GnRH will result in extremely low levels of circulating LH and FSH, causing testicular atrophy, as in Kallmann's syndrome. Hypersecretion of LH and FSH, increased activin, and an increased number of FSH receptors all lead to hyperfunction of the testis, not hypofunction. A failure of the hypothalamus to respond to testosterone increases LH, leading to increased Leydig cell androgens and testicular hypertrophy.
2. **The answer is E.** Follistatin is a binding protein for activin. Activin cannot increase FSH secretion when follistatin is bound to it, so FSH secretion decreases. Follistatin does not bind FSH, does not inhibit seminal fluid production and Leydig cell testosterone secretion, and does not stimulate the production of spermatogonia.
3. **The answer is A.** The epididymis and vas deferens are major storage sites of spermatozoa. Spermatozoa develop in the seminiferous tubules. Sertoli cells, not the epididymis, secrete estrogens and inhibin. The prostate gland, seminal vesicles, and bulbourethral glands secrete the seminal fluids.
4. **The answer is B.** It takes approximately 65 to 70 days to develop spermatozoa from the earliest stages of spermatogonia. Because the production of sperm depends on LH and FSH, a lack of GnRH (Kallmann's syndrome) will reduce the production of LH, FSH, and sperm. Temperature is important in regulating sperm production. Optimal sperm production occurs at 2 to 3°C lower than body temperature.
5. **The answer is A.** The initial reaction and the rate-limiting step in the production of testosterone is the conversion of cholesterol to pregnenolone, which is regulated by LH-stimulated cAMP in the Leydig cells. The cholesterol side-chain cleavage enzyme is located in mitochondria. All other sex hormone synthesis occurs outside of the mitochondria. Aromatization is the last reaction, the conversion of testosterone to estradiol. Pregnenolone is the immediate derivative of cholesterol, not progesterone. The initial reaction is stimulated by LH, not FSH. LH receptors are on Leydig cells, the site of testosterone synthesis.
6. **The answer is C.** The enzyme 5α-reductase is found in

the prostate and converts testosterone to dihydrotestosterone. Testosterone does not bind HDL; HDL is a source of cholesterol. Activin does not bind testosterone. Testosterone cannot be converted directly to 17-hydroxyprogesterone, which is derived from progesterone and is converted to androstenedione. The side-chain cleavage enzyme converts cholesterol to pregnenolone.

7. **The answer is A.** Sex hormone-binding globulin binds to both testosterone and estradiol, but it binds with higher affinity to testosterone. The bioactivity of testosterone is reduced by SHBG because testosterone cannot bind to its receptor when bound by SHBG. SHBG increases the circulating half-life of testosterone by slowing the clearance and metabolism of testosterone. SHBG does not alter the secretion of inhibin or androgen-binding protein.
8. **The answer is D.** The production of estradiol requires Leydig cells, under the influence of LH, which stimulates androgen production. The androgen diffuses to Sertoli cells, which contain aromatase, the enzyme that converts androgens to estrogens under the influence of FSH. Therefore, Leydig cells, Sertoli cells, LH, and FSH are required. Follistatin binds activin and would reduce FSH secretion, an essential component for estradiol production. Estradiol is not produced by Leydig cells. Activin would increase the secretion of FSH, which is a necessary component for estradiol, but other cells and hormones are required. Similarly, Leydig cells would need LH to stimulate the production of the androgen precursor of estrogen. Sertoli cells, under the influence of FSH, are needed to aromatize androgen from Leydig cells.
9. **The answer is C.** Androgens and estrogens are known to stimulate the closure of the epiphyses at puberty. Because eunuchs are castrated, they have no testicular source of androgen and estrogen, and the closure of the epiphyses is delayed. In eunuchs, long bones continue to grow, resulting in a tall stature. Estrogens do have a positive effect in maintaining bone; however, eunuchs have little or no estrogen because the testes are absent. Choice B is incorrect, although eunuchs may have elevated circulating LH (as a result of the lack of androgen negative feedback). LH has no effect on bone. The absence of testes delays the closure of the epiphyses, and androgen levels are low in eunuchs because of the lack of testes.

Chapter 38

1. **The answer is B.** Aromatase is present only in granulosa cells and is regulated mainly by FSH. Although LH may stimulate aromatase in granulosa cells, granulosa cells do not produce androgens. Estradiol synthesis in the graafian follicle is unrelated to progesterone synthesis in the corpus luteum and does not increase LH during this phase. Estradiol increases LH secretion during the LH surge. There is no evidence for synergy between FSH and progesterone in regulating estradiol secretion by the graafian follicle.
2. **The answer is A.** Granulosa cells do not have the enzyme called 17α-hydroxylase, which converts progesterone to 17α-hydroxyprogesterone. Aromatase is the enzyme that converts androgens to estrogens. 5α-Reductase converts testosterone to dihydrotestosterone. Sulfatase is an enzyme that conjugates steroids with sulfate for subsequent excretion in the urine. Steroidogenic acute regulatory protein transports cholesterol from the outer to the inner mitochondrial membrane.
3. **The answer is B.** One of the first clinical measures for menopause is an increase in the serum concentration of FSH (and LH), indicative of the lack of ovarian function. Menses starts at age 12, not age 50, and its onset at this time would not indicate menopause. Excessive corpora lutea would likely indicate multiple ovulations or a failure of luteal regression. Increased vaginal cornification is an indicator of estrogen secretion, which does not occur in menopause. Menstrual cycles become irregular at menopause.
4. **The answer is D.** Progesterone has a thermogenic effect on the hypothalamus, increasing the basal body temperature for a few days after ovulation. Women who, because of ovulatory problems, are having trouble getting pregnant are sometimes asked to record their daily oral temperatures and look for the increase in basal body temperature, indicating an increase in progesterone (which indicates ovulation). Progesterone induces a secretory type of endometrium, whereas estrogens induce a proliferative type. During the luteal phase, when progesterone is increasing, graafian follicles are not present. Progesterone levels decrease during luteal regression. FSH decreases when progesterone is rising.
5. **The answer is A.** Theca interna cells produce androgens under the influence of LH, whereas granulosa cells do not produce androgens. Theca interna cells do contain cholesterol side-chain cleavage enzyme, which converts cholesterol to pregnenolone. Because theca cells do not express aromatase, they cannot convert testosterone to estradiol. The theca interna has a rich blood supply. Granulosa cells produce inhibin.
6. **The answer is A.** Disruption of the hypothalamic-pituitary portal system leads to a lack of dopamine and GnRH reaching the pituitary. Because dopamine inhibits PRL secretion, PRL levels will increase. In addition, the lack of GnRH will lead to reduced secretion of LH and FSH, reduced ovarian function, and eventual ovarian atrophy. PRL will have no effect on the ovary or inhibit ovarian follicle development. Disruption of the hypothalamic-pituitary axis will lead to reduced follicular development, lack of ovulation, and low circulating progesterone. Inhibin levels will decrease, but FSH will not increase because there is no GnRH reaching the pituitary from the disrupted axis. Excessive ovarian androgen usually occurs in the presence of excessive LH secretion or an androgen-producing tumor in the ovary. LH secretion is reduced by the lack of GnRH.
7. **The answer is B.** Inhibin is produced by granulosa cells and inhibits the secretion of FSH. Inhibin does not inhibit the secretion of LH and PRL. Although inhibin can have local ovarian effects, it has profound in-

hibitory effects on FSH secretion. Inhibin has two forms, A and B; the α subunits are the same, whereas the β subunits are different. Inhibin binds activin and decreases FSH secretion.

8. **The answer is D.** Estrogen induces the formation of a stringy vaginal secretion that is called spinnbarkeit, observed in the late follicular phase. The secretory endometrium is under the influence of progesterone; therefore, spinnbarkeit would not be present. Spinnbarkeit is not produced in response to progesterone, androgen, or prolactin.
9. **The answer is A.** Fertilization occurs in the oviduct. The oocyte must have entered a second meiotic division to reduce the chromosome number of the oocyte to a haploid state (n) so that it may fuse with the sperm (also haploid), producing a 2n zygote. Fertilization does not occur in the uterus, especially not after the first meiotic division when the chromosome number is 2n. In the adult ovary, oocytes do not undergo mitosis. Graafian follicles do not enter the oviduct and are not fertilized. Fertilization does not occur in the uterus, and the oocyte does not implant. The blastocyst will implant in the uterus. In addition, extrusion of the polar body is associated with fertilization, but this event occurs within the oviduct.
10. **The answer is B.** 5α-Reductase is the enzyme that converts testosterone to dihydrotestosterone. 5α-Reductase is associated with increasing the most potent androgen, dihydrotestosterone, and reducing LH secretion. Estrogens are associated with female secondary sex characteristics, although some androgens regulate pubic hair development.

Chapter 39

1. **The answer is D.** Suckling involves hormonal and neuronal components, but the hormonal component is efferent and the neuronal component is afferent. When the baby suckles, neural signals from the nipple travel via nerves to the spinal cord and up to the brain (afferent component), which triggers the release of oxytocin from the supraoptic nuclei (efferent component). Oxytocin enters the circulation, enters the breasts, and causes contraction of the myoepithelial cells. Placental lactogen is no longer present after parturition; it is a placental hormone. Dopamine release is decreased by suckling, and as a result, PRL secretion is increased.
2. **The answer is D.** Under normal circumstances, the uterus must be primed with both progesterone and estrogen for successful implantation. Implantation occurs on day 7 after fertilization. The decidual reaction occurs as the result of the implanting blastocyst. The embryo is in the blastocyst stage of development at the time of implantation. A morula does not implant. The developing embryo enters the uterus on day 3 or 4, it remains suspended in the uterus for 3 or 4 more days, and implantation occurs on day 7.
3. **The answer is A.** The acrosome reaction causes a fusion of the plasma membrane and the acrosomal membrane of the sperm, with subsequent release of proteolytic enzymes that help the sperm enter the ovum. The zona reaction and pronuclei formation occur after the sperm has entered the ovum. Sperm enter the perivitelline space after penetration; there is no evidence that this space has any role in penetration. Cumulus expansion assists in movement of the sperm through the mass of granulosa cells for the sperm to get to the surface of the zona pellucida. However, the cumulus cells do not assist in actual penetration of the zona.
4. **The answer is B.** The production of hCG by trophoblast cells stimulates the corpus luteum to continue to produce progesterone so that luteal regression does not occur at the end of the anticipated cycle. Although PRL levels increase throughout pregnancy, PRL is not responsible for maintenance of the corpus luteum of pregnancy. Prostaglandins are generally luteolytic, causing regression of the corpus luteum, termination of the luteal phase, and return to the next cycle; they do not prolong the cycle or postpone it. Oocytes are not depleted after implantation. In fact, pregnancy tends to preserve oocytes, as ovulation ceases during pregnancy. Plasma progesterone levels are high during pregnancy as a result of activation of the corpus luteum and placental production of progesterone. Elevated progesterone blocks follicular development and the ensuing LH surge; low levels of progesterone would permit a return to cyclicity.
5. **The answer is A.** Fertilization by more than one sperm is prevented by the cortical reaction. Cortical granules containing proteolytic enzymes fuse just beneath the entire surface of the oolemma. The proteolytic enzymes are released to the perivitelline space, destroy the sperm receptors, and harden the zona, preventing other sperm from penetrating the fertilized ovum. Enzyme reaction is a nonspecific term with little meaning for polyspermy. The acrosome reaction allows the sperm to penetrate the zona. The decidual reaction is an inflammatory-like reaction that occurs simultaneously with implantation of the blastocyst into the uterine endometrium.
6. **The answer is A.** Oral steroidal contraceptives generally contain progesterone and estrogen-like molecules, which feed back negatively on the hypothalamic-pituitary axis and reduce the secretion of LH and FSH; this is the primary mechanism of action in preventing pregnancy. Choices B, C, and D are not the best answers, although oral contraceptives do alter the uterine environment, thicken the cervical mucus, and reduce sperm motility. If ovulation were not blocked, the other parameters would not be effective in blocking pregnancy. Oral contraceptives block the LH surge; they do not induce a premature surge.
7. **The answer is D.** hCG is produced by trophoblast cells prior to implantation of the embryo and binds to luteal LH receptors, signaling them to produce progesterone, which is necessary for the maintenance of pregnancy. Therefore, hCG signals the mother that she is pregnant via stimulation of luteal LH receptors. Placental lactogen is not produced until after pregnancy is well established. Progesterone is a common hormone associated with the menstrual cycle and pregnancy. In humans, the inflammatory reaction at implantation does

not signal the mother that she is pregnant and it follows secretion of hCG.

8. **The answer is C.** The placenta cannot make androgens from progestin precursors because it lacks 17α-hydroxylase. DHEAS from the fetal adrenal glands is converted to 16OH-DHEAS by the fetal liver and then to estriol by the placenta; this reaction is substantial and is an indicator of fetal stress (estriol low) or well-being (estriol high). The mother's adrenal can also make DHEAS, which can be converted to 16OH-DHEAS by 16-hydroxylase in the fetal liver, but this reaction is limited (10%). Androgens cannot be produced from cholesterol in the placenta; the placenta lacks 17α-hydroxylase. Estradiol is generally not converted to estriol. Androgens from the ovary are generally not converted to estriol.
9. **The answer is C.** Insulin resistance is associated with reduced utilization of glucose by the mother and this glucose is spared for the fetus. Plasma glucose is not lower but higher with insulin resistance. Insulin moves glucose into cells. During pregnancy, the development of insulin resistance may be a predictor of diabetes later in life. Reduced pituitary function occurs because of the high levels of steroids and PRL, all independent of insulin resistance. Progesterone, not insulin, increases appetite during pregnancy.
10. **The answer is C.** The female phenotype can develop in an XY male if the biological action of testosterone is absent. This absence can be due to a lack of testosterone secretion caused by enzyme deficiencies or a lack of the testosterone (DHT) receptor. In this process, called testicular feminization, a phenotypic female develops in the presence of an XY karyotype. There is a lack of pubic and axillary hair, well-developed breasts (as a result of the conversion of testosterone to estrogen), with inguinal or abdominal testes, no uterus (because AMH is secreted), underdeveloped male accessory ducts (lack of testosterone action), and the vagina ends in a blind pouch. Progesterone has no effect on phenotype. There is no evidence that adrenal insufficiency (low cortisol and androgens from the adrenals) have any effect on inducing female phenotype in a male. Inhibin would reduce FSH secretion and ultimately reduce adult testis size, but in the fetus there is no effect on the development of the female phenotype. AMH will prevent formation of the oviducts, uterus, and upper vagina; it does not increase female characteristics in the male.

APPENDIX B

Common Abbreviations in Physiology

A	amount; area
A	alveolar
a	arterial; ambient
ABP	androgen-binding protein
AC	adenylyl cyclase
ACE	angiotensin-converting enzyme
ACh	acetylcholine
AChE	acetylcholinesterase
ACTH	adrenocorticotropic hormone (corticotropin)
ADH	antidiuretic hormone
ADP	adenosine diphosphate
AE	anion exchanger
AMH	antimüllerian hormone (müllerian-inhibiting substance)
AMP	adenosine monophosphate
ANP	atrial natriuretic peptide
ANS	autonomic nervous system
AQP	aquaporin
A_r	effective radiating surface area
ARDS	adult respiratory distress syndrome
ATP	adenosine triphosphate
AV	atrioventricular
A-V	arteriovenous
AVP	arginine vasopressin (antidiuretic hormone or ADH)
aw	airway
B	barometric
b	body; blood
BF	blood flow
BMR	basal metabolic rate
BS	urinary space of Bowman's capsule
BSC	bumetanide-sensitive (Na^+-K^+-$2Cl^-$) cotransporter
BUN	blood urea nitrogen
C	concentration; compliance; capacitance; capacity; clearance; kilocalorie; conductance
C	heat loss by convection
c	core
CA	carbonic anhydrase
CaBP	calcium-binding protein (calbindin)
CaM	calmodulin
CAM	cell adhesion molecule
cAMP	cyclic AMP (cyclic adenosine 3′,5′-monophosphate)
CBG	corticosteroid-binding globulin
CCK	cholecystokinin
CFC	capillary filtration coefficient
CFTR	cystic fibrosis transmembrane conductance regulator
cGMP	cyclic GMP (cyclic guanosine monophosphate [guanosine 3′,5′-monophosphate])
CGRP	calcitonin-gene-related peptide
CIA	central inspiratory activity
CMK	calmodulin-dependent protein kinase
CNS	central nervous system
CO	cardiac output; carbon monoxide
COP	colloid osmotic pressure (oncotic pressure)
COPD	chronic obstructive pulmonary disease
COX	cyclooxygenase
CRH	corticotropin-releasing hormone
CSF	cerebrospinal fluid
CT	computed tomography; calcitonin (thyrocalcitonin)
CYP	cytochrome P450 enzyme
D	diffusion coefficient; diffusing capacity
D	dead space
DAG	diacylglycerol
DHEAS	dehydroepiandrosterone sulfate
DHT	dihydrotestosterone
DIT	diiodotyrosine
DMT	divalent metal transporter
DNA	deoxyribonucleic acid
DPG	diphosphoglycerate
DPPC	dipalmitoylphosphatidylcholine
E	extraction (extraction ratio)
E	evaporative heat loss
E	expiratory
e	emissivity
EABV	effective arterial blood volume
ECaC	epithelial calcium channel
ECF	extracellular fluid
ECG	electrocardiogram
ECL cell	enterochromaffin-like cell
ED_{50}	median effective dose
EDRF	endothelium-derived relaxing factor (NO)
EDV	end-diastolic volume
EEG	electroencephalogram
EF	ejection fraction
EGF	epidermal growth factor
E_{ion}	equilibrium potential for an ionic species
EJP	excitatory junction potential

ELISA enzyme-linked immunosorbent assay
E_m membrane potential
ENaC epithelial sodium channel
ENS enteric nervous system
EPP equal pressure point
EPSP excitatory postsynaptic potential
ER endoplasmic reticulum
ERV expiratory reserve volume
ESR erythrocyte sedimentation rate
ESV end-systolic volume
F fractional concentration of gas; farad; Faraday constant
f frequency
FEF forced expiratory flow
FEV forced expiratory volume
FGF fibroblast growth factor
FL focal length
FM frequency modulation
FRC functional residual capacity
FSH follicle-stimulating hormone
FVC forced vital capacity
g ionic conductance
G protein guanine nucleotide-binding protein
GABA γ-aminobutyric acid
GAP GnRH-associated peptide
GC glomerular capillary
GDP guanosine diphosphate
GFR glomerular filtration rate
GH growth hormone (somatotropin)
GHRH growth hormone-releasing hormone
GI gastrointestinal
GIP gastric inhibitory peptide (glucose-dependent insulinotropic peptide)
GLP glucagon-like peptide
GLUT glucose transporter
GnRH gonadotropin-releasing hormone (LHRH)
GPCR G protein-coupled receptor
GRE glucocorticoid response element
GRP gastrin-releasing peptide
GTO Golgi tendon organ
GTP guanosine triphosphate
Hb hemoglobin
h_c convective heat transfer coefficient
hCG human chorionic gonadotropin
HDL high-density lipoprotein
h_e evaporative heat transfer coefficient
HF heat flow
HGF hepatocyte growth factor
hPL human placental lactogen (human chorionic somatomammotropin)
HR heart rate
HRE hormone response element
HSL hormone-sensitive lipase
5-HT 5-hydroxytryptamine (serotonin)
5-HTP 5-hydroxytryptophan
I inspiratory
IC inspiratory capacity
ICC interstitial cell of Cajal
ICF intracellular fluid
IGF-I insulin-like growth factor I (somatomedin C)
I/G ratio insulin/glucagon ratio
I_{ion} ionic current
IJP inhibitory junction potential
IL interleukin
IP_3 inositol 1,4,5-trisphosphate
IPSP inhibitory postsynaptic potential
IRDS infant respiratory distress syndrome
IRV inspiratory reserve volume
J flow (or flux) of a solute or water; joule
K heat loss by conduction
K_f ultrafiltration coefficient
K_h hydraulic conductivity
L lung
LDL low-density lipoprotein
L-DOPA L-3,4-dihydroxyphenylalanine
LH luteinizing hormone
LHRH luteinizing hormone-releasing hormone
M metabolic rate
MAP microtubule-associated protein
MAP kinase mitogen-activated protein kinase
MCH mean cell (corpuscular) hemoglobin
MCHC mean cell (corpuscular) hemoglobin concentration
MCR metabolic clearance rate
M-CSF macrophage-colony stimulating factor
MCV mean cell (corpuscular) volume
MIT monoiodotyrosine
MLCK myosin light-chain kinase
MLCP myosin light-chain phosphatase
MMC migrating motor complex
MSH melanocyte-stimulating hormone
MVC maximal voluntary contraction
NANC nonadrenergic noncholinergic
NaPi sodium-coupled phosphate transporter
NE norepinephrine
NHE Na^+/H^+ exchanger
NK natural killer
NMDA *N*-methyl-D-aspartate
NO nitric oxide
NOS nitric oxide synthase
N_R Reynolds number
OAT organic anion transporter
OCT organic cation transporter
P pressure; partial pressure; permeability; permeability coefficient; plasma; plasma concentration
P_{50} PO_2 at which 50% of hemoglobin is saturated
PAH *p*-aminohippurate
Pc′ pulmonary end-capillary
PCr phosphocreatine (creatine phosphate)
PDE phosphodiesterase
PEF peak expiratory flow

PG	prostaglandin
PI	phosphatidylinositol
P_i	inorganic phosphate
PIF	peak inspiratory flow
PIP_2	phosphatidylinositol 4,5-bisphosphate
PKA	protein kinase A
PKC	protein kinase C
PKG	cGMP-dependent protein kinase
pl	pleural
PLC	phospholipase C
PNS	peripheral nervous system
POMC	proopiomelanocortin
PRL	prolactin
PRU	peripheral resistance unit
PT	prothrombin time
PTH	parathyroid hormone (parathormone)
PTT	partial thromboplastin time
PVC	premature ventricular complex
pw	pulmonary wedge
$\dot{Q}$	blood flow
R	respiratory exchange ratio; resistance; universal gas constant
R	heat loss by radiation
r	radius; radiant environment
RAAS	renin-angiotensin-aldosterone system
RBF	renal blood flow
RBP	retinol-binding protein
RDA	recommended daily allowance
REM	rapid eye movement
rh	relative humidity
RIA	radioimmunoassay
RISA	radioiodinated serum albumin
ROS	reactive oxygen species
RPF	renal plasma flow
RQ	respiratory quotient
RV	residual volume
RVD	regulatory volume decrease
RVI	regulatory volume increase
S	saturation; siemens
S	rate of heat storage in the body
s	shunt
SA	sinoatrial
SF-1	steroidogenic factor-1
SGLT	Na^+-glucose cotransporter
SH2	src homology domain
SHBG	sex hormone-binding globulin
SIADH	syndrome of inappropriate ADH
sk	skin
SN	single nephron
SPCA	serum prothrombin conversion accelerator
SRIF	somatotropin release inhibiting factor (somatostatin)
SRY	sex-determining region, Y chromosome
SSRI	selective serotonin reuptake inhibitor
StAR	steroidogenic acute regulatory protein
SV	stroke volume
SVR	systemic vascular resistance (total peripheral resistance)
SW	stroke work
T	tension; temperature; time
T	tidal
t	time
T_3	triiodothyronine
T_4	thyroxine
ta	transairway
TBG	thyroxine-binding globulin
TBW	total body water
TEA	tetraethylammonium
TF	tubular fluid
TGF	transforming growth factor
TLC	total lung capacity
tm	transmural
Tm	tubular transport maximum
Tn-C	calcium-binding troponin
TNF	tumor necrosis factor
Tn-I	troponin that inhibits actin-myosin interactions
Tn-T	tropomyosin-binding troponin
tp	transpulmonary
TPA	tissue plasminogen activator
TR	thyroid hormone receptor
TRE	thyroid hormone response element
TRH	thyrotropin-releasing hormone
TSC	thiazide-sensitive (Na^+-Cl^-) cotransporter
TSH	thyroid-stimulating hormone
T tubule	transverse tubule
U	urine concentration
UCP	uncoupling protein
UDP	uridine diphosphate
UP	ultrafiltration pressure gradient
V	volume; volt; vasopressin; vacuolar
v	velocity; venous
$\dot{V}$	gas volume per unit time (airflow); minute ventilation; urine flow rate
$\dot{V}_A$	alveolar ventilation
VC	vital capacity
VEGF	vascular endothelial growth factor
VIP	vasoactive intestinal peptide
VLDL	very low density lipoprotein
$\dot{V}O_2$	oxygen uptake
W	heat loss as mechanical work
w	wettedness
z	valence of an ion
φ	osmotic coefficient
η	viscosity
λ	length (space) constant; wavelength
μ	electrochemical potential
π	osmotic pressure; 3.14 (pi)
ρ	density
σ	reflection coefficient
τ	time constant

INDEX

Page numbers followed by *b* denote boxes, those followed by *f* denote figures, and those followed by *t* denote tables.